Table of Integrals, Series, and Products

Table of Integrals, Series, and Products

By I. S. GRADSHTEYN and I. M. RYZHIK

Fourth Edition Prepared by

YU. V. GERONIMUS/M. YU. TSEYTLIN

Translated from the Russian by Scripta Technica, Inc.

Translation edited by

ALAN JEFFREY
PROFESSOR OF ENGINEERING MATHEMATICS
UNIVERSITY OF NEWCASTLE UPON TYNE, ENGLAND

1965

ACADEMIC PRESS New York and London

A Subsidiary of Harcourt Brace Jovanovich, Publishers

ACADEMIC PRESS INC.
111 Fifth Avenue, New York, New York 10003

United Kingdom Edition published by
ACADEMIC PRESS, INC. (LONDON) LTD.
24/28 Oval Road, London NW1

Originally published as:
Tablitsy Integralov, Summ, Ryadov I Proievedeniy

by: Gosudarstvennoe Izdatel'stvo
Fiziko-Matematicheskoy Literatury

Moscow, 1963

Library of Congress Catalog Card Number: 65-29097

PRINTED IN THE UNITED STATES OF AMERICA.

EDITOR'S FOREWORD

This book is a translation of the fourth and most recent edition of Ryzhik and Gradshteyn's Tables of Integrals. It represents a very considerable enlargement of the third edition to which many new sections have been added. Its thorough coverage of the functions of mathematical physics and its careful arrangement of material will make it invaluable to mathematicians, engineers and physicists working in all fields.

In preparing this translation the opportunity has been taken to add a short note on the use of the tables. After a brief reference to the method of arrangement of the tables, the introduction then summarizes some of the more common differences in notation and definition of higher transcendental functions. A short classified bibliography has also been added at the end of the book to supplement the original Russian bibliography—much of which is likely to be inaccessible to the user of these tables.

A. Jeffrey

ACKNOWLEDGEMENT

The Publisher and Editor would like to take this opportunity to express their gratitude to the following users of the Tables of Integrals, Series and Products, who have generously supplied us with a number of corrections to the original printing:

Dr. J. E. Bowcock, Dr. T. H. Boyer, Dr. D. J. Buch,
Dr. F. Calogero, Dr. Herman W. Chew, Dr. R. W. Cleary,
Dr. D. K. Cohoon, Dr. J. W. Criss, Dr. A. Degasperis,
Dr. K. Evans, Dr. G. R. Gamertsfelder, Dr. I. J. Good,
Dr. J. Good, Dr. T. Hagfors, Dr. D. O. Harris,
Dr. R. E. Hise, Dr. Y. Iksbe, Dr. B. Jancovici,
Dr. D. Monowalow, Dr. L. P. Kok, Dr. S. L. Levie,
Dr. I. Manning, Dr. J. Marmur, Dr. C. Muhlhausen,
Dr. P. Noerdlinger, Dr. A. H. Nuttall, Dr. J. R. Roth,
Dr. G. K. Tannahill.

PREFACE TO THE FIRST EDITION

The number of formulas for integrals, sums, series, and products available in the existing mathematical reference books, is definitely insufficient for mathematicians working in scientific fields and for scientists and engineers doing theoretical and experimental research. The present tables have been compiled with the purpose of filling this gap. More than five thousand formulas have been collected in the present volume from various sources.

The book is primarily intended for scientists, technicians, and engineers carrying out investigations in the physical-mathematical sciences. Therefore, exposition takes up only a small portion of thc book, and basically, it is a collection of formulas.

Considerable attention has been given to special functions, in particular, elliptic, Bessel, and Legendre functions. Many formulas dealing with these functions are included.

I wish to take this occasion to express my deep gratitude to Professors V. V. Stepanov, A. I. Markushevich, and I. N. Bronshteyn for valuable advice and for suggestions made by them in the preparation of this work.

I. Ryzhik

PREFACE TO THE THIRD EDITION

I. M. Ryzhik, the author of the first and second editions of these tables, died during World War II. These tables have been considerably revised in preparation for printing.

All the formulas, definitions, and theorems have been placed into numbered divisions. The numbering procedure has some similarity with the decimal system of classification and can easily be understood from the arrangement of the table of contents. Only the main divisions are included in the table of contents. These have numbers containing one, two, or three digits. The smallest divisions of the book are numbered with four digits. These divisions contain one or several formulas, theorems, or definitions, which are numbered in light type with successive integers. The digit 0 is reserved for divisions of a general character: introductions, definitions, etc. The number 0 is also used for the first chapter of the book, which includes a number of theorems of a general nature and which is itself something of of an introduction.

I. Gradshteyn

PREFACE TO THE FOURTH EDITION

In preparing the fourth edition, I. S. Gradshteyn intended to expand this reference work considerably. His death prevented completion of this plan. He had compiled new tables of integrals of the elementary functions and had collected material for preparing tables of integrals of special functions.

The publishing authorities entrusted us with preparing the manuscript left by Gradshteyn and with adding supplementary material where it was needed.

In carrying out this work, we have tried to follow the plan of the manuscript and of the preceding edition. Throughout, we have kept its main feature: the order of the formulas. We have included the divisions dealing with sums, series, products, and elementary functions without change from the preceding edition. The other divisions were somewhat revised. The tables of definite integrals of elementary and special functions, in particular, were considerably extended. We have added a few new divisions; the integrals of Mathieu functions, Struve functions, Lommel functions, and a number of other functions that were entirely omitted from the older editions. In general, the number of special functions considered in the fourth edition is much greater than in the third. As a consequence, the chapters dealing with special functions were supplemented with the corresponding divisions.

The majority of definitions of special functions that were used in the preceding edition have been kept. We changed to new definitions only in certain cases when we were following references that contain the major source of material on integrals of the special functions in question.

Certain notations have also been changed. The chapter in the third edition that is devoted to integral transformations was dropped from the fourth edition. The material included in it appears in other portions of the book.

We wish to express our deep gratitude to A. F. Lapko, who read the manuscript carefully and made a number of useful comments.

Yu. Geronimus, M. Tseytlin

CONTENTS

THE ORDER OF PRESENTATION OF THE FORMULAS

The question of the most expedient order in which to give the formulas, in particular, in what division to include particular formulas such as the definite integrals, turned out to be quite complicated. The thought naturally occurs to set up an order analogous to that of a dictionary. However, it is almost impossible to create such a system for the formulas of integral calculus. Indeed, in an arbitrary formula of the form

$$\int_a^b f(x)\,dx = A$$

one may make a large number of substitutions of the form $x=\varphi(t)$ and thus obtain a number of "synonyms" of the given formula. We must point out that the table of definite integrals by Bierens de Haan and the earlier editions of the present reference both sin in the plethora of such "synonyms" and the formulas of complicated form. In the present edition, we have tried to keep only the simplest of the "synonym" formulas. Basically, we judged the simplicity of a formula from the standpoint of the simplicity of the arguments of the "outer" functions that appear in the integrand. Where possible, we have replaced a complicated formula with a simpler one. Sometimes, several complicated formulas were thereby reduced to a single simpler one. We then kept only the simplest formula. As a result of such substitutions, we sometimes obtained an integral that could be evaluated by use of the formulas of chapter two and the Newton-Leibnitz formula, or to an integral of the form

$$\int_{-a}^{a} f(x)\,dx,$$

where $f(x)$ is an odd function. In such cases the complicated integrals have been omitted.

Let us give an example using the expression

$$\int_0^{\frac{\pi}{4}} \frac{(\operatorname{ctg} x - 1)^{p-1}}{\sin^2 x} \ln \operatorname{tg} x\,dx = -\frac{\pi}{p} \operatorname{cosec} p\pi. \qquad (1)$$

By making the natural substitution $\operatorname{ctg} x - 1 = u$, we obtain

$$\int_0^{\infty} u^{p-1} \ln(1+u)\,du = \frac{\pi}{p} \operatorname{cosec} p\pi. \qquad (2)$$

Integrals similar to formula (1) are omitted in this new edition. Instead, we have formula (2) and the formula obtained from the integral (1) by making the substitution $\operatorname{ctg} x = v$.

As a second example, let us take

$$I = \int_0^{\frac{\pi}{2}} \ln(\operatorname{tg}^p x + \operatorname{ctg}^p x) \ln \operatorname{tg} x \, dx = 0.$$

The substitution $\operatorname{tg} x = u$ yields

$$I = \int_0^{\infty} \frac{\ln(u^p + u^{-p}) \ln u}{1 + u^2} \, du.$$

If we now set $v = \ln u$, we obtain

$$I = \int_{-\infty}^{\infty} \frac{ve^v}{1 + e^{2v}} \ln(e^{pv} + e^{-pv}) \, dv = \int_{-\infty}^{\infty} v \frac{\ln 2 \operatorname{ch} pv}{2 \operatorname{ch} v} \, dv.$$

The integrand is odd and, consequently, the integral is equal to 0.

Thus, before looking for an integral in the tables, the user should simplify as much as possible the arguments (the "inner" functions) of the functions in the integrand.

The functions are ordered as follows:

First we have the elementary functions:

1. The function $f(x) = x$.
2. The exponential function.
3. The hyperbolic functions.
4. The trigonometric functions.
5. The logarithmic function.
6. The inverse hyperbolic functions. (These are replaced with the corresponding logarithms in the formulas containing definite integrals.)
7. The inverse trigonometric functions.

Then follow the special functions:

8. Elliptic integrals.
9. Elliptic functions.
10. The logarithm integral, the exponential integral, the sine integral, and the cosine integral functions.
11. Probability integrals and Fresnel's integrals.
12. The Gamma function and related functions.
13. Bessel functions.
14. Mathieu functions.
15. Legendre functions.
16. Orthogonal polynomials.
17. Hypergeometric functions.
18. Degenerate hypergeometric functions.
19. Functions of a parabolic cylinder.
20. Meijer's and MacRobert's functions.
21. Riemann's zeta function.

The integrals are arranged in order of outer function according to the above scheme: the farther down in the list a function occurs, (i.e. the more complex it is) the later will the corresponding formula appear in the tables. Suppose that several expressions have the same outer function. For example, consider $\sin e^x$, $\sin x$, $\sin \ln x$. Here, the outer function is the sine in all three cases. Such expressions are then arranged in order of the inner function. In the present work, these functions are therefore arranged in the following order: $\sin x$, $\sin e^x$, $\sin \ln x$.

Our list does not include polynomials, rational functions, powers, or other algebraic functions. An algebraic function that is included in tables of definite integrals can usually be reduced to a finite combination of roots of rational power. Therefore, for classifying our formulas, we can conditionally treat a power function as a generalization of an algebraic and, consequently, of a rational function.* We shall distinguish between all these functions and those listed above and we shall treat them as operators. Thus, in the expression $\sin^2 e^x$, we shall think of the squaring operator as applied to the outer function, namely, the sine. In the expression $\frac{\sin x + \cos x}{\sin x - \cos x}$, we shall think of the rational operator as applied to the trigonometric functions sine and cosine. We shall arrange the operators according to the following order:

1. Polynomials (listed in order of their degree).
2. Rational operators.
3. Algebraic operators (expressions of the form $A^{\frac{p}{q}}$, where q and p are rational, and $q > 0$; these are listed according to the size of q).
4. Power operators.

Expressions with the same outer and inner functions are arranged in the order of complexity of the operators. For example, the following functions (whose outer functions are all trigonometric, and whose inner functions are all $f(x) = x$) are arranged in the order shown:

$$\sin x,\ \sin x \cos x,\ \frac{1}{\sin x} = \operatorname{cosec} x,\ \frac{\sin x}{\cos x} = \operatorname{tg} x,\ \frac{\sin x + \cos x}{\sin x - \cos x},\ \sin^m x, \sin^m x \cos x.$$

Furthermore, if two outer functions $\varphi_1(x)$ and $\varphi_2(x)$, where $\varphi_1(x)$ is more complex than $\varphi_2(x)$, appear in an integrand and if any of the operations mentioned are performed on them, then the corresponding integral will appear (in the order determined by the position of $\varphi_2(x)$ in the list) after all integrals containing only the function $\varphi_1(x)$. Thus, following the trigonometric functions are the trigonometric and power functions (that is, $\varphi_2(x) = x$). Then come

combinations of trigonometric and exponential functions,

combinations of trigonometric functions, exponential functions, and powers, etc.,

combinations of trigonometric and hyperbolic functions, etc.

---

*For any natural number n, the involution $(a+bx)^n$ of the binomial $a+bx$ is a polynomial. If n is a negative integer, $(a+bx)^n$ is a rational function. If n is irrational, the function $(a+bx)^n$ is not even an algebraic function.

Integrals containing two functions $\varphi_1(x)$ and $\varphi_2(x)$ are located in the division and order corresponding to the more complicated function of the two. However, if the positions of several integrals coincide because they contain the same complicated function, these integrals are put in the position defined by the complexity of the second function.

To these rules of a general nature, we need to add certain particular considerations that will be easily understood from the tables. For example, according to the above remarks, the function $e^{\frac{1}{x}}$ comes after e^x as regards complexity, but $\ln x$ and $\ln \frac{1}{x}$ are equally complex since $\ln \frac{1}{x} = -\ln x$. In the section on "powers and algebraic functions", polynomials, rational functions, and powers of powers are formed from power functions of the form $(a+bx)^n$ and $(\alpha+\beta x)^\nu$.

USE OF THE TABLES*

For the effective use of the tables contained in this book it is necessary that the user should first become familiar with the classification system for integrals devised by the authors Ryzhik and Gradshteyn. This classification is described in detail in the section entitled The Order of Presentation of the Formulas and essentially involves the separation of the integrand into *inner* and *outer* functions. The principal function involved in the integrand is called the *outer* function and its argument, which is itself usually another function, is called the *inner* function. Thus, if the integrand comprised the expression $\ln \sin x$, the *outer* function would be the logarithmic function while its argument, the *inner* function, would be the trigonometric function $\sin x$. The desired integral would then be found in the section dealing with logarithmic functions, its position within that section being determined by the position of the *inner* function (here a trigonometric function) in Ryzhik and Gradshteyn's list of functional forms.

It is inevitable that some duplication of symbols will occur within such a large collection of integrals and this happens most frequently in the first part of the book dealing with algebraic and trigonometric integrands. The symbols most frequently involved are α, β, γ, δ, t, u, z, z_k, and Δ. The expressions associated with these symbols are used consistently within each section and are defined at the start of each new section in which they occur. Consequently, reference should be made to the beginning of the section being used in order to verify the meaning of the substitutions involved.

Integrals of algebraic functions are expressed as combinations of roots with rational power indices, and definite integrals of such functions are frequently expressed in terms of the Legendre elliptic integrals $F(\phi, k)$, $E(\phi, k)$ and $\Pi(\phi, n, k)$, respectively, of the first, second and third kinds.

An abbreviated notation is used for the trigonometric and hyperbolic functions $\tan x$, $\cot x$, $\sinh x$, $\cosh x$, $\tanh x$ and $\coth x$ which are denoted, respectively, by $\operatorname{tg} x$, $\operatorname{ctg} x$, $\operatorname{sh} x$, $\operatorname{ch} x$, $\operatorname{th} x$ and $\operatorname{cth} x$. Also the four inverse hyperbolic functions $\operatorname{Arsh} z$, $\operatorname{Arch} z$, $\operatorname{Arth} z$ and $\operatorname{Arcth} z$ are introduced through the definitions

$$\arcsin z = \frac{1}{i} \operatorname{Arsh}(iz)$$

$$\arccos z = \frac{1}{i} \operatorname{Arch} z$$

$$\operatorname{arctg} z = \frac{1}{i} \operatorname{Arth}(iz)$$

---

*Prepared by Alan Jeffrey for the English language edition.

$$\operatorname{arcctg} z = i \operatorname{Arcth}(iz)$$

or,

$$\operatorname{Arsh} z = \frac{1}{i} \arcsin(iz)$$

$$\operatorname{Arch} z = i \arccos z$$

$$\operatorname{Arth} z = \frac{1}{i} \operatorname{arctg}(iz)$$

$$\operatorname{Arcth} z = \frac{1}{i} \operatorname{arcctg}(-iz)$$

The numerical constants **C** and **G** which often appear in the definite integrals denote Euler's constant and Catalan's constant, respectively. Euler's constant **C** is defined by the limit

$$\mathbf{C} = \lim_{s \to \infty} \left(\sum_{m=1}^{s} \frac{1}{m} - \ln s \right) = 0.577215\ldots .$$

On occasions other writers denote Euler's constant by the symbol γ, but this is also often used instead to denote the constant

$$\gamma = e^{\mathbf{C}} = 1.781072\ldots .$$

Catalan's constant **G** is related to the complete elliptic integral

$$\mathbf{K} \equiv \mathbf{K}(k) \equiv \int_0^{\pi/2} \frac{da}{\sqrt{1 - k^2 \sin^2 a}}$$

by the expression

$$\mathbf{G} = \frac{1}{2} \int_0^1 \mathbf{K}\, dk = \sum_{m=0}^{\infty} \frac{(-1)^m}{(2m+1)^2} = 0.915965\ldots .$$

Since the notations and definitions for higher transcendental functions that are used by different authors are by no means uniform, it is advisable to check the definitions of the functions that occur in these tables. This can be done by identifying the required function by symbol and name in the Index of Special Functions and Notations that appears at the front of the book on page **xxxix**, and by then referring to the defining formula or section number listed there. We now present a brief discussion of some of the most commonly used alternative notations and definitions for higher transcendental functions.

Bernoulli and Euler Polynomials and Numbers

Extensive use is made throughout the book of the Bernoulli and Euler numbers B_n and E_n that are defined in terms of the n-th Bernoulli and Euler polynomials $B_n(x)$ and $E_n(x)$, respectively. These polynomials are defined by the generating functions

$$\frac{te^{xt}}{e^t - 1} = \sum_{n=0}^{\infty} B_n(x) \frac{t^n}{n!} \qquad \text{for} \qquad |t| < 2\pi$$

and

$$\frac{2e^{xt}}{e^t + 1} = \sum_{n=0}^{\infty} E_n(x) \frac{t^n}{n!} \qquad \text{for} \qquad |t| < \pi.$$

The Bernoulli numbers are always denoted by B_n and are defined by the relation

$$B_n = B_n(0) \qquad \text{for} \qquad n = 0,1, \ldots ,$$

when

$$B_0 = 1, \; B_1 = -\tfrac{1}{2}, \; B_2 = 1/6, \; B_4 = -1/30, \ldots .$$

The Euler numbers E_n are defined by setting

$$E_n = 2^n E_n\left(\frac{1}{2}\right) \qquad \text{for} \qquad n = 0,1, \ldots .$$

The E_n are all integral and $E_0 = 1$, $E_2 = -1$, $E_4 = 5$, $E_6 = -61, \ldots$.

An alternative definition of Bernoulli numbers, which we shall denote by the symbol B_n^*, uses the same generating function but identifies the B_n^* differently in the following manner:

$$\frac{t}{e^t - 1} = 1 - \frac{1}{2}t + B_1^* \frac{t^2}{2!} - B_2^* \frac{t^4}{4!} + \cdots .$$

This definition then gives rise to the alternative set of Bernoulli numbers

$B_1^* = 1/6$, $B_2^* = 1/30$, $B_3^* = 1/42$, $B_4^* = 1/30$, $B_5^* = 5/66$,
$B_6^* = 691/2730$, $B_7^* = 7/6$, $B_8^* = 3617/510, \ldots$.

These differences in notation must also be taken into account when using the following relationships that exist between the Bernoulli and Euler polynomials:

$$B_n(x) = \frac{1}{2^n} \sum_{k=0}^{n} \binom{n}{k} B_{n-k} E_k(2x) \qquad n = 0,1, \ldots$$

$$E_{n-1}(x) = \frac{2^n}{n}\left\{B_n\left(\frac{x+1}{2}\right) - B_n\left(\frac{x}{2}\right)\right\}$$

or

$$E_{n-1}(x) = \frac{2}{n}\left\{B_n(x) - 2^n B_n\left(\frac{x}{2}\right)\right\} \qquad n = 1,2,\ldots$$

and

$$E_{n-2}(x) = 2\binom{n}{2}^{-1}\sum_{k=0}^{n-2}\binom{n}{k}(2^{n-k}-1)B_{n-k}B_k(x) \qquad n = 2,3,\ldots .$$

There are also alternative definitions of the Euler polynomials and it should be noted that some authors, using a modification of the third expression above, call

$$\left(\frac{2}{n+1}\right)\left\{B_n(x) - 2^n B_n\left(\frac{x}{2}\right)\right\}$$

the n-th Euler polynomial.

Elliptic Functions and Elliptic Integrals

The following notations are often used in connection with the inverse elliptic functions $\mathrm{sn}\,u$, $\mathrm{cn}\,u$ and $\mathrm{dn}\,u$:

$$\mathrm{ns}\,u = \frac{1}{\mathrm{sn}\,u} \qquad \mathrm{nc}\,u = \frac{1}{\mathrm{cn}\,u} \qquad \mathrm{nd}\,u = \frac{1}{\mathrm{dn}\,u}$$

$$\mathrm{sc}\,u = \frac{\mathrm{sn}\,u}{\mathrm{cn}\,u} \qquad \mathrm{cs}\,u = \frac{\mathrm{cn}\,u}{\mathrm{sn}\,u} \qquad \mathrm{ds}\,u = \frac{\mathrm{dn}\,u}{\mathrm{sn}\,u}$$

$$\mathrm{sd}\,u = \frac{\mathrm{sn}\,u}{\mathrm{dn}\,u} \qquad \mathrm{cd}\,u = \frac{\mathrm{cn}\,u}{\mathrm{dn}\,u} \qquad \mathrm{dc}\,u = \frac{\mathrm{dn}\,u}{\mathrm{cn}\,u}$$

The following elliptic integral of the third kind is defined by Ryzhik and Gradshteyn to be

$$\Pi(\varphi,n,k) = \int_0^{\varphi}\frac{da}{(1+n\sin^2 a)\sqrt{1-k^2\sin^2 a}} = \int_0^{\sin\varphi}\frac{dx}{(1+nx^2)\sqrt{(1-x^2)(1-k^2x^2)}}.$$

Other authors use the definition

$$\Pi(\varphi,n^2,k) = \int_0^{\varphi}\frac{da}{(1-n^2\sin^2 a)\sqrt{1-k^2\sin^2 a}}$$

$$= \int_0^{\sin\varphi}\frac{dx}{(1-n^2x^2)\sqrt{(1-x^2)(1-k^2x^2)}} \qquad (-\infty < n^2 < \infty).$$

The Jacobi Zeta Function and Theta Functions

The Jacobi zeta function $\operatorname{zn}(u,k)$, frequently written $Z(u)$, is defined by the relation

$$\operatorname{zn}(u,k) = Z(u) = \int_0^u \left\{\operatorname{dn}^2 v - \frac{E}{K}\right\} dv = E(u) - \frac{E}{K}\, u .$$

This is related to the theta functions by the relationship

$$\operatorname{zn}(u,k) = \frac{\partial}{\partial u} \ln \Theta(u)$$

giving

$$\text{(i)} \quad \operatorname{zn}(u,k) = \frac{\pi}{2K} \frac{\vartheta_1'\left(\frac{\pi u}{2K}\right)}{\vartheta_1\left(\frac{\pi u}{2K}\right)} - \frac{\operatorname{cn} u \operatorname{dn} u}{\operatorname{sn} u}$$

$$\text{(ii)} \quad \operatorname{zn}(u,k) = \frac{\pi}{2K} \frac{\vartheta_2'\left(\frac{\pi u}{2K}\right)}{\vartheta_2\left(\frac{\pi u}{2K}\right)} + \frac{\operatorname{dn} u \operatorname{sn} u}{\operatorname{cn} u}$$

$$\text{(iii)} \quad \operatorname{zn}(u,k) = \frac{\pi}{2K} \frac{\vartheta_3'\left(\frac{\pi u}{2K}\right)}{\vartheta_3\left(\frac{\pi u}{2K}\right)} - k^2 \frac{\operatorname{sn} u \operatorname{cn} u}{\operatorname{dn} u}$$

$$\text{(iv)} \quad \operatorname{zn}(u,k) = \frac{\pi}{2K} \frac{\vartheta_4'\left(\frac{\pi u}{2K}\right)}{\vartheta_4\left(\frac{\pi u}{2K}\right)} .$$

Many different notations for the theta function are in current use. The most common variants are the replacement of the argument u by the argument u/π and, occasionally, a permutation of the identification of the functions ϑ_1 to ϑ_4 with the function ϑ_4 replaced by ϑ.

The Factorial (Gamma) Function

In older reference texts the gamma function $\Gamma(z)$, defined by the Euler integral

$$\Gamma(z) = \int_0^\infty t^{z-1} e^{-t}\, dt ,$$

is sometimes expressed in the alternative notation

$$\Gamma(1 + z) = z! = \Pi(z).$$

On occasions the related derivative of the logarithmic factorial function $\Psi(z)$ is used where

$$\frac{d(\ln z!)}{dz} = \frac{(z!)'}{z!} = \Psi(z+1).$$

This function satisfies the recurrence relation

$$\Psi(z) = \Psi(z-1) + \frac{1}{z-1}$$

and is defined by the series

$$\Psi(z) = -C + \sum_{n=0}^{\infty}\left(\frac{1}{n+1} - \frac{1}{z+n}\right).$$

The derivative $\Psi'(z)$ satisfies the recurrence relation

$$-\Psi'(z-1) = -\Psi'(z) + \frac{1}{z^2}$$

and is defined by the series

$$\Psi'(z) = \sum_{n=0}^{\infty} \frac{1}{(z+n)^2}.$$

Exponential and Related Integrals

The exponential integrals $E_n(z)$ have been defined by Schloemilch using the integral

$$E_n(z) = \int_0^{\infty} e^{-zt}\, t^{-n}\, dt \qquad (n = 0,1,\ldots,\ \mathrm{Re}\, z > 0).$$

They should not be confused with the Euler polynomials already mentioned. The function $E_1(z)$ is related to the exponential integral $\mathrm{Ei}(z)$ and to the logarithmic integral $\mathrm{li}(z)$ through the expressions

$$E_1(z) = -\mathrm{Ei}(-z) = \int_z^{\infty} e^{-t}\, t^{-1}\, dt$$

and

$$\mathrm{li}(z) = \int_0^z \frac{dt}{\ln t} = \mathrm{Ei}(\ln z) \qquad (z > 1).$$

The functions $E_n(z)$ satisfy the recurrence relations

$$E_n(z) = \frac{1}{n-1}\left\{e^{-z} - zE_{n-1}(z)\right\}, \qquad (n > 1)$$

and

$$E_n'(z) = -E_{n-1}(z)$$

with

$$E_0(z) = e^{-z}/z.$$

The function $E_n(z)$ has the asymptotic expansion

$$E_n(z) \sim \frac{e^{-z}}{z}\left\{1 - \frac{n}{z} + \frac{n(n+1)}{z^2} - \frac{n(n+1)(n+2)}{z^3} + \cdots\right\}$$

$$(z \gg 1 \text{ and } n > 0),$$

while for large n,

$$E_n(x) = \frac{e^{-x}}{x+n}\left\{1 + \frac{n}{(x+n)^2} + \frac{n(n-2x)}{(x+n)^4} + \frac{n(6x^2 - 8nx + n^2)}{(x+n)^6} + R(n,x)\right\},$$

where

$$-0.36\,n^{-4} \le R(n,x) \le \left(1 + \frac{1}{x+n-1}\right)n^{-4} \qquad (x > 0).$$

The sine and cosine integrals $\mathrm{si}(x)$ and $\mathrm{ci}(x)$ are related to the functions $\mathrm{Si}(x)$ and $\mathrm{Ci}(x)$ by the integrals

$$\mathrm{Si}(x) = \int_0^x \frac{\sin t}{t}\,dt = \mathrm{si}(x) + \pi/2$$

and

$$\mathrm{Ci}(x) = \mathbf{C} + \ln x + \int_0^x \frac{(\cos t - 1)}{t}\,dt.$$

The hyperbolic sine and cosine integrals Shi (x) and Chi (x) are defined by the relations

$$\operatorname{Shi}(x) = \int_0^x \frac{\sinh t}{t}\,dt$$

and

$$\operatorname{Chi}(x) = \mathbf{C} + \ln x + \int_0^x \frac{(\cosh t - 1)}{t}\,dt.$$

Some authors write

$$\operatorname{Cin}(x) = \int_0^x \frac{(1 - \cos t)}{t}\,dt$$

when

$$\operatorname{Cin}(x) = -\operatorname{Ci}(x) + \ln x + \mathbf{C}.$$

The error function erf (x) is defined by the relation

$$\operatorname{erf}(x) = \Phi(x) = \frac{2}{\sqrt{\pi}} \int_0^x e^{-t^2}\,dt$$

and the complementary error function erfc (x) is related to the error function erf (x) and to $\Phi(x)$ by the expression

$$\operatorname{erfc}(x) = 1 - \operatorname{erf}(x).$$

The Fresnel integrals $S(x)$ and $C(x)$ are defined by Ryzhik and Gradshteyn as

$$S(x) = \frac{2}{\sqrt{2\pi}} \int_0^x \sin t^2\,dt$$

and

$$C(x) = \frac{2}{\sqrt{2\pi}} \int_0^x \cos t^2\,dt.$$

Other definitions that are in use are

$$S_1(x) = \int_0^x \sin \frac{\pi t^2}{2}\,dt, \qquad C_1(x) = \int_0^x \cos \frac{\pi t^2}{2}\,dt$$

and

$$S_2(x) = \frac{1}{\sqrt{2\pi}} \int_0^x \frac{\sin t}{\sqrt{t}}\, dt\,, \qquad C_2(x) = \frac{1}{\sqrt{2\pi}} \int_0^x \frac{\cos t}{\sqrt{t}}\, dt$$

These are related by the expressions

$$S(x) = S_1\left(x\sqrt{\frac{2}{\pi}}\right) = S_2\left(x^2\sqrt{\frac{\pi}{2}}\right)$$

and

$$C(x) = C_1\left(x\sqrt{\frac{2}{\pi}}\right) = C_2\left(x^2\sqrt{\frac{\pi}{2}}\right)$$

Hermite and Chebyshev Orthogonal Polynomials

The Hermite polynomials $H_n(x)$ are related to the Hermite polynomials $He_n(x)$ by the relations

$$He_n(x) = 2^{-n/2} H_n\left(\frac{x}{\sqrt{2}}\right)$$

and

$$H_n(x) = 2^{n/2} He_n(x\sqrt{2})\,.$$

These functions satisfy the differential equations

$$\frac{d^2 H_n}{dx^2} - 2x\frac{dH_n}{dx} + 2n\, H_n = 0$$

and

$$\frac{d^2 He_n}{dx^2} - x\frac{dHe_n}{dx} + n He_n = 0\,.$$

They obey the recurrence relations

$$H_{n+1} = 2x\, H_n - 2n\, H_{n-1}$$

and

$$He_{n+1} = x\, He_n - n\, He_{n-1}\,.$$

The first six orthogonal polynomials He_n are

$He_0 = 1$, $He_1 = x$, $He_2 = x^2 - 1$, $He_3 = x^3 - 3x$, $He_4 = x^4 - 6x^2 + 3$, $He_5 = x^5 - 10x^3 + 15x$.

Sometimes the Chebyshev polynomial $U_n(x)$ of the second kind is defined as a solution of the equation

$$(1 - x^2)\frac{d^2y}{dx^2} - 3x\frac{dy}{dx} + n(n+2)y = 0.$$

Bessel Functions

A variety of different notations for Bessel functions are in use. Some common ones involve the replacement of $Y_n(z)$ by $N_n(z)$ and the introduction of the symbol

$$\Lambda_n(z) = \left(\frac{1}{2}z\right)^{-n}\Gamma(n+1)\,J_n(z).$$

In the book by Gray, Mathews and MacRobert the symbol $Y_n(z)$ is used to denote $\frac{1}{2}\pi Y_n(z) + (\ln 2 - C)J_n(z)$ while Neumann uses the symbol $Y^{(n)}(z)$ for the identical quantity.

The Hankel functions $H_\nu^{(1)}(z)$ and $H_\nu^{(2)}(z)$ are sometimes denoted by $Hs_\nu(z)$ and $Hi_\nu(z)$ and some authors write $G_\nu(z) = \left(\frac{1}{2}\right)\pi i\, H_\nu^{(1)}(z)$.

The Neumann polynomial $O_n(t)$ is a polynomial of degree $n+1$ in $1/t$, with $O_o(t) = 1/t$. The polynomials $O_n(t)$ are defined by the generating function

$$\frac{1}{t-z} = J_o(z)\,O_o(t) + 2\sum_{k=1}^{\infty} J_k(z)\,O_k(t),$$

giving

$$O_n(t) = \frac{1}{4}\sum_{k=0}^{\left[\frac{1}{2}n\right]} \frac{n(n-k-1)!}{k!}\left(\frac{2}{t}\right)^{n-2k+1} \qquad \text{for } n = 1,2,\ldots,$$

where $[\frac{1}{2}n]$ signifies the integral part of $\frac{1}{2}n$. The following relationship holds between three successive polynomials:

$$(n-1)\,O_{n+1}(t) + (n+1)\,O_{n-1}(t) - \frac{2(n^2-1)}{t}\,O_n(t) = \frac{2n}{t}\sin^2\frac{n\pi}{2}.$$

The Airy functions $\mathrm{Ai}(z)$ and $\mathrm{Bi}(z)$ are independent solutions of the equation

$$\frac{d^2u}{dz^2} - zu = 0.$$

These solutions can be represented in terms of Bessel functions by the expressions

$$Ai(z) = \frac{1}{3}\sqrt{z}\left\{I_{-1/3}\left(\frac{2}{3}z^{3/2}\right) - I_{1/3}\left(\frac{2}{3}z^{3/2}\right)\right\},$$

$$Ai(-z) = \frac{1}{3}\sqrt{z}\left\{J_{1/3}\left(\frac{2}{3}z^{3/2}\right) + J_{-1/3}\left(\frac{2}{3}z^{3/2}\right)\right\}$$

and by

$$Bi(z) = \sqrt{\frac{z}{3}}\left\{I_{-1/3}\left(\frac{2}{3}z^{3/2}\right) + I_{1/3}\left(\frac{2}{3}z^{3/2}\right)\right\},$$

$$Bi(-z) = \sqrt{\frac{z}{3}}\left\{J_{-1/3}\left(\frac{2}{3}z^{3/2}\right) - J_{1/3}\left(\frac{2}{3}z^{3/2}\right)\right\}.$$

Parabolic Cylinder Functions and Whittaker Functions

The differential equation

$$\frac{d^2y}{dz^2} + (az^2 + bz + c)y = 0$$

has associated with it the two equations

$$\frac{d^2y}{dz^2} + \left(\frac{1}{4}z^2 + a\right)y = 0 \qquad \text{and} \qquad \frac{d^2y}{dz^2} - \left(\frac{1}{4}z^2 + a\right)y = 0$$

the solutions of which are parabolic cylinder functions. The first equation can be derived from the second by replacing z by $ze^{i\pi/4}$ and a by $-ia$.

The solutions of the equation

$$\frac{d^2y}{dz^2} - \left(\frac{1}{4}z^2 + a\right)y = 0$$

are sometimes written $U(a,z)$ and $V(a,z)$. These solutions are related to Whittaker's function $D_p(z)$ by the expressions

$$U(a,z) = D_{-a-\frac{1}{2}}(z)$$

and

$$V(a,z) = \frac{1}{\pi}\Gamma\left(\frac{1}{2} + a\right)\left\{D_{-a-\frac{1}{2}}(-z) + (\sin\pi a)D_{-a-\frac{1}{2}}(z)\right\}.$$

Mathieu Functions

There are several accepted notations for Mathieu functions and for their associated parameters. The defining equation used by Ryzhik and Gradshteyn is

$$\frac{d^2y}{dz^2} + (a - 2k^2 \cos 2z)y = 0 \qquad \text{with } k^2 = q.$$

Different notations involve the replacement of a and q in this equation by h and θ, λ and h^2 and b and $c = 2\sqrt{q}$, respectively. The periodic solutions $se_n(z,q)$ and $ce_n(z,q)$ and the modified periodic solutions $Se_n(z,q)$ and $Ce_n(z,q)$ are suitably altered and, sometimes, re-normalized. A description of these relationships together with the normalizing factors is contained in: Tables relating to Mathieu functions. National Bureau of Standards, Columbia University Press, New York, 1951.

INDEX OF SPECIAL FUNCTIONS AND NOTATIONS

Notation	Name of the function and the number of the formula containing its definition	
$\mathrm{am}\,(u, k)$	Amplitude (of an elliptic function)	8.141
B_n	Bernoulli numbers	9.61, 9.71
$B_n(x)$	Bernoulli polynomials	9.620
$\mathrm{B}(x, y)$	Beta functions	8.38
$\mathrm{B}_x(p, q)$	Incomplete beta functions	8.39
$\beta(x)$		8.37
$\mathrm{bei}\,(z)$, $\mathrm{ber}\,(z)$	Thomson's functions	8.56
$\boldsymbol{C}$	Euler's constant	9.73, 8.367
$C(x)$	Fresnel's cosine-integral	8.25
$C_n^{\lambda}(t)$	Gegenbauer's polynomials	8.93
$C_{\nu}^{\lambda}(x)$	Gegenbauer's function	8.932 1
$\mathrm{ce}_{2n}(z, q)$, $\mathrm{ce}_{2n+1}(z, q)$	Periodic Mathieu functions (Mathieu functions of the first kind)	8.61
$\mathrm{Ce}_{2n}(z, q)$, $\mathrm{Ce}_{2n+1}(z, q)$	Associated (modified) Mathieu functions of the first kind	8.63
$\mathrm{chi}\,(x)$	Hyperbolic-cosine-integral function	8.22
$\mathrm{ci}\,(x)$	Cosine-integral	8.23
$\mathrm{cn}\,(u)$	Cosine-amplitude	8.14
$D(k) \equiv D$		8.112
$D(\varphi, k)$		8.111
$\mathrm{D}_n(z)$, $\mathrm{D}_p(z)$	Parabolic cylinder functions	9.24—9.25
$\mathrm{dn}\,u$	Delta amplitude	8.14
e_1, e_2, e_3		8.162
E_n	Euler numbers	9.63, 9.72
$E(\varphi, k)$	Elliptic integral of the second kind	8.11—8.12
$\left.\begin{matrix} \boldsymbol{E}(k) = \boldsymbol{E} \\ \boldsymbol{E}(k') = \boldsymbol{E}' \end{matrix}\right\}$	Complete elliptic integral of the second kind	8.11—8.12
$E(p;\ a_r:\ q;\ \varrho_s:\ x)$	MacRobert's function	9.4
$\mathrm{E}_{\nu}(z)$	Weber's function	8.58
$\mathrm{Ei}\,(z)$	Exponential-integral function	8.21
$\mathrm{Erfc}\,(x) = 1 - \Phi(x)$	See probability integral	8.25
$\zeta(u)$	Weierstrass's zeta function	8.17
$\left.\begin{matrix} \zeta(s) \\ \zeta(z, a) \end{matrix}\right\}$	Riemann's zeta function	9.51—9.54
$F(\varphi, k)$	Elliptic integral of the first kind	8.11—8.12
${}_pF_q(\alpha_1, \ldots, \alpha_p;\ \beta_1, \ldots, \beta_q;\ z)$	Generalized hypergeometric series	9.14
${}_2F_1(\alpha, \beta;\ \gamma;\ z) = F(\alpha, \beta;\ \gamma;\ z)$	Gauss' hypergeometric function	9.10—9.13
${}_1F_1(\alpha;\ \gamma;\ z) = \Phi(\alpha;\ \gamma;\ z)$	Degenerate hypergeometric function	9.21

continued

Notation	Name of the function and the number of the formula containing its definition	
$F_A(\alpha; \beta_1, \ldots, \beta_n; \gamma_1, \ldots, \gamma_n; z_1, \ldots, z_n)$	Hypergeometric function of several variables	9.19
F_1, F_2, F_3, F_4	Hypergeometric functions of two variables	9.18
$\mathrm{fe}_n(z, q), \mathrm{Fe}_n(z, q) \ldots$ $\mathrm{Fey}_n(z, q), \mathrm{Fek}_n(z, q) \ldots$	Other nonperiodic solutions of Mathieu's equation	8.64 8.663
$\boldsymbol{G}$	Catalan's constant	9.73
g_2, g_3	Invariants of the $\wp(u)$-function	8.161
$\mathrm{gd}\, x$	Gudermannian	1.49
$\mathrm{ge}_n(z, q), \mathrm{Ge}_n(z, q)$ $\mathrm{Gey}_n(z, q), \mathrm{Gek}_n(z, q)$	Other nonperiodic solutions of Mathieu's equation	8.64 8.663
$\Gamma(z)$	Gamma function	8.31—8.33
$\gamma(a, x), \Gamma(a, x)$	Incomplete gamma function	8.35
$G^{m, n}_{p, q}\left(x \middle\vert \begin{matrix} a_1, \ldots, a_p \\ b_1, \ldots, b_q \end{matrix}\right)$	Meijer's functions	9.3
$\mathrm{hei}_\nu(z)$ $\mathrm{her}_\nu(z)$	Thomson's functions	8.56
$H_0^{(1)}(z)$ $H_0^{(2)}(z)$		8.473, 8.531
$H_1^{(1)}(z), H_1^{(2)}(z)$		8.473
$H_\nu^{(1)}(z), H_\nu^{(2)}(z)$	Hankel's functions of the first and second kinds	8.405, 8.42
$H(u) = \vartheta_1\left(\frac{\pi u}{2\boldsymbol{K}}\right)$		8.192
$H_1(u) = \vartheta_2\left(\frac{\pi u}{2\boldsymbol{K}}\right)$		8.192
$H_n(z)$	Hermite polynomials	8.95
$\mathbf{H}_\nu(z)$	Struve functions	8.55
$I_\nu(z)$	Bessel functions of an imaginary argument	8.406, 8.43
$I_x(p, q)$	Incomplete beta function	8.39
$J_\nu(z)$	Bessel function	8.402, 8.41
$\mathbf{J}_\nu(z)$	Anger's function	8.58
$\boldsymbol{K}(k) = \boldsymbol{K}, \boldsymbol{K}(k') = \boldsymbol{K}'$	Complete elliptic integral of the first kind	8.11—8.12
$K_\nu(z)$	Bessel functions of imaginary argument	8.407, 8.43
$\mathrm{kei}(z), \mathrm{ker}(z)$	Thomson's functions	8.56
$\xi(s)$		9.56
$L(x)$	Lobachevskiy's function	8.26
$\mathbf{L}_\nu(z)$	Modified Struve function	8.55
$L_n^\alpha(z)$	Laguerre polynomials	8.97
$\mathrm{li}(x)$	Logarithm-integral	8.24
$\lambda(x, y)$		9.640
$M_{\lambda, \mu}(z)$	Whittaker's functions	9.22,9.23
$\mu(x, \beta)$		9.640
$N_\nu(z)$	Neumann's functions	8.403, 8.41
$\nu(x)$		9.640
$\nu(x, \alpha)$		9.640
$O_n(x)$	Neumann's polynomials	8.59
$\wp(u)$	Weierstrass elliptic function	8.16
$P_\nu^\mu(z), \mathrm{P}_\nu^\mu(x)$	Associated Legendre functions of the first kind	8.7, 8.8
$P_\nu(z), \mathrm{P}_n(x)$	Legendre functions and polynomials	8.82, 8.83, 8.91
$P\begin{Bmatrix} a & b & c & \\ \alpha & \beta & \gamma & z \\ \alpha' & \beta' & \gamma' & \end{Bmatrix}$	Riemann's differential equation (diagram)	9.160

(continued)

Notation	Name of the function and the number of the formula containing its definition	
$P_n^{(\alpha,\beta)}(x)$	Jacobi's polynomials	8.96
$\Pi(x)$	Lobachevskiy's angle of parallelism	1.48
$\Pi(\varphi, n, k)$	Elliptic integral of the third kind	8.11
$\Phi(x)$	Probability integral	8.25
$\Phi(z, s, v)$		9.55
$\Phi(\alpha, \gamma; x) = {}_1F_1(\alpha; \gamma; x)$		9.21
$\Phi_1(\alpha, \beta, \gamma, x, y)$, $\Phi_2(\beta, \beta', \gamma, x, y)$, $\Phi_3(\beta, \gamma, x, y)$	Degenerate hypergeometric series in two variables	9.26
$\psi(x)$	Euler's psi function	8.36
$\Psi(a, c; x)$	Degenerate hypergeometric function	9.21
$Q_\nu^\mu(z)$, $\mathrm{Q}_\nu^\mu(x)$	Ass. Legendre func. of the 2nd kind	8.7, 8.8
$Q_\nu(z)$, $\mathrm{Q}_\nu(x)$	Legendre functions of the second kind	8.82, 8.83
$S(x)$	Fresnel's sine-integral	8.25
$S_n(x)$	Schläfli's polynomials	8.59
$s_{\mu,\nu}(z)$, $S_{\mu,\nu}(z)$	Lommel's functions	8.57
$\mathrm{se}_{2n+1}(z, q)$, $\mathrm{se}_{2n+2}(z, q)$	Periodic Mathieu functions	8.61
$\mathrm{Se}_{2n+1}(z, q)$, $\mathrm{Se}_{2n+2}(z, q)$	Mathieu functions of an imaginary argument	8.63
$\mathrm{shi}(x)$	Hyperbolic-sine-integral	8.22
$\mathrm{si}(x)$	Sine-integral	8.23
$\mathrm{sn}\, u$	Sine-amplitude	8.14
$\sigma(u)$	Weierstrass' sigma function	8.17
$T_n(x)$	Chebyshev polynom. of the 1st kind	8.94
$\Theta(u)$, $\Theta_1(u)$	Jacobi's theta function	8.191—8.196
$\Theta_1(u) = \vartheta_3\left(\frac{\pi u}{2\boldsymbol{K}}\right)$		8.192
$\Theta(u) = \vartheta_4\left(\frac{\pi u}{2\boldsymbol{K}}\right)$		8.192
$\vartheta_0(v \mid \tau) = \vartheta_4(v \mid \tau)$, $\vartheta_1(v \mid \tau)$, $\vartheta_2(v \mid \tau)$, $\vartheta_3(v \mid \tau)$	Elliptic theta functions	8.18, 8.19
$U_n(x)$	Chebyshev polynom. of the 2nd kind	8.94
$U_\nu(w, z)$, $V_\nu(w, z)$	Lommel's functions of two variables	8.57
$W_{\lambda,\mu}(z)$	Whittaker's function	9.22, 9.23
$Z_\nu(z)$	Bessel functions	8.401

NOTATIONS

The letter k (when not used as an index of summation) denotes a number in the interval $[0, 1]$. This notation is used in integrals that lead to elliptic integrals. In such a connection, the number $\sqrt{1-k^2}$ is denoted by k'.

$R(x)$ — A rational function

$\operatorname{Re} z \equiv x, \ \operatorname{Im} z \equiv y$ — The real and imaginary parts of the complex number $z = x + iy$.

$\bar{z} = x - iy$ — The complex conjugate of $z = x + iy$.

$\arg z$ — The argument of the complex number $z = x + iy$.

$\operatorname{sign} x$ — The sign (signum) of the real number x; $\operatorname{sign} x = +1$ for $x > 0$; $\operatorname{sign} x = -1$ for $x < 0$.

$E(x)$ — The integral part of the real number x.

$\int_a^{(b+)}$ $\int_a^{(b-)}$ — Contour integrals; the path of integration starting at the point a extends to the point b (along a straight line unless there is an indication to the contrary), encircles the point b along a small circle in the positive (negative) direction, and returns to the point a, proceeding along the original path in the opposite direction.

$\int_C$ — Line integral along the curve C.

$n!$ $= 1 \cdot 2 \cdot 3 \ldots n, \quad 0! = 1.$

$(2n+1)!!$ $= 1 \cdot 3 \ldots (2n+1).$

$(2n)!!$ $= 2 \cdot 4 \ldots (2n).$

$\binom{p}{n}$ $= \dfrac{p(p-1)\ldots(p-n+1)}{1 \cdot 2 \ldots n}, \quad \binom{p}{0} = 1.$

$(a)_n$ $= a(a+1)\ldots(a+n-1) = \dfrac{\Gamma(a+n)}{\Gamma(a)}.$

$\sum_{k=m}^{n} u_k$ $= u_m + u_{m+1} + \ldots + u_n$. If $n < m$, we define $\sum_{k=m}^{n} u_k = 0$.

$\sum_n{}'$, $\sum_{m,n}{}'$ — Summation over all integral values of n excluding $n = 0$, and summation over all integral values of n and m excluding $m = n = 0$, respectively.

$O(f(z))$ — The order of the function $f(z)$. Suppose that the point z approaches z_0. If there exists an $M > 0$ such that $|g(z)| \leqslant M|f(z)|$ in some sufficiently small neighborhood of the point z_0, we write $g(z) = O(f(z))$.

NOTE ON THE BIBLIOGRAPHIC REFERENCES

The letters and numbers following equations refer to the sources used by Russian editors. The key to the letters will be found preceding each entry in the Bibliography beginning on page 1081. Roman numerals indicate the volume number of a multivolume work. Numbers without parentheses indicate page numbers, numbers in single parentheses refer to equation numbers in the original sources, and numbers in double parentheses denote the number of a table in the source.

Some formulas were changed from their form in the source material. In such cases, the letter *a* appears at the end of the bibliographic references.

As an example we may use the reference to equation 3.354-5:

ET I 118 (1) *a*.

The key on page 1081 indicates that the book referred to is:

Erdelyi, A. et al., *Tables of Integral Transforms.*

The Roman numeral denotes volume one of the work, 118 is the page on which the formula will be found, (1) refers to the number of the formula in this source, and the *a* indicates that the expression appearing in the source differs in some respect from the formula in this book.

In several cases the editors have used Russian editions of works published in other languages, under such circumstances, because the pagination and numbering of equations may be altered, we have referred the reader only to the original sources and dispensed with page and equation numbers.

0. INTRODUCTION

0.1 Finite Sums

0.11 Progressions

0.111 Arithmetic progression.

$$\sum_{k=0}^{n-1} (a+kr) = \frac{n}{2}\left[2a+(n-1)r\right] = \frac{n}{2}(a+l) \quad [l\text{—the last term}]$$

0.112 Geometric progression.

$$\sum_{k=1}^{n} aq^{k-1} = \frac{a(q^n-1)}{q-1}.$$

0.113 Arithmetico-geometric progression.

$$\sum_{k=0}^{n-1} (a+kr)q^k = \frac{a-[a+(n-1)r]q^n}{1-q} + \frac{rq(1-q^{n-1})}{(1-q)^2}.$$ JO (5)

0.12 Sums of powers of natural numbers

0.121

$$\sum_{k=1}^{n} k^q = \frac{n^{q+1}}{q+1} + \frac{n^q}{2} + \frac{1}{2}\binom{q}{1}B_2n^{q-1} + \frac{1}{4}\binom{q}{3}B_4n^{q-3} + \frac{1}{6}\binom{q}{5}B_6n^{q-5} + \ldots =$$

$$= \frac{n^{q+1}}{q+1} + \frac{n^q}{2} + \frac{qn^{q-1}}{12} - \frac{q(q-1)(q-2)}{720}n^{q-3} + \frac{q(q-1)(q-2)(q-3)(q-4)}{30{,}240}n^{q-5} - \ldots$$

[last term contains either n or n^2]. CE 332

1. $\sum_{k=1}^{n} k = \frac{n(n+1)}{2}.$ CE 333

2. $\sum_{k=1}^{n} k^2 = \frac{n(n+1)(2n+1)}{6}.$ CE 333

3. $\sum_{k=1}^{n} k^3 = \left[\frac{n(n+1)}{2}\right]^2.$ CE 333

4. $\sum_{k=1}^{n} k^4 = \frac{1}{30}n(n+1)(2n+1)(3n^2+3n-1).$ CE 333

5. $\sum_{k=1}^{n} k^5 = \frac{1}{12} n^2 (n+1)^2 (2n^2 + 2n - 1).$ CE 333

6. $\sum_{k=1}^{n} k^6 = \frac{1}{42} n (n+1)(2n+1)(3n^4 + 6n^3 - 3n + 1).$ CE 333

7. $\sum_{k=1}^{n} k^7 = \frac{1}{24} n^2 (n+1)^2 (3n^4 + 6n^3 - n^2 - 4n + 2).$ CE 333

0.122 $$\sum_{k=1}^{n} (2k-1)^q = \frac{2^q}{q+1} n^{q+1} - \frac{1}{2}\binom{q}{1} 2^{q-1} B_2 n^{q-1} - \\ - \frac{1}{4}\binom{q}{3} 2^{q-3}(2^3 - 1) B_4 n^{q-3} - \ldots$$

[last term contains either n or n^2.]

1. $\sum_{k=1}^{n} (2k-1) = n^2.$

2. $\sum_{k=1}^{n} (2k-1)^2 = \frac{1}{3} n (4n^2 - 1).$ JO (32a)

3. $\sum_{k=1}^{n} (2k-1)^3 = n^2 (2n^2 - 1).$ JO (32b)

0.123 $\sum_{k=1}^{n} k(k+1)^2 = \frac{1}{12} n (n+1)(n+2)(3n+5).$

0.124 $\sum_{k=1}^{q} k(n^2 - k^2) = \frac{1}{4} q (q+1)(2n^2 - q^2 - q).$

0.125 $\sum_{k=1}^{n} k! \cdot k = (n+1)! - 1.$ AD (188.1)

0.126 $\sum_{k=1}^{n} \frac{(n+k)!}{k!\,(n-k)!} = \sqrt{\frac{e}{\pi}}\, K_{n+\frac{1}{2}}\left(\frac{1}{2}\right).$ WA 94

0.13 Sums of reciprocals of natural numbers

0.131 $$\sum_{k=1}^{n} \frac{1}{k} = C + \ln n + \frac{1}{2n} - \sum_{k=2}^{\infty} \frac{A_k}{n(n+1)\ldots(n+k-1)},$$

where

$$A_k = \frac{1}{k} \int_0^1 x(1-x)(2-x)(3-x)\ldots(k-1-x)\,dx.$$

$$A_2 = \frac{1}{12}, \qquad A_3 = \frac{1}{12},$$

$$A_4 = \frac{19}{80}, \qquad A_5 = \frac{9}{20},$$ JO (59), AD (1876)

0.132 $$\sum_{k=1}^{n} \frac{1}{2k-1} = \frac{1}{2}(C + \ln n) + \ln 2 + \frac{B_2}{8n^2} + \frac{(2^3-1)B_4}{64n^4} + \ldots$$ JO (71a)a

0.133 $$\sum_{k=2}^{n} \frac{1}{k^2-1} = \frac{3}{4} - \frac{2n+1}{2n(n+1)}.$$ JO (184f)

0.14 Sums of products of reciprocals of natural numbers

0.141

1. $$\sum_{k=1}^{n} \frac{1}{[p+(k-1)q](p+kq)} = \frac{n}{p(p+nq)}.$$ GI III (64)a

2. $$\sum_{k=1}^{n} \frac{1}{[p+(k-1)q](p+kq)[p+(k+1)q]} = \frac{n(2p+nq+q)}{2p(p+q)(p+nq)[p+(n+1)q]}.$$ GI III (65)a

3. $$\sum_{k=1}^{n} \frac{1}{[p+(k-1)q](p+kq)\ldots[p+(k+l)q]} = \frac{1}{(l+1)q}\left\{\frac{1}{p(p+q)\ldots(p+lq)} - \right.$$
$$\left. - \frac{1}{(p+nq)[p+(n+1)q]\ldots[p+(n+l)q]}\right\}.$$ AD (1856)a

4. $$\sum_{k=1}^{n} \frac{1}{[1+(k-1)q][1+(k-1)q+p]} =$$
$$= \frac{1}{p}\left[\sum_{k=1}^{n} \frac{1}{1+(k-1)q} - \sum_{k=1}^{n} \frac{1}{1+(k-1)q+p}\right].$$ GI III (66)a

0.142 $$\sum_{k=1}^{n} \frac{k^2+k-1}{(k+2)!} = \frac{1}{2} - \frac{n+1}{(n+2)!}.$$ JO (157)

0.15 Sums of the binomial coefficients

(n is a natural number)

0.151

1. $$\sum_{k=0}^{m} \binom{n+k}{n} = \binom{n+m+1}{n+1}.$$ KR 64 (70.1)

2. $$1 + \binom{n}{2} + \binom{n}{4} + \ldots = 2^{n-1}.$$ KR 62 (58.1)

3. $$\binom{n}{1} + \binom{n}{3} + \binom{n}{5} + \ldots = 2^{n-1}.$$ KR 62 (58.1)

4. $$\sum_{k=0}^{m} (-1)^k \binom{n}{k} = (-1)^m \binom{n-1}{m}.$$ KR 64 (70.2)

0.152

1. $$\binom{n}{0} + \binom{n}{3} + \binom{n}{6} + \ldots = \frac{1}{3}\left(2^n + 2\cos\frac{n\pi}{3}\right).$$ KR 62 (59.1)

2. $\binom{n}{1}+\binom{n}{4}+\binom{n}{7}+\ldots=\frac{1}{3}\left(2^n+2\cos\frac{(n-2)\pi}{3}\right).$ KR 62 (59.2)

3. $\binom{n}{2}+\binom{n}{5}+\binom{n}{8}+\ldots=\frac{1}{3}\left(2^n+2\cos\frac{(n-4)\pi}{3}\right).$ KR 62 (59.3)

0.153

1. $\binom{n}{0}+\binom{n}{4}+\binom{n}{8}+\ldots=\frac{1}{2}\left(2^{n-1}+2^{\frac{n}{2}}\cos\frac{n\pi}{4}\right).$ KR 63 (60.1)

2. $\binom{n}{1}+\binom{n}{5}+\binom{n}{9}+\ldots=\frac{1}{2}\left(2^{n-1}+2^{\frac{n}{2}}\sin\frac{n\pi}{4}\right).$ KR 63 (60.2)

3. $\binom{n}{2}+\binom{n}{6}+\binom{n}{10}+\ldots=\frac{1}{2}\left(2^{n-1}-2^{\frac{n}{2}}\cos\frac{n\pi}{4}\right).$ KR 63 (60.3)

4. $\binom{n}{3}+\binom{n}{7}+\binom{n}{11}+\ldots=\frac{1}{2}\left(2^{n-1}-2^{\frac{n}{2}}\sin\frac{n\pi}{4}\right).$ KR 63 (60.4)

0.154

1. $\sum_{k=0}^{n}(k+1)\binom{n}{k}=2^{n-1}(n+2).$ KR 63 (66.1)

2. $\sum_{k=1}^{n}(-1)^{k+1}k\binom{n}{k}=0.$ KR 63 (66.2)

0.155

1. $\sum_{k=1}^{n}\frac{(-1)^{k+1}}{k+1}\binom{n}{k}=\frac{n}{n+1}.$ KR 63 (67)

2. $\sum_{k=0}^{n}\frac{1}{k+1}\binom{n}{k}=\frac{2^{n+1}-1}{n+1}.$ KR 63 (68.1)

3. $\sum_{k=0}^{n}\frac{\alpha^{k+1}}{k+1}\binom{n}{k}=\frac{(\alpha+1)^{n+1}-1}{n+1}.$ KR 63 (68.2)

4. $\sum_{k=1}^{n}\frac{(-1)^{k+1}}{k}\binom{n}{k}=\sum_{m=1}^{n}\frac{1}{m}.$ KR 64 (69)

0.156

1. $\sum_{k=0}^{p}\binom{n}{k}\binom{m}{p-k}=\binom{n+m}{p}$ [m is a natural number]. KR 64 (71.1)

2. $\sum_{k=0}^{n-p}\binom{n}{k}\binom{n}{p+k}=\frac{(2n)!}{(n-p)!\,(n+p)!}.$ KR 64 (71.2)

0.157

1. $\sum_{k=0}^{n}\binom{n}{k}^2=\binom{2n}{n}.$ KR 64 (72.1)

2. $\sum_{k=0}^{2n}(-1)^k\binom{2n}{k}^2=(-1)^n\binom{2n}{n}.$ KR 64 (72.2)

3. $\sum_{k=0}^{2n+1} (-1)^k \binom{2n+1}{k}^2 = 0.$ KR 64 (72.3)

4. $\sum_{k=1}^{n} k \binom{n}{k}^2 = \frac{(2n-1)!}{[(n-1)!]^2}.$ KR 64 (72.4)

0.2 Numerical Series and Infinite Products

0.21 The convergence of numerical series

The series

0.211 $$\sum_{k=1}^{\infty} u_k = u_1 + u_2 + u_3 + \dots$$

is said to *converge absolutely* if the series

0.212 $$\sum_{k=1}^{\infty} |u_k| = |u_1| + |u_2| + |u_3| + \dots,$$

composed of the absolute values of its terms converges. If the series 0.211 converges and the series 0.212 diverges, the series 0.211 is said to *converge conditionally*. Every absolutely convergent series converges.

0.22 Convergence tests

0.221 Suppose that

$$\lim_{k\to\infty} |u_k|^{\frac{1}{k}} = q.$$

If $q < 1$, the series 0.211 converges absolutely. On the other hand, if $q > 1$, the series 0.211 diverges. (Cauchy)

0.222 Suppose that

$$\lim_{k\to\infty} \left| \frac{u_{k+1}}{u_k} \right| = q.$$

Here, if $q < 1$, the series 0.211 converges absolutely. If $q > 1$, the series 0.211 diverges. If $\left| \frac{u_{k+1}}{u_k} \right|$ approaches 1 but remains greater than unity, the series 0.211 diverges. (d'Alembert)

0.223 Suppose that

$$\lim_{k\to\infty} k \left\{ \left| \frac{u_k}{u_{k+1}} \right| - 1 \right\} = q.$$

Here, if $q > 1$, the series 0.211 converges absolutely. If $q < 1$, the series 0.211 diverges. (Raabe)

0.224 Suppose that $f(x)$ is a positive decreasing function and that

$$\lim_{k\to\infty} \frac{e^k f(e^k)}{f(k)} = q.$$

for natural k. If $q < 1$, the series $\sum_{k=1}^{\infty} f(k)$ converges. If $q > 1$, this series diverges. (Ermakov)

0.225 Suppose that

$$\left| \frac{u_k}{u_{k+1}} \right| = 1 + \frac{q}{k} + \frac{|v_k|}{k^p},$$

where $p > 1$ and the $|v_k|$ are bounded, that is, the $|v_k|$ are all less than some M, which is independent of k. Here, if $q > 1$, the series **0.211** converges absolutely. If $q \leqslant 1$, this series diverges. (Gauss)

0.226 Suppose that a function $f(x)$ defined for $x \geqslant q \geqslant 1$ is continuous, positive, and decreasing. Under these conditions, the series

$$\sum_{k=1}^{\infty} f(k)$$

converges or diverges according as the integral

$$\int_q^{\infty} f(x)\, dx.$$

converges or diverges (the Cauchy integral test.)

0.227 Suppose that all terms of a sequence $u_1, u_2, \ldots, u_n$ are positive. In such a case, the series

1. $$\sum_{k=1}^{\infty} (-1)^{k+1} u_k = u_1 - u_2 + u_3 - \ldots$$

is called an *alternating series*.

If the terms of an alternating series decrease monotonically in absolute value and approach zero, that is, if

2. $$u_{k+1} < u_k \text{ and } \lim_{k \to \infty} u_k = 0,$$

the series **0.227** 1. converges. Here, the remainder of the series is

3. $$\sum_{k=n+1}^{\infty} (-1)^{k-n+1} u_k = \left| \sum_{k=1}^{\infty} (-1)^{k+1} u_k - \sum_{k=1}^{n} (-1)^{k+1} u_k \right| < u_{n+1}.$$

(Leibnitz)

0.228 If the series

1. $$\sum_{k=1}^{\infty} v_k = v_1 + v_2 + \ldots + v_k + \ldots$$

converges and the numbers u_k form a monotonic bounded sequence, that is, if $|u_k| < M$ for some number M and for all k, the series

2. $$\sum_{k=1}^{\infty} u_k v_k = u_1 v_1 + u_2 v_2 + \ldots + u_k v_k + \ldots$$ **FI II 354**

converges. (Abel)

0.229 If the partial sums of the series **0.228** 1. are bounded and if the numbers u_k constitute a monotonic sequence that approaches zero, that is, if

$$\left| \sum_{k=1}^{n} v_k \right| < M \; [n = 1, 2, 3, \ldots] \text{ and } \lim_{k \to \infty} u_k = 0,$$ **FI II 355**

then the series **0.228** 2. converges. (Dirichlet)

0.23-0.24 Examples of numerical series

0.231 Progressions

1. $\sum_{k=0}^{\infty} aq^k = \frac{a}{1-q} \qquad [|q|<1].$

2. $\sum_{k=0}^{\infty} (a+kr)\,q^k = \frac{a}{1-q} + \frac{rq}{(1-q)^2} \qquad [|q|<1]$ (cf. **0.113**).

0.232

1. $\sum_{k=1}^{\infty} (-1)^{k+1}\frac{1}{k} = \ln 2$ (cf. **1.511**).

2. $\sum_{k=1}^{\infty} (-1)^{k+1}\frac{1}{2k-1} = 1 - 2\sum_{k=1}^{\infty}\frac{1}{(4k-1)(4k+1)} = \frac{\pi}{4}$ (cf. **1.643**).

0.233

1. $\sum_{k=1}^{\infty} \frac{1}{k^p} = 1 + \frac{1}{2^p} + \frac{1}{3^p} + \ldots = \zeta(p) \qquad [\operatorname{Re} p > 1]$ WH

2. $\sum_{k=1}^{\infty} (-1)^{k+1}\frac{1}{k^p} = (1-2^{1-p})\,\zeta(p) \qquad [\operatorname{Re} p > 0]$ WH

3. $\sum_{k=1}^{\infty} \frac{1}{k^{2n}} = \frac{2^{2n-1}\pi^{2n}}{(2n)!}\,|B_{2n}|.$ FI II 721

4. $\sum_{k=1}^{\infty} (-1)^{k+1}\frac{1}{k^{2n}} = \frac{(2^{2n-1}-1)\,\pi^{2n}}{(2n)!}\,|B_{2n}|.$ JO (165)

5. $\sum_{k=1}^{\infty} \frac{1}{(2k-1)^{2n}} = \frac{(2^{2n}-1)\,\pi^{2n}}{2\cdot(2n)!}\,|B_{2n}|.$ JO (184b)

6. $\sum_{k=1}^{\infty} (-1)^{k+1}\frac{1}{(2k-1)^{2n+1}} = \frac{\pi^{2n+1}}{2^{2n+2}\,(2n)!}\,|E_{2n}|.$ JO (184d)

0.234

1. $\sum_{k=1}^{\infty} (-1)^{k+1}\frac{1}{k^2} = \frac{\pi^2}{12}.$ EU

2. $\sum_{k=1}^{\infty} \frac{1}{(2k-1)^2} = \frac{\pi^2}{8}.$ EU

3. $\sum_{k=0}^{\infty} \frac{(-1)^k}{(2k+1)^2} = G.$ FI II 482

4. $\sum_{k=1}^{\infty} \frac{(-1)^{k+1}}{(2k-1)^3} = \frac{\pi^3}{32}.$ EU

5. $\sum_{k=1}^{\infty} \frac{1}{(2k-1)^4} = \frac{\pi^4}{96}.$ EU

6. $\sum_{k=1}^{\infty} \frac{(-1)^{k+1}}{(2k-1)^5} = \frac{5\pi^5}{1536}$. EU

7. $\sum_{k=1}^{\infty} (-1)^{k+1} \frac{k}{(k+1)^2} = \frac{\pi^2}{12} - \ln 2.$

0.235 $S_n = \sum_{k=1}^{\infty} \frac{1}{(4k^2-1)^n}$,

$S_1 = \frac{1}{2}, \quad S_2 = \frac{\pi^2-8}{16}, \quad S_3 = \frac{32-3\pi^2}{64}, \quad S_4 = \frac{\pi^4+30\pi^2-384}{768}$. JO (186)

0.236

1. $\sum_{k=1}^{\infty} \frac{1}{k(4k^2-1)} = 2 \ln 2 - 1.$ BR 51a

2. $\sum_{k=1}^{\infty} \frac{1}{k(9k^2-1)} = \frac{3}{2}(\ln 3 - 1).$ Br 51a

3. $\sum_{k=1}^{\infty} \frac{1}{k(36k^2-1)} = -3 + \frac{3}{2} \ln 3 + 2 \ln 2.$ BR 52, AD (6913.3)

4. $\sum_{k=1}^{\infty} \frac{k}{(4k^2-1)^2} = \frac{1}{8}$. BR 52

5. $\sum_{k=1}^{\infty} \frac{1}{k(4k^2-1)^2} = \frac{3}{2} - 2 \ln 2.$ BR 52

6. $\sum_{k=1}^{\infty} \frac{12k^2-1}{k(4k^2-1)^2} = 2 \ln 2.$ AD (6917.3), BR 52

0.237

1. $\sum_{k=1}^{\infty} \frac{1}{(2k-1)(2k+1)} = \frac{1}{2}$. AD (6917.2), BR 52

2. $\sum_{k=1}^{\infty} \frac{1}{(4k-1)(4k+1)} = \frac{1}{2} - \frac{\pi}{8}$.

3. $\sum_{k=2}^{\infty} \frac{1}{(k-1)(k+1)} = \frac{3}{4}$ (cf. 0.133).

4. $\sum_{k=1,\, k\neq m}^{\infty}{}' \frac{1}{(m+k)(m-k)} = -\frac{3}{4m^2}$ [m is an integer]. AD (6916.1)

5. $\sum_{k=1,\, k\neq m}^{\infty}{}' \frac{(-1)^{k-1}}{(m-k)(m+k)} = \frac{3}{4m^2}$ [m is an even number]. AD (6916.2)

0.238

1. $\sum_{k=1}^{\infty} \frac{1}{(2k-1)2k(2k+1)} = \ln 2 - \frac{1}{2}$. GI III (93)

2. $\sum_{k=1}^{\infty} \frac{(-1)^{k+1}}{(2k-1)\,2k\,(2k+1)} = \frac{1}{2}(1-\ln 2).$ GI III (94)a

3. $\sum_{k=0}^{\infty} \frac{1}{(3k+1)(3k+2)(3k+3)(3k+4)} = \frac{1}{6} - \frac{1}{4}\ln 3 + \frac{\pi}{12\sqrt{3}}.$ GI III (95)

0.239

1. $\sum_{k=1}^{\infty} (-1)^{k+1}\frac{1}{3k-2} = \frac{1}{3}\left(\frac{\pi}{\sqrt{3}} + \ln 2\right).$ GI III (85), BR* 161 (1)

2. $\sum_{k=1}^{\infty} (-1)^{k+1}\frac{1}{3k-1} = \frac{1}{3}\left(\frac{\pi}{\sqrt{3}} - \ln 2\right).$ BR* 161 (1)

3. $\sum_{k=1}^{\infty} (-1)^{k+1}\frac{1}{4k-3} = \frac{1}{4\sqrt{2}}\left[\pi + 2\ln\left(\sqrt{2}+1\right)\right].$ BR* 161 (1)

4. $\sum_{k=1}^{\infty} (-1)^{E\left(\frac{k+3}{2}\right)}\frac{1}{k} = \frac{\pi}{4} + \frac{1}{2}\ln 2.$ GI III (87)

5. $\sum_{k=1}^{\infty} (-1)^{E\left(\frac{k+3}{2}\right)}\frac{1}{2k-1} = \frac{\pi}{2\sqrt{2}}.$

6. $\sum_{k=1}^{\infty} (-1)^{E\left(\frac{k+5}{3}\right)}\frac{1}{2k-1} = \frac{5\pi}{12}.$ GI III (88)

7. $\sum_{k=1}^{\infty} \frac{1}{(8k-1)(8k+1)} = \frac{1}{2} - \frac{\pi}{16}\left(\sqrt{2}+1\right).$

0.241

1. $\sum_{k=1}^{\infty} \frac{1}{2^k k} = \ln 2.$ JO (172g)

2. $\sum_{k=1}^{\infty} \frac{1}{2^k k^2} = \frac{\pi^2}{12} - \frac{1}{2}(\ln 2)^2.$ JO (174)

0.242 $\sum_{k=0}^{\infty} (-1)^k \frac{1}{n^{2k}} = \frac{n^2}{n^2+1}.$

0.243

1. $\sum_{k=1}^{\infty} \frac{1}{[p+(k-1)q](p+kq)\dots[p+(k+l)q]} = \frac{1}{(l+1)q}\,\frac{1}{p(p+q)\dots(p+lq)}$

(see also 0.141 3.)

2. $\sum_{k=1}^{\infty} \frac{x^{k-1}}{[p+(k-1)q][p+(k-1)q+1][p+(k-1)q+2]\dots[p+(k-1)q+l-1]} =$

$$= \frac{1}{l!}\int_0^1 \frac{t^{p-1}(1-t)^l}{1-xt^q}\,dt \quad [q>0,\ x^2 \leqslant 1].$$ BR* 161 (2), AD (6.704)

0.244

1. $$\sum_{k=1}^{\infty} \frac{1}{(k+p)(k+q)} = \frac{1}{q-p} \int_0^1 \frac{x^p - x^q}{1-x}\, dx \quad [p > -1,\ q > -1,\ p \neq q].$$ GI III (90)

2. $$\sum_{k=1}^{\infty} (-1)^{k+1} \frac{1}{p+(k-1)q} = \int_0^1 \frac{t^{p-1}}{1+t^q}\, dt \quad [p > 0,\ q > 0].$$ BR* 161 (1)

Summations of reciprocals of factorials

0.245

1. $$\sum_{k=0}^{\infty} \frac{1}{k!} = e = 2.71828 \ldots$$

2. $$\sum_{k=0}^{\infty} \frac{(-1)^k}{k!} = \frac{1}{e} = 0.36787 \ldots$$

3. $$2 \sum_{k=1}^{\infty} \frac{k}{(2k+1)!} = \frac{1}{e} = 0.36787 \ldots$$

4. $$\sum_{k=1}^{\infty} \frac{k}{(k+1)!} = 1.$$

5. $$\sum_{k=0}^{\infty} \frac{1}{(2k)!} = \frac{1}{2}\left(e + \frac{1}{e}\right) = 1.54308 \ldots$$

6. $$\sum_{k=0}^{\infty} \frac{1}{(2k+1)!} = \frac{1}{2}\left(e - \frac{1}{e}\right) = 1.17520 \ldots$$

7. $$\sum_{k=0}^{\infty} \frac{(-1)^k}{(2k)!} = \cos 1 = \cos 57^\circ 17' 45'' = 0.54030 \ldots$$

8. $$\sum_{k=1}^{\infty} \frac{(-1)^{k-1}}{(2k-1)!} = \sin 1 = \sin 57^\circ 17' 45'' = 0.84147 \ldots$$

0.246

1. $$\sum_{k=0}^{\infty} \frac{1}{(k!)^2} = I_0(2) = 2.27958530 \ldots$$

2. $$\sum_{k=0}^{\infty} \frac{1}{k!\,(k+1)!} = I_1(2) = 1.590636855 \ldots$$

3. $$\sum_{k=0}^{\infty} \frac{1}{k!\,(k+n)!} = I_n(2).$$

4. $$\sum_{k=0}^{\infty} \frac{(-1)^k}{(k!)^2} = J_0(2) = 0.22389078 \ldots$$

5. $\sum_{k=0}^{\infty} \frac{(-1)^k}{k!\,(k+1)!} = J_1(2) = 0.57672481\ldots$

6. $\sum_{k=0}^{\infty} \frac{(-1)^k}{k!\,(k+n)!} = J_n(2).$

0.247 $\sum_{k=1}^{\infty} \frac{k!}{(n+k-1)!} = \frac{1}{(n-2)\cdot(n-1)!}.$ JE (159)

0.248 $\sum_{k=1}^{\infty} \frac{k^n}{k!} = S_n,$

$S_1 = e, \quad S_2 = 2e, \quad S_3 = 5e, \quad S_4 = 15e,$
$S_5 = 52e, \quad S_6 = 203e, \quad S_7 = 877e, \quad S_8 = 4140e.$ JE (185)

0.249 $\sum_{k=1}^{\infty} \frac{(k+1)^3}{k!} = 15e.$ JE (76)

0.25 Infinite products

0.250 Suppose that a sequence of numbers $a_1, a_2, \ldots, a_k, \ldots$ is given. If the limit $\lim_{n\to\infty} \prod_{k=1}^{n} (1+a_k)$ exists, whether finite or infinite (but of definite sign), this limit is called the value of the *infinite product* $\prod_{k=1}^{\infty} (1+a_k)$ and we write

1. $$\lim_{n\to\infty} \prod_{k=1}^{n} (1+a_k) = \prod_{k=1}^{\infty} (1+a_k).$$

If an infinite product has a finite *nonzero* value, it is said to converge. Otherwise, the infinite product is said to diverge. FI II 400

0.251 For the infinite product **0.250** 1. to converge, it is necessary that $\lim_{k\to\infty} a_k = 0.$ FI II 403

0.252 If $a_k > 0$ or $a_k < 0$ for all values of the index k starting with some particular value, then, for the product 0.250 1. to converge, it is necessary and sufficient that the series $\sum_{k=1}^{\infty} a_k$ converges.

0.253 The product $\prod_{k=1}^{\infty} (1+a_k)$ is said to converge absolutely if the product $\prod_{k=1}^{\infty} (1+|a_k|)$ converges. FI II 403

0.254 Absolute convergence of an infinite product implies its convergence.

0.255 The product $\prod_{k=1}^{\infty} (1+a_k)$ converges absolutely if, and only if, the series $\sum_{k=1}^{\infty} a_k$ converges absolutely. FI II 406

0.26 Examples of infinite products

0.261 $\prod_{k=1}^{\infty}\left(1+\frac{(-1)^{k+1}}{2k-1}\right)=\sqrt{2}.$ EU

0.262

1. $\prod_{k=2}^{\infty}\left(1-\frac{1}{k^2}\right)=\frac{1}{2}.$ FI II 401

2. $\prod_{k=1}^{\infty}\left(1-\frac{1}{(2k)^2}\right)=\frac{2}{\pi}.$ FI II 401

3. $\prod_{k=1}^{\infty}\left(1-\frac{1}{(2k+1)^2}\right)=\frac{\pi}{4}.$ FI II 401

0.263 $\frac{2}{1}\cdot\left(\frac{4}{3}\right)^{\frac{1}{2}}\left(\frac{6\cdot 8}{5\cdot 7}\right)^{\frac{1}{4}}\left(\frac{10\cdot 12\cdot 14\cdot 16}{9\cdot 11\cdot 13\cdot 15}\right)^{\frac{1}{8}}\ldots=e.$

0.264 $\prod_{k=1}^{\infty}\frac{\sqrt[k]{e}}{1+\frac{1}{k}}=e^{C}.$ FI II 402

0.265 $\sqrt{\frac{1}{2}}\cdot\sqrt{\frac{1}{2}+\frac{1}{2}\sqrt{\frac{1}{2}}}\cdot\sqrt{\frac{1}{2}+\frac{1}{2}\sqrt{\frac{1}{2}+\frac{1}{2}\sqrt{\frac{1}{2}}}}\cdots=\frac{2}{\pi}.$ FI II 402

0.266 $\prod_{k=0}^{\infty}(1+x^{2^k})=\frac{1}{1-x}\qquad[|x|<1].$ FI II 401

0.3 Functional Series

0.30 Definitions and theorems

0.301 The series

1. $$\sum_{k=1}^{\infty} f_k(x),$$

the terms of which are functions, is called a *functional series*. The set of values of the independent variable x for which the series 0.301 1. converges constitutes what is called the *region of convergence* of that series.

0.302 A series that converges for all values of x in a region M is said to *converge uniformly* in that region if, for every $\varepsilon>0$, there exists a number N such that, for $n>N$, the inequality

$$\left|\sum_{k=n+1}^{\infty} f_k(x)\right|<\varepsilon$$

holds for *all* x in M.

0.303 If the terms of the functional series 0.301 1. satisfy the inequalities

$$|f_k(x)|<u_k\quad(k=1, 2, 3, \ldots),$$

throughout the region M, where the u_k are the terms of some *convergent* numerical series

$$\sum_{k=1}^{\infty} u_k=u_1+u_2+\ldots+u_k+\ldots,$$

the series 0.301 1. converges uniformly in M. (Weierstrass)

0.304 Suppose that the series 0.301 1. converges uniformly in a region M and that a set of functions $g_k(x)$ constitutes (for each x) a monotonic sequence, and that these functions are uniformly bounded, that is, suppose that a number L exists such that the inequalities

1. $$|g_n(x)| \leqslant L;$$

hold for all n and x. Then, the series

2. $$\sum_{k=1}^{\infty} f_k(x)\, g_k(x)$$

converges uniformly in the region M. (Abel) FI II 451

0.305 Suppose that the partial sums of the series 0.301 1. are uniformly bounded; that is, suppose that, for some L and for all n and x in M, the inequalities

$$\left|\sum_{k=1}^{n} f_k(x)\right| < L;$$

hold. Suppose also that for each x the functions $g_n(x)$ constitute a monotonic sequence that approaches zero uniformly in the region M. Then, the series 0.304 2. converges uniformly in the region M. (Dirichlet) FI II 451

0.306 If the functions $f_k(x)$ $(k = 1, 2, 3, \ldots)$ are integrable on the interval $[a, b]$ and if the series 0.301 1. made up of these functions converges uniformly on that interval, this series may be integrated *termwise*; that is,

$$\int_a^x \left(\sum_{k=1}^{\infty} f_k(x)\right) dx = \sum_{k=1}^{\infty} \int_a^x f_k(x)\, dx \qquad [a \leqslant x \leqslant b].$$ FI II 459

0.307 Suppose that the functions $f_k(x)$ (for $(k = 1, 2, 3, \ldots)$ have continuous derivatives $f_k'(x)$ on the interval $[a, b]$. If the series 0.301 1. converges on this interval and if the series $\sum_{k=1}^{\infty} f_k'(x)$ of these derivatives converges uniformly, the series 0.301 1. may be differentiated termwise; that is,

$$\left\{\sum_{k=1}^{\infty} f_k(x)\right\}' = \sum_{k=1}^{\infty} f_k'(x).$$ FI II 460

0.31 Power series

0.311 A functional series of the form

1. $$\sum_{k=0}^{\infty} a_k (x-\xi)^k = a_0 + a_1(x-\xi) + a_2(x-\xi)^2 + \ldots$$

is called a *power series*. The following is true of any power series: if it is not everywhere convergent, the region of convergence is a circle with its center at the point ξ and a radius equal to R; at every interior point of this circle, the power series 0.311 1. converges absolutely and outside this circle, it diverges. This circle is called the *circle of convergence* and its radius is called the *radius of convergence*. If the series converges at all points of the complex plane, we say that the radius of convergence is infinite $(R = +\infty)$.

0.312 Power series may be integrated and differentiated termwise inside the circle of convergence; that is,

$$\int_{\xi}^{x}\left\{\sum_{k=0}^{\infty} a_k (x-\xi)^k\right\} dx = \sum_{k=0}^{\infty} \frac{a_k}{k+1}(x-\xi)^{k+1},$$

$$\frac{d}{dx}\left\{\sum_{k=0}^{\infty} a_k (x-\xi)^k\right\} = \sum_{k=1}^{\infty} k a_k (x-\xi)^{k-1}.$$

The radius of convergence of a series that is obtained from termwise integration or differentiation of another power series coincides with the radius of convergence of the original series.

Operations on power series

0.313 Division of power series.

$$\frac{\sum_{k=0}^{\infty} b_k x^k}{\sum_{k=0}^{\infty} a_k x^k} = \frac{1}{a_0}\sum_{k=0}^{\infty} c_k x^k,$$

where

$$c_n + \frac{1}{a_0}\sum_{k=1}^{n} c_{n-k} a_k - b_n = 0,$$

or

$$c_n = \frac{(-1)^n}{a_0^n}\begin{vmatrix} a_1 b_0 - a_0 b_1 & a_0 & 0 & \dots & 0 \\ a_2 b_0 - a_0 b_2 & a_1 & a_0 & \dots & 0 \\ a_3 b_0 - a_0 b_3 & a_2 & a_1 & \dots & 0 \\ \dots & \dots & \dots & \dots & \dots \\ \dots & \dots & \dots & \dots & \dots \\ a_{n-1} b_0 - a_0 b_{n-1} & a_{n-2} & a_{n-3} & \dots & a_0 \\ a_n b_0 - a_0 b_n & a_{n-1} & a_{n-2} & \dots & a_1 \end{vmatrix}.$$ AD (6360)

0.314 Power series raised to powers.

$$\left(\sum_{k=0}^{\infty} a_k x^k\right)^n = \sum_{k=0}^{\infty} c_k x^k,$$

where

$$c_0 = a_0^n, \quad c_m = \frac{1}{m a_0}\sum_{k=1}^{m} (kn - m + k) a_k c_{m-k} \quad \text{for} \quad m \geqslant 1$$

[n is a natural number]. AD (6361)

0.315 The substitution of one series into another.

$$\sum_{k=1}^{\infty} b_k y^k = \sum_{k=1}^{\infty} c_k x^k, \quad y = \sum_{k=1}^{\infty} a_k x^k;$$

$$c_1 = a_1 b_1, \quad c_2 = a_2 b_1 + a_1^2 b_2, \quad c_3 = a_3 b_1 + 2a_1 a_2 b_2 + a_1^3 b_3,$$

$$c_4 = a_4 b_1 + a_2^2 b_2 + 2a_1 a_3 b_2 + 3a_1^2 a_2 b_3 + a_1^4 b_4, \ \dots$$ AD (6362)

0.316 Multiplication of power series

$$\sum_{k=0}^{\infty} a_k x^k \sum_{k=0}^{\infty} b_k x^k = \sum_{k=0}^{\infty} c_k x^k; \quad c_n = \sum_{k=0}^{n} a_k b_{n-k}.$$ FI II 372

Taylor series

0.317 If a function $f(x)$ has derivatives of all orders throughout a neighborhood of a point ξ, then we may write the series

1. $$f(\xi) + \frac{(x-\xi)}{1!} f'(\xi) + \frac{(x-\xi)^2}{2!} f''(\xi) + \frac{(x-\xi)^3}{3!} f'''(\xi) + \ldots,$$

which is known as the Taylor series of the function $f(x)$.

The Taylor series converges to the function $f(x)$ if the remainder

2. $$R_n(x) = f(x) - f(\xi) - \sum_{k=1}^{n} \frac{(x-\xi)^k}{k!} f^{(k)}(\xi)$$

approaches zero as $n \to \infty$.

The following are different forms for the remainder of a Taylor series.

3. $$R_n(x) = \frac{(x-\xi)^{n+1}}{(n+1)!} f^{(n+1)}(\xi + \theta(x-\xi)) \quad [0 < \theta < 1]. \text{ (Lagrange)}$$

4. $$R_n(x) = \frac{(x-\xi)^{n+1}}{n!} (1-\theta)^n f^{(n+1)}(\xi + \theta(x-\xi)) \quad [0 < \theta < 1]. \quad \text{(Cauchy)}$$

5. $$R_n(x) = \frac{\psi(x-\xi) - \psi(0)}{\psi'[(x-\xi)(1-\theta)]} \frac{(x-\xi)^n (1-\theta)^n}{n!} f^{(n+1)}(\xi + \theta(x-\xi)) \quad [0 < \theta < 1],$$ (Schlömilch)

where $\psi(x)$ is an arbitrary function satisfying the following two conditions: 1) It and its derivative $\psi'(x)$ are continuous in the interval $(0, x-\xi)$; 2) the derivative $\psi'(x)$ does not change sign in that interval. If we set $\psi(x) = x^{p+1}$, we obtain the following form for the remainder:

$$R_n(x) = \frac{(x-\xi)^{n+1}(1-\theta)^{n-p-1}}{(p+1)\, n!} f^{(n+1)}(\xi + \theta(x-\xi)) \quad [0 < p \leqslant n;\ 0 < \theta < 1].$$ (Rouché)

6. $$R_n(x) = \frac{1}{n!} \int_{\xi}^{x} f^{(n+1)}(t)(x-t)^n \, dt.$$

0.318 Other forms in which a Taylor series may be written:

1. $$f(a+x) = \sum_{k=0}^{\infty} \frac{x^k}{k!} f^{(k)}(a) = f(a) + \frac{x}{1!} f'(a) + \frac{x^2}{2!} f''(a) + \ldots$$

2. $$f(x) = \sum_{k=0}^{\infty} \frac{x^k}{k!} f^{(k)}(0) = f(0) + \frac{x}{1!} f'(0) + \frac{x^2}{2!} f''(0) + \ldots$$

[Maclaurin series]

0.319 The Taylor series of functions of several variables:

$$f(x, y) = f(\xi, \eta) + (x-\xi) \frac{\partial f(\xi, \eta)}{\partial x} + (y-\eta) \frac{\partial f(\xi, \eta)}{\partial y} +$$
$$+ \frac{1}{2!} \left\{ (x-\xi)^2 \frac{\partial^2 f(\xi, \eta)}{\partial x^2} + 2(x-\xi)(y-\eta) \frac{\partial^2 f(\xi, \eta)}{\partial x \, \partial y} + (y-\eta)^2 \frac{\partial^2 f(\xi, \eta)}{\partial y^2} \right\} + \ldots$$

0.32 Fourier series

0.320 Suppose that $f(x)$ is a *periodic* function of period $2l$ and that it is absolutely integrable (possibly improperly) over the interval $(-l, l)$. The following trigonometric series is called the *Fourier series* of $f(x)$:

1. $$\frac{a_0}{2}+\sum_{k=1}^{\infty} a_k \cos\frac{k\pi x}{l}+b_k \sin\frac{k\pi x}{l},$$

the coefficients of which (the Fourier coefficients) are given by the formulas

2. $$a_k=\frac{1}{l}\int_{-l}^{l} f(t)\cos\frac{k\pi t}{l}\,dt=\frac{1}{l}\int_{\alpha}^{\alpha+2l} f(t)\cos\frac{k\pi t}{l}\,dt \quad (k=0,\ 1,\ 2,\ \ldots),$$

3. $$b_k=\frac{1}{l}\int_{-l}^{l} f(t)\sin\frac{k\pi t}{l}\,dt=\frac{1}{l}\int_{\alpha}^{\alpha+2l} f(t)\sin\frac{k\pi t}{l}\,dt \quad (k=1,\ 2,\ \ldots).$$

Convergence tests

0.321 The Fourier series of a function $f(x)$ at a point x_0 converges to the number

$$\frac{f(x_0+0)+f(x_0-0)}{2},$$

if, for some $h>0$, the integral

$$\int_0^h \frac{|f(x_0+t)+f(x_0-t)-f(x_0+0)-f(x_0-0)|}{t}\,dt$$

exists. Here, it is assumed that the function $f(x)$ either is continuous at the point x_0 or has a discontinuity of the first kind (a *saltus*) at that point and that both one-sided limits $f(x_0+0)$ and $f(x_0-0)$ exist. (Dini) **FI III 524**

0.322 The Fourier series of a periodic function $f(x)$ that satisfies the Dirichlet conditions on the interval $[a, b]$ converges at every point x_0 to the value $\frac{1}{2}\{f(x_0+0)+f(x_0-0)\}$. (Dirichlet)

We say that a function $f(x)$ satisfies the Dirichlet conditions on the interval $[a, b]$ if it is bounded on that interval and if the interval $[a, b]$ can be partitioned into a finite number of subintervals inside each of which the function $f(x)$ is continuous and monotonic.

0.323 The Fourier series of a function $f(x)$ at a point x_0 converges to $\frac{1}{2}\{f(x_0+0)+f(x_0-0)\}$ if $f(x)$ is of bounded variation in some interval $(x_0-h,\ x_0+h)$ with its center at x_0. (Jordan-Dirichlet) **FI III 528**

The definition of a function of bounded variation. Suppose that a function $f(x)$ is defined on some interval $[a, b]$, where $a<b$. Let us partition this interval in an arbitrary manner into subintervals with the dividing points

$$a=x_0<x_1<x_2<\ldots<x_{n-1}<x_n=b$$

and let us form the sum

$$\sum_{k=1}^{n} |f(x_k)-f(x_{k-1})|.$$

Different partitions of the interval $[a, b]$ (that is, different choices of points of division x_i) yield, generally speaking, different sums. If the set of these sums is bounded above, we say that the function $f(x)$ is *of bounded variation* on the interval $[a, b]$. The least upper bound of these sums is called the *total variation* of the function $f(x)$ on the interval $[a, b]$.

0.324 Suppose that a function $f(x)$ is piecewise-continuous on the interval $[a, b]$ and that in each interval of continuity it has a piecewise-continuous derivative. Then, at every point x_0 of the interval $[a, b]$, the Fourier series of the function $f(x)$ converges to $\frac{1}{2}\{f(x_0+0)+f(x_0-0)\}$.

0.325 A function $f(x)$ defined in the interval $(0, l)$ can be expanded in a cosine series of the form

1. $$\frac{a_0}{2}+\sum_{k=1}^{\infty} a_k \cos\frac{k\pi x}{l},$$

where

2. $$a_k=\frac{2}{l}\int_0^l f(t)\cos\frac{k\pi t}{l}\,dt.$$

0.326 A function $f(x)$ defined in the interval $(0, l)$ can be expanded in a sine series of the form

1. $$\sum_{k=1}^{\infty} b_k \sin\frac{k\pi x}{l},$$

where

2. $$b_k=\frac{2}{l}\int_0^l f(t)\sin\frac{k\pi t}{l}\,dt.$$

The convergence tests for the series 0.325 1. and 0.326 1. are analogous to the convergence tests for the series 0.320 1. (see 0.321—0.324).

0.327 The Fourier coefficients a_k and b_k (given by formulas 0.320 2. and 0.320 3.) of an absolutely integrable function approach zero as $k \to \infty$.

If a function $f(x)$ is square-integrable on the interval $(-l, l)$, the equation of closure is satisfied:

$$\frac{a_0^2}{2}+\sum_{k=1}^{\infty}(a_k^2+b_k^2)=\frac{1}{l}\int_{-l}^{l} f^2(x)\,dx. \qquad \text{(A. M. Lyapunov)}$$ FI III 705

0.328 Suppose that $f(x)$ and $\varphi(x)$ are two functions that are square-integrable on the interval $(-l, l)$ and that a_k, b_k and α_k, β_k are their Fourier coefficients. For such functions, the generalized equation of closure (Parseval's equation) holds:

$$\frac{a_0\alpha_0}{2}+\sum_{k=1}^{\infty}(a_k\alpha_k+b_k\beta_k)=\frac{1}{l}\int_{-l}^{l} f(x)\varphi(x)\,dx.$$ FI III 709

For examples of Fourier series, see 1.44 and 1.45.

0.33 Asymptotic series

0.330 Included in the collection of all divergent series is the broad class of series known as *asymptotic* or *semiconvergent* series. *Despite the fact that these series diverge*, the values of the functions that they represent can be calculated with a high degree of accuracy if we take the sum of a suitable number of terms of such series. In the case of alternating asymptotic series, we obtain greatest accuracy if we break off the series in question at whatever term is of lowest absolute value. In this case, the error (in absolute value) does not exceed the absolute value of the first of the discarded terms (cf. 0.227 3.).

Asymptotic series have many properties that are analogous to the properties of convergent series and, for that reason, they play a significant role in analysis.

The asymptotic expansion of a function is denoted as follows:

$$f(z) \sim \sum_{n=0}^{\infty} A_n z^{-n}.$$

The definition of an asymptotic expansion. The divergent series $\sum_{n=0}^{\infty} \frac{A_n}{z^n}$ is called the *asymptotic expansion* of a function $f(z)$ in a given region of values of arg z if the expression $R_n(z) = z^n [f(z) - S_n(z)]$, where $S_n(z) = \sum_{k=0}^{n} \frac{A_k}{z^k}$, satisfies the condition $\lim_{|z| \to \infty} R_n(z) = 0$ for fixed n.

FI II 820

A divergent series that represents the asymptotic expansion of some function is called an *asymptotic series*.

0.331 Properties of asymptotic series

1. The operations of addition, subtraction, multiplication, and raising to a power can be performed on asymptotic series just as on absolutely convergent series. The series obtained as a result of these operations will also be asymptotic.

2. One asymptotic series can be divided by another provided that the first term A_0 of the divisor is not equal to zero. The series obtained as a result of division will also be asymptotic. **FI II 823-825**

3. An asymptotic series can be integrated termwise, and the resultant series will also be asymptotic. In contrast, differentiation of an asymptotic series is, in general, not permissible. **FI II 824**

4. A single asymptotic expansion can represent different functions. On the other hand, a given function can be expanded in an asymptotic series in only one manner. **WH**

0.4 Certain Formulas from Differential Calculus

0.41 Differentiation of a definite integral with respect to a parameter

0.410 $$\frac{d}{da}\int_{\psi(a)}^{\varphi(a)} f(x, a)\,dx = f(\varphi(a), a)\frac{d\varphi(a)}{da} - f(\psi(a), a)\frac{d\psi(a)}{da} + \int_{\psi(a)}^{\varphi(a)} \frac{d}{da} f(x, a)\,dx.$$ **FI II 680**

0.411 In particular,

1. $$\frac{d}{da}\int_b^a f(x)\,dx = f(a).$$

2. $$\frac{d}{db}\int_b^a f(x)\,dx = -f(b).$$

0.42 The *n*th derivative of a product

(Leibnitz' rule)

Suppose that u and v are n-times-differentiable functions of x. Then,

$$\frac{d^n(uv)}{dx^n} = u\frac{d^n v}{dx^n} + \binom{n}{1}\frac{du}{dx}\frac{d^{n-1}v}{dx^{n-1}} + \binom{n}{2}\frac{d^2u}{dx^2}\frac{d^{n-2}v}{dx^{n-2}} + \binom{n}{3}\frac{d^3u}{dx^3}\frac{d^{n-3}v}{dx^{n-3}} + \ldots + v\frac{d^n u}{dx^n}$$

or, symbolically,

$$\frac{d^n(uv)}{dx^n} = (u+v)^{(n)}.$$ FI I 272

0.43 The *n*th derivative of a composite function

0.430 If $f(x) = F(y)$ and $y = \varphi(x)$, then

1. $$\frac{d^n}{dx^n} f(x) = \frac{U_1}{1!}F'(y) + \frac{U_2}{2!}F''(y) + \frac{U_3}{3!}F'''(y) + \ldots + \frac{U_n}{n!}F^{(n)}(y),$$

where

$$U_k = \frac{d^n}{dx^n}y^k - \frac{k}{1!}y\frac{d^n}{dx^n}y^{k-1} + \frac{k(k-1)}{2!}y^2\frac{d^n}{dx^n}y^{k-2} - \ldots + (-1)^{k-1}ky^{k-1}\frac{d^n y}{dx^n}.$$

AD (7361) GO

2. $$\frac{d^n}{dx^n} f(x) = \sum \frac{n!}{i!\,j!\,h!\ldots k!}\frac{d^mF}{dy^m}\left(\frac{y'}{1!}\right)^i\left(\frac{y''}{2!}\right)^j\left(\frac{y'''}{3!}\right)^h\cdots\left(\frac{y^{(l)}}{l!}\right)^k,$$

Here, the symbol $\sum$ indicates summation over all solutions in non negative integers of the equation $i + 2j + 3h + \ldots + lk = n$ and $m = i + j + h + \ldots + k$.

0.431

1. $$(-1)^n\frac{d^n}{dx^n}F\left(\frac{1}{x}\right) = \frac{1}{x^{2n}}F^{(n)}\left(\frac{1}{x}\right) + \frac{n-1}{x^{2n-1}}\frac{n}{1!}F^{(n-1)}\left(\frac{1}{x}\right) + \frac{(n-1)(n-2)}{x^{2n-2}}\frac{n(n-1)}{2!}F^{(n-2)}\left(\frac{1}{x}\right) + \ldots$$ AD (7362.1)

2. $$(-1)^n\frac{d^n}{dx^n}e^{\frac{a}{x}} = \frac{1}{x^n}e^{\frac{a}{x}}\left\{\left(\frac{a}{x}\right)^n + (n-1)\binom{n}{1}\left(\frac{a}{x}\right)^{n-1} + (n-1)(n-2)\binom{n}{2}\left(\frac{a}{x}\right)^{n-2} + (n-1)(n-2)(n-3)\binom{n}{3}\left(\frac{a}{x}\right)^{n-3} + \ldots\right\}.$$

AD (7362.2)

0.432

1. $$\frac{d^n}{dx^n} F(x^2) = (2x)^n F^{(n)}(x^2) + \frac{n(n-1)}{1!}(2x)^{n-2} F^{(n-1)}(x^2) + \\ + \frac{n(n-1)(n-2)(n-3)}{2!}(2x)^{n-4} F^{(n-2)}(x^2) + \\ + \frac{n(n-1)(n-2)(n-3)(n-4)(n-5)}{3!}(2x)^{n-6} F^{(n-3)}(x^2) + \dots$$ AD (7363.1)

2. $$\frac{d^n}{dx^n} e^{ax^2} = (2ax)^n e^{ax^2} \left\{ 1 + \frac{n(n-1)}{1!\,(4ax^2)} + \frac{n(n-1)(n-2)(n-3)}{2!\,(4ax^2)^2} + \right. \\ \left. + \frac{n(n-1)(n-2)(n-3)(n-4)(n-5)}{3!\,(4ax^2)^3} + \dots \right\}.$$ AD (7363.2)

3. $$\frac{d^n}{dx^n}(1+ax^2)^p = \frac{p(p-1)(p-2)\dots(p-n+1)(2ax)^n}{(1+ax^2)^{n-p}} \times \\ \times \left\{ 1 + \frac{n(n-1)}{1!\,(p-n+1)} \frac{1+ax^2}{4ax^2} + \frac{n(n-1)(n-2)(n-3)}{2!\,(p-n+1)(p-n+2)} \left(\frac{1+ax^2}{4ax^2} \right)^2 + \dots \right\}.$$ AD (7363.3)

4. $$\frac{d^{m-1}}{dx^{m-1}} (1-x^2)^{m-\frac{1}{2}} = (-1)^{m-1} \frac{(2m-1)!!}{m} \sin(m \arccos x).$$ AD (7363.4)

0.433

1. $$\frac{d^n}{dx^n} F\left(\sqrt{x}\right) = \frac{F^{(n)}\left(\sqrt{x}\right)}{\left(2\sqrt{x}\right)^n} - \frac{n(n-1)}{1!} \frac{F^{(n-1)}\left(\sqrt{x}\right)}{\left(2\sqrt{x}\right)^{n+1}} + \\ + \frac{(n+1)\,n(n-1)(n-2)}{2!} \frac{F^{(n-2)}\left(\sqrt{x}\right)}{\left(2\sqrt{x}\right)^{n+2}} - \dots$$ AD (7364.1)

2. $$\frac{d^n}{dx^n}\left(1+a\sqrt{x}\right)^{2n-1} = \frac{(2n-1)!!}{2^n} \frac{a}{\sqrt{x}} \left(a^2 - \frac{1}{x}\right)^{n-1}.$$ AD (7364.2)

0.434 $$\frac{d^n}{dx^n} y^p = p\binom{n-p}{n}\left\{ -\binom{n}{1}\frac{1}{p-1} y^{p-1} \frac{d^n y}{dx^n} + \binom{n}{2} \frac{1}{p-2} y^{p-2} \frac{d^n y^2}{dx^n} - \dots \right\}.$$ AD (737.1)

0.435 $$\frac{d^n}{dx^n} \ln y = \binom{n}{1}\left\{ \frac{1}{1\cdot y}\frac{d^n y}{dx^n} - \binom{n}{2} \frac{1}{2\cdot y^2} \frac{d^n y^2}{dx^n} + \binom{n}{3} \frac{1}{3\cdot y^3} \frac{d^n y^3}{dx^n} - \dots \right\}.$$ AD (737.2)

0.44 Integration by substitution

0.440 Let $f(x)$ and $g(x)$ be continuous in $[a, b]$. Further, let $g'(x)$ exist and be continuous there. Then

$$\int_a^b f[g(x)]\, g'(x)\, dx = \int_{g(a)}^{g(b)} f(u)\, du.$$

1. ELEMENTARY FUNCTIONS

1.1 Power of Binomials

1.11 Power series

1.110 $$(1+x)^q = 1 + qx + \frac{q(q-1)}{2!}x^2 + \ldots + \frac{q(q-1)\ldots(q-k+1)}{k!}x^k + \ldots$$

If q is neither a natural number nor zero, the series converges absolutely for $|x| < 1$ and diverges for $|x| > 1$. For $x = 1$, the series converges for $q > -1$ and diverges for $q \leqslant -1$. For $x = 1$, the series converges absolutely for $q > 0$. For $x = -1$, it converges absolutely for $q > 0$ and diverges for $q < 0$. If $q = n$ is a natural number, the series **1.110** is reduced to the finite sum **1.111**.

FI II 425

1.111

$$(a+x)^n = \sum_{k=0}^{n} \binom{n}{k} x^k a^{n-k}.$$

1.112

1. $$(1+x)^{-1} = 1 - x + x^2 - x^3 + \ldots = \sum_{k=1}^{\infty} (-1)^{k-1} x^{k-1}$$

(see also **1.121** 2.).

2. $$(1+x)^{-2} = 1 - 2x + 3x^2 - 4x^3 + \ldots = \sum_{k=1}^{\infty} (-1)^{k-1} kx^{k-1}.$$

3. $$(1+x)^{\frac{1}{2}} = 1 + \frac{1}{2}x - \frac{1\cdot 1}{2\cdot 4}x^2 + \frac{1\cdot 1\cdot 3}{2\cdot 4\cdot 6}x^3 - \frac{1\cdot 1\cdot 3\cdot 5}{2\cdot 4\cdot 6\cdot 8}x^4 + \ldots$$

4. $$(1+x)^{-\frac{1}{2}} = 1 - \frac{1}{2}x + \frac{1\cdot 3}{2\cdot 4}x^2 - \frac{1\cdot 3\cdot 5}{2\cdot 4\cdot 6}x^3 + \ldots$$

1.113 $$\frac{x}{(1-x)^2} = \sum_{k=1}^{\infty} kx^k \qquad [x^2 < 1].$$

1.114

1. $$(1+\sqrt{1+x})^q = 2^q\left\{1 + \frac{q}{1!}\left(\frac{x}{4}\right) + \frac{q(q-3)}{2!}\left(\frac{x}{4}\right)^2 + \frac{q(q-4)(q-5)}{3!}\left(\frac{x}{4}\right)^3 + \ldots\right\} \quad [x^2 < 1,\ q \text{ is a real number}].$$

AD (6351.1)

2. $$(x+\sqrt{1+x^2})^q = 1+\sum_{k=0}^{\infty}\frac{q^2(q^2-2^2)(q^2-4^2)\dots[q^2-(2k)^2]\,x^{2k+2}}{(2k+2)!}+$$
$$+qx+q\sum_{k=1}^{\infty}\frac{(q^2-1^2)(q^2-3^2)\dots[q^2-(2k-1)^2]}{(2k+1)!}x^{2k+1}$$
$[x^2<1,\ q$ is a real number]. AD (6351.2)

1.12 Series of rational fractions

1.121

1. $$\frac{x}{1-x}=\sum_{k=1}^{\infty}\frac{2^{k-1}x^{2^{k-1}}}{1+x^{2^{k-1}}}=\sum_{k=1}^{\infty}\frac{x^{2^{k-1}}}{1-x^{2^k}}\qquad [x^2<1].$$ AD (6350.3)

2. $$\frac{1}{x-1}=\sum_{k=1}^{\infty}\frac{2^{k-1}}{x^{2^{k-1}}+1}\qquad [x^2>1].$$ AD (6350.3)

1.2 The Exponential Function

1.21 Series representations

1.211

1. $$e^x=\sum_{k=0}^{\infty}\frac{x^k}{k!}.$$

2. $$a^x=\sum_{k=0}^{\infty}\frac{(x\ln a)^k}{k!}.$$

3. $$e^{-x^2}=\sum_{k=0}^{\infty}(-1)^k\frac{x^{2k}}{k!}.$$

1.212 $$e^x(1+x)=\sum_{k=0}^{\infty}\frac{x^k(k+1)}{k!}.$$

1.213 $$\frac{x}{e^x-1}=1-\frac{x}{2}+\sum_{k=1}^{\infty}\frac{B_{2k}x^{2k}}{(2k)!}\qquad [x<2\pi].$$ FI II 520

1.214 $$e^{e^x}=e\left(1+x+\frac{2x^2}{2!}+\frac{5x^3}{3!}+\frac{15x^4}{4!}+\dots\right).$$ AD (6460.3)

1.215

1. $$e^{\sin x}=1+x+\frac{x^2}{2!}-\frac{3x^4}{4!}-\frac{8x^5}{5!}-\frac{3x^6}{6!}+\frac{56x^7}{7!}+\dots$$ AD (6460.4)

2. $$e^{\cos x}=e\left(1-\frac{x^2}{2!}+\frac{4x^4}{4!}-\frac{31x^6}{6!}+\dots\right).$$ AD (6460.5)

3. $$e^{\operatorname{tg} x}=1+x+\frac{x^2}{2!}+\frac{3x^3}{3!}+\frac{9x^4}{4!}+\frac{37x^5}{5!}+\dots$$ AD (6460.6)

1.216

1. $$e^{\arcsin x}=1+x+\frac{x^2}{2!}+\frac{2x^3}{3!}+\frac{5x^4}{4!}+\dots$$ AD (6460.7)

2. $$e^{\operatorname{arctg} x}=1+x+\frac{x^2}{2!}-\frac{x^3}{3!}-\frac{7x^4}{4!}-\dots$$ AD (6460.8)

1.217

1. $\pi \frac{e^{\pi x}+e^{-\pi x}}{e^{\pi x}-e^{-\pi x}} = \frac{1}{x} + 2x \sum_{k=1}^{\infty} \frac{1}{x^2+k^2}$ (cf. **1.421** 3.). **AD (6707.1)**

2. $\frac{2\pi}{e^{\pi x}-e^{-\pi x}} = \frac{1}{x} + 2x \sum_{k=1}^{\infty} (-1)^k \frac{1}{x^2+k^2}$ (cf. **1.422** 3.). **AD (6707.2)**

1.22 Functional relations

1.221

1. $a^x = e^{x \ln a}$.

2. $a^{\log_a x} = a^{\frac{1}{\log_x a}} = x$.

1.222

1. $e^x = \operatorname{ch} x + \operatorname{sh} x$.

2. $e^{ix} = \cos x + i \sin x$.

1.223 $e^{ax} - e^{bx} = (a-b)\, x \exp\left[\frac{1}{2}(a+b)\, x\right] \prod_{k=1}^{\infty} \left[1 + \frac{(a-b)^2 x^2}{4k^2\pi^2}\right]$. **MO 216**

1.23 Series of exponentials

1.231 $\sum_{k=0}^{\infty} a^{kx} = \frac{1}{1-a^x}$ $[a > 1$ and $x < 0$ or $0 < a < 1$ and $x > 0]$.

1.232

1. $\operatorname{th} x = 1 + 2 \sum_{k=1}^{\infty} (-1)^k e^{-2kx}$ $[x > 0]$.

2. $\operatorname{sech} x = 2 \sum_{k=0}^{\infty} (-1)^k e^{-(2k+1)x}$ $[x > 0]$.

3. $\operatorname{cosech} x = 2 \sum_{k=0}^{\infty} e^{-(2k+1)x}$ $[x > 0]$.

1.3-1.4 Trigonometric and Hyperbolic Functions

1.30 Introduction

The trigonometric and hyperbolic sines are related by the identities

$$\operatorname{sh} x = \frac{1}{i} \sin ix, \qquad \sin x = \frac{1}{i} \operatorname{sh} ix.$$

The trigonometric and hyperbolic cosines are related by the identities

$$\operatorname{ch} x = \cos ix, \qquad \cos x = \operatorname{ch} ix.$$

Because of this duality, every relation involving trigonometric functions has its formal counterpart involving the corresponding hyperbolic functions, and vice-versa. In many (though not all) cases, both pairs of relationships are meaningful.

The idea of matching the relationships is carried out in the list of formulas given below. However, not all the meaningful "pairs" are included in the list.

1.31 The basic functional relations

1.311

1. $\sin x = \frac{1}{2i}(e^{ix} - e^{-ix});$
$= -i \operatorname{sh} ix.$

2. $\operatorname{sh} x = \frac{1}{2}(e^{x} - e^{-x});$
$= -i \sin(ix).$

3. $\cos x = \frac{1}{2}(e^{ix} + e^{-ix});$
$= \operatorname{ch} ix.$

4. $\operatorname{ch} x = \frac{1}{2}(e^{x} + e^{-x});$
$= \cos ix.$

5. $\operatorname{tg} x = \frac{\sin x}{\cos x} = \frac{1}{i} \operatorname{th} ix.$

6. $\operatorname{th} x = \frac{\operatorname{sh} x}{\operatorname{ch} x} = \frac{1}{i} \operatorname{tg} ix.$

7. $\operatorname{ctg} x = \frac{\cos x}{\sin x} = \frac{1}{\operatorname{tg} x} = i \operatorname{cth} ix.$

8. $\operatorname{cth} x = \frac{\operatorname{ch} x}{\operatorname{sh} x} = \frac{1}{\operatorname{th} x} = i \operatorname{ctg} ix.$

1.312

1. $\cos^2 x + \sin^2 x = 1.$

2. $\operatorname{ch}^2 x - \operatorname{sh}^2 x = 1.$

1.313

1. $\sin(x \pm y) = \sin x \cos y \pm \sin y \cos x.$
2. $\operatorname{sh}(x \pm y) = \operatorname{sh} x \operatorname{ch} y \pm \operatorname{sh} y \operatorname{ch} x.$
3. $\sin(x \pm iy) = \sin x \operatorname{ch} y \pm i \operatorname{sh} y \cos x.$
4. $\operatorname{sh}(x \pm iy) = \operatorname{sh} x \cos y \pm i \sin y \operatorname{ch} x.$
5. $\cos(x \pm y) = \cos x \cos y \mp \sin x \sin y.$
6. $\operatorname{ch}(x \pm y) = \operatorname{ch} x \operatorname{ch} y \pm \operatorname{sh} x \operatorname{sh} y.$
7. $\cos(x \pm iy) = \cos x \operatorname{ch} y \mp i \sin x \operatorname{sh} y.$
8. $\operatorname{ch}(x \pm iy) = \operatorname{ch} x \cos y \pm i \operatorname{sh} x \sin y.$
9. $\operatorname{tg}(x \pm y) = \frac{\operatorname{tg} x \pm \operatorname{tg} y}{1 \mp \operatorname{tg} x \operatorname{tg} y}.$
10. $\operatorname{th}(x \pm y) = \frac{\operatorname{th} x \pm \operatorname{th} y}{1 \pm \operatorname{th} x \operatorname{th} y}.$
11. $\operatorname{tg}(x \pm iy) = \frac{\operatorname{tg} x \pm i \operatorname{th} y}{1 \mp i \operatorname{tg} x \operatorname{th} y}.$
12. $\operatorname{th}(x \pm iy) = \frac{\operatorname{th} x \pm i \operatorname{tg} y}{1 \pm i \operatorname{th} x \operatorname{tg} y}.$

1.314

1. $\sin x \pm \sin y = 2 \sin \frac{1}{2}(x \pm y) \cos \frac{1}{2}(x \mp y).$
2. $\operatorname{sh} x \pm \operatorname{sh} y = 2 \operatorname{sh} \frac{1}{2}(x \pm y) \operatorname{ch} \frac{1}{2}(x \mp y).$
3. $\cos x + \cos y = 2 \cos \frac{1}{2}(x + y) \cos \frac{1}{2}(x - y).$
4. $\operatorname{ch} x + \operatorname{ch} y = 2 \operatorname{ch} \frac{1}{2}(x + y) \operatorname{ch} \frac{1}{2}(x - y).$

5. $\cos x - \cos y = 2 \sin \frac{1}{2}(x+y) \sin \frac{1}{2}(y-x).$

6. $\operatorname{ch} x - \operatorname{ch} y = 2 \operatorname{sh} \frac{1}{2}(x+y) \operatorname{sh} \frac{1}{2}(x-y).$

7. $\operatorname{tg} x \pm \operatorname{tg} y = \frac{\sin(x \pm y)}{\cos x \cos y}.$ 8. $\operatorname{th} x \pm \operatorname{th} y = \frac{\operatorname{sh}(x \pm y)}{\operatorname{ch} x \operatorname{ch} y}.$

1.315

1. $\sin^2 x - \sin^2 y = \sin(x+y)\sin(x-y) = \cos^2 y - \cos^2 x.$
2. $\operatorname{sh}^2 x - \operatorname{sh}^2 y = \operatorname{sh}(x+y)\operatorname{sh}(x-y) = \operatorname{ch}^2 x - \operatorname{ch}^2 y.$
3. $\cos^2 x - \sin^2 y = \cos(x+y)\cos(x-y) = \cos^2 y - \sin^2 x.$
4. $\operatorname{sh}^2 x + \operatorname{ch}^2 y = \operatorname{ch}(x+y)\operatorname{ch}(x-y) = \operatorname{ch}^2 x + \operatorname{sh}^2 y.$

1.316

1. $(\cos x + i \sin x)^n = \cos nx + i \sin nx.$ 2. $(\operatorname{ch} x + \operatorname{sh} x)^n = \operatorname{sh} nx + \operatorname{ch} ny$

[n is an integer].

1.317

1. $\sin \frac{x}{2} = \pm \sqrt{\frac{1}{2}(1 - \cos x)}.$ 2. $\operatorname{sh} \frac{x}{2} = \pm \sqrt{\frac{1}{2}(\operatorname{ch} x - 1)}.$

3. $\cos \frac{x}{2} = \pm \sqrt{\frac{1}{2}(1 + \cos x)}.$ 4. $\operatorname{ch} \frac{x}{2} = \sqrt{\frac{1}{2}(\operatorname{ch} x + 1)}.$

5. $\operatorname{tg} \frac{x}{2} = \frac{1-\cos x}{\sin x} = \frac{\sin x}{1+\cos x}.$ 6. $\operatorname{th} \frac{x}{2} = \frac{\operatorname{ch} x - 1}{\operatorname{sh} x} = \frac{\operatorname{sh} x}{\operatorname{ch} x + 1}.$

The signs in front of the radical in formulas **1.317** 1., **1.317** 2., and **1.317** 3. are taken so as to agree with the signs of the left hand members. The sign of the left hand members depends in turn on the value of x.

1.32 The representation of powers of trigonometric and hyperbolic functions in terms of functions of multiples of the argument (angle)

1.320

1. $\sin^{2n} x = \frac{1}{2^{2n}} \left\{ \sum_{k=0}^{n-1} (-1)^{n-k}\, 2 \binom{2n}{k} \cos 2(n-k)x + \binom{2n}{n} \right\}.$ KR 56 (10, 2)

2. $\operatorname{sh}^{2n} x = \frac{(-1)^n}{2^{2n}} \left\{ \sum_{k=0}^{n-1} (-1)^{n-k}\, 2 \binom{2n}{k} \operatorname{ch} 2(n-k)x + \binom{2n}{n} \right\}.$

3. $\sin^{2n-1} x = \frac{1}{2^{2n-2}} \sum_{k=0}^{n-1} (-1)^{n+k-1} \binom{2n-1}{k} \sin(2n-2k-1)x.$ KR 56 (10, 4)

4. $\operatorname{sh}^{2n-1} x = \frac{(-1)^{n-1}}{2^{2n-2}} \sum_{k=0}^{n-1} (-1)^{n+k-1} \binom{2n-1}{k} \operatorname{sh}(2n-2k-1)x.$

5. $\cos^{2n} x = \frac{1}{2^{2n}} \left\{ \sum_{k=0}^{n-1} 2 \binom{2n}{k} \cos 2(n-k)x + \binom{2n}{n} \right\}.$ KR 56 (10, 1)

6. $\operatorname{ch}^{2n} x = \frac{1}{2^{2n}} \left\{ \sum_{k=0}^{n-1} 2 \binom{2n}{k} \operatorname{ch} 2(n-k)x + \binom{2n}{n} \right\}.$

7. $\cos^{2n-1} x = \frac{1}{2^{2n-2}} \sum_{k=0}^{n-1} \binom{2n-1}{k} \cos(2n-2k-1)x.$ KR 56 (10, 3)

8. $\operatorname{ch}^{2n-1} x = \frac{1}{2^{2n-2}} \sum_{k=0}^{n-1} \binom{2n-1}{k} \operatorname{ch}(2n-2k-1)x.$

Special cases

1.321

1. $\sin^2 x = \frac{1}{2}(-\cos 2x + 1).$

2. $\sin^3 x = \frac{1}{4}(-\sin 3x + 3\sin x).$

3. $\sin^4 x = \frac{1}{8}(\cos 4x - 4\cos 2x + 3).$

4. $\sin^5 x = \frac{1}{16}(\sin 5x - 5\sin 3x + 10\sin x).$

5. $\sin^6 x = \frac{1}{32}(-\cos 6x + 6\cos 4x - 15\cos 2x + 10).$

6. $\sin^7 x = \frac{1}{64}(-\sin 7x + 7\sin 5x - 21\sin 3x + 35\sin x).$

1.322

1. $\operatorname{sh}^2 x = \frac{1}{2}(\operatorname{ch} 2x - 1).$

2. $\operatorname{sh}^3 x = \frac{1}{4}(\operatorname{sh} 3x - 3\operatorname{sh} x).$

3. $\operatorname{sh}^4 x = \frac{1}{8}(\operatorname{ch} 4x - 4\operatorname{ch} 2x + 3).$

4. $\operatorname{sh}^5 x = \frac{1}{16}(\operatorname{sh} 5x - 5\operatorname{sh} 3x + 10\operatorname{sh} x).$

5. $\operatorname{sh}^6 x = \frac{1}{32}(\operatorname{ch} 6x - 6\operatorname{ch} 4x + 15\operatorname{ch} 2x - 10).$

6. $\operatorname{sh}^7 x = \frac{1}{64}(\operatorname{sh} 7x - 7\operatorname{sh} 5x + 21\operatorname{sh} 3x - 35\operatorname{sh} x).$

1.323

1. $\cos^2 x = \frac{1}{2}(\cos 2x + 1).$

2. $\cos^3 x = \frac{1}{4}(\cos 3x + 3\cos x).$

3. $\cos^4 x = \frac{1}{8}(\cos 4x + 4\cos 2x + 3).$

4. $\cos^5 x = \frac{1}{16}(\cos 5x + 5\cos 3x + 10\cos x).$

5. $\cos^6 x = \frac{1}{32}(\cos 6x + 6\cos 4x + 15\cos 2x + 10).$

6. $\cos^7 x = \frac{1}{64}(\cos 7x + 7\cos 5x + 21\cos 3x + 35\cos x).$

1.324

1. $\operatorname{ch}^2 x = \frac{1}{2}(\operatorname{ch} 2x + 1).$

2. $\operatorname{ch}^3 x = \frac{1}{4}(\operatorname{ch} 3x + 3 \operatorname{ch} x).$

3. $\operatorname{ch}^4 x = \frac{1}{8}(\operatorname{ch} 4x + 4 \operatorname{ch} 2x + 3).$

4. $\operatorname{ch}^5 x = \frac{1}{16}(\operatorname{ch} 5x + 5 \operatorname{ch} 3x + 10 \operatorname{ch} x).$

5. $\operatorname{ch}^6 x = \frac{1}{32}(\operatorname{ch} 6x + 6 \operatorname{ch} 4x + 15 \operatorname{ch} 2x + 10).$

6. $\operatorname{ch}^7 x = \frac{1}{64}(\operatorname{ch} 7x + 7 \operatorname{ch} 5x + 21 \operatorname{ch} 3x + 35 \operatorname{ch} x).$

1.33 The representation of trigonometric and hyperbolic functions of multiples of the argument (angle) in terms of powers of these functions

1.331

1. $$\sin nx = n \cos^{n-1} x \sin x - \binom{n}{3} \cos^{n-3} x \sin^3 x + \binom{n}{5} \cos^{n-5} x \sin^5 x - \ldots;$$
$$= \sin x \left\{ 2^{n-1} \cos^{n-1} x - \binom{n-2}{1} 2^{n-3} \cos^{n-3} x + \right.$$
$$\left. + \binom{n-3}{2} 2^{n-5} \cos^{n-5} x - \binom{n-4}{3} 2^{n-7} \cos^{n-7} x + \ldots \right\}.$$ AD (3.175)

2. $$\operatorname{sh} nx = \operatorname{sh} x \sum_{k=1}^{E\left(\frac{n+1}{2}\right)} \binom{n}{2k-1} \operatorname{sh}^{2k-2} x \operatorname{ch}^{n-2k+1} x;$$
$$= \operatorname{sh} x \sum_{k=0}^{E\left(\frac{n-1}{2}\right)} (-1)^k \binom{n-k-1}{k} 2^{n-2k-1} \operatorname{ch}^{n-2k-1} x.$$

3. $$\cos nx = \cos^n x - \binom{n}{2} \cos^{n-2} x \sin^2 x + \binom{n}{4} \cos^{n-4} x \sin^4 x - \ldots;$$
$$= 2^{n-1} \cos^n x - \frac{n}{1} 2^{n-3} \cos^{n-2} x +$$
$$+ \frac{n}{2} \binom{n-3}{1} 2^{n-5} \cos^{n-4} x - \frac{n}{3} \binom{n-4}{2} 2^{n-7} \cos^{n-6} x + \ldots$$ AD (3.175)

4. $$\operatorname{ch} nx = \sum_{k=0}^{E\left(\frac{n}{2}\right)} \binom{n}{2k} \operatorname{sh}^{2k} x \operatorname{ch}^{n-2k} x =$$
$$= 2^{n-1} \operatorname{ch}^n x + n \sum_{k=1}^{E\left(\frac{n}{2}\right)} (-1)^k \frac{1}{k} \binom{n-k-1}{k-1} 2^{n-2k-1} \operatorname{ch}^{n-2k} x,$$

1.332

1. $\sin 2nx = 2n\cos x\left\{\sin x - \frac{4n^2-2^2}{3!}\sin^3 x + \frac{(4n^2-2^2)(4n^2-4^2)}{5!}\sin^5 x - \dots\right\};$ AD (3.171)

$= (-1)^{n-1}\cos x\left\{2^{2n-1}\sin^{2n-1}x - \frac{2n-2}{1!}2^{2n-3}\sin^{2n-3}x + \right.$

$+ \frac{(2n-3)(2n-4)}{2!}2^{2n-5}\sin^{2n-5}x -$

$\left. - \frac{(2n-4)(2n-5)(2n-6)}{3!}2^{2n-7}\sin^{2n-7}x + \dots\right\}.$ AD (3.173)

2. $\sin(2n-1)x = (2n-1)\left\{\sin x - \frac{(2n-1)^2-1^2}{3!}\sin^3 x + \right.$

$\left. + \frac{[(2n-1)^2-1^2][(2n-1)^2-3^2]}{5!}\sin^5 x - \dots\right\};$ AD (3.172)

$= (-1)^{n-1}\left\{2^{2n-2}\sin^{2n-1}x - \frac{2n-1}{1!}2^{2n-4}\sin^{2n-3}x + \right.$

$+ \frac{(2n-1)(2n-4)}{2!}2^{2n-6}\sin^{2n-5}x -$

$\left. - \frac{(2n-1)(2n-5)(2n-6)}{3!}2^{2n-8}\sin^{2n-7}x + \dots\right\}.$ AD (3.174)a

3. $\cos 2nx = 1 - \frac{4n^2}{2!}\sin^2 x +$

$+ \frac{4n^2(4n^2-2^2)}{4!}\sin^4 x - \frac{4n^2(4n^2-2^2)(4n^2-4^2)}{6!}\sin^6 x + \dots;$ AD (3.171)

$= (-1)^n\left\{2^{2n-1}\sin^{2n}x - \frac{2n}{1!}2^{2n-3}\sin^{2n-2}x + \right.$

$\left. + \frac{2n(2n-3)}{2!}2^{2n-5}\sin^{2n-4} - \frac{2n(2n-4)(2n-5)}{3!}2^{2n-7}\sin^{2n-6} + \dots\right\}.$ AD (3.173)a

4. $\cos(2n-1)x = \cos x\left\{1 - \frac{(2n-1)^2-1^2}{2!}\sin^2 x + \right.$

$\left. + \frac{[(2n-1)^2-1^2][(2n-1)^2-3^2]}{4!}\sin^4 x - \dots\right\};$ AD (3.172)

$= (-1)^{n-1}\cos x\left\{2^{2n-2}\sin^{2n-2}x - \frac{2n-3}{1!}2^{2n-4}\sin^{2n-4}x + \right.$

$+ \frac{(2n-4)(2n-5)}{2!}2^{2n-6}\sin^{2n-6}x -$

$\left. - \frac{(2n-5)(2n-6)(2n-7)}{3!}2^{2n-8}\sin^{2n-8}x + \dots\right\}.$ AD (3.174)

By using the formulas and values of **1.30**, we can write formulas for $\operatorname{sh} 2nx$, $\operatorname{sh}(2n-1)x$, $\operatorname{ch} 2nx$, and $\operatorname{ch}(2n-1)x$ that are analogous to those of **1.332**, just as was done in the formulas in **1.331**.

Special cases

1.333

1. $\sin 2x = 2\sin x\cos x.$
2. $\sin 3x = 3\sin x - 4\sin^3 x.$
3. $\sin 4x = \cos x(4\sin x - 8\sin^3 x).$

4. $\sin 5x = 5 \sin x - 20 \sin^3 x + 16 \sin^5 x.$

5. $\sin 6x = \cos x (6 \sin x - 32 \sin^3 x + 32 \sin^5 x).$

6. $\sin 7x = 7 \sin x - 56 \sin^3 x + 112 \sin^5 x - 64 \sin^7 x.$

1.334

1. $\operatorname{sh} 2x = 2 \operatorname{sh} x \operatorname{ch} x.$

2. $\operatorname{sh} 3x = 3 \operatorname{sh} x + 4 \operatorname{sh}^3 x.$

3. $\operatorname{sh} 4x = \operatorname{ch} x (4 \operatorname{sh} x + 8 \operatorname{sh}^3 x).$

4. $\operatorname{sh} 5x = 5 \operatorname{sh} x + 20 \operatorname{sh}^3 x + 16 \operatorname{sh}^5 x.$

5. $\operatorname{sh} 6x = \operatorname{ch} x (6 \operatorname{sh} x + 32 \operatorname{sh}^3 x + 32 \operatorname{sh}^5 x).$

6. $\operatorname{sh} 7x = 7 \operatorname{sh} x + 56 \operatorname{sh}^3 x + 112 \operatorname{sh}^5 x + 64 \operatorname{sh}^7 x.$

1.335

1. $\cos 2x = 2 \cos^2 x - 1.$

2. $\cos 3x = 4 \cos^3 x - 3 \cos x.$

3. $\cos 4x = 8 \cos^4 x - 8 \cos^2 x + 1.$

4. $\cos 5x = 16 \cos^5 x - 20 \cos^3 x + 5 \cos x.$

5. $\cos 6x = 32 \cos^6 x - 48 \cos^4 x + 18 \cos^2 x - 1.$

6. $\cos 7x = 64 \cos^7 x - 112 \cos^5 x + 56 \cos^3 x - 7 \cos x.$

1.336

1. $\operatorname{ch} 2x = 2 \operatorname{ch}^2 x - 1.$

2. $\operatorname{ch} 3x = 4 \operatorname{ch}^3 x - 3 \operatorname{ch} x.$

3. $\operatorname{ch} 4x = 8 \operatorname{ch}^4 x - 8 \operatorname{ch}^2 x + 1.$

4. $\operatorname{ch} 5x = 16 \operatorname{ch}^5 x - 20 \operatorname{ch}^3 x + 5 \operatorname{ch} x.$

5. $\operatorname{ch} 6x = 32 \operatorname{ch}^6 x - 48 \operatorname{ch}^4 x + 18 \operatorname{ch}^2 x - 1.$

6. $\operatorname{ch} 7x = 64 \operatorname{ch}^7 x - 112 \operatorname{ch}^5 x + 56 \operatorname{ch}^3 x - 7 \operatorname{ch} x.$

1.34 Certain sums of trigonometric and hyperbolic functions

1.341

1. $\sum_{k=0}^{n-1} \sin (x + ky) = \sin \left(x + \frac{n-1}{2} y \right) \sin \frac{ny}{2} \operatorname{cosec} \frac{y}{2}.$ AD (361.8)

2. $\sum_{k=0}^{n-1} \operatorname{sh} (x + ky) = \operatorname{sh} \left(x + \frac{n-1}{2} y \right) \operatorname{sh} \frac{ny}{2} \frac{1}{\operatorname{sh} \frac{y}{2}}.$

3. $\sum_{k=0}^{n-1} \cos (x + ky) = \cos \left(x + \frac{n-1}{2} y \right) \sin \frac{ny}{2} \operatorname{cosec} \frac{y}{2}.$ AD (361.9)

4. $\sum_{k=0}^{n-1} \operatorname{ch} (x + ky) = \operatorname{ch} \left(x + \frac{n-1}{2} y \right) \operatorname{sh} \frac{ny}{2} \frac{1}{\operatorname{sh} \frac{y}{2}}.$

5. $\sum_{k=0}^{2n-1} (-1)^k \cos (x + ky) = \sin \left(x + \frac{2n-1}{2} y \right) \sin ny \sec \frac{y}{2}.$ JO (202)

6. $$\sum_{k=0}^{n-1} (-1)^k \sin(x+ky) = \sin\left\{x+\frac{n-1}{2}(y+\pi)\right\} \sin\frac{n(y+\pi)}{2} \sec\frac{y}{2}.$$ JO (202a)

Special cases

1.342

1. $$\sum_{k=1}^{n} \sin kx = \sin\frac{n+1}{2}x \sin\frac{nx}{2} \operatorname{cosec}\frac{x}{2}.$$ AD (361.1)

2. $$\sum_{k=0}^{n} \cos kx = \cos\frac{n+1}{2}x \sin\frac{nx}{2} \operatorname{cosec}\frac{x}{2} + 1 = \cos\frac{nx}{2} \sin\frac{n+1}{2}x \operatorname{cosec}\frac{x}{2}.$$ AD (361.2)

3. $$\sum_{k=1}^{n} \sin(2k-1)x = \sin^2 nx \operatorname{cosec} x.$$ AD (361.7)

4. $$\sum_{k=1}^{n} \cos(2k-1)x = \frac{1}{2}\sin 2nx \operatorname{cosec} x.$$ JO (207)

1.343

1. $$\sum_{k=1}^{n} (-1)^k \cos kx = -\frac{1}{2} + \frac{(-1)^n \cos\left(\frac{2n+1}{2}x\right)}{2\cos\frac{x}{2}}.$$ AD (361.11)

2. $$\sum_{k=1}^{n} (-1)^{k+1} \sin(2k-1)x = (-1)^{n+1} \frac{\sin 2nx}{2\cos x}.$$ AD (361.10)

3. $$\sum_{k=1}^{n} \cos(4k-3)x + \sum_{k=1}^{n} \sin(4k-1)x =$$
$$= \sin 2nx (\cos 2nx + \sin 2nx)(\cos x + \sin x) \operatorname{cosec} 2x.$$ JO (208)

1.344

1. $$\sum_{k=1}^{n-1} \sin\frac{\pi k}{n} = \operatorname{ctg}\frac{\pi}{2n}.$$ AD (361.19)

2. $$\sum_{k=1}^{n-1} \sin\frac{2\pi k^2}{n} = \frac{\sqrt{n}}{2}\left(1+\cos\frac{n\pi}{2} - \sin\frac{n\pi}{2}\right).$$ AD (361.18)

3. $$\sum_{k=0}^{n-1} \cos\frac{2\pi k^2}{n} = \frac{\sqrt{n}}{2}\left(1+\cos\frac{n\pi}{2} + \sin\frac{n\pi}{2}\right).$$ AD (361.17)

1.35 Sums of powers of trigonometric functions of multiple angles

1.351

1. $$\sum_{k=1}^{n} \sin^2 kx = \frac{1}{4}[(2n+1)\sin x - \sin(2n+1)x] \operatorname{cosec} x;$$
$$= \frac{n}{2} - \frac{\cos(n+1)x \sin nx}{2\sin x}.$$ AD (361.3)

2. $$\sum_{k=1}^{n} \cos^2 kx = \frac{n-1}{2} + \frac{1}{2}\cos nx \sin(n+1)x \operatorname{cosec} x;$$
$$= \frac{n}{2} + \frac{\cos(n+1)x \sin nx}{2\sin x}.$$ AD (361.4)a

3. $$\sum_{k=1}^{n} \sin^3 kx = \frac{3}{4}\sin\frac{n+1}{2}x \sin\frac{nx}{2}\operatorname{cosec}\frac{x}{2} - \frac{1}{4}\sin\frac{3(n+1)x}{2}\sin\frac{3nx}{2}\operatorname{cosec}\frac{3x}{2}.$$ JO (210)

4. $$\sum_{k=1}^{n} \cos^3 kx = \frac{3}{4}\cos\frac{n+1}{2}x \sin\frac{nx}{2}\operatorname{cosec}\frac{x}{2} + \frac{1}{4}\cos\frac{3(n+1)}{2}x \sin\frac{3nx}{2}\operatorname{cosec}\frac{3x}{2}.$$ JO (211)a

5. $$\sum_{k=1}^{n} \sin^4 kx = \frac{1}{8}[3n - 4\cos(n+1)x \sin nx \operatorname{cosec} x + \cos 2(n+1)x \sin 2nx \operatorname{cosec} 2x].$$ JO (212)

6. $$\sum_{k=1}^{n} \cos^4 kx = \frac{1}{8}[3n + 4\cos(n+1)x \sin nx \operatorname{cosec} x + \cos 2(n+1)x \sin 2nx \operatorname{cosec} 2x].$$ JO (213)

1.352

1. $$\sum_{k=1}^{n-1} k \sin kx = \frac{\sin nx}{4\sin^2\frac{x}{2}} - \frac{n\cos\frac{2n-1}{2}x}{2\sin\frac{x}{2}}.$$ AD (361.5)

2. $$\sum_{k=1}^{n-1} k \cos kx = \frac{n\sin\frac{2n-1}{2}x}{2\sin\frac{x}{2}} - \frac{1-\cos nx}{4\sin^2\frac{x}{2}}.$$ AD (361.6)

1.353

1. $$\sum_{k=1}^{n-1} p^k \sin kx = \frac{p\sin x - p^n \sin nx + p^{n+1}\sin(n-1)x}{1-2p\cos x + p^2}.$$ AD (361.12)a

2. $$\sum_{k=1}^{n-1} p^k \operatorname{sh} kx = \frac{p\operatorname{sh} x - p^n \operatorname{sh} nx + p^{n+1}\operatorname{sh}(n-1)x}{1-2p\operatorname{ch} x + p^2}.$$

3. $$\sum_{k=0}^{n-1} p^k \cos kx = \frac{1-p\cos x - p^n\cos nx + p^{n+1}\cos(n-1)x}{1-2p\cos x + p^2}.$$ AD (361.13)a

4. $$\sum_{k=0}^{n-1} p^k \operatorname{ch} kx = \frac{1-p\operatorname{ch} x - p^n\operatorname{ch} nx + p^{n+1}\operatorname{ch}(n-1)x}{1-2p\operatorname{ch} x + p^2}.$$ JO (396)

1.36 Sums of products of trigonometric functions of multiple angles

1.361

1. $\sum_{k=1}^{n} \sin kx \sin (k+1)x = \frac{1}{4}[(n+1)\sin 2x - \sin 2(n+1)x] \operatorname{cosec} x.$ JO (214)

2. $\sum_{k=1}^{n} \sin kx \sin (k+2)x = \frac{n}{2}\cos 2x - \frac{1}{2}\cos (n+3)x \sin nx \operatorname{cosec} x.$ JO (216)

3. $2\sum_{k=1}^{n} \sin kx \cos (2k-1)y = \sin\left\{ny + \frac{n+1}{2}x\right\} \sin \frac{n(x+2y)}{2} \operatorname{cosec} \frac{x+2y}{2} -$
$- \sin\left\{ny - \frac{n+1}{2}x\right\} \sin \frac{n(2y-x)}{2} \operatorname{cosec} \frac{2y-x}{2}.$ JO (217)

1.362

1. $\sum_{k=1}^{n} \left(2^k \sin^2 \frac{x}{2^k}\right)^2 = \left(2^n \sin \frac{x}{2^n}\right)^2 - \sin^2 x.$ AD (361.15)

2. $\sum_{k=1}^{n} \left(\frac{1}{2^k} \sec \frac{x}{2^k}\right)^2 = \operatorname{cosec}^2 x - \left(\frac{1}{2^n} \operatorname{cosec} \frac{x}{2^n}\right)^2.$ AD (361.14)

1.37 Sums of tangents of multiple angles

1.371

1. $\sum_{k=0}^{n} \frac{1}{2^k} \operatorname{tg} \frac{x}{2^k} = \frac{1}{2^n} \operatorname{ctg} \frac{x}{2^n} - 2 \operatorname{ctg} 2x.$ AD (361.16)

2. $\sum_{k=0}^{n} \frac{1}{2^{2k}} \operatorname{tg}^2 \frac{x}{2^k} = \frac{2^{2n+2}-1}{3 \cdot 2^{2n-1}} + 4 \operatorname{ctg}^2 2x - \frac{1}{2^{2n}} \operatorname{ctg}^2 \frac{x}{2^n}.$ AD (361.20)

1.38 Sums leading to hyperbolic tangents and cotangents

1.381

1. $\sum_{k=0}^{n-1} \frac{\operatorname{th} x \dfrac{1}{n \sin^2 \dfrac{2k+1}{4n}\pi}}{1 + \dfrac{\operatorname{th}^2 x}{\operatorname{tg}^2 \dfrac{2k+1}{4n}\pi}} = \operatorname{th} 2nx.$ JO (402)a

2. $\sum_{k=1}^{n-1} \frac{\operatorname{th} x \dfrac{1}{n \sin^2 \dfrac{k\pi}{2n}}}{1 + \dfrac{\operatorname{th}^2 x}{\operatorname{tg}^2 \dfrac{k\pi}{2n}}} = \operatorname{cth} 2nx - \frac{1}{2n}(\operatorname{th} x + \operatorname{cth} x).$ JO (403)

3. $\sum_{k=0}^{n-1} \frac{\operatorname{th} x \dfrac{2}{(2n+1) \sin^2 \dfrac{2k+1}{2(2n+1)}\pi}}{1 + \dfrac{\operatorname{th}^2 x}{\operatorname{tg}^2 \dfrac{2k+1}{2(2n+1)}\pi}} = \operatorname{th}(2n+1)x - \frac{\operatorname{th} x}{2n+1}.$ JO (404)

4. $$\sum_{k=1}^{n} \frac{\operatorname{th} x \dfrac{2}{(2n+1)\sin^2 \dfrac{k\pi}{(2n+1)}}}{1+\dfrac{\operatorname{th}^2 x}{\operatorname{tg}^2 \dfrac{k\pi}{(2n+1)}}} = \operatorname{cth}(2n+1)x - \frac{\operatorname{cth} x}{2n+1}.$$ JO (405)

1.382

1. $$\sum_{k=0}^{n-1} \frac{1}{\dfrac{\sin^2 \dfrac{2k+1}{4n}\pi}{\operatorname{sh} x} + \dfrac{1}{2}\operatorname{th}\dfrac{x}{2}} = 2n \operatorname{th} nx.$$ JO (406)

2. $$\sum_{k=1}^{n-1} \frac{1}{\dfrac{\sin^2 \dfrac{k\pi}{2n}}{\operatorname{sh} x} + \dfrac{1}{2}\operatorname{th}\dfrac{x}{2}} = 2n \operatorname{cth} nx - 2\operatorname{cth} x.$$ JO (407)

3. $$\sum_{k=0}^{n-1} \frac{1}{\dfrac{\sin^2 \dfrac{2k+1}{2(2n+1)}\pi}{\operatorname{sh} x} + \dfrac{1}{2}\operatorname{th}\dfrac{x}{2}} = (2n+1)\operatorname{th}\frac{(2n+1)x}{2} - \operatorname{th}\frac{x}{2}.$$ JO (408)

4. $$\sum_{k=1}^{n} \frac{1}{\dfrac{\sin^2 \dfrac{k\pi}{2n+1}}{\operatorname{sh} x} + \dfrac{1}{2}\operatorname{th}\dfrac{x}{2}} = (2n+1)\operatorname{cth}\frac{(2n+1)x}{2} - \operatorname{cth}\frac{x}{2}.$$ JO (409)

1.39 The representation of cosines and sines of multiples of the angle as finite products

1.391

1. $$\sin nx = n \sin x \cos x \prod_{k=1}^{\frac{n-2}{2}} \left(1 - \frac{\sin^2 x}{\sin^2 \dfrac{k\pi}{n}}\right) \qquad [n\text{-even}].$$ JO (568)

2. $$\cos nx = \prod_{k=1}^{\frac{n}{2}} \left(1 - \frac{\sin^2 x}{\sin^2 \dfrac{(2k-1)\pi}{2n}}\right) \qquad [n\text{-even}].$$ JO (569)

3. $$\sin nx = n \sin x \prod_{k=1}^{\frac{n-1}{2}} \left(1 - \frac{\sin^2 x}{\sin^2 \dfrac{k\pi}{n}}\right) \qquad [n\text{-odd}].$$ JO (570)

4. $$\cos nx = \cos x \prod_{k=1}^{\frac{n-1}{2}} \left(1 - \frac{\sin^2 x}{\sin^2 \dfrac{(2k-1)\pi}{2n}}\right) \qquad [n\text{- odd}].$$ JO (571)a

1.392

1. $$\sin nx = 2^{n-1} \prod_{k=0}^{n-1} \sin\left(x + \frac{k\pi}{n}\right).$$ JO (548)

2. $$\cos nx = 2^{n-1} \prod_{k=1}^{n} \sin\left(x + \frac{2k-1}{2n}\pi\right).$$ JO (549)

1.393

1. $$\prod_{k=0}^{n-1} \cos\left(x+\frac{2k}{n}\pi\right)=\frac{1}{2^{n-1}}\cos nx \qquad [n\text{-odd}].$$

$$=\frac{1}{2^{n-1}}\left[(-1)^{\frac{n}{2}}-\cos nx\right] \quad [n\text{-even}].$$ JO (543)

2. $$\prod_{k=0}^{n-1} \sin\left(x+\frac{2k}{n}\pi\right)=\frac{(-1)^{\frac{n-1}{2}}}{2^{n-1}}\sin nx \qquad [n\text{-odd}].$$

$$=\frac{(-1)^{\frac{n}{2}}}{2^{n-1}}(1-\cos nx) \qquad [n\text{-even}].$$ JO (544)

1.394 $$\prod_{k=0}^{n-1}\left\{x^2-2xy\cos\left(\alpha+\frac{2k\pi}{n}\right)+y^2\right\}=x^{2n}-2x^ny^n\cos n\alpha+y^{2n}.$$ JO (573)

1.395

1. $$\cos nx-\cos ny=2^{n-1}\prod_{k=0}^{n-1}\left\{\cos x-\cos x\left(y+\frac{2k\pi}{n}\right)\right\}.$$ JO (573)

2. $$\operatorname{ch} nx-\cos ny=2^{n-1}\prod_{k=0}^{n-1}\left\{\operatorname{ch} x-\cos\left(y+\frac{2k\pi}{n}\right)\right\}.$$ JO (538)

1.396

1. $$\prod_{k=1}^{n-1}\left(x^2-2x\cos\frac{k\pi}{n}+1\right)=\frac{x^{2n}-1}{x^2-1}.$$ KR 58 (28.1)

2. $$\prod_{k=1}^{n}\left(x^2-2x\cos\frac{2k\pi}{2n+1}+1\right)=\frac{x^{2n+1}-1}{x-1}.$$ KR 58(28.2)

3. $$\prod_{k=1}^{n}\left(x^2+2x\cos\frac{2k\pi}{2n+1}+1\right)=\frac{x^{2n+1}-1}{x+1}.$$ KR 58 (28.3)

4. $$\prod_{k=0}^{n-1}\left(x^2-2x\cos\frac{(2k+1)\pi}{2n}+1\right)=x^{2n}+1.$$ KR 58 (28.4)

1.41 The expansion of trigonometric and hyperbolic functions in power series

1.411

1. $$\sin x=\sum_{k=0}^{\infty}(-1)^k\frac{x^{2k+1}}{(2k+1)!}.$$

2. $$\operatorname{sh} x=\sum_{k=0}^{\infty}\frac{x^{2k+1}}{(2k+1)!}.$$

3. $$\cos x=\sum_{k=0}^{\infty}(-1)^k\frac{x^{2k}}{(2k)!}.$$

4. $$\operatorname{ch} x=\sum_{k=0}^{\infty}\frac{x^{2k}}{(2k)!}.$$

5. $\operatorname{tg} x = \sum_{k=1}^{\infty} \frac{2^{2k}(2^{2k}-1)}{(2k)!} |B_{2k}| x^{2k-1} \left[x^2 < \frac{\pi^2}{4}\right].$ FI II 523

6. $\operatorname{th} x = x - \frac{x^3}{3} + \frac{2x^5}{15} - \frac{17}{315}x^7 + \ldots = \sum_{k=1}^{\infty} \frac{2^{2k}(2^{2k}-1)}{(2k)!} B_{2k} x^{2k-1} \left[x^2 < \frac{\pi^2}{4}\right].$

7. $\operatorname{ctg} x = \frac{1}{x} - \sum_{k=1}^{\infty} \frac{2^{2k}|B_{2k}|}{(2k)!} x^{2k-1} \quad [x^2 < \pi^2].$ FI II 523a

8. $\operatorname{cth} x = \frac{1}{x} + \frac{x}{3} - \frac{x^3}{45} + \frac{2x^5}{945} - \ldots = \frac{1}{x} + \sum_{k=1}^{\infty} \frac{2^{2k} B_{2k}}{(2k)!} x^{2k-1} [x^2 < \pi^2].$ FI II 522a

9. $\sec x = \sum_{k=0}^{\infty} \frac{|E_{2k}|}{(2k)!} x^{2k} \quad \left[x^2 < \frac{\pi^2}{4}\right].$ CE 330a

10. $\operatorname{sech} x = 1 - \frac{x^2}{2} + \frac{5x^4}{24} - \frac{61x^6}{720} + \ldots = 1 + \sum_{k=1}^{\infty} \frac{E_{2k}}{(2k)!} x^{2k} \quad \left[x^2 < \frac{\pi^2}{4}\right]$ CE 330

11. $\operatorname{cosec} x = \frac{1}{x} + \sum_{k=1}^{\infty} \frac{2(2^{2k-1}-1)|B_{2k}| x^{2k-1}}{(2k)!} \quad [x^2 < \pi^2].$ CE 329a

12. $\operatorname{cosech} x = \frac{1}{x} - \frac{1}{6}x + \frac{7x^3}{360} - \frac{31x^5}{15120} + \ldots = \frac{1}{x} - \sum_{k=1}^{\infty} \frac{2(2^{2k-1}-1) B_{2k}}{(2k)!} x^{2k-1}$
$[x^2 < \pi^2].$ JO (418)

1.412

1. $\sin^2 x = \sum_{k=1}^{\infty} (-1)^{k+1} \frac{2^{2k-1} x^{2k}}{(2k)!}.$ JO (452)a

2. $\cos^2 x = 1 - \sum_{k=1}^{\infty} (-1)^{k+1} \frac{2^{2k-1} x^{2k}}{(2k)!}.$ JO (443)

3. $\sin^3 x = \frac{1}{4} \sum_{k=1}^{\infty} (-1)^{k+1} \frac{3^{2k+1}-3}{(2k+1)!} x^{2k+1}.$ JO (452a)a

4. $\cos^3 x = \frac{1}{4} \sum_{k=0}^{\infty} (-1)^k \frac{(3^{2k}+3) x^{2k}}{(2k)!}.$ JO (443a)

1.413

1. $\operatorname{sh} x = \operatorname{cosec} x \sum_{k=1}^{\infty} (-1)^{k+1} \frac{2^{2k-1} x^{4k-2}}{(4k-2)!}.$ JO (508)

2. $\operatorname{ch} x = \sec x + \sec x \sum_{k=1}^{\infty} (-1)^k \frac{2^{2k} x^{4k}}{(4k)!}.$ JO (507)

3. $\operatorname{sh} x = \sec x \sum_{k=1}^{\infty} (-1)^{E\left(\frac{k}{2}\right)} \frac{2^{k-1} x^{2k-1}}{(2k-1)!}.$ JO (510)

4. $\operatorname{ch} x = \operatorname{cosec} x \sum_{k=1}^{\infty} (-1)^{E\left(\frac{k-1}{2}\right)} \frac{2^{k-1} x^{2k-1}}{(2k-1)!}$. JO (509)

1.414

1. $\cos [n \ln (x + \sqrt{1+x^2})] =$

$$= 1 - \sum_{k=0}^{\infty} (-1)^k \frac{(n^2+0^2)(n^2+2^2)\dots[n^2+(2k)^2]}{(2k+2)!} x^{2k+2} \quad [x^2 < 1].$$ AD (6456.1)

2. $\sin [n \ln (x + \sqrt{1+x^2})] =$

$$= nx - n^2 \sum_{k=1}^{\infty} (-1)^{k+1} \frac{(n^2+1^2)(n^2+3^2)\dots[n^2+(2k-1)^2]\, x^{2k+1}}{(2k+1)!} \quad [x^2 < 1].$$ AD (6456.2)

Power series for $\ln \sin x$, $\ln \cos x$ and $\ln \operatorname{tg} x$ see 1.518.

1.42 Expansion in series of simple fractions

1.421

1. $\operatorname{tg} \frac{\pi x}{2} = \frac{4x}{\pi} \sum_{k=1}^{\infty} \frac{1}{(2k-1)^2 - x^2}$. BR* (191), AD (6495.1)

2. $\operatorname{th} \frac{\pi x}{2} = \frac{4x}{\pi} \sum_{k=1}^{\infty} \frac{1}{(2k-1)^2 + x^2}$.

3. $\operatorname{ctg} \pi x = \frac{1}{\pi x} + \frac{2x}{\pi} \sum_{k=1}^{\infty} \frac{1}{x^2-k^2} = \frac{1}{\pi x} + \frac{x}{\pi} \sum_{\substack{k=-\infty \\ k \neq 0}}^{\infty} \frac{1}{k(x-k)}$.

AD (6495.2), JO (450a)

4. $\operatorname{cth} \pi x = \frac{1}{\pi x} + \frac{2x}{\pi} \sum_{k=1}^{\infty} \frac{1}{x^2+k^2}$ (cf. 1.217 1.).

5. $\operatorname{tg}^2 \frac{\pi x}{2} = x^2 \sum_{k=1}^{\infty} \frac{2(2k-1)^2 - x^2}{(1^2-x^2)^2(3^2-x^2)^2 \dots [(2k-1)^2 - x^2]^2}$. JO (450)

1.422

1. $\sec \frac{\pi x}{2} = \frac{4}{\pi} \sum_{k=1}^{\infty} (-1)^{k+1} \frac{2k-1}{(2k-1)^2 - x^2}$. AD (6495.3)a

2. $\sec^2 \frac{\pi x}{2} = \frac{4}{\pi^2} \sum_{k=1}^{\infty} \left\{ \frac{1}{(2k-1-x)^2} + \frac{1}{(2k-1+x)^2} \right\}$. JO (451)a

3. $\operatorname{cosec} \pi x = \frac{1}{\pi x} + \frac{2x}{\pi} \sum_{k=1}^{\infty} \frac{(-1)^k}{x^2-k^2}$ (see also 1.217 2.). AD (6495.4)a

4. $\operatorname{cosec}^2 \pi x = \frac{1}{\pi^2} \sum_{k=-\infty}^{\infty} \frac{1}{(x-k)^2} = \frac{1}{\pi^2 x^2} + \frac{2}{\pi^2} \sum_{k=1}^{\infty} \frac{x^2+k^2}{(x^2-k^2)^2}$. JO (446)

5. $$\frac{1+x\operatorname{cosec} x}{2x^2} = \frac{1}{x^2} - \sum_{k=1}^{\infty} \frac{(-1)^{k+1}}{(x^2-k^2\pi^2)}.$$ JO (449)

6. $$\operatorname{cosec} \pi x = \frac{1}{\pi x} + \frac{1}{\pi} \sum_{k=-\infty}^{\infty} (-1)^k \left(\frac{1}{x-k} + \frac{1}{k}\right).$$ JO (450b)

1.423 $$\frac{\pi^2}{4m^2} \operatorname{cosec}^2 \frac{\pi}{m} + \frac{\pi}{4m} \operatorname{ctg} \frac{\pi}{m} - \frac{1}{2} = \sum_{k=1}^{\infty} \frac{1}{(1-k^2m^2)^2}.$$ JO (477)

1.43 Representation in the form of an infinite product

1.431

1. $$\sin x = x \prod_{k=1}^{\infty} \left(1 - \frac{x^2}{k^2\pi^2}\right).$$ EU

2. $$\operatorname{sh} x = x \prod_{k=1}^{\infty} \left(1 + \frac{x^2}{k^2\pi^2}\right).$$ EU

3. $$\cos x = \prod_{k=0}^{\infty} \left(1 - \frac{4x^2}{(2k+1)^2\pi^2}\right).$$ EU

4. $$\operatorname{ch} x = \prod_{k=0}^{\infty} \left(1 + \frac{4x^2}{(2k+1)^2\pi^2}\right).$$ EU

1.432

1. $\cos x - \cos y =$

$$= 2\left(1 - \frac{x^2}{y^2}\right) \sin^2 \frac{y}{2} \prod_{k=1}^{\infty} \left(1 - \frac{x^2}{(2k\pi+y)^2}\right)\left(1 - \frac{x^2}{(2k\pi-y)^2}\right).$$ AD (653.2)

2. $\operatorname{ch} x - \cos y =$

$$= 2\left(1 + \frac{x^2}{y^2}\right) \sin^2 \frac{y}{2} \prod_{k=1}^{\infty} \left(1 + \frac{x^2}{(2k\pi+y)^2}\right)\left(1 + \frac{x^2}{(2k\pi-y)^2}\right).$$ AD (653.1)

1.433 $$\cos \frac{\pi x}{4} - \sin \frac{\pi x}{4} = \prod_{k=1}^{\infty} \left[1 + \frac{(-1)^k x}{2k-1}\right].$$ BR* 189

1.434 $$1 - \cos^2 x = \frac{1}{4}(\pi + 2x)^2 \prod_{k=1}^{\infty} \left[1 - \left(\frac{\pi+2x}{2k\pi}\right)^2\right]^2.$$ MO 216

1.435 $$\frac{\sin \pi(x+a)}{\sin \pi a} = \frac{x+a}{a} \prod_{k=1}^{\infty} \left(1 - \frac{x}{k-a}\right)\left(1 + \frac{x}{k+a}\right).$$ MO 216

1.436 $$1 - \frac{\sin^2 \pi x}{\sin^2 \pi a} = \prod_{k=-\infty}^{\infty} \left[1 - \left(\frac{x}{k-a}\right)^2\right].$$ MO 216

1.437 $$\frac{\sin 3x}{\sin x} = -\prod_{k=-\infty}^{\infty} \left[1 - \left(\frac{2x}{x+k\pi}\right)^2\right].$$ MO 216

1.438 $\dfrac{\operatorname{ch} x - \cos a}{1 - \cos a} = \prod_{k=-\infty}^{\infty} \left[1 + \left(\dfrac{x}{2k\pi + a}\right)^2\right].$ MO 216

1.439

1. $\sin x = x \prod_{k=1}^{\infty} \cos \dfrac{x}{2^k}$ $[|x| < 1].$ AD (615), MO 216

2. $\dfrac{\sin x}{x} = \prod_{k=1}^{\infty} \left[1 - \dfrac{4}{3}\sin^2\left(\dfrac{x}{3^k}\right)\right].$ MO 216

1.44-1.45 Trigonometric (Fourier) series

1.441

1. $\sum_{k=1}^{\infty} \dfrac{\sin kx}{k} = \dfrac{\pi - x}{2}$ $[0 < x < 2\pi].$ FI III 539

2. $\sum_{k=1}^{\infty} \dfrac{\cos kx}{k} = \dfrac{1}{2} \ln \dfrac{1}{2(1 - \cos x)}$ $[0 < x < 2\pi].$ FI III 550a, AD (6814)

3. $\sum_{k=1}^{\infty} \dfrac{(-1)^{k-1} \sin kx}{k} = \dfrac{x}{2}$ $[-\pi < x < \pi].$ FI III 542

4. $\sum_{k=1}^{\infty} (-1)^{k-1} \dfrac{\cos kx}{k} = \ln\left(2 \cos \dfrac{x}{2}\right)$ $[-\pi < x < \pi].$ FI III 550

1.442

1. $\sum_{k=1}^{\infty} \dfrac{\sin (2k-1) x}{2k-1} = \dfrac{\pi}{4}$ $[0 < x < \pi].$ FI III 541

2. $\sum_{k=1}^{\infty} \dfrac{\cos (2k-1) x}{2k-1} = \dfrac{1}{2} \ln \operatorname{ctg} \dfrac{x}{2}$ $[0 < x < \pi].$ BR* 168, JO (266), GI III (195)

3. $\sum_{k=1}^{\infty} (-1)^{k-1} \dfrac{\sin (2k-1) x}{2k-1} = \dfrac{1}{2} \ln \operatorname{tg}\left(\dfrac{\pi}{4} + \dfrac{x}{2}\right)$ $\left[-\dfrac{\pi}{2} < x < \dfrac{\pi}{2}\right].$ BR* 168, JO (268)a

4. $\sum_{k=1}^{\infty} (-1)^{k-1} \dfrac{\cos (2k-1) x}{2k-1} = -\dfrac{\pi}{4}$ $\left[\dfrac{\pi}{2} < x < \pi\right].$ BR* 168, JO (269)

1.443

1. $\sum_{k=1}^{\infty} \dfrac{\cos k\pi x}{k^{2n}} = (-1)^{n-1} 2^{2n-1} \dfrac{\pi^{2n}}{(2n)!} \sum_{k=0}^{2n} (-1)^k \binom{2n}{k} B_{2n-k} \varrho^k;$

$= (-1)^{n-1} \dfrac{1}{2} \dfrac{(2\pi)^{2n}}{(2n)!} B_{2n}\left(\dfrac{x}{2}\right)$

$\left[0 < x < 1, \varrho = \dfrac{x}{2} - E\left(\dfrac{x}{2}\right)\right].$ CE 340, GE 71

2. $\sum_{k=1}^{\infty} \frac{\sin k\pi x}{k^{2n+1}} = (-1)^{n-1} 2^{2n} \frac{\pi^{2n+1}}{(2n+1)!} \sum_{k=0}^{2n+1} \binom{2n+1}{k} B_{2n-k+1} \varrho^k;$

$= (-1)^{n-1} \frac{1}{2} \frac{(2\pi)^{2n+1}}{(2n+1)!} B_{2n+1}\left(\frac{x}{2}\right)$

$\left[0 < x < 1;\ \varrho = \frac{x}{2} - E\left(\frac{x}{2}\right)\right].$ CE 340

3. $\sum_{k=1}^{\infty} \frac{\cos kx}{k^2} = \frac{\pi^2}{6} - \frac{\pi x}{2} + \frac{x^2}{4}$ $[0 \leqslant x \leqslant 2\pi].$ FI III 547

4. $\sum_{k=1}^{\infty} (-1)^{k-1} \frac{\cos kx}{k^2} = \frac{\pi^2}{12} - \frac{x^2}{4}$ $[-\pi \leqslant x \leqslant \pi].$ FI III 544

5. $\sum_{k=1}^{\infty} \frac{\sin kx}{k^3} = \frac{\pi^2 x}{6} - \frac{\pi x^2}{4} + \frac{x^3}{12}$ AD (6816)

6. $\sum_{k=1}^{\infty} \frac{\cos kx}{k^4} = \frac{\pi^4}{90} - \frac{\pi^2 x^2}{12} + \frac{\pi x^3}{12} - \frac{x^4}{48}$ $[0 \leqslant x \leqslant 2\pi].$ AD (6817)

7. $\sum_{k=1}^{\infty} \frac{\sin kx}{k^5} = \frac{\pi^4 x}{90} - \frac{\pi^2 x^3}{36} + \frac{\pi x^4}{48} - \frac{x^5}{240}$ AD (6818)

1.444

1. $\sum_{k=1}^{\infty} \frac{\sin 2(k+1)x}{k(k+1)} = \sin 2x - (\pi - 2x)\sin^2 x - \sin x \cos x \ln(4\sin^2 x)$

$[0 \leqslant x \leqslant \pi].$ BR* 168, GI III (190)

2. $\sum_{k=1}^{\infty} \frac{\cos 2(k+1)x}{k(k+1)} = \cos 2x - \left(\frac{\pi}{2} - x\right)\sin 2x + \sin^2 x \ln(4\sin^2 x)$

$[0 \leqslant x \leqslant \pi].$ BR* 168

3. $\sum_{k=1}^{\infty} (-1)^k \frac{\sin(k+1)x}{k(k+1)} = \sin x - \frac{x}{2}(1 + \cos x) - \sin x \ln\left|2\cos\frac{x}{2}\right|.$ MO 213

4. $\sum_{k=1}^{\infty} (-1)^k \frac{\cos(k+1)x}{k(k+1)} = \cos x - \frac{x}{2}\sin x - (1 + \cos x)\ln\left|2\cos\frac{x}{2}\right|.$

MO 213

5. $\sum_{k=0}^{\infty} (-1)^k \frac{\sin(2k+1)x}{(2k+1)^2} = \frac{\pi}{4} x$ $\left[-\frac{\pi}{2} \leqslant x \leqslant \frac{\pi}{2}\right];$

$= \frac{\pi}{4}(\pi - x)$ $\left[\frac{\pi}{2} \leqslant x \leqslant \frac{3}{2}\pi\right].$ MO 213

6. $\sum_{k=1}^{\infty} \frac{\cos(2k-1)x}{(2k-1)^2} = \frac{\pi}{4}\left(\frac{\pi}{2} - |x|\right)$ $[-\pi \leqslant x \leqslant \pi].$ FI III 546

7. $\sum_{k=1}^{\infty} \frac{\cos 2kx}{(2k-1)(2k+1)} = \frac{1}{2} - \frac{\pi}{4}\sin x$ $\left[0 \leqslant x \leqslant \frac{\pi}{2}\right].$ JO (591)

1.445

1. $$\sum_{k=1}^{\infty} \frac{k \sin kx}{k^2+\alpha^2} = \frac{\pi}{2} \frac{\operatorname{sh} \alpha(\pi - x)}{\operatorname{sh} \alpha\pi} \qquad [0 < x < 2\pi].$$ BR* 157, JO (411)

2. $$\sum_{k=1}^{\infty} \frac{\cos kx}{k^2+\alpha^2} = \frac{\pi}{2\alpha} \frac{\operatorname{ch} \alpha(\pi - x)}{\operatorname{sh} \alpha\pi} - \frac{1}{2\alpha^2} \qquad [0 < x < 2\pi].$$ BR* 257, JO (410)

3. $$\sum_{k=1}^{\infty} \frac{(-1)^k \cos kx}{k^2+\alpha^2} = \frac{\pi}{2\alpha} \frac{\operatorname{ch} \alpha x}{\operatorname{sh} \alpha\pi} - \frac{1}{2\alpha^2} \qquad [-\pi \leqslant x \leqslant \pi].$$ FI III 546

4. $$\sum_{k=1}^{\infty} (-1)^{k-1} \frac{k \sin kx}{k^2+\alpha^2} = \frac{\pi}{2} \frac{\operatorname{sh} \alpha x}{\operatorname{sh} \alpha\pi} \qquad [-\pi < x < \pi].$$ FI III 546

5. $$\sum_{k=0}^{\infty} \frac{k \sin kx}{k^2-\alpha^2} = \pi \frac{\sin\{\alpha[(2m+1)\pi - x]\}}{2 \sin \alpha\pi}$$

$[2m\pi \leqslant x \leqslant (2m+2)\pi$, α-not an integer$]$. MO 213

6. $$\sum_{k=1}^{\infty} \frac{\cos kx}{k^2-\alpha^2} = \frac{1}{2\alpha^2} - \frac{\pi}{2} \frac{\cos[\alpha\{(2m+1)\pi - x\}]}{\alpha \sin \alpha\pi}$$

$[2m\pi \leqslant x \leqslant (2m+2)\pi$, α-not an integer$]$. MO 213

7. $$\sum_{k=0}^{\infty} (-1)^k \frac{k \sin kx}{k^2-\alpha^2} = \pi \frac{\sin[\alpha(2m\pi - x)]}{2 \sin \alpha\pi}$$

$[(2m-1)\pi \leqslant x \leqslant (2m+1)\pi$, α-not an integer$]$. FI III 545a

8. $$\sum_{k=1}^{\infty} (-1)^k \frac{\cos kx}{k^2-\alpha^2} = \frac{1}{2\alpha^2} - \frac{\pi}{2} \frac{\cos[\alpha(2m\pi - x)]}{\alpha \sin \alpha\pi}$$

$[(2m-1)\pi \leqslant x \leqslant (2m+1)\pi$, α-not an integer$]$. FI III 545a

1.446

$$\sum_{k=1}^{\infty} \frac{(-1)^{k+1} \cos(2k+1)x}{(2k-1)(2k+1)(2k+3)} = \frac{\pi}{8} \cos^2 x - \frac{1}{3} \cos x$$

$$\left[-\frac{\pi}{2} \leqslant x \leqslant \frac{\pi}{2}\right].$$ BR* 256, GI III (189)

1.447

1. $$\sum_{k=1}^{\infty} p^k \sin kx = \frac{p \sin x}{1 - 2p\cos x + p^2}$$ FI II 559

2. $$\sum_{k=0}^{\infty} p^k \cos kx = \frac{1 - p\cos x}{1 - 2p\cos x + p^2}$$ FI II 559a

3. $$1 + 2\sum_{k=1}^{\infty} p^k \cos kx = \frac{1 - p^2}{1 - 2p\cos x + p^2}$$ FI II 559a, MO 213

$[|p| < 1]$ (applies to 1–3).

1.448

1. $$\sum_{k=1}^{\infty} \frac{p^k \sin kx}{k} = \operatorname{arctg} \frac{p \sin x}{1 - p \cos x}$$ FI II 559

2. $$\sum_{k=1}^{\infty} \frac{p^k \cos kx}{k} = \ln \frac{1}{\sqrt{1 - 2p \cos x + p^2}}$$ FI II 559

3. $$\sum_{k=1}^{\infty} \frac{p^{2k-1} \sin (2k-1) x}{2k-1} = \frac{1}{2} \operatorname{arctg} \frac{2p \sin x}{1 - p^2}$$ JO (594)

4. $$\sum_{k=1}^{\infty} \frac{p^{2k-1} \cos (2k-1) x}{2k-1} = \frac{1}{4} \ln \frac{1 + 2p \cos x + p^2}{1 - 2p \cos x + p^2}$$ JO (259)

$[0 < x < 2\pi, \ p^2 \leqslant 1]$.

5. $$\sum_{k=1}^{\infty} \frac{(-1)^{k-1} p^{2k-1} \sin (2k-1) x}{2k-1} = \frac{1}{4} \ln \frac{1 + 2p \sin x + p^2}{1 - 2p \sin x + p^2}$$ JO (261)

6. $$\sum_{k=1}^{\infty} \frac{(-1)^{k-1} p^{2k-1} \cos (2k-1) x}{2k-1} = \frac{1}{2} \operatorname{arctg} \frac{2p \cos x}{1 - p^2}$$ JO (597)

$[0 < x < \pi, \ p^2 \leqslant 1]$.

1.449

1. $$\sum_{k=1}^{\infty} \frac{p^k \sin kx}{k!} = e^{p \cos x} \sin (p \sin x)$$ JO (486)

2. $$\sum_{k=0}^{\infty} \frac{p^k \cos kx}{k!} = e^{p \cos x} \cos (p \sin x)$$ JO (485)

$[p^2 \leqslant 1]$.

Fourier expansions of hyperbolic functions

1.451

1. $$\operatorname{sh} x = \cos x \sum_{k=0}^{\infty} \frac{(1^2+0^2)(1^2+2^2) \ldots [1^2+(2k)^2]}{(2k+1)!} \sin^{2k+1} x.$$ JO (504)

2. $$\operatorname{ch} x = \cos x + \cos x \sum_{k=1}^{\infty} \frac{(1^2+1^2)(1^2+3^2) \ldots [1^2+(2k-1)^2]}{(2k)!} \sin^{2k} x.$$ JO (503)

1.452

1. $\operatorname{sh}(x\cos\theta)=\sec(x\sin\theta)\sum_{k=0}^{\infty}\frac{x^{2k+1}\cos(2k+1)\theta}{(2k+1)!}$ JO (391)

2. $\operatorname{ch}(x\cos\theta)=\sec(x\sin\theta)\sum_{k=0}^{\infty}\frac{x^{2k}\cos 2k\theta}{(2k)!}$ JO (390)

3. $\operatorname{sh}(x\cos\theta)=\operatorname{cosec}(x\sin\theta)\sum_{k=1}^{\infty}\frac{x^{2k}\sin 2k\theta}{(2k)!}$ JO (393)

4. $\operatorname{ch}(x\cos\theta)=\operatorname{cosec}(x\sin\theta)\sum_{k=0}^{\infty}\frac{x^{2k+1}\sin(2k+1)\theta}{(2k+1)!}$ JO (392)

$[x^2<1]$.

1.46 Series of products of exponential and trigonometric functions

1.461

1. $\sum_{k=0}^{\infty}e^{-kt}\sin kx=\frac{1}{2}\frac{\sin x}{\operatorname{ch}t-\cos x}\quad[t>0].$ MO 213

2. $1+2\sum_{k=1}^{\infty}e^{-kt}\cos kx=\frac{\operatorname{sh}t}{\operatorname{ch}t-\cos x}\quad[t>0].$ MO 213

1.462 $\sum_{k=1}^{\infty}\frac{\sin kx\sin ky}{k}e^{-2k|t|}=\frac{1}{4}\ln\left[\frac{\sin^2\frac{x+y}{2}+\operatorname{sh}^2 t}{\sin^2\frac{x-y}{2}+\operatorname{sh}^2 t}\right].$ MO 214

1.463

1. $e^{x\cos\varphi}\cos(x\sin\varphi)=\sum_{n=0}^{\infty}\frac{x^n\cos n\varphi}{n!}\quad[x^2<1].$ AD (6476.1)

2. $e^{x\cos\varphi}\sin(x\sin\varphi)=\sum_{n=1}^{\infty}\frac{x^n\sin n\varphi}{n!}\quad[x^2<1].$ AD (6476.2)

1.47 Series of hyperbolic functions

1.471

1. $\sum_{k=1}^{\infty}\frac{\operatorname{sh}kx}{k!}=e^{\operatorname{ch}x}\operatorname{sh}(\operatorname{sh}x).$ JO (395)

2. $\sum_{k=0}^{\infty}\frac{\operatorname{ch}kx}{k!}=e^{\operatorname{ch}x}\operatorname{ch}(\operatorname{sh}x).$ JO (394)

1.472

1. $\sum_{k=1}^{\infty}p^k\operatorname{sh}kx=\frac{p\operatorname{sh}x}{1-2p\operatorname{ch}x+p^2}\quad[p^2<1].$ JO (396)

2. $\sum_{k=0}^{\infty}p^k\operatorname{ch}kx=\frac{1-p\operatorname{ch}x}{1-2p\operatorname{ch}x+p^2}\quad[p^2<1].$ JO (397)a

1.48 Lobachevskiy's "Angle of parallelism" $\Pi(x)$

1.480 Definition.

1. $\Pi(x) = 2 \operatorname{arcctg} e^{x} = 2 \operatorname{arctg} e^{-x} \qquad [x \geqslant 0].$ LO III 297, LO I 120

2. $\Pi(x) = \pi - \Pi(-x) \qquad [x < 0].$ LO III 183, LO I 193

1.481 Functional relations.

1. $\sin \Pi(x) = \dfrac{1}{\operatorname{ch} x}.$ LO III 297

2. $\cos \Pi(x) = \operatorname{th} x.$ LO III 297

3. $\operatorname{tg} \Pi(x) = \dfrac{1}{\operatorname{sh} x}.$ LO III 297

4. $\operatorname{ctg} \Pi(x) = \operatorname{sh} x.$ LO III 297

5. $\sin \Pi(x+y) = \dfrac{\sin \Pi(x) \sin \Pi(y)}{1+\cos \Pi(x) \cos \Pi(y)}.$ LO III 297

6. $\cos \Pi(x+y) = \dfrac{\cos \Pi(x) + \cos \Pi(y)}{1+\cos \Pi(x) \cos \Pi(y)}.$ LO III 183

1.482 Connection with the gudermannian.

$$\operatorname{gd}(-x) = \Pi(x) - \frac{\pi}{2}.$$

(Definite) integral of the angle of parallelism; cf. 4.581 and 4.561

1.49 The hyperbolic amplitude (the Gudermannian) $\operatorname{gd} x$

1.490 Definition.

1. $\operatorname{gd} x = \displaystyle\int_0^x \frac{dt}{\operatorname{ch} t} = 2 \operatorname{arctg} e^{x} - \frac{\pi}{2}.$ JA

2. $x = \displaystyle\int_0^{\operatorname{gd} x} \frac{dt}{\cos t} = \ln \operatorname{tg}\left(\frac{\operatorname{gd} x}{2} + \frac{\pi}{4}\right).$ JA

1.491 Functional relations.

1. $\operatorname{ch} x = \sec(\operatorname{gd} x).$ AD (343.1), JA

2. $\operatorname{sh} x = \operatorname{tg}(\operatorname{gd} x).$ AD (343.2), JA

3. $e^{x} = \sec(\operatorname{gd} x) + \operatorname{tg}(\operatorname{gd} x) = \operatorname{tg}\left(\frac{\pi}{4} + \frac{\operatorname{gd} x}{2}\right) = \dfrac{1+\sin(\operatorname{gd} x)}{\cos(\operatorname{gd} x)}.$ AD (343.5), JA

4. $\operatorname{th} x = \sin(\operatorname{gd} x).$ AD (344.3), JA

5. $\operatorname{th} \frac{x}{2} = \operatorname{tg}\left(\frac{1}{2} \operatorname{gd} x\right).$ AD (344.4), JA

6. $\operatorname{arctg}(\operatorname{th} x) = \frac{1}{2} \operatorname{gd} 2x.$ AD (344.6)a

1.492 If $\gamma = \operatorname{gd} x$, then $ix = \operatorname{gd} i\gamma$. JA

1.493 Series expansion.

1. $\frac{\operatorname{gd} x}{2} = \sum_{k=0}^{\infty} \frac{(-1)^k}{2k+1} \operatorname{th}^{2k+1} \frac{x}{2}.$ JA

2. $\frac{x}{2} = \sum_{k=0}^{\infty} \frac{1}{2k+1} \operatorname{tg}^{2k+1}\left(\frac{1}{2} \operatorname{gd} x\right).$ JA

3. $\operatorname{gd} x = x - \frac{x^3}{6} + \frac{x^5}{24} - \frac{61x^7}{5040} + \dots$ JA

4. $x = \operatorname{gd} x + \frac{(\operatorname{gd} x)^3}{6} + \frac{(\operatorname{gd} x)^5}{24} + \frac{61 (\operatorname{gd} x)^7}{5040} + \dots \quad \left[\operatorname{gd} x < \frac{\pi}{2}\right].$ JA

1.5 The Logarithm

1.51 Series representation

1.511 $\ln(1+x) = x - \frac{1}{2}x^2 + \frac{1}{3}x^3 - \frac{1}{4}x^4 + \dots =$

$$= \sum_{k=1}^{\infty} (-1)^{k+1} \frac{x^k}{k} \quad [-1 < x \leqslant 1].$$

1.512

1. $\ln x = (x-1) - \frac{1}{2}(x-1)^2 + \frac{1}{3}(x-1)^3 - \dots =$

$$= \sum_{k=1}^{\infty} (-1)^{k+1} \frac{(x-1)^k}{k} \quad [0 < x \leqslant 2].$$

2. $\ln x = 2\left[\frac{x-1}{x+1} + \frac{1}{3}\left(\frac{x-1}{x+1}\right)^3 + \frac{1}{5}\left(\frac{x-1}{x+1}\right)^5 + \dots\right] =$

$$= 2\sum_{k=1}^{\infty} \frac{1}{2k-1}\left(\frac{x-1}{x+1}\right)^{2k-1} \quad [0 < x].$$

3. $\ln x = \frac{x-1}{x} + \frac{1}{2}\left(\frac{x-1}{x}\right)^2 + \frac{1}{3}\left(\frac{x-1}{x}\right)^3 + \dots =$

$$= \sum_{k=1}^{\infty} \frac{1}{k}\left(\frac{x-1}{x}\right)^k \quad \left[x \geqslant \frac{1}{2}\right].$$ AD (644.6)

1.513

1. $\ln \frac{1+x}{1-x} = 2\sum_{k=1}^{\infty} \frac{1}{2k-1} x^{2k-1} \quad [x^2 < 1].$ FI II 421

2. $\ln \frac{x+1}{x-1} = 2\sum_{k=1}^{\infty} \frac{1}{(2k-1)x^{2k-1}} \quad [x^2 > 1].$ AD (644.9)

3. $\ln \frac{x}{x-1} = \sum_{k=1}^{\infty} \frac{1}{kx^k} \quad [x^2 > 1].$ JO (88a)

4. $\ln \frac{1}{1-x} = \sum_{k=1}^{\infty} \frac{x^k}{k} \quad [x^2 < 1].$ JO (88b)

5. $\frac{1-x}{x} \ln \frac{1}{1-x} = 1 - \sum_{k=1}^{\infty} \frac{x^k}{k(k+1)}$ $[x^2 < 1]$. JO (102)

6. $\frac{1}{1-x} \ln \frac{1}{1-x} = \sum_{k=1}^{\infty} x^k \sum_{n=1}^{k} \frac{1}{n}$ $[x^2 < 1]$. JO (88e)

7. $\frac{(1-x)^2}{2x^3} \ln \frac{1}{1-x} = \frac{1}{2x^2} - \frac{3}{4x} + \sum_{k=1}^{\infty} \frac{x^{k-1}}{k(k+1)(k+2)}$ $[x^2 < 1]$. AD (6445.1)

1.514 $\ln(1 - 2x\cos\varphi + x^2) = -2 \sum_{k=1}^{\infty} \frac{\cos k\varphi}{k} x^k.$ MO 98, FI II 485

$\ln(x + \sqrt{1+x^2}) = \operatorname{Arsh} x$ see **1.631**, **1.641**, **1.642**, **1.646**.

1.515

1. $\ln(1 + \sqrt{1+x^2}) = \ln 2 + \frac{1 \cdot 1}{2 \cdot 2} x^2 - \frac{1 \cdot 1 \cdot 3}{2 \cdot 4 \cdot 4} x^4 - \frac{1 \cdot 1 \cdot 3 \cdot 5}{2 \cdot 4 \cdot 6 \cdot 6} x^6 + \ldots;$

$$= \ln 2 - \sum_{k=1}^{\infty} (-1)^k \frac{(2k-1)!}{2^{2k}(k!)^2} x^{2k} \qquad [x^2 \leqslant 1].$$ JO (91)

2. $\ln(1 + \sqrt{1+x^2}) = \ln x + \frac{1}{x} - \frac{1}{2 \cdot 3x^3} + \frac{1 \cdot 3}{2 \cdot 4 \cdot 5x^5} - \ldots;$

$$= \ln x + \frac{1}{x} + \sum_{k=1}^{\infty} (-1)^k \frac{(2k-1)!}{2^{2k-1} \cdot k!\,(k-1)!\,(2k+1)\,x^{2k+1}}$$

$[x^2 \geqslant 1]$. AD (644.4)

3. $\sqrt{1+x^2} \ln(x + \sqrt{1+x^2}) =$

$$= x - \sum_{k=1}^{\infty} (-1)^k \frac{2^{2k-1}(k-1)!\,k!}{(2k+1)!} x^{2k+1} \qquad [x^2 < 1].$$ JO (93)

4. $\frac{\ln(x + \sqrt{1+x^2})}{\sqrt{1+x^2}} = \sum_{k=0}^{\infty} (-1)^k \frac{2^{2k}(k!)^2}{(2k+1)!} x^{2k+1}$ $[x^2 < 1]$. JO (94)

1.516

1. $\frac{1}{2}\{\ln(1 \pm x)\}^2 = \sum_{k=1}^{\infty} \frac{(\mp 1)^{k+1} x^{k+1}}{k+1} \sum_{n=1}^{k} \frac{1}{n}$ $[x^2 < 1]$. JO (86), JO (85)

2. $\frac{1}{6}\{\ln(1+x)\}^3 = \sum_{k=1}^{\infty} \frac{(-1)^{k+1} x^{k+2}}{k+2} \sum_{n=1}^{k} \frac{1}{n+1} \sum_{m=1}^{n} \frac{1}{m}$ $[x^2 < 1]$. AD (644.14)

3. $-\ln(1+x) \cdot \ln(1-x) = \sum_{k=1}^{\infty} \frac{x^{2k}}{k} \sum_{n=1}^{2k-1} \frac{(-1)^{n+1}}{n}$ $[x^2 < 1]$. | JO (87)

4. $\frac{1}{4x}\left\{\frac{1+x}{\sqrt{x}} \ln \frac{1+\sqrt{x}}{1-\sqrt{x}} + 2\ln(1-x)\right\} = \frac{1}{2x} + \sum_{k=1}^{\infty} \frac{x^{k-1}}{(2k-1)\,2k\,(2k+1)}$

$[0 < x < 1]$. AD (6445.2)

1.517

1. $$\frac{1}{2x}\left\{1-\ln(1+x)-\frac{1-x}{\sqrt{x}}\operatorname{arctg}x\right\}=\sum_{k=1}^{\infty}\frac{(-1)^{k+1}x^{k-1}}{(2k-1)\,2k\,(2k+1)}\quad [0<x\leqslant 1].$$
AD (6445.3)

2. $$\frac{1}{2}\operatorname{arctg}x\ln\frac{1+x}{1-x}=\sum_{k=1}^{\infty}\frac{x^{4k-2}}{2k-1}\sum_{n=1}^{2k-1}\frac{(-1)^{n-1}}{2n-1}\quad [x^2<1].$$
BR* 163

3. $$\frac{1}{2}\operatorname{arctg}x\ln(1+x^2)=\sum_{k=1}^{\infty}\frac{(-1)^{k+1}x^{2k+1}}{2k+1}\sum_{n=1}^{2k}\frac{1}{n}\quad [x^2<1].$$
AD (6455.3)

1.518

1. $$\ln\sin x=\ln x-\frac{x^2}{6}-\frac{x^4}{180}-\frac{x^6}{2835}-\ldots;$$
$$=\ln x+\sum_{k=1}^{\infty}\frac{(-1)^k 2^{2k-1}B_{2k}x^{2k}}{k\,(2k)!}\quad [x^2<\pi^2].$$
AD (643.1)a

2. $$\ln\cos x=-\frac{x^2}{2}-\frac{x^4}{12}-\frac{x^6}{45}-\frac{17x^8}{2520}-\ldots,$$
$$=\sum_{k=1}^{\infty}(-1)^k\frac{2^{2k-1}(2^{2k}-1)B_{2k}}{k\,(2k)!}x^{2k}=-\frac{1}{2}\sum_{k=1}^{\infty}\frac{\sin^{2k}x}{k}\left[x^2<\frac{\pi^2}{4}\right].$$
FI II 524

3. $$\ln\operatorname{tg}x=\ln x+\frac{x^2}{3}+\frac{7}{90}x^4+\frac{62}{2835}x^6+\frac{127}{18{,}900}x^8+\ldots;$$
$$=\ln x+\sum_{k=1}^{\infty}(-1)^{k+1}\frac{(2^{2k-1}-1)\,2^{2k}B_{2k}x^{2k}}{k\,(2k)!}\left[x^2<\frac{\pi^2}{4}\right].$$
AD (643.3)a

Power series for $\frac{\cos}{\sin}\left\{n\ln(x+\sqrt{1+x^2})\right\}$ cf. **1.414**.

1.52 Series of logarithms (cf. 1.431)

1.521

1. $$\sum_{k=1}^{\infty}\ln\left(1-\frac{4x^2}{(2k-1)^2\pi^2}\right)=\ln\cos x.$$

2. $$\sum_{k=1}^{\infty}\ln\left(1-\frac{x^2}{k^2\pi^2}\right)=\ln\sin x-\ln x.$$

1.6 The Inverse Trigonometric and Hyperbolic Functions

1.61 The domain of definition

The principal values of the inverse trigonometric functions are defined by the inequalities:

$$-\frac{\pi}{2} \leqslant \arcsin x \leqslant \frac{\pi}{2}; \quad 0 \leqslant \arccos x \leqslant \pi \qquad [-1 \leqslant x \leqslant 1]. \qquad \text{FI II 553}$$

$$-\frac{\pi}{2} < \operatorname{arctg} x < \frac{\pi}{2}; \ 0 < \operatorname{arcctg} x < \pi \qquad [-\infty < x < +\infty]. \qquad \text{FI II 552}$$

1.62-1.63 Functional relations

1.621 The relationship between the inverse and the direct trigonometric functions.

1. $\arcsin(\sin x) = x - 2n\pi \quad \left[2n\pi - \frac{\pi}{2} \leqslant x \leqslant 2n\pi + \frac{\pi}{2}\right];$
   $= -x + (2n+1)\pi \quad \left[(2n+1)\pi - \frac{\pi}{2} \leqslant x \leqslant (2n+1)\pi + \frac{\pi}{2}\right].$
2. $\arccos(\cos x) = x - 2n\pi \quad [2n\pi \leqslant x \leqslant (2n+1)\pi];$
   $= -x + 2(n+1)\pi \quad [(2n+1)\pi \leqslant x \leqslant 2(n+1)\pi].$
3. $\operatorname{arctg}(\operatorname{tg} x) = x - n\pi \quad \left[n\pi - \frac{\pi}{2} < x < n\pi + \frac{\pi}{2}\right].$
4. $\operatorname{arcctg}(\operatorname{ctg} x) = x - n\pi \quad [n\pi < x < (n+1)\pi].$

1.622 The relationship between the inverse trigonometric functions, the inverse hyperbolic functions, and the logarithm.

1. $\arcsin z = \frac{1}{i}\ln\left(iz + \sqrt{1-z^2}\right) = \frac{1}{i}\operatorname{Arsh}(iz)$
2. $\arccos z = \frac{1}{i}\ln\left(z + \sqrt{z^2-1}\right) = \frac{1}{i}\operatorname{Arch} z.$
3. $\operatorname{arctg} z = \frac{1}{2i}\ln\frac{1+iz}{1-iz} = \frac{1}{i}\operatorname{Arth}(iz).$
4. $\operatorname{arcctg} z = \frac{1}{2i}\ln\frac{iz-1}{iz+1} = i\operatorname{Arcth}(iz).$
5. $\operatorname{Arsh} z = \ln\left(z + \sqrt{z^2+1}\right) = \frac{1}{i}\arcsin(iz).$
6. $\operatorname{Arch} z = \ln\left(z + \sqrt{z^2-1}\right) = i\arccos z.$
7. $\operatorname{Arth} z = \frac{1}{2}\ln\frac{1+z}{1-z} = \frac{1}{i}\operatorname{arctg} iz.$
8. $\operatorname{Arcth} z = \frac{1}{2}\ln\frac{z+1}{z-1} = \frac{1}{i}\operatorname{arcctg}(-iz).$

Relations between different inverse trigonometric functions

1.623

1. $\arcsin x + \arccos x = \frac{\pi}{2}.$ NV 43
2. $\operatorname{arctg} x + \operatorname{arcctg} x = \frac{\pi}{2}.$ NV 43

1.624

1. $\arcsin x = \arccos \sqrt{1-x^2} \qquad [0 \leqslant x \leqslant 1];$

$= -\arccos \sqrt{1-x^2} \qquad [-1 \leqslant x \leqslant 0].$ NV 47 (5)

2. $\arcsin x = \operatorname{arctg} \dfrac{x}{\sqrt{1-x^2}} \qquad [x^2 < 1].$ NV 46 (2)

3. $\arcsin x = \operatorname{arcctg} \dfrac{\sqrt{1-x^2}}{x} \qquad [0 < x \leqslant 1];$

$= \operatorname{arcctg} \dfrac{\sqrt{1-x^2}}{x} - \pi \qquad [-1 \leqslant x < 0].$ NV 49 (10)

4. $\arccos x = \arcsin \sqrt{1-x^2} \qquad [0 \leqslant x \leqslant 1];$

$= \pi - \arcsin \sqrt{1-x^2} \qquad [-1 \leqslant x \leqslant 0].$ NV 48 (6)

5. $\arccos x = \operatorname{arctg} \dfrac{\sqrt{1-x^2}}{x} \qquad [0 < x \leqslant 1];$

$= \pi + \operatorname{arctg} \dfrac{\sqrt{1-x^2}}{x} \qquad [-1 \leqslant x < 0].$ NV 48 (8)

6. $\arccos x = \operatorname{arcctg} \dfrac{x}{\sqrt{1-x^2}} \qquad [-1 \leqslant x < 1].$ NV 46 (4)

7. $\operatorname{arctg} x = \arcsin \dfrac{x}{\sqrt{1+x^2}}.$ NV 6 (3)

8. $\operatorname{arctg} x = \arccos \dfrac{1}{\sqrt{1+x^2}} \qquad [x \geqslant 0];$

$= -\arccos \dfrac{1}{\sqrt{1+x^2}} \qquad [x \leqslant 0].$ NV 48 (7)

9. $\operatorname{arctg} x = \operatorname{arcctg} \dfrac{1}{x} \qquad [x > 0],$

$= \operatorname{arcctg} \dfrac{1}{x} - \pi \qquad [x < 0].$ NV 49 (9)

10. $\operatorname{arcctg} x = \arcsin \dfrac{1}{\sqrt{1+x^2}} \qquad [x > 0];$

$= \pi - \arcsin \dfrac{1}{\sqrt{1+x^2}} \qquad [x < 0].$ NV 49 (11)

11. $\operatorname{arcctg} x = \arccos \dfrac{x}{\sqrt{1+x^2}}.$ NV 46 (4)

12. $\operatorname{arcctg} x = \operatorname{arctg} \dfrac{1}{x} \qquad [x > 0];$

$= \pi + \operatorname{arctg} \dfrac{1}{x} \qquad [x < 0].$ NV 49 (12)

1.625.

1. $\arcsin x + \arcsin y = \arcsin \left(x\sqrt{1-y^2} + y\sqrt{1-x^2}\right)$

$[xy \leqslant 0 \quad \text{or} \quad x^2+y^2 \leqslant 1];$

$= \pi - \arcsin \left(x\sqrt{1-y^2} + y\sqrt{1-x^2}\right)$

$[x > 0,\ y > 0 \ \text{and} \ x^2+y^2 > 1];$

$= -\pi - \arcsin \left(x\sqrt{1-y^2} + y\sqrt{1-x^2}\right)$

$[x < 0,\ y < 0, \ \text{and} \ x^2+y^2 > 1].$ NV 54(1), GI I (880)

2. $\arcsin x + \arcsin y = \arccos\left(\sqrt{1-x^2}\sqrt{1-y^2} - xy\right) \quad [x \geqslant 0,\ y \geqslant 0];$

$= -\arccos\left(\sqrt{1-x^2}\sqrt{1-y^2} - xy\right) \quad [x < 0,\ y < 0].$ NV 55

3. $\arcsin x + \arcsin y = \operatorname{arctg} \dfrac{x\sqrt{1-y^2} + y\sqrt{1-x^2}}{\sqrt{1-x^2}\sqrt{1-y^2} - xy}$

$[xy \leqslant 0 \quad \text{or} \quad x^2 + y^2 < 1];$

$= \operatorname{arctg} \dfrac{x\sqrt{1-y^2} + y\sqrt{1-x^2}}{\sqrt{1-x^2}\sqrt{1-y^2} - xy} + \pi$

$[x > 0,\ y > 0 \ \text{and}\ x^2 + y^2 > 1];$

$= \operatorname{arctg} \dfrac{x\sqrt{1-y^2} + y\sqrt{1-x^2}}{\sqrt{1-x^2}\sqrt{1-y^2} - xy} - \pi$

$[x < 0,\ y < 0 \ \text{and}\ x^2 + y^2 > 1].$ NV 56

4. $\arcsin x - \arcsin y = \arcsin\left(x\sqrt{1-y^2} - y\sqrt{1-x^2}\right)$

$[xy \geqslant 0 \quad \text{or} \quad x^2 + y^2 \leqslant 1];$

$= \pi - \arcsin\left(x\sqrt{1-y^2} - y\sqrt{1-x^2}\right)$

$[x > 0,\ y < 0 \ \text{and}\ x^2 + y^2 > 1];$

$= -\pi - \arcsin\left(x\sqrt{1-y^2} - y\sqrt{1-x^2}\right)$

$[x < 0,\ y > 0 \ \text{and}\ x^2 + y^2 > 1].$ NV 55(2)

5. $\arcsin x - \arcsin y = \arccos\left(\sqrt{1-x^2}\sqrt{1-y^2} + xy\right) \quad [x > y];$

$= -\arccos\left(\sqrt{1-x^2}\sqrt{1-y^2} + xy\right) \quad [x < y].$ NV 56

6. $\arccos x + \arccos y = \arccos\left(xy - \sqrt{1-x^2}\sqrt{1-y^2}\right) \quad [x + y \geqslant 0];$

$= 2\pi - \arccos\left(xy - \sqrt{1-x^2}\sqrt{1-y^2}\right) \quad [x + y < 0].$ NV 57 (3)

7. $\arccos x - \arccos y = -\arccos\left(xy + \sqrt{1-x^2}\sqrt{1-y^2}\right) \quad [x \geqslant y];$

$= \arccos\left(xy + \sqrt{1-x^2}\sqrt{1-y^2}\right) \quad [x < y].$ NV 57 (4)

8. $\operatorname{arctg} x + \operatorname{arctg} y = \operatorname{arctg} \dfrac{x+y}{1-xy} \quad [xy < 1];$

$= \pi + \operatorname{arctg} \dfrac{x+y}{1-xy} \quad [x > 0,\ xy > 1];$

$= -\pi + \operatorname{arctg} \dfrac{x+y}{1-xy} \quad [x < 0,\ xy > 1].$ NV 59(5), GI I (879)

9. $\operatorname{arctg} x - \operatorname{arctg} y = \operatorname{arctg} \dfrac{x-y}{1+xy} \quad [xy > -1];$

$= \pi + \operatorname{arctg} \dfrac{x-y}{1+xy} \quad [x > 0,\ xy < -1];$

$= -\pi + \operatorname{arctg} \dfrac{x-y}{1+xy} \quad [x < 0,\ xy < -1].$ NV 59 (6)

1.626

1. $2 \arcsin x = \arcsin(2x\sqrt{1-x^2}) \quad \left[|x| \leqslant \frac{1}{\sqrt{2}}\right];$

$= \pi - \arcsin(2x\sqrt{1-x^2}) \quad \left[\frac{1}{\sqrt{2}} < x \leqslant 1\right];$

$= -\pi - \arcsin(2x\sqrt{1-x^2}) \quad \left[-1 \leqslant x < -\frac{1}{\sqrt{2}}\right].$

NV 61 (7)

2. $2 \arccos x = \arccos(2x^2-1) \quad [0 \leqslant x \leqslant 1];$

$= 2\pi - \arccos(2x^2-1) \quad [-1 \leqslant x < 0].$ NV 61 (8)

3. $2 \operatorname{arctg} x = \operatorname{arctg} \frac{2x}{1-x^2} \quad [|x| < 1];$

$= \operatorname{arctg} \frac{2x}{1-x^2} + \pi \quad [x > 1];$

$= \operatorname{arctg} \frac{2x}{1-x^2} - \pi \quad [x < -1].$ NV 61 (9)

1.627

1. $\operatorname{arctg} x + \operatorname{arctg} \frac{1}{x} = \frac{\pi}{2} \quad [x > 0];$

$= -\frac{\pi}{2} \quad [x < 0].$ GI I (878)

2. $\operatorname{arctg} x + \operatorname{arctg} \frac{1-x}{1+x} = \frac{\pi}{4} \quad [x > -1];$

$= -\frac{3}{4}\pi \quad [x < -1].$ NV 62, GI I (881)

1.628

1. $\arcsin \frac{2x}{1+x^2} = -\pi - 2 \operatorname{arctg} x \quad [x < -1];$

$= 2 \operatorname{arctg} x \quad [-1 \leqslant x \leqslant 1];$

$= \pi - 2 \operatorname{arctg} x \quad [x > 1].$ NV 65

2. $\arccos \frac{1-x^2}{1+x^2} = 2 \operatorname{arctg} x \quad [x \geqslant 0];$

$= -2 \operatorname{arctg} x \quad [x \leqslant 0].$ NV 66

1.629 $\frac{2x-1}{2} - \frac{1}{\pi} \operatorname{arctg}\left(\operatorname{tg} \frac{2x-1}{2}\pi\right) = E(x).$ GI (886)

1.631 Relations between the inverse hyperbolic functions.

1. $\operatorname{Arsh} x = \operatorname{Arch} \sqrt{x^2+1} = \operatorname{Arth} \frac{x}{\sqrt{x^2+1}}.$ JA

2. $\operatorname{Arch} x = \operatorname{Arsh} \sqrt{x^2-1} = \operatorname{Arth} \frac{\sqrt{x^2-1}}{x}.$ JA

3. $\operatorname{Arth} x = \operatorname{Arsh} \frac{x}{\sqrt{1-x^2}} = \operatorname{Arch} \frac{1}{\sqrt{1-x^2}} = \operatorname{Arcth} \frac{1}{x}.$ JA

4. $\operatorname{Arsh} x \pm \operatorname{Arsh} y = \operatorname{Arsh}(x\sqrt{1+y^2} \pm y\sqrt{1+x^2}).$ JA

5. $\operatorname{Arch} x \pm \operatorname{Arch} y = \operatorname{Arch}(xy \pm \sqrt{(x^2-1)(y^2-1)}).$ JA

6. $\operatorname{Arth} x \pm \operatorname{Arth} y = \operatorname{Arth} \frac{x \pm y}{1 \pm xy}.$ JA

1.64 Series representations

1.641

1. $$\arcsin x = \frac{\pi}{2} - \arccos x = x + \frac{1}{2\cdot 3}x^3 + \frac{1\cdot 3}{2\cdot 4\cdot 5}x^5 + \frac{1\cdot 3\cdot 5}{2\cdot 4\cdot 6\cdot 7}x^7 + \ldots;$$
$$= \sum_{k=0}^{\infty} \frac{(2k)!}{2^{2k}(k!)^2(2k+1)} x^{2k+1} = xF\left(\frac{1}{2}, \frac{1}{2}; \frac{3}{2}; x^2\right) \quad [x^2 < 1].$$
FI II 479

2. $$\operatorname{Arsh} x = x - \frac{1}{2\cdot 3}x^3 + \frac{1\cdot 3}{2\cdot 4\cdot 5}x^5 - \ldots;$$
$$= \sum_{k=0}^{\infty} (-1)^k \frac{(2k)!}{2^{2k}(k!)^2(2k+1)} x^{2k+1};$$
$$= xF\left(\frac{1}{2}, \frac{1}{2}; \frac{3}{2}; -x^2\right) \quad [x^2 < 1].$$
FI II 480

1.642

1. $$\operatorname{Arsh} x = \ln 2x + \frac{1}{2}\frac{1}{2x^2} - \frac{1\cdot 3}{2\cdot 4}\frac{1}{4x^4} + \ldots;$$
$$= \ln 2x + \sum_{k=1}^{\infty} (-1)^{k+1} \frac{(2k)!\, x^{-2k}}{2^{2k}(k!)^2\, 2k} \quad [x^2 > 1].$$
AD (6480. 2)a

2. $$\operatorname{Arch} x = \ln 2x - \sum_{k=1}^{\infty} \frac{(2k)!}{2^{2k}(k!)^2\, 2k} x^{-2k} \quad [x^2 > 1].$$
AD (6480. 3)a

1.643

1. $$\operatorname{arctg} x = x - \frac{x^3}{3} + \frac{x^5}{5} - \frac{x^7}{7} + \ldots;$$
$$= \sum_{k=0}^{\infty} \frac{(-1)^k x^{2k+1}}{2k+1} \quad [x^2 \leqslant 1].$$
FI II 479

2. $$\operatorname{Arth} x = x + \frac{x^3}{3} + \frac{x^5}{5} + \ldots = \sum_{k=0}^{\infty} \frac{x^{2k+1}}{2k+1} \quad [x^2 < 1].$$
AD (6480. 4)

1.644

1. $$\operatorname{arctg} x = \frac{x}{\sqrt{1+x^2}} \sum_{k=0}^{\infty} \frac{(2k)!}{2^{2k}(k!)^2(2k+1)} \left(\frac{x^2}{1+x^2}\right)^k;$$
$$= \frac{x}{\sqrt{1+x^2}} F\left(\frac{1}{2}, \frac{1}{2}; \frac{3}{2}; \frac{x^2}{1+x^2}\right) \quad [x^2 < \infty].$$
AD (641.3)

2. $$\operatorname{arctg} x = \frac{\pi}{2} - \frac{1}{x} + \frac{1}{3x^3} - \frac{1}{5x^5} + \frac{1}{7x^7} - \ldots;$$
$$= \frac{\pi}{2} - \sum_{k=0}^{\infty} (-1)^k \frac{1}{(2k+1)\, x^{2k+1}} \quad [x^2 \geqslant 1]$$
(see also **1.643**).
AD (641.4)

1.645

1. $$\operatorname{arcsec} x = \frac{\pi}{2} - \frac{1}{x} - \frac{1}{2\cdot 3x^3} - \frac{1\cdot 3}{2\cdot 4\cdot 5x^5} - \dots;$$

$$= \frac{\pi}{2} - \sum_{k=0}^{\infty} \frac{(2k)!\, x^{-(2k+1)}}{(k!)^2\, 2^{2k}\, (2k+1)};$$

$$= \frac{\pi}{2} - \frac{1}{x} F\left(\frac{1}{2}, \frac{1}{2}; \frac{3}{2}; \frac{1}{x^2}\right) \quad [x^2 > 1].$$

AD (641.5)

2. $$(\arcsin x)^2 = \sum_{k=0}^{\infty} \frac{2^{2k}\, (k!)^2 x^{2k+2}}{(2k+1)!\, (k+1)} \quad [x^2 \leqslant 1].$$

AD (642.2), GI III (152)a

3. $$(\arcsin x)^3 = x^3 + \frac{3!}{5!} 3^2 \left(1 + \frac{1}{3^2}\right) x^5 + \frac{3!}{7!} 3^2 \cdot 5^2 \left(1 + \frac{1}{3^2} + \frac{1}{5^2}\right) x^7 + \dots \quad [x^2 \leqslant 1].$$

BR* 188, AD (642.2), GI III (153)a

1.646

1. $$\operatorname{Arsh} \frac{1}{x} = \operatorname{Arcosech} x = \sum_{k=0}^{\infty} \frac{(-1)^k\, (2k)!}{2^{2k}\, (k!)^2\, (2k+1)} x^{-2k-1} \quad [x^2 > 1].$$

AD (6480.5)

2. $$\operatorname{Arch} \frac{1}{x} = \operatorname{Arsech} x = \ln \frac{2}{x} - \sum_{k=1}^{\infty} \frac{(2k)!}{2^{2k}\, (k!)^2 2k} x^{2k} \quad [0 < x < 1].$$

AD (6480.6)

3. $$\operatorname{Arsh} \frac{1}{x} = \operatorname{Arcosech} x = \ln \frac{2}{x} + \sum_{k=1}^{\infty} \frac{(-1)^{k+1}\, (2k)!}{2^{2k}\, (k!)^2\, 2k} x^{2k} \quad [0 < x < 1].$$

AD (6480.7)a

4. $$\operatorname{Arth} \frac{1}{x} = \operatorname{Arcth} x = \sum_{k=0}^{\infty} \frac{x^{-(2k+1)}}{2k+1} \quad [x^2 > 1].$$

AD (6480.8)

2. INDEFINITE INTEGRALS OF ELEMENTARY FUNCTIONS

2.0 Introduction

2.00 General remarks

We omit the constant of integration in all the formulas of this chapter. Therefore, the equality sign (=) means that the functions on the left and right of this symbol differ by a constant. For example (see 201 15.), we write

$$\int \frac{dx}{1+x^2} = \operatorname{arctg} x = -\operatorname{arcctg} x,$$

although

$$\operatorname{arctg} x = -\operatorname{arcctg} x + \frac{\pi}{2}.$$

When we integrate certain functions, we obtain the logarithm of the absolute value $\left(\text{for example, } \int \frac{dx}{\sqrt{1+x^2}} = \ln\left|x+\sqrt{1+x^2}\right|\right)$. In such formulas, the absolute-value bars in the argument of the logarithm are omitted for simplicity in writing.

In certain cases, it is important to give the complete form of the primitive function. Such primitive functions, written in the form of definite integrals, are given in Chapter 2 and in other chapters.

Closely related to these formulas are formulas in which the limits of integration and the integrand depend on the same parameter.

A number of formulas lose their meaning for certain values of the constants (parameters) or for certain relationships between these constants (for example, formula 2.02 8. for $n = -1$ or formula 2.02 15. for $a = b$). These values of the constants and the relationships between them are for the most part completely clear from the very structure of the right hand member of the formula (the one not containing an integral sign). Therefore, throughout the chapter, we omit remarks to this effect. However, if the value of the integral is given by means of some other formula for those values of the parameters for which the formula in question loses meaning, we accompany this second formula with the appropriate explanation.

The letters $x, y, t, \ldots$ denote independent variables; $f, g, \varphi, \ldots$ denote functions of $x, y, t, \ldots$; $f', g', \varphi', \ldots, f'', g'', \varphi'', \ldots$ denote their first, second, etc., derivatives; $a, b, m, p, \ldots$ denote constants, by which we generally mean arbitrary real numbers. If a particular formula is valid only for certain values of the constants (for example, only for positive numbers or only for integers), an appropriate remark is made provided the restriction that we make does not follow from the form of the formula itself. Thus, in formulas 2.148 4. and 2.424 6., we make no remark since it is clear from the form of these formulas themselves that n must be a natural number (that is, a positive integer).

2.01 The basic integrals

1. $\int x^n\,dx = \frac{x^{n+1}}{n+1} \quad (n \neq -1).$

For $n = -1$

2. $\int \frac{dx}{x} = \ln x.$

3. $\int e^x\,dx = e^x.$

4. $\int a^x\,dx = \frac{a^x}{\ln a}.$

5. $\int \sin x\,dx = -\cos x.$

6. $\int \cos x\,dx = \sin x.$

7. $\int \frac{dx}{\sin^2 x} = -\operatorname{ctg} x.$

8. $\int \frac{dx}{\cos^2 x} = \operatorname{tg} x.$

9. $\int \frac{\sin x}{\cos^2 x}\,dx = \sec x.$

10. $\int \frac{\cos x}{\sin^2 x}\,dx = -\operatorname{cosec} x.$

11. $\int \operatorname{tg} x\,dx = -\ln\cos x.$

12. $\int \operatorname{ctg} x\,dx = \ln\sin x.$

13. $\int \frac{dx}{\sin x} = \ln\operatorname{tg}\frac{x}{2}.$

14. $\int \frac{dx}{\cos x} = \ln\operatorname{tg}\left(\frac{\pi}{4} + \frac{x}{2}\right) = \ln(\sec x + \operatorname{tg} x).$

15. $\int \frac{dx}{1+x^2} = \operatorname{arctg} x = -\operatorname{arcctg} x.$

16. $\int \frac{dx}{1-x^2} = \operatorname{Arth} x = \frac{1}{2}\ln\frac{1+x}{1-x}.$

17. $\int \frac{dx}{\sqrt{1-x^2}} = \arcsin x = -\arccos x.$

18. $\int \frac{dx}{\sqrt{x^2+1}} = \operatorname{Arsh} x = \ln(x + \sqrt{x^2+1}).$

19. $\int \frac{dx}{\sqrt{x^2-1}} = \operatorname{Arch} x = \ln(x + \sqrt{x^2-1}).$

20. $\int \operatorname{sh} x\,dx = \operatorname{ch} x.$

21. $\int \operatorname{ch} x\,dx = \operatorname{sh} x.$

22. $\int \frac{dx}{\operatorname{sh}^2 x} = -\operatorname{cth} x.$

23. $\int \frac{dx}{\operatorname{ch}^2 x} = \operatorname{th} x.$

24. $\int \operatorname{th} x\,dx = \ln\operatorname{ch} x.$

25 $\int \operatorname{cth} x\,dx = \ln\operatorname{sh} x.$

26. $\int \frac{dx}{\operatorname{sh} x} = \ln\operatorname{th}\frac{x}{2}.$

2.02 General formulas

1. $\int af\,dx = a\int f\,dx.$

2. $\int [af \pm b\varphi \pm c\psi \pm \ldots]\,dx = a\int f\,dx \pm b\int \varphi\,dx \pm c\int \psi\,dx \pm \ldots$

3. $\frac{d}{dx}\int f\,dx = f.$

4. $\int f'\,dx = f.$

5. $\int f'\varphi\,dx = f\varphi - \int f\varphi'\,dx$ [integration by parts].

6. $\int f^{(n+1)}\varphi\,dx = \varphi f^{(n)} - \varphi' f^{(n-1)} + \varphi'' f^{(n-2)} - \ldots + (-1)^n \varphi^{(n)} f +$

$$+(-1)^{n+1}\int \varphi^{(n+1)} f\,dx.$$

7. $\int f(x)\,dx = \int f[\varphi(y)]\,\varphi'(y)\,dy \quad [x = \varphi(y)]$ [change of variable].

8. $\int (f)^n f'\,dx = \frac{(f)^{n+1}}{n+1}.$

For $n = -1$

$$\int \frac{f'\,dx}{f} = \ln f.$$

9. $\int (af+b)^n f'\,dx = \frac{(af+b)^{n+1}}{a(n+1)}.$

10. $\int \frac{f'\,dx}{\sqrt{af+b}} = \frac{2\sqrt{af+b}}{a}.$

11. $\int \frac{f'\varphi - \varphi' f}{\varphi^2}\,dx = \frac{f}{\varphi}.$

12. $\int \frac{f'\varphi - \varphi' f}{f\varphi}\,dx = \ln\frac{f}{\varphi}.$

13. $\int \frac{dx}{f(f\pm\varphi)} = \pm\int \frac{dx}{f\varphi} \mp \int \frac{dx}{\varphi(f\pm\varphi)}.$

14. $\int \frac{f'\,dx}{\sqrt{f^2+a}} = \ln(f + \sqrt{f^2+a}).$

15. $\int \frac{f\,dx}{(f+a)(f+b)} = \frac{a}{a-b}\int \frac{dx}{(f+a)} - \frac{b}{a-b}\int \frac{dx}{(f+b)}.$

For $a = b$

$$\int \frac{f\,dx}{(f+a)^2} = \int \frac{dx}{f+a} - a\int \frac{dx}{(f+a)^2}.$$

16. $\int \frac{f\,dx}{(f+\varphi)^n} = \int \frac{dx}{(f+\varphi)^{n-1}} - \int \frac{\varphi\,dx}{(f+\varphi)^n}.$

17. $\int \frac{f'\,dx}{p^2+q^2f^2} = \frac{1}{pq}\operatorname{arctg}\frac{qf}{p}.$

18. $\int \frac{f'\,dx}{q^2f^2-p^2} = \frac{1}{2pq}\ln\frac{qf-p}{qf+p}$

19. $\int \frac{f\,dx}{1-f} = -x + \int \frac{dx}{1-f}.$

20. $\int \frac{f^2\,dx}{f^2-a^2} = \frac{1}{2}\int \frac{f\,dx}{f-a} + \frac{1}{2}\int \frac{f\,dx}{f+a}.$

21. $\int \frac{f'\,dx}{\sqrt{a^2-f^2}} = \arcsin \frac{f}{a}.$

22. $\int \frac{f'\,dx}{af^2+bf} = \frac{1}{b}\ln \frac{f}{af+b}.$

23. $\int \frac{f'\,dx}{f\sqrt{f^2-a^2}} = \frac{1}{a}\operatorname{arcsec}\frac{f}{a}.$

24. $\int \frac{(f'\varphi - f\varphi')\,dx}{f^2+\varphi^2} = \operatorname{arctg}\frac{f}{\varphi}.$

25. $\int \frac{(f'\varphi - f\varphi')\,dx}{f^2-\varphi^2} = \frac{1}{2}\ln \frac{f-\varphi}{f+\varphi}.$

2.1 Rational Functions

2.10 *General integration rules*

2.101 To integrate an arbitrary rational function $\frac{F(x)}{f(x)}$, where $F(x)$ and $f(x)$ are polynomials with no common factors, we first need to separate out the integral part $E(x)$ (where $E(x)$ is a polynomial), if there is an integral part, and then to integrate separately the integral part and the remainder, thus:

$$\int \frac{F(x)\,dx}{f(x)} = \int E(x)\,dx + \int \frac{\varphi(x)}{f(x)}\,dx.$$

Integration of the remainder, which is then a proper rational function (that is, one in which the degree of the numerator is less than the degree of the denominator) is based on the decomposition of the fraction into elementary fractions, the so-called *partial fractions.*

2.102 If $a, b, c, \ldots, m$ are roots of the equation $f(x)=0$ and if $\alpha, \beta, \gamma, \ldots, \mu$ are their corresponding multiplicities, so that $f(x)=(x-a)^\alpha (x-b)^\beta \ldots (x-m)^\mu$ then, $\frac{\varphi(x)}{f(x)}$ can be decomposed into the following partial fractions:

$$\begin{aligned}\frac{\varphi(x)}{f(x)} = {} & \frac{A_\alpha}{(x-a)^\alpha} + \frac{A_{\alpha-1}}{(x-a)^{\alpha-1}} + \ldots + \frac{A_1}{x-a} + \\ & + \frac{B_\beta}{(x-b)^\beta} + \frac{B_{\beta-1}}{(x-b)^{\beta-1}} + \ldots + \frac{B_1}{x-b} + \\ & + \ldots\ldots\ldots\ldots + \\ & + \frac{M_\mu}{(x-m)^\mu} + \frac{M_{\mu-1}}{(x-m)^{\mu-1}} + \ldots + \frac{M_1}{x-m},\end{aligned}$$

where the numerators of the individual fractions are determined by the following formulas:

$$A_{\alpha-k+1} = \frac{\psi_1^{(k-1)}(a)}{(k-1)!}, \quad B_{\beta-k+1} = \frac{\psi_2^{(k-1)}(b)}{(k-1)!}, \ \ldots, \ M_{\mu-k+1} = \frac{\psi_m^{(k-1)}(m)}{(k-1)!},$$

$$\psi_1(x) = \frac{\varphi(x)(x-a)^\alpha}{f(x)}, \quad \psi_2(x) = \frac{\varphi(x)(x-b)^\beta}{f(x)}, \ \ldots, \ \psi_m(x) = \frac{\varphi(x)(x-m)^\mu}{f(x)}. \qquad \text{TI 51a}$$

If $a, b, \ldots, m$ are simple roots, that is, if $\alpha=\beta=\ldots=\mu=1$, then

$$\frac{\varphi(x)}{f(x)} = \frac{A}{x-a} + \frac{B}{x-b} + \ldots + \frac{M}{x-m},$$

where

$$A=\frac{\varphi(a)}{f'(a)}\qquad B=\frac{\varphi(b)}{f'(b)},\ \ldots,\ \ M=\frac{\varphi(m)}{f'(m)}.$$

If some of the roots of the equation $f(x)=0$ are imaginary, we group together the fractions that represent conjugate roots of the equation. Then, after certain manipulations, we represent the corresponding pairs of fractions in the form of real fractions of the form

$$\frac{M_1x+N_1}{x^2+2Bx+C}+\frac{M_2x+N_2}{(x^2+2Bx+C)^2}+\ldots+\frac{M_px+N_p}{(x^2+2Bx+C)^p}.$$

2.103 Thus, the integration of a proper rational fraction $\frac{\varphi(x)}{f(x)}$ reduces to integrals of the form $\int\frac{g\,dx}{(x-a)^{\alpha}}$ or $\int\frac{Mx+N}{(A+2Bx+Cx^2)^p}dx$. Fractions of the first form yield rational functions for $\alpha>1$ and logarithms for $\alpha=1$. Fractions of the second form yield rational functions and logarithms or arctangents:

1. $\int\frac{g\,dx}{(x-a)^{\alpha}}=g\int\frac{d(x-a)}{(x-a)^{\alpha}}=-\frac{g}{(\alpha-1)(x-a)^{\alpha-1}}.$

2. $\int\frac{g\,dx}{x-a}=g\int\frac{d(x-a)}{x-a}=g\ln|x-a|.$

3. $$\int\frac{Mx+N}{(A+2Bx+Cx^2)^p}dx=\frac{NB-MA+(NC-MB)x}{2(p-1)(AC-B^2)(A+2Bx+Cx^2)^{p-1}}+\frac{(2p-3)(NC-MB)}{2(p-1)(AC-B^2)}\int\frac{dx}{(A+2Bx+Cx^2)^{p-1}}.$$

4. $$\int\frac{dx}{A+2Bx+Cx^2}=\frac{1}{\sqrt{AC-B^2}}\operatorname{arctg}\frac{Cx+B}{\sqrt{AC-B^2}}\qquad [AC>B^2];$$
$$=\frac{1}{2\sqrt{B^2-AC}}\ln\left|\frac{Cx+B-\sqrt{B^2-AC}}{Cx+B+\sqrt{B^2-AC}}\right|\qquad [AC<B^2].$$

5. $$\int\frac{(Mx+N)\,dx}{A+2Bx+Cx^2}=\frac{M}{2C}\ln|A+2Bx+Cx^2|+\frac{NC-MB}{C\sqrt{AC-B^2}}\operatorname{arctg}\frac{Cx+B}{\sqrt{AC-B^2}}\qquad [AC>B^2];$$
$$=\frac{M}{2C}\ln|A+2Bx+Cx^2|+\frac{NC-MB}{2C\sqrt{B^2-AC}}\ln\left|\frac{Cx+B-\sqrt{B^2-AC}}{Cx+B+\sqrt{B^2-AC}}\right|\qquad [AC<B^2].$$

The Ostrogradskiy-Hermite method

2.104 By means of the Ostrogradskiy-Hermite method, we can find the rational part of $\int\frac{\varphi(x)}{f(x)}dx$ without finding the roots of the equation $f(x)=0$ and without decomposing the integrand into partial fractions:

$$\int\frac{\varphi(x)}{f(x)}dx=\frac{M}{D}+\int\frac{N\,dx}{Q}.$$ FI II 49

Here, M, N, D, and Q are rational functions of x. Specifically, D is the greatest common divisor of the function $f(x)$ and its derivative $f'(x)$; $Q=\frac{f(x)}{D}$; M is a polynomial of degree no higher than $m-1$, where m is the degree of the polynomial D; N is a polynomial of degree no higher than $n-1$, where n is the degree of the polynomial Q. The coefficients of the polynomials M and N are determined by equating the coefficients of like powers of x in the following identity:

$$\varphi(x) = M'Q - M(T - Q') + ND$$

where $T = \frac{f'(x)}{D}$ and M' and Q' are the derivatives of the polynomials M and Q.

2.11-2.13 Forms containing the binomial $a + bx^k$

2.110 Reduction formulas for $z_k = a + bx^k$.

1. $$\int x^n z_k^m\,dx = \frac{x^{n+1} z_k^m}{km+n+1} + \frac{amk}{km+n+1}\int x^n z_k^{m-1}\,dx =$$ LA 126 (4)

$$= \frac{x^{n+1}}{m+1}\sum_{s=0}^{p} \frac{(ak)^s (m+1)m(m-1)\ldots(m-s+1) z_k^{m-s}}{[mk+n+1][(m-1)k+n+1]\ldots[(m-s)k+n+1]} +$$

$$+ \frac{(ak)^{p+1} m(m-1)\ldots(m-p+1)(m-p)}{[mk+n+1][(m-1)k+n+1]\ldots[(m-p)k+n+1]}\int x^n z_k^{m-p-1}\,dx.$$

2. $$\int x^n z_k^m\,dx = \frac{-x^{n+1} z_k^{m+1}}{ak(m+1)} + \frac{km+k+n+1}{ak(m+1)}\int x^n z_k^{m+1}\,dx.$$ LA 126 (6)

3. $$\int x^n z_k^m\,dx = \frac{x^{n+1} z_k^m}{n+1} - \frac{bkm}{n+1}\int x^{n+k} z_k^{m-1}\,dx.$$ LA 125 (1)

4. $$\int x^n z_k^m\,dx = \frac{x^{n+1-k} z_k^{m+1}}{bk(m+1)} - \frac{n+1-k}{bk(m+1)}\int x^{n-k} z_k^{m+1}\,dx.$$ LA 125 (2)

5. $$\int x^n z_k^m\,dx = \frac{x^{n+1-k} z_k^{m+1}}{b(km+n+1)} - \frac{a(n+1-k)}{b(km+n+1)}\int x^{n-k} z_k^m\,dx.$$ LA 126 (3)

6. $$\int x^n z_k^m\,dx = \frac{x^{n+1} z_k^{m+1}}{a(n+1)} - \frac{b(km+k+n+1)}{a(n+1)}\int x^{n+k} z_k^m\,dx.$$ LA 126 (5)

Forms containing the binomial $z_1 = a + bx$

2.111

1. $$\int z_1^m\,dx = \frac{z_1^{m+1}}{b(m+1)}.$$

For $m = -1$

$$\int \frac{dx}{z_1} = \frac{1}{b}\ln z_1.$$

2. $$\int \frac{x^n\,dx}{z_1^m} = \frac{x^n}{z_1^{m-1}(n+1-m)b} - \frac{na}{(n+1-m)b}\int \frac{x^{n-1}\,dx}{z_1^m}.$$

For $n = m-1$, we may use the formula

3. $$\int \frac{x^{m-1}\,dx}{z_1^m} = -\frac{x^{m-1}}{z_1^{m-1}(m-1)b} + \frac{1}{b}\int \frac{x^{m-2}\,dx}{z_1^{m-1}}.$$

For $m = 1$

$$\int \frac{x^n\,dx}{z_1} = \frac{x^n}{nb} - \frac{ax^{n-1}}{(n-1)b^2} + \frac{a^2 x^{n-2}}{(n-2)b^3} - \ldots + (-1)^{n-1}\frac{a^{n-1}x}{1\cdot b^n} + \frac{(-1)^n a^n}{b^{n+1}}\ln z_1.$$

4. $$\int \frac{x^n\,dx}{z_1^2} = \sum_{k=1}^{n-1} (-1)^{k-1}\frac{ka^{k-1}x^{n-k}}{(n-k)b^{k+1}} +$$

$$+ (-1)^{n-1}\frac{a^n}{b^{n+1}z_1} + (-1)^{n+1}\frac{na^{n-1}}{b^{n+1}}\ln z_1.$$

2.112

1. $\int \frac{x\,dx}{z_1} = \frac{x}{b} - \frac{a}{b^2} \ln z_1.$

2. $\int \frac{x^2\,dx}{z_1} = \frac{x^2}{2b} - \frac{ax}{b^2} + \frac{a^2}{b^3} \ln z_1.$

2.113

1. $\int \frac{dx}{z_1^2} = -\frac{1}{bz_1}.$

2. $\int \frac{x\,dx}{z_1^2} = -\frac{x}{bz_1} + \frac{1}{b^2} \ln z_1 = \frac{a}{b^2 z_1} + \frac{1}{b^2} \ln z_1.$

3. $\int \frac{x^2\,dx}{z_1^2} = \frac{x}{b^2} - \frac{a^2}{b^3 z_1} - \frac{2a}{b^3} \ln z_1.$

2.114

1. $\int \frac{dx}{z_1^3} = -\frac{1}{2bz_1^2}.$

2. $\int \frac{x\,dx}{z_1^3} = -\left[\frac{x}{b} + \frac{a}{2b^2}\right] \frac{1}{z_1^2}.$

3. $\int \frac{x^2\,dx}{z_1^3} = \left[\frac{2ax}{b^2} + \frac{3a^2}{2b^3}\right] \frac{1}{z_1^2} + \frac{1}{b^3} \ln z_1.$

4. $\int \frac{x^3\,dx}{z_1^3} = \left[\frac{x}{b} + 2\frac{a}{b^2} x^2 - 2\frac{a^2}{b^3} x - \frac{5}{2}\frac{a^3}{b^4}\right] \frac{1}{z_1^2} - 3\frac{a}{b^4} \ln z_1.$

2.115

1. $\int \frac{dx}{z_1^4} = -\frac{1}{3bz_1^3}.$

2. $\int \frac{x\,dx}{z_1^4} = -\left[\frac{x}{2b} + \frac{a}{6b^2}\right] \frac{1}{z_1^3}.$

3. $\int \frac{x^2\,dx}{z_1^4} = -\left[\frac{x^2}{b} + \frac{ax}{b^2} + \frac{a^2}{3b^3}\right] \frac{1}{z_1^3}.$

4. $\int \frac{x^3 dx}{z_1^4} = \left[\frac{3ax^2}{b^2} + \frac{9a^2x}{2b^3} + \frac{11a^3}{6b^4}\right] \frac{1}{z_1^3} + \frac{1}{b^4} \ln z_1.$

2.116

1. $\int \frac{dx}{z_1^5} = -\frac{1}{4bz_1^4}.$

2. $\int \frac{x\,dx}{z_1^5} = -\left[\frac{x}{3b} + \frac{a}{12b^2}\right] \frac{1}{z_1^4}.$

3. $\int \frac{x^2\,dx}{z_1^5} = -\left[\frac{x^2}{2b} + \frac{ax}{3b^2} + \frac{a^2}{12b^3}\right] \frac{1}{z_1^4}.$

4. $\int \frac{x^3\,dx}{z_1^5} = -\left[\frac{x^3}{b} + \frac{3ax^2}{2b^2} + \frac{a^2x}{b^3} + \frac{a^3}{4b^4}\right] \frac{1}{z_1^4}.$

2.117

1. $\int \frac{dx}{x^n z_1^m} = \frac{-1}{(n-1)\,ax^{n-1}z_1^{m-1}} + \frac{b\,(2-n-m)}{a\,(n-1)} \int \frac{dx}{x^{n-1}z_1^m}.$

2. $\int \frac{dx}{z_1^m} = -\frac{1}{(m-1)\,bz_1^{m-1}}.$

3. $\int \frac{dx}{xz_1^m} = \frac{1}{z_1^{m-1}a\,(m-1)} + \frac{1}{a} \int \frac{dx}{xz_1^{m-1}}.$

4. $\int \frac{dx}{x^n z_1} = \sum_{k=1}^{n-1} \frac{(-1)^k b^{k-1}}{(n-k)\,a^k x^{n-k}} + \frac{(-1)^n b^{n-1}}{a^n} \ln \frac{z_1}{x}.$

2.118

1. $\int \frac{dx}{xz_1} = -\frac{1}{a}\ln\frac{z_1}{x}$.

2. $\int \frac{dx}{x^2 z_1} = -\frac{1}{ax} + \frac{b}{a^2}\ln\frac{z_1}{x}$.

3. $\int \frac{dx}{x^3 z_1} = -\frac{1}{2ax^2} + \frac{b}{a^2 x} - \frac{b^2}{a^3}\ln\frac{z_1}{x}$.

2.119

1. $\int \frac{dx}{xz_1^2} = \frac{1}{az_1} - \frac{1}{a^2}\ln\frac{z_1}{x}$.

2. $\int \frac{dx}{x^2 z_1^2} = -\left[\frac{1}{ax} + \frac{2b}{a^2}\right]\frac{1}{z_1} + \frac{2b}{a^3}\ln\frac{z_1}{x}$.

3. $\int \frac{dx}{x^3 z_1^2} = \left[-\frac{1}{2ax^2} + \frac{3b}{2a^2 x} + \frac{3b^2}{a^3}\right]\frac{1}{z_1} - \frac{3b^2}{a^4}\ln\frac{z_1}{x}$.

2.121

1. $\int \frac{dx}{xz_1^3} = \left[\frac{3}{2a} + \frac{bx}{a^2}\right]\frac{1}{z_1^2} - \frac{1}{a^3}\ln\frac{z_1}{x}$.

2. $\int \frac{dx}{x^2 z_1^3} = -\left[\frac{1}{ax} + \frac{9b}{2a^2} + \frac{3b^2 x}{a^3}\right]\frac{1}{z_1^2} + \frac{3b}{a^4}\ln\frac{z_1}{x}$.

3. $\int \frac{dx}{x^3 z_1^3} = \left[-\frac{1}{2ax^2} + \frac{2b}{a^2 x} + \frac{9b^2}{a^3} + \frac{6b^3 x}{a^4}\right]\frac{1}{z_1^2} - \frac{6b^2}{a^5}\ln\frac{z_1}{x}$.

2.122

1. $\int \frac{dx}{xz_1^4} = \left[\frac{11}{6a} + \frac{5bx}{2a^2} + \frac{b^2 x^2}{a^3}\right]\frac{1}{z_1^3} - \frac{1}{a^4}\ln\frac{z_1}{x}$.

2. $\int \frac{dx}{x^2 z_1^4} = -\left[\frac{1}{ax} + \frac{22b}{3a^2} + \frac{10b^2 x}{a^3} + \frac{4b^3 x^2}{a^4}\right]\frac{1}{z_1^3} + \frac{4b}{a^5}\ln\frac{z_1}{x}$.

3. $\int \frac{dx}{x^3 z_1^4} = \left[-\frac{1}{2ax^2} + \frac{5b}{2a^2 x} + \frac{55b^2}{3a^3} + \frac{25b^3 x}{a^4} + \frac{10b^4 x^2}{a^5}\right]\frac{1}{z_1^3} - \frac{10b^2}{a^6}\ln\frac{z_1}{x}$.

2.123

1. $\int \frac{dx}{xz^5} = \left[\frac{25}{12a} + \frac{13bx}{3a^2} + \frac{7b^2 x^2}{2a^3} + \frac{b^3 x^3}{a^4}\right]\frac{1}{z_1^4} - \frac{1}{a^5}\ln\frac{z_1}{x}$.

2. $\int \frac{dx}{x^2 z_1^5} = \left[-\frac{1}{ax} - \frac{125b}{12a^2} - \frac{65b^2 x}{3a^3} - \frac{35b^3 x^2}{2a^4} - \frac{5b^4 x^3}{a^5}\right]\frac{1}{z_1^4} + \frac{5b}{a^6}\ln\frac{z_1}{x}$.

3. $\int \frac{dx}{x^3 z_1^5} = \left[-\frac{1}{2ax^2} + \frac{3b}{a^2 x} + \frac{125b^2}{4a^3} + \frac{65b^3 x}{a^4} + \frac{105b^4 x^2}{2a^5} + \frac{15b^5 x^3}{a^6}\right]\frac{1}{z_1^4} - \frac{15b^2}{a^7}\ln\frac{z_1}{x}$.

2.124 Forms containing the binomial $z_2 = a + bx^2$.

1. $\int \frac{dx}{z_2} = \frac{1}{\sqrt{ab}}\operatorname{arctg} x\sqrt{\frac{b}{a}}$ $[ab > 0]$ (see also **2.141** 2.);

$= \frac{1}{2i\sqrt{ab}}\ln\frac{a + xi\sqrt{ab}}{a - xi\sqrt{ab}}$ $[ab < 0]$ (see also **2.143** 2. and **2.143** 3.).

2. $\int \frac{x\,dx}{z_2^m} = -\frac{1}{2b(m-1)\,z_2^{m-1}}$ (see also **2.145** 2., **2.145** 6, and **2.18**).

Forms containing the binomial $z_3 = a + bx^3$

Notation: $\alpha = \sqrt[3]{\frac{a}{b}}$

2.125

1. $\int \frac{x^n\,dx}{z_3^m} = \frac{x^{n-2}}{z_3^{m-1}(n+1-3m)\,b} - \frac{(n-2)\,a}{b\,(n+1-3m)} \int \frac{x^{n-3}\,dx}{z_3^m}$.

2. $\int \frac{x^n\,dx}{z_3^m} = \frac{x^{n+1}}{3a\,(m-1)\,z_3^{m-1}} - \frac{n+4-3m}{3a\,(m-1)} \int \frac{x^n\,dx}{z_3^{m-1}}$. LA 133 (1)

2.126

1. $\int \frac{dx}{z_3} = \frac{\alpha}{3a}\left\{\frac{1}{2}\ln\frac{(x+\alpha)^2}{x^2-\alpha x+\alpha^2} + \sqrt{3}\operatorname{arctg}\frac{x\sqrt{3}}{2\alpha - x}\right\}$;

   $= \frac{\alpha}{3a}\left\{\frac{1}{2}\ln\frac{(x+\alpha)^2}{x^2-\alpha x+\alpha^2} + \sqrt{3}\operatorname{arctg}\frac{2x-\alpha}{\alpha\sqrt{3}}\right\}$

   (see also **2.141** 3. and **2.143** 4.).

2. $\int \frac{x\,dx}{z_3} = -\frac{1}{3b\alpha}\left\{\frac{1}{2}\ln\frac{(x+\alpha)^2}{x^2-\alpha x+\alpha^2} - \sqrt{3}\operatorname{arctg}\frac{2x-\alpha}{\alpha\sqrt{3}}\right\}$

   (see also **2.145** 3. and **2.145** 7.).

3. $\int \frac{x^2\,dx}{z_3} = \frac{1}{3b}\ln\left(1 + x^3\alpha^{-3}\right) = \frac{1}{3b}\ln z_3$.

4. $\int \frac{x^3\,dx}{z_3} = \frac{x}{b} - \frac{a}{b}\int \frac{dx}{z_3}$ (see **2.126** 1.).

5. $\int \frac{x^4\,dx}{z_3} = \frac{x^2}{2b} - \frac{a}{b}\int \frac{x\,dx}{z_3}$ (see **2.126** 2.).

2.127

1. $\int \frac{dx}{z_3^2} = \frac{x}{3az_3} + \frac{2}{3a}\int \frac{dx}{z_3}$ (see **2.126** 1.).

2. $\int \frac{x\,dx}{z_3^2} = \frac{x^2}{3az_3} + \frac{1}{3a}\int \frac{x\,dx}{z_3}$ (see **2.126** 2.).

3. $\int \frac{x^2\,dx}{z_3^2} = -\frac{1}{3bz_3}$.

4. $\int \frac{x^3\,dx}{z_3^2} = -\frac{x}{3bz_3} + \frac{1}{3b}\int \frac{dx}{z_3}$ (see **2.126** 1.).

2.128

1. $\int \frac{dx}{x^n z_3^m} = -\frac{1}{(n-1)\,ax^{n-1}z_3^{m-1}} - \frac{b\,(3m+n-4)}{a\,(n-1)} \int \frac{dx}{x^{n-3}z_3^m}$.

2. $\int \frac{dx}{x^n z_3^m} = \frac{1}{3a\,(m-1)\,x^{n-1}z_3^{m-1}} + \frac{n+3m-4}{3a\,(m-1)} \int \frac{dx}{x^n z_3^{m-1}}$. LA 133 (2)

2.129

1. $\int \frac{dx}{xz_3} = \frac{1}{3a}\ln\frac{x^3}{z_3}$.

2. $\int \frac{dx}{x^2z_3} = -\frac{1}{ax} - \frac{b}{a}\int \frac{x\,dx}{z_3}$ (see **2.126** 2.).

3. $\int \frac{dx}{x^3z_3} = -\frac{1}{2ax^2} - \frac{b}{a}\int \frac{dx}{z_3}$ (see **2.126** 1.).

2.131

1. $\int \frac{dx}{xz_3^2} = \frac{1}{3az_3} + \frac{1}{3a^2} \ln \frac{x^3}{z_3}$

2. $\int \frac{dx}{x^2 z_3^2} = -\left[\frac{1}{ax} + \frac{4bx^2}{3a^2}\right] \frac{1}{z_3} - \frac{4b}{3a^2} \int \frac{x\,dx}{z_3}$ (see 2.126 2.).

3. $\int \frac{dx}{x^3 z_3^2} = -\left[\frac{1}{2ax^2} + \frac{5bx}{6a^2}\right] \frac{1}{z_3} - \frac{5b}{3a^2} \int \frac{dx}{z_3}$ (see 2.126 1.).

Forms containing the binomial $z_4 = a + bx^4$

Notations: $\alpha = \sqrt[4]{\frac{a}{b}}$ $\alpha' = \sqrt[4]{\frac{-a}{b}}$

2.132

1. $\int \frac{dx}{z_4} = \frac{\alpha}{4a\sqrt{2}} \left\{ \ln \frac{x^2 + \alpha x\sqrt{2} + \alpha^2}{x^2 - \alpha x\sqrt{2} + \alpha^2} + 2 \operatorname{arctg} \frac{\alpha x\sqrt{2}}{\alpha^2 - x^2} \right\}$ $[ab > 0]$

(see also 2.141 4.).

$= \frac{\alpha'}{4a} \left\{ \ln \frac{x + \alpha}{x - \alpha'} + 2 \operatorname{arctg} \frac{x}{\alpha'} \right\}$ $[ab < 0]$ (see also 2.143 5.).

2. $\int \frac{x\,dx}{z} = \frac{1}{2\sqrt{ab}} \operatorname{arctg} x^2 \sqrt{\frac{b}{a}}$ $[ab > 0]$ (see also 2.145 4.).

$= \frac{1}{4i\sqrt{ab}} \ln \frac{a + x^2 i\sqrt{ab}}{a - x^2 i\sqrt{ab}}$ $[ab < 0]$ (see also 2.145 8.).

3. $\int \frac{x^2\,dx}{z_4} = \frac{1}{4b\alpha\sqrt{2}} \left\{ \ln \frac{x^2 - \alpha x\sqrt{2} + \alpha^2}{x^2 + \alpha x\sqrt{2} + \alpha^2} + 2 \operatorname{arctg} \frac{\alpha x\sqrt{2}}{\alpha^2 - x^2} \right\}$ $[ab > 0]$;

$= -\frac{1}{4b\alpha'} \left\{ \ln \frac{x + \alpha'}{x - \alpha'} - 2 \operatorname{arctg} \frac{x}{\alpha'} \right\}$ $[ab < 0]$.

4. $\int \frac{x^3\,dx}{z_4} = \frac{1}{4b} \ln z_4$

2.133

1. $\int \frac{x^n\,dx}{z_4^m} = \frac{x^{n+1}}{4a(m-1)z_4^{m-1}} + \frac{4m - n - 5}{4a(m-1)} \int \frac{x^n\,dx}{z_4^{m-1}}$ LA 134 (1)

2. $\int \frac{x^n\,dx}{z_4^m} = \frac{x^{n-3}}{z_4^{m-1}(n+1-4m)b} - \frac{(n-3)a}{b(n+1-4m)} \int \frac{x^{n-4}\,dx}{z_4^m}$.

2.134

1. $\int \frac{dx}{z_4^2} = \frac{x}{4az_4} + \frac{3}{4a} \int \frac{dx}{z_4}$ (see 2.132 1.).

2. $\int \frac{x\,dx}{z_4^2} = \frac{x^2}{4az_4} + \frac{1}{2a} \int \frac{x\,dx}{z_4}$ (see 2.132 2.).

3. $\int \frac{x^2\,dx}{z_4^2} = \frac{x^3}{4az_4} + \frac{1}{4a} \int \frac{x^2\,dx}{z_4}$ (see 2.132 3.).

4. $\int \frac{x^3\,dx}{z_4^2} = \frac{x^4}{4az_4} = -\frac{1}{4bz_4}$.

2.135 $\int \frac{dx}{x^n z_4^m} = -\frac{1}{(n-1)ax^{n-1}z_4^{m-1}} - \frac{b(4m+n-5)}{(n-1)a} \int \frac{dx}{x^{n-4} z_4^m}$.

For $n = 1$

$$\int \frac{dx}{xz_4^m} = \frac{1}{a} \int \frac{dx}{xz_4^{m-1}} - \frac{b}{a} \int \frac{dx}{x^{-3} z_4^m}.$$

2.136

1. $\int \frac{dx}{xz_4} = \frac{\ln x}{a} - \frac{\ln z_4}{4a} = \frac{1}{4a} \ln \frac{x^4}{z_4}.$

2. $\int \frac{dx}{x^2 z_4} = -\frac{1}{ax} - \frac{b}{a} \int \frac{x^2\,dx}{z_4}$ (see 2.132 3.).

2.14 Forms containing the binomial $1 \pm x^n$

2.141

1. $\int \frac{dx}{1+x} = \ln(1+x).$

2. $\int \frac{dx}{1+x^2} = \operatorname{arctg} x = -\operatorname{arcctg} x$ (see also 2.124 1.).

3. $\int \frac{dx}{1+x^3} = \frac{1}{3} \ln \frac{1+x}{\sqrt{1-x+x^2}} + \frac{1}{\sqrt{3}} \operatorname{arctg} \frac{x\sqrt{3}}{2-x}$ (see also 2.126 1.)

4. $\int \frac{dx}{1+x^4} = \frac{1}{4\sqrt{2}} \ln \frac{1+x\sqrt{2}+x^2}{1-x\sqrt{2}+x^2} + \frac{1}{2\sqrt{2}} \operatorname{arctg} \frac{x\sqrt{2}}{1-x^2}$ (see also 2.132 1.).

2.142 $\int \frac{dx}{1+x^n} = -\frac{2}{n} \sum_{k=0}^{\frac{n}{2}-1} P_k \cos \frac{2k+1}{n}\pi + \frac{2}{n} \sum_{k=0}^{\frac{n}{2}-1} Q_k \sin \frac{2k+1}{n}\pi$

[n-a positive even number]; TI (43)a

$= \frac{1}{n} \ln(1+x) - \frac{2}{n} \sum_{k=0}^{\frac{n-3}{2}} P_k \cos \frac{2k+1}{n}\pi + \frac{2}{n} \sum_{k=0}^{\frac{n-3}{2}} Q_k \sin \frac{2k+1}{n}\pi$

[n-a positive odd number]. TI (45)

$$P_k = \frac{1}{2} \ln\left(x^2 - 2x\cos\frac{2k+1}{n}\pi + 1\right).$$

$$Q_k = \operatorname{arctg} \frac{x \sin \frac{2k+1}{n}\pi}{1 - x\cos\frac{2k+1}{n}\pi} = \operatorname{arctg} \frac{x - \cos\frac{2k+1}{n}\pi}{\sin\frac{2k+1}{n}\pi}.$$

2.143

1. $\int \frac{dx}{1-x} = -\ln(1-x).$

2. $\int \frac{dx}{1-x^2} = \frac{1}{2} \ln \frac{1+x}{1-x} = \operatorname{Arth} x \quad [-1 < x < 1]$ (see also 2.141 1.).

3. $\int \frac{dx}{x^2-1} = \frac{1}{2} \ln \frac{x-1}{x+1} = -\operatorname{Arcth} x \quad [x > 1, \quad x < -1].$

4. $\int \frac{dx}{1-x^3} = \frac{1}{3} \ln \frac{\sqrt{1+x+x^2}}{1-x} + \frac{1}{\sqrt{3}} \operatorname{arctg} \frac{x\sqrt{3}}{2+x}$ (see also 2.126 1.).

5. $\int \frac{dx}{1-x^4} = \frac{1}{4} \ln \frac{1+x}{1-x} + \frac{1}{2} \operatorname{arctg} x = \frac{1}{2}(\operatorname{Arth} x + \operatorname{arctg} x)$

(see also 2.132 1.).

2.144

1. $$\int \frac{dx}{1-x^n} = \frac{1}{n} \ln \frac{1+x}{1-x} - \frac{2}{n} \sum_{k=1}^{\frac{n}{2}-1} P_k \cos \frac{2k}{n} \pi + \frac{2}{n} \sum_{k=1}^{\frac{n}{2}-1} Q_k \sin \frac{2k}{n} \pi$$

[n-a positive even number].

$$P_k = \frac{1}{2} \ln \left(x^2 - 2x \cos \frac{2k}{n} \pi + 1 \right), \quad Q_k = \operatorname{arctg} \frac{x - \cos \frac{2k}{n} \pi}{\sin \frac{2k}{n} \pi}.$$ TI (47)

2. $$\int \frac{dx}{1-x^n} = -\frac{1}{n} \ln (1-x) + \frac{2}{n} \sum_{k=0}^{\frac{n-3}{2}} P_k \cos \frac{2k+1}{n} \pi +$$

$$+ \frac{2}{n} \sum_{k=0}^{\frac{n-3}{2}} Q_k \sin \frac{2k+1}{n} \pi$$ [n-a positive odd number]. TI (49)

$$P_k = \frac{1}{2} \ln \left(x^2 + 2x \cos \frac{2k+1}{n} \pi + 1 \right), \quad Q_k = \operatorname{arctg} \frac{x + \cos \frac{2k+1}{n} \pi}{\sin \frac{2k+1}{n} \pi}.$$

2.145

1. $$\int \frac{x\,dx}{1+x} = x - \ln (1+x).$$
2. $$\int \frac{x\,dx}{1+x^2} = \frac{1}{2} \ln (1+x^2).$$
3. $$\int \frac{x\,dx}{1+x^3} = -\frac{1}{6} \ln \frac{(1+x)^2}{1-x+x^2} + \frac{1}{\sqrt{3}} \operatorname{arctg} \frac{2x-1}{\sqrt{3}}$$ (see also **2.126** 2.).
4. $$\int \frac{x\,dx}{1+x^4} = \frac{1}{2} \operatorname{arctg} x^2.$$
5. $$\int \frac{x\,dx}{1-x} = -\ln (1-x) - x.$$
6. $$\int \frac{x\,dx}{1-x^2} = -\frac{1}{2} \ln (1-x^2).$$
7. $$\int \frac{x\,dx}{1-x^3} = -\frac{1}{6} \ln \frac{(1-x)^2}{1+x+x^2} - \frac{1}{\sqrt{3}} \operatorname{arctg} \frac{2x+1}{\sqrt{3}}$$ (see also **2.126** 2.).
8. $$\int \frac{x\,dx}{1-x^4} = \frac{1}{4} \ln \frac{1+x^2}{1-x^2}$$ (see also **2.132** 2.).

2.146 For m and n-natural numbers.

1. $$\int \frac{x^{m-1}dx}{1+x^{2n}} = -\frac{1}{2n} \sum_{k=1}^{n} \cos \frac{m\pi(2k-1)}{2n} \ln \left\{ 1 - 2x \cos \frac{2k-1}{2n} \pi + x^2 \right\} +$$

$$+ \frac{1}{n} \sum_{k=1}^{n} \sin \frac{m\pi(2k-1)}{2n} \operatorname{arctg} \frac{x - \cos \frac{2k-1}{2n} \pi}{\sin \frac{2k-1}{2n} \pi}$$ [$m < 2n$]. TI (44)a

2. $$\int \frac{x^{m-1}\,dx}{1+x^{2n+1}} = (-1)^{m+1} \frac{\ln(1+x)}{2n+1} - \frac{1}{2n+1} \sum_{k=1}^{n} \cos\frac{m\pi(2k-1)}{2n+1} \ln\left\{1 - 2x\cos\frac{2k-1}{2n+1}\pi + x^2\right\} + \frac{2}{2n+1} \sum_{k=1}^{n} \sin\frac{m\pi(2k-1)}{2n+1} \operatorname{arctg} \frac{x - \cos\frac{2k-1}{2n+1}\pi}{\sin\frac{2k-1}{2n+1}\pi} \qquad [m \leqslant 2n].$$ TI (46)a

3. $$\int \frac{x^{m-1}\,dx}{1-x^{2n}} = \frac{1}{2n}\{(-1)^{m+1}\ln(1+x) - \ln(1-x)\} - \frac{1}{2n}\sum_{k=1}^{n-1} \cos\frac{km\pi}{n} \ln\left(1 - 2x\cos\frac{k\pi}{n} + x^2\right) + \frac{1}{n}\sum_{k=1}^{n-1} \sin\frac{km\pi}{n} \operatorname{arctg} \frac{x - \cos\frac{k\pi}{n}}{\sin\frac{k\pi}{n}} \qquad [m < 2n].$$ TI (48)

4. $$\int \frac{x^{m-1}\,dx}{1-x^{2n+1}} = -\frac{1}{2n+1}\ln(1-x) + (-1)^{m+1}\frac{1}{2n+1}\sum_{k=1}^{n} \cos\frac{m\pi(2k-1)}{2n+1} \ln\left(1 + 2x\cos\frac{2k-1}{2n+1}\pi + x^2\right) + (-1)^{m+1}\frac{2}{2n+1}\sum_{k=1}^{n} \sin\frac{m\pi(2k-1)}{2n+1} \operatorname{arctg} \frac{x + \cos\frac{2k-1}{2n+1}\pi}{\sin\frac{2k-1}{2n+1}\pi} \qquad [m \leqslant 2n].$$ TI (50)

2.147

1. $$\int \frac{x^m\,dx}{1-x^{2n}} = \frac{1}{2}\int \frac{x^m\,dx}{1-x^n} + \frac{1}{2}\int \frac{x^m\,dx}{1+x^n}.$$

2. $$\int \frac{x^m\,dx}{(1+x^2)^n} = -\frac{1}{2n-m-1}\cdot\frac{x^{m-1}}{(1+x^2)^{n-1}} + \frac{m-1}{2n-m-1}\int \frac{x^{m-2}\,dx}{(1+x^2)^n}.$$ LA 139 (28)

3. $$\int \frac{x^m}{1+x^2}\,dx = \frac{x^{m-1}}{m-1} - \int \frac{x^{m-2}}{1+x^2}\,dx.$$

4. $$\int \frac{x^m\,dx}{(1-x^2)^n} = \frac{1}{2n-m-1}\,\frac{x^{m-1}}{(1-x^2)^{n-1}} - \frac{m-1}{2n-m-1}\int \frac{x^{m-2}\,dx}{(1-x^2)^n};$$
$$= \frac{1}{2n-2}\,\frac{x^{m-1}}{(1-x^2)^{n-1}} - \frac{m-1}{2n-2}\int \frac{x^{m-2}\,dx}{(1-x^2)^{n-1}}.$$ LA 139 (33)

5. $$\int \frac{x^m\,dx}{1-x^2} = -\frac{x^{m-1}}{m-1} + \int \frac{x^{m-2}\,dx}{1-x^2}.$$

2.148

1. $$\int \frac{dx}{x^m(1+x^2)^n} = -\frac{1}{m-1}\,\frac{1}{x^{m-1}(1+x^2)^{n-1}} - \frac{2n+m-3}{m-1}\int \frac{dx}{x^{m-2}(1+x^2)^n}.$$ LA 139 (29)

For $m = 1$

$$\int \frac{dx}{x(1+x^2)^n} = \frac{1}{2n-2}\,\frac{1}{(1+x^2)^{n-1}} + \int \frac{dx}{x(1+x^2)^{n-1}}.$$ LA 139 (31)

For $m=1$ and $n=1$

$$\int \frac{dx}{x(1+x^2)} = \ln \frac{x}{\sqrt{1+x^2}}.$$

2. $$\int \frac{dx}{x^m(1+x^2)} = -\frac{1}{(m-1)x^{m-1}} - \int \frac{dx}{x^{m-2}(1+x^2)}.$$

3. $$\int \frac{dx}{(1+x^2)^n} = \frac{1}{2n-2}\frac{x}{(1+x^2)^{n-1}} + \frac{2n-3}{2n-2}\int \frac{dx}{(1+x^2)^{n-1}}.$$ FI II 40

4. $$\int \frac{dx}{(1+x^2)^n} = \frac{x}{2n-1}\sum_{k=1}^{n-1} \frac{(2n-1)(2n-3)(2n-5)\dots(2n-2k+1)}{2^k(n-1)(n-2)\dots(n-k)(1+x^2)^{n-k}} + \frac{(2n-3)!!}{2^{n-1}(n-1)!}\operatorname{arctg} x.$$ TI (91)

2.149

1. $$\int \frac{dx}{x^m(1-x^2)^n} = -\frac{1}{(m-1)x^{m-1}(1-x^2)^{n-1}} + \frac{2n+m-3}{m-1}\int \frac{dx}{x^{m-2}(1-x^2)^n}.$$ LA 139 (34)

For $m=1$

$$\int \frac{dx}{x(1-x^2)^n} = \frac{1}{2(n-1)(1-x^2)^{n-1}} + \int \frac{dx}{x(1-x^2)^{n-1}}.$$ LA 139 (36)

For $m=1$ and $n=1$

$$\int \frac{dx}{x(1-x^2)} = \ln \frac{x}{\sqrt{1-x^2}}.$$

2. $$\int \frac{dx}{(1-x^2)^n} = \frac{1}{2n-2}\frac{x}{(1-x^2)^{n-1}} + \frac{2n-3}{2n-2}\int \frac{dx}{(1-x^2)^{n-1}}.$$ LA 139 (35)

3. $$\int \frac{dx}{(1-x^2)^n} = \frac{x}{2n-1}\sum_{k=1}^{n-1} \frac{(2n-1)(2n-3)(2n-5)\dots(2n-2k+1)}{2^k(n-1)(n-2)\dots(n-k)(1-x^2)^{n-k}} + \frac{(2n-3)!!}{2^n\cdot(n-1)!}\ln\frac{1+x}{1-x}.$$ TI (91)

2.15 Forms containing pairs of binomials: $a+bx$ and $\alpha+\beta x$

Notations: $z=a+bx$; $t=\alpha+\beta x$; $\Delta=a\beta-\alpha b$

2.151 $$\int z^n t^m\,dx = \frac{z^{n+1}t^m}{(m+n+1)b} - \frac{m\Delta}{(m+n+1)b}\int z^n t^{m-1}\,dx.$$

2.152

1. $$\int \frac{z}{t}\,dx = \frac{bx}{\beta} + \frac{\Delta}{\beta^2}\ln t.$$

2. $$\int \frac{t}{z}\,dx = \frac{\beta x}{b} - \frac{\Delta}{b^2}\ln z.$$

2.153 $$\int \frac{t^m\,dx}{z^n} = \frac{1}{(m-n+1)b}\frac{t^m}{z^{n-1}} - \frac{m\Delta}{(m-n+1)b}\int \frac{t^{m-1}\,dx}{z^n};$$

$$= \frac{1}{(n-1)\Delta}\frac{t^{m+1}}{z^{n-1}} - \frac{(m-n+2)\beta}{(n-1)\Delta}\int \frac{t^m\,dx}{z^{n-1}};$$

$$= -\frac{1}{(n-1)b}\frac{t^m}{z^{n-1}} + \frac{m\beta}{(n-1)b}\int \frac{t^{m-1}}{z^{n-1}}\,dx.$$

2.154 $\int \frac{dx}{zt} = \frac{1}{\Delta} \ln \frac{t}{z}.$

2.155 $\int \frac{dx}{z^n t^m} = -\frac{1}{(m-1)\Delta} \frac{1}{t^{m-1} z^{n-1}} - \frac{(m+n-2)b}{(m-1)\Delta} \int \frac{dx}{t^{m-1} z^n};$

$$= \frac{1}{(n-1)\Delta} \frac{1}{t^{m-1} z^{n-1}} + \frac{(m+n-2)\beta}{(n-1)\Delta} \int \frac{dx}{t^m z^{n-1}}.$$

2.156 $\int \frac{x\,dx}{zt} = \frac{1}{\Delta}\left(\frac{a}{b} \ln z - \frac{\alpha}{\beta} \ln t\right).$

2.16 Forms containing the trinomial $a + bx^k + cx^{2k}$

2.160 Reduction formulas for $R_k = a + bx^k + cx^{2k}$.

1. $\int x^{m-1} R_k^n\,dx = \frac{x^m R_k^{n+1}}{ma} - \frac{(m+k+nk)b}{ma} \int x^{m+k-1} R_k^n\,dx -$

$$- \frac{(m+2k+2kn)c}{ma} \int x^{m+2k-1} R_k^n\,dx.$$

2. $\int x^{m-1} R_k^n\,dx = \frac{x^m R_k^n}{m} - \frac{bkn}{m} \int x^{m+k-1} R_k^{n-1}\,dx - \frac{2ckn}{m} \int x^{m+2k-1} R_k^{n-1}\,dx.$

3. $\int x^{m-1} R_k^n\,dx = \frac{x^{m-2k} R_k^{n+1}}{(m+2kn)c} - \frac{(m-2k)a}{(m+2kn)c} \int x^{m-2k-1} R_k^n\,dx -$

$$- \frac{(m-k+kn)b}{(m+2kn)c} \int x^{m-k-1} R_k^n\,dx;$$

$$= \frac{x^m R_k^n}{m+2kn} + \frac{2kna}{m+2kn} \int x^{m-1} R_k^{n-1}\,dx + \frac{bkn}{m+2kn} \int x^{m+k-1} R_k^{n-1}\,dx.$$

2.161 Forms containing the trinomial $R_2 = a + bx^2 + cx^4$.

Notations: $f = \frac{b}{2} - \frac{1}{2}\sqrt{b^2 - 4ac}, \quad g = \frac{b}{2} + \frac{1}{2}\sqrt{b^2 - 4ac},$

$h = \sqrt{b^2 - 4ac}, \quad q = \sqrt[4]{\frac{a}{c}}, \quad l = 2a(n-1)(b^2 - 4ac), \quad \cos\alpha = -\frac{b}{2\sqrt{ac}}.$

1. $\int \frac{dx}{R_2} = \frac{c}{h}\left\{\int \frac{dx}{cx^2+f} - \int \frac{dx}{cx^2+g}\right\} \quad [h^2 > 0];$ LA 146 (5)

$$= \frac{1}{4cq^3 \sin\alpha}\left\{\sin\frac{\alpha}{2} \ln \frac{x^2 + 2qx\cos\frac{\alpha}{2} + q^2}{x^2 - 2qx\cos\frac{\alpha}{2} + q^2} + 2\cos\frac{\alpha}{2} \operatorname{arctg} \frac{x^2 - q^2}{2qx \sin\frac{\alpha}{2}}\right\} \quad [h^2 < 0].$$

LA 146 (8)a

2. $\int \frac{x\,dx}{R_2} = \frac{1}{2h} \ln \frac{cx^2+f}{cx^2+g} \quad [h^2 > 0];$ LA 146 (6)

$$= \frac{1}{2cq^2 \sin\alpha} \operatorname{arctg} \frac{x^2 - q^2\cos\alpha}{q^2 \sin\alpha} \quad [h^2 < 0].$$ LA 146 (9)a

3. $\int \frac{x^2\,dx}{R_2} = \frac{g}{h}\int \frac{dx}{cx^2+g} - \frac{f}{h}\int \frac{dx}{cx^2+f} \quad [h^2 > 0].$ LA 146 (7)

4. $\int \frac{dx}{R_2^2} = \frac{bcx^3 + (b^2 - 2ac)x}{lR_2} + \frac{b^2 - 6ac}{l}\int \frac{dx}{R_2} + \frac{bc}{l}\int \frac{x^2\,dx}{R_2}.$

5. $\int \frac{dx}{R_2^n} = \frac{bcx^3 + (b^2 - 2ac)x}{lR_2^{n-1}} + \frac{(4n-7)bc}{l}\int \frac{x^2\,dx}{R_2^{n-1}} +$

$$+ \frac{2(n-1)h^2 + 2ac - b^2}{l}\int \frac{dx}{R_2^{n-1}} \quad [n > 1].$$ LA 146 (10)

6. $$\int \frac{dx}{x^m R_2^n} = -\frac{1}{(m-1)\,a x^{m-1} R_2^{n-1}} - \frac{(m+2n-3)\,b}{(m-1)\,a} \int \frac{dx}{x^{m-2} R^n} - \frac{(m+4n-5)\,b}{(m-1)\,a} \int \frac{dx}{x^{m-4} R_2^n}.$$ LA 147 (12)a

2.17 Forms containing the quadratic trinomial $a+bx+cx^2$ and powers of x

Notations: $R = a + bx + cx^2$; $\Delta = 4ac - b^2$

2.171

1. $$\int x^{m+1} R^n\,dx = \frac{x^m R^{n+1}}{c\,(m+2n+2)} - \frac{am}{c\,(m+2n+2)} \int x^{m-1} R^n\,dx - \frac{b\,(m+n+1)}{c\,(m+2n+2)} \int x^m R^n\,dx.$$ TI (97)

2. $$\int \frac{R^n\,dx}{x^{m+1}} = -\frac{R^{n+1}}{amx^m} + \frac{b\,(n-m+1)}{am} \int \frac{R^n\,dx}{x^m} + \frac{c\,(2n-m+2)}{am} \int \frac{R^n\,dx}{x^{m-1}}.$$ LA 142(3), TI (98)a

3. $$\int \frac{dx}{R^{n+1}} = \frac{b+2cx}{n\Delta R^n} + \frac{(4n-2)\,c}{n\Delta} \int \frac{dx}{R^n}.$$ TI (94)a

4. $$\int \frac{dx}{R^{n+1}} = \frac{(2cx+b)}{2n+1} \sum_{k=0}^{n-1} \frac{2^k\,(2n+1)\,(2n-1)\,(2n-3)\ldots(2n-2k+1)\,c^k}{n\,(n-1)\ldots(n-k)\,\Delta^{k+1} R^{n-k}} + 2^n \frac{(2n-1)!!\,c^n}{n!\,\Delta^n} \int \frac{dx}{R}.$$ TI (96)a

2.172 $$\int \frac{dx}{R} = \frac{1}{\sqrt{-\Delta}} \ln \frac{b+2cx-\sqrt{-\Delta}}{b+2cx+\sqrt{-\Delta}} = \frac{-2}{\sqrt{-\Delta}} \operatorname{Arth} \frac{b+2cx}{\sqrt{-\Delta}} \qquad [\Delta < 0];$$
$$= \frac{-2}{b+2cx} \qquad [\Delta = 0];$$
$$= \frac{2}{\sqrt{\Delta}} \operatorname{arctg} \frac{b+2cx}{\sqrt{\Delta}} \qquad [\Delta > 0].$$

2.173

1. $$\int \frac{dx}{R^2} = \frac{b+2cx}{\Delta R} + \frac{2c}{\Delta} \int \frac{dx}{R}$$ (see **2.172**).

2. $$\int \frac{dx}{R^3} = \frac{b+2cx}{\Delta} \left\{ \frac{1}{2R^2} + \frac{3c}{\Delta R} \right\} + \frac{6c^2}{\Delta^2} \int \frac{dx}{R}$$ (see **2.172**).

2.174

1. $$\int \frac{x^m\,dx}{R^n} = -\frac{x^{m-1}}{(2n-m-1)\,cR^{n-1}} - \frac{(n-m)\,b}{(2n-m-1)\,c} \int \frac{x^{m-1}\,dx}{R^n} + \frac{(m-1)\,a}{(2n-m-1)\,c} \int \frac{x^{m-2}\,dx}{R^n}.$$

For $m = 2n-1$, this formula is inapplicable. Instead, we may use

2. $$\int \frac{x^{2n-1}\,dx}{R^n} = \frac{1}{c} \int \frac{x^{2n-3}\,dx}{R^{n-1}} - \frac{a}{c} \int \frac{x^{2n-3}\,dx}{R^n} - \frac{b}{c} \int \frac{x^{2n-2}\,dx}{R^n}.$$

2.175

1. $$\int \frac{x\,dx}{R} = \frac{1}{2c} \ln R - \frac{b}{2c} \int \frac{dx}{R}$$ (see **2.172**).

2. $$\int \frac{x\,dx}{R^2} = -\frac{2a+bx}{\Delta R} - \frac{b}{\Delta} \int \frac{dx}{R}$$ (see **2.172**).

3. $$\int \frac{x\,dx}{R^3} = -\frac{2a+bx}{2\Delta R^2} - \frac{3b(b+2cx)}{2\Delta^2 R} - \frac{3bc}{\Delta^2}\int \frac{dx}{R}$$ (see **2.172**).

4. $$\int \frac{x^2\,dx}{R} = \frac{x}{c} - \frac{b}{2c^2}\ln R + \frac{b^2-2ac}{2c^2}\int \frac{dx}{R}$$ (see **2.172**).

5. $$\int \frac{x^2\,dx}{R^2} = \frac{ab+(b^2-2ac)x}{c\Delta R} + \frac{2a}{\Delta}\int \frac{dx}{R}$$ (see **2.172**).

6. $$\int \frac{x^2\,dx}{R^3} = \frac{ab+(b^2-2ac)x}{2c\Delta R^2} + \frac{(2ac+b^2)(b+2cx)}{2c\Delta^2 R} + \frac{2ac+b^2}{\Delta^2}\int \frac{dx}{R}$$ (see **2.172**).

7. $$\int \frac{x^3\,dx}{R} = \frac{x^2}{2c} - \frac{bx}{c^2} + \frac{b^2-ac}{2c^3}\ln R - \frac{b(b^2-3ac)}{2c^3}\int \frac{dx}{R}$$ (see **2.172**).

8. $$\int \frac{x^3\,dx}{R^2} = \frac{1}{2c^2}\ln R + \frac{a(2ac-b^2)+b(3ac-b^2)x}{c^2\Delta R} - \frac{b(6ac-b^2)}{2c^2\Delta}\int \frac{dx}{R}$$ (see **2.172**).

9. $$\int \frac{x^3\,dx}{R^3} = -\left(\frac{x^2}{c} + \frac{abx}{c\Delta} + \frac{2a^2}{c\Delta}\right)\frac{1}{2R^2} - \frac{3ab}{2c\Delta}\int \frac{dx}{R^2}$$ (see **2.173** 1.).

2.176 $$\int \frac{dx}{x^m R^n} = \frac{-1}{(m-1)ax^{m-1}R^{n-1}} - \frac{b(m+n-2)}{a(m-1)}\int \frac{dx}{x^{m-1}R^n} - \frac{c(m+2n-3)}{a(m-1)}\int \frac{dx}{x^{m-2}R^n}.$$

2.177

1. $$\int \frac{dx}{xR} = \frac{1}{2a}\ln\frac{x^2}{R} - \frac{b}{2a}\int \frac{dx}{R}$$ (see **2.172**).

2. $$\int \frac{dx}{xR^2} = \frac{1}{2a^2}\ln\frac{x^2}{R} + \frac{1}{2aR}\left\{1 - \frac{b(b+2cx)}{\Delta}\right\} - \frac{b}{2a^2}\left(1+\frac{2ac}{\Delta}\right)\int \frac{dx}{R}$$ (see **2.172**).

3. $$\int \frac{dx}{xR^3} = \frac{1}{4aR^2} + \frac{1}{2a^2R} + \frac{1}{2a^3}\ln\frac{x^2}{R} - \frac{b}{2a}\int \frac{dx}{R^3} - \frac{b}{2a^2}\int \frac{dx}{R^2} - \frac{b}{2a^3}\int \frac{dx}{R}$$ (see **2.172**, **2.173**).

4. $$\int \frac{dx}{x^2R} = -\frac{b}{2a^2}\ln\frac{x^2}{R} - \frac{1}{ax} + \frac{b^2-2ac}{2a^2}\int \frac{dx}{R}$$ (see **2.172**).

5. $$\int \frac{dx}{x^2R^2} = -\frac{b}{a^3}\ln\frac{x^2}{R} - \frac{a+bx}{a^2xR} + \frac{(b^2-3ac)(b+2cx)}{a^2\Delta R} - \frac{1}{\Delta}\left(\frac{b^4}{a^3} - \frac{6b^2c}{a^2} + \frac{6c^2}{a}\right)\int \frac{dx}{R}$$ (see **2.172**).

6. $$\int \frac{dx}{x^2R^3} = -\frac{1}{axR^2} - \frac{3b}{a}\int \frac{dx}{xR^3} - \frac{5c}{a}\int \frac{dx}{R^3}$$ (see **2.173** and **2.177** 3.).

7. $$\int \frac{dx}{x^3R} = -\frac{ac-b^2}{2a^3}\ln\frac{x^2}{R} + \frac{b}{a^2x} - \frac{1}{2ax^2} + \frac{b(3ac-b^2)}{2a^3}\int \frac{dx}{R}$$ (see **2.172**).

8. $$\int \frac{dx}{x^3R^2} = \left(-\frac{1}{2ax^2} + \frac{3b}{2a^2x}\right)\frac{1}{R} + \left(\frac{3b^2}{a^2} - \frac{2c}{a}\right)\int \frac{dx}{xR^2} + \frac{9bc}{2a^2}\int \frac{dx}{R^2}$$ (see **2.173** 1. and **2.177** 2.).

9. $$\int \frac{dx}{x^3R^3} = \left(\frac{-1}{2ax^2} + \frac{2b}{a^2x}\right)\frac{1}{R^2} + \left(\frac{6b^2}{a^2} - \frac{3c}{a}\right)\int \frac{dx}{xR^3} + \frac{10bc}{a^2}\int \frac{dx}{R^3}$$ (see **2.173** 2., **2.177** 3.).

2.18 Forms containing the quadratic trinomial $a+bx+cx^2$ and the binomial $\alpha+\beta x$

Notations: $R=a+bx+cx^2$; $z=\alpha+\beta x$; $A=a\beta^2-\alpha b\beta+c\alpha^2$; $B=b\beta-2c\alpha$; $\Delta=4ac-b^2$.

1. $$\int z^m R^n\,dx=\frac{\beta z^{m-1}R^{n+1}}{(m+2n+1)c}-\frac{(m+n)B}{(m+2n+1)c}\int z^{m-1}R^n\,dx-\frac{(m-1)A}{(m+2n+1)c}\int z^{m-2}R^n\,dx.$$

2. $$\int\frac{R^n\,dx}{z^m}=-\frac{1}{(m-2n-1)\beta}\frac{R^n}{z^{m-1}}-\frac{2nA}{(m-2n-1)\beta^2}\int\frac{R^{n-1}\,dx}{z^m}-\frac{nB}{(m-2n-1)\beta^2}\int\frac{R^{n-1}\,dx}{z^{m-1}};$$ LA 184 (4)a

$$=\frac{-\beta}{(m-1)A}\frac{R^{n+1}}{z^{m-1}}-\frac{(m-n-2)B}{(m-1)A}\int\frac{R^n\,dx}{z^{m-1}}-\frac{(m-2n-3)c}{(m-1)A}\int\frac{R^n\,dx}{z^{m-2}};$$ LA 148 (5)

$$=-\frac{1}{(m-1)\beta}\frac{R^n}{z^{m-1}}+\frac{nB}{(m-1)\beta^2}\int\frac{R^{n-1}\,dx}{z^{m-1}}+\frac{2nc}{(m-1)\beta^2}\int\frac{R^{n-1}\,dx}{z^{m-2}}$$ LA 148 (6)

3. $$\int\frac{z^m\,dx}{R^n}=\frac{\beta}{(m-2n+1)c}\frac{z^{m-1}}{R^{n-1}}-\frac{(m-n)B}{(m-2n+1)c}\int\frac{z^{m-1}\,dx}{R^n}-\frac{(m-1)A}{(m-2n+1)c}\int\frac{z^{m-2}\,dx}{R^n};$$ LA 147 (1)

$$=\frac{b+2cx}{(n-1)\Delta}\frac{z^m}{R^{n-1}}-\frac{2(m-2n+3)c}{(n-1)\Delta}\int\frac{z^m\,dx}{R^{n-1}}-\frac{Bm}{(n-1)\Delta}\int\frac{z^{m-1}\,dx}{R^{n-1}}$$ LA 148 (3)

4. $$\int\frac{dx}{z^mR^n}=-\frac{\beta}{(m-1)A}\frac{1}{z^{m-1}R^{n-1}}-\frac{(m+n-2)B}{(m-1)A}\int\frac{dx}{z^{m-1}R^n}-\frac{(m+2n-3)c}{(m-1)A}\int\frac{dx}{z^{m-2}R^n};$$ LA 148 (7)

$$=\frac{\beta}{2(n-1)A}\frac{1}{z^{m-1}R^{n-1}}-\frac{B}{2A}\int\frac{dx}{z^{m-1}R^n}+\frac{(m+2n-3)\beta^2}{2(n-1)A}\int\frac{dx}{z^mR^{n-1}}.$$ LA 148 (8)

For $m=1$ and $n=1$

$$\int\frac{dx}{zR}=\frac{\beta}{2A}\ln\frac{z^2}{R}-\frac{B}{2A}\int\frac{dx}{R}.$$

For $A=0$

$$\int\frac{dx}{z^mR^n}=-\frac{\beta}{(m+n-1)B}\frac{1}{z^mR^{n-1}}-\frac{(m+2n-2)c}{(m+n-1)B}\int\frac{dx}{z^{m-1}R^n}.$$ LA 148 (9)

2.2 Algebraic Functions

2.20 Introduction

2.201 The integrals $\int R\left(x,\left(\frac{\alpha x+\beta}{\gamma x+\delta}\right)^r,\left(\frac{\alpha x+\beta}{\gamma x+\delta}\right)^s,\ldots\right)dx$, where $r, s, \ldots$ are rational numbers, can be reduced to integrals of rational functions by means of the substitution

$$\frac{\alpha x+\beta}{\gamma x+\delta}=t^m,$$ FI II 57

where m is the common denominator of the fractions $r, s, \ldots$.

2.202 Integrals of the form $\int x^m (a+bx^n)^p \, dx$,† where m, n, and p are rational numbers, can be expressed in terms of elementary functions only in the following cases:

(a) When p is an integer; then, this integral takes the form of the sum of the integrals shown in **2.201**;

(b) When $\frac{m+1}{n}$ is an integer: by means of the substitution $x^n = z$, this integral can be transformed to the form $\frac{1}{n}\int (a+bz)^p z^{\frac{m+1}{n}-1} dz$, which we considered in **2.201**;

(c) When $\frac{m+1}{n}+p$ is an integer: by means of the same substitution $x^n = z$, this integral can be reduced to an integral of the form $\frac{1}{n}\int \left(\frac{a+bz}{z}\right)^p z^{\frac{m+1}{n}+p-1} dz$, considered in **2.201**.

For reduction formulas for integrals of binomial differentials, see **2.110**.

2.21 Forms containing the binomial $a+bx^k$ and $\sqrt{x}$

Notation: $z_1 = a + bx$.

2.211
$$\int \frac{dx}{z_1 \sqrt{x}} = \frac{2}{\sqrt{ab}} \operatorname{arctg} \sqrt{\frac{bx}{a}} \qquad [ab > 0];$$
$$= \frac{1}{i\sqrt{ab}} \ln \frac{a - bx + 2i\sqrt{xab}}{z_1} \qquad [ab < 0].$$

2.212
$$\int \frac{x^m \sqrt{x}}{z_1} dx = 2\sqrt{x} \sum_{k=0}^{m} \frac{(-1)^k a^k x^{m-k}}{(2m-2k+1) b^{k+1}} + (-1)^{m+1} \frac{a^{m+1}}{b^{m+1}} \int \frac{dx}{z_1 \sqrt{x}}$$
(see **2.211**).

2.213

1. $$\int \frac{\sqrt{x}\, dx}{z_1} = \frac{2\sqrt{x}}{b} - \frac{a}{b} \int \frac{dx}{z_1 \sqrt{x}}$$ (see **2.211**).

2. $$\int \frac{x\sqrt{x}\, dx}{z_1} = \left(\frac{x}{3b} - \frac{a}{b^2}\right) 2\sqrt{x} + \frac{a^2}{b^2} \int \frac{dx}{z_1 \sqrt{x}}$$ (see **2.211**).

3. $$\int \frac{x^2\sqrt{x}\, dx}{z_1} = \left(\frac{x^2}{5b} - \frac{xa}{3b^2} + \frac{a^2}{b^3}\right) 2\sqrt{x} - \frac{a^3}{b^3} \int \frac{dx}{z_1 \sqrt{x}}$$ (see **2.211**).

4. $$\int \frac{dx}{z_1^2 \sqrt{x}} = \frac{\sqrt{x}}{az_1} + \frac{1}{2a} \int \frac{dx}{z_1 \sqrt{x}}$$ (see **2.211**).

5. $$\int \frac{\sqrt{x}\, dx}{z_1^2} = -\frac{\sqrt{x}}{bz_1} + \frac{1}{2b} \int \frac{dx}{z_1 \sqrt{x}}$$ (see **2.211**).

6. $$\int \frac{x\sqrt{x}\, dx}{z_1^2} = \frac{2x\sqrt{x}}{bz_1} - \frac{3a}{b} \int \frac{\sqrt{x}\, dx}{z_1^2}$$ (see **2.213** 5.).

7. $$\int \frac{x^2\sqrt{x}\, dx}{z_1^2} = \left(\frac{x^2}{3b} - \frac{5ax}{3b^2}\right) \frac{2\sqrt{x}}{z_1} + \frac{5a^2}{b^2} \int \frac{\sqrt{x}\, dx}{z_1^2}$$ (see **2.213** 5.).

8. $$\int \frac{dx}{z_1^3 \sqrt{x}} = \left(\frac{1}{2az_1^2} + \frac{3}{4a^2 z_1}\right)\sqrt{x} + \frac{3}{8a^2} \int \frac{dx}{z_1 \sqrt{x}}$$ (see **2.211**).

---

† Transl. The authors term such integrals "integrals of binomial differentials".

9. $\int \frac{\sqrt{x}\,dx}{z_1^3} = \left(-\frac{1}{2bz_1^2} + \frac{1}{4abz_1}\right)\sqrt{x} + \frac{1}{8ab}\int \frac{dx}{z_1\sqrt{x}}$ (see **2.211**).

10. $\int \frac{x\sqrt{x}\,dx}{z_1^3} = -\frac{2x\sqrt{x}}{bz_1^2} + \frac{3a}{b}\int \frac{\sqrt{x}\,dx}{z_1^3}$ (see **2.213** 9.).

11. $\int \frac{x^2\sqrt{x}\,dx}{z_1^3} = \left(\frac{x^2}{b} + \frac{5ax}{b^2}\right)\frac{2\sqrt{x}}{z_1^2} - \frac{15a^2}{b^2}\int \frac{\sqrt{x}\,dx}{z_1^3}$ (see **2.213** 9.).

Notations: $z_2 = a + bx^2$, $\alpha = \sqrt[4]{\frac{a}{b}}$, $\alpha' = \sqrt[4]{-\frac{a}{b}}$.

2.214 $\int \frac{dx}{z_2\sqrt{x}} = \frac{1}{b\alpha^3\sqrt{2}}\left[\ln\frac{x+\alpha\sqrt{2x}+\alpha^2}{\sqrt{z_2}} + \operatorname{arctg}\frac{\alpha\sqrt{2x}}{\alpha^2-x}\right]$ $\left[\frac{a}{b} > 0\right]$;

$= \frac{1}{2b\alpha'^3}\left(\ln\frac{\alpha'-\sqrt{x}}{\alpha'+\sqrt{x}} - 2\operatorname{arctg}\frac{\sqrt{x}}{\alpha'}\right)$ $\left[\frac{a}{b} < 0\right]$.

2.215 $\int \frac{\sqrt{x}\,dx}{z_2} = \frac{1}{b\alpha\sqrt{2}}\left[-\ln\frac{x+\alpha\sqrt{2x}+\alpha^2}{\sqrt{z_2}} + \operatorname{arctg}\frac{\alpha\sqrt{2x}}{\alpha^2-x}\right]$ $\left[\frac{a}{b} > 0\right]$;

$= \frac{1}{2b\alpha'}\left[\ln\frac{\alpha'-\sqrt{x}}{\alpha'+\sqrt{x}} + 2\operatorname{arctg}\frac{\sqrt{x}}{\alpha'}\right]$ $\left[\frac{a}{b} < 0\right]$.

2.216

1. $\int \frac{x\sqrt{x}\,dx}{z_2} = \frac{2\sqrt{x}}{b} - \frac{a}{b}\int \frac{dx}{z_2\sqrt{x}}$ (see **2.214**).

2. $\int \frac{x^2\sqrt{x}\,dx}{z_2} = \frac{2x\sqrt{x}}{3b} - \frac{a}{b}\int \frac{\sqrt{x}\,dx}{z_2}$ (see **2.215**).

3. $\int \frac{dx}{z_2^2\sqrt{x}} = \frac{\sqrt{x}}{2az_2} + \frac{3}{4a}\int \frac{dx}{z_2\sqrt{x}}$ (see **2.214**).

4. $\int \frac{\sqrt{x}\,dx}{z_2^2} = \frac{x\sqrt{x}}{2az_2} + \frac{1}{4a}\int \frac{\sqrt{x}\,dx}{z_2}$ (see **2.215**).

5. $\int \frac{x\sqrt{x}\,dx}{z_2^2} = -\frac{\sqrt{x}}{2bz_2} + \frac{1}{4b}\int \frac{dx}{z_2\sqrt{x}}$ (see **2.214**).

6. $\int \frac{x^2\sqrt{x}\,dx}{z_2^2} = -\frac{x\sqrt{x}}{2bz_2} + \frac{3}{4b}\int \frac{\sqrt{x}\,dx}{z_2}$ (see **2.215**).

7. $\int \frac{dx}{z_2^3\sqrt{x}} = \left(\frac{1}{4az_2^2} + \frac{7}{16a^2z_2}\right)\sqrt{x} + \frac{21}{32a^2}\int \frac{dx}{z_2\sqrt{x}}$ (see **2.214**).

8. $\int \frac{\sqrt{x}\,dx}{z_2^3} = \left(\frac{1}{4az_2^2} + \frac{5}{16a^2z_2}\right)x\sqrt{x} + \frac{5}{32a^2}\int \frac{\sqrt{x}\,dx}{z_2}$ (see **2.215**).

9. $\int \frac{x\sqrt{x}\,dx}{z_2^3} = \frac{(bx^2-3a)\sqrt{x}}{16abz_2^2} + \frac{3}{32ab}\int \frac{dx}{z_2\sqrt{x}}$ (see **2.214**).

10. $\int \frac{x^2\sqrt{x}\,dx}{z_2^3} = -\frac{2x\sqrt{x}}{5bz_2^2} + \frac{3a}{5b}\int \frac{\sqrt{x}\,dx}{z_2^3}$ (see **2.216** 8.).

2.22-2.23 Forms containing $\sqrt[n]{(a+bx)^k}$

Notation: $z = a + bx$.

2.220 $\int x^n \sqrt[l]{z^{lm+f}}\,dx = \left\{\sum_{k=0}^{n} \frac{(-1)^k \binom{n}{k} z^{n-k}a^k}{ln - lk + l(m+1) + f}\right\} \frac{l\sqrt[l]{z^{l(m+1)+f}}}{b^{n+1}}$.

The square root

2.221 $$\int x^n \sqrt{z^{2m-1}}\,dx = \left\{ \sum_{k=0}^{n} \frac{(-1)^k \binom{n}{k} z^{n-k} a^k}{2n-2k+2m+1} \right\} \frac{2\sqrt{z^{2m+1}}}{b^{n+1}}.$$

2.222

1. $$\int \frac{dx}{\sqrt{z}} = \frac{2}{b}\sqrt{z}.$$
2. $$\int \frac{x\,dx}{\sqrt{z}} = \left(\frac{1}{3}z - a\right)\frac{2\sqrt{z}}{b^2}.$$
3. $$\int \frac{x^2\,dx}{\sqrt{z}} = \left(\frac{1}{5}z^2 - \frac{2}{3}az + a^2\right)\frac{2\sqrt{z}}{b^3}.$$

2.223

1. $$\int \frac{dx}{\sqrt{z^3}} = -\frac{2}{b\sqrt{z}}.$$
2. $$\int \frac{x\,dx}{\sqrt{z^3}} = (z+a)\frac{2}{b^2\sqrt{z}}.$$
3. $$\int \frac{x^2\,dx}{\sqrt{z^3}} = \left(\frac{z^2}{3} - 2az - a^2\right)\frac{2}{b^3\sqrt{z}}.$$

2.224

1. $$\int \frac{z^m\,dx}{x^n\sqrt{z}} = -\frac{z^m\sqrt{z}}{(n-1)\,ax^{n-1}} + \frac{2m-2n+3}{2(n-1)}\,\frac{b}{a}\int \frac{z^m\,dx}{x^{n-1}\sqrt{z}}.$$
2. $$\int \frac{z^m\,dx}{x^n\sqrt{z}} = -z^m\sqrt{z}\left\{\frac{1}{(n-1)\,ax^{n-1}} + \right.$$
$$+\sum_{k=1}^{n-2} \frac{(2m-2n+3)(2m-2n+5)\dots(2m-2n+2k+1)}{2^k(n-1)(n-2)\dots(n-k-1)\,x^{n-k-1}}\,\frac{b^k}{a^{k+1}} +$$
$$+\frac{(2m-2n+3)(2m-2n+5)\dots(2m-3)(2m-1)}{2^{n-1}(n-1)!\,x}\,\frac{b^{n-1}}{a^{n-1}}\int \frac{z^m\,dx}{x\sqrt{z}}.$$

For $n = 1$

3. $$\int \frac{z^m}{x\sqrt{z}}\,dx = \frac{2z^m}{(2m-1)\sqrt{z}} + a\int \frac{z^{m-1}}{x\sqrt{z}}\,dx.$$
4. $$\int \frac{z^m}{x\sqrt{z}}\,dx = \sum_{k=1}^{m} \frac{2a^{m-k}z^k}{(2k-1)\sqrt{z}} + a^m\int \frac{dx}{x\sqrt{z}}.$$
5. $$\int \frac{dx}{x\sqrt{z}} = \frac{1}{\sqrt{a}}\ln\frac{\sqrt{z}-\sqrt{a}}{\sqrt{z}+\sqrt{a}} \qquad [a > 0];$$
$$= \frac{2}{\sqrt{-a}}\operatorname{arctg}\frac{\sqrt{z}}{\sqrt{-a}} \qquad [a < 0].$$

2.225

1. $$\int \frac{\sqrt{z}\,dx}{x} = 2\sqrt{z} + a\int \frac{dx}{x\sqrt{z}}$$ (see **2.224** 4.).
2. $$\int \frac{\sqrt{z}\,dx}{x^2} = -\frac{\sqrt{z}}{x} + \frac{b}{2}\int \frac{dx}{x\sqrt{z}}$$ (see **2.224** 4.).
3. $$\int \frac{\sqrt{z}\,dx}{x^3} = -\frac{\sqrt{z^3}}{2ax^2} + \frac{b\sqrt{z}}{4ax} - \frac{b^2}{8a}\int \frac{dx}{x\sqrt{z}}$$ (see **2.224** 4.).

2.226

1. $\int \frac{\sqrt{z^3}\,dx}{x} = \left(\frac{z}{3}+a\right) 2\sqrt{z} + a^2 \int \frac{dx}{x\sqrt{z}}$ (see **2.224** 4.).

2. $\int \frac{\sqrt{z^3}\,dx}{x^2} = -\frac{\sqrt{z^5}}{ax} + \frac{3b}{2a} \int \frac{\sqrt{z^3}\,dx}{x}$ (see **2.226** 1.).

3. $\int \frac{\sqrt{z^3}\,dx}{x^3} = -\left(\frac{1}{2ax^2}+\frac{b}{4a^2x}\right)\sqrt{z^5} + \frac{3b^2}{8a^2} \int \frac{\sqrt{z^3}\,dx}{x}$ (see **2.226** 1.).

2.227 $\int \frac{dx}{xz^m\sqrt{z}} = \sum_{k=0}^{m-1} \frac{2}{(2k+1)\,a^{m-k}z^k\sqrt{z}} + \frac{1}{a^m} \int \frac{dx}{x\sqrt{z}}$. (see **2.224** 4.).

2.228

1. $\int \frac{dx}{x^2\sqrt{z}} = -\frac{\sqrt{z}}{ax} - \frac{b}{2a} \int \frac{dx}{x\sqrt{z}}$ (see **2.224** 4.).

2. $\int \frac{dx}{x^3\sqrt{z}} = \left(-\frac{1}{2ax^2}+\frac{3b}{4a^2x}\right)\sqrt{z} + \frac{3b^2}{8a^2} \int \frac{dx}{x\sqrt{z}}$ (see **2.224** 4.).

2.229

1. $\int \frac{dx}{x\sqrt{z^3}} = \frac{2}{a\sqrt{z}} + \frac{1}{a} \int \frac{dx}{x\sqrt{z}}$ (see **2.224** 4.).

2. $\int \frac{dx}{x^2\sqrt{z^3}} = \left(-\frac{1}{ax}-\frac{3b}{a^2}\right)\frac{1}{\sqrt{z}} - \frac{3b}{2a^2} \int \frac{dx}{x\sqrt{z}}$ (see **2.224** 4.).

3. $\int \frac{dx}{x^3\sqrt{z^3}} = \left(-\frac{1}{2ax^2}+\frac{5b}{4a^2x}+\frac{15b^2}{4a^3}\right)\frac{1}{\sqrt{z}} + \frac{15b^2}{8a^3} \int \frac{dx}{x\sqrt{z}}$ (see **2.224** 4.).

Cube root

2.231

1. $\int \sqrt[3]{z^{3m+1}}x^n dx = \left\{\sum_{k=0}^{n} \frac{(-1)^k \binom{n}{k} z^{n-k}a^k}{3n-3k+3\,(m+1)+1}\right\} \frac{3\sqrt[3]{z^{3(m+1)+1}}}{b^{n+1}}$.

2. $\int \frac{x^n\,dx}{\sqrt[3]{z^{3m+2}}} = \left\{\sum_{k=0}^{n} \frac{(-1)^k \binom{n}{k} z^{n-k}a^k}{3n-3k-3\,(m-1)-2}\right\} \frac{3}{b^{n+1}\sqrt[3]{z^{3(m-1)+2}}}$.

3. $\int \sqrt[3]{z^{3m+2}}x^n\,dx = \left\{\sum_{k=0}^{n} \frac{(-1)^k \binom{n}{k} z^{n-k}a^k}{3n-3k+3\,(m+1)+2}\right\} \frac{3\sqrt[3]{z^{3(m+1)+2}}}{b^{n+1}}$.

4. $\int \frac{x^n\,dx}{\sqrt[3]{z^{3m+1}}} = \left\{\sum_{k=0}^{n} \frac{(-1)^k \binom{n}{k} z^{n-k}a^k}{3n-3k-3\,(m-1)-1}\right\} \frac{3}{b^{n+1}\sqrt[3]{z^{3(m-1)+1}}}$.

5. $\int \frac{z^n\,dx}{x^m\sqrt[3]{z^2}} = -\frac{z^{n+\frac{1}{3}}}{(m-1)\,ax^{m-1}} + \frac{3n-3m+4}{3\,(m-1)}\,\frac{b}{a} \int \frac{z^n\,dx}{x^{m-1}\sqrt[3]{z^2}}$.

For $m=1$

$$\int \frac{z^n\,dx}{x\sqrt[3]{z^2}} = \frac{3z^n}{(3n-2)\sqrt[3]{z^2}} + a \int \frac{z^{n-1}\,dx}{x\sqrt[3]{z^2}}.$$

6. $\int \frac{dx}{xz^n\sqrt[3]{z^2}} = \frac{3\sqrt[3]{z}}{(3n-1)\,az^n} + \frac{1}{a} \int \frac{\sqrt[3]{z}\,dx}{xz^n}$.

2.232 $\int \frac{dx}{x\sqrt[3]{z^2}} = \frac{1}{\sqrt[3]{a^2}}\left\{\frac{3}{2}\ln\frac{\sqrt[3]{z}-\sqrt[3]{a}}{\sqrt[3]{x}} - \sqrt{3}\operatorname{arctg}\frac{\sqrt{3}\sqrt[3]{z}}{\sqrt[3]{z}+2\sqrt[3]{a}}\right\}.$

2.233

1. $\int \frac{\sqrt[3]{z}\,dx}{x} = 3\sqrt[3]{z} + a\int\frac{dx}{x\sqrt[3]{z^2}}$ (see 2.232).
2. $\int \frac{\sqrt[3]{z}\,dx}{x^2} = -\frac{z\sqrt[3]{z}}{ax} + \frac{b}{a}\sqrt[3]{z} + \frac{b}{3}\int\frac{dx}{x\sqrt[3]{z^2}}$ (see 2.232).
3. $\int \frac{\sqrt[3]{z}\,dx}{x^3} = \left(-\frac{1}{2ax^2} + \frac{b}{3a^2x}\right)z\sqrt[3]{z} - \frac{b^2}{3a^2}\sqrt[3]{z} - \frac{b^2}{9a}\int\frac{dx}{x\sqrt[3]{z^2}}$ (see 2.232).
4. $\int \frac{dx}{x^2\sqrt[3]{z^2}} = -\frac{\sqrt[3]{z}}{ax} - \frac{2b}{3a}\int\frac{dx}{x\sqrt[3]{z^2}}$ (see 2.232).
5. $\int \frac{dx}{x^3\sqrt[3]{z^2}} = \left[-\frac{1}{2ax^2} + \frac{5b}{6a^2x}\right]\sqrt[3]{z} + \frac{5b^2}{9a^2}\int\frac{dx}{x\sqrt[3]{z^2}}$ (see 2.232).

2.234

1. $\int \frac{z^n\,dx}{x^m\sqrt[3]{z}} = -\frac{z^n\sqrt[3]{z^2}}{(m-1)ax^{m-1}} + \frac{3n-3m+5}{3(m-1)}\,\frac{b}{a}\int\frac{z^n\,dx}{x^{m-1}\sqrt[3]{z}}.$

For $m=1$:

2. $\int \frac{z^n\,dx}{x\sqrt[3]{z}} = \frac{3z^n}{(3n-1)\sqrt[3]{z}} + a\int\frac{z^{n-1}\,dx}{x\sqrt[3]{z}}.$
3. $\int \frac{dx}{xz^n\sqrt[3]{z}} = \frac{3\sqrt[3]{z^2}}{(3n-2)az^n} + \frac{1}{a}\int\frac{\sqrt[3]{z^2}\,dx}{xz^n}.$

2.235 $\int \frac{dx}{x\sqrt[3]{z}} = \frac{1}{\sqrt[3]{a^2}}\left\{\frac{3}{2}\ln\frac{\sqrt[3]{z}-\sqrt[3]{a}}{\sqrt[3]{x}} + \sqrt{3}\operatorname{arctg}\frac{\sqrt{3}\sqrt[3]{z}}{\sqrt[3]{z}+2\sqrt[3]{a}}\right\}.$

2.236

1. $\int \frac{\sqrt[3]{z^2}\,dx}{x} = \frac{3}{2}\sqrt[3]{z^2} + a\int\frac{dx}{x\sqrt[3]{z}}$ (see 2.235).
2. $\int \frac{\sqrt[3]{z^2}\,dx}{x^2} = -\frac{\sqrt[3]{z^5}}{ax} + \frac{b}{a}\sqrt[3]{z^2} + \frac{2b}{3}\int\frac{dx}{x\sqrt[3]{z}}$ (see 2.235).
3. $\int \frac{\sqrt[3]{z^2}\,dx}{x^3} = \left[-\frac{1}{2ax^2} + \frac{b}{6a^2x}\right]z^{\frac{5}{3}} - \frac{b^2}{6a^2}\sqrt[3]{z^2} - \frac{b^2}{9a}\int\frac{dx}{x\sqrt[3]{z}}$ (see 2.235).
4. $\int \frac{dx}{x^2\sqrt[3]{z}} = -\frac{\sqrt[3]{z^2}}{ax} - \frac{b}{3a}\int\frac{dx}{x\sqrt[3]{z}}$ (see 2.235).
5. $\int \frac{dx}{x^3\sqrt[3]{z}} = \left[-\frac{1}{2ax^2} + \frac{2b}{3a^2x}\right]\sqrt[3]{z^2} + \frac{2b^2}{9a^2}\int\frac{dx}{x\sqrt[3]{z}}$ (see 2.235).

2.24 Forms containing $\sqrt{a+bx}$ and the binomial $\alpha+\beta x$

Notation: $z = a+bx,\ t = \alpha+\beta x,\ \Delta = a\beta - b\alpha.$

2.241

1. $\int \frac{z^m t^n\,dx}{\sqrt{z}} = \frac{2}{(2n+2m+1)\beta}t^{n+1}z^{m-1}\sqrt{z} + \frac{(2m-1)\Delta}{(2n+2m+1)\beta}\int\frac{z^{m-1}t^n\,dx}{\sqrt{z}}.$

LA 176 (1)

2. $\int \frac{t^n z^m\,dx}{\sqrt{z}} = 2\sqrt{z^{2m+1}}\sum_{k=0}^{n}\binom{n}{k}\frac{\alpha^{n-k}\beta^k}{b^{k+1}}\sum_{p=0}^{k}(-1)^p\binom{k}{p}\frac{z^{k-p}a^p}{2k-2p+2m+1}.$

2.242

1. $$\int \frac{t\,dx}{\sqrt{z}} = \frac{2\alpha\sqrt{z}}{b} + \beta\left(\frac{z}{3} - a\right)\frac{2\sqrt{z}}{b^2}.$$

2. $$\int \frac{t^2\,dx}{\sqrt{z}} = \frac{2\alpha^2\sqrt{z}}{b} + 2\alpha\beta\left(\frac{z}{3} - a\right)\frac{2\sqrt{z}}{b^2} + \beta^2\left(\frac{z^2}{5} - \frac{2}{3}za + a^2\right)\frac{2\sqrt{z}}{b^3}.$$

3. $$\int \frac{t^3\,dx}{\sqrt{z}} = \frac{2\alpha^3\sqrt{z}}{b} + 3\alpha^2\beta\left(\frac{z}{3} - a\right)\frac{2\sqrt{z}}{b^2} +$$
$$+ 3\alpha\beta^2\left(\frac{z^2}{5} - \frac{2}{3}za + a^2\right)\frac{2\sqrt{z}}{b^3} + \beta^3\left(\frac{z^3}{7} - \frac{3z^2a}{5} + za^2 - a^3\right)\frac{2\sqrt{z}}{b^4}.$$

4. $$\int \frac{tz\,dx}{\sqrt{z}} = \frac{2\alpha\sqrt{z^3}}{3b} + \beta\left(\frac{z}{5} - \frac{a}{3}\right)\frac{2\sqrt{z^3}}{b^2}.$$

5. $$\int \frac{t^2z\,dx}{\sqrt{z}} = \frac{2\alpha^2\sqrt{z^3}}{3b} + 2\alpha\beta\left(\frac{z}{5} - \frac{a}{3}\right)\frac{2\sqrt{z^3}}{b^2} + \beta^2\left(\frac{z^2}{7} - \frac{2za}{5} + \frac{a^2}{3}\right)\frac{2\sqrt{z^3}}{b^3}.$$

6. $$\int \frac{t^3z\,dx}{\sqrt{z}} = \frac{2\alpha^3\sqrt{z^3}}{3b} + 3\alpha^2\beta\left(\frac{z}{5} - \frac{a}{3}\right)\frac{2\sqrt{z^3}}{b^2} +$$
$$+ 3\alpha\beta^2\left(\frac{z^2}{7} - \frac{2za}{5} + \frac{a^2}{3}\right)\frac{2\sqrt{z^3}}{b^3} + \beta^3\left(\frac{z^3}{9} - \frac{3z^2a}{7} + \frac{3za^2}{5} - \frac{a^3}{3}\right)\frac{2\sqrt{z^3}}{b^4}.$$

7. $$\int \frac{tz^2\,dx}{\sqrt{z}} = \frac{2\alpha\sqrt{z^5}}{5b} + \beta\left(\frac{z}{7} - \frac{a}{5}\right)\frac{2\sqrt{z^5}}{b^2}.$$

8. $$\int \frac{t^2z^2\,dx}{\sqrt{z}} = \frac{2\alpha^2\sqrt{z^5}}{5b} + 2\alpha\beta\left(\frac{z}{7} - \frac{a}{5}\right)\frac{2\sqrt{z^5}}{b^2} + \beta^2\left(\frac{z^2}{9} - \frac{2za}{7} + \frac{a^2}{5}\right)\frac{2\sqrt{z^5}}{b^3}.$$

9. $$\int \frac{t^3z^2\,dx}{\sqrt{z}} = \frac{2\alpha^3\sqrt{z^5}}{5b} + 3\alpha^2\beta\left(\frac{z}{7} - \frac{a}{5}\right)\frac{2\sqrt{z^5}}{b^2} +$$
$$+ 3\alpha\beta^2\left(\frac{z^2}{9} - \frac{2za}{7} + \frac{a^2}{5}\right)\frac{2\sqrt{z^5}}{b^3} + \beta^3\left(\frac{z^3}{11} - \frac{3z^2a}{9} + \frac{3za^2}{7} - \frac{a^3}{5}\right)\frac{2\sqrt{z^5}}{b^4}$$

10. $$\int \frac{tz^3\,dx}{\sqrt{z}} = \frac{2\alpha\sqrt{z^7}}{7b} + \beta\left(\frac{z}{9} - \frac{a}{7}\right)\frac{2\sqrt{z^7}}{b^2}.$$

11. $$\int \frac{t^2z^3\,dx}{\sqrt{z}} = \frac{2\alpha^2\sqrt{z^7}}{7b} + 2\alpha\beta\left(\frac{z}{9} - \frac{a}{7}\right)\frac{2\sqrt{z^7}}{b^2} + \beta^2\left(\frac{z^2}{11} - \frac{2za}{9} + \frac{a^2}{7}\right)\frac{2\sqrt{z^7}}{b^3}.$$

12. $$\int \frac{t^3z^3\,dx}{\sqrt{z}} = \frac{2\alpha^3\sqrt{z^7}}{7b} + 3\alpha^2\beta\left(\frac{z}{9} - \frac{a}{7}\right)\frac{2\sqrt{z^7}}{b^2} +$$
$$+ 3\alpha\beta^2\left(\frac{z^2}{11} - \frac{2za}{9} + \frac{a^2}{7}\right)\frac{2\sqrt{z^7}}{b^3} + \beta^3\left(\frac{z^3}{13} - \frac{3z^2a}{11} + \frac{3za^2}{9} - \frac{a^3}{7}\right)\frac{2\sqrt{z^7}}{b^4}.$$

2.243

1. $$\int \frac{t^n\,dx}{z^m\sqrt{z}} = \frac{2}{(2m-1)\Delta}\,\frac{t^{n+1}}{z^m}\sqrt{z} - \frac{(2n-2m+3)\beta}{(2m-1)\Delta}\int \frac{t^n\,dx}{z^{m-1}\sqrt{z}};$$
$$= -\frac{2}{(2m-1)b}\,\frac{t^n}{z^m}\sqrt{z} + \frac{2n\beta}{(2m-1)b}\int \frac{t^{n-1}\,dx}{z^{m-1}\sqrt{z}}.$$ LA 176 (2)

2. $$\int \frac{t^n\,dx}{z^m\sqrt{z}} = \frac{2}{\sqrt{z^{2m-1}}}\sum_{k=0}^{n}\binom{n}{k}\frac{\alpha^{n-k}\beta^k}{b^{k+1}}\sum_{p=0}^{k}(-1)^p\binom{k}{p}\frac{z^{k-p}a^p}{2k-2p-2m+1}.$$

2.244

1. $$\int \frac{t\,dx}{z\sqrt{z}} = -\frac{2\alpha}{b\sqrt{z}} + \frac{2\beta(z+a)}{b^2\sqrt{z}}.$$

2. $$\int \frac{t^2\,dx}{z\sqrt{z}} = -\frac{2\alpha^2}{b\sqrt{z}} + \frac{4\alpha\beta(z+a)}{b^2\sqrt{z}} + \frac{2\beta^2\left(\frac{z^2}{3} - 2za - a^2\right)}{b^3\sqrt{z}}.$$

3. $$\int \frac{t^3\,dx}{z\sqrt{z}} = -\frac{2\alpha^3}{b\sqrt{z}} + \frac{6\alpha^2\beta(z+a)}{b^2\sqrt{z}} + \frac{6\alpha\beta^2\left(\frac{z^2}{3} - 2za - a^2\right)}{b^3\sqrt{z}} + \frac{2\beta^3\left(\frac{z^3}{5} - z^2a + 3za^2 + a^3\right)}{b^4\sqrt{z}}.$$

4. $$\int \frac{t\,dx}{z^2\sqrt{z}} = -\frac{2\alpha}{3b\sqrt{z^3}} - \frac{2\beta\left(z - \frac{a}{3}\right)}{b^2\sqrt{z^3}}.$$

5. $$\int \frac{t^2\,dx}{z^2\sqrt{z}} = -\frac{2\alpha^2}{3b\sqrt{z^3}} - \frac{4\alpha\beta\left(z - \frac{a}{3}\right)}{b^2\sqrt{z^3}} + \frac{2\beta^2\left(z^2 + 2az - \frac{a^2}{3}\right)}{b^3\sqrt{z^3}}.$$

6. $$\int \frac{t^3\,dx}{z^2\sqrt{z}} = -\frac{2\alpha^3}{3b\sqrt{z^3}} - \frac{6\alpha^2\beta\left(z - \frac{a}{3}\right)}{b^2\sqrt{z^3}} + \frac{6\alpha\beta^2\left(z^2 + 2za - \frac{a^2}{3}\right)}{b^3\sqrt{z^3}} + \frac{2\beta^3\left(\frac{z^3}{3} - 3z^2a - 3za^2 + \frac{a^3}{3}\right)}{b^4\sqrt{z^3}}.$$

7. $$\int \frac{t\,dx}{z^3\sqrt{z}} = -\frac{2\alpha}{5b\sqrt{z^5}} - \frac{2\beta\left(\frac{z}{3} - \frac{a}{5}\right)}{b^2\sqrt{z^5}}.$$

8. $$\int \frac{t^2\,dx}{z^3\sqrt{z}} = -\frac{2\alpha^2}{5b\sqrt{z^5}} - \frac{4\alpha\beta\left(\frac{z}{3} - \frac{a}{5}\right)}{b^2\sqrt{z^5}} - \frac{2\beta^2\left(z^2 - \frac{2za}{3} + \frac{a^2}{5}\right)}{b^3\sqrt{z^5}}.$$

9. $$\int \frac{t^3\,dx}{z^3\sqrt{z}} = -\frac{2\alpha^3}{5b\sqrt{z^5}} - \frac{6\alpha^2\beta\left(\frac{z}{3} - \frac{a}{5}\right)}{b^2\sqrt{z^5}} - \frac{6\alpha\beta^2\left(z^2 - \frac{2za}{3} + \frac{a^2}{5}\right)}{b^3\sqrt{z^5}} + \frac{2\beta^3\left(z^3 + 3z^2a - za^2 + \frac{a^3}{5}\right)}{b^4\sqrt{z^5}}.$$

2.245

1. $$\int \frac{z^m\,dx}{t^n\sqrt{z}} = -\frac{2}{(2n-2m-1)\beta}\frac{z^{m-1}}{t^{n-1}}\sqrt{z} - \frac{(2m-1)\Delta}{(2n-2m-1)\beta}\int \frac{z^{m-1}\,dx}{t^n\sqrt{z}};$$ LA 176 (3)

$$= -\frac{1}{(n-1)\beta}\frac{z^{m-1}}{t^{n-1}}\sqrt{z} + \frac{(2m-1)b}{2(n-1)\beta}\int \frac{z^{m-1}}{t^{n-1}\sqrt{z}}\,dx;$$

$$= -\frac{1}{(n-1)\Delta}\frac{z^m}{t^{n-1}}\sqrt{z} - \frac{(2n-2m-3)b}{2(n-1)\Delta}\int \frac{z^m\,dx}{t^{n-1}\sqrt{z}}.$$

2. $$\int \frac{z^m\,dx}{t^n\sqrt{z}} = -z^m\sqrt{z}\left\{\frac{1}{(n-1)\Delta}\frac{1}{t^{n-1}} + \sum_{k=2}^{n-1}\frac{(2n-2m-3)(2n-2m-5)\dots(2n-2m-2k+1)b^{k-1}}{2^{k-1}(n-1)(n-2)\dots(n-k)\Delta^k}\frac{1}{t^{n-k}}\right\} - \frac{(2n-2m-3)(2n-2m-5)\dots(-2m+3)(-2m+1)b^{n-1}}{2^{n-1}\cdot(n-1)!\,\Delta^n}\int \frac{z^m\,dx}{t\sqrt{z}}.$$

For $n=1$

3. $$\int \frac{z^m\,dx}{t\sqrt{z}} = \frac{2}{(2m-1)\beta}\frac{z^m}{\sqrt{z}} + \frac{\Delta}{\beta}\int \frac{z^{m-1}\,dx}{t\sqrt{z}}.$$

4. $$\int \frac{z^m\,dx}{t\sqrt{z}} = 2\sum_{k=0}^{m-1} \frac{\Delta^k}{(2m-2k-1)\beta^{k+1}}\frac{z^{m-k}}{\sqrt{z}} + \frac{\Delta^m}{\beta^m}\int \frac{dx}{t\sqrt{z}}.$$

2.246 $$\int \frac{dx}{t\sqrt{z}} = \frac{1}{\sqrt{\beta\Delta}} \ln \frac{\beta\sqrt{z}-\sqrt{\beta\Delta}}{\beta\sqrt{z}+\sqrt{\beta\Delta}} \qquad [\beta\Delta > 0];$$
$$= \frac{2}{\sqrt{-\beta\Delta}} \operatorname{arctg} \frac{\beta\sqrt{z}}{\sqrt{-\beta\Delta}} \qquad [\beta\Delta < 0];$$
$$= -\frac{2\sqrt{z}}{bt} \qquad [\Delta = 0].$$

2.247 $$\int \frac{dx}{tz^m\sqrt{z}} = \frac{2}{z^{m-1}\sqrt{z}} + \sum_{k=1}^{m} \frac{\beta^{k-1}z^k}{\Delta^k(2m-2k+1)} + \frac{\beta^m}{\Delta^m}\int \frac{dx}{t\sqrt{z}} \qquad \text{(see 2.246).}$$

2.248

1. $$\int \frac{dx}{tz\sqrt{z}} = \frac{2}{\Delta\sqrt{z}} + \frac{\beta}{\Delta}\int \frac{dx}{t\sqrt{z}} \quad \text{(see 2.246).}$$

2. $$\int \frac{dx}{tz^2\sqrt{z}} = \frac{2}{3\Delta z\sqrt{z}} + \frac{2\beta}{\Delta^2\sqrt{z}} + \frac{\beta^2}{\Delta^2}\int \frac{dx}{t\sqrt{z}} \quad \text{(see 2.246).}$$

3. $$\int \frac{dx}{tz^3\sqrt{z}} = \frac{2}{5\Delta z^2\sqrt{z}} + \frac{2\beta}{3\Delta^2 z\sqrt{z}} + \frac{2\beta^2}{\Delta^3\sqrt{z}} + \frac{\beta^3}{\Delta^3}\int \frac{dx}{t\sqrt{z}} \quad \text{(see 2.246).}$$

4. $$\int \frac{dx}{t^2\sqrt{z}} = -\frac{\sqrt{z}}{\Delta t} - \frac{b}{2\Delta}\int \frac{dx}{t\sqrt{z}} \quad \text{(see 2.246).}$$

5. $$\int \frac{dx}{t^2z\sqrt{z}} = -\frac{1}{\Delta t\sqrt{z}} - \frac{3b}{\Delta^2\sqrt{z}} - \frac{3b\beta}{2\Delta^2}\int \frac{dx}{t\sqrt{z}} \quad \text{(see 2.246).}$$

6. $$\int \frac{dx}{t^2z^2\sqrt{z}} = -\frac{1}{\Delta tz^2\sqrt{z}} - \frac{5b}{3\Delta^2 z\sqrt{z}} - \frac{5b\beta}{\Delta^3\sqrt{z}} - \frac{5b\beta^2}{2\Delta^3}\int \frac{dx}{t\sqrt{z}} \quad \text{(see 2.246).}$$

7. $$\int \frac{dx}{t^2z^3\sqrt{z}} = -\frac{1}{\Delta tz^2\sqrt{z}} - \frac{7b}{5\Delta^2 z^2\sqrt{z}} - \frac{7b\beta}{3\Delta^3 z\sqrt{z}} - \frac{7b\beta^2}{\Delta^4\sqrt{z}} - \frac{7b\beta^3}{2\Delta^4}\int \frac{dx}{t\sqrt{z}} \quad \text{(see 2.246).}$$

8. $$\int \frac{dx}{t^3\sqrt{z}} = -\frac{\sqrt{z}}{2\Delta t^2} + \frac{3b\sqrt{z}}{4\Delta^2 t} + \frac{3b^2}{8\Delta^2}\int \frac{dx}{t\sqrt{z}} \quad \text{(see 2.246).}$$

9. $$\int \frac{dx}{t^3z\sqrt{z}} = -\frac{1}{2\Delta t^2\sqrt{z}} + \frac{5b}{4\Delta^2 t\sqrt{z}} + \frac{15b^2}{4\Delta^3\sqrt{z}} + \frac{15b^2\beta}{8\Delta^3}\int \frac{dx}{t\sqrt{z}} \quad \text{(see 2.246).}$$

10. $$\int \frac{dx}{t^3z^2\sqrt{z}} = -\frac{1}{2\Delta t^2z\sqrt{z}} + \frac{7b\sqrt{z}}{4\Delta^2 tz\sqrt{z}} + \frac{35b^2}{12\Delta^3 z\sqrt{z}} + \frac{35b^2\beta}{4\Delta^4\sqrt{z}} + \frac{35b^2\beta^2}{8\Delta^4}\int \frac{dx}{t\sqrt{z}} \quad \text{(see 2.246).}$$

11. $$\int \frac{dx}{t^3z^3\sqrt{z}} = -\frac{1}{2\Delta t^2z^2\sqrt{z}} + \frac{9b}{4\Delta^2 tz^2\sqrt{z}} + \frac{63b^2}{20\Delta^3 z^2\sqrt{z}} + \frac{21b^2\beta}{4\Delta^4 z\sqrt{z}} + \frac{63b^2\beta^2}{4\Delta^5\sqrt{z}} + \frac{63b^2\beta^3}{8\Delta^5}\int \frac{dx}{t\sqrt{z}} \quad \text{(see 2.246).}$$

12. $\int \frac{z\,dx}{t\sqrt{z}} = \frac{2\sqrt{z}}{\beta} + \frac{\Delta}{\beta}\int \frac{dx}{t\sqrt{z}}$ (see **2.246**).

13. $\int \frac{z^2\,dx}{t\sqrt{z}} = \frac{2z\sqrt{z}}{3\beta} + \frac{2\Delta\sqrt{z}}{\beta^2} + \frac{\Delta^2}{\beta^2}\int \frac{dx}{t\sqrt{z}}$ (see **2.246**).

14. $\int \frac{z^3\,dx}{t\sqrt{z}} = \frac{2z^2\sqrt{z}}{5\beta} + \frac{2\Delta z\sqrt{z}}{3\beta^2} + \frac{2\Delta^2\sqrt{z}}{\beta^3} + \frac{\Delta^3}{\beta^3}\int \frac{dx}{t\sqrt{z}}$ (see **2.246**).

15. $\int \frac{z\,dx}{t^2\sqrt{z}} = -\frac{z\sqrt{z}}{\Delta t} + \frac{b\sqrt{z}}{\beta\Delta} + \frac{b}{2\beta}\int \frac{dx}{t\sqrt{z}}$ (see **2.246**).

16. $\int \frac{z^2\,dx}{t^2\sqrt{z}} = -\frac{z^2\sqrt{z}}{\Delta t} + \frac{bz\sqrt{z}}{\beta\Delta} + \frac{3b\sqrt{z}}{\beta^2} + \frac{3b\Delta}{2\beta^2}\int \frac{dx}{t\sqrt{z}}$ (see **2.246**).

17. $\int \frac{z^3\,dx}{t^2\sqrt{z}} = -\frac{z^3\sqrt{z}}{\Delta t} + \frac{bz^2\sqrt{z}}{\beta\Delta} + \frac{5bz\sqrt{z}}{3\beta^2} + \frac{5b\Delta\sqrt{z}}{\beta^3} +$

$$+ \frac{5\Delta^2 b}{2\beta^3}\int \frac{dx}{t\sqrt{z}}$$ (see **2.246**).

18. $\int \frac{z\,dx}{t^3\sqrt{z}} = -\frac{z\sqrt{z}}{2\Delta t^2} - \frac{bz\sqrt{z}}{4\Delta^2 t} + \frac{b^2\sqrt{z}}{4\beta\Delta^2} + \frac{b^2}{8\beta\Delta}\int \frac{dx}{t\sqrt{z}}$ (see **2.246**).

19. $\int \frac{z^2\,dx}{t^3\sqrt{z}} = -\frac{z^2\sqrt{z}}{2\Delta t^2} + \frac{bz^2\sqrt{z}}{4\Delta^2 t} + \frac{b^2 z\sqrt{z}}{4\beta\Delta^2} +$

$$+ \frac{3b^2\sqrt{z}}{4\beta^2\Delta} + \frac{3b^2}{8\beta^2}\int \frac{dx}{t\sqrt{z}}$$ (see **2.246**).

20. $\int \frac{z^3\,dx}{t^3\sqrt{z}} = -\frac{z^3\sqrt{z}}{2\Delta t^2} + \frac{3bz^3\sqrt{z}}{\Delta^2 t} + \frac{3b^2z^2\sqrt{z}}{4\beta\Delta^2} + \frac{5b^2 z\sqrt{z}}{4\beta^2\Delta} +$

$$+ \frac{15b^2\sqrt{z}}{4\beta^3} + \frac{15b^2\Delta}{8\beta^3}\int \frac{dx}{t\sqrt{z}}$$ (see **2.246**).

2.249

1. $\int \frac{dx}{z^m t^n\sqrt{z}} = \frac{2}{(2m-1)\Delta}\,\frac{\sqrt{z}}{t^{n-1}z^m} + \frac{(2n+2m-3)\beta}{(2m-1)\Delta}\int \frac{dx}{t^n z^{m-1}\sqrt{z}};$ LA 177 (4)

$$= -\frac{1}{(n-1)\Delta}\,\frac{\sqrt{z}}{z^m t^{n-1}} - \frac{(2n+2m-3)\,b}{2(n-1)\Delta}\int \frac{dx}{t^{n-1}z^m\sqrt{z}}.$$

2. $\int \frac{dx}{z^m t^n\sqrt{z}} = \frac{\sqrt{z}}{z^m}\left\{\frac{-1}{(n-1)\Delta}\,\frac{1}{t^{n-1}} +\right.$

$$\left.+\sum_{k=2}^{n-1}(-1)^k\,\frac{(2n+2m-3)(2n+2m-5)\ldots(2n+2m-2k+1)\,b^{k-1}}{2^{k-1}(n-1)(n-2)\ldots(n-k)\,\Delta^k}\cdot\frac{1}{t^{n-k}}\right\}+$$

$$+(-1)^{n-1}\,\frac{(2n+2m-3)(2n+2m-5)\ldots(-2m+3)(-2m+1)\,b^{n-1}}{2^{n-1}(n-1)!\,\Delta^{n-1}}\int \frac{dx}{tz^m\sqrt{z}}.$$

For $n=1$

$$\int \frac{dx}{z^m t\sqrt{z}} = \frac{2}{(2m-1)\Delta}\,\frac{1}{z^{m-1}\sqrt{z}} + \frac{\beta}{\Delta}\int \frac{dx}{tz^{m-1}\sqrt{z}}.$$

2.25 Forms containing $\sqrt{a+bx+cx^2}$

Integration techniques

2.251 It is possible to rationalize the integrand in integrals of the form $\int R(x, \sqrt{a+bx+cx^2})\,dx$ by using one or more of the following three substitutions, known as the "Euler substitutions".

1) $\sqrt{a+bx+cx^2} = xt \pm \sqrt{a}$ for $a > 0$;
2) $\sqrt{a+bx+cx^2} = t \pm x\sqrt{c}$ for $c > 0$;
3) $\sqrt{c(x-x_1)(x-x_2)} = t(x-x_1)$ when x_1 and x_2 are real roots of the equation $a+bx+cx^2=0$.

2.252 Besides the Euler substitutions, there is also the following method of calculating integrals of the form $\int R\left(x, \sqrt{a+bx+cx^2}\right)dx$. By removing the irrational expressions in the denominator and performing simple algebraic operations, we can reduce the integrand to the sum of some rational function of x and an expression of the form $\frac{P_1(x)}{P_2(x)\sqrt{a+bx+cx^2}}$, where $P_1(x)$ and $P_2(x)$ are both polynomials. By separating the integral portion of the rational function $\frac{P_1(x)}{P_2(x)}$ from the remainder and decomposing the latter into partial fractions, we can reduce the integral of these partial fractions to the sum of integrals each of which is in one of the following three forms:

I. $\int \frac{P(x)\,dx}{\sqrt{a+bx+cx^2}}$, where $P(x)$ is a polynomial of some degree r;

II. $\int \frac{dx}{(x+p)^k\sqrt{a+bx+cx^2}}$;

III. $\int \frac{(Mx+N)\,dx}{(\alpha+\beta x+x^2)^m\sqrt{c(a_1+b_1x+x^2)}}$, $\left(a_1=\frac{a}{c},\ b_1=\frac{b}{c}\right)$.

I. $\int \frac{P(x)\,dx}{\sqrt{a+bx+cx^2}} = Q(x)\sqrt{a+bx+cx^2} + \lambda \int \frac{dx}{\sqrt{a+bx+cx^2}}$, where $Q(x)$ is a polynomial of degree $(r-1)$. Its coefficients, and also the number λ, can be calculated by the method of undetermined coefficients from the identity

$$P(x) = Q'(x)(a+bx+cx^2) + \frac{1}{2}Q(x)(b+2cx) + \lambda. \qquad \text{LI II 77}$$

Integrals of the form $\int \frac{P(x)\,dx}{\sqrt{a+bx+cx^2}}$ (where $r \leqslant 3$) can also be calculated by use of formulas **2.26**.

II. Integrals of the form $\int \frac{P(x)\,dx}{(x+p)^k\sqrt{a+bx+cx^2}}$, where the degree n of the polynomial $P(x)$ is lower than k can, by means of the substitution $t=\frac{1}{x+p}$, be reduced to an integral of the form $\int \frac{P(t)\,dt}{\sqrt{\alpha+\beta t+\gamma t^2}}$. (See also **2.281**).

III. Integrals of the form $\int \frac{(Mx+N)\,dx}{(\alpha+\beta x+x^2)^m\sqrt{c(a_1+b_1x+x^2)}}$ can be calculated by the following procedure.

If $b_1 \neq \beta$, by using the substitution

$$x = \frac{a_1 - \alpha}{\beta - b_1} + \frac{t-1}{t+1} \frac{\sqrt{(a_1 - \alpha)^2 - (\alpha b_1 - a_1 \beta)(\beta - b_1)}}{\beta - b_1}$$

we can reduce this integral to an integral of the form $\int \frac{P(t)\,dt}{(t^2+p)^m \sqrt{c(t^2+q)}}$, where $P(t)$ is a polynomial of degree no higher than $2m-1$. The integral $\int \frac{P(t)\,dt}{(t^2+p)^m \sqrt{t^2+q}}$ can be reduced to the sum of integrals of the forms $\int \frac{t\,dt}{(t^2+p)^k \sqrt{t^2+q}}$ and $\int \frac{dt}{(t^2+p)^k \sqrt{t^2+q}}$.

If $b_1 = \beta$, we can reduce it to integrals of the form $\int \frac{P(t)\,dt}{(t^2+p)^m \sqrt{c(t^2+q)}}$ by means of the substitution $t = x + \frac{b_1}{2}$.

The integral $\int \frac{t\,dt}{(t^2+p)^k \sqrt{c(t^2+q)}}$ can be evaluated by means of the substitution $t^2 + q = u^2$.

The integral $\int \frac{dt}{(t^2+p)^k \sqrt{c(t^2+q)}}$ can be evaluated by means of the substitution $\frac{t}{\sqrt{t^2+q}} = v$ (see also **2.283**). FI II 78-82

2.26 Forms containing $\sqrt{a+bx+cx^2}$ and integral powers of x

Notation: $R = a + bx + cx^2$, $\Delta = 4ac - b^2$

Simplified formulas for the case $b = 0$. See **2.27**.

2.260

1. $$\int x^m \sqrt{R^{2n+1}}\,dx = \frac{x^{m-1}\sqrt{R^{2n+3}}}{(m+2n+2)c} - \frac{(2m+2n+1)b}{2(m+2n+2)c} \int x^{m-1}\sqrt{R^{2n+1}}\,dx - \frac{(m-1)a}{(m+2n+2)c} \int x^{m-2}\sqrt{R^{2n+1}}\,dx.$$ TI (192)a

2. $$\int \sqrt{R^{2n+1}}\,dx = \frac{2cx+b}{4(n+1)c}\sqrt{R^{2n+1}} + \frac{2n+1}{8(n+1)} \frac{\Delta}{c} \int \sqrt{R^{2n-1}}\,dx.$$ TI (188)

3. $$\int \sqrt{R^{2n+1}}\,dx = \frac{(2cx+b)\sqrt{R}}{4(n+1)c} \left\{ R^n + \sum_{k=0}^{n-1} \frac{(2n+1)(2n-1)\dots(2n-2k+1)}{8^{k+1}n(n-1)\dots(n-k)} \left(\frac{\Delta}{c}\right)^{k+1} R^{n-k-1} \right\} + \frac{(2n+1)!!}{8^{n+1}(n+1)!} \left(\frac{\Delta}{c}\right)^{n+1} \int \frac{dx}{\sqrt{R}}.$$ TI (190)

2.261 For $n = -1$

$$\int \frac{dx}{\sqrt{R}} = \frac{1}{\sqrt{c}} \ln (2\sqrt{cR} + 2cx + b) \quad [c > 0];$$ TI 127)

$$= \frac{1}{\sqrt{c}} \operatorname{Arsh} \frac{2cx+b}{\sqrt{\Delta}} \quad [c > 0,\ \Delta > 0];$$ DW

$$= \frac{-1}{\sqrt{-c}} \arcsin \frac{2cx+b}{\sqrt{-\Delta}} \quad [c < 0,\ \Delta < 0];$$ TI (128)

$$= \frac{1}{\sqrt{c}} \ln (2cx+b) \quad [c > 0,\ \Delta = 0].$$ DW

2.262

1. $\int \sqrt{R}\,dx = \frac{(2cx+b)\sqrt{R}}{4c} + \frac{\Delta}{8c}\int \frac{dx}{\sqrt{R}}$ (see 2.261).

2. $\int x\sqrt{R}\,dx = \frac{\sqrt{R^3}}{3c} - \frac{(2cx+b)\,b}{8c^2}\sqrt{R} - \frac{b\Delta}{16c^2}\int \frac{dx}{\sqrt{R}}$ (see 2.261).

3. $\int x^2\sqrt{R}\,dx = \left(\frac{x}{4c} - \frac{5b}{24c^2}\right)\sqrt{R^3} + \left(\frac{5b^2}{16c^2} - \frac{a}{4c}\right)\frac{(2cx+b)\sqrt{R}}{4c} + \left(\frac{5b^2}{16c^2} - \frac{a}{4c}\right)\frac{\Delta}{8c}\int \frac{dx}{\sqrt{R}}$ (see 2.261).

4. $\int x^3\sqrt{R}\,dx = \left(\frac{x^2}{5c} - \frac{7bx}{40c^2} + \frac{7b^2}{48c^3} - \frac{2a}{15c^2}\right)\sqrt{R^3} - \left(\frac{7b^3}{32c^3} - \frac{3ab}{8c^2}\right)\frac{(2cx+b)\sqrt{R}}{4c} - \left(\frac{7b^3}{32c^3} - \frac{3ab}{8c^2}\right)\frac{\Delta}{8c}\int \frac{dx}{\sqrt{R}}$ (see 2.261).

5. $\int \sqrt{R^3}\,dx = \left(\frac{R}{8c} + \frac{3\Delta}{64c^2}\right)(2cx+b)\sqrt{R} + \frac{3\Delta^2}{128c^2}\int \frac{dx}{\sqrt{R}}$ (see 2.261).

6. $\int x\sqrt{R^3}\,dx = \frac{\sqrt{R^5}}{5c} - (2cx+b)\left(\frac{b}{16c^2}\sqrt{R^3} + \frac{3\Delta b}{128c^3}\sqrt{R}\right) - \frac{3\Delta^2 b}{256c^3}\int \frac{dx}{\sqrt{R}}$ (see 2.261).

7. $\int x^2\sqrt{R^3}\,dx = \left(\frac{x}{6c} + \frac{7b}{60c^2}\right)\sqrt{R^5} + \left(\frac{7b^2}{24c^2} - \frac{a}{6c}\right)\left(2x + \frac{b}{c}\right)\left(\frac{\sqrt{R^3}}{8} + \frac{3\Delta}{64c}\sqrt{R}\right) + \left(\frac{7b^2}{4c} - a\right)\frac{\Delta^2}{256c^3}\int \frac{dx}{\sqrt{R}}$ (see 2.261).

8. $\int x^3\sqrt{R^3}\,dx = \left(\frac{x^2}{7c} - \frac{3bx}{28c^2} + \frac{3b^2}{40c^3} - \frac{2a}{35c^2}\right)\sqrt{R^5} - \left(\frac{3b^3}{16c^3} - \frac{ab}{4c^2}\right)\left(2x + \frac{b}{c}\right)\left(\frac{\sqrt{R^3}}{8} + \frac{3\Delta}{64c}\sqrt{R}\right) - \left(\frac{3b^2}{4c} - a\right)\frac{3\Delta^2 b}{512c^4}\int \frac{dx}{\sqrt{R}}$ (see 2.261).

2.263

1. $\int \frac{x^m\,dx}{\sqrt{R^{2n+1}}} = \frac{x^{m-1}}{(m-2n)\,c\sqrt{R^{2n-1}}} - \frac{(2m-2n-1)\,b}{2(m-2n)\,c}\int \frac{x^{m-1}\,dx}{\sqrt{R^{2n+1}}} - \frac{(m-1)\,a}{(m-2n)\,c}\int \frac{x^{m-2}\,dx}{\sqrt{R^{2n+1}}}.$ TI (193)a

For $m = 2n$

2. $\int \frac{x^{2n}\,dx}{\sqrt{R^{2n+1}}} = -\frac{x^{2n-1}}{(2n-1)\,c\sqrt{R^{2n-1}}} - \frac{b}{2c}\int \frac{x^{2n-1}}{\sqrt{R^{2n+1}}}dx + \frac{1}{c}\int \frac{x^{2n-2}}{\sqrt{R^{2n-1}}}dx.$ TI (194)a

3. $$\int \frac{dx}{\sqrt{R^{2n+1}}} = \frac{2(2cx+b)}{(2n-1)\Delta\sqrt{R^{2n-1}}} + \frac{8(n-1)c}{(2n-1)\Delta}\int \frac{dx}{\sqrt{R^{2n-1}}}.$$ TI (189)

4. $$\int \frac{dx}{\sqrt{R^{2n+1}}} = \frac{2(2cx+b)}{(2n-1)\Delta\sqrt{R^{2n-1}}} \times$$
$$\times\left\{1 + \sum_{k=1}^{n-1} \frac{8^k(n-1)(n-2)\ldots(n-k)}{(2n-3)(2n-5)\ldots(2n-2k-1)} \frac{c^k}{\Delta^k} R^k\right\} \quad [n \geqslant 1].$$ TI (191)

2.264

1. $$\int \frac{dx}{\sqrt{R}} \quad \text{(see 2.261).}$$

2. $$\int \frac{x\,dx}{\sqrt{R}} = \frac{\sqrt{R}}{c} - \frac{b}{2c}\int \frac{dx}{\sqrt{R}} \quad \text{(see 2.261).}$$

3. $$\int \frac{x^2\,dx}{\sqrt{R}} = \left(\frac{x}{2c} - \frac{3b}{4c^2}\right)\sqrt{R} + \left(\frac{3b^2}{8c^2} - \frac{a}{2c}\right)\int \frac{dx}{\sqrt{R}} \quad \text{(see 2.261).}$$

4. $$\int \frac{x^3\,dx}{\sqrt{R}} = \left(\frac{x^2}{3c} - \frac{5bx}{12c^2} + \frac{5b^2}{8c^3} - \frac{2a}{3c^2}\right)\sqrt{R} -$$
$$-\left(\frac{5b^3}{16c^3} - \frac{3ab}{4c^2}\right)\int \frac{dx}{\sqrt{R}} \quad \text{(see 2.261).}$$

5. $$\int \frac{dx}{\sqrt{R^3}} = \frac{2(2cx+b)}{\Delta\sqrt{R}}.$$

6. $$\int \frac{x\,dx}{\sqrt{R^3}} = -\frac{2(2a+bx)}{\Delta\sqrt{R}}.$$

7. $$\int \frac{x^2\,dx}{\sqrt{R^3}} = -\frac{(\Delta-b^2)x-2ab}{c\Delta\sqrt{R}} + \frac{1}{c}\int \frac{dx}{\sqrt{R}} \quad \text{(see 2.261).}$$

8. $$\int \frac{x^3\,dx}{\sqrt{R^3}} = \frac{c\Delta x^2 + b(10ac-3b^2)x + a(8ac-3b^2)}{c^2\Delta\sqrt{R}} - \frac{3b}{2c^2}\int \frac{dx}{\sqrt{R}} \quad \text{(see 2.261).}$$

2.265 $$\int \frac{\sqrt{R^{2n+1}}}{x^m}\,dx =$$
$$= -\frac{\sqrt{R^{2n+3}}}{(m-1)ax^{m-1}} + \frac{(2n-2m+5)b}{2(m-1)a}\int \frac{\sqrt{R^{2n+1}}}{x^{m-1}}\,dx +$$
$$+ \frac{(2n-m+4)c}{(m-1)a}\int \frac{\sqrt{R^{2n+1}}}{x^{m-2}}\,dx.$$ TI (195)

For $m=1$

$$\int \frac{\sqrt{R^{2n+1}}}{x}\,dx = \frac{\sqrt{R^{2n+1}}}{2n+1} + \frac{b}{2}\int \sqrt{R^{2n-1}}\,dx + a\int \frac{\sqrt{R^{2n-1}}}{x}\,dx.$$ TI (198)

For $a=0$

$$\int \frac{\sqrt{(bx+cx^2)^{2n+1}}}{x^m}\,dx = \frac{2\sqrt{(bx+cx^2)^{2n+3}}}{(2n-2m+3)bx^m} +$$
$$+ \frac{2(m-2n-3)c}{(2n-2m+3)b}\int \frac{\sqrt{(bx+cx^2)^{2n+1}}}{x^{m-1}}.$$ LA 169 (3)

For $m=0$ see **2.260** 2. and **2.260** 3.

For $n=-1$ and $m=1$:

2.266 $$\int \frac{dx}{x\sqrt{R}} = -\frac{1}{\sqrt{a}} \ln \frac{2a+bx+2\sqrt{aR}}{x} \qquad [a>0];$$ TI (137)

$$= \frac{1}{\sqrt{-a}} \arcsin \frac{2a+bx}{x\sqrt{b^2-4ac}} \qquad [a<0,\ \Delta<0];$$ TI (138)

$$= \frac{1}{\sqrt{-a}} \operatorname{arctg} \frac{2a+bx}{2\sqrt{-a}\sqrt{R}} \qquad [a<0];$$ LA 178 (6)a

$$= -\frac{1}{\sqrt{a}} \operatorname{Arsh} \frac{2a+bx}{x\sqrt{\Delta}} \qquad [a>0,\ \Delta>0];$$ DW

$$= -\frac{1}{\sqrt{a}} \operatorname{Arth} \frac{2a+bx}{2\sqrt{a}\sqrt{R}} \qquad [a>0];$$

$$= \frac{1}{\sqrt{a}} \ln \frac{x}{2a+bx} \qquad [a>0,\ \Delta=0];$$

$$= -\frac{2\sqrt{bx+cx^2}}{bx} \qquad [a=0,\ b\neq 0].$$ La 170 (16)

2.267

1. $$\int \frac{\sqrt{R}\,dx}{x} = \sqrt{R} + a\int \frac{dx}{x\sqrt{R}} + \frac{b}{2}\int \frac{dx}{\sqrt{R}}$$ (see 2.261 and 2.266).

2. $$\int \frac{\sqrt{R}\,dx}{x^2} = -\frac{\sqrt{R}}{x} + \frac{b}{2}\int \frac{dx}{x\sqrt{R}} + c\int \frac{dx}{\sqrt{R}}$$ (see 2.261 and 2.266).

For $a=0$

$$\int \frac{\sqrt{bx+cx^2}}{x^2}\,dx = -\frac{2\sqrt{bx+cx^2}}{x} + c\int \frac{dx}{\sqrt{bx+cx^2}}$$ (see 2.261).

3. $$\int \frac{\sqrt{R}\,dx}{x^3} = -\left(\frac{1}{2x^2} + \frac{b}{4ax}\right)\sqrt{R} - \left(\frac{b^2}{8a} - \frac{c}{2}\right)\int \frac{dx}{x\sqrt{R}}$$ (see 2.266).

For $a=0$

$$\int \frac{\sqrt{bx+cx^2}}{x^3}\,dx = -\frac{2\sqrt{(bx+cx^2)^3}}{3bx^3}$$

4. $$\int \frac{\sqrt{R^3}}{x}\,dx = \frac{\sqrt{R^3}}{3} + \frac{2bcx+b^2+8ac}{8c}\sqrt{R} + a^2\int \frac{dx}{x\sqrt{R}} + \frac{b(12ac-b^2)}{16c}\int \frac{dx}{\sqrt{R}}$$ (see 2.261 and 2.266).

5. $$\int \frac{\sqrt{R^3}}{x^2}\,dx = -\frac{\sqrt{R^5}}{ax} + \frac{cx+b}{a}\sqrt{R^3} + \frac{3}{4}(2cx+3b)\sqrt{R} + \frac{3}{2}ab\int \frac{dx}{x\sqrt{R}} + \frac{3(4ac+b^2)}{8}\int \frac{dx}{\sqrt{R}}$$ (see 2.261 and 2.266).

For $a=0$

$$\int \frac{\sqrt{(bx+cx^2)^3}}{x^2} = \frac{\sqrt{(bx+cx^2)^3}}{2x} + \frac{3b}{4}\sqrt{bx+cx^2} + \frac{3b^2}{8}\int \frac{dx}{\sqrt{bx+cx^2}}$$ (see 2.261).

6. $$\int \frac{\sqrt{R^3}}{x^3}\,dx = -\left(\frac{1}{2ax^2} + \frac{b}{4a^2x}\right)\sqrt{R^5} + \frac{bcx+2ac+b^2}{4a^2}\sqrt{R^3} + \frac{3(bcx+2ac+b^2)}{4a}\sqrt{R} + \frac{3}{8}(4ac+b^2)\int \frac{dx}{x\sqrt{R}} + \frac{3}{2}bc\int \frac{dx}{\sqrt{R}}$$ (see 2.261 and 2.266).

For $a=0$

$$\int \frac{\sqrt{(bx+cx^2)^3}}{x^3}dx=\left(c-\frac{2b}{x}\right)\sqrt{bx+cx^2}+\frac{3bc}{2}\int\frac{dx}{\sqrt{bx+cx^2}}$$ (see **2.261**).

2.268

$$\int\frac{dx}{x^m\sqrt{R^{2n+1}}}=-\frac{1}{(m-1)\,ax^{m-1}\sqrt{R^{2n-1}}}-$$
$$-\frac{(2n+2m-3)\,b}{2\,(m-1)\,a}\int\frac{dx}{x^{m-1}\sqrt{R^{2n+1}}}-\frac{(2n+2m-2)\,c}{(m-1)\,a}\int\frac{dx}{x^{m-2}\sqrt{R^{2n+1}}}.$$ TI (196)

For $m=1$

$$\int\frac{dx}{x\sqrt{R^{2n+1}}}=\frac{1}{(2n-1)\,a\sqrt{R^{2n-1}}}-\frac{b}{2a}\int\frac{dx}{\sqrt{R^{2n+1}}}+\frac{1}{a}\int\frac{dx}{x\sqrt{R^{2n-1}}}.$$ TI (199)

For $a=0$

$$\int\frac{dx}{x^m\sqrt{(bx+cx^2)^{2n+1}}}=-\frac{2}{(2n+2m-1)\,bx^m\sqrt{(bx+cx^2)^{2n-1}}}-$$
$$-\frac{(4n+2m-2)\,c}{(2n+2m-1)\,b}\int\frac{dx}{x^{m-1}\sqrt{(bx+cx^2)^{2n+1}}}$$ (cf. **2.265**).

2.269

1. $\int\frac{dx}{x\sqrt{R}}$ (see **2.266**).

2. $\int\frac{dx}{x^2\sqrt{R}}=-\frac{\sqrt{R}}{ax}-\frac{b}{2a}\int\frac{dx}{x\sqrt{R}}$ (see **2.266**).

For $a=0$

$$\int\frac{dx}{x^2\sqrt{bx+cx^2}}=\frac{2}{3}\left(-\frac{1}{bx^2}+\frac{2c}{b^2x}\right)\sqrt{bx+cx^2}.$$

3. $\int\frac{dx}{x^3\sqrt{R}}=\left(-\frac{1}{2ax^2}+\frac{3b}{4a^2x}\right)\sqrt{R}+\left(\frac{3b^2}{8a^2}-\frac{c}{2a}\right)\int\frac{dx}{x\sqrt{R}}$ (see **2.266**).

For $a=0$

$$\int\frac{dx}{x^3\sqrt{bx+cx^2}}=\frac{2}{5}\left(-\frac{1}{bx^3}+\frac{4c}{3b^2x^2}-\frac{8c^2}{3b^3x}\right)\sqrt{bx+cx^2}.$$

4. $\int\frac{dx}{x\sqrt{R^3}}=-\frac{2\,(bcx-2ac+b^2)}{a\Delta\sqrt{R}}+\frac{1}{a}\int\frac{dx}{x\sqrt{R}}$ (see **2.266**).

For $a=0$

$$\int\frac{dx}{x\sqrt{(bx+cx^2)^3}}=\frac{2}{3}\left(-\frac{1}{bx}+\frac{4c}{b^2}-\frac{8c^2x}{b^3}\right)\frac{1}{\sqrt{bx+cx^2}}.$$

5. $\int\frac{dx}{x^2\sqrt{R^3}}=\left(-\frac{1}{ax}+\frac{2bc}{a\Delta}+\frac{c\,(3b^2-3ac)\,x}{a^2\Delta}\right)\frac{1}{\sqrt{R}}-\frac{3b}{2a^2}\int\frac{dx}{x\sqrt{R}}$

(see **2.266**).

For $a=0$

$$\int\frac{dx}{x^2\sqrt{(bx+cx^2)^3}}=\frac{2}{5}\left(-\frac{1}{bx^2}+\frac{2c}{b^2x}-\frac{8c^2}{b^3}-\frac{16c^3x}{b^4}\right)\frac{1}{\sqrt{bx+cx^2}}.$$

6. $\int\frac{dx}{x^3\sqrt{R^3}}=\left(-\frac{1}{ax^2}+\frac{5b}{2a^2x}-\frac{15b^4-62acb^2+24a^2c^2}{2a^3\Delta}-\right.$
$\left.-\frac{bc\,(15b^2-52ac)\,x}{2a^3\Delta}\right)\frac{1}{2\sqrt{R}}+\frac{15b^2-12ac}{8a^3}\int\frac{dx}{x\sqrt{R}}$ (see **2.266**).

For $a=0$

$$\int\frac{dx}{x^3\sqrt{(bx+cx^2)^3}}=\frac{2}{7}\left(-\frac{1}{bx^3}+\frac{8c}{5b^2x^2}-\frac{16c^2}{5b^3x}+\frac{64c^3}{5b^4}+\frac{128c^4x}{5b^5}\right)\frac{1}{\sqrt{bx+cx^2}}.$$

2.27 Forms containing $\sqrt{a+cx^2}$ and integral powers of x

Notations: $u=\sqrt{a+cx^2}$.

$$I_1=\frac{1}{\sqrt{c}}\ln\left(x\sqrt{c}+u\right) \qquad [c>0];$$

$$=\frac{1}{\sqrt{-c}}\arcsin x\sqrt{-\frac{c}{a}} \qquad [c<0 \text{ and } a>0].$$

$$I_2=\frac{1}{2\sqrt{a}}\ln\frac{u-\sqrt{a}}{u+\sqrt{a}} \qquad [a>0 \text{ and } c>0];$$

$$=\frac{1}{2\sqrt{a}}\ln\frac{\sqrt{a}-u}{\sqrt{a}+u} \qquad [a>0 \text{ and } c<0];$$

$$=\frac{1}{\sqrt{-a}}\operatorname{arcsec} x\sqrt{-\frac{c}{a}}=\frac{1}{\sqrt{-a}}\arccos\frac{1}{x}\sqrt{-\frac{a}{c}} \quad [a<0 \text{ and } c>0].$$

2.271

1. $\int u^5\,dx=\frac{1}{6}xu^5+\frac{5}{24}axu^3+\frac{5}{16}a^2xu+\frac{5}{16}a^3I_1.$ DW

2. $\int u^3\,dx=\frac{1}{4}xu^3+\frac{3}{8}axu+\frac{3}{8}a^2I_1.$ DW

3. $\int u\,dx=\frac{1}{2}xu+\frac{1}{2}aI_1.$ DW

4. $\int\frac{dx}{u}=I_1.$ DW

5. $\int\frac{dx}{u^3}=\frac{1}{a}\frac{x}{u}$ DW

6. $\int\frac{dx}{u^{2n+1}}=\frac{1}{a^n}\sum_{k=0}^{n-1}\frac{(-1)^k}{2k+1}\binom{n-1}{k}\frac{c^kx^{2k+1}}{u^{2k+1}}.$

7. $\int\frac{x\,dx}{u^{2n+1}}=-\frac{1}{(2n-1)cu^{2n-1}}.$ DW

2.272

1. $\int x^2u^3\,dx=\frac{1}{6}\frac{xu^5}{c}-\frac{1}{24}\frac{axu^3}{c}-\frac{1}{16}\frac{a^2xu}{c}-\frac{1}{16}\frac{a^3}{c}I_1.$ DW

2. $\int x^2u\,dx=\frac{1}{4}\frac{xu^3}{c}-\frac{1}{8}\frac{axu}{c}-\frac{1}{8}\frac{a^2}{c}I_1.$ DW

3. $\int\frac{x^2}{u}\,dx=\frac{1}{2}\frac{xu}{c}-\frac{1}{2}\frac{a}{c}I_1.$ DW

4. $\int\frac{x^2}{u^3}\,dx=-\frac{x}{cu}+\frac{1}{c}I_1.$ DW

5. $\int\frac{x^2}{u^5}\,dx=\frac{1}{3}\frac{x^3}{au^3}.$ DW

6. $\int\frac{x^2\,dx}{u^{2n+1}}=\frac{1}{a^{n-1}}\sum_{k=0}^{n-2}\frac{(-1)^k}{2k+3}\binom{n-2}{k}\frac{c^kx^{2k+3}}{u^{2k+3}}.$

7. $\int\frac{x^3\,dx}{u^{2n+1}}=-\frac{1}{(2n-3)c^2u^{2n-3}}+\frac{a}{(2n-1)c^2u^{2n-1}}.$ DW

2.273

1. $\int x^4 u^3\,dx = \frac{1}{8}\frac{x^3u^5}{c} - \frac{axu^5}{16c^2} + \frac{a^2xu^3}{64c^2} + \frac{3a^3xu}{128c^2} + \frac{3a^4}{128c^2} I_1.$ DW

2. $\int x^4 u\,dx = \frac{1}{6}\frac{x^3u^3}{c} - \frac{axu^3}{8c^2} + \frac{a^2xu}{16c^2} + \frac{a^3}{16c^2} I_1.$ DW

3. $\int \frac{x^4}{u}\,dx = \frac{1}{4}\frac{x^3u}{c} - \frac{3}{8}\frac{axu}{c^2} + \frac{3}{8}\frac{a^2}{c^2} I_1.$ DW

4. $\int \frac{x^4}{u^3}\,dx = \frac{1}{2}\frac{xu}{c^2} + \frac{ax}{c^2u} - \frac{3}{2}\frac{a}{c^2} I_1.$ DW

5. $\int \frac{x^4}{u^5}\,dx = -\frac{x}{c^2u} - \frac{1}{3}\frac{x^3}{cu^3} + \frac{1}{c^2} I_1.$ DW

6. $\int \frac{x^4}{u^7}\,dx = \frac{1}{5}\frac{x^5}{au^5}.$ DW

7. $\int \frac{x^4\,dx}{u^{2n+1}} = \frac{1}{a^{n-2}} \sum_{k=0}^{n-3} \frac{(-1)^k}{2k+5}\binom{n-3}{k}\frac{c^k x^{2k+5}}{u^{2k+5}}.$

8. $\int \frac{x^5\,dx}{u^{2n+1}} = -\frac{1}{(2n-5)\,c^3u^{2n-5}} + \frac{2a}{(2n-3)\,c^3u^{2n-3}} - \frac{a^2}{(2n-1)\,c^3u^{2n-1}}.$ DW

2.274

1. $\int x^6u^3\,dx = \frac{1}{10}\frac{x^5u^5}{c} - \frac{ax^3u^5}{16c^2} + \frac{a^2xu^5}{32c^3} - \frac{a^3xu^3}{128c^3} - \frac{3a^4xu}{256c^3} - \frac{3}{256}\frac{a^5}{c^3} I_1.$

2. $\int x^6u\,dx = \frac{1}{8}\frac{x^5u^3}{c} - \frac{5}{48}\frac{ax^3u^3}{c^2} + \frac{5a^2xu^3}{64c^3} - \frac{5a^3xu}{128c^3} - \frac{5}{128}\frac{a^4}{c^3} I_1.$

3. $\int \frac{x^6}{u}\,dx = \frac{1}{6}\frac{x^5u}{c} - \frac{5}{24}\frac{ax^3u}{c^2} + \frac{5}{16}\frac{a^2xu}{c^3} - \frac{5}{16}\frac{a^3}{c^3} I_1.$ DW

4. $\int \frac{x^6}{u^3}\,dx = \frac{1}{4}\frac{x^5}{cu} - \frac{5}{8}\frac{ax^3}{c^2u} - \frac{15}{8}\frac{a^2x}{c^3u} + \frac{15}{8}\frac{a^2}{c^3} I_1.$ DW

5. $\int \frac{x^6}{u^5}\,dx = \frac{1}{2}\frac{x^5}{cu^3} + \frac{10}{3}\frac{ax^3}{c^2u^3} + \frac{5}{2}\frac{a^2x}{c^3u^3} - \frac{5}{2}\frac{a}{c^3} I_1.$ DW

6. $\int \frac{x^6}{u^7}\,dx = -\frac{23}{15}\frac{x^5}{cu^5} - \frac{7}{3}\frac{ax^3}{c^2u^5} - \frac{a^2x}{c^3u^5} + \frac{1}{c^3} I_1.$ DW

7. $\int \frac{x^6}{u^9}\,dx = \frac{1}{7}\frac{x^7}{au^7}.$ DW

8. $\int \frac{x^6\,dx}{u^{2n+1}} = \frac{1}{a^{n-3}} \sum_{k=0}^{n-4} \frac{(-1)^k}{2k+7}\binom{n-4}{k}\frac{c^k x^{2k+7}}{u^{2k+7}}.$

9. $\int \frac{x^7\,dx}{u^{2n+1}} = -\frac{1}{(2n-7)\,c^4u^{2n-7}} + \frac{3a}{(2n-5)\,c^4u^{2n-5}} - \frac{3a^2}{(2n-3)\,c^4u^{2n-3}} + \frac{a^3}{(2n-1)\,c^4u^{2n-1}}.$ DW

2.275

1. $\int \frac{u^5}{x}\,dx = \frac{u^5}{5} + \frac{1}{3}au^3 + a^2u + a^3 I_2.$ DW

2. $\int \frac{u^3}{x}\,dx = \frac{u^3}{3} + au + a^2 I_2.$ DW

3. $\int \frac{u}{x}\,dx = u + aI_2.$ DW

4. $\int \frac{dx}{xu} = I_2.$ DW

5. $\int \frac{dx}{xu^{2n+1}} = \frac{1}{a^n} I_2 + \sum_{k=0}^{n-1} \frac{1}{(2k+1)\, a^{n-k} u^{2k+1}}$.

6. $\int \frac{u^5}{x^2} dx = -\frac{u^5}{x} + \frac{5}{4} cxu^3 + \frac{15}{8} acxu + \frac{15}{8} a^2 I_1$. DW

7. $\int \frac{u^3}{x^2} dx = -\frac{u^3}{x} + \frac{3}{2} cxu + \frac{3}{2} a I_1$. DW

8. $\int \frac{u}{x^2} dx = -\frac{u}{x} + I_1$. DW

9. $\int \frac{dx}{x^2 u^{2n+1}} = -\frac{1}{a^{n+1}} \left\{ \frac{u}{x} + \sum_{k=1}^{n} \frac{(-1)^{k+1}}{2k-1} \binom{n}{k} c^k \left(\frac{x}{u}\right)^{2k-1} \right\}$.

2.276

1. $\int \frac{u^5}{x^3} dx = -\frac{u^5}{2x^2} + \frac{5}{6} cu^3 + \frac{5}{2} acu + \frac{5}{2} a^2 c I_2$. DW

2. $\int \frac{u^3}{x^3} dx = -\frac{u^3}{2x^2} + \frac{3}{2} cu + \frac{3}{2} ac I_2$. DW

3. $\int \frac{u}{x^3} dx = -\frac{u}{2x^2} + \frac{c}{2} I_2$. DW

4. $\int \frac{dx}{x^3 u} = -\frac{u}{2ax^2} - \frac{c}{2a} I_2$. DW

5. $\int \frac{dx}{x^3 u^3} = -\frac{1}{2ax^2 u} - \frac{3c}{2a^2 u} - \frac{3c}{2a^2} I_2$. DW

6. $\int \frac{dx}{x^3 u^5} = -\frac{1}{2ax^2 u^3} - \frac{5}{6} \frac{c}{a^2 u^3} - \frac{5}{2} \frac{c}{a^3 u} - \frac{5}{2} \frac{c}{a^3} I_2$. DW

7. $\int \frac{u^5}{x^4} dx = -\frac{au^3}{3x^3} - \frac{2acu}{x} + \frac{c^2 xu}{2} + \frac{5}{2} ac I_1$. DW

8. $\int \frac{u^3}{x^4} dx = -\frac{u^3}{3x^3} - \frac{cu}{x} + c I_1$. DW

9. $\int \frac{u}{x^4} dx = -\frac{u^3}{3ax^3}$. DW

10. $\int \frac{dx}{x^4 u^{2n+1}} = \frac{1}{a^{n+2}} \left\{ -\frac{u^3}{3x^3} + (n+1) \frac{cu}{x} + \sum_{k=2}^{n+1} \frac{(-1)^k}{2k-3} \binom{n+1}{k} c^k \left(\frac{x}{u}\right)^{2k-3} \right\}$.

2.277

1. $\int \frac{u^3}{x^5} dx = -\frac{u^3}{4x^4} - \frac{3}{8} \frac{cu^3}{ax^2} + \frac{3}{8} \frac{c^2 u}{a} + \frac{3}{8} c^2 I_2$. DW

2. $\int \frac{u}{x^5} dx = -\frac{u}{4x^4} - \frac{1}{8} \frac{cu}{ax^2} - \frac{1}{8} \frac{c^2}{a} I_2$. DW

3. $\int \frac{dx}{x^5 u} = -\frac{u}{4ax^4} + \frac{3}{8} \frac{cu}{a^2 x^2} + \frac{3}{8} \frac{c^2}{a^2} I_2$. DW

4. $\int \frac{dx}{x^5 u^3} = -\frac{1}{4ax^4 u} + \frac{5}{8} \frac{c}{a^2 x^2 u} + \frac{15}{8} \frac{c^2}{a^3 u} + \frac{15}{8} \frac{c^2}{a^3} I_2$. DW

2.278

1. $\int \frac{u^3}{x^6} dx = -\frac{u^5}{5ax^5}$. DW

2. $\int \frac{u}{x^6} dx = -\frac{u^3}{5ax^5} + \frac{2}{15} \frac{cu^3}{a^2 x^3}$. DW

3. $$\int \frac{dx}{x^6 u} = \frac{1}{a^3}\left(-\frac{u^5}{5x^5} + \frac{2}{3}\frac{cu^3}{x^3} - \frac{c^2 u}{x}\right).$$ DW

4. $$\int \frac{dx}{x^6 u^{2n+1}} = \frac{1}{a^{n+3}}\left\{-\frac{u^5}{5x^5} + \frac{1}{3}\binom{n+2}{1}\frac{cu^3}{x^3} - \binom{n+2}{2}\frac{c^2 u}{x} + \right.$$
$$\left. + \sum_{k=3}^{n+2} \frac{(-1)^k}{2k-5}\binom{n+2}{k} c^k \left(\frac{x}{u}\right)^{2k-5}\right\}.$$

2.28 Forms containing $\sqrt{a+bx+cx^2}$ and first- and second-degree polynomials

Notation: $R = a + bx + cx^2$

See also **2.252.**

2.281 $$\int \frac{dx}{(x+p)^n \sqrt{R}} = -\int \frac{t^{n-1}dt}{\sqrt{c + (b-2pc)\,t + (a-bp+cp^2)\,t^2}} \qquad \left[t = \frac{1}{x+p}\right].$$

2.282

1. $$\int \frac{\sqrt{R}\,dx}{x+p} = c\int \frac{x\,dx}{\sqrt{R}} + (b-cp)\int \frac{dx}{\sqrt{R}} + (a-bp+cp^2)\int \frac{dx}{(x+p)\sqrt{R}}.$$

2. $$\int \frac{dx}{(x+p)(x+q)\sqrt{R}} = \frac{1}{q-p}\int \frac{dx}{(x+p)\sqrt{R}} + \frac{1}{p-q}\int \frac{dx}{(x+q)\sqrt{R}}.$$

3. $$\int \frac{\sqrt{R}\,dx}{(x+p)(x+q)} = \frac{1}{q-p}\int \frac{\sqrt{R}dx}{x+p} + \frac{1}{p-q}\int \frac{\sqrt{R}dx}{x+q}.$$

4. $$\int \frac{(x+p)\sqrt{R}\,dx}{x+q} = \int \sqrt{R}\,dx + (p-q)\int \frac{\sqrt{R}\,dx}{x+q}.$$

5. $$\int \frac{(rx+s)\,dx}{(x+p)(x+q)\sqrt{R}} = \frac{s-pr}{q-p}\int \frac{dx}{(x+p)\sqrt{R}} + \frac{s-qr}{p-q}\int \frac{dx}{(x+q)\sqrt{R}}.$$

2.283 $$\int \frac{(Ax+B)\,dx}{(p+R)^n\sqrt{R}} = \frac{A}{c}\int \frac{du}{(p+u^2)^n} + \frac{2Bc-Ab}{2c}\int \frac{(1-cv^2)^{n-1}\,dv}{\left[p+a-\frac{b^2}{4c}-cpv^2\right]^n},$$

where $u = \sqrt{R}$ and $v = \frac{b+2cx}{2c\sqrt{R}}$.

2.284 $$\int \frac{Ax+B}{(p+R)\sqrt{R}}dx = \frac{A}{c}I_1 + \frac{2Bc-Ab}{\sqrt{c^2p\,[b^2-4(a+p)c]}}I_2,$$

where

$$I_1 = \frac{1}{\sqrt{p}}\operatorname{arctg}\sqrt{\frac{R}{p}} \qquad [p>0];$$

$$= \frac{1}{2\sqrt{-p}}\ln\frac{\sqrt{-p}-\sqrt{R}}{\sqrt{-p}+\sqrt{R}} \qquad [p<0].$$

$$I_2 = \operatorname{arctg} \sqrt{\frac{p}{b^2-4(a+p)c}}\,\frac{b+2cx}{\sqrt{R}} \qquad [p\{b^2-4(a+p)c\}>0,\ p<0];$$

$$= -\operatorname{arctg} \sqrt{\frac{p}{b^2-4(a+p)c}}\,\frac{b+2cx}{\sqrt{R}} \qquad [p\{b^2-4(a+p)c\}>0,\ p>0];$$

$$= \frac{1}{2i}\ln\frac{\sqrt{4(a+p)c-b^2}\,\sqrt{R}+\sqrt{p}(b+2cx)}{\sqrt{4(a+p)c-b^2}\,\sqrt{R}-\sqrt{p}(b+2cx)} \qquad [p\{b^2-4(a+p)c\}<0,\ p>0];$$

$$= \frac{1}{2i}\ln\frac{\sqrt{b^2-4(a+p)c}\,\sqrt{R}-\sqrt{-p}\,(b+2cx)}{\sqrt{b^2-4(a+p)c}\,\sqrt{R}+\sqrt{-p}\,(b+2cx)} \qquad [p\{b^2-4(a+p)c\}<0,\ p<0].$$

2.29 Integrals that can be reduced to elliptic or pseudo-elliptic integrals

2.290 Integrals of the form $\int R\left(x, \sqrt{P(x)}\right)dx$, where $P(x)$ is a third- or fourth-degree polynomial can, by means of algebraic transformations, be reduced to a sum of integrals expressed in terms of elementary functions and elliptic integrals (see 8.11). Since the substitutions that transform the given integral into an elliptic integral in the normal Legendre form are different for different intervals of integration, the corresponding formulas are given in the chapter on definite integrals (see 3.13, 3.17).

2.291 Certain integrals of the form $\int R(x, \sqrt{P(x)})\,dx$, where $k \geqslant 2$ and $P_n(x)$ is a polynomial of not more than fourth degree, can be reduced to integrals of the form $\int R\left(x, \sqrt[k]{P_n(x)}\right)dx$. Below are examples of this procedure.

1. $$\int \frac{dx}{\sqrt{1-x^6}} = -\int \frac{dz}{\sqrt{3+3z^2+z^4}} \qquad \left[x^2 = \frac{1}{1+z^2}\right].$$

2. $$\int \frac{dx}{\sqrt{a+bx^2+cx^4+dx^6}} = \frac{1}{2}\int \frac{dz}{\sqrt{az+bz^2+cz^3+dz^4}} \qquad [x^2 = z].$$

3. $$\int (a+2bx+cx^2+gx^3)^{\pm\frac{1}{3}}\,dx = \frac{3}{2}\int \frac{z^2 A^{\pm\frac{1}{3}}dz}{B}$$

$$\left[a+2bx+cx^2 = z^3,\ A = g\left(\frac{-b+\sqrt{b^2+(z^3-a)c}}{c}\right)^3 + z^3,\ B = \sqrt{b^2+(z^3-a)c}\right]$$

4. $$\int \frac{dx}{\sqrt{a+bx+cx^2+dx^3+cx^4+bx^5+ax^6}} =$$

$$= -\frac{1}{\sqrt{2}}\int \frac{dx}{\sqrt{(z+1)p}} - \frac{1}{\sqrt{2}}\int \frac{dz}{\sqrt{(z-1)p}} \qquad \left[x = z+\sqrt{z^2-1}\right];$$

$$= -\frac{1}{\sqrt{2}}\int \frac{d}{\sqrt{(z+1)p}} + \frac{1}{\sqrt{2}}\int \frac{dz}{\sqrt{(z-1)p}} \qquad \left[x = z-\sqrt{z^2-1}\right],$$

where

$$p = 2a(4z^3-3z)+2b(2z^2-1)+2cz+d.$$

5. $$\int \frac{dx}{\sqrt{a+bx^2+cx^4+bx^6+ax^8}} = \frac{1}{2}\int \frac{dy}{\sqrt{y}\sqrt{a+by+cy^2+by^3+ay^4}} \quad [x=\sqrt{y}];$$
$$= -\frac{1}{2\sqrt{2}}\int \frac{dz}{\sqrt{(z+1)\,p}} - \frac{1}{2\sqrt{2}}\int \frac{dz}{\sqrt{(z-1)\,p}} \quad [y=z+\sqrt{z^2-1}];$$
$$= -\frac{1}{2\sqrt{2}}\int \frac{dz}{\sqrt{(z+1)\,p}} + \frac{1}{2\sqrt{2}}\int \frac{dz}{\sqrt{(z-1)\,p}} \quad [y=z-\sqrt{z^2-1}],$$

where $p = 2a(2z^2-1) + 2bz + c$.

6. $$\int \frac{dx}{\sqrt{a+bx^4+cx^8}} = \frac{1}{2}\sqrt[8]{\frac{a}{c}}\int \frac{dt}{\sqrt{t}\sqrt{a+b_1t^2+at^4}} \quad \left[x=\sqrt[8]{\frac{a}{c}}\sqrt{t}\right];$$
$$= -\frac{1}{2\sqrt{2}}\sqrt[8]{\frac{a}{c}}\left\{\int \frac{dz}{\sqrt{(z+1)\,p}} - \int \frac{dz}{\sqrt{(z-1)\,p}}\right\} \quad [t=z+\sqrt{z^2-1}];$$
$$= -\frac{1}{2\sqrt{2}}\sqrt[8]{\frac{a}{c}}\left\{\int \frac{dz}{\sqrt{(z+1)\,p}} + \int \frac{dz}{\sqrt{(z-1)\,p}}\right\} \quad [t=z-\sqrt{z^2-1}],$$

where $p = 2a(2z^2-1) + b_1$; $b_1 = b\sqrt{\frac{a}{c}}$.

7. $$\int \frac{x\,dx}{\sqrt[4]{a+bx^2+cx^4}} = 2\int \frac{z^2dz}{\sqrt{A+Bz^4}}$$
$$[a+bx^2+cx^4 = z^4, \qquad A = b^2-4ac, \qquad B = 4c].$$

8. $$\int \frac{dx}{\sqrt[4]{a+2bx^2+cx^4}} = \int \frac{\sqrt{b^2-a(c-z^4)}+b}{(c-z^4)\sqrt{b^2-a(c-z^4)}}\,z^2\,dz =$$
$$= \int R_1(z^4)\,z^2\,dz + \int \frac{R_2(z^4)\,z^2\,dz}{\sqrt{b^2-a(c-z^4)}},$$

where $R_1(z^4)$ and $R_2(z^4)$ are rational functions of z^4; $a+2bx^2+cx^4 = x^4z^4$.

2.292 In certain cases, integrals of the form $\int R(x, \sqrt{P(x)})\,dx$, where $P(x)$ is a third- or fourth-degree polynomial, can be expressed in terms of elementary functions. Such integrals are called *pseudo-elliptic* integrals.

Thus, if the relations

$$f_1(x) = -f_1\left(\frac{1}{k^2x}\right), \quad f_2(x) = -f_2\left(\frac{1-k^2x}{k^2(1-x)}\right), \quad f_3(x) = -f_3\left(\frac{1-x}{1-k^2x}\right),$$

hold, then

1. $$\int \frac{f_1(x)\,dx}{\sqrt{x(1-x)(1-k^2x)}} = \int R_1(z)\,dz \quad \left[zx = \sqrt{x(1-x)(1-k^2x)}\right];$$

2. $$\int \frac{f_2(x)\,dx}{\sqrt{x(1-x)(1-k^2x)}} = \int R_2(z)\,dz \quad \left[z = \frac{\sqrt{x(1-k^2x)}}{\sqrt{1-x}}\right];$$

3. $$\int \frac{f_3(x)\,dx}{\sqrt{x(1-x)(1-k^2x)}} = \int R_3(z)\,dz \quad \left[z = \frac{\sqrt{x(1-x)}}{\sqrt{1-k^2x}}\right],$$

where $R_1(z)$, $R_2(z)$, and $R_3(z)$ are rational functions of z.

2.3 The Exponential Function

2.31 Forms containing e^{ax}

2.311 $\int e^{ax}\,dx = \frac{e^{ax}}{a}$.

2.312 a^x in the integrands should be replaced with $e^{x \ln a} = a^x$.

2.313

1. $\int \frac{dx}{a+be^{mx}} = \frac{1}{am}[mx - \ln(a+be^{mx})]$. PE (410)

2. $\int \frac{dx}{1+e^x} = \ln\frac{e^x}{1+e^x} = x - \ln(1+e^x)$. PE (409)

2.314 $\int \frac{dx}{ae^{mx}+be^{-mx}} = \frac{1}{m\sqrt{ab}}\operatorname{arctg}\left(e^{mx}\sqrt{\frac{a}{b}}\right)$ $[ab>0]$; PE (411)

$= \frac{1}{2m\sqrt{-ab}}\ln\frac{b+e^{mx}\sqrt{-ab}}{b-e^{mx}\sqrt{-ab}}$ $[ab<0]$.

2.315 $\int \frac{dx}{\sqrt{a+be^{mx}}} = \frac{1}{m\sqrt{a}}\ln\frac{\sqrt{a+be^{mx}}-\sqrt{a}}{\sqrt{a+be^{mx}}+\sqrt{a}}$ $[a>0]$;

$= \frac{2}{m\sqrt{-a}}\operatorname{arctg}\frac{\sqrt{a+be^{mx}}}{\sqrt{-a}}$ $[a<0]$.

2.32 The exponential combined with rational functions of x

2.321

1. $\int x^m e^{ax}\,dx = \frac{x^m e^{ax}}{a} - \frac{m}{a}\int x^{m-1}e^{ax}\,dx$.

2. $\int x^n e^{ax}\,dx = e^{ax}\left(\frac{x^n}{a} + \sum_{k=1}^{n}(-1)^k \frac{n(n-1)\dots(n-k+1)}{a^{k+1}}x^{n-k}\right)$.

2.322

1. $\int xe^{ax}\,dx = e^{ax}\left(\frac{x}{a} - \frac{1}{a^2}\right)$.

2. $\int x^2 e^{ax}\,dx = e^{ax}\left(\frac{x^2}{a} - \frac{2x}{a^2} + \frac{2}{a^3}\right)$.

3. $\int x^3 e^{ax}\,dx = e^{ax}\left(\frac{x^3}{a} - \frac{3x^2}{a^2} + \frac{6x}{a^3} - \frac{6}{a^4}\right)$.

2.323 $\int P_m(x)\,e^{ax}\,dx = \frac{e^{ax}}{a}\sum_{k=0}^{m}(-1)^k\frac{P^{(k)}(x)}{a^k}$,

where $P_m(x)$ is a polynomial in x of degree m and $P^{(k)}(x)$ is the k-th derivative of $P_m(x)$ with respect to x.

2.324

1. $\int \frac{e^{ax}\,dx}{x^m} = \frac{1}{m-1}\left[-\frac{e^{ax}}{x^{m-1}} + a\int\frac{e^{ax}\,dx}{x^{m-1}}\right]$.

2. $\int \frac{e^{ax}}{x^n}\,dx = -e^{ax}\sum_{k=1}^{n-1}\frac{a^{k-1}}{(n-1)(n-2)\dots(n-k)\,x^{n-k}} + \frac{a^{n-1}}{(n-1)!}\operatorname{Ei}(ax)$.

2.325

1. $\int \frac{e^{ax}}{x}\,dx = \mathrm{Ei}\,(ax).$

2. $\int \frac{e^{ax}}{x^2}\,dx = -\frac{e^{ax}}{x} + a\,\mathrm{Ei}\,(ax).$

3. $\int \frac{e^{ax}}{x^3}\,dx = -\frac{e^{ax}}{2x^2} - \frac{ae^{ax}}{2x} + \frac{a^2}{2}\,\mathrm{Ei}\,(ax).$

2.326 $\int \frac{xe^{ax}\,dx}{(1+ax)^2} = \frac{e^{ax}}{a^2(1+ax)}.$

2.4 Hyperbolic Functions

2.41-2.43 Powers of $\mathrm{sh}\,x$, $\mathrm{ch}\,x$, $\mathrm{th}\,x$ and $\mathrm{cth}\,x$

2.411

$$\int \mathrm{sh}^p x\,\mathrm{ch}^q x\,dx = \frac{\mathrm{sh}^{p+1} x\,\mathrm{ch}^{q-1} x}{p+q} + \frac{q-1}{p+q}\int \mathrm{sh}^p x\,\mathrm{ch}^{q-2} x\,dx;$$

$$= \frac{\mathrm{sh}^{p-1} x\,\mathrm{ch}^{q+1} x}{p+q} - \frac{p-1}{p+q}\int \mathrm{sh}^{p-2} x\,\mathrm{ch}^{q} x\,dx;$$

$$= \frac{\mathrm{sh}^{p-1} x\,\mathrm{ch}^{q+1} x}{q+1} - \frac{p-1}{q+1}\int \mathrm{sh}^{p-2} x\,\mathrm{ch}^{q+2} x\,dx;$$

$$= \frac{\mathrm{sh}^{p+1} x\,\mathrm{ch}^{q-1} x}{p+1} - \frac{q-1}{p+1}\int \mathrm{sh}^{p+2} x\,\mathrm{ch}^{q-2} x\,dx;$$

$$= \frac{\mathrm{sh}^{p+1} x\,\mathrm{ch}^{q+1} x}{p+1} - \frac{p+q+2}{p+1}\int \mathrm{sh}^{p+2} x\,\mathrm{ch}^{q} x\,dx;$$

$$= -\frac{\mathrm{sh}^{p+1} x\,\mathrm{ch}^{q+1} x}{q+1} + \frac{p+q+2}{q+1}\int \mathrm{sh}^{p} x\,\mathrm{ch}^{q+2} x\,dx.$$

2.412

1. $$\int \mathrm{sh}^p x\,\mathrm{ch}^{2n} x\,dx = \frac{\mathrm{sh}^{p+1} x}{2n+p}\left\{\mathrm{ch}^{2n-1} x + \sum_{k=1}^{n-1} \frac{(2n-1)(2n-3)\ldots(2n-2k+1)}{(2n+p-2)(2n+p-4)\ldots(2n+p-2k)}\,\mathrm{ch}^{2n-2k-1} x\right\} + \frac{(2n-1)!!}{(2n+p)(2n+p-2)\ldots(p+2)}\int \mathrm{sh}^p x\,dx.$$

This formula is applicable for arbitrary real p except for the following negative even integers: $-2, -4, \ldots, -2n$. If p is a natural number and $n=0$, we have

2. $$\int \mathrm{sh}^{2m} x\,dx = (-1)^m \binom{2m}{m}\frac{x}{2^{2m}} + \frac{1}{2^{2m-1}}\sum_{k=0}^{m-1}(-1)^k\binom{2m}{k}\frac{\mathrm{sh}\,(2m-2k)\,x}{2m-2k}.$$ TI (543)

3. $$\int \mathrm{sh}^{2m+1} x\,dx = \frac{1}{2^{2m}}\sum_{k=0}^{m}(-1)^k\binom{2m+1}{k}\frac{\mathrm{ch}\,(2m-2k+1)\,x}{2m-2k+1};$$ TI (544)

$$= (-1)^n\sum_{k=0}^{m}(-1)^k\binom{m}{k}\frac{\mathrm{ch}^{2k+1} x}{2k+1}.$$ GU ((351)) (5)

4. $$\int \operatorname{sh}^p x \operatorname{ch}^{2n+1} x\,dx = \frac{\operatorname{sh}^{p+1} x}{2n+p+1}\left\{\operatorname{ch}^{2n} x + \sum_{k=1}^{n} \frac{2^k n(n-1)\dots(n-k+1)\operatorname{ch}^{2n-2k} x}{(2n+p-1)(2n+p-3)\dots(2n+p-2k+1)}\right\}.$$

This formula is applicable for arbitrary real p except for the following negative odd integers: $-1,\ -3,\ \dots,\ -(2n+1)$.

2.413

1. $$\int \operatorname{ch}^p x \operatorname{sh}^{2n} x\,dx = \frac{\operatorname{ch}^{p+1} x}{2n+p}\left\{\operatorname{sh}^{2n-1} x + \sum_{k=1}^{n-1} (-1)^k \frac{(2n-1)(2n-3)\dots(2n-2k+1)\operatorname{sh}^{2n-2k-1} x}{(2n+p-2)(2n+p-4)\dots(2n+p-2k)}\right\} + (-1)^n \frac{(2n-1)!!}{(2n+p)(2n+p-2)\dots(p+2)}\int \operatorname{ch}^p x\,dx.$$

This formula is applicable for arbitrary real p except for the following negative even integers: $-2,\ -4,\ \dots,\ -2n$. If p is a natural number and $n=0$, we have

2. $$\int \operatorname{ch}^{2m} x\,dx = \binom{2m}{m}\frac{x}{2^{2m}} + \frac{1}{2^{2m-1}}\sum_{k=0}^{m-1}\binom{2m}{k}\frac{\operatorname{sh}(2m-2k)x}{2m-2k}.$$ TI (541)

3. $$\int \operatorname{ch}^{2m+1} x\,dx = \frac{1}{2^{2m}}\sum_{k=0}^{m}\binom{2m+1}{k}\frac{\operatorname{sh}(2m-2k+1)x}{2m-2k+1};$$ TI (542)

$$= \sum_{k=0}^{m}\binom{m}{k}\frac{\operatorname{sh}^{2k+1} x}{2k+1}.$$ GU ((351)) (8)

4. $$\int \operatorname{ch}^p x \operatorname{sh}^{2n+1} x\,dx = \frac{\operatorname{ch}^{p+1} x}{2n+p+1}\left\{\operatorname{sh}^{2n} x + \sum_{k=1}^{n}(-1)^k \frac{2^k n(n-1)\dots(n-k+1)\operatorname{sh}^{2n-2k} x}{(2n+p-1)(2n+p-3)\dots(2n+p-2k+1)}\right\}.$$

This formula is applicable for arbitrary real p except for the following negative odd integers: $-1,\ -3,\ \dots,\ -(2n+1)$.

2.414

1. $$\int \operatorname{sh} ax\,dx = \frac{1}{a}\operatorname{ch} ax.$$

2. $$\int \operatorname{sh}^2 ax\,dx = \frac{1}{4a}\operatorname{sh} 2ax - \frac{x}{2}.$$

3. $$\int \operatorname{sh}^3 x\,dx = -\frac{3}{4}\operatorname{ch} x + \frac{1}{12}\operatorname{ch} 3x = \frac{1}{3}\operatorname{ch}^3 x - \operatorname{ch} x.$$

4. $$\int \operatorname{sh}^4 x\,dx = \frac{3}{8}x - \frac{1}{4}\operatorname{sh} 2x + \frac{1}{32}\operatorname{sh} 4x = \frac{3}{8}x - \frac{3}{8}\operatorname{sh} x \operatorname{ch} x + \frac{1}{4}\operatorname{sh}^3 x \operatorname{ch} x.$$

5. $$\int \operatorname{sh}^5 x\,dx = \frac{5}{8}\operatorname{ch} x - \frac{5}{48}\operatorname{ch} 3x + \frac{1}{80}\operatorname{ch} 5x;$$

$$= \frac{4}{5}\operatorname{ch} x + \frac{1}{5}\operatorname{sh}^4 x \operatorname{ch} x - \frac{4}{15}\operatorname{ch}^3 x.$$

6. $\int \operatorname{sh}^6 x\,dx = -\frac{5}{16}x + \frac{15}{64}\operatorname{sh} 2x - \frac{3}{64}\operatorname{sh} 4x + \frac{1}{192}\operatorname{sh} 6x;$

$= -\frac{5}{16}x + \frac{1}{6}\operatorname{sh}^5 x \operatorname{ch} x - \frac{5}{24}\operatorname{sh}^3 x \operatorname{ch} x + \frac{5}{16}\operatorname{sh} x \operatorname{ch} x.$

7. $\int \operatorname{sh}^7 x\,dx = -\frac{35}{64}\operatorname{ch} x + \frac{7}{64}\operatorname{ch} 3x - \frac{7}{320}\operatorname{ch} 5x + \frac{1}{448}\operatorname{ch} 7x;$

$= -\frac{24}{35}\operatorname{ch} x + \frac{8}{35}\operatorname{ch}^3 x - \frac{6}{35}\operatorname{ch} x \operatorname{sh}^4 x + \frac{1}{7}\operatorname{ch} x \operatorname{sh}^6 x.$

8. $\int \operatorname{ch} ax\,dx = \frac{1}{a}\operatorname{sh} ax.$

9. $\int \operatorname{ch}^2 ax\,dx = \frac{x}{2} + \frac{1}{4a}\operatorname{sh} 2ax.$

10. $\int \operatorname{ch}^3 x\,dx = \frac{3}{4}\operatorname{sh} x + \frac{1}{12}\operatorname{sh} 3x = \operatorname{sh} x + \frac{1}{3}\operatorname{sh}^3 x.$

11. $\int \operatorname{ch}^4 x\,dx = \frac{3}{8}x + \frac{1}{4}\operatorname{sh} 2x + \frac{1}{32}\operatorname{sh} 4x = \frac{3}{8}x + \frac{3}{8}\operatorname{sh} x \operatorname{ch} x + \frac{1}{4}\operatorname{sh} x \operatorname{ch}^3 x.$

12. $\int \operatorname{ch}^5 x\,dx = \frac{5}{8}\operatorname{sh} x + \frac{5}{48}\operatorname{sh} 3x + \frac{1}{80}\operatorname{sh} 5x;$

$= \frac{4}{5}\operatorname{sh} x + \frac{1}{5}\operatorname{ch}^4 x \operatorname{sh} x + \frac{4}{15}\operatorname{sh}^3 x.$

13. $\int \operatorname{ch}^6 x\,dx = \frac{5}{16}x + \frac{15}{64}\operatorname{sh} 2x + \frac{3}{64}\operatorname{sh} 4x + \frac{1}{192}\operatorname{sh} 6x;$

$= \frac{5}{16}x + \frac{5}{16}\operatorname{sh} x \operatorname{ch} x + \frac{5}{24}\operatorname{sh} x \operatorname{ch}^3 x + \frac{1}{6}\operatorname{sh} x \operatorname{ch}^5 x.$

14. $\int \operatorname{ch}^7 x\,dx = \frac{35}{64}\operatorname{sh} x + \frac{7}{64}\operatorname{sh} 3x + \frac{7}{320}\operatorname{sh} 5x + \frac{1}{448}\operatorname{sh} 7x;$

$= \frac{24}{35}\operatorname{sh} x + \frac{8}{35}\operatorname{sh}^3 x + \frac{6}{35}\operatorname{sh} x \operatorname{ch}^4 x + \frac{1}{7}\operatorname{sh} x \operatorname{ch}^6 x.$

2.415

1. $\int \operatorname{sh} ax \operatorname{ch} bx\,dx = \frac{\operatorname{ch}(a+b)x}{2(a+b)} + \frac{\operatorname{ch}(a-b)x}{2(a-b)}.$

2. $\int \operatorname{sh} ax \operatorname{ch} ax\,dx = \frac{1}{4a}\operatorname{ch} 2ax.$

3. $\int \operatorname{sh}^2 x \operatorname{ch} x\,dx = \frac{1}{3}\operatorname{sh}^3 x.$

4. $\int \operatorname{sh}^3 x \operatorname{ch} x\,dx = \frac{1}{4}\operatorname{sh}^4 x.$

5. $\int \operatorname{sh}^4 x \operatorname{ch} x\,dx = \frac{1}{5}\operatorname{sh}^5 x.$

6. $\int \operatorname{sh} x \operatorname{ch}^2 x\,dx = \frac{1}{3}\operatorname{ch}^3 x.$

7. $\int \operatorname{sh}^2 x \operatorname{ch}^2 x\,dx = -\frac{x}{8} + \frac{1}{32}\operatorname{sh} 4x.$

8. $\int \operatorname{sh}^3 x \operatorname{ch}^2 x\,dx = \frac{1}{5}\left(\operatorname{sh}^2 x - \frac{2}{3}\right)\operatorname{ch}^3 x.$

9. $\int \operatorname{sh}^4 x \operatorname{ch}^2 x\,dx = \frac{x}{16} - \frac{1}{64}\operatorname{sh} 2x - \frac{1}{64}\operatorname{sh} 4x + \frac{1}{192}\operatorname{sh} 6x.$

10. $\int \operatorname{sh} x \operatorname{ch}^3 x\, dx = \frac{1}{4} \operatorname{ch}^4 x.$

11. $\int \operatorname{sh}^2 x \operatorname{ch}^3 x\, dx = \frac{1}{5}\left(\operatorname{ch}^2 x + \frac{2}{3}\right) \operatorname{sh}^3 x.$

12. $\int \operatorname{sh}^3 x \operatorname{ch}^3 x\, dx = -\frac{3}{64} \operatorname{ch} 2x + \frac{1}{192} \operatorname{ch} 6x = \frac{1}{48} \operatorname{ch}^3 2x - \frac{1}{16} \operatorname{ch} 2x;$

$$= \frac{\operatorname{sh}^6 x}{6} + \frac{\operatorname{sh}^4 x}{4} = \frac{\operatorname{ch}^6 x}{6} - \frac{\operatorname{ch}^4 x}{4}.$$

13. $\int \operatorname{sh}^4 x \operatorname{ch}^3 x\, dx = \frac{1}{7} \operatorname{sh}^3 x \left(\operatorname{ch}^4 x - \frac{3}{5} \operatorname{ch}^2 x - \frac{2}{5}\right) = \frac{1}{7}\left(\operatorname{ch}^2 x + \frac{2}{5}\right) \operatorname{sh}^5 x.$

14. $\int \operatorname{sh} x \operatorname{ch}^4 x\, dx = \frac{1}{5} \operatorname{ch}^5 x.$

15. $\int \operatorname{sh}^2 x \operatorname{ch}^4 x\, dx = -\frac{x}{16} - \frac{1}{64} \operatorname{sh} 2x + \frac{1}{64} \operatorname{sh} 4x + \frac{1}{192} \operatorname{sh} 6x.$

16. $\int \operatorname{sh}^3 x \operatorname{ch}^4 x\, dx = \frac{1}{7} \operatorname{ch}^3 x \left(\operatorname{sh}^4 x + \frac{3}{5} \operatorname{sh}^2 x - \frac{2}{5}\right) = \frac{1}{7}\left(\operatorname{sh}^2 x - \frac{2}{5}\right) \operatorname{ch}^5 x.$

17. $\int \operatorname{sh}^4 x \operatorname{ch}^4 x\, dx = \frac{3x}{128} - \frac{1}{128} \operatorname{sh} 4x + \frac{1}{1024} \operatorname{sh} 8x.$

2.416

1. $$\int \frac{\operatorname{sh}^p x}{\operatorname{ch}^{2n} x} dx = \frac{\operatorname{sh}^{p+1}}{2n-1} \left\{ \operatorname{sech}^{2n-1} x + \right.$$

$$+ \sum_{k=1}^{n-1} \frac{(2n-p-2)(2n-p-4)\dots(2n-p-2k)}{(2n-3)(2n-5)\dots(2n-2k-1)} \left. \operatorname{sech}^{2n-2k-1} x \right\} +$$

$$+ \frac{(2n-p-2)(2n-p-4)\dots(-p+2)(-p)}{(2n-1)!!} \int \operatorname{sh}^p x\, dx.$$

This formula is applicable for arbitrary real p. For $\int \operatorname{sh}^p x\, dx$, where p is a natural number, see **2.412** 2. and **2.412** 3. For $n=0$ and p a negative integer, we have for this integral:

2. $$\int \frac{dx}{\operatorname{sh}^{2m} x} = \frac{\operatorname{ch} x}{2m-1} \left\{ -\operatorname{cosech}^{2m-1} x + \right.$$

$$+ \sum_{k=1}^{m-1} (-1)^{k-1} \cdot \frac{2^k (m-1)(m-2)\dots(m-k)}{(2m-3)(2m-5)\dots(2m-2k-1)} \left. \operatorname{cosech}^{2m-2k-1} x \right\}.$$

3. $$\int \frac{dx}{\operatorname{sh}^{2m+1} x} = \frac{\operatorname{ch} x}{2m} \left\{ -\operatorname{cosech}^{2m} x + \right.$$

$$+ \sum_{k=1}^{m-1} (-1)^{k-1} \cdot \frac{(2m-1)(2m-3)\dots(2m-2k+1)}{2^k (m-1)(m-2)\dots(m-k)} \left. \operatorname{cosech}^{2m-2k} x \right\} +$$

$$+ (-1)^m \frac{(2m-1)!!}{(2m)!!} \ln \operatorname{th} \frac{x}{2}.$$

2.417

1. $$\int \frac{\operatorname{sh}^p x}{\operatorname{ch}^{2n+1} x}\,dx = \frac{\operatorname{sh}^{p+1} x}{2n}\left\{\operatorname{sech}^{2n} x + \sum_{k=1}^{n-1} \frac{(2n-p-1)(2n-p-3)\dots(2n-p-2k+1)}{2^k (n-1)(n-2)\dots(n-k)} \operatorname{sech}^{2n-2k} x\right\} + \frac{(2n-p-1)(2n-p-3)\dots(3-p)(1-p)}{2^n n!}\int \frac{\operatorname{sh}^p x}{\operatorname{ch} x}\,dx.$$

This formula is applicable for arbitrary real p. For $n=0$ and p integral, we have

2. $$\int \frac{\operatorname{sh}^{2m+1} x}{\operatorname{ch} x}\,dx = \sum_{k=1}^{m} \frac{(-1)^{m+k}}{2k} \operatorname{sh}^{2k} x + (-1)^m \ln \operatorname{ch} x;$$
$$= \sum_{k=1}^{m} \frac{(-1)^{m+k}}{2k}\binom{m}{k} \operatorname{ch}^{2k} x + (-1)^m \ln \operatorname{ch} x \quad [m \geqslant 1].$$

3. $$\int \frac{\operatorname{sh}^{2m} x}{\operatorname{ch} x}\,dx = \sum_{k=1}^{m} \frac{(-1)^{m+k}}{2k-1} \operatorname{sh}^{2k-1} x + (-1)^m \operatorname{arctg}(\operatorname{sh} x) \quad [m \geqslant 1].$$

4. $$\int \frac{dx}{\operatorname{sh}^{2m+1} x \operatorname{ch} x} = \sum_{k=1}^{m} \frac{(-1)^k \operatorname{cosech}^{2m-2k+2} x}{2m-2k+2} + (-1)^m \ln \operatorname{th} x.$$

5. $$\int \frac{dx}{\operatorname{sh}^{2m} x \operatorname{ch} x} = \sum_{k=1}^{m} \frac{(-1)^k \operatorname{cosech}^{2m-2k+1} x}{2m-2k+1} + (-1)^m \operatorname{arctg} \operatorname{sh} x.$$

2.418

1. $$\int \frac{\operatorname{ch}^p x}{\operatorname{sh}^{2n} x}\,dx = -\frac{\operatorname{ch}^{p+1} x}{2n-1}\left\{\operatorname{cosech}^{2n-1} x + \sum_{k=1}^{n-1} \frac{(-1)^k (2n-p-2)(2n-p-4)\dots(2n-p-2k)}{(2n-3)(2n-5)\dots(2n-2k-1)} \operatorname{cosech}^{2n-2k-1} x\right\} + \frac{(-1)^n (2n-p-2)(2n-p-4)\dots(-p+2)(-p)}{(2n-1)!!}\int \operatorname{ch}^p x\,dx.$$

This formula is applicable for arbitrary real p. For the integral $\int \operatorname{ch}^p x\,dx$, where p is a natural number, see **2.413** 2. and **2.413** 3. If p is a negative integer, we have for this integral:

2. $$\int \frac{dx}{\operatorname{ch}^{2m} x} = \frac{\operatorname{sh} x}{2m-1}\left\{\operatorname{sech}^{2m-1} x + \sum_{k=1}^{m-1} \frac{2^k (m-1)(m-2)\dots(m-k)}{(2m-3)(2m-5)\dots(2m-2k-1)} \operatorname{sech}^{2m-2k-1} x\right\}.$$

3. $$\int \frac{dx}{\operatorname{ch}^{2m+1} x} = \frac{\operatorname{sh} x}{2m}\left\{\operatorname{sech}^{2m} x + \sum_{k=1}^{m-1} \frac{(2m-1)(2m-3)\dots(2m-2k+1)}{2^k (m-1)(m-2)\dots(m-k)} \operatorname{sech}^{2m-2k} x\right\} + \frac{(2m-1)!!}{(2m)!!} \operatorname{arctg} \operatorname{sh} x.$$

2.419

1. $$\int \frac{\mathrm{ch}^p x}{\mathrm{sh}^{2n+1} x} dx = -\frac{\mathrm{ch}^{p+1} x}{2n} \left\{ \mathrm{cosech}^{2n} x + \right.$$
$$\left. + \sum_{k=1}^{n-1} \frac{(-1)^k (2n-p-1)(2n-p-3) \dots (2n-p-2k+1)}{2^k (n-1)(n-2) \dots (n-k)} \mathrm{cosech}^{2n-2k} x \right\} +$$
$$+ \frac{(-1)^n (2n-p-1)(2n-p-3) \dots (3-p)(1-p)}{2^n n!} \int \frac{\mathrm{ch}^p x}{\mathrm{sh}\, x} dx.$$

This formula is applicable for arbitrary real p. For $n=0$ and p an integer

2. $$\int \frac{\mathrm{ch}^{2m} x}{\mathrm{sh}\, x} dx = \sum_{k=1}^{m} \frac{\mathrm{ch}^{2k-1} x}{2k-1} + \ln \mathrm{th} \frac{x}{2}.$$

3. $$\int \frac{\mathrm{ch}^{2m+1} x}{\mathrm{sh}\, x} dx = \sum_{k=1}^{m} \frac{\mathrm{ch}^{2k} x}{2k} + \ln \mathrm{sh}\, x;$$
$$= \sum_{k=1}^{m} \binom{m}{k} \frac{\mathrm{sh}^{2k} x}{2k} + \ln \mathrm{sh}\, x.$$

4. $$\int \frac{dx}{\mathrm{sh}\, x\, \mathrm{ch}^{2m} x} = \sum_{k=1}^{m} \frac{\mathrm{sech}^{2m-2k+1} x}{2m-2k+1} + \ln \mathrm{th} \frac{x}{2}.$$

5. $$\int \frac{dx}{\mathrm{sh}\, x\, \mathrm{ch}^{2m+1} x} = \sum_{k=1}^{m} \frac{\mathrm{sech}^{2m-2k+2} x}{2m-2k+2} + \ln \mathrm{th}\, x.$$

2.421

1. $$\int \frac{\mathrm{sh}^{2n+1} x}{\mathrm{ch}^m x} dx = \sum_{\substack{k=0 \\ k \neq \frac{m-1}{2}}}^{n} (-1)^{n+k} \binom{n}{k} \frac{\mathrm{ch}^{2k-m+1} x}{2k-m+1} +$$
$$+ s(-1)^{n+\frac{m-1}{2}} \binom{n}{\frac{m-1}{2}} \ln \mathrm{ch}\, x.$$

2. $$\int \frac{\mathrm{ch}^{2n+1} x}{\mathrm{sh}^m x} dx = \sum_{\substack{k=0 \\ k \neq \frac{m-1}{2}}}^{n} \binom{n}{k} \frac{\mathrm{sh}^{2k-m+1} x}{2k-m+1} + s \binom{n}{\frac{m-1}{2}} \ln \mathrm{sh}\, x.$$

[In formulas **2.421** 1. and **2.421** 2., $s=1$ for m odd and $m < 2n+1$; in all other cases, $s=0$.]

GI ((351))(11, 13)

2.422

1. $$\int \frac{dx}{\mathrm{sh}^{2m} x\, \mathrm{ch}^{2n} x} = \sum_{k=0}^{m+n-1} \frac{(-1)^{k+1}}{2m-2k-1} \binom{m+n-1}{k} \mathrm{th}^{2k-2m+1} x.$$

2. $$\int \frac{dx}{\mathrm{sh}^{2m+1} x\, \mathrm{ch}^{2n+1} x} = \sum_{\substack{k=0 \\ k \neq m}}^{m+n} \frac{(-1)^{k+1}}{2m-2k} \binom{m+n}{k} \mathrm{th}^{2k-2m} x +$$
$$+ (-1)^m \binom{m+n}{m} \ln \mathrm{th}\, x.$$

GI ((351))(15)

2.423

1. $\int \frac{dx}{\operatorname{sh} x} = \ln \operatorname{th} \frac{x}{2} = \frac{1}{2} \ln \frac{\operatorname{ch} x - 1}{\operatorname{ch} x + 1}.$

2. $\int \frac{dx}{\operatorname{sh}^2 x} = -\operatorname{cth} x.$

3. $\int \frac{dx}{\operatorname{sh}^3 x} = -\frac{\operatorname{ch} x}{2\operatorname{sh}^2 x} - \frac{1}{2} \ln \operatorname{th} \frac{x}{2}.$

4. $\int \frac{dx}{\operatorname{sh}^4 x} = -\frac{\operatorname{ch} x}{3\operatorname{sh}^3 x} + \frac{2}{3} \operatorname{cth} x = -\frac{1}{3} \operatorname{cth}^3 x + \operatorname{cth} x.$

5. $\int \frac{dx}{\operatorname{sh}^5 x} = -\frac{\operatorname{ch} x}{4\operatorname{sh}^4 x} + \frac{3}{8} \frac{\operatorname{ch} x}{\operatorname{sh}^2 x} + \frac{3}{8} \ln \operatorname{th} \frac{x}{2}.$

6. $\int \frac{dx}{\operatorname{sh}^6 x} = -\frac{\operatorname{ch} x}{5\operatorname{sh}^5 x} + \frac{4}{15} \operatorname{cth}^3 x - \frac{4}{5} \operatorname{cth} x;$

$= -\frac{1}{5} \operatorname{cth}^5 x + \frac{2}{3} \operatorname{cth}^3 x - \operatorname{cth} x.$

7. $\int \frac{dx}{\operatorname{sh}^7 x} = -\frac{\operatorname{ch} x}{6\operatorname{sh}^2 x} \left(\frac{1}{\operatorname{sh}^4 x} - \frac{5}{4\operatorname{sh}^2 x} + \frac{15}{8} \right) - \frac{5}{16} \ln \operatorname{th} \frac{x}{2}.$

8. $\int \frac{dx}{\operatorname{sh}^8 x} = \operatorname{cth} x - \operatorname{cth}^3 x + \frac{3}{5} \operatorname{cth}^5 x - \frac{1}{7} \operatorname{cth}^7 x.$

9. $\int \frac{dx}{\operatorname{ch} x} = \operatorname{arctg} (\operatorname{sh} x) = 2\operatorname{arctg} (e^x);$

$= \arcsin (\operatorname{th} x);$

$= \operatorname{gd} x.$

10. $\int \frac{dx}{\operatorname{ch}^2 x} = \operatorname{th} x.$

11. $\int \frac{dx}{\operatorname{ch}^3 x} = \frac{\operatorname{sh} x}{2\operatorname{ch}^2 x} + \frac{1}{2} \operatorname{arctg} (\operatorname{sh} x).$

12. $\int \frac{dx}{\operatorname{ch}^4 x} = \frac{\operatorname{sh} x}{3\operatorname{ch}^3 x} + \frac{2}{3} \operatorname{th} x;$

$= -\frac{1}{3} \operatorname{th}^3 x + \operatorname{th} x.$

13. $\int \frac{dx}{\operatorname{ch}^5 x} = \frac{\operatorname{sh} x}{4\operatorname{ch}^4 x} + \frac{3}{8} \frac{\operatorname{sh} x}{\operatorname{ch}^2 x} + \frac{3}{8} \operatorname{arctg} (\operatorname{sh} x).$

14. $\int \frac{dx}{\operatorname{ch}^6 x} = \frac{\operatorname{sh} x}{5\operatorname{ch}^5 x} - \frac{4}{15} \operatorname{th}^3 x + \frac{4}{5} \operatorname{th} x;$

$= \frac{1}{5} \operatorname{th}^5 x - \frac{2}{3} \operatorname{th}^3 x + \operatorname{th} x.$

15. $\int \frac{dx}{\operatorname{ch}^7 x} = \frac{\operatorname{sh} x}{6\operatorname{ch}^6 x} \left(\frac{1}{\operatorname{ch}^4 x} + \frac{5}{4\operatorname{ch}^2 x} + \frac{15}{8} \right) + \frac{5}{16} \operatorname{arctg} (\operatorname{sh} x).$

16. $\int \frac{dx}{\operatorname{ch}^8 x} = -\frac{1}{7} \operatorname{th}^7 x + \frac{3}{5} \operatorname{th}^5 x - \operatorname{th}^3 x + \operatorname{th} x.$

17. $\int \frac{\operatorname{sh} x}{\operatorname{ch} x} dx = \ln \operatorname{ch} x.$

18. $\int \frac{\operatorname{sh}^2 x}{\operatorname{ch} x} dx = \operatorname{sh} x - \operatorname{arctg} (\operatorname{sh} x).$

19. $\int \frac{\operatorname{sh}^3 x}{\operatorname{ch} x} dx = \frac{1}{2} \operatorname{sh}^2 x - \ln \operatorname{ch} x;$

$= \frac{1}{2} \operatorname{ch}^2 x - \ln \operatorname{ch} x.$

20. $\int \frac{\mathrm{sh}^4 x}{\mathrm{ch}\, x} dx = \frac{1}{3} \mathrm{sh}^3 x - \mathrm{sh}\, x + \mathrm{arctg}\,(\mathrm{sh}\, x).$

21. $\int \frac{\mathrm{sh}\, x}{\mathrm{ch}^2 x} dx = -\frac{1}{\mathrm{ch}\, x}.$

22. $\int \frac{\mathrm{sh}^2 x}{\mathrm{ch}^2 x} dx = x - \mathrm{th}\, x.$

23. $\int \frac{\mathrm{sh}^3 x}{\mathrm{ch}^2 x} dx = \mathrm{ch}\, x + \frac{1}{\mathrm{ch}\, x}.$

24. $\int \frac{\mathrm{sh}^4 x}{\mathrm{ch}^2 x} dx = -\frac{3}{2} x + \frac{1}{4} \mathrm{sh}\, 2x + \mathrm{th}\, x.$

25. $\int \frac{\mathrm{sh}\, x}{\mathrm{ch}^3 x} dx = -\frac{1}{2\mathrm{ch}^2 x};$

$= \frac{1}{2} \mathrm{th}^2 x.$

26. $\int \frac{\mathrm{sh}^2 x}{\mathrm{ch}^3 x} dx = -\frac{\mathrm{sh}\, x}{2\mathrm{ch}^2 x} + \frac{1}{2} \mathrm{arctg}\,(\mathrm{sh}\, x).$

27. $\int \frac{\mathrm{sh}^3 x}{\mathrm{ch}^3 x} dx = -\frac{1}{2} \mathrm{th}^2 x + \ln \mathrm{ch}\, x;$

$= \frac{1}{2\mathrm{ch}^2 x} + \ln \mathrm{ch}\, x.$

28. $\int \frac{\mathrm{sh}^4 x}{\mathrm{ch}^3 x} dx = \frac{\mathrm{sh}\, x}{2\mathrm{ch}\, x} + \mathrm{sh}\, x - \frac{3}{2} \mathrm{arctg}\,(\mathrm{sh}\, x).$

29. $\int \frac{\mathrm{sh}\, x}{\mathrm{ch}^4 x} dx = -\frac{1}{3\mathrm{ch}^3 x}.$

30. $\int \frac{\mathrm{sh}^2 x}{\mathrm{ch}^4 x} dx = \frac{1}{3} \mathrm{th}^3 x.$

31. $\int \frac{\mathrm{sh}^3 x}{\mathrm{ch}^4 x} dx = -\frac{1}{\mathrm{ch}\, x} + \frac{1}{3\mathrm{ch}^3 x}.$

32. $\int \frac{\mathrm{sh}^4 x}{\mathrm{ch}^4 x} dx = -\frac{1}{3} \mathrm{th}^3 x - \mathrm{th}\, x + x.$

33. $\int \frac{\mathrm{ch}\, x}{\mathrm{sh}\, x} dx = \ln \mathrm{sh}\, x.$

34. $\int \frac{\mathrm{ch}^2 x}{\mathrm{sh}\, x} dx = \mathrm{ch}\, x + \ln \mathrm{th}\, \frac{x}{2}.$

35. $\int \frac{\mathrm{ch}^3 x}{\mathrm{sh}\, x} dx = \frac{1}{2} \mathrm{ch}^2 x + \ln \mathrm{sh}\, x.$

36. $\int \frac{\mathrm{ch}^4 x}{\mathrm{sh}\, x} dx = \frac{1}{3} \mathrm{ch}^3 x + \mathrm{ch}\, x + \ln \mathrm{th}\, \frac{x}{2}.$

37. $\int \frac{\mathrm{ch}\, x}{\mathrm{sh}^2 x} dx = -\frac{1}{\mathrm{sh}\, x}.$

38. $\int \frac{\mathrm{ch}^2 x}{\mathrm{sh}^2 x} dx = x - \mathrm{cth}\, x.$

39. $\int \frac{\mathrm{ch}^3 x}{\mathrm{sh}^2 x} dx = \mathrm{sh}\, x - \frac{1}{\mathrm{sh}\, x}.$

40. $\int \frac{\mathrm{ch}^4 x}{\mathrm{sh}^2 x} dx = \frac{3}{2} x + \frac{1}{4} \mathrm{sh}\, 2x - \mathrm{cth}\, x.$

41. $\int \frac{\mathrm{ch}\, x}{\mathrm{sh}^3 x} dx = -\frac{1}{2\mathrm{sh}^2 x};$

$= -\frac{1}{2} \mathrm{cth}^2 x.$

42. $\int \frac{\operatorname{ch}^2 x}{\operatorname{sh}^3 x}\,dx = -\frac{\operatorname{ch} x}{2\operatorname{sh}^2 x} + \ln \operatorname{th}\frac{x}{2}.$

43. $\int \frac{\operatorname{ch}^3 x}{\operatorname{sh}^3 x}\,dx = -\frac{1}{2\operatorname{sh}^2 x} + \ln \operatorname{sh} x;$

$= -\frac{1}{2}\operatorname{cth}^2 x + \ln \operatorname{sh} x.$

44. $\int \frac{\operatorname{ch}^4 x}{\operatorname{sh}^3 x}\,dx = -\frac{\operatorname{ch} x}{2\operatorname{sh}^2 x} + \operatorname{ch} x + \frac{3}{2}\ln \operatorname{th}\frac{x}{2}.$

45. $\int \frac{\operatorname{ch} x}{\operatorname{sh}^4 x}\,dx = -\frac{1}{3\operatorname{sh}^3 x}.$

46. $\int \frac{\operatorname{ch}^2 x}{\operatorname{sh}^4 x}\,dx = -\frac{1}{3}\operatorname{cth}^3 x.$

47. $\int \frac{\operatorname{ch}^3 x}{\operatorname{sh}^4 x}\,dx = -\frac{1}{\operatorname{sh} x} - \frac{1}{3\operatorname{sh}^3 x}.$

48. $\int \frac{\operatorname{ch}^4 x}{\operatorname{sh}^4 x}\,dx = -\frac{1}{3}\operatorname{cth}^3 x - \operatorname{cth} x + x.$

49. $\int \frac{dx}{\operatorname{sh} x \operatorname{ch} x} = \ln \operatorname{th} x.$

50. $\int \frac{dx}{\operatorname{sh} x \operatorname{ch}^2 x} = \frac{1}{\operatorname{ch} x} + \ln \operatorname{th}\frac{x}{2}.$

51. $\int \frac{dx}{\operatorname{sh} x \operatorname{ch}^3 x} = \frac{1}{2\operatorname{ch}^2 x} + \ln \operatorname{th} x;$

$= -\frac{1}{2}\operatorname{th}^2 x + \ln \operatorname{th} x.$

52. $\int \frac{dx}{\operatorname{sh} x \operatorname{ch}^4 x} = \frac{1}{\operatorname{ch} x} + \frac{1}{3\operatorname{ch}^3 x} + \ln \operatorname{th}\frac{x}{2}.$

53. $\int \frac{dx}{\operatorname{sh}^2 x \operatorname{ch} x} = -\frac{1}{\operatorname{sh} x} - \operatorname{arctg} \operatorname{sh} x.$

54. $\int \frac{dx}{\operatorname{sh}^2 x \operatorname{ch}^2 x} = -2\operatorname{cth} 2x.$

55. $\int \frac{dx}{\operatorname{sh}^2 x \operatorname{ch}^3 x} = -\frac{\operatorname{sh} x}{2\operatorname{ch}^2 x} - \frac{1}{\operatorname{sh} x} - \frac{3}{2}\operatorname{arctg} \operatorname{sh} x.$

56. $\int \frac{dx}{\operatorname{sh}^2 x \operatorname{ch}^4 x} = \frac{1}{3\operatorname{sh} x \operatorname{ch}^3 x} - \frac{8}{3}\operatorname{cth} 2x.$

57. $\int \frac{dx}{\operatorname{sh}^3 x \operatorname{ch} x} = -\frac{1}{2\operatorname{sh}^2 x} - \ln \operatorname{th} x;$

$= -\frac{1}{2}\operatorname{cth}^2 x + \ln \operatorname{cth} x.$

58. $\int \frac{dx}{\operatorname{sh}^3 x \operatorname{ch}^2 x} = -\frac{1}{\operatorname{ch} x} - \frac{\operatorname{ch} x}{2\operatorname{sh}^2 x} - \frac{3}{2}\ln \operatorname{th}\frac{x}{2}.$

59. $\int \frac{dx}{\operatorname{sh}^3 x \operatorname{ch}^3 x} = -\frac{2\operatorname{ch} 2x}{\operatorname{sh}^2 2x} - 2\ln \operatorname{th} x;$

$= \frac{1}{2}\operatorname{th}^2 x - \frac{1}{2}\operatorname{cth}^2 x - 2\ln \operatorname{th} x.$

60. $\int \frac{dx}{\operatorname{sh}^3 x \operatorname{ch}^4 x} = -\frac{2}{\operatorname{ch} x} - \frac{1}{3\operatorname{ch}^2 x} - \frac{\operatorname{ch} x}{2\operatorname{sh}^2 x} - \frac{5}{2}\ln \operatorname{th}\frac{x}{2}.$

61. $\int \frac{dx}{\operatorname{sh}^4 x \operatorname{ch} x} = \frac{1}{\operatorname{sh} x} - \frac{1}{3\operatorname{sh}^3 x} + \operatorname{arctg} \operatorname{sh} x.$

62. $\int \frac{dx}{\operatorname{sh}^4 x \operatorname{ch}^2 x} = -\frac{1}{3\operatorname{ch} x \operatorname{sh}^3 x} + \frac{8}{3} \operatorname{cth} 2x.$

63. $\int \frac{dx}{\operatorname{sh}^4 x \operatorname{ch}^3 x} = \frac{2}{\operatorname{sh} x} - \frac{1}{3\operatorname{sh}^3 x} + \frac{\operatorname{sh} x}{2\operatorname{ch}^2 x} + \frac{5}{2} \operatorname{arctg} \operatorname{sh} x.$

64. $\int \frac{dx}{\operatorname{sh}^4 x \operatorname{ch}^4 x} = 8\operatorname{cth} 2x - \frac{8}{3} \operatorname{cth}^3 2x.$

2.424

1. $\int \operatorname{th}^p x\, dx = -\frac{\operatorname{th}^{p-1} x}{p-1} + \int \operatorname{th}^{p-2} x\, dx \qquad [p \neq 1].$

2. $\int \operatorname{th}^{2n+1} x\, dx = \sum_{k=1}^{n} \frac{(-1)^{k-1}}{2k} \binom{n}{k} \frac{1}{\operatorname{ch}^{2k} x} + \ln \operatorname{ch} x;$

$$= -\sum_{k=1}^{n} \frac{\operatorname{th}^{2n-2k+2} x}{2n-2k+2} + \ln \operatorname{ch} x.$$

3. $\int \operatorname{th}^{2n} x\, dx = -\sum_{k=1}^{n} \frac{\operatorname{th}^{2n-2k+1} x}{2n-2k+1} + x.$ GU ((351))(12)

4. $\int \operatorname{cth}^p x\, dx = -\frac{\operatorname{cth}^{p-1} x}{p-1} + \int \operatorname{cth}^{p-2} x\, dx \qquad [p \neq 1].$

5. $\int \operatorname{cth}^{2n+1} x\, dx = -\sum_{k=1}^{n} \frac{1}{2n} \binom{n}{k} \frac{1}{\operatorname{sh}^{2k} x} + \ln \operatorname{sh} x;$

$$= -\sum_{k=1}^{n} \frac{\operatorname{cth}^{2n-2k+2} x}{2n-2k+2} + \ln \operatorname{sh} x.$$

6. $\int \operatorname{cth}^{2n} x\, dx = -\sum_{k=1}^{n} \frac{\operatorname{cth}^{2n-2k+1} x}{2n-2k+1} + x.$ GU ((351))(14)

For formulas containing powers of $\operatorname{th} x$ and $\operatorname{cth} x$ equal to $n = 1, 2, 3, 4$, see **2.423** 17., **2.423** 22., **2.423** 27., **2.423** 32., **2.423** 33., **2.423** 38., **2.423** 43., **2.423** 48..

Powers of hyperbolic functions and hyperbolic functions of linear functions of the argument

2.425

1. $\int \operatorname{sh}(ax+b) \operatorname{sh}(cx+d)\, dx = \frac{1}{2(a+c)} \operatorname{sh}[(a+c)x+b+d] - \frac{1}{2(a-c)} \operatorname{sh}[(a-c)x+b-d] \qquad [a^2 \neq c^2].$ GU ((352))(2a)

2. $\int \operatorname{sh}(ax+b) \operatorname{ch}(cx+d)\, dx = \frac{1}{2(a+c)} \operatorname{ch}[(a+c)x+b+d] + \frac{1}{2(a-c)} \operatorname{ch}[(a-c)x+b-d] \qquad [a^2 \neq c^2].$ GU ((352))(2c)

3. $\int \operatorname{ch}(ax+b) \operatorname{ch}(cx+d)\, dx = \frac{1}{2(a+c)} \operatorname{sh}[(a+c)x+b+d] + \frac{1}{2(a-c)} \operatorname{sh}[(a-c)x+b-d] \qquad [a^2 \neq c^2].$ GU ((352))(2b)

When $a = c$:

4. $$\int \operatorname{sh}(ax+b)\operatorname{sh}(ax+d)\,dx = -\frac{x}{2}\operatorname{ch}(b-d)+\frac{1}{4a}\operatorname{sh}(2ax+b+d).$$ GU ((352))(3a)

5. $$\int \operatorname{sh}(ax+b)\operatorname{ch}(ax+d)\,dx = \frac{x}{2}\operatorname{sh}(b-d)+\frac{1}{4a}\operatorname{ch}(2ax+b+d).$$ GU ((352))(3c)

6. $$\int \operatorname{ch}(ax+b)\operatorname{ch}(ax+d)\,dx = \frac{x}{2}\operatorname{ch}(b-d)+\frac{1}{4a}\operatorname{sh}(2ax+b+d).$$ GU ((352))(3b)

2.426

1. $$\int \operatorname{sh} ax \operatorname{sh} bx \operatorname{sh} cx\,dx = \frac{\operatorname{ch}(a+b+c)x}{4(a+b+c)} - \frac{\operatorname{ch}(-a+b+c)x}{4(-a+b+c)} - \frac{\operatorname{ch}(a-b+c)x}{4(a-b+c)} - \frac{\operatorname{ch}(a+b-c)x}{4(a+b-c)}.$$ GU ((352))(4a)

2. $$\int \operatorname{sh} ax \operatorname{sh} bx \operatorname{ch} cx\,dx = \frac{\operatorname{sh}(a+b+c)x}{4(a+b+c)} - \frac{\operatorname{sh}(-a+b+c)x}{4(-a+b+c)} - \frac{\operatorname{sh}(a-b+c)x}{4(a-b+c)} + \frac{\operatorname{sh}(a+b-c)x}{4(a+b-c)}.$$ GU ((352))(4b)

3. $$\int \operatorname{sh} ax \operatorname{ch} bx \operatorname{ch} cx\,dx = \frac{\operatorname{ch}(a+b+c)x}{4(a+b+c)} - \frac{\operatorname{ch}(-a+b+c)x}{4(-a+b+c)} + \frac{\operatorname{ch}(a-b+c)x}{4(a-b+c)} + \frac{\operatorname{ch}(a+b-c)x}{4(a+b-c)}.$$ GU ((352))(4c)

4. $$\int \operatorname{ch} ax \operatorname{ch} bx \operatorname{ch} cx\,dx = \frac{\operatorname{sh}(a+b+c)x}{4(a+b+c)} + \frac{\operatorname{sh}(-a+b+c)x}{4(-a+b+c)} + \frac{\operatorname{sh}(a-b+c)x}{4(a-b+c)} + \frac{\operatorname{sh}(a+b-c)x}{4(a+b-c)}.$$ GU ((352))(4d)

2.427

1. $$\int \operatorname{sh}^p x \operatorname{sh} ax\,dx = \frac{1}{p+a}\left\{\operatorname{sh}^p x \operatorname{ch} ax - p\int \operatorname{sh}^{p-1} x \operatorname{ch}(a-1)x\,dx\right\}.$$

2. $$\int \operatorname{sh}^p x \operatorname{sh}(2n+1)x\,dx = \frac{\Gamma(p+1)}{\Gamma\left(\frac{p+3}{2}+n\right)} \times \left\{\sum_{k=0}^{n-1}\left[\frac{\Gamma\left(\frac{p+1}{2}+n-2k\right)}{2^{2k+1}\Gamma(p-2k+1)}\operatorname{sh}^{p-2k} x \operatorname{ch}(2n-2k+1)x - \frac{\Gamma\left(\frac{p-1}{2}+n-2k\right)}{2^{2k+2}\Gamma(p-2k)}\operatorname{sh}^{p-2k-1} x \operatorname{sh}(2n-2k)x\right] + \frac{\Gamma\left(\frac{p+3}{2}-n\right)}{2^{2n}\Gamma(p+1-2n)}\int \operatorname{sh}^{p-2n} x \operatorname{sh} x\,dx\right\}$$

[p is not a negative integer].

3. $$\int \operatorname{sh}^p x \operatorname{sh} 2nx\,dx = \frac{\Gamma(p+1)}{\Gamma\left(\frac{p}{2}+n+1\right)} \times$$
$$\times \sum_{k=0}^{n-1}\left[\frac{\Gamma\left(\frac{p}{2}+n-2k\right)}{2^{2k+1}\Gamma(p-2k+1)} \operatorname{sh}^{p-2k} x \operatorname{ch}(2n-2k)x - \right.$$
$$\left. - \frac{\Gamma\left(\frac{p}{2}+n-2k-1\right)}{2^{2k+2}\Gamma(p-2k)} \operatorname{sh}^{p-2k-1} x \operatorname{sh}(2n-2k-1)x\right]$$

[p is not a negative integer]. GU ((352))(5)a

2.428

1. $$\int \operatorname{sh}^p x \operatorname{ch} ax\,dx = \frac{1}{p+a}\left\{\operatorname{sh}^p x \operatorname{sh} ax - p\int \operatorname{sh}^{p-1} x \operatorname{sh}(a-1)x\,dx\right\}.$$

2. $$\int \operatorname{sh}^p x \operatorname{ch}(2n+1)x\,dx = \frac{\Gamma(p+1)}{\Gamma\left(\frac{p+3}{2}+n\right)} \times$$
$$\times \left\{\sum_{k=0}^{n-1}\left[\frac{\Gamma\left(\frac{p+1}{2}+n-2k\right)}{2^{2k+1}\Gamma(p-2k+1)} \operatorname{sh}^{p-2k} x \operatorname{sh}(2n-2k+1)x - \right.\right.$$
$$\left. - \frac{\Gamma\left(\frac{p-1}{2}+n-2k\right)}{2^{2k+2}\Gamma(p-2k)} \operatorname{sh}^{p-2k-1} x \operatorname{ch}(2n-2k)x\right] +$$
$$\left. + \frac{\Gamma\left(\frac{p+3}{2}-n\right)}{2^{2n}\Gamma(p+1-2n)}\int \operatorname{sh}^{p-2n} x \operatorname{ch} x\,dx\right\}$$

[p is not a negative integer].

3. $$\int \operatorname{sh}^p x \operatorname{ch} 2nx\,dx = \frac{\Gamma(p+1)}{\Gamma\left(\frac{p}{2}+n+1\right)} \times$$
$$\times \left\{\sum_{k=0}^{n-1}\left[\frac{\Gamma\left(\frac{p}{2}+n-2k\right)}{2^{2k+1}\Gamma(p-2k+1)} \operatorname{sh}^{p-2k} x \operatorname{sh}(2n-2k)x - \right.\right.$$
$$\left.\left. - \frac{\Gamma\left(\frac{p}{2}+n-2k-1\right)}{2^{2k+2}\Gamma(p-2k)} \operatorname{sh}^{p-2k-1} x \operatorname{ch}(2n-2k-1)x\right] + \frac{\Gamma\left(\frac{p}{2}-n+1\right)}{2^{2n}\Gamma(p+1-2n)}\int \operatorname{sh}^{p-2n} x\,dx\right\}$$

[p is not a negative integer]. GU ((352))(6)a

2.429

1. $$\int \operatorname{ch}^p x \operatorname{sh} ax\,dx = \frac{1}{p+a}\left\{\operatorname{ch}^p x \operatorname{ch} ax + p\int \operatorname{ch}^{p-1} x \operatorname{sh}(a-1)x\,dx\right\}.$$

2. $$\int \operatorname{ch}^p x \operatorname{sh}(2n+1)x\,dx = \frac{\Gamma(p+1)}{\Gamma\left(\frac{p+3}{2}+n\right)}\left\{\sum_{k=0}^{n-1}\frac{\Gamma\left(\frac{p+1}{2}+n-k\right)}{2^{k+1}\Gamma(p-k+1)} \times\right.$$
$$\left. \times \operatorname{ch}^{p-k} x \operatorname{ch}(2n-k+1)x + \frac{\Gamma\left(\frac{p+3}{2}\right)}{2^{n}\Gamma(p-n+1)}\int \operatorname{ch}^{p-n} x \operatorname{sh}(n+1)x\,dx\right\}$$

[p is not a negative integer].

3. $$\int \operatorname{ch}^p x \operatorname{sh} 2nx\, dx = \frac{\Gamma(p+1)}{\Gamma\left(\frac{p}{2}+n+1\right)} \left\{ \sum_{k=0}^{n-1} \frac{\Gamma\left(\frac{p}{2}+n-k\right)}{2^{k+1}\Gamma(p-k+1)} \times \right.$$
$$\left. \times \operatorname{ch}^{p-k} x \operatorname{ch}(2n-k)x + \frac{\Gamma\left(\frac{p}{2}+1\right)}{2^n\Gamma(p-n+1)} \int \operatorname{ch}^{p-n} x \operatorname{sh} nx\, dx \right\}$$

[p is not a negative integer]. GU ((352))(7)a

2.431

1. $$\int \operatorname{ch}^p x \operatorname{ch} ax\, dx = \frac{1}{p+a} \left\{ \operatorname{ch}^p x \operatorname{sh} ax + p \int \operatorname{ch}^{p-1} x \operatorname{ch}(a-1)x\, dx \right\}.$$

2. $$\int \operatorname{ch}^p x \operatorname{ch}(2n+1)x\, dx = \frac{\Gamma(p+1)}{\Gamma\left(\frac{p+3}{2}+n\right)} \left\{ \sum_{k=0}^{n-1} \frac{\Gamma\left(\frac{p+1}{2}+n-k\right)}{2^{k+1}\Gamma(p-k+1)} \times \right.$$
$$\left. \times \operatorname{ch}^{p-k} x \operatorname{sh}(2n-k+1)x + \frac{\Gamma\left(\frac{p+3}{2}\right)}{2^n\Gamma(p-n+1)} \int \operatorname{ch}^{p-n} x \operatorname{ch}(n+1)x\, dx \right\}$$

[p is not a negative integer].

3. $$\int \operatorname{ch}^p x \operatorname{ch} 2nx\, dx = \frac{\Gamma(p+1)}{\Gamma\left(\frac{p}{2}+n+1\right)} \left\{ \sum_{k=0}^{n-1} \frac{\Gamma\left(\frac{p}{2}+n-k\right)}{2^{k+1}\Gamma(p-k+1)} \times \right.$$
$$\left. \times \operatorname{ch}^{p-k} x \operatorname{sh}(2n-k)x + \frac{\Gamma\left(\frac{p}{2}+1\right)}{2^n\Gamma(p-n+1)} \int \operatorname{ch}^{p-n} x \operatorname{ch} nx\, dx \right\}$$

[p is not a negative integer]. GU ((352))(8)a

2.432

1. $$\int \operatorname{sh}(n+1)x \operatorname{sh}^{n-1} x\, dx = \frac{1}{n} \operatorname{sh}^n x \operatorname{sh} nx.$$

2. $$\int \operatorname{sh}(n+1)x \operatorname{ch}^{n-1} x\, dx = \frac{1}{n} \operatorname{ch}^n x \operatorname{ch} nx.$$

3. $$\int \operatorname{ch}(n+1)x \operatorname{sh}^{n-1} x\, dx = \frac{1}{n} \operatorname{sh}^n x \operatorname{ch} nx.$$

4. $$\int \operatorname{ch}(n+1)x \operatorname{ch}^{n-1} x\, dx = \frac{1}{n} \operatorname{ch}^n x \operatorname{sh} nx.$$

2.433

1. $$\int \frac{\operatorname{sh}(2n+1)x}{\operatorname{sh} x}\, dx = 2\sum_{k=0}^{n-1} \frac{\operatorname{sh}(2n-2k)x}{2n-2k} + x.$$

2. $$\int \frac{\operatorname{sh} 2nx}{\operatorname{sh} x}\, dx = 2\sum_{k=0}^{n-1} \frac{\operatorname{sh}(2n-2k-1)x}{2n-2k-1}.$$ GU ((352))(5d)

3. $$\int \frac{\operatorname{ch}(2n+1)x}{\operatorname{sh} x}\, dx = 2\sum_{k=0}^{n-1} \frac{\operatorname{ch}(2n-2k)x}{2n-2k} + \ln \operatorname{sh} x.$$

4. $\int \frac{\operatorname{ch} 2nx}{\operatorname{sh} x} dx = 2 \sum_{k=0}^{n-1} \frac{\operatorname{ch}(2n-2k-1)x}{2n-2k-1} + \ln \operatorname{th} \frac{x}{2}.$ GU ((352))(6d)

5. $\int \frac{\operatorname{sh}(2n+1)x}{\operatorname{ch} x} dx = 2 \sum_{k=0}^{n-1} (-1)^k \frac{\operatorname{ch}(2n-2k)x}{2n-2k} + (-1)^n \ln \operatorname{ch} x.$

6. $\int \frac{\operatorname{sh} 2nx}{\operatorname{ch} x} dx = 2 \sum_{k=0}^{n-1} (-1)^k \frac{\operatorname{ch}(2n-2k-1)x}{2n-2k-1}.$ GU ((352))(7d)

7. $\int \frac{\operatorname{ch}(2n+1)x}{\operatorname{ch} x} dx = 2 \sum_{k=0}^{n-1} (-1)^k \frac{\operatorname{sh}(2n-2k)x}{2n-2k} + (-1)^n x.$

8. $\int \frac{\operatorname{ch} 2nx}{\operatorname{ch} x} dx = 2 \sum_{k=0}^{n-1} (-1)^k \frac{\operatorname{sh}(2n-2k-1)x}{2n-2k-1} + (-1)^n \arcsin(\operatorname{th} x).$

GU ((352))(8d)

9. $\int \frac{\operatorname{sh} 2x}{\operatorname{sh}^n x} dx = -\frac{2}{(n-2)\operatorname{sh}^{n-2} x}.$

For $n = 2$:

10. $\int \frac{\operatorname{sh} 2x}{\operatorname{sh}^2 x} dx = 2 \ln \operatorname{sh} x.$

11. $\int \frac{\operatorname{sh} 2x\, dx}{\operatorname{ch}^n x} = \frac{2}{(2-n)\operatorname{ch}^{n-2} x}.$

For $n = 2$:

12. $\int \frac{\operatorname{sh} 2x}{\operatorname{ch}^2 x} dx = 2 \ln \operatorname{ch} x.$

13. $\int \frac{\operatorname{ch} 2x}{\operatorname{sh} x} dx = 2 \operatorname{ch} x + \ln \operatorname{th} \frac{x}{2}.$

14. $\int \frac{\operatorname{ch} 2x}{\operatorname{sh}^2 x} dx = -\operatorname{cth} x + 2x.$

15. $\int \frac{\operatorname{ch} 2x}{\operatorname{sh}^3 x} dx = -\frac{\operatorname{ch} x}{2 \operatorname{sh}^2 x} + \frac{3}{2} \ln \operatorname{th} \frac{x}{2}.$

16. $\int \frac{\operatorname{ch} 2x}{\operatorname{ch} x} dx = 2 \operatorname{sh} x - \arcsin(\operatorname{th} x).$

17. $\int \frac{\operatorname{ch} 2x}{\operatorname{ch}^2 x} dx = -\operatorname{th} x + 2x.$

18. $\int \frac{\operatorname{ch} 2x}{\operatorname{ch}^3 x} dx = -\frac{\operatorname{sh} x}{2 \operatorname{ch}^2 x} + \frac{3}{2} \arcsin(\operatorname{th} x).$

19. $\int \frac{\operatorname{sh} 3x}{\operatorname{sh} x} dx = x + \operatorname{sh} 2x.$

20. $\int \frac{\operatorname{sh} 3x}{\operatorname{sh}^2 x} dx = 3 \ln \operatorname{th} \frac{x}{2} + 4 \operatorname{ch} x.$

21. $\int \frac{\operatorname{sh} 3x}{\operatorname{sh}^3 x} dx = -3 \operatorname{cth} x + 4x.$

22. $\int \frac{\operatorname{sh} 3x}{\operatorname{ch}^n x} dx = \frac{4}{(3-n)\operatorname{ch}^{n-3} x} - \frac{1}{(1-n)\operatorname{ch}^{n-1} x}.$

For $n=1$ and $n=3$:

23. $$\int \frac{\operatorname{sh} 3x}{\operatorname{ch} x}\,dx = 2\operatorname{sh}^2 x - \ln \operatorname{ch} x.$$

24. $$\int \frac{\operatorname{sh} 3x}{\operatorname{ch}^3 x}\,dx = \frac{1}{2\operatorname{ch}^2 x} + 4\ln \operatorname{ch} x.$$

25. $$\int \frac{\operatorname{ch} 3x}{\operatorname{sh}^n x}\,dx = \frac{4}{(3-n)\operatorname{sh}^{n-3} x} + \frac{1}{(1-n)\operatorname{sh}^{n-1} x}.$$

For $n=1$ and $n=3$:

26. $$\int \frac{\operatorname{ch} 3x}{\operatorname{sh} x}\,dx = 2\operatorname{sh}^2 x + \ln \operatorname{sh} x.$$

27. $$\int \frac{\operatorname{ch} 3x}{\operatorname{sh}^3 x}\,dx = -\frac{1}{2\operatorname{sh}^2 x} + 4\ln \operatorname{sh} x.$$

28. $$\int \frac{\operatorname{ch} 3x}{\operatorname{ch} x}\,dx = \operatorname{sh} 2x - x.$$

29. $$\int \frac{\operatorname{ch} 3x}{\operatorname{ch}^2 x}\,dx = 4\operatorname{sh} x - 3\arcsin(\operatorname{th} x).$$

30. $$\int \frac{\operatorname{ch} 3x}{\operatorname{ch}^3 x}\,dx = 4x - 3\operatorname{th} x.$$

2.44-2.45 Rational functions of hyperbolic functions

2.441

1. $$\int \frac{A+B\operatorname{sh} x}{(a+b\operatorname{sh} x)^n}\,dx = \frac{aB-bA}{(n-1)(a^2+b^2)}\cdot\frac{\operatorname{ch} x}{(a+b\operatorname{sh} x)^{n-1}} + \frac{1}{(n-1)(a^2+b^2)}\int \frac{(n-1)(aA+bB)+(n-2)(aB-bA)\operatorname{sh} x}{(a+b\operatorname{sh} x)^{n-1}}\,dx.$$

For $n=1$:

2. $$\int \frac{A+B\operatorname{sh} x}{a+b\operatorname{sh} x}\,dx = \frac{B}{b}x - \frac{aB-bA}{b}\int \frac{dx}{a+b\operatorname{sh} x}$$ (see **2.441** 3.).

3. $$\int \frac{dx}{a+b\operatorname{sh} x} = \frac{1}{\sqrt{a^2+b^2}}\ln \frac{a\operatorname{th}\frac{x}{2} - b + \sqrt{a^2+b^2}}{a\operatorname{th}\frac{x}{2} - b - \sqrt{a^2+b^2}};$$

$$= \frac{2}{\sqrt{a^2+b^2}}\operatorname{Arth}\frac{a\operatorname{th}\frac{x}{2} - b}{\sqrt{a^2+b^2}}.$$

2.442

1. $$\int \frac{A+B\operatorname{ch} x}{(a+b\operatorname{sh} x)^n}\,dx = -\frac{B}{(n-1)b(a+b\operatorname{sh} x)^{n-1}} + A\int \frac{dx}{(a+b\operatorname{sh} x)^n}.$$

For $n=1$:

2. $$\int \frac{A+B\operatorname{ch} x}{a+b\operatorname{sh} x}\,dx = \frac{B}{b}\ln(a+b\operatorname{sh} x) + A\int \frac{dx}{a+b\operatorname{sh} x}$$ (see **2.441** 3.)

2.443

1. $$\int \frac{A+B\operatorname{ch} x}{(a+b\operatorname{ch} x)^n}\,dx = \frac{aB-bA}{(n-1)(a^2-b^2)}\cdot\frac{\operatorname{sh} x}{(a+b\operatorname{ch} x)^{n-1}} + \frac{1}{(n-1)(a^2-b^2)}\int \frac{(n-1)(aA-bB)+(n-2)(aB-bA)\operatorname{ch} x}{(a+b\operatorname{ch} x)^{n-1}}\,dx.$$

For $n=1$:

2. $\int \frac{A+B \operatorname{ch} x}{a+b \operatorname{ch} x} dx = \frac{B}{b} x - \frac{aB-bA}{b} \int \frac{dx}{a+b \operatorname{ch} x}$ (see **2.443** 3.).

3. $\int \frac{dx}{a+b \operatorname{ch} x} = \frac{1}{\sqrt{b^2-a^2}} \arcsin \frac{b+a \operatorname{ch} x}{a+b \operatorname{ch} x} \quad [b^2 > a^2,\ x < 0];$

$$= -\frac{1}{\sqrt{b^2-a^2}} \arcsin \frac{b+a \operatorname{ch} x}{a+b \operatorname{ch} x} \quad [b^2 > a^2,\ x > 0];$$

$$= \frac{1}{\sqrt{a^2-b^2}} \ln \frac{a+b+\sqrt{a^2-b^2} \operatorname{th} \frac{x}{2}}{a+b-\sqrt{a^2-b^2} \operatorname{th} \frac{x}{2}} \quad [a^2 > b^2].$$

2.444

1. $\int \frac{dx}{\operatorname{ch} a + \operatorname{ch} x} = \operatorname{cosech} a \left[\ln \operatorname{ch} \frac{x+a}{2} - \ln \operatorname{ch} \frac{x-a}{2} \right];$

$$= 2 \operatorname{cosech} a \operatorname{Arth} \left(\operatorname{th} \frac{x}{2} \operatorname{th} \frac{a}{2} \right).$$

2. $\int \frac{dx}{\cos a + \operatorname{ch} x} = 2 \operatorname{cosec} a \operatorname{arctg} \left(\operatorname{th} \frac{x}{2} \operatorname{tg} \frac{a}{2} \right).$

2.445

1. $\int \frac{A+B \operatorname{sh} x}{(a+b \operatorname{ch} x)^n} dx = -\frac{B}{(n-1) b (a+b \operatorname{ch} x)^{n-1}} + A \int \frac{dx}{(a+b \operatorname{ch} x)^n}.$

For $n=1$:

2. $\int \frac{A+B \operatorname{sh} x}{a+b \operatorname{ch} x} dx = \frac{B}{b} \ln (a+b \operatorname{ch} x) + A \int \frac{dx}{a+b \operatorname{ch} x}$ (see **2.443** 3.)

In evaluating definite integrals by use of formulas **2.441** – **2.443** and **2.445**, one may not take the integral over points at which the integrand becomes infinite, that is, over the points

$$x = \operatorname{Arsh} \left(-\frac{a}{b} \right)$$

in formulas **2.441** or **2.442** or over the points

$$x = \operatorname{Arch} \left(-\frac{a}{b} \right)$$

in formulas **2.443** or **2.445**. Formulas **2.443** are not applicable for $a^2 = b^2$. Instead, we may use the following formulas in these cases:

2.446

1. $\int \frac{A+B \operatorname{ch} x}{(\varepsilon + \operatorname{ch} x)^n} dx = \frac{B \operatorname{sh} x}{(1-n)(\varepsilon + \operatorname{ch} x)^n} +$

$$+ \left(\varepsilon A + \frac{n}{n-1} B \right) \frac{(n-1)!}{(2n-1)!!} \operatorname{sh} x \sum_{k=0}^{n-1} \frac{(2n-2k-3)!!}{(n-k-1)!} \cdot \frac{\varepsilon^k}{(\varepsilon + \operatorname{ch} x)^{n-k}} \quad [\varepsilon = \pm 1,\ n > 1].$$

For $n=1$:

2. $\int \frac{A+B \operatorname{ch} x}{\varepsilon + \operatorname{ch} x} dx = Bx + (\varepsilon A - B) \frac{\operatorname{ch} x - \varepsilon}{\operatorname{sh} x} \quad [\varepsilon = \pm 1].$

2.447

1. $$\int \frac{\operatorname{sh} x\, dx}{a \operatorname{ch} x + b \operatorname{sh} x} = \frac{a \ln \operatorname{ch}\left(x + \operatorname{Arth} \frac{b}{a}\right) - bx}{a^2 - b^2} \quad [a > |b|];$$
$$= \frac{bx - a \ln \operatorname{sh}\left(x + \operatorname{Arth} \frac{a}{b}\right)}{b^2 - a^2} \quad [b > |a|].$$ MZ 215

For $a = b = 1$:

2. $$\int \frac{\operatorname{sh} x\, dx}{\operatorname{ch} x + \operatorname{sh} x} = \frac{x}{2} + \frac{1}{4} e^{-2x}.$$

For $a = -b = 1$:

3. $$\int \frac{\operatorname{sh} x\, dx}{\operatorname{ch} x - \operatorname{sh} x} = -\frac{x}{2} + \frac{1}{4} e^{2x}.$$ MZ 215

2.448

1. $$\int \frac{\operatorname{ch} x\, dx}{a \operatorname{ch} x + b \operatorname{sh} x} = \frac{ax - b \ln \operatorname{ch}\left(x + \operatorname{Arth} \frac{b}{a}\right)}{a^2 - b^2} \quad [a > |b|];$$
$$= \frac{-ax + b \ln \operatorname{sh}\left(x + \operatorname{Arth} \frac{a}{b}\right)}{b^2 - a^2} \quad [b > |a|].$$

For $a = b = 1$:

2. $$\int \frac{\operatorname{ch} x\, dx}{\operatorname{ch} x + \operatorname{sh} x} = \frac{x}{2} - \frac{1}{4} e^{-2x}.$$

For $a = -b = 1$:

3. $$\int \frac{\operatorname{ch} x\, dx}{\operatorname{ch} x - \operatorname{sh} x} = \frac{x}{2} + \frac{1}{4} e^{2x}.$$ MZ 214, 215

2.449

1. $$\int \frac{dx}{(a \operatorname{ch} x + b \operatorname{sh} x)^n} = \frac{1}{\sqrt{(a^2 - b^2)^n}} \int \frac{dx}{\operatorname{ch}^n \left(x + \operatorname{Arth} \frac{b}{a}\right)} \quad [a > |b|];$$
$$= \frac{1}{\sqrt{(b^2 - a^2)^n}} \int \frac{dx}{\operatorname{ch}^n \left(x + \operatorname{Arth} \frac{a}{b}\right)} \quad [b > |a|].$$

For $n = 1$:

2. $$\int \frac{dx}{a \operatorname{ch} x + b \operatorname{sh} x} = \frac{1}{\sqrt{a^2 - b^2}} \operatorname{arctg} \left| \operatorname{sh}\left(x + \operatorname{Arth} \frac{b}{a}\right) \right| \quad [a > |b|];$$
$$= \frac{1}{\sqrt{b^2 - a^2}} \ln \left| \operatorname{th} \frac{x + \operatorname{Arth} \frac{a}{b}}{2} \right| \quad [b > |a|].$$

For $a = b = 1$:

3. $$\int \frac{dx}{\operatorname{ch} x + \operatorname{sh} x} = -e^{-x} = \operatorname{sh} x - \operatorname{ch} x.$$

For $a = -b = 1$:

4. $$\int \frac{dx}{\operatorname{ch} x - \operatorname{sh} x} = e^{x} = \operatorname{sh} x + \operatorname{ch} x.$$ MZ 214

2.451

1. $$\int \frac{A+B\operatorname{ch}x+C\operatorname{sh}x}{(a+b\operatorname{ch}x+c\operatorname{sh}x)^n}\,dx = \frac{Bc-Cb+(Ac-Ca)\operatorname{ch}x+(Ab-Ba)\operatorname{sh}x}{(1-n)(a^2-b^2+c^2)(a+b\operatorname{ch}x+c\operatorname{sh}x)^{n-1}} + \frac{1}{(n-1)(a^2-b^2+c^2)}\times$$
$$\times\int \frac{(n-1)(Aa-Bb+Cc)-(n-2)(Ab-Ba)\operatorname{ch}x-(n-2)(Ac-Ca)\operatorname{sh}x}{(a+b\operatorname{ch}x+c\operatorname{sh}x)^{n-1}}\,dx$$
$[a^2+c^2\neq b^2]$;
$$= \frac{Bc-Cb-Ca\operatorname{ch}x-Ba\operatorname{sh}x}{(n-1)a(a+b\operatorname{ch}x+c\operatorname{sh}x)^n} + \left[\frac{A}{a}+\frac{n(Bb-Cc)}{(n-1)a^2}\right](c\operatorname{ch}x+b\operatorname{sh}x)\frac{(n-1)!}{(2n-1)!!}\times$$
$$\times\sum_{k=0}^{n-1}\frac{(2n-2k-3)!!}{(n-k-1)!\,a^k}\,\frac{1}{(a+b\operatorname{ch}x+c\operatorname{sh}x)^{n-k}}$$
$[a^2+c^2=b^2]$.

2. $$\int \frac{A+B\operatorname{ch}x+C\operatorname{sh}x}{a+b\operatorname{ch}x+c\operatorname{sh}x}\,dx = \frac{Cb-Bc}{b^2-c^2}\ln(a+b\operatorname{ch}x+c\operatorname{sh}x) + \frac{Bb-Cc}{b^2-c^2}x + \left(A-a\frac{Bb-Cc}{b^2-c^2}\right)\int\frac{dx}{a+b\operatorname{ch}x+c\operatorname{sh}x}$$
$[b^2\neq c^2]$ (see **2.451 4**).

3. $$\int \frac{A+B\operatorname{ch}x+C\operatorname{sh}x}{a+b\operatorname{ch}x\pm b\operatorname{sh}x}\,dx = \frac{C\mp B}{2a}(\operatorname{ch}x\mp\operatorname{sh}x) + \left[\frac{A}{a}-\frac{(B\mp C)b}{2a^2}\right]x + \left[\frac{C\pm B}{2b}\pm\frac{A}{a}-\frac{(C\mp B)b}{2a^2}\right]\ln(a+b\operatorname{ch}x\pm b\operatorname{sh}x)$$
$[ab\neq 0]$.

4. $$\int \frac{dx}{a+b\operatorname{ch}x+c\operatorname{sh}x} = \frac{2}{\sqrt{b^2-a^2-c^2}}\operatorname{arctg}\frac{(b-a)\operatorname{th}\frac{x}{2}+c}{\sqrt{b^2-a^2-c^2}}$$
$[b^2>a^2+c^2$ and $a\neq b]$;
$$= \frac{1}{\sqrt{a^2-b^2+c^2}}\ln\frac{(a-b)\operatorname{th}\frac{x}{2}-c+\sqrt{a^2-b^2+c^2}}{(a-b)\operatorname{th}\frac{x}{2}-c-\sqrt{a^2-b^2+c^2}}$$
$[b^2<a^2+c^2$ and $a\neq b]$;
$$= \frac{1}{c}\ln\left(a+c\operatorname{th}\frac{x}{2}\right) \qquad [a=b,\ c\neq 0];$$
$$= \frac{2}{(a-b)\operatorname{th}\frac{x}{2}+c} \qquad [b^2=a^2+c^2].$$
GU ((351))(18)

2.452

1. $$\int \frac{A+B\operatorname{ch}x+C\operatorname{sh}x}{(a_1+b_1\operatorname{ch}x+c_1\operatorname{sh}x)(a_2+b_2\operatorname{ch}x+c_2\operatorname{sh}x)}\,dx = A_0\ln\frac{a_1+b_1\operatorname{ch}x+c_1\operatorname{sh}x}{a_2+b_2\operatorname{ch}x+c_2\operatorname{sh}x} + A_1\int\frac{dx}{a_1+b_1\operatorname{ch}x+c_1\operatorname{sh}x} + A_2\int\frac{dx}{a_2+b_2\operatorname{ch}x+c_2\operatorname{sh}x},$$

where

$$A_0 = \frac{\begin{vmatrix} a_1 & b_1 & c_1 \\ A & B & C \\ a_2 & b_2 & c_2 \end{vmatrix}}{\begin{vmatrix} a_1 & b_1 \\ a_2 & b_2 \end{vmatrix}^2 + \begin{vmatrix} b_1 & c_1 \\ b_2 & c_2 \end{vmatrix}^2 - \begin{vmatrix} c_1 & a_1 \\ c_2 & a_2 \end{vmatrix}^2}, \qquad A_1 = \frac{\begin{vmatrix} a_1 & b_1 & c_1 \\ \begin{vmatrix} b_1 & c_1 \\ B & C \end{vmatrix} & \begin{vmatrix} c_1 & a_1 \\ C & A \end{vmatrix} & \begin{vmatrix} a_1 & b_1 \\ A & B \end{vmatrix} \\ a_2 & b_2 & c_2 \end{vmatrix}}{\begin{vmatrix} a_1 & b_1 \\ a_2 & b_2 \end{vmatrix}^2 + \begin{vmatrix} b_1 & c_1 \\ b_2 & c_2 \end{vmatrix}^2 - \begin{vmatrix} c_1 & a_1 \\ c_2 & a_2 \end{vmatrix}^2},$$

$$A_2 = \frac{\begin{vmatrix} a_1 & b_1 & c_1 \\ \begin{vmatrix} C & B \\ c_2 & b_2 \end{vmatrix} & \begin{vmatrix} C & A \\ c_2 & a_2 \end{vmatrix} & \begin{vmatrix} B & A \\ b_2 & a_2 \end{vmatrix} \\ a_2 & b_2 & c_2 \end{vmatrix}}{\begin{vmatrix} a_1 & b_1 \\ a_2 & b_2 \end{vmatrix}^2 + \begin{vmatrix} b_1 & c_1 \\ b_2 & c_2 \end{vmatrix}^2 - \begin{vmatrix} c_1 & a_1 \\ c_2 & a_2 \end{vmatrix}^2},$$

$$\left[\begin{vmatrix} a_1 & b_1 \\ a_2 & b_2 \end{vmatrix}^2 + \begin{vmatrix} b_1 & c_1 \\ b_2 & c_2 \end{vmatrix}^2 \neq \begin{vmatrix} c_1 & a_1 \\ c_2 & a_2 \end{vmatrix}^2 \right].$$ GU ((351))(19)

2. $$\int \frac{A \operatorname{ch}^2 x + 2B \operatorname{sh} x \operatorname{ch} x + C \operatorname{sh}^2 x}{a \operatorname{ch}^2 x + 2b \operatorname{sh} x \operatorname{ch} x + c \operatorname{sh}^2 x} dx =$$
$$= \frac{1}{4b^2 - (a+c)^2} \{[4Bb - (A+C)(a+c)] x +$$
$$+ [(A+C) b - B(a+c)] \ln (a \operatorname{ch}^2 x + 2b \operatorname{sh} x \operatorname{ch} x + c \operatorname{sh}^2 x) +$$
$$+ [2(A-C) b^2 + 2Bb(a-c) + (Ca - Ac)(a+c)] f(x)\},$$

where

$$f(x) = \frac{1}{2\sqrt{b^2 - ac}} \ln \frac{c \operatorname{th} x + b - \sqrt{b^2 - ac}}{c \operatorname{th} x + b + \sqrt{b^2 - ac}} \qquad [b^2 > ac];$$
$$= \frac{1}{\sqrt{ac - b^2}} \operatorname{arctg} \frac{c \operatorname{th} x + b}{\sqrt{ac - b^2}} \qquad [b^2 < ac];$$
$$= -\frac{1}{c \operatorname{th} x + b} \qquad [b^2 = ac].$$

GU ((351))(24)

2.453

1. $$\int \frac{(A + B \operatorname{sh} x)\, dx}{\operatorname{sh} x (a + b \operatorname{sh} x)} = \frac{1}{a} \left[A \ln \left| \operatorname{th} \frac{x}{2} \right| + (aB - bA) \int \frac{dx}{a + b \operatorname{sh} x} \right]$$
(see **2.441** 3.).

2. $$\int \frac{(A + B \operatorname{sh} x)\, dx}{\operatorname{sh} x (a + b \operatorname{ch} x)} = \frac{A}{a^2 - b^2} \left(a \ln \left| \operatorname{th} \frac{x}{2} \right| + b \ln \left| \frac{a + b \operatorname{ch} x}{\operatorname{sh} x} \right| \right) +$$
$$+ B \int \frac{dx}{a + b \operatorname{ch} x}$$ (see **2.443** 3.).

For $a^2 = b^2 (= 1)$:

3. $$\int \frac{(A + B \operatorname{sh} x)\, dx}{\operatorname{sh} x (1 + \operatorname{ch} x)} = \frac{A}{2} \left(\ln \left| \operatorname{th} \frac{x}{2} \right| - \frac{1}{2} \operatorname{th}^2 \frac{x}{2} \right) + B \operatorname{th} \frac{x}{2}.$$

4. $$\int \frac{(A + B \operatorname{sh} x)\, dx}{\operatorname{sh} x (1 - \operatorname{ch} x)} = \frac{A}{2} \left(-\ln \left| \operatorname{cth} \frac{x}{2} \right| + \frac{1}{2} \operatorname{cth}^2 \frac{x}{2} \right) + B \operatorname{cth} \frac{x}{2}.$$

2.454

1. $$\int \frac{(A + B \operatorname{sh} x)\, dx}{\operatorname{ch} x (a + b \operatorname{sh} x)} = \frac{1}{a^2 + b^2} \left[(Aa + Bb) \operatorname{arctg} (\operatorname{sh} x) + \right.$$
$$\left. + (Ab - Ba) \ln \left| \frac{a + b \operatorname{sh} x}{\operatorname{ch} x} \right| \right].$$

2. $$\int \frac{(A + B \operatorname{ch} x)\, dx}{\operatorname{sh} x (a + b \operatorname{sh} x)} = \frac{1}{a} \left(A \ln \left| \operatorname{th} \frac{x}{2} \right| + B \ln \left| \frac{\operatorname{sh} x}{a + b \operatorname{sh} x} \right| - Ab \int \frac{dx}{a + b \operatorname{sh} x} \right)$$
(see **2.441** 3.).

2.455

1. $$\int \frac{(A + B \operatorname{ch} x)\, dx}{\operatorname{sh} x (a + b \operatorname{ch} x)} = \frac{1}{a^2 - b^2} \left[(Aa + Bb) \ln \left| \operatorname{th} \frac{x}{2} \right| + \right.$$
$$\left. + (Ab - Ba) \ln \left| \frac{a + b \operatorname{ch} x}{\operatorname{sh} x} \right| \right].$$

For $a^2 = b^2 (= 1)$:

2. $$\int \frac{(A+B \operatorname{ch} x)\, dx}{\operatorname{sh} x (1+\operatorname{ch} x)} = \frac{A+B}{2} \ln \left| \operatorname{th} \frac{x}{2} \right| - \frac{A-B}{4} \operatorname{th}^2 \frac{x}{2}.$$

3. $$\int \frac{(A+B \operatorname{ch} x)\, dx}{\operatorname{sh} x (1-\operatorname{ch} x)} = \frac{A+B}{4} \operatorname{cth}^2 \frac{x}{2} - \frac{A-B}{2} \ln \operatorname{cth} \frac{x}{2}.$$

2.456 $$\int \frac{(A+B \operatorname{ch} x)\, dx}{\operatorname{ch} x (a+b \operatorname{sh} x)} = \frac{A}{a^2+b^2} \left[a \operatorname{arctg} (\operatorname{sh} x) + b \ln \left| \frac{a+b \operatorname{sh} x}{\operatorname{ch} x} \right| \right] + B \int \frac{dx}{a+b \operatorname{sh} x}$$ (see 2.441 3.).

2.457

$$\int \frac{(A+B \operatorname{ch} x)\, dx}{\operatorname{ch} x (a+b \operatorname{ch} x)} = \frac{1}{a} \left[A \operatorname{arctg} \operatorname{sh} x - (Ab - Ba) \int \frac{dx}{a+b \operatorname{ch} x} \right]$$

(see 2.443 3.).

2.458

1. $$\int \frac{dx}{a+b \operatorname{sh}^2 x} = \frac{1}{\sqrt{a(b-a)}} \operatorname{arctg} \left(\sqrt{\frac{b}{a} - 1} \operatorname{th} x \right) \qquad \left[\frac{b}{a} > 1 \right]$$

$$= \frac{1}{\sqrt{a(a-b)}} \operatorname{Arth} \left(\sqrt{1 - \frac{b}{a}} \operatorname{th} x \right)$$

$$\left[0 < \frac{b}{a} < 1 \quad \text{or} \quad \frac{b}{a} < 0 \text{ and } \operatorname{sh}^2 x < -\frac{a}{b} \right];$$

$$= \frac{1}{\sqrt{a(a-b)}} \operatorname{Arcth} \left(\sqrt{1 - \frac{b}{a}} \operatorname{th} x \right) \left[\frac{b}{a} < 0 \text{ and } \operatorname{sh}^2 x > -\frac{a}{b} \right].$$

MZ 195

2. $$\int \frac{dx}{a+b \operatorname{ch}^2 x} = \frac{1}{\sqrt{-a(a+b)}} \operatorname{arctg} \left(\sqrt{-\left(1 + \frac{b}{a}\right)} \operatorname{cth} x \right) \left[\frac{b}{a} < -1 \right];$$

$$= \frac{1}{\sqrt{a(a+b)}} \operatorname{Arth} \left(\sqrt{1 + \frac{b}{a}} \operatorname{cth} x \right)$$

$$\left[-1 < \frac{b}{a} < 0 \text{ and } \operatorname{ch}^2 x > -\frac{a}{b} \right];$$

$$= \frac{1}{\sqrt{a(a+b)}} \operatorname{Arcth} \left(\sqrt{1 + \frac{b}{a}} \operatorname{cth} x \right)$$

$$\left[\frac{b}{a} > 0 \quad \text{or} \quad -1 < \frac{b}{a} < 0 \text{ and } \operatorname{ch}^2 x < -\frac{a}{b} \right].$$ MZ 202

For $a^2 = b^2 = 1$:

3. $$\int \frac{dx}{1+\operatorname{sh}^2 x} = \operatorname{th} x.$$

4. $$\int \frac{dx}{1-\operatorname{sh}^2 x} = \frac{1}{\sqrt{2}} \operatorname{Arth} (\sqrt{2} \operatorname{th} x) \qquad [\operatorname{sh}^2 x < 1];$$

$$= \frac{1}{\sqrt{2}} \operatorname{Arcth} (\sqrt{2} \operatorname{th} x) \qquad [\operatorname{sh}^2 x > 1].$$

5. $$\int \frac{dx}{1+\operatorname{ch}^2 x} = \frac{1}{\sqrt{2}} \operatorname{Arcth} (\sqrt{2} \operatorname{cth} x).$$

6. $$\int \frac{dx}{1-\operatorname{ch}^2 x} = \operatorname{cth} x.$$

2.459

1. $$\int \frac{dx}{(a+b \operatorname{sh}^2 x)^2} = \frac{1}{2a(b-a)} \left[\frac{b \operatorname{sh} x \operatorname{ch} x}{a+b \operatorname{sh}^2 x} + (b-2a) \int \frac{dx}{a+b \operatorname{sh}^2 x} \right]$$

(see 2.458 1.). MZ 196

2. $$\int \frac{dx}{(a+b\,\mathrm{ch}^2 x)^2} = \frac{1}{2a(a+b)}\left[-\frac{b\,\mathrm{sh}\,x\,\mathrm{ch}\,x}{a+b\,\mathrm{ch}^2 x} + (2a+b)\int \frac{dx}{a+b\,\mathrm{ch}^2 x}\right]$$ (see 2.458 2.). MZ 203

3. $$\int \frac{dx}{(a+b\,\mathrm{sh}^2 x)^3} = \frac{1}{8pa^3}\left[\left(3-\frac{2}{p^2}+\frac{3}{p^4}\right)\mathrm{arctg}\,(p\,\mathrm{th}\,x) + \left(3-\frac{2}{p^2}-\frac{3}{p^4}\right)\frac{p\,\mathrm{th}\,x}{1+p^2\,\mathrm{th}^2 x} + \left(1+\frac{2}{p^2}-\frac{1}{p^2}\mathrm{th}^2 x\right)\frac{2p\,\mathrm{th}\,x}{(1+p^2\,\mathrm{th}^2 x)^2}\right]$$ $\left[p^2 = \frac{b}{a} - 1 > 0\right]$;

$$= \frac{1}{8qa^3}\left[\left(3+\frac{2}{q^2}+\frac{3}{q^4}\right)\mathrm{Arth}\,(q\,\mathrm{th}\,x) + \left(3+\frac{2}{q^2}-\frac{3}{q^4}\right)\frac{q\,\mathrm{th}\,x}{1-q^2\,\mathrm{th}^2 x} + \left(1-\frac{2}{q^2}+\frac{1}{q^2}\mathrm{th}^2 x\right)\frac{2q\,\mathrm{th}\,x}{(1-q^2\,\mathrm{th}^2 x)^2}\right]$$ $\left[q^2 = 1 - \frac{b}{a} > 0\right]$. MZ 196

4. $$\int \frac{dx}{(a+b\,\mathrm{ch}^2 x)^3} = \frac{1}{8pa^3}\left[\left(3-\frac{2}{p^2}+\frac{3}{p^4}\right)\mathrm{arctg}\,(p\,\mathrm{cth}\,x) + \left(3-\frac{2}{p^2}-\frac{3}{p^4}\right)\frac{p\,\mathrm{cth}\,x}{1+p^2\,\mathrm{cth}^2 x} + \left(1+\frac{2}{p^2}-\frac{1}{p^2}\mathrm{cth}^2 x\right)\frac{2p\,\mathrm{cth}\,x}{(1+p^2\,\mathrm{cth}^2 x)^2}\right]$$ $\left[p^2 = -1 - \frac{b}{a} > 0\right]$;

$$= \frac{1}{8qa^3}\left[\left(3+\frac{2}{q^2}+\frac{3}{q^4}\right)\varphi(x)\,^{*)} + \left(3+\frac{2}{q^2}-\frac{3}{q^4}\right)\frac{q\,\mathrm{cth}\,x}{1-q^2\,\mathrm{cth}^2 x} + \left(1-\frac{2}{q^2}+\frac{1}{q^2}\mathrm{cth}^2 x\right)\frac{2q\,\mathrm{cth}\,x}{(1-q^2\,\mathrm{cth}^2 x)^2}\right]$$ $\left[q^2 = 1 + \frac{b}{a} > 0\right]$.

2.46 Algebraic functions of hyperbolic functions

2.461

1. $$\int \sqrt{\mathrm{th}\,x}\,dx = \mathrm{Arth}\sqrt{\mathrm{th}\,x} - \mathrm{arctg}\sqrt{\mathrm{th}\,x}.$$ MZ 221

2. $$\int \sqrt{\mathrm{cth}\,x}\,dx = \mathrm{Arcth}\sqrt{\mathrm{cth}\,x} - \mathrm{arctg}\sqrt{\mathrm{cth}\,x}.$$ MZ 222

2.462

1. $$\int \frac{\mathrm{sh}\,x\,dx}{\sqrt{a^2+\mathrm{sh}^2 x}} = \mathrm{Arsh}\frac{\mathrm{ch}\,x}{\sqrt{a^2-1}} = \ln\left(\mathrm{ch}\,x + \sqrt{a^2+\mathrm{sh}^2 x}\right) \qquad [a^2 > 1];$$
$$= \mathrm{Arch}\frac{\mathrm{ch}\,x}{\sqrt{1-a^2}} = \ln\left(\mathrm{ch}\,x + \sqrt{a^2+\mathrm{sh}^2 x}\right) \qquad [a^2 < 1];$$
$$= \ln\,\mathrm{ch}\,x \qquad [a^2 = 1].$$

2. $$\int \frac{\mathrm{sh}\,x\,dx}{\sqrt{a^2-\mathrm{sh}^2 x}} = \arcsin\frac{\mathrm{ch}\,x}{\sqrt{a^2+1}} \qquad [\mathrm{sh}^2 x < a^2].$$

3. $$\int \frac{\mathrm{sh}\,x\,dx}{\sqrt{\mathrm{sh}^2 x - a^2}} = \mathrm{Arch}\frac{\mathrm{ch}\,x}{\sqrt{a^2+1}} = \ln\left(\mathrm{ch}\,x + \sqrt{\mathrm{sh}^2 x - a^2}\right) \qquad [\mathrm{sh}^2 x > a^2].$$ MZ 199

---

*If $\frac{b}{a} < 0$ and $\mathrm{ch}^2 x > -\frac{a}{b}$, then $\varphi(x) = \mathrm{Arth}\,(q\,\mathrm{cth}\,x)$. If $\frac{b}{a} < 0$, but $\mathrm{ch}^2 x < -\frac{a}{b}$, or if $\frac{b}{a} > 0$, then $\varphi(x) = \mathrm{Arcth}\,(q\,\mathrm{cth}\,x)$.

4. $\int \frac{\operatorname{ch} x\,dx}{\sqrt{a^2+\operatorname{sh}^2 x}} = \operatorname{Arsh}\frac{\operatorname{sh} x}{a} = \ln\left(\operatorname{sh} x + \sqrt{a^2+\operatorname{sh}^2 x}\right).$

5. $\int \frac{\operatorname{ch} x\,dx}{\sqrt{a^2-\operatorname{sh}^2 x}} = \arcsin\frac{\operatorname{sh} x}{a} \qquad [\operatorname{sh}^2 x < a^2].$

6. $\int \frac{\operatorname{ch} x\,dx}{\sqrt{\operatorname{sh}^2 x-a^2}} = \operatorname{Arch}\frac{\operatorname{sh} x}{a} = \ln\left(\operatorname{sh} x + \sqrt{\operatorname{sh}^2 x-a^2}\right) \qquad [\operatorname{sh}^2 x > a^2].$

7. $\int \frac{\operatorname{sh} x\,dx}{\sqrt{a^2+\operatorname{ch}^2 x}} = \operatorname{Arsh}\frac{\operatorname{ch} x}{a} = \ln\left(\operatorname{ch} x + \sqrt{a^2+\operatorname{ch}^2 x}\right).$

8. $\int \frac{\operatorname{sh} x\,dx}{\sqrt{a^2-\operatorname{ch}^2 x}} = \arcsin\frac{\operatorname{ch} x}{a} \qquad [\operatorname{ch}^2 x < a^2].$

9. $\int \frac{\operatorname{sh} x\,dx}{\sqrt{\operatorname{ch}^2 x-a^2}} = \operatorname{Arch}\frac{\operatorname{ch} x}{a} = \ln\left(\operatorname{ch} x + \sqrt{\operatorname{ch}^2 x-a^2}\right) \qquad [\operatorname{ch}^2 x > a^2].$

MZ 215, 216

10. $\int \frac{\operatorname{ch} x\,dx}{\sqrt{a^2+\operatorname{ch}^2 x}} = \operatorname{Arsh}\frac{\operatorname{sh} x}{\sqrt{a^2+1}} = \ln\left(\operatorname{sh} x + \sqrt{a^2+\operatorname{ch}^2 x}\right).$

11. $\int \frac{\operatorname{ch} x\,dx}{\sqrt{a^2-\operatorname{ch}^2 x}} = \arcsin\frac{\operatorname{sh} x}{\sqrt{a^2-1}} \qquad [\operatorname{ch}^2 x < a^2].$

12. $\int \frac{\operatorname{ch} x\,dx}{\sqrt{\operatorname{ch}^2 x-a^2}} = \operatorname{Arch}\frac{\operatorname{sh} x}{\sqrt{a^2-1}} \qquad [a^2 > 1];$

$= \ln \operatorname{sh} x \qquad [a^2 = 1].$ **MZ 206**

13. $\int \frac{\operatorname{cth} x\,dx}{\sqrt{a+b\operatorname{sh} x}} = 2\sqrt{a}\operatorname{Arcth}\sqrt{1+\frac{b}{a}\operatorname{sh} x} \qquad [b\operatorname{sh} x > 0.\ a > 0];$

$= 2\sqrt{a}\operatorname{Arth}\sqrt{1+\frac{b}{a}\operatorname{sh} x} \qquad [b\operatorname{sh} x < 0,\ a > 0];$

$= 2\sqrt{-a}\operatorname{Arth}\sqrt{-\left(1+\frac{b}{a}\operatorname{sh} x\right)} \qquad a < 0.$

14. $\int \frac{\operatorname{th} x\,dx}{\sqrt{a+b\operatorname{ch} x}} = 2\sqrt{a}\operatorname{Arcth}\sqrt{1+\frac{b}{a}\operatorname{ch} x} \qquad [b\operatorname{ch} x > 0,\ a > 0];$

$= 2\sqrt{a}\operatorname{Arth}\sqrt{1+\frac{b}{a}\operatorname{ch} x} \qquad [b\operatorname{ch} x < 0,\ a > 0];$

$= 2\sqrt{-a}\operatorname{Arth}\sqrt{-\left(1+\frac{b}{a}\operatorname{ch} x\right)} \qquad [a<0].$ **MZ 220, 221**

2.463

1. $\int \frac{\operatorname{sh} x\sqrt{a+b\operatorname{ch} x}}{p+q\operatorname{ch} x}\,dx = 2\sqrt{\frac{aq-bp}{q}}\operatorname{Arcth}\sqrt{\frac{q(a+b\operatorname{ch} x)}{aq-bp}}$

$\left[b\operatorname{ch} x > 0,\ \frac{aq-bp}{q} > 0\right];$

$= 2\sqrt{\frac{aq-bp}{q}}\operatorname{Arth}\sqrt{\frac{q(a+b\operatorname{ch} x)}{aq-bp}}$

$\left[b\operatorname{ch} x < 0,\ \frac{aq-bp}{q} > 0\right];$

$= 2\sqrt{\frac{bp-aq}{q}}\operatorname{Arth}\sqrt{\frac{q(a+b\operatorname{ch} x)}{bp-aq}}$

$\left[\frac{aq-bp}{q} < 0\right].$ **MZ 220**

2. $\int \frac{\operatorname{ch} x\sqrt{a+b\operatorname{sh} x}}{p+q\operatorname{sh} x}\,dx = 2\sqrt{\frac{aq-bp}{q}}\operatorname{Arcth}\sqrt{\frac{q(a+b\operatorname{sh} x)}{aq-bp}}$

$\left[b\operatorname{sh} x > 0,\ \frac{aq-bp}{q} > 0\right];$

$$= 2\sqrt{\frac{aq-bp}{q}}\,\text{Arth}\sqrt{\frac{q(a+b\,\text{sh}\,x)}{aq-bp}} \qquad \left[b\,\text{sh}\,x<0,\ \frac{aq-bp}{q}>0\right];$$

$$= 2\sqrt{\frac{bp-aq}{q}}\,\text{Arth}\sqrt{\frac{q(a+b\,\text{sh}\,x)}{bp-aq}} \qquad \left[\frac{aq-bp}{q}<0\right].$$ MZ 221

2.464

1. $\int\frac{dx}{\sqrt{k^2+k'^2\,\text{ch}^2\,x}}=\int\frac{dx}{\sqrt{1+k'^2\,\text{sh}^2\,x}}=F(\arcsin(\text{th}\,x),\ k)\quad[x>0].$ BY (295.00)(295.10)

2. $\int\frac{dx}{\sqrt{\text{ch}^2\,x-k^2}}=\int\frac{dx}{\sqrt{\text{sh}^2\,x+k'^2}}=F\left(\arcsin\left(\frac{1}{\text{ch}\,x}\right),\ k\right)\quad[x>0].$ BY (295.40)(295.30)

3. $\int\frac{dx}{\sqrt{1-k'^2\,\text{ch}^2\,x}}=F\left(\arcsin\left(\frac{\text{th}\,x}{k}\right),\ k\right)\left[0<x<\text{Arch}\,\frac{1}{k'}\right].$ BY (295.20)

In **2.464** 4. — **2.464** 8., we set $\alpha=\arccos\frac{1-\text{sh}\,2ax}{1+\text{sh}\,2ax},\ r=\frac{1}{\sqrt{2}}\quad[ax>0]$:

4. $\int\frac{dx}{\sqrt{\text{sh}\,2ax}}=\frac{1}{2a}F(\alpha,\ r).$ BY (296.50)

5. $\int\sqrt{\text{sh}\,2ax}\,dx=\frac{1}{2a}[F(\alpha,\ r)-2E(\alpha,\ r)]+\frac{1}{a}\frac{\sqrt{\text{sh}\,2ax\,(1+\text{sh}^2\,2ax)}}{1+\text{sh}\,2ax}.$ BY (296.53)

6. $\int\frac{\text{ch}^2\,2ax\,dx}{(1+\text{sh}\,2ax)^2\sqrt{\text{sh}\,2ax}}=\frac{1}{2a}E(\alpha,\ r).$ BY (296.51)

7. $\int\frac{(1-\text{sh}\,2ax)^2\,dx}{(1+\text{sh}\,2ax)^2\sqrt{\text{sh}\,2ax}}=\frac{1}{2a}[2E(\alpha,\ r)-F(\alpha,\ r)].$ BY (296.55)

8. $\int\frac{\sqrt{\text{sh}\,2ax}\,dx}{(1+\text{sh}\,2ax)^2}=\frac{1}{4a}[F(\alpha,\ r)-E(\alpha,\ r)].$ BY (296.54)

In **2.464** 9. — **2.464** 15., we set $\alpha=\arcsin\sqrt{\frac{\text{ch}\,2ax-1}{\text{ch}\,2ax}},\ r=\frac{1}{\sqrt{2}}\quad[x\neq 0]$:

9. $\int\frac{dx}{\sqrt{\text{ch}\,2ax}}=\frac{1}{a\sqrt{2}}F(\alpha,\ r).$ BY (296.00)

10. $\int\sqrt{\text{ch}\,2ax}\,dx=\frac{1}{a\sqrt{2}}[F(\alpha,\ r)-2E(\alpha,\ r)]+\frac{\text{sh}\,2ax}{a\sqrt{\text{ch}\,2ax}}.$ BY (296.03)

11. $\int\frac{dx}{\sqrt{\text{ch}^3\,2ax}}=\frac{1}{a\sqrt{2}}[2E(\alpha,\ r)-F(\alpha,\ r)].$ BY (296.04)

12. $\int\frac{dx}{\sqrt{\text{ch}^5\,2ax}}=\frac{1}{3\sqrt{2}\,a}F(\alpha,\ r)+\frac{\text{th}\,2ax}{3a\sqrt{\text{ch}\,2ax}}.$ BY (296.04)

13. $\int\frac{\text{sh}^2\,2ax\,dx}{\sqrt{\text{ch}\,2ax}}=-\frac{\sqrt{2}}{3a}F(\alpha,\ r)+\frac{1}{3a}\text{sh}\,2ax\sqrt{\text{ch}\,2ax}.$ BY (296.07)

14. $\int\frac{\text{th}^2\,2ax\,dx}{\sqrt{\text{ch}\,2ax}}=\frac{\sqrt{2}}{3a}F(\alpha,\ r)-\frac{\text{th}\,2ax}{3a\sqrt{\text{ch}\,2ax}}.$ BY (296.05)

15. $\int\frac{\sqrt{\text{ch}\,2ax}\,dx}{p^2+(1-p^2)\,\text{ch}\,2ax}=\frac{1}{a\sqrt{2}}\Pi(\alpha,\ p^2,\ r).$ BY (296.02)

In **2.464** 16. — **2.464** 20., we set $\alpha = \arccos \dfrac{\sqrt{a^2+b^2}-a-b\operatorname{sh} x}{\sqrt{a^2+b^2}+a+b\operatorname{sh} x}$,
$r = \sqrt{\dfrac{a+\sqrt{a^2+b^2}}{2\sqrt{a^2+b^2}}}$ $\left[a>0,\ b>0,\ x>-\operatorname{Arsh}\frac{a}{b}\right]$:

16. $$\int \frac{dx}{\sqrt{a+b\operatorname{sh} x}} = \frac{1}{\sqrt[4]{a^2+b^2}} F(\alpha, r).$$ BY (298.00)

17. $$\int \sqrt{a+b\operatorname{sh} x}\, dx = \sqrt[4]{a^2+b^2}\,[F(\alpha, r) - 2E(\alpha, r)] + \frac{2b\operatorname{ch} x\sqrt{a+b\operatorname{sh} x}}{\sqrt{a^2+b^2}+a+b\operatorname{sh} x}.$$ BY (298.02)

18. $$\int \frac{\sqrt{a+b\operatorname{sh} x}}{\operatorname{ch}^2 x}\, dx = \sqrt[4]{a^2+b^2}\, E(\alpha, r) - \frac{\sqrt{a^2+b^2}-a}{2\sqrt[4]{a^2+b^2}} F(\alpha, r) - \frac{a+\sqrt{a^2+b^2}}{b}\cdot\frac{\sqrt{a^2+b^2}-a-b\operatorname{sh} x}{\sqrt{a^2+b^2}+a+b\operatorname{sh} x}\cdot\frac{\sqrt{a+b\operatorname{sh} x}}{\operatorname{ch} x}.$$ BY (298.03)

19. $$\int \frac{\operatorname{ch}^2 x\, dx}{\left[\sqrt{a^2+b^2}+a+b\operatorname{sh} x\right]^2\sqrt{a+b\operatorname{sh} x}} = \frac{1}{b^2\sqrt[4]{a^2+b^2}} E(\alpha, r).$$ BY (298.01)

20. $$\int \frac{\sqrt{a+b\operatorname{sh} x}\, dx}{\left[\sqrt{a^2+b^2}-a-b\operatorname{sh} x\right]^2} = -\frac{1}{\sqrt[4]{a^2+b^2}\left(\sqrt{a^2+b^2}-a\right)} E(\alpha, r) + \frac{b}{\sqrt{a^2+b^2}-a}\cdot\frac{\operatorname{ch} x\sqrt{a+b\operatorname{sh} x}}{a^2+b^2-(a+b\operatorname{sh} x)^2}.$$ BY (298.04)

In **2.464** 21. — **2.464** 31, we set $\alpha = \arcsin\left(\operatorname{th}\frac{x}{2}\right)$, $r = \sqrt{\dfrac{a-b}{a+b}}$ $[0<b<a,\ x>0]$:

21. $$\int \frac{dx}{\sqrt{a+b\operatorname{ch} x}} = \frac{2}{\sqrt{a+b}} F(\alpha, r).$$ BY (297.25)

22. $$\int \sqrt{a+b\operatorname{ch} x}\, dx = 2\sqrt{a+b}\,[F(\alpha, r) - E(\alpha, r)] + 2\operatorname{th}\frac{x}{2}\sqrt{a+b\operatorname{ch} x}.$$ BY (297.29)

23. $$\int \frac{\operatorname{ch} x\, dx}{\sqrt{a+b\operatorname{ch} x}} = \frac{2}{\sqrt{a+b}} F(\alpha, r) - \frac{2\sqrt{a+b}}{b} E(\alpha, r) + \frac{2}{b}\operatorname{th}\frac{x}{2}\sqrt{a+b\operatorname{ch} x}.$$ BY (297.33)

24. $$\int \frac{\operatorname{th}^2\frac{x}{2}}{\sqrt{a+b\operatorname{ch} x}}\, dx = \frac{2\sqrt{a+b}}{a-b}[F(\alpha, r) - E(\alpha, r)].$$ BY (297.28)

25. $$\int \frac{\operatorname{th}^4\frac{x}{2}}{\sqrt{a+b\operatorname{ch} x}}\, dx = \frac{2\sqrt{a+b}}{3(a-b)^2}[(3a+b)F(\alpha, r) - 4aE(\alpha, r)] + \frac{2}{3(a-b)}\,\frac{\operatorname{sh}\frac{x}{2}\sqrt{a+b\operatorname{ch} x}}{\operatorname{ch}^3\frac{x}{2}}.$$ BY (297.28)

26. $$\int \frac{\operatorname{ch} x - 1}{\sqrt{a+b\operatorname{ch} x}}\, dx = \frac{2}{b}\left[\operatorname{th}\frac{x}{2}\sqrt{a+b\operatorname{ch} x} - \sqrt{a+b}\, E(\alpha, r)\right].$$ BY (297.31)

27. $\int \frac{(\operatorname{ch} x-1)^2}{\sqrt{a+b \operatorname{ch} x}}\,dx = \frac{4\sqrt{a+b}}{3b^2}[(a+3b)E(\alpha, r) - bF(\alpha, r)] + \frac{4}{3b^2}\left[b \operatorname{ch}^2 \frac{x}{2} - (a+3b)\right] \operatorname{th}\frac{x}{2}\sqrt{a+b \operatorname{ch} x}.$ BY (297.31)

28. $\int \frac{\sqrt{a+b \operatorname{ch} x}}{\operatorname{ch} x+1}\,dx = \sqrt{a+b}\,E(\alpha, r).$ BY (297.26)

29. $\int \frac{dx}{(\operatorname{ch} x+1)\sqrt{a+b \operatorname{ch} x}} = \frac{\sqrt{a+b}}{a-b}E(\alpha, r) - \frac{2b}{(a-b)\sqrt{a+b}}F(\alpha, r).$ BY (297.30)

30. $\int \frac{dx}{(\operatorname{ch} x+1)^2\sqrt{a+b \operatorname{ch} x}} = \frac{1}{3(a-b)^2\sqrt{a+b}}[b(5b-a)F(\alpha, r) + (a-3b)(a+b)E(\alpha, r)] + \frac{1}{6(a-b)}\cdot\frac{\operatorname{sh}\frac{x}{2}}{\operatorname{ch}^3\frac{x}{2}}\sqrt{a+b \operatorname{ch} x}.$ BY (297.30)

31. $\int \frac{(1+\operatorname{ch} x)\,dx}{[1+p^2+(1-p^2)\operatorname{ch} x]\sqrt{a+b \operatorname{ch} x}} = \frac{2}{\sqrt{a+b}}\Pi(\alpha, p^2, r).$ BY (297.27)

In **2.464** 32. — **2.464** 40., we set $\alpha = \arcsin\sqrt{\frac{a-b \operatorname{ch} x}{a-b}}$, $r = \sqrt{\frac{a-b}{a+b}}$ $\left[0 < b < a,\ 0 < x < \operatorname{Arch}\frac{a}{b}\right]$:

32. $\int \frac{dx}{\sqrt{a-b \operatorname{ch} x}} = \frac{2}{\sqrt{a+b}}F(\alpha, r).$ BY (297.50)

33. $\int \sqrt{a-b \operatorname{ch} x}\,dx = 2\sqrt{a+b}[F(\alpha, r) - E(\alpha, r)].$ BY (297.54)

34. $\int \frac{\operatorname{ch} x\,dx}{\sqrt{a-b \operatorname{ch} x}} = \frac{2\sqrt{a+b}}{b}E(\alpha, r) - \frac{2}{\sqrt{a+b}}F(\alpha, r).$ BY (297.56)

35. $\int \frac{\operatorname{ch}^2 x\,dx}{\sqrt{a-b \operatorname{ch} x}} = \frac{2(b-2a)}{3b\sqrt{a+b}}F(\alpha, r) + \frac{4a\sqrt{a+b}}{3b^2}E(\alpha, r) + \frac{2}{3b}\operatorname{sh} x\sqrt{a-b \operatorname{ch} x}.$ BY (297.56)

36. $\int \frac{(1+\operatorname{ch} x)\,dx}{\sqrt{a-b \operatorname{ch} x}} = \frac{2\sqrt{a+b}}{b}E(\alpha, r).$ BY (297.51)

37. $\int \frac{dx}{\operatorname{ch} x\sqrt{a-b \operatorname{ch} x}} = \frac{2b}{a\sqrt{a+b}}\Pi\left(\alpha, \frac{a-b}{a}, r\right).$ BY (297.57)

38. $\int \frac{dx}{(1+\operatorname{ch} x)\sqrt{a-b \operatorname{ch} x}} = \frac{1}{\sqrt{a+b}}E(\alpha, r) - \frac{1}{a+b}\operatorname{th}\frac{x}{2}\sqrt{a-b \operatorname{ch} x}.$ BY (297.58)

39. $\int \frac{dx}{(1+\operatorname{ch} x)^2\sqrt{a-b \operatorname{ch} x}} = \frac{1}{3\sqrt{(a+b)^3}}[(a+3b)E(\alpha, r) - bF(\alpha, r)] - \frac{1}{3(a+b)^2}\,\frac{\operatorname{th}\frac{x}{2}\sqrt{a-b \operatorname{ch} x}}{\operatorname{ch} x+1}[2a+4b+(a+3b)\operatorname{ch} x].$ BY (297.58)

40. $\int \frac{dx}{(a-b-ap^2+bp^2 \operatorname{ch} x)\sqrt{a-b \operatorname{ch} x}} = \frac{2}{(a-b)\sqrt{a+b}}\Pi(\alpha, p^2, r).$ BY (297.52)

In **2.464** 41. — **2.464** 47., we set $\alpha = \arcsin \sqrt{\frac{b(\operatorname{ch} x - 1)}{b \operatorname{ch} x - a}}$, $r = \sqrt{\frac{a+b}{2b}}$ $[0 < a < b,\ x > 0]$:

41. $\int \frac{dx}{\sqrt{b \operatorname{ch} x - a}} = \sqrt{\frac{2}{b}} F(\alpha, r).$ BY (297.00)

42. $\int \sqrt{b \operatorname{ch} x - a}\, dx = (b-a)\sqrt{\frac{2}{b}}\, F(\alpha, r) - 2\sqrt{2b}\, E(\alpha, r) + \frac{2b \operatorname{sh} x}{\sqrt{b \operatorname{ch} x - a}}.$ BY (297.05)

43. $\int \frac{dx}{\sqrt{(b \operatorname{ch} x - a)^3}} = \frac{1}{b^2 - a^2} \cdot \sqrt{\frac{2}{b}} [2bE(\alpha, r) - (b-a) F(\alpha, r)].$ BY (297.06)

44. $\int \frac{dx}{\sqrt{(b \operatorname{ch} x - a)^5}} = \frac{1}{3(b^2 - a^2)^2} \sqrt{\frac{2}{b}} [(b - 3a)(b-a) F(\alpha, r) + 8abE(\alpha, r)] + \frac{2b}{3(b^2 - a^2)} \cdot \frac{\operatorname{sh} x}{\sqrt{(b \operatorname{ch} x - a)^3}}.$ BY (297.06)

45. $\int \frac{\operatorname{ch} x\, dx}{\sqrt{b \operatorname{ch} x - a}} = \sqrt{\frac{2}{b}} [F(\alpha, r) - 2E(\alpha, r)] + \frac{2 \operatorname{sh} x}{\sqrt{b \operatorname{ch} x - a}}.$ BY (297.03)

46. $\int \frac{(\operatorname{ch} x + 1)\, dx}{\sqrt{(b \operatorname{ch} x - a)^3}} = \frac{2}{b-a} \sqrt{\frac{2}{b}} E(\alpha, r).$ BY (297.01)

47. $\int \frac{\sqrt{b \operatorname{ch} x - a}\, dx}{p^2 b - a + b(1 - p^2) \operatorname{ch} x} = \sqrt{\frac{2}{b}} \Pi(\alpha, p^2, r).$ BY (297.02)

In **2.464** 48. — **2.464** 55., we set $\alpha = \arcsin \sqrt{\frac{b \operatorname{ch} x - a}{b(\operatorname{ch} x - 1)}}$, $r = \sqrt{\frac{2b}{a+b}}$ $\left[0 < b < a,\ x > \operatorname{Arch} \frac{a}{b}\right]$:

48. $\int \frac{dx}{\sqrt{b \operatorname{ch} x - a}} = \frac{2}{\sqrt{a+b}} F(\alpha, r).$ BY (297.75)

49. $\int \sqrt{b \operatorname{ch} x - a}\, dx = -2\sqrt{a+b}\, E(\alpha, r) + 2 \operatorname{cth} \frac{x}{2} \sqrt{b \operatorname{ch} x - a}.$ BY (297.79)

50. $\int \frac{\operatorname{cth}^2 \frac{x}{2}\, dx}{\sqrt{b \operatorname{ch} x - a}} = \frac{2\sqrt{a+b}}{a-b} E(\alpha, r).$ BY (297.76)

51. $\int \frac{\sqrt{b \operatorname{ch} x - a}}{\operatorname{ch} x - 1} dx = \sqrt{a+b}\, [F(\alpha, r) - E(\alpha, r)].$ BY (297.77)

52. $\int \frac{dx}{(\operatorname{ch} x - 1)\sqrt{b \operatorname{ch} x - a}} = \frac{\sqrt{a+b}}{a-b} E(\alpha, r) - \frac{1}{\sqrt{a+b}} F(\alpha, r).$ BY (297.78)

53. $\int \frac{dx}{(\operatorname{ch} x - 1)^2 \sqrt{b \operatorname{ch} x - a}} = \frac{1}{3(a-b)^2 \sqrt{a+b}} [(a - 2b)(a-b) F(\alpha, r) + (3a - b)(a+b) E(\alpha, r)] + \frac{a+b}{6b(a-b)} \cdot \frac{\operatorname{ch} \frac{x}{2}}{\operatorname{sh}^3 \frac{x}{2}} \sqrt{b \operatorname{ch} x - a}.$ BY (297.78)

54. $\int \frac{dx}{(\operatorname{ch} x + 1)\sqrt{b \operatorname{ch} x - a}} = \frac{1}{\sqrt{a+b}} [F(\alpha, r) - E(\alpha, r)] + \frac{2\sqrt{b \operatorname{ch} x - a}}{(a+b) \operatorname{sh} x}.$ BY (297.80)

55. $$\int \frac{dx}{(\operatorname{ch} x+1)^2 \sqrt{b \operatorname{ch} x - a}} = \frac{1}{3\sqrt{(a+b)^3}} [(a+2b) F(\alpha, r) - (a+3b) E(\alpha, r)] + \frac{\sqrt{b \operatorname{ch} x - a}}{3(a+b) \operatorname{sh} x} \left(2 \frac{a+3b}{a+b} - \operatorname{th}^2 \frac{x}{2}\right).$$ BY (297.80)

In **2.464** 56. — **2.464** 60., we set $\alpha = \arccos \frac{\sqrt[4]{b^2-a^2}}{\sqrt{a \operatorname{sh} x + b \operatorname{ch} x}}$, $r = \frac{1}{\sqrt{2}}$ $\left[0 < a < b, \quad -\operatorname{Arsh} \frac{a}{\sqrt{b^2-a^2}} < x\right]$:

56. $$\int \frac{dx}{\sqrt{a \operatorname{sh} x + b \operatorname{ch} x}} = \sqrt[4]{\frac{4}{b^2-a^2}} F(\alpha, r).$$ BY (299.00)

57. $$\int \sqrt{a \operatorname{sh} x + b \operatorname{ch} x}\, dx = \sqrt[4]{4(b^2-a^2)} [F(\alpha, r) - 2E(\alpha, r)] + \frac{2(a \operatorname{ch} x + b \operatorname{sh} x)}{\sqrt{a \operatorname{sh} x + b \operatorname{ch} x}}.$$ BY (299.02)

58. $$\int \frac{dx}{\sqrt{(a \operatorname{sh} x + b \operatorname{ch} x)^3}} = \sqrt[4]{\frac{4}{(b^2-a^2)^3}} [2E(\alpha, r) - F(\alpha, r)].$$ BY (299.03)

59. $$\int \frac{dx}{\sqrt{(a \operatorname{sh} x + b \operatorname{ch} x)^5}} = \frac{1}{3} \sqrt[4]{\frac{4}{(b^2-a^2)^5}} F(\alpha, r) + \frac{2}{3(b^2-a^2)} \cdot \frac{a \operatorname{ch} x + b \operatorname{sh} x}{\sqrt{(a \operatorname{sh} x + b \operatorname{ch} x)^3}}.$$ BY (299.03)

60. $$\int \frac{(\sqrt{b^2-a^2} + a \operatorname{sh} x + b \operatorname{ch} x)\, dx}{\sqrt{(a \operatorname{sh} x + b \operatorname{ch} x)^3}} = 2 \sqrt[4]{\frac{4}{b^2-a^2}} E(\alpha, r).$$ BY (299.01)

2.47 Combinations of hyperbolic functions and powers

2.471

1. $$\int x^r \operatorname{sh}^p x \operatorname{ch}^q x\, dx = \frac{1}{(p+q)^2} \Big[(p+q) x^r \operatorname{sh}^{p+1} x \operatorname{ch}^{q-1} x - r x^{r-1} \operatorname{sh}^p x \operatorname{ch}^q x + r(r+1) \int x^{r-2} \operatorname{sh}^p x \operatorname{ch}^q x\, dx + rp \int x^{r-1} \operatorname{sh}^{p-1} x \operatorname{ch}^{q-1} x\, dx + (q-1)(p+q) \int x^r \operatorname{sh}^p x \operatorname{ch}^{q-2} x\, dx \Big];$$
$$= \frac{1}{(p+q)^2} \Big[(p+q) x^r \operatorname{sh}^{p-1} x \operatorname{ch}^{q+1} x - r x^{r-1} \operatorname{sh}^p x \operatorname{ch}^q x + r(r-1) \int x^{r-2} \operatorname{sh}^p x \operatorname{ch}^q x\, dx - rq \int x^{r-1} \operatorname{sh}^{p-1} x \operatorname{ch}^{q-1} x\, dx - (p-1)(p+q) \int x^r \operatorname{sh}^{p-2} x \operatorname{ch}^q x\, dx \Big].$$ GU ((353))(1)

2. $$\int x^n \operatorname{sh}^{2m} x\, dx = (-1)^m \binom{2m}{m} \frac{x^{n+1}}{2^{2m}(n+1)} + \frac{1}{2^{2m-1}} \sum_{k=0}^{m-1} (-1)^k \binom{2m}{k} \int x^n \operatorname{ch}(2m-2k) x\, dx.$$

3. $$\int x^n \operatorname{sh}^{2m+1} x\, dx = \frac{1}{2^{2m}} \sum_{k=0}^{m} (-1)^k \binom{2m+1}{k} \int x^n \operatorname{sh}(2m-2k+1) x\, dx.$$

4. $$\int x^n \operatorname{ch}^{2m} x\,dx = \binom{2m}{m} \frac{x^{n+1}}{2^{2m}(n+1)} + \frac{1}{2^{2m-1}} \sum_{k=0}^{m-1} \binom{2m}{k} \int x^n \operatorname{ch}(2m-2k)x\,dx.$$

5. $$\int x^n \operatorname{ch}^{2m+1} x\,dx = \frac{1}{2^{2m}} \sum_{k=0}^{m} \binom{2m+1}{k} \int x^n \operatorname{ch}(2m-2k+1)x\,dx.$$

2.472

1. $$\int x^n \operatorname{sh} x\,dx = x^n \operatorname{ch} x - n \int x^{n-1} \operatorname{ch} x\,dx = x^n \operatorname{ch} x - nx^{n-1} \operatorname{sh} x + n(n-1) \int x^{n-2} \operatorname{sh} x\,dx.$$

2. $$\int x^n \operatorname{ch} x\,dx = x^n \operatorname{sh} x - n \int x^{n-1} \operatorname{sh} x\,dx = x^n \operatorname{sh} x - nx^{n-1} \operatorname{ch} x + n(n-1) \int x^{n-2} \operatorname{ch} x\,dx.$$

3. $$\int x^{2n} \operatorname{sh} x\,dx = (2n)! \left\{ \sum_{k=0}^{n} \frac{x^{2k}}{(2k)!} \operatorname{ch} x - \sum_{k=1}^{n} \frac{x^{2k-1}}{(2k-1)!} \operatorname{sh} x \right\}.$$

4. $$\int x^{2n+1} \operatorname{sh} x\,dx = (2n+1)! \sum_{k=0}^{n} \left\{ \frac{x^{2k+1}}{(2k+1)!} \operatorname{ch} x - \frac{x^{2k}}{(2k)!} \operatorname{sh} x \right\}.$$

5. $$\int x^{2n} \operatorname{ch} x\,dx = (2n)! \left\{ \sum_{k=1}^{n} \frac{x^{2k}}{(2k)!} \operatorname{sh} x - \sum_{k=1}^{n} \frac{x^{2k-1}}{(2k-1)!} \operatorname{ch} x \right\}.$$

6. $$\int x^{2n+1} \operatorname{ch} x\,dx = (2n+1)! \sum_{k=0}^{n} \left\{ \frac{x^{2k+1}}{(2k+1)!} \operatorname{sh} x - \frac{x^{2k}}{(2k)!} \operatorname{ch} x \right\}.$$

7. $$\int x \operatorname{sh} x\,dx = x \operatorname{ch} x - \operatorname{sh} x.$$

8. $$\int x^2 \operatorname{sh} x\,dx = (x^2+2) \operatorname{ch} x - 2x \operatorname{sh} x.$$

9. $$\int x \operatorname{ch} x\,dx = x \operatorname{sh} x - \operatorname{ch} x.$$

10. $$\int x^2 \operatorname{ch} x\,dx = (x^2+2) \operatorname{sh} x - 2x \operatorname{ch} x.$$

2.473 Notation: $z_1 = a + bx$

1. $$\int z_1 \operatorname{sh} kx\,dx = \frac{1}{k} z_1 \operatorname{ch} kx - \frac{b}{k^2} \operatorname{sh} kx.$$

2. $$\int z_1 \operatorname{ch} kx\,dx = \frac{1}{k} z_1 \operatorname{sh} kx - \frac{b}{k^2} \operatorname{ch} kx.$$

3. $$\int z_1^2 \operatorname{sh} kx\,dx = \frac{1}{k}\left(z_1^2 + \frac{2b^2}{k^2} \right) \operatorname{ch} kx - \frac{2bz_1}{k^2} \operatorname{sh} kx.$$

4. $$\int z_1^2 \operatorname{ch} kx\,dx = \frac{1}{k}\left(z_1^2 + \frac{2b^2}{k^2} \right) \operatorname{sh} kx - \frac{2bz_1}{k^2} \operatorname{ch} kx.$$

5. $$\int z_1^3 \operatorname{sh} kx\,dx = \frac{z_1}{k}\left(z_1^2 + \frac{6b^2}{k^2} \right) \operatorname{ch} kx - \frac{3b}{k^2}\left(z_1^2 + \frac{2b^2}{k^2} \right) \operatorname{sh} kx.$$

6. $\int z_1^3 \operatorname{ch} kx\, dx = \frac{z_1}{k}\left(z_1^2 + \frac{6b^2}{k^2}\right)\operatorname{sh} kx - \frac{3b}{k^2}\left(z_1^2 + \frac{2b^2}{k^2}\right)\operatorname{ch} kx.$

7. $\int z_1^4 \operatorname{sh} kx\, dx = \frac{1}{k}\left(z_1^4 + \frac{12b^2}{k^2} z_1^2 + \frac{24b^4}{k^4}\right)\operatorname{ch} kx - \frac{4bz_1}{k^2}\left(z_1^2 + \frac{6b^2}{k^2}\right)\operatorname{sh} kx.$

8. $\int z_1^4 \operatorname{ch} kx\, dx = \frac{1}{k}\left(z_1^4 + \frac{12b^2}{k^2} z_1^2 + \frac{24b^4}{k^4}\right)\operatorname{sh} kx - \frac{4bz_1}{k^2}\left(z_1^2 + \frac{6b^2}{k^2}\right)\operatorname{ch} kx.$

9. $\int z_1^5 \operatorname{sh} kx\, dx = \frac{z_1}{k}\left(z_1^4 + \frac{20b^2}{k^2} z_1^2 + 120\frac{b^4}{k^4}\right)\operatorname{ch} kx -$

$- \frac{5b}{k^2}\left(z_1^4 + 12\frac{b^2}{k^2} z_1^2 + 24\frac{b^4}{k^4}\right)\operatorname{sh} kx.$

10. $\int z_1^5 \operatorname{ch} kx\, dx = \frac{z_1}{k}\left(z_1^4 + 20\frac{b^2}{k^2} z_1^2 + 120\frac{b^4}{k^4}\right)\operatorname{sh} kx -$

$- \frac{5b}{k^2}\left(z_1^4 + 12\frac{b^2}{k^2} z_1^2 + 24\frac{b^4}{k^4}\right)\operatorname{ch} kx.$

11. $\int z_1^6 \operatorname{sh} kx\, dx = \frac{1}{k}\left(z_1^6 + 30\frac{b^2}{k^2} z_1^4 + 360\frac{b^4}{k^4} z_1^2 + 720\frac{b^6}{k^6}\right)\operatorname{ch} kx -$

$- \frac{6bz_1}{k^2}\left(z_1^4 + 20\frac{b^2}{k^2} z_1^2 + 120\frac{b^4}{k^4}\right)\operatorname{sh} kx.$

12. $\int z_1^6 \operatorname{ch} kx\, dx = \frac{1}{k}\left(z_1^6 + 30\frac{b^2}{k^2} z_1^4 + 360\frac{b^4}{k^4} z_1^2 + 720\frac{b^6}{k^6}\right)\operatorname{sh} kx -$

$- \frac{6bz_1}{k^2}\left(z_1^4 + 20\frac{h^2}{k^2} z_1 + 120\frac{b^4}{k^4}\right)\operatorname{ch} kx.$

2.474

1. $\int x^n \operatorname{sh}^2 x\, dx = -\frac{x^{n+1}}{2(n+1)} +$

$+ \frac{n!}{4}\sum_{k=0}^{E\left(\frac{n}{2}\right)}\left\{\frac{x^{n-2k}}{2^{2k}(n-2k)!}\operatorname{sh} 2x - \frac{x^{n-2k-1}}{2^{2k+1}(n-2k-1)!}\operatorname{ch} 2x\right\}.$ GU ((353))(2b)

2. $\int x^n \operatorname{ch}^2 x\, dx = \frac{x^{n+1}}{2(n+1)} +$

$+ \frac{n!}{4}\sum_{k=0}^{E\left(\frac{n}{2}\right)}\left\{\frac{x^{n-2k}}{2^{2k}(n-2k)!}\operatorname{sh} 2x - \frac{x^{n-2k-1}}{2^{2k+1}(n-2k-1)!}\operatorname{ch} 2x\right\}.$ GU ((353))(3e)

3. $\int x \operatorname{sh}^2 x\, dx = \frac{1}{4} x \operatorname{sh} 2x - \frac{1}{8}\operatorname{ch} 2x - \frac{x^2}{4}.$

4. $\int x^2 \operatorname{sh}^2 x\, dx = \frac{1}{4}\left(x^2 + \frac{1}{2}\right)\operatorname{sh} 2x - \frac{x}{4}\operatorname{ch} 2x - \frac{x^3}{6}.$ MZ 257

5. $\int x \operatorname{ch}^2 x\, dx = \frac{x}{4}\operatorname{sh} 2x - \frac{1}{8}\operatorname{ch} 2x + \frac{x^2}{4}.$

6. $\int x^2 \operatorname{ch}^2 x\, dx = \frac{1}{4}\left(x^2 + \frac{1}{2}\right)\operatorname{sh} 2x - \frac{x}{4}\operatorname{ch} 2x + \frac{x^3}{6}.$ MZ 261

7. $\int x^n \operatorname{sh}^3 x\, dx = \frac{n!}{4}\sum_{k=0}^{E\left(\frac{n}{2}\right)}\left\{\frac{x^{n-2k}}{(n-2k)!}\left(\frac{\operatorname{ch} 3x}{3^{2k+1}} - 3\operatorname{ch} x\right) -\right.$

$\left. - \frac{x^{n-2k-1}}{(n-2k-1)!}\left(\frac{\operatorname{sh} 3x}{3^{2k+2}} - 3\operatorname{sh} x\right)\right\}.$ GU ((353))(2f)

8. $$\int x^n \operatorname{ch}^3 x\, dx = \frac{n!}{4} \sum_{k=0}^{E\left(\frac{n}{2}\right)} \left\{ \frac{x^{n-2k}}{(n-2k)!} \left(\frac{\operatorname{sh} 3x}{3^{2k+1}} + 3 \operatorname{sh} x \right) - \frac{x^{n-2k-1}}{(n-2k-1)!} \left(\frac{\operatorname{ch} 3x}{3^{2k+2}} + 3 \operatorname{ch} x \right) \right\}.$$ GU ((353))(3f)

9. $$\int x \operatorname{sh}^3 x\, dx = \frac{3}{4} \operatorname{sh} x - \frac{1}{36} \operatorname{sh} 3x - \frac{3}{4} x \operatorname{ch} x - \frac{x}{12} \operatorname{ch} 3x.$$

10. $$\int x^2 \operatorname{sh}^3 x\, dx = -\left(\frac{3x^2}{4} + \frac{3}{2} \right) \operatorname{ch} x + \left(\frac{x^2}{12} + \frac{1}{54} \right) \operatorname{ch} 3x + \frac{3x}{2} \operatorname{sh} x - \frac{x}{18} \operatorname{sh} 3x.$$ MZ 257

11. $$\int x \operatorname{ch}^3 x\, dx = -\frac{3}{4} \operatorname{ch} x - \frac{1}{36} \operatorname{ch} 3x + \frac{3}{4} x \operatorname{sh} x + \frac{x}{12} \operatorname{sh} 3x.$$

12. $$\int x^2 \operatorname{ch}^3 x\, dx = \left(\frac{3}{4} x^2 + \frac{3}{2} \right) \operatorname{sh} x + \left(\frac{x^2}{12} + \frac{1}{54} \right) \operatorname{sh} 3x - \frac{3}{2} x \operatorname{ch} x - \frac{x}{18} \operatorname{ch} 3x.$$ MZ 262

2.475

1. $$\int \frac{\operatorname{sh}^q x}{x^p} dx = -\frac{(p-2) \operatorname{sh}^q x + qx \operatorname{sh}^{q-1} x \operatorname{ch} x}{(p-1)(p-2) x^{p-1}} + \frac{q(q-1)}{(p-1)(p-2)} \int \frac{\operatorname{sh}^{q-2} x}{x^{p-2}} dx + \frac{q^2}{(p-1)(p-2)} \int \frac{\operatorname{sh}^q x}{x^{p-2}} dx \quad [p > 2].$$ GU ((353))(6a)

2. $$\int \frac{\operatorname{ch}^q x}{x^p} dx = -\frac{(p-2) \operatorname{ch}^q x + qx \operatorname{ch}^{q-1} x \operatorname{sh} x}{(p-1)(p-2) x^{p-1}} - \frac{q(q-1)}{(p-1)(p-2)} \int \frac{\operatorname{ch}^{q-2} x}{x^{p-2}} dx + \frac{q^2}{(p-1)(p-2)} \int \frac{\operatorname{ch}^q x}{x^{p-2}} dx \quad [p > 2].$$ GU ((353))(7a)

3. $$\int \frac{\operatorname{sh} x}{x^{2n}} dx = -\frac{1}{x(2n-1)!} \left\{ \sum_{k=0}^{n-2} \frac{(2k+1)!}{x^{2k+1}} \operatorname{ch} x + \sum_{k=0}^{n-1} \frac{(2k)!}{x^{2k}} \operatorname{sh} x \right\} + \frac{1}{(2n-1)!} \operatorname{chi}(x).$$ GU ((353))(6b)

4. $$\int \frac{\operatorname{sh} x}{x^{2n+1}} dx = -\frac{1}{x(2n)!} \left\{ \sum_{k=0}^{n-1} \frac{(2k)!}{x^{2k}} \operatorname{ch} x + \sum_{k=0}^{n-1} \frac{(2k+1)!}{x^{2k+1}} \operatorname{sh} x \right\} + \frac{1}{(2n)!} \operatorname{shi}(x).$$ GU ((353))(6b)

5. $$\int \frac{\operatorname{ch} x}{x^{2n}} dx = -\frac{1}{x(2n-1)!} \left\{ \sum_{k=0}^{n-2} \frac{(2k+1)!}{x^{2k+1}} \operatorname{sh} x + \sum_{k=0}^{n-1} \frac{(2k)!}{x^{2k}} \operatorname{ch} x \right\} + \frac{1}{(2n-1)!} \operatorname{shi}(x).$$ GU ((353))(7b)

6. $$\int \frac{\operatorname{ch} x}{x^{2n+1}} dx = -\frac{1}{(2n)!\, x} \left\{ \sum_{k=0}^{n-1} \frac{(2k)!}{x^{2k}} \operatorname{sh} x + \sum_{k=0}^{n-1} \frac{(2k+1)!}{x^{2k+1}} \operatorname{ch} x \right\} + \frac{1}{(2n)!} \operatorname{chi}(x).$$ GU ((353))(7b)

7. $$\int \frac{\operatorname{sh}^{2m} x}{x} dx = \frac{1}{2^{2m-1}} \sum_{k=0}^{m-1} (-1)^k \binom{2m}{k} \operatorname{chi}(2m-2k) x + \frac{(-1)^m}{2^{2m}} \binom{2m}{m} \ln x.$$ GU ((353))(6c)

8. $$\int \frac{\operatorname{sh}^{2m+1} x}{x} dx = \frac{1}{2^{2m}} \sum_{k=0}^{m} (-1)^k \binom{2m+1}{k} \operatorname{shi}(2m-2k+1) x.$$ GU ((353))(6d)

9. $$\int \frac{\operatorname{ch}^{2m} x}{x} dx = \frac{1}{2^{2m-1}} \sum_{k=0}^{m-1} \binom{2m}{k} \operatorname{chi}(2m-2k) x + \frac{1}{2^{2m}} \binom{2m}{m} \ln x.$$ GU ((353))(7c)

10. $$\int \frac{\operatorname{ch}^{2m+1} x}{x} dx = \frac{1}{2^{2m}} \sum_{k=0}^{m} \binom{2m+1}{k} \operatorname{chi}(2m-2k+1) x.$$ GU ((353))(7c)

11. $$\int \frac{\operatorname{sh}^{2m} x}{x^2} dx = \frac{(-1)^{m-1}}{2^{2m} x} \binom{2m}{m} + \frac{1}{2^{2m-1}} \sum_{k=0}^{m-1} (-1)^{k+1} \binom{2m}{k} \left\{ \frac{\operatorname{ch}(2m-2k) x}{x} - (2m-2k) \operatorname{shi}(2m-2k) x \right\}.$$

12. $$\int \frac{\operatorname{sh}^{2m+1} x}{x^2} dx = \frac{1}{2^{2m}} \sum_{k=0}^{m} (-1)^{k+1} \binom{2m+1}{k} \times \left\{ \frac{\operatorname{sh}(2m-2k+1) x}{x} - (2m-2k+1) \operatorname{chi}(2m-2k+1) x \right\}.$$

13. $$\int \frac{\operatorname{ch}^{2m} x}{x^2} dx = -\frac{1}{2^{2m} x} \binom{2m}{m} - \frac{1}{2^{2m-1}} \sum_{k=0}^{m-1} \binom{2m}{k} \left\{ \frac{\operatorname{ch}(2m-2k) x}{x} - (2m-2k) \operatorname{shi}(2m-2k) x \right\}.$$

14. $$\int \frac{\operatorname{ch}^{2m+1} x}{x^2} dx = -\frac{1}{2^{2m}} \sum_{k=0}^{m} \binom{2m+1}{k} \times \left\{ \frac{\operatorname{ch}(2m-2k+1) x}{x} - (2m-2k+1) \operatorname{shi}(2m-2k+1) x \right\}.$$

2.476

1. $$\int \frac{\operatorname{sh} kx}{a+bx} dx = \frac{1}{b} \left[\operatorname{ch} \frac{ka}{b} \operatorname{shi}(u) - \operatorname{sh} \frac{ka}{b} \operatorname{chi}(u) \right];$$
$$= \frac{1}{2b} \left[\exp\left(-\frac{ka}{b}\right) \operatorname{Ei}(u) - \exp\left(\frac{ka}{b}\right) \operatorname{Ei}(-u) \right]$$
$$\left[u = \frac{k}{b}(a+bx) \right].$$

2. $$\int \frac{\operatorname{ch} kx}{a+bx} dx = \frac{1}{b} \left[\operatorname{ch} \frac{ka}{b} \operatorname{chi}(u) - \operatorname{sh} \frac{ka}{b} \operatorname{shi}(u) \right];$$
$$= \frac{1}{2b} \left[\exp\left(-\frac{ka}{b}\right) \operatorname{Ei}(u) + \exp\left(\frac{ka}{b}\right) \operatorname{Ei}(-u) \right]$$
$$\left[u = \frac{k}{b}(a+bx) \right].$$

3. $\int \frac{\operatorname{sh} kx}{(a+bx)^2} dx = -\frac{1}{b} \cdot \frac{\operatorname{sh} kx}{a+bx} + \frac{k}{b} \int \frac{\operatorname{ch} kx}{a+bx} dx$ (see **2.476** 2.).

4. $\int \frac{\operatorname{ch} kx}{(a+bx)^2} dx = -\frac{1}{b} \cdot \frac{\operatorname{ch} kx}{a+bx} + \frac{k}{b} \int \frac{\operatorname{sh} kx}{a+bx} dx$ (see **2.476** 1.).

5. $\int \frac{\operatorname{sh} kx}{(a+bx)^3} dx = -\frac{\operatorname{sh} kx}{2b(a+bx)^2} - \frac{k \operatorname{ch} kx}{2b^2(a+bx)} + \frac{k^2}{2b^2} \int \frac{\operatorname{sh} kx}{a+bx} dx$ (see **2.476** 1.).

6. $\int \frac{\operatorname{ch} kx}{(a+bx)^3} dx = -\frac{\operatorname{ch} kx}{2b(a+bx)^2} - \frac{k \operatorname{sh} kx}{2b^2(a+bx)} + \frac{k^2}{2b^2} \int \frac{\operatorname{ch} kx}{a+bx} dx$ (see **2.476** 2.).

7. $\int \frac{\operatorname{sh} kx}{(a+bx)^4} dx = -\frac{\operatorname{sh} kx}{3b(a+bx)^3} - \frac{k \operatorname{ch} kx}{6b^2(a+bx)^2} - \frac{k^2 \operatorname{sh} kx}{6b^3(a+bx)} + \frac{k^3}{6b^3} \int \frac{\operatorname{ch} kx}{a+bx} dx$ (see **2.476** 2.).

8. $\int \frac{\operatorname{ch} kx}{(a+bx)^4} dx = -\frac{\operatorname{ch} kx}{3b(a+bx)^3} - \frac{k \operatorname{sh} kx}{6b^2(a+bx)^2} - \frac{k^2 \operatorname{ch} kx}{6b^3(a+bx)} + \frac{k^3}{6b^3} \int \frac{\operatorname{sh} kx}{a+bx} dx$ (see **2.476** 1.).

9. $\int \frac{\operatorname{sh} kx}{(a+bx)^5} dx = -\frac{\operatorname{sh} kx}{4b(a+bx)^4} - \frac{k \operatorname{ch} kx}{12b^2(a+bx)^3} - \frac{k^2 \operatorname{sh} kx}{24b^3(a+bx)^2} - \frac{k^3 \operatorname{ch} kx}{24b^4(a+bx)} + \frac{k^4}{24b^4} \int \frac{\operatorname{sh} kx}{a+bx} dx$ (see **2.476** 1.).

10. $\int \frac{\operatorname{ch} kx}{(a+bx)^5} dx = -\frac{\operatorname{ch} kx}{4b(a+bx)^4} - \frac{k \operatorname{sh} kx}{12b^2(a+bx)^3} - \frac{k^2 \operatorname{ch} kx}{24b^3(a+bx)^2} - \frac{k^3 \operatorname{sh} kx}{24b^4(a+bx)} + \frac{k^4}{24b^4} \int \frac{\operatorname{ch} kx}{a+bx} dx$ (see **2.476** 2.).

11. $\int \frac{\operatorname{sh} kx}{(a+bx)^6} dx = -\frac{\operatorname{sh} kx}{5b(a+bx)^5} - \frac{k \operatorname{ch} kx}{20b^2(a+bx)^4} - \frac{k^2 \operatorname{sh} kx}{60b^3(a+bx)^3} - \frac{k^3 \operatorname{ch} kx}{120b^4(a+bx)^2} - \frac{k^4 \operatorname{sh} kx}{120b^5(a+bx)} + \frac{k^5}{120b^5} \int \frac{\operatorname{ch} kx}{a+bx} dx$ (see **2.476** 2.).

12. $\int \frac{\operatorname{ch} kx}{(a+bx)^6} dx = -\frac{\operatorname{ch} kx}{5b(a+bx)^5} - \frac{k \operatorname{sh} kx}{20b^2(a+bx)^4} - \frac{k^2 \operatorname{ch} kx}{60b^3(a+bx)^3} - \frac{k^3 \operatorname{sh} kx}{120b^4(a+bx)^2} - \frac{k^4 \operatorname{ch} kx}{120b^5(a+bx)} + \frac{k^5}{120b^5} \int \frac{\operatorname{sh} kx}{a+bx} dx$ (see **2.476** 1.).

2.477

1. $\int \frac{x^p\, dx}{\operatorname{sh}^q x} = \frac{-px^{p-1} \operatorname{sh} x - (q-2) x^p \operatorname{ch} x}{(q-1)(q-2) \operatorname{sh}^{q-1} x} + \frac{p(p-1)}{(q-1)(q-2)} \int \frac{x^{p-2}}{\operatorname{sh}^{q-2} x} dx - \frac{q-2}{q-1} \int \frac{x^p\, dx}{\operatorname{sh}^{q-2} x}$ $[q > 2]$. GU ((353))(8a)

2. $\int \frac{x^p\, dx}{\operatorname{ch}^q x} = \frac{px^{p-1} \operatorname{ch} x + (q-2) x^p \operatorname{sh} x}{(q-1)(q-2) \operatorname{ch}^{q-1} x} - \frac{p(p-1)}{(q-1)(q-2)} \int \frac{x^{p-2}\, dx}{\operatorname{ch}^{q-2} x} + \frac{q-2}{q-1} \int \frac{x^p\, dx}{\operatorname{ch}^{q-2} x}$ $[q > 2]$. GU ((353))(10a)

3. $\int \frac{x^n}{\operatorname{sh} x} dx = \sum_{k=0}^{\infty} \frac{(2-2^{2k}) B_{2k}}{(n+2k)(2k)!} x^{n+2k}$ $[|x| < \pi, n > 0]$. GU ((353))(8b)

4. $$\int \frac{x^n}{\operatorname{ch} x}\,dx = \sum_{k=0}^{\infty} \frac{E_{2k} x^{n+2k+1}}{(n+2k+1)(2k)!} \qquad \left[|x| < \frac{\pi}{2},\ n \geqslant 0\right].$$ GU ((353))(10b)

5. $$\int \frac{dx}{x^n \operatorname{sh} x} = -[1+(-1)^n]\frac{2^{n-1}-1}{n!} B_n \ln x + \sum_{\substack{k=0\\ k\neq \frac{n}{2}}}^{\infty} \frac{2-2^{2k}}{(2k-n)(2k)!} B_{2k} x^{2k-n} \qquad [|x| < \pi,\ n \geqslant 1].$$ GU ((353))(9b)

6. $$\int \frac{dx}{x^n \operatorname{ch} x} = \sum_{\substack{k=0\\ k\neq \frac{n-1}{2}}}^{\infty} \frac{E_{2k}}{(2k-n+1)(2k)!} x^{2k-n+1} + \frac{1}{2}[1-(-1)^{n-1}] + \frac{E_{n-1}}{(n-1)!} \ln x \qquad \left[|x| < \frac{\pi}{2}\right].$$ GU ((353))(11b)

7. $$\int \frac{x^n}{\operatorname{sh}^2 x}\,dx = -x^n \operatorname{cth} x + n\sum_{k=0}^{\infty} \frac{2^{2k} B_{2k}}{(n+2k-1)(2k)!} x^{n+2k-1} \qquad [n > 1,\quad |x| < \pi].$$ GU ((353))(8c)

8. $$\int \frac{x^n}{\operatorname{ch}^2 x}\,dx = x^n \operatorname{th} x - n\sum_{k=1}^{\infty} \frac{2^{2k}(2^{2k}-1) B_{2k}}{(n+2k-1)(2k)!} x^{n+2k-1} \qquad \left[n > 1,\quad |x| < \frac{\pi}{2}\right].$$ GU ((353))(10c)

9. $$\int \frac{dx}{x^n \operatorname{sh}^2 x} = -\frac{\operatorname{cth} x}{x^n} - [1-(-1)^n]\frac{2^n n}{(n+1)!} B_{n+1} \ln x - \frac{n}{x^{n+1}} \sum_{\substack{k=0\\ k\neq \frac{n+1}{2}}}^{\infty} \frac{B_{2k}}{(2k-n-1)(2k)!}(2x)^{2k} \qquad [|x| < \pi].$$ GU ((353))(9c)

10. $$\int \frac{dx}{x^n \operatorname{ch}^2 x} = \frac{\operatorname{th} x}{x^n} + [1-(-1)^n]\frac{2^n(2^{n+1}-1)n'}{(n+1)!} B_{n+1} \ln x + \frac{n}{x^{n+1}} \sum_{\substack{k=1\\ k\neq \frac{n+1}{2}}} \frac{(2^{2k}-1) B_{2k}}{(2k-n-1)(2k)!}(2x)^{2k} \qquad \left[|x| < \frac{\pi}{2}\right].$$ GU ((353))(11c)

11. $$\int \frac{x}{\operatorname{sh}^{2n} x}\,dx = \sum_{k=1}^{n-1} (-1)^k \frac{(2n-2)(2n-4)\dots(2n-2k+2)}{(2n-1)(2n-3)\dots(2n-2k+1)} \times \left\{\frac{x \operatorname{ch} x}{\operatorname{sh}^{2n-2k+1} x} + \frac{1}{(2n-2k)\operatorname{sh}^{2n-2k} x}\right\} + (-1)^{n-1}\frac{(2n-2)!!}{(2n-1)!!}\int \frac{x\,dx}{\operatorname{sh}^2 x}$$

(see **2.477** 17.). GU ((353))(8e)

12. $$\int \frac{x}{\operatorname{sh}^{2n-1} x}\,dx = \sum_{k=1}^{n-1} (-1)^k \frac{(2n-3)(2n-5)\dots(2n-2k+1)}{(2n-2)(2n-4)\dots(2n-2k)} \times \left\{\frac{x \operatorname{ch} x}{\operatorname{sh}^{2n-2k} x} + \frac{1}{(2n-2k-1)\operatorname{sh}^{2n-2k-1} x}\right\} + (-1)^{n-1}\frac{(2n-3)!!}{(2n-2)!!}\int \frac{x\,dx}{\operatorname{sh} x}$$

(see **2.477** 15.). GU ((353))(8e)

13. $$\int \frac{x}{\operatorname{ch}^{2n} x}\,dx = \sum_{k=1}^{n-1} \frac{(2n-2)(2n-4)\ldots(2n-2k+2)}{(2n-1)(2n-3)\ldots(2n-2k+1)} \times \left\{\frac{x \operatorname{sh} x}{\operatorname{ch}^{2n-2k+1} x} + \frac{1}{(2n-2k)\operatorname{ch}^{2n-2k} x}\right\} + \frac{(2n-2)!!}{(2n-1)!!}\int \frac{x\,dx}{\operatorname{ch}^2 x}$$
(see 2.477 18.). GU ((353))(10e)

14. $$\int \frac{x}{\operatorname{ch}^{2n-1} x}\,dx = \sum_{k=1}^{n-1} \frac{(2n-3)(2n-5)\ldots(2n-2k+1)}{(2n-2)(2n-4)\ldots(2n-2k)} \times \left\{\frac{x \operatorname{sh} x}{\operatorname{ch}^{2n-2k} x} + \frac{1}{(2n-2k-1)\operatorname{ch}^{2n-2k-1} x}\right\} + \frac{(2n-3)!!}{(2n-2)!!}\int \frac{x\,dx}{\operatorname{ch} x}$$
(see 2.477 16.). GU ((353))(10e)

15. $$\int \frac{x\,dx}{\operatorname{sh} x} = \sum_{k=0}^{\infty} \frac{2-2^{2k}}{(2k+1)(2k)!} B_{2k} x^{2k+1} \qquad |x| < \pi.$$ GU ((353))(8b)a

16. $$\int \frac{x\,dx}{\operatorname{ch} x} = \sum_{k=0}^{\infty} \frac{E_{2k} x^{2k+2}}{(2k+2)(2k)!} \qquad |x| < \frac{\pi}{2}.$$ GU ((353))(10b)a

17. $$\int \frac{x\,dx}{\operatorname{sh}^2 x} = -x \operatorname{cth} x + \ln \operatorname{sh} x.$$ MZ 257

18. $$\int \frac{x\,dx}{\operatorname{ch}^2 x} = x \operatorname{th} x - \ln \operatorname{ch} x.$$ MZ 262

19. $$\int \frac{x\,dx}{\operatorname{sh}^3 x} = -\frac{x \operatorname{ch} x}{2 \operatorname{sh}^2 x} - \frac{1}{2 \operatorname{sh} x} - \frac{1}{2}\int \frac{x\,dx}{\operatorname{sh} x}$$ (see 2.477 15.). MZ 257

20. $$\int \frac{x\,dx}{\operatorname{ch}^3 x} = \frac{x \operatorname{sh} x}{2 \operatorname{ch}^2 x} + \frac{1}{2 \operatorname{ch} x} + \frac{1}{2}\int \frac{x\,dx}{\operatorname{ch} x}$$ (see 2.477 16.). MZ 262

21. $$\int \frac{x\,dx}{\operatorname{sh}^4 x} = -\frac{x \operatorname{ch} x}{3 \operatorname{sh}^3 x} - \frac{1}{6 \operatorname{sh}^2 x} + \frac{2}{3} x \operatorname{cth} x - \frac{2}{3} \ln \operatorname{sh} x.$$ MZ 258

22. $$\int \frac{x\,dx}{\operatorname{ch}^4 x} = \frac{x \operatorname{sh} x}{3 \operatorname{ch}^3 x} + \frac{1}{6 \operatorname{ch}^2 x} + \frac{2}{3} x \operatorname{th} x - \frac{2}{3} \ln \operatorname{ch} x.$$ MZ 262

23. $$\int \frac{x\,dx}{\operatorname{sh}^5 x} = -\frac{x \operatorname{ch} x}{4 \operatorname{sh}^4 x} - \frac{1}{12 \operatorname{sh}^3 x} + \frac{3x \operatorname{ch} x}{8 \operatorname{sh}^2 x} + \frac{3}{8 \operatorname{sh} x} + \frac{3}{8}\int \frac{x\,dx}{\operatorname{sh} x}$$ (see 2.477 15.). MZ 258

24. $$\int \frac{x\,dx}{\operatorname{ch}^5 x} = \frac{x \operatorname{sh} x}{4 \operatorname{ch}^4 x} + \frac{1}{12 \operatorname{ch}^3 x} + \frac{3x \operatorname{sh} x}{8 \operatorname{ch}^2 x} + \frac{3}{8 \operatorname{ch} x} + \frac{3}{8}\int \frac{x\,dx}{\operatorname{ch} x}$$ (see 2.477 16.). MZ 262

2.478

1. $$\int \frac{x^n \operatorname{ch} x\,dx}{(a+b \operatorname{sh} x)^m} = -\frac{x^n}{(m-1)b(a+b \operatorname{sh} x)^{m-1}} + \frac{n}{(m-1)b}\int \frac{x^{n-1}\,dx}{(a+b \operatorname{sh} x)^{m-1}} \qquad [m \neq 1].$$ MZ 263

2. $$\int \frac{x^n \operatorname{sh} x\,dx}{(a+b \operatorname{ch} x)^m} = -\frac{x^n}{(m-1)b(a+b \operatorname{ch} x)^{m-1}} + \frac{n}{(m-1)b}\int \frac{x^{n-1}\,dx}{(a+b \operatorname{ch} x)^{m-1}} \qquad [m \neq 1].$$ MZ 263

3. $$\int \frac{x\,dx}{1+\operatorname{ch} x} = x \operatorname{th} \frac{x}{2} - 2 \ln \operatorname{ch} \frac{x}{2}.$$

4. $\int \frac{x\,dx}{1-\operatorname{ch} x} = x \operatorname{cth} \frac{x}{2} - 2 \ln \operatorname{sh} \frac{x}{2}$.

5. $\int \frac{x \operatorname{sh} x\,dx}{(1+\operatorname{ch} x)^2} = -\frac{x}{1+\operatorname{ch} x} + \operatorname{th} \frac{x}{2}$.

6. $\int \frac{x \operatorname{sh} x\,dx}{(1-\operatorname{ch} x)^2} = \frac{x}{1-\operatorname{ch} x} - \operatorname{cth} \frac{x}{2}$. MZ 262-264

7. $\int \frac{x\,dx}{\operatorname{ch} 2x - \cos 2t} = \frac{1}{2 \sin 2t} [L(u+t) - L(u-t) - 2L(t)]$

$[u = \operatorname{arctg}(\operatorname{th} x \operatorname{ctg} t),\ t \neq \pm n\pi]$. LO III 402

8. $\int \frac{x \operatorname{ch} x\,dx}{\operatorname{ch} 2x - \cos 2t} = \frac{1}{2 \sin t} \left[L\left(\frac{u+t}{2}\right) - L\left(\frac{u-t}{2}\right) + \right.$

$\left. + L\left(\pi - \frac{v+t}{2}\right) + L\left(\frac{v-t}{2}\right) - 2L\left(\frac{t}{2}\right) - 2L\left(\frac{\pi-t}{2}\right) \right]$

$\left[u = 2 \operatorname{arctg}\left(\operatorname{th} \frac{x}{2} \cdot \operatorname{ctg} \frac{t}{2}\right),\quad v = 2 \operatorname{arctg}\left(\operatorname{cth} \frac{x}{2} \cdot \operatorname{ctg} \frac{t}{2}\right);\ t \neq \pm n\pi \right]$. LO III 403

2.479

1. $\int x^p \frac{\operatorname{sh}^{2m} x}{\operatorname{ch}^n x} dx = \sum_{k=0}^{m} (-1)^{m+k} \binom{m}{k} \int \frac{x^p\,dx}{\operatorname{ch}^{n-2k} x}$ (see 2.477 2.).

2. $\int x^p \frac{\operatorname{sh}^{2m+1} x}{\operatorname{ch}^n x} dx = \sum_{k=0}^{m} (-1)^{m+k} \binom{m}{k} \int x^p \frac{\operatorname{sh} x}{\operatorname{ch}^{n-2k} x} dx$

$[n > 1]$ (see 2.479 3.)

3. $\int x^p \frac{\operatorname{sh} x}{\operatorname{ch}^n x} dx = -\frac{x^p}{(n-1) \operatorname{ch}^{n-1} x} + \frac{p}{n-1} \int \frac{x^{p-1}\,dx}{\operatorname{ch}^{n-1} x}$

$[n > 1]$ (see 2.477 2.). GU ((353))(12)

4. $\int x^p \frac{\operatorname{ch}^{2m} x}{\operatorname{sh}^n x} dx = \sum_{k=0}^{m} \binom{m}{k} \int \frac{x^p\,dx}{\operatorname{sh}^{n-2k} x}$ (see 2.477 1.).

5. $\int x^p \frac{\operatorname{ch}^{2m+1} x}{\operatorname{sh}^n x} dx = \sum_{k=0}^{m} \binom{m}{k} \int \frac{x^p \operatorname{ch} x}{\operatorname{sh}^{n-2k} x} dx$ (see 2.479 6.).

6. $\int x^p \frac{\operatorname{ch} x}{\operatorname{sh}^n x} dx = -\frac{x^p}{(n-1) \operatorname{sh}^{n-1} x} + \frac{p}{n-1} \int \frac{x^{p-1}\,dx}{\operatorname{sh}^{n-1} x}$

$[n > 1]$ (see 2.477 1.). GU ((353))(13c)

7. $\int x^p \operatorname{th} x\,dx = \sum_{k=1}^{\infty} \frac{2^{2k}(2^{2k}-1) B_{2k}}{(2k+p)(2k)!} x^{p+2k}$ $\left[p \geqslant -1,\ |x| < \frac{\pi}{2}\right]$.

GU ((353))(12d)

8. $\int x^p \operatorname{cth} x\,dx = \sum_{k=0}^{\infty} \frac{2^{2k} B_{2k}}{(p+2k)(2k)!} x^{p+2k}$ $[p \geqslant +1,\ |x| < \pi]$.

GU ((353))(13d)

9. $\int \frac{x \operatorname{ch} x}{\operatorname{sh}^2 x} dx = \ln \operatorname{th} \frac{x}{2} - \frac{x}{\operatorname{sh} x}$.

10. $\int \frac{x \operatorname{sh} x}{\operatorname{ch}^2 x} dx = -\frac{x}{\operatorname{ch} x} + \operatorname{arctg}(\operatorname{sh} x)$. MZ 263

2.48 Combinations of hyperbolic functions, exponentials, and powers

2.481

1. $\int e^{ax} \operatorname{sh}(bx+c)\,dx = \frac{e^{ax}}{a^2-b^2}[a \operatorname{sh}(bx+c) - b \operatorname{ch}(bx+c)]$ $[a^2 \neq b^2]$.

2. $\int e^{ax} \operatorname{ch}(bx+c)\,dx = \frac{e^{ax}}{a^2-b^2}[a \operatorname{ch}(bx+c) - b \operatorname{sh}(bx+c)]$ $[a^2 \neq b^2]$.

For $a^2 = b^2$:

3. $\int e^{ax} \operatorname{sh}(ax+c)\,dx = -\frac{1}{2} xe^{-c} + \frac{1}{4a} e^{2ax+c}$.

4. $\int e^{-ax} \operatorname{sh}(ax+c)\,dx = \frac{1}{2} xe^{c} + \frac{1}{4a} e^{-(2ax+c)}$.

5. $\int e^{ax} \operatorname{ch}(ax+c)\,dx = \frac{1}{2} xe^{-c} + \frac{1}{4a} e^{2ax+c}$.

6. $\int e^{-ax} \operatorname{ch}(ax+c)\,dx = \frac{1}{2} xe^{c} - \frac{1}{4a} e^{-(2ax+c)}$. MZ 275-277

2.482

1. $\int x^p e^{ax} \operatorname{sh} bx\,dx = \frac{1}{2}\left\{\int x^p e^{(a+b)x}\,dx - \int x^p e^{(a-b)x}\,dx\right\}$ $[a^2 \neq b^2]$ (see 2.321).

2. $\int x^p e^{ax} \operatorname{ch} bx\,dx = \frac{1}{2}\left\{\int x^p e^{(a+b)x}\,dx + \int x^p e^{(a-b)x}\,dx\right\}$ $[a^2 \neq b^2]$ (see 2.321).

For $a^2 = b^2$:

3. $\int x^p e^{ax} \operatorname{sh} ax\,dx = \frac{1}{2}\int x^p e^{2ax}\,dx - \frac{x^{p+1}}{2(p+1)}$ (see 2.321).

4. $\int x^p e^{-ax} \operatorname{sh} ax\,dx = \frac{x^{p+1}}{2(p+1)} - \frac{1}{2}\int x^p e^{-2ax}\,dx$ (see 2.321).

5. $\int x^p e^{ax} \operatorname{ch} ax\,dx = \frac{x^{p+1}}{2(p+1)} + \frac{1}{2}\int x^p e^{2ax}\,dx$ (see 2.321).

MZ 276, 278

2.483

1. $\int xe^{ax} \operatorname{sh} bx\,dx = \frac{e^{ax}}{a^2-b^2}\left[\left(ax - \frac{a^2+b^2}{a^2-b^2}\right)\operatorname{sh} bx - \left(bx - \frac{2ab}{a^2-b^2}\right)\operatorname{ch} bx\right]$ $[a^2 \neq b^2]$.

2. $\int xe^{ax} \operatorname{ch} bx\,dx = \frac{e^{ax}}{a^2-b^2}\left[\left(ax - \frac{a^2+b^2}{a^2-b^2}\right)\operatorname{ch} bx - \left(bx - \frac{2ab}{a^2-b^2}\right)\operatorname{sh} bx\right]$ $[a^2 \neq b^2]$.

3. $\int x^2 e^{ax} \operatorname{sh} bx\,dx = \frac{e^{ax}}{a^2-b^2}\left\{\left[ax^2 - \frac{2(a^2+b^2)}{a^2-b^2}x + \frac{2a(a^2+3b^2)}{(a^2-b^2)^2}\right]\operatorname{sh} bx - \left[bx^2 - \frac{4ab}{a^2-b^2}x + \frac{2b(3a^2+b^2)}{(a^2-b^2)^2}\right]\operatorname{ch} x\right\}$ $[a^2 \neq b^2]$.

4. $\int x^2 e^{ax} \operatorname{ch} bx\,dx = \frac{e^{ax}}{a^2-b^2}\left\{\left[ax^2 - \frac{2(a^2+b^2)}{a^2-b^2}x + \frac{2a(a^2+3b^2)}{(a^2-b^2)^2}\right]\operatorname{ch} bx - \left[bx^2 - \frac{4ab}{a^2-b^2}x + \frac{2b(3a^2+b^2)}{(a^2-b^2)^2}\right]\operatorname{sh} x\right\}$ $[a^2 \neq b^2]$.

For $a^2 = b^2$:

5. $\int xe^{ax} \operatorname{sh} ax\, dx = \frac{e^{2ax}}{4a}\left(x - \frac{1}{2a}\right) - \frac{x^2}{4}.$

6. $\int xe^{-ax} \operatorname{sh} ax\, dx = \frac{e^{-2ax}}{4a}\left(x + \frac{1}{2a}\right) + \frac{x^2}{4}.$ MZ 276, 278

7. $\int xe^{ax} \operatorname{ch} ax\, dx = \frac{x^2}{4} + \frac{e^{2ax}}{4a}\left(x - \frac{1}{2a}\right).$

8. $\int xe^{-ax} \operatorname{ch} ax\, dx = \frac{x^2}{4} - \frac{e^{-2ax}}{4a}\left(x + \frac{1}{2a}\right).$

9. $\int x^2 e^{ax} \operatorname{sh} ax\, dx = \frac{e^{2ax}}{4a}\left(x^2 - \frac{x}{a} + \frac{1}{2a^2}\right) - \frac{x^3}{6}.$

10. $\int x^2 e^{-ax} \operatorname{sh} ax\, dx = \frac{e^{-2ax}}{4a}\left(x^2 + \frac{x}{a} + \frac{1}{2a^2}\right) + \frac{x^3}{6}.$

11. $\int x^2 e^{ax} \operatorname{ch} ax\, dx = \frac{x^3}{6} + \frac{e^{2ax}}{4a}\left(x^2 - \frac{x}{a} + \frac{1}{2a^2}\right).$

2.484

1. $\int e^{ax} \operatorname{sh} bx \frac{dx}{x} = \frac{1}{2}\{\operatorname{Ei}[(a+b)x] - \operatorname{Ei}[(a-b)x]\}$ $[a^2 \neq b^2].$

2. $\int e^{ax} \operatorname{ch} bx \frac{dx}{x} = \frac{1}{2}\{\operatorname{Ei}[(a+b)x] + \operatorname{Ei}[(a-b)x]\}$ $[a^2 \neq b^2].$

3. $\int e^{ax} \operatorname{sh} bx \frac{dx}{x^2} = -\frac{e^{ax} \operatorname{sh} bx}{2x} + \frac{1}{2}\{(a+b)\operatorname{Ei}[(a+b)x] - (a-b)\operatorname{Ei}[(a-b)x]\}$ $[a^2 \neq b^2].$

4. $\int e^{ax} \operatorname{ch} bx \frac{dx}{x^2} = -\frac{e^{ax} \operatorname{ch} bx}{2x} + \frac{1}{2}\{(a+b)\operatorname{Ei}[(a+b)x] + (a-b)\operatorname{Ei}[(a-b)x]\}$ $[a^2 \neq b^2].$

For $a^2 = b^2$:

5. $\int e^{ax} \operatorname{sh} ax \frac{dx}{x} = \frac{1}{2}[\operatorname{Ei}(2ax) - \ln x].$

6. $\int e^{-ax} \operatorname{sh} ax \frac{dx}{x} = \frac{1}{2}[\ln x - \operatorname{Ei}(-2ax)].$

7. $\int e^{ax} \operatorname{ch} ax \frac{dx}{x} = \frac{1}{2}[\ln x + \operatorname{Ei}(2ax)].$

8. $\int e^{ax} \operatorname{sh} ax \frac{dx}{x^2} = -\frac{1}{2x}(e^{2ax} - 1) + a\operatorname{Ei}(2ax).$

9. $\int e^{-ax} \operatorname{sh} ax \frac{dx}{x^2} = -\frac{1}{2x}(1 - e^{-2ax}) + a\operatorname{Ei}(-2ax).$

10. $\int e^{ax} \operatorname{ch} ax \frac{dx}{x^2} = -\frac{1}{2x}(e^{2ax} + 1) + a\operatorname{Ei}(2ax).$ MZ 276, 278

2.5-2.6 Trigonometric Functions

2.50 Introduction

2.501 Integrals of the form $\int R(\sin x, \cos x)\, dx$ can always be reduced to integrals of rational functions by means of the substitution $t = \operatorname{tg} \frac{x}{2}$.

2.502 If $R(\sin x, \cos x)$ satisfies the relation

$$R(\sin x, \cos x) = -R(-\sin x, \cos x),$$

it is convenient to make the substitution $t = \cos x$.

2.503 If this function satisfies the relation

$$R(\sin x, \cos x) = -R(\sin x, -\cos x),$$

it is convenient to make the substitution $t = \sin x$.

2.504 If this function satisfies the relation

$$R(\sin x, \cos x) = R(-\sin x, -\cos x),$$

it is convenient to make the substitution $t = \operatorname{tg} x$.

2.51-2.52 Powers of trigonometric functions

2.510

$$\int \sin^p x \cos^q x\,dx = -\frac{\sin^{p-1} x \cos^{q+1} x}{q+1} + \frac{p-1}{q+1}\int \sin^{p-2} x \cos^{q+2} x\,dx;$$

$$= -\frac{\sin^{p-1} x \cos^{q+1} x}{p+q} + \frac{p-1}{p+q}\int \sin^{p-2} x \cos^{q} x\,dx;$$

$$= \frac{\sin^{p+1} x \cos^{q+1} x}{p+1} + \frac{p+q+2}{p+1}\int \sin^{p+2} x \cos^{q} x\,dx;$$

$$= \frac{\sin^{p+1} x \cos^{q-1} x}{p+1} + \frac{q-1}{p+1}\int \sin^{p+2} x \cos^{q-2} x\,dx;$$

$$= \frac{\sin^{p+1} x \cos^{q-1} x}{p+q} + \frac{q-1}{p+q}\int \sin^{p} x \cos^{q-2} x\,dx;$$

$$= -\frac{\sin^{p+1} x \cos^{q+1} x}{q+1} + \frac{p+q+2}{q+1}\int \sin^{p} x \cos^{q+2} x\,dx;$$

$$= \frac{\sin^{p-1} x \cos^{q-1} x}{p+q}\left\{\sin^2 x - \frac{q-1}{p+q-2}\right\} +$$

$$+ \frac{(p-1)(q-1)}{(p+q)(p+q-2)}\int \sin^{p-2} x \cos^{q-2} x\,dx.$$ FI II 89, TI 214

2.511

1. $$\int \sin^p x \cos^{2n} x\,dx = \frac{\sin^{p+1} x}{2n+p}\left\{\cos^{2n-1} x + \right.$$

$$\left. + \sum_{k=1}^{n-1} \frac{(2n-1)(2n-3)\ldots(2n-2k+1)\cos^{2n-2k-1} x}{(2n+p-2)(2n+p-4)\ldots(2n+p-2k)}\right\} +$$

$$+ \frac{(2n-1)!!}{(2n+p)(2n+p-2)\ldots(p+2)}\int \sin^p x\,dx.$$

This formula is applicable for arbitrary real p except for the following negative even integers: $-2, -4, \ldots, -2n$. If p is a natural number and $n = 0$, we have:

2. $$\int \sin^{2l} x\,dx = -\frac{\cos x}{2l}\left\{\sin^{2l-1} x + \right.$$

$$\left. + \sum_{k=1}^{l-1} \frac{(2l-1)(2l-3)\ldots(2l-2k+1)}{2^k(l-1)(l-2)\ldots(l-k)}\sin^{2l-2k-1} x\right\} + \frac{(2l-1)!!}{2^l\, l!}\,x$$

(see also 2.513 1.). TI (232)

3. $$\int \sin^{2l+1} x\,dx = -\frac{\cos x}{2l+1}\left\{\sin^{2l} x + \right.$$

$$\left. + \sum_{k=0}^{l-1} \frac{2^{k+1}\, l(l-1)\ldots(l-k)}{(2l-1)(2l-3)\ldots(2l-2k-1)}\sin^{2l-2k-2} x\right\}$$

(see also 2.513 2.). TI (233)

4. $$\int \sin^p x \cos^{2n+1} x\,dx = \frac{\sin^{p+1} x}{2n+p+1}\left\{\cos^{2n} x + \sum_{k=1}^{n} \frac{2^k n(n-1)\dots(n-k+1)\cos^{2n-2k} x}{(2n+p-1)(2n+p-3)\dots(2n+p-2k+1)}\right\}.$$

This formula is applicable for arbitrary real p except for the negative odd integers: $-1, -3, \dots, -(2n+1)$.

2.512

1. $$\int \cos^p x \sin^{2n} x\,dx = -\frac{\cos^{p+1} x}{2n+p}\left\{\sin^{2n-1} x + \sum_{k=1}^{n-1} \frac{(2n-1)(2n-3)\dots(2n-2k+1)\sin^{2n-2k-1} x}{(2n+p-2)(2n+p-4)\dots(2n+p-2k)}\right\} + \frac{(2n-1)!!}{(2n+p)(2n+p-2)\dots(p+2)}\int \cos^p x\,dx.$$

This formula is applicable for arbitrary real p except for the following negative even integers: $-2, -4, \dots, -2n$. If p is a natural number and $n=0$, we have

2. $$\int \cos^{2l} x\,dx = \frac{\sin x}{2l}\left\{\cos^{2l-1} x + \sum_{k=1}^{l-1} \frac{(2l-1)(2l-3)\dots(2l-2k+1)}{2^k(l-1)(l-2)\dots(l-k)}\cos^{2l-2k-1} x\right\} + \frac{(2l-1)!!}{2^l\, l!}x$$

(see also **2.513** 3.). TI (230)

3. $$\int \cos^{2l+1} x\,dx = \frac{\sin x}{2l+1}\left\{\cos^{2l} x + \sum_{k=0}^{l-1} \frac{2^{k+1} l(l-1)\dots(l-k)}{(2l-1)(2l-3)\dots(2l-2k-1)}\cos^{2l-2k-2} x\right\}$$

(see also **2.513** 4.). TI (231)

4. $$\int \cos^p x \sin^{2n+1} x\,dx = -\frac{\cos^{p+1} x}{2n+p+1}\left\{\sin^{2n} x + \sum_{k=1}^{n} \frac{2^k n(n-1)\dots(n-k+1)\sin^{2n-2k} x}{(2n+p-1)(2n+p-3)\dots(2n+p-2k+1)}\right\}.$$

This formula is applicable for arbitrary real p except for the following negative odd integers: $-1, -3, \dots, -(2n+1)$.

2.513

1. $$\int \sin^{2n} x\,dx = \frac{1}{2^{2n}}\binom{2n}{n}x + \frac{(-1)^n}{2^{2n-1}}\sum_{k=0}^{n-1}(-1)^k\binom{2n}{k}\frac{\sin(2n-2k)x}{2n-2k}$$

(see also **2.511** 2.). TI (226)

2. $$\int \sin^{2n+1} x\,dx = \frac{1}{2^{2n}}(-1)^{n+1}\sum_{k=0}^{n}(-1)^k\binom{2n+1}{k}\frac{\cos(2n+1-2k)x}{2n+1-2k}$$

(see also **2.511** 3.). TI (227)

3. $\int \cos^{2n} x\,dx = \frac{1}{2^{2n}}\binom{2n}{n}x + \frac{1}{2^{2n-1}}\sum_{k=0}^{n-1}\binom{2n}{k}\frac{\sin(2n-2k)x}{2n-2k}$

(see also **2.512** 2.). TI (224)

4. $\int \cos^{2n+1} x\,dx = \frac{1}{2^{2n}}\sum_{k=0}^{n}\binom{2n+1}{k}\frac{\sin(2n-2k+1)x}{2n-2k+1}$

(see also **2.512** 3.). TI (225)

5. $\int \sin^2 x\,dx = -\frac{1}{4}\sin 2x + \frac{1}{2}x = -\frac{1}{2}\sin x\cos x + \frac{1}{2}x.$

6. $\int \sin^3 x\,dx = \frac{1}{12}\cos 3x - \frac{3}{4}\cos x = \frac{1}{3}\cos^3 x - \cos x.$

7. $\int \sin^4 x\,dx = \frac{3x}{8} - \frac{\sin 2x}{4} + \frac{\sin 4x}{32} =$

$= -\frac{3}{8}\sin x\cos x - \frac{1}{4}\sin^3 x\cos x + \frac{3}{8}x.$

8. $\int \sin^5 x\,dx = -\frac{5}{8}\cos x + \frac{5}{48}\cos 3x - \frac{1}{80}\cos 5x =$

$= -\frac{1}{5}\sin^4 x\cos x + \frac{4}{15}\cos^3 x - \frac{4}{5}\cos x.$

9. $\int \sin^6 x\,dx = \frac{5}{16}x - \frac{15}{64}\sin 2x + \frac{3}{64}\sin 4x - \frac{1}{192}\sin 6x =$

$= -\frac{1}{6}\sin^5 x\cos x - \frac{5}{24}\sin^3 x\cos x - \frac{5}{16}\sin x\cos x + \frac{5}{16}x.$

10. $\int \sin^7 x\,dx = -\frac{35}{64}\cos x + \frac{7}{64}\cos 3x - \frac{7}{320}\cos 5x + \frac{1}{448}\cos 7x =$

$= -\frac{1}{7}\sin^6 x\cos x - \frac{6}{35}\sin^4 x\cos x + \frac{8}{35}\cos^3 x - \frac{24}{35}\cos x.$

11. $\int \cos^2 x\,dx = \frac{1}{4}\sin 2x + \frac{x}{2} = \frac{1}{2}\sin x\cos x + \frac{1}{2}x.$

12. $\int \cos^3 x\,dx = \frac{1}{12}\sin 3x + \frac{3}{4}\sin x = \sin x - \frac{1}{3}\sin^3 x.$

13. $\int \cos^4 x\,dx = \frac{3}{8}x + \frac{1}{4}\sin 2x + \frac{1}{32}\sin 4x =$

$= \frac{3}{8}x + \frac{3}{8}\sin x\cos x + \frac{1}{4}\sin x\cos^3 x.$

14. $\int \cos^5 x\,dx = \frac{5}{8}\sin x + \frac{5}{48}\sin 3x + \frac{1}{80}\sin 5x =$

$= \frac{4}{5}\sin x - \frac{4}{15}\sin^3 x + \frac{1}{5}\cos^4 x\sin x.$

15. $\int \cos^6 x\,dx = \frac{5}{16}x + \frac{15}{64}\sin 2x + \frac{3}{64}\sin 4x + \frac{1}{192}\sin 6x =$

$= \frac{5}{16}x + \frac{5}{16}\sin x\cos x + \frac{5}{24}\sin x\cos^3 x + \frac{1}{6}\sin x\cos^5 x.$

16. $\int \cos^7 x\,dx = \frac{35}{64}\sin x + \frac{7}{64}\sin 3x + \frac{7}{320}\sin 5x + \frac{1}{448}\sin 7x =$

$= \frac{24}{35}\sin x - \frac{8}{35}\sin^3 x + \frac{6}{35}\sin x\cos^4 x + \frac{1}{7}\sin x\cos^6 x.$

17. $\int \sin x \cos^2 x \, dx = -\frac{1}{4}\left\{\frac{1}{3}\cos 3x + \cos x\right\} = -\frac{\cos^3 x}{3}.$

18. $\int \sin x \cos^3 x \, dx = -\frac{\cos^4 x}{4}.$

19. $\int \sin x \cos^4 x \, dx = -\frac{\cos^5 x}{5}.$

20. $\int \sin^2 x \cos x \, dx = -\frac{1}{4}\left\{\frac{1}{3}\sin 3x - \sin x\right\} = \frac{\sin^3 x}{3}.$

21. $\int \sin^2 x \cos^2 x \, dx = -\frac{1}{8}\left\{\frac{1}{4}\sin 4x - x\right\}.$

22. $\int \sin^2 x \cos^3 x \, dx = -\frac{1}{16}\left\{\frac{1}{5}\sin 5x + \frac{1}{3}\sin 3x - 2\sin x\right\} =$
$= \frac{\sin^3 x}{5}\left(\cos^2 x + \frac{2}{3}\right) = \frac{\sin^3 x}{5}\left(\frac{5}{3} - \sin^2 x\right).$

23. $\int \sin^2 x \cos^4 x \, dx = \frac{x}{16} + \frac{1}{64}\sin 2x - \frac{1}{64}\sin 4x - \frac{1}{192}\sin 6x.$

24. $\int \sin^3 x \cos x \, dx = \frac{1}{8}\left(\frac{1}{4}\cos 4x - \cos 2x\right) = \frac{\sin^4 x}{4}.$

25. $\int \sin^3 x \cos^2 x \, dx = \frac{1}{16}\left(\frac{1}{5}\cos 5x - \frac{1}{3}\cos 3x - 2\cos x\right) =$
$= \frac{1}{5}\cos^5 x - \frac{1}{3}\cos^3 x.$

26. $\int \sin^3 x \cos^3 x \, dx = \frac{1}{32}\left(\frac{1}{6}\cos 6x - \frac{3}{2}\cos 2x\right).$

27. $\int \sin^3 x \cos^4 x \, dx = \frac{1}{7}\cos^3 x\left(-\frac{2}{5} - \frac{3}{5}\sin^2 x + \sin^4 x\right).$

28. $\int \sin^4 x \cos x \, dx = \frac{\sin^5 x}{5}.$

29. $\int \sin^4 x \cos^2 x \, dx = \frac{1}{16}x - \frac{1}{64}\sin 2x - \frac{1}{64}\sin 4x + \frac{1}{192}\sin 6x.$

30. $\int \sin^4 x \cos^3 x \, dx = \frac{1}{7}\sin^3 x\left(\frac{2}{5} + \frac{3}{5}\cos^2 x - \cos^4 x\right).$

31. $\int \sin^4 x \cos^4 x \, dx = \frac{3}{128}x - \frac{1}{128}\sin 4x + \frac{1}{1024}\sin 8x.$

2.514 $\int \frac{\sin^p x}{\cos^{2n} x} dx = \frac{\sin^{p+1} x}{2n-1}\left\{\sec^{2n-1} x + \right.$
$$+ \sum_{k=1}^{n-1} \frac{(2n-p-2)(2n-p-4)\dots(2n-p-2k)}{(2n-3)(2n-5)\dots(2n-2k-1)}\sec^{2n-2k-1}x\Big\} +$$
$$+ \frac{(2n-p-2)(2n-p-4)\dots(-p+2)(-p)}{(2n-1)!!}\int \sin^p x \, dx.$$

This formula is applicable for arbitrary real p. For $\int \sin^p x \, dx$, where p is a natural number, see **2.511** 2., 3. and **2.513** 1., 2. If $n = 0$ and p is a negative integer, we have for this integral:

2.515

1. $$\int \frac{dx}{\sin^{2l} x} = -\frac{\cos x}{2l-1}\left\{\operatorname{cosec}^{2l-1} x + \sum_{k=1}^{l-1} \frac{2^k (l-1)(l-2)\ldots(l-k)}{(2l-3)(2l-5)\ldots(2l-2k-1)} \operatorname{cosec}^{2l-2k-1} x\right\}.$$ TI (242)

2. $$\int \frac{dx}{\sin^{2l+1} x} = -\frac{\cos x}{2l}\left\{\operatorname{cosec}^{2l} x + \sum_{k=1}^{l-1} \frac{(2l-1)(2l-3)\ldots(2l-2k+1)}{2^k (l-1)(l-2)\ldots(l-k)} \operatorname{cosec}^{2l-2k} x\right\} + \frac{(2l-1)!!}{2^l l!} \ln \operatorname{tg} \frac{x}{2}.$$ TI (243)

2.516

1. $$\int \frac{\sin^p x\, dx}{\cos^{2n+1} x} = \frac{\sin^{p+1}}{2n}\left\{\sec^{2n} x + \sum_{k=1}^{n-1} \frac{(2n-p-1)(2n-p-3)\ldots(2n-p-2k+1)}{2^k (n-1)(n-2)\ldots(n-k)} \sec^{2n-2k} x\right\} + \frac{(2n-p-1)(2n-p-3)\ldots(3-p)(1-p)}{2^n n!} \int \frac{\sin^p x}{\cos x}\, dx.$$

This formula is applicable for arbitrary real p. For $n=0$ and p a natural number, we have

2. $$\int \frac{\sin^{2l+1} x\, dx}{\cos x} = -\sum_{k=1}^{l} \frac{\sin^{2k} x}{2k} - \ln \cos x.$$

3. $$\int \frac{\sin^{2l} x\, dx}{\cos x} = -\sum_{k=1}^{l} \frac{\sin^{2k-1} x}{2k-1} + \ln \operatorname{tg}\left(\frac{\pi}{4} + \frac{x}{2}\right).$$

2.517

1. $$\int \frac{dx}{\sin^{2m+1} x \cos x} = -\sum_{k=1}^{m} \frac{1}{(2m-2k+2)\sin^{2m-2k+2} x} + \ln \operatorname{tg} x.$$

2. $$\int \frac{dx}{\sin^{2m} x \cos x} = -\sum_{k=1}^{m} \frac{1}{(2m-2k+1)\sin^{2m-2k+1} x} + \ln \operatorname{tg}\left(\frac{\pi}{4} - \frac{x}{2}\right).$$

2.518

1. $$\int \frac{\sin^p x}{\cos^2 x}\, dx = \frac{\sin^{p-1} x}{\cos x} - (p-1)\int \sin^{p-2} x\, dx.$$

2. $$\int \frac{\cos^p x\, dx}{\sin^{2n} x} = -\frac{\cos^{p+1} x}{2n-1}\left\{\operatorname{cosec}^{2n-1} x + \sum_{k=1}^{n-1} \frac{(2n-p-2)(2n-p-4)\ldots(2n-p-2k)}{(2n-3)(2n-5)\ldots(2n-2k-1)} \operatorname{cosec}^{2n-2k-1} x\right\} + \frac{(2n-p-2)(2n-p-4)\ldots(2-p)(-p)}{(2n-1)!!} \int \cos^p x\, dx.$$

This formula is applicable for arbitrary real p. For $\int \cos^p x\,dx$ where p is a natural number, see **2.512** 2., 3. and **2.513** 3., 4. If $n=0$ and p is a negative integer, we have for this integral:

2.519

1. $$\int \frac{dx}{\cos^{2l} x} = \frac{\sin x}{2l-1}\left\{\sec^{2l-1} x + \sum_{k=1}^{l-1} \frac{2^k (l-1)(l-2)\ldots(l-k)}{(2l-3)(2l-5)\ldots(2l-2k-1)} \sec^{2l-2k-1} x\right\}.$$ TI (240)

2. $$\int \frac{dx}{\cos^{2l+1} x} = \frac{\sin x}{2l}\left\{\sec^{2l} x + \sum_{k=1}^{l-1} \frac{(2l-1)(2l-3)\ldots(2l-2k+1)}{2^k (l-1)(l-2)\ldots(l-k)} \sec^{2l-2k} x\right\} + \frac{(2l-1)!!}{2^l l!} \ln \operatorname{tg}\left(\frac{\pi}{4}+\frac{x}{2}\right).$$ TI (241)

2.521

1. $$\int \frac{\cos^p x\,dx}{\sin^{2n+1} x} = -\frac{\cos^{p+1} x}{2n}\left\{\operatorname{cosec}^{2n} x + \sum_{k=1}^{n-1} \frac{(2n-p-1)(2n-p-3)\ldots(2n-p-2k+1)}{2^k (n-1)(n-2)\ldots(n-k)} \operatorname{cosec}^{2n-2k} x\right\} + \frac{(2n-p-1)(2n-p-3)\ldots(3-p)(1-p)}{2^n\cdot n!} \int \frac{\cos^p x}{\sin x}\,dx.$$

This formula is applicable for arbitrary real p. For $n=0$ and p a natural number, we have

2. $$\int \frac{\cos^{2l+1} x\,dx}{\sin x} = \sum_{k=1}^{l} \frac{\cos^{2k} x}{2k} + \ln \sin x.$$

3. $$\int \frac{\cos^{2l} x\,dx}{\sin x} = \sum_{k=1}^{l} \frac{\cos^{2k-1} x}{2k-1} + \ln \operatorname{tg} \frac{x}{2}.$$

2.522

1. $$\int \frac{dx}{\sin x \cos^{2m+1} x} = \sum_{k=1}^{m} \frac{1}{(2m-2k+2)\cos^{2m-2k+2} x} + \ln \operatorname{tg} x.$$

2. $$\int \frac{dx}{\sin x \cos^{2m} x} = \sum_{k=1}^{m} \frac{1}{(2m-2k+1)\cos^{2m-2k+1} x} + \ln \operatorname{tg} \frac{x}{2}.$$ GW ((331))(15)

2.523 $$\int \frac{\cos^m x}{\sin^2 x}\,dx = -\frac{\cos^{m-1} x}{\sin x} - (m-1)\int \cos^{m-2} x\,dx.$$

2.524

1. $$\int \frac{\sin^{2n+1} x}{\cos^m x} dx = \sum_{\substack{k=0 \\ k \neq \frac{m-1}{2}}}^{n} (-1)^{k+1} \binom{n}{k} \frac{\cos^{2k-m+1} x}{2k-m+1} + s(-1)^{\frac{m+1}{2}} \binom{n}{\frac{m-1}{2}} \ln \cos x.$$ GU ((331))(11d)

2. $$\int \frac{\cos^{2n+1} x}{\sin^m x} dx = \sum_{\substack{k=0 \\ k \neq \frac{m-1}{2}}}^{n} (-1)^{k} \binom{n}{k} \frac{\sin^{2k-m+1} x}{2k-m+1} + s(-1)^{\frac{m-1}{2}} \binom{n}{\frac{m-1}{2}} \ln \sin x.$$

[In formulas 2.524 1. and 2.524 2., $s=1$ for m odd and $m < 2n+1$; in other cases, $s=0$.] GU ((331))(13d)

2.525

1. $$\int \frac{dx}{\sin^{2m} x \cos^{2n} x} = \sum_{k=0}^{m+n-1} \binom{m+n-1}{k} \frac{\operatorname{tg}^{2k-2m+1} x}{2k-2m+1}.$$ TI (267)

2. $$\int \frac{dx}{\sin^{2m+1} x \cos^{2n+1} x} = \sum_{k=0}^{m+n} \binom{m+n}{k} \frac{\operatorname{tg}^{2k-2m} x}{2k-2m} + \binom{m+n}{m} \ln \operatorname{tg} x.$$

TI (268), GU ((331))(15f)

2.526

1. $$\int \frac{dx}{\sin x} = \ln \operatorname{tg} \frac{x}{2}.$$

2. $$\int \frac{dx}{\sin^2 x} = -\operatorname{ctg} x.$$

3. $$\int \frac{dx}{\sin^3 x} = -\frac{1}{2} \frac{\cos x}{\sin^2 x} + \frac{1}{2} \ln \operatorname{tg} \frac{x}{2}.$$

4. $$\int \frac{dx}{\sin^4 x} = -\frac{\cos x}{3 \sin^3 x} - \frac{2}{3} \operatorname{ctg} x = -\frac{1}{3} \operatorname{ctg}^3 x - \operatorname{ctg} x.$$

5. $$\int \frac{dx}{\sin^5 x} = -\frac{\cos x}{4 \sin^4 x} - \frac{3}{8} \frac{\cos x}{\sin^2 x} + \frac{3}{8} \ln \operatorname{tg} \frac{x}{2}.$$

6. $$\int \frac{dx}{\sin^6 x} = -\frac{\cos x}{5 \sin^5 x} - \frac{4}{15} \operatorname{ctg}^3 x - \frac{4}{5} \operatorname{ctg} x;$$
$$= -\frac{1}{5} \operatorname{ctg}^5 x - \frac{2}{3} \operatorname{ctg}^3 x - \operatorname{ctg} x.$$

7. $$\int \frac{dx}{\sin^7 x} = -\frac{\cos x}{6 \sin^2 x} \left(\frac{1}{\sin^4 x} + \frac{5}{4 \sin^2 x} + \frac{15}{8} \right) + \frac{5}{16} \ln \operatorname{tg} \frac{x}{2}.$$

8. $$\int \frac{dx}{\sin^8 x} = -\left(\frac{1}{7} \operatorname{ctg}^7 x + \frac{3}{5} \operatorname{ctg}^5 x + \operatorname{ctg}^3 x + \operatorname{ctg} x \right).$$

9. $$\int \frac{dx}{\cos x} = \ln \operatorname{tg} \left(\frac{\pi}{4} + \frac{x}{2} \right) = \ln \operatorname{ctg} \left(\frac{\pi}{4} - \frac{x}{2} \right) = \ln \sqrt{\frac{1+\sin x}{1-\sin x}}.$$

10. $\int \frac{dx}{\cos^2 x} = \operatorname{tg} x.$

11. $\int \frac{dx}{\cos^3 x} = \frac{1}{2} \frac{\sin x}{\cos^2 x} + \frac{1}{2} \ln \operatorname{tg}\left(\frac{\pi}{4} + \frac{x}{2}\right).$

12. $\int \frac{dx}{\cos^4 x} = \frac{\sin x}{3\cos^3 x} + \frac{2}{3} \operatorname{tg} x = \frac{1}{3} \operatorname{tg}^3 x + \operatorname{tg} x.$

13. $\int \frac{dx}{\cos^5 x} = \frac{\sin x}{4\cos^4 x} + \frac{3}{8} \frac{\sin x}{\cos^2 x} + \frac{3}{8} \ln \operatorname{tg}\left(\frac{x}{2} + \frac{\pi}{4}\right).$

14. $\int \frac{dx}{\cos^6 x} = \frac{\sin x}{5\cos^5 x} + \frac{4}{15} \operatorname{tg}^3 x + \frac{4}{5} \operatorname{tg} x = \frac{1}{5} \operatorname{tg}^5 x + \frac{2}{3} \operatorname{tg}^3 x + \operatorname{tg} x.$

15. $\int \frac{dx}{\cos^7 x} = \frac{\sin x}{6\cos^6 x} + \frac{5 \sin x}{24 \cos^4 x} + \frac{5 \sin x}{16 \cos^2 x} + \frac{5}{16} \ln \operatorname{tg}\left(\frac{x}{2} + \frac{\pi}{4}\right).$

16. $\int \frac{dx}{\cos^8 x} = \frac{1}{7} \operatorname{tg}^7 x + \frac{3}{5} \operatorname{tg}^5 x + \operatorname{tg}^3 x + \operatorname{tg} x.$

17. $\int \frac{\sin x}{\cos x} dx = -\ln \cos x.$

18. $\int \frac{\sin^2 x}{\cos x} dx = -\sin x + \ln \operatorname{tg}\left(\frac{\pi}{4} + \frac{x}{2}\right).$

19. $\int \frac{\sin^3 x}{\cos x} dx = -\frac{\sin^2 x}{2} - \ln \cos x = -\frac{1}{2} \cos^2 x - \ln \cos x.$

20. $\int \frac{\sin^4 x}{\cos x} dx = -\frac{1}{3} \sin^3 x - \sin x + \ln \operatorname{tg}\left(\frac{x}{2} + \frac{\pi}{4}\right).$

21. $\int \frac{\sin x \, dx}{\cos^2 x} = \frac{1}{\cos x}.$

22. $\int \frac{\sin^2 x \, dx}{\cos^2 x} = \operatorname{tg} x - x.$

23. $\int \frac{\sin^3 x \, dx}{\cos^2 x} = \cos x + \frac{1}{\cos x}.$

24. $\int \frac{\sin^4 x \, dx}{\cos^2 x} = \operatorname{tg} x + \frac{1}{2} \sin x \cos x - \frac{3}{2} x.$

25. $\int \frac{\sin x \, dx}{\cos^3 x} = \frac{1}{2\cos^2 x} = \frac{1}{2} \operatorname{tg}^2 x.$

26. $\int \frac{\sin^2 x \, dx}{\cos^3 x} = \frac{\sin x}{2\cos^2 x} - \frac{1}{2} \ln \operatorname{tg}\left(\frac{\pi}{4} + \frac{x}{2}\right).$

27. $\int \frac{\sin^3 x \, dx}{\cos^3 x} = \frac{1}{2\cos^2 x} + \ln \cos x.$

28. $\int \frac{\sin^4 x \, dx}{\cos^3 x} = \frac{1}{2} \frac{\sin x}{\cos^2 x} + \sin x - \frac{3}{2} \ln \operatorname{tg}\left(\frac{x}{2} + \frac{\pi}{4}\right).$

29. $\int \frac{\sin x \, dx}{\cos^4 x} = \frac{1}{3\cos^3 x}.$

30. $\int \frac{\sin^2 x \, dx}{\cos^4 x} = \frac{1}{3} \operatorname{tg}^3 x.$

31. $\int \frac{\sin^3 x \, dx}{\cos^4 x} = -\frac{1}{\cos x} + \frac{1}{3\cos^3 x}.$

32. $\int \frac{\sin^4 x \, dx}{\cos^4 x} = \frac{1}{3} \operatorname{tg}^3 x - \operatorname{tg} x + x.$

33. $\int \frac{\cos x \, dx}{\sin x} = \ln \sin x.$

34. $\int \frac{\cos^2 x \, dx}{\sin x} = \cos x + \ln \operatorname{tg} \frac{x}{2}.$

35. $\int \frac{\cos^3 x\,dx}{\sin x} = \frac{\cos^2 x}{2} + \ln \sin x.$

36. $\int \frac{\cos^4 x\,dx}{\sin x} = \frac{1}{3}\cos^3 x + \cos x + \ln \operatorname{tg}\left(\frac{x}{2}\right).$

37. $\int \frac{\cos x}{\sin^2 x}\,dx = -\frac{1}{\sin x}.$

38. $\int \frac{\cos^2 x}{\sin^2 x}\,dx = -\operatorname{ctg} x - x.$

39. $\int \frac{\cos^3 x}{\sin^2 x}\,dx = -\sin x - \frac{1}{\sin x}.$

40. $\int \frac{\cos^4 x}{\sin^2 x}\,dx = -\operatorname{ctg} x - \frac{1}{2}\sin x \cos x - \frac{3}{2}x.$

41. $\int \frac{\cos x}{\sin^3 x}\,dx = -\frac{1}{2\sin^2 x}.$

42. $\int \frac{\cos^2 x}{\sin^3 x}\,dx = -\frac{\cos x}{2\sin^2 x} - \frac{1}{2}\ln \operatorname{tg}\frac{x}{2}.$

43. $\int \frac{\cos^3 x}{\sin^3 x}\,dx = -\frac{1}{2\sin^2 x} - \ln \sin x.$

44. $\int \frac{\cos^4 x}{\sin^3 x}\,dx = -\frac{1}{2}\frac{\cos x}{\sin^2 x} - \cos x - \frac{3}{2}\ln \operatorname{tg}\frac{x}{2}.$

45. $\int \frac{\cos x}{\sin^4 x}\,dx = -\frac{1}{3\sin^3 x}.$

46. $\int \frac{\cos^2 x}{\sin^4 x}\,dx = -\frac{1}{3}\operatorname{ctg}^3 x.$

47. $\int \frac{\cos^3 x}{\sin^4 x}\,dx = \frac{1}{\sin x} - \frac{1}{3\sin^3 x}.$

48. $\int \frac{\cos^4 x}{\sin^4 x}\,dx = -\frac{1}{3}\operatorname{ctg}^3 x + \operatorname{ctg} x + x.$

49. $\int \frac{dx}{\sin x \cos x} = \ln \operatorname{tg} x.$

50. $\int \frac{dx}{\sin x \cos^2 x} = \frac{1}{\cos x} + \ln \operatorname{tg}\frac{x}{2}.$

51. $\int \frac{dx}{\sin x \cos^3 x} = \frac{1}{2\cos^2 x} + \ln \operatorname{tg} x.$

52. $\int \frac{dx}{\sin x \cos^4 x} = \frac{1}{\cos x} + \frac{1}{3\cos^3 x} + \ln \operatorname{tg}\frac{x}{2}.$

53. $\int \frac{dx}{\sin^2 x \cos x} = \ln \operatorname{tg}\left(\frac{\pi}{4} + \frac{x}{2}\right) - \operatorname{cosec} x.$

54. $\int \frac{dx}{\sin^2 x \cos^2 x} = -2\operatorname{ctg} 2x.$

55. $\int \frac{dx}{\sin^2 x \cos^3 x} = \left(\frac{1}{2\cos^2 x} - \frac{3}{2}\right)\frac{1}{\sin x} + \frac{3}{2}\ln\left(\frac{\pi}{4} + \frac{x}{2}\right).$

56. $\int \frac{dx}{\sin^2 x \cos^4 x} = \frac{1}{3\sin x \cos^3 x} - \frac{8}{3}\operatorname{ctg} 2x.$

57. $\int \frac{dx}{\sin^3 x \cos x} = -\frac{1}{2\sin^2 x} + \ln \operatorname{tg} x.$

58. $\int \frac{dx}{\sin^3 x \cos^2 x} = -\frac{1}{\cos x}\left(\frac{1}{2\sin^2 x} - \frac{3}{2}\right) + \frac{3}{2}\ln \operatorname{tg}\frac{x}{2}.$

59. $\int \frac{dx}{\sin^3 x \cos^3 x} = -\frac{2\cos 2x}{\sin^2 2x} + 2\ln \operatorname{tg} x.$

60. $\int \frac{dx}{\sin^3 x \cos^4 x} = \frac{2}{\cos x} + \frac{1}{3\cos^3 x} - \frac{\cos x}{2\sin^2 x} + \frac{5}{2} \ln \operatorname{tg} \frac{x}{2}$.

61. $\int \frac{dx}{\sin^4 x \cos x} = -\frac{1}{\sin x} - \frac{1}{3\sin^3 x} + \ln \operatorname{tg}\left(\frac{x}{2} + \frac{\pi}{4}\right)$.

62. $\int \frac{dx}{\sin^4 x \cos^2 x} = -\frac{1}{3\cos x \sin^3 x} - \frac{8}{3} \operatorname{ctg} 2x$.

63. $\int \frac{dx}{\sin^4 x \cos^3 x} = -\frac{2}{\sin x} - \frac{1}{3\sin^3 x} + \frac{\sin x}{2\cos^2 x} + \frac{5}{2} \ln \operatorname{tg}\left(\frac{x}{2} + \frac{\pi}{4}\right)$.

64. $\int \frac{dx}{\sin^4 x \cos^4 x} = -8 \operatorname{ctg} 2x - \frac{8}{3} \operatorname{ctg}^3 2x$.

2.527

1. $\int \operatorname{tg}^p x \, dx = \frac{\operatorname{tg}^{p-1} x}{p-1} - \int \operatorname{tg}^{p-2} x \, dx \quad [p \neq 1]$.

2. $\int \operatorname{tg}^{2n+1} x \, dx = \sum_{k=1}^{n} (-1)^{n+k} \binom{n}{k} \frac{1}{2k \cos^{2k} x} - (-1)^n \ln \cos x =$

$$= \sum_{k=1}^{n} \frac{(-1)^{k-1} \operatorname{tg}^{2n-2k+2} x}{2n-2k+2} - (-1)^n \ln \cos x.$$

3. $\int \operatorname{tg}^{2n} x \, dx = \sum_{k=1}^{n} (-1)^{k-1} \frac{\operatorname{tg}^{2n-2k+1} x}{2n-2k+1} + (-1)^n x$. GU ((331))(12)

4. $\int \operatorname{ctg}^p x \, dx = -\frac{\operatorname{ctg}^{p-1} x}{p-1} - \int \operatorname{ctg}^{p-2} x \, dx \quad [p \neq 1]$.

5. $\int \operatorname{ctg}^{2n+1} x \, dx = \sum_{k=1}^{n} (-1)^{n+k+1} \binom{n}{k} \frac{1}{2k \sin^{2k} x} + (-1)^n \ln \sin x =$

$$= \sum_{k=1}^{n} (-1)^k \frac{\operatorname{ctg}^{2n-2k+2} x}{2n-2k+2} + (-1)^n \ln \sin x.$$

6. $\int \operatorname{ctg}^{2n} x \, dx = \sum_{k=1}^{n} (-1)^k \frac{\operatorname{ctg}^{2n-2k+1} x}{2n-2k+1} + (-1)^n x$. GU((331))(14)

For special formulas for $p = 1, 2, 3, 4$, see **2.526** 17., **2.526** 33., **2.526** 22., **2.526** 38., **2.526** 27., **2.526** 43., **2.526** 32., **2.526** 48..

2.53-2.54 Sines and cosines of multiple angles and of linear and more complicated functions of the argument

2.531

1. $\int \sin(ax + b) \, dx = -\frac{1}{a} \cos(ax + b)$.

2. $\int \cos(ax + b) \, dx = \frac{1}{a} \sin(ax + b)$.

2.532

1. $\int \sin(ax + b) \sin(cx + d) \, dx = \frac{\sin[(a-c)x + b - d]}{2(a-c)} -$

$$- \frac{\sin[(a+c)x + b + d]}{2(a+c)} \quad [a^2 \neq c^2].$$

2. $$\int \sin(ax+b)\cos(cx+d)\,dx = -\frac{\cos[(a-c)x+b-d]}{2(a-c)} - \frac{\cos[(a+c)x+b+d]}{2(a+c)} \quad [a^2 \neq c^2].$$

3. $$\int \cos(ax+b)\cos(cx+d)\,dx = \frac{\sin[(a-c)x+b-d]}{2(a-c)} + \frac{\sin[(a+c)x+b+d]}{2(a+c)} \quad [a^2 \neq c^2].$$

For $c=a$:

4. $$\int \sin(ax+b)\sin(ax+d)\,dx = \frac{x}{2}\cos(b-d) - \frac{\sin(2ax+b+d)}{4a}.$$

5. $$\int \sin(ax+b)\cos(ax+d)\,dx = \frac{x}{2}\sin(b-d) - \frac{\cos(2ax+b+d)}{4a}.$$

6. $$\int \cos(ax+b)\cos(ax+d)\,dx = \frac{x}{2}\cos(b-d) + \frac{\sin(2ax+b+d)}{4a}.$$

GU ((332))(3)

2.533

1. $$\int \sin ax \cos bx\,dx = -\frac{\cos(a+b)x}{2(a+b)} - \frac{\cos(a-b)x}{2(a-b)} \quad [a^2 \neq b^2].$$

2. $$\int \sin ax \sin bx \sin cx\,dx = -\frac{1}{4}\left\{\frac{\cos(a-b+c)x}{a-b+c} + \frac{\cos(b+c-a)x}{b+c-a} + \frac{\cos(a+b-c)x}{a+b-c} - \frac{\cos(a+b+c)x}{a+b+c}\right\}.$$ PE (376)

3. $$\int \sin ax \cos bx \cos cx\,dx = -\frac{1}{4}\left\{\frac{\cos(a+b+c)x}{a+b+c} - \frac{\cos(b+c-a)x}{b+c-a} + \frac{\cos(a+b-c)x}{a+b-c} + \frac{\cos(a+c-b)x}{a+c-b}\right\}.$$ PE (378)

4. $$\int \cos ax \sin bx \sin cx\,dx = \frac{1}{4}\left\{\frac{\sin(a+b-c)x}{a+b-c} + \frac{\sin(a+c-b)x}{a+c-b} - \frac{\sin(a+b+c)x}{a+b+c} - \frac{\sin(b+c-a)x}{b+c-a}\right\}.$$ PE (379)

5. $$\int \cos ax \cos bx \cos cx\,dx = \frac{1}{4}\left\{\frac{\sin(a+b+c)x}{a+b+c} + \frac{\sin(b+c-a)x}{b+c-a} + \frac{\sin(a+c-b)x}{a+c-b} + \frac{\sin(a+b-c)x}{a+b-c}\right\}.$$ PE (377)

2.534

1. $$\int \frac{\cos px + i\sin px}{\sin nx}\,dx = -2\int \frac{z^{p+n-1}}{1-z^{2n}}\,dz$$ PE (374)

2. $$\int \frac{\cos px + i\sin px}{\cos nx}\,dx = -2i\int \frac{z^{p+n-1}}{1+z^{2n}}\,dz$$ PE (373)

$[z = \cos x + i \sin x]$.

2.535

1. $$\int \sin^p x \sin ax\,dx = \frac{1}{p+a}\left\{-\sin^p x \cos ax + p\int \sin^{p-1} x \cos(a-1)x\,dx\right\}.$$

GU ((332))(5a)

2. $$\int \sin^p x \sin(2n+1)x\,dx = (2n+1)\left\{\int \sin^{p+1} x\,dx + \right.$$

$$\left. + \sum_{k=1}^{n} (-1)^k \frac{[(2n+1)^2-1^2][(2n+1)^2-3^2]\dots[(2n+1)^2-(2k-1)^2]}{(2k+1)!} \int \sin^{2k+p+1} x\,dx\right\};$$

TI (299)

$$= \frac{\Gamma(p+1)}{\Gamma\left(\frac{p+3}{2}+n\right)} \left\{\sum_{k=0}^{n-1} \left[\frac{(-1)^{k-1}\Gamma\left(\frac{p+1}{2}+n-2k\right)}{2^{2k+1}\Gamma(p-2k+1)} \sin^{p-2k} x \cos(2n-2k+1)x + \right.\right.$$

$$\left. + (-1)^k \frac{\Gamma\left(\frac{p-1}{2}+n-2k\right)}{2^{2k+2}\Gamma(p-2k)} \sin^{p-2k-1} x \sin(2n-2k)x\right] +$$

$$\left. + \frac{(-1)^n \Gamma\left(\frac{p+3}{2}-n\right)}{2^{2n}\Gamma(p-2n+1)} \int \sin^{p-2n+1} x\,dx\right\}.$$

GU ((332))(5c)

3. $$\int \sin^p x \sin 2nx\,dx = 2n\left\{\frac{\sin^{p+2} x}{p+2} + \right.$$

$$\left. + \sum_{k=1}^{n-1} (-1)^k \frac{(4n^2-2^2)(4n^2-4^2)\dots[4n^2-(2k)^2]}{(2k+1)!\,(2k+p+2)} \sin^{2k+p+2} x\right\};$$

TI (303)

$$= \frac{\Gamma(p+1)}{\Gamma\left(\frac{p}{2}+n+1\right)} \left\{\sum_{k=0}^{n-1} \frac{(-1)^{k-1}\Gamma\left(\frac{p}{2}+n-2k\right)}{2^{2k+1}\Gamma(p-2k+1)} \sin^{p-2k} x \cos(2n-2k)x - \right.$$

$$\left. - \frac{(-1)^k \Gamma\left(\frac{p}{2}+n-2k-1\right)}{2^{2k+2}\Gamma(p-2k)} \sin^{p-2k-1} x \sin(2n-2k-1)x\right\};$$

$[p$ is not equal to $-2, -4, \dots, -2n]$. GU ((332))(5c)

2.536

1. $$\int \sin^p x \cos ax\,dx = \frac{1}{p+a}\left\{\sin^p x \sin ax - p\int \sin^{p-1} x \sin(a-1)x\,dx\right\}.$$

GU ((332))(6a)

2. $$\int \sin^p x \cos(2n+1)x\,dx = \frac{\sin^{p+1} x}{p+1} +$$

$$+ \sum_{k=1}^{n} (-1)^k \frac{[(2n+1)^2-1^2][(2n+1)^2-3^2]\dots[(2n+1)^2-(2k-1)^2]}{(2k)!\,(2k+p+1)} \sin^{2k+p+1} x;$$

TI (301)

$$= \frac{\Gamma(p+1)}{\Gamma\left(\frac{p+3}{2}+n\right)} \left\{\sum_{k=0}^{n-1} \left[\frac{(-1)^{k}\Gamma\left(\frac{p+1}{2}+n-2k\right)}{2^{2k+1}\Gamma(p-2k+1)} \sin^{p-2k} x \sin(2n-2k+1)x + \right.\right.$$

$$\left. + \frac{(-1)^k \Gamma\left(\frac{p-1}{2}+n-2k\right)}{2^{2k+2}\Gamma(p-2k)} \sin^{p-2k-1} x \cos(2n-2k)x\right] +$$

$$\left. + \frac{(-1)^n \Gamma\left(\frac{p+3}{2}-n\right)}{2^{2n}\Gamma(p-2n+1)} \int \sin^{p-2n} x \cos x\,dx\right\};$$

$[p$ is not equal to $-3, -5, \dots, -(2n+1)]$. GU ((332))(6c)

3. $$\int \sin^p x \cos 2nx\,dx = \int \sin^p x\,dx + \sum_{k=1}^{n} (-1)^k \frac{4n^2 \cdot (4n^2-2^2) \ldots [4n^2-(2k-2)^2]}{(2k)!} \int \sin^{2k+p} x\,dx;$$ TI (300)

$$= \frac{\Gamma(p+1)}{\Gamma\left(\frac{p}{2}+n+1\right)} \left\{ \sum_{k=0}^{n-1} \left[\frac{(-1)^k \Gamma\left(\frac{p}{2}+n-2k\right)}{2^{2k+1}\Gamma(p-2k+1)} \sin^{p-2k} x \sin(2n-2k)x + \right.\right.$$
$$\left. + \frac{(-1)^k \Gamma\left(\frac{p}{2}+n-2k-1\right)}{2^{2k+2}\Gamma(p-2k)} \sin^{p-2k-1} x \cos(2n-2k-1)x \right] +$$
$$\left. + \frac{(-1)^n \Gamma\left(\frac{p}{2}-n+1\right)}{2^{2n}\Gamma(p-2n+1)} \int \sin^{p-2n} x\,dx \right\}.$$ GU ((332))(6c)

2.537

1. $$\int \cos^p x \sin ax\,dx = \frac{1}{p+a}\left\{ -\cos^p x \cos ax + p \int \cos^{p-1} x \sin(a-1)x\,dx \right\}.$$ GU ((332))(7a)

2. $$\int \cos^p x \sin(2n+1)x\,dx = (-1)^{n+1} \left\{ \frac{\cos^{p+1} x}{p+1} + \right.$$
$$\left. + \sum_{k=1}^{n} (-1)^k \frac{[(2n+1)^2-1^2][(2n+1)^2-3^2] \ldots [(2n+1)^2-(2k-1)^2]}{(2k)!\,(2k+p+1)} \cos^{2k+p+1} x \right\};$$ TI 295)

$$= \frac{\Gamma(p+1)}{\Gamma\left(\frac{p+3}{2}+n\right)} \left\{ -\sum_{k=0}^{n-1} \frac{\Gamma\left(\frac{p+1}{2}+n-k\right)}{2^{k+1}\Gamma(p-k+1)} \cos^{p-k} x \cos(2n-k+1)x + \right.$$
$$\left. + \frac{\Gamma\left(\frac{p+3}{2}\right)}{2^n\Gamma(p-n+1)} \int \cos^{p-n} x \sin(n+1)x\,dx \right\};$$

[p is not equal to $-3,\ -5,\ \ldots,\ -(2n+1)$]. GU ((332))(7b)a

3. $$\int \cos^p x \sin 2nx\,dx = (-1)^n \left\{ \frac{\cos^{p+2} x}{p+2} + \sum_{k=1}^{n-1} (-1)^k \frac{(4n^2-2^2)(4n^2-4^2) \ldots [4n^2-(2k)^2]}{(2k+1)!\,(2k+p+2)} \cos^{2k+p+2} x \right\};$$ TI (297)

$$= \frac{\Gamma(p+1)}{\Gamma\left(\frac{p}{2}+n+1\right)} \left\{ -\sum_{k=0}^{n-1} \frac{\Gamma\left(\frac{p}{2}+n-k\right)}{2^{k+1}\Gamma(p-k+1)} \cos^{p-k} x \cos(2n-k)x + \right.$$
$$\left. + \frac{\Gamma\left(\frac{p}{2}+1\right)}{2^n\Gamma(p-n+1)} \int \cos^{p-n} x \sin nx\,dx \right\};$$

[p is not equal to $-2,\ -4,\ \ldots,\ -2n$]. GU ((332))(7b)a

2.538

1. $$\int \cos^p x \cos ax\,dx = \frac{1}{p+a}\left\{ \cos^p x \sin ax + p \int \cos^{p-1} x \cos(a-1)x\,dx \right\}.$$ GU ((332))(8a)

2. $$\int \cos^p x \cos(2n+1)x\,dx = (-1)^n (2n+1) \left\{ \int \cos^{p+1} x\,dx + \right.$$

$$\left. + \sum_{k=1}^{n} (-1)^k \frac{[(2n+1)^2-1^2]\,[(2n+1)^2-3^2]\,\ldots\,[(2n+1)^2-(2k-1)^2]}{(2k+1)!} \int \cos^{2k+p+1} x\,dx \right\};$$

TI (293)

$$= \frac{\Gamma(p+1)}{\Gamma\left(\frac{p+3}{2}+n\right)} \left\{ \sum_{k=0}^{n-1} \frac{\Gamma\left(\frac{p+1}{2}+n-k\right)}{2^{k+1}\Gamma(p-k+1)} \cos^{p-k} x \sin(2n-k+1)x + \right.$$

$$\left. + \frac{\Gamma\left(\frac{p+3}{2}\right)}{2^n\Gamma(p-n+1)} \int \cos^{p-n} x \cos(n+1)x\,dx \right\}.$$

GU ((332))(8b)a

3. $$\int \cos^p x \cos 2nx\,dx = (-1)^n \left\{ \int \cos^p x\,dx + \right.$$

$$\left. + \sum_{k=1}^{n} (-1)^k \frac{4n^2\,[4n^2-2^2]\,\ldots\,[4n^2-(2k-2)^2]}{(2k)!} \int \cos^{2k+p} x\,dx \right\};$$

TI (294)

$$= \frac{\Gamma(p+1)}{\Gamma\left(\frac{p}{2}+n+1\right)} \left\{ \sum_{k=0}^{n-1} \frac{\Gamma\left(\frac{p}{2}+n-k\right)}{2^{k+1}\Gamma(p-k+1)} \cos^{p-k} x \sin(2n-k)x + \right.$$

$$\left. + \frac{\Gamma\left(\frac{p}{2}+1\right)}{2^n\Gamma(p-n+1)} \int \cos^{p-n} x \cos nx\,dx \right\}.$$

GU ((332))(8b)a

2.539

1. $$\int \frac{\sin(2n+1)x}{\sin x}\,dx = 2\sum_{k=1}^{n} \frac{\sin 2kx}{2k} + x.$$

2. $$\int \frac{\sin 2nx}{\sin x}\,dx = 2\sum_{k=1}^{n} \frac{\sin(2k-1)x}{2k-1}.$$

GU ((332))(5e)

3. $$\int \frac{\cos(2n+1)x}{\sin x}\,dx = 2\sum_{k=1}^{n} \frac{\cos 2kx}{2k} + \ln \sin x.$$

4. $$\int \frac{\cos 2nx}{\sin x}\,dx = 2\sum_{k=1}^{n} \frac{\cos(2k-1)x}{2k-1} + \ln \operatorname{tg}\frac{x}{2}.$$

GU((332))(6e)

5. $$\int \frac{\sin(2n+1)x}{\cos x}\,dx = 2\sum_{k=1}^{n} (-1)^{n-k+1} \frac{\cos 2kx}{2k} + (-1)^{n+1} \ln \cos x.$$

6. $$\int \frac{\sin 2nx}{\cos x}\,dx = 2\sum_{k=1}^{n} (-1)^{n-k+1} \frac{\cos(2k-1)x}{2k-1}.$$

GU ((332))(7d)

7. $$\int \frac{\cos(2n+1)x}{\cos x}\,dx = 2\sum_{k=1}^{n} (-1)^{n-k} \frac{\sin 2kx}{2k} + (-1)^n x.$$

8. $$\int \frac{\cos 2nx}{\cos x}\,dx = 2\sum_{k=1}^{n} (-1)^{n-k} \frac{\sin(2k-1)x}{2k-1} + (-1)^n \ln \operatorname{tg}\left(\frac{\pi}{4}+\frac{x}{2}\right).$$

GU ((332))(8d)

2.541

1. $\int \sin(n+1)x \sin^{n-1} x\, dx = \frac{1}{n} \sin^n x \sin nx.$ BI ((71))(1)a

2. $\int \sin(n+1)x \cos^{n-1} x\, dx = -\frac{1}{n} \cos^n x \cos nx.$ BI ((71))(2)a

3. $\int \cos(n+1)x \sin^{n-1} x\, dx = \frac{1}{n} \sin^n x \cos nx.$ BI ((71))(3)a

4. $\int \cos(n+1)x \cos^{n-1} x\, dx = \frac{1}{n} \cos^n x \sin nx.$ BI ((71))(4)a

5. $\int \sin\left[(n+1)\left(\frac{\pi}{2} - x\right)\right] \sin^{n-1} x\, dx = \frac{1}{n} \sin^n x \cos n\left(\frac{\pi}{2} - x\right).$ BI ((71))(5)a

6. $\int \cos\left[(n+1)\left(\frac{\pi}{2} - x\right)\right] \sin^{n-1} x\, dx = -\frac{1}{n} \sin^n x \sin n\left(\frac{\pi}{2} - x\right).$ BI ((71))(6)a

2.542

1. $\int \frac{\sin 2x}{\sin^n x}\, dx = -\frac{2}{(n-2)\sin^{n-2} x}.$

For $n = 2$:

2. $\int \frac{\sin 2x}{\sin^2 x}\, dx = 2 \ln \sin x.$

2.543

1. $\int \frac{\sin 2x\, dx}{\cos^n x} = \frac{2}{(n-2)\cos^{n-2} x}.$

For $n = 2$:

2. $\int \frac{\sin 2x}{\cos^2 x}\, dx = -2 \ln \cos x.$

2.544

1. $\int \frac{\cos 2x\, dx}{\sin x} = 2\cos x + \ln \operatorname{tg} \frac{x}{2}.$

2. $\int \frac{\cos 2x\, dx}{\sin^2 x} = -\operatorname{ctg} x - 2x.$

3. $\int \frac{\cos 2x\, dx}{\sin^3 x} = -\frac{\cos x}{2\sin^2 x} - \frac{3}{2} \ln \operatorname{tg} \frac{x}{2}.$

4. $\int \frac{\cos 2x\, dx}{\cos x} = 2\sin x - \ln \operatorname{tg}\left(\frac{\pi}{4} + \frac{x}{2}\right).$

5. $\int \frac{\cos 2x\, dx}{\cos^2 x} = 2x - \operatorname{tg} x.$

6. $\int \frac{\cos 2x\, dx}{\cos^3 x} = -\frac{\sin x}{2\cos^2 x} + \frac{3}{2} \ln \operatorname{tg}\left(\frac{\pi}{4} + \frac{x}{2}\right).$

7. $\int \frac{\sin 3x\, dx}{\sin x} = x + \sin 2x.$

8. $\int \frac{\sin 3x}{\sin^2 x}\, dx = 3 \ln \operatorname{tg} \frac{x}{2} + 4\cos x.$

9. $\int \frac{\sin 3x}{\sin^3 x}\, dx = -3 \operatorname{ctg} x - 4x.$

2.545

1. $\int \frac{\sin 3x}{\cos^n x}\, dx = \frac{4}{(n-3)\cos^{n-3} x} - \frac{1}{(n-1)\cos^{n-1} x}.$

For $n=1$ and $n=3$:

2. $\int \frac{\sin 3x}{\cos x}\,dx = 2\sin^2 x + \ln\cos x.$

3. $\int \frac{\sin 3x}{\cos^3 x}\,dx = -\frac{1}{2\cos^2 x} - 4\ln\cos x.$

2.546

1. $\int \frac{\cos 3x}{\sin^n x}\,dx = \frac{4}{(n-3)\sin^{n-3} x} - \frac{1}{(n-1)\sin^{n-1} x}.$

For $n=1$ and $n=3$:

2. $\int \frac{\cos 3x}{\sin x}\,dx = -2\sin^2 x + \ln\sin x.$

3. $\int \frac{\cos 3x}{\sin^3 x}\,dx = -\frac{1}{2\sin^2 x} - 4\ln\sin x.$

2.547

1. $\int \frac{\sin nx}{\cos^p x}\,dx = 2\int \frac{\sin(n-1)x\,dx}{\cos^{p-1} x} - \int \frac{\sin(n-2)x\,dx}{\cos^p x}.$

2. $\int \frac{\cos 3x}{\cos x}\,dx = \sin 2x - x.$

3. $\int \frac{\cos 3x}{\cos^2 x}\,dx = 4\sin x - 3\ln\operatorname{tg}\left(\frac{\pi}{4}+\frac{x}{2}\right).$

4. $\int \frac{\cos 3x}{\cos^3 x}\,dx = 4x - 3\operatorname{tg} x.$

2.548

1. $$\int \frac{\sin^m x\,dx}{\sin(2n+1)x} = \frac{1}{2n+1}\sum_{k=0}^{2n}(-1)^{n+k}\cos^m\left[\frac{2k+1}{2(2n+1)}\pi\right]\ln\frac{\sin\left[\frac{(k-n)\pi}{2(2n+1)}+\frac{x}{2}\right]}{\sin\left[\frac{k+n+1}{2(2n+1)}\pi-\frac{x}{2}\right]}$$

[m-a natural number $\leqslant 2n$]. TI (378)

2. $$\int \frac{\sin^{2m} x\,dx}{\sin 2nx} = \frac{(-1)^n}{2n}\left\{\ln\cos x + \sum_{k=1}^{n-1}(-1)^k\cos^{2m}\frac{k\pi}{2n}\ln\left(\cos^2 x - \sin^2\frac{k\pi}{2n}\right)\right\}$$

[m-a natural number $\leqslant n$]. TI (379)

3. $$\int \frac{\sin^{2m+1} x}{\sin 2nx}\,dx = \frac{(-1)^n}{2n}\left\{\ln\operatorname{tg}\left(\frac{\pi}{4}-\frac{x}{2}\right) + \sum_{k=1}^{n-1}(-1)^k\cos^{2m+1}\frac{k\pi}{2n}\ln\left[\operatorname{tg}\left(\frac{n+k}{4n}\pi-\frac{x}{2}\right)\operatorname{tg}\left(\frac{n-k}{4n}\pi-\frac{x}{2}\right)\right]\right\}$$

[m-a natural number $< n$]. TI (380)

4. $$\int \frac{\sin^{2m} x\,dx}{\cos(2n+1)x} = \frac{(-1)^{n+1}}{2n+1}\left\{\ln\operatorname{tg}\left(\frac{\pi}{4}-\frac{x}{2}\right) + \sum_{k=1}^{n}(-1)^k\cos^{2m}\frac{k\pi}{2n+1}\ln\left[\operatorname{tg}\left(\frac{2n+2k+1}{4(2n+1)}\pi-\frac{x}{2}\right)\operatorname{tg}\left(\frac{2n-2k+1}{4(2n+1)}\pi-\frac{x}{2}\right)\right]\right\}$$

[m-a natural number $\leqslant n$]. TI (381)

5. $$\int \frac{\sin^{2m+1} x\,dx}{\cos(2n+1)x} = \frac{(-1)^{n+1}}{2n+1}\left\{\ln\cos x + \sum_{k=1}^{n}(-1)^k \cos^{2m+1}\frac{k\pi}{2n+1}\ln\left(\cos^2 x - \sin^2\frac{k\pi}{2n+1}\right)\right\}$$

[m-a natural number $\leqslant n$]. TI (382)a

6. $$\int \frac{\sin^m x\,dx}{\cos 2nx} = \frac{1}{2n}\sum_{k=0}^{2n-1}(-1)^{n+k}\cos^m\left[\frac{2k+1}{4n}\pi\right]\ln\frac{\sin\left[\frac{2k-2n+1}{8n}\pi + \frac{x}{2}\right]}{\sin\left[\frac{2k+2n+1}{8n}\pi - \frac{x}{2}\right]}$$

[m-a natural number $< 2n$]. TI (377)

7. $$\int \frac{\cos^{2m+1} x\,dx}{\sin(2n+1)x} = \frac{1}{2n+1}\left\{\ln\sin x + \sum_{k=1}^{n}(-1)^k \cos^{2m+1}\frac{k\pi}{2n+1}\ln\left(\sin^2 x - \sin^2\frac{k\pi}{2n+1}\right)\right\}$$

[m-a natural number $\leqslant n$]. TI (376)

8. $$\int \frac{\cos^{2m} x\,dx}{\sin(2n+1)x} = \frac{1}{2n+1}\left\{\ln\operatorname{tg}\frac{x}{2} + \sum_{k=1}^{n}(-1)^k \cos^{2m}\frac{k\pi}{2n+1}\ln\left[\operatorname{tg}\left(\frac{x}{2}+\frac{k\pi}{4n+2}\right)\operatorname{tg}\left(\frac{x}{2}-\frac{k\pi}{4n+2}\right)\right]\right\}$$

[m-a natural number $\leqslant n$]. TI (375)

9. $$\int \frac{\cos^{2m+1} x}{\sin 2nx}\,dx = \frac{1}{2n}\left\{\ln\operatorname{tg}\frac{x}{2} + \sum_{k=1}^{n-1}(-1)^k \cos^{2m+1}\frac{k\pi}{2n}\ln\left[\operatorname{tg}\left(\frac{x}{2}+\frac{k\pi}{4n}\right)\operatorname{tg}\left(\frac{x}{2}-\frac{k\pi}{4n}\right)\right]\right\}$$

[m-a natural number $< n$]. TI (374)

10. $$\int \frac{\cos^{2m} x}{\sin 2nx}\,dx = \frac{1}{2n}\left\{\ln\sin x + \sum_{k=1}^{n-1}(-1)^k \cos^{2m}\frac{k\pi}{2n}\ln\left(\sin^2 x - \sin^2\frac{k\pi}{2n}\right)\right\}$$

[m-a natural number $\leqslant n$]. TI (373)

11. $$\int \frac{\cos^m x}{\cos nx}\,dx = \frac{1}{n}\sum_{k=0}^{n-1}(-1)^k \cos^m\frac{2k+1}{2n}\pi\ln\frac{\sin\left[\frac{2k+1}{4n}\pi + \frac{x}{2}\right]}{\sin\left[\frac{2k+1}{4n}\pi - \frac{x}{2}\right]}$$

[m-a natural number $\leqslant n$]. TI (372)

2.549

1. $$\int \sin x^2\,dx = \sqrt{\frac{\pi}{2}}\,S(x).$$

2. $$\int \cos x^2\,dx = \sqrt{\frac{\pi}{2}}\,C(x).$$

3. $$\int \sin(ax^2+2bx+c)\,dx = \sqrt{\frac{\pi}{2a}}\left\{\cos\frac{ac-b^2}{a}\,S\left(\frac{ax+b}{\sqrt{a}}\right)+ \sin\frac{ac-b^2}{a}\,C\left(\frac{ax+b}{\sqrt{a}}\right)\right\}.$$

4. $$\int \cos(ax^2+2bx+c)\,dx = \sqrt{\frac{\pi}{2a}}\left\{\cos\frac{ac-b^2}{a}\,C\left(\frac{ax+b}{\sqrt{a}}\right)- \sin\frac{ac-b^2}{a}\,S\left(\frac{ax+b}{\sqrt{a}}\right)\right\}.$$

5. $$\int \sin\ln x\,dx = \frac{x}{2}(\sin\ln x - \cos\ln x).$$ PE (444)

6. $$\int \cos\ln x\,dx = \frac{x}{2}(\sin\ln x + \cos\ln x).$$ PE (445)

2.55-2.56 Rational functions of the sine and cosine

2.551

1. $$\int \frac{A+B\sin x}{(a+b\sin x)^n}\,dx = \frac{1}{(n-1)(a^2-b^2)}\left[\frac{(Ab-aB)\cos x}{(a+b\sin x)^{n-1}} + \int \frac{(Aa-Bb)(n-1)+(aB-bA)(n-2)\sin x}{(a+b\sin x)^{n-1}}\,dx\right].$$ TI (358)a

For $n=1$:

2. $$\int \frac{A+B\sin x}{a+b\sin x}\,dx = \frac{B}{b}x + \frac{Ab-aB}{b}\int\frac{dx}{a+b\sin x}$$ (see **2.551** 3.). TI (342)

3. $$\int \frac{dx}{a+b\sin x} = \frac{2}{\sqrt{a^2-b^2}}\operatorname{arctg}\frac{a\operatorname{tg}\frac{x}{2}+b}{\sqrt{a^2-b^2}} \qquad [a^2>b^2];$$

$$= \frac{1}{\sqrt{b^2-a^2}}\ln\frac{a\operatorname{tg}\frac{x}{2}+b-\sqrt{b^2-a^2}}{a\operatorname{tg}\frac{x}{2}+b+\sqrt{b^2-a^2}} \qquad [a^2<b^2].$$

2.552

1. $$\int \frac{A+B\cos x}{(a+b\sin x)^n}\,dx = -\frac{B}{(n-1)b(a+b\sin x)^{n-1}} + A\int\frac{dx}{(a+b\sin x)^n}$$ (see **2.552** 3.). TI (361)

For $n=1$:

2. $$\int \frac{A+B\cos x}{a+b\sin x}\,dx = \frac{B}{b}\ln(a+b\sin x) + A\int\frac{dx}{a+b\sin x}$$ (see **2.551** 3.). TI (344)

3. $$\int \frac{dx}{(a+b\sin x)^n} = \frac{1}{(n-1)(a^2-b^2)}\left\{\frac{b\cos x}{(a+b\sin x)^{n-1}} + \int\frac{(n-1)a-(n-2)b\sin x}{(a+b\sin x)^{n-1}}\,dx\right\}$$ (see **2.551** 1.). TI (359)

2.553

1. $$\int \frac{A+B\sin x}{(a+b\cos x)^n}\,dx = \frac{B}{(n-1)b(a+b\cos x)^{n-1}} + A\int\frac{dx}{(a+b\cos x)^n}$$ (see **2.554** 3.). TI (355)

For $n=1$:

2. $$\int \frac{A+B\sin x}{a+b\cos x}\,dx = -\frac{B}{b}\ln(a+b\cos x)+A\int\frac{dx}{a+b\cos x}$$ (see 2.553 3.). TI (343)

3. $$\int\frac{dx}{a+b\cos x}=\frac{2}{\sqrt{a^2-b^2}}\operatorname{arctg}\frac{\sqrt{a^2-b^2}\operatorname{tg}\frac{x}{2}}{a+b} \qquad [a^2>b^2];$$

$$=\frac{1}{\sqrt{b^2-a^2}}\ln\frac{\sqrt{b^2-a^2}\operatorname{tg}\frac{x}{2}+a+b}{\sqrt{b^2-a^2}\operatorname{tg}\frac{x}{2}-a-b} \qquad [a^2<b^2].$$ TI II 93, 94, TI (305)

2.554

1. $$\int\frac{A+B\cos x}{(a+b\cos x)^n}\,dx=\frac{1}{(n-1)(a^2-b^2)}\left[\frac{(aB-Ab)\sin x}{(a+b\cos x)^{n-1}}+\right.$$
$$\left.+\int\frac{(Aa-bB)(n-1)+(n-2)(aB-bA)\cos x}{(a+b\cos x)^{n-1}}\,dx\right].$$ TI (353)

For $n=1$:

2. $$\int\frac{A+B\cos x}{a+b\cos x}\,dx=\frac{B}{b}x+\frac{Ab-aB}{b}\int\frac{dx}{a+b\cos x}$$ (see 2.553 3.). TI (341)

3. $$\int\frac{dx}{(a+b\cos x)^n}=-\frac{1}{(n-1)(a^2-b^2)}\left\{\frac{b\sin x}{(a+b\cos x)^{n-1}}-\right.$$
$$\left.-\int\frac{(n-1)a-(n-2)b\cos x}{(a+b\cos x)^{n-1}}\,dx\right\}$$ (see 2.554 1.). TI (354)

In integrating the functions in formulas 2.551 3. and 2.553 3, we may not take the integration over points at which the integrand becomes infinite, that is, over the points $x=\arcsin\left(-\frac{a}{b}\right)$ in formula 2.551 3. or over the points $x=\arccos\left(-\frac{a}{b}\right)$ in formula 2.553 3.

2.555 Formulas 2.551 3. and 2.553 3 are not applicable for $a^2=b^2$. Instead, we may use the following formulas in these cases:

1. $$\int\frac{A+B\sin x}{(1\pm\sin x)^n}\,dx=-\frac{1}{2^{n-1}}\left\{2B\sum_{k=0}^{n-2}\binom{n-2}{k}\frac{\operatorname{tg}^{2k+1}\left(\frac{\pi}{4}\mp\frac{x}{2}\right)}{2k+1}\pm\right.$$
$$\left.\pm(A\mp B)\sum_{k=0}^{n-1}\binom{n-1}{k}\frac{\operatorname{tg}^{2k+1}\left(\frac{\pi}{4}\mp\frac{x}{2}\right)}{2k+1}\right\}.$$ TI (361)a

2. $$\int\frac{A+B\cos x}{(1\pm\cos x)^n}\,dx=\frac{1}{2^{n-1}}\left\{2B\sum_{k=0}^{n-2}\binom{n-2}{k}\frac{\operatorname{tg}^{2k+1}\left[\frac{\pi}{4}\mp\left(\frac{\pi}{4}-\frac{x}{2}\right)\right]}{2k+1}\pm\right.$$
$$\left.\pm(A\mp B)\sum_{k=0}^{n-1}\binom{n-1}{k}\frac{\operatorname{tg}^{2k+1}\left[\frac{\pi}{4}\mp\left(\frac{\pi}{4}-\frac{x}{2}\right)\right]}{2k+1}\right\}.$$ TI (356)

For $n=1$:

3. $$\int\frac{A+B\sin x}{1\pm\sin x}\,dx=\pm Bx+(A\mp B)\operatorname{tg}\left(\frac{\pi}{4}\mp\frac{x}{2}\right).$$ TI (250)

4. $\int \frac{A+B\cos x}{1\pm\cos x}\,dx = \pm Bx \pm (A\mp B)\operatorname{tg}\left[\frac{\pi}{4}\mp\left(\frac{\pi}{4}-\frac{x}{2}\right)\right].$ TI (248)

2.556

1. $\int \frac{(1-a^2)\,dx}{1-2a\cos x+a^2} = 2\operatorname{arctg}\left(\frac{1+a}{1-a}\operatorname{tg}\frac{x}{2}\right) \quad [0<a<1,\ |x|<\pi].$ FI II 93

2. $\int \frac{(1-a\cos x)\,dx}{1-2a\cos x+a^2} = \frac{x}{2}+\operatorname{arctg}\left(\frac{1+a}{1-a}\operatorname{tg}\frac{x}{2}\right) \quad [0<a<1,\ |x|<\pi].$

FI II 93

2.557

1. $\int \frac{dx}{(a\cos x+b\sin x)^n} = \frac{1}{\sqrt{(a^2+b^2)^n}}\int \frac{dx}{\sin^n\left(x+\operatorname{arctg}\frac{a}{b}\right)}$

(see 2.515). MZ 173a

2. $\int \frac{\sin x\,dx}{a\cos x+b\sin x} = \frac{ax-b\ln\sin\left(x+\operatorname{arctg}\frac{a}{b}\right)}{a^2+b^2}.$

3. $\int \frac{\cos x\,dx}{a\cos x+b\sin x} = \frac{ax+b\ln\sin\left(x+\operatorname{arctg}\frac{a}{b}\right)}{a^2+b^2}.$ MZ 174a

4. $\int \frac{dx}{a\cos x+b\sin x} = \frac{\ln\operatorname{tg}\left[\frac{1}{2}\left(x+\operatorname{arctg}\frac{a}{b}\right)\right]}{\sqrt{a^2+b^2}}.$

5. $\int \frac{dx}{(a\cos x+b\sin x)^2} = -\frac{\operatorname{ctg}\left(x+\operatorname{arctg}\frac{a}{b}\right)}{a^2+b^2} =$

$= +\frac{1}{a^2+b^2}\cdot\frac{a\sin x-b\cos x}{a\cos x+b\sin x}.$ MZ 174a

2.558

1. $\int \frac{A+B\cos x+C\sin x}{(a+b\cos x+c\sin x)^n}\,dx = \frac{(Bc-Cb)+(Ac-Ca)\cos x-(Ab-Ba)\sin x}{(n-1)(a^2-b^2-c^2)(a+b\cos x+c\sin x)^{n-1}}+$

$+\frac{1}{(n-1)(a^2-b^2-c^2)}\int \frac{(n-1)(Aa-Bb-Cc)-(n-2)[(Ab-Ba)\cos x-(Ac-Ca)\sin x]}{(a+b\cos x+c\sin x)^{n-1}}\,dx$

$[n\neq 1,\quad a^2\neq b^2+c^2];$

$=\frac{Cb-Bc+Ca\cos x-Ba\sin x}{(n-1)a(a+b\cos x+c\sin x)^n}+\left(\frac{A}{a}+\frac{n(Bb+Cc)}{(n-1)a^2}\right)(-c\cos x+b\sin x)\times$

$\times\frac{(n-1)!}{(2n-1)!!}\sum_{k=0}^{n-1}\frac{(2n-2k-3)!!}{(n-k-1)!\,a^k}\cdot\frac{1}{(a+b\cos x+c\sin x)^{n-k}} \qquad [n\neq 1,\quad a^2=b^2+c^2].$

For $n=1$:

2. $\int \frac{A+B\cos x+C\sin x}{a+b\cos x+c\sin x}\,dx = \frac{Bc-Cb}{b^2+c^2}\ln(a+b\cos x+c\sin x)+\frac{Bb+Cc}{b^2+c^2}x+$

$+\left(A-\frac{Bb+Cc}{b^2+c^2}a\right)\int\frac{dx}{a+b\cos x+c\sin x}$ (see 2.558 4.). GU ((331))(18)

3. $\int \frac{dx}{(a+b\cos x+c\sin x)^n} = \int \frac{d(x-\alpha)}{[a+r\cos(x-\alpha)]^n},$

where $b=r\cos\alpha,\quad c=r\sin\alpha$ (see 2.554 3.).

4. $\int \frac{dx}{a+b\cos x+c\sin x}=$

$$=\frac{2}{\sqrt{a^2-b^2-c^2}}\operatorname{arctg}\frac{(a-b)\operatorname{tg}\frac{x}{2}+c}{\sqrt{a^2-b^2-c^2}}\qquad [a^2>b^2+c^2];$$ TI (253), FI II 94

$$=\frac{1}{\sqrt{b^2+c^2-a^2}}\ln\frac{(a-b)\operatorname{tg}\frac{x}{2}+c-\sqrt{b^2+c^2-a^2}}{(a-b)\operatorname{tg}\frac{x}{2}+c+\sqrt{b^2+c^2-a^2}}\qquad [a^2<b^2+c^2];$$ TI (253)a

$$=\frac{1}{c}\ln\left(a+c\cdot\operatorname{tg}\frac{x}{2}\right)\qquad [a=b];$$

$$=\frac{-2}{c+(a-b)\operatorname{tg}\frac{x}{2}}\qquad [a^2=b^2+c^2].$$ TI (253)a

2.559

1. $$\int\frac{dx}{[a(1+\cos x)+c\sin x]^2}=\frac{1}{c^3}\left[\frac{c(a\sin x-c\cos x)}{a(1+\cos x)+c\sin x}-a\ln\left(a+c\operatorname{tg}\frac{x}{2}\right)\right].$$

2. $$\int\frac{A+B\cos x+C\sin x}{(a_1+b_1\cos x+c_1\sin x)(a_2+b_2\cos x+c_2\sin x)}dx=$$

$$=A_0\ln\frac{a_1+b_1\cos x+c_1\sin x}{a_2+b_2\cos x+c_2\sin x}+A_1\int\frac{dx}{a_1+b_1\cos x+c_1\sin x}+$$

$$+A_2\int\frac{dx}{a_2+b_2\cos x+c_2\sin x},$$

where

$$A_0=\frac{\begin{vmatrix}A&B&C\\a_1&b_1&c_1\\a_2&b_2&c_2\end{vmatrix}}{\begin{vmatrix}a_1&b_1\\a_2&b_2\end{vmatrix}^2-\begin{vmatrix}b_1&c_1\\b_2&c_2\end{vmatrix}^2+\begin{vmatrix}c_1&a_1\\c_2&a_2\end{vmatrix}^2},\qquad A_1=\frac{\begin{vmatrix}\begin{vmatrix}B&C\\b_1&c_1\end{vmatrix}&\begin{vmatrix}A&C\\a_1&c_1\end{vmatrix}&\begin{vmatrix}B&A\\b_1&a_1\end{vmatrix}\\a_1&b_1&c_1\\a_2&b_2&c_2\end{vmatrix}}{\begin{vmatrix}a_1&b_1\\a_2&b_2\end{vmatrix}^2-\begin{vmatrix}b_1&c_1\\b_2&c_2\end{vmatrix}^2+\begin{vmatrix}c_1&a_1\\c_2&a_2\end{vmatrix}^2},$$

$$A_2=\frac{\begin{vmatrix}\begin{vmatrix}C&B\\c_2&b_2\end{vmatrix}&\begin{vmatrix}C&A\\c_2&a_2\end{vmatrix}&\begin{vmatrix}A&B\\a_2&b_2\end{vmatrix}\\a_1&b_1&c_1\\a_2&b_2&c_2\end{vmatrix}}{\begin{vmatrix}a_1&b_1\\a_2&b_2\end{vmatrix}^2-\begin{vmatrix}b_1&c_1\\b_2&c_2\end{vmatrix}^2+\begin{vmatrix}c_1&a_1\\c_2&a_2\end{vmatrix}^2};$$

$$\left[\begin{vmatrix}a_1&b_1\\a_2&b_2\end{vmatrix}^2+\begin{vmatrix}c_1&a_1\\c_2&a_2\end{vmatrix}^2\neq\begin{vmatrix}b_1&c_1\\b_2&c_2\end{vmatrix}^2\right]$$ (see **2.558** 4.). GU ((331))(19)

3. $$\int\frac{A\cos^2 x+2B\sin x\cos x+C\sin^2 x}{a\cos^2 x+2b\sin x\cos x+c\sin^2 x}dx=$$

$$=\frac{1}{4b^2+(a-c)^2}\{[4Bb+(A-C)(a-c)]x+[(A-C)b-B(a-c)]\times$$

$$\times\ln(a\cos^2 x+2b\sin x\cos x+c\sin^2 x)+$$

$$+[2(A+C)b^2-2Bb(a+c)+(aC-Ac)(a-c)]f(x)\},$$

where

$$f(x) = \frac{1}{2\sqrt{b^2-ac}} \ln \frac{c \operatorname{tg} x + b - \sqrt{b^2-ac}}{c \operatorname{tg} x + b + \sqrt{b^2-ac}} \qquad [b^2 > ac];$$

$$= \frac{1}{\sqrt{ac-b^2}} \operatorname{arctg} \frac{c \operatorname{tg} x + b}{\sqrt{ac-b^2}} \qquad [b^2 < ac];$$

$$= -\frac{1}{c \operatorname{tg} x + b} \qquad [b^2 = ac].$$

GU ((331))(24)

2.561

1. $$\int \frac{(A+B\sin x)\,dx}{\sin x\,(a+b\sin x)} = \frac{A}{a} \ln \operatorname{tg} \frac{x}{2} + \frac{Ba-Ab}{a} \int \frac{dx}{a+b\sin x}$$
(see 2.551 3.). TI (348)

2. $$\int \frac{(A+B\sin x)\,dx}{\sin x\,(a+b\cos x)} = \frac{A}{a^2-b^2} \left\{ a \ln \operatorname{tg} \frac{x}{2} + b \ln \frac{a+b\cos x}{\sin x} \right\} + B \int \frac{dx}{a+b\cos x}$$
(see 2.553 3.). TI (349)

For $a^2 = b^2\ (=1)$:

3. $$\int \frac{(A+B\sin x)\,dx}{\sin x\,(1+\cos x)} = \frac{A}{2} \left\{ \ln \operatorname{tg} \frac{x}{2} + \frac{1}{1+\cos x} \right\} + B \operatorname{tg} \frac{x}{2}.$$

4. $$\int \frac{(A+B\sin x)\,dx}{\sin x\,(1-\cos x)} = \frac{A}{2} \left\{ \ln \operatorname{tg} \frac{x}{2} - \frac{1}{1-\cos x} \right\} - B \operatorname{ctg} \frac{x}{2}.$$

5. $$\int \frac{(A+B\sin x)\,dx}{\cos x\,(a+b\sin x)} = \frac{1}{a^2-b^2} \left\{ (Aa-Bb) \ln \operatorname{tg}\left(\frac{\pi}{4}+\frac{x}{2}\right) - (Ab-aB) \ln \frac{a+b\sin x}{\cos x} \right\}.$$
TI (346)

For $a^2 = b^2\ (=1)$:

6. $$\int \frac{(A+B\sin x)\,dx}{\cos x\,(1\pm\sin x)} = \frac{A\pm B}{2} \ln \operatorname{tg}\left(\frac{\pi}{4}+\frac{x}{2}\right) \mp \frac{A\mp B}{2(1\pm\sin x)}.$$

7. $$\int \frac{(A+B\sin x)\,dx}{\cos x\,(a+b\cos x)} = \frac{A}{a} \ln \operatorname{tg}\left(\frac{\pi}{4}+\frac{x}{2}\right) + \frac{B}{a} \ln \frac{a+b\cos x}{\cos x} - \frac{Ab}{a} \int \frac{dx}{a+b\cos x}$$
(see 2.553 3.). TI (351)a

8. $$\int \frac{(A+B\cos x)\,dx}{\sin x\,(a+b\sin x)} = \frac{A}{a} \ln \operatorname{tg} \frac{x}{2} - \frac{B}{a} \ln \frac{a+b\sin x}{\sin x} - \frac{Ab}{a} \int \frac{dx}{a+b\sin x}$$
(see 2.551 3.). TI (352)

9. $$\int \frac{(A+B\cos x)\,dx}{\sin x\,(a+b\cos x)} = \frac{1}{a^2-b^2} \left\{ (Aa-Bb) \ln \operatorname{tg} \frac{x}{2} + (Ab-Ba) \ln \frac{a+b\cos x}{\sin x} \right\}.$$
TI (345)

For $a^2 = b^2\ (=1)$:

10. $$\int \frac{(A+B\cos x)\,dx}{\sin x\,(1\pm\cos x)} = \pm \frac{A\mp B}{2(1\pm\cos x)} + \frac{A\pm B}{2} \ln \operatorname{tg} \frac{x}{2}.$$

11. $$\int \frac{(A+B\cos x)\,dx}{\cos x\,(a+b\sin x)} = \frac{A}{a^2-b^2} \left\{ a \ln \operatorname{tg}\left(\frac{\pi}{4}+\frac{x}{2}\right) - b \ln \frac{a+b\sin x}{\cos x} \right\} + B \int \frac{dx}{a+b\sin x}$$
(see 2.551 3.) TI (350)

For $a^2 = b^2 (= 1)$:

12. $$\int \frac{(A+B\sin x)\,dx}{\cos x\,(1\pm\sin x)} = \frac{A\pm B}{2}\ln\operatorname{tg}\left(\frac{\pi}{4}+\frac{x}{2}\right) \mp \frac{A\mp B}{2(1\pm\sin x)}.$$

13. $$\int \frac{(A+B\cos x)\,dx}{\cos x\,(a+b\cos x)} = \frac{A}{a}\ln\operatorname{tg}\left(\frac{\pi}{4}+\frac{x}{2}\right) + \frac{Ba-Ab}{a}\int\frac{dx}{a+b\cos x}$$

(see 2.553 3.). TI (347)

2.562

1. $$\int\frac{dx}{a+b\sin^2 x} = \frac{\operatorname{sign} a}{\sqrt{a(a+b)}}\operatorname{arctg}\left(\sqrt{\frac{a+b}{a}}\operatorname{tg} x\right) \quad \left[\frac{b}{a} > -1\right];$$

$$= \frac{\operatorname{sign} a}{\sqrt{-a(a+b)}}\operatorname{Arth}\left(\sqrt{-\frac{a+b}{a}}\operatorname{tg} x\right)$$

$$\left[\frac{b}{a} < -1,\ \sin^2 x < -\frac{a}{b}\right];$$

$$= \frac{\operatorname{sign} a}{\sqrt{-a(a+b)}}\operatorname{Arcth}\left(\sqrt{-\frac{a+b}{a}}\operatorname{tg} x\right)$$

$$\left[\frac{b}{a} < -1,\ \sin^2 x > -\frac{a}{b}\right].$$ MZ 155

2. $$\int\frac{dx}{a+b\cos^2 x} = \frac{-\operatorname{sign} a}{\sqrt{a(a+b)}}\operatorname{arctg}\left(\sqrt{\frac{a+b}{a}}\operatorname{ctg} x\right) \quad \left[\frac{b}{a} > -1\right];$$

$$= \frac{-\operatorname{sign} a}{\sqrt{-a(a+b)}}\operatorname{Arth}\left(\sqrt{-\frac{a+b}{a}}\operatorname{ctg} x\right)$$

$$\left[\frac{b}{a} < -1,\quad \cos^2 x < -\frac{a}{b}\right];$$

$$= \frac{-\operatorname{sign} a}{\sqrt{-a(a+b)}}\operatorname{Arcth}\left(\sqrt{-\frac{a+b}{a}}\operatorname{ctg} x\right)$$

$$\left[\frac{b}{a} < -1,\ \cos^2 x > -\frac{a}{b}\right].$$ MZ 162

3. $$\int\frac{dx}{1+\sin^2 x} = \frac{1}{\sqrt{2}}\operatorname{arctg}(\sqrt{2}\operatorname{tg} x).$$

4. $$\int\frac{dx}{1-\sin^2 x} = \operatorname{tg} x.$$

5. $$\int\frac{dx}{1+\cos^2 x} = -\frac{1}{\sqrt{2}}\operatorname{arctg}(\sqrt{2}\operatorname{ctg} x).$$

6. $$\int\frac{dx}{1-\cos^2 x} = -\operatorname{ctg} x.$$

2.563

1. $$\int\frac{dx}{(a+b\sin^2 x)^2} = \frac{1}{2a(a+b)}\left[(2a+b)\int\frac{dx}{a+b\sin^2 x} + \frac{b\sin x\cos x}{a+b\sin^2 x}\right]$$ (see 2.562 1.). MZ 155

2. $$\int\frac{dx}{(a+b\cos^2 x)^2} = \frac{1}{2a(a+b)}\left[(2a+b)\int\frac{dx}{a+b\cos^2 x} - \frac{b\sin x\cos x}{a+b\cos^2 x}\right]$$ (see 2.562 2.). MZ 163

3. $$\int \frac{dx}{(a+b\sin^2 x)^3} = \frac{1}{8pa^3}\left[\left(3+\frac{2}{p^2}+\frac{3}{p^4}\right)\operatorname{arctg}(p\operatorname{tg}x) + \left(3+\frac{2}{p^2}-\frac{3}{p^4}\right)\frac{p\operatorname{tg}x}{1+p^2\operatorname{tg}^2 x} + \left(1-\frac{2}{p^2}-\frac{1}{p^2}\operatorname{tg}^2 x\right)\frac{2p\operatorname{tg}x}{(1+p^2\operatorname{tg}^2 x)^2}\right]$$
$$\left[p^2 = 1+\frac{b}{a} > 0\right];$$
$$= \frac{1}{8qa^3}\left[\left(3-\frac{2}{q^2}+\frac{3}{q^4}\right)\operatorname{Arth}(q\operatorname{tg}x) + \left(3-\frac{2}{q^2}-\frac{3}{q^4}\right)\frac{q\operatorname{tg}x}{1-q^2\operatorname{tg}^2 x} + \left(1+\frac{2}{q^2}+\frac{1}{q^2}\operatorname{tg}^2 x\right)\frac{2q\operatorname{tg}x}{(1-q^2\operatorname{tg}^2 x)^2}\right]$$
$\left[q^2 = -1-\frac{b}{a} > 0,\quad \sin^2 x < -\frac{a}{b}\right.$; for $\sin^2 x > -\frac{a}{b}$, one should replace $\operatorname{Arth}(q\operatorname{tg}x)$ with $\left.\operatorname{Arcth}(q\operatorname{tg}x)\right]$. MZ 156

4. $$\int \frac{dx}{(a+b\cos^2 x)^3} = -\frac{1}{8pa^3}\left[\left(3+\frac{2}{p^2}+\frac{3}{p^4}\right)\operatorname{arctg}(p\operatorname{ctg}x) + \left(3+\frac{2}{p^2}-\frac{3}{p^4}\right)\frac{p\operatorname{ctg}x}{1+p^2\operatorname{ctg}^2 x} + \left(1-\frac{2}{p^2}-\frac{1}{p^2}\operatorname{ctg}^2 x\right)\frac{2p\operatorname{ctg}x}{(1+p^2\operatorname{ctg}^2 x)^2}\right]$$
$$\left[p^2 = 1+\frac{b}{a} > 0\right];$$
$$= -\frac{1}{8qa^3}\left[\left(3-\frac{2}{q^2}+\frac{3}{q^4}\right)\operatorname{Arth}(q\operatorname{ctg}x) + \left(3-\frac{2}{q^2}-\frac{3}{q^4}\right)\frac{q\operatorname{ctg}x}{1-q^2\operatorname{ctg}^2 x} + \left(1+\frac{2}{q^2}+\frac{1}{q^2}\operatorname{ctg}^2 x\right)\frac{2q\operatorname{ctg}x}{(1-q^2\operatorname{ctg}^2 x)^2}\right]$$
$\left[q^2 = -1-\frac{b}{a} > 0,\quad \cos^2 x < -\frac{a}{b}\right.$, for $\cos^2 x > -\frac{a}{b}$, one should replace $\operatorname{Arth}(q\operatorname{ctg}x)$ with $\left.\operatorname{Arcth}(q\operatorname{ctg}x)\right]$. MZ 163a

2.564

1. $$\int \frac{\operatorname{tg}x\,dx}{1+m^2\operatorname{tg}^2 x} = \frac{\ln(\cos^2 x + m^2\sin^2 x)}{2(m^2-1)}.$$ LA 210(10)

2. $$\int \frac{\operatorname{tg}\alpha - \operatorname{tg}x}{\operatorname{tg}\alpha + \operatorname{tg}x}\,dx = \sin 2\alpha \ln\sin(x+\alpha) - x\cos 2\alpha.$$ LA 210(11)a

3. $$\int \frac{\operatorname{tg}x\,dx}{a+b\operatorname{tg}x} = \frac{1}{a^2+b^2}\{bx - a\ln(a\cos x + b\sin x)\}.$$ PE (335)

4. $$\int \frac{dx}{a+b\operatorname{tg}^2 x} = \frac{1}{a-b}\left[x - \sqrt{\frac{b}{a}}\operatorname{arctg}\left(\sqrt{\frac{b}{a}}\operatorname{tg}x\right)\right].$$ PE (334)

2.57 Forms containing $\sqrt{a \pm b\sin x}$, or $\sqrt{a \pm b\cos x}$ and forms reducible to such expressions

Notations: $\alpha = \arcsin\sqrt{\frac{1-\sin x}{2}}$, $\beta = \arcsin\sqrt{\frac{b(1-\sin x)}{a+b}}$,

$\gamma = \arcsin\sqrt{\frac{b(1-\cos x)}{a+b}}$, $\delta = \arcsin\sqrt{\frac{(a+b)(1-\cos x)}{2(a-b\cos x)}}$, $r = \sqrt{\frac{2b}{a+b}}$.

2.571

1. $$\int \frac{dx}{\sqrt{a+b\sin x}} = \frac{-2}{\sqrt{a+b}}F(\alpha, r) \quad \left[a > b > 0,\ -\frac{\pi}{2} \leqslant x < \frac{\pi}{2}\right];$$
$$= -\sqrt{\frac{2}{b}}F\left(\beta, \frac{1}{r}\right) \quad \left[0 < |a| < b,\ -\arcsin\frac{a}{b} < x < \frac{\pi}{2}\right].$$

BY (288.00, 288.50)

2. $$\int \frac{\sin x\,dx}{\sqrt{a+b\sin x}} = \frac{2a}{b\sqrt{a+b}} F(\alpha, r) - \frac{2\sqrt{a+b}}{b} E(\alpha, r)$$

$$\left[a > b > 0,\ -\frac{\pi}{2} \leqslant x < \frac{\pi}{2}\right];$$ BY (288.03)

$$= \sqrt{\frac{2}{b}} \left\{F\left(\beta, \frac{1}{r}\right) - 2E\left(\beta, \frac{1}{r}\right)\right\}$$

$$\left[0 < |a| < b,\ -\arcsin\frac{a}{b} < x < \frac{\pi}{2}\right].$$ BY (288.54)

3. $$\int \frac{\sin^2 x\,dx}{\sqrt{a+b\sin x}} = \frac{4a\sqrt{a+b}}{3b^2} E(\alpha, r) - \frac{2(2a^2+b^2)}{3b^2\sqrt{a+b}} F(\alpha, r) -$$

$$- \frac{2}{3b}\cos x\sqrt{a+b\sin x} \qquad \left[a > b > 0,\ -\frac{\pi}{2} \leqslant x < \frac{\pi}{2}\right];$$

$$= \sqrt{\frac{2}{b}} \left\{\frac{4a}{3b} E\left(\beta, \frac{1}{r}\right) - \frac{2a+b}{3b} F\left(\beta, \frac{1}{r}\right)\right\} - \frac{2}{3b}\cos x\sqrt{a+b\sin x}$$

$$\left[0 < |a| < b,\ -\arcsin\frac{a}{b} < x < \frac{\pi}{2}\right].$$ BY (288.03, 288.54)

4. $$\int \frac{dx}{\sqrt{a+b\cos x}} = \frac{2}{\sqrt{a+b}} F\left(\frac{x}{2}, r\right) \qquad [a > b > 0,\ 0 \leqslant x \leqslant \pi];$$

BY (289.00)

$$= \sqrt{\frac{2}{b}} F\left(\gamma, \frac{1}{r}\right)$$

$$\left[b \geqslant |a| > 0,\ 0 \leqslant x < \arccos\left(-\frac{a}{b}\right)\right].$$ BY (290.00)

5. $$\int \frac{dx}{\sqrt{a-b\cos x}} = \frac{2}{\sqrt{a+b}} F(\delta, r) \qquad [a > b > 0,\ 0 \leqslant x \leqslant \pi].$$

BY (291.00)

6. $$\int \frac{\cos x\,dx}{\sqrt{a+b\cos x}} = \frac{2}{b\sqrt{a+b}} \left\{(a+b)E\left(\frac{x}{2}, r\right) - aF\left(\frac{x}{2}, r\right)\right\}$$

$$[a > b > 0,\ 0 \leqslant x \leqslant \pi];$$ BY (289.03)

$$= \sqrt{\frac{2}{b}} \left\{2E\left(\gamma, \frac{1}{r}\right) - F\left(\gamma, \frac{1}{r}\right)\right\}$$

$$\left[b > |a| > 0,\ 0 \leqslant x < \arccos\left(-\frac{a}{b}\right)\right].$$ BY (290.04)

7. $$\int \frac{\cos x\,dx}{\sqrt{a-b\cos x}} = \frac{2}{b\sqrt{a+b}} \{(b-a)\Pi(\delta, r^2, r) + aF(\delta, r)\}$$

$$[a > b > 0,\ 0 \leqslant x \leqslant \pi].$$ BY (291.03)

8. $$\int \frac{\cos^2 x\,dx}{\sqrt{a+b\cos x}} = \frac{2}{3b^2\sqrt{a+b}} \left\{(2a^2+b^2) F\left(\frac{x}{2}, r\right) -\right.$$

$$\left. -2a(a+b) E\left(\frac{x}{2}, r\right)\right\} + \frac{2}{3b}\sin x\sqrt{a+b\cos x}$$

$$[a > b > 0,\ 0 \leqslant x \leqslant \pi];$$ BY (289.03)

$$= \frac{1}{3b}\sqrt{\frac{2}{b}} \left\{(2a+b) F\left(\gamma, \frac{1}{r}\right) - 4aE\left(\gamma, \frac{1}{r}\right)\right\} +$$

$$+ \frac{2}{3b}\sin x\sqrt{a+b\cos x} \qquad \left[b \geqslant |a| > 0,\ 0 \leqslant x < \arccos\left(-\frac{a}{b}\right)\right].$$ BY (290.04)

9. $$\int \frac{\cos^2 x\,dx}{\sqrt{a-b\cos x}} = \frac{2}{3b^2\sqrt{a+b}} \left\{(2a^2+b^2) F(\delta, r) - 2a(a+b) E(\delta, r)\right\} +$$

$$+ \frac{2}{3b}\sin x \frac{a+b\cos x}{\sqrt{a-b\cos x}} \qquad [a > b > 0,\ 0 \leqslant x < \pi].$$ BY (291.04)a

2.572

$$\int \frac{\operatorname{tg}^2 x\,dx}{\sqrt{a+b\sin x}} = \frac{1}{\sqrt{a+b}} F(\alpha, r) + \frac{a}{(a-b)\sqrt{a+b}} E(\alpha, r) - \frac{b-a\sin x}{(a^2-b^2)\cos x}\sqrt{a+b\sin x} \quad \left[0<b<a,\ -\frac{\pi}{2}<x<\frac{\pi}{2}\right];$$

$$= \sqrt{\frac{2}{b}}\left\{\frac{2a+b}{2(a+b)}F\left(\beta, \frac{1}{r}\right) + \frac{ab}{a^2-b^2}E\left(\beta, \frac{1}{r}\right)\right\} - \frac{b-a\sin x}{(a^2-b^2)\cos x}\sqrt{a+b\sin x} \quad \left[0<|a|<b,\ -\arcsin\frac{a}{b}<x<\frac{\pi}{2}\right].$$

BY (288.08, 288.58)

2.573

1. $$\int \frac{1-\sin x}{1+\sin x}\cdot\frac{dx}{\sqrt{a+b\sin x}} = \frac{2}{a-b}\left\{\sqrt{a+b}\,E(\alpha, r) - \operatorname{tg}\left(\frac{\pi}{4}-\frac{x}{2}\right)\sqrt{a+b\sin x}\right\} \quad \left[0<b<a,\ -\frac{\pi}{2}\leqslant x<\frac{\pi}{2}\right].$$ BY (288.07)

2. $$\int \frac{1-\cos x}{1+\cos x}\,\frac{dx}{\sqrt{a+b\cos x}} = \frac{2}{a-b}\operatorname{tg}\frac{x}{2}\sqrt{a+b\cos x} - \frac{2\sqrt{a+b}}{a-b}E\left(\frac{x}{2}, r\right) \quad [a>b>0,\ 0\leqslant x<\pi].$$ BY (289.07)

2.574

1. $$\int \frac{dx}{(2-p^2+p^2\sin x)\sqrt{a+b\sin x}} = -\frac{1}{a+b}\Pi(\alpha, p^2, r) \quad \left[0<b<a,\ -\frac{\pi}{2}\leqslant x<\frac{\pi}{2}\right].$$ BY (288.02)

2. $$\int \frac{dx}{(a+b-p^2b+p^2b\sin x)\sqrt{a+b\sin x}} = -\frac{1}{a+b}\sqrt{\frac{2}{b}}\,\Pi\left(\beta, p^2, \frac{1}{r}\right) \quad \left[0<|a|<b,\ -\arcsin\frac{a}{b}<x<\frac{\pi}{2}\right].$$ BY (288.52)

3. $$\int \frac{dx}{(2-p^2+p^2\cos x)\sqrt{a+b\cos x}} = \frac{1}{\sqrt{a+b}}\Pi\left(\frac{x}{2}, p^2, r\right) \quad [a>b>0,\ 0\leqslant x<\pi].$$ BY (289.02)

4. $$\int \frac{dx}{(a+b-p^2b+p^2b\cos x)\sqrt{a+b\cos x}} = \frac{\sqrt{2}}{(a+b)\sqrt{b}}\Pi\left(\gamma, p^2, \frac{1}{r}\right) \quad \left[b\geqslant|a|>0,\ 0\leqslant x<\arccos\left(-\frac{a}{b}\right)\right].$$ BY (290.02)

2.575

1. $$\int \frac{dx}{\sqrt{(a+b\sin x)^3}} = \frac{2b\cos x}{(a^2-b^2)\sqrt{a+b\sin x}} - \frac{2}{(a-b)\sqrt{a+b}}E(\alpha, r) \quad \left[0<b<a,\ -\frac{\pi}{2}\leqslant x<\frac{\pi}{2}\right];$$ BY (288.05)

$$= \sqrt{\frac{2}{b}}\left\{\frac{2b}{b^2-a^2}E\left(\beta, \frac{1}{r}\right) - \frac{1}{a+b}F\left(\beta, \frac{1}{r}\right)\right\} + \frac{2b}{b^2-a^2}\cdot\frac{\cos x}{\sqrt{a+b\sin x}} \quad \left[0<|a|<b,\ -\arcsin\frac{a}{b}<x<\frac{\pi}{2}\right].$$ BY (288.56)

2. $$\int \frac{dx}{\sqrt{(a+b\sin x)^5}} = \frac{2}{3(a^2-b^2)^2\sqrt{a+b}}\left\{(a^2-b^2)F(\alpha, r) - 4a(a+b)E(\alpha, r)\right\} + \frac{2b(5a^2-b^2+4ab\sin x)}{3(a^2-b^2)^2\sqrt{(a+b\sin x)^3}}\cos x$$

$$\left[0 < b < a,\ -\frac{\pi}{2} \leqslant x < \frac{\pi}{2}\right];$$ BY (288.05)

$$= -\frac{1}{3(a^2-b^2)^2}\sqrt{\frac{2}{b}}\left\{(3a-b)(a-b)F\left(\beta, \frac{1}{r}\right) + 8abE\left(\beta, \frac{1}{r}\right)\right\} + \frac{2b[a^2-b^2+4a(a+b\sin x)]}{3(a^2-b^2)^2\sqrt{(a+b\sin x)^3}}\cos x$$

$$\left[0 < |a| < b,\ -\arcsin\frac{a}{b} < x < \frac{\pi}{2}\right].$$ BY (288.56)

3. $$\int \frac{dx}{\sqrt{(a+b\cos x)^3}} = \frac{2}{(a-b)\sqrt{a+b}}E\left(\frac{x}{2}, r\right) - \frac{2b}{a^2-b^2}\cdot\frac{\sin x}{\sqrt{a+b\cos x}}$$

$$[a > b > 0,\ 0 \leqslant x \leqslant \pi];$$ BY (289.05)

$$= \frac{1}{a^2-b^2}\sqrt{\frac{2}{b}}\left\{(a-b)F\left(\gamma, \frac{1}{r}\right) + 2bE\left(\gamma, \frac{1}{r}\right)\right\} + \frac{2b}{b^2-a^2}\cdot\frac{\sin x}{\sqrt{a+b\cos x}}$$

$$\left[b \geqslant |a| > 0,\ 0 \leqslant x < \arccos\left(-\frac{a}{b}\right)\right].$$ BY (290.06)

4. $$\int \frac{dx}{\sqrt{(a-b\cos x)^3}} = \frac{2}{(a-b)\sqrt{a+b}}E(\delta, r) \qquad [a>b>0,\ 0 \leqslant x \leqslant \pi].$$

BY (291.01)

5. $$\int \frac{dx}{\sqrt{(a+b\cos x)^5}} = \frac{2\sqrt{a+b}}{3(a^2-b^2)^2}\left\{4aE\left(\frac{x}{2}, r\right) - (a-b)F\left(\frac{x}{2}, r\right)\right\} - \frac{2b}{3(a^2-b^2)^2}\cdot\frac{5a^2-b^2+4ab\cos x}{\sqrt{(a+b\cos x)^3}}\sin x \quad [a > b > 0,\ 0 \leqslant x \leqslant \pi];$$ BY (289.05)

$$= \frac{1}{3(a^2-b^2)^2}\sqrt{\frac{2}{b}}\left\{(a-b)(3a-b)F\left(\gamma, \frac{1}{r}\right) + 8abE\left(\gamma, \frac{1}{r}\right)\right\} + \frac{2b(5a^2-b^2+4ab\cos x)\sin x}{3(a^2-b^2)^2\sqrt{(a+b\cos x)^3}}$$

$$\left[b \geqslant |a| > 0,\ 0 \leqslant x < \arccos\left(-\frac{a}{b}\right)\right].$$ BY (290.06)

2.576

1. $$\int \sqrt{a+b\cos x}\,dx = 2\sqrt{a+b}\,E\left(\frac{x}{2}, r\right) \quad [a > b > 0,\ 0 \leqslant x \leqslant \pi];$$

BY (289.01)

$$= \sqrt{\frac{2}{b}}\left\{(a-b)F\left(\gamma, \frac{1}{r}\right) + 2bE\left(\gamma, \frac{1}{r}\right)\right\}$$

$$\left[b \geqslant |a| > 0,\ 0 \leqslant x < \arccos\left(-\frac{a}{b}\right)\right].$$ BY (290.03)

2. $$\int \sqrt{a-b\cos x}\,dx = 2\sqrt{a+b}\,E(\delta, r) - \frac{2b\sin x}{\sqrt{a-b\cos x}}$$

$$[a > b > 0,\ 0 \leqslant x \leqslant \pi].$$ BY (291.05)

2.577 $$\int \sqrt{\frac{a-b\cos x}{1+p\cos x}}\,dx = \frac{2(a-b)}{(1+p)\sqrt{a+b}}\,\Pi\left(\delta, \frac{2ap}{(a+b)(1+p)}, r\right)$$
$$[a > b > 0,\ 0 \leqslant x \leqslant \pi,\ p \neq -1].$$ BY (291.02)

2.578 $$\int \frac{\operatorname{tg} x\,dx}{\sqrt{a+b\operatorname{tg}^2 x}} = \frac{1}{\sqrt{b-a}} \arccos\left(\frac{\sqrt{b-a}}{\sqrt{b}}\cos x\right) \quad [b > a,\ b > 0].$$ PE (333)

2.58-2.62 Integrals reducible to elliptic and pseudo-elliptic integrals

2.580

1. $$\int \frac{d\varphi}{\sqrt{a+b\cos\varphi+c\sin\varphi}} = 2\int \frac{d\psi}{\sqrt{a-p+2p\cos^2\psi}}$$
$$\left[\varphi = 2\psi+\alpha,\ \operatorname{tg}\alpha = \frac{c}{b},\ p = \sqrt{b^2+c^2}\right]$$

2. $$\int \frac{d\varphi}{\sqrt{a+b\cos\varphi+c\sin\varphi+d\cos^2\varphi+e\sin\varphi\cos\varphi+f\sin^2\varphi}} =$$
$$= 2\int \frac{dx}{\sqrt{A+Bx+Cx^2+Dx^3+Ex^4}}$$
$$\left[\operatorname{tg}\frac{\varphi}{2} = x,\ A = a+b+d,\ B = 2c+2e,\ C = 2a-2d+4f,\ D = 2c-2e,\ E = a-b+d\right]$$

Forms containing $\sqrt{1-k^2\sin^2 x}$

Notations: $\Delta = \sqrt{1-k^2\sin^2 x}$, $\quad k' = \sqrt{1-k^2}$

2.581

1. $$\int \sin^m x\cos^n x\,\Delta^r\,dx =$$
$$= \frac{1}{(m+n+r)k^2}\left\{\sin^{m-3} x\cos^{n+1} x\Delta^{r+2} + [m+n-2(m+r-1)k^2]\times\right.$$
$$\left.\times\int \sin^{m-2} x\cos^n x\Delta^r\,dx - (m-3)\int \sin^{m-4} x\cos^n x\,\Delta^r\,dx\right\} =$$
$$= \frac{1}{(m+n+r)k^2}\left\{\sin^{m+1} x\cos^{n-3} x\Delta^{r+2} + [(n+r-1)k^2-(m+n-2)k'^2]\times\right.$$
$$\left.\times\int \sin^m x\cos^{n-2} x\,\Delta^r\,dx + (n-3)k'^2\int \sin^m x\cos^{n-4} x\,\Delta^r\,dx\right\}$$
$$[m+n+r \neq 0]$$

For $r = -3$ and $r = -5$:

2. $$\int \frac{\sin^m x\cos^n x}{\Delta^3}\,dx = \frac{\sin^{m-1} x\cos^{n-1} x}{k^2\Delta} -$$
$$-\frac{m-1}{k^2}\int \frac{\sin^{m-2} x\cos^n x}{\Delta}\,dx + \frac{n-1}{k^2}\int \frac{\sin^m x\cos^{n-2} x}{\Delta}\,dx.$$

3. $$\int \frac{\sin^m x\cos^n x}{\Delta^5}\,dx = \frac{\sin^{m-1} x\cos^{n-1} x}{3k^2\Delta^3} -$$
$$-\frac{m-1}{3k^2}\int \frac{\sin^{m-2} x\cos^n x}{\Delta^3}\,dx + \frac{n-1}{3k^2}\int \frac{\sin^m x\cos^{n-2} x}{\Delta^3}\,dx.$$

For $m=1$ or $n=1$:

4. $\int \sin x \cos^n x \Delta^r \, dx = -\frac{\cos^{n-1} x \Delta^{r+2}}{(n+r+1)k^2} - \frac{(n-1)k'^2}{(n+r+1)k^2} \int \cos^{n-2} x \sin x \Delta^r \, dx.$

5. $\int \sin^m x \cos x \, \Delta^r \, dx = -\frac{\sin^{m-1} x \Delta^{r+2}}{(m+r+1)k^2} + \frac{m-1}{(m+r+1)k^2} \int \sin^{m-2} x \cos x \, \Delta^r \, dx.$

For $m=3$ or $n=3$:

6. $$\int \sin^3 x \cos^n x \Delta^r \, dx = \frac{(n+r+1)k^2 \cos^2 x - [(r+2)k^2 + n + 1]}{(n+r+1)(n+r+3)k^4} \cos^{n-1} x \Delta^{r+2} - \\ - \frac{[(r+2)k^2+n+1](n-1)k'^2}{(n+r+1)(n+r+3)k^4} \int \cos^{n-2} x \sin x \, \Delta^r \, dx.$$

7. $$\int \sin^m x \cos^3 x \, \Delta^r \, dx = \frac{(m+r+1)k^2 \sin^2 x - [(r+2)k^2 - (m+1)k'^2]}{(m+r+1)(m+r+3)k^4} \times \\ \times \sin^{m-1} x \Delta^{r+2} + \frac{[(r+2)k^2 - (m+1)k'^2](m-1)}{(m+r+1)(m+r+3)k^4} \int \sin^{m-2} x \cos x \Delta^r \, dx.$$

2.582

1. $$\int \Delta^n \, dx = \frac{n-1}{n}(2-k^2) \int \Delta^{n-2} \, dx - \frac{n-2}{n}(1-k^2) \int \Delta^{n-4} \, dx + \\ + \frac{k^2}{n} \sin x \cos x \cdot \Delta^{n-2}.$$ LA 316(1)a

2. $$\int \frac{dx}{\Delta^{n+1}} = -\frac{k^2 \sin x \cos x}{(n-1)k'^2 \Delta^{n-1}} + \frac{n-2}{n-1} \frac{2-k^2}{k'^2} \int \frac{dx}{\Delta^{n-1}} - \frac{n-3}{n-1} \frac{1}{k'^2} \int \frac{dx}{\Delta^{n-3}}$$ LA 317(8)a

3. $$\int \frac{\sin^n x}{\Delta} \, dx = \frac{\sin^{n-3} x}{(n-1)k^2} \cos x \cdot \Delta + \frac{n-2}{n-1} \frac{1+k^2}{k^2} \int \frac{\sin^{n-2} x}{\Delta} \, dx - \\ - \frac{n-3}{(n-1)k^2} \int \frac{\sin^{n-4} x}{\Delta} \, dx.$$ LA 316(1)a

4. $$\int \frac{\cos^n x}{\Delta} \, dx = \frac{\cos^{n-3} x}{(n-1)k^2} \sin x \cdot \Delta + \frac{n-2}{n-1} \frac{2k^2-1}{k^2} \int \frac{\cos^{n-2} x}{\Delta} \, dx + \\ + \frac{n-3}{n-1} \frac{k'^2}{k^2} \int \frac{\cos^{n-4} x}{\Delta} \, dx.$$ LA 316(2)a

5. $$\int \frac{\operatorname{tg}^n x}{\Delta} \, dx = \frac{\operatorname{tg}^{n-3} x}{(n-1)k'^2} \frac{\Delta}{\cos^2 x} - \frac{(n-2)(2-k^2)}{(n-1)k'^2} \int \frac{\operatorname{tg}^{n-2} x}{\Delta} \, dx - \\ - \frac{n-3}{(n-1)k'^2} \int \frac{\operatorname{tg}^{n-4} x}{\Delta} \, dx.$$ LA 317(3)

6. $$\int \frac{\operatorname{ctg}^n x}{\Delta} \, dx = -\frac{\operatorname{ctg}^{n-1} x}{n-1} \frac{\Delta}{\cos^2 x} - \frac{n-2}{n-1}(2-k^2) \int \frac{\operatorname{ctg}^{n-2} x}{\Delta} \, dx - \\ - \frac{n-3}{n-1} k'^2 \int \frac{\operatorname{ctg}^{n-4} x}{\Delta} \, dx.$$ LA 317(6)

2.583

1. $\int \Delta \, dx = E(x, k).$

2. $\int \Delta \sin x \, dx = -\frac{\Delta \cos x}{2} - \frac{k'^2}{2k} \ln(k \cos x + \Delta).$

3. $\int \Delta \cos x \, dx = \frac{\Delta \sin x}{2} + \frac{1}{2k} \arcsin(k \sin x).$

4. $\int \Delta \sin^2 x \, dx = -\frac{\Delta}{3} \sin x \cos x + \frac{k'^2}{3k^2} F(x, k) + \frac{2k^2-1}{3k^2} E(x, k).$

5. $\int \Delta \sin x \cos x\, dx = -\frac{\Delta^3}{3k^2}$.

6. $\int \Delta \cos^2 x\, dx = \frac{\Delta}{3} \sin x \cos x - \frac{k'^2}{3k^2} F(x, k) + \frac{k^2+1}{3k^2} E(x, k)$.

7. $\int \Delta \sin^3 x\, dx = -\frac{2k^2 \sin^2 x + 3k^2 - 1}{8k^2} \Delta \cos x + \frac{3k^4 - 2k^2 - 1}{8k^3} \ln(k \cos x + \Delta)$.

8. $\int \Delta \sin^2 x \cos x\, dx = \frac{2k^2 \sin^2 x - 1}{8k^2} \Delta \sin x + \frac{1}{8k^3} \arcsin(k \sin x)$.

9. $\int \Delta \sin x \cos^2 x\, dx = -\frac{2k^2 \cos^2 x + k'^2}{8k^2} \Delta \cos x + \frac{k'^4}{8k^3} \ln(k \cos x + \Delta)$.

10. $\int \Delta \cos^3 x\, dx = \frac{2k^2 \cos^2 x + 2k^2 + 1}{8k^2} \Delta \sin x + \frac{4k^2 - 1}{8k^3} \arcsin(k \sin x)$.

11. $\int \Delta \sin^4 x\, dx = -\frac{3k^2 \sin^2 x + 4k^2 - 1}{15k^2} \Delta \sin x \cos x +$
$+ \frac{2(2k^4 - k^2 - 1)}{15k^4} F(x, k) + \frac{8k^4 - 3k^2 - 2}{15k^4} E(x, k)$.

12. $\int \Delta \sin^3 x \cos x\, dx = \frac{3k^4 \sin^4 x - k^2 \sin^2 x - 2}{15k^4} \Delta$.

13. $\int \Delta \sin^2 x \cos^2 x\, dx = -\frac{3k^2 \cos^2 x - 2k^2 + 1}{15k^2} \Delta \sin x \cos x -$
$- \frac{k'^2 (1 + k'^2)}{15k^4} F(x, k) + \frac{2(k^4 - k^2 + 1)}{15k^4} E(x, k)$.

14. $\int \Delta \sin x \cos^3 x\, dx = -\frac{3k^4 \sin^4 x - k^2 (5k^2 + 1) \sin^2 x + 5k^2 - 2}{15k^4} \Delta$.

15. $\int \Delta \cos^4 x\, dx = \frac{3k^2 \cos^2 x + 3k^2 + 1}{15k^2} \Delta \sin x \cos x +$
$+ \frac{2k'^2 (k'^2 - 2k^2)}{15k^4} F(x, k) + \frac{3k^4 + 7k^2 - 2}{15k^4} E(x, k)$.

16. $\int \Delta \sin^5 x\, dx = -\frac{8k^4 \sin^4 x - 2k^2 (5k^2 - 1) \sin^2 x - 15k^4 + 4k^2 + 3}{48k^4} \Delta \cos x +$
$+ \frac{5k^6 - 3k^4 - k^2 - 1}{16k^5} \ln(k \cos x + \Delta)$.

17. $\int \Delta \sin^4 x \cos x\, dx = \frac{8k^4 \sin^4 x - 2k^2 \sin^2 x - 3}{48k^4} \Delta \sin x +$
$+ \frac{1}{16k^5} \arcsin(k \sin x)$.

18. $\int \Delta \sin^3 x \cos^2 x\, dx = \frac{8k^4 \sin^4 x - 2k^2 (k^2 + 1) \sin^2 x - 3k^4 + 2k^2 - 3}{48k^4} \Delta \cos x +$
$+ \frac{k'^4 (k^2 + 1)}{16k^5} \ln(k \cos x + \Delta)$.

19. $\int \Delta \sin^2 x \cos^3 x\, dx = \frac{-8k^4 \sin^4 x + 2k^2 (6k^2 + 1) \sin^2 x - 6k^2 + 3}{48k^4} \Delta \sin x +$
$+ \frac{2k^2 - 1}{16k^5} \arcsin(k \sin x)$.

20. $\int \Delta \sin x \cos^4 x\, dx = \frac{-8k^4 \sin^4 x + 2k^2 (7k^2 + 1) \sin^2 x - 3k^4 - 8k^2 + 3}{48k^4} \Delta \cos x -$
$- \frac{k'^6}{16k^5} \ln(k \cos x + \Delta)$.

21. $\int \Delta \cos^5 x\, dx = \frac{8k^4 \sin^4 x - 2k^2 (12k^2+1) \sin^2 x + 24k^4 + 12k^2 - 3}{48k^4} \Delta \sin x + $
$+ \frac{8k^4 - 4k^2 + 1}{16k^5} \arcsin (k \sin x).$

22. $\int \Delta^3\, dx = \frac{2}{3} (1 + k'^2) E(x, k) - \frac{k'^2}{3} F(x, k) + \frac{k^2}{3} \Delta \sin x \cos x.$

23. $\int \Delta^3 \sin x\, dx = \frac{2k^2 \sin^2 x + 3k^2 - 5}{8} \Delta \cos x - \frac{3k'^4}{8k} \ln (k \cos x + \Delta).$

24. $\int \Delta^3 \cos x\, dx = \frac{-2k^2 \sin^2 x + 5}{8} \Delta \sin x + \frac{3}{8k} \arcsin (k \sin x).$

25. $\int \Delta^3 \sin^2 x\, dx = \frac{3k^2 \sin^2 x + 4k^2 - 6}{15} \Delta \sin x \cos x + \frac{k'^2 (3 - 4k^2)}{15k^2} F(x, k) -$
$- \frac{8k^4 - 13k^2 + 3}{15k^2} E(x, k).$

26. $\int \Delta^3 \sin x \cos x\, dx = -\frac{\Delta^5}{5k^2}.$

27. $\int \Delta^3 \cos^2 x\, dx = \frac{-3k^2 \sin^2 x + k^2 + 6}{15} \Delta \sin x \cos x - \frac{k'^2 (k^2+3)}{15k^2} F(x, k) -$
$- \frac{2k^4 - 7k^2 - 3}{15k^2} E(x, k).$

28. $\int \Delta^3 \sin^3 x\, dx = \frac{8k^4 \sin^4 x + 2k^2 (5k^2 - 7) \sin^2 x + 15k^4 - 22k^2 + 3}{48k^2} \Delta \cos x -$
$- \frac{5k^6 - 9k^4 + 3k^2 + 1}{16k^3} \ln (k \cos x + \Delta).$

29. $\int \Delta^3 \sin^2 x \cos x\, dx = \frac{-8k^4 \sin^4 x + 14k^2 \sin^2 x - 3}{48k^2} \Delta \sin x +$
$+ \frac{1}{16k^3} \arcsin (k \sin x).$

30. $\int \Delta^3 \sin x \cos^2 x\, dx = \frac{-8k^4 \sin^4 x + 2k^2 (k^2+7) \sin^2 x + 3k^4 - 8k^2 - 3}{48k^2} \times$
$\times \Delta \cos x + \frac{k'^6}{16k^3} \ln (k \cos x + \Delta).$

31. $\int \Delta^3 \cos^3 x\, dx = \frac{8k^4 \sin^4 x - 2k^2 (6k^2 + 7) \sin^2 x + 30k^2 + 3}{48k^2} \Delta \sin x +$
$+ \frac{6k^2 - 1}{16k^3} \arcsin (k \sin x).$

32. $\int \frac{\Delta\, dx}{\sin x} = -\frac{1}{2} \ln \frac{\Delta + \cos x}{\Delta - \cos x} + k \ln (k \cos x + \Delta).$

33. $\int \frac{\Delta\, dx}{\cos x} = \frac{k'}{2} \ln \frac{\Delta + k' \sin x}{\Delta - k' \sin x} + k \arcsin (k \sin x).$

34. $\int \frac{\Delta dx}{\sin^2 x} = k'^2 F(x, k) - E(x, k) - \Delta \operatorname{ctg} x.$

35. $\int \frac{\Delta\, dx}{\sin x \cos x} = \frac{1}{2} \ln \frac{1 - \Delta}{1 + \Delta} + \frac{k'}{2} \ln \frac{\Delta + k'}{\Delta - k'}.$

36. $\int \frac{\Delta\, dx}{\cos^2 x} = F(x, k) - E(x, k) + \Delta \operatorname{tg} x.$

37. $\int \frac{\sin x}{\cos x} \Delta\, dx = \int \Delta \operatorname{tg} x\, dx = -\Delta + \frac{k'}{2} \ln \frac{\Delta + k'}{\Delta - k'}.$

38. $\int \frac{\cos x}{\sin x} \Delta\, dx = \int \Delta \operatorname{ctg} x\, dx = \Delta + \frac{1}{2} \ln \frac{1-\Delta}{1+\Delta}$.

39. $\int \frac{\Delta\, dx}{\sin^3 x} = -\frac{\Delta \cos x}{2 \sin^2 x} + \frac{k'^2}{4} \ln \frac{\Delta + \cos x}{\Delta - \cos x}$.

40. $\int \frac{\Delta\, dx}{\sin^2 x \cos x} = \frac{-\Delta}{\sin x} - \frac{1+k^2}{2k'} \ln \frac{\Delta - k' \sin x}{\Delta + k' \sin x}$.

41. $\int \frac{\Delta\, dx}{\sin x \cos^2 x} = \frac{\Delta}{\cos x} + \frac{1}{2} \ln \frac{\Delta + \cos x}{\Delta - \cos x}$.

42. $\int \frac{\Delta\, dx}{\cos^3 x} = \frac{\Delta \sin x}{2 \cos^2 x} + \frac{1}{4k'} \ln \frac{\Delta + k' \sin x}{\Delta - k' \sin x}$.

43. $\int \frac{\Delta \sin x\, dx}{\cos^2 x} = \frac{\Delta}{\cos x} - k \ln (k \cos x + \Delta)$.

44. $\int \frac{\Delta \cos x\, dx}{\sin^2 x} = -\frac{\Delta}{\sin x} - k \arcsin (k \sin x)$.

45. $\int \frac{\Delta \sin^2 x\, dx}{\cos x} = -\frac{\Delta \sin x}{2} + \frac{2k^2 - 1}{2k} \arcsin (k \sin x) + \frac{k'}{2} \ln \frac{\Delta + k' \sin x}{\Delta - k' \sin x}$.

46. $\int \frac{\Delta \cos^2 x\, dx}{\sin x} = \frac{\Delta \cos x}{2} + \frac{k^2+1}{2k} \ln (k \cos x + \Delta) + \frac{1}{2} \ln \frac{\Delta + \cos x}{\Delta - \cos x}$.

47. $\int \frac{\Delta\, dx}{\sin^4 x} = \frac{1}{3} \{ -\Delta \operatorname{ctg}^3 x + (k^2 - 3) \Delta \operatorname{ctg} x + 2k'^2 F(x, k) + (k^2 - 2) E(x, k) \}$.

48. $\int \frac{\Delta\, dx}{\sin^3 x \cos x} = -\frac{\Delta}{2 \sin^2 x} + \frac{k'}{2} \ln \frac{\Delta + k'}{\Delta - k'} + \frac{k^2 - 2}{4} \ln \frac{1+\Delta}{1-\Delta}$.

49. $\int \frac{\Delta\, dx}{\sin^2 x \cos^2 x} = \left(\frac{1}{k'^2} \operatorname{tg} x - \operatorname{ctg} x \right) \Delta + 2F(x, k) - \frac{1+k'^2}{k'^2} E(x, k)$.

50. $\int \frac{\Delta\, dx}{\sin x \cos^3 x} = \frac{\Delta}{2 \cos^2 x} - \frac{1}{2} \ln \frac{1+\Delta}{1-\Delta} + \frac{2-k^2}{4k'} \ln \frac{\Delta + k'}{\Delta - k'}$.

51. $\int \frac{\Delta\, dx}{\cos^4 x} = \frac{1}{3k'^2} \{ [k'^2 \operatorname{tg}^3 x - (2k^2 - 3) \operatorname{tg} x] \Delta + 2k'^2 F(x, k) + (k^2 - 2) E(x, k) \}$.

52. $\int \frac{\sin x}{\cos^3 x} \Delta\, dx = \frac{\Delta}{2 \cos^2 x} + \frac{k^2}{4k'} \ln \frac{\Delta + k'}{\Delta - k'}$.

53. $\int \frac{\cos x}{\sin^3 x} \Delta\, dx = -\frac{\Delta}{2 \sin^2 x} + \frac{k^2}{4} \ln \frac{1+\Delta}{1-\Delta}$.

54. $\int \frac{\sin^2 x}{\cos^2 x} \Delta\, dx = \int \operatorname{tg}^2 x \Delta\, dx = \Delta \operatorname{tg} x + F(x, k) - 2E(x, k)$.

55. $\int \frac{\cos^2 x}{\sin^2 x} \Delta\, dx = \int \operatorname{ctg}^2 x\, \Delta\, dx = -\Delta \operatorname{ctg} x + k'^2 F(x, k) - 2E(x, k)$.

56. $\int \frac{\sin^3 x}{\cos x} \Delta\, dx = -\frac{k^2 \sin^2 x + 3k^2 - 1}{3k^2} \Delta + \frac{k'}{2} \ln \frac{\Delta + k'}{\Delta - k'}$.

57. $\int \frac{\cos^3 x}{\sin x} \Delta\, dx = -\frac{k^2 \sin^2 x - 3k^2 - 1}{3k^2} \Delta + \frac{1}{2} \ln \frac{1-\Delta}{1+\Delta}$.

58. $\int \frac{\Delta\, dx}{\sin^5 x} = \frac{(k^2 - 3) \sin^2 x + 2}{8 \sin^4 x} \cos x\, \Delta + \frac{k'^2 (k^2 + 3)}{16} \ln \frac{\Delta + \cos x}{\Delta - \cos x}$.

59. $\int \frac{\Delta\, dx}{\sin^4 x \cos x} = -\frac{(3 - k^2) \sin^2 x + 1}{3 \sin^3 x} \Delta - \frac{k'}{2} \ln \frac{\Delta - k' \sin x}{\Delta + k' \sin x}$.

60. $\int \frac{\Delta\, dx}{\sin^3 x \cos^2 x} = \frac{3 \sin^2 x - 1}{2 \sin^2 x \cos x} \Delta + \frac{k^2 - 3}{4} \ln \frac{\Delta - \cos x}{\Delta + \cos x}$.

61. $\int \frac{\Delta\, dx}{\sin^2 x \cos^3 x} = \frac{3 \sin^2 x - 2}{2 \sin x \cos^2 x} \Delta - \frac{2k^2 - 3}{4k'} \ln \frac{\Delta + k' \sin x}{\Delta - k' \sin x}$.

62. $\int \frac{\Delta\, dx}{\sin x \cos^4 x} = \frac{(2k^2-3)\sin^2 x - 3k^2+4}{3k'^2 \cos^3 x}\Delta + \frac{1}{2}\ln\frac{\Delta+\cos x}{\Delta-\cos x}$.

63. $\int \frac{\Delta\, dx}{\cos^5 x} = \frac{(2k^2-3)\sin^2 x - 4k^2+5}{8k'^2\cos^4 x}\sin x\,\Delta - \frac{4k^2-3}{16k'^3}\ln\frac{\Delta+k'\sin x}{\Delta-k'\sin x}$.

64. $\int \frac{\sin x}{\cos^4 x}\Delta\, dx = \frac{-(2k^2+1)k^2\sin^2 x + 3k^4 - k^2+1}{3k'^2\cos^3 x}\Delta$.

65. $\int \frac{\cos x}{\sin^4 x}\Delta\, dx = -\frac{\Delta^3}{3\sin^3 x}$.

66. $\int \frac{\sin^2 x}{\cos^3 x}\Delta\, dx = \frac{\sin x}{2\cos^2 x}\Delta + \frac{2k^2-1}{4k'}\ln\frac{\Delta+k'\sin x}{\Delta-k'\sin x} - k\arcsin(k\sin x)$.

67. $\int \frac{\cos^2 x}{\sin^3 x}\Delta\, dx = -\frac{\cos x}{2\sin^2 x}\Delta - \frac{k^2+1}{4}\ln\frac{\Delta+\cos x}{\Delta-\cos x} - k\ln(k\cos x+\Delta)$.

68. $\int \frac{\sin^3 x}{\cos^2 x}\Delta\, dx = -\frac{\sin^2 x - 3}{2\cos x}\Delta - \frac{3k^2-1}{2k}\ln(k\cos x+\Delta)$.

69. $\int \frac{\cos^3 x}{\sin^2 x}\Delta\, dx = -\frac{\sin^2 x+2}{2\sin x}\Delta - \frac{2k^2+1}{2k}\arcsin(k\sin x)$.

70. $\int \frac{\sin^4 x}{\cos x}\Delta\, dx = -\frac{2k^2\sin^2 x+4k^2-1}{8k^2}\sin x\Delta +$
$$+\frac{8k^4-4k^2-1}{8k^3}\arcsin(k\sin x) + \frac{k'}{2}\ln\frac{\Delta+k'\sin x}{\Delta-k'\sin x}.$$

71. $\int \frac{\cos^4 x}{\sin x}\Delta\, dx = \frac{-2k^2\sin^2 x+5k^2+1}{8k^2}\cos x\Delta +$
$$+\frac{1}{2}\ln\frac{\Delta+\cos x}{\Delta-\cos x} + \frac{3k^4+6k^2-1}{8k^3}\ln(k\cos x+\Delta).$$

2.584

1. $\int \frac{dx}{\Delta} = F(x, k)$.

2. $\int \frac{\sin x\, dx}{\Delta} = \frac{1}{2k}\ln\frac{\Delta-k\cos x}{\Delta+k\cos x} = -\frac{1}{k}\ln(k\cos x+\Delta)$.

3. $\int \frac{\cos x\, dx}{\Delta} = \frac{1}{k}\arcsin(k\sin x) = \frac{1}{k}\operatorname{arctg}\frac{k\sin x}{\Delta}$.

4. $\int \frac{\sin^2 x\, dx}{\Delta} = \frac{1}{k^2}F(x, k) - \frac{1}{k^2}E(x, k)$.

5. $\int \frac{\sin x\cos x\, dx}{\Delta} = -\frac{\Delta}{k^2}$.

6. $\int \frac{\cos^2 x\, dx}{\Delta} = \frac{1}{k^2}E(x, k) - \frac{k'^2}{k^2}F(x, k)$.

7. $\int \frac{\sin^3 x\, dx}{\Delta} = \frac{\cos x\,\Delta}{2k^2} - \frac{1+k^2}{2k^3}\ln(k\cos x+\Delta)$.

8. $\int \frac{\sin^2 x\cos x\, dx}{\Delta} = -\frac{\sin x\,\Delta}{2k^2} + \frac{\arcsin(k\sin x)}{2k^3}$.

9. $\int \frac{\sin x\cos^2 x\, dx}{\Delta} = -\frac{\cos x\,\Delta}{2k^2} + \frac{k'^2}{2k^3}\ln(k\cos x+\Delta)$.

10. $\int \frac{\cos^3 x\, dx}{\Delta} = \frac{\sin x\,\Delta}{2k^2} + \frac{2k^2-1}{2k^3}\arcsin(k\sin x)$.

11. $\int \frac{\sin^4 x\, dx}{\Delta} = \frac{\sin x\cos x\,\Delta}{3k^2} + \frac{2+k^2}{3k^4}F(x, k) - \frac{2(1+k^2)}{3k^4}E(x, k)$.

12. $\int \frac{\sin^3 x\cos x\, dx}{\Delta} = -\frac{1}{3k^4}(2+k^2\sin^2 x)\Delta$.

13. $\int \frac{\sin^2 x \cos^2 x\, dx}{\Delta} = -\frac{\sin x \cos x\, \Delta}{3k^2} + \frac{2-k^2}{3k^4} E(x, k) + \frac{2k^2-2}{3k^4} F(x, k).$

14. $\int \frac{\sin x \cos^3 x\, dx}{\Delta} = -\frac{1}{3k^4}(k^2 \cos^2 x - 2k'^2)\, \Delta.$

15. $\int \frac{\cos^4 x\, dx}{\Delta} = \frac{\sin x \cos x\, \Delta}{3k^2} + \frac{4k^2-2}{3k^4} E(x, k) + \frac{3k^4-5k^2+2}{3k^4} F(x, k).$

16. $\int \frac{\sin^5 x\, dx}{\Delta} = \frac{2k^2 \sin^2 x + 3k^2 + 3}{8k^4} \cos x\, \Delta - \frac{3+2k^2+3k^4}{8k^5} \ln(k \cos x + \Delta).$

17. $\int \frac{\sin^4 x \cos x\, dx}{\Delta} = -\frac{2k^2 \sin^2 x + 3}{8k^4} \sin x\, \Delta + \frac{3}{8k^5} \arcsin(k \sin x).$

18. $\int \frac{\sin^3 x \cos^2 x\, dx}{\Delta} = \frac{2k^2 \cos^2 x - k^2 - 3}{8k^4} \cos x\, \Delta - \frac{k^4+2k^2-3}{8k^5} \ln(k \cos x + \Delta).$

19. $\int \frac{\sin^2 x \cos^3 x\, dx}{\Delta} = -\frac{2k^2 \cos^2 x + 2k^2 - 3}{8k^4} \sin x\, \Delta + \frac{4k^2-3}{8k^5} \arcsin(k \sin x).$

20. $\int \frac{\sin x \cos^4 x\, dx}{\Delta} = \frac{3-5k^2+2k^2 \sin^2 x}{8k^4} \cos x\, \Delta -$
$$- \frac{3k^4-6k^2+3}{8k^5} \ln(k \cos x + \Delta).$$

21. $\int \frac{\cos^5 x\, dx}{\Delta} = \frac{2k^2 \cos^2 x + 6k^2 - 3}{8k^4} \sin x\, \Delta + \frac{8k^4-8k^2+3}{8k^5} \arcsin(k \sin x).$

22. $\int \frac{\sin^6 x\, dx}{\Delta} = \frac{3k^2 \sin^2 x + 4k^2 + 4}{15k^4} \sin x \cos x\, \Delta +$
$$+ \frac{4k^4+3k^2+8}{15k^6} F(x, k) - \frac{8k^4+7k^2+8}{15k^6} E(x, k).$$

23. $\int \frac{\sin^5 x \cos x\, dx}{\Delta} = -\frac{3k^4 \sin^4 x + 4k^2 \sin^2 x + 8}{15k^6}\, \Delta.$

24. $\int \frac{\sin^4 x \cos^2 x\, dx}{\Delta} = \frac{3k^2 \cos^2 x - 2k^2 - 4}{15k^4} \sin x \cos x\, \Delta +$
$$+ \frac{k^4+7k^2-8}{15k^6} F(x, k) - \frac{2k^4+3k^2-8}{15k^6} E(x, k).$$

25. $\int \frac{\sin^3 x \cos^3 x\, dx}{\Delta} = \frac{3k^4 \sin^4 x - (5k^4-4k^2) \sin^2 x - 10k^2 + 8}{15k^6}\, \Delta.$

26. $\int \frac{\sin^2 x \cos^4 x\, dx}{\Delta} = -\frac{3k^2 \cos^2 x + 3k^2 - 4}{15k^4} \sin x \cos x\, \Delta +$
$$+ \frac{9k^4-17k^2+8}{15k^6} F(x, k) - \frac{3k^4-13k^2+8}{15k^6} E(x, k).$$

27. $\int \frac{\sin x \cos^5 x\, dx}{\Delta} = \frac{-3k^4 \cos^4 x + 4k^2 k'^2 \cos^2 x - 8k^4 + 16k^2 - 8}{15k^6}\, \Delta.$

28. $\int \frac{\cos^6 x\, dx}{\Delta} = \frac{3k^2 \cos^2 x + 8k^2 - 4}{15k^4} \sin x \cos x\, \Delta +$
$$+ \frac{15k^6-34k^4+27k^2-8}{15k^6} F(x, k) + \frac{23k^4-23k^2+8}{15k^6} E(x, k).$$

29. $\int \frac{\sin^7 x\, dx}{\Delta} = \frac{8k^4 \sin^4 x + 10k^2 (k^2+1) \sin^2 x + 15k^4 + 14k^2 + 15}{48k^6} \cos x\, \Delta -$
$$- \frac{(5k^4-2k^2+5)(k^2+1)}{16k^7} \ln(k \cos x + \Delta).$$

30. $\int \frac{\sin^6 x \cos x\, dx}{\Delta} = -\frac{8k^4 \sin^4 x + 10k^2 \sin^2 x + 15}{48k^6} \sin x\, \Delta + \frac{5}{16k^7} \arcsin(k \sin x).$

31. $$\int \frac{\sin^5 x \cos^2 x \, dx}{\Delta} = \frac{-8k^4 \sin^4 x + 2k^2(k^2-5)\sin^2 x + 3k^4 + 4k^2 - 15}{48k^6} \cos x \, \Delta - \frac{k^6 + k^4 + 3k^2 - 5}{16k^7} \ln (k \cos x + \Delta).$$

32. $$\int \frac{\sin^4 x \cos^3 x \, dx}{\Delta} = \frac{8k^4 \sin^4 x - 2k^2 (6k^2-5) \sin^2 x - 18k^2 + 15}{48k^6} \sin x \, \Delta + \frac{6k^2 - 5}{16k^7} \arcsin (k \sin x).$$

33. $$\int \frac{\sin^3 x \cos^4 x \, dx}{\Delta} = \frac{8k^4 \sin^4 x - 2k^2 (7k^2-5) \sin^2 x + 3k^4 - 22k^2 + 15}{48k^6} \cos x \, \Delta - \frac{k^6 + 3k^4 - 9k^2 + 5}{16k^7} \ln (k \cos x + \Delta).$$

34. $$\int \frac{\sin^2 x \cos^5 x \, dx}{\Delta} = \frac{-8k^4 \sin^4 x + 2k^2 (12k^2-5) \sin^2 x - 24k^4 + 36k^2 - 15}{48k^6} \sin x \, \Delta + \frac{8k^4 - 12k^2 + 5}{16k^7} \arcsin (k \sin x).$$

35. $$\int \frac{\sin x \cos^6 x \, dx}{\Delta} = \frac{-8k^4 \sin^4 x + 2k^2 (13k^2-5) \sin^2 x - 33k^4 + 40k^2 - 15}{48k^6} \cos x \, \Delta + \frac{5k'^6}{16k^7} \ln (k \cos x + \Delta).$$

36. $$\int \frac{\cos^7 x \, dx}{\Delta} = \frac{8k^4 \sin^4 x - 2k^2 (18k^2-5) \sin^2 x + 72k^4 - 54k^2 + 15}{48k^6} \sin x \, \Delta + \frac{16k^6 - 24k^4 + 18k^2 - 5}{16k^7} \arcsin (k \sin x).$$

37. $$\int \frac{dx}{\Delta^3} = \frac{1}{k'^2} E(x, k) - \frac{k^2}{k'^2} \frac{\sin x \cos x}{\Delta}.$$

38. $$\int \frac{\sin x \, dx}{\Delta^3} = -\frac{\cos x}{k'^2 \Delta}.$$

39. $$\int \frac{\cos x \, dx}{\Delta^3} = \frac{\sin x}{\Delta}.$$

40. $$\int \frac{\sin^2 x \, dx}{\Delta^3} = \frac{1}{k'^2 k^2} E(x, k) - \frac{1}{k^2} F(x, k) - \frac{1}{k'^2} \frac{\sin x \cos x}{\Delta}.$$

41. $$\int \frac{\sin x \cos x \, dx}{\Delta^3} = \frac{1}{k^2 \Delta}.$$

42. $$\int \frac{\cos^2 x \, dx}{\Delta^3} = \frac{1}{k^2} F(x, k) - \frac{1}{k^2} E(x, k) + \frac{\sin x \cos x}{\Delta}.$$

43. $$\int \frac{\sin^3 x \, dx}{\Delta^3} = -\frac{\cos x}{k^2 k'^2 \Delta} + \frac{1}{k^3} \ln (k \cos x + \Delta).$$

44. $$\int \frac{\sin^2 x \cos x \, dx}{\Delta^3} = \frac{\sin x}{k^2 \Delta} - \frac{1}{k^3} \arcsin (k \sin x).$$

45. $$\int \frac{\sin x \cos^2 x \, dx}{\Delta^3} = \frac{\cos x}{k^2 \Delta} - \frac{1}{k^3} \ln (k \cos x + \Delta).$$

46. $$\int \frac{\cos^3 x \, dx}{\Delta^3} = -\frac{k'^2 \sin x}{k^2 \Delta} + \frac{1}{k^3} \arcsin (k \sin x).$$

47. $$\int \frac{\sin^4 x \, dx}{\Delta^3} = \frac{k'^2 + 1}{k'^2 k^4} E(x, k) - \frac{2}{k^4} F(x, k) - \frac{\sin x \cos x}{k^2 k'^2 \Delta}.$$

48. $$\int \frac{\sin^3 x \cos x \, dx}{\Delta^3} = \frac{2 - k^2 \sin^2 x}{k^4 \Delta}.$$

49. $\int \frac{\sin^2 x \cos^2 x\, dx}{\Delta^3} = \frac{2-k^2}{k^4} F(x, k) - \frac{2}{k^4} E(x, k) + \frac{\sin x \cos x}{k^2 \Delta}.$

50. $\int \frac{\sin x \cos^3 x\, dx}{\Delta^3} = \frac{k^2 \sin^2 x + k^2 - 2}{k^4 \Delta}.$

51. $\int \frac{\cos^4 x\, dx}{\Delta^3} = \frac{k'^2 + 1}{k^4} E(x, k) - \frac{2k'^2}{k^4} F(x, k) - \frac{k'^2 \sin x \cos x}{k^2 \Delta}.$

52. $\int \frac{\sin^5 x\, dx}{\Delta^3} = \frac{k^2 k'^2 \sin^2 x + k^2 - 3}{2k^4 k'^2 \Delta} \cos x + \frac{k^2 + 3}{2k^5} \ln(k \cos x + \Delta).$

53. $\int \frac{\sin^4 x \cos x\, dx}{\Delta^3} = \frac{-k^2 \sin^2 x + 3}{2k^4 \Delta} \sin x - \frac{3}{2k^5} \arcsin(k \sin x).$

54. $\int \frac{\sin^3 x \cos^2 x\, dx}{\Delta} = \frac{-k^2 \sin^2 x + 3}{2k^4 \Delta} \cos x + \frac{k^2 - 3}{2k^5} \ln(k \cos x + \Delta).$

55. $\int \frac{\sin^2 x \cos^3 x\, dx}{\Delta^3} = \frac{k^2 \sin^2 x + 2k^2 - 3}{2k^4 \Delta} \sin x - \frac{2k^2 - 3}{2k^5} \arcsin(k \sin x).$

56. $\int \frac{\sin x \cos^4 x\, dx}{\Delta^3} = \frac{k^2 \sin^2 x + 2k^2 - 3}{2k^4 \Delta} \cos x + \frac{3k'^2}{2k^5} \ln(k \cos x + \Delta).$

57. $\int \frac{\cos^5 x\, dx}{\Delta^3} = \frac{-k^2 \sin^2 x + 2k^4 - 4k^2 + 3}{2k^4 \Delta} \sin x + \frac{4k^2 - 3}{2k^5} \arcsin(k \sin x).$

58. $\int \frac{dx}{\Delta^5} = \frac{-k^2 \sin x \cos x}{3k'^2 \Delta^3} - \frac{2k^2 (k'^2 + 1) \sin x \cos x}{3k'^4 \Delta} - \frac{1}{3k'^2} F(x, k) + \frac{2(k'^2 + 1)}{3k'^4} E(x, k).$

59. $\int \frac{\sin x\, dx}{\Delta^5} = \frac{2k^2 \sin^2 x + k^2 - 3}{3k'^4 \Delta^3} \cos x.$

60. $\int \frac{\cos x\, dx}{\Delta^5} = \frac{-2k^2 \sin^2 x + 3}{3\Delta^3} \sin x.$

61. $\int \frac{\sin^2 x\, dx}{\Delta^5} = \frac{k^2 + 1}{3k'^4 k^2} E(x, k) - \frac{1}{3k'^2 k^2} F(x, k) + \frac{k^2 (k^2 + 1) \sin^2 x - 2}{3k'^4 \Delta^3} \sin x \cos x.$

62. $\int \frac{\sin x \cos x\, dx}{\Delta^5} = \frac{1}{3k^2 \Delta^3}.$

63. $\int \frac{\cos^2 x\, dx}{\Delta^5} = \frac{1}{3k^2} F(x, k) + \frac{2k^2 - 1}{3k^2 k'^2} E(x, k) + \frac{k^2 (2k^2 - 1) \sin^2 x - 3k^2 + 2}{3k'^2 \Delta} \sin x \cos x.$

64. $\int \frac{\sin^3 x}{\Delta^5} dx = \frac{(3k^2 - 1) \sin^2 x - 2}{3k'^4 \Delta^3} \cos x.$

65. $\int \frac{\sin^2 x \cos x}{\Delta^5} dx = \frac{\sin^3 x}{3\Delta^3}.$

66. $\int \frac{\sin x \cos^2 x}{\Delta^5} dx = -\frac{\cos^3 x}{3k'^2 \Delta^3}.$

67. $\int \frac{\cos^3 x\, dx}{\Delta^5} = \frac{-(2k^2 + 1) \sin^2 x + 3}{3\Delta^3} \sin x.$

68. $\int \frac{dx}{\Delta \sin x} = -\frac{1}{2} \ln \frac{\Delta + \cos x}{\Delta - \cos x}.$

69. $\int \frac{dx}{\Delta \cos x} = -\frac{1}{2k'} \ln \frac{\Delta - k' \sin x}{\Delta + k' \sin x}.$

70. $\int \frac{dx}{\Delta \sin^2 x} = \int \frac{1+\operatorname{ctg}^2 x}{\Delta} dx = F(x, k) - E(x, k) - \Delta \operatorname{ctg} x.$

71. $\int \frac{dx}{\Delta \sin x \cos x} = \int (\operatorname{tg} x + \operatorname{ctg} x) \frac{dx}{\Delta} = \frac{1}{2} \ln \frac{1-\Delta}{1+\Delta} + \frac{1}{2k'} \ln \frac{\Delta + k'}{\Delta - k'}.$

72. $\int \frac{dx}{\Delta \cos^2 x} = \int (1+\operatorname{tg}^2 x) \frac{dx}{\Delta} = F(x, k) - \frac{1}{k'^2} E(x, k) + \frac{1}{k'^2} \Delta \operatorname{tg} x.$

73. $\int \frac{\sin x}{\cos x} \frac{dx}{\Delta} = \int \operatorname{tg} x \frac{dx}{\Delta} = \frac{1}{2k'} \ln \frac{\Delta + k'}{\Delta - k'}.$

74. $\int \frac{\cos x}{\sin x} \frac{dx}{\Delta} = \int \operatorname{ctg} x \frac{dx}{\Delta} = \frac{1}{2} \ln \frac{1-\Delta}{1+\Delta}.$

75. $\int \frac{dx}{\Delta \sin^3 x} = -\frac{\Delta \cos x}{2 \sin^2 x} - \frac{1+k^2}{4} \ln \frac{\Delta + \cos x}{\Delta - \cos x}.$

76. $\int \frac{dx}{\Delta \sin^2 x \cos x} = -\frac{\Delta}{\sin x} - \frac{1}{2k'} \ln \frac{\Delta - k' \sin x}{\Delta + k' \sin x}.$

77. $\int \frac{dx}{\Delta \sin x \cos^2 x} = \frac{\Delta}{k'^2 \cos x} + \frac{1}{2} \ln \frac{\Delta - \cos x}{\Delta + \cos x}.$

78. $\int \frac{dx}{\Delta \cos^3 x} = \frac{\Delta \sin x}{2k'^2 \cos^2 x} + \frac{2k^2 - 1}{4k'^3} \ln \frac{\Delta - k' \sin x}{\Delta + k' \sin x}.$

79. $\int \frac{\sin x}{\cos^2 x} \frac{dx}{\Delta} = \frac{\Delta}{k'^2 \cos x}.$

80. $\int \frac{\cos x}{\sin^2 x} \frac{dx}{\Delta} = -\frac{\Delta}{\sin x}.$

81. $\int \frac{\sin^2 x}{\cos x} \frac{dx}{\Delta} = \frac{1}{2k'} \ln \frac{\Delta + k' \sin x}{\Delta - k' \sin x} - \frac{1}{k} \arcsin (k \sin x).$

82. $\int \frac{\cos^2 x}{\sin x} \frac{dx}{\Delta} = \frac{1}{2} \ln \frac{\Delta + \cos x}{\Delta - \cos x} + \frac{1}{k} \ln (k \cos x + \Delta).$

83. $\int \frac{dx}{\Delta \sin^4 x} = \frac{1}{3} \{ -\Delta \operatorname{ctg}^3 x - \Delta (2k^2 + 3) \operatorname{ctg} x + (k^2 + 2) F(x, k) -$
$- 2(k^2 + 1) E(x, k) \}.$

84. $\int \frac{dx}{\Delta \sin^3 x \cos x} = \int (\operatorname{tg} x + 2 \operatorname{ctg} x + \operatorname{ctg}^3 x) \frac{dx}{\Delta} =$
$= -\frac{\Delta}{2 \sin^2 x} + \frac{1}{2k'} \ln \frac{\Delta + k'}{\Delta - k'} - \frac{k^2 + 2}{4} \ln \frac{1+\Delta}{1-\Delta}.$

85. $\int \frac{dx}{\Delta \sin^2 x \cos^2 x} = \int (\operatorname{tg}^2 x + 2 + \operatorname{ctg}^2 x) \frac{dx}{\Delta} =$
$= \left(\frac{\operatorname{tg} x}{k'^2} - \operatorname{ctg} x \right) \Delta + \frac{k^2 - 2}{k'^2} E(x, k) + 2F(x, k).$

86. $\int \frac{dx}{\Delta \sin x \cos^3 x} = \int (\operatorname{ctg} x + 2 \operatorname{tg} x + \operatorname{tg}^3 x) \frac{dx}{\Delta} =$
$= \frac{\Delta}{2k'^2 \cos^2 x} - \frac{1}{2} \ln \frac{1+\Delta}{1-\Delta} + \frac{2 - 3k^2}{4k'^3} \ln \frac{\Delta + k'}{\Delta - k'}.$

87. $\int \frac{dx}{\Delta \cos^4 x} = \frac{1}{3k'^2} \left\{ \Delta \operatorname{tg}^3 x - \frac{5k^2 - 3}{k'^2} \Delta \operatorname{tg} x - (3k^2 - 2) F(x, k) + \right.$
$\left. + \frac{2(2k^2 - 1)}{k'^2} E(x, k) \right\}.$

88. $\int \frac{\sin x}{\cos^3 x} \frac{dx}{\Delta} = \int \operatorname{tg} x (1 + \operatorname{tg}^2 x) \frac{dx}{\Delta} = \frac{\Delta}{2k'^2 \cos^2 x} - \frac{k^2}{4k'^3} \ln \frac{\Delta + k'}{\Delta - k'}.$

89. $\int \frac{\cos x}{\sin^3 x} \frac{dx}{\Delta} = -\frac{\Delta}{2 \sin^2 x} - \frac{k^2}{4} \ln \frac{1+\Delta}{1-\Delta}.$

90. $\int \frac{\sin^2 x}{\cos^2 x} \frac{dx}{\Delta} = \int \frac{\operatorname{tg}^2 x}{\Delta} dx = \frac{\Delta}{k'^2} \operatorname{tg} x - \frac{1}{k'^2} E(x, k).$

91. $\int \frac{\cos^2 x}{\sin^2 x} \frac{dx}{\Delta} = \int \frac{\operatorname{ctg}^2 x}{\Delta} dx = -\Delta \operatorname{ctg} x - E(x, k).$

92. $\int \frac{\sin^3 x}{\cos x} \frac{dx}{\Delta} = \frac{\Delta}{k^2} + \frac{1}{2k'} \ln \frac{\Delta + k'}{\Delta - k'}.$

93. $\int \frac{\cos^3 x}{\sin x} \frac{dx}{\Delta} = \frac{\Delta}{k^2} - \frac{1}{2} \ln \frac{1+\Delta}{1-\Delta}.$

94. $\int \frac{dx}{\Delta \sin^5 x} = -\frac{[3(1+k^2)\sin^2 x + 2]}{8 \sin^4 x} \Delta \cos x + \frac{3k^4 + 2k^2 + 3}{16} \ln \frac{\Delta + \cos x}{\Delta - \cos x}.$

95. $\int \frac{dx}{\Delta \sin^4 x \cos x} = -\frac{(3+2k^2)\sin^2 x + 1}{3 \sin^3 x} \Delta - \frac{1}{2k'} \ln \frac{\Delta - k' \sin x}{\Delta + k' \sin x}.$

96. $\int \frac{dx}{\Delta \sin^3 x \cos^2 x} = \frac{(3-k^2)\sin^2 x - k'^2}{2k'^2 \sin^2 x \cos x} \Delta + \frac{k^2+3}{4} \ln \frac{\Delta - \cos x}{\Delta + \cos x}.$

97. $\int \frac{dx}{\Delta \sin^2 x \cos^3 x} = \frac{(3-2k^2)\sin^2 x - 2k'^2}{2k'^2 \sin x \cos^2 x} \Delta - \frac{4k^2-3}{4k'^3} \ln \frac{\Delta + k' \sin x}{\Delta - k' \sin x}.$

98. $\int \frac{dx}{\Delta \sin x \cos^4 x} = \frac{(5k^2-3)\sin^2 x - 6k^2 + 4}{3k'^4 \cos^3 x} \Delta - \frac{1}{2} \ln \frac{\Delta + \cos x}{\Delta - \cos x}.$

99. $\int \frac{dx}{\Delta \cos^5 x} = \frac{3(2k^2-1)\sin^2 x - 8k^2 + 5}{8k'^4 \cos^4 x} \Delta \sin x +$

$$+ \frac{8k^4 - 8k^2 + 3}{16k'^5} \ln \frac{\Delta + k' \sin x}{\Delta - k' \sin x}.$$

100. $\int \frac{\sin x}{\cos^4 x} \frac{dx}{\Delta} = -\frac{2k^2 \cos^2 x - k'^2}{3k'^4 \cos^3 x} \Delta.$

101. $\int \frac{\cos x}{\sin^4 x} \frac{dx}{\Delta} = -\frac{2k^2 \sin^2 x + 1}{3 \sin^3 x} \Delta.$

102. $\int \frac{\sin^2 x}{\cos^3 x} \frac{dx}{\Delta} = \frac{\Delta \sin x}{2k'^2 \cos^2 x} - \frac{1}{4k'^3} \ln \frac{\Delta + k' \sin x}{\Delta - k' \sin x}.$

103. $\int \frac{\cos^2 x}{\sin^3 x} \frac{dx}{\Delta} = -\frac{\Delta \cos x}{2 \sin^2 x} + \frac{k'^2}{4} \ln \frac{\Delta + \cos x}{\Delta - \cos x}.$

104. $\int \frac{\sin^3 x}{\cos^2 x} \frac{dx}{\Delta} = \frac{\Delta}{k'^2 \cos x} + \frac{1}{k} \ln (k \cos x + \Delta).$

105. $\int \frac{\cos^3 x}{\sin^2 x} \frac{dx}{\Delta} = \frac{-\Delta}{\sin x} - \frac{1}{k} \arcsin (k \sin x).$

106. $\int \frac{\sin^4 x}{\cos x} \frac{dx}{\Delta} = \frac{\Delta \sin x}{2k^2} + \frac{1}{2k'} \ln \frac{\Delta + k' \sin x}{\Delta - k' \sin x} - \frac{2k^2+1}{2k^3} \arcsin (k \sin x).$

107. $\int \frac{\cos^4 x}{\sin x} \frac{dx}{\Delta} = \frac{\Delta \cos x}{2k^2} + \frac{1}{2} \ln \frac{\Delta + \cos x}{\Delta - \cos x} + \frac{3k^2-1}{2k^3} \ln (k \cos x + \Delta).$

2.585

1. $\int \frac{(a + \sin x)^{p+3} dx}{\Delta} = \frac{1}{(p+2)k^2} \Big[(a + \sin x)^p \cos x \, \Delta +$

$$+ 2(2p+3) ak^2 \int \frac{(a + \sin x)^{p+2} dx}{\Delta} + (p+1)(1 + k^2 - 6a^2k^2) \int \frac{(a + \sin x)^{p+1} dx}{\Delta} -$$

$$- a(2p+1)(1 + k^2 - 2a^2k^2) \int \frac{(a + \sin x)^p dx}{\Delta} -$$

$$- p(1-a^2)(1-a^2k^2) \int \frac{(a + \sin x)^{p-1} dx}{\Delta} \Big]$$

$$\left[p \neq -2, \quad a \neq \pm 1, \quad a \neq \pm \frac{1}{k} \right].$$

For $p=n$ a natural number, this integral can be reduced to the following three integrals:

2. $$\int \frac{a+\sin x}{\Delta} dx = aF(x, k) + \frac{1}{2k} \ln \frac{\Delta - k\cos x}{\Delta + k\cos x}.$$

3. $$\int \frac{(a+\sin x)^2}{\Delta} dx = \frac{1+k^2a^2}{k^2} F(x, k) - \frac{1}{k^2} E(x, k) + \frac{a}{k} \ln \frac{\Delta - k\cos x}{\Delta + k\cos x}.$$

4. $$\int \frac{dx}{(a+\sin x)\Delta} = \frac{1}{a}\Pi\left(x, -\frac{1}{a^2}, k\right) - \int \frac{\sin x\, dx}{(a^2-\sin^2 x)\Delta},$$

where

5. $$\int \frac{\sin x\, dx}{(a^2-\sin^2 x)\Delta} = \frac{-1}{2\sqrt{(1-a^2)(1-a^2k^2)}} \ln \frac{\sqrt{1-a^2}\,\Delta - \sqrt{1-k^2a^2}\cos x}{\sqrt{1-a^2}\,\Delta + \sqrt{1-k^2a^2}\cos x}.$$

2.586

1. $$\int \frac{dx}{(a+\sin x)^n \Delta} = \frac{1}{(n-1)(1-a^2)(1-a^2k^2)} \left[-\frac{\cos x\,\Delta}{(a+\sin x)^{n-1}} - \right.$$
$$-(2n-3)(1+k^2-2a^2k^2)\,a \int \frac{dx}{(a+\sin x)^{n-1}\Delta} -$$
$$-(n-2)(6a^2k^2-k^2-1) \int \frac{dx}{(a+\sin x)^{n-2}\Delta} -$$
$$\left. -(10-4n)\,ak^2 \int \frac{dx}{(a+\sin x)^{n-3}\Delta} - (n-3)\,k^2 \int \frac{dx}{(a+\sin x)^{n-4}\Delta} \right]$$
$$\left[n \neq 1, \quad a \neq \pm 1, \quad a \neq \pm \frac{1}{k}\right].$$

This integral can be reduced to the integrals:

2. $$\int \frac{dx}{(a+\sin x)^2 \Delta} = \frac{1}{(1-a^2)(1-a^2k^2)} \left[-\frac{\cos x\,\Delta}{a+\sin x} - \right.$$
$$-a(1+k^2-2a^2k^2) \int \frac{dx}{(a+\sin x)\Delta} - 2ak^2 \int \frac{(a+\sin x)\,dx}{\Delta} +$$
$$\left. + k^2 \int \frac{(a+\sin x)^2\,dx}{\Delta} \right]$$
(see 2.585 2., 3., 4.).

3. $$\int \frac{dx}{(a+\sin x)^3 \Delta} = \frac{1}{2(1-a^2)(1-a^2k^2)} \left[-\frac{\cos x\,\Delta}{(a+\sin x)^2} - \right.$$
$$\left. -3a(1+k^2-2a^2k^2) \int \frac{dx}{(a+\sin x)^2\Delta} - (6a^2k^2-k^2-1) \int \frac{dx}{(a+\sin x)\Delta} + 2ak^2F(x, k) \right]$$
(see 2.585 4. and 2.586 2.).

For $a = \pm 1$, we have:

4. $$\int \frac{dx}{(1\pm\sin x)^n \Delta} = \frac{1}{(2n-1)\,k'^2} \left[\mp \frac{\cos x\,\Delta}{(1\pm\sin x)^n} + \right.$$
$$+ (n-1)(1-5k^2) \int \frac{dx}{(1\pm\sin x)^{n-1}\Delta} + 2(2n-3)\,k^2 \int \frac{dx}{(1\pm\sin x)^{n-2}\Delta} -$$
$$\left. -(n-2)\,k^2 \int \frac{dx}{(1\pm\sin x)^{n-3}\Delta} \right].$$
GU ((241))(6a)

This integral can be reduced to the integrals

5. $$\int \frac{dx}{(1\pm\sin x)\Delta} = \frac{\mp\cos x\Delta}{k'^2(1\pm\sin x)} + F(x, k) - \frac{1}{k'^2} E(x, k).$$
GU ((241))(6c)

6. $$\int \frac{dx}{(1\pm\sin x)^2\Delta} = \frac{1}{3k'^4} \left[\mp \frac{k'^2\cos x\,\Delta}{(1\pm\sin x)^2} \mp \frac{(1-5k^2)\cos x\,\Delta}{1\pm\sin x} + \right.$$
$$\left. + (1-3k^2)\,k'^2 F(x, k) - (1-5k^2)\,E(x, k) \right].$$
GU ((241))(6b)

For $a = \pm \frac{1}{k}$, we have

7. $$\int \frac{dx}{(1 \pm k \sin x)^n \Delta} = \frac{1}{(2n-1) k'^2} \left[\pm \frac{k \cos x \Delta}{(1 \pm k \sin x)^n} + (n-1)(5-k^2) \int \frac{dx}{(1 \pm k \sin x)^{n-1} \Delta} - 2(2n-3) \int \frac{dx}{(1 \pm k \sin x)^{n-2} \Delta} + (n-2) \int \frac{dx}{(1 \pm k \sin x)^{n-3} \Delta} \right].$$ GU ((241))(7a)

This integral can be reduced to the integrals

8. $$\int \frac{dx}{(1 \pm k \sin x) \Delta} = \pm \frac{k \cos x \Delta}{k'^2 (1 \pm \sin x)} + \frac{1}{k'^2} E(x, k).$$ GU ((241))(7b)

9. $$\int \frac{dx}{(1 \pm k \sin x)^2 \Delta} = \frac{1}{3k'^4} \left[\pm \frac{kk'^2 \cos x \Delta}{(1 \pm k \sin x)^2} \pm \frac{k(5-k^2) \cos x \Delta}{1 \pm k \sin x} - 2k'^2 F(x, k) + (5-k^2) E(x, k) \right].$$ GU ((241))(7c)

2.587

1. $$\int \frac{(b + \cos x)^{p+3} dx}{\Delta} = \frac{1}{(p+2) k^2} \left[(b + \cos x)^p \sin x \Delta + 2(2p+3) bk^2 \int \frac{(b+\cos x)^{p+2} dx}{\Delta} - (p+1)(k'^2 - k^2 + 6b^2k^2) \int \frac{(b+\cos x)^{p+1} dx}{\Delta} + (2p+1) b (k'^2 - k^2 + b^2k^2) \int \frac{(b+\cos x)^p dx}{\Delta} + p(1-b^2)(k'^2 + k^2b^2) \int \frac{(b+\cos x)^{p-1} dx}{\Delta} \right]$$
$$\left[p \neq -2, \quad b \neq \pm 1, \quad b \neq \frac{ik'}{k} \right].$$

For $p = n$ a natural number, this integral can be reduced to the following three integrals:

2. $$\int \frac{b + \cos x}{\Delta} dx = bF(x, k) + \frac{1}{k} \arcsin(k \sin x).$$

3. $$\int \frac{(b+\cos x)^2}{\Delta} dx = \frac{b^2k^2 - k'^2}{k^2} F(x, k) + \frac{1}{k^2} E(x, k) + \frac{2b}{k} \arcsin(k \sin x).$$

4. $$\int \frac{dx}{(b+\cos x) \Delta} = \frac{b}{b^2-1} \Pi\left(x, \frac{b}{b^2-1}, k\right) + \int \frac{\cos x \, dx}{(1 - b^2 - \sin^2 x) \Delta},$$

where

5. $$\int \frac{\cos x \, dx}{(1-b^2-\sin^2 x) \Delta} = \frac{1}{2\sqrt{(1-b^2)(k'^2+k^2b^2)}} \ln \frac{\sqrt{1-b^2}\, \Delta + k \sqrt{k'^2+k^2b^2} \sin x}{\sqrt{1-b^2}\, \Delta - k \sqrt{k'^2+k^2b^2} \sin x}.$$

2.588

1. $$\int \frac{dx}{(b+\cos x)^n \Delta} = \frac{1}{(n-1)(1-b^2)(k'^2+b^2k^2)} \left[\frac{-k'^2 \sin x \Delta}{(b+\cos x)^{n-1}} - (2n-3)(1-2k^2+2b^2k^2) b \int \frac{dx}{(b+\cos x)^{n-1} \Delta} - (n-2)(2k^2 - 1 - 6b^2k^2) \int \frac{dx}{(b+\cos x)^{n-2} \Delta} - (4n-10) bk^2 \int \frac{dx}{(b+\cos x)^{n-3} \Delta} + (n-3) k^2 \int \frac{dx}{(b+\cos x)^{n-4} \Delta} \right]$$
$$\left[n \neq 1, \quad b \neq \pm 1, \quad b \neq \pm \frac{ik'}{k} \right].$$

This integral can be reduced to the following integrals:

2. $$\int \frac{dx}{(b+\cos x)^2 \Delta} =$$
$$= \frac{1}{(1-b^2)(k'^2+b^2k^2)} \left[\frac{-k'^2 \sin x \, \Delta}{b+\cos x} - (1-2k^2+2b^2k^2)\, b \int \frac{dx}{(b+\cos x)\Delta} + \right.$$
$$\left. + 2bk^2 \int \frac{b+\cos x}{\Delta} dx - k^2 \int \frac{(b+\cos x)^2}{\Delta} dx \right] \quad \text{(see 2.587 2., 3., 4.).}$$

3. $$\int \frac{dx}{(b+\cos x)^3 \Delta} = \frac{1}{2(1-b^2)(k'^2+b^2k^2)} \left[\frac{-k'^2 \sin x \, \Delta}{(b+\cos x)^2} - \right.$$
$$- 3b(1-2k^2+2k^2b^2) \int \frac{dx}{(b+\cos x)^2 \Delta} -$$
$$\left. -(2k^2-1-6b^2k^2) \int \frac{dx}{(b+\cos x)\Delta} - 2bk^2 F(x, k) \right] \quad \text{(see 2.588 2. and 2.587 4.).}$$

2.589

1. $$\int \frac{(c+\operatorname{tg} x)^{p+3} dx}{\Delta} =$$
$$= \frac{1}{(p+2)k'^2} \left[\frac{(c+\operatorname{tg} x)^p \Delta}{\cos^2 x} + 2(2n+3)ck'^2 \int \frac{(c+\operatorname{tg} x)^{p+2} dx}{\Delta} - \right.$$
$$- (p+1)(1+k'^2+6c^2k'^2) \int \frac{(c+\operatorname{tg} x)^{p+1} dx}{\Delta} +$$
$$+ (2p+1)c(1+k'^2+2c^2k'^2) \int \frac{(c+\operatorname{tg} x)^p dx}{\Delta} -$$
$$\left. - p(1+c^2)(1+k'^2c^2) \int \frac{(c+\operatorname{tg} x)^{p-1} dx}{\Delta} \right] \quad [p \neq -2].$$

For $p = n$ a natural number, this integral can be reduced to the following three integrals:

2. $$\int \frac{c+\operatorname{tg} x}{\Delta} dx = cF(x, k) + \frac{1}{2k'} \ln \frac{\Delta+k'}{\Delta-k'}.$$

3. $$\int \frac{(c+\operatorname{tg} x)^2}{\Delta} dx = \frac{1}{k'^2} \operatorname{tg} x \, \Delta + c^2 F(x, k) - \frac{1}{k'^2} E(x, k) + \frac{c}{k'} \ln \frac{\Delta+k'}{\Delta-k'}.$$

4. $$\int \frac{dx}{(c+\operatorname{tg} x)\Delta} = \frac{c}{1+c^2} F(x, k) + \frac{1}{c(1+c^2)} \Pi\left(x, -\frac{1+c^2}{c^2}, k\right) -$$
$$- \int \frac{\sin x \cos x \, dx}{[c^2-(1+c^2)\sin^2 x]\Delta},$$

where

5. $$\int \frac{\sin x \cos x \, dx}{[c^2-(1+c^2)\sin^2 x]\Delta} = \frac{1}{2\sqrt{(1+c^2)(1+c^2k'^2)}} \ln \frac{\sqrt{1+c^2k'^2}+\sqrt{1+c^2}\,\Delta}{\sqrt{1+c^2k'^2}-\sqrt{1+c^2}\,\Delta}.$$

2.591

1. $$\int \frac{dx}{(c+\operatorname{tg} x)^n \Delta} = \frac{1}{(n-1)(1+c^2)(1+k'^2c^2)} \left[-\frac{\Delta}{(c+\operatorname{tg} x)^{n-1}\cos^2 x} + \right.$$
$$+ (2n-3)c(1+k'^2+2c^2k'^2) \int \frac{dx}{(c+\operatorname{tg} x)^{n-1}\Delta} -$$
$$- (n-2)(1+k'^2+6c^2k'^2) \int \frac{dx}{(c+\operatorname{tg} x)^{n-2}\Delta} +$$
$$\left. + (4n-10)ck'^2 \int \frac{dx}{(c+\operatorname{tg} x)^{n-3}\Delta} - (n-3)k'^2 \int \frac{dx}{(c+\operatorname{tg} x)^{n-4}\Delta} \right].$$

This integral can be reduced to the integrals:

2. $$\int \frac{dx}{(c+\operatorname{tg} x)^2 \Delta} = \frac{1}{(1+c^2)(1+k'^2c^2)} \left[\frac{-\Delta}{(c+\operatorname{tg} x)\cos^2 x} + \right.$$
$$\left. + c(1+k'^2+2c^2k'^2) \int \frac{dx}{(c+\operatorname{tg} x)\Delta} - 2ck'^2 \int \frac{c+\operatorname{tg} x}{\Delta} dx + k'^2 \int \frac{(c+\operatorname{tg} x)^2}{\Delta} dx \right]$$
(see **2.589** 2., 3., 4.).

3. $$\int \frac{dx}{(c+\operatorname{tg} x)^3 \Delta} = \frac{1}{2(1+c^2)(1+k'^2c^2)} \left[\frac{-\Delta}{(c+\operatorname{tg} x)^2 \cos^2 x} + \right.$$
$$+ 3c(1+k'^2+2c^2k'^2) \int \frac{dx}{(c+\operatorname{tg} x)^2 \Delta} -$$
$$\left. -(1+k'^2+6c^2k'^2) \int \frac{dx}{(c+\operatorname{tg} x)\Delta} + 2ck'^2 F(x, k) \right]$$ (see **2.591** 2. and **2.589** 4.).

2.592

1. $$P_n = \int \frac{(a+\sin^2 x)^n}{\Delta} dx.$$

The recursion formula

$$P_{n+2} = \frac{1}{(2n+3)k^2} \{(a+\sin^2 x)^n \sin x \cos x \Delta + (2n+2)(1+k^2+3ak^2) P_{n+1} -$$
$$- (2n+1)[1+2a(1+k^2)+3a^2k^2] P_n + 2na(1+a)(1+k^2a) P_{n-1}\}$$

reduces this integral (for n an integer) to the integrals

2. P_1 see **2.584** 1. and **2.584** 4.

3. P_0 see **2.584** 1.

4. $$P_{-1} = \int \frac{dx}{(a+\sin^2 x)\Delta} = \frac{1}{a} \Pi\left(x, \frac{1}{a}, k\right).$$

For $a = 0$

5. $$\int \frac{dx}{\sin^2 x \Delta}$$ see **2.584** 70. ZH (124)a

6. $$T_n = \int \frac{dx}{(h+g\sin^2 x)^n \Delta}$$

can be calculated by means of the recursion formula:

$$T_{n-3} = \frac{1}{(2n-5)k^2} \left\{ \frac{-g^2 \sin x \cos x \Delta}{(h+g\sin^2 x)^{n-1}} + 2(n-2)[g(1+k^2)+3hk^2] T_{n-2} - \right.$$
$$\left. -(2n-3)[g^2+2hg(1+k^2)+3h^2k^2] T_{n-1} + 2(n-1)h(g+h)(g+hk^2) T_n \right\}.$$

2.593

1. $$Q_n = \int \frac{(b+\cos^2 x)^n}{\Delta} dx.$$

The recursion formula

$$Q_{n+2} = \frac{1}{(2n+3)k^2} \left\{ (b+\cos^2 x)^n \sin x \cos x \Delta - (2n+2)(1-2k^2-3bk^2) Q_{n+1} + \right.$$
$$\left. + (2n+1)[k'^2+2b(k'^2-k^2)-3b^2k^2] Q_n - 2nb(1-b)(k'^2-k^2b) Q_{n-1} \right\}$$

reduces this integral (for n an integer) to the integrals:

2. Q_1 see **2.584** 1. and **2.584** 6.

3. Q_0 see **2.584** 1.

4. $$Q_{-1} = \int \frac{dx}{(b+\cos^2 x)\Delta} = \frac{1}{b+1} \Pi\left(x, -\frac{1}{b+1}, k\right).$$

For $b=0$

5. $\int \frac{dx}{\cos^2 x\,\Delta}$ see **2.584** 72. ZH (123)

2.594

1. $R_n = \int \frac{(c+\operatorname{tg}^2 x)^n\,dx}{\Delta}$.

The recursion formula

$$R_{n+2} = \frac{1}{(2n+3)k'^2}\left\{\frac{(c+\operatorname{tg}^2 x)^n \operatorname{tg} x\,\Delta}{\cos^2 x} - (2n+2)(1+k'^2-3ck'^2)R_{n+1} + \right.$$
$$\left. + (2n+1)[1-2c(1+k'^2)+3c^2k'^2]R_n + 2nc(1-c)(1-k'^2c)R_{n-1}\right\}$$

reduces this integral (for n an integer) to the integrals:

2. R_1 see **2.584** 1. and **2.584** 90.

3. R_0 see **2.584** 1.

4. $R_{-1} = \int \frac{dx}{(c+\operatorname{tg}^2 x)\,\Delta} = \frac{1}{c-1} F(x, k) + \frac{1}{c(1-c)} \Pi\left(x, \frac{1-c}{c}, k\right)$.

For $c=0$ see **2.582** 5.

2.595 Integrals of the type $\int R(\sin x, \cos x, \sqrt{1-p^2\sin^2 x})\,dx$ for $p^2>1$.

Notation: $\alpha = \arcsin(p\sin x)$.

Basic formulas

1. $\int \frac{dx}{\sqrt{1-p^2\sin^2 x}} = \frac{1}{p} F\left(\alpha, \frac{1}{p}\right) \quad [p^2>1]$. BY (283.00)

2. $\int \sqrt{1-p^2\sin^2 x}\,dx = pE\left(\alpha, \frac{1}{p}\right) - \frac{p^2-1}{p} F\left(\alpha, \frac{1}{p}\right) \quad [p^2>1]$.

BY (283.03)

3. $\int \frac{dx}{(1-r^2\sin^2 x)\sqrt{1-p^2\sin^2 x}} = \frac{1}{p} \Pi\left(\alpha, \frac{r^2}{p^2}, \frac{1}{p}\right) \quad [p^2>1]$.

BY (283.02)

To evaluate integrals of the form $\int R(\sin x, \cos x, \sqrt{1-p^2\sin^2 x})\,dx$ for $p^2>1$, we may use formulas **2.583** and **2.584**, making the following modifications in them.

We replace (1) k with p, (2) k'^2 with $1-p^2$, (3) $F(x, k)$ with $\frac{1}{p} F\left(\alpha, \frac{1}{p}\right)$, and (4) $E(x, k)$ with $pE\left(\alpha, \frac{1}{p}\right) - \frac{p^2-1}{p} F\left(\alpha, \frac{1}{p}\right)$.

For example (see **2.584** 15.):

2.596

1. $$\int \frac{\cos^4 x\,dx}{\sqrt{1-p^2\sin^2 x}} = \frac{\sin x\cos x\sqrt{1-p^2\sin^2 x}}{3p^2} + \frac{4p^2-2}{3p^4}\left[pE\left(\alpha; \frac{1}{p}\right) - \right.$$
$$\left. - \frac{p^2-1}{p} F\left(\alpha, \frac{1}{p}\right)\right] + \frac{2-5p^2+3p^4}{3p^4} \cdot \frac{1}{p} F\left(\alpha, \frac{1}{p}\right) =$$
$$= \frac{\sin x\cos x\sqrt{1-p^2\sin^2 x}}{3p^2} - \frac{p^2-1}{3p^3} F\left(\alpha, \frac{1}{p}\right) + \frac{4p^2-2}{3p^3} E\left(\alpha, \frac{1}{p}\right) \quad [p^2>1];$$

(see **2.583** 36.):

2. $$\int \frac{\sqrt{1-p^2\sin^2 x}}{\cos^2 x}\,dx = \operatorname{tg} x\sqrt{1-p^2\sin^2 x} + \frac{1}{p}F\left(\alpha, \frac{1}{p}\right) - \left[pE\left(\alpha, \frac{1}{p}\right) - \frac{p^2-1}{p}F\left(\alpha, \frac{1}{p}\right)\right] = p\left[F\left(\alpha, \frac{1}{p}\right) - E\left(\alpha, \frac{1}{p}\right)\right] + \operatorname{tg} x\sqrt{1-p^2\sin^2 x} \quad [p^2>1];$$

(see **2.584** 37.):

3. $$\int \frac{dx}{\sqrt{(1-p^2\sin^2 x)^3}} = \frac{-1}{p^2-1}\left[pE\left(\alpha, \frac{1}{p}\right) - \frac{p^2-1}{p}F\left(\alpha, \frac{1}{p}\right)\right] - \frac{p^2}{1-p^2}\cdot\frac{\sin x\cos x}{\sqrt{1-p^2\sin^2 x}} = \frac{p^2}{p^2-1}\cdot\frac{\sin x\cos x}{\sqrt{1-p^2\sin^2 x}} + \frac{1}{p}F\left(\alpha, \frac{1}{p}\right) - \frac{p}{p^2-1}E\left(\alpha, \frac{1}{p}\right) \quad [p^2>1].$$

2.597 Integrals of the form $\int R(\sin x, \cos x, \sqrt{1+p^2\sin^2 x})\,dx$.

Notation: $\alpha = \arcsin\frac{\sqrt{1+p^2}\sin x}{\sqrt{1+p^2\sin^2 x}}$.

Basic formulas

1. $$\int \frac{dx}{\sqrt{1+p^2\sin^2 x}} = \frac{1}{\sqrt{1+p^2}}F\left(\alpha, \frac{p}{\sqrt{1+p^2}}\right).$$ BY (282.00)

2. $$\int \sqrt{1+p^2\sin^2 x}\,dx = \sqrt{1+p^2}E\left(\alpha, \frac{p}{\sqrt{1+p^2}}\right) - p^2\frac{\sin x\cos x}{\sqrt{1+p^2\sin^2 x}}.$$ BY (282.03)

3. $$\int \frac{\sqrt{1+p^2\sin^2 x}\,dx}{1+(p^2-r^2p^2-r^2)\sin^2 x} = \frac{1}{\sqrt{1+p^2}}\Pi\left(\alpha, r^2, \frac{p}{\sqrt{1+p^2}}\right).$$ BY (282.02)

4. $$\int \frac{\sin x\,dx}{\sqrt{1+p^2\sin^2 x}} = -\frac{1}{p}\arcsin\left(\frac{p\cos x}{\sqrt{1+p^2}}\right).$$

5. $$\int \frac{\cos x\,dx}{\sqrt{1+p^2\sin^2 x}} = \frac{1}{p}\ln\left(p\sin x + \sqrt{1+p^2\sin^2 x}\right).$$

6. $$\int \frac{dx}{\sin x\sqrt{1+p^2\sin^2 x}} = \frac{1}{2}\ln\frac{\sqrt{1+p^2\sin^2 x}-\cos x}{\sqrt{1+p^2\sin^2 x}+\cos x}.$$

7. $$\int \frac{dx}{\cos x\sqrt{1+p^2\sin^2 x}} = \frac{1}{2\sqrt{1+p^2}}\ln\frac{\sqrt{1+p^2\sin^2 x}+\sqrt{1+p^2}\sin x}{\sqrt{1+p^2\sin^2 x}-\sqrt{1+p^2}\sin x}.$$

8. $$\int \frac{\operatorname{tg} x\,dx}{\sqrt{1+p^2\sin^2 x}} = \frac{1}{2\sqrt{1+p^2}}\ln\frac{\sqrt{1+p^2\sin^2 x}+\sqrt{1+p^2}}{\sqrt{1+p^2\sin^2 x}-\sqrt{1+p^2}}.$$

9. $$\int \frac{\operatorname{ctg} x\,dx}{\sqrt{1+p^2\sin^2 x}} = \frac{1}{2}\ln\frac{1-\sqrt{1+p^2\sin^2 x}}{1+\sqrt{1+p^2\sin^2 x}}.$$

2.598 To calculate integrals of the form $\int R\left(\sin x, \cos x, \sqrt{1+p^2\sin^2 x}\right)dx$, we may use formulas **2.583** and **2.584**, making the following modifications in them:

We replace 1) k^2 with $-p^2$; 2) k'^2 with $1+p^2$;

3) $F(x, k)$ with $\frac{1}{\sqrt{1+p^2}} F\left(\alpha, \frac{p}{\sqrt{1+p^2}}\right)$;

4) $E(x, k)$ with $\sqrt{1+p^2}\, E\left(\alpha, \frac{p}{\sqrt{1+p^2}}\right) - p^2 \frac{\sin x \cos x}{\sqrt{1+p^2 \sin^2 x}}$;

5) $\frac{1}{k} \ln (k \cos x + \Delta)$ with $\frac{1}{p} \arcsin \frac{p \cos x}{\sqrt{1+p^2}}$;

6) $\frac{1}{k} \arcsin (k \sin x)$ with $\frac{1}{p} \ln \left(p \sin x + \sqrt{1+p^2 \sin^2 x}\right)$.

For example (see **2.584** 90.):

1. $$\int \frac{\operatorname{tg}^2 x\, dx}{\sqrt{1+p^2 \sin^2 x}} = \frac{1}{(1+p^2)} \left[\operatorname{tg} x \sqrt{1+p^2 \sin^2 x} - \right.$$
$$\left. - \sqrt{1+p^2}\, E\left(\alpha, \frac{p}{\sqrt{1+p^2}}\right) + p^2 \frac{\sin x \cos x}{\sqrt{1+p^2 \sin^2 x}} \right] =$$
$$= -\frac{1}{\sqrt{1+p^2}} E\left(\alpha, \frac{p}{\sqrt{1+p^2}}\right) + \frac{\operatorname{tg} x}{\sqrt{1+p^2 \sin^2 x}}$$

(see **2.584** 37.):

2. $$\int \frac{dx}{\sqrt{(1+p^2 \sin^2 x)^3}} = \frac{1}{\sqrt{1+p^2}} E\left(\alpha, \frac{p}{\sqrt{1+p^2}}\right).$$

2.599 Integrals of the form $\int R\left(\sin x, \cos x, \sqrt{a^2 \sin^2 x - 1}\right) dx$ $[a^2 > 1]$

Notation: $\alpha = \arcsin \frac{a \cos x}{\sqrt{a^2 - 1}}$.

Basic formulas:

1. $$\int \frac{dx}{\sqrt{a^2 \sin^2 x - 1}} = -\frac{1}{a} F\left(\alpha, \frac{\sqrt{a^2-1}}{a}\right) \quad [a^2 > 1].$$ BY (285.00)a

2. $$\int \sqrt{a^2 \sin^2 x - 1}\, dx = \frac{1}{a} F\left(\alpha, \frac{\sqrt{a^2-1}}{a}\right) - aE\left(\alpha, \frac{\sqrt{a^2-1}}{a}\right) [a^2 > 1].$$ BY (285.06)a

3. $$\int \frac{dx}{(1 - r^2 \sin^2 x)\sqrt{a^2 \sin^2 x - 1}} = \frac{1}{a(r^2-1)} \Pi\left(\alpha, \frac{r^2(a^2-1)}{a^2(r^2-1)}, \frac{\sqrt{a^2-1}}{a}\right)$$
$$[a^2 > 1,\ r^2 > 1].$$ BY (285.02)a

4. $$\int \frac{\sin x\, dx}{\sqrt{a^2 \sin^2 x - 1}} = -\frac{\alpha}{a} \quad [a^2 > 1].$$

5. $$\int \frac{\cos x\, dx}{\sqrt{a^2 \sin^2 x - 1}} = \frac{1}{a} \ln \left(a \sin x + \sqrt{a^2 \sin^2 x - 1}\right) \quad [a^2 > 1].$$

6. $$\int \frac{dx}{\sin x \sqrt{a^2 \sin^2 x - 1}} = -\operatorname{arctg} \frac{\cos x}{\sqrt{a^2 \sin^2 x - 1}} \quad [a^2 > 1].$$

7. $$\int \frac{dx}{\cos x \sqrt{a^2 \sin^2 x - 1}} = \frac{1}{2\sqrt{a^2-1}} \ln \frac{\sqrt{a^2-1}\, \sin x + \sqrt{a^2 \sin^2 x - 1}}{\sqrt{a^2-1}\, \sin x - \sqrt{a^2 \sin^2 x - 1}}$$
$$[a^2 > 1].$$

8. $\int \frac{\operatorname{tg} x\, dx}{\sqrt{a^2 \sin^2 x - 1}} = \frac{1}{2\sqrt{a^2-1}} \ln \frac{\sqrt{a^2-1} + \sqrt{a^2 \sin^2 x - 1}}{\sqrt{a^2-1} - \sqrt{a^2 \sin^2 x - 1}} \quad [a^2 > 1].$

9. $\int \frac{\operatorname{ctg} x\, dx}{\sqrt{a^2 \sin^2 x - 1}} = -\arcsin \frac{1}{a \sin x} \quad [a^2 > 1].$

2.611 To calculate integrals of the type $\int R\left(\sin x, \cos x, \sqrt{a^2 \sin^2 x - 1}\right) dx$ (for $a^2 > 1$), we may use formulas **2.583** and **2.584**. In doing so, we should follow the procedure outlined below:

1) In the right members of these formulas, the following functions should be replaced with integrals equal to them:

$F(x, k)$ should be replaced with $\int \frac{dx}{\Delta}$,

$E(x, k)$ should be replaced with $\int \Delta\, dx$,

$-\frac{1}{k} \ln (k \cos x + \Delta)$ should be replaced with $\int \frac{\sin x\, dx}{\Delta}$,

$\frac{1}{k} \arcsin (k \sin x)$ should be replaced with $\int \frac{\cos x\, dx}{\Delta}$,

$\frac{1}{2} \ln \frac{\Delta - \cos x}{\Delta + \cos x}$ should be replaced with $\int \frac{dx}{\Delta \sin x}$,

$\frac{1}{2k'} \ln \frac{\Delta + k' \sin x}{\Delta - k' \sin x}$ should be replaced with $\int \frac{dx}{\Delta \cos x}$,

$\frac{1}{2k'} \ln \frac{\Delta + k'}{\Delta - k'}$ should be replaced with $\int \frac{\operatorname{tg} x}{\Delta}\, dx$,

$\frac{1}{2} \ln \frac{1-\Delta}{1+\Delta}$ should be replaced with $\int \frac{\operatorname{ctg} x}{\Delta}\, dx$.

2) Then, on both sides of the equations, we should replace Δ with $i\sqrt{a^2 \sin^2 x - 1}$, k with a and k'^2 with $1 - a^2$.

3) Both sides of the resulting equations should be multiplied by i, as a result of which only real functions ($a^2 > 1$) should appear on both sides of the equations.

4) The integrals on the right sides of the equations should be replaced with their values found from formulas **2.599**.

Examples:

1. We rewrite equation **2.584** 4. in the form

$$\int \frac{\sin^2 x}{i\sqrt{a^2 \sin^2 x - 1}}\, dx = \frac{1}{a^2} \int \frac{dx}{i\sqrt{a^2 \sin^2 x - 1}} - \frac{1}{a^2} \int i\sqrt{a^2 \sin^2 x - 1}\, dx,$$

from which we get

$$\int \frac{\sin^2 x\, dx}{\sqrt{a^2 \sin^2 x - 1}} = \frac{1}{a^2} \left\{ \int \frac{dx}{\sqrt{a^2 \sin^2 x - 1}} + \int \sqrt{a^2 \sin^2 x - 1}\, dx \right\} =$$

$$= -\frac{1}{a} E\left(\alpha, \frac{\sqrt{a^2-1}}{a}\right) \quad [a^2 > 1].$$

2. We rewrite equation **2.584** 58. as follows:

$$\int \frac{dx}{i^5 \sqrt{(a^2 \sin^2 x - 1)^5}} = -\frac{2a^4 (a^2 - 2) \sin^2 x - (3a^2 - 5) a^2}{3(1 - a^2)^2 i^3 \sqrt{(a^2 \sin^2 x - 1)^3}} \sin x \cos x -$$

$$- \frac{1}{3(1 - a^2)} \int \frac{dx}{i\sqrt{a^2 \sin^2 x - 1}} - \frac{2a^2 - 4}{3(1 - a^2)^2} \int i\sqrt{a^2 \sin^2 x - 1}\, dx,$$

from which we obtain

$$\int \frac{dx}{\sqrt{(a^2 \sin^2 x - 1)^5}} = \frac{2a^4(a^2-2)\sin^2 x - (3a^2-5)a^2}{3(1-a^2)^2\sqrt{(a^2\sin^2 x - 1)^3}} \sin x \cos x + \frac{1}{3(1-a^2)^2 a} \times$$

$$\times \left\{ (a^2-3) F\left(\alpha, \frac{\sqrt{a^2-1}}{a}\right) - 2a^2(a^2-2) E\left(\alpha, \frac{\sqrt{a^2-1}}{a}\right) \right\} \quad [a^2 > 1].$$

3. We rewrite equation **2.584** 71. in the form

$$\int \frac{dx}{\sin x \cos x \, i\sqrt{a^2 \sin^2 x - 1}} = \int \frac{\operatorname{ctg} x \, dx}{i\sqrt{a^2\sin^2 x - 1}} + \int \frac{\operatorname{tg} x \, dx}{i\sqrt{a^2 \sin^2 x - 1}},$$

from which we obtain

$$\int \frac{dx}{\sin x \cos x \sqrt{a^2 \sin^2 x - 1}} = \frac{1}{2\sqrt{a^2-1}} \ln \frac{\sqrt{a^2-1} + \sqrt{a^2\sin^2 x - 1}}{\sqrt{a^2-1} - \sqrt{a^2 \sin^2 x - 1}} -$$

$$- \arcsin \frac{1}{a \sin x} \quad [a^2 > 1].$$

2.612 Integrals of the form $\int R\left(\sin x, \cos x, \sqrt{1-k^2\cos^2 x}\right) dx$.

To find integrals of the form $\int R\left(\sin x, \cos x, \sqrt{1-k^2\cos^2 x}\right) dx$ we make the substitution $x = \frac{\pi}{2} - y$, which yields

$$\int R\left(\sin x, \cos x, \sqrt{1-k^2\cos^2 x}\right) dx = -\int R\left(\cos y, \sin y, \sqrt{1-k^2\sin^2 y}\right) dy.$$

The integrals $\int R\left(\cos y, \sin y, \sqrt{1-k^2\sin^2 y}\right) dy$ are found from formulas **2.583** and **2.584**. As a result of the use of these formulas (where it is assumed that the original integral can be reduced only to integrals of the first and second Legendre forms), when we replace the functions $F(x, k)$ and $E(x, k)$ with the corresponding integrals, we obtain an expression of the form

$$-g(\cos y, \sin y) - A\int \frac{dy}{\sqrt{1-k^2\sin^2 y}} - B\int \sqrt{1-k^2 \sin^2 y} \, dy.$$

Returning now to the original variable x, we obtain

$$\int R\left(\sin x, \cos x, \sqrt{1-k^2\cos^2 x}\right) dx =$$

$$= -g(\sin x, \cos x) - A\int \frac{dx}{\sqrt{1-k^2\cos^2 x}} - B\int \sqrt{1-k^2\cos^2 x} \, dx.$$

The integrals appearing in this expression are found from the formulas

1. $\displaystyle\int \frac{dx}{\sqrt{1-k^2\cos^2 x}} = F\left(\arcsin \frac{\sin x}{\sqrt{1-k^2\cos^2 x}}, k\right).$

2. $\displaystyle\int \sqrt{1-k^2\cos^2 x} \, dx = E\left(\arcsin \frac{\sin x}{\sqrt{1-k^2\cos^2 x}}, k\right) - \frac{k^2 \sin x \cos x}{\sqrt{1-k^2\cos^2 x}}.$

2.613 Integrals of the form $\int R\left(\sin x, \cos x, \sqrt{1-p^2\cos^2 x}\right) dx$ $[p > 1]$.

To find integrals of the type $\int R\left(\sin x, \cos x, \sqrt{1-p^2\cos^2 x}\right) dx$, where $[p > 1]$, we proceed as in **2.612**. Here, we use the formulas

1. $\displaystyle\int \frac{dx}{\sqrt{1-p^2\cos^2 x}} = -\frac{1}{p} F\left(\arcsin(p\cos x), \frac{1}{p}\right) \quad [p > 1].$

2. $\int \sqrt{1-p^2\cos^2 x}\, dx = \frac{p^2-1}{p} F\left(\arcsin (p\cos x), \frac{1}{p}\right) - pE\left(\arcsin (p\cos x), \frac{1}{p}\right)$.

2.614 Integrals of the form $\int R\left(\sin x, \cos x, \sqrt{1+p^2\cos^2 x}\right) dx$.

To find integrals of the type $\int R\left(\sin x, \cos x, \sqrt{1+p^2\cos^2 x}\right) dx$, we need to make the substitution $x = \frac{\pi}{2} - y$. This yields

$$\int R\left(\sin x, \cos x, \sqrt{1+p^2\cos^2 x}\right) dx = -\int R\left(\cos y, \sin y, \sqrt{1+p^2\sin^2 y}\right) dy.$$

To calculate the integrals $-\int R\left(\cos y, \sin y, \sqrt{1+p^2\sin^2 y}\right) dy$, we need to use first what was said in **2.598** and **2.612** and then, after returning to the variable x, the formulas

1. $\int \frac{dx}{\sqrt{1+p^2\cos^2 x}} = \frac{1}{\sqrt{1+p^2}} F\left(x, \frac{p}{\sqrt{1+p^2}}\right)$.

2. $\int \sqrt{1+p^2\cos^2 x}\, dx = \sqrt{1+p^2}\, E\left(x, \frac{p}{\sqrt{1+p^2}}\right)$.

2.615 Integrals of the form $\int R\left(\sin x, \cos x, \sqrt{a^2\cos^2 x - 1}\right) dx$ $[a > 1]$.

To find integrals of the type $\int R\left(\sin x, \cos x, \sqrt{a^2\cos^2 x - 1}\right) dx$, we need to make the substitution $x = \frac{\pi}{2} - y$. This yields

$$\int R\left(\sin x, \cos x, \sqrt{a^2\cos^2 x - 1}\right) dx = -\int R\left(\cos y, \sin y, \sqrt{a^2\sin^2 y - 1}\right) dy.$$

To calculate the integrals $-\int R\left(\cos y, \sin y, \sqrt{a^2\sin^2 y - 1}\right) dy$, we use what was said in **2.611** and then, after returning to the variable x, we use the formulas

1. $\int \frac{dx}{\sqrt{a^2\cos^2 x - 1}} = \frac{1}{a} F\left(\arcsin \frac{a\sin x}{\sqrt{a^2-1}}, \frac{\sqrt{a^2-1}}{a}\right)$ $[a > 1]$.

2. $\int \sqrt{a^2\cos^2 x - 1}\, dx = aE\left(\arcsin \frac{a\sin x}{\sqrt{a^2-1}}, \frac{\sqrt{a^2-1}}{a}\right) - \frac{1}{a} F\left(\arcsin \frac{a\sin x}{\sqrt{a^2-1}}, \frac{\sqrt{a^2-1}}{a}\right)$ $[a > 1]$.

2.616 Integrals of the form $\int R\left(\sin x, \cos x, \sqrt{1-p^2\sin^2 x}, \sqrt{1-q^2\sin^2 x}\right) dx$.

Notation: $\alpha = \arcsin \frac{\sqrt{1-p^2}\sin x}{\sqrt{1-p^2\sin^2 x}}$.

1. $\int \frac{dx}{\sqrt{(1-p^2\sin^2 x)(1-q^2\sin^2 x)}} = \frac{1}{\sqrt{1-p^2}} F\left(\alpha, \sqrt{\frac{q^2-p^2}{1-p^2}}\right)$

$\left[0 < p^2 < q^2 < 1,\ 0 < x \leqslant \frac{\pi}{2}\right]$. BY (284.00)

2. $$\int \frac{\operatorname{tg}^2 x\,dx}{\sqrt{(1-p^2\sin^2 x)(1-q^2\sin^2 x)}} = \frac{\operatorname{tg} x\sqrt{1-q^2\sin^2 x}}{(1-q^2)\sqrt{1-p^2\sin^2 x}} - \frac{1}{(1-q^2)\sqrt{1-p^2}}E\left(\alpha, \sqrt{\frac{q^2-p^2}{1-p^2}}\right) \quad \left[0<p^2<q^2<1,\ 0<x\leqslant\frac{\pi}{2}\right].$$

BY (284.07)

3. $$\int \frac{\operatorname{tg}^4 x\,dx}{\sqrt{(1-p^2\sin^2 x)(1-q^2\sin^2 x)}} = \frac{1}{3(1-q^2)^2(1-p^2)^{3/2}}\times$$
$$\times\left[2(2-p^2-q^2)E\left(\alpha, \sqrt{\frac{q^2-p^2}{1-p^2}}\right)-(1-q^2)F\left(\alpha, \sqrt{\frac{q^2-p^2}{1-p^2}}\right)\right]+$$
$$+\frac{2p^2+q^2-3+\sin^2 x(4-3p^2-2q^2+p^2q^2)}{3(1-p^2)(1-q^2)^2}\frac{\sin x}{\cos^3 x}\sqrt{\frac{1-q^2\sin^2 x}{1-p^2\sin^2 x}}$$
$$\left[0<p^2<q^2<1,\ 0<x\leqslant\frac{\pi}{2}\right].$$

BY (284.07)

4. $$\int \frac{\sin^2 x\,dx}{\sqrt{(1-p^2\sin^2 x)(1-q^2\sin^2 x)^3}} = \frac{\sqrt{1-p^2}}{(1-q^2)(q^2-p^2)}E\left(\alpha, \sqrt{\frac{q^2-p^2}{1-p^2}}\right)-$$
$$-\frac{1}{(q^2-p^2)\sqrt{1-p^2}}F\left(\alpha, \sqrt{\frac{q^2-p^2}{1-p^2}}\right)-$$
$$-\frac{\sin x\cos x}{(1-q^2)\sqrt{(1-p^2\sin^2 x)(1-q^2\sin^2 x)}} \quad \left[0<p^2<q^2<1,\ 0<x\leqslant\frac{\pi}{2}\right].$$

BY (284.06)

5. $$\int \frac{\cos^2 x\,dx}{\sqrt{(1-p^2\sin^2 x)^3(1-q^2\sin^2 x)}} =$$
$$=\frac{\sqrt{1-p^2}}{q^2-p^2}E\left(\alpha, \sqrt{\frac{q^2-p^2}{1-p^2}}\right)-\frac{1-q^2}{(q^2-p^2)\sqrt{1-p^2}}F\left(\alpha, \sqrt{\frac{q^2-p^2}{1-p^2}}\right)$$
$$\left[0<p^2<q^2<1,\ 0<x\leqslant\frac{\pi}{2}\right].$$

BY (284.05)

6. $$\int \frac{\cos^4 x\,dx}{\sqrt{(1-p^2\sin^2 x)^5(1-q^2\sin^2 x)}} =$$
$$=\frac{(1-p^2)^{3/2}}{3(q^2-p^2)^2}\left[\frac{(2+p^2-3q^2)(1-q^2)}{(1-p^2)^2}F\left(\alpha, \sqrt{\frac{q^2-p^2}{1-p^2}}\right)+\right.$$
$$\left.+2\,\frac{2q^2-p^2-1}{1-p^2}E\left(\alpha, \sqrt{\frac{q^2-p^2}{1-p^2}}\right)\right]+\frac{(1-p^2)\sin x\cos x\sqrt{1-q^2\sin^2 x}}{3(q^2-p^2)\sqrt{(1-p^2\sin^2 x)^3}}$$
$$\left[0<p^2<q^2<1,\ 0<x\leqslant\frac{\pi}{2}\right].$$

BY (284.05)

7. $$\int \frac{dx}{1-p^2\sin^2 x}\sqrt{\frac{1-q^2\sin^2 x}{1-p^2\sin^2 x}} = \frac{1}{\sqrt{1-p^2}}E\left(\alpha, \sqrt{\frac{q^2-p^2}{1-p^2}}\right)$$
$$\left[0<p^2<q^2<1,\ 0<x\leqslant\frac{\pi}{2}\right].$$

BY (284.01)

8. $$\int \sqrt{\frac{1-p^2\sin^2 x}{(1-q^2\sin^2 x)^3}}\,dx = \frac{\sqrt{1-p^2}}{1-q^2}E\left(\alpha, \sqrt{\frac{q^2-p^2}{1-p^2}}\right)-$$
$$-\frac{q^2-p^2}{1-q^2}\frac{\sin x\cos x}{\sqrt{(1-p^2\sin^2 x)(1-q^2\sin^2 x)}} \quad \left[0<p^2<q^2<1,\ 0<x\leqslant\frac{\pi}{2}\right].$$

BY (284.04)

9. $$\int \frac{dx}{1+(p^2r^2-p^2-r^2)\sin^2 x}\sqrt{\frac{1-p^2\sin^2 x}{1-q^2\sin^2 x}} =$$
$$= \frac{1}{\sqrt{1-p^2}}\,\Pi\left(\alpha, r^2, \sqrt{\frac{q^2-p^2}{1-p^2}}\right) \quad \left[0 < p^2 < q^2 < 1,\ 0 < x \leqslant \frac{\pi}{2}\right].$$
BY (284.02)

2.617 Notation: $\alpha = \arcsin\sqrt{\frac{\sqrt{b^2+c^2} - b\sin x - c\cos x}{2\sqrt{b^2+c^2}}}$,

$$r = \sqrt{\frac{2\sqrt{b^2+c^2}}{a+\sqrt{b^2+c^2}}}.$$

1. $$\int \frac{dx}{\sqrt{a+b\sin x+c\cos x}} = -\frac{2}{\sqrt{a+\sqrt{b^2+c^2}}}F(\alpha, r)$$
$$\left[0 < \sqrt{b^2+c^2} < a,\ \arcsin\frac{b}{\sqrt{b^2+c^2}} - \pi \leqslant x < \arcsin\frac{b}{\sqrt{b^2+c^2}}\right];$$ BY (294.00)
$$= -\frac{\sqrt{2}}{\sqrt[4]{b^2+c^2}}F(\alpha, r)$$
$$\left[0 < |a| < \sqrt{b^2+c^2},\ \arcsin\frac{b}{\sqrt{b^2+c^2}} - \arccos\left(-\frac{a}{\sqrt{b^2+c^2}}\right) \leqslant x < \arcsin\frac{b}{\sqrt{b^2+c^2}}\right].$$ BY (293.00)

2. $$\int \frac{\sin x\,dx}{\sqrt{a+b\sin x+c\cos x}} = -\frac{\sqrt{2}b}{\sqrt[4]{(b^2+c^2)^3}}\{2E(\alpha, r) - F(\alpha, r)\} + \frac{2c}{b^2+c^2}\sqrt{a+b\sin x+c\cos x}$$
$$\left[0 < |a| < \sqrt{b^2+c^2},\ \arcsin\frac{b}{\sqrt{b^2+c^2}} - \arccos\left(-\frac{a}{\sqrt{b^2+c^2}}\right) \leqslant x < \arcsin\frac{b}{\sqrt{b^2+c^2}}\right].$$ BY (293.05)

3. $$\int \frac{(b\cos x - c\sin x)\,dx}{\sqrt{a+b\sin x+c\cos x}} = 2\sqrt{a+b\sin x+c\cos x}.$$

4. $$\int \frac{\sqrt{b^2+c^2}+b\sin x+c\cos x}{\sqrt{a+b\sin x+c\cos x}}dx = -2\sqrt{a+\sqrt{b^2+c^2}}E(\alpha, r) + \frac{2(a-\sqrt{b^2+c^2})}{\sqrt{a+\sqrt{b^2+c^2}}}F(\alpha, r)$$
$$\left[0 < \sqrt{b^2+c^2} < a,\ \arcsin\frac{b}{\sqrt{b^2+c^2}} - \pi \leqslant x < \arcsin\frac{b}{\sqrt{b^2+c^2}}\right];$$ BY (294.04)
$$= -2\sqrt{2}\sqrt[4]{b^2+c^2}E(\alpha, r)$$
$$\left[0 < |a| < \sqrt{b^2+c^2},\ \arcsin\frac{b}{\sqrt{b^2+c^2}} - \arccos\left(-\frac{a}{\sqrt{b^2+c^2}}\right) \leqslant x < \arcsin\frac{b}{\sqrt{b^2+c^2}}\right].$$ BY (293.01)

5. $\int \sqrt{a+b\sin x+c\cos x}\,dx=-2\sqrt{a+\sqrt{b^2+c^2}}\,E(\alpha, r)$

$\left[0<\sqrt{b^2+c^2}<a,\ \arcsin\frac{b}{\sqrt{b^2+c^2}}-\pi\leqslant x<\arcsin\frac{b}{\sqrt{b^2+c^2}}\right];$ BY (294.01)

$=-2\sqrt{2}\sqrt[4]{b^2+c^2}\,E(\alpha, r)+$

$+\frac{\sqrt{2}(\sqrt{b^2+c^2}-a)}{\sqrt[4]{b^2+c^2}}F(\alpha, r)\ \left[0<|a|<\sqrt{b^2+c^2},\right.$

$\left.\arcsin\frac{b}{\sqrt{b^2+c^2}}-\arccos\left(\frac{-a}{\sqrt{b^2+c^2}}\right)\leqslant x<\arcsin\frac{b}{\sqrt{b^2+c^2}}\right].$ BY (293.03)

2.618 Integrals of the form $\int R(\sin ax,\ \cos ax,\ \sqrt{\cos 2ax})\,dx=$

$=\frac{1}{a}\int R(\sin t,\ \cos t,\ \sqrt{1-2\sin^2 t})\,dt\quad (t=ax).$

Notation: $\alpha=\arcsin(\sqrt{2}\sin ax)$.

The integrals $\int R(\sin ax,\ \cos ax,\ \sqrt{\cos 2ax})\,dx$ are special cases of the integrals **2.595.** for $(p=2)$. We give some formulas:

1. $\int\frac{dx}{\sqrt{\cos 2ax}}=\frac{1}{a\sqrt{2}}F\left(\alpha,\frac{1}{\sqrt{2}}\right)\quad\left[0<ax\leqslant\frac{\pi}{4}\right].$

2. $\int\frac{\cos^2 ax}{\sqrt{\cos 2ax}}dx=\frac{1}{a\sqrt{2}}E\left(\alpha,\frac{1}{\sqrt{2}}\right)\quad\left[0<ax\leqslant\frac{\pi}{4}\right].$

3. $\int\frac{dx}{\cos^2 ax\sqrt{\cos 2ax}}=\frac{\sqrt{2}}{a}E\left(\alpha,\frac{1}{\sqrt{2}}\right)-\frac{\operatorname{tg} x}{a}\sqrt{\cos 2ax}\quad\left[0<ax\leqslant\frac{\pi}{4}\right].$

4. $\int\frac{dx}{\cos^4 ax\sqrt{\cos 2ax}}=\frac{2\sqrt{2}}{a}E\left(\alpha,\frac{1}{\sqrt{2}}\right)-$

$-\frac{\sqrt{2}}{3a}F\left(\alpha,\frac{1}{\sqrt{2}}\right)-\frac{(6\cos^2 ax+1)\sin ax}{3a\cos^3 ax}\sqrt{\cos 2ax}\quad\left[0<x\leqslant\frac{\pi}{4}\right].$

5. $\int\frac{\operatorname{tg}^2 ax\,dx}{\sqrt{\cos 2ax}}=\frac{\sqrt{2}}{a}E\left(\alpha,\frac{1}{\sqrt{2}}\right)-\frac{1}{a\sqrt{2}}F\left(\alpha,\frac{1}{\sqrt{2}}\right)-$

$-\frac{1}{a}\operatorname{tg} ax\sqrt{\cos 2ax}\quad\left[0<x\leqslant\frac{\pi}{2}\right].$

6. $\int\frac{\operatorname{tg}^4 ax\,dx}{\sqrt{\cos 2ax}}=\frac{1}{3a\sqrt{2}}F\left(\alpha,\frac{1}{\sqrt{2}}\right)-\frac{\sin ax}{3a\cos^3 ax}\sqrt{\cos 2ax}\quad\left[0<ax\leqslant\frac{\pi}{4}\right].$

7. $\int\frac{dx}{(1-2r^2\sin^2 ax)\sqrt{\cos 2ax}}=\frac{1}{a\sqrt{2}}\Pi\left(\alpha,r^2,\frac{1}{\sqrt{2}}\right)\quad\left[0<ax\leqslant\frac{\pi}{4}\right].$

8. $\int\frac{dx}{\sqrt{\cos^3 2ax}}=\frac{1}{a\sqrt{2}}F\left(\alpha,\frac{1}{\sqrt{2}}\right)-\frac{\sqrt{2}}{a}E\left(\alpha,\frac{1}{\sqrt{2}}\right)+\frac{\sin 2ax}{a\sqrt{\cos 2ax}}$

$\left[0<ax\leqslant\frac{\pi}{4}\right].$

9. $\int\frac{\sin^2 ax\,dx}{\sqrt{\cos^3 2ax}}=\frac{\sin 2ax}{2a\sqrt{\cos 2ax}}-\frac{1}{a\sqrt{2}}E\left(\alpha,\frac{1}{\sqrt{2}}\right)\quad\left[0<ax\leqslant\frac{\pi}{4}\right].$

10. $\int\frac{dx}{\sqrt{\cos^5 2ax}}=\frac{1}{3a\sqrt{2}}F\left(\alpha,\frac{1}{\sqrt{2}}\right)+\frac{\sin 2ax}{3a\sqrt{\cos^3 2ax}}\quad\left[0<ax\leqslant\frac{\pi}{4}\right].$

11. $\int \sqrt{\cos 2ax}\, dx = \frac{\sqrt{2}}{a} E\left(\alpha, \frac{1}{\sqrt{2}}\right) - \frac{1}{a\sqrt{2}} F\left(\alpha, \frac{1}{\sqrt{2}}\right) \quad \left[0 < ax \leqslant \frac{\pi}{4}\right]$

12. $\int \frac{\sqrt{\cos 2ax}}{\cos^2 ax}\, dx = \frac{\sqrt{2}}{a} \left\{F\left(\alpha, \frac{1}{\sqrt{2}}\right) - E\left(\alpha, \frac{1}{\sqrt{2}}\right)\right\} + \frac{1}{a} \operatorname{tg} ax \sqrt{\cos 2ax} \quad \left[0 < x \leqslant \frac{\pi}{4}\right].$

2.619

Integrals of the form $\int R(\sin ax, \cos ax, \sqrt{-\cos 2ax})\, dx = \frac{1}{a} \int R(\sin x, \cos x, \sqrt{2\sin^2 x - 1})\, dx.$

Notation: $\alpha = \arcsin(\sqrt{2} \cos ax)$.

The integrals $\int R(\sin x, \cos x, \sqrt{2\sin^2 x - 1})\, dx$ are special cases of the integrals 2.599 and 2.611 for $(a = 2)$. We give some formulas:

1. $\int \frac{dx}{\sqrt{-\cos 2ax}} = -\frac{1}{a\sqrt{2}} F\left(\alpha, \frac{1}{\sqrt{2}}\right).$

2. $\int \frac{\cos^2 ax\, dx}{\sqrt{-\cos 2ax}} = \frac{1}{a\sqrt{2}} \left[E\left(\alpha, \frac{1}{\sqrt{2}}\right) - F\left(\alpha, \frac{1}{\sqrt{2}}\right)\right].$

3. $\int \frac{\cos^4 ax\, dx}{\sqrt{-\cos 2ax}} = \frac{1}{3a\sqrt{2}} \left[3F\left(\alpha, \frac{1}{\sqrt{2}}\right) - \frac{5}{2} E\left(\alpha, \frac{1}{\sqrt{2}}\right)\right] - \frac{1}{12a} \sin 2ax \sqrt{-\cos 2ax}.$

4. $\int \frac{dx}{\sin^2 ax \sqrt{-\cos 2ax}} = \frac{1}{a} \operatorname{ctg} ax \sqrt{-\cos 2ax} - \frac{\sqrt{2}}{a} E\left(\alpha, \frac{1}{\sqrt{2}}\right).$

5. $\int \frac{dx}{\sin^4 ax \sqrt{-\cos 2ax}} = \frac{2}{3a\sqrt{2}} \left[F\left(\alpha, \frac{1}{\sqrt{2}}\right) - 6E\left(\alpha, \frac{1}{\sqrt{2}}\right)\right] + \frac{1}{3a} \frac{\cos ax}{\sin^3 ax} (6\sin^2 ax + 1) \sqrt{-\cos 2ax}.$

6. $\int \frac{\operatorname{ctg}^2 ax\, dx}{\sqrt{-\cos 2ax}} = \frac{1}{a\sqrt{2}} \left[F\left(\alpha, \frac{1}{\sqrt{2}}\right) - 2E\left(\alpha, \frac{1}{\sqrt{2}}\right)\right] + \frac{1}{a} \operatorname{ctg} ax \sqrt{-\cos 2ax}.$

7. $\int \frac{dx}{(1 - 2r^2 \cos^2 ax) \sqrt{-\cos 2ax}} = -\frac{1}{a\sqrt{2}} \Pi\left(\alpha, r^2, \frac{1}{\sqrt{2}}\right).$

8. $\int \frac{dx}{\sqrt{-\cos^3 2ax}} = \frac{1}{a\sqrt{2}} \left[F\left(\alpha, \frac{1}{\sqrt{2}}\right) - 2E\left(\alpha, \frac{1}{\sqrt{2}}\right)\right] + \frac{\sin 2ax}{a\sqrt{-\cos 2ax}}.$

9. $\int \frac{\cos^2 ax\, dx}{\sqrt{-\cos^3 2ax}} = \frac{\sin 2ax}{2a\sqrt{-\cos 2ax}} - \frac{1}{a\sqrt{2}} E\left(\alpha, \frac{1}{\sqrt{2}}\right).$

10. $\int \frac{dx}{\sqrt{-\cos^5 2ax}} = -\frac{1}{3a\sqrt{2}} F\left(\alpha, \frac{1}{\sqrt{2}}\right) - \frac{\sin 2ax}{3a\sqrt{-\cos^3 2ax}}.$

11. $\int \sqrt{-\cos 2ax}\, dx = \frac{1}{a\sqrt{2}} \left[F\left(\alpha, \frac{1}{\sqrt{2}}\right) - 2E\left(\alpha, \frac{1}{\sqrt{2}}\right)\right].$

Integrals of the form $\int R(\sin ax, \cos ax, \sqrt{\sin 2ax})\,dx$.

Notation: $\alpha = \arcsin\sqrt{\frac{2\sin ax}{1+\sin ax+\cos ax}}$.

2.621

1. $\int \frac{dx}{\sqrt{\sin 2ax}} = \frac{\sqrt{2}}{a} F\left(\alpha, \frac{1}{\sqrt{2}}\right)$. BY (287.50)

2. $\int \frac{\sin ax\,dx}{\sqrt{\sin 2ax}} = \frac{\sqrt{2}}{a}\left\{\frac{1+i}{2}\Pi\left(\alpha, \frac{1+i}{2}, \frac{1}{\sqrt{2}}\right) + \right.$
$\left. + \frac{1-i}{2}\Pi\left(\alpha, \frac{1-i}{2}, \frac{1}{\sqrt{2}}\right) + F\left(\alpha, \frac{1}{\sqrt{2}}\right) - 2E\left(\alpha, \frac{1}{\sqrt{2}}\right)\right\}$. BY (287.57)

3. $\int \frac{\sin ax\,dx}{(1+\sin ax+\cos ax)\sqrt{\sin 2ax}} = \frac{\sqrt{2}}{a}\left[F\left(\alpha, \frac{1}{\sqrt{2}}\right) - E\left(\alpha, \frac{1}{\sqrt{2}}\right)\right]$. BY (287.54)

4. $\int \frac{\sin ax\,dx}{(1-\sin ax+\cos ax)\sqrt{\sin 2ax}} = \frac{\sqrt{2}}{a}\left\{\sqrt{\operatorname{tg} ax} - E\left(\alpha, \frac{1}{\sqrt{2}}\right)\right\} \quad \left[ax \neq \frac{\pi}{2}\right]$. BY (287.55)

5. $\int \frac{(1+\cos ax)\,dx}{(1+\sin ax+\cos ax)\sqrt{\sin 2ax}} = \frac{\sqrt{2}}{a} E\left(\alpha, \frac{1}{\sqrt{2}}\right)$. BY (287.51)

6 $\int \frac{(1+\cos ax)\,dx}{(1-\sin ax+\cos ax)\sqrt{\sin 2ax}} =$
$= \frac{\sqrt{2}}{a}\left\{F\left(\alpha, \frac{1}{\sqrt{2}}\right) - E\left(\alpha, \frac{1}{\sqrt{2}}\right) + \sqrt{\operatorname{tg} ax}\right\} \quad \left[ax \neq \frac{\pi}{2}\right]$. BY (287.56)

7. $\int \frac{(1-\sin ax+\cos ax)\,dx}{(1+\sin ax+\cos ax)\sqrt{\sin 2ax}} = \frac{\sqrt{2}}{a}\left\{2E\left(\alpha, \frac{1}{\sqrt{2}}\right) - F\left(\alpha, \frac{1}{\sqrt{2}}\right)\right\}$. BY (287.53)

8. $\int \frac{(1+\sin ax+\cos ax)\,dx}{[1+\cos ax+(1-2r^2)\sin ax]\sqrt{\sin 2ax}} = \frac{\sqrt{2}}{a}\Pi\left(\alpha, r^2, \frac{1}{\sqrt{2}}\right)$. BY (287.52)

2.63-2.65 Products of trigonometric functions and powers

2.631

1. $\int x^r \sin^p x \cos^q x\,dx = \frac{1}{(p+q)^2}\left[(p+q)x^r \sin^{p+1} x \cos^{q-1} x + \right.$
$+ rx^{r-1}\sin^p x \cos^q x - r(r-1)\int x^{r-2}\sin^p x \cos^q x\,dx -$
$\left. - rp\int x^{r-1}\sin^{p-1} x \cos^{q-1} x\,dx + (q-1)(p+q)\int x^r \sin^p x \cos^{q-2} x\,dx\right];$
$= \frac{1}{(p+q)^2}\left[-(p+q)x^r \sin^{p-1} x \cos^{q+1} x + \right.$
$+ rx^{r-1}\sin^p x \cos^q x - r(r-1)\int x^{r-2}\sin^p x \cos^q x\,dx +$
$\left. + rq\int x^{r-1}\sin^{p-1} x \cos^{q-1} x\,dx + (p-1)(p+q)\int x^r \sin^{p-2} x \cos^q x\,dx\right]$. GU ((331))(1)

2. $$\int x^m \sin^n x\,dx = \frac{x^{m-1}\sin^{n-1}x}{n^2}\{m\sin x - nx\cos x\} + \frac{n-1}{n}\int x^m \sin^{n-2}x\,dx - \frac{m(m-1)}{n^2}\int x^{m-2}\sin^n x\,dx.$$

3. $$\int x^m \cos^n x\,dx = \frac{x^{m-1}\cos^{n-1}x}{n^2}\{m\cos x + nx\sin x\} + \frac{n-1}{n}\int x^m \cos^{n-2}x\,dx - \frac{m(m-1)}{n^2}\int x^{m-2}\cos^n x\,dx.$$

4. $$\int x^n \sin^{2m} x\,dx = \binom{2m}{m}\frac{x^{n+1}}{2^{2m}(n+1)} + \frac{(-1)^m}{2^{2m-1}}\sum_{k=0}^{m-1}(-1)^k\binom{2m}{k}\int x^n\cos(2m-2k)x\,dx$$ (see 2.633 2.). TI 333

5. $$\int x^n \sin^{2m+1} x\,dx = \frac{(-1)^m}{2^{2m}}\sum_{k=0}^{m}(-1)^k\binom{2m+1}{k}\int x^n\sin(2m-2k+1)x\,dx$$ (see 2.633 1.). TI 333

6. $$\int x^n \cos^{2m} x\,dx = \binom{2m}{m}\frac{x^{n+1}}{2^{2m}(n+1)} + \frac{1}{2^{2m-1}}\sum_{k=0}^{m-1}\binom{2m}{k}\int x^n\cos(2m-2k)x\,dx$$ (see 2.633 2.). TI 333

7. $$\int x^n \cos^{2m+1} x\,dx = \frac{1}{2^{2m}}\sum_{k=0}^{m}\binom{2m+1}{k}\int x^n\cos(2m-2k+1)x\,dx.$$ (see 2.633 2.). TI 333

2.632

1. $$\int x^{\mu-1}\sin\beta x\,dx = \frac{i}{2}(i\beta)^{-\mu}\gamma(\mu, i\beta x) - \frac{i}{2}(-i\beta)^{-\mu}\gamma(\mu, -i\beta x)$$ $[\operatorname{Re}\mu > -1,\ x > 0]$. ET I 317(2)

2. $$\int x^{\mu-1}\sin ax\,dx = -\frac{1}{2a^\mu}\left\{\exp\left[\frac{\pi i}{2}(\mu-1)\right]\Gamma(\mu, -iax) + \exp\left[\frac{\pi i}{2}(1-\mu)\right]\Gamma(\mu, iax)\right\}$$ $[\operatorname{Re}\mu < 1,\ a > 0,\ x > 0]$. ET I 317(3)

3. $$\int x^{\mu-1}\cos\beta x\,dx = \frac{1}{2}\{(i\beta)^{-\mu}\gamma(\mu, i\beta x) + (-i\beta)^{-\mu}\gamma(\mu, -i\beta x)\}$$ $[\operatorname{Re}\mu > 0,\ x > 0]$. ET I 319(22)

4. $$\int x^{\mu-1}\cos ax\,dx = -\frac{1}{2a^\mu}\left\{\exp\left(i\mu\frac{\pi}{2}\right)\Gamma(\mu, -iax) + \exp\left(-i\mu\frac{\pi}{2}\right)\Gamma(\mu, iax)\right\}.$$ ET I 319(23)

2.633

1. $$\int x^n\sin ax\,dx = -\sum_{k=0}^{n}k!\binom{n}{k}\frac{x^{n-k}}{a^{k+1}}\cos\left(ax + \frac{1}{2}k\pi\right).$$ TI (487)

2. $$\int x^n \cos ax\,dx = \sum_{k=0}^{n} k! \binom{n}{k} \frac{x^{n-k}}{a^{k+1}} \sin\left(ax + \frac{1}{2}k\pi\right).$$ TI (486)

3. $$\int x^{2n} \sin x\,dx = (2n)! \left\{ \sum_{k=0}^{n} (-1)^{k+1} \frac{x^{2n-2k}}{(2n-2k)!} \cos x + \sum_{k=0}^{n-1} (-1)^k \frac{x^{2n-2k-1}}{(2n-2k-1)!} \sin x \right\}.$$

4. $$\int x^{2n+1} \sin x\,dx = (2n+1)! \left\{ \sum_{k=0}^{n} (-1)^{k+1} \frac{x^{2n-2k+1}}{(2n-2k+1)!} \cos x + \sum_{k=0}^{n} (-1)^k \frac{x^{2n-2k}}{(2n-2k)!} \sin x \right\}.$$

5. $$\int x^{2n} \cos x\,dx = (2n)! \left\{ \sum_{k=0}^{n} (-1)^k \frac{x^{2n-2k}}{(2n-2k)!} \sin x + \sum_{k=0}^{n-1} (-1)^k \frac{x^{2n-2k-1}}{(2n-2k-1)!} \cos x \right\}.$$

6. $$\int x^{2n+1} \cos x\,dx = (2n+1)! \left\{ \sum_{k=0}^{n} (-1)^k \frac{x^{2n-2k+1}}{(2n-2k+1)!} \sin x + \sum_{k=0}^{n} \frac{x^{2n-2k}}{(2n-2k)!} \cos x \right\}.$$

2.634

1. $$\int P_n(x) \sin mx\,dx = -\frac{\cos mx}{m} \sum_{k=0}^{E\left(\frac{n}{2}\right)} (-1)^k \frac{P_n^{(2k)}(x)}{m^{2k}} + \frac{\sin mx}{m} \sum_{k=1}^{E\left(\frac{n+1}{2}\right)} (-1)^{k-1} \frac{P_n^{(2k-1)}(x)}{m^{2k-1}}.$$

2. $$\int P_n(x) \cos mx\,dx = \frac{\sin mx}{m} \sum_{k=0}^{E\left(\frac{n}{2}\right)} (-1)^k \frac{P_n^{(2k)}(x)}{m^{2k}} + \frac{\cos mx}{m} \sum_{k=1}^{E\left(\frac{n+1}{2}\right)} (-1)^{k-1} \frac{P_n^{(2k-1)}(x)}{m^{2k-1}}.$$

In formulas **2.634**, $P_n(x)$ is an nth-degree polynomial and $P_n^{(k)}(x)$ is its kth derivative with respect to x.

Notation: $z_1 = a + bx$.

2.635

1. $$\int z_1 \sin kx\,dx = -\frac{1}{k} z_1 \cos kx + \frac{b}{k^2} \sin kx.$$

2. $$\int z_1 \cos kx\,dx = \frac{1}{k} z_1 \sin kx + \frac{b}{k^2} \cos kx.$$

3. $\int z_1^2 \sin kx\,dx = \frac{1}{k}\left(\frac{2b^2}{k^2} - z_1^2\right)\cos kx + \frac{2bz_1}{k^2}\sin kx.$

4. $\int z_1^2 \cos kx\,dx = \frac{1}{k}\left(z_1^2 - \frac{2b^2}{k^2}\right)\sin kx + \frac{2bz_1}{k^2}\cos kx.$

5. $\int z_1^3 \sin kx\,dx = \frac{z_1}{k}\left(\frac{6b^2}{k^2} - z_1^2\right)\cos kx + \frac{3b}{k^2}\left(z_1^2 - \frac{2b^2}{k^2}\right)\sin kx.$

6. $\int z_1^3 \cos kx\,dx = \frac{z_1}{k}\left(z_1^2 - \frac{6b^2}{k^2}\right)\sin kx + \frac{3b}{k^2}\left(z_1^2 - \frac{2b^2}{k^2}\right)\cos kx.$

7. $\int z_1^4 \sin kx\,dx = -\frac{1}{k}\left(z_1^4 - \frac{12b^2}{k^2}z_1^2 + \frac{24b^4}{k^4}\right)\cos kx + \frac{4bz_1}{k^2}\left(z_1^2 - \frac{6b^2}{k^2}\right)\sin kx.$

8. $\int z_1^4 \cos kx\,dx = \frac{1}{k}\left(z_1^4 - \frac{12b^2}{k^2}z_1^2 + \frac{24b^4}{k^4}\right)\sin kx + \frac{4bz_1}{k^2}\left(z_1^2 - \frac{6b^2}{k^2}\right)\cos kx.$

9. $\int z_1^5 \sin kx\,dx = \frac{5b}{k^2}\left(z_1^4 - \frac{12b^2}{k^2}z_1^2 + \frac{24b^4}{k^4}\right)\sin kx - \frac{z_1}{k}\left(z_1^4 - \frac{20b^2}{k^2}z_1^2 + \frac{120b^4}{k^4}\right)\cos kx.$

10. $\int z_1^5 \cos kx\,dx = \frac{5b}{k^2}\left(z_1^4 - \frac{12b^2}{k^2}z_1^2 + \frac{24b^4}{k^4}\right)\cos kx + \frac{z_1}{k}\left(z_1^4 - \frac{20b^2}{k^2}z_1^2 + \frac{120b^4}{k^4}\right)\sin kx.$

11. $\int z_1^6 \sin kx\,dx = \frac{6bz_1}{k^2}\left(z_1^4 - \frac{20b^2}{k^2}z_1^2 + \frac{120b^4}{k^4}\right)\sin kx - \frac{1}{k}\left(z_1^6 - \frac{30b^2}{k^2}z_1^4 + \frac{360b^4}{k^4}z_1^2 - \frac{720b^6}{k^6}\right)\cos kx.$

12. $\int z_1^6 \cos kx\,dx = \frac{6bz_1}{k^2}\left(z_1^4 - \frac{20b^2}{k^2}z_1^2 + \frac{120b^4}{k^4}\right)\cos kx + \frac{1}{k}\left(z_1^6 - \frac{30b^2}{k^2}z_1^4 + \frac{360b^4}{k^4}z_1^2 - \frac{720b^6}{k^6}\right)\sin kx.$

2.636

1. $$\int x^n \sin^2 x\,dx = \frac{x^{n+1}}{2(n+1)} + \frac{n!}{4}\left\{\sum_{k=0}^{E\left(\frac{n}{2}\right)} \frac{(-1)^{k+1}x^{n-2k}}{2^{2k}(n-2k)!}\sin 2x + \sum_{k=0}^{E\left(\frac{n-1}{2}\right)} \frac{(-1)^{k+1}x^{n-2k-1}}{2^{2k+1}(n-2k-1)!}\cos 2x\right\}.$$
GU ((333))(2e)

2. $$\int x^n \cos^2 x\,dx = \frac{x^{n+1}}{2(n+1)} - \frac{n!}{4}\left\{\sum_{k=0}^{E\left(\frac{n}{2}\right)} \frac{(-1)^{k+1}x^{n-2k}}{2^{2k}(n-2k)!}\sin 2x + \sum_{k=0}^{E\left(\frac{n-1}{2}\right)} \frac{(-1)^{k+1}x^{n-2k-1}}{2^{2k+1}(n-2k-1)!}\cos 2x\right\}.$$
GU ((333))(3e)

3. $\int x \sin^2 x\,dx = \frac{x^2}{4} - \frac{x}{4}\sin 2x - \frac{1}{8}\cos 2x.$

4. $\int x^2 \sin^2 x \, dx = \frac{x^3}{6} - \frac{x}{4} \cos 2x - \frac{1}{4}\left(x^2 - \frac{1}{2}\right) \sin 2x.$ MZ 241

5. $\int x \cos^2 x \, dx = \frac{x^2}{4} + \frac{x}{4} \sin 2x + \frac{1}{8} \cos 2x.$

6. $\int x^2 \cos^2 x \, dx = \frac{x^3}{6} + \frac{x}{4} \cos 2x + \frac{1}{4}\left(x^2 - \frac{1}{2}\right) \sin 2x.$ MZ 245

2.637

1. $$\int x^n \sin^3 x \, dx = \frac{n!}{4} \left\{ \sum_{k=0}^{E\left(\frac{n}{2}\right)} \frac{(-1)^k x^{n-2k}}{(n-2k)!} \left(\frac{\cos 3x}{3^{2k+1}} - 3 \cos x\right) - \right.$$
$$\left. - \sum_{k=0}^{E\left(\frac{n-1}{2}\right)} (-1)^k \frac{x^{n-2k-1}}{(n-2k-1)!} \left(\frac{\sin 3x}{3^{2k+2}} - 3 \sin x\right)\right\}.$$ GU ((333))(2f)

2. $$\int x^n \cos^3 x \, dx = \frac{n!}{4} \left\{ \sum_{k=0}^{E\left(\frac{n}{2}\right)} \frac{(-1)^k x^{n-2k}}{(n-2k)!} \left(\frac{\sin 3x}{3^{2k+1}} + 3 \sin x\right) + \right.$$
$$\left. + \sum_{k=0}^{E\left(\frac{n-1}{2}\right)} (-1)^k \frac{x^{n-2k-1}}{(n-2k-1)!} \left(\frac{\cos 3x}{3^{2k+2}} + 3 \cos x\right)\right\}.$$ GU ((333))(3f)

3. $\int x \sin^3 x \, dx = \frac{3}{4} \sin x - \frac{1}{36} \sin 3x - \frac{3}{4} x \cos x + \frac{x}{12} x \sin x.$

4. $$\int x^2 \sin^3 x \, dx = -\left(\frac{3}{4} x^2 + \frac{3}{2}\right) \cos x + \left(\frac{x^2}{12} + \frac{1}{54}\right) \cos 3x +$$
$$+ \frac{3}{2} x \sin x - \frac{x}{18} \sin 3x.$$ MZ 241

5. $\int x \cos^3 x \, dx = \frac{3}{4} \cos x + \frac{1}{36} \cos 3x + \frac{3}{4} x \sin x + \frac{x}{12} \sin 3x.$

6. $$\int x^2 \cos^3 x \, dx = \left(\frac{3}{4} x^2 - \frac{3}{2}\right) \sin x + \left(\frac{x^2}{12} - \frac{1}{54}\right) \sin 3x +$$
$$+ \frac{3}{2} x \cos x + \frac{x}{18} \cos 3x.$$ MZ 245, 246

2.638

1. $$\int \frac{\sin^q x}{x^p} dx = -\frac{\sin^{q-1} x \left[(p-2) \sin x + q x \cos x\right]}{(p-1)(p-2) x^{p-1}} -$$
$$- \frac{q^2}{(p-1)(p-2)} \int \frac{\sin^q x \, dx}{x^{p-2}} + \frac{q(q-1)}{(p-1)(p-2)} \int \frac{\sin^{q-2} x \, dx}{x^{p-2}}$$
$[p \neq 1,\ p \neq 2].$ TI (496)

2. $$\int \frac{\cos^q x}{x^p} dx = -\frac{\cos^{q-1} x \left[(p-2) \cos x - q x \sin x\right]}{(p-1)(p-2) x^{p-1}} -$$
$$- \frac{q^2}{(p-1)(p-2)} \int \frac{\cos^q x \, dx}{x^{p-2}} + \frac{q(q-1)}{(p-1)(p-2)} \int \frac{\cos^{q-2} x \, dx}{x^{p-2}}$$
$[p \neq 1,\ p \neq 2].$ TI (495)

3. $$\int \frac{\sin x \, dx}{x^p} = -\frac{\sin x}{(p-1) x^{p-1}} + \frac{1}{p-1} \int \frac{\cos x \, dx}{x^{p-1}};$$
$$= -\frac{\sin x}{(p-1) x^{p-1}} - \frac{\cos x}{(p-1)(p-2) x^{p-2}} - \frac{1}{(p-1)(p-2)} \int \frac{\sin x \, dx}{x^{p-2}}$$
$(n > 2).$ TI (492)

4. $$\int \frac{\cos x\,dx}{x^p} = -\frac{\cos x}{(p-1)x^{p-1}} - \frac{1}{p-1}\int \frac{\sin x\,dx}{x^{p-1}};$$
$$= -\frac{\cos x}{(p-1)x^{p-1}} + \frac{\sin x}{(p-1)(p-2)x^{p-2}} - \frac{1}{(p-1)(p-2)}\int \frac{\cos x\,dx}{x^{p-2}}$$
$(n > 2)$. TI (491)

2.639

1. $$\int \frac{\sin x\,dx}{x^{2n}} = \frac{(-1)^{n+1}}{x(2n-1)!}\left\{\sum_{k=0}^{n-2} \frac{(-1)^k(2k+1)!}{x^{2k+1}}\cos x + \right.$$
$$\left. + \sum_{k=0}^{n-1} \frac{(-1)^{k+1}(2k)!}{x^{2k}}\sin x\right\} + \frac{(-1)^{n+1}}{(2n-1)!}\operatorname{ci}(x).$$ GU ((333))(6b)a

2. $$\int \frac{\sin x}{x^{2n+1}}\,dx = \frac{(-1)^{n+1}}{x(2n)!}\left\{\sum_{k=0}^{n-1} \frac{(-1)^{k+1}(2k)!}{x^{2k}}\cos x + \right.$$
$$\left. + \sum_{k=0}^{n-1} \frac{(-1)^{k+1}(2k+1)!}{x^{2k+1}}\sin x\right\} + \frac{(-1)^n}{(2n)!}\operatorname{si}(x).$$ GU ((333))(6b)a

3. $$\int \frac{\cos x}{x^{2n}}\,dx = \frac{(-1)^{n+1}}{x(2n-1)!}\left\{\sum_{k=0}^{n-1} \frac{(-1)^{k+1}(2k)!}{x^{2k}}\cos x - \right.$$
$$\left. - \sum_{k=0}^{n-2} \frac{(-1)^k(2k+1)!}{x^{2k+1}}\sin x\right\} + \frac{(-1)^n}{(2n-1)!}\operatorname{si}(x).$$ GU ((333))(7b)

4. $$\int \frac{\cos x\,dx}{x^{2n+1}} = \frac{(-1)^{n+1}}{x(2n)!}\left\{\sum_{k=0}^{n-1} \frac{(-1)^{k+1}(2k+1)!}{x^{2k+1}}\cos x - \right.$$
$$\left. - \sum_{k=0}^{n-1} \frac{(-1)^{k+1}(2k)!}{x^{2k}}\sin x\right\} + \frac{(-1)^n}{(2n)!}\operatorname{ci}(x).$$ GU ((333))(7b)

2.641

1. $$\int \frac{\sin kx}{a+bx}\,dx = \frac{1}{b}\left[\cos\frac{ka}{b}\operatorname{si}(u) - \sin\frac{ka}{b}\operatorname{ci}(u)\right] \quad \left[u = \frac{k}{b}(a+bx)\right].$$

2. $$\int \frac{\cos kx}{a+bx}\,dx = \frac{1}{b}\left[\cos\frac{ka}{b}\operatorname{ci}(u) + \sin\frac{ka}{b}\operatorname{si}(u)\right] \quad \left[u = \frac{k}{b}(a+bx)\right].$$

3. $$\int \frac{\sin kx}{(a+bx)^2}\,dx = -\frac{1}{b}\frac{\sin kx}{a+bx} + \frac{k}{b}\int \frac{\cos kx}{a+bx}\,dx$$ (see **2.641** 2.).

4. $$\int \frac{\cos kx}{(a+bx)^2}\,dx = -\frac{1}{b}\frac{\cos kx}{a+bx} - \frac{k}{b}\int \frac{\sin kx}{a+bx}\,dx$$ (see **2.641** 1.).

5. $$\int \frac{\sin kx}{(a+bx)^3}\,dx = -\frac{\sin kx}{2b(a+bx)^2} - \frac{k\cos kx}{2b^2(a+bx)} - \frac{k^2}{2b^2}\int \frac{\sin kx}{a+bx}\,dx$$ (see **2.641** 1.).

6. $$\int \frac{\cos kx}{(a+bx)^3}\,dx = -\frac{\cos kx}{2b(a+bx)^2} + \frac{k\sin kx}{2b^2(a+bx)} - \frac{k^2}{2b^2}\int \frac{\cos kx}{a+bx}\,dx$$ (see **2.641** 2.).

7. $$\int \frac{\sin kx}{(a+bx)^4}\,dx = -\frac{\sin kx}{3b(a+bx)^3} - \frac{k\cos kx}{6b^2(a+bx)^2} +$$
$$+ \frac{k^2\sin kx}{6b^3(a+bx)} - \frac{k^3}{6b^3}\int \frac{\cos kx}{a+bx}\,dx$$ (see **2.641** 2.).

8. $$\int \frac{\cos kx}{(a+bx)^4}\,dx = -\frac{\cos kx}{3b(a+bx)^3} + \frac{k\sin kx}{6b^2(a+bx)^2} + \frac{k^2\cos kx}{6b^3(a+bx)} + \frac{k^3}{6b^3}\int \frac{\sin kx}{a+bx}\,dx$$ (see **2.641** 1.).

9. $$\int \frac{\sin kx}{(a+bx)^5}\,dx = -\frac{\sin kx}{4b(a+bx)^4} - \frac{k\cos kx}{12b^2(a+bx)^3} + \frac{k^2\sin kx}{24b^3(a+bx)^2} + \frac{k^3\cos kx}{24b^4(a+bx)} + \frac{k^4}{24b^4}\int \frac{\sin kx}{a+bx}\,dx$$ (see **2.641** 1.).

10. $$\int \frac{\cos kx}{(a+bx)^5}\,dx = -\frac{\cos kx}{4b(a+bx)^4} + \frac{k\sin kx}{12b^2(a+bx)^3} + \frac{k^2\cos kx}{24b^3(a+bx)^2} - \frac{k^3\sin kx}{24b^4(a+bx)} + \frac{k^4}{24b^4}\int \frac{\cos kx}{a+bx}\,dx$$ (see **2.641** 2.).

11. $$\int \frac{\sin kx}{(a+bx)^6}\,dx = -\frac{\sin kx}{5b(a+bx)^5} - \frac{k\cos kx}{20b^2(a+bx)^4} + \frac{k^2\sin kx}{60b^3(a+bx)^3} + \frac{k^3\cos kx}{120b^4(a+bx)^2} - \frac{k^4\sin kx}{120b^5(a+bx)} + \frac{k^5}{120b^5}\int \frac{\cos kx}{a+bx}\,dx$$ (see **2.641** 2.).

12. $$\int \frac{\cos kx}{(a+bx)^6}\,dx = -\frac{\cos kx}{5b(a+bx)^5} + \frac{k\sin kx}{20b^2(a+bx)^4} + \frac{k^2\cos kx}{60b^3(a+bx)^3} - \frac{k^3\sin kx}{120b^4(a+bx)^2} - \frac{k^4\cos kx}{120b^5(a+bx)} - \frac{k^5}{120b^5}\int \frac{\sin kx}{a+bx}\,dx$$ (see **2.641** 1.).

2.642

1. $$\int \frac{\sin^{2m} x}{x}\,dx = \binom{2m}{m}\frac{\ln x}{2^{2m}} + \frac{(-1)^m}{2^{2m-1}}\sum_{k=0}^{m-1}(-1)^k\binom{2m}{k}\operatorname{ci}[(2m-2k)x].$$

2. $$\int \frac{\sin^{2m+1} x}{x}\,dx = \frac{(-1)^m}{2^{2m}}\sum_{k=0}^{m}(-1)^k\binom{2m+1}{k}\operatorname{si}[(2m-2k+1)x].$$

3. $$\int \frac{\cos^{2m} x}{x}\,dx = \binom{2m}{m}\frac{\ln x}{2^{2m}} + \frac{1}{2^{2m-1}}\sum_{k=0}^{m-1}\binom{2m}{k}\operatorname{ci}[(2m-2k)x].$$

4. $$\int \frac{\cos^{2m+1} x}{x}\,dx = \frac{1}{2^{2m}}\sum_{k=0}^{m}\binom{2m+1}{k}\operatorname{ci}[(2m-2k+1)x].$$

5. $$\int \frac{\sin^{2m} x}{x^2}\,dx = -\binom{2m}{m}\frac{1}{2^{2m}x} + \frac{(-1)^m}{2^{2m-1}}\sum_{k=0}^{m-1}(-1)^{k+1}\binom{2m}{k}\left\{\frac{\cos(2m-2k)x}{x} + (2m-2k)\operatorname{si}[(2m-2k)x]\right\}.$$

6. $$\int \frac{\sin^{2m+1} x}{x^2}\,dx = \frac{(-1)^m}{2^{2m}}\sum_{k=0}^{m}(-1)^{k+1}\binom{2m+1}{k}\times\left\{\frac{\sin(2m-2k+1)x}{x} - (2m-2k+1)\operatorname{ci}[(2m-2k+1)x]\right\}.$$

7. $$\int \frac{\cos^{2m} x}{x^2}\,dx = -\binom{2m}{m}\frac{1}{2^{2m}x} - \frac{1}{2^{2m-1}}\sum_{k=0}^{m-1}\binom{2m}{k}\left\{\frac{\cos(2m-2k)x}{x} + (2m-2k)\operatorname{si}[(2m-2k)x]\right\}.$$

8. $$\int \frac{\cos^{2m+1} x}{x^2}\,dx = -\frac{1}{2^{2m}} \sum_{k=0}^{m} \binom{2m+1}{k} \left\{ \frac{\cos(2m-2k+1)x}{x} + (2m-2k+1)\operatorname{si}[(2m-2k+1)x] \right\}.$$

2.643

1. $$\int \frac{x^p\,dx}{\sin^q x} = -\frac{x^{p-1}[p\sin x + (q-2)x\cos x]}{(q-1)(q-2)\sin^{q-1} x} + \frac{q-2}{q-1}\int \frac{x^p\,dx}{\sin^{q-2} x} + \frac{p(p-1)}{(q-1)(q-2)}\int \frac{x^{p-2}\,dx}{\sin^{q-2} x}.$$

2. $$\int \frac{x^p\,dx}{\cos^q x} = -\frac{x^{p-1}[p\cos x - (q-2)x\sin x]}{(q-1)(q-2)\cos^{q-1} x} + \frac{q-2}{q-1}\int \frac{x^p\,dx}{\cos^{q-2} x} + \frac{p(p-1)}{(q-1)(q-2)}\int \frac{x^{p-2}\,dx}{\cos^{q-2} x}.$$

3. $$\int \frac{x^n}{\sin x}\,dx = \frac{x^n}{n} + \sum_{k=1}^{\infty} (-1)^{k+1} \frac{2(2^{k-1}-1)}{(n+2k)(2k)!} B_{2k} x^{n+2k} \quad [|x| < \pi,\ n > 0].$$ TU ((333))(8b)

4. $$\int \frac{dx}{x^n \sin x} = -\frac{1}{nx^n} - [1+(-1)^n](-1)^{\frac{n}{2}} \frac{2^{n-1}-1}{n!} B_n \ln x - \sum_{\substack{k=1 \\ k\neq \frac{n}{2}}}^{\infty} (-1)^k \frac{2(2^{2n-1}-1)}{(2k-n)\cdot(2k)!} B_{2k} x^{2k-n} \quad [n > 1,\ |x| < \pi].$$ GU ((333))(9b)

5. $$\int \frac{x^n\,dx}{\cos x} = \sum_{k=0}^{\infty} \frac{|E_{2k}|\, x^{n+2k+1}}{(n+2k+1)(2k)!} \quad \left[|x| < \frac{\pi}{2},\ n > 0\right].$$ GU ((333))(10b)

6. $$\int \frac{dx}{x^n \cos x} = \frac{1}{2}[1-(-1)^n] \frac{|E_{n-1}|}{(n-1)!} \ln x + \sum_{\substack{k=0 \\ k\neq \frac{n-1}{2}}}^{\infty} \frac{|E_{2k}|\, x^{2k-n+1}}{(2k-n+1)\cdot(2k)!} \quad \left[|x| < \frac{\pi}{2}\right].$$ GU ((333))(11b)

7. $$\int \frac{x^n\,dx}{\sin^2 x} = -x^n \operatorname{ctg} x + \frac{n}{n-1} x^{n-1} + n \sum_{k=1}^{\infty} (-1)^k \frac{2^{2k} x^{n+2k-1}}{(n+2k-1)(2k)!} B_{2k} \quad [|x| < \pi,\ n > 1].$$ GU ((333))(8c)

8. $$\int \frac{dx}{x^n \sin^2 x} = -\frac{\operatorname{ctg} x}{x^n} + \frac{n}{(n+1)x^{n+1}} - [1-(-1)^n](-1)^{\frac{n+1}{2}} \frac{2^n n}{(n+1)!} B_{n+1} \ln x - \frac{n}{x^{n+1}} \sum_{\substack{k=1 \\ k\neq \frac{n+1}{2}}}^{\infty} \frac{(-1)^k (2x)^{2k}}{(2k-n-1)(2k)!} B_{2k} \quad [|x| < \pi].$$ GU ((333))(9c)

9. $$\int \frac{x^n\,dx}{\cos^2 x} = x^n \operatorname{tg} x + n \sum_{k=1}^{\infty} (-1)^k \frac{2^{2k}(2^{2k}-1)\,x^{n+2k-1}}{(n+2k-1)\cdot(2k)!} B_{2k}$$

$$\left[n > 1,\ |x| < \frac{\pi}{2}\right].$$ GU ((333))(10c)

10. $$\int \frac{dx}{x^n \cos^2 x} = \frac{\operatorname{tg} x}{x^n} - [1-(-1)^n](-1)^{\frac{n+1}{2}} \frac{2^n n}{(n+1)!}(2^{n+1}-1) B_{n+1} \ln x -$$

$$- \frac{n}{x^{n+1}} \sum_{\substack{k=1 \\ k \neq \frac{n+1}{2}}}^{\infty} \frac{(-1)^k (2^{2k}-1)(2x)^{2k}}{(2k-n-1)(2k)!} B_{2k}$$

$$\left[|x| < \frac{\pi}{2}\right].$$ GU ((333))(11c)

2.644

1. $$\int \frac{x\,dx}{\sin^{2n} x} =$$

$$= -\sum_{k=0}^{n-1} \frac{(2n-2)(2n-4)\dots(2n-2k+2)}{(2n-1)(2n-3)\dots(2n-2k+3)} \frac{\sin x + (2n-2k)\,x\cos x}{(2n-2k+1)(2n-2k)\sin^{2n-2k+1} x} +$$

$$+ \frac{2^{n-1}(n-1)!}{(2n-1)!!}(\ln \sin x - x \operatorname{ctg} x).$$

2. $$\int \frac{x\,dx}{\sin^{2n+1} x} =$$

$$= -\sum_{k=0}^{n-1} \frac{(2n-1)(2n-3)\dots(2n-2k+1)}{2n(2n-2)\dots(2n-2k+2)} \frac{\sin x + (2n-2k-1)\,x\cos x}{(2n-2k)(2n-2k-1)\sin^{2n-2k} x} +$$

$$+ \frac{(2n-1)!!}{2^n n!} \int \frac{x\,dx}{\sin x}$$ (see **2.644** 5.).

3. $$\int \frac{x\,dx}{\cos^{2n} x} =$$

$$= \sum_{k=0}^{n-1} \frac{(2n-2)(2n-4)\dots(2n-2k+2)}{(2n-1)(2n-3)\dots(2n-2k+3)} \frac{(2n-2k)\,x\sin x - \cos x}{(2n-2k+1)(2n-2k)\cos^{2n-2k+1} x} +$$

$$+ \frac{2^{n-1}(n-1)!}{(2n-1)!!}(x \operatorname{tg} x + \ln \cos x).$$

4. $$\int \frac{x\,dx}{\cos^{2n+1} x} =$$

$$= \sum_{k=0}^{n-1} \frac{(2n-1)(2n-3)\dots(2n-2k+1)}{2n(2n-2)\dots(2n-2k+2)} \frac{(2n-2k+1)\,x\sin x - \cos x}{(2n-2k)(2n-2k-1)\cos^{2n-2k} x} +$$

$$+ \frac{(2n-1)!!}{2^n n!} \int \frac{x\,dx}{\cos x}$$ (see **2.644** 6.).

5. $$\int \frac{x\,dx}{\sin x} = x + \sum_{k=1}^{\infty} (-1)^{k+1} \frac{2(2^{2k-1}-1)}{(2k+1)!} B_{2k} x^{2k+1}.$$

6. $$\int \frac{x\,dx}{\cos x} = \sum_{k=0}^{\infty} \frac{|E_{2k}|\,x^{2k+2}}{(2k+2)(2k)!}.$$

7. $\int \frac{x\,dx}{\sin^2 x} = -x \operatorname{ctg} x + \ln \sin x.$

8. $\int \frac{x\,dx}{\cos^2 x} = x \operatorname{tg} x + \ln \cos x.$

9. $\int \frac{x\,dx}{\sin^3 x} = -\frac{\sin x + x\cos x}{2\sin^2 x} + \frac{1}{2}\int \frac{x}{\sin x}\,dx$ (see **2.644** 5.).

10. $\int \frac{x\,dx}{\cos^3 x} = \frac{x\sin x - \cos x}{2\cos^2 x} + \frac{1}{2}\int \frac{x\,dx}{\cos x}$ (see **2.644** 6.).

11. $\int \frac{x\,dx}{\sin^4 x} = -\frac{x\cos x}{3\sin^3 x} - \frac{1}{6\sin^2 x} - \frac{2}{3} x \operatorname{ctg} x + \frac{2}{3}\ln(\sin x).$

12. $\int \frac{x\,dx}{\cos^4 x} = \frac{x\sin x}{3\cos^3 x} - \frac{1}{6\cos^2 x} + \frac{2}{3} x \operatorname{tg} x - \frac{2}{3}\ln(\cos x).$

13. $\int \frac{x\,dx}{\sin^5 x} = -\frac{x\cos x}{4\sin^4 x} - \frac{1}{12\sin^3 x} - \frac{3x\cos x}{8\sin^2 x} - \frac{3}{8\sin x} + \frac{3}{8}\int \frac{x\,dx}{\sin x}$ (see **2.644** 5.).

14. $\int \frac{x\,dx}{\cos^5 x} = \frac{x\sin x}{4\cos^4 x} - \frac{1}{12\cos^3 x} + \frac{3x\sin x}{8\cos^2 x} - \frac{3}{8\cos x} + \frac{3}{8}\int \frac{x\,dx}{\cos x}$ (see **2.644** 6.).

2.645

1. $\int x^p \frac{\sin^{2m} x}{\cos^n x}\,dx = \sum_{k=0}^{m} (-1)^k \binom{m}{k} \int \frac{x^p\,dx}{\cos^{n-2k} x}$ (see **2.643** 2.).

2. $\int x^p \frac{\sin^{2m+1} x}{\cos^n x}\,dx = \sum_{k=0}^{m} (-1)^k \binom{m}{k} \int \frac{x^p \sin x}{\cos^{n-2k} x}\,dx$ (see **2.645** 3.).

3. $\int x^p \frac{\sin x\,dx}{\cos^n x} = \frac{x^p}{(n-1)\cos^{n-1} x} - \frac{p}{n-1}\int \frac{x^{p-1}}{\cos^{n-1} x}\,dx$
$[n > 1]$ (see **2.643** 2.). GU ((333))(12)

4. $\int x^p \frac{\cos^{2m} x}{\sin^n x}\,dx = \sum_{k=0}^{m} (-1)^k \binom{m}{k} \int \frac{x^p\,dx}{\sin^{n-2k} x}$ (see **2.643** 1.).

5. $\int x^p \frac{\cos^{2m+1} x}{\sin^n x}\,dx = \sum_{k=0}^{m} (-1)^k \binom{m}{k} \int \frac{x^p \cos x}{\sin^{n-2k} x}\,dx$ (see **2.645** 6.).

6. $\int x^p \frac{\cos x}{\sin^n x}\,dx = -\frac{x^p}{(n-1)\sin^{n-1} x} + \frac{p}{n-1}\int \frac{x^{p-1}\,dx}{\sin^{n-1} x}$
$[n > 1]$ (see **2.643** 1.). GU ((333))(13)

7. $\int \frac{x\cos x}{\sin^2 x}\,dx = -\frac{x}{\sin x} + \ln \operatorname{tg} \frac{x}{2}.$

8. $\int \frac{x\sin x}{\cos^2 x}\,dx = \frac{x}{\cos x} - \ln \operatorname{tg}\left(\frac{x}{2} + \frac{\pi}{4}\right).$

2.646

1. $\int x^p \operatorname{tg} x\,dx = \sum_{k=1}^{\infty} (-1)^{k+1} \frac{2^{2k}(2^{2k-1}-1)}{(p+2k)\cdot(2k)!} B_{2k} x^{p+2k}$
$\left[p \geqslant -1,\ |x| < \frac{\pi}{2}\right].$ GU ((333))(12d)

2. $\int x^p \operatorname{ctg} x\, dx = \sum_{k=0}^{\infty} (-1)^k \frac{2^{2k} B_{2k}}{(p+2k)(2k)!} x^{p+2k}$

$[p \geqslant 1,\ |x| < \pi].$ GU ((333))(13d)

3. $\int x \operatorname{tg}^2 x\, dx = x \operatorname{tg} x + \ln \cos x - \frac{x^2}{2}.$

4. $\int x \operatorname{ctg}^2 x\, dx = -x \operatorname{ctg} x + \ln \sin x - \frac{x^2}{2}.$

2.647

1. $\int \frac{x^n \cos x\, dx}{(a+b \sin x)^m} = -\frac{x^n}{(m-1)\, b\, (a+b \sin x)^{m-1}} + \frac{n}{(m-1)\, b} \int \frac{x^{n-1}\, dx}{(a+b \sin x)^{m-1}}$ $[m \neq 1].$ MZ 247

2. $\int \frac{x^n \sin x\, dx}{(a+b \cos x)^m} = \frac{x^n}{(m-1)\, b\, (a+b \cos x)^{m-1}} - \frac{n}{(m-1)\, b} \int \frac{x^{n-1}\, dx}{(a+b \cos x)^{m-1}}$ $[m \neq 1].$ MZ 247

3. $\int \frac{x\, dx}{1+\sin x} = -x \operatorname{tg}\left(\frac{\pi}{4} - \frac{x}{2}\right) + 2 \ln \cos\left(\frac{\pi}{4} - \frac{x}{2}\right).$ PE (329)

4. $\int \frac{x\, dx}{1-\sin x} = x \operatorname{ctg}\left(\frac{\pi}{4} - \frac{x}{2}\right) + 2 \ln \sin\left(\frac{\pi}{4} - \frac{x}{2}\right).$ PE (330)

5. $\int \frac{x\, dx}{1+\cos x} = x \operatorname{tg} \frac{x}{2} + 2 \ln \cos \frac{x}{2}.$ PE (331)

6. $\int \frac{x\, dx}{1-\cos x} = -x \operatorname{ctg} \frac{x}{2} + 2 \ln \sin \frac{x}{2}.$ PE (332)

7. $\int \frac{x \cos x}{(1+\sin x)^2}\, dx = -\frac{x}{1+\sin x} + \operatorname{tg}\left(\frac{x}{2} - \frac{\pi}{4}\right).$

8. $\int \frac{x \cos x}{(1-\sin x)^2}\, dx = \frac{x}{1-\sin x} + \operatorname{tg}\left(\frac{x}{2} + \frac{\pi}{4}\right).$

9. $\int \frac{x \sin x}{(1+\cos x)^2}\, dx = \frac{x}{1+\cos x} - \operatorname{tg} \frac{x}{2}.$

10. $\int \frac{x \sin x}{(1-\cos x)^2}\, dx = -\frac{x}{1-\cos x} - \operatorname{ctg} \frac{x}{2}.$ MZ 247a

2.648

1. $\int \frac{x+\sin x}{1+\cos x}\, dx = x \operatorname{tg} \frac{x}{2}.$

2. $\int \frac{x-\sin x}{1-\cos x}\, dx = -x \operatorname{ctg} \frac{x}{2}.$ GU ((333))(16)

2.649 $\int \frac{x^2\, dx}{[(ax-b) \sin x + (a+bx) \cos x]^2} = \frac{x \sin x + \cos x}{b\, [(ax-b) \sin x + (a+bx) \cos x]}.$

GU ((333))(17)

2.651 $\int \frac{dx}{[a+(ax+b) \operatorname{tg} x]^2} = \frac{\operatorname{tg} x}{a\, [a+(ax+b) \operatorname{tg} x]}.$ GU ((333))(18)

2.652 $$\int \frac{x\,dx}{\cos(x+t)\cos(x-t)} = \operatorname{cosec} 2t \left\{ x \ln \frac{\cos(x-t)}{\cos(x+t)} - L(x+t) + L(x-t) \right\}$$

$$\left[t \neq n\pi;\ |x| < \left| \frac{\pi}{2} - |t_0| \right| \right],$$

where t_0 is the value of the argument t, which is reduced by multiples of the argument π to lie in the interval $\left(-\frac{\pi}{2}, \frac{\pi}{2}\right)$. LO III 288

2.653

1. $$\int \frac{\sin x}{\sqrt{x}}\,dx = \sqrt{2\pi}\,S\left(\sqrt{x}\right) \qquad \text{(cf. 2.528 1.).}$$

2. $$\int \frac{\cos x}{\sqrt{x}}\,dx = \sqrt{2\pi}\,C\left(\sqrt{x}\right) \qquad \text{(cf. 2.528 2.).}$$

2.654 Notation: $\Delta = \sqrt{1-k^2\sin^2 x}$, $k' = \sqrt{1-k^2}$:

1. $$\int \frac{x\sin x\cos x}{\Delta}\,dx = -\frac{x\Delta}{k^2} + \frac{1}{k^2}E(x, k).$$

2. $$\int \frac{x\sin^3 x\cos x}{\Delta}\,dx = \frac{k'^2}{9k^4}F(x, k) + \frac{2k^2+5}{9k^4}E(x, k) - \frac{1}{9k^4}\left[3(3-\Delta^2)x + k^2\sin x\cos x\right]\Delta.$$

3. $$\int \frac{x\sin x\cos^3 x}{\Delta}\,dx = -\frac{k'^2}{9k^4}F(x, k) + \frac{7k^2-5}{9k^4}E(x, k) - \frac{1}{9k^4}\left[3(\Delta^2-3k'^2)x - k^2\sin x\cos x\right]\Delta.$$

4. $$\int \frac{x\sin x\,dx}{\Delta^3} = -\frac{x\cos x}{k'^2\Delta} + \frac{1}{kk'^2}\arcsin(k\sin x).$$

5. $$\int \frac{x\cos x\,dx}{\Delta^3} = \frac{x\sin x}{\Delta} + \frac{1}{k}\ln(k\cos x + \Delta).$$

6. $$\int \frac{x\sin x\cos x\,dx}{\Delta^3} = \frac{x}{k^2\Delta} - \frac{1}{k^2}F(x, k).$$

7. $$\int \frac{x\sin^3 x\cos x\,dx}{\Delta^3} = x\frac{2-k^2\sin^2 x}{k^4\Delta} - \frac{1}{k^4}\left[E(x, k) + F(x, k)\right].$$

8. $$\int \frac{x\sin x\cos^3 x\,dx}{\Delta^3} = x\frac{k^2\sin^2 x + k^2 - 2}{k^4\Delta} + \frac{k'^2}{k^4}F(x, k) + \frac{1}{k^4}E(x, k).$$

Integrals containing $\sin x^2$ and $\cos x^2$

In integrals containing $\sin x^2$ and $\cos x^2$, it is expedient to make the substitution $x^2 = u$.

2.655

1. $$\int x^p \sin x^2\,dx = -\frac{x^{p-1}}{2}\cos x^2 + \frac{p-1}{2}\int x^{p-2}\cos x^2\,dx.$$

2. $$\int x^p \cos x^2\,dx = \frac{x^{p-1}}{2}\sin x^2 - \frac{p-1}{2}\int x^{p-2}\sin x^2\,dx.$$

3. $$\int x^n \sin x^2\,dx = (n-1)!!\left\{\sum_{k=1}^{r}(-1)^k\left[\frac{x^{n-4k+3}\cos x^2}{2^{2k-1}(n-4k+3)!!}-\frac{x^{n-4k+1}\sin x^2}{2^{2k}(n-4k+1)!!}\right]+\frac{(-1)^r}{2^{2r}(n-4r-1)!!}\int x^{n-4r}\sin x^2\,dx\right\}$$
$$\left[r=E\left(\frac{n}{4}\right)\right].$$ GU ((336))(4a)

4. $$\int x^n \cos x^2\,dx = (n-1)!!\left\{\sum_{k=1}^{r}(-1)^{k-1}\left[\frac{x^{n-4k+3}\sin x^2}{2^{2k-1}(n-4k+3)!!}+\frac{x^{n-4k+1}\cos x^2}{2^{2k}(n-4k+1)!!}\right]+\frac{(-1)^r}{2^{2r}(n-4r-1)!!}\int x^{n-4r}\cos x^2\,dx\right\}$$
$$\left[r=E\left(\frac{n}{4}\right)\right].$$ GU ((336))(5a)

5. $$\int x\sin x^2\,dx = -\frac{\cos x^2}{2}.$$

6. $$\int x\cos x^2\,dx = \frac{\sin x^2}{2}.$$

7. $$\int x^2\sin x^2\,dx = -\frac{x}{2}\cos x^2+\frac{1}{2}\sqrt{\frac{\pi}{2}}\,C(x).$$

8. $$\int x^2\cos x^2\,dx = \frac{x}{2}\sin x^2-\frac{1}{2}\sqrt{\frac{\pi}{2}}\,S(x).$$

9. $$\int x^3\sin x^2\,dx = -\frac{x^2}{2}\cos x^2+\frac{1}{2}\sin x^2.$$

10. $$\int x^3\cos x^2\,dx = \frac{x^2}{2}\sin x^2+\frac{1}{2}\cos x^2.$$

2.66 Combinations of trigonometric functions and exponentials

2.661 $$\int e^{ax}\sin^p x\cos^q x\,dx =$$

$$=\frac{1}{a^2+(p+q)^2}\left\{e^{ax}\sin^p x\cos^{q-1}x\,[a\cos x+(p+q)\sin x]-pa\int e^{ax}\sin^{p-1}x\cos^{q-1}x\,dx+(q-1)(p+q)\int e^{ax}\sin^p x\cos^{q-2}x\,dx\right\};$$
TI (523)

$$=\frac{1}{a^2+(p+q)^2}\left\{e^{ax}\sin^{p-1}x\cos^q x\,[a\sin x-(p+q)\cos x]+qa\int e^{ax}\sin^{p-1}x\cos^{q-1}x\,dx+(p-1)(p+q)\int e^{ax}\sin^{p-2}x\cos^q x\,dx\right\};$$
TI (524)

$$=\frac{1}{a^2+(p+q)^2}\left\{e^{ax}\sin^{p-1}x\cos^{q-1}x\,[a\sin x\cos x+q\sin^2 x-p\cos^2 x]+q(q-1)\int e^{ax}\sin^p x\cos^{q-2}x\,dx+p(p-1)\int e^{ax}\sin^{p-2}x\cos^q x\,dx\right\};$$ TI (525)

$$= \frac{1}{a^2+(p+q)^2}\left\{e^{ax}\sin^{p-1}x\cos^{q-1}x\,(a\sin x\cos x + q\sin^2 x - p\cos^2 x) + \right.$$
$$+ q(q-1)\int e^{ax}\sin^{p-2}x\cos^{q-2}x\,dx -$$
$$\left. -(q-p)(p+q-1)\int e^{ax}\sin^{p-2}x\cos^{q}x\,dx\right\}; \qquad \text{TI (526)}$$

$$= \frac{1}{a^2+(p+q)^2}\left\{e^{ax}\sin^{p-1}x\cos^{q-1}x\,(a\sin x\cos x + q\sin^2 x - p\cos^2 x) + \right.$$
$$+ p(p-1)\int e^{ax}\sin^{p-2}x\cos^{q-2}x\,dx +$$
$$\left. +(q-p)(p+q-1)\int e^{ax}\sin^{p}x\cos^{q-2}x\,dx\right\}. \qquad \text{GU ((334))(1a)}$$

For $p=m$ and $q=n$ even integers, the integral $\int e^{ax}\sin^m x\cos^n x\,dx$ can be reduced by means of these formulas to the integral $\int e^{ax}\,dx$. However, when only m or only n is even, they can be reduced to integrals of the form $\int e^{ax}\cos^n x\,dx$ or $\int e^{ax}\sin^m x\,dx$ respectively.

2.662

1. $$\int e^{ax}\sin^n bx\,dx = \frac{1}{a^2+n^2b^2}\left[(a\sin bx - nb\cos bx)\,e^{ax}\sin^{n-1}bx + \right.$$
$$\left. + n(n-1)\,b^2\int e^{ax}\sin^{n-2}bx\,dx\right].$$

2. $$\int e^{ax}\cos^n bx\,dx = \frac{1}{a^2+n^2b^2}\left[(a\cos bx + nb\sin bx)\,e^{ax}\cos^{n-1}bx + \right.$$
$$\left. + n(n-1)\,b^2\int e^{ax}\cos^{n-2}bx\,dx\right].$$

3. $$\int e^{ax}\sin^{2m} bx\,dx =$$
$$= \sum_{k=0}^{m-1}\frac{(2m)!\,b^{2k}e^{ax}\sin^{2m-2k-1}bx}{(2m-2k)!\,[a^2+(2m)^2b^2]\,[a^2+(2m-2)^2b^2]\ldots[a^2+(2m-2k)^2b^2]}\times$$
$$\times[a\sin bx-(2m-2k)\,b\cos bx] + \frac{(2m)!\,b^{2m}e^{ax}}{[a^2+(2m)^2b^2]\,[a^2+(2m-2)^2b^2]\ldots[a^2+4b^2]\,a} =$$
$$= \binom{2m}{m}\frac{e^{ax}}{2^{2m}a} + \frac{e^{ax}}{2^{2m-1}}\sum_{k=1}^{m}(-1)^k\binom{2m}{m-k}\frac{1}{a^2+4b^2k^2}(a\cos 2bkx + 2bk\sin 2bkx).$$

4. $$\int e^{ax}\sin^{2m+1} bx\,dx =$$
$$= \sum_{k=0}^{m}\frac{(2m+1)!\,b^{2k}e^{ax}\sin^{2m-2k}bx\,[a\sin bx-(2m-2k+1)\,b\cos bx]}{(2m-2k+1)!\,[a^2+(2m+1)^2b^2]\,[a^2+(2m-1)^2b^2]\ldots[a^2+(2m-2k+1)^2b^2]} =$$
$$= \frac{e^{ax}}{2^{2m}}\sum_{k=0}^{m}\frac{(-1)^k}{a^2+(2k+1)^2b^2}\binom{2m+1}{m-k}[a\sin(2k+1)\,bx-(2k+1)\,b\cos(2k+1)\,bx].$$

5. $\int e^{ax}\cos^{2m} bx\,dx =$

$$= \sum_{k=0}^{m-1} \frac{(2m)!\, b^{2k} e^{ax} \cos^{2m-2k-1} bx\,[a\cos bx + (2m-2k)\, b \sin bx]}{(2m-2k)!\,[a^2+(2m)^2 b^2]\,[a^2+(2m-2)^2 b^2]\ldots[a^2+(2m-2k)^2 b^2]} +$$

$$+ \frac{(2m)!\, b^{2m} e^{ax}}{[a^2+(2m)^2 b^2]\,[a^2+(2m-2)^2 b^2]\ldots[a^2+4b^2]\,a} =$$

$$= \binom{2m}{m} \frac{e^{ax}}{2^{2m} a} + \frac{e^{ax}}{2^{2m-1}} \sum_{k=1}^{m} \binom{2m}{m-k} \frac{1}{a^2+4b^2k^2}\,[a\cos 2kbx + 2kb \sin 2kbx].$$

6. $\int e^{ax}\cos^{2m+1} bx\,dx =$

$$= \sum_{k=0}^{m} \frac{(2m+1)!\, b^{2k} e^{ax} \cos^{2m-2k} bx\,[a\cos bx + (2m-2k+1)\, b \sin bx]}{(2m-2k+1)!\,[a^2+(2m+1)^2 b^2]\,[a^2+(2m-1)^2 b^2]\ldots[a^2+(2m-2k+1)^2 b^2]} =$$

$$= \frac{e^{ax}}{2^{2m}} \sum_{k=0}^{m} \binom{2m+1}{m-k} \frac{1}{a^2+(2k+1)^2 b^2}\,[a\cos(2k+1)\,bx + (2k+1)\,b\sin(2k+1)\,bx].$$

2.663

1. $$\int e^{ax}\sin bx\,dx = \frac{e^{ax}(a\sin bx - b\cos bx)}{a^2+b^2}.$$

2. $$\int e^{ax}\sin^2 bx\,dx = \frac{e^{ax}\sin bx\,(a\sin bx - 2b\cos bx)}{4b^2+a^2} + \frac{2b^2 e^{ax}}{(4b^2+a^2)\,a} =$$
$$= \frac{e^{ax}}{2a} - \frac{e^{ax}}{a^2+4b^2}\left(\frac{a}{2}\cos 2bx + b\sin 2bx\right).$$

3. $$\int e^{ax}\cos bx\,dx = \frac{e^{ax}(a\cos bx + b\sin bx)}{a^2+b^2}.$$

4. $$\int e^{ax}\cos^2 bx\,dx = \frac{e^{ax}\cos bx\,(a\cos bx + 2b\sin bx)}{4b^2+a^2} + \frac{2b^2 e^{ax}}{(4b^2+a^2)\,a} =$$
$$= \frac{e^{ax}}{2a} + \frac{e^{ax}}{a^2+4b^2}\left(\frac{a}{2}\cos 2bx + b\sin 2bx\right).$$

2.664

1. $$\int e^{ax}\sin bx\cos cx\,dx = \frac{e^{ax}}{2}\left[\frac{a\sin(b+c)\,x - (b+c)\cos(b+c)\,x}{a^2+(b+c)^2} + \right.$$
$$\left. + \frac{a\sin(b-c)\,x - (b-c)\cos(b-c)\,x}{a^2+(b-c)^2}\right].$$ GU ((334))(6b)

2. $$\int e^{ax}\sin^2 bx\cos cx\,dx = \frac{e^{ax}}{4}\left[2\,\frac{a\cos cx + c\sin cx}{a^2+c^2} - \right.$$
$$- \frac{a\cos(2b+c)\,x + (2b+c)\sin(2b+c)\,x}{a^2+(2b+c)^2} -$$
$$\left. - \frac{a\cos(2b-c)\,x + (2b-c)\sin(2b-c)\,x}{a^2+(2b-c)^2}\right].$$ GU ((334))(6c)

3. $$\int e^{ax}\sin bx\cos^2 cx\,dx = \frac{e^{ax}}{4}\left[2\,\frac{a\sin bx - b\cos bx}{a^2+b^2} + \right.$$
$$+ \frac{a\sin(b+2c)\,x - (b+2c)\cos(b+2c)\,x}{a^2+(b+2c)^2} +$$
$$\left. + \frac{a\sin(b-2c)\,x - (b-2c)\cos(b-2c)\,x}{a^2+(b-2c)^2}\right].$$ GU ((334))(6d)

2.665

1. $$\int \frac{e^{ax}\,dx}{\sin^p bx} = -\frac{e^{ax}\,[a \sin bx + (p-2)\, b \cos bx]}{(p-1)(p-2)\, b^2 \sin^{p-1} bx} + \frac{a^2 + (p-2)^2 b^2}{(p-1)(p-2)\, b^2} \int \frac{e^{ax}\,dx}{\sin^{p-2} bx}.$$ TI (530)a

2. $$\int \frac{e^{ax}\,dx}{\cos^p bx} = -\frac{e^{ax}\,[a \cos bx - (p-2)\, b \sin bx]}{(p-1)(p-2)\, b^2 \cos^{p-1} bx} + \frac{a^2 + (p-2)^2 b^2}{(p-1)(p-2)\, b^2} \int \frac{e^{ax}\,dx}{\cos^{p-2} bx}.$$ TI (529)a

By successive applications of formulas **2.665** for p a natural number, we obtain integrals of the form $\int \frac{e^{ax}\,dx}{\sin bx}$, $\int \frac{e^{ax}\,dx}{\sin^2 bx}$, $\int \frac{e^{ax}\,dx}{\cos bx}$, $\int \frac{e^{ax}\,dx}{\cos^2 bx}$, which are not expressible in terms of a finite combination of elementary functions.

2.666

1. $$\int e^{ax} \operatorname{tg}^p x\,dx = \frac{e^{ax}}{p-1} \operatorname{tg}^{p-1} x - \frac{a}{p-1} \int e^{ax} \operatorname{tg}^{p-1} x\,dx - \int e^{ax} \operatorname{tg}^{p-2} x\,dx.$$ TI (527)

2. $$\int e^{ax} \operatorname{ctg}^p x\,dx = -\frac{e^{ax} \operatorname{ctg}^{p-1} x}{p-1} + \frac{a}{p-1} \int e^{ax} \operatorname{ctg}^{p-1} x\,dx - \int e^{ax} \operatorname{ctg}^{p-2} x\,dx.$$ TI (528)

3. $$\int e^{ax} \operatorname{tg} x\,dx = \frac{e^{ax} \operatorname{tg} x}{a} - \frac{1}{a} \int \frac{e^{ax}\,dx}{\cos^2 x}$$ (see remark following **2.665**).

4. $$\int e^{ax} \operatorname{tg}^2 x\,dx = \frac{e^{ax}}{a} (a \operatorname{tg} x - 1) - a \int e^{ax} \operatorname{tg} x\,dx$$ (see **2.666** 3.). TI 355

5. $$\int e^{ax} \operatorname{ctg} x\,dx = \frac{e^{ax} \operatorname{ctg} x}{a} + \frac{1}{a} \int \frac{e^{ax}\,dx}{\sin^2 x}$$ (see remark following **2.665**).

6. $$\int e^{ax} \operatorname{ctg}^2 x\,dx = -\frac{e^{ax}}{a} (a \operatorname{ctg} x + 1) + a \int e^{ax} \operatorname{ctg} x\,dx$$ (see **2.666** 5.).

Integrals of the type $\int R(x, e^{ax}, \sin bx, \cos cx)\,dx$

Notation: $\sin t = -\frac{b}{\sqrt{a^2+b^2}}$; $\cos t = \frac{a}{\sqrt{a^2+b^2}}$.

2.667

1. $$\int x^p e^{ax} \sin bx\,dx = \frac{x^p e^{ax}}{a^2+b^2} (a \sin bx - b \cos bx) - \frac{p}{a^2+b^2} \int x^{p-1} e^{ax} (a \sin bx - b \cos bx)\,dx;$$
$$= \frac{x^p e^{ax}}{\sqrt{a^2+b^2}} \sin (bx + t) - \frac{p}{\sqrt{a^2+b^2}} \int x^{p-1} e^{ax} \sin (bx + t)\,dx.$$

2. $$\int x^p e^{ax} \cos bx\,dx = \frac{x^p e^{ax}}{a^2+b^2} (a \cos bx + b \sin bx) - \frac{p}{a^2+b^2} \int x^{p-1} e^{ax} (a \cos bx + b \sin bx)\,dx;$$
$$= \frac{x^p e^{ax}}{\sqrt{a^2+b^2}} \cos (bx + t) - \frac{p}{\sqrt{a^2+b^2}} \int x^{p-1} e^{ax} \cos (bx + t)\,dx.$$

3. $\int x^n e^{ax} \sin bx\,dx = e^{ax} \sum_{k=1}^{n+1} \frac{(-1)^{k+1} n!\, x^{n-k+1}}{(n-k+1)!\,(a^2+b^2)^{k/2}} \sin(bx+kt).$

4. $\int x^n e^{ax} \cos bx\,dx = e^{ax} \sum_{k=1}^{n+1} \frac{(-1)^{k+1} n!\, x^{n-k+1}}{(n-k+1)!\,(a^2+b^2)^{k/2}} \cos(bx+kt).$

5. $\int xe^{ax} \sin bx\,dx = \frac{e^{ax}}{a^2+b^2}\left[\left(ax - \frac{a^2-b^2}{a^2+b^2}\right)\sin bx - \right.$

$\left. - \left(bx - \frac{2ab}{a^2+b^2}\right)\cos bx\right].$

6. $\int xe^{ax} \cos bx\,dx = \frac{e^{ax}}{a^2+b^2}\left[\left(ax - \frac{a^2-b^2}{a^2+b^2}\right)\cos bx + \right.$

$\left. + \left(bx - \frac{2ab}{a^2+b^2}\right)\sin bx\right].$

7. $\int x^2 e^{ax} \sin bx\,dx =$

$= \frac{e^{ax}}{a^2+b^2}\left\{\left[ax^2 - \frac{2(a^2-b^2)}{a^2+b^2}x + \frac{2a(a^2-3b^2)}{(a^2+b^2)^2}\right]\sin bx - \right.$

$\left. - \left[bx^2 - \frac{4ab}{a^2+b^2}x + \frac{2b(3a^2-b^2)}{(a^2+b^2)^2}\right]\cos bx\right\}.$

8. $\int x^2 e^{ax} \cos bx\,dx =$

$= \frac{e^{ax}}{a^2+b^2}\left\{\left[ax^2 - \frac{2(a^2-b^2)}{a^2+b^2}x + \frac{2a(a^2-3b^2)}{(a^2+b^2)^2}\right]\cos bx + \right.$

$\left. + \left[bx^2 - \frac{4ab}{a^2+b^2}x + \frac{2b(3a^2-b^2)}{(a^2+b^2)^2}\right]\sin bx\right\}.$ GU ((335)), MZ 274-275

2.67 Combinations of trigonometric and hyperbolic functions

2.671

1. $\int \operatorname{sh}(ax+b)\sin(cx+d)\,dx = \frac{a}{a^2+c^2}\operatorname{ch}(ax+b)\sin(cx+d) -$

$- \frac{c}{a^2+c^2}\operatorname{sh}(ax+b)\cos(cx+d).$

2. $\int \operatorname{sh}(ax+b)\cos(cx+d)\,dx = \frac{a}{a^2+c^2}\operatorname{ch}(ax+b)\cos(cx+d) +$

$+ \frac{c}{a^2+c^2}\operatorname{sh}(ax+b)\sin(cx+d).$

3. $\int \operatorname{ch}(ax+b)\sin(cx+d)\,dx = \frac{a}{a^2+c^2}\operatorname{sh}(ax+b)\sin(cx+d) -$

$- \frac{c}{a^2+c^2}\operatorname{ch}(ax+b)\cos(cx+d).$

4. $\int \operatorname{ch}(ax+b)\cos(cx+d)\,dx = \frac{a}{a^2+c^2}\operatorname{sh}(ax+b)\cos(cx+d) +$

$+ \frac{c}{a^2+c^2}\operatorname{ch}(ax+b)\sin(cx+d).$ GU ((354))(1)

2.672

1. $\int \operatorname{sh} x \sin x\,dx = \frac{1}{2}(\operatorname{ch} x \sin x - \operatorname{sh} x \cos x).$

2. $\int \operatorname{sh} x \cos x \, dx = \frac{1}{2} (\operatorname{ch} x \cos x + \operatorname{sh} x \sin x).$

3. $\int \operatorname{ch} x \sin x \, dx = \frac{1}{2} (\operatorname{sh} x \sin x - \operatorname{ch} x \cos x).$

4. $\int \operatorname{ch} x \cos x \, dx = \frac{1}{2} (\operatorname{sh} x \cos x + \operatorname{ch} x \sin x).$

2.673

1. $$\int \operatorname{sh}^{2m} (ax+b) \sin^{2n} (cx+d) \, dx = \frac{(-1)^m}{2^{2m+2n}} \binom{2m}{m} \binom{2n}{n} x +$$
$$+ \frac{(-1)^{m+n}}{2^{2m+2n-1}} \binom{2m}{m} \sum_{k=0}^{n-1} \frac{(-1)^k}{(2n-2k)c} \binom{2n}{k} \sin[(2n-2k)(cx+d)] +$$
$$+ \frac{(-1)^n}{2^{2m+2n-2}} \sum_{j=0}^{m-1} \sum_{k=0}^{n-1} \frac{(-1)^{j+k} \binom{2m}{j} \binom{2n}{k}}{(2m-2j)^2 a^2 + (2n-2k)^2 c^2} \times$$
$$\times \{(2m-2j) a \operatorname{sh} [(2m-2j)(ax+b)] \cos [(2n-2k)(cx+d)] +$$
$$+ (2n-2k) c \operatorname{ch} [(2m-2j)(ax+b)] \sin [(2n-2k)(cx+d)]\}.$$ GU ((354))(3a)

2. $$\int \operatorname{sh}^{2m} (ax+b) \sin^{2n-1} (cx+d) \, dx =$$
$$= \frac{(-1)^{m+n}}{2^{2m+2n-2}} \binom{2m}{m} \sum_{k=0}^{n-1} \frac{(-1)^k}{(2n-2k-1)c} \binom{2n-1}{k} \cos[(2n-2k-1)(cx+d)] +$$
$$+ \frac{(-1)^{n-1}}{2^{2m+2n-3}} \sum_{j=0}^{m-1} \sum_{k=0}^{n-1} \frac{(-1)^{j+k} \binom{2m}{j} \binom{2n-1}{k}}{(2m-2j)^2 a^2 + (2n-2k-1)^2 c^2} \times$$
$$\times \{(2m-2j) a \operatorname{sh} [(2m-2j)(ax+b)] \sin [(2n-2k-1)(cx+d)] -$$
$$- (2n-2k-1) c \operatorname{ch} [(2m-2j)(ax+b)] \cos [(2n-2k-1)(cx+d)]\}.$$ GU ((354))(3b)

3. $$\int \operatorname{sh}^{2m-1} (ax+b) \sin^{2n} (cx+d) \, dx =$$
$$= \frac{\binom{2n}{n}}{2^{2m+2n-2}} \sum_{j=0}^{m-1} \frac{(-1)^j \binom{2m-1}{j}}{(2m-2j-1)a} \operatorname{ch} [(2m-2j-1)(ax+b)] +$$
$$+ \frac{(-1)^n}{2^{2m+2n-3}} \sum_{j=0}^{m-1} \sum_{k=0}^{n-1} \frac{(-1)^{j+k} \binom{2m-1}{j} \binom{2n}{k}}{(2m-2j-1)^2 a^2 + (2n-2k)^2 c^2} \times$$
$$\times \{(2m-2j-1) a \operatorname{ch} [(2m-2j-1)(ax+b)] \cos [(2n-2k)(cx+d)] +$$
$$+ (2n-2k) c \operatorname{sh} [(2m-2j-1)(ax+b)] \sin [(2n-2k)(cx+d)]\}.$$ GU ((354))(3c)

4. $$\int \operatorname{sh}^{2m-1} (ax+b) \sin^{2n-1} (cx+d) \, dx =$$
$$= \frac{(-1)^{n-1}}{2^{2m-2n-4}} \sum_{j=0}^{m-1} \sum_{k=0}^{n-1} \frac{(-1)^{j+k} \binom{2m-1}{j} \binom{2n-1}{k}}{(2m-2j-1)^2 a^2 + (2n-2k-1)^2 c^2} \times$$
$$\times \{(2m-2j-1) a \operatorname{ch} [(2m-2j-1)(ax+b)] \sin [(2n-2k-1)(cx+d)] -$$
$$- (2n-2k-1) c \operatorname{sh} [(2m-2j-1)(ax+b)] \cos [(2n-2k-1)(cx+d)]\}.$$

GU ((354))(3d)

5. $$\int \operatorname{sh}^{2m}(ax+b)\cos^{2n}(cx+d)\,dx = \frac{(-1)^m}{2^{2m+2n}}\binom{2m}{m}\binom{2n}{n}x + \frac{\binom{2n}{n}}{2^{2m+2n-1}}\sum_{j=0}^{m-1}\frac{(-1)^j\binom{2m}{j}}{(2m-2j)\,a}\operatorname{sh}[(2m-2j)(ax+b)] + \frac{(-1)^m\binom{2m}{m}}{2^{2m+2n-1}}\sum_{k=0}^{n-1}\frac{\binom{2n}{k}}{(2n-2k)\,c}\sin[(2n-2k)(cx+d)] + \frac{1}{2^{2m+2n-2}}\sum_{j=0}^{m-1}\sum_{k=0}^{n-1}\frac{(-1)^j\binom{2m}{j}\binom{2n}{k}}{(2m-2j)^2a^2+(2n-2k)^2c^2}\times$$
$$\times\{(2m-2j)\,a\operatorname{sh}[(2m-2j)(ax+b)]\cos[(2n-2k)(cx+d)] + (2n-2k)\,c\operatorname{ch}[(2m-2j)(ax+b)]\sin[(2n-2k)(cx+d)]\}.$$

GU ((354))(4a)

6. $$\int \operatorname{sh}^{2m}(ax+b)\cos^{2n-1}(cx+d)\,dx = \frac{(-1)^m\binom{2m}{m}}{2^{2m+2n-2}}\sum_{k=0}^{n-1}\frac{\binom{2n-1}{k}}{(2n-2k-1)\,c}\sin[(2n-2k-1)(cx+d)] + \frac{1}{2^{2m+2n-3}}\sum_{j=0}^{m-1}\sum_{k=0}^{n-1}\frac{(-1)^j\binom{2m}{j}\binom{2n-1}{k}}{(2m-2j)^2a^2+(2n-2k-1)^2c^2}\times$$
$$\times\{(2m-2j)\,a\operatorname{sh}[(2m-2j)(ax+b)]\cos[(2n-2k-1)(cx+d)] + (2n-2k-1)\,c\operatorname{ch}[(2m-2j)(ax+b)]\sin[(2n-2k-1)(cx+d)]\}.$$

GU ((354))(4a)

7. $$\int \operatorname{sh}^{2m-1}(ax+b)\cos^{2n}(cx+d)\,dx = \frac{\binom{2n}{n}}{2^{2m+2n-2}}\sum_{j=0}^{m-1}\frac{(-1)^j\binom{2m-1}{j}}{(2m-2j-1)\,a}\operatorname{ch}[(2m-2j-1)(ax+b)] + \frac{1}{2^{2m-2n-3}}\sum_{j=0}^{m-1}\sum_{k=0}^{n-1}\frac{(-1)^j\binom{2m-1}{j}\binom{2n}{k}}{(2m-2j-1)^2a^2+(2n-2k)^2c^2}\times$$
$$\times\{(2m-2j-1)\,a\operatorname{ch}[(2m-2j-1)(ax+b)]\cos[(2n-2k)(cx+d)] + (2n-2k)\,c\operatorname{sh}[(2m-2j-1)(ax+b)]\sin[(2n-2k)(cx+d)]\}.$$

GU ((354))(4b)

8. $$\int \operatorname{sh}^{2m-1}(ax+b)\cos^{2n-1}(cx+d)\,dx = \frac{1}{2^{2m+2n-4}}\sum_{j=0}^{m-1}\sum_{k=0}^{n-1}\frac{(-1)^j\binom{2m-1}{j}\binom{2n-1}{k}}{(2m-2j-1)^2a^2+(2n-2k-1)^2c^2}\times$$
$$\times\{(2m-2j-1)\,a\operatorname{ch}[(2m-2j-1)(ax+b)]\cos[(2n-2k-1)(cx+d)] + (2n-2k-1)\,c\operatorname{sh}[(2m-2j-1)(ax+b)]\sin[(2n-2k-1)(cx+d)]\}.$$

GU ((354))(4b)

9. $$\int \operatorname{ch}^{2m}(ax+b)\sin^{2n}(cx+d)\,dx = \frac{\binom{2m}{m}\binom{2n}{n}}{2^{2m+2n}}x + \frac{(-1)^n\binom{2m}{m}}{2^{2m+2n-1}}\sum_{k=0}^{m-1}\frac{(-1)^k\binom{2n}{k}}{(2n-2k)c}\sin[(2n-2k)(cx+d)] + \frac{\binom{2n}{n}}{2^{2m+2n-1}}\sum_{j=0}^{m-1}\frac{\binom{2m}{j}}{(2m-2j)a}\operatorname{sh}[(2m-2j)(ax+b)] + \frac{(-1)^n}{2^{2m+2n-2}}\sum_{j=0}^{m-1}\sum_{k=0}^{n-1}\frac{(-1)^k\binom{2m}{j}\binom{2n}{k}}{(2m-2j)^2a^2+(2n-2k)^2c^2}\times$$
$$\times\{(2m-2j)a\operatorname{sh}[(2m-2j)(ax+b)]\cos[(2n-2k)(cx+d)] + (2n-2k)c\operatorname{ch}[(2m-2j)(ax+b)]\sin[(2n-2k)(cx+d)]\}.$$

GU (354))(5a)

10. $$\int \operatorname{ch}^{2m-1}(ax+b)\sin^{2n}(cx+d)\,dx = \frac{\binom{2n}{n}}{2^{2m+2n-2}}\sum_{j=0}^{m-1}\frac{\binom{2m-1}{j}}{(2m-2j-1)a}\operatorname{sh}[(2m-2j-1)(ax+b)] + \frac{(-1)^n}{2^{2m+2n-3}}\sum_{j=0}^{m-1}\sum_{k=0}^{n-1}\frac{(-1)^k\binom{2m-1}{j}\binom{2n}{k}}{(2m-2j-1)^2a^2+(2n-2k)^2c^2}\times$$
$$\times\{(2m-2j-1)a\operatorname{sh}[(2m-2j-1)(ax+b)]\cos[(2n-2k)(cx+d)] + (2n-2k)c\operatorname{ch}[(2m-2j-1)(ax+b)]\sin[(2n-2k)(cx+d)]\}.$$

GU ((354))(5a)

11. $$\int \operatorname{ch}^{2m}(ax+b)\sin^{2n-1}(cx+d)\,dx = \frac{(-1)^{n-1}\binom{2m}{m}}{2^{2m+2n-2}}\sum_{k=0}^{n-1}\frac{(-1)^{k+1}\binom{2n-1}{k}}{(2n-2k-1)c}\cos[(2n-2k-1)(cx+d)] + \frac{(-1)^{n-1}}{2^{2m+2n-3}}\sum_{j=0}^{m-1}\sum_{k=0}^{n-1}\frac{(-1)^k\binom{2m}{j}\binom{2n-1}{k}}{(2m-2j)^2a^2+(2n-2k-1)^2c^2}\times$$
$$\times\{(2m-2j)a\operatorname{sh}[(2m-2j)(ax+b)]\sin[(2n-2k-1)(cx+d)] - (2n-2k-1)c\operatorname{ch}[(2m-2j)(ax+b)]\cos[(2n-2k-1)(cx+d)]\}.$$

GU ((354))(5b)

12. $$\int \operatorname{ch}^{2m-1}(ax+b)\sin^{2n-1}(cx+d)\,dx = \frac{(-1)^{n-1}}{2^{2m+2n-4}}\sum_{j=0}^{m-1}\sum_{k=0}^{n-1}\frac{(-1)^k\binom{2m-1}{j}\binom{2n-1}{k}}{(2m-2j-1)^2a^2+(2n-2k-1)^2c^2}\times$$
$$\times\{(2m-2j-1)a\operatorname{sh}[(2m-2j-1)(ax+b)]\sin[(2n-2k-1)(cx+d)] - (2n-2k-1)c\operatorname{ch}[(2m-2j-1)(ax+b)]\cos[(2n-2k-1)(cx+d)]\}.$$

GU ((354))(5b)

13. $$\int \operatorname{ch}^{2m}(ax+b)\cos^{2n}(cx+d)\,dx = \frac{\binom{2m}{m}\binom{2n}{n}}{2^{2m+2n}}x + \frac{\binom{2m}{m}}{2^{2m+2n-1}}\sum_{k=0}^{n-1}\frac{\binom{2n}{k}}{(2n-2k)c}\sin[(2n-2k)(cx+d)] + \frac{\binom{2n}{n}}{2^{2m+2n-1}}\sum_{j=0}^{m-1}\frac{\binom{2m}{j}}{(2m-2j)a}\operatorname{sh}[(2m-2j)(ax+b)] + \frac{1}{2^{2m+2n-2}}\sum_{j=0}^{m-1}\sum_{k=0}^{n-1}\frac{\binom{2m}{j}\binom{2n}{k}}{(2m-2j)^2a^2+(2n-2k)^2c^2}\times$$
$$\times\{(2m-2j)a\operatorname{sh}[(2m-2j)(ax+b)]\cos[(2n-2k)(cx+d)] + (2n-2k)c\operatorname{ch}[(2m-2j)(ax+b)]\sin[(2n-2k)(cx+d)]\}.$$

GU ((354))(6)

14. $$\int \operatorname{ch}^{2m-1}(ax+b)\cos^{2n}(cx+d)\,dx = \frac{\binom{2n}{n}}{2^{2m+2n-2}}\sum_{j=0}^{m-1}\frac{\binom{2m-1}{j}}{(2m-2j-1)a}\operatorname{sh}[(2m-2j-1)(ax+b)] + \frac{1}{2^{2m+2n-3}}\sum_{j=0}^{m-1}\sum_{k=0}^{n-1}\frac{\binom{2m-1}{j}\binom{2n}{k}}{(2m-2j-1)^2a^2+(2n-2k)^2c^2}\times$$
$$\times\{(2m-2j-1)a\operatorname{sh}[(2m-2j-1)(ax+b)]\cos[(2n-2k)(cx+d)] + (2n-2k)c\operatorname{ch}[(2m-2j-1)(ax+b)]\sin[(2n-2k)(cx+d)]\}.$$

GU ((354))(6)

15. $$\int \operatorname{ch}^{2m}(ax+b)\cos^{2n-1}(cx+d)\,dx = \frac{\binom{2m}{m}}{2^{2m+2n-2}}\sum_{k=0}^{n-1}\frac{\binom{2n-1}{k}}{(2n-2k-1)c}\sin[(2n-2k-1)(cx+d)] + \frac{1}{2^{2m+2n-3}}\sum_{j=0}^{m-1}\sum_{k=0}^{n-1}\frac{\binom{2m}{j}\binom{2n-1}{k}}{(2m-2j)^2a^2+(2n-2k-1)^2c^2}\times$$
$$\times\{(2m-2j)a\operatorname{sh}[(2m-2j)(ax+b)]\cos[(2n-2k-1)(cx+d)] + (2n-2k-1)c\operatorname{ch}[(2m-2j)(ax+b)]\sin[(2n-2k-1)(cx+d)]\}.$$

GU ((354))(6)

16. $$\int \operatorname{ch}^{2m-1}(ax+b)\cos^{2n-1}(cx+d)\,dx = \frac{1}{2^{2m+2n-4}}\sum_{j=0}^{m-1}\sum_{k=0}^{n-1}\frac{\binom{2m-1}{j}\binom{2n-1}{k}}{(2m-2j-1)^2a^2+(2n-2k-1)^2c^2}\times$$
$$\times\{(2m-2j-1)a\operatorname{sh}[(2m-2j-1)(ax+b)]\cos[(2n-2k-1)(cx+d)] + (2n-2k-1)c\operatorname{ch}[(2m-2j-1)(ax+b)]\sin[(2n-2k-1)(cx+d)]\}.$$

GU ((354))(6)

2.674

1. $$\int e^{ax} \operatorname{sh} bx \sin cx\, dx = \frac{e^{(a+b)x}}{2[(a+b)^2+c^2]}[(a+b)\sin cx - c\cos cx] - \frac{e^{(a-b)x}}{2[(a-b)^2+c^2]}[(a-b)\sin cx - c\cos cx].$$

2. $$\int e^{ax} \operatorname{sh} bx \cos cx\, dx = \frac{e^{(a+b)x}}{2[(a+b)^2+c^2]}[(a+b)\cos cx + c\sin cx] - \frac{e^{(a-b)x}}{2[(a-b)^2+c^2]}[(a-b)\cos cx + c\sin cx].$$

3. $$\int e^{ax} \operatorname{ch} bx \sin cx\, dx = \frac{e^{(a+b)x}}{2[(a+b)^2+c^2]}[(a+b)\sin cx - c\cos cx] + \frac{e^{(a-b)x}}{2[(a-b)^2+c^2]}[(a-b)\sin cx - c\cos cx].$$

4. $$\int e^{ax} \operatorname{ch} bx \cos cx\, dx = \frac{e^{(a+b)x}}{2[(a+b)^2+c^2]}[(a+b)\cos cx + c\sin cx] + \frac{e^{(a-b)x}}{2[(a-b)^2+c^2]}[(a-b)\cos cx + c\sin cx]$$

MZ 379

2.7 Logarithms and Inverse-Hyperbolic Functions

2.71 The logarithm

2.711 $$\int \ln^m x\, dx = x\ln^m x - m\int \ln^{m-1} x\, dx = \frac{x}{m+1}\sum_{k=0}^{m}(-1)^k (m+1)m(m-1)\ldots(m-k+1)\ln^{m-k} x \quad (m > 0).$$ TI (603)

2.72-2.73 Combinations of logarithms and algebraic functions

2.721

1. $$\int x^n \ln^m x\, dx = \frac{x^{n+1}\ln^m x}{n+1} - \frac{m}{n+1}\int x^n \ln^{m-1} x\, dx \quad \text{(see } \mathbf{2.722}\text{)}.$$

For $n = -1$

2. $$\int \frac{\ln^m x\, dx}{x} = \frac{\ln^{m+1} x}{m+1}.$$

For $n = -1$ and $m = -1$

3. $$\int \frac{dx}{x\ln x} = \ln(\ln x).$$

2.722 $$\int x^n \ln^m x\, dx = \frac{x^{n+1}}{m+1}\sum_{k=0}^{m}(-1)^k (m+1)m(m-1)\ldots(m-k+1)\frac{\ln^{m-k} x}{(n+1)^{k+1}}.$$

TI (604)

2.723

1. $$\int x^n \ln x\, dx = x^{n+1}\left[\frac{\ln x}{n+1} - \frac{1}{(n+1)^2}\right].$$ TI 375

2. $\int x^n \ln^2 x\,dx = x^{n+1}\left[\frac{\ln^2 x}{n+1} - \frac{2\ln x}{(n+1)^2} + \frac{2}{(n+1)^3}\right]$. TI 375

3. $\int x^n \ln^3 x\,dx = x^{n+1}\left[\frac{\ln^3 x}{n+1} - \frac{3\ln^2 x}{(n+1)^2} + \frac{6\ln x}{(n+1)^3} - \frac{6}{(n+1)^4}\right]$.

2.724

1. $\int \frac{x^n\,dx}{(\ln x)^m} = -\frac{x^{n+1}}{(m-1)(\ln x)^{m-1}} + \frac{n+1}{m-1}\int \frac{x^n\,dx}{(\ln x)^{m-1}}$.

For $m=1$

2. $\int \frac{x^n\,dx}{\ln x} = \operatorname{li}(x^{n+1})$.

2.725

1. $\int (a+bx)^m \ln x\,dx =$

$$= \frac{1}{(m+1)b}\left[(a+bx)^{m+1}\ln x - \int \frac{(a+bx)^{m+1}\,dx}{x}\right].$$ TI 374

2. $$\int (a+bx)^m \ln x\,dx = \frac{1}{(m+1)b}\left[(a+bx)^{m+1} - a^{m+1}\right]\ln x - \\ - \sum_{k=0}^{m} \frac{\binom{m}{k} a^{m-k} b^k x^{k+1}}{(k+1)^2}.$$

For $m=-1$ see **2.727** 2.

2.726

1. $\int (a+bx)\ln x\,dx = \left[\frac{(a+bx)^2}{2b} - \frac{a^2}{2b}\right]\ln x - \left(ax + \frac{1}{4}bx^2\right)$.

2. $\int (a+bx)^2 \ln x\,dx = \frac{1}{3b}\left[(a+bx)^3 - a^3\right]\ln x - \left(a^2x + \frac{abx^2}{2} + \frac{b^2x^3}{9}\right)$.

3. $$\int (a+bx)^3 \ln x\,dx = \frac{1}{4b}\left[(a+bx)^4 - a^4\right]\ln x - \\ - \left(a^3x + \frac{3}{4}a^2bx^2 + \frac{1}{3}ab^2x^3 + \frac{1}{16}b^3x^4\right).$$

2.727

1. $\int \frac{\ln x\,dx}{(a+bx)^m} = \frac{1}{b(m-1)}\left[-\frac{\ln x}{(a+bx)^{m-1}} + \int \frac{dx}{x(a+bx)^{m-1}}\right]$. TI 376

For $m=1$

2. $\int \frac{\ln x\,dx}{a+bx} = \frac{1}{b}\ln x \ln(a+bx) - \frac{1}{b}\int \frac{\ln(a+bx)\,dx}{x}$ (see **2.728** 2.).

3. $\int \frac{\ln x\,dx}{(a+bx)^2} = -\frac{\ln x}{b(a+bx)} + \frac{1}{ab}\ln\frac{x}{a+bx}$.

4. $\int \frac{\ln x\,dx}{(a+bx)^3} = -\frac{\ln x}{2b(a+bx)^2} + \frac{1}{2ab(a+bx)} + \frac{1}{2a^2b}\ln\frac{x}{a+bx}$.

5. $$\int \frac{\ln x\,dx}{\sqrt{a+bx}} = \frac{2}{b}\left\{(\ln x - 2)\sqrt{a+bx} + \sqrt{a}\ln\frac{\sqrt{a+bx}+\sqrt{a}}{\sqrt{a+bx}-\sqrt{a}}\right\} \quad [a>0];$$

$$= \frac{2}{b}\left\{(\ln x - 2)\sqrt{a+bx} + 2\sqrt{-a}\operatorname{arctg}\sqrt{\frac{a+bx}{-a}}\right\} \quad [a<0].$$

2.728

1. $$\int x^m \ln(a+bx)\,dx = \frac{1}{m+1}\left[x^{m+1}\ln(a+bx) - b\int \frac{x^{m+1}\,dx}{a+bx}\right].$$

2. $\int \frac{\ln(a+bx)}{x}\,dx$ cannot be expressed as a finite combination of elementary functions. See **1.511** and **0.312**.

2.729

1. $$\int x^m \ln(a+bx)\,dx = \frac{1}{m+1}\left[x^{m+1} - \frac{(-a)^{m+1}}{b^{m+1}}\right]\ln(a+bx) + \\ + \frac{1}{m+1}\sum_{k=1}^{m+1} \frac{(-1)^k x^{m-k+2} a^{k-1}}{(m-k+2)\,b^{k-1}}.$$

2. $$\int x\ln(a+bx)\,dx = \frac{1}{2}\left[x^2 - \frac{a^2}{b^2}\right]\ln(a+bx) - \frac{1}{2}\left[\frac{x^2}{2} - \frac{ax}{b}\right].$$

3. $$\int x^2\ln(a+bx)\,dx = \frac{1}{3}\left[x^3 + \frac{a^3}{b^3}\right]\ln(a+bx) - \frac{1}{3}\left[\frac{x^3}{3} - \frac{ax^2}{2b} + \frac{a^2x}{b^2}\right].$$

4. $$\int x^3\ln(a+bx)\,dx = \frac{1}{4}\left[x^4 - \frac{a^4}{b^4}\right]\ln(a+bx) - \\ - \frac{1}{4}\left[\frac{x^4}{4} - \frac{ax^3}{3b} + \frac{a^2x^2}{2b^2} - \frac{a^3x}{b^3}\right].$$

2.731 $$\int x^{2n}\ln(x^2+a^2)\,dx = \frac{1}{2n+1}\left\{x^{2n+1}\ln(x^2+a^2) + (-1)^n\,2a^{2n+1}\operatorname{arctg}\frac{x}{a} - \\ - 2\sum_{k=0}^{n}\frac{(-1)^{n-k}}{2k+1}\,a^{2n-2k}x^{2k+1}\right\}.$$

2.732 $$\int x^{2n+1}\ln(x^2+a^2)\,dx = \frac{1}{2n+1}\left\{(x^{2n+2} + (-1)^n a^{2n+2})\ln(x^2+a^2) + \\ + \sum_{k=1}^{n+1}\frac{(-1)^{n-k}}{k}\,a^{2n-2k+2}x^{2k}\right\}.$$

2.733

1. $$\int \ln(x^2+a^2)\,dx = x\ln(x^2+a^2) - 2x + 2a\operatorname{arctg}\frac{x}{a}.$$ DW

2. $$\int x\ln(x^2+a^2)\,dx = \frac{1}{2}\left[(x^2+a^2)\ln(x^2+a^2) - x^2\right].$$ DW

3. $$\int x^2\ln(x^2+a^2)\,dx = \frac{1}{3}\left[x^3\ln(x^2+a^2) - \frac{2}{3}x^3 + 2a^2x - 2a^3\operatorname{arctg}\frac{x}{a}\right].$$ DW

4. $$\int x^3\ln(x^2+a^2)\,dx = \frac{1}{4}\left[(x^4-a^4)\ln(x^2+a^2) - \frac{x^4}{2} + a^2x^2\right].$$ DW

5. $$\int x^4\ln(x^2+a^2)\,dx = \frac{1}{5}\left[x^5\ln(x^2+a^2) - \frac{2}{5}x^5 + \frac{2}{3}a^2x^3 - 2a^4x + \\ + 2a^5\operatorname{arctg}\frac{x}{a}\right].$$ DW

2.734 $$\int x^{2n} \ln|x^2-a^2|\,dx = \frac{1}{2n+1}\left\{x^{2n+1}\ln|x^2-a^2| + a^{2n+1}\ln\left|\frac{x+a}{x-a}\right| - 2\sum_{k=0}^{n}\frac{1}{2k+1}a^{2n-2k}x^{2k+1}\right\}.$$

2.735 $$\int x^{2n+1}\ln|x^2-a^2|\,dx = \frac{1}{2n+2}\left\{(x^{2n+2}-a^{2n+2})\ln|x^2-a^2| - \sum_{k=1}^{n+1}\frac{1}{k}a^{2n-2k+2}x^{2k}\right\}.$$

2.736

1. $$\int \ln|x^2-a^2|\,dx = x\ln|x^2-a^2| - 2x + a\ln\left|\frac{x+a}{x-a}\right|.$$ DW

2. $$\int x\ln|x^2-a^2|\,dx = \frac{1}{2}\{(x^2-a^2)\ln|x^2-a^2| - x^2\}.$$ DW

3. $$\int x^2\ln|x^2-a^2|\,dx = \frac{1}{3}\left\{x^3\ln|x^2-a^2| - \frac{2}{3}x^3 - 2a^2x + a^3\ln\left|\frac{x+a}{x-a}\right|\right\}.$$ DW

4. $$\int x^3\ln|x^2-a^2|\,dx = \frac{1}{4}\left\{(x^4-a^4)\ln|x^2-a^2| - \frac{x^4}{2} - a^2x^2\right\}.$$ DW

5. $$\int x^4\ln|x^2-a^2|\,dx = \frac{1}{5}\left\{x^5\ln|x^2-a^2| - \frac{2}{5}x^5 - \frac{2}{3}a^2x^3 - 2a^4x + a^5\ln\left|\frac{x+a}{x-a}\right|\right\}.$$ DW

2.74 Inverse hyperbolic functions

2.741

1. $$\int \operatorname{Arsh}\frac{x}{a}\,dx = x\operatorname{Arsh}\frac{x}{a} - \sqrt{x^2+a^2}.$$ DW

2. $$\int \operatorname{Arch}\frac{x}{a}\,dx = x\operatorname{Arch}\frac{x}{a} - \sqrt{x^2-a^2} \qquad \left[\operatorname{Arch}\frac{x}{a} > 0\right];$$
$$= x\operatorname{Arch}\frac{x}{a} + \sqrt{x^2-a^2} \qquad \left[\operatorname{Arch}\frac{x}{a} < 0\right].$$ DW

3. $$\int \operatorname{Arth}\frac{x}{a}\,dx = x\operatorname{Arth}\frac{x}{a} + \frac{a}{2}\ln(a^2-x^2).$$ DW

4. $$\int \operatorname{Arcth}\frac{x}{a}\,dx = x\operatorname{Arcth}\frac{x}{a} + \frac{a}{2}\ln(x^2-a^2).$$ DW

2.742

1. $$\int x\operatorname{Arsh}\frac{x}{a}\,dx = \left(\frac{x^2}{2}+\frac{a^2}{4}\right)\operatorname{Arsh}\frac{x}{a} - \frac{x}{4}\sqrt{x^2+a^2}.$$ DW

2. $$\int x\operatorname{Arch}\frac{x}{a}\,dx = \left(\frac{x^2}{2}-\frac{a^2}{4}\right)\operatorname{Arch}\frac{x}{a} - \frac{x}{4}\sqrt{x^2-a^2} \qquad \left[\operatorname{Arch}\frac{x}{a} > 0\right];$$
$$= \left(\frac{x^2}{2}-\frac{a^2}{4}\right)\operatorname{Arch}\frac{x}{a} + \frac{x}{4}\sqrt{x^2-a^2} \qquad \left[\operatorname{Arch}\frac{x}{a} < 0\right].$$ DW

2.8 Inverse Trigonometric Functions

2.81 Arcsines and arccosines

2.811 $$\int \left(\arcsin\frac{x}{a}\right)^n dx = x\sum_{k=0}^{E\left(\frac{n}{2}\right)} (-1)^k \binom{n}{2k}\cdot(2k)!\left(\arcsin\frac{x}{a}\right)^{n-2k} + \\ + \sqrt{a^2-x^2}\sum_{k=1}^{E\left(\frac{n+1}{2}\right)} (-1)^{k-1}\binom{n}{2k-1}\cdot(2k-1)!\left(\arcsin\frac{x}{a}\right)^{n-2k+1}.$$

2.812 $$\int \left(\arccos\frac{x}{a}\right)^n dx = x\sum_{k=0}^{E\left(\frac{n}{2}\right)} (-1)^k \binom{n}{2k}\cdot(2k)!\left(\arccos\frac{x}{a}\right)^{n-2k} + \\ + \sqrt{a^2-x^2}\sum_{k=1}^{E\left(\frac{n+1}{2}\right)} (-1)^{k}\binom{n}{2k-1}\cdot(2k-1)!\left(\arccos\frac{x}{a}\right)^{n-2k+1}.$$

2.813

1. $$\int \arcsin\frac{x}{a}\,dx = x\arcsin\frac{x}{a} + \sqrt{a^2-x^2}.$$
2. $$\int \left(\arcsin\frac{x}{a}\right)^2 dx = x\left(\arcsin\frac{x}{a}\right)^2 + 2\sqrt{a^2-x^2}\arcsin\frac{x}{a} - 2x.$$
3. $$\int \left(\arcsin\frac{x}{a}\right)^3 dx = x\left(\arcsin\frac{x}{a}\right)^3 + 3\sqrt{a^2-x^2}\left(\arcsin\frac{x}{a}\right)^2 - \\ - 6x\arcsin\frac{x}{a} - 6\sqrt{a^2-x^2}.$$

2.814

1. $$\int \arccos\frac{x}{a}\,dx = x\arccos\frac{x}{a} - \sqrt{a^2-x^2}.$$
2. $$\int \left(\arccos\frac{x}{a}\right)^2 dx = x\left(\arccos\frac{x}{a}\right)^2 - 2\sqrt{a^2-x^2}\arccos\frac{x}{a} - 2x.$$
3. $$\int \left(\arccos\frac{x}{a}\right)^3 dx = x\left(\arccos\frac{x}{a}\right)^3 - 3\sqrt{a^2-x^2}\left(\arccos\frac{x}{a}\right)^2 - \\ - 6x\arccos\frac{x}{a} + 6\sqrt{a^2-x^2}.$$

2.82 The arcsecant, the arccosecant, the arctangent and the arccotangent

2.821

1. $$\int \operatorname{arccosec}\frac{x}{a}\,dx = \int \arcsin\frac{a}{x}\,dx = \\ = x\arcsin\frac{a}{x} + a\ln\left(x+\sqrt{x^2-a^2}\right) \quad \left[0 < \arcsin\frac{a}{x} < \frac{\pi}{2}\right]; \\ = x\arcsin\frac{a}{x} - a\ln\left(x+\sqrt{x^2-a^2}\right) \quad \left[-\frac{\pi}{2} < \arcsin\frac{a}{x} < 0\right].$$ DW
2. $$\int \operatorname{arcsec}\frac{x}{a}\,dx = \int \arccos\frac{a}{x}\,dx = \\ = x\arccos\frac{a}{x} - a\ln\left(x+\sqrt{x^2-a^2}\right) \quad \left[0 < \arccos\frac{a}{x} < \frac{\pi}{2}\right]; \\ = x\arccos\frac{a}{x} + a\ln\left(x+\sqrt{x^2-a^2}\right) \quad \left[-\frac{\pi}{2} < \arccos\frac{a}{x} < 0\right].$$ DW

2.822

1. $\int \operatorname{arctg}\frac{x}{a}\,dx = x\operatorname{arctg}\frac{x}{a} - \frac{a}{2}\ln(a^2+x^2).$ DW

2. $\int \operatorname{arcctg}\frac{x}{a}\,dx = x\operatorname{arcctg}\frac{x}{a} + \frac{a}{2}\ln(a^2+x^2).$ DW

2.83 Combinations of arcsine or arccosine and algebraic functions

2.831 $\int x^n \arcsin\frac{x}{a}\,dx = \frac{x^{n+1}}{n+1}\arcsin\frac{x}{a} - \frac{1}{n+1}\int \frac{x^{n+1}\,dx}{\sqrt{a^2-x^2}}$

(see **2.263** 1., **2.264**, **2.27**).

2.832 $\int x^n \arccos\frac{x}{a}\,dx = \frac{x^{n+1}}{n+1}\arccos\frac{x}{a} + \frac{1}{n+1}\int \frac{x^{n+1}\,dx}{\sqrt{a^2-x^2}}$

(see **2.263** 1., **2.264**, **2.27**).

1. For $n=-1$, these integrals $\left(\text{that is, } \int \frac{\arcsin x}{x}\,dx \text{ and} \int \frac{\arccos x}{x}\,dx\right)$ cannot be expressed as a finite combination of elementary functions.

2. $\int \frac{\arccos x}{x}\,dx = -\frac{\pi}{2}\ln\frac{1}{x} - \int \frac{\arcsin x}{x}\,dx.$

2.833

1. $\int x\arcsin\frac{x}{a}\,dx = \left(\frac{x^2}{2}-\frac{a^2}{4}\right)\arcsin\frac{x}{a} + \frac{x}{4}\sqrt{a^2-x^2}.$

2. $\int x\arccos\frac{x}{a}\,dx = \left(\frac{x^2}{2}-\frac{a^2}{4}\right)\arccos\frac{x}{a} - \frac{x}{4}\sqrt{a^2-x^2}.$

2.834

1. $\int \frac{1}{x^2}\arcsin\frac{x}{a}\,dx = -\frac{1}{x}\arcsin\frac{x}{a} - \frac{1}{a}\ln\frac{a+\sqrt{a^2-x^2}}{x}.$

2. $\int \frac{1}{x^2}\arccos\frac{x}{a}\,dx = -\frac{1}{x}\arccos\frac{x}{a} + \frac{1}{a}\ln\frac{a+\sqrt{a^2-x^2}}{x}.$

2.835 $\int \frac{\arcsin x}{(a+bx)^2}\,dx = -\frac{\arcsin x}{b(a+bx)} - \frac{2}{b\sqrt{a^2-b^2}}\operatorname{arctg}\sqrt{\frac{(a-b)(1-x)}{(a+b)(1+x)}}$

$[a^2 > b^2];$

$= -\frac{\arcsin x}{b(a+bx)} - \frac{1}{b\sqrt{b^2-a^2}}\ln\frac{\sqrt{(a+b)(1+x)}+\sqrt{(b-a)(1-x)}}{\sqrt{(a+b)(1+x)}-\sqrt{(b-a)(1-x)}}$ $[a^2 < b^2].$

2.836 $\int \frac{x\arcsin x}{(1+cx^2)^2}\,dx = \frac{\arcsin x}{2c(1+cx^2)} + \frac{1}{2c\sqrt{c+1}}\operatorname{arctg}\frac{\sqrt{c+1}x}{\sqrt{1-x^2}}$ $[c > -1];$

$= -\frac{\arcsin x}{2c(1+cx^2)} + \frac{1}{4c\sqrt{-(c+1)}}\ln\frac{\sqrt{1-x^2}+x\sqrt{-(c+1)}}{\sqrt{1-x^2}-x\sqrt{-(c+1)}}$ $[c < -1].$

2.837

1. $\int \frac{x\arcsin x}{\sqrt{1-x^2}}\,dx = x - \sqrt{1-x^2}\arcsin x.$

2. $\int \frac{x^2 \arcsin x}{\sqrt{1-x^2}} dx = \frac{x^2}{4} - \frac{x}{2}\sqrt{1-x^2} \arcsin x + \frac{1}{4}(\arcsin x)^2.$

3. $\int \frac{x^3 \arcsin x}{\sqrt{1-x^2}} dx = \frac{x^3}{9} + \frac{2x}{3} - \frac{1}{3}(x^2+2)\sqrt{1-x^2} \arcsin x.$

2.838

1. $\int \frac{\arcsin x}{\sqrt{(1-x^2)^3}} dx = \frac{x \arcsin x}{\sqrt{1-x^2}} + \frac{1}{2}\ln(1-x^2).$

2. $\int \frac{x \arcsin x}{\sqrt{(1-x^2)^3}} dx = \frac{\arcsin x}{\sqrt{1-x^2}} + \frac{1}{2}\ln\frac{1-x}{1+x}.$

2.84 Combinations of the arcsecant and arccosecant with powers of x

2.841

1. $\int x \operatorname{arcsec}\frac{x}{a} dx = \int x \arccos\frac{a}{x} dx =$

$= \frac{1}{2}\left\{x^2 \arccos\frac{a}{x} - a\sqrt{x^2-a^2}\right\} \left[0 < \arccos\frac{a}{x} < \frac{\pi}{2}\right];$

$= \frac{1}{2}\left\{x^2 \arccos\frac{a}{x} + a\sqrt{x^2-a^2}\right\} \left[\frac{\pi}{2} < \arccos\frac{a}{x} < \pi\right].$ DW

2. $\int x^2 \operatorname{arcsec}\frac{x}{a} dx = \int x^2 \arccos\frac{a}{x} dx =$

$= \frac{1}{3}\left\{x^3 \arccos\frac{a}{x} - \frac{a}{2}x\sqrt{x^2-a^2} - \frac{a^3}{2}\ln(x+\sqrt{x^2-a^2})\right\}$

$\left[0 < \arccos\frac{a}{x} < \frac{\pi}{2}\right];$

$= \frac{1}{3}\left\{x^3 \arccos\frac{a}{x} + \frac{a}{2}x\sqrt{x^2-a^2} + \frac{a^3}{2}\ln(x+\sqrt{x^2-a^2})\right\}$

$\left[\frac{\pi}{2} < \arccos\frac{a}{x} < \pi\right].$ DW

3. $\int x \operatorname{arccosec}\frac{x}{a} dx = \int x \arcsin\frac{a}{x} dx =$

$= \frac{1}{2}\left\{x^2 \arcsin\frac{a}{x} + a\sqrt{x^2-a^2}\right\} \left[0 < \arcsin\frac{a}{x} < \frac{\pi}{2}\right];$

$= \frac{1}{2}\left\{x^2 \arcsin\frac{a}{x} - a\sqrt{x^2-a^2}\right\} \left[-\frac{\pi}{2} < \arcsin\frac{a}{x} < 0\right].$ DW

2.85 Combinations of the arctangent and arccotangent with algebraic functions

2.851 $\int x^n \operatorname{arctg}\frac{x}{a} dx = \frac{x^{n+1}}{n+1}\operatorname{arctg}\frac{x}{a} - \frac{a}{n+1}\int \frac{x^{n+1} dx}{a^2+x^2}.$

2.852

1. $\int x^n \operatorname{arcctg}\frac{x}{a} dx = \frac{x^{n+1}}{n+1}\operatorname{arcctg}\frac{x}{a} + \frac{a}{n+1}\int \frac{x^{n+1} dx}{a^2+x^2}.$

For $n = -1$

$\int \frac{\operatorname{arctg} x}{x} dx$ cannot be expressed as a finite combination of elementary functions.

2. $\int \frac{\operatorname{arcctg} x}{x} dx = \frac{\pi}{2} \ln x - \int \frac{\operatorname{arctg} x}{x} dx.$

2.853

1. $\int x \operatorname{arctg} \frac{x}{a} dx = \frac{1}{2}(x^2 + a^2) \operatorname{arctg} \frac{x}{a} - \frac{ax}{2}.$

2. $\int x \operatorname{arcctg} \frac{x}{a} dx = \frac{1}{2}(x^2 + a^2) \operatorname{arcctg} \frac{x}{a} + \frac{ax}{2}.$

2.854 $\int \frac{1}{x^2} \operatorname{arctg} \frac{x}{a} dx = -\frac{1}{x} \operatorname{arctg} \frac{x}{a} - \frac{1}{2a} \ln \frac{a^2 + x^2}{x^2}.$

2.855 $\int \frac{\operatorname{arctg} x}{(\alpha+\beta x)^2} dx = \frac{1}{\alpha^2+\beta^2} \left\{ \ln \frac{\alpha+\beta x}{\sqrt{1+x^2}} - \frac{\beta - \alpha x}{\alpha+\beta x} \operatorname{arctg} x \right\}.$

2.856

1. $\int \frac{x \operatorname{arctg} x}{1+x^2} dx = \frac{1}{2} \operatorname{arctg} x \ln(1+x^2) - \frac{1}{2} \int \frac{\ln(1+x^2)\, dx}{1+x^2}.$ TI (689)

2. $\int \frac{x^2 \operatorname{arctg} x}{1+x^2} dx = x \operatorname{arctg} x - \frac{1}{2} \ln(1+x^2) - \frac{1}{2} (\operatorname{arctg} x)^2.$ TI (405)

3. $\int \frac{x^3 \operatorname{arctg} x}{1+x^2} dx = -\frac{1}{2} x + \frac{1}{2}(1+x^2) \operatorname{arctg} x - \int \frac{x \operatorname{arctg} x}{1+x^2} dx.$

(see 2.8511.)

4. $\int \frac{x^4 \operatorname{arctg} x}{1+x^2} dx = -\frac{1}{6} x^2 + \frac{2}{3} \ln(1+x^2) + \left(\frac{x^3}{3} - x\right) \operatorname{arctg} x + \frac{1}{2} (\operatorname{arctg} x)^2.$

2.857
$$\int \frac{\operatorname{arctg} x\, dx}{(1+x^2)^{n+1}} = \left[\sum_{k=1}^{n} \frac{(2n-2k)!!\,(2n-1)!!}{(2n)!!\,(2n-2k+1)!!} \frac{x}{(1+x^2)^{n-k+1}} + \frac{1}{2} \frac{(2n-1)!!}{(2n)!!} \operatorname{arctg} x \right] \operatorname{arctg} x + \frac{1}{2} \sum_{k=1}^{n} \frac{(2n-1)!!\,(2n-2k)!!}{(2n)!!\,(2n-2k+1)!!\,(n-k+1)} \frac{1}{(1+x^2)^{n-k+1}}.$$

2.858 $\int \frac{x \operatorname{arctg} x}{\sqrt{1-x^2}} dx = -\sqrt{1-x^2} \operatorname{arctg} x + \sqrt{2} \operatorname{arctg} \frac{x\sqrt{2}}{\sqrt{1-x^2}} - \arcsin x.$

2.859
$$\int \frac{\operatorname{arctg} x}{\sqrt{(a+bx^2)^3}} dx = \frac{x \operatorname{arctg} x}{a\sqrt{a+bx^2}} - \frac{1}{a\sqrt{b-a}} \operatorname{arctg} \sqrt{\frac{a+bx^2}{b-a}} \quad [a < b];$$
$$= \frac{x \operatorname{arctg} x}{a\sqrt{a+bx^2}} - \frac{1}{2a\sqrt{a-b}} \ln \frac{\sqrt{a+bx^2} - \sqrt{a-b}}{\sqrt{a+bx^2} + \sqrt{a-b}} \quad [a > b].$$

3.-4. DEFINITE INTEGRALS OF ELEMENTARY FUNCTIONS

3.0 Introduction*

3.01 Theorems of a general nature

3.011 Suppose that $f(x)$ is integrable** over the largest of the intervals (p, q), (p, r), (r, q). Then (depending on the relative positions of the points p, q, and r) it is also integrable over the other two intervals and we have

$$\int_p^q f(x)\,dx = \int_p^r f(x)\,dx + \int_r^q f(x)\,dx.$$ **FI II 126**

3.012 *The first mean-value theorem.* Suppose (1) that $f(x)$ is continuous and that $g(x)$ is integrable over the interval (p, q), (2) that $m \leqslant f(x) \leqslant M$ and (3) that $g(x)$ does not change sign anywhere in the interval (p, q). Then, there exists at least one point $\xi\ (p \leqslant \xi \leqslant q)$ such that

$$\int_p^q f(x)\,g(x)\,dx = f(\xi)\int_p^q g(x)\,dx.$$ **FI II 132**

3.013 *The second mean-value theorem.* If $f(x)$ is monotonic and non-negative throughout the interval (p, q), where $p < q$, and if $g(x)$ is integrable over that interval, then there exists at least one point $\xi\,[p \leqslant \xi \leqslant q]$ such that

1. $$\int_p^q f(x)\,g(x)\,dx = f(p)\int_p^\xi g(x)\,dx.$$

Under the conditions of Theorem **3.013** 1, if $f(x)$ is nondecreasing, then

2. $$\int_p^q f(x)\,g(x)\,dx = f(q)\int_\xi^q g(x)\,dx \quad [p \leqslant \xi \leqslant q].$$

---

* We omit the definition of definite and multiple integrals since they are widely known and can easily be found in any textbook on the subject. Here we give only certain theorems of a general nature which provide estimates, or which reduce the given integral to a simpler one.

** A function $f(x)$ is said to be integrable over the interval (p, q), if the integral $\int_p^q f(x)\,dx$ exists. Here, we usually mean the existence of the integral in the sense of Riemann. When it is a matter of the existence of the integral in the sense of Stieltjes or Lebesgue, etc., we shall speak of integrability in the sense of Stieltjes or Lebesgue.

If $f(x)$ is monotonic in the interval (p, q), where $p<q$, and if $g(x)$ is integrable over that interval, then

3. $$\int_p^q f(x) g(x)\, dx = f(p) \int_p^\xi g(x)\, dx + f(q) \int_\xi^q g(x)\, dx \quad [p \leqslant \xi \leqslant q],$$

or

4. $$\int_p^q f(x) g(x)\, dx = A \int_p^\xi g(x)\, dx + B \int_\xi^q g(x)\, dx \quad [p \leqslant \xi \leqslant q],$$

where A and B are any two numbers satisfying the conditions

$$A \geqslant f(p+0) \text{ and } B \leqslant f(q-0) \quad [\text{if } f \text{ decreases}],$$
$$A \leqslant f(p+0) \text{ and } B \geqslant f(q-0) \quad [\text{if } f \text{ increases}].$$

In particular,

5. $$\int_p^q f(x) g(x)\, dx = f(p+0) \int_p^\xi g(x)\, dx + f(q-0) \int_\xi^q g(x)\, dx.$$ FI II 138

3.02 Change of variable in a definite integral

3.020 $$\int_\alpha^\beta f(x)\, dx = \int_\varphi^\psi f[g(t)]\, g'(t)\, dt; \quad x = g(t).$$

This formula is valid under the following conditions:

1. $f(x)$ is continuous on some interval $A \leqslant x \leqslant B$ containing the original limits of integration α and β.
2. The equalities $\alpha = g(\varphi)$ and $\beta = g(\psi)$ hold.
3. $g(t)$ and its derivative $g'(t)$ are continuous on the interval $\varphi \leqslant t \leqslant \psi$.
4. As t varies from φ to ψ, the function $g(t)$ always varies in the same direction from $g(\varphi) = \alpha$ to $g(\psi) = \beta$.*

3.021 The integral $\int_\alpha^\beta f(x)\, dx$ can be transformed into another integral with given limits φ and ψ by means of the linear substitution

$$x = \frac{\beta - \alpha}{\psi - \varphi} t + \frac{\alpha\psi - \beta\varphi}{\psi - \varphi}:$$

1. $$\int_\alpha^\beta f(x)\, dx = \frac{\beta - \alpha}{\psi - \varphi} \int_\varphi^\psi f\left(\frac{\beta - \alpha}{\psi - \varphi} t + \frac{\alpha\psi - \beta\varphi}{\psi - \varphi} \right) dt.$$

In particular, for $\varphi = 0$ and $\psi = 1$,

2. $$\int_\alpha^\beta f(x)\, dx = (\beta - \alpha) \int_0^1 f((\beta - \alpha) t + \alpha)\, dt.$$

---

* If this last condition is not satisfied, the interval $\varphi \leqslant t \leqslant \psi$ should be partitioned into subintervals throughout each of which the condition is satisfied:

$$\int_\alpha^\beta f(x)\, dx = \int_\varphi^{\varphi_1} f[g(t)]\, g'(t)\, dt + \int_{\varphi_1}^{\varphi_2} f[g(t)]\, g'(t)\, dt + \ldots + \int_{\varphi_{n-1}}^{\psi} f[g(t)]\, g'(t)\, dt.$$

For $\varphi=0$ and $\psi=\infty$,

$$3.\quad \int_{\alpha}^{\beta} f(x)\,dx=(\beta-\alpha)\int_{0}^{\infty} f\left(\frac{\alpha+\beta t}{1+t}\right)\frac{dt}{(1+t)^2}.$$

3.022 The following formulas also hold:

$$1.\quad \int_{\alpha}^{\beta} f(x)\,dx=\int_{\alpha}^{\beta} f(\alpha+\beta-x)\,dx.$$

$$2.\quad \int_{0}^{\beta} f(x)\,dx=\int_{0}^{\beta} f(\beta-x)\,dx.$$

$$3.\quad \int_{-\alpha}^{\alpha} f(x)\,dx=\int_{-\alpha}^{\alpha} f(-x)\,dx.$$

3.03 General formulas

3.031

1. Suppose that a function $f(x)$ is integrable over the interval $(-p, p)$ and satisfies the relation $f(-x)=f(x)$ on that interval. (A function satisfying the latter condition is called an *even* function.) Then,

$$\int_{-p}^{p} f(x)\,dx=2\int_{0}^{p} f(x)\,dx.$$ FI II 159

2. Suppose that $f(x)$ is a function that is integrable on the interval $(-p, p)$ and satisfies the relation $f(-x)=-f(x)$ on that interval. (A function satisfying the latter condition is called an *odd* function). Then,

$$\int_{-p}^{p} f(x)\,dx=0.$$ FI II 159

3.032

$$1.\quad \int_{0}^{\frac{\pi}{2}} f(\sin x)\,dx=\int_{0}^{\frac{\pi}{2}} f(\cos x)\,dx,$$

where $f(x)$ is a function that is integrable on the interval $(0, 1)$. FI II 159

$$2.\quad \int_{0}^{2\pi} f(p\cos x+q\sin x)\,dx=2\int_{0}^{\pi} f\left(\sqrt{p^2+q^2}\cos x\right)dx,$$

where $f(x)$ is integrable on the interval $\left(-\sqrt{p^2+q^2},\ \sqrt{p^2+q^2}\right)$. FI II 160

$$3.\quad \int_{0}^{\frac{\pi}{2}} f(\sin 2x)\cos x\,dx=\int_{0}^{\frac{\pi}{2}} f(\cos^2 x)\cos x\,dx,$$

where $f(x)$ is integrable on the interval $(0, 1)$. FI II 161

3.033

1. If $f(x+\pi)=f(x)$ and $f(-x)=f(x)$, then

$$\int_0^\infty f(x)\frac{\sin x}{x}\,dx=\int_0^{\frac{\pi}{2}} f(x)\,dx.$$ LO V 277(3)

2. If $f(x+\pi)=-f(x)$ and $f(-x)=f(x)$, then

$$\int_0^\infty f(x)\frac{\sin x}{x}\,dx=\int_0^{\frac{\pi}{2}} f(x)\cos x\,dx.$$ LO V 279(4)

In formulas **3.033**, it is assumed that the integrals in the left members of the formulas exist.

3.034 $\displaystyle\int_0^\infty \frac{f(px)-f(qx)}{x}\,dx=[f(0)-f(+\infty)]\ln\frac{q}{p},$

if $f(x)$ is continuous for $x\geqslant 0$ and if there exists a finite limit $f(+\infty)=\lim\limits_{x\to+\infty} f(x)$.

FI II 633

3.035

1. $$\int_0^\pi \frac{f(\alpha+e^{xi})+f(\alpha+e^{-xi})}{1+2p\cos x+p^2}\,dx=\frac{2\pi}{1-p^2}f(\alpha+p)\quad[|p|<1].$$ LA 230(16)

2. $$\int_0^\pi \frac{1-p\cos x}{1-2p\cos x+p^2}\{f(\alpha+e^{xi})+f(\alpha+e^{-xi})\}\,dx=\pi\{f(\alpha+p)+f(\alpha)\}$$
$$[|p|<1].$$ BE 169

3. $$\int_0^\pi \frac{f(\alpha+e^{-xi})-f(\alpha+e^{xi})}{1-2p\cos x+p^2}\sin x\,dx=\frac{\pi}{pi}\{f(\alpha+p)-f(\alpha)\}\quad[|p|<1].$$

BE 169

In formulas **3.035**, it is assumed that the function f is analytic in the closed unit circle with its center at the point α.

3.036

1. $$\int_0^\pi f\left(\frac{\sin^2 x}{1+2p\cos x+p^2}\right)dx=\int_0^\pi f(\sin^2 x)\,dx\quad[p^2\geqslant 1];$$
$$=\int_0^\pi f\left(\frac{\sin^2 x}{p^2}\right)dx\quad[p^2<1].$$ LA 228(6)

2. $$\int_0^\pi F^{(n)}(\cos x)\sin^{2n}x\,dx=(2n-1)!!\int_0^\pi F(\cos x)\cos nx\,dx.$$ B 174

3.037 If f is analytic in the circle of radius r and if

$$f[r(\cos x+i\sin x)]=f_1(r,\ x)+if_2(r,\ x),$$

then

1. $$\int_0^\infty \frac{f_1(r,\ x)}{p^2+x^2}\,dx = \frac{\pi}{2p} f(re^{-p}).$$ LA 230(19)

2. $$\int_0^\infty f_2(r,\ x) \frac{x\,dx}{p^2+x^2} = \frac{\pi}{2}\left[f(re^{-p}) - f(0)\right].$$ LA 230(20)

3. $$\int_0^\infty \frac{f_2(r,\ x)}{x}\,dx = \frac{\pi}{2}\left[f(r) - f(0)\right].$$ LA 230(21)

4. $$\int_0^\infty \frac{f_2(r,\ x)}{x(p^2+x^2)}\,dx = \frac{\pi}{2p^2}\left[f(r) - f(re^{-p})\right].$$ LA 230(22)

3.038 $$\int_{-\infty}^{\infty} \frac{x\,dx}{\sqrt{1+x^2}} F\left(qx + p\sqrt{1+x^2}\right) = \int_{-\infty}^{\infty} F(p \operatorname{ch} x + q \operatorname{sh} x) \operatorname{sh} x\,dx =$$
$$= 2q \int_0^\infty F'\left(\operatorname{sign} p \cdot \sqrt{p^2-q^2} \operatorname{ch} x\right) \operatorname{sh}^2 x\,dx$$

[F is a function with a continuous derivative in the interval $(-\infty,\ \infty)$; all these integrals converge.]

3.04 Improper integrals

3.041 Suppose that a function $f(x)$ is defined on an interval $(p,\ +\infty)$ and that it is integrable over an arbitrary finite subinterval of the form $(p,\ P)$. Then, by definition

$$\int_p^{+\infty} f(x)\,dx = \lim_{P\to+\infty} \int_p^P f(x)\,dx,$$

if this limit exists. If it does exist, we say that the integral $\int_p^{+\infty} f(x)\,dx$ exists or that it converges. Otherwise, we say that the integral diverges.

3.042 Suppose that a function $f(x)$ is bounded and integrable in an arbitrary interval $(p,\ q-\eta)$ (for $0 < \eta < q-p$) but is unbounded in every interval $(q-\eta,\ q)$ to the left of the point q. The point q is then called a *singular point.* Then, by definition,

$$\int_p^q f(x)\,dx = \lim_{\eta\to 0} \int_p^{q-\eta} f(x)\,dx,$$

if this limit exists. In this case, we say that the integral $\int_p^q f(x)\,dx$ *exists* or that it *converges*.

3.043 If not only the integral of $f(x)$ but also the integral of $|f(x)|$ exists, we say that the integral of $f(x)$ converges *absolutely*.

3.044 The integral $\int_p^{+\infty} f(x)\,dx$ converges absolutely if there exists a number

$\alpha > 1$ such that the limit

$$\lim_{x\to+\infty} \{x^{\alpha} |f(x)|\}$$

exists. On the other hand, if

$$\lim_{x\to+\infty} \{x |f(x)|\} = L > 0,$$

the integral $\int_p^{+\infty} |f(x)| \, dx$ diverges.

3.045 Suppose that the upper limit q of the integral $\int_p^q f(x) \, dx$ is a singular point. Then, this integral converges absolutely if there exists a number $\alpha < 1$ such that the limit

$$\lim_{x\to q} [(q-x)^{\alpha} |f(x)|]$$

exists. On the other hand, if

$$\lim_{x\to q} [(q-x) |f(x)|] = L > 0,$$

the integral $\int_p^q f(x) \, dx$ diverges.

3.046 Suppose that the functions $f(x)$ and $g(x)$ are defined on the interval $(p, +\infty)$, that $f(x)$ is integrable over every finite interval of the form (p, P), that the integral

$$\int_p^P f(x) \, dx$$

is a bounded function of P, that $g(x)$ is monotonic, and that $g(x) \to 0$ as $x \to +\infty$. Then, the integral

$$\int_p^{+\infty} f(x) g(x) \, dx$$

converges.

FI II 577

3.05 *The principal values of improper integrals*

3.051 Suppose that a function $f(x)$ has a singular point r somewhere inside the interval (p, q), that $f(x)$ is defined at r, and that $f(x)$ is integrable over every portion of this interval that does not contain the point r. Then, by definition

$$\int_p^q f(x) \, dx = \lim_{\substack{\eta\to 0 \\ \eta'\to 0}} \left\{ \int_p^{r-\eta} f(x) \, dx + \int_{r+\eta'}^{q} f(x) \, dx \right\},$$

Here, the limit must exist for *independent* modes of approach of η and η' to zero. If this limit does not exist but the limit

$$\lim_{\eta\to 0} \left\{ \int_p^{r-\eta} f(x) \, dx + \int_{r+\eta}^{q} f(x) \, dx \right\},$$

does exist, we say that this latter limit is the *principal value* of the improper integral $\int_p^q f(x)\,dx$ and we say that the integral $\int_p^q f(x)\,dx$ exists in the sense of principal value. FI II 603

3.052 Suppose that the function $f(x)$ is continuous over the interval (p, q) and vanishes at only one point r inside this interval. Suppose that the first derivative $f'(x)$ exists in a neighborhood of the point r. Suppose that $f'(r) \neq 0$ and that the second derivative $f''(r)$ exists at the point r itself. Then,

$$\int_p^q \frac{dx}{f(x)}$$

FI II 605

diverges, but exists in the sense of principal values.

3.053 A divergent integral of a positive function cannot exist in the sense of principal values. FI II 605

3.054 Suppose that the function $f(x)$ has no singular points in the interval $(-\infty, +\infty)$. Then, by definition

$$\int_{-\infty}^{+\infty} f(x)\,dx = \lim_{\substack{P\to-\infty \\ Q\to+\infty}} \int_P^Q f(x)\,dx,$$

Here, the limit must exist for independent approach of P and Q to $\pm\infty$. If this limit does not exist but the limit

$$\lim_{P\to+\infty} \int_{-P}^{+P} f(x)\,dx,$$

does exist, this last limit is called the principal value of the improper integral

$$\int_{-\infty}^{+\infty} f(x)\,dx.$$

FI II 607

3.055 The principal value of an improper integral of an even function exists only when this integral converges (in the ordinary sense). FI II 607

3.1-3.2 Power and Algebraic Functions

3.11 Rational functions

3.111 $$\int_{-\infty}^{\infty} \frac{p+qx}{r^2+2rx\cos\lambda+x^2}\,dx = \frac{\pi}{r\sin\lambda}(p-qr\cos\lambda) \quad \text{(principal value*)}$$

(see also **3.194** 8. and **3.252** 1. and 2.). BI ((22))(14)

*We give the values of proper and improper convergent integrals and also the principal values of divergent integrals (see 3.05) if the latter exist. Henceforth, we make no special indication of principal values.

3.112 Integrals of the form $\int\limits_{-\infty}^{\infty} \frac{g_n(x)\,dx}{h_n(x)\,h_n(-x)}$,

where

$$g_n(x) = b_0 x^{2n-2} + b_1 x^{2n-4} + \ldots + b_{n-1},$$
$$h_n(x) = a_0 x^n + a_1 x^{n-1} + \ldots + a_n$$

[All roots of $h_n(x)$ lie in the upper half-plane.]

1. $$\int\limits_{-\infty}^{\infty} \frac{g_n(x)\,dx}{h_n(x)\,h_n(-x)} = \frac{\pi i}{a_0}\frac{M_n}{\Delta_n},$$ JE

where

$$\Delta_n = \begin{vmatrix} a_1 & a_3 & a_5 & \ldots & 0 \\ a_0 & a_2 & a_4 & \ldots & 0 \\ 0 & a_1 & a_3 & \ldots & 0 \\ \cdot & \cdot & \cdot & \cdot & \cdot \\ \cdot & \cdot & \cdot & \cdot & \cdot \\ \cdot & \cdot & \cdot & \cdot & \cdot \\ 0 & 0 & 0 & \ldots & a_n \end{vmatrix},$$

$$M_n = \begin{vmatrix} b_0 & b_1 & b_2 & \ldots & b_{n-1} \\ a_0 & a_2 & a_4 & \ldots & 0 \\ 0 & a_1 & a_3 & \ldots & 0 \\ \cdot & \cdot & \cdot & \cdot & \cdot \\ \cdot & \cdot & \cdot & \cdot & \cdot \\ \cdot & \cdot & \cdot & \cdot & \cdot \\ 0 & 0 & 0 & \ldots & a_n \end{vmatrix}.$$

2. $$\int\limits_{-\infty}^{\infty} \frac{g_1(x)\,dx}{h_1(x)\,h_1(-x)} = \frac{\pi i b_0}{a_0 a_1}.$$ JE

3. $$\int\limits_{-\infty}^{\infty} \frac{g_2(x)\,dx}{h_2(x)\,h_2(-x)} = \pi i \frac{-b_0 + \frac{a_0 b_1}{a_2}}{a_0 a_1}.$$

4. $$\int\limits_{-\infty}^{\infty} \frac{g_3(x)\,dx}{h_3(x)\,h_3(-x)} = \pi i \frac{-a_2 b_0 + a_0 b_1 - \frac{a_0 a_1 b_2}{a_3}}{a_0(a_0 a_3 - a_1 a_2)}.$$ JE

5. $$\int\limits_{-\infty}^{\infty} \frac{g_4(x)\,dx}{h_4(x)\,h_4(-x)} =$$

$$= \pi i \frac{b_0(-a_1 a_4 + a_2 a_3) - a_0 a_3 b_1 + a_0 a_1 b_2 + \frac{a_0 b_3}{a_4}(a_0 a_3 - a_1 a_2)}{a_0(a_0 a_3^2 + a_1^2 a_4 - a_1 a_2 a_3)}.$$ JE

6. $$\int\limits_{-\infty}^{\infty} \frac{g_5(x)\,dx}{h_5(x)\,h_5(-x)} = \pi i \frac{M_5}{a_0 \Delta_5},$$

where

$$M_5 = b_0(-a_0a_4a_5 + a_1a_4^2 + a_2^2a_5 - a_2a_3a_4) + a_0b_1(-a_2a_5 + a_3a_4) +$$
$$+ a_0b_2(a_0a_5 - a_1a_4) + a_0b_3(-a_0a_3 + a_1a_2) + \frac{a_0b_4}{a_5}(-a_0a_1a_5 + a_0a_3^2 + a_1^2a_4 - a_1a_2a_3),$$
$$\Delta_5 = a_0^2a_5^2 - 2a_0a_1a_4a_5 - a_0a_2a_3a_5 + a_0a_3^2a_4 + a_1^2a_4^2 + a_1a_2^2a_5 - a_1a_2a_3a_4.$$ JE

3.12 Products of rational functions and expressions that can be reduced to square roots of first- and second-degree polynomials

3.121

1. $$\int_0^1 \frac{1}{1-2x\cos\lambda+x^2}\frac{dx}{\sqrt{x}} = 2\operatorname{cosec}\lambda\sum_{k=1}^{\infty}\frac{\sin k\lambda}{2k-1}.$$ BI ((10))(17)

2. $$\int_0^1 \frac{1}{q-px}\frac{dx}{\sqrt{x(1-x)}} = \frac{\pi}{\sqrt{q(q-p)}} \qquad [0 < p < q].$$ BI ((10))(9)

3. $$\int_0^1 \frac{dx}{1-2rx+r^2}\sqrt{\frac{1\mp x}{1\pm x}} = \pm\frac{\pi}{4r} \mp \frac{1}{r}\frac{1\mp r}{1\pm r}\operatorname{arctg}\frac{1+r}{1-r}.$$ LI ((14))(5, 16)

3.13-3.17 Expressions that can be reduced to square roots of third- and fourth-degree polynomials and their products with rational functions

In 3.131 — 3.137 we set: $\alpha = \arcsin\sqrt{\frac{a-c}{a-u}}$, $\beta = \arcsin\sqrt{\frac{c-u}{b-u}}$,

$\gamma = \arcsin\sqrt{\frac{u-c}{b-c}}$, $\delta = \arcsin\sqrt{\frac{(a-c)(b-u)}{(b-c)(a-u)}}$,

$\varkappa = \arcsin\sqrt{\frac{(a-c)(u-b)}{(a-b)(u-c)}}$, $\lambda = \arcsin\sqrt{\frac{a-u}{a-b}}$,

$\mu = \arcsin\sqrt{\frac{u-a}{u-b}}$, $\nu = \arcsin\sqrt{\frac{a-c}{u-c}}$, $p = \sqrt{\frac{a-b}{a-c}}$, $q = \sqrt{\frac{b-c}{a-c}}$.

3.131

1. $$\int_{-\infty}^{u} \frac{dx}{\sqrt{(a-x)(b-x)(c-x)}} = \frac{2}{\sqrt{a-c}}F(\alpha, p) \quad [a > b > c \geqslant u].$$ BY (231.00)

2. $$\int_u^c \frac{dx}{\sqrt{(a-x)(b-x)(c-x)}} = \frac{2}{\sqrt{a-c}}F(\beta, p) \quad [a > b > c > u].$$ BY (232.00)

3. $$\int_c^u \frac{dx}{\sqrt{(a-x)(b-x)(x-c)}} = \frac{2}{\sqrt{a-c}}F(\gamma, q) \quad [a > b \geqslant u > c].$$ BY (233.00)

4. $$\int_u^b \frac{dx}{\sqrt{(a-x)(b-x)(x-c)}} = \frac{2}{\sqrt{a-c}}F(\delta, q) \quad [a > b > u \geqslant c].$$ BY (234.00)

5. $$\int_b^u \frac{dx}{\sqrt{(a-x)(x-b)(x-c)}} = \frac{2}{\sqrt{a-c}}F(\varkappa, p) \quad [a \geqslant u > b > c].$$ BY (235.00)

6. $$\int_u^a \frac{dx}{\sqrt{(a-x)(x-b)(x-c)}} = \frac{2}{\sqrt{a-c}} F(\lambda, p) \quad [a > u \geqslant b > c].$$ BY (236.00)

7. $$\int_a^u \frac{dx}{\sqrt{(x-a)(x-b)(x-c)}} = \frac{2}{\sqrt{a-c}} F(\mu, q) \quad [u > a > b > c].$$ BY (237.00)

8. $$\int_u^\infty \frac{dx}{\sqrt{(x-a)(x-b)(x-c)}} = \frac{2}{\sqrt{a-c}} F(\nu, q) \quad [u \geqslant a > b > c].$$ BY (238.00)

3.132

1. $$\int_u^c \frac{x\,dx}{\sqrt{(a-x)(b-x)(c-x)}} = \frac{2}{\sqrt{a-c}} [cF(\beta, p) + (a-c) E(\beta, p)] - 2\sqrt{\frac{(a-u)(c-u)}{b-u}} \quad [a > b > c > u].$$ BY (232.19)

2. $$\int_c^u \frac{x\,dx}{\sqrt{(a-x)(b-x)(x-c)}} = \frac{2a}{\sqrt{a-c}} F(\gamma, q) - 2\sqrt{a-c}\, E(\gamma, q) \quad [a > b \geqslant u > c].$$ BY (233.17)

3. $$\int_u^b \frac{x\,dx}{\sqrt{(a-x)(b-x)(x-c)}} = \frac{2}{\sqrt{a-c}} [(b-a)\Pi(\delta, q^2, q) + aF(\delta, q)] \quad [a > b > u \geqslant c].$$ BY (234.16)

4. $$\int_b^u \frac{x\,dx}{\sqrt{(a-x)(x-b)(x-c)}} = \frac{2}{\sqrt{a-c}} [(b-c)\Pi(\varkappa, p^2, p) + cF(\varkappa, p)] \quad [a \geqslant u > b > c].$$ BY (235.16)

5. $$\int_u^a \frac{x\,dx}{\sqrt{(a-x)(x-b)(x-c)}} = \frac{2c}{\sqrt{a-c}} F(\lambda, p) + 2\frac{a}{b}\sqrt{a-c}\, E(\lambda, p) \quad [a > u \geqslant b > c].$$ BY (236.16)

6. $$\int_a^u \frac{x\,dx}{\sqrt{(x-a)(x-b)(x-c)}} = \frac{2}{b\sqrt{a-c}} [a(a-b)\Pi(\mu, 1, q) + b^2 F(\mu, q)] \quad [u > a > b > c].$$ BY (237.16)

3.133

1. $$\int_{-\infty}^u \frac{dx}{\sqrt{(a-x)^3(b-x)(c-x)}} = \frac{2}{(a-b)\sqrt{a-c}} [F(\alpha, p) - E(\alpha, p)] \quad [a > b > c \geqslant u].$$ BY (231.08)

2. $$\int_u^c \frac{dx}{\sqrt{(a-x)^3(b-x)(c-x)}} = \frac{2}{(a-b)\sqrt{a-c}} [F(\beta, p) - E(\beta, p)] + \frac{2}{a-c}\sqrt{\frac{c-u}{(a-u)(b-u)}} \quad [a > b > c > u].$$ BY (232.13)

3. $$\int_c^u \frac{dx}{\sqrt{(a-x)^3(b-x)(x-c)}} = \frac{2}{(a-b)\sqrt{a-c}} E(\gamma, q) - \frac{2}{(a-b)(a-c)} \sqrt{\frac{(b-u)(u-c)}{a-u}} \quad [a > b \geqslant u > c].$$ BY (233.09)

4. $$\int_u^b \frac{dx}{\sqrt{(a-x)^3(b-x)(x-c)}} = \frac{2}{(a-b)\sqrt{a-c}} E(\delta, q) \quad [a > b > u \geqslant c].$$ BY (234.05)

5. $$\int_b^u \frac{dx}{\sqrt{(a-x)^3(x-b)(x-c)}} = \frac{2}{(a-b)\sqrt{a-c}} [F(\varkappa, p) - E(\varkappa, p)] + \frac{2}{a-b} \sqrt{\frac{u-b}{(a-u)(u-c)}} \quad [a > u > b > c].$$ BY (235.04)

6. $$\int_u^\infty \frac{dx}{\sqrt{(x-a)^3(x-b)(x-c)}} = \frac{2}{(b-a)\sqrt{a-c}} E(\nu, q) + \frac{2}{a-b} \sqrt{\frac{u-b}{(u-a)(u-c)}} \quad [u > a > b > c].$$ BY (238.05)

7. $$\int_{-\infty}^u \frac{dx}{\sqrt{(a-x)(b-x)^3(c-x)}} = \frac{2\sqrt{a-c}}{(a-b)(b-c)} E(\alpha, p) - \frac{2}{(a-b)\sqrt{a-c}} F(\alpha, p) - \frac{2}{b-c} \sqrt{\frac{c-u}{(a-u)(b-u)}} \quad [a > b > c \geqslant u].$$ BY (231.09)

8. $$\int_u^c \frac{dx}{\sqrt{(a-x)(b-x)^3(c-x)}} = \frac{2\sqrt{a-c}}{(a-b)(b-c)} E(\beta, p) - \frac{2}{(a-b)\sqrt{a-c}} F(\beta, p) \quad [a > b > c > u].$$ BY (232.14)

9. $$\int_c^u \frac{dx}{\sqrt{(a-x)(b-x)^3(x-c)}} = \frac{2}{(b-c)\sqrt{a-c}} F(\gamma, q) - \frac{2\sqrt{a-c}}{(a-b)(b-c)} E(\gamma, q) + \frac{2}{(a-b)(b-c)} \sqrt{\frac{(a-u)(u-c)}{b-u}} \quad [a > b > u > c].$$ BY (233.10)

10. $$\int_u^a \frac{dx}{\sqrt{(a-x)(x-b)^3(x-c)}} = \frac{2}{(a-b)\sqrt{a-c}} F(\lambda, p) - \frac{2\sqrt{a-c}}{(a-b)(b-c)} E(\lambda, p) + \frac{2}{(a-b)(b-c)} \sqrt{\frac{(a-u)(u-c)}{u-b}} \quad [a > u > b > c].$$ BY (236.09)

11. $$\int_a^u \frac{dx}{\sqrt{(x-a)(x-b)^3(x-c)}} = \frac{2\sqrt{a-c}}{(a-b)(b-c)} E(\mu, q) - \frac{2}{(b-c)\sqrt{a-c}} F(\mu, q) \quad [u > a > b > c].$$ BY (237.12)

12. $$\int_u^\infty \frac{dx}{\sqrt{(x-a)(x-b)^3(x-c)}} = \frac{2\sqrt{a-c}}{(a-b)(b-c)} E(\nu, q) - \frac{2}{(b-c)\sqrt{a-c}} F(\nu, q) - \frac{2}{a-b}\sqrt{\frac{u-a}{(u-b)(u-c)}} \quad [u \geqslant a > b > c].$$ BY (238.04)

13. $$\int_{-\infty}^u \frac{dx}{\sqrt{(a-x)(b-x)(c-x)^3}} = \frac{2}{(c-b)\sqrt{a-c}} E(\alpha, p) + \frac{2}{b-c}\sqrt{\frac{b-u}{(a-u)(c-u)}} \quad [a > b > c > u].$$ BY (231.10)

14. $$\int_u^b \frac{dx}{\sqrt{(a-x)(b-x)(x-c)^3}} = \frac{2}{(b-c)\sqrt{a-c}} [F(\delta, q) - E(\delta, q)] + \frac{2}{b-c}\sqrt{\frac{b-u}{(a-u)(u-c)}} \quad [a > b > u > c].$$ BY (234.04)

15. $$\int_b^u \frac{dx}{\sqrt{(a-x)(x-b)(x-c)^3}} = \frac{2}{(b-c)\sqrt{a-c}} E(\varkappa, p) \quad [a \geqslant u > b > c].$$ BY (235.01)

16. $$\int_u^a \frac{dx}{\sqrt{(a-x)(x-b)(x-c)^3}} = \frac{2}{(b-c)\sqrt{a-c}} E(\lambda, p) - \frac{2}{(b-c)(a-c)}\sqrt{\frac{(a-u)(u-b)}{u-c}} \quad [a > u \geqslant b > c].$$ BY (236.10)

17. $$\int_a^u \frac{dx}{\sqrt{(x-a)(x-b)(x-c)^3}} = \frac{2}{(b-c)\sqrt{a-c}} [F(\mu, q) - E(\mu, q)] + \frac{2}{a-c}\sqrt{\frac{u-a}{(u-b)(u-c)}} \quad [u > a > b > c].$$ BY (237.13)

18. $$\int_u^\infty \frac{dx}{\sqrt{(x-a)(x-b)(x-c)^3}} = \frac{2}{(b-c)\sqrt{a-c}} [F(\nu, q) - E(\nu, q)] \quad [u \geqslant a > b > c].$$ BY (238.03)

3.134

1. $$\int_{-\infty}^u \frac{dx}{\sqrt{(a-x)^5(b-x)(c-x)}} = \frac{2}{3(a-b)^2\sqrt{(a-c)^3}} \times [(3a-b-2c)F(\alpha, p) - 2(2a-b-c)E(\alpha, p)] + \frac{2}{3(a-c)(a-b)}\sqrt{\frac{(c-u)(b-u)}{(a-u)^3}} \quad [a > b > c \geqslant u].$$ BY (231.08)

2. $$\int_u^c \frac{dx}{\sqrt{(a-x)^5(b-x)(c-x)}} = \frac{2}{3(a-b)^2\sqrt{(a-c)^3}} \times [(3a-b-2c)F(\beta, p) - 2(2a-b-c)E(\beta, p)] + \frac{2[4a^2-3ab-2ac+bc-u(3a-2b-c)]}{3(a-b)(a-c)^2}\sqrt{\frac{c-u}{(a-u)^3(b-u)}} \quad [a > b > c > u].$$ BY (232.13)

3. $$\int_c^u \frac{dx}{\sqrt{(a-x)^5(b-x)(x-c)}} = \frac{2}{3(a-b)^2\sqrt{(a-c)^3}} \times$$
$$\times [2(2a-b-c)E(\gamma, q)-(a-b)F(\gamma, q)] -$$
$$-\frac{2[5a^2-3ab-3ac+bc-2u(2a-b-c)]}{3(a-b)^2(a-c)^2}\sqrt{\frac{(b-u)(u-c)}{(a-u)^3}} \quad [a>b\geqslant u>c].$$

BY (233.09)

4. $$\int_u^b \frac{dx}{\sqrt{(a-x)^5(b-x)(x-c)}} = \frac{2}{3(a-b)^2\sqrt{(a-c)^3}} \times$$
$$\times [2(2a-b-c)E(\delta, q)-(a-b)F(\delta, q)] -$$
$$-\frac{2}{3(a-b)(a-c)}\sqrt{\frac{(b-u)(u-c)}{(a-u)^3}} \quad [a>b>u\geqslant c].$$

BY (234.05)

5. $$\int_b^u \frac{dx}{\sqrt{(a-x)^5(x-b)(x-c)}} = \frac{2}{3(a-b)^2\sqrt{(a-c)^3}} \times$$
$$\times [(3a-b-2c)F(\varkappa, p)-2(2a-b-c)E(\varkappa, p)] +$$
$$+\frac{2[4a^2-2ab-3ac+bc-u(3a-b-2c)]}{3(a-b)^2(a-c)}\sqrt{\frac{u-b}{(a-u)^3(u-c)}} \quad [a>u>b>c].$$

BY (235.04)

6. $$\int_u^\infty \frac{dx}{\sqrt{(x-a)^5(x-b)(x-c)}} = \frac{2}{3(a-b)^2\sqrt{(a-c)^3}} \times$$
$$\times [2(2a-b-c)E(\nu, q)-(a-b)F(\nu, q)] +$$
$$+\frac{2[4a^2-2ab-3ac+bc+u(b+2c-3a)]}{3(a-b)^2(a-c)}\sqrt{\frac{u-b}{(u-a)^3(u-c)}} \quad [u>a>b>c].$$

BY (238.05)

7. $$\int_{-\infty}^u \frac{dx}{\sqrt{(a-x)(b-x)^5(c-x)}} = \frac{2}{3(a-b)^2(b-c)^2\sqrt{a-c}} \times$$
$$\times [2(a-c)(a+c-2b)E(\alpha, p)+(b-c)(3b-a-2c)F(\alpha, p)] -$$
$$-\frac{2[3ab-ac+2bc-4b^2-u(2a-3b+c)]}{3(a-b)(b-c)^2}\sqrt{\frac{c-u}{(a-u)(b-u)^3}} \quad [a>b>c\geqslant u].$$

BY (231.09)

8. $$\int_u^c \frac{dx}{\sqrt{(a-x)(b-x)^5(c-x)}} = \frac{2}{3(a-b)^2(b-c)^2\sqrt{a-c}} \times$$
$$\times [(b-c)(3b-a-2c)F(\beta, p)+2(a-c)(a-2b+c)E(\beta, p)] +$$
$$+\frac{2}{3(a-b)(b-c)}\sqrt{\frac{(a-u)(c-u)}{(b-u)^3}} \quad [a>b>c>u].$$

BY (232.14)

9. $$\int_c^u \frac{dx}{\sqrt{(a-x)(b-x)^5(x-c)}} = \frac{2}{3(a-b)^2(b-c)^2\sqrt{a-c}} \times$$
$$\times [(a-b)(2a-3b+c)F(\gamma, q)+2(a-c)(2b-a-c)E(\gamma, q)] +$$
$$+\frac{2[3ab+3bc-ac-5b^2-2u(a-2b+c)]}{3(a-b)^2(b-c)^2}\sqrt{\frac{(a-u)(u-c)}{(b-u)^3}} \quad [a>b>u>c].$$

BY (233.10)

10. $$\int_u^a \frac{dx}{\sqrt{(a-x)(x-b)^5(x-c)}} = \frac{2}{3(a-b)^2(b-c)^2\sqrt{a-c}} \times$$
$$\times [(b-c)(3b-2c-a)F(\lambda, p) + 2(a-c)(a+c-2b)E(\lambda, p)] +$$
$$+ \frac{2[3ab+3bc-ac-5b^2+2u(2b-a-c)]}{3(a-b)^2(b-c)^2}\sqrt{\frac{(a-u)(u-c)}{(u-b)^3}} \quad [a > u > b > c].$$

BY (236.09)

11. $$\int_a^u \frac{dx}{\sqrt{(x-a)(x-b)^5(x-c)}} = \frac{2}{3(a-b)^2(b-c)^2\sqrt{a-c}} \times$$
$$\times [(a-b)(2a+c-3b)F(\mu, q) + 2(a-c)(2b-a-c)E(\mu, q)] +$$
$$+ \frac{2}{3(a-b)(b-c)}\sqrt{\frac{(u-a)(u-c)}{(u-b)^3}} \quad [u > a > b > c].$$ BY (237.12)

12. $$\int_u^\infty \frac{dx}{\sqrt{(x-a)(x-b)^5(x-c)}} = \frac{2}{3(a-b)^2(b-c)^2\sqrt{a-c}} \times$$
$$\times [(a-b)(2a+c-3b)F(\nu, q) + 2(a-c)(2b-c-a)E(\nu, q)] -$$
$$- \frac{2[3bc+2ab-ac-4b^2+u(3b-a-2c)]}{3(a-b)^2(b-c)}\sqrt{\frac{u-a}{(u-b)^3(u-c)}} \quad [u \geqslant a > b > c].$$

BY (238.04)

13. $$\int_{-\infty}^u \frac{dx}{\sqrt{(a-x)(b-x)(c-x)^5}} = \frac{2}{3(b-c)^2\sqrt{(a-c)^3}} \times$$
$$\times [2(a+b-2c)E(\alpha, p) - (b-c)F(\alpha, p)] +$$
$$+ \frac{2[ab-3ac-2bc+4c^2+u(2a+b-3c)]}{3(a-c)(b-c)^2}\sqrt{\frac{b-u}{(a-u)(c-u)^3}}$$
$$[a > b > c > u].$$ BY (231.10)

14. $$\int_u^b \frac{dx}{\sqrt{(a-x)(b-x)(x-c)^5}} = \frac{2}{3(b-c)^2\sqrt{(a-c)^3}} \times$$
$$\times [(2a+b-3c)F(\delta, q) - 2(a+b-2c)E(\delta, q)] +$$
$$+ \frac{2[ab-3ac-2bc+4c^2+u(2a+b-3c)]}{3(b-c)^2(a-c)}\sqrt{\frac{b-u}{(a-u)(u-c)^3}}$$
$$[a > b > u > c].$$ BY (234.04)

15. $$\int_b^u \frac{dx}{\sqrt{(a-x)(x-b)(x-c)^5}} = \frac{2}{3(b-c)^2\sqrt{(a-c)^3}} \times$$
$$\times [2(a+b-2c)E(\varkappa, p) - (b-c)F(\varkappa, p)] +$$
$$+ \frac{2}{3(a-c)(b-c)}\sqrt{\frac{(a-u)(u-b)}{(u-c)^3}} \quad [a \geqslant u > b > c].$$ BY (235.20)

16. $$\int_u^a \frac{dx}{\sqrt{(a-x)(x-b)(x-c)^5}} = \frac{2}{3(b-c)^2\sqrt{(a-c)^3}} \times$$
$$\times [2(a+b-2c)E(\lambda, p) - (b-c)F(\lambda, p)] -$$
$$- \frac{2[ab-3ac-3bc+5c^2+2u(a+b-2c)]}{3(b-c)^2(a-c)^2}\sqrt{\frac{(a-u)(u-b)}{(u-c)^3}}$$
$$[a > u \geqslant b > c].$$ BY (236.10)

17. $$\int_a^u \frac{dx}{\sqrt{(x-a)(x-b)(x-c)^5}} = \frac{2}{3(b-c)^2\sqrt{(a-c)^3}} \times$$
$$\times [(2a+b-3c)F(\mu, q) - 2(a+b-2c)E(\mu, q)] +$$
$$+ \frac{2[4c^2-ab-2ac-bc+u(3a+2b-5c)]}{3(b-c)(a-c)^2}\sqrt{\frac{u-a}{(u-b)(u-c)^3}}$$
$$[u > a > b > c].$$ BY (237.13)

18. $$\int_u^\infty \frac{dx}{\sqrt{(x-a)(x-b)(x-c)^5}} = \frac{2}{3(b-c)^2\sqrt{(a-c)^3}} \times$$
$$\times [(2a+b-3c)F(\nu, q) - 2(a+b-2c)E(\nu, q)] +$$
$$+ \frac{2}{3(a-c)(b-c)}\sqrt{\frac{(u-a)(u-b)}{(u-c)^3}} \quad [u \geqslant a > b > c].$$ BY (238.03)

3.135

1. $$\int_{-\infty}^u \frac{dx}{\sqrt{(a-x)(b-x)^3(c-x)^3}} = \frac{2}{(a-b)(b-c)^2\sqrt{a-c}} \times$$
$$\times [(b-c)F(\alpha, p) - (a+b-2c)E(\alpha, p)] +$$
$$+ \frac{2(b+c-2u)}{(b-c)^2\sqrt{(a-u)(b-u)(c-u)}} \quad [a > b > c > u].$$ BY (231.13)

2. $$\int_u^a \frac{dx}{\sqrt{(a-x)(x-b)^3(x-c)^3}} = \frac{2}{(a-b)(b-c)^2\sqrt{a-c}} \times$$
$$\times [(b-c)F(\lambda, p) - 2(2a-b-c)E(\lambda, p)] +$$
$$+ \frac{2(a-b-c+u)}{(a-b)(b-c)(a-c)}\sqrt{\frac{a-u}{(u-b)(u-c)}} \quad [a > u > b > c].$$ BY (236.15)

3. $$\int_a^u \frac{dx}{\sqrt{(x-a)(x-b)^3(x-c)^3}} = \frac{2}{(a-b)(b-c)^2\sqrt{a-c}} \times$$
$$\times [(2a-b-c)E(\mu, q) - 2(a-b)F(\mu, q)] +$$
$$+ \frac{2}{(a-c)(b-c)}\sqrt{\frac{u-a}{(u-b)(u-c)}} \quad [u > a > b > c].$$ BY (236.14)

4. $$\int_u^\infty \frac{dx}{\sqrt{(x-a)(x-b)^3(x-c)^3}} = \frac{2}{(a-b)(b-c)^2\sqrt{a-c}} \times$$
$$\times [(2a-b-c)E(\nu, q) - 2(a-b)F(\nu, q)] -$$
$$- \frac{2}{(a-b)(b-c)}\sqrt{\frac{u-a}{(u-b)(u-c)}} \quad [u \geqslant a > b > c].$$ BY (238.13)

5. $$\int_{-\infty}^u \frac{dx}{\sqrt{(a-x)^3(b-x)(c-x)^3}} = \frac{2}{(a-b)(b-c)\sqrt{(a-c)^3}} \times$$
$$\times [(2b-a-c)E(\alpha, p) - (b-c)F(\alpha, p)] +$$
$$+ \frac{2}{(b-c)(a-c)}\sqrt{\frac{b-u}{(a-u)(c-u)}} \quad [a > b > c > u].$$ BY (231.12)

6. $$\int_u^b \frac{dx}{\sqrt{(a-x)^3(b-x)(x-c)^3}} = \frac{2}{(b-c)(a-b)\sqrt{(a-c)^3}} \times$$
$$\times [(a-b)F(\delta, q) + (2b-a-c)E(\delta, q)] +$$
$$+\frac{2}{(b-c)(a-c)}\sqrt{\frac{b-u}{(a-u)(u-c)}} \quad [a > b > u > c].$$ BY (234.03)

7. $$\int_b^u \frac{dx}{\sqrt{(a-x)^3(x-b)(x-c)^3}} = \frac{2}{(a-b)(b-c)\sqrt{(a-c)^3}} \times$$
$$\times [(b-c)F(\varkappa, p) - (2b-a-c)E(\varkappa, p)] +$$
$$+\frac{2}{(a-b)(a-c)}\sqrt{\frac{u-b}{(a-u)(u-c)}} \quad [a > u > b > c].$$ BY (235.15)

8. $$\int_u^\infty \frac{dx}{\sqrt{(x-a)^3(x-b)(x-c)^3}} = \frac{2}{(a-b)(b-c)\sqrt{(a-c)^3}} \times$$
$$\times [(a+c-2b)E(\nu, q) - (a-b)F(\nu, q)] +$$
$$+\frac{2}{(a-b)(a-c)}\sqrt{\frac{u-b}{(u-a)(u-c)}} \quad [u > a > b > c].$$ BY (238.14)

9. $$\int_{-\infty}^u \frac{dx}{\sqrt{(a-x)^3(b-x)^3(c-x)}} = \frac{2}{(b-c)(a-b)^2\sqrt{a-c}} \times$$
$$\times [(a+b-2c)E(\alpha, p) - 2(b-c)F(\alpha, p)] -$$
$$-\frac{2}{(a-b)(b-c)}\sqrt{\frac{c-u}{(a-u)(b-u)}} \quad [a > b > c \geqslant u].$$ BY (231.11)

10. $$\int_u^c \frac{dx}{\sqrt{(a-x)^3(b-x)^3(c-x)}} = \frac{2}{(a-b)^2(b-c)\sqrt{a-c}} \times$$
$$\times [(a+b-2c)E(\beta, p) - 2(b-c)F(\beta, p)] +$$
$$+\frac{2}{(a-b)(a-c)}\sqrt{\frac{c-u}{(a-u)(b-u)}} \quad [a > b > c > u].$$ BY (232.15)

11. $$\int_c^u \frac{dx}{\sqrt{(a-x)^3(b-x)^3(x-c)}} = \frac{2}{(a-b)^2(b-c)\sqrt{a-c}} \times$$
$$\times [(a-b)F(\gamma, q) - (a+b-2c)E(\gamma, q)] +$$
$$+\frac{2[a^2+b^2-ac-bc-u(a+b-2c)]}{(a-b)^2(b-c)(a-c)}\sqrt{\frac{u-c}{(a-u)(b-u)}} \quad [a > b > u > c].$$
BY (233.11)

12. $$\int_u^\infty \frac{dx}{\sqrt{(x-a)^3(x-b)^3(x-c)}} = \frac{2}{(a-b)^2(b-c)\sqrt{a-c}} \times$$
$$\times [(a-b)F(\nu, q) - (a+b-2c)E(\nu, q)] +$$
$$+\frac{2u-a-b}{(a-b)^2\sqrt{(u-a)(u-b)(u-c)}} \quad [u > a > b > c].$$ BY (238.15)

3.136

1. $$\int_{-\infty}^{u} \frac{dx}{\sqrt{(a-x)^3 (b-x)^3 (c-x)^3}} = \frac{2}{(a-b)^2 (b-c)^2 \sqrt{(a-c)^3}} \times$$
$$\times [(b-c)(a+b-2c) F(\alpha, p) - 2(c^2+a^2+b^2-ab-ac-bc) E(\alpha, p)] +$$
$$+ \frac{2[c(a-c)+b(a-b)-u(2a-c-b)]}{(a-b)(a-c)(b-c)^2 \sqrt{(a-u)(b-u)(c-u)}} \quad [a > b > c > u].$$ BY (231.14)

2. $$\int_{u}^{\infty} \frac{dx}{\sqrt{(x-a)^3 (x-b)^3 (x-c)^3}} = \frac{2}{(a-b)^2 (b-c)^2 \sqrt{(a-c)^3}} \times$$
$$\times [(a-b)(2a-b-c) F(\nu, q) - 2(a^2+b^2+c^2-ab-ac-bc) E(\nu, q)] +$$
$$+ \frac{2[u(a+b-2c)-a(a-c)-b(b-c)]}{(a-b)^2 (a-c)(b-c) \sqrt{(u-a)(u-b)(u-c)}} \quad [u > a > b > c].$$ BY (238.16)

3.137

1. $$\int_{-\infty}^{u} \frac{dx}{(r-x)\sqrt{(a-x)(b-x)(c-x)}} = \frac{2}{(a-r)\sqrt{a-c}} \times$$
$$\times \left[\Pi\left(\alpha, \frac{a-r}{a-c}, p\right) - F(\alpha, p)\right] \quad [a > b > c \geqslant u].$$ BY (231.15)

2. $$\int_{u}^{c} \frac{dx}{(r-x)\sqrt{(a-x)(b-x)(c-x)}} = \frac{2(c-b)}{(r-b)(r-c)\sqrt{a-c}} \times$$
$$\times \Pi\left(\beta, \frac{r-b}{r-c}, p\right) + \frac{2}{(r-b)\sqrt{a-c}} F(\beta, p) \quad [a > b > c > u, r \neq 0].$$ BY (232.17)

3. $$\int_{c}^{u} \frac{dx}{(r-x)\sqrt{(a-x)(b-x)(x-c)}} = \frac{2}{(r-c)\sqrt{a-c}} \Pi\left(\gamma, \frac{b-c}{r-c}, q\right)$$
$$[a > b \geqslant u > c, \ r \neq c].$$ BY (233.02)

4. $$\int_{u}^{b} \frac{dx}{(r-x)\sqrt{(a-x)(b-x)(x-c)}} = \frac{2}{(r-a)(r-b)\sqrt{a-c}} \times$$
$$\times \left[(b-a)\Pi\left(\delta, q^2\frac{r-a}{r-b}, q\right) + (r-b)F(\delta, q)\right]$$
$$[a > b > u \geqslant c, \ r \neq b].$$ BY (234.18)

5. $$\int_{b}^{u} \frac{dx}{(x-r)\sqrt{(a-x)(x-b)(x-c)}} = \frac{2}{(c-r)(b-r)\sqrt{a-c}} \times$$
$$\times \left[(c-b)\Pi\left(\varkappa, p^2\frac{c-r}{b-r}, p\right) + (b-r)F(\varkappa, p)\right]$$
$$[a \geqslant u > b > c, \ r \neq b].$$ BY (235.17)

6. $$\int_{u}^{a} \frac{dx}{(x-r)\sqrt{(a-x)(x-b)(x-c)}} = \frac{2}{(a-r)\sqrt{a-c}} \Pi\left(\lambda, \frac{a-b}{a-r}, p\right)$$
$$[a > u \geqslant b > c, \ r \neq a].$$ BY (236.02)

7. $$\int_a^u \frac{dx}{(x-r)\sqrt{(x-a)(x-b)(x-c)}} = \frac{2}{(b-r)(a-r)\sqrt{a-c}} \times$$
$$\times \left[(b-a)\Pi\left(\mu, \frac{b-r}{a-b}, q\right) + (a-p)F(\mu, q)\right]$$
$$[u > a > b > c,\ r \neq a].$$ BY (237.17)

8. $$\int_u^\infty \frac{dx}{(x-r)\sqrt{(x-a)(x-b)(x-c)}} = \frac{2}{(r-c)\sqrt{a-c}} \times$$
$$\times \left[\Pi\left(\nu, \frac{r-c}{a-c}, q\right) - F(\nu, q)\right] \quad [u \geqslant a > b > c].$$ BY (238.06)

3.138

1. $$\int_0^u \frac{dx}{\sqrt{x(1-x)(1-k^2x)}} = 2F\left(\arcsin\sqrt{u},\ k\right) \quad [0<u<1].$$ PE (532), JA

2. $$\int_u^1 \frac{dx}{\sqrt{x(1-x)(k'^2+k^2x)}} = 2F\left(\arccos\sqrt{u},\ k\right) \quad [0<u<1].$$ PE (533)

3. $$\int_u^1 \frac{dx}{\sqrt{x(1-x)(x-k'^2)}} = 2F\left(\arcsin\frac{\sqrt{1-u}}{k},\ k\right) \quad [0<u<1].$$ PE (534)

4. $$\int_0^u \frac{dx}{\sqrt{x(1+x)(1+k'^2x)}} = 2F\left(\operatorname{arctg}\sqrt{u},\ k\right) \quad [0<u<1].$$ PE (535)

5. $$\int_0^u \frac{dx}{\sqrt{x[1+x^2+2(k'^2-k^2)x]}} = F\left(2\operatorname{arctg}\sqrt{u},\ k\right) \quad [0<u<1].$$ JA

6. $$\int_u^1 \frac{dx}{\sqrt{x[k'^2(1+x^2)+2(1+k^2)x]}} = F\left(\frac{\pi}{2} - 2\operatorname{arctg}\sqrt{u},\ k\right) \quad [0<u<1].$$ JA

7. $$\int_\alpha^u \frac{dx}{\sqrt{(x-\alpha)[(x-m)^2+n^2]}} = \frac{1}{\sqrt{p}} F\left(2\operatorname{arctg}\sqrt{\frac{u-\alpha}{p}},\ \sqrt{\frac{p+m-\alpha}{2p}}\right)$$
$$[\alpha < u],$$

8. $$\int_u^\alpha \frac{dx}{\sqrt{(\alpha-x)[(x-m)^2+n^2]}} = \frac{1}{\sqrt{p}} F\left(2\operatorname{arcctg}\sqrt{\frac{\alpha-u}{p}},\ \sqrt{\frac{p-m+\alpha}{2p}}\right)$$
$$[u < \alpha],$$

where $p = \sqrt{(m-\alpha)^2+n^2}$.

3.139 Notation: $\alpha = \arccos\frac{1-\sqrt{3}-u}{1+\sqrt{3}-u}$, $\beta = \arccos\frac{\sqrt{3}-1+u}{\sqrt{3}+1-u}$,
$\gamma = \arccos\frac{\sqrt{3}+1-u}{\sqrt{3}-1+u}$, $\delta = \arccos\frac{u-1-\sqrt{3}}{u-1+\sqrt{3}}$.

1. $\int_{-\infty}^{u} \frac{dx}{\sqrt{1-x^3}} = \frac{1}{\sqrt[4]{3}} F(\alpha, \sin 75^\circ).$ ZH 66 (285)

2. $\int_{u}^{1} \frac{dx}{\sqrt{1-x^3}} = \frac{1}{\sqrt[4]{3}} F(\beta, \sin 75^\circ).$ ZH 65 (284)

3. $\int_{1}^{u} \frac{dx}{\sqrt{x^3-1}} = \frac{1}{\sqrt[4]{3}} F(\gamma, \sin 15^\circ).$ ZH 65 (283)

4. $\int_{u}^{\infty} \frac{dx}{\sqrt{x^3-1}} = \frac{1}{\sqrt[4]{3}} F(\delta, \sin 15^\circ).$ ZH 65 (282)

5. $\int_{0}^{1} \frac{dx}{\sqrt{1-x^3}} = \frac{1}{2\pi \sqrt{3} \sqrt[3]{2}} \left\{\Gamma\left(\frac{1}{3}\right)\right\}^2.$ MO 9

6. $\int_{0}^{1} \frac{x\,dx}{\sqrt{1-x^3}} = \frac{1}{\pi} \frac{\sqrt{3}}{\sqrt[3]{4}} \left\{\Gamma\left(\frac{2}{3}\right)\right\}^2.$ MO 9

7. $\int_{u}^{1} \sqrt{1-x^3}\,dx = \frac{1}{5}\{\sqrt[4]{27}\,F(\beta, \sin 75^\circ) - 2u\sqrt{1-u^3}\}.$ BY (244.01)

8. $\int_{u}^{1} \frac{x\,dx}{\sqrt{1-x^3}} = (3^{-\frac{1}{4}} - 3^{\frac{1}{4}}) F(\beta, \sin 75^\circ) +$

$+ 2\sqrt[4]{3} E(\beta, \sin 75^\circ) - \frac{2\sqrt{1-u^3}}{\sqrt{3}+1-u}.$ BY (244.05)

9. $\int_{u}^{1} \frac{x^m\,dx}{\sqrt{1-x^3}} = \frac{2u^{m-2}\sqrt{1-u^3}}{2m-1} + \frac{2(m-2)}{2m-1} \int_{u}^{1} \frac{x^{m-3}\,dx}{\sqrt{1-x^3}}.$ BY (244.07)

10. $\int_{1}^{u} \frac{x\,dx}{\sqrt{x^3-1}} = (3^{-\frac{1}{4}} + 3^{\frac{1}{4}}) F(\gamma, \sin 15^\circ) -$

$- 2\sqrt[4]{3} E(\gamma, \sin 15^\circ) + \frac{2\sqrt{u^3-1}}{\sqrt{3}-1+u}.$ BY (240.05)

11. $\int_{-\infty}^{u} \frac{dx}{(1-x)\sqrt{1-x^3}} = \frac{1}{\sqrt[4]{27}} [F(\alpha, \sin 75^\circ) - 2E(\alpha, \sin 75^\circ)] +$

$+ \frac{2}{\sqrt{3}} \frac{\sqrt{1+u+u^2}}{(1+\sqrt{3}-u)\sqrt{1-u}} \quad [u \neq 1].$ BY (246.06)

12. $\int_{u}^{\infty} \frac{dx}{(x-1)\sqrt{x^3-1}} = \frac{1}{\sqrt[4]{27}} [F(\delta, \sin 15^\circ) - 2E(\delta, \sin 15^\circ)] +$

$+ \frac{2}{\sqrt{3}} \frac{\sqrt{1+u+u^2}}{(u-1+\sqrt{3})\sqrt{u-1}} \quad [u \neq 1].$ BY (242.03)

13. $$\int_{-\infty}^{u} \frac{(1-x)\,dx}{(1+\sqrt{3}-x)^2\sqrt{1-x^3}} =$$
$$= \frac{2-\sqrt{3}}{\sqrt[4]{27}}\,[F(\alpha, \sin 75^\circ) - E(\alpha, \sin 75^\circ)].$$ BY (246.07)

14. $$\int_{u}^{1} \frac{(1-x)\,dx}{(1+\sqrt{3}-x)^2\sqrt{1-x^3}} = \frac{2-\sqrt{3}}{\sqrt[4]{27}}\,[F(\beta, \sin 75^\circ) - E(\beta, \sin 75^\circ)].$$ BY (244.04)

15. $$\int_{1}^{u} \frac{(x-1)\,dx}{(1+\sqrt{3}-x)^2\sqrt{x^3-1}} = \frac{2(\sqrt{3}-2)}{\sqrt{3}}\,\frac{\sqrt{u^3-1}}{u^2-2u-2} - \frac{2-\sqrt{3}}{\sqrt[4]{27}}\,E(\gamma, \sin 15^\circ).$$ BY (240.08)

16. $$\int_{u}^{\infty} \frac{(x-1)\,dx}{(1+\sqrt{3}-x)^2\sqrt{x^3-1}} = \frac{2(2-\sqrt{3})}{\sqrt{3}}\,\frac{\sqrt{u^3-1}}{u^2-2u-2} - \frac{2-\sqrt{3}}{\sqrt[4]{27}}\,E(\delta, \sin 15^\circ).$$ BY (242.07)

17. $$\int_{-\infty}^{u} \frac{(1-x)\,dx}{(1-\sqrt{3}-x)^2\sqrt{1-x^3}} = \frac{2+\sqrt{3}}{\sqrt[4]{27}}\left[\frac{2\sqrt[4]{3}\sqrt{1-u^3}}{u^2-2u-2} - E(\alpha, \sin 75^\circ)\right].$$ BY (246.08)

18. $$\int_{1}^{u} \frac{(x-1)\,dx}{(1-\sqrt{3}-x)^2\sqrt{x^3-1}} = \frac{2+\sqrt{3}}{\sqrt[4]{27}}\,[F(\gamma, \sin 15^\circ) - E(\gamma, \sin 15^\circ)].$$ BY (240.04)

19. $$\int_{u}^{\infty} \frac{(x-1)\,dx}{(1-\sqrt{3}-x)^2\sqrt{x^3-1}} = \frac{2+\sqrt{3}}{\sqrt[4]{27}}\,[F(\delta, \sin 15^\circ) - E(\delta, \sin 15^\circ)].$$ BY (242.05)

20. $$\int_{-\infty}^{u} \frac{(x^2+x+1)\,dx}{(1+\sqrt{3}-x)^2\sqrt{1-x^3}} = \frac{1}{\sqrt[4]{3}}\,E(\alpha, \sin 75^\circ).$$ BY (246.01)

21. $$\int_{u}^{1} \frac{(x^2+x+1)\,dx}{(x-1+\sqrt{3})^2\sqrt{1-x^3}} = \frac{1}{\sqrt[4]{3}}\,E(\beta, \sin 75^\circ).$$ BY (244.02)

22. $$\int_{1}^{u} \frac{(x^2+x+1)\,dx}{(\sqrt{3}+x-1)^2\sqrt{x^3-1}} = \frac{1}{\sqrt[4]{3}}\,E(\gamma, \sin 15^\circ).$$ BY (240.01)

23. $$\int_{u}^{\infty} \frac{(x^2+x+1)\,dx}{(x-1+\sqrt{3})^2\sqrt{x^3-1}} = \frac{1}{\sqrt[4]{3}}\,E(\delta, \sin 15^\circ).$$ BY (242.01)

24. $$\int_{1}^{u} \frac{(x-1)\,dx}{(x^2+x+1)\sqrt{x^3-1}} = \frac{4}{\sqrt[4]{27}}\,E(\gamma, \sin 15^\circ) -$$
$$- \frac{2+\sqrt{3}}{\sqrt[4]{27}}\,F(\gamma, \sin 15^\circ) - \frac{2-\sqrt{3}}{\sqrt{3}}\,\frac{2(u-1)(\sqrt{3}+1-u)}{(\sqrt{3}-1+u)\sqrt{u^3-1}}$$ BY (240.09)

25. $$\int_{-\infty}^{u} \frac{(1+\sqrt{3}-x)^2\,dx}{[(1+\sqrt{3}-x)^2-4\sqrt{3}p^2(1-x)]\sqrt{1-x^3}} = \frac{1}{\sqrt[4]{3}}\Pi(\alpha,\ p^2,\ \sin 75^\circ).$$

BY (246.02)

26. $$\int_{u}^{1} \frac{(1+\sqrt{3}-x)^2\,dx}{[(1+\sqrt{3}-x)^2-4\sqrt{3}p^2(1-x)]\sqrt{1-x^3}} = \frac{1}{\sqrt[4]{3}}\Pi(\beta,\ p^2,\ \sin 75^\circ).$$

BY (244.03)

27. $$\int_{1}^{u} \frac{(1-\sqrt{3}-x)^2\,dx}{[(1-\sqrt{3}-x)^2-4\sqrt{3}p^2(x-1)]\sqrt{x^3-1}} = \frac{1}{\sqrt[4]{3}}\Pi(\gamma,\ p^2,\ \sin 15^\circ).$$

BY (240.02)

28. $$\int_{u}^{\infty} \frac{(1-\sqrt{3}-x)^2\,dx}{[(1-\sqrt{3}-x)^2-4\sqrt{3}p^2(x-1)]\sqrt{x^3-1}} = \frac{1}{\sqrt[4]{3}}\Pi(\delta,\ p^2,\ \sin 15^\circ).$$

BY (242.02)

In **3.141** and **3.142** we set: $\alpha=\arcsin\sqrt{\frac{a-c}{a-u}}$, $\beta=\arcsin\sqrt{\frac{c-u}{b-u}}$, $\gamma=\arcsin\sqrt{\frac{u-c}{b-c}}$, $\delta=\arcsin\sqrt{\frac{(a-c)(b-u)}{(b-c)(a-u)}}$, $\varkappa=\arcsin\sqrt{\frac{(a-c)(u-b)}{(a-b)(u-c)}}$, $\lambda=\arcsin\sqrt{\frac{a-u}{a-b}}$, $\mu=\arcsin\sqrt{\frac{u-a}{u-b}}$, $\nu=\arcsin\sqrt{\frac{a-c}{u-c}}$, $p=\sqrt{\frac{a-b}{a-c}}$, $q=\sqrt{\frac{b-c}{a-c}}$.

3.141

1. $$\int_{u}^{c} \sqrt{\frac{a-x}{(b-x)(c-x)}}\,dx = 2\sqrt{a-c}\,[F(\beta,\ p)-E(\beta,\ p)] + 2\sqrt{\frac{(a-u)(c-u)}{b-u}} \qquad [a>b>c>u].$$

BY (232.06)

2. $$\int_{c}^{u} \sqrt{\frac{a-x}{(b-x)(x-c)}}\,dx = 2\sqrt{a-c}\,E(\gamma,\ q) \qquad [a>b\geqslant u>c].$$

BY (233.01)

3. $$\int_{u}^{b} \sqrt{\frac{a-x}{(b-x)(x-c)}}\,dx = 2\sqrt{a-c}\,E(\delta,\ q) - 2\sqrt{\frac{(b-u)(u-c)}{a-u}} \qquad [a>b>u\geqslant c].$$

BY (234.06)

4. $$\int_{b}^{u} \sqrt{\frac{a-x}{(x-b)(x-c)}}\,dx = 2\sqrt{a-c}\,[F(\varkappa,\ p)-E(\varkappa,\ p)] + 2\sqrt{\frac{(a-u)(u-b)}{u-c}} \qquad [a\geqslant u>b>c].$$

BY (235.07)

5. $$\int_u^a \sqrt{\frac{a-x}{(x-b)(x-c)}}\,dx = 2\sqrt{a-c}\,[F(\lambda, p) - E(\lambda, p)]$$

$$[a > u \geqslant b > c]$$ BY (236.04)

6. $$\int_a^u \sqrt{\frac{x-a}{(x-b)(x-c)}}\,dx = -2\sqrt{a-c}\,E(\mu, q) + 2\sqrt{\frac{(u-a)(u-c)}{u-b}}$$

$$[u > a > b > c].$$ BY (237.03)

7. $$\int_u^c \sqrt{\frac{b-x}{(a-x)(c-x)}}\,dx = \frac{2(b-c)}{\sqrt{a-c}}F(\beta, p) - 2\sqrt{a-c}\,E(\beta, p) +$$

$$+2\sqrt{\frac{(a-u)(c-u)}{b-u}} \qquad [a > b > c > u].$$ BY (232.07)

8. $$\int_c^u \sqrt{\frac{b-x}{(a-x)(x-c)}}\,dx = 2\sqrt{a-c}\,E(\gamma, q) - \frac{2(a-b)}{\sqrt{a-c}}F(\gamma, q)$$

$$[a > b \geqslant u > c].$$ BY (233.04)

9. $$\int_u^b \sqrt{\frac{b-x}{(a-x)(x-c)}}\,dx = 2\sqrt{a-c}\,E(\delta, q) - \frac{2(a-b)}{\sqrt{a-c}}F(\delta, q) -$$

$$-2\sqrt{\frac{(b-u)(u-c)}{a-u}} \qquad [a > b > u \geqslant c].$$ BY (234.07)

10. $$\int_b^u \sqrt{\frac{x-b}{(a-x)(x-c)}}\,dx = 2\sqrt{a-c}\,E(\varkappa, p) - \frac{2(b-c)}{\sqrt{a-c}}F(\varkappa, p) -$$

$$-2\sqrt{\frac{(a-u)(u-b)}{u-c}} \qquad [a \geqslant u > b > c].$$ BY (235.06)

11. $$\int_u^a \sqrt{\frac{x-b}{(a-x)(x-c)}}\,dx = 2\sqrt{a-c}\,E(\lambda, p) - \frac{2(b-c)}{\sqrt{a-c}}F(\lambda, p)$$

$$[a > u \geqslant b > c].$$ BY (236.03)

12. $$\int_a^u \sqrt{\frac{x-b}{(x-a)(x-c)}}\,dx = \frac{2(a-b)}{\sqrt{a-c}}F(\mu, q) - 2\sqrt{a-c}\,E(\mu, q) +$$

$$+2\sqrt{\frac{(u-a)(u-c)}{u-b}} \qquad [u > a > b > c].$$ BY (237.04)

13. $$\int_u^c \sqrt{\frac{c-x}{(a-x)(b-x)}}\,dx = -2\sqrt{a-c}\,E(\beta, p) +$$

$$+2\sqrt{\frac{(a-u)(c-u)}{b-u}} \qquad [a > b > c > u].$$ BY (232.08)

14. $$\int_c^u \sqrt{\frac{x-c}{(a-x)(b-x)}}\,dx = 2\sqrt{a-c}\,[F(\gamma, q) - E(\gamma, q)] \qquad [a > b \geqslant u > c].$$

BY (233.03)

15. $\int_u^b \sqrt{\frac{x-c}{(a-x)(b-x)}}\,dx = 2\sqrt{a-c}\,[F(\delta, q) - E(\delta, q)] +$

$+2\sqrt{\frac{(b-u)(u-c)}{a-u}} \qquad [a > b > u \geqslant c].$ BY (234.08)

16. $\int_b^u \sqrt{\frac{x-c}{(a-x)(x-b)}}\,dx = 2\sqrt{a-c}\,E(\varkappa, p) - 2\sqrt{\frac{(a-u)(u-b)}{u-c}}$

$[a \geqslant u > b > c].$ BY (235.07)

17. $\int_u^a \sqrt{\frac{x-c}{(a-x)(x-b)}}\,dx = 2\sqrt{a-c}\,E(\lambda, p) \qquad [a > u \geqslant b > c].$

BY (236.01)

18. $\int_a^u \sqrt{\frac{x-c}{(x-a)(x-b)}}\,dx = 2\sqrt{a-c}\,[F(\mu, q) - E(\mu, q)] +$

$+2\sqrt{\frac{(u-a)(u-c)}{u-b}} \qquad [u > a > b > c].$ BY (237.05)

19. $\int_u^c \sqrt{\frac{(b-x)(c-x)}{a-x}}\,dx = \frac{2}{3}\sqrt{a-c}\,[(2a-b-c)\,E(\beta, p) -$

$-(b-c)\,F(\beta, p)] + \frac{2}{3}(2b-2a+c-u)\sqrt{\frac{(a-u)(c-u)}{b-u}}$

$[a > b > c > u].$ BY (232.11)

20. $\int_c^u \sqrt{\frac{(x-c)(b-x)}{a-x}}\,dx = \frac{2}{3}\sqrt{a-c}\,[(2a-b-c)\,E(\gamma, q) -$

$-2(a-b)\,F(\gamma, q)] - \frac{2}{3}\sqrt{(a-u)(b-u)(u-c)}$

$[a > b \geqslant u > c].$ BY (233.06)

21. $\int_u^b \sqrt{\frac{(x-c)(b-x)}{a-x}}\,dx = \frac{2}{3}\sqrt{a-c}\,[2(b-a)\,F(\delta, q) +$

$+(2a-b-c)\,E(\delta, q)] + \frac{2}{3}(2c-b-u)\sqrt{\frac{(b-u)(u-c)}{a-u}}$

$[a > b > u \geqslant c].$ BY (234.11)

22. $\int_b^u \sqrt{\frac{(x-b)(x-c)}{a-x}}\,dx = \frac{2}{3}\sqrt{a-c}\,[(2a-b-c)\,E(\varkappa, p) -$

$-(b-c)\,F(\varkappa, p)] + \frac{2}{3}(b+2c-2a-u)\sqrt{\frac{(a-u)(u-b)}{u-c}}$

$[a \geqslant u > b > c].$ BY (235.10)

23. $\int_u^a \sqrt{\frac{(x-b)(x-c)}{a-x}}\,dx = \frac{2}{3}\sqrt{a-c}\,[(2a-b-c)\,E(\lambda, p) -$

$-(b-c)\,F(\lambda, p)] + \frac{2}{3}\sqrt{(a-u)(u-b)(u-c)} \qquad [a > u \geqslant b > c].$

BY (236.07)

24. $$\int_a^u \sqrt{\frac{(x-b)(x-c)}{x-a}}\,dx = \frac{2}{3}\sqrt{a-c}\,[2(a-b)F(\mu, q) + (b+c-2a)E(\mu, q)] + \frac{2}{3}(u+2a-2b-c)\sqrt{\frac{(u-a)(u-b)}{u-c}}$$

$[u > a > b > c]$. BY (237.08)

25. $$\int_u^c \sqrt{\frac{(a-x)(c-x)}{b-x}}\,dx = \frac{2}{3}\sqrt{a-c}\,[(2b-a-c)E(\beta, p) - (b-c)F(\beta, p)] + \frac{2}{3}(a+c-b-u)\sqrt{\frac{(a-u)(c-u)}{b-u}}$$

$[a > b > c > u]$. BY (232.10)

26. $$\int_c^u \sqrt{\frac{(a-x)(x-c)}{b-x}}\,dx = \frac{2}{3}\sqrt{a-c}\,[(2b-a-c)E(\gamma, q) + (a-b)F(\gamma, q)] - \frac{2}{3}\sqrt{(a-u)(b-u)(u-c)}$$

$[a > b \geqslant u > c]$. BY (233.05)

27. $$\int_u^b \sqrt{\frac{(a-x)(x-c)}{b-x}}\,dx = \frac{2}{3}\sqrt{a-c}\,[(a-b)F(\delta, q) + (2b-a-c)E(\delta, q)] + \frac{2}{3}(2a+c-2b-u)\sqrt{\frac{(b-u)(u-c)}{a-u}}$$

$[a > b > u \geqslant c]$. BY (234.10)

28. $$\int_b^u \sqrt{\frac{(a-x)(x-c)}{x-b}}\,dx = \frac{2}{3}\sqrt{a-c}\,[(b-c)F(\varkappa, p) + (a+c-2b)E(\varkappa, p)] + \frac{2}{3}(2b-a-2c+u)\sqrt{\frac{(a-u)(u-b)}{u-c}}$$

$[a \geqslant u > b > c]$. BY (235.11)

29. $$\int_u^a \sqrt{\frac{(a-x)(x-c)}{x-b}}\,dx = \frac{2}{3}\sqrt{a-c}\,[(a+c-2b)E(\lambda, p) + (b-c)F(\lambda, p)] - \frac{2}{3}\sqrt{(a-u)(u-b)(u-c)}$$

$[a > u \geqslant b > c]$. BY (236.06)

30. $$\int_a^u \sqrt{\frac{(x-a)(x-c)}{x-b}}\,dx = \frac{2}{3}\frac{\sqrt{(a-c)^3}}{b-c}\,[(a+c-2b)E(\mu, q) - (a-b)F(\mu, q)] + \frac{2}{3}\frac{a-c}{b-c}(u+b-a-c)\sqrt{\frac{(u-a)(u-c)}{u-b}}$$

$[u > a > b > c]$. BY (237.06)

31. $$\int_u^c \sqrt{\frac{(a-x)(b-x)}{c-x}}\,dx = \frac{2}{3}\sqrt{a-c}\,[2(b-c)F(\beta, p) + (2c-a-b)E(\beta, p)] + \frac{2}{3}(a+2b-2c-u)\sqrt{\frac{(a-u)(c-u)}{b-u}}$$
$[a > b > c > u]$. BY (232.09)

32. $$\int_c^u \sqrt{\frac{(a-x)(b-x)}{x-c}}\,dx = \frac{2}{3}\sqrt{a-c}\,[(a+b-2c)E(\gamma, q) - (a-b)F(\gamma, q)] + \frac{2}{3}\sqrt{(a-u)(b-u)(u-c)}$$
$[a > b \geqslant u > c]$. BY (233.07)

33. $$\int_u^b \sqrt{\frac{(a-x)(b-x)}{x-c}}\,dx = \frac{2}{3}\sqrt{a-c}\,[(a+b-2c)E(\delta, q) - (a-b)F(\delta, q)] + \frac{2}{3}(2c-2a-b+u)\sqrt{\frac{(b-u)(u-c)}{a-u}}$$
$[a > b > u \geqslant c]$. BY (234.09)

34. $$\int_b^u \sqrt{\frac{(a-x)(x-b)}{x-c}}\,dx = \frac{2}{3}\sqrt{a-c}\,[(a+b-2c)E(\varkappa, p) - 2(b-c)F(\varkappa, p)] + \frac{2}{3}(u+c-a-b)\sqrt{\frac{(a-u)(u-b)}{u-c}}$$
$[a \geqslant u > b > c]$. BY (235.09)

35. $$\int_u^a \sqrt{\frac{(a-x)(x-b)}{x-c}}\,dx = \frac{2}{3}\sqrt{a-c}\,[(a+b-2c)E(\lambda, p) - 2(b-c)F(\lambda, p)] - \frac{2}{3}\sqrt{(a-u)(u-b)(u-c)}$$
$[a > u \geqslant b > c]$. BY (236.05)

36. $$\int_a^u \sqrt{\frac{(x-a)(x-b)}{x-c}}\,dx = \frac{2}{3}\sqrt{a-c}\,[(a+b-2c)E(\mu, q) - (a-b)F(\mu, q)] + \frac{2}{3}(u+2c-a-2b)\sqrt{\frac{(u-a)(u-c)}{u-b}}$$
$[u > a > b > c]$. BY (237.07)

3.142

1. $$\int_{-\infty}^u \sqrt{\frac{a-x}{(b-x)(c-x)^3}}\,dx = \frac{2}{\sqrt{a-c}}F(\alpha, p) - \frac{2\sqrt{a-c}}{b-c}E(\alpha, p) + \frac{2(a-c)}{b-c}\sqrt{\frac{b-u}{(a-u)(c-u)}}$$
$[a > b > c > u]$. BY (231.05)

2. $$\int_u^b \sqrt{\frac{a-x}{(b-x)(x-c)^3}}\,dx = 2\frac{a-b}{(b-c)\sqrt{a-c}}F(\delta, q) - \frac{2\sqrt{a-c}}{b-c}E(\delta, q) + 2\frac{a-c}{b-c}\sqrt{\frac{b-u}{(a-u)(u-c)}}$$
$[a > b > u > c]$. BY (234.13)

3. $$\int_b^u \sqrt{\frac{a-x}{(x-b)(x-c)^3}}\,dx = \frac{2\sqrt{a-c}}{b-c}E(\varkappa, p) - \frac{2}{\sqrt{a-c}}F(\varkappa, p)$$
$$[a \geqslant u > b > c].$$ BY (235.12)

4. $$\int_u^a \sqrt{\frac{a-x}{(x-b)(x-c)^3}}\,dx = \frac{2\sqrt{a-c}}{b-c}E(\lambda, p) - \frac{2}{\sqrt{a-c}}F(\lambda, p) - \frac{2}{b-c}\sqrt{\frac{(a-u)(u-b)}{u-c}}$$
$$[a > u \geqslant b > c].$$ BY (236.12)

5. $$\int_a^u \sqrt{\frac{x-a}{(x-b)(x-c)^3}}\,dx = \frac{2\sqrt{a-c}}{b-c}E(\mu, q) - \frac{2(a-b)}{(b-c)\sqrt{a-c}}F(\mu, q) - 2\sqrt{\frac{u-a}{(u-b)(u-c)}}$$
$$[u > a > b > c].$$ BY (237.10)

6. $$\int_u^\infty \sqrt{\frac{x-a}{(x-b)(x-c)^3}}\,dx = \frac{2\sqrt{a-c}}{b-c}E(\nu, q) - \frac{2(a-b)}{(b-c)\sqrt{a-c}}F(\nu, q)$$
$$[u \geqslant a > b > c].$$ BY (238.09)

7. $$\int_{-\infty}^u \sqrt{\frac{a-x}{(b-x)^3(c-x)}}\,dx = \frac{2\sqrt{a-c}}{b-c}E(\alpha, p) - 2\frac{a-b}{b-c}\sqrt{\frac{c-u}{(a-u)(b-u)}}$$
$$[a > b > c \geqslant u].$$ BY (231.03)

8. $$\int_u^c \sqrt{\frac{a-x}{(b-x)^3(c-x)}}\,dx = \frac{2\sqrt{a-c}}{b-c}E(\beta, p) \qquad [a > b > c > u].$$
BY (232.01)

9. $$\int_c^u \sqrt{\frac{a-x}{(b-x)^3(x-c)}}\,dx = \frac{2\sqrt{a-c}}{b-c}[F(\gamma, q) - E(\gamma, q)] + \frac{2}{b-c}\sqrt{\frac{(a-u)(u-c)}{b-u}}$$
$$[a > b > u > c].$$ BY (233.15)

10. $$\int_u^a \sqrt{\frac{a-x}{(x-b)^3(x-c)}}\,dx = \frac{2\sqrt{a-c}}{c-b}E(\lambda, p) + \frac{2}{b-c}\sqrt{\frac{(a-u)(u-c)}{u-b}}$$
$$[a > u > b > c].$$ BY (236.11)

11. $$\int_a^u \sqrt{\frac{x-a}{(x-b)^3(x-c)}}\,dx = \frac{2\sqrt{a-c}}{b-c}[F(\mu, q) - E(\mu, q)] \qquad [u > a > b > c].$$
BY (237.09)

12. $$\int_u^\infty \sqrt{\frac{x-a}{(x-b)^3(x-c)}}\,dx = \frac{2\sqrt{a-c}}{b-c}[F(\nu, q) - E(\nu, q)] + 2\sqrt{\frac{u-a}{(u-b)(u-c)}}$$
$$[u \geqslant a > b > c].$$ BY (238.10)

13. $$\int_{-\infty}^{u} \sqrt{\frac{b-x}{(a-x)^3(c-x)}}\,dx = \frac{2}{\sqrt{a-c}} E(\alpha, p) \qquad [a > b > c \geqslant u].$$

BY (231.01)

14. $$\int_{u}^{c} \sqrt{\frac{b-x}{(a-x)^3(c-x)}}\,dx = \frac{2}{\sqrt{a-c}} E(\beta, p) - \frac{2(a-b)}{a-c}\sqrt{\frac{c-u}{(a-u)(b-u)}} \qquad [a > b > c > u].$$

BY (232.05)

15. $$\int_{c}^{u} \sqrt{\frac{b-x}{(a-x)^3(x-c)}}\,dx = \frac{2}{\sqrt{a-c}} [F(\gamma, q) - E(\gamma, q)] + \frac{2}{a-c}\sqrt{\frac{(b-u)(u-c)}{a-u}} \qquad [a > b \geqslant u > c].$$

BY (233.13)

16. $$\int_{u}^{b} \sqrt{\frac{b-x}{(a-x)^3(x-c)}}\,dx = \frac{2}{\sqrt{a-c}} [F(\delta, q) - E(\delta, q)] \qquad [a > b > u \geqslant c].$$

BY (234.15)

17. $$\int_{b}^{u} \sqrt{\frac{x-b}{(a-x)^3(x-c)}}\,dx = -\frac{2}{\sqrt{a-c}} E(\varkappa, p) + 2\sqrt{\frac{u-b}{(a-u)(u-c)}} \qquad [a > u > b > c].$$

BY (235.08)

18. $$\int_{u}^{\infty} \sqrt{\frac{x-b}{(x-a)^3(x-c)}}\,dx = \frac{2}{\sqrt{a-c}} [F(\nu, q) - E(\nu, q)] + 2\sqrt{\frac{u-b}{(u-a)(u-c)}} \qquad [u > a > b > c].$$

BY (238.07)

19. $$\int_{-\infty}^{u} \sqrt{\frac{b-x}{(a-x)(c-x)^3}}\,dx = \frac{2}{\sqrt{a-c}} [F(\alpha, p) - E(\alpha, p)] + 2\sqrt{\frac{b-u}{(a-u)(c-u)}} \qquad [a > b > c > u].$$

BY (231.04)

20. $$\int_{u}^{b} \sqrt{\frac{b-x}{(a-x)(x-c)^3}}\,dx = -\frac{2}{\sqrt{a-c}} E(\delta, q) + 2\sqrt{\frac{b-u}{(a-u)(u-c)}} \qquad [a > b > u > c].$$

BY (234.14)

21. $$\int_{b}^{u} \sqrt{\frac{x-b}{(a-x)(x-c)^3}}\,dx = \frac{2}{\sqrt{a-c}} [F(\varkappa, p) - E(\varkappa, p)] \qquad [a \geqslant u > b > c].$$

BY (235.03)

22. $$\int_{u}^{a} \sqrt{\frac{x-b}{(a-x)(x-c)^3}}\,dx = \frac{2}{\sqrt{a-c}} [F(\lambda, p) - E(\lambda, p)] + \frac{2}{a-c}\sqrt{\frac{(a-u)(u-b)}{u-c}} \qquad [a > u \geqslant b > c].$$

BY (236.14)

23. $$\int_a^u \sqrt{\frac{x-b}{(x-a)(x-c)^3}}\,dx = \frac{2}{\sqrt{a-c}} E(\mu, q) - 2\,\frac{b-c}{a-c}\sqrt{\frac{u-a}{(u-b)(u-c)}} \qquad [u > a > b > c].$$ BY (237.11)

24. $$\int_u^\infty \sqrt{\frac{x-b}{(x-a)(x-c)^3}}\,dx = \frac{2}{\sqrt{a-c}} E(\nu, q) \qquad [u \geqslant a > b > c].$$ BY (238.01)

25. $$\int_{-\infty}^u \sqrt{\frac{c-x}{(a-x)^3(b-x)}}\,dx = \frac{2\sqrt{a-c}}{a-b} E(\alpha, p) - \frac{2(b-c)}{(a-b)\sqrt{a-c}} F(\alpha, p) \qquad [a > b > c \geqslant u].$$ BY (231.07)

26. $$\int_u^c \sqrt{\frac{c-x}{(a-x)^3(b-x)}}\,dx = \frac{2\sqrt{a-c}}{a-b} E(\beta, p) - \frac{2(b-c)}{(a-b)\sqrt{a-c}} F(\beta, p) - 2\sqrt{\frac{c-u}{(a-u)(b-u)}} \qquad [a > b > c > u].$$ BY (232.03)

27. $$\int_c^u \sqrt{\frac{x-c}{(a-x)^3(b-x)}}\,dx = \frac{2\sqrt{a-c}}{a-b} E(\gamma, q) - \frac{2}{\sqrt{a-c}} F(\gamma, q) - \frac{2}{a-b}\sqrt{\frac{(b-u)(u-c)}{a-u}} \qquad [a > b \geqslant u > c].$$ BY (233.14)

28. $$\int_u^b \sqrt{\frac{x-c}{(a-x)^3(b-x)}}\,dx = \frac{2\sqrt{a-c}}{a-b} E(\delta, q) - \frac{2}{\sqrt{a-c}} F(\delta, q) \qquad [a > b > u \geqslant c].$$ BY (234.20)

29. $$\int_b^u \sqrt{\frac{x-c}{(a-x)^3(x-b)}}\,dx = \frac{2(b-c)}{(a-b)\sqrt{a-c}} F(\varkappa, p) - \frac{2\sqrt{a-c}}{a-b} E(\varkappa, p) + 2\,\frac{a-c}{a-b}\sqrt{\frac{u-b}{(a-u)(u-c)}} \qquad [a > u > b > c].$$ BY (235.13)

30. $$\int_u^\infty \sqrt{\frac{x-c}{(x-a)^3(x-b)}}\,dx = \frac{2}{\sqrt{a-c}} F(\nu, q) - \frac{2\sqrt{a-c}}{a-b} E(\nu, q) + \frac{2(a-c)}{a-b}\sqrt{\frac{u-b}{(u-a)(u-c)}} \qquad [u > a > b > c].$$ BY (238.08)

31. $$\int_{-\infty}^u \sqrt{\frac{c-x}{(a-x)(b-x)^3}}\,dx = \frac{2\sqrt{a-c}}{a-b}[F(\alpha, p) - E(\alpha, p)] + 2\sqrt{\frac{c-u}{(a-u)(b-u)}} \qquad [a > b > c \geqslant u].$$ BY (231.06)

32. $$\int_u^c \sqrt{\frac{c-x}{(a-x)(b-x)^3}}\,dx = \frac{2\sqrt{a-c}}{a-b}[F(\beta, p) - E(\beta, p)] \qquad [a > b > c > u].$$ BY (232.04)

33. $$\int_c^u \sqrt{\frac{x-c}{(a-x)(b-x)^3}}\,dx = -\frac{2\sqrt{a-c}}{a-b}E(\gamma, q) + \frac{2}{a-b}\sqrt{\frac{(a-u)(u-c)}{b-u}} \quad [a > b > u > c].$$ BY (233.16)

34. $$\int_u^a \sqrt{\frac{x-c}{(a-x)(x-b)^3}}\,dx = \frac{2\sqrt{a-c}}{a-b}[F(\lambda, p) - E(\lambda, p)] + \frac{2}{a-b}\sqrt{\frac{(a-u)(u-c)}{u-b}} \quad [a > u > b > c].$$ BY (236.13)

35. $$\int_a^u \sqrt{\frac{x-c}{(x-a)(x-b)^3}}\,dx = \frac{2\sqrt{a-c}}{a-b}E(\mu, q) \quad [u > a > b > c].$$ BY (237.01)

36. $$\int_u^\infty \sqrt{\frac{x-c}{(x-a)(x-b)^3}}\,dx = \frac{2\sqrt{a-c}}{a-b}E(\nu, q) - 2\frac{b-c}{a-b}\sqrt{\frac{u-a}{(u-b)(u-c)}} \quad [u \geqslant a > b > c].$$ BY (238.11)

3.143

1. $$\int_u^1 \frac{dx}{\sqrt{1+x^4}} = \frac{1}{2}F\left(\arccos\frac{u\sqrt{2}}{\sqrt{1+u^4}},\ \sin 80°7'15''\right)^*$$ ZH 66 (286)

2. $$\int_u^\infty \frac{dx}{\sqrt{1+x^4}} = \frac{1}{2}F\left(\arccos\frac{u^2-1}{u^2+1},\ \frac{\sqrt{2}}{2}\right).$$ ZH 66 (287)

3.144 Notation: $\alpha = \arcsin\frac{1}{\sqrt{u^2-u+1}}$.

1. $$\int_u^\infty \frac{dx}{\sqrt{x(x-1)(x^2-x+1)}} = F\left(\alpha, \frac{\sqrt{3}}{2}\right) \quad [u \geqslant 1].$$ BY (261.50)

2. $$\int_u^\infty \frac{dx}{\sqrt{x^3(x-1)^3(x^2-x+1)}} = \frac{2(2u-1)}{\sqrt{u(u-1)(u^2-u+1)}} - 4E\left(\alpha, \frac{\sqrt{3}}{2}\right) \quad [u > 1].$$ BY (261.54)

3. $$\int_u^\infty \frac{(2x-1)^2\,dx}{\sqrt{x^3(x-1)^3(x^2-x+1)}} = 4\left[F\left(\alpha, \frac{\sqrt{3}}{2}\right) - E\left(\alpha, \frac{\sqrt{3}}{2}\right) + \frac{2u-1}{2\sqrt{u(u-1)(u^2-u+1)}}\right] \quad [u > 1].$$ BY (261.56)

4. $$\int_u^\infty \frac{dx}{\sqrt{x(x-1)(x^2-x+1)^3}} = \frac{4}{3}\left[F\left(\alpha, \frac{\sqrt{3}}{2}\right) - E\left(\alpha, \frac{\sqrt{3}}{2}\right)\right] \quad [u \geqslant 1].$$ BY (261.52)

* $\sin 80°7'15'' = 2\sqrt[4]{2}(\sqrt{2}-1) = 0.985171\ldots$

5. $$\int_u^\infty \frac{(2x-1)^2\,dx}{\sqrt{x(x-1)(x^2-x+1)^3}} = 4E\left(\alpha, \frac{\sqrt{3}}{2}\right) \qquad [u>1].$$ BY (261.51)

6. $$\int_u^\infty \sqrt{\frac{x(x-1)}{(x^2-x+1)^3}}\,dx = \frac{4}{3}E\left(\alpha, \frac{\sqrt{3}}{2}\right) - \frac{1}{3}F\left(\alpha, \frac{\sqrt{3}}{2}\right) \qquad [u>1].$$ BY (261.53)

7. $$\int_u^\infty \frac{dx}{(2x-1)^2}\sqrt{\frac{x(x-1)}{x^2-x+1}} = \frac{1}{3}\left[F\left(\alpha, \frac{\sqrt{3}}{2}\right) - E\left(\alpha, \frac{\sqrt{3}}{2}\right)\right] + \frac{1}{2(2u-1)}\sqrt{\frac{u(u-1)}{u^2-u+1}} \qquad [u>1].$$ BY (261.57)

8. $$\int_u^\infty \frac{dx}{(2x-1)^2}\sqrt{\frac{x^2-x+1}{x(x-1)}} = E\left(\alpha, \frac{\sqrt{3}}{2}\right) - \frac{3}{2(2u-1)}\sqrt{\frac{u(u-1)}{u^2-u+1}} \qquad [u>1].$$ BY (261.58)

9. $$\int_u^\infty \frac{dx}{(2x-1)^2\sqrt{x(x-1)(x^2-x+1)}} = \frac{4}{3}E\left(\alpha, \frac{\sqrt{3}}{2}\right) - \frac{1}{3}F\left(\alpha, \frac{\sqrt{3}}{2}\right) - \frac{2}{2u-1}\sqrt{\frac{u(u-1)}{u^2-u+1}} \qquad [u>1].$$ BY (261.55)

10. $$\int_u^\infty \frac{dx}{\sqrt{x^5(x-1)^5(x^2-x+1)}} = \frac{40}{3}E\left(\alpha, \frac{\sqrt{3}}{2}\right) - \frac{4}{3}F\left(\alpha, \frac{\sqrt{3}}{2}\right) - \frac{2(2u-1)(9u^2-9u-1)}{3\sqrt{u^3(u-1)^3(u^2-u+1)}} \qquad [u>1].$$ BY (261.54)

11. $$\int_u^\infty \frac{dx}{\sqrt{x(x-1)(x^2-x+1)^5}} = \frac{44}{27}F\left(\alpha, \frac{\sqrt{3}}{2}\right) - \frac{56}{27}E\left(\alpha, \frac{\sqrt{3}}{2}\right) + \frac{2(2u-1)\sqrt{u(u-1)}}{9\sqrt{(u^2-u+1)^3}} \qquad [u>1].$$ BY (261.52)

12. $$\int_u^\infty \frac{dx}{(2x-1)^4\sqrt{x(x-1)(x^2-x+1)}} = \frac{16}{27}E\left(\alpha, \frac{\sqrt{3}}{2}\right) - \frac{1}{27}F\left(\alpha, \frac{\sqrt{3}}{2}\right) - \frac{8(5u^2-5u+2)}{9(2u-1)^3}\sqrt{\frac{u(u-1)}{u^2-u+1}} \qquad [u>1].$$ BY (261.55)

3.145

1. $$\int_\alpha^u \frac{dx}{\sqrt{(x-\alpha)(x-\beta)[(x-m)^2+n^2]}} = \frac{1}{\sqrt{pq}}F\left(2\operatorname{arctg}\sqrt{\frac{q(u-\alpha)}{p(u-\beta)}}, \frac{1}{2}\sqrt{\frac{(p+q)^2+(\alpha-\beta)^2}{pq}}\right) \qquad [\beta<\alpha<u].$$

2. $$\int_\beta^u \frac{dx}{\sqrt{(\alpha-x)(x-\beta)[(x-m)^2+n^2]}} = \frac{1}{\sqrt{pq}}F\left(2\operatorname{arcctg}\sqrt{\frac{q(\alpha-u)}{p(u-\beta)}}, \frac{1}{2}\sqrt{\frac{-(p-q)^2+(\alpha-\beta)^2}{pq}}\right) \qquad [\beta<u<\alpha].$$

3. $\int_u^\beta \frac{dx}{\sqrt{(x-\alpha)(x-\beta)[(x-m)^2+n^2]}} =$

$$= \frac{1}{\sqrt{pq}} F\left(2 \operatorname{arctg} \sqrt{\frac{q(\beta-u)}{p(\alpha-u)}}, \frac{1}{2}\sqrt{\frac{(p+q)^2+(\alpha-\beta)^2}{pq}}\right) \quad [u<\beta<\alpha],$$

where $(m-\alpha)^2+n^2=p^2$, $(m-\beta)^2+n^2=q^2$ *).

4. Set

$$(m_1-m)^2+(n_1+n)^2=p^2, \ (m_1-m)^2+(n_1-n)^2=p_1^2,$$
$$\operatorname{ctg}\alpha = \sqrt{\frac{(p+p_1)^2-4n^2}{4n^2-(p-p_1)^2}};$$

then

$$\int_{m-n \operatorname{tg}\alpha}^{u} \frac{dx}{\sqrt{[(x-m)^2+n^2][(x-m_1)^2+n_1^2]}} =$$
$$= \frac{2}{p+p_1} F\left(\alpha + \operatorname{arctg}\frac{u-m}{n}, \frac{2\sqrt{pp_1}}{p+p_1}\right) \quad [m-n\operatorname{tg}\alpha < u < m+n\operatorname{ctg}\alpha].$$

3.146

1. $\int_0^1 \frac{1}{1+x^4}\frac{dx}{\sqrt{1-x^4}} = \frac{\pi}{8}+\frac{1}{4}\sqrt{2}\boldsymbol{K}\left(\frac{\sqrt{2}}{2}\right).$ BI ((13))(6)

2. $\int_0^1 \frac{x^2}{1+x^4}\frac{dx}{\sqrt{1-x^4}} = \frac{\pi}{8}.$ BI ((13))(7)

3. $\int_0^1 \frac{x^4}{1+x^4}\frac{dx}{\sqrt{1-x^4}} = -\frac{\pi}{8}+\frac{1}{4}\sqrt{2}\boldsymbol{K}\left(\frac{\sqrt{2}}{2}\right).$ BI ((13))(8)

In **3.147 — 3.151** we set: $\alpha = \arcsin\sqrt{\frac{(a-c)(d-u)}{(a-d)(c-u)}}$,

$$\beta = \arcsin\sqrt{\frac{(a-c)(u-d)}{(c-d)(a-u)}}, \quad \gamma = \arcsin\sqrt{\frac{(b-d)(c-u)}{(c-d)(b-u)}},$$
$$\delta = \arcsin\sqrt{\frac{(b-d)(u-c)}{(b-c)(u-d)}}, \quad \varkappa = \arcsin\sqrt{\frac{(a-c)(b-u)}{(b-c)(a-u)}},$$
$$\lambda = \arcsin\sqrt{\frac{(a-c)(u-b)}{(a-b)(u-c)}}, \quad \mu = \arcsin\sqrt{\frac{(b-d)(a-u)}{(a-b)(u-d)}},$$
$$\nu = \arcsin\sqrt{\frac{(b-d)(u-a)}{(a-d)(u-b)}}, \quad q = \sqrt{\frac{(b-c)(a-d)}{(a-c)(b-d)}}, \quad r = \sqrt{\frac{(a-b)(c-d)}{(a-c)(b-d)}}.$$

3.147

1. $\int_u^d \frac{dx}{\sqrt{(a-x)(b-x)(c-x)(d-x)}} = \frac{2}{\sqrt{(a-c)(b-d)}} F(\alpha, q)$

$$[a>b>c>d>u].$$ BY (251.00)

---

* Formulas 3.145 are not valid for $\alpha+\beta=2m$. In this case, we make the substitution $x-m=z$, which leads to one of the formulas 3.152.

2. $$\int_d^u \frac{dx}{\sqrt{(a-x)(b-x)(c-x)(x-d)}} = \frac{2}{\sqrt{(a-c)(b-d)}} F(\beta, r)$$
$[a > b > c \geqslant u > d]$. BY (254.00)

3. $$\int_u^c \frac{dx}{\sqrt{(a-x)(b-x)(c-x)(x-d)}} = \frac{2}{\sqrt{(a-c)(b-d)}} F(\gamma, r)$$
$[a > b > c > u \geqslant d]$. BY (253.00)

4. $$\int_c^u \frac{dx}{\sqrt{(a-x)(b-x)(x-c)(x-d)}} = \frac{2}{\sqrt{(a-c)(b-d)}} F(\delta, q)$$
$[a > b \geqslant u > c > d]$. BY (254.00)

5. $$\int_u^b \frac{dx}{\sqrt{(a-x)(b-x)(x-c)(x-d)}} = \frac{2}{\sqrt{(a-c)(b-d)}} F(\varkappa, q)$$
$[a > b > u \geqslant c > d]$. BY (255.00)

6. $$\int_b^u \frac{dx}{\sqrt{(a-x)(x-b)(x-c)(x-d)}} = \frac{2}{\sqrt{(a-c)(b-d)}} F(\lambda, r)$$
$[a \geqslant u > b > c > d]$. BY (256.00)

7. $$\int_u^a \frac{dx}{\sqrt{(a-x)(x-b)(x-c)(x-d)}} = \frac{2}{\sqrt{(a-c)(b-d)}} F(\mu, r)$$
$[a > u \geqslant b > c > d]$. BY (257.00)

8. $$\int_a^u \frac{dx}{\sqrt{(x-a)(x-b)(x-c)(x-d)}} = \frac{2}{\sqrt{(a-c)(b-d)}} F(\nu, q)$$
$[u > a > b > c > d]$. BY (258.00)

3.148

1. $$\int_u^d \frac{x\,dx}{\sqrt{(a-x)(b-x)(c-x)(d-x)}} = \frac{2}{\sqrt{(a-c)(b-d)}} \left\{ c\Pi\left(\alpha, \frac{a-d}{a-c}, q\right) - (c-d) F(\alpha, q) \right\}$$
$[a > b > c > d > u]$. BY (251.03)

2. $$\int_d^u \frac{x\,dx}{\sqrt{(a-x)(b-x)(c-x)(x-d)}} = \frac{2}{\sqrt{(a-c)(b-d)}} \left\{ (d-a)\Pi\left(\beta, \frac{d-c}{a-c}, r\right) + aF(\beta, r) \right\}$$
$[a > b > c \geqslant u > d]$. BY (252.11)

3. $$\int_u^c \frac{x\,dx}{\sqrt{(a-x)(b-x)(c-x)(x-d)}} = \frac{2}{\sqrt{(a-c)(b-d)}} \left\{ (c-b)\Pi\left(\gamma, \frac{c-d}{b-d}, r\right) + bF(\gamma, r) \right\}$$
$[a > b > c > u \geqslant d]$. BY (253.11)

4. $$\int_c^u \frac{x\,dx}{\sqrt{(a-x)(b-x)(x-c)(x-d)}} = \frac{2}{\sqrt{(a-c)(b-d)}}\left\{(c-d)\Pi\left(\delta, \frac{b-c}{b-d}, q\right) + dF(\delta, q)\right\} \quad [a > b \geqslant u > c > d].$$ BY (254.10)

5. $$\int_u^b \frac{x\,dx}{\sqrt{(a-x)(b-x)(x-c)(x-d)}} = \frac{2}{\sqrt{(a-c)(b-d)}}\left\{(b-a)\Pi\left(\varkappa, \frac{b-c}{a-c}, q\right) + aF(\varkappa, q)\right\} \quad [a > b > u \geqslant c > d].$$ BY (255.17)

6. $$\int_b^u \frac{x\,dx}{\sqrt{(a-x)(x-b)(x-c)(x-d)}} = \frac{2}{\sqrt{(a-c)(b-d)}}\left\{(b-c)\Pi\left(\lambda, \frac{a-b}{a-c}, r\right) + cF(\lambda, r)\right\} \quad [a \geqslant u > b > c > d].$$ BY (256.11)

7. $$\int_u^a \frac{x\,dx}{\sqrt{(a-x)(x-b)(x-c)(x-d)}} = \frac{2}{\sqrt{(a-c)(b-d)}}\left\{(a-d)\Pi\left(\mu, \frac{b-a}{b-d}, r\right) + dF(\mu, r)\right\} \quad [a > u \geqslant b > c > d].$$ BY (257.11)

8. $$\int_a^u \frac{x\,dx}{\sqrt{(x-a)(x-b)(x-c)(x-d)}} = \frac{2}{\sqrt{(a-c)(b-d)}}\left\{(a-b)\Pi\left(\nu, \frac{a-d}{b-d}, q\right) + bF(\nu, q)\right\} \quad [u > a > b > c > d].$$ BY (258.11)

3.149

1. $$\int_u^d \frac{dx}{x\sqrt{(a-x)(b-x)(c-x)(d-x)}} = \frac{2}{cd\sqrt{(a-c)(b-d)}}\left\{(c-d)\Pi\left(\alpha, \frac{c(a-d)}{d(a-c)}, q\right) + dF(\alpha, q)\right\} \quad [a > b > c > d > u].$$ BY (251.04)

2. $$\int_d^u \frac{dx}{x\sqrt{(a-x)(b-x)(c-x)(x-d)}} = \frac{2}{ad\sqrt{(a-c)(b-d)}}\left\{(a-d)\Pi\left(\beta, \frac{a(d-c)}{d(a-c)}, r\right) + dF(\beta, r)\right\} \quad [a > b > c \geqslant u > d].$$ BY (252.12)

3. $$\int_u^c \frac{dx}{x\sqrt{(a-x)(b-x)(c-x)(x-d)}} = \frac{2}{bc\sqrt{(a-c)(b-d)}}\left\{(b-c)\Pi\left(\gamma, \frac{b(c-d)}{c(b-d)}, r\right) + cF(\gamma, r)\right\} \quad [a > b > c > u \geqslant d].$$ BY (253.12)

4. $$\int_c^u \frac{dx}{x\sqrt{(a-x)(b-x)(x-c)(x-d)}} = \frac{2}{cd\sqrt{(a-c)(b-d)}}\left\{(d-c)\Pi\left(\delta, \frac{d(b-c)}{c(b-d)}, q\right) + cF(\delta, q)\right\} \quad [a > b \geqslant u > c > d].$$ BY (254.11)

5. $$\int_u^b \frac{dx}{x\sqrt{(a-x)(b-x)(x-c)(x-d)}} = \frac{2}{ab\sqrt{(a-c)(b-d)}} \times$$
$$\times \left\{(a-b)\Pi\left(\varkappa, \frac{a(b-c)}{b(a-c)}, q\right) + bF(\varkappa, q)\right\} \quad [a > b > u \geqslant c > d].$$ BY (255.18)

6. $$\int_b^u \frac{dx}{x\sqrt{(a-x)(x-b)(x-c)(x-d)}} = \frac{2}{bc\sqrt{(a-c)(b-d)}} \times$$
$$\times \left\{(c-b)\Pi\left(\lambda, \frac{c(a-b)}{b(a-c)}, r\right) + bF(\lambda, r)\right\} \quad [a \geqslant u > b > c > d].$$ BY (256.12)

7. $$\int_u^a \frac{dx}{x\sqrt{(a-x)(x-b)(x-c)(x-d)}} = \frac{2}{ad\sqrt{(a-c)(b-d)}} \times$$
$$\times \left\{(d-a)\Pi\left(\mu, \frac{d(b-a)}{a(b-d)}, r\right) + aF(\mu, r)\right\} \quad [a > u \geqslant b > c > d].$$ BY (257.12)

8. $$\int_a^u \frac{dx}{x\sqrt{(x-a)(x-b)(x-c)(x-d)}} =$$
$$= \frac{2}{ab\sqrt{(a-c)(b-d)}} \left\{(b-a)\Pi\left(\nu, \frac{b(a-d)}{a(b-d)}, q\right) + aF(\nu\ q)\right\}$$
$$[u > a > b > c > d].$$ BY (258.12)

3.151

1. $$\int_u^d \frac{dx}{(p-x)\sqrt{(a-x)(b-x)(c-x)(d-x)}} = \frac{2}{(p-c)(p-d)\sqrt{(a-c)(b-d)}} \times$$
$$\times \left[(d-c)\Pi\left(\alpha, \frac{(a-d)(p-c)}{(a-c)(p-d)}, q\right) + (p-d)F(\alpha, q)\right]$$
$$[a > b > c > d > u, \quad p \neq d].$$ BY (251.39)

2. $$\int_d^u \frac{dx}{(p-x)\sqrt{(a-x)(b-x)(c-x)(x-d)}} = \frac{2}{(p-a)(p-d)\sqrt{(a-c)(b-d)}} \times$$
$$\times \left[(d-a)\Pi\left(\beta, \frac{(d-c)(p-a)}{(a-c)(p-d)}, r\right) + (p-d)F(\beta, r)\right]$$
$$[a > b > c \geqslant u > d, \quad p \neq d].$$ BY (252.39)

3. $$\int_u^c \frac{dx}{(p-x)\sqrt{(a-x)(b-x)(c-x)(x-d)}} = \frac{2}{(p-b)(p-c)\sqrt{(a-c)(b-d)}} \times$$
$$\times \left[(c-b)\Pi\left(\gamma, \frac{(c-d)(p-b)}{(b-d)(p-c)}, r\right) + (p-c)F(\gamma, r)\right]$$
$$[a > b > c > u \geqslant d, \quad p \neq c].$$ BY (253.39)

4. $$\int_c^u \frac{dx}{(p-x)\sqrt{(a-x)(b-x)(x-c)(x-d)}} = \frac{2}{(p-c)(p-d)\sqrt{(a-c)(b-d)}} \times$$
$$\times \left[(c-d)\Pi\left(\delta, \frac{(b-c)(p-d)}{(b-d)(p-c)}, q\right) + (p-c)F(\delta, q)\right]$$
$$[a > b \geqslant u > c > d, \quad p \neq c].$$ BY (254.39)

5. $$\int_u^b \frac{dx}{(p-x)\sqrt{(a-x)(b-x)(x-c)(x-d)}} = \frac{2}{(p-a)(p-b)\sqrt{(a-c)(b-d)}} \times$$
$$\times \left[(b-a)\,\Pi\left(\varkappa, \frac{(b-c)(p-a)}{(a-c)(p-b)}, q\right) + (p-b)F(\varkappa, q)\right]$$
$$[a > b > u \geqslant c > d,\ p \neq b].$$ BY (255.38)

6. $$\int_b^u \frac{dx}{(x-p)\sqrt{(a-x)(x-b)(x-c)(x-d)}} = \frac{2}{(b-p)(p-c)\sqrt{(a-c)(b-d)}} \times$$
$$\times \left[(b-c)\,\Pi\left(\lambda, \frac{(a-b)(p-c)}{(a-c)(p-b)}, r\right) + (p-b)F(\lambda, r)\right]$$
$$[a \geqslant u > b > c > d,\ p \neq b].$$ BY (256.39)

7 $$\int_u^a \frac{dx}{(p-x)\sqrt{(a-x)(x-b)(x-c)(x-d)}} = \frac{2}{(p-a)(p-d)\sqrt{(a-c)(b-d)}} \times$$
$$\times \left[(a-d)\,\Pi\left(\mu, \frac{(b-a)(p-d)}{(b-d)(p-a)}, r\right) + (p-a)F(\mu, r)\right]$$
$$[a > u \geqslant b > c > d,\ p \neq a].$$ BY (257.39)

8. $$\int_a^u \frac{dx}{(p-x)\sqrt{(x-a)(x-b)(x-c)(x-d)}} = \frac{2}{(p-a)(p-b)\sqrt{(a-c)(b-d)}} \times$$
$$\times \left[(a-b)\,\Pi\left(\nu, \frac{(a-d)(p-b)}{(b-d)(p-a)}, q\right) + (p-a)F(\nu, q)\right]$$
$$[u > a > b > c > d,\ p \neq a].$$ BY (258.39)

In **3.152 — 3.163** we set: $\alpha = \operatorname{arctg}\frac{u}{b}$, $\beta = \operatorname{arcctg}\frac{u}{a}$,

$$\gamma = \arcsin\frac{u}{b}\sqrt{\frac{a^2+b^2}{a^2+u^2}}, \quad \delta = \arccos\frac{u}{b}, \quad \varepsilon = \arccos\frac{b}{u}, \quad \xi = \arcsin\sqrt{\frac{a^2+b^2}{a^2+u^2}},$$

$$\eta = \arcsin\frac{u}{b}, \quad \zeta = \arcsin\frac{a}{b}\sqrt{\frac{b^2-u^2}{a^2-u^2}}, \quad \varkappa = \arcsin\frac{a}{u}\sqrt{\frac{u^2-b^2}{a^2-b^2}},$$

$$\lambda = \arcsin\sqrt{\frac{a^2-u^2}{a^2-b^2}}, \quad \mu = \arcsin\sqrt{\frac{u^2-a^2}{u^2-b^2}}, \quad \nu = \arcsin\frac{a}{u}, \quad q = \frac{\sqrt{a^2-b^2}}{a},$$

$$r = \frac{b}{\sqrt{a^2+b^2}}, \quad s = \frac{a}{\sqrt{a^2+b^2}}, \quad t = \frac{b}{a}.$$

3.152

1. $$\int_0^u \frac{dx}{\sqrt{(x^2+a^2)(x^2+b^2)}} = \frac{1}{a}F(\alpha, q) \quad [a > b > 0].$$ ZH 62(258), BY (221.00)

2. $$\int_u^\infty \frac{dx}{\sqrt{(x^2+a^2)(x^2+b^2)}} = \frac{1}{a}F(\beta, q) \quad [a > b > 0].$$ ZH 63(259), BY (222.00)

3. $$\int_0^u \frac{dx}{\sqrt{(x^2+a^2)(b^2-x^2)}} = \frac{1}{\sqrt{a^2+b^2}}F(\gamma, r) \quad [b \geqslant u > 0].$$ ZH 63(260)

4. $$\int_u^b \frac{dx}{\sqrt{(x^2+a^2)(b^2-x^2)}} = \frac{1}{\sqrt{a^2+b^2}}F(\delta, r) \quad [b > u \geqslant 0].$$ ZH 63(261), BY (213.00)

5. $$\int_b^u \frac{dx}{\sqrt{(x^2+a^2)(x^2-b^2)}} = \frac{1}{\sqrt{a^2+b^2}} F(\varepsilon, s) \quad [u > b > 0].$$

ZH 63(262), BY(211.00)

6. $$\int_u^\infty \frac{dx}{\sqrt{(x^2+a^2)(x^2-b^2)}} = \frac{1}{\sqrt{a^2+b^2}} F(\xi, s) \quad [u > b > 0].$$

ZH 63(263), BY(212.00)

7. $$\int_0^u \frac{dx}{\sqrt{(a^2-x^2)(b^2-x^2)}} = \frac{1}{a} F(\eta, t) \quad [a > b \geqslant u > 0].$$

ZH 63(264), BY(219.00)

8. $$\int_u^b \frac{dx}{\sqrt{(a^2-x^2)(b^2-x^2)}} = \frac{1}{a} F(\zeta, t) \quad [a > b > u \geqslant 0].$$

ZH 63(265), BY(220.00)

9. $$\int_b^u \frac{dx}{\sqrt{(a^2-x^2)(x^2-b^2)}} = \frac{1}{a} F(\varkappa, q) \quad [a \geqslant u > b > 0].$$

ZH 63(266), BY(217.00)

10. $$\int_u^a \frac{dx}{\sqrt{(a^2-x^2)(x^2-b^2)}} = \frac{1}{a} F(\lambda, q) \quad [a > u \geqslant b > 0].$$

ZH 63(257), BY(218.00)

11. $$\int_a^u \frac{dx}{\sqrt{(x^2-a^2)(x^2-b^2)}} = \frac{1}{a} F(\mu, t) \quad [u > a > b > 0].$$

ZH 63(268), BY(216.00)

12. $$\int_u^\infty \frac{dx}{\sqrt{(x^2-a^2)(x^2-b^2)}} = \frac{1}{a} F(\nu, t) \quad [u \geqslant a > b > 0].$$

ZH 64(269), BY(215.00)

3.153

1. $$\int_0^u \frac{x^2\,dx}{\sqrt{(x^2+a^2)(x^2+b^2)}} = u\sqrt{\frac{a^2+u^2}{b^2+u^2}} - aE(\alpha, q) \quad [u>0,\ a>b].$$ BY (221.09)

2. $$\int_0^u \frac{x^2\,dx}{\sqrt{(a^2+x^2)(b^2-x^2)}} = \sqrt{a^2+b^2}\,E(\gamma, r) - \frac{a^2}{\sqrt{a^2+b^2}} F(\gamma, r) - u\sqrt{\frac{b^2-u^2}{a^2+u^2}} \quad [b \geqslant u > 0].$$ BY (214.05)

3. $$\int_u^b \frac{x^2\,dx}{\sqrt{(a^2+x^2)(b^2-x^2)}} = \sqrt{a^2+b^2}\,E(\delta, r) - \frac{a^2}{\sqrt{a^2+b^2}} F(\delta, r) \quad [b > u \geqslant 0].$$

BY (213.06)

4. $$\int_b^u \frac{x^2\,dx}{\sqrt{(a^2+x^2)(x^2-b^2)}} = \frac{b^2}{\sqrt{a^2+b^2}} F(\varepsilon, s) - \sqrt{a^2+b^2}\,E(\varepsilon, s) + \frac{1}{u}\sqrt{(u^2+a^2)(u^2-b^2)} \quad [u > b > 0].$$ BY (211.09)

5. $$\int_0^u \frac{x^2\,dx}{\sqrt{(a^2-x^2)(b^2-x^2)}} = a\{F(\eta, t) - E(\eta, t)\} \qquad [a > b \geqslant u > 0].$$

BY (219.05)

6. $$\int_u^b \frac{x^2\,dx}{\sqrt{(a^2-x^2)(b^2-x^2)}} = a\{F(\zeta, t) - E(\zeta, t)\} + u\sqrt{\frac{b^2-u^2}{a^2-u^2}}$$

$$[a > b > u \geqslant 0].$$ BY (220.06)

7. $$\int_b^u \frac{x^2\,dx}{\sqrt{(a^2-x^2)(x^2-b^2)}} = aE(\varkappa, q) - \frac{1}{u}\sqrt{(a^2-u^2)(u^2-b^2)}$$

$$[a \geqslant u > b > 0].$$ BY (217.05)

8. $$\int_u^a \frac{x^2\,dx}{\sqrt{(a^2-x^2)(x^2-b^2)}} = aE(\lambda, q) \qquad [a > u \geqslant b > 0].$$ BY (218.06)

9. $$\int_a^u \frac{x^2\,dx}{\sqrt{(x^2-a^2)(x^2-b^2)}} = a\{F(\nu, t) - E(\nu, t)\} + u\sqrt{\frac{u^2-a^2}{u^2-b^2}}$$

$$[u > a > b > 0].$$ BY (216.06)

10. $$\int_0^1 \frac{x^2\,dx}{\sqrt{(1+x^2)(1+k^2x^2)}} = \frac{1}{k^2}\left\{\sqrt{\frac{1+k^2}{2}} - E\left(\frac{\pi}{4}, \sqrt{1-k^2}\right)\right\}.$$

BI ((14))(9)

3.154

1. $$\int_0^u \frac{x^4\,dx}{\sqrt{(x^2+a^2)(x^2+b^2)}} = \frac{a}{3}\{2(a^2+b^2)E(\alpha, q) - b^2F(\alpha. q)\} +$$

$$+\frac{u}{3}(u^2-2a^2-b^2)\sqrt{\frac{a^2+u^2}{b^2+u^2}} \qquad [a > b, \quad u > 0].$$ BY (221.09)

2. $$\int_0^u \frac{x^4\,dx}{\sqrt{(a^2+x^2)(b^2-x^2)}} = \frac{1}{3\sqrt{a^2+b^2}}\{(2a^2-b^2)a^2F(\gamma, r) -$$

$$-2(a^4-b^4)E(\gamma, r)\} - \frac{u}{3}(2b^2-a^2+u^2)\sqrt{\frac{b^2-u^2}{a^2+u^2}}$$

$$[a \geqslant u > 0].$$ BY (214.05)

3. $$\int_u^b \frac{x^4\,dx}{\sqrt{(a^2+x^2)(b^2-x^2)}} = \frac{1}{3\sqrt{a^2+b^2}}\{(2a^2-b^2)a^2F(\delta, r) -$$

$$-2(a^4-b^4)E(\delta, r)\} + \frac{u}{3}\sqrt{(a^2+u^2)(b^2-u^2)}$$

$$[b > u \geqslant 0].$$ BY (213.06)

4. $$\int_b^u \frac{x^4\,dx}{\sqrt{(a^2+x^2)(x^2-b^2)}} = \frac{1}{3\sqrt{a^2+b^2}}\{(2b^2-a^2)b^2F(\varepsilon, s) +$$

$$+2(a^4-b^4)E(\varepsilon, s)\} + \frac{2b^2-2a^2+u^2}{3u}\sqrt{(u^2+a^2)(u^2-b^2)}$$

$$[u > b > 0].$$ BY (211.09)

5. $$\int_0^u \frac{x^4\,dx}{\sqrt{(a^2-x^2)(b^2-x^2)}} = \frac{a}{3}\{(2a^2+b^2)F(\eta, t) - 2(a^2+b^2)E(\eta, t)\} + \frac{u}{3}\sqrt{(a^2-u^2)(b^2-u^2)} \quad [a > b \geqslant u > 0].$$ BY (219.05)

6. $$\int_u^b \frac{x^4\,dx}{\sqrt{(a^2-x^2)(b^2-x^2)}} = \frac{a}{3}\{(2a^2+b^2)F(\zeta, t) - 2(a^2+b^2)E(\zeta, t)\} + \frac{u}{3}(u^2+a^2+2b^2)\sqrt{\frac{b^2-u^2}{a^2-u^2}} \quad [a > b > u \geqslant 0].$$ BY (220.06)

7. $$\int_b^u \frac{x^4\,dx}{\sqrt{(a^2-x^2)(x^2-b^2)}} = \frac{a}{3}\{2(a^2+b^2)E(\varkappa, q) - b^2F(\varkappa, q)\} - \frac{u^2+2a^2+2b^2}{3u}\sqrt{(a^2-u^2)(u^2-b^2)} \quad [a \geqslant u > b > 0].$$ BY (217.05)

8. $$\int_u^a \frac{x^4\,dx}{\sqrt{(a^2-x^2)(x^2-b^2)}} = \frac{a}{3}\{2(a^2+b^2)E(\lambda, q) - b^2F(\lambda, q)\} + \frac{u}{3}\sqrt{(a^2-u^2)(u^2-b^2)} \quad [a > u \geqslant b > 0].$$ BY (218.06)

9. $$\int_a^u \frac{x^4\,dx}{\sqrt{(x^2-a^2)(x^2-b^2)}} = \frac{a}{3}\{(2a^2+b^2)F(\mu, t) - 2(a^2+b^2)E(\mu, t)\} + \frac{u}{3}(u^2+2a^2+b^2)\sqrt{\frac{u^2-a^2}{u^2-b^2}} \quad [u > a > b > 0].$$ BY (216.06)

3.155

1. $$\int_u^a \sqrt{(a^2-x^2)(x^2-b^2)}\,dx = \frac{a}{3}\{(a^2+b^2)E(\lambda, q) - 2b^2F(\lambda, q)\} - \frac{u}{3}\sqrt{(a^2-u^2)(u^2-b^2)} \quad [a > u \geqslant b > 0].$$ BY (218.11)

2. $$\int_a^u \sqrt{(x^2-a^2)(x^2-b^2)}\,dx = \frac{a}{3}\{(a^2+b^2)E(\mu, t) - (a^2-b^2)F(\mu, t)\} + \frac{u}{3}(u^2-a^2-2b^2)\sqrt{\frac{u^2-a^2}{u^2-b^2}} \quad [u > a > b > 0].$$ BY (216.10)

3. $$\int_0^u \sqrt{(x^2+a^2)(x^2+b^2)}\,dx = \frac{a}{3}\{2b^2F(\alpha, q) - (a^2+b^2)E(\alpha, q)\} + \frac{u}{3}(u^2+a^2+2b^2)\sqrt{\frac{a^2+u^2}{b^2+u^2}} \quad [a > b, \quad u > 0].$$ BY (221.08)

4 $$\int_0^u \sqrt{(a^2+x^2)(b^2-x^2)}\,dx = \frac{1}{3}\sqrt{a^2+b^2}\{a^2F(\gamma, r) - (a^2-b^2)E(\gamma, r)\} + \frac{u}{3}(u^2+2a^2-b^2)\sqrt{\frac{b^2-u^2}{a^2+u^2}} \quad [a \geqslant u > 0].$$ BY (214.12)

5. $$\int_u^b \sqrt{(a^2+x^2)(b^2-x^2)}\,dx = \frac{1}{3}\sqrt{a^2+b^2}\,\{a^2F(\delta, r) + 2(b^2-a^2)E(\delta, r)\} + \frac{u}{3}\sqrt{(a^2+u^2)(b^2-u^2)} \qquad [b > u \geqslant 0].$$ BY (213.13)

6. $$\int_b^u \sqrt{(a^2+x^2)(x^2-b^2)}\,dx = \frac{1}{3}\sqrt{a^2+b^2}\,\{(b^2-a^2)E(\varepsilon, s) - b^2F(\varepsilon, s)\} + \frac{u^2+a^2-b^2}{3u}\sqrt{(a^2+u^2)(u^2-b^2)} \qquad [u > b > 0].$$ BY (211.08)

7. $$\int_0^u \sqrt{(a^2-x^2)(b^2-x^2)}\,dx = \frac{a}{3}\{(a^2+b^2)E(\eta, t) - (a^2-b^2)F(\eta, t)\} + \frac{u}{3}\sqrt{(a^2-u^2)(b^2-u^2)} \qquad [a > b \geqslant u > 0].$$ BY (219.11)

8. $$\int_u^b \sqrt{(a^2-x^2)(b^2-x^2)}\,dx = \frac{a}{3}\{(a^2+b^2)E(\zeta, t) - (a^2-b^2)F(\zeta, t)\} + \frac{u}{3}(u^2-2a^2-b^2)\sqrt{\frac{b^2-u^2}{a^2-u^2}} \qquad [a > b > u \geqslant 0].$$ BY (220.05)

9. $$\int_b^u \sqrt{(a^2-x^2)(x^2-b^2)}\,dx = \frac{a}{3}\{(a^2+b^2)E(\varkappa, q) - 2b^2F(\varkappa, q)\} + \frac{u^2-a^2-b^2}{3u}\sqrt{(a^2-u^2)(u^2-b^2)} \qquad [a \geqslant u > b > 0].$$ BY (217.09)

3.156

1. $$\int_u^\infty \frac{ux}{x^2\sqrt{(x^2+a^2)(x^2+b^2)}} = \frac{1}{ub^2}\sqrt{\frac{b^2+u^2}{a^2+u^2}} - \frac{1}{ab^2}E(\alpha, q) \qquad [a \geqslant b, \quad u > 0].$$ BY (222.04)

2. $$\int_u^b \frac{dx}{x^2\sqrt{(x^2+a^2)(b^2-x^2)}} = \frac{1}{a^2b^2\sqrt{a^2+b^2}}\{a^2F(\delta, r) - (a^2+b^2)E(\delta, r)\} + \frac{1}{a^2b^2u}\sqrt{(a^2+u^2)(b^2-u^2)} \qquad [b > u > 0].$$ BY (213.09)

3. $$\int_b^u \frac{dx}{x^2\sqrt{(x^2+a^2)(x^2-b^2)}} = \frac{1}{a^2b^2\sqrt{a^2+b^2}}\{(a^2+b^2)E(\varepsilon, s) - b^2F(\varepsilon, s)\} \qquad [u > b > 0].$$ BY (211.11)

4. $$\int_u^\infty \frac{dx}{x^2\sqrt{(x^2+a^2)(x^2-b^2)}} = \frac{1}{a^2b^2\sqrt{a^2+b^2}}\{(a^2+b^2)E(\gamma, s) - b^2F(\gamma, s)\} - \frac{1}{b^2u}\sqrt{\frac{u^2-b^2}{a^2+u^2}} \qquad [u \geqslant b > 0].$$ BY (212.06)

5. $$\int\limits_u^b \frac{dx}{x^2\sqrt{(a^2-x^2)(b^2-x^2)}} = \frac{1}{ab^2}\{F(\zeta, t)-E(\zeta, t)\}+ + \frac{1}{b^2u}\sqrt{\frac{b^2-u^2}{a^2-u^2}} \quad [a>b>u>0].$$ BY (220.09)

6. $$\int\limits_b^u \frac{dx}{x^2\sqrt{(a^2-x^2)(x^2-b^2)}} = \frac{1}{ab^2}E(\varkappa, q) \quad [a \geqslant u > b > 0].$$ BY (217.01)

7. $$\int\limits_u^a \frac{dx}{x^2\sqrt{(a^2-x^2)(x^2-b^2)}} = \frac{1}{ab^2}E(\lambda, q) - \frac{1}{a^2b^2u}\sqrt{(a^2-u^2)(u^2-b^2)} \quad [a>u\geqslant b>0].$$ BY (218.12)

8. $$\int\limits_a^u \frac{dx}{x^2\sqrt{(x^2-a^2)(x^2-b^2)}} = \frac{1}{ab^2}\{F(\mu, t)-E(\mu, t)\}+ + \frac{1}{a^2u}\sqrt{\frac{u^2-a^2}{u^2-b^2}} \quad [u>a>b>0].$$ BY (216.09)

9. $$\int\limits_u^\infty \frac{dx}{x^2\sqrt{(x^2-a^2)(x^2-b^2)}} = \frac{1}{ab^2}\{F(\nu, t)-E(\nu, t)\} \quad [u\geqslant a>b>0].$$ BY (215.07)

3.157

1. $$\int\limits_0^u \frac{dx}{(p-x^2)\sqrt{(x^2+a^2)(x^2+b^2)}} = = \frac{1}{a(p+b^2)}\left\{\frac{b^2}{p}\Pi\left(\alpha, \frac{p+b^2}{p}, q\right)+F(\alpha, q)\right\} \quad [p\neq 0].$$ BY (221.13)

2. $$\int\limits_u^\infty \frac{dx}{(p-x^2)\sqrt{(x^2+a^2)(x^2+b^2)}} = = -\frac{1}{a(a^2+p)}\left\{\Pi\left(\beta, \frac{a^2+p}{a^2}, q\right)-F(\beta, q)\right\}.$$ BY (222.11)

3. $$\int\limits_0^u \frac{dx}{(p-x^2)\sqrt{(a^2+x^2)(b^2-x^2)}} = = \frac{1}{p(p+a^2)\sqrt{a^2+b^2}}\left\{a^2\Pi\left(\gamma, \frac{b^2(p+a^2)}{p(a^2+b^2)}, r\right)+pF(\gamma, r)\right\} \quad [b\geqslant u>0, \ p\neq 0].$$ BY (214.13)a

4. $$\int\limits_u^b \frac{dx}{(p-x^2)\sqrt{(a^2+x^2)(b^2-x^2)}} = = \frac{1}{(p-b^2)\sqrt{a^2+b^2}}\Pi\left(\delta, \frac{b^2}{b^2-p}, r\right) \quad [b>u\geqslant 0, \ p\neq b^2].$$ BY (213.02)

5. $$\int\limits_b^u \frac{dx}{(p-x^2)\sqrt{(a^2+x^2)(x^2-b^2)}} = = \frac{1}{p(p-b^2)\sqrt{a^2+b^2}}\left\{b^2\Pi\left(\varepsilon, \frac{p}{p-b^2}, s\right)+(p-b^2)F(\varepsilon, s)\right\} \quad [u>b>0, \ p\neq b^2].$$ BY (211.14)

6. $$\int_u^\infty \frac{dx}{(x^2-p)\sqrt{(a^2+x^2)(x^2-b^2)}} =$$
$$= \frac{1}{(a^2+p)\sqrt{a^2+b^2}}\left\{\Pi\left(\xi, \frac{a^2+p}{a^2+b^2}, s\right) - F(\xi, s)\right\} \quad [u \geqslant b > 0].$$ BY (212.12)

7. $$\int_0^u \frac{dx}{(p-x^2)\sqrt{(a^2-x^2)(b^2-x^2)}} = \frac{1}{ap}\Pi\left(\eta, \frac{b^2}{p}, t\right)$$
$[a > b \geqslant u > 0;\ p \neq b]$. BY (219.02)

8. $$\int_u^b \frac{dx}{(p-x^2)\sqrt{(a^2-x^2)(b^2-x^2)}} =$$
$$= \frac{1}{a(p-a^2)(p-b^2)}\left\{(b^2-a^2)\Pi\left(\zeta, \frac{b^2(p-a^2)}{a^2(p-b^2)}, t\right) + (p-b^2)F(\zeta, t)\right\}$$
$[a > b > u \geqslant 0;\ p \neq b^2]$. BY (220.13)

9. $$\int_b^u \frac{dx}{(p-x^2)\sqrt{(a^2-x^2)(x^2-b^2)}} =$$
$$= \frac{1}{ap(p-b^2)}\left\{b^2\Pi\left(\varkappa, \frac{p(a^2-b^2)}{a^2(p-b^2)}, q\right) + (p-b^2)F(\varkappa, q)\right\}$$
$[a \geqslant u > b > 0;\ p \neq b^2]$. BY (217.12)

10. $$\int_u^a \frac{dx}{(x^2-p)\sqrt{(a^2-x^2)(x^2-b^2)}} = \frac{1}{a(a^2-p)}\Pi\left(\lambda, \frac{a^2-b^2}{a^2-p}, q\right)$$
$[a > u \geqslant b > 0;\ p \neq a^2]$. BY (218.02)

11. $$\int_a^u \frac{dx}{(p-x^2)\sqrt{(x^2-a^2)(x^2-b^2)}} =$$
$$= \frac{1}{a(p-a^2)(p-b^2)}\left\{(a^2-b^2)\Pi\left(\mu, \frac{p-b^2}{p-a^2}, t\right) + (p-a^2)F(\mu, t)\right\}$$
$[u > a > b > 0;\ p \neq a^2,\ p \neq b^2]$. BY (216.12)

12. $$\int_u^\infty \frac{dx}{(x^2-p)\sqrt{(x^2-a^2)(x^2-b^2)}} = \frac{1}{ap}\left\{\Pi\left(\nu, \frac{p}{a^2}, t\right) - F(\nu, t)\right\}$$
$[u \geqslant a > b > 0;\ p \neq 0]$. BY (215.12)

3.158

1. $$\int_0^u \frac{dx}{\sqrt{(x^2+a^2)(x^2+b^2)^3}} = \frac{1}{ab^2(a^2-b^2)}\{a^2E(\alpha, q) - b^2F(\alpha, q)\}$$
$[a > b;\ u > 0]$. BY (221.05)

2. $$\int_u^\infty \frac{dx}{\sqrt{(x^2+a^2)(x^2+b^2)^3}} = \frac{1}{ab^2(a^2-b^2)}\{a^2E(\beta, q) - b^2F(\beta, q)\} -$$
$$- \frac{u}{b^2\sqrt{(a^2+u^2)(b^2+u^2)}} \quad [a > b,\ u \geqslant 0].$$ BY (222.05)

3. $$\int_0^u \frac{dx}{\sqrt{(x^2+a^2)^3(x^2+b^2)}} = \frac{1}{a(a^2-b^2)}\{F(\alpha, q) - E(\alpha, q)\} + \frac{u}{a^2\sqrt{(u^2+a^2)(u^2+b^2)}} \quad [a > b;\ u > 0].$$ BY (221.06)

4. $$\int_u^\infty \frac{dx}{\sqrt{(a^2+x^2)^3(x^2+b^2)}} = \frac{1}{a(a^2-b^2)}\{F(\beta, q) - E(\beta, q)\} \quad [a > b,\ u \geqslant 0].$$ BY (222.03)

5. $$\int_0^u \frac{dx}{\sqrt{(a^2+x^2)^3(b^2-x^2)}} = \frac{1}{a^2\sqrt{a^2+b^2}} E(\gamma, r) \quad [b \geqslant u > 0].$$ BY (214.01)a

6. $$\int_u^b \frac{dx}{\sqrt{(a^2+x^2)^3(b^2-x^2)}} = \frac{1}{a^2\sqrt{a^2+b^2}} E(\delta, r) - \frac{u}{a^2(a^2+b^2)}\sqrt{\frac{b^2-u^2}{a^2+u^2}} \quad [b > u \geqslant 0].$$ BY (213.08)

7. $$\int_b^u \frac{dx}{\sqrt{(a^2+x^2)^3(x^2-b^2)}} = \frac{1}{a^2\sqrt{a^2+b^2}}\{F(\varepsilon, s) - E(\varepsilon, s)\} + \frac{1}{(a^2+b^2)u}\sqrt{\frac{u^2-b^2}{u^2+a^2}} \quad [u > b > 0].$$ BY (211.05)

8. $$\int_u^\infty \frac{dx}{\sqrt{(a^2+x^2)^3(x^2-b^2)}} = \frac{1}{a^2\sqrt{a^2+b^2}}\{F(\xi, s) - E(\xi, s)\} \quad [u \geqslant b > 0].$$ BY (212.03)

9. $$\int_0^u \frac{dx}{\sqrt{(a^2+x^2)(b^2-x^2)^3}} = \frac{1}{b^2\sqrt{a^2+b^2}}\{F(\gamma, r) - E(\gamma, r)\} + \frac{u}{b^2\sqrt{(a^2+u^2)(b^2-u^2)}} \quad [b > u > 0].$$ BY (214.10)

10. $$\int_u^\infty \frac{dx}{\sqrt{(a^2+x^2)(x^2-b^2)^3}} = \frac{u}{b^2\sqrt{(a^2+u^2)(u^2-b^2)}} - \frac{1}{b^2\sqrt{a^2+b^2}} E(\xi, s) \quad [u \geqslant b > 0].$$ BY (212.04)

11. $$\int_0^u \frac{dx}{\sqrt{(a^2-x^2)^3(b^2-x^2)}} = \frac{1}{a^2(a^2-b^2)}\left\{aE(\eta, t) - u\sqrt{\frac{b^2-u^2}{a^2-u^2}}\right\} \quad [a > b \geqslant u > 0].$$ BY (219.07)

12. $$\int_u^b \frac{dx}{\sqrt{(a^2-x^2)^3(b^2-x^2)}} = \frac{1}{a(a^2-b^2)} E(\zeta, t) \quad [a > b > u \geqslant 0].$$ BY (220.10)

13. $$\int_b^u \frac{dx}{\sqrt{(a^2-x^2)^3(x^2-b^2)}} = \frac{1}{a(a^2-b^2)}\left\{F(\varkappa, q) - E(\varkappa, q) + \frac{a}{u}\sqrt{\frac{u^2-b^2}{a^2-u^2}}\right\}$$

$[a > u > b > 0]$. BY (217.10)

14. $$\int_u^\infty \frac{dx}{\sqrt{(x^2-a^2)^3(x^2-b^2)}} = \frac{1}{a(b^2-a^2)}\left\{E(\nu, t) - \frac{a}{u}\sqrt{\frac{u^2-b^2}{u^2-a^2}}\right\}$$

$[u > a > b > 0]$. BY (215.04)

15. $$\int_0^u \frac{dx}{\sqrt{(a^2-x^2)(b^2-x^2)^3}} = \frac{1}{ab^2}F(\eta, t) - \frac{1}{b^2(a^2-b^2)} \times$$

$$\times\left\{aE(\eta, t) - u\sqrt{\frac{a^2-u^2}{b^2-u^2}}\right\} \quad [a > b > u > 0].$$ BY (219.06)

16. $$\int_u^a \frac{dx}{\sqrt{(a^2-x^2)(x^2-b^2)^3}} = \frac{1}{ab^2(a^2-b^2)}\left\{b^2F(\lambda, q) - a^2E(\lambda, q) + \right.$$

$$\left. + au\sqrt{\frac{a^2-u^2}{u^2-b^2}}\right\} \quad [a > u > b > 0].$$ BY (218.04)

17. $$\int_a^u \frac{dx}{\sqrt{(x^2-a^2)(x^2-b^2)^3}} = \frac{a}{b^2(a^2-b^2)}E(\mu, t) - \frac{1}{ab^2}F(\mu, t)$$

$[u > a > b > 0]$. BY (216.11)

18. $$\int_u^\infty \frac{dx}{\sqrt{(x^2-a^2)(x^2-b^2)^3}} = \frac{1}{b^2(a^2-b^2)}\left\{aE(\nu, t) - \frac{b^2}{u}\sqrt{\frac{u^2-a^2}{u^2-b^2}}\right\} -$$

$$-\frac{1}{ab^2}F(\nu, t) \quad [u \geqslant a > b > 0].$$ BY (215.06)

3.159

1. $$\int_0^u \frac{x^2\,dx}{\sqrt{(x^2+a^2)(x^2+b^2)^3}} = \frac{a}{a^2-b^2}\{F(\alpha, q) - E(\alpha, q)\}$$

$[a > b,\ u > 0]$. BY (221.12)

2. $$\int_u^\infty \frac{x^2\,dx}{\sqrt{(x^2+a^2)(x^2+b^2)^3}} = \frac{a}{a^2-b^2}\{F(\beta, q) - E(\beta, q)\} +$$

$$+\frac{u}{\sqrt{(a^2+u^2)(b^2+u^2)}} \quad [a > b,\ u \geqslant 0].$$ BY (222.10)

3. $$\int_0^u \frac{x^2\,dx}{\sqrt{(x^2+a^2)^3(x^2+b^2)}} = \frac{1}{a(a^2-b^2)}\{a^2E(\alpha, q) - b^2F(\alpha, q)\} -$$

$$-\frac{u}{\sqrt{(a^2+u^2)(b^2+u^2)}} \quad [a > b,\ u > 0].$$ BY (221.11)

4. $$\int_u^\infty \frac{x^2\,dx}{\sqrt{(x^2+a^2)^3(x^2+b^2)}} = \frac{1}{a(a^2-b^2)}\{a^2E(\beta, q) - b^2F(\beta, q)\}$$

$[a > b,\ u \geqslant 0]$. BY (222.07)

5. $$\int_0^u \frac{x^2\,dx}{\sqrt{(a^2+x^2)^3(b^2-x^2)}} = \frac{1}{\sqrt{a^2+b^2}}\{F(\gamma, r) - E(\gamma, r)\}$$

$[b \geqslant u > 0]$. BY (214.04)

6. $$\int_u^b \frac{x^2\,dx}{\sqrt{(a^2+x^2)^3(b^2-x^2)}} = \frac{1}{\sqrt{a^2+b^2}}\{F(\delta, r) - E(\delta, r)\} + \frac{u}{a^2+b^2}\sqrt{\frac{b^2-u^2}{a^2+u^2}}$$

$[b > u \geqslant 0]$. BY (213.07)

7. $$\int_b^u \frac{x^2\,dx}{\sqrt{(a^2+x^2)^3(x^2-b^2)}} = \frac{1}{\sqrt{a^2+b^2}} E(\varepsilon, s) - \frac{a^2}{u(a^2+b^2)}\sqrt{\frac{u^2-b^2}{u^2+a^2}}$$

$[u > b > 0]$. BY (211.13)

8. $$\int_u^\infty \frac{x^2\,dx}{\sqrt{(a^2+x^2)^3(x^2-b^2)}} = \frac{1}{\sqrt{a^2+b^2}} E(\xi, s)$$ $[u \geqslant b > 0]$. BY (212.01)

9. $$\int_0^u \frac{x^2\,dx}{\sqrt{(a^2+x^2)(b^2-x^2)^3}} = \frac{u}{\sqrt{(a^2+u^2)(b^2-u^2)}} - \frac{1}{\sqrt{a^2+b^2}} E(\gamma, r)$$

$[b > u > 0]$. BY (214.07)

10. $$\int_u^\infty \frac{x^2\,dx}{\sqrt{(a^2+x^2)(x^2-b^2)^3}} = \frac{1}{\sqrt{a^2+b^2}}\{F(\xi, s) - E(\xi, s)\} + \frac{u}{\sqrt{(a^2+u^2)(u^2-b^2)}}$$

$[u > b > 0]$. BY (212.10)

11. $$\int_0^u \frac{x^2\,dx}{\sqrt{(a^2-x^2)^3(b^2-x^2)}} = \frac{1}{a^2-b^2}\left\{aE(\eta, t) - u\sqrt{\frac{b^2-u^2}{a^2-u^2}}\right\} - \frac{1}{a}F(\eta, t)$$

$[a > b \geqslant u > 0]$. BY (219.04)

12. $$\int_u^b \frac{x^2\,dx}{\sqrt{(a^2-x^2)^3(b^2-x^2)}} = \frac{a}{a^2-b^2} E(\zeta, t) - \frac{1}{a} F(\zeta, t)$$ $[a > b > u \geqslant 0]$.

BY (220.08)

13. $$\int_b^u \frac{x^2\,dx}{\sqrt{(a^2-x^2)^3(x^2-b^2)}} = \frac{1}{a(a^2-b^2)}\left\{b^2F(\varkappa, q) - a^2E(\varkappa, q) + \frac{a^3}{u}\sqrt{\frac{u^2-b^2}{a^2-u^2}}\right\}$$

$[a > u > b > 0]$. BY (217.06)

14. $$\int_u^\infty \frac{x^2\,dx}{\sqrt{(x^2-a^2)^3(x^2-b^2)}} = \frac{a}{a^2-b^2}\left\{\frac{a}{u}\sqrt{\frac{u^2-b^2}{u^2-a^2}} - E(\nu, t)\right\} + \frac{1}{a}F(\nu, t)$$

$[u > a > b > 0]$. BY (215.09)

15. $$\int_0^u \frac{x^2\,dx}{\sqrt{(a^2-x^2)(b^2-x^2)^3}} = \frac{1}{a^2-b^2}\left\{u\sqrt{\frac{a^2-u^2}{b^2-u^2}} - aE(\eta, t)\right\}$$

$[a > b > u > 0]$. BY (219.12)

16. $\int\limits_u^a \frac{x^2\,dx}{\sqrt{(a^2-x^2)(x^2-b^2)^3}} = \frac{1}{a^2-b^2}\left\{aF(\lambda, q) - aE(\lambda, q) + u\sqrt{\frac{a^2-u^2}{u^2-b^2}}\right\}$

$[a > u > b > 0]$. BY (218.07)

17. $\int\limits_a^u \frac{x^2\,dx}{\sqrt{(x^2-a^2)(x^2-b^2)^3}} = \frac{a}{a^2-b^2}E(\mu, t) \quad [u > a > b > 0]$. BY (216.01)

18. $\int\limits_u^\infty \frac{x^2\,dx}{\sqrt{(x^2-a^2)(x^2-b^2)^3}} = \frac{1}{a^2-b^2}\left\{aE(\nu, t) - \frac{b^2}{u}\sqrt{\frac{u^2-a^2}{u^2-b^2}}\right\}$

$[u \geqslant a > b > 0]$. BY (215.11)

3.161

1. $\int\limits_u^\infty \frac{dx}{x^4\sqrt{(x^2+a^2)(x^2+b^2)}} = \frac{1}{3a^3b^4}\{2(a^2+b^2)E(\beta, q) - b^2F(\beta, q)\} + $

$+\frac{a^2b^2-u^2(2a^2+b^2)}{3a^2b^4u^3} \quad [a > b,\ u > 0]$. BY (222.04)

2. $\int\limits_u^b \frac{dx}{x^4\sqrt{(x^2+a^2)(b^2-x^2)}} =$

$= \frac{1}{3a^4b^4\sqrt{a^2+b^2}}\{a^2(2a^2-b^2)F(\delta, r) - 2(a^4-b^4)E(\delta, r)\} +$

$+\frac{a^2b^2+2u^2(a^2-b^2)}{3a^4b^4u^3}\sqrt{(b^2-u^2)(a^2+u^2)} \quad [b > u > 0]$. BY (213.09)

3. $\int\limits_b^u \frac{dx}{x^4\sqrt{(x^2+a^2)(x^2-b^2)}} = \frac{2b^2-a^2}{3a^4b^2\sqrt{a^2+b^2}}F(\varepsilon, s) +$

$+\frac{2}{3}\frac{(a^2-b^2)\sqrt{a^2+b^2}}{a^4b^4}E(\varepsilon, s) + \frac{1}{3a^2b^2u^3}\sqrt{(u^2+a^2)(u^2-b^2)}$

$[u > b > 0]$. BY (211.11)

4. $\int\limits_u^\infty \frac{dx}{x^4\sqrt{(x^2+a^2)(x^2-b^2)}} =$

$= \frac{1}{3a^4b^4\sqrt{a^2+b^2}}\{2(a^4-b^4)E(\xi, s) + b^2(2b^2-a^2)F(\xi, s)\} -$

$-\frac{a^2b^2+u^2(2a^2-b^2)}{3a^2b^4u^3}\sqrt{\frac{u^2-b^2}{u^2+a^2}} \quad [u \geqslant b > 0]$. BY (212.06)

5. $\int\limits_u^b \frac{dx}{x^4\sqrt{(a^2-x^2)(b^2-x^2)}} = \frac{1}{3a^3b^4}\left\{(2a^2+b^2)F(\zeta, t) - 2(a^2+b^2)E(\zeta, t) + \right.$

$\left.+\frac{[(2a^2+b^2)u^2+a^2b^2]a}{u^3}\sqrt{\frac{b^2-u^2}{a^2-u^2}}\right\} \quad [a > b > u > 0]$. BY (220.09)

6. $\int\limits_b^u \frac{dx}{x^4\sqrt{(a^2-x^2)(x^2-b^2)}} = \frac{1}{3a^3b^4}\{2(a^2+b^2)E(\varkappa, q) - b^2F(\varkappa, q)\} +$

$+\frac{1}{3a^2b^2u^3}\sqrt{(a^2-u^2)(u^2-b^2)} \quad [a \geqslant u > b > 0]$. BY (217.14)

7. $$\int_u^a \frac{dx}{x^4\sqrt{(a^2-x^2)(x^2-b^2)}} = \frac{1}{3a^3b^4}\left\{2(a^2+b^2)E(\lambda, q) - b^2F(\lambda, q) - \right.$$
$$\left. - \frac{2(a^2+b^2)u^2+a^2b^2}{au^3}\sqrt{(a^2-u^2)(u^2-b^2)}\right\} \qquad [a > u \geqslant b > 0].$$

BY (218.12)

8. $$\int_a^u \frac{dx}{x^4\sqrt{(x^2-a^2)(x^2-b^2)}} = \frac{1}{3a^3b^4}\left\{(2a^2+b^2)F(\mu, t) - \right.$$
$$\left. - 2(a^2+b^2)E(\mu, t) + \frac{[(a^2+2b^2)u^2+a^2b^2]b^2}{au^3}\sqrt{\frac{u^2-a^2}{u^2-b^2}}\right\}$$
$$[u > a > b > 0].$$

BY (216.09)

9. $$\int_u^\infty \frac{dx}{x^4\sqrt{(x^2-a^2)(x^2-b^2)}} = \frac{1}{3a^3b^4}\left\{(2a^2+b^2)F(\nu, t) - 2(a^2+b^2)E(\nu, t) + \right.$$
$$\left. + \frac{ab^2}{u^3}\sqrt{(u^2-a^2)(u^2-b^2)}\right\} \qquad [u \geqslant a > b > 0].$$

BY (215.07)

3.162

1. $$\int_0^u \frac{dx}{\sqrt{(x^2+a^2)^5(x^2+b^2)}} =$$
$$= \frac{1}{3a^3(a^2-b^2)^2}\{(3a^2-b^2)F(\alpha, q) - 2(2a^2-b^2)E(\alpha, q)\} +$$
$$+ \frac{u[a^2(4a^2-3b^2)+u^2(3a^2-2b^2)]}{3a^4(a^2-b^2)\sqrt{(u^2+a^2)^3(u^2+b^2)}} \qquad [a > b, \ u > 0].$$

BY (221.06)

2 $$\int_u^\infty \frac{dx}{\sqrt{(x^2+a^2)^5(x^2+b^2)}} = \frac{1}{3a^3(a^2-b^2)^2}\{(3a^2-b^2)F(\beta, q) -$$
$$- 2(2a^2-b^2)E(\beta, q)\} + \frac{u}{3a^2(a^2-b^2)}\sqrt{\frac{u^2+b^2}{(a^2+u^2)^3}}$$
$$[a > b, \ u \geqslant 0].$$

BY (222.03)

3. $$\int_0^u \frac{dx}{\sqrt{(x^2+a^2)(x^2+b^2)^5}} = \frac{3b^2-a^2}{3ab^2(a^2-b^2)^2}F(\alpha, q) + \frac{a(2a^2-4b^2)}{3b^4(a^2-b^2)^2}E(\alpha, q) +$$
$$+ \frac{u}{3b^2(a^2-b^2)}\sqrt{\frac{u^2+a^2}{(u^2+b^2)^3}} \qquad [a > b, \ u > 0].$$

BY (221.05)

4. $$\int_u^\infty \frac{dx}{\sqrt{(x^2+a^2)(x^2+b^2)^5}} =$$
$$= \frac{1}{3ab^4(a^2-b^2)^2}\{2a^2(a^2-2b^2)E(\beta, q) + b^2(3b^2-a^2)F(\beta, q)\} -$$
$$- \frac{u[b^2(3a^2-4b^2)+u^2(2a^2-3b^2)]}{3b^4(a^2-b^2)\sqrt{(u^2+a^2)(u^2+b^2)^3}} \qquad [a > b, \ u \geqslant 0].$$

BY (222.05)

5. $$\int_0^u \frac{dx}{\sqrt{(a^2+x^2)^5(b^2-x^2)}} = \frac{1}{3a^4\sqrt{(a^2+b^2)^3}}\{2(b^2+2a^2)E(\gamma, r) - a^2F(\gamma, r)\} +$$
$$+ \frac{u}{3a^2(a^2+b^2)}\sqrt{\frac{b^2-u^2}{(a^2+u^2)^3}} \qquad [b \geqslant u > 0].$$

BY (214.15)

6. $$\int_u^b \frac{dx}{\sqrt{(a^2+x^2)^5 (b^2-x^2)}} = \frac{1}{3a^4 \sqrt{(a^2+b^2)^3}} \{(4a^2+2b^2) E(\delta, r) - a^2 F(\delta, r)\} - \frac{u[a^2(5a^2+3b^2)+u^2(4a^2+2b^2)]}{3a^4(a^2+b^2)^2} \sqrt{\frac{b^2-u^2}{(a^2+u^2)^3}} \quad [b > u > 0].$$ BY (213.08)

7. $$\int_b^u \frac{dx}{\sqrt{(a^2+x^3)^5 (x^2-b^2)}} = \frac{1}{3a^4 \sqrt{(a^2+b^2)^3}} \{(3a^2+2b^2) F(\varepsilon, s) - (4a^2+2b^2) E(\varepsilon, s)\} + \frac{(3a^2+b^2)u^2+2(2a^2+b^2)a^2}{3a^2(a^2+b^2)^2 u} \sqrt{\frac{u^2-b^2}{(u^2+a^2)^3}}$$ $[u > b > 0]$. BY (211.05)

8. $$\int_u^\infty \frac{dx}{\sqrt{(a^2+x^2)^5 (x^2-b^2)}} = \frac{1}{3a^4 \sqrt{(a^2+b^2)^3}} \{(3a^2+2b^2) F(\xi, s) - (4a^2+2b^2) E(\xi, s)\} + \frac{u}{3a^2(a^2+b^2)} \sqrt{\frac{u^2-b^2}{(a^2+u^2)^3}} \quad [u > b > 0].$$ BY (212.03)

9. $$\int_0^u \frac{dx}{\sqrt{(a^2+x^2)(b^2-x^2)^5}} = \frac{1}{3b^4 \sqrt{(a^2+b^2)^3}} \{(2a^2+3b^2) F(\gamma, r) - (2a^2+4b^2) E(\gamma, r)\} + \frac{u[(3a^2+4b^2)b^2-(2a^2+3b^2)u^2]}{3b^4(a^2+b^2)\sqrt{(a^2+u^2)(b^2-u^2)^3}} \quad [b > u > 0].$$ BY (214.10)

10. $$\int_u^\infty \frac{dx}{\sqrt{(a^2+x^2)(x^2-b^2)^5}} = \frac{1}{3b^4 \sqrt{(a^2+b^2)^3}} \{(2a^2+4b^2) E(\xi, s) - b^2 F(\xi, s)\} + \frac{u[(3a^2+4b^2)b^2-(2a^2+3b^2)u^2]}{3b^4(a^2+b^2)\sqrt{(a^2+u^2)(u^2-b^2)^3}} \quad [u > b > 0].$$ BY (212.04)

11. $$\int_0^u \frac{dx}{\sqrt{(a^2-x^2)(b^2-x^2)^5}} = \frac{2a^2-3b^2}{3ab^4(a^2-b^2)} F(\eta, t) + \frac{2a(2b^2-a^2)}{3b^4(a^2-b^2)^2} E(\eta, t) + \frac{u[(3a^2-5b^2)b^2-2(a^2-2b^2)u^2]}{3b^4(a^2-b^2)^2(b^2-u^2)} \sqrt{\frac{a^2-u^2}{b^2-u^2}}$$ $[a > b > a > 0]$. BY (219.06)

12 $$\int_u^a \frac{dx}{\sqrt{(a^2-x^2)(x^2-b^2)^5}} = \frac{3b^2-a^2}{3ab^2(a^2-b^2)^2} F(\lambda, q) + \frac{2a(a^2-2b^2)}{3b^4(a^2-b^2)^2} E(\lambda, q) + \frac{u[2(2b^2-a^2)u^2+(3a^2-5b^2)b^2]}{3b^4(a^2-b^2)^2(u^2-b^2)} \sqrt{\frac{a^2-u^2}{u^2-b^2}}$$ $[a > u > b > 0]$. BY (218.04)

13. $$\int_a^u \frac{dx}{\sqrt{(x^2-a^2)(x^2-b^2)^5}} = \frac{2a^2-3b^2}{3ab^4(a^2-b^2)} F(\mu, t) + \frac{2a(2b^2-a^2)}{3b^4(a^2-b^2)^2} E(\mu, t) + \frac{u}{3b^2(a^2-b^2)(u^2-b^2)} \sqrt{\frac{u^2-a^2}{u^2-b^2}}$$ $[u > a > b > 0]$. BY (216.11)

14. $$\int_u^\infty \frac{dx}{\sqrt{(x^2-a^2)(x^2-b^2)^5}} = \frac{(4b^2-2a^2)\,a}{3b^4(a^2-b^2)^2} E(\nu, t) + \frac{2a^2-3b^2}{3ab^4(a^2-b^2)} F(\nu, t) - \frac{(3b^2-a^2)\,u^2-(4b^2-2a^2)\,b^2}{3b^2u(a^2-b^2)^2(u^2-b^2)} \sqrt{\frac{u^2-a^2}{u^2-b^2}}$$
$[u \geqslant a > b > 0]$. BY (215.06)

15. $$\int_0^u \frac{dx}{\sqrt{(a^2-x^2)^5(b^2-x^2)}} = \frac{1}{3a^3(a^2-b^2)^2}\left\{(4a^2-2b^2)E(\eta, t) - (a^2-b^2)F(\eta, t) - \frac{u[(5a^2-3b^2)\,a^2-(4a^2-2b^2)\,u^2]}{a(a^2-u^2)}\sqrt{\frac{b^2-u^2}{a^2-u^2}}\right\}$$
$[a > b \geqslant u > 0]$. BY (219.07)

16. $$\int_u^b \frac{dx}{\sqrt{(a^2-x^2)^5(b^2-x^2)}} = \frac{2(2a^2-b^2)}{3a^3(a^2-b^2)^2} E(\zeta, r) - \frac{1}{3a^3(a^2-b^2)} F(\zeta, t) + \frac{u}{3a^2(a^2-b^2)(a^2-u^2)}\sqrt{\frac{b^2-u^2}{a^2-u^2}}$$
$[a > b > u \geqslant 0]$. BY (220.10)

17. $$\int_b^u \frac{dx}{\sqrt{(a^2-x^2)^5(x^2-b^2)}} = \frac{1}{3a^3(a^2-b^2)^2}\{(3a^2-b^2)F(\varkappa, q) - (4a^2-2b^2)E(\varkappa, q)\} + \frac{2(2a^2-b^2)\,a^2+(b^2-3a^2)\,u^2}{3a^2u(a^2-b^2)^2(a^2-u^2)}\sqrt{\frac{u^2-b^2}{a^2-u^2}},$$
$[a > u > b > 0]$. BY (217.10)

18. $$\int_u^\infty \frac{dx}{\sqrt{(x^2-a^2)^5(x^2-b^2)}} = \frac{1}{3a^3(a^2-b^2)^2}\{(4a^2-2b^2)E(\nu, t) - (a^2-b^2)F(\nu, t)\} + \frac{(4a^2-2b^2)\,a^2+(b^2-3a^2)\,u^2}{3a^2u(a^2-b^2)^2(u^2-a^2)}\sqrt{\frac{u^2-b^2}{u^2-a^2}}$$
$[u > a > b > 0]$. BY (215.04)

3.163

1. $$\int_0^u \frac{dx}{\sqrt{(x^2+a^2)^3(x^2+b^2)^3}} = \frac{1}{ab^2(a^2-b^2)^2}\{(a^2+b^2)E(\alpha, q) - 2b^2F(\alpha, q)\} - \frac{u}{a^2(a^2-b^2)\sqrt{(a^2+u^2)(b^2+u^2)}}$$
$[a > b,\ u > 0]$. BY (221.07)

2. $$\int_u^\infty \frac{dx}{\sqrt{(x^2+a^2)^3(x^2+b^2)^3}} = \frac{1}{ab^2(a^2-b^2)^2}\{(a^2+b^2)E(\beta, q) - 2b^2F(\beta, q)\} - \frac{u}{b^2(a^2-b^2)\sqrt{(a^2+u^2)(b^2+u^2)}}$$
$[a > b,\ u \geqslant 0]$. BY (222.12)

3. $$\int_0^u \frac{dx}{\sqrt{(x^2+a^2)^3(b^2-x^2)^3}} = \frac{1}{a^2b^2\sqrt{(a^2+b^2)^3}}\{a^2F(\gamma, r) - (a^2-b^2)E(\gamma, r)\} + \frac{u}{b^2(a^2+b^2)\sqrt{(a^2+u^2)(b^2-u^2)}}$$
$[b > u > 0]$. BY (214.15)

4. $$\int_u^\infty \frac{dx}{\sqrt{(x^2+a^2)^3(x^2-b^2)^3}} = \frac{b^2-a^2}{a^2b^2\sqrt{(a^2+b^2)^3}} E(\xi, s) - \frac{1}{a^2\sqrt{(a^2+b^2)^3}} F(\xi, s) + \frac{u}{b^2(a^2+b^2)\sqrt{(u^2+a^2)(u^2-b^2)}} \quad [u > b > 0].$$ BY (212.05)

5. $$\int_0^u \frac{dx}{\sqrt{(a^2-x^2)^3(b^2-x^2)^3}} = \frac{1}{ab^2(a^2-b^2)} F(\eta, t) - \frac{a^2+b^2}{ab^2(a^2-b^2)^2} E(\eta, t) + \frac{[a^4+b^4-(a^2+b^2)u^2]u}{a^2b^2(a^2-b^2)^2\sqrt{(a^2-u^2)(b^2-u^2)}} \quad [a > b > u > 0].$$ BY (279.08)

6. $$\int_u^\infty \frac{dx}{\sqrt{(x^2-a^2)^3(x^2-b^2)^3}} = \frac{1}{ab^2(a^2-b^2)} F(\nu, t) - \frac{a^2+b^2}{ab^2(a^2-b^2)^2} E(\nu, t) + \frac{1}{u(a^2-b^2)\sqrt{(u^2-a^2)(u^2-b^2)}} \quad [u > a > b > 0].$$ BY (215.10)

3.164 Notations: $\alpha = \arccos \frac{u^2-\varrho\bar{\varrho}}{u^2+\varrho\bar{\varrho}}$, $r = \frac{1}{2}\sqrt{-\frac{(\varrho-\bar{\varrho})^2}{\varrho\bar{\varrho}}}$.

1. $$\int_u^\infty \frac{dx}{\sqrt{(x^2+\varrho^2)(x^2+\bar{\varrho}^2)}} = \frac{1}{\sqrt{\varrho\bar{\varrho}}} F(\alpha, r).$$ BY (225.00)

2. $$\int_u^\infty \frac{x^2\,dx}{(x^2-\varrho\bar{\varrho})^2\sqrt{(x^2+\varrho^2)(x^2+\bar{\varrho}^2)}} = \frac{2u\sqrt{(u^2+\varrho^2)(u^2+\bar{\varrho}^2)}}{(\varrho+\bar{\varrho})^2(u^4-\varrho^2\bar{\varrho}^2)} - \frac{1}{(\varrho+\bar{\varrho})^2\sqrt{\varrho\bar{\varrho}}} E(\alpha, r).$$ BY (225.03)

3. $$\int_u^\infty \frac{x^2\,dx}{(x^2+\varrho\bar{\varrho})^2\sqrt{(x^2+\varrho^2)(x^2+\bar{\varrho}^2)}} = -\frac{1}{(\varrho-\bar{\varrho})^2\sqrt{\varrho\bar{\varrho}}}[F(\alpha, r) - E(\alpha, r)].$$ BY (225.07)

38 4. $$\int_u^\infty \frac{x^2\,dx}{\sqrt{(x^2+\varrho^2)^3(x^2+\bar{\varrho}^2)^3}} = -\frac{4\sqrt{\varrho\bar{\varrho}}}{(\varrho^2-\bar{\varrho}^2)^2} E(\alpha, r) + \frac{1}{(\varrho-\bar{\varrho})^2\sqrt{\varrho\bar{\varrho}}} F(\alpha, r) - \frac{2u(u^2-\varrho\bar{\varrho})}{(\varrho+\bar{\varrho})^2(u^2+\varrho\bar{\varrho})\sqrt{(u^2+\varrho^2)(u^2+\bar{\varrho}^2)}}.$$ BY (225.05)

5. $$\int_u^\infty \frac{(x^2-\varrho\bar{\varrho})^2\,dx}{\sqrt{(x^2+\varrho^2)^3(x^2+\bar{\varrho}^2)^3}} = -\frac{4\sqrt{\varrho\bar{\varrho}}}{(\varrho-\bar{\varrho})^2}[F(\alpha, r) - E(\alpha, r)] + \frac{2u(u^2-\varrho\bar{\varrho})}{(u^2+\varrho\bar{\varrho})\sqrt{(u^2+\varrho^2)(u^2+\bar{\varrho}^2)}}.$$ BY (225.06)

6. $$\int_u^\infty \frac{\sqrt{(x^2+\varrho^2)(x^2+\bar{\varrho}^2)}}{(x^2+\varrho\bar{\varrho})^2}\,dx = \frac{1}{\sqrt{\varrho\bar{\varrho}}} E(\alpha, r).$$ BY (225.01)

7. $$\int_u^\infty \frac{(x^2-\varrho\bar{\varrho})^2\,dx}{(x^2+\varrho\bar{\varrho})^2\sqrt{(x^2+\varrho^2)(x^2+\bar{\varrho}^2)}} = -\frac{4\sqrt{\varrho\bar{\varrho}}}{(\varrho-\bar{\varrho})^2}E(\alpha, r) + \frac{(\varrho+\bar{\varrho})^2}{(\varrho-\bar{\varrho})^2\sqrt{\varrho\bar{\varrho}}}F(\alpha, r).$$ BY (225.08)

8. $$\int_u^\infty \frac{(x^2+\varrho\bar{\varrho})^2\,dx}{[(x^2+\varrho\bar{\varrho})^2-4p^2\varrho\bar{\varrho}x^2]\sqrt{(x^2+\varrho^2)(x^2+\bar{\varrho}^2)}} = \frac{1}{\sqrt{\varrho\bar{\varrho}}}\Pi(\alpha, p^2, r).$$ BY (225.02)

3.165 Notations: $\alpha = \arccos\frac{u^2-a^2}{u^2+a^2}$, $r = \frac{\sqrt{a^2-b^2}}{a\sqrt{2}}$.

1. $$\int_u^a \frac{dx}{\sqrt{x^4+2b^2x^2+a^4}} = \frac{\sqrt{2}}{a\sqrt{2}+\sqrt{a^2+b^2}} \times F\left[\operatorname{arctg}\left(\frac{a\sqrt{2}+\sqrt{a^2-b^2}}{\sqrt{a^2+b^2}}\,\frac{a-u}{a+u}\right),\ \frac{2\sqrt{a\sqrt{2(a^2-b^2)}}}{a\sqrt{2}+\sqrt{a^2-b^2}}\right]$$ $[a > b,\ a > u \geqslant 0]$. BY (264.00)

2. $$\int_u^\infty \frac{dx}{\sqrt{x^4+2b^2x^2+a^4}} = \frac{1}{2a}F(\alpha, r)$$ $[a^2 > b^2 > -\infty,\ a^2 > 0,\ u \geqslant 0]$. BY (263.00, 266.00)

3. $$\int_u^\infty \frac{dx}{x^2\sqrt{x^4+2b^2x^2+a^4}} = \frac{1}{2a^3}[F(\alpha, r)-2E(\alpha, r)] + \frac{\sqrt{u^4+2b^2u^2+a^4}}{a^2u(u^2+a^2)}$$ $[a > b > 0,\ u > 0]$. BY (263.06)

4. $$\int_u^\infty \frac{x^2\,dx}{(x^2+a^2)^2\sqrt{x^4+2b^2x^2+a^4}} = \frac{1}{4a(a^2-b^2)}[F(\alpha, r)-E(\alpha, r)]$$ $[a^2 > b^2 > -\infty,\ a^2 > 0,\ u \geqslant 0]$. BY (263.03, 266.05)

5. $$\int_u^\infty \frac{x^2\,dx}{(x^2-a^2)^2\sqrt{x^4+2b^2x^2+a^4}} = \frac{u\sqrt{u^4+2b^2u^2+a^4}}{2(a^2+b^2)(u^4-a^4)} - \frac{1}{4a(a^2+b^2)}E(\alpha, r)$$ $[a^2 > b^2 > -\infty,\ u^2 > a^2 > 0]$. BY (263.05, 266.02)

6. $$\int_u^\infty \frac{x^2\,dx}{\sqrt{(x^4+2b^2x^2+a^4)^3}} = \frac{a}{2(a^4-b^4)}E(\alpha, r) - \frac{1}{4a(a^2-b^2)}F(\alpha, r) - \frac{u(u^2-a^2)}{2(a^2+b^2)(u^2+a^2)\sqrt{u^4+2b^2u^2+a^4}}$$ $[a^2 > b^2 > -\infty,\ a^2 > 0,\ u \geqslant 0]$. BY (263.08, 266.03)

7. $$\int_u^\infty \frac{(x^2-a^2)^2\,dx}{\sqrt{(x^4+2b^2x^2+a^4)^3}} = \frac{a}{a^2-b^2}[F(\alpha, r)-E(\alpha, r)] + \frac{u^2-a^2}{u^2+a^2}\,\frac{u}{\sqrt{u^4+2b^2u^2+a^4}}$$ $[|b^2| < a^2,\ u \geqslant 0]$. BY (266.08)

8. $$\int_u^\infty \frac{(x^2+a^2)^2\,dx}{\sqrt{(x^2+2b^2x^2+a^4)^3}} = \frac{a}{a^2+b^2}E(\alpha, r) - \frac{a^2-b^2}{a^2+b^2}\cdot\frac{u^2-a^2}{u^2+a^2}\cdot\frac{u}{\sqrt{u^4+2b^2u^2+a^4}}$$
$$[|b^2| < a^2,\ u \geqslant 0].$$ BY (266.06)a

9. $$\int_u^\infty \frac{(x^2-a^2)^2\,dx}{(x^2+a^2)^2\sqrt{x^4+2b^2x^2+a^4}} = \frac{a}{a^2-b^2}E(\alpha, r) - \frac{a^2+b^2}{2a(a^2-b^2)}F(\alpha, r)$$
$$[a^2 > b^2 > -\infty,\ a^2 > 0,\ u \geqslant 0].$$ BY (263.04, 266.07)

10. $$\int_u^\infty \frac{\sqrt{x^4+2b^2x^2+a^4}}{(x^2+a^2)^2}\,dx = \frac{1}{2a}E(\alpha, r) \quad [a^2 > b^2 > -\infty,\ a^2 > 0,\ u \geqslant 0].$$
BY (263.01, 266.01)

11. $$\int_u^\infty \frac{\sqrt{x^4+2b^2x^2+a^4}}{(x^2-a^2)^2}\,dx = \frac{1}{2a}[F(\alpha, r) - E(\alpha, r)] +$$
$$+\frac{u}{u^4-a^4}\sqrt{u^4+2b^2u^2+a^4} \quad [a > b > 0,\ u > a].$$ BY (263)

12. $$\int_u^\infty \frac{(x^2+a^2)^2\,dx}{[(x^2+a^2)^2-4a^2p^2x^2]\sqrt{x^4+2b^2x^2+a^4}} = \frac{1}{2a}\Pi(\alpha, p^2, r) \quad [a > b > 0, u \geqslant 0].$$
BY (263.02)

3.166 Notations: $\alpha = \arccos\frac{u^2-1}{u^2+1}$, $\beta = \operatorname{arctg}\left\{\left(1+\sqrt{2}\right)\frac{1-u}{1+u}\right\}$, $\gamma = \arccos u$, $\delta = \arccos\frac{1}{u}$, $\varepsilon = \arccos\frac{1-u^2}{1+u^2}$, $r = \frac{\sqrt{2}}{2}$, $q = 2\sqrt{3\sqrt{2}-4} = 2\sqrt[4]{2}\left(\sqrt{2}-1\right) = \sin 80°7'15'' \approx 0{,}985171$.

1. $$\int_u^\infty \frac{dx}{\sqrt{x^4+1}} = \frac{1}{2}F(\alpha, r) \quad [u \geqslant 0].$$ ZH (287, BY (263.50)

2. $$\int_u^\infty \frac{dx}{x^2\sqrt{x^4+1}} = \frac{1}{2}[F(\alpha, r) - 2E(\alpha, r)] + \frac{\sqrt{u^4+1}}{u(u^2+1)} \quad [u > 0].$$ BY (263.57)

3. $$\int_u^\infty \frac{x^2\,dx}{(x^4+1)\sqrt{x^4+1}} = \frac{1}{2}E(\alpha, r) - \frac{1}{4}F(\alpha, r) - \frac{u(u^2-1)}{2(u^2+1)\sqrt{u^4+1}} \quad [u \geqslant 0].$$
BY (263.59)

4. $$\int_u^\infty \frac{x^2\,dx}{(x^2+1)^2\sqrt{x^4+1}} = \frac{1}{4}[F(\alpha, r) - E(\alpha, r)] \quad [u \geqslant 0].$$ BY (263.53)

5. $$\int_u^\infty \frac{x^2\,dx}{(x^2-1)^2\sqrt{x^4+1}} = \frac{u\sqrt{u^4+1}}{2(u^4-1)} - \frac{1}{4}E(\alpha, r) \quad [u > 1].$$ BY (263.55)

6. $$\int_u^\infty \frac{\sqrt{x^4+1}}{(x^2-1)^2}\,dx = \frac{1}{2}[F(\alpha, r) - E(\alpha, r)] + \frac{u\sqrt{u^4+1}}{u^4-1} \quad [u > 1].$$
BY (263.58)

7. $\int_u^\infty \frac{(x^2-1)^2\,dx}{(x^2+1)^2\sqrt{x^4+1}} = E(\alpha, r) - \frac{1}{2}F(\alpha, r) \quad [u \geqslant 0].$ BY (263.54)

8. $\int_u^\infty \frac{\sqrt{x^4+1}\,dx}{(x^2+1)^2} = \frac{1}{2}E(\alpha, r) \quad [u \geqslant 0].$ BY (263.51)

9. $\int_u^\infty \frac{(x^2+1)^2\,dx}{[(x^2+1)^2-4p^2x^2]\sqrt{x^4+1}} = \frac{1}{2}\Pi(\alpha, p^2, r) \quad [u \geqslant 0].$ BY (263.52)

10. $\int_0^u \frac{dx}{\sqrt{x^4+1}} = \frac{1}{2}F(\varepsilon, r).$ ZH 66(288)

11. $\int_u^1 \frac{dx}{\sqrt{x^4+1}} = \left(2-\sqrt{2}\right)F(\beta, q) \quad [0 \leqslant u < 1].$ BY (264.50)

12. $\int_u^1 \frac{(x^2+x\sqrt{2}+1)\,dx}{(x^2-x\sqrt{2}+1)\sqrt{x^4+1}} = \left(2+\sqrt{2}\right)E(\beta, q) \quad [0 \leqslant u < 1].$ BY (264.51)

13. $\int_u^1 \frac{(1-x)^2\,dx}{(x^2-x\sqrt{2}+1)\sqrt{x^4+1}} = \frac{1}{\sqrt{2}}[F(\beta, q) - E(\beta, q)] \quad [0 \leqslant u < 1].$ BY (264.55)

14. $\int_u^1 \frac{(1+x)^2\,dx}{(x^2-x\sqrt{2}+1)\sqrt{x^4+1}} = \frac{3\sqrt{2}+4}{2}E(\beta, q) - \frac{3\sqrt{2}-4}{2}F(\beta, q) \quad [0 \leqslant u < 1].$ BY (264.56)

15. $\int_u^1 \frac{dx}{\sqrt{1-x^4}} = \frac{1}{\sqrt{2}}F(\gamma, r) \quad [u < 1].$ ZH 66(290), BY(259.75)

16. $\int_0^1 \frac{dx}{\sqrt{1-x^4}} = \frac{1}{4\sqrt{2\pi}}\left\{\Gamma\left(\frac{1}{4}\right)\right\}^2.$

17. $\int_1^u \frac{dx}{\sqrt{x^4-1}} = \frac{1}{\sqrt{2}}F(\delta, r) \quad [u > 1].$ ZH 66(289), BY(260.75)

18. $\int_u^1 \frac{x^2\,dx}{\sqrt{1-x^4}} = \sqrt{2}\,E(\gamma, r) - \frac{1}{\sqrt{2}}F(\gamma, r) \quad [u < 1].$ BY (259.76)

19. $\int_1^u \frac{x^2\,dx}{\sqrt{x^4-1}} = \frac{1}{\sqrt{2}}F(\delta, r) - \sqrt{2}\,E(\delta, r) + \frac{1}{u}\sqrt{u^4-1} \quad [u > 1].$ BY (260.77)

20. $\int_u^1 \frac{x^4\,dx}{\sqrt{1-x^4}} = \frac{1}{3\sqrt{2}}F(\gamma, r) + \frac{u}{3}\sqrt{1-u^4} \quad [u < 1].$ BY (259.76)

21. $\int_1^u \frac{x^4\,dx}{\sqrt{x^4-1}} = \frac{1}{3}F(\delta, r) + \frac{\sqrt{2}}{3}u\sqrt{u^4-1} \quad [u > 1].$ BY (260.77)

22. $\int_0^u \frac{dx}{\sqrt{x(1+x^3)}} = \frac{1}{\sqrt[4]{3}} F\left(\arccos \frac{1+(1-\sqrt{3})u}{1+(1+\sqrt{3})u}, \frac{\sqrt{2+\sqrt{3}}}{2}\right)$

$[u > 0]$. BY (260.50)

23. $\int_0^u \frac{dx}{\sqrt{x(1-x^3)}} = \frac{1}{\sqrt[4]{3}} F\left(\arccos \frac{1-(1+\sqrt{3})u}{1+(\sqrt{3}-1)u}, \frac{\sqrt{2-\sqrt{3}}}{2}\right)$

$[1 \geqslant u > 0]$. BY (259.50)

In **3.167** and **3.168** we set: $\alpha = \arcsin \sqrt{\frac{(a-c)(d-u)}{(a-d)(c-u)}}$,

$\beta = \arcsin \sqrt{\frac{(a-c)(u-d)}{(c-d)(a-u)}}$, $\gamma = \arcsin \sqrt{\frac{(b-d)(c-u)}{(c-d)(b-u)}}$,

$\delta = \arcsin \sqrt{\frac{(b-d)(u-c)}{(b-c)(u-d)}}$, $\varkappa = \arcsin \sqrt{\frac{(a-c)(b-u)}{(b-c)(a-u)}}$,

$\lambda = \arcsin \sqrt{\frac{(a-c)(u-b)}{(a-b)(u-c)}}$, $\mu = \arcsin \sqrt{\frac{(b-d)(a-u)}{(a-b)(u-d)}}$,

$\nu = \arcsin \sqrt{\frac{(b-d)(u-a)}{(a-d)(u-b)}}$, $q = \sqrt{\frac{(b-c)(a-d)}{(a-c)(b-d)}}$, $r = \sqrt{\frac{(a-b)(c-d)}{(a-c)(b-d)}}$.

3.167

1. $\int_u^d \sqrt{\frac{d-x}{(a-x)(b-x)(c-x)}}\, dx = \frac{2(c-d)}{\sqrt{(a-c)(b-d)}} \left\{\Pi\left(\alpha, \frac{a-d}{a-c}, q\right) - F(\alpha, q)\right\}$

$[a > b > c > d > u]$. BY (251.05)

2. $\int_d^u \sqrt{\frac{x-d}{(a-x)(b-x)(c-x)}}\, dx = \frac{2(d-a)}{\sqrt{(a-c)(b-d)}} \left\{\Pi\left(\beta, \frac{d-c}{a-c}, r\right) - F(\beta, r)\right\}$

$[a > b > c \geqslant u > d]$. BY (252.14)

3. $\int_u^c \sqrt{\frac{x-d}{(a-x)(b-x)(c-x)}}\, dx =$

$= \frac{2}{\sqrt{(a-c)(b-d)}} \left\{(c-b)\Pi\left(\gamma, \frac{c-d}{b-d}, r\right) + (b-d)F(\gamma, r)\right\}$

$[a > b > c > u \geqslant d]$. BY (253.14)

4. $\int_c^u \sqrt{\frac{x-d}{(a-x)(b-x)(x-c)}}\, dx = \frac{2(c-d)}{\sqrt{(a-c)(b-d)}} \Pi\left(\delta, \frac{b-c}{b-d}, q\right)$

$[a > b \geqslant u > c > d]$. BY (254.02)

5. $\int_u^b \sqrt{\frac{x-d}{(a-x)(b-x)(x-c)}}\, dx =$

$= \frac{2}{\sqrt{(a-c)(b-d)}} \left\{(b-a)\Pi\left(\varkappa, \frac{b-c}{a-c}, q\right) + (a-d)F(\varkappa, q)\right\}$

$[a > b > u \geqslant c > d]$. BY (255.20)

6. $$\int_b^u \sqrt{\frac{x-d}{(a-x)(x-b)(x-c)}}\,dx = \frac{2}{\sqrt{(a-c)(b-d)}}\left\{(b-c)\,\Pi\left(\lambda, \frac{a-b}{a-c}, r\right) + (c-d)\,F(\lambda, r)\right\}$$

$[a \geqslant u > b > c > d]$. BY (256.13)

7. $$\int_u^a \sqrt{\frac{x-d}{(a-x)(x-b)(x-c)}}\,dx = \frac{2(a-d)}{\sqrt{(a-c)(b-d)}}\,\Pi\left(\mu, \frac{b-a}{b-d}, r\right)$$

$[a > u \geqslant b > c > d]$. BY (257.02)

8. $$\int_a^u \sqrt{\frac{x-d}{(x-a)(x-b)(x-c)}}\,dx = \frac{2}{\sqrt{(a-c)(b-d)}}\left\{(a-b)\,\Pi\left(\nu, \frac{a-d}{b-d}, q\right) + (b-d)\,F(\nu, q)\right\}$$

$[u > a > b > c > d]$. BY (258.14)

9. $$\int_u^d \sqrt{\frac{c-x}{(a-x)(b-x)(d-x)}}\,dx = \frac{2(c-d)}{\sqrt{(a-c)(b-d)}}\,\Pi\left(\alpha, \frac{a-d}{a-c}, q\right)$$

$[a > b > c > d > u]$. BY (251.02)

10. $$\int_d^u \sqrt{\frac{c-x}{(a-x)(b-x)(x-d)}}\,dx = \frac{2}{\sqrt{(a-c)(b-d)}}\left[(a-d)\,\Pi\left(\beta, \frac{d-c}{a-c}, r\right) - (a-c)\,F(\beta, r)\right]$$

$[a > b > c \geqslant u > d]$. BY (252.13)

11. $$\int_u^c \sqrt{\frac{c-x}{(a-x)(b-x)(x-d)}}\,dx = \frac{2(b-c)}{\sqrt{(a-c)(b-d)}}\left[\Pi\left(\gamma, \frac{c-d}{b-d}, r\right) - F(\gamma, r)\right]$$

$[a > b > c > u \geqslant d]$. BY (253.13)

12. $$\int_c^u \sqrt{\frac{x-c}{(a-x)(b-x)(x-d)}}\,dx = \frac{2(c-d)}{\sqrt{(a-c)(b-d)}}\left[\Pi\left(\delta, \frac{b-c}{b-d}, q\right) - F(\delta, q)\right]$$

$[a > b \geqslant u > c > d]$. BY (254.12)

13. $$\int_u^b \sqrt{\frac{x-c}{(a-x)(b-x)(x-d)}}\,dx = \frac{2}{\sqrt{(a-c)(b-d)}}\left[(b-a)\,\Pi\left(\varkappa, \frac{b-c}{a-c}, q\right) + (a-c)\,F(\varkappa, q)\right]$$

$[a > b > u \geqslant c > d]$. BY (259.19)

14. $$\int_b^u \sqrt{\frac{x-c}{(a-x)(x-b)(x-d)}}\,dx = \frac{2(b-c)}{\sqrt{(a-c)(b-d)}}\,\Pi\left(\lambda, \frac{a-b}{a-c}, r\right)$$

$[a \geqslant u > b > c > d]$. BY (256.02)

15. $$\int_u^a \sqrt{\frac{x-c}{(a-x)(x-b)(x-d)}}\,dx = \frac{2}{\sqrt{(a-c)(b-d)}}\left[(a-d)\Pi\left(\mu, \frac{b-a}{b-d}, r\right)+\right.$$
$$\left. +(d-c)F(\mu, r)\right] \qquad [a > u \geqslant b > c > d].$$ BY (257.13)

16. $$\int_a^u \sqrt{\frac{x-c}{(x-a)(x-b)(x-d)}}\,dx =$$
$$= \frac{2}{\sqrt{(a-c)(b-d)}}\left[(a-b)\Pi\left(\nu, \frac{a-d}{b-d}, q\right) + (b-c)F(\nu, q)\right]$$
$$[u > a > b > c > d].$$ BY (258.13)

17. $$\int_u^d \sqrt{\frac{b-x}{(a-x)(c-x)(d-x)}}\,dx =$$
$$= \frac{2}{\sqrt{(a-c)(b-d)}}\left[(c-d)\Pi\left(\alpha, \frac{a-d}{a-c}, q\right) + (b-c)F(\alpha, q)\right]$$
$$[a > b > c > d > u].$$ BY (251.07)

18. $$\int_d^u \sqrt{\frac{b-x}{(a-x)(c-x)(x-d)}}\,dx =$$
$$= \frac{2}{\sqrt{(a-c)(b-d)}}\left[(a-d)\Pi\left(\beta, \frac{d-c}{a-c}, r\right) - (a-b)F(\beta, r)\right]$$
$$[a > b > c \geqslant u > d].$$ BY (252.15)

19. $$\int_u^c \sqrt{\frac{b-x}{(a-x)(c-x)(x-d)}}\,dx = \frac{2(b-c)}{\sqrt{(a-c)(b-d)}}\Pi\left(\gamma, \frac{c-d}{b-d}, r\right)$$
$$[a > b > c > u \geqslant d].$$ BY (253.02)

20. $$\int_c^u \sqrt{\frac{b-x}{(a-x)(x-c)(x-d)}}\,dx =$$
$$= \frac{2}{\sqrt{(a-c)(b-d)}}\left[(d-c)\Pi\left(\delta, \frac{b-c}{b-d}, q\right) + (b-d)F(\delta, q)\right]$$
$$[a > b \geqslant u > c > d].$$ BY (254.14)

21. $$\int_u^b \sqrt{\frac{b-x}{(a-x)(x-c)(x-d)}}\,dx =$$
$$= \frac{2(a-b)}{\sqrt{(a-c)(b-d)}}\left[\Pi\left(\varkappa, \frac{b-c}{a-c}, q\right) - F(\varkappa, q)\right]$$
$$[a > b > u \geqslant c > d].$$ BY (255.21)

22. $$\int_b^u \sqrt{\frac{x-b}{(a-x)(x-c)(x-d)}}\,dx = \frac{2(b-c)}{\sqrt{(a-c)(b-d)}}\left[\Pi\left(\lambda, \frac{a-b}{a-c}, r\right) - F(\lambda, r)\right]$$
$$[a \geqslant u > b > c > d].$$ BY (256.15)

23. $$\int_u^a \sqrt{\frac{x-b}{(a-x)(x-c)(x-d)}}\,dx = \frac{2}{\sqrt{(a-c)(b-d)}}\left[(d-a)\,\Pi\left(\mu, \frac{b-a}{b-d}, r\right) - (b-d)\,F(\mu, r)\right]$$
$[a > u \geqslant b > c > d]$. BY (257.15)

24. $$\int_a^u \sqrt{\frac{x-b}{(x-a)(x-c)(x-d)}}\,dx = \frac{2(a-b)}{\sqrt{(a-c)(b-d)}}\,\Pi\left(\nu, \frac{a-d}{b-d}, q\right)$$
$[u > a > b > c > d]$. BY (258.02)

25. $$\int_u^d \sqrt{\frac{a-x}{(b-x)(c-x)(d-x)}}\,dx = \frac{2}{\sqrt{(a-c)(b-d)}}\left[(c-d)\,\Pi\left(\alpha, \frac{a-d}{a-c}, q\right) + (a-c)\,F(\alpha, q)\right]$$
$[a > b > c > d > u]$. BY (251.06)

26. $$\int_d^u \sqrt{\frac{a-x}{(b-x)(c-x)(x-d)}}\,dx = \frac{2(a-d)}{\sqrt{(a-c)(b-d)}}\,\Pi\left(\beta, \frac{d-c}{a-c}, r\right)$$
$[a > b > c \geqslant u > d]$. BY (252.02)

27. $$\int_u^c \sqrt{\frac{a-x}{(b-x)(c-x)(x-d)}}\,dx = \frac{2}{\sqrt{(a-c)(b-d)}}\left[(b-c)\,\Pi\left(\gamma, \frac{c-d}{b-d}, r\right) + (a-b)\,F(\gamma, r)\right]$$
$[a > b > c > u \geqslant d]$. BY (253.15)

28. $$\int_c^u \sqrt{\frac{a-x}{(b-x)(x-c)(x-d)}}\,dx = \frac{2}{\sqrt{(a-c)(b-d)}}\left[(d-c)\,\Pi\left(\delta, \frac{b-c}{b-d}, q\right) + (a-d)\,F(\delta, q)\right]$$
$[a > b \geqslant u > c > d]$. BY (254.13)

29. $$\int_u^b \sqrt{\frac{a-x}{(b-x)(x-c)(x-d)}}\,dx = \frac{2(a-b)}{\sqrt{(a-c)(b-d)}}\,\Pi\left(\varkappa, \frac{b-c}{a-c}, q\right)$$
$[a > b > u \geqslant c > d]$. BY (255.02)

30. $$\int_b^u \sqrt{\frac{a-x}{(x-b)(x-c)(x-d)}}\,dx = \frac{2}{\sqrt{(a-c)(b-d)}}\left[(c-b)\,\Pi\left(\lambda, \frac{a'-b}{a-c}, r\right) + (a-c)\,F(\lambda, r)\right]$$
$[a \geqslant u > b > c > d]$. BY (256.14)

31. $\int\limits_u^a \sqrt{\frac{a-x}{(x-b)(x-c)(x-d)}}\,dx = \frac{2(d-a)}{\sqrt{(a-c)(b-d)}}\left[\Pi\left(\mu, \frac{b-a}{b-d}, r\right) - F(\mu, r)\right]$

$[a > u \geqslant b > c > d]$. BY (257.14)

32. $\int\limits_a^u \sqrt{\frac{x-a}{(x-b)(x-c)(x-d)}}\,dx = \frac{2(a-b)}{\sqrt{(a-c)(b-d)}}\left[\Pi\left(\nu, \frac{a-d}{b-d}, q\right) - F(\nu, q)\right]$

$[u > a > b > c > d]$. BY (258.15)

3.168

1. $\int\limits_u^c \sqrt{\frac{c-x}{(a-x)(b-x)(x-d)^3}}\,dx =$

$= \frac{2}{d-a}\left[\sqrt{\frac{a-c}{b-d}}\,E(\gamma, r) - \sqrt{\frac{(a-u)(c-u)}{(b-u)(u-d)}}\right]$

$[a > b > c > u > d]$. BY (253.06)

2. $\int\limits_c^u \sqrt{\frac{x-c}{(a-x)(b-x)(x-d)^3}}\,dx = \frac{2}{a-d}\sqrt{\frac{a-c}{b-d}}\,[F(\delta, q) - E(\delta, q)]$

$[a > b \geqslant u > c > d]$. BY (254.04)

3. $\int\limits_u^b \sqrt{\frac{x-c}{(a-x)(b-x)(x-d)^3}}\,dx = \frac{2}{a-d}\sqrt{\frac{a-c}{b-d}}\,[F(\varkappa, q) - E(\varkappa, q)] +$

$+ \frac{2}{b-d}\sqrt{\frac{(b-u)(u-c)}{(a-u)(u-d)}}$ $[a > b > u \geqslant c > d]$. BY (255.09)

4. $\int\limits_b^u \sqrt{\frac{x-c}{(a-x)(x-b)(x-d)^3}}\,dx =$

$= \frac{2}{a-d}\left[\sqrt{\frac{a-c}{b-d}}\,E(\lambda, r) - \frac{c-d}{b-d}\sqrt{\frac{(a-u)(u-b)}{(u-c)(u-d)}}\right]$

$[a \geqslant u > b > c > d]$. BY (256.06)

5. $\int\limits_u^a \sqrt{\frac{x-c}{(a-x)(x-b)(x-d)^3}}\,dx = \frac{2}{a-d}\sqrt{\frac{a-c}{b-d}}\,E(\mu, r)$

$[a > u \geqslant b > c > d]$. BY (257.01)

6. $\int\limits_a^u \sqrt{\frac{x-c}{(x-a)(x-b)(x-d)^3}}\,dx =$

$= \frac{2}{a-d}\sqrt{\frac{a-c}{b-d}}\,[F(\nu, q) - E(\nu, q)] + \frac{2}{a-d}\sqrt{\frac{(u-a)(u-c)}{(u-b)(u-d)}}$

$[u > a > b > c > d]$. BY (258.10)

7. $\int\limits_u^c \sqrt{\frac{b-x}{(a-x)(c-x)(x-d)^3}}\,dx =$

$= \frac{2}{(a-d)(c-d)\sqrt{(a-c)(b-d)}}\,[(b-c)(a-d)F(\gamma, r) - (a-c)(b-d)E(\gamma, r)] +$

$+ \frac{2(b-d)}{(a-d)(c-d)}\sqrt{\frac{(a-u)(c-u)}{(b-u)(u-d)}}$ $[a > b > c > u > d]$. BY (253.03)

8. $$\int_c^u \sqrt{\frac{b-x}{(a-x)(x-c)(x-d)^3}}\,dx = \frac{2}{(a-d)(c-d)\sqrt{(a-c)(b-d)}} \times$$
$$\times [(a-c)(b-d)E(\delta, q) - (a-b)(c-d)F(\delta, q)]$$
$$[a > b \geqslant u > c > d].$$ BY (254.15)

9. $$\int_u^b \sqrt{\frac{b-x}{(a-x)(x-c)(x-d)^3}}\,dx = \frac{2}{(a-d)(c-d)\sqrt{(a-c)(b-d)}} \times$$
$$\times [(a-c)(b-d)E(\varkappa, q) - (a-b)(c-d)F(\varkappa, q)] -$$
$$- \frac{2}{c-d}\sqrt{\frac{(b-u)(u-c)}{(a-u)(u-d)}} \qquad [a > b > u \geqslant c > d].$$ BY (255.06)

10. $$\int_b^u \sqrt{\frac{x-b}{(a-x)(x-c)(x-d)^3}}\,dx = \frac{2}{(a-d)(c-d)\sqrt{(a-c)(b-d)}} \times$$
$$\times [(a-c)(b-d)E(\lambda, r) - (a-d)(b-c)F(\lambda, r)] -$$
$$- \frac{2}{a-d}\sqrt{\frac{(a-u)(u-b)}{(u-c)(u-d)}} \qquad [a \geqslant u > b > c > d].$$ BY (256.03)

11. $$\int_u^a \sqrt{\frac{x-b}{(a-x)(x-c)(x-d)^3}}\,dx = 2\frac{\sqrt{(a-c)(b-d)}}{(a-d)(c-d)} E(\mu, r) -$$
$$- \frac{2(b-c)}{(c-d)\sqrt{(a-c)(b-d)}} F(\mu, r) \quad [a > u \geqslant b > c > d].$$ BY (257.09)

12. $$\int_a^u \sqrt{\frac{x-b}{(x-a)(x-c)(x-d)^3}}\,dx = \frac{2(b-d)}{(a-d)(c-d)}\sqrt{\frac{(u-a)(u-c)}{(u-b)(u-d)}} +$$
$$+ \frac{2(a-b)}{(a-d)\sqrt{(a-c)(b-d)}} F(\nu, q) + 2\frac{\sqrt{(a-c)(b-d)}}{(a-d)(c-d)} E(\nu, q)$$
$$[u > a > b > c > d].$$ BY (258.09)

13. $$\int_u^c \sqrt{\frac{a-x}{(b-x)(c-x)(x-d)^3}}\,dx = \frac{2}{c-d}\sqrt{\frac{a-c}{b-d}}[F(\gamma, r) - E(\gamma, r)] +$$
$$+ \frac{2}{c-d}\sqrt{\frac{(a-u)(c-u)}{(b-u)(u-d)}} \qquad [a > b > c > u > d].$$ BY (253.04)

14. $$\int_c^u \sqrt{\frac{a-x}{(b-x)(x-c)(x-d)^3}}\,dx = \frac{2}{c-d}\sqrt{\frac{a-c}{b-d}} E(\delta, q)$$
$$[a > b \geqslant u > c > d].$$ BY (254.01)

15. $$\int_u^b \sqrt{\frac{a-x}{(b-x)(x-c)(x-d)^3}}\,dx = \frac{2}{c-d}\sqrt{\frac{a-c}{b-d}} E(\varkappa, q) -$$
$$- \frac{2(a-d)}{(b-d)(c-d)}\sqrt{\frac{(b-u)(u-c)}{(a-u)(u-d)}} \qquad [a > b > u \geqslant c > d].$$ BY (255.08)

16. $$\int_b^u \sqrt{\frac{a-x}{(x-b)(x-c)(x-d)^3}}\,dx =$$

$$= \frac{2}{c-d}\sqrt{\frac{a-c}{b-d}}\,[F(\lambda, r) - E(\lambda, r)] + \frac{2}{b-d}\sqrt{\frac{(a-u)(u-b)}{(u-c)(u-d)}}$$

$[a \geqslant u > b > c > d]$. BY (256.05)

17. $$\int_u^a \sqrt{\frac{a-x}{(x-b)(x-c)(x-d)^3}}\,dx = \frac{2}{c-d}\sqrt{\frac{a-c}{b-d}}\,[F(\mu, r) - E(\mu, r)]$$

$[a > u \geqslant b > c > d]$. BY (257.06)

18. $$\int_a^u \sqrt{\frac{x-a}{(x-b)(x-c)(x-d)^3}}\,dx = \frac{-2}{c-d}\sqrt{\frac{a-c}{b-d}}\,E(\nu, q) +$$

$$+ \frac{2}{c-d}\sqrt{\frac{(u-a)(u-c)}{(u-b)(u-d)}}$$

$[u > a > b > c > d]$. BY (258.05)

19. $$\int_u^d \sqrt{\frac{d-x}{(a-x)(b-x)(c-x)^3}}\,dx = \frac{2}{b-c}\sqrt{\frac{b-d}{a-c}}\,[F(\alpha, q) - E(\alpha, q)]$$

$[a > b > c > d > u]$. BY (251.01)

20. $$\int_d^u \sqrt{\frac{x-d}{(a-x)(b-x)(c-x)^3}}\,dx = \frac{-2}{b-c}\sqrt{\frac{b-d}{a-c}}\,E(\beta, r) +$$

$$+ \frac{2}{b-c}\sqrt{\frac{(b-u)(u-d)}{(a-u)(c-u)}}$$

$[a > b > c \geqslant u > d]$. BY (252.06)

21. $$\int_u^b \sqrt{\frac{x-d}{(a-x)(b-x)(x-c)^3}}\,dx =$$

$$= \frac{2}{b-c}\sqrt{\frac{b-d}{a-c}}\,[F(\varkappa, q) - E(\varkappa, q)] + \frac{2}{b-c}\sqrt{\frac{(b-u)(u-d)}{(a-u)(u-c)}}$$

$[a > b > u > c > d]$. BY (255.05)

22. $$\int_b^u \sqrt{\frac{x-d}{(a-x)(x-b)(x-c)^3}}\,dx = \frac{2}{b-c}\sqrt{\frac{b-d}{a-c}}\,E(\lambda, r)$$

$[a \geqslant u > b > c > d]$. BY (256.01)

23. $$\int_u^a \sqrt{\frac{x-d}{(a-x)(x-b)(x-c)^3}}\,dx =$$

$$= \frac{2}{b-c}\sqrt{\frac{b-d}{a-c}}\,E(\mu, r) - \frac{2(c-d)}{(a-c)(b-c)}\sqrt{\frac{(a-u)(u-b)}{(u-c)(u-d)}}$$

$[a > u \geqslant b > c > d]$. BY (257.06)

24. $$\int_a^u \sqrt{\frac{x-d}{(x-a)(x-b)(x-c)^3}}\,dx =$$

$$= \frac{2}{b-c}\sqrt{\frac{b-d}{a-c}}\,[F(\nu, q) - E(\nu, q)] + \frac{2}{a-c}\sqrt{\frac{(u-a)(u-d)}{(u-b)(u-c)}}$$

$[u > a > b > c > d]$. BY (258.06)

25. $$\int_u^a \sqrt{\frac{b-x}{(a-x)(c-x)^3(d-x)}}\,dx = \frac{2}{c-d}\sqrt{\frac{b-d}{a-c}}E(\alpha, q)$$

$[a > b > c > d > u]$. BY (251.01)

26. $$\int_d^u \sqrt{\frac{b-x}{(a-x)(c-x)^3(x-d)}}\,dx =$$

$$= \frac{2}{c-d}\sqrt{\frac{b-d}{a-c}}[F(\beta, r) - E(\beta, r)] + \frac{2}{c-d}\sqrt{\frac{(b-u)(u-d)}{(a-u)(c-u)}}$$

$[a > b > c > u > d]$. BY (252.03)

27. $$\int_u^b \sqrt{\frac{b-x}{(a-x)(x-c)^3(x-d)}}\,dx =$$

$$= \frac{2}{d-c}\sqrt{\frac{b-d}{a-c}}E(\varkappa, q) + \frac{2}{c-d}\sqrt{\frac{(b-u)(u-d)}{(a-u)(u-c)}}$$

$[a > b > u > c > d]$. BY (255.03)

28. $$\int_b^u \sqrt{\frac{x-b}{(a-x)(x-c)^3(x-d)}}\,dx = \frac{2}{c-d}\sqrt{\frac{b-d}{a-c}}[F(\lambda, r) - E(\lambda, r)]$$

$[a \geqslant u > b > c > d]$. BY (256.08)

29. $$\int_u^a \sqrt{\frac{x-b}{(a-x)(x-c)^3(x-d)}}\,dx =$$

$$= \frac{2}{c-d}\sqrt{\frac{b-d}{a-c}}[F(\mu, r) - E(\mu, r)] + \frac{2}{a-c}\sqrt{\frac{(a-u)(u-b)}{(u-c)(u-d)}}$$

$[a > u \geqslant b > c > d]$. BY (257.03)

30. $$\int_a^u \sqrt{\frac{x-b}{(x-a)(x-c)^3(x-d)}}\,dx =$$

$$= \frac{2}{c-d}\sqrt{\frac{b-d}{a-c}}E(\nu, q) - \frac{2(b-c)}{(a-c)(c-d)}\sqrt{\frac{(u-a)(u-d)}{(u-b)(u-c)}}$$

$[u > a > b > c > d]$. BY (258.03)

31. $$\int_u^d \sqrt{\frac{a-x}{(b-x)(c-x)^3(d-x)}}\,dx =$$

$$= \frac{2\sqrt{(a-c)(b-d)}}{(b-c)(c-d)}E(\alpha, q) - \frac{a-b}{b-c}\frac{2}{\sqrt{(a-c)(b-d)}}F(\alpha, q)$$

$[a > b > c > d > u]$. BY (251.08)

32. $$\int_d^u \sqrt{\frac{a-x}{(b-x)(c-x)^3(x-d)}}\,dx = \frac{2(a-d)}{(c-d)\sqrt{(a-c)(b-d)}}F(\beta, r) -$$

$$-2\frac{\sqrt{(a-c)(b-d)}}{(b-c)(c-d)}E(\beta, r) + 2\frac{a-c}{(b-c)(c-d)}\sqrt{\frac{(b-u)(u-d)}{(a-u)(c-u)}}$$

$[a > b > c > u > d]$. BY (252.04)

33. $$\int_u^b \sqrt{\frac{a-x}{(b-x)(x-c)^3(x-d)}}\,dx = \frac{2(a-b)}{(b-c)\sqrt{(a-c)(b-d)}}F(\varkappa, q) - 2\frac{\sqrt{(a-c)(b-d)}}{(b-c)(c-d)}E(\varkappa, q) + \frac{2(a-c)}{(b-c)(c-d)}\sqrt{\frac{(b-u)(u-d)}{(a-u)(u-c)}}$$

$[a > b > u > c > d]$. BY (255.04)

34. $$\int_b^u \sqrt{\frac{a-x}{(x-b)(x-c)^3(x-d)}}\,dx = \frac{2\sqrt{(a-c)(b-d)}}{(b-c)(c-d)}E(\lambda, r) - \frac{2(a-d)}{(c-d)\sqrt{(a-c)(b-d)}}F(\lambda, r)$$

$[a \geqslant u > b > c > d]$. BY (256.09)

35. $$\int_u^a \sqrt{\frac{a-x}{(x-b)(x-c)^3(x-d)}}\,dx = \frac{2\sqrt{(a-c)(b-d)}}{(b-c)(c-d)}E(\mu, r) - \frac{2(a-d)}{(c-d)\sqrt{(a-c)(b-d)}}F(\mu, r) - \frac{2}{b-c}\sqrt{\frac{(a-u)(u-b)}{(u-c)(u-d)}}$$

$[a > u \geqslant b > c > d]$. BY (257.04)

36. $$\int_a^u \sqrt{\frac{x-a}{(x-b)(x-c)^3(x-d)}}\,dx = \frac{2\sqrt{(a-c)(b-d)}}{(b-c)(c-d)}E(\nu, q) - \frac{2(a-b)}{(b-c)\sqrt{(a-c)(b-d)}}F(\nu, q) - \frac{2}{c-d}\sqrt{\frac{(u-a)(u-d)}{(u-b)(u-c)}}$$

$[u > a > b > c > d]$. BY (258.04)

37. $$\int_u^d \sqrt{\frac{d-x}{(a-x)(b-x)^3(c-x)}}\,dx = \frac{2\sqrt{(a-c)(b-d)}}{(a-b)(b-c)}E(\alpha, q) - \frac{2(c-d)}{(b-c)\sqrt{(a-c)(b-d)}}F(\alpha, q) - \frac{2}{a-b}\sqrt{\frac{(a-u)(d-u)}{(b-u)(c-u)}}$$

$[a > b > c > d > u]$. BY (251.11)

38. $$\int_d^u \sqrt{\frac{x-d}{(a-x)(b-x)^3(c-x)}}\,dx = \frac{2\sqrt{(a-c)(b-d)}}{(a-b)(b-c)}E(\beta, r) - \frac{2(a-d)}{(a-b)\sqrt{(a-c)(b-d)}}F(\beta, r) + \frac{2}{b-c}\sqrt{\frac{(c-u)(u-d)}{(a-u)(b-u)}}$$

$[a > b > c \geqslant u > d]$. BY (252.07)

39. $$\int_u^c \sqrt{\frac{x-d}{(a-x)(b-x)^3(c-x)}}\,dx = \frac{2\sqrt{(a-c)(b-d)}}{(a-b)(b-c)}E(\gamma, r) - \frac{2(a-d)}{(a-b)\sqrt{(a-c)(b-d)}}F(\gamma, r)$$

$[a > b > c > u \geqslant d]$. BY (253.07)

40. $$\int_c^u \sqrt{\frac{x-d}{(a-x)(b-x)^3(x-c)}}\,dx = \frac{2(c-d)}{(b-c)\sqrt{(a-c)(b-d)}}F(\delta, q) - \frac{2\sqrt{(a-c)(b-d)}}{(a-b)(b-c)}E(\delta, q) + \frac{2(b-d)}{(a-b)(b-c)}\sqrt{\frac{(a-u)(u-c)}{(b-u)(u-d)}}$$
$[a > b > u > c > d]$. BY (254.05)

41. $$\int_u^a \sqrt{\frac{x-d}{(a-x)(x-b)^3(x-c)}}\,dx = \frac{2(a-d)}{(a-b)\sqrt{(a-c)(b-d)}}F(\mu, r) - \frac{2\sqrt{(a-c)(b-d)}}{(a-b)(b-c)}E(\mu, r) + \frac{2(b-d)}{(a-b)(b-c)}\sqrt{\frac{(a-u)(u-c)}{(u-b)(u-d)}}$$
$[a > u > b > c > d]$. BY (257.07)

42. $$\int_a^u \sqrt{\frac{x-d}{(x-a)(x-b)^3(x-c)}}\,dx = \frac{2\sqrt{(a-c)(b-d)}}{(a-b)(b-c)}E(\nu, q) - \frac{2(c-d)}{(b-c)\sqrt{(a-c)(b-d)}}F(\nu, q)$$
$[u > a > b > c > d]$. BY (258.07)

43. $$\int_u^d \sqrt{\frac{c-x}{(a-x)(b-x)^3(d-x)}}\,dx = \frac{2}{a-b}\sqrt{\frac{a-c}{b-d}}E(\alpha, q) - \frac{2(b-c)}{(a-b)(b-d)}\sqrt{\frac{(a-u)(d-u)}{(b-u)(c-u)}}$$
$[a > b > c > d > u]$. BY (251.14)

44. $$\int_d^u \sqrt{\frac{c-x}{(a-x)(b-x)^3(x-d)}}\,dx = \frac{2}{a-b}\sqrt{\frac{a-c}{b-d}}[F(\beta, r) - E(\beta, r)] + \frac{2}{b-d}\sqrt{\frac{(c-u)(u-d)}{(a-u)(b-u)}}$$
$[a > b > c \geqslant u > d]$. BY (252.10)

45. $$\int_u^c \sqrt{\frac{c-x}{(a-x)(b-x)^3(x-d)}}\,dx = \frac{2}{a-b}\sqrt{\frac{a-c}{b-d}}[F(\gamma, r) - E(\gamma, r)]$$
$[a > b > c > u \geqslant d]$. BY (254.08)

46. $$\int_c^u \sqrt{\frac{x-c}{(a-x)(b-x)^3(x-d)}}\,dx = \frac{2}{b-a}\sqrt{\frac{a-c}{b-d}}E(\delta, q) + \frac{2}{a-b}\sqrt{\frac{(a-u)(u-c)}{(b-u)(u-d)}}$$
$[a > b \geqslant u > c > d]$. BY (254.08)

47. $$\int_u^a \sqrt{\frac{x-c}{(a-x)(x-b)^3(x-d)}}\,dx = \frac{2}{a-b}\sqrt{\frac{a-c}{b-d}}[F(\mu, r) - E(\mu, r)] + \frac{2}{a-b}\sqrt{\frac{(a-u)(u-c)}{(u-b)(u-d)}}$$
$[a > u \geqslant b > c > d]$. BY (257.10)

48. $$\int_a^u \sqrt{\frac{x-c}{(x-a)(x-b)^3(x-d)}}\,dx = \frac{2}{a-b}\sqrt{\frac{a-c}{b-d}}E(\nu, q)$$
$[u > a > b > c > d]$. BY (258.01)

49. $$\int_u^d \sqrt{\frac{a-x}{(b-x)^3(c-x)(d-x)}}\,dx = \frac{2}{b-c}\sqrt{\frac{a-c}{b-d}}\,[F(\alpha, q)-E(\alpha, q)]+\frac{2}{b-d}\sqrt{\frac{(a-u)(d-u)}{(b-u)(c-u)}}$$
$[a>b>c>d>u]$. BY (251.12)

50. $$\int_d^u \sqrt{\frac{a-x}{(b-x)^3(c-x)(x-d)}}\,dx = \frac{2}{b-c}\sqrt{\frac{a-c}{b-d}}\,E(\beta, r)-\frac{2(a-b)}{(b-c)(b-d)}\sqrt{\frac{(u-d)(c-u)}{(a-u)(b-u)}}$$
$[a>b>c\geqslant u>d]$. BY (252.09)

51. $$\int_u^c \sqrt{\frac{a-x}{(b-x)^3(c-x)(x-d)}}\,dx = \frac{2}{b-c}\sqrt{\frac{a-c}{b-d}}\,E(\gamma, r)$$
$[a>b>c>u\geqslant d]$. BY (253.01)

52. $$\int_c^u \sqrt{\frac{a-x}{(b-x)^3(x-c)(x-d)}}\,dx = \frac{2}{b-c}\sqrt{\frac{a-c}{b-d}}\,[F(\delta, q)-E(\delta, q)]+\frac{2}{b-c}\sqrt{\frac{(a-u)(u-c)}{(b-u)(u-d)}}$$
$[a>b>u>c>d]$. BY (254.06)

53. $$\int_u^a \sqrt{\frac{a-x}{(x-b)^3(x-c)(x-d)}}\,dx = \frac{2}{c-b}\sqrt{\frac{a-c}{b-d}}\,E(\mu, r)+\frac{2}{b-c}\sqrt{\frac{(a-u)(u-c)}{(u-b)(u-d)}}$$
$[a>u>b>c>d]$. BY (257.08)

54. $$\int_a^u \sqrt{\frac{x-a}{(x-b)^3(x-c)(x-d)}}\,dx = \frac{2}{b-c}\sqrt{\frac{a-c}{b-d}}\,[F(\nu, q)-E(\nu, q)]$$
$[u>a>b>c>d]$. BY (258.08)

55. $$\int_u^d \sqrt{\frac{d-x}{(a-x)^3(b-x)(c-x)}}\,dx = \frac{2}{b-a}\sqrt{\frac{b-d}{a-c}}\,E(\alpha, q)+\frac{2}{a-b}\sqrt{\frac{(b-u)(d-u)}{(a-u)(c-u)}}$$
$[a>b>c>d>u]$. BY (251.09)

56. $$\int_d^u \sqrt{\frac{x-d}{(a-x)^3(b-x)(c-x)}}\,dx = \frac{2}{a-b}\sqrt{\frac{b-d}{a-c}}\,[F(\beta, q)-E(\beta, q)]$$
$[a>b>c\geqslant u>d]$. BY (252.05)

57. $$\int_u^c \sqrt{\frac{x-d}{(a-x)^3(b-x)(c-x)}}\,dx = \frac{2}{a-b}\sqrt{\frac{b-d}{a-c}}\,[F(\gamma, r)-E(\gamma, r)]+\frac{2}{a-c}\sqrt{\frac{(c-u)(u-d)}{(a-u)(b-u)}}$$
$[a>b>c>u\geqslant d]$. BY (253.05)

58. $$\int_c^u \sqrt{\frac{x-d}{(a-x)^3(b-x)(x-c)}}\,dx = \frac{2}{a-b}\sqrt{\frac{b-d}{a-c}}\,E(\delta, q) - \frac{2(a-d)}{(a-b)(a-c)}\sqrt{\frac{(b-u)(u-c)}{(a-u)(u-d)}}$$
$[a > b \geqslant u > c > d]$. BY (254.03)

59. $$\int_u^b \sqrt{\frac{x-d}{(a-x)^3(b-x)(x-c)}}\,dx = \frac{2}{a-b}\sqrt{\frac{b-d}{a-c}}\,E(\varkappa, q)$$
$[a > b > u \geqslant c > d]$. BY (255.01)

60. $$\int_b^u \sqrt{\frac{x-d}{(a-x)^3(x-b)(x-c)}}\,dx = \frac{2}{a-b}\sqrt{\frac{b-d}{a-c}}\,[F(\lambda, r) - E(\lambda, r)] + \frac{2}{a-b}\sqrt{\frac{(u-b)(u-d)}{(a-u)(u-c)}}$$
$[a > u > b > c > d]$. BY (256.10)

61. $$\int_u^d \sqrt{\frac{c-x}{(a-x)^3(b-x)(d-x)}}\,dx = \frac{2(c-d)}{(a-d)\sqrt{(a-c)(b-d)}}\,F(\alpha, q) - \frac{2\sqrt{(a-c)(b-d)}}{(a-b)(a-d)}\,E(\alpha, q) + \frac{2(a-c)}{(a-b)(a-d)}\sqrt{\frac{(b-u)(d-u)}{(a-u)(c-u)}}$$
$[a > b > c > d > u]$. BY (251.15)

62. $$\int_d^u \sqrt{\frac{c-x}{(a-x)^3(b-x)(x-d)}}\,dx = \frac{2\sqrt{(a-c)(b-d)}}{(a-b)(a-d)}\,E(\beta, r) - \frac{2(b-c)}{(a-b)\sqrt{(a-c)(b-d)}}\,F(\beta, r)$$
$[a > b > c \geqslant u > d]$. BY (252.08)

63. $$\int_u^c \sqrt{\frac{c-x}{(a-x)^3(b-x)(x-d)}}\,dx = \frac{2\sqrt{(a-c)(b-d)}}{(a-b)(a-d)}\,E(\gamma, r) - \frac{2(b-c)}{(a-b)\sqrt{(a-c)(b-d)}}\,F(\gamma, r) - \frac{2}{a-d}\sqrt{\frac{(c-u)(u-d)}{(a-u)(b-u)}}$$
$[a > b > c > u \geqslant d]$. BY (253.10)

64. $$\int_c^u \sqrt{\frac{x-c}{(a-x)^3(b-x)(x-d)}}\,dx = \frac{2\sqrt{(a-c)(b-d)}}{(a-b)(a-d)}\,E(\delta, q) - \frac{2(c-d)}{(a-d)\sqrt{(a-c)(b-d)}}\,F(\delta, q) - \frac{2}{a-b}\sqrt{\frac{(b-u)(u-c)}{(a-u)(u-d)}}$$
$[a > b \geqslant u > c > d]$. BY (254.09)

65. $$\int_u^b \sqrt{\frac{x-c}{(a-x)^3(b-x)(x-d)}}\,dx = \frac{2\sqrt{(a-c)(b-d)}}{(a-b)(a-d)}\,E(\varkappa, q) - \frac{2(c-d)}{(a-d)\sqrt{(a-c)(b-d)}}\,F(\varkappa, q)$$
$[a > b > u \geqslant c > d]$. BY (255.10)

66. $$\int_b^u \sqrt{\frac{x-c}{(a-x)^3(x-b)(x-d)}}\,dx =$$

$$= \frac{2(b-c)}{(a-b)\sqrt{(a-c)(b-d)}} F(\lambda, r) - \frac{2\sqrt{(a-c)(b-d)}}{(a-b)(a-d)} E(\lambda, r) +$$

$$+ \frac{2(a-c)}{(a-b)(a-d)} \sqrt{\frac{(u-b)(u-d)}{(a-u)(u-c)}}$$

$[a > u > b > c > d]$. BY (256.07)

67. $$\int_u^d \sqrt{\frac{b-x}{(a-x)^3(c-x)(d-x)}}\,dx =$$

$$= \frac{2}{a-d}\sqrt{\frac{b-d}{a-c}}[F(\alpha, q) - E(\alpha, q)] + \frac{2}{a-d}\sqrt{\frac{(b-u)(d-u)}{(a-u)(c-u)}}$$

$[a > b > c > d > u]$. BY (251.13)

68. $$\int_d^u \sqrt{\frac{b-x}{(a-x)^3(c-x)(x-d)}}\,dx = \frac{2}{a-d}\sqrt{\frac{b-d}{a-c}} E(\beta, r)$$

$[a > b > c \geqslant u > d]$. BY (252.01)

69. $$\int_u^c \sqrt{\frac{b-x}{(a-x)^3(c-x)(x-d)}}\,dx =$$

$$= \frac{2}{a-d}\sqrt{\frac{b-d}{a-c}} E(\gamma, r) - \frac{2(a-b)}{(a-c)(a-d)}\sqrt{\frac{(c-u)(u-d)}{(a-u)(b-u)}}$$

$[a > b > c > u \geqslant d]$. BY (253.08)

70. $$\int_c^u \sqrt{\frac{b-x}{(a-x)^3(x-c)(x-d)}}\,dx =$$

$$= \frac{2}{a-d}\sqrt{\frac{b-d}{a-c}}[F(\delta, q) - E(\delta, q)] + \frac{2}{a-c}\sqrt{\frac{(b-u)(u-c)}{(a-u)(u-d)}}$$

$[a > b \geqslant u > c > d]$. BY (254.07)

71. $$\int_u^b \sqrt{\frac{b-x}{(a-x)^3(x-c)(x-d)}}\,dx = \frac{2}{a-d}\sqrt{\frac{b-d}{a-c}}[F(\varkappa, q) - E(\varkappa, q)]$$

$[a > b > u \geqslant c > d]$. BY (255.07)

72. $$\int_b^u \sqrt{\frac{x-b}{(a-x)^3(x-c)(x-d)}}\,dx =$$

$$= \frac{-2}{a-d}\sqrt{\frac{b-d}{a-c}} E(\lambda, r) + \frac{2}{a-d}\sqrt{\frac{(u-b)(u-d)}{(a-u)(u-c)}}$$

$[a \geqslant u > b > c > d]$. BY (256.04)

In **3.169—3.172**, we set: $\alpha = \operatorname{arctg}\frac{u}{b}$, $\beta = \operatorname{arctg}\frac{a}{u}$,

$$\gamma = \arcsin\frac{u}{b}\sqrt{\frac{a^2+b^2}{a^2+u^2}},\ \delta = \arccos\frac{u}{b},\ \varepsilon = \arccos\frac{b}{u},\ \xi = \arcsin\sqrt{\frac{a^2+b^2}{a^2+u^2}}.$$

$$\eta = \arcsin\frac{u}{b},\quad \zeta = \arcsin\frac{a}{b}\sqrt{\frac{b^2-u^2}{a^2-u^2}},\quad \varkappa = \arcsin\frac{a}{u}\sqrt{\frac{u^2-b^2}{a^2-b^2}},$$

$$\lambda = \arcsin\sqrt{\frac{a^2-u^2}{a^2-b^2}},\ \mu = \arcsin\sqrt{\frac{u^2-a^2}{u^2-b^2}},\ \nu = \arcsin\frac{a}{u},\ q = \frac{\sqrt{a^2-b^2}}{a},$$

$$r = \frac{b}{\sqrt{a^2+b^2}},\quad s = \frac{a}{\sqrt{a^2+b^2}}.\quad t = \frac{b}{a}.$$

3.169

1. $$\int_0^u \sqrt{\frac{x^2+a^2}{x^2+b^2}}\,dx = a\{F(\alpha, q) - E(\alpha, q)\} + u\sqrt{\frac{a^2+u^2}{b^2+u^2}}$$
$[a > b, \quad u > 0]$. BY (221.03)

2. $$\int_0^u \sqrt{\frac{x^2+b^2}{x^2+a^2}}\,dx = \frac{b^2}{a}F(\beta, q) - aE(\beta, q) + u\sqrt{\frac{a^2+u^2}{b^2+u^2}}$$
$[a > b, \quad u > 0]$. BY (221.04)

3. $$\int_0^u \sqrt{\frac{x^2+a^2}{b^2-x^2}}\,dx = \sqrt{a^2+b^2}\,E(\gamma, r) - u\sqrt{\frac{b^2-u^2}{a^2+u^2}}$$
$[b \geqslant u > 0]$. BY (214.11)

4. $$\int_u^b \sqrt{\frac{a^2+x^2}{b^2-x^2}}\,dx = \sqrt{a^2+b^2}\,E(\delta, r)$$
$[b > u \geqslant 0]$. BY(213.01), ZH 64(273)

5. $$\int_b^u \sqrt{\frac{a^2+x^2}{x^2-b^2}}\,dx = \sqrt{a^2+b^2}\{F(\varepsilon, s) - E(\varepsilon, s)\} + \frac{1}{u}\sqrt{(u^2+a^2)(u^2-b^2)}$$
$[u > b > 0]$. BY (211.03)

6. $$\int_0^u \sqrt{\frac{b^2-x^2}{a^2+x^2}}\,dx = \sqrt{a^2+b^2}\{F(\gamma, r) - E(\gamma, r)\} + u\sqrt{\frac{b^2-u^2}{a^2+u^2}}$$
$[b \geqslant u > 0]$. BY (214.03)

7. $$\int_u^b \sqrt{\frac{b^2-x^2}{a^2+x^2}}\,dx = \sqrt{a^2+b^2}\{F(\delta, r) - E(\delta, r)\}$$
$[b > u \geqslant 0]$. BY (213.03)

8. $$\int_b^u \sqrt{\frac{x^2-b^2}{a^2+x^2}}\,dx = \frac{1}{u}\sqrt{(a^2+u^2)(u^2-b^2)} - \sqrt{a^2+b^2}\,E(\varepsilon, s)$$
$[u > b > 0]$. BY (211.04)

9. $$\int_0^u \sqrt{\frac{b^2-x^2}{a^2-x^2}}\,dx = aE(\eta, t) - \frac{a^2-b^2}{a}F(\eta, t)$$
$[a > b \geqslant u > 0]$. BY (219.03)

10. $\int_u^b \sqrt{\frac{b^2-x^2}{a^2-x^2}}\,dx = aE(\zeta, t) - \frac{a^2-b^2}{a} F(\zeta, t) - u\sqrt{\frac{b^2-u^2}{a^2-u^2}}$

$[a > b > u \geqslant 0]$. BY (220.04)

11. $\int_b^u \sqrt{\frac{x^2-b^2}{a^2-x^2}}\,dx = aE(\varkappa, q) - \frac{b^2}{a} F(\varkappa, q) - \frac{1}{u}\sqrt{(a^2-u^2)(u^2-b^2)}$ $[a \geqslant u > b > 0]$. BY (217.04)

12. $\int_u^a \sqrt{\frac{x^2-b^2}{a^2-x^2}}\,dx = aE(\lambda, q) - \frac{b^2}{a} F(\lambda, q)$ $[a > u \geqslant b > 0]$. BY (218.03)

13. $\int_a^u \sqrt{\frac{x^2-b^2}{x^2-a^2}}\,dx = \frac{a^2-b^2}{a} F(\mu, t) - aE(\mu, t) + u\sqrt{\frac{u^2-a^2}{u^2-b^2}}$

$[u > a > b > 0]$. BY (216.03)

14. $\int_0^u \sqrt{\frac{a^2-x^2}{b^2-x^2}}\,dx = aE(\eta, t)$ $[a > b \geqslant u > 0]$. ZH 64(276), BY(219.01)

15. $\int_u^b \sqrt{\frac{a^2-x^2}{b^2-x^2}}\,dx = a\left\{E(\zeta, t) - \frac{u}{a}\sqrt{\frac{b^2-u^2}{a^2-u^2}}\right\}$

$[a > b > u \geqslant 0]$. BY (220.03)

16. $\int_b^u \sqrt{\frac{a^2-x^2}{x^2-b^2}}\,dx = a\{F(\varkappa, q) - E(\varkappa, q)\} + \frac{1}{u}\sqrt{(a^2-u^2)(u^2-b^2)}$ $[a \geqslant u > b > 0]$. BY (217.03)

17. $\int_u^a \sqrt{\frac{a^2-x^2}{x^2-b^2}}\,dx = a\{F(\lambda, q) - E(\lambda, q)\}$ $[a > u \geqslant b > 0]$. BY (218.09)

18. $\int_a^u \sqrt{\frac{x^2-a^2}{x^2-b^2}}\,dx = u\sqrt{\frac{u^2-a^2}{u^2-b^2}} - aE(\mu, t)$ $[u > a > b > 0]$. BY (216.04)

3.171

1. $\int_b^u \frac{dx}{x^2}\sqrt{\frac{a^2+x^2}{x^2-b^2}} = \frac{\sqrt{a^2+b^2}}{b^2} E(\varepsilon, s)$

$[u > b > 0]$. BY(211.01), ZH 64(274)

2. $\int_u^\infty \frac{dx}{x^2}\sqrt{\frac{a^2+x^2}{x^2-b^2}} = \frac{\sqrt{a^2+b^2}}{b^2} E(\xi, s) - \frac{a^2}{b^2 u}\sqrt{\frac{u^2-b^2}{a^2+u^2}}$

$[u \geqslant b > 0]$. BY (212.09)

3. $\int_u^b \frac{dx}{x^2}\sqrt{\frac{a^2-x^2}{b^2-x^2}} = \frac{a^2-b^2}{ab^2} F(\zeta, t) - \frac{a}{b^2} E(\zeta, t) + \frac{a^2}{b^2 u}\sqrt{\frac{b^2-u^2}{a^2-u^2}}$

$[a > b > u > 0]$. BY (220.12)

4. $$\int_b^u \frac{dx}{x^2}\sqrt{\frac{a^2-x^2}{x^2-b^2}}=\frac{a}{b^2}E(\varkappa, q)-\frac{1}{a}F(\varkappa, q)$$

$[a \geqslant u > b > 0]$. BY (217.11)

5. $$\int_u^a \frac{dx}{x^2}\sqrt{\frac{a^2-x^2}{x^2-b^2}}=\frac{a}{b^2}E(\lambda, q)-\frac{1}{a}F(\lambda, q)-\frac{\sqrt{(a^2-u^2)(u^2-b^2)}}{b^2u}$$

$[a > u \geqslant b > 0]$. BY (218.10)

6. $$\int_a^u \frac{dx}{x^2}\sqrt{\frac{x^2-a^2}{x^2-b^2}}=\frac{a}{b^2}E(\mu, t)-\frac{a^2-b^2}{ab^2}F(\mu, t)-\frac{1}{u}\sqrt{\frac{u^2-a^2}{u^2-b^2}}$$

$[u > a > b > 0]$. BY (216.08)

7. $$\int_u^\infty \frac{dx}{x^2}\sqrt{\frac{x^2+a^2}{x^2+b^2}}=\frac{1}{a}F(\beta, q)-\frac{a}{b^2}E(\beta, q)+\frac{a^2}{b^2u}\sqrt{\frac{b^2+u^2}{a^2+u^2}}$$

$[a > b,\ u > 0]$. BY (222.08)

8. $$\int_u^\infty \frac{dx}{x^2}\sqrt{\frac{x^2+b^2}{x^2+a^2}}=\frac{1}{a}\{F(\beta, q)-E(\beta, q)\}+\frac{1}{u}\sqrt{\frac{b^2+u^2}{a^2+u^2}}$$

$[a > b,\ u > 0]$. BY (222.09)

9. $$\int_u^b \frac{dx}{x^2}\sqrt{\frac{b^2-x^2}{a^2+x^2}}=\frac{\sqrt{(b^2-u^2)(a^2+u^2)}}{a^2u}-\frac{\sqrt{a^2+b^2}}{a^2}E(\delta, r)$$

$[b > u > 0]$. BY (213.10)

10. $$\int_b^u \frac{dx}{x^2}\sqrt{\frac{x^2-b^2}{a^2+x^2}}=\frac{\sqrt{a^2+b^2}}{a^2}\{F(\varepsilon, s)-E(\varepsilon, s)\}$$

$[u > b > 0]$. BY (211.07)

11. $$\int_u^\infty \frac{dx}{x^2}\sqrt{\frac{x^2-b^2}{a^2+x^2}}=\frac{\sqrt{a^2+b^2}}{a^2}\{F(\xi, s)-E(\xi, s)\}+\frac{1}{u}\sqrt{\frac{u^2-b^2}{a^2+u^2}}$$

$[u \geqslant b > 0]$. BY (212.11)

12. $$\int_u^b \frac{dx}{x^2}\sqrt{\frac{a^2+x^2}{b^2-x^2}}=\frac{\sqrt{a^2+b^2}}{b^2}\{F(\delta, r)-E(\delta, r)\}+\frac{\sqrt{(b^2-u^2)(a^2+u^2)}}{b^2u}$$

$[b > u > 0]$. BY (213.05)

13. $$\int_u^\infty \frac{dx}{x^2}\sqrt{\frac{x^2-a^2}{x^2-b^2}}=\frac{a}{b^2}E(\nu, t)-\frac{a^2-b^2}{ab^2}F(\nu, t)$$

$[u \geqslant a > b > 0]$. BY (215.08)

14. $$\int_u^b \frac{dx}{x^2}\sqrt{\frac{b^2-x^2}{a^2-x^2}}=\frac{1}{u}\sqrt{\frac{b^2-u^2}{a^2-u^2}}-\frac{1}{a}E(\zeta, t)$$

$[a > b > u > 0]$. BY (220.11)

15. $\int_b^u \frac{dx}{x^2}\sqrt{\frac{x^2-b^2}{a^2-x^2}}=\frac{1}{a}\{F(\varkappa, q)-E(\varkappa, q)\}$

$[a \geqslant u > b > 0]$. BY (217.08)

16. $\int_u^a \frac{dx}{x^2}\sqrt{\frac{x^2-b^2}{a^2-x^2}}=\frac{1}{a}\{F(\lambda, q)-E(\lambda, q)\}+\frac{\sqrt{(a^2-u^2)(u^2-b^2)}}{a^2u}$

$[a > u \geqslant b > 0]$. BY (218.08)

17. $\int_a^u \frac{dx}{x^2}\sqrt{\frac{x^2-b^2}{x^2-a^2}}=\frac{1}{a}E(\mu, t)-\frac{1}{u}\sqrt{\frac{u^2-a^2}{u^2-b^2}}$

$[u > a > b > 0]$. BY (216.07)

18. $\int_u^\infty \frac{dx}{x^2}\sqrt{\frac{x^2-b^2}{x^2-a^2}}=\frac{1}{a}E(\nu, t) \quad [u \geqslant a > b > 0]$. BY(215.01), ZH 65(281)

3.172

1. $\int_0^u \sqrt{\frac{x^2+b^2}{(x^2+a^2)^3}}\,dx=\frac{1}{a}E(\alpha, q)-\frac{a^2-b^2}{a^2}\frac{u}{\sqrt{(a^2+u^2)(b^2+u^2)}}$

$[a > b,\ u > 0]$. BY (221.10)

2. $\int_u^\infty \sqrt{\frac{x^2+b^2}{(x^2+a^2)^3}}\,dx=\frac{1}{a}E(\beta, q) \quad [a > b,\ u \geqslant 0]$. ZH 64 (271)

3. $\int_0^u \sqrt{\frac{x^2+a^2}{(x^2+b^2)^3}}\,dx=\frac{a}{b^2}E(\alpha, q) \quad [a > b,\ u > 0]$. ZH 64 (270)

4. $\int_u^\infty \sqrt{\frac{x^2+a^2}{(x^2+b^2)^3}}\,dx=\frac{a}{b^2}E(\beta, q)-\frac{a^2-b^2}{b^2}\frac{u}{\sqrt{(a^2+u^2)(b^2+u^2)}}$

$[a > b,\ u \geqslant 0]$. BY (222.06)

5. $\int_0^u \sqrt{\frac{b^2-x^2}{(a^2+x^2)^3}}\,dx=\frac{\sqrt{a^2+b^2}}{a^2}E(\gamma, r)-\frac{1}{\sqrt{a^2+b^2}}F(\gamma, r)$

$[b \geqslant u > 0]$. BY (214.08)

6. $\int_u^b \sqrt{\frac{b^2-x^2}{(a^2+x^2)^3}}\,dx=\frac{\sqrt{a^2+b^2}}{a^2}E(\delta, r)-\frac{1}{\sqrt{a^2+b^2}}F(\delta, r)-$

$-\frac{u}{a^2}\sqrt{\frac{b^2-u^2}{a^2+u^2}} \quad [b > u \geqslant 0]$. BY (213.04)

7. $\int_b^u \sqrt{\frac{x^2-b^2}{(a^2+x^2)^3}}\,dx=\frac{\sqrt{a^2+b^2}}{a^2}E(\varepsilon, s)-\frac{b^2}{a^2\sqrt{a^2+b^2}}F(\varepsilon, s)-$

$-\frac{1}{u}\sqrt{\frac{u^2-b^2}{u^2+a^2}} \quad [u > b > 0]$. BY (211.06)

8. $$\int_u^\infty \sqrt{\frac{x^2-b^2}{(a^2+x^2)^3}}\,dx = \frac{\sqrt{a^2+b^2}}{a^2}E(\xi, s) - \frac{b^2}{a^2\sqrt{a^2+b^2}}F(\xi, s)$$
$$[u \geqslant b > 0].$$ BY (212.08)

9. $$\int_0^u \sqrt{\frac{x^2+a^2}{(b^2-x^2)^3}}\,dx = \frac{a^2}{b^2\sqrt{a^2+b^2}}F(\gamma, r) - \frac{\sqrt{a^2+b^2}}{b^2}E(\gamma, r) + \frac{(a^2+b^2)\,u}{b^2\sqrt{(a^2+u^2)(b^2-u^2)}}$$
$$[b > u > 0].$$ BY (214.09)

10. $$\int_u^\infty \sqrt{\frac{x^2+a^2}{(x^2-b^2)^3}}\,dx = \frac{1}{\sqrt{a^2+b^2}}F(\xi, s) - \frac{\sqrt{a^2+b^2}}{b^2}E(\xi, s) + \frac{(a^2+b^2)\,u}{b^2\sqrt{(a^2+u^2)(u^2-b^2)}}$$
$$[u > b > 0].$$ BY (212.07)

11. $$\int_0^u \sqrt{\frac{b^2-x^2}{(a^2-x^2)^3}}\,dx = \frac{1}{a}\left\{F(\eta, t) - E(\eta, t) + \frac{u}{a}\sqrt{\frac{b^2-u^2}{a^2-u^2}}\right\}$$
$$[a > b \geqslant u > 0].$$ BY (219.09)

12. $$\int_u^b \sqrt{\frac{b^2-x^2}{(a^2-x^2)^3}}\,dx = \frac{1}{a}\{F(\zeta, t) - E(\zeta, t)\}$$
$$[a > b > u \geqslant 0].$$ BY (220.07)

13. $$\int_b^u \sqrt{\frac{x^2-b^2}{(a^2-x^2)^3}}\,dx = \frac{1}{u}\sqrt{\frac{u^2-b^2}{a^2-u^2}} - \frac{1}{a}E(\varkappa, q)$$
$$[a > u > b > 0].$$ BY (217.07)

14. $$\int_u^\infty \sqrt{\frac{x^2-b^2}{(x^2-a^2)^3}}\,dx = \frac{1}{a}[F(\nu, t) - E(\nu, t)] + \frac{1}{u}\sqrt{\frac{u^2-b^2}{u^2-a^2}}$$
$$[u > a > b > 0].$$ BY (215.05)

15. $$\int_0^u \sqrt{\frac{a^2-x^2}{(b^2-x^2)^3}}\,dx = \frac{a}{b^2}[F(\eta, t) - E(\eta, t)] + \frac{u}{b^2}\sqrt{\frac{a^2-u^2}{b^2-u^2}}$$
96
$$[a > b > u > 0].$$ BY (219.10)

16. $$\int_u^a \sqrt{\frac{a^2-x^2}{(x^2-b^2)^3}}\,dx = \frac{u}{b^2}\sqrt{\frac{a^2-u^2}{u^2-b^2}} - \frac{a}{b^2}E(\lambda, q)$$
$$[a > u > b > 0].$$ BY (218.05)

17. $$\int_a^u \sqrt{\frac{x^2-a^2}{(x^2-b^2)^3}}\,dx = \frac{a}{b^2}[F(\mu, t) - E(\mu, t)]$$
$$[u > a > b > 0].$$ BY (216.05)

18. $$\int_u^\infty \sqrt{\frac{x^2-a^2}{(x^2-b^2)^3}}\,dx = \frac{a}{b^2}[F(\nu, t) - E(\nu, t)] + \frac{1}{u}\sqrt{\frac{u^2-a^2}{u^2-b^2}}$$
$$[u \geqslant a > b > 0].$$ BY (215.03)

3.173

1. $$\int_u^1 \frac{dx}{x^2}\sqrt{\frac{x^2+1}{1-x^2}} = \sqrt{2}\left[F\left(\arccos u, \frac{\sqrt{2}}{2}\right) - E\left(\arccos u, \frac{\sqrt{2}}{2}\right)\right] + \frac{\sqrt{1-u^4}}{u} \qquad [u<1].$$ BY (259.77)

2. $$\int_1^u \frac{dx}{x^2}\sqrt{\frac{x^2+1}{x^2-1}} = \sqrt{2}\,E\left(\arccos\frac{1}{u}, \frac{\sqrt{2}}{2}\right) \qquad [u>1].$$ BY (260.76)

In **3.174** and **3.175**, we take: $\alpha = \arccos\dfrac{1+(1-\sqrt{3})u}{1+(1+\sqrt{3})u}$,

$$\beta = \arccos\frac{1-(1+\sqrt{3})u}{1+(\sqrt{3}-1)u}, \quad p = \frac{\sqrt{2+\sqrt{3}}}{2}, \quad q = \frac{\sqrt{2-\sqrt{3}}}{2}.$$

3.174

1. $$\int_0^u \frac{dx}{[1+(1+\sqrt{3})x]^2}\sqrt{\frac{1-x+x^2}{x(1+x)}} = \frac{1}{\sqrt[4]{3}}E(\alpha, p) \qquad [u>0].$$ BY (260.51)

2. $$\int_0^u \frac{dx}{[1+(\sqrt{3}-1)x]^2}\sqrt{\frac{1+x+x^2}{x(1-x)}} = \frac{1}{\sqrt[4]{3}}E(\beta, q) \qquad [1 \geqslant u > 0].$$ BY (259.51)

3. $$\int_0^u \frac{dx}{1-x+x^2}\sqrt{\frac{x(1+x)}{1-x+x^2}} = \frac{1}{\sqrt[4]{27}}E(\alpha, p) - \frac{2-\sqrt{3}}{\sqrt[4]{27}}F(\alpha, p) - \frac{2(2+\sqrt{3})}{\sqrt{3}}\,\frac{1+(1-\sqrt{3})u}{1+(1+\sqrt{3})u}\sqrt{\frac{u(1+u)}{1-u+u^2}} \qquad [u>0].$$ BY (260.54)

4. $$\int_0^u \frac{dx}{1+x+x^2}\sqrt{\frac{x(1-x)}{1+x+x^2}} = \frac{4}{\sqrt[4]{27}}E(\beta, q) - \frac{2+\sqrt{3}}{\sqrt[4]{27}}F(\beta, q) - \frac{2(2-\sqrt{3})}{\sqrt{3}}\,\frac{1-(1+\sqrt{3})u}{1+(\sqrt{3}-1)u}\sqrt{\frac{u(1-u)}{1+u+u^2}} \qquad [1 \geqslant u > 0].$$ BY (259.55)

3.175

1. $$\int_0^u \frac{dx}{1+x}\sqrt{\frac{x}{1+x^3}} = \frac{1}{\sqrt[4]{27}}[F(\alpha, p) - 2E(\alpha, p)] + \frac{2}{\sqrt{3}}\,\frac{\sqrt{u(1-u+u^2)}}{\sqrt{1+u}\,[1+(1+\sqrt{3})u]} \qquad [u>0].$$ BY (260.55)

2. $$\int_0^u \frac{dx}{1-x}\sqrt{\frac{x}{1-x^3}} = \frac{1}{\sqrt[4]{27}}[F(\beta, q) - 2E(\beta, q)] + \frac{2}{\sqrt{3}}\,\frac{\sqrt{u(1+u+u^2)}}{\sqrt{1-u}\,[1+(\sqrt{3}-1)u]} \qquad [0<u<1].$$ BY (259.52)

3.18 Expressions that can be reduced to fourth roots of second-degree polynomials and their products with rational functions

3.181

1. $$\int_b^u \frac{dx}{\sqrt[4]{(a-x)(x-b)}} = \sqrt{a-b}\left\{2\left[E\left(\frac{1}{\sqrt{2}}\right) + E\left(\arccos\sqrt[4]{\frac{4(a-u)(u-b)}{(a-b)^2}}, \frac{1}{\sqrt{2}}\right)\right] - \left[K\left(\frac{1}{\sqrt{2}}\right) + F\left(\arccos\sqrt[4]{\frac{4(a-u)(u-b)}{(a-b)^2}}, \frac{1}{\sqrt{2}}\right)\right]\right\} \quad [a \geqslant u > b].$$ BY (271.05)

2. $$\int_a^u \frac{dx}{\sqrt[4]{(x-a)(x-b)}} = \sqrt{\frac{a-b}{2}} F\left[\left(\arccos\frac{a-b-2\sqrt{(u-a)(u-b)}}{a-b+2\sqrt{(u-a)(u-b)}}, \frac{1}{\sqrt{2}}\right) - 2E\left(\arccos\frac{a-b-2\sqrt{(u-a)(u-b)}}{a-b+2\sqrt{(u-a)(u-b)}}, \frac{1}{\sqrt{2}}\right)\right] + \frac{2(2u-a-b)\sqrt[4]{(u-a)(u-b)}}{a-b+2\sqrt{(u-a)(u-b)}} \quad [u > a > b].$$ BY (272.05)

3.182

1. $$\int_b^u \frac{dx}{\sqrt[4]{[(a-x)(x-b)]^3}} = \frac{2}{\sqrt{a-b}}\left[K\left(\frac{1}{\sqrt{2}}\right) + F\left(\arccos\sqrt{\frac{4(a-u)(u-b)}{(a-b)^2}}, \frac{1}{\sqrt{2}}\right)\right] \quad [a \geqslant u > b].$$ BY (271.01)

2. $$\int_a^u \frac{dx}{\sqrt[4]{[(x-a)(x-b)]^3}} = \frac{\sqrt{2}}{\sqrt{a-b}} F\left(\arccos\frac{a-b-2\sqrt{(u-a)(u-b)}}{a-b+2\sqrt{(u-a)(u-b)}}, \frac{1}{\sqrt{2}}\right) \quad [u > a > b].$$ BY (272.00)

In **3.183—3.186** we set: $\alpha = \arccos\frac{1}{\sqrt[4]{u^2+1}}$,

$$\beta = \arccos\sqrt[4]{1-u^2}, \quad \gamma = \arccos\frac{1-\sqrt{u^2-1}}{1+\sqrt{u^2-1}}.$$

3.183

1. $$\int_0^u \frac{dx}{\sqrt[4]{x^2+1}} = \sqrt{2}\left[F\left(\alpha, \frac{1}{\sqrt{2}}\right) - 2E\left(\alpha, \frac{1}{\sqrt{2}}\right)\right] + \frac{2u}{\sqrt[4]{u^2+1}} \quad [u > 0].$$ BY (273.55)

2. $$\int_0^u \frac{dx}{\sqrt[4]{1-x^2}} = \sqrt{2}\left[2E\left(\beta, \frac{1}{\sqrt{2}}\right) - F\left(\beta, \frac{1}{\sqrt{2}}\right)\right] \quad [0 < u \leqslant 1].$$ BY (271.55)

3. $$\int_1^u \frac{dx}{\sqrt[4]{x^2-1}} = F\left(\gamma, \frac{1}{\sqrt{2}}\right) - 2E\left(\gamma, \frac{1}{\sqrt{2}}\right) + \frac{2u\sqrt[4]{u^2-1}}{1+\sqrt{u^2-1}} \quad [u > 1].$$ BY (272.55)

3.184

1. $$\int_0^u \frac{x^2\,dx}{\sqrt[4]{1-x^2}} = \frac{2\sqrt{2}}{5}\left[2E\left(\beta, \frac{1}{\sqrt{2}}\right) - F\left(\beta, \frac{1}{\sqrt{2}}\right)\right] - \frac{2u}{5}\sqrt[4]{(1-u^2)^3}$$
$[0 < u \leqslant 1]$. BY (271.59)

2. $$\int_1^u \frac{dx}{x^2\sqrt[4]{x^2-1}} = E\left(\gamma, \frac{1}{\sqrt{2}}\right) - \frac{1}{2}F\left(\gamma, \frac{1}{\sqrt{2}}\right) - \frac{1-\sqrt{u^2-1}}{1+\sqrt{u^2-1}} \cdot \frac{\sqrt{u^2-1}}{u}$$
$[u > 1]$. BY (272.54)

3.185

1. $$\int_0^u \frac{dx}{\sqrt[4]{(x^2+1)^3}} = \sqrt{2}F\left(\alpha, \frac{1}{\sqrt{2}}\right) \quad [u > 0].$$ BY (273.50)

2. $$\int_0^u \frac{dx}{\sqrt[4]{(1-x^2)^3}} = \sqrt{2}F\left(\beta, \frac{1}{\sqrt{2}}\right) \quad [0 < u \leqslant 1].$$ BY (271.51)

3. $$\int_1^u \frac{dx}{\sqrt[4]{(x^2-1)^3}} = F\left(\gamma, \frac{1}{\sqrt{2}}\right) \quad [u > 1].$$ BY (272.50)

4. $$\int_0^u \frac{x^2\,dx}{\sqrt[4]{(1-x^2)^3}} = \frac{2\sqrt{2}}{3}F\left(\beta, \frac{1}{\sqrt{2}}\right) - \frac{2}{3}u\sqrt[4]{1-u^2}$$
$[0 < u \leqslant 1]$. BY (271.54)

5. $$\int_0^u \frac{dx}{\sqrt[4]{(x^2+1)^5}} = 2\sqrt{2}E\left(\alpha, \frac{1}{\sqrt{2}}\right) - \sqrt{2}F\left(\alpha, \frac{1}{\sqrt{2}}\right)$$
$[u > 0]$. BY (273.54)

6. $$\int_0^u \frac{x^2\,dx}{\sqrt[4]{(x^2+1)^5}} = 2\sqrt{2}\left[F\left(\alpha, \frac{1}{\sqrt{2}}\right) - 2E\left(\alpha, \frac{1}{\sqrt{2}}\right)\right] + \frac{2u}{\sqrt[4]{u^2+1}}$$
$[u > 0]$. BY (273.56)

7. $$\int_0^u \frac{x^2\,dx}{\sqrt[4]{(x^2+1)^7}} = \frac{1}{3\sqrt{2}}F\left(\alpha, \frac{1}{\sqrt{2}}\right) - \frac{u}{6\sqrt[4]{(u^2+1)^3}}$$
$[u > 0]$. BY (273.53)

3.186

1. $$\int_0^u \frac{1+\sqrt{x^2+1}}{(x^2+1)\sqrt[4]{x^2+1}}\,dx = 2\sqrt{2}E\left(\alpha, \frac{1}{\sqrt{2}}\right) \quad [u > 0].$$ BY (273.51)

2. $$\int_0^u \frac{dx}{(1+\sqrt{1-x^2})\sqrt[4]{1-x^2}} = \sqrt{2}\left[F\left(\beta, \frac{1}{\sqrt{2}}\right) - E\left(\beta, \frac{1}{\sqrt{2}}\right)\right] + \frac{u\sqrt[4]{1-u^2}}{1+\sqrt{1-u^2}}$$
$[0 < u \leqslant 1]$. BY (271.58)

3. $$\int_1^u \frac{dx}{(x^2+2\sqrt{x^2-1})\sqrt[4]{x^2-1}} = \frac{1}{2}\left[F\left(\gamma, \frac{1}{\sqrt{2}}\right) - E\left(\gamma, \frac{1}{\sqrt{2}}\right)\right]$$
$[u > 1]$. BY (272.53)

4. $\int_0^u \frac{1-\sqrt{1-x^2}}{1+\sqrt{1-x^2}} \cdot \frac{dx}{\sqrt[4]{(1-x^2)^3}} = \sqrt{2}\left[2E\left(\beta, \frac{1}{\sqrt{2}}\right) - F\left(\beta, \frac{1}{\sqrt{2}}\right)\right] - \frac{2u\sqrt[4]{1-u^2}}{1+\sqrt{1-u^2}}$ $[0 < u \leqslant 1]$. BY (271.57)

5. $\int_1^u \frac{x^2\,dx}{(x^2+2\sqrt{x^2-1})\sqrt[4]{(x^2-1)^3}} = E\left(\gamma, \frac{1}{\sqrt{2}}\right)$ $[u > 1]$. BY (272.51)

3.19-3.23 Combinations of powers of x and powers of binomials of the form $(\alpha+\beta x)$

3.191

1. $\int_0^u x^{\nu-1}(u-x)^{\mu-1}\,dx = u^{\mu+\nu-1}\mathrm{B}(\mu, \nu)$ $[\operatorname{Re}\mu>0, \operatorname{Re}\nu>0]$. ET II 185(7)

2. $\int_u^\infty x^{-\nu}(x-u)^{\mu-1}\,dx = u^{\mu-\nu}\mathrm{B}(\nu-\mu, \mu)$ $[\operatorname{Re}\nu > \operatorname{Re}\mu > 0]$. ET II 201(6)

3. $\int_0^1 x^{\nu-1}(1-x)^{\mu-1}\,dx = \int_0^1 x^{\mu-1}(1-x)^{\nu-1}\,dx = \mathrm{B}(\mu, \nu)$ $[\operatorname{Re}\mu > 0, \quad \operatorname{Re}\nu > 0]$. FI II 774(1)

3.192

1. $\int_0^1 \frac{x^p\,dx}{(1-x)^p} = p\pi \operatorname{cosec} p\pi$ $[p^2 < 1]$. BI ((3))(4)

2. $\int_0^1 \frac{x^p\,dx}{(1-x)^{p+1}} = -\pi \operatorname{cosec} p\pi$ $[-1 < p < 0]$. BI ((3))(5)

3. $\int_0^1 \frac{(1-x)^p}{x^{p+1}}\,dx = -\pi \operatorname{cosec} p\pi$ $[-1 < p < 0]$. BI ((4))(6)

4. $\int_1^\infty (x-1)^{p-\frac{1}{2}}\frac{dx}{x} = \pi \sec p\pi$ $\left[-\frac{1}{2} < p < \frac{1}{2}\right]$. BI ((23))(7)

3.193 $\int_0^n x^{\nu-1}(n-x)^n\,dx = \frac{n!\,n^{\nu+n}}{\nu(\nu+1)(\nu+2)\ldots(\nu+n)}$ $[\operatorname{Re}\nu > 0]$. EH I 2

3.194

1. $\int_0^u \frac{x^{\mu-1}\,dx}{(1+\beta x)^\nu} = \frac{u^\mu}{\mu}\,{}_2F_1(\nu, \mu; 1+\mu; -\beta u)$ $[|\arg(1+\beta u)| < \pi, \operatorname{Re}\mu > 0]$. ET I 310(20)

2. $$\int_u^\infty \frac{x^{\mu-1}\,dx}{(1+\beta x)^\nu} = \frac{u^{\mu-\nu}}{\beta^\nu(\nu-\mu)}\,{}_2F_1\left(\nu,\ \nu-\mu;\quad \nu-\mu+1;\ -\frac{1}{\beta u}\right)$$

$[\operatorname{Re}\mu > \operatorname{Re}\nu]$. ET I 310(21)

3. $$\int_0^\infty \frac{x^{\mu-1}\,dx}{(1+\beta x)^\nu} = \beta^{-\mu}\mathrm{B}(\mu,\ \nu-\mu) \quad [|\arg\beta| < \pi,\ \operatorname{Re}\nu > \operatorname{Re}\mu > 0].$$

FI II 775a, ET I 310(19)

4. $$\int_0^\infty \frac{x^{\mu-1}\,dx}{(1+\beta x)^{n+1}} = (-1)^n \frac{\pi}{\beta^\mu}\binom{\mu-1}{n}\operatorname{cosec}(\mu\pi)$$

$[|\arg\beta| < \pi,\quad 0 < \operatorname{Re}\nu < n+1]$. ET I 308(6)

5. $$\int_0^u \frac{x^{\mu-1}\,dx}{1+\beta x} = \frac{u^\mu}{\mu}\,{}_2F_1(1,\ \mu;\ 1+\mu;\ -u\beta)$$

$[|\arg(1-u\beta)| < \pi,\quad \operatorname{Re}\mu > 0]$. ET I 308(5)

6. $$\int_0^\infty \frac{x^{\mu-1}\,dx}{(1+\beta x)^2} = \frac{(1-\mu)\pi}{\beta^\mu}\operatorname{cosec}\mu\pi \qquad [0 < \operatorname{Re}\mu < 2].$$ BI ((16))(4)

7. $$\int_0^\infty \frac{x^m\,dx}{(a+bx)^{n+\frac{1}{2}}} = 2^{m+1}m!\,\frac{(2n-2m-3)!!}{(2n-1)!!}\,\frac{a^{m-n+\frac{1}{2}}}{b^{m+1}}$$

$\left[m < n-\frac{1}{2},\ a > 0,\ b > 0\right]$. BI ((21))(2)

8. $$\int_0^1 \frac{x^{n-1}\,dx}{(1+x)^m} = 2^{-n}\sum_{k=0}^\infty \binom{m-n-1}{k}\frac{(-2)^{-k}}{n+k}.$$ BI ((3))(1)

3.195 $$\int_0^\infty \frac{(1+x)^{p-1}}{(x+a)^{p+1}}\,dx = \frac{1-a^{-p}}{p(a-1)} \qquad [a > 0].$$ LI ((19))(6)

3.196

1. $$\int_0^u (x+\beta)^\nu (u-x)^{\mu-1}\,dx = \frac{\beta^\nu u^\mu}{\mu}\,{}_2F_1\left(1,\ -\nu;\ 1+\mu;\ -\frac{u}{\beta}\right)$$

$\left[\left|\arg\frac{u}{\beta}\right| < \pi\right]$. ET II 185(8)

2. $$\int_u^\infty (x+\beta)^{-\nu}(x-u)^{\mu-1}\,dx = (u+\beta)^{\mu-\nu}\,\mathrm{B}(\nu-\mu,\ \mu)$$

$\left[\left|\arg\frac{u}{\beta}\right| < \pi,\ \operatorname{Re}\nu > \operatorname{Re}\mu > 0\right]$. ET II 201(7)

3. $$\int_a^b (x-a)^{\mu-1}(b-x)^{\nu-1}\,dx = (b-a)^{\mu+\nu-1}\,\mathrm{B}(\mu,\ \nu)$$

$[b > a,\ \operatorname{Re}\mu > 0,\ \operatorname{Re}\nu > 0]$. EH I 10(13)

4. $\int_1^\infty \frac{dx}{(a-bx)(x-1)^\nu} = -\frac{\pi}{b} \operatorname{cosec} \nu\pi \left(\frac{b}{b-a}\right)^\nu$

$[a < b,\ b > 0,\ 0 < \nu < 1]$. LI ((23))(5)

5. $\int_{-\infty}^1 \frac{dx}{(a-bx)(1-x)^\nu} = \frac{\pi}{b} \operatorname{cosec} \nu\pi \left(\frac{b}{a-b}\right)^\nu$

$[a > b > 0,\ 0 < \nu < 1]$. LI ((24))(10)

3.197

1. $\int_0^\infty x^{\nu-1} (\beta+x)^{-\mu} (x+\gamma)^{-\varrho}\, dx = \beta^{-\mu}\gamma^{\nu-\varrho} \mathrm{B}(\nu, \mu-\nu+\varrho) \times$

$\times\, {}_2F_1\left(\mu, \nu;\ \mu+\varrho;\ 1-\frac{\gamma}{\beta}\right)$ $[|\arg\beta| < \pi,\ |\arg\gamma| < \pi,$

$\operatorname{Re}\nu > 0,\ \operatorname{Re}\mu > \operatorname{Re}(\nu-\varrho)]$. ET II 233(9)

2. $\int_u^\infty x^{-\lambda} (x+\beta)^\nu (x-u)^{\mu-1}\, dx = u^{\mu+\nu-\lambda} \mathrm{B}(\lambda-\mu-\nu, \mu) \times$

$\times\, {}_2F_1\left(-\nu,\ \lambda-\mu-\nu;\ \lambda-\nu;\ -\frac{\beta}{u}\right)$

$\left[\left|\arg\frac{u}{\beta}\right| < \pi \text{ or } \left|\frac{\beta}{u}\right| < 1,\quad 0 < \operatorname{Re}\mu < \operatorname{Re}(\lambda-\nu)\right]$. ET II 201(8)

3. $\int_0^1 x^{\lambda-1} (1-x)^{\mu-1} (1-\beta x)^{-\nu}\, dx = \mathrm{B}(\lambda, \mu)\, {}_2F_1(\nu, \lambda;\ \lambda+\mu;\ \beta)$

$[\operatorname{Re}\lambda > 0,\ \operatorname{Re}\mu > 0,\ |\beta| < 1]$. WH

4. $\int_0^1 x^{\mu-1} (1-x)^{\nu-1} (1+ax)^{-\mu-\nu}\, dx = (1+a)^{-\mu} \mathrm{B}(\mu, \nu)$

$[\operatorname{Re}\mu > 0,\ \operatorname{Re}\nu > 0,\ a > -1]$. BI((5))4, EH I 10(11)

5. $\int_0^\infty x^{\lambda-1} (1+x)^\nu (1+\alpha x)^\mu\, dx = \mathrm{B}(\lambda, -\mu-\nu-\lambda) \times$

$\times\, {}_2F_1(-\mu, \lambda;\ -\mu-\nu;\ 1-\alpha)$ $[|\arg\alpha| < \pi,\ -\operatorname{Re}(\mu+\nu) > \operatorname{Re}\lambda > 0]$.

EH I 60(12), ET I 310(23)

6. $\int_1^\infty x^{\lambda-\nu} (x-1)^{\nu-\mu-1} (\alpha x-1)^{-\lambda}\, dx = \alpha^{-\lambda} \mathrm{B}(\mu, \nu-\mu)\, {}_2F_1(\nu, \mu;\ \lambda;\ \alpha^{-1})$

$[1+\operatorname{Re}\nu > \operatorname{Re}\lambda > \operatorname{Re}\mu,\ |\arg(\alpha-1)| < \pi]$. EH I 115(6)

7. $\int_0^\infty x^{\mu-\frac{1}{2}} (x+a)^{-\mu} (x+b)^{-\mu}\, dx = \sqrt{\pi}\left(\sqrt{a}+\sqrt{b}\right)^{1-2\mu} \frac{\Gamma\left(\mu-\frac{1}{2}\right)}{\Gamma(\mu)}$

$[\operatorname{Re}\mu > 0]$. BI 19(5)

8. $$\int_0^u x^{\nu-1}(x+\alpha)^{\lambda}(u-x)^{\mu-1}\,dx=\alpha^{\lambda}u^{\mu+\nu-1}\mathrm{B}(\mu,\nu)\,{}_2F_1\left(-\lambda,\nu;\mu+\nu;-\frac{u}{\alpha}\right)$$

$$\left[\left|\arg\left(\frac{u}{\alpha}\right)\right|<\pi,\ \mathrm{Re}\,\mu>0,\ \mathrm{Re}\,\nu>0\right].$$ ET II 186(9)

9. $$\int_0^\infty x^{\lambda-1}(1+x)^{-\mu+\nu}(x+\beta)^{-\nu}\,dx=\mathrm{B}(\mu-\lambda,\lambda)\,{}_2F_1(\nu,\mu-\lambda;\mu;1-\beta)$$

$$[\mathrm{Re}\,\mu>\mathrm{Re}\,\lambda>0].$$ EH I 205

10. $$\int_0^1\frac{x^{q-1}\,dx}{(1-x)^q(1+px)}=\frac{\pi}{(1+p)^q}\operatorname{cosec}q\pi\quad[0<q<1,\ p>-1].$$ BI ((5))(1)

11. $$\int_0^1\frac{x^{p-\frac{1}{2}}\,dx}{(1-x)^p(1+qx)^p}=$$

$$=\frac{2\Gamma\left(p+\frac{1}{2}\right)\Gamma(1-p)}{\sqrt{\pi}}\cos^{2p}\left(\operatorname{arctg}\sqrt{q}\right)\frac{\sin\left[(2p-1)\operatorname{arctg}(\sqrt{q})\right]}{(2p-1)\sin\left[\operatorname{arctg}(\sqrt{q})\right]}$$

$$\left[-\frac{1}{2}<p<1,\ q>0\right].$$ BI ((11))(1)

12. $$\int_0^1\frac{x^{p-\frac{1}{2}}\,dx}{(1-x)^p(1-qx)^p}=\frac{\Gamma\left(p+\frac{1}{2}\right)\Gamma(1-p)}{\sqrt{\pi}}\,\frac{(1-\sqrt{q})^{1-2p}-(1+\sqrt{q})^{1-2p}}{(2p-1)\sqrt{q}}$$

$$\left[-\frac{1}{2}<p<1,\ 0<q<1\right].$$ BI ((11))(2)

3.198 $$\int_0^1 x^{\mu-1}(1-x)^{\nu-1}\left[ax+b(1-x)+c\right]^{-(\mu+\nu)}\,dx=$$

$$=(a+c)^{-\mu}(b+c)^{-\nu}\mathrm{B}(\mu,\nu)$$

$$[a\geqslant 0,\ b\geqslant 0,\ c>0,\ \mathrm{Re}\,\mu>0,\ \mathrm{Re}\,\nu>0].$$ FI II 787

3.199 $$\int_a^b (x-a)^{\mu-1}(b-x)^{\nu-1}(x-c)^{-\mu-\nu}\,dx=$$

$$=(b-a)^{\mu+\nu-1}(b-c)^{-\mu}(a-c)^{-\nu}\mathrm{B}(\mu,\nu)$$

$$[\mathrm{Re}\,\mu>0,\ \mathrm{Re}\,\nu>0,\ c<a<b].$$ EH I 10(14)

3.211 $$\int_0^1 x^{\lambda-1}(1-x)^{\mu-1}(1-ux)^{-\varrho}(1-vx)^{-\sigma}\,dx=$$

$$=\mathrm{B}(\mu,\lambda)\,F_1(\lambda,\varrho,\sigma,\lambda+\mu;\ u,v)$$

$$[\mathrm{Re}\,\lambda>0,\ \mathrm{Re}\,\mu>0].$$ EH I 231(5)

3.212 $$\int_0^\infty\left[(1+ax)^{-p}+(1+bx)^{-p}\right]x^{q-1}\,dx=$$

$$=2(ab)^{-\frac{q}{2}}\mathrm{B}(q,p-q)\cos\left\{q\arccos\left[\frac{a+b}{2\sqrt{ab}}\right]\right\}$$

$$[p>q>0].$$ BI ((19))(9)

3.213 $\int_0^\infty [(1+ax)^{-p} - (1+bx)^{-p}] x^{q-1} dx =$

$$= -2i(ab)^{-\frac{q}{2}} \mathrm{B}(q, p-q) \sin\left\{q \arccos\left[\frac{a+b}{2\sqrt{ab}}\right]\right\}$$

$[p > q > 0]$. BI ((19))(10)

3.214 $\int_0^1 [(1+x)^{\mu-1}(1-x)^{\nu-1} + (1+x)^{\nu-1}(1-x)^{\mu-1}] dx = 2^{\mu+\nu-1} \mathrm{B}(\mu, \nu)$

$[\mathrm{Re}\,\mu > 0, \ \mathrm{Re}\,\nu > 0]$. LI(1))(15), EH I 10(10)

3.215 $\int_0^1 \{a^\mu x^{\mu-1}(1-ax)^{\nu-1} + (1-a)^\nu x^{\nu-1}[1-(1-a)x]^{\mu-1}\} dx = \mathrm{B}(\mu, \nu)$

$[\mathrm{Re}\,\mu > 0, \ \mathrm{Re}\,\nu > 0, \ |a| < 1]$. BI ((1))(16)

3.216

1. $\int_0^1 \frac{x^{\mu-1} + x^{\nu-1}}{(1+x)^{\mu+\nu}} dx = \mathrm{B}(\mu, \nu)$ $[\mathrm{Re}\,\mu > 0, \ \mathrm{Re}\,\nu > 0]$. FI II 775

2. $\int_1^\infty \frac{x^{\mu-1} + x^{\nu-1}}{(1+x)^{\mu+\nu}} dx = \mathrm{B}(\mu, \nu)$ $[\mathrm{Re}\,\mu > 0, \ \mathrm{Re}\,\nu > 0]$. FI II 775

3.217 $\int_0^\infty \left\{\frac{b^p x^{p-1}}{(1+bx)^p} - \frac{(1+bx)^{p-1}}{b^{p-1} x^p}\right\} dx = \pi \operatorname{ctg} p\pi$ $[0 < p < 1, \ b > 0]$.

BI ((18))(13)

3.218 $\int_0^\infty \frac{x^{2p-1} - (a+x)^{2p-1}}{(a+x)^p x^p} dx = \pi \operatorname{ctg} p\pi$ $[p < 1]$ (cf. 3.217).

BI ((18))(7)

3.219 $\int_0^\infty \left\{\frac{x^\nu}{(x+1)^{\nu+1}} - \frac{x^\mu}{(x+1)^{\mu+1}}\right\} dx = \psi(\mu+1) - \psi(\nu+1)$

$[\mathrm{Re}\,\mu > 1, \ \mathrm{Re}\,\nu > 1]$. BI ((19))(13)

3.221

1. $\int_a^\infty \frac{(x-a)^{p-1}}{x-b} dx = \pi(a-b)^{p-1} \operatorname{cosec} p\pi$ $[a > b, \ 0 < p < 1]$. LI ((24))(8)

2. $\int_{-\infty}^a \frac{(a-x)^{p-1}}{x-b} dx = -\pi(b-a)^{p-1} \operatorname{cosec} p\pi$ $[a < b, \ 0 < p < 1]$.

LI ((24))(8)

3.222

1. $$\int_0^1 \frac{x^{\mu-1}\,dx}{1+x} = \beta(\mu) \qquad [\operatorname{Re}\mu > 0].$$ WH

2. $$\int_0^\infty \frac{x^{\mu-1}\,dx}{x+a} = \begin{cases} \pi \operatorname{cosec}(\mu\pi)\, a^{\mu-1} & \text{for } a>0, \\ -\pi \operatorname{ctg}(\mu\pi)\,(-a)^{\mu-1} & \text{for } a<0, \end{cases}$$ FI II 718, FI II 737, BI((18))(2), ET II 249(28)

$$[0 < \operatorname{Re}\mu < 1].$$

3.223

1. $$\int_0^\infty \frac{x^{\mu-1}\,dx}{(\beta+x)(\gamma+x)} = \frac{\pi}{\gamma-\beta}(\beta^{\mu-1} - \gamma^{\mu-1}) \operatorname{cosec}(\mu\pi)$$

$$[|\arg\beta| < \pi, \ |\arg\gamma| < \pi, \quad 0 < \operatorname{Re}\mu < 2].$$ ET I 309(7)

2. $$\int_0^\infty \frac{x^{\mu-1}\,dx}{(\beta+x)(\alpha-x)} = \frac{\pi}{\alpha+\beta}[\beta^{\mu-1} \operatorname{cosec}(\mu\pi) + \alpha^{\mu-1} \operatorname{ctg}(\mu\pi)]$$

$$[|\arg\beta| < \pi, \ \alpha > 0, \ 0 < \operatorname{Re}\mu < 2].$$ ET I 309(8)

3. $$\int_0^\infty \frac{x^{\mu-1}\,dx}{(a-x)(b-x)} = \pi \operatorname{ctg}(\mu\pi) \frac{a^{\mu-1} - b^{\mu-1}}{b-a}$$

$$[a > b > 0, \ 0 < \operatorname{Re}\mu < 2].$$ ET I 309(9)

3.224 $$\int_0^\infty \frac{(x+\beta)\, x^{\mu-1}\,dx}{(x+\gamma)(x+\delta)} = \pi \operatorname{cosec}(\mu\pi) \left\{ \frac{\gamma-\beta}{\gamma-\delta}\gamma^{\mu-1} + \frac{\delta-\beta}{\delta-\gamma}\delta^{\mu-1} \right\}$$

$$[|\arg\gamma| < \pi, \ |\arg\delta| < \pi, \ 0 < \operatorname{Re}\mu < 1].$$ ET I 309(10)

3.225

1. $$\int_1^\infty \frac{(x-1)^{p-1}}{x^2}\,dx = (1-p)\,\pi \operatorname{cosec} p\pi \qquad [-1 < p < 1].$$ BI ((23))(8)

2. $$\int_1^\infty \frac{(x-1)^{1-p}}{x^3}\,dx = \frac{1}{2}\,p(1-p)\,\pi \operatorname{cosec} p\pi \qquad [0 < p < 1].$$ BI ((23))(1)

3. $$\int_0^\infty \frac{x^p\,dx}{(1+x)^3} = \frac{\pi}{2}\,p(1-p) \operatorname{cosec} p\pi \qquad [-1 < p < 2].$$ BI ((16))(5)

3.226

1. $$\int_0^1 \frac{x^n\,dx}{\sqrt{1-x}} = 2\,\frac{(2n)!!}{(2n+1)!!}$$ BI ((8))(1)

2. $$\int_0^1 \frac{x^{n-\frac{1}{2}}\,dx}{\sqrt{1-x}} = \frac{(2n-1)!!}{(2n)!!}\,\pi.$$ BI ((8))(2)

3.227

1. $$\int_0^\infty \frac{x^{\nu-1}(\beta+x)^{1-\mu}}{\gamma+x}\,dx = \beta^{1-\mu}\gamma^{\nu-1}\mathrm{B}(\nu,\ \mu-\nu)\,{}_2F_1\left(\mu-1,\ \nu;\ \mu;\ 1-\frac{\gamma}{\beta}\right)$$

$[|\arg\beta|<\pi,\ |\arg\gamma|<\pi,\ 0<\operatorname{Re}\nu<\operatorname{Re}\mu].$ ET II 217(9)

2. $$\int_0^\infty \frac{x^{-\varrho}(\beta+x)^{-\sigma}}{\gamma+x}\,dx = \pi\gamma^{-\varrho}(\beta-\gamma)^{-\sigma}\operatorname{cosec}(\varrho\pi)\, I_{1-\gamma/\beta}(\sigma,\ \varrho)$$

$[|\arg\beta|<\pi,\ |\arg\gamma|<\pi,\ -\operatorname{Re}\sigma<\operatorname{Re}\varrho<1].$ ET II 217(10)

3.228

1. $$\int_a^b \frac{(x-a)^{\nu}(b-x)^{-\nu}}{x-c}\,dx = \pi\operatorname{cosec}(\nu\pi)\left[1-\left(\frac{a-c}{b-c}\right)^{\nu}\right] \quad \text{for } c<a;$$

$$= \pi\operatorname{cosec}(\nu\pi)\left[1-\cos(\nu\pi)\left(\frac{c-a}{b-c}\right)^{\nu}\right] \quad \text{for } a<c<b;$$

$$= \pi\operatorname{cosec}(\nu\pi)\left[1-\left(\frac{c-a}{c-b}\right)^{\nu}\right] \quad \text{for } c>b$$

$[|\operatorname{Re}\nu|<1].$ ET II 250(31)

2. $$\int_a^b \frac{(x-a)^{\nu-1}(b-x)^{-\nu}}{x-c}\,dx = \frac{\pi\operatorname{cosec}(\nu\pi)}{b-c}\left|\frac{a-c}{b-c}\right|^{\nu-1} \quad \text{for } c<a \text{ or } c>b;$$

$$= -\frac{\pi(c-a)^{\nu-1}}{(b-c)^{\nu}}\operatorname{ctg}(\nu\pi) \quad \text{for } a<c<b$$

$[0<\operatorname{Re}\nu<1].$ ET II 250(32)

3. $$\int_a^b \frac{(x-a)^{\nu-1}(b-x)^{\mu-1}}{x-c}\,dx = \frac{(b-a)^{\mu+\nu-1}}{b-c}\mathrm{B}(\mu,\ \nu)\,{}_2F_1\left(1,\ \mu;\ \mu+\nu;\ \frac{b-a}{b-c}\right)$$

for $c<a$ or $c>b$;

$$= (c-a)^{\nu-1}(b-c)^{\mu-1}\operatorname{ctg}\mu\pi-(b-a)^{\mu+\nu-2}\,\mathrm{B}(\mu-1,\ \nu)\times$$
$$\times\,{}_2F_1\left(2-\mu-\nu,\ 1;\ 2-\mu;\ \frac{b-c}{b-a}\right) \quad \text{for } a<c<b$$

$[\operatorname{Re}\mu>0,\ \operatorname{Re}\nu>0].$ ET II 250(33)

4. $$\int_0^1 \frac{(1-x)^{\nu-1}x^{-\nu}}{a-bx}\,dx = \frac{\pi(a-b)^{\nu-1}}{a^{\nu}}\operatorname{cosec}(\nu\pi)$$

$[0<\operatorname{Re}\nu<1,\ 0<b<a].$ BI ((5))(8)

5. $$\int_0^\infty \frac{x^{\nu-1}(x+a)^{1-\mu}}{x-c}\,dx = a^{1-\mu}(-c)^{\nu-1}\,\mathrm{B}(\mu-\nu,\ \nu)\,{}_2F_1\left(\mu-1,\ \nu;\ \mu;\ 1+\frac{c}{a}\right)$$

for $c<0$;

$$= c^{\nu-1}(a+c)^{1-\mu}\operatorname{ctg}[(\mu-\nu)\pi] -$$

$$-\frac{a^{1-\mu+\nu}}{\pi(a+c)}\mathrm{B}(\mu-\nu-1,\ \nu)\,{}_2F_1\left(2-\mu,\ 1;\ 2-\mu+\nu;\ \frac{a}{a+c}\right) \quad \text{for} \quad c>0$$

$[a > 0,\ 0 < \operatorname{Re}\nu < \operatorname{Re}\mu]$. ET II 251(34)

3.229 $$\int_0^1 \frac{x^{\mu-1}\,dx}{(1-x)^{\mu}(1+ax)(1+bx)} = \frac{\pi\operatorname{cosec}\mu\pi}{a-b}\left[\frac{a}{(1+a)^{\mu}} - \frac{b}{(1+b)^{\mu}}\right]$$

$[0 < \operatorname{Re}\mu < 1]$. BI ((5))(7)

3.231

1. $$\int_0^1 \frac{x^{p-1}-x^{-p}}{1-x}\,dx = \pi\operatorname{ctg}p\pi \qquad [p^2 < 1].$$ BI ((4))(4)

2. $$\int_0^1 \frac{x^{p-1}-x^{-p}}{1+x}\,dx = \pi\operatorname{cosec}p\pi \qquad [p^2 < 1].$$ BI ((4))(1)

3. $$\int_0^1 \frac{x^{p}-x^{-p}}{x-1}\,dx = \frac{1}{p} - \pi\operatorname{ctg}p\pi \qquad [p^2 < 1].$$ BI ((4))(3)

4. $$\int_0^1 \frac{x^{p}-x^{-p}}{1+x}\,dx = \frac{1}{p} - \pi\operatorname{cosec}p\pi \qquad [p^2 < 1].$$ BI ((4))(2)

5. $$\int_0^1 \frac{x^{\mu-1}-x^{\nu-1}}{1-x}\,dx = \psi(\nu) - \psi(\mu) \qquad [\operatorname{Re}\mu > 0,\ \operatorname{Re}\nu > 0].$$

FI II 815, BI((4))(5)

6. $$\int_0^\infty \frac{x^{p-1}-x^{q-1}}{1-x}\,dx = \pi(\operatorname{ctg}p\pi - \operatorname{ctg}q\pi) \qquad [p > 0,\ q > 0].$$ FI II 718

3.232 $$\int_0^\infty \frac{(c+ax)^{-\mu}-(c+bx)^{-\mu}}{x}\,dx = c^{-\mu}\ln\frac{b}{a}$$

$[\operatorname{Re}\mu > -1;\ a > 0;\ b > 0;\ c > 0]$. BI ((18))(14)

3.233 $$\int_0^\infty \left\{\frac{1}{1+x} - (1+x)^{-\nu}\right\}\frac{dx}{x} = \psi(\nu) + C \qquad [\operatorname{Re}\nu > 0].$$ EH I 17, WH

3.234

1. $$\int_0^1 \left(\frac{x^{q-1}}{1-ax} - \frac{x^{-q}}{a-x}\right)dx = \pi a^{-q}\operatorname{ctg}q\pi \qquad [0 < q < 1,\ a > 0].$$ BI ((55))(11)

2. $$\int_0^1 \left(\frac{x^{q-1}}{1+ax} + \frac{x^{-q}}{a+x}\right)dx = \pi a^{-q}\operatorname{cosec}q\pi \qquad [0 < q < 1,\ a > 0].$$ BI ((5))(10)

3.235 $\int_0^\infty \frac{(1+x)^\mu - 1}{(1+x)^\nu} \frac{dx}{x} = \psi(\nu) - \psi(\nu - \mu) \qquad [\operatorname{Re} \nu > \operatorname{Re} \mu > 0].$ BI ((18))(5)

3.236 $\int_0^1 \frac{x^{\frac{\mu}{2}} \, dx}{[(1-x)(1-a^2x)]^{\frac{\mu+1}{2}}} = \frac{(1-a)^{-\mu} - (1+a)^{-\mu}}{2a\mu\sqrt{\pi}} \Gamma\left(1 + \frac{\mu}{2}\right) \Gamma\left(\frac{1-\mu}{2}\right)$

$[-2 < \mu < 1, \ |a| < 1].$ BI ((12))(32)

3.237 $\sum_{n=0}^{\infty} (-1)^{n+1} \int_n^{n+1} \frac{dx}{x+u} = \ln \frac{u\left[\Gamma\left(\frac{u}{2}\right)\right]^2}{2\left[\Gamma\left(\frac{u+1}{2}\right)\right]^2} \qquad [|\arg u| < \pi].$ ET II 216(1)

3.238

1. $\int_{-\infty}^{\infty} \frac{|x|^{\nu-1}}{x-u} \, dx = -\pi \operatorname{ctg} \frac{\nu\pi}{2} |u|^{\nu-1} \operatorname{sign} u \qquad [0 < \operatorname{Re} \nu < 1].$

ET II 249(29)

2. $\int_{-\infty}^{\infty} \frac{|x|^{\nu-1}}{x-u} \operatorname{sign} x \, dx = \pi \operatorname{tg} \frac{\nu\pi}{2} |u|^{\nu-1} \qquad [0 < \operatorname{Re} \nu < 1].$ ET II 249(30)

3. $\int_a^b \frac{(b-x)^{\mu-1}(x-a)^{\nu-1}}{|x-u|^{\mu+\nu}} \, dx = \frac{(b-a)^{\mu+\nu-1}}{|a-u|^\mu |b-u|^\nu} \frac{\Gamma(\mu)\Gamma(\nu)}{\Gamma(\mu+\nu)}$

$[\operatorname{Re} \mu > 0, \ \operatorname{Re} \nu > 0, \ 0 < u < a < b$ or $0 < a < b < u].$ MO 7

3.24-3.27 Powers of x, of binomials of the form $\alpha + \beta x^p$ and of polynomials in x

3.241

1. $\int_0^1 \frac{x^{\mu-1} \, dx}{1+x^p} = \frac{1}{p} \beta\left(\frac{\mu}{p}\right) \qquad [\operatorname{Re} \mu > 0, \ p > 0].$ WH, BI ((2))(13)

2. $\int_0^\infty \frac{x^{\mu-1} \, dx}{1+x^\nu} = \frac{\pi}{\nu} \operatorname{cosec} \frac{\mu\pi}{\nu} = \frac{1}{\nu} \mathrm{B}\left(\frac{\mu}{\nu}, \frac{\nu-\mu}{\nu}\right) \qquad [\operatorname{Re} \nu \geqslant \operatorname{Re} \mu > 0].$

ET I 309(15)a, BI ((17))(10)

3. $\int_0^\infty \frac{x^{p-1} \, dx}{1-x^q} = \frac{\pi}{q} \operatorname{ctg} \frac{p\pi}{q} \qquad [p < q].$ BI ((17))(11)

4. $\int_0^\infty \frac{x^{\mu-1} \, dx}{(p+qx^\nu)^{n+1}} = \frac{1}{\nu p^{n+1}} \left(\frac{p}{q}\right)^{\frac{\mu}{\nu}} \frac{\Gamma\left(\frac{\mu}{\nu}\right) \Gamma\left(1+n-\frac{\mu}{\nu}\right)}{\Gamma(1+n)}$

$\left[0 < \frac{\mu}{\nu} < n+1\right].$ BI ((17))(22)a

5. $\int_0^\infty \frac{x^{p-1}\,dx}{(1+x^q)^2} = \frac{(p-q)\pi}{q^2} \operatorname{cosec} \frac{(p-q)\pi}{q} \quad [p < 2q].$ BI ((17))(18)

3.242 $\int_{-\infty}^\infty \frac{x^{2m}\,dx}{x^{4n}+2x^{2n}\cos t+1} = \frac{\pi}{n} \sin\left[\frac{(2n-2m-1)}{2n} t\right] \operatorname{cosec} t \operatorname{cosec} \frac{(2m+1)\pi}{2n}$

$[m < n, \ t^2 < \pi^2].$ FI II 642

3.243 $\int_0^\infty \frac{x^{\mu-1}\,dx}{(1+x^{2\nu})(1+x^{3\nu})} = -\frac{\pi}{8\nu} \frac{\operatorname{cosec}\left(\frac{\mu\pi}{3\nu}\right)}{1-4\cos^2\left(\frac{\mu\pi}{3\nu}\right)}$

$[0 < \operatorname{Re}\mu < 5\operatorname{Re}\nu].$ ET 312(34)

3.244

1. $\int_0^1 \frac{x^{p-1}+x^{q-p-1}}{1+x^q}\,dx = \frac{\pi}{q} \operatorname{cosec} \frac{p\pi}{q} \quad [q > p > 0].$ BI ((2))(14)

2. $\int_0^1 \frac{x^{p-1}-x^{q-p-1}}{1-x^q}\,dx = \frac{\pi}{q} \operatorname{ctg} \frac{p\pi}{q} \quad [q > p > 0].$ BI ((2))(16)

3. $\int_0^1 \frac{x^{\nu-1}-x^{\mu-1}}{1-x^\nu}\,dx = \frac{1}{\nu}\left[C+\psi\left(\frac{\mu}{\nu}\right)\right] \quad [\operatorname{Re}\mu > \operatorname{Re}\nu > 0].$ BI ((2))(17)

4. $\int_{-\infty}^\infty \frac{x^{2m}-x^{2n}}{1-x^{2l}}\,dx = \frac{\pi}{l}\left[\operatorname{ctg}\left(\frac{2m+1}{2l}\pi\right) - \operatorname{ctg}\left(\frac{2n+1}{2l}\pi\right)\right]$

$[m < l, \ n < l].$ FI II 640

3.245 $\int_0^\infty [x^{\nu-\mu} - x^\nu (1+x)^{-\mu}]\,dx = \frac{\nu}{\nu-\mu+1} \mathrm{B}(\nu, \ \mu-\nu)$

$[\operatorname{Re}\mu > \operatorname{Re}\nu > 0].$ BI ((16))(13)

3.246 $\int_0^\infty \frac{1-x^q}{1-x^r} x^{p-1}\,dx = \frac{\pi}{r} \sin\frac{q\pi}{r} \operatorname{cosec} \frac{p\pi}{r} \operatorname{cosec} \frac{(p+q)\pi}{r}$

$[p+q < r, \ p > 0].$ ET I 311(33), BI ((17))(12)

Integrals of the form $\int f(x^p \pm x^{-p}, \ x^q \pm x^{-q}, \ \dots)\frac{dx}{x}$ can be transformed by the substitution $x = e^t$ or $x = e^{-t}$. For example, instead of $\int_0^1 (x^{1+p} + x^{1-p})^{-1}\,dx$, we should seek to evaluate $\int_0^\infty \operatorname{sech} px\,dx$ and, instead of $\int_0^1 \frac{x^{n-m-1}+x^{n+m-1}}{1+2x^n\cos a + x^{2n}}\,dx$, we should seek to evaluate $\int_0^\infty \operatorname{ch} mx (\operatorname{ch} nx - \cos a)^{-1}\,dx$ (see 3.514 2.).

3.247

1. $$\int_0^1 \frac{x^{\alpha-1}(1-x)^{n-1}}{1-\xi x^b}\,dx = (n-1)!\sum_{k=0}^{\infty}\frac{\xi^k}{(\alpha+kb)(\alpha+kb+1)\dots(\alpha+kb+k-1)}$$
$[b > 0,\ |\xi| < 1]$. AD 6.704

2. $$\int_0^\infty \frac{(1-x^p)\,x^{\nu-1}}{1-x^{np}}\,dx = \frac{\pi}{np}\sin\left(\frac{\pi}{n}\right)\operatorname{cosec}\frac{(p+\nu)\pi}{np}\operatorname{cosec}\frac{\pi\nu}{np}$$
$[0 < \operatorname{Re}\nu < (n-1)p]$. ET 311(33)

3.248

1. $$\int_0^\infty \frac{x^{\mu-1}\,dx}{\sqrt{1+x^\nu}} = 2^{\frac{2\mu}{\nu}}\,B(\nu-2\mu,\mu) \qquad [\nu > 2\mu].$$ BI ((21))(9)

2. $$\int_0^1 \frac{x^{2n+1}\,dx}{\sqrt{1-x^2}} = \frac{(2n)!!}{(2n+1)!!}.$$ BI ((8))(14)

3. $$\int_0^1 \frac{x^{2n}\,dx}{\sqrt{1-x^2}} = \frac{(2n-1)!!}{(2n)!!}\frac{\pi}{2}.$$ BI ((8))(13)

3.249

1. $$\int_0^\infty \frac{dx}{(x^2+a^2)^n} = \frac{(2n-3)!!}{2\cdot(2n-2)!!}\frac{\pi}{a^{2n-1}}.$$ FI II 743

2. $$\int_0^a (a^2-x^2)^{n-\frac{1}{2}}\,dx = a^{2n}\frac{(2n-1)!!}{2\,(2n)!!}\pi.$$ FI II 156

3. $$\int_{-1}^1 \frac{(1-x^2)^n\,dx}{(a-x)^{n+1}} = 2^{n+1}Q_n(a).$$ EH II 181(31)

4. $$\int_0^1 \frac{x^\mu\,dx}{1+x^2} = \frac{1}{2}\beta\left(\frac{\mu+1}{2}\right) \qquad [\operatorname{Re}\mu > -1].$$ BI ((2))(7)

5. $$\int_0^1 (1-x^2)^{\mu-1}\,dx = 2^{2\mu-2}B(\mu,\mu) = \frac{1}{2}B\left(\frac{1}{2},\mu\right)$$
$[\operatorname{Re}\mu > 0]$. FI II 784

6. $$\int_0^1 \left(1-\sqrt{x}\right)^{p-1}dx = \frac{2}{p(p+1)} \qquad [p > 0].$$ BI ((7))(7)

7. $$\int_0^1 (1-x^\mu)^{-\frac{1}{\nu}}\,dx = \frac{1}{\mu}B\left(\frac{1}{\mu},1-\frac{1}{\nu}\right) \qquad [\operatorname{Re}\mu > 0,\ |\nu| > 1].$$

3.251

1. $$\int_0^1 x^{\mu-1}(1-x^\lambda)^{\nu-1}\,dx = \frac{1}{\lambda}B\left(\frac{\mu}{\lambda},\nu\right) \qquad [\operatorname{Re}\mu > 0,\ \operatorname{Re}\nu > 0,\ \lambda > 0].$$
FI II 787

2. $$\int_0^\infty x^{\mu-1}(1+x^2)^{\nu-1}\,dx=\frac{1}{2}\mathrm{B}\left(\frac{\mu}{2},\ 1-\nu-\frac{\mu}{2}\right)$$
$$\left[\operatorname{Re}\mu>0,\ \operatorname{Re}\left(\nu+\frac{1}{2}\mu\right)<1\right].$$

3. $$\int_1^\infty x^{\mu-1}(x^p-1)^{\nu-1}\,dx=\frac{1}{p}\mathrm{B}\left(1-\nu-\frac{\mu}{p},\ \nu\right)$$
$$[p>0,\ \operatorname{Re}\nu>0,\ \operatorname{Re}\mu<p-p\operatorname{Re}\nu].$$ ET 311(32)

4. $$\int_0^\infty \frac{x^{2m}\,dx}{(ax^2+c)^n}=\frac{(2m-1)!!\,(2n-2m-3)!!\,\pi}{2\cdot(2n-2)!!\,a^m c^{n-m-1}\sqrt{ac}}$$
$$[a>0,\ c>0,\ n>m+1].$$ GU ((141))(8a)

5. $$\int_0^\infty \frac{x^{2m+1}\,dx}{(ax^2+c)^n}=\frac{m!\,(n-m-2)!}{2\,(n-1)!\,a^{m+1}c^{n-m-1}}$$
$$[ac>0,\ n>m+1\geqslant 1].$$ GU ((141))(8b)

6. $$\int_0^\infty \frac{x^{\mu+1}}{(1+x^2)^2}\,dx=\frac{\mu\pi}{4\sin\frac{\mu\pi}{2}}\qquad[-2<\operatorname{Re}\mu<2].$$ WH

7. $$\int_0^1 \frac{x^\mu\,dx}{(1+x^2)^2}=-\frac{1}{4}+\frac{\mu-1}{4}\beta\left(\frac{\mu-1}{2}\right)\qquad[\operatorname{Re}\mu>1].$$ LI ((3))(11)

8. $$\int_0^1 x^{q+p-1}(1-x^q)^{-\frac{p}{q}}\,dx=\frac{p\pi}{q^2}\operatorname{cosec}\frac{p\pi}{q}\qquad[q>p].$$ BI ((9))(22)

9. $$\int_0^1 x^{\frac{q}{p}-1}(1-x^q)^{-\frac{1}{p}}\,dx=\frac{\pi}{q}\operatorname{cosec}\frac{\pi}{p}\qquad[p>1,\ q>0].$$ BI ((9))(23)a

10. $$\int_0^1 x^{p-1}(1-x^q)^{-\frac{p}{q}}\,dx=\frac{\pi}{q}\operatorname{cosec}\frac{p\pi}{q}\qquad[q>p>0].$$ BI ((9))(20)

11. $$\int_0^\infty x^{\mu-1}(1+\beta x^p)^{-\nu}\,dx=\frac{1}{p}\beta^{-\frac{\mu}{p}}\mathrm{B}\left(\frac{\mu}{p},\ \nu-\frac{\mu}{p}\right)$$
$$[|\arg\beta|<\pi,\ p>0,\ 0<\operatorname{Re}\mu<p\operatorname{Re}\nu].$$ BI ((17))(20, EH I 10(16)

3.252

1. $$\int_0^\infty \frac{dx}{(ax^2+2bx+c)^n}=\frac{(-1)^{n-1}}{(n-1)!}\frac{\partial^{n-1}}{\partial c^{n-1}}\left[\frac{1}{\sqrt{ac-b^2}}\operatorname{arcctg}\frac{b}{\sqrt{ac-b^2}}\right]$$
$$[a>0,\quad ac>b^2].$$ GW ((131))(4)

2. $$\int_{-\infty}^\infty \frac{dx}{(ax^2+2bx+c)^n}=\frac{(2n-3)!!\,\pi a^{n-1}}{(2n-2)!!\,(ac-b^2)^{n-\frac{1}{2}}}\qquad[a>0,\quad ac>b^2].$$ GW ((131))(5)

3. $$\int_0^\infty \frac{dx}{(ax^2+2bx+c)^{n+\frac{3}{2}}} = \frac{(-2)^n}{(2n+1)!!} \frac{\partial^n}{\partial c^n} \left\{ \frac{1}{\sqrt{c}(\sqrt{ac}+b)} \right\}$$

$[a \geqslant 0,\ c > 0,\ b > -\sqrt{ac}]$. GW ((213))(4)

4. $$\int_0^\infty \frac{x\,dx}{(ax^2+2bx+c)^n} = \frac{(-1)^n}{(n-1)!} \frac{\partial^{n-2}}{\partial c^{n-2}} \left\{ \frac{1}{2(ac-b^2)} - \right.$$

$$\left. - \frac{b}{2(ac-b^2)^{\frac{3}{2}}} \operatorname{arcctg} \frac{b}{\sqrt{ac-b^2}} \right\} \quad \text{for} \quad ac > b^2;$$

$$= \frac{(-1)^n}{(n-1)!} \frac{\partial^{n-2}}{\partial c^{n-2}} \left\{ \frac{1}{2(ac-b^2)} + \frac{b}{4(b^2-ac)^{\frac{3}{2}}} \ln \frac{b+\sqrt{b^2-ac}}{b-\sqrt{b^2-ac}} \right\}$$

for $b^2 > ac > 0$;

$$= \frac{a^{n-2}}{2(n-1)(2n-1)b^{2n-2}} \quad \text{for} \quad ac = b^2 \qquad [a > 0,\ b > 0,\ n \geqslant 2].$$

GW ((141))(5)

5. $$\int_{-\infty}^\infty \frac{x\,dx}{(ax^2+2bx+c)^n} = -\frac{(2n-3)!!\,\pi b a^{n-2}}{(2n-2)!!\,(ac-b^2)^{\frac{(2n-1)}{2}}}$$

$[ac > b^2,\ a > 0,\quad n \geqslant 2]$. GW ((141))(6)

6. $$\int_{-\infty}^\infty \frac{x^m\,dx}{(ax^2+2bx+c)^n} = \frac{(-1)^m \pi a^{n-m-1} b^m}{(2n-2)!!\,(ac-b^2)^{n-\frac{1}{2}}} \times$$

$$\times \sum_{k=0}^{E\left(\frac{m}{2}\right)} \binom{m}{2k} (2k-1)!!\,(2n-2k-3)!! \left(\frac{ac-b^2}{b^2}\right)^k$$

$[ac > b^2,\quad 0 \leqslant m \leqslant 2n-2]$. GW ((141))(17)

7. $$\int_0^\infty \frac{x^n\,dx}{(ax^2+2bx+c)^{n+\frac{3}{2}}} = \frac{n!}{(2n+1)!!\,\sqrt{c}(\sqrt{ac}+b)^{n+1}}$$

$[a \geqslant 0,\quad c > 0,\quad b > -\sqrt{ac}]$. GW ((213))(5a)

8. $$\int_0^\infty \frac{x^{n+1}\,dx}{(ax^2+2bx+c)^{n+\frac{3}{2}}} = \frac{n!}{(2n+1)!!\,\sqrt{a}(\sqrt{ac}+b)^{n+1}}$$

$[a > 0,\ c \geqslant 0,\ b > -\sqrt{ac}]$. GW ((213))(5b)

9. $$\int_0^\infty \frac{x^{n+\frac{1}{2}}\,dx}{(ax^2+2bx+c)^{n+1}} = \frac{(2n-1)!!\,\pi}{2^{2n+\frac{1}{2}}(b+\sqrt{ac})^{n+\frac{1}{2}} n!\,\sqrt{a}}$$

$[a > 0,\ c > 0,\ b+\sqrt{ac} > 0]$. LI ((21))(19)

10. $\int_0^\infty \frac{x^{\mu-1}\,dx}{(1+2x\cos t+x^2)^\nu} =$

$$= 2^{\nu-\frac{1}{2}} \sin^{\nu-\frac{1}{2}} t\,\Gamma\left(\nu+\frac{1}{2}\right) \mathrm{B}(\mu,\ 2\nu-\mu)\, P^{\frac{1}{2}-\nu}_{\mu-\nu-\frac{1}{2}}(\cos t)$$

$[-\pi < t < \pi, \quad 0 < \operatorname{Re}\mu < \operatorname{Re} 2\nu]$. ET I 310(22)

11. $\int_0^\infty (1+2\beta x+x^2)^{\mu-\frac{1}{2}} x^{-\nu-1}\,dx = 2^{-\mu}(\beta^2-1)^{\frac{\mu}{2}}\Gamma(1-\mu)\times$

$$\times \mathrm{B}(\nu-2\mu+1,\ -\nu)\, P^{\mu}_{\nu-\mu}(\beta)$$

$[\operatorname{Re}\nu < 0, \quad \operatorname{Re}(2\mu-\nu) < 1, \quad |\arg(\beta\pm 1)| < \pi]$; EH I 160(33)

$$= -\pi \operatorname{cosec}\nu\pi\, C_\nu^{\frac{1}{2}-\mu}(\beta)$$

$\left[-2 < \operatorname{Re}\left(\frac{1}{2}-\mu\right) < \operatorname{Re}\nu < 0, \quad |\arg(\beta\pm 1)| < \pi\right]$. EH I 178(24)

12. $\int_0^\infty \frac{x^{\mu-1}\,dx}{x^2+2ax\cos t+a^2} = -\pi a^{\mu-2}\operatorname{cosec} t \operatorname{cosec}(\mu\pi)\sin[(\mu-1)t]$

$[a > 0, \ |t| < \pi, \ 0 < \operatorname{Re}\mu < 2]$. FI II 738, BI((20))(3)

13. $\int_0^\infty \frac{x^{\mu-1}\,dx}{(x^2+2ax\cos t+a^2)^2} = \frac{\pi a^{\mu-4}}{2}\operatorname{cosec}\mu\pi \operatorname{cosec}^3 t\times$

$$\times\{(\mu-1)\sin t\cos[(\mu-2)t] - \sin[(\mu-1)t]\}$$

$[a > 0, \ |t| < \pi, \ 0 < \operatorname{Re}\mu < 4]$. LI((20))(8)a, ET I 309(13)

14. $\int_0^\infty \frac{x^{\mu-1}\,dx}{\sqrt{1+2x\cos t+x^2}} = \pi\operatorname{cosec}(\mu\pi)\,\mathrm{P}_{\mu-1}(\cos t) \quad [|t| < \pi, \ 0 < \operatorname{Re}\mu < 1]$.

ET 310(17)

3.253 $\int_{-1}^{1} \frac{(1+x)^{2\mu-1}(1-x)^{2\nu-1}}{(1+x^2)^{\mu+\nu}}\,dx = 2^{\mu+\nu-2}\mathrm{B}(\mu,\ \nu) \qquad [\operatorname{Re}\mu > 0, \ \operatorname{Re}\nu > 0]$.

FI II 787

3.254

1. $\int_0^u x^{\lambda-1}(u-x)^{\mu-1}(x^2+\beta^2)^\nu\,dx = \beta^{2\nu}u^{\lambda+\mu-1}\mathrm{B}(\lambda,\ \mu)\times$

$$\times {}_3F_2\left(-\nu,\ \frac{\lambda}{2},\ \frac{\lambda+1}{2};\ \frac{\lambda+\mu}{2},\ \frac{\lambda+\mu+1}{2};\ \frac{-u^2}{\beta^2}\right)$$

$\left[\operatorname{Re}\left(\frac{u}{\beta}\right) > 0, \quad \operatorname{Re}\lambda > 0, \quad \operatorname{Re}\mu > 0\right]$. ET II 186(10)

2. $$\int_u^\infty x^{-\lambda}(x-u)^{\mu-1}(x^2+\beta^2)^\nu\,dx = u^{\mu-\lambda+2\nu}\frac{\Gamma(\mu)\,\Gamma(\lambda-\mu-2\nu)}{\Gamma(\lambda-\mu)}\times$$
$$\times\,{}_3F_2\left(-\nu,\ \frac{\lambda-\mu}{2}-\nu,\ \frac{1+\lambda-\mu}{2}-\nu;\ \frac{\lambda}{2}-\nu,\ \frac{1+\lambda}{2}-\nu;\ -\frac{\beta^2}{u^2}\right)$$
$$\left[|u|>|\beta| \text{ or } \operatorname{Re}\left(\frac{\beta}{u}\right)>0,\ 0<\operatorname{Re}\mu<\operatorname{Re}(\lambda-2\nu)\right].$$

ET II 202(9)

3.255 $$\int_0^1 \frac{x^{\mu+\frac{1}{2}}(1-x)^{\mu-\frac{1}{2}}}{(c+2bx-ax^2)^{\mu+1}}\,dx =$$
$$= \frac{\sqrt{\pi}}{\{a+(\sqrt{c+2b-a}+\sqrt{c})^2\}^{\mu+\frac{1}{2}}\sqrt{c+2b-a}}\,\frac{\Gamma\left(\mu+\frac{1}{2}\right)}{\Gamma(\mu+1)}$$
$$\left[a+(\sqrt{c+2b-a}+\sqrt{c})^2>0,\quad c+2b-a>0,\quad \operatorname{Re}\mu>-\frac{1}{2}\right].$$

BI ((14))(2)

3.256

1. $$\int_0^1 \frac{x^{p-1}+x^{q-1}}{(1-x^2)^{\frac{p+q}{2}}}\,dx = \frac{1}{2}\cos\left(\frac{q-p}{4}\pi\right)\sec\left(\frac{q+p}{4}\pi\right)\mathrm{B}\left(\frac{p}{2},\frac{q}{2}\right)$$
$$[p>0,\quad q>0,\quad p+q<2].$$ BI ((8))(25)

2. $$\int_0^1 \frac{x^{p-1}-x^{q-1}}{(1-x^2)^{\frac{p+q}{2}}}\,dx = \frac{1}{2}\sin\left(\frac{q-p}{4}\pi\right)\operatorname{cosec}\left(\frac{q+p}{4}\pi\right)\mathrm{B}\left(\frac{p}{2},\frac{q}{2}\right)$$
$$[p>0,\quad q>0,\quad p+q<2].$$ BI ((8))(26)

3.257 $$\int_0^\infty \left[\left(ax+\frac{b}{x}\right)^2+c\right]^{-p-1}dx = \frac{2\sqrt{\pi}\,\Gamma\left(p+\frac{1}{2}\right)}{ac^{p+\frac{1}{2}}\,\Gamma(p+1)}.$$ BI ((20))(4)

3.258

1. $$\int_b^\infty \left(x-\sqrt{x^2-a^2}\right)^n dx = \frac{a^2}{2(n-1)}\left(b-\sqrt{b^2-a^2}\right)^{n-1}-$$
$$-\frac{1}{2(n+1)}\left(b-\sqrt{b^2-a^2}\right)^{n+1}\qquad [0<a\leqslant b,\quad n\geqslant 2].$$ (GW ((215))(5)

2. $$\int_b^\infty \left(\sqrt{x^2+1}-x\right)^n dx = \frac{(\sqrt{b^2+1}-b)^{n-1}}{2(n-1)}+\frac{(\sqrt{b^2+1}-b)^{n+1}}{2(n+1)}$$
$$[n\geqslant 2].$$ GW ((214))(7)

3. $$\int_0^\infty \left(\sqrt{x^2+a^2}-x\right)^n dx = \frac{na^{n+1}}{n^2-1}\qquad [n\geqslant 2].$$ GW ((214))(6a)

4. $$\int_0^\infty \frac{dx}{(x+\sqrt{x^2+a^2})^n} = \frac{n}{a^{n-1}(n^2-1)} \qquad [n \geqslant 2].$$ GW ((214))(5a)

5. $$\int_0^\infty x^m (\sqrt{x^2+a^2}-x)^n \, dx = \frac{n \cdot m! \, a^{m+n+1}}{(n-m-1)(n-m+1)\dots(m+n+1)}$$
$$[a>0, \quad 0 \leqslant m \leqslant n-2].$$ GW ((214))(6)

6. $$\int_0^\infty \frac{x^m \, dx}{(x+\sqrt{x^2+a^2})^n} = \frac{n \cdot m!}{(n-m-1)(n-m+1)\dots(m+n+1)\, a^{n-m-1}}$$
$$[a>0, \quad 0 \leqslant m \leqslant n-2].$$ GW ((214))(5)

7. $$\int_a^\infty (x-a)^m (x-\sqrt{x^2-a^2})^n \, dx = \frac{n \cdot (n-m-2)! \, (2m+1)! \, a^{m+n+1}}{2^m (n+m+1)!}$$
$$[a>0, \quad n \geqslant m+2].$$ GW ((215))(6)

3.259

1. $$\int_0^1 x^{p-1}(1-x)^{n-1}(1+bx^m)^l \, dx = (n-1)! \sum_{k=0}^{\infty} \binom{l}{k} \frac{b^k \Gamma(p+km)}{\Gamma(p+n+km)}$$
$$[b^2>1].$$ BI ((1))(14)

2. $$\int_0^u x^{\nu-1}(u-x)^{\mu-1}(x^m+\beta^m)^\lambda \, dx = \beta^{m\lambda} u^{\mu+\nu+1} \mathrm{B}(\mu, \nu) \times$$
$$\times {}_{m+1}F_m\left(-\lambda, \frac{\nu}{m}, \frac{\nu+1}{m}, \dots, \frac{\nu+m-1}{m}; \frac{\mu+\nu}{m}, \frac{\mu+\nu+1}{m}, \dots \right.$$
$$\left. \dots, \frac{\mu+\nu+m-1}{m}; \frac{-u^m}{\beta^m}\right)$$
$$\left[\operatorname{Re}\mu>0, \quad \operatorname{Re}\nu>0, \quad \left|\arg\left(\frac{u}{\beta}\right)\right| < \frac{\pi}{m}\right].$$ ET II 186(11)

3. $$\int_0^\infty x^{\lambda-1}(1+\alpha x^p)^{-\mu}(1+\beta x^p)^{-\nu} \, dx = \frac{1}{p} \alpha^{-\frac{\lambda}{p}} \mathrm{B}\left(\frac{\lambda}{p}, \mu+\nu-\frac{\lambda}{p}\right) \times$$
$$\times {}_2F_1\left(\nu, \frac{\lambda}{p}; \mu+\nu; 1-\frac{\beta}{\alpha}\right)$$
$$[|\arg\alpha|<\pi, \ |\arg\beta|<\pi, \ p>0, \ 0<\operatorname{Re}\lambda<2\operatorname{Re}(\mu+\nu)].$$ ET I 312(35)

3.261

1. $$\int_0^1 \frac{(1-x\cos t)\, x^{\mu-1} \, dx}{1-2x\cos t+x^2} = \sum_{k=0}^{\infty} \frac{\cos kt}{\mu+k}$$
$$[\operatorname{Re}\mu>0, \quad t \neq 2n\pi].$$ BI ((6))(9)

2. $$\int_0^1 \frac{(x^\nu+x^{-\nu})\, dx}{1+2x\cos t+x^2} = \frac{\pi \sin \nu t}{\sin t \sin \nu\pi}$$
$$[\nu^2<1, \quad t \neq (2n+1)\pi].$$ BI ((6))(8)

3. $$\int_0^1 \frac{(x^{1+p}+x^{1-p})\, dx}{(1+2x\cos t+x^2)^2} = \frac{\pi(p\sin t\cos pt-\cos t\sin pt)}{2\sin^3 t \sin p\pi}$$
$$[p^2<1, \quad t \neq (2n+1)\pi].$$ BI ((6))(18)

4. $$\int_0^1 \frac{x^{\mu-1}}{1+2ax\cos t+a^2x^2}\cdot\frac{dx}{(1-x)^{\mu}}=$$

$$=\frac{\pi\,\mathrm{cosec}\,t\,\mathrm{cosec}\,\mu\pi}{(1+2a\cos t+a^2)^{\frac{\mu}{2}}}\sin\left(t-\mu\,\mathrm{arctg}\frac{a\sin t}{1+a\cos t}\right)$$

$[a>0,\quad 0<\mathrm{Re}\,\mu<1]$. BI ((6))(21)

3.262 $$\int_0^\infty \frac{x^{-p}\,dx}{1+x^3}=\frac{\pi}{3}\,\mathrm{cosec}\,\frac{(1-p)\pi}{3}\quad [-2<p<1].$$ LI ((18))(3)

3.263 $$\int_0^\infty \frac{x^{\nu}\,dx}{(x+\gamma)(x^2+\beta^2)}=\frac{\pi}{2(\beta^2+\gamma^2)}\left[\gamma\beta^{\nu-1}\sec\frac{\nu\pi}{2}+\beta^{\nu}\,\mathrm{cosec}\,\frac{\nu\pi}{2}-\right.$$

$$\left.-2\gamma^{\nu}\,\mathrm{cosec}\,(\nu\pi)\right]\quad [\mathrm{Re}\,\beta>0,\quad |\arg\gamma|<\pi,\quad -1<\mathrm{Re}\,\nu<2].$$ ET II 216(7)

3.264

1. $$\int_0^\infty \frac{x^{p-1}\,dx}{(a^2+x^2)(b^2-x^2)}=\frac{\pi}{2}\,\frac{a^{p-2}+b^{p-2}\cos\frac{p\pi}{2}}{a^2+b^2}\,\mathrm{cosec}\frac{p\pi}{2}\quad [p<4].$$

BI ((19))(14)

2. $$\int_0^\infty \frac{x^{\mu-1}\,dx}{(\beta+x^2)(\gamma+x^2)}=\frac{\pi}{2}\,\frac{\gamma^{\frac{\mu}{2}-1}-\beta^{\frac{\mu}{2}-1}}{\beta-\gamma}\,\mathrm{cosec}\,\frac{\mu\pi}{2}$$

$[|\arg\beta|<\pi,\quad |\arg\gamma|<\pi,\ 0<\mathrm{Re}\,\mu<4]$. ET I 309(14)

3.265 $$\int_0^1 \frac{1-x^{\mu-1}}{1-x}\,dx=\psi(\mu)+C\quad [\mathrm{Re}\,\mu>0];$$ FI II 796, WH, ET I 16(13)

$$=\psi(1-\mu)+C-\pi\,\mathrm{ctg}\,(\mu\pi)\quad [\mathrm{Re}\,\mu>0].$$ EH I 16(15)a

3.266 $$\int_0^\infty \frac{(x^{\nu}-a^{\nu})\,dx}{(x-a)(\beta+x)}=\frac{\pi}{a+\beta}\left\{\beta^{\nu}\,\mathrm{cosec}\,(\nu\pi)-a^{\nu}\,\mathrm{ctg}\,(\nu\pi)-\frac{a^{\nu}}{\pi}\ln\frac{\beta}{a}\right\}$$

$[|\arg\beta|<\pi,\quad |\mathrm{Re}\,\nu|<1]$. ET II 216(8)

3.267

1. $$\int_0^1 \frac{x^{3n}\,dx}{\sqrt[3]{1-x^3}}=\frac{2\pi}{3\sqrt{3}}\,\frac{\Gamma\left(n+\frac{1}{3}\right)}{\Gamma\left(\frac{1}{3}\right)\Gamma(n+1)}.$$ BI ((9))(6)

2. $$\int_0^1 \frac{x^{3n-1}\,dx}{\sqrt[3]{1-x^3}}=\frac{(n-1)!\,\Gamma\left(\frac{2}{3}\right)}{3\Gamma\left(n+\frac{2}{3}\right)}.$$ BI ((9))(7)

3.268

1. $$\int_0^1 \left(\frac{1}{1-x} - \frac{px^{p-1}}{1-x^p} \right) dx = \ln p.$$ BI ((5))(14)

2. $$\int_0^1 \frac{1-x^\mu}{1-x} x^{\nu-1}\, dx = \psi(\mu+\nu) - \psi(\nu) \qquad [\operatorname{Re}\nu > 0,\ \operatorname{Re}\mu > 0].$$ BI ((2))(3)

3. $$\int_0^1 \left[\frac{n}{1-x} - \frac{x^{\mu-1}}{1-\sqrt[n]{x}} \right] dx = nC + \sum_{k=1}^{n} \psi\left(\mu + \frac{n-k}{n}\right)$$ $$[\operatorname{Re}\mu > 0].$$ BI ((13))(10)

3.269

1. $$\int_0^1 \frac{x^p - x^{-p}}{1-x^2} x\, dx = \frac{\pi}{2} \operatorname{ctg} \frac{p\pi}{2} - \frac{1}{p} \qquad [p^2 < 1].$$ BI ((4))(12)

2. $$\int_0^1 \frac{x^p - x^{-p}}{1+x^2} x\, dx = \frac{1}{p} - \frac{\pi}{2} \operatorname{cosec} \frac{p\pi}{2} \qquad [p^2 < 1].$$ BI ((4))(8)

3. $$\int_0^1 \frac{x^\mu - x^\nu}{1-x^2}\, dx = \frac{1}{2}\psi\left(\frac{\nu+1}{2}\right) - \frac{1}{2}\psi\left(\frac{\mu+1}{2}\right)$$ $$[\operatorname{Re}\mu > -1, \quad \operatorname{Re}\nu > -1].$$ BI ((2))(9)

3.271

1. $$\int_0^\infty \frac{x^p - x^q}{x-1} \frac{dx}{x+a} = \frac{\pi}{1+a} \left(\frac{a^p - \cos p\pi}{\sin p\pi} - \frac{a^q - \cos q\pi}{\sin q\pi} \right)$$ $$[p^2 < 1, \quad q^2 < 1, \quad a > 0].$$ BI ((19))(2)

2. $$\int_0^\infty \frac{x^p - a^p}{x-a} \frac{x^p - 1}{x-1}\, dx = \frac{\pi}{a-1} \left\{ \frac{a^{2p} - 1}{\sin(2p\pi)} - \frac{1}{\pi} a^p \ln a \right\}$$ $$\left[p^2 < \frac{1}{4}\right].$$ BI ((19))(3)

3. $$\int_0^\infty \frac{x^p - a^p}{x-a} \frac{x^{-p} - 1}{x-1}\, dx = \frac{\pi}{a-1} \left\{ 2(a^p - 1) \operatorname{ctg} p\pi - \frac{1}{\pi}(a^p + 1) \ln a \right\}$$ $$[p^2 < 1].$$ BI ((18))(9)

4. $$\int_0^\infty \frac{x^p - a^p}{x-a} \frac{1 - x^{-p}}{1-x} x^q\, dx = \frac{\pi}{a-1} \left\{ \frac{a^{p+q} - 1}{\sin[(p+q)\pi]} + \right.$$ $$\left. + \frac{a^p - a^q}{\sin[(q-p)\pi]} \right\} \frac{\sin p\pi}{\sin q\pi} \qquad [(p+q)^2 < 1, \quad (p-q)^2 < 1].$$ BI ((19))(4)

5. $$\int_0^\infty \left(\frac{x^p - x^{-p}}{1-x} \right)^2 dx = 2(1 - 2p\pi \operatorname{ctg} 2p\pi) \qquad \left[p^2 < \frac{1}{4}\right].$$ BI ((16))(3)

3.272

1. $$\int_0^1 \frac{x^{n-1}+x^{n-\frac{1}{2}}-2x^{2n-1}}{1-x}\,dx = 2\ln 2.$$ BI ((8))(8)

2. $$\int_0^1 \frac{x^{n-1}+x^{n-\frac{2}{3}}+x^{n-\frac{1}{3}}-3x^{3n-1}}{1-x}\,dx = 3\ln 3.$$ BI ((8))(9)

3.273

1. $$\int_0^1 \frac{\sin t - a^n x^n \sin[(n+1)t] + a^{n+1}x^{n+1}\sin nt}{1-2ax\cos t + a^2x^2}(1-x)^{p-1}\,dx = \Gamma(p)\sum_{k=1}^{n}\frac{(k-1)!\,a^{k-1}\sin kt}{\Gamma(p+k)} \quad [p>0].$$ BI ((6))(13)

2. $$\int_0^1 \frac{\cos t - ax - a^n x^n \cos[(n+1)t] + a^{n+1}x^{n+1}\cos nt}{1-2ax\cos t + a^2x^2}(1-x)^{p-1}\,dx = \Gamma(p)\sum_{k=1}^{n}\frac{(k-1)!\,a^{k-1}\cos kt}{\Gamma(p+k)} \quad [p>0].$$ BI ((6))(14)

3. $$\int_0^1 x\,\frac{\sin t - x^n \sin[(n+1)t] + x^{n+1}\sin nt}{1-2x\cos t + x^2}\,dx = \sum_{k=1}^{n}\frac{\sin kt}{k+1}.$$ BI ((6))(12)

4. $$\int_0^1 \frac{1 - x\cos t - x^{n+1}\cos[(n+1)t] + x^{n+2}\cos nt}{1-2x\cos t + x^2}\,dx = \sum_{k=0}^{n}\frac{\cos kt}{k+1}.$$ BI ((6))(11)

3.274

1. $$\int_0^\infty \frac{x^{\mu-1}(1-x)}{1-x^n}\,dx = \frac{\pi}{n}\sin\frac{\pi}{n}\operatorname{cosec}\frac{\mu\pi}{n}\operatorname{cosec}\frac{(\mu+1)\pi}{n} \quad [0<\operatorname{Re}\mu<n-1].$$ BI ((20))(13)

2. $$\int_0^1 \frac{1-x^n}{(1+x)^{n+1}}\frac{dx}{1-x} = \frac{1}{2^{n+1}}\sum_{k=1}^{n}\frac{2^k}{k}.$$ BI ((5))(3)

3. $$\int_0^\infty \frac{x^q-1}{x^p-x^{-p}}\frac{dx}{x} = \frac{\pi}{2p}\operatorname{tg}\frac{q\pi}{2p} \quad [p>q].$$ BI ((18))(6)

3.275

1. $$\int_0^1 \left\{\frac{x^{n-1}}{1-x^{\frac{1}{p}}} - \frac{px^{np-1}}{1-x}\right\}dx = p\ln p \quad [p>0].$$ BI ((13))(9)

2. $$\int_0^1 \left\{\frac{nx^{n-1}}{1-x^n} - \frac{x^{mn-1}}{1-x}\right\}dx = C + \frac{1}{n}\sum_{k=1}^{n}\psi\left(m+\frac{n-k}{n}\right).$$ BI ((5))(13)

3. $$\int_0^1 \left(\frac{x^{p-1}}{1-x} - \frac{qx^{pq-1}}{1-x^q}\right)dx = \ln q \quad [q>0].$$ BI ((5))(12)

4. $$\int_0^\infty \left\{\frac{1}{1+x^{2^n}} - \frac{1}{1+x^{2^m}}\right\} \frac{dx}{x} = 0.$$ BI ((18))(17)

3.276

1. $$\int_0^\infty \frac{\left[\left(ax+\frac{b}{x}\right)^2+c\right]^{-p-1} dx}{x^2} = \frac{\sqrt{\pi}}{2bc^{p+\frac{1}{2}}} \frac{\Gamma\left(p+\frac{1}{2}\right)}{\Gamma(p+1)} \quad \left[p > -\frac{1}{2}\right].$$ BI ((20))(19)

2. $$\int_0^\infty \left(a+\frac{b}{x^2}\right)\left[\left(ax+\frac{b}{x}\right)^2+c\right]^{-p-1} dx = \frac{\sqrt{\pi}}{c^{p+\frac{1}{2}}} \frac{\Gamma\left(p+\frac{1}{2}\right)}{\Gamma(p+1)}$$
$$\left[p > -\frac{1}{2}\right].$$ BI ((20))(5)

3.277

1. $$\int_0^\infty \frac{x^{\mu-1}\left[\sqrt{1+x^2}+\beta\right]^\nu}{\sqrt{1+x^2}} dx = 2^{\frac{\mu}{2}-1}(\beta^2-1)^{\frac{\nu}{2}+\frac{\mu}{4}}\Gamma\left(\frac{\mu}{2}\right)\Gamma(1-\mu-\nu)P_{\frac{\mu}{2}-1}^{\frac{\nu+\mu}{2}}(\beta)$$
$$[\operatorname{Re}\beta > -1, \quad 0 < \operatorname{Re}\mu < 1-\operatorname{Re}\nu].$$ ET I 310(25)

2. $$\int_0^\infty \frac{x^{\mu-1}\left[\sqrt{\beta^2+x^2}+x\right]^\nu}{\sqrt{\beta^2+x^2}} dx = \frac{\beta^{\mu+\nu-1}}{2^\mu} \mathrm{B}\left(\mu, \quad \frac{1-\mu-\nu}{2}\right)$$
$$[\operatorname{Re}\beta > 0, \quad 0 < \operatorname{Re}\mu < 1-\operatorname{Re}\nu].$$ ET I 311(28)

3. $$\int_0^\infty \frac{x^{\mu-1}\left[\cos t \pm i \sin t \sqrt{1+x^2}\right]^\nu}{\sqrt{1+x^2}} dx = 2^{\frac{\mu-1}{2}} \sin^{\frac{1-\mu}{2}} t \frac{\Gamma\left(\frac{\mu}{2}\right)\Gamma(1-\mu-\nu)}{\Gamma(-\nu)} \times$$
$$\times \left[\pi^{-\frac{1}{2}} Q_{-\frac{\mu+1}{2}-\nu}^{\frac{\mu-1}{2}}(\cos t) \mp \frac{i}{2}\pi^{\frac{1}{2}} P_{-\frac{\mu+1}{2}-\nu}^{\frac{\mu-1}{2}}(\cos t)\right] \quad [\operatorname{Re}\mu > 0].$$ ET I 311 (27)

4. $$\int_0^\infty \frac{x^{\mu-1}\left[\sqrt{(\beta^2-1)(x^2+1)}+\beta\right]^\nu}{\sqrt{x^2+1}} dx = \frac{2^{\frac{\mu-1}{2}}}{\sqrt{\pi}} \times$$
$$\times e^{-\frac{1}{2}i\pi(\mu-1)} \frac{\Gamma\left(\frac{\mu}{2}\right)\Gamma(1-\mu-\nu)}{\Gamma(-\nu)} (\beta^2-1)^{\frac{1-\mu}{4}} Q_{-\nu-\frac{\mu+1}{2}}^{\frac{\mu-1}{2}}(\beta)$$
$$[\operatorname{Re}\beta > 1, \quad \operatorname{Re}\nu < 0, \quad \operatorname{Re}\mu < 1-\operatorname{Re}\nu].$$ ET I 311(26)

5. $$\int_u^\infty \frac{(x-u)^{\mu-1}\left(\sqrt{x+1}-\sqrt{x-1}\right)^{2\nu}}{\sqrt{x^2-1}} dx =$$
$$= \frac{2^{\nu+\frac{1}{2}}}{\sqrt{\pi}} e^{\left(\mu-\frac{1}{2}\right)\pi i}(u^2-1)^{\frac{2\mu-1}{4}} Q_{\nu-\frac{1}{2}}^{\frac{1}{2}-\mu}(u)$$
$$[|\arg(u-1)| < \pi, \quad 0 < \operatorname{Re}\mu < 1+\operatorname{Re}\nu].$$ ET II 202(10)

6. $\int\limits_1^\infty \frac{x^{\mu-1}[(x-\sqrt{x^2-1})^\nu+(x-\sqrt{x^2-1})^{-\nu}]}{\sqrt{x^2-1}}\,dx =$

$= 2^{-\mu}\mathrm{B}\left(\frac{1-\mu+\nu}{2},\ \frac{1-\mu-\nu}{2}\right)$ [Re $\mu < 1 +$ Re ν]. ET I 311(29)

7. $\int\limits_0^u \frac{(u-x)^{\mu-1}[(\sqrt{x+2}+\sqrt{x})^{2\nu}+(\sqrt{x+2}-\sqrt{x})^{2\nu}]}{\sqrt{x(x+2)}}\,dx =$

$= 2^{\frac{2\mu+1}{2}}\sqrt{\pi}\,[u(u+2)]^{\mu-\frac{1}{2}}P_{\nu-\frac{1}{2}}^{\frac{1}{2}-\mu}(u+1)$

[$|\arg u| < \pi$, Re $\mu > 0$]. ET II 186(12)

3.278 $\int\limits_0^\infty \left(\frac{x^p}{1+x^{2p}}\right)^q \frac{dx}{1-x^2} = 0.$

3.3-3.4 Exponential Functions

3.31 Exponential functions

3.310 $\int\limits_0^\infty e^{-px}\,dx = \frac{1}{p}$ [Re $p > 0$].

3.311

1. $\int\limits_0^\infty \frac{dx}{1+e^{px}} = \frac{\ln 2}{p}.$ LO III 284a

2. $\int\limits_0^\infty \frac{e^{-\mu x}}{1+e^{-x}}\,dx = \beta(\mu)$ [Re $\mu > 0$]. EH I 20(3), ET I 144(7)

3. $\int\limits_{-\infty}^\infty \frac{e^{-px}}{1+e^{-qx}}\,dx = \frac{\pi}{q}\operatorname{cosec}\frac{p\pi}{q}$ [$q > p > 0$ or $0 > p > q$] (cf. **3.241** 2.). BI ((28))(7)

4. $\int\limits_0^\infty \frac{e^{-qx}\,dx}{1-ae^{-px}} = \sum_{k=0}^\infty \frac{a^k}{q+kp}$ [$0 < a < 1$]. BI ((27))(7)

5. $\int\limits_0^\infty \frac{1-e^{\nu x}}{e^x-1}\,dx = \psi(\nu) + C + \pi\operatorname{ctg}(\pi\nu)$ [Re $\nu < 1$] (cf. **3.266**). EH I 16(16)

6. $\int\limits_0^\infty \frac{e^{-x}-e^{-\nu x}}{1-e^{-x}}\,dx = \psi(\nu) + C$ [Re $\nu > 0$]. WH, EH 16(14)

7. $\int\limits_0^\infty \frac{e^{-\mu x}-e^{-\nu x}}{1-e^{-x}}\,dx = \psi(\nu) - \psi(\mu)$ [Re $\mu > 0$, Re $\nu > 0$] (cf. **3.231** 5.). BI ((27))(8)

8. $\int\limits_{-\infty}^{\infty} \frac{e^{-\mu x}\,dx}{b-e^{-x}} = \pi b^{\mu-1} \operatorname{ctg}(\mu\pi) \qquad [b>0,\ 0<\operatorname{Re}\mu<1].$ ET I 120(14)a

9. $\int\limits_{-\infty}^{\infty} \frac{e^{-\mu x}\,dx}{b+e^{-x}} = \pi b^{\mu-1} \operatorname{cosec}(\mu\pi) \qquad [|\arg b|<\pi,\ 0<\operatorname{Re}\mu<1].$ ET I 120(15)a

10. $\int\limits_{0}^{\infty} \frac{e^{-px}-e^{-qx}}{1-e^{-(p+q)x}}\,dx = \frac{\pi}{p+q} \operatorname{ctg}\frac{p\pi}{p+q} \qquad [p>0,\ q>0].$ GW ((311))(16c)

11. $\int\limits_{0}^{\infty} \frac{e^{px}-e^{qx}}{e^{rx}-e^{sx}}\,dx = \frac{1}{r-s}\left[\psi\left(\frac{r-q}{r-s}\right)-\psi\left(\frac{r-p}{r-s}\right)\right]$
$[r>s,\ r>p,\ r>q].$ GW ((311))(16)

12. $\int\limits_{0}^{\infty} \frac{a^x-b^x}{c^x-d^x}\,dx = \frac{1}{\ln\frac{c}{d}}\left\{\psi\left(\frac{\ln\frac{c}{b}}{\ln\frac{c}{d}}\right)-\psi\left(\frac{\ln\frac{c}{a}}{\ln\frac{c}{d}}\right)\right\}$
$[c>a>0,\ b>0,\ d>0].$ GW ((311))(16a)

3.312

1. $\int\limits_{0}^{\infty} \left(1-e^{-\frac{x}{\beta}}\right)^{\nu-1} e^{-\mu x}\,dx = \beta\,\mathrm{B}(\beta\mu,\ \nu) \qquad [\operatorname{Re}\beta>0,\ \operatorname{Re}\nu>0,\ \operatorname{Re}\mu>0].$

LI((25))(13), EH I 11(24)

2. $\int\limits_{0}^{\infty} (1-e^{-x})^{-1}(1-e^{-\alpha x})(1-e^{-\beta x})\,e^{-px}\,dx =$
$= \psi(p+\alpha)+\psi(p+\beta)-\psi(p+\alpha+\beta)-\psi(p)$
$[\operatorname{Re}p>0,\ \operatorname{Re}p>-\operatorname{Re}\alpha,\ \operatorname{Re}p>-\operatorname{Re}\beta,\ \operatorname{Re}p>-\operatorname{Re}(\alpha+\beta)].$ ET I 145(15)

3. $\int\limits_{0}^{\infty} (1-e^{-x})^{\nu-1}(1-\beta e^{-x})^{-\varrho}\,e^{-\mu x}\,dx = \mathrm{B}(\mu,\ \nu)\,{}_2F_1(\varrho,\ \mu;\ \mu+\nu;\ \beta)$
$[\operatorname{Re}\mu>0,\ \operatorname{Re}\nu>0,\ |\arg(1-\beta)|<\pi].$ EH I 116(15)

3.313 $\int\limits_{-\infty}^{\infty} \frac{e^{-\mu x}\,dx}{(1-e^{-x})^n} = \pi \operatorname{cosec}\mu\pi \prod_{k=1}^{n-1} \frac{k-\mu}{(n-1)!} \qquad [0<\operatorname{Re}\mu<n].$ ET I 120(20)

3.314 $\int\limits_{-\infty}^{\infty} \frac{e^{-\mu x}\,dx}{(e^{\beta/\gamma}+e^{-x/\gamma})^{\nu}} = \gamma\exp\left[\beta\left(\mu-\frac{\nu}{\gamma}\right)\right]\mathrm{B}(\gamma\mu,\ \nu-\gamma\mu)$
$\left[\operatorname{Re}\left(\frac{\nu}{\gamma}\right)>\operatorname{Re}\mu>0,\ |\operatorname{Im}\beta|<\pi\operatorname{Re}\gamma\right].$ ET I 120(21)

3 315

1. $\int\limits_{-\infty}^{\infty} \frac{e^{-\mu x}\,dx}{(e^{\beta}+e^{-x})^{\nu}(e^{\gamma}+e^{-x})^{\varrho}} = \exp[\gamma(\mu-\varrho)-\beta\nu]\times$
$\times\mathrm{B}(\mu,\ \nu+\varrho-\mu)\,{}_2F_1(\nu,\ \mu;\ \nu+\varrho;\ 1-e^{\gamma-\beta})$
$[|\operatorname{Im}\beta|<\pi,\ |\operatorname{Im}\gamma|<\pi,\ 0<\operatorname{Re}\mu<\operatorname{Re}(\nu+\varrho)].$ ET I 121(22)

2. $$\int_{-\infty}^{\infty} \frac{e^{-\mu x}\,dx}{(\beta+e^{-x})(\gamma+e^{-x})} = \frac{\pi(\beta^{\mu-1}-\gamma^{\mu-1})}{\gamma-\beta}\operatorname{cosec}(\mu\pi)$$

$[|\arg\beta|<\pi,\ |\arg\gamma|<\pi,\ \beta\neq\gamma,\ 0<\operatorname{Re}\mu<2]$. ET I 120(18)

3.316 $$\int_{-\infty}^{\infty} \frac{(1+e^{-x})^{\nu}-1}{(1+e^{-x})^{\mu}}\,dx = \psi(\mu)-\psi(\mu-\nu) \qquad [\operatorname{Re}\mu>\operatorname{Re}\nu>0]$$

(cf. **3.235**). BI ((28))(8)

3.317

1. $$\int_{-\infty}^{\infty} \left\{\frac{1}{1+e^{-x}}-\frac{1}{(1+e^{-x})^{\mu}}\right\}dx = C+\psi(\mu) \qquad [\operatorname{Re}\mu>0]$$

(cf. **3.233** 2.). BI ((28))(10)

2. $$\int_{-\infty}^{\infty} \left\{\frac{1}{(1+e^{-x})^{\nu}}-\frac{1}{(1+e^{-x})^{\mu}}\right\}dx = \psi(\mu)-\psi(\nu)$$

$[\operatorname{Re}\mu>0,\quad \operatorname{Re}\nu>0]$ (cf. **3.219**). BI ((28))(11)

3.318

1. $$\int_{0}^{\infty} \frac{[\beta+\sqrt{1-e^{-x}}]^{-\nu}+[\beta-\sqrt{1-e^{-x}}]^{-\nu}}{\sqrt{1-e^{-x}}}e^{-\mu x}\,dx =$$

$$= \frac{2^{\mu+1}e^{(\mu-\nu)\pi i}(\beta^2-1)^{(\mu-\nu)/2}\Gamma(\mu)\,Q_{\mu-1}^{\nu-\mu}(\beta)}{\Gamma(\nu)}$$

$[\operatorname{Re}\mu>0]$. ET I 145(18)

2. $$\int_{u}^{\infty} \frac{1}{\sqrt{1-e^{-2x}}}\left\{e^{-u}\sqrt{1-e^{-2x}}-e^{-x}\sqrt{1-e^{-2u}}\right\}^{\nu} e^{-\mu x}\,dx =$$

$$= \frac{2^{-\frac{1}{2}(\mu+\nu)}\sqrt{\pi}\,e^{-\frac{u}{2}(\mu+\nu)}\Gamma(\mu)\Gamma(\nu+1)\,P_{-\frac{1}{2}(\mu-\nu)}^{-\frac{1}{2}(\mu+\nu)}\left(\sqrt{1-e^{-2u}}\right)}{\Gamma[(\mu+\nu+1)/2]}$$

$[u>0,\ \operatorname{Re}\mu>0,\ \operatorname{Re}\nu>-1]$. ET I 145(19)

3.32-3.34 Exponentials of more complicated arguments

3.321

1. $$\int_{0}^{u} e^{-x^2}\,dx = \sum_{k=0}^{\infty}\frac{(-1)^k u^{2k+1}}{k!\,(2k+1)};$$

$$= e^{-u^2}\sum_{k=0}^{\infty}\frac{2^k u^{2k+1}}{(2k+1)!!}.$$ AD 6.700

2. $$\int_{0}^{u} e^{-q^2x^2}\,dx = \frac{\sqrt{\pi}}{2q}\Phi(qu) \qquad [q>0].$$

3. $\int_0^\infty e^{-q^2x^2}\,dx = \frac{\sqrt{\pi}}{2q}$ $[q>0]$. FI II 624

3.322

1. $\int_u^\infty \exp\left(-\frac{x^2}{4\beta}-\gamma x\right)dx = \sqrt{\pi\beta}\,e^{\beta\gamma^2}\left[1-\Phi\left(\gamma\sqrt{\beta}+\frac{u}{2\sqrt{\beta}}\right)\right]$

$[\operatorname{Re}\beta>0,\ u>0]$. ET I 146(21)

2. $\int_0^\infty \exp\left(-\frac{x^2}{4\beta}-\gamma x\right)dx = \sqrt{\pi\beta}\exp(\beta\gamma^2)\left[1-\Phi\left(\gamma\sqrt{\beta}\right)\right]$ $[\operatorname{Re}\beta>0]$.

NT 27(1)a

3.323

1. $\int_1^\infty \exp(-qx-x^2)\,dx = \frac{e^{-q-1}}{q+2}\sum_{k=0}^\infty (-1)^k 2^k \frac{(2k-1)!!}{(q+2)^{2k}}$. BI ((29))(4)

2. $\int_{-\infty}^\infty \exp(-p^2x^2 \pm qx)\,dx = \exp\left(\frac{q^2}{4p^2}\right)\frac{\sqrt{\pi}}{p}$ $[p>0]$. BI ((28))(1)

3. $\int_0^\infty \exp(-\beta^2x^4-2\gamma^2x^2)\,dx = 2^{-\frac{3}{2}}\frac{\gamma}{\beta}e^{\frac{\gamma^4}{2\beta^2}}K_{\frac{1}{4}}\left(\frac{\gamma^4}{2\beta^2}\right)$

$\left[|\arg\gamma|<\frac{\pi}{4},\quad |\arg\beta|<\frac{\pi}{4}\right]$. ET I 147(34)a

3.324

1. $\int_0^\infty \exp\left(-\frac{\beta}{4x}-\gamma x\right)dx = \sqrt{\frac{\beta}{\gamma}}K_1\left(\sqrt{\beta\gamma}\right)$ $[\operatorname{Re}\beta\geqslant 0,\ \operatorname{Re}\gamma>0]$.

ET I 146(25)

2. $\int_{-\infty}^\infty \exp\left[-\left(x-\frac{b}{x}\right)^{2n}\right]dx = \frac{1}{n}\Gamma\left(\frac{1}{2n}\right)$.

3.325 $\int_0^\infty \exp\left(-ax^2-\frac{b}{x^2}\right)dx = \frac{1}{2}\sqrt{\frac{\pi}{a}}\exp\left(-2\sqrt{ab}\right)$ $[a>0,\ b>0]$.

FI II 644

3.326 $\int_0^\infty \exp(-x^\mu)\,dx = \frac{1}{\mu}\Gamma\left(\frac{1}{\mu}\right)$ $[\operatorname{Re}\mu>0]$. BI ((26))(4)

Exponentials of exponentials

3.327 $\int_0^\infty \exp(-ae^{nx})\,dx = -\frac{1}{n}\operatorname{Ei}(-a)$. LI ((26))(5)

3.328 $\int_{-\infty}^{\infty} \exp(-e^x) e^{\mu x}\, dx = \Gamma(\mu) \qquad [\operatorname{Re}\mu > 0].$ NH 145(14)

3.329 $\int_0^{\infty} \left[\frac{a \exp(-ce^{ax})}{1-e^{-ax}} - \frac{b \exp(-ce^{bx})}{1-e^{-bx}} \right] dx = e^{-c} \ln \frac{b}{a}$

$[a > 0, \quad b > 0, \quad c > 0].$ BI ((27))(12)

3.331

1. $\int_0^{\infty} \exp(-\beta e^{-x} - \mu x)\, dx = \beta^{-\mu} \gamma(\mu, \beta) \qquad [\operatorname{Re}\mu > 0].$ ET I 147(36)

2. $\int_0^{\infty} \exp(-\beta e^{x} - \mu x)\, dx = \beta^{\mu} \Gamma(-\mu, \beta) \qquad [\operatorname{Re}\beta > 0].$ ET I 147(37)

3. $\int_0^{\infty} (1-e^{-x})^{\nu-1} \exp(\beta e^{-x} - \mu x)\, dx = \mathrm{B}(\mu, \nu)\, \beta^{-\frac{\mu+\nu}{2}} e^{\frac{\beta}{2}} M_{\frac{\nu-\mu}{2}, \frac{\nu+\mu-1}{2}}(\beta)$

$[\operatorname{Re}\mu > 0, \ \operatorname{Re}\nu > 0].$ ET I 147(38)

4. $\int_0^{\infty} (1-e^{-x})^{\nu-1} \exp(-\beta e^{x} - \mu x)\, dx = \Gamma(\nu)\, \beta^{\frac{\mu-1}{2}} e^{-\frac{\beta}{2}} W_{\frac{1-\mu-2\nu}{2}, \frac{-\mu}{2}}(\beta)$

$[\operatorname{Re}\beta > 0, \ \operatorname{Re}\nu > 0].$ ET I 147(39)

3.332 $\int_0^{\infty} (1-e^{-x})^{\nu-1} (1-\lambda e^{-x})^{-\varrho} \exp(\beta e^{-x} - \mu x)\, dx =$

$= \mathrm{B}(\mu, \nu)\, \Phi_1(\mu, \varrho, \nu, \lambda, \beta)$

$[\operatorname{Re}\mu > 0, \ \operatorname{Re}\nu > 0, \ |\arg(1-\lambda)| < \pi].$ ET I 147(40)

3.333

1. $\int_{-\infty}^{\infty} \frac{e^{-\mu x}\, dx}{\exp(e^{-x}) - 1} = \Gamma(\mu)\, \zeta(\mu) \qquad [\operatorname{Re}\mu < 1].$ ET I 121(24)

2. $\int_{-\infty}^{\infty} \frac{e^{-\mu x}\, dx}{\exp(e^{-x}) + 1} = (1 - 2^{1-\mu})\, \Gamma(\mu)\, \zeta(\mu) \qquad [\operatorname{Re}\mu > 0].$ ET I 121(25)

3.334 $\int_0^{\infty} (e^x - 1)^{\nu-1} \exp\left[-\frac{\beta}{e^x - 1} - \mu x \right] dx =$

$= \Gamma(\mu - \nu + 1)\, e^{\frac{\beta}{2}} \beta^{\frac{\nu-1}{2}} W_{\frac{\nu-2\mu-1}{2}, \frac{\nu}{2}}(\beta) \qquad [\operatorname{Re}\beta > 0, \ \operatorname{Re}\mu > \operatorname{Re}\nu - 1].$

ET I 137(41)

Exponentials of hyperbolic functions

3.335 $$\int_0^\infty (e^{\nu x} + e^{-\nu x}\cos\nu\pi)\exp(-\beta\,\mathrm{sh}\,x)\,dx = -\pi[\mathbf{E}_\nu(\beta) + N_\nu(\beta)]$$

$[\mathrm{Re}\,\beta > 0]$. EH II 35(34)

3.336

1. $$\int_0^\infty \exp(-\nu x - \beta\,\mathrm{sh}\,x)\,dx = \pi\,\mathrm{cosec}\,\nu\pi\,[\mathbf{J}_\nu(\beta) - J_\nu(\beta)]$$

$\left[|\arg\beta| < \frac{\pi}{2} \text{ and } |\arg\beta| = \frac{\pi}{2} \text{ for } \mathrm{Re}\,\nu > 0;\ \nu\text{—not an integer}\right]$. WA 341(2)

2. $$\int_0^\infty \exp(nx - \beta\,\mathrm{sh}\,x)\,dx = \frac{1}{2}[S_n(\beta) - \pi\mathbf{E}_n(\beta) - \pi N_n(\beta)]$$

$[\mathrm{Re}\,\beta > 0;\ n = 0, 1, 2, \ldots]$. WA 342(6)

3. $$\int_0^\infty \exp(-nx - \beta\,\mathrm{sh}\,x)\,dx = \frac{1}{2}(-1)^{n+1}[S_n(\beta) + \pi\mathbf{E}_n(\beta) + \pi N_n(\beta)]$$

$[\mathrm{Re}\,\beta > 0;\ n = 0, 1, 2, \ldots]$. EH II 84(47)

3.337

1. $$\int_{-\infty}^\infty \exp(-\alpha x - \beta\,\mathrm{ch}\,x)\,dx = 2K_\alpha(\beta) \qquad \left[|\arg\beta| < \frac{\pi}{2}\right].$$ WA 201(7)

2. $$\int_{-\infty}^\infty \exp(-\nu x + i\beta\,\mathrm{ch}\,x)\,dx = i\pi e^{\frac{i\nu\pi}{2}} H_\nu^1(\beta) \qquad [0 < \arg z < \pi].$$ EH II 21(27)

3. $$\int_{-\infty}^\infty \exp(-\nu x - i\beta\,\mathrm{ch}\,x)\,dx = -i\pi e^{-\frac{i\nu\pi}{2}} H_\nu^2(\beta) \qquad [-\pi < \arg z < 0].$$ EH II 21(30)

Exponentials of trigonometric functions and logarithms

3.338

1. $$\int_0^\pi \{\exp i[(\nu-1)x - \beta\sin x] - \exp i[(\nu+1)x - \beta\sin x]\}\,dx =$$
$$= 2\pi[\mathbf{J}'_\nu(\beta) + i\mathbf{E}'_\nu(\beta)] \qquad [\mathrm{Re}\,\beta > 0].$$ EH II 36

2. $$\int_0^\pi \exp[\pm i(\nu x - \beta\sin x)]\,dx = \pi[\mathbf{J}_\nu(\beta) \pm i\mathbf{E}_\nu(\beta)]$$

$[\mathrm{Re}\,\beta > 0]$. EH II 35(32)

3. $\int_0^\infty \exp[-\gamma(x-\beta\sin x)]\,dx = \frac{1}{\gamma} + 2\sum_{k=1}^\infty \frac{\gamma J_k(k\beta)}{\gamma^2+k^2}$

[$\operatorname{Re}\gamma > 0$]. WA 619(4)

3.339 $\int_0^\pi \exp(2\cos x)\,dx = \pi I_0(2)$. BI ((277))(2)a

3.341 $\int_0^{\frac{\pi}{2}} \exp(-p\,\operatorname{tg} x)\,dx = \operatorname{ci}(p)\sin p - \operatorname{si}(p)\cos(p)$ [$p > 0$].

BI ((271))(2)a

3.342 $\int_0^1 \exp(-px\ln x)\,dx = \int_0^1 x^{-px}\,dx = \sum_{k=1}^\infty \frac{p^{k-1}}{k^k}$. BI ((29))(1)

3.35 Combinations of exponentials and rational functions

3.351

1. $\int_0^u x^n e^{-\mu x}\,dx = \frac{n!}{\mu^{n+1}} - e^{-u\mu}\sum_{k=0}^n \frac{n!}{k!}\frac{u^k}{\mu^{n-k+1}}$

[$u > 0$, $\operatorname{Re}\mu > 0$]. ET I 134(5)

2. $\int_u^\infty x^n e^{-\mu x}\,dx = e^{-u\mu}\sum_{k=0}^n \frac{n!}{k!}\frac{u^k}{\mu^{n-k+1}}$

[$u > 0$, $\operatorname{Re}\mu > 0$]. ET I 133(4)

3. $\int_0^\infty x^n e^{-\mu x}\,dx = n!\,\mu^{-n-1}$ [$\operatorname{Re}\mu > 0$]. ET I 133(3)

4. $\int_u^\infty \frac{e^{-px}\,dx}{x^{n+1}} = (-1)^{n+1}\frac{p^n \operatorname{Ei}(-pu)}{n!} + \frac{e^{-pu}}{u^n}\sum_{k=0}^{n-1}\frac{(-1)^k p^k u^k}{n(n-1)\dots(n-k)}$

[$p > 0$]. NT 21(3)

5. $\int_1^\infty \frac{e^{-\mu x}\,dx}{x} = -\operatorname{Ei}(-\mu)$ [$\operatorname{Re}\mu > 0$]. BI ((104))(10)

6. $\int_{-\infty}^u \frac{e^x}{x}\,dx = \operatorname{li}(e^u) = \operatorname{Ei}(u)$ [$u < 0$].

3.352

1. $\int_0^u \frac{e^{-\mu x}\,dx}{x+\beta} = e^{\mu\beta}[\operatorname{Ei}(-\mu u - \mu\beta) - \operatorname{Ei}(-\mu\beta)]$

[$|\arg\beta| < \pi$]. ET II 217(12)

2. $$\int_u^\infty \frac{e^{-\mu x}\,dx}{x+\beta} = -e^{\beta\mu}\,\mathrm{Ei}\,(-\mu u-\mu\beta)$$

$[u \geqslant 0,\ |\arg(u+\beta)| < \pi,\ \mathrm{Re}\,\mu > 0]$. ET I 134(6), JA

3. $$\int_u^v \frac{e^{-\mu x}\,dx}{x+\alpha} = e^{\alpha\mu}\{\mathrm{Ei}\,[-(\alpha+v)\mu]-\mathrm{Ei}\,[-(\alpha+u)\mu]\}$$

$[-\alpha < u,$ or $-\alpha > v,\ \mathrm{Re}\,\mu > 0]$. ET I 134 (7)

4. $$\int_0^\infty \frac{e^{-\mu x}\,dx}{x+\beta} = -e^{\beta\mu}\,\mathrm{Ei}\,(-\mu\beta) \qquad [|\arg\beta| < \pi,\ \mathrm{Re}\,\mu > 0].$$

ET II 217(11)

5. $$\int_u^\infty \frac{e^{-px}\,dx}{a-x} = e^{-pa}\,\mathrm{Ei}\,(pa-pu)$$

$[p > 0,\ a < u;$ for $a > u$, one should replace $\mathrm{Ei}\,(pa-pu)$ in this formula with $\overline{\mathrm{Ei}}\,(pa-pu)]$. ET II 251(37)

6. $$\int_0^\infty \frac{e^{-\mu x}\,dx}{a-x} = e^{-\mu a}\,\mathrm{Ei}\,(a\mu) \qquad [a > 0,\ \mathrm{Re}\,\mu > 0].$$ BI ((91))(4)

7. $$\int_{-\infty}^\infty \frac{e^{ipx}\,dx}{x-a} = i\pi e^{iap} \qquad [p > 0].$$ ET II 251(38)

3.353

1. $$\int_u^\infty \frac{e^{-\mu x}\,dx}{(x+\beta)^n} = e^{-u\mu}\sum_{k=1}^{n-1}\frac{(k-1)!\,(-\mu)^{n-k-1}}{(n-1)!\,(u+\beta)^k} - \frac{(-\mu)^{n-1}}{(n-1)!}\,e^{\beta\mu}\,\mathrm{Ei}\,[-(u+\beta)\mu]$$

$[n \geqslant 2,\ |\arg(u+\beta)| < \pi,\ \mathrm{Re}\,\mu > 0]$. ET I 134(10)

2. $$\int_0^\infty \frac{e^{-\mu x}\,dx}{(x+\beta)^n} = \frac{1}{(n-1)!}\sum_{k=1}^{n-1}(k-1)!\,(-\mu)^{n-k-1}\beta^{-k} - \frac{(-\mu)^{n-1}}{(n-1)!}\,e^{\beta\mu}\,\mathrm{Ei}\,(-\beta\mu)$$

$[n > 2,\ |\arg\beta| < \pi,\ \mathrm{Re}\,\mu > 0]$. ET I 134(9), BI((92))(2)

3. $$\int_0^\infty \frac{e^{-px}\,dx}{(a\pm x)^2} = pe^{\pm ap}\,\mathrm{Ei}\,(\mp ap) \pm \frac{1}{a} \qquad [p > 0].$$

LI ((281))(28), LI (281))(29)

4. $$\int_0^1 \frac{xe^x}{(1+x)^2}\,dx = \frac{e}{2}-1.$$ BI ((80))(6)

5. $\int\limits_0^\infty \frac{x^n e^{-\mu x}}{x+\beta}\,dx = (-1)^{n-1}\beta^n e^{\beta\mu}\operatorname{Ei}(-\beta\mu) + \sum_{k=1}^{n}(k-1)!\,(-\beta)^{n-k}\mu^{-k}$

$[|\arg\beta| < \pi,\ \operatorname{Re}\mu > 0].$ BI ((91))(3)a, ET I 135(11)

3.354

1. $\int\limits_0^\infty \frac{e^{-\mu x}\,dx}{\beta^2+x^2} = \frac{1}{\beta}[\operatorname{ci}(\beta\mu)\sin\beta\mu - \operatorname{si}(\beta\mu)\cos\beta\mu]$

$[\operatorname{Re}\beta > 0,\ \operatorname{Re}\mu > 0].$ BI ((91))(7)

2. $\int\limits_0^\infty \frac{xe^{-\mu x}\,dx}{\beta^2+x^2} = -\operatorname{ci}(\beta\mu)\cos\beta\mu - \operatorname{si}(\beta\mu)\sin\beta\mu$

$[\operatorname{Re}\beta > 0,\ \operatorname{Re}\mu > 0].$ BI ((91))(8)

3. $\int\limits_0^\infty \frac{e^{-\mu x}\,dx}{\beta^2-x^2} = \frac{1}{2\beta}[e^{-\beta\mu}\operatorname{Ei}(\beta\mu) - e^{\beta\mu}\operatorname{Ei}(-\beta\mu)]$

$[|\arg(\pm\beta)| < \pi,\ \operatorname{Re}\mu > 0;$ for $\beta > 0$, one should replace $\operatorname{Ei}(\beta\mu)$ in this formula with $\overline{\operatorname{Ei}}(\beta\mu)]$. BI ((91))(14)

4. $\int\limits_0^\infty \frac{xe^{-\mu x}\,dx}{\beta^2-x^2} = \frac{1}{2}[e^{-\beta\mu}\operatorname{Ei}(\beta\mu) + e^{\beta\mu}\operatorname{Ei}(-\beta\mu)]$

$[|\arg(\pm\beta)| < \pi,\ \operatorname{Re}\mu > 0;$ for $\beta > 0$ one should replace $\operatorname{Ei}(\beta\mu)$ in this formula with $\overline{\operatorname{Ei}}(\beta\mu)]$. BI ((91))(15)

5. $\int\limits_{-\infty}^\infty \frac{e^{-ipx}\,dx}{a^2+x^2} = \frac{\pi}{a}e^{-|ap|}$ $[a > 0]$. p real. ET I 118(1)a

3.355

1. $\int\limits_0^\infty \frac{e^{-\mu x}\,dx}{(\beta^2+x^2)^2} = \frac{1}{2\beta^3}\{\operatorname{ci}(\beta\mu)\sin\beta\mu - \operatorname{si}(\beta\mu)\cos\beta\mu -$

$- \beta\mu[\operatorname{ci}(\beta\mu)\cos\beta\mu + \operatorname{si}(\beta\mu)\sin\beta\mu]\}.$ LI ((92))(6)

2. $\int\limits_0^\infty \frac{xe^{-\mu x}\,dx}{(\beta^2+x^2)^2} = \frac{1}{2\beta^2}\{1 - \beta\mu[\operatorname{ci}(\beta\mu)\sin\beta\mu - \operatorname{si}(\beta\mu)\cos\beta\mu]\}$

$[\operatorname{Re}\beta > 0,\ \operatorname{Re}\mu > 0].$ BI ((92))(7)

3. $\int\limits_0^\infty \frac{e^{-px}\,dx}{(a^2-x^2)^2} = \frac{1}{4a^3}[(ap-1)e^{ap}\operatorname{Ei}(-ap) + (1+ap)e^{-ap}\operatorname{Ei}(ap)]$

$[a > 0,\ p > 0].$ BI ((92))(8)

4. $\int\limits_0^\infty \frac{xe^{-px}\,dx}{(a^2-x^2)^2} = \frac{1}{4a^2}\{-2 + ap[e^{-ap}\operatorname{Ei}(ap) - e^{ap}\operatorname{Ei}(-ap)]\}.$ LI ((92))(9)

3.356

1. $$\int_0^\infty \frac{x^{2n+1}e^{-px}}{a^2+x^2}\,dx = (-1)^{n-1}a^{2n}[\operatorname{ci}(ap)\cos ap + \operatorname{si}(ap)\sin ap] + \frac{1}{p^{2n}}\sum_{k=1}^{n}(2n-2k+1)!\,(-a^2p^2)^{k-1} \qquad [p>0].$$ BI ((91))(12)

2. $$\int_0^\infty \frac{x^{2n}e^{-px}}{a^2+x^2}\,dx = (-1)^n a^{2n-1}[\operatorname{ci}(ap)\sin ap - \operatorname{si}(ap)\cos ap] + \frac{1}{p^{2n-1}}\sum_{k=1}^{n}(2n-2k)!\,(-a^2p^2)^{k-1} \qquad [p>0].$$ BI ((91))(11)

3. $$\int_0^\infty \frac{x^{2n+1}e^{-px}}{a^2-x^2}\,dx = \frac{1}{2}a^{2n}[e^{ap}\operatorname{Ei}(-ap) + e^{-ap}\operatorname{Ei}(ap)] - \frac{1}{p^{2n}}\sum_{k=1}^{n}(2n-2k+1)!\,(a^2p^2)^{k-1} \qquad [p>0].$$ BI ((91))(17)

4. $$\int_0^\infty \frac{x^{2n}e^{-px}}{a^2-x^2}\,dx = \frac{1}{2}a^{2n-1}[e^{-ap}\operatorname{Ei}(ap) - e^{ap}\operatorname{Ei}(-ap)] - \frac{1}{p^{2n-1}}\sum_{k=1}^{n}(2n-2k)!\,(a^2p^2)^{k-1} \qquad [p>0].$$ BI ((91))(16)

3.357

1. $$\int_0^\infty \frac{e^{-\mu x}\,dx}{a^3+a^2x+ax^2+x^3} = \frac{1}{2a^2}\{\operatorname{ci}(a\mu)(\sin a\mu + \cos a\mu) + \operatorname{si}(a\mu)(\sin a\mu - \cos a\mu) - e^{a\mu}\operatorname{Ei}(-a\mu)\} \qquad [\operatorname{Re}\mu>0,\ a>0].$$ BI ((92))(18)

2. $$\int_0^\infty \frac{xe^{-\mu x}\,dx}{a^3+a^2x+ax^2+x^3} = \frac{1}{2a}\{\operatorname{ci}(a\mu)(\sin a\mu - \cos a\mu) - \operatorname{si}(a\mu)(\sin a\mu + \cos a\mu) + e^{a\mu}\operatorname{Ei}(-a\mu)\} \qquad [\operatorname{Re}\mu>0,\ a>0].$$ BI ((92))(19)

3. $$\int_0^\infty \frac{x^2e^{-\mu x}\,dx}{a^3+a^2x+ax^2+x^3} = \frac{1}{2}\{-\operatorname{ci}(a\mu)(\sin a\mu + \cos a\mu) - \operatorname{si}(a\mu)(\sin a\mu - \cos a\mu) - e^{a\mu}\operatorname{Ei}(-a\mu)\} \qquad [\operatorname{Re}\mu>0,\ a>0].$$ BI ((92))(20)

4. $$\int_0^\infty \frac{e^{-\mu x}\,dx}{a^3-a^2x+ax^2-x^3} = \frac{1}{2a^2}\{\operatorname{ci}(a\mu)(\sin a\mu - \cos a\mu) - \operatorname{si}(a\mu)(\sin a\mu + \cos a\mu) + e^{-a\mu}\operatorname{Ei}(a\mu)\} \qquad [\operatorname{Re}\mu>0,\ a>0].$$ BI ((92))(21)

5. $$\int_0^\infty \frac{xe^{-\mu x}\,dx}{a^3-a^2x+ax^2-x^3} = \frac{1}{2a}\{-\operatorname{ci}(a\mu)(\sin a\mu + \cos a\mu) - \operatorname{si}(a\mu)(\sin a\mu - \cos a\mu) + e^{-a\mu}\operatorname{Ei}(a\mu)\} \qquad [\operatorname{Re}\mu>0,\ a>0].$$ BI ((92))(22)

6. $$\int_0^\infty \frac{x^2 e^{-\mu x}\,dx}{a^3-a^2x+ax^2-x^3} = \frac{1}{2}\{\operatorname{ci}(a\mu)(\cos a\mu - \sin a\mu) +$$
$$+ \operatorname{si}(a\mu)(\cos a\mu + \sin a\mu) + e^{-a\mu}\operatorname{Ei}(a\mu)\} \quad [\operatorname{Re}\mu > 0,\ a > 0].$$ BI ((92))(23)

3.358

1. $$\int_0^\infty \frac{e^{-px}}{a^4-x^4}\,dx = \frac{1}{4a^3}\{e^{-ap}\operatorname{Ei}(ap) - e^{ap}\operatorname{Ei}(-ap) +$$
$$+ 2\operatorname{ci}(ap)\sin ap - 2\operatorname{si}(ap)\cos ap\} \quad [p > 0,\quad a > 0].$$ BI ((91))(18)

2. $$\int_0^\infty \frac{xe^{-px}\,dx}{a^4-x^4} = \frac{1}{4a^2}\{e^{ap}\operatorname{Ei}(-ap) + e^{-ap}\operatorname{Ei}(ap) -$$
$$- 2\operatorname{ci}(ap)\cos ap - 2\operatorname{si}(ap)\sin ap\} \quad [p > 0,\quad a > 0].$$ BI ((91))(19)

3. $$\int_0^\infty \frac{x^2e^{-px}\,dx}{a^4-x^4} = \frac{1}{4a}\{e^{-ap}\operatorname{Ei}(ap) - e^{ap}\operatorname{Ei}(-ap) -$$
$$- 2\operatorname{ci}(ap)\sin ap + 2\operatorname{si}(ap)\cos ap\} \quad [p > 0,\quad a > 0].$$ BI ((91))(20)

4. $$\int_0^\infty \frac{x^3e^{-px}\,dx}{a^4-x^4} = \frac{1}{4}\{e^{ap}\operatorname{Ei}(-ap) + e^{-ap}\operatorname{Ei}(ap) +$$
$$+ 2\operatorname{ci}(ap)\cos ap + 2\operatorname{si}(ap)\sin ap\} \quad [p > 0,\quad a > 0].$$ BI ((91))(21)

5. $$\int_0^\infty \frac{x^{4n}e^{-px}}{a^4-x^4}\,dx = \frac{1}{4}a^{4n-3}[e^{-ap}\operatorname{Ei}(ap) - e^{ap}\operatorname{Ei}(-ap) +$$
$$+ 2\operatorname{ci}(ap)\sin ap - 2\operatorname{si}(ap)\cos ap] - \frac{1}{p^{4n-3}}\sum_{k=1}^{n}(4n-4k)!\,(a^4p^4)^{k-1}$$
$$[p > 0,\quad a > 0].$$ BI ((91))(22)

6. $$\int_0^\infty \frac{x^{4n+1}e^{-px}}{a^4-x^4}\,dx = \frac{1}{4}a^{4n-2}[e^{ap}\operatorname{Ei}(-ap) + e^{-ap}\operatorname{Ei}(ap) -$$
$$- 2\operatorname{ci}(ap)\cos ap - 2\operatorname{si}(ap)\sin ap] - \frac{1}{p^{4n-2}}\sum_{k=1}^{n}(4n-4k+1)!\,(a^4p^4)^{k-1}$$
$$[p > 0,\quad a > 0].$$ BI ((91))(23)

7. $$\int_0^\infty \frac{x^{4n+2}e^{-px}}{a^4-x^4}\,dx = \frac{1}{4}a^{4n-1}[e^{-ap}\operatorname{Ei}(ap) - e^{ap}\operatorname{Ei}(-ap) -$$
$$- 2\operatorname{ci}(ap)\sin ap + 2\operatorname{si}(ap)\cos ap] - \frac{1}{p^{4n-1}}\sum_{k=1}^{n}(4n-4k+2)!\,(a^4p^4)^{k-1}$$
$$[p > 0,\quad a > 0].$$ BI ((91))(24)

8. $\int_0^\infty \frac{x^{4n+3}e^{-px}}{a^4-x^4}\,dx = \frac{1}{4}a^{4n}\left[e^{ap}\operatorname{Ei}(-ap) + e^{-ap}\operatorname{Ei}(ap) + \right.$

$\left. + 2\operatorname{ci}(ap)\cos ap + 2\operatorname{si}(ap)\sin ap\right] - \frac{1}{p^{4n}}\sum_{k=1}^{n}(4n-4k+3)!\,(a^4p^4)^{k-1}$

$[p>0, \quad a>0].$ BI ((91))(25)

3.359 $\int_{-\infty}^{\infty} \frac{(i-x)^n}{(i+x)^n}\frac{e^{-ipx}}{1+x^2}\,dx = (-1)^{n-1}\,2\pi p e^{-p}L_{n-1}(2p)$ for $p>0$;

$= 0$ for $p<0$.

ET I 118(2)

3.36-3.37 Combinations of exponentials and algebraic functions

3.361

1. $\int_0^u \frac{e^{-qx}}{\sqrt{qx}}\,dx = \sqrt{\frac{\pi}{q}}\,\Phi(\sqrt{qu}).$

2. $\int_0^\infty \frac{e^{-qx}}{\sqrt{x}}\,dx = \sqrt{\frac{\pi}{q}} \quad [q>0].$ BI ((98))(10)

3. $\int_{-1}^\infty \frac{e^{-qx}}{\sqrt{1+x}}\,dx = e^q\sqrt{\frac{\pi}{q}} \quad [q>0].$ BI ((104))(16)

3.362

1. $\int_1^\infty \frac{e^{-\mu x}\,dx}{\sqrt{x-1}} = \sqrt{\frac{\pi}{\mu}}\,e^{-\mu} \quad [\operatorname{Re}\mu>0].$ BI ((104))(11)a

2. $\int_0^\infty \frac{e^{-\mu x}\,dx}{\sqrt{x+\beta}} = \sqrt{\frac{\pi}{\mu}}\,e^{\beta\mu}\left[1-\Phi(\sqrt{\beta\mu})\right] \quad [\operatorname{Re}\mu>0, \ |\arg\beta|<\pi].$

ET I 135(18)

3.363

1. $\int_u^\infty \frac{\sqrt{x-u}}{x}\,e^{-\mu x}\,dx = \sqrt{\frac{\pi}{\mu}}\,e^{-u\mu} - \pi\sqrt{u}\left[1-\Phi(\sqrt{u\mu})\right]$

$[u>0, \quad \operatorname{Re}\mu>0].$ ET I 136(23)

2. $\int_u^\infty \frac{e^{-\mu x}\,dx}{x\sqrt{x-u}} = \frac{\pi}{\sqrt{u}}\left[1-\Phi(\sqrt{u\mu})\right] \quad [u>0, \quad \operatorname{Re}\mu \geqslant 0].$ ET I 136(26)

3.364

1. $\int_0^2 \frac{e^{-px}\,dx}{\sqrt{x(2-x)}} = \pi e^{-p}I_0(p) \quad [p>0].$ GW ((312))(7a)

2. $$\int_{-1}^{1} \frac{e^{2x}\,dx}{\sqrt{1-x^2}} = \pi I_0(2).$$ BI ((277))(2)a

3. $$\int_0^\infty \frac{e^{-px}\,dx}{\sqrt{x(x+a)}} = e^{\frac{ap}{2}} K_0\left(\frac{ap}{2}\right) \quad [a>0, \quad p>0].$$ GW ((312))(8a)

3.365

1. $$\int_0^u \frac{xe^{-\mu x}\,dx}{\sqrt{u^2-x^2}} = \frac{\pi u}{2}\left[\mathbf{L}_1(\mu u) - I_1(\mu u)\right] + u$$
$$[u>0, \quad \operatorname{Re}\mu>0].$$ ET I 136(28)

2. $$\int_u^\infty \frac{xe^{-\mu x}\,dx}{\sqrt{x^2-u^2}} = uK_1(u\mu) \quad [u>0, \quad \operatorname{Re}\mu>0].$$ ET I 136(29)

3.366

1. $$\int_0^{2u} \frac{(u-x)\,e^{-\mu x}\,dx}{\sqrt{2ux-x^2}} = \pi u e^{-u\mu} I_1(u\mu) \quad [\operatorname{Re}\mu>0].$$ ET I 136(31)

2. $$\int_0^\infty \frac{(x+\beta)\,e^{-\mu x}\,dx}{\sqrt{x^2+2\beta x}} = \beta e^{\beta\mu} K_1(\beta\mu) \; [\operatorname{Re}\mu>0, \quad |\arg\beta|<\pi].$$ ET I (136(30)

3. $$\int_0^\infty \frac{xe^{-\mu x}\,dx}{\sqrt{x^2+\beta^2}} = \frac{\beta\pi}{2}\left[\mathbf{H}_1(\beta\mu) - N_1(\beta\mu)\right] - \beta$$
$$\left[|\arg\beta|<\frac{\pi}{2}, \quad \operatorname{Re}\mu>0\right].$$ ET I 136(27)

3.367 $$\int_0^\infty \frac{e^{-\mu x}\,dx}{(1+\cos t+x)\sqrt{x^2+2x}} = \frac{\exp\left(2\mu\cos^2\frac{t}{2}\right)}{\sin t} \times$$
$$\times\left(t - \sin t \int_0^\mu K_0(v)\,e^{-v\cos t}\,dv\right) \quad [\operatorname{Re}\mu>0].$$ ET I 136(33)

3.368 $$\int_0^\infty \frac{e^{-\mu x}\,dx}{x+\sqrt{x^2+\beta^2}} = \frac{\pi}{2\beta\mu}\left[\mathbf{H}_1(\beta\mu) - N_1(\beta\mu)\right] - \frac{1}{\beta^2\mu^2}$$
$$\left[|\arg\beta|<\frac{\pi}{2}, \quad \operatorname{Re}\mu>0\right].$$ ET I 136(32)

3.369 $$\int_0^\infty \frac{e^{-\mu x}\,dx}{\sqrt{(x+a)^3}} = \frac{2}{\sqrt{a}} - 2\sqrt{\pi\mu}\,e^{a\mu}\left(1-\Phi\left(\sqrt{a\mu}\right)\right)$$
$$[|\arg a|<\pi, \quad \operatorname{Re}\mu>0].$$ ET I 135(20)

3.371 $$\int_0^\infty x^{n-\frac{1}{2}} e^{-\mu x}\,dx = \sqrt{\pi}\cdot\frac{1}{2}\cdot\frac{3}{2}\ldots\frac{2n-1}{2}\mu^{-n-\frac{1}{2}}$$

$[\operatorname{Re}\mu > 0]$. ET I 135(17)

3.372 $$\int_0^\infty x^{n-\frac{1}{2}}(2+x)^{n-\frac{1}{2}} e^{-px}\,dx = \frac{(2n-1)!!}{p^n} e^p K_n(p)$$

$[p > 0, \quad n = 0, 1, 2, \ldots]$. GW ((312))(8)

3.373 $$\int_0^\infty [(x+\sqrt{x^2+\beta^2})^n + (x-\sqrt{x^2+\beta^2})^n]\, e^{-\mu x}\,dx = 2\beta^{n+1} O_n(\beta\mu)$$

$[\operatorname{Re}\mu > 0]$. WA 305(1)

3.374

1. $$\int_0^\infty \frac{(x+\sqrt{1+x^2})^n}{\sqrt{1+x^2}} e^{-\mu x}\,dx = \frac{1}{2}[S_n(\mu) - \pi \mathbf{E}_n(\mu) - \pi N_n(\mu)]$$

$[\operatorname{Re}\mu > 0]$. ET I 137(35)

2. $$\int_0^\infty \frac{(x-\sqrt{1+x^2})^n}{\sqrt{1+x^2}} e^{-\mu x}\,dx = -\frac{1}{2}[S_n(\mu) + \pi \mathbf{E}_n(\mu) + \pi N_n(\mu)]$$

$[\operatorname{Re}\mu > 0]$. ET I 137(36)

3.38-3.39 Combinations of exponentials and arbitrary powers

3.381

1. $$\int_0^u x^{\nu-1} e^{-\mu x}\,dx = \mu^{-\nu}\gamma(\nu, \mu u)$$

$[\operatorname{Re}\nu > 0]$ EH I 266(22), EH II 133(1)

2. $$\int_0^u x^{p-1} e^{-x}\,dx = \sum_{k=0}^\infty (-1)^k \frac{u^{p+k}}{k!\,(p+k)};$$

$$= e^{-u}\sum_{k=0}^\infty \frac{u^{p+k}}{p(p+1)\ldots(p+k)}.$$ AD 6.705

3. $$\int_u^\infty x^{\nu-1} e^{-\mu x}\,dx = \mu^{-\nu}\Gamma(\nu, \mu u)$$

$[u > 0, \operatorname{Re}\mu > 0]$. EH I 256(21), EH II 133(2)

4. $$\int_0^\infty x^{\nu-1} e^{-\mu x}\,dx = \frac{1}{\mu^\nu}\Gamma(\nu) \quad [\operatorname{Re}\mu > 0, \quad \operatorname{Re}\nu > 0].$$ FI II 779

5. $$\int_0^\infty x^{\nu-1}e^{-(p+iq)x}\,dx = \Gamma(\nu)\,(p^2+q^2)^{-\frac{\nu}{2}}\exp\left(-i\nu\operatorname{arctg}\frac{q}{p}\right)$$

$[p>0,\quad \operatorname{Re}\nu>0 \quad \text{or} \quad p=0,\quad 0<\operatorname{Re}\nu<1]$. EH I 12(32)

6. $$\int_u^\infty \frac{e^{-x}}{x^\nu}\,dx = u^{-\frac{\nu}{2}}e^{-\frac{u}{2}}W_{-\frac{\nu}{2},\frac{(1-\nu)}{2}}(u) \quad [u>0].$$ WH

3.382

1. $$\int_0^u (u-x)^\nu e^{-\mu x}\,dx = \mu^{-\nu-1}e^{-u\mu}\gamma(\nu+1,-u\mu)$$

$[\operatorname{Re}\nu>-1,\quad u>0]$. ET I 137(6)

2. $$\int_u^\infty (x-u)^\nu e^{-\mu x}\,dx = \mu^{-\nu-1}e^{-u\mu}\Gamma(\nu+1)$$

$[u>0,\ \operatorname{Re}\nu>-1,\ \operatorname{Re}\mu>0]$. ET I 137(5), ET II 202(11)

3. $$\int_0^\infty (1+x)^{-\nu}e^{-\mu x}\,dx = \mu^{\frac{\nu}{2}-1}e^{\frac{\mu}{2}}W_{-\frac{\nu}{2},\frac{(1-\nu)}{2}}(\mu)$$

$[\operatorname{Re}\mu>0]$. WH

4. $$\int_0^\infty (x+\beta)^\nu e^{-\mu x}\,dx = \mu^{-\nu-1}e^{\beta\mu}\Gamma(\nu+1,\ \beta\mu)$$

$[|\arg\beta|<\pi,\ \operatorname{Re}\mu>0]$. ET I 137(4), ET II 233(10)

5. $$\int_0^u (a+x)^{\mu-1}e^{-x}\,dx = e^a[\gamma(\mu,\ a+u)-\gamma(\mu,\ a)]$$

$[\operatorname{Re}\mu>0]$. EH II 139

6. $$\int_{-\infty}^\infty (\beta+ix)^{-\nu}e^{-ipx}\,dx = 0 \qquad \text{for}\quad p>0;$$

$$= \frac{2\pi(-p)^{\nu-1}e^{\beta p}}{\Gamma(\nu)} \qquad \text{for}\quad p<0$$

$[\operatorname{Re}\nu>0,\ \operatorname{Re}\beta>0]$. ET I 118(4)

7. $$\int_{-\infty}^\infty (\beta-ix)^{-\nu}e^{-ipx}\,dx = \frac{2\pi p^{\nu-1}e^{-\beta p}}{\Gamma(\nu)} \qquad \text{for}\quad p>0;$$

$$=0 \qquad \text{for}\quad p<0$$

$[\operatorname{Re}\nu>0,\ \operatorname{Re}\beta>0]$. ET I 118(3)

3.383

1. $$\int_0^u x^{\nu-1}(u-x)^{\mu-1}e^{\beta x}\,dx = \mathrm{B}(\mu,\ \nu)\,u^{\mu+\nu-1}{}_1F_1(\nu;\ \mu+\nu;\ \beta u)$$

$[\operatorname{Re}\mu>0,\ \operatorname{Re}\nu>0]$. ET II 187(14)

2. $$\int_0^u x^{\mu-1}(u-x)^{\mu-1}e^{\beta x}\,dx = \sqrt{\pi}\left(\frac{u}{\beta}\right)^{\mu-\frac{1}{2}}\exp\left(\frac{\beta u}{2}\right)\Gamma(\mu)\,I_{\mu-\frac{1}{2}}\left(\frac{\beta u}{2}\right)$$

$[\operatorname{Re}\mu > 0]$. ET II 187(13)

3. $$\int_u^\infty x^{\mu-1}(x-u)^{\mu-1}e^{-\beta x}\,dx = \frac{1}{\sqrt{\pi}}\left(\frac{u}{\beta}\right)^{\mu-\frac{1}{2}}\Gamma(\mu)\exp\left(-\frac{\beta u}{2}\right)K_{\mu-\frac{1}{2}}\left(\frac{\beta u}{2}\right)$$

$[\operatorname{Re}\mu > 0,\ \operatorname{Re}\beta u > 0]$. ET II 202(12)

4. $$\int_u^\infty x^{\nu-1}(x-u)^{\mu-1}e^{-\beta x}\,dx =$$

$$= \beta^{-\frac{\mu+\nu}{2}}u^{\frac{\mu+\nu-2}{2}}\Gamma(\mu)\exp\left(-\frac{\beta u}{2}\right)W_{\frac{\nu-\mu}{2},\frac{1-\mu-\nu}{2}}(\beta u)$$

$[\operatorname{Re}\mu > 0,\ \operatorname{Re}\beta u > 0]$. ET II 202(13)

5. $$\int_0^\infty \frac{x^{q-1}e^{-px}}{(1+ax)^n}\,dx = p^{-q}\Gamma(q)\sum_{k=0}^\infty \binom{n+k-1}{k}\frac{\Gamma(q+k)}{\Gamma(q)}\left(\frac{a}{p}\right)^n$$

$[q > 0,\ p > 0,\ a > 0]$. BI ((92))(3)

6. $$\int_0^\infty x^{\nu-1}(x+\beta)^{-\nu+\frac{1}{2}}e^{-\mu x}\,dx = 2^{\nu-\frac{1}{2}}\Gamma(\nu)\,\mu^{-\frac{1}{2}}e^{\frac{\beta\mu}{2}}D_{1-2\nu}\left(\sqrt{2\beta\mu}\right)$$

2 $[|\arg\beta| < \pi,\ \operatorname{Re}\nu > 0,\ \operatorname{Re}\mu \geqslant 0]$. ET I 139(20), EH II 119(2)a

7. $$\int_0^\infty x^{\nu-1}(x+\beta)^{-\nu-\frac{1}{2}}e^{-\mu x}\,dx = 2^{\nu}\Gamma(\nu)\,\beta^{-\frac{1}{2}}e^{\frac{\beta\mu}{2}}D_{-2\nu}\left(\sqrt{2\beta\mu}\right)$$

$[|\arg\beta| < \pi,\ \operatorname{Re}\nu > 0,\ \operatorname{Re}\mu \geqslant 0]$. ET I 139(21), EH II 119(1)a

8. $$\int_0^\infty x^{\nu-1}(x+\beta)^{-\varrho}e^{-\mu x}\,dx = \beta^{\frac{\nu-\varrho-1}{2}}\mu^{\frac{\varrho-\nu-1}{2}}e^{\frac{\beta\mu}{2}}\Gamma(\nu)\,W_{\frac{1-\nu-\varrho}{2},\frac{\nu-\varrho}{2}}(\beta\mu)$$

$[|\arg\beta| < \pi,\ \operatorname{Re}\mu > 0,\ \operatorname{Re}\nu > 0]$;

WH, ET II 234(12), EH I 255(2)a

$$= \frac{1}{\sqrt{\pi}}\left(\frac{\beta}{\mu}\right)^{\nu-\frac{1}{2}}e^{\beta\mu}\Gamma(\nu)\,K_{\frac{1}{2}-\nu}\left(\frac{\beta\mu}{2}\right)$$

$[|\arg\beta| < \pi,\ \operatorname{Re}\mu > 0,\ \operatorname{Re}\nu > 0]$.

ET II 233(11), EH II 19(16)a, EH II 82(22)a

9. $$\int_u^\infty \frac{(x-u)^\nu e^{-\mu x}}{x}\,dx = u^\nu\Gamma(\nu+1)\,\Gamma(-\nu,\ u\mu)$$

$[u > 0,\ \operatorname{Re}\nu > -1,\ \operatorname{Re}\mu > 0]$. ET I 138(8)

10. $$\int_0^\infty \frac{x^{\nu-1}e^{-\mu x}}{x+\beta}\,dx = \beta^{\nu-1}e^{\beta\mu}\Gamma(\nu)\,\Gamma(1-\nu,\ \beta\mu)$$

$[|\arg\beta| < \pi,\ \operatorname{Re}\mu > 0,\ \operatorname{Re}\nu > 0]$. EH II 137(3)

3.384

1. $\int_{-1}^{1} (1-x)^{\nu-1}(1+x)^{\mu-1}e^{-ipx}\,dx = 2^{\mu+\nu-1}\mathrm{B}(\mu, \nu)\, e^{ip}\,{}_1F_1(\mu;\ \nu+\mu;\ -2ip)$

$[\operatorname{Re}\nu > 0,\ \operatorname{Re}\mu > 0].$ ET I 119(13)

2. $\int_{u}^{v} (x-u)^{2\mu-1}(v-x)^{2\nu-1}e^{-px}\,dx = \mathrm{B}(2\mu,\ 2\nu)(v-u)^{\mu+\nu-1}\times$

$\times p^{-\mu-\nu}\exp\left(-p\frac{u+v}{2}\right)M_{\mu-\nu,\ \mu+\nu-\frac{1}{2}}(vp-up)$

$[v > u > 0,\ \operatorname{Re}\mu > 0,\ \operatorname{Re}\nu > 0].$ ET I 139(23)

3. $\int_{u}^{\infty} (x+\beta)^{2\nu-1}(x-u)^{2\varrho 1}e^{-\mu x}\,dx = \frac{(u+\beta)^{\nu+\varrho-1}}{\mu^{\nu+\varrho}}\exp\left[\frac{(\beta-u)\mu}{2}\right]\times$

$\times\Gamma(2\varrho)\,W_{\nu-\varrho,\ \nu+\varrho-\frac{1}{2}}(u\mu+\beta\mu)$

$[u > 0,\ |\arg(\beta+u)| < \pi,\ \operatorname{Re}\mu > 0,\ \operatorname{Re}\varrho > 0].$ ET I 139(22)

4. $\int_{u}^{\infty} (x+\beta)^{\nu}(x-u)^{-\nu}e^{-\mu x}\,dx = \frac{1}{\mu}\nu\pi\operatorname{cosec}(\nu\pi)\quad e^{-\frac{(\beta+u)\mu}{2}}k_{2\nu}\left[\frac{(\beta+u)\mu}{2}\right]$

2 $[u > 0,\ |\arg(u+\beta)| < \pi,\ \operatorname{Re}\mu > 0,\ \operatorname{Re}\nu < 1].$ ET I 139(17)

5. $\int_{u}^{\infty} (x-u)^{\nu-1}(x+u)^{-\nu+\frac{1}{2}}e^{-\mu x}\,dx = \frac{1}{\sqrt{\mu}}2^{\nu-\frac{1}{2}}\Gamma(\nu)D_{1-2\nu}\left(2\sqrt{u\mu}\right)$

$[u > 0,\ \operatorname{Re}\mu > 0,\ \operatorname{Re}\nu > 0].$ ET I 139(18)

6. $\int_{u}^{\infty} (x-u)^{\nu-1}(x+u)^{-\nu-\frac{1}{2}}e^{-\mu x}\,dx = \frac{1}{\sqrt{u}}2^{\nu-\frac{1}{2}}\Gamma(\nu)D_{-2\nu}(2\sqrt{u\mu})$

$[u > 0,\ \operatorname{Re}\mu \geqslant 0,\ \operatorname{Re}\nu > 0].$ ET I 139(19)

7. $\int_{-\infty}^{\infty} (\beta-ix)^{-\mu}(\gamma-ix)^{-\nu}e^{-ipx}\,dx =$

$= \frac{2\pi e^{-\beta p}p^{\mu+\nu-1}}{\Gamma(\mu+\nu)}\,{}_1F_1(\nu;\ \mu+\nu;\ (\beta-\gamma)p)$ for $p > 0;$

$= 0$ for $p < 0$

$[\operatorname{Re}\beta > 0,\ \operatorname{Re}\gamma > 0,\ \operatorname{Re}(\mu+1) > \nu].$ ET I 119(10)

8. $\int_{-\infty}^{\infty} (\beta+ix)^{-\mu}(\gamma+ix)^{-\nu}e^{-ipx}\,dx = 0$ for $p > 0;$

$= -\frac{2\pi e^{\beta p}(-p)^{\mu+\nu-1}}{\Gamma(\mu+\nu)}\,{}_1F_1[\mu;\ \mu+\nu;\ (\beta-\gamma)p]$ for $p < 0$

$[\operatorname{Re}\beta > 0,\ \operatorname{Re}\gamma > 0,\ \operatorname{Re}(\mu+\nu) > 1].$ ET I 119(11)

9. $$\int_{-\infty}^{\infty} (\beta + ix)^{-2\mu} (\gamma - ix)^{-2\nu} e^{-ipx}\, dx =$$

$$= -2\pi (\beta + \gamma)^{-\mu-\nu} \frac{p^{\mu+\nu-1}}{\Gamma(2\nu)} \exp\left(\frac{\gamma-\beta}{2} p\right) \times$$
$$\times W_{\nu-\mu, \frac{1}{2}-\nu-\mu} (\beta p + \gamma p) \quad \text{for} \quad p > 0;$$

$$= 2\pi (\beta + \gamma)^{-\mu-\nu} \frac{(-p)^{\mu+\nu-1}}{\Gamma(2\mu)} \exp\left(\frac{\beta-\gamma}{2} p\right) \times$$
$$\times W_{\mu-\nu, \frac{1}{2}-\nu-\mu} (-\beta p - \gamma p) \quad \text{for} \quad p < 0$$

$$\left[\operatorname{Re}\beta > 0,\ \operatorname{Re}\gamma > 0,\ \operatorname{Re}(\mu+\nu) > \frac{1}{2}\right].$$ ET I 119(12)

3.385 $$\int_0^1 x^{\nu-1} (1-x)^{\lambda-1} (1-\beta x)^{-\varrho} e^{-\mu x}\, dx = \mathrm{B}(\nu, \lambda)\, \Phi_1(\nu, \varrho, \lambda+\nu; \beta, -\mu)$$

$$[\operatorname{Re}\lambda > 0,\ \operatorname{Re}\nu > 0,\ |\arg(1-\beta)| < \pi].$$ ET I 139(24)

3.386

1. $$\int_{-\infty}^{\infty} \frac{(ix)^{\nu_0} \prod_{k=1}^{n} (\beta_k + ix)^{\nu_k} e^{-ipx}\, dx}{\beta_0 - ix} = 2\pi e^{-\beta_0 p} \beta_0^{\nu_0} \prod_{k=1}^{n} (\beta_0 + \beta_k)^{\nu_k}$$

$$\left[\operatorname{Re}\nu_0 > -1,\quad \operatorname{Re}\beta_k > 0,\ \sum_{k=0}^{n} \operatorname{Re}\nu_k < 1,\ \arg ix = \frac{\pi}{2} \operatorname{sign} x,\ p > 0\right].$$

ET I 118(8)

2. $$\int_{-\infty}^{\infty} \frac{(ix)^{\nu_0} \prod_{k=1}^{n} (\beta_k + ix)^{\nu_k} e^{-ipx}\, dx}{\beta_0 + ix} = 0$$

$$\left[\operatorname{Re}\nu_0 > -1,\ \operatorname{Re}\beta_k > 0,\ \sum_{k=0}^{n} \operatorname{Re}\nu_k < 1,\ \arg ix = \frac{\pi}{2} \operatorname{sign} x,\ p > 0\right].$$

ET I 119(9)

3.387

1. $$\int_{-1}^{1} (1-x^2)^{\nu-1} e^{-\mu x}\, dx = \sqrt{\pi} \left(\frac{2}{\mu}\right)^{\nu-\frac{1}{2}} \Gamma(\nu)\, I_{\nu-\frac{1}{2}}(\mu)$$

$$\left[\operatorname{Re}\nu \geqslant 0,\ |\arg\mu| < \frac{\pi}{2}\right].$$ WA 190(2)a

2. $$\int_{-1}^{1} (1-x^2)^{\nu-1} e^{i\mu x}\, dx = \sqrt{\pi} \left(\frac{2}{\mu}\right)^{\nu-\frac{1}{2}} \Gamma(\nu)\, J_{\nu-\frac{1}{2}}(\mu) \qquad [\operatorname{Re}\nu > 0].$$

WA 34(3)a, WA 60(4)a

3. $$\int_1^\infty (x^2-1)^{\nu-1} e^{-\mu x}\,dx = \frac{1}{\sqrt{\pi}}\left(\frac{2}{\mu}\right)^{\nu-\frac{1}{2}} \Gamma(\nu) K_{\nu-\frac{1}{2}}(\mu)$$

$$\left[|\arg\mu| < \frac{\pi}{2},\quad \operatorname{Re}\nu > 0\right].$$ WA 190(4)a

4. $$\int_1^\infty (x^2-1)^{\nu-1} e^{i\mu x}\,dx = i\frac{\sqrt{\pi}}{2}\left(\frac{2}{\mu}\right)^{\nu-\frac{1}{2}} \Gamma(\nu) H^{(1)}_{\frac{1}{2}-\nu}(\mu)$$

$$[\operatorname{Im}\mu > 0,\quad \operatorname{Re}\nu > 0];$$ EH II 83(28)a

$$= -i\frac{\sqrt{\pi}}{2}\left(-\frac{2}{\mu}\right)^{\nu-\frac{1}{2}} \Gamma(\nu) H^{(2)}_{\frac{1}{2}-\nu}(-\mu)$$

$$[\operatorname{Im}\mu < 0,\ \operatorname{Re}\nu > 0].$$ EH II 83(29)a

5. $$\int_0^u (u^2-x^2)^{\nu-1} e^{\mu x}\,dx = \frac{\sqrt{\pi}}{2}\left(\frac{2u}{\mu}\right)^{\nu-\frac{1}{2}} \Gamma(\nu)\left[I_{\nu-\frac{1}{2}}(u\mu) + \mathbf{L}_{\nu-\frac{1}{2}}(u\mu)\right].$$

$$[u > 0,\quad \operatorname{Re}\nu > 0].$$ ET II 188(20)a

6. $$\int_u^\infty (x^2-u^2)^{\nu-1} e^{-\mu x}\,dx = \frac{1}{\sqrt{\pi}}\left(\frac{2u}{\mu}\right)^{\nu-\frac{1}{2}} \Gamma(\nu) K_{\nu-\frac{1}{2}}(u\mu)$$

$$[u > 0,\ \operatorname{Re}\mu > 0,\ \operatorname{Re}\nu > 0].$$ ET II 203(17)a

7. $$\int_0^\infty (x^2+u^2)^{\nu-1} e^{-\mu x}\,dx = \frac{\sqrt{\pi}}{2}\left(\frac{2u}{\mu}\right)^{\nu-\frac{1}{2}} \Gamma(\nu)\left[\mathbf{H}_{\nu-\frac{1}{2}}(u\mu) - N_{\nu-\frac{1}{2}}(u\mu)\right]$$

$$[|\arg u| < \pi,\quad \operatorname{Re}\mu > 0].$$ ET I 138(10)

3.388

1. $$\int_0^{2u} (2ux-x^2)^{\nu-1} e^{-\mu x}\,dx = \sqrt{\pi}\left(\frac{2u}{\mu}\right)^{\nu-\frac{1}{2}} e^{-u\mu}\Gamma(\nu) I_{\nu-\frac{1}{2}}(u\mu)$$

$$[u > 0,\quad \operatorname{Re}\nu > 0].$$ ET I 138(14)

2. $$\int_0^\infty (2\beta x+x^2)^{\nu-1} e^{-\mu x}\,dx = \frac{1}{\sqrt{\pi}}\left(\frac{2\beta}{\mu}\right)^{\nu-\frac{1}{2}} e^{\beta\mu}\Gamma(\nu) K_{\nu-\frac{1}{2}}(\beta\mu)$$

$$[|\arg\beta| < \pi;\quad \operatorname{Re}\nu > 0,\quad \operatorname{Re}\mu > 0].$$ ET I 138(13)

3. $$\int_0^\infty (x^2+ix)^{\nu-1} e^{-\mu x}\,dx = -\frac{i\sqrt{\pi}\,e^{\frac{i\mu}{2}}}{2\mu^{\nu-\frac{1}{2}}}\Gamma(\nu) H^{(2)}_{\nu-\frac{1}{2}}\left(\frac{\mu}{2}\right)$$

$$[\operatorname{Re}\mu > 0,\quad \operatorname{Re}\nu > 0].$$ ET I 138(15)

4. $$\int_0^\infty (x^2-ix)^{\nu-1} e^{-\mu x}\,dx = \frac{i\sqrt{\pi}\,e^{-\frac{i\mu}{2}}}{2\mu^{\nu-\frac{1}{2}}}\Gamma(\nu) H^{(1)}_{\nu-\frac{1}{2}}\left(\frac{\mu}{2}\right)$$

$$[\operatorname{Re}\mu > 0,\quad \operatorname{Re}\nu > 0].$$ ET I 138(16)

3.389

1. $$\int_0^u x^{2\nu-1}(u^2-x^2)^{\varrho-1}e^{\mu x}\,dx =$$
$$= \frac{1}{2}\mathrm{B}(\nu,\ \varrho)\,u^{2\nu+2\varrho-2}\,{}_1F_2\left(\nu;\frac{1}{2},\ \nu+\varrho;\frac{\mu^2u^2}{4}\right)+$$
$$+\frac{\mu}{2}\mathrm{B}\left(\nu+\frac{1}{2},\ \varrho\right)u^{2\nu+2\varrho-1}\,{}_1F_2\left(\nu+\frac{1}{2};\ \frac{3}{2},\ \nu+\varrho+\frac{1}{2};\ \frac{\mu^2u^2}{4}\right)$$
$[\operatorname{Re}\varrho>0,\quad \operatorname{Re}\nu>0].$ ET II 188(21)

2. $$\int_0^\infty x^{2\nu-1}(u^2+x^2)^{\varrho-1}e^{-\mu x}\,dx = \frac{u^{2\nu+2\varrho-2}}{2\sqrt{\pi}\,\Gamma(1-\varrho)}G_{13}^{31}\left(\frac{\mu^2u^2}{4}\middle|\begin{matrix}1-\nu\\ 1-\varrho-\nu,\ 0,\ \frac{1}{2}\end{matrix}\right)$$
$\left[|\arg u|<\frac{\pi}{2},\ \operatorname{Re}\mu>0,\ \operatorname{Re}\nu>0\right].$ ET II 234(15)a

3. $$\int_0^u x(u^2-x^2)^{\nu-1}e^{\mu x}\,dx = \frac{u^{2\nu}}{2\nu}+\frac{\sqrt{\pi}}{2}\left(\frac{\mu}{2}\right)^{\frac{1}{2}-\nu}u^{\nu+\frac{1}{2}}\,\Gamma(\nu)\times$$
$$\times\left[I_{\nu+\frac{1}{2}}(\mu u)+\mathbf{L}_{\nu+\frac{1}{2}}(\mu u)\right]\quad[\operatorname{Re}\nu>0].$$ ET II 188(19)a

4. $$\int_u^\infty x(x^2-u^2)^{\nu-1}e^{-\mu x}\,dx = 2^{\nu-\frac{1}{2}}\left(\sqrt{\pi}\right)^{-1}\mu^{\frac{1}{2}-\nu}u^{\nu+\frac{1}{2}}\Gamma(\nu)\,K_{\nu+\frac{1}{2}}(u\mu)$$
$[\operatorname{Re}(u\mu)>0].$ ET II 203(16)a

5. $$\int_{-\infty}^\infty \frac{(ix)^{-\nu}e^{-ipx}\,dx}{\beta^2+x^2}=\pi\beta^{-\nu-1}e^{-|p|\beta}$$
$\left[|\nu|<1,\quad \operatorname{Re}\beta>0,\quad \arg ix=\frac{\pi}{2}\operatorname{sign}x\right].$ ET I 118(5)

6. $$\int_0^\infty \frac{x^\nu e^{-\mu x}}{\beta^2+x^2}\,dx=\frac{1}{2}\Gamma(\nu)\,\beta^{\nu-1}\left[\exp\left(i\mu\beta+i\frac{(\nu-1)\pi}{2}\right)\times\right.$$
$$\left.\times\Gamma(1-\nu,\ i\beta\mu)+\exp\left(-i\beta\mu-i\frac{(\nu-1)\pi}{2}\right)\Gamma(1-\nu,\ -i\beta\mu)\right]$$
$[\operatorname{Re}\beta>0,\ \operatorname{Re}\mu>0,\ \operatorname{Re}\nu>-1].$ ET II 218(22)

7. $$\int_0^\infty \frac{x^{\nu-1}e^{-\mu x}\,dx}{1+x^2}=\pi\operatorname{cosec}(\nu\pi)\,V_\nu(2\mu,\ 0)\qquad[\operatorname{Re}\mu>0,\quad \operatorname{Re}\nu>0].$$
ET I 138(9)

8. $$\int_{-\infty}^\infty \frac{(\beta+ix)^{-\nu}e^{-ipx}}{\gamma^2+x^2}\,dx=\frac{\pi}{\gamma}(\beta+\gamma)^{-\nu}e^{-p\gamma}$$
$[\operatorname{Re}\nu>-1,\ p>0,\ \operatorname{Re}\beta>0,\ \operatorname{Re}\gamma>0].$ ET I 118(6)

9. $$\int_{-\infty}^\infty \frac{(\beta-ix)^{-\nu}e^{-ipx}}{\gamma^2+x^2}\,dx=\frac{\pi}{\gamma}(\beta-\gamma)^{-\nu}e^{\gamma p}$$
$[p>0,\ \operatorname{Re}\beta>0,\ \operatorname{Re}\gamma>0,\ \beta\neq\gamma,\ \operatorname{Re}\nu>-1].$ ET I 118(7)

3.391 $\int_0^\infty [(\sqrt{x+2\beta}+\sqrt{x})^{2\nu}-(\sqrt{x+2\beta}-\sqrt{x})^{2\nu}]\,e^{-\mu x}\,dx =$

$$= 2^{\nu+1}\frac{\nu}{\mu}\beta^\nu e^{\beta\mu}K_\nu(\beta\mu) \qquad [|\arg\beta| < \pi, \quad \operatorname{Re}\mu > 0]$$

ET I 140(30)

3.392

1. $\int_0^\infty (x+\sqrt{1+x^2})^\nu e^{-\mu x}\,dx = \frac{1}{\mu}S_{1,\nu}(\mu)+\frac{\nu}{\mu}S_{0,\nu}(\mu) \qquad [\operatorname{Re}\mu > 0].$

ET I 140(25)

2. $\int_0^\infty (\sqrt{1+x^2}-x)^\nu e^{-\mu x}\,dx = \frac{1}{\mu}S_{1,\nu}(\mu)-\frac{\nu}{\mu}S_{0,\nu}(\mu) \qquad [\operatorname{Re}\mu > 0].$

ET I 140(26)

3. $\int_0^\infty \frac{(x+\sqrt{1+x^2})^\nu}{\sqrt{1+x^2}}e^{-\mu x}\,dx = \pi\operatorname{cosec}\nu\pi\,[\mathbf{J}_{-\nu}(\mu)-J_{-\nu}(\mu)]$

$[\operatorname{Re}\mu > 0].$ ET I 140(27), EH II 35(33)

4. $\int_0^\infty \frac{(\sqrt{1+x^2}-x)^\nu}{\sqrt{1+x^2}}e^{-\mu x}\,dx = S_{0,\nu}(\mu)-\nu S_{-1,\nu}(\mu) \qquad [\operatorname{Re}\mu > 0].$

ET I 140(28)

3.393 $\int_0^\infty \frac{(x+\sqrt{x^2+4\beta^2})^{2\nu}}{\sqrt{x^3+4\beta^2 x}}e^{-\mu x}\,dx =$

$$= \frac{\sqrt{\mu\pi^3}}{2^{2\nu+\frac{3}{2}}\beta^{2\nu}}\left[J_{\nu+\frac{1}{4}}(\beta\mu)N_{\nu-\frac{1}{4}}(\beta\mu)-J_{\nu-\frac{1}{4}}(\beta\mu)N_{\nu+\frac{1}{4}}(\beta\mu)\right]$$

$[\operatorname{Re}\beta > 0, \quad \operatorname{Re}\mu > 0].$ ET I 140(33)

3.394 $\int_0^\infty \frac{(1+\sqrt{1+x^2})^{\nu+\frac{1}{2}}}{x^{\nu+1}\sqrt{1+x^2}}e^{-\mu x}\,dx = \sqrt{2}\,\Gamma(-\nu)\,D_\nu(\sqrt{2i\mu})\,D_\nu(\sqrt{-2i\mu})$

$[\operatorname{Re}\mu \geqslant 0, \ \operatorname{Re}\nu > 0].$ ET I 140(32)

3.395

1. $\int_1^\infty \frac{(\sqrt{x^2-1}+x)^\nu+(\sqrt{x^2-1}+x)^{-\nu}}{\sqrt{x^2-1}}e^{-\mu x}\,dx = 2K_\nu(\mu) \qquad [\operatorname{Re}\mu > 0].$

ET I 140(29)

2. $\int_1^\infty \frac{(x+\sqrt{x^2-1})^{2\nu}+(x-\sqrt{x^2-1})^{2\nu}}{\sqrt{x(x^2-1)}}e^{-\mu x}\,dx =$

$$= \sqrt{\frac{2\mu}{\pi}}K_{\nu+\frac{1}{4}}\left(\frac{\mu}{2}\right)K_{\nu-\frac{1}{4}}\left(\frac{\mu}{2}\right) \qquad [\operatorname{Re}\mu > 0].$$

ET I 140(34)

3. $$\int_0^\infty \frac{(x+\sqrt{x^2+1})^\nu+\cos\nu\pi\,(x+\sqrt{x^2+1})^{-\nu}}{\sqrt{x^2+1}} e^{-\mu x}\,dx =$$
$$= -\pi\left[\mathbf{E}_\nu(\mu)+N_\nu(\mu)\right] \qquad [\operatorname{Re}\mu>0].$$ EH II 35(34)

3.41-3.44 Combinations of rational functions of powers and exponentials

3.411

1. $$\int_0^\infty \frac{x^{\nu-1}\,dx}{e^{\mu x}-1} = \frac{1}{\mu^\nu}\Gamma(\nu)\zeta(\nu) \qquad [\operatorname{Re}\mu>0,\ \operatorname{Re}\nu>1].$$ FI II 792a

2. $$\int_0^\infty \frac{x^{2n-1}\,dx}{e^{px}-1} = (-1)^{n-1}\left(\frac{2\pi}{p}\right)^{2n}\frac{B_{2n}}{4n}.$$ FI II 721a

3. $$\int_0^\infty \frac{x^{\nu-1}\,dx}{e^{\mu x}+1} = \frac{1}{\mu^\nu}(1-2^{1-\nu})\Gamma(\nu)\zeta(\nu) \qquad [\operatorname{Re}\mu>0,\ \operatorname{Re}\nu>0].$$ FI II 792a, WH

4. $$\int_0^\infty \frac{x^{2n-1}\,dx}{e^{px}+1} = (1-2^{1-2n})\left(\frac{2\pi}{p}\right)^{2n}\frac{|B_{2n}|}{4n}.$$ BI((83))(2), EH I 39(25)

5. $$\int_0^{\ln 2} \frac{x\,dx}{1-e^{-x}} = \frac{\pi^2}{12}.$$ BI ((104))(5)

6. $$\int_0^\infty \frac{x^{\nu-1}e^{-\mu x}}{1-\beta e^{-x}}\,dx = \Gamma(\nu)\,\Phi(\beta;\nu;\mu)$$
$$[\operatorname{Re}\mu>0 \text{ and either } |\beta|\leqslant 1,\ \beta\neq 1,\ \operatorname{Re}\nu>0;\ \text{or } \beta=1,\ \operatorname{Re}\nu>1].$$ EH I 27(3)

7. $$\int_0^\infty \frac{x^{\nu-1}e^{-\mu x}}{1-e^{-\beta x}}\,dx = \frac{1}{\beta^\nu}\Gamma(\nu)\zeta\left(\nu,\frac{\mu}{\beta}\right) \qquad [\operatorname{Re}\mu>0,\ \operatorname{Re}\nu>1].$$ ET I 144(10)

8. $$\int_0^\infty \frac{x^{n-1}e^{-px}}{1+e^x}\,dx = (n-1)!\sum_{k=1}^\infty \frac{(-1)^{k-1}}{(p+k)^n} \qquad [p>-1;\ n=1,2,\ldots].$$ BI ((83))(9)

9. $$\int_0^\infty \frac{xe^{-x}\,dx}{e^x-1} = \frac{\pi^2}{6}-1$$ (cf. 4.231 2.). BI ((82))(1)

10. $$\int_0^\infty \frac{xe^{-2x}\,dx}{e^{-x}+1} = 1-\frac{\pi^2}{12}$$ (cf. 4.251 6.). BI ((82))(2)

11. $$\int_0^\infty \frac{xe^{-3x}}{e^{-x}+1}\,dx = \frac{\pi^2}{12}-\frac{3}{4}$$ (cf. 4.251 5.). BI ((82))(3)

12. $$\int_0^\infty \frac{xe^{-2nx}}{1+e^x}\,dx = -\frac{\pi^2}{12} + \sum_{k=1}^{2n-1} \frac{(-1)^{k-1}}{k^2}$$ (cf. 4.251 6.) BI ((82))(5)

13. $$\int_0^\infty \frac{xe^{-(2n-1)x}}{1+e^x}\,dx = \frac{\pi^2}{12} + \sum_{k=1}^{2n} \frac{(-1)^k}{k^2}$$ (cf. 4.251 5.). BI ((82))(4)

14. $$\int_0^\infty \frac{x^2e^{-nx}}{1-e^{-x}}\,dx = 2\sum_{k=n}^{\infty} \frac{1}{k^3}$$ (cf. 4.261 12.). BI ((82))(9)

15. $$\int_0^\infty \frac{x^2e^{-nx}}{1+e^{-x}}\,dx = 2\sum_{k=n}^{\infty} \frac{(-1)^{n+k}}{k^3}$$ (cf. 4.261 11.). LI ((82))(10)

16. $$\int_{-\infty}^\infty \frac{x^2e^{-\mu x}}{1+e^{-x}}\,dx = \pi^3 \cos^3 \mu\pi\,(2-\sin^2\mu\pi)$$

$[0 < \operatorname{Re}\mu < 1]$. ET I 120(17)a

17. $$\int_0^\infty \frac{x^3e^{-nx}}{1-e^{-x}}\,dx = \frac{\pi^4}{15} - 6\sum_{k=1}^{n-1} \frac{1}{k^4}$$ (cf. 4.262 5.). BI ((82))(12)

18. $$\int_0^\infty \frac{x^3e^{-nx}}{1+e^{-x}}\,dx = 6\sum_{k=n}^{\infty} \frac{(-1)^{n+k}}{k^4}$$ (cf. 4.262 4.). LI ((82))(13)

19. $$\int_0^\infty e^{-px}(e^{-x}-1)^n \frac{dx}{x} = -\sum_{k=0}^{n} (-1)^k n_k \ln(p+n-k)$$

$$\left[\binom{n}{k} = n(n+1)\ldots(n+k-1);\ n_0 = 1\right].$$ LI ((89))(10)

20. $$\int_0^\infty e^{-px}(e^{-x}-1)^n \frac{dx}{x^2} = \sum_{k=0}^{n} (-1)^k n_k (p+n-k)\ln(p+n-k)$$

$$\left[\binom{n}{k} = n(n+1)\ldots(n+k-1);\ n_0 = 1\right].$$ LI ((89))(15)

21. $$\int_0^\infty x^{n-1}\frac{1-e^{-mx}}{1-e^x}\,dx = (n-1)!\sum_{k=1}^{m} \frac{1}{k^n}$$ (cf. 4.272 11.). LI ((83))(8)

22. $$\int_0^\infty \frac{x^{p-1}}{e^{rx}-q}\,dx = \frac{1}{qr^p}\Gamma(p)\sum_{k=1}^{\infty} \frac{q^k}{k^p}$$ $[p > 0]$. BI ((83))(5)

23. $$\int_{-\infty}^\infty \frac{xe^{\mu x}\,dx}{\beta+e^x} = \pi\beta^{\mu-1}\operatorname{cosec}(\mu\pi)\,[\ln\beta - \pi\operatorname{ctg}(\mu\pi)]$$

$[|\arg\beta| < \pi,\ 0 < \operatorname{Re}\mu < 1]$. BI ((101))(5), ET I 120(16)a

24. $$\int_{-\infty}^\infty \frac{xe^{\mu x}}{e^{\nu x}-1}\,dx = \left(\frac{\pi}{\nu}\operatorname{cosec}\frac{\mu\pi}{\nu}\right)^2$$ $[\operatorname{Re}\nu > \operatorname{Re}\mu > 0]$

(cf. 4.254 2.). LI ((101))(3)

25. $$\int_0^\infty x\,\frac{1+e^{-x}}{e^x-1}\,dx = \frac{\pi^2}{3} - 1$$ (cf. 4.231 3.). BI ((82))(6)

26. $\int_0^\infty x \frac{1-e^{-x}}{1+e^{-3x}} e^{-x} dx = \frac{2\pi^2}{27}$. LI ((82))(7)a

27. $\int_0^\infty \frac{1-e^{-\mu x}}{1+e^x} \frac{dx}{x} = \ln\left[\frac{\Gamma\left(\frac{\mu}{2}+1\right)}{\Gamma\left(\frac{\mu+1}{2}\right)}\sqrt{\pi}\right]$ $[\operatorname{Re}\mu > -1]$. BI ((93))(4)

28. $\int_0^\infty \frac{e^{-\nu x}-e^{-\mu x}}{e^{-x}+1} \frac{dx}{x} = \ln \frac{\Gamma\left(\frac{\nu}{2}\right)\Gamma\left(\frac{\mu+1}{2}\right)}{\Gamma\left(\frac{\mu}{2}\right)\Gamma\left(\frac{\nu+1}{2}\right)}$

$[\operatorname{Re}\mu > 0, \ \operatorname{Re}\nu > 0]$. BI ((93))(6)

29. $\int_{-\infty}^\infty \frac{e^{px}-e^{qx}}{1+e^{rx}} \frac{dx}{x} = \ln\left[\operatorname{tg}\frac{p\pi}{2r}\operatorname{ctg}\frac{q\pi}{2r}\right]$

$[|r| > |p|, \ |r| > |q|, \ rp > 0, \ rq > 0]$

(cf. **4.267** 18.). BI ((103))(3)

30. $\int_{-\infty}^\infty \frac{e^{px}-e^{qx}}{1-e^{rx}} \frac{dx}{x} = \ln\left[\sin\frac{p\pi}{r}\operatorname{cosec}\frac{q\pi}{r}\right]$

$[|r| > |p|, \ |r| > |q|, \ rp > 0, \ rq > 0]$

(cf. **4.267** 19.). BI ((103))(4)

31. $\int_0^\infty \frac{e^{-qx}+e^{(q-p)x}}{1-e^{-px}} x\, dx = \left(\frac{\pi}{p}\operatorname{cosec}\frac{q\pi}{p}\right)^2$ $[0 < q < p]$. BI ((82))(8)

32. $\int_0^\infty \frac{e^{-px}-e^{(p-q)x}}{e^{-qx}+1} \frac{dx}{x} = \ln\operatorname{ctg}\frac{p\pi}{2q}$ $[0 < p < q]$. BI (93))(7)

3.412 $\int_0^\infty \left\{\frac{a+be^{-px}}{ce^{px}+g+he^{-px}} - \frac{a+be^{-qx}}{ce^{qx}+g+he^{-qx}}\right\} \frac{dx}{x} =$

$= \frac{a+b}{c+g+h}\ln\frac{p}{q}$ $[p > 0, \ q > 0]$. BI ((96))(7)

3.413

1. $\int_0^\infty \frac{(1-e^{-\beta x})(1-e^{-\nu x})e^{-\mu x}}{1-e^{-x}} \frac{dx}{x} = \ln\frac{\Gamma(\mu)\Gamma(\beta+\gamma+\mu)}{\Gamma(\mu+\beta)\Gamma(\mu+\gamma)}$

$[\operatorname{Re}\mu > 0, \ \operatorname{Re}\mu > -\operatorname{Re}\beta, \ \operatorname{Re}\mu > -\operatorname{Re}\gamma, \ \operatorname{Re}\mu > -\operatorname{Re}(\beta+\gamma)]$

(cf. **4.267** 25.). BI ((93))(13)

2. $\int_0^\infty \frac{\{1-e^{(q-p)x}\}^2}{e^{qx}-e^{(q-2p)x}} \frac{dx}{x} = \ln\operatorname{cosec}\frac{q\pi}{2p}$ $[0 < q < p]$. BI ((95))(6)

3. $$\int_0^\infty \frac{e^{-px}-e^{-qx}}{1+e^{-x}} \frac{1+e^{-(2n+1)x}}{x}\,dx =$$
$$= \ln\left\{\frac{q(q+2)(q+4)\dots(q+2n)}{p(p+2)(p+4)\dots(p+2n)} \frac{(p+1)(p+3)\dots(p+2n-1)}{(q+1)(q+3)\dots(q+2n-1)}\right\}$$
$[\operatorname{Re} p > -2n,\ \operatorname{Re} q > -2n]$ (cf. **4.267** 14.). BI((93))(11)

3.414 $$\int_0^\infty \frac{(1-e^{-\beta x})(1-e^{-\gamma x})(1-e^{-\delta x})e^{-\mu x}}{1-e^{-x}} \frac{dx}{x} =$$
$$= \ln \frac{\Gamma(\mu)\Gamma(\mu+\beta+\gamma)\Gamma(\mu+\beta+\delta)\Gamma(\mu+\gamma+\delta)}{\Gamma(\mu+\beta)\Gamma(\mu+\gamma)\Gamma(\mu+\delta)\Gamma(\mu+\beta+\gamma+\delta)}$$
$[2\operatorname{Re}\mu > |\operatorname{Re}\beta| + |\operatorname{Re}\gamma| + |\operatorname{Re}\delta|]$
(cf. **4.267** 31.). BI((93))(14), ET I 145(17)

3.415

1. $$\int_0^\infty \frac{x\,dx}{(x^2+\beta^2)(e^{\mu x}-1)} = \frac{1}{2}\left[\ln\left(\frac{\beta\mu}{2\pi}\right) - \frac{\pi}{\beta\mu} - \psi\left(\frac{\beta\mu}{2\pi}\right)\right]$$
$[\operatorname{Re}\beta > 0,\ \operatorname{Re}\mu > 0]$. BI((97))(20), EH I 18(27)

2. $$\int_0^\infty \frac{x\,dx}{(x^2+\beta^2)^2(e^{2\pi x}-1)} = -\frac{1}{8\beta^3} - \frac{1}{4\beta^2} + \frac{1}{4\beta}\psi'(\beta);$$
$$= \frac{1}{4\beta^4}\sum_{k=0}^{\infty} \frac{|B_{2k+2}|}{\beta^{2k}} \qquad [\operatorname{Re}\beta > 0].$$
BI((97))(22), ET I 22(12)

3.416

1. $$\int_0^\infty \frac{(1+ix)^{2n}-(1-ix)^{2n}}{i} \frac{dx}{e^{2\pi x}-1} = \frac{1}{2}\,\frac{2n-1}{2n+1}.$$ BI ((88))(4)

2. $$\int_0^\infty \frac{(1+ix)^{2n}-(1-ix)^{2n}}{i} \frac{dx}{e^{\pi x}+1} = \frac{1}{2n+1}.$$ BI ((87))(1)

3. $$\int_0^\infty \frac{(1+ix)^{2n-1}-(1-ix)^{2n-1}}{i} \frac{dx}{e^{\pi x}+1} = \frac{1}{2n}[1-2^{2n}B_{2n}].$$ BI ((87))(2)

3.417

1. $$\int_{-\infty}^\infty \frac{x\,dx}{a^2e^x+b^2e^{-x}} = \frac{\pi}{2ab}\ln\frac{b}{a} \qquad [ab>0]$$
(cf. **4.231** 6.). BI ((101))(1)

2. $$\int_{-\infty}^\infty \frac{x\,dx}{a^2e^x-b^2e^{-x}} = \frac{\pi^2}{4ab}$$ (cf. **4.231** 8.). LI ((101))(2)

3.418

1. $$\int_0^\infty \frac{x\,dx}{e^x+e^{-x}-1} = 1.171\,953\,6193\ldots$$ LI ((88))(1)

2. $$\int_0^\infty \frac{xe^{-x}\,dx}{e^x+e^{-x}-1} = 0.311\,821\,1319\ldots$$ LI ((88))(2)

3. $$\int_0^{\ln 2} \frac{x\,dx}{e^x+2e^{-x}-2} = \frac{\pi}{8}\ln 2.$$ BI ((104))(7)

3.419

1. $$\int_{-\infty}^\infty \frac{x\,dx}{(\beta+e^x)(1+e^{-x})} = \frac{(\ln\beta)^2}{2(\beta-1)} \qquad [|\arg\beta|<\pi]$$ (cf. 4.232 2.). BI ((101))(16)

2. $$\int_{-\infty}^\infty \frac{x\,dx}{(\beta+e^x)(1-e^{-x})} = \frac{\pi^2+(\ln\beta)^2}{2(\beta+1)} \qquad [|\arg\beta|<\pi]$$ (cf. 4.232 3.). BI ((101))(17)

3. $$\int_{-\infty}^\infty \frac{x^2\,dx}{(\beta+e^x)(1-e^{-x})} = \frac{[\pi^2+(\ln\beta)^2]\ln\beta}{3(\beta+1)} \qquad [|\arg\beta|<\pi]$$ (cf. 4.261 4.). BI ((102))(6)

4. $$\int_{-\infty}^\infty \frac{x^3\,dx}{(\beta+e^x)(1-e^{-x})} = \frac{\pi^2+(\ln\beta)^2}{4(\beta+1)} \qquad [|\arg\beta|<\pi]$$ (cf. 4.262 3.). BI ((102))(9)

5. $$\int_{-\infty}^\infty \frac{x^4\,dx}{(\beta+e^x)(1-e^{-x})} = \frac{[\pi^2+(\ln\beta)^2]^2}{15(\beta+1)}[7\pi^2+3(\ln\beta)^2]\ln\beta$$ (cf. 4.263 1.). BI ((102))(10)

6. $$\int_{-\infty}^\infty \frac{x^5\,dx}{(\beta+e^x)(1-e^{-x})} = \frac{[\pi^2+(\ln\beta)^2]^2}{6(\beta+1)}[3\pi^2+(\ln\beta)^2]^2$$ (cf. 4.264 3). BI ((102))(7)

7. $$\int_{-\infty}^\infty \frac{(x-\ln\beta)\,x\,dx}{(\beta-e^x)(1-e^{-x})} = \frac{-[4\pi^2+(\ln\beta)^2]\ln\beta}{6(\beta-1)} \qquad [|\arg\beta|<\pi]$$ (cf. 4.257 4.). BI ((102))(7)

3.421

1. $$\int_0^\infty (e^{-\nu x}-1)^n(e^{-\varrho x}-1)^m e^{-\mu x}\frac{dx}{x^2} = \sum_{k=0}^n(-1)^k\binom{n}{k}\sum_{l=0}^m(-1)^l\binom{m}{l}\times$$
$$\times\{(m-l)\varrho+(n-k)\nu+\mu\}\ln[(m-l)\varrho+(n-k)\nu+\mu]$$
$$[\operatorname{Re}\nu>0,\ \operatorname{Re}\mu>0,\ \operatorname{Re}\varrho>0].$$ BI ((89))(17)

2. $$\int_0^\infty (1-e^{-\nu x})^n (1-e^{-\varrho x}) e^{-x} \frac{dx}{x^3} = \frac{1}{2}\sum_{k=0}^{n}(-1)^k \binom{n}{k} (\varrho + k\nu + 1)^2 \times$$

$$\times \ln(\varrho + k\nu + 1) + \frac{1}{2}\sum_{k=1} (-1)^{k-1}\binom{n}{k}(k\nu+1)^2 \ln(k\nu+1)$$

$[\operatorname{Re}\mu > 0,\ \operatorname{Re}\nu > 0,\ \operatorname{Re}\varrho > 0]$. BI ((89))(31)

3. $$\int_{-\infty}^\infty \frac{xe^{-\mu x}\,dx}{(\beta+e^{-x})(\gamma+e^{-x})} = \frac{\pi(\beta^{\mu-1}\ln\beta - \gamma^{\mu-1}\ln\gamma)}{(\beta-\gamma)\sin\mu\pi} + \frac{\pi^2(\beta^{\mu-1}-\gamma^{\mu-1})\cos\mu\pi}{(\gamma-\beta)\sin^2\mu\pi}$$

$[|\arg\beta| < \pi,\ |\arg\gamma| < \pi,\ \beta \neq \gamma,\ 0 < \operatorname{Re}\mu < 2]$. ET I 120(19)

4. $$\int_0^\infty (e^{-px} - e^{-qx})(e^{-rx} - e^{-sx}) e^{-x}\frac{dx}{x} = \ln\frac{(p+s+1)(q+r+1)}{(p+r+1)(q+s+1)}$$

$[p+s > -1,\ p+r > -1,\ q > p]$ (cf. 4.267 24.). BI ((89))(11)

5. $$\int_0^\infty (1-e^{-px})(1-e^{-qx})(1-e^{-rx})e^{-x}\frac{dx}{x^2} = (p+q+1)\ln(p+q+1) +$$

$$+(p+r+1)\ln(p+r+1) + (q+r+1)\ln(q+r+1) - (p+1)\ln(p+1) -$$

$$-(q+1)\ln(q+1) - (r+1)\ln(r+1) - (p+q+r)\ln(p+q+r)$$

$[p > 0,\ q > 0,\ r > 0]$ (cf. 4.268 3.). BI ((89))(14)

3.422 $$\int_{-\infty}^\infty \frac{x(x-a)e^{\mu x}\,dx}{(\beta-e^x)(1-e^{-x})} = \frac{-\pi^2}{e^a-1}\operatorname{cosec}^2\mu\pi\,[(e^{a\mu}+1)\ln\mu - 2\pi\operatorname{ctg}\mu\pi\,(e^{a\mu}-1)]$$

$[a > 0,\ |\arg\beta| < \pi,\ |\operatorname{Re}\mu| < 1]$ (cf. 4.257 5.). BI ((102))(8)a

3.423

1. $$\int_0^\infty \frac{x^{\nu-1}}{(e^x-1)^2}\,dx = \Gamma(\nu)[\zeta(\nu-1) - \zeta(\nu)] \qquad [\operatorname{Re}\nu > 2].$$

ET I 313(10)

2. $$\int_0^\infty \frac{x^{\nu-1}e^{-\mu x}}{(e^x-1)^2}\,dx = \Gamma(\nu)[\zeta(\nu-1,\ \mu+1) - (\mu+1)\zeta(\nu,\ \mu+1)]$$

$[\operatorname{Re}\mu > -2,\ \operatorname{Re}\nu > 2]$. ET I 313(11)

3. $$\int_0^\infty \frac{x^q e^{-px}\,dx}{(1-ae^{-px})^2} = \frac{\Gamma(q+1)}{ap^{q+1}}\sum_{k=1}^\infty \frac{a^k}{k^q} \qquad [a < 1,\ q > -1,\ p > 0].$$

BI ((85))(13)

4. $$\int_0^\infty \frac{x^{\nu-1}e^{-\mu x}}{(1-\beta e^{-x})^2}\,dx = \Gamma(\nu)[{}_1F_1(\beta;\ \nu-1;\ \mu-1) - (\mu-1)\,{}_1F_1(\beta;\ \nu;\ \mu-1)]$$

$[\operatorname{Re}\nu > 0,\ \operatorname{Re}\mu > 0,\ |\arg(1-\beta)| < \pi]$. ET I 313(12)

5. $$\int_{-\infty}^\infty \frac{xe^x\,dx}{(\beta+e^x)^2} = \frac{1}{\beta}\ln\beta \qquad [|\arg\beta| < \pi]$$ (cf. 4.231 3.).

BI ((101))(10)

3.424

1. $$\int_0^\infty \frac{(1+a)\,e^x-a}{(1-e^x)^2}\,e^{-ax}x^n\,dx = n!\,\zeta(n,\ a).$$ BI ((85))(15)

2. $$\int_0^\infty \frac{(1+a)\,e^x+a}{(1+e^x)^2}\,e^{-ax}x^n\,dx = n!\sum_{k=1}^\infty \frac{(-1)^k}{(a+k)^n}.$$ BI ((85))(14)

3. $$\int_{-\infty}^\infty \frac{a^2e^x+b^2e^{-x}}{(a^2e^x-b^2e^{-x})^2}\,x^2\,dx = \frac{\pi^2}{2ab} \qquad [ab>0].$$ BI ((102))(3)a

4. $$\int_{-\infty}^\infty \frac{a^2e^x-b^2e^{-x}}{(a^2e^x+b^2e^{-x})^2}\,x^2\,dx = \frac{\pi}{ab}\ln\frac{b}{a} \qquad [ab>0].$$ BI ((102))(1)

5. $$\int_0^\infty \frac{e^x-e^{-x}+2}{(e^x-1)^2}\,x^2\,dx = \frac{2}{3}\pi^2-2.$$ BI ((85))(7)

3.425

1. $$\int_{-\infty}^\infty \frac{xe^x\,dx}{(a^2+b^2e^{2x})^n} = \frac{\sqrt{\pi}\,\Gamma\left(n-\frac{1}{2}\right)}{4a^{2n-1}b\Gamma(n)}\left[2\ln\frac{a}{2b}-C-\psi\left(n-\frac{1}{2}\right)\right]$$

$[ab>0,\ n>0]$ (cf. 4.231 5.).

BI((101))(13), LI((101))(13)

2. $$\int_{-\infty}^\infty \frac{(a^2e^x-e^{-x})\,x^2\,dx}{(a^2e^x+e^{-x})^{p+1}} = -\frac{1}{a^{p+1}}\,\mathrm{B}\left(\frac{p}{2},\ \frac{p}{2}\right)\ln a$$

$[a>0,\ p>0]$. BI ((102))(5)

3.426

1. $$\int_{-\infty}^\infty \frac{(e^x-ae^{-x})\,x^2\,dx}{(a+e^x)^2\,(1+e^{-x})^2} = \frac{(\ln a)^2}{a-1}.$$ BI ((102))(12)

2. $$\int_{-\infty}^\infty \frac{(e^x-ae^{-x})\,x^2\,dx}{(a+e^x)^2\,(1-e^{-x})^2} = \frac{\pi^2+(\ln a)^2}{a+1}.$$ BI ((102))(13)

3.427

1. $$\int_0^\infty \left(\frac{e^{-x}}{x}+\frac{e^{-\mu x}}{e^{-x}-1}\right)dx = \psi(\mu) \qquad [\operatorname{Re}\mu>0]$$

(cf. 4.281 4.). WH

2. $$\int_0^\infty \left(\frac{1}{1-e^{-x}}-\frac{1}{x}\right)e^{-x}\,dx = C$$ (cf. 4.281 1.). BI ((94))(1)

3. $$\int_0^\infty \left(\frac{1}{2}-\frac{1}{1+e^{-x}}\right)\frac{e^{-2x}}{x}\,dx = \frac{1}{2}\ln\frac{\pi}{4}.$$ BI ((94))(5)

4. $$\int_0^\infty \left(\frac{1}{2} - \frac{1}{x} + \frac{1}{e^x - 1}\right) \frac{e^{-\mu x}}{x}\, dx = \ln \Gamma(\mu) - \left(\mu - \frac{1}{2}\right) \ln \mu + \mu - \frac{1}{2} \ln(2\pi)$$

$[\mathrm{Re}\, \mu > 0]$. WH

5. $$\int_0^\infty \left(\frac{1}{2} e^{-2x} - \frac{1}{e^x + 1}\right) \frac{dx}{x} = -\frac{1}{2} \ln \pi.$$ BI ((94))(6)

6. $$\int_0^\infty \left(\frac{e^{\mu x} - 1}{1 - e^{-x}} - \mu\right) \frac{e^{-x}}{x}\, dx = -\ln \Gamma(\mu) - \ln \sin(\pi\mu) + \ln \pi$$

$[\mathrm{Re}\, \mu < 1]$. EH I 21(6)

7. $$\int_0^\infty \left(\frac{e^{-\nu x}}{1 - e^{-x}} - \frac{e^{-\mu x}}{x}\right) dx = \ln \mu - \psi(\nu)$$ (cf. 4.281 5.). BI ((94))(3)

8. $$\int_0^\infty \left(\frac{n}{x} - \frac{e^{-\mu x}}{1 - e^{-x/n}}\right) e^{-x}\, dx = n\psi(n\mu + n) - n \ln n \quad [\mathrm{Re}\, \mu > 0].$$ BI ((94))(4)

9. $$\int_0^\infty \left(\mu - \frac{1 - e^{-\mu x}}{1 - e^{-x}}\right) \frac{e^{-x}}{x}\, dx = \ln \Gamma(\mu + 1) \quad [\mathrm{Re}\, \mu > -1].$$ WH

10. $$\int_0^\infty \left(\nu e^{-x} - \frac{e^{-\mu x} - e^{-(\mu+\nu)x}}{e^x - 1}\right) \frac{dx}{x} = \ln \frac{\Gamma(\mu + \nu + 1)}{\Gamma(\mu + 1)} \quad [\mathrm{Re}\, \mu > -1,\ \mathrm{Re}\, \nu > 0]$$

(cf. 4.267 33.). BI ((94))(8)

11. $$\int_0^\infty [(1 - e^x)^{-1} + x^{-1} - 1]\, e^{-xz}\, dx = \psi(z) - \ln z \quad [\mathrm{Re}\, z > 0].$$ EH I 18(24)

3.428

1. $$\int_0^\infty \left(\nu e^{-\mu x} - \frac{1}{\mu} e^{-x} - \frac{1}{\mu} \frac{e^{-x} - e^{-\mu\nu x}}{1 - e^{-x}}\right) \frac{dx}{x} = \frac{1}{\mu} \ln \Gamma(\mu\nu) - \nu \ln \mu$$

$[\mathrm{Re}\, \mu > 0,\ \mathrm{Re}\, \nu > 0]$. BI ((94))(18)

2. $$\int_0^\infty \left(\frac{n-1}{2} + \frac{n-1}{1 - e^{-x}} + \frac{e^{(1-\mu)x}}{1 - e^{x/n}} + \frac{e^{-n\mu x}}{1 - e^{-x}}\right) e^{-x} \frac{dx}{x} =$$
$$= \frac{n-1}{2} \ln 2\pi - \left(n\mu + \frac{1}{2}\right) \ln n \quad [\mathrm{Re}\, \mu > 0].$$ BI ((94))(14)

3. $$\int_0^\infty \left(n\mu - \frac{n-1}{2} - \frac{n}{1 - e^{-x}} - \frac{e^{(1-\mu)x}}{1 - e^{\frac{x}{n}}}\right) \frac{e^{-x}}{x}\, dx = \sum_0^{n-1} \ln \Gamma\left(\mu - \frac{k}{n} + 1\right)$$

$[\mathrm{Re}\, \mu > 0]$. BI ((94))(13)

4. $$\int_0^\infty \left(\frac{e^{-\nu x}}{1 - e^x} - \frac{e^{-\mu\nu x}}{1 - e^{\mu x}} - \frac{e^x}{1 - e^x} + \frac{e^{\mu x}}{1 - e^{\mu x}}\right) \frac{dx}{x} = \nu \ln \mu$$

$[\mathrm{Re}\, \mu > 0,\ \mathrm{Re}\, \nu > 0]$. LI ((94))(15)

5. $$\int_0^\infty \left[\frac{1}{e^x-1} - \frac{\mu e^{-\mu x}}{1-e^{-\mu x}} + \left(a\mu - \frac{\mu+1}{2}\right)e^{-\mu x} + (1-a\mu)e^{-x}\right]\frac{dx}{x} =$$
$$= \frac{\mu-1}{2}\ln(2\pi) + \left(\frac{1}{2} - a\mu\right)\ln\mu \quad [\operatorname{Re}\mu > 0].$$ BI ((94))(16)

6. $$\int_0^\infty \left[\frac{e^{-\nu x}}{1-e^{-x}} - \frac{e^{-\mu\nu x}}{1-e^{-\mu x}} - \frac{(\mu-1)e^{-\mu x}}{1-e^{-\mu x}} - \frac{\mu-1}{2}e^{-\mu x}\right]\frac{dx}{x} =$$
$$= \frac{\mu-1}{2}\ln(2\pi) + \left(\frac{1}{2} - \mu\nu\right)\ln\mu$$
$$[\operatorname{Re}\mu > 0, \ \operatorname{Re}\nu > 0]$$ (cf. 4.267 37.). BI ((94))(17)

7. $$\int_0^\infty \left[1 - e^{-x} - \frac{(1-e^{-\nu x})(1-e^{-\mu x})}{1-e^{-x}}\right]\frac{dx}{x} = \ln B(\mu, \nu)$$
$$[\operatorname{Re}\mu > 0, \ \operatorname{Re}\nu > 0]$$ (cf. 4.267 35.). BI ((94))(12)

3.429 $$\int_0^\infty [e^{-x} - (1+x)^{-\mu}]\frac{dx}{x} = \psi(\mu) \quad [\operatorname{Re}\mu > 0].$$ NH 184(7)

3.431

1. $$\int_0^\infty \left(e^{-\mu x} - 1 + \mu x - \frac{1}{2}\mu^2x^2\right)x^{\nu-1}\,dx = \frac{-1}{\nu(\nu+1)(\nu+2)\mu^\nu}\Gamma(\nu+3)$$
$$[\operatorname{Re}\mu > 0, \ -2 > \operatorname{Re}\nu > -3].$$ LI ((90))(5)

2. $$\int_0^\infty \left[x^{-1} - \frac{1}{2}x^{-2}(x+2)(1-e^{-x})\right]e^{-px}\,dx = -1 + \left(p + \frac{1}{2}\right)\ln\left(1 + \frac{1}{p}\right)$$
$$[\operatorname{Re} p > 0].$$ ET I 144(6)

3.432

1. $$\int_0^\infty x^{\nu-1}e^{-mx}(e^{-x}-1)^n\,dx = \Gamma(\nu)\sum_{k=0}^{n}(-1)^k\binom{n}{k}\frac{1}{(n+m-k)^\nu}$$
$$[\operatorname{Re}\nu > 0].$$ LI ((90))(10)

2. $$\int_0^\infty [x^{\nu-1}e^{-x} - e^{-\mu x}(1-e^{-x})^{\nu-1}]\,dx = \Gamma(\nu) - \frac{\Gamma(\mu)}{\Gamma(\mu+\nu)}$$
$$[\operatorname{Re}\mu > 0, \ \operatorname{Re}\nu > 0].$$ LI ((81))(14)

3.433 $$\int_0^\infty x^{p-1}\left[e^{-x} + \sum_{k=1}^{n}(-1)^k\frac{x^{k-1}}{(k-1)!}\right]dx = \Gamma(p) \quad [-n < p < -n+1].$$
FI II 805

3.434

1. $$\int_0^\infty \frac{e^{-\nu x} - e^{-\mu x}}{x^{\varrho+1}}\,dx = \frac{\mu^\varrho - \nu^\varrho}{\varrho}\Gamma(1-\varrho) \quad [\operatorname{Re}\mu > 0, \ \operatorname{Re}\nu > 0, \ \operatorname{Re}\varrho < 1].$$
BI ((90))(6)

2. $\int\limits_0^\infty \frac{e^{-\mu x}-e^{-\nu x}}{x}\,dx = \ln\frac{\nu}{\mu}$ $[\operatorname{Re}\mu > 0,\ \operatorname{Re}\nu > 0]$. FI II 634

3.435

1. $\int\limits_0^\infty \left\{(x+1)e^{-x} - e^{-\frac{x}{2}}\right\}\frac{dx}{x} = 1 - \ln 2.$ LI ((89))(19)

2. $\int\limits_0^\infty \frac{1-e^{-\mu x}}{x(x+\beta)}\,dx = \frac{1}{\beta}\left[\ln(\beta\mu) - e^{\beta\mu}\operatorname{Ei}(-\beta\mu)\right]$ $[|\arg\beta| < \pi,\ \operatorname{Re}\mu > 0]$.

ET II 217(18)

3. $\int\limits_0^\infty \left(\frac{1}{1+x} - e^{-x}\right)\frac{dx}{x} = C.$ FI II 795, 802

4. $\int\limits_0^\infty \left(e^{-\mu x} - \frac{1}{1+ax}\right)\frac{dx}{x} = \ln\frac{a}{\mu} - C$ $[a > 0,\ \operatorname{Re}\mu > 0]$. BI ((92))(10)

3.436 $\int\limits_0^\infty \left\{\frac{e^{-npx}-e^{-nqx}}{n} - \frac{e^{-mpx}-e^{-mqx}}{m}\right\}\frac{dx}{x^2} = (q-p)\ln\frac{m}{n}$ $[p > 0,\ q > 0]$.

BI ((89))(28)

3.437 $\int\limits_0^\infty \left\{pe^{-x} - \frac{1-e^{-px}}{x}\right\}\frac{dx}{x} = p\ln p - p$ $[p > 0]$. BI ((89))(24)

3.438

1. $\int\limits_0^\infty \left\{\left(\frac{1}{2}+\frac{1}{x}\right)e^{-x} - \frac{1}{x}e^{-\frac{x}{2}}\right\}\frac{dx}{x} = \frac{\ln 2 - 1}{2}.$ BI ((89))(19)

2. $\int\limits_0^\infty \left\{\frac{p^2}{6}e^{-x} - \frac{p^2}{2x} - \frac{p}{x^2} - \frac{1-e^{-px}}{x^3}\right\}\frac{dx}{x} = \frac{p^2}{6}\ln p - \frac{11}{36}p^3$ $[p > 0]$.

BI ((89))(33)

3. $\int\limits_0^\infty \left(e^{-x} - e^{-2x} - \frac{1}{x}e^{-2x}\right)\frac{dx}{x} = 1 - \ln 2.$ BI ((89))(25)

4. $\int\limits_0^\infty \left\{\left(p-\frac{1}{2}\right)e^{-x} + \frac{x+2}{2x}\left(e^{-px} - e^{-\frac{x}{2}}\right)\right\}\frac{dx}{x} = \left(p-\frac{1}{2}\right)(\ln p - 1)$

$[p > 0]$. BI ((89))(22)

3.439 $\int\limits_0^\infty \left\{(p-q)e^{-rx} + \frac{1}{mx}\left(e^{-mpx} - e^{-mqx}\right)\right\}\frac{dx}{x} =$

$= p\ln p - q\ln q - (p-q)\left(1 + \ln\frac{r}{m}\right)$ $[p > 0,\ q > 0,\ r > 0]$.

LI((89))(26), LI((89))(27)

3.441 $\int_0^\infty \{(p-r)e^{-qx}+(r-q)e^{-px}+(q-p)e^{-rx}\}\frac{dx}{x^2}=(r-q)p\ln p+$

$+(p-r)q\ln q+(q-p)r\ln r$

$[p>0,\ q>0,\ r>0]$ (cf. **4.268** 6.). BI ((89))(18)

3.442

1. $\int_0^\infty \left\{1-\frac{x+2}{2x}(1-e^{-x})\right\}e^{-qx}\frac{dx}{x}=-1+\left(q+\frac{1}{2}\right)\ln\frac{q+1}{q}$ $[q>0]$.

BI ((89))(23)

2. $\int_0^\infty \left(\frac{e^{-x}-1}{x}+\frac{1}{1+x}\right)\frac{dx}{x}=C-1.$ BI ((92))(16)

3. $\int_0^\infty \left(e^{-px}-\frac{1}{1+a^2x^2}\right)\frac{dx}{x}=-C+\ln\frac{a}{p}$ $[p>0]$. BI ((92))(11)

3.443

1. $\int_0^\infty \left\{\frac{e^{-x}p^2}{2}-\frac{p}{x}+\frac{1-e^{-px}}{x^2}\right\}\frac{dx}{x}=\frac{p^2}{2}\ln p-\frac{3}{4}\ p^2$ $[p>0]$. BI ((89))(32)

2. $\int_0^\infty \frac{(1-e^{-px})^n e^{-qx}}{x^3}dx=\frac{1}{2}\sum_{k=2}^{n}(-1)^{k-1}\binom{n}{k}(q+kp)^2\ln(q+kp)$

$[n>2,\ q>0,\ pn+q>0]$ (cf. **4.268** 4.). BI ((89))(30)

3. $\int_0^\infty (1-e^{-px})^2 e^{-qx}\frac{dx}{x^2}=(2p+q)\ln(2p+q)-2(p+q)\ln(p+q)+q\ln q$

$[q>0,\ 2p>-q]$ (cf. **4.268** 2.). BI ((89))(13)

3.45 Combinations of powers and algebraic functions of exponentials

3.451

1. $\int_0^\infty xe^{-x}\sqrt{1-e^{-x}}\,dx=\frac{4}{3}\left(\frac{4}{3}-\ln 2\right).$ BI ((99))(1)

2. $\int_0^\infty xe^{-x}\sqrt{1-e^{-2x}}\,dx=\frac{\pi}{4}\left(\frac{1}{2}+\ln 2\right)$ (cf. **4.241** 9.). BI ((99))(2)

3.452

1. $\int_0^\infty \frac{x\,dx}{\sqrt{e^x-1}}=2\pi\ln 2.$ FI II 643a, BI((99))(4)

2. $\int_0^\infty \frac{x^2\,dx}{\sqrt{e^x-1}}=4\pi\left\{(\ln 2)^2+\frac{\pi^2}{12}\right\}.$ BI ((99))(5)

3. $\int_0^\infty \frac{xe^{-x}\,dx}{\sqrt{e^x-1}}=\frac{\pi}{2}[2\ln 2-1].$ BI ((99))(6)

4. $\int\limits_0^\infty \frac{xe^{-x}\,dx}{\sqrt{e^{2x}-1}} = 1 - \ln 2.$ BI ((99))(8)

5. $\int\limits_0^\infty \frac{xe^{-2x}\,dx}{\sqrt{e^{x}-1}} = \frac{3}{4}\pi\left(\ln 2 - \frac{7}{12}\right).$ BI ((99))(7)

3.453

1. $\int\limits_0^\infty \frac{xe^{x}}{a^2e^{x}-(a^2-b^2)}\frac{dx}{\sqrt{e^{x}-1}} = \frac{2\pi}{ab}\ln\left(1+\frac{b}{a}\right) \quad [ab > 0]$ (cf. 4.298 18.). BI ((99))(16)

2. $\int\limits_0^\infty \frac{xe^{x}\,dx}{[a^2e^{x}-(a^2+b^2)]\sqrt{e^{x}-1}} = \frac{2\pi}{ab}\operatorname{arctg}\frac{b}{a} \quad [ab > 0]$ (cf. 4.298 19.). BI ((99(17)

3.454

1. $\int\limits_0^\infty \frac{xe^{-2nx}\,dx}{\sqrt{e^{2x}+1}} = \frac{(2n-1)!!}{(2n)!!}\frac{\pi}{2}\left\{\ln 2 + \sum_{k=1}^{2n}\frac{(-1)^k}{k}\right\}.$ LI ((99))(10)

2. $\int\limits_0^\infty \frac{xe^{-(2n-1)x}\,dx}{\sqrt{e^{2x}-1}} = -\frac{(2n-2)!!}{(2n-1)!!}\left\{\ln 2 + \sum_{k=1}^{2n-1}\frac{(-1)^k}{k}\right\}.$ LI ((99))(9)

3.455

1. $\int\limits_0^\infty \frac{x^2e^{x}\,dx}{\sqrt{(e^{x}-1)^3}} = 8\pi\ln 2.$ BI ((99))(11)

2. $\int\limits_0^\infty \frac{x^3e^{x}\,dx}{\sqrt{(e^{x}-1)^3}} = 24\pi\left[(\ln 2)^2 + \frac{\pi^2}{12}\right].$ BI ((99))(12)

3.456

1. $\int\limits_0^\infty \frac{x\,dx}{\sqrt[3]{e^{3x}-1}} = \frac{\pi}{3\sqrt{3}}\left[\ln 3 + \frac{\pi}{3\sqrt{3}}\right].$ BI ((99))(13)

2. $\int\limits_0^\infty \frac{x\,dx}{\sqrt[3]{(e^{3x}-1)^2}} = \frac{\pi}{3\sqrt{3}}\left[\ln 3 - \frac{\pi}{3\sqrt{3}}\right]$ (cf. 4.244 3.). BI ((99))(14)

3.457

1. $\int\limits_0^\infty xe^{-x}\left(1-e^{-2x}\right)^{n-\frac{1}{2}}dx = \frac{(2n-1)!!}{4\cdot(2n)!!}\pi\left[\boldsymbol{C} + \psi(n+1) + 2\ln 2\right]$ (cf. 4.241 5.). BI ((99))(3)

2. $\int\limits_{-\infty}^\infty \frac{xe^{x}\,dx}{(a+e^{x})^{n+\frac{3}{2}}} = \frac{2}{(2n+1)\,a^{n+\frac{1}{2}}}\left[\ln(4a) - 3\boldsymbol{C} - 2\psi(2n) - \psi(n)\right].$ BI ((101))(12)

3. $\int\limits_{-\infty}^\infty \frac{x\,dx}{(a^2e^{x}+e^{-x})^{\mu}} = \frac{-1}{2a^{\mu}}\,\mathrm{B}\left(\frac{\mu}{2}, \frac{\mu}{2}\right)\ln a$ $[a > 0, \ \operatorname{Re}\mu > 0].$ BI ((101))(14)

3.458

1. $$\int_0^{\ln 2} xe^x (e^x - 1)^{p-1} dx = \frac{1}{p}\left[\ln 2 + \sum_{k=0}^{\infty} \frac{(-1)^{k-1}}{p+k+1}\right].$$ BI ((104))(4)

2. $$\int_{-\infty}^{\infty} \frac{xe^x\, dx}{(a+e^x)^{\nu+1}} = \frac{1}{\nu a^\nu}[\ln a - C - \psi(\nu)] \qquad [a > 0];$$

$$= \frac{1}{\nu a^\nu}\left[\ln a - \sum_{k=1}^{\nu-1} \frac{1}{k}\right] \qquad [\nu\text{-an integer}].$$ BI ((101))(11)

3.46-3.48 Combinations of exponentials of more complicated arguments and powers

3.461

1. $$\int_u^\infty \frac{e^{-p^2x^2}}{x^{2n}} dx = \frac{(-1)^n 2^{n-1} p^{2n-1}\sqrt{\pi}}{(2n-1)!!}[1 - \Phi(pu)] +$$

$$+ \frac{e^{-p^2u^2}}{2u^{2n-1}} \sum_{k=0}^{n-1} \frac{(-1)^k 2^{k+1}(pu)^{2k}}{(2n-1)(2n-3)\ldots(2n-2k-1)} \qquad [p > 0].$$ NT 21(4)

2.* $$\int_0^\infty x^{2n} e^{-px^2} dx = \frac{(2n-1)!!}{2(2p)^n}\sqrt{\frac{\pi}{p}} \qquad [p > 0].$$ FI II 743

3. $$\int_0^\infty x^{2n+1} e^{-px^2} dx = \frac{n!}{2p^{n+1}} \qquad [p > 0].$$ BI ((81))(7)

4. $$\int_{-\infty}^\infty (x + ai)^{2n} e^{-x^2} dx = \frac{(2n-1)!!}{2^n}\sqrt{\pi}\sum_{k=0}^{n}(-1)^k \frac{(2a)^{2k} n!}{(2k)!(n-k)!}.$$ BI ((100))(12)

5. $$\int_u^\infty e^{-\mu x^2} \frac{dx}{x^2} = \frac{1}{u} e^{-\mu u^2} - \sqrt{\mu\pi}[1 - \Phi(\sqrt{\mu} u)]$$

$$\left[|\arg\mu| < \frac{\pi}{4},\ u > 0\right].$$ ET I 135(19)a

3.462

1. $$\int_0^\infty x^{\nu-1} e^{-\beta x^2 - \gamma x} dx = (2\beta)^{-\frac{\nu}{2}}\Gamma(\nu)\exp\left(\frac{\gamma^2}{8\beta}\right) D_{-\nu}\left(\frac{\gamma}{\sqrt{2\beta}}\right)$$

$$[\operatorname{Re}\beta > 0,\ \operatorname{Re}\nu > 0].$$ EH II 119(3)a, ET I 313(13)

2. $$\int_{-\infty}^\infty x^n e^{-px^2 + 2qx} dx = \frac{1}{2^{n-1}p}\sqrt{\frac{\pi}{p}} \frac{d^{n-1}}{dq^{n-1}}\left(qe^{\frac{q^2}{p}}\right) \qquad [p > 0];$$ BI ((100))(8)

$$= n!\, e^{\frac{q^2}{p}} \sqrt{\frac{\pi}{p}}\left(\frac{q}{p}\right)^n \sum_{k=0}^{E\left(\frac{n}{2}\right)} \frac{1}{(n-2k)!(k)!}\left(\frac{p}{4q^2}\right)^k \qquad [p > 0].$$ LI ((100))(8)

3. $$\int_{-\infty}^{\infty} (ix)^{\nu} e^{-\beta^2 x^2 - iqx}\,dx = 2^{-\frac{\nu}{2}} \sqrt{\pi}\,\beta^{-\nu-1} \exp\left(-\frac{q^2}{8\beta^2}\right) D_{\nu}\left(\frac{q}{\beta\sqrt{2}}\right)$$

$$\left[\operatorname{Re}\beta > 0,\ \operatorname{Re}\nu > -1,\ \arg ix = \frac{\pi}{2}\operatorname{sign} x\right].$$ ET I 121(23)

4. $$\int_{-\infty}^{\infty} x^n \exp[-(x-\beta)^2]\,dx = (2i)^{-n}\sqrt{\pi}\,H_n(i\beta).$$ EH II 195(31)

5. $$\int_0^{\infty} x e^{-\mu x^2 - 2\nu x}\,dx = \frac{1}{2\mu} - \frac{\nu}{2\mu}\sqrt{\frac{\pi}{\mu}}\,e^{\frac{\nu^2}{\mu}}\left[1 - \Phi\left(\frac{\nu}{\sqrt{\mu}}\right)\right]$$

$$\left[|\arg\nu| < \frac{\pi}{2},\ \operatorname{Re}\mu > 0\right].$$ ET I 146(31)a

6. $$\int_{-\infty}^{\infty} x e^{-px^2 + 2qx}\,dx = \frac{q}{p}\sqrt{\frac{\pi}{p}}\exp\left(\frac{q^2}{p}\right) \qquad [\operatorname{Re} p > 0].$$ BI ((100))(7)

7. $$\int_0^{\infty} x^2 e^{-\mu x^2 - 2\nu x}\,dx = -\frac{\nu}{2\mu^2} + \sqrt{\frac{\pi}{\mu^5}}\,\frac{2\nu^2 + \mu}{4}\,e^{\frac{\nu^2}{\mu}}\left[1 - \Phi\left(\frac{\nu}{\sqrt{\mu}}\right)\right]$$

$$\left[|\arg\nu| < \frac{\pi}{2},\ \operatorname{Re}\mu > 0\right].$$ ET I 146(32)

8. $$\int_{-\infty}^{\infty} x^2 e^{-\mu x^2 + 2\nu x}\,dx = \frac{1}{2\mu}\sqrt{\frac{\pi}{\mu}}\left(1 + 2\frac{\nu^2}{\mu}\right)e^{\frac{\nu^2}{\mu}}$$

$$[|\arg\nu| < \pi,\ \operatorname{Re}\mu > 0].$$ BI ((100))(8)a

3.463 $$\int_0^{\infty} \left(e^{-x^2} - e^{-x}\right)\frac{dx}{x} = \frac{1}{2}\,\boldsymbol{C}.$$ BI ((89))(5)

3.464 $$\int_0^{\infty} \left(e^{-\mu x^2} - e^{-\nu x^2}\right)\frac{dx}{x^2} = \sqrt{\pi}\left(\sqrt{\nu} - \sqrt{\mu}\right)$$

$$[\operatorname{Re}\mu > 0,\ \operatorname{Re}\nu > 0].$$ FI II 645

3.465 $$\int_0^{\infty} (1 + 2\beta x^2)\,e^{-\mu x^2}\,dx = \frac{\mu + \beta}{2}\sqrt{\frac{\pi}{\mu^3}} \qquad [\operatorname{Re}\mu > 0].$$ ET I 136(24)a

3.466

1. $$\int_0^{\infty} \frac{e^{-\mu^2 x^2}}{x^2 + \beta^2}\,dx = [1 - \Phi(\beta\mu)]\,\frac{\pi}{2\beta}\,e^{\beta^2\mu^2}$$

$$\left[\operatorname{Re}\beta > 0,\ |\arg\mu| < \frac{\pi}{4}\right].$$ NT 19(13)

2. $$\int_0^{\infty} \frac{x^2 e^{-\mu^2 x^2}}{x^2 + \beta^2}\,dx = \frac{\sqrt{\pi}}{2\mu} - \frac{\pi\beta}{2}\,e^{\mu^2\beta^2}\,[1 - \Phi(\beta\mu)]$$

$$\left[\operatorname{Re}\beta > 0,\ |\arg\mu| < \frac{\pi}{4}\right].$$ ET II 217(16)

3. $\int_0^1 \frac{e^{x^2}-1}{x^2}dx = \sum_{k=1}^{\infty} \frac{1}{k!(2k-1)}$. FI II 683

3.467 $\int_0^\infty \left(e^{-x^2} - \frac{1}{1+x^2}\right)\frac{dx}{x} = -\frac{1}{2}C$. BI ((92))(12)

3.468

1. $\int_{u\sqrt{2}}^\infty \frac{e^{-x^2}}{\sqrt{x^2-u^2}}\frac{dx}{x} = \frac{\pi}{4u}[1-\Phi(u)]^2 \qquad [u>0]$. NT 33(17)

2. $\int_0^\infty \frac{xe^{-\mu x^2}dx}{\sqrt{a^2+x^2}} = \frac{1}{2}\sqrt{\frac{\pi}{\mu}}e^{a^2\mu}\left[1-\Phi\left(a\sqrt{\mu}\right)\right]$

$[\operatorname{Re}\mu>0,\ a>0]$. NT 19(11)

3.469

1. $\int_0^\infty e^{-\mu x^4-2\nu x^2}dx = \frac{1}{4}\sqrt{\frac{2\nu}{\mu}}\exp\left(\frac{\nu^2}{2\mu}\right)K_{\frac{1}{4}}\left(\frac{\nu^2}{2\mu}\right)$

$[\operatorname{Re}\mu>0]$. ET I 146(23)

2. $\int_0^\infty (e^{-x^4}-e^{-x})\frac{dx}{x} = \frac{3}{4}C$. BI ((89))(7)

3. $\int_0^\infty (e^{-x^4}-e^{-x^2})\frac{dx}{x} = \frac{1}{4}C$. BI ((89))(6)

3.471

1. $\int_0^u \exp\left(-\frac{\beta}{x}\right)\frac{dx}{x^2} = \frac{1}{\beta}\exp\left(-\frac{\beta}{u}\right)$. ET II 188(22)

2. $\int_0^u x^{\nu-1}(u-x)^{\mu-1}e^{-\frac{\beta}{x}}dx = \beta^{\frac{\nu-1}{2}}u^{\frac{2\mu+\nu-1}{2}}\exp\left(-\frac{\beta}{2u}\right)\times$

$\times\Gamma(\mu)W_{\frac{1-2\mu-\nu}{2},\frac{\nu}{2}}\left(\frac{\beta}{u}\right) \qquad [\operatorname{Re}\mu>0,\ \operatorname{Re}\beta>0,\ u>0]$.

ET II 187(18)

3. $\int_0^u x^{-\mu-1}(u-x)^{\mu-1}e^{-\frac{\beta}{x}}dx = \beta^{-\mu}u^{\mu-1}\Gamma(\mu)\exp\left(-\frac{\beta}{u}\right)$

$[\operatorname{Re}\mu>0,\ u>0]$. ET II 187(16)

4. $\int_0^u x^{-2\mu}(u-x)^{\mu-1}e^{-\frac{\beta}{x}}dx = \frac{1}{\sqrt{\pi u}}\beta^{\frac{1}{2}-\mu}e^{-\frac{\beta}{2u}}\Gamma(\mu)K_{\mu-\frac{1}{2}}\left(\frac{\beta}{2u}\right)$

$[u>0,\ \operatorname{Re}\beta>0,\ \operatorname{Re}\mu>0]$. ET II 187(17)

5. $$\int_u^\infty x^{\nu-1}(x-u)^{\mu-1}e^{\frac{\beta}{x}}\,dx = \mathrm{B}(1-\mu-\nu,\mu)\,u^{\mu+\nu-1}{}_1F_1\left(1-\mu-\nu;\ 1-\nu;\ \frac{\beta}{u}\right)$$
$[0 < \operatorname{Re}\mu < \operatorname{Re}(1-\nu),\ u > 0]$. ET II 203(15)

6. $$\int_u^\infty x^{-2\mu}(x-u)^{\mu-1}e^{\frac{\beta}{x}}\,dx = \sqrt{\frac{\pi}{u}}\,\beta^{\frac{1}{2}-\mu}\,\Gamma(\mu)\exp\left(\frac{\beta}{2u}\right)I_{\mu-\frac{1}{2}}\left(\frac{\beta}{2u}\right)$$
$[\operatorname{Re}\mu > 0,\ u > 0]$. ET II 202(14)

7. $$\int_0^\infty x^{\nu-1}(x+\gamma)^{\mu-1}e^{-\frac{\beta}{x}}\,dx = \beta^{\frac{\nu-1}{2}}\gamma^{\frac{\nu-1}{2}+\mu}\,\Gamma(1-\mu-\nu)\,e^{\frac{\beta}{2\gamma}}W_{\frac{\nu-1}{2}+\mu,\,-\frac{\nu}{2}}\left(\frac{\beta}{\gamma}\right)$$
$[|\arg\gamma| < \pi,\ \operatorname{Re}(1-\mu) > \operatorname{Re}\nu > 0]$. ET II 234(13)a

8. $$\int_0^u x^{-2\mu}(u^2-x^2)^{\mu-1}e^{-\frac{\beta}{x}}\,dx = \frac{1}{\sqrt{\pi}}\left(\frac{2}{\beta}\right)^{\mu-\frac{1}{2}}u^{\mu-\frac{3}{2}}\Gamma(\mu)\,K_{\mu-\frac{1}{2}}\left(\frac{\beta}{u}\right)$$
$[\operatorname{Re}\beta > 0,\ u > 0,\ \operatorname{Re}\mu > 0]$. ET II 188(23)a

9. $$\int_0^\infty x^{\nu-1}e^{-\frac{\beta}{x}-\gamma x}\,dx = 2\left(\frac{\beta}{\gamma}\right)^{\frac{\nu}{2}}K_\nu\left(2\sqrt{\beta\gamma}\right) \qquad [\operatorname{Re}\beta > 0,\ \operatorname{Re}\gamma > 0].$$
EH II 82(23)a, ET I 146(29)

10. $$\int_0^\infty x^{\nu-1}\exp\left[\frac{i\mu}{2}\left(x-\frac{\beta^2}{x}\right)\right]dx = 2\beta^\nu e^{\frac{i\nu\pi}{2}}K_{-\nu}(\beta\mu)$$
$[\operatorname{Im}\mu > 0,\ \operatorname{Im}(\beta^2\mu) > 0]$. EH II 82(24)

11. $$\int_0^\infty x^{\nu-1}\exp\left[\frac{i\mu}{2}\left(x+\frac{\beta^2}{x}\right)\right]dx = i\pi\beta^\nu e^{-\frac{i\nu\pi}{2}}H_{-\nu}^{(1)}(\beta\mu)$$
$[\operatorname{Im}\mu > 0,\ \operatorname{Im}(\beta^2\mu) > 0]$. EH II 21(33)

12. $$\int_{-\infty}^\infty x^{\nu-1}\exp\left(-x-\frac{\mu^2}{4x}\right)dx = 2\left(\frac{\mu}{2}\right)^\nu K_{-\nu}(\mu)$$
$\left[|\arg\mu| < \frac{\pi}{2},\ \operatorname{Re}\mu^2 > 0\right]$. WA 203(15)

13. $$\int_0^\infty \frac{x^{\nu-1}e^{-\frac{\beta}{x}}}{x+\gamma}\,dx = \gamma^{\nu-1}e^{\frac{\beta}{\gamma}}\,\Gamma(1-\nu)\,\Gamma\left(\nu,\ \frac{\beta}{\gamma}\right)$$
$[|\arg\gamma| < \pi,\ \operatorname{Re}\beta > 0,\ \operatorname{Re}\nu < 1]$. ET II 218(19)

14. $$\int_0^1 \frac{\exp\left(1-\frac{1}{x}\right)-x^\nu}{x(1-x)}\,dx = \psi(\nu) \qquad [\operatorname{Re}\nu > 0].$$
BI ((80))(7)

3.472

1. $$\int_0^\infty \left(\exp\left(-\frac{a}{x^2}\right)-1\right) e^{-\mu x^2}\,dx = \frac{1}{2}\sqrt{\frac{\pi}{\mu}}\left[\exp\left(-2\sqrt{a\mu}\right)-1\right]$$

$[\operatorname{Re}\mu > 0,\ \operatorname{Re} a > 0]$. ET I 146(30)

2. $$\int_0^\infty x^2 \exp\left(-\frac{a}{x^2}-\mu x^2\right) dx = \frac{1}{4}\sqrt{\frac{\pi}{\mu^3}}\left(1+2\sqrt{a\mu}\right)\exp\left(-2\sqrt{a\mu}\right)$$

$[\operatorname{Re}\mu > 0,\ \operatorname{Re} a > 0]$. ET I 146(26)

3. $$\int_0^\infty \exp\left(-\frac{a}{x^2}-\mu x^2\right)\frac{dx}{x^2} = \frac{1}{2}\sqrt{\frac{\pi}{a}}\exp\left(-2\sqrt{a\mu}\right)$$

$[\operatorname{Re}\mu > 0,\ a > 0]$. ET I 146(28)a

4. $$\int_0^\infty \exp\left[-\frac{1}{2a}\left(x^2+\frac{1}{x^2}\right)\right]\frac{dx}{x^4} = \sqrt{\frac{a\pi}{2}}\,(1+a)\,e^{-\frac{1}{a}}$$

$[a > 0]$. BI ((98))(14)

3.473 $$\int_0^\infty \exp\left(-x^n\right) x^{\left(m+\frac{1}{2}\right)n-1}\,dx = \frac{(2m-1)!!}{2^m n}\sqrt{\pi}.$$ BI ((98))(6)

3.474

1. $$\int_0^1 \left\{\frac{n\exp\left(1-x^{-n}\right)}{1-x^n}-\frac{x^{np}}{1-x}\right\}\frac{dx}{x} = \frac{1}{n}\sum_{k=1}^{n}\psi\left(p+\frac{k-1}{n}\right)$$

$[p > 0]$. BI ((80))(8)

2. $$\int_0^1 \left\{\frac{n\exp\left(1-x^{-n}\right)}{1-x^n}-\frac{\exp\left(1-\frac{1}{x}\right)}{1-x}\right\}\frac{dx}{x} = -\ln n.$$ BI ((80))(9)

3.475

1. $$\int_0^\infty \left\{\exp\left(-x^{2^n}\right)-\frac{1}{1+x^{2^{n+1}}}\right\}\frac{dx}{x} = -\frac{1}{2^n}\,C.$$ BI ((92))(14)

2. $$\int_0^\infty \left\{\exp\left(-x^{2^n}\right)-\frac{1}{1+x^2}\right\}\frac{dx}{x} = -2^{-n}C.$$ BI ((92))(13)

3. $$\int_0^\infty \left\{\exp\left(-x^{2^n}\right)-e^{-x}\right\}\frac{dx}{x} = \left(1-2^{-n}\right)C.$$ BI ((89))(8)

3.476

1. $$\int_0^\infty \left[\exp\left(-\nu x^p\right)-\exp\left(-\mu x^p\right)\right]\frac{dx}{x} = \frac{1}{p}\ln\frac{\mu}{\nu}$$

$[\operatorname{Re}\mu > 0,\ \operatorname{Re}\nu > 0]$. BI ((89))(3)

2. $$\int_0^\infty [\exp(-x^p) - \exp(-x^q)] \frac{dx}{x} = \frac{p-q}{pq} C$$

$[p > 0, q > 0]$. BI ((89))(9)

3.477

1. $$\int_{-\infty}^\infty \frac{\exp(-a|x|)}{x-u} dx = \frac{\operatorname{sign} u}{\pi} [\exp(a|u|) \operatorname{Ei}(-a|u|) - \exp(-a|u|) \overline{\operatorname{Ei}}(a|u|)] \quad [a > 0].$$ ET II 251(35)

2. $$\int_{-\infty}^\infty \frac{\operatorname{sign} x \exp(-a|x|)}{x-u} dx = -[\exp(a|u|) \operatorname{Ei}(-a|u|) - \exp(-a|u|) \overline{\operatorname{Ei}}(a|u|)] \quad [a > 0].$$ ET II 251(36)

3.478

1. $$\int_0^\infty x^{\nu-1} \exp(-\mu x^p) dx = \frac{1}{|p|} \mu^{-\frac{\nu}{p}} \Gamma\left(\frac{\nu}{p}\right)$$

$[\operatorname{Re} \mu > 0, \operatorname{Re} \nu > 0]$. BI((81))(8)a, ET I 313(15, 16)

2. $$\int_0^\infty x^{\nu-1} [1 - \exp(-\mu x^p)] dx = -\frac{1}{|p|} \mu^{-\frac{\nu}{p}} \Gamma\left(\frac{\nu}{p}\right)$$

$[\operatorname{Re} \mu > 0$ and $-p < \operatorname{Re} \nu < 0$ for $p > 0$, $0 < \operatorname{Re} \nu < -p$ for $p < 0]$. ET I 313(18, 19)

3. $$\int_0^u x^{\nu-1} (u-x)^{\mu-1} \exp(\beta x^n) dx = \mathrm{B}(\mu, \nu) u^{\mu+\nu-1} \times {}_nF_n\left(\frac{\nu}{n}, \frac{\nu+1}{n}, \ldots, \frac{\nu+n-1}{n}; \frac{\mu+\nu}{n}, \frac{\mu+\nu+1}{n}, \ldots, \frac{\mu+\nu+n-1}{n}; \beta u^n\right)$$

$[\operatorname{Re} \mu > 0, \operatorname{Re} \nu > 0, n = 2, 3, \ldots]$. ET II 187(15)

4. $$\int_0^\infty x^{\nu-1} \exp(-\beta x^p - \gamma x^{-p}) dx = \frac{2}{p} \left(\frac{\gamma}{\beta}\right)^{\frac{\nu}{2p}} K_{\frac{\nu}{p}}(2\sqrt{\beta\gamma})$$

$[\operatorname{Re} \beta > 0, \operatorname{Re} \gamma > 0]$. ET I 313(17)

3.479

1. $$\int_0^\infty \frac{x^{\nu-1} \exp(-\beta\sqrt{1+x})}{\sqrt{1+x}} dx = \frac{2}{\sqrt{\pi}} \left(\frac{\beta}{2}\right)^{\frac{1}{2}-\nu} \Gamma(\nu) K_{\frac{1}{2}-\nu}(\beta)$$

$[\operatorname{Re} \beta > 0, \operatorname{Re} \nu > 0]$. ET I 313(14)

2. $$\int_0^\infty \frac{x^{\nu-1}\exp(i\mu\sqrt{1+x^2})}{\sqrt{1+x^2}}\,dx = i\frac{\sqrt{\pi}}{2}\left(\frac{\mu}{2}\right)^{\frac{1-\nu}{2}}\Gamma\left(\frac{\mu}{2}\right)H^{(1)}_{\frac{1-\nu}{2}}(\mu)$$

$[\operatorname{Im}\mu > 0,\ \operatorname{Re}\nu > 2].$ EH II 83(30)

3.481

1. $$\int_{-\infty}^\infty xe^x\exp(-\mu e^x)\,dx = -\frac{1}{\mu}(C+\ln\mu) \qquad [\operatorname{Re}\mu > 0].$$ BI ((100))(13)

2. $$\int_{-\infty}^\infty xe^x\exp(-\mu e^{2x})\,dx = -\frac{1}{4}[C+\ln(4\mu)]\sqrt{\frac{\pi}{\mu}}$$

$[\operatorname{Re}\mu > 0].$ BI ((100))(14)

3.482

1. $$\int_0^\infty \exp(nx-\beta\operatorname{sh}x)\,dx = \frac{1}{2}[S_n(\beta)-\pi\mathbf{E}_n(\beta)+\pi N_n(\beta)]$$

$[\operatorname{Re}\beta > 0].$ ET I 168(11)

2. $$\int_0^\infty \exp(-nx-\beta\operatorname{sh}x)\,dx = (-1)^{n+1}\frac{1}{2}[S_n(\beta)+\pi\mathbf{E}_n(\beta)+\pi N_n(\beta)]$$

$[\operatorname{Re}\beta > 0].$ ET I 168(12)

3. $$\int_0^\infty \exp(-\nu x-\beta\operatorname{sh}x)\,dx = \frac{\pi}{\sin\nu\pi}[\mathbf{J}_\nu(\beta)-J_\nu(\beta)]$$

$[\operatorname{Re}\beta > 0].$ ET I 168(13)

3.483 $$\int_{-\infty}^\infty \frac{\exp(\nu\operatorname{Arsh}x-iax)}{\sqrt{1+x^2}}\,dx = \begin{cases} -2\exp\left(-\frac{i\nu\pi}{2}\right)K_\nu(a) & \text{for } a>0,\\ -2\exp\left(\frac{i\nu\pi}{2}\right)K_\nu(a) & \text{for } a<0\end{cases}$$

$[|\operatorname{Re}\nu| < 1].$ ET I 122(32)

3.484 $$\int_0^\infty \left[\left(1+\frac{a}{qx}\right)^{qx}-\left(1+\frac{a}{px}\right)^{px}\right]\frac{dx}{x} = (e^a-1)\ln\frac{q}{p}$$

$[p>0,\ q>0].$ BI ((89))(34)

3.485 $$\int_0^{\frac{\pi}{2}} \exp(-\operatorname{tg}^2x)\,dx = \frac{\pi e}{2}[1-\Phi(1)].$$

3.486 $$\int_0^1 x^{-x}\,dx = \int_0^1 e^{-x\ln x}\,dx = \sum_{k=1}^\infty k^{-k}.$$ FI II 483

3.5 Hyperbolic Functions

3.51 Hyperbolic functions

3.511

1. $\int_0^\infty \frac{dx}{\operatorname{ch} ax} = \frac{\pi}{2a} \quad [a > 0].$

2. $\int_0^\infty \frac{\operatorname{sh} ax}{\operatorname{sh} bx} dx = \frac{\pi}{2b} \operatorname{tg} \frac{a\pi}{2b} \quad [b > |a|].$ BI ((27))(10)a

3. $\int_0^\infty \frac{\operatorname{sh} ax}{\operatorname{ch} bx} dx = \frac{\pi}{2b} \sec \frac{a\pi}{2b} - \frac{1}{b} \beta \left(\frac{a+b}{2b}\right) \quad [b > |a|].$ GW ((351))(3b)

4. $\int_0^\infty \frac{\operatorname{ch} ax}{\operatorname{ch} bx} dx = \frac{\pi}{2b} \sec \frac{a\pi}{2b} \quad [b > |a|].$ BI ((4))(14)a

5. $\int_0^\infty \frac{\operatorname{sh} ax \operatorname{ch} bx}{\operatorname{sh} cx} dx = \frac{\pi}{2c} \frac{\sin \frac{a\pi}{c}}{\cos \frac{a\pi}{c} + \cos \frac{b\pi}{c}} \quad [c > |a| + |b|].$ BI ((27))(11)

6. $\int_0^\infty \frac{\operatorname{ch} ax \operatorname{ch} bx}{\operatorname{ch} cx} dx = \frac{\pi}{c} \frac{\cos \frac{a\pi}{2c} \cos \frac{b\pi}{2c}}{\cos \frac{a\pi}{c} + \cos \frac{b\pi}{c}} \quad [c > |a| + |b|].$ BI ((27))(5)a

7. $\int_0^\infty \frac{\operatorname{sh} ax \operatorname{sh} bx}{\operatorname{ch} cx} dx = \frac{\pi}{c} \frac{\sin \frac{a\pi}{2c} \sin \frac{b\pi}{2c}}{\cos \frac{a\pi}{c} + \cos \frac{b\pi}{c}} \quad [c > |a| + |b|].$ BI ((27))(6)a

8. $\int_0^\infty \frac{dx}{\operatorname{ch} x^2} = \sqrt{\pi} \sum_{k=0}^\infty \frac{(-1)^k}{\sqrt{2k+1}}$ BI ((98))(25)

9. $\int_{-\infty}^\infty \frac{\operatorname{sh}^2 ax}{\operatorname{sh}^2 x} dx = 1 - a\pi \operatorname{ctg} a\pi \quad [a^2 < 1].$ BI ((16))(3)a

10. $\int_0^\infty \frac{\operatorname{sh} ax \operatorname{sh} bx}{\operatorname{ch}^2 bx} dx = \frac{a\pi}{2b^2} \sec \frac{a\pi}{2b} \quad [b > |a|].$ BI ((27))(16)a

3.512

1. $\int_0^\infty \frac{\operatorname{ch} 2\beta x}{\operatorname{ch}^{2\nu} ax} dx = \frac{4^{\nu-1}}{a} \mathrm{B}\left(\nu + \frac{\beta}{a}, \nu - \frac{\beta}{a}\right)$
$[\operatorname{Re}(\nu \pm \beta) > 0, \ a > 0].$ LI((27))(17)a, EH I 11(26)

2. $\int_0^\infty \frac{\operatorname{sh}^\mu x}{\operatorname{ch}^\nu x} dx = \frac{1}{2} \mathrm{B}\left(\frac{\mu+1}{2}, \frac{\nu-1}{2}\right) \quad [\operatorname{Re} \mu > 0, \ \operatorname{Re}(\mu - \nu) > 0].$ EH I 11(23)

3.513

1. $$\int_0^\infty \frac{dx}{a+b\operatorname{sh}x} = \frac{1}{\sqrt{a^2+b^2}} \ln \frac{a+b+\sqrt{a^2+b^2}}{a+b-\sqrt{a^2+b^2}} \qquad [ab \neq 0].$$ GW ((351))(8)

2. $$\int_0^\infty \frac{dx}{a+b\operatorname{ch}x} = \frac{2}{\sqrt{b^2-a^2}} \operatorname{arctg} \frac{\sqrt{b^2-a^2}}{a+b} \qquad [b^2 > a^2];$$

$$= \frac{1}{\sqrt{a^2-b^2}} \ln \frac{a+b+\sqrt{a^2-b^2}}{a+b-\sqrt{a^2-b^2}} \qquad [b^2 < a^2].$$ GW ((351))(7)

3. $$\int_0^\infty \frac{dx}{a\operatorname{sh}x+b\operatorname{ch}x} = \frac{2}{\sqrt{b^2-a^2}} \operatorname{arctg} \frac{\sqrt{b^2-a^2}}{a+b} \qquad [b^2 > a^2];$$

$$= \frac{1}{\sqrt{a^2-b^2}} \ln \frac{a+b+\sqrt{a^2-b^2}}{a+b-\sqrt{a^2-b^2}} \qquad [a^2 > b^2].$$ GW ((351))(9)

4. $$\int_0^\infty \frac{dx}{a+b\operatorname{ch}x+c\operatorname{sh}x} = \frac{2}{\sqrt{b^2-a^2-c^2}} \left[\operatorname{arctg} \frac{\sqrt{b^2-a^2-c^2}}{a+b+c} + \varepsilon\pi \right]$$

$[b^2 > a^2+c^2$; $\varepsilon = 0$ for $(b-a)(a+b+c) > 0$,
$|\varepsilon| = 1$ for $(b-a)(a+b+c) < 0$, also $\varepsilon = 1$ for $a < b+c$
and $\varepsilon = -1$ for $a > b+c]$;

$$= \frac{1}{\sqrt{a^2-b^2+c^2}} \ln \frac{a+b+c+\sqrt{a^2-b^2+c^2}}{a+b+c-\sqrt{a^2-b^2+c^2}}$$
$[b^2 < a^2+c^2,\ a^2 \neq b^2]$;

$$= \frac{1}{c} \ln \frac{a+c}{a} \qquad [a = b \neq 0,\ c \neq 0];$$

$$= \frac{2(a-b)}{c(a-b-c)} \qquad [b^2 = a^2+c^2,\ c(a-b-c) < 0].$$

GW ((351))(6)

3.514

1. $$\int_0^\infty \frac{dx}{\operatorname{ch}ax+\cos t} = \frac{t}{a} \operatorname{cosec} t \qquad [0 < t < \pi].$$ BI ((27))(22)a

2. $$\int_0^\infty \frac{\operatorname{ch}ax-\cos t_1}{\operatorname{ch}bx-\cos t_2} dx = \frac{\pi}{b} \frac{\sin \frac{a(\pi-t_2)}{b}}{\sin t_2 \sin \frac{a}{b}\pi} - \frac{\pi-t_2}{b\sin t_2} \cos t_1 \qquad [b > |a|,\ 0 < t < \pi].$$

BI ((6))(20)a

3. $$\int_0^\infty \frac{\operatorname{ch}ax\,dx}{(\operatorname{ch}x+\cos t)^2} = \frac{\pi(-\cos t \sin at + a \sin t \cos at)}{\sin^3 t \sin a\pi}$$

$[a^2 < 1,\ 0 < t < \pi].$ BI ((6))(18)a

4. $$\int_0^\infty \frac{\operatorname{sh}ax\operatorname{sh}bx}{(\operatorname{ch}ax+\cos t)^2} dx = \frac{b\pi}{a^2} \operatorname{cosec} t \operatorname{cosec} \frac{b\pi}{a} \sin \frac{bt}{a}$$

$[a > |b|,\ 0 < t < \pi].$ BI ((27))(27)a

3.515 $\int_{-\infty}^{\infty}\left(1-\frac{\sqrt{2}\operatorname{ch}x}{\sqrt{\operatorname{ch}2x}}\right)dx = -\ln 2.$ BI ((21))(12)a

3.516

1. $$\int_0^\infty \frac{dx}{(z+\sqrt{z^2-1}\operatorname{ch}x)^\mu} = \frac{1}{2}\int_{-\infty}^{\infty}\frac{dx}{(z+\sqrt{z^2-1}\operatorname{ch}x)^\mu} = Q_{\mu-1}(z)$$
$[\operatorname{Re}\mu > -1]$.

For a suitable choice of a single-valued branch of the integrand, this formula is valid for arbitrary values of z in the z-plane cut from -1 to $+1$ provided $\mu < 0$. If $\mu > 0$, this formula ceases to be valid for points at which the denominator vanishes.

CO, WH

2. $$\int_0^\infty \frac{dx}{(\beta+\sqrt{\beta^2-1}\operatorname{ch}x)^{n+1}} = Q_n(\beta).$$ EH II 181(32)

3. $$\int_0^\infty \frac{\operatorname{ch}\gamma x\,dx}{(\beta+\sqrt{\beta^2-1}\operatorname{ch}x)^{\nu+1}} = \frac{e^{-i\gamma\pi}\Gamma(\nu-\gamma+1)Q_\nu^\gamma(\beta)}{\Gamma(\nu+1)}$$
$[\operatorname{Re}(\nu\pm\gamma) > -1,\ \nu \neq -1, -2, -3, \ldots]$. EH I 157(12)

4. $$\int_0^\infty \frac{\operatorname{sh}^{2\mu}x\,dx}{(\beta+\sqrt{\beta^2-1}\operatorname{ch}x)^{\nu+1}} = \frac{2^\mu e^{-i\mu\pi}\Gamma(\nu-2\mu+1)\Gamma\left(\mu+\frac{1}{2}\right)}{\sqrt{\pi}(\beta^2-1)^{\frac{\mu}{2}}\Gamma(\nu+1)} Q_{\nu-\mu}^\mu(\beta)$$
$[\operatorname{Re}(\nu-2\mu+1) > 0,\quad \operatorname{Re}(\nu+1) > 0]$. EH I 155(2)

3.517

1. $$\int_0^\infty \frac{\operatorname{ch}\left(\gamma+\frac{1}{2}\right)x\,dx}{(\beta+\operatorname{ch}x)^{\nu+\frac{1}{2}}} = \sqrt{\frac{\pi}{2}}(\beta^2-1)^{-\frac{\nu}{2}}\frac{\Gamma(\nu+\gamma+1)\Gamma(\nu-\gamma)P_\gamma^{-\nu}(\beta)}{\Gamma\left(\nu+\frac{1}{2}\right)}$$
$[\operatorname{Re}(\nu-\gamma) > 0,\quad \operatorname{Re}(\nu+\gamma+1) > 0]$. EH I 156(11)

2. $$\int_0^a \frac{\operatorname{ch}\left(\gamma+\frac{1}{2}\right)x\,dx}{(\operatorname{ch}a-\operatorname{ch}x)^{\nu+\frac{1}{2}}} = \sqrt{\frac{\pi}{2}}\frac{\Gamma\left(\frac{1}{2}-\nu\right)}{\operatorname{sh}^\nu a}P_\gamma^\nu(\operatorname{ch}a)$$
$\left[\operatorname{Re}\nu < \frac{1}{2},\ a > 0\right]$. EH I 156(8)

3.518

1. $$\int_0^\infty \frac{\operatorname{sh}^{2\mu}x\,dx}{(\operatorname{ch}a+\operatorname{sh}a\operatorname{ch}x)^{\nu+1}} = \frac{2^\mu e^{-i\mu\pi}}{\sqrt{\pi}\operatorname{sh}^\mu a}\,\frac{\Gamma(\nu-2\mu+1)\Gamma\left(\mu+\frac{1}{2}\right)}{\Gamma(\nu+1)}Q_{\nu-\mu}^\mu(\operatorname{ch}a)$$
$[\operatorname{Re}(\nu+1) > 0,\quad \operatorname{Re}(\nu-2\mu+1) > 0,\quad a > 0]$. EH I 155(3)a

2. $$\int_0^\infty \frac{\operatorname{sh}^{2\mu+1} x\, dx}{(\beta+\operatorname{ch} x)^{\nu+1}} = 2^\mu (\beta^2-1)^{\frac{\mu-\nu}{2}} \Gamma(\nu-2\mu)\Gamma(\mu+1) P_\mu^{\mu-\nu}(\beta)$$

[Re $(-\mu-\nu) >$ Re $\mu > -1$, β does not lie on the ray $(-1, +\infty)$ of the real axis]. EH I 155(1)

3. $$\int_0^\infty \frac{\operatorname{sh}^{2\mu-1} x \operatorname{ch} x\, dx}{(1+a \operatorname{sh}^2 x)^\nu} = \frac{1}{2} a^{-\mu} \mathrm{B}(\mu, \nu-\mu) \qquad [\operatorname{Re} \nu > \operatorname{Re} \mu > 0,\ a>0].$$

EH I 11(22)

4. $$\int_0^\infty \frac{\operatorname{sh}^{\mu-1} x (\operatorname{ch} x+1)^{\nu-1}\, dx}{(\beta+\operatorname{ch} x)^\varrho} = \frac{2^{-\varrho+\frac{\mu}{2}} {}_2F_1\left(\varrho, \varrho-\frac{\mu}{2}; 1+\varrho-\frac{\mu}{4}-\frac{\nu}{2}; \frac{1-\beta}{2}\right)}{\mathrm{B}\left(\varrho-\frac{\mu}{2}, 1+\frac{\mu}{4}-\frac{\nu}{2}\right)}$$

$\left[\operatorname{Re} 2\varrho > \operatorname{Re} \mu;\quad \operatorname{Re}\left(1+\frac{\mu}{4}\right) > \operatorname{Re}\left(\frac{\nu}{2}\right)\right]$. EH I 115(11)

5. $$\int_0^\infty \frac{\operatorname{sh}^{\mu-1} x (\operatorname{ch} x-1)^{\nu-1}\, dx}{(\beta+\operatorname{ch} x)^\varrho} =$$

$$= \frac{2^{-(2-\mu-\nu+\varrho)} {}_2F_1\left(\varrho, 2-\mu-\nu+\varrho; 1+\varrho-\frac{\mu}{2}; \frac{1-\beta}{2}\right)}{\mathrm{B}\left(2-\mu-\nu+\varrho, -1+\nu+\frac{\mu}{2}\right)}$$

$[\operatorname{Re}(1+\varrho) > \operatorname{Re}(\mu+\nu),\ \operatorname{Re}(4\varrho+2\nu+\mu) > 0]$. EH I 115(10)

6. $$\int_0^\infty \frac{\operatorname{sh}^{\mu-1} x \operatorname{ch}^{\nu-1} x}{(\operatorname{ch}^2 x-\beta)^\varrho}\, dx = \frac{{}_2F_1\left(\varrho, 1+\varrho-\frac{\mu+\nu}{2}; 1+\varrho-\frac{\nu}{2}; \beta\right)}{2\mathrm{B}\left(\frac{\mu}{2}, 1+\varrho-\frac{\mu+\nu}{2}\right)}$$

$[2\operatorname{Re}(1+\varrho) > \operatorname{Re} \nu,\quad 2\operatorname{Re}(1+\varrho) > \operatorname{Re}(\mu+\nu)]$. EH I 115(9)

3.519 $$\int_0^{\frac{\pi}{2}} \frac{\operatorname{sh}[(r-p)\operatorname{tg} x]}{\operatorname{sh}(r \operatorname{tg} x)}\, dx = \pi \sum_{k=1}^\infty \frac{1}{k\pi+r} \sin\frac{pk\pi}{r} \qquad [p^2 < r^2].$$ BI ((274))(13)

3.52-3.53 Combinations of hyperbolic functions and algebraic functions

3.521

1. $$\int_0^\infty \frac{x\, dx}{\operatorname{sh} ax} = \frac{\pi^2}{2a^2} \qquad [a>0].$$ GW ((352))(2b)

2. $$\int_0^\infty \frac{x\, dx}{\operatorname{ch} x} = 2G = \pi \ln 2 - 4L\left(\frac{\pi}{4}\right) = 1.831931188\ldots$$

LI III 225(103a), BI((84))(1)a

3. $$\int_1^\infty \frac{dx}{x \operatorname{sh} ax} = -2\sum_{k=0}^\infty \operatorname{Ei}[-(2k+1)a].$$ LI ((104))(14)

4. $\int\limits_1^\infty \frac{dx}{x \operatorname{ch} ax} = 2 \sum_{k=0}^{\infty} (-1)^{k+1} \operatorname{Ei}[-(2k+1)a].$ LI ((104))(13)

3.522

1. $\int\limits_0^\infty \frac{x\,dx}{(b^2+x^2)\operatorname{sh} ax} = \frac{\pi}{2ab} + \pi \sum_{k=1}^{\infty} \frac{(-1)^k}{ab+k\pi} \quad [a>0,\ b>0].$

2. $\int\limits_0^\infty \frac{x\,dx}{(b^2+x^2)\operatorname{sh} \pi x} = \frac{1}{2b} - \beta(b+1) \quad [b>0].$ BI((97))(16), GW((352))(8)

3. $\int\limits_0^\infty \frac{dx}{(b^2+x^2)\operatorname{ch} ax} = \frac{2\pi}{b} \sum_{k=1}^{\infty} \frac{(-1)^{k-1}}{2ab+(2k-1)\pi} \quad [a>0,\ b>0].$ BI ((97))(5)

4. $\int\limits_0^\infty \frac{dx}{(b^2+x^2)\operatorname{ch} \pi x} = \frac{1}{b} \beta\left(b+\frac{1}{2}\right) \quad [b>0].$ BI ((97))(4)

5. $\int\limits_0^\infty \frac{x\,dx}{(1+x^2)\operatorname{sh} \pi x} = \ln 2 - \frac{1}{2}.$ BI ((97))(7)

6. $\int\limits_0^\infty \frac{dx}{(1+x^2)\operatorname{ch} \pi x} = 2 - \frac{\pi}{2}.$ BI ((97))(1)

7. $\int\limits_0^\infty \frac{x\,dx}{(1+x^2)\operatorname{sh} \frac{\pi x}{2}} = \frac{\pi}{2} - 1.$ BI ((97))(8)

8. $\int\limits_0^\infty \frac{dx}{(1+x^2)\operatorname{ch} \frac{\pi x}{2}} = \ln 2.$ BI ((97))(2)

9. $\int\limits_0^\infty \frac{x\,dx}{(1+x^2)\operatorname{sh} \frac{\pi x}{4}} = \frac{1}{\sqrt{2}}[\pi + 2\ln(\sqrt{2}+1)] - 2.$ BI ((97))(9)

10. $\int\limits_0^\infty \frac{dx}{(1+x^2)\operatorname{ch} \frac{\pi x}{4}} = \frac{1}{\sqrt{2}}[\pi - 2\ln(\sqrt{2}+1)].$ BI ((97))(3)

3.523

1. $\int\limits_0^\infty \frac{x^{\beta-1}}{\operatorname{sh} ax}\,dx = \frac{2^\beta - 1}{2^{\beta-1} a^\beta} \Gamma(\beta)\zeta(\beta) \quad [\operatorname{Re}\beta > 1,\ a>0].$ WH

2. $\int\limits_0^\infty \frac{x^{2n-1}}{\operatorname{sh} ax}\,dx = \frac{2^{2n}-1}{2n}\left(\frac{\pi}{a}\right)^{2n} |B_{2n}| \quad [a>0].$ WH, GW((352))(2a)

3. $\int\limits_0^\infty \frac{x^{\beta-1}}{\operatorname{ch} ax}\,dx = \frac{2}{(2a)^\beta} \Gamma(\beta)\, {}_1F_1\left(-1;\ \beta;\ \frac{1}{2}\right) =$

$= \frac{2}{(2a)^\beta} \Gamma(\beta) \sum_{k=0}^{\infty} (-1)^k \left(\frac{2}{2k+1}\right)^\beta \quad [\operatorname{Re}\beta > 0,\ a>0].$

EH I 35, ET 322(1)

4. $$\int_0^\infty \frac{x^{2n}}{\operatorname{ch} ax} dx = \left(\frac{\pi}{2a}\right)^{2n+1} |E_{2n}| \qquad [a > 0].$$ BI((84))(12)a, GW((352))(1a)

5. $$\int_0^\infty \frac{x^2 \, dx}{\operatorname{ch} x} = \frac{\pi^3}{8}$$ (cf. **4.261** 6.). BI ((84))(3)

6. $$\int_0^\infty \frac{x^3 \, dx}{\operatorname{sh} x} = \frac{\pi^4}{8}$$ (cf. **4.262** 1. and 2.). BI ((84))(5)

7. $$\int_0^\infty \frac{x^4 \, dx}{\operatorname{ch} x} = \frac{5}{32}\pi^5.$$ BI ((84))(7)

8. $$\int_0^\infty \frac{x^5}{\operatorname{sh} x} dx = \frac{\pi^6}{4}.$$ BI ((84))(8)

9. $$\int_0^\infty \frac{x^6}{\operatorname{ch} x} dx = \frac{61}{128}\pi^7.$$ BI ((84))(9)

10. $$\int_0^\infty \frac{x^7}{\operatorname{sh} x} dx = \frac{17}{16}\pi^8.$$ BI ((84))(10)

11. $$\int_0^\infty \frac{\sqrt{x}\, dx}{\operatorname{ch} x} = \sqrt{\pi} \sum_{k=0}^\infty (-1)^k \frac{1}{\sqrt{(2k+1)^3}}.$$ BI ((98))(7)a

12. $$\int_0^\infty \frac{dx}{\sqrt{x} \operatorname{ch} x} = 2\sqrt{\pi} \sum_{k=0}^\infty \frac{(-1)^k}{\sqrt{2k+1}}.$$ BI ((98))(25)a

3.524

1. $$\int_0^\infty x^{\mu-1} \frac{\operatorname{sh} \beta x}{\operatorname{sh} \gamma x} dx = \frac{\Gamma(\mu)}{(2\gamma)^\mu} \left\{ \zeta\left[\mu, \frac{1}{2}\left(1 - \frac{\beta}{\gamma}\right)\right] - \right.$$
$$\left. - \zeta\left[\mu, \frac{1}{2}\left(1 + \frac{\beta}{\gamma}\right)\right]\right\} \qquad [\operatorname{Re}\gamma > |\operatorname{Re}\beta|, \ \operatorname{Re}\mu > -1].$$
ET 323(10)

2 $$\int_0^\infty x^{2m} \frac{\operatorname{sh} ax}{\operatorname{sh} bx} dx = \frac{\pi}{2b} \frac{d^{2m}}{da^{2m}} \operatorname{tg} \frac{a\pi}{2b} \qquad [b > |a|].$$ BI ((112))(20)a

3. $$\int_0^\infty \frac{\operatorname{sh} ax \, dx}{\operatorname{sh} bx \, x^p} = \Gamma(1-p) \sum_{k=0}^\infty \left\{ \frac{1}{[b(2k+1)-a]^{1-p}} - \frac{1}{[b(2k+1)+a]^{1-p}} \right\}$$
$$[b > |a|, \ p < 1].$$ BI ((131))(2)a

4. $$\int_0^\infty x^{2m+1} \frac{\operatorname{sh} ax}{\operatorname{ch} bx} dx = \frac{\pi}{2b} \frac{d^{2m+1}}{da^{2m+1}} \sec \frac{a\pi}{2b} \qquad [b > |a|].$$ BI ((112))(18)a

5. $$\int_0^\infty x^{\mu-1} \frac{\operatorname{ch} \beta x}{\operatorname{sh} \gamma x} dx = \frac{\Gamma(\mu)}{(2\gamma)^\mu} \left\{ \zeta\left[\mu, \frac{1}{2}\left(1 - \frac{\beta}{\gamma}\right)\right] + \right.$$
$$\left. + \zeta\left[\mu, \frac{1}{2}\left(1 + \frac{\beta}{\gamma}\right)\right]\right\} \qquad [\operatorname{Re}\gamma > |\operatorname{Re}\beta|, \ \operatorname{Re}\mu > 1].$$
ET I 323(12)

6. $$\int_0^\infty x^{2m} \frac{\operatorname{ch} ax}{\operatorname{ch} bx} dx = \frac{\pi}{2b} \frac{d^{2m}}{da^{2m}} \sec \frac{a\pi}{2b} \qquad [b > |a|].$$ BI ((112))(17)

7. $$\int_0^\infty \frac{\operatorname{ch} ax}{\operatorname{ch} bx} \cdot \frac{dx}{x^p} = \Gamma(1-p) \sum_{k=0}^\infty (-1)^k \left\{ \frac{1}{[b(2k+1)-a]^{1-p}} + \frac{1}{[b(2k+1)+a]^{1-p}} \right\} \qquad [b > |a|,\ p < 1].$$ BI ((131))(1)a

8. $$\int_0^\infty x^{2m+1} \frac{\operatorname{ch} ax}{\operatorname{sh} bx} dx = \frac{\pi}{2b} \frac{d^{2m+1}}{da^{2m+1}} \operatorname{tg} \frac{a\pi}{2b} \qquad [b > |a|].$$ BI ((112))(19)a

9. $$\int_0^\infty x^{2m-1} \operatorname{cth} ax\, dx = \frac{2^{2m-1}-1}{m} \left(\frac{\pi}{2a} \right)^{2m} |B_{2m}| \qquad [a > 0].$$ BI ((83))(11)

10. $$\int_0^\infty x^2 \frac{\operatorname{sh} ax}{\operatorname{sh} bx} dx = \frac{\pi^3}{4b^3} \sin \frac{a\pi}{2b} \sec^3 \frac{a\pi}{2b} \qquad [b > |a|].$$ BI ((84))(18)

11. $$\int_0^\infty x^4 \frac{\operatorname{sh} ax}{\operatorname{sh} bx} dx = 8 \left(\frac{\pi}{2b} \sec \frac{a\pi}{2b} \right)^5 \cdot \sin \frac{a\pi}{2b} \cdot \left(2 + \sin^2 \frac{a\pi}{2b} \right) \qquad [b > |a|].$$ BI ((82))(17)a

12. $$\int_0^\infty x^6 \frac{\operatorname{sh} ax}{\operatorname{sh} bx} dx = 16 \left(\frac{\pi}{2b} \sec \frac{a\pi}{2b} \right)^7 \sin \frac{a\pi}{2b} \left(45 - 30 \cos^2 \frac{a\pi}{2b} + 2 \cos^4 \frac{a\pi}{2b} \right) \qquad [b > |a|].$$ BI ((82))(21)a

13. $$\int_0^\infty x \frac{\operatorname{sh} ax}{\operatorname{ch} bx} dx = \frac{\pi^2}{4b^2} \sin \frac{a\pi}{2b} \sec^2 \frac{a\pi}{2b} \qquad [b > |a|].$$ BI ((84))(15)a

14. $$\int_0^\infty x^3 \frac{\operatorname{sh} ax}{\operatorname{ch} bx} dx = \left(\frac{\pi}{2b} \sec \frac{a\pi}{2b} \right)^4 \sin \frac{a\pi}{2b} \cdot \left(6 - \cos^2 \frac{a\pi}{2b} \right) \qquad [b > |a|].$$ BI ((82))(14)a

15. $$\int_0^\infty x^5 \frac{\operatorname{sh} ax}{\operatorname{ch} bx} dx = \left(\frac{\pi}{2b} \sec \frac{a\pi}{2b} \right)^6 \sin \frac{a\pi}{2b} \left(120 - 60 \cos^2 \frac{a\pi}{2b} + \cos^4 \frac{a\pi}{2b} \right) \qquad [b > |a|].$$ BI ((82))(18)a

16. $$\int_0^\infty x^7 \frac{\operatorname{sh} ax}{\operatorname{ch} bx} dx = \left(\frac{\pi}{2b} \sec \frac{a\pi}{2b} \right)^8 \sin \frac{a\pi}{2b} \times \left(5040 - 4200 \cos^2 \frac{a\pi}{2b} + 546 \cos^4 \frac{a\pi}{2b} - \cos^6 \frac{a\pi}{2b} \right) \qquad [b > |a|].$$ BI ((82))(22)a

17. $$\int_0^\infty x \frac{\operatorname{ch} ax}{\operatorname{sh} bx} dx = \left(\frac{\pi}{2b} \sec \frac{a\pi}{2b} \right)^2 \qquad [b > |a|].$$ BI ((84))(16)a

18. $\int_0^\infty x^3 \frac{\operatorname{ch} ax}{\operatorname{sh} bx} dx = 2\left(\frac{\pi}{2b} \sec \frac{a\pi}{2b}\right)^4 \left(1 + 2 \sin^2 \frac{a\pi}{2b}\right)$

$[b > |a|]$. BI ((82))(15)a

19. $\int_0^\infty x^5 \frac{\operatorname{ch} ax}{\operatorname{sh} bx} dx = 8\left(\frac{\pi}{2b} \sec \frac{a\pi}{2b}\right)^6 \left(15 - 15 \cos^2 \frac{a\pi}{2b} + 2 \cos^4 \frac{a\pi}{2b}\right)$

$[b > |a|]$. BI ((82))(19)a

20. $\int_0^\infty x^7 \frac{\operatorname{ch} ax}{\operatorname{sh} bx} dx = 16\left(\frac{\pi}{2b} \sec \frac{a\pi}{2b}\right)^8 \times$

$\times \left(315 - 420 \cos^2 \frac{a\pi}{2b} + 126 \cos^4 \frac{a\pi}{2b} - 4 \cos^6 \frac{a\pi}{2b}\right)$

$[b > |a|]$. BI ((82))(23)a

21. $\int_0^\infty x^2 \frac{\operatorname{ch} ax}{\operatorname{ch} bx} dx = \frac{\pi^3}{8b^3}\left(2 \sec^3 \frac{a\pi}{2b} - \sec \frac{a\pi}{2b}\right) \quad [b > |a|]$. BI ((84))(17)a

22. $\int_0^\infty x^4 \frac{\operatorname{ch} ax}{\operatorname{ch} bx} dx = \left(\frac{\pi}{2b} \sec \frac{a\pi}{2b}\right)^5 \left(24 - 20 \cos^2 \frac{a\pi}{2b} + \cos^4 \frac{a\pi}{2b}\right)$

$[b > |a|]$. BI ((82))(16)a

23. $\int_0^\infty x^6 \frac{\operatorname{ch} ax}{\operatorname{ch} bx} dx = \left(\frac{\pi}{2b} \sec \frac{a\pi}{2b}\right)^7 \left(720 - 840 \cos^2 \frac{a\pi}{2b} + \right.$

$\left. + 182 \cos^4 \frac{a\pi}{2b} - \cos^6 \frac{a\pi}{2b}\right) \quad [b > |a|]$. BI ((82))(20)a

24. $\int_0^\infty \frac{\operatorname{sh} ax}{\operatorname{ch} bx} \cdot \frac{dx}{x} = \ln \operatorname{tg}\left(\frac{a\pi}{4b} + \frac{\pi}{4}\right) \quad [b > |a|]$. BI ((95))(3)a

3.525

1. $\int_0^\infty \frac{\operatorname{sh} ax}{\operatorname{sh} \pi x} \cdot \frac{dx}{1+x^2} = -\frac{a}{2} \cos a + \frac{1}{2} \sin a \ln [2(1 + \cos a)]$

$[\pi \geqslant |a|]$. BI ((97))(10)a

2. $\int_0^\infty \frac{\operatorname{sh} ax}{\operatorname{sh} \frac{\pi}{2} x} \cdot \frac{dx}{1+x^2} = \frac{\pi}{2} \sin a + \frac{1}{2} \cos a \ln \frac{1-\sin a}{1+\sin a}$

$[\pi \geqslant 2|a|]$. BI ((97))(11)a

3. $\int_0^\infty \frac{\operatorname{ch} ax}{\operatorname{sh} \pi x} \cdot \frac{x\, dx}{1+x^2} = \frac{1}{2}(a \sin a - 1) + \frac{1}{2} \cos a \ln [2(1 + \cos a)]$

$[\pi > |a|]$. BI ((97))(12)a

4. $$\int_0^\infty \frac{\operatorname{ch} ax}{\operatorname{sh} \frac{\pi}{2} x} \cdot \frac{x\,dx}{1+x^2} = \frac{\pi}{2} \cos a - 1 + \frac{1}{2} \sin a \ln \frac{1+\sin a}{1-\sin a}$$

$$\left[\frac{\pi}{2} > |a|\right].$$ BI ((97))(13)a

5. $$\int_0^\infty \frac{\operatorname{sh} ax}{\operatorname{ch} \pi x} \cdot \frac{x\,dx}{1+x^2} = -2 \sin \frac{a}{2} + \frac{\pi}{2} \sin a - \cos a \ln \operatorname{tg} \frac{a+\pi}{4}$$

$$[\pi > |a|].$$ GW ((352))(12)

6. $$\int_0^\infty \frac{\operatorname{ch} ax}{\operatorname{ch} \pi x} \cdot \frac{dx}{1+x^2} = 2 \cos \frac{a}{2} - \frac{\pi}{2} \cos a - \sin a \ln \operatorname{tg} \frac{a+\pi}{4}$$

$$[\pi > |a|].$$ GW ((352))(11)

7. $$\int_0^\infty \frac{\operatorname{sh} ax}{\operatorname{sh} bx} \cdot \frac{dx}{c^2+x^2} = \frac{\pi}{c} \sum_{k=1}^\infty \frac{\sin \frac{k(b-a)}{b} \pi}{bc+k\pi} \qquad [b \geqslant |a|].$$ BI ((97))(18)

8. $$\int_0^\infty \frac{\operatorname{ch} ax}{\operatorname{sh} bx} \cdot \frac{x\,dx}{c^2+x^2} = \frac{\pi}{2bc} + \pi \sum_{k=1}^\infty \frac{\cos \frac{k(b-a)}{b} \pi}{bc+k\pi} \qquad [b > |a|].$$ BI ((97))(19)

3.526

1. $$\int_0^\infty \frac{\operatorname{sh} ax \operatorname{ch} bx}{\operatorname{ch} cx} \cdot \frac{dx}{x} = \frac{1}{2} \ln \left\{ \operatorname{tg} \frac{(a+b+c)\pi}{4c} \operatorname{ctg} \frac{(b+c-a)\pi}{4c} \right\}$$

$$[c > |a| + |b|].$$ BI ((93))(10)a

2. $$\int_0^\infty \frac{\operatorname{sh}^2 ax}{\operatorname{sh} bx} \cdot \frac{dx}{x} = \frac{1}{2} \ln \sec \frac{a}{b} \pi \qquad [b > |2a|].$$ BI ((95))(5)a

3. $$\int_0^\infty \frac{x^{\mu-1}}{\operatorname{sh} \beta x \operatorname{ch} \gamma x}\,dx = \frac{\Gamma(\mu)}{(2\gamma)^\mu} \left\{ {}_1F_1\left[-1;\ \mu;\ \frac{1}{2}\left(1+\frac{\beta}{\gamma}\right)\right] + \right.$$

$$\left. + {}_1F_1\left[-1;\ \mu;\ \frac{1}{2}\left(1-\frac{\beta}{\gamma}\right)\right]\right\} \qquad [\operatorname{Re} \gamma > |\operatorname{Re} \beta|,\ \operatorname{Re} \mu > 0].$$

ET I 323(11)

3.527

1. $$\int_0^\infty \frac{x^{\mu-1}}{\operatorname{sh}^2 ax}\,dx = \frac{4}{(2a)^\mu} \Gamma(\mu) \zeta(\mu-1) \qquad [\operatorname{Re} a > 0,\ \operatorname{Re} \mu > 2].$$ BI ((86))(7)a

2. $$\int_0^\infty \frac{x^{2m}}{\operatorname{sh}^2 ax}\,dx = \frac{\pi^{2m}}{a^{2m+1}} |B_{2m}| \qquad [a > 0].$$ BI ((86))(5)a

3. $$\int_0^\infty \frac{x^{\mu-1}}{\operatorname{ch}^2 ax}\,dx = \frac{4}{(2a)^\mu} (1-2^{2-\mu}) \Gamma(\mu) \zeta(\mu-1)$$

$$[\operatorname{Re} a > 0,\ \operatorname{Re} \mu > 0].$$ BI ((86))(6)a

4. $\int_0^\infty \frac{x\,dx}{\operatorname{ch}^2 ax} = \frac{\ln 2}{a^2}.$ LO III 396

5. $\int_0^\infty \frac{x^{2m}}{\operatorname{ch}^2 ax}\,dx = \frac{(2^{2m}-2)\,\pi^{2m}}{(2a)^{2m}\,a}\,|B_{2m}| \qquad [a>0].$ BI ((86))(2)a

6. $\int_0^\infty x^{\mu-1}\frac{\operatorname{sh} ax}{\operatorname{ch}^2 ax}\,dx = \frac{2\Gamma(\mu)}{a^\mu}\sum_{k=0}^\infty \frac{(-1)^k}{(2k+1)^{\mu-1}}$

$[\operatorname{Re}\mu>0,\ a>0].$ BI ((86))(15)a

7. $\int_0^\infty \frac{x \operatorname{sh} ax}{\operatorname{ch}^2 ax}\,dx = \frac{\pi}{2a^2}.$ BI ((86))(8)a

8. $\int_0^\infty x^{2m+1}\frac{\operatorname{sh} ax}{\operatorname{ch}^2 ax}\,dx = \frac{2m+1}{a}\left(\frac{\pi}{2a}\right)^{2m+1}|E_{2m}| \qquad [a>0].$ BI ((86))(12)a

9. $\int_0^\infty x^{2m+1}\frac{\operatorname{ch} ax}{\operatorname{sh}^2 ax}\,dx = \frac{2^{2m+1}-1}{a^2(2a)^{2m}}\,(2m+1)!\,\zeta(2m+1).$ BI ((86))(13)a

10. $\int_0^\infty x^{2m}\frac{\operatorname{ch} ax}{\operatorname{sh}^2 ax}\,dx = \frac{2^{2m}-1}{a}\left(\frac{\pi}{a}\right)^{2m}|B_{2m}| \qquad [a>0].$ BI ((86))(14)a

11. $\int_0^\infty \frac{x \operatorname{sh} ax}{\operatorname{ch}^{2\mu+1} ax}\,dx = \frac{\sqrt{\pi}}{4\mu a^2}\,\frac{\Gamma(\mu)}{\Gamma\left(\mu+\frac{1}{2}\right)} \qquad [\mu>0].$ LI ((86))(9)

12. $\int_{-\infty}^\infty \frac{x^2\,dx}{\operatorname{sh}^2 x} = \frac{\pi^2}{3}.$ BI ((102))(2)a

13. $\int_0^\infty x^2\frac{\operatorname{ch} ax}{\operatorname{sh}^2 ax}\,dx = \frac{\pi^2}{2a^3}.$ BI ((86))(11)a

14. $\int_0^\infty x^2\frac{\operatorname{sh} ax}{\operatorname{ch}^2 ax}\,dx = \frac{\ln 2}{2a^3}.$ BI ((86))(10)a

15. $\int_0^\infty \frac{\operatorname{th}\frac{x}{2}}{\operatorname{ch} x}\,\frac{dx}{x} = \ln 2.$ BI ((93))(17)a

3.528

1. $\int_0^\infty \frac{(1+xi)^{2n-1}-(1-xi)^{2n-1}}{i \operatorname{sh}\frac{\pi x}{2}}\,dx = 2.$ BI ((87))(8)

2. $\int_0^\infty \frac{(1+xi)^{2n}-(1-xi)^{2n}}{i \operatorname{sh}\frac{\pi x}{2}}\,dx = (-1)^{n+1}\,2\,|E_{2n}|+2.$ BI ((87))(7)

3.529

1. $$\int_0^\infty \left(\frac{1}{\operatorname{sh} x} - \frac{1}{x}\right)\frac{dx}{x} = -\ln 2.$$ BI ((94))(10)a

2. $$\int_0^\infty \frac{\operatorname{ch} ax - 1}{\operatorname{sh} bx}\cdot\frac{dx}{x} = -\ln\cos\frac{a\pi}{2b} \quad [b > |a|].$$ GW ((352))(66)

3. $$\int_0^\infty \left(\frac{a}{\operatorname{sh} ax} - \frac{b}{\operatorname{sh} bx}\right)\frac{dx}{x} = (b-a)\ln 2.$$ BI ((94))(11)a

3.531

1. $$\int_0^\infty \frac{x\,dx}{2\operatorname{ch} x - 1} = 1.1719536194\ldots$$ LI ((88))(1)

2. $$\int_0^\infty \frac{x\,dx}{\operatorname{ch} 2x + \cos 2t} = \frac{t\ln 2 - L(t)}{\sin t\cos t}.$$ LO III 402

3. $$\int_0^\infty \frac{x^2\,dx}{\operatorname{ch} x + \cos t} = \frac{t}{3}\cdot\frac{\pi^2 - t^2}{\sin t} \quad [0 < t < \pi].$$ BI ((88))(3)a

4. $$\int_0^\infty \frac{x^4\,dx}{\operatorname{ch} x + \cos t} = \frac{t}{15}\cdot\frac{(\pi^2 - t^2)(7\pi^2 - 3t^2)}{\sin t} \quad [0 < t < \pi].$$ BI ((88))(4)a

5. $$\int_0^\infty \frac{x^{2m}\,dx}{\operatorname{ch} x - \cos 2a\pi} = 2\cdot(2m)!\operatorname{cosec} 2a\pi\sum_{k=1}^\infty \frac{\sin 2ka\pi}{k^{2m+1}}.$$ BI (88))(5)a

6. $$\int_0^\infty \frac{x^{\mu-1}\,dx}{\operatorname{ch} x - \cos t} = \frac{i\Gamma(\mu)}{\sin t}\left[e^{-it}{}_1F_1(e^{-it}; \mu; 1) - e^{it}{}_1F_1(e^{it}; \mu; 1)\right]$$
$$[\operatorname{Re}\mu > 0,\ 0 < t < 2\pi].$$ ET I 323(5)

7. $$\int_0^\infty \frac{x^\mu\,dx}{\operatorname{ch} x + \cos t} = \frac{2\Gamma(\mu+1)}{\sin t}\sum_{k=1}^\infty (-1)^{k-1}\frac{\sin kt}{k^{\mu+1}} \quad [\mu > -1].$$ BI ((96))(14)a

8. $$\int_0^u \frac{x\,dx}{\operatorname{ch} 2x - \cos 2t} = \frac{1}{2}\operatorname{cosec} 2t\,[L(\theta + t) - L(\theta - t) - 2L(t)]$$
$$[\theta = \operatorname{arctg}(\operatorname{th} u \operatorname{ctg} t),\ t \neq n\pi].$$ LO III 402

3.532

1. $$\int_0^\infty \frac{x^n\,dx}{a\operatorname{ch} x + b\operatorname{sh} x} = \frac{(2n)!}{a+b}\sum_{k=0}^\infty \frac{1}{(2k+1)^{n+1}}\left(\frac{b-a}{b+a}\right)^k$$
$$[a > 0,\ b > 0,\ n > -1].$$ GW ((352))(5)

2. $$\int_0^u \frac{x \operatorname{ch} x\, dx}{\operatorname{ch} 2x - \cos 2t} = \frac{1}{2} \operatorname{cosec} t \left\{ L\left(\frac{\theta+t}{2}\right) - L\left(\frac{\theta-t}{2}\right) + \right.$$
$$\left. + L\left(\pi - \frac{\psi+t}{2}\right) + L\left(\frac{\psi-t}{2}\right) - 2L\left(\frac{t}{2}\right) - 2L\left(\frac{\pi-t}{2}\right) \right\}$$
$$\left[\operatorname{tg} \frac{\theta}{2} = \operatorname{th} \frac{u}{2} \operatorname{ctg} \frac{t}{2},\ \operatorname{tg} \frac{\psi}{2} = \operatorname{cth} \frac{u}{2} \operatorname{ctg} \frac{t}{2};\ t \neq n\pi \right].$$ LO III 288a

3.533

1. $$\int_0^\infty \frac{x \operatorname{ch} x\, dx}{\operatorname{ch} 2x - \cos 2t} = \operatorname{cosec} t \left[\frac{\pi}{2} \ln 2 - L\left(\frac{t}{2}\right) - L\left(\frac{\pi-t)}{2}\right) \right]$$
$$[t \neq m\pi].$$ LO III 403

2. $$\int_0^\infty x \frac{\operatorname{sh} ax\, dx}{(\operatorname{ch} ax - \cos t)^2} = \frac{t}{a^2} \operatorname{cosec} t$$
$[0 < t < \pi]$ (cf. **3.514** 1.). BI ((88))(11)a

3. $$\int_0^\infty x^3 \frac{\operatorname{sh} x\, dx}{(\operatorname{ch} x + \cos t)^2} = \frac{t(\pi^2 - t^2)}{\sin t}$$
$[0 < t < \pi]$ (cf. **3.531** 3.). BI ((88))(13)

4. $$\int_0^\infty x^{2m+1} \frac{\operatorname{sh} x\, dx}{(\operatorname{ch} x - \cos 2a\pi)^2} = 2(2m+1)! \operatorname{cosec} 2a\pi \sum_{k=1}^\infty \frac{\cos 2ka\pi}{k^{2m+1}}$$
$[0 < a < \pi]$. BI ((88))(14)

3.534

1. $$\int_0^1 \sqrt{1-x^2} \operatorname{ch} ax\, dx = \frac{\pi}{2a} I_1(a).$$ WA 94(9)

2. $$\int_0^1 \frac{\operatorname{ch} ax}{\sqrt{1-x^2}} dx = \frac{\pi}{2} I_0(a).$$ WA 94(9)

3.535 $$\int_0^1 \frac{x}{\sqrt{\operatorname{ch} 2a - \operatorname{ch} 2ax}} \cdot \frac{dx}{\operatorname{sh} ax} = \frac{\pi}{2\sqrt{2a^2}} \cdot \frac{\arcsin(\operatorname{th} a)}{\operatorname{sh} a}.$$ BI ((80))(11)

3.536

1. $$\int_0^\infty \frac{x^2}{\operatorname{ch} x^2} dx = \frac{\sqrt{\pi}}{2} \sum_{k=0}^\infty \frac{(-1)^k}{\sqrt{(2k+1)^3}}.$$ BI ((98))(7)

2. $$\int_0^\infty \frac{x^2 \operatorname{th} x^2\, dx}{\operatorname{ch} x^2} = \frac{\sqrt{\pi}}{2} \sum_{k=0}^\infty \frac{(-1)^k}{\sqrt{2k+1}}.$$ BI ((98))(8)

3. $$\int_0^\infty \operatorname{sh}(\nu \operatorname{Arsh} x) \frac{x^{\mu-1}}{\sqrt{1+x^2}} dx = \frac{\sin \frac{\mu\pi}{2} \sin \frac{\nu\pi}{2}}{2^\mu \pi} \Gamma(\mu) \Gamma\left(\frac{1-\mu-\nu}{2}\right) \times$$
$$\times \Gamma\left(\frac{1-\mu+\nu)}{2}\right) \quad [-1 < \operatorname{Re} \mu < 1 - |\operatorname{Re} \nu|].$$ ET I 324(14)

4. $\int_0^\infty \operatorname{ch}(\nu \operatorname{Arch} x) \frac{x^{\mu-1}}{\sqrt{1+x^2}} dx = \frac{\cos\frac{\mu\pi}{2}\cos\frac{\nu\pi}{2}}{2^\mu \pi} \Gamma(\mu)\Gamma\left(\frac{1-\mu-\nu}{2}\right) \times$
$\times \Gamma\left(\frac{1-\mu+\nu}{2}\right) \quad [0 < \operatorname{Re}\mu < 1 - |\operatorname{Re}\nu|].$ ET I 324(15)

3.54 Combinations of hyperbolic functions and exponentials

3.541

1. $\int_0^\infty e^{-\mu x} \operatorname{sh}^\nu \beta x \, dx = \frac{1}{2^{\nu+1}\beta} B\left(\frac{\mu}{2\beta} - \frac{\nu}{2}, \nu+1\right)$
$[\operatorname{Re}\beta > 0, \operatorname{Re}\nu > -1, \operatorname{Re}\mu > \operatorname{Re}\beta\nu].$ EH I 11(25, ET I 163(5)

2. $\int_0^\infty e^{-\mu x} \frac{\operatorname{sh}\beta x}{\operatorname{sh} bx} dx = \frac{1}{2b}\left[\psi\left(\frac{1}{2} + \frac{\mu+\beta}{2b}\right) - \psi\left(\frac{1}{2} + \frac{\mu-\beta}{2b}\right)\right]$
$[\operatorname{Re}(\mu + b \pm \beta) > 0].$ EH I 16

3. $\int_{-\infty}^\infty e^{-\mu x} \frac{\operatorname{sh}\mu x}{\operatorname{sh}\beta x} dx = \frac{\pi}{2\beta} \operatorname{tg}\frac{\mu\pi}{\beta} \quad [\operatorname{Re}\beta > 2|\operatorname{Re}\mu|].$ BI ((18))(6)

4. $\int_0^\infty e^{-x} \frac{\operatorname{sh} ax}{\operatorname{sh} x} dx = \frac{1}{a} - \frac{\pi}{2}\operatorname{ctg}\frac{a\pi}{2}.$ BI ((4))(3)

5. $\int_0^\infty \frac{e^{-px}\, dx}{(\operatorname{ch} px)^{2q+1}} = \frac{2^{2q-2}}{p} B(q, q) - \frac{1}{2qp}.$ LI ((27))(19)

6. $\int_0^\infty e^{-\mu x} \frac{dx}{\operatorname{ch} x} = \beta\left(\frac{\mu+1}{2}\right) \quad [\operatorname{Re}\mu > -1].$ ET I 163(7)

7. $\int_0^\infty e^{-\mu x} \operatorname{th} x \, dx = \beta\left(\frac{\mu}{2}\right) - \frac{1}{\mu} \quad [\operatorname{Re}\mu > 0].$ ET I 163(9)

8. $\int_0^\infty \frac{e^{-\mu x}}{\operatorname{ch}^2 x} dx = \beta\left(\frac{\mu}{2}\right) - 1 \quad [\operatorname{Re}\mu > 0].$ ET I 163(8)

9. $\int_0^\infty e^{-\mu x} \frac{\operatorname{sh}\mu x}{\operatorname{ch}^2 \mu x} dx = \frac{1}{\mu}(1 - \ln 2) \quad [\operatorname{Re}\mu > 0].$ LI ((27))(15)

10. $\int_0^\infty e^{-qx} \frac{\operatorname{sh} px}{\operatorname{sh} qx} dx = \frac{1}{p} - \frac{\pi}{2q}\operatorname{ctg}\frac{p\pi}{2q} \quad [0 < p < 2q].$ BI ((27))(9)a

3.542

1. $\int_0^\infty e^{-\mu x} (\operatorname{ch}\beta x - 1)^\nu dx = \frac{1}{2^\nu \beta} B\left(\frac{\mu}{\beta} - \nu, 2\nu + 1\right)$
$\left[\operatorname{Re}\beta > 0, \operatorname{Re}\nu > -\frac{1}{2}, \operatorname{Re}\mu > \operatorname{Re}\beta\nu\right].$ ET I 163(6)

2. $$\int_0^\infty e^{-\mu x}(\operatorname{ch} x - \operatorname{ch} u)^{\nu-1}\,dx = -i\sqrt{\frac{2}{\pi}}\,e^{i\pi\nu}\Gamma(\nu)\,\operatorname{sh}^{\nu-\frac{1}{2}} u\,Q^{\frac{1}{2}-\nu}_{\mu-\frac{1}{2}}(\operatorname{ch} u)$$

$[\operatorname{Re}\nu > 0,\ \operatorname{Re}\mu > \operatorname{Re}\nu - 1].$ EH I 155(4), ET I 164(23)

3.543

1. $$\int_{-\infty}^\infty \frac{e^{-ibx}\,dx}{\operatorname{sh} x + \operatorname{sh} t} = -\frac{i\pi e^{itb}}{\operatorname{sh}\pi b\operatorname{ch} t}(\operatorname{ch}\pi b - e^{-2itb}) \quad [t > 0].$$ ET I 121(30)

2. $$\int_0^\infty \frac{e^{-\mu x}}{\operatorname{ch} x - \cos t}\,dx = 2\operatorname{cosec} t\sum_{k=1}^\infty \frac{\sin kt}{\mu + k} \quad [\operatorname{Re}\mu > -1,\ t \neq 2n\pi].$$ BI ((6))(10)a

3. $$\int_0^\infty \frac{1 - e^{-x}\cos t}{\operatorname{ch} x - \cos t}\,e^{-(\mu-1)x}\,dx = 2\sum_{k=0}^\infty \frac{\cos kt}{\mu + k}$$

$[\operatorname{Re}\mu > 0,\ \ t \neq 2n\pi].$ BI ((6))(9)a

4. $$\int_0^\infty \frac{e^{px} + \cos t}{(\operatorname{ch} px + \cos t)^2}\,dx = \frac{1}{p}\left(t\operatorname{cosec} t + \frac{1}{1 + \cos t}\right) \quad [p > 0].$$ BI ((27))(26)a

3.544 $$\int_u^\infty \frac{\exp\left[-\left(n+\frac{1}{2}\right)x\right]}{\sqrt{2(\operatorname{ch} x - \operatorname{ch} u)}}\,dx = Q_n(\operatorname{ch} u).$$ EH II 181(33)

3.545

1. $$\int_0^\infty \frac{\operatorname{sh} ax}{e^{px} + 1}\,dx = \frac{\pi}{2p}\operatorname{cosec}\frac{a\pi}{p} - \frac{1}{2a} \quad [p > a,\ p > 0].$$ BI ((27))(3)

2. $$\int_0^\infty \frac{\operatorname{sh} ax}{e^{px} - 1}\,dx = \frac{1}{2a} - \frac{\pi}{2p}\operatorname{ctg}\frac{a\pi}{p} \quad [p > a,\ p > 0].$$ BI ((27))(9)

3.546

1. $$\int_0^\infty e^{-\beta x^2}\operatorname{sh} ax\,dx = \frac{1}{2}\frac{\sqrt{\pi}}{\sqrt{\beta}}\exp\frac{a^2}{4\beta}\,\Phi\left(\frac{a}{2\sqrt{\beta}}\right) \quad [\operatorname{Re}\beta > 0].$$ ET I 166(38)a

2. $$\int_0^\infty e^{-\beta x^2}\operatorname{ch} ax\,dx = \frac{1}{2}\sqrt{\frac{\pi}{\beta}}\exp\frac{a^2}{4\beta} \quad [\operatorname{Re}\beta > 0].$$ FI II 720a

3. $$\int_0^\infty e^{-\beta x^2}\operatorname{sh}^2 ax\,dx = \frac{1}{4}\sqrt{\frac{\pi}{\beta}}\left(\exp\frac{a^2}{\beta} - 1\right) \quad [\operatorname{Re}\beta > 0].$$ ET I 166(40)

4. $$\int_0^\infty e^{-\beta x^2}\operatorname{ch}^2 ax\,dx = \frac{1}{4}\sqrt{\frac{\pi}{\beta}}\left(\exp\frac{a^2}{\beta} + 1\right) \quad [\operatorname{Re}\beta > 0].$$ ET I 166(41)

3.547

1. $$\int_0^\infty \exp(-\beta \operatorname{sh} x) \operatorname{sh} \gamma x \, dx = \frac{\pi}{2} \operatorname{ctg} \frac{\gamma\pi}{2} [J_\gamma(\beta) - \mathbf{J}_\gamma(\beta)] - \frac{\pi}{2} [\mathbf{E}_\gamma(\beta) + N_\gamma(\beta)] = \gamma S_{-1,\gamma}(\beta) \quad [\operatorname{Re} \beta > 0].$$ WA 341(5), ET I 168(14)a

2. $$\int_0^\infty \exp(-\beta \operatorname{ch} x) \operatorname{sh} \gamma x \operatorname{sh} x \, dx = \frac{\gamma}{\beta} K_\gamma(\beta).$$

3. $$\int_0^\infty \exp(-\beta \operatorname{sh} x) \operatorname{ch} \gamma x \, dx = \frac{\pi}{2} \operatorname{tg} \frac{\pi\gamma}{2} [\mathbf{J}_\gamma(\beta) - J_\gamma(\beta)] - \frac{\pi}{2} [\mathbf{E}_\gamma(\beta) + N_\gamma(\beta)] = S_{0,\gamma}(\beta) \quad [\operatorname{Re} \beta > 0, \ \gamma \text{ not an integer}].$$ ET I 168(16)a, WA 341(4), EH II 84(50)

4. $$\int_0^\infty \exp(-\beta \operatorname{ch} x) \operatorname{ch} \gamma x \, dx = K_\gamma(\beta) \quad [\operatorname{Re} \beta > 0].$$ ET I 168(16)a, WA 201(5)

5. $$\int_0^\infty \exp(-\beta \operatorname{sh} x) \operatorname{sh} \gamma x \operatorname{ch} x \, dx = \frac{\gamma}{\beta} S_{0,\gamma}(\beta) \quad [\operatorname{Re} \beta > 0].$$ ET I 168(7), EH II 85(51)

6. $$\int_0^\infty \exp(-\beta \operatorname{sh} x) \operatorname{sh} [(2n+1)x] \operatorname{ch} x \, dx = O_{2n+1}(\beta) \quad [\operatorname{Re} \beta > 0].$$ ET I 167(5)

7. $$\int_0^\infty \exp(-\beta \operatorname{sh} x) \operatorname{ch} \gamma x \operatorname{ch} x \, dx = \frac{1}{\beta} S_{1,\gamma}(\beta) \quad [\operatorname{Re} \beta > 0].$$

8. $$\int_0^\infty \exp(-\beta \operatorname{sh} x) \operatorname{ch} 2nx \operatorname{ch} x \, dx = O_{2n}(\beta) \quad [\operatorname{Re} \beta > 0].$$ ET I 168(6)

9. $$\int_0^\infty \exp(-\beta \operatorname{ch} x) \operatorname{sh}^{2\nu} x \, dx = \frac{1}{\sqrt{\pi}} \left(\frac{2}{\beta}\right)^\nu \Gamma\left(\nu + \frac{1}{2}\right) K_\nu(\beta) \quad \left[\operatorname{Re} \beta > 0, \ \operatorname{Re} \nu > -\frac{1}{2}\right].$$ EH II 82(20)

10. $$\int_0^\infty \exp[-2(\beta \operatorname{cth} x + \mu x)] \operatorname{sh}^{2\nu} x \, dx = \frac{1}{4} \beta^{\frac{\nu-1}{2}} \Gamma(\mu - \nu) \times [W_{-\mu+\frac{1}{2},\nu}(4\beta) - (\mu - \nu) W_{-\mu-\frac{1}{2},\nu}(4\beta)] \quad [\operatorname{Re} \beta > 0, \ \operatorname{Re} \mu > \operatorname{Re} \nu].$$ ET I 165(31)

11. $$\int_0^\infty \exp\left(-\frac{\beta^2}{2} \operatorname{sh} x\right) \operatorname{sh}^{\nu-1} x \operatorname{ch}^\nu x \, dx = -\pi D_\nu\left(\beta e^{\frac{i\pi}{4}}\right) D_\nu\left(\beta e^{-\frac{i\pi}{4}}\right) \quad \left[\operatorname{Re} \nu > 0, \ |\arg \beta| \leqslant \frac{\pi}{4}\right].$$ EH II 120(10)

12. $$\int_0^\infty \frac{\exp(2\nu x - 2\beta \operatorname{sh} x)}{\sqrt{\operatorname{sh} x}}\,dx = \frac{1}{2}\sqrt{\pi^3\beta}\left[J_{\nu+\frac{1}{4}}(\beta)\,J_{\nu-\frac{1}{4}}(\beta) + N_{\nu+\frac{1}{4}}(\beta)\,N_{\nu-\frac{1}{4}}(\beta)\right] \quad [\operatorname{Re}\beta > 0].$$ EH I 169(20)

13. $$\int_0^\infty \frac{\exp(-2\nu x - 2\beta \operatorname{sh} x)}{\sqrt{\operatorname{sh} x}}\,dx = \frac{1}{2}\sqrt{\pi^3\beta}\left[J_{\nu+\frac{1}{4}}(\beta)\,N_{\nu-\frac{1}{4}}(\beta) - J_{\nu-\frac{1}{4}}(\beta)\,N_{\nu+\frac{1}{4}}(\beta)\right] \quad [\operatorname{Re}\beta > 0].$$ ET I 169(21)

14. $$\int_0^\infty \frac{\exp(-2\beta \operatorname{sh} x)\operatorname{sh} 2\nu x}{\sqrt{\operatorname{sh} x}}\,dx = \frac{1}{4i}\sqrt{\frac{\pi^3\beta}{2}}\left[e^{\nu\pi i}H^{(1)}_{\frac{1}{2}+\nu}(\beta)\,H^{(2)}_{\frac{1}{2}-\nu}(\beta) - e^{-\nu\pi i}H^{(1)}_{\frac{1}{2}-\nu}(\beta)\,H^{(2)}_{\frac{1}{2}+\nu}(\beta)\right] \quad [\operatorname{Re}\beta > 0].$$ ET I 170(24)

15. $$\int_0^\infty \frac{\exp(-2\beta \operatorname{sh} x)\operatorname{ch} 2\nu x}{\sqrt{\operatorname{sh} x}}\,dx = \frac{1}{4}\sqrt{\frac{\pi^3\beta}{2}}\left[e^{\nu\pi i}H^{(1)}_{\frac{1}{2}+\nu}(\beta)\,H^{(2)}_{\frac{1}{2}-\nu}(\beta) + e^{-\nu\pi i}H^{(1)}_{\frac{1}{2}-\nu}(\beta)\,H^{(2)}_{\frac{1}{2}+\nu}(\beta)\right] \quad [\operatorname{Re}\beta > 0].$$ ET I 170(25)

16. $$\int_0^\infty \frac{\exp(-2\beta \operatorname{ch} x)\operatorname{ch} 2\nu x}{\sqrt{\operatorname{ch} x}}\,dx = \sqrt{\frac{\beta}{\pi}}\,K_{\nu+\frac{1}{4}}(\beta)\,K_{\nu-\frac{1}{4}}(\beta) \quad [\operatorname{Re}\beta > 0].$$ ET I 170(26)

17. $$\int_0^\infty \frac{\exp[-2\beta(\operatorname{ch} x - 1)]\operatorname{ch} 2\nu x}{\sqrt{\operatorname{ch} x}}\,dx = \sqrt{\frac{\beta}{\pi}}\cdot e^{2\beta}K_{\nu+\frac{1}{2}}(\beta)\,K_{\nu-\frac{1}{2}}(\beta) \quad [\operatorname{Re}\beta > 0].$$ ET I 170(27)

18. $$\int_0^\infty \frac{\cos\left[\left(\nu+\frac{1}{4}\right)\pi\right]\exp(-2\nu x - 2\beta \operatorname{sh} x) + \sin\left[\left(\nu+\frac{1}{4}\right)\pi\right]\exp(2\nu x - 2\beta \operatorname{sh} x)}{\sqrt{\operatorname{sh} x}}\,dx = \frac{1}{2}\sqrt{\pi^3\beta}\left[J_{\frac{1}{4}+\nu}(\beta)\,J_{\frac{1}{4}-\nu}(\beta) + N_{\frac{1}{4}+\nu}(\beta)\,N_{\frac{1}{4}-\nu}(\beta)\right] \quad [\operatorname{Re}\beta > 0].$$ ET I 169(22)

19. $$\int_0^\infty \frac{\sin\left[\left(\nu+\frac{1}{4}\right)\pi\right]\exp(-2\nu x - 2\beta \operatorname{sh} x) - \cos\left[\left(\nu+\frac{1}{4}\right)\pi\right]\exp(2\nu x - 2\beta \operatorname{sh} x)}{\sqrt{\operatorname{sh} x}}\,dx = \frac{1}{2}\sqrt{\pi^3\beta}\left[J_{\frac{1}{4}+\nu}(\beta)\,N_{\frac{1}{4}-\nu}(\beta) - J_{\frac{1}{4}-\nu}(\beta)\,N_{\frac{1}{4}+\nu}(\beta)\right] \quad [\operatorname{Re}\beta > 0].$$ ET I 169(23)

20. $$\int_0^\infty \frac{\exp[-\beta(\operatorname{ch} x - 1)]\operatorname{ch}\nu x \operatorname{sh} x}{\sqrt{\operatorname{ch} x(\operatorname{ch} x - 1)}}\,dx = e^{\beta}K_\nu(\beta) \quad [\operatorname{Re}\beta > 0].$$ ET I 169(19)

3.548

1. $$\int_0^\infty e^{-\mu x^4} \operatorname{sh} ax^2\,dx = \frac{\pi}{4}\sqrt{\frac{a}{2\mu}} \exp\left(\frac{a^2}{8\mu}\right) I_{\frac{1}{4}}\left(\frac{a^2}{8\mu}\right)$$
$[\operatorname{Re}\mu > 0]$. ET I 166(42)

2. $$\int_0^\infty e^{-\mu x^4} \operatorname{ch} ax^2\,dx = \frac{\pi}{4}\sqrt{\frac{a}{2\mu}} \exp\left(\frac{a^2}{8\mu}\right) I_{-\frac{1}{4}}\left(\frac{a^2}{8\mu}\right)$$
$[\operatorname{Re}\mu > 0]$. ET I 166(43)

3.549

1. $$\int_0^\infty e^{-\beta x} \operatorname{sh}\left[(2n+1)\operatorname{Arsh} x\right] dx = O_{2n+1}(\beta)$$
$[\operatorname{Re}\beta > 0]$ (cf. 3.547 6.). ET I 167(5)

2. $$\int_0^\infty e^{-\beta x} \operatorname{ch}(2n \operatorname{Arsh} x)\,dx = O_{2n}(\beta)$$
$[\operatorname{Re}\beta > 0]$ (cf. 3.547 8.). ET I 168(6)

3. $$\int_0^\infty e^{-\beta x} \operatorname{sh}(\nu \operatorname{Arsh} x)\,dx = \frac{\nu}{\beta} S_{0,\nu}(\beta) \quad [\operatorname{Re}\beta > 0]$$ (cf. 3.547 5.).
ET I 168(7)

4. $$\int_0^\infty e^{-\beta x} \operatorname{ch}(\nu \operatorname{Arsh} x)\,dx = \frac{1}{\beta} S_{1,\nu}(\beta) \quad [\operatorname{Re}\beta > 0]$$ (cf. 3.547 7.).

A number of other integrals containing hyperbolic functions and exponentials, depending on $\operatorname{Arsh} x$ or $\operatorname{Arch} x$ can be found by first making the substitution $x = \operatorname{sh} t$ or $x = \operatorname{ch} t$.

3.55-3.56 Combinations of hyperbolic functions, exponentials and powers

3.551

1. $$\int_0^\infty x^{\mu-1} e^{-\beta x} \operatorname{sh} \gamma x\,dx = \frac{1}{2}\Gamma(\mu)\left[(\beta-\gamma)^{-\mu} - (\beta+\gamma)^{-\mu}\right]$$
$[\operatorname{Re}\mu > -1, \quad \operatorname{Re}\beta > |\operatorname{Re}\gamma|]$. ET I 164(18)

2. $$\int_0^\infty x^{\mu-1} e^{-\beta x} \operatorname{ch} \gamma x\,dx = \frac{1}{2}\Gamma(\mu)\left[(\beta-\gamma)^{-\mu} + (\beta+\gamma)^{-\mu}\right]$$
$[\operatorname{Re}\mu > 0, \quad \operatorname{Re}\beta > |\operatorname{Re}\gamma|]$. ET I 164(19)

3. $$\int_0^\infty x^{\mu-1} e^{-\beta x} \operatorname{cth} x\,dx = \Gamma(\mu)\left[2^{1-\mu}\zeta\left(\mu, \frac{\beta}{2}\right) - \beta^{-\mu}\right]$$
$[\operatorname{Re}\mu > 1, \quad \operatorname{Re}\beta > 0]$. ET I 164(21)

4. $$\int_0^\infty x^n e^{-(p+mq)x} \operatorname{sh}^m qx\,dx = 2^{-m} n! \sum_{k=0}^{m} \binom{m}{k} \frac{(-1)^k}{(p+2kq)^{n+1}}$$
$[p > 0, \quad q > 0, \quad m < p+qm]$. LI ((81))(4)

5. $$\int_0^1 \frac{e^{-\beta x}}{x} \operatorname{sh} \gamma x \, dx = \frac{1}{2} \left[\ln \frac{\beta+\gamma}{\beta-\gamma} + \operatorname{Ei}(\gamma-\beta) - \operatorname{Ei}(-\gamma-\beta) \right].$$

BI ((80))(4)

6. $$\int_0^\infty \frac{e^{-\beta x}}{x} \operatorname{sh} \gamma x \, dx = \frac{1}{2} \ln \frac{\beta+\gamma}{\beta-\gamma} \quad [\operatorname{Re} \beta > |\operatorname{Re} \gamma|].$$

ET I 163(12)

7. $$\int_1^\infty \frac{e^{-\beta x}}{x} \operatorname{ch} \gamma x \, dx = \frac{1}{2} \left[-\operatorname{Ei}(\gamma-\beta) - \operatorname{Ei}(-\gamma-\beta) \right]$$
$$[\operatorname{Re} \beta > |\operatorname{Re} \gamma]|.$$

ET I 164(15)

8. $$\int_0^\infty x e^{-x} \operatorname{cth} x \, dx = \frac{\pi^2}{3} - 1.$$

BI ((82))(6)

9. $$\int_0^\infty e^{-\beta x} \operatorname{th} x \frac{dx}{x} = \ln \frac{\beta}{4} + 2 \ln \frac{\Gamma\left(\frac{\beta}{4}\right)}{\Gamma\left(\frac{\beta}{4}+\frac{1}{2}\right)} \quad [\operatorname{Re} \beta > 0].$$

ET I 164(16)

3.552

1. $$\int_0^\infty \frac{x^{\mu-1} e^{-\beta x}}{\operatorname{sh} x} dx = 2^{1-\mu} \Gamma(\mu) \zeta\left[\mu, \frac{1}{2}(\beta+1)\right]$$
$$[\operatorname{Re} \mu > 1, \quad \operatorname{Re} \beta > -1].$$

ET I 164(20)

2. $$\int_0^\infty \frac{x^{2m-1} e^{-ax}}{\operatorname{sh} ax} dx = \frac{1}{2m} \left| B_{2m} \right| \left(\frac{\pi}{a}\right)^{2m}$$

EH I 38(24)a

3. $$\int_0^\infty \frac{x^{\mu-1} e^{-x}}{\operatorname{ch} x} dx = 2^{1-\mu} (1 - 2^{1-\mu}) \Gamma(\mu) \zeta(\mu) \quad [\operatorname{Re} \mu > 0].$$

EH I 32(5)

4. $$\int_0^\infty \frac{x^{2m-1} e^{-ax}}{\operatorname{ch} ax} dx = \frac{1-2^{1-2m}}{2m} \left| B_{2m} \right| \left(\frac{\pi}{a}\right)^{2m}.$$

EH I 39(25)a

5. $$\int_0^\infty \frac{x^2 e^{-2nx}}{\operatorname{sh} x} dx = 4 \sum_{k=n}^{\infty} \frac{1}{(2k+1)^3} \quad (\text{cf. } \mathbf{4.261} \ 13.).$$

BI ((84))(4)

6. $$\int_0^\infty \frac{x^3 e^{-2nx}}{\operatorname{sh} x} dx = \frac{\pi^4}{8} - 12 \sum_{k=1}^{n} \frac{1}{(2k-1)^4} \quad (\text{cf. } \mathbf{4.262} \ 6.).$$

BI ((84))(6)

3.553

1. $$\int_0^\infty \frac{\operatorname{sh}^2 ax}{\operatorname{sh} x} \frac{e^{-x} \, dx}{x} = \frac{1}{2} \ln(a\pi \operatorname{cosec} a\pi) \quad [a < 1].$$

BI ((95))(7)

2. $$\int_0^\infty \frac{\operatorname{sh}^2 \frac{x}{2}}{\operatorname{ch} x} \cdot \frac{e^{-x} \, dx}{x} = \frac{1}{2} \ln \frac{4}{\pi} \quad (\text{cf. } \mathbf{4.267} \ 2.).$$

BI ((95))(4)

3.554

1. $$\int_0^\infty e^{-\beta x}(1-\operatorname{sech} x)\frac{dx}{x} = 2\ln\frac{\Gamma\left(\frac{\beta+3}{4}\right)}{\Gamma\left(\frac{\beta+1}{4}\right)} - \ln\frac{\beta}{4} \quad [\operatorname{Re}\beta > 0].$$

ET I 164(17)

2. $$\int_0^\infty e^{-\beta x}\left(\frac{1}{x} - \operatorname{cosech} x\right)dx = \psi\left(\frac{\beta+1}{2}\right) - \ln\frac{\beta}{2} \quad [\operatorname{Re}\beta > 0].$$

ET I 163(10)

3. $$\int_0^\infty \left[\frac{\operatorname{sh}\left(\frac{1}{2}-\beta\right)x}{\operatorname{sh}\frac{x}{2}} - (1-2\beta)e^{-x}\right]\frac{dx}{x} = 2\ln\Gamma(\beta) - \ln\pi + \ln(\sin\pi\beta)$$

$[0 < \operatorname{Re}\beta < 1]$. EH I 21(7)

4. $$\int_0^\infty e^{-\beta x}\left(\frac{1}{x} - \operatorname{cth} x\right)dx = \psi\left(\frac{\beta}{2}\right) - \ln\frac{\beta}{2} + \frac{1}{\beta} \quad [\operatorname{Re}\beta > 0].$$

ET I 163(11)

5. $$\int_0^\infty \left\{-\frac{\operatorname{sh} qx}{\operatorname{sh}\frac{x}{2}} + 2qe^{-x}\right\}\frac{dx}{x} = 2\ln\Gamma\left(q+\frac{1}{2}\right) + \ln\cos\pi q - \ln\pi$$

$\left[q^2 < \frac{1}{2}\right]$. WH

6. $$\int_0^\infty x^{\mu-1}e^{-\beta x}(\operatorname{cth} x - 1)\,dx = 2^{1-\mu}\Gamma(\mu)\,\zeta\left(\mu, \frac{\beta}{2}+1\right)$$

$[\operatorname{Re}\beta > 0; \quad \operatorname{Re}\mu > 1]$. ET I 164(22)

3.555

1. $$\int_0^\infty \frac{\operatorname{sh}^2 ax}{1-e^{px}}\cdot\frac{dx}{x} = \frac{1}{4}\ln\left(\frac{p}{2a\pi}\sin\frac{2a\pi}{p}\right) \quad [2a < p] \qquad (\text{cf. } 3.545\ 2.).$$

BI ((93))(15)

2. $$\int_0^\infty \frac{\operatorname{sh}^2 ax}{e^x+1}\cdot\frac{dx}{x} = -\frac{1}{4}\ln(a\pi\operatorname{ctg} a\pi) \quad \left[a < \frac{1}{2}\right] \qquad (\text{cf. } 3.545\ 1.).$$

BI ((93))(9)

3.556

1. $$\int_{-\infty}^\infty x\frac{1-e^{px}}{\operatorname{sh} x}\,dx = -\frac{\pi^2}{2}\operatorname{tg}^2\frac{p\pi}{2} \quad [p < 1] \qquad (\text{cf. } 4.255\ 3.).$$

BI ((101))(4)

2. $$\int_0^\infty \frac{1-e^{-px}}{\operatorname{sh} x}\cdot\frac{1-e^{-(p+1)x}}{x}\,dx = 2p\ln 2 \quad [p > -1].$$ BI ((95))(8)

3.557

1. $$\int_0^\infty \frac{e^{-px}-e^{-qx}}{\operatorname{ch} x-\cos\frac{m}{n}\pi}\cdot\frac{dx}{x}=$$

$$=2\operatorname{cosec}\frac{m}{n}\pi\sum_{k=1}^{n-1}(-1)^{k-1}\sin\left(\frac{km}{n}\pi\right)\ln\frac{\Gamma\left(\frac{n+q+k}{2n}\right)\Gamma\left(\frac{p+k}{2n}\right)}{\Gamma\left(\frac{n+p+k}{2n}\right)\Gamma\left(\frac{q+k}{2n}\right)}$$

$[m+n \text{ odd}]$;

$$=2\operatorname{cosec}\frac{m}{n}\pi\sum_{k=1}^{\frac{n-1}{2}}(-1)^{k-1}\sin\left(\frac{km}{n}\pi\right)\ln\frac{\Gamma\left(\frac{n+q-k}{n}\right)\Gamma\left(\frac{p+k}{n}\right)}{\Gamma\left(\frac{n+p-k}{n}\right)\Gamma\left(\frac{q+k}{n}\right)}$$

$[m+n \text{ even}]$; $[p>-1, \quad q>-1]$. BI ((96))(1)

2. $$\int_0^\infty \frac{(1-e^{-x})^2}{\operatorname{ch} x+\cos\frac{m}{n}\pi}\cdot\frac{dx}{x}=$$

$$=2\operatorname{cosec}\frac{m}{n}\pi\sum_{k=1}^{n-1}(-1)^{k-1}\sin\left(\frac{km}{n}\pi\right)\times$$

$$\times\ln\frac{\left[\Gamma\left(\frac{n+k+1}{2n}\right)\right]^2\Gamma\left(\frac{k+2}{2n}\right)\Gamma\left(\frac{k}{2n}\right)}{\left[\Gamma\left(\frac{k+1}{2n}\right)\right]^2\Gamma\left(\frac{n+k}{2n}\right)\Gamma\left(\frac{n+k+2}{2n}\right)}$$ $[m+n \text{ odd}]$;

$$=2\operatorname{cosec}\frac{m}{n}\pi\sum_{k=1}^{\frac{n-1}{2}}(-1)^{k-1}\sin\left(\frac{km}{n}\pi\right)\times$$

$$\times\ln\frac{\left[\Gamma\left(\frac{n-k+1}{n}\right)\right]^2\Gamma\left(\frac{k+2}{n}\right)\Gamma\left(\frac{k}{n}\right)}{\left[\Gamma\left(\frac{k+1}{n}\right)\right]^2\Gamma\left(\frac{n-k}{n}\right)\Gamma\left(\frac{n-k+2}{n}\right)}$$ $[m+n \text{ even}]$.

BI ((96))(2)

3. $$\int_0^\infty\left[e^{-x}\operatorname{tg}\frac{m}{2n}\pi-\frac{e^{-px}\sin\frac{m}{n}\pi}{\operatorname{ch} x+\cos\frac{m}{n}\pi}\right]\cdot\frac{dx}{x}=$$

$$=\operatorname{tg}\left(\frac{m}{2n}\pi\right)\ln(2n)+2\sum_{k=1}^{n-1}(-1)^{k-1}\sin\left(\frac{km}{n}\pi\right)\ln\frac{\Gamma\left(\frac{p+n+k}{2n}\right)}{\Gamma\left(\frac{p+k}{2n}\right)}$$

$[m+n \text{ odd}]$;

$$=\operatorname{tg}\left(\frac{m}{2n}\pi\right)\ln n+2\sum_{k=1}^{\frac{n-1}{2}}(-1)^{k-1}\sin\left(\frac{km}{n}\pi\right)\ln\frac{\Gamma\left(\frac{p+n-k}{n}\right)}{\Gamma\left(\frac{p+k}{n}\right)}$$

$[m+n \text{ even}]$. BI ((96))(3)

4. $$\int_0^\infty \frac{1+e^{-x}}{\operatorname{ch} x+\cos a}\cdot\frac{dx}{x^{1-p}} = 2\sec\frac{a}{2}\,\Gamma(p)\sum_{k=1}^\infty(-1)^{k-1}\frac{\cos\left(k-\frac{1}{2}\right)a}{k^p} \qquad [p>0].$$ LI ((96))(5)

5. $$\int_0^\infty \frac{x^q e^{-\frac{x}{2}}\operatorname{ch}\frac{x}{2}}{\operatorname{ch} x+\cos\lambda}\,dx = \frac{\Gamma(q+1)}{\cos\frac{\lambda}{2}}\sum_{k=1}^\infty(-1)^{k-1}\frac{\cos\left(k-\frac{1}{2}\right)\lambda}{k^{q+1}} \qquad [q>-1].$$ LI ((96))(5)a

6. $$\int_0^\infty x\,\frac{e^{-x}-\cos a}{\operatorname{ch} x-\cos a}\,dx = a\pi-\frac{a^2}{2}-\frac{\pi^2}{3}.$$ BI ((88))(8)

7. $$\int_0^\infty x^{2m+1}\,\frac{e^{-x}-\cos a\pi}{\operatorname{ch} x-\cos a\pi}\,dx = 2\cdot(2m+1)!\sum_{k=1}^\infty\frac{\cos ka\pi}{k^{2m+2}}.$$ BI ((88))(6)

3.558

1. $$\int_0^\infty x\,\frac{1-e^{-nx}}{\operatorname{sh}^2\frac{x}{2}}\,dx = \frac{2n\pi^2}{3}-4\sum_{k=1}^{n-1}\frac{n-k}{k^2}.$$ BI ((85))(3)

2. $$\int_0^\infty x\,\frac{1-(-1)^n e^{-nx}}{\operatorname{ch}^2\frac{x}{2}}\,dx = \frac{n\pi^2}{3}+4\sum_{k=1}^{n-1}(-1)^k\frac{n-k}{k^2}.$$ LI ((85))(1)

3. $$\int_0^\infty x^2\,\frac{1-e^{-nx}}{\operatorname{sh}^2\frac{x}{2}}\,dx = 8n\zeta(3)-8\sum_{k=1}^{n-1}\frac{n-k}{k^3}.$$ BI ((85))(5)

4. $$\int_0^\infty x^2 e^x\,\frac{1-e^{-2nx}}{\operatorname{sh}^2 x}\,dx = 8n\sum_{k=1}^\infty\frac{1}{(2k-1)^3}-8\sum_{k=1}^{n-1}\frac{n-k}{(2k-1)^3}.$$ LI ((85))(6)

5. $$\int_0^\infty x^2\,\frac{1+(-1)^n e^{-nx}}{\operatorname{ch}^2\frac{x}{2}}\,dx = 6n\zeta(3)-8\sum_{k=1}^{n-1}\frac{n-k}{k^3}.$$ LI ((85))(4)

6. $$\int_0^\infty x^3\,\frac{1-e^{-nx}}{\operatorname{sh}^2\frac{x}{2}}\,dx = \frac{4}{15}n\pi^4-24\sum_{k=1}^{n-1}\frac{n-k}{k^4}.$$ BI ((85))(9)

7. $$\int_0^\infty x^3\,\frac{1+(-1)^n e^{-nx}}{\operatorname{ch}^2\frac{x}{2}}\,dx = \frac{7}{30}n\pi^4+24\sum_{k=1}^{n-1}(-1)^k\frac{n-k}{k^4}.$$ BI ((85))(8)

3.559 $$\int_0^\infty e^{-x}\left[a-\frac{1}{2}+\frac{(1-e^{-x})(1-ax)-xe^{-x}}{4\operatorname{sh}^2\frac{x}{2}}e^{(2-a)x}\right]\frac{dx}{x} =$$
$$= a-\frac{1}{2}+\ln\Gamma(a)-\frac{1}{2}\ln(2\pi) \qquad [a>0].$$ BI ((96))(6)

3.561 $\int_0^\infty \frac{e^{-2x} \operatorname{th} \frac{x}{2}}{x \operatorname{ch} x} dx = 2 \ln \frac{\pi}{2\sqrt{2}}$. BI ((93))(18)

3.562

1. $\int_0^\infty x^{2\mu-1} e^{-\beta x^2} \operatorname{sh} \gamma x \, dx = \frac{1}{2} \Gamma(2\mu) (2\beta)^{-\mu} \exp\left(\frac{\gamma^2}{8\beta}\right) \times$

$\times \left[D_{-2\mu}\left(\frac{\gamma}{\sqrt{2\beta}}\right) - D_{-2\mu}\left(-\frac{\gamma}{\sqrt{2\beta}}\right) \right] \quad \left[\operatorname{Re} \mu > -\frac{1}{2}, \operatorname{Re} \beta > 0\right]$. ET I 166(44)

2. $\int_0^\infty x^{2\mu-1} e^{-\beta x^2} \operatorname{ch} \gamma x \, dx = \frac{1}{2} \Gamma(2\mu) (2\beta)^{-\mu} \exp\left(\frac{\gamma^2}{8\beta}\right) \times$

$\times \left[D_{-2\mu}\left(-\frac{\gamma}{\sqrt{2\beta}}\right) + D_{-2\mu}\left(\frac{\gamma}{\sqrt{2\beta}}\right) \right] \quad [\operatorname{Re} \mu > 0, \operatorname{Re} \beta > 0]$. ET I 166(45)

3. $\int_0^\infty x e^{-\beta x^2} \operatorname{sh} \gamma x \, dx = \frac{\gamma}{4\beta} \sqrt{\frac{\pi}{\beta}} \exp \frac{\gamma^2}{4\beta} \quad [\operatorname{Re} \beta > 0]$.

BI((81))(12)a, ET I 165(34)

4. $\int_0^\infty x e^{-\beta x^2} \operatorname{ch} \gamma x \, dx = \frac{\gamma}{4\beta} \sqrt{\frac{\pi}{\beta}} \exp \frac{\gamma^2}{4\beta} \Phi\left(\frac{\gamma}{2\sqrt{\beta}}\right) + \frac{1}{2\beta} \quad [\operatorname{Re} \beta > 0]$.

ET I 166(35)

5. $\int_0^\infty x^2 e^{-\beta x^2} \operatorname{sh} \gamma x \, dx = \frac{\sqrt{\pi}(2\beta + \gamma^2)}{8\beta^2 \sqrt{\beta}} \exp\left(\frac{\gamma^2}{4\beta}\right) \Phi\left(\frac{\gamma}{2\sqrt{\beta}}\right) + \frac{\gamma}{4\beta^2}$

$[\operatorname{Re} \beta > 0]$. ET I 166(36)

6. $\int_0^\infty x^2 e^{-\beta x^2} \operatorname{ch} \gamma x \, dx = \frac{\sqrt{\pi}(2\beta + \gamma^2)}{8\beta^2 \sqrt{\beta}} \exp\left(\frac{\gamma^2}{4\beta}\right) \quad [\operatorname{Re} \beta > 0]$. ET I 166(37)

3.6-4.1 Trigonometric Functions

3.61 Rational functions of sines and cosines and trigonometric functions of multiple angles

3.611

1. $\int_0^{2\pi} (1 - \cos x)^n \sin nx \, dx = 0$. BI ((68))(10)

2. $\int_0^{2\pi} (1 - \cos x)^n \cos nx \, dx = (-1)^n \frac{\pi}{2^{n-1}}$. BI ((68))(11)

3. $\int_0^\pi (\cos t + i \sin t \cos x)^n \, dx = \int_0^\pi (\cos t + i \sin t \cos x)^{-n-1} \, dx = \pi P_n(\cos t)$.

EH I 158(23)a

3.612

1. $$\int_0^{\pi} \frac{\sin nx \cos mx}{\sin x}\, dx = 0 \quad \text{for} \quad n \leqslant m;$$
$$= \pi \quad \text{for} \quad n > m, \quad \text{if} \quad m+n \quad \text{is odd;}$$
$$= 0 \quad \text{for} \quad n > m, \quad \text{if} \quad m+n \quad \text{is even.}$$
LI ((64))(3)

2. $$\int_0^{\pi} \frac{\sin nx}{\sin x}\, dx = 0 \quad \text{for} \quad n \quad \text{even;}$$
$$= \pi \quad \text{for} \quad n \quad \text{odd.}$$
BI ((64))(1, 2)

3. $$\int_0^{\frac{\pi}{2}} \frac{\sin (2n-1) x}{\sin x}\, dx = \frac{\pi}{2}.$$
FI II 145

4. $$\int_0^{\frac{\pi}{2}} \frac{\sin 2nx}{\sin x}\, dx = 2\left(1 - \frac{1}{3} + \frac{1}{5} - \ldots + \frac{(-1)^{n-1}}{2n-1}\right).$$
GW ((332))(21b)

5. $$\int_0^{\pi} \frac{\sin 2nx}{\cos x}\, dx = 2\int_0^{\frac{\pi}{2}} \frac{\sin 2nx}{\cos x}\, dx = (-1)^{n-1} 4\left(1 - \frac{1}{3} + \frac{1}{5} - \ldots + \frac{(-1)^{n-1}}{2n-1}\right).$$
GW ((332))(22a)

6. $$\int_0^{\pi} \frac{\cos (2n+1) x}{\cos x}\, dx = 2\int_0^{\frac{\pi}{2}} \frac{\cos (2n+1) x}{\cos x}\, dx = (-1)^n \pi.$$
GW ((332))(22b)

7. $$\int_0^{\frac{\pi}{2}} \frac{\sin 2nx \cos x}{\sin x}\, dx = \frac{\pi}{2}.$$
LI ((45))(17)

3.613

1. $$\int_0^{\pi} \frac{\cos nx\, dx}{1 + a \cos x} = \frac{\pi}{\sqrt{1-a^2}} \left(\frac{\sqrt{1-a^2}-1}{a}\right)^n \qquad [a^2 < 1].$$
BI ((64))(12)

2. $$\int_0^{\pi} \frac{\cos nx\, dx}{1 - 2a \cos x + a^2} = \frac{\pi a^n}{1-a^2} \qquad [a^2 < 1];$$
$$= \frac{\pi}{(a^2-1)\, a^n} \qquad [a^2 > 1].$$
BI ((65))(3)

3. $$\int_0^{\pi} \frac{\sin nx \sin x\, dx}{1 - 2a \cos x + a^2} = \frac{\pi}{2} a^{n-1} \qquad [a^2 < 1];$$
$$= \frac{\pi}{2a^{n+1}} \qquad [a^2 > 1].$$
BI((65))(4), GW((332))(34a)

4. $$\int_0^\pi \frac{\cos nx \cos x \, dx}{1-2a\cos x+a^2} = \frac{\pi}{2} \cdot \frac{1+a^2}{1-a^2} a^{n-1} \qquad [a^2 < 1];$$
$$= \frac{\pi}{2a^{n+1}} \cdot \frac{a^2+1}{a^2-1} \qquad [a^2 > 1].$$

BI((65))(5), GW((332))(34b)

5. $$\int_0^\pi \frac{\cos(2n-1)x \, dx}{1-2a\cos 2x+a^2} = \int_0^\pi \frac{\cos 2nx \cos x \, dx}{1-2a\cos 2x+a^2} = 0 \qquad [a^2 \neq 1].$$

BI ((65))(9, 10)

6. $$\int_0^\pi \frac{\cos(2n-1)x \cos 2x \, dx}{1-2a\cos 2x+a^2} = 0 \qquad [a^2 \neq 1].$$ BI ((65))(12)

7. $$\int_0^\pi \frac{\sin 2nx \sin x \, dx}{1-2a\cos 2x+a^2} = \int_0^\pi \frac{\sin(2n-1)x \sin 2x \, dx}{1-2a\cos 2x+a^2} = 0 \qquad [a^2 \neq 1].$$

BI ((65))(6, 7)

8. $$\int_0^\pi \frac{\sin(2n-1)x \sin x \, dx}{1-2a\cos 2x+a^2} = \frac{\pi}{2} \cdot \frac{a^{n-1}}{1+a} \qquad [a^2 < 1];$$
$$= \frac{\pi}{2} \cdot \frac{1}{(1+a)a^n} \qquad [a^2 > 1].$$ BI ((65))(8)

9. $$\int_0^\pi \frac{\cos(2n-1)x \cos x \, dx}{1-2a\cos 2x+a^2} = \frac{\pi}{2} \cdot \frac{a^{n-1}}{1-a} \qquad [a^2 < 1];$$
$$= \frac{\pi}{2} \cdot \frac{1}{(a-1)a^n} \qquad [a^2 > 1].$$ BI ((65))(11)

10. $$\int_0^\pi \frac{\sin nx - a\sin(n-1)x}{1-2a\cos x+a^2} \sin mx \, dx = 0 \qquad \text{for} \quad m < n;$$
$$= \frac{\pi}{2} a^{m-n} \qquad \text{for} \quad m \geqslant n;$$
$$[a^2 < 1].$$ LI ((65))(13)

11. $$\int_0^\pi \frac{\cos nx - a\cos(n-1)x}{1-2a\cos x+a^2} \cos mx \, dx = \frac{\pi}{2}(a^{m-n}-1) \qquad [a^2 < 1].$$ BI ((65))(14)

12. $$\int_0^\pi \frac{\sin nx - a\sin[(n+1)x]}{1-2a\cos x+a^2} dx = 0 \qquad [a^2 < 1].$$ BI ((68))(13)

13. $$\int_0^\pi \frac{\cos nx - a\cos[(n+1)x]}{1-2a\cos x+a^2} dx = 2\pi a^n \qquad [a^2 < 1].$$ BI ((68))(14)

3.614 $$\int_0^\pi \frac{\sin x}{a^2-2ab\cos x+b^2} \cdot \frac{\sin px \cdot dx}{1-2a^p\cos px+a^{2p}} = \frac{\pi b^{p-1}}{2a^{p+1}(1-b^p)}$$
$$[0 < a < 1, \ 0 < a < b, \ p > 0].$$ BI ((66))(9)

3.615

1. $$\int_0^{\frac{\pi}{2}} \frac{\cos 2nx\, dx}{1-a^2 \sin^2 x} = \frac{(-1)^n \pi}{2\sqrt{1-a^2}} \left(\frac{1-\sqrt{1-a^2}}{a}\right)^{2n} \quad [a^2 < 1].$$ BI ((47))(27)

2. $$\int_0^{\pi} \frac{\cos x \sin 2nx\, dx}{1+(a+b\sin x)^2} = -\frac{\pi}{b} \sin\left\{2n \operatorname{arctg} \sqrt{\frac{s}{2}}\right\} \operatorname{tg}^{2n}\left(\frac{1}{2} \arccos \sqrt{\frac{s}{2a^2}}\right).$$

3. $$\int_0^{\pi} \frac{\cos x \cos(2n+1)x\, dx}{1+(a+b\sin x)^2} =$$
$$= \frac{\pi}{b} \cos\left\{(2n+1) \operatorname{arctg} \sqrt{\frac{s}{2}}\right\} \operatorname{tg}^{2n+1}\left(\frac{1}{2} \arccos \sqrt{\frac{s}{2a^2}}\right),$$
where $s = -(1+b^2-a^2) + \sqrt{(1+b^2-a^2)^2+4a^2}$. BI ((65))(21, 22)

3.616

1. $$\int_0^{\pi} (1-2a\cos x+a^2)^n\, dx = \pi \sum_{k=0}^{n} \binom{n}{k}^2 a^{2k}.$$ BI ((63))(1)

2. $$\int_0^{\pi} \frac{dx}{(1-2a\cos x+a^2)^n} = \frac{1}{2}\int_0^{2\pi} \frac{dx}{(1-2a\cos x+a^2)^n} =$$
$$= \frac{\pi}{(1-a^2)^n} \sum_{k=0}^{n-1} \frac{(n+k-1)!}{(k!)^2 (n-k-1)!} \cdot \left(\frac{a^2}{1-a^2}\right)^k \quad [a^2 < 1];$$
$$= \frac{\pi}{(a^2-1)^n} \sum_{k=0}^{n-1} \frac{(n+k-1)!}{(k!)^2 (n-k-1)!} \cdot \frac{1}{(a^2-1)^k} \quad [a^2 > 1].$$ BI ((331))(63)

3. $$\int_0^{\pi} (1-2a\cos x+a^2)^n \cos nx\, dx = (-1)^n \pi a^n.$$ BI ((63))(2)

4. $$\int_0^{\pi} (1-2a\cos x+a^2)^n \cos mx\, dx = \frac{1}{2}\int_0^{2\pi} (1-2a\cos x+a^2)^n \cos mx\, dx =$$
$$= 0 \quad [n < m];$$
$$= \pi(-a)^m (1+a^2)^{n-m} \sum_{k=0}^{E\left(\frac{n-m}{2}\right)} \binom{n}{k} \binom{n-k}{m+k} \left(\frac{a}{1+a^2}\right)^{2k} \quad [n \geqslant m].$$
GW ((332))(35a)

5. $$\int_0^{2\pi} \frac{\sin nx\, dx}{(1-2a\cos 2x+a^2)^m} = 0.$$ GW ((332))(32a)

6. $$\int_0^{\pi} \frac{\sin x\, dx}{(1-2a\cos 2x+a^2)^m} = \frac{1}{2(m-1)a}\left[\frac{1}{(1-a)^{2m-2}} - \frac{1}{(1+a)^{2m-2}}\right]$$
$[a \neq 0, \pm 1],$ GW ((332))(32c)

7. $$\int_0^{\pi} \frac{\cos nx\,dx}{(1-2a\cos x+a^2)^m} = \frac{1}{2}\int_0^{2\pi} \frac{\cos nx\,dx}{(1-2a\cos x+a^2)^m} =$$

$$= \frac{a^{2m+n-2}\pi}{(1-a^2)^{2m-1}} \sum_{k=0}^{m-1} \binom{m+n-1}{k}\binom{2m-k-2}{m-1}\left(\frac{1-a^2}{a^2}\right)^k \qquad [a^2<1];$$

$$= \frac{\pi}{a^n(a^2-1)^{2m-1}} \sum_{k=0}^{m-1} \binom{m+n-1}{k}\binom{2m-k-2}{m-1}(a^2-1)^k \qquad [a^2>1].$$

GW ((332))(31)

8. $$\int_0^{\frac{\pi}{2}} \frac{\cos 2nx\,dx}{(a^2\cos^2 x+b^2\sin^2 x)^{n+1}} = \binom{2n}{n}\frac{(b^2-a^2)^n}{(2ab)^{2n+1}}\pi \qquad [a>0,\ b>0].$$

GW ((332))(30b)

3.62 Powers of trigonometric functions

3.621

1. $$\int_0^{\frac{\pi}{2}} \sin^{\mu-1} x\,dx = \int_0^{\frac{\pi}{2}} \cos^{\mu-1} x\,dx = 2^{\mu-2}\,\mathrm{B}\left(\frac{\mu}{2},\ \frac{\mu}{2}\right).$$ FI II 789

2. $$\int_0^{\frac{\pi}{2}} \sin^{\frac{3}{2}} x\,dx = \int_0^{\frac{\pi}{2}} \cos^{\frac{3}{2}} x\,dx = \frac{1}{6\sqrt{2\pi}}\,\Gamma\left(\frac{1}{4}\right).$$

3. $$\int_0^{\frac{\pi}{2}} \sin^{2m} x\,dx = \int_0^{\frac{\pi}{2}} \cos^{2m} x\,dx = \frac{(2m-1)!!}{(2m)!!}\,\frac{\pi}{2}.$$ FI II 151

4. $$\int_0^{\frac{\pi}{2}} \sin^{2m+1} x\,dx = \int_0^{\frac{\pi}{2}} \cos^{2m+1} x\,dx = \frac{(2m)!!}{(2m+1)!!}.$$ FI II 151

5. $$\int_0^{\frac{\pi}{2}} \sin^{\mu-1} x \cos^{\nu-1} x\,dx = \frac{1}{2}\,\mathrm{B}\left(\frac{\mu}{2},\ \frac{\nu}{2}\right) \qquad [\mathrm{Re}\,\mu>0,\ \mathrm{Re}\,\nu>0].$$

LO V 113(50), LO V 122, FI II 788

3.622

1. $$\int_0^{\frac{\pi}{2}} \mathrm{tg}^{\pm\mu} x\,dx = \frac{\pi}{2}\sec\frac{\mu\pi}{2} \qquad [|\mathrm{Re}\,\mu|<1].$$ BI ((42))(1)

2. $$\int_0^{\frac{\pi}{4}} \mathrm{tg}^{\mu} x\,dx = \frac{1}{2}\,\beta\left(\frac{\mu+1}{2}\right) \qquad [\mathrm{Re}\,\mu>-1].$$ BI ((34))(1)

3. $\int_0^{\frac{\pi}{4}} \operatorname{tg}^{2n} x\,dx = (-1)^n \frac{\pi}{4} + \sum_{k=0}^{n-1} \frac{(-1)^k}{2n-2k-1}$. BI ((34))(2)

4. $\int_0^{\frac{\pi}{4}} \operatorname{tg}^{2n+1} x\,dx = (-1)^{n+1} \frac{\ln 2}{2} + \sum_{k=0}^{n-1} \frac{(-1)^k}{2n-2k}$. BI ((34))(3)

3.623

1. $\int_0^{\frac{\pi}{2}} \operatorname{tg}^{\mu-1} x \cos^{2\nu-2} x\,dx = \int_0^{\frac{\pi}{2}} \operatorname{ctg}^{\mu-1} x \sin^{2\nu-2} x\,dx =$

$= \frac{1}{2} \mathrm{B}\left(\frac{\mu}{2}, \nu - \frac{\mu}{2}\right) \quad [0 < \operatorname{Re}\mu < 2\operatorname{Re}\nu]$. BI((42))(6), BI((45))(22)

2. $\int_0^{\frac{\pi}{4}} \operatorname{tg}^{\mu} x \sin^2 x\,dx = \frac{1+\mu}{4} \beta\left(\frac{\mu+1}{2}\right) \quad [\operatorname{Re}\mu > -1]$. BI ((34))(4)

3. $\int_0^{\frac{\pi}{4}} \operatorname{tg}^{\mu} x \cos^2 x\,dx = \frac{1-\mu}{4} \beta\left(\frac{\mu+1}{2}\right) \quad [\operatorname{Re}\mu > -1]$. BI ((34))(5)

3.624

1. $\int_0^{\frac{\pi}{4}} \frac{\sin^p x}{\cos^{p+2} x}\,dx = \frac{1}{p+1} \quad [p > -1]$. GW ((331))(34b)

2. $\int_0^{\frac{\pi}{2}} \frac{\sin^{\mu-\frac{1}{2}} x}{\cos^{2\mu-1} x}\,dx = \int_0^{\frac{\pi}{2}} \frac{\cos^{\mu-\frac{1}{2}} x}{\sin^{2\mu-1} x}\,dx = \frac{\Gamma\left(\frac{\mu}{2}+\frac{1}{4}\right)\Gamma(1-\mu)}{\Gamma\left(\frac{5}{4}-\frac{\mu}{2}\right)}$

$\left[-\frac{1}{2} < \operatorname{Re}\mu < 1\right]$. LI ((55))(12)

3. $\int_0^{\frac{\pi}{4}} \frac{\cos^{n-\frac{1}{2}} 2x}{\cos^{2n+1} x}\,dx = \frac{(2n-1)!!}{2\cdot(2n)!!}\pi$. BI ((38))(3)

4. $\int_0^{\frac{\pi}{4}} \frac{\cos^{\mu} 2x}{\cos^{2(\mu+1)} x}\,dx = 2^{2\mu} \mathrm{B}(\mu+1, \mu+1) \quad [\operatorname{Re}\mu > -1]$. BI ((35))(1)

5. $\int_0^{\frac{\pi}{4}} \frac{\sin^{2\mu-2} x}{\cos^{\mu} 2x}\,dx = 2^{1-2\mu} \mathrm{B}(2\mu-1, 1-\mu) = \frac{\Gamma\left(\mu-\frac{1}{2}\right)\Gamma(1-\mu)}{2\sqrt{\pi}}$

$\left[\frac{1}{2} < \operatorname{Re}\mu < 1\right]$. BI ((35))(4)

6. $$\int_0^{\frac{\pi}{2}} \left(\frac{\sin ax}{\sin x}\right)^2 dx = \frac{a\pi}{2}.$$ FI II 145

3.625

1. $$\int_0^{\frac{\pi}{4}} \frac{\sin^{2n-1} x \cos^p 2x}{\cos^{2p+2n+1} x} dx = \frac{(n-1)!}{2} \cdot \frac{\Gamma(p+1)}{\Gamma(p+n+1)} =$$
$$= \frac{(n-1)!}{2(p+n)(p+n-1)\ldots(p+1)} = \frac{1}{2} \mathrm{B}(n, p+1)$$
$$[p > -1], \qquad \text{(cf. } \mathbf{3.251}\ 1.).$$ BI ((35))(2)

2. $$\int_0^{\frac{\pi}{4}} \frac{\sin^{2n} x \cos^p 2x}{\cos^{2p+2n+2} x} dx = \frac{1}{2} \mathrm{B}\left(n + \frac{1}{2}, p+1\right)$$
$$[p > -1], \qquad \text{(cf. } \mathbf{3.251}\ 1.).$$ BI ((35))(3)

3. $$\int_0^{\frac{\pi}{4}} \frac{\sin^{2n-1} x \cos^{m-\frac{1}{2}} 2x}{\cos^{2n+2m} x} dx = \frac{(2n-2)!!\,(2m-1)!!}{(2n+2m-1)!!}.$$ BI ((38))(6)

4. $$\int_0^{\frac{\pi}{4}} \frac{\sin^{2n} x \cos^{m-\frac{1}{2}} 2x}{\cos^{2n+2m+1} x} dx = \frac{(2n-1)!!\,(2m-1)!!}{(2n+2m)!!} \cdot \frac{\pi}{2}.$$ BI ((38))(7)

3.626

1. $$\int_0^{\frac{\pi}{4}} \frac{\sin^{2n-1} x}{\cos^{2n+2} x} \sqrt{\cos 2x}\, dx = \frac{(2n-2)!!}{(2n+1)!!} \qquad \text{(cf. } \mathbf{3.251}\ 1.).$$ BI ((38))(4)

2. $$\int_0^{\frac{\pi}{4}} \frac{\sin^{2n} x}{\cos^{2n+3} x} \sqrt{\cos 2x}\, dx = \frac{(2n-1)!!}{(2n+2)!!} \cdot \frac{\pi}{2} \qquad \text{(cf. } \mathbf{3.251}\ 1.).$$ BI ((38))(5)

3.627 $$\int_0^{\frac{\pi}{2}} \frac{\operatorname{tg}^{\mu} x}{\cos^{\mu} x} dx = \int_0^{\frac{\pi}{2}} \frac{\operatorname{ctg}^{\mu} x}{\sin^{\mu} x} dx = \frac{\Gamma(\mu)\,\Gamma\left(\frac{1}{2} - \mu\right)}{2^{\mu} \sqrt{\pi}} \sin \frac{\mu\pi}{2}$$
$$\left[-1 < \operatorname{Re} \mu < \frac{1}{2}\right].$$ BI ((55))(12)a

3.628 $$\int_0^{\frac{\pi}{2}} \sec^{2p+1} x \frac{d \sin^{2p} x}{dx} dx = \frac{1}{\sqrt{\pi}} \Gamma(p+1)\, \Gamma\left(\frac{1}{2} - p\right)$$
$$\left[\frac{1}{2} > p > 0\right].$$ WA 691

3.63 Powers of trigonometric functions and trigonometric functions of linear functions

3.631

1. $$\int_0^{\pi} \sin^{\nu-1} x \sin ax\, dx = \frac{\pi \sin \frac{a\pi}{2}}{2^{\nu-1} \nu B\left(\frac{\nu+a+1}{2}, \frac{\nu-a+1}{2}\right)}$$
$$[\operatorname{Re} \nu > 0].$$ LO V 121(67)a, WA 337a

2. $$\int_0^{\frac{\pi}{2}} \sin^{\nu-2} x \sin \nu x\, dx = \frac{-1}{\nu-1} \cos \frac{\nu\pi}{2} \qquad [\operatorname{Re} \nu > 1].$$
GW((332))(16d), FI II 152

3. $$\int_0^{\pi} \sin^{\nu} x \sin \nu x\, dx = 2^{-\nu} \pi \sin \frac{\nu\pi}{2} \qquad [\operatorname{Re} \nu > -1].$$ LO V 121(69)

4. $$\int_0^{\pi} \sin^{n} x \sin 2mx\, dx = 0.$$ GW ((332))(11a)

5. $$\int_0^{\pi} \sin^{2n} x \sin (2m+1) x\, dx = 2 \int_0^{\frac{\pi}{2}} \sin^{2n} x \sin (2m+1) x\, dx =$$
$$= \frac{(-1)^m 2^{n+1} n!\,(2n-1)!!}{(2n-2m-1)!!\,(2m+2n+1)!!} \qquad [m \leqslant n]^{*};$$
$$= \frac{(-1)^n 2^{n+1} n!\,(2m-2n-1)!!\,(2n-1)!!}{(2m+2n+1)!!} \qquad [m \geqslant n]^{*}.$$ GW ((332))(11b)

6. $$\int_0^{\pi} \sin^{2n+1} x \sin (2m+1) x\, dx = 2 \int_0^{\frac{\pi}{2}} \sin^{2n+1} x \sin (2m+1) x\, dx =$$
$$= \frac{(-1)^m \pi}{2^{2n+1}} \binom{2n+1}{n-m} \qquad [n \geqslant m];$$
$$= 0 \qquad [n < m].$$ BI((40))(12), GW((332))(11c)

7. $$\int_0^{\pi} \sin^{n} x \cos (2m+1) x\, dx = 0.$$ GW ((332))(12a)

8. $$\int_0^{\pi} \sin^{\nu-1} x \cos ax\, dx = \frac{\pi \cos \frac{a\pi}{2}}{2^{\nu-1} \nu B\left(\frac{\nu+a+1}{2}, \frac{\nu-a+1}{2}\right)}$$
$$[\operatorname{Re} \nu > 0].$$ LO V 121(68)a, WA 337a

9. $$\int_0^{\frac{\pi}{2}} \cos^{\nu-1} x \cos ax\, dx = \frac{\pi}{2^{\nu} \nu B\left(\frac{\nu+a+1}{2}, \frac{\nu-a+1}{2}\right)} \qquad [\operatorname{Re} \nu > 0].$$
GW ((332))(9c)

---

* For $m=n$ we should set $(2n-2m-1)!!=1$.

10. $$\int_0^{\frac{\pi}{2}} \sin^{\nu-2} x \cos \nu x \, dx = \frac{1}{\nu-1} \sin \frac{\nu\pi}{2} \qquad [\operatorname{Re} \nu > 1].$$

GW((332))(16b), FI II 152

11. $$\int_0^{\pi} \sin^{\nu} x \cos \nu x \, dx = \frac{\pi}{2^{\nu}} \cos \frac{\nu\pi}{2} \qquad [\operatorname{Re} \nu > -1].$$ LO V 121(70)a

12. $$\int_0^{\pi} \sin^{2n} x \cos 2mx \, dx = 2 \int_0^{\frac{\pi}{2}} \sin^{2n} x \cos 2mx \, dx =$$

$$= \frac{(-1)^m}{2^{2n}} \binom{2n}{n-m} \qquad [n \geqslant m];$$

$$= 0 \qquad [n < m].$$ BI((40))(16), GW((332))(12b)

13. $$\int_0^{\pi} \sin^{2n+1} x \cos 2mx \, dx = 2 \int_0^{\frac{\pi}{2}} \sin^{2n+1} x \cos 2mx \, dx =$$

$$= \frac{(-1)^m \, 2^{n+1} \, n! \, (2n+1)!!}{(2n-2m+1)!! \, (2m+2n+1)!!} \qquad [n \geqslant m-1];$$

$$= \frac{(-1)^{n+1} \, 2^{n+1} \, n! \, (2m-2n+1)!! \, (2n+1)!!}{(2m+2n+1)!!} \qquad [n < m-1].$$

GW ((332))(12c)

14. $$\int_0^{\frac{\pi}{2}} \cos^{\nu-2} x \sin \nu x \, dx = \frac{1}{\nu-1} \qquad [\operatorname{Re} \nu > 1].$$ GW((332))(16c), FI II 152

15. $$\int_0^{\pi} \cos^m x \sin nx \, dx = [1-(-1)^{m+n}] \int_0^{\frac{\pi}{2}} \cos^m x \sin nx \, dx =$$

$$= [1-(-1)^{m+n}] \left\{ \sum_{k=0}^{r-1} \frac{m!}{(m-k)\,!} \, \frac{(m+n-2k-2)!!}{(m+n)!!} + s \, \frac{m! \, (n-m-2)!!}{(m+n)!!} \right\}$$

$$\left[r = \begin{cases} m & [m \leqslant n], \\ n & [m \geqslant n], \end{cases} \qquad s = \begin{cases} 2 & [n-m = 4l+2 > 0], \\ 1 & [n-m = 2l+1 > 0], \\ 0 & [n-m = 4l \text{ or } n-m < 0] \end{cases} \right].$$

GW ((332))(13a)

16. $$\int_0^{\frac{\pi}{2}} \cos^n x \sin nx \, dx = \frac{1}{2^{n+1}} \sum_{k=1}^{n} \frac{2^k}{k}.$$ FI II 153

17. $\int_0^{\pi} \cos^n x \cos mx\, dx = [1+(-1)^{m+n}] \int_0^{\frac{\pi}{2}} \cos^n x \cos mx\, dx =$

$$= [1+(-1)^{m+n}] \begin{cases} s \dfrac{n!}{(m-n)(m-n+2)\ldots(m+n)} & [n < m]; \\ \dfrac{\pi}{2^{n+1}} \dbinom{n}{k} & [m \leqslant n \text{ and } n-m=2k]; \\ \dfrac{n!}{(2k+1)!!\,(2m+2k+1)!!} & [m < n \text{ and } n-m=2k+1]; \end{cases}$$

$$\text{where } s = \begin{cases} 0 & [m-n=2k], \\ 1 & [m-n=4k+1], \\ -1 & [m-n=4k-1]. \end{cases}$$

GW ((332))(15a)

18. $\int_0^{\pi} \cos^m x \cos ax\, dx = \frac{(-1)^m \sin a\pi}{2^m (m+a)}\, {}_2F_1\left(-m, \frac{a+m}{2};\ 1-\frac{a+m}{2};\ -1\right)$

$[a \neq 0,\ \pm 1,\ \pm 2, \ldots]$. WA 342

19. $\int_0^{\frac{\pi}{2}} \cos^{\nu-2} x \cos \nu x\, dx = 0 \quad [\operatorname{Re} \nu > 1]$. GW((332))(16a), FI II 152

20. $\int_0^{\frac{\pi}{2}} \cos^n x \cos nx\, dx = \frac{\pi}{2^{n+1}}$. LO V 122(78), FI II 153

3.632

1. $\int_0^{\pi} \sin^{p-1} x \cos\left[a\left(\frac{\pi}{2}-x\right)\right] dx = 2^{p-1} \frac{\Gamma\left(\frac{p-a}{2}\right)\Gamma\left(\frac{p+a}{2}\right)}{\Gamma(p-a)\,\Gamma(p+a)}\, \Gamma(p)$

$[p^2 < a^2]$. BI ((62))(11)

2. $\int_{-\frac{\pi}{2}}^{\frac{\pi}{2}} \cos^{\nu-1} x \sin\left[a\left(x+\frac{\pi}{2}\right)\right] dx = \frac{\pi \sin \frac{a\pi}{2}}{2^{\nu-1}\nu B\left(\frac{\nu+a+1}{2}, \frac{\nu-a+1}{2}\right)}$

$[\operatorname{Re} \nu > 0]$. WA 337a

3. $\int_0^{\frac{\pi}{2}} \cos^p x \sin[(p+2n)x]\, dx = (-1)^{n-1} \sum_{k=0}^{n-1} \frac{(-1)^k 2^k}{p+k+1} \binom{n-1}{k}$. LI ((41))(12)

4. $\int_{-\pi}^{\pi} \cos^{n-1} x \cos[m(x-a)]\, dx = [1-(-1)^{n+m}] = \int_{-\frac{\pi}{2}}^{\frac{\pi}{2}} \cos^{n-1} x \cos[m(x-a)]\, dx =$

$$= \frac{[1-(-1)^{n+m}]\, \pi \cos ma}{2^{n-1} n B\left(\frac{n+m+1}{2}, \frac{n-m+1}{2}\right)} \qquad [n \geqslant m].$$

LO V 123(80), LO V 139(94a)

5. $$\int_0^{\frac{\pi}{2}} \cos^{p+q-2} x \cos[(p-q)x]\,dx = \frac{\pi}{2^{p+q-1}(p+q-1)\,\mathrm{B}(p,q)}$$

$[p+q>1]$. WH

3.633

1. $$\int_0^{\frac{\pi}{2}} \cos^{p-1} x \sin ax \sin x\,dx = \frac{a\pi}{2^{p+1}\,p(p+1)\,\mathrm{B}\left(\frac{p+a}{2}+1,\ \frac{p-a}{2}+1\right)}.$$

LO V 150(110)

2. $$\int_0^{\frac{\pi}{2}} \cos^n x \sin nx \sin 2mx\,dx = \int_0^{\frac{\pi}{2}} \cos^n x \cos nx \cos 2mx\,dx = \frac{\pi}{2^{n+2}}\binom{n}{m}.$$

BI ((42))(19, 20)

3. $$\int_0^{\frac{\pi}{2}} \cos^{n-1} x \cos[(n+1)x] \cos 2mx\,dx = \frac{\pi}{2^{n+1}}\binom{n-1}{m-1} \quad [n>m-1].$$

BI ((42))(21)

4. $$\int_0^{\frac{\pi}{2}} \cos^{p+q} x \cos px \cos qx\,dx = \frac{\pi}{2^{p+q+2}}\left[1+\frac{1}{(p+q+1)\,\mathrm{B}(p+1,\ q+1)}\right]$$

$[p+q>-1]$. GW ((332))(10c)

5. $$\int_0^{\frac{\pi}{2}} \cos^{p+q} x \sin px \sin qx\,dx = \frac{\pi}{2^{p+q+2}}\sum_{k=1}^{\infty}\binom{p}{k}\binom{q}{k}$$

$[p+q>-1]$. BI ((42))(16)

3.634

1. $$\int_0^{\frac{\pi}{2}} \sin^{\mu-1} x \cos^{\nu-1} x \sin(\mu+\nu)x\,dx = \sin\frac{\mu\pi}{2}\,\mathrm{B}(\mu,\nu)$$

$[\operatorname{Re}\mu>0,\ \operatorname{Re}\nu>0]$. BI((42))(23), FI II 814a

2. $$\int_0^{\frac{\pi}{2}} \sin^{\mu-1} x \cos^{\nu-1} x \cos(\mu+\nu)x\,dx = \cos\frac{\mu\pi}{2}\,\mathrm{B}(\mu,\nu)$$

$[\operatorname{Re}\mu>0,\ \operatorname{Re}\nu>0]$. BI((42))(24), FI II 814a

3. $$\int_0^{\frac{\pi}{2}} \cos^{p+n-1} x \sin px \cos[(n+1)x] \sin x\,dx = \frac{\pi}{2^{p+n+1}}\frac{\Gamma(p+n)}{n!\,\Gamma(p)} \quad [p>-n].$$

BI ((42))(15)

3.635

1. $$\int_0^{\frac{\pi}{4}} \cos^{\mu-1} 2x \operatorname{tg} x\, dx = \frac{1}{4}\left[\psi\left(\frac{\mu+1}{2}\right) - \psi\left(\frac{\mu}{2}\right)\right] \quad [\operatorname{Re}\mu > 0].$$ BI ((34))(7)

2. $$\int_0^{\frac{\pi}{2}} \cos^{p+2n} x \sin px \operatorname{tg} x\, dx =$$
$$= \frac{\pi}{2^{p+2n+1}\Gamma(p)} \sum_{k=0}^{\infty} \binom{n}{k} \frac{\Gamma(p+n-k)}{(n-k)!} \quad [p > -2n].$$ BI ((42))(22)

3. $$\int_0^{\frac{\pi}{2}} \cos^{n-1} x \sin[(n+1)x] \operatorname{ctg} x\, dx = \frac{\pi}{2}.$$ BI ((45))(18)

3.636

1. $$\int_0^{\frac{\pi}{2}} \operatorname{tg}^{\pm\mu} x \sin 2x\, dx = \frac{\mu\pi}{2} \operatorname{cosec} \frac{\mu\pi}{2} \quad [0 < \operatorname{Re}\mu < 2].$$ BI ((45))(20)a

2. $$\int_0^{\frac{\pi}{2}} \operatorname{tg}^{\pm\mu} x \cos 2x\, dx = \mp \frac{\mu\pi}{2} \sec \frac{\mu\pi}{2} \quad [|\operatorname{Re}\mu| < 1].$$ BI ((45))(21)

3. $$\int_0^{\frac{\pi}{2}} \frac{\operatorname{tg}^{2\mu} x}{\cos x}\, dx = \int_0^{\frac{\pi}{2}} \frac{\operatorname{ctg}^{2\mu} x}{\sin x}\, dx = \frac{\Gamma\left(\mu+\frac{1}{2}\right)\Gamma(-\mu)}{2\sqrt{\pi}}$$
$$\left[-\frac{1}{2} < \operatorname{Re}\mu < 1\right], \qquad \text{(cf. 3.251 1.).}$$ BI ((45))(13, 14)

3.637

1. $$\int_0^{\frac{\pi}{2}} \operatorname{tg}^p x \sin^{q-2} x \sin qx\, dx = -\cos\frac{(p+q)\pi}{2} \mathrm{B}(p+q-1, 1-p)$$
$$[p+q > 1 > p].$$ GW ((332))(15d)

2. $$\int_0^{\frac{\pi}{2}} \operatorname{tg}^p x \sin^{q-2} x \cos qx\, dx = \sin\frac{(p+q)\pi}{2} \mathrm{B}(p+q-1, 1-p)$$
$$[p+q > 1 > p].$$ GW ((332))(15b)

3. $$\int_0^{\frac{\pi}{2}} \operatorname{ctg}^p x \cos^{q-2} x \sin qx\, dx = \cos\frac{p\pi}{2} \mathrm{B}(p+q-1, 1-p)$$
$$[p+q > 1 > p].$$ GW ((332))(15c)

4. $$\int_0^{\frac{\pi}{2}} \operatorname{ctg}^p x \cos^{q-2} x \cos qx \, dx = \sin \frac{p\pi}{2} \mathrm{B}(p+q-1,\ 1-p) \quad [p+q>1>p].$$ GW ((332))(15a)

3.638

1. $$\int_0^{\frac{\pi}{4}} \frac{\sin^{2\mu} x \, dx}{\cos^{\mu+\frac{1}{2}} 2x \cos x} = \frac{\pi}{2} \sec \mu\pi \quad \left[|\operatorname{Re} \mu| < \frac{1}{2} \right],$$ (cf. **3.192** 2.). BI ((38))(8)

2. $$\int_0^{\frac{\pi}{4}} \frac{\sin^{\mu-\frac{1}{2}} 2x \, dx}{\cos^{\mu} 2x \cos x} = \frac{2}{2\mu-1} \cdot \frac{\Gamma\left(\mu+\frac{1}{2}\right) \Gamma(1-\mu)}{\sqrt{\pi}} \sin\left(\frac{2\mu-1}{4}\pi\right) \quad \left[-\frac{1}{2} < \operatorname{Re} \mu < 1 \right].$$ BI ((38))(17)

3. $$\int_0^{\frac{\pi}{2}} \frac{\cos^{p-1} x \sin px}{\sin x} dx = \frac{\pi}{2} \quad [p>0].$$ GW((332))(17), BI((45))(5)

3.64-3.65 Powers and rational functions of trigonometric functions

3.641

1. $$\int_0^{\frac{\pi}{2}} \frac{\sin^{p-1} x \cos^{-p} x}{a \cos x + b \sin x} dx = \int_0^{\frac{\pi}{2}} \frac{\sin^{-p} x \cos^{p-1} x}{a \sin x + b \cos x} dx = \frac{\pi \operatorname{cosec} p\pi}{a^{1-p} b^p} \quad [ab>0,\ 0<p<1].$$ GW ((331))(62)

2. $$\int_0^{\frac{\pi}{2}} \frac{\sin^{1-p} x \cos^p x}{(\sin x + \cos x)^3} dx = \int_0^{\frac{\pi}{2}} \frac{\sin^p x \cos^{1-p} x}{(\sin x + \cos x)^3} dx = \frac{(1-p)\, p}{2} \pi \operatorname{cosec} p\pi \quad [-1<p<2].$$ BI ((48))(5)

3.642

1. $$\int_0^{\frac{\pi}{2}} \frac{\sin^{2\mu-1} x \cos^{2\nu-1} x \, dx}{(a^2 \sin^2 x + b^2 \cos^2 x)^{\mu+\nu}} = \frac{1}{2a^{2\mu} b^{2\nu}} \mathrm{B}(\mu, \nu) \quad [\operatorname{Re} \mu > 0,\ \operatorname{Re} \nu > 0].$$ BI ((48))(28)

2. $$\int_0^{\frac{\pi}{2}} \frac{\sin^{n-1} x \cos^{n-1} x \, dx}{(a^2 \cos^2 x + b^2 \sin^2 x)^n} = \frac{\mathrm{B}\left(\frac{n}{2}, \frac{n}{2}\right)}{2(ab)^n} \quad [ab>0].$$ GW ((331))(59a)

3. $$\int_0^{\frac{\pi}{2}} \frac{\sin^{2n} x\, dx}{(a^2\cos^2 x + b^2 \sin^2 x)^{n+1}} = \frac{1}{2}\int_0^{\pi} \frac{\sin^{2n} x\, dx}{(a^2\cos^2 x + b^2 \sin^2 x)^{n+1}} =$$
$$= \int_0^{\frac{\pi}{2}} \frac{\cos^{2n} x\, dx}{(a^2\sin^2 x + b^2 \cos^2 x)^{n+1}} = \frac{1}{2}\int_0^{\pi} \frac{\cos^{2n} x\, dx}{(a^2\sin^2 x + b^2 \cos^2 x)^{n+1}} =$$
$$= \frac{(2n-1)!!\,\pi}{2^{n+1} n!\, ab^{2n+1}} \qquad [ab > 0].$$ GW ((331))(58)

4. $$\int_0^{\frac{\pi}{2}} \frac{\cos^{p+2n} x \cos px\, dx}{(a^2\cos^2 x + b^2 \sin^2 x)^{n+1}} = \pi \sum_{k=0}^{n} \binom{2n-k}{n}\binom{p+k-1}{k} \frac{b^{p-1}}{(2a)^{2n-k+1}(a+b)^{p+k}}$$
$$[a > 0,\ b > 0,\ p > -2n-1].$$ GW ((332))(30)

3.643

1. $$\int_0^{\frac{\pi}{2}} \frac{\cos^p x \cos px\, dx}{1 - 2a\cos 2x + a^2} = \frac{\pi}{2^{p+1}} \cdot \frac{(1+a)^{p-1}}{1-a} \qquad [a^2 < 1,\ p > -1].$$ GW ((332))(33c)

2. $$\int_0^{\frac{\pi}{2}} \frac{\sin^{2n} x \cos^{\mu} x \cos \beta x}{(1 - 2a\cos 2x + a^2)^m}\, dx =$$
$$= \frac{(-1)^n \pi (1-a)^{2n-2m+1}}{2^{2m-\beta-1}(1+a)^{2m+\beta+1}} \sum_{k=0}^{m-1} \sum_{l=0}^{m-k-1} \binom{\beta}{k}\binom{2n}{l}\binom{2m-k-l-2}{m-1}(-2)^l (a-1)^k$$
$$[a^2 < 1,\ \beta = 2m - 2n - \mu - 2,\ \mu > -1].$$ GW ((332))(33)

3.644 *

1. $$\int_0^{\pi} \frac{\sin^m x}{p + q\cos x}\, dx =$$
$$= 2^{m-2} \frac{p}{q^2} \sum_{\nu=1}^{k} \left(\frac{p^2-q^2}{-4q^2}\right)^{\nu-1} \mathrm{B}\left(\frac{m+1-2\nu}{2},\ \frac{m+1-2\nu}{2}\right) + \left(\frac{p^2-q^2}{-q^2}\right)^k A;$$
$$A = \frac{\pi p}{q^2}\left(1 - \sqrt{1 - \frac{q^2}{p^2}}\right) \qquad [m = 2k+2];$$
$$A = \frac{1}{q} \ln \frac{p+q}{p-q} \qquad [m = 2k+1]$$
$$[k \geqslant 1,\ q \neq 0,\ p^2 - q^2 \geqslant 0].$$

2. $$\int_0^{\pi} \frac{\sin^m x}{1 + \cos x}\, dx = 2^{m-1} \mathrm{B}\left(\frac{m-1}{2},\ \frac{m+1}{2}\right) \qquad [m \geqslant 2].$$

---

* The integrals 3.644 appear in the article by K. V. Brodovitskiy "Ob integrale $\int_0^{\pi} \frac{\sin^m x}{p+q\cos x}\, dx$". (On the integral $\int_0^{\pi} \frac{\sin^m x}{p+q\cos x}\, dx$), *Doklady Akad. nauk*, 120, No. 6 (1958).

3. $$\int_0^{\pi} \frac{\sin^m x}{1-\cos x}\,dx = 2^{m-1}\mathrm{B}\left(\frac{m-1}{2}, \frac{m+1}{2}\right) \quad [m \geqslant 2].$$

4. $$\int_0^{\pi} \frac{\sin^2 x}{p+q\cos x}\,dx = \frac{p\pi}{q^2}\left(1-\sqrt{1-\frac{q^2}{p^2}}\right).$$

5. $$\int_0^{\pi} \frac{\sin^3 x}{p+q\cos x}\,dx = 2\frac{p}{q^2}+\frac{1}{q}\left(1-\frac{p^2}{q^2}\right)\ln\frac{p+q}{p-q}.$$

3.645 $$\int_0^{\pi} \frac{\cos^n x\,dx}{(a+b\cos x)^{n+1}} = \frac{\pi}{2^n(a+b)^n\sqrt{a^2-b^2}}\times$$
$$\times\sum_{k=0}^{n}(-1)^k\frac{(2n-2k-1)!!\,(2k-1)!!}{(n-k)!\,k!}\left(\frac{a+b}{a-b}\right)^k \quad [a^2>b^2].$$ LI ((64))(16)

3.646

1. $$\int_0^{\frac{\pi}{2}} \frac{\cos^n x\sin nx\sin 2x}{1-2a\cos 2x+a^2}\,dx = \frac{\pi}{4a}\left[\left(\frac{1+a}{2}\right)^n-\frac{1}{2^n}\right] \quad [a^2<1].$$ BI ((50))(6)

2. $$\int_0^{\frac{\pi}{2}} \frac{1-a\cos 2nx}{1-2a\cos 2nx+a^2}\cos^m x\cos mx\,dx =$$
$$= \frac{\pi}{2^{m+2}}\sum_{k=1}^{\infty}\binom{m}{kn}a^k+\frac{\pi}{2^{m+1}} \quad [a^2<1].$$ LI ((50))(7)

3.647

$$\int_0^{\frac{\pi}{2}} \frac{\cos^p x\cos px\,dx}{a^2\sin^2 x+b^2\cos^2 x} = \frac{\pi}{2b}\cdot\frac{a^{p-1}}{(a+b)^p} \quad [p>-1,\ a>0,\ b>0].$$ BI ((47))(20)

3.648

1. $$\int_0^{\frac{\pi}{4}} \frac{\operatorname{tg}^l x\,dx}{1+\cos\frac{m}{n}\pi\sin 2x} = \frac{1}{2n}\operatorname{cosec}\frac{m}{n}\pi\sum_{k=0}^{n-1}(-1)^{k-1}\sin\frac{km}{n}\pi\times$$
$$\times\left[\psi\left(\frac{n+l+k}{2n}\right)-\psi\left(\frac{l+k}{2n}\right)\right] \quad [m+n \text{ is odd}];$$
$$= \frac{1}{n}\operatorname{cosec}\frac{m}{n}\pi\sum_{k=0}^{\frac{n-1}{2}}(-1)^{k-1}\sin\frac{km}{n}\pi\times$$
$$\times\left[\psi\left(\frac{n+l-k}{n}\right)-\psi\left(\frac{l+k}{n}\right)\right] \quad [m+n \text{ is even}]$$
$[l$ is a natural number$]$. BI ((36))(5)

2. $\int_0^{\frac{\pi}{2}} \frac{\operatorname{tg}^{\pm\mu} x \, dx}{1+\cos t \sin 2x} = \pi \operatorname{cosec} t \sin \mu t \operatorname{cosec}(\mu\pi) \quad [|\operatorname{Re}\mu| < 1, \; t^2 < \pi^2].$

BI ((47))(4)

3.649

1. $\int_0^{\frac{\pi}{2}} \frac{\operatorname{tg}^{\pm\mu} x \sin 2x \, dx}{1 \mp 2a\cos 2x + a^2} = \frac{\pi}{4a} \operatorname{cosec} \frac{\mu\pi}{2} \left[1 - \left(\frac{1-a}{1+a}\right)^{\mu}\right] \quad [a^2 < 1];$

$= \frac{\pi}{4a} \operatorname{cosec} \frac{\mu\pi}{2} \left[1 + \left(\frac{a-1}{a+1}\right)^{\mu}\right] \quad [a^2 > 1]$

$[-2 < \operatorname{Re}\mu < 1].$ BI ((50))(3)

2. $\int_0^{\frac{\pi}{2}} \frac{\operatorname{tg}^{\pm\mu} x (1 \mp a\cos 2x)}{1 \mp 2a\cos 2x + a^2} dx = \frac{\pi}{4} \sec \frac{\mu\pi}{2} \left[1 + \left(\frac{1-a}{1+a}\right)^{\mu}\right] \quad [a^2 < 1];$

$= \frac{\pi}{4} \sec \frac{\mu\pi}{2} \left[1 - \left(\frac{a-1}{a+1}\right)^{\mu}\right] \quad [a^2 > 1]$

$[|\operatorname{Re}\mu| < 1].$ BI ((50))(4)

3.651

1. $\int_0^{\frac{\pi}{4}} \frac{\operatorname{tg}^{\mu} x \, dx}{1+\sin x \cos x} = \frac{1}{3}\left[\psi\left(\frac{\mu+2}{3}\right) - \psi\left(\frac{\mu+1}{3}\right)\right] \quad [\operatorname{Re}\mu > -1].$ BI ((36))(3)

2. $\int_0^{\frac{\pi}{4}} \frac{\operatorname{tg}^{\mu} x \, dx}{1-\sin x \cos x} = \frac{1}{3}\left[\beta\left(\frac{\mu+2}{3}\right) + \beta\left(\frac{\mu+1}{3}\right)\right] \quad [\operatorname{Re}\mu > -1].$

BI ((36))(4)a

3.652

1. $\int_0^{\frac{\pi}{2}} \frac{\operatorname{tg}^{\mu} x \, dx}{(\sin x + \cos x)\sin x} = \int_0^{\frac{\pi}{2}} \frac{\operatorname{ctg}^{\mu} x \, dx}{(\sin x + \cos x)\cos x} = \pi \operatorname{cosec} \mu\pi \quad [0 < \operatorname{Re}\mu < 1].$

BI ((49))(1)

2. $\int_0^{\frac{\pi}{2}} \frac{\operatorname{tg}^{\mu} x \, dx}{(\sin x - \cos x)\sin x} = \int_0^{\frac{\pi}{2}} \frac{\operatorname{ctg}^{\mu} x \, dx}{(\cos x - \sin x)\cos x} = -\pi \operatorname{ctg} \mu\pi \quad [0 < \operatorname{Re}\mu < 1]$

BI ((49))(2)

3. $\int_0^{\frac{\pi}{2}} \frac{\operatorname{ctg}^{\mu+\frac{1}{2}} x \, dx}{(\sin x + \cos x)\cos x} = \int_0^{\frac{\pi}{2}} \frac{\operatorname{tg}^{\mu-\frac{1}{2}} x \, dx}{(\sin x + \cos x)\cos x} = \pi \sec \mu\pi \quad \left[|\operatorname{Re}\mu| < \frac{1}{2}\right].$

BI ((61))(1, 2)

3.653

1. $$\int_0^{\frac{\pi}{2}} \frac{\operatorname{tg}^{1-2\mu} x\, dx}{a^2\cos^2 x + b^2 \sin^2 x} = \int_0^{\frac{\pi}{2}} \frac{\operatorname{ctg}^{1-2\mu} x\, dx}{a^2 \sin^2 x + b^2\cos^2 x} = \frac{\pi}{2a^{2\mu} b^{2-2\mu}\sin \mu\pi}$$
$[0 < \operatorname{Re}\mu < 1]$. GW ((331))(59b)

2. $$\int_0^{\frac{\pi}{2}} \frac{\operatorname{tg}^{\mu} x\, dx}{1 - a\sin^2 x} = \int_0^{\frac{\pi}{2}} \frac{\operatorname{ctg}^{\mu} x\, dx}{1 - a\cos^2 x} = \frac{\pi \sec \frac{\mu\pi}{2}}{2\sqrt{(1-a)^{\mu+1}}} \qquad [|\operatorname{Re}\mu| < 1,\ a < 1].$$
BI ((49))(6)

3. $$\int_0^{\frac{\pi}{2}} \frac{\operatorname{tg}^{\pm\mu} x\, dx}{1 - \cos^2 t \sin^2 2x} = \frac{\pi}{2}\operatorname{cosec} t \sec\frac{\mu\pi}{2} \cos\left[\left(\frac{\pi}{2} - t\right)\mu\right]$$
$[|\operatorname{Re}\mu| < 1,\ t^2 < \pi^2]$. BI((49))(7), BI((47))(21)

4. $$\int_0^{\frac{\pi}{2}} \frac{\operatorname{tg}^{\pm\mu} x \sin 2x}{1 - \cos^2 t \sin^2 2x}\, dx = \pi \operatorname{cosec} 2t \operatorname{cosec}\frac{\mu\pi}{2} \sin\left[\left(\frac{\pi}{2} - t\right)\mu\right]$$
$[|\operatorname{Re}\mu| < 1,\ t^2 < \pi^2]$. BI ((47))(22)a

5. $$\int_0^{\frac{\pi}{2}} \frac{\operatorname{tg}^{\mu} x \sin^2 x\, dx}{1 - \cos^2 t \sin^2 2x} = \int_0^{\frac{\pi}{2}} \frac{\operatorname{ctg}^{\mu} x \cos^2 x\, dx}{1 - \cos^2 t \sin^2 2x} =$$
$$= \frac{\pi}{2} \operatorname{cosec} 2t \sec\frac{\mu\pi}{2} \cos\left[\frac{\mu\pi}{2} - (\mu+1)t\right] \qquad [|\operatorname{Re}\mu| < 1\quad t^2 < \pi^2].$$
BI((47))(23)a, BI((49))(10)

6. $$\int_0^{\frac{\pi}{2}} \frac{\operatorname{tg}^{\mu} x \cos^2 x\, dx}{1 - \cos^2 t \sin^2 2x} = \int_0^{\frac{\pi}{2}} \frac{\operatorname{ctg}^{\mu} x \sin^2 x\, dx}{1 - \cos^2 t \sin^2 2x} =$$
$$= \frac{\pi}{2} \operatorname{cosec} 2t \sec\frac{\mu\pi}{2} \cos\left[\frac{\mu\pi}{2} - (\mu-1)t\right] \qquad [|\operatorname{Re}\mu| < 1,\ t^2 < \pi^2].$$
BI((47))(24)a, BI((49))(9)

3.654

1. $$\int_0^{\frac{\pi}{2}} \frac{\operatorname{tg}^{\mu+1} x \cos^2 x\, dx}{(1+\cos t \sin 2x)^2} = \int_0^{\frac{\pi}{2}} \frac{\operatorname{ctg}^{\mu+1} x \sin^2 x\, dx}{(1+\cos t\sin 2x)^2} = \frac{\pi(\mu \sin t \cos \mu t - \cos t \sin \mu t)}{2\sin\mu\pi \sin^3 t}$$
$[|\operatorname{Re}\mu| < 1,\ t^2 < \pi^2]$. BI((48))(3), BI((49))(22)

2. $$\int_0^{\frac{\pi}{2}} \frac{\operatorname{tg}^{\pm\mu} x\, dx}{(\sin x + \cos x)^2} = \frac{\mu\pi}{\sin\mu\pi} \qquad [0 < \operatorname{Re}\mu < 1].$$
BI ((56))(9)a

3. $$\int_0^{\frac{\pi}{2}} \frac{\operatorname{tg}^{\pm(\mu-1)} x\, dx}{\cos^2 x - \sin^2 x} = \pm\frac{\pi}{2}\operatorname{ctg}\frac{\mu\pi}{2} \qquad [0 < \operatorname{Re}\mu < 2].$$
BI ((45))(27, 29)

3.655 $$\int_0^{\frac{\pi}{2}} \frac{\operatorname{tg}^{2\mu-1} x \, dx}{1-2a(\cos t_1 \sin^2 x + \cos t_2 \cos^2 x) + a^2} =$$

$$= \int_0^{\frac{\pi}{2}} \frac{\operatorname{ctg}^{2\mu-1} x \, dx}{1-2a(\cos t_1 \cos^2 x + \cos t_2 \sin^2 x) + a^2} =$$

$$= \frac{\pi \operatorname{cosec} \mu\pi}{(1-2a\cos t_2 + a^2)^{\mu} (1-2a \cos t_1 + a^2)^{1-\mu}}$$

$[0 < \operatorname{Re}\mu < 1, \quad t_1^2 < \pi^2, \quad t_2^2 < \pi^2]$. BI ((50))(18)

3.656

1. $$\int_0^{\frac{\pi}{4}} \frac{\operatorname{tg}^{\mu} x \, dx}{1-\sin^2 x \cos^2 x} = \frac{1}{12}\left\{-\psi\left(\frac{\mu+1}{6}\right) - \psi\left(\frac{\mu+2}{6}\right) + \right.$$

$$\left. + \psi\left(\frac{\mu+4}{6}\right) + \psi\left(\frac{\mu+5}{6}\right) + 2\psi\left(\frac{\mu+2}{3}\right) - 2\psi\left(\frac{\mu+1}{3}\right)\right\}$$

$[\operatorname{Re}\mu > -1]$, (cf. 3.651 1. and 2.). LI ((36))(10)

2. $$\int_0^{\frac{\pi}{2}} \frac{\operatorname{tg}^{\mu-1} x \cos^2 x \, dx}{1-\sin^2 x \cos^2 x} = \int_0^{\frac{\pi}{2}} \frac{\operatorname{ctg}^{\mu-1} x \sin^2 x \, dx}{1-\sin^2 x \cos^2 x} =$$

$$= \frac{\pi}{4\sqrt{3}} \operatorname{cosec} \frac{\mu\pi}{6} \operatorname{cosec}\left(\frac{2+\mu}{6}\pi\right) \quad [0 < \operatorname{Re}\mu < 4].$$ LI ((47))(26)

3.66 Forms containing powers of linear functions of trigonometric functions

3.661

1. $$\int_0^{2\pi} (a \sin x + b \cos x)^{2n+1} \, dx = 0.$$ BI ((68))(9)

2. $$\int_0^{2\pi} (a \sin x + b \cos x)^{2n} \, dx = \frac{(2n-1)!!}{(2n)!!} \cdot 2\pi (a^2+b^2)^n.$$ BI ((68))(8)

3. $$\int_0^{\pi} (a + b \cos x)^n \, dx = \frac{1}{2} \int_0^{2\pi} (a + b\cos x)^n \, dx =$$

$$= \pi (a^2-b^2)^{\frac{n}{2}} P_n\left(\frac{a}{\sqrt{a^2-b^2}}\right) =$$

$$= \frac{\pi}{2^n} \sum_{k=0}^{E\left(\frac{n}{2}\right)} \frac{(-1)^k (2n-2k)!}{k!\,(n-k)!\,(n-2k)!} a^{n-2k} (a^2-b^2)^k \quad [a^2 > b^2].$$ GW ((332))(37a)

4. $$\int_0^{\pi} \frac{dx}{(a+b\cos x)^{n+1}} = \frac{1}{2}\int_0^{2\pi} \frac{dx}{(a+b\cos x)^{n+1}} =$$

$$= \frac{\pi}{(a^2-b^2)^{\frac{n+1}{2}}} P_n\left(\frac{a}{\sqrt{a^2-b^2}}\right) =$$

$$= \frac{\pi}{2^n (a+b)^n \sqrt{a^2-b^2}} \sum_{k=0}^{n} \frac{(2n-2k-1)!!\,(2k-1)!!}{(n-k)!\,k!} \cdot \left(\frac{a+b}{a-b}\right)^k$$

$[a > |b|]$. GW((332))(38), LI((64))(14)

3.662

1. $$\int_0^{\frac{\pi}{2}} (\sec x - 1)^{\mu} \sin x \, dx = \int_0^{\frac{\pi}{2}} (\operatorname{cosec} x - 1)^{\mu} \cos x \, dx =$$

$$= \mu\pi \operatorname{cosec} \mu\pi \quad [|\operatorname{Re}\mu| < 1].$$ BI ((55))(13)

2. $$\int_0^{\frac{\pi}{2}} (\operatorname{cosec} x - 1)^{\mu} \sin 2x \, dx = (1-\mu)\,\mu\pi \operatorname{cosec} \mu\pi$$

$[-1 < \operatorname{Re}\mu < 2]$. BI ((48))(7)

3. $$\int_0^{\frac{\pi}{2}} (\sec x - 1)^{\mu} \operatorname{tg} x \, dx = \int_0^{\frac{\pi}{2}} (\operatorname{cosec} x - 1)^{\mu} \operatorname{ctg} x \, dx = -\pi \operatorname{cosec} \mu\pi$$

$[-1 < \operatorname{Re}\mu < 0]$, (cf. **3.192** 2.). BI ((46))(4, 6)

4. $$\int_0^{\frac{\pi}{4}} (\operatorname{ctg} x - 1)^{\mu} \frac{dx}{\sin 2x} = -\frac{\pi}{2} \operatorname{cosec} \mu\pi \quad [-1 < \operatorname{Re}\mu < 0].$$ BI ((38))(22)a

5. $$\int_0^{\frac{\pi}{4}} (\operatorname{ctg} x - 1)^{\mu} \frac{dx}{\cos^2 x} = \mu\pi \operatorname{cosec} \mu\pi \quad [|\operatorname{Re}\mu| < 1].$$ BI ((38))(11)a

3.663

1. $$\int_0^{u} (\cos x - \cos u)^{\nu - \frac{1}{2}} \cos ax \, dx = \sqrt{\frac{\pi}{2}} \sin^{\nu} u \, \Gamma\left(\nu + \frac{1}{2}\right) P_{a-\frac{1}{2}}^{-\nu}(\cos u)$$

$\left[\operatorname{Re}\nu > -\frac{1}{2};\ a > 0,\ 0 < u < \pi\right]$. EH I 159(27), ET I 22(28)

2. $$\int_0^{u} (\cos x - \cos u)^{\nu - 1} \cos[(\nu+\beta)x] \, dx =$$

$$= \frac{\sqrt{\pi}\,\Gamma(\beta+1)\,\Gamma(\nu)\,\Gamma(2\nu)\sin^{2\nu-1} u}{2^{\nu}\,\Gamma(\beta+2\nu)\,\Gamma\left(\nu+\frac{1}{2}\right)} C_{\beta}^{\nu}(\cos u)$$

$[\operatorname{Re}\nu > 0,\quad \operatorname{Re}\beta > -1,\quad 0 < u < \pi]$. EH I 178(23)

3.664

1. $$\int_0^\pi (z+\sqrt{z^2-1}\cos x)^q\,dx = \pi P_q(z)$$

$\left[\operatorname{Re} z>0, \quad \arg(z+\sqrt{z^2-1}\cos x)=\arg z \text{ for } x=\frac{\pi}{2}\right].$ SM 482

2. $$\int_0^\pi \frac{dx}{(z+\sqrt{z^2-1}\cos x)^q} = \pi P_{q-1}(z)$$

$\left[\operatorname{Re} z>0, \quad \arg(z+\sqrt{z^2-1}\cos x)=\arg z \quad \text{for} \quad x=\frac{\pi}{2}\right].$ WH

3. $$\int_0^\pi (z+\sqrt{z^2-1}\cos x)^q \cos nx\,dx = \frac{\pi}{(q+1)(q+2)\dots(q+n)} P_q^n(z)\ .$$

$[\operatorname{Re} z>0, \quad \arg(z+\sqrt{z^2-1}\cos x)=\arg z \text{ for } x=\frac{\pi}{2}$, z lies outside the interval $(-1, 1)$ of the real axis].

WH, SM 483(15)

4. $$\int_0^\pi (z+\sqrt{z^2-1}\cos x)^\mu \sin^{2\nu-1} x\,dx = \frac{2^{2\nu-1}\Gamma(\mu+1)[\Gamma(\nu)]^2}{\Gamma(2\nu+\mu)} C_\mu^\nu(z) =$$

$$= \frac{\sqrt{\pi}\,\Gamma(\nu)\,\Gamma(2\nu)\Gamma(\mu+1)}{\Gamma(2\nu+\mu)\,\Gamma\left(\nu+\frac{1}{2}\right)} C_\mu^\nu(z) = 2^\nu \sqrt{\frac{\pi}{2}}\,(z^2-1)^{\frac{1}{4}-\frac{\nu}{2}}\,\Gamma(\nu)\,P_{\mu+\nu-\frac{1}{2}}^{\frac{1}{2}-\nu}(z)$$

$[\operatorname{Re}\nu>0].$ EH I 155(6)a, EH I 178(22)

5. $$\int_0^{2\pi} [\beta+\sqrt{\beta^2-1}\cos(a-x)]^\nu (\gamma+\sqrt{\gamma^2-1}\cos x)^{\nu-1}\,dx =$$

$$= 2\pi P_\nu[\beta\gamma-\sqrt{\beta^2-1}\sqrt{\gamma^2-1}\cos a] \quad [\operatorname{Re}\beta>0, \ \operatorname{Re}\gamma>0].$$ EH I 157(18)

3.665

1. $$\int_0^\pi \frac{\sin^{\mu-1} x\,dx}{(a+b\cos x)^\mu} = \frac{2^{\mu-1}}{\sqrt{(a^2-b^2)^\mu}} \mathrm{B}\left(\frac{\mu}{2}, \frac{\mu}{2}\right)$$

$[\operatorname{Re}\mu>0, \quad 0<b<a].$ FI II 790a

2. $$\int_0^\pi \frac{\sin^{2\mu-1} x\,dx}{(1+2a\cos x+a^2)^\nu} = \mathrm{B}\left(\mu, \frac{1}{2}\right) F\left(\nu, \nu-\mu+\frac{1}{2}; \mu+\frac{1}{2}; a^2\right)$$

$[\operatorname{Re}\mu>0, \quad |a|<1].$ EH I 81(9)

3.666

1. $$\int_0^\pi (\beta+\cos x)^{\mu-\nu-\frac{1}{2}} \sin^{2\nu} x\,dx =$$

$$= \frac{2^{\nu+\frac{1}{2}} e^{-i\mu\pi} (z^2-1)^{\frac{\mu}{2}}\,\Gamma\left(\nu+\frac{1}{2}\right) Q_{\nu-\frac{1}{2}}^{\mu}(\beta)}{\Gamma\left(\nu+\mu+\frac{1}{2}\right)}$$

$\left[\operatorname{Re}\left(\nu+\mu+\frac{1}{2}\right)>0, \quad \operatorname{Re}\nu>-\frac{1}{2}\right].$ EH I 155(5)a

2. $$\int_0^{\frac{\pi}{2}} (\operatorname{ch}\beta + \operatorname{sh}\beta\cos x)^{\mu+\nu}\sin^{-2\nu} x\,dx = \frac{\sqrt{\pi}}{2^\nu}\operatorname{sh}^\nu(\beta)\,\Gamma\left(\frac{1}{2}-\nu\right)P_\mu^\nu(\operatorname{ch}\beta) \quad \left[\operatorname{Re}\nu < \frac{1}{2}\right].$$ EH I 156(7)

3. $$\int_0^{\pi} (\cos t + i\sin t\cos x)^{\mu}\sin^{2\nu-1} x\,dx = 2^{\nu-\frac{1}{2}}\sqrt{\pi}\sin^{\frac{1}{2}-\nu} t\,\Gamma(\nu)\,\mathrm{P}^{\frac{1}{2}-\nu}_{\mu+\nu-\frac{1}{2}}(\cos t) \quad [\operatorname{Re}\nu > 0,\ t^2 < \pi^2].$$ EH I 158(23)

4. $$\int_0^{2\pi} [\cos t + i\sin t\cos(a-x)]^{\nu}\cos mx\,dx = \frac{i^{3m}2\pi\Gamma(\nu+1)}{\Gamma(\nu+m+1)}\cos ma\mathrm{P}_\nu^m(\cos t) \quad \left[0 < t < \frac{\pi}{2}\right].$$ EH I 159(25)

5. $$\int_0^{2\pi} [\cos t + i\sin t\cos(a-x)]^{\nu}\sin mx\,dx = \frac{i^{3m}2\pi\Gamma(\nu+1)}{\Gamma(\nu+m+1)}\sin ma\mathrm{P}_\nu^m(\cos t) \quad \left[0 < t < \frac{\pi}{2}\right].$$ EH I 159(26)

3.667

1. $$\int_0^{\frac{\pi}{4}} \frac{\sin^{\mu-1} 2x\,dx}{(\cos x + \sin x)^{2\mu}} = \frac{\sqrt{\pi}}{2^{\mu+1}}\frac{\Gamma(\mu)}{\Gamma\left(\mu+\frac{1}{2}\right)} \quad [\operatorname{Re}\mu > 0].$$ BI ((37))(1)

2. $$\int_0^{\frac{\pi}{4}} \frac{\sin^{\mu} x\,dx}{(\cos x - \sin x)^{\mu+1}\cos x} = -\pi\operatorname{cosec}\mu\pi \quad [-1 < \operatorname{Re}\mu < 0],$$ (cf. 3.192 2.). BI ((37))(16)

3. $$\int_0^{\frac{\pi}{4}} \frac{(\cos x - \sin x)^{\mu}}{\sin^{\mu} x\sin 2x}\,dx = -\frac{\pi}{2}\operatorname{cosec}\mu\pi \quad [-1 < \operatorname{Re}\mu < 0].$$ BI ((35))(27)

4. $$\int_0^{\frac{\pi}{4}} \frac{\sin^{\mu} x\,dx}{(\cos x - \sin x)^{\mu}\sin 2x} = \frac{\pi}{2}\operatorname{cosec}\mu\pi \quad [0 < \operatorname{Re}\mu < 1].$$ LI ((37))(20)a

5. $$\int_0^{\frac{\pi}{4}} \frac{\sin^{\mu} x\,dx}{(\cos x - \sin x)^{\mu}\cos^2 x} = \mu\pi\operatorname{cosec}\mu\pi \quad [|\operatorname{Re}\mu| < 1].$$ BI ((37))(17)

6. $$\int_0^{\frac{\pi}{4}} \frac{\sin^{\mu} x\,dx}{(\cos x - \sin x)^{\mu-1}\cos^3 x} = \frac{1-\mu}{2}\mu\pi\operatorname{cosec}\mu\pi \quad [|\operatorname{Re}\mu| < 1].$$ BI((35))(24), BI((37))(18)

7. $$\int_0^{\frac{\pi}{2}} \frac{\sin^{\mu-1} x \cos^{\nu-1} x}{(\sin x+\cos x)^{\mu+\nu}}\, dx = \mathrm{B}(\mu, \nu) \quad [\operatorname{Re}\mu > 0, \quad \operatorname{Re}\nu > 0].$$

BI ((48))(8)

3.668

1. $$\int_{-\frac{\pi}{4}}^{\frac{\pi}{4}} \left(\frac{\cos x+\sin x}{\cos x-\sin x}\right)^{\cos 2t} dx = \frac{\pi}{2\sin(\pi\cos^2 t)}.$$

FI II 788

2. $$\int_u^v \frac{(\cos u-\cos x)^{\mu-1}}{(\cos x-\cos v)^{\mu}} \cdot \frac{\sin x\, dx}{1-2a\cos x+a^2} =$$
$$= \frac{(1-2a\cos u+a^2)^{\mu-1}}{(1-2a\cos v+a^2)^{\mu}} \cdot \frac{\pi}{\sin\mu\pi} \qquad [0 < \operatorname{Re}\mu < 1,\ a^2 < 1].$$

BI ((73))(2)

3.669 $$\int_0^{\frac{\pi}{2}} \frac{\sin^{p-1} x \cos^{q-p-1} x\, dx}{(a\cos x+b\sin x)^q} =$$
$$= \int_0^{\frac{\pi}{2}} \frac{\sin^{q-p-1} x \cos^{p-1} x}{(a\sin x+b\cos x)^q}\, dx = \frac{\mathrm{B}(p, q-p)}{a^{q-p}b^p} \qquad [q > p > 0, \quad ab > 0].$$

BI ((331))(90)

3.67 Square roots of expressions containing trigonometric functions

3.671

1. $$\int_0^{\frac{\pi}{2}} \sin^\alpha x \cos^\beta x \sqrt{1-k^2\sin^2 x}\, dx =$$
$$= \frac{1}{2}\mathrm{B}\left(\frac{\alpha+1}{2}, \frac{\beta+1}{2}\right) F\left(\frac{\alpha+1}{2}, -\frac{1}{2}; \frac{\alpha+\beta+2}{2}; k^2\right)$$
$$[\alpha > -1,\ \beta > -1,\ |k| < 1].$$

GW ((331))(93)

2. $$\int_0^{\frac{\pi}{2}} \frac{\sin^\alpha x \cos^\beta x}{\sqrt{1-k^2\sin^2 x}}\, dx = \frac{1}{2}\mathrm{B}\left(\frac{\alpha+1}{2}, \frac{\beta+1}{2}\right) F\left(\frac{\alpha+1}{2}, \frac{1}{2}; \frac{\alpha+\beta+2}{2}; k^2\right)$$
$$[\alpha > -1,\ \beta > -1,\ |k| < 1].$$

GW ((331))(92)

3. $$\int_0^{\pi} \frac{\sin^{2n} x\, dx}{\sqrt{1-k^2\sin^2 x}} = \frac{\pi}{2^n} \sum_{j=0}^{\infty} \frac{(2j-1)!!\,(2n+2j-1)!!}{2^{2j} j!\,(n+j)!} k^{2j} \quad [k^2 < 1];$$
$$= \frac{(2n-1)!!\,\pi}{2^n\sqrt{1-k^2}} \sum_{j=0}^{\infty} \frac{[(2j-1)!!]^2}{2^{2j} j!\,(n+j)!} \left(\frac{k^2}{k^2-1}\right)^j \left[k^2 < \frac{1}{2}\right].$$

LI ((67))(2)

3.672

1. $$\int_0^{\frac{\pi}{4}} \frac{\sin^n x}{\cos^{n+1} x} \cdot \frac{dx}{\sqrt{\cos x\,(\cos x - \sin x)}} = 2 \cdot \frac{(2n)!!}{(2n+1)!!}.$$ BI ((39))(5)

2. $$\int_0^{\frac{\pi}{4}} \frac{\sin^n x}{\cos^{n+1} x} \cdot \frac{dx}{\sqrt{\sin x\,(\cos x - \sin x)}} = \frac{(2n-1)!!}{(2n)!!}\pi.$$ BI ((39))(6)

3.673 $$\int_u^{\frac{\pi}{2}} \frac{dx}{\sqrt{\sin x - \sin u}} = \sqrt{2}\,\boldsymbol{K}\left(\sin \frac{\pi - 2u}{4}\right).$$ BI ((74))(11)

3.674

1. $$\int_0^{\pi} \frac{dx}{\sqrt{1 \pm 2p\cos x + p^2}} = 2\boldsymbol{K}(p) \qquad [p^2 < 1].$$ BI ((67))(5)

2. $$\int_0^{\pi} \frac{\sin x\,dx}{\sqrt{1 - 2p\cos x + p^2}} = 2 \qquad [p^2 \leqslant 1];$$
$$= \frac{2}{p} \qquad [p^2 \geqslant 1].$$ BI ((67))(6)

3. $$\int_0^{\pi} \frac{\cos x\,dx}{\sqrt{1 - 2p\cos x + p^2}} = \frac{2}{p}[\boldsymbol{K}(p) - \boldsymbol{E}(p)] \qquad [p^2 < 1].$$ BI ((67))(7)

3.675

1. $$\int_u^{\pi} \frac{\sin\left(n + \frac{1}{2}\right)x\,dx}{\sqrt{2(\cos u - \cos x)}} = \frac{\pi}{2} P_n(\cos u).$$ WH

2. $$\int_0^{u} \frac{\cos\left(n + \frac{1}{2}\right)x\,dx}{\sqrt{2(\cos x - \cos u)}} = \frac{\pi}{2} P_n(\cos u).$$ FI II 684, WH

3.676

1. $$\int_0^{\frac{\pi}{2}} \frac{\sin x\,dx}{\sqrt{1 + p^2\sin^2 x}} = \frac{1}{p}\operatorname{arctg} p.$$ BI ((60))(5)

2. $$\int_0^{\frac{\pi}{2}} \operatorname{tg}^2 x \sqrt{1 - p^2 \sin^2 x}\,dx = \infty.$$ BI ((53))(8)

3. $$\int_0^{\frac{\pi}{2}} \frac{dx}{\sqrt{p^2\cos^2 x + q^2\sin^2 x}} = \frac{1}{p}\boldsymbol{K}\left(\frac{\sqrt{p^2 - q^2}}{p}\right) \qquad [0 < q < p].$$ FI II 165

3.677

1. $$\int_0^{\frac{\pi}{2}} \frac{\sin^2 x\,dx}{\sqrt{1+\sin^2 x}} = \sqrt{2}\,\boldsymbol{E}\left(\frac{\sqrt{2}}{2}\right) - \frac{1}{\sqrt{2}}\boldsymbol{K}\left(\frac{\sqrt{2}}{2}\right).$$ BI ((60))(2)

2. $$\int_0^{\frac{\pi}{2}} \frac{\cos^2 x\,dx}{\sqrt{1+\sin^2 x}} = \sqrt{2}\left[\boldsymbol{K}\left(\frac{\sqrt{2}}{2}\right) - \boldsymbol{E}\left(\frac{\sqrt{2}}{2}\right)\right].$$ BI ((60))(3)

3.678

1. $$\int_0^{\frac{\pi}{4}} \left(\sec^{\frac{1}{2}} 2x - 1\right)\frac{dx}{\operatorname{tg} x} = \ln 2.$$ BI ((38))(23)

2. $$\int_0^{\frac{\pi}{4}} \frac{\operatorname{tg}^2 x\,dx}{\sqrt{1-k^2\sin^2 2x}} = \sqrt{1-k^2} - \boldsymbol{E}(k) + \frac{1}{2}\boldsymbol{K}(k).$$ BI ((39))(2)

3. $$\int_0^{u} \sqrt{\frac{\cos 2x - \cos 2u}{\cos 2x + 1}}\,dx = \frac{\pi}{2}(1-\cos u) \qquad \left[u^2 < \frac{\pi^2}{4}\right].$$ LI ((74))(6)

4. $$\int_0^{\frac{\pi}{4}} \frac{(\cos x - \sin x)^{n-\frac{1}{2}}}{\cos^{n+1} x}\sqrt{\operatorname{cosec} x}\,dx = \frac{(2n-1)!!}{(2n)!!}\pi.$$ BI ((38))(24)

5. $$\int_0^{\frac{\pi}{4}} \frac{(\cos x - \sin x)^{n-\frac{1}{2}}}{\cos^{n+1} x}\operatorname{tg}^m x\sqrt{\operatorname{cosec} x}\,dx = \frac{(2n-1)!!\,(2m-1)!!}{(2n+2m)!!}\pi.$$

BI ((38))(25)

3.679

1. $$\int_0^{\frac{\pi}{2}} \frac{\cos^2 x}{1-\cos^2\beta\cos^2 x}\cdot\frac{dx}{\sqrt{1-k^2\sin^2 x}} =$$
$$= \frac{1}{\sin\beta\cos\beta\sqrt{1-k'^2\sin^2\beta}}\left\{\frac{\pi}{2} - \boldsymbol{K}E(\beta, k') - \boldsymbol{E}F(\beta, k') + \boldsymbol{K}F(\beta, k')\right\}.$$

MO 138

2. $$\int_0^{\frac{\pi}{2}} \frac{\sin^2 x}{1-(1-k'^2\sin^2\beta)\sin^2 x}\cdot\frac{dx}{\sqrt{1-k^2\sin^2 x}} =$$
$$= \frac{1}{k'^2\sin\beta\cos\beta\sqrt{1-k'^2\sin^2\beta}}\left\{\frac{\pi}{2} - \boldsymbol{K}E(\beta, k') - \boldsymbol{E}F(\beta, k') + \boldsymbol{K}F(\beta, k')\right\}.$$

MO 138

3. $$\int_0^{\frac{\pi}{2}} \frac{\sin^2 x}{1-k^2\sin^2\beta\sin^2 x}\cdot\frac{dx}{\sqrt{1-k^2\sin^2 x}} =$$
$$= \frac{\boldsymbol{K}E(\beta, k) - \boldsymbol{E}F(\beta, k)}{k^2\sin\beta\cos\beta\sqrt{1-k^2\sin^2\beta}}.$$ MO 138

3.68 Various forms of powers of trigonometric functions

3.681

1. $$\int_0^{\frac{\pi}{2}} \frac{\sin^{2\mu-1}x\cos^{2\nu-1}x\,dx}{(1-k^2\sin^2 x)^{\varrho}} = \frac{1}{2}\,\mathrm{B}(\mu, \nu)F(\varrho, \mu; \mu+\nu; k^2)$$
$[\operatorname{Re}\mu > 0, \quad \operatorname{Re}\nu > 0]$. EH I 115(7)

2. $$\int_0^{\frac{\pi}{2}} \frac{\sin^{2\mu-1}x\cos^{2\nu-1}x\,dx}{(1-k^2\sin^2 x)^{\mu+\nu}} = \frac{\mathrm{B}(\mu, \nu)}{2(1-k^2)^{\mu}} \qquad [\operatorname{Re}\mu > 0, \quad \operatorname{Re}\nu > 0].$$
EH I 10(20)

3. $$\int_0^{\frac{\pi}{2}} \frac{\sin^{\mu}x\,dx}{\cos^{\mu-3}x\,(1-k^2\sin^2 x)^{\frac{\mu}{2}-1}} =$$
$$= \frac{\Gamma\left(\frac{\mu+1}{2}\right)\Gamma\left(2-\frac{\mu}{2}\right)}{k^3\sqrt{\pi}(\mu-1)(\mu-3)(\mu-5)}\left\{\frac{1+(\mu-3)k+k^2}{(1+k)^{\mu-3}} - \frac{1-(\mu-3)k+k^2}{(1-k)^{\mu-3}}\right\}$$
$[-1 < \operatorname{Re}\mu < 4]$. BI ((54))(10)

4. $$\int_0^{\frac{\pi}{2}} \frac{\sin^{\mu+1}x\,dx}{\cos^{\mu}x\,(1-k^2\sin^2 x)^{\frac{\mu+1}{2}}} = \frac{(1-k)^{-\mu}-(1+k)^{-\mu}}{4k\mu\sqrt{\pi}}\,\Gamma\left(1+\frac{\mu}{2}\right)\Gamma\left(\frac{1-\mu}{2}\right)$$
$[-2 < \operatorname{Re}\mu < 1]$. BI ((61))(5)

3.682 $$\int_0^{\frac{\pi}{2}} \frac{\sin^{\mu}x\cos^{\nu}x}{(a-b\cos^2 x)^{\varrho}}\,dx =$$
$$= \frac{1}{2a^{\varrho}}\,\mathrm{B}\left(\frac{\mu+1}{2}, \frac{\nu+1}{2}\right)F\left(\frac{\nu+1}{2}, \varrho; \frac{\mu+\nu}{2}+1; \frac{b}{a}\right)$$
$[\operatorname{Re}\mu > -1, \operatorname{Re}\nu > -1, a > |b| \geqslant 0]$. GW ((331))(64)

3.683

1. $$\int_0^{\frac{\pi}{4}} (\sin^n 2x - 1)\operatorname{tg}\left(\frac{\pi}{4}+x\right)dx = \int_0^{\frac{\pi}{4}} (\cos^n 2x - 1)\operatorname{ctg}x\,dx =$$
$$= -\frac{1}{2}\sum_{k=1}^{n}\frac{1}{k} = -\frac{1}{2}[C+\psi(n+1)].$$ BI((34))(8), BI((35))(11)

2. $$\int_0^{\frac{\pi}{4}} (\sin^{\mu} 2x - 1) \operatorname{cosec}^{\mu} 2x \operatorname{tg}\left(\frac{\pi}{4} + x\right) dx =$$

$$= \int_0^{\frac{\pi}{4}} (\cos^{\mu} 2x - 1) \sec^{\mu} 2x \operatorname{ctg} x \, dx = \frac{1}{2} [C + \psi(1 - \mu)];$$

$[\operatorname{Re} \mu < 1]$. BI ((35))(20)

3. $$\int_0^{\frac{\pi}{4}} (\sin^{2\mu} 2x - 1) \operatorname{cosec}^{\mu} 2x \operatorname{tg}\left(\frac{\pi}{4} + x\right) dx =$$

$$= \int_0^{\frac{\pi}{4}} (\cos^{2\mu} 2x - 1) \sec^{\mu} 2x \operatorname{ctg} x \, dx = -\frac{1}{2\mu} + \frac{\pi}{2} \operatorname{ctg} \mu\pi.$$ BI ((35))(21)

4. $$\int_0^{\frac{\pi}{4}} (1 - \sec^{\mu} 2x) \operatorname{ctg} x \, dx = \int_0^{\frac{\pi}{4}} (1 - \operatorname{cosec}^{\mu} 2x) \operatorname{tg}\left(\frac{\pi}{4} + x\right) dx =$$

$$= \frac{1}{2} [C + \psi(1 - \mu)] \qquad [\operatorname{Re} \mu < 1].$$ BI ((35))(13)

3.684 $$\int_0^{\frac{\pi}{4}} \frac{(\operatorname{ctg}^{\mu} x - 1)\, dx}{(\cos x - \sin x) \sin x} = \int_0^{\frac{\pi}{2}} \frac{(\operatorname{tg}^{\mu} x - 1)\, dx}{(\sin x - \cos x) \cos x} =$$

$$= -C - \psi(1 - \mu) \qquad [\operatorname{Re} \mu < 1].$$ BI ((37))(9)

3.685

1. $$\int_0^{\frac{\pi}{4}} (\sin^{\mu-1} 2x - \sin^{\nu-1} 2x) \operatorname{tg}\left(\frac{\pi}{4} + x\right) dx =$$

$$= \int_0^{\frac{\pi}{4}} (\cos^{\mu-1} 2x - \cos^{\nu-1} 2x) \operatorname{ctg} x \, dx = \frac{1}{2} [\psi(\nu) - \psi(\mu)]$$

$[\operatorname{Re} \mu > 0, \ \operatorname{Re} \nu > 0]$. BI((34))(9), BI((35))(12)

2. $$\int_0^{\frac{\pi}{2}} (\sin^{\mu-1} x - \sin^{\nu-1} x) \frac{dx}{\cos x} = \int_0^{\frac{\pi}{2}} (\cos^{\mu-1} x - \cos^{\nu-1} x) \frac{dx}{\sin x} =$$

$$= \frac{1}{2} \left[\psi\left(\frac{\nu}{2}\right) - \psi\left(\frac{\mu}{2}\right) \right] \qquad [\operatorname{Re} \mu > 0, \ \operatorname{Re} \nu > 0].$$ BI ((46))(2)

3. $$\int_0^{\frac{\pi}{2}} (\sin^{\mu} x - \operatorname{cosec}^{\mu} x) \frac{dx}{\cos x} = \int_0^{\frac{\pi}{2}} (\cos^{\mu} x - \sec^{\mu} x) \frac{dx}{\sin x} =$$

$$= -\frac{\pi}{2} \operatorname{tg} \frac{\mu\pi}{2} \qquad [|\operatorname{Re} \mu| < 1].$$ BI ((46))(1, 3)

4. $\int_0^{\frac{\pi}{4}} (\sin^{\mu} 2x - \operatorname{cosec}^{\mu} 2x) \operatorname{ctg}\left(\frac{\pi}{4}+x\right) dx =$

$$= \int_0^{\frac{\pi}{4}} (\cos^{\mu} 2x - \sec^{\mu} 2x) \operatorname{tg} x \, dx = \frac{1}{2\mu} - \frac{\pi}{2} \operatorname{cosec} \mu\pi$$

$[|\operatorname{Re}\mu| < 1]$. BI ((35))(19, 22)

5 $\int_0^{\frac{\pi}{4}} (\sin^{\mu} 2x - \operatorname{cosec}^{\mu} 2x) \operatorname{tg}\left(\frac{\pi}{4}+x\right) dx =$

$$= \int_0^{\frac{\pi}{4}} (\cos^{\mu} 2x - \sec^{\mu} 2x) \operatorname{ctg} x \, dx = -\frac{1}{2\mu} + \frac{\pi}{2} \operatorname{ctg} \mu\pi$$

$[|\operatorname{Re}\mu| < 1]$. BI ((35))(14)

6 $\int_0^{\frac{\pi}{4}} (\sin^{\mu-1} 2x + \operatorname{cosec}^{\mu} 2x) \operatorname{ctg}\left(\frac{\pi}{4}+x\right) dx =$

$$= \int_0^{\frac{\pi}{4}} (\cos^{\mu-1} 2x + \sec^{\mu} 2x) \operatorname{tg} x \, dx = \frac{\pi}{2} \operatorname{cosec} \mu\pi$$

$[0 < \operatorname{Re}\mu < 1]$. BI ((35))(18, 8)

7 $\int_0^{\frac{\pi}{4}} (\sin^{\mu-1} 2x - \operatorname{cosec}^{\mu} 2x) \operatorname{tg}\left(\frac{\pi}{4}+x\right) dx =$

$$= \int_0^{\frac{\pi}{4}} (\cos^{\mu-1} 2x - \sec^{\mu} 2x) \operatorname{ctg} x \, dx = \frac{\pi}{2} \operatorname{ctg} \mu\pi$$

$[0 < \operatorname{Re}\mu < 1]$. BI((35))(7), LI((34))(10)

3.686 $$\int_0^{\frac{\pi}{2}} \frac{\operatorname{tg} x \, dx}{\cos^{\mu} x + \sec^{\mu} x} = \int_0^{\frac{\pi}{2}} \frac{\operatorname{ctg} x \, dx}{\sin^{\mu} x + \operatorname{cosec}^{\mu} x} = \frac{\pi}{4\mu}.$$

BI((47))(28), BI((49))(14)

3.687

1. $\int_0^{\frac{\pi}{2}} \frac{\sin^{\mu-1} x + \sin^{\nu-1} x}{\cos^{\mu+\nu-1} x} dx = \int_0^{\frac{\pi}{2}} \frac{\cos^{\mu-1} x + \cos^{\nu-1} x}{\sin^{\mu+\nu-1} x} dx =$

$$= \frac{\cos\left(\frac{\nu-\mu}{4}\pi\right)}{2\cos\left(\frac{\nu+\mu}{4}\pi\right)} \mathrm{B}\left(\frac{\mu}{2}, \frac{\nu}{2}\right)$$

$[\operatorname{Re}\mu > 0, \ \operatorname{Re}\nu > 0]$. BI ((46))(7)

2. $$\int_0^{\frac{\pi}{2}} \frac{\sin^{\mu-1} x - \sin^{\nu-1} x}{\cos^{\mu+\nu-1} x}\, dx = \int_0^{\frac{\pi}{2}} \frac{\cos^{\mu-1} x - \cos^{\nu-1} x}{\sin^{\mu+\nu-1} x}\, dx = \frac{\sin\left(\frac{\nu-\mu}{4}\pi\right)}{2\sin\left(\frac{\nu+\mu}{4}\pi\right)} \mathrm{B}\left(\frac{\mu}{2}, \frac{\nu}{2}\right) \qquad [\operatorname{Re}\mu > 0,\ \operatorname{Re}\nu > 0].$$ BI((46))(8)

3. $$\int_0^{\frac{\pi}{2}} \frac{\sin^{\mu} x + \sin^{\nu} x}{\sin^{\mu+\nu} x + 1} \operatorname{ctg} x\, dx = \int_0^{\frac{\pi}{2}} \frac{\cos^{\mu} x + \cos^{\nu} x}{\cos^{\mu+\nu} x + 1} \operatorname{tg} x\, dx = \frac{\pi}{\mu+\nu} \sec\left(\frac{\mu-\nu}{\mu+\nu}\cdot\frac{\pi}{2}\right) \qquad [\operatorname{Re}\mu > 0,\ \operatorname{Re}\nu > 0].$$ BI((49))(15)a, BI((47))(29)

4. $$\int_0^{\frac{\pi}{2}} \frac{\sin^{\mu} x - \sin^{\nu} x}{\sin^{\mu+\nu} x - 1} \operatorname{ctg} x\, dx = \int_0^{\frac{\pi}{2}} \frac{\cos^{\mu} x - \cos^{\nu} x}{\cos^{\mu+\nu} x - 1} \operatorname{tg} x\, dx = \frac{\pi}{\mu+\nu} \operatorname{tg}\left(\frac{\mu-\nu}{\mu+\nu}\cdot\frac{\pi}{2}\right) \qquad [\operatorname{Re}\mu > 0,\ \operatorname{Re}\nu > 0].$$ BI((149))(16)a, BI((47))(30)

5. $$\int_0^{\frac{\pi}{2}} \frac{\cos^{\mu} x + \sec^{\mu} x}{\cos^{\nu} x + \sec^{\nu} x} \operatorname{tg} x\, dx = \frac{\pi}{2\nu} \sec\left(\frac{\mu}{\nu}\cdot\frac{\pi}{2}\right) \qquad [|\operatorname{Re}\nu| > |\operatorname{Re}\mu|].$$ BI ((49))(12)

6. $$\int_0^{\frac{\pi}{2}} \frac{\cos^{\mu} x - \sec^{\mu} x}{\cos^{\nu} x - \sec^{\nu} x} \operatorname{tg} x\, dx = \frac{\pi}{2\nu} \operatorname{tg}\left(\frac{\mu}{\nu}\cdot\frac{\pi}{2}\right) \qquad [|\operatorname{Re}\nu| > |\operatorname{Re}\mu|].$$ BI ((49))(13)

3.688

1. $$\int_0^{\frac{\pi}{4}} \frac{\operatorname{tg}^{\nu} x - \operatorname{tg}^{\mu} x}{\cos x - \sin x} \cdot \frac{dx}{\sin x} = \psi(\mu) - \psi(\nu) \qquad [\operatorname{Re}\mu > 0,\ \operatorname{Re}\nu > 0].$$ BI ((37))(10)

2. $$\int_0^{\frac{\pi}{4}} \frac{\operatorname{tg}^{\mu} x - \operatorname{tg}^{1-\mu} x}{\cos x - \sin x} \cdot \frac{dx}{\sin x} = \pi \operatorname{ctg} \mu\pi \qquad [0 < \operatorname{Re}\mu < 1].$$ BI ((37))(11)

3. $$\int_0^{\frac{\pi}{4}} (\operatorname{tg}^{\mu} x + \operatorname{ctg}^{\mu} x)\, dx = \frac{\pi}{2} \sec \frac{\mu\pi}{2} \qquad [|\operatorname{Re}\mu| < 1].$$ BI ((35))(9)

4. $$\int_0^{\frac{\pi}{4}} (\operatorname{tg}^{\mu} x - \operatorname{ctg}^{\mu} x) \operatorname{tg} x \, dx = \frac{1}{\mu} - \frac{\pi}{2} \operatorname{cosec} \frac{\mu\pi}{2}$$

$[0 < \operatorname{Re} \mu < 2]$. BI ((35))(15)

5. $$\int_0^{\frac{\pi}{4}} \frac{\operatorname{tg}^{\mu-1} x - \operatorname{ctg}^{\mu-1} x}{\cos 2x} \, dx = \frac{\pi}{2} \operatorname{ctg} \frac{\mu\pi}{2}$$ $[|\operatorname{Re} \mu| < 2]$. BI ((35))(10)

6 $$\int_0^{\frac{\pi}{4}} \frac{\operatorname{tg}^{\mu} x - \operatorname{ctg}^{\mu} x}{\cos 2x} \operatorname{tg} x \, dx = -\frac{1}{\mu} + \frac{\pi}{2} \operatorname{ctg} \frac{\mu\pi}{2}$$

$[-2 < \operatorname{Re} \mu < 0]$. BI ((35))(23)

7 $$\int_0^{\frac{\pi}{4}} \frac{\operatorname{tg}^{\mu} x + \operatorname{ctg}^{\mu} x}{1 + \cos t \sin 2x} \, dx = \pi \operatorname{cosec} t \operatorname{cosec} \mu\pi \sin \mu t$$

$[t \neq n\pi, \ |\operatorname{Re} \mu| < 1]$. BI ((36))(6)

8. $$\int_0^{\frac{\pi}{4}} \frac{\operatorname{tg}^{\mu-1} x + \operatorname{ctg}^{\mu} x}{(\sin x + \cos x) \cos x} \, dx = \pi \operatorname{cosec} \mu\pi$$

$[0 < \operatorname{Re} \mu < 1]$. BI ((37))(3)

9. $$\int_0^{\frac{\pi}{4}} \frac{\operatorname{tg}^{\mu} x - \operatorname{ctg}^{\mu} x}{(\sin x + \cos x) \cos x} \, dx = -\pi \operatorname{cosec} \mu\pi + \frac{1}{\mu}$$

$[0 < \operatorname{Re} \mu < 1]$. BI ((37))(4)

10. $$\int_0^{\frac{\pi}{4}} \frac{\operatorname{tg}^{\nu} x - \operatorname{ctg}^{\mu} x}{(\cos x - \sin x) \cos x} \, dx = \psi(1 - \mu) - \psi(1 + \nu)$$

$[\operatorname{Re} \mu < 1, \ \operatorname{Re} \nu > -1]$. BI ((37))(5)

11. $$\int_0^{\frac{\pi}{4}} \frac{\operatorname{tg}^{\mu-1} x - \operatorname{ctg}^{\mu} x}{(\cos x - \sin x) \cos x} \, dx = \pi \operatorname{ctg} \mu\pi$$

$[0 < \operatorname{Re} \mu < 1]$. BI ((37))(7)

12 $$\int_0^{\frac{\pi}{4}} \frac{\operatorname{tg}^{\mu} x - \operatorname{ctg}^{\mu} x}{(\cos x - \sin x) \cos x} \, dx = \pi \operatorname{ctg} \mu\pi - \frac{1}{\mu}$$

$[0 < \operatorname{Re} \mu < 1]$. BI ((37))(8)

13. $$\int_0^{\frac{\pi}{4}} \frac{1}{\operatorname{tg}^{\mu} x + \operatorname{ctg}^{\mu} x} \cdot \frac{dx}{\sin 2x} = \frac{\pi}{8\mu}$$ $[\operatorname{Re} \mu \neq 0]$. BI ((37))(12)

14. $$\int_0^{\frac{\pi}{2}} \frac{1}{(\mathrm{tg}^{\mu} x+\mathrm{ctg}^{\mu} x)^{\nu}} \cdot \frac{dx}{\mathrm{tg}\, x} = \int_0^{\frac{\pi}{2}} \frac{1}{(\mathrm{tg}^{\mu} x+\mathrm{ctg}^{\mu} x)^{\nu}} \cdot \frac{dx}{\sin 2x} =$$
$$= \frac{\sqrt{\pi}}{2^{2\nu+1}\mu} \frac{\Gamma(\nu)}{\Gamma\left(\nu+\frac{1}{2}\right)} \qquad [\nu > 0].$$ BI((49))(25), BI((49))(26)

15. $$\int_0^{\frac{\pi}{4}} (\mathrm{tg}^{\mu} x - \mathrm{ctg}^{\mu} x)(\mathrm{tg}^{\nu} x - \mathrm{ctg}^{\nu} x)\, dx = \frac{2\pi \sin\frac{\mu\pi}{2} \sin\frac{\nu\pi}{2}}{\cos \mu\pi + \cos \nu\pi}$$
$$[|\mathrm{Re}\, \mu| < 1, \ |\mathrm{Re}\, \nu| < 1].$$ BI ((35))(17)

16. $$\int_0^{\frac{\pi}{4}} (\mathrm{tg}^{\mu} x + \mathrm{ctg}^{\mu} x)(\mathrm{tg}^{\nu} x + \mathrm{ctg}^{\nu} x)\, dx = \frac{2\pi \cos\frac{\mu\pi}{2} \cos\frac{\nu\pi}{2}}{\cos \mu\pi + \cos \nu\pi}$$
$$[|\mathrm{Re}\, \mu| < 1, \ |\mathrm{Re}\, \nu| < 1].$$ BI ((35))(16)

17. $$\int_0^{\frac{\pi}{4}} \frac{(\mathrm{tg}^{\mu} x - \mathrm{ctg}^{\mu} x)(\mathrm{tg}^{\nu} x + \mathrm{ctg}^{\nu} x)}{\cos 2x}\, dx = -\pi \frac{\sin \mu\pi}{\cos \mu\pi + \cos \nu\pi}$$
$$[|\mathrm{Re}\, \mu| < 1, \ |\mathrm{Re}\, \nu| < 1].$$ BI ((35))(25)

18. $$\int_0^{\frac{\pi}{4}} \frac{\mathrm{tg}^{\nu} x - \mathrm{ctg}^{\nu} x}{\mathrm{tg}^{\mu} x - \mathrm{ctg}^{\mu} x} \cdot \frac{dx}{\sin 2x} = \frac{\pi}{4\mu} \mathrm{tg} \frac{\nu\pi}{2\mu}$$
$$[0 < \mathrm{Re}\, \nu < 1].$$ BI ((37))(14)

19. $$\int_0^{\frac{\pi}{4}} \frac{\mathrm{tg}^{\nu} x + \mathrm{ctg}^{\nu} x}{\mathrm{tg}^{\mu} x + \mathrm{ctg}^{\mu} x} \cdot \frac{dx}{\sin 2x} = \frac{\pi}{4\mu} \sec \frac{\nu\pi}{2\mu}$$
$$[0 < \mathrm{Re}\, \nu < 1].$$ BI ((37))(13)

20. $$\int_0^{\frac{\pi}{2}} \frac{(1+\mathrm{tg}\, x)^{\nu} - 1}{(1+\mathrm{tg}\, x)^{\mu+\nu}} \frac{dx}{\sin x \cos x} = \psi(\mu+\nu) - \psi(\mu)$$
$$[\mu > 0, \ \nu > 0].$$ BI ((49))(29)

3.689

1. $$\int_0^{\frac{\pi}{2}} \frac{(\sin^{\mu} x + \mathrm{cosec}^{\mu} x)\, \mathrm{ctg}\, x\, dx}{\sin^{\nu} x - 2\cos t + \mathrm{cosec}^{\nu} x} = \frac{\pi}{\nu} \mathrm{cosec}\, t\, \mathrm{cosec} \frac{\mu\pi}{\nu} \sin \frac{\mu t}{\nu}$$
$$[\mu < \nu]$$ LI ((50))(14)

2. $$\int_0^{\frac{\pi}{2}} \frac{\sin^{\mu} x - 2\cos t_1 + \mathrm{cosec}^{\mu} x}{\sin^{\nu} x + 2\cos t_2 + \mathrm{cosec}^{\nu} x} \cdot \mathrm{ctg}\, x \cdot dx =$$
$$= \frac{\pi}{\nu} \mathrm{cosec}\, t_2\, \mathrm{cosec} \frac{\mu\pi}{\nu} \sin \frac{\mu t_2}{\nu} - \frac{t_2}{\nu} \mathrm{cosec}\, t_2 \cos t_1$$
$$[\nu > \mu > 0 \quad \text{or} \quad \nu < \mu < 0 \quad \text{or} \quad \mu > 0, \ \nu < 0 \text{ and } \mu+\nu < 0$$
$$\text{or} \quad \mu < 0, \ \nu > 0 \text{ and } \mu+\nu > 0].$$ BI ((50))(15)

3.69-3.71 Trigonometric functions of more complicated arguments

3.691

1. $$\int_0^\infty \sin(ax^2)\,dx = \int_0^\infty \cos ax^2\,dx = \frac{1}{2}\sqrt{\frac{\pi}{2a}} \quad [a>0].$$

FI II 743a, ET I 64(7)a

2. $$\int_0^1 \sin(ax^2)\,dx = \sqrt{\frac{\pi}{2a}}\,S(\sqrt{a}) \quad [a>0].$$

3. $$\int_0^1 \cos(ax^2)\,dx = \sqrt{\frac{\pi}{2a}}\,C(\sqrt{a}) \quad [a>0].$$

ET I 8(5)a

4. $$\int_0^\infty \sin(ax^2)\sin 2bx\,dx = \sqrt{\frac{\pi}{2a}}\left\{\cos\frac{b^2}{a}\,C\left(\frac{b}{\sqrt{a}}\right) + \sin\frac{b^2}{a}\,S\left(\frac{b}{\sqrt{a}}\right)\right\}$$
$$[a>0,\ b>0].$$

ET I 82(1)a

5. $$\int_0^\infty \sin(ax^2)\cos 2bx\,dx = \frac{1}{2}\sqrt{\frac{\pi}{2a}}\left\{\cos\frac{b^2}{a} - \sin\frac{b^2}{a}\right\} =$$
$$= \frac{1}{2}\sqrt{\frac{\pi}{a}}\cos\left(\frac{b^2}{a}+\frac{\pi}{4}\right) \qquad [a>0,\ b>0].$$

ET I 82(18), BI((70))(13) GW((334))(5a)

6. $$\int_0^\infty \cos ax^2 \sin 2bx\,dx = \sqrt{\frac{\pi}{2a}}\left\{\sin\frac{b^2}{a}\,C\left(\frac{b}{\sqrt{a}}\right) - \cos\frac{b^2}{a}\,S\left(\frac{b}{\sqrt{a}}\right)\right\}$$
$$[a>0,\ b>0].$$

ET I 83(3)a

7. $$\int_0^\infty \cos ax^2 \cos 2bx\,dx = \frac{1}{2}\sqrt{\frac{\pi}{2a}}\left\{\cos\frac{b^2}{a} + \sin\frac{b^2}{a}\right\}$$
$$[a>0,\ b>0].$$

GW((334))(5a), BI((70))(14), ET I 24(7)

8. $$\int_0^\infty (\cos ax + \sin ax)\sin(b^2x^2)\,dx = \frac{1}{2b}\sqrt{\frac{\pi}{2}}\exp\left(-\frac{a^2}{2b}\right)$$
$$[a>0,\ b>0].$$

ET I 85(22)

9. $$\int_0^\infty (\cos ax + \sin ax)\cos(b^2x^2)\,dx = \frac{1}{2b}\sqrt{\frac{\pi}{2}}\exp\left(-\frac{a^2}{2b}\right)$$
$$[a>0,\ b>0].$$

ET I 25(21)

10. $$\int_0^\infty \sin(a^2x^2)\sin 2bx \sin 2cx\,dx = \frac{\sqrt{\pi}}{2a}\sin\frac{2bc}{a^2}\cos\left(\frac{b^2+c^2}{a^2}-\frac{\pi}{4}\right)$$
$$[a>0,\ b>0,\ c>0].$$

ET I 84(15)

11. $$\int_0^\infty \sin(a^2x^2)\cos 2bx \cos 2cx\, dx = \frac{\sqrt{\pi}}{2a}\cos\frac{2bc}{a^2}\cos\left(\frac{b^2+c^2}{a^2}+\frac{\pi}{4}\right)$$

$[a>0,\ b>0,\ c>0]$. ET I 84(21)

12. $$\int_0^\infty \cos(a^2x^2)\sin 2bx \sin 2cx\, dx = \frac{\sqrt{\pi}}{2a}\sin\frac{2bc}{a^2}\sin\left(\frac{b^2+c^2}{a^2}-\frac{\pi}{4}\right)$$

$[a>0,\ b>0,\ c>0]$. ET I 25(19)

13. $$\int_0^\infty \sin(ax^2)\cos(bx^2)\, dx = \frac{1}{4}\sqrt{\frac{\pi}{2}}\left(\frac{1}{\sqrt{a+b}}+\frac{1}{\sqrt{a-b}}\right) \quad [a>b>0];$$

$$= \frac{1}{4}\sqrt{\frac{\pi}{2}}\left(\frac{1}{\sqrt{b+a}}-\frac{1}{\sqrt{b-a}}\right) \quad [b>a>0].$$

BI ((177))(21)

14. $$\int_0^\infty (\sin^2 ax^2 - \sin^2 bx^2)\, dx = \frac{1}{8}\left(\sqrt{\frac{\pi}{b}}-\sqrt{\frac{\pi}{a}}\right) \quad [a>0,\ b>0].$$

BI ((178))(1)

15. $$\int_0^\infty (\cos^2 ax^2 - \sin^2 bx^2)\, dx = \frac{1}{8}\left(\sqrt{\frac{\pi}{b}}+\sqrt{\frac{\pi}{a}}\right) \quad [a>0,\ b>0].$$

BI ((178))(3)

16. $$\int_0^\infty (\cos^2 ax^2 - \cos^2 bx^2)\, dx = \frac{1}{8}\left(\sqrt{\frac{\pi}{a}}-\sqrt{\frac{\pi}{b}}\right) \quad [a>0,\ b>0].$$

BI ((178))(5)

17. $$\int_0^\infty (\sin^4 ax^2 - \sin^4 bx^2)\, dx = \frac{1}{64}(8-\sqrt{2})\left(\sqrt{\frac{\pi}{b}}-\sqrt{\frac{\pi}{a}}\right)$$

$[a>0,\ b>0]$. BI ((178))(2)

18. $$\int_0^\infty (\cos^4 ax^2 - \sin^4 bx^2)\, dx = \frac{1}{8}\left(\sqrt{\frac{\pi}{a}}+\sqrt{\frac{\pi}{b}}\right)+\frac{1}{32}\left(\sqrt{\frac{\pi}{2a}}-\sqrt{\frac{\pi}{2b}}\right)$$

$[a>0,\ b>0]$. BI ((178))(4)

19. $$\int_0^\infty (\cos^4 ax^2 - \cos^4 bx^2)\, dx = \frac{1}{64}(8+\sqrt{2})\left(\sqrt{\frac{\pi}{a}}-\sqrt{\frac{\pi}{b}}\right)$$

$[a>0,\ b>0]$. BI ((178))(6)

20. $$\int_0^\infty \sin^{2n} ax^2\, dx = \int_0^\infty \cos^{2n} ax^2\, dx = \infty.$$

BI ((177))(5, 6)

21. $$\int_0^\infty \sin^{2n+1}(ax^2)\, dx = \frac{1}{2^{2n+1}}\sum_{k=0}^{n}(-1)^{n+k}\binom{2n+1}{k}\sqrt{\frac{\pi}{2(2n-2k+1)a}}$$

$[a>0]$. BI ((70))(9)

22. $\int_0^\infty \cos^{2n+1}(ax^2)\,dx = \frac{1}{2^{2n+1}} \sum_{k=0}^{n} \binom{2n+1}{k} \sqrt{\frac{\pi}{2(2n-2k+1)a}}$

$[a > 0]$. BI((177))(7)a, BI((70))(10)

3.692

1. $\int_0^\infty [\sin(a-x^2)+\cos(a-x^2)]\,dx = \sqrt{\frac{\pi}{2}}\sin a.$

GW((333))(30c), BI((178))(7)a

2. $\int_0^\infty \cos\left(\frac{x^2}{2}-\frac{\pi}{8}\right)\cos ax\,dx = \sqrt{\frac{\pi}{2}}\cos\left(\frac{a^2}{2}-\frac{\pi}{8}\right)$

$[a > 0]$. ET I 24(8)

3. $\int_0^\infty \sin[a(1-x^2)]\cos bx\,dx = -\frac{1}{2}\sqrt{\frac{\pi}{a}}\cos\left(a+\frac{b^2}{4a}+\frac{\pi}{4}\right)$

$[a > 0]$. ET I 23(2)

4. $\int_0^\infty \cos[a(1-x^2)]\cos bx\,dx = \frac{1}{2}\sqrt{\frac{\pi}{a}}\sin\left(a+\frac{b^2}{4a}+\frac{\pi}{4}\right)$

$[a > 0]$. ET I 24(10)

5. $\int_0^\infty \sin\left(ax^2+\frac{b^2}{a}\right)\cos 2bx\,dx = \int_0^\infty \cos\left(ax^2+\frac{b^2}{a}\right)\cos 2bx\,dx = \frac{1}{2}\sqrt{\frac{\pi}{2a}}$

$[a > 0]$. BI ((70))(19, 20)

3.693

1. $\int_0^\infty \sin(ax^2+2bx)\,dx = \frac{1}{2}\left(\cos\frac{b^2}{a}-\sin\frac{b^2}{a}\right)\sqrt{\frac{\pi}{2a}}$

$[a > 0]$. BI ((70))(3)

2. $\int_0^\infty \cos(ax^2+2bx)\,dx = \frac{1}{2}\left(\cos\frac{b^2}{a}+\sin\frac{b^2}{a}\right)\sqrt{\frac{\pi}{2a}}$

$[a > 0]$. BI ((70))(4)

3.694

1. $\int_0^\infty \sin(ax^2+2bx+c)\,dx = \frac{1}{2}\int_{-\infty}^\infty \sin(ax^2+2bx+c)\,dx =$

$= \frac{1}{2}\sqrt{\frac{\pi}{a}}\sin\left(\frac{\pi}{4}+\frac{ac-b^2}{a}\right) \quad [a > 0].$ GW ((334))(4a)

2. $\int_0^\infty \cos(ax^2+2bx+c)\,dx = \frac{1}{2}\int_{-\infty}^\infty \cos(ax^2+2bx+c)\,dx =$

$= \frac{1}{2}\sqrt{\frac{\pi}{a}}\cos\left(\frac{\pi}{4}+\frac{ac-b^2}{a}\right) \quad [a > 0].$ GW ((334))(4b)

3.695

1. $$\int_0^\infty \sin(a^3x^3)\sin(bx)\,dx = \frac{\pi}{6a}\sqrt{\frac{b}{3a}}\left\{J_{\frac{1}{3}}\left(\frac{2b}{3a}\sqrt{\frac{b}{3a}}\right)+\right.$$
$$\left.+J_{-\frac{1}{3}}\left(\frac{2b}{3a}\sqrt{\frac{b}{3a}}\right)-\frac{\sqrt{3}}{\pi}K_{\frac{1}{3}}\left(\frac{2b}{3a}\sqrt{\frac{b}{3a}}\right)\right\} \quad [a>0,\ b>0].$$ ET I 83(5)

2. $$\int_0^\infty \cos(a^3x^3)\cos(bx)\,dx = \frac{\pi}{6a}\sqrt{\frac{b}{3a}}\left\{J_{\frac{1}{3}}\left(\frac{2b}{3a}\sqrt{\frac{b}{3a}}\right)+\right.$$
$$\left.+J_{-\frac{1}{3}}\left(\frac{2b}{3a}\sqrt{\frac{b}{3a}}\right)+\frac{\sqrt{3}}{\pi}K_{\frac{1}{3}}\left(\frac{2b}{3a}\sqrt{\frac{b}{3a}}\right)\right\} \quad [a>0,\ b>0].$$ ET I 24(11)

3.696

1. $$\int_0^\infty \sin(ax^4)\sin(bx^2)\,dx = -\frac{\pi}{4}\sqrt{\frac{b}{2a}}\sin\left(\frac{b^2}{8a}-\frac{3}{8}\pi\right)J_{\frac{1}{4}}\left(\frac{b^2}{8a}\right)$$
$$[a>0,\ b>0].$$ ET I 83(2)

2. $$\int_0^\infty \sin(ax^4)\cos(bx^2)\,dx = -\frac{\pi}{4}\sqrt{\frac{b}{2a}}\sin\left(\frac{b^2}{8a}-\frac{\pi}{8}\right)J_{-\frac{1}{4}}\left(\frac{b^2}{8a}\right)$$
$$[a>0,\ b>0].$$ ET I 84(19)

3. $$\int_0^\infty \cos(ax^4)\sin(bx^2)\,dx = \frac{\pi}{4}\sqrt{\frac{b}{2a}}\cos\left(\frac{b^2}{8a}-\frac{3}{8}\pi\right)J_{\frac{1}{4}}\left(\frac{b^2}{8a}\right)$$
$$[a>0,\ b>0].$$ ET I 83(4), ET I 25(24)

4. $$\int_0^\infty \cos(ax^4)\cos(bx^2)\,dx = \frac{\pi}{4}\sqrt{\frac{b}{2a}}\cos\left(\frac{b^2}{8a}-\frac{\pi}{8}\right)J_{-\frac{1}{4}}\left(\frac{b^2}{8a}\right)$$
$$[a>0,\ b>0].$$ ET I 25(25)

3.697 $$\int_0^\infty \sin\left(\frac{a^2}{x}\right)\sin(bx)\,dx = \frac{a\pi}{2\sqrt{b}}J_1(2a\sqrt{b}) \quad [a>0,\ b>0].$$ ET I 83(6)

3.698

1. $$\int_0^\infty \sin\left(\frac{a^2}{x^2}\right)\sin(b^2x^2)\,dx = \frac{1}{4b}\sqrt{\frac{\pi}{2}}[\sin 2ab-\cos 2ab+e^{-2ab}]$$
$$[a>0,\ b>0].$$ ET I 83(9)

2. $$\int_0^\infty \sin\left(\frac{a^2}{x^2}\right)\cos(b^2x^2)\,dx = \frac{1}{4b}\sqrt{\frac{\pi}{2}}[\sin 2ab+\cos 2ab+e^{-2ab}]$$
$$[a>0,\ b>0].$$ ET I 24(13)

3. $\int_0^\infty \cos\left(\frac{a^2}{x^2}\right)\sin(b^2x^2)\,dx = \frac{1}{4b}\sqrt{\frac{\pi}{2}}\left[\sin 2ab + \cos 2ab + e^{-2ab}\right]$

$[a > 0,\ b > 0]$. ET I 84(12)

4. $\int_0^\infty \cos\left(\frac{a^2}{x^2}\right)\cos(b^2x^2)\,dx = \frac{1}{4b}\sqrt{\frac{\pi}{2}}\left[\cos 2ab - \sin 2ab + e^{-2ab}\right]$

$[a > 0,\ b > 0]$. ET I 24(14)

3.699

1. $\int_0^\infty \sin\left(a^2x^2 + \frac{b^2}{x^2}\right)dx = \frac{\sqrt{2\pi}}{4a}(\cos 2ab + \sin 2ab)$

$[a > 0,\ b > 0]$. BI ((70))(27)

2. $\int_0^\infty \cos\left(a^2x^2 + \frac{b^2}{x^2}\right)dx = \frac{\sqrt{2\pi}}{4a}(\cos 2ab - \sin 2ab)$

$[a > 0,\ b > 0]$. BI ((70))(28)

3. $\int_0^\infty \sin\left(a^2x^2 - 2ab + \frac{b^2}{x^2}\right)dx = \int_0^\infty \cos\left(a^2x^2 - 2ab + \frac{b^2}{x^2}\right)dx = \frac{\sqrt{2\pi}}{4a}$

$[a > 0,\quad b > 0]$. BI((179))(11, 12)a, ET I 83(6)

4. $\int_0^\infty \sin\left(a^2x^2 - \frac{b^2}{x^2}\right)dx = \frac{\sqrt{2\pi}}{4a}e^{-2ab}$ $[a > 0,\quad b > 0]$. GW ((334))(9b)a

5. $\int_0^\infty \cos\left(a^2x^2 - \frac{b^2}{x^2}\right)dx = \frac{\sqrt{2\pi}}{4a}e^{-2ab}$ $[a > 0,\ b > 0]$. GW ((334))(9b)a

3.711 $\int_0^u \sin\left(a\sqrt{u^2 - x^2}\right)\cos bx\,dx = \frac{\pi a u}{2\sqrt{a^2+b^2}} J_1\left(u\sqrt{a^2+b^2}\right)$

$[a > 0,\quad b > 0,\quad u > 0]$. ET I 27(37)

3.712

1. $\int_0^\infty \sin(ax^p)\,dx = \frac{\Gamma\left(\frac{1}{p}\right)\sin\frac{\pi}{2p}}{pa^{\frac{1}{p}}}$ $[a > 0,\quad p > 1]$. EH I 13(40)

2. $\int_0^\infty \cos(ax^p)\,dx = \frac{\Gamma\left(\frac{1}{p}\right)\cos\frac{\pi}{2p}}{pa^{\frac{1}{p}}}$ $[a > 0,\quad p > 1]$. EH I 13(39)

3.713

1. $\int_0^\infty \sin(ax^p + bx^q)\,dx = \frac{1}{p}\sum_{k=0}^\infty \frac{(-b)^k}{k!} a^{-\frac{kq+1}{p}}\Gamma\left(\frac{kq+1}{p}\right)\times$

$\times \sin\left[\frac{k(q-p)+1}{2p}\pi\right]$ $[a > 0,\quad b > 0,\quad p > 0,\quad q > 0]$. BI ((70))(7)

2. $$\int_0^\infty \cos(ax^p + bx^q)\,dx = \frac{1}{p}\sum_{k=0}^\infty \frac{(-b)^k}{k!} a^{-\frac{kq+1}{p}} \Gamma\left(\frac{kq+1}{p}\right) \times$$
$$\times \cos\left[\frac{k(q-p)+1}{2p}\pi\right] \quad [a>0, \quad b>0, \quad p>0, \quad q>0].$$ BI ((70))(8)

3.714

1. $$\int_0^\infty \cos(z \operatorname{sh} x)\,dx = K_0(z) \quad [\operatorname{Re} z > 0].$$ WA 202(14)

2. $$\int_0^\infty \sin(z \operatorname{ch} x)\,dx = \frac{\pi}{2} J_0(z) \quad [\operatorname{Re} z > 0].$$ MO 36

3. $$\int_0^\infty \cos(z \operatorname{ch} x)\,dx = -\frac{\pi}{2} N_0(z) \quad [\operatorname{Re} z > 0].$$ MO 37

4. $$\int_0^\infty \cos(z \operatorname{sh} x) \operatorname{ch} \mu x\,dx = \cos\frac{\mu\pi}{2} K_\mu(z) \quad [\operatorname{Re} z > 0, \quad |\operatorname{Re}\mu| < 1].$$ WA 202(13)

5. $$\int_0^\pi \cos(z \operatorname{ch} x) \sin^{2\mu} x\,dx = \sqrt{\pi}\left(\frac{2}{z}\right)^\mu \Gamma\left(\mu+\frac{1}{2}\right) I_\mu(z)$$
$$\left[\operatorname{Re} z > 0, \quad \operatorname{Re}\mu > -\frac{1}{2}\right].$$ WH

3.715

1. $$\int_0^\pi \sin(z\sin x)\sin ax\,dx = \sin a\pi\, s_{0,a}(z) =$$
$$= \sin a\pi \sum_{k=1}^\infty \frac{(-1)^{k-1} z^{2k-1}}{(1^2-a^2)(3^2-a^2)\dots[(2k-1)^2-a^2]} \quad [a>0].$$ WA 338(13)

2. $$\int_0^\pi \sin(z\sin x)\sin nx\,dx = \frac{1}{2}\int_{-\pi}^{\pi} \sin(z\sin x)\sin nx\,dx =$$
$$= [1-(-1)^n]\int_0^{\frac{\pi}{2}} \sin(z\sin x)\sin nx\,dx =$$
$$= [1-(-1)^n]\frac{\pi}{2} J_n(z) \quad [n = 0, \pm 1, \pm 2, \dots].$$ WA 30(6), GW((334))(153a)

3. $$\int_0^{\frac{\pi}{2}} \sin(z\sin x)\sin 2x\,dx = \frac{2}{z^2}(\sin z - z\cos z).$$ LI ((43))(14)

4. $$\int_0^{\pi} \sin(z \sin x) \cos ax \, dx = (1 + \cos a\pi)\, s_{0,\,a}(z) =$$
$$= (1 + \cos a\pi) \sum_{k=1}^{\infty} \frac{(-1)^{k-1} z^{2k-1}}{(1^2 - a^2)(3^2 - a^2) \ldots [(2k-1)^2 - a^2]} \quad [a > 0].$$

WA 338(14)

5. $$\int_0^{\pi} \sin(z \sin x) \cos[(2n+1)x] \, dx = 0.$$

GW ((334))(53b)

6. $$\int_0^{\pi} \cos(z \sin x) \sin ax \, dx = -a(1 - \cos a\pi)\, s_{-1,\,a}(z) =$$
$$= -a(1 - \cos a\pi) \left\{ -\frac{1}{a^2} + \sum_{k=1}^{\infty} \frac{(-1)^{k-1} z^{2k}}{a^2 (2^2 - a^2)(4^2 - a^2) \ldots [(2k)^2 - a^2]} \right\}$$
$$[a > 0].$$

WA 338(12)

7. $$\int_0^{\pi} \cos(z \sin x) \sin 2nx \, dx = 0.$$

GW ((334))(54a)

8. $$\int_0^{\pi} \cos(z \sin x) \cos ax \, dx = -a \sin a\pi \, s_{-1,\,a}(z) =$$
$$= -a \sin a\pi \left\{ -\frac{1}{a^2} + \sum_{k=1}^{\infty} \frac{(-1)^{k-1} z^{2k}}{a^2 (2^2 - a^2)(4^2 - a^2) \ldots [(2k)^2 - a^2]} \right\} \quad [a > 0].$$

WA 338(11)

9. $$\int_0^{\pi} \cos(z \sin x) \cos nx \, dx = \frac{1}{2} \int_{-\pi}^{\pi} \cos(z \sin x) \cos nx \, dx =$$
$$= [1 + (-1)^n] \int_0^{\frac{\pi}{2}} \cos(z \sin x) \cos nx \, dx = [1 + (-1)^n] \frac{\pi}{2} J_n(z).$$

GW ((334))(54b)

10. $$\int_0^{\frac{\pi}{2}} \cos(z \sin x) \cos^{2n} x \, dx = \frac{\pi}{2} \frac{(2n-1)!!}{z^n} J_n(z) \quad \left[\operatorname{Re} n > -\frac{1}{2}\right].$$

FI II 486, WA 35a

11. $$\int_0^{\frac{\pi}{2}} \sin(z \cos x) \sin 2x \, dx = \frac{2}{z^2} (\sin z - z \cos z).$$

LI ((43))(15)

12. $$\int_0^{\frac{\pi}{2}} \sin(z\cos x)\cos ax\,dx = \cos\frac{a\pi}{2}\, s_{0,a}(z) =$$

$$= \frac{\pi}{4}\operatorname{cosec}\frac{a\pi}{2}\left[\mathbf{J}_\nu(z) - \mathbf{J}_{-\nu}(z)\right] =$$

$$= -\frac{\pi}{4}\sec\frac{a\pi}{4}\left[\mathbf{E}_\nu(z) + \mathbf{E}_{-\nu}(z)\right] =$$

$$= \cos\frac{a\pi}{2}\sum_{k=1}^{\infty}\frac{(-1)^{k-1}z^{2k-1}}{(1^2-a^2)(3^2-a^2)\ldots[(2k-1)^2-a^2]} \quad [a>0].$$ WA 339

13. $$\int_0^{\pi} \sin(z\cos x)\cos nx\,dx = \frac{1}{2}\int_{-\pi}^{\pi}\sin(z\cos x)\cos nx\,dx =$$

$$= \pi\sin\frac{n\pi}{2} J_n(z).$$ GW ((334))(55b)

14. $$\int_0^{\frac{\pi}{2}} \sin(z\cos x)\cos[(2n+1)x]\,dx = (-1)^n\frac{\pi}{2}J_{2n+1}(z).$$ WA 30(8)

15. $$\int_0^{\frac{\pi}{2}} \sin(a\cos x)\operatorname{tg} x\,dx = \operatorname{si}(a) + \frac{\pi}{2} \quad [a>0].$$ BI ((43))(17)

16. $$\int_0^{\frac{\pi}{2}} \sin(z\cos x)\sin^{2\nu} x\,dx = \frac{\sqrt{\pi}}{2}\left(\frac{2}{z}\right)^{\nu}\Gamma\left(\nu+\frac{1}{2}\right)\mathbf{H}_\nu(z)$$

$$\left[\operatorname{Re}\nu > -\frac{1}{2}\right].$$ WA 358(1)

17. $$\int_0^{\frac{\pi}{2}} \cos(z\cos x)\cos ax\,dx = -a\sin\frac{a\pi}{2}\, s_{-1,a}(z) =$$

$$= \frac{\pi}{4}\sec\frac{a\pi}{2}\left[\mathbf{J}_\nu(z) + \mathbf{J}_{-\nu}(z)\right] = \frac{\pi}{4}\operatorname{cosec}\frac{a\pi}{2}\left[\mathbf{E}_\nu(z) - \mathbf{E}_{-\nu}(z)\right] =$$

$$= -a\sin\frac{a\pi}{2}\left\{-\frac{1}{a^2} + \sum_{k=1}^{\infty}\frac{(-1)^{k-1}z^{2k}}{a^2(2^2-a^2)(4^2-a^2)\ldots[(2k)^2-a^2]}\right\} \quad [a>0].$$

WA 339

18. $$\int_0^{\pi} \cos(z\cos x)\cos nx\,dx = \frac{1}{2}\int_{-\pi}^{\pi}\cos(z\cos x)\cos nx\,dx =$$

$$= \pi\cos\frac{n\pi}{2} J_n(z).$$ GW ((334))(56b)

19. $$\int_0^{\frac{\pi}{2}} \cos(z\cos x)\cos 2nx\,dx = (-1)^n\cdot\frac{\pi}{2}J_{2n}(z).$$ WA 30(9)

20. $$\int_0^{\frac{\pi}{2}} \cos(z\cos x)\sin^{2\nu}x\,dx = \frac{\sqrt{\pi}}{2}\left(\frac{2}{z}\right)^{\nu}\Gamma\left(\nu+\frac{1}{2}\right)J_\nu(z)$$

$$\left[\operatorname{Re}\nu > -\frac{1}{2}\right].$$ WA 35, WH

21. $$\int_0^{\pi} \cos(z\cos x)\sin^{2\mu}x\,dx = \sqrt{\pi}\left(\frac{2}{z}\right)^{\mu}\Gamma\left(\mu+\frac{1}{2}\right)J_\mu(z)$$

$$\left[\operatorname{Re}\mu > -\frac{1}{2}\right].$$ WH

3.716

1. $$\int_0^{\frac{\pi}{2}} \sin(a\operatorname{tg}x)\,dx = \frac{1}{2}\left[e^{-a}\overline{\operatorname{Ei}}(a) - e^{a}\operatorname{Ei}(-a)\right]$$ (cf. 3.723 1.).

BI ((43))(1)

2. $$\int_0^{\frac{\pi}{2}} \cos(a\operatorname{tg}x)\,dx = \frac{\pi}{2}e^{-a}.$$ BI ((43))(2)

3. $$\int_0^{\frac{\pi}{2}} \sin(a\operatorname{tg}x)\sin 2x\,dx = \frac{a\pi}{2}e^{-a}.$$ BI ((43))(7)

4. $$\int_0^{\frac{\pi}{2}} \cos(a\operatorname{tg}x)\sin^2 x\,dx = \frac{1-a}{4}\pi e^{-a}.$$ BI ((43))(8)

5. $$\int_0^{\frac{\pi}{2}} \cos(a\operatorname{tg}x)\cos^2 x\,dx = \frac{1+a}{4}\pi e^{-a}.$$ BI ((43))(9)

6. $$\int_0^{\frac{\pi}{2}} \sin(a\operatorname{tg}x)\operatorname{tg}x\,dx = \frac{\pi}{2}e^{-a}.$$ BI ((43))(5)

7. $$\int_0^{\frac{\pi}{2}} \cos(a\operatorname{tg}x)\operatorname{tg}x\,dx = -\frac{1}{2}\left[e^{-a}\overline{\operatorname{Ei}}(a) + e^{a}\operatorname{Ei}(-a)\right]$$ (cf. 3.723 5.).

BI ((43))(6)

8. $$\int_0^{\frac{\pi}{2}} \sin(a\operatorname{tg}x)\sin^2 x\operatorname{tg}x\,dx = \frac{2-a}{4}\pi e^{-a}.$$ BI ((43))(11)

9. $$\int_0^{\frac{\pi}{2}} \sin^2(a\operatorname{tg}x)\,dx = \frac{\pi}{4}\left(1-e^{-2a}\right)$$ (cf. 3.742 1.). BI ((43))(3)

10. $\int_0^{\frac{\pi}{2}} \cos^2(a \operatorname{tg} x)\, dx = \frac{\pi}{4}(1 + e^{-2a})$ (cf. 3.742 3.). BI ((43))(4)

11. $\int_0^{\frac{\pi}{2}} \sin^2(a \operatorname{tg} x) \operatorname{ctg}^2 x\, dx = \frac{\pi}{4}(e^{-2a} + 2a - 1).$ BI ((43))(19)

12. $\int_0^{\frac{\pi}{2}} [1 - \sec^2 x \cos(\operatorname{tg} x)] \frac{dx}{\operatorname{tg} x} = C.$ BI ((51))(14)

13. $\int_0^{\frac{\pi}{2}} \sin(a \operatorname{ctg} x) \sin 2x\, dx = \frac{a\pi}{2} e^{-a}$ (cf. 3.716 3.),

and in general, formulas **3.716** remain valid if we replace $\operatorname{tg} x$ in the argument of the sine or cosine with $\operatorname{ctg} x$, if we also replace $\sin x$ with $\cos x$, $\cos x$ with $\sin x$, hence $\operatorname{tg} x$ with $\operatorname{ctg} x$, $\operatorname{ctg} x$ with $\operatorname{tg} x$, $\sec x$ with $\operatorname{cosec} x$, and $\operatorname{cosec} x$ with $\sec x$ in the factors. Analogously,

3.717 $\int_0^{\frac{\pi}{2}} \sin(a \operatorname{cosec} x) \sin(a \operatorname{ctg} x) \frac{dx}{\cos x} =$

$$= \int_0^{\frac{\pi}{2}} \sin(a \sec x) \sin(a \operatorname{tg} x) \frac{dx}{\sin x} = \frac{\pi}{2} \sin a \qquad [a > 0].$$

BI ((52))(11, 12)

3.718

1. $\int_0^{\frac{\pi}{2}} \sin\left(\frac{\pi}{2} p - a \operatorname{tg} x\right) \operatorname{tg}^{p-1} x\, dx =$

$$= \int_0^{\frac{\pi}{2}} \cos\left(\frac{\pi}{2} p - a \operatorname{tg} x\right) \operatorname{tg}^p x\, dx = \frac{\pi}{2} e^{-a} \qquad [p^2 < 1].$$

BI ((44))(5, 6)

2. $\int_0^{\frac{\pi}{2}} \sin(a \operatorname{tg} x - \nu x) \sin^{\nu-2} x\, dx = 0$ $[\operatorname{Re} \nu > 0]$. NH 157(15)

3. $\int_0^{\frac{\pi}{2}} \sin(n \operatorname{tg} x + \nu x) \frac{\cos^{\nu-1} x}{\sin x}\, dx = \frac{\pi}{2}$ $[\operatorname{Re} \nu > 0]$. BI ((51))(15)

4. $$\int_0^{\frac{\pi}{2}} \cos(a \operatorname{tg} x - \nu x) \cos^{\nu-2} x \, dx = \frac{\pi e^{-a} a^{\nu-1}}{\Gamma(\nu)} \qquad [\operatorname{Re} \nu > 1].$$

LO V 153(112), NT 157(14)

5. $$\int_0^{\frac{\pi}{2}} \cos(a \operatorname{tg} x + \nu x) \cos^{\nu} x \, dx = 2^{-\nu-1} \pi e^{-a} \qquad [\operatorname{Re} \nu > -1].$$

BI ((44))(4)

6. $$\int_0^{\frac{\pi}{2}} \cos(a \operatorname{tg} x - \gamma x) \cos^{\nu} x \, dx =$$
$$= \frac{\pi a^{\frac{\nu}{2}}}{2^{\frac{\nu}{2}+1}} \cdot \frac{W_{\frac{\gamma}{2}, -\frac{\nu+1}{2}}(2a)}{\Gamma\left(1+\frac{\gamma+\nu}{2}\right)} \qquad \left[a > 0, \ \operatorname{Re} \nu > -1, \ \frac{\nu+\gamma}{2} \neq -1, -2, \ldots\right].$$

EH I 274(13)

7. $$\int_0^{\frac{\pi}{2}} \frac{\sin nx - \sin(nx - a \operatorname{tg} x)}{\sin x} \cos^{n-1} x \, dx = \pi.$$ LO V 153(114)

3.719

1. $$\int_0^{\frac{\pi}{2}} \sin(\nu x - z \sin x) \, dx = \pi \mathbf{E}_\nu(z).$$ WA 336(2)

2. $$\int_0^{\pi} \cos(nx - z \sin x) \, dx = \pi J_n(z).$$ WH

3. $$\int_0^{\pi} \cos(\nu x - z \sin x) \, dx = \pi \mathbf{J}_\nu(z).$$ WA 336(1)

3.72-3.74 Combinations of trigonometric and rational functions

3.721

1. $$\int_0^{\infty} \frac{\sin(ax)}{x} dx = \frac{\pi}{2} \operatorname{sign} a.$$ FI II 645

2. $$\int_1^{\infty} \frac{\sin(ax)}{x} dx = -\operatorname{si}(a).$$ BI 203(1)

3. $$\int_1^{\infty} \frac{\cos(ax)}{x} dx = -\operatorname{ci}(a).$$ BI 203(5)

3.722

1. $\int_0^\infty \frac{\sin(ax)}{x+\beta}\,dx = \operatorname{ci}(a\beta)\sin(a\beta) - \cos(a\beta)\operatorname{si}(a\beta)$

$[|\arg\beta| < \pi,\ a > 0]$. BI((160))(1), FI II 646a

2. $\int_{-\infty}^\infty \frac{\sin(ax)}{x+\beta}\,dx = \pi\cos(a\beta)$ $[|\arg\beta| < \pi,\ a > 0]$. BI ((202))(1)

3. $\int_0^\infty \frac{\cos(ax)}{x+\beta}\,dx = -\sin(a\beta)\operatorname{si}(a\beta) - \cos(a\beta)\operatorname{ci}(a\beta)$

$[|\arg\beta| < \pi,\ a > 0]$. ET I 8(7), BI((160))(2)

4. $\int_{-\infty}^\infty \frac{\cos(ax)}{x+\beta}\,dx = \pi\sin(a\beta)$ $[|\arg\beta| < \pi,\ a > 0]$. BI ((202))(4)

5. $\int_0^\infty \frac{\sin(ax)}{\beta-x}\,dx = \sin(\beta a)\operatorname{ci}(\beta a) - \cos(\beta a)[\operatorname{si}(\beta a)+\pi]$

$[a > 0]$. FI II 646, BI((161))(1)

6. $\int_{-\infty}^\infty \frac{\sin(ax)}{\beta-x}\,dx = -\pi\cos(a\beta)$ $[a > 0]$. BI ((202))(3)

7. $\int_0^\infty \frac{\cos(ax)}{\beta-x}\,dx = \cos(a\beta)\operatorname{ci}(a\beta) + \sin(a\beta)[\operatorname{si}(a\beta)+\pi]$

$[a > 0]$. ET I 8(8), BI((161))(2)a

8. $\int_{-\infty}^\infty \frac{\cos(ax)}{\beta-x}\,dx = \pi\sin(a\beta)$ $[a > 0]$. BI ((202))(6)

3.723

1. $\int_0^\infty \frac{\sin(ax)}{\beta^2+x^2}\,dx = \frac{1}{2\beta}\left[e^{-a\beta}\overline{\operatorname{Ei}}(a\beta) - e^{a\beta}\operatorname{Ei}(-a\beta)\right]$

$[a > 0,\ \operatorname{Re}\beta > 0]$. ET I 65(15), BI((160))(3)

2. $\int_0^\infty \frac{\cos(ax)}{\beta^2+x^2}\,dx = \frac{\pi}{2\beta}e^{-a\beta}$ $[a > 0,\ \operatorname{Re}\beta > 0]$.

FI II 741, 750, ET I 8(11), WH

3. $\int_0^\infty \frac{x\sin(ax)}{\beta^2+x^2}\,dx = \frac{\pi}{2}e^{-a\beta}$ $[a > 0,\ \operatorname{Re}\beta > 0]$.

FI II 741, 750, ET I 65(15), WH

4. $\int_{-\infty}^\infty \frac{x\sin(ax)}{\beta^2+x^2}\,dx = \pi e^{-a\beta}$ $[a > 0,\ \operatorname{Re}\beta > 0]$.

BI ((202))(10)

5. $$\int_0^\infty \frac{x\cos(ax)}{\beta^2+x^2}\,dx = -\frac{1}{2}\left[e^{-a\beta}\,\overline{\mathrm{Ei}}\,(a\beta)+e^{a\beta}\,\mathrm{Ei}\,(-a\beta)\right]$$

$[a>0,\ \mathrm{Re}\,\beta>0]$. BI ((160))(6)

6. $$\int_{-\infty}^\infty \frac{\sin[a(b-x)]}{c^2+x^2}\,dx = \frac{\pi}{c}\,e^{-ac}\sin(ab) \qquad [a>0,\ b>0,\ c>0].$$

LI ((202))(9)

7. $$\int_{-\infty}^\infty \frac{\cos[a(b-x)]}{c^2+x^2}\,dx = \frac{\pi}{c}\,e^{-ac}\cos(ab) \qquad [a>0,\ b>0,\ c>0].$$

LI ((202))(11)a

8. $$\int_0^\infty \frac{\sin(ax)}{\beta^2-x^2}\,dx = \frac{1}{\beta}\left[\sin(a\beta)\,\mathrm{ci}\,(a\beta)-\cos(a\beta)\left(\mathrm{si}\,(a\beta)+\frac{\pi}{2}\right)\right]$$

$[|\arg\beta|<\pi,\ a>0]$. BI ((161))(3)

9. $$\int_0^\infty \frac{\cos(ax)}{b^2-x^2}\,dx = \frac{\pi}{2b}\sin(ab) \qquad [a>0,\ b>0].$$

BI((161))(5), ET I 9(15)

10. $$\int_0^\infty \frac{x\sin(ax)}{b^2-x^2}\,dx = -\frac{\pi}{2}\cos(ab) \qquad [a>0].$$ FI II 647, ET II 252(45)

11. $$\int_0^\infty \frac{x\cos(ax)}{\beta^2-x^2}\,dx = \cos(a\beta)\,\mathrm{ci}\,(a\beta)+\sin(a\beta)\left[\mathrm{si}\,(a\beta)+\frac{\pi}{2}\right]$$

$[|\arg\beta|<\pi,\ a>0]$. BI ((161))(6)

12. $$\int_{-\infty}^\infty \frac{\sin(ax)}{x(x-b)}\,dx = \pi\,\frac{\cos(ab)-1}{b} \qquad [a>0,\ b>0].$$ ET II 252(44)

3.724

1. $$\int_{-\infty}^\infty \frac{b+cx}{p+2qx+x^2}\sin(ax)\,dx = \left(\frac{cq-b}{\sqrt{p-q^2}}\sin(aq)+c\cos(aq)\right)\pi e^{-a\sqrt{p-q^2}}$$

$[a>0,\ p>q^2]$. BI ((202))(12)

2. $$\int_{-\infty}^\infty \frac{b+cx}{p+2qx+x^2}\cos(ax)\,dx = \left(\frac{b-cq}{\sqrt{p-q^2}}\cos(aq)+c\sin(aq)\right)\pi e^{-a\sqrt{p-q^2}}$$

$[a>0,\ p>q^2]$. BI ((202))(13)

3. $$\int_{-\infty}^\infty \frac{\cos[(b-1)t]-x\cos(bt)}{1-2x\cos t+x^2}\cos(ax)\,dx = \pi e^{-a\sin t}\sin(bt+a\cos t)$$

$[a>0,\ t^2<\pi^2]$. BI ((202))(14)

3.725

1. $$\int_0^\infty \frac{\sin(ax)\,dx}{x(\beta^2+x^2)} = \frac{\pi}{2\beta^2}\left(1-e^{-a\beta}\right) \qquad [\operatorname{Re}\beta > 0,\ a > 0].$$ BI ((172))(1)

2. $$\int_0^\infty \frac{\sin(ax)\,dx}{x(b^2-x^2)} = \frac{\pi}{2b^2}\left(1-\cos(ab)\right) \qquad [a > 0].$$ BI ((172))(4)

3. $$\int_0^\infty \frac{\sin(ax)\cos(bx)}{x(x^2+\beta^2)}\,dx = \frac{\pi}{2\beta^2} e^{-\beta b}\operatorname{sh}(a\beta) \qquad [0 < a < b];$$
$$= -\frac{\pi}{2\beta^2} e^{-a\beta}\operatorname{ch}(b\beta) + \frac{\pi}{2\beta^2} \qquad [a > b > 0].$$ ET I 19(4)

3.726

1. $$\int_0^\infty \frac{x\sin(ax)\,dx}{b^3 \pm b^2x + bx^2 \pm x^3} = \pm\frac{1}{4b}\left[e^{-ab}\,\overline{\operatorname{Ei}}(ab) - e^{ab}\operatorname{Ei}(-ab) - \right.$$
$$\left. - 2\operatorname{ci}(ab)\sin(ab) + 2\cos(ab)\left(\operatorname{si}(ab)+\frac{\pi}{2}\right)\right] + \frac{\pi e^{-ab} - \pi\cos(ab)}{4b}$$

$[a > 0,\ b > 0$; if the lower sign is taken, the above expression indicates the principal value]. ET I 65(21)a, BI((176))(10, 13)

2. $$\int_0^\infty \frac{x^2\sin(ax)\,dx}{b^3 \pm b^2x + bx^2 \pm x^3} = \frac{1}{4}\left[e^{ab}\operatorname{Ei}(-ab) - e^{-ab}\,\overline{\operatorname{Ei}}(ab) + \right.$$
$$\left. + 2\operatorname{ci}(ab)\sin(ab) - 2\cos(ab)\left(\operatorname{si}(ab)+\frac{\pi}{2}\right)\right] \pm \pi\left(e^{-ab} + \cos(ab)\right)$$

$[a > 0,\ b > 0$; if the lower sign is taken, the above expression indicates the principal value]. ET I 66(22), BI((176))(11, 14)

3.727

1. $$\int_0^\infty \frac{\cos(ax)\,dx}{b^4+x^4} = \frac{\pi\sqrt{2}}{4b^3}\exp\left(-\frac{ab}{\sqrt{2}}\right)\left(\cos\frac{ab}{\sqrt{2}} + \sin\frac{ab}{\sqrt{2}}\right)$$
$$[a > 0,\ b > 0].$$ BI((160))(25)a, ET I 9(19)

2. $$\int_0^\infty \frac{\sin(ax)\,dx}{b^4-x^4} = \frac{1}{4b^3}\left[2\sin(ab)\operatorname{ci}(ab) - 2\cos(ab)\left(\operatorname{si}(ab)+\frac{\pi}{2}\right) + \right.$$
$$\left. + e^{-ab}\,\overline{\operatorname{Ei}}(ab) - e^{ab}\operatorname{Ei}(-ab)\right] \qquad [a > 0,\ b > 0],$$
(cf. 3.723 1. and 3.723 8.). BI ((161))(12)

3. $$\int_0^\infty \frac{\cos(ax)\,dx}{b^4-x^4} = \frac{\pi}{4b^3}\left[e^{-ab} + \sin(ab)\right] \qquad [a > 0,\ b > 0]$$
(cf. 3.723 2. and 3.723 9.). BI ((161))(16)

4. $$\int_0^\infty \frac{x\sin(ax)\,dx}{b^4+x^4} = \frac{\pi}{2b^2}\exp\left(-\frac{ab}{\sqrt{2}}\right)\sin\frac{ab}{\sqrt{2}} \qquad [a > 0,\ b > 0].$$ BI ((160))(23)a

5. $$\int_0^\infty \frac{x \sin(ax)}{b^4 - x^4} dx = \frac{\pi}{4b^2} [e^{-ab} - \cos(ab)] \qquad [a > 0,\ b > 0],$$

(cf. **3.723** 3. and **3.723** 10.). BI ((161))(13)

6. $$\int_0^\infty \frac{x \cos(ax)\, dx}{b^4 - x^4} = \frac{1}{4b^2} \Big[2 \cos(ab) \operatorname{ci}(ab) + 2 \sin(ab) \Big(\operatorname{si}(ab) + \frac{\pi}{2} \Big) - e^{-ab} \overline{\operatorname{Ei}}(ab) - e^{ab} \operatorname{Ei}(-ab) \Big] \qquad [a > 0,\ b > 0],$$

(cf. **3.723** 5. and **3.723** 11.). BI ((161))(17)

7. $$\int_0^\infty \frac{x^2 \cos(ax)\, dx}{b^4 + x^4} = \frac{\pi \sqrt{2}}{4b} \exp\Big(-\frac{ab}{\sqrt{2}} \Big) \Big(\cos \frac{ab}{\sqrt{2}} - \sin \frac{ab}{\sqrt{2}} \Big)$$

$[a > 0,\ b > 0]$. BI ((160))(26)a

8. $$\int_0^\infty \frac{x^2 \sin(ax)\, dx}{b^4 - x^4} = \frac{1}{4b} \Big[2 \sin(ab) \operatorname{ci}(ab) - 2 \cos(ab) \Big(\operatorname{si}(ab) + \frac{\pi}{2} \Big) - e^{-ab} \overline{\operatorname{Ei}}(ab) + e^{ab} \operatorname{Ei}(-ab) \Big]$$

$[a > 0,\ b > 0]$, (cf. **3.723** 1. and **3.723** 8.). BI ((161))(14)

9. $$\int_0^\infty \frac{x^2 \cos(ax)\, dx}{b^4 - x^4} = \frac{\pi}{4b} (\sin(ab) - e^{-ab}) \qquad [a > 0,\ b > 0],$$

(cf. **3.723** 2. and **3.723** 9.). BI ((161))(18)

10. $$\int_0^\infty \frac{x^3 \sin(ax)}{b^4 + x^4} dx = \frac{\pi}{2} \exp\Big(-\frac{ab}{\sqrt{2}} \Big) \cos \frac{ab}{\sqrt{2}}$$

$[a > 0,\ b > 0]$. BI ((160))(24)

11. $$\int_0^\infty \frac{x^3 \sin(ax)}{b^4 - x^4} dx = \frac{-\pi}{4} [e^{-ab} + \cos(ab)] \qquad [a > 0,\ b > 0],$$

(cf. **3.723** 4. and **3.723** 10.). BI ((161))(15)

12. $$\int_0^\infty \frac{x^3 \cos(ax)\, dx}{b^4 - x^4} = \frac{1}{4} \Big[2 \cos(ab) \operatorname{ci}(ab) + 2 \sin(ab) \Big(\operatorname{si}(ab) + \frac{\pi}{2} \Big) + e^{-ab} \overline{\operatorname{Ei}}(ab) + e^{ab} \operatorname{Ei}(-ab) \Big] \qquad [a > 0,\ b > 0],$$

(cf. **3.723** 5. and **3.723** 11.). BI ((161))(19)

3.728

1. $$\int_0^\infty \frac{\cos(ax)\, dx}{(\beta^2 + x^2)(\gamma^2 + x^2)} = \frac{\pi (\beta e^{-a\gamma} - \gamma e^{-a\beta})}{2\beta\gamma (\beta^2 - \gamma^2)}$$

$[a > 0,\ \operatorname{Re} \beta > 0,\ \operatorname{Re} \gamma > 0]$. BI ((175))(1)

2. $\int_0^\infty \frac{x \sin(ax)\,dx}{(\beta^2+x^2)(\gamma^2+x^2)} = \frac{\pi(e^{-a\beta}-e^{-a\gamma})}{2(\gamma^2-\beta^2)}$

$[a>0, \ \operatorname{Re}\beta>0, \ \operatorname{Re}\gamma>0]$. BI ((174))(1)

3. $\int_0^\infty \frac{x^2 \cos(ax)\,dx}{(\beta^2+x^2)(\gamma^2+x^2)} = \frac{\pi(\beta e^{-a\beta}-\gamma e^{-a\gamma})}{2(\beta^2-\gamma^2)}$

$[a>0, \ \operatorname{Re}\beta>0, \ \operatorname{Re}\gamma>0]$. BI ((175))(2)

4. $\int_0^\infty \frac{x^3 \sin(ax)\,dx}{(\beta^2+x^2)(\gamma^2+x^2)} = \frac{\pi(\beta^2 e^{-a\beta}-\gamma^2 e^{-a\gamma})}{2(\beta^2-\gamma^2)}$

$[a>0, \ \operatorname{Re}\beta>0, \ \operatorname{Re}\gamma>0]$. BI ((174))(2)

5. $\int_0^\infty \frac{\cos(ax)\,dx}{(b^2-x^2)(c^2-x^2)} = \frac{\pi(b\sin(ac)-c\sin(ab))}{2bc(b^2-c^2)}$

$[a>0, \ b>0, \ c>0]$. BI ((175))(3)

6. $\int_0^\infty \frac{x\sin(ax)\,dx}{(b^2-x^2)(c^2-x^2)} = \frac{\pi(\cos(ab)-\cos(ac))}{2(b^2-c^2)}$ $[a>0]$. BI ((174))(3)

7. $\int_0^\infty \frac{x^2\cos(ax)\,dx}{(b^2-x^2)(c^2-x^2)} = \frac{\pi(c\sin(ac)-b\sin(ab))}{2(b^2-c^2)}$

$[a>0, \ b>0, \ c>0]$. BI ((175))(4)

8. $\int_0^\infty \frac{x^3\sin(ax)\,dx}{(b^2-x^2)(c^2-x^2)} = \frac{\pi(b^2\cos(ab)-c^2\cos(ac))}{2(b^2-c^2)}$

$[a>0, \ b>0, \ c>0]$. BI ((174))(4)

3.729

1. $\int_0^\infty \frac{\cos(ax)\,dx}{(b^2+x^2)^2} = \frac{\pi}{4b^3}(1+ab)e^{-ab}$ $[a>0, \ b>0]$ BI ((170))(7)

2. $\int_0^\infty \frac{x\sin(ax)\,dx}{(b^2+x^2)^2} = \frac{\pi}{4b}ae^{-ab}$ $[a>0, \ b>0]$. BI ((170))(3)

3. $\int_0^\infty \cos(px)\frac{1-x^2}{(1+x^2)^2}\,dx = \frac{\pi p}{2}e^{-p}$. BI ((43))(10)a

4. $\int_0^\infty \frac{x^3\sin(ax)\,dx}{(b^2+x^2)^2} = \frac{\pi}{4}(2-ab)e^{-ab}$ $[a>0, \ b>0]$. BI ((170))(4)

3.731 Notations: $2A^2=\sqrt{b^4+c^2}+b^2, \ 2B^2=\sqrt{b^4+c^2}-b^2$,

1. $\int_0^\infty \frac{\cos(ax)\,dx}{(x^2+b^2)^2+c^2} = \frac{\pi}{2c}\,\frac{e^{-aA}(B\cos(aB)+A\sin(aB))}{\sqrt{b^4+c^2}}$

$[a>0, \ b>0, \ c>0]$. BI ((176))(3)

2. $$\int_0^\infty \frac{x \sin(ax)\,dx}{(x^2+b^2)^2+c^2} = \frac{\pi}{2c} e^{-aA} \sin(aB) \qquad [a>0,\ b>0,\ c>0].$$ BI ((176))(1)

3. $$\int_0^\infty \frac{(x^2+b^2)\cos(ax)\,dx}{(x^2+b^2)^2+c^2} = \frac{\pi}{2} \frac{e^{-aA}(A\cos(aB)-B\sin(aB))}{\sqrt{b^4+c^2}}$$
$$[a>0,\ b>0,\ c>0].$$ BI ((176))(4)

4. $$\int_0^\infty \frac{x(x^2+b^2)\sin(ax)\,dx}{(x^2+b^2)^2+c^2} = \frac{\pi}{2} e^{-aA}\cos(aB)$$
$$[a>0,\ b>0,\ c>0].$$ BI ((176))(2)

3.732

1. $$\int_0^\infty \left[\frac{1}{\beta^2+(\gamma-x)^2} - \frac{1}{\beta^2+(\gamma+x)^2}\right]\sin(ax)\,dx = \frac{\pi}{\beta} e^{-a\beta}\sin(a\gamma)$$
$[a>0,\ \mathrm{Re}\,\beta>0,\ \gamma+i\beta$ is not real]. ET I 65(16)

2. $$\int_0^\infty \left[\frac{1}{\beta^2+(\gamma-x)^2} + \frac{1}{\beta^2+(\gamma+x)^2}\right]\cos(ax)\,dx =$$
$$= \frac{\pi}{\beta} e^{-a\beta}\cos(a\gamma) \qquad [a>0,\ |\mathrm{Im}\,\gamma|<\mathrm{Re}\,\beta].$$ ET I 8(13)

3. $$\int_0^\infty \left[\frac{\gamma+x}{\beta^2+(\gamma+x)^2} - \frac{\gamma-x}{\beta^2+(\gamma-x)^2}\right]\sin(ax)\,dx = \pi e^{-a\beta}\cos(a\gamma)$$
$[a>0,\ \mathrm{Re}\,\beta>0,\ \gamma+i\beta$ is not real]. LI ((175))(17)

4. $$\int_0^\infty \left[\frac{\gamma+x}{\beta^2+(\gamma+x)^2} + \frac{\gamma-x}{\beta^2+(\gamma-x)^2}\right]\cos(ax)\,dx = \pi e^{-a\beta}\sin(a\gamma)$$
$$[a>0,\ |\mathrm{Im}\,a|<\mathrm{Re}\,\beta].$$ LI ((176))(21)

3.733

1. $$\int_0^\infty \frac{\cos(ax)\,dx}{x^4+2b^2x^2\cos 2t+b^4} = \frac{\pi}{2b^3}\exp(-ab\cos t)\frac{\sin(t+ab\sin t)}{\sin 2t}$$
$$\left[a>0,\ b>0,\ |t|<\frac{\pi}{2}\right].$$ BI ((176))(7)

2. $$\int_0^\infty \frac{x\sin(ax)\,dx}{x^4+2b^2x^2\cos 2t+b^4} = \frac{\pi}{2b^2}\exp(-ab\cos t)\frac{\sin(ab\sin t)}{\sin 2t}$$
$$\left[a>0,\ b>0,\ |t|<\frac{\pi}{2}\right].$$ BI((176))(5), ET I 66(23)

3. $$\int_0^\infty \frac{x^2\cos(ax)\,dx}{x^4+2b^2x^2\cos 2t+b^4} = \frac{\pi}{2b}\exp(-ab\cos t)\frac{\sin(t-ab\sin t)}{\sin 2t}$$
$$\left[a>0,\ b>0,\ |t|<\frac{\pi}{2}\right].$$ BI ((176))(8)

4. $$\int_0^\infty \frac{x^3 \sin(ax)\, dx}{x^4 + 2b^2x^2 \cos 2t + b^4} = \frac{\pi}{2} \exp(-ab\cos t) \frac{\sin(2t - ab\sin t)}{\sin 2t} \qquad \left[a > 0,\ b > 0,\ |t| < \frac{\pi}{2}\right].$$ BI ((176))(6)

5. $$\int_0^\infty \frac{\sin(ax)\, dx}{x(x^4 + 2b^2x^2 \cos 2t + b^4)} = \frac{\pi}{2b^4}\left[1 - \exp(-ab\cos t)\frac{\sin(2t + ab\sin t)}{\sin 2t}\right] \qquad \left[a > 0,\ b > 0,\ |t| < \frac{\pi}{2}\right].$$ BI ((176))(22)

3.734

1. $$\int_0^\infty \frac{\sin(ax)\, dx}{x(b^4 + x^4)} = \frac{\pi}{2b^4}\left[1 - \exp\left(-\frac{ab}{\sqrt{2}}\right)\cos\frac{ab}{\sqrt{2}}\right] \qquad [a > 0,\ b > 0].$$ BI ((172))(7)

2. $$\int_0^\infty \frac{\sin(ax)\, dx}{x(b^4 - x^4)} = \frac{\pi}{4b^4}[2 - e^{-ab} - \cos(ab)] \qquad [a > 0,\ b > 0].$$ BI ((172))(10)

3.735 $$\int_0^\infty \frac{\sin(ax)\, dx}{x(b^2 + x^2)^2} = \frac{\pi}{2b^4}\left[1 - \frac{1}{2}e^{-ab}(2 + ab)\right] \qquad [a > 0,\ b > 0].$$ WH, BI ((172))(22)

3.736

1. $$\int_0^\infty \frac{\cos(ax)\, dx}{(b^2 + x^2)(b^4 - x^4)} = \frac{\pi}{8b^5}[\sin(ab) + (2 + ab)e^{-ab}] \qquad [a > 0,\ b > 0],$$ (cf. **3.723** 2. and 9. and **3.729** 1.). BI ((176))(5)

2. $$\int_0^\infty \frac{x \sin(ax)\, dx}{(b^2 + x^2)(b^4 - x^4)} = \frac{\pi}{8b^4}[(1 + ab)e^{-ab} - \cos(ab)] \qquad [a > 0,\ b > 0],$$ (cf. **3.723** 3. and 10. and **3.729** 2.). BI ((174))(5)

3. $$\int_0^\infty \frac{x^2 \cos(ax)\, dx}{(b^2 + x^2)(b^4 - x^4)} = \frac{\pi}{8b^3}[\sin(ab) - abe^{-ab}] \qquad [a > 0,\ b > 0],$$ (cf. **3.723** 2. and 9. and **3.729** 1.). BI ((175))(6)

4. $$\int_0^\infty \frac{x^3 \sin(ax)\, dx}{(b^2 + x^2)(b^4 - x^4)} = \frac{\pi}{8b^2}[(1 - ab)e^{-ab} - \cos(ab)] \qquad [a > 0,\ b > 0],$$ (cf. **3.723** 3. and 10. and **3.729** 2.). BI ((174))(6)

5. $$\int_0^\infty \frac{x^4 \cos(ax)\,dx}{(b^2+x^2)(b^4-x^4)} = \frac{\pi}{8b}\left[\sin(ab) + (ab-2)e^{-ab}\right] \quad [a>0,\ b>0],$$

(cf. **3.723** 2. and 9. and **3.729** 1.). BI ((175))(7)

6. $$\int_0^\infty \frac{x^5 \sin(ax)\,dx}{(b^2+x^2)(b^4-x^4)} = \frac{\pi}{8}\left[(ab-3)e^{-ab} - \cos(ab)\right] \quad [a>0,\ b>0],$$

(cf. **3.723** 3. and 10. and **3.729** 2.). BI ((174))(7)

3.737

1. $$\int_0^\infty \frac{\cos(ax)\,dx}{(b^2+x^2)^n} = \frac{\pi e^{-ab}}{(2b)^{2n-1}(n-1)!}\sum_{k=0}^{n-1}\frac{(2n-k-2)!\,(2ab)^k}{k!\,(n-k-1)!};$$

$$= \frac{(-1)^{n-1}\pi}{b^{2n-1}(n-1)!}\left[\frac{d^{n-1}}{dp^{n-1}}\left(\frac{e^{-ab\sqrt{p}}}{\sqrt{p}}\right)\right]_{p=1};$$

$$= \frac{(-1)^{n-1}\pi}{2b^{2n-1}(n-1)!}\left[\frac{d^{n-1}}{dp^{n-1}}\frac{e^{-abp}}{(1+p)^n}\right]_{p=1} \quad [a>0,\ b>0].$$

GW((333))(67b), WA 209, WA 192

2 $$\int_0^\infty \frac{x\sin(ax)\,dx}{(x^2+\beta^2)^{n+1}} = \frac{\pi a e^{-a\beta}}{2^{2n}n!\beta^{2n-1}}\sum_{k=0}^{n-1}\frac{(2n-k-2)!\,(2a\beta)^k}{k!\,(n-k-1)!} \quad [a>0,\ \operatorname{Re}\beta>0].$$

GW ((333))(66c)

3. $$\int_0^\infty \frac{\sin(ax)\,dx}{x(\beta^2+x^2)^{n+1}} = \frac{\pi}{2\beta^{2n+2}}\left[1 - \frac{e^{-a\beta}}{2^n n!}F_n(a\beta)\right]$$

$[a>0,\ \operatorname{Re}\beta>0,\ F_0(z)=1,\ F_1(z)=z+2,\ \ldots,\ F_n(z)= = (z+2n)F_{n-1}(z) - zF'_{n-1}(z)].$ GW ((333))(66e)

4. $$\int_0^\infty \frac{x\sin(ax)\,dx}{(b^2+x^2)^3} = \frac{\pi a}{16b^3}(1+ab)e^{-ab} \quad [a>0,\ b>0].$$

BI((170))(5), ET I 67(35)a

5. $$\int_0^\infty \frac{x\sin(ax)\,dx}{(b^2+x^2)^4} = \frac{\pi a}{96b^5}(3+3ab+a^2b^2)e^{-ab} \quad [a>0,\ b>0].$$

BI((170))(6), ET I 67(35)a

3.738

1. $$\int_0^\infty \frac{x^{m-1}\sin(ax)}{x^{2n}+\beta^{2n}}\,dx = 0 \quad [m \text{ is odd}];$$

$$= -\frac{\pi\beta^{m-2n}}{2n}\sum_{k=1}^{n}\exp\left[-a\beta\sin\frac{(2k-1)\pi}{2n}\right]\times$$

$$\times\left\{\cos\frac{(2k-1)m\pi}{n} + a\beta\cos\frac{(2k-1)\pi}{2n}\right\} \quad [m \text{ is even}];$$

$$\left[a>0,\ |\arg\beta| < \frac{\pi}{2n},\ 0 \leqslant m < 2n\right].$$ ET I 67(38)

2. $$\int_0^\infty \frac{x^{m-1}\cos(ax)}{x^{2n}+\beta^{2n}}dx = 0 \qquad [m \text{ is even}];$$

$$= \frac{\pi\beta^{m-2n}}{2n}\sum_{k=1}^{n}\exp\left[-a\beta\sin\frac{(2k-1)\pi}{2n}\right]\times$$

$$\times\left\{\sin\frac{(2k-1)m\pi}{2n}+a\beta\cos\frac{(2k-1)\pi}{2n}\right\} \qquad [m \text{ is odd}];$$

$$\left[a>0,\ |\arg\beta|<\frac{\pi}{2n},\ 0<m<2n+1\right].$$ BI((161))(20)a, ET I 10(29)

3.739

1. $$\int_0^\infty \frac{\sin(ax)\,dx}{x(x^2+2^2)(x^2+4^2)\dots(x^2+4n^2)} =$$

$$= \frac{\pi(-1)^n}{(2n)!\,2^{2n+1}}\left[2\sum_{k=0}^{n-1}(-1)^k\binom{2n}{k}e^{2(k-n)a}+(-1)^n\binom{2n}{n}\right].$$ LI ((174))(8)

2. $$\int_0^\infty \frac{\cos(ax)\,dx}{(x^2+1^2)(x^2+3^2)\dots[x^2+(2n+1)^2]} =$$

$$= \frac{(-1)^n}{(2n+1)!}\frac{\pi}{2^{2n+1}}\sum_{k=0}^{n}(-1)^k\binom{2n+1}{k}e^{(2k-2n-1)a}.$$ BI((175))(8)

3. $$\int_0^\infty \frac{x\sin(ax)\,dx}{(x^2+1^2)(x^2+3^2)\dots[x^2+(2n+1)^2]} =$$

$$= \frac{\pi(-1)^n}{(2n+1)!\,2^{2n+1}}\sum_{k=0}^{n}(-1)^k\binom{2n+1}{k}(2n-2k+1)e^{(2k-2n-1)a}.$$ LI ((174))(9)

3.741

1. $$\int_0^\infty \frac{\sin(ax)\sin(bx)}{x}dx = \frac{1}{4}\ln\left(\frac{a+b}{a-b}\right)^2 \qquad [a>0,\ b>0,\ a\neq b].$$ FI II 647

2. $$\int_0^\infty \frac{\sin(ax)\cos(bx)}{x}dx = \frac{\pi}{2} \qquad [a>b\geqslant 0];$$

$$= \frac{\pi}{4} \qquad [a=b>0];$$

$$= 0 \qquad [b>a\geqslant 0].$$ FI II 645

3. $$\int_0^\infty \frac{\sin(ax)\sin(bx)}{x^2}dx = \frac{a\pi}{2} \qquad [0<a\leqslant b];$$

$$= \frac{b\pi}{2} \qquad [0<b\leqslant a].$$ BI ((157))(1)

3.742

1. $$\int_0^\infty \frac{\sin(ax)\sin(bx)}{\beta^2+x^2}\,dx = \frac{\pi}{4\beta}\left(e^{-|a-b|\beta} - e^{-(a+b)\beta}\right)$$

$$[a>0,\ b>0,\ \operatorname{Re}\beta>0].$$ BI((162))(1)a, GW((333))(71a)

2. $$\int_0^\infty \frac{\sin(ax)\cos(bx)}{\beta^2+x^2}\,dx = \frac{1}{4\beta}e^{-a\beta}\{e^{b\beta}\operatorname{Ei}[\beta(a-b)] + e^{-b\beta}\operatorname{Ei}[\beta(a+b)]\} - \frac{1}{4\beta}e^{a\beta}\{e^{b\beta}\operatorname{Ei}[-\beta(a+\beta)] + e^{-b\beta}\operatorname{Ei}[\beta(b-a)]\}.$$ BI ((162))(3)

3. $$\int_0^\infty \frac{\cos(ax)\cos(bx)}{\beta^2+x^2}\,dx = \frac{\pi}{4\beta}\left[e^{-|a-b|\beta} + e^{-(a+b)\beta}\right]$$

$$[a>0,\ b>0,\ \operatorname{Re}\beta>0].$$ BI((163))(1)a, GW((333))(71c)

4. $$\int_0^\infty \frac{x\cos(ax)\cos(bx)}{\beta^2+x^2}\,dx = -\frac{1}{4}e^{a\beta}\{e^{b\beta}\operatorname{Ei}[-\beta(a+b)] + e^{-b\beta}\operatorname{Ei}[\beta(b-a)]\} - \frac{1}{4}e^{-a\beta}\{e^{b\beta}\operatorname{Ei}[\beta(a-b)] + e^{-b\beta}\operatorname{Ei}[\beta(a+b)]\} \quad [a\neq b];$$

$$= \infty \quad [a=b].$$ BI ((163))(2)

5. $$\int_0^\infty \frac{x\sin(ax)\cos(bx)}{x^2+\beta^2}\,dx = \frac{\pi}{2}e^{-a\beta}\operatorname{ch}(b\beta) \qquad [0<b<a];$$

$$= \frac{\pi}{4}e^{-2a\beta} \quad [0<b=a];$$

$$= -\frac{\pi}{2}e^{-b\beta}\operatorname{sh}(a\beta) \qquad [0<a<b].$$

BI ((162))(4)

6. $$\int_0^\infty \frac{\sin(ax)\sin(bx)}{p^2-x^2}\,dx = -\frac{\pi}{2p}\cos(ap)\sin(bp) \qquad [a>b>0];$$

$$= -\frac{\pi}{4p}\sin(2ap) \qquad [a=b>0];$$

$$= -\frac{\pi}{2p}\sin(ap)\cos(bp) \qquad [0<b<a].$$

BI ((166))(1)

7. $$\int_0^\infty \frac{\sin(ax)\cos(bx)}{p^2-x^2}\,x\,dx = -\frac{\pi}{2}\cos(ap)\cos(bp) \qquad [a>b>0];$$

$$= -\frac{\pi}{4}\cos(2ap) \qquad [a=b>0];$$

$$= \frac{\pi}{2}\sin(ap)\sin(bp) \qquad [b>a>0].$$

BI ((166))(2)

8. $\int_0^\infty \frac{\cos(ax)\cos(bx)}{p^2-x^2}\,dx = \frac{\pi}{2p}\sin(ap)\cos(bp) \quad [a > b > 0];$

$= \frac{\pi}{4p}\sin(2ap) \quad [a = b > 0];$

$= \frac{\pi}{2p}\cos(ap)\sin(bp) \quad [b > a > 0].$

BI ((166))(3)

3.743

1. $\int_0^\infty \frac{\sin(ax)}{\sin(bx)}\cdot\frac{dx}{x^2+\beta^2} = \frac{\pi}{2\beta}\cdot\frac{\operatorname{sh}(a\beta)}{\operatorname{sh}(b\beta)} \quad [0 < a < b,\ \operatorname{Re}\beta > 0].$ ET I 80(21)

2. $\int_0^\infty \frac{\sin(ax)}{\cos(bx)}\cdot\frac{x\,dx}{x^2+\beta^2} = -\frac{\pi}{2}\cdot\frac{\operatorname{sh}(a\beta)}{\operatorname{ch}(b\beta)} \quad [0 < a < b,\ \operatorname{Re}\beta > 0].$

ET I 81(30)

3. $\int_0^\infty \frac{\cos(ax)}{\sin(bx)}\cdot\frac{x\,dx}{x^2+\beta^2} = \frac{\pi}{2}\cdot\frac{\operatorname{ch}(a\beta)}{\operatorname{sh}(b\beta)} \quad [0 < a < b,\ \operatorname{Re}\beta > 0].$ ET I 23(37)

4. $\int_0^\infty \frac{\cos(ax)}{\cos(bx)}\cdot\frac{dx}{x^2+\beta^2} = \frac{\pi}{2\beta}\cdot\frac{\operatorname{ch}(a\beta)}{\operatorname{ch}(b\beta)} \quad [0 < a < b,\ \operatorname{Re}\beta > 0].$ ET I 23(36)

5. $\int_0^\infty \frac{\sin(2ax)}{\sin x}\cdot\frac{dx}{b^2-x^2} = \frac{\pi}{b}\cdot\frac{\sin^2(ab)}{\sin b} \quad [0 < a < 1,\ b > 0].$ BI ((191))(18)

3.744

1. $\int_0^\infty \frac{\sin(ax)}{\cos(bx)}\cdot\frac{dx}{x(x^2+\beta^2)} = \frac{\pi}{2\beta^2}\cdot\frac{\operatorname{sh}(a\beta)}{\operatorname{ch}(b\beta)} \quad [0 < a < b,\ \operatorname{Re}\beta > 0].$

ET I 82(32)

2. $\int_0^\infty \frac{\sin(ax)}{\cos(bx)}\cdot\frac{dx}{x(c^2-x^2)} = 0 \quad [0 < a < b,\ c > 0].$ ET I 82(31)

3.745

1. $\int_0^\infty \frac{\sin(2ax)}{\sin x}\cdot\frac{dx}{(b^2-x^2)^2} = \frac{\pi}{4b^3}\left[2\frac{\sin^2(ab)}{\sin b} - ab\frac{\sin(2ab)}{\sin b} + \right.$

$\left. + 2b\frac{\cos b}{\sin^2 b}\sin^2(ab)\right] \quad [0 < a < 1,\ b > 0].$ BI ((199))(1)a

2. $\int_0^\infty \frac{\sin(2ax)}{\sin x}\cdot\frac{x^2\,dx}{(b^2-x^2)^2} = \frac{\pi}{4b}\left[-2\frac{\sin^2(ab)}{\sin b} - ab\frac{\sin(2ab)}{\sin b} + \right.$

$\left. + 2b\frac{\cos b}{\sin^2 b}\sin^2(ab)\right] \quad [0 < a < 1,\ b > 0].$ BI ((199))(2)

3.746

1. $$\int_0^\infty \frac{dx}{x^{n+1}} \prod_{k=0}^{n} \sin(a_k x) = \frac{\pi}{2} \prod_{k=1}^{n} a_k \qquad \left[a_0 > \sum_{k=1}^{n} a_k,\ a_k > 0 \right].$$

FI II 646

2. $$\int_0^\infty \frac{\sin(ax)}{x^{n+1}}\, dx \prod_{k=1}^{n} \sin(a_k x) \prod_{j=1}^{m} \cos(b_j x) = \frac{\pi}{2} \prod_{k=1}^{n} a_k$$
$$\left[a > \sum_{k=1}^{n} |a_k| + \sum_{j=1}^{m} |b_j| \right].$$

WH

3.747

1. $$\int_0^{\frac{\pi}{2}} \frac{x^m}{\sin x}\, dx = \left(\frac{\pi}{2}\right)^m \left[\frac{1}{m} + \sum_{k=1}^{\infty} \frac{2^{2k-1}-1}{4^{2k-1}(m+2k)} \zeta(2k) \right].$$

LI ((206))(2)

2. $$\int_0^{\frac{\pi}{2}} \frac{x\, dx}{\sin x} = \int_0^{\frac{\pi}{2}} \frac{\left(\frac{\pi}{2} - x\right) dx}{\cos x} = 2G.$$

BI((204))(18), BI((206))(1), GW((333))(32)

3. $$\int_0^\infty \frac{x\, dx}{(x^2+b^2)\sin(ax)} = \frac{\pi}{2\,\mathrm{sh}(ab)} \qquad [b > 0].$$

GW ((333))(79c)

4. $$\int_0^\pi x \,\mathrm{tg}\, x\, dx = -\pi \ln 2.$$

BI ((218))(4)

5. $$\int_0^{\frac{\pi}{2}} x \,\mathrm{tg}\, x\, dx = \infty.$$

BI ((205))(2)

6. $$\int_0^{\frac{\pi}{4}} x \,\mathrm{tg}\, x\, dx = -\frac{\pi}{8} \ln 2 + \frac{1}{2} G = 0.1857845358 \ldots$$

BI ((204))(1)

7. $$\int_0^{\frac{\pi}{2}} x \,\mathrm{ctg}\, x\, dx = \frac{\pi}{2} \ln 2.$$

FI II 623

8. $$\int_0^{\frac{\pi}{4}} x \,\mathrm{ctg}\, x\, dx = \frac{\pi}{8} \ln 2 + \frac{1}{2} G = 0.730\,181\,0584 \ldots$$

BI ((204))(2)

9. $$\int_0^{\frac{\pi}{2}} \left(\frac{\pi}{2} - x\right) \mathrm{tg}\, x\, dx = \frac{1}{2} \int_0^{\pi} \left(\frac{\pi}{2} - x\right) \mathrm{tg}\, x\, dx = \frac{\pi}{2} \ln 2.$$

GW((333))(33b), BI((218))(12)

10. $$\int_0^\infty \operatorname{tg} ax \frac{dx}{x} = \frac{\pi}{2} \qquad [a > 0].$$ LO V 279(5)

11. $$\int_0^{\frac{\pi}{2}} \frac{x \operatorname{ctg} x}{\cos 2x} dx = \frac{\pi}{4} \ln 2.$$ BI ((206))(12)

3.748

1. $$\int_0^{\frac{\pi}{4}} x^m \operatorname{tg} x \, dx = \frac{1}{2}\left(\frac{\pi}{4}\right)^m \sum_{k=1}^{\infty} \frac{(4^k - 1)\,\zeta(2k)}{4^{2k-1}(m+2k)}.$$ LI ((204))(5)

2. $$\int_0^{\frac{\pi}{2}} x^p \operatorname{ctg} x \, dx = \left(\frac{\pi}{2}\right)^p \left\{\frac{1}{p} - 2\sum_{k=1}^{\infty} \frac{1}{4^k(p+2k)} \zeta(2k)\right\}.$$ LI ((205))(7)

3. $$\int_0^{\frac{\pi}{4}} x^m \operatorname{ctg} x \, dx = \frac{1}{2}\left(\frac{\pi}{4}\right)^m \left[\frac{2}{m} - \sum_{k=1}^{\infty} \frac{\zeta(2k)}{4^{2k-1}(m+2k)}\right].$$ LI ((204))(6)

3.749

1. $$\int_0^\infty \frac{x \operatorname{tg}(ax)\,dx}{x^2+b^2} = \frac{\pi}{e^{2ab}+1} \qquad [a > 0, \ b > 0].$$ GW ((333))(79a)

2. $$\int_0^\infty \frac{x \operatorname{ctg}(ax)\,dx}{x^2+b^2} = \frac{\pi}{e^{2ab}-1} \qquad [a > 0, \ b > 0].$$ GW ((333))(79b)

3. $$\int_0^\infty \frac{x \operatorname{tg}(ax)\,dx}{b^2-x^2} = \int_0^\infty \frac{x \operatorname{ctg}(ax)\,dx}{b^2-x^2} = \int_0^\infty \frac{x \operatorname{cosec}(ax)\,dx}{b^2-x^2} = \infty.$$ BI ((161))(7, 8, 9)

3.75 Combinations of trigonometric and algebraic functions

3.751

1. $$\int_0^\infty \frac{\sin(ax)\,dx}{\sqrt{x+\beta}} = \sqrt{\frac{\pi}{2a}}\left[\cos(a\beta) - \sin(a\beta) + 2C\left(\sqrt{a\beta}\right)\sin(a\beta) - 2S\left(\sqrt{a\beta}\right)\cos(a\beta)\right] \qquad [a > 0, \ |\arg\beta| < \pi].$$ ET I 65(12)a

2. $$\int_0^\infty \frac{\cos(ax)\,dx}{\sqrt{x+\beta}} = \sqrt{\frac{\pi}{2a}}\left[\cos a\beta + \sin(a\beta) - 2C\left(\sqrt{a\beta}\right)\cos(a\beta) - 2S\left(\sqrt{a\beta}\right)\sin(a\beta)\right] \qquad [a > 0, \ |\arg\beta| < \pi].$$ ET I 8(9)a

3. $$\int_u^\infty \frac{\sin(ax)}{\sqrt{x-u}}\,dx = \sqrt{\frac{\pi}{2a}}\,[\sin(au)+\cos(au)] \qquad [a>0,\ u>0].$$

ET I 65(13)

4. $$\int_u^\infty \frac{\cos(ax)}{\sqrt{x-u}}\,dx = \sqrt{\frac{\pi}{2a}}\,[\cos(au)-\sin(au)] \qquad [a>0,\ u>0].$$

ET I 8(10)

3.752

1. $$\int_0^1 \sin(ax)\sqrt{1-x^2}\,dx = \sum_{k=0}^\infty \frac{(-1)^k a^{2k+1}}{(2k+1)!!\,(2k+3)!!} \qquad [a>0].$$

BI ((149))(6)

2. $$\int_0^1 \cos(ax)\sqrt{1-x^2}\,dx = \frac{\pi}{2a} J_1(a).$$ KU 65(6)a

3.753

1. $$\int_0^1 \frac{\sin(ax)\,dx}{\sqrt{1-x^2}} = \sum_{k=0}^\infty \frac{(-1)^k a^{2k+1}}{[(2k+1)!!]^2} \qquad [a>0].$$ BI ((149))(9)

2. $$\int_0^1 \frac{\cos(ax)\,dx}{\sqrt{1-x^2}} = \frac{\pi}{2} J_0(a).$$ WA 30(7)a

3. $$\int_1^\infty \frac{\sin(ax)\,dx}{\sqrt{x^2-1}} = \frac{\pi}{2} J_0(a). \qquad [a>0].$$ WA 200(14)

4. $$\int_1^\infty \frac{\cos(ax)}{\sqrt{x^2-1}}\,dx = -\frac{\pi}{2} N_0(a).$$ WA 200(15)

5. $$\int_0^1 \frac{x\sin(ax)}{\sqrt{1-x^2}}\,dx = \frac{\pi}{2} J_1(a) \qquad [a>0].$$ WA 30(6)

3.754

1. $$\int_0^\infty \frac{\sin(ax)\,dx}{\sqrt{\beta^2+x^2}} = \frac{\pi}{2}[I_0(a\beta)-\mathbf{L}_0(a\beta)] \qquad [a>0,\ \operatorname{Re}\beta>0].$$

ET I 66(26)

2. $$\int_0^\infty \frac{\cos(ax)\,dx}{\sqrt{\beta^2+x^2}} = K_0(a\beta) \qquad [a>0,\ \operatorname{Re}\beta>0].$$

WA 191(1), GW((333))(78a)

3. $$\int_0^\infty \frac{x\sin(ax)}{\sqrt{(\beta^2+x^2)^3}}\,dx = aK_0(a\beta) \qquad [a>0,\ \operatorname{Re}\beta>0].$$

ET I 66(27)

3.755

1. $$\int_0^\infty \frac{\sqrt{\sqrt{x^2+\beta^2}-\beta}\,\sin(ax)\,dx}{\sqrt{x^2+\beta^2}} = \sqrt{\frac{\pi}{2a}}\,e^{-a\beta} \qquad [a>0].$$

ET I 66(31)

2. $$\int_0^\infty \frac{\sqrt{\sqrt{x^2+\beta^2}+\beta}\,\cos(ax)\,dx}{\sqrt{x^2+\beta^2}} = \sqrt{\frac{\pi}{2a}}\,e^{-a\beta} \qquad [a>0,\ \operatorname{Re}\beta>0].$$

ET I 10(25)

3.756

1. $$\int_0^\infty \frac{\sin(ax)}{x^{\frac{n}{2}-1}} \prod_{k=2}^{n} \sin(a_k x)\,dx = 0 \qquad \left[a_k>0,\ a>\sum_{k=2}^{n} a_k\right].$$

ET I 80(22)

2. $$\int_0^\infty x^{\frac{n}{2}-1} \cos(ax) \prod_{k=1}^{n} \cos(a_k x)\,dx = 0 \qquad \left[a_k>0,\ a>\sum_{k=1}^{n} a_k\right].$$

ET I 22(26)

3.757

1. $$\int_0^\infty \frac{\sin(ax)}{\sqrt{x}}\,dx = \sqrt{\frac{\pi}{2a}}.$$ BI ((177))(1)

2. $$\int_0^\infty \frac{\cos(ax)}{\sqrt{x}}\,dx = \sqrt{\frac{\pi}{2a}}.$$ BI ((177))(2)

3.76-3.77 Combinations of trigonometric functions and powers

3.761

1. $$\int_0^1 x^{\mu-1}\sin(ax)\,dx = \frac{-i}{2\mu}\left[{}_1F_1(\mu;\ \mu+1;\ ia) - {}_1F_1(\mu;\ \mu+1;\ -ia)\right]$$
$$[a>0,\ \operatorname{Re}\mu>-1].$$ ET I 68(2)a

2. $$\int_u^\infty x^{\mu-1}\sin x\,dx = \frac{i}{2}\left[e^{-\frac{\pi}{2}i\mu}\,\Gamma(\mu,\ iu) - e^{\frac{\pi}{2}i\mu}\,\Gamma(\mu,\ -iu)\right]$$
$$[\operatorname{Re}\mu>-1].$$ EH II 149(2)

3. $$\int_1^\infty \frac{\sin(ax)}{x^{2n}}\,dx = \frac{a^{2n-1}}{(2n-1)!}\left[\sum_{k=1}^{2n-1} \frac{(2n-k-1)!}{a^{2n-k}} \sin\left(a+(k-1)\frac{\pi}{2}\right) + (-1)^n \operatorname{ci}(a)\right] \qquad [a>0].$$ LI ((203))(15)

4. $$\int_0^\infty x^{\mu-1}\sin(ax)\,dx = \frac{\Gamma(\mu)}{a^\mu}\sin\frac{\mu\pi}{2} = \frac{\pi\sec\frac{\mu\pi}{2}}{2a^\mu\,\Gamma(1-\mu)}$$
$$[a>0;\ 0<|\operatorname{Re}\mu|<1].$$ FI II 809a, BI((150))(1)

5. $$\int_0^\pi x^m \sin(nx)\,dx = \frac{(-1)^{n+1}}{n^{m+1}} \sum_{k=0}^{E\left(\frac{m}{2}\right)} (-1)^k \frac{m!}{(m-2k)!}(n\pi)^{m-2k} - -(-1)^{E\left(\frac{m}{2}\right)} \frac{m!\left[m-2E\left(\frac{m}{2}\right)-1\right]}{n^{m+1}}$$ GW((333))(6)

6. $$\int_0^1 x^{\mu-1}\cos(ax)\,dx = \frac{1}{2\mu}\left[{}_1F_1(\mu;\ \mu+1;\ ia) + {}_1F_1(\mu,\ \mu+1;\ -ia)\right]$$ $[a > 0,\ \operatorname{Re}\mu > 0]$. ET I 11(2)

7. $$\int_u^\infty x^{\mu-1}\cos x\,dx = \frac{1}{2}\left[e^{-\frac{\pi}{2}i\mu}\,\Gamma(\mu,\ iu) + e^{\frac{\pi}{2}i\mu}\,\Gamma(\mu,\ -iu)\right]$$ $[\operatorname{Re}\mu < 1]$. EH II 149(1)

8. $$\int_1^\infty \frac{\cos(ax)}{x^{2n+1}}\,dx = \frac{a^{2n}}{(2n)!}\left[\sum_{k=1}^{2n} \frac{(2n-k)!}{a^{2n-k+1}}\cos\left(a+(k-1)\frac{\pi}{2}\right) + (-1)^{n+1}\operatorname{ci}(a)\right]$$ $[a > 0]$. LI ((203))(16)

9. $$\int_0^\infty x^{\mu-1}\cos(ax)\,dx = \frac{\Gamma(\mu)}{a^\mu}\cos\frac{\mu\pi}{2} = \frac{\pi\operatorname{cosec}\frac{\mu\pi}{2}}{2a^\mu\Gamma(1-\mu)}$$ $[a > 0,\ 0 < \operatorname{Re}\mu < 1]$. FI II 809a, BI((150))(2)

10. $$\int_0^\pi x^m\cos(nx)\,dx = \frac{(-1)^n}{n^{m+1}} \sum_{k=0}^{E\left(\frac{m-1}{2}\right)} (-1)^k \frac{m!}{(m-2k-1)!}(n\pi)^{m-2k-1} + +(-1)^{E\left(\frac{m+1}{2}\right)} \frac{2E\left(\frac{m+1}{2}\right)-m}{n^{m+1}}\cdot m!$$ GW ((333))(7)

11. $$\int_0^{\frac{\pi}{2}} x^m\cos x\,dx = \sum_{k=0}^{E\left[\frac{m}{2}\right]} (-1)^k \frac{m!}{(m-2k)!}\left(\frac{\pi}{2}\right)^{m-2k} + +(-1)^{E\left(\frac{m}{2}\right)}\left[2E\left(\frac{m}{2}\right)-m\right]m!.$$ GW ((333))(9c)

12. $$\int_0^{2n\pi} x^m\cos kx\,dx = -\sum_{j=0}^{m-1} \frac{j!}{k^{j+1}}\binom{m}{j}(2n\pi)^{m-j}\cos\frac{j+1}{2}\pi.$$ BI ((226))(2)

3.762

1. $$\int_0^\infty x^{\mu-1}\sin(ax)\sin(bx)\,dx = \frac{1}{2}\cos\frac{\mu\pi}{2}\,\Gamma(\mu)\left[|b-a|^{-\mu} - (b+a)^{-\mu}\right]$$ $[a > 0,\ b > 0,\ a \neq b,\ -2 < \operatorname{Re}\mu < 1]$

(for $\mu=0$, see 3.741 1., for $\mu=-1$, see 3.741 3.). BI((149))(7), ET I 321(40)

2. $$\int_0^\infty x^{\mu-1}\sin(ax)\cos(bx)\,dx = \frac{1}{2}\sin\frac{\mu\pi}{2}\Gamma(\mu)[(a+b)^{-\mu} + |a-b|^{-\mu}\operatorname{sign}(a-b)] \qquad [a>0,\ b>0,\ |\operatorname{Re}\mu|<1]$$

(for $\mu=0$ see 3.741 2.). BI((159))(8)a, ET I 321(41)

3. $$\int_0^\infty x^{\mu-1}\cos(ax)\cos(bx)\,dx = \frac{1}{2}\cos\frac{\mu\pi}{2}\Gamma(\mu)[(a+b)^{-\mu}+|a-b|^{-\mu}]$$

$[a>0,\ b>0,\ 0<\operatorname{Re}\mu<1]$. ET I 20(17)

3.763

1. $$\int_0^\infty \frac{\sin(ax)\sin(bx)\sin(cx)}{x^\nu}dx = \frac{1}{4}\cos\frac{\nu\pi}{2}\Gamma(1-\nu)[(c+a-b)^{\nu-1} - (c+a+b)^{\nu-1} - |c-a+b|^{\nu-1}\operatorname{sign}(a-b-c) + |c-a-b|^{\nu-1}\operatorname{sign}(a+b-c)]$$

$[c>0,\ 0<\operatorname{Re}\nu<4,\ \nu\neq 1, 2, 3,\ a\geqslant b>0]$. GW(333))(26a)a, ET I 79(13)

2. $$\int_0^\infty \frac{\sin(ax)\sin(bx)\sin(cx)}{x}dx = 0 \qquad [c<a-b \text{ and } c>a+b];$$

$$= \frac{\pi}{8} \qquad [c=a-b \text{ and } c=a+b];$$

$$= \frac{\pi}{4} \qquad [a-b<c<a+b]$$

$[a\geqslant b>0,\ c>0]$. FI II 645

3. $$\int_0^\infty \frac{\sin(ax)\sin(bx)\sin(cx)}{x^2}dx = \frac{1}{4}(c+a+b)\ln(c+a+b) - \frac{1}{4}(c+a-b)\ln(c+a-b) - \frac{1}{4}|c-a-b|\ln|c-a-b|\times$$
$$\times\operatorname{sign}(a+b-c) + \frac{1}{4}|c-a+b|\ln|c-a+b|\operatorname{sign}(a-b-c)$$

$[a\geqslant b>0,\ c>0]$. BI((157))(8)a, ET I 79(11)

4. $$\int_0^\infty \frac{\sin(ax)\sin(bx)\sin(cx)}{x^3}dx = \frac{\pi bc}{2} \qquad [0<c<a-b \text{ and } c>a+b];$$

$$= \frac{\pi bc}{2} - \frac{\pi(a-b-c)^2}{8} \qquad [a-b<c<a+b];$$

$[a\geqslant b>0,\ c>0]$. BI((157))(20), ET I 79(12)

3.764

1. $$\int_0^\infty x^p\sin(ax+b)\,dx = \frac{1}{a^{p+1}}\Gamma(1+p)\cos\left(b+\frac{p\pi}{2}\right)$$

$[a>0,\ -1<p<0]$. GW ((333))(30a)

2. $$\int_0^\infty x^p\cos(ax+b)\,dx = -\frac{1}{a^{p+1}}\Gamma(1+p)\sin\left(b+\frac{\pi p}{2}\right)$$

$[a>0,\ -1<p<0]$. GW ((333))(30b)

3.765

1. $$\int_0^\infty \frac{\sin(ax)\,dx}{x^\nu(x+\beta)} = \frac{i}{2\beta^\nu}\Gamma(1-\nu)\left[e^{-ia\beta}\Gamma(\nu, -ia\beta) - e^{ia\beta}\Gamma(\nu, ia\beta)\right]$$
$$[a>0, \quad -1<\operatorname{Re}\nu<2, \quad |\arg\beta|<\pi].$$
ET I 219(34)

2. $$\int_0^\infty \frac{\cos(ax)\,dx}{x^\nu(x+\beta)} = \frac{\Gamma(1-\nu)}{2\beta^\nu}\left[e^{ia\beta}\Gamma(\nu, ia\beta) + e^{-ia\beta}\Gamma(\nu, -ia\beta)\right]$$
$$[a>0, \quad |\operatorname{Re}\nu|<1, \quad |\arg\beta|<\pi].$$
ET II 221(52)

3.766

1. $$\int_0^\infty \frac{x^{\mu-1}\sin(ax)}{1+x^2}\,dx = \frac{\pi}{2}\sec\frac{\mu\pi}{2}\operatorname{sh}a +$$
$$+\frac{1}{2}\sin\frac{\mu\pi}{2}\Gamma(\mu)\left\{\exp\left[-a+i\pi(1-\mu)\right]\gamma(1-\mu, -a) - e^a\gamma(1-\mu, a)\right\}$$
$$[a>0, \quad -1<\operatorname{Re}\mu<3].$$
ET I 317(4)

2. $$\int_0^\infty \frac{x^{\mu-1}\cos(ax)}{1+x^2}\,dx = \frac{\pi}{2}\operatorname{cosec}\frac{\mu\pi}{2}\operatorname{ch}a +$$
$$+\frac{1}{2}\cos\frac{\mu\pi}{2}\Gamma(\mu)\left\{\exp\left[-a+i\pi(1-\mu)\right]\gamma(1-\mu, -a) - e^a\gamma(1-\mu, a)\right\}$$
$$[a>0, \quad 0<\operatorname{Re}\mu<3].$$
ET I 319(24)

3. $$\int_0^\infty \frac{x^{2\mu+1}\sin(ax)\,dx}{x^2+b^2} = -\frac{\pi}{2}b^{2\mu}\sec(\mu\pi)\operatorname{sh}(ab) -$$
$$-\frac{\sin(\mu\pi)}{a^{2\mu}}\Gamma(2\mu)\left[{}_1F_1(1;\ 1-2\mu;\ ab) + {}_1F_1(1;\ 1-2\mu;\ -ab)\right]$$
$$\left[a>0, \quad -\frac{3}{2}<\operatorname{Re}\mu<\frac{1}{2}\right].$$
ET II 220(39)

4. $$\int_0^\infty \frac{x^{2\mu+1}\cos(ax)\,dx}{x^2+b^2} = -\frac{\pi}{2}b^{2\mu}\operatorname{cosec}(\mu\pi)\operatorname{ch}(ab) -$$
$$-\frac{\cos(\mu\pi)}{2a^{2\mu}}\Gamma(2\mu)\left[{}_1F_1(1;\ 1-2\mu;\ ab) + {}_1F_1(1;\ 1-2\mu;\ -ab)\right]$$
$$\left[a>0, \quad -1<\operatorname{Re}\mu<\frac{1}{2}\right].$$
ET II 221(56)

3.767

1. $$\int_0^\infty \frac{x^{\beta-1}\sin\left(ax-\frac{\beta\pi}{2}\right)}{\gamma^2+x^2}\,dx = -\frac{\pi}{2}\gamma^{\beta-2}e^{-a\gamma}$$
$$[a>0, \quad \operatorname{Re}\gamma>0, \quad 0<\operatorname{Re}\beta<2].$$
BI ((160))(20)

2. $$\int_0^\infty \frac{x^{\beta}\cos\left(ax-\frac{\beta\pi}{2}\right)}{\gamma^2+x^2}\,dx = \frac{\pi}{2}\gamma^{\beta-1}e^{-a\gamma}$$
$$[a>0, \quad \operatorname{Re}\gamma>0, \quad |\operatorname{Re}\beta|<1].$$
BI ((160))(21)

3. $$\int_0^\infty \frac{x^{\beta-1}\sin\left(ax-\frac{\beta\pi}{2}\right)}{x^2-b^2}\,dx = \frac{\pi}{2}\,b^{\beta-2}\cos\left(ab-\frac{\pi\beta}{2}\right)$$

$[a>0,\ b>0,\ 0<\operatorname{Re}\beta<2]$. BI ((161))(11)

4. $$\int_0^\infty \frac{x^{\beta}\cos\left(ax-\frac{\beta\pi}{2}\right)}{x^2-b^2}\,dx = -\frac{\pi}{2}\,b^{\beta-1}\sin\left(ab-\frac{\beta\pi}{2}\right)$$

$[a>0,\ b>0,\ |\beta|<1]$. GW ((333))(82)

3.768

1. $$\int_u^\infty (x-u)^{\mu-1}\sin(ax)\,dx = \frac{\Gamma(\mu)}{a^\mu}\sin\left(au+\frac{\mu\pi}{2}\right)$$

$[a>0,\ 0<\operatorname{Re}\mu<1]$. ET II 203(19)

2. $$\int_u^\infty (x-u)^{\mu-1}\cos(ax)\,dx = \frac{\Gamma(\mu)}{a^\mu}\cos\left(au+\frac{\mu\pi}{2}\right)$$

$[a>0,\ 0<\operatorname{Re}\mu<1]$. ET II 204(24)

3. $$\int_0^1 (1-x)^\nu\sin(ax)\,dx = \frac{1}{a}-\frac{\Gamma(\nu+1)}{a^{\nu+1}}C_\nu(a)$$

$[a>0,\ \operatorname{Re}\nu>-1]$. ET I 68(3)

4. $$\int_0^1 (1-u)^\nu\cos(ax)\,dx = \frac{i}{2}a^{-\nu-1}\left\{\exp\left[\frac{i}{2}(\nu\pi-2a)\right]\gamma(\nu+1,-ia)-\right.$$
$$\left.-\exp\left[-\frac{i}{2}(\nu\pi-2a)\right]\gamma(\nu+1,\ ia)\right\}$$

$[a>0,\ \operatorname{Re}\nu>-1]$. ET I 11(3)a

5. $$\int_0^u x^{\nu-1}(u-x)^{\mu-1}\sin(ax)\,dx =$$
$$=\frac{u^{\mu+\nu-1}}{2i}\mathrm{B}(\mu,\nu)[{}_1F_1(\nu;\ \mu+\nu;\ iau)-{}_1F_1(\nu;\ \mu+\nu;\ -iau)]$$

$[a>0,\ \operatorname{Re}\mu>0,\ \operatorname{Re}\nu>-1]$. ET II 189(26)

6. $$\int_0^u x^{\nu-1}(u-x)^{\mu-1}\cos(ax)\,dx =$$
$$=\frac{u^{\mu+\nu-1}}{2}\mathrm{B}(\mu,\nu)[{}_1F_1(\nu;\ \mu+\nu;\ iau)+{}_1F_1(\nu;\ \mu+\nu;\ -iau)]$$

$[a>0,\ \operatorname{Re}\mu>0,\ \operatorname{Re}\nu>0]$. ET II 189(32)

7. $$\int_0^u x^{\mu-1}(u-x)^{\mu-1}\sin(ax)\,dx = \sqrt{\pi}\left(\frac{u}{a}\right)^{\mu-\frac{1}{2}}\sin\frac{au}{2}\,\Gamma(\mu)\,J_{\mu-\frac{1}{2}}\left(\frac{au}{2}\right)$$

$[\operatorname{Re}\mu>0]$. ET II 189(25)

8. $$\int_u^\infty x^{\mu-1}(x-u)^{\mu-1}\sin(ax)\,dx =$$
$$=\frac{\sqrt{\pi}}{2}\left(\frac{u}{a}\right)^{\mu-\frac{1}{2}}\Gamma(\mu)\left[\cos\frac{au}{2}J_{\frac{1}{2}-\mu}\left(\frac{au}{2}\right)-\sin\frac{au}{2}N_{\frac{1}{2}-\mu}\left(\frac{au}{2}\right)\right]$$

$\left[a>0,\ 0<\operatorname{Re}\mu<\frac{1}{2}\right]$. ET II 203(20)

9. $$\int_0^u x^{\mu-1}(u-x)^{\mu-1}\cos(ax)\,dx = \sqrt{\pi}\left(\frac{u}{a}\right)^{\mu-\frac{1}{2}}\cos\frac{au}{2}\,\Gamma(\mu)\,J_{\mu-\frac{1}{2}}\left(\frac{au}{2}\right)$$

$[\operatorname{Re}\mu > 0]$. ET II 189(31)

10. $$\int_u^\infty x^{\mu-1}(x-u)^{\mu-1}\cos(ax)\,dx =$$
$$= -\frac{\sqrt{\pi}}{2}\left(\frac{u}{a}\right)^{\mu-\frac{1}{2}}\Gamma(\mu)\left[\sin\frac{au}{2}\,J_{\frac{1}{2}-\mu}\left(\frac{au}{2}\right) - \cos\frac{au}{2}\,N_{\frac{1}{2}-\mu}\left(\frac{au}{2}\right)\right]$$

$\left[a > 0,\ 0 < \operatorname{Re}\mu < \frac{1}{2}\right]$. ET II 204(25)

11. $$\int_0^1 x^{\nu-1}(1-x)^{\mu-1}\sin(ax)\,dx = -\frac{i}{2}\,\mathrm{B}(\mu,\nu)\,[{}_1F_1(\nu,\nu+\mu;ia) -$$
$$- {}_1F_1(\nu;\nu+\mu;-ia)]$$

$[a > 0;\ \operatorname{Re}\mu > 0,\ \operatorname{Re}\nu > 0]$.

ET I 58(5)a, ET I 317(5)

12. $$\int_0^1 x^{\nu-1}(1-x)^{\mu-1}\cos(ax)\,dx = \frac{1}{2}\,\mathrm{B}(\mu,\nu)\,[{}_1F_1(\nu;\nu+\mu;ia) +$$
$$+ {}_1F_1(\nu;\nu+\mu;-ia)]$$

$[a > 0,\ \operatorname{Re}\mu > 0,\ \operatorname{Re}\nu > 0]$.

ET I 11(5)

13. $$\int_0^1 x^\mu(1-x)^\mu\sin(ax)\,dx = \frac{\sqrt{\pi}}{(2a)^{\mu+\frac{1}{2}}}\Gamma(\mu+1)\sin a\,J_{\mu+\frac{1}{2}}(a)$$

$[a > 0,\ \operatorname{Re}\mu > -1]$. ET I 68(4)

14. $$\int_0^1 x^\mu(1-x)^\mu\cos(ax)\,dx = \frac{\sqrt{\pi}}{(2a)^{\mu+\frac{1}{2}}}\Gamma(\mu+1)\cos a\,J_{\mu+\frac{1}{2}}(a)$$

$[a > 0,\ \operatorname{Re}\mu > -1]$. ET I 11(4)

3.769

1. $$\int_0^\infty [(\beta+ix)^{-\nu} - (\beta-ix)^{-\nu}]\sin(ax)\,dx =$$
$$= \frac{\pi i a^{\nu-1}e^{-a\beta}}{\Gamma(\nu)}$$

$[a > 0,\ \operatorname{Re}\beta > 0,\ \operatorname{Re}\nu > 0]$. ET I 70(15)

2. $$\int_0^\infty [(\beta+ix)^{-\nu} + (\beta-ix)^{-\nu}]\cos(ax)\,dx = \frac{\pi a^{\nu-1}e^{-a\beta}}{\Gamma(\nu)}$$

$[a > 0,\ \operatorname{Re}\beta > 0,\ \operatorname{Re}\nu > 0]$. ET I 13(19)

3. $$\int_0^\infty x[(\beta+ix)^{-\nu} + (\beta-ix)^{-\nu}]\sin(ax)\,dx =$$
$$= -\frac{\pi a^{\nu-2}(1-a\beta)}{\Gamma(\nu)}e^{-a\beta}$$

$[a > 0,\ \operatorname{Re}\beta > 0,\ \operatorname{Re}\nu > 0]$. ET I 70(16)

4. $$\int_0^\infty x^{2n}\left[(\beta - ix)^{-\nu} - (\beta + ix)^{-\nu}\right]\sin(ax)\,dx =$$
$$= \frac{(-1)^{n+1} i}{\Gamma(\nu)}(2n)!\,\pi a^{\nu-2n-1} e^{-a\beta} L_{2n}^{\nu-2n-1}(a\beta)$$
$$[a > 0,\ \operatorname{Re}\beta > 0,\ 0 \leqslant 2n \leqslant \operatorname{Re}\nu].$$
ET I 70(17)

5. $$\int_0^\infty x^{2n}\left[(\beta + ix)^{-\nu} + (\beta - ix)^{-\nu}\right]\cos(ax)\,dx =$$
$$= \frac{(-1)^{n}}{\Gamma(\nu)}(2n)!\,\pi a^{\nu-2n-1} e^{-a\beta} L_{2n}^{\nu-2n-1}(a\beta)$$
$$[a > 0,\ \operatorname{Re}\beta > 0,\ 0 \leqslant 2n < \operatorname{Re}\nu].$$
ET I 13(20)

6. $$\int_0^\infty x^{2n+1}\left[(\beta + ix)^{-\nu} + (\beta - ix)^{-\nu}\right]\sin(ax)\,dx =$$
$$= \frac{(-1)^{n+1}}{\Gamma(\nu)}(2n+1)!\,\pi a^{\nu-2n-2} e^{-a\beta} L_{2n+1}^{\nu-2n-2}(a\beta)$$
$$[a > 0,\ \operatorname{Re}\beta > 0,\ -1 \leqslant 2n+1 < \operatorname{Re}\nu].$$
ET I 70(18)

7. $$\int_0^\infty x^{2n+1}\left[(\beta + ix)^{-\nu} - (\beta - ix)^{-\nu}\right]\cos(ax)\,dx =$$
$$= \frac{(-1)^{n+1} i}{\Gamma(\nu)}(2n+1)!\,\pi a^{\nu-2n-2} e^{-a\beta} L_{2n+1}^{\nu-2n-2}(a\beta)$$
$$[a > 0,\ \operatorname{Re}\beta > 0,\ 0 \leqslant 2n < \operatorname{Re}\nu - 1].$$
ET I 13(21)

3.771

1. $$\int_0^\infty (\beta^2 + x^2)^{\nu - \frac{1}{2}} \sin(ax)\,dx = \frac{\sqrt{\pi}}{2}\left(\frac{2\beta}{a}\right)^\nu \Gamma\left(\nu + \frac{1}{2}\right)\left[I_{-\nu}(a\beta) - \mathbf{L}_\nu(a\beta)\right]$$
$$\left[a > 0,\ \operatorname{Re}\beta > 0,\ \operatorname{Re}\nu < \frac{1}{2},\nu \neq -\frac{1}{2},\ -\frac{3}{2},\ -\frac{5}{2},\ \dots\right].$$
EH II 38a, ET I 68(6)

2. $$\int_0^\infty (\beta^2 + x^2)^{\nu - \frac{1}{2}} \cos(ax)\,dx = \frac{1}{\sqrt{\pi}}\left(\frac{2\beta}{a}\right)^\nu \cos(\pi\nu)\ \Gamma\left(\nu + \frac{1}{2}\right) K_{-\nu}(a\beta)$$
$$\left[a > 0,\ \operatorname{Re}\beta > 0,\ \operatorname{Re}\nu < \frac{1}{2}\right].$$
WA 191(1)a, GW((333))(78)a

3. $$\int_0^u x^{2\nu-1}(u^2 - x^2)^{\mu-1}\sin(ax)\,dx =$$
$$= \frac{a}{2} u^{2\mu+2\nu-1}\,\mathrm{B}\left(\mu,\ \nu + \frac{1}{2}\right) {}_1F_2\left(\nu + \frac{1}{2};\ \frac{3}{2},\ \mu + \nu + \frac{1}{2};\ -\frac{a^2u^2}{4}\right)$$
$$\left[\operatorname{Re}\mu > 0,\ \operatorname{Re}\nu > -\frac{1}{2}\right].$$
ET II 189(29)

4. $$\int_0^u x^{2\nu-1}(u^2-x^2)^{\mu-1}\cos(ax)\,dx=\frac{1}{2}u^{2\mu+2\nu-2}\mathrm{B}(\mu,\ \nu)\times$$
$$\times{}_1F_2\left(\nu;\ \frac{1}{2},\ \mu+\nu;\ -\frac{a^2u^2}{4}\right)\quad [\operatorname{Re}\mu>0,\ \operatorname{Re}\nu>0].$$ ET II 190(35)

5. $$\int_0^\infty x(x^2+\beta^2)^{\nu-\frac{1}{2}}\sin(ax)\,dx=\frac{1}{\sqrt{\pi}}\beta\left(\frac{2\beta}{a}\right)^\nu\cos\nu\pi\Gamma\left(\nu+\frac{1}{2}\right)K_{\nu+1}(a\beta)$$
$$[a>0,\ \operatorname{Re}\beta>0,\ \operatorname{Re}\nu>-2].$$ ET I 69(11)

6. $$\int_0^u (u^2-x^2)^{\nu-\frac{1}{2}}\sin(ax)\,dx=\frac{\sqrt{\pi}}{2}\left(\frac{2u}{a}\right)^\nu\Gamma\left(\nu+\frac{1}{2}\right)\mathbf{H}_\nu(au)$$
$$\left[a>0,\ u>0,\ \operatorname{Re}\nu>-\frac{1}{2}\right].$$ ET I 69(7), WA 358(1)a

7. $$\int_u^\infty (x^2-u^2)^{\nu-\frac{1}{2}}\sin(ax)\,dx=\frac{\sqrt{\pi}}{2}\left(\frac{2u}{a}\right)^\nu\Gamma\left(\nu+\frac{1}{2}\right)J_{-\nu}(au)$$
$$\left[a>0,\ u>0,\ |\operatorname{Re}\nu|<\frac{1}{2}\right].$$ EH II 81(12)a, ET I 69(8), WA 187(3)a

8. $$\int_0^u (u^2-x^2)^{\nu-\frac{1}{2}}\cos(ax)\,dx=\frac{\sqrt{\pi}}{2}\left(\frac{2u}{a}\right)^\nu\Gamma\left(\nu+\frac{1}{2}\right)J_\nu(au)$$
$$\left[a>0,\ u>0,\ \operatorname{Re}\nu>-\frac{1}{2}\right].$$ ET I 11(8)

9. $$\int_u^\infty (x^2-u^2)^{\nu-\frac{1}{2}}\cos(ax)\,dx=-\frac{\sqrt{\pi}}{2}\left(\frac{2u}{a}\right)^\nu\Gamma\left(\nu+\frac{1}{2}\right)N_{-\nu}(au)$$
$$\left[a>0,\ u>0,\ |\operatorname{Re}\nu|<\frac{1}{2}\right].$$ WA 187(4)a, EH II 82(13)a, ET I 11(9)

10. $$\int_0^u x(u^2-x^2)^{\nu-\frac{1}{2}}\sin(ax)\,dx=\frac{\sqrt{\pi}}{2}u\left(\frac{2u}{a}\right)^\nu\Gamma\left(\nu+\frac{1}{2}\right)J_{\nu+1}(au)$$
$$\left[a>0,\ u>0,\ \operatorname{Re}\nu>-\frac{1}{2}\right].$$ ET I 69(9)

11 $$\int_u^\infty x(x^2-u^2)^{\nu-\frac{1}{2}}\sin(ax)\,dx=\frac{\sqrt{\pi}}{2}u\left(\frac{2u}{a}\right)^\nu\Gamma\left(\nu+\frac{1}{2}\right)N_{-\nu-1}(au)$$
$$\left[a>0,\ u>0,\ -\frac{1}{2}<\operatorname{Re}\nu<0\right].$$ ET I 69(10)

12. $$\int_0^u x(u^2-x^2)^{\nu-\frac{1}{2}}\cos(ax)\,dx=-\frac{u^{\nu+1}}{a^\nu}s_{\nu,\,\nu+1}(au)=$$
$$=\frac{1}{2}\left(\nu+\frac{1}{2}\right)u^{2\nu+1}-\frac{\sqrt{\pi}}{2}u\left(\frac{2u}{a}\right)^\nu\Gamma\left(\nu+\frac{1}{2}\right)\mathbf{H}_{\nu+1}(au)$$
$$\left[a>0,\ u>0,\ \operatorname{Re}\nu>-\frac{1}{2}\right].$$ ET I 12(10)

13. $\int\limits_u^\infty x(x^2-u^2)^{\nu-\frac{1}{2}}\cos(ax)\,dx = \frac{\sqrt{\pi}\,u}{2}\left(\frac{2u}{a}\right)^\nu \Gamma\left(\nu+\frac{1}{2}\right) J_{-\nu-1}(au)$

$\left[a>0,\quad u>0,\ 0<\operatorname{Re}\nu<\frac{1}{2}\right].$ ET I 12(11)

3.772

1. $\int\limits_0^\infty (x^2+2\beta x)^{\nu-\frac{1}{2}}\sin(ax)\,dx =$

$= \frac{\sqrt{\pi}}{2}\left(\frac{2\beta}{a}\right)^\nu \Gamma\left(\nu+\frac{1}{2}\right)[J_{-\nu}(a\beta)\cos(a\beta)+N_{-\nu}(a\beta)\sin(a\beta)]$

$\left[a>0,\ |\arg\beta|<\pi,\ \frac{1}{2}>\operatorname{Re}\nu>-\frac{3}{2}\right].$ ET I 69(12)

2. $\int\limits_0^\infty (x^2+2\beta x)^{\nu-\frac{1}{2}}\cos(ax)\,dx =$

$= -\frac{\sqrt{\pi}}{2}\left(\frac{2\beta}{a}\right)^\nu \Gamma\left(\nu+\frac{1}{2}\right)[N_{-\nu}(a\beta)\cos(a\beta)-J_{-\nu}(a\beta)\sin(a\beta)]$

$\left[a>0,\ |\operatorname{Re}\nu|<\frac{1}{2}\right].$ ET I 12(13)

3. $\int\limits_0^{2u} (2ux-x^2)^{\nu-\frac{1}{2}}\sin(ax)\,dx = \sqrt{\pi}\left(\frac{2u}{a}\right)^\nu \Gamma\left(\nu+\frac{1}{2}\right)\sin(au)\,J_\nu(au)$

$\left[a>0,\ u>0,\ \operatorname{Re}\nu>-\frac{1}{2}\right].$ ET I 69(13)a

4. $\int\limits_{2u}^\infty (x^2-2ux)^{\nu-\frac{1}{2}}\sin(ax)\,dx =$

$= \frac{\sqrt{\pi}}{2}\left(\frac{2u}{a}\right)^\nu \Gamma\left(\nu+\frac{1}{2}\right)[J_{-\nu}(au)\cos(au)-N_{-\nu}(au)\sin(au)]$

$\left[a>0,\ u>0,\ |\operatorname{Re}\nu|<\frac{1}{2}\right].$ ET I 70(14)

5. $\int\limits_0^{2u} (2ux-x^2)^{\nu-\frac{1}{2}}\cos(ax)\,dx = \sqrt{\pi}\left(\frac{2u}{a}\right)^\nu \Gamma\left(\nu+\frac{1}{2}\right) J_\nu(au)\cos(au)$

$\left[a>0,\ u>0,\ \operatorname{Re}\nu>-\frac{1}{2}\right].$ ET I 12(14)

6. $\int\limits_{2u}^\infty (x^2-2ux)^{\nu-\frac{1}{2}}\cos(ax)\,dx =$

$= -\frac{\sqrt{\pi}}{2}\left(\frac{2u}{a}\right)^\nu \Gamma\left(\nu+\frac{1}{2}\right)[J_{-\nu}(au)\sin(au)+N_{-\nu}(au)\cos(au)]$

$\left[a>0,\ u>0,\ |\operatorname{Re}\nu|<\frac{1}{2}\right].$ ET I 12(12)

3.773

1. $$\int_0^\infty \frac{x^{2\nu}}{(x^2+\beta^2)^{\mu+1}} \sin(ax)\,dx =$$

$$= \frac{1}{2}\beta^{2\nu-2\mu} a \mathrm{B}(1+\nu, \mu-\nu)_1 = F_2\left(\nu+1;\ \nu+1-\mu,\ \frac{3}{2};\ \frac{\beta^2 a^2}{4}\right) +$$

$$+ \frac{\sqrt{\pi}\, a^{2\mu-2\nu+1}}{4^{\mu-\nu+1}} + \frac{\Gamma(\nu-\mu)}{\Gamma\left(\mu-\nu+\frac{3}{2}\right)}\, {}_1F_2\left(\mu+1;\ \mu-\nu+\frac{3}{2},\ \mu-\nu+1;\ \frac{\beta^2 a^2}{4}\right) =$$

$$= \frac{\sqrt{\pi}}{2\Gamma(\mu+1)} \beta^{2\nu-2\mu-1} G_{13}^{21}\left(\frac{a^2\beta^2}{4}\middle|\begin{matrix}-\nu+\frac{1}{2}\\ \mu-\nu+\frac{1}{2},\ \frac{1}{2},\ 0\end{matrix}\right)$$

$[a > 0,\ \operatorname{Re}\beta > 0,\ -1 < \operatorname{Re}\nu < \operatorname{Re}\mu + 1].$ ET I 71(28)a, ET II 234(17)

2. $$\int_0^\infty \frac{x^{2m+1}\sin(ax)}{(z+x^2)^{n+1}}\,dx = \frac{(-1)^{n+m}}{n!}\cdot\frac{\pi}{2}\frac{d^n}{dz^n}\left(z^m e^{-a\sqrt{z}}\right)$$

$[a > 0,\ 0 \leqslant m \leqslant n,\ |\arg z| < \pi].$ ET I 68(39)

3. $$\int_0^\infty \frac{x^{2m+1}\sin(ax)\,dx}{(\beta^2+x^2)^{n+\frac{1}{2}}} = \frac{(-1)^{m+1}\sqrt{\pi}}{2^n\beta^n\Gamma\left(n+\frac{1}{2}\right)}\frac{d^{2m+1}}{da^{2m+1}}\left[a^n K_n(a\beta)\right]$$

$[a > 0,\ \operatorname{Re}\beta > 0,\ -1 \leqslant m \leqslant n].$ ET I 67(37)

4. $$\int_0^\infty \frac{x^{2\nu}\cos(ax)\,dx}{(x^2+\beta^2)^{\mu+1}} =$$

$$= \frac{1}{2}\beta^{2\nu-2\mu-1}\mathrm{B}\left(\nu+\frac{1}{2},\ \mu-\nu+\frac{1}{2}\right){}_1F_2\left(\nu+\frac{1}{2};\ \nu-\mu+\frac{1}{2},\ \frac{1}{2};\ \frac{\beta^2 a^2}{4}\right) +$$

$$+ \frac{\sqrt{\pi}\, a^{2\mu-2\nu+1}}{4^{\mu-\nu+1}}\frac{\Gamma\left(\nu-\mu-\frac{1}{2}\right)}{\Gamma(\mu-\nu+1)}\,{}_1F_2\left(\mu+1;\ \mu-\nu+1,\ \mu-\nu+\frac{3}{2};\ \frac{\beta^2 a^2}{4}\right) =$$

$$= \frac{\sqrt{\pi}}{2\Gamma(\mu+1)}\beta^{2\nu-2\mu-1} G_{13}^{21}\left(\frac{a^2\beta^2}{4}\middle|\begin{matrix}-\nu+\frac{1}{2}\\ \mu-\nu+\frac{1}{2},\ 0,\ \frac{1}{2}\end{matrix}\right)$$

$\left[a > 0,\ \operatorname{Re}\beta > 0,\ -\frac{1}{2} < \operatorname{Re}\nu < \operatorname{Re}\mu + 1\right].$ ET I 14(29)a, ET II 235(19)

5. $$\int_0^\infty \frac{x^{2m}\cos(ax)\,dx}{(z+x^2)^{n+1}} = (-1)^{m+n}\frac{\pi}{2\cdot n!}\cdot\frac{d^n}{dz^n}\left(z^{m-\frac{1}{2}}e^{-a\sqrt{z}}\right)$$

$[a > 0,\ n+1 > m \geqslant 0,\ |\arg z| < \pi].$ ET I 10(28)

6. $$\int_0^\infty \frac{x^{2m}\cos(ax)\,dx}{(\beta^2+x^2)^{n+\frac{1}{2}}} = \frac{(-1)^m\sqrt{\pi}}{2^n\beta^n\Gamma\left(n+\frac{1}{2}\right)}\cdot\frac{d^{2m}}{da^{2m}}\left\{a^n K_n(a\beta)\right\}$$

$\left[a > 0,\ \operatorname{Re}\beta > 0,\ 0 \leqslant m < n+\frac{1}{2}\right].$ ET I 14(28)

3.774

1. $$\int_0^\infty \frac{\sin(ax)\,dx}{\sqrt{x^2+b^2}\left(x+\sqrt{x^2+b^2}\right)^\nu} = \frac{\pi}{b^\nu\sin(\nu\pi)}\left[\sin\frac{\nu\pi}{2} I_\nu(ab) + \frac{i}{2}\mathbf{J}_\nu(iab) - \right.$$
$$\left. - \frac{i}{2}\mathbf{J}_\nu(-iab)\right]$$

$[a > 0,\ b > 0,\ \operatorname{Re}\nu > -1].$ ET I 70(19)

2. $$\int_0^\infty \frac{\cos(ax)\,dx}{\sqrt{x^2+b^2}\,(x+\sqrt{x^2+b^2})^\nu} = \frac{\pi}{b^\nu \sin\nu\pi}\left[\frac{1}{2}\mathbf{J}_\nu(iab)+\frac{1}{2}\mathbf{J}_\nu(-iab)-\right.$$
$$\left.-\cos\frac{\nu\pi}{2} I_\nu(ab)\right] \quad [a>0,\ b>0,\ \operatorname{Re}\nu>-1].$$ ET I 12(15)

3. $$\int_0^\infty \frac{(x+\sqrt{x^2+\beta^2})^\nu}{\sqrt{x(x^2+\beta^2)}}\sin(ax)\,dx = \sqrt{\frac{a\pi}{2}}\,\beta^\nu I_{\frac{1}{4}-\frac{\nu}{2}}\left(\frac{a\beta}{2}\right)K_{\frac{1}{4}+\frac{\nu}{2}}\left(\frac{a\beta}{2}\right)$$
$$\left[a>0,\ \operatorname{Re}\beta>0,\ \operatorname{Re}\nu<\frac{3}{2}\right].$$ ET I 71(23)

4. $$\int_0^\infty \frac{(\sqrt{x^2+\beta^2}-x)^\nu}{\sqrt{x(x^2+\beta^2)}}\cos(ax)\,dx = \sqrt{\frac{a\pi}{2}}\,\beta^\nu I_{-\frac{1}{4}+\frac{\nu}{2}}\left(\frac{a\beta}{2}\right)K_{-\frac{1}{4}-\frac{\nu}{2}}\left(\frac{a\beta}{2}\right)$$
$$\left[a>0,\ \operatorname{Re}\beta>0,\ \operatorname{Re}\nu>-\frac{3}{2}\right].$$ ET I 12(17)

5. $$\int_0^\infty \frac{(\beta+\sqrt{x^2+\beta^2})^\nu}{x^{\nu+\frac{1}{2}}\sqrt{x^2+\beta^2}}\sin(ax)\,dx = \frac{1}{\beta}\sqrt{\frac{2}{a}}\,\Gamma\left(\frac{3}{4}-\frac{\nu}{2}\right)W_{\frac{\nu}{2},\frac{1}{4}}(a\beta)\,M_{-\frac{\nu}{2},\frac{1}{4}}(a\beta)$$
$$\left[a>0,\ \operatorname{Re}\beta>0,\ \operatorname{Re}\nu<\frac{3}{2}\right].$$ ET I 71(27)

6. $$\int_0^\infty \frac{(\beta+\sqrt{x^2+\beta^2})^\nu}{x^{\nu+\frac{1}{2}}\sqrt{\beta^2+x^2}}\cos(ax)\,dx = \frac{1}{\beta\sqrt{2a}}\,\Gamma\left(\frac{1}{4}-\frac{\nu}{2}\right)W_{\frac{\nu}{2},-\frac{1}{4}}(a\beta)\,M_{-\frac{\nu}{2},-\frac{1}{4}}(a\beta)$$
$$\left[a>0,\ \operatorname{Re}\beta>0,\ \operatorname{Re}\nu<\frac{1}{2}\right].$$ ET I 12(18)

3.775

1. $$\int_0^\infty \frac{(\sqrt{x^2+\beta^2}+x)^\nu-(\sqrt{x^2+\beta^2}-x)^\nu}{\sqrt{x^2+\beta^2}}\sin(ax)\,dx = 2\beta^\nu\sin\frac{\nu\pi}{2}K_\nu(a\beta)$$
$$[a>0,\ \operatorname{Re}\beta>0,\ |\operatorname{Re}\nu|<1].$$ ET I 70(20)

2. $$\int_0^\infty \frac{(\sqrt{x^2+\beta^2}+x)^\nu+(\sqrt{x^2+\beta^2}-x)^\nu}{\sqrt{x^2+\beta^2}}\cos(ax)\,dx = 2\beta^\nu\cos\frac{\nu\pi}{2}K_\nu(a\beta)$$
$$[a>0,\ \operatorname{Re}\beta>0,\ |\operatorname{Re}\nu|<1].$$ ET I 13(22)

3. $$\int_u^\infty \frac{(x+\sqrt{x^2-u^2})^\nu+(x-\sqrt{x^2-u^2})^\nu}{\sqrt{x^2-u^2}}\sin(ax)\,dx =$$
$$= \pi u^\nu\left[J_\nu(au)\cos\frac{\nu\pi}{2}-N_\nu(au)\sin\frac{\nu\pi}{2}\right]$$
$$[a>0,\ u>0,\ |\operatorname{Re}\nu|<1].$$ ET I 70(22)

4. $$\int_u^\infty \frac{(x+\sqrt{x^2-u^2})^\nu+(x-\sqrt{x^2-u^2})^\nu}{\sqrt{x^2-u^2}}\cos(ax)\,dx =$$
$$= -\pi u^\nu\left[N_\nu(au)\cos\frac{\nu\pi}{2}+J_\nu(au)\sin\frac{\nu\pi}{2}\right]$$
$$[a>0,\ u>0,\ |\operatorname{Re}\nu|<1].$$ ET I 13(25)

5. $$\int_0^u \frac{(x+i\sqrt{u^2-x^2})^\nu+(x-i\sqrt{u^2-x^2})^\nu}{\sqrt{u^2-x^2}}\sin(ax)\,dx=$$
$$=\frac{\pi}{2}u^\nu\operatorname{cosec}\frac{\nu\pi}{2}[\mathbf{J}_\nu(au)-\mathbf{J}_{-\nu}(au)] \qquad [a>0,\ u>0].$$
ET I 70(21)

6. $$\int_0^u \frac{(x+i\sqrt{u^2-x^2})^\nu+(x-i\sqrt{u^2-x^2})^\nu}{\sqrt{u^2-x^2}}\cos(ax)\,dx=$$
$$=\frac{\pi}{2}u^\nu\sec\frac{\nu\pi}{2}[\mathbf{J}_\nu(au)+\mathbf{J}_{-\nu}(au)] \qquad [a>0,\ u>0,\ |\operatorname{Re}\nu|<1].$$
ET I 13(24)

7. $$\int_u^\infty \frac{(x+\sqrt{x^2-u^2})^\nu+(x-\sqrt{x^2-u^2})^\nu}{\sqrt{x(x^2-u^2)}}\sin(ax)\,dx=$$
$$=-\sqrt{\left(\frac{\pi}{2}\right)^3 a}\,u^\nu\left[J_{\frac{1}{4}+\frac{\nu}{2}}\left(\frac{au}{2}\right)N_{\frac{1}{4}-\frac{\nu}{2}}\left(\frac{au}{2}\right)+\right.$$
$$\left.+J_{\frac{1}{4}-\frac{\nu}{2}}\left(\frac{au}{2}\right)N_{\frac{1}{4}+\frac{\nu}{2}}\left(\frac{au}{2}\right)\right] \qquad \left[a>0,\ u>0,\ \operatorname{Re}\nu<\frac{3}{2}\right]$$
ET I 71(25)

8. $$\int_u^\infty \frac{(x+\sqrt{x^2-u^2})^\nu+(x-\sqrt{x^2-u^2})^\nu}{\sqrt{x(x^2-u^2)}}\cos(ax)\,dx=$$
$$=-\sqrt{\left(\frac{\pi}{2}\right)^3 a}\,u^\nu\left[J_{-\frac{1}{4}+\frac{\nu}{2}}\left(\frac{au}{2}\right)N_{-\frac{1}{4}-\frac{\nu}{2}}\left(\frac{au}{2}\right)+\right.$$
$$\left.+J_{-\frac{1}{4}-\frac{\nu}{2}}\left(\frac{au}{2}\right)N_{-\frac{1}{4}+\frac{\nu}{2}}\left(\frac{au}{2}\right)\right] \qquad \left[a>0,\ u>0,\ \operatorname{Re}\nu<\frac{3}{2}\right].$$
ET I 13(26)

9. $$\int_0^\infty \frac{(x+\beta+\sqrt{x^2+2\beta x})^\nu+(x+\beta-\sqrt{x^2+2\beta x})^\nu}{\sqrt{x^2+2\beta x}}\sin(ax)\,dx=$$
$$=\pi\beta^\nu\left[N_\nu(\beta a)\sin\left(\beta a-\frac{\nu\pi}{2}\right)+J_\nu(\beta a)\cos\left(\beta a-\frac{\nu\pi}{2}\right)\right]$$
$$[a>0,\ |\arg\beta|<\pi,\ |\operatorname{Re}\nu|<1].$$
ET I 71(26)

10. $$\int_0^\infty \frac{(x+\beta+\sqrt{x^2+2\beta x})^\nu+(x+\beta-\sqrt{x^2+2\beta x})^\nu}{\sqrt{x^2+2\beta x}}\cos(ax)\,dx=$$
$$=\pi\beta^\nu\left[J_\nu(\beta a)\sin\left(\beta a-\frac{\nu\pi}{2}\right)-N_\nu(\beta a)\cos\left(\beta a-\frac{\nu\pi}{2}\right)\right]$$
$$[a>0,\ |\arg\beta|<\pi,\ |\operatorname{Re}\nu|<1].$$
ET I 13(23)

11. $$\int_0^{2u} \frac{(\sqrt{2u+x}+i\sqrt{2u-x})^{4\nu}+(\sqrt{2u+x}-i\sqrt{2u-x})^{4\nu}}{\sqrt{4u^2x-x^3}}\cos(ax)\,dx=$$
$$=(4u)^{2\nu}\pi^{\frac{3}{2}}\sqrt{\frac{a}{2}}\,J_{\nu-\frac{1}{4}}(au)\,J_{-\nu-\frac{1}{4}}(au) \qquad [a>0,\ u>0].$$
ET I 14(27)

3.776

1. $$\int_0^\infty \frac{a^2 (b+x)^2 + p(p+1)}{(b+x)^{p+2}} \sin(ax)\, dx = \frac{a}{b^p} \qquad [a>0,\ b>0,\ p>0].$$ BI ((170))(1)

2. $$\int_0^\infty \frac{a^2 (b+x)^2 + p(p+1)}{(b+x)^{p+2}} \cos(ax)\, dx = \frac{p}{b^{p+1}} \qquad [a>0,\ b>0,\ p>0].$$ BI ((170))(2)

3.78-3.81 Rational functions of x and of trigonometric functions

3.781

1. $$\int_0^\infty \left(\frac{\sin x}{x} - \frac{1}{1+x}\right)\frac{dx}{x} = 1 - C$$ (cf. 3.784 4. and 3.781 2.). BI ((173))(7)

2. $$\int_0^\infty \left(\cos x - \frac{1}{1+x}\right)\frac{dx}{x} = -C.$$ BI ((173))(8)

3.782

1. $$\int_0^u \frac{1-\cos x}{x}\, dx - \int_u^\infty \frac{\cos x}{x}\, dx = C + \ln u \qquad [u>0].$$ GW ((333))(31)

2. $$\int_0^\infty \frac{1-\cos ax}{x^2}\, dx = \frac{a\pi}{2}.$$ BI ((158))(1)

3. $$\int_{-\infty}^\infty \frac{1-\cos ax}{x(x-b)}\, dx = \frac{\sin ab}{b} \qquad [a>0].$$ ET II 253(48)

3.783

1. $$\int_0^\infty \left[\frac{\cos x - 1}{x^2} + \frac{1}{2(1+x)}\right]\frac{dx}{x} = \frac{1}{2} C - \frac{3}{4}.$$ BI ((173))(19)

2. $$\int_0^\infty \left(\cos x - \frac{1}{1+x^2}\right)\frac{dx}{x} = -C.$$ EH I 17, BI((273))(21)

3.784

1. $$\int_0^\infty \frac{\cos ax - \cos bx}{x}\, dx = \ln \frac{b}{a} \qquad [ab \neq 0].$$ FI II 635, GW((333))(20)

2. $$\int_0^\infty \frac{a \sin bx - b \sin ax}{x^2}\, dx = ab \ln \frac{a}{b} \qquad [a>0,\ b>0].$$ FI II 647

3. $$\int_0^\infty \frac{\cos ax - \cos bx}{x^2}\, dx = \frac{(b-a)\pi}{2} \qquad [a \geqslant 0,\ b \geqslant 0].$$ BI((158))(12), FI II 645

4. $$\int_0^\infty \frac{\sin x - x\cos x}{x^2}\,dx = 1.$$ BI ((158))(3)

5. $$\int_0^\infty \frac{\cos ax - \cos bx}{x(x+\beta)}\,dx = \frac{1}{\beta}\left[\operatorname{ci}(a\beta)\cos a\beta + \operatorname{si}(a\beta)\sin a\beta - \right.$$
$$\left. - \operatorname{ci}(b\beta)\cos b\beta - \operatorname{si}(b\beta)\sin b\beta + \ln\frac{b}{a}\right] \quad [a>0,\ b>0,\ |\arg\beta|<\pi].$$ ET II 221(49)

6. $$\int_0^\infty \frac{\cos ax + x\sin ax}{1+x^2}\,dx = \pi e^{-a} \qquad [a>0].$$ GW ((333))(73)

7. $$\int_0^\infty \frac{\sin ax - ax\cos ax}{x^3}\,dx = \frac{\pi}{4}a^2\operatorname{sign} a.$$ LI ((158))(5)

8. $$\int_0^\infty \frac{\cos ax - \cos bx}{x^2(x^2+\beta^2)}\,dx = \frac{\pi\left[(b-a)\beta + e^{-b\beta} - e^{-a\beta}\right]}{2\beta^3}$$
$$[a>0,\ b>0,\ |\arg\beta|<\pi].$$ BI((173))(20)a, ET II 222(59)

3.785 $$\int_0^\infty \frac{1}{x}\sum_{k=1}^n a_k\cos b_k x\,dx = -\sum_{k=1}^n a_k\ln b_k \qquad \left[b_k>0,\ \sum_{k=1}^n a_k = 0\right].$$ FI II 649

3.786

1. $$\int_0^\infty \frac{(1-\cos ax)\sin bx}{x^2}\,dx = \frac{b}{2}\ln\frac{b^2-a^2}{b^2} + \frac{a}{2}\ln\frac{a+b}{a-b} \qquad [a>0,\ b>0].$$ ET I 81(29)

2. $$\int_0^\infty \frac{(1-\cos ax)\cos bx}{x}\,dx = \ln\frac{\sqrt{|a^2-b^2|}}{b} \qquad [a>0,\ b>0,\ a\neq b].$$ FI II 647

3. $$\int_0^\infty \frac{(1-\cos ax)\cos bx}{x^2}\,dx = \frac{\pi}{2}(a-b) \qquad [0<b\leqslant a];$$
$$=0 \qquad [0<a\leqslant b].$$ ET I 20(16)

3.787

1. $$\int_0^\infty \frac{(\cos a - \cos nax)\sin mx}{x}\,dx = \frac{\pi}{2}(\cos a - 1) \qquad [m>na>0];$$
$$=\frac{\pi}{2}\cos a \qquad [na>m].$$ BI ((155))(7)

2. $$\int_0^\infty \frac{\sin^2 ax - \sin^2 bx}{x}\,dx = \frac{1}{2}\ln\frac{a}{b} \qquad [ab\neq 0].$$ GW ((333))(20b)

3. $$\int_0^\infty \frac{x^3-\sin^3 x}{x^5}\,dx = \frac{13}{32}\pi.$$ BI ((158))(6)

4. $$\int_0^\infty \frac{(3-4\sin^2 ax)\sin^2 ax}{x}\,dx = \frac{1}{2}\ln 2.$$ BI ((155))(6)

3.788 $$\int_0^{\frac{\pi}{2}} \left(\frac{1}{x}-\operatorname{ctg} x\right)dx = \ln\frac{\pi}{2}.$$ GW ((333))(61)a

3.789 $$\int_0^{\frac{\pi}{2}} \frac{4x^2\cos x+(\pi-x)x}{\sin x}\,dx = \pi^2\ln 2.$$ LI ((206))(10)

3.791

1. $$\int_0^{\frac{\pi}{2}} \frac{x\,dx}{1+\sin x} = \ln 2.$$ GW ((333))(55a)

2. $$\int_0^{\pi} \frac{x\cos x}{1+\sin x}\,dx = \pi\ln 2-4G.$$ GW ((333))(55c)

3. $$\int_0^{\frac{\pi}{2}} \frac{x\cos x}{1+\sin x}\,dx = \pi\ln 2-2G.$$ GW ((333))(55b)

4. $$\int_0^{\pi} \frac{\left(\frac{\pi}{2}-x\right)\cos x}{1-\sin x}\,dx = 2\int_0^{\frac{\pi}{2}} \frac{\left(\frac{\pi}{2}-x\right)\cos x}{1-\sin x}\,dx =$$
$$=\pi\ln 2+4G=5.8414484669\ldots$$ BI((207))(3), GW((333))(56c)

5. $$\int_0^{\frac{\pi}{2}} \frac{x^2\,dx}{1-\cos x} = -\frac{\pi^2}{4}+\pi\ln 2+4G=3.3740473667\ldots$$ BI ((207))(3)

6. $$\int_0^{\pi} \frac{x^2\,dx}{1-\cos x} = 4\pi\ln 2.$$ BI ((219))(1)

7. $$\int_0^{\frac{\pi}{2}} \frac{x^{p+1}\,dx}{1-\cos x} = -\left(\frac{\pi}{2}\right)^{p+1}+\left(\frac{\pi}{2}\right)^{p}(p+1)\left\{\frac{2}{p}-\sum_{k=1}^{\infty}\frac{1}{4^{2k-1}(p+2k)}\zeta(2k)\right\}$$
$[p>0]$. LI ((207))(4)

8. $\int\limits_0^{\frac{\pi}{2}} \frac{x\,dx}{1+\cos x} = \frac{\pi}{2} - \ln 2.$ GW ((333))(55a)

9. $\int\limits_0^{\frac{\pi}{2}} \frac{x \sin x\,dx}{1-\cos x} = \frac{\pi}{2} \ln 2 + 2G.$ GW ((333))(56a)

10. $\int\limits_0^{\pi} \frac{x \sin x\,dx}{1-\cos x} = 2\pi \ln 2.$ GW ((333))(56b)

11. $\int\limits_0^{\pi} \frac{x-\sin x}{1-\cos x}\,dx = \frac{\pi}{2} + \int\limits_0^{\frac{\pi}{2}} \frac{x-\sin x}{1-\cos x}\,dx = 2.$ GW ((333))(57a)

12. $\int\limits_0^{\frac{\pi}{2}} \frac{x \sin x}{1+\cos x}\,dx = -\frac{\pi}{2} \ln 2 + 2G.$ GW ((333))(55b)

3.792

1. $\int\limits_{-\pi}^{\pi} \frac{dx}{1-2a\cos x + a^2} = \frac{2\pi}{1-a^2}$ $[a^2 < 1].$ FI II 485

2. $\int\limits_0^{\frac{\pi}{2}} \frac{x \cos x\,dx}{1+2a\sin x + a^2} = \frac{\pi}{2a} \ln(1+a) - \sum\limits_{k=0}^{\infty} (-1)^k \frac{a^{2k}}{(2k+1)^2}$ $[a^2 < 1].$ LI ((241))(2)

3. $\int\limits_0^{\pi} \frac{x \sin x\,dx}{1-2a\cos x + a^2} = \frac{\pi}{a} \ln(1+a)$ $[a^2 < 1];$

$= \frac{\pi}{a} \ln\left(1+\frac{1}{a}\right)$ $[a^2 > 1].$ BI ((221))(2)

4. $\int\limits_0^{2\pi} \frac{x \sin x\,dx}{1-2a\cos x + a^2} = \frac{2\pi}{a} \ln(1-a)$ $[a^2 < 1];$

$= \frac{2\pi}{a} \ln\left(1-\frac{1}{a}\right)$ $[a^2 > 1].$ BI ((223))(4)

5. $\int\limits_0^{2\pi} \frac{x \sin nx\,dx}{1-2a\cos x + a^2} = \frac{2\pi}{1-a^2}\left[(a^{-n}-a^n)\ln(1-a) + \sum\limits_{k=1}^{n-1} \frac{a^{-k}-a^k}{n-k}\right]$ $[a^2 < 1].$ BI ((223))(5)

6. $\int\limits_0^{\infty} \frac{\sin x}{1-2a\cos x + a^2} \cdot \frac{dx}{x} = \frac{\pi}{4a}\left[\left|\frac{1+a}{1-a}\right| - 1\right].$ GW ((333))(62b)

7. $$\int_0^\infty \frac{\sin bx}{1-2a\cos x+a^2}\cdot\frac{dx}{x}=\frac{\pi}{2}\frac{1+a-2a^{E(b)+1}}{(1-a^2)(1-a)} \qquad [b\neq 0, 1, 2, \ldots];$$

$$=\frac{\pi}{2}\frac{1+a-a^b-a^{b+1}}{(1-a^2)(1-a)} \qquad [b=0, 1, 2, \ldots]$$

$[0<a<1]$. ET I 181(26)

8. $$\int_0^\infty \frac{\sin x\cos bx}{1-2a\cos x+a^2}\cdot\frac{dx}{x}=\frac{\pi}{2(1-a)}a^{E(b)} \qquad [b\neq 0, 1, 2, \ldots];$$

$$=\frac{\pi}{2(1-a)}a^b+\frac{\pi}{4}a^{b-1} \qquad [b=0, 1, 2, \ldots]$$

$[0<a<1,\ b>0]$. ET I 19(5)

9. $$\int_0^\infty \frac{(1-a\cos x)\sin bx}{1-2a\cos x+a^2}\cdot\frac{dx}{x}=\frac{\pi}{2}\cdot\frac{1-a^{E(b)+1}}{1-a} \qquad [b\neq 1, 2, 3, \ldots];$$

$$=\frac{\pi}{2}\cdot\frac{1-a^b}{1-a}+\frac{\pi a^b}{4} \qquad [b=1, 2, 3, \ldots]$$

$[0<a<1]$. ET I 82(33)

10. $$\int_0^\infty \frac{1}{1-2a\cos bx+a^2}\,\frac{dx}{\beta^2+x^2}=\frac{\pi}{2\beta(1-a^2)}\,\frac{1+ae^{-b\beta}}{1-ae^{-b\beta}} \qquad [a^2<1].$$

BI ((192))(1)

11. $$\int_0^\infty \frac{1}{1-2a\cos bx+a^2}\,\frac{dx}{\beta^2-x^2}=\frac{a\pi}{\beta(1-a^2)}\,\frac{\sin b\beta}{1-2a\cos b\beta+a^2} \qquad [a^2<1].$$

BI ((193))(1)

12. $$\int_0^\infty \frac{\sin bcx}{1-2a\cos bx+a^2}\,\frac{x\,dx}{\beta^2+x^2}=\frac{\pi}{2}\,\frac{e^{-\beta bc}-a^c}{(1-ae^{-b\beta})(1-ae^{b\beta})} \qquad [a^2<1].$$

BI ((192))(8)

13. $$\int_0^\infty \frac{\sin bx}{1-2a\cos bx+a^2}\,\frac{x\,dx}{\beta^2+x^2}=\frac{\pi}{2}\,\frac{1}{e^{b\beta}-a} \qquad [a^2<1];$$

$$=\frac{\pi}{2a}\,\frac{1}{ae^{b\beta}-1} \qquad [a^2>1].$$

BI ((192))(2)

14. $$\int_0^\infty \frac{\sin bcx}{1-2a\cos bx+a^2}\,\frac{x\,dx}{\beta^2-x^2}=\frac{\pi}{2}\,\frac{a^c-\cos\beta bc}{1-2a\cos\beta b+a^2} \qquad [a^2<1].$$

BI ((193))(5)

15. $$\int_0^\infty \frac{\cos bcx}{1-2a\cos bx+a^2}\,\frac{dx}{\beta^2-x^2}=\frac{\pi}{2\beta(1-a^2)}\,\frac{(1-a^2)\sin\beta bc+2a^{c+1}\sin\beta b}{1-2a\cos\beta b+a^2}$$

$[a^2<1]$. BI ((193))(9)

16. $$\int_0^\infty \frac{1-a\cos bx}{1-2a\cos bx+a^2}\,\frac{dx}{1+x^2}=\frac{\pi}{2}\,\frac{e^b}{e^b-a}.$$

FI II 719

17. $$\int_0^\infty \frac{\cos bx}{1-2a\cos x+a^2}\cdot\frac{dx}{x^2+\beta^2}=\frac{\pi\left(e^{\beta-\beta b}+ae^{\beta b}\right)}{2\beta\left(1-a^2\right)\left(e^\beta-a\right)}$$
$[0\leqslant b<1,\quad |a|<1,\ \operatorname{Re}\beta>0]$. ET I 21(21)

18. $$\int_0^\infty \frac{\sin bx\sin x}{1-2a\cos x+a^2}\cdot\frac{dx}{x^2+\beta^2}=$$
$$=\frac{\pi}{2\beta}\,\frac{\operatorname{sh} b\beta}{e^\beta-a}\qquad [0\leqslant b<1];$$
$$=\frac{\pi}{4\beta\left(ae^\beta-1\right)}\left[a^m e^{\beta(m+1-b)}-e^{(1-b)\beta}\right]-$$
$$-\frac{\pi}{4\beta\left(ae^{-\beta}-1\right)}\left[a^m e^{-(m+1-b)\beta}-e^{-(1-b)\beta}\right]\quad [m\leqslant b\leqslant m+1]$$
$[0<a<1,\ \operatorname{Re}\beta>0]$. ET I 81(27)

19. $$\int_0^\infty \frac{(\cos x-a)\cos bx}{1-2a\cos x+a^2}\cdot\frac{dx}{x^2+\beta^2}=\frac{\pi\operatorname{ch}\beta b}{2\beta\left(e^\beta-a\right)}$$
$[0\leqslant b<1,\ |a|<1,\ \operatorname{Re}\beta>0]$. ET I 21(23)

20. $$\int_0^\infty \frac{\sin x}{\left(1-2a\cos 2x+a^2\right)^{n+1}}\,\frac{dx}{x}=\int_0^\infty \frac{\operatorname{tg} x}{\left(1-2a\cos 2x+a^2\right)^{n+1}}\,\frac{dx}{x}=$$
$$=\int_0^\infty \frac{\operatorname{tg} x}{\left(1-2a\cos 4x+a^2\right)^{n+1}}\,\frac{dx}{x}=\frac{\pi}{2\left(1-a^2\right)^{2n+1}}\sum_{k=0}^{n}\binom{n}{k}^2 a^{2k}.$$
BI ((187))(14-16)

3.793

1. $$\int_0^{2\pi} \frac{\sin nx-a\sin[(n+1)x]}{1-2a\cos x+a^2}\,x\,dx=2\pi a^n\left[\ln(1-a)+\sum_{k=1}^{n}\frac{1}{ka^k}\right]$$
$[|a|<1]$. BI ((223))(9)

2. $$\int_0^{2\pi} \frac{\cos nx-a\cos[(n+1)x]}{1-2a\cos x+a^2}\,x\,dx=2\pi^2a^n$$
$[a^2<1]$. BI ((223))(13)

3.794

1. $$\int_0^{\pi} \frac{x\,dx}{a\pm\cos x}=\frac{\pi^2}{2\sqrt{a^2-1}}\pm\frac{4}{\sqrt{a^2-1}}\sum_{k=1}^{\infty}\frac{\left(a-\sqrt{a^2-1}\right)^{2k+1}}{(2k+1)^2}$$
$[a>1]$. LI ((219))(2)

2. $$\int_0^{2\pi} \frac{x\sin nx}{1\pm a\cos x}\,dx=\frac{2\pi}{\sqrt{1-a^2}}\left[(\mp 1)^n\frac{\left(1+\sqrt{1-a^2}\right)^n-\left(1-\sqrt{1-a^2}\right)^n}{a^n}\times\right.$$
$$\left.\times\ln\frac{2\sqrt{1\pm a}}{\sqrt{1+a}+\sqrt{1-a}}+\sum_{k=1}^{n-1}\frac{(\mp 1)^k}{n-k}\,\frac{\left(1+\sqrt{1-a^2}\right)^k-\left(1-\sqrt{1-a^2}\right)^k}{a^k}\right]$$
$[a^2<1]$. BI ((223))(2)

3. $\int_0^{2\pi} \frac{x \cos nx}{1 \pm a \cos x}\, dx = \frac{2\pi^2}{\sqrt{1-a^2}} \left(\frac{1-\sqrt{1-a^2}}{\pm a} \right)^n$ $[a^2 < 1]$. BI ((223))(3)

4. $\int_0^{\pi} \frac{x \sin x\, dx}{a + b \cos x} = \frac{\pi}{b} \ln \frac{a + \sqrt{a^2 - b^2}}{2(a-b)}$ $[a > |b| > 0]$. GW ((333))(53a)

5. $\int_0^{2\pi} \frac{x \sin x\, dx}{a + b \cos x} = \frac{2\pi}{b} \ln \frac{a + \sqrt{a^2 - b^2}}{2(a+b)}$ $[a > |b| > 0]$. GW ((333))(53b)

6. $\int_0^{\infty} \frac{\sin x}{a \pm b \cos 2x} \cdot \frac{dx}{x} = \frac{\pi}{2\sqrt{a^2 - b^2}}$ $[a^2 > b^2]$;

$= 0$ $[a^2 < b^2]$. BI ((181))(1)

3.795 $\int_{-\infty}^{\infty} \frac{(b^2 + c^2 + x^2)\, x \sin ax - (b^2 - c^2 - x^2)\, c \operatorname{sh} ac}{[x^2 + (b-c)^2]\,[x^2 + (b+c)^2]\,(\cos ax + \operatorname{ch} ac)}\, dx = \pi$ $[c > b > 0]$;

$= \frac{2\pi}{e^{ab} + 1}$ $[b > c > 0]$;

$[a > 0]$. BI ((202))(18)

3.796

1. $\int_0^{\frac{\pi}{2}} \frac{\cos x \pm \sin x}{\cos x \mp \sin x}\, x\, dx = \mp \frac{\pi}{4} \ln 2 - G.$ BI ((207))(8, 9)

2. $\int_0^{\frac{\pi}{4}} \frac{\cos x - \sin x}{\cos x + \sin x}\, x\, dx = \frac{\pi}{4} \ln 2 - \frac{1}{2} G.$ BI ((204))(23)

3.797

1. $\int_0^{\frac{\pi}{4}} \left(\frac{\pi}{4} - x \operatorname{tg} x \right) \operatorname{tg} x\, dx = \frac{1}{2} \ln 2 + \frac{\pi^2}{32} - \frac{\pi}{4} + \frac{\pi}{8} \ln 2.$ BI ((204))(8)

2. $\int_0^{\frac{\pi}{4}} \frac{\left(\frac{\pi}{4} - x \right) \operatorname{tg} x\, dx}{\cos 2x} = -\frac{\pi}{8} \ln 2 + \frac{1}{2} G.$ BI ((204))(19)

3. $\int_0^{\frac{\pi}{4}} \frac{\frac{\pi}{4} - x \operatorname{tg} x}{\cos 2x}\, dx = \frac{\pi}{8} \ln 2 + \frac{1}{2} G.$ BI ((204))(20)

3.798

1. $\int_0^{\infty} \frac{\operatorname{tg} x}{a + b \cos 2x} \cdot \frac{dx}{x} = \frac{\pi}{2\sqrt{a^2 - b^2}}$ $[a^2 > b^2]$;

$= 0$ $[a^2 < b^2]$; $[a > 0]$. BI ((181))(2)

2. $\int_0^\infty \frac{\operatorname{tg} x}{a+b\cos 4x}\cdot\frac{dx}{x} = \frac{\pi}{2\sqrt{a^2-b^2}}$ $[a^2 > b^2]$;

$= 0$ $[a^2 < b^2]$; $[a > 0]$. BI ((181))(3)

3.799

1. $\int_0^{\frac{\pi}{2}} \frac{x\,dx}{(\sin x + a\cos x)^2} = \frac{a}{1+a^2}\frac{\pi}{2} - \frac{\ln a}{1+a^2}$ $[a > 0]$. BI ((208))(5)

2. $\int_0^{\frac{\pi}{4}} \frac{x\,dx}{(\cos x + a\sin x)^2} = \frac{1}{1+a^2}\ln\frac{1+a}{\sqrt{2}} + \frac{\pi}{4}\cdot\frac{1-a}{(1+a)(1+a^2)}$ $[a > 0]$.

BI ((204))(24)

3. $\int_0^{\pi} \frac{a\cos x + b}{(a+b\cos x)^2} x^2\,dx = \frac{2\pi}{b}\ln\frac{2(a-b)}{a+\sqrt{a^2-b^2}}$ $[a > |b| > 0]$. GW ((333))(58a)

3.811

1. $\int_0^{\pi} \frac{\sin x}{1-\cos t_1\cos x}\cdot\frac{x\,dx}{1-\cos t_2\cos x} = \pi\operatorname{cosec}\frac{t_1+t_2}{2}\operatorname{cosec}\frac{t_1-t_2}{2}\ln\frac{1+\operatorname{tg}\frac{t_1}{2}}{1+\operatorname{tg}\frac{t_2}{2}}$

(cf. **3.794** 4.). BI ((222))(5)

2. $\int_0^{\frac{\pi}{2}} \frac{x\,dx}{(\cos x \pm \sin x)\sin x} = \frac{\pi}{4}\ln 2 \pm G$. BI ((208))(16, 17)

3. $\int_0^{\frac{\pi}{4}} \frac{x\,dx}{(\cos x + \sin x)\sin x} = -\frac{\pi}{8}\ln 2 + G$. BI ((204))(29)

4. $\int_0^{\frac{\pi}{4}} \frac{x\,dx}{(\cos x + \sin x)\cos x} = \frac{\pi}{8}\ln 2$. BI ((204))(28)

5. $\int_0^{\frac{\pi}{4}} \frac{\sin x}{\sin x + \cos x}\,\frac{x\,dx}{\cos^2 x} = -\frac{\pi}{8}\ln 2 + \frac{\pi}{4} - \frac{1}{2}\ln 2$. BI ((204))(30)

3.812

1. $\int_0^{\pi} \frac{x\sin x\,dx}{a+b\cos^2 x} = \frac{\pi}{\sqrt{ab}}\operatorname{arctg}\sqrt{\frac{b}{a}}$ $[a > 0,\ b > 0]$;

$= \frac{\pi}{2\sqrt{-ab}}\ln\frac{\sqrt{a}+\sqrt{-b}}{\sqrt{a}-\sqrt{-b}}$ $[a > -b > 0]$. GW ((333))(60a)

2. $\int_0^{\frac{\pi}{2}} \frac{x \sin 2x \, dx}{1+a \cos^2 x} = \frac{\pi}{a} \ln \frac{1+\sqrt{1+a}}{2} \quad [a > -1, \; a \neq 0].$ BI ((207))(10)

3. $\int_0^{\frac{\pi}{2}} \frac{x \sin 2x \, dx}{1+a \sin^2 x} = \frac{\pi}{a} \ln \frac{2(1+a-\sqrt{1+a})}{a} \quad [a > -1, \; a \neq 0].$ BI ((207))(2)

4. $\int_0^{\pi} \frac{x \, dx}{a^2 - \cos^2 x} = \frac{\pi^2}{2a\sqrt{a^2-1}} \quad [a^2 > 1];$

$= 0 \quad [a^2 < 1].$ BI ((219))(10)

5. $\int_0^{\pi} \frac{x \sin x \, dx}{a^2 - \cos^2 x} = \frac{\pi}{2a} \ln \frac{1+a}{1-a} \quad [a \neq 1].$ BI ((219))(13)

6. $\int_0^{\pi} \frac{x \sin 2x \, dx}{a^2 - \cos^2 x} = \pi \ln \{4(1-a^2)\} \quad [a^2 < 1];$

$= 2\pi \ln [2(1-a^2+a\sqrt{a^2-1})] \quad [a^2 > 1].$ BI ((219))(19)

7. $\int_0^{\frac{\pi}{2}} \frac{x \sin x \, dx}{\cos^2 t - \sin^2 x} = -2 \operatorname{cosec} t \sum_{k=0}^{\infty} \frac{\sin(2k+1)t}{(2k+1)^2}.$ BI ((207))(1)

8. $\int_0^{\pi} \frac{x \sin x \, dx}{1-\cos^2 t \sin^2 x} = \pi(\pi - 2t) \operatorname{cosec} 2t.$ BI ((219))(12)

9. $\int_0^{\pi} \frac{x \cos x \, dx}{\cos^2 t - \cos^2 x} = 4 \operatorname{cosec} t \sum_{k=0}^{\infty} \frac{\sin(2k+1)t}{(2k+1)^2}.$ BI ((219))(17)

10. $\int_0^{\pi} \frac{x \sin x \, dx}{\operatorname{tg}^2 t + \cos^2 x} = \frac{\pi}{2}(\pi - 2t) \operatorname{ctg} t.$ BI ((219))(14)

11. $\int_0^{\infty} \frac{x(a \cos x + b) \sin x \, dx}{\operatorname{ctg}^2 t + \cos^2 x} = 2a\pi \ln \cos \frac{t}{2} + \pi bt \operatorname{tg} t.$ BI ((219))(18)

3.813

1. $\int_0^{\pi} \frac{x \, dx}{a^2 \cos^2 x + b^2 \sin^2 x} = \frac{1}{4} \int_0^{2\pi} \frac{x \, dx}{a^2 \cos^2 x + b^2 \sin^2 x} = \frac{\pi^2}{2ab}$

$[a > 0, \; b > 0].$ GW ((333))(36)

2. $\int_0^{\infty} \frac{1}{\beta^2 \sin^2 ax + \gamma^2 \cos^2 ax} \cdot \frac{dx}{x^2+\delta^2} = \frac{\pi \operatorname{sh}(2a\delta)}{4\delta(\beta^2 \operatorname{sh}^2(a\delta) - \gamma^2 \operatorname{ch}^2(a\delta))} \left[\frac{\beta}{\gamma} - \frac{\gamma}{\beta} - \frac{2}{\operatorname{sh}(2a\delta)}\right]$

$\left[\left|\arg \frac{\beta}{\gamma}\right| < \pi, \; \operatorname{Re} \delta > 0, \; a > 0\right].$ GW((333))(81), ET II 222(63)

3. $\int_0^{\infty} \frac{\sin x \, dx}{x(a^2 \sin^2 x + b^2 \cos^2 x)} = \frac{\pi}{2ab} \quad [ab > 0].$ BI ((181))(8)

4. $\int_0^\infty \frac{\sin^2 x\, dx}{x(a^2\cos^2 x + b^2 \sin^2 x)} = \frac{\pi}{2b(a+b)} \quad [a>0,\ b>0].$ BI ((181))(11)

5. $\int_0^{\frac{\pi}{2}} \frac{x \sin 2x\, dx}{a^2\cos^2 x + b^2\sin^2 x} = \frac{\pi}{a^2-b^2} \ln \frac{a+b}{2b} \quad [a>0,\ b>0,\ a \neq b].$

GW ((333))(52a)

6. $\int_0^{\pi} \frac{x \sin 2x\, dx}{a^2\cos^2 x + b^2\sin^2 x} = \frac{2\pi}{a^2-b^2} \ln \frac{a+b}{2a} \quad [a>0,\ b>0,\ a \neq b].$

GW ((333))(52b)

7. $\int_0^\infty \frac{\sin 2x}{a^2\cos^2 x + b^2\sin^2 x} \cdot \frac{dx}{x} = \frac{\pi}{a(a+b)} \quad [a>0,\ b>0].$ BI ((182))(3)

8. $\int_0^\infty \frac{\sin 2ax}{\beta^2 \sin^2 ax + \gamma^2 \cos^2 ax} \cdot \frac{x\, dx}{x^2+\delta^2} = \frac{\pi}{2(\beta^2 \operatorname{sh}^2(a\delta) - \gamma^2 \operatorname{ch}^2(a\delta))} \left[\frac{\beta-\gamma}{\beta+\gamma} - e^{-2a\delta}\right]$

$\left[a>0,\ \left|\arg\frac{\beta}{\gamma}\right| < \pi,\ \operatorname{Re}\delta > 0\right].$

ET II 222(64), GW((333))(80)

9. $\int_0^\infty \frac{(1-\cos x)\sin x}{a^2\cos^2 x + b^2\sin^2 x} \cdot \frac{dx}{x} = \frac{\pi}{2b(a+b)} \quad [a>0,\ b>0].$ BI ((182))(7)a

10. $\int_0^\infty \frac{\sin x \cos^2 x}{a^2\cos^2 x + b^2\sin^2 x} \cdot \frac{dx}{x} = \frac{\pi}{2a(a+b)} \quad [a>0,\ b>0].$ BI ((182))(4)

11. $\int_0^\infty \frac{\sin^3 x}{a^2\cos^2 x + b^2\sin^2 x} \cdot \frac{dx}{x} = \frac{\pi}{2b} \cdot \frac{1}{a+b} \quad [a>0,\ b>0].$ BI ((182))(1)

3.814

1. $\int_0^{\frac{\pi}{2}} \frac{(1 - x \operatorname{ctg} x)\, dx}{\sin^2 x} = \frac{\pi}{4}.$ BI ((206))(9)

2. $\int_0^{\frac{\pi}{4}} \frac{x \operatorname{tg} x\, dx}{(\sin x + \cos x)\cos x} = -\frac{\pi}{8}\ln 2 + \frac{\pi}{4} - \frac{1}{2}\ln 2.$ BI ((204))(30)

3. $\int_0^\infty \frac{\operatorname{tg} x}{a^2\cos^2 x + b^2\sin^2 x} \frac{dx}{x} = \frac{\pi}{2ab} \quad [a>0,\ b>0].$ BI ((181))(9)

4. $\int_0^{\frac{\pi}{2}} \frac{x \operatorname{ctg} x\, dx}{a^2\cos^2 x + b^2\sin^2 x} = \frac{\pi}{2a^2} \ln \frac{a+b}{b} \quad [a>0,\ b>0].$ LI ((208))(20)

5. $$\int_0^{\frac{\pi}{2}} \frac{\left(\frac{\pi}{2}-x\right)\operatorname{tg} x\,dx}{a^2\cos^2 x+b^2\sin^2 x}=\frac{1}{2}\int_0^{\pi}\frac{\left(\frac{\pi}{2}-x\right)\operatorname{tg} x\,dx}{a^2\cos^2 x+b^2\sin^2 x}=$$
$$=\frac{\pi}{2b^2}\ln\frac{a+b}{a} \quad [a>0,\ b>0].$$ GW ((333))(59)

6. $$\int_0^{\infty}\frac{\sin^2 x\operatorname{tg} x}{a^2\cos^2 x+b^2\sin^2 x}\cdot\frac{dx}{x}=\frac{\pi}{2b(a+b)} \quad [a>0,\ b>0].$$ BI ((182))(6)

7. $$\int_0^{\infty}\frac{\operatorname{tg} x}{a^2\cos^2 2x+b^2\sin^2 2x}\cdot\frac{dx}{x}=\frac{\pi}{2ab} \quad [a>0,\ b>0].$$ BI ((181))(10)a

8. $$\int_0^{\infty}\frac{\sin^2 2x\operatorname{tg} x}{a^2\cos^2 2x+b^2\sin^2 2x}\cdot\frac{dx}{x}=\frac{\pi}{2b}\cdot\frac{1}{a+b} \quad [a>0,\ b>0].$$ BI ((182))(2)a

9. $$\int_0^{\infty}\frac{\cos^2 2x\operatorname{tg} x}{a^2\cos^2 2x+b^2\sin^2 2x}\cdot\frac{dx}{x}=\frac{\pi}{2a}\cdot\frac{1}{a+b} \quad [a>0,\ b>0].$$ BI ((182))(5)a

10. $$\int_0^{\infty}\frac{\sin^2 x\cos x}{a^2\cos^2 2x+b^2\sin^2 2x}\cdot\frac{dx}{x\cos 4x}=-\frac{\pi}{8b}\,\frac{a}{a^2+b^2}$$
$$[a>0,\ b>0].$$ BI ((186))(12)a

11. $$\int_0^{\infty}\frac{\sin x}{a^2\cos^2 x+b^2\sin^2 x}\cdot\frac{dx}{x\cos 2x}=\frac{\pi}{2ab}\cdot\frac{b^2-a^2}{b^2+a^2}$$
$$[a>0,\ b>0].$$ BI ((186))(4)a

12. $$\int_0^{\infty}\frac{\sin x\cos x}{a^2\cos^2 x+b^2\sin^2 x}\cdot\frac{dx}{x\cos 2x}=\frac{\pi}{2a}\cdot\frac{b}{a^2+b^2} \quad [a>0,\ b>0].$$ BI ((186))(7)a

13. $$\int_0^{\infty}\frac{\sin x\cos^2 x}{a^2\cos^2 x+b^2\sin^2 x}\cdot\frac{dx}{x\cos 2x}=\frac{\pi}{2ab}\cdot\frac{b^2}{a^2+b^2} \quad [a>0,\ b>0].$$ BI ((186))(8)a

14. $$\int_0^{\infty}\frac{\sin^3 x}{a^2\cos^2 x+b^2\sin^2 x}\cdot\frac{dx}{x\cos 2x}=-\frac{\pi}{2b}\cdot\frac{a}{a^2+b^2} \quad [a>0,\ b>0].$$ BI ((186))(10)

15. $$\int_0^{\infty}\frac{1-\cos x}{a^2\cos^2 x+b^2\sin^2 x}\cdot\frac{dx}{x\sin x}=\frac{\pi}{2ab} \quad [a>0,\ b>0].$$ BI ((186))(3)a

3.815

1. $$\int_0^{\frac{\pi}{2}}\frac{x\sin 2x\,dx}{(1+a\sin^2 x)(1+b\sin^2 x)}=\frac{\pi}{a-b}\ln\left\{\frac{1+\sqrt{1+b}}{1+\sqrt{1+a}}\cdot\frac{\sqrt{1+a}}{\sqrt{1+b}}\right\}$$
$$[a>0,\ b>0],$$ (cf. **3.812** 3.). BI ((208))(22)

2. $$\int_0^{\frac{\pi}{2}} \frac{x \sin 2x \, dx}{(1+a\sin^2 x)(1+b\cos^2 x)} = \frac{\pi}{a+ab+b} \ln \frac{(1+\sqrt{1+b})\sqrt{1+a}}{1+\sqrt{1+a}}$$
$[a>0,\ b>0]$, (cf. **3.812** 2. and 3.). BI ((208))(24)

3. $$\int_0^{\frac{\pi}{2}} \frac{x \sin 2x \, dx}{(1+a\cos^2 x)(1+b\cos^2 x)} = \frac{\pi}{a-b} \ln \frac{1+\sqrt{1+a}}{1+\sqrt{1+b}}$$
$[a>0,\ b>0]$, (cf. **3.812** 2.). BI ((208))(23)

4. $$\int_0^{\frac{\pi}{2}} \frac{x \sin 2x \, dx}{(1-\sin^2 t_1 \cos^2 x)(1-\sin^2 t_2 \cos^2 x)} =$$
$$= \frac{2\pi}{\cos^2 t_1 - \cos^2 t_2} \ln \frac{\cos \frac{t_1}{2}}{\cos \frac{t_2}{2}} \quad [-\pi < t_1 < \pi,\ -\pi < t_2 < \pi].$$ BI ((208))(21)

3.816

1. $$\int_0^{\pi} \frac{x^2 \sin 2x}{(a^2-\cos^2 x)^2} dx = \pi^2 \frac{\sqrt{a^2-1}-a}{a(a^2-1)} \quad [a>1].$$ LI ((220))(9)

2. $$\int_0^{\pi} \frac{(a^2-1-\sin^2 x)\cos x}{(a^2-\cos^2 x)^2} x^2 \, dx = \frac{\pi}{a} \ln \frac{1-a}{1+a} \quad [a>0,\ a \neq 1],$$
(cf. **3.812** 5.). BI ((220))(12)

3. $$\int_0^{\pi} \frac{a\cos 2x - \sin^2 x}{(a+\sin^2 x)^2} x^2 \, dx = -2\pi \ln\left[2\left(-a+\sqrt{a(a+1)}\right)\right] \quad [a>0].$$
LI ((220))(10)

4. $$\int_0^{\pi} \frac{a\cos 2x + \sin^2 x}{(a-\sin^2 x)^2} x^2 \, dx = \pi \ln(4a) \quad [a>1],$$ (cf. **3.812** 6.).
LI ((220))(11)

5. $$\int_0^{\frac{\pi}{2}} \frac{(\cos^2 t + \sin^2 x)\cos x}{(\cos^2 t - \sin^2 x)^2} \cdot x^2 dx = -\frac{\pi^2}{4\sin^2 t} + \frac{4}{\sin t} \sum_{k=0}^{\infty} \frac{\sin[(2k+1)t]}{(2k+1)^2}$$
(cf. **3.812** 7.). BI ((208))(14)

3.817

1. $$\int_0^{\infty} \frac{\sin x}{(a^2\cos^2 x + b^2 \sin^2 x)^2} \cdot \frac{dx}{x} = \frac{\pi}{4} \cdot \frac{a^2+b^2}{a^3 b^3} \quad [ab>0].$$ BI ((181))(12)

2. $$\int_0^{\infty} \frac{\sin x \cos x}{(a^2\cos^2 x + b^2 \sin^2 x)^2} \cdot \frac{dx}{x} = \frac{\pi}{4a^3 b} \quad [ab>0].$$ BI ((182))(8)

3. $$\int_0^{\infty} \frac{\sin^3 x}{(a^2\cos^2 x + b^2 \sin^2 x)^2} \cdot \frac{dx}{x} = \frac{\pi}{4ab^3} \quad [ab>0].$$ BI ((181))(15)

4. $\int_0^\infty \frac{\sin x \cos^2 x}{(a^2 \cos^2 x + b^2 \sin^2 x)^2} \cdot \frac{dx}{x} = \frac{\pi}{4a^3 b} \quad [ab > 0].$ BI ((182))(9)

5. $\int_0^\infty \frac{\operatorname{tg} x}{(a^2 \cos^2 x + b^2 \sin^2 x)^2} \cdot \frac{dx}{x} = \frac{\pi}{4} \cdot \frac{a^2 + b^2}{a^3 b^3} \quad [ab > 0].$ BI ((181))(13)

6. $\int_0^\infty \frac{\operatorname{tg} x}{(a^2 \cos^2 2x + b^2 \sin^2 2x)^2} \cdot \frac{dx}{x} = \frac{\pi}{4} \frac{a^2 + b^2}{a^3 b^3} \quad [ab > 0].$ BI ((181))(14)

7. $\int_0^\infty \frac{\sin^2 x \operatorname{tg} x}{(a^2 \cos^2 x + b^2 \sin^2 x)^2} \cdot \frac{dx}{x} = \frac{\pi}{4ab^3} \quad [ab > 0].$ BI ((182))(11)

8. $\int_0^\infty \frac{\operatorname{tg} x \cos^2 2x}{(a^2 \cos^2 2x + b^2 \sin^2 2x)^2} \cdot \frac{dx}{x} = \frac{\pi}{4a^3 b} \quad [ab > 0].$ BI ((182))(10)

3.818

1. $\int_0^\infty \frac{\sin x}{(a^2 \cos^2 x + b^2 \sin^2 x)^3} \cdot \frac{dx}{x} = \frac{\pi}{16} \cdot \frac{3a^4 + 2a^2 b^2 + 3b^4}{a^5 b^5} \quad [ab > 0].$ BI ((181))(16)

2. $\int_0^\infty \frac{\sin x \cos x}{(a^2 \cos^2 x + b^2 \sin^2 x)^3} \cdot \frac{dx}{x} = \frac{\pi}{16} \cdot \frac{a^2 + 3b^2}{a^5 b^3} \quad [ab > 0].$ BI ((182))(13)

3. $\int_0^\infty \frac{\sin x \cos^2 x}{(a^2 \cos^2 x + b^2 \sin^2 x)^3} \cdot \frac{dx}{x} = \frac{\pi}{16} \cdot \frac{a^2 + 3b^2}{a^5 b^3} \quad [ab > 0].$ BI ((182))(14)

4. $\int_0^\infty \frac{\sin^3 x}{(a^2 \cos^2 x + b^2 \sin^2 x)^3} \cdot \frac{dx}{x} = \frac{\pi}{16} \cdot \frac{3a^2 + b^2}{a^3 b^5} \quad [ab > 0].$ LI ((181))(19)

5. $\int_0^\infty \frac{\sin^3 x \cos x}{(a^2 \cos^2 2x + b^2 \sin^2 2x)^3} \cdot \frac{dx}{x} = \frac{\pi}{64} \cdot \frac{3a^2 + b^2}{a^3 b^5} \quad [ab > 0].$ BI ((182))(17)

6. $\int_0^\infty \frac{\operatorname{tg} x}{(a^2 \cos^2 x + b^2 \sin^2 x)^3} \cdot \frac{dx}{x} = \frac{\pi}{16} \frac{3a^4 + 2a^2 b^2 + 3b^4}{a^5 b^5} \quad [ab > 0].$ BI ((181))(17)

7. $\int_0^\infty \frac{\sin^2 x \operatorname{tg} x}{(a^2 \cos^2 x + b^2 \sin^2 x)^3} \cdot \frac{dx}{x} = \frac{\pi}{16} \cdot \frac{3a^2 + b^2}{a^3 b^5} \quad [ab > 0].$ BI((182))(16)

8. $\int_0^\infty \frac{\operatorname{tg} x}{(a^2 \cos^2 2x + b^2 \sin^2 2x)^3} \cdot \frac{dx}{x} = \frac{\pi}{16} \cdot \frac{3a^4 + 2a^2 b^2 + 3b^4}{a^5 b^5} \quad [ab > 0].$ BI ((181))(18)

9. $\int_0^\infty \frac{\operatorname{tg} x \cos^2 2x}{(a^2 \cos^2 2x + b^2 \sin^2 2x)^3} \cdot \frac{dx}{x} = \frac{\pi}{16} \cdot \frac{a^2 + 3b^2}{a^5 b^3} \quad [ab > 0].$ BI ((182))(15)

3.819

1. $\int_0^\infty \frac{\sin x}{(a^2 \cos^2 x + b^2 \sin^2 x)^4} \cdot \frac{dx}{x} = \frac{\pi}{32} \cdot \frac{5a^6 + 3a^4 b^2 + 3a^2 b^4 + 5b^6}{a^7 b^7} \quad [ab > 0].$

BI ((181))(20)

2. $$\int_0^\infty \frac{\sin x \cos x}{(a^2\cos^2 x+b^2\sin^2 x)^4}\cdot\frac{dx}{x}=\frac{\pi}{32}\cdot\frac{a^4+2a^2b^2+5b^4}{a^7b^5} \quad [ab>0].$$ BI ((182))(18)

3. $$\int_0^\infty \frac{\sin x \cos^2 x}{(a^2\cos^2 x+b^2\sin^2 x)^4}\cdot\frac{dx}{x}=\frac{\pi}{32}\cdot\frac{a^4+2a^2b^2+5b^4}{a^7b^5} \quad [ab>0].$$ BI ((182))(19)

4. $$\int_0 \frac{\sin^3 x}{(a^2\cos^2 x+b^2\sin^2 x)^4}\cdot\frac{dx}{x}=\frac{\pi}{32}\cdot\frac{5a^4+a^2b^2+b^4}{a^5b^7} \quad [ab>0].$$ BI ((181))(23)

5. $$\int_0^\infty \frac{\sin^3 x \cos x}{(a^2\cos^2 x+b^2\sin^2 x)^4}\cdot\frac{dx}{x}=\frac{\pi}{32}\cdot\frac{a^2+b^2}{a^5b^5} \quad [ab>0].$$ BI ((182))(26)

6. $$\int_0^\infty \frac{\sin x \cos^3 x}{(a^2\cos^2 x+b^2\sin^2 x)^4}\cdot\frac{dx}{x}=\frac{\pi}{32}\cdot\frac{a^2+5b^2}{a^7b^3} \quad [ab>0].$$ BI ((182))(23)

7. $$\int_0^\infty \frac{\sin^3 x \cos^2 x}{(a^2\cos^2 x+b^2\sin^2 x)^4}\cdot\frac{dx}{x}=\frac{\pi}{32}\cdot\frac{a^2+b^2}{a^5b^5} \quad [ab>0].$$ BI ((182))(27)

8. $$\int_0^\infty \frac{\sin x \cos^4 x}{(a^2\cos^2 x+b^2\sin^2 x)^4}\cdot\frac{dx}{x}=\frac{\pi}{32}\cdot\frac{a^2+5b^2}{a^7b^3} \quad [ab>0].$$ BI ((182))(24)

9. $$\int_0^\infty \frac{\sin^5 x}{(a^2\cos^2 x+b^2\sin^2 x)^4}\cdot\frac{dx}{x}=\frac{\pi}{32}\cdot\frac{5a^2+b^2}{a^3b^7} \quad [ab>0].$$ BI ((181))(24)

10. $$\int_0^\infty \frac{\sin^3 x \cos x}{(a^2\cos^2 2x+b^2\sin^2 2x)^4}\cdot\frac{dx}{x}=\frac{\pi}{128}\cdot\frac{5a^4+2a^2b^2+b^4}{a^5b^7} \quad [ab>0].$$ BI ((182))(22)

11. $$\int_0^\infty \frac{\sin^5 x \cos^3 x}{(a^2\cos^2 2x+b^2\sin^2 2x)^4}\cdot\frac{dx}{x}=\frac{\pi}{512}\cdot\frac{5a^2+b^2}{a^3b^7} \quad [ab>0].$$ BI ((182))(30)

12. $$\int_0^\infty \frac{\sin^2 x \operatorname{tg} x}{(a^2\cos^2 x+b^2\sin^2 x)^4}\cdot\frac{dx}{x}=\frac{\pi}{32}\cdot\frac{5a^4+2a^2b^2+b^4}{a^5b^7} \quad [ab>0].$$ BI ((182))(21)

13. $$\int_0^\infty \frac{\sin^4 x \operatorname{tg} x}{(a^2\cos^2 x+b^2\sin^2 x)^4}\cdot\frac{dx}{x}=\frac{\pi}{32}\cdot\frac{5a^2+b^2}{a^3b^7} \quad [ab>0].$$ BI ((182))(29)

14. $$\int_0^\infty \frac{\cos^2 2x \operatorname{tg} x}{(a^2\cos^2 2x+b^2\sin^2 2x)^4}\cdot\frac{dx}{x}=\frac{\pi}{32}\cdot\frac{a^4+2a^2b^2+5b^4}{a^7b^5} \quad [ab>0].$$ BI ((182))(29)

15. $$\int_0^\infty \frac{\sin^2 4x \operatorname{tg} x}{(a^2\cos^2 2x+b^2\sin^2 2x)^4}\cdot\frac{dx}{x}=\frac{\pi}{8}\cdot\frac{a^2+b^2}{a^5b^5} \quad [ab>0].$$ BI ((182))(28)

16. $$\int_0^\infty \frac{\cos^4 2x \operatorname{tg} x}{(a^2\cos^2 2x+b^2\sin^2 2x)^4}\cdot\frac{dx}{x}=\frac{\pi}{32}\cdot\frac{a^2+5b^2}{a^7b^3} \quad [ab>0].$$ BI ((182))(25)

3.82-3.83 Powers of trigonometric functions combined with other powers

3.821

1. $$\int_0^{\pi} x \sin^p x \, dx = \frac{\pi^2}{2^{p+1}} \frac{\Gamma(p+1)}{\left[\Gamma\left(\frac{p}{2}+1\right)\right]^2} \quad [p > -1].$$ BI((218))(7), LO V 121(71)

2. $$\int_0^{r\pi} x \sin^n x \, dx = \frac{\pi^2}{2} \cdot \frac{(2m-1)!!}{(2m)!!} r^2 \quad [n = 2m];$$
$$= (-1)^{r+1} \pi \frac{(2m)!!}{(2m+1)!!} r \quad [n = 2m+1],$$
[r is a natural number]. GW ((333))(8c)

3. $$\int_0^{\frac{\pi}{2}} x \cos^n x \, dx = -\sum_{k=0}^{m-1} \frac{(n-2k+1)(n-2k+3)\ldots(n-1)}{(n-2k)(n-2k+2)\ldots n} \cdot \frac{1}{n-2k} +$$
$$+ \begin{cases} \frac{\pi}{2} \cdot \frac{(2m-2)!!}{(2m-1)!!} & [n = 2m-1]; \\ \frac{\pi^2}{8} \cdot \frac{(2m-1)!!}{(2m)!!} & [n = 2m]. \end{cases}$$ GW ((333))(9b)

4. $$\int_0^{\pi} x \cos^{2m} x \, dx = \frac{\pi^2}{2} \frac{(2m-1)!!}{(2m)!!}.$$ BI ((218))(10)

5. $$\int_{r\pi}^{s\pi} x \cos^{2m} x \, dx = \frac{\pi^2}{2} (s^2 - r^2) \cdot \frac{(2m-1)!!}{(2m)!!}.$$ BI ((226))(3)

6. $$\int_0^{\infty} \frac{\sin^p x}{x} \, dx = \frac{\sqrt{\pi}}{2} \cdot \frac{\Gamma\left(\frac{p}{2}\right)}{\Gamma\left(\frac{p+1}{2}\right)} = 2^{p-2} \mathrm{B}\left(\frac{p}{2}, \frac{p}{2}\right);$$
[p is a fraction with odd numerator and denominator].

LO V 278, FI II 808

7. $$\int_0^{\infty} \frac{\sin^{2n+1} x}{x} \, dx = \frac{(2n-1)!!}{(2n)!!} \cdot \frac{\pi}{2}.$$ BI ((151))(4)

8. $$\int_0^{\infty} \frac{\sin^{2n} x}{x} \, dx = \infty.$$ BI ((151))(3)

9. $$\int_0^{\infty} \frac{\sin^2 ax}{x^2} \, dx = \frac{a\pi}{2} \quad [a > 0].$$ LO V 307, 312, FI II 632

10. $$\int_0^{\infty} \frac{\sin^{2m} ax}{x^2} \, dx = \frac{(2m-3)!!}{(2m-2)!!} \cdot \frac{a\pi}{2} \quad [a > 0].$$ GW ((333))(14b)

11. $\int_0^\infty \frac{\sin^{2m+1} ax}{x^3}\,dx = \frac{(2m-3)!!}{(2m)!!}(2m+1)\frac{a^2\pi}{4} \quad [a>0].$ GW ((333))(14d)

12. $\int_0^\infty \frac{\sin^p x}{x^m}\,dx = \frac{p}{m-1}\int_0^\infty \frac{\sin^{p-1} x}{x^{m-1}}\cos x\,dx \quad [p>m-1>0];$

$$= \frac{p(p-1)}{(m-1)(m-2)}\int_0^\infty \frac{\sin^{p-2} x}{x^{m-2}}\,dx - \frac{p^2}{(m-1)(m-2)}\int_0^\infty \frac{\sin^p x}{x^{m-2}}\,dx$$

$[p>m-1>1].$ GW ((333))(17)

13. $\int_0^\infty \frac{\sin^{2n} px}{\sqrt{x}}\,dx = \infty.$ BI ((177))(5)

14. $\int_0^\infty \sin^{2n+1} px \frac{dx}{\sqrt{x}} = \frac{1}{2^{2n}}\sqrt{\frac{\pi}{2p}}\sum_{k=0}^{n}(-1)^k \binom{2n+1}{n+k+1}\frac{1}{\sqrt{2k+1}}$

BI ((177))(7)

3.822

1. $\int_0^{\frac{\pi}{2}} x^p \cos^m x\,dx = -\frac{p(p-1)}{m^2}\int_0^{\frac{\pi}{2}} x^{p-2}\cos^m x\,dx + \frac{m-1}{m}\int_0^{\frac{\pi}{2}} x^p \cos^{m-2} x\,dx$

$[m>1,\ p>1].$ GW ((333))(9a)

2. $\int_0^\infty x^{-\frac{1}{2}}\cos^{2n+1}(px)\,dx = \frac{1}{2^{2n}}\sqrt{\frac{\pi}{2p}}\sum_{k=0}^{n}\binom{2n+1}{n+k+1}\frac{1}{\sqrt{2k+1}}.$

BI ((177))(8)

3.823

$$\int_0^\infty x^{\mu-1}\sin^2 ax\,dx = -\frac{\Gamma(\mu)\cos\frac{\mu\pi}{2}}{2^{\mu+1}a^\mu} \qquad [a>0,\quad -2<\operatorname{Re}\mu<0].$$

ET I 319(15), GW((333))(19c)a

3.824

1. $\int_0^\infty \frac{\sin^2 ax}{x^2+\beta^2}\,dx = \frac{\pi}{4\beta}(1-e^{-2a\beta}) \qquad [a>0,\ \operatorname{Re}\beta>0].$ BI ((160))(10)

2. $\int_0^\infty \frac{\cos^2 ax}{x^2+\beta^2}\,dx = \frac{\pi}{4\beta}(1+e^{-2a\beta}) \qquad [a>0,\ \operatorname{Re}\beta>0].$ BI ((160))(11)

3. $\int_0^\infty \sin^{2m} x \frac{dx}{a^2+x^2} = \frac{(-1)^m}{2^{2m+1}}\cdot\frac{\pi}{a}\Big\{2^{2m}\operatorname{sh}^{2m} a -$

$-2\sum_{k=0}^{m}(-1)^k\binom{2m}{k}\operatorname{sh}[2(m-k)a]\Big\} \qquad [a>0].$ BI ((160))(12)

4. $$\int_0^\infty \sin^{2m+1} x \frac{dx}{a^2+x^2} =$$

$$= \frac{(-1)^{m-1}}{2^{2m+2}a}\left\{e^{(2m+1)a} \sum_{k=0}^{2m+1} (-1)^k \binom{2m+1}{k} e^{-2ka} \operatorname{Ei}[(2k-2m-1)a] + \right.$$

$$\left. + e^{-(2m+1)a} \sum_{k=1}^{2m+1} (-1)^{k-1} \binom{2m+1}{k} e^{2ka} \operatorname{Ei}[(2m+1-2k)a]\right\} \quad [a>0].$$

BI ((160))(14)

5. $$\int_0^\infty \sin^{2m+1} x \frac{x\,dx}{a^2+x^2} = \frac{(-1)^{m-1}}{2^{2m+2}} e^{-(2m+1)a} \left\{(1-e^{2(2m+1)a})(1-e^{-2a})^{2m+1} - \right.$$

$$\left. -2\sum_{k=0}^{m} (-1)^k \binom{2m+1}{k} e^{2ka}\right\} \quad [a>0].$$ BI ((160))(15)

6. $$\int_0^\infty \cos^{2m} x \frac{dx}{a^2+x^2} = \frac{\pi}{2^{2m+1}a}\binom{2m}{m} + \frac{\pi}{2^{2m}} \sum_{k=1}^{m} \binom{2m}{m+k} e^{-2ka} \quad [a>0].$$

BI ((160))(16)

7. $$\int_0^\infty \cos^{2m+1} x \frac{dx}{a^2+x^2} = \frac{\pi}{2^{2m+1}a} \sum_{k=1}^{m} \binom{2m+1}{m+k+1} e^{-(2k+1)a} \quad [a>0].$$

BI ((160))(17)

8. $$\int_0^\infty \cos^{2m+1} x \frac{x\,dx}{a^2+x^2} = -\frac{e^{-(2m+1)a}}{2^{2m+2}} \sum_{k=0}^{2m+1} \binom{2m+1}{k} e^{2ka} \operatorname{Ei}[(2m-2k+1)a] -$$

$$-\frac{e^{(2m+1)a}}{2^{2m+2}} \sum_{k=0}^{2m+1} \binom{2m+1}{k} e^{-2ka} \operatorname{Ei}[(2k-2m-1)a].$$ BI ((160))(18)

9. $$\int_0^\infty \frac{\cos^2 ax}{b^2-x^2}\,dx = \frac{\pi}{4b} \sin 2ab \quad [a>0,\ b>0].$$ BI ((161))(10)

10. $$\int_0^\infty \frac{\sin^2 ax \cos^2 bx}{\beta^2+x^2}\,dx = \frac{\pi}{8\beta}\left[1 - \frac{1}{2}e^{-2(a+b)\beta} + e^{-2b\beta} - \frac{1}{2}e^{2(b-a)\beta} - e^{-2a\beta}\right]$$

$$[a>b];$$

$$= \frac{\pi}{16\beta}[1-e^{-4a\beta}] \quad [a=b];$$

$$= \frac{\pi}{8\beta}\left[1 - \frac{1}{2}e^{-2(a+b)\beta} + e^{-2b\beta} - \frac{1}{2}e^{2(a-b)\beta} - e^{-2a\beta}\right] \quad [a<b];$$

$[a>0,\ b>0]$, (cf. 3.824 1. and 3.). BI ((162))(6)

11. $$\int_0^\infty \frac{x \sin 2ax \cos^2 bx}{\beta^2+x^2}\,dx = \frac{\pi}{8}[2e^{-2a\beta} + e^{-2(a+b)\beta} + e^{2(b-a)\beta}] \quad [a>b];$$

$$= \frac{\pi}{8}[e^{-4a\beta} + 2e^{-2a\beta}] \quad [a=b];$$

$$= \frac{\pi}{8}[2e^{-2a\beta} + e^{-2(a+b)\beta} - e^{2(a-b)\beta}] \quad [a<b].$$

LI ((162))(5)

3.825

1. $\int_0^\infty \frac{\sin^2 ax\,dx}{(b^2+x^2)(c^2+x^2)} = \frac{\pi(b-c+ce^{-2ab}-be^{-2ac})}{4bc(b^2-c^2)}$ $[a>0, \quad b>0, \quad c>0].$

BI ((174))(15)

2. $\int_0^\infty \frac{\cos^2 ax\,dx}{(b^2+x^2)(c^2+x^2)} = \frac{\pi(b-c+be^{-2ac}-ce^{-2ab})}{4bc(b^2-c^2)}$ $[a>0, \quad b>0, \quad c>0].$

BI ((175))(14)

3. $\int_0^\infty \frac{\sin^2 ax\,dx}{(b^2-x^2)(c^2-x^2)} = \frac{\pi(c\sin 2ab - b\sin 2ac)}{4bc(b^2-c^2)}$ $[a>0, \quad b>0, \quad c>0].$

LI ((174))(16)

4. $\int_0^\infty \frac{\cos^2 ax\,dx}{(b^2-x^2)(c^2-x^2)} = \frac{\pi(b\sin 2ac - c\sin 2ab)}{4bc(b^2-c^2)}$ $[a>0, \quad b>0, \quad c>0].$

LI ((175))(15)

3.826

1. $\int_0^\infty \frac{\sin^2 ax\,dx}{x^2(b^2+x^2)} = \frac{\pi}{4b^2}\left[2a - \frac{1}{b}(1-e^{-2ab})\right]$ $[a>0, \quad b>0].$

BI ((172))(13)

2. $\int_0^\infty \frac{\sin^2 ax\,dx}{x^2(b^2-x^2)} = \frac{\pi}{4b^2}\left(2a - \frac{1}{b}\sin 2ab\right)$ $[a>0, \ b>0].$

BI ((172))(14)

3.827

1. $\int_0^\infty \frac{\sin^3 ax}{x^\nu}\,dx = \frac{3-3^{\nu-1}}{4} a^{\nu-1} \cos\frac{\nu\pi}{2}\,\Gamma(1-\nu)$ $[a>0, \quad 0<\operatorname{Re}\nu<2].$

GW ((333))(19f)

2. $\int_0^\infty \frac{\sin^3 ax}{x}\,dx = \frac{\pi}{4}\operatorname{sign} a.$ LO V 277

3. $\int_0^\infty \frac{\sin^3 ax}{x^2}\,dx = \frac{3}{4} a \ln 3.$ BI ((156))(2)

4. $\int_0^\infty \frac{\sin^3 ax}{x^3}\,dx = \frac{3}{8} a^2\pi \operatorname{sign} a.$ BI((156))(7)a, LO V 312

5. $\int_0^\infty \frac{\sin^4 ax}{x^2}\,dx = \frac{a\pi}{4}$ $[a>0].$ BI ((156))(3)

6. $\int_0^\infty \frac{\sin^4 ax}{x^3}\,dx = a^2 \ln 2.$ BI ((156))(8)

7. $\int_0^\infty \frac{\sin^4 ax}{x^4} dx = \frac{a^3\pi}{3} \quad [a > 0].$ BI((156))(11), LO V 312

8. $\int_0^\infty \frac{\sin^5 ax}{x^2} dx = \frac{5}{16} a (3 \ln 3 - \ln 5).$ BI ((156))(4)

9. $\int_0^\infty \frac{\sin^5 ax}{x^3} dx = \frac{5}{32} a^2\pi \quad [a > 0].$ BI ((156))(9)

10. $\int_0^\infty \frac{\sin^5 ax}{x^4} dx = \frac{5}{96} a^3 (25 \ln 5 - 27 \ln 3) \quad [a > 0].$ BI ((156))(12)

11. $\int_0^\infty \frac{\sin^5 ax}{x^5} dx = \frac{115}{384} a^4\pi \quad [a > 0].$ BI((156))(13), LO V 312

12. $\int_0^\infty \frac{\sin^6 ax}{x^2} dx = \frac{3}{16} a\pi \quad [a > 0].$ BI ((156))(5)

13. $\int_0^\infty \frac{\sin^6 ax}{x^3} dx = \frac{3}{16} a^2 (8 \ln 2 - 3 \ln 3).$ BI ((156))(10)

14. $\int_0^\infty \frac{\sin^6 ax}{x^5} dx = \frac{1}{16} a^4 (27 \ln 3 - 32 \ln 2).$ BI ((156))(14)

15. $\int_0^\infty \frac{\sin^6 ax}{x^6} dx = \frac{11}{40} a^5\pi \quad [a > 0].$ LO V 312

3.828

1. $\int_0^\infty \frac{\sin px \sin qx}{x} dx = \ln \sqrt{\frac{p+q}{|p-q|}} \quad [p \neq q].$ FI II 647

2. $\int_0^\infty \sin qx \sin px \frac{dx}{x^2} = \frac{1}{2} p\pi \quad [p \leqslant q];$

$= \frac{1}{2} q\pi \quad [p \geqslant q].$ BI ((157))(1)

3. $\int_0^\infty \frac{\sin^2 ax \sin bx}{x} dx = \frac{\pi}{4} \quad [0 < b < 2a];$

$= \frac{\pi}{8} \quad [b = 2a];$

$= 0 \quad [b > 2a].$ BI ((151))(10)

4. $\int_0^\infty \frac{\sin^2 ax \cos bx}{x} dx = \frac{1}{4} \ln \frac{4a^2 - b^2}{b^2}$ BI ((151))(12)

5. $\int_0^\infty \frac{\sin^2 ax \cos 2bx}{x^2} dx = \frac{\pi}{2} (a - b) \quad [b < a];$

$= 0 \quad [b > a].$ FI III 648a, BI((157))(5)a

6. $$\int_0^\infty \frac{\sin 2ax \cos^2 bx}{x}\,dx = \frac{\pi}{2} \quad [a > b];$$
$$= \frac{3}{8}\pi \quad [a = b];$$
$$= \frac{\pi}{4} \quad [a < b].$$ BI ((151))(9)

7. $$\int_0^\infty \frac{\sin^2 ax \sin bx \sin cx}{x^2}\,dx = \frac{\pi}{16}(|b-2a-c| - |2a-b-c| + 2c)$$
$$[a > 0,\ b > 0,\ c > 0].$$ BI((157))(9)a, ET I 79(15)

8. $$\int_0^\infty \frac{\sin^2 ax \sin bx \sin cx}{x}\,dx = \frac{1}{4}\ln\frac{b+c}{b-c} + $$
$$+ \frac{1}{8}\ln\frac{(2a-b+c)(2a+b-c)}{(2a+b+c)(2a-b-c)} \quad [a > 0,\ b > 0,\ c > 0,\ b \neq c].$$ LI ((152))(2)

9. $$\int_0^\infty \frac{\sin^2 ax \sin^2 bx}{x^2}\,dx = \frac{\pi}{4}a \quad [0 \leqslant a \leqslant b];$$
$$= \frac{\pi}{4}b \quad [0 \leqslant b \leqslant a].$$ BI ((157))(3)

10. $$\int_0^\infty \frac{\sin^2 ax \sin^2 bx}{x^4}\,dx = \frac{1}{6}a^2\pi(3b-a) \quad [0 \leqslant a \leqslant b];$$
$$= \frac{1}{6}b^2\pi(3a-b) \quad [0 \leqslant b \leqslant a].$$ BI ((157))(27)

11. $$\int_0^\infty \frac{\sin^2 ax \cos^2 bx}{x^2}\,dx = \frac{2a-b}{4}\pi \quad [a \geqslant b > 0];$$
$$= \frac{a\pi}{4} \quad [0 < a \leqslant b].$$ BI ((157))(6)

12. $$\int_0^\infty \frac{\sin^3 ax \sin 3bx}{x^4}\,dx = \frac{a^3\pi}{2} \quad [b > a];$$
$$= \frac{\pi}{16}[8a^3 - 9(a-b)^3] \quad [a \leqslant 3b \leqslant 3a];$$ BI ((157))(28)
$$= \frac{9b\pi}{8}(a^2 - b^2) \quad [3b \leqslant a].$$ LI ((157))(28)

13. $$\int_0^\infty \frac{\sin^3 ax \cos bx}{x}\,dx = 0 \quad [b > 3a];$$
$$= -\frac{\pi}{16} \quad [b = 3a];$$
$$= -\frac{\pi}{8} \quad [3a > b > a];$$
$$= \frac{\pi}{16} \quad [b = a];$$
$$= \frac{\pi}{4} \quad [a > b] \quad [a > 0,\ b > 0].$$ BI ((151))(15)

14. $$\int_0^\infty \frac{\sin^3 ax \cos 3bx}{x^2}\,dx = \frac{3}{8}\Big\{(a+b)\ln[3(a+b)] + (b-a)\ln[3(b-a)] - \frac{1}{3}(a+3b)\ln(a+3b) - \frac{1}{3}(3b-a)\ln(3b-a)\Big\}$$
$[a > 0,\ b > 0]$. BI((157))(7)a, ET I 19(9)

15. $$\int_0^\infty \frac{\sin^3 ax \cos bx}{x^3}\,dx = \frac{\pi}{8}(3a^2 - b^2) \quad [b < a];$$
$$= \frac{\pi b^2}{4} \quad [a = b];$$
$$= \frac{\pi}{16}(3a - b)^2 \quad [a < b < 3a];$$
$$= 0 \quad [3a < b];\ [a > 0,\ b > 0].$$
BI((157))(19), ET I 19(10)

16. $$\int_0^\infty \frac{\sin^3 ax \sin bx}{x^4}\,dx = \frac{b\pi}{24}(9a^2 - b^2) \quad [0 < b \leqslant a];$$
$$= \frac{\pi}{48}[24a^3 - (3a-b)^3] \quad [0 < a \leqslant b \leqslant 3a];$$
$$= \frac{\pi a^3}{2} \quad [0 < 3a \leqslant b].$$
ET I 79(16)

17. $$\int_0^\infty \frac{\sin^3 ax \sin^2 bx}{x}\,dx = \frac{\pi}{8} \quad [2b > 3a];$$
$$= \frac{5\pi}{32} \quad [2b = 3a];$$
$$= \frac{3\pi}{16} \quad [3a > 2b > a];$$
$$= \frac{3\pi}{32} \quad [2b = a];$$
$$= 0 \quad [a > 2b]; \qquad [a > 0,\ b > 0].$$
BI ((151))(14)

18. $$\int_0^\infty \frac{\sin^2 ax \cos^3 bx}{x}\,dx = \frac{1}{16}\ln\frac{(2a+b)^3(b-2a)^3(2a+3b)(3b-2a)}{9b^8}$$
$[b > 2a > 0$ or $2a > 3b > 0]$;
$$= \frac{1}{16}\ln\frac{(2a+b)^3(2a-b)^3(2a+3b)(3b-2a)}{9b^8}$$
$[3b > 2a > b]$. BI ((151))(13)

19. $$\int_0^\infty \frac{\sin^2 ax \sin^2 bx \sin 2cx}{x}\,dx =$$
$$= \frac{\pi}{16}[1 + \operatorname{sign}(c - a + b) + \operatorname{sign}(c + a - b) - 2\operatorname{sign}(c - a) - 2\operatorname{sign}(c - b)]$$
$[a > 0,\ b > 0,\ c > 0]$. ET I 80(17)

20. $$\int_0^\infty \frac{\sin^2 ax \sin^2 bx \sin 2cx\, dx}{x^2} = \frac{a-b-c}{16} \ln 4(a-b-c)^2 - \frac{a+b+c}{16} \ln 4(a+b+c)^2 + \frac{a+b-c}{16} \ln 4(a+b-c)^2 - \frac{a-b+c}{16} \ln 4(a-b+c)^2 + \frac{a+c}{8} \ln 4(a+c)^2 - \frac{a-c}{8} \ln 4(a-c)^2 + \frac{b+c}{8} \ln 4(b+c)^2 - \frac{b-c}{8} \ln 4(b-c)^2 - \frac{1}{2} c \ln 2c$$
$[a > 0,\ b > 0,\ c > 0]$. BI ((157))(10)

21 $$\int_0^\infty \frac{\sin^2 ax \sin^3 bx}{x^3} dx = \frac{3b^2\pi}{16} \quad [2a > 3b];$$
$$= \frac{a^2\pi}{12} \quad [2a = 3b];$$
$$= \frac{6b^2 - (3b-2a)^2}{32} \pi \quad [3b > 2a > b];$$
$$= \frac{a^2\pi}{4} \quad [b > 2a]; \qquad [a > 0,\ b > 0].$$ BI ((157))(18)

3.829

1. $$\int_0^\infty \frac{x^n - \sin^n x}{x^{n+2}} dx = \frac{\pi}{2^n (n+1)!} \sum_{k=0}^{E\left(\frac{n-1}{2}\right)} (-1)^k \binom{n}{k} (n-2k)^{n+1}$$ GW ((333))(63)

2. $$\int_0^\infty (1 - \cos^{2m-1} x) \frac{dx}{x^2} = \int_0^\infty (1 - \cos^{2m} x) \frac{dx}{x^2} = \frac{m\pi}{2^{2m}} \binom{2m}{m}.$$ BI ((158))(7, 8)

3.831

1. $$\int_0^\infty \frac{\sin^{2n} ax - \sin^{2n} bx}{x} dx = \frac{(2n-1)!!}{(2n)!!} \ln \frac{b}{a} \quad [a > 0,\ b > 0].$$ FI II 651

2. $$\int_0^\infty \frac{\cos^{2n} ax - \cos^{2n} bx}{x} dx = \left[1 - \frac{(2n-1)!!}{(2n)!!}\right] \ln \frac{b}{a} \quad [a > 0,\ b > 0].$$ FI II 651

3. $$\int_0^\infty \frac{\cos^{2m+1} ax - \cos^{2m+1} bx}{x} dx = \ln \frac{b}{a} \quad [a > 0,\ b > 0].$$ FI II

4. $$\int_0^\infty \frac{\cos^m ax \cos max - \cos^m bx \cos mbx}{x} dx = \left(1 - \frac{1}{2^m}\right) \ln \frac{b}{a}$$
$[ab > 0]$. LI ((155))(8)

3.832

1. $$\int_0^{\frac{\pi}{2}} x \cos^{p-1} x \sin ax\, dx = \frac{\pi}{2^{p+1}} \Gamma(p) \frac{\psi\left(\frac{p+a+1}{2}\right) - \psi\left(\frac{p-a+1}{2}\right)}{\Gamma\left(\frac{p+a+1}{2}\right) \Gamma\left(\frac{p-a+1}{2}\right)}$$
$[p > 0,\ -(p+1) < a < p+1]$. BI ((205))(6)

2. $\int_0^\infty \sin^{2m+1} x \sin 2mx \frac{dx}{a^2+x^2} = \frac{(-1)^m \pi}{2^{2m+1}a} [(1-e^{-2a})^{2m} - 1] \operatorname{sh} a \quad [a > 0].$

BI ((162))(17)

3. $\int_0^\infty \sin^{2m-1} x \sin [(2m-1)x] \frac{dx}{a^2+x^2} = \frac{(-1)^{m+1} \pi}{2^{2m}a} (1-e^{-2a})^{2m-1} \quad [a > 0].$

BI ((162))(11)

4. $\int_0^\infty \sin^{2m-1} x \sin [(2m+1)x] \frac{dx}{a^2+x^2} = \frac{(-1)^{m-1} \pi}{2^{2m}a} e^{-2a}(1-e^{-2a})^{2m-1} \quad [a > 0].$

BI ((162))(12)

5. $\int_0^\infty \sin^{2m+1} x \sin [3(2m+1)x] \frac{dx}{a^2+x^2} = \frac{(-1)^m \pi}{2a} e^{-3(2m+1)a} \operatorname{sh}^{2m+1} a$

$[a > 0].$ BI ((162))(18)

6. $\int_0^\infty \sin^{2m} x \sin [(2m-1)x] \frac{x\,dx}{a^2+x^2} = \frac{(-1)^m \pi}{2^{2m+1}} e^a [(1-e^{-2a})^{2m} - 1] \quad [a > 0].$

BI ((162))(13)

7. $\int_0^\infty \sin^{2m} x \sin (2mx) \frac{x\,dx}{a^2+x^2} = \frac{(-1)^m \pi}{2^{2m+1}} [(1-e^{-2a})^{2m} - 1] \quad [a > 0].$

BI ((162))(14)

8. $\int_0^\infty \sin^{2m} x \sin [(2m+2)x] \frac{x\,dx}{a^2+x^2} = \frac{(-1)^m \pi}{2^{2m+1}} e^{-2a} (1-e^{-2a})^{2m} \quad [a > 0].$

BI ((162))(15)

9. $\int_0^\infty \sin^{2m} x \sin 4mx \frac{x\,dx}{a^2+x^2} = \frac{(-1)^m \pi}{2} e^{-4ma} \operatorname{sh}^{2m} a \quad [a > 0].$ BI ((162))(16)

10. $\int_0^\infty \sin^{2m} x \cos x \frac{dx}{x^2} = \frac{(2m-3)!!}{(2m)!!} \cdot \frac{\pi}{2}.$ GW ((333))(15a)

11. $\int_0^\infty \sin^{2m} x \cos [(2m-1)x] \frac{dx}{a^2+x^2} = \frac{(-1)^m \pi}{2^{2m}a} [(1-e^{-2a})^{2m-1} - 1] \operatorname{sh} a$

$[a > 0].$ BI ((162))(25)

12. $\int_0^\infty \sin^{2m} x \cos (2mx) \frac{dx}{a^2+x^2} = \frac{(-1)^m \pi}{2^{2m+1}a} (1-e^{-2a})^{2m} \quad [a > 0].$

BI ((162))(26)

13. $\int_0^\infty \sin^{2m} x \cos [(2m+2)x] \frac{dx}{a^2+x^2} = \frac{(-1)^m \pi}{2^{2m+1}a} e^{-2a} (1-e^{-2a})^{2m}$

$[a > 0].$ BI ((162))(27)

14. $\int_0^\infty \sin^{2m} x \cos 4mx \frac{dx}{a^2+x^2} = \frac{(-1)^m \pi}{2a} e^{-4ma} \operatorname{sh}^{2m} a \quad [a > 0].$ BI ((162))(28)

15. $\int_0^\infty \sin^{2m+1} x \cos x \frac{dx}{x} = \frac{(2m-1)!!}{(2m+2)!!} \cdot \frac{\pi}{2}.$ GW ((333))(15)

16. $\int_0^\infty \sin^{2m+1} x \cos x \frac{dx}{x^3} = \frac{(2m-3)!!}{(2m)!!} \cdot \frac{\pi}{2}.$ GW ((333))(15b)

17. $\int_0^\infty \sin^{2m-1} x \cos [(2m-1) x] \frac{x\,dx}{a^2+x^2} = \frac{(-1)^m \pi}{2^{2m}} [(1 - e^{-2a})^{2m-1} - 1]$

$[a > 0].$ BI ((162))(23)

18. $\int_0^\infty \sin^{2m+1} x \cos 2mx \frac{x\,dx}{a^2+x^2} = \frac{(-1)^{m-1} \pi}{2^{2m+2}} e^{-a}[(1 - e^{-2a})^{2m+1} - 1]$

$[a > 0].$ BI ((162))(29)

19. $\int_0^\infty \sin^{2m-1} x \cos [(2m+1) x] \frac{x\,dx}{a^2+x^2} = \frac{(-1)^m \pi}{2^{2m}} e^{-2a} (1 - e^{-2a})^{2m-1}$

$[a > 0].$ BI ((162))(24)

20. $\int_0^\infty \sin^{2m+1} x \cos [2(2m+1) x] \frac{x\,dx}{a^2+x^2} = \frac{(-1)^{m-1} \pi}{2} e^{-2(2m+1)a} \operatorname{sh}^{2m+1} a$

$[a > 0].$ BI ((162))(30)

21. $\int_0^\infty \cos^m x \sin mx \frac{dx}{a^2+x^2} = \frac{1}{2^{m+1}a} \sum_{k=1}^{m} \binom{m}{k} [e^{-2ka} \operatorname{Ei}(2ka) - e^{2ka} \operatorname{Ei}(-2ka)]$

$[a > 0].$ BI ((162))(8)

22. $\int_0^\infty \cos^n sx \sin nsx \frac{x\,dx}{a^2+x^2} = \frac{\pi}{2^{n+1}} [(1 + e^{-2as})^n - 1].$ BI ((163))(9)

23. $\int_0^\infty \cos^n sx \sin nsx \frac{x\,dx}{a^2-x^2} = \frac{\pi}{2} (2^{-n} - \cos^n as \cos nas).$ BI ((166))(10)

24. $\int_0^\infty \cos^{m-1} x \sin [(m+1) x] \frac{x\,dx}{a^2+x^2} = \frac{\pi}{2^m} e^{-2a} (1 + e^{-2a})^{m-1} \quad [a > 0].$

BI ((163))(6)

25. $\int_0^\infty \cos^m x \sin [(m+1) x] \frac{x\,dx}{a^2+x^2} = \frac{\pi}{2^{m+1}} e^{-a} (1 + e^{-2a})^m$

$[a > 0].$ BI ((163))(10)

26. $\int_0^\infty \cos^m x \sin [(m-1) x] \frac{x\,dx}{a^2+x^2} = \frac{\pi}{2^{m+1}} e^{a} (1 + e^{-2a})^m$

$[a > 0].$ BI ((163))(7)

27. $\int_0^\infty \cos^m x \sin(3mx) \frac{x\,dx}{a^2+x^2} = \frac{\pi}{2} e^{-3a} \operatorname{ch}^m a \quad [a>0].$ BI ((163))(11)

28. $\int_0^\infty \cos^n sx \cos nsx \frac{dx}{a^2+x^2} = \frac{\pi}{2^{n+1}a} (1+e^{-2as})^n.$ BI ((163))(16)

29. $\int_0^\infty \cos^n sx \cos nsx \frac{dx}{a^2-x^2} = \frac{\pi}{2a} \cos^n as \sin nas.$

30. $\int_0^\infty \cos^{m-1} x \cos[(m+1)x] \frac{dx}{a^2+x^2} = \frac{\pi}{2^m a} e^{-2a} (1+e^{-2a})^{m-1}$

$[a>0].$ BI ((163))(14)

31. $\int_0^\infty \cos^m x \cos[(m-1)x] \frac{dx}{a^2+x^2} = \frac{\pi}{2^{m+1}a} e^a (1+e^{-2a})^m$

$[a>0].$ BI ((163))(15)

32. $\int_0^\infty \cos^m x \cos[(m+1)x] \frac{dx}{a^2+x^2} = \frac{\pi}{2^{m+1}a} e^{-a} (1+e^{-2a})^m$

$[a>0].$ BI ((163))(17)

33. $\int_0^\infty \sin^p x \cos x \frac{dx}{x^q} = \frac{p}{q-1} \int_0^\infty \frac{\sin^{p-1} x}{x^{q-1}}\,dx - \frac{p+1}{q-1} \int_0^\infty \frac{\sin^{p+1} x}{x^{q-1}}\,dx \quad [p>q-1>0];$

$= \frac{p(p-1)}{(q-1)(q-2)} \int_0^\infty \sin^{p-2} x \cos x \frac{dx}{x^{q-2}} -$

$- \frac{(p+1)^2}{(q-1)(q-2)} \int_0^\infty \sin^p x \cos x \frac{dx}{x^{q-2}} \quad [p>q-1>1].$ GW ((333))(18)

34. $\int_0^\infty \cos^{2m} x \cos 2nx \sin x \frac{dx}{x} = \int_0^\infty \cos^{2m-1} x \cos 2nx \sin x \frac{dx}{x} = \frac{\pi}{2^{2m+1}} \binom{2m}{m+n}.$

BI ((152))(5, 6)

35. $\int_0^\infty \cos^p ax \sin bx \cos x \frac{dx}{x} = \frac{\pi}{2} \quad [b>ap,\ p>-1].$ BI ((153))(12)

36. $\int_0^\infty \cos^p ax \sin pax \cos x \frac{dx}{x} = \frac{\pi}{2^{p+1}} (2^p-1) \quad [p>-1].$ BI ((153))(2)

37. $\int_0^\infty \frac{dx}{x^2} \prod_{k=1}^n \cos^{p_k} a_k x \cdot \sin bx \sin x = \frac{\pi}{2}$

$\left[b > \sum_{k=1}^n a_k p_k;\ a_k>0,\ p_k>0\right].$ BI ((157))(15)

3.833

1. $$\int_0^\infty \sin^{2m+1} x \cos^{2n} x \frac{dx}{x} = \int_0^\infty \sin^{2m+1} x \cos^{2n-1} x \frac{dx}{x} = \frac{(2m-1)!!\,(2n-1)!!}{2^{m+n+1}\,(m+n)!}\pi.$$ BI ((151))(24, 25)

$$= \frac{1}{2} \mathrm{B}\left(m+\frac{1}{2},\, n+\frac{1}{2}\right).$$ GW ((333))(24)

2 $$\int_0^\infty \sin^{2m+1} 2x \cos^{2n-1} 2x \cos^2 x \frac{dx}{x} = \frac{\pi}{2} \cdot \frac{(2m-1)!!\,(2n-1)!!}{(2m+2n)!!}.$$ LI ((152))(4)

3.834

1. $$\int_0^\infty \frac{\sin^{2m+1} x}{1-2a\cos x+a^2} \cdot \frac{dx}{x} = \frac{(-1)^m \pi (1+a)^{4m}}{2^{2m+2} a^{2m+1}} \left\{ \left|\frac{1-a}{1+a}\right|^{2m-1} - \right.$$

$$\left. - \sum_{k=0}^{2m} (-1)^k \binom{m-\frac{1}{2}}{k} \left(\frac{4a}{(1+a)^2}\right)^k \right\} \quad [\,|a| \neq 1].$$ GW ((333))(62a)

2 $$\int_0^\infty \frac{\sin^{2m+1} x \cos^n x}{(1-2a\cos x+a^2)^p} \cdot \frac{dx}{x} =$$

$$= \frac{n!\,\pi}{2^{n+1}(2m+n+1)!\,(1+a)^{2p}} \sum_{k=0}^{n} \frac{(-1)^k (2m+2n-2k+1)!!\,(2m+2k-1)!!}{k!\,(n-k)!} \times$$

$$\times F\left(m+n-k+\frac{3}{2},\, p;\, 2m+n+2;\, \frac{4a}{(1+a)^2}\right) \quad [a \neq \pm 1].$$ GW ((333))(62)

3.835

1. $$\int_0^\infty \frac{\cos^{2m} x \cos 2mx \sin x}{a^2\cos^2 x+b^2\sin^2 x} \cdot \frac{dx}{x} = \frac{\pi}{2} \frac{b^{2m-1}}{a(a+b)^{2m}} \quad [ab > 0].$$ BI ((182))(31)a

2 $$\int_0^\infty \frac{\cos^{2m-1} x \cos 2mx \sin x}{a^2\cos^2 x+b^2\sin^2 x} \cdot \frac{dx}{x} = \frac{\pi}{2a} \frac{b^{2m-1}}{(a+b)^{2m}} \quad [ab > 0].$$ LI ((182))(32)a

3.836

1 $$\int_0^\infty \left(\frac{\sin x}{x}\right)^n \frac{\sin mx}{x} dx = \frac{\pi}{2} \quad [m \geqslant n].$$ LI ((159))(12)

2 $$\int_0^\infty \left(\frac{\sin x}{x}\right)^n \cos mx\, dx = \frac{n\pi}{2^n} \sum_{0<k<\frac{m+n}{2}} \frac{(-1)^k (n+m-2k)^{n-1}}{k!\,(n-k)!} \quad [0 < m < n];$$

$$= 0 \quad [m \geqslant n] \quad [n \geqslant 2].$$ GI((159))(14), ET I 20(11)

3. $$\int_0^\infty \left(\frac{\sin x}{x}\right)^{n-1} \sin nx \cos x \frac{dx}{x} = \frac{\pi}{2}.$$ BI ((159))(20)

4. $$\int_0^\infty \left(\frac{\sin x}{x}\right)^n \frac{\sin(anx)}{x}\,dx = \frac{\pi}{2}\left[1-\frac{1}{2^{n-1}n!}\sum_{0\leqslant k<\frac{n}{2}(1-a)}(-1)^k\binom{n}{k}(n-an-2k)^n\right] \qquad [0<a\leqslant 1].$$ LO V 341(15)

5. $$\int_0^\infty \left(\frac{\sin x}{x}\right)^n \cos(anx)\,dx = \frac{\pi}{2^n}\sum_{0\leqslant k<(1\pm a)\frac{n}{2}}(-1)^k\binom{n}{k}\frac{\Gamma(n\pm an-2k+1)}{(n-1)!\,\Gamma(2\pm an-2k)}$$

[For $0<a\leqslant 1$, the sign in the binomials $1\pm a$ and $2\pm an$ can be chosen arbitrarily but they must be the same throughout the formula]. LO V 340(14)

3.837

1. $$\int_0^{\frac{\pi}{2}} \frac{x^2\,dx}{\sin^2 x} = \pi\ln 2.$$ BI ((206))(9)

2. $$\int_0^{\frac{\pi}{4}} \frac{x^2\,dx}{\sin^2 x} = -\frac{\pi^2}{16}+\frac{\pi}{4}\ln 2+G = 0.8435118417\ldots$$ BI ((204))(10)

3. $$\int_0^{\frac{\pi}{4}} \frac{x^2\,dx}{\cos^2 x} = \frac{\pi^2}{16}+\frac{\pi}{4}\ln 2-G.$$ GW ((333))(35a)

4. $$\int_0^{\frac{\pi}{4}} \frac{x^{p+1}}{\sin^2 x}\,dx = -\left(\frac{\pi}{4}\right)^{p+1}+(p+1)\left(\frac{\pi}{4}\right)^p\left\{\frac{1}{p}-\frac{1}{2}\sum_{k=1}^{\infty}\frac{1}{4^{2k-1}(p+2k)}\zeta(2k)\right\}$$ $$[p>0].$$ LI ((204))(14)

5. $$\int_0^{\frac{\pi}{2}} \frac{x^2\cos x}{\sin^2 x}\,dx = -\frac{\pi^2}{4}+4G = 1.1964612764\ldots$$ BI ((206))(7)

6. $$\int_0^{\frac{\pi}{2}} \frac{x^3\cos x}{\sin^3 x}\,dx = -\frac{\pi^3}{16}+\frac{3}{2}\pi\ln 2.$$ BI ((206))(8)

7. $$\int_0^\infty \frac{\cos 2nx}{\cos x}\sin^{2n}x\,\frac{dx}{x^m} = 0 \qquad \left[n>\frac{m-1}{2},\ m>0\right].$$ BI ((180))(16)

8. $$\int_0^\infty \frac{\cos 2nx}{\cos x}\sin^{2n+1}x\,\frac{dx}{x^m} = 0 \qquad \left[n>\frac{m-2}{2},\ m>0\right].$$ BI ((180))(17)

9. $\int_0^1 \frac{x\,dx}{\cos ax \cos[a(1-x)]} = \frac{1}{a} \operatorname{cosec} a \cdot \ln \sec a \quad \left[a < \frac{\pi}{2}\right].$ BI ((149))(20)

3.838

1. $\int_0^{\frac{\pi}{2}} \frac{x \cos^{p-1} x}{\sin^{p+1} x} dx = \frac{\pi}{2p} \sec \frac{\pi p}{2} \quad [p < 1].$ BI ((206))(13)a

2. $\int_0^{\frac{\pi}{4}} \frac{x \sin^{p-1} x}{\cos^{p+1} x} dx = \frac{\pi}{4p} - \frac{1}{2p} \beta\left(\frac{p+1}{2}\right) \quad [p > -1].$ LI ((204))(15)

3. $\int_0^{\frac{\pi}{4}} \frac{x \sin^{2m-1} x}{\cos^{2m+1} x} dx = \frac{\pi}{8m}(1 - \cos m\pi) + \frac{1}{2m} \sum_{k=0}^{m-1} \frac{(-1)^{k-1}}{2m-2k-1}.$ BI ((204))(17)

4. $\int_0^{\frac{\pi}{4}} \frac{x \sin^{2m} x}{\cos^{2m+2} x} dx = \frac{1}{2(2m+1)} \left[\frac{\pi}{2} + (-1)^{m-1} \ln 2 + \sum_{k=0}^{m-1} \frac{(-1)^{k-1}}{m-k}\right].$

BI ((204))(16)

3.839

1. $\int_0^{\frac{\pi}{4}} x \operatorname{tg}^2 x\, dx = \frac{\pi}{4} - \frac{\pi^2}{32} - \frac{1}{2} \ln 2.$ BI ((204))(3)

2. $\int_0^{\frac{\pi}{4}} x \operatorname{tg}^3 x\, dx = \frac{\pi}{4} - \frac{1}{2} + \frac{\pi}{8} \ln 2 - \frac{1}{2} \boldsymbol{G}.$ BI ((204))(7)

3. $\int_0^{\frac{\pi}{4}} \frac{x^2 \operatorname{tg} x}{\cos^2 x} dx = \frac{1}{2} \ln 2 - \frac{\pi}{4} + \frac{\pi^2}{16}$ (cf. 3.839 1.). BI ((204))(13)

4. $\int_0^{\frac{\pi}{4}} \frac{x^2 \operatorname{tg}^2 x}{\cos^2 x} dx = \frac{1}{3}\left(1 - \frac{\pi}{4} \ln 2 - \frac{\pi}{2} + \frac{\pi^2}{16} + \boldsymbol{G}\right)$

(cf. 3.839 2.). BI ((204))(12)

5. $\int_0^{\frac{\pi}{2}} x \cos^p x \operatorname{tg} x\, dx = \frac{\pi}{2^{p+1} p} \cdot \frac{\Gamma(p+1)}{\left[\Gamma\left(\frac{p}{2}+1\right)\right]^2} \quad [p > -1].$ BI ((205))(3)

6. $\int_0^{\frac{\pi}{2}} x \sin^p x \operatorname{ctg} x\, dx = \frac{\pi}{2p} - \frac{2^{p-1}}{p} \mathrm{B}\left(\frac{p+1}{2}, \frac{p+1}{2}\right)$

$[p > -1].$ BI ((206))(11)

7. $\int_0^\infty \sin^{2n} x \operatorname{tg} x \frac{dx}{x} = \frac{\pi}{2} \cdot \frac{(2n-1)!!}{(2n)!!}$. GW ((333))(16)

8. $\int_0^\infty \cos^s rx \operatorname{tg} qx \frac{dx}{x} = \frac{\pi}{2} \quad [s > -1]$. BI ((151))(26)

9. $\int_0^\infty \frac{\cos [(2n-1)x]}{\cos x} \cdot \left(\frac{\sin x}{x}\right)^{2n} dx = (-1)^{n-1} \frac{2^{2n}-1}{(2n)!} \cdot 2^{2n-1}\pi |B_{2n}|$.

BI ((180))(15)

10. $\int_0^\infty \operatorname{tg}^r px \frac{dx}{q^2+x^2} = \frac{\pi}{2q} \sec \frac{r\pi}{2} \operatorname{th}^r pq \quad [r^2 < 1]$. BI ((160))(19)

3.84 Integrals containing the expressions $\sqrt{1-k^2\sin^2 x}$, $\sqrt{1-k^2\cos^2 x}$ and similar expressions

3.841

1. $\int_0^\infty \sin x \sqrt{1-k^2 \sin^2 x} \frac{dx}{x} = \boldsymbol{E}(k)$. BI ((154))(8)

2. $\int_0^\infty \sin x \sqrt{1-k^2 \cos^2 x} \frac{dx}{x} = \boldsymbol{E}(k)$. BI ((154))(20)

3. $\int_0^\infty \operatorname{tg} x \sqrt{1-k^2 \sin^2 x} \frac{dx}{x} = \boldsymbol{E}(k)$. BI ((154))(9)

4. $\int_0^\infty \operatorname{tg} x \sqrt{1-k^2 \cos^2 x} \frac{dx}{x} = \boldsymbol{E}(k)$. BI ((154))(21)

3.842

1. $\int_0^\infty \frac{\sin x}{\sqrt{1+\sin^2 x}} \frac{dx}{x} = \int_0^\infty \frac{\operatorname{tg} x}{\sqrt{1+\sin^2 x}} \cdot \frac{dx}{x} =$

$$= \int_0^\infty \frac{\sin x}{\sqrt{1+\cos^2 x}} \frac{dx}{x} = \int_0^\infty \frac{\operatorname{tg} x}{\sqrt{1+\cos^2 x}} \frac{dx}{x} = \frac{1}{\sqrt{2}} \boldsymbol{K}\left(\frac{1}{\sqrt{2}}\right).$$

BI ((183))(4, 5, 9, 10)

2. $\int_u^{\frac{\pi}{2}} \frac{x \cos x \, dx}{\sqrt{\sin^2 x - \sin^2 u}} = \frac{\pi}{2} \ln(1+\cos u)$. BI ((226))(4)

3. $\int_0^\infty \frac{\sin x}{\sqrt{1-k^2\sin^2 x}} \frac{dx}{x} = \int_0^\infty \frac{\operatorname{tg} x}{\sqrt{1-k^2 \sin^2 x}} \frac{dx}{x} =$

$$= \int_0^\infty \frac{\sin x}{\sqrt{1-k^2\cos^2 x}} \frac{dx}{x} = \int_0^\infty \frac{\operatorname{tg} x}{\sqrt{1-k^2\cos^2 x}} \frac{dx}{x} = \boldsymbol{K}(k).$$

BI ((183))(12, 13, 21, 22)

4. $$\int_0^{\frac{\pi}{2}} \frac{x \sin x \cos x}{\sqrt{1-k^2\sin^2 x}} dx = \frac{1}{2k^2}[-\pi k' + 2\boldsymbol{E}(k)].$$ BI ((211))(1)

5. $$\int_0^{\frac{\pi}{2}} \frac{x \sin x \cos x}{\sqrt{1-k^2\cos^2 x}} dx = \frac{1}{2k^2}[\pi - 2\boldsymbol{E}(k)].$$ BI ((214))(1)

6. $$\int_0^{\alpha} \frac{x \sin x\, dx}{\cos^2 x \sqrt{\sin^2\alpha - \sin^2 x}} = \frac{\pi \sin^2 \frac{\alpha}{2}}{\cos^2\alpha}.$$ LO III 284

7. $$\int_0^{\beta} \frac{x \sin x\, dx}{(1-\sin^2\alpha\sin^2 x)\sqrt{\sin^2\beta - \sin^2 x}} = \frac{\pi \ln \dfrac{\cos\alpha + \sqrt{1-\sin^2\alpha\sin^2\beta}}{2\cos\beta\cos^2\frac{\alpha}{2}}}{2\cos\alpha\sqrt{1-\sin^2\alpha\sin^2\beta}}.$$

LO III 284

3.843

1. $$\int_0^\infty \operatorname{tg} x \sqrt{1-k^2\sin^2 2x}\, \frac{dx}{x} = \boldsymbol{E}(k).$$ BI ((154))(10)

2. $$\int_0^\infty \operatorname{tg} x \sqrt{1-k^2\cos^2 2x}\, \frac{dx}{x} = \boldsymbol{E}(k).$$ BI ((154))(22)

3. $$\int_0^\infty \frac{\operatorname{tg} x}{\sqrt{1+\sin^2 2x}} \frac{dx}{x} = \int_0^\infty \frac{\operatorname{tg} x}{\sqrt{1+\cos^2 2x}} \frac{dx}{x} = \frac{1}{\sqrt{2}} \boldsymbol{K}\left(\frac{1}{\sqrt{2}}\right).$$

BI ((183))(6, 11)

4. $$\int_0^\infty \frac{\operatorname{tg} x}{\sqrt{1-k^2\sin^2 2x}} \frac{dx}{x} = \int_0^\infty \frac{\operatorname{tg} x}{\sqrt{1-k^2\cos^2 2x}} \frac{dx}{x} = \boldsymbol{K}(k).$$

BI ((183))(14, 23)

3.844

1. $$\int_0^\infty \frac{\sin x \cos x}{\sqrt{1-k^2\cos^2 x}} \frac{dx}{x} = \frac{1}{k^2}[\boldsymbol{K}(k) - \boldsymbol{E}(k)].$$ BI ((185))(20)

2. $$\int_0^\infty \frac{\sin x \cos^2 x}{\sqrt{1-k^2\cos^2 x}} \cdot \frac{dx}{x} = \frac{1}{k^2}[\boldsymbol{K}(k) - \boldsymbol{E}(k)].$$ BI ((185))(21)

3. $$\int_0^\infty \frac{\sin x \cos^3 x}{\sqrt{1-k^2\cos^2 x}} \cdot \frac{dx}{x} = \frac{1}{3k^4}[(2+k^2)\boldsymbol{K}(k) - 2(1+k^2)\boldsymbol{E}(k)].$$

BI ((185))(22)

4. $\int\limits_0^\infty \frac{\sin x \cos^4 x}{\sqrt{1-k^2\cos^2 x}} \cdot \frac{dx}{x} = \frac{1}{3k^4}[(2+k^2)\boldsymbol{K}(k) - 2(1+k^2)\boldsymbol{E}(k)].$ BI ((185))(23)

5. $\int\limits_0^\infty \frac{\sin^3 x \cos x}{\sqrt{1-k^2\cos^2 x}} \cdot \frac{dx}{x} = \frac{1}{3k^4}[(1+k'^2)\boldsymbol{E}(k) - 2k'^2\boldsymbol{K}(k)].$ BI ((185))(24)

6. $\int\limits_0^\infty \frac{\sin^3 x \cos^2 x}{\sqrt{1-k^2\cos^2 x}} \cdot \frac{dx}{x} = \frac{1}{3k^4}[(1+k'^2)\boldsymbol{E}(k) - 2k'^2\boldsymbol{K}(k)].$ BI ((185))(25)

7. $\int\limits_0^\infty \frac{\sin^2 x \operatorname{tg} x}{\sqrt{1-k^2\cos^2 x}} \cdot \frac{dx}{x} = \frac{1}{k^2}[\boldsymbol{E}(k) - k'^2\boldsymbol{K}(k)].$ BI ((184))(16)

8. $\int\limits_0^\infty \frac{\sin^4 x \operatorname{tg} x}{\sqrt{1-k^2\cos^2 x}} \cdot \frac{dx}{x} = \frac{1}{3k^4}[(2+3k^2)k'^2\boldsymbol{K}(k) - 2(k'^2-k^2)\boldsymbol{E}(k)].$ BI ((184))(18)

3.845

1. $\int\limits_0^\infty \frac{\sin x \cos x}{\sqrt{1+\cos^2 x}} \cdot \frac{dx}{x} = \sqrt{2}\left[\boldsymbol{E}\left(\frac{\sqrt{2}}{2}\right) - \frac{1}{2}\boldsymbol{K}\left(\frac{\sqrt{2}}{2}\right)\right].$ BI ((185))(6)

2. $\int\limits_0^\infty \frac{\sin x \cos^2 x}{\sqrt{1+\cos^2 x}} \cdot \frac{dx}{x} = \sqrt{2}\left[\boldsymbol{E}\left(\frac{\sqrt{2}}{2}\right) - \frac{1}{2}\boldsymbol{K}\left(\frac{\sqrt{2}}{2}\right)\right].$ BI ((185))(7)

3. $\int\limits_0^\infty \frac{\sin^2 x \operatorname{tg} x}{\sqrt{1+\cos^2 x}} \cdot \frac{dx}{x} = \sqrt{2}\left[\boldsymbol{K}\left(\frac{\sqrt{2}}{2}\right) - \boldsymbol{E}\left(\frac{\sqrt{2}}{2}\right)\right].$ BU ((184))(8)

3.846

1. $\int\limits_0^\infty \frac{\sin x \cos x}{\sqrt{1-k^2\sin^2 x}} \cdot \frac{dx}{x} = \frac{1}{k^2}[\boldsymbol{E}(k) - k'^2\boldsymbol{K}(k)].$ BI ((185))(9)

2. $\int\limits_0^\infty \frac{\sin x \cos^2 x}{\sqrt{1-k^2\sin^2 x}} \cdot \frac{dx}{x} = \frac{1}{k^2}[\boldsymbol{E}(k) - k'^2\boldsymbol{K}(k)].$ BI ((185))(10)

3. $\int\limits_0^\infty \frac{\sin x \cos^3 x}{\sqrt{1-k^2\sin^2 x}} \cdot \frac{dx}{x} = \frac{1}{3k^4}[(2-3k^2)k'^2\boldsymbol{K}(k) - 2(k'^2-k^2)\boldsymbol{E}(k)].$ BI ((185))(11)

4. $\int\limits_0^\infty \frac{\sin x \cos^4 x}{\sqrt{1-k^2\sin^2 x}} \cdot \frac{dx}{x} = \frac{1}{3k^4}[(2-3k^2)k'^2\boldsymbol{K}(k) - 2(k'^2-k^2)\boldsymbol{E}(k)].$ BI ((185))(12)

5. $\int\limits_0^\infty \frac{\sin^3 x \cos x}{\sqrt{1-k^2\sin^2 x}} \cdot \frac{dx}{x} = \frac{1}{3k^4}[(1+k'^2)\boldsymbol{E}(k) - 2k'^2\boldsymbol{K}(k)].$ BI ((185))(13)

6. $$\int_0^\infty \frac{\sin^3 x \cos^2 x}{\sqrt{1-k^2\sin^2 x}} \cdot \frac{dx}{x} = \frac{1}{3k^4}\left[(1+k'^2)\,\boldsymbol{E}(k) - 2k'^2\boldsymbol{K}(k)\right].$$ BI ((185))(14)

7. $$\int_0^\infty \frac{\sin^2 x \operatorname{tg} x}{\sqrt{1-k^2\sin^2 x}} \cdot \frac{dx}{x} = \frac{1}{k^2}\left[\boldsymbol{K}(k) - \boldsymbol{E}(k)\right].$$ BI ((184))(9)

8. $$\int_0^\infty \frac{\sin^4 x \operatorname{tg} x}{\sqrt{1-k^2\sin^2 x}} \cdot \frac{dx}{x} = \frac{1}{3k^4}\left[(2+k^2)\,\boldsymbol{K}(k) - 2(1+k^2)\,\boldsymbol{E}(k)\right].$$ BI ((184))(11)

3.847 $$\int_0^\infty \frac{\sin x \cos x}{\sqrt{1+\sin^2 x}} \cdot \frac{dx}{x} = \int_0^\infty \frac{\sin x \cos^2 x}{\sqrt{1+\sin^2 x}} \cdot \frac{dx}{x} = \sqrt{2}\left[\boldsymbol{K}\left(\frac{\sqrt{2}}{2}\right) - \boldsymbol{E}\left(\frac{\sqrt{2}}{2}\right)\right].$$ BI ((185))(3, 4)

3.848

1 $$\int_0^\infty \frac{\sin^3 x \cos x}{\sqrt{1-k^2\sin^2 2x}} \cdot \frac{dx}{x} = \frac{1}{4k^2}\left[\boldsymbol{K}(k) - \boldsymbol{E}(k)\right].$$ BI ((185))(15)

2 $$\int_0^\infty \frac{\cos^2 2x \operatorname{tg} x}{\sqrt{1-k^2\sin^2 2x}} \cdot \frac{dx}{x} = \frac{1}{k^2}\left[\boldsymbol{E}(k) - k'^2\boldsymbol{K}(k)\right].$$ BI ((184))(12)

3 $$\int_0^\infty \frac{\cos^4 2x \operatorname{tg} x}{\sqrt{1-k^2\sin^2 2x}} \cdot \frac{dx}{x} = \frac{1}{3k^4}\left[(2-3k^2)\,k'^2\boldsymbol{K}(k) - 2(k'^2-k^2)\,\boldsymbol{E}(k)\right].$$ BI ((184))(13)

4. $$\int_0^\infty \frac{\sin^2 4x \operatorname{tg} x}{\sqrt{1-k^2\sin^2 2x}} \cdot \frac{dx}{x} = \frac{4}{3k^4}\left[(1+k'^2)\,\boldsymbol{E}(k) - 2k'^2\boldsymbol{K}(k)\right].$$ BI ((184))(17)

5 $$\int_0^\infty \frac{\sin^3 x \cos x}{\sqrt{1-k^2\cos^2 2x}} \cdot \frac{dx}{x} = \frac{1}{4k^2}\left[\boldsymbol{E}(k) - k'^2\boldsymbol{K}(k)\right].$$ BI ((185))(26)

6. $$\int_0^\infty \frac{\cos^2 2x \operatorname{tg} x}{\sqrt{1-k^2\cos^2 2x}} \cdot \frac{dx}{x} = \frac{1}{k^2}\left[\boldsymbol{K}(k) - \boldsymbol{E}(k)\right].$$ BI ((184))(19)

7. $$\int_0^\infty \frac{\cos^4 2x \operatorname{tg} x}{\sqrt{1-k^2\cos^2 2x}} \cdot \frac{dx}{x} = \frac{1}{3k^4}\left[(2+k^2)\,\boldsymbol{K}(k) - 2(1+k^2)\,\boldsymbol{E}(k)\right].$$ BI ((184))(20)

3.849

1. $$\int_0^\infty \frac{\sin^3 x \cos x}{\sqrt{1+\cos^2 2x}} \cdot \frac{dx}{x} = \frac{1}{2\sqrt{2}}\left[\boldsymbol{K}\left(\frac{\sqrt{2}}{2}\right) - \boldsymbol{E}\left(\frac{\sqrt{2}}{2}\right)\right].$$ BI ((185))(8)

2. $$\int_0^\infty \frac{\sin^3 x \cos x}{\sqrt{1+\sin^2 2x}} \cdot \frac{dx}{x} = \frac{\sqrt{2}}{8}\left[2\boldsymbol{E}\left(\frac{\sqrt{2}}{2}\right) - \boldsymbol{K}\left(\frac{\sqrt{2}}{2}\right)\right].$$ BI ((185))(5)

3. $\int\limits_0^\infty \frac{\cos^2 2x \operatorname{tg} x}{\sqrt{1+\sin^2 2x}} \cdot \frac{dx}{x} = \sqrt{2}\left[\boldsymbol{K}\left(\frac{\sqrt{2}}{2}\right) - \boldsymbol{E}\left(\frac{\sqrt{2}}{2}\right)\right].$ BI ((184))(7)

3.85-3.88 Trigonometric functions of more complicated arguments combined with powers

3.851

1. $\int\limits_0^\infty x \sin(ax^2) \sin(2bx)\, dx = \frac{b}{2a}\sqrt{\frac{\pi}{2a}}\left(\cos\frac{b^2}{a} + \sin\frac{b^2}{a}\right)$

$[a > 0,\ b > 0].$ BI ((150))(4)

2. $\int\limits_0^\infty x \sin(ax^2) \cos(2bx)\, dx =$

$= \frac{1}{2a} - \frac{b}{a}\sqrt{\frac{\pi}{2a}}\left[\sin\frac{b^2}{a}\, C\left(\frac{b}{\sqrt{a}}\right) - \cos\frac{b^2}{a}\, S\left(\frac{b}{\sqrt{a}}\right)\right].$ BI ((150))(5)a

3. $\int\limits_0^\infty x \cos(ax^2) \sin(2bx)\, dx = \frac{b}{2a}\sqrt{\frac{\pi}{2a}}\left(\sin\frac{b^2}{a} - \cos\frac{b^2}{a}\right)$

$[a > 0,\ b > 0],$ (cf. **3.691** 7.). BI ((150))(7)

4. $\int\limits_0^\infty x \cos(ax^2) \cos(2bx)\, dx =$

$= \frac{b}{a}\sqrt{\frac{\pi}{2a}}\left[\cos\frac{b^2}{a}\, C\left(\frac{b}{\sqrt{a}}\right) + \sin\frac{b^2}{a}\, S\left(\frac{b}{\sqrt{a}}\right)\right].$ BI ((150))(6)a

5. $\int\limits_0^\infty \sin(ax^2) \cos(bx)\, \frac{dx}{x^2} = \frac{b\pi}{2}\left\{S\left(\frac{b}{2\sqrt{a}}\right) - C\left(\frac{b}{2\sqrt{a}}\right) +\right.$

$\left. + \sqrt{a\pi}\sin\left(\frac{b^2}{4a} + \frac{\pi}{4}\right)\right\}$ $[a > 0,\ b > 0],$ (cf. **3.691** 7.).

ET I 23(3)a

3.852

1. $\int\limits_0^\infty \frac{\sin(ax^2)}{x^2}\, dx = \sqrt{\frac{a\pi}{2}}.$ BI ((177))(10)a

2. $\int\limits_0^\infty \sin(ax^2) \cos(bx^2)\, \frac{dx}{x^2} = \frac{1}{2}\sqrt{\frac{\pi}{2}}\left(\sqrt{a+b} + \sqrt{a-b}\right)$ $[a > b > 0];$

$= \frac{1}{2}\sqrt{\pi a}$ $[b = a \geqslant 0];$

$= \frac{1}{2}\sqrt{\frac{\pi}{2}}\left(\sqrt{a+b} - \sqrt{b-a}\right)$ $[b > a > 0],$

(cf. **3.852** 1.). BI ((177))(23)

3. $\int\limits_0^\infty \frac{\sin^2(a^2x^2)}{x^4}\, dx = \frac{2\sqrt{\pi}}{3}\, a^3$ $[a \geqslant 0].$ GW ((333))(19e)

4. $\int_0^\infty \frac{\sin^3(a^2x^2)}{x^2}\,dx = \frac{3-\sqrt{3}}{8}\sqrt{\pi}a \qquad [a \geqslant 0].$ GW ((333))(19g)

5. $\int_0^\infty (\sin x^2 - x^2\cos x^2)\frac{dx}{x^4} = \frac{1}{3}\sqrt{\frac{\pi}{2}}.$ BI ((178))(8)

6. $\int_0^\infty \left\{\cos x^2 - \frac{1}{1+x^2}\right\}\frac{dx}{x} = -\frac{1}{2}C.$ BI ((173))(22)

3.853

1. $\int_0^\infty \frac{\sin(ax^2)}{\beta^2+x^2}\,dx = \frac{\pi}{2\beta}\left[\sqrt{2}\sin\left(a\beta^2+\frac{\pi}{4}\right)C(\sqrt{a}\,\beta) - \right.$

$\left. -\sqrt{2}\cos\left(a\beta^2+\frac{\pi}{4}\right)S(\sqrt{a}\,\beta) - \sin(a\beta^2)\right] \qquad [a>0,\ \operatorname{Re}\beta>0].$ ET II 219(33)a

2. $\int_0^\infty \frac{\cos(ax^2)}{\beta^2+x^2}\,dx = \frac{\pi}{2\beta}\left[\cos(a\beta^2) - \sqrt{2}\cos\left(a\beta^2+\frac{\pi}{4}\right)C(\sqrt{a}\,\beta) - \right.$

$\left. -\sqrt{2}\sin\left(a\beta^2+\frac{\pi}{4}\right)S(\sqrt{a}\,\beta)\right] \qquad [a>0,\ \operatorname{Re}\beta>0].$ ET II 221(51)a

3. $\int_0^\infty \frac{x^2\sin(ax^2)}{\beta^2+x^2}\,dx = \frac{\beta\pi}{2}\left[\sin(a\beta^2) - \sqrt{2}\sin\left(a\beta^2+\frac{\pi}{4}\right)C(\sqrt{a}\,\beta) + \right.$

$\left. +\sqrt{2}\cos\left(a\beta^2+\frac{\pi}{4}\right)S(\sqrt{a}\,\beta)\right] - \frac{1}{2}\sqrt{\frac{\pi}{2a}} \qquad [a>0,\ \operatorname{Re}\beta>0].$ ET II 219(32)a

4. $\int_0^\infty \frac{x^2\cos(ax^2)}{\beta^2+x^2}\,dx = \frac{1}{2}\sqrt{\frac{\pi}{2a}} - \frac{\beta\pi}{2}\left[\cos(a\beta^2) - \right.$

$\left. -\sqrt{2}\cos\left(a\beta^2+\frac{\pi}{4}\right)C(\sqrt{a}\,\beta) - \sqrt{2}\sin\left(a\beta^2+\frac{\pi}{4}\right)S(\sqrt{a}\,\beta)\right]$

$[a>0,\ \operatorname{Re}\beta>0].$ ET II 221(50)a

3.854

1. $\int_0^\infty (\cos(ax^2) - \sin(ax^2))\frac{dx}{x^4+b^4} = \frac{\pi e^{-ab^2}}{2b^3\sqrt{2}} \qquad [a>0,\ b>0].$

LI((178))(11)a, BI((168))(25)

2. $\int_0^\infty (\cos(ax^2) + \sin(ax^2))\frac{x^2\,dx}{x^4+b^4} = \frac{\pi e^{-ab^2}}{2b\sqrt{2}} \qquad [a>0,\ b>0].$

LI ((178))(12)

3. $\int_0^\infty (\cos(ax^2) + \sin(ax^2))\frac{x^2\,dx}{(x^4+b^4)^2} = \frac{\pi e^{-ab^2}}{4\sqrt{2}b^3}\left(a+\frac{1}{2b^2}\right)$

$[a>0,\ b>0].$ LI ((178))(14)

4. $\int_0^\infty (\cos(ax^2) - \sin(ax^2))\frac{x^4\,dx}{(x^4+b^4)^2} = \frac{\pi e^{-ab^2}}{4\sqrt{2}\,b}\left(\frac{1}{2b^2}-a\right)$

$[a>0,\ b>0].$ BI ((178))(15)

3.855

1. $$\int_0^\infty \frac{\sin(ax^2)}{\sqrt{\beta^2+x^4}}\,dx = \frac{1}{2}\sqrt{\frac{a\pi}{2}}\, I_{\frac{1}{4}}\left(\frac{a\beta}{2}\right) K_{\frac{1}{4}}\left(\frac{a\beta}{2}\right)$$
$[a>0,\ \operatorname{Re}\beta>0]$. ET I 66(28)

2. $$\int_0^\infty \frac{\cos(ax^2)}{\sqrt{\beta^2+x^4}}\,dx = \frac{1}{2}\sqrt{\frac{a\pi}{2}}\, I_{-\frac{1}{4}}\left(\frac{a\beta}{2}\right) K_{\frac{1}{4}}\left(\frac{a\beta}{2}\right) \qquad [a>0,\ \operatorname{Re}\beta>0].$$
ET I 9(22)

3. $$\int_0^u \frac{\sin(a^2x^2)}{\sqrt{u^4-x^4}}\,dx = \frac{a}{4}\sqrt{\frac{\pi^3}{2}}\left[J_{\frac{1}{4}}\left(\frac{a^2u^2}{2}\right)\right]^2 \qquad [a>0].$$
ET I 66(29)

4. $$\int_u^\infty \frac{\sin(a^2x^2)}{\sqrt{x^4-u^4}}\,dx = -\frac{a}{4}\sqrt{\frac{\pi^3}{2}}\, J_{\frac{1}{4}}\left(\frac{a^2u^2}{2}\right) N_{\frac{1}{4}}\left(\frac{a^2u^2}{2}\right) \qquad [a>0].$$
ET I 66(30)

5. $$\int_0^u \frac{\cos(a^2x^2)}{\sqrt{u^4-x^4}}\,dx = \frac{a}{4}\sqrt{\frac{\pi^3}{2}}\left[J_{-\frac{1}{4}}\left(\frac{a^2u^2}{2}\right)\right]^2.$$
ET I 9(23)

6. $$\int_u^\infty \frac{\cos(a^2x^2)}{\sqrt{x^4-u^4}}\,dx = -\frac{a}{4}\sqrt{\frac{\pi^3}{2}}\, J_{-\frac{1}{4}}\left(\frac{a^2u^2}{2}\right) N_{-\frac{1}{4}}\left(\frac{a^2u^2}{2}\right).$$
ET I 10(24)

3.856

1. $$\int_0^\infty \frac{\left(\sqrt{\beta^4+x^4}+x^2\right)^\nu}{\sqrt{\beta^4+x^4}}\sin(a^2x^2)\,dx =$$
$$= \frac{a}{2}\sqrt{\frac{\pi}{2}}\,\beta^{2\nu} I_{\frac{1}{4}-\frac{\nu}{2}}\left(\frac{a^2\beta^2}{2}\right) K_{\frac{1}{4}+\frac{\nu}{2}}\left(\frac{a^2\beta^2}{2}\right)$$
$\left[\operatorname{Re}\nu<\frac{3}{2},\ |\arg\beta|<\frac{\pi}{4}\right]$. ET I 71(23)

2. $$\int_0^\infty \frac{\left(\sqrt{\beta^4+x^4}+x^2\right)^\nu}{\sqrt{\beta^4+x^4}}\cos(a^2x^2)\,dx =$$
$$= \frac{a}{2}\sqrt{\frac{\pi}{2}}\,\beta^{2\nu} I_{-\frac{1}{4}-\frac{\nu}{2}}\left(\frac{a^2\beta^2}{2}\right) K_{-\frac{1}{4}+\frac{\nu}{2}}\left(\frac{a^2\beta^2}{2}\right)$$
$\left[\operatorname{Re}\nu<\frac{3}{2},\ |\arg\beta|<\frac{\pi}{4}\right]$. ET I 12(16)

3. $$\int_0^\infty \frac{\left(\sqrt{\beta^4+x^4}-x^2\right)^\nu}{\sqrt{\beta^4+x^4}}\cos(a^2x^2)\,dx =$$
$$= \frac{a}{2}\sqrt{\frac{\pi}{2}}\,\beta^{2\nu} I_{-\frac{1}{4}+\frac{\nu}{2}}\left(\frac{a^2\beta^2}{2}\right) K_{-\frac{1}{4}-\frac{\nu}{2}}\left(\frac{a^2\beta^2}{2}\right)$$
$\left[\operatorname{Re}\nu>-\frac{3}{2},\ |\arg\beta|<\frac{\pi}{4}\right]$. ET I 12(17)

4. $$\int_0^\infty \frac{\sin(a^2x^2)\,dx}{\sqrt{\beta^4+x^4}\sqrt{x^2+\sqrt{\beta^4+x^4}}} = \frac{\operatorname{sh}\frac{a^2\beta^2}{2}}{\sqrt{2}\,\beta^2} K_0\left(\frac{a^2\beta^2}{2}\right) \qquad \left[|\arg\beta| < \frac{\pi}{4}\right].$$

74

ET I 66(32)

5. $$\int_0^\infty \frac{\cos(a^2x^2)\,dx}{\sqrt{\beta^4+x^4}\sqrt{(x^2+\sqrt{\beta^4+x^4})^3}} = \frac{\operatorname{sh}\frac{a^2\beta^2}{2}}{2\sqrt{2}\,\beta^4} K_1\left(\frac{a^2\beta^2}{2}\right) \qquad \left[|\arg\beta| < \frac{\pi}{4}\right].$$

ET I 10(27)

6. $$\int_0^\infty \frac{\sqrt{\sqrt{\beta^4+x^4}+x^2}}{\sqrt{\beta^4+x^4}} \sin(a^2x^2)\,dx = \frac{\pi}{2\sqrt{2}} e^{-\frac{a^2\beta^2}{2}} I_0\left(\frac{a^2\beta^2}{2}\right)$$

$$\left[|\arg\beta| < \frac{\pi}{4}\right],$$

ET I 67(33)

3.857

1. $$\int_0^\infty \frac{x^2}{R_1R_2}\sqrt{\frac{R_2-R_1}{R_2+R_1}} \sin(ax^2)\,dx = \frac{1}{2\sqrt{b}} K_0(ac)\sin ab$$

$$\left[R_1 = \sqrt{c^2+(b-x^2)^2}, \quad R_2 = \sqrt{c^2+(b+x^2)^2}, \quad a > 0, \ c > 0\right].$$

ET I 67(34)

2. $$\int_0^\infty \frac{x^2}{R_1R_2}\sqrt{\frac{R_2+R_1}{R_2-R_1}} \cos(ax^2)\,dx = \frac{1}{2\sqrt{b}} K_0(ac)\cos ab$$

$$\left[R_1 = \sqrt{c^2+(b-x^2)^2}, \quad R_2 = \sqrt{c^2+(b+x^2)^2}, \quad a > 0, \ c > 0\right].$$

ET I 10(26)

3.858

1. $$\int_u^\infty \frac{(x^2+\sqrt{x^4-u^4})^\nu + (x^2-\sqrt{x^4-u^4})^\nu}{\sqrt{x^4-u^4}} \sin(a^2x^2)\,dx =$$

$$= -\frac{a}{4}\sqrt{\frac{\pi^3}{2}}\, u^{2\nu}\left[J_{\frac{1}{4}+\frac{\nu}{2}}\left(\frac{a^2u^2}{2}\right) N_{\frac{1}{4}-\frac{\nu}{2}}\left(\frac{a^2u^2}{2}\right) + \right.$$

$$\left. + J_{\frac{1}{4}-\frac{\nu}{2}}\left(\frac{a^2u^2}{2}\right) N_{\frac{1}{4}+\frac{\nu}{2}}\left(\frac{a^2u^2}{2}\right)\right] \qquad \left[\operatorname{Re}\nu < \frac{3}{2}\right].$$

ET I 71(25)

2. $$\int_u^\infty \frac{(x^2+\sqrt{x^4-u^4})^\nu + (x^2-\sqrt{x^4-u^4})^\nu}{\sqrt{x^4-u^4}} \cos(a^2x^2)\,dx =$$

$$= -\frac{a}{4}\sqrt{\frac{\pi^3}{2}}\, u^{2\nu}\left[J_{-\frac{1}{4}+\frac{\nu}{2}}\left(\frac{a^2u^2}{2}\right) N_{-\frac{1}{4}-\frac{\nu}{2}}\left(\frac{a^2u^2}{2}\right) + \right.$$

$$\left. + J_{-\frac{1}{4}-\frac{\nu}{2}}\left(\frac{a^2u^2}{2}\right) N_{-\frac{1}{4}+\frac{\nu}{2}}\left(\frac{a^2u^2}{2}\right)\right] \quad \left[\operatorname{Re}\nu < \frac{3}{2}\right].$$

ET I 13(26)

3.859 $$\int_0^\infty \left[\cos(x^{2^n}) - \frac{1}{1+x^{2^{n+1}}}\right]\frac{dx}{x} = -\frac{1}{2^n} C.$$

BI ((173))(24)

3.861

1. $$\int_0^\infty \sin^{2n+1}(ax^2)\frac{dx}{x^{2m}} = \pm\frac{\sqrt{\pi}\,a^{m-\frac{1}{2}}}{2^{2n-m+\frac{1}{2}}(2m-1)!!}\times$$
$$\times\sum_{k=1}^{n+1}(-1)^{k-1}\binom{2n+1}{n+k}(2k-1)^{m-\frac{1}{2}}$$

[the sign + is taken when $m \equiv 0 \pmod 4$ or $m \equiv 1 \pmod 4$, the sign − is taken when $m \equiv 2 \pmod 4$ or $m \equiv 3 \pmod 4$].

BI ((177))(19)a

2. $$\int_0^\infty \sin^{2n}(ax^2)\frac{dx}{x^{2m}} =$$
$$= \pm\frac{\sqrt{\pi}\,a^{m-\frac{1}{2}}}{2^{2n-2m+1}(2m-1)!!}\sum_{k=1}^{n}(-1)^k\binom{2n}{n+k}k^{m-\frac{1}{2}}$$

[the sign + is taken when $m \equiv 0 \pmod 4$ or $m \equiv 3 \pmod 4$, the sign − is taken when $m \equiv 2 \pmod 4$ or $m \equiv 1 \pmod 4$].

BI((177))(18)a, LI((177))(18)

3.862 $$\int_0^\infty \left[\cos\left(ax^2\sqrt{n}\right)+\sin\left(ax^2\sqrt{n}\right)\right]\left(\frac{\sin x^2}{x^2}\right)^n dx =$$
$$= \frac{\sqrt{\pi}}{(2n-1)!!\sqrt{2}}\sum_{k=0}^{n}(-1)^k\binom{n}{k}\left(n-2k+a\sqrt{n}\right)^{n-\frac{1}{2}} \quad \left[a > \sqrt{n} > 0\right].$$

BI ((178))(9)

3.863

1. $$\int_0^\infty x^2\cos(ax^4)\sin(2bx^2)\,dx = -\frac{\pi}{8}\sqrt{\frac{b^3}{a^3}}\left[\sin\left(\frac{b^2}{2a}-\frac{\pi}{8}\right)J_{-\frac{1}{4}}\left(\frac{b^2}{2a}\right)+\right.$$
$$\left.+\cos\left(\frac{b^2}{2a}-\frac{\pi}{8}\right)J_{\frac{3}{4}}\left(\frac{b^2}{2a}\right)\right] \quad [a > 0, \quad b > 0].$$

ET I 25(22)

2. $$\int_0^\infty x^2\cos(ax^4)\cos(2bx^2)\,dx =$$
$$= -\frac{\pi}{8}\sqrt{\frac{b^3}{a^3}}\left[\sin\left(\frac{b^2}{2a}+\frac{\pi}{8}\right)J_{-\frac{3}{4}}\left(\frac{b^2}{2a}\right)+\right.$$
$$\left.+\cos\left(\frac{b^2}{2a}+\frac{\pi}{8}\right)J_{-\frac{1}{4}}\left(\frac{b^2}{2a}\right)\right] \quad [a > 0, \ b > 0].$$

ET I 25(23)

3.864

1. $$\int_0^\infty \sin\frac{b}{x}\sin ax\frac{dx}{x} = \frac{\pi}{2}N_0\left(2\sqrt{ab}\right)+K_0\left(2\sqrt{ab}\right) \quad [a > 0, \ b > 0].$$

WA 204(3)a

2. $$\int_0^\infty \cos\frac{b}{x}\cos ax\frac{dx}{x} = -\frac{\pi}{2}N_0\left(2\sqrt{ab}\right)+K_0\left(2\sqrt{ab}\right) \qquad [a>0, \quad b>0].$$

WA 204(4)a, ET I 24(12)

3.865

1. $$\int_0^u \frac{(u^2-x^2)^{\mu-1}}{x^{2\mu}}\sin\frac{a}{x}\,dx = \frac{\sqrt{\pi}}{2}\left(\frac{2}{a}\right)^{\mu-\frac{1}{2}}u^{\mu-\frac{3}{2}}\Gamma(\mu)J_{\frac{1}{2}-\mu}\left(\frac{a}{u}\right)$$

$$[a>0,\ u>0,\ 0<\operatorname{Re}\mu<1].$$ ET II 189(30)

2. $$\int_u^\infty \frac{(x-u)^{\mu-1}}{x^{2\mu}}\sin\frac{a}{x}\,dx = \sqrt{\frac{\pi}{u}}a^{\frac{1}{2}-\mu}\Gamma(\mu)\sin\frac{a}{2u}J_{\mu-\frac{1}{2}}\left(\frac{a}{2u}\right)$$

$$[a>0,\ u>0,\ \operatorname{Re}\mu>0].$$ ET II 203(21)

3. $$\int_0^u \frac{(u^2-x^2)^{\mu-1}}{x^{2\mu}}\cos\frac{a}{x}\,dx = -\frac{\sqrt{\pi}}{2}\left(\frac{2}{a}\right)^{\mu-\frac{1}{2}}\Gamma(\mu)u^{\mu-\frac{3}{2}}N_{\frac{1}{2}-\mu}\left(\frac{a}{u}\right)$$

$$[a>0,\ u>0,\ 0<\operatorname{Re}\mu<1].$$ ET II 190(36)

4. $$\int_u^\infty \frac{(x-u)^{\mu-1}}{x^{2\mu}}\cos\frac{a}{x}\,dx = \sqrt{\frac{\pi}{u}}a^{\frac{1}{2}-\mu}\Gamma(\mu)\cos\frac{a}{2u}J_{\mu-\frac{1}{2}}\left(\frac{a}{2u}\right)$$

$$[a>0,\ u>0,\ \operatorname{Re}\mu>0].$$ ET II 204(26)

3.866

1. $$\int_0^\infty x^{\mu-1}\sin\frac{b^2}{x}\sin(a^2x)\,dx = \frac{\pi}{4}\left(\frac{b}{a}\right)^\mu\operatorname{cosec}\frac{\mu\pi}{2}\times$$
$$\times[J_\mu(2ab)-J_{-\mu}(2ab)+I_{-\mu}(2ab)-I_\mu(2ab)]$$
$$[a>0,\ b>0,\ |\operatorname{Re}\mu|<1].$$ ET I 322(42)

2. $$\int_0^\infty x^{\mu-1}\sin\frac{b^2}{x}\cos(a^2x)\,dx = \frac{\pi}{4}\left(\frac{b}{a}\right)^\mu\sec\frac{\mu\pi}{2}\times$$
$$\times[J_\mu(2ab)+J_{-\mu}(2ab)+I_\mu(2ab)-I_{-\mu}(2ab)]$$
$$[a>0,\ b>0,\ |\operatorname{Re}\mu|<1].$$ ET I 322(43)

3. $$\int_0^\infty x^{\mu-1}\cos\frac{b^2}{x}\cos(a^2x)\,dx = \frac{\pi}{4}\left(\frac{b}{a}\right)^\mu\operatorname{cosec}\frac{\mu\pi}{2}\times$$
$$\times[J_{-\mu}(2ab)-J_\mu(2ab)+I_{-\mu}(2ab)-I_\mu(2ab)]$$
$$[a>0,\ b>0,\ |\operatorname{Re}\mu|<1].$$ ET I 322(44)

3.867

1. $$\int_0^1 \frac{\cos ax-\cos\frac{a}{x}}{1-x^2}\,dx = \frac{1}{2}\int_0^\infty \frac{\cos ax-\cos\frac{a}{x}}{1-x^2}\,dx = \frac{\pi}{2}\sin a$$

$$[a>0].$$ GW ((334))(7a)

2. $$\int_0^1 \frac{\cos ax + \cos \frac{a}{x}}{1+x^2}\,dx = \frac{1}{2}\int_0^\infty \frac{\cos ax + \cos \frac{a}{x}}{1+x^2}\,dx = \frac{\pi}{2}e^{-a} \qquad [a>0].$$

GW ((334))(7b)

3.868

1. $$\int_0^\infty \sin\left(a^2x+\frac{b^2}{x}\right)\frac{dx}{x} = \pi J_0(2ab) \qquad [a>0,\ b>0].$$

GW ((334))(11a), WA 200(16)

2. $$\int_0^\infty \cos\left(a^2x+\frac{b^2}{x}\right)\frac{dx}{x} = -\pi N_0(2ab). \qquad [a>0,\ b>0].$$

GW ((334))(11a)

3. $$\int_0^\infty \sin\left(a^2x-\frac{b^2}{x}\right)\frac{dx}{x} = 0 \qquad [a>0,\ b>0].$$ GW ((334))(11b)

4. $$\int_0^\infty \cos\left(a^2x-\frac{b^2}{x}\right)\frac{dx}{x} = 2K_0(2ab) \qquad [a>0,\ b>0].$$ GW ((334))(11b)

3.869

1. $$\int_0^\infty \sin\left(ax-\frac{b}{x}\right)\frac{x\,dx}{\beta^2+x^2} = \frac{\pi}{2}\exp\left(-a\beta-\frac{b}{\beta}\right)$$

$[a>0,\ b>0,\ \operatorname{Re}\beta>0]$. ET II 220(42)

2. $$\int_0^\infty \cos\left(ax-\frac{b}{x}\right)\frac{dx}{\beta^2+x^2} = \frac{\pi}{2\beta}\exp\left(-a\beta-\frac{b}{\beta}\right)$$

$[a>0,\ b>0,\ \operatorname{Re}\beta>0]$. ET II 222(58)

3.871

1. $$\int_0^\infty x^{\mu-1}\sin\left[a\left(x+\frac{b^2}{x}\right)\right]dx = \pi b^\mu\left[J_\mu(2ab)\cos\frac{\mu\pi}{2} - N_\mu(2ab)\sin\frac{\mu\pi}{2}\right]$$

$[a>0,\ b>0,\ \operatorname{Re}\mu<1]$. ET I 319(17)

2. $$\int_0^\infty x^{\mu-1}\cos\left[a\left(x+\frac{b^2}{x}\right)\right]dx = -\pi b^\mu\left[J_\mu(2ab)\sin\frac{\mu\pi}{2} + N_\mu(2ab)\cos\frac{\mu\pi}{2}\right]$$

$[a>0,\ b>0,\ |\operatorname{Re}\mu|<1]$. ET I 321(35)

3. $$\int_0^\infty x^{\mu-1}\sin\left[a\left(x-\frac{b^2}{x}\right)\right]dx = 2b^\mu K_\mu(2ab)\sin\frac{\mu\pi}{2}$$

$[a>0,\ b>0,\ |\operatorname{Re}\mu|<1]$. ET I 319(16)

4. $$\int_0^\infty x^{\mu-1}\cos\left[a\left(x-\frac{b^2}{x}\right)\right]dx = 2b^\mu K_\mu(2ab)\cos\frac{\mu\pi}{2}$$

$[a>0,\ b>0,\ |\operatorname{Re}\mu|<1]$. ET I 321(36)

3.872

1. $$\int_0^1 \sin\left[a\left(x+\frac{1}{x}\right)\right]\sin\left[a\left(x-\frac{1}{x}\right)\right]\frac{dx}{1-x^2}=$$
$$=\frac{1}{2}\int_0^\infty \sin\left[a\left(x+\frac{1}{x}\right)\right]\sin\left[a\left(x-\frac{1}{x}\right)\right]\frac{dx}{1-x^2}=-\frac{\pi}{4}\sin 2a$$
$[a \geqslant 0]$ BI((149))(15), GW((334))(8a)

2. $$\int_0^1 \cos\left[a\left(x+\frac{1}{x}\right)\right]\cos\left[a\left(x-\frac{1}{x}\right)\right]\frac{dx}{1+x^2}=$$
$$=\frac{1}{2}\int_0^\infty \cos\left[a\left(x+\frac{1}{x}\right)\right]\cos\left[a\left(x-\frac{1}{x}\right)\right]\frac{dx}{1+x^2}=\frac{\pi}{4}e^{-2a}$$
$[a \geqslant 0]$. GW ((334))(8b)

3.873

1. $$\int_0^\infty \sin\frac{a^2}{x^2}\cos b^2x^2\,\frac{dx}{x^2}=\frac{\sqrt{\pi}}{4\sqrt{2a}}\left[\sin(2ab)+\cos(2ab)+e^{-2ab}\right]$$
$[a>0,\ b>0]$. ET I 24(15)

2. $$\int_0^\infty \cos\frac{a^2}{x^2}\cos b^2x^2\,\frac{dx}{x^2}=\frac{\sqrt{\pi}}{4\sqrt{2a}}\left[\cos(2ab)-\sin(2ab)+e^{-2ab}\right]$$
$[a>0,\ b>0]$. ET I 24(16)

3 874

1. $$\int_0^\infty \sin\left(a^2x^2+\frac{b^2}{x^2}\right)\frac{dx}{x^2}=\frac{\sqrt{\pi}}{2b}\sin\left(2ab+\frac{\pi}{4}\right)$$
$[a>0,\ b>0]$. BI((179))(6)a, GW((334))(10a)

2. $$\int_0^\infty \cos\left(a^2x^2+\frac{b^2}{x^2}\right)\frac{dx}{x^2}=\frac{\sqrt{\pi}}{2b}\cos\left(2ab+\frac{\pi}{4}\right)\qquad [a>0,\ b>0].$$
GI((179))(8)a, GW((334))(10a)

3. $$\int_0^\infty \sin\left(a^2x^2-\frac{b^2}{x^2}\right)\frac{dx}{x^2}=-\frac{\sqrt{\pi}}{2\sqrt{2}b}e^{-2ab}$$
$[a \geqslant 0,\ b>0]$. GW ((334))(10b)

4. $$\int_0^\infty \cos\left(a^2x^2-\frac{b^2}{x^2}\right)\frac{dx}{x^2}=\frac{\sqrt{\pi}}{2\sqrt{2}b}e^{-2ab}\qquad [a \geqslant 0,\ b>0].$$ GW ((334))(10b)

5 $$\int_0^\infty \sin\left(ax-\frac{b}{x}\right)^2\frac{dx}{x^2}=\frac{\sqrt{2\pi}}{4b}\qquad [a>0,\ b>0].$$ BI ((179))(13)a

6. $$\int_0^\infty \cos\left(ax-\frac{b}{x}\right)^2\frac{dx}{x^2}=\frac{\sqrt{2\pi}}{4b}\qquad [a>0,\ b>0].$$ BI ((179))(14)a

3.875

1. $$\int_u^\infty \frac{x\sin\left(p\sqrt{x^2-u^2}\right)}{x^2+a^2}\cos bx\,dx = \frac{\pi}{2}\exp\left(-p\sqrt{a^2+u^2}\right)\operatorname{ch} ab$$

$[0<b<p].$ ET I 27(39)

2. $$\int_u^\infty \frac{x\sin\left(p\sqrt{x^2-u^2}\right)}{a^2+x^2-u^2}\cos bx\,dx = \frac{\pi}{2}e^{-ap}\cos\left(b\sqrt{u^2-a^2}\right)$$

$[0<b<p,\ a>0].$ ET I 27(38)

3. $$\int_0^\infty \frac{\sin\left(p\sqrt{a^2+x^2}\right)}{(a^2+x^2)^2}\cos bx\,dx = \frac{\pi p}{2a}e^{-ab} \qquad [b>a>0].$$ ET I 26(29)

3.876

1. $$\int_0^\infty \frac{\sin\left(p\sqrt{x^2+a^2}\right)}{\sqrt{x^2+a^2}}\cos bx\,dx = \frac{\pi}{2}J_0\left(a\sqrt{p^2-b^2}\right) \qquad [0<b<p];$$

$$= 0 \qquad [b>p>0];$$

$[a>0].$ ET I 26(30)

2. $$\int_0^\infty \frac{\cos\left(p\sqrt{x^2+a^2}\right)}{\sqrt{x^2+a^2}}\cos bx\,dx = -\frac{\pi}{2}N_0\left(a\sqrt{p^2-b^2}\right) \qquad [0<b<p];$$

$$= K_0\left(a\sqrt{b^2-p^2}\right) \qquad [b>p>0];$$

$[a>0].$ ET I 26(34)

3. $$\int_0^\infty \frac{\cos\left(p\sqrt{x^2+a^2}\right)}{x^2+c^2}\cos bx\,dx = \frac{\pi}{2c}e^{-bc}\cos\left(p\sqrt{a^2-c^2}\right)$$

$[c>0,\ b>p].$ ET I 26(33)

4. $$\int_0^\infty \frac{\sin\left(p\sqrt{x^2+a^2}\right)}{(x^2+c^2)\sqrt{x^2+a^2}}\cos bx\,dx = \frac{\pi}{2c}\,\frac{e^{-bc}\sin\left(p\sqrt{a^2-c^2}\right)}{\sqrt{a^2-c^2}} \qquad [c\neq a];$$

$$= \frac{\pi}{2}e^{-ba}\frac{p}{a} \qquad [c=a];$$

$[b>p,\ c>0].$ ET I 26(31)a

5. $$\int_0^\infty \frac{\cos\left(p\sqrt{x^2+a^2}\right)}{(x^2+a^2)^{\frac{3}{2}}}\cos bx\,dx = \frac{\pi}{2a}e^{-ab} \qquad [b>a>0].$$ ET I 27(35)a

6. $$\int_0^\infty \frac{x\cos\left(p\sqrt{x^2+a^2}\right)}{(x^2+a^2)^{\frac{3}{2}}}\sin bx\,dx = \frac{\pi}{2}e^{-ab}$$

$[a>0,\ b>p>0].$ ET I 85(29)a

7. $$\int_0^u \frac{\cos\left(p\sqrt{u^2-x^2}\right)}{\sqrt{u^2-x^2}} \cos bx\,dx = \frac{\pi}{2} J_0\left(u\sqrt{b^2+p^2}\right).$$ ET I 28(42)

8 $$\int_u^\infty \frac{\cos\left(p\sqrt{x^2-u^2}\right)}{\sqrt{x^2-u^2}} \cos bx\,dx = K_0\left(u\sqrt{p^2-b^2}\right) \qquad [0 < b < |p|].$$

$$= -\frac{\pi}{2} N_0\left(u\sqrt{b^2-p^2}\right) \qquad [b > |p|].$$

ET I 28(43)

3.877

1 $$\int_0^u \frac{\sin\left(p\sqrt{u^2-x^2}\right)}{\sqrt[4]{(u^2-x^2)^3}} \cos bx\,dx =$$

$$= \sqrt{\frac{\pi^3 p}{8}} J_{\frac{1}{4}}\left[\frac{u}{2}\left(\sqrt{b^2+p^2}-b\right)\right] J_{\frac{1}{4}}\left[\frac{u}{2}\left(\sqrt{b^2+p^2}+b\right)\right]$$

$[b > 0,\ p > 0]$. ET I 27(40)

2. $$\int_u^\infty \frac{\sin\left(p\sqrt{x^2-u^2}\right)}{\sqrt[4]{(x^2-u^2)^3}} \cos bx\,dx =$$

$$= -\sqrt{\frac{\pi^3 p}{8}} J_{\frac{1}{4}}\left[\frac{u}{2}\left(b-\sqrt{b^2-p^2}\right)\right] N_{\frac{1}{4}}\left[\frac{u}{2}\left(b+\sqrt{b^2-p^2}\right)\right]$$

$[b > p > 0]$. ET I 27(41)

3. $$\int_0^u \frac{\cos\left(p\sqrt{u^2-x^2}\right)}{\sqrt[4]{(u^2-x^2)^3}} \cos bx\,dx =$$

$$= \sqrt{\frac{\pi^3 p}{8}} J_{-\frac{1}{4}}\left[\frac{u}{2}\left(\sqrt{p^2+b^2}-b\right)\right] J_{-\frac{1}{4}}\left[\frac{u}{2}\left(\sqrt{p^2+b^2}+b\right)\right]$$

$[u > 0,\ p > 0]$. ET I 28(44)

4 $$\int_u^\infty \frac{\cos\left(p\sqrt{x^2-u^2}\right)}{\sqrt[4]{(x^2-u^2)^3}} \cos bx\,dx =$$

$$= -\sqrt{\frac{\pi^3 p}{8}} J_{-\frac{1}{4}}\left[\frac{u}{2}\left(b-\sqrt{b^2-p^2}\right)\right] N_{\frac{1}{4}}\left[\frac{u}{2}\left(b+\sqrt{b^2-p^2}\right)\right]$$

$[b > p > 0]$. ET I 28(45)

3.878

1 $$\int_0^\infty \frac{\sin\left(p\sqrt{x^4+a^4}\right)}{\sqrt{x^4+a^4}} \cos bx^2\,dx =$$

$$= \frac{1}{2}\sqrt{\left(\frac{\pi}{2}\right)^3 b}\, J_{-\frac{1}{4}}\left[\frac{a^2}{2}\left(p-\sqrt{p^2-b^2}\right)\right] J_{\frac{1}{4}}\left[\frac{a^2}{2}\left(p+\sqrt{p^2-b^2}\right)\right]$$

$[p > b > 0]$. ET I 26(32)

2. $\int\limits_0^\infty \frac{\cos(p\sqrt{x^4+a^4})}{\sqrt{x^4+a^4}} \cos bx^2\,dx =$

$$= -\frac{1}{2}\sqrt{\left(\frac{\pi}{2}\right)^3 b}\, J_{-\frac{1}{4}}\left[\frac{a^2}{2}\left(p-\sqrt{p^2-b^2}\right)\right] N_{\frac{1}{4}}\left[\frac{a^2}{2}\left(p+\sqrt{p^2-b^2}\right)\right]$$

$[a > 0,\ p > b > 0]$. ET I 27(36)

3. $\int\limits_0^u \frac{\cos(p\sqrt{u^4-x^4})}{\sqrt{u^4-x^4}} \cos bx^2\,dx =$

$$= \frac{1}{2}\sqrt{\left(\frac{\pi}{2}\right)^3 b}\, J_{-\frac{1}{4}}\left[\frac{u^2}{2}\left(\sqrt{p^2+b^2}-p\right)\right] J_{-\frac{1}{4}}\left[\frac{u^2}{2}\left(\sqrt{p^2+b^2}+p\right)\right]$$

$[p > 0,\ b > 0]$. ET I 28(46)

3.879 $\int\limits_0^\infty \sin ax^p \frac{dx}{x} = \frac{\pi}{2p}$ $[a > 0,\ p > 0]$. GW ((334))(6)

3.881

1. $\int\limits_0^{\frac{\pi}{2}} x \sin(a \operatorname{tg} x)\,dx = \frac{\pi}{4} e^{-a}\left[\boldsymbol{C} + \ln 2a - e^{2a}\operatorname{Ei}(-2a)\right]$ $[a > 0]$. BI ((205))(9)

2. $\int\limits_0^\infty \sin(a \operatorname{tg} x) \frac{dx}{x} = \frac{\pi}{2}(1-e^{-a})$ $[a > 0]$. BI ((151))(6)

3. $\int\limits_0^\infty \sin(a \operatorname{tg} x)\cos x \frac{dx}{x} = \frac{\pi}{2}(1-e^{-a})$ $[a > 0]$. BI ((151))(19)

4. $\int\limits_0^\infty \cos(a \operatorname{tg} x)\sin x \frac{dx}{x} = \frac{\pi}{2} e^{-a}$ $[a > 0]$. BI ((151))(20)

5. $\int\limits_0^\infty \sin(a \operatorname{tg} x)\sin 2x \frac{dx}{x} = \frac{1+a}{2}\pi e^{-a}$ $[a > 0]$. BI ((152))(11)

6. $\int\limits_0^\infty \cos(a \operatorname{tg} x)\sin^3 x \frac{dx}{x} = \frac{1-a}{4}\pi e^{-a}$ $[a > 0]$. BI ((151))(23)

7. $\int\limits_0^\infty \sin(a \operatorname{tg} x)\operatorname{tg}\frac{x}{2}\cos^2 x \frac{dx}{x} = \frac{1+a}{4}\pi e^{-a}$ $[a > 0]$. BI ((152))(13)

8. $\int\limits_0^{\frac{\pi}{2}} \cos(a \operatorname{tg} x) \frac{x\,dx}{\sin 2x} = -\frac{\pi}{4}\operatorname{Ei}(-a)$ $[a > 0]$. BI ((206))(15)

9. $\int\limits_0^{\frac{\pi}{2}} \sin(a \operatorname{ctg} x) \frac{x\,dx}{\sin^2 x} = \frac{1-e^{-a}}{2a}\pi$ $[a > 0]$. LI ((206))(14)

10. $\int_0^{\frac{\pi}{2}} x \cos(a \operatorname{tg} x) \operatorname{tg} x \, dx = -\frac{\pi}{4} e^{-a} [C + \ln 2a + e^{2a} \operatorname{Ei}(-2a)] \quad [a > 0].$ BI ((205))(10)

11. $\int_0^{\infty} \cos(a \operatorname{tg} x) \operatorname{tg} x \frac{dx}{x} = \frac{\pi}{2} e^{-a} \quad [a > 0].$ BI ((151))(21)

12. $\int_0^{\infty} \cos(a \operatorname{tg} x) \sin^2 x \operatorname{tg} x \frac{dx}{x} = \frac{1-a}{16} \pi e^{-a} \quad [a > 0].$ BI ((152))(15)

13. $\int_0^{\infty} \sin(a \operatorname{tg} x) \operatorname{tg}^2 x \frac{dx}{x} = \frac{\pi}{2} e^{-a} \quad [a > 0].$ BI ((152))(9)

14. $\int_0^{\infty} \cos(a \operatorname{tg} 2x) \operatorname{tg} x \frac{dx}{x} = \frac{\pi}{2} e^{-a} \quad [a > 0].$ BI ((151))(22)

15. $\int_0^{\infty} \sin(a \operatorname{tg} 2x) \cos^2 2x \operatorname{tg} x \frac{dx}{x} = \frac{1+a}{4} \pi e^{-a} \quad [a > 0].$ BI ((152))(13)

16. $\int_0^{\infty} \sin(a \operatorname{tg} 2x) \operatorname{tg} x \operatorname{tg} 2x \frac{dx}{x} = \frac{\pi}{2} e^{-a} \quad [a > 0].$ BI ((152))(10)

17. $\int_0^{\infty} \sin(a \operatorname{tg} 2x) \operatorname{tg} x \operatorname{ctg} 2x \frac{dx}{x} = \frac{\pi}{2} (1 - e^{-a}) \quad [a > 0].$ BI ((180))(6)

3.882

1. $\int_0^{\infty} \sin(a \operatorname{tg}^2 x) \frac{x \, dx}{b^2 + x^2} = \frac{\pi}{2} [\exp(-a \operatorname{th} b) - e^{-a}] \quad [a > 0, \ b > 0].$ BI ((160))(22)

2. $\int_0^{\infty} \cos(a \operatorname{tg}^2 x) \cos x \frac{dx}{b^2 + x^2} = \frac{\pi}{2b} [\operatorname{ch} b \exp(-a \operatorname{th} b) - e^{-a} \operatorname{sh} b]$ $[a > 0, b > 0].$ BI ((163))(3)

3. $\int_0^{\infty} \cos(a \operatorname{tg}^2 x) \operatorname{cosec} 2x \frac{x \, dx}{b^2 + x^2} = \frac{\pi}{2 \operatorname{sh} 2b} \exp(-a \operatorname{th} b) \quad [a > 0, \ b > 0].$ BI ((191))(10)

4. $\int_0^{\infty} \cos(a \operatorname{tg}^2 x) \operatorname{tg} x \frac{x \, dx}{b^2 + x^2} = \frac{\pi}{2 \operatorname{ch} b} [e^{-a} \operatorname{ch} b - \exp(-a \operatorname{th} b) \operatorname{sh} b]$ $[a > 0, \ b > 0].$ BI ((163))(4)

5. $\int_0^{\infty} \cos(a \operatorname{tg}^2 x) \operatorname{ctg} x \frac{x \, dx}{b^2 + x^2} = \frac{\pi}{2} [\operatorname{cth} b \exp(-a \operatorname{th} b) - e^{-a}]$ $[a > 0, \ b > 0].$ BI ((163))(5)

6. $\int_0^\infty \cos(a \operatorname{tg}^2 x) \operatorname{ctg} 2x \frac{x\,dx}{b^2+x^2} = \frac{\pi}{2}[\operatorname{cth} 2b \exp(-a \operatorname{th} b) - e^{-a}]$

$[a>0, \quad b>0]$. BI ((191))(11)

3.883

1. $\int_0^1 \cos(a \ln x) \frac{dx}{(1+x)^2} = \frac{a\pi}{2 \operatorname{sh} a\pi}$. BI ((404))(4)

2. $\int_0^1 x^{\mu-1} \sin(\beta \ln x)\,dx = -\frac{\beta}{\beta^2+\mu^2}$ $[\operatorname{Re}\mu > |\operatorname{Im}\beta|]$. ET I 319(19)

3. $\int_0^1 x^{\mu-1} \cos(\beta \ln x)\,dx = \frac{\mu}{\beta^2+\mu^2}$ $[\operatorname{Re}\mu > |\operatorname{Im}\beta|]$. ET I 321(38)

3.884 $\int_{-\infty}^\infty \frac{\sin a\sqrt{|x|}}{x-b} \operatorname{sign} x\,dx = \cos a\sqrt{|b|} + \exp\left(-a\sqrt{|b|}\right)$

$[a>0]$. ET II 253(46)

3.89-3.91 Trigonometric functions and exponentials

3.891

1. $\int_0^{2\pi} e^{imx} \sin nx\,dx = 0$ $[m \neq n; \; m=n=0]$;

$= \pi i$ $[m=n\neq 0]$.

2. $\int_0^{2\pi} e^{imx} \cos nx\,dx = 0$ $[m \neq n]$;

$= \pi$ $[m=n\neq 0]$;

$= 2\pi$ $[m=n=0]$.

3.892

1. $\int_0^\pi e^{i\beta x} \sin^{\nu-1} x\,dx = \frac{\pi e^{i\beta\frac{\pi}{2}}}{2^{\nu-1}\nu \mathrm{B}\left(\frac{\nu+\beta+1}{2}, \frac{\nu-\beta+1}{2}\right)}$

$[\operatorname{Re}\nu > -1]$. NH 158, EH I 12(29)

2. $\int_{-\frac{\pi}{2}}^{\frac{\pi}{2}} e^{i\beta x} \cos^{\nu-1} x\,dx = \frac{\pi}{2^{\nu-1}\nu \mathrm{B}\left(\frac{\nu+\beta+1}{2}, \frac{\nu-\beta+1}{2}\right)}$

$[\operatorname{Re}\nu > -1]$. GW ((335))(19)

3. $\int_0^{\frac{\pi}{2}} e^{i2\beta x} \sin^{2\mu} x \cos^{2\nu} x\,dx =$

$$= \frac{1}{2^{2\mu+2\nu+1}} \left\{ \exp\left[i\pi \left(\beta - \nu - \frac{1}{2} \right) \right] \mathrm{B}(\beta-\mu-\nu,\ 2\nu+1) \times \right.$$
$$\times F(-2\mu,\ \beta-\mu-\nu;\ 1+\beta-\mu+\nu;\ -1) + \exp\left[i\pi \left(\mu + \frac{1}{2} \right) \right] \times$$
$$\left. \times \mathrm{B}(\beta-\mu-\nu,\ 2\mu+1) F(-2\nu,\ \beta-\mu-\nu;\ 1+\beta+\mu-\nu;\ -1) \right\}$$
$$\left[\operatorname{Re}\mu > -\frac{1}{2},\ \operatorname{Re}\nu > -\frac{1}{2} \right].$$ ET I 80(6)

4. $$\int_0^{\pi} e^{i2\beta x} \sin^{2\mu} x \cos^{2\nu} x\, dx = \frac{\pi \exp[i\pi(\beta-\nu)] F(-2\nu,\ \beta-\mu-\nu;\ 1+\beta+\mu-\nu;\ -1)}{4^{\mu+\nu}(2\mu+1)\mathrm{B}(1-\beta+\mu+\nu,\ 1+\beta+\mu-\nu)}.$$ EH I 80(8)

5. $$\int_0^{\frac{\pi}{2}} e^{i(\mu+\nu)x} \sin^{\mu-1} x \cos^{\nu-1} x\, dx = e^{i\mu\frac{\pi}{2}} \mathrm{B}(\mu,\ \nu) =$$
$$= \frac{1}{2^{\mu+\nu-1}} e^{i\mu\frac{\pi}{2}} \left\{ \frac{1}{\mu} F(1-\nu,\ 1;\ \mu+1;\ -1) + \frac{1}{\nu} F(1-\mu,\ 1,\ \nu+1;\ -1) \right\}$$
$$[\operatorname{Re}\mu > 0, \quad \operatorname{Re}\nu > 0].$$ EH I 80(7)

3.893

1. $$\int_0^{\infty} e^{-px} \sin(qx+\lambda)\, dx = \frac{1}{p^2+q^2}(q\cos\lambda + p\sin\lambda) \qquad [p > 0].$$ BI ((261))(3)

2. $$\int_0^{\infty} e^{-px} \cos(qx+\lambda)\, dx = \frac{1}{p^2+q^2}(p\cos\lambda - q\sin\lambda) \qquad [p > 0].$$ BI ((261))(4)

3 $$\int_0^{\infty} e^{-x\cos t} \cos(t - x\sin t)\, dx = 1.$$ BI ((261))(7)

4. $$\int_0^{\infty} \frac{e^{-\beta x}\sin ax}{\sin bx}\, dx = \frac{1}{2bi}\left[\psi\left(\frac{a+b}{2b} - i\frac{\beta}{2b} \right) - \psi\left(\frac{b-a}{2b} - i\frac{\beta}{2b} \right) \right] \quad [\operatorname{Re}\beta > 0, \quad b \neq 0].$$ GW ((335))(15)

5. $$\int_0^{\infty} \frac{e^{-2px}\sin[(2n+1)x]}{\sin x}\, dx = \frac{1}{2p} + \sum_{k=1}^{n} \frac{p}{p^2+k^2} \quad [p > 0].$$ BI ((267))(15)

6. $$\int_0^{\infty} \frac{e^{-px}\sin 2nx}{\sin x}\, dx = 2p\sum_{k=0}^{n-1} \frac{1}{p^2+(2k+1)^2} \quad [p > 0].$$ GW ((335))(15c)

7. $$\int_0^{\infty} e^{-px}\cos[(2n+1)x] \operatorname{tg} x\, dx = \frac{2n+1}{p^2+(2n+1)^2} + (-1)^n 2\sum_{k=0}^{n-1} \frac{(-1)^k(2k+1)}{p^2+(2k+1)^2} \quad [p > 0].$$ LI ((267))(16)

3.894 $$\int_{-\pi}^{\pi} \left[\beta + \sqrt{\beta^2 - 1} \cos x\right]^{\nu} e^{inx}\, dx = \frac{2\pi\Gamma(\nu+1) P_{\nu}^{m}(\beta)}{\Gamma(\nu+m+1)}$$

$[\operatorname{Re}\beta > 0]$. ET I 157(15)

3.895

1. $$\int_0^{\infty} e^{-\beta x} \sin^{2m} x\, dx = \frac{(2m)!}{\beta(\beta^2+2^2)(\beta^2+4^2)\dots[\beta^2+(2m)^2]};$$

$[\operatorname{Re}\beta > 0]$. FI II 615, WA 620a

2. $$\int_0^{\pi} e^{-px} \sin^{2m} x\, dx = \frac{(2m)!\,(1-e^{-p\pi})}{p(p^2+2^2)(p^2+4^2)\dots[p^2+(2m)^2]}$$

$[p \neq 0]$. GW ((335))(4a)

3. $$\int_0^{\frac{\pi}{2}} e^{-px} \sin^{2m} x\, dx = \frac{(2m)!}{p(p^2+2^2)(p^2+4^2)\dots[p^2+(2m)^2]} \times$$

$$\times \left\{1 - e^{-\frac{p\pi}{2}} \left[1 + \frac{p^2}{2!} + \frac{p^2(p^2+2^2)}{4!} + \dots + \frac{p^2(p^2+2^2)\dots[p^2+(2m-2)^2]}{(2m)!}\right]\right\}$$

$[p \neq 0]$. BI ((270))(4)

4. $$\int_0^{\infty} e^{-\beta x} \sin^{2m+1} x\, dx = \frac{(2m+1)!}{(\beta^2+1^2)(\beta^2+3^2)\dots[\beta^2+(2m+1)^2]}$$

$[\operatorname{Re}\beta > 0]$. FI II 615, WA 620a

5. $$\int_0^{\pi} e^{-px} \sin^{2m+1} x\, dx = \frac{(2m+1)!\,(1+e^{-p\pi})}{(p^2+1^2)(p^2+3^2)\dots[p^2+(2m+1)^2]}$$

$[p \neq 0]$. GW ((335))(4b)

6. $$\int_0^{\frac{\pi}{2}} e^{-px} \sin^{2m+1} x\, dx = \frac{(2m+1)!}{(p^2+1^2)(p^2+3^2)\dots[p^2+(2m+1)^2]} \times$$

$$\times \left\{1 - pe^{\frac{p\pi}{2}} \left[1 + \frac{p^2+1^2}{3!} + \dots + \frac{(p^2+1^2)(p^2+3^2)\dots[p^2+(2m-1)^2]}{(2m+1)!}\right]\right\}$$

$[p \neq 0]$. BI ((270))(5)

7. $$\int_0^{\infty} e^{-px} \cos^{2m} x\, dx = \frac{(2m)!}{p(p^2+2^2)\dots[p^2+(2m)^2]} \times$$

$$\times \left\{1 + \frac{p^2}{2!} + \frac{p^2(p^2+2^2)}{4!} + \dots + \frac{p^2(p^2+2^2)\dots[p^2+(2m-2)^2]}{(2m)!}\right\}$$

$[p > 0]$. BI ((262))(3)

8. $$\int_0^{\frac{\pi}{2}} e^{-px} \cos^{2m} x\, dx = \frac{(2m)!}{p\,(p^2+2^2)\,\ldots\,[p^2+(2m)^2]} \times$$

$$\times \left\{ -e^{-p\frac{\pi}{2}} + 1 + \frac{p^2}{2!} + \frac{p^2\,(p^2+2^2)}{4!} + \ldots + \frac{p^2\,(p^2+2^2)\,\ldots\,[p^2+(2m-2)^2]}{(2m)!} \right\}$$

$[p \neq 0]$. BI ((270))(6)

9. $$\int_0^{\infty} e^{-px} \cos^{2m+1} x\, dx = \frac{(2m+1)!\,p}{(p^2+1^2)\,(p^2+3^2)\,\ldots\,[p^2+(2m+1)^2]} \times$$

$$\times \left\{ 1 + \frac{p^2+1^2}{3!} + \frac{(p^2+1^2)\,(p^2+3^2)}{5!} + \ldots + \frac{(p^2+1^2)\,(p^2+3^2)\,\ldots\,[p^2+(2m-1)^2]}{(2m+1)!} \right\}$$

$[p > 0]$. BI ((262))(4)

10. $$\int_0^{\frac{\pi}{2}} e^{-px} \cos^{2m+1} x\, dx = \frac{(2m+1)!}{(p^2+1^2)\,(p^2+3^2)\,\ldots\,[p^2+(2m+1)^2]} \times$$

$$\times \left\{ e^{-p\frac{\pi}{2}} + p\left[1 + \frac{p^2+1^2}{3!} + \ldots + \frac{(p^2+1)\,(p^2+3^2)\,\ldots\,[p^2+(2m-1)^2]}{(2m+1)!} \right] \right\}$$

$[p \neq 0]$. BI ((270))(7)

11. $$\int_0^{\infty} e^{-\beta x} \sin^{2n} x \sin ax\, dx =$$

$$= -\frac{1}{(-4)^{n+1}\,(2n+1)} \left\{ \frac{1}{\binom{\frac{a}{2}+i\frac{\beta}{2}+n}{2n+1}} + \frac{1}{\binom{\frac{a}{2}-i\frac{\beta}{2}+n}{2n+1}} \right\}$$

$[\operatorname{Re}\beta > 0,\ a > 0]$. ET I 80(19)

12. $$\int_0^{\infty} e^{-\beta x} \sin^{2n-1} x \sin ax\, dx =$$

$$= \frac{-i}{(-4)^{n+1} n} \left\{ \frac{1}{\binom{\frac{a}{2}-i\frac{\beta}{2}+n-\frac{1}{2}}{2n}} - \frac{1}{\binom{\frac{a}{2}+i\frac{\beta}{2}+n-\frac{1}{2}}{2n}} \right\}$$

$[\operatorname{Re}\beta > 0,\ a > 0]$. ET I 80(20)a

$$\int_0^{\infty} e^{-\beta x} \sin^{2n} x \cos ax\, dx = \frac{(-1)^n\, i}{(2n+1)\,2^{2n+2}} \left\{ \frac{1}{\binom{\frac{a}{2}+i\frac{\beta}{2}+n}{2n+1}} - \frac{1}{\binom{\frac{a}{2}-i\frac{\beta}{2}+n}{2n+1}} \right\}$$

$[\operatorname{Re}\beta > 0,\ a > 0]$. ET I 20(12)a

14. $$\int_0^{\infty} e^{-\beta x} \sin^{2n-1} x \cos ax\, dx =$$

$$= \frac{(-1)^n}{2^{2n+2} n} \left\{ \frac{1}{\binom{\frac{a}{2}-i\frac{\beta}{2}+n-\frac{1}{2}}{2n}} + \frac{1}{\binom{\frac{a}{2}+i\frac{\beta}{2}+n-\frac{1}{2}}{2n}} \right\}$$

$[\operatorname{Re}\beta > 0,\ a > 0]$. ET I 20(13)a

3.896

1. $$\int_{-\infty}^{\infty} e^{-q^2x^2} \sin [p(x+\lambda)]\, dx = \frac{\sqrt{\pi}}{q} e^{-\frac{p^2}{4q^2}} \sin p\lambda.$$ BI ((269))(2)

2. $$\int_{-\infty}^{\infty} e^{-q^2x^2} \cos [p(x+\lambda)]\, dx = \frac{\sqrt{\pi}}{q} e^{-\frac{p^2}{4q^2}} \cos p\lambda.$$ BI ((269))(3)

3. $$\int_0^{\infty} e^{-ax^2} \sin bx\, dx = \frac{b}{2a} \exp\left(-\frac{b^2}{4a}\right) {}_1F_1\left(\frac{1}{2}; \frac{3}{2}; \frac{b^2}{4a}\right) =$$
$$= \frac{b}{2a} {}_1F_1\left(1; \frac{3}{2}; -\frac{b^2}{4a}\right);$$ ET I 73(18)
$$= \frac{b}{2a} \sum_{k=1}^{\infty} \frac{1}{(2k-1)!!} \left(-\frac{b^2}{2a}\right)^{k-1} \quad [a > 0].$$ FI II 720

4. $$\int_0^{\infty} e^{-\beta x^2} \cos bx\, dx = \frac{1}{2} \sqrt{\frac{\pi}{\beta}} \exp\left(-\frac{b^2}{4\beta}\right) \quad [\operatorname{Re} \beta > 0].$$ BI ((263))(2)

3.897

1. $$\int_0^{\infty} e^{-\beta x^2 - \gamma x} \sin bx\, dx = -\frac{i}{4} \sqrt{\frac{\pi}{\beta}} \left\{\exp \frac{(\gamma - ib)^2}{4\beta} \left[1 - \Phi\left(\frac{\gamma - ib}{2\sqrt{\beta}}\right)\right] - \right.$$
$$\left. - \exp \frac{(\gamma + ib)^2}{4\beta} \left[1 - \Phi\left(\frac{\gamma + ib}{2\sqrt{\beta}}\right)\right]\right\} \quad [\operatorname{Re} \beta > 0,\ b > 0].$$ ET I 74(27)

2. $$\int_0^{\infty} e^{-\beta x^2 - \gamma x} \cos bx\, dx = \frac{1}{4} \sqrt{\frac{\pi}{\beta}} \left\{\exp \frac{(\gamma - ib)^2}{4\beta} \left[1 - \Phi\left(\frac{\gamma - ib}{2\sqrt{\beta}}\right)\right] + \right.$$
$$\left. + \exp \frac{(\gamma + ib)^2}{4\beta} \left[1 - \Phi\left(\frac{\gamma + ib}{2\sqrt{\beta}}\right)\right]\right\} \quad [\operatorname{Re} \beta > 0,\ b > 0].$$ ET I 15(16)

3.898

1. $$\int_0^{\infty} e^{-\beta x^2} \sin ax \sin bx\, dx = \frac{1}{4} \sqrt{\frac{\pi}{\beta}} \{e^{-\frac{(a-b)^2}{4\beta}} - e^{-\frac{(a+b)^2}{4\beta}}\}$$
$$[a > 0,\ b > 0,\ \operatorname{Re} \beta > 0].$$ BI ((263))(4)

2. $$\int_0^{\infty} e^{-\beta x^2} \cos ax \cos bx\, dx = \frac{1}{4} \sqrt{\frac{\pi}{\beta}} \{e^{-\frac{(a-b)^2}{4\beta}} + e^{-\frac{(a+b)^2}{4\beta}}\} \quad [\operatorname{Re} \beta > 0].$$ BI ((263))(5)

3. $$\int_0^{\infty} e^{-px^2} \sin^2 ax\, dx = \frac{1}{2} \sqrt{\frac{\pi}{p}} (1 - e^{-\frac{a^2}{p}}) \quad [p > 0].$$ BI ((263))(6)

3.899

1. $$\int_0^{\infty} \frac{e^{-p^2x^2} \sin [(2n+1)x]}{\sin x}\, dx = \frac{\sqrt{\pi}}{p} \left[\frac{1}{2} + \sum_{k=1}^{n} e^{-\left(\frac{k}{p}\right)^2}\right] \quad [p > 0].$$ BI ((267))(17)

2. $$\int_0^\infty \frac{e^{-p^2x^2}\sin[(4n+1)x]}{\cos x}\,dx = \frac{\sqrt{\pi}}{p}\left[\frac{1}{2}+\sum_{k=1}^{2n}(-1)^k e^{\left(\frac{k}{p}\right)^2}\right] \quad [p>0].$$

BI ((267))(18)

3. $$\int_0^\infty \frac{e^{-px^2}\,dx}{1-2a\cos x+a^2} = \frac{\sqrt{\frac{\pi}{p}}}{1-a^2}\left\{\frac{1}{2}+\sum_{k=1}^{\infty} a^k \exp\left(-\frac{k^2}{4p}\right)\right\} \quad [a^2<1,\ p>0];$$

BI ((266))(1)

$$= \frac{\sqrt{\frac{\pi}{p}}}{a^2-1}\left\{\frac{1}{2}+\sum_{k=1}^{\infty} a^{-k} \exp\left(-\frac{k^2}{4p}\right)\right\} \quad [a^2>1,\ p>0].$$

LI ((266))(1)

3.911

1. $$\int_0^\infty \frac{\sin ax}{e^{\beta x}+1}\,dx = \frac{1}{2a} - \frac{\pi}{2\beta \operatorname{sh}\frac{a\pi}{\beta}} \quad [a>0,\ \operatorname{Re}\beta>0].$$ BI ((264))(1)

2. $$\int_0^\infty \frac{\sin ax}{e^{\beta x}-1}\,dx = \frac{\pi}{2\beta}\operatorname{cth}\left(\frac{\pi a}{\beta}\right) - \frac{1}{2a} \quad [a>0,\ \operatorname{Re}\beta>0].$$

BI ((264))(2), WH

3. $$\int_0^\infty \frac{\sin ax}{e^x-1} e^{\frac{x}{2}}\,dx = -\frac{1}{2}\operatorname{th}(a\pi) \quad [a>0].$$ ET I 73(13)

4. $$\int_0^\infty \frac{\sin ax}{1-e^{-x}} e^{-nx}\,dx = \frac{\pi}{2} - \frac{1}{2a} + \frac{\pi}{e^{2\pi a}-1} - \sum_{k=1}^{n-1}\frac{a}{a^2+k^2} \quad [a>0].$$

BI ((264))(8)

5. $$\int_0^\infty \frac{\sin ax}{e^{\beta x}-e^{\gamma x}}\,dx = \frac{1}{2i(\beta-\gamma)}\left[\psi\left(\frac{\beta+ia}{\beta-\gamma}\right)-\psi\left(\frac{\beta-ia}{\beta-\gamma}\right)\right]$$

$[\operatorname{Re}\beta>0,\ \operatorname{Re}\gamma>0].$ GW ((335))(8)

6. $$\int_0^\infty \frac{\sin ax\,dx}{e^{\beta x}(e^{-x}-1)} = \frac{i}{2}[\psi(\beta+ia)-\psi(\beta-ia)] \quad [\operatorname{Re}\beta>-1].$$ ET 73(15)

3.912

1. $$\int_0^\infty e^{-\beta x}(1-e^{-\gamma x})^{\nu-1}\sin ax\,dx = -\frac{i}{2\gamma}\left[\mathrm{B}\left(\nu,\frac{\beta-ia}{\gamma}\right)-\mathrm{B}\left(\nu,\frac{\beta+ia}{\gamma}\right)\right]$$

$[\operatorname{Re}\beta>0,\quad \operatorname{Re}\gamma>0,\quad \operatorname{Re}\nu>0,\ a>0].$ ET I 73(17)

2. $$\int_0^\infty e^{-\beta x}(1-e^{-\gamma x})^{\nu-1}\cos ax\,dx = \frac{1}{2\gamma}\left[\mathrm{B}\left(\nu,\frac{\beta-ia}{\gamma}\right)+\mathrm{B}\left(\nu,\frac{\beta+ia}{\gamma}\right)\right]$$

$[\operatorname{Re}\beta>0,\ \operatorname{Re}\gamma>0,\ \operatorname{Re}\nu>0,\ a>0].$ ET I 15(10)

3.913

1. $$\int_{-\frac{\pi}{2}}^{\frac{\pi}{2}} e^{i\beta x} \cos^{\nu} x \, (\beta^2 e^{ix} + \nu^2 e^{-ix})^{\mu} \, dx =$$
$$= \frac{\pi {}_2F_1\left(-\mu, \frac{\beta}{2} - \frac{\nu}{2} - \frac{\mu}{2}; 1 + \frac{\beta}{2} + \frac{\nu}{2} - \frac{\mu}{2}; \frac{\beta^2}{\nu^2}\right)}{2^{\nu} (\nu+1) \, \mathrm{B}\left(1 + \frac{\beta}{2} + \frac{\nu}{2} - \frac{\mu}{2}, 1 - \frac{\beta}{2} + \frac{\nu}{2} + \frac{\mu}{2}\right)}$$
$[\operatorname{Re} \nu > -1, \ |\nu| > |\beta|]$. EH I 81(11)a

2. $$\int_{-\frac{\pi}{2}}^{\frac{\pi}{2}} e^{-iux} \cos^{\mu} x \, (a^2 e^{ix} + b^2 e^{-ix})^{\nu} \, dx =$$
$$= \frac{\pi b^{2\nu} {}_2F_1\left(-\nu, \frac{u+\mu+\nu}{2}; 1 + \frac{\mu-\nu-u}{2}, \frac{a^2}{b^2}\right)}{2^{\mu} (\mu+1) \, \mathrm{B}\left(1 - \frac{u+\nu-\mu}{2}, 1 + \frac{u+\mu+\nu}{2}\right)} \quad \text{for } a^2 < b^2;$$
$$= \frac{\pi a^{2\nu} {}_2F_1\left(-\nu, \frac{\mu+\nu-u}{2}; 1 + \frac{\mu-\nu+u}{2}; \frac{b^2}{a^2}\right)}{2^{\mu} (\mu+1) \, \mathrm{B}\left(1 + \frac{u+\mu-\nu}{2}, 1 + \frac{\mu+\nu-u}{2}\right)} \quad \text{for } b^2 < a^2$$
$[\operatorname{Re} \mu > -1]$. ET I 122(31)a

3.914 $$\int_0^{\infty} e^{-\beta \sqrt{\gamma^2 + x^2}} \cos bx \, dx = \frac{\beta\gamma}{\sqrt{\beta^2 + b^2}} K_1\left(\gamma \sqrt{\beta^2 + b^2}\right)$$
$[\operatorname{Re} \beta > 0, \ \operatorname{Re} \gamma > 0]$. ET I 16(26)

3.915

1 $$\int_0^{\pi} e^{a \cos x} \sin x \, dx = \frac{2}{a} \operatorname{sh} a.$$ GW ((337))(15c)

2. $$\int_0^{\pi} e^{i\beta \cos x} \cos nx \, dx = i^n \pi J_n(\beta).$$ EH II 81(2)

3. $$\int_{-\frac{\pi}{2}}^{\frac{\pi}{2}} e^{i\beta \cos x} \cos^{2\nu} x \, dx = \sqrt{\pi} \left(\frac{2}{\beta}\right)^{\nu} \Gamma\left(\nu + \frac{1}{2}\right) J_{\nu}(\beta) \qquad \left[\operatorname{Re} \nu > -\frac{1}{2}\right].$$ EH II 81(6)

4 $$\int_0^{\pi} e^{\pm\beta \cos x} \sin^{2\nu} x \, dx = \sqrt{\pi} \left(\frac{2}{\beta}\right)^{\nu} \Gamma\left(\nu + \frac{1}{2}\right) I_{\nu}(\beta) \qquad \left[\operatorname{Re} \nu > -\frac{1}{2}\right].$$ GW ((337))(15b)

5. $$\int_0^{\pi} e^{i\beta \cos x} \sin^{2\nu} x \, dx = \sqrt{\pi} \left(\frac{2}{\beta}\right)^{\nu} \Gamma\left(\nu + \frac{1}{2}\right) J_{\nu}(\beta) \qquad \left[\operatorname{Re} \nu > -\frac{1}{2}\right].$$ WA 34(2), WA 60(6)

3.916

1. $$\int_0^{\frac{\pi}{2}} e^{-p^2 \operatorname{tg} x} \frac{\sin \frac{x}{2} \sqrt{\cos x}}{\sin 2x} dx = \left[C(p) - \frac{1}{2} \right]^2 + \left[S(p) - \frac{1}{2} \right]^2.$$

NT 33(18)a

2. $$\int_0^{\frac{\pi}{2}} \frac{\exp(-p \operatorname{tg} x)\, dx}{\sin 2x + a \cos 2x + a} = -\frac{1}{2} e^{ap} \operatorname{Ei}(-ap) \quad [p > 0], \quad \text{(cf. 3.552 4. and 6.)}.$$

BI ((273))(11)

3. $$\int_0^{\frac{\pi}{2}} \frac{\exp(-p \operatorname{ctg} x)\, dx}{\sin 2x + a \cos 2x - a} = -\frac{1}{2} e^{-ap} \operatorname{Ei}(ap) \quad [p > 0], \quad \text{(cf. 3.552 4. and 6.)}.$$

BI ((273))(12)

4. $$\int_0^{\frac{\pi}{2}} \frac{\exp(-p \operatorname{tg} x) \sin 2x\, dx}{(1-a^2) - 2a^2 \cos 2x - (1+a^2) \cos^2 2x} = -\frac{1}{4} [e^{-ap} \operatorname{Ei}(ap) + e^{ap} \operatorname{Ei}(-ap)]$$

$[p > 0]$. BI ((273))(13)

5. $$\int_0^{\frac{\pi}{2}} \frac{\exp(-p \operatorname{ctg} x) \sin 2x\, dx}{(1-a^2) + 2a^2 \cos 2x - (1+a^2) \cos^2 2x} = -\frac{1}{4} [e^{-ap} \operatorname{Ei}(ap) + e^{ap} \operatorname{Ei}(-ap)]$$

$[p > 0]$. BI ((273))(14)

3.917

38 1. $$\int_0^{\frac{\pi}{2}} e^{-2\beta \operatorname{ctg} x} \cos^{\nu - \frac{1}{2}} x \sin^{-(\nu+1)} x \sin\left[\beta - \left(\nu - \frac{1}{2}\right) x\right] dx =$$

$$= \frac{\sqrt{\pi}}{2 \cdot (2\beta)^{\nu}} \Gamma\left(\nu + \frac{1}{2}\right) J_{\nu}(\beta) \quad \left[\operatorname{Re} \nu > -\frac{1}{2}\right].$$

WA 186(7)

2. $$\int_0^{\frac{\pi}{2}} e^{-2\beta \operatorname{ctg} x} \cos^{\nu - \frac{1}{2}} x \sin^{-(\nu+1)} x \cos\left[\beta - \left(\nu - \frac{1}{2}\right) x\right] dx =$$

$$= \frac{\sqrt{\pi}}{2 \cdot (2\beta)^{\nu}} \Gamma\left(\nu + \frac{1}{2}\right) N_{\nu}(\beta) \quad \left[\operatorname{Re} \nu > -\frac{1}{2}\right].$$

WA 186(8)

3.918

1. $$\int_0^{\frac{\pi}{2}} \frac{\cos^{\mu} x}{\sin^{2\mu+2} x} e^{i\gamma(\beta - \mu x) - 2 \cdot \beta \operatorname{ctg} x} dx = \frac{i\gamma}{2} \sqrt{\frac{\pi}{2\beta}} (2\beta)^{-\mu} \Gamma(\mu + 1) H^{(\varepsilon)}_{\mu + \frac{1}{2}}(\beta)$$

$[\varepsilon = 1, 2;\ \gamma = (-1)^{\varepsilon+1};\quad \operatorname{Re} \beta > 0,\ \operatorname{Re} \mu > -1]$. GW ((337))(16)

2. $$\int_0^{\frac{\pi}{2}} \frac{\cos^{\mu} x \sin(\beta - \mu x)}{\sin^{2\mu+2} x} e^{-2\beta \operatorname{ctg} x} dx = \frac{1}{2} \sqrt{\frac{\pi}{2\beta}} (2\beta)^{-\mu} \Gamma(\mu + 1) J_{\mu + \frac{1}{2}}(\beta)$$

$[\operatorname{Re} \beta > 0,\ \operatorname{Re} \mu > -1]$. WH

3. $$\int_0^{\frac{\pi}{2}} \frac{\cos^{\mu} x \cos(\beta - \mu x)}{\sin^{2\mu+2} x} e^{-2\beta \operatorname{ctg} x}\, dx = -\frac{1}{2}\sqrt{\frac{\pi}{2\beta}}(2\beta)^{-\mu}\Gamma(\mu+1) N_{\mu+\frac{1}{2}}(\beta)$$

$[\operatorname{Re}\beta > 0, \ \operatorname{Re}\mu > -1]$. GW ((337))(17b)

3.919

1. $$\int_0^{\frac{\pi}{2}} \frac{\sin 2nx}{\sin^{2n+2} x} \cdot \frac{dx}{\exp(2\pi \operatorname{ctg} x) - 1} = (-1)^{n-1} \frac{2n-1}{4(2n+1)}.$$ BI ((275))(6), LI ((275))(6)

2. $$\int_0^{\frac{\pi}{2}} \frac{\sin 2nx}{\sin^{2n+2} x} \frac{dx}{\exp(\pi \operatorname{ctg} x) - 1} = (-1)^{n-1} \frac{n}{2n+1}.$$ BI ((275))(7), LI ((275))(7)

3.92 Trigonometric functions of more complicated arguments combined with exponentials

3.921 $$\int_0^{\infty} e^{-\beta x} \cos ax^2 (\cos \gamma x - \sin \gamma x)\, dx = \sqrt{\frac{\pi}{8a}} \exp\left(-\frac{\gamma^2}{2a}\right)$$

$[\operatorname{Re}\gamma \geqslant |\operatorname{Im}\beta|]$. ET I 26(28)

3.922

1. $$\int_0^{\infty} e^{-\beta x^2} \sin ax^2\, dx = \frac{1}{2}\int_{-\infty}^{\infty} e^{-\beta x^2} \sin ax^2\, dx = \sqrt{\frac{\pi}{8}} \sqrt{\frac{\sqrt{\beta^2+a^2} - \beta}{\beta^2+a^2}} =$$

$$= \frac{\sqrt{\pi}}{2\sqrt[4]{\beta^2+a^2}} \sin\left(\frac{1}{2} \operatorname{arctg} \frac{a}{\beta}\right)$$ $[\operatorname{Re}\beta > 0, \ a > 0]$. FI II 750, BI ((263))(8)

2. $$\int_0^{\infty} e^{-\beta x^2} \cos ax^2\, dx = \frac{1}{2}\int_{-\infty}^{\infty} e^{-\beta x^2} \cos ax^2\, dx =$$

$$= \sqrt{\frac{\pi}{8}} \sqrt{\frac{\sqrt{\beta^2+a^2}+\beta}{\beta^2+a^2}} = \frac{\sqrt{\pi}}{2\sqrt[4]{\beta^2+a^2}} \cos\left(\frac{1}{2} \operatorname{arctg} \frac{a}{\beta}\right)$$

$[\operatorname{Re}\beta > 0, \ a > 0]$. FI II 750, BI ((263))(9)

3. $$\int_0^{\infty} e^{-\beta x^2} \sin ax^2 \cos bx\, dx = -\frac{1}{2}\sqrt{\frac{\pi}{\beta^2+a^2}} e^{-A\beta}(B \sin Aa - C \cos Aa) =$$

$$= \frac{\sqrt{\pi}}{2\sqrt[4]{\beta^2+a^2}} \exp\left(-\frac{\beta b^2}{4(\beta^2+a^2)}\right) \sin\left\{\frac{1}{2} \operatorname{arctg} \frac{a}{\beta} - \frac{ab^2}{4(\beta^2+a^2)}\right\}.$$

LI ((263))(10), GW ((337))(5)

4. $$\int_0^{\infty} e^{-\beta x^2} \cos ax^2 \cos bx\, dx = \frac{1}{2}\sqrt{\frac{\pi}{\beta^2+a^2}} e^{-A\beta}(B \cos Aa + C \sin Aa) =$$

$$= \frac{\sqrt{\pi}}{2\sqrt[4]{\beta^2+a^2}} \exp\left(-\frac{\beta b^2}{4(\beta^2+a^2)}\right) \cos\left\{\frac{1}{2} \operatorname{arctg} \frac{a}{\beta} - \frac{ab^2}{4(\beta^2+a^2)}\right\}.$$

LI ((263))(11), GW ((337))(5)

[In formulas **3.922** 3 and 4. $a > 0,\ b > 0,\ \operatorname{Re}\beta > 0,\quad A = \frac{b^2}{4(a^2+\beta^2)}$,
$B = \sqrt{\tfrac{1}{2}\left(\sqrt{\beta^2+a^2}+\beta\right)},\quad C = \sqrt{\tfrac{1}{2}\left(\sqrt{\beta^2+a^2}-\beta\right)}$.
If a is complex, $\operatorname{Re}\beta > |\operatorname{Im} a|$.]

3.923

1. $$\int_{-\infty}^{\infty} \exp[-(ax^2+2bx+c)] \sin(px^2+2qx+r)\,dx =$$
$$= \frac{\sqrt{\pi}}{\sqrt[4]{a^2+p^2}} \exp \frac{a(b^2-ac)-(aq^2-2bpq+cp^2)}{a^2+p^2} \times$$
$$\times \sin\left\{\frac{1}{2}\operatorname{arctg}\frac{p}{a} - \frac{p(q^2-pr)-(b^2p-2abq+a^2r)}{a^2+p^2}\right\} \qquad [a > 0].$$
GW ((337))(3), BI ((296))(6)

2. $$\int_{-\infty}^{\infty} \exp[-(ax^2+2bx+c)] \cos(px^2+2qx+r)\,dx =$$
$$= \frac{\sqrt{\pi}}{\sqrt[4]{a^2+p^2}} \exp \frac{a(b^2-ac)-(aq^2-2bpq+cp^2)}{a^2+p^2} \times$$
$$\times \cos\left\{\frac{1}{2}\operatorname{arctg}\frac{p}{a} - \frac{p(q^2-pr)-(b^2p-2abq+a^2r)}{a^2+p^2}\right\} \qquad [a > 0].$$
GW ((337))(3), BI ((269))(7)

3.924

1. $$\int_0^{\infty} e^{-\beta x^4} \sin bx^2\,dx = \frac{\pi}{4}\sqrt{\frac{b}{2\beta}} \exp\left(-\frac{b^2}{8\beta}\right) I_{\frac{1}{4}}\left(\frac{b^2}{8\beta}\right) \qquad [\operatorname{Re}\beta > 0,\ b > 0].$$
ET 73(22)

2. $$\int_0^{\infty} e^{-\beta x^4} \cos bx^2\,dx = \frac{\pi}{4}\sqrt{\frac{b}{2\beta}} \exp\left(-\frac{b^2}{8\beta}\right) I_{-\frac{1}{4}}\left(\frac{b^2}{8\beta}\right) \qquad [\operatorname{Re}\beta > 0,\ b > 0].$$
ET I 15(12)

3.925

1. $$\int_0^{\infty} e^{-\frac{p^2}{x^2}} \sin 2a^2x^2\,dx = \frac{1}{2}\int_{-\infty}^{\infty} e^{-\frac{p^2}{x^2}} \sin 2a^2x^2\,dx =$$
$$= \frac{\sqrt{\pi}}{4a} e^{-2ap}(\cos 2ap + \sin 2ap) \qquad [a > 0,\ b > 0].$$
BI ((268))(12)

2. $$\int_0^{\infty} e^{-\frac{p^2}{x^2}} \cos 2a^2x^2\,dx = \frac{1}{2}\int_{-\infty}^{\infty} e^{-\frac{p^2}{x^2}} \cos 2a^2x^2\,dx =$$
$$= \frac{\sqrt{\pi}}{4a} e^{-2ap}(\cos 2ap - \sin 2ap) \qquad [a > 0,\ b > 0].$$
BI ((268))(13)

3.926

1. $$\int_0^{\infty} e^{-\left(\beta x^2+\frac{\gamma}{x^2}\right)} \sin ax^2\,dx = \frac{1}{2}\sqrt{\frac{\pi}{a^2+\beta^2}} e^{-2u\sqrt{\gamma}} \times$$
$$\times \left[v\cos\left(2v\sqrt{\gamma}\right) + u\sin\left(2v\sqrt{\gamma}\right)\right] \qquad [\operatorname{Re}\beta > 0,\ \operatorname{Re}\gamma > 0].$$
BI ((268))(14)

2. $$\int_0^\infty e^{-\left(\beta x^2+\frac{\gamma}{x^2}\right)} \cos ax^2\, dx = \frac{1}{2}\sqrt{\frac{\pi}{a^2+\beta^2}}\, e^{-2u\sqrt{\gamma}} \times$$
$$\times \left[u \cos\left(2v\sqrt{\gamma}\right) - v \sin\left(2v\sqrt{\gamma}\right)\right] \quad [\operatorname{Re}\beta > 0,\ \operatorname{Re}\gamma > 0].$$ BI ((268))(15)

[In formulas 3.926 1., 3.926 2.
$$u = \sqrt{\frac{\sqrt{a^2+\beta^2}+\beta}{2}}, \quad v = \sqrt{\frac{\sqrt{a^2+\beta^2}-\beta}{2}}.\ ]$$

3.927 $$\int_0^\infty e^{-\frac{p}{x}} \sin^2 \frac{a}{x}\, dx = a \operatorname{arctg} \frac{2a}{p} + \frac{p}{4} \ln \frac{p^2}{p^2+4a^2} \quad [a > 0,\ p > 0].$$ LI ((268))(4)

3.928

1. $$\int_0^\infty \exp\left[-\left(p^2x^2+\frac{q^2}{x^2}\right)\right] \sin\left(a^2x^2+\frac{b^2}{x^2}\right) dx =$$
$$= \frac{\sqrt{\pi}}{2r} e^{-2rs\cos(A+B)} \sin\{A+2rs\sin(A+B)\}.$$ BI ((268))(22)

2. $$\int_0^\infty \exp\left[-\left(p^2x^2+\frac{q^2}{x^2}\right)\right] \cos\left(a^2x^2+\frac{b^2}{x^2}\right) dx =$$
$$= \frac{\sqrt{\pi}}{2r} e^{-2rs\cos(A+B)} \cos\{A+2rs\sin(A+B)\}.$$ BI ((268))(23)

[In formulas 3.928 1., 3.928 2. $a^2+p^2>0$ and
$$r = \sqrt[4]{a^4+p^4}, \quad s = \sqrt[4]{b^4+q^4}, \quad A = \frac{1}{2}\operatorname{arctg}\frac{a^2}{p^2}, \quad B = \frac{1}{2}\operatorname{arctg}\frac{b^2}{q^2}.\ ]$$

3.929 $$\int_0^\infty \left[e^{-x}\cos\left(p\sqrt{x}\right) + pe^{-x^2}\sin px\right] dx = 1.$$ LI ((268))(3)

3.93 Trigonometric and exponential functions of trigonometric functions

3.931

1. $$\int_0^{\frac{\pi}{2}} e^{-p\cos x} \sin(p\sin x)\, dx = \operatorname{Ei}(-p) - \operatorname{ci}(p).$$ NT 13(27)

2. $$\int_0^{\pi} e^{-p\cos x} \sin(p\sin x)\, dx = -\int_{-\pi}^{0} e^{-p\cos x} \sin(p\sin x)\, dx = -2\operatorname{shi}(p).$$ GW ((337))(11b)

3. $$\int_0^{\frac{\pi}{2}} e^{-p\cos x} \cos(p\sin x)\, dx = -\operatorname{si}(p).$$ NT 13(26)

4. $$\int_0^{\pi} e^{-p\cos x}\cos(p\sin x)\,dx = \frac{1}{2}\int_0^{2\pi} e^{-p\cos x}\cos(p\sin x)\,dx = \pi.$$

GW ((337))(11a)

3.932

1. $$\int_0^{\pi} e^{p\cos x}\sin(p\sin x)\sin mx\,dx =$$
$$= \frac{1}{2}\int_0^{2\pi} e^{p\cos x}\sin(p\sin x)\sin mx\,dx = \frac{\pi}{2}\cdot\frac{p^m}{m!}.$$

BI ((277))(7), GW((337))(13a)

2. $$\int_0^{\pi} e^{p\cos x}\cos(p\sin x)\cos mx\,dx = \frac{1}{2}\int_0^{2\pi} e^{p\cos x}\cos(p\sin x)\cos mx\,dx =$$
$$= \frac{\pi}{2}\cdot\frac{p^m}{m!}.$$

BI ((227))(8), GW((337))(13b)

3.933 $$\int_0^{\pi} e^{p\cos x}\sin(p\sin x)\operatorname{cosec} x\,dx = \pi \operatorname{sh} p.$$ BI ((278))(1)

3.934

1. $$\int_0^{\pi} e^{p\cos x}\sin(p\sin x)\operatorname{tg}\frac{x}{2}\,dx = \pi(1-e^p).$$ BI ((271))(8)

2. $$\int_0^{\pi} e^{p\cos x}\sin(p\sin x)\operatorname{ctg}\frac{x}{2}\,dx = \pi(e^p-1).$$ BI ((272))(5)

3.935 $$\int_0^{\pi} e^{p\cos x}\cos(p\sin x)\frac{\sin 2nx}{\sin x}\,dx = \pi\sum_{k=0}^{n-1}\frac{p^{2k+1}}{(2k+1)!} \quad [p>0].$$ LI ((278))(3)

3.936

1 $$\int_0^{2\pi} e^{p\cos x}\cos(p\sin x - mx)\,dx = 2\int_0^{\pi} e^{p\cos x}\cos(p\sin x - mx)\,dx = \frac{2\pi p^m}{m!}.$$

BI ((277))(9), GW ((337))(14a)

2 $$\int_0^{2\pi} e^{p\sin x}\sin(p\cos x + mx)\,dx = \frac{2\pi p^m}{m!}\sin\frac{m\pi}{2} \quad [p>0].$$ GW ((337))(14b)

3 $$\int_0^{2\pi} e^{p\sin x}\cos(p\cos x + mx)\,dx = \frac{2\pi p^m}{m!}\cos\frac{m\pi}{2} \quad [p>0].$$ GW ((337))(14b)

4. $$\int_0^{2\pi} e^{\cos x}\sin(mx - \sin x)\,dx = 0.$$ WH

5. $\int_0^\pi e^{\beta\cos x}\cos(ax+\beta\sin x)\,dx=\beta^{-a}\sin(a\pi)\,\gamma(a,\beta).$ EH II 137(2)

3.937

1. $\int_0^{2\pi}\exp(p\cos x+q\sin x)\sin(a\cos x+b\sin x-mx)\,dx=$

$$=i\pi\left[(b-p)^2+(a+q)^2\right]^{-\frac{m}{2}}\left\{(A+iB)^{\frac{m}{2}}I_m\left(\sqrt{C-iD}\right)-\right.$$
$$\left.-(A-iB)^{\frac{m}{2}}I_m\left(\sqrt{C+iD}\right)\right\}.$$ GW ((337))(9b)

2. $\int_0^{2\pi}\exp(p\cos x+q\sin x)\cos(a\cos x+b\sin x-mx)\,dx=$

$$=\pi\left[(b-p)^2+(a+q)^2\right]^{-\frac{m}{2}}\left\{(A+iB)^{\frac{m}{2}}I_m\left(\sqrt{C-iD}\right)+\right.$$
$$\left.+(A-iB)^{\frac{m}{2}}I_m\left(\sqrt{C+iD}\right)\right\}.$$

[In formulas **3.937** 1. and **3.937** 2. $(b-p)^2+(a+q)^2>0$, $m=0, 1, 2, \ldots$, $A=p^2-q^2+a^2-b^2$, $B=2(pq+ab)$, $C=p^2+q^2-a^2-b^2$, $D=-2(ap+bq)$.] GW ((337))(9a)

3. $\int_0^{2\pi}\exp(p\cos x+q\sin x)\sin(q\cos x-p\sin x+mx)\,dx=$

$$=\frac{2\pi}{m!}(p^2+q^2)^{\frac{m}{2}}\sin\left(m\operatorname{arctg}\frac{q}{p}\right).$$ GW ((337))(12)

4. $\int_0^{2\pi}\exp(p\cos x+q\sin x)\cos(q\cos x-p\sin x+mx)\,dx=$

$$=\frac{2\pi}{m!}(p^2+q^2)^{\frac{m}{2}}\cos\left(m\operatorname{arctg}\frac{q}{p}\right).$$ GW ((337))(12)

3.938

1. $\int_0^\pi e^{r(\cos px+\cos qx)}\sin(r\sin px)\sin(r\sin qx)\,dx=$

$$=\frac{\pi}{2}\sum_{k=1}^\infty\frac{1}{\Gamma(pk+1)\Gamma(qk+1)}r^{(p+q)k}.$$ BI ((277))(14)

2. $\int_0^\pi e^{r(\cos px+\cos qx)}\cos(r\sin px)\cos(r\sin qx)\,dx=$

$$=\frac{\pi}{2}\left(2+\sum_{k=1}^\infty\frac{r^{(p+q)k}}{\Gamma(pk+1)\Gamma(qk+1)}\right).$$ BI ((277))(15)

3.939

1. $\int_0^\pi e^{q\cos x}\frac{\sin rx}{1-2p^r\cos rx+p^{2r}}\sin(q\sin x)\,dx=\frac{\pi}{2pr}\sum_{k=1}^\infty\frac{(pq)^{kr}}{\Gamma(kr+1)}\quad[p^r<1].$

BI ((278))(15)

2. $$\int_0^{\pi} e^{q\cos x}\frac{1-p^r\cos rx}{1-2p^r\cos rx+p^{2r}}\cos(q\sin x)\,dx=\frac{\pi}{2}\left[2+\sum_{k=1}^{\infty}\frac{(pq)^{kr}}{\Gamma(kr+1)}\right]$$
$[p^2<1]$. BI ((278))(16)

3. $$\int_0^{\frac{\pi}{2}}\frac{e^{p\cos 2x}\cos(p\sin 2x)\,dx}{\cos^2 x+q^2\sin^2 x}=\frac{\pi}{2q}\exp\left(p\frac{q-1}{q+1}\right).$$ BI ((273))(8)

3.94-3.97 Combinations involving trigonometric functions, exponentials, and powers

3.941

1. $$\int_0^{\infty} e^{-px}\sin qx\frac{dx}{x}=\operatorname{arctg}\frac{q}{p}\qquad [p>0].$$ BI ((365))(1)

2. $$\int_0^{\infty} e^{-px}\cos qx\frac{dx}{x}=\infty.$$ BI ((365))(2)

3.942

1. $$\int_0^{\infty} e^{-px}\cos px\frac{x\,dx}{b^4+x^4}=\frac{\pi}{4b^2}\exp\left(-bp\sqrt{2}\right)\quad [p>0,\ b>0].$$ BI ((386))(6)a

2. $$\int_0^{\infty} e^{-px}\cos px\frac{x\,dx}{b^4-x^4}=\frac{\pi}{4b^2}e^{-bp}\sin bp\quad [p>0,\ b>0].$$ BI ((386))(7)a

3.943 $$\int_0^{\infty} e^{-\beta x}(1-\cos ax)\frac{dx}{x}=\frac{1}{2}\ln\frac{a^2+\beta^2}{\beta^2}\qquad [\operatorname{Re}\beta>0].$$ BI ((367))(6)

3.944

1. $$\int_0^{u} x^{\mu-1}e^{-\beta x}\sin\delta x\,dx=\frac{i}{2}(\beta+i\delta)^{-\mu}\gamma[\mu,\ (\beta+i\delta)u]-$$
$$-\frac{i}{2}(\beta-i\delta)^{-\mu}\gamma[\mu,\ (\beta-i\delta)u]\quad [\operatorname{Re}\mu>-1].$$ ET I 318(8)

2. $$\int_u^{\infty} x^{\mu-1}e^{-\beta x}\sin\delta x\,dx=\frac{i}{2}(\beta+i\delta)^{-\mu}\Gamma[\mu,\ (\beta+i\delta)u]-$$
$$-\frac{i}{2}(\beta-i\delta)^{-\mu}\Gamma[\mu,\ (\beta-i\delta)u]\quad [\operatorname{Re}\beta>|\operatorname{Im}\delta|].$$ ET I 318(9)

3. $$\int_0^{u} x^{\mu-1}e^{-\beta x}\cos\delta x\,dx=\frac{1}{2}(\beta+i\delta)^{-\mu}\gamma[\mu,\ (\beta+i\delta)u]+$$
$$+\frac{1}{2}(\beta-i\delta)^{-\mu}\gamma[\mu,\ (\beta-i\delta)u]\quad [\operatorname{Re}\mu>0].$$ ET I 320(28)

4. $$\int_u^{\infty} x^{\mu-1}e^{-\beta x}\cos\delta x\,dx=\frac{1}{2}(\beta+i\delta)^{-\mu}\Gamma[\mu,\ (\beta+i\delta)u]+$$
$$+\frac{1}{2}(\beta-i\delta)^{-\mu}\Gamma[\mu,\ (\beta-i\delta)u]\quad [\operatorname{Re}\beta>|\operatorname{Im}\delta|].$$ ET I 320(29)

5. $$\int_0^\infty x^{\mu-1}e^{-\beta x}\sin\delta x\,dx=\frac{\Gamma(\mu)}{(\beta^2+\delta^2)^{\frac{\mu}{2}}}\sin\left(\mu\operatorname{arctg}\frac{\delta}{\beta}\right)$$

$[\operatorname{Re}\mu>-1,\ \operatorname{Re}\beta>|\operatorname{Im}\delta|]$. FI II 812, BI((361))(9)

6. $$\int_0^\infty x^{\mu-1}e^{-\beta x}\cos\delta x\,dx=\frac{\Gamma(\mu)}{(\delta^2+\beta^2)^{\frac{\mu}{2}}}\cos\left(\mu\operatorname{arctg}\frac{\delta}{\beta}\right)$$

$[\operatorname{Re}\mu>0,\ \operatorname{Re}\beta>|\operatorname{Im}\delta|]$. FI II 812, BI((361))(10)

7. $$\int_0^\infty x^{\mu-1}\exp(-ax\cos t)\sin(ax\sin t)\,dx=\Gamma(\mu)a^{-\mu}\sin(\mu t)$$

$\left[\operatorname{Re}\mu>-1,\ a>0,\ |t|<\frac{\pi}{2}\right]$. EH I 13(36)

8. $$\int_0^\infty x^{\mu-1}\exp(-ax\cos t)\cos(ax\sin t)\,dx=\Gamma(\mu)a^{-\mu}\cos(\mu t)$$

$\left[\operatorname{Re}\mu>-1,\ a>0,\ |t|<\frac{\pi}{2}\right]$. EH I 13(35)

9. $$\int_0^\infty x^{p-1}e^{-qx}\sin(qx\operatorname{tg}t)\,dx=\frac{1}{q^p}\Gamma(p)\cos^p t\sin pt\left[|t|<\frac{\pi}{2},\ q>0\right].$$

LO V 288(16)

10. $$\int_0^\infty x^{p-1}e^{-qx}\cos(qx\operatorname{tg}t)\,dx=\frac{1}{q^p}\Gamma(p)\cos^p(t)\cos pt\quad\left[|t|<\frac{\pi}{2},\ q>0\right].$$

LO V 288(15)

11. $$\int_0^\infty x^n e^{-\beta x}\sin bx\,dx=n!\left(\frac{\beta}{\beta^2+b^2}\right)^{n+1}\sum_{0\leqslant 2k\leqslant n}(-1)^k\binom{n+1}{2k+1}\left(\frac{b}{\beta}\right)^{2k+1}=$$

$$=(-1)^n\frac{\partial^n}{\partial\beta^n}\left(\frac{b}{b^2+\beta^2}\right)\ [\operatorname{Re}\beta>0,\ b>0].$$ GW ((336))(3), ET I 72(3)

12. $$\int_0^\infty x^n e^{-\beta x}\cos bx\,dx=n!\left(\frac{\beta}{\beta^2+b^2}\right)^{n+1}\sum_{0\leqslant 2k\leqslant n+1}(-1)^k\binom{n+1}{2k}\left(\frac{b}{\beta}\right)^{2k}=$$

$$=(-1)^n\frac{\partial^n}{\partial\beta^n}\left(\frac{\beta}{b^2+\beta^2}\right)\quad[\operatorname{Re}\beta>0,\ b>0].$$ GW ((336))(4), ET I 14(5)

13. $$\int_0^\infty x^{n-\frac{1}{2}}e^{-\beta x}\sin bx\,dx=(-1)^n\sqrt{\frac{\pi}{2}}\frac{d^n}{d\beta^n}\frac{\sqrt{\sqrt{\beta^2+b^2}-\beta}}{\sqrt{\beta^2+b^2}}$$

$[\operatorname{Re}\beta>0,\ b>0]$. ET I 72(6)

14. $$\int_0^\infty x^{n-\frac{1}{2}}e^{-\beta x}\cos bx\,dx=(-1)^n\sqrt{\frac{\pi}{2}}\frac{d^n}{d\beta^n}\frac{\sqrt{\sqrt{\beta^2+b^2}+\beta}}{\sqrt{\beta^2+b^2}}$$

$[\operatorname{Re}\beta>0,\ b>0]$. ET I 15(6)

3.945

1. $$\int_0^\infty (e^{-\beta x}\sin ax - e^{-\gamma x}\sin bx)\frac{dx}{x^r} =$$

$$= \Gamma(1-r)\left\{(b^2+\gamma^2)^{\frac{r-1}{2}}\sin\left[(r-1)\operatorname{arctg}\frac{b}{\gamma}\right] - \right.$$

$$\left. -(a^2+\beta^2)^{\frac{r-1}{2}}\sin\left[(r-1)\operatorname{arctg}\frac{a}{\beta}\right]\right\}$$

$[\operatorname{Re}\beta > 0,\ \operatorname{Re}\gamma > 0,\ r < 2,\ r \neq 1]$. BI ((371))(6)

2. $$\int_0^\infty (e^{-\beta x}\cos ax - e^{-\gamma x}\cos bx)\frac{dx}{x^r} =$$

$$= \Gamma(1-r)\left\{(a^2+\beta^2)^{\frac{r-1}{2}}\cos\left[(r-1)\operatorname{arctg}\frac{a}{\beta}\right] - (b^2+\gamma^2)^{\frac{r-1}{2}} \times \right.$$

$$\left. \times \cos\left[(r-1)\operatorname{arctg}\frac{b}{\gamma}\right]\right\}$$ $[\operatorname{Re}\beta > 0,\ \operatorname{Re}\gamma > 0,\ r < 2,\ r \neq 1]$.

BI ((371))(7)

3. $$\int_0^\infty (ae^{-\beta x}\sin bx - be^{-\gamma x}\sin ax)\frac{dx}{x^2} =$$

$$= ab\left[\frac{1}{2}\ln\frac{a^2+\gamma^2}{b^2+\beta^2} + \frac{\gamma}{a}\operatorname{arcctg}\frac{\gamma}{a} - \frac{\beta}{b}\operatorname{arcctg}\frac{\beta}{b}\right]$$

$[\operatorname{Re}\beta > 0,\ \operatorname{Re}\gamma > 0]$. BI ((368))(22)

3 946

1. $$\int_0^\infty e^{-px}\sin^{2m+1} ax\,\frac{dx}{x} = \frac{(-1)^m}{2^{2m}}\sum_{k=0}^{m}(-1)^k\binom{2m+1}{k}\operatorname{arctg}\frac{(2m-2k+1)a}{p}$$

$[p > 0]$. GW ((336))(9a)

2. $$\int_0^\infty e^{-px}\sin^{2m} ax\,\frac{dx}{x} = \frac{(-1)^{m+1}}{2^{2m}}\sum_{k=0}^{m-1}(-1)^k\binom{2m}{k}\ln\left[p^2+(2m-2k)^2a^2\right] -$$

$$-\frac{1}{2^{2m}}\binom{2m}{m}\ln p$$ $[p > 0]$. GW ((336))(9b)

3.947

1. $$\int_0^\infty e^{-\beta x}\sin\gamma x\sin ax\,\frac{dx}{x} = \frac{1}{4}\ln\frac{\beta^2+(a+\gamma)^2}{\beta^2+(a-\gamma)^2}$$

$[\operatorname{Re}\beta > |\operatorname{Im}\gamma|,\ a > 0]$. BI ((365))(5)

2. $$\int_0^\infty e^{-px}\sin ax\sin bx\,\frac{dx}{x^2} = \frac{a}{2}\operatorname{arctg}\frac{2pb}{p^2+a^2-b^2} + \frac{b}{2}\operatorname{arctg}\frac{2pa}{p^2+b^2-a^2} +$$

$$+\frac{p}{4}\ln\frac{p^2+(a-b)^2}{p^2+(a+b)^2}$$ $[p > 0]$. BI ((368))(1), FI II 744

3. $\int_0^\infty e^{-px} \sin ax \cos bx \frac{dx}{x} = \frac{1}{2} \operatorname{arctg} \frac{2pa}{p^2-a^2+b^2} + s\frac{\pi}{2}$

$[a \geqslant 0,\ p > 0,\ s = 0 \text{ for } p^2-a^2+b^2 \geqslant 0 \text{ and } s = 1 \text{ for } p^2-a^2+b^2 < 0].$

GW ((336))(10b)

3.948

1. $\int_0^\infty e^{-\beta x} (\sin ax - \sin bx) \frac{dx}{x} = \operatorname{arctg} \frac{(a-b)\beta}{ab+\beta^2}$

$[\operatorname{Re} \beta > 0]$, (cf. **3.951** 2.). BI ((367))(7)

2. $\int_0^\infty e^{-\beta x} (\cos ax - \cos bx) \frac{dx}{x} = \frac{1}{2} \ln \frac{b^2+\beta^2}{a^2+\beta^2}$

$[\operatorname{Re} \beta > 0]$, (cf. **3.951** 3.). BI ((367))(8), FI II 748a

3. $\int_0^\infty e^{-\beta x} (\cos ax - \cos bx) \frac{dx}{x^2} = \frac{\beta}{2} \ln \frac{a^2+\beta^2}{b^2+\beta^2} +$

$+ b \operatorname{arctg} \frac{b}{\beta} - a \operatorname{arctg} \frac{a}{\beta} \qquad [\operatorname{Re} p > 0].$ BI ((368))(20)

4. $\int_0^\infty e^{-px} (\sin^2 ax - \sin^2 bx) \frac{dx}{x^2} = a \operatorname{arctg} \frac{2a}{p} -$

$- b \operatorname{arctg} \frac{2b}{p} - \frac{p}{4} \ln \frac{p^2+4a^2}{p^2+4b^2} \qquad [p > 0].$ BI ((368))(25)

5. $\int_0^\infty e^{-px} (\cos^2 ax - \cos^2 bx) \frac{dx}{x^2} = -a \operatorname{arctg} \frac{2a}{p} +$

$+ b \operatorname{arctg} \frac{2b}{p} + \frac{p}{4} \ln \frac{p^2+4a^2}{p^2+4b^2} \qquad [p > 0].$ BI ((368))(26)

3.949

1. $\int_0^\infty e^{-px} \sin ax \sin bx \sin cx \frac{dx}{x} = -\frac{1}{4} \operatorname{arctg} \frac{a+b+c}{p} +$

$+ \frac{1}{4} \operatorname{arctg} \frac{a+b-c}{p} + \frac{1}{4} \operatorname{arctg} \frac{a-b+c}{p} + \frac{1}{4} \operatorname{arctg} \frac{-a+b+c}{p} \qquad [p > 0].$

BI ((365))(11)

2. $\int_0^\infty e^{-px} \sin^2 ax \sin bx \frac{dx}{x} = \frac{1}{2} \operatorname{arctg} \frac{b}{p} -$

$- \frac{1}{4} \operatorname{arctg} \frac{2pb}{p^2+a^2-b^2} \qquad [p > 0].$ BI ((365))(8)

3. $\int_0^\infty e^{-px} \sin^2 ax \cos bx \frac{dx}{x} = \frac{1}{8} \ln \frac{[p^2+(2a+b)^2][p^2+(2a-b)^2]}{(p^2+b^2)^2}$

$[p > 0].$ BI ((365))(9)

4. $$\int_0^\infty e^{-px} \sin ax \cos^2 bx \frac{dx}{x} = \frac{1}{2} \operatorname{arctg} \frac{a}{p} + \frac{1}{4} \operatorname{arctg} \frac{2pa}{p^2+b^2-a^2}$$
$[p > 0]$. BI ((365))(10)

5. $$\int_0^\infty e^{-px} \sin^2 ax \sin bx \sin cx \frac{dx}{x} = \frac{1}{8} \ln \frac{p^2+(b+c)^2}{p^2+(b-c)^2} + \frac{1}{16} \ln \frac{[p^2+(2a-b+c)^2][p^2+(2a+b-c)^2]}{[p^2+(2a+b+c)^2][p^2+(2a-b-c)^2]}$$
$[p > 0]$. BI ((365))(15)

3.951

1. $$\int_0^\infty (1-e^{-x}) \cos x \frac{dx}{x} = \ln \sqrt{2}.$$
FI II 745

2. $$\int_0^\infty \frac{e^{-\gamma x} - e^{-\beta x}}{x} \sin bx \, dx = \operatorname{arctg} \frac{(\beta-\gamma) b}{b^2+\beta\gamma}$$
$[\operatorname{Re} \beta > 0, \ \operatorname{Re} \gamma \geqslant 0]$. BI ((367))(3)

3. $$\int_0^\infty \frac{e^{-\gamma x} - e^{-\beta x}}{x} \cos bx \, dx = \frac{1}{2} \ln \frac{b^2+\beta^2}{b^2+\gamma^2}$$
$[\operatorname{Re} \beta > 0, \ \operatorname{Re} \gamma \geqslant 0]$. BI ((367))(4)

4. $$\int_0^\infty \frac{e^{-\gamma x} - e^{-\beta x}}{x^2} \sin bx \, dx = \frac{b}{2} \ln \frac{b^2+\beta^2}{b^2+\gamma^2} + \beta \operatorname{arctg} \frac{b}{\beta} - \gamma \operatorname{arctg} \frac{b}{\gamma}$$
$[\operatorname{Re} \beta > 0, \ \operatorname{Re} \gamma > 0]$. BI ((368))(21)a

5. $$\int_0^\infty \frac{x}{e^{\beta x}-1} \cos bx \, dx = \frac{1}{2b^2} - \frac{\pi^2}{2\beta^2} \operatorname{cosech}^2 \frac{b\pi}{\beta}$$
$[\operatorname{Re} \beta > 0]$. ET I 15(18)

6. $$\int_0^\infty \left(\frac{1}{e^x-1} - \frac{1}{x} \right) \cos bx \, dx = \ln b - \frac{1}{2} [\psi(ib) + \psi(-ib)]$$
$[b > 0]$. ET I 15(9)

7. $$\int_0^\infty \frac{1-\cos ax}{e^{2\pi x}-1} \cdot \frac{dx}{x} = \frac{a}{4} + \frac{1}{2} \ln \frac{1-e^{-a}}{a}$$
$[a > 0]$. BI ((387))(10)

8. $$\int_0^\infty (e^{-\beta x} - e^{-\gamma x} \cos ax) \frac{dx}{x} = \frac{1}{2} \ln \frac{a^2+\gamma^2}{\beta^2}$$
$[\operatorname{Re} \beta > 0, \ \operatorname{Re} \gamma > 0]$. BI ((367))(10)

9. $$\int_0^\infty \frac{\cos px - e^{-px}}{b^4+x^4} \frac{dx}{x} = \frac{\pi}{2b^4} \exp\left(-\frac{1}{2} bp\sqrt{2} \right) \sin \left(\frac{1}{2} bp\sqrt{2} \right)$$
$[p > 0]$. BI ((390))(6)

10. $\int_0^\infty \left(\frac{1}{e^x-1}-\frac{\cos x}{x}\right) dx = \mathbf{C}.$ NT 65(8)

11. $\int_0^\infty \left(ae^{-px}-\frac{e^{-qx}}{x}\sin ax\right)\frac{dx}{x}=\frac{a}{2}\ln\frac{a^2+q^2}{p^2}+q\operatorname{arctg}\frac{a}{q}-a$

$[p>0,\ q>0].$ BI ((368))(24)

12. $\int_0^\infty \frac{x^{2m}\sin bx}{e^x-1}dx=(-1)^m\frac{\partial^{2m}}{\partial b^{2m}}\left[\frac{\pi}{2}\operatorname{cth} b\pi-\frac{1}{2b}\right]\quad [b>0].$ GW ((336))(15a)

13. $\int_0^\infty \frac{x^{2m+1}\cos bx}{e^x-1}dx=(-1)^m\frac{\partial^{2m+1}}{\partial b^{2m+1}}\left[\frac{\pi}{2}\operatorname{cth} b\pi-\frac{1}{2b}\right]\quad [b>0].$

GW ((336))(15b)

14. $\int_0^\infty \frac{x^{2m}\sin bx\,dx}{e^{(2n+1)cx}-e^{(2n-1)cx}}=(-1)^m\frac{\partial^{2m}}{\partial b^{2m}}\left[\frac{\pi}{4c}\operatorname{th}\frac{b\pi}{2c}-\right.$

$\left.-\sum_{k=1}^{n}\frac{b}{b^2+(2k-1)^2c^2}\right]^{*}\quad [b>0].$ GW ((336))(14a)

15. $\int_0^\infty \frac{x^{2m+1}\cos bx\,dx}{e^{(2n+1)cx}-e^{(2n-1)cx}}=(-1)^m\frac{\partial^{2m+1}}{\partial b^{2m+1}}\left[\frac{\pi}{4c}\operatorname{th}\frac{b\pi}{2c}-\right.$

$\left.-\sum_{k=1}^{n}\frac{b}{b^2+(2k-1)^2c^2}\right]^{*}\quad [b>0].$ GW ((336))(14b)

16. $\int_0^\infty \frac{x^{2m}\sin bx\,dx}{e^{2ncx}-e^{(2n-2)cx}}=(-1)^m\frac{\partial^{2m}}{\partial b^{2m}}\left[\frac{\pi}{4c}\operatorname{cth}\frac{b\pi}{2c}-\right.$

$\left.-\frac{1}{2b}-\sum_{k=1}^{n-1}\frac{b}{b^2+(2k)^2c^2}\right]^{**}\quad [b>0,\ c>0].$ GW ((336))(14c)

17. $\int_0^\infty \frac{x^{2m+1}\cos bx\,dx}{e^{2ncx}-e^{(2n-2)cx}}=(-1)^m\frac{\partial^{2m+1}}{\partial b^{2m+1}}\left[\frac{\pi}{4c}\operatorname{cth}\frac{b\pi}{2c}-\right.$

$\left.-\frac{1}{2b}-\sum_{k=1}^{n-1}\frac{b}{b^2+(2k)^2c^2}\right]^{**}\quad [b>0,\ c>0].$ GW ((336))(14d)

18. $\int_0^\infty \frac{\cos ax-\cos bx}{e^{(2m+1)px}-e^{(2m-1)px}}\frac{dx}{x}=\frac{1}{2}\ln\frac{\operatorname{ch}\frac{b\pi}{2p}}{\operatorname{ch}\frac{a\pi}{2p}}-$

$-\frac{1}{2}\sum_{k=1}^{m}\ln\frac{b^2+(2k-1)^2p^2}{a^2+(2k-1)^2p^2}{}^{***}\quad [p>0].$ GW ((336))(16a)

---

* For $n=0$ the sum vanishes.
** For $n=1$ the sum vanishes.
*** For $m=0$ the sum vanishes.

19. $\int_0^\infty \frac{\cos ax - \cos bx}{e^{2mpx} - e^{(2m-2)px}} \frac{dx}{x} = \frac{1}{2} \ln \frac{a \operatorname{sh} \frac{b\pi}{2p}}{b \operatorname{sh} \frac{a\pi}{2p}} -$

$- \frac{1}{2} \sum_{k=1}^{m-1} \ln \frac{b^2 + 4k^2p^2}{a^2 + 4k^2p^2}$ * $[p > 0]$. GW ((336))(16b)

20. $\int_0^\infty \frac{\sin x \sin bx}{1 - e^x} \cdot \frac{dx}{x} = \frac{1}{4} \ln \frac{(b+1) \operatorname{sh} [(b-1)\pi]}{(b-1) \operatorname{sh} [(b+1)\pi]}$ $[b^2 \neq 1]$. LO V 305

21. $\int_0^\infty \frac{\sin^2 ax}{1 - e^x} \cdot \frac{dx}{x} = \frac{1}{4} \ln \frac{2a\pi}{\operatorname{sh} 2a\pi}$. LO V 306, BI ((387))(5)

3.952

1. $\int_0^\infty xe^{-p^2x^2} \sin ax \, dx = \frac{a\sqrt{\pi}}{4p^3} \exp\left(-\frac{a^2}{4p^2}\right)$. BI ((362))(1)

2 $\int_0^\infty xe^{-p^2x^2} \cos ax \, dx = \frac{1}{2p^2} - \frac{a}{4p^3} \sum_{k=0}^{\infty} \frac{(-1)^k k!}{(2k+1)!} \left(\frac{a}{p}\right)^{2k+1}$ $[a > 0]$. BI ((362))(2)

3. $\int_0^\infty x^2 e^{-p^2x^2} \sin ax \, dx = \frac{a}{4p^4} + \frac{2p^2 - a^2}{8p^5} \sum_{k=0}^{\infty} \frac{(-1)^k k!}{(2k+1)!} \left(\frac{a}{p}\right)^{2k+1}$ $[a > 0]$. BI ((362))(4)

4. $\int_0^\infty x^2 e^{-p^2x^2} \cos ax \, dx = \sqrt{\pi} \frac{2p^2 - a^2}{8p^5} \exp\left(-\frac{a^2}{4p^2}\right)$. BI ((362))(5)

5. $\int_0^\infty x^3 e^{-p^2x^2} \sin ax \, dx = \sqrt{\pi} \frac{6ap^2 - a^3}{16p^7} \exp\left(-\frac{a^2}{4p^2}\right)$. BI ((362))(6)

6. $\int_0^\infty e^{-p^2x^2} \sin ax \frac{dx}{x} = \frac{a\sqrt{\pi}}{2p} \sum_{k=0}^{\infty} \frac{(-1)^k}{k!(2k+1)} \left(\frac{a}{2p}\right)^{2k}$. BI ((365))(21)

7. $\int_0^\infty x^{\mu-1} e^{-\beta x^2} \sin \gamma x \, dx = \frac{\gamma e^{-\frac{\gamma^2}{4\beta}}}{2\beta^{\frac{\mu+1}{2}}} \Gamma\left(\frac{1+\mu}{2}\right) {}_1F_1\left(1 - \frac{\mu}{2}; \frac{3}{2}; \frac{\gamma^2}{4\beta}\right)$

$[\operatorname{Re}\beta > 0, \operatorname{Re}\mu > -1]$. ET I 318(10)

8. $\int_0^\infty x^{\mu-1} e^{-\beta x^2} \cos ax \, dx = \frac{\Gamma\left(\frac{\mu}{2}\right)}{2\beta^{\frac{\mu}{2}}} {}_1F_1\left(\frac{\mu}{2}; \frac{1}{2}; -\frac{a^2}{4\beta}\right)$

$[\operatorname{Re}\beta > 0, \operatorname{Re}\mu > 0, a > 0]$. ET I 15(14)

* For $m = 1$ the sum vanishes.

9. $$\int_0^\infty x^{2n} e^{-\beta^2 x^2} \cos ax\, dx = (-1)^n \frac{\sqrt{\pi}}{2^{n+1}\beta^{2n+1}} \exp\left(-\frac{a^2}{8\beta^2}\right) D_{2n}\left(\frac{a}{\beta\sqrt{2}}\right) =$$
$$= (-1)^n \frac{\sqrt{\pi}}{(2\beta)^{2n+1}} \exp\left(-\frac{a^2}{4\beta^2}\right) H_{2n}\left(\frac{a}{2\beta}\right)$$
$$\left[|\arg\beta| < \frac{\pi}{4},\ a > 0\right].$$ WH, ET I 15(13)

10. $$\int_0^\infty x^{2n+1} e^{-\beta^2 x^2} \sin ax\, dx = (-1)^n \frac{\sqrt{\pi}}{2^{n+\frac{3}{2}}\beta^{2n+2}} \exp\left(-\frac{a^2}{8\beta^2}\right) D_{2n+1}\left(\frac{a}{\beta\sqrt{2}}\right) =$$
$$= (-1)^n \frac{\sqrt{\pi}}{(2\beta)^{2n+2}} \exp\left(-\frac{a^2}{4\beta^2}\right) H_{2n+1}\left(\frac{a}{2\beta}\right)$$
$$\left[|\arg\beta| < \frac{\pi}{4},\ a > 0\right].$$ WH, ET I 74(23)

3.953

1. $$\int_0^\infty x^{\mu-1} e^{-\gamma x - \beta x^2} \sin ax\, dx = -\frac{i}{2(2\beta)^{\frac{\mu}{2}}} \exp\frac{\gamma^2 - a^2}{8\beta} \times$$
$$\times \Gamma(\mu)\left\{\exp\left(-\frac{ia\gamma}{4\beta}\right) D_{-\mu}\left(\frac{\gamma - ia}{\sqrt{2\beta}}\right) - \exp\frac{ia\gamma}{4\beta} D_{-\mu}\left(\frac{\gamma + ia}{\sqrt{2\beta}}\right)\right\}$$
$$[\operatorname{Re}\mu > -1,\ \operatorname{Re}\beta > 0,\ a > 0].$$ ET I 318(11)

2. $$\int_0^\infty x^{\mu-1} e^{-\gamma x - \beta x^2} \cos ax\, dx = \frac{1}{2(2\beta)^{\frac{\mu}{2}}} \exp\frac{\gamma^2 - a^2}{8\beta} \times$$
$$\times \Gamma(\mu)\left\{\exp\left(-\frac{ia\gamma}{4\beta}\right) D_{-\mu}\left(\frac{\gamma - ia}{\sqrt{2\beta}}\right) + \exp\frac{ia\gamma}{4\beta} D_{-\mu}\left(\frac{\gamma + ia}{\sqrt{2\beta}}\right)\right\}$$
$$[\operatorname{Re}\mu > 0,\ \operatorname{Re}\beta > 0,\ a > 0].$$ ET I 16(18)

3. $$\int_0^\infty x e^{-\gamma x - \beta x^2} \sin ax\, dx =$$
$$= \frac{i\sqrt{\pi}}{8\sqrt{\beta^3}}\left\{(\gamma - ia)\exp\left[-\frac{(\gamma - ia)^2}{4\beta}\right]\left[1 - \Phi\left(\frac{\gamma - ia}{2\sqrt{\beta}}\right)\right] -\right.$$
$$\left. - (\gamma + ia)\exp\left[-\frac{(\gamma + ia)^2}{4\beta}\right]\left[1 - \Phi\left(\frac{\gamma + ia}{2\sqrt{\beta}}\right)\right]\right\} \quad [\operatorname{Re}\beta > 0,\ a > 0]$$ ET I 74(28)

4. $$\int_0^\infty x e^{-\gamma x - \beta x^2} \cos ax\, dx =$$
$$= -\frac{\sqrt{\pi}}{8\sqrt{\beta^3}}\left\{(\gamma - ia)\exp\frac{(\gamma - ia)^2}{4\beta}\left[1 - \Phi\left(\frac{\gamma - ia}{2\sqrt{\beta}}\right)\right] +\right.$$
$$\left. + (\gamma + ia)\exp\frac{(\gamma + ia)^2}{4\beta}\left[1 - \Phi\left(\frac{\gamma + ia}{2\sqrt{\beta}}\right)\right]\right\} + \frac{1}{2\beta} \quad [\operatorname{Re}\beta > 0,\ a > 0].$$ ET I 16(17)

3.954

1. $$\int_0^\infty e^{-\beta x^2} \sin ax \frac{x\,dx}{\gamma^2+x^2} =$$
$$= -\frac{\pi^{\frac{3}{2}}}{8} e^{\beta\gamma^2}\left[2\,\mathrm{sh}\, a\gamma + e^{-\gamma a}\Phi\left(\gamma\sqrt{\beta} - \frac{a}{2\sqrt{\beta}}\right) - e^{\gamma a}\Phi\left(\gamma\sqrt{\beta} + \frac{a}{2\sqrt{\beta}}\right)\right]$$
$[\mathrm{Re}\,\beta > 0,\ \mathrm{Re}\,\gamma > 0,\ a > 0]$. ET I 74(26)a

2. $$\int_0^\infty e^{-\beta x^2} \cos ax \frac{dx}{\gamma^2+x^2} = \frac{\pi^{\frac{3}{2}}}{8\gamma} e^{\beta\gamma^2}\left[2\,\mathrm{ch}\, a\gamma - e^{-\gamma a}\Phi\left(\gamma\sqrt{\beta} - \frac{a}{2\sqrt{\beta}}\right) - \right.$$
$$\left. - e^{\gamma a}\Phi\left(\gamma\sqrt{\beta} + \frac{a}{2\sqrt{\beta}}\right)\right]$$
$[\mathrm{Re}\,\beta > 0,\ \mathrm{Re}\,\gamma > 0,\ a > 0]$. ET I 15(15)

3.955 $$\int_0^\infty x^\nu e^{-\frac{x^2}{2}} \cos\left(\beta x - \nu\frac{\pi}{2}\right) dx =$$
$$= \sqrt{\frac{\pi}{2}} e^{-\frac{\beta^2}{4}} D_\nu(\beta)$$
$[\mathrm{Re}\,\nu > -1]$. EH II 120(4)

3.956 $$\int_0^\infty e^{-x^2}(2x\cos x - \sin x)\sin x \frac{dx}{x^2} = \sqrt{\pi}\,\frac{e-1}{2e}.$$
BI ((369))(19)

3.957

1. $$\int_0^\infty x^{\mu-1} \exp\left(\frac{-\beta^2}{4x}\right) \sin ax\, dx = \frac{i}{2^\mu} \beta^\mu a^{-\frac{\mu}{2}} \times$$
$$\times \left[\exp\left(-\frac{i}{4}\mu\pi\right) K_\mu\left(\beta e^{\frac{\pi i}{4}}\sqrt{a}\right) - \exp\left(\frac{i}{4}\mu\pi\right) K_\mu\left(\beta e^{-\frac{\pi i}{4}}\sqrt{a}\right)\right]$$
$[\mathrm{Re}\,\beta > 0,\ \mathrm{Re}\,\mu < 1,\ a > 0]$. ET I 318(12)

2. $$\int_0^\infty x^{\mu-1} \exp\left(\frac{-\beta^2}{4x}\right) \cos ax\, dx =$$
$$= \frac{1}{2^\mu} \beta^\mu a^{-\frac{\mu}{2}} \left[\exp\left(-\frac{i}{4}\mu\pi\right) K_\mu\left(\beta e^{\frac{\pi i}{4}}\sqrt{a}\right) + \right.$$
$$\left. + \exp\left(\frac{i}{4}\mu\pi\right) K_\mu\left(\beta e^{-\frac{\pi i}{4}}\sqrt{a}\right)\right]$$
$[\mathrm{Re}\,\beta > 0,\ \mathrm{Re}\,\mu < 1,\ a > 0]$. ET I 320(32)a

3.958

1. $$\int_{-\infty}^\infty x^n e^{-(ax^2+bx+c)} \sin(px+q)\, dx =$$
$$= -\left(\frac{-1}{2a}\right)^n \sqrt{\frac{\pi}{a}} \exp\left(\frac{b^2-p^2}{4a} - c\right) \sum_{k=0}^{E\left(\frac{n}{2}\right)} \frac{n!}{(n-2k)!\,k!} a^k \times$$
$$\times \sum_{j=0}^{n-2k} \binom{n-2k}{j} b^{n-2k-j} p^j \sin\left(\frac{pb}{2a} - q + \frac{\pi}{2}j\right)$$
$[a > 0]$. GW ((337))(1b)

2. $$\int_{-\infty}^{\infty} x^n e^{-(ax^2+bx+c)} \cos(px+q)\, dx =$$

$$= \left(\frac{-1}{2a}\right)^n \sqrt{\frac{\pi}{a}} \exp\left(\frac{b^2-p^2}{4a}-c\right) \sum_{k=0}^{E\left(\frac{n}{2}\right)} \frac{n!}{(n-2k)!\,k!} a^k \times$$

$$\times \sum_{j=0}^{n-2k} \binom{n-2k}{j} b^{n-2k-j} p^j \cos\left(\frac{pb}{2a}-q+\frac{\pi}{2}j\right)$$

$[a > 0]$. GW ((337))(1a)

3.959 $$\int_0^{\infty} xe^{-p^2x^2} \operatorname{tg} ax\, dx = \frac{a\sqrt{\pi}}{p^3} \sum_{k=1}^{\infty} (-1)^k k \exp\left(-\frac{a^2k^2}{p^2}\right)$$

$[p > 0]$. BI ((362))(15)

3.961

1. $$\int_0^{\infty} \exp\left(-\beta\sqrt{\gamma^2+x^2}\right) \sin ax \frac{x\, dx}{\sqrt{\gamma^2+x^2}} =$$

$$= \frac{a\gamma}{\sqrt{a^2+\beta^2}} K_1\left(\gamma\sqrt{a^2+\beta^2}\right) \qquad [\operatorname{Re}\beta > 0, \quad \operatorname{Re}\gamma > 0, \quad a > 0].$$

ET I 75(36)

2. $$\int_0^{\infty} \exp\left[-\beta\sqrt{\gamma^2+x^2}\right] \cos ax \frac{dx}{\sqrt{\gamma^2+x^2}} = K_0\left(\gamma\sqrt{a^2+\beta^2}\right)$$

$[\operatorname{Re}\beta > 0,\ \operatorname{Re}\gamma > 0,\ a > 0]$. ET I 17(27)

3.962

1. $$\int_0^{\infty} \frac{\sqrt{\sqrt{\gamma^2+x^2}-\gamma}\exp\left(-\beta\sqrt{\gamma^2+x^2}\right)}{\sqrt{\gamma^2+x^2}} \sin ax\, dx =$$

$$= \sqrt{\frac{\pi}{2}} \frac{a\exp\left(-\gamma\sqrt{a^2+\beta^2}\right)}{\sqrt{\beta^2+a^2}\sqrt{\beta+\sqrt{a^2+\beta^2}}}$$

$[\operatorname{Re}\beta > 0,\ \operatorname{Re}\gamma > 0,\ a > 0]$. ET I 75(38)

2. $$\int_0^{\infty} \frac{x\exp\left(-\beta\sqrt{\gamma^2+x^2}\right)}{\sqrt{\gamma^2+x^2}\sqrt{\sqrt{\gamma^2+x^2}-\gamma}} \cos ax\, dx =$$

$$= \sqrt{\frac{\pi}{2}} \frac{\sqrt{\beta+\sqrt{a^2+\beta^2}}}{\sqrt{a^2+\beta^2}} \exp\left[-\gamma\sqrt{a^2+\beta^2}\right]$$

$[\operatorname{Re}\beta > 0,\ \operatorname{Re}\gamma > 0,\ a > 0]$. ET I 17(29)

3.963

1. $$\int_0^{\infty} e^{-\operatorname{tg}^2 x} \frac{\sin x}{\cos^2 x} \frac{dx}{x} = \frac{\sqrt{\pi}}{2}.$$ BI ((391))(1)

2. $$\int_0^{\frac{\pi}{2}} e^{-p \operatorname{tg} x} \frac{x\,dx}{\cos^2 x} = \frac{1}{p}[\operatorname{ci}(p)\sin p - \cos p \operatorname{si}(p)]$$

$[p > 0]$; (cf. **3.339**). BI ((396))(3)

3. $$\int_0^{\frac{\pi}{2}} xe^{-\operatorname{tg}^2 x} \sin 4x \frac{dx}{\cos^2 x} = -\frac{3}{2}\sqrt{\pi}.$$ BI ((396))(5)

4. $$\int_0^{\frac{\pi}{2}} xe^{-\operatorname{tg}^2 x} \sin^2 2x \frac{dx}{\cos^2 x} = 2\sqrt{\pi}.$$ BI ((396))(6)

3.964

1. $$\int_0^{\frac{\pi}{2}} xe^{-p \operatorname{tg} x} \frac{p \sin x - \cos x}{\cos^3 x}\,dx = -\sin p \operatorname{si}(p) - \operatorname{ci}(p)\cos p \quad [p > 0].$$

LI ((396))(4)

2. $$\int_0^{\frac{\pi}{2}} xe^{-p \operatorname{tg}^2 x} \frac{p - \cos^2 x}{\cos^4 x \operatorname{ctg} x}\,dx = \frac{1}{4}\sqrt{\frac{\pi}{p}} \quad [p > 0].$$ BI ((396))(7)

3. $$\int_0^{\frac{\pi}{2}} xe^{-p \operatorname{tg}^2 x} \frac{p - 2\cos^2 x}{\cos^6 x \operatorname{ctg} x}\,dx = \frac{1+2p}{8}\sqrt{\frac{\pi}{p}} \quad [p > 0].$$ BI ((396))(8)

3.965

1. $$\int_0^{\infty} xe^{-\beta x} \sin ax^2 \sin \beta x\,dx = \frac{\beta}{4}\sqrt{\frac{\pi}{2a^3}}\, e^{-\frac{\beta^2}{2a}} \quad \left[|\arg \beta| < \frac{\pi}{4},\ a > 0\right].$$

ET I 84(17)

2. $$\int_0^{\infty} xe^{-\beta x} \cos ax^2 \cos \beta x\,dx = \frac{\beta}{4}\sqrt{\frac{\pi}{2a^3}}\, e^{-\frac{\beta^2}{2a}} \quad [a > 0,\ \operatorname{Re}\beta > |\operatorname{Im}\beta|].$$

ET 26(27)

3.966

1. $$\int_0^{\infty} xe^{-px} \cos(2x^2 + px)\,dx = 0 \quad [p > 0].$$ BI ((361))(16)

2. $$\int_0^{\infty} xe^{-px} \cos(2x^2 - px)\,dx = \frac{p\sqrt{\pi}}{8} \exp\left(-\frac{1}{4}p^2\right) \quad [p > 0].$$ BI ((361))(17)

3. $$\int_0^{\infty} x^2 e^{-px} [\sin(2x^2 + px) + \cos(2x^2 + px)]\,dx = 0 \quad [p > 0].$$ BI ((361))(18)

4. $$\int_0^{\infty} x^2 e^{-px} [\sin(2x^2 - px) - \cos(2x^2 - px)]\,dx =$$

$$= \frac{\sqrt{\pi}}{16}(2 - p^2) \exp\left(-\frac{1}{4}p^2\right).$$ BI ((361))(19)

5. $$\int_0^\infty x^{\mu-1}e^{-x}\cos(x+ax^2)\,dx = \frac{e^{+\frac{1}{4a}}\Gamma(\mu)}{(2a)^{\frac{\mu}{2}}}\cos\frac{\mu\pi}{2}D_{-\mu}\left(\frac{1}{\sqrt{a}}\right)$$
$[\operatorname{Re}\mu > 0,\ a > 0]$. ET I 321(37)

6. $$\int_0^\infty x^{\mu-1}e^{-x}\sin(x+ax^2)\,dx = \frac{e^{+\frac{1}{4a}}\Gamma(\mu)}{(2a)^{\frac{\mu}{2}}}\sin\frac{\mu\pi}{4}D_{-\mu}\left(\frac{1}{\sqrt{a}}\right)$$
$[\operatorname{Re}\mu > 0,\ a > 0]$. ET I 319(18)

3.967

1. $$\int_0^\infty e^{-\frac{\beta^2}{x^2}}\sin a^2x^2\,\frac{dx}{x^2} = \frac{\sqrt{\pi}}{2\beta}e^{-\sqrt{2}a\beta}\sin\left(\sqrt{2}a\beta\right) \quad [\operatorname{Re}\beta > 0,\ a > 0].$$
ET I 75(30)a, BI((369))(3)a

2. $$\int_0^\infty e^{-\frac{\beta^2}{x^2}}\cos a^2x^2\,\frac{dx}{x^2} = \frac{\sqrt{\pi}}{2\beta}e^{-\sqrt{2}a\beta}\cos\left(\sqrt{2}a\beta\right) \quad [\operatorname{Re}\beta > 0,\ a > 0].$$
BI ((369))(4), ET I 16(20)

3. $$\int_0^\infty x^2e^{-\beta x^2}\cos ax^2\,dx = \frac{\sqrt{\pi}}{4\sqrt[4]{(a^2+\beta^2)^3}}\cos\left(\frac{3}{2}\operatorname{arctg}\frac{a}{\beta}\right) \quad [\operatorname{Re}\beta > 0].$$
ET I 14(3)a

3.968

1. $$\int_0^\infty e^{-\beta x^2}\sin ax^4\,dx = -\frac{\pi}{8}\sqrt{\frac{\beta}{a}}\left[J_{\frac{1}{4}}\left(\frac{\beta^2}{8a}\right)\cos\left(\frac{\beta^2}{8a}+\frac{\pi}{8}\right)+\right.$$
$$\left.+N_{\frac{1}{4}}\left(\frac{\beta^2}{8a}\right)\sin\left(\frac{\beta^2}{8a}+\frac{\pi}{8}\right)\right] \quad [\operatorname{Re}\beta > 0,\ a > 0].$$
ET I 75(34)

2. $$\int_0^\infty e^{-\beta x^2}\cos ax^4\,dx = \frac{\pi}{8}\sqrt{\frac{\beta}{a}}\left[J_{\frac{1}{4}}\left(\frac{\beta^2}{8a}\right)\sin\left(\frac{\beta^2}{8a}+\frac{\pi}{8}\right)-\right.$$
$$\left.-N_{\frac{1}{4}}\left(\frac{\beta^2}{8a}\right)\cos\left(\frac{\beta^2}{8a}+\frac{\pi}{8}\right)\right] \quad [\operatorname{Re}\beta > 0,\ > 0].$$
ET I 16(24)

3.969

1. $$\int_0^\infty e^{-p^2x^4+q^2x^2}\left[2px\cos(2pqx^3)+q\sin(2pqx^3)\right]dx = \frac{\sqrt{\pi}}{2}.$$
BI ((363))(7)

2. $$\int_0^\infty e^{-p^2x^4+q^2x^2}\left[2px\sin(2pqx^3)-q\cos(2pqx^3)\right]dx = 0.$$
BI ((363))(8)

3.971

1. $$\int_0^\infty \exp\left(-px^2-\frac{q}{x^2}\right)\sin\left(ax^2+\frac{b}{x^2}\right)\frac{dx}{x^2} =$$
$$= \frac{1}{2}\int_{-\infty}^\infty \exp\left(-px^2-\frac{q}{x^2}\right)\sin\left(ax^2+\frac{b}{x^2}\right)\frac{dx}{x^2} =$$
$$= \frac{\sqrt{\pi}}{2s}\exp\left[-2rs\cos(A+B)\right]\sin\left[A+2rs\sin(A+B)\right].$$
BI ((369))(16, 17)

2. $$\int_0^\infty \exp\left(-px^2-\frac{q}{x^2}\right)\cos\left(ax^2+\frac{b}{x^2}\right)\frac{dx}{x^2}=$$
$$=\frac{1}{2}\int_{-\infty}^\infty \exp\left(-px^2-\frac{q}{x^2}\right)\cos\left(ax^2+\frac{b}{x^2}\right)\frac{dx}{x^2}=$$
$$=\frac{\sqrt{\pi}}{2s}\exp\left[-2rs\cos(A+B)\right]\cos\left[A+2rs\sin(A+B)\right].$$

$\left[\text{In formulas 3.971 1. and 2. } p\geqslant 0,\ q\geqslant 0,\ r=\sqrt[4]{a^2+p^2},\ s=\sqrt[4]{b^2+q^2},\ A=\operatorname{arctg}\frac{a}{p},\ B=\operatorname{arctg}\frac{b}{q}.\right]$ BI ((369))(15, 18)

3.972

1. $$\int_0^\infty \exp\left[-\beta\sqrt{\gamma^4+x^4}\right]\sin ax^2\frac{dx}{\sqrt{\gamma^4+x^4}}=$$
$$=\sqrt{\frac{a\pi}{8}}\,I_{\frac{1}{4}}\left[\frac{\gamma^2}{2}\left(\sqrt{\beta^2+a^2}-\beta\right)\right]K_{\frac{1}{4}}\left[\frac{\gamma^2}{4}\left(\sqrt{\beta^2+a^2}+\beta\right)\right]$$
$$\left[\operatorname{Re}\beta>0,\ |\arg\gamma|<\frac{\pi}{4},\ a>0\right].$$ ET I 75(37)

2. $$\int_0^\infty \exp\left[-\beta\sqrt{\gamma^4+x^4}\right]\cos ax^2\frac{dx}{\sqrt{\gamma^4+x^4}}=$$
$$=\sqrt{\frac{a\pi}{8}}\,I_{-\frac{1}{4}}\left[\frac{\gamma^2}{2}\left(\sqrt{\beta^2+a^2}-\beta\right)\right]K_{\frac{1}{4}}\left[\frac{\gamma^2}{4}\left(\sqrt{\beta^2+a^2}+\beta\right)\right]$$
$$\left[\operatorname{Re}\beta>0,\ |\arg\gamma|<\frac{\pi}{4},\ a>0\right].$$ ET I 17(28)

3.973

1. $$\int_0^\infty \exp(p\cos ax)\sin(p\sin ax)\frac{dx}{x}=\frac{\pi}{2}(e^p-1)$$
$$[p>0,\ a>0].$$ WH, FI II 725

2. $$\int_0^\infty \exp(p\cos ax)\sin(p\sin ax+bx)\frac{x\,dx}{c^2+x^2}=$$
$$=\frac{\pi}{2}\exp\left(-cb+pe^{-ac}\right)\qquad [a>0,\ b>0,\ c>0,\ p>0].$$ BI ((372))(3)

3. $$\int_0^\infty \exp(p\cos ax)\cos(p\sin ax+bx)\frac{dx}{c^2+x^2}=$$
$$=\frac{\pi}{2c}\exp\left(-cb+pe^{-ac}\right)\qquad [a>0,\ b>0,\ c>0,\ p>0].$$ BI ((372))(4)

4. $$\int_0^\infty \exp(p\cos x)\sin(p\sin x+nx)\frac{dx}{x}=\frac{\pi}{2}e^p \quad [p>0].$$ BI ((366))(2)

5. $$\int_0^\infty \exp(p\cos x)\sin(p\sin x)\cos nx\frac{dx}{x}=$$
$$=\frac{p^n}{n!}\cdot\frac{\pi}{4}+\frac{\pi}{2}\sum_{k=n+1}^{\infty}\frac{p^k}{k!} \quad [p>0].$$ LI ((366))(3)

6. $$\int_0^\infty \exp(p\cos x)\cos(p\sin x)\sin nx\frac{dx}{x}=$$
$$=\frac{\pi}{2}\sum_{k=0}^{n-1}\frac{p^k}{k!}+\frac{p^n}{n!}\frac{\pi}{4} \quad [p>0].$$ LI ((366))(4)

3.974

1. $$\int_0^\infty \exp(p\cos ax)\sin(p\sin ax)\operatorname{cosec} ax\frac{dx}{b^2+x^2}=\frac{\pi[e^p-\exp(pe^{-ab})]}{2b\operatorname{sh} ab}$$
$$[a>0,\ b>0,\ p>0].$$ BI ((391))(4)

2. $$\int_0^\infty [1-\exp(p\cos ax)\cos(p\sin ax)]\operatorname{cosec} ax\frac{x\,dx}{b^2+x^2}=\frac{\pi[e^p-\exp(pe^{-ab})]}{2\operatorname{sh} ab}$$
$$[a>0,\ b>0,\ p>0].$$ BI ((391))(5)

3. $$\int_0^\infty \exp(p\cos ax)\sin(p\sin ax+ax)\operatorname{cosec} ax\frac{dx}{b^2+x^2}=$$
$$=\frac{\pi[e^p-\exp(pe^{-ab}-ab)]}{2b\operatorname{sh} ab} \quad [a>0,\ b>0,\ p>0].$$ BI ((391))(6)

4. $$\int_0^\infty \exp(p\cos ax)\cos(p\sin ax+ax)\operatorname{cosec} ax\frac{x\,dx}{b^2+x^2}=$$
$$=\frac{\pi[e^p-\exp(pe^{-ab}-ab)]}{2\operatorname{sh} ab} \quad [a>0,\ b>0,\ p>0].$$ BI ((391))(7)

5. $$\int_0^\infty \exp(p\cos ax)\sin(p\sin ax)\frac{x\,dx}{b^2-x^2}=$$
$$=\frac{\pi}{2}[1-\exp(p\cos ab)\cos(p\sin ab)] \quad [p>0,\ a>0].$$ BI ((378))(1)

6. $$\int_0^\infty \exp(p\cos ax)\cos(p\sin ax)\frac{dx}{b^2-x^2}=$$
$$=\frac{\pi}{2b}\exp(p\cos ab)\sin(p\sin ab) \quad [a>0,\ b>0,\ p>0].$$ BI ((378))(2)

7. $\int\limits_0^\infty \exp(p\cos ax)\sin(p\sin ax)\operatorname{tg} ax \frac{dx}{b^2+x^2} =$

$= \frac{\pi}{2b}\cdot \operatorname{th} ab\,[\exp(pe^{-ab}) - e^p] \quad [a>0,\ b>0,\ p>0].$ BI ((372))(14)

8. $\int\limits_0^\infty \exp(p\cos ax)\sin(p\sin ax)\operatorname{ctg} ax \frac{dx}{b^2+x^2} =$

$= \frac{\pi}{2b}\operatorname{cth} ab\,[e^p - \exp(pe^{-ab})] \quad [a>0,\ b>0,\ p>0].$ BI ((372))(15)

9. $\int\limits_0^\infty \exp(p\cos ax)\sin(p\sin ax)\operatorname{cosec} ax \frac{dx}{b^2-x^2} =$

$= \frac{\pi}{2b}\operatorname{cosec} ab\,[e^p - \exp(p\cos ab)\cos(p\sin ab)]$

$[a>0,\ b>0,\ p>0].$ BI ((391))(12)

10. $\int\limits_0^\infty [1-\exp(p\cos ax)\cos(p\sin ax)]\operatorname{cosec} ax \frac{x\,dx}{b^2-x^2} =$

$= -\frac{\pi}{2}\exp(p\cos ab)\sin(p\sin ab)\operatorname{cosec} ab$

$[a>0,\ b>0,\ p>0].$ BI ((391))(13)

3.975

1. $\int\limits_0^\infty \frac{\sin\left(\beta\operatorname{arctg}\frac{x}{\gamma}\right)}{(\gamma^2+x^2)^{\frac{\beta}{2}}}\cdot\frac{dx}{e^{2\pi x}-1} = \frac{1}{2}\zeta(\beta,\ \gamma) - \frac{1}{4\gamma^\beta} - \frac{\gamma^{1-\beta}}{2(\beta-1)}$

$[\operatorname{Re}\beta>1,\ \operatorname{Re}\gamma>0].$ WH, ET I 26(7)

2. $\int\limits_0^\infty \frac{\sin(\beta\operatorname{arctg} x)}{(1+x^2)^{\frac{\beta}{2}}}\cdot\frac{dx}{e^{2\pi x}+1} = \frac{1}{2(\beta-1)} - \frac{\zeta(\beta)}{2^\beta} \quad [\operatorname{Re}\beta>1].$ EH I 33(13)

3.976 $\int\limits_0^\infty (1+x^2)^{\beta-\frac{1}{2}} e^{-px^2}\cos[2px+(2\beta-1)\operatorname{arctg} x]\,dx = \frac{e^{-p}}{2p^\beta}\sin\pi\beta\,\Gamma(\beta)$

$[\operatorname{Re}\beta>0,\ p>0].$ WH

3.98-3.99 Combinations of trigonometric and hyperbolic functions

3.981

1. $\int\limits_0^\infty \frac{\sin ax}{\operatorname{sh}\beta x}\,dx = \frac{\pi}{2\beta}\operatorname{th}\frac{a\pi}{2\beta} \quad [\operatorname{Re}\beta>0,\ a>0].$ BI ((264))(16)

2. $\int\limits_0^\infty \frac{\sin ax}{\operatorname{ch}\beta x}\,dx = -\frac{\pi}{2\beta}\operatorname{th}\frac{a\pi}{2\beta} - \frac{i}{2\beta}\left[\psi\left(\frac{\beta+ai}{4\beta}\right) - \psi\left(\frac{\beta-ai}{4\beta}\right)\right]$

$[\operatorname{Re}\beta>0,\ a>0].$ GW ((335))(12), ET I 88(1)

3. $\int\limits_0^\infty \frac{\cos ax}{\operatorname{ch}\beta x}\,dx = \frac{\pi}{2\beta}\operatorname{sech}\frac{a\pi}{2\beta} \quad [\operatorname{Re}\beta>0,\ a>0].$ BI ((264))(14)

4. $$\int_0^\infty \sin ax \frac{\operatorname{sh} \beta x}{\operatorname{sh} \gamma x}\, dx = \frac{\pi}{2\gamma} \frac{\operatorname{sh} \frac{a\pi}{\gamma}}{\operatorname{ch} \frac{a\pi}{\gamma} + \cos \frac{\beta\pi}{\gamma}} + \frac{i}{2\gamma}\left[\psi\left(\frac{\beta+\gamma+ia}{2\gamma}\right) - \psi\left(\frac{\beta+\gamma-ia}{2\gamma}\right)\right] \quad [|\operatorname{Re}\beta| < \operatorname{Re}\gamma,\ a > 0].$$ ET I 88(5)

5. $$\int_0^\infty \cos ax \frac{\operatorname{sh} \beta x}{\operatorname{sh} \gamma x}\, dx = \frac{\pi}{2\gamma} \frac{\sin \frac{\pi\beta}{\gamma}}{\operatorname{ch} \frac{a\pi}{\gamma} + \cos \frac{\beta\pi}{\gamma}} \quad [|\operatorname{Re}\beta| < \operatorname{Re}\gamma,\ a > 0].$$ BI ((265))(7)

6. $$\int_0^\infty \sin ax \frac{\operatorname{sh} \beta x}{\operatorname{ch} \gamma x}\, dx = \frac{\pi}{\gamma} \frac{\sin \frac{\beta\pi}{2\gamma} \operatorname{sh} \frac{a\pi}{2\gamma}}{\operatorname{ch} \frac{a\pi}{\gamma} + \cos \frac{\beta\pi}{\gamma}} \quad [|\operatorname{Re}\beta| < \operatorname{Re}\gamma;\ a > 0].$$ BI ((265))(2)

7. $$\int_0^\infty \cos ax \frac{\operatorname{sh} \beta x}{\operatorname{ch} \gamma x}\, dx = \frac{1}{4\gamma}\left\{\psi\left(\frac{3\gamma-\beta+ia}{4\gamma}\right) + \psi\left(\frac{3\gamma-\beta-ia}{4\gamma}\right) - \psi\left(\frac{3\gamma+\beta-ia}{4\gamma}\right) - \psi\left(\frac{3\gamma+\beta+ia}{4\gamma}\right) + \frac{2\pi \sin \frac{\pi\beta}{\gamma}}{\cos \frac{\pi\beta}{\gamma} + \operatorname{ch} \frac{\pi a}{\gamma}}\right\} \quad [|\operatorname{Re}\beta| < \operatorname{Re}\gamma,\ a > 0].$$ ET I 31(13)

8. $$\int_0^\infty \sin ax \frac{\operatorname{ch} \beta x}{\operatorname{sh} \gamma x}\, dx = \frac{\pi}{2\gamma} \cdot \frac{\operatorname{sh} \frac{\pi a}{\gamma}}{\operatorname{ch} \frac{\pi a}{\gamma} + \cos \frac{\pi\beta}{\gamma}} \quad [|\operatorname{Re}\beta| < \operatorname{Re}\gamma,\ a > 0].$$ BI ((265))(4)

9. $$\int_0^\infty \sin ax \frac{\operatorname{ch} \beta x}{\operatorname{ch} \gamma x}\, dx = \frac{i}{4\gamma}\left[\psi\left(\frac{3\gamma+\beta+ai}{4\gamma}\right) - \psi\left(\frac{3\gamma+\beta-ai}{4\gamma}\right) + \psi\left(\frac{3\gamma-\beta+ia}{4\gamma}\right) - \psi\left(\frac{3\gamma-\beta-ai}{4\gamma}\right) - \frac{2\pi i \operatorname{sh} \frac{\pi a}{\gamma}}{\operatorname{ch} \frac{a\pi}{\gamma} + \cos \frac{\beta\pi}{\gamma}}\right] \quad [|\operatorname{Re}\beta| < \operatorname{Re}\gamma,\ a > 0].$$ ET I 88(6)

10. $$\int_0^\infty \cos ax \frac{\operatorname{ch} \beta x}{\operatorname{ch} \gamma x}\, dx = \frac{\pi}{\gamma} \frac{\cos \frac{\beta\pi}{2\gamma} \operatorname{ch} \frac{a\pi}{2\gamma}}{\operatorname{ch} \frac{a\pi}{\gamma} + \cos \frac{\beta\pi}{\gamma}} \quad [|\operatorname{Re}\beta| < \operatorname{Re}\gamma,\ a > 0].$$ BI ((265))(6)

11. $$\int_0^{\frac{\pi}{2}} \cos^{2m} x \operatorname{ch} \beta x\, dx = \frac{(2m)!\, \operatorname{sh} \frac{\pi\beta}{2}}{\beta(\beta^2+2^2) \ldots [\beta^2+(2m)^2]} \quad [\operatorname{Re}\beta > 0].$$ WA 620a

12. $$\int_0^{\frac{\pi}{2}} \cos^{2m-1} x \operatorname{ch} \beta x \, dx = \frac{(2m-1)! \operatorname{ch} \frac{\pi\beta}{2}}{(\beta^2+1^2)(\beta^2+3^2)\ldots[\beta^2+(2m+1)^2]}$$

$[\operatorname{Re}\beta > 0]$. WA 620a

3.982

1. $$\int_0^\infty \frac{\cos ax}{\operatorname{ch}^2 \beta x} dx = \frac{a\pi}{2\beta^2 \operatorname{sh} \frac{a\pi}{2\beta}} \quad [\operatorname{Re}\beta > 0, \ a > 0].$$ BI ((264))(16)

2. $$\int_0^\infty \sin ax \frac{\operatorname{sh} \beta x}{\operatorname{ch}^2 \gamma x} dx = \frac{\pi \left(a \sin \frac{\beta\pi}{2\gamma} \operatorname{ch} \frac{a\pi}{2\gamma} - \beta \cos \frac{\beta\pi}{2\gamma} \operatorname{sh} \frac{a\pi}{2\gamma} \right)}{\gamma^2 \left(\operatorname{ch} \frac{a\pi}{\gamma} - \cos \frac{\beta\pi}{\gamma} \right)}$$

$[|\operatorname{Re}\beta| < 2\operatorname{Re}\gamma, \ a > 0]$. ET I 88(9)

3.983

1. $$\int_0^\infty \frac{\cos ax \, dx}{b \operatorname{ch} \beta x + c} = \frac{\pi \sin\left(\frac{a}{\beta} \operatorname{arch} \frac{c}{b} \right)}{\beta \sqrt{c^2 - b^2} \operatorname{sh} \frac{a\pi}{\beta}} \quad [c > b > 0];$$

$$= \frac{\pi \operatorname{sh}\left(\frac{a}{\beta} \arccos \frac{c}{b} \right)}{\beta \sqrt{b^2 - c^2} \operatorname{sh} \frac{a\pi}{\beta}} \quad [b > |c| > 0];$$

$[\operatorname{Re}\beta > 0, \ a > 0]$. GW ((335))(13a)

2. $$\int_0^\infty \frac{\cos ax \, dx}{\operatorname{ch} \beta x + \cos \gamma} = \frac{\pi}{\beta} \frac{\operatorname{sh} \frac{a\gamma}{\beta}}{\sin \gamma \operatorname{sh} \frac{a\pi}{\beta}} \quad [\pi \operatorname{Re}\beta < \operatorname{Im} \bar{\beta}\gamma, \ a > 0].$$ BI ((267))(3)

3. $$\int_0^\infty \frac{\cos ax \, dx}{\operatorname{ch} x - \operatorname{ch} b} = -\pi \operatorname{ch} a\pi \frac{\sin ab}{\operatorname{sh} b} \quad [a > 0, \ b > 0].$$

BI ((267))(4), ET I 30(8)

4. $$\int_0^\infty \frac{\cos ax \, dx}{1 + 2 \operatorname{ch}\left(\sqrt{\frac{2}{3}} \pi x \right)} = \frac{\sqrt{\frac{\pi}{2}}}{1 + 2 \operatorname{ch}\left(\sqrt{\frac{2}{3}} \pi a \right)} \quad [a > 0].$$ ET I 30(9)

5. $$\int_0^\infty \frac{\sin ax \operatorname{sh} \beta x}{\operatorname{ch} \gamma x + \cos \delta} dx =$$

$$= \frac{\pi \left\{ \sin\left[\frac{\beta}{\gamma}(\pi - \delta) \right] \operatorname{sh}\left[\frac{a}{\gamma}(\pi + \delta) \right] - \sin\left[\frac{\beta}{\gamma}(\pi + \delta) \right] \operatorname{sh}\left[\frac{a}{\gamma}(\pi - \delta) \right] \right\}}{\gamma \sin \delta \left(\operatorname{ch} \frac{2\pi a}{\gamma} - \cos \frac{2\pi\beta}{\gamma} \right)}$$

$[\pi \operatorname{Re}\gamma > |\operatorname{Re} \bar{\gamma}\delta|, \ |\operatorname{Re}\beta| < \operatorname{Re}\gamma, \ a > 0]$. BI ((267))(2)

6. $$\int_0^\infty \frac{\cos ax \operatorname{ch} \beta x}{\operatorname{ch} \gamma x + \cos b} dx =$$
$$= \frac{\pi \left\{\cos\left[\frac{\beta}{\gamma}(\pi - b)\right] \operatorname{ch}\left[\frac{a}{\gamma}(\pi + b)\right] - \cos\left[\frac{\beta}{\gamma}(\pi + b)\right] \operatorname{ch}\left[\frac{a}{\gamma}(\pi - b)\right]\right\}}{\gamma \sin b \left(\operatorname{ch} \frac{2\pi a}{\gamma} - \cos \frac{2\pi\beta}{\gamma}\right)}$$
$[|\operatorname{Re}\beta| < \operatorname{Re}\gamma,\ 0 < b < \pi,\ a > 0]$. BI ((267))(6)

7. $$\int_0^\infty \frac{\cos ax\, dx}{(\beta + \sqrt{\beta^2 - 1} \operatorname{ch} x)^{\nu+1}} = \Gamma(\nu + 1 - ai)\, e^{a\pi} \frac{Q_\nu^{ai}(\beta)}{\Gamma(\nu + 1)}$$
$[\operatorname{Re}\nu > -1,\ |\arg(\beta \pm 1)| < \pi,\ a > 0]$. ET I 30(10)

3.984

1. $$\int_0^\infty \frac{\sin ax \operatorname{sh} x}{\operatorname{ch} x + \cos b} dx = \pi \frac{\operatorname{ch} ab}{\operatorname{ch} a\pi} \qquad [b \leqslant \pi,\ a > 0].$$ BI ((267))(1)

2. $$\int_0^\infty \frac{\cos ax \operatorname{ch} x}{\operatorname{ch} x + \cos b} dx = -\pi \operatorname{ctg} b \frac{\operatorname{sh} ab}{\operatorname{sh} a\pi} \qquad [b \leqslant \pi].$$ BI ((267))(5)

3. $$\int_0^\infty \frac{\sin ax \operatorname{sh} \frac{x}{2}}{\operatorname{ch} x + \cos \beta} dx = \frac{\operatorname{sh} a\beta}{2 \sin \frac{\beta}{2} \operatorname{ch} a\pi} \qquad [\operatorname{Re}\beta < \pi,\ a > 0].$$ ET I 80(10)

4. $$\int_0^\infty \frac{\cos ax \operatorname{ch} \frac{\beta}{2} x}{\operatorname{ch} \beta x + \operatorname{ch} \gamma} dx = \frac{\pi \cos \frac{a\gamma}{\beta}}{2\beta \operatorname{ch} \frac{\gamma}{2} \operatorname{ch} \frac{a\pi}{\beta}} \qquad [\pi \operatorname{Re}\beta > |\operatorname{Im}(\bar{\beta}\gamma)|].$$ ET I 31(16)

5. $$\int_0^\infty \frac{\sin ax \operatorname{sh} \beta x}{\operatorname{ch} 2\beta x + \cos 2ax} dx = \frac{a\pi}{4(a^2 + \beta^2)} \qquad [a > 0,\ \operatorname{Re}\beta > 0].$$ BI ((267))(7)

6. $$\int_0^\infty \frac{\cos ax \operatorname{ch} \beta x}{\operatorname{ch} 2\beta x + \cos 2ax} dx = \frac{\beta\pi}{4(a^2 + \beta^2)} \qquad [\operatorname{Re}\beta > 0,\ a > 0].$$ BI ((267))(8)

7. $$\int_0^\infty \frac{\operatorname{sh}^{2\mu-1} x \operatorname{ch}^{2\varrho - 2\nu + 1} x}{(\operatorname{ch}^2 x - \beta \operatorname{sh}^2 x)^\varrho} dx = \frac{1}{2} \mathrm{B}(\mu, \nu - \mu)\, {}_2F_1(\varrho, \mu; \nu; \beta)$$
$[\operatorname{Re}\nu > \operatorname{Re}\mu > 0]$. EH I 115(12)

3.985

1. $$\int_0^\infty \frac{\cos ax\, dx}{\operatorname{ch}^\nu \beta x} = \frac{2^{\nu-2}}{\beta \Gamma(\nu)} \Gamma\left(\frac{\nu}{2} + \frac{ai}{2\beta}\right) \Gamma\left(\frac{\nu}{2} - \frac{ai}{2\beta}\right)$$
$[\operatorname{Re}\beta > 0,\ \operatorname{Re}\nu > 0,\ a > 0]$. ET I 30(5)

2. $$\int_0^\infty \frac{\cos ax\, dx}{\operatorname{ch}^{2n} \beta x} = \frac{4^{n-1}\pi a}{2(2n-1)!\, \beta^2 \operatorname{sh} \frac{a\pi}{2\beta}} \prod_{k=1}^{n-1} \left(\frac{a^2}{4\beta^2} + k^2\right);$$
$$= \frac{\pi a (a^2 + 2^2\beta^2)(a^2 + 4^2\beta^2) \ldots [a^2 + (2n-2)^2\beta^2]}{2(2n-1)!\, \beta^{2n} \operatorname{sh} \frac{a\pi}{2\beta}} \qquad [n \geqslant 2,\ a > 0].$$
ET I 30(3)

3. $$\int_0^\infty \frac{\cos ax\,dx}{\operatorname{ch}^{2n+1}\beta x} = \frac{\pi\cdot 2^{2n-1}}{(2n)!\,\beta\operatorname{ch}\frac{a\pi}{2\beta}} \prod_{k=1}^{n}\left[\frac{a^2}{4\beta^2}+\left(\frac{2k-1}{2}\right)^2\right];$$
$$= \frac{\pi(a^2+\beta^2)(a^2+3^2\beta^2)\ldots[a^2+(2n-1)^2\beta^2]}{2(2n)!\,\beta^{2n+1}\operatorname{ch}\frac{a\pi}{2\beta}} \qquad [a>0].$$ ET I 30(4)

3.986

1. $$\int_0^\infty \frac{\sin\beta x\sin\gamma x}{\operatorname{ch}\delta x}\,dx = \frac{\pi}{\delta}\cdot\frac{\operatorname{sh}\frac{\beta\pi}{2\delta}\operatorname{sh}\frac{\gamma\pi}{2\delta}}{\operatorname{ch}\frac{\beta}{\delta}\pi+\operatorname{ch}\frac{\gamma}{\delta}\pi}$$
$[|\operatorname{Im}(\beta+\gamma)|<\operatorname{Re}\delta]$. BI ((264))(19)

2. $$\int_0^\infty \frac{\sin\alpha x\cos\beta x}{\operatorname{sh}\gamma x}\,dx = \frac{\pi\operatorname{sh}\frac{\pi\alpha}{\gamma}}{2\gamma\left(\operatorname{ch}\frac{\alpha\pi}{\gamma}+\operatorname{ch}\frac{\beta\pi}{\gamma}\right)}$$
$[|\operatorname{Im}(\alpha+\beta)|<\operatorname{Re}\gamma]$. LI ((264))(20)

3. $$\int_0^\infty \frac{\cos\beta x\cos\gamma x}{\operatorname{ch}\delta x}\,dx = \frac{\pi}{\delta}\cdot\frac{\operatorname{ch}\frac{\beta\pi}{2\delta}\operatorname{ch}\frac{\gamma\pi}{2\delta}}{\operatorname{ch}\frac{\beta\pi}{\delta}+\operatorname{ch}\frac{\gamma\pi}{\delta}} \qquad [|\operatorname{Im}(\beta+\gamma)|<\operatorname{Re}\delta].$$
BI ((264))(21)

4 $$\int_0^\infty \frac{\sin^2\beta x}{\operatorname{sh}^2\pi x}\,dx = \frac{\beta}{\pi(e^{2\beta}-1)}+\frac{\beta-1}{2\pi} \qquad [|\operatorname{Im}\beta|<\pi].$$ EH I 44(3)

3.987

1. $$\int_0^\infty \sin ax\,(1-\operatorname{th}\beta x)\,dx = \frac{1}{a}-\frac{\pi}{2\beta\operatorname{sh}\frac{a\pi}{2\beta}} \qquad [\operatorname{Re}\beta>0].$$ ET I 88(4)a

2. $$\int_0^\infty \sin ax\,(\operatorname{cth}\beta x-1)\,dx = \frac{\pi}{2\beta}\operatorname{cth}\frac{a\pi}{2\beta}-\frac{1}{a} \qquad [\operatorname{Re}\beta>0].$$ ET I 88(3)

3.988

1. $$\int_0^{\frac{\pi}{2}} \frac{\cos ax\operatorname{sh}(2b\cos x)}{\sqrt{\cos x}}\,dx = \frac{\pi}{2}\sqrt{\pi b}\,I_{\frac{a}{2}+\frac{1}{4}}(b)\,I_{-\frac{a}{2}+\frac{1}{4}}(b) \qquad [a>0].$$
ET I 37(66)

2. $$\int_0^{\frac{\pi}{2}} \frac{\cos ax\operatorname{ch}(2b\cos x)}{\sqrt{\cos x}}\,dx = \frac{\pi}{2}\sqrt{\pi b}\,I_{\frac{a}{2}-\frac{1}{4}}(b)\,I_{-\frac{a}{2}-\frac{1}{4}}(b) \qquad [a>0].$$
ET I 37(67)

3. $$\int_0^\infty \frac{\cos ax\,dx}{\sqrt{\operatorname{ch}x+\cos b}} = \frac{\pi P_{-\frac{1}{2}+ia}(\cos b)}{\sqrt{2}\operatorname{ch}a\pi} \qquad [a>0,\ b>0].$$ ET I 30(7)

3.989

1. $$\int_0^\infty \frac{\sin\frac{a^2x^2}{\pi}\sin bx}{\operatorname{sh} ax}\,dx = \frac{\pi}{2a}\sin\frac{\pi b^2}{4a^2}\operatorname{cosech}\frac{\pi b}{2a} \qquad [a>0,\ b>0].$$

ET I 93(44)

2. $$\int_0^\infty \frac{\cos\frac{a^2x^2}{\pi}\sin bx}{\operatorname{sh} ax}\,dx = \frac{\pi}{2a}\,\frac{\operatorname{ch}\frac{\pi b}{a}-\cos\frac{\pi b^2}{4a^2}}{\operatorname{sh}\frac{\pi b}{2a}} \qquad [a>0,\ b>0].$$

ET I 93(45)

3. $$\int_0^\infty \frac{\sin\frac{x^2}{\pi}\cos ax}{\operatorname{ch} x}\,dx = \frac{\pi}{2}\,\frac{\cos\frac{a^2\pi}{4}-\frac{1}{\sqrt{2}}}{\operatorname{ch}\frac{a\pi}{2}} \qquad [a>0].$$ ET I 36(54)

4. $$\int_0^\infty \frac{\cos\frac{x^2}{\pi}\cos ax}{\operatorname{ch} x}\,dx = \frac{\pi}{2}\cdot\frac{\sin\frac{a^2\pi}{4}+\frac{1}{\sqrt{2}}}{\operatorname{ch}\frac{a\pi}{2}} \qquad [a>0].$$ ET I 36(55)

5. $$\int_0^\infty \frac{\sin(\pi a x^2)\cos bx}{\operatorname{ch}\pi x}\,dx = -\sum_{k=0}^\infty \exp\left[-\left(k+\frac{1}{2}\right)b\right]\sin\left[\left(k+\frac{1}{2}\right)^2\pi a\right]+$$
$$+\frac{1}{\sqrt{a}}\sum_{k=0}^\infty \exp\left[-\frac{b\left(k+\frac{1}{2}\right)}{a}\right]\sin\left[\frac{\pi}{4}-\frac{b^2}{4\pi a}+\frac{\left(k+\frac{1}{2}\right)^2\pi}{a}\right]$$
$$[a>0,\ b>0].$$ ET I 36(56)

6. $$\int_0^\infty \frac{\cos(\pi a x^2)\cos bx}{\operatorname{ch}\pi x}\,dx =$$
$$=\sum_{k=0}^\infty (-1)^k \exp\left[-\left(k+\frac{1}{2}\right)b\right]\cos\left[\left(k+\frac{1}{2}\right)^2\pi a\right]+$$
$$+\frac{1}{\sqrt{a}}\sum_{k=0}^\infty \exp\left[-\frac{b\left(k+\frac{1}{2}\right)}{a}\right]\cos\left[\frac{\pi}{4}-\frac{b^2}{4\pi a}+\frac{\left(k+\frac{1}{2}\right)^2\pi}{a}\right]$$
$$[a>0,\ b>0].$$ ET I 36(57)

3.991

1. $$\int_0^\infty \sin\pi x^2 \sin ax \operatorname{cth}\pi x\,dx = \frac{1}{2}\operatorname{th}\frac{a}{2}\sin\left(\frac{\pi}{4}+\frac{a^2}{4\pi}\right) \qquad [a>0].$$

ET I 93(42)

2. $$\int_0^\infty \cos\pi x^2 \sin ax \operatorname{cth}\pi x\,dx = \frac{1}{2}\operatorname{th}\frac{a}{2}\left[1-\cos\left(\frac{\pi}{4}+\frac{a^2}{4\pi}\right)\right] \qquad [a>0].$$

ET I 93(43)

3.992

1. $$\int_0^\infty \frac{\sin\pi x^2\cos ax}{1+2\operatorname{ch}\left(\frac{2}{\sqrt{3}}\pi x\right)}\,dx = -\sqrt{3}+\frac{\cos\left(\frac{\pi}{12}-\frac{a^2}{4\pi}\right)}{4\operatorname{ch}\frac{a}{\sqrt{3}}-2} \qquad [a>0].$$

ET I 37(60)

2. $$\int_0^\infty \frac{\cos \pi x^2 \cos ax}{1+2\operatorname{ch}\left(\frac{2}{\sqrt{3}}\pi x\right)}\,dx = 1-\frac{\sin\left(\frac{\pi}{12}-\frac{a^2}{4\pi}\right)}{4\operatorname{ch}\frac{a}{\sqrt{3}}-2} \qquad [a>0].$$ ET I 37(61)

3.993 $$\int_0^\infty \frac{\sin x^2+\cos x^2}{\operatorname{ch}(\sqrt{\pi}\,x)}\cos ax\,dx = \frac{\sqrt{\pi}}{2}\cdot\frac{\sin a^2+\cos a^2}{\operatorname{ch}(\sqrt{\pi}\,a)} \qquad [a>0].$$ ET I 37(58)

3.994

1. $$\int_0^\infty \frac{\sin(2a\operatorname{ch}x)\cos bx}{\sqrt{\operatorname{ch}x}}\,dx = -\frac{\pi}{4}\sqrt{a\pi}\left[J_{\frac{1}{4}+\frac{ib}{2}}(a)\,N_{\frac{1}{4}-\frac{ib}{2}}(a)+J_{\frac{1}{4}-\frac{ib}{2}}(a)\,N_{\frac{1}{4}+\frac{ib}{2}}(a)\right] \qquad [a>0,\ b>0].$$ ET I 37(62)

2. $$\int_0^\infty \frac{\cos(2a\operatorname{ch}x)\cos bx}{\sqrt{\operatorname{ch}x}}\,dx = -\frac{\pi}{4}\sqrt{a\pi}\left[J_{-\frac{1}{4}+\frac{ib}{2}}(a)\,N_{-\frac{1}{4}-\frac{ib}{2}}(a)+J_{-\frac{1}{4}-\frac{ib}{2}}(a)\,N_{-\frac{1}{4}+\frac{ib}{2}}(a)\right] \qquad [a>0,\ b>0].$$ ET I 37(63)

3. $$\int_0^\infty \frac{\sin(2a\operatorname{sh}x)\sin bx}{\sqrt{\operatorname{sh}x}}\,dx = -\frac{i}{2}\sqrt{\pi a}\left[I_{\frac{1}{4}-\frac{ib}{2}}(a)\,K_{-\frac{1}{4}+\frac{ib}{2}}(a)-I_{\frac{1}{4}+\frac{ib}{2}}(a)\,K_{\frac{1}{4}-\frac{ib}{2}}(a)\right] \qquad [a>0,\ b>0].$$ ET I 93(47)

4. $$\int_0^\infty \frac{\cos(2a\operatorname{sh}x)\sin bx}{\sqrt{\operatorname{sh}x}}\,dx =$$
$$= -\frac{i}{2}\sqrt{\pi a}\left[I_{-\frac{1}{4}-\frac{ib}{2}}(a)\,K_{-\frac{1}{4}+\frac{ib}{2}}(a)-I_{-\frac{1}{4}+\frac{ib}{2}}(a)\,K_{-\frac{1}{4}-\frac{ib}{2}}(a)\right]$$
$$[a>0,\ b>0].$$ ET I 93(48)

5. $$\int_0^\infty \frac{\sin(2a\operatorname{sh}x)\cos bx}{\sqrt{\operatorname{sh}x}}\,dx = \frac{\sqrt{\pi a}}{2}\cdot\left[I_{\frac{1}{4}-\frac{ib}{2}}(a)\,K_{\frac{1}{4}+\frac{ib}{2}}(a)+I_{\frac{1}{4}+\frac{ib}{2}}(a)\,K_{\frac{1}{4}-\frac{ib}{2}}(a)\right] \qquad [a>0,\ b>0].$$ ET I 37(64)

6. $$\int_0^\infty \frac{\cos(2a\operatorname{sh}x)\cos bx}{\sqrt{\operatorname{sh}x}}\,dx = \frac{\sqrt{\pi a}}{2}\left[I_{-\frac{1}{4}-\frac{ib}{2}}(a)\,K_{-\frac{1}{4}+\frac{ib}{2}}(a)+I_{-\frac{1}{4}+\frac{ib}{2}}(a)\,K_{-\frac{1}{4}-\frac{ib}{2}}(a)\right] \qquad [a>0,\ b>0].$$ ET I 37(65)

7. $$\int_0^\infty \sin(a\operatorname{ch}x)\sin(a\operatorname{sh}x)\frac{dx}{\operatorname{sh}x} = \frac{\pi}{2}\sin a \qquad [a>0].$$ BI ((264))(22)

3.995

1. $$\int_0^{\frac{\pi}{2}} \frac{\sin(2a\cos^2 x)\,\mathrm{ch}\,(a\sin 2x)}{b^2\cos^2 x + c^2\sin^2 x}\,dx = \frac{\pi}{2bc}\sin\frac{2ac}{b+c}$$

$[b > 0,\ c > 0]$. BI ((273))(9)

2. $$\int_0^{\frac{\pi}{2}} \frac{\cos(2a\cos^2 x)\,\mathrm{ch}\,(a\sin 2x)}{b^2\cos^2 x + c^2\sin^2 x}\,dx = \frac{\pi}{2bc}\cos\frac{2ac}{b+c}$$

$[b > 0,\ c > 0]$. BI ((273))(10)

3.996

1. $$\int_0^{\infty} \sin(a\,\mathrm{sh}\,x)\,\mathrm{sh}\,\beta x\,dx = \sin\frac{\beta\pi}{2}K_\beta(a)$$

$[|\mathrm{Re}\,\beta| < 1,\ a > 0]$. EH II 82(26)

2. $$\int_0^{\infty} \cos(a\,\mathrm{sh}\,x)\,\mathrm{ch}\,\beta x\,dx = \cos\frac{\beta\pi}{2}K_\beta(a)$$

$[|\mathrm{Re}\,\beta| < 1,\ a > 0]$. WA 202(13)

3. $$\int_0^{\frac{\pi}{2}} \cos(a\sin x)\,\mathrm{ch}\,(\beta\cos x)\,dx = \frac{\pi}{2}J_0\left(\sqrt{a^2-\beta^2}\right).$$

MO 40

4. $$\int_0^{\infty} \sin\left(a\,\mathrm{ch}\,x - \frac{1}{2}\beta\pi\right)\mathrm{ch}\,\beta x\,dx = \frac{\pi}{2}J_\beta(a)$$

$[|\mathrm{Re}\,\beta| < 1,\ a > 0]$. WA 199(12)

5. $$\int_0^{\infty} \cos\left(a\,\mathrm{ch}\,x - \frac{1}{2}\beta\pi\right)\mathrm{ch}\,\beta x\,dx = -\frac{\pi}{2}N_\beta(a)$$

$[|\mathrm{Re}\,\beta| < 1,\ a > 0]$. WA 199(13)

3.997

1. $$\int_0^{\frac{\pi}{2}} \sin^\nu x\,\mathrm{sh}\,(\beta\cos x)\,dx = \frac{\sqrt{\pi}}{2}\left(\frac{2}{\beta}\right)^{\frac{\nu}{2}}\Gamma\left(\frac{\nu+1}{2}\right)\mathbf{L}_{\frac{\nu}{2}}(\beta)$$

$[\mathrm{Re}\,\nu > -1]$. EH II 38(53)

2. $$\int_0^{\pi} \sin^\nu x\,\mathrm{ch}\,(\beta\cos x)\,dx = \sqrt{\pi}\left(\frac{2}{\beta}\right)^{\frac{\nu}{2}}\Gamma\left(\frac{\nu+1}{2}\right)I_{\frac{\nu}{2}}(\beta)$$

$[\mathrm{Re}\,\nu > -1]$. WH

3. $$\int_0^{\frac{\pi}{2}} \frac{dx}{\operatorname{ch}(\operatorname{tg} x)\cos x\sqrt{\sin 2x}} = \sqrt{2\pi}\sum_{k=0}^{\infty}\frac{(-1)^k}{\sqrt{2k+1}}.$$ BI ((276))(13)

4. $$\int_0^{\frac{\pi}{2}} \frac{\operatorname{tg}^q x}{\operatorname{ch}(\operatorname{tg} x)+\cos\lambda}\,\frac{dx}{\sin 2x} = \frac{\Gamma(q)}{\sin\lambda}\sum_{k=1}^{\infty}(-1)^{k-1}\frac{\sin k\lambda}{k^q}$$

$[q > 0]$. BI ((275))(20)

4.11-4.12 Combinations involving trigonometric and hyperbolic functions and powers

4.111

1. $$\int_0^{\infty} \frac{\sin ax}{\operatorname{sh}\beta x}\cdot x^{2m}\,dx = (-1)^m\frac{\pi}{2\beta}\cdot\frac{\partial^{2m}}{\partial a^{2m}}\left(\operatorname{th}\frac{a\pi}{2\beta}\right)$$

$[\operatorname{Re}\beta > 0]$ (cf. 3.981 1.). GW ((336))(17a)

2. $$\int_0^{\infty} \frac{\cos ax}{\operatorname{sh}\beta x}\cdot x^{2m+1}\,dx = (-1)^m\frac{\pi}{2\beta}\,\frac{\partial^{2m+1}}{\partial a^{2m+1}}\left(\operatorname{th}\frac{a\pi}{2\beta}\right)$$

$[\operatorname{Re}\beta > 0]$ (cf. 3.981 1.). GW ((336))(17b)

3. $$\int_0^{\infty} \frac{\sin ax}{\operatorname{ch}\beta x}\cdot x^{2m+1}\,dx = (-1)^{m+1}\frac{\pi}{2\beta}\cdot\frac{\partial^{2m+1}}{\partial a^{2m+1}}\left(\frac{1}{\operatorname{ch}\frac{a\pi}{2\beta}}\right)$$

$[\operatorname{Re}\beta > 0]$ (cf. 3.981 3.). GW ((336))(18b)

4. $$\int_0^{\infty} \frac{\cos ax}{\operatorname{ch}\beta x}\cdot x^{2m}\,dx = (-1)^m\frac{\pi}{2\beta}\cdot\frac{\partial^{2m}}{\partial a^{2m}}\left(\frac{1}{\operatorname{ch}\frac{a\pi}{2\beta}}\right)$$

$[\operatorname{Re}\beta > 0]$ (cf. 3.981 3.). GW ((336))(18a)

5. $$\int_0^{\infty} x\,\frac{\sin 2ax}{\operatorname{ch}\beta x}\,dx = \frac{\pi^2}{4\beta^2}\cdot\frac{\operatorname{sh}\frac{a\pi}{\beta}}{\operatorname{ch}^2\frac{a\pi}{\beta}}$$ $[\operatorname{Re}\beta > 0,\ a > 0]$.

BI ((364))(6)a

6. $$\int_0^{\infty} x\,\frac{\cos 2ax}{\operatorname{sh}\beta x}\,dx = \frac{\pi^2}{4\beta^2}\cdot\frac{1}{\operatorname{ch}^2\frac{a\pi}{\beta}}$$ $[\operatorname{Re}\beta > 0,\ a > 0]$. BI ((364))(1)a

7. $$\int_0^{\infty} \frac{\sin ax}{\operatorname{ch}\beta x}\cdot\frac{dx}{x} = 2\operatorname{arctg}\left(\exp\frac{\pi a}{2\beta}\right)-\frac{\pi}{2}$$

$[\operatorname{Re}\beta > 0,\ a > 0]$. BI ((387))(1), ET I 89(13), LI((298))(17)

4.112

1. $$\int_0^{\infty} (x^2+\beta^2)\,\frac{\cos ax}{\operatorname{ch}\frac{\pi x}{2\beta}}\,dx = \frac{2\beta^3}{\operatorname{ch}^3 a\beta}$$ $[\operatorname{Re}\beta > 0,\ a > 0]$. ET I 32(19)

2. $$\int_0^\infty x\,(x^2+4\beta^2)\frac{\cos ax}{\operatorname{sh}\frac{\pi x}{2\beta}}\,dx=\frac{6\beta^4}{\operatorname{ch}^4 a\beta}\qquad [\operatorname{Re}\beta>0,\ a>0].$$ ET I 32(20)

4.113

1. $$\int_0^\infty \frac{\sin ax}{\operatorname{sh}\pi x}\cdot\frac{dx}{x^2+\beta^2}=-\frac{1}{2\beta^2}-\frac{\pi e^{-a\beta}}{\beta\sin\pi\beta}+$$
$$+\frac{1}{2\beta^2}\left[{}_2F_1(1,\ -\beta;\ 1-\beta;\ -e^{-a})+{}_2F_1(1,\ \beta;\ 1+\beta;\ -e^{-a})\right]=$$
$$=\frac{1}{2\beta^2}-\frac{\pi e^{-a\beta}}{2\beta\sin\pi\beta}-\sum_{k=1}^\infty\frac{(-1)^k e^{-ak}}{k^2-\beta^2}\qquad[\operatorname{Re}\beta>0,\ \beta\neq 0,\ 1,\ 2,\ \dots,\ a>0].$$ ET I 90(18)

2. $$\int_0^\infty \frac{\sin ax}{\operatorname{sh}\pi x}\cdot\frac{dx}{x^2+m^2}=\frac{(-1)^m a e^{-ma}}{2m}+$$
$$+\frac{1}{2m}\sum_{k=1}^{m-1}\frac{(-1)^k e^{-ka}}{m-k}+\frac{(-1)^m e^{-ma}}{2m}\ln(1+e^{-a})+$$
$$+\frac{1}{2m!}\frac{d^{m-1}}{dz^{m-1}}\left[\frac{(1+z)^{m-1}}{z}\ln(1+z)\right]_{z=e^{-a}}\qquad[a>0].$$ ET I 89(17)

3. $$\int_0^\infty \frac{\sin ax}{\operatorname{sh}\pi x}\cdot\frac{dx}{1+x^2}=\frac{1}{2}\int_{-\infty}^\infty\frac{\sin ax}{\operatorname{sh}\pi x}\cdot\frac{dx}{1+x^2}=$$
$$=-\frac{a}{2}\operatorname{ch}a+\operatorname{sh}a\ln\left(2\operatorname{ch}\frac{a}{2}\right).$$ GW ((336))(21b)

4. $$\int_0^\infty \frac{\sin ax}{\operatorname{sh}\frac{\pi}{2}x}\cdot\frac{dx}{1+x^2}=\frac{1}{2}\int_{-\infty}^\infty\frac{\sin ax}{\operatorname{sh}\frac{\pi}{2}x}\cdot\frac{dx}{1+x^2}=$$
$$=\frac{\pi}{2}\operatorname{sh}a-\operatorname{ch}a\operatorname{arctg}(\operatorname{sh}a).$$ GW ((336))(21a)

5. $$\int_0^\infty \frac{\sin ax}{\operatorname{sh}\frac{\pi}{4}x}\cdot\frac{dx}{1+x^2}=-\frac{\pi}{\sqrt{2}}e^{-a}+\frac{\operatorname{sh}a}{\sqrt{2}}\ln\frac{2\operatorname{ch}a+\sqrt{2}}{2\operatorname{ch}a-\sqrt{2}}+$$
$$+\sqrt{2}\operatorname{ch}a\operatorname{arctg}\frac{\sqrt{2}}{2\operatorname{sh}a}\qquad[a>0].$$ LI ((389))(1)

6. $$\int_0^\infty \frac{\sin ax}{\operatorname{ch}\frac{\pi}{4}x}\cdot\frac{x\,dx}{1+x^2}=\frac{\pi}{\sqrt{2}}e^{-a}+\frac{\operatorname{sh}a}{\sqrt{2}}\ln\frac{2\operatorname{ch}a+\sqrt{2}}{2\operatorname{ch}a-\sqrt{2}}-$$
$$-\sqrt{2}\operatorname{ch}a\operatorname{arctg}\left(\frac{1}{\sqrt{2}\operatorname{sh}a}\right)\qquad[a>0].$$ BI ((388))(1)

7. $$\int_0^\infty \frac{\cos ax}{\operatorname{sh}\pi x}\cdot\frac{x\,dx}{1+x^2}=-\frac{1}{2}+\frac{a}{2}e^{-a}+\operatorname{ch}a\ln(1+e^{-a})$$
$$[a>0].$$ BI ((389))(14), ET I 32(24)

8. $$\int_0^\infty \frac{\cos ax}{\operatorname{sh}\frac{\pi}{2}x}\cdot\frac{x\,dx}{1+x^2} = 2\operatorname{sh} a \operatorname{arctg}(e^{-a}) + \frac{\pi}{2}e^{-a} - 1$$

$[a > 0]$. BI ((389))(11)

9. $$\int_0^\infty \frac{\cos ax}{\operatorname{ch}\pi x}\cdot\frac{dx}{x^2+\beta^2} = \sum_{k=0}^{\infty}(-1)^k \frac{\left(k+\frac{1}{2}\right)^2 e^{-a\beta} - \beta e^{-\left(k+\frac{1}{2}\right)a}}{\beta\left[\left(k+\frac{1}{2}\right)^2 - \beta^2\right]}$$

$[\operatorname{Re}\beta > 0,\ a > 0]$. ET I 32(26)

10. $$\int_0^\infty \frac{\cos ax}{\operatorname{ch}\pi x}\cdot\frac{dx}{\left(m+\frac{1}{2}\right)^2+x^2} = \frac{(-1)^m e^{-\left(m+\frac{1}{2}\right)a}}{2m+1}[a+\ln(1+e^{-a})] +$$
$$+\frac{e^{-\frac{a}{2}}}{2m+1}\sum_{k=0}^{m-1}\frac{(-1)^k e^{-ak}}{k-m} + \frac{e^{-\frac{a}{2}}}{(2m+1)(m+1)}\cdot {}_2F_1(1,\ m+1;\ m+2; -e^{-a})$$

$[a > 0]$. ET I 32(25)

11. $$\int_0^\infty \frac{\cos ax}{\operatorname{ch}\pi x}\cdot\frac{dx}{1+x^2} = 2\operatorname{ch}\frac{a}{2} - [e^a \operatorname{arctg}(e^{-\frac{a}{2}}) + e^{-a}\operatorname{arctg}(e^{\frac{a}{2}})]$$

$[a > 0]$. ET I 32(21)

12. $$\int_0^\infty \frac{\cos ax}{\operatorname{ch}\frac{\pi}{2}x}\cdot\frac{dx}{1+x^2} = ae^{-a} + \operatorname{ch} a \ln(1+e^{-2a})$$

$[a > 0]$. BI ((388))(6)

13. $$\int_0^\infty \frac{\cos ax}{\operatorname{ch}\frac{\pi}{4}x}\cdot\frac{dx}{1+x^2} = \frac{\pi}{\sqrt{2}}e^{-a} + \frac{2\operatorname{sh} a}{\sqrt{2}}\operatorname{arctg}\left(\frac{1}{\sqrt{2}\operatorname{sh} a}\right) -$$
$$-\frac{\operatorname{ch} a}{\sqrt{2}}\ln\frac{2\operatorname{ch} a+\sqrt{2}}{2\operatorname{ch} a-\sqrt{2}}$$

$[a > 0]$. BI ((388))(5)

4.114

1. $$\int_0^\infty \frac{\sin ax}{x}\cdot\frac{\operatorname{sh}\beta x}{\operatorname{sh}\gamma x}\,dx = \operatorname{arctg}\left(\operatorname{tg}\frac{\beta\pi}{2\gamma}\operatorname{th}\frac{a\pi}{2\gamma}\right)$$

$[|\operatorname{Re}\beta| < \operatorname{Re}\gamma,\ a > 0]$. BI ((387))(6)a

2. $$\int_0^\infty \frac{\cos ax}{x}\cdot\frac{\operatorname{sh}\beta x}{\operatorname{ch}\gamma x}\,dx = \frac{1}{2}\ln\frac{\operatorname{ch}\frac{a\pi}{2\gamma}+\sin\frac{\beta\pi}{2\gamma}}{\operatorname{ch}\frac{a\pi}{2\gamma}-\sin\frac{\beta\pi}{2\gamma}}$$

$[|\operatorname{Re}\beta| < \operatorname{Re}\gamma]$. ET I 33(34)

4.115

1. $$\int_0^\infty \frac{x\sin ax}{x^2+b^2}\cdot\frac{\operatorname{sh}\beta x}{\operatorname{sh}\pi x}\,dx = \frac{\pi}{2}\frac{e^{-ab}\sin b\beta}{\sin b\pi} + \sum_{k=1}^{\infty}(-1)^k\frac{ke^{-ak}\sin k\beta}{k^2-b^2}$$

$[0 < \operatorname{Re}\beta < \pi,\ a > 0,\ b > 0]$. BI ((389))(23)

2. $$\int_0^\infty \frac{x \sin ax}{x^2+1} \cdot \frac{\operatorname{sh} \beta x}{\operatorname{sh} \pi x} dx = \frac{1}{2} e^{-a} (a \sin \beta - \beta \cos \beta) - \frac{1}{2} \operatorname{sh} a \sin \beta \ln [1 + 2e^{-a} \cos \beta + e^{-2a}] + \operatorname{ch} a \cos \beta \operatorname{arctg} \frac{\sin \beta}{e^a + \cos \beta}$$
$[|\operatorname{Re} \beta| < \pi, \ a > 0]$. LI ((389))(10)

3. $$\int_0^\infty \frac{x \sin ax}{x^2+1} \cdot \frac{\operatorname{sh} \beta x}{\operatorname{sh} \frac{\pi}{2} x} dx = \frac{\pi}{2} e^{-a} \sin \beta + \frac{1}{2} \cos \beta \operatorname{sh} a \ln \frac{\operatorname{ch} a + \sin \beta}{\operatorname{ch} a - \sin \beta} - \sin \beta \operatorname{ch} a \operatorname{arctg} \left(\frac{\cos \beta}{\operatorname{sh} a} \right)$$
$\left[|\operatorname{Re} \beta| < \frac{\pi}{2}, \ a > 0 \right]$. BI ((389))(8)

4. $$\int_0^\infty \frac{\cos ax}{x^2+b^2} \cdot \frac{\operatorname{sh} \beta x}{\operatorname{sh} \pi x} dx = \frac{\pi}{2b} \cdot \frac{e^{-ab} \sin b\beta}{\sin b\pi} + \sum_{k=1}^\infty (-1)^k \frac{e^{-ak} \sin k\beta}{k^2 - b^2}$$
$[0 < \operatorname{Re} \beta < \pi, \ a > 0, \ b > 0]$. BI ((389))(22)

5. $$\int_0^\infty \frac{\cos ax}{x^2+1} \cdot \frac{\operatorname{sh} \beta x}{\operatorname{sh} \pi x} dx = \frac{1}{2} e^{-a} (a \sin \beta - \beta \cos \beta) + \frac{1}{2} \operatorname{ch} a \sin \beta \ln (1 + 2e^{-a} \cos \beta + e^{-2a}) - \operatorname{sh} a \cos \beta \operatorname{arctg} \frac{\sin \beta}{e^a + \cos \beta}$$
$[|\operatorname{Re} \beta| < \pi, \ a > 0 \ \ b > 0]$

BI ((389))(20)a

6. $$\int_0^\infty \frac{\cos ax}{x^2+1} \cdot \frac{\operatorname{sh} \beta x}{\operatorname{sh} \frac{\pi}{2} x} dx = \frac{\pi}{2} e^{-a} \sin \beta - \frac{1}{2} \operatorname{ch} a \cos \beta \ln \frac{\operatorname{ch} a + \sin \beta}{\operatorname{ch} a - \sin \beta} + \operatorname{sh} a \sin \beta \operatorname{arctg} \frac{\cos \beta}{\operatorname{sh} a}$$
$\left[|\operatorname{Re} \beta| < \frac{\pi}{2}, \ a > 0, \ b > 0 \right]$. BI ((389))(18)

7. $$\int_0^\infty \frac{\sin ax}{x^2 + \frac{1}{4}} \cdot \frac{\operatorname{sh} \beta x}{\operatorname{ch} \pi x} dx = e^{-\frac{a}{2}} \left(a \sin \frac{\beta}{2} - \beta \cos \frac{\beta}{2} \right) - \operatorname{sh} \frac{a}{2} \sin \frac{\beta}{2} \ln (1 + 2e^{-a} \cos \beta + e^{-2a}) + \operatorname{ch} \frac{a}{2} \cos \frac{\beta}{2} \operatorname{arctg} \frac{\sin \beta}{1 + e^{-a} \cos \beta}$$
$[|\operatorname{Re} \beta| < \pi, \ a > 0]$.

ET I 91(26)

8. $$\int_0^\infty \frac{\sin ax}{x^2+\beta^2} \cdot \frac{\operatorname{ch} \gamma x}{\operatorname{sh} \pi x} dx = \frac{1}{2\beta^2} - \frac{\pi}{2\beta} \cdot \frac{e^{-a\beta} \cos \beta\gamma}{\sin \beta\pi} + \sum_{k=1}^\infty (-1)^{k-1} \frac{e^{-ak} \cos k\gamma}{k^2 - \beta^2}$$
$[0 \leqslant \operatorname{Re} \beta, \ |\operatorname{Re} \gamma| < \pi, \ a > 0]$.

BI ((389))(21)

9. $$\int_0^\infty \frac{\sin ax}{x^2+1} \cdot \frac{\operatorname{ch} \beta x}{\operatorname{sh} \pi x} dx = -\frac{1}{2} e^{-a} (a \cos\beta + \beta \sin\beta) + \frac{1}{2} \operatorname{sh} a \cos\beta \ln(1 + 2e^{-a} \cos\beta + e^{-2a}) + \operatorname{ch} a \sin\beta \operatorname{arctg} \frac{\sin\beta}{e^a + \cos\beta} \qquad [|\operatorname{Re}\beta| < \pi,\ a > 0].$$

ET I 91(25), LI((389))(9)

10. $$\int_0^\infty \frac{\sin ax}{x^2+1} \cdot \frac{\operatorname{ch} \beta x}{\operatorname{sh} \frac{\pi}{2} x} dx = -\frac{\pi}{2} e^{-a} \cos\beta + \frac{1}{2} \operatorname{sh} a \sin\beta \ln \frac{\operatorname{ch} a + \sin\beta}{\operatorname{ch} a - \sin\beta} + \operatorname{ch} a \cos\beta \operatorname{arctg} \frac{\cos\beta}{\operatorname{sh} a} \qquad \left[|\operatorname{Re}\beta| < \frac{\pi}{2},\ a > 0\right],$$

BI ((389))(7)

11. $$\int_0^\infty \frac{x \cos ax}{x^2+b^2} \cdot \frac{\operatorname{ch} \beta x}{\operatorname{sh} \pi x} dx = \frac{\pi}{2} \cdot \frac{e^{-ab} \cos b\beta}{\sin b\pi} + \sum_{k=1}^\infty (-1)^k \frac{k e^{-ak} \cos k\beta}{k^2 - b^2} \qquad [|\operatorname{Re}\beta| < \pi,\ a > 0].$$

BI ((389))(24)

12. $$\int_0^\infty \frac{x \cos ax}{x^2+1} \cdot \frac{\operatorname{ch} \beta x}{\operatorname{sh} \pi x} dx = \frac{1}{2} e^{-a} (a \cos\beta + \beta \sin\beta) - \frac{1}{2} + \frac{1}{2} \operatorname{ch} a \cos\beta \ln[1 + 2e^{-a} \cos\beta + e^{-2a}] + \operatorname{sh} a \sin\beta \operatorname{arctg} \frac{\sin\beta}{e^a + \cos\beta} \qquad [|\operatorname{Re}\beta| < \pi,\ a > 0].$$

BI ((389))(19)

13. $$\int_0^\infty \frac{x \cos ax}{x^2+1} \cdot \frac{\operatorname{ch} \beta x}{\operatorname{sh} \frac{\pi}{2} x} dx = -1 + \frac{\pi}{2} e^{-a} \cos\beta + \frac{1}{2} \operatorname{ch} a \sin\beta \ln \frac{\operatorname{ch} a + \sin\beta}{\operatorname{ch} a - \sin\beta} + \operatorname{sh} a \cos\beta \operatorname{arctg} \frac{\cos\beta}{\operatorname{sh} a} \qquad \left[|\operatorname{Re}\beta| < \frac{\pi}{2},\ a > 0\right].$$

BI ((389))(17)

14. $$\int_0^\infty \frac{\cos ax}{x^2+1} \cdot \frac{\operatorname{ch} \beta x}{\operatorname{ch} \frac{\pi}{2} x} dx = a e^{-a} \cos\beta + \beta e^{-a} \sin\beta + \operatorname{sh} a \sin\beta \operatorname{arctg} \frac{e^{-2a} \sin 2\beta}{1 + e^{-2a} \cos 2\beta} + \frac{1}{2} \operatorname{ch} a \cos\beta \ln(1 + 2e^{-2a} \cos 2\beta + e^{-4a}) \qquad \left[|\operatorname{Re}\beta| < \frac{\pi}{2},\ a > 0\right].$$

ET I 34(37)

4.116

1. $$\int_0^\infty x \cos 2ax \operatorname{th} x\, dx = -\frac{\pi^2}{4} \cdot \frac{\operatorname{ch} a\pi}{\operatorname{sh}^2 a\pi} \qquad [a > 0].$$

BI ((364))(2)

2. $\int_0^\infty \cos ax \operatorname{th} \beta x \frac{dx}{x} = \ln \operatorname{cth} \frac{a\pi}{4\beta}$ $[\operatorname{Re} \beta > 0,\ a > 0]$. BI ((387))(8)

3. $\int_0^\infty \cos ax \operatorname{cth} \beta x \frac{dx}{x} = -\ln\left(2 \operatorname{sh} \frac{a\pi}{2\beta}\right)$ $[\operatorname{Re} \beta > 0,\ a > 0]$. BI ((387))(9)

4.117

1. $\int_0^\infty \frac{\sin ax}{1+x^2} \operatorname{th} \frac{\pi x}{2} dx = a \operatorname{ch} a - \operatorname{sh} a \ln (2 \operatorname{sh} a)$ $[a > 0]$. BI ((388))(3)

2. $\int_0^\infty \frac{\sin ax}{1+x^2} \operatorname{th} \frac{\pi x}{4} dx = -\frac{\pi}{2} e^a + \operatorname{sh} a \ln \operatorname{cth} \frac{a}{2} + 2 \operatorname{ch} a \operatorname{arctg}(e^a)$. BI ((388))(4)

3. $\int_0^\infty \frac{\sin ax}{1+x^2} \operatorname{cth} \pi x \, dx = \frac{a}{2} e^{-a} - \operatorname{sh} a \ln (1 - e^{-a})$ $[a > 0]$. BI ((389))(5)

4. $\int_0^\infty \frac{\sin ax}{1+x^2} \operatorname{cth} \frac{\pi}{2} x \, dx = \operatorname{sh} a \ln \operatorname{cth} \frac{a}{2}$ $[a > 0]$. BI ((389))(6)

5. $\int_0^\infty \frac{x \cos ax}{1+x^2} \operatorname{th} \frac{\pi}{2} x \, dx = -a e^{-a} + \operatorname{sh} a \ln (1 - e^{-2a})$ $[a > 0]$. BI ((388))(7)

6. $\int_0^\infty \frac{x \cos ax}{1+x^2} \operatorname{th} \frac{\pi}{4} x \, dx = -\frac{\pi}{2} e^a + \operatorname{ch} a \ln \operatorname{cth} \frac{a}{2} + 2 \operatorname{sh} a \operatorname{arctg}(e^a)$ $[a > 0]$. BI ((388))(8)

7. $\int_0^\infty \frac{x \cos ax}{1+x^2} \operatorname{cth} \pi x \, dx = -\frac{a}{2} e^{-a} - \frac{1}{2} - \operatorname{ch} a \ln (1 - e^{-a})$. BI ((389))(15)a, ET I 33(31)a

8. $\int_0^\infty \frac{x \cos ax}{1+x^2} \operatorname{cth} \frac{\pi}{2} x \, dx = -1 + \operatorname{ch} a \ln \operatorname{cth} \frac{a}{2}$ $[a > 0]$. BI ((389))(12)

9. $\int_0^\infty \frac{x \cos ax}{1+x^2} \operatorname{cth} \frac{\pi}{4} x \, dx = -2 + \frac{\pi}{2} e^{-a} + \operatorname{ch} a \ln \operatorname{cth} \frac{a}{2} + 2 \operatorname{sh} a \operatorname{arctg}(e^{-a})$ $[a > 0]$. BI ((389))(13)

4.118 $\int_0^\infty \frac{x \sin ax}{\operatorname{ch}^2 x} dx = -\frac{d}{da}\left(\frac{\pi a}{2 \operatorname{sh} \frac{\pi a}{2}}\right)$ $[a > 0]$. ET I 89(14)

4.119 $\int_0^\infty \frac{1-\cos px}{\operatorname{sh} qx} \cdot \frac{dx}{x} = \ln\left(\operatorname{ch}\frac{p\pi}{2q}\right).$ BI ((387))(2)a

4.121

1. $\int_0^\infty \frac{\sin ax - \sin bx}{\operatorname{ch} \beta x} \cdot \frac{dx}{x} = 2 \operatorname{arctg} \frac{\exp\frac{a\pi}{2\beta} - \exp\frac{b\pi}{2\beta}}{1+\exp\frac{(a+b)\pi}{2\beta}}$ $[\operatorname{Re}\beta > 0]$. GW ((336))(19b)

2. $\int_0^\infty \frac{\cos ax - \cos bx}{\operatorname{sh} \beta x} \cdot \frac{dx}{x} = \ln \frac{\operatorname{ch}\frac{b\pi}{2\beta}}{\operatorname{ch}\frac{a\pi}{2\beta}}$ $[\operatorname{Re}\beta > 0]$. GW ((336))(19a)

4.122

1. $\int_0^\infty \frac{\cos \beta x \sin \gamma x}{\operatorname{ch} \delta x} \cdot \frac{dx}{x} = \operatorname{arctg} \frac{\operatorname{sh}\frac{\gamma\pi}{2\delta}}{\operatorname{ch}\frac{\beta\pi}{2\delta}}$ $[\operatorname{Re}\delta > |\operatorname{Im}(\beta+\gamma)|]$. ET I 93(46)a

2. $\int_0^\infty \sin^2 ax \frac{\operatorname{ch} \beta x}{\operatorname{sh} x} \cdot \frac{dx}{x} = \frac{1}{4} \ln \frac{\operatorname{ch} 2a\pi + \cos \beta\pi}{1+\cos \beta\pi}$ $[|\operatorname{Re}\beta| < 1]$. BI ((387))(7)

4.123

1. $\int_0^\infty \frac{\sin x}{\operatorname{ch} ax + \cos x} \cdot \frac{x\,dx}{x^2-\pi^2} = \operatorname{arctg}\frac{1}{a} - \frac{1}{a}.$ BI ((390))(1)

2. $\int_0^\infty \frac{\sin x}{\operatorname{ch} ax - \cos x} \cdot \frac{x\,dx}{x^2-\pi^2} = \frac{a}{1+a^2} - \operatorname{arctg}\frac{1}{a}.$ BI ((390))(2)

3. $\int_0^\infty \frac{\sin 2x}{\operatorname{ch} 2ax - \cos 2x} \cdot \frac{x\,dx}{x^2-\pi^2} = \frac{1}{2a} \cdot \frac{1+2a^2}{1+a^2} - \operatorname{arctg}\frac{1}{a}.$ BI ((390))(4)

4. $\int_0^\infty \frac{\operatorname{ch} ax \sin x}{\operatorname{ch} 2ax - \cos 2x} \cdot \frac{x\,dx}{x^2-\pi^2} = \frac{-1}{2a(1+a^2)}.$ LI ((390))(3)

5. $\int_0^\infty \frac{\cos ax}{\operatorname{ch} \pi x + \cos \pi\beta} \cdot \frac{dx}{x^2+\gamma^2} = \frac{\pi e^{-a\gamma}}{2\gamma(\cos\gamma\pi + \cos\beta\pi)} +$

$$+ \frac{1}{\operatorname{sh}\beta\pi} \sum_{k=0}^\infty \left\{ \frac{\exp[-(2k+1-\beta)a]}{\gamma^2-(2k+1-\beta)^2} - \frac{\exp[-(2k+1+\beta)a]}{\gamma^2-(2k+1+\beta)^2} \right\}$$

$[0 < \operatorname{Re}\beta < 1, \quad \operatorname{Re}\gamma > 0, \quad a > 0]$. ET I 33(27)

6. $\int_0^\infty \frac{\sin ax \operatorname{sh} bx}{\cos 2ax + \operatorname{ch} 2bx} x^{p-1}\,dx =$

$$= \frac{\Gamma(p)}{(a^2+b^2)^{\frac{p}{2}}} \sin\left(p \operatorname{arctg}\frac{a}{b}\right) \sum_{k=0}^\infty \frac{(-1)^k}{(2k+1)^p}$$

$[p > 0]$. BI ((364))(8)

7. $$\int_0^\infty \sin ax^2 \frac{\sin\frac{\pi x}{2}\operatorname{sh}\frac{\pi x}{2}}{\cos\pi x+\operatorname{ch}\pi x}\cdot x\,dx=\frac{1}{4}\left[\frac{\partial\theta_1(z,\ q)}{\partial z}\right]_{z=0,\ q=e^{-2a}} \qquad [a>0].$$

ET I 93(49)

4.124

1. $$\int_0^1 \frac{\cos px\operatorname{ch}\left(q\sqrt{1-x^2}\right)}{\sqrt{1-x^2}}\,dx=\frac{\pi}{2}J_0\left(\sqrt{p^2-q^2}\right).$$

MO (40)

2. $$\int_u^\infty \cos ax\operatorname{ch}\sqrt{\beta(u^2-x^2)}\cdot\frac{dx}{\sqrt{u^2-x^2}}=\frac{\pi}{2}J_0\left(\frac{u}{\sqrt{a^2-\beta^2}}\right).$$

ET I 34(38)

4.125

1. $$\int_0^\infty \operatorname{sh}(a\sin x)\cos(a\cos x)\sin x\sin 2nx\frac{dx}{x}=$$
$$=\frac{(-1)^{n-1}a^{2n-1}}{(2n-1)!}\frac{\pi}{8}\left[1+\frac{a^2}{2n(2n+1)}\right].$$

LI ((367))(14)

2. $$\int_0^\infty \operatorname{ch}(a\sin x)\cos(a\cos x)\sin x\cos(2n-1)x\frac{dx}{x}=$$
$$=\frac{(-1)^{n-1}a^{2(n-1)}}{[2(n-1)]!}\frac{\pi}{8}\left[1-\frac{a^2}{2n(2n-1)}\right].$$

LI ((367))(15)

3. $$\int_0^\infty \operatorname{sh}(a\sin x)\cos(a\cos x)\cos x\cos 2nx\frac{dx}{x}=$$
$$=\frac{\pi}{2}\sum_{k=n+1}^{\infty}\frac{(-1)^k a^{2k+1}}{(2k+1)!}+\frac{(-1)^n a^{2n+1}}{(2n+1)!}\frac{3\pi}{8}+\frac{(-1)^{n-1}a^{2n-1}}{(2n-1)!}\frac{\pi}{8}.$$

LI ((367))(21)

4.126

1. $$\int_0^\infty \sin(a\cos bx)\operatorname{sh}(a\sin bx)\frac{x\,dx}{c^2-x^2}=$$
$$=\frac{\pi}{2}[\cos(a\cos bc)\operatorname{ch}(a\sin bc)-1] \qquad [b>0].$$

BI ((381))(2)

2. $$\int_0^\infty \sin(a\cos bx)\operatorname{ch}(a\sin bx)\frac{dx}{c^2-x^2}=\frac{\pi}{2c}\cos(a\cos bc)\operatorname{sh}(a\sin bc)$$
$$[b>0,\ c>0].$$

BI ((381))(1)

3. $$\int_0^\infty \cos(a\cos bx)\operatorname{sh}(a\sin bx)\frac{x\,dx}{c^2-x^2}=\frac{\pi}{2}[a\cos bc-\sin(a\cos bc)\operatorname{ch}(a\sin bc)]$$
$$[b>0].$$

BI ((381))(4)

4. $$\int_0^\infty \cos(a\cos bx)\operatorname{ch}(a\sin bx)\frac{dx}{c^2-x^2}=-\frac{\pi}{2c}\sin(a\cos bc)\operatorname{sh}(a\sin bc)$$
$$[b>0].$$

BI ((381))(3)

4.13 Combinations of trigonometric and hyperbolic functions and exponentials

4.131

1. $$\int_0^\infty \sin ax \operatorname{sh}^\nu \gamma x \cdot e^{-\beta x}\, dx =$$
$$= -\frac{i\Gamma(\nu+1)}{2^{\nu+2}\gamma}\left\{\frac{\Gamma\left(\frac{\beta-\nu\gamma-ai}{2\gamma}\right)}{\Gamma\left(\frac{\beta+\nu\gamma-ai}{2\gamma}+1\right)} - \frac{\Gamma\left(\frac{\beta-\nu\gamma+ai}{2\gamma}\right)}{\Gamma\left(\frac{\beta+\gamma\nu+ai}{2\gamma}+1\right)}\right\}$$
$[\operatorname{Re}\nu > -2,\ \operatorname{Re}\gamma > 0,\ |\operatorname{Re}(\gamma\nu)| < \operatorname{Re}\beta].$ ET I 91(30)a

2. $$\int_0^\infty \cos ax \operatorname{sh}^\nu \gamma x \cdot e^{-\beta x}\, dx =$$
$$= \frac{\Gamma(\nu+1)}{2^{\nu+2}\gamma}\left\{\frac{\Gamma\left(\frac{\beta-\nu\gamma-ai}{2\gamma}\right)}{\Gamma\left(\frac{\beta+\gamma\nu-ai}{2\gamma}+1\right)} + \frac{\Gamma\left(\frac{\beta-\nu\gamma+ai}{2\gamma}\right)}{\Gamma\left(\frac{\beta+\nu\gamma+ai}{2\gamma}+1\right)}\right\}$$
$[\operatorname{Re}\nu > -1,\ \operatorname{Re}\gamma > 0,\ |\operatorname{Re}(\gamma\nu)| < \operatorname{Re}\beta].$ ET I 34(40)a

3. $$\int_0^\infty e^{-\beta x}\frac{\sin ax}{\operatorname{sh}\gamma x}\, dx = \sum_{k=1}^\infty \frac{2a}{a^2+[\beta+(2k-1)\gamma]^2};$$ BI ((264))(9)a
$$= \frac{1}{2\gamma i}\left[\psi\left(\frac{\beta+\gamma+ia}{2\gamma}\right) - \psi\left(\frac{\beta+\gamma-ia}{2\gamma}\right)\right]$$
$[\operatorname{Re}\beta > |\operatorname{Re}\gamma|].$ ET I 91(28)

4. $$\int_0^\infty e^{-x}\frac{\sin ax}{\operatorname{sh} x}\, dx = \frac{\pi}{2}\operatorname{cth}\frac{a\pi}{2} - \frac{1}{a}.$$ ET I 91(29)

4.132

1. $$\int_0^\infty \frac{\sin ax \operatorname{sh}\beta x}{e^{\gamma x}-1}\, dx = -\frac{a}{2(a^2+\beta^2)} + \frac{\pi}{2\gamma}\cdot\frac{\operatorname{sh}\frac{2\pi a}{\gamma}}{\operatorname{ch}\frac{2\pi a}{\gamma} - \cos\frac{2\pi\beta}{\gamma}} +$$
$$+\frac{i}{2\gamma}\left[\psi\left(\frac{\beta}{\gamma}+i\frac{a}{\gamma}+1\right) - \psi\left(\frac{\beta}{\gamma}-i\frac{a}{\gamma}+1\right)\right]$$ $[\operatorname{Re}\gamma > |\operatorname{Re}\beta|,\ a > 0].$ ET I 92(33)

2. $$\int_0^\infty \frac{\sin ax \operatorname{ch}\beta x}{e^{\gamma x}-1}\, dx = -\frac{a}{2(a^2+\beta^2)} + \frac{\pi}{2\gamma}\cdot\frac{\operatorname{sh}\frac{2\pi a}{\gamma}}{\operatorname{ch}\frac{2\pi a}{\gamma} - \cos\frac{2\pi\beta}{\gamma}}$$ $[\operatorname{Re}\gamma > |\operatorname{Re}\beta|].$ BI ((265))(5)a, ET I 92(34)

3. $$\int_0^\infty \frac{\sin ax \operatorname{ch}\beta x}{e^{\gamma x}+1}\, dx = \frac{a}{2(a^2+\beta^2)} - \frac{\pi}{\gamma}\cdot\frac{\operatorname{sh}\frac{a\pi}{\gamma}\cos\frac{\beta\pi}{\gamma}}{\operatorname{ch}\frac{2a\pi}{\gamma} - \cos\frac{2\beta\pi}{\gamma}}$$ $[\operatorname{Re}\gamma > |\operatorname{Re}\beta|].$ ET I 92(35)

4. $$\int_0^\infty \frac{\cos ax \operatorname{sh}\beta x}{e^{\gamma x}-1}\, dx = \frac{\beta}{2(a^2+\beta^2)} - \frac{\pi}{2\gamma}\cdot\frac{\sin\frac{2\pi\beta}{\gamma}}{\operatorname{ch}\frac{2a\pi}{\gamma} - \cos\frac{2\beta\pi}{\gamma}}$$ $[\operatorname{Re}\gamma > |\operatorname{Re}\beta|].$ LI ((265))(8)

5. $\int\limits_0^\infty \frac{\cos ax \operatorname{sh} \beta x}{e^{\gamma x}+1} dx = -\frac{\beta}{2(a^2+\beta^2)} + \frac{\pi}{\gamma} \cdot \frac{\sin \frac{\pi\beta}{\gamma} \operatorname{ch} \frac{\pi a}{\gamma}}{\operatorname{ch} \frac{2a\pi}{\gamma} - \cos \frac{2\beta\pi}{\gamma}}$ $[\operatorname{Re} \gamma > |\operatorname{Re} \beta|]$. ET I 34(39)

4.133

1. $\int\limits_0^\infty \sin ax \operatorname{sh} \beta x \exp\left(-\frac{x^2}{4\gamma}\right) dx =$

$= \sqrt{\pi\gamma} \exp \gamma (\beta^2 - a^2) \sin (2a\beta\gamma)$ $[\operatorname{Re} \gamma > 0]$. ET I 92(37)

2. $\int\limits_0^\infty \cos ax \operatorname{ch} \beta x \exp\left(-\frac{x^2}{4\gamma}\right) dx =$

$= \sqrt{\pi\gamma} \exp \gamma (\beta^2 - a^2) \cos (2a\beta\gamma)$ $[\operatorname{Re} \gamma > 0]$. ET I 35(41)

4.134

1. $\int\limits_0^\infty e^{-\beta x^2} (\operatorname{ch} x + \cos x) dx = \sqrt{\frac{\pi}{\beta}} \operatorname{ch} \frac{1}{4\beta}$ $[\operatorname{Re} \beta > 0]$. ME 24

2. $\int\limits_0^\infty e^{-\beta x^2} (\operatorname{ch} x - \cos x) dx = \sqrt{\frac{\pi}{\beta}} \operatorname{sh} \frac{1}{4\beta}$ $[\operatorname{Re} \beta > 0]$. ME 24

4.135

1. $\int\limits_0^\infty \sin ax^2 \operatorname{ch} 2\gamma x \cdot e^{-\beta x^2} dx = \frac{1}{2} \sqrt[4]{\frac{\pi^2}{a^2+\beta^2}} \exp\left(-\frac{\beta\gamma^2}{a^2+\beta^2}\right) \times$

$\times \sin\left(\frac{a\gamma^2}{a^2+\beta^2} + \frac{1}{2} \operatorname{arctg} \frac{a}{\beta}\right)$ $[\operatorname{Re} \beta > 0]$. LI ((268))(7)

2. $\int\limits_0^\infty \cos ax^2 \operatorname{ch} 2\gamma x \cdot e^{-\beta x^2} dx = \frac{1}{2} \sqrt[4]{\frac{\pi^2}{a^2+\beta^2}} \exp\left(-\frac{\beta\gamma^2}{a^2+\beta^2}\right) \times$

$\times \cos\left(\frac{a\gamma^2}{a^2+\beta^2} + \frac{1}{2} \operatorname{arctg} \frac{a}{\beta}\right)$ $[\operatorname{Re} \beta > 0]$. LI ((268))(8)

4.136

1. $\int\limits_0^\infty (\operatorname{sh} x^2 + \sin x^2) e^{-\beta x^4} dx = \frac{\sqrt{2}\pi}{4\sqrt{\beta}} I_{\frac{1}{4}}\left(\frac{1}{8\beta}\right) \operatorname{ch} \frac{1}{8\beta}$ $[\operatorname{Re} \beta > 0]$. ME 24

2. $\int\limits_0^\infty (\operatorname{sh} x^2 - \sin x^2) e^{-\beta x^4} dx = \frac{\sqrt{2}\pi}{4\sqrt{\beta}} I_{\frac{1}{4}}\left(\frac{1}{8\beta}\right) \operatorname{sh} \frac{1}{8\beta}$ $[\operatorname{Re} \beta > 0]$. ME 24

3. $\int\limits_0^\infty (\operatorname{ch} x^2 + \cos x^2) e^{-\beta x^4} dx = \frac{\sqrt{2}\pi}{4\sqrt{\beta}} I_{-\frac{1}{4}}\left(\frac{1}{8\beta}\right) \operatorname{ch} \frac{1}{8\beta}$ $[\operatorname{Re} \beta > 0]$. ME 24

4. $\int\limits_0^\infty (\operatorname{ch} x^2 - \cos x^2) e^{-\beta x^4} dx = \frac{\sqrt{2}\pi}{4\sqrt{\beta}} I_{-\frac{1}{4}}\left(\frac{1}{8\beta}\right) \operatorname{sh} \frac{1}{8\beta}$ $[\operatorname{Re} \beta > 0]$. ME 24

4.137

1. $$\int_0^\infty \sin 2x^2 \operatorname{sh} 2x^2 e^{-\beta x^4}\,dx = \frac{\pi}{\sqrt[4]{128\beta^2}} J_{-\frac{1}{4}}\left(\frac{1}{\beta}\right)\cos\left(\frac{1}{\beta}+\frac{\pi}{4}\right) \qquad [\operatorname{Re}\beta > 0].$$ MI 32

2. $$\int_0^\infty \sin 2x^2 \operatorname{ch} 2x^2 e^{-\beta x^4}\,dx = \frac{\pi}{\sqrt[4]{128\beta^2}} J_{\frac{1}{4}}\left(\frac{1}{\beta}\right)\cos\left(\frac{1}{\beta}-\frac{\pi}{4}\right) \qquad [\operatorname{Re}\beta > 0].$$ MI 32

3. $$\int_0^\infty \cos 2x^2 \operatorname{sh} 2x^2 e^{-\beta x^4}\,dx = \frac{-\pi}{\sqrt[4]{128\beta^2}} J_{\frac{1}{4}}\left(\frac{1}{\beta}\right)\sin\left(\frac{1}{\beta}-\frac{\pi}{4}\right) \qquad [\operatorname{Re}\beta > 0].$$ MI 32

4. $$\int_0^\infty \cos 2x^2 \operatorname{ch} 2x^2 e^{-\beta x^4}\,dx = \frac{\pi}{\sqrt[4]{128\beta^2}} J_{-\frac{1}{4}}\left(\frac{1}{\beta}\right)\sin\left(\frac{1}{\beta}+\frac{\pi}{4}\right) \qquad [\operatorname{Re}\beta > 0].$$ MI 32

4.138

1. $$\int_0^\infty (\sin 2x^2 \operatorname{ch} 2x^2 + \cos 2x^2 \operatorname{sh} 2x^2)\, e^{-\beta x^4}\,dx = \frac{\pi}{\sqrt[4]{32\beta^2}} J_{\frac{1}{4}}\left(\frac{1}{\beta}\right)\cos\left(\frac{1}{\beta}\right) \qquad [\operatorname{Re}\beta > 0].$$ MI 32

2. $$\int_0^\infty (\sin 2x^2 \operatorname{ch} 2x^2 - \cos 2x^2 \operatorname{sh} 2x^2)\, e^{-\beta x^4}\,dx = \frac{\pi}{\sqrt[4]{32\beta^2}} J_{\frac{1}{4}}\left(\frac{1}{\beta}\right)\sin\left(\frac{1}{\beta}\right) \qquad [\operatorname{Re}\beta > 0].$$ MI 32

3. $$\int_0^\infty (\cos 2x^2 \operatorname{ch} 2x^2 + \sin 2x^2 \operatorname{sh} 2x^2)\, e^{-\beta x^4}\,dx = \frac{\pi}{\sqrt[4]{32\beta^2}} J_{-\frac{1}{4}}\left(\frac{1}{\beta}\right)\cos\left(\frac{1}{\beta}\right) \qquad [\operatorname{Re}\beta > 0].$$ MI 32

4. $$\int_0^\infty (\cos 2x^2 \operatorname{ch} 2x^2 - \sin 2x^2 \operatorname{sh} 2x^2)\, e^{-\beta x^4}\,dx = \frac{\pi}{\sqrt[4]{32\beta^2}} J_{-\frac{1}{4}}\left(\frac{1}{\beta}\right)\sin\left(\frac{1}{\beta}\right) \qquad [\operatorname{Re}\beta > 0].$$ MI 32

4.14 Combinations of trigonometric and hyperbolic functions, exponentials, and powers

4.141

1. $$\int_0^\infty x e^{-\beta x^2} \operatorname{ch} x \sin x\,dx = \frac{1}{4}\sqrt{\frac{\pi}{\beta^3}}\left(\cos\frac{1}{2\beta}+\sin\frac{1}{2\beta}\right) \qquad [\operatorname{Re}\beta > 0].$$ MI 32

2. $$\int_0^\infty x e^{-\beta x^2} \operatorname{sh} x \cos x\,dx = \frac{1}{4}\sqrt{\frac{\pi}{\beta^3}}\left(\cos\frac{1}{2\beta}-\sin\frac{1}{2\beta}\right) \qquad [\operatorname{Re}\beta > 0].$$ MI 32

3. $\int\limits_0^\infty x^2 e^{-\beta x^2} \operatorname{ch} x \cos x \, dx = \frac{1}{4} \sqrt{\frac{\pi}{\beta^3}} \left(\cos \frac{1}{2\beta} - \frac{1}{\beta} \sin \frac{1}{2\beta} \right) \quad [\operatorname{Re} \beta > 0].$ MI 32

4. $\int\limits_0^\infty x^2 e^{-\beta x^2} \operatorname{sh} x \sin x \, dx = \frac{1}{4} \sqrt{\frac{\pi}{\beta^3}} \left(\sin \frac{1}{2\beta} + \frac{1}{\beta} \cos \frac{1}{2\beta} \right) \quad [\operatorname{Re} \beta > 0].$ MI 32

4.142

1. $\int\limits_0^\infty x e^{-\beta x^2} (\operatorname{sh} x + \sin x) \, dx = \frac{1}{2} \sqrt{\frac{\pi}{\beta^3}} \operatorname{ch} \frac{1}{4\beta} \quad [\operatorname{Re} \beta > 0].$ ME 24

2. $\int\limits_0^\infty x e^{-\beta x^2} (\operatorname{sh} x - \sin x) \, dx = \frac{1}{2} \sqrt{\frac{\pi}{\beta^3}} \operatorname{sh} \frac{1}{4\beta} \quad [\operatorname{Re} \beta > 0].$ ME 24

3. $\int\limits_0^\infty x^2 e^{-\beta x^2} (\operatorname{ch} x + \cos x) \, dx = \frac{1}{2} \sqrt{\frac{\pi}{\beta^3}} \left(\operatorname{ch} \frac{1}{4\beta} + \frac{1}{2\beta} \operatorname{sh} \frac{1}{4\beta} \right) \quad [\operatorname{Re} \beta > 0].$ ME 24

4. $\int\limits_0^\infty x^2 e^{-\beta x^2} (\operatorname{ch} x - \cos x) \, dx = \frac{1}{2} \sqrt{\frac{\pi}{\beta^3}} \left(\operatorname{sh} \frac{1}{4\beta} + \frac{1}{2\beta} \operatorname{ch} \frac{1}{4\beta} \right) \quad [\operatorname{Re} \beta > 0].$ ME 24

4.143

1. $\int\limits_0^\infty x e^{-\beta x^2} (\operatorname{ch} x \sin x + \operatorname{sh} x \cos x) \, dx = \frac{1}{2\beta} \sqrt{\frac{\pi}{\beta}} \cos \frac{1}{2\beta} \quad [\operatorname{Re} \beta > 0].$ MI 32

2. $\int\limits_0^\infty x e^{-\beta x^2} (\operatorname{ch} x \sin x - \operatorname{sh} x \cos x) \, dx = \frac{1}{2\beta} \sqrt{\frac{\pi}{\beta}} \sin \frac{1}{2\beta} \quad [\operatorname{Re} \beta > 0].$ MI 32

4.144 $\int\limits_0^\infty e^{-x^2} \operatorname{sh} x^2 \cos ax \frac{dx}{x^2} = \sqrt{\frac{\pi}{2}} e^{-\frac{a^2}{8}} - \frac{\pi a}{4} \left[1 - \Phi\left(\frac{a}{\sqrt{8}} \right) \right] \quad [a > 0].$ ET I 35(44)

4.145

1. $\int\limits_0^\infty x e^{-\beta x^2} \operatorname{ch}(2ax \sin t) \sin(2ax \cos t) \, dx =$

$$= \frac{a}{2} \sqrt{\frac{\pi}{\beta^3}} \exp\left(-\frac{a^2}{\beta} \cos 2t \right) \cos\left(t - \frac{a^2}{\beta} \sin 2t \right) \quad [\operatorname{Re} \beta > 0].$$ BI ((363))(5)

2. $\int\limits_0^\infty x e^{-\beta x^2} \operatorname{sh}(2ax \sin t) \cos(2ax \cos t) \, dx =$

$$= \frac{a}{2} \sqrt{\frac{\pi}{\beta^3}} \exp\left(-\frac{a^2}{\beta} \cos 2t \right) \sin\left(t - \frac{a^2}{\beta} \sin 2t \right) \quad [\operatorname{Re} \beta > 0].$$ BI ((363))(6)

4.2-4.4 Logarithmic Functions

4.21 Logarithmic Functions

4.211

1. $$\int_e^\infty \frac{dx}{\ln\frac{1}{x}} = -\infty.$$ BI ((33))(9)

2. $$\int_0^u \frac{dx}{\ln x} = \operatorname{li} u.$$ FI III 653, FI II 606

4.212

1. $$\int_0^1 \frac{dx}{a+\ln x} = e^{-a}\overline{\operatorname{Ei}}(a).$$ BI ((31))(4)

2. $$\int_0^1 \frac{dx}{a-\ln x} = -e^a \operatorname{Ei}(-a).$$ BI ((31))(5)

3. $$\int_0^1 \frac{dx}{(a+\ln x)^2} = -\frac{1}{a} + e^{-a}\overline{\operatorname{Ei}}(a) \quad [a>0].$$ BI ((31))(14)

4. $$\int_0^1 \frac{dx}{(a-\ln x)^2} = \frac{1}{a} + e^a \operatorname{Ei}(-a) \quad [a>0].$$ BI ((31))(16)

5. $$\int_0^1 \frac{\ln x\, dx}{(a+\ln x)^2} = 1+(1-a)\, e^{-a}\overline{\operatorname{Ei}}(a) \quad [a>0].$$ BI ((31))(15)

6. $$\int_0^1 \frac{\ln x\, dx}{(a-\ln x)^2} = 1+(1+a)\, e^a \operatorname{Ei}(-a) \quad [a>0].$$ BI ((31))(17)

7. $$\int_1^e \frac{\ln x\, dx}{(1+\ln x)^2} = \frac{e}{2}-1.$$ BI ((33))(10)

8. $$\int_0^1 \frac{dx}{(a+\ln x)^n} = \frac{1}{(n-1)!}\, a^{-a}\overline{\operatorname{Ei}}(a) - \frac{1}{(n-1)!}\sum_{k=1}^{n-1}(n-k-1)!\, a^{k-n}$$
$$[a>0].$$ BI ((31))(22)

9 $$\int_0^1 \frac{dx}{(a-\ln x)^n} = \frac{(-1)^n}{(n-1)!}\, e^a \operatorname{Ei}(-a) + \frac{(-1)^{n-1}}{(n-1)!}\sum_{k=1}^{n-1}(n-k-1)!\,(-a)^{k-n}$$
$$[a>0].$$ BI ((31))(23)

In integrals of the form $\int \frac{(\ln x)^m}{[a^n+(\ln x)^n]^l}\, dx$ it is convenient to make the substitution $x=e^{-t}$.

4.213

1. $$\int_0^1 \frac{dx}{a^2+(\ln x)^2} = \frac{1}{a}[\operatorname{ci}(a)\sin a - \operatorname{si}(a)\cos a] \qquad [a>0].$$ BI ((31))(6)

2. $$\int_0^1 \frac{dx}{a^2-(\ln x)^2} = \frac{1}{2a}[e^{-a}\,\overline{\operatorname{Ei}}(a) - e^{a}\operatorname{Ei}(-a)] \qquad [a>0],$$
(cf. **4.212** 1. and 2.). BI ((31))(8)

3. $$\int_0^1 \frac{\ln x\,dx}{a^2+(\ln x)^2} = \operatorname{ci}(a)\cos(a) + \operatorname{si}(a)\sin a \qquad [a>0].$$ BI ((31))(7)

4. $$\int_0^1 \frac{\ln x\,dx}{a^2-(\ln x)^2} = -\frac{1}{2}[e^{-a}\,\overline{\operatorname{Ei}}(a) + e^{a}\operatorname{Ei}(-a)] \qquad [a>0],$$
(cf. **4.212** 1. and 2.). BI ((31))(9)

5. $$\int_0^1 \frac{dx}{[a^2+(\ln x)^2]^2} = \frac{1}{2a^3}[\operatorname{ci}(a)\sin a - \operatorname{si}(a)\cos a] - \frac{1}{2a^2}[\operatorname{ci}(a)\cos a + \operatorname{si}(a)\sin a] \qquad [a>0].$$ LI ((31))(18)

6. $$\int_0^1 \frac{dx}{[a^2-(\ln x)^2]^2} = \frac{1}{4a^3}[(a-1)e^{a}\operatorname{Ei}(-a) + (1+a)e^{-a}\overline{\operatorname{Ei}}(a)] \qquad [a>0].$$ BI ((31))(20)

7. $$\int_0^1 \frac{\ln x\,dx}{[a^2+(\ln x)^2]^2} = \frac{1}{2a}[\operatorname{ci}(a)\sin a - \operatorname{si}(a)\cos a] - \frac{1}{2a^2} \qquad [a>0].$$ BI ((31))(19)

8. $$\int_0^1 \frac{\ln x\,dx}{[a^2-(\ln x)^2]^2} = \frac{1}{4a^2}\{2 + a[e^{a}\operatorname{Ei}(-a) - e^{-a}\overline{\operatorname{Ei}}(a)]\} \qquad [a>0].$$ LI ((31))(21)

4.214

1. $$\int_0^1 \frac{dx}{a^4-(\ln x)^4} = -\frac{1}{4a^3}[e^{a}\operatorname{Ei}(-a) - e^{-a}\,\overline{\operatorname{Ei}}(a) - 2\operatorname{ci}(a)\sin a + 2\operatorname{si}(a)\cos a] \qquad [a>0].$$ BI ((31))(10)

2. $$\int_0^1 \frac{\ln x\,dx}{a^4-(\ln x)^4} = -\frac{1}{4a^2}[e^{a}\operatorname{Ei}(-a) + e^{-a}\,\overline{\operatorname{Ei}}(a) - 2\operatorname{ci}(a)\cos a - 2\operatorname{si}(a)\sin a] \qquad [a>0].$$ BI ((31))(11)

3. $$\int_0^1 \frac{(\ln x)^2\,dx}{a^4-(\ln x)^4} = -\frac{1}{4a}[e^{a}\operatorname{Ei}(-a) - e^{-a}\,\overline{\operatorname{Ei}}(a) + 2\operatorname{ci}(a)\sin a - 2\operatorname{si}(a)\cos a] \qquad [a>0].$$ BI ((31))(12)

4. $$\int_0^1 \frac{(\ln x)^3\,dx}{a^4-(\ln x)^4} = -\frac{1}{4}\left[e^a \operatorname{Ei}(-a) + e^{-a}\,\overline{\operatorname{Ei}}(a) + 2\operatorname{ci}(a)\cos a + 2\operatorname{si}(a)\sin a\right] \qquad [a > 0].$$ BI ((31))(13)

4.215

1. $$\int_0^1 \left(\ln\frac{1}{x}\right)^{\mu-1} dx = \Gamma(\mu) \qquad [\operatorname{Re}\mu > 0].$$ FI II 778

2. $$\int_0^1 \frac{dx}{\left(\ln\frac{1}{x}\right)^{\mu}} = \frac{\pi}{\Gamma(\mu)}\operatorname{cosec}\mu\pi \qquad [\operatorname{Re}\mu < 1].$$ BI ((31))(1)

3. $$\int_0^1 \sqrt{\ln\frac{1}{x}}\,dx = \frac{\sqrt{\pi}}{2}.$$ BI ((32))(1)

4. $$\int_0^1 \frac{dx}{\sqrt{\ln\frac{1}{x}}} = \sqrt{\pi}.$$ BI ((32))(3)

4.216 $$\int_0^{\frac{1}{e}} \frac{dx}{\sqrt{(\ln x)^2-1}} = K_0(1).$$ GW ((321))(2)

4.22 Logarithms of more complicated arguments

4.221

1. $$\int_0^1 \ln x \ln(1-x)\,dx = 2 - \frac{\pi^2}{6}.$$ BI ((30))(7)

2. $$\int_0^1 \ln x \ln(1+x)\,dx = 2 - \frac{\pi^2}{12} - 2\ln 2.$$ BI ((30))(8)

3. $$\int_0^1 \ln\frac{1-ax}{1-a}\,\frac{dx}{\ln x} = -\sum_{k=1}^{\infty} a^k \frac{\ln(1+k)}{k} \qquad [a < 1].$$ BI ((31))(3)

4.222

1. $$\int_0^\infty \ln\frac{a^2+x^2}{b^2+x^2}\,dx = (a-b)\pi \qquad [a > 0,\ b > 0].$$ GW ((322))(20)

2. $$\int_0^\infty \ln x \ln\frac{a^2+x^2}{b^2+x^2}\,dx = \pi(b-a) + \pi\ln\frac{a^a}{b^b} \qquad [a > 0,\ b > 0].$$ BI ((33))(1)

3. $$\int_0^\infty \ln x \ln\left(1+\frac{b^2}{x^2}\right) dx = \pi b(\ln b - 1) \qquad [b > 0].$$ BI ((33))(2)

4. $\int_0^\infty \ln(1+a^2x^2)\ln\left(1+\frac{b^2}{x^2}\right)dx = 2\pi\left[\frac{1+ab}{a}\ln(1+ab)-b\right]$

$[a>0,\ b>0].$ BI ((33))(3)

5. $\int_0^\infty \ln(a^2+x^2)\ln\left(1+\frac{b^2}{x^2}\right)dx = 2\pi[(a+b)\ln(a+b)-a\ln a-b]$

$[a>0,\ b>0].$ BI ((33))(4)

6. $\int_0^\infty \ln\left(1+\frac{a^2}{x^2}\right)\ln\left(1+\frac{b^2}{x^2}\right)dx = 2\pi[(a+b)\ln(a+b)-a\ln a-b\ln b]$

$[a>0,\ b>0].$ BI ((33))(5)

7. $\int_0^\infty \ln\left(a^2+\frac{1}{x^2}\right)\ln\left(1+\frac{b^2}{x^2}\right)dx = 2\pi\left[\frac{1+ab}{a}\ln(1+ab)-b\ln b\right]$

$[a>0,\ b>0].$ BI ((33))(7)

4.223

1. $\int_0^\infty \ln(1+e^{-x})\,dx = \frac{\pi^2}{12}.$ BI ((256))(10)

2. $\int_0^\infty \ln(1-e^{-x})\,dx = -\frac{\pi^2}{6}.$ BI ((256))(11)

3. $\int_0^\infty \ln(1+2e^{-x}\cos t+e^{-2x})\,dx = \frac{\pi^2}{6}-\frac{t^2}{2}$ $[|t|<\pi].$ BI ((256))(18)

4.224

1. $\int_0^u \ln\sin x\,dx = L\left(\frac{\pi}{2}-u\right)-L\left(\frac{\pi}{2}\right).$ LO III 186(15)

2. $\int_0^{\frac{\pi}{4}} \ln\sin x\,dx = -\frac{\pi}{4}\ln 2-\frac{1}{2}G.$ BI ((285))(1)

3. $\int_0^{\frac{\pi}{2}} \ln\sin x\,dx = \frac{1}{2}\int_0^{\pi}\ln\sin x\,dx = -\frac{\pi}{2}\ln 2.$ FI II 629, 643

4. $\int_0^u \ln\cos x\,dx = -L(u).$ LO III 184(10)

5. $\int_0^{\frac{\pi}{4}} \ln\cos x\,dx = -\frac{\pi}{4}\ln 2+\frac{1}{2}G.$ BI ((286))(1)

6. $\int_0^{\frac{\pi}{2}} \ln\cos x\,dx = -\frac{\pi}{2}\ln 2.$ BI 306(1)

7. $$\int_0^{\frac{\pi}{2}} (\ln \sin x)^2 \, dx = \frac{\pi}{2}\left[(\ln 2)^2 + \frac{\pi^2}{12}\right].$$ BI ((305))(19)

8. $$\int_0^{\frac{\pi}{2}} (\ln \cos x)^2 \, dx = \frac{\pi}{2}\left[(\ln 2)^2 + \frac{\pi^2}{12}\right].$$ BI ((306))(14)

9. $$\int_0^{\pi} \ln (a + b \cos x) \, dx = \pi \ln \frac{a + \sqrt{a^2 - b^2}}{2} \qquad [a \geqslant |b| > 0].$$ SW ((322))(15)

10. $$\int_0^{\pi} \ln (1 \pm \sin x) \, dx = -\pi \ln 2 \pm 4\boldsymbol{G}.$$ GW ((322))(16a)

11. $$\int_0^{\frac{\pi}{2}} \ln (1 + a \sin x)^2 \, dx = \int_0^{\frac{\pi}{2}} \ln (1 + a \cos x)^2 \, dx =$$
$$= \pi \ln \frac{1 + \sqrt{1 - a^2}}{2} \qquad [a^2 < 1];$$
$$= -\pi \ln 2a \qquad [a^2 > 1];$$
$$= -\pi \ln 2 + 4\boldsymbol{G} \qquad [a = 1];$$
$$= -\pi \ln 2 - 4\boldsymbol{G} \qquad [a = -1].$$
BI ((308))(5, 6, 7, 8)

12. $$\int_0^{\pi} \ln (1 + a \cos x)^2 \, dx = 2\pi \ln \frac{1 + \sqrt{1 - a^2}}{2} \qquad [a^2 \leqslant 1].$$ BI ((330))(1)

13. $$\int_0^{\frac{\pi}{2}} \ln (1 + 2a \sin x + a^2) \, dx = \sum_{k=0}^{\infty} \frac{2^k k!}{(2k+1)\cdot(2k+1)!!}\left(\frac{2a}{1+a^2}\right)^{2k+1}$$
$$[a^2 \leqslant 1].$$ BI ((308))(24)

14. $$\int_0^{n\pi} \ln (1 - 2a \cos x + a^2) \, dx = 0 \qquad [a^2 < 1];$$
$$= n\pi \ln a^2 \qquad [a^2 > 1].$$
FI II 142, 163, 688

4.225

1. $$\int_0^{\frac{\pi}{4}} \ln (\cos x - \sin x) \, dx = -\frac{\pi}{8} \ln 2 - \frac{1}{2}\boldsymbol{G}.$$ GW ((322))(9b)

2. $$\int_0^{\frac{\pi}{4}} \ln (\cos x + \sin x) \, dx = \frac{1}{2}\int_0^{\frac{\pi}{2}} \ln (\cos x + \sin x) \, dx = -\frac{\pi}{8} \ln 2 + \frac{1}{2}\boldsymbol{G}.$$
GW ((322))(9a)

3. $$\int_0^{2\pi} \ln(1 + a\sin x + b\cos x)\,dx = 2\pi \ln \frac{1+\sqrt{1-a^2-b^2}}{2}$$

$$[a^2 + b^2 < 1].$$ BI ((332))(2)

4. $$\int_0^{2\pi} \ln(1 + a^2 + b^2 + 2a\sin x + 2b\cos x)\,dx =$$

$$= 0 \qquad [a^2 + b^2 \leqslant 1];$$
$$= 2\pi \ln(a^2 + b^2) \qquad [a^2 + b^2 \geqslant 1].$$ BI ((322))(3)

4.226

1. $$\int_0^{\frac{\pi}{2}} \ln(a^2 - \sin^2 x)^2\,dx = -2\pi \ln 2 \qquad [a^2 \leqslant 1];$$

$$= 2\pi \ln \frac{a+\sqrt{a^2-1}}{2} = 2\pi(\operatorname{Arch} a - \ln 2) \qquad [a > 1].$$

FI II 644, 687

2. $$\int_0^{\frac{\pi}{2}} \ln(1 + a\sin^2 x)\,dx = \frac{1}{2}\int_0^{\pi} \ln(1 + a\sin^2 x)\,dx =$$

$$= \int_0^{\frac{\pi}{2}} \ln(1 + a\cos^2 x)\,dx = \frac{1}{2}\int_0^{\pi} \ln(1 + a\cos^2 x)\,dx = \pi \ln \frac{1+\sqrt{1+a}}{2}$$

$$[a \geqslant -1].$$ BI ((308))(15), GW((322))(12)

3. $$\int_0^{u} \ln(1 - \sin^2\alpha \sin^2 x)\,dx = (\pi - 2\theta)\ln \operatorname{ctg}\frac{\alpha}{2} +$$

$$+ 2u \ln\left(\frac{1}{2}\sin\alpha\right) - \frac{\pi}{2}\ln 2 + L(\theta + u) - L(\theta - u) + L\left(\frac{\pi}{2} - 2u\right)$$

$$\left[\operatorname{ctg}\theta = \cos\alpha \operatorname{tg} u; \quad -\pi \leqslant \alpha \leqslant \pi, \quad -\frac{\pi}{2} \leqslant u \leqslant \frac{\pi}{2}\right].$$ LO III 287

4. $$\int_0^{\frac{\pi}{2}} \ln[1 - \cos^2 x(\sin^2\alpha - \sin^2\beta \sin^2 x)]\,dx =$$

$$= \pi \ln\left[\frac{1}{2}\left(\cos^2\frac{\alpha}{2} + \sqrt{\cos^4\frac{\alpha}{2} + \sin^2\frac{\beta}{2}\cos^2\frac{\beta}{2}}\right)\right]$$

$$[\alpha > \beta > 0].$$ LO III 283

5. $$\int_0^{u} \ln\left(1 - \frac{\sin^2 x}{\sin^2\alpha}\right)dx = -u\ln\sin^2\alpha - L\left(\frac{\pi}{2} - \alpha + u\right) + L\left(\frac{\pi}{2} - \alpha - u\right)$$

$$\left[-\frac{\pi}{2} \leqslant u \leqslant \frac{\pi}{2}, \quad |\sin u| \leqslant |\sin\alpha|\right].$$ LO III 287

6. $$\int_0^{\frac{\pi}{2}} \ln(a^2\cos^2 x + b^2\sin^2 x)\,dx = \frac{1}{2}\int_0^{\pi} \ln(a^2\cos^2 x + b^2\sin^2 x)\,dx = \pi \ln\frac{a+b}{2} \quad [a>0,\ b>0].$$ GW ((322))(13)

7. $$\int_0^{\frac{\pi}{2}} \ln\frac{1+\sin t\cos^2 x}{1-\sin t\cos^2 x}\,dx = \pi \ln\frac{1+\sin\frac{t}{2}}{\cos\frac{t}{2}} = \pi\ln\operatorname{ctg}\frac{\pi-t}{4} \quad \left[|t|<\frac{\pi}{2}\right].$$ LO III 283

4.227

1. $$\int_0^{u} \ln\operatorname{tg} x\,dx = L(u) + L\left(\frac{\pi}{2}-u\right) - L\left(\frac{\pi}{2}\right).$$ LO III 186(16)

2. $$\int_0^{\frac{\pi}{4}} \ln\operatorname{tg} x\,dx = -\int_{\frac{\pi}{4}}^{\frac{\pi}{2}} \ln\operatorname{tg} x\,dx = -\boldsymbol{G}.$$ BI ((286))(11)

3. $$\int_0^{\frac{\pi}{2}} \ln(a\operatorname{tg} x)\,dx = \frac{\pi}{2}\ln a \quad [a>0].$$ BI ((307))(2)

4. $$\int_0^{\frac{\pi}{4}} (\ln\operatorname{tg} x)^n\,dx = n!\,(-1)^n \sum_{k=0}^{\infty} \frac{(-1)^k}{(2k+1)^{n+1}}.$$ BI ((286))(21)

5. $$\int_0^{\frac{\pi}{2}} (\ln\operatorname{tg} x)^{2n}\,dx = 2\,(2n)! \sum_{k=0}^{\infty} \frac{(-1)^k}{(2k+1)^{2n+1}}.$$ BI ((307))(15)

6. $$\int_0^{\frac{\pi}{2}} (\ln\operatorname{tg} x)^{2n+1}\,dx = 0.$$ BI ((307))(14)

7. $$\int_0^{\frac{\pi}{4}} (\ln\operatorname{tg} x)^2\,dx = \frac{\pi^3}{16}.$$ BI ((286))(16)

8. $$\int_0^{\frac{\pi}{4}} (\ln\operatorname{tg} x)^4\,dx = \frac{5}{64}\pi^5.$$ BI ((286))(19)

9. $$\int_0^{\frac{\pi}{4}} \ln(1+\operatorname{tg} x)\,dx = \frac{\pi}{8}\ln 2.$$ BI ((287))(1)

10. $\int_0^{\frac{\pi}{2}} \ln(1+\operatorname{tg} x)\,dx = \frac{\pi}{4}\ln 2 + G.$ BI ((308))(9)

11. $\int_0^{\frac{\pi}{4}} \ln(1-\operatorname{tg} x)\,dx = \frac{\pi}{8}\ln 2 - G.$ BI ((287))(2)

12. $\int_0^{\frac{\pi}{2}} \ln(1-\operatorname{tg} x)^2\,dx = \frac{\pi}{2}\ln 2 - 2G.$ BI ((308))(10)

13. $\int_0^{\frac{\pi}{4}} \ln(1+\operatorname{ctg} x)\,dx = \frac{\pi}{8}\ln 2 + G.$ BI ((287))(3)

14. $\int_0^{\frac{\pi}{4}} \ln(\operatorname{ctg} x - 1)\,dx = \frac{\pi}{8}\ln 2.$ BI ((287))(4)

15. $\int_0^{\frac{\pi}{4}} \ln(\operatorname{tg} x + \operatorname{ctg} x)\,dx = \frac{1}{2}\int_0^{\frac{\pi}{2}} \ln(\operatorname{tg} x + \operatorname{ctg} x)\,dx = \frac{\pi}{2}\ln 2.$

BI ((287))(5), BI ((308))(11)

16. $\int_0^{\frac{\pi}{4}} \ln(\operatorname{ctg} x - \operatorname{tg} x)^2\,dx = \frac{1}{2}\int_0^{\frac{\pi}{2}} \ln(\operatorname{ctg} x - \operatorname{tg} x)^2\,dx = \frac{\pi}{2}\ln 2.$

BI ((287))(6), BI ((308))(12)

17. $\int_0^{\frac{\pi}{2}} \ln(a^2 + b^2\operatorname{tg}^2 x)\,dx = \frac{1}{2}\int_0^{\pi} \ln(a^2 + b^2\operatorname{tg}^2 x)\,dx = \pi\ln(a+b)$

$[a > 0,\ b > 0].$ GW ((322))(17)

4.228

1 $\int_0^{\frac{\pi}{2}} \ln\left(\sin t \sin x + \sqrt{1-\cos^2 t \sin^2 x}\right)dx =$

$= \frac{\pi}{2}\ln 2 - 2L\left(\frac{t}{2}\right) - 2L\left(\frac{\pi - t}{2}\right).$ LO III 290

2. $\int_0^{u} \ln\left(\cos x + \sqrt{\cos^2 x - \cos^2 t}\right)dx = -\left(\frac{\pi}{2} - t - \varphi\right)\ln\cos t +$

$+ \frac{1}{2}L(u+\varphi) - \frac{1}{2}L(u-\varphi) - L(\varphi)$

$\left[\cos\varphi = \frac{\sin u}{\sin t};\ 0 \leqslant u \leqslant t \leqslant \frac{\pi}{2}\right].$ LO III 290

3. $\int_0^t \ln\left(\cos x + \sqrt{\cos^2 x - \cos^2 t}\right) dx = -\left(\frac{\pi}{2} - t\right) \ln \cos t.$ LO III 285

4. $\int_0^u \ln \frac{\sin u + \sin t \cos x \sqrt{\sin^2 u - \sin^2 x}}{\sin u - \sin t \cos x \sqrt{\sin^2 u - \sin^2 x}} dx =$

$= \pi \ln\left[\operatorname{tg}\frac{t}{2} \sin u + \sqrt{\operatorname{tg}^2 \frac{t}{2} \sin^2 u + 1}\right] \quad [t > 0,\ u > 0].$ LO III 283

5. $\int_0^{\frac{\pi}{4}} \sqrt{\ln \operatorname{ctg} x}\, dx = \frac{\sqrt{\pi}}{2} \sum_{k=0}^{\infty} \frac{(-1)^k}{\sqrt{(2k+1)^3}}.$ BI ((297))(9)

6. $\int_0^{\frac{\pi}{4}} \frac{dx}{\sqrt{\ln \operatorname{ctg} x}} = \sqrt{\pi} \sum_{k=0}^{\infty} \frac{(-1)^k}{\sqrt{2k+1}}.$ BI ((304))(24)

7. $\int_0^{\frac{\pi}{4}} \ln\left(\sqrt{\operatorname{tg} x} + \sqrt{\operatorname{ctg} x}\right) dx = \frac{1}{2} \int_0^{\frac{\pi}{2}} \ln\left(\sqrt{\operatorname{tg} x} + \sqrt{\operatorname{ctg} x}\right) dx =$

$= \frac{\pi}{8} \ln 2 + \frac{1}{2} G.$ BI ((287))(7), BI ((308))(22)

8 $\int_0^{\frac{\pi}{4}} \ln\left(\sqrt{\operatorname{ctg} x} - \sqrt{\operatorname{tg} x}\right)^2 dx = \frac{1}{2} \int_0^{\frac{\pi}{2}} \ln\left(\sqrt{\operatorname{ctg} x} - \sqrt{\operatorname{tg} x}\right)^2 dx =$

$= \frac{\pi}{4} \ln 2 - G.$ BI ((287))(8), BI ((308))(23)

4.229

1 $\int_0^1 \ln\left(\ln \frac{1}{x}\right) dx = -C.$ FI II 807

2. $\int_0^1 \frac{dx}{\ln\left(\ln \frac{1}{x}\right)} = 0.$ BI ((31))(2)

3. $\int_0^1 \ln\left(\ln \frac{1}{x}\right) \frac{dx}{\sqrt{\ln \frac{1}{x}}} = -(C + 2 \ln 2)\sqrt{\pi}.$ BI ((32))(4)

4 $\int_0^1 \ln\left(\ln \frac{1}{x}\right)\left(\ln \frac{1}{x}\right)^{\mu-1} dx = \psi(\mu)\,\Gamma(\mu) \quad [\operatorname{Re} \mu > 0].$ BI ((30))(10)

If the integrand contains $\ln\left(\ln \frac{1}{x}\right)$, it is convenient to make the substitution $\ln \frac{1}{x} = u$, i. e., $x = e^{-u}$.

5. $\int_0^1 \ln(a + \ln x)\, dx = \ln a - e^{-a} \overline{\operatorname{Ei}}(a) \quad [a > 0].$ BI ((30))(5)

6. $\int_0^1 \ln(a - \ln x)\, dx = \ln a - e^a \operatorname{Ei}(-a) \quad [a > 0].$ BI ((30))(6)

7. $\int_{\frac{\pi}{4}}^{\frac{\pi}{2}} \ln \ln \operatorname{tg} x\, dx = \frac{\pi}{2} \ln \left\{ \frac{\Gamma\left(\frac{3}{4}\right)}{\Gamma\left(\frac{1}{4}\right)} \sqrt{2\pi} \right\}.$ BI ((308))(28)

4.23 Combinations of logarithms and rational functions

4.231

1. $\int_0^1 \frac{\ln x}{1+x}\, dx = -\frac{\pi^2}{12}.$ FI II 483a

2. $\int_0^1 \frac{\ln x}{1-x}\, dx = -\frac{\pi^2}{6}.$ FI II 714

3. $\int_0^1 \frac{x \ln x}{1-x}\, dx = 1 - \frac{\pi^2}{6}.$ BI ((108))(7)

4. $\int_0^1 \frac{1+x}{1-x} \ln x\, dx = 1 - \frac{\pi^2}{3}.$ BI ((108))(9)

5. $\int_0^\infty \frac{\ln x\, dx}{(x+a)^2} = \frac{\ln a}{a} \quad [0 < a < 1].$ BI ((139))(1)

6. $\int_0^1 \frac{\ln x}{(1+x)^2}\, dx = -\ln 2.$ BI ((111))(1)

7. $\int_0^\infty \ln x \frac{dx}{(a^2 + b^2 x^2)^n} = \frac{\Gamma\left(n - \frac{1}{2}\right)\sqrt{\pi}}{4 \cdot (n-1)!\, a^{2n-1} b} \left[2 \ln \frac{a}{2b} - C - \psi\left(n - \frac{1}{2}\right) \right]$
$[a > 0,\ b > 0].$ LI ((139))(3)

8. $\int_0^\infty \frac{\ln x\, dx}{a^2 + b^2 x^2} = \frac{\pi}{2ab} \ln \frac{a}{b} \quad [ab > 0].$ BI ((135))(6)

9. $\int_0^\infty \frac{\ln px}{q^2 + x^2}\, dx = \frac{\pi}{2q} \ln pq \quad [p > 0,\ q > 0].$ BI ((135))(4)

10. $\int_0^\infty \frac{\ln x\, dx}{a^2 - b^2 x^2} = -\frac{\pi^2}{4ab} \quad [ab > 0].$ LI ((324))(7b)

11. $\int_0^a \frac{\ln x\, dx}{x^2 + a^2} = \frac{\pi \ln a}{4a} - \frac{G}{a} \quad [a > 0].$ GW ((324))(7b)

12. $\int_0^1 \frac{\ln x}{1+x^2}\, dx = -\int_1^\infty \frac{\ln x}{1+x^2}\, dx = -G.$ FI II 482, 614

13. $$\int_0^1 \frac{\ln x\, dx}{1-x^2} = -\frac{\pi^2}{8}.$$ BI ((108))(11)

14. $$\int_0^1 \frac{x \ln x}{1+x^2}\, dx = -\frac{\pi^2}{48}.$$ GW ((324))(7b)

15. $$\int_0^1 \frac{x \ln x}{1-x^2}\, dx = -\frac{\pi^2}{24}.$$

16. $$\int_0^1 \ln x \frac{1-x^{2n+2}}{(1-x^2)^2}\, dx = -\frac{(n+1)\pi^2}{8} + \sum_{k=1}^{n} \frac{n-k+1}{(2k-1)^2}.$$ BI ((111))(5)

17. $$\int_0^1 \ln x \frac{1+(-1)^n x^{n+1}}{(1+x)^2}\, dx = -\frac{(n+1)\pi^2}{12} - \sum_{k=1}^{n} (-1)^k \frac{n-k+1}{k^2}.$$ BI ((111))(2)

18. $$\int_0^1 \ln x \frac{1-x^{n+1}}{(1-x)^2}\, dx = -\frac{(n+1)\pi^2}{6} + \sum_{k=1}^{n} \frac{n-k+1}{k^2}.$$ BI ((111))(3)

4.232

1. $$\int_u^v \frac{\ln x\, dx}{(x+u)(x+v)} = \frac{\ln uv}{2(v-u)} \ln \frac{(u+v)^2}{4uv}.$$ BI ((145))(32)

2. $$\int_0^\infty \frac{\ln x\, dx}{(x+\beta)(x+\gamma)} = \frac{(\ln \beta)^2 - (\ln \gamma)^2}{2(\beta-\gamma)} \qquad [|\arg \beta| < \pi, \ |\arg \gamma| < \pi].$$ ET II 218(24)

3. $$\int_0^\infty \frac{\ln x}{x+a} \frac{dx}{x-1} = \frac{\pi^2 + (\ln a)^2}{2(a+1)} \qquad [a > 0].$$ BI ((140))(10)

4.233

1. $$\int_0^1 \frac{\ln x\, dx}{1+x+x^2} = -0.781\,302\,412\,9\ \ldots$$ LI ((113))(1)

2. $$\int_0^1 \frac{\ln x\, dx}{1-x+x^2} = -1.171\,953\,619\,34\ \ldots$$ LI ((113))(2)

3. $$\int_0^1 \frac{x \ln x\, dx}{1+x+x^2} = -0.157\,660\,149\,17\ \ldots$$ LI ((113))(2)

4. $$\int_0^1 \frac{x \ln x\, dx}{1-x+x^2} = -0.311\,821\,131\,9\ \ldots$$ LI ((113))(4)

5. $$\int_0^\infty \frac{\ln x\, dx}{x^2+2xa\cos t+a^2} = \frac{t \ln a}{a \sin t} \qquad [a > 0, \ 0 < t < \pi].$$ GW ((324))(13c)

4.234

1. $\int_1^\infty \frac{\ln x\,dx}{(1+x^2)^2} = \ln 2.$ BI ((144))(18)a

2. $\int_0^1 \frac{x \ln x\,dx}{(1+x^2)^2} = -\frac{1}{4}\ln 2.$ BI ((111))(4)

3. $\int_0^\infty \frac{1+x^2}{(1-x^2)^2}\ln x\,dx = 0.$ BI ((142))(2)a

4. $\int_0^\infty \frac{1-x^2}{(1+x^2)^2}\ln x\,dx = -\frac{\pi}{2}.$ BI ((142))(1)a

5. $\int_0^1 \frac{x^2 \ln x\,dx}{(1-x^2)(1+x^4)} = -\frac{\pi^2}{16(2+\sqrt{2})}.$ BI ((112))(21)

6. $\int_0^\infty \frac{\ln x\,dx}{(a^2+b^2x^2)(1+x^2)} = \frac{b\pi}{2a(b^2-a^2)}\ln\frac{a}{b} \quad [ab>0].$ BI ((317))(16)a

7. $\int_0^\infty \frac{\ln x}{x^2+a^2}\cdot\frac{dx}{1+b^2x^2} = \frac{\pi}{2(1-a^2b^2)}\left(\frac{1}{a}\ln a + b\ln b\right)$
$[a>0, \quad b>0].$ LI ((140))(12)

8. $\int_0^\infty \frac{x^2 \ln x\,dx}{(a^2+b^2x^2)(1+x^2)} = \frac{a\pi}{2b(b^2-a^2)}\ln\frac{b}{a} \quad [ab>0].$

LI ((140))(12), BI((317))(15)a

4.235

1. $\int_0^\infty \ln x \frac{(1-x)x^{n-2}}{1-x^{2n}}\,dx = -\frac{\pi^2}{4n^2}\operatorname{tg}^2\frac{\pi}{2n} \quad [n>1].$ BI ((135))(10)

2. $\int_0^\infty \ln x \frac{(1-x^2)x^{m-1}}{1-x^{2n}}\,dx = -\frac{\pi^2 \sin\frac{m+1}{n}\pi \sin\frac{\pi}{n}}{4n^2\sin^2\frac{m\pi}{2n}\sin^2\left(\frac{m+2}{2n}\pi\right)}$ LI ((135))(12)

3. $\int_0^\infty \ln x \frac{(1-x^2)x^{n-2}}{1-x^{2n}}\,dx = -\frac{\pi^2}{4n^2}\operatorname{tg}^2\frac{\pi}{n} \quad [n>2].$ BI ((135))(11)

4. $\int_0^1 \ln x \frac{x^{m-1}+x^{n-m-1}}{1-x^n}\,dx = -\frac{\pi^2}{n^2\sin^2\left(\frac{m}{n}\pi\right)} \quad [n>m].$ BI ((108))(15)

4.236

1. $\int_0^1 \left\{\frac{1+(p-1)\ln x}{1-x} + \frac{x\ln x}{(1-x)^2}\right\} x^{p-1}\,dx = -1+\psi'(p) \quad [p>0].$

BI ((111))(6)a, GW ((326))(13)

2. $\int_0^1 \left[\frac{1}{1-x} + \frac{x\ln x}{(1-x)^2}\right] dx = \frac{\pi^2}{6} - 1.$ GW ((326))(13a)

4.24 Combinations of logarithms and algebraic functions

4.241

1. $$\int_0^1 \frac{x^{2n}\ln x}{\sqrt{1-x^2}}\,dx = \frac{(2n-1)!!}{(2n)!!}\cdot\frac{\pi}{2}\left(\sum_{k=1}^{2n}\frac{(-1)^{k-1}}{k}-\ln 2\right).$$

BI ((118))(5)a

2. $$\int_0^1 \frac{x^{2n+1}\ln x}{\sqrt{1-x^2}}\,dx = \frac{(2n)!!}{(2n+1)!!}\left(\ln 2+\sum_{k=1}^{2n+1}\frac{(-1)^{k}}{k}\right).$$ BI ((118))(5)a

3. $$\int_0^1 x^{2n}\sqrt{1-x^2}\ln x\,dx = \frac{(2n-1)!!}{(2n+2)!!}\cdot\frac{\pi}{2}\left(\sum_{k=1}^{2n}\frac{(-1)^{k-1}}{k}-\frac{1}{2n+2}-\ln 2\right).$$

LI ((117))(4), GW ((324))(53a)

4. $$\int_0^1 x^{2n+1}\sqrt{1-x^2}\ln x\,dx = \frac{(2n)!!}{(2n+3)!!}\left(\ln 2+\sum_{k=1}^{2n+1}\frac{(-1)^{k}}{k}-\frac{1}{2n+3}\right).$$

BI ((117))(5), GW ((324))(53b)

5. $$\int_0^1 \ln x\cdot\sqrt{(1-x^2)^{2n-1}}\,dx = -\frac{(2n-1)!!}{4\cdot(2n)!!}\pi\left[\psi(n+1)+C+\ln 4\right].$$

BI ((117))(3)

6. $$\int_0^{\sqrt{\frac{1}{2}}} \frac{\ln x\,dx}{\sqrt{1-x^2}} = -\frac{\pi}{4}\ln 2-\frac{1}{2}G.$$ BI ((145))(1)

7. $$\int_0^1 \frac{\ln x\,dx}{\sqrt{1-x^2}} = -\frac{\pi}{2}\ln 2.$$ FI II 614, 643

8. $$\int_1^\infty \frac{\ln x\,dx}{x^2\sqrt{x^2-1}} = 1-\ln 2.$$ BI ((144))(17)

9. $$\int_0^1 \sqrt{1-x^2}\ln x\,dx = -\frac{\pi}{8}-\frac{\pi}{4}\ln 2.$$ BI((117))(1), GW ((324))(53c)

10. $$\int_0^1 x\sqrt{1-x^2}\ln x\,dx = \frac{1}{3}\ln 2-\frac{4}{9}.$$ BI ((117))(2)

11. $$\int_0^1 \frac{\ln x\,dx}{\sqrt{x(1-x^2)}} = -\frac{\sqrt{2\pi}}{8}\left[\Gamma\left(\frac{1}{4}\right)\right]^2.$$ GW ((324))(54a)

4.242

1. $$\int_0^\infty \frac{\ln x\,dx}{\sqrt{(a^2+x^2)(x^2+b^2)}} = \frac{1}{2a}K\left(\frac{\sqrt{a^2-b^2}}{a}\right)\ln ab \quad [a>b>0].$$

BY (800.04)

2. $$\int_0^b \frac{\ln x\,dx}{\sqrt{(a^2+x^2)(b^2-x^2)}} = \frac{1}{2\sqrt{a^2+b^2}}\left[\boldsymbol{K}\left(\frac{b}{\sqrt{a^2+b^2}}\right)\ln ab - \frac{\pi}{2}\boldsymbol{K}\left(\frac{a}{\sqrt{a^2+b^2}}\right)\right] \quad [a>0, \quad b>0].$$ BY (800.02)

3. $$\int_b^\infty \frac{\ln x\,dx}{\sqrt{(x^2+a^2)(x^2-b^2)}} = \frac{1}{2\sqrt{a^2+b^2}}\left[\boldsymbol{K}\left(\frac{a}{\sqrt{a^2+b^2}}\right)\ln ab + \frac{\pi}{2}\boldsymbol{K}\left(\frac{b}{\sqrt{a^2+b^2}}\right)\right] \quad [a>0, \quad b>0].$$ BY (800.06)

4. $$\int_0^b \frac{\ln x\,dx}{\sqrt{(a^2-x^2)(b^2-x^2)}} = \frac{1}{2a}\left[\boldsymbol{K}\left(\frac{b}{a}\right)\ln ab - \frac{\pi}{2}\boldsymbol{K}\left(\frac{\sqrt{a^2-b^2}}{a}\right)\right]$$ $[a>b>0]$. BY (800.01)

5. $$\int_b^a \frac{\ln x\,dx}{\sqrt{(a^2-x^2)(x^2-b^2)}} = \frac{1}{2a}\boldsymbol{K}\left(\frac{\sqrt{a^2-b^2}}{a}\right)\ln ab.$$ BY (800.03)

6. $$\int_a^\infty \frac{\ln x\,dx}{\sqrt{(x^2-a^2)(x^2-b^2)}} = \frac{1}{2a}\left[\boldsymbol{K}\left(\frac{b}{a}\right)\ln ab + \frac{\pi}{2}\boldsymbol{K}\left(\frac{\sqrt{a^2-b^2}}{a}\right)\right]$$ $[a>b>0]$. BY (800.05)

4.243 $$\int_0^1 \frac{x\ln x}{\sqrt{1-x^4}}\,dx = -\frac{\pi}{8}\ln 2.$$ GW ((324))(56b)

4.244

1. $$\int_0^1 \frac{\ln x\,dx}{\sqrt[3]{x(1-x^2)^2}} = -\frac{1}{8}\left[\Gamma\left(\frac{1}{3}\right)\right]^3.$$ GW ((324))(54b)

2. $$\int_0^1 \frac{\ln x\,dx}{\sqrt[3]{1-x^3}} = -\frac{\pi}{3\sqrt{3}}\left(\ln 3 + \frac{\pi}{3\sqrt{3}}\right).$$ BI ((118))(7)

3. $$\int_0^1 \frac{x\ln x\,dx}{\sqrt[3]{(1-x^3)^2}} = \frac{\pi}{3\sqrt{3}}\left(\frac{\pi}{3\sqrt{3}} - \ln 3\right).$$ BI ((118))(8)

4.245

1. $$\int_0^1 \frac{x^{4n+1}\ln x}{\sqrt{1-x^4}}\,dx = \frac{(2n-1)!!}{(2n)!!}\cdot\frac{\pi}{8}\left(\sum_{k=1}^{2n}\frac{(-1)^{k-1}}{k} - \ln 2\right).$$ GW ((324))(56a)

2. $$\int_0^1 \frac{x^{4n+3}\ln x}{\sqrt{1-x^4}}\,dx = \frac{(2n)!!}{4\cdot(2n+1)!!}\left(\ln 2 + \sum_{k=1}^{2n+1}\frac{(-1)^k}{k}\right).$$ GW ((324))(56c)

4.246 $$\int_0^1 (1-x^2)^{n-\frac{1}{2}}\ln x\,dx = -\frac{(2n-1)!!}{(2n)!!}\cdot\frac{\pi}{4}\left[2\ln 2 + \sum_{k=1}^{n}\frac{1}{k}\right].$$ GW ((324))(55)

4.247

1. $$\int_0^1 \frac{\ln x}{\sqrt[n]{1-x^{2n}}}\,dx = -\frac{\pi B\left(\frac{1}{2n}, \frac{1}{2n}\right)}{8n^2 \sin\frac{\pi}{2n}} \qquad [n > 1].$$ GW ((324))(54c)a

2. $$\int_0^1 \frac{\ln x\,dx}{\sqrt[n]{x^{n-1}(1-x^2)}} = -\frac{\pi B\left(\frac{1}{2n}, \frac{1}{2n}\right)}{8 \sin\frac{\pi}{2n}}.$$ GW ((324))(54)

4.25 Combinations of logarithms and powers

4.251

1. $$\int_0^\infty \frac{x^{\mu-1}\ln x}{\beta+x}\,dx = \frac{\pi\beta^{\mu-1}}{\sin\mu\pi}(\ln\beta - \pi\operatorname{ctg}\mu\pi)$$
$$[|\arg\beta| < \pi, \quad 0 < \operatorname{Re}\mu < 1].$$ BI ((135))(1)

2. $$\int_0^\infty \frac{x^{\mu-1}\ln x}{a-x}\,dx = \pi a^{\mu-1}\left(\operatorname{ctg}\mu\pi\ln a - \frac{\pi}{\sin^2\mu\pi}\right)$$
$$[a > 0, \quad 0 < \operatorname{Re}\mu < 1].$$ ET I 314(5)

3. $$\int_0^1 \frac{x^{\mu-1}\ln x}{x+1}\,dx = \frac{1}{2}\beta'(\mu) \qquad [\operatorname{Re}\mu > 0].$$ GW((324))(6), ET I 314(3)

4. $$\int_0^1 \frac{x^{\mu-1}\ln x}{1-x}\,dx = -\psi'(\mu) = -\zeta(2, \mu) \qquad [\operatorname{Re}\mu > 0].$$ BI ((108))(8)

5. $$\int_0^1 \ln x \frac{x^{2n}\,dx}{1+x} = -\frac{\pi^2}{12} + \sum_{k=1}^{2n} \frac{(-1)^{k-1}}{k^2}.$$ BI ((108))(4)

6. $$\int_0^1 \ln x \frac{x^{2n-1}\,dx}{1+x} = \frac{\pi^2}{12} + \sum_{k=1}^{2n-1} \frac{(-1)^k}{k^2}.$$ BI ((108))(5)

4.252

1. $$\int_0^\infty \frac{x^{\mu-1}\ln x}{(x+\beta)(x+\gamma)}\,dx = \frac{\pi}{(\gamma-\beta)\sin\mu\pi}[\beta^{\mu-1}\ln\beta - \gamma^{\mu-1}\ln\gamma -$$
$$-\pi\operatorname{ctg}\mu\pi(\beta^{\mu-1} - \gamma^{\mu-1})] \quad [|\arg\beta| < \pi, \quad |\arg\gamma| < \pi,$$
$$0 < \operatorname{Re}\mu < 2, \ \mu \neq 1].$$ BI ((140))(9)a, ET 314(6)

2. $$\int_0^\infty \frac{x^{\mu-1}\ln x\,dx}{(x+\beta)(x-1)} = \frac{\pi}{(\beta+1)\sin^2\mu\pi}[\pi - \beta^{\mu-1}(\sin\mu\pi\ln\beta - \pi\cos\mu\pi)]$$
$$[|\arg\beta| < \pi, \quad 0 < \operatorname{Re}\mu < 2, \quad \mu \neq 1].$$ BI ((140))(11)

3. $$\int_0^\infty \frac{x^{p-1}\ln x}{1-x^2}\,dx = -\frac{\pi^2}{4}\operatorname{cosec}^2\frac{p\pi}{2} \qquad [0 < p < 2]$$
(see also **4.254** 2.).

4. $\int_0^\infty \frac{x^{\mu-1}\ln x}{(x+a)^2}dx = \frac{(1-\mu)a^{\mu-2}\pi}{\sin\mu\pi}\left(\ln a - \pi\operatorname{ctg}\mu\pi + \frac{1}{\mu-1}\right)$

$[a>0,\ 0<\operatorname{Re}\mu<2\ (\mu\neq 1)].$ GW ((324))(13b)

4.253

1. $\int_0^1 x^{\mu-1}(1-x^r)^{\nu-1}\ln x\,dx = \frac{1}{r^2}\mathrm{B}\left(\frac{\mu}{r},\nu\right)\left[\psi\left(\frac{\mu}{r}\right)-\psi\left(\frac{\mu}{r}+\nu\right)\right]$

$[\operatorname{Re}\mu>0,\ \operatorname{Re}\nu>0,\ r>0].$ GW ((324))(3b)a, BI ((107))(5)a

2. $\int_0^1 \frac{x^{p-1}}{(1-x)^{p+1}}\ln x\,dx = -\frac{\pi}{p}\operatorname{cosec}p\pi \quad [0<p<1].$ BI ((319))(10)a

3. $\int_u^\infty \frac{(x-u)^{\mu-1}\ln x\,dx}{x^\lambda} = u^{\mu-\lambda}\mathrm{B}(\lambda-\mu,\ \mu)\left[\ln u+\psi(\lambda)-\psi(\lambda-\mu)\right]$

$[0<\operatorname{Re}\mu<\operatorname{Re}\lambda].$ ET II 203(18)

4. $\int_0^1 \ln x\left(\frac{x}{a^2+x^2}\right)^p\frac{dx}{x} = \frac{\ln a}{2a^p}\mathrm{B}\left(\frac{p}{2},\frac{p}{2}\right)$

$[a>0,\ p>0].$ BI ((140))(6)

5. $\int_1^\infty (x-1)^{p-1}\ln x\,dx = \frac{\pi}{p}\operatorname{cosec}\pi p \quad [-1<p<0].$ BI ((289))(12)a

6. $\int_0^\infty \ln x\frac{dx}{(a+x)^{\mu+1}} = \frac{1}{\mu a^\mu}(\ln a - C - \psi(\mu))$

$[\operatorname{Re}\mu>0,\ a\neq 0,\ \mu-a$ is not a natural number$].$ NT 68(7)

7. $\int_0^\infty \ln x\frac{dx}{(a+x)^{n+\frac{1}{2}}} = \frac{2}{(2n-1)a^{n-\frac{1}{2}}}\left(\ln a+2\ln 2-\sum_{k=1}^{n-2}\frac{1}{k}-2\sum_{k=n-1}^{2n-3}\frac{1}{k}\right)$

$[a>0].$ BI ((142))(5)

4.254

1. $\int_0^1 \frac{x^{p-1}\ln x}{1-x^q}dx = -\frac{1}{q^2}\psi'\left(\frac{p}{q}\right) \quad [p>0,\ q>0].$ GW ((324))(5)

2. $\int_0^\infty \frac{x^{p-1}\ln x}{1-x^q}dx = -\frac{\pi^2}{q^2\sin^2\frac{p\pi}{q}} \quad [0<p<q].$ BI ((135))(8)

3. $\int_0^\infty \frac{\ln x}{x^q-1}\frac{dx}{x^p} = \frac{\pi^2}{q^2\sin^2\frac{p-1}{q}\pi} \quad [p<1,\ p+q>1].$ BI ((140))(2)

4. $\int_0^1 \frac{x^{p-1}\ln x}{1+x^q}dx = \frac{1}{2q^2}\beta\left(\frac{p}{q}\right) \quad [p>0,\ q>0].$ GW ((324))(7)

5. $\int_0^\infty \frac{x^{p-1}\ln x}{1+x^q}\,dx = -\frac{\pi^2}{q^2}\frac{\cos\frac{p\pi}{q}}{\sin^2\frac{p\pi}{q}}$ $[0<p<q]$. BI ((135))(7)

6. $\int_0^1 \frac{x^{q-1}\ln x}{1-x^{2q}}\,dx = -\frac{\pi^2}{8q^2}$ $[q>0]$. BI ((108))(12)

4.255

1. $\int_0^1 \ln x \frac{(1-x^2)\,x^{p-2}}{1+x^{2p}}\,dx = -\left(\frac{\pi}{2p}\right)^2 \frac{\sin\frac{\pi}{2p}}{\cos^2\frac{\pi}{2p}}$ $[p>1]$. BI ((108))(13)

2. $\int_0^1 \ln x \frac{(1+x^2)\,x^{p-2}}{1-x^{2p}}\,dx = -\left(\frac{\pi}{2p}\right)^2 \sec^2\frac{\pi}{2p}$ $[p>1]$. BI ((108))(14)

3. $\int_0^\infty \ln x \frac{1-x^p}{1-x^2}\,dx = \frac{\pi^2}{4}\operatorname{tg}^2\frac{p\pi}{2}$ $[p<1]$. BI ((140))(3)

4.256 $\int_0^1 \ln\frac{1}{x}\frac{x^{\mu-1}\,dx}{\sqrt[n]{(1-x^n)^{n-m}}} = \frac{1}{n^2}\mathrm{B}\left(\frac{\mu}{n},\frac{m}{n}\right)\left[\psi\left(\frac{\mu+m}{n}\right)-\psi\left(\frac{\mu}{n}\right)\right]$

$[\operatorname{Re}\mu>0]$. LI ((118))(12)

4.257

1. $\int_0^\infty \frac{x^\nu \ln\frac{x}{\beta}\,dx}{(x+\beta)(x+\gamma)} = \frac{\pi\left[\gamma^\nu\ln\frac{\gamma}{\beta}+\pi(\beta^\nu-\gamma^\nu)\operatorname{ctg}\nu\pi\right]}{\sin\nu\pi\,(\gamma-\beta)}$

$[|\arg\beta|<\pi,\ |\arg\gamma|<\pi,\ |\operatorname{Re}\nu|<1]$. ET II 219(30)

2. $\int_0^\infty \ln\frac{x}{q}\left(\frac{x^p}{q^{2p}+x^{2p}}\right)\frac{dx}{x} = 0$ $[q>0]$. BI ((140))(4)a

3. $\int_0^\infty \ln\frac{x}{q}\left(\frac{x^p}{q^{2p}+x^{2p}}\right)^r\frac{dx}{q^2+x^2} = 0$ $[q>0]$. BI ((140))(4)a

4. $\int_0^\infty \ln x\ln\frac{x}{a}\frac{dx}{(x-1)(x-a)} = \frac{[4\pi^2+(\ln a)^2]\ln a}{6(a-1)}$ $[a>0]$,

$[a=1$ see **4.261** 5.]. BI ((141))(5)

5. $\int_0^\infty \ln x\ln\frac{x}{a}\frac{x^p\,dx}{(x-1)(x-a)} = \frac{\pi^2[(a^p+1)\ln a-2\pi(a^p-1)\operatorname{ctg}p\pi]}{(a-1)\sin^2 p\pi}$

$[p^2<1,\ a>0]$. BI ((141))(6)

4.26-4.27 Combinations involving powers of the logarithm and other powers

4.261

1. $\int_0^1 (\ln x)^2\frac{dx}{1+2x\cos t+x^2} = \frac{t(\pi^2-t^2)}{6\sin t}$. BI ((113))(7)

2. $$\int_0^1 \frac{(\ln x)^2\, dx}{x^2-x+1} = \frac{1}{2}\int_0^\infty \frac{(\ln x)^2\, dx}{x^2-x+1} = \frac{10\pi^3}{81\sqrt{3}}.$$ GW ((324))(16c)

3. $$\int_0^1 \frac{(\ln x)^2\, dx}{x^2+x+1} = \frac{1}{2}\int_0^\infty \frac{(\ln x)^2\, dx}{x^2+x+1} = \frac{8\pi^3}{81\sqrt{3}}.$$ GW ((324))(16b)

4. $$\int_0^\infty (\ln x)^2 \frac{dx}{(x-1)(x+a)} = \frac{[\pi^2+(\ln a)^2]\ln a}{3(1+a)}.$$ BI ((141))(1)

5. $$\int_0^\infty (\ln x)^2 \frac{dx}{(1-x)^2} = \frac{2}{3}\pi^2.$$ BI ((139))(4)

6. $$\int_0^1 (\ln x)^2 \frac{dx}{1+x^2} = \frac{\pi^3}{16}.$$ BI ((109))(3)

7. $$\int_0^1 (\ln x)^2 \frac{1+x^2}{1+x^4}\, dx = \frac{1}{2}\int_0^\infty (\ln x)^2 \frac{1+x^2}{1+x^4}\, dx = \frac{3\sqrt{2}}{64}\pi^3.$$ BI ((109))(5), BI ((135))(13)

8. $$\int_0^1 (\ln x)^2 \frac{1-x}{1-x^6}\, dx = \frac{\sqrt{3}}{27}\pi^3.$$ BI ((109))(6)

9. $$\int_0^1 (\ln x)^2 \frac{dx}{\sqrt{1-x^2}} = \frac{\pi}{2}\left[(\ln 2)^2 + \frac{\pi^2}{12}\right].$$ BI ((118))(13)

10. $$\int_0^\infty (\ln x)^2 \frac{x^{\mu-1}}{1+x}\, dx = \frac{\pi^3(2-\sin^2\mu\pi)}{\sin^3\mu\pi} \qquad [0 < \operatorname{Re}\mu < 1].$$ ET I 315(10)

11. $$\int_0^1 (\ln x)^2 \frac{x^n\, dx}{1+x} = 2\sum_{k=n}^\infty \frac{(-1)^{n+k}}{(k+1)^3}.$$ BI ((109)(1)

12. $$\int_0^1 (\ln x)^2 \frac{x^n\, dx}{1-x} = 2\sum_{k=n}^\infty \frac{1}{(k+1)^3}.$$ BI ((109))(2)

13. $$\int_0^1 (\ln x)^2 \frac{x^{2n}\, dx}{1-x^2} = 2\sum_{k=n}^\infty \frac{1}{(2k+1)^3}.$$ BI ((109))(4)

14. $$\int_0^\infty (\ln x)^2 \frac{x^{p-1}\, dx}{x^2+2x\cos t+1} = \frac{\pi\sin(1-p)t}{\sin t\sin p\pi} \times$$
$$\times \{\pi^2 - t^2 + 2\pi\operatorname{ctg} p\pi[\pi\operatorname{ctg} p\pi + t\operatorname{ctg}(1-p)t]\}$$
$$[0 < t < \pi,\ 0 < p < 2\ (p \neq 1)].$$ GW ((324))(17)

15. $$\int_0^1 (\ln x)^2 \frac{x^{2n}\, dx}{\sqrt{1-x^2}} = \frac{(2n-1)!!}{2\cdot(2n)!!}\pi\left\{\frac{\pi^2}{12} + \sum_{k=1}^{2n}\frac{(-1)^k}{k^2} + \left[\sum_{k=1}^{2n}\frac{(-1)^k}{k} + \ln 2\right]^2\right\}.$$ GW ((324))(60a)

16 $$\int_0^1 (\ln x)^2 \frac{x^{2n+1}\,dx}{\sqrt{1-x^2}} = \frac{(2n)!!}{(2n+1)!!}\left\{-\frac{\pi^2}{12} - \sum_{k=1}^{2n+1}\frac{(-1)^k}{k^2} + \left[\sum_{k=1}^{2n+1}\frac{(-1)^k}{k} + \ln 2\right]^2\right\}.$$ GW ((324))(60b)

17. $$\int_0^1 (\ln x)^2 x^{\mu-1}(1-x)^{\nu-1}\,dx = \mathrm{B}(\mu,\nu)\{[\psi(\mu)-\psi(\nu+\mu)]^2 + \psi'(\mu) - \psi'(\mu+\nu)\} \qquad [\operatorname{Re}\mu > 0,\ \operatorname{Re}\nu > 0].$$ ET I 315(11)

18 $$\int_0^1 (\ln x)^2 \frac{1-x^{n+1}}{(1-x)^2}\,dx = 2(n+1)\zeta(3) - 2\sum_{k=1}^{n}\frac{n-k+1}{k^3}.$$ LI ((111))(8)

19 $$\int_0^1 (\ln x)^2 \frac{1+(-1)^n x^{n+1}}{(1+x)^2}\,dx = \frac{3}{2}(n+1)\zeta(3) - 2\sum_{k=1}^{n}(-1)^{k-1}\frac{n-k+1}{k^3}.$$ LI ((111))(7)

20 $$\int_0^1 (\ln x)^2 \frac{1-x^{2n+2}}{(1-x)^2}\,dx = 2(n+1)\sum_{k=0}^{\infty}\frac{1}{(2k+1)^3} - 2\sum_{k=1}^{n}\frac{n-k+1}{(2k-1)^3}.$$ LI ((111))(9)

21 $$\int_0^1 (\ln x)^2 x^{p-1}(1-x^r)^{q-1}\,dx = \frac{1}{r^3}\mathrm{B}\left(\frac{p}{r}, q\right)\left\{\psi'\left(\frac{p}{r}\right) - \psi'\left(\frac{p}{r}+q\right) + \left[\psi\left(\frac{p}{r}\right) - \psi\left(\frac{p}{r}+q\right)\right]^2\right\} \qquad [p>0,\ q>0,\ r>0].$$ GW ((324))(8a)

4.262

1 $$\int_0^1 (\ln x)^3 \frac{dx}{1+x} = -\frac{7}{120}\pi^4.$$ BI ((109))(9)

2. $$\int_0^1 (\ln x)^3 \frac{dx}{1-x} = -\frac{\pi^4}{15}.$$ BI ((109))(11)

3. $$\int_0^\infty (\ln x)^3 \frac{dx}{(x+a)(x-1)} = \frac{[\pi^2+(\ln a)^2]^2}{4(a+1)} \qquad [a>0].$$ BI ((141))(2)

4 $$\int_0^1 (\ln x)^3 \frac{x^n\,dx}{1+x} = (-1)^{n+1}\left[\frac{7\pi^4}{120} - 6\sum_{k=0}^{n-1}\frac{(-1)^k}{(k+1)^4}\right].$$ BI ((109))(10)

5. $$\int_0^1 (\ln x)^3 \frac{x^n\,dx}{1-x} = -\frac{\pi^4}{15} + 6\sum_{k=0}^{n-1}\frac{1}{(k+1)^4}.$$ BI ((109))(12)

6. $$\int_0^1 (\ln x)^3 \frac{x^{2n}\,dx}{1-x^2} = -\frac{\pi^4}{16} + 6\sum_{k=0}^{n-1}\frac{1}{(2k+1)^4}.$$ BI ((109))(14)

7. $\int_0^1 (\ln x)^3 \frac{1-x^{n+1}}{(1-x)^2}\,dx = -\frac{(n+1)\pi^4}{15} + 6\sum_{k=1}^{n} \frac{n-k+1}{k^4}$. BI ((111))(11)

8. $\int_0^1 (\ln x)^3 \frac{1+(-1)^n x^{n+1}}{(1+x)^2}\,dx = -\frac{7(n+1)\pi^4}{120} + 6\sum_{k=1}^{n} (-1)^{k-1} \frac{n-k+1}{k^4}$. BI ((111))(10)

9. $\int_0^1 (\ln x)^3 \frac{1-x^{2n+2}}{(1-x^2)^2}\,dx = -\frac{(n+1)\pi^4}{16} + 6\sum_{k=1}^{n} \frac{n-k+1}{(2k-1)^4}$. BI ((111))(12)

4.263

1. $\int_0^\infty (\ln x)^4 \frac{dx}{(x-1)(x+a)} = \frac{\ln a\,[\pi^2+(\ln a)^2]^2\,[7\pi^2+3(\ln a)^2]}{15(1+a)}$ $[a > 0]$. BI ((141))(3)

2. $\int_0^1 (\ln x)^4 \frac{dx}{1+x^2} = \frac{5\pi^5}{64}$. BI ((109))(17)

3. $\int_0^1 (\ln x)^4 \frac{dx}{1+2x\cos t+x^2} = \frac{t(\pi^2-t^2)(7\pi^2-3t^2)}{30\sin t}$ $[|t| < \pi]$. BI ((113))(8)

4.264

1. $\int_0^1 (\ln x)^5 \frac{dx}{1+x} = -\frac{31\pi^6}{252}$. BI ((109))(20)

2. $\int_0^1 (\ln x)^5 \frac{dx}{1-x} = -\frac{8\pi^6}{63}$. BI ((109))(21)

3. $\int_0^\infty (\ln x)^5 \frac{dx}{(x-1)(x+a)} = \frac{[\pi^2+(\ln a)^2]^2\,[3\pi^2+(\ln a)^2]}{6(1+a)}$ $[a > 0]$. BI ((141))(4)

4.265 $\int_0^1 (\ln x)^6 \frac{dx}{1+x^2} = \frac{61\pi^7}{256}$. BI ((109))(25)

4.266

1. $\int_0^1 (\ln x)^7 \frac{dx}{1+x} = -\frac{127\pi^8}{240}$. BI ((109))(28)

2. $\int_0^1 (\ln x)^7 \frac{dx}{1-x} = -\frac{8\pi^8}{15}$. BI ((109))(29)

4.267

1. $\int_0^1 \frac{1-x}{1+x}\frac{dx}{\ln x} = \ln\frac{2}{\pi}$. BI ((127))(3)

2. $$\int_0^1 \frac{(1-x)^2}{1+x^2} \frac{dx}{\ln x} = \ln \frac{\pi}{4}.$$ BI ((128))(2)

3. $$\int_0^1 \frac{(1-x)^2}{1+2x\cos\frac{mx}{n}+x^2} \cdot \frac{dx}{\ln x} = \frac{1}{\sin\frac{m\pi}{n}} \sum_{k=1}^{n-1} (-1)^k \sin\frac{km\pi}{n} \times$$
$$\times \ln \frac{\left\{\Gamma\left(\frac{n+k+1}{2n}\right)\right\}^2 \Gamma\left(\frac{k+2}{2n}\right) \Gamma\left(\frac{k}{2n}\right)}{\left\{\Gamma\left(\frac{k+1}{2n}\right)\right\}^2 \Gamma\left(\frac{n+k}{2n}\right) \Gamma\left(\frac{n+k+2}{2n}\right)} \qquad [m+n \text{ is odd}];$$
$$= \frac{1}{\sin\frac{m\pi}{n}} \sum_{k=1}^{\frac{1}{2}(n-1)} (-1)^k \sin\frac{km\pi}{n} \times$$
$$\times \ln \frac{\left\{\Gamma\left(\frac{n-k+1}{n}\right)\right\}^2 \Gamma\left(\frac{k+2}{n}\right) \Gamma\left(\frac{k}{n}\right)}{\left\{\Gamma\left(\frac{k+1}{n}\right)\right\}^2 \Gamma\left(\frac{n-k}{n}\right) \Gamma\left(\frac{n-k+2}{n}\right)} \qquad [m+n \text{ is even}];$$
$[m < n]$. BI ((130))(3)

4. $$\int_0^1 \frac{1-x}{1+x} \cdot \frac{1}{1+x^2} \cdot \frac{dx}{\ln x} = -\frac{\ln 2}{2}.$$ BI ((130))(16)

5. $$\int_0^1 \frac{1-x}{1+x} \cdot \frac{x^2}{1+x^2} \cdot \frac{dx}{\ln x} = \ln \frac{2\sqrt{2}}{\pi}.$$ BI ((130))(17)

6. $$\int_0^1 (1-x)^p \frac{dx}{\ln x} = \sum_{k=1}^{\infty} (-1)^k \binom{p}{k} \ln(1+k) \qquad [p \geqslant 1].$$ BI ((123))(2)

7. $$\int_0^1 \left(\frac{1-x^p}{1-x} - p\right) \frac{dx}{\ln x} = \ln \Gamma(p+1).$$ GW ((326))(10)

8. $$\int_0^1 \frac{x^{p-1}-x^{q-1}}{\ln x} dx = \ln \frac{p}{q} \qquad [p > 0,\ q > 0].$$ FI II 647

9. $$\int_0^1 \frac{x^{p-1}-x^{q-1}}{\ln x} \cdot \frac{dx}{1+x} = \ln \frac{\Gamma\left(\frac{q}{2}\right) \Gamma\left(\frac{p+1}{2}\right)}{\Gamma\left(\frac{p}{2}\right) \Gamma\left(\frac{q+1}{2}\right)} \qquad [p > 0,\ q > 0].$$ FI II 186

10. $$\int_0^1 \frac{x^{p-1}-x^{-p}}{(1+x)\ln x} dx = \frac{1}{2} \int_0^{\infty} \frac{x^{p-1}-x^{-p}}{(1+x)\ln x} dx = \ln\left(\operatorname{tg}\frac{p\pi}{2}\right) \qquad [0 < p < 1].$$ FI II 816

11. $$\int_0^1 (x^p - x^q) x^{r-1} \frac{dx}{\ln x} = \ln \frac{p+r}{r+q} \qquad [r > 0,\ p > 0,\ q > 0].$$ LI ((123))(5)

12. $$\int_0^1 \frac{x^p - x^q}{(1-ax)^n} \frac{dx}{x \ln x} = \ln \frac{p}{q} + \sum_{k=1}^{\infty} \binom{n+k-1}{k} a^k \ln \frac{p+k}{q+k}$$
$[p > 0,\ q > 0,\ a^2 < 1]$. BI ((130))(15)

13. $\int_0^1 (x^p-1)(x^q-1)\frac{dx}{\ln x} = \ln\frac{p+q+1}{(p+1)(q+1)}$ $[p>-1,\ q>-1,\ p+q>-1]$

GW ((324))(19b)

14. $\int_0^1 \frac{x^p-x^q}{1+x}\cdot\frac{1+x^{2n+1}}{x\ln x}\,dx = \ln\frac{\Gamma\left(\frac{p}{2}+n+1\right)\Gamma\left(\frac{q+1}{2}+n\right)\Gamma\left(\frac{p+1}{2}\right)\Gamma\left(\frac{q}{2}\right)}{\Gamma\left(\frac{q}{2}+n+1\right)\Gamma\left(\frac{p+1}{2}+n\right)\Gamma\left(\frac{q+1}{2}\right)\Gamma\left(\frac{p}{2}\right)}$

$[p>0,\ q>0].$ BI ((127))(7)

15. $\int_0^1 \frac{x^p-x^q}{1-x}\cdot\frac{1-x^r}{\ln x}\,dx = \ln\frac{\Gamma(q+1)\,\Gamma(p+r+1)}{\Gamma(p+1)\,\Gamma(q+r+1)}$

$[p>-1,\ q>-1,\ p+r>-1,\ q+r>-1].$ GW ((324))(23)

16. $\int_0^1 \frac{x^{p-1}-x^{q-1}}{(1+x^r)\ln x}\,dx = \ln\frac{\Gamma\left(\frac{p+r}{2r}\right)\Gamma\left(\frac{q}{2r}\right)}{\Gamma\left(\frac{q+r}{2r}\right)\Gamma\left(\frac{p}{2r}\right)}$ $[p>0,\ q>0,\ r>0].$

GW ((324))(21)

17. $\int_0^1 \frac{1-x^{2p-2q}}{1+x^{2p}}\,\frac{x^{q-1}\,dx}{\ln x} = \ln \operatorname{tg}\frac{q\pi}{4p}$ $[0<q<p]$ (see also **3.524** 27.).

BI ((128))(6)

18. $\int_0^\infty \frac{x^{p-1}-x^{q-1}}{(1+x^r)\ln x}\,dx = \ln\left(\operatorname{tg}\frac{p\pi}{2r}\operatorname{ctg}\frac{q\pi}{2r}\right)$ $[0<p<r,\ 0<q<r].$

GW ((324))(22), BI ((143))(2)

19. $\int_0^\infty \frac{x^{p-1}-x^{q-1}}{(1-x^r)\ln x}\,dx = \ln\left(\frac{\sin\frac{p\pi}{r}}{\sin\frac{q\pi}{r}}\right)$ $[0<p<r,\ 0<q<r].$ BI ((143))(4)

20. $\int_0^1 \frac{x^{p-1}-x^{q-1}}{1-x^{2n}}\cdot\frac{1-x^2}{\ln x}\,dx = \ln\frac{\Gamma\left(\frac{p+2}{2n}\right)\Gamma\left(\frac{q}{2n}\right)}{\Gamma\left(\frac{q+2}{2n}\right)\Gamma\left(\frac{p}{2n}\right)}$ $[p>0,\ q>0].$

BI ((128))(11)

21. $\int_0^1 \frac{x^{p-1}-x^{q-1}}{1+x^{2(2n+1)}}\,\frac{1+x^2}{\ln x}\,dx =$

$$= \ln\frac{\Gamma\left(\frac{p+4n+4}{4(2n+1)}\right)\Gamma\left(\frac{q+2}{4(2n+1)}\right)\Gamma\left(\frac{p+4n+2}{4(2n+1)}\right)\Gamma\left(\frac{q}{4(2n+1)}\right)}{\Gamma\left(\frac{q+4n+4}{4(2n+1)}\right)\Gamma\left(\frac{p+2}{4(2n+1)}\right)\Gamma\left(\frac{q+4n+2}{4(2n+1)}\right)\Gamma\left(\frac{p}{4(2n+1)}\right)}$$

$[p>0,\ q>0].$ BI ((128))(7)

22. $\int_0^\infty \frac{x^{p-1}-x^{q-1}}{1+x^{2(2n+1)}}\cdot\frac{1+x^2}{\ln x}\,dx =$

$$= \ln\left\{\operatorname{tg}\frac{p\pi}{4(2n+1)}\cdot\operatorname{tg}\frac{(p+2)\pi}{4(2n+1)}\cdot\operatorname{ctg}\frac{q\pi}{4(2n+1)}\cdot\operatorname{ctg}\frac{(q+2)\pi}{4(2n+1)}\right\}$$

$[0<p<4n,\ 0<q<4n].$ BI ((143))(5)

23. $$\int_0^\infty \frac{x^{p-1}-x^{q-1}}{1-x^{2n}} \frac{1-x^2}{\ln x} dx = \ln \frac{\sin \frac{p\pi}{2n} \cdot \sin \frac{(q+2)\pi}{2n}}{\sin \frac{q\pi}{2n} \cdot \sin \frac{(p+2)\pi}{2n}}$$
$[0 < p < 2n,\ 0 < q < 2n]$. BI ((143))(6)

24. $$\int_0^1 (1-x^p)(1-x^q) \frac{x^{r-1}\,dx}{\ln x} = \ln \frac{(p+q+r)r}{(p+r)(q+r)}$$
$[p > 0,\ q > 0,\ r > 0]$. BI ((123))(8)

25. $$\int_0^1 (1-x^p)(1-x^q) \frac{x^{r-1}\,dx}{(1-x)\ln x} = \ln \frac{\Gamma(p+r)\Gamma(q+r)}{\Gamma(p+q+r)\Gamma(r)}$$
$[r > 0,\ r+p > 0,\ r+q > 0,\ r+p+q > 0]$. FI II 815a

26. $$\int_0^1 (1-x^p)(1-x^q)(1-x^r) \frac{dx}{\ln x} = \ln \frac{(p+q+1)(q+r+1)(r+p+1)}{(p+q+r+1)(p+1)(q+1)(r+1)}$$
$[p > -1,\ q > -1,\ r > -1,\ p+q > -1,\ p+r > -1,\ q+r > -1,$
$p+q+r > -1]$. GW ((324))(19c)

27. $$\int_0^1 (1-x^p)(1-x^q)(1-x^r) \frac{dx}{(1-x)\ln x} =$$
$$= \ln \frac{\Gamma(p+1)\Gamma(q+1)\Gamma(r+1)\Gamma(p+q+r+1)}{\Gamma(p+q+1)\Gamma(p+r+1)\Gamma(q+r+1)}$$
$[p > -1,\ q > -1,\ r > -1,\ p+q > -1,\ p+r > -1,\ q+r > -1,$
$p+q+r > -1]$. FI II 815

28. $$\int_0^1 (1-x^p)(1-x^q)(1-x^r) \frac{x^{s-1}\,dx}{\ln x} = \ln \frac{(p+q+s)(p+r+s)(q+r+s)s}{(p+s)(q+s)(r+s)(p+q+r+s)}$$
$[p > 0,\ q > 0,\ r > 0,\ s > 0]$. BI ((123))(10)

29. $$\int_0^1 (1-x^p)(1-x^q) \frac{x^{s-1}\,dx}{(1-x^r)\ln x} = \ln \frac{\Gamma\left(\frac{p+s}{r}\right)\Gamma\left(\frac{q+s}{r}\right)}{\Gamma\left(\frac{s}{r}\right)\Gamma\left(\frac{p+q+s}{r}\right)}$$
$[p > 0,\ q > 0,\ r > 0,\ s > 0]$. GW ((324))(23a)

30. $$\int_0^\infty (1-x^p)(1-x^q) \frac{x^{s-1}\,dx}{(1-x^{p+q+2s})\ln x} = 2\int_0^1 (1-x^p)(1-x^q) \frac{x^{s-1}\,dx}{(1-x^{p+q+2s})\ln x} =$$
$$= 2\ln\left\{\sin \frac{s\pi}{p+q+2s} \operatorname{cosec} \frac{(p+s)\pi}{p+q+2s}\right\}$$ $[s > 0,\ s+p > 0,\ s+p+q > 0]$.
GW ((324))(23b)a

31. $$\int_0^1 (1-x^p)(1-x^q)(1-x^r) \frac{x^{s-1}\,dx}{(1-x)\ln x} =$$
$$= \ln \frac{\Gamma(p+s)\Gamma(q+s)\Gamma(r+s)\Gamma(p+q+r+s)}{\Gamma(p+q+s)\Gamma(p+r+s)\Gamma(q+r+s)\Gamma(s)}$$
$[p > 0,\ q > 0,\ r > 0,\ s > 0]$*. BI ((127))(11)

---

*These restrictions can be somewhat weakened by writing, for example, $s > 0,\ p+s > 0,\ q+s > 0,\ r+s > 0,\ p+q+s > 0,\ p+r+s > 0,\ q+r+s > 0,\ p+q+r+s > 0$, in 4.267 31. and 32.

32. $$\int_0^1 (1-x^p)(1-x^q)(1-x^r)\frac{x^{s-1}\,dx}{(1-x^t)\ln x} =$$
$$= \ln \frac{\Gamma\left(\frac{p+s}{t}\right)\Gamma\left(\frac{q+s}{t}\right)\Gamma\left(\frac{r+s}{t}\right)\Gamma\left(\frac{p+q+r+s}{t}\right)}{\Gamma\left(\frac{p+q+s}{t}\right)\Gamma\left(\frac{q+r+s}{t}\right)\Gamma\left(\frac{p+r+s}{t}\right)\Gamma\left(\frac{s}{t}\right)}$$
$[p > 0,\ q > 0,\ r > 0,\ s > 0,\ t > 0]$*. GW ((324))(23b)

33. $$\int_0^1 \left\{\frac{x^p - x^{p+q}}{1-x} - q\right\}\frac{dx}{\ln x} = \ln \frac{\Gamma(p+q+1)}{\Gamma(p+1)}$$
$[p > -1,\ p+q > -1]$. BI ((127))(19)

34. $$\int_0^1 \left\{\frac{x^\mu - x}{x-1} - x(\mu-1)\right\}\frac{dx}{x\ln x} = \ln\Gamma(\mu) \quad [\operatorname{Re}\mu > 0].$$
WH, BI ((127))(18)

35. $$\int_0^1 \left\{1 - x - \frac{(1-x^p)(1-x^q)}{1-x}\right\}\frac{dx}{x\ln x} = -\ln\{B(p,\ q)\} \quad [p > 0,\ q > 0].$$
BI ((130))(18)

36. $$\int_0^1 \left\{\frac{x^{p-1}}{1-x} - \frac{x^{pq-1}}{1-x^q} - \frac{1}{x(1-x)} + \frac{1}{x(1-x^q)}\right\}\frac{dx}{\ln x} = q\ln p \quad [p > 0].$$
BI ((130))(20)

37. $$\int_0^1 \left\{\frac{x^{q-1}}{1-x} - \frac{x^{pq-1}}{1-x^p} - \frac{p-1}{1-x^p}x^{p-1} - \frac{p-1}{2}x^{p-1}\right\}\frac{dx}{\ln x} =$$
$$= \frac{1-p}{2}\ln(2\pi) + \left(pq - \frac{1}{2}\right)\ln p \quad [p > 0,\ q > 0].$$
BI ((130))(22)

38. $$\int_0^1 \frac{(1-x^p)(1-x^q) - (1-x)^2}{x(1-x)\ln x}\,dx = \ln B(p,\ q) \quad [p > 0,\ q > 0].$$
GW ((324))(24)

39. $$\int_0^1 (x^p - 1)^n \frac{dx}{\ln x} = \sum_{k=0}^{n} \binom{n}{n-k}(-1)^{n-k}\ln(pk+1) \quad \left[p > -\frac{1}{n}\right]$$
GW ((324))(19d), BI ((123))(12)a

40. $$\int_0^1 \frac{(1-x^p)^n}{1-x}\frac{dx}{\ln x} = \sum_{k=0}^{n}(-1)^{k-1}\ln\Gamma[(n-k)p+1] \quad \left[p > -\frac{1}{n}\right].$$
BI ((127))(12)

41. $$\int_0^1 (x^p - 1)^n x^{q-1}\frac{dx}{\ln x} = \sum_{k=0}^{n}(-1)^k\binom{n}{k}\ln[q + (n-k)p]$$
$\left[p > -\frac{1}{n},\ q > 0\right]$. BI ((123))(12)

*See footnote on preceding page.

42. $$\int_0^1 (1-x^p)^n x^{q-1} \frac{dx}{(1-x)\ln x} = \sum_{k=0}^{n} (-1)^{k-1} \ln \Gamma[(n-k)p+q]$$

$$\left[p > -\frac{1}{n},\ q > 0\right].$$ BI ((127))(13)

43. $$\int_0^1 (x^p-1)^n (x^q-1)^m \frac{x^{r-1}\,dx}{\ln x} =$$

$$= \sum_{j=0}^{n} (-1)^j \binom{n}{j} \sum_{k=0}^{m} (-1)^k \binom{m}{k} [r+(m-k)q+(n-j)p]$$

$$\left[p > -\frac{1}{n},\ q > -\frac{1}{m},\ r > 0\right].$$ BI ((123))(16)

4.268.

1. $$\int_0^1 \frac{(x^p-x^q)(1-x^r)}{(\ln x)^2}\,dx = (p+1)\ln(p+1) - (q+1)\ln(q+1) -$$

$$-(p+r+1)\ln(p+r+1) + (q+r+1)\ln(q+r+1)$$

$$[p > -1,\ q > -1,\ p+r > -1,\ q+r > -1].$$ GW ((324))(26)

2. $$\int_0^1 (x^p-x^q)^2 \frac{dx}{(\ln x)^2} = (2p+1)\ln(2p+1) + (2q+1)\ln(2q+1) -$$

$$-2(p+q+1)\ln(p+q+1) \quad \left[p > -\frac{1}{2},\ q > -\frac{1}{2}\right]$$ GW ((324))(26a)

3. $$\int_0^1 (1-x^p)(1-x^q)(1-x^r)\frac{dx}{(\ln x)^2} = (p+q+1)\ln(p+q+1) +$$

$$+(q+r+1)\ln(q+r+1) + (p+r+1)\ln(p+r+1) - (p+1)\ln(p+1) -$$

$$-(q+1)\ln(q+1) - (r+1)\ln(r+1) - (p+q+r)\ln(p+q+r)$$

$$[p > -1,\ q > -1,\ r > -1,\ p+q > -1,\ p+r > -1,\ q+r > -1,$$

$$p+q+r > 0].$$ BI ((124))(4)

4. $$\int_0^1 (1-x^p)^n x^{q-1} \frac{dx}{(\ln x)^2} = \frac{1}{2}\sum_{k=0}^{n} (-1)^k \binom{n}{k} (pk+q)^2 \ln(pk+q)$$

$$\left[q > 0,\ p > -\frac{q}{n}\right].$$ BI ((124))(14)

5. $$\int_0^1 (1-x^p)^n (1-x^q)^m x^{r-1} \frac{dx}{(\ln x)^2} = \sum_{j=0}^{n} (-1)^j \binom{n}{j} \sum_{k=0}^{m} (-1)^k \binom{m}{k} \times$$

$$\times [(m-k)q+(n-j)p+r] \ln[(m-k)q+(n-j)p+r]$$

$$[r > 0,\ mq+r > 0,\ np+r > 0,\ mq+np+r > 0].$$ BI ((124))(8)

6. $$\int_0^1 [(q-r)x^{p-1}+(r-p)x^{q-1}+(p-q)x^{r-1}]\frac{dx}{(\ln x)^2}=$$
$$=(q-r)p\ln p+(r-p)q\ln q+(p-q)r\ln r$$
$$[p>0,\ q>0,\ r>0].$$ BI ((124))(9)

7. $$\int_0^1\left[\frac{x^{p-1}}{(p-q)(p-r)(p-s)}+\frac{x^{q-1}}{(q-p)(q-r)(q-s)}+\frac{x^{r-1}}{(r-p)(r-q)(r-s)}+\right.$$
$$\left.+\frac{x^{s-1}}{(s-p)(s-q)(s-r)}\right]\frac{dx}{(\ln x)^2}=\frac{1}{2}\left[\frac{p^2\ln p}{(p-q)(p-r)(p-s)}+\right.$$
$$\left.+\frac{q^2\ln q}{(q-p)(q-r)(q-s)}+\frac{r^2\ln r}{(r-p)(r-q)(r-s)}+\frac{s^2\ln s}{(s-p)(s-q)(s-r)}\right]$$
$$[p>0,\ q>0,\ r>0,\ s>0].$$ BI ((124))(16)

4.269

1. $$\int_0^1\sqrt{\ln\frac{1}{x}}\cdot\frac{dx}{1+x^2}=\frac{\sqrt{\pi}}{2}\sum_{k=0}^{\infty}\frac{(-1)^k}{\sqrt{(2k+1)^3}}.$$ BI ((115))(33)

2. $$\int_0^1\frac{dx}{\sqrt{\ln\frac{1}{x}\cdot(1+x)^2}}=\sqrt{\pi}\sum_{k=0}^{\infty}\frac{(-1)^k}{\sqrt{2k+1}}.$$ BI ((133))(2)

3. $$\int_0^1\sqrt{\ln\frac{1}{x}}\cdot x^{p-1}\,dx=\frac{1}{2}\sqrt{\frac{\pi}{p^3}}\qquad[p>0].$$ GW ((324))(1c)

4. $$\int_0^1\frac{x^{p-1}}{\sqrt{\ln\frac{1}{x}}}\,dx=\sqrt{\frac{\pi}{p}}\qquad[p>0].$$ BI ((133))(1)

5. $$\int_0^1\frac{\sin t-x^n\sin[(n+1)t]+x^{n+1}\sin nt}{1-2x\cos t+x^2}\cdot\frac{dx}{\sqrt{\ln\frac{1}{x}}}=$$
$$=\sqrt{\pi}\sum_{k=1}^{n}\frac{\sin kt}{\sqrt{k}}\qquad[|t|<\pi].$$ BI ((133))(5)

6. $$\int_0^1\frac{\cos t-x-x^{n-1}\cos nt+x^n\cos[(n-1)t]}{1-2x\cos t+x^2}\cdot\frac{dx}{\sqrt{\ln\frac{1}{x}}}=$$
$$=\sqrt{\pi}\sum_{k=1}^{n-1}\frac{\cos kt}{\sqrt{k}}\qquad[|t|<\pi].$$ BI ((133))(6)

7. $$\int_u^v\frac{dx}{x\cdot\sqrt{\ln\frac{x}{u}\ln\frac{v}{x}}}=\pi\qquad[uv>0].$$ BI ((145))(37)

4.271

1. $$\int_0^1(\ln x)^{2n}\frac{dx}{1+x}=\frac{2^{2n}-1}{2^{2n}}\cdot(2n)!\,\zeta(2n+1).$$ BI ((110))(1)

2. $\int_0^1 (\ln x)^{2n-1} \frac{dx}{1+x} = \frac{1-2^{2n-1}}{2n} \pi^{2n} |B_{2n}|.$ BI ((110))(2)

3. $\int_0^1 (\ln x)^{2n-1} \frac{dx}{1-x} = -\frac{1}{n} 2^{2n-2}\pi^{2n} |B_{2n}|.$ BI((110))(5), GW((324))(9a)

4. $\int_0^1 (\ln x)^{p-1} \frac{dx}{1-x} = e^{i(p-1)\pi}\Gamma(p)\zeta(p) \qquad [p>1].$ GW ((324))(9b)

5. $\int_0^1 (\ln x)^n \frac{dx}{1+x^2} = (-1)^n n! \sum_{k=0}^{\infty} \frac{(-1)^k}{(2k+1)^{n+1}}.$ BI ((110))(11)

6. $\int_0^1 (\ln x)^{2n} \frac{dx}{1+x^2} = \frac{1}{2}\int_0^{\infty} (\ln x)^{2n} \frac{dx}{1+x^2} = \frac{\pi^{2n+1}}{2^{2n+2}} |E_{2n}|.$ GW ((324))(10)a

7. $\int_0^{\infty} \frac{(\ln x)^{2n+1}}{1+bx+x^2} dx = 0 \qquad [|b|<2].$ BI ((135))(2)

8. $\int_0^1 (\ln x)^{2n} \frac{dx}{1-x^2} = \frac{2^{2n+1}-1}{2^{2n+1}} \cdot (2n)!\, \zeta(2n+1).$ BI ((110))(12)

9. $\int_0^{\infty} (\ln x)^{2n} \frac{dx}{1-x^2} = 0.$ BI ((312))(7)a

10. $\int_0^1 (\ln x)^{2n-1} \frac{dx}{1-x^2} = \frac{1}{2}\int_0^{\infty} (\ln x)^{2n-1} \frac{dx}{1-x^2} = \frac{1-2^{2n}}{4n} \pi^{2n} |B_{2n}|.$

BI ((290))(17)a, BI((312))(6)a

11. $\int_0^1 (\ln x)^{2n-1} \frac{x\,dx}{1-x^2} = -\frac{1}{4n} \pi^{2n} |B_{2n}|.$ BI ((290))(19)a

12. $\int_0^1 (\ln x)^{2n} \frac{1+x^2}{(1-x^2)^2} dx = \frac{2^{2n}-1}{2} \pi^{2n} |B_{2n}|.$ BI ((296))(17)a

13. $\int_0^1 (\ln x)^{2n+1} \frac{(\cos 2a\pi - x)\,dx}{1-2x\cos 2a\pi + x^2} = -(2n+1)! \sum_{k=1}^{\infty} \frac{\cos 2ak\pi}{k^{2n+2}}$

[a is not an integer]. LI ((113))(10)

14. $\int_0^1 (\ln x)^n \frac{x^{\nu-1}\,dx}{a^2+2ax\cos t + x^2} = -\pi \cos t \frac{d^n}{d\nu^n}\left[a^{\nu-2}\frac{\sin(\nu-1)t}{\sin\nu\pi}\right]$

$[a>0,\ 0<\operatorname{Re}\nu<2,\ |t|<\pi].$ ET I 315(12)

15. $\int_0^1 (\ln x)^n \frac{x^{p-1}}{1-x^q} dx = -\frac{1}{q^{n+1}} \psi^{(n)}\left(\frac{p}{q}\right) \qquad [p>0,\ q>0].$ GW ((324))(9)

16. $\int_0^1 (\ln x)^n \frac{x^{p-1}}{1+x^q} dx = \frac{1}{2^n q^{n+1}} \beta\left(\frac{p}{q}\right) \qquad [p>0,\ q>0].$ GW ((324))(10)

4.272

1. $$\int_0^1 \frac{\left[\ln\left(\frac{1}{x}\right)\right]^{q-1} dx}{1+2x\cos t+x^2} = \operatorname{cosec} t\,\Gamma(q)\sum_{k=1}^{\infty}(-1)^{k-1}\frac{\sin kt}{k^q}$$
$[|t| < \pi,\ q < 1]$. LI ((130))(1)

2. $$\int_0^1 \left(\ln\frac{1}{x}\right)^{q-1}\frac{(1+x)\,dx}{1+2x\cos t+x^2} = \sec\frac{t}{2}\cdot\Gamma(q)\sum_{k=1}^{\infty}(-1)^{k-1}\frac{\cos\left[\left(k-\frac{1}{2}\right)t\right]}{k^q}$$
$\left[|t| < \pi,\ q < \frac{1}{2}\right]$. LI ((130))(5)

3. $$\int_0^{\infty}\left[\ln\left(\frac{1}{x}\right)\right]^{\mu}\frac{x^{\nu-1}\,dx}{1-2ax\cos t+x^2a^2} = \frac{\Gamma(\mu+1)}{a\sin t}\sum_{k=0}^{\infty}\frac{a^k\sin kt}{(\nu+k-1)^{\mu+1}}$$
$[a > 0,\ \operatorname{Re}\mu > 0,\ 0 < \operatorname{Re}\nu < 2,\ |t| < \pi]$. BI ((140))(14)a

4. $$\int_0^1\left(\ln\frac{1}{x}\right)^{r-1}\frac{\cos\lambda - px}{1+p^2x^2-2px\cos\lambda}x^{q-1}\,dx =$$
$$= \Gamma(r)\sum_{k=1}^{\infty}\frac{p^{k-1}\cos k\lambda}{(q+k-1)^r}$$
$[r > 0,\ q > 0]$. BI ((113))(11)

5. $$\int_1^{\infty}(\ln x)^p\frac{dx}{x^2} = \Gamma(1+p)$$
$[p > -1]$. BI ((149))(1)

6. $$\int_0^1\left(\ln\frac{1}{x}\right)^{\mu-1}x^{\nu-1}\,dx = \frac{1}{\nu^{\mu}}\Gamma(\mu)$$
$[\operatorname{Re}\mu > 0,\ \operatorname{Re}\nu > 0]$. BI ((107))(3)

7. $$\int_0^1\left(\ln\frac{1}{x}\right)^{n-\frac{1}{2}}x^{\nu-1}\,dx = \frac{(2n-1)!!}{(2\nu)^n}\sqrt{\frac{\pi}{\nu}}$$
$[\operatorname{Re}\nu > 0]$. BI ((107))(2)

8. $$\int_0^1\left(\ln\frac{1}{x}\right)^{n-1}\frac{x^{\nu-1}}{1+x}\,dx = (n-1)!\sum_{k=0}^{\infty}\frac{(-1)^k}{(\nu+k)^n}$$
$[\operatorname{Re}\nu > 0]$. BI ((110))(4)

9. $$\int_0^1\left(\ln\frac{1}{x}\right)^{n-1}\frac{x^{\nu-1}}{1-x}\,dx = (n-1)!\,\zeta(n,\nu)$$
$[\operatorname{Re}\nu > 0]$. BI ((110))(7)

10. $$\int_0^1\left(\ln\frac{1}{x}\right)^{\mu-1}(x-1)^n\left(a+\frac{nx}{x-1}\right)x^{a-1}\,dx =$$
$$= \Gamma(\mu)\sum_{k=0}^{n}\frac{(-1)^k n(n-1)\ldots(n-k+1)}{(a+n-k)^{\mu-1}\,k!}$$
$[\operatorname{Re}\mu > 0]$. LI ((110))(10)

11. $$\int_0^1\left(\ln\frac{1}{x}\right)^{n-1}\frac{1-x^m}{1-x}\,dx = (n-1)!\sum_{k=1}^{m}\frac{1}{k^n}.$$
LI ((110))(9)

12. $$\int_0^1 \left(\ln\frac{1}{x}\right)^{\mu-1} \frac{x^{\nu-1}\,dx}{1-x^2} = \Gamma(\mu) \sum_{k=0}^{\infty} \frac{1}{(\nu+2k)^{\mu}} = \frac{1}{2^{\mu}}\Gamma(\mu)\,\zeta\left(\mu, \frac{\nu}{2}\right) \qquad [\mathrm{Re}\,\mu > 0,\ \mathrm{Re}\,\nu > 0].$$ BI ((110))(13)

13. $$\int_0^1 \frac{x^q - x^{-q}}{1-x^2}\left(\ln\frac{1}{x}\right)^p dx = \Gamma(p+1) \sum_{k=1}^{\infty} \left\{\frac{1}{(2k+q-1)^{p+1}} - \frac{1}{(2k-q-1)^{p+1}}\right\} \qquad [p > -1,\ q^2 < 1].$$ LI ((326))(12)a

14. $$\int_0^1 \left(\ln\frac{1}{x}\right)^{r-1} \frac{x^{p-1}\,dx}{(1+x^q)^s} = \Gamma(r) \sum_{k=0}^{\infty} \binom{-s}{k} \frac{1}{(p+kq)^r} \qquad [p > 0,\ q > 0,\ r > 0,\ 0 < s < r+2].$$ GW ((324))(11)

15. $$\int_0^1 \left(\ln\frac{1}{x}\right)^n (1+x^q)^m x^{p-1}\,dx = n! \sum_{k=0}^{m} \binom{m}{k} \frac{1}{(p+kq)^{n+1}} \qquad [p > 0,\ q > 0].$$ BI ((107))(6)

16. $$\int_0^1 \left(\ln\frac{1}{x}\right)^n (1-x^q)^m x^{p-1}\,dx = n! \sum_{k=0}^{m} \binom{m}{k} \frac{(-1)^k}{(p+kq)^{n+1}} \qquad [p > 0,\ q > 0].$$ BI ((107))(7)

17. $$\int_0^1 \left(\ln\frac{1}{x}\right)^{p-1} \frac{x^{q-1}\,dx}{1-ax^q} = \frac{1}{aq^p}\Gamma(p) \sum_{k=1}^{\infty} \frac{a^k}{k^p} \qquad [p > 0,\ q > 0,\ a < 1].$$ LI ((110))(8)

18. $$\int_0^1 \left(\ln\frac{1}{x}\right)^{2-\frac{1}{n}} (x^{p-1} - x^{q-1})\,dx = \frac{n}{n-1}\Gamma\left(\frac{1}{n}\right)\left(q^{1-\frac{1}{n}} - p^{1-\frac{1}{n}}\right) \qquad [q > p > 0].$$ BI ((133))(4)

19 $$\int_0^1 \left(\ln\frac{1}{x}\right)^{2n-1} \frac{x^p - x^{-p}}{1-x^q} x^{q-1}\,dx = \frac{1}{p^{2n}} \sum_{k=n}^{\infty} \left(\frac{2p\pi}{q}\right)^k \frac{|B_{2k}|}{2k\cdot(2k-2n)!} \qquad \left[p < \frac{q}{2}\right].$$ LI ((110))(16)

4.273 $$\int_u^v \left(\ln\frac{x}{u}\right)^{p-1}\left(\ln\frac{v}{x}\right)^{q-1}\frac{dx}{x} = \mathrm{B}(p, q)\left(\ln\frac{v}{u}\right)^{p+q-1} \qquad [p > 0,\ q > 0,\ uv > 0].$$ BI ((145))(36)

4.274 $$\int_0^{\frac{1}{e}} \frac{\sqrt[q]{x}\,dx}{x\sqrt{-(1+\ln x)}} = \frac{\sqrt{q\pi}}{\sqrt[q]{e}} \qquad [q > 0].$$ BI ((145))(4)

4.275

1. $$\int_0^1 \left[\left(\ln \frac{1}{x}\right)^{q-1} - x^{p-1}(1-x)^{q-1}\right] dx = \frac{\Gamma(q)}{\Gamma(p+q)}\left[\Gamma(p+q) - \Gamma(p)\right] \qquad [p > 0,\ q > 0].$$ BI ((107))(8)

2. $$\int_0^1 \left[x - \left(\frac{1}{1-\ln x}\right)^q\right] \frac{dx}{x \ln x} = -\psi(q) \qquad [q > 0].$$ BI ((126))(5)

4.28 Combinations of rational functions of $\ln x$ and powers

4.281

1. $$\int_0^1 \left[\frac{1}{\ln x} + \frac{1}{1-x}\right] dx = C.$$ BI ((127))(15)

2. $$\int_1^\infty \frac{dx}{x^2 (\ln p - \ln x)} = \frac{1}{p} \operatorname{li}(p).$$ LA 281(30)

3. $$\int_0^1 \frac{x^{p-1}\, dx}{q \pm \ln x} = \pm e^{\mp pq} \operatorname{Ei}(\pm pq) \qquad [p > 0,\ q > 0].$$ LI ((144))(11, 12)

4. $$\int_0^1 \left[\frac{1}{\ln x} + \frac{x^{\mu-1}}{1-x}\right] dx = -\psi(\mu) \qquad [\operatorname{Re} \mu > 0].$$ WH

5. $$\int_0^1 \left[\frac{x^{p-1}}{\ln x} + \frac{x^{q-1}}{1-x}\right] dx = \ln p - \psi(q) \qquad [p > 0,\ q > 0].$$ BI ((127))(17)

6. $$\int_0^1 \left[\frac{1}{1-x^2} + \frac{1}{2x \ln x}\right] \frac{dx}{\ln x} = \frac{\ln 2}{2}.$$ LI ((130))(19)

7. $$\int_0^1 \left[q - \frac{1}{2} + \frac{(1-x)(1+q\ln x) + x \ln x}{(1-x)^2} x^{q-1}\right] \frac{dx}{\ln x} = \frac{1}{2} - q - \ln \Gamma(q) + \frac{\ln 2\pi}{2} \qquad [q > 0].$$ BI ((128))(15)

4.282

1. $$\int_0^1 \frac{\ln x}{4\pi^2 + (\ln x)^2} \cdot \frac{dx}{1-x} = \frac{1}{4} - \frac{1}{2} C.$$ BI ((129))(1)

2. $$\int_0^1 \frac{1}{a^2 + (\ln x)^2} \cdot \frac{dx}{1+x^2} = \frac{1}{2a} \beta\left(\frac{2a+\pi}{4\pi}\right) \qquad \left[a > -\frac{\pi}{2}\right].$$ BI ((129))(9)

3. $$\int_0^1 \frac{1}{\pi^2 + (\ln x)^2} \frac{dx}{1+x^2} = \frac{4-\pi}{4\pi}.$$ BI ((129))(6)

4. $$\int_0^1 \frac{\ln x}{\pi^2 + (\ln x)^2} \cdot \frac{dx}{1-x^2} = \frac{1}{2}\left(\frac{1}{2} - \ln 2\right).$$ BI ((129))(10)

5. $\int_0^1 \frac{\ln x}{a^2+(\ln x)^2} \cdot \frac{x\,dx}{1-x^2} = \frac{1}{2}\left[\frac{\pi}{2a} + \ln\frac{\pi}{a} + \psi\left(\frac{a}{\pi}\right)\right]$ $[a > 0]$. BI ((129))(14)

6. $\int_0^1 \frac{\ln x}{\pi^2+(\ln x)^2} \cdot \frac{x\,dx}{1-x^2} = \frac{1}{2}\left(\frac{1}{2} - \boldsymbol{C}\right)$. BI ((129))(13)

7. $\int_0^1 \frac{1}{\pi^2+4(\ln x)^2} \cdot \frac{dx}{1+x^2} = \frac{\ln 2}{4\pi}$. BI ((129))(7)

8. $\int_0^1 \frac{\ln x}{\pi^2+4(\ln x)^2} \cdot \frac{dx}{1-x^2} = \frac{2-\pi}{16}$. BI ((129))(11)

9. $\int_0^1 \frac{1}{\pi^2+16(\ln x)^2} \cdot \frac{dx}{1+x^2} = \frac{1}{8\pi\sqrt{2}}\,[\pi + 2\ln(\sqrt{2}-1)]$. BI ((129))(8)

10. $\int_0^1 \frac{\ln x}{\pi^2+16(\ln x)^2} \cdot \frac{dx}{1-x^2} = -\frac{\pi}{32\sqrt{2}} + \frac{1}{16} + \frac{1}{16\sqrt{2}}\ln(\sqrt{2}-1)$. BI ((129))(12)

11. $\int_0^1 \frac{\ln x}{[a^2+(\ln x)^2]^2}\,\frac{dx}{1-x} = -\frac{\pi^2}{a^4}\sum_{k=1}^{\infty} |B_{2k}| \left(\frac{2\pi}{a}\right)^{2k-2}$ BI ((129))(4)

12. $\int_0^1 \frac{\ln x}{[a^2+(\ln x)^2]^2}\,\frac{x\,dx}{1-x^2} = -\frac{\pi^2}{4a^4}\sum_{k=1}^{\infty} |B_{2k}| \left(\frac{\pi}{a}\right)^{2k-2}$ BI ((129))(16)

13. $\int_0^1 \frac{x^p - x^{-p}}{x^2-1}\,\frac{dx}{q^2+\ln^2 x} = \frac{2\pi}{q}\sum_{k=1}^{\infty} (-1)^{k-1}\frac{\sin kp\pi}{2q+k\pi}$ $[p^2 < 1]$. BI ((132))(13)a

4.283

1. $\int_0^1 \left(\frac{x-1}{\ln x} - x\right)\frac{dx}{\ln x} = \ln 2 - 1$. BI ((132))(17)a

2. $\int_0^1 \left(\frac{1}{\ln x} + \frac{1}{1-x} - \frac{1}{2}\right)\frac{dx}{\ln x} = \frac{\ln 2\pi}{2} - 1$. BI ((127))(20)

3. $\int_0^1 \left(\frac{1}{\ln x} + \frac{x}{1-x} + \frac{x}{2}\right)\frac{dx}{x\ln x} = \frac{\ln 2\pi}{2}$. BI ((127))(23)

4. $\int_0^1 \left[\frac{1}{(\ln x)^2} - \frac{x}{(1-x)^2}\right] dx = \boldsymbol{C} - \frac{1}{2}$. GW ((326))(8a)

5. $\int_0^1 \left(\frac{1}{1-x^2} + \frac{1}{2\ln x} - \frac{1}{2}\right)\frac{dx}{\ln x} = \frac{\ln 2 - 1}{2}$. BI ((128))(14)

6. $\int_0^1 \left(\frac{1}{\ln x} + \frac{1}{2}\cdot\frac{1+x}{1-x} - \ln x\right)\frac{dx}{\ln x} = \frac{\ln 2\pi}{2}$. BI ((127))(22)

7. $\int_0^1 \left[\frac{1}{1-\ln x} - x\right] \frac{dx}{x \ln x} = -C.$ GW ((326))(11a)

8. $\int_0^1 \left[\frac{x^q-1}{x(\ln x)^2} - \frac{q}{\ln x}\right] dx = q \ln q - q \quad [q > 0].$ BI ((126))(2)

9. $\int_0^1 \left[x + \frac{1}{a \ln x - 1}\right] \frac{dx}{x \ln x} = \ln \frac{a}{q} + C \quad [a > 0,\ q > 0].$ BI ((126))(8)

10. $\int_0^1 \left[\frac{1}{\ln x} + \frac{1+x}{2(1-x)}\right] \frac{x^{p-1}}{\ln x} dx = -\ln \Gamma(p) + \left(p - \frac{1}{2}\right) \ln p - p + \frac{\ln 2\pi}{2} \quad [p > 0].$ GW ((326))(9)

11. $\int_0^1 \left[p - 1 - \frac{1}{1-x} + \left(\frac{1}{2} - \frac{1}{\ln x}\right) x^{p-1}\right] \frac{dx}{\ln x} = \left(\frac{1}{2} - p\right) \ln p + p - \frac{\ln 2\pi}{2} \quad [p > 0].$ BI ((127))(25)

12. $\int_0^1 \left[-\frac{1}{(\ln x)^2} + \frac{(p-2)x^p - (p-1)x^{p-1}}{(1-x)^2}\right] dx = -\psi(p) + p - \frac{3}{2} \quad [p > 0]$ GW ((326))(8)

13. $\int_0^1 \left[\left(p - \frac{1}{2}\right) x^3 + \frac{1}{2}\left(1 - \frac{1}{\ln x}\right)(x^{2p-1} - 1)\right] \frac{dx}{\ln x} = \left(\frac{1}{2} - p\right)(\ln p - 1) \quad [p > 0].$ BI ((132))(23)a

14. $\int_0^1 \left[\left(q - \frac{1}{2}\right) \frac{x^{p-1} - x^{r-1}}{\ln x} + \frac{p x^{pq-1}}{1-x^p} - \frac{r x^{rq-1}}{1-x^r}\right] \frac{dx}{\ln x} = (p - r)\left[\frac{1}{2} - q - \ln \Gamma(q) + \frac{\ln 2\pi}{2}\right] \quad [q > 0].$ BI ((132))(13)

4.284

1. $\int_0^1 \left[\frac{x^q-1}{x(\ln x)^3} - \frac{q}{x(\ln x)^2} - \frac{q^2}{2 \ln x}\right] dx = \frac{q^2}{2} \ln q - \frac{3}{4} q^2 \quad [q > 0].$ BI ((126))(3)

2. $\int_0^1 \left[\frac{x^q-1}{x(\ln x)^4} - \frac{q}{x(\ln x)^3} - \frac{q^2}{2x(\ln x)^2} - \frac{q^3}{6 \ln x}\right] dx = \frac{q^3}{6} \ln q - \frac{11}{36} q^3 \quad [q > 0].$ BI ((126))(4)

4.285 $\int_0^1 \frac{x^{p-1}\, dx}{(q + \ln x)^n} = \frac{p^{n-1}}{(n-1)!} e^{-pq} \operatorname{Ei}(pq) - \frac{1}{(n-1)!\, q^{n-1}} \sum_{k=1}^{n-1} (n-k-1)!\, (pq)^{k-1} \quad [p > 0,\ q < 0].$ BI ((125))(21)

In integrals of the form $\int \frac{x^a (\ln x)^n\, dx}{[b \pm (\ln x)^m]^l}$, we should make the substitution $x = e^t$ or $x = e^{-t}$ and then seek the resulting integrals in 3.351 – 3.356.

4.29-4.32 Combinations of logarithmic functions of more complicated arguments and powers

4.291

1. $\int_0^1 \frac{\ln(1+x)}{x}\,dx = \frac{\pi^2}{12}.$ FI II 483

2. $\int_0^1 \frac{\ln(1-x)}{x}\,dx = -\frac{\pi^2}{6}.$ FI II 714

3. $\int_0^{\frac{1}{2}} \frac{\ln(1-x)}{x}\,dx = \frac{1}{2}(\ln 2)^2 - \frac{\pi^2}{12}.$ BI ((145))(2)

4. $\int_0^1 \ln\left(1-\frac{x}{2}\right)\frac{dx}{x} = \frac{1}{2}(\ln 2)^2 - \frac{\pi^2}{12}.$ BI ((114))(18)

5. $\int_0^1 \frac{\ln\frac{1+x}{2}}{1-x}\,dx = \frac{1}{2}(\ln 2)^2 - \frac{\pi^2}{12}.$ BI ((115))(1)

6. $\int_0^1 \frac{\ln(1+x)}{1+x}\,dx = \frac{1}{2}(\ln 2)^2.$ BI ((114))(14)a

7. $\int_0^\infty \frac{\ln(1+ax)}{1+x}\,dx = \frac{\pi}{4}\ln(1+a^2) - \int_0^a \frac{\ln u\,du}{1+u^2} \quad [a>0].$ GI II (2209)

8. $\int_0^1 \frac{\ln(1+x)}{1+x^2}\,dx = \frac{\pi}{8}\ln 2.$ FI II 157

9. $\int_0^\infty \frac{\ln(1+x)}{1+x^2}\,dx = \frac{\pi}{4}\ln 2 + G.$ BI ((136))(1)

10. $\int_0^1 \frac{\ln(1-x)}{1+x^2}\,dx = \frac{\pi}{8}\ln 2 - G.$ BI ((114))(17)

11. $\int_1^\infty \frac{\ln(x-1)}{1+x^2}\,dx = \frac{\pi}{8}\ln 2.$ BI ((144))(4)

12. $\int_0^1 \frac{\ln(1+x)}{x(1+x)}\,dx = \frac{\pi^2}{12} - \frac{1}{2}(\ln 2)^2.$ BI ((144))(4)

13. $\int_0^\infty \frac{\ln(1+x)}{x(1+x)}\,dx = \frac{\pi^2}{6}.$ BI ((141))(9)a

14. $\int_0^1 \frac{\ln(1+x)}{(ax+b)^2}\,dx = \frac{1}{a(a-b)} \ln \frac{a+b}{b} + \frac{2\ln 2}{b^2-a^2} \quad [a \neq b,\ ab > 0];$

$= \frac{1}{2a^2}(1-\ln 2) \quad [a = b].$ LI ((114))(5)a

15. $\int_0^\infty \frac{\ln(1+x)}{(ax+b)^2}\,dx = \frac{\ln \frac{a}{b}}{a(a-b)} \quad [ab > 0].$ BI ((139))(5)

16. $\int_0^1 \ln(a+x) \frac{dx}{a+x^2} = \frac{1}{2\sqrt{a}} \operatorname{arcctg} \sqrt{a} \ln[(1+a)a] \quad [a > 0].$

BI ((114))(20)

17. $\int_0^\infty \ln(a+x) \frac{dx}{(b+x)^2} = \frac{a \ln a - b \ln b}{b(a-b)} \quad [a > 0,\ b > 0,\ a \neq b].$

LI ((139))(6)

18. $\int_0^a \frac{\ln(1+ax)}{1+x^2}\,dx = \frac{1}{2} \operatorname{arctg} a \ln(1+a^2).$ GI II (2195)

19. $\int_0^1 \frac{\ln(1+ax)}{1+ax^2}\,dx = \frac{1}{2\sqrt{a}} \operatorname{arctg} \sqrt{a} \ln(1+a) \quad [a > 0].$ BI ((114))(21)

20. $\int_0^1 \frac{\ln(ax+b)}{(1+x)^2}\,dx = \frac{1}{a-b}\left[\frac{1}{2}(a+b)\ln(a+b) - b\ln b - a\ln 2\right]$

$[a > 0,\ b > 0,\ a \neq b].$ BI ((114))(22)

21. $\int_0^\infty \frac{\ln(ax+b)}{(1+x)^2}\,dx = \frac{1}{a-b}[a \ln a - b \ln b] \quad [a > 0,\ b > 0].$

BI ((139))(8)

22. $\int_0^\infty \ln(a+x) \frac{x\,dx}{(b^2+x^2)^2} = \frac{1}{2(a^2+b^2)}\left(\ln b + \frac{a\pi}{2b} + \frac{a^2}{b^2}\ln a\right) \quad [a > 0,\ b > 0].$

BI ((139))(9)

23. $\int_0^1 \ln(1+x) \frac{1+x^2}{(1+x)^4}\,dx = -\frac{1}{3}\ln 2 + \frac{23}{72}.$ LI ((114))(12)

24. $\int_0^1 \ln(1+x) \frac{1+x^2}{a^2+x^2} \cdot \frac{dx}{1+a^2x^2} = \frac{1}{2a(1+a^2)}\left[\frac{\pi}{2}\ln(1+a^2) -\right.$

$\left. - 2 \operatorname{arctg} a \cdot \ln a\right] \quad [a > 0].$ LI ((114))(11)

25. $\int_0^1 \ln(1+x) \frac{1-x^2}{(ax+b)^2} \frac{dx}{(bx+a)^2} = \frac{1}{a^2-b^2}\left\{\frac{1}{a-b}\left[\frac{a+b}{ab}\ln(a+b) -\right.\right.$

$\left.\left. - \frac{1}{a}\ln b - \frac{1}{b}\ln a\right] + \frac{4\ln 2}{b^2-a^2}\right\} \quad [a > 0,\ b > 0,\ a^2 \neq b^2].$

LI ((114))(13)

26. $$\int_0^\infty \ln(1+x)\frac{1-x^2}{(ax+b)^2}\cdot\frac{dx}{(bx+a)^2}=\frac{1}{ab(a^2-b^2)}\ln\frac{b}{a} \qquad [a>0,\ b>0].$$

LI ((139))(14)

27. $$\int_0^1 \ln(1+ax)\frac{1-x^2}{(1+x^2)^2}dx=\frac{1}{2}\frac{(1+a)^2}{1+a^2}\ln(1+a)-$$
$$-\frac{1}{2}\cdot\frac{a}{1+a^2}\ln 2-\frac{\pi}{4}\cdot\frac{a^2}{1+a^2} \qquad [a>-1].$$

BI ((114))(23)

28. $$\int_0^\infty \ln(a+x)\frac{b^2-x^2}{(b^2+x^2)^2}dx=\frac{1}{a^2+b^2}\left(a\ln\frac{b}{a}-\frac{b\pi}{2}\right) \qquad [a>0,\ b>0].$$

BI ((139))(11)

29. $$\int_0^\infty \ln(a-x)^2\frac{b^2-x^2}{(b^2+x^2)^2}dx=\frac{2}{a^2+b^2}\left(a\ln\frac{a}{b}-\frac{b\pi}{2}\right) \qquad [a>0,\ b>0].$$

BI ((139))(12)

30. $$\int_0^\infty \ln(a-x)^2\frac{x\,dx}{(b^2+x^2)^2}=\frac{1}{a^2+b^2}\left(\ln b-\frac{a\pi}{2b}+\frac{a^2}{b^2}\ln a\right) \qquad [a>0,\ b>0].$$

BI ((139))(10)

4.292

1. $$\int_0^1 \frac{\ln(1\pm x)}{\sqrt{1-x^2}}dx=-\frac{\pi}{2}\ln 2\pm 2G.$$

GW ((325))(20)

2. $$\int_0^1 \frac{x\ln(1\pm x)}{\sqrt{1-x^2}}dx=-1\pm\frac{\pi}{2}.$$

GW ((325))(22c)

3. $$\int_{-a}^a \frac{\ln(1+bx)}{\sqrt{a^2-x^2}}dx=\pi\ln\frac{1+\sqrt{1-a^2b^2}}{2} \qquad \left[0\leqslant|b|\leqslant\frac{1}{a}\right].$$

BI ((145))(16, 17)a, GW ((325))(21e)

4. $$\int_0^1 \frac{x\ln(1+ax)}{\sqrt{1-x^2}}dx=-1+\frac{\pi}{2}\cdot\frac{1-\sqrt{1-a^2}}{a}+\frac{\sqrt{1-a^2}}{a}\arcsin a \qquad [|a|\leqslant 1];$$
$$=-1+\frac{\pi}{2a}+\frac{\sqrt{a^2-1}}{a}\ln\left(a+\sqrt{a^2-1}\right) \qquad [a\geqslant 1].$$

GW ((325))(22)

5. $$\int_0^1 \frac{\ln(1+ax)}{x\sqrt{1-x^2}}dx=\frac{1}{2}\arcsin a\,(\pi-\arcsin a)=$$
$$=\frac{\pi^2}{8}-\frac{1}{2}(\arccos a)^2 \qquad [|a|\leqslant 1].$$

BI ((120))(4), GW ((325))(21a)

4.293

1. $$\int_0^1 x^{\mu-1}\ln(1+x)\,dx=\frac{1}{\mu}[\ln 2-\beta(\mu+1)] \quad [\operatorname{Re}\mu>-1].$$

BI ((106))(4)a

2. $\int_1^\infty x^{\mu-1}\ln(1+x)\,dx = \frac{1}{\mu}[\beta(-\mu) - \ln 2]$ $[\operatorname{Re}\mu < 0]$. ET I 315(17)

3. $\int_0^\infty x^{\mu-1}\ln(1+x)\,dx = \frac{\pi}{\mu\sin\mu\pi}$ $[-1 < \operatorname{Re}\mu < 0]$. GW ((325))(3)a

4. $\int_0^1 x^{2n-1}\ln(1+x)\,dx = \frac{1}{2n}\sum_{k=1}^{2n}\frac{(-1)^{k-1}}{k}$ GW ((325))(2b)

5. $\int_0^1 x^{2n}\ln(1+x)\,dx = \frac{1}{2n+1}\left[\ln 4 + \sum_{k=1}^{2n+1}\frac{(-1)^k}{k}\right]$. GW ((325))(2c)

6. $\int_0^1 x^{n-\frac{1}{2}}\ln(1+x)\,dx = \frac{2\ln 2}{2n+1} + \frac{(-1)^n\cdot 4}{2n+1}\left[\pi - \sum_{k=0}^{n}\frac{(-1)^k}{2k+1}\right]$. GW ((325))(2f)

7. $\int_0^\infty x^{\mu-1}\ln|1-x|\,dx = \frac{\pi}{\mu}\operatorname{ctg}(\mu\pi)$ $[-1 < \operatorname{Re}\mu < 0]$.

BI((134))(4), ET I 315(18)

8. $\int_0^1 x^{\mu-1}\ln(1-x)\,dx = -\frac{1}{\mu}[\psi(\mu+1) - \psi(1)]$ $[\operatorname{Re}\mu > -1]$.

ET I 316(19)

9. $\int_1^\infty x^{\mu-1}\ln(x-1)\,dx = \frac{1}{\mu}[\pi\operatorname{ctg}(\mu\pi) + \psi(\mu+1) - \psi(1)]$ $[\operatorname{Re}\mu < 0]$.

ET I 316(20)

10. $\int_0^\infty x^{\mu-1}\ln(1+\gamma x)\,dx = \frac{\pi}{\mu\gamma^\mu\sin\mu\pi}$ $[-1 < \operatorname{Re}\mu < 0, \quad |\arg\gamma| < \pi]$.

BI ((134))(3)

11. $\int_0^\infty \frac{x^{\mu-1}\ln(1+x)}{1+x}\,dx = \frac{\pi}{\sin\mu\pi}[C + \psi(1-\mu)]$ $[-1 < \operatorname{Re}\mu < 1]$.

ET I 316(21)

12. $\int_0^1 \frac{\ln(1+x)}{(1+x)^{\mu+1}}\,dx = \frac{-\ln 2}{2^\mu\mu} + \frac{2^\mu - 1}{2^\mu\mu^2}$. BI ((114))(6)

13. $\int_0^1 \frac{x^{\mu-1}\ln(1-x)}{(1-x)^{1-\nu}}\,dx = \mathrm{B}(\mu, \nu)[\psi(\nu) - \psi(\mu+\nu)]$

$[\operatorname{Re}\mu > 0, \quad \operatorname{Re}\nu > 0]$. ET I 316(122)

14. $\int_0^\infty \frac{x^{\mu-1}\ln(\gamma+x)}{(\gamma+x)^\nu}\,dx = \gamma^{\mu-\nu}\mathrm{B}(\mu, \nu-\mu)[\psi(\nu) - \psi(\nu-\mu) + \ln\gamma]$

$[0 < \operatorname{Re}\mu < \operatorname{Re}\nu]$. ET I 316(23)

4.294

1. $$\int_0^1 \ln(1+x)\frac{(p-1)x^{p-1}-px^{-p}}{x}\,dx = 2\ln 2 - \frac{\pi}{\sin p\pi} \qquad [0<p<1].$$

BI ((114))(2)

2. $$\int_0^1 \ln(1+x)\frac{1+x^{2n+1}}{1+x}\,dx = 2\ln 2\sum_{k=0}^{n}\frac{1}{2k+1} - \sum_{j=1}^{2n+1}\frac{1}{j}\sum_{k=1}^{j}\frac{(-1)^{k-1}}{k}.$$

BI ((114))(7)

3. $$\int_0^1 \ln(1+x)\frac{1-x^{2n}}{1+x}\,dx = 2\ln 2\cdot\sum_{k=0}^{n-1}\frac{1}{2k+1} - \sum_{j=1}^{2n}\frac{1}{j}\sum_{k=1}^{j}\frac{(-1)^{k-1}}{k}.$$

BI ((114))(8)

4. $$\int_0^1 \ln(1+x)\frac{1-x^{2n}}{1-x}\,dx = 2\ln 2\cdot\sum_{k=0}^{n-1}\frac{1}{2k+1} + \sum_{i=1}^{2n}\frac{(-1)^j}{j}\sum_{k=1}^{j}\frac{(-1)^{k-1}}{k}.$$

BI ((114))(9)

5. $$\int_0^1 \ln(1+x)\frac{1-x^{2n+1}}{1-x}\,dx = 2\ln 2\sum_{k=0}^{n}\frac{1}{2k+1} + \sum_{j=1}^{2n+1}\frac{(-1)^j}{j}\sum_{k=1}^{j}\frac{(-1)^{k-1}}{k}.$$

BI ((114))(10)

6. $$\int_0^1 \ln(1-x)\frac{1-(-1)^n x^n}{1-x}\,dx = \sum_{j=1}^{n}\frac{(-1)^j}{j}\sum_{k=1}^{j}\frac{1}{k}.$$ BI ((114))(15)

7 $$\int_0^1 \ln(1-x)\frac{1-x^n}{1-x}\,dx = -\sum_{j=1}^{n}\frac{1}{j}\sum_{k=1}^{j}\frac{1}{k}.$$ BI ((114))(16)

8 $$\int_0^\infty \ln(1-x)^2 x^p\,dx = \frac{2\pi}{p+1}\operatorname{ctg} p\pi \qquad [-2<p<-1].$$ BI ((134))(13)a

9. $$\int_0^1 [\ln(1+x)]^n (1+x)^r\,dx = (-1)^{n-1}\frac{n!}{(r+1)^{n+1}} + 2^{r+1}\sum_{k=0}^{n}\frac{(-1)^k n!(\ln 2)^{n-k}}{(n-k)!(r+1)^{k+1}}.$$ LI ((106))(34)a

10. $$\int_0^1 [\ln(1-x)]^n (1-x)^r\,dx = (-1)^n\frac{n!}{(r+1)^{n+1}} \qquad [r>-1].$$ BI ((106))(35)a

11. $$\int_0^1 \left(\ln\frac{1}{1-x^2}\right)^n x^{2q-1}\,dx = \frac{n!}{2}\zeta(n+1,\ q+1) \qquad [-1<q<0].$$

BI ((311))(15)a

12. $$\int_0^1 (\ln x)^{2n}\ln(1-x^2)\frac{dx}{x} = -\frac{\pi^{2n+2}}{2(n+1)(2n+1)}|B_{2n+2}|.$$ BI ((309))(5)a

4.295

1. $\int_0^\infty \ln(\mu x^2+\beta)\frac{dx}{\gamma+x^2}=\frac{\pi}{\sqrt{\gamma}}\ln\left(\sqrt{\mu\gamma}+\sqrt{\beta}\right)$

$[\operatorname{Re}\beta>0,\ \operatorname{Re}\mu>0,\ \ |\arg\gamma|<\pi]$. ET II 218(27)

2. $\int_0^1 \ln(1+x^2)\frac{dx}{x^2}=\frac{\pi}{2}-\ln 2.$ GW ((325))(2g)

3. $\int_0^\infty \ln(1+x^2)\frac{dx}{x^2}=\pi.$ GW ((325))(4c)

4. $\int_0^\infty \ln(1+x^2)\frac{dx}{(a+x)^2}=\frac{2a}{1+a^2}\left(\frac{\pi}{2a}+\ln a\right)\quad [a>0].$ BI ((319))(6)a

5. $\int_0^1 \ln(1+x^2)\frac{dx}{1+x^2}=\frac{\pi}{2}\ln 2-\boldsymbol{G}.$ BI ((114))(24)

6. $\int_1^\infty \ln(1+x^2)\frac{dx}{1+x^2}=\frac{\pi}{2}\ln 2+\boldsymbol{G}.$ BI ((114))(5)

7. $\int_0^\infty \ln(a^2+b^2x^2)\frac{dx}{c^2+g^2x^2}=\frac{\pi}{cg}\ln\frac{ag+bc}{g}$

$[a>0,\ b>0,\ c>0,\ g>0].$ BI ((136))(11-14)a

8. $\int_0^\infty \ln(a^2+b^2x^2)\frac{dx}{c^2-g^2x^2}=-\frac{\pi}{cg}\operatorname{arctg}\frac{bc}{ag}$

$[a>0,\ b>0,\ c>0,\ g>0].$ BI ((136))(15)a

9. $\int_0^\infty \frac{\ln(1+p^2x^2)-\ln(1+q^2x^2)}{x^2}\,dx=\pi(p-q)\quad [p>0,\ q>0].$ FI II 645

10. $\int_0^1 \ln\frac{1+a^2x^2}{1+a^2}\,\frac{dx}{1-x^2}=-(\operatorname{arctg} a)^2.$ BI ((115))(2)

11. $\int_0^1 \ln(1-x^2)\frac{dx}{x}=-\frac{\pi^2}{12}.$

12. $\int_0^\infty \ln(1-x^2)^2\frac{dx}{x^2}=0.$ BI ((142))(9)a

13. $\int_0^1 \ln(1-x^2)\frac{dx}{1+x^2}=\frac{\pi}{4}\ln 2-\boldsymbol{G}.$ GW ((325))(17)

14. $\int_1^\infty \ln(x^2-1)\frac{dx}{1+x^2}=\frac{\pi}{4}\ln 2+\boldsymbol{G}$ BI ((144))(6)

15. $\int_0^\infty \ln (a^2 - x^2)^2 \frac{dx}{b^2+x^2} = \frac{\pi}{b} \ln (a^2 + b^2) \quad [b > 0].$ BI ((136))(16)

16. $\int_0^\infty \ln (a^2 - x^2)^2 \frac{b^2-x^2}{(b^2+x^2)^2} dx = -\frac{2b\pi}{a^2+b^2} \quad [b > 0].$ BI ((136))(20)

17. $\int_0^1 \ln (1+x^2) \frac{dx}{x(1+x^2)} = \frac{1}{2}\left[\frac{\pi^2}{12} - \frac{1}{2}(\ln 2)^2\right].$ BI ((114))(25)

18. $\int_0^\infty \ln (1+x^2) \frac{dx}{x(1+x^2)} = \frac{\pi^2}{12}.$ BI ((141))(9)

19. $\int_0^1 \ln (\cos^2 t + x^2 \sin^2 t) \frac{dx}{1-x^2} = -t^2.$ BI ((114))(27)a

20. $\int_0^\infty \ln (a^2 + b^2x^2) \frac{dx}{(c+gx)^2} = \frac{2 \ln b}{cg} + \frac{b^2}{a^2g^2+b^2c^2}\left(\frac{a}{b}\pi + 2\frac{c}{g}\ln\frac{c}{g} + \right.$

$\left. + 2\frac{a^2g}{b^2c}\ln\frac{a}{b}\right) \quad [a > 0,\ b > 0,\ c > 0,\ g > 0].$ BI ((139))(16)a

21. $\int_0^1 \ln (a^2 + b^2x^2) \frac{dx}{(c+gx)^2} = \frac{2}{c(c+g)} \ln a + \frac{b^2}{a^2g^2+b^2c^2}\left[\frac{2a}{b} \operatorname{arcctg} \frac{a}{b} + \right.$

$\left. + \frac{cb^2 - ga^2}{b^2(c+g)} \ln \frac{a^2+b^2}{a^2} - 2\frac{c}{g} \ln \frac{c+g}{c}\right] \quad [a > 0,\ b > 0,\ c > 0,\ g > 0].$

BI ((114))(28)a

22. $\int_0^\infty \frac{\ln (1+p^2x^2)}{r^2+q^2x^2} dx = \int_0^\infty \frac{\ln (p^2+x^2)}{q^2+r^2x^2} dx = \frac{\pi}{qr} \ln \frac{q+pr}{q} \quad [qr > 0,\ p > 0].$

FI II 745a, BI((318))(1)a, BI((318))(4)a

23. $\int_0^\infty \frac{\ln (1+a^2x^2)}{b^2+c^2x^2} \frac{dx}{d^2+g^2x^2} = \frac{\pi}{b^2g^2 - c^2d^2}\left[\frac{g}{d} \ln \left(1 + \frac{ad}{g}\right) - \right.$

$\left. - \frac{c}{b} \ln \left(1 + \frac{ab}{c}\right)\right] \quad [a > 0,\ b > 0,\ c > 0,\ d > 0,\ g > 0 \quad b^2g^2 \neq c^2d^2].$

BI ((141))(10)

24. $\int_0^\infty \frac{\ln (1+a^2x^2)}{b^2+c^2x^2} \frac{x^2\, dx}{d^2+g^2x^2} = \frac{\pi}{b^2g^2 - c^2d^2}\left[\frac{b}{c} \ln \left(1 + \frac{ab}{c}\right) - \right.$

$\left. - \frac{d}{g} \ln \left(1 + \frac{ad}{g}\right)\right] [a > 0,\ b > 0,\ c > 0,\ d > 0,\ g > 0,\ b^2g^2 \neq c^2d^2].$

BI ((141))(11)

25. $\int_0^\infty \ln (a^2 + b^2x^2) \frac{dx}{(c^2+g^2x^2)^2} = \frac{\pi}{2c^3g}\left(\ln \frac{ag+bc}{g} - \frac{bc}{ag+bc}\right)$

$[a > 0,\ b > 0,\ c > 0,\ g > 0].$ GW ((325))(18a)

26. $\int_0^\infty \ln(a^2+b^2x^2)\frac{x^2\,dx}{(c^2+g^2x^2)^2} = \frac{\pi}{2cg^3}\left(\ln\frac{ag+bc}{g}+\frac{bc}{ag+bc}\right)$

$[a>0,\ b>0,\ c>0,\ g>0]$. GW ((325))(18b)

27. $\int_0^1 \ln(1+ax^2)\sqrt{1-x^2}\,dx = \frac{\pi}{2}\left\{\ln\frac{1+\sqrt{1+a}}{2}+\frac{1}{2}\,\frac{1-\sqrt{1+a}}{1+\sqrt{1+a}}\right\}$

$[a>0]$. BI ((117))(6)

28. $\int_0^1 \ln(1+a-ax^2)\sqrt{1-x^2}\,dx = \frac{\pi}{2}\left\{\ln\frac{1+\sqrt{1+a}}{2}-\frac{1}{2}\,\frac{1-\sqrt{1+a}}{1+\sqrt{1+a}}\right\}$

$[a>0]$. BI ((117))(7)

29. $\int_0^1 \ln(1-a^2x^2)\frac{dx}{\sqrt{1-x^2}} = \pi\ln\frac{1+\sqrt{1-a^2}}{2}\quad [a^2<1]$. BI ((119))(1)

30. $\int_0^1 \ln(1-a^2x^2)\frac{dx}{x\sqrt{1-x^2}} = \frac{\pi^2}{4}-(\arccos a)^2\quad [a^2<1]$. LI ((120))(11)

31. $\int_0^1 \ln(1-x^2)\frac{dx}{\sqrt{(1-x^2)(1-k^2x^2)}} = \ln\frac{k'}{k}\,\boldsymbol{K}(k)-\frac{\pi}{2}\,\boldsymbol{K}(k')$. BI ((120))(12)

32. $\int_0^1 \ln(1\pm kx^2)\frac{dx}{\sqrt{(1-x^2)(1-k^2x^2)}} = \frac{1}{2}\ln\frac{2\pm 2k}{\sqrt{k}}\,\boldsymbol{K}(k)-\frac{\pi}{8}\,\boldsymbol{K}(k')$.

BI ((120))(8), BI ((120))(14)

33. $\int_0^1 \frac{\ln(1-k^2x^2)}{\sqrt{(1-x^2)(1-k^2x^2)}}\,dx = \ln k'\,\boldsymbol{K}(k)$. BI ((119))(27)

34. $\int_0^1 \ln(1-k^2x^2)\sqrt{\frac{1-k^2x^2}{1-x^2}}\,dx = (2-k^2)\boldsymbol{K}(k)-(2-\ln k')\,\boldsymbol{E}(k)$.

BI ((119))(3)

35. $\int_0^1 \sqrt{\frac{1-x^2}{1-k^2x^2}}\ln(1-k^2x^2)\,dx = \frac{1}{k^2}(1+k'^2-k'^2\ln k')\,\boldsymbol{K}(k)-$

$-(2-\ln k')\,\boldsymbol{E}(k)$. BI ((119))(7)

36. $\int_{-1}^1 \ln(1-x^2)\frac{dx}{(a+bx)\sqrt{1-x^2}} = \frac{2\pi}{\sqrt{a^2-b^2}}\ln\frac{\sqrt{a^2-b^2}}{a+\sqrt{a^2-b^2}}$

$[a>0,\ b>0,\ a\neq b]$. BI ((145))(15)

37. $\int_0^1 \ln(1-x^2)(px^{p-1}-qx^{q-1})\,dx = \psi\left(\frac{p}{2}+1\right)-\psi\left(\frac{q}{2}+1\right)$

$[p>-2,\ q>-2]$. BI ((106))(15)

38. $\int_0^1 \ln(1+ax^2)\frac{dx}{\sqrt{1-x^2}} = \pi\ln\frac{1+\sqrt{1+a}}{2}\quad [a\geqslant -1]$. GW ((325))(21b)

39. $\int_0^1 \ln(1+x^2)\, x^{\mu-1}\, dx = \frac{1}{\mu}\left[\ln 2 - \beta\left(\frac{\mu}{2}+1\right)\right]$

$[\operatorname{Re}\mu > -2].$ BI ((106))(12)

40. $\int_0^\infty \ln(1+x^2)\, x^{\mu-1}\, dx = \frac{\pi}{\mu \sin\frac{\mu\pi}{2}} \quad [-2 < \operatorname{Re}\mu < 0].$

BI ((311))(4)a, ET I 315(15)

41. $\int_0^\infty \ln(1+x^2)\, \frac{x^{\mu-1}\, dx}{1+x} = \frac{\pi}{\sin\mu\pi}\left\{\ln 2 - (1-\mu)\sin\frac{\mu\pi}{2}\beta\left(\frac{1-\mu}{2}\right) -\right.$

$\left. -(2-\mu)\cos\frac{\mu\pi}{2}\beta\left(\frac{2-\mu}{2}\right)\right\} \quad [-2 < \operatorname{Re}\mu < 1].$ ET I 316(25)

4.296

1. $\int_0^1 \ln(1+2x\cos t + x^2)\frac{dx}{x} = \frac{\pi^2}{6} - \frac{t^2}{2}.$ BI ((114))(34)

2. $\int_{-\infty}^\infty \ln(a^2 - 2ax\cos t + x^2)\frac{dx}{1+x^2} = \pi\ln(1+2a\sin t + a^2).$ BI ((145))(28)

3. $\int_0^\infty \ln(1+2x\cos t + x^2)\, x^{\mu-1}\, dx = \frac{2\pi}{\mu}\,\frac{\cos\mu t}{\sin\mu\pi}$

$[|t| < \pi, \ -1 < \operatorname{Re}\mu < 0].$ ET I 316(27)

4.297

1. $\int_0^1 \ln\frac{ax+b}{bx+a}\,\frac{dx}{(1+x)^2} = \frac{1}{a-b}\left[(a+b)\ln\frac{a+b}{2} - a\ln a - b\ln b\right]$

$[a > 0, \ b > 0].$ BI ((115))(16)

2. $\int_0^\infty \ln\frac{ax+b}{bx+a}\,\frac{dx}{(1+x)^2} = 0 \quad [ab > 0].$ BI ((139))(23)

3. $\int_0^1 \ln\frac{1-x}{x}\,\frac{dx}{1+x^2} = \frac{\pi}{8}\ln 2.$ BI ((115)(5)

4. $\int_0^1 \ln\frac{1+x}{1-x}\,\frac{dx}{1+x^2} = G.$ BI ((115))(17)

5. $\int_0^\infty \ln\left(\frac{1+x}{1-x}\right)^2\frac{dx}{x(1+x^2)} = \frac{\pi^2}{2}.$ BI ((141))(13)

6. $\int_u^v \ln\frac{v+x}{u+x}\,\frac{dx}{x} = \frac{1}{2}\left(\ln\frac{v}{u}\right)^2 \quad [uv > 0].$ BI ((145))(33)

7. $\int_0^\infty \frac{b\ln(1+ax) - a\ln(1+bx)}{x^2}\, dx = ab\ln\frac{b}{a} \quad [a > 0, \ b > 0].$ FI II 647

8. $\int\limits_0^1 \ln \frac{1+ax}{1-ax} \frac{dx}{x\sqrt{1-x^2}} = \pi \arcsin a \quad [|a| \leqslant 1]$ GW((325))(21c), BI((122))(2)

9. $\int\limits_u^v \ln\left(\frac{1+ax}{1-ax}\right) \frac{dx}{\sqrt{(x^2-u^2)(v^2-x^2)}} = \frac{\pi}{v} F\left(\arcsin av, \frac{u}{v}\right)$

$[|av| < 1].$ BI((145))(35)

4.298

1. $\int\limits_0^\infty \ln \frac{1+x^2}{x} \frac{x^{2n-1}}{1+x} dx = \frac{\ln 2}{2n} + \frac{1}{4n^2} - \frac{1}{2n} \beta(2n+1).$ BI((137))(1)

2. $\int\limits_0^\infty \ln \frac{1+x^2}{x} \frac{x^{2n}}{1+x} dx = \frac{\ln 2}{2n} + \frac{1}{4n^2} - \frac{1}{2n} \beta(2n+1).$ BI((137))(3)

3. $\int\limits_0^\infty \ln \frac{1+x^2}{x} \frac{x^{2n-1}}{1-x} dx = \frac{\ln 2}{2n} + \frac{1}{4n^2} - \frac{1}{2n} \beta(2n+1).$ BI((137))(2)

4. $\int\limits_0^\infty \ln \frac{1+x^2}{x} \frac{x^{2n}}{1-x} dx = -\frac{\ln 2}{2n} - \frac{1}{4n^2} + \frac{1}{2n} \beta(2n+1).$ BI((137))(4)

5. $\int\limits_0^\infty \ln \frac{1+x^2}{x} \frac{x^{2n-1}}{1+x^2} dx = \frac{\ln 2}{2n} + \frac{1}{4n^2} - \frac{1}{2n} \beta(2n+1).$ BI((137))(10)

6. $\int\limits_0^1 \ln \frac{1+x^2}{x} x^{2n} dx = \frac{1}{2n+1} \left\{(-1)^n \frac{\pi}{2} + \ln 2 - \frac{1}{2n+1} + 2 \sum_{k=0}^{n-1} \frac{(-1)^k}{2n-2k-1}\right\}.$

BI((294))(8)

7 $\int\limits_0^1 \ln \frac{1+x^2}{x} x^{2n-1} dx = \frac{1}{2n} \left\{(-1)^{n+1} \ln 2 + \ln 2 - \frac{1}{2n} + (-1)^{n+1} \sum_{k=1}^{n-1} \frac{(-1)^k}{k}\right\}.$

BI((294))(9)a

8. $\int\limits_0^1 \ln \frac{1+x^2}{x} \frac{dx}{1+x^2} = \frac{\pi}{2} \ln 2.$ BI((115))(7)

9. $\int\limits_0^\infty \ln \frac{1+x^2}{x} \frac{dx}{1+x^2} = \pi \ln 2.$ BI((137))(8)

10. $\int\limits_0^\infty \ln \frac{1+x^2}{x} \frac{dx}{1-x^2} = 0.$ BI((137))(9)

11. $\int\limits_0^1 \ln \frac{1-x^2}{x} \frac{dx}{1+x^2} = \frac{\pi}{4} \ln 2.$ BI((115))(9)

12. $\int\limits_1^\infty \ln \frac{1+x^2}{x+1} \frac{dx}{1+x^2} = \frac{3\pi}{8} \ln 2.$ BI((144))(8)

13. $\int_0^1 \ln \frac{1+x^2}{x+1} \frac{dx}{1+x^2} = \frac{3\pi}{8} \ln 2 - G.$ BI ((115))(18)

14. $\int_1^\infty \ln \frac{1+x^2}{x-1} \frac{dx}{1+x^2} = \frac{3\pi}{8} \ln 2 + G.$ BI ((144))(9)

15. $\int_0^1 \ln \frac{1+x^2}{1-x} \frac{dx}{1+x^2} = \frac{3\pi}{8} \ln 2.$ BI ((115))(19)

16. $\int_0^\infty \ln \frac{1+x^2}{x^2} \frac{x\,dx}{1+x^2} = \frac{\pi^2}{12}.$ BI ((138))(3)

17. $\int_0^\infty \ln \frac{a^2+b^2x^2}{x^2} \frac{dx}{c^2+g^2x^2} = \frac{\pi}{cg} \ln \frac{ag+bc}{c}$

$[a > 0,\ b > 0,\ c > 0,\ g > 0].$ BI ((138))(6, 7, 9, 10)a

18. $\int_0^\infty \ln \frac{a^2+b^2x^2}{x^2} \frac{dx}{c^2-g^2x^2} = \frac{1}{cg} \operatorname{arctg} \frac{ag}{bc}$

$[a > 0\ \ b > 0,\ c > 0,\ g > 0].$ BI ((138))(8, 11)a

19. $\int_0^\infty \ln \frac{1+x^2}{x^2} \frac{x^2\,dx}{(1+x^2)^2} = \frac{\pi}{4} (\ln 4 - 1).$ BI ((139))(21)

20. $\int_0^1 \ln \left(\frac{1-x^2}{x^2}\right)^2 \sqrt{1-x^2}\,dx = \pi.$ FI II 643a

21. $\int_0^1 \ln \frac{1+2x\cos t + x^2}{(1+x)^2} \frac{dx}{x} = \frac{1}{2} \int_0^\infty \ln \frac{1+2x\cos t + x^2}{(1+x)^2} \frac{dx}{x} = -\frac{t^2}{2}$

$[|t| < \pi].$ BI ((115))(23), BI ((134))(15)

22. $\int_0^\infty \ln \frac{1+2x\cos t + x^2}{(1+x)^2} x^{p-1}\,dx = -\frac{2\pi(1-\cos p\pi)}{p \sin p\pi}$

$[|p| < 1,\ |t| < \pi].$ BI ((134))(17)

23. $\int_0^1 \ln \frac{1+x^2 \sin t}{1-x^2 \sin t} \frac{dx}{\sqrt{1-x^2}} = \pi \ln \operatorname{ctg} \left(\frac{\pi - t}{4}\right)$

$[|t| < \pi].$ GW ((325))(21d)

4.299

1. $\int_0^\infty \ln \frac{(x+1)(x+a^2)}{(x+a)^2} \frac{dx}{x} = (\ln a)^2 \qquad [a > 0].$ BI ((134))(14)

2. $\int_0^1 \ln \frac{(1-ax)(1+ax^2)}{(1-ax^2)^2} \frac{dx}{1+ax^2} = \frac{1}{2\sqrt{a}} \operatorname{arctg} \sqrt{a} \ln(1+a)$

$[a > 0].$ BI ((115))(25)

3. $\int_0^1 \ln\frac{(1-a^2x^2)(1+ax^2)}{(1-ax^2)^2}\frac{dx}{1+ax^2} = \frac{1}{\sqrt{a}}\operatorname{arctg}\sqrt{a}\ln(1+a)$

$[a > 0]$. BI ((115))(26)

4. $\int_0^1 \ln\frac{(x+1)(x+a^2)}{(x+a)^2} x^{\mu-1}\,dx = \frac{\pi(a^\mu-1)^2}{\mu\sin\mu\pi}$

$[a > 0,\ \operatorname{Re}\mu > 0]$. BI ((134))(16)

4.311

1. $\int_0^\infty \ln(a^3-x^3)\frac{dx}{x^3} = \frac{\pi}{4a^2}\sqrt{3}.$ BI ((134))(7)

2. $\int_0^\infty \ln(1+x^3)\frac{dx}{1-x+x^2} = \frac{2\pi}{\sqrt{3}}\ln 3.$ LI ((136))(8)

3. $\int_0^\infty \ln(1+x^3)\frac{dx}{1+x^3} = \frac{\pi}{\sqrt{3}}\ln 3 - \frac{\pi^2}{9}.$ LI ((136))(6)

4. $\int_0^\infty \ln(1+x^3)\frac{x\,dx}{1+x^3} = \frac{\pi}{\sqrt{3}}\ln 3 + \frac{\pi^2}{9}.$ LI ((136))(7)

5. $\int_0^\infty \ln(1+x^3)\frac{1-x}{1+x^3}dx = -\frac{2}{9}\pi^2.$ BI ((136))(9)

4.312

1. $\int_0^\infty \ln\frac{1+x^3}{x^3}\frac{dx}{1+x^3} = \frac{\pi}{\sqrt{3}}\ln 3 + \frac{\pi^2}{9}.$ BI ((138))(12)

2. $\int_0^\infty \ln\frac{1+x^3}{x^3}\frac{x\,dx}{1+x^3} = \frac{\pi}{\sqrt{3}}\ln 3 - \frac{\pi^2}{9}.$ BI ((138))(13)

4.313

1. $\int_0^\infty \ln x\ln(1+a^2x^2)\frac{dx}{x^2} = \pi a(1-\ln a)$

$[a > 0]$. BI ((134))(18)

2. $\int_0^\infty \ln(1+c^2x^2)\ln(a^2+b^2x^2)\frac{dx}{x^2} =$

$$= 2\pi\left[\left(c+\frac{b}{a}\right)\ln(b+ac) - \frac{b}{a}\ln b - c\ln c\right]$$

$[a > 0,\ b > 0,\ c > 0]$. BI ((134))(20, 21)a

3. $\int_0^\infty \ln(1+c^2x^2)\ln\left(a^2+\frac{b^2}{x^2}\right)\frac{dx}{x^2} = 2\pi\left[\frac{a+bc}{b}\ln(a+bc) - \frac{a}{b}\ln a - c\right]$

$[a > 0,\ a+bc > 0]$. BI ((134))(22, 23)a

4. $$\int_0^\infty \ln x \ln \frac{1+a^2x^2}{1+b^2x^2} \frac{dx}{x^2} = \pi(a-b) + \pi \ln \frac{b^b}{a^a}$$

$[a > 0, \ b > 0]$. BI ((134))(24)

5. $$\int_0^\infty \ln x \ln \frac{a^2+2bx+x^2}{a^2-2bx+x^2} \frac{dx}{x} = 2\pi \ln a \arcsin \frac{b}{a}$$

$[a \geqslant |b|]$. BI ((134))(25)

6. $$\int_0^\infty \ln(1+x) \frac{x \ln x - x - a}{(x+a)^2} \frac{dx}{x} = \frac{(\ln a)^2}{2(a-1)} \qquad [a > 0].$$ BI ((141))(7)

7. $$\int_0^\infty \ln(1-x)^2 \frac{x \ln x - x - a}{(x+a)^2} \frac{dx}{x} = \frac{\pi^2 + (\ln a)^2}{1+a}$$

$[a > 0]$. LI ((141))(8)

4.314

1. $$\int_0^1 \ln(1+ax) \frac{x^{p-1} - x^{q-1}}{\ln x} dx = \sum_{k=1}^\infty \frac{a^k}{k} \ln \frac{p+k}{q+k} + \ln \frac{p}{q}$$

$[a > 0, \ p > 0, \ q > 0]$. BI ((123))(18)

2. $$\int_0^\infty \left[\frac{(q-1)x}{(1+x)^2} - \frac{1}{x+1} + \frac{1}{(1+x)^q} \right] \frac{dx}{x \ln(1+x)} = \ln \Gamma(q)$$

$[q > 0]$. BI ((143))(7)

3. $$\int_0^1 \frac{x \ln x + 1 - x}{x(\ln x)^2} \ln(1+x)\, dx = \ln \frac{4}{\pi}.$$ BI ((126))(12)

4. $$\int_0^1 \frac{\ln(1-x^2)\, dx}{x(q^2 + \ln^2 x)} = -\frac{\pi}{q} \ln \Gamma\left(\frac{q+\pi}{\pi}\right) + \frac{\pi}{2q} \ln 2q + \ln \frac{q}{\pi} - 1$$

$[q > \quad]$. LI ((327))(12)a

4.315

1. $$\int_0^1 \ln(1+x)(\ln x)^{n-1} \frac{dx}{x} = (-1)^{n-1}(n-1)! \left(1 - \frac{1}{2^n}\right) \zeta(n+1).$$

BI ((116))(3)

2. $$\int_0^1 \ln(1+x)(\ln x)^{2n} \frac{dx}{x} = \frac{2^{2n+1}-1}{(2n+1)(2n+2)} \pi^{2n+2} |B_{2n+2}|.$$

BI ((116))(1)

3. $$\int_0^1 \ln(1-x)(\ln x)^{n-1} \frac{dx}{x} = (-1)^n (n-1)!\, \zeta(n+1).$$ BI ((116))(4)

4. $$\int_0^1 \ln(1-x)(\ln x)^{2n} \frac{dx}{x} = -\frac{2^{2n}}{(n+1)(2n+1)} \pi^{2n+2} |B_{2n+2}|.$$

BI ((116))(2)

4.316

1. $$\int_0^1 \ln(1-ax^r)\left(\ln\frac{1}{x}\right)^p \frac{dx}{x} = -\frac{1}{r^{p+1}}\Gamma(p+1)\sum_{k=1}^{\infty}\frac{a^k}{k^{p+2}}$$
$[p>-1,\ a<1,\ r>0]$. BI ((116))(7)

2. $$\int_0^1 \ln(1-2ax\cos t+a^2x^2)\left(\ln\frac{1}{x}\right)^p \frac{dx}{x} =$$
$$= -2\Gamma(p+1)\sum_{k=1}^{\infty}\frac{a^k\cos kt}{k^{p+2}}.$$ LI ((116))(8)

4.317

1. $$\int_0^\infty \ln\frac{\sqrt{1+x^2}+a}{\sqrt{1+x^2}-a}\,\frac{dx}{\sqrt{1+x^2}} = \pi\arcsin a$$
$[|a|<1]$. BI ((142))(11)

2. $$\int_0^1 \ln\frac{\sqrt{1-a^2x^2}-x\sqrt{1-a^2}}{1-x}\,\frac{dx}{x} = \frac{1}{2}(\arcsin a)^2.$$ BI ((115))(32)

3. $$\int_0^1 \ln\frac{1+\cos t\sqrt{1-x^2}}{1-\cos t\sqrt{1-x^2}}\,\frac{dx}{x^2+\operatorname{tg}^2 v} = \pi\operatorname{ctg} t\,\frac{\cos\frac{v-t}{2}}{\sin\frac{v+t}{2}}.$$ BI ((115))(30)

4. $$\int_0^1 \ln\left(\frac{x+\sqrt{1-x^2}}{x-\sqrt{1-x^2}}\right)^2 \frac{x\,dx}{1-x^2} = \frac{\pi^2}{2}.$$ BI ((115))(31)

5. $$\int_0^1 \ln\{\sqrt{1+kx}+\sqrt{1-kx}\}\,\frac{dx}{\sqrt{(1-x^2)(1-k^2x^2)}} =$$
$$= \frac{1}{4}\ln(4k)\,\boldsymbol{K}(k)+\frac{\pi}{8}\boldsymbol{K}(k').$$ BI ((121))(8)

6. $$\int_0^1 \ln\{\sqrt{1+kx}-\sqrt{1-kx}\}\,\frac{dx}{\sqrt{(1-x^2)(1-k^2x^2)}} =$$
$$= \frac{1}{4}\ln(4k)\,\boldsymbol{K}(k)+\frac{3}{8}\pi\boldsymbol{K}(k').$$ BI ((121))(9)

7 $$\int_0^1 \ln\{1+\sqrt{1-k^2x^2}\}\,\frac{dx}{\sqrt{(1-x^2)(1-k^2x^2)}} =$$
$$= \frac{1}{2}\ln k\boldsymbol{K}(k)+\frac{\pi}{4}\boldsymbol{K}(k').$$ BI ((121))(6)

8. $$\int_0^1 \ln\{1-\sqrt{1-k^2x^2}\}\,\frac{dx}{\sqrt{(1-x^2)(1-k^2x^2)}} =$$
$$= \frac{1}{2}\ln k\,\boldsymbol{K}(k)-\frac{3}{4}\pi\boldsymbol{K}(k').$$ BI ((121))(7)

9. $\int_0^1 \ln \frac{1+p\sqrt{1-x^2}}{1-p\sqrt{1-x^2}} \frac{dx}{1-x} = \pi \arcsin p \qquad [p^2 < 1].$ BI ((115))(29)

10. $\int_0^1 \ln \frac{1+q\sqrt{1-k^2x^2}}{1-q\sqrt{1-k^2x^2}} \frac{dx}{\sqrt{(1-x^2)(1-k^2x^2)}} = \pi F(\arcsin q, k')$

$[q^2 < 1].$ BI ((122))(15)

4.318

1. $\int_0^1 \frac{\ln(1-x^q)}{1+(\ln x)^2} \frac{dx}{x} = \pi \left[\ln \Gamma\left(\frac{q}{2\pi}+1\right) - \frac{\ln q}{2} + \frac{q}{2\pi}\left(\ln \frac{q}{2\pi} - 1\right)\right]$

$[q > 0].$ BI ((126))(11)

2. $\int_0^\infty \ln(1+x^r)\left[\frac{(p-r)x^p-(q-r)x^q}{\ln x} + \frac{x^q-x^p}{(\ln x)^2}\right]\frac{dx}{x^{r+1}} =$

$= r \ln\left(\operatorname{tg}\frac{q\pi}{2r} \operatorname{ctg}\frac{p\pi}{2r}\right) \qquad [p < r, \ q < r].$ BI ((143))(9)

In integrals containing $\ln(a+bx^r)$, it is useful to make the substitution $x^r = t$ and then to seek the resulting integral in the tables. For example,

$$\int_0^\infty x^{p-1} \ln(1+x^r)\, dx = \frac{1}{r}\int_0^\infty t^{\frac{p}{r}-1} \ln(1+t)\, dt = \frac{\pi}{p \sin \frac{p\pi}{r}}$$

(see 4.293 3.).

4.319

1. $\int_0^\infty \ln(1-e^{-2a\pi x}) \frac{dx}{1+x^2} = -\pi\left[\frac{1}{2}\ln 2a\pi + a(\ln a - 1) - \ln\Gamma(a+1)\right]$

$[a > 0].$ BI ((354))(6)

2. $\int_0^\infty \ln(1+e^{-2a\pi x})\frac{dx}{1+x^2} = \pi\left[\ln\Gamma(2a) - \ln\Gamma(a) + \right.$

$\left. + a(1-\ln a) - \left(2a - \frac{1}{2}\right)\ln 2\right] \qquad [a > 0].$ BI ((354))(7)

3. $\int_0^\infty \ln \frac{a+be^{-px}}{a+be^{-qx}} \frac{dx}{x} = \ln \frac{a}{a+b} \ln \frac{p}{q} \qquad \left[\frac{b}{a} > -1, \ pq > 0\right].$

FI II 635, BI ((354))(1)

4.321

1. $\int_{-\infty}^\infty x \ln \operatorname{ch} x\, dx = 0.$ BI ((358))(2)a

2. $\int_0^\infty \ln \operatorname{ch} x \frac{dx}{1-x^2} = 0.$ BI ((138))(20)a

4.322

1. $$\int_0^{\pi} \ln \sin x \, x \, dx = \frac{1}{2} \int_0^{\pi} \ln \cos^2 x \, x \, dx = -\frac{\pi^2}{2} \ln 2.$$ BI ((432))(1, 2) FI II 643

2. $$\int_0^{\infty} \frac{\ln \sin^2 ax}{b^2+x^2} dx = \frac{\pi}{b} \ln \frac{1-e^{-2ab}}{2} \qquad [a > 0,\ b > 0].$$ GW ((338))(28b)

3. $$\int_0^{\infty} \frac{\ln \cos^2 ax}{b^2+x^2} dx = \frac{\pi}{b} \ln \frac{1+e^{-2ab}}{2} \qquad [a > 0,\ b > 0].$$ GW ((338))(28a)

4. $$\int_0^{\infty} \frac{\ln \sin^2 ax}{b^2-x^2} dx = -\frac{\pi^2}{2b} + a\pi \qquad [a > 0,\ b > 0].$$ BI ((418))(1)

5. $$\int_0^{\infty} \frac{\ln \cos^2 ax}{b^2-x^2} dx = a\pi \qquad [a > 0].$$ BI ((418))(2)

6. $$\int_0^{\infty} \frac{\ln \cos^2 x}{x^2} dx = -\pi.$$ FI II 686

7. $$\int_0^{\frac{\pi}{4}} \ln \sin x \, x^{\mu-1} \, dx = -\frac{1}{2\mu} \left(\frac{\pi}{4}\right)^{\mu} \left[\ln 2 + \frac{2}{\mu} - \sum_{k=1}^{\infty} \frac{\zeta(2k)}{4^{2k-1}(\mu+2k)} \right]$$
$$[\operatorname{Re} \mu > 0].$$ LI ((425))(1)

8. $$\int_0^{\frac{\pi}{2}} \ln \sin x \, x^{\mu-1} \, dx = -\frac{1}{\mu} \left(\frac{\pi}{2}\right)^{\mu} \left[\frac{1}{\mu} - \sum_{k=1}^{\infty} \frac{\zeta(2k)}{4^{k}(\mu+2k)} \right]$$
$$[\operatorname{Re} \mu > 0].$$ LI ((430))(1)

9. $$\int_0^{\frac{\pi}{2}} \ln (1-\cos x) \, x^{\mu-1} \, dx = \frac{-1}{\mu} \left(\frac{\pi}{2}\right)^{\mu} \left[\frac{2}{\mu} + \sum_{k=1}^{\infty} \frac{\zeta(2k)}{4^{2k-1}(\mu+2k)} \right]$$
$$[\operatorname{Re} \mu > 0].$$ LI ((430))(2)

10. $$\int_0^{\infty} \ln (1 \pm 2p \cos \beta x + p^2) \frac{dx}{q^2+x^2} = \frac{\pi}{q} \ln (1 \pm p e^{-\beta q}) \qquad [p^2 < 1];$$
$$= \frac{\pi}{q} \ln (p \pm e^{-\beta q}) \qquad [p^2 > 1].$$ FI II 718a

4.323

1. $$\int_0^{\pi} \ln \operatorname{tg}^2 x \, x \, dx = 0.$$ BI ((432))(3)

2. $$\int_0^{\infty} \frac{\ln \operatorname{tg}^2 ax}{b^2+x^2} dx = \frac{\pi}{b} \ln \operatorname{th} ab \qquad [a > 0,\ b > 0].$$ GW ((338))(28c)

3. $\int\limits_0^\infty \ln\left(\frac{1+\operatorname{tg} x}{1-\operatorname{tg} x}\right)^2 \frac{dx}{x} = \frac{\pi^2}{2}.$ GW ((338))(26)

4.324

1. $\int\limits_0^\infty \ln\left(\frac{1+\sin x}{1-\sin x}\right)^2 \frac{dx}{x} = \pi^2.$ GW ((338))(25)

2. $\int\limits_0^\infty \ln \frac{1+2a\cos px+a^2}{1+2a\cos qx+a^2}\frac{dx}{x} = \ln(1+a)\ln\frac{q^2}{p^2} \qquad [-1<a\leqslant 1];$

$= \ln\left(1+\frac{1}{a}\right)\ln\frac{q^2}{p^2} \qquad [a<-1 \text{ or } a\geqslant 1].$

GW ((338))(27)

3. $\int\limits_0^\infty \ln(a^2\sin^2 px+b^2\cos^2 px)\frac{dx}{c^2+x^2} =$

$= \frac{\pi}{c}[\ln(a\operatorname{sh} cp+b\operatorname{ch} cp)-cp] \qquad [a>0,\ b>0,\ c>0,\ p>0].$

GW ((338))(29)

4.325

1. $\int\limits_0^1 \ln\ln\left(\frac{1}{x}\right)\frac{dx}{1+x} = -C\ln 2+\sum_{k=2}^{\infty}(-1)^k\frac{\ln k}{k} =$

$= -C\ln 2+0.159\,868\,905\ldots$ GW ((325))(25a)

2. $\int\limits_0^1 \ln\ln\left(\frac{1}{x}\right)\frac{dx}{x+e^{i\lambda}} = \sum_{k=1}^{\infty}\frac{(-1)^k}{k}e^{-ik\lambda}(C+\ln k).$ GW ((325))(26)

3. $\int\limits_0^1 \ln\ln\left(\frac{1}{x}\right)\frac{dx}{(1+x)^2} = \int\limits_1^\infty \ln\ln x\,\frac{dx}{(1+x)^2} =$

$= \frac{1}{2}\left[\psi\left(\frac{1}{2}\right)+\ln 2\pi\right] = \frac{1}{2}\left(\ln\frac{\pi}{2}-C\right).$ BI ((147))(7)

4. $\int\limits_0^1 \ln\ln\left(\frac{1}{x}\right)\frac{dx}{1+x^2} = \int\limits_1^\infty \ln\ln x\,\frac{dx}{1+x^2} =$

$= \frac{\pi}{2}\ln\frac{\sqrt{2\pi}\,\Gamma\left(\frac{3}{4}\right)}{\Gamma\left(\frac{1}{4}\right)}.$ BI ((148))(1)

5. $\int\limits_0^1 \ln\ln\left(\frac{1}{x}\right)\frac{dx}{1+x+x^2} = \int\limits_1^\infty \ln\ln x\,\frac{dx}{1+x+x^2} =$

$= \frac{\pi}{\sqrt{3}}\ln\frac{\sqrt[3]{2\pi}\,\Gamma\left(\frac{2}{3}\right)}{\Gamma\left(\frac{1}{3}\right)}.$ BI ((148))(2)

6. $$\int_0^1 \ln\ln\left(\frac{1}{x}\right)\frac{dx}{1-x+x^2} = \int_1^\infty \ln\ln x \frac{dx}{1-x+x^2} = \frac{2\pi}{\sqrt{3}}\left[\frac{5}{6}\ln 2\pi - \ln\Gamma\left(\frac{1}{6}\right)\right].$$ BI ((148))(5)

7. $$\int_0^1 \ln\ln\left(\frac{1}{x}\right)\frac{dx}{1+2x\cos t+x^2} = \int_1^\infty \ln\ln x \frac{dx}{1+2x\cos t+x^2} = \frac{\pi}{2\sin t}\ln\frac{(2\pi)^{t/\pi}\,\Gamma\left(\frac{1}{2}+\frac{t}{2\pi}\right)}{\Gamma\left(\frac{1}{2}-\frac{t}{2\pi}\right)}.$$ BI ((147))(9)

8 $$\int_0^1 \ln\ln\frac{1}{x}\, x^{\mu-1}\, dx = -\frac{1}{\mu}(\boldsymbol{C} + \ln\mu)$$

$[\operatorname{Re}\mu > 0]$. BI ((147))(1)

9. $$\int_1^\infty \ln\ln x \frac{x^{n-2}\, dx}{1+x^2+x^4+\ldots+x^{2n-2}} =$$

$$= \frac{\pi}{2n}\operatorname{tg}\frac{\pi}{2n}\ln 2\pi + \frac{\pi}{n}\sum_{k=1}^{n-1}(-1)^{k-1}\sin\frac{k\pi}{n}\ln\frac{\Gamma\left(\frac{n+k}{2n}\right)}{\Gamma\left(\frac{k}{2n}\right)}$$ [n is even].

$$= \frac{\pi}{2n}\operatorname{tg}\frac{\pi}{2n}\ln\pi + \frac{\pi}{n}\sum_{k=1}^{\frac{n-1}{2}}(-1)^{k-1}\sin\frac{k\pi}{n}\ln\frac{\Gamma\left(\frac{n-k}{n}\right)}{\Gamma\left(\frac{k}{n}\right)}$$ [n is odd].

DI ((148))(4)

10. $$\int_0^1 \ln\ln\left(\frac{1}{x}\right)\frac{dx}{(1+x^2)\sqrt{\ln\frac{1}{x}}} = \int_1^\infty \ln\ln x \frac{dx}{(1+x^2)\sqrt{\ln x}} = \sqrt{\pi}\sum_{k=0}^\infty \frac{(-1)^{k+1}}{\sqrt{2k+1}}[\ln(2k+1) + 2\ln 2 + \boldsymbol{C}].$$ BI ((147))(4)

11. $$\int_0^1 \ln\ln\left(\frac{1}{x}\right)\frac{x^{\mu-1}\, dx}{\sqrt{\ln\frac{1}{x}}} = -(\boldsymbol{C} + \ln 4\mu)\sqrt{\frac{\pi}{\mu}}$$

$[\operatorname{Re}\mu > 0]$. BI ((147))(3)

12. $$\int_0^1 \ln\ln\left(\frac{1}{x}\right)\left(\ln\frac{1}{x}\right)^{\mu-1} x^{\nu-1}\, dx = \frac{1}{\nu^\mu}\Gamma(\mu)\,[\psi(\mu) - \ln(\nu)]$$

$[\operatorname{Re}\mu > 0,\ \operatorname{Re}\nu > 0]$. BI ((147))(2)

4.326

1. $$\int_0^1 \ln(a - \ln x)\, x^{\mu-1}\, dx = \frac{1}{\mu}[\ln a - e^{a\mu}\operatorname{Ei}(-a\mu)]$$

$[\operatorname{Re}\mu > 0,\ a > 0]$. BI ((107))(23)

2. $\int_0^{\frac{1}{e}} \ln\left(2\ln\frac{1}{x}-1\right)\frac{x^{2\mu-1}}{\ln x}\,dx = -\frac{1}{2}[\mathrm{Ei}(-\mu)]^2$

$[\operatorname{Re}\mu > 0]$. BI ((145))(5)

4.327

1. $\int_0^1 \ln[a^2+(\ln x)^2]\frac{dx}{1+x^2} = \pi\ln\frac{2\Gamma\left(\frac{2a+3\pi}{4\pi}\right)}{\Gamma\left(\frac{2a+\pi}{4\pi}\right)} + \frac{\pi}{2}\ln\frac{\pi}{2}$

$\left[a > -\frac{\pi}{2}\right]$. BI ((147))(10)

2. $\int_0^1 \ln[a^2+4(\ln x)^2]\frac{dx}{1+x^2} = \pi\ln\frac{2\Gamma\left(\frac{a+3\pi}{4\pi}\right)}{\Gamma\left(\frac{a+\pi}{4\pi}\right)} + \frac{\pi}{2}\ln\pi$

$[a > -\pi]$. BI ((147))(16)a

3. $\int_0^\infty \ln[a^2+(\ln x)^2]\,x^{\mu-1}\,dx = \frac{2}{\mu}[-\cos a\mu\,\mathrm{ci}(a\mu) -$

$-\sin a\mu\,\mathrm{si}(a\mu)+\ln a]$ $[a > 0,\ \operatorname{Re}\mu > 0]$. GW ((325))(28)

If the integrand contains a logarithm whose argument also contains a logarithm, for example, if the integrand contains $\ln\ln\frac{1}{x}$, it is useful to make the substitution $\ln x = t$ and then seek the transformed integral in the tables.

4.33-4.34 Combinations of logarithms and exponentials

4.331

1. $\int_0^\infty e^{-\mu x}\ln x\,dx = -\frac{1}{\mu}(C+\ln\mu)$ $[\operatorname{Re}\mu > 0]$. BI ((256))(2)

2. $\int_1^\infty e^{-\mu x}\ln x\,dx = -\frac{1}{\mu}\mathrm{Ei}(-\mu)$ $[\operatorname{Re}\mu > 0]$. BI ((260))(5)

3. $\int_0^1 e^{\mu x}\ln x\,dx = -\frac{1}{\mu}\int_0^1\frac{e^{\mu x}-1}{x}\,dx$ $[\mu \neq 0]$. GW ((324))(81a)

4.332

1. $\int_0^\infty \frac{\ln x\,dx}{e^x+e^{-x}-1} = \frac{2\pi}{\sqrt{3}}\left[\frac{5}{6}\ln 2\pi - \ln\Gamma\left(\frac{1}{6}\right)\right]$

(cf. **4.325** 6.). BI ((257))(6)

2. $\int_0^\infty \frac{\ln x\,dx}{e^x+e^{-x}+1} = \frac{\pi}{\sqrt{3}}\ln\left[\frac{\Gamma\left(\frac{2}{3}\right)}{\Gamma\left(\frac{1}{3}\right)}\sqrt{2\pi}\right]$

(cf. **4.325** 5.). BI ((257))(7)a, LI ((260))(3)

4.333 $$\int_0^\infty e^{-\mu x^2} \ln x \, dx = -\frac{1}{4}(C + \ln 4\mu)\sqrt{\frac{\pi}{\mu}}$$

$[\operatorname{Re}\mu > 0]$. BI ((256))(8), FI II 807a

4.334 $$\int_0^\infty \frac{\ln x \, dx}{e^{x^2}+1+e^{-x^2}} = \frac{1}{2}\sqrt{\frac{\pi}{3}} \sum_{k=1}^{\infty} (-1)^k \frac{C+\ln 4k}{\sqrt{k}} \sin\frac{k\pi}{3}.$$ BI ((357))(13)

4.335

1. $$\int_0^\infty e^{-\mu x} (\ln x)^2 \, dx = \frac{1}{\mu}\left[\frac{\pi^2}{6} + (C + \ln\mu)^2\right] \qquad [\operatorname{Re}\mu > 0].$$ ET I 149(13)

2. $$\int_0^\infty e^{-x^2} (\ln x)^2 \, dx = \frac{\sqrt{\pi}}{8}\left[(C + 2\ln 2)^2 + \frac{\pi^2}{2}\right].$$ FI II 808

3. $$\int_0^\infty e^{-\mu x} (\ln x)^3 \, dx = -\frac{1}{\mu}\left[(C + \ln\mu)^3 + \frac{\pi^2}{2}(C + \ln\mu) - \psi''(1)\right].$$ MI 26

4.336

1. $$\int_0^\infty \frac{e^{-x}}{\ln x} \, dx = 0.$$ BI ((260))(9)

2. $$\int_0^\infty \frac{e^{-\mu x} \, dx}{\pi^2 + (\ln x)^2} = \nu'(\mu) - e^{\mu} \qquad [\operatorname{Re}\mu > 0].$$ MI 26

4.337

1. $$\int_0^\infty e^{-\mu x} \ln(\beta + x) \, dx = \frac{1}{\mu}[\ln\beta - e^{\mu\beta} \operatorname{Ei}(-\beta\mu)]$$

$[|\arg\beta| < \pi, \ \operatorname{Re}\mu > 0]$. BI ((256))(3)

2. $$\int_0^\infty e^{-\mu x} \ln(1 + \beta x) \, dx = -\frac{1}{\mu} e^{\frac{\mu}{\beta}} \operatorname{Ei}\left(-\frac{\mu}{\beta}\right)$$

$[|\arg\beta| < \pi, \ \operatorname{Re}\mu > 0]$. ET I 148(4)

3. $$\int_0^\infty e^{-\mu x} \ln|a - x| \, dx = \frac{1}{\mu}[\ln a - e^{-a\mu} \overline{\operatorname{Ei}}(a\mu)] \qquad [a > 0, \ \operatorname{Re}\mu > 0].$$

BI ((256))(4)

4. $$\int_0^\infty e^{-\mu x} \ln\frac{\beta}{\beta - x} \, dx = \frac{1}{\mu}[e^{-\beta\mu} \operatorname{Ei}(\beta\mu)]$$

[β cannot be a real positive number, $\operatorname{Re}\mu > 0$].

MI 26

4.338

1. $$\int_0^\infty e^{-\mu x} \ln(\beta^2 + x^2)\, dx = \frac{2}{\mu}[\ln\beta - \operatorname{ci}(\beta\mu)\cos(\beta\mu) - \operatorname{si}(\beta\mu)\sin(\beta\mu)]$$

$[\operatorname{Re}\beta > 0, \ \operatorname{Re}\mu > 0]$. BI ((256))(6)

2 $$\int_0^\infty e^{-\mu x} \ln(x^2 - \beta^2)^2\, dx = \frac{2}{\mu}[\ln\beta^2 - e^{\beta\mu}\operatorname{Ei}(-\beta\mu) - e^{-\beta\mu}\operatorname{Ei}(\beta\mu)]$$

$[\operatorname{Im}\beta > 0, \ \operatorname{Re}\mu > 0]$. BI ((256))(5)

4.339 $$\int_0^\infty e^{-\mu x} \ln\left|\frac{x+1}{x-1}\right| dx = \frac{1}{\mu}[e^{-\mu}(\ln 2\mu + \gamma) - e^{\mu}\operatorname{Ei}(-2\mu)]$$

$[\operatorname{Re}\mu > 0]$. MI 27

4.341 $$\int_0^\infty e^{-\mu x} \ln\frac{\sqrt{x+ai}+\sqrt{x-ai}}{\sqrt{2a}}\, dx = \frac{\pi}{4\mu}[\mathbf{H}_0(a\mu) - N_0(a\mu)]$$

$[a > 0, \ \operatorname{Re}\mu > 0]$. ET I 149(20)

4.342

1. $$\int^\infty e^{-2nx} \ln(\operatorname{sh} x)\, dx = \frac{1}{2n}\left[\frac{1}{n} + \ln 2 - 2\beta(2n+1)\right].$$ BI ((256))(17)

2 $$\int_0^\infty e^{-\mu x} \ln(\operatorname{ch} x)\, dx = \frac{1}{\mu}\left[\beta\left(\frac{\mu}{2}\right) - \frac{1}{\mu}\right] \quad [\operatorname{Re}\mu > 0].$$ ET I 165(32)

3. $$\int_0^\infty e^{-\mu x}[\ln(\operatorname{sh} x) - \ln x]\, dx = \frac{1}{\mu}\left[\ln\frac{\mu}{2} - \frac{1}{2\mu} - \psi\left(\frac{\mu}{2}\right)\right]$$

$[\operatorname{Re}\mu > 0]$. ET I 165(33)

4.343 $$\int_0^\pi e^{\mu\cos x}[\ln(2\mu\sin^2 x) + C]\, dx = -\pi K_0(\mu).$$ WA 95(16)

4.35-4.36 Combinations of logarithms, exponentials, and powers

4.351

1. $$\int_0^1 (1-x)e^{-x}\ln x\, dx = \frac{1-e}{e}.$$ BI ((352))(1)

2. $$\int_0^1 e^{\mu x}(\mu x^2 + 2x)\ln x\, dx = \frac{1}{\mu^2}[(1-\mu)e^{\mu} - 1].$$ BI ((352))(2)

3. $$\int_1^\infty \frac{e^{-\mu x}\ln x}{1+x}\, dx = \frac{1}{2}e^{\mu}[\operatorname{Ei}(-\mu)]^2 \quad [\operatorname{Re}\mu > 0].$$ NT 32(10)

4.352

1. $$\int_0^\infty x^{\nu-1}e^{-\mu x}\ln x\,dx=\frac{1}{\mu^\nu}\Gamma(\nu)[\psi(\nu)-\ln\mu] \qquad [\operatorname{Re}\mu>0,\ \operatorname{Re}\nu>0].$$
BI ((353))(3), ET I 315(10)a

2. $$\int_0^\infty x^n e^{-\mu x}\ln x\,dx=\frac{n!}{\mu^{n+1}}\left[1+\frac{1}{2}+\frac{1}{3}+\ldots+\frac{1}{n}-C-\ln\mu\right]$$
$$[\operatorname{Re}\mu>0].$$
ET I 148(7)

3. $$\int_0^\infty x^{n-\frac{1}{2}}e^{-\mu x}\ln x\,dx=\sqrt{\pi}\frac{(2n-1)!!}{2^n\mu^{n+\frac{1}{2}}}\left[2\left(1+\frac{1}{3}+\frac{1}{5}+\ldots+\frac{1}{2n-1}\right)-\right.$$
$$\left.-C-\ln 4\mu\right] \qquad [\operatorname{Re}\mu>0].$$
ET I 148(10)

4. $$\int_0^\infty x^{\mu-1}e^{-x}\ln x\,dx=\Gamma'(\mu) \qquad [\operatorname{Re}\mu>0].$$
GW ((324))(83a)

4.353

1. $$\int_0^\infty (x-\nu)x^{\nu-1}e^{-x}\ln x\,dx=\Gamma(\nu) \qquad [\operatorname{Re}\nu>0].$$
GW ((324))(84)

2. $$\int_0^\infty\left(\mu x-n-\frac{1}{2}\right)x^{n-\frac{1}{2}}e^{-\mu x}\ln x\,dx=\frac{(2n-1)!!}{(2\mu)^n}\sqrt{\frac{\pi}{\mu}}$$
$$[\operatorname{Re}\mu>0].$$
BI ((357))(2)

3. $$\int_0^1(\mu x+n+1)x^n e^{\mu x}\ln x\,dx=e^\mu\sum_{k=0}^n(-1)^{k-1}\frac{n!}{(n-k)!\,\mu^{k+1}}+(-1)^n\frac{n!}{\mu^{n+1}}$$
$$[\mu\neq 0].$$
GW ((324))(82)

4.354

1. $$\int_0^\infty\frac{x^{\nu-1}\ln x}{e^x+1}dx=\Gamma(\nu)\sum_{k=1}^\infty\frac{(-1)^{k-1}}{k^\nu}[\psi(\nu)-\ln k] \qquad [\operatorname{Re}\nu>0].$$
GW ((324))(86a)

2. $$\int_0^\infty\frac{x^{\nu-1}\ln x}{(e^x+1)^2}dx=\Gamma(\nu)\sum_{k=2}^\infty\frac{(-1)^k(k-1)}{k^\nu}[\psi(\nu)-\ln k] \qquad [\operatorname{Re}\nu>0].$$
GW ((324))(86b)

3. $$\int_0^\infty\frac{(x-\nu)e^x-\nu}{(e^x+1)^2}x^{\nu-1}\ln x\,dx=\Gamma(\nu)\sum_{k=1}^\infty\frac{(-1)^{k-1}}{k^\nu} \qquad [\operatorname{Re}\nu>0].$$
GW ((324))(87a)

4. $$\int_0^\infty\frac{(x-2n)e^x-2n}{(e^x+1)^2}x^{2n-1}\ln x\,dx=\frac{2^{2n-1}-1}{2n}\pi^{2n}|B_{2n}|.$$
GW ((324))(87b)

5. $$\int_0^\infty \frac{x^{\nu-1}\ln x}{(e^x+1)^n}\,dx = (-1)^n \frac{\Gamma(\nu)}{(n-1)!} \sum_{k=n}^{\infty} \frac{(-1)^k (k-1)!}{(k-n)!\,k^\nu} [\psi(\nu) - \ln k]$$
$$[\operatorname{Re} \nu > 0].$$ GW ((324))(86c)

4.355

1. $$\int_0^\infty x^2 e^{-\mu x^2} \ln x\,dx = \frac{1}{8\mu}(2 - \ln 4\mu - C)\sqrt{\frac{\pi}{\mu}} \qquad [\operatorname{Re}\mu > 0].$$ BI ((357))(1)a

2. $$\int_{-\infty}^\infty x(\mu x^2 - \nu x - 1)\, e^{-\mu x^2 + 2\nu x} \ln x\,dx = \frac{\nu}{2\mu}\sqrt{\frac{\pi}{\mu}} \exp\left(\frac{\nu^2}{\mu}\right)$$
$$[\operatorname{Re}\mu > 0].$$ BI ((358))(1)

3. $$\int_0^\infty (\mu x^2 - n)\, x^{2n-1} e^{-\mu x^2} \ln x\,dx = \frac{(n-1)!}{4\mu^n} \qquad [\operatorname{Re}\mu > 0].$$ BI ((353))(4)

4. $$\int_0^\infty (2\mu x^2 - 2n - 1)\, x^{2n} e^{-\mu x^2} \ln x\,dx = \frac{(2n-1)!!}{2(2\mu)^n}\sqrt{\frac{\pi}{\mu}} \qquad [\operatorname{Re}\mu > 0].$$ BI ((353))(5)

4.356

1. $$\int_0^\infty \exp\left[-\mu\left(\frac{x}{a} + \frac{a}{x}\right)\right] \ln x \frac{dx}{x} = 2 \ln a\, K_0(2\mu) \qquad [a > 0,\ \operatorname{Re}\mu > 0].$$ GW ((324))(91)

2. $$\int_0^\infty \exp\left(-ax - \frac{b}{x}\right) \ln x\, [2ax^2 - (2n+1)x - 2b]\, x^{n-\frac{1}{2}}\,dx =$$
$$= 2\left(\frac{b}{a}\right)^{\frac{n}{2}} \sqrt{\frac{\pi}{a}}\, e^{-2\sqrt{ab}} \sum_{k=0}^{\infty} \frac{(n+k)!}{(n-k)!\,(2k)!!\,(2\sqrt{ab})^k} \qquad [a > 0,\ b > 0].$$ BI ((357))(4)

3. $$\int_0^\infty \exp\left(-ax - \frac{b}{x}\right) \ln x\, [2ax^2 + (2n-1)x - 2b] \frac{dx}{x^{n+\frac{3}{2}}} =$$
$$= 2\left(\frac{a}{b}\right)^{\frac{n}{2}} \sqrt{\frac{\pi}{a}}\, e^{-2\sqrt{ab}} \sum_{k=0}^{\infty} \frac{(n+k-1)!}{(n-k-1)!\,(2k)!!\,(2\sqrt{ab})^k} \qquad [a > 0,\ b > 0].$$ BI ((357))(11)

For $n = \frac{1}{2}$:

4. $$\int_0^\infty \exp\left(-ax - \frac{b}{x}\right) \ln x \frac{ax^2 - b}{x^2}\,dx = 2K_0\left(2\sqrt{ab}\right) \qquad [a > 0,\ b > 0].$$ GW ((324))(92c)

For $n=0$:

5. $$\int_0^\infty \exp\left(-ax-\frac{b}{x}\right)\ln x\,\frac{2ax^2-x-2b}{x\sqrt{x}}\,dx=2\sqrt{\frac{\pi}{a}}\,e^{-2\sqrt{ab}}.$$

$[a>0,\ b>0]$. BI((357))(7), GW((324))(92a)

For $n=-1$:

6. $$\int_0^\infty \exp\left(-ax-\frac{b}{x}\right)\ln x\,\frac{2ax^2-3x-2b}{\sqrt{x}}\,dx=\frac{1+2\sqrt{ab}}{a}\sqrt{\frac{\pi}{a}}\,e^{-2\sqrt{ab}}$$

$[a>0,\ b>0]$. LI((357))(6), GW((324))(92b)

4.357

1. $$\int_0^\infty \exp\left(-\frac{1+x^4}{2ax^2}\right)\ln x\,\frac{1+ax^2-x^4}{x^2}\,dx=-\frac{\sqrt{2a^3\pi}}{2\sqrt[a]{e}}$$

$[a>0]$. BI((357))(8)

2. $$\int_0^\infty \exp\left(-\frac{1+x^4}{2ax^2}\right)\ln x\,\frac{x^4+ax^2-1}{x^4}\,dx=\frac{\sqrt{2a^3\pi}}{2\sqrt[a]{e}}$$

$[a>0]$. BI((357))(9)

3. $$\int_0^\infty \exp\left(-\frac{1+x^4}{2ax^2}\right)\ln x\cdot\frac{x^4+3ax-1}{x^6}\,dx=\frac{(1+a)\sqrt{2a^3\pi}}{2\sqrt[a]{e}}$$

$[a>0]$. BI((357))(10)

4.358

1. $$\int_1^\infty x^{\nu-1}e^{-\mu x}(\ln x)^m\,dx=\frac{1}{\mu}\frac{\partial^m}{\partial\nu^m}\{\mu^{1-\nu}\Gamma(\mu,\nu)\}$$

$[\operatorname{Re}\mu>0,\ \operatorname{Re}\nu>0]$. MI 26

2. $$\int_0^\infty x^{\nu-1}e^{-\mu x}(\ln x)^2\,dx=\frac{\Gamma(\nu)}{\mu^\nu}\{[\psi(\nu)-\ln\mu]^2+\zeta(2,\nu-1)\}$$

$[\operatorname{Re}\mu>0,\ \operatorname{Re}\nu>0]$. MI 26

3. $$\int_0^\infty x^{\nu-1}e^{-\mu x}(\ln x)^3\,dx=\frac{\Gamma(\nu)}{\mu^\nu}\{[\psi(\nu)-\ln\mu]^3+$$
$$+[2\psi(\nu)-3\ln\mu]\,\zeta(2,\nu-1)-2\zeta(3,\nu-1)\}$$

$[\operatorname{Re}\mu>0,\ \operatorname{Re}\nu>0]$. MI 26

4.359

1. $$\int_0^\infty e^{-\mu x}\frac{x^{p-1}-x^{q-1}}{\ln x}\,dx=\frac{1}{\mu}[\lambda(\mu,p-1)-\lambda(\mu,q-1)]$$

$[\operatorname{Re}\mu>0,\ p>0,\ q>0]$. MI 27

2. $$\int_0^1 e^{\mu x}\frac{x^{p-1}-x^{q-1}}{\ln x}\,dx=\sum_{k=0}^\infty\frac{\mu^k}{k!}\ln\frac{p+k}{q+k}$$

$[\operatorname{Re}\mu>0,\ p>0,\ q>0]$. BI((352))(9)

4.361

1. $$\int_0^\infty \frac{(x+1)\,e^{-\mu x}}{\pi^2+(\ln x)^2}\,dx = \nu'(\mu)-\nu''(\mu) \quad [\operatorname{Re}\mu>0].$$ MI 27

2. $$\int_0^\infty \frac{e^{-\mu x}\,dx}{x\,[\pi^2+(\ln x)^2]} = e^{\mu}-\nu(\mu) \quad [\operatorname{Re}\mu>0].$$ MI 27

4.362

1. $$\int_0^1 xe^x \ln(1-x)\,dx = 1-e.$$ BI ((352))(5)a

2. $$\int_1^\infty e^{-\mu x}\ln(2x-1)\frac{dx}{x} = \frac{1}{2}\left[\operatorname{Ei}\left(-\frac{\mu}{2}\right)\right]^2 \quad [\operatorname{Re}\mu>0].$$ ET I 148(8)

4.363

1. $$\int_0^\infty e^{-\mu x}\ln(a+x)\,\frac{\mu(x+a)\ln(x+a)-2}{x+a}\,dx =$$
$$= \frac{1}{4}\int_0^\infty e^{-\mu x}\ln(a-x)^2\,\frac{\mu(x-a)\ln(x-a)^2-4}{x-a}\,dx = (\ln a)^2$$
$$[\operatorname{Re}\mu>0,\ a>0].$$ BI ((354))(4, 5)

2. $$\int_0^1 x(1-x)(2-x)\,e^{-(1-x)^2}\ln(1-x)\,dx = \frac{1-e}{4e}.$$ BI ((352))(4)

4.364

1. $$\int_0^\infty e^{-\mu x}\ln[(x+a)(x+b)]\,\frac{dx}{x+a+b} =$$
$$= e^{(a+b)\mu}\{\operatorname{Ei}(-a\mu)\operatorname{Ei}(-b\mu)-\ln(ab)\operatorname{Ei}[-(a+b)\mu]\}$$
$$[a>0,\ b>0,\ \operatorname{Re}\mu>0].$$ BI ((354))(11)

2. $$\int_0^\infty e^{-\mu x}\ln(x+a+b)\left(\frac{1}{x+a}+\frac{1}{x+b}\right)dx =$$
$$= (1+\ln a\ln b)\ln(a+b)+e^{-(a+b)\mu}\{\operatorname{Ei}(-a\mu)\operatorname{Ei}(-b\mu)+$$
$$+(1-\ln(ab))\operatorname{Ei}[-(a+b)\mu]\} \quad [a>0,\ b>0,\ \operatorname{Re}\mu>0].$$ BI ((354))(12)

4.365 $$\int_0^\infty \left[e^{-x}-\frac{x}{(1+x)^{p+1}\ln(1+x)}\right]\frac{dx}{x} = \ln p \quad [p>0].$$ BI ((354))(15)

4.366

1. $$\int_0^\infty e^{-\mu x}\ln\left(1+\frac{x^2}{a^2}\right)\frac{dx}{x} = [\operatorname{ci}(a\mu)]^2+[\operatorname{si}(a\mu)]^2$$
$$[\operatorname{Re}\mu>0].$$ NT 32(11)a

2. $$\int_0^\infty e^{-\mu x}\ln\left|1-\frac{x^2}{a^2}\right|\frac{dx}{x} = \operatorname{Ei}(a\mu)\operatorname{Ei}(-a\mu) \quad [\operatorname{Re}\mu>0].$$ ME 18

3. $\int\limits_0^\infty xe^{-\mu x^2}\ln\left|\frac{1+x^2}{1-x^2}\right|dx=\frac{1}{\mu}\left[\operatorname{ch}\mu\operatorname{sh}i(\mu)-\operatorname{sh}\mu\operatorname{ch}i(\mu)\right]$

$[\operatorname{Re}\mu>0]$; (cf. **4.339**). MI 27

4.367 $\int\limits_0^\infty xe^{-\mu x^2}\ln\frac{x+\sqrt{x^2+2\beta}}{\sqrt{2\beta}}\,dx=\frac{e^{\beta\mu}}{4\mu}K_0(\beta\mu)$

$[|\arg\beta|<\pi,\ \operatorname{Re}\mu>0]$. ET I 149(19)

4.368 $\int\limits_0^{2u} e^{-\mu x^2}\ln\frac{x^2(4u^2-x^2)}{u^4}\frac{dx}{\sqrt{4u^2-x^2}}=\frac{\pi}{2}e^{-2u^2\mu}\left[\frac{\pi}{2}N_0(2iu^2\mu)-\right.$

$\left.-(\boldsymbol{C}-\ln 2)J_0(2iu^2\mu)\right]\quad[\operatorname{Re}\mu>0]$. ET I 149(21)a

4.369

1. $\int\limits_0^\infty x^{\nu-1}e^{-\mu x}[\psi(\nu)-\ln x]\,dx=\frac{\Gamma(\nu)\ln\mu}{\mu^\nu}\quad[\operatorname{Re}\nu>0]$. ET I 149(12)

2. $\int\limits_0^\infty x^n e^{-\mu x}\left\{\left[\ln x-\frac{1}{2}\psi(n+1)\right]^2-\frac{1}{2}\psi'(n+1)\right\}dx=$

$=\frac{n!}{\mu^{n+1}}\left\{\left[\ln\mu-\frac{1}{2}\psi(n+1)\right]^2+\frac{1}{2}\psi'(n+1)\right\}\quad[\operatorname{Re}\mu>0]$. MI 26

4.37 Combinations of logarithms and hyperbolic functions

4.371

1. $\int\limits_0^\infty \frac{\ln x}{\operatorname{ch}x}\,dx=\pi\ln\left[\frac{\sqrt{2\pi}\,\Gamma\left(\frac{3}{4}\right)}{\Gamma\left(\frac{1}{4}\right)}\right]$. LI ((260))(1)a

2. $\int\limits_0^\infty \frac{\ln x\,dx}{\operatorname{ch}x+\cos t}=\frac{\pi}{\sin t}\ln\frac{(2\pi)^{\frac{t}{\pi}}\,\Gamma\left(\frac{\pi+t}{2\pi}\right)}{\Gamma\left(\frac{\pi-t}{2\pi}\right)}\quad[t^2<\pi^2]$. BI ((257))(7)a

3. $\int\limits_0^\infty \frac{\ln x\,dx}{\operatorname{ch}^2 x}=\psi\left(\frac{1}{2}\right)+\ln\pi=\ln\pi-2\ln 2-\boldsymbol{C}$. BI ((257))(4)a

4.372

1. $\int\limits_1^\infty \ln x\frac{\operatorname{sh}mx}{\operatorname{sh}nx}\,dx=$

$=\frac{\pi}{2n}\operatorname{tg}\frac{m\pi}{2n}\ln 2\pi+\frac{\pi}{n}\sum_{k=1}^{n-1}(-1)^{k-1}\sin\frac{km\pi}{n}\ln\frac{\Gamma\left(\frac{n+k}{2n}\right)}{\Gamma\left(\frac{k}{2n}\right)}\quad[m+n \text{ is odd}];$

$=\frac{\pi}{2n}\operatorname{tg}\frac{m\pi}{2n}\ln\pi+\frac{\pi}{n}\sum_{k=1}^{\frac{n-1}{2}}(-1)^{k-1}\sin\frac{km\pi}{n}\ln\frac{\Gamma\left(\frac{n-k}{n}\right)}{\Gamma\left(\frac{k}{n}\right)}\quad[m+n \text{ is even}].$

BI ((148))(3)a

2. $$\int_1^\infty \ln x \frac{\operatorname{ch} mx}{\operatorname{ch} nx}\,dx =$$

$$= \frac{\pi}{2n}\frac{\ln 2\pi}{\cos\frac{m\pi}{2n}} + \frac{\pi}{n}\sum_{k=1}^{n}(-1)^{k-1}\cos\frac{(2k-1)m\pi}{2n}\ln\frac{\Gamma\left(\frac{2n+2k-1}{4n}\right)}{\Gamma\left(\frac{2k-1}{4n}\right)}$$

$[m+n$ is odd];

$$= \frac{\pi}{2n}\frac{\ln \pi}{\cos\frac{m\pi}{2n}} + \frac{\pi}{n}\sum_{k=1}^{\frac{n-1}{2}}(-1)^{k-1}\cos\frac{(2k-1)m\pi}{2n}\ln\frac{\Gamma\left(\frac{2n-2k+1}{2n}\right)}{\Gamma\left(\frac{2k-1}{2n}\right)}$$

$[m+n$ is even]. BI ((148))(6)a

4.373

1. $$\int_0^\infty \frac{\ln(a^2+x^2)}{\operatorname{ch} bx}\,dx = \frac{\pi}{b}\left[2\ln\frac{2\Gamma\left(\frac{2ab+3\pi}{4\pi}\right)}{\Gamma\left(\frac{2ab+\pi}{4\pi}\right)} - \ln\frac{2b}{\pi}\right]$$

$\left[b>0,\ a>-\frac{\pi}{2b}\right]$. BI ((258))(11)a

2. $$\int_0^\infty \ln(1+x^2)\frac{dx}{\operatorname{ch}\frac{\pi x}{2}} = 2\ln\frac{4}{\pi}.$$ BI ((258))(1)a

3. $$\int_0^\infty \ln(a^2+x^2)\frac{\operatorname{sh}\left(\frac{2}{3}\pi x\right)}{\operatorname{sh}\pi x}\,dx = 2\sin\frac{\pi}{3}\ln\frac{6\Gamma\left(\frac{a+4}{6}\right)\Gamma\left(\frac{a+5}{6}\right)}{\Gamma\left(\frac{a+1}{6}\right)\Gamma\left(\frac{a+2}{6}\right)}$$

$[a>-1]$. BI ((258))(12)

4. $$\int_0^\infty \ln(1+x^2)\frac{dx}{\operatorname{sh}^2 ax} = \frac{2}{a}\left[\ln\frac{a}{\pi}+\frac{\pi}{2a}-\psi\left(\frac{\pi+a}{\pi}\right)\right] \quad [a>0].$$

BI ((258))(5)

5. $$\int_0^\infty \ln(1+x^2)\frac{\operatorname{ch}\frac{\pi}{2}x}{\operatorname{sh}^2\frac{\pi}{2}x}\,dx = \frac{2\pi-4}{\pi}.$$ BI ((258))(3)

6. $$\int_0^\infty \ln(1+x^2)\frac{\operatorname{ch}\frac{\pi}{4}x}{\operatorname{sh}^2\frac{\pi}{4}x}\,dx = 4\sqrt{2}-\frac{16}{\pi}+\frac{8\sqrt{2}}{\pi}\ln\left(\sqrt{2}+1\right).$$ BI ((258))(2)

4.374

1. $$\int_0^\infty \ln(\cos^2 t + e^{-2x}\sin^2 t)\frac{dx}{\operatorname{sh} x} = -2t^2.$$ BI ((259))(10)a

2. $$\int_0^\infty \ln(a+be^{-2x})\frac{dx}{\operatorname{ch}^2 x} = \frac{2}{(b-a)}\left[\frac{a+b}{2}\ln(a+b) - a\ln a - b\ln 2\right]$$

$[a>0,\ a+b>0]$. LI ((259))(14)

4.375

1. $\int_0^\infty \ln \operatorname{ch} \frac{x}{2} \frac{dx}{\operatorname{ch} x} = G - \frac{\pi}{4} \ln 2.$ BI ((259))(11)

2. $\int_0^\infty \ln \operatorname{cth} x \frac{dx}{\operatorname{ch} x} = \frac{\pi}{2} \ln 2.$ BI ((259))(16)

4.376

1. $\int_0^\infty \frac{\ln x}{\sqrt{x} \operatorname{ch} x} dx = 2\sqrt{\pi} \sum_{k=0}^\infty \frac{(-1)^{k+1}}{\sqrt{2k+1}} \{\ln(2k+1) + 2\ln 2 + C\}.$ BI ((147))(4)

2. $\int_0^\infty \ln x \frac{(\mu+1) \operatorname{ch} x - x \operatorname{sh} x}{\operatorname{ch}^2 x} x^\mu dx = 2\Gamma(\mu+1) \sum_{k=0}^\infty \frac{(-1)^{k+1}}{(2k+1)^{\mu+1}}$ $[\operatorname{Re} \mu > -1].$ BI ((356))(10)

3. $\int_0^\infty \ln x \frac{(n+1) \operatorname{ch} x - x \operatorname{sh} x}{\operatorname{ch}^2 x} x^n dx = \frac{(-1)^n}{2^n} \beta^{(n)}\left(\frac{1}{2}\right).$

4. $\int_0^\infty \ln 2x \frac{n \operatorname{sh} 2ax - ax}{\operatorname{sh}^2 ax} x^{2n-1} dx = -\frac{1}{n} \left(\frac{\pi}{a}\right)^{2n} |B_{2n}|.$ BI ((356))(9)a

5. $\int_0^\infty \ln x \frac{ax \operatorname{ch} ax - (2n+1) \operatorname{sh} ax}{\operatorname{sh}^2 ax} x^{2n} dx = 2 \frac{2^{2n+1}-1}{(2a)^{2n+1}} (2n)! \, \zeta(2n+1).$ BI ((356))(14)

6. $\int_0^\infty \ln x \frac{ax \operatorname{ch} ax - 2n \operatorname{sh} ax}{\operatorname{sh}^2 ax} x^{2n-1} dx = \frac{2^{2n-1}-1}{2n} |B_{2n}| \left(\frac{\pi}{a}\right)^{2n}$ $[a > 0].$ BI ((356))(15)

7. $\int_0^\infty \ln x \frac{(2n+1) \operatorname{ch} ax - ax \operatorname{sh} ax}{\operatorname{ch}^2 ax} x^{2n} dx = -\left(\frac{\pi}{2a}\right)^{2n+1} |E_{2n}|$ $[a > 0].$ BI ((356))(11)

8. $\int_0^\infty \ln x \frac{2ax \operatorname{sh} ax - (2n+1) \operatorname{ch} ax}{\operatorname{ch}^3 ax} x^{2n} dx = \frac{2}{a} (2^{2n-1} - 1) \left(\frac{\pi}{2a}\right)^{2n} |B_{2n}|$ $[a > 0].$ BI ((356))(2)

9. $\int_0^\infty \ln x \frac{2ax \operatorname{ch} ax - (2n+1) \operatorname{sh} ax}{\operatorname{sh}^3 ax} x^{2n} dx = \frac{1}{a} \left(\frac{\pi}{a}\right)^{2n} |B_{2n}|.$ BI ((356))(6)a

10. $\int_0^\infty \ln x \frac{x \operatorname{sh} x - 6 \operatorname{sh}^2\left(\frac{x}{2}\right) - 6 \cos^2 \frac{t}{2}}{(\operatorname{ch} x + \cos t)^2} x^2 dx = \frac{(\pi^2 - t^2) t}{3 \sin t}$ $[0 < t < \pi].$ BI ((356))(16)a

11. $\int_0^\infty \ln(1+x^2)\frac{\operatorname{ch}\pi x+\pi x\operatorname{sh}\pi x}{\operatorname{ch}^2\pi x}\frac{dx}{x^2}=4-\pi.$ BI ((356))(12)

12. $\int_0^\infty \ln(1+4x^2)\frac{\operatorname{ch}\pi x+\pi x\operatorname{sh}\pi x}{\operatorname{ch}^2\pi x}\frac{dx}{x^2}=4\ln 2.$ BI ((356))(13)

4.377 $\int_0^\infty \ln 2x\,\frac{ax-n(1-e^{-2ax})}{\operatorname{sh}^2 ax}x^{2n-1}\,dx=\frac{1}{2n}\left(\frac{\pi}{a}\right)^{2n}|B_{2n}|.$ LI ((356))(8)a

4.38-4.41 Logarithms and trigonometric functions

4.381

1. $\int_0^1 \ln x\sin ax\,dx=-\frac{1}{a}[C+\ln a-\operatorname{ci}(a)]\quad [a>0].$ GW ((338))(2a)

2. $\int_0^1 \ln x\cos ax\,dx=-\frac{1}{a}\left[\operatorname{si}(a)+\frac{\pi}{2}\right]\quad [a>0].$ BI ((284))(2)

3. $\int_0^{2\pi} \ln x\sin nx\,dx=-\frac{1}{n}[C+\ln(2n\pi)-\operatorname{ci}(2n\pi)].$ GW ((338))(1a)

4. $\int_0^{2\pi} \ln x\cos nx\,dx=-\frac{1}{n}\left[\operatorname{si}(2n\pi)+\frac{\pi}{2}\right].$ GW ((338))(1b)

4.382

1. $\int_0^\infty \ln\left|\frac{x+a}{x-a}\right|\sin bx\,dx=\frac{\pi}{b}\sin ab\quad [a>0,\ b>0].$ ET I 77(11)

2. $\int_0^\infty \ln\left|\frac{x+a}{x-a}\right|\cos bx\,dx=\frac{2}{b}[\cos ab\operatorname{si}(ab)-\sin ab\operatorname{ci}(ab)]$

$[a>0,\ b>0].$ ET I 18(9)

3. $\int_0^\infty \ln\frac{a^2+x^2}{b^2+x^2}\cos cx\,dx=\frac{\pi}{c}(e^{-bc}-e^{-ac})\quad [a>0,\ b>0,\ c>0].$

FI III 648a, BI ((337))(5)

4. $\int_0^\infty \ln\frac{x^2+x+a^2}{x^2-x+a^2}\sin bx\,dx=\frac{2\pi}{b}\exp\left(-b\sqrt{a^2-\frac{1}{4}}\right)\sin\frac{b}{2}$

$[b>0].$ ET I 77(12)

5. $\int_0^\infty \ln\frac{(x+\beta)^2+\gamma^2}{(x-\beta)^2+\gamma^2}\sin bx\,dx=\frac{2\pi}{b}e^{-\gamma b}\sin\beta b$

$[\operatorname{Re}\gamma>0,\ |\operatorname{Im}\beta|\leqslant\operatorname{Re}\gamma,\ b>0].$ ET I 77(13)

4.383

1. $$\int_0^\infty \ln(1+e^{-\beta x})\cos bx\,dx = \frac{\beta}{2b^2} - \frac{\pi}{2b\,\operatorname{sh}\left(\frac{\pi b}{\beta}\right)}$$

$[\operatorname{Re}\beta > 0,\ b > 0]$. ET I 18(13)

2. $$\int_0^\infty \ln(1-e^{-\beta x})\cos bx\,dx = \frac{\beta}{2b^2} - \frac{\pi}{2b}\operatorname{cth}\left(\frac{\pi b}{\beta}\right)$$

$[\operatorname{Re}\beta > 0,\ b > 0]$. ET I 18(14)

4.384

1. $$\int_0^1 \ln(\sin\pi x)\sin 2n\pi x\,dx = 0.$$ GW ((338))(3a)

2. $$\int_0^1 \ln(\sin\pi x)\sin(2n+1)\pi x\,dx = 2\int_0^{\frac{1}{2}} \ln(\sin\pi x)\sin(2n+1)\pi x\,dx =$$
$$= -\frac{1}{(2n+1)\pi}\left[2C + 2\ln 2 + \psi\left(\frac{1}{2}+n\right) + \psi\left(-\frac{1}{2}-n\right)\right] =$$
$$= \frac{2}{(2n+1)\pi}\left[\ln 2 - 2 - \frac{2}{3} - \ldots - \frac{2}{2n-1} - \frac{1}{2n+1}\right].$$ GW ((338))(3b)

3. $$\int_0^1 \ln(\sin\pi x)\cos 2n\pi x\,dx = 2\int_0^{\frac{1}{2}} \ln(\sin\pi x)\cos 2n\pi x\,dx =$$
$$= -\ln 2 \qquad [n=0];$$
$$= -\frac{1}{2n} \qquad [n>0].$$ GW ((338))(3c)

4. $$\int_0^1 \ln(\sin\pi x)\cos(2n+1)\pi x\,dx = 0.$$ GW ((338))(3d)

5. $$\int_0^{\frac{\pi}{2}} \ln\sin x\sin x\,dx = \ln 2 - 1.$$ BI ((305))(4)

6. $$\int_0^{\frac{\pi}{2}} \ln\sin x\cos x\,dx = -1.$$ BI ((305))(5)

7. $$\int_0^{\frac{\pi}{2}} \ln\sin x\cos 2nx\,dx = -\frac{\pi}{4n}.$$ LI ((305))(6)

8. $$\int_0^{\pi} \ln\sin x\cos[2m(x-n)]\,dx = -\frac{\pi\cos 2mn}{2m}.$$ LI ((330))(8)

9. $$\int_0^{\frac{\pi}{2}} \ln\sin x\sin^2 x\,dx = \frac{\pi}{8}(1-\ln 4).$$ BI ((305))(7)

10. $\int_0^{\frac{\pi}{2}} \ln \sin x \cos^2 x \, dx = -\frac{\pi}{8}(1 + \ln 4).$ BI ((305))(8)

11. $\int_0^{\frac{\pi}{2}} \ln \sin x \sin x \cos^2 x \, dx = \frac{1}{9}(\ln 8 - 4).$ BI ((305))(9)

12. $\int_0^{\frac{\pi}{2}} \ln \sin x \operatorname{tg} x \, dx = -\frac{\pi^2}{24}.$ BI ((305))(11)

13. $\int_0^{\frac{\pi}{2}} \ln \sin 2x \sin x \, dx = \int_0^{\frac{\pi}{2}} \ln \sin 2x \cos x \, dx = 2(\ln 2 - 1).$

BI ((305))(16, 17)

14. $\int_0^{\pi} \frac{\ln(1 + p \cos x)}{\cos x} dx = \pi \arcsin p \qquad [p^2 < 1].$ FI II 484

15. $\int_0^{\pi} \ln \sin x \frac{dx}{1 - 2a \cos x + a^2} = \frac{\pi}{1 - a^2} \ln \frac{1 - a^2}{2} \qquad [a^2 < 1];$

$= \frac{\pi}{a^2 - 1} \ln \frac{a^2 - 1}{2a^2} \qquad [a^2 > 1].$ BI ((331))(8)

16. $\int_0^{\pi} \ln \sin bx \frac{dx}{1 - 2a \cos x + a^2} = \frac{\pi}{1 - a^2} \ln \frac{1 - a^{2b}}{2} \qquad [a^2 < 1].$ BI ((331))(10)

17. $\int_0^{\pi} \ln \cos bx \frac{dx}{1 - 2a \cos x + a^2} = \frac{\pi}{1 - a^2} \ln \frac{1 + a^{2b}}{2} \qquad [a^2 < 1].$ BI ((331))(11)

18. $\int_0^{\frac{\pi}{2}} \ln \sin x \frac{dx}{1 - 2a \cos 2x + a^2} = \frac{1}{2} \int_0^{\pi} \ln \sin x \frac{dx}{1 - 2a \cos 2x + a^2} =$

$= \frac{\pi}{2(1 - a^2)} \ln \frac{1 - a}{2} \qquad [a^2 < 1];$

$= \frac{\pi}{2(a^2 - 1)} \ln \frac{a - 1}{2a} \qquad [a^2 > 1].$ BI ((321))(1), BI ((331))(13)

19. $\int_0^{\pi} \ln \sin bx \frac{dx}{1 - 2a \cos 2x + a^2} = \frac{\pi}{1 - a^2} \ln \frac{1 - a^b}{2} \qquad [a^2 < 1].$

BI ((331))(18)

20. $\int_0^{\pi} \ln \cos bx \frac{dx}{1 - 2a \cos 2x + a^2} = \frac{\pi}{1 - a^2} \ln \frac{1 + a^b}{2} \qquad [a^2 < 1].$

BI ((331))(21)

21. $\int_0^{\frac{\pi}{2}} \frac{\ln \cos x \, dx}{1-2p \cos 2x+p^2} = \frac{\pi}{2(1-p^2)} \ln \frac{1+p}{2} \quad [p^2 < 1];$

$= \frac{\pi}{2(p^2-1)} \ln \frac{p+1}{2p} \quad [p^2 > 1].$ BI ((321))(8)

22. $\int_0^{\pi} \ln \sin x \frac{\cos x \, dx}{1-2a \cos x+a^2} = \frac{\pi}{2a} \frac{1+a^2}{1-a^2} \ln (1-a^2) - \frac{a\pi \ln 2}{1-a^2} \quad [a^2 < 1];$

$= \frac{\pi}{2a} \frac{a^2+1}{a^2-1} \ln \frac{a^2-1}{a^2} - \frac{\pi \ln 2}{a(a^2-1)} \quad [a^2 > 1].$

LI ((331))(9)

23. $\int_0^{\pi} \ln \sin bx \frac{\cos x \, dx}{1-2a \cos 2x+a^2} = \int_0^{\pi} \ln \cos bx \frac{\cos x \, dx}{1-2a \cos 2x+a^2} = 0$

$[0 < a < 1].$ BI ((331))(19, 22)

24. $\int_0^{\pi} \ln \sin x \frac{\cos^2 x \, dx}{1-2a \cos 2x+a^2} = \frac{\pi}{4a} \frac{1+a}{1-a} \ln (1-a) - \frac{\pi \ln 2}{2(1-a)} \quad [0 < a < 1];$

$= \frac{\pi}{4a} \frac{a+1}{a-1} \ln \frac{a-1}{a} - \frac{\pi \ln 2}{2a(a-1)} \quad [a > 1].$ BI ((331))(16)

25. $\int_0^{\frac{\pi}{2}} \ln \sin x \frac{\cos 2x \, dx}{1-2a \cos 2x+a^2} = \frac{1}{2} \int_0^{\pi} \ln \sin x \frac{\cos 2x \, dx}{1-2a \cos 2x+a^2} =$

$= \frac{\pi}{2a(1-a^2)} \left\{ \frac{1+a^2}{2} \ln (1-a) - a^2 \ln 2 \right\} \quad [a^2 < 1];$

$= \frac{\pi}{2a(a^2-1)} \left\{ \frac{1+a^2}{2} \ln \frac{a-1}{a} - \ln 2 \right\} \quad [a^2 > 1].$

BI ((321))(2), BI ((331))(15), LI ((321))(2)

26. $\int_0^{\frac{\pi}{2}} \ln \cos x \frac{\cos 2x \, dx}{1-2a \cos 2x+a^2} =$

$= \frac{\pi}{2a(1-a^2)} \left\{ \frac{1+a^2}{2} \ln (1+a) - a^2 \ln 2 \right\} \quad [a^2 < 1];$

$= \frac{\pi}{2a(a^2-1)} \left\{ \frac{1+a^2}{2} \ln \frac{1+a}{a} - \ln 2 \right\} \quad [a^2 > 1].$ BI ((321))(9)

4.385

1. $\int_0^{\pi} \ln \sin x \frac{dx}{a+b \cos x} = \frac{\pi}{\sqrt{a^2-b^2}} \ln \frac{\sqrt{a^2-b^2}}{a+\sqrt{a^2-b^2}} \quad [a > 0, \ a > b].$

BI ((331))(6)

2. $\int_0^{\frac{\pi}{2}} \ln \sin x \frac{dx}{(a \sin x \pm b \cos x)^2} = \int_0^{\frac{\pi}{2}} \ln \cos x \frac{dx}{(a \cos x \pm b \sin x)^2} =$

$= \frac{1}{b(a^2+b^2)} \left(\mp a \ln \frac{a}{b} - \frac{b\pi}{2} \right) \quad [a > 0, \ b > 0].$ BI ((319))(1, 6)a

3. $$\int_0^{\frac{\pi}{2}} \frac{\ln \sin x \, dx}{a^2 \sin^2 x + b^2 \cos^2 x} = \int_0^{\frac{\pi}{2}} \frac{\ln \cos x \, dx}{b^2 \sin^2 x + a^2 \cos^2 x} = \frac{\pi}{2ab} \ln \frac{b}{a+b}$$
$$[a > 0, \quad b > 0].$$ BI ((317))(4, 10)

4. $$\int_0^{\frac{\pi}{2}} \ln \sin x \frac{\sin 2x \, dx}{(a \sin^2 x + b \cos^2 x)^2} =$$
$$= \int_0^{\frac{\pi}{2}} \ln \cos x \frac{\sin 2x \, dx}{(b \sin^2 x + a \cos^2 x)^2} = \frac{1}{2b(b-a)} \ln \frac{a}{b}$$
$$[a > 0, \ b > 0].$$ BI ((319))(3, 7), LI ((319))(3)

5. $$\int_0^{\frac{\pi}{2}} \ln \sin x \frac{a^2 \sin^2 x - b^2 \cos^2 x}{(a^2 \sin^2 x + b^2 \cos^2 x)^2} dx =$$
$$= \int_0^{\frac{\pi}{2}} \ln \cos x \frac{a^2 \cos^2 x - b^2 \sin^2 x}{(a^2 \cos^2 x + b^2 \sin^2 x)^2} dx = \frac{\pi}{2b(a+b)} \quad [a > 0, \ b > 0].$$
LI ((319))(2, 8)

4.386

1. $$\int_0^{\frac{\pi}{2}} \ln \sin x \frac{\sin x}{\sqrt{1 + \sin^2 x}} dx = \int_0^{\frac{\pi}{2}} \frac{\cos x \ln \cos x}{\sqrt{1 + \cos^2 x}} dx = -\frac{\pi}{8} \ln 2.$$
BI ((322))(1, 6)

2. $$\int_0^{\frac{\pi}{2}} \frac{\sin^3 x \ln \sin x}{\sqrt{1 + \sin^2 x}} dx = \int_0^{\frac{\pi}{2}} \frac{\cos^3 x \ln \cos x}{\sqrt{1 + \cos^2 x}} dx = \frac{\ln 2 - 1}{4}.$$ BI ((322))(2, 7)

3. $$\int_0^{\frac{\pi}{2}} \ln \sin x \frac{dx}{\sqrt{1 - k^2 \sin^2 x}} = -\frac{1}{2} \boldsymbol{K}(k) \ln k - \frac{\pi}{4} \boldsymbol{K}(k').$$ BI ((322))(3)

4. $$\int_0^{\frac{\pi}{2}} \frac{\ln \cos x \, dx}{\sqrt{1 - k^2 \sin^2 x}} = \frac{1}{2} \boldsymbol{K}(k) \ln \frac{k'}{k} - \frac{\pi}{4} \boldsymbol{K}(k').$$ BI ((322))(9)

4.387

1. $$\int_0^{\frac{\pi}{2}} \ln \sin x \sin^{\mu} x \cos^{\nu} x \, dx = \int_0^{\frac{\pi}{2}} \ln \cos x \cos^{\mu} x \sin^{\nu} x \, dx =$$
$$= \frac{1}{4} \mathrm{B}\left(\frac{\mu+1}{2}, \frac{\nu+1}{2}\right) \left[\psi\left(\frac{\mu+1}{2}\right) - \psi\left(\frac{\mu+\nu+2}{2}\right)\right]$$
$$[\operatorname{Re} \mu > -1, \ \operatorname{Re} \nu > -1].$$ GW ((338))(6c)

2. $\int_0^{\frac{\pi}{2}} \ln \sin x \sin^{\mu-1} x\, dx = \frac{\sqrt{\pi}\, \Gamma\left(\frac{\mu}{2}\right)}{4\Gamma\left(\frac{\mu+1}{2}\right)} \left[\psi\left(\frac{\mu}{2}\right) - \psi\left(\frac{\mu+1}{2}\right)\right]$ [Re $\mu > 0$].

GW ((338))(6a)

3. $\int_0^{\frac{\pi}{2}} \ln \sin x \cos^{\nu-1} x\, dx = \frac{\sqrt{\pi}\, \Gamma\left(\frac{\nu}{2}\right)}{4\Gamma\left(\frac{\nu+1}{2}\right)} \left[\psi\left(\frac{\nu}{2}\right) - \psi\left(\frac{\nu+1}{2}\right)\right]$ [Re $\nu > 0$].

GW ((338))(6b)

4. $\int_0^{\frac{\pi}{2}} \ln \sin x \sin^{2n} x\, dx = \frac{(2n-1)!!}{(2n)!!} \frac{\pi}{2} \left\{\sum_{k=1}^{2n} \frac{(-1)^{k+1}}{k} - \ln 2\right\}.$ FI II 811

5. $\int_0^{\frac{\pi}{2}} \ln \sin x \sin^{2n+1} x\, dx = \frac{(2n)!!}{(2n+1)!!} \left\{\sum_{k=1}^{2n+1} \frac{(-1)^k}{k} + \ln 2\right\}.$ BI ((305))(13)

6. $\int_0^{\frac{\pi}{2}} \ln \sin x \cos^{2n} x\, dx = -\frac{(2n-1)!!}{(2n)!!} \frac{\pi}{4} \left[\sum_{k=1}^{n} \frac{1}{k} + \ln 4\right] =$

$= -\frac{(2n-1)!!}{(2n)!!} \frac{\pi}{4} [C + \psi(n+1) + \ln 4].$ BI ((305))(14)

7. $\int_0^{\frac{\pi}{2}} \ln \sin x \cos^{2n+1} x\, dx = -\frac{(2n)!!}{(2n+1)!!} \sum_{k=0}^{n} \frac{1}{2k+1} =$

$= -\frac{(2n)!!}{2(2n+1)!!} \left[\psi\left(n+\frac{3}{2}\right) - \psi\left(\frac{1}{2}\right)\right].$ GW ((338))(7b)

8. $\int_0^{\frac{\pi}{2}} \ln \cos x \sin^{2n} x\, dx = -\frac{(2n-1)!!}{2^{n+1} \cdot n!} \frac{\pi}{2} \{C + 2\ln 2 + \psi(n+1)\}.$

BI ((306))(8)

9. $\int_0^{\frac{\pi}{2}} \ln \cos x \cos^{2n} x\, dx = -\frac{(2n-1)!!}{2^n\, n!} \frac{\pi}{2} \left\{\ln 2 + \sum_{k=1}^{2n}{}' \frac{(-1)^k}{k}\right\}.$ BI ((306))(10)

10. $\int_0^{\frac{\pi}{2}} \ln \cos x \cos^{2n-1} x\, dx = \frac{2^{n-1}(n-1)!}{(2n-1)!!} \left[\ln 2 + \sum_{k=1}^{2n-1} \frac{(-1)^k}{k}\right].$

BI ((306))(9)

4.388

1. $\int_0^{\frac{\pi}{4}} \ln \sin x \frac{\sin^{2n} x}{\cos^{2n+2} x}\, dx =$

$= \frac{1}{2n+1}\left[\frac{1}{2}\ln 2 + (-1)^n \frac{\pi}{4} + \sum_{k=0}^{n-1}{}' \frac{(-1)^k}{2n-2k-1}\right].$ BI ((288))(1)

2. $$\int_0^{\frac{\pi}{4}} \ln \sin x \frac{\sin^{2n-1} x}{\cos^{2n+1} x} dx = \frac{1}{4n}\left[-\ln 2 + (-1)^n \ln 2 + \sum_{k=1}^{n-1} \frac{(-1)^k}{n-k}\right].$$ LI ((288))(2)

3. $$\int_0^{\frac{\pi}{4}} \ln \cos x \frac{\sin^{2n} x}{\cos^{2n+2} x} dx = \frac{1}{2n+1}\left[-\frac{1}{2}\ln 2 + (-1)^{n+1}\frac{\pi}{4} + \sum_{k=0}^{n} \frac{(-1)^{k-1}}{2n-2k+1}\right].$$ BI ((288))(10)

4. $$\int_0^{\frac{\pi}{4}} \ln \cos x \frac{\sin^{2n-1} x}{\cos^{2n+1} x} dx = \frac{1}{4n}\left[-\ln 2 + (-1)^n \ln 2 + \sum_{k=0}^{n-1} \frac{(-1)^k}{n-k}\right].$$ BI ((288))(11)

5. $$\int_0^{\frac{\pi}{2}} \ln \sin x \frac{\sin^{p-1} x}{\cos^{p+1} x} dx = -\frac{\pi}{2p} \operatorname{cosec} \frac{p\pi}{2} \qquad [0 < p < 2].$$ BI ((310))(4)

6. $$\int_0^{\frac{\pi}{2}} \ln \sin x \frac{dx}{\operatorname{tg}^{p-1} x \sin 2x} = \frac{1}{4}\frac{\pi}{p-1} \sec \frac{p\pi}{2} \qquad [p^2 < 1].$$ BI ((310))(3)

4.389

1. $$\int_0^{\pi} \ln \sin x \sin^{2n} 2x \cos 2x \, dx = -\frac{(2n-1)!!}{(2n)!!}\frac{\pi}{4n+2}.$$ BI ((330))(9)

2. $$\int_0^{\frac{\pi}{4}} \ln \sin x \cos^n 2x \sin 2x \, dx = -\frac{1}{4(n+1)}\{C + \psi(n+2) + \ln 2\}.$$ BI ((285))(2)

3. $$\int_0^{\frac{\pi}{4}} \ln \cos x \cos^{\mu-1} 2x \operatorname{tg} 2x \, dx = \frac{1}{4(1-\mu)} \beta(\mu) \qquad [\operatorname{Re} \mu > 0].$$ BI ((286))(2)

4. $$\int_0^{\frac{\pi}{2}} \ln \sin x \sin^{\mu-1} x \cos x \, dx = \int_0^{\frac{\pi}{2}} \ln \cos x \cos^{\mu-1} x \sin x \, dx = -\frac{1}{\mu^2} \qquad [\operatorname{Re} \mu > 0].$$ BI ((306))(11)

5. $$\int_{-\frac{\pi}{2}}^{\frac{\pi}{2}} \ln \cos x \cos^p x \cos px \, dx = -\frac{\pi}{2^p} \ln 2 \qquad [p > -1].$$ BI ((337))(6)

6. $$\int_0^{\frac{\pi}{2}} \ln \cos x \cos^{p-1} x \sin px \sin x \, dx = \frac{\pi}{2^{p+2}}\left[C + \psi(p) - \frac{1}{p} - 2\ln 2\right] \qquad [p > 0].$$ BI ((306))(12)

4.391

1. $$\int_0^{\frac{\pi}{4}} (\ln \cos 2x)^n \cos^{p-1} 2x \operatorname{tg} x \, dx = \int_0^{\frac{\pi}{4}} (\ln \sin 2x)^n \sin^{p-1} 2x \operatorname{tg}\left(\frac{\pi}{4} - x\right) dx = \frac{1}{2} \beta^{(n)}(p) \quad [p > 0].$$

BI((286))(10), BI((285))(18)

2. $$\int_0^{\frac{\pi}{4}} (\ln \sin 2x)^n \sin^{p-1} 2x \operatorname{tg}\left(\frac{\pi}{4} + x\right) dx = \frac{(-1)^n n!}{2} \zeta(n+1, p).$$

BI ((285))(17)

3. $$\int_0^{\frac{\pi}{4}} (\ln \cos 2x)^{2n-1} \operatorname{tg} x \, dx = \frac{1 - 2^{2n-1}}{4n} \pi^{2n} |B_{2n}|.$$ BI ((286))(7)

4. $$\int_0^{\frac{\pi}{4}} (\ln \cos 2x)^{2n} \operatorname{tg} x \, dx = \frac{2^{2n} - 1}{2^{2n+1}} (2n)! \, \zeta(2n+1).$$ BI ((286))(8)

4.392

1. $$\int_0^{\frac{\pi}{4}} \ln(\sin x \cos x) \frac{\sin^{2n} x}{\cos^{2n+2} x} dx = \frac{1}{2n+1} \left[(-1)^{n+1} \frac{\pi}{2} - \ln 2 + \frac{1}{2n+1} + 2 \sum_{k=0}^{n-1} \frac{(-1)^{k-1}}{2n-2k-1} \right].$$ BI ((294))(8)

2. $$\int_0^{\frac{\pi}{4}} \ln(\sin x \cos x) \frac{\sin^{2n-1} x}{\cos^{2n+1} x} dx = \frac{1}{2n} \left[(-1)^n \ln 2 - \ln 2 + \frac{1}{2n} + (-1)^n \sum_{k=1}^{n-1} \frac{(-1)^k}{k} \right].$$ BI ((294))(9)

4.393

1. $$\int_0^{\frac{\pi}{2}} \ln \operatorname{tg} x \sin x \, dx = \ln 2.$$ BI ((307))(3)

2. $$\int_0^{\frac{\pi}{2}} \ln \operatorname{tg} x \cos x \, dx = -\ln 2.$$ BI ((307))(4)

3. $$\int_0^{\frac{\pi}{2}} \ln \operatorname{tg} x \sin^2 x \, dx = -\int_0^{\frac{\pi}{2}} \ln \operatorname{tg} x \cos^2 x \, dx = \frac{\pi}{4}.$$ BI ((307))(5, 6)

4. $$\int_0^{\frac{\pi}{4}} \frac{\ln \operatorname{tg} x}{\cos 2x}\, dx = -\frac{\pi^2}{8}.$$ GW ((338))(10b)a

5. $$\int_0^{\frac{\pi}{2}} \sin x \ln \operatorname{ctg} \frac{x}{2}\, dx = \ln 2.$$ LO III 290

4.394

1. $$\int_0^{\frac{\pi}{2}} \frac{\ln \operatorname{tg} x\, dx}{1-2a\cos 2x+a^2} = \frac{\pi}{2(1-a^2)} \ln \frac{1-a}{1+a} \qquad [a^2<1];$$
$$= \frac{\pi}{2(a^2-1)} \ln \frac{a-1}{a+1} \qquad [a^2>1].$$
BI ((321))(15)

2. $$\int_0^{\frac{\pi}{2}} \frac{\ln \operatorname{tg} x \cos 2x\, dx}{1-2a\cos 2x+a^2} = \frac{\pi}{4a}\, \frac{1+a^2}{1-a^2} \ln \frac{1-a}{1+a} \qquad [a^2<1];$$
$$= \frac{\pi}{4a}\, \frac{a^2+1}{a^2-1} \ln \frac{a-1}{a+1} \qquad [a^2>1].$$ BI ((321))(16)

3. $$\int_0^{\pi} \frac{\ln \operatorname{tg} bx\, dx}{1-2a\cos 2x+a^2} = \frac{\pi}{1-a^2} \ln \frac{1-a^b}{1+a^b} \qquad [0<a<1,\ b>0].$$ BI ((331))(24)

4. $$\int_0^{\pi} \frac{\ln \operatorname{tg} bx \cos x\, dx}{1-2a\cos 2x+a^2} = 0 \qquad [0<a<1].$$ BI ((331))(25)

5. $$\int_0^{\frac{\pi}{4}} \ln \operatorname{tg} x \frac{\cos 2x\, dx}{1-a\sin 2x} = -\frac{\arcsin a}{4a}(\pi+\arcsin a) \qquad [a^2 \leqslant 1].$$
BI ((291))(2, 3)

6. $$\int_0^{\frac{\pi}{4}} \ln \operatorname{tg} x \frac{\cos 2x\, dx}{1-a^2\sin^2 2x} = -\frac{\pi}{4a} \arcsin a \qquad [a^2<1].$$ BI ((291))(9)

7. $$\int_0^{\frac{\pi}{4}} \ln \operatorname{tg} x \frac{\cos 2x\, dx}{1+a^2\sin^2 x} = -\frac{\pi}{4a} \operatorname{Arsh} a = -\frac{\pi}{4a} \ln\left(a+\sqrt{1+a^2}\right)$$
$$[a^2<1].$$ BI ((291))(10)

8. $$\int_0^{u} \frac{\sin x \ln \operatorname{ctg} \frac{x}{2}}{1-\cos^2\alpha \sin^2 x}\, dx =$$
$$= \operatorname{cosec} 2\alpha \left\{\frac{\pi}{2} \ln 2 + L(\varphi-\alpha) - L(\varphi+\alpha) - L\left(\frac{\pi}{2}-2\alpha\right)\right\}$$
$$[\operatorname{tg}\varphi = \operatorname{ctg}\alpha \cos u;\ 0<u<\pi].$$ LO III 290

9. $$\int_0^{\frac{\pi}{4}} \frac{\ln \operatorname{tg} x \sin 2x \, dx}{1-\cos^2 t \sin^2 2x} = \operatorname{cosec} 2t \left[L\left(\frac{\pi}{2}-t\right) - \left(\frac{\pi}{2}-t\right) \ln 2 \right].$$

LO III 290a

4.395

1. $$\int_0^{\frac{\pi}{2}} \frac{\ln \operatorname{tg} x \, dx}{\sqrt{1-k^2 \sin^2 x}} = -\ln k' \boldsymbol{K}(k).$$ BI ((322))(11)

2. $$\int_u^{\frac{\pi}{4}} \frac{\ln \operatorname{tg} x \sin 4x \, dx}{(\sin^2 u + \operatorname{tg}^2 v \sin^2 2x) \sqrt{\sin^2 2x - \sin^2 u}} =$$

$$= -\frac{\pi}{2} \frac{\cos^2 v}{\sin u \sin v} \ln \frac{\sin v + \sqrt{1-\cos^2 u \cos^2 v}}{\sin u (1+\sin v)}$$

$$\left[0 < u < \frac{\pi}{2},\ 0 < v < \frac{\pi}{2}\right].$$ LO III 285a

4.396

1. $$\int_0^{\frac{\pi}{2}} \ln (a \operatorname{tg} x) \sin^{\mu-1} 2x \, dx = 2^{\mu-2} \ln a \frac{\left\{\Gamma\left(\frac{a}{2}\right)\right\}^2}{\Gamma(a)}$$

$$[a > 0,\ \operatorname{Re} \mu > 0]$$ LI ((307))(8)

2. $$\int_0^{\frac{\pi}{2}} \ln \operatorname{tg} x \cos^{2(\mu-1)} x \, dx = -\frac{\sqrt{\pi}}{4} \frac{\Gamma\left(\mu-\frac{1}{2}\right)}{\Gamma(\mu)} \left[\boldsymbol{C} + \psi\left(\frac{2\mu-1}{2}\right) + \ln 4 \right]$$

$$\left[\operatorname{Re} \mu > \frac{1}{2}\right].$$ BI ((307))(9)

3. $$\int_0^{\frac{\pi}{2}} \ln \operatorname{tg} x \cos^{q-1} x \operatorname{ctg} x \sin [(q+1)x] \, dx = -\frac{\pi}{2} [\boldsymbol{C} + \psi(q+1)]$$

$$[q > -1].$$ BI ((307))(11)

4. $$\int_0^{\frac{\pi}{2}} \ln \operatorname{tg} x \cos^{q-1} x \cos [(q+1)x] \, dx = -\frac{\pi}{2q} \qquad [q > 0].$$ BI ((307))(10)

5. $$\int_0^{\frac{\pi}{4}} (\ln \operatorname{tg} x)^n \operatorname{tg}^p x \, dx = \frac{1}{2^{n+1}} \beta^{(n)}\left(\frac{p+1}{2}\right) \qquad [p > -1].$$ LI ((286))(22)

6. $$\int_0^{\frac{\pi}{2}} (\ln \operatorname{tg} x)^{2n-1} \frac{dx}{\cos 2x} = \frac{1-2^{2n}}{2n} \pi^{2n} |B_{2n}|.$$ BI ((312))(6)

7. $$\int_0^{\frac{\pi}{4}} \ln \operatorname{tg} x \operatorname{tg}^{2n+1} x \, dx = \frac{(-1)^{n+1}}{4} \left[\frac{\pi^2}{12} + \sum_{k=1}^{n} \frac{(-1)^k}{k^2} \right].$$ GW ((338))(8a)

4.397

1. $$\int_0^{\frac{\pi}{2}} \ln(1+p\sin x)\frac{dx}{\sin x} = \frac{\pi^2}{8} - \frac{1}{2}(\arccos p)^2 \quad [p^2 < 1].$$ BI ((313))(1)

2. $$\int_0^{\frac{\pi}{2}} \ln(1+p\cos x)\frac{dx}{\cos x} = \frac{\pi^2}{8} - \frac{1}{2}(\arccos p)^2 \quad [p^2 < 1].$$ BI ((313))(8)

3. $$\int_0^{\pi} \ln(1+p\cos x)\frac{dx}{\cos x} = \pi \arcsin p \quad [p^2 < 1].$$ BI ((331))(1)

4. $$\int_0^{\frac{\pi}{2}} \frac{\cos x \ln(1+\cos\alpha\cos x)}{1-\cos^2\alpha\cos^2 x}\,dx = \frac{L\left(\frac{\pi}{2}-\alpha\right) - \alpha\ln\sin\alpha}{\sin\alpha\cos\alpha} \quad \left[0 < \alpha < \frac{\pi}{2}\right].$$ LO III 291

5. $$\int_0^{\frac{\pi}{2}} \frac{\cos x \ln(1-\cos\alpha\cos x)}{1-\cos^2\alpha\cos^2 x}\,dx = \frac{L\left(\frac{\pi}{2}-\alpha\right) + (\pi-\alpha)\ln\sin\alpha}{\sin\alpha\cos\alpha} \quad \left[0 < \alpha < \frac{\pi}{2}\right].$$ LO III 291

6. $$\int_0^{\pi} \ln(1-2a\cos x + a^2)\cos nx\,dx = \frac{1}{2}\int_0^{2\pi} \ln(1-2a\cos x+a^2)\cos nx\,dx =$$
$$= -\frac{\pi}{n}a^n \quad [a^2 < 1];$$ BI((330))(11), BI((332))(5)
$$= -\frac{\pi}{na^n} \quad [a^2 > 1].$$ GW ((338))(13a)

7. $$\int_0^{\pi} \ln(1-2a\cos x+a^2)\sin nx \sin x\,dx =$$
$$= \frac{1}{2}\int_0^{2\pi} \ln(1-2a\cos x+a^2)\sin nx\sin x\,dx = \frac{\pi}{2}\left(\frac{a^{n+1}}{n+1} - \frac{a^{n-1}}{n-1}\right)$$
$$[a^2 < 1].$$ BI((330))(10), BI((332))(4)

8. $$\int_0^{\pi} \ln(1-2a\cos x+a^2)\cos nx\cos x\,dx =$$
$$= \frac{1}{2}\int_0^{2\pi} \ln(1-2a\cos x+a^2)\cos nx\cos x\,dx = -\frac{\pi}{2}\left(\frac{a^{n+1}}{n+1} + \frac{a^{n-1}}{n-1}\right)$$
$$[a^2 < 1].$$ BI((330))(12), BI((332))(6)

9. $$\int_0^{\pi} \ln(1-2a\cos 2x+a^2)\cos(2n-1)x\,dx = 0 \quad [a^2 < 1].$$ BI ((330))(15)

10. $$\int_0^{\pi} \ln(1-2a\cos 2x+a^2)\sin 2nx\sin x\,dx = 0 \quad [a^2 < 1].$$ BI ((330))(13)

11. $\int_0^\pi \ln(1-2a\cos 2x+a^2)\sin(2n-1)x\sin x\,dx =$

$$= \frac{\pi}{2}\left(\frac{a^n}{n}-\frac{a^{n-1}}{n-1}\right) \quad [a^2<1].$$ BI ((330))(14)

12. $\int_0^\pi \ln(1-2a\cos 2x+a^2)\cos 2nx\cos x\,dx = 0 \quad [a^2<1].$ BI ((330))(16)

13. $\int_0^\pi \ln(1-2a\cos 2x+a^2)\cos(2n-1)x\cos x\,dx =$

$$= -\frac{\pi}{2}\left(\frac{a^n}{n}+\frac{a^{n-1}}{n-1}\right) \quad [a^2<1].$$ BI ((330))(17)

14. $\int_0^{\frac{\pi}{2}} \ln(1+2a\cos 2x+a^2)\sin^2 x\,dx = -\frac{a\pi}{4} \quad [a^2<1];$

$$= \frac{\pi\ln a^2}{4}-\frac{\pi}{4a} \quad [a^2>1].$$ BI((309))(22), LI((309))(22)

15. $\int_0^{\frac{\pi}{2}} \ln(1+2a\cos 2x+a^2)\cos^2 x\,dx = \frac{a\pi}{4} \quad [a^2<1];$

$$= \frac{\pi\ln a^2}{4}+\frac{\pi}{4a} \quad [a^2>1].$$ BI((309))(23), LI((309))(23)

16. $\int_0^\pi \frac{\ln(1-2a\cos x+a^2)}{1-2b\cos x+b^2}\,dx = \frac{2\pi\ln(1-ab)}{1-b^2} \quad [a^2\leqslant 1,\ b^2<1].$ BI((331))(26)

4.398

1. $\int_0^\pi \ln\frac{1+2a\cos x+a^2}{1-2a\cos x+a^2}\sin(2n+1)x\,dx = (-1)^n\frac{2\pi a^{2n+1}}{2n+1}$

$$[a^2<1].$$ BI ((330))(18)

2. $\int_0^{2\pi} \ln\frac{1-2a\cos x+a^2}{1-2a\cos nx+a^2}\cos mx\,dx = 2\pi\left(\frac{n}{m}a^{\frac{m}{n}}-\frac{a^m}{m}\right) \quad [a^2\leqslant 1];$

$$= 2\pi\left(\frac{n}{m}a^{-\frac{m}{n}}-\frac{a^{-m}}{m}\right) \quad [a^2\geqslant 1].$$ BI ((332))(9)

3. $\int_0^\pi \ln\frac{1+2a\cos 2x+a^2}{1+2a\cos 2nx+a^2}\operatorname{ctg} x\,dx = 0.$ BI((331))(5), LI((331))(5)

4.399

1. $\int_0^{\frac{\pi}{2}} \ln(1+a\sin^2 x)\sin^2 x\,dx = \frac{\pi}{2}\left(\ln\frac{1+\sqrt{1+a}}{2}-\frac{1}{2}\,\frac{1-\sqrt{1+a}}{1+\sqrt{1+a}}\right)$

$$[a>-1].$$ BI ((309))(14)

2. $$\int_0^{\frac{\pi}{2}} \ln(1+a\sin^2 x)\cos^2 x\,dx = \frac{\pi}{2}\left(\ln\frac{1+\sqrt{1+a}}{2} + \frac{1}{2}\frac{1-\sqrt{1+a}}{1+\sqrt{1+a}}\right) \quad [a > -1].$$ BI ((309))(15)

3. $$\int_0^{\frac{\pi}{2}} \frac{\ln(1-\cos^2\beta\cos^2 x)}{1-\cos^2\alpha\cos^2 x}\,dx = -\frac{\pi}{\sin\alpha}\ln\frac{1+\sin\alpha}{\sin\alpha+\sin\beta}$$
$$\left[0<\beta<\frac{\pi}{2},\ 0<\alpha<\frac{\pi}{2}\right].$$ LO III 285

4.411

1. $$\int_0^{\pi} \ln\frac{1+\sin x}{1+\cos\lambda\sin x}\frac{dx}{\sin x} = \lambda^2 \quad [\lambda^2 < \pi^2].$$ BI ((331))(2)

2. $$\int_0^{\frac{\pi}{2}} \ln\frac{p+q\sin ax}{p-q\sin ax}\frac{dx}{\sin ax} = \int_0^{\frac{\pi}{2}} \ln\frac{p+q\cos ax}{p-q\cos ax}\frac{dx}{\cos ax} =$$
$$= \int_0^{\frac{\pi}{2}} \ln\frac{p+q\operatorname{tg} ax}{p-q\operatorname{tg} ax}\frac{dx}{\operatorname{tg} ax} = \pi\arcsin\frac{q}{p} \quad [p > q > 0].$$
FI II 695a, BI ((315))(5, 13, 17)a

3. $$\int_0^{\frac{\pi}{2}} \frac{\cos x}{1-\cos^2\alpha\cos^2 x}\ln\frac{1+\cos\beta\cos x}{1-\cos\beta\cos x}\,dx = \frac{2\pi}{\sin 2\alpha}\ln\frac{\cos\frac{\alpha-\beta}{2}}{\sin\frac{\alpha+\beta}{2}}$$
$$\left[0<\alpha\leqslant\beta<\frac{\pi}{2}\right].$$ LO III 284

4.412

1. $$\int_0^{\frac{\pi}{4}} \ln\operatorname{tg}\left(\frac{\pi}{4}\pm x\right)\frac{dx}{\sin 2x} = \pm\frac{\pi^2}{8}.$$ BI ((293))(1)

2. $$\int_0^{\frac{\pi}{4}} \ln\operatorname{tg}\left(\frac{\pi}{4}\pm x\right)\frac{dx}{\operatorname{tg} 2x} = \pm\frac{\pi^2}{16}.$$ BI ((293))(2)

3. $$\int_0^{\frac{\pi}{4}} \ln\operatorname{tg}\left(\frac{\pi}{4}\pm x\right)(\ln\operatorname{tg} x)^{2n}\frac{dx}{\sin 2x} = \pm\frac{2^{2n+2}-1}{4(n+1)(2n+1)}\pi^{2n+2}|B_{2n+2}|.$$
BI ((294))(24)

4. $$\int_0^{\frac{\pi}{4}} \ln\operatorname{tg}\left(\frac{\pi}{4}\pm x\right)(\ln\operatorname{tg} x)^{2n-1}\frac{dx}{\sin 2x} = \pm\frac{1-2^{2n+1}}{2^{2n+2}n}(2n)!\,\zeta(2n+1).$$
BI ((294))(25)

5. $\int_0^{\frac{\pi}{4}} \ln \operatorname{tg}\left(\frac{\pi}{4} \pm x\right)(\ln \sin 2x)^{n-1} \frac{dx}{\operatorname{tg} 2x} = \frac{(-1)^{n-1}}{2}(n-1)!\,\zeta(n+1).$ LI ((294))(20)

4.413

1. $\int_0^{\frac{\pi}{2}} \ln(p^2 + q^2 \operatorname{tg}^2 x) \frac{dx}{a^2 \sin^2 x + b^2 \cos^2 x} = \frac{\pi}{ab} \ln \frac{ap+bq}{a}$

$[a > 0,\ b > 0,\ p > 0,\ q > 0].$ BI ((318))(1-4)a

2. $\int_0^{\frac{\pi}{2}} \ln(1 + q^2 \operatorname{tg}^2 x) \frac{1}{p^2 \sin^2 x + r^2 \cos^2 x} \frac{dx}{s^2 \sin^2 x + t^2 \cos^2 x} =$

$= \frac{\pi}{p^2t^2 - s^2r^2} \left\{ \frac{p^2 - r^2}{pr} \ln\left(1 + \frac{qr}{p}\right) + \frac{t^2 - s^2}{st} \ln\left(1 + \frac{qt}{s}\right) \right\}$

$[q > 0,\ p > 0,\ r > 0,\ s > 0,\ t > 0].$ BI ((320))(18)

3. $\int_0^{\frac{\pi}{2}} \ln(1 + q^2 \operatorname{tg}^2 x) \frac{\sin^2 x}{p^2 \sin^2 x + r^2 \cos^2 x} \frac{dx}{s^2 \sin^2 x + t^2 \cos^2 x} =$

$= \frac{\pi}{p^2t^2 - s^2r^2} \left\{ \frac{t}{s} \ln\left(1 + \frac{qt}{s}\right) - \frac{r}{p} \ln\left(1 + \frac{qr}{p}\right) \right\}$

$[q > 0,\ p > 0,\ r > 0,\ s > 0,\ t > 0].$ BI ((320))(20)

4. $\int_0^{\frac{\pi}{2}} \ln(1 + q^2 \operatorname{tg}^2 x) \frac{\cos^2 x}{p^2 \sin^2 x + r^2 \cos^2 x} \frac{dx}{s^2 \sin^2 x + t^2 \cos^2 x} =$

$= \frac{\pi}{p^2t^2 - s^2r^2} \left\{ \frac{p}{r} \ln\left(1 + \frac{qr}{p}\right) - \frac{s}{t} \ln\left(1 + \frac{qt}{s}\right) \right\}$

$[q > 0,\ p > 0,\ r > 0,\ s > 0,\ t > 0].$ BI ((320))(21)

5. $\int_0^{\pi} \frac{\ln \operatorname{tg} rx\, dx}{1 - 2p\cos x + p^2} = \frac{\pi}{1-p^2} \ln \frac{1-p^{2r}}{1+p^{2r}} \qquad [p^2 < 1].$ BI ((331))(12)

4.414

1. $\int_0^{\frac{\pi}{2}} \ln(1 - k^2 \sin^2 x) \frac{dx}{\sqrt{1 - k^2 \sin^2 x}} = \ln k' \boldsymbol{K}(k).$ BI ((323))(1)

2. $\int_0^{\frac{\pi}{2}} \ln(1 - k^2 \sin^2 x) \frac{\sin^2 x\, dx}{\sqrt{1 - k^2 \sin^2 x}} = \frac{1}{k^2}\{(k^2 - 2 + \ln k') \boldsymbol{K}(k) +$

$+ (2 - \ln k') \boldsymbol{E}(k)\}.$ BI ((323))(3)

3. $\int_0^{\frac{\pi}{2}} \ln(1 - k^2 \sin^2 x) \frac{\cos^2 x\, dx}{\sqrt{1 - k^2 \sin^2 x}} = \frac{1}{k^2}[(1 + k'^2 - k'^2 \ln k') \boldsymbol{K}(k) -$

$- (2 - \ln k') \boldsymbol{E}(k)].$ BI ((323))(6)

4. $$\int_0^{\frac{\pi}{2}} \ln(1-k^2\sin^2 x)\frac{dx}{\sqrt{(1-k^2\sin^2 x)^3}} = \frac{1}{k'^2}[(k^2-2)\boldsymbol{K}(k) + (2+\ln k')\boldsymbol{E}(k)].$$ BI ((323))(9)

5. $$\int_0^{\frac{\pi}{2}} \ln(1-k^2\sin^2 x)\frac{\sin^2 x\,dx}{\sqrt{(1-k^2\sin^2 x)^3}} = \frac{1}{k^2k'^2}[(2+\ln k')\boldsymbol{E}(k) - (1+k'^2+k'^2\ln k')\boldsymbol{K}(k)].$$ BI ((323))(10)

6. $$\int_0^{\frac{\pi}{2}} \ln(1-k^2\sin^2 x)\frac{\cos^2 x\,dx}{\sqrt{(1-k^2\sin^2 x)^3}} = \frac{1}{k^2}[(1+k'^2+\ln k')\boldsymbol{K}(k) - (2+\ln k')\boldsymbol{E}(k)].$$ BI ((323))(16)

7. $$\int_0^{\frac{\pi}{2}} \ln(1-k^2\sin^2 x)\sqrt{1-k^2\sin^2 x}\,dx = (1+k'^2)\boldsymbol{K}(k) - (2-\ln k')\boldsymbol{E}(k).$$ BI ((324))(18)

8. $$\int_0^{\frac{\pi}{2}} \ln(1-k^2\sin^2 x)\sin^2 x\sqrt{1-k^2\sin^2 x}\,dx = \frac{1}{9k^2}\{(-2+11k^2-6k^4+3k'^2\ln k')\boldsymbol{K}(k) + [2-10k^2 - 3(1-2k^2)\ln k']\boldsymbol{E}(k)\}.$$ BI ((324))(20)

9. $$\int_0^{\frac{\pi}{2}} \ln(1-k^2\sin^2 x)\cos^2 x\sqrt{1-k^2\sin^2 x}\,dx = \frac{1}{9k^2}\{(2+7k^2-3k^4-3k'^2\ln k')\boldsymbol{K}(k) - [2+8k^2-3(1+k^2)\ln k']\boldsymbol{E}(k)\}.$$ BI ((324))(21), LI ((324))(21)

10. $$\int_0^{\frac{\pi}{2}} \ln(1-k^2\sin^2 x)\frac{\sin x\cos x\,dx}{\sqrt{(1-k^2\sin^2 x)^{2n+1}}} = \frac{2}{(2n-1)^2k^2}\{[1+(2n-1)\ln k']k'^{1-2n} - 1\}.$$ BI ((324))(17)

4.415

1. $$\int_0^{\infty} \ln x \sin ax^2\,dx = -\frac{1}{4}\sqrt{\frac{\pi}{2a}}\left(\ln 4a + \boldsymbol{C} - \frac{\pi}{2}\right) \quad [a>0].$$ GW ((338))(19)

2. $$\int_0^{\infty} \ln x \cos ax^2\,dx = -\frac{1}{4}\sqrt{\frac{\pi}{2a}}\left(\ln 4a + \boldsymbol{C} + \frac{\pi}{2}\right) \quad [a>0].$$ GW ((338))(19)

4.416

1. $$\int_0^{\frac{\pi}{2}} \frac{\cos x \ln\left(1+\sqrt{\sin^2\beta-\cos^2\beta \operatorname{tg}^2\alpha \sin^2 x}\right)}{1-\sin^2\alpha\cos^2 x}\,dx =$$
$$=\operatorname{cosec} 2\alpha\,\{(2\alpha+2\gamma-\pi)\ln\cos\beta+2L(\alpha)-2L(\gamma)+L(\alpha+\gamma)-L(\alpha-\gamma)\}$$
$$\left[\cos\gamma=\frac{\sin\alpha}{\sin\beta};\ 0<\alpha<\beta<\frac{\pi}{2}\right].$$ LO III 291

2. $$\int_0^{\frac{\pi}{2}} \frac{\cos x \ln\left(1-\sqrt{\sin^2\beta-\cos^2\beta \operatorname{tg}^2\alpha \sin^2 x}\right)}{1-\sin^2\alpha\cos^2 x}\,dx =$$
$$=\operatorname{cosec} 2\alpha\,\{(\pi+2\alpha-2\gamma)\ln\cos\beta+2L(\alpha)+2L(\gamma)-L(\alpha+\gamma)+L(\alpha-\gamma)\}$$
$$\left[\cos\gamma=\frac{\sin\alpha}{\sin\beta};\ 0<\alpha<\beta<\frac{\pi}{2}\right].$$ LO III 291

3. $$\int_\beta^{\frac{\pi}{2}} \frac{\ln\left(\sin x+\sqrt{\sin^2 x-\sin^2\beta}\right)}{1-\cos^2\alpha\cos^2 x}\,dx =$$
$$=-\operatorname{cosec}\alpha\left\{\operatorname{arctg}\left(\frac{\operatorname{tg}\beta}{\sin\alpha}\right)\ln\sin\beta+\frac{\pi}{2}\ln\frac{1+\sin\alpha}{\sin\alpha+\sqrt{1-\cos^2\alpha\cos^2\beta}}\right\}$$
$$\left[0<\alpha<\pi,\ 0<\beta<\frac{\pi}{2}\right].$$ LO III 285

4. $$\int_0^{\frac{\pi}{4}} \ln\operatorname{tg} x\,(\ln\cos 2x)^{n-1}\operatorname{tg} 2x\,dx=(-1)^{n-1}\frac{(n-1)!}{4}\sum_{k=0}^{\infty}\frac{1}{(1+2k)^{n+1}}=$$
$$=(-1)^{n-1}\frac{(n-1)!}{2^{n+3}}\zeta\left(n+1,\frac{1}{2}\right).$$ BI ((287))(20)

4.42-4.43 Combinations of logarithms, trigonometric functions, and powers

4.421

1. $$\int_0^\infty \ln x\sin ax\frac{dx}{x}=-\frac{\pi}{2}(C+\ln a)\qquad [a>0].$$ FI II 810a

2. $$\int_0^\infty \ln ax\sin bx\frac{x\,dx}{\beta^2+x^2}=\frac{\pi}{2}e^{-b\beta'}\ln(a\beta')-$$
$$-\frac{\pi}{4}\left[e^{b\beta'}\operatorname{Ei}(-b\beta')+e^{-b\beta'}\operatorname{Ei}(b\beta')\right]\quad [\beta'=\beta\operatorname{sign}\beta;\ a>0,\ b>0].$$ ET I 76(5), NT 27(10)a

3. $$\int_0^\infty \ln ax\cos bx\frac{\beta'\,dx}{\beta^2+x^2}=\frac{\pi}{2}e^{-b\beta'}\ln(a\beta')+$$
$$+\frac{\pi}{4}\left[e^{b\beta'}\operatorname{Ei}(-b\beta')-e^{-b\beta'}\operatorname{Ei}(b\beta')\right]\quad [\beta'=\beta\operatorname{sign}\beta;\ a>0,\ b>0].$$ ET I 17(3), NT 27(11)a

4. $$\int_0^\infty \ln ax \sin bx \frac{x\,dx}{x^2-c^2} = \frac{\pi}{2}\{-\operatorname{si}(bc)\sin bc + \cos bc\,[\ln ac - \operatorname{ci}(bc)]\} \qquad [a>0,\ b>0,\ c>0].$$ BI ((422))(5)

5. $$\int_0^\infty \ln ax \cos bx \frac{dx}{x^2-c^2} = \frac{\pi}{2c}\{\sin bc\,[\operatorname{ci}(bc) - \ln ac] - \cos bc \operatorname{si}(bc)\} \qquad [a>0,\ b>0,\ c>0].$$ BI ((422))(6)

4.422

1. $$\int_0^\infty \ln x \sin ax\, x^{\mu-1}\,dx = \frac{\Gamma(\mu)}{a^\mu}\sin\frac{\mu\pi}{2}\left[\psi(\mu) - \ln a + \frac{\pi}{2}\operatorname{ctg}\frac{\mu\pi}{2}\right] \qquad [a>0,\ |\operatorname{Re}\mu|<1].$$ BI ((411))(5)

2. $$\int_0^\infty \ln x \cos ax\, x^{\mu-1}\,dx = \frac{\Gamma(\mu)}{a^\mu}\cos\frac{\mu\pi}{2}\left[\psi(\mu) - \ln a - \frac{\pi}{2}\operatorname{tg}\frac{\mu\pi}{2}\right] \qquad [a>0,\ 0<\operatorname{Re}\mu<1].$$ BI ((411))(6)

4.423

1. $$\int_0^\infty \ln x \frac{\cos ax - \cos bx}{x}\,dx = \ln\frac{a}{b}\left(C + \frac{1}{2}\ln ab\right) \qquad [a>0,\ b>0].$$ GW ((338))(21a)

2. $$\int_0^\infty \ln x \frac{\cos ax - \cos bx}{x^2}\,dx = \frac{\pi}{2}[(a-b)(C-1) + a\ln a - b\ln b] \qquad [a>0,\ b>0].$$ GW ((338))(21b)

3. $$\int_0^\infty \ln x \frac{\sin^2 ax}{x^2}\,dx = -\frac{a\pi}{2}(C + \ln 2a - 1) \qquad [a>0].$$ GW ((338))(20b)

4.424

1. $$\int_0^\infty (\ln x)^2 \sin ax \frac{dx}{x} = \frac{\pi}{2}C^2 + \frac{\pi^3}{24} + \pi C\ln a + \frac{\pi}{2}(\ln a)^2 \qquad [a>0].$$ ET I 77(9), FI II 810a

2. $$\int_0^\infty (\ln x)^2 \sin ax\, x^{\mu-1}\,dx = \frac{\Gamma(\mu)}{a^\mu}\sin\frac{\mu\pi}{2}\Big[\psi'(\mu) + \psi^2(\mu) + \pi\psi(\mu)\operatorname{ctg}\frac{\mu\pi}{2} - 2\psi(\mu)\ln a - \pi\ln a\operatorname{ctg}\frac{\mu\pi}{2} + (\ln a)^2 - \pi^2\Big] \qquad [a>0,\ 0<\operatorname{Re}\mu<1].$$ ET I 77(10)

4.425

1. $$\int_0^\infty \ln(1+x)\cos ax\frac{dx}{x} = \frac{1}{2}\{[\operatorname{si}(a)]^2 + [\operatorname{ci}(a)]^2\} \qquad [a>0].$$ ET I 18(8)

2. $$\int_0^\infty \ln\left(\frac{b+x}{b-x}\right)^2 \cos ax\frac{dx}{x} = -2\pi\operatorname{si}(ab) \qquad [a>0,\ b>0].$$ ET I 18(11)

3. $\int_0^\infty \ln(1+b^2x^2)\sin ax \frac{dx}{x} = -\pi \operatorname{Ei}\left(-\frac{a}{b}\right) \qquad [a>0,\ b>0].$

GW ((338))(24), ET I 77(14)

4. $\int_0^1 \ln(1-x^2)\cos(p\ln x)\frac{dx}{x} = \frac{1}{2p^2} + \frac{\pi}{2p}\operatorname{cth}\frac{p\pi}{2}.$ LI ((309))(1)a

4.426

1. $\int_0^\infty \ln\frac{b^2+x^2}{c^2+x^2}\sin ax\, x\,dx = \frac{\pi}{a^2}\left[(1+ac)e^{-ac} - (1+ab)e^{-ab}\right]$

$[b \geqslant 0,\ c \geqslant 0,\ a>0].$ GW ((338))(23)

2. $\int_0^\infty \ln\frac{b^2x^2+p^2}{c^2x^2+p^2}\sin ax\frac{dx}{x} = \pi\left[\operatorname{Ei}\left(-\frac{ap}{c}\right) - \operatorname{Ei}\left(-\frac{ap}{b}\right)\right]$

$[b>0,\ c>0,\ p>0,\ a>0].$ ET I 77(15)

4.427 $\int_0^\infty \ln\left(x+\sqrt{\beta^2+x^2}\right)\frac{\sin ax}{\sqrt{\beta^2+x^2}}dx = \frac{\pi}{2}K_0(a\beta) +$

$+\frac{\pi}{2}\ln(\beta)\left[I_0(a\beta) - \mathbf{L}(a\beta)\right] \qquad [\operatorname{Re}\beta>0,\ a>0].$ ET I 77(16)

4.428

1. $\int_0^\infty \ln\cos^2 ax\frac{\cos bx}{x^2}dx = \pi b\ln 2 - a\pi \qquad [a>0,\ b>0].$ ET I 22(29)

2. $\int_0^\infty \ln(4\cos^2 ax)\frac{\cos bx}{x^2+c^2}dx = \frac{\pi}{c}\operatorname{ch}(bc)\ln(1+e^{-2ac})$

$\left[0<b<2a<\frac{\pi}{c}\right].$ ET I 22(30)

3. $\int_0^\infty \ln\cos^2 ax\frac{\sin bx}{x(1+x^2)}dx = \pi\ln(1+e^{-2a})\operatorname{sh} b -$

$-\pi\ln 2(1-e^{-b}) \qquad [a>0,\ b>0].$ ET I 82(36)

4. $\int_0^\infty \ln\cos^2 ax\frac{\cos bx}{x^2(1+x^2)}dx = -\pi\ln(1+e^{-2a})\operatorname{ch} b +$

$+(b+e^{-b})\pi\ln 2 - a\pi \qquad [a>0,\ b>0].$ ET I 22(31)

4.429 $\int_0^1 \frac{(1+x)x}{\ln x}\sin(\ln x)\,dx = \frac{\pi}{4}.$ BI ((326))(2)a

4.431

1. $\int_0^\infty \ln(2\pm 2\cos x)\frac{\sin bx}{x^2+c^2}x\,dx = -\pi\operatorname{sh}(bc)\ln(1\pm e^{-c})$

$[b>0,\ c>0].$ ET I 22(32)

2. $$\int_0^\infty \ln(2 \pm 2\cos x)\frac{\cos bx}{x^2+c^2}\,dx = \frac{\pi}{c}\operatorname{ch}(bc)\ln(1 \pm e^{-c})$$
$$[b > 0,\ c > 0].$$ ET I 22(32)

3. $$\int_0^\infty \ln(1+2a\cos x+a^2)\frac{\sin bx}{x}\,dx =$$
$$= -\frac{\pi}{2}\sum_{k=1}^{E(b)}\frac{(-a)^k}{k}[1+\operatorname{sign}(b-k)] \quad [0<a<1,\ b>0].$$ ET I 82(35)

4. $$\int_0^\infty \ln(1-2a\cos x+a^2)\frac{\cos bx}{x^2+c^2}\,dx =$$
$$= \frac{\pi}{c}\ln(1-ae^{-c})\operatorname{ch}(bc) + \frac{\pi}{c}\sum_{k=1}^{E(b)}{}' \frac{a^k}{k}\operatorname{sh}[c(b-k)]$$
$$[|a|<1,\ b>0,\ c>0].$$ ET I 22(33)

4.432

1. $$\int_0^\infty \ln(1-k^2\sin^2 x)\frac{\sin x}{\sqrt{1-k^2\sin^2 x}}\frac{dx}{x} =$$
$$= \int_0^\infty \ln(1-k^2\cos^2 x)\frac{\sin x}{\sqrt{1-k^2\cos^2 x}}\frac{dx}{x} = \ln k'\,\boldsymbol{K}(k).$$ BI ((412, 414))(4)

2. $$\int_0^{\frac{\pi}{2}} \ln(1-k^2\sin^2 x)\frac{\sin x\cos x}{\sqrt{1-k^2\sin^2 x}}\,x\,dx =$$
$$= \frac{1}{k^2}\{\pi k'(1-\ln k') + (2-k^2)\boldsymbol{K}(k) - (4-\ln k')\boldsymbol{E}(k)\}.$$ BI ((426))(3)

3. $$\int_0^{\frac{\pi}{2}} \ln(1-k^2\cos^2 x)\frac{\sin x\cos x}{\sqrt{1-k^2\cos^2 x}}\,x\,dx =$$
$$= \frac{1}{k^2}\{-\pi - (2-k^2)\boldsymbol{K}(k) + (4-\ln k')\boldsymbol{E}(k)\}.$$ BI ((426))(6)

4. $$\int_0^\infty \ln(1-k^2\sin^2 x)\frac{\sin x\cos x}{\sqrt{1-k^2\sin^2 x}}\frac{dx}{x} =$$
$$= \frac{1}{k^2}\{(2-k^2-k'^2\ln k')\boldsymbol{K}(k) - (2-\ln k')\boldsymbol{E}(k)\}.$$ BI ((412))(5)

5. $$\int_0^\infty \ln(1-k^2\cos^2 x)\frac{\sin x\cos x}{\sqrt{1-k^2\cos^2 x}}\frac{dx}{x} =$$
$$= \frac{1}{k^2}\{(k^2-2+\ln k')\boldsymbol{K}(k) + (2-\ln k')\boldsymbol{E}(k)\}.$$ BI ((414))(5)

6. $\int\limits_0^\infty \ln(1 \pm k\sin^2 x)\frac{\sin x}{\sqrt{1-k^2\sin^2 x}}\frac{dx}{x} =$

$= \int\limits_0^\infty \ln(1 \pm k\cos^2 x)\frac{\sin x}{\sqrt{1-k^2\cos^2 x}}\frac{dx}{x} =$

$= \int\limits_0^\infty \ln(1 \pm k\sin^2 x)\frac{\operatorname{tg} x}{\sqrt{1-k^2\sin^2 x}}\frac{dx}{x} =$

$= \int\limits_0^\infty \ln(1 \pm k\cos^2 x)\frac{\operatorname{tg} x}{\sqrt{1-k^2\cos^2 x}}\frac{dx}{x} =$

$= \int\limits_0^\infty \ln(1 \pm k\sin^2 2x)\frac{\operatorname{tg} x}{\sqrt{1-k^2\sin^2 2x}}\frac{dx}{x} =$

$= \int\limits_0^\infty \ln(1 \pm k^2\cos^2 2x)\frac{\operatorname{tg} x}{\sqrt{1-k^2\cos^2 2x}}\frac{dx}{x} =$

$= \frac{1}{2}\ln\frac{2(1 \pm k)}{\sqrt{k}}\boldsymbol{K}(k) - \frac{\pi}{8}\boldsymbol{K}(k').$

BI ((413))(1-6), BI ((415))(1-6)

7. $\int\limits_0^\infty \ln(1-k^2\sin^2 x)\frac{\sin^3 x}{\sqrt{1-k^2\sin^2 x}}\frac{dx}{x} =$

$= \frac{1}{k^2}\{(k^2-2+\ln k')\boldsymbol{K}(k) + (2-\ln k')\boldsymbol{E}(k)\}.$ BI ((412))(6)

8. $\int\limits_0^\infty \ln(1-k^2\cos^2 x)\frac{\sin^3 x}{\sqrt{1-k^2\cos^2 x}}\frac{dx}{x} =$

$= \frac{1}{k^2}\{(2-k^2-k'^2\ln k')\boldsymbol{K}(k) - (2-\ln k')\boldsymbol{E}(k)\}.$ BI ((414))(6)a

9. $\int\limits_0^\infty \ln(1-k^2\sin^2 x)\frac{\sin x\cos^2 x}{\sqrt{1-k^2\sin^2 x}}\frac{dx}{x} =$

$= \frac{1}{k^2}\{(2-k^2-k'^2\ln k')\boldsymbol{K}(k) - (2-\ln k')\boldsymbol{E}(k)\}.$ BI ((412))(7)

10. $\int\limits_0^\infty \ln(1-k^2\cos^2 x)\frac{\sin x\cos^2 x}{\sqrt{1-k^2\cos^2 x}}\frac{dx}{x} =$

$= \frac{1}{k^2}\{(k^2-2+\ln k')\boldsymbol{K}(k) + (2-\ln k')\boldsymbol{E}(k)\}.$ BI ((414))(7)

11. $\int\limits_0^\infty \ln(1-k^2\sin^2 x)\frac{\operatorname{tg} x}{\sqrt{1-k^2\sin^2 x}}\frac{dx}{x} =$

$= \int\limits_0^\infty \ln(1-k^2\cos^2 x)\frac{\operatorname{tg} x}{\sqrt{1-k^2\cos^2 x}}\frac{dx}{x} = \ln k'\boldsymbol{K}(k).$ BI ((412, 414))(9)

12. $$\int_0^\infty \ln(1-k^2\sin^2 x)\frac{\sin^2 x \operatorname{tg} x}{\sqrt{1-k^2\sin^2 x}}\frac{dx}{x} =$$
$$= \frac{1}{k^2}\{(k^2-2+\ln k')\boldsymbol{K}(k)+(2-\ln k')\boldsymbol{E}(k)\}.$$ BI ((412))(8)

13. $$\int_0^\infty \ln(1-k^2\cos^2 x)\frac{\sin^2 x \operatorname{tg} x}{\sqrt{1-k^2\cos^2 x}}\frac{dx}{x} =$$
$$= \frac{1}{k^2}\{(2-k^2-k'^2\ln k')\boldsymbol{K}(k)-(2-\ln k')\boldsymbol{E}(k)\}.$$ BI ((414))(8)

14. $$\int_0^\infty \ln(1-k^2\sin^2 x)\frac{\sin x}{\sqrt{(1-k^2\sin^2 x)^3}}\frac{dx}{x} =$$
$$= \int_0^\infty \ln(1-k^2\cos^2 x)\frac{\sin x}{\sqrt{(1-k^2\cos^2 x)^3}}\frac{dx}{x} =$$
$$= \frac{1}{k'^2}\{(k^2-2)\boldsymbol{K}(k)+(2+\ln k')\boldsymbol{E}(k)\}.$$ BI ((412, 414))(13)

15. $$\int_0^{\frac{\pi}{2}} \ln(1-k^2\sin^2 x)\frac{\sin x\cos x}{\sqrt{(1-k^2\sin^2 x)^3}}x\,dx =$$
$$= \frac{1}{k^2}\left\{(1+\ln k')\frac{\pi}{k'}-(2+\ln k')\boldsymbol{K}(k)\right\}.$$ BI ((426))(9)

16. $$\int_0^{\frac{\pi}{2}} \ln(1-k^2\cos^2 x)\frac{\sin x\cos x}{\sqrt{(1-k^2\cos^2 x)^3}}x\,dx =$$
$$= \frac{1}{k^2}\{-\pi+(2+\ln k')\boldsymbol{K}(k)\}.$$ BI ((426))(15)

17. $$\int_0^\infty \ln(1-k^2\sin^2 x)\frac{\sin x\cos x}{\sqrt{(1-k^2\sin^2 x)^3}}\frac{dx}{x} =$$
$$= \int_0^\infty \ln(1-k^2\cos^2 x)\frac{\sin^3 x}{\sqrt{(1-k^2\cos^2 x)^3}}\frac{dx}{x} =$$
$$= \frac{1}{k^2}\{(2-k^2+\ln k')\boldsymbol{K}(k)-(2+\ln k')\boldsymbol{E}(k)\}.$$
BI((412))(14), BI((414))(15)

18. $$\int_0^\infty \ln(1-k^2\sin^2 x)\frac{\sin^3 x}{\sqrt{(1-k^2\sin^2 x)^3}}\frac{dx}{x} =$$
$$= \int_0^\infty \ln(1-k^2\cos^2 x)\frac{\sin x\cos x}{\sqrt{(1-k^2\cos^2 x)^3}}\frac{dx}{x} =$$
$$= \frac{1}{k^2k'^2}\{(2+\ln k')\boldsymbol{E}(k)-(2-k^2+k'^2\ln k')\boldsymbol{K}(k)\}.$$
BI((412))(15), BI((414))(14)

19. $$\int_0^\infty \ln(1-k^2\sin^2 x)\frac{\sin x\cos^2 x}{\sqrt{(1-k^2\sin^2 x)^3}}\frac{dx}{x}=$$

$$=\int_0^\infty \ln(1-k^2\cos^2 x)\frac{\sin^2 x\operatorname{tg} x}{\sqrt{(1-k^2\cos^2 x)^3}}\frac{dx}{x}=$$

$$=\frac{1}{k^2}\{(2-k^2+\ln k')\boldsymbol{K}(k)-(2+\ln k')\boldsymbol{E}(k)\}.$$ BI((412))(16), BI((414)(17)

20. $$\int_0^\infty \ln(1-k^2\sin^2 x)\frac{\sin^2 x\operatorname{tg} x}{\sqrt{(1-k^2\sin^2 x)^3}}\frac{dx}{x}=$$

$$=\int_0^\infty \ln(1-k^2\cos^2 x)\frac{\sin x\cos^2 x}{\sqrt{(1-k^2\cos^2 x)^3}}\frac{dx}{x}=$$

$$=\frac{1}{k^2k'^2}\{(2+\ln k')\boldsymbol{E}(k)-(2-k^2+k'^2\ln k')\boldsymbol{K}(k)\}.$$

BI((412))(17), BI((414))(16)

21. $$\int_0^\infty \ln(1-k^2\sin^2 x)\frac{\operatorname{tg} x}{\sqrt{(1-k^2\sin^2 x)^3}}\frac{dx}{x}=$$

$$=\int_0^\infty \ln(1-k^2\cos^2 x)\frac{\operatorname{tg} x}{\sqrt{(1-k^2\cos^2 x)^3}}\frac{dx}{x}=$$

$$=\frac{1}{k'^2}\{(k^2-2)\boldsymbol{K}(k)+(2+\ln k')\boldsymbol{E}(k)\}.$$ BI ((412, 414))(18)

22. $$\int_0^\infty \ln(1-k^2\sin^2 x)\sqrt{1-k^2\sin^2 x}\,\sin x\,\frac{dx}{x}=$$

$$=\int_0^\infty \ln(1-k^2\cos^2 x)\sqrt{1-k^2\cos^2 x}\,\sin x\,\frac{dx}{x}=$$

$$=(2-k^2)\boldsymbol{K}(k)-(2-\ln k')\boldsymbol{E}(k).$$ BI ((412, 414))(1)

23. $$\int_0^{\frac{\pi}{2}} \ln(1-k^2\sin^2 x)\sqrt{1-k^2\sin^2 x}\,\sin x\cos x\cdot x\,dx=$$

$$=\frac{1}{27k^2}\{3\pi k'^3(1-3\ln k')+(22k'^2+6k^4-3k'^2\ln k')\boldsymbol{K}(k)-$$
$$-(2-k^2)(14-6\ln k')\boldsymbol{E}(k)\}.$$ BI ((426))(1)

24. $$\int_0^{\frac{\pi}{2}} \ln(1-k^2\cos^2 x)\sqrt{1-k^2\cos^2 x}\,\sin x\cos x\cdot x\,dx=\frac{1}{27k^2}\{-3\pi-$$

$$-(22k'^2+6k^4-3k'^2\ln k')\boldsymbol{K}(k)+(2-k^2)(14-6\ln k')\boldsymbol{E}(k)\}.$$

BI ((426))(2)

25. $\int_0^\infty \ln(1-k^2\sin^2 x)\sqrt{1-k^2\sin^2 x}\ \operatorname{tg} x \frac{dx}{x} =$

$= \int_0^\infty \ln(1-k^2\cos^2 x)\sqrt{1-k^2\cos^2 x}\ \operatorname{tg} x \frac{dx}{x} =$

$= (2-k^2)\, \boldsymbol{K}(k) - (2-\ln k')\, \boldsymbol{E}(k).$ BI ((412, 414))(2)

26. $\int_0^\infty \ln(\sin^2 x + k' \cos^2 x) \frac{\sin x}{\sqrt{1-k^2\cos^2 x}} \frac{dx}{x} =$

$= \int_0^\infty \ln(\sin^2 x + k' \cos^2 x) \frac{\operatorname{tg} x}{\sqrt{1-k^2\cos^2 x}} \frac{dx}{x} =$

$= \int_0^\infty \ln(\sin^2 2x + k' \cos^2 2x) \frac{\operatorname{tg} x}{\sqrt{1-k^2\cos^2 2x}} \frac{dx}{x} =$

$= \frac{1}{2} \ln\left[\frac{2(\sqrt{k'})^3}{1+k'}\right] \boldsymbol{K}(k).$ BI ((415))(19-21)

4.44 Combinations of logarithms, trigonometric functions, and exponentials

4.441

1. $\int_0^\infty e^{-qx} \sin px \ln x\, dx = \frac{1}{p^2+q^2}\left[q \operatorname{arctg}\frac{p}{q} - p\boldsymbol{C} + \frac{p}{2}\ln(p^2+q^2)\right]$

$[q>0,\ p>0].$ BI ((467))(1)

2. $\int_0^\infty e^{-qx} \cos px \ln x\, dx = -\frac{1}{p^2+q^2}\left[\frac{q}{2}\ln(p^2+q^2) + p \operatorname{arctg}\frac{p}{q} + q\boldsymbol{C}\right]$

$[q>0].$ BI ((467))(2)

4.442 $\int_0^{\frac{\pi}{2}} \frac{e^{-p \operatorname{tg} x} \ln \cos x\, dx}{\sin x \cos x} = -\frac{1}{2}[\operatorname{ci}(p)]^2 + \frac{1}{2}[\operatorname{si}(p)]^2 \quad [\operatorname{Re} p > 0].$ NT 32(11)

4.5 Inverse Trigonometric Functions

4.51 Inverse trigonometric functions

4.511 $\int_0^\infty \operatorname{arcctg} px \operatorname{arcctg} qx\, dx =$

$= \frac{\pi}{2}\left\{\frac{1}{p}\ln\left(1+\frac{p}{q}\right) + \frac{1}{q}\ln\left(1+\frac{q}{p}\right)\right\} \quad [p>0,\ q>0].$ BI ((77))(8)

4.512 $\int_0^\pi \operatorname{arctg}(\cos x)\, dx = 0.$ BI ((345))(1)

4.52 Combinations of arcsines, arccosines, and powers

4.521

1. $\int_0^1 \frac{\arcsin x}{x} dx = \frac{\pi}{2} \ln 2.$ FI II 614, 623

2. $\int_0^1 \frac{\arccos x}{1 \pm x} dx = \mp \frac{\pi}{2} \ln 2 + 2G.$ BI ((231))(7, 8)

3. $\int_0^1 \arcsin x \frac{x}{1+qx^2} dx = \frac{\pi}{2q} \ln \frac{2\sqrt{1+q}}{1+\sqrt{1+q}} \quad [q > -1].$ BI ((231))(1)

4. $\int_0^1 \arcsin x \frac{x}{1-p^2x^2} dx = \frac{\pi}{2p^2} \ln \frac{1+\sqrt{1-p^2}}{2\sqrt{1-p^2}} \quad [p^2 < 1].$ LI ((231))(3)

5. $\int_0^1 \arccos x \frac{dx}{\sin^2 \lambda - x^2} = 2 \operatorname{cosec} \lambda \sum_{k=0}^{\infty} \frac{\sin[(2k+1)\lambda]}{(2k+1)^2}.$ BI ((231))(10)

6. $\int_0^1 \arcsin x \frac{dx}{x(1+qx^2)} = \frac{\pi}{2} \ln \frac{1+\sqrt{1+q}}{\sqrt{1+q}} \quad [q > -1].$ BI ((235))(10)

7. $\int_0^1 \arcsin x \frac{x}{(1+qx^2)^2} dx = \frac{\pi}{4q} \frac{\sqrt{1+q}-1}{1+q} \quad [q > -1].$ BI ((234))(2)

8. $\int_0^1 \arccos x \frac{x}{(1+qx^2)^2} dx = \frac{\pi}{4q} \frac{\sqrt{1+q}-1}{\sqrt{1+q}} \quad [q > -1].$ BI ((234))(4)

4.522

1. $\int_0^1 x\sqrt{1-k^2x^2} \arccos x \, dx = \frac{1}{9k^2}\left[\frac{3}{2}\pi + k'^2 K(k) - 2(1+k'^2) E(k)\right].$ BI ((236))(9)

2. $\int_0^1 x\sqrt{1-k^2x^2} \arcsin x \, dx =$
$= \frac{1}{9k^2}\left[-\frac{3}{2}\pi k'^3 - k'^2 K(k) + 2(1+k'^2) E(k)\right].$ BI ((236))(1)

3. $\int_0^1 x\sqrt{k'^2+k^2x^2} \arcsin x \, dx = \frac{1}{9k^2}\left[\frac{3}{2}\pi + k'^2 K(k) - 2(1+k'^2) E(k)\right].$ BI ((236))(5)

4. $\int_0^1 \frac{x \arcsin x}{\sqrt{1-k^2x^2}} dx = \frac{1}{k^2}\left[-\frac{\pi}{2} k' + E(k)\right].$ BI ((237))(1)

5. $\int_0^1 \frac{x \arccos x}{\sqrt{1-k^2x^2}} dx = \frac{1}{k^2}\left[\frac{\pi}{2} - E(k)\right].$ BI ((240))(1)

6. $$\int_0^1 \frac{x \arcsin x}{\sqrt{k'^2+k^2x^2}}\,dx = \frac{1}{k^2}\left[\frac{\pi}{2} - \boldsymbol{E}(k)\right].$$ BI ((238))(1)

7. $$\int_0^1 \frac{x \arccos x}{\sqrt{k'^2+k^2x^2}}\,dx = \frac{1}{k^2}\left[-\frac{\pi}{2}k' + \boldsymbol{E}(k)\right].$$ BI ((241))(1)

8. $$\int_0^1 \frac{x \arcsin x\,dx}{(x^2-\cos^2\lambda)\sqrt{1-x^2}} = \frac{2}{\sin\lambda}\sum_{k=0}^{\infty}\frac{\sin[(2k+1)\lambda]}{(2k+1)^2}.$$ BI ((243))(11)

9. $$\int_0^1 \frac{x \arcsin kx}{\sqrt{(1-x^2)(1-k^2x^2)}}\,dx = -\frac{\pi}{2k}\ln k'.$$ BI ((239))(1)

10. $$\int_0^1 \frac{x \arccos kx}{\sqrt{(1-x^2)(1-k^2x^2)}}\,dx = \frac{\pi}{2k}\ln(1+k).$$ BI ((242))(1)

4.523

1. $$\int_0^1 x^{2n}\arcsin x\,dx = \frac{1}{2n+1}\left[\frac{\pi}{2} - \frac{2^n n!}{(2n+1)!!}\right].$$ BI ((229))(1)

2. $$\int_0^1 x^{2n-1}\arcsin x\,dx = \frac{\pi}{4n}\left[1 - \frac{(2n-1)!!}{2^n n!}\right].$$ BI ((229))(2)

3. $$\int_0^1 x^{2n}\arccos x\,dx = \frac{2^n n!}{(2n+1)(2n+1)!!}.$$ BI ((229))(4)

4. $$\int_0^1 x^{2n-1}\arccos x\,dx = \frac{\pi}{4n}\frac{(2n-1)!!}{2^n n!}.$$ BI ((229))(5)

5. $$\int_{-1}^1 (1-x^2)^n \arccos x\,dx = \pi\frac{2^n n!}{(2n+1)!!}.$$ BI ((254))(2)

6. $$\int_{-1}^1 (1-x^2)^{n-\frac{1}{2}}\arccos x\,dx = \frac{\pi^2}{2}\frac{(2n-1)!!}{2^n n!}.$$ BI ((254))(3)

4.524

1. $$\int_0^1 (\arcsin x)^2\frac{dx}{x^2\sqrt{1-x^2}} = \pi\ln 2.$$ BI ((243))(13)

2. $$\int_0^1 (\arccos x)^2\frac{dx}{(\sqrt{1-x^2})^3} = \pi\ln 2.$$ BI ((244))(9)

4.53-4.54 Combinations of arctangents, arccotangents, and powers

4.531

1. $$\int_0^1 \frac{\operatorname{arctg} x}{x}\,dx = \int_1^\infty \frac{\operatorname{arcctg} x}{x}\,dx = \boldsymbol{G}.$$ FI II 482, BI ((253))(8)

2. $$\int_0^\infty \frac{\operatorname{arcctg} x}{1 \pm x}\,dx = \pm \frac{\pi}{4} \ln 2 + G.$$ BI ((248))(6, 7)

3. $$\int_0^1 \frac{\operatorname{arctg} x}{x(1+x)}\,dx = -\frac{\pi}{8} \ln 2 + G.$$ BI ((235))(11)

4. $$\int_0^\infty \frac{\operatorname{arctg} x}{1-x^2}\,dx = -G.$$ BI ((248))(2)

5. $$\int_0^1 \operatorname{arctg} qx \frac{dx}{(1+px)^2} = \frac{1}{2} \frac{q}{p^2+q^2} \ln \frac{(1+p)^2}{1+q^2} + \frac{q^2-p}{(1+p)(p^2+q^2)} \operatorname{arctg} q$$ $$[p > -1].$$ BI ((243))(7)

6. $$\int_0^1 \operatorname{arcctg} qx \frac{dx}{(1+px)^2} = \frac{1}{2} \frac{q}{p^2+q^2} \ln \frac{1+q^2}{(1+p)^2} + \frac{p}{p^2+q^2} \operatorname{arctg} q + \frac{1}{1+p} \operatorname{arcctg} q \quad [p > -1].$$ BI ((234))(10)

7. $$\int_0^1 \frac{\operatorname{arctg} x}{x(1+x^2)}\,dx = \frac{\pi}{8} \ln 2 + \frac{1}{2} G.$$ BI ((235))(12)

8. $$\int_0^\infty \frac{x \operatorname{arctg} x}{1+x^4}\,dx = \frac{\pi^2}{16}.$$ BI ((248))(3)

9. $$\int_0^\infty \frac{x \operatorname{arctg} x}{1-x^4}\,dx = -\frac{\pi}{8} \ln 2.$$ BI ((248))(4)

10. $$\int_0^\infty \frac{x \operatorname{arcctg} x}{1-x^4}\,dx = \frac{\pi}{8} \ln 2.$$ BI ((248))(12)

11. $$\int_0^\infty \frac{\operatorname{arctg} x}{x\sqrt{1+x^2}}\,dx = \int_0^\infty \frac{\operatorname{arcctg} x}{\sqrt{1+x^2}}\,dx = 2G.$$ BI ((251))(3, 10)

12. $$\int_0^1 \frac{\operatorname{arctg} x}{x\sqrt{1-x^2}}\,dx = \frac{\pi}{2} \ln\left(1+\sqrt{2}\right).$$ FI II 694

13. $$\int_0^1 \frac{x \operatorname{arctg} x\,dx}{\sqrt{(1+x^2)(1+k'^2x^2)}} = \frac{1}{k^2}\left[F\left(\frac{\pi}{4}, k\right) - \frac{\pi}{2\sqrt{2(1+k'^2)}}\right].$$ BI ((244))(14)

4.532

1. $$\int_0^1 x^p \operatorname{arctg} x\,dx = \frac{1}{2(p+1)}\left[\frac{\pi}{2} - \beta\left(\frac{p}{2}+1\right)\right] \quad [p > -2].$$ BI ((229))(7)

2. $$\int_0^\infty x^p \operatorname{arctg} x\,dx = \frac{\pi}{2(p+1)} \operatorname{cosec} \frac{p\pi}{2} \quad [-1 > p > -2].$$ BI ((246))(1)

3. $\int_0^1 x^p \operatorname{arcctg} x \, dx = \frac{1}{2(p+1)} \left[\frac{\pi}{2} + \beta\left(\frac{p}{2}+1\right) \right]$

$[p > -1]$. BI ((229))(8)

4. $\int_0^\infty x^p \operatorname{arcctg} x \, dx = -\frac{\pi}{2(p+1)} \operatorname{cosec} \frac{p\pi}{2} \quad [-1 < p < 0]$. BI ((246))(2)

5. $\int_0^\infty \left(\frac{x^p}{1+x^{2p}} \right)^{2q} \operatorname{arctg} x \frac{dx}{x} = \frac{\sqrt{\pi^3}}{2^{2q+2} p} \frac{\Gamma(q)}{\Gamma\left(q+\frac{1}{2}\right)} \quad [q > 0]$. BI ((250))(10)

4.533

1. $\int_0^\infty (1 - x \operatorname{arcctg} x) \, dx = \frac{\pi}{4}$. BI ((246))(3)

2. $\int_0^1 \left(\frac{\pi}{4} - \operatorname{arctg} x \right) \frac{dx}{1-x} = -\frac{\pi}{8} \ln 2 + G$. BI ((232))(2)

3. $\int_0^1 \left(\frac{\pi}{4} - \operatorname{arctg} x \right) \frac{1+x}{1-x} \frac{dx}{1+x^2} = \frac{\pi}{8} \ln 2 + \frac{1}{2} G$. BI ((235))(25)

4. $\int_0^1 \left(x \operatorname{arcctg} x - \frac{1}{x} \operatorname{arctg} x \right) \frac{dx}{1-x^2} = -\frac{\pi}{4} \ln 2$. BI ((232))(1)

4.534 $\int_0^\infty (\operatorname{arctg} x)^2 \frac{dx}{x^2 \sqrt{1+x^2}} = \int_0^\infty (\operatorname{arcctg} x)^2 \frac{x \, dx}{\sqrt{1+x^2}} = -\frac{\pi^2}{4} + 4G$.

BI ((251))(9, 17)

4.535

1. $\int_0^1 \frac{\operatorname{arctg} px}{1+p^2 x} dx = \frac{1}{2p^2} \operatorname{arctg} p \ln(1+p^2)$. BI ((231))(19)

2. $\int_0^1 \frac{\operatorname{arcctg} px}{1+p^2 x} dx = \frac{1}{p^2} \left\{ \frac{\pi}{4} + \frac{1}{2} \operatorname{arcctg} p \right\} \ln(1+p^2) \quad [p > 0]$. BI ((231))(24)

3. $\int_0^\infty \frac{\operatorname{arctg} qx}{(p+x)^2} dx = -\frac{q}{1+p^2 q^2} \left(\ln pq - \frac{\pi}{2} pq \right)$

$[p > 0, \ q > 0]$. BI ((249))(1)

4. $\int_0^\infty \frac{\operatorname{arcctg} qx}{(p+x)^2} dx = \frac{q}{1+p^2 q^2} \left(\ln pq + \frac{\pi}{2pq} \right) \quad [p > 0, \ q > 0]$. BI ((249))(8)

5. $\int_0^\infty \frac{x \operatorname{arcctg} px}{q^2 + x^2} dx = \frac{\pi}{2} \ln \frac{1+pq}{pq} \quad [p > 0, \ q > 0]$. BI ((248))(9)

6. $\int_0^\infty \frac{x \operatorname{arcctg} px \, dx}{x^2 - q^2} = \frac{\pi}{4} \ln \frac{1+p^2 q^2}{p^2 q^2} \quad [p > 0, \ q > 0]$. BI ((248))(10)

7. $\int_0^\infty \frac{\operatorname{arctg} px}{x(1+x^2)}\,dx = \frac{\pi}{2}\ln(1+p) \quad [p \geqslant 0].$ FI II 745

8. $\int_0^\infty \frac{\operatorname{arctg} px}{x(1-x^2)}\,dx = \frac{\pi}{4}\ln(1+p^2) \quad [p \geqslant 0].$ BI ((250))(6)

9. $\int_0^\infty \operatorname{arctg} qx \frac{dx}{x(p^2+x^2)} = \frac{\pi}{2p^2}\ln(1+pq) \quad [p>0,\ q \geqslant 0].$ BI ((250))(3)

10. $\int_0^\infty \operatorname{arctg} qx \frac{dx}{x(1-p^2x^2)} = \frac{\pi}{4}\ln\frac{p^2+q^2}{p^2} \quad [q \geqslant 0].$ BI ((250))(6)

11. $\int_0^\infty \frac{x \operatorname{arctg} qx}{(p^2+x^2)^2}\,dx = \frac{\pi q}{4p(1+pq)} \quad [p>0,\ q \geqslant 0].$ BI ((252))(12)a

12. $\int_0^\infty \frac{x \operatorname{arcctg} qx}{(p^2+x^2)^2}\,dx = \frac{\pi}{4p^2(1+pq)} \quad [p>0,\ q \geqslant 0].$ BI ((252))(20)a

13. $\int_0^1 \frac{\operatorname{arctg} qx}{x\sqrt{1-x^2}}\,dx = \frac{\pi}{2}\ln\left(q+\sqrt{1+q^2}\right).$ BI ((244))(11)

4.536

1. $$\int_0^\infty \operatorname{arctg} qx \arcsin x \frac{dx}{x^2} = \frac{1}{2} q\pi \ln\frac{1+\sqrt{1+q^2}}{\sqrt{1+q^2}} + \frac{\pi}{2}\ln\left(q+\sqrt{1+q^2}\right) - \frac{\pi}{2}\operatorname{arctg} q.$$ BI ((230))(7)

2. $\int_0^\infty \frac{\operatorname{arctg} px - \operatorname{arctg} qx}{x}\,dx = \frac{\pi}{2}\ln\frac{p}{q} \quad [p>0,\ q>0].$ FI II 635

3. $\int_0^\infty \frac{\operatorname{arctg} px \operatorname{arctg} qx}{x^2}\,dx = \frac{\pi}{2}\ln\frac{(p+q)^{p+q}}{p^p q^q} \quad [p>0,\ q>0].$ FI II 745

4.537

1. $$\int_0^1 \operatorname{arctg}\left(\sqrt{1-x^2}\right)\frac{dx}{1-x^2\cos^2\lambda} = \frac{\pi}{\cos\lambda}\ln\left[\cos\left(\frac{\pi-4\lambda}{8}\right)\operatorname{cosec}\left(\frac{\pi+4\lambda}{8}\right)\right].$$ BI ((245))(9)

2. $$\int_0^1 \operatorname{arctg}\left(p\sqrt{1-x^2}\right)\frac{dx}{1-x^2} = \frac{1}{2}\pi\ln\left(p+\sqrt{1+p^2}\right) \quad [p>0].$$ BI ((245))(10)

3. $$\int_0^1 \operatorname{arctg}\left(\operatorname{tg}\lambda\sqrt{1-k^2x^2}\right)\sqrt{\frac{1-x^2}{1-k^2x^2}}\,dx = \frac{\pi}{2k^2}\left[E(\lambda, k) - k'^2 F(\lambda, k)\right] - \frac{\pi}{2k^2}\operatorname{ctg}\lambda\left(1-\sqrt{1-k^2\sin^2\lambda}\right).$$ BI ((245))(12)

4. $\int_0^1 \operatorname{arctg}\left(\operatorname{tg}\lambda\sqrt{1-k^2x^2}\right)\sqrt{\frac{1-k^2x^2}{1-x^2}}\,dx =$

$= \frac{\pi}{2}E(\lambda, k) - \frac{\pi}{2}\operatorname{ctg}\lambda\left(1-\sqrt{1-k^2\sin^2\lambda}\right).$ BI ((245))(11)

5. $\int_0^1 \frac{\operatorname{arctg}\left(\operatorname{tg}\lambda\sqrt{1-k^2x^2}\right)}{\sqrt{(1-x^2)(1-k^2x^2)}}\,dx = \frac{\pi}{2}F(\lambda, k).$ BI ((245))(13)

4.538

1. $\int_0^\infty \operatorname{arctg} x^2 \frac{dx}{1+x^2} = \int_0^\infty \operatorname{arctg} x^3 \frac{dx}{1+x^2};$ BI ((252))(10, 11)

$= \int_0^\infty \operatorname{arcctg} x^2 \frac{dx}{1+x^2} = \int_0^\infty \operatorname{arcctg} x^3 \frac{dx}{1+x^2} = \frac{\pi^2}{8}.$

BI ((252))(18, 19)

2. $\int_0^1 \frac{1-x^2}{x^2}\operatorname{arctg} x^2\,dx = \frac{\pi}{2}\left(\sqrt{2}-1\right).$ BI ((244))(10)a

4.539 $\int_0^\infty x^{s-1}\operatorname{arctg}(ae^{-x})\,dx = 2^{-s-1}\Gamma(s)\,a\Phi\left(-a^2, s+1, \frac{1}{2}\right).$ ET I 222(47)

4.541 $\int_0^\infty \operatorname{arctg}\left(\frac{p\sin qx}{1+p\cos qx}\right)\frac{x\,dx}{1+x^2} = \frac{\pi}{2}\ln\left(1+pe^{-q}\right) \quad [p > -e^q].$

BI ((341))(14)a

4.55 Combinations of inverse trigonometric functions and exponentials

4.551

1. $\int_0^1 (\arcsin x)\,e^{-bx}\,dx = \frac{\pi}{2b}\left[I_0(b) - \mathbf{L}_0(b)\right].$ ET I 160(1)

2. $\int_0^1 x(\arcsin x)\,e^{-bx}\,dx = \frac{\pi}{2b^2}\left[\mathbf{L}_0(b) - I_0(b) + b\mathbf{L}_1(b) - bI_1(b)\right] + \frac{1}{b}.$

ET I 161(2)

3. $\int_0^\infty \left(\operatorname{arctg}\frac{x}{a}\right)e^{-bx}\,dx = \frac{1}{b}\left[-\operatorname{ci}(ab)\sin(ab) - \operatorname{si}(ab)\cos(ab)\right]$

$[\operatorname{Re} b > 0].$ ET I 161(3)

4. $\int_0^\infty \left(\operatorname{arcctg}\frac{x}{a}\right)e^{-bx}\,dx = \frac{1}{b}\left[\frac{\pi}{2} + \operatorname{ci}(ab)\sin(ab) + \operatorname{si}(ab)\cos(ab)\right]$

$[\operatorname{Re} b > 0].$ ET I 161(4)

4.552 $\int_0^\infty \frac{\operatorname{arctg}\frac{x}{q}}{e^{2\pi x}-1}\,dx = \frac{1}{2}\left[\ln\Gamma(q) - \left(q-\frac{1}{2}\right)\ln q + q - \frac{1}{2}\ln 2\pi\right] \quad [q > 0].$

WH

4.553 $\int\limits_0^\infty \left(\frac{2}{\pi}\operatorname{arcctg} x - e^{-px}\right)\frac{dx}{x} = C + \ln p \quad [p>0].$ NT 66(12)

4.56 A combination of the arctangent and an inverse hyperbolic function

4.561 $\int\limits_{-\infty}^{\infty} \frac{\operatorname{arctg} e^{-x}}{\operatorname{ch}^{2q} px}\,dx = \frac{1}{2}\int\limits_{-\infty}^{\infty}\frac{\Pi(x)}{\operatorname{ch}^{2q} px}\,dx = \frac{\sqrt{\pi^3}}{4p}\frac{\Gamma(q)}{\Gamma\left(q+\frac{1}{2}\right)}$

$[q>0].$ LI ((282))(10)

4.57 Combinations of inverse and direct trigonometric functions

4.571 $\int\limits_0^{\frac{\pi}{2}} \arcsin(k\sin x)\frac{\sin x\,dx}{\sqrt{1-k^2\sin^2 x}} = -\frac{\pi}{2k}\ln k'.$ BI ((344))(2)

4.572 $\int\limits_0^\infty \left(\frac{2}{\pi}\operatorname{arcctg} x - \cos px\right)dx = C + \ln p \quad [p>0].$ NT 66(12)

4.573

1. $\int\limits_0^\infty \operatorname{arcctg} qx \sin px\,dx = \frac{\pi}{2p}\left(1-e^{-\frac{p}{q}}\right) \quad [p>0,\ q>0]$ BI ((347))(1)a

2. $\int\limits_0^\infty \operatorname{arcctg} qx\cos px\,dx = \frac{1}{2p}\left[e^{-\frac{p}{q}}\operatorname{Ei}\left(\frac{p}{q}\right) - e^{\frac{p}{q}}\operatorname{Ei}\left(-\frac{p}{q}\right)\right]$

$[p>0,\quad q>0].$ BI ((347))(2)a

3. $\int\limits_0^\infty \operatorname{arcctg} rx\frac{\sin px\,dx}{1\pm 2q\cos px+q^2} =$

$= \pm\frac{\pi}{2pq}\ln\frac{1\pm q}{1\pm qe^{-\frac{p}{r}}} \quad [q^2<1,\quad r>0,\quad p>0];$

$= \pm\frac{\pi}{2pq}\ln\frac{q\pm 1}{q\pm e^{-\frac{p}{r}}} \quad [q^2>1,\quad r>0,\quad p>0].$ BI ((347))(10)

4. $\int\limits_0^\infty \operatorname{arcctg} px\frac{\operatorname{tg} x\,dx}{q^2\cos^2 x+r^2\sin^2 x} = \frac{\pi}{2r^2}\ln\left(1+\frac{r}{q}\operatorname{th}\frac{1}{p}\right)$

$[p>0,\quad q>0,\quad r>0].$ BI ((347))(9)

4.574

1. $\int\limits_0^\infty \operatorname{arctg}\left(\frac{2a}{x}\right)\sin(bx)\,dx = \frac{\pi}{b}e^{-ab}\operatorname{sh}(ab)$

$[\operatorname{Re} a>0,\quad b>0].$ ET I 87(8)

2. $$\int_0^\infty \operatorname{arctg}\frac{a}{x}\cos(bx)\,dx = \frac{1}{2b}\left[e^{-ab}\,\overline{\mathrm{Ei}}(ab) - e^{ab}\,\mathrm{Ei}(-ab)\right]$$

$[a > 0, \quad b > 0].$ ET I 29(7)

3. $$\int_0^\infty \operatorname{arctg}\left[\frac{2ax}{x^2+c^2}\right]\sin(bx)\,dx = \frac{\pi}{b}e^{-b\sqrt{a^2+c^2}}\operatorname{sh}(ab)$$

$[b > 0].$ ET I 87(9)

4. $$\int_0^\infty \operatorname{arctg}\left(\frac{2}{x^2}\right)\cos(bx)\,dx = \frac{\pi}{b}e^{-b}\sin b \quad [b > 0].$$ ET I 29(8)

4.575

1. $$\int_0^\pi \operatorname{arctg}\frac{p\sin x}{1-p\cos x}\sin nx\,dx = \frac{\pi}{2n}p^n \quad [p^2 < 1].$$ BI ((345))(4)

2. $$\int_0^\pi \operatorname{arctg}\frac{p\sin x}{1-p\cos x}\sin nx\cos x\,dx = \frac{\pi}{4}\left(\frac{p^{n+1}}{n+1} + \frac{p^{n-1}}{n-1}\right)$$

$[p^2 < 1].$ BI ((345))(5)

3. $$\int_0^\pi \operatorname{arctg}\frac{p\sin x}{1-p\cos x}\cos nx\sin x\,dx = \frac{\pi}{4}\left(\frac{p^{n+1}}{n+1} - \frac{p^{n-1}}{n-1}\right)$$

$[p^2 < 1].$ BI ((345))(6)

4.576

1. $$\int_0^\pi \operatorname{arctg}\frac{p\sin x}{1-p\cos x}\frac{dx}{\sin x} = \frac{\pi}{2}\ln\frac{1+p}{1-p} \quad [p^2 < 1].$$ BI ((346))(1)

2. $$\int_0^\pi \operatorname{arctg}\frac{p\sin x}{1-p\cos x}\frac{dx}{\operatorname{tg} x} = -\frac{\pi}{2}\ln(1-p^2) \quad [p^2 < 1].$$ BI ((346))(3)

4.577

1. $$\int_0^{\frac{\pi}{2}} \operatorname{arctg}\left(\operatorname{tg}\lambda\sqrt{1-k^2\sin^2 x}\right)\frac{\sin^2 x\,dx}{\sqrt{1-k^2\sin^2 x}} =$$

$$= \frac{\pi}{2k^2}\left[F(\lambda, k) - E(\lambda, k) + \operatorname{ctg}\lambda\left(1-\sqrt{1-k^2\sin^2\lambda}\right)\right].$$ BI ((344))(4)

2. $$\int_0^{\frac{\pi}{2}} \operatorname{arctg}\left(\operatorname{tg}\lambda\sqrt{1-k^2\sin^2 x}\right)\frac{\cos^2 x\,dx}{\sqrt{1-k^2\sin^2 x}} =$$

$$= \frac{\pi}{2k^2}\left[E(\lambda, k) - k'^2F(\lambda, k) + \operatorname{ctg}\lambda\left(\sqrt{1-k^2\sin^2\lambda} - 1\right)\right].$$

BI ((344))(5)

4.58 A combination involving an inverse and a direct trigonometric function and a power

4.581 $$\int_0^\infty \operatorname{arctg} x \cos px \frac{dx}{x} = \int_0^\infty \operatorname{arctg} \frac{x}{p} \cos x \frac{dx}{x} = -\frac{\pi}{2} \operatorname{Ei}(-p) \quad [\operatorname{Re}(p) > 0].$$ ET III 654, NT 25(13)

4.59 Combinations of inverse trigonometric functions and logarithms

4.591

1. $$\int_0^1 \arcsin x \ln x \, dx = 2 - \ln 2 - \frac{1}{2}\pi.$$ BI ((339))(1)

2. $$\int_0^1 \arccos x \ln x \, dx = \ln 2 - 2.$$ BI ((339))(2)

4.592 $$\int_0^1 \arccos x \frac{dx}{\ln x} = -\sum_{k=0}^\infty \frac{(2k-1)!!}{2^k k!} \frac{\ln(2k+2)}{2k+1}.$$ BI ((339))(8)

4.593

1. $$\int_0^1 \operatorname{arctg} x \ln x \, dx = \frac{1}{2}\ln 2 - \frac{\pi}{4} + \frac{1}{48}\pi^2.$$ BI ((339))(3)

2. $$\int_0^1 \operatorname{arcctg} x \ln x \, dx = -\frac{1}{48}\pi^2 - \frac{\pi}{4} - \frac{1}{2}\ln 2.$$ BI ((339))(4)

4.594 $$\int_0^1 \operatorname{arctg} x (\ln x)^{n-1} (\ln x + n) \, dx = \frac{n!}{(-2)^{n+1}} (2^{-n} - 1) \zeta(n+1).$$ BI ((339))(7)

4.6 Multiple Integrals

4.60 Change of variables in multiple integrals

4.601

1. $$\iint_{(\sigma)} f(x, y) \, dx \, dy = \iint_{(\sigma')} f[\varphi(u, v), \psi(u, v)] \, |\Delta| \, du \, dv,$$

where $x = \varphi(u, v)$, $y = \psi(u, v)$, and $\Delta = \frac{\partial\varphi}{\partial u}\frac{\partial\psi}{\partial v} - \frac{\partial\psi}{\partial u}\frac{\partial\varphi}{\partial v} \equiv \frac{D(\varphi, \psi)}{D(u, v)}$ is the Jacobian determinant of the functions φ and ψ.

2. $$\iiint_{(V)} f(x, y, z) \, dx \, dy \, dz = \iiint_{(V')} f[\varphi(u, v, w), \psi(u, v, w), \chi(u, v, w)] \, |\Delta| \, du \, dv \, dw,$$

where $x = \varphi(u, v, w)$, $y = \psi(u, v, w)$, and $z = \chi(u, v, w)$ and where

$$\Delta = \begin{vmatrix} \frac{\partial\varphi}{\partial u} & \frac{\partial\varphi}{\partial v} & \frac{\partial\varphi}{\partial w} \\ \frac{\partial\psi}{\partial u} & \frac{\partial\psi}{\partial v} & \frac{\partial\psi}{\partial w} \\ \frac{\partial\chi}{\partial u} & \frac{\partial\chi}{\partial v} & \frac{\partial\chi}{\partial w} \end{vmatrix} \equiv \frac{D(\varphi, \psi, \chi)}{D(u, v, w)}$$

is the Jacobian determinant of the functions φ, ψ, and χ.

Here, we assume, both in (4.601 2.) and in (4.601 1.) that

(a) the functions φ, ψ, and χ and also their first partial derivatives are continuous in the region of integration;

(b) the Jacobian does not change sign in this region;

(c) there exists a one-to-one correspondence between the old variables x, y, z and the new ones u, v, w in the region of integration;

(d) when we change from the variables x, y, z to the variables u, v, w, the region V (resp. σ) is mapped into the region V' (resp. σ').

4.602 Transformation to polar coordinates:

$$x = r\cos\varphi, \qquad y = r\sin\varphi; \qquad \frac{D(x, y)}{D(r, \varphi)} = r.$$

4.603 Transformation to spherical coordinates:

$$x = r\sin\theta\cos\varphi, \quad y = r\sin\theta\sin\varphi, \quad z = r\cos\theta, \quad \frac{D(x, y, z)}{D(r, \theta, \varphi)} = r^2\sin\theta.$$

4.61 Change of the order of integration and change of variables

4.611

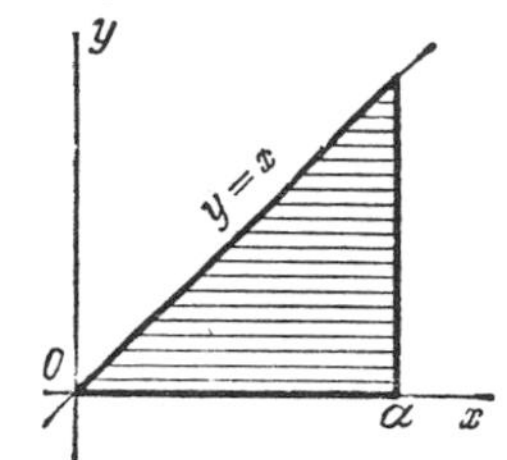

1. $$\int_0^a dx \int_0^x f(x, y)\,dy = \int_0^a dy \int_y^a f(x, y)\,dx.$$

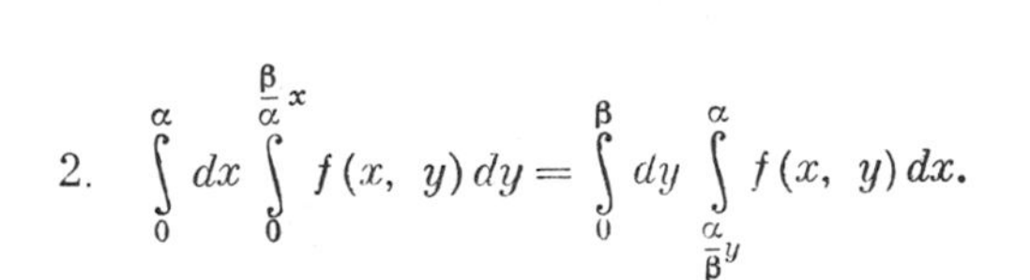

2. $$\int_0^a dx \int_0^{\frac{\beta}{\alpha}x} f(x, y)\,dy = \int_0^{\beta} dy \int_{\frac{\alpha}{\beta}y}^{\alpha} f(x, y)\,dx.$$

4.612

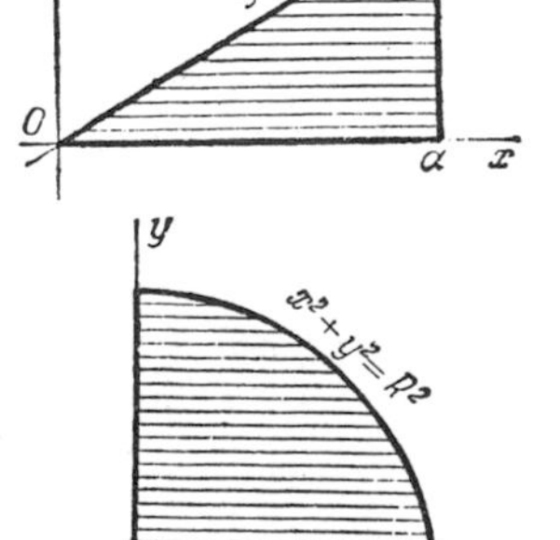

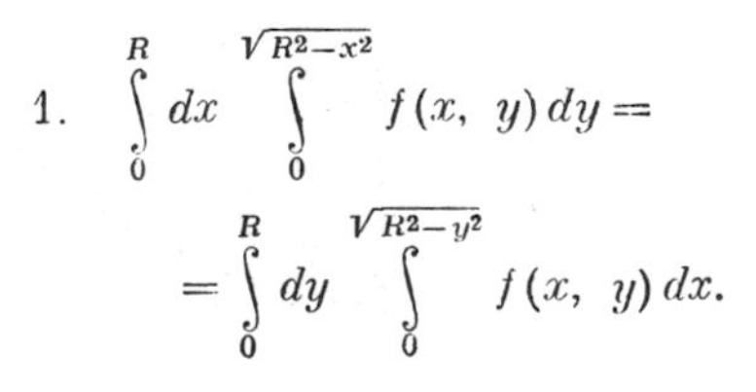

1. $$\int_0^R dx \int_0^{\sqrt{R^2-x^2}} f(x, y)\,dy = \int_0^R dy \int_0^{\sqrt{R^2-y^2}} f(x, y)\,dx.$$

2. $$\int_0^{2p} dx \int_0^{\frac{q}{p}\sqrt{2px-x^2}} f(x, y)\, dy = \int_0^q dy \int_{p\left[1-\sqrt{1-\left(\frac{y}{q}\right)^2}\right]}^{p\left[1+\sqrt{1-\left(\frac{y}{q}\right)^2}\right]} f(x, y)\, dx.$$

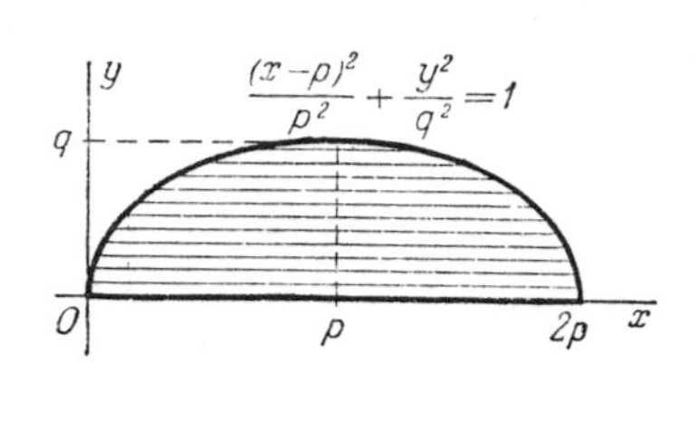

4.613

1. $$\int_0^{\alpha} dx \int_0^{\frac{\beta}{\beta+x}} f(x, y)\, dy = \int_0^{\frac{\beta}{\beta+\alpha}} dy \int_0^{\alpha} f(x, y)\, dx + \int_{\frac{\beta}{\beta+\alpha}}^{1} dy \int_0^{\frac{\beta}{y}(1-y)} f(x, y)\, dx.$$

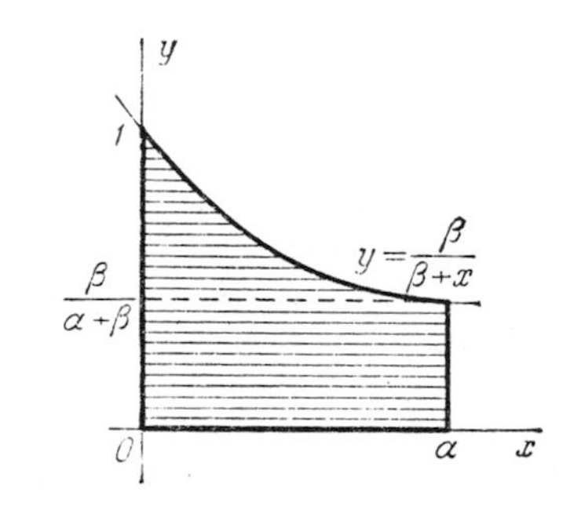

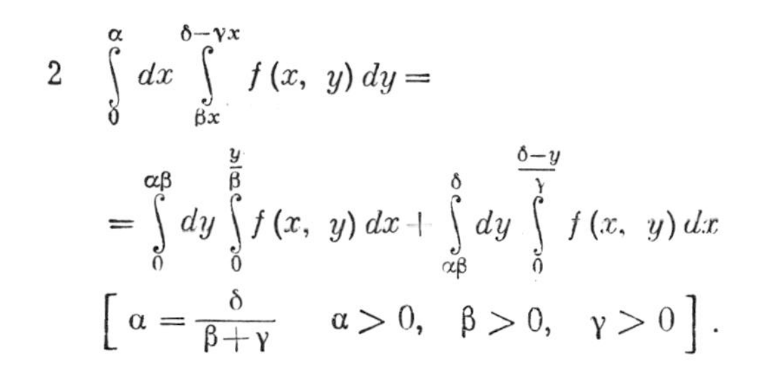

2 $$\int_0^{\alpha} dx \int_{\beta x}^{\delta-\gamma x} f(x, y)\, dy = \int_0^{\alpha\beta} dy \int_0^{\frac{y}{\beta}} f(x, y)\, dx + \int_{\alpha\beta}^{\delta} dy \int_0^{\frac{\delta-y}{\gamma}} f(x, y)\, dx$$

$$\left[\alpha = \frac{\delta}{\beta+\gamma} \quad \alpha > 0, \quad \beta > 0, \quad \gamma > 0\right].$$

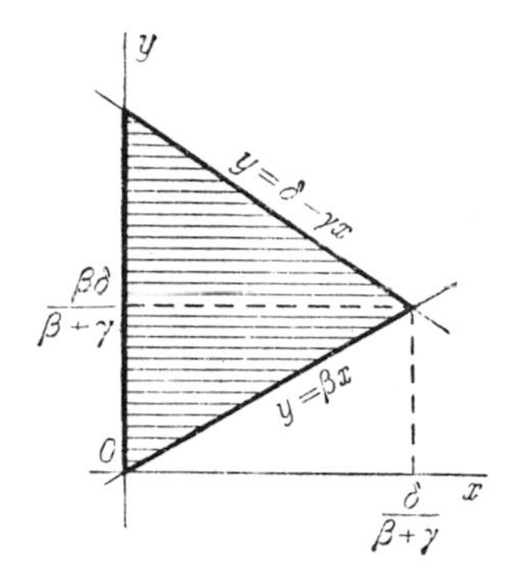

3. $$\int_0^{2\alpha} dx \int_{\frac{x^2}{4\alpha}}^{3\alpha-x} f(x, y)\, dy = \int_0^{\alpha} dy \int_0^{2\sqrt{\alpha y}} f(x, y)\, dx + \int_{\alpha}^{3\alpha} dy \int_0^{3\alpha-y} f(x, y)\, dx.$$

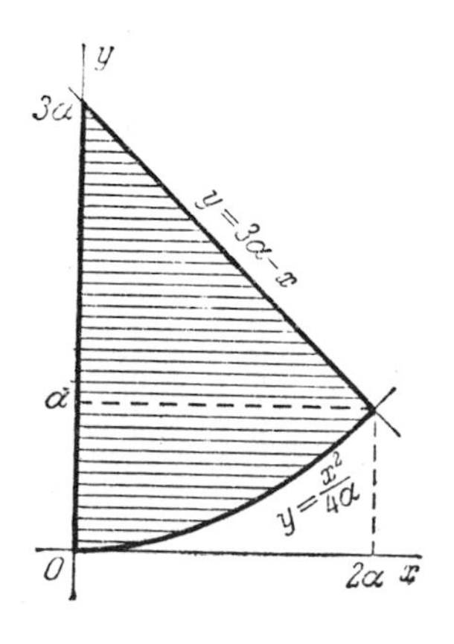

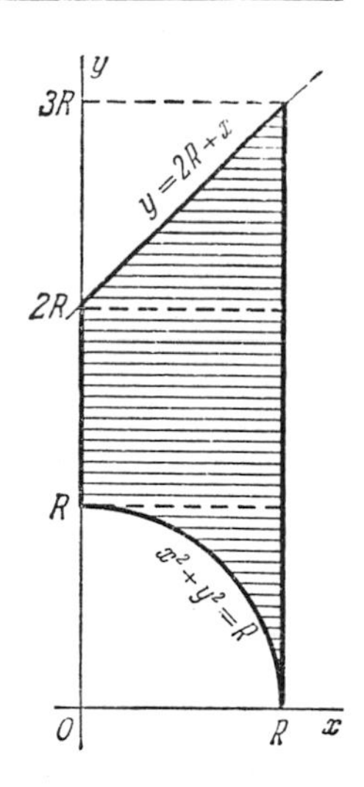

4. $$\int_0^R dx \int_{\sqrt{R^2-x^2}}^{x+2R} f(x, y)\,dy = \int_0^R dy \int_{\sqrt{R^2-y^2}}^{R} f(x, y)\,dx + \int_R^{2R} dy \int_0^R f(x, y)\,dx + \int_{2R}^{3R} dy \int_{y-2R}^{R} f(x, y)\,dx.$$

4.614 $$\int_0^{\frac{\pi}{2}} d\varphi \int_0^{2R\cos\varphi} f(r, \varphi)\,dr = \int_0^{2R} dr \int_0^{\arccos\frac{r}{2R}} f(r, \varphi)\,d\varphi.$$

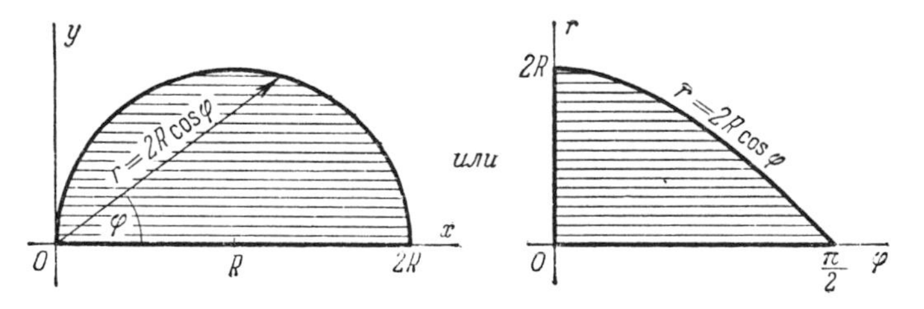

4.615 $$\int_0^R dx \int_0^{\sqrt{R^2-x^2}} f(x, y)\,dy = \int_0^{\frac{\pi}{2}} d\varphi \int_0^R f(r\cos\varphi, r\sin\varphi)\,r\,dr.$$

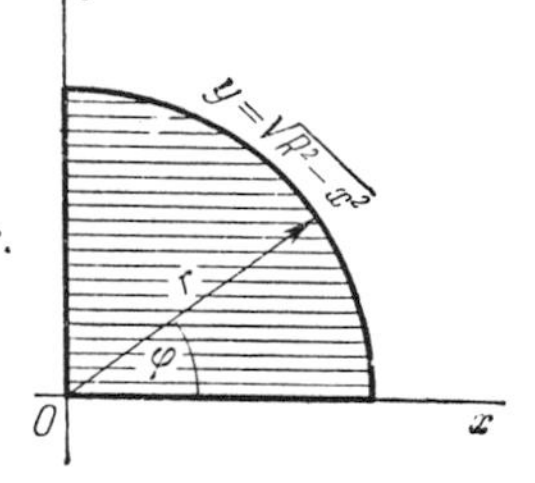

4.616 $$\int_0^{2R} dx \int_0^{\sqrt{2Rx-x^2}} f(x, y)\,dy = \int_0^{\frac{\pi}{2}} d\varphi \int_0^{2R\cos\varphi} f(r\cos\varphi\, r,\ \sin\varphi)\,r\,dr.$$

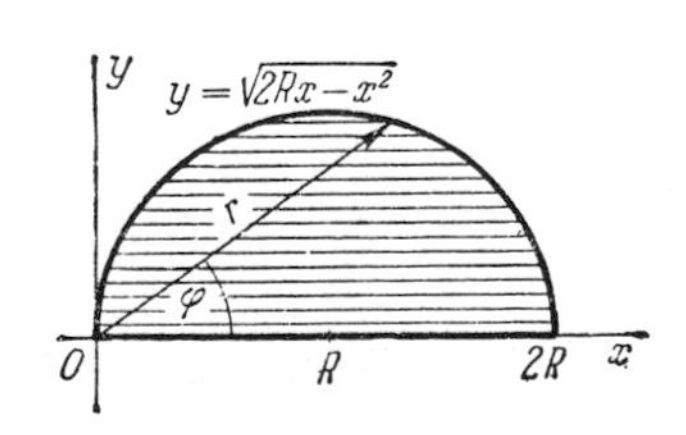

4.617 $$\int_{\alpha}^{\beta} dx \int_{\varphi_1(x)}^{\varphi_2(x)} f(x, y)\, dy = \int_0^{\beta} dx \int_0^{\varphi_2(x)} f(x, y)\, dy - \int_0^{\beta} dx \int_0^{\varphi_1(x)} f(x, y)\, dy - \int_0^{\alpha} dx \int_0^{\varphi_2(x)} f(x, y)\, dy + \int_0^{\alpha} dx \int_0^{\varphi_1(x)} f(x, y)\, dy$$

$[\varphi_1(x) \leqslant \varphi_2(x) \quad \text{for} \quad \alpha \leqslant x \leqslant \beta]$.

4.618 $$\int_0^{\gamma} dx \int_0^{\varphi(x)} f(x, y)\, dy = \int_0^{\gamma} dx \int_0^1 f[x, z\varphi(x)]\, \varphi(x)\, dz \qquad [y = z\varphi(x)];$$

$$= \gamma \int_0^1 dz \int_0^{\varphi(\gamma z)} f(\gamma z, y)\, dy \qquad [x = \gamma z].$$

4.619 $$\int_{x_0}^{x_1} dx \int_{y_0}^{y_1} f(x, y)\, dy = \int_{x_0}^{x_1} dx \int_0^1 (y_1 - y_0) f[x, y_0 + (y_1 - y_0) t]\, dt$$

$[y = y_0 + (y_1 - y_0) t]$.

4.62 Double and triple integrals with constant limits

4.620 General formulas

1. $$\int_0^{\pi} d\omega \int_0^{\infty} f'(p \operatorname{ch} x + q \cos\omega \operatorname{sh} x) \operatorname{sh} x\, dx = -\frac{\pi \operatorname{sign} p}{\sqrt{p^2 - q^2}} f(\operatorname{sign} p \sqrt{p^2 - q^2}) \qquad [p^2 > q^2, \lim_{x \to +\infty} f(x) = 0]$$

LO III 389

2. $$\int_0^{2\pi} d\omega \int_0^{\infty} f'[p \operatorname{ch} x + (q \cos\omega + r \sin\omega) \operatorname{sh} x] \operatorname{sh} x\, dx = -\frac{2\pi \operatorname{sign} p}{\sqrt{p^2 - q^2 - r^2}} f(\operatorname{sign} p \sqrt{p^2 - q^2 - r^2}) \quad [p^2 > q^2 + r^2, \lim_{x \to +\infty} f(x) = 0].$$

LO III 390

3. $$\int_0^{\pi} \int_0^{\pi} \frac{dx\, dy}{\sin x \sin^2 y} f'\left[\frac{p - q\cos x}{\sin x \sin y} + r \operatorname{ctg} y\right] = -\frac{2\pi \operatorname{sign} p}{\sqrt{p^2 - q^2 - r^2}} f(\operatorname{sign} p \sqrt{p^2 - q^2 - r^2}) \quad [p^2 > q^2 + r^2, \lim_{x \to +\infty} f(x) = 0].$$

LO III 280

4. $$\int_{-\infty}^{\infty} dx \int_{-\infty}^{\infty} f'(p \operatorname{ch} x \operatorname{ch} y + q \operatorname{sh} x \operatorname{ch} y + r \operatorname{sh} y) \operatorname{ch} y\, dy = -\frac{2\pi \operatorname{sign} p}{\sqrt{p^2 - q^2 - r^2}} f(\operatorname{sign} p \sqrt{p^2 - q^2 - r^2}) \quad [p^2 > q^2 + r^2, \lim_{x \to +\infty} f(x) = 0].$$

LO III 390

5. $$\int_0^\infty dx \int_0^\pi f(p \operatorname{ch} x + q \cos\omega \operatorname{sh} x) \operatorname{sh}^2 x \sin\omega \, d\omega =$$
$$= 2\int_0^\infty f(\operatorname{sign} p \sqrt{p^2-q^2} \operatorname{ch} x) \operatorname{sh}^2 x \, dx \qquad \left[\lim_{x\to+\infty} f(x) = 0\right].$$ LO III 391

6. $$\int_0^\infty dx \int_0^{2\pi} d\omega \int_0^\pi f[p \operatorname{ch} x + (q\cos\omega + r\sin\omega)\sin\theta \operatorname{sh} x] \operatorname{sh}^2 x \sin\theta \, d\theta =$$
$$= 4\int_0^\infty f(\operatorname{sign} p \sqrt{p^2-q^2-r^2} \operatorname{ch} x) \operatorname{sh}^2 x \, dx$$
$$\left[p^2 > q^2 + r^2, \quad \lim_{x\to+\infty} f(x) = 0\right].$$ LO III 390

7. $$\int_0^\infty dx \int_0^{2\pi} d\omega \int_0^\pi f\{p \operatorname{ch} x + [(q\cos\omega + r\sin\omega)\sin\theta + s \operatorname{ch}\theta] \operatorname{sh} x\} \times$$
$$\times \operatorname{sh}^2 x \sin\theta \, d\theta = 4\pi \int_0^\infty f(\operatorname{sign} p \sqrt{p^2-q^2-r^2-s^2} \operatorname{ch} x) \operatorname{sh}^2 x \, dx$$
$$\left[p^2 > q^2 + r^2 + s^2, \quad \lim_{x\to+\infty} f(x) = 0\right].$$ LO III 391

4.621

1. $$\int_0^{\frac{\pi}{2}} \int_0^{\frac{\pi}{2}} \frac{\sin y \sqrt{1-k^2 \sin^2 x \sin^2 y}}{1-k^2\sin^2 y} \, dx \, dy = \frac{\pi}{2\sqrt{1-k^2}}.$$ LO I 252(90)

2. $$\int_0^{\frac{\pi}{2}} \int_0^{\frac{\pi}{2}} \frac{\cos y \sqrt{1-k^2 \sin^2 x \sin^2 y}}{1-k^2\sin^2 y} \, dx \, dy = \boldsymbol{K}(k).$$ LO I 252(91)

3. $$\int_0^{\frac{\pi}{2}} \int_0^{\frac{\pi}{2}} \frac{\sin\alpha \sin y \, dx \, dy}{\sqrt{1-\sin^2\alpha \sin^2 x \sin^2 y}} = \frac{\pi\alpha}{2}.$$ LO I 253

4.622

1. $$\int_0^\pi \int_0^\pi \int_0^\pi \frac{dx \, dy \, dz}{1-\cos x \cos y \cos z} = 4\pi \boldsymbol{K}^2\left(\frac{\sqrt{2}}{2}\right).$$ MO 137

2. $$\int_0^\pi \int_0^\pi \int_0^\pi \frac{dx \, dy \, dz}{3-\cos y \cos z - \cos x \cos z - \cos x \cos y} = \sqrt{3}\pi \boldsymbol{K}^2\left(\sin\frac{\pi}{12}\right).$$ MO 137

3. $$\int_0^\pi \int_0^\pi \int_0^\pi \frac{dx \, dy \, dz}{3-\cos x - \cos y - \cos z} =$$
$$= 4\pi[18 + 12\sqrt{2} - 10\sqrt{3} - 7\sqrt{6}] \boldsymbol{K}^2[(2-\sqrt{3})(\sqrt{3}-\sqrt{2})].$$ MO 137

4.623 $$\int_0^\infty \int_0^\infty \varphi(a^2x^2 + b^2y^2) \, dx \, dy = \frac{\pi}{4ab} \int_0^\infty \varphi(x) \, x \, dx.$$

4.624
$$\int_0^{\pi}\int_0^{2\pi} f(\alpha\cos\theta+\beta\sin\theta\cos\psi+\gamma\sin\theta\sin\psi)\sin\theta\,d\theta\,d\psi =$$
$$= 2\pi\int_0^{\pi} f(R\cos p)\sin p\,dp = 2\pi\int_{-1}^{1} f(Rt)\,dt \qquad \left[R=\sqrt{\alpha^2+\beta^2+\gamma^2}\right].$$

4.63-4.64 Multiple integrals

4.631
$$\int_p^x dt_{n-1}\int_p^{t_{n-1}} dt_{n-2}\ldots\int_p^{t_1} f(t)\,dt = \frac{1}{(n-1)!}\int_p^x (x-t)^{n-1} f(t)\,dt,$$
where $f(t)$ is continuous on the interval $[p, q]$ and $p\leqslant x\leqslant q$. FI II 692

4.632

1. $$\int\int\limits_{\substack{x_1\geqslant 0,\ x_2\geqslant 0,\ \ldots,\ x_n\geqslant 0\\ x_1+x_2+\ldots+x_n\leqslant h}}\ldots\int dx_1\,dx_2\ldots dx_n = \frac{h^n}{n!}$$ [the volume of an n-dimensional simplex] FI III 472

2. $$\int\int\limits_{x_1^2+x_2^2+\ldots+x_n^2\leqslant R^2}\ldots\int dx_1\,dx_2\ldots dx_n = \frac{\sqrt{\pi^n}}{\Gamma\left(\frac{n}{2}+1\right)}R^n$$ [the volume of an n-dimensional sphere] FI III 473

4.633
$$\int\int\limits_{x_1^2+x_2^2+\ldots+x_n^2\leqslant 1}\ldots\int \frac{dx_1\,dx_2\ldots dx_n}{\sqrt{1-x_1^2-x_2^2-\ldots-x_n^2}} = \frac{\pi^{\frac{n+1}{2}}}{\Gamma\left(\frac{n+1}{2}\right)} \qquad [n>1]$$
[Half-area of the surface of an $(n+1)$-dimensional sphere $x_1^2+x_2^2+\ldots+x_{n+1}^2=1$]. FI III 474

4.634
$$\int\int\limits_{\substack{x_1\geqslant 0,\ x_2\geqslant 0,\ \ldots,\ x_n\geqslant 0\\ \left(\frac{x_1}{q_1}\right)^{\alpha_1}+\left(\frac{x_2}{q_2}\right)^{\alpha_2}+\ldots+\left(\frac{x_n}{q_n}\right)^{\alpha_n}\leqslant 1}}\ldots\int x_1^{p_1-1}x_2^{p_2-1}\ldots x_n^{p_n-1}\,dx_1\,dx_2\ldots dx_n =$$
$$= \frac{q_1^{p_1}q_2^{p_2}\ldots q_n^{p_n}}{\alpha_1\alpha_2\ldots\alpha_n}\,\frac{\Gamma\left(\frac{p_1}{\alpha_1}\right)\Gamma\left(\frac{p_2}{\alpha_2}\right)\ldots\Gamma\left(\frac{p_n}{\alpha_n}\right)}{\Gamma\left(\frac{p_1}{\alpha_1}+\frac{p_2}{\alpha_2}+\ldots+\frac{p_n}{\alpha_n}+1\right)}$$
$[\alpha_i>0,\ p_i>0,\ q_i>0,\ i=1, 2, \ldots, n]$. FI III 477

4.635

1. $$\int\int\limits_{\substack{x_1\geqslant 0,\ x_2\geqslant 0,\ \ldots,\ x_n\geqslant 0\\ \left(\frac{x_1}{q_1}\right)^{\alpha_1}+\left(\frac{x_2}{q_2}\right)^{\alpha_2}+\ldots+\left(\frac{x_n}{q_n}\right)^{\alpha_n}\geqslant 1}}\ldots\int f\left[\left(\frac{x_1}{q_1}\right)^{\alpha_1}+\left(\frac{x_2}{q_2}\right)^{\alpha_2}+\ldots+\left(\frac{x_n}{q_n}\right)^{\alpha_n}\right]\times$$
$$\times x_1^{p_1-1}x_2^{p_2-1}\ldots x_n^{p_n-1}\,dx_1\,dx_2\ldots dx_n =$$
$$= \frac{q_1^{p_1}q_2^{p_2}\ldots q_n^{p_n}}{\alpha_1\alpha_2\ldots\alpha_n}\,\frac{\Gamma\left(\frac{p_1}{\alpha_1}\right)\Gamma\left(\frac{p_2}{\alpha_2}\right)\ldots\Gamma\left(\frac{p_n}{\alpha_n}\right)}{\Gamma\left(\frac{p_1}{\alpha_1}+\frac{p_2}{\alpha_2}+\ldots+\frac{p_n}{\alpha_n}\right)}\int_1^{\infty} f(x)\,x^{\frac{p_1}{\alpha_1}+\frac{p_2}{\alpha_2}+\ldots+\frac{p_n}{\alpha_n}-1}\,dx$$

under the assumption that the integral on the right converges absolutely.

FI III 487

2. $$\underset{\substack{x_1\geqslant 0,\ x_2\geqslant 0,\ \dots,\ x_n\geqslant 0\\ \left(\frac{x_1}{q_1}\right)^{\alpha_1}+\left(\frac{x_2}{q_2}\right)^{\alpha_2}+\dots+\left(\frac{x_n}{q_n}\right)^{\alpha_n}\leqslant 1}}{\int\int\dots\int} f\left[\left(\frac{x_1}{q_1}\right)^{\alpha_1}+\left(\frac{x_2}{q_2}\right)^{\alpha_2}+\dots+\left(\frac{x_n}{q_n}\right)^{\alpha_n}\right]\times$$

$$\times x_1^{p_1-1}x_2^{p_2-1}\dots x_n^{p_n-1}\,dx_1\,dx_2\dots dx_n=$$

$$=\frac{q_1^{p_1}q_2^{p_2}\dots q_n^{p_n}}{\alpha_1\alpha_2\dots\alpha_n}\frac{\Gamma\left(\frac{p_1}{\alpha_1}\right)\Gamma\left(\frac{p_2}{\alpha_2}\right)\dots\Gamma\left(\frac{p_n}{\alpha_n}\right)}{\Gamma\left(\frac{p_1}{\alpha_1}+\frac{p_2}{\alpha_2}+\dots+\frac{p_n}{\alpha_n}\right)}\int_0^1 f(x)\,x^{\frac{p_1}{\alpha_1}+\frac{p_2}{\alpha_2}+\dots+\frac{p_n}{\alpha_n}-1}\,dx$$

under the assumptions that the one-dimensional integral on the right converges absolutely and that the numbers q_i, α_i, and p_i are positive.

FI III 479

In particular,

3. $$\underset{\substack{x_1\geqslant 0,\ x_2\geqslant 0,\ \dots,\ x_n\geqslant 0\\ x_1+x_2+\dots+x_n\leqslant 1}}{\int\int\dots\int} x_1^{p_1-1}x_2^{p_2-1}\dots x_n^{p_n-1}e^{-q(x_1+x_2+\dots+x_n)}\,dx_1\,dx_2\dots dx_n=$$

$$=\frac{\Gamma(p_1)\Gamma(p_2)\dots\Gamma(p_n)}{\Gamma(p_1+p_2+\dots+p_n)}\int_0^1 x^{p_1+p_2+\dots+p_n-1}e^{-qx}\,dx$$

$$[n>0,\ p_1>0,\ p_2>0,\ \dots,\ p_n>0].$$

4. $$\underset{\substack{x_1\geqslant 0,\ x_2\geqslant 0,\ \dots,\ x_n\geqslant 0\\ x_1^{\alpha_1}+x_2^{\alpha_2}+\dots+x_n^{\alpha_n}\leqslant 1}}{\int\int\dots\int} \frac{x_1^{p_1-1}x_2^{p_2-1}\dots x_n^{p_n-1}}{(1-x_1^{\alpha_1}-x_2^{\alpha_2}-\dots-x_n^{\alpha_n})^{\mu}}\,dx_1\,dx_2\dots dx_n=$$

$$=\frac{1}{\alpha_1\alpha_2\dots\alpha_n}\frac{\Gamma\left(\frac{p_1}{\alpha_1}\right)\Gamma\left(\frac{p_2}{\alpha_2}\right)\dots\Gamma\left(\frac{p_n}{\alpha_n}\right)}{\Gamma\left(1-\mu+\frac{p_1}{\alpha_1}+\frac{p_2}{\alpha_2}+\dots+\frac{p_n}{\alpha_n}\right)}\Gamma(1-\mu)$$

$$[p_1>0,\ p_2>0,\ \dots,\ p_n>0,\ \mu<1].$$

FI III 480

4.636

1. $$\underset{\substack{x_1\geqslant 0,\ x_2\geqslant 0,\ \dots,\ x_n\geqslant 0\\ x_1^{\alpha_1}+x_2^{\alpha_2}+\dots+x_n^{\alpha_n}\geqslant 1}}{\int\int\dots\int} \frac{x_1^{p_1-1}x_2^{p_2-1}\dots x_n^{p_n-1}}{(x_1^{\alpha_1}+x_2^{\alpha_2}+\dots+x_n^{\alpha_n})^{\mu}}\,dx_1\,dx_2\dots dx_n=$$

$$=\frac{1}{\alpha_1\alpha_2\dots\alpha_n\left(\mu-\frac{p_1}{\alpha_1}-\frac{p_2}{\alpha_2}-\dots-\frac{p_n}{\alpha_n}\right)}\frac{\Gamma\left(\frac{p_1}{\alpha_1}\right)\Gamma\left(\frac{p_2}{\alpha_2}\right)\dots\Gamma\left(\frac{p_n}{\alpha_n}\right)}{\Gamma\left(\frac{p_1}{\alpha_1}+\frac{p_2}{\alpha_2}+\dots+\frac{p_n}{\alpha_n}\right)}$$

$$\left[p_1>0,\ p_2>0,\ \dots,\ p_n>0;\ \mu>\frac{p_1}{\alpha_1}+\frac{p_2}{\alpha_2}+\dots+\frac{p_n}{\alpha_n}\right].$$

FI III 488

2. $$\underset{\substack{x_1\geqslant 0,\, x_2\geqslant 0,\, \dots,\, x_n\geqslant 0\\ x_1^{\alpha_1}+x_2^{\alpha_2}+\dots+x_n^{\alpha_n}\leqslant 1}}{\int\int\cdots\int}\frac{x_1^{p_1-1}x_2^{p_2-1}\dots x_n^{p_n-1}}{(x_1^{\alpha_1}+x_2^{\alpha_2}+\dots+x_n^{\alpha_n})^{\mu}}\,dx_1\,dx_2\dots dx_n=$$

$$=\frac{1}{\alpha_1\alpha_2\dots\alpha_n\left(\frac{p_1}{\alpha_1}+\frac{p_2}{\alpha_2}+\dots+\frac{p_n}{\alpha_n}-\mu\right)}\frac{\Gamma\left(\frac{p_1}{\alpha_1}\right)\Gamma\left(\frac{p_2}{\alpha_2}\right)\dots\Gamma\left(\frac{p_n}{\alpha_n}\right)}{\Gamma\left(\frac{p_1}{\alpha_1}+\frac{p_2}{\alpha_2}+\dots+\frac{p_n}{\alpha_n}\right)}$$

$$\left[\mu<\frac{p_1}{\alpha_1}+\frac{p_2}{\alpha_2}+\dots+\frac{p_n}{\alpha_n}\right].$$ FI III 480

3. $$\underset{\substack{x_1\geqslant 0,\, x_2\geqslant 0,\, \dots,\, x_n\geqslant 0\\ x_1^{\alpha_1}+x_2^{\alpha_2}+\dots+x_n^{\alpha_n}\leqslant 1}}{\int\int\cdots\int} x_1^{p_1-1}x_2^{p_2-1}\dots x_n^{p_n-1}\sqrt{\frac{1-x_1^{\alpha_1}-x_2^{\alpha_2}-\dots-x_n^{\alpha_n}}{1+x_1^{\alpha_1}+x_2^{\alpha_2}+\dots+x_n^{\alpha_n}}}\times$$

$$\times\,dx_1\,dx_2\dots dx_n=\frac{\sqrt{\pi}}{2}\frac{\Gamma\left(\frac{p_1}{\alpha_1}\right)\Gamma\left(\frac{p_2}{\alpha_2}\right)\dots\Gamma\left(\frac{p_n}{\alpha_n}\right)}{\alpha_1\alpha_2\dots\alpha_n}\frac{1}{\Gamma(m)}\times$$

$$\times\left\{\frac{\Gamma\left(\frac{m}{2}\right)}{\Gamma\left(\frac{m+1}{2}\right)}-\frac{\Gamma\left(\frac{m+1}{2}\right)}{\Gamma\left(\frac{m+2}{2}\right)}\right\},$$

where $m=\frac{p_1}{\alpha_1}+\frac{p_2}{\alpha_2}+\dots+\frac{p_n}{\alpha_n}$. FI III 480

4.637 $$\underset{\substack{x_1\geqslant 0,\, x_2\geqslant 0,\, \dots,\, x_n\geqslant 0\\ x_1+x_2+\dots+x_n\leqslant 1}}{\int\int\cdots\int} f(x_1+x_2+\dots+x_n)\times$$

$$\times\frac{x_1^{p_1-1}x_2^{p_2-1}\dots x_n^{p_n-1}\,dx_1\,dx_2\dots dx_n}{(q_1x_1+q_2x_2+\dots+q_nx_n+r)^{p_1+p_2+\dots+p_n}}=$$

$$=\frac{\Gamma(p_1)\Gamma(p_2)\dots\Gamma(p_n)}{\Gamma(p_1+p_2+\dots+p_n)}\int_0^{\infty}f(x)\frac{x^{p_1+p_2+\dots+p_n-1}}{(q_1x+r)^{p_1}(q_2x+r)^{p_2}\dots(q_nx+r)^{p_n}}\,dx,$$

where $f(x)$ is continuous on the interval $(0, 1)$

$$[q_1\geqslant 0,\ q_2\geqslant 0,\ \dots,\ q_n\geqslant 0;\ r>0].$$

4.638

1. $$\int_0^{\infty}\int_0^{\infty}\cdots\int_0^{\infty}\frac{x_1^{p_1-1}x_2^{p_2-1}\dots x_n^{p_n-1}e^{-(q_1x_1+q_2x_2+\dots+q_nx_n)}}{(r_0+r_1x_1+r_2x_2+\dots+r_nx_n)^s}\,dx_1\,dx_2\dots dx_n=$$

$$=\frac{\Gamma(p_1)\Gamma(p_2)\dots\Gamma(p_n)}{\Gamma(s)}\int_0^{\infty}\frac{e^{-r_0x}x^{s-1}\,dx}{(q_1+r_1x)^{p_1}(q_2+r_2x)^{p_2}\dots(q_n+r_nx)^{p_n}}$$

where p_i, q_i, r_i, and s are positive. This result is also valid for $r_0=0$ provided $p_1+p_2+\dots+p_n>s$.

2. $$\int_0^{\infty}\int_0^{\infty}\cdots\int_0^{\infty}\frac{x_1^{p_1-1}x_2^{p_2-1}\dots x_n^{p_n-1}}{(r_0+r_1x_1+r_2x_2+\dots+r_nx_n)^s}\,dx_1\,dx_2\dots dx_n=$$

$$=\frac{\Gamma(p_1)\Gamma(p_2)\dots\Gamma(p_n)\Gamma(s-p_1-p_2-\dots-p_n)}{r_1^{p_1}r_2^{p_2}\dots r_n^{p_n}r_0^{s-p_1-p_2-\dots-p_n}\Gamma(s)}\qquad[p_i>0,\ r_i>0,\ s>0].$$

3. $$\int_0^\infty\int_0^\infty\cdots\int_0^\infty \frac{x_1^{p_1-1}x_2^{p_2-1}\cdots x_n^{p_n-1}}{[1+(r_1x_1)^{q_1}+(r_2x_2)^{q_2}+\cdots+(r_nx_n)^{q_n}]^s}\,dx_1\,dx_2\ldots dx_n =$$
$$= \frac{\Gamma\left(\frac{p_1}{q_1}\right)\Gamma\left(\frac{p_2}{q_2}\right)\cdots\Gamma\left(\frac{p_n}{q_n}\right)}{q_1q_2\cdots q_n r_1^{p_1q_1}r_2^{p_2q_2}\cdots r_n^{p_nq_n}}\,\frac{\Gamma\left(s-\frac{p_1}{q_1}-\frac{p_2}{q_2}-\cdots-\frac{p_n}{q_n}\right)}{\Gamma(s)}$$
$[p_i>0,\ q_i>0,\ r_i>0,\ s>0]$.

4.639

1. $$\int\int\cdots\int_{x_1^2+x_2^2+\cdots+x_n^2\le 1} (p_1x_1+p_2x_2+\cdots+p_nx_n)^{2m}\,dx_1\,dx_2\ldots dx_n =$$
$$= \frac{(2m-1)!!}{2^m}\,\frac{\sqrt{\pi^n}}{\Gamma\left(\frac{n}{2}+m+1\right)}\,(p_1^2+p_2^2+\cdots+p_n^2)^m.$$ FI III 482

2. $$\int\int\cdots\int_{x_1^2+x_2^2+\cdots+x_n^2\le 1} (p_1x_1+p_2x_2+\cdots+p_nx_n)^{2m+1}\,dx_1\,dx_2\ldots dx_n = 0.$$ FI III 483

4.641

1. $$\int\int\cdots\int_{x_1^2+x_2^2+\cdots+x_n^2\le 1} e^{p_1x_1+p_2x_2+\cdots+p_nx_n}\,dx_1\,dx_2\ldots dx_n =$$
$$= \sqrt{\pi^n}\sum_{k=0}^\infty \frac{1}{k!\,\Gamma\left(\frac{n}{2}+k+1\right)}\left(\frac{p_1^2+p_2^2+\cdots+p_n^2}{4}\right)^k.$$ FI III 483

2. $$\int\int\cdots\int_{x_1^2+x_2^2+\cdots+x_{2n}^2\le 1} e^{p_1x_1+p_2x_2+\cdots+p_{2n}x_{2n}}\,dx_1\,dx_2\ldots dx_{2n} =$$
$$= \frac{(2\pi)^n I_n\left(\sqrt{p_1^2+p_2^2+\cdots+p_{2n}^2}\right)}{(p_1^2+p_2^2+\cdots+p_{2n}^2)^{n/2}}.$$ FI III 483a

4.642 $$\int\int\cdots\int_{x_1^2+x_2^2+\cdots+x_n^2\le R^2} f\left(\sqrt{x_1^2+x_2^2+\cdots+x_n^2}\right)dx_1\,dx_2\ldots dx_n =$$
$$= \frac{2\sqrt{\pi^n}}{\Gamma\left(\frac{n}{2}\right)}\int_0^R x^{n-1}f(x)\,dx,$$
where $f(x)$ is a function that is continuous on the interval $(0, R)$. FI III 485

4.643 $$\int_0^1\int_0^1\cdots\int_0^1 f(x_1x_2\ldots x_n)(1-x_1)^{p_1-1}(1-x_2)^{p_2-1}\ldots(1-x_n)^{p_n-1}\times$$
$$\times x_2^{p_1}x_3^{p_1+p_2}\ldots x_n^{p_1+p_2+\cdots+p_{n-1}}\,dx_1\,dx_2\ldots dx_n =$$
$$= \frac{\Gamma(p_1)\Gamma(p_2)\ldots\Gamma(p_n)}{\Gamma(p_1+p_2+\cdots+p_n)}\int_0^1 f(x)(1-x)^{p_1+p_2+\cdots+p_n-1}\,dx$$
under the assumption that the integral on the right converges absolutely. FI III 488

4.644 $$\overbrace{\int\int\cdots\int}^{n-1}_{x_1^2+x_2^2+\ldots+x_n^2=1} f(p_1x_1+p_2x_2+\ldots+p_nx_n)\frac{dx_1\,dx_2\ldots dx_{n-1}}{|x_n|}=$$

$$=2\int\int\cdots\int_{x_1^2+x_2^2+\ldots+x_{n-1}^2\leqslant 1} f(p_1x_1+p_2x_2+\ldots+p_nx_n)\frac{dx_1\,dx_2\ldots dx_{n-1}}{\sqrt{1-x_1^2-x_2^2-\ldots-x_{n-1}^2}}=$$

$$=\frac{2\sqrt{\pi^{n-1}}}{\Gamma\left(\frac{n-1}{2}\right)}\int_0^\pi f\left(\sqrt{p_1^2+p_2^2+\ldots+p_n^2}\cos x\right)\sin^{n-2}x\,dx \qquad [n\geqslant 3],$$

where $f(x)$ is continuous on the interval $\left[-\sqrt{p_1^2+p_2^2+\ldots+p_n^2},\ \sqrt{p_1^2+p_2^2+\ldots+p_n^2}\right]$.

FI III 489

4.645 Suppose that two functions $f(x_1, x_2, \ldots, x_n)$ and $g(x_1, x_2, \ldots, x_n)$ are continuous in a closed bounded region D and that the smallest and greatest values of the function g in D are m and M respectively. Let $\varphi(u)$ denote a function that is continuous for $m\leqslant u\leqslant M$. We denote by $\psi(u)$ the integral

1. $$\psi(u)=\int\int\cdots\int_{m\leqslant g(x_1,x_2,\ldots,x_n)\leqslant u} f(x_1, x_2, \ldots, x_n)\,dx_1\,dx_2\ldots dx_n,$$

over that portion of the region D on which the inequality $m\leqslant g(x_1, x_2, \ldots, x_n)\leqslant u$ is satisfied. Then

2. $$\int\int\cdots\int_{m\leqslant g(x_1,x_2,\ldots x_n)\leqslant M} f(x_1, x_2, \ldots, x_n)\,\varphi[g(x_1, x_2, \ldots, x_n)]\,dx_1\,dx_2\ldots dx_n=$$

$$=(S)\int_m^M\varphi(u)\,d\psi(u)=(R)\int_m^M\varphi(u)\frac{d\psi(u)}{du}\,du,$$

where the middle integral must be understood in the sense of Stieltjes. If the derivative $\frac{d\psi}{du}$ exists and is continuous, the Riemann integral on the right exists.

M may be $+\infty$ in formula 4.645 2., in which case $\int_m^{+\infty}$ should be understood to mean $\lim\limits_{M\to+\infty}\int_m^M$.

4.646 $$\int\int\cdots\int_{\substack{x_1\geqslant 0,\ x_2\geqslant 0,\ \ldots,\ x_n\geqslant 0\\ x_1+x_2+\ldots+x_n\leqslant 1}}\frac{x_1^{p_1-1}x_2^{p_2-1}\ldots x_n^{p_n-1}}{(q_1x_1+q_2x_2+\ldots q_nx_n)^n}\,dx_1\,dx_2\ldots dx_n=$$

$$=\frac{\Gamma(p_1)\Gamma(p_2)\ldots\Gamma(p_n)}{\Gamma(p_1+p_2+\ldots+p_n-r+1)\Gamma(r)}\int_0^\infty\frac{x^{r-1}\,dx}{(1+q_1x)^{p_1}(1+q_2x)^{p_2}\ldots(1+q_nx)^{p_n}}$$

$[p_1>0,\ p_2>0,\ \ldots,\ p_n>0,\ \ q_1>0,\ q_2>0,\ \ldots,\ q_n>0,$

$p_1+p_2+\ldots+p_n>r>0]$. FI III 493

4.647 $$\underset{0\leq x_1^2+x_2^2+\ldots+x_n^2\leq 1}{\int\int\ldots\int} \exp\left\{\frac{p_1x_1+p_2x_2+\ldots+p_nx_n}{\sqrt{x_1^2+x_2^2+\ldots+x_n^2}}\right\} dx_1\,dx_2\,\ldots\,dx_n =$$

$$= \frac{2\sqrt{\pi^n}}{n\,(p_1^2+p_2^2+\ldots+p_n^2)^{\frac{n}{4}-\frac{1}{2}}} I_{\frac{n}{2}-1}\left(\sqrt{p_1^2+p_2^2+\ldots+p_n^2}\right).$$ FI III 495

4.648 $$\int_0^\infty\int_0^\infty\ldots\int_0^\infty \exp\left[-\left(x_1+x_2+\ldots+x_n+\frac{\lambda^{n+1}}{x_1x_2\ldots x_n}\right)\right]\times$$

$$\times\, x_1^{\frac{1}{n+1}-1} x_2^{\frac{2}{n+1}-1}\ldots x_n^{\frac{n}{n+1}-1}\,dx_1\,dx_2\,\ldots\,dx_n =$$

$$= \frac{1}{\sqrt{n+1}}(2\pi)^{\frac{n}{2}} e^{-(n+1)\lambda}.$$ FI III 496

5. INDEFINITE INTEGRALS OF SPECIAL FUNCTIONS

5.1 Elliptic Integrals and Functions

5.11 Complete elliptic integrals

5.111

1. $$\int \boldsymbol{K}(k)\, k^{2p+3}\, dk = \frac{1}{(2p+3)^2}\left\{4(p+1)^2 \int \boldsymbol{K}(k)\, k^{2p+1}\, dk + \right.$$
$$\left. + k^{2p+2}\left[\boldsymbol{E}(k) - (2p+3)\,\boldsymbol{K}(k)\, k'^2\right]\right\}.$$ BY (610.04)

2. $$\int \boldsymbol{E}(k)\, k^{2p+3}\, dk = \frac{1}{4p^2+16p+15}\left\{4(p+1)^2 \int \boldsymbol{E}(k)\, k^{2p+1}\, dk - \right.$$
$$\left. - \boldsymbol{E}(k)\, k^{2p+2}\left[(2p+3)\, k'^2 - 2\right] - k^{2p+2} k'^2 \boldsymbol{K}(k)\right\}$$ BY (611.04)

5.112

1. $$\int \boldsymbol{K}(k)\, dk = \frac{\pi k}{2}\left[1 + \sum_{j=1}^{\infty} \frac{[(2j)!]^2\, k^{2j}}{(2j+1)\, 2^{4j}\, (j!)^4}\right].$$ BY (610.00)

2. $$\int \boldsymbol{E}(k)\, dk = \frac{\pi k}{2}\left[1 - \sum_{j=1}^{\infty} \frac{[2j]^2\, k^{2j}}{(4j^2-1)\, 2^{4j}\, (j!)^4}\right].$$ BY (611.00)

3. $$\int \boldsymbol{K}(k)\, k\, dk = \boldsymbol{E}(k) - k'^2 \boldsymbol{K}(k).$$ BY (610.01)

4. $$\int \boldsymbol{E}(k)\, k\, dk = \frac{1}{3}\left[(1+k^2)\,\boldsymbol{E}(k) - k'^2 \boldsymbol{K}(k)\right].$$ BY (611.01)

5. $$\int \boldsymbol{K}(k)\, k^3\, dk = \frac{1}{9}\left[(4+k^2)\,\boldsymbol{E}(k) - k'^2(4+3k^2)\,\boldsymbol{K}(k)\right].$$ BY (610.02)

6. $$\int \boldsymbol{E}(k)\, k^3\, dk = \frac{1}{45}\left[(4+k^2+9k^4)\,\boldsymbol{E}(k) - k'^2(4+3k^2)\,\boldsymbol{K}(k)\right].$$ BY (611.02)

7. $$\int \boldsymbol{K}(k)\, k^5\, dk = \frac{1}{225}\left[(64+16k^2+9k^4)\,\boldsymbol{E}(k) - \right.$$
$$\left. - k'^2(64+48k^2+45k^4)\,\boldsymbol{K}(k)\right].$$ BY (610.03)

8. $$\int \boldsymbol{E}(k)\, k^5\, dk = \frac{1}{1575}\left[(64+16k^2+9k^4+225k^6)\,\boldsymbol{E}(k) - \right.$$
$$\left. - k'^2(64+48k^2+45k^4)\,\boldsymbol{K}(k)\right].$$ BY (611.03)

9. $\int \frac{\boldsymbol{K}(k)}{k^2}\,dk = -\frac{\boldsymbol{E}(k)}{k}$ BY (612.05)

10. $\int \frac{\boldsymbol{E}(k)}{k^2}\,dk = \frac{1}{k}\left[k'^2\boldsymbol{K}(k) - 2\boldsymbol{E}(k)\right].$ BY (612.02)

11. $\int \frac{\boldsymbol{E}(k)}{k'^2}\,dk = k\boldsymbol{K}(k).$ BY (612.01)

12. $\int \frac{\boldsymbol{E}(k)}{k^4}\,dk = \frac{1}{9k^3}\left[2(k^2-2)\boldsymbol{E}(k) + k'^2\boldsymbol{K}(k)\right].$ BY (612.03)

13. $\int \frac{k\boldsymbol{E}(k)}{k'^2}\,dk = \boldsymbol{K}(k) - \boldsymbol{E}(k).$ BY (612.04)

5.113

1. $\int \left[\boldsymbol{K}(k) - \boldsymbol{E}(k)\right]\frac{dk}{k} = -\boldsymbol{E}(k).$ BY (612.06)

2. $\int \left[\boldsymbol{E}(k) - k'^2\boldsymbol{K}(k)\right]\frac{dk}{k} = 2\boldsymbol{E}(k) - k'^2\boldsymbol{K}(k).$ BY (612.09)

3. $\int \left[(1+k^2)\boldsymbol{K}(k) - \boldsymbol{E}(k)\right]\frac{dk}{k} = -k'^2\boldsymbol{K}(k).$ BY (612.12)

4. $\int \left[\boldsymbol{K}(k) - \boldsymbol{E}(k)\right]\frac{dk}{k^2} = \frac{1}{k}\left[\boldsymbol{E}(k) - k'^2\boldsymbol{K}(k)\right].$ BY (612.07)

5. $\int \left[\boldsymbol{E}(k) - k'^2\boldsymbol{K}(k)\right]\frac{dk}{k^2k'^2} = \frac{1}{k}\left[\boldsymbol{K}(k) - \boldsymbol{E}(k)\right].$

6. $\int \left[(1+k^2)\boldsymbol{E}(k) - k'^2\boldsymbol{K}(k)\right]\frac{dk}{kk'^4} = \frac{\boldsymbol{E}(k)}{k'^2}.$ BY (612.13)

5.114 $\int \frac{k\boldsymbol{K}(k)\,dk}{\left[\boldsymbol{E}(k) - k'^2\boldsymbol{K}(k)\right]^2} = \frac{1}{k'^2\boldsymbol{K}(k) - \boldsymbol{E}(k)}.$ BY (612.11)

5.115

1. $\int \Pi\left(\frac{\pi}{2}, r^2, k\right)k\,dk = (k^2 - r^2)\,\Pi\left(\frac{\pi}{2}, r^2, k\right) - \boldsymbol{K}(k) + \boldsymbol{E}(k).$ BY (612.14)

2. $\int \left[\boldsymbol{K}(k) - \Pi\left(\frac{\pi}{2}, r^2, k\right)\right]k\,dk = k^2\boldsymbol{K}(k) - (k^2 - r^2)\,\Pi\left(\frac{\pi}{2}, r^2, k\right).$ BY (612.15)

3. $\int \left[\frac{\boldsymbol{E}(k)}{k'^2} + \Pi\left(\frac{\pi}{2}, r^2, k\right)\right]k\,dk = (k^2 - r^2)\,\Pi\left(\frac{\pi}{2}, r^2, k\right).$ BY (612.16)

5.12 Elliptic integrals

5.121 $\int_0^x \frac{F(x, k)\,dx}{\sqrt{1-k^2\sin^2 x}} = \frac{[F(x, k)]^2}{2} \qquad \left[0 < x \leqslant \frac{\pi}{2}\right].$ BY (630.01)

5.122 $\int_0^x E(x, k)\sqrt{1-k^2\sin^2 x}\,dx = \frac{[E(x, k)]^2}{2}.$ BY (630.32)

5.123

1. $$\int_0^x F(x, k)\sin x\,dx = -\cos x\,F(x, k) + \frac{1}{k}\arcsin(k\sin x).$$

BY (630.11)

2. $$\int_0^x F(x, k)\cos x\,dx = \sin x\,F(x, k) + \frac{1}{k}\operatorname{Arch}\sqrt{\frac{1-k^2\sin^2 x}{k'^2}} - \frac{1}{k}\operatorname{Arch}\left(\frac{1}{k'}\right).$$

BY (630.21)

5.124

1. $$\int_0^x E(x, k)\sin x\,dx = -\cos x\,E(x, k) + \frac{1}{2k}\left[k\sin x\sqrt{1-k^2\sin^2 x} + \arcsin(k\sin x)\right].$$

BY (630.12)

2. $$\int_0^x E(x, k)\cos x\,dx = \sin x\,E(x, k) + \frac{1}{2k}\left[k\cos x\sqrt{1-k^2\sin^2 x} - k'^2\operatorname{Arch}\sqrt{\frac{1-k^2\sin^2 x}{k'^2}} - k + k'^2\operatorname{Arch}\left(\frac{1}{k'}\right)\right].$$

BY (630.22)

5.125

1. $$\int_0^x \Pi(x, \alpha^2, k)\sin x\,dx = -\cos x\,\Pi(x, \alpha^2, k) + \frac{1}{\sqrt{k^2-\alpha^2}}\operatorname{arctg}\left[\sqrt{\frac{k^2-\alpha^2}{1-k^2\sin^2 x}}\sin x\right] \qquad [\alpha^2 < k^2];$$

$$= -\cos x\,\Pi(x, \alpha^2, k) + \frac{1}{\sqrt{\alpha^2-k^2}}\operatorname{Arth}\left[\sqrt{\frac{\alpha^2-k^2}{1-k^2\sin^2 x}}\sin x\right] \qquad [\alpha^2 > k^2].$$

BY (630.13)

2. $$\int_0^x \Pi(x, \alpha^2, k)\cos x\,dx = \sin x\,\Pi(x, \alpha^2, k) - f + f_0,$$

where

$$f = \frac{1}{2\sqrt{(1-\alpha^2)(\alpha^2-k^2)}}\operatorname{arctg}\left[\frac{2(1-\alpha^2)(\alpha^2-k^2)+(1-\alpha^2\sin^2 x)(2k^2-\alpha^2-\alpha^2k^2)}{2\alpha^2\sqrt{(1-\alpha^2)(\alpha^2-k^2)}\cos x\sqrt{1-k^2\sin^2 x}}\right]$$

for $(1-\alpha^2)(\alpha^2-k^2) > 0$;

$$= \frac{1}{2\sqrt{(\alpha^2-1)(\alpha^2-k^2)}}\ln\left[\frac{2(\alpha^2-1)(\alpha^2-k^2)+(1-\alpha^2\sin^2 x)(\alpha^2+\alpha^2k^2-2k^2)}{1-\alpha^2\sin^2 x} + \frac{2\alpha^2\sqrt{(\alpha^2-1)(\alpha^2-k^2)}\cos x\sqrt{1-k^2\sin^2 x}}{1-\alpha^2\sin^2 x}\right]$$

for $(1-\alpha^2)(\alpha^2-k^2) < 0$,

f_0 is the value of f at $x = 0$.

BY (630.23)

Integration with respect to the modulus

5.126 $\int F(x, k)\, k\, dk = E(x, k) - k'^2 F(x, k) + \left(\sqrt{1-k^2\sin^2 x} - 1\right)\operatorname{ctg} x.$ BY (613.01)

5.127 $\int E(x, k)\, k\, dk = \frac{1}{3}\left[(1+k^2)E(x, k) - k'^2 F(x, k) + \right.$
$\left. + \left(\sqrt{1-k^2\sin^2 x} - 1\right)\operatorname{ctg} x\right].$ BY (613.02)

5.128 $\int \Pi(x, r^2, k)\, k\, dk = (k^2 - r^2)\Pi(x, r^2, k) - F(x, k) + E(x, k) +$
$+ \left(\sqrt{1-k^2\sin^2 x} - 1\right)\operatorname{ctg} x.$ BY (613.03)

5.13 Jacobian elliptic functions

5.131

1. $\int \operatorname{sn}^m u\, du = \frac{1}{m+1}\left[\operatorname{sn}^{m+1} u \operatorname{cn} u \operatorname{dn} u + (m+2)(1+k^2)\int \operatorname{sn}^{m+2} u\, du - \right.$
$\left. - (m+3)k^2 \int \operatorname{sn}^{m+4} u\, du\right].$ SI 259, PE (567)

2. $\int \operatorname{cn}^m u\, du = \frac{1}{(m+1)k'^2}\left[-\operatorname{cn}^{m+1} u \operatorname{sn} u \operatorname{dn} u + \right.$
$\left. + (m+2)(1-2k^2)\int \operatorname{cn}^{m+2} u\, du + (m+3)k^2 \int \operatorname{cn}^{m+4} u\, du\right].$ PE (568)

3. $\int \operatorname{dn}^m u\, du = \frac{1}{(m+1)k'^2}\left[k^2 \operatorname{dn}^{m+1} u \operatorname{sn} u \operatorname{cn} u + \right.$
$\left. + (m+2)(2-k^2)\int \operatorname{dn}^{m+2} u\, du - (m+3)\int \operatorname{dn}^{m+4} u\, du\right].$ PE (569)

By using formulas **5.131**, we can reduce the integrals $\int \operatorname{sn}^m u\, du$, $\int \operatorname{cn}^m u\, du$, $\int \operatorname{dn}^m u\, du$ to the integrals **5.132**, **5.133** and **5.134**.

5.132

1. $\int \frac{du}{\operatorname{sn} u} = \ln \frac{\operatorname{sn} u}{\operatorname{cn} u + \operatorname{dn} u};$ ZH 87(164)
$= \ln \frac{\operatorname{dn} u - \operatorname{cn} u}{\operatorname{sn} u}.$ SI 266(4)

2. $\int \frac{du}{\operatorname{cn} u} = \frac{1}{k'} \ln \frac{k' \operatorname{sn} u + \operatorname{dn} u}{\operatorname{cn} u}.$ SI 266(5)

3. $\int \frac{du}{\operatorname{dn} u} = \frac{1}{k'} \operatorname{arctg} \frac{k' \operatorname{sn} u - \operatorname{cn} u}{k' \operatorname{sn} u + \operatorname{cn} u};$ ZH 88(166)
$= \frac{1}{k'} \arccos \frac{\operatorname{cn} u}{\operatorname{dn} u};$ JA
$= \frac{1}{ik'} \ln \frac{\operatorname{cn} u + ik' \operatorname{sn} u}{\operatorname{dn} u};$ SI 266(6)
$= \frac{1}{k'} \arcsin \frac{k' \operatorname{sn} u}{\operatorname{dn} u}.$ JA

5.133

1. $\int \operatorname{sn} u\, du = \frac{1}{k} \ln (\operatorname{dn} u - k \operatorname{cn} u);$ ZH 87(161)

$= \frac{1}{k} \operatorname{Arch} \frac{\operatorname{dn} u - k^2 \operatorname{cn} u}{1-k^2};$ JA

$= \frac{1}{k} \operatorname{Arsh} \left(k \frac{\operatorname{dn} u - \operatorname{cn} u}{1-k^2} \right);$ JA

$= -\frac{1}{k} \ln (\operatorname{dn} u + k \operatorname{cn} u).$ SI 365(1)

2. $\int \operatorname{cn} u\, du = \frac{1}{k} \arccos (\operatorname{dn} u);$ ZH 87(162)

$= \frac{i}{k} \ln (\operatorname{dn} u - ik \operatorname{sn} u);$ SI 265(2)a, ZH 87(162)

$= \frac{1}{k} \arcsin (k \operatorname{sn} u).$ JA

3. $\int \operatorname{dn} u\, du = \arcsin (\operatorname{sn} u);$ ZH 87(163)

$= \operatorname{am} u = i \ln (\operatorname{cn} u - i \operatorname{sn} u).$ SI 266(3), ZH 87(163)

5.134

1. $\int \operatorname{sn}^2 u\, du = \frac{1}{k^2} [u - E (\operatorname{am} u, k)].$ PE (564)

2. $\int \operatorname{cn}^2 u\, du = \frac{1}{k^2} [E (\operatorname{am} u, k) - k'^2 u].$ PE (565)

3. $\int \operatorname{dn}^2 u\, du = E (\operatorname{am} u, k).$ PE (566)

5.135

1. $\int \frac{\operatorname{sn} u}{\operatorname{cn} u} du = \frac{1}{k'} \ln \frac{\operatorname{dn} u + k'}{\operatorname{cn} u};$ SI 266(7)

$= \frac{1}{2k'} \ln \frac{\operatorname{dn} u + k'}{\operatorname{dn} u - k'}.$ ZH 88(167)

2. $\int \frac{\operatorname{sn} u}{\operatorname{dn} u} du = \frac{i}{kk'} \ln \frac{ik' - k \operatorname{cn} u}{\operatorname{dn} u};$ SI 266(8)

$= \frac{1}{kk'} \operatorname{arcctg} \frac{k \operatorname{cn} u}{k'}.$ ZH 88(169)

3. $\int \frac{\operatorname{cn} u}{\operatorname{sn} u} du = \ln \frac{1 - \operatorname{dn} u}{\operatorname{sn} u};$ SI 266(10)

$= \frac{1}{2} \ln \frac{1 - \operatorname{dn} u}{1 + \operatorname{dn} u}.$ ZH 88(168)

4. $\int \frac{\operatorname{cn} u}{\operatorname{dn} u} du = -\frac{1}{k} \ln \frac{1 - k \operatorname{sn} u}{\operatorname{dn} u};$ SI 266(9)

$= \frac{1}{2k} \ln \frac{1 + k \operatorname{sn} u}{1 - k \operatorname{sn} u}.$ ZH 88(171)

5. $\int \frac{\operatorname{dn} u}{\operatorname{cn} u} du = \frac{1}{2} \ln \frac{1 + \operatorname{sn} u}{1 - \operatorname{sn} u};$ ZH 88(172)

$= \ln \frac{1 + \operatorname{sn} u}{\operatorname{cn} u}.$ JA

6. $\int \frac{\operatorname{dn} u}{\operatorname{sn} u} du = \frac{1}{2} \ln \frac{1 - \operatorname{cn} u}{1 + \operatorname{cn} u}.$ ZH 87(170)

5.136

1. $\int \operatorname{sn} u \operatorname{cn} u \, du = -\frac{1}{k^2} \operatorname{dn} u.$

2. $\int \operatorname{sn} u \operatorname{dn} u \, du = -\operatorname{cn} u.$

3. $\int \operatorname{cn} u \operatorname{dn} u \, du = \operatorname{sn} u.$

5.137

1. $\int \frac{\operatorname{sn} u}{\operatorname{cn}^2 u} du = \frac{1}{k'^2} \frac{\operatorname{dn} u}{\operatorname{cn} u}.$ ZH 88(173)

2. $\int \frac{\operatorname{sn} u}{\operatorname{dn}^2 u} du = -\frac{1}{k'^2} \frac{\operatorname{cn} u}{\operatorname{dn} u}.$ ZH 88(175)

3. $\int \frac{\operatorname{cn} u}{\operatorname{sn}^2 u} du = -\frac{\operatorname{dn} u}{\operatorname{sn} u}.$ ZH 88(174)

4. $\int \frac{\operatorname{cn} u}{\operatorname{dn}^2 u} du = \frac{\operatorname{sn} u}{\operatorname{dn} u}.$ ZH 88(177)

5. $\int \frac{\operatorname{dn} u}{\operatorname{sn}^2 u} du = -\frac{\operatorname{cn} u}{\operatorname{sn} u}.$ ZH 88(176)

6. $\int \frac{\operatorname{dn} u}{\operatorname{cn}^2 u} du = \frac{\operatorname{sn} u}{\operatorname{cn} u}.$ ZH 88(178)

5.138

1. $\int \frac{\operatorname{cn} u}{\operatorname{sn} u \operatorname{dn} u} du = \ln \frac{\operatorname{sn} u}{\operatorname{dn} u}.$ ZH 88(183)

2. $\int \frac{\operatorname{sn} u}{\operatorname{cn} u \operatorname{dn} u} du = \frac{1}{k'^2} \ln \frac{\operatorname{dn} u}{\operatorname{cn} u}.$ ZH 88(182)

3. $\int \frac{\operatorname{dn} u}{\operatorname{sn} u \operatorname{cn} u} du = \ln \frac{\operatorname{sn} u}{\operatorname{cn} u}.$ ZH 88(184)

5.139

1. $\int \frac{\operatorname{cn} u \operatorname{dn} u}{\operatorname{sn} u} du = \ln \operatorname{sn} u.$ ZH 88(179)

2. $\int \frac{\operatorname{sn} u \operatorname{dn} u}{\operatorname{cn} u} du = \ln \frac{1}{\operatorname{cn} u}.$ ZH 88(180)

3. $\int \frac{\operatorname{sn} u \operatorname{cn} u}{\operatorname{dn} u} du = -\frac{1}{k^2} \ln \operatorname{dn} u.$ ZH 88(181)

5.14 Weierstrass elliptic functions

5.141

1. $\int \wp(u) \, du = -\zeta(u).$

2. $\int \wp^2(u) \, du = \frac{1}{6} \wp'(u) + \frac{1}{12} g_2 u.$ ZH 120(192)

3. $\int \wp^3(u) \, du = \frac{1}{120} \wp'''(u) - \frac{3}{20} g_2 \zeta(u) + \frac{1}{10} g_3 u.$ ZH 120(193)

4. $$\int \frac{du}{\wp(u)-\wp(v)} = \frac{1}{\wp'(v)}\left[2u\zeta(v) + \ln\frac{\sigma(u-v)}{\sigma(u+v)}\right].$$ ZH 120(194)

5. $$\int \frac{\alpha\wp(u)+\beta}{\gamma\wp(u)+\delta}\,du = \frac{\alpha u}{\gamma} - \frac{\alpha\delta-\beta\gamma}{\gamma^2\wp'(v)}\left[\ln\frac{\sigma(u+v)}{\sigma(u-v)} - 2u\zeta(v)\right],$$

where $\wp'(v) = -\frac{\delta}{\gamma}$. ZH 120(195)

5.2 The Exponential-Integral Function

5.21 The exponential-integral function

5.211 $$\int_x^\infty \mathrm{Ei}(-\beta x)\,\mathrm{Ei}(-\gamma x)\,dx = \left(\frac{1}{\beta}+\frac{1}{\gamma}\right)\mathrm{Ei}[-(\beta+\gamma)x] -$$

$$- x\,\mathrm{Ei}(-\beta x)\,\mathrm{Ei}(-\gamma x) - \frac{e^{-\beta x}}{\beta}\mathrm{Ei}(-\gamma x) - \frac{e^{-\gamma x}}{\gamma}\mathrm{Ei}(-\beta x)$$

$[\mathrm{Re}(\beta+\gamma) > 0]$. NT 53(2)

5.22 Combinations of the exponential-integral function and powers

5.221

1. $$\int_x^\infty \frac{\mathrm{Ei}[-a(x+b)]}{x^{n+1}}\,dx = \left[\frac{1}{x^n} - \frac{(-1)^n}{b^n}\right]\frac{\mathrm{Ei}[-a(x+b)]}{n} +$$

$$+ \frac{e^{-ab}}{n}\sum_{k=0}^{n-1}\frac{(-1)^{n-k-1}}{b^{n-k}}\int_x^\infty \frac{e^{-ax}}{x^{k+1}}\,dx \qquad [a>0,\ b>0].$$ NT 52(3)

2. $$\int_x^\infty \frac{\mathrm{Ei}[-a(x+b)]}{x^2}\,dx = \left(\frac{1}{x}+\frac{1}{b}\right)\mathrm{Ei}[-a(x+b)] - \frac{e^{-ab}\,\mathrm{Ei}(-ax)}{b}$$

$[a>0,\ b>0]$. NT 52(4)

5.23 Combinations of the exponential-integral and the exponential

5.231

1. $$\int_0^x e^x\,\mathrm{Ei}(-x)\,dx = -\ln x - \boldsymbol{C} + e^x\,\mathrm{Ei}(-x).$$ ET II 308(11)

2. $$\int_0^x e^{-\beta x}\,\mathrm{Ei}(-\alpha x)\,dx = -\frac{1}{\beta}\left\{e^{-\beta x}\,\mathrm{Ei}(-\alpha x) + \ln\left(1+\frac{\beta}{\alpha}\right) - \right.$$

$$\left. - \mathrm{Ei}[-(\alpha+\beta)x]\right\}.$$ ET II 308(12)

5.3 The Sine-Integral and the Cosine-Integral

5.31

1. $$\int \cos\alpha x\,\mathrm{ci}(\beta x)\,dx = \frac{\sin\alpha x\,\mathrm{ci}(\beta x)}{\alpha} - \frac{\mathrm{si}(\alpha x+\beta x)+\mathrm{si}(\alpha x-\beta x)}{2\alpha}.$$ NT 49(1)

2. $$\int \sin\alpha x\,\mathrm{ci}(\beta x)\,dx = -\frac{\cos\alpha x\,\mathrm{ci}(\beta x)}{\alpha} + \frac{\mathrm{ci}(\alpha x+\beta x)+\mathrm{ci}(\alpha x-\beta x)}{2\alpha}.$$ NT 49(2)

5.32

1. $$\int \cos \alpha x \operatorname{si}(\beta x)\, dx = \frac{\sin \alpha x \operatorname{si}(\beta x)}{\alpha} + \frac{\operatorname{ci}(\alpha x + \beta x) - \operatorname{ci}(\alpha x - \beta x)}{2\alpha}.$$ NT 49(3)

2. $$\int \sin \alpha x \operatorname{si}(\beta x)\, dx = -\frac{\cos \alpha x \operatorname{si}(\beta x)}{\alpha} + \frac{\operatorname{si}(\alpha x + \beta x) - \operatorname{si}(\alpha x - \beta x)}{2\alpha}.$$ NT 49(4)

5.33

1. $$\int \operatorname{ci}(\alpha x) \operatorname{ci}(\beta x)\, dx = x \operatorname{ci}(\alpha x) \operatorname{ci}(\beta x) + \frac{1}{2\alpha}(\operatorname{si}(\alpha x + \beta x) + \operatorname{si}(\alpha x - \beta x)) + $$
$$+ \frac{1}{2\beta}(\operatorname{si}(\alpha x + \beta x) + \operatorname{si}(\beta x - \alpha x)) - \frac{1}{\alpha} \sin \alpha x \operatorname{ci}(\beta x) - \frac{1}{\beta} \sin \beta x \operatorname{ci}(\alpha x).$$ NT 53(5)

2. $$\int \operatorname{si}(\alpha x) \operatorname{si}(\beta x)\, dx = x \operatorname{si}(\alpha x) \operatorname{si}(\beta x) - \frac{1}{2\beta}(\operatorname{si}(\alpha x + \beta x) + \operatorname{si}(\alpha x - \beta x)) - $$
$$- \frac{1}{2\alpha}(\operatorname{si}(\alpha x + \beta x) + \operatorname{si}(\beta x - \alpha x)) + \frac{1}{\alpha} \cos \alpha x \operatorname{si}(\beta x) + \frac{1}{\beta} \cos \beta x \operatorname{si}(\alpha x).$$ NT 54(6)

3. $$\int \operatorname{si}(\alpha x) \operatorname{ci}(\beta x)\, dx = x \operatorname{si}(\alpha x) \operatorname{ci}(\beta x) + \frac{1}{\alpha} \cos \alpha x \operatorname{ci}(\beta x) - $$
$$- \frac{1}{\beta} \sin \beta x \operatorname{si}(\alpha x) - \left(\frac{1}{2\alpha} + \frac{1}{2\beta}\right) \operatorname{ci}(\alpha x + \beta x) - \left(\frac{1}{2\alpha} - \frac{1}{2\beta}\right) \operatorname{ci}(\alpha x - \beta x).$$ NT 54(10)

5.34

1. $$\int_x^\infty \operatorname{si}[a(x+b)] \frac{dx}{x^2} = \left(\frac{1}{x} + \frac{1}{b}\right) \operatorname{si}[a(x+b)] - $$
$$- \frac{\cos ab \operatorname{si}(ax) + \sin ab \operatorname{ci}(ax)}{b} \qquad [a > 0,\ b > 0].$$ NT 52(6)

2. $$\int_x^\infty \operatorname{ci}[a(x+b)] \frac{dx}{x^2} = \left(\frac{1}{x} + \frac{1}{b}\right) \operatorname{ci}[a(x+b)] + $$
$$+ \frac{\sin ab \operatorname{si}(ax) - \cos ab \operatorname{ci}(ax)}{b} \qquad [a > 0,\ b > 0].$$ NT 52(5)

5.4 The Probability Integral and Fresnel Integrals

5.41 $$\int \Phi(\alpha x)\, dx = x\Phi(\alpha x) + \frac{e^{-\alpha^2 x^2}}{\alpha\sqrt{\pi}}.$$ NT 12(20)a

5.42 $$\int S(\alpha x)\, dx = xS(\alpha x) + \frac{\cos \alpha^2 x^2}{\alpha\sqrt{2\pi}}.$$ NT 12(22)a

5.43 $$\int C(\alpha x)\, dx = xC(\alpha x) - \frac{\sin \alpha^2 x^2}{\alpha\sqrt{2\pi}}.$$ NT 12(21)a

5.5 Bessel Functions

5.51 $$\int J_p(x)\, dx = 2 \sum_{k=0}^{\infty} J_{p+2k+1}(x).$$ JA, MO 30

5.52

1. $\int x^{p+1} Z_p(x)\,dx = x^{p+1} Z_{p+1}(x)$ * WA 146(1)

2. $\int x^{-p+1} Z_p(x)\,dx = -x^{-p+1} Z_{p-1}(x)$* . WA 146(2)

5.53 $\int \left[(\alpha^2-\beta^2)x - \frac{p^2-q^2}{x}\right] Z_p(\alpha x)\,\mathfrak{Z}_p(\beta x)\,dx = \beta x Z_p(\alpha x)\,\mathfrak{Z}_{q-1}(\beta x) - $
$- \alpha x Z_{p-1}(\alpha x)\,\mathfrak{Z}_q(\beta x) + (p-q)\,Z_p(\alpha x)\,\mathfrak{Z}_q(\beta x)$ *

JA, MO 30, WA 148(7)a

5.54

1. $\int x Z_p(\alpha x)\,\mathfrak{Z}_p(\beta x)\,dx = \frac{\beta x Z_p(\alpha x)\,\mathfrak{Z}_{p-1}(\beta x) - \alpha x Z_{p-1}(\alpha x)\,\mathfrak{Z}_p(\beta x)}{\alpha^2-\beta^2}$ * .

WA 148(8)a

2. $\int x\,[Z_p(\alpha x)]^2\,dx = \frac{x^2}{2}\{[Z_p(\alpha x)]^2 - Z_{p-1}(\alpha x)\,Z_{p+1}(\alpha x)\}$ * . WA 149(11)

5.55 $\int \frac{1}{x} Z_p(\alpha x)\,\mathfrak{Z}_q(\alpha x)\,dx = \alpha x\,\frac{Z_{p-1}(\alpha x)\,\mathfrak{Z}_q(\alpha x) - Z_p(\alpha x)\,\mathfrak{Z}_{q-1}(\alpha x)}{p^2-q^2} - $
$- \frac{Z_p(\alpha x)\,\mathfrak{Z}_q(\alpha x)}{p+q}$ * WA 149(13)

5.56

1. $\int Z_1(x)\,dx = -Z_0(x)$ * . JA

2. $\int x Z_0(x)\,dx = x Z_1(x)$ * . JA

---

*In formulas 5.52—5.56, $Z_p(x)$ and $\mathfrak{Z}_p(x)$ are arbitrary Bessel functions.

6.-7. DEFINITE INTEGRALS OF SPECIAL FUNCTIONS

6.1 Elliptic Integrals and Functions

6.11 Forms containing $F(x, k)$

6.111 $$\int_0^{\frac{\pi}{2}} F(x, k) \operatorname{ctg} x \, dx = \frac{\pi}{4} \boldsymbol{K}(k') + \frac{1}{2} \ln k \, \boldsymbol{K}(k).$$ BI ((350))(1)

6.112

1. $$\int_0^{\frac{\pi}{2}} F(x, k) \frac{\sin x \cos x}{1 + k \sin^2 x} dx = \frac{1}{4k} \boldsymbol{K}(k) \ln \frac{(1+k)\sqrt{k}}{2} + \frac{\pi}{16k} \boldsymbol{K}(k').$$ BI ((350))(6)

2. $$\int_0^{\frac{\pi}{2}} F(x, k) \frac{\sin x \cos x}{1 - k \sin^2 x} dx = \frac{1}{4k} \boldsymbol{K}(k) \ln \frac{2}{(1-k)\sqrt{k}} - \frac{\pi}{16k} \boldsymbol{K}(k').$$ BI ((350))(7)

3. $$\int_0^{\frac{\pi}{2}} F(x, k) \frac{\sin x \cos x}{1 - k^2 \sin^2 x} dx = -\frac{1}{2k^2} \ln k' \, \boldsymbol{K}(k).$$ BI ((350))(2)a, BY (802.12)a

6.113

1. $$\int_0^{\frac{\pi}{2}} F(x, k') \frac{\sin x \cos x \, dx}{\cos^2 x + k \sin^2 x} = \frac{1}{4(1-k)} \ln \frac{2}{(1+k)\sqrt{k}} \boldsymbol{K}(k').$$ BI ((350))(5)

2. $$\int_0^{\frac{\pi}{2}} F(x, k) \frac{\sin x \cos x}{1 - k^2 \sin^2 t \sin^2 x} \cdot \frac{dx}{\sqrt{1 - k^2 \sin^2 x}} = -\frac{1}{k^2 \sin t \cos t} \times$$ $$\times \left[\boldsymbol{K}(k) \operatorname{arctg}(k' \operatorname{tg} t) - \frac{\pi}{2} F(t, k) \right].$$ BI ((350))(12)

6.114 $$\int_u^v F(x, k)\frac{dx}{\sqrt{(\sin^2 x-\sin^2 u)(\sin^2 v-\sin^2 x)}} =$$
$$= \frac{1}{2\cos u \sin v} \boldsymbol{K}(k) \boldsymbol{K}\left(\sqrt{1-\operatorname{tg}^2 u \operatorname{ctg}^2 v}\right)$$
$$[k^2 = 1-\operatorname{ctg}^2 u \cdot \operatorname{ctg}^2 v].$$ BI ((351))(9)

6.115 $$\int_0^1 F(\arcsin x, k)\frac{x\,dx}{1+kx^2} = \frac{1}{4k}\boldsymbol{K}(k)\ln\frac{(1+k)\sqrt{k}}{2} + \frac{\pi}{16k}\boldsymbol{K}(k')$$
(cf. **6.112** 2.). BI ((466))(1)

This and similar formulas can be obtained from formulas **6.111** — **6.113** by means of the substitution $x = \arcsin t$.

6.12 Forms containing $\boldsymbol{E}(x, k)$

6.121 $$\int_0^{\frac{\pi}{2}} E(x, k)\frac{\sin x \cos x}{1-k^2\sin^2 x}\,dx = \frac{1}{2k^2}\{(1+k'^2)\boldsymbol{K}(k) - (2+\ln k')\boldsymbol{E}(k)\}.$$
BI ((350))(4)

6.122 $$\int_0^{\frac{\pi}{2}} E(x, k)\frac{dx}{\sqrt{1-k^2\sin^2 x}} = \frac{1}{2}\{\boldsymbol{E}(k)\boldsymbol{K}(k) - \ln k'\}.$$
BI ((350))(10), BY (630.02)

6.123 $$\int_0^{\frac{\pi}{2}} E(x, k)\frac{\sin x\cos x}{1-k^2\sin^2 t\sin^2 x}\cdot\frac{dx}{\sqrt{1-k^2\sin^2 x}} = -\frac{1}{k^2\sin t\cos t}\times$$
$$\times\left[\boldsymbol{E}(k)\operatorname{arctg}(k'\operatorname{tg} t) - \frac{\pi}{2}E(t, k) + \frac{\pi}{2}\operatorname{ctg} t\left(1-\sqrt{1-k^2\sin^2 t}\right)\right].$$
BI ((350))(13)

6.124 $$\int_u^v E(x, k)\frac{dx}{\sqrt{(\sin^2 x-\sin^2 u)(\sin^2 v-\sin^2 x)}} =$$
$$= \frac{1}{2\cos u\sin v}\boldsymbol{E}(k)\boldsymbol{K}\left(\sqrt{1-\frac{\operatorname{tg}^2 u}{\operatorname{tg}^2 v}}\right) + \frac{k^2\sin v}{2\cos u}\boldsymbol{K}\left(\sqrt{1-\frac{\sin^2 2u}{\sin^2 2v}}\right)$$
$$[k^2 = 1-\operatorname{ctg}^2 u\operatorname{ctg}^2 v].$$ BI ((351))(10)

6.13 Integration of elliptic integrals with respect to the modulus

6.131 $$\int_0^1 F(x, k)\,k\,dk = \frac{1-\cos x}{\sin x} = \operatorname{tg}\frac{x}{2}.$$ BY (616.03)

6.132 $\int_0^1 E(x, k)\, k\, dk = \frac{\sin^2 x + 1 - \cos x}{3 \sin x}$. BY (616.04)

6.133 $\int_0^1 \Pi(x, r^2, k)\, k\, dk = \operatorname{tg}\frac{x}{2} - r \ln \sqrt{\frac{1+r\sin x}{1-r\sin x}} - r^2\Pi(x, r^2, 0)$. BY (616.05)

6.14-6.15 Complete elliptic integrals

6.141

1. $\int_0^1 \boldsymbol{K}(k)\, dk = 2\boldsymbol{G}$. FI II 755

2. $\int_0^1 \boldsymbol{K}(k')\, dk = \frac{\pi^2}{4}$. BY (615.03)

6.142 $\int_0^1 \left(\boldsymbol{K}(k) - \frac{\pi}{2}\right)\frac{dk}{k} = \pi \ln 2 - 2\boldsymbol{G}$. BY (615.05)

6.143 $\int_0^1 \boldsymbol{K}(k)\frac{dk}{k'} = \boldsymbol{K}^2\left(\frac{\sqrt{2}}{2}\right)$. BY (615.08)

6.144 $\int_0^1 \boldsymbol{K}(k)\frac{dk}{1+k} = \frac{\pi^2}{8}$. BY (615.09)

6.145 $\int_0^1 \left(\boldsymbol{K}(k') - \ln\frac{4}{k}\right)\frac{dk}{k} = \frac{1}{12}[24(\ln 2)^2 - \pi^2]$. BY (615.13)

6.146 $n^2 \int_0^1 k^n \boldsymbol{K}(k)\, dk = (n-1)^2 \int_0^1 k^{n-2}\boldsymbol{K}(k)\, dk + 1$. BY (615.12)

6.147 $n \int_0^1 k^n \boldsymbol{K}(k')\, dk = (n-1)\int_0^1 k^{n-2}\boldsymbol{E}(k)\, dk \quad [n > 1]$

(see 6.152). BY (615.11)

6.148

1. $\int_0^1 \boldsymbol{E}(k)\, dk = \frac{1}{2} + \boldsymbol{G}$. BY (615.02)

2. $\int_0^1 \boldsymbol{E}(k')\, dk = \frac{\pi^2}{8}$. BY (615.04)

6.149

1. $\int_0^1 \left(\boldsymbol{E}(k) - \frac{\pi}{2}\right)\frac{dk}{k} = \pi \ln 2 - 2\boldsymbol{G} + 1 - \frac{\pi}{2}$. BY (615.06)

2. $\int\limits_0^1 (\boldsymbol{E}(k') - 1)\frac{dk}{k} = 2\ln 2 - 1.$ BY (615.07)

6.151 $\int\limits_0^1 \boldsymbol{E}(k)\frac{dk}{k'} = \frac{1}{8}\left[4\boldsymbol{K}^2\left(\frac{\sqrt{2}}{2}\right) + \frac{\pi^2}{\boldsymbol{K}^2\left(\frac{\sqrt{2}}{2}\right)}\right].$ BY (615.10)

6.152 $(n+2)\int\limits_0^1 k^n \boldsymbol{E}(k')\,dk = (n+1)\int\limits_0^1 k^n \boldsymbol{K}(k')\,dk \quad [n > 1]$

(see 6.147). BY (615.14)

6.153 $\int\limits_0^a \frac{\boldsymbol{K}(k)\,k\,dk}{k'^2\sqrt{a^2-k^2}} = \frac{\pi a}{2\sqrt{1-a^2}} \quad [a^2 < 1].$ LO I 252

6.154 $\int\limits_0^{\frac{\pi}{2}} \frac{\boldsymbol{E}(p\sin x)}{1-p^2\sin^2 x}\sin x\,dx = \frac{\pi}{2\sqrt{1-p^2}} \quad [p^2 < 1].$ FI II 489

6.16 The theta function

6.161

1. $\int\limits_0^\infty x^{s-1}\theta_2(0 \mid ix^2)\,dx = 2^s(1-2^{-s})\pi^{-\frac{s}{2}}\Gamma\left(\frac{1}{2}s\right)\zeta(s)$

$[\operatorname{Re} s > 2].$ ET I 339(20)

2. $\int\limits_0^\infty x^{s-1}[\theta_3(0 \mid ix^2) - 1]\,dx = \pi^{-\frac{s}{2}}\Gamma\left(\frac{1}{2}s\right)\zeta(s)$

$[\operatorname{Re} s > 2].$ ET I 339(21)

3. $\int\limits_0^\infty x^{s-1}[1 - \theta_4(0 \mid ix^2)]\,dx = (1-2^{1-s})\pi^{-\frac{1}{2}s}\Gamma\left(\frac{1}{2}s\right)\zeta(s)$

$[\operatorname{Re} s > 2].$ ET I 339(22)

4. $\int\limits_0^\infty x^{s-1}[\theta_4(0 \mid ix^2) + \theta_2(0 \mid ix^2) - \theta_3(0 \mid ix^2)]\,dx =$

$= -(2^s-1)(2^{1-s}-1)\pi^{-\frac{1}{2}s}\Gamma\left(\frac{1}{2}s\right)\zeta(s).$ ET I 339(24)

6.162

1. $\int\limits_0^\infty e^{-ax}\theta_4\left(\frac{b\pi}{2l}\,\middle|\,\frac{i\pi x}{l^2}\right)dx = \frac{l}{\sqrt{a}}\operatorname{ch}(b\sqrt{a})\operatorname{cosech}(l\sqrt{a})$

$[\operatorname{Re} a > 0, |b| \leqslant l].$ ET I 224(1)a

2. $\int\limits_0^\infty e^{-ax}\theta_1\left(\frac{b\pi}{2l}\,\middle|\,\frac{i\pi x}{l^2}\right)dx = -\frac{l}{\sqrt{a}}\operatorname{sh}(b\sqrt{a})\operatorname{sech}(l\sqrt{a})$

$[\operatorname{Re} a > 0, |b| \leqslant l].$ ET I 224(2)a

3. $\int_0^\infty e^{-ax}\theta_2\left(\frac{(1+b)\pi}{2l}\middle|\frac{i\pi x}{l^2}\right)dx = -\frac{l}{\sqrt{a}}\operatorname{sh}(b\sqrt{a})\operatorname{sech}(l\sqrt{a})$

$[\operatorname{Re} a > 0,\ |b| \leqslant l]$. ET I 224(3)a

4. $\int_0^\infty e^{-ax}\theta_3\left(\frac{(1+b)\pi}{2l}\middle|\frac{i\pi x}{l^2}\right)dx = \frac{l}{\sqrt{a}}\operatorname{ch}(b\sqrt{a})\operatorname{cosech}(l\sqrt{a})$

$[\operatorname{Re} a > 0,\ |b| \leqslant l]$. ET I 224(4)a

6.163 $\int_0^\infty e^{-(a-\mu)x}\theta_3(\pi\sqrt{\mu}\,x\,|\,i\pi x)\,dx = \frac{1}{2\sqrt{a}}\left[\operatorname{th}(\sqrt{a}+\sqrt{\mu}) + \operatorname{th}(\sqrt{a}-\sqrt{\mu})\right]$

$[\operatorname{Re} a > 0]$. ET I 224(7)a

6.164 $\int_0^\infty [\theta_4(0\,|\,ie^{2x}) + \theta_2(0\,|\,ie^{2x}) - \theta_3(0\,|\,ie^{2x})]\,e^{\frac{1}{2}x}\cos(ax)\,dx =$

$= \frac{1}{2}\left(2^{\frac{1}{2}+ia} - 1\right)\left(1 - 2^{\frac{1}{2}-ia}\right)\pi^{-\frac{1}{4}-\frac{1}{2}ia}\Gamma\left(\frac{1}{4}+\frac{1}{2}ia\right)\zeta\left(\frac{1}{2}+ia\right)$

$[a > 0]$. ET I 61(11)

6.165 $\int_0^\infty e^{\frac{1}{2}x}[\theta_3(0\,|\,ie^{2x}) - 1]\cos(ax)\,dx =$

$= 2(1+4a^2)^{-1}\left\{1 + \left[\left(a^2+\frac{1}{4}\right)\pi^{-\frac{1}{2}ia-\frac{1}{4}}\Gamma\left(\frac{1}{2}ia+\frac{1}{4}\right)\zeta\left(ia+\frac{1}{2}\right)\right]\right\}$

$[a > 0]$. ET I 62(12)

6.2-6.3 The Exponential-Integral Function and Functions Generated by It

6.21 The logarithm-integral

6.211 $\int_0^1 \operatorname{li}(x)\,dx = -\ln 2$. BI ((79))(5)

6.212

1. $\int_0^1 \operatorname{li}\left(\frac{1}{x}\right)x\,dx = 0$. BI ((255))(1)

2. $\int_0^1 \operatorname{li}(x)\,x^{p-1}\,dx = -\frac{1}{p}\ln(p+1) \quad [p > -1]$. BI ((255))(2)

3. $\int_0^1 \operatorname{li}(x)\frac{dx}{x^{q+1}} = \frac{1}{q}\ln(1-q) \quad [q < 1]$. BI ((255))(3)

4. $\int_1^\infty \operatorname{li}(x)\frac{dx}{x^{q+1}} = -\frac{1}{q}\ln(q-1) \quad [q > 1]$. BI ((255))(4)

6.213

1. $\int_0^1 \operatorname{li}\left(\frac{1}{x}\right) \sin(a \ln x)\, dx = \frac{1}{1+a^2}\left(a \ln a - \frac{\pi}{2}\right) \quad [a > 0].$ BI ((475))(1)

2. $\int_1^\infty \operatorname{li}\left(\frac{1}{x}\right) \sin(a \ln x)\, dx = -\frac{1}{1+a^2}\left(\frac{\pi}{2} + a \ln a\right) \quad [a > 0].$ BI ((475))(9)

3. $\int_0^1 \operatorname{li}\left(\frac{1}{x}\right) \cos(a \ln x)\, dx = -\frac{1}{1+a^2}\left(\ln a + \frac{\pi}{2} a\right) \quad [a > 0].$ BI ((475))(2)

4. $\int_1^\infty \operatorname{li}\left(\frac{1}{x}\right) \cos(a \ln x)\, dx = \frac{1}{1+a^2}\left(\ln a - \frac{\pi}{2} a\right) \quad [a > 0].$ BI ((475))(10)

5. $\int_0^1 \operatorname{li}(x) \sin(a \ln x) \frac{dx}{x} = \frac{\ln(1+a^2)}{2a} \quad [a > 0].$ BI((479))(1), ET I 98(20)a

6. $\int_0^1 \operatorname{li}(x) \cos(a \ln x) \frac{dx}{x} = -\frac{\operatorname{arctg} a}{a}.$ BI ((479))(2)

7. $\int_0^1 \operatorname{li}(x) \sin(a \ln x) \frac{dx}{x^2} = \frac{1}{1+a^2}\left(a \ln a + \frac{\pi}{2}\right) \quad [a > 0].$ BI ((479))(3)

8. $\int_1^\infty \operatorname{li}(x) \sin(a \ln x) \frac{dx}{x^2} = \frac{1}{1+a^2}\left(\frac{\pi}{2} - a \ln a\right) \quad [a > 0].$ BI ((479))(13)

9. $\int_0^1 \operatorname{li}(x) \cos(a \ln x) \frac{dx}{x^2} = \frac{1}{1+a^2}\left(\ln a - \frac{\pi}{2} a\right) \quad [a > 0].$ BI ((479))(4)

10. $\int_1^\infty \operatorname{li}(x) \cos(a \ln x) \frac{dx}{x^2} = -\frac{1}{1+a^2}\left(\ln a + \frac{\pi}{2} a\right) \quad [a > 0].$ BI ((479))(14)

11. $\int_0^1 \operatorname{li}(x) \sin(a \ln x) x^{p-1}\, dx = \frac{1}{a^2+p^2}\left\{\frac{a}{2} \ln[(1+p)^2 + a^2] - p \operatorname{arctg} \frac{a}{1+p}\right\}$

$[p > 0].$ BI ((477))(1)

12. $\int_0^1 \operatorname{li}(x) \cos(a \ln x) x^{p-1}\, dx = -\frac{1}{a^2+p^2}\left\{a \operatorname{arctg} \frac{a}{1+p} + \right.$

$\left. + \frac{p}{2} \ln[(1+p)^2 + a^2]\right\} \quad [p > 0].$ BI ((477))(2)

6.214

1. $\int_0^1 \operatorname{li}\left(\frac{1}{x}\right)\left(\ln \frac{1}{x}\right)^{p-1} dx = -\pi \operatorname{ctg} p\pi \cdot \Gamma(p) \quad [0 < p < 1].$ BI ((340))(1)

2. $\int_1^\infty \operatorname{li}\left(\frac{1}{x}\right)(\ln x)^{p-1}\, dx = -\frac{\pi}{\sin p\pi} \Gamma(p) \quad [p > 0].$ BI ((340))(9)

6.215

1. $$\int_0^1 \mathrm{li}(x) \frac{x^{p-1}}{\sqrt{\ln\left(\frac{1}{x}\right)}} dx = -2\sqrt{\frac{\pi}{p}} \operatorname{Arsh}\sqrt{p} =$$
$$= -2\sqrt{\frac{\pi}{p}} \ln\left(\sqrt{p}+\sqrt{p+1}\right) \quad [p>0].$$ BI ((444))(3)

2. $$\int_0^1 \mathrm{li}(x) \frac{dx}{x^{p+1}\sqrt{\ln\left(\frac{1}{x}\right)}} = -2\sqrt{\frac{\pi}{p}} \arcsin\sqrt{p} \quad [1>p>0].$$ BI ((444))(4)

6.216

1. $$\int_0^1 \mathrm{li}(x) \left[\ln\left(\frac{1}{x}\right)\right]^{p-1} \frac{dx}{x} = -\frac{1}{p}\Gamma(p) \quad [0<p\leqslant 1].$$ BI ((444))(1)

2. $$\int_0^1 \mathrm{li}(x) \left[\ln\left(\frac{1}{x}\right)\right]^{p-1} \frac{dx}{x^2} = -\frac{\pi\Gamma(p)}{\sin p\pi} \quad [0<p\leqslant 1].$$ BI ((444))(2)

6.22-6.23 The exponential-integral function

6.221 $$\int_0^p \mathrm{Ei}(\alpha x)\, dx = p\,\mathrm{Ei}(\alpha p) + \frac{1-e^{\alpha p}}{\alpha}.$$ NT 11(7)

6.222 $$\int_0^\infty \mathrm{Ei}(-px)\,\mathrm{Ei}(-qx)\, dx = \left(\frac{1}{p}+\frac{1}{q}\right)\ln(p+q) - \frac{\ln q}{p} - \frac{\ln p}{q}$$
$$[p>0,\ q>0].$$ FI II 653, NT 53(3)

6.223 $$\int_0^\infty \mathrm{Ei}(-\beta x)\, x^{\mu-1}\, dx = -\frac{\Gamma(\mu)}{\mu\beta^\mu} \quad [\operatorname{Re}\beta\geqslant 0,\ \operatorname{Re}\mu>0].$$ NT 55(7), ET I 325(10)

6.224

1. $$\int_0^\infty \mathrm{Ei}(-\beta x)\, e^{-\mu x}\, dx = -\frac{1}{\mu}\ln\left(1+\frac{\mu}{\beta}\right) \quad [\operatorname{Re}(\beta+\mu)\geqslant 0,\ \mu>0];$$
$$= 1 \quad [\mu = 0].$$ FI II 652, NT 48(8)

2. $$\int_0^\infty \mathrm{Ei}(ax)\, e^{-\mu x}\, dx = -\frac{1}{\mu}\ln\left(\frac{\mu}{a}-1\right) \quad [a>0,\ \operatorname{Re}\mu>0,\ \mu>a].$$ ET I 178(23)a, BI ((283))(3)

6.225

1. $$\int_0^\infty \mathrm{Ei}(-x^2)\, e^{-\mu x^2}\, dx = -\sqrt{\frac{\pi}{\mu}} \operatorname{Arsh}\sqrt{\mu} = -\sqrt{\frac{\pi}{\mu}} \ln\left(\sqrt{\mu}+\sqrt{1+\mu}\right)$$
$$[\operatorname{Re}\mu>0].$$ BI ((283))(5), ET I 178(25)a

2. $$\int_0^\infty \mathrm{Ei}(-x^2)\, e^{px^2}\, dx = -\sqrt{\frac{\pi}{p}} \arcsin\sqrt{p} \quad [1>p>0].$$ NT 59(9)a

6.226

1. $\int_0^\infty \mathrm{Ei}\left(-\frac{1}{4x}\right) e^{-\mu x}\, dx = -\frac{2}{\mu} K_0(\sqrt{\mu}) \quad [\mathrm{Re}\,\mu > 0].$ MI 34

2. $\int_0^\infty \mathrm{Ei}\left(\frac{a^2}{4x}\right) e^{-\mu x}\, dx = -\frac{2}{\mu} K_0(a\sqrt{\mu}) \quad [a > 0,\ \mathrm{Re}\,\mu > 0].$ MI 34

3. $\int_0^\infty \mathrm{Ei}\left(-\frac{1}{4x^2}\right) e^{-\mu x^2}\, dx = \sqrt{\frac{\pi}{\mu}}\,\mathrm{Ei}(-\sqrt{\mu}) \quad [\mathrm{Re}\,\mu > 0].$ MI 34

4. $\int_0^\infty \mathrm{Ei}\left(-\frac{1}{4x^2}\right) e^{-\mu x^2 + \frac{1}{4x^2}}\, dx =$

$= \sqrt{\frac{\pi}{\mu}}\left[\cos\sqrt{\mu}\,\mathrm{ci}\sqrt{\mu} - \sin\sqrt{\mu}\,\mathrm{si}\sqrt{\mu}\right] \quad [\mathrm{Re}\,\mu > 0].$ MI 34

6.227

1. $\int_0^\infty \mathrm{Ei}(-x)\, e^{-\mu x} x\, dx = \frac{1}{\mu(\mu+1)} - \frac{1}{\mu^2}\ln(1+\mu) \quad [\mathrm{Re}\,\mu > 0].$ MI 34

2. $\int_0^\infty \left[\frac{e^{-ax}\,\mathrm{Ei}(ax)}{x-b} - \frac{e^{ax}\,\mathrm{Ei}(-ax)}{x+b}\right] dx = 0 \quad [a > 0,\ b < 0];$

$= \pi^2 e^{-ab} \quad [a > 0,\ b > 0].$ ET II 253(1)a

6.228

1. $\int_0^\infty \mathrm{Ei}(-x)\, e^{x} x^{\nu-1}\, dx = -\frac{\pi\Gamma(\nu)}{\sin\nu\pi} \quad [0 < \mathrm{Re}\,\nu < 1].$ ET II 308(13)

2. $\int_0^\infty \mathrm{Ei}(-\beta x)\, e^{-\mu x} x^{\nu-1}\, dx = -\frac{\Gamma(\nu)}{\nu(\beta+\mu)^\nu}\, {}_2F_1\left(1,\ \nu;\ \nu+1;\ \frac{\mu}{\beta+\mu}\right)$

$[|\arg\beta| < \pi,\ \mathrm{Re}(\beta+\mu) > 0,\ \mathrm{Re}\,\nu > 0].$ ET II 308(14)

6.229 $\int_0^\infty \mathrm{Ei}\left(-\frac{1}{4x^2}\right) \exp\left(-\mu x^2 + \frac{1}{4x^2}\right) \frac{dx}{x^2} =$

$= 2\sqrt{\pi}\left(\cos\sqrt{\mu}\,\mathrm{si}\sqrt{\mu} - \sin\sqrt{\mu}\,\mathrm{ci}\sqrt{\mu}\right) \quad [\mathrm{Re}\,\mu > 0].$ MI 34

6.231 $\int_{-\ln a}^\infty [\mathrm{Ei}(-a) - \mathrm{Ei}(-e^{-x})]\, e^{-\mu x}\, dx = \frac{1}{\mu}\gamma(\mu,\ a) \quad [a < 1,\ \mathrm{Re}\,\mu > 0].$

MI 34

6.232

1. $\int_0^\infty \mathrm{Ei}(-ax)\sin bx\, dx = -\frac{\ln\left(1+\frac{b^2}{a^2}\right)}{2b} \quad [a > 0,\ b > 0].$ BI ((473))(1)a

2. $\int_0^\infty \mathrm{Ei}(-ax)\cos bx\, dx = -\frac{1}{b}\,\mathrm{arctg}\,\frac{b}{a} \quad [a > 0,\ b > 0].$ BI ((473))(2)a

6.233

1. $$\int_0^\infty \operatorname{Ei}(-x)\, e^{-\mu x} \sin \beta x\, dx = -\frac{1}{\beta^2+\mu^2} \times$$
$$\times \left\{\frac{\beta}{2} \ln[(1+\mu)^2+\beta^2] - \mu \operatorname{arctg} \frac{\beta}{1+\mu}\right\} \quad [\operatorname{Re}\mu > |\operatorname{Im}\beta|].$$ BI ((473))(7)a

2. $$\int_0^\infty \operatorname{Ei}(-x)\, e^{-\mu x} \cos \beta x\, dx = -\frac{1}{\beta^2+\mu^2} \times$$
$$\times \left\{\frac{\mu}{2} \ln[(1+\mu)^2+\beta^2] + \beta \operatorname{arctg} \frac{\beta}{1+\mu}\right\} \quad [\operatorname{Re}\mu > |\operatorname{Im}\beta|].$$ BI ((473))(8)a

6.234 $$\int_0^\infty \operatorname{Ei}(-x) \ln x\, dx = C + 1.$$ NT 56(10)

6.24-6.26 The sine- and cosine-integral functions

6.241

1. $$\int_0^\infty \operatorname{si}(px) \operatorname{si}(qx)\, dx = \frac{\pi}{2p} \quad [p \geqslant q].$$ FI II 653, NT 54(8)

2. $$\int_0^\infty \operatorname{ci}(px) \operatorname{ci}(qx)\, dx = \frac{\pi}{2p} \quad [p \geqslant q].$$ FI II 653, NT 54(7)

3. $$\int_0^\infty \operatorname{si}(px) \operatorname{ci}(qx)\, dx = \frac{1}{4q} \ln\left(\frac{p+q}{p-q}\right)^2 + \frac{1}{4p} \ln \frac{(p^2-q^2)^2}{q^4} \quad [p \neq q];$$
$$= \frac{1}{q} \ln 2 \quad [p = q].$$ FI II 653, NT 54(10, 12)

6.242 $$\int_0^\infty \frac{\operatorname{ci}(ax)}{\beta+x}\, dx = -\frac{1}{2} \{[\operatorname{si}(a\beta)]^2 + [\operatorname{ci}(a\beta)]^2\} \quad [a > 0, \ |\arg\beta| < \pi].$$ ET II 224(1)

6.243

1. $$\int_{-\infty}^\infty \frac{\operatorname{si}(a|x|)}{x-b} \operatorname{sign} x\, dx = \pi \operatorname{ci}(a|b|) \quad [a > 0, \ b > 0].$$ ET II 253(3)

2. $$\int_{-\infty}^\infty \frac{\operatorname{ci}(a|x|)}{x-b}\, dx = -\pi \operatorname{sign} b \cdot \operatorname{si}(a|b|) \quad [a > 0].$$ ET II 253(2)

6.244

1. $$\int_0^\infty \left[\operatorname{si}(px) + \frac{\pi}{2}\right] \frac{x\, dx}{q^2+x^2} = \frac{\pi}{2} \operatorname{Ei}(-pq) \quad [p > 0, \ q > 0].$$ BI ((255))(6)

2. $$\int_0^\infty \left[\operatorname{si}(px) + \frac{\pi}{2}\right] \frac{x\, dx}{q^2-x^2} = -\frac{\pi}{2} \operatorname{ci}(pq) \quad [p > 0, \ q > 0].$$ BI ((255))(6)

6.245

1. $\int_0^\infty \operatorname{ci}(px) \frac{dx}{q^2+x^2} = \frac{\pi}{2q} \operatorname{Ei}(-pq) \quad [p>0,\ q>0].$ BI ((255))(7)

2. $\int_0^\infty \operatorname{ci}(px) \frac{dx}{q^2-x^2} = \frac{\pi}{2q} \operatorname{si}(pq) \quad [p>0,\ q>0].$ BI ((255))(8)

6.246

1. $\int_0^\infty \operatorname{si}(ax)\, x^{\mu-1}\, dx = -\frac{\Gamma(\mu)}{\mu a^\mu} \sin\frac{\mu\pi}{2} \quad [a>0,\ 0<\operatorname{Re}\mu<1].$

NT 56(9), ET I 325 (12)a

2. $\int_0^\infty \operatorname{ci}(ax)\, x^{\mu-1}\, dx = -\frac{\Gamma(\mu)}{\mu a^\mu} \cos\frac{\mu\pi}{2} \quad [a>0,\ 0<\operatorname{Re}\mu<1].$

NT 56(8), ET I 325(13)a

6.247

1. $\int_0^\infty \operatorname{si}(\beta x)\, e^{-\mu x}\, dx = -\frac{1}{\mu} \operatorname{arctg}\frac{\mu}{\beta} \quad [\operatorname{Re}\mu>0].$ NT 49(12), ET I 177(18)

2. $\int_0^\infty \operatorname{ci}(\beta x)\, e^{-\mu x}\, dx = -\frac{1}{\mu} \ln\sqrt{1+\frac{\mu^2}{\beta^2}} \quad [\operatorname{Re}\mu>0].$

NT 49(11), ET I 178(19)a

6.248

1. $\int_0^\infty \operatorname{si}(x)\, e^{-\mu x^2} x\, dx = \frac{\pi}{4\mu}\left[1-\Phi\left(\frac{1}{2\sqrt{\mu}}\right)\right] \quad [\operatorname{Re}\mu>0].$ MI 34

2. $\int_0^\infty \operatorname{ci}(x)\, e^{-\mu x^2}\, dx = \frac{1}{4}\sqrt{\frac{\pi}{\mu}} \operatorname{Ei}\left(-\frac{1}{4\mu}\right) \quad [\operatorname{Re}\mu>0].$ MI 34

6.249 $\int_0^\infty \left[\operatorname{si}(x^2)+\frac{\pi}{2}\right] e^{-\mu x}\, dx = \frac{\pi}{\mu}\left\{\left[S\left(\frac{\mu^2}{4}\right)-\frac{1}{2}\right]^2+\left[C\left(\frac{\mu^2}{4}\right)-\frac{1}{2}\right]^2\right\}$

$[\operatorname{Re}\mu>0].$ ME 26

6.251

1. $\int_0^\infty \operatorname{si}\left(\frac{1}{x}\right) e^{-\mu x}\, dx = \frac{2}{\mu} \operatorname{kei}(2\sqrt{\mu}) \quad [\operatorname{Re}\mu>0].$ MI 34

2. $\int_0^\infty \operatorname{ci}\left(\frac{1}{x}\right) e^{-\mu x}\, dx = -\frac{2}{\mu} \operatorname{ker}(2\sqrt{\mu}) \quad [\operatorname{Re}\mu>0].$ MI 34

6.252

1. $\int_0^\infty \sin px \operatorname{si}(qx)\, dx = -\frac{\pi}{2p} \quad [p^2>q^2];$

$= -\frac{\pi}{4p} \quad [p^2=q^2];$

$= 0 \quad [p^2<q^2].$ FI II 652, NT 50(8)

2. $$\int_0^\infty \cos px \operatorname{si}(qx)\,dx = -\frac{1}{4p}\ln\left(\frac{p+q}{p-q}\right)^2 \quad [p \neq 0,\ p^2 \neq q^2];$$
$$= 1 \quad [p = 0].$$ FI II 652, NT 50(10)

3. $$\int_0^\infty \sin px \operatorname{ci}(qx)\,dx = -\frac{1}{4p}\ln\left(\frac{p^2}{q^2}-1\right)^2 \quad [p \neq 0,\ p^2 \neq q^2];$$
$$= 0 \quad [p = 0].$$ FI II 652, NT 50(9)

4. $$\int_0^\infty \cos px \operatorname{ci}(qx)\,dx = -\frac{\pi}{2p} \quad [p^2 > q^2];$$
$$= -\frac{\pi}{4p} \quad [p^2 = q^2];$$
$$= 0 \quad [p^2 < q^2].$$ FI II 654, NT 50(7)

6.253 $$\int_0^\infty \frac{\operatorname{si}(ax)\sin bx}{1-2r\cos x+r^2}\,dx = -\frac{\pi(r^m+r^{m+1})}{4b(1-r)(1-r^2)} \quad [b = a-m];$$
$$= -\frac{\pi(2+2r-r^m-r^{m+1})}{4b(1-r)(1-r^2)} \quad [b = a+m];$$
$$= -\frac{\pi r^{m+1}}{2b(1-r)(1-r^2)} \quad [a-m-1 < b < a-m];$$
$$= -\frac{\pi(1+r-r^{m+1})}{2b(1-r)(1-r^2)} \quad [a+m < b < a+m+1].$$
ET I 97(10)

6.254

1. $$\int_0^\infty \left[\operatorname{si}(ax)+\frac{\pi}{2}\right]\sin bx\frac{dx}{x} = \frac{1}{2}\left[L_2\left(\frac{a}{b}\right)-L_2\left(-\frac{a}{b}\right)\right]$$
$$[a > 0,\ b > 0].$$ ET I 97(12)

2. $$\int_0^\infty \left[\operatorname{si}(ax)+\frac{\pi}{2}\right]\cos bx\cdot\frac{dx}{x} = \frac{\pi}{2}\ln\frac{a}{b} \quad [a > 0,\ b > 0].$$ ET I 41(11)

6.255

1. $$\int_{-\infty}^\infty [\cos ax \operatorname{ci}(a|x|)+\sin(a|x|)\operatorname{si}(a|x|)]\frac{dx}{x-b} =$$
$$= -\pi[\operatorname{sign} b\cos ab\operatorname{si}(a|b|)-\sin ab\operatorname{ci}(a|b|)] \quad [a > 0].$$ ET II 253(4)

2. $$\int_{-\infty}^\infty [\sin ax\operatorname{ci}(a|x|)-\operatorname{sign} x\cos ax\operatorname{si}(a|x|)]\frac{dx}{x-b} =$$
$$= -\pi[\sin(a|b|)\operatorname{si}(a|b|)+\cos ab\operatorname{ci}(a|b|)] \quad [a > 0].$$ ET II 253(5)

6.256 $$\int_0^\infty [\operatorname{si}^2(x)+\operatorname{ci}^2(x)]\cos ax\,dx = \frac{\pi}{a}\ln(1+a) \quad [a > 0].$$ ET I 42(18)

6.257 $$\int_0^\infty \operatorname{si}\left(\frac{a}{x}\right)\sin bx\,dx = -\frac{\pi}{2b}J_0\left(2\sqrt{ab}\right) \quad [b > 0].$$ ET I 96(9)

6.258

1. $$\int_0^\infty \left[\operatorname{si}(ax)+\frac{\pi}{2}\right]\sin bx \frac{dx}{x^2+c^2} =$$

$$= \frac{\pi}{4c}\{e^{-bc}[\operatorname{Ei}(bc)-\operatorname{Ei}(-ac)]+e^{bc}[\operatorname{Ei}(-ac)-\operatorname{Ei}(-bc)]\} \quad [0<b\leqslant a,\ c>0];$$

$$= \frac{\pi}{4c}e^{-bc}[\operatorname{Ei}(ac)-\operatorname{Ei}(-ac)] \quad [0<a\leqslant b,\ c>0].$$ BI ((460))(1)

2. $$\int_0^\infty \left[\operatorname{si}(ax)+\frac{\pi}{2}\right]\cos bx \frac{x\,dx}{x^2+c^2} =$$

$$= -\frac{\pi}{4}\{e^{-bc}[\operatorname{Ei}(bc)-\operatorname{Ei}(-ac)]+e^{bc}[\operatorname{Ei}(-bc)-\operatorname{Ei}(-ac)]\}$$
$$[0<b\leqslant a,\ c>0];$$

$$= \frac{\pi}{4}e^{-bc}[\operatorname{Ei}(-ac)-\operatorname{Ei}(ac)] \quad [0<a\leqslant b,\ c>0].$$ BI ((460))(2, 5)

6.259

1. $$\int_0^\infty \operatorname{si}(ax)\sin bx\frac{dx}{x^2+c^2} = \frac{\pi}{2c}\operatorname{Ei}(-ac)\operatorname{sh}(bc) \quad [0<b\leqslant a,\ c>0];$$

$$= \frac{\pi}{4c}e^{-cb}[\operatorname{Ei}(-bc)+\operatorname{Ei}(bc)-\operatorname{Ei}(-ac)-$$
$$-\operatorname{Ei}(ac)]+\frac{\pi}{2c}\operatorname{Ei}(-bc)\operatorname{sh}(bc) \quad [0<a\leqslant b,\ c>0].$$ ET I 96(8)

2. $$\int_0^\infty \operatorname{ci}(ax)\sin bx\frac{x\,dx}{x^2+c^2} = -\frac{\pi}{2}\operatorname{sh}(bc)\operatorname{Ei}(-ac) \quad [0<b\leqslant a,\ c>0];$$

$$= -\frac{\pi}{2}\operatorname{sh}(bc)\operatorname{Ei}(-bc)+\frac{\pi}{4}e^{-bc}[\operatorname{Ei}(-bc)+\operatorname{Ei}(bc)-$$
$$-\operatorname{Ei}(-ac)-\operatorname{Ei}(ac)] \quad [0<a\leqslant b,\ c>0].$$ BI ((460))(3)a, ET I 97(15)a

3. $$\int_0^\infty \operatorname{ci}(ax)\cos bx\frac{dx}{x^2+c^2} = \frac{\pi}{2c}\operatorname{ch} bc\operatorname{Ei}(-ac) \quad [0<b\leqslant a,\ c>0];$$

$$= \frac{\pi}{4c}\{e^{-bc}[\operatorname{Ei}(ac)+\operatorname{Ei}(-ac)-\operatorname{Ei}(bc)]+e^{bc}\operatorname{Ei}(-bc)\}$$
$$[0<a\leqslant b,\ c>0].$$ BI ((460))(4), ET I 41(15)

6.261

1. $$\int_0^\infty \operatorname{si}(bx)\cos ax\, e^{-px}\,dx = -\frac{1}{2(a^2+p^2)}\left[\frac{a}{2}\ln\frac{p^2+(a+b)^2}{p^2+(a-b)^2}+\right.$$
$$\left.+p\operatorname{arctg}\frac{2bp}{b^2-a^2-p^2}\right] \quad [a>0,\ b>0,\ p>0].$$ ET I 40(8)

2. $$\int_0^\infty \operatorname{si}(\beta x)\cos ax\, e^{-\mu x}\,dx = -\frac{\operatorname{arctg}\frac{\mu+ai}{\beta}}{2(\mu+ai)}-\frac{\operatorname{arctg}\frac{\mu-ai}{\beta}}{2(\mu-ai)}$$
$$[a>0,\ \operatorname{Re}\mu>|\operatorname{Im}\beta|].$$ ET I 40(9)

6.262

1. $$\int_0^\infty \operatorname{ci}(bx)\sin ax\, e^{-\mu x}\,dx = \frac{1}{2(a^2+\mu^2)} \times \left\{\mu \operatorname{arctg}\frac{2a\mu}{\mu^2+b^2-a^2} - \frac{a}{2}\ln\frac{(\mu^2+b^2-a^2)^2+4a^2\mu^2}{b^4}\right\}$$
$[a>0,\ b>0,\ \operatorname{Re}\mu>0]$. ET I 98(16)a

2. $$\int_0^\infty \operatorname{ci}(bx)\cos ax\, e^{-px}\,dx = \frac{-1}{2(a^2+p^2)} \times \left\{\frac{p}{2}\ln\frac{[(b^2+p^2-a^2)^2+4a^2p^2]}{b^4} + a\operatorname{arctg}\frac{2ap}{b^2+p^2-a^2}\right\}$$
$[a>0,\ b>0,\ \operatorname{Re}p>0]$. ET I 41(16)

3. $$\int_0^\infty \operatorname{ci}(\beta x)\cos ax e^{-\mu x}\,dx = \frac{-\ln\left[1+\frac{(\mu+ai)^2}{\beta^2}\right]}{4(\mu+ai)} - \frac{\ln\left[1+\frac{(\mu-ai)^2}{\beta^2}\right]}{4(\mu-ai)}$$
$[a>0,\ \operatorname{Re}\mu>|\operatorname{Im}\beta|]$. ET I 41(17)

6.263

1. $$\int_0^\infty [\operatorname{ci}(x)\cos x + \operatorname{si}(x)\sin x]\, e^{-\mu x}\,dx = \frac{-\frac{\pi}{2}-\mu\ln\mu}{1+\mu^2}$$
$[\operatorname{Re}\mu>0]$. ME 26a, ET I 178(21)a

2. $$\int_0^\infty [\operatorname{si}(x)\cos x - \operatorname{ci}(x)\sin x]\, e^{-\mu x}\,dx = \frac{-\frac{\pi}{2}\mu+\ln\mu}{1+\mu^2}$$
$[\operatorname{Re}\mu>0]$. ME 26a, ET I 178(20)a

3. $$\int_0^\infty [\sin x - x\operatorname{ci}(x)]\, e^{-\mu x}\,dx = \frac{\ln(1+\mu^2)}{2\mu^2} \quad [\operatorname{Re}\mu>0].$$ ME 26

6.264

1. $$\int_0^\infty \operatorname{si}(x)\ln x\,dx = \boldsymbol{C}+1.$$ NT 46(10)

2. $$\int_0^\infty \operatorname{ci}(x)\ln x\,dx = \frac{\pi}{2}.$$ NT 56(11)

6.27 The hyperbolic-sine- and -cosine-integral functions

6.271

1. $$\int_0^\infty \operatorname{shi}(x)\, e^{-\mu x}\,dx = \frac{1}{2\mu}\ln\frac{\mu+1}{\mu-1} = \frac{1}{\mu}\operatorname{Arcth}\mu \quad [\operatorname{Re}\mu>1].$$ MI 34

2. $$\int_0^\infty \operatorname{chi}(x)\, e^{-\mu x}\,dx = -\frac{1}{2\mu}\ln(\mu^2-1) \quad [\operatorname{Re}\mu>1].$$ MI 34

6.272 $\int_0^\infty \operatorname{chi}(x) e^{-px^2} dx = \frac{1}{4}\sqrt{\frac{\pi}{p}} \operatorname{Ei}\left(\frac{1}{4p}\right) \quad [p > 0].$ MI 35

6.273

1. $\int_0^\infty [\operatorname{ch} x \operatorname{shi}(x) - \operatorname{sh} x \operatorname{chi}(x)] e^{-\mu x} dx = \frac{\ln \mu}{\mu^2 - 1} \quad [\operatorname{Re} \mu > 0].$ MI 35

2. $\int_0^\infty [\operatorname{ch} x \operatorname{chi}(x) + \operatorname{sh} x \operatorname{shi}(x)] e^{-\mu x} dx = \frac{\mu \ln \mu}{1 - \mu^2} \quad [\operatorname{Re} \mu > 2].$ MI 35

6.274 $\int_0^\infty [\operatorname{ch} x \operatorname{shi}(x) - \operatorname{sh} x \operatorname{chi}(x)] e^{-\mu x^2} dx = \frac{1}{4}\sqrt{\frac{\pi}{\mu}} e^{\frac{1}{4\mu}} \operatorname{Ei}\left(-\frac{1}{4\mu}\right)$

$[\operatorname{Re} \mu > 0].$ MI 35

6.275 $\int_0^\infty [x \operatorname{chi}(x) - \operatorname{sh} x] e^{-\mu x} dx = -\frac{\ln(\mu^2 - 1)}{2\mu^2} \quad [\operatorname{Re} \mu > 1].$ MI 35

6.276 $\int_0^\infty [\operatorname{ch} x \operatorname{chi}(x) + \operatorname{sh} x \operatorname{shi}(x)] e^{-\mu x^2} x \, dx =$

$= \frac{1}{8}\sqrt{\frac{\pi}{\mu^3}} \exp\left(\frac{1}{4\mu}\right) \operatorname{Ei}\left(-\frac{1}{4\mu}\right) \quad [\operatorname{Re} \mu > 0].$ MI 35

6.277

1. $\int_0^\infty [\operatorname{chi}(x) + \operatorname{ci}(x)] e^{-\mu x} dx = -\frac{\ln(\mu^4 - 1)}{2\mu} \quad [\operatorname{Re} \mu > 1].$ MI 34

2. $\int_0^\infty [\operatorname{chi}(x) - \operatorname{ci}(x)] e^{-\mu x} dx = \frac{1}{2\mu} \ln \frac{\mu^2 + 1}{\mu^2 - 1} \quad [\operatorname{Re} \mu > 1].$ MI 35

6.28-6.31 The probability integral

6.281 $\int_0^\infty [1 - \Phi(px)] x^{2q-1} dx = \frac{\Gamma\left(q + \frac{1}{2}\right)}{2\sqrt{\pi}\, q p^{2q}} \quad [\operatorname{Re} q > 0, \ \operatorname{Re} p > 0].$

NT 56(12), ET II 306(1)a

6.282

1. $\int_0^\infty \Phi(qt) e^{-pt} dt = \frac{1}{p}\left[1 - \Phi\left(\frac{p}{2q}\right)\right] \exp\left(\frac{p^2}{4q^2}\right).$

MO 175, EH II 148(11)

2. $\int_0^\infty \left[\Phi\left(x + \frac{1}{2}\right) - \Phi\left(\frac{1}{2}\right)\right] e^{-\mu x + \frac{1}{4}} dx =$

$= \frac{1}{(\mu + 1)(\mu + 2)} \exp \frac{(\mu + 1)^2}{4} \left[1 - \Phi\left(\frac{\mu + 1}{2}\right)\right].$ ME 27

6.283

1. $$\int_0^\infty e^{\beta x}\left[1-\Phi\left(\sqrt{\alpha x}\right)\right]dx=\frac{1}{\beta}\left[\frac{\sqrt{\alpha}}{\sqrt{\alpha-\beta}}-1\right]$$
$[\operatorname{Re}\alpha>0,\ \operatorname{Re}\beta<\operatorname{Re}\alpha]$. ET II 307(5)

2. $$\int_0^\infty \Phi\left(\sqrt{qt}\right)e^{-pt}\,dt=\frac{\sqrt{q}}{p}\,\frac{1}{\sqrt{p+q}}$$
$[\operatorname{Re}p>0,\ \operatorname{Re}(q+p)>0]$. EH II 148(12)

6.284 $$\int_0^\infty\left[1-\Phi\left(\frac{q}{2\sqrt{x}}\right)\right]e^{-px}\,dx=\frac{1}{p}\,e^{-q\sqrt{p}}$$
$\left[\operatorname{Re}p>0,\ |\arg q|<\frac{\pi}{4}\right]$. EF 147(235), EH II 148(13)

6.285

1. $$\int_0^\infty[1-\Phi(x)]\,e^{-\mu^2x^2}\,dx=\frac{\operatorname{arctg}\mu}{\sqrt{\pi}\,\mu}$$ $[\operatorname{Re}\mu>0]$. MI 37

2. $$\int_0^\infty\Phi(iat)\,e^{-a^2t^2-st}\,dt=\frac{-1}{2ai\sqrt{\pi}}\exp\left(\frac{s^2}{4a^2}\right)\operatorname{Ei}\left(-\frac{s^2}{4a^2}\right)$$
$\left[\operatorname{Re}s>0,\ |\arg a|<\frac{\pi}{4}\right]$. EH II 148(14)a

6.286

1. $$\int_0^\infty[1-\Phi(\beta x)]\,e^{\mu^2x^2}x^{\nu-1}\,dx=\frac{\Gamma\left(\frac{\nu+1}{2}\right)}{\sqrt{\pi}\,\nu\beta^\nu}\times$$
$$\times\,{}_2F_1\left(\frac{\nu}{2},\ \frac{\nu+1}{2};\ \frac{\nu}{2}+1;\ \frac{\mu^2}{\beta^2}\right)$$ $[\operatorname{Re}\beta^2>\operatorname{Re}\mu^2,\ \operatorname{Re}\nu>0]$. ET II 306(2)

2. $$\int_0^\infty\left[1-\Phi\left(\frac{\sqrt{2}\,x}{2}\right)\right]e^{\frac{x^2}{2}}x^{\nu-1}\,dx=2^{\frac{\nu}{2}-1}\sec\frac{\nu\pi}{2}\,\Gamma\left(\frac{\nu}{2}\right)$$
$[0<\operatorname{Re}\nu<1]$. ET I 325(9)

6.287

1. $$\int_0^\infty\Phi(\beta x)\,e^{-\mu x^2}x\,dx=\frac{\beta}{2\mu\sqrt{\mu+\beta^2}}$$ $[\operatorname{Re}\mu>-\operatorname{Re}\beta^2,\ \operatorname{Re}\mu>0]$.
ME 27a, ET I 176(4)

2. $$\int_0^\infty[1-\Phi(\beta x)]\,e^{-\mu x^2}x\,dx=\frac{1}{2\mu}\left(1-\frac{\beta}{\sqrt{\mu+\beta^2}}\right)$$
$[\operatorname{Re}\mu>-\operatorname{Re}\beta^2,\ \operatorname{Re}\mu>0]$. NT 49(14), ET I 177(9)

6.288 $$\int_0^\infty\Phi(iax)\,e^{-\mu x^2}x\,dx=\frac{ai}{2\mu\sqrt{\mu-a^2}}$$ $[a>0,\ \operatorname{Re}\mu>\operatorname{Re}a^2]$. MI 37a

6.289

1. $\int_0^\infty \Phi(\beta x)\, e^{(\beta^2-\mu^2)x^2} x\, dx = \frac{\beta}{2\mu(\mu^2-\beta^2)} \quad \left[\operatorname{Re}\mu^2 > \operatorname{Re}\beta^2,\ |\arg\mu| < \frac{\pi}{4}\right].$ ET I 176(5)

2. $\int_0^\infty [1-\Phi(\beta x)]\, e^{(\beta^2-\mu^2)x^2} x\, dx = \frac{1}{2\mu(\mu+\beta)}$

$\left[\operatorname{Re}\mu^2 > \operatorname{Re}\beta^2,\quad \arg\mu < \frac{\pi}{4}\right]$ ET I 177(10)

3. $\int_0^\infty \Phi\left(\sqrt{b-a}\, x\right) e^{-(a+\mu)x^2} x\, dx = \frac{\sqrt{b-a}}{2(\mu+a)\sqrt{\mu+b}}$

$[\operatorname{Re}\mu > -a > 0,\ b > a].$ ME 27

6.291 $\int_0^\infty \Phi(ix)\, e^{-(\mu x+x^2)} x\, dx = \frac{i}{\sqrt{\pi}}\left[\frac{1}{\mu} + \frac{\mu}{4}\operatorname{Ei}\left(-\frac{\mu^2}{4}\right)\right]$

$[\operatorname{Re}\mu > 0].$ MI 37

6.292 $\int_0^\infty [1-\Phi(x)]\, e^{-\mu^2x^2} x^2\, dx = \frac{1}{2\sqrt{\pi}}\left\{\frac{\operatorname{arctg}\mu}{\mu^3} - \frac{1}{\mu^2(\mu^2+1)}\right\}$

$\left[|\arg\mu| < \frac{\pi}{4}\right].$ MI 37

6.293 $\int_0^\infty \Phi(x)\, e^{-\mu x^2} \frac{dx}{x} = \frac{1}{2}\ln\frac{\sqrt{\mu+1}+1}{\sqrt{\mu+1}-1} = \operatorname{Arcth}\sqrt{\mu+1}$

$[\operatorname{Re}\mu > 0].$ MI 37a

6.294

1. $\int_0^\infty \left[1-\Phi\left(\frac{\beta}{x}\right)\right] e^{-\mu^2x^2} x\, dx = \frac{1}{2\mu^2}\exp(-2\beta\mu)$

$\left[|\arg\beta| < \frac{\pi}{4},\ |\arg\mu| < \frac{\pi}{4}\right].$ ET I 177(11)

2. $\int_0^\infty \left[1-\Phi\left(\frac{1}{x}\right)\right] e^{-\mu^2x^2} \frac{dx}{x} = -\operatorname{Ei}(-2\mu) \quad \left[|\arg\mu| < \frac{\pi}{4}\right].$ MI 37

6.295

1. $\int_0^\infty \left[1-\Phi\left(\frac{1}{x}\right)\right] \exp\left(-\mu^2x^2 + \frac{1}{x^2}\right) dx =$

$= \frac{1}{\sqrt{\pi\mu}}[\sin 2\mu\, \operatorname{ci}(2\mu) - \cos 2\mu\, \operatorname{si}(2\mu)] \quad \left[|\arg\mu| < \frac{\pi}{4}\right].$ MI 37

2. $\int_0^\infty \left[1-\Phi\left(\frac{1}{x}\right)\right] \exp\left(-\mu^2x^2 + \frac{1}{x^2}\right) x\, dx =$

$= \frac{\pi}{2\mu}[\mathbf{H}_1(2\mu) - N_1(2\mu)] - \frac{1}{\mu^2} \quad \left[|\arg\mu| < \frac{\pi}{4}\right].$ MI 37

3. $$\int_0^\infty \left[1-\Phi\left(\frac{1}{x}\right)\right] \exp\left(-\mu^2 x^2 + \frac{1}{x^2}\right) \frac{dx}{x} =$$
$$= \frac{\pi}{2} [\mathbf{H}_0(2\mu) - N_0(2\mu)] \qquad \left[|\arg\mu| < \frac{\pi}{4}\right].$$ MI 37

6.296 $$\int_0^\infty \left\{(x^2+a^2)\left[1-\Phi\left(\frac{a}{\sqrt{2}\,x}\right)\right] - \sqrt{\frac{2}{\pi}}\, ax \cdot e^{-\frac{a^2}{2x^2}}\right\} e^{-\mu^2 x^2} x\, dx =$$
$$= \frac{1}{2\mu^4} e^{-a\mu\sqrt{2}} \qquad \left[|\arg\mu| < \frac{\pi}{4},\ a > 0\right].$$ MI 38a

6.297

1. $$\int_0^\infty \left[1-\Phi\left(\gamma x + \frac{\beta}{x}\right)\right] e^{(\gamma^2-\mu)x^2} x\, dx =$$
$$= \frac{1}{2\sqrt{\mu}(\sqrt{\mu}+\gamma)} \exp[-2(\beta\gamma + \beta\sqrt{\mu})]$$
$$[\operatorname{Re}\beta > 0,\ \operatorname{Re}\mu > 0].$$ ET I 177(12)a

2. $$\int_0^\infty \left[1-\Phi\left(\frac{b+2ax^2}{2x}\right)\right] \exp[-(\mu^2-a^2)x^2 + ab]\, x\, dx =$$
$$= \frac{e^{-b\mu}}{2\mu(\mu+a)} \qquad [a > 0,\ b > 0,\ \operatorname{Re}\mu > 0].$$ MI 38

3. $$\int_0^\infty \left\{\left[1-\Phi\left(\frac{b-2ax^2}{2x}\right)\right] e^{-ab} + \left[1-\Phi\left(\frac{b+2ax^2}{2x}\right)\right] e^{ab}\right\} e^{-\mu x^2} x\, dx =$$
$$= \frac{1}{\mu} \exp\left(-b\sqrt{a^2+\mu}\right) \qquad [a > 0,\ b > 0,\ \operatorname{Re}\mu > 0].$$ MI 38

6.298 $$\int_0^\infty \left\{2\operatorname{ch} ab - e^{-ab}\Phi\left(\frac{b-2ax^2}{2x}\right) - e^{ab}\Phi\left(\frac{b+2ax^2}{2x}\right)\right\} e^{-(\mu-a^2)x^2} x\, dx =$$
$$= \frac{1}{\mu-a^2} \exp\left(-b\sqrt{\mu}\right)$$
$$[a > 0,\ b > 0,\ \operatorname{Re}\mu > 0].$$ MI 38

6.299 $$\int_0^\infty \operatorname{ch}(2\nu t) \exp[(a \operatorname{ch} t)^2]\, [1-\Phi(a \operatorname{ch} t)]\, dt =$$
$$= \frac{1}{2\cos(\nu\pi)} \exp\left(\frac{1}{2}a^2\right) K_\nu(a^2)$$
$$\left[\operatorname{Re} a > 0,\ -\frac{1}{2} < \operatorname{Re}\nu < \frac{1}{2}\right].$$ ET II 308(10)

6.311 $$\int_0^\infty [1-\Phi(ax)] \sin bx\, dx = \frac{1}{b}\left(1 - e^{-\frac{b^2}{4a^2}}\right)$$
$$[a > 0,\ b > 0].$$ ET I 96(4)

6.312 $$\int_0^\infty \Phi(ax) \sin bx^2\, dx = \frac{1}{4\sqrt{2\pi b}} \left(\ln \frac{b+a^2+a\sqrt{2b}}{b+a^2-a\sqrt{2b}} + 2\operatorname{arctg} \frac{a\sqrt{2b}}{b-a^2}\right)$$
$$[a > 0,\ b > 0].$$ ET I 96(3)

6.313

1. $$\int_0^\infty \sin(\beta x)\left[1-\Phi\left(\sqrt{\alpha x}\right)\right]dx = \frac{1}{\beta}-\left(\frac{\frac{\alpha}{2}}{\alpha^2+\beta^2}\right)^{\frac{1}{2}}\left[(\alpha^2+\beta^2)^{\frac{1}{2}}-\alpha\right]^{-\frac{1}{2}}$$
$[\operatorname{Re}\alpha > |\operatorname{Im}\beta|]$. ET II 307(6)

2. $$\int_0^\infty \cos(\beta x)\left[1-\Phi\left(\sqrt{\alpha x}\right)\right]dx = \left(\frac{\frac{\alpha}{2}}{\alpha^2+\beta^2}\right)^{\frac{1}{2}}\left[(\alpha^2+\beta^2)^{\frac{1}{2}}+\alpha\right]^{-\frac{1}{2}}$$
$[\operatorname{Re}\alpha > |\operatorname{Im}\beta|]$. ET II 307(7)

6.314

1. $$\int_0^\infty \sin(bx)\left[1-\Phi\left(\sqrt{\frac{a}{x}}\right)\right]dx = b^{-1}\exp\left[-(2ab)^{\frac{1}{2}}\right]\cos\left[(2ab)^{\frac{1}{2}}\right]$$
$[\operatorname{Re}a > 0,\ b > 0]$. ET II 307(8)

2. $$\int_0^\infty \cos(bx)\left[1-\Phi\left(\sqrt{\frac{a}{x}}\right)\right]dx = -b^{-1}\exp\left[-(2ab)^{\frac{1}{2}}\right]\sin\left[(2ab)^{\frac{1}{2}}\right]$$
$[\operatorname{Re}a > 0,\ b > 0]$. ET II 307(9)

6.315

1. $$\int_0^\infty x^{\nu-1}\sin(\beta x)\left[1-\Phi(\alpha x)\right]dx = \frac{\Gamma\left(1+\frac{1}{2}\nu\right)\beta}{\sqrt{\pi}(\nu+1)\alpha^{\nu+1}}\,{}_2F_2\left(\frac{\nu+1}{2},\ \frac{\nu}{2}+1;\ \frac{3}{2},\ \frac{\nu+3}{2};\ -\frac{\beta^2}{4\alpha^2}\right)$$
$[\operatorname{Re}\alpha > 0,\ \operatorname{Re}\nu > -1]$. ET II 307(3)

2. $$\int_0^\infty x^{\nu-1}\cos(\beta x)\left[1-\Phi(\alpha x)\right]dx = \frac{\Gamma\left(\frac{1}{2}+\frac{1}{2}\nu\right)}{\sqrt{\pi}\,\nu\alpha^{\nu}}\,{}_2F_2\left(\frac{\nu}{2},\ \frac{\nu+1}{2};\ \frac{1}{2},\ \frac{\nu}{2}+1;\ -\frac{\beta^2}{4\alpha^2}\right)$$
$[\operatorname{Re}\alpha > 0,\ \operatorname{Re}\nu > 0]$. ET II 307(4)

3. $$\int_0^\infty \left[1-\Phi(ax)\right]\cos bx\cdot x\,dx = \frac{1}{2a^2}\exp\left(-\frac{b^2}{4a^2}\right)-\frac{1}{b^2}\left[1-\exp\left(-\frac{b^2}{4a^2}\right)\right]$$
$[a > 0,\ b > 0]$. ET I 40(5)

4. $$\int_0^\infty [\Phi(ax) - \Phi(bx)] \cos px \frac{dx}{x} = \frac{1}{2}\left[\operatorname{Ei}\left(-\frac{p^2}{4b^2}\right) - \operatorname{Ei}\left(\frac{p^2}{4a^2}\right)\right]$$

$[a > 0,\ b > 0,\ p > 0].$ ET I 40(6)

5. $$\int_0^\infty x^{-\frac{1}{2}} \Phi\left(a\sqrt{x}\right) \sin bx\, dx =$$

$$= \frac{1}{2\sqrt{2\pi b}}\left\{\ln\left[\frac{b + a\sqrt{2b} + a^2}{b - a\sqrt{2b} + a^2}\right] + 2\operatorname{arctg}\left[\frac{a\sqrt{2b}}{b - a^2}\right]\right\}$$

$[a > 0,\ b > 0].$ ET I 96(3)

6.316 $$\int_0^\infty e^{\frac{1}{2}x^2}\left[1 - \Phi\left(\frac{x}{\sqrt{2}}\right)\right] \sin bx\, dx =$$

$$= \sqrt{\frac{\pi}{2}}\, e^{\frac{b^2}{2}}\left[1 - \Phi\left(\frac{b}{\sqrt{2}}\right)\right] \qquad [b > 0].$$ ET I 96(5)

6.317 $$\int_0^\infty e^{-a^2x^2} \Phi(iax) \sin bx\, dx = \frac{\pi i}{4a} e^{-\frac{b^2}{4a^2}} \qquad [b > 0].$$ ET I 96(2)

6.318 $$\int_0^\infty [1 - \Phi(x)] \operatorname{si}(2px)\, dx = \frac{2}{\pi p}\left(1 - e^{-p^2}\right) - \frac{2}{\sqrt{\pi}}\left(1 - \Phi(p)\right)$$

$[p > 0].$ NT 61(13)a

6.32 Fresnel integrals

6.321

1. $$\int_0^\infty \left[\frac{1}{2} - S(px)\right] x^{2q-1}\, dx =$$

$$= \frac{\sqrt{2}\,\Gamma\left(q + \frac{1}{2}\right) \sin \frac{2q+1}{4}\pi}{4\sqrt{\pi}\, q p^{2q}} \qquad \left[0 < \operatorname{Re} q < \frac{3}{2},\ p > 0\right].$$ NT 56(14)a

2. $$\int_0^\infty \left[\frac{1}{2} - C(px)\right] x^{2q-1}\, dx =$$

$$= \frac{\sqrt{2}\,\Gamma\left(q + \frac{1}{2}\right) \cos \frac{2q+1}{4}\pi}{4\sqrt{\pi}\, q p^{2q}} \qquad \left[0 < \operatorname{Re} q < \frac{3}{2},\ p > 0\right].$$ NT 56(13)a

6.322

1. $$\int_0^\infty S(t)\, e^{-pt}\, dt = \frac{1}{p}\left\{\cos\frac{p^2}{4}\left[\frac{1}{2} - C\left(\frac{p}{2}\right)\right] + \right.$$

$$\left. + \sin\frac{p^2}{4}\left[\frac{1}{2} - S\left(\frac{p}{2}\right)\right]\right\}.$$ MO 173a

2. $$\int_0^\infty C(t) e^{-pt}\,dt = \frac{1}{p}\left\{\cos\frac{p^2}{4}\left[\frac{1}{2} - S\left(\frac{p}{2}\right)\right] - \sin\frac{p^2}{4}\left[\frac{1}{2} - C\left(\frac{p}{2}\right)\right]\right\}.$$ MO 172a

6.323

1. $$\int_0^\infty S(\sqrt{t})\,e^{-pt}\,dt = \frac{(\sqrt{p^2+1}-p)^{\frac{1}{2}}}{2p\sqrt{p^2+1}}.$$ EF 122(58)a

2. $$\int_0^\infty C(\sqrt{t})\,e^{-pt}\,dt = \frac{(\sqrt{p^2+1}+p)^{\frac{1}{2}}}{2p\sqrt{p^2+1}}.$$ EF 122(58)a

6.324

1. $$\int_0^\infty \left[\frac{1}{2} - S(x)\right]\sin 2px\,dx = -\frac{2\sqrt{2}\cos\frac{\pi}{8}}{\pi}\,\frac{\sin\frac{p^2}{2}}{p} \qquad [p > 0].$$ NT 61(12)a

2. $$\int_0^\infty \left[\frac{1}{2} - C(x)\right]\sin 2px\,dx = -\frac{2\sqrt{2}\sin\frac{\pi}{8}}{\pi}\,\frac{\sin\frac{p^2}{2}}{p} \qquad [p > 0].$$ NT 61(11)a

6.325

1. $$\int_0^\infty S(x)\sin b^2x^2\,dx = \frac{1}{b}\sqrt{\pi}\,2^{-\frac{5}{2}} \qquad [0 < b^2 < 1];$$
$$= 0 \qquad [b^2 > 1].$$ ET I 98(21)a

2. $$\int_0^\infty C(x)\cos b^2x^2\,dx = \frac{\sqrt{\pi}}{b}\,2^{-\frac{5}{2}} \qquad [0 < b^2 < 1];$$
$$= 0 \qquad [b^2 > 1].$$ ET I 42(22)

6.326

1. $$\int_0^\infty \left[\frac{1}{2} - S(x)\right]\operatorname{si}(2px)\,dx = \frac{\sqrt{8}\cos\frac{\pi}{8}}{\sqrt{\pi}}\left[\frac{1}{2} - S(p\sqrt{2})\right] \qquad [p > 0].$$ NT 61(15)a

2. $$\int_0^\infty \left[\frac{1}{2} - C(x)\right]\operatorname{si}(2px)\,dx = \frac{\sqrt{8}\sin\frac{\pi}{8}}{\sqrt{\pi}}\left[\frac{1}{2} - S(p\sqrt{2})\right] \qquad [p > 0].$$ NT 61(14)a

6.4 The Gamma Function and Functions Generated by It

6.41 The gamma function

6.411 $$\int_{-\infty}^{\infty} \Gamma(\alpha+x)\Gamma(\beta-x)\,dx = -i\pi 2^{1-\alpha-\beta}\Gamma(\alpha+\beta)$$

$[\operatorname{Re}(\alpha+\beta) < 1,\ \operatorname{Im}\alpha,\ \operatorname{Im}\beta > 0]$; ET II 297(3)

$$= i\pi 2^{1-\alpha-\beta}\Gamma(\alpha+\beta)$$

$[\operatorname{Re}(\alpha+\beta) < 1,\ \operatorname{Im}\alpha,\ \operatorname{Im}\beta < 0]$; ET II 297(2)

$$= 0$$

$[\operatorname{Re}(\alpha+\beta) < 1,\ \operatorname{Im}\alpha\cdot\operatorname{Im}\beta < 0]$. ET II 297(1)

6.412 $$\int_{-i\infty}^{i\infty} \Gamma(\alpha+s)\Gamma(\beta+s)\Gamma(\gamma-s)\Gamma(\delta-s)\,ds = 2\pi i\frac{\Gamma(\alpha+\gamma)\Gamma(\alpha+\delta)\Gamma(\beta+\gamma)\Gamma(\beta+\delta)}{\Gamma(\alpha+\beta+\gamma+\delta)}$$

$[\operatorname{Re}\alpha,\ \operatorname{Re}\beta,\ \operatorname{Re}\gamma,\ \operatorname{Re}\delta > 0]$. ET II 302(32)

6.413

1. $$\int_0^{\infty} |\Gamma(a+ix)\Gamma(b+ix)|^2\,dx = \frac{\sqrt{\pi}\,\Gamma(a)\Gamma\left(a+\frac{1}{2}\right)\Gamma(b)\Gamma\left(b+\frac{1}{2}\right)\Gamma(a+b)}{2\Gamma\left(a+b+\frac{1}{2}\right)}$$

$[a > 0,\ b > 0]$. ET II 302(27)

2. $$\int_0^{\infty} \left|\frac{\Gamma(a+ix)}{\Gamma(b+ix)}\right|^2 dx = \frac{\sqrt{\pi}\,\Gamma(a)\Gamma\left(a+\frac{1}{2}\right)\Gamma\left(b-a-\frac{1}{2}\right)}{2\Gamma(b)\Gamma\left(b-\frac{1}{2}\right)\Gamma(b-a)}$$

$\left[0 < a < b-\frac{1}{2}\right]$. ET II 302(28)

6.414

1. $$\int_{-\infty}^{\infty} \frac{\Gamma(\alpha+x)}{\Gamma(\beta+x)}\,dx = 0 \qquad [\operatorname{Im}\alpha \neq 0,\ \operatorname{Re}(\alpha-\beta) < -1].$$ ET II 297(4)

2. $$\int_{-\infty}^{\infty} \frac{dx}{\Gamma(\alpha+x)\Gamma(\beta-x)} = \frac{2^{\alpha+\beta-2}}{\Gamma(\alpha+\beta-1)} \qquad [\operatorname{Re}(\alpha+\beta) > 1].$$ ET II 297(5)

3. $$\int_{-\infty}^{\infty} \frac{\Gamma(\gamma+x)\Gamma(\delta+x)}{\Gamma(\alpha+x)\Gamma(\beta+x)}\,dx = 0$$

$[\operatorname{Re}(\alpha+\beta-\gamma-\delta) > 1,\ \operatorname{Im}\gamma,\ \operatorname{Im}\delta > 0]$. ET II 299(18)

4. $$\int_{-\infty}^{\infty} \frac{\Gamma(\gamma+x)\Gamma(\delta+x)}{\Gamma(\alpha+x)\Gamma(\beta+x)}dx =$$
$$= \frac{\pm 2\pi^2 i\Gamma(\alpha+\beta-\gamma-\delta-1)}{\sin[\pi(\gamma-\delta)]\Gamma(\alpha-\gamma)\Gamma(\alpha-\delta)\Gamma(\beta-\gamma)\Gamma(\beta-\delta)}$$

[$\mathrm{Re}(\alpha+\beta-\gamma-\delta) > 1$, $\mathrm{Im}\,\gamma$, $\mathrm{Im}\,\delta < 0$. In the numerator, we take the plus sign if $\mathrm{Im}\,\gamma > \mathrm{Im}\,\delta$ and the minus sign if $\mathrm{Im}\,\gamma < \mathrm{Im}\,\delta$.] ET II 300(19)

5. $$\int_{-\infty}^{\infty} \frac{\Gamma(\alpha-\beta-\gamma+x+1)\,dx}{\Gamma(\alpha+x)\Gamma(\beta-x)\Gamma(\gamma+x)} =$$
$$= \frac{\pi\exp\left[\pm\frac{1}{2}\pi(\delta-\gamma)i\right]}{\Gamma(\beta+\gamma-1)\Gamma\left[\frac{1}{2}(\alpha+\beta)\right]\Gamma\left[\frac{1}{2}(\gamma-\delta+1)\right]}$$

[$\mathrm{Re}(\beta+\gamma) > 1$, $\delta = \alpha-\beta-\gamma+1$, $\mathrm{Im}\,\delta \neq 0$. The sign is plus in the argument of the exponential for $\mathrm{Im}\,\delta > 0$ and minus for $\mathrm{Im}\,\delta < 0$.] ET II 300(20)

6. $$\int_{-\infty}^{\infty} \frac{dx}{\Gamma(\alpha+x)\Gamma(\beta-x)\Gamma(\gamma+x)\Gamma(\delta-x)} =$$
$$= \frac{\Gamma(\alpha+\beta+\gamma+\delta-3)}{\Gamma(\alpha+\beta-1)\Gamma(\beta+\gamma-1)\Gamma(\gamma+\delta-1)\Gamma(\delta+\alpha-1)}$$
[$\mathrm{Re}(\alpha+\beta+\gamma+\delta) > 3$]. ET II 300(21)

6.415

1. $$\int_{-\infty}^{\infty} \frac{R(x)\,dx}{\Gamma(\alpha+x)\Gamma(\beta-x)\Gamma(\gamma+x)\Gamma(\delta-x)} =$$
$$= \frac{\Gamma(\alpha+\beta+\gamma+\delta-3)}{\Gamma(\alpha+\beta-1)\Gamma(\beta+\gamma-1)\Gamma(\gamma+\delta-1)\Gamma(\delta+\alpha-1)} \int_0^1 R(t)\,dt$$
[$\mathrm{Re}(\alpha+\beta+\gamma+\delta) > 3$, $R(x+1) = R(x)$]. ET II 301(24)

2. $$\int_{-\infty}^{\infty} \frac{R(x)\,dx}{\Gamma(\alpha+x)\Gamma(\beta-x)\Gamma(\gamma+x)\Gamma(\delta-x)} =$$
$$= \frac{\int_0^1 R(t)\cos\left[\frac{1}{2}\pi(2t+\alpha-\beta)\right]dt}{\Gamma\left(\frac{\alpha+\beta}{2}\right)\Gamma\left(\frac{\gamma+\delta}{2}\right)\Gamma(\alpha+\delta-1)}$$
[$\alpha+\delta = \beta+\gamma$, $\mathrm{Re}(\alpha+\beta+\gamma+\delta) > 2$, $R(x+1) = -R(x)$]. ET II 301(25)

6.42 Combinations of the gamma function, the exponential, and powers

6.421

1. $$\int_{-\infty}^{\infty} \Gamma(\alpha+x)\Gamma(\beta-x)\exp[2(\pi n+\theta)xi]\,dx =$$
$$= 2\pi i\Gamma(\alpha+\beta)(2\cos\theta)^{-\alpha-\beta}\exp[(\beta-\alpha)i\theta]\times$$

$$\times [\eta_n(\beta)\exp(2n\pi\beta i) - \eta_n(-\alpha)\exp(-2n\pi\alpha i)]$$

$\left[\operatorname{Re}(\alpha+\beta) < 1;\ -\frac{\pi}{2} < \theta < \frac{\pi}{2};\ n \text{ — an integer};\ \eta_n(\zeta) = 0,\right.$

if $\left(\frac{1}{2} - n\right)\operatorname{Im}\zeta > 0,\ \eta_n(\zeta) = \operatorname{sign}\left(\frac{1}{2} - n\right),$

if $\left.\left(\frac{1}{2} - n\right)\operatorname{Im}\zeta < 0\right].$ ET II 298(7)

2. $$\int_{-\infty}^{\infty} \frac{e^{\pi icx}\,dx}{\Gamma(\alpha+x)\,\Gamma(\beta-x)\,\Gamma(\gamma+kx)\,\Gamma(\delta-kx)} = 0$$

$[\operatorname{Re}(\alpha+\beta+\gamma+\delta) > 2,\ c,\ k \text{ — are real:}$
$|c| > |k| + 1].$ ET II 301(26)

3. $$\int_{-\infty}^{\infty} \frac{\Gamma(\alpha+x)}{\Gamma(\beta+x)} \exp[(2\pi n + \pi - 2\theta)xi]\,dx =$$

$$= 2\pi i \operatorname{sign}\left(n + \frac{1}{2}\right) \frac{(2\cos\theta)^{\beta-\alpha-1}}{\Gamma(\beta-\alpha)} \exp[-(2\pi n + \pi - \theta)\alpha i + \theta i(\beta - 1)]$$

$\left[\operatorname{Re}(\beta-\alpha) > 0,\ -\frac{\pi}{2} < \theta < \frac{\pi}{2},\ n \text{ — an integer},\ \left(n + \frac{1}{2}\right)\operatorname{Im}\alpha < 0\right].$

ET II 298(8)

4. $$\int_{-\infty}^{\infty} \frac{\Gamma(\alpha+x)}{\Gamma(\beta+x)} \exp[(2\pi n + \pi - 2\theta)xi]\,dx = 0$$

$\left[\operatorname{Re}(\beta-\alpha) > 0,\ -\frac{\pi}{2} < \theta < \frac{\pi}{2},\ n \text{ — an integer},\ \left(n + \frac{1}{2}\right)\operatorname{Im}\alpha > 0\right].$

ET II 297(6)

6.422

1. $$\int_{-i\infty}^{i\infty} \Gamma(s-k-\lambda)\,\Gamma\left(\lambda+\mu-s+\frac{1}{2}\right)\Gamma\left(\lambda-\mu-s+\frac{1}{2}\right) z^s\,ds =$$

$$= 2\pi i \Gamma\left(\frac{1}{2} - k - \mu\right)\Gamma\left(\frac{1}{2} - k + \mu\right) z^\lambda e^{\frac{z}{2}} W_{k,\mu}(z)$$

$$\left[\operatorname{Re}(k+\lambda) < 0,\ \operatorname{Re}\lambda > |\operatorname{Re}\mu| - \frac{1}{2},\ |\arg z| < \frac{3\pi}{2}\right].$$

ET II 302(29)

2. $$\int_{\gamma-i\infty}^{\gamma+i\infty} \Gamma(\alpha+s)\,\Gamma(-s)\,\Gamma(1-c-s)\,x^s\,ds =$$

$$= 2\pi i \Gamma(\alpha)\,\Gamma(\alpha-c+1)\,\Psi(\alpha, c; x)$$

$$\left[-\operatorname{Re}\alpha < \gamma < \min(0, 1 - \operatorname{Re}c),\ -\frac{3\pi}{2} < \arg x < \frac{3\pi}{2}\right].$$

EH I 256(5)

3. $$\int_{\gamma-i\infty}^{\gamma+i\infty} \Gamma(-s)\,\Gamma(\beta+s)\,t^s\,ds = 2\pi i \Gamma(\beta)(1+t)^{-\beta}$$

$[0 > \gamma > \operatorname{Re}(1-\beta),\ |\arg t| < \pi].$ EH I 256, BU 75

4. $$\int_{-\infty i}^{\infty i} \Gamma\left(\frac{t-p}{2}\right) \Gamma(-t) \left(\sqrt{2}\right)^{t-p-2} z^t \, dt =$$

$$= 2\pi i e^{\frac{1}{4} z^2} \Gamma(-p) D_p(z)$$

$\left[|\arg z| < \frac{3}{4}\pi;\ p \text{ — not a positive integer}\right]$. WH

5. $$\int_{-i\infty}^{i\infty} \Gamma(s) \Gamma\left(\frac{1}{2}\nu + \frac{1}{4} - s\right) \Gamma\left(\frac{1}{2}\nu - \frac{1}{4} - s\right) \left(\frac{z^2}{2}\right)^s ds =$$

$$= 2\pi i \cdot 2^{\frac{1}{4} - \frac{1}{2}\nu} z^{-\frac{1}{2}} e^{\frac{3}{4} z^2} \Gamma\left(\frac{1}{2}\nu + \frac{1}{4}\right) \Gamma\left(\frac{1}{2}\nu - \frac{1}{4}\right) D_\nu(z)$$

$\left[|\arg z| < \frac{3}{4}\pi,\quad \nu \neq \frac{1}{2},\ -\frac{1}{2},\ -\frac{3}{2}, \ldots\right]$. EH II 120

6. $$\int_{c-i\infty}^{c+i\infty} \left(\frac{1}{2}x\right)^{-s} \Gamma\left(\frac{1}{2}\nu + \frac{1}{2}s\right) \left[\Gamma\left(1 + \frac{1}{2}\nu - \frac{1}{2}s\right)\right]^{-1} ds = -4\pi i J_\nu(x)$$

$[x > 0,\ -\operatorname{Re}\nu < c < 1]$. EH II 21(34)

7. $$\int_{-c-i\infty}^{-c+i\infty} \Gamma(-\nu - s) \Gamma(-s) \left(-\frac{1}{2} iz\right)^{\nu+2s} ds = -2\pi^2 e^{\frac{1}{2} i\nu\pi} H_\nu^{(1)}(z)$$

$\left[|\arg(-iz)| < \frac{\pi}{2}\quad 0 < \operatorname{Re}\nu < c\right]$. EH II 83(34)

8. $$\int_{-c-i\infty}^{-c+i\infty} \Gamma(-\nu - s) \Gamma(-s) \left(\frac{1}{2} iz\right)^{\nu+2s} ds = 2\pi^2 e^{-\frac{1}{2} i\nu\pi} H_\nu^{(2)}(z)$$

$\left[|\arg(iz)| < \frac{\pi}{2},\ 0 < \operatorname{Re}\nu < c\right]$. EH II 83(35)

9. $$\int_{-i\infty}^{i\infty} \Gamma(-s) \frac{\left(\frac{1}{2}x\right)^{\nu+2s}}{\Gamma(\nu+s+1)} ds = 2\pi i J_\nu(x) \qquad [x > 0,\ \operatorname{Re}\nu > 0].$$

EH II 83(36)

10. $$\int_{-i\infty}^{i\infty} \Gamma(-s) \Gamma(-2\nu - s) \Gamma\left(\nu + s + \frac{1}{2}\right) (-2iz)^s ds =$$

$$= -\pi^{\frac{5}{2}} e^{-i(z-\nu\pi)} \sec(\nu\pi) (2z)^{-\nu} H_\nu^{(1)}(z)$$

$\left[|\arg(-iz)| < \frac{3}{2}\pi,\quad 2\nu \neq \pm 1, \pm 3 \ldots\right]$. EH II 83(37)

11. $$\int_{-i\infty}^{i\infty} \Gamma(-s) \Gamma(-2\nu - s) \Gamma\left(\nu + s + \frac{1}{2}\right) (2iz)^s ds =$$

$$= \pi^{\frac{5}{2}} e^{i(z-\nu\pi)} \sec(\nu\pi)\ (2z)^{-\nu} H_\nu^{(2)}(z)$$

$\left[|\arg(iz)| < \frac{3}{2}\pi,\ 2\nu \neq \pm 1,\ \pm 3 \ldots\right]$. EH II 84(38)

12. $$\int_{-i\infty}^{i\infty} \Gamma(s)\,\Gamma\left(\frac{1}{2}-s-\nu\right)\Gamma\left(\frac{1}{2}-s+\nu\right)(2z)^s\,ds =$$

$$= 2^{\frac{3}{2}}\pi^{\frac{3}{2}} i z^{\frac{1}{2}} e^z \sec(\nu\pi)\,K_\nu(z)$$

$$\left[|\arg z| < \frac{3\pi}{2},\ 2\nu \neq \pm 1,\ \pm 3,\ \ldots\right].$$ EH II 84(39)

13. $$\int_{-\frac{1}{2}-i\infty}^{-\frac{1}{2}+i\infty} \frac{\Gamma(-s)}{s\Gamma(1+s)} x^{2s}\,ds = 4\pi \int_{2x}^{\infty} \frac{J_0(t)}{t}\,dt \qquad [x > 0].$$ MO 41

14 $$\int_{-i\infty}^{i\infty} \frac{\Gamma(\alpha+s)\,\Gamma(\beta+s)\,\Gamma(-s)}{\Gamma(\gamma+s)}(-z)^s\,ds =$$

$$= 2\pi i\,\frac{\Gamma(\alpha)\,\Gamma(\beta)}{\Gamma(\gamma)} F(\alpha,\beta;\gamma;z)$$

For $\arg(-z) < \pi$, the path of integration must separate the poles of the integrand at the points $s = 0, 1, 2, 3, \ldots$ from the poles $s = -\alpha - n$ and $s = -\beta - n$ for $n = 0, 1, 2, \ldots)$].

15. $$\int_{\delta-i\infty}^{\delta+i\infty} \frac{\Gamma(\alpha+s)\,\Gamma(-s)}{\Gamma(\gamma+s)}(-z)^s\,ds = \frac{2\pi i\Gamma(\alpha)}{\Gamma(\gamma)}\,{}_1F_1(\alpha;\gamma;z)$$ EH I 62(15)

$$\left[-\frac{\pi}{2} < \arg(-z) < \frac{\pi}{2},\ 0 > \delta > -\operatorname{Re}\alpha,\ \gamma \neq 0, 1, 2, \ldots\right].$$ EH I 256(4)

16. $$\int_{-i\infty}^{i\infty} \left[\frac{\Gamma\left(\frac{1}{2}-s\right)}{\Gamma(s)}\right]^2 z^s\,ds = 2\pi i z^{\frac{1}{2}}\left[2\pi^{-1}K_0\left(4z^{\frac{1}{4}}\right) - N_0\left(4z^{\frac{1}{4}}\right)\right] \qquad [z > 0].$$

ET II 303(33)

17. $$\int_{-i\infty}^{i\infty} \frac{\Gamma\left(\lambda+\mu-s+\frac{1}{2}\right)\Gamma\left(\lambda-\mu-s+\frac{1}{2}\right)}{\Gamma(\lambda-k-s+1)} z^s\,ds =$$

$$= 2\pi i z^\lambda e^{-\frac{z}{2}} W_{k,\mu}(z) \qquad \left[\operatorname{Re}\lambda > |\operatorname{Re}\mu| - \frac{1}{2},\quad |\arg z| < \frac{\pi}{2}\right].$$

ET II 302(30)

18. $$\int_{-i\infty}^{i\infty} \frac{\Gamma(k-\lambda+s)\,\Gamma\left(\lambda+\mu-s+\frac{1}{2}\right)}{\Gamma\left(\mu-\lambda+s+\frac{1}{2}\right)} z^s\,ds =$$

$$= 2\pi i\,\frac{\Gamma\left(k+\mu+\frac{1}{2}\right)}{\Gamma(2\mu+1)} z^\lambda e^{-\frac{z}{2}} M_{k,\mu}(z)$$

$$\left[\operatorname{Re}(k-\lambda) > 0,\ \operatorname{Re}(\lambda+\mu) > -\frac{1}{2},\ |\arg z| < \frac{\pi}{2}\right].$$ ET II 302(31)

19. $$\int_{-i\infty}^{i\infty} \frac{\prod_{j=1}^{m} \Gamma(b_j - s) \prod_{j=1}^{n} \Gamma(1 - a_j + s)}{\prod_{j=m+1}^{q} \Gamma(1 - b_j + s) \prod_{j=n+1}^{p} \Gamma(a_j - s)} z^s \, ds = 2\pi i G_{pq}^{mn}\left(z \left| \begin{matrix} a_1, \ldots, a_p \\ b_1, \ldots, b_q \end{matrix} \right.\right)$$

$\left[p + q < 2(m+n); \ |\arg z| < \left(m + n - \frac{1}{2}p - \frac{1}{2}q\right)\pi;\right.$
$\left.\operatorname{Re} a_k < 1, \ k = 1, \ldots, n; \ \operatorname{Re} b_j > 0, \ j = 1, \ldots, m\right].$ ET II 303(34)

6.423

1. $$\int_0^\infty e^{-\alpha x} \frac{dx}{\Gamma(1+x)} = \nu(e^{-\alpha}).$$ MI 39, EH III 222(16)

2. $$\int_0^\infty e^{-\alpha x} \frac{dx}{\Gamma(x+\beta+1)} = e^{\beta\alpha} \nu(e^{-\alpha}, \beta).$$ MI 39, EH III 222(16)

3. $$\int_0^\infty e^{-\alpha x} \frac{x^m}{\Gamma(x+1)} dx = \mu(e^{-\alpha}, m)$$

$[\operatorname{Re} m > -1]$. MI 39, EH III 222(17)

4. $$\int_0^\infty e^{-\alpha x} \frac{x^m}{\Gamma(x+n+1)} dx = e^{n\alpha} \mu(e^{-\alpha}, m, n).$$ MI 39, EH III 222(17)

6.424 $$\int_{-\infty}^{\infty} \frac{R(x) \exp[(2\pi n + \theta) xi] \, dx}{\Gamma(\alpha + x) \Gamma(\beta - x)} = \frac{\left[2 \cos\left(\frac{\theta}{2}\right)\right]^{\alpha+\beta-2}}{\Gamma(\alpha+\beta-1)} \exp\left[\frac{1}{2}\theta(\beta - \alpha) i\right] \int_0^1 R(t) \exp(2\pi n t i) \, dt$$

$[\operatorname{Re}(\alpha + \beta) > 1, \ -\pi < \theta < \pi, \ n$ — an integer, $R(x+1) = R(x)]$.

ET II 299(16)

6.43 Combinations of the gamma function and trigonometric functions

6.431

1. $$\int_{-\infty}^{\infty} \frac{\sin rx \, dx}{\Gamma(p+x) \Gamma(q-x)} = \frac{\left(2 \cos \frac{r}{2}\right)^{p+q-2} \sin \frac{r(q-p)}{2}}{\Gamma(p+q-1)} \quad [|r| < \pi];$$

$$= 0 \quad [|r| > \pi];$$

$[r$ — real; $\operatorname{Re}(p+q) > 1]$. MO 10a, ET II 298(9, 10)

2. $$\int_{-\infty}^{\infty} \frac{\cos rx \, dx}{\Gamma(p+x) \Gamma(q-x)} = \frac{\left(2 \cos \frac{r}{2}\right)^{p+q-2} \cos \frac{r(q-p)}{2}}{\Gamma(p+q-1)} \quad [|r| < \pi];$$

$$= 0 \quad [|r| > \pi];$$

$[r$ — real; $\operatorname{Re}(p+q) > 1]$. MO 10a, ET II 299(13, 14)

6.432 $$\int_{-\infty}^{\infty} \frac{\sin(m\pi x)}{\sin(\pi x)} \frac{dx}{\Gamma(\alpha+x)\Gamma(\beta-x)} = 0 \qquad [m \text{ — an even integer}];$$

$$= \frac{2^{\alpha+\beta-2}}{\Gamma(\alpha+\beta-1)} \qquad [m \text{ — an odd integer}]$$

$$[\operatorname{Re}(\alpha+\beta) > 1].$$ ET II 298(11, 12)

6.433

1. $$\int_{-\infty}^{\infty} \frac{\sin \pi x \, dx}{\Gamma(\alpha+x)\Gamma(\beta-x)\Gamma(\gamma+x)\Gamma(\delta-x)} =$$

$$= \frac{\sin\left[\frac{\pi}{2}(\beta-\alpha)\right]}{2\Gamma\left(\frac{\alpha+\beta}{2}\right)\Gamma\left(\frac{\gamma+\delta}{2}\right)\Gamma(\alpha+\delta-1)}$$

$$[\alpha+\delta = \beta+\gamma, \ \operatorname{Re}(\alpha+\beta+\gamma+\delta) > 2].$$ ET II 300(22)

2. $$\int_{-\infty}^{\infty} \frac{\cos \pi x \, dx}{\Gamma(\alpha+x)\Gamma(\beta-x)\Gamma(\gamma+x)\Gamma(\delta-x)} =$$

$$= \frac{\cos\left[\frac{\pi}{2}(\beta-\alpha)\right]}{2\Gamma\left(\frac{\alpha+\beta}{2}\right)\Gamma\left(\frac{\gamma+\delta}{2}\right)\Gamma(\alpha+\delta-1)}$$

$$[\alpha+\delta = \beta+\gamma, \ \operatorname{Re}(\alpha+\beta+\gamma+\delta) > 2].$$ ET II 301(23)

6.44 The logarithm of the gamma function*

6.441

1. $$\int_{p}^{p+1} \ln \Gamma(x) \, dx = \frac{1}{2} \ln 2\pi + p \ln p - p.$$ FI II 784

2. $$\int_{0}^{1} \ln \Gamma(x) \, dx = \int_{0}^{1} \ln \Gamma(1-x) \, dx = \frac{1}{2} \ln 2\pi.$$ FI II 783

3. $$\int_{0}^{1} \ln \Gamma(x+q) \, dx = \frac{1}{2} \ln 2\pi + q \ln q - q \qquad [q \geqslant 0].$$

NH 89(17), ET II 304(40)

4. $$\int_{0}^{z} \ln \Gamma(x+1) \, dx = \frac{z}{2} \ln 2\pi - \frac{z(z+1)}{2} + z \ln \Gamma(z+1) - \ln G(z+1),$$

where $G(z+1) = (2\pi)^{\frac{z}{2}} \exp\left(-\frac{z(z+1)}{2} - \frac{Cz^2}{2}\right) \prod_{k=1}^{\infty} \left\{\left(1+\frac{z}{k}\right)^k \exp\left(-z+\frac{z^2}{2k}\right)\right.$

WH

*Here, we are violating our usual order of presentation of the formulas in order to m it easier to examine the integrals involving the gamma function.

5. $\int_0^n \ln\Gamma(a+x)\,dx = \sum_{k=0}^{n-1}(a+k)\ln(a+k) - na + \frac{1}{2}n\ln(2\pi) - \frac{1}{2}n(n-1) \quad [a \geqslant 0;\ n=1, 2, \ldots].$ ET II 304(41)

6.442 $\int_0^1 \exp(2\pi nxi)\ln\Gamma(a+x)\,dx = (2\pi ni)^{-1}[\ln a - \exp(-2\pi nai)\,\mathrm{Ei}(2\pi nai)] \quad [a>0;\ n=\pm 1, \pm 2, \ldots].$ ET II 304(38)

6.443

1. $\int_0^1 \ln\Gamma(x)\sin 2\pi nx\,dx = \frac{1}{2\pi n}[\ln(2\pi n) + C].$ NH 203(5), ET II 304(42)

2. $\int_0^1 \ln\Gamma(x)\sin(2n+1)\pi x\,dx = \frac{1}{(2n+1)\pi}\left[\ln\left(\frac{\pi}{2}\right) + 2\left(1 + \frac{1}{3} + \ldots + \frac{1}{2n-1}\right) + \frac{1}{2n+1}\right].$ ET II 305(43)

3. $\int_0^1 \ln\Gamma(x)\cos 2\pi nx\,dx = \frac{1}{4n}.$ NH 203(6), ET II 305(44)

4. $\int_0^1 \ln\Gamma(x)\cos(2n+1)\pi x\,dx = 0.$ NH 203(6)

5. $\int_0^1 \sin(2\pi nx)\ln\Gamma(a+x)\,dx = -(2\pi n)^{-1}[\ln a + \cos(2\pi na)\,\mathrm{ci}(2\pi na) - \sin(2\pi na)\,\mathrm{si}(2\pi na)] \quad [a>0;\ n=1, 2, \ldots].$ ET II 304(36)

6. $\int_0^1 \cos(2\pi nx)\ln\Gamma(a+x)\,dx = -(2\pi n)^{-1}[\sin(2\pi na)\,\mathrm{ci}(2\pi na) + \cos(2\pi na)\,\mathrm{si}(2\pi na)] \quad [a>0;\ n=1, 2, \ldots].$ ET II 304(37)

6.45 The incomplete gamma function

6.451

1. $\int_0^\infty e^{-\alpha x}\gamma(\beta, x)\,dx = \frac{1}{\alpha}\Gamma(\beta)(1+\alpha)^{-\beta} \quad [\beta>0].$ MI 39

2. $\int_0^\infty e^{-\alpha x}\Gamma(\beta, x)\,dx = \frac{1}{\alpha}\Gamma(\beta)\left[1 - \frac{1}{(\alpha+1)^\beta}\right] \quad [\beta>0].$ MI 39

6.452

1. $$\int_0^\infty e^{-\mu x}\gamma\left(\nu, \frac{x^2}{8a^2}\right)dx = \frac{1}{\mu}2^{-\nu-1}\Gamma(2\nu)e^{(a\mu)^2}D_{-2\nu}(2a\mu)$$
$$\left[|\arg a| < \frac{\pi}{4},\ \operatorname{Re}\nu > -\frac{1}{2},\ \operatorname{Re}\mu > 0\right].$$ ET I 179(36)

2. $$\int_0^\infty e^{-\mu x}\gamma\left(\frac{1}{4}, \frac{x^2}{8a^2}\right)dx = \frac{2^{\frac{3}{4}}\sqrt{a}}{\sqrt{\mu}}e^{(a\mu)^2}K_{\frac{1}{4}}(a^2\mu^2)$$
$$\left[|\arg a| < \frac{\pi}{4},\ \operatorname{Re}\mu > 0\right].$$ ET I 179(35)

6.453 $$\int_0^\infty e^{-\mu x}\Gamma\left(\nu, \frac{a}{x}\right)dx = 2a^{\frac{1}{2}\nu}\mu^{\frac{1}{2}\nu-1}K_\nu(2\sqrt{\mu a})$$
$$\left[|\arg a| < \frac{\pi}{2},\ \operatorname{Re}\mu > 0\right].$$ ET I 179(32)

6.454 $$\int_0^\infty e^{-\beta x}\gamma(\nu, \alpha x^{\frac{1}{2}})dx = 2^{-\frac{1}{2}\nu}\alpha^\nu\beta^{-\frac{1}{2}\nu-1}\Gamma(\nu)\exp\left(\frac{\alpha^2}{8\beta}\right)D_{-\nu}\left(\frac{\alpha}{\sqrt{2\beta}}\right)$$
$$\left[\operatorname{Re}\beta > 0,\ \operatorname{Re}\nu > 0\right].$$ ET II 309(19), MI 39a

6.455

1. $$\int_0^\infty x^{\mu-1}e^{-\beta x}\Gamma(\nu, \alpha x)dx = \frac{\alpha^\nu\Gamma(\mu+\nu)}{\mu(\alpha+\beta)^{\mu+\nu}}\,{}_2F_1\left(1, \mu+\nu; \mu+1; \frac{\beta}{\alpha+\beta}\right)$$
$$[\operatorname{Re}(\alpha+\beta) > 0,\ \operatorname{Re}\mu > 0,\ \operatorname{Re}(\mu+\nu) > 0].$$ ET II 309(16)

2. $$\int_0^\infty x^{\mu-1}e^{-\beta x}\gamma(\nu, \alpha x)dx = \frac{\alpha^\nu\Gamma(\mu+\nu)}{\nu(\alpha+\beta)^{\mu+\nu}}\,{}_2F_1\left(1, \mu+\nu; \nu+1; \frac{\alpha}{\alpha+\beta}\right)$$
$$[\operatorname{Re}(\alpha+\beta) > 0,\ \operatorname{Re}\beta > 0,\ \operatorname{Re}(\mu+\nu) > 0].$$ ET II 308(15)

6.456

1. $$\int_0^\infty e^{-\alpha x}(4x)^{\nu-\frac{1}{2}}\gamma\left(\nu, \frac{1}{4x}\right)dx = \sqrt{\pi}\frac{\gamma(2\nu, \sqrt{\alpha})}{\alpha^{\nu+\frac{1}{2}}}.$$ MI 39a

2. $$\int_0^\infty e^{-\alpha x}(4x)^{\nu-\frac{1}{2}}\Gamma\left(\nu, \frac{1}{4x}\right)dx = \frac{\sqrt{\pi}\,\Gamma(2\nu, \sqrt{\alpha})}{\alpha^{\nu+\frac{1}{2}}}.$$ MI 39a

6.457

1. $$\int_0^\infty e^{-\alpha x}\frac{(4x)^\nu}{\sqrt{x}}\gamma\left(\nu+1, \frac{1}{4x}\right)dx = \sqrt{\pi}\frac{\gamma(2\nu+1, \sqrt{\alpha})}{\alpha^{\nu+\frac{1}{2}}}.$$ MI 39

2. $$\int_0^\infty e^{-\alpha x}\frac{(4x)^\nu}{\sqrt{x}}\Gamma\left(\nu+1, \frac{1}{4x}\right)dx = \sqrt{\pi}\frac{\Gamma(2\nu+1, \sqrt{\alpha})}{\alpha^{\nu+\frac{1}{2}}}.$$ MI 39

6.458 $\int_0^\infty x^{1-2\nu} \exp(\alpha x^2) \sin(bx) \Gamma(\nu, \alpha x^2)\, dx =$

$$= \pi^{\frac{1}{2}} 2^{-\nu} \alpha^{\nu-1} \Gamma\left(\frac{3}{2} - \nu\right) \exp\left(\frac{b^2}{8\alpha}\right) D_{2\nu-2}\left[\frac{b}{(2\alpha)^{\frac{1}{2}}}\right]$$

$\left[|\arg \alpha| < \frac{3\pi}{2},\ 0 < \operatorname{Re} \nu < 1\right]$. ET II 309(18)

6.46-6.47 The function $\psi(x)$

6.461 $\int_1^x \psi(x)\, dx = \ln \Gamma(x)$.

6.462 $\int_0^1 \psi(\alpha + x)\, dx = \ln \alpha \qquad [\alpha > 0]$. ET II 305(1)

6.463 $\int_0^\infty x^{-\alpha} [C + \psi(1 + x)] = -\pi \operatorname{cosec}(\pi\alpha) \zeta(\alpha) \qquad [1 < \operatorname{Re} \alpha < 2]$.

ET II 305(6)

6.464 $\int_0^1 e^{2\pi n x i} \psi(\alpha + x)\, dx = e^{-2\pi n \alpha i} \operatorname{Ei}(2\pi n \alpha i)$

$[\alpha > 0;\ n = \pm 1,\ \pm 2,\ \dots]$. ET II 305(2)

6.465

1. $\int_0^1 \psi(x) \sin \pi x\, dx = 0$. NH 204

2. $\int_0^1 \psi(x) \sin(2\pi n x)\, dx = -\frac{1}{2}\pi \qquad [n = 1,\ 2,\ \dots]$. ET II 305(3)

6.466 $\int_0^\infty [\psi(\alpha + ix) - \psi(\alpha - ix)] \sin xy\, dx = i\pi e^{-\alpha y} (1 - e^{-y})^{-1}$

$[\alpha > 0,\ y > 0]$. ET I 96(1)

6.467

1. $\int_0^1 \sin(2\pi n x) \psi(\alpha + x)\, dx = \sin(2\pi n \alpha) \operatorname{ci}(2\pi n \alpha) + \cos(2\pi n \alpha) \operatorname{si}(2\pi n \alpha)$

$[\alpha \geqslant 0;\ n = 1,\ 2,\ \dots]$. ET II 305(4)

2. $\int_0^1 \cos(2\pi n x) \psi(\alpha + x)\, dx = \sin(2\pi n \alpha) \operatorname{si}(2\pi n \alpha) - \cos(2\pi n \alpha) \operatorname{ci}(2\pi n \alpha)$

$[\alpha > 0;\ n = 1,\ 2,\ \dots]$. ET II 305(5)

6.468 $\int_0^1 \psi(x) \sin^2 \pi x\, dx = -\frac{1}{2}[C + \ln(2\pi)]$. NH 204

6.469

1. $$\int_0^1 \psi(x) \sin \pi x \cos \pi x \, dx = -\frac{\pi}{4}.$$ NH 204

2. $$\int_0^1 \psi(x) \sin \pi x \sin(n\pi x) \, dx = 0 \qquad [n \text{ — even}],$$
$$= \frac{1}{2} \ln \frac{n-1}{n+1} \qquad [n \text{ — odd}].$$ NH 204(8)a

6.471

1. $$\int_0^\infty x^{-\alpha} [\ln x - \psi(1+x)] \, dx = \pi \operatorname{cosec}(\pi\alpha) \zeta(\alpha) \qquad [0 < \operatorname{Re} \alpha < 1].$$ ET II 306(7)

2. $$\int_0^\infty x^{-\alpha} [\ln(1+x) - \psi(1+x)] \, dx = \pi \operatorname{cosec}(\pi\alpha)[\zeta(\alpha) - (\alpha-1)^{-1}]$$
$$[0 < \operatorname{Re} \alpha < 1].$$ ET II 306(8)

3. $$\int_0^\infty [\psi(x+1) - \ln x] \cos(2\pi x y) \, dx = \frac{1}{2} [\psi(y+1) - \ln y].$$ ET II 306(12)

6.472

1. $$\int_0^\infty x^{-\alpha} [(1+x)^{-1} - \psi'(1+x)] \, dx = -\pi\alpha \operatorname{cosec}(\pi\alpha) [\zeta(1+\alpha) - \alpha^{-1}]$$
$$[|\operatorname{Re} \alpha| < 1].$$ ET II 306(9)

2. $$\int_0^\infty x^{-\alpha} [x^{-1} - \psi'(1+x)] \, dx = -\pi\alpha \operatorname{cosec}(\pi\alpha) \zeta(1+\alpha)$$
$$[-2 < \operatorname{Re} \alpha < 0].$$ ET II 306(10)

6.473 $$\int_0^\infty x^{-\alpha} \psi^{(n)}(1+x) \, dx = (-1)^{n-1} \frac{\pi \Gamma(\alpha+n)}{\Gamma(\alpha) \sin \pi\alpha} \zeta(\alpha+n)$$
$$[n = 1, 2, \ldots; \; 0 < \operatorname{Re} \alpha < 1].$$ ET II 306(11)

6.5-6.7 Bessel Functions

6.51 Bessel functions

6.511

1. $$\int_0^\infty J_\nu(bx) \, dx = \frac{1}{b} \qquad [\operatorname{Re} \nu > -1, \; b > 0].$$ ET II 22(3)

2. $$\int_0^\infty N_\nu(bx) \, dx = -\frac{1}{b} \operatorname{tg}\left(\frac{\nu\pi}{2}\right) \qquad [|\operatorname{Re} \nu| < 1, \; b > 0].$$ WA 432(7), ET II 96(1)

3. $\int_0^a J_\nu(x)\,dx = 2\sum_{k=0}^{\infty} J_{\nu+2k+1}(a)$ $[\operatorname{Re}\nu > -1]$. ET II 333(1)

4. $\int_0^a J_{\frac{1}{2}}(t)\,dt = 2S(\sqrt{a})$. WA 599(4)

5. $\int_0^a J_{-\frac{1}{2}}(t)\,dt = 2C(\sqrt{a})$. WA 599(3)

6. $\int_0^a J_0(x)\,dx = aJ_0(a) + \frac{\pi a}{2}[J_1(a)\mathbf{H}_0(a) - J_0(a)\mathbf{H}_1(a)]$ $[a > 0]$. ET II 7(2)

7. $\int_0^a J_1(x)\,dx = 1 - J_0(a)$ $[a > 0]$. ET II 18(1)

8. $\int_a^\infty J_0(x)\,dx = 1 - aJ_0(a) + \frac{\pi a}{2}[J_0(a)\mathbf{H}_1(a) - J_1(a)\mathbf{H}_0(a)]$ $[a > 0]$. ET II 7(3)

9. $\int_a^\infty J_1(x)\,dx = J_0(a)$ $[a > 0]$. ET II 18(2)

10. $\int_a^b N_\nu(x)\,dx = 2\sum_{n=0}^{\infty}[N_{\nu+2n+1}(b) - N_{\nu+2n+1}(a)]$. ET II 339(46)

11. $\int_0^a I_\nu(x)\,dx = 2\sum_{n=0}^{\infty}(-1)^n I_{\nu+2n+1}(a)$ $[\operatorname{Re}\nu > -1]$. ET II 364(1)

6.512

1. $\int_0^\infty J_\mu(ax)J_\nu(bx)\,dx = b^\nu a^{-\nu-1} \times$

$$\times \frac{\Gamma\left(\frac{\mu+\nu+1}{2}\right)}{\Gamma(\nu+1)\Gamma\left(\frac{\mu-\nu+1}{2}\right)} F\left(\frac{\mu+\nu+1}{2}, \frac{\nu-\mu+1}{2}; \nu+1; \frac{b^2}{a^2}\right)$$

$[a > 0,\ b > 0,\ \operatorname{Re}(\mu+\nu) > -1,\ b < a.$ For $a < b$, the positions of μ and ν should be reversed]. ET II 48(6)

2. $\int_0^\infty J_{\nu+n}(\alpha t)J_{\nu-n-1}(\beta t)\,dt = \frac{\beta^{\nu-n-1}\Gamma(\nu)}{\alpha^{\nu-n}n!\,\Gamma(\nu-n)} F\left(\nu, -n; \nu-n; \frac{\beta^2}{\alpha^2}\right)$

$[0 < \beta < \alpha]$;

$= (-1)^n \frac{1}{2\alpha}$ $[0 < \beta = \alpha]$;

$= 0$ $[0 < \alpha < \beta]$ $[\operatorname{Re}(\nu) > 0]$. MO 50

3. $$\int_0^\infty J_\nu(ax) J_{\nu-1}(\beta x)\,dx = \frac{\beta^{\nu-1}}{\alpha^\nu} \quad [\beta < \alpha];$$
$$= \frac{1}{2\beta} \quad [\beta = \alpha];$$
$$= 0 \quad [\beta > \alpha]; \quad [\operatorname{Re}\nu > 0].$$

WA 444(8), KU (40)a

4. $$\int_0^\infty J_{\nu+2n+1}(ax) J_\nu(bx)\,dx = b^\nu a^{-\nu-1} P_n^{(\nu,\,0)}\left(1 - \frac{2b^2}{a^2}\right)$$
$[\operatorname{Re}\nu > -1-n,\ 0 < b < a];$
$$= 0 \quad [\operatorname{Re}\nu > -1-n,\ 0 < a < b].$$

ET II 47(5)

5. $$\int_0^\infty J_{\nu+n}(ax) N_{\nu-n}(ax)\,dx = (-1)^{n+1}\frac{1}{2a}$$
$$\left[\operatorname{Re}\nu > -\frac{1}{2};\ a > 0;\ n = 0, 1, 2, \ldots\right].$$

ET II 347(57)

6. $$\int_0^\infty J_1(bx) N_0(ax)\,dx = -\frac{b^{-1}}{\pi}\ln\left(1 - \frac{b^2}{a^2}\right)$$
$[0 < b < a].$

ET II 21(31)

7. $$\int_0^a J_\nu(x) J_{\nu+1}(x)\,dx = \sum_{n=0}^\infty [J_{\nu+n+1}(a)]^2$$
$[\operatorname{Re}\nu > -1].$

ET II 338(37)

6.513

1. $$\int_0^\infty [J_\mu(ax)]^2 J_\nu(bx)\,dx =$$
$$= a^{2\mu} b^{-2\mu-1} \frac{\Gamma\left(\frac{1+\nu+2\mu}{2}\right)}{[\Gamma(\mu+1)]^2\,\Gamma\left(\frac{1+\nu-2\mu}{2}\right)} \times$$
$$\times \left[F\left(\frac{1-\nu+2\mu}{2}, \frac{1+\nu+2\mu}{2}; \mu+1; \frac{1-\sqrt{1-\frac{4a^2}{b^2}}}{2}\right)\right]^2$$
$[\operatorname{Re}\nu + \operatorname{Re}2\mu > -1,\ 0 < 2a < b].$

ET II 52(33)

2. $$\int_0^\infty [J_\mu(ax)]^2 K_\nu(bx)\,dx =$$
$$= \frac{b^{-1}}{2}\Gamma\left(\frac{2\mu+\nu+1}{2}\right)\Gamma\left(\frac{2\mu-\nu+1}{2}\right)\left[P^{-\mu}_{\frac{1}{2}\nu-\frac{1}{2}}\left(\sqrt{1+\frac{4a^2}{b^2}}\right)\right]^2$$
$[2\operatorname{Re}\mu > |\operatorname{Re}\nu| - 1,\ \operatorname{Re}b > 2|\operatorname{Im}a|].$

ET II 138(18)

3. $\int_0^\infty I_\mu(ax) K_\mu(ax) J_\nu(bx)\,dx =$

$$= \frac{e^{\mu\pi i}\Gamma\left(\frac{\nu+2\mu+1}{2}\right)}{b\Gamma\left(\frac{\nu-2\mu+1}{2}\right)} P^{-\mu}_{\frac{1}{2}\nu-\frac{1}{2}}\left(\sqrt{1+\frac{4a^2}{b^2}}\right) Q^{-\mu}_{\frac{1}{2}\nu-\frac{1}{2}}\left(\sqrt{1+\frac{4a^2}{b^2}}\right)$$

$[\operatorname{Re} a > 0,\ b > 0,\ \operatorname{Re}\nu > -1,\ \operatorname{Re}(\nu+2\mu) > -1]$ ET II 65(20)

4. $\int_0^\infty J_\mu(ax) J_{-\mu}(ax) K_\nu(bx)\,dx =$

$$= \frac{\pi}{2b}\sec\left(\frac{\nu\pi}{2}\right) P^{\mu}_{\frac{1}{2}\nu-\frac{1}{2}}\left(\sqrt{1+\frac{4a^2}{b^2}}\right) P^{-\mu}_{\frac{1}{2}\nu-\frac{1}{2}}\left(\sqrt{1+\frac{4a^2}{b^2}}\right)$$

$[|\operatorname{Re}\nu| < 1,\ \operatorname{Re} b > 2|\operatorname{Im} a|]$. ET II 138(21)

5. $\int_0^\infty [K_\mu(ax)]^2 J_\nu(bx)\,dx =$

$$= \frac{e^{2\mu\pi i}\Gamma\left(\frac{1+\nu+2\mu}{2}\right)}{b\Gamma\left(\frac{1+\nu-2\mu}{2}\right)}\left[Q^{-\mu}_{\frac{1}{2}\nu-\frac{1}{2}}\left(\sqrt{1+\frac{4a^2}{b^2}}\right)\right]^2$$

$\left[\operatorname{Re} a > 0,\ b > 0,\ \operatorname{Re}\left(\frac{1}{2}\nu \pm \mu\right) > -\frac{1}{2}\right]$. ET II 66(28)

6. $\int_0^z J_\mu(x) J_\nu(z-x)\,dx = 2\sum_{k=0}^\infty (-1)^k J_{\mu+\nu+2k+1}(z)$

$[\operatorname{Re}\mu > -1,\ \operatorname{Re}\nu > -1]$ (see also 6.683 3.). WA 414(2)

7. $\int_0^z J_\mu(x) J_{-\mu}(z-x)\,dx = \sin z \qquad [-1 < \operatorname{Re}\mu < 1]$. WA 415(4)

8. $\int_0^z J_\mu(x) J_{1-\mu}(z-x)\,dx = J_0(z) - \cos(z)$

$[-1 < \operatorname{Re}\mu < 2]$. WA 415(4)

6.514

1. $\int_0^\infty J_\nu\left(\frac{a}{x}\right) J_\nu(bx)\,dx = b^{-1} J_{2\nu}(2\sqrt{ab})$

$\left[a > 0,\ b > 0,\ \operatorname{Re}\nu > -\frac{1}{2}\right]$. ET II 57(9)

2. $\int_0^\infty J_\nu\left(\frac{a}{x}\right) N_\nu(bx)\,dx = b^{-1}\left[N_{2\nu}(2\sqrt{ab}) + \frac{2}{\pi} K_{2\nu}(\sqrt{2ab})\right]$

$\left[a > 0,\ b > 0,\ -\frac{1}{2} < \operatorname{Re}\nu < \frac{3}{2}\right]$. ET II 110(12)

3. $$\int_0^\infty J_\nu\left(\frac{a}{x}\right) K_\nu(bx)\,dx =$$

$$= b^{-1} e^{\frac{1}{2} i(\nu+1)\pi} K_{2\nu}\left[2e^{\frac{1}{4} i\pi}\sqrt{ab}\right] + b^{-1} e^{-\frac{1}{2} i(\nu+1)\pi} K_{2\nu}\left[2e^{-\frac{1}{4}\pi i}\sqrt{ab}\right]$$

$$\left[a > 0,\ \operatorname{Re} b > 0,\ |\operatorname{Re}\nu| < \frac{5}{2}\right].$$ ET II 141(31)

4. $$\int_0^\infty N_\nu\left(\frac{a}{x}\right) J_\nu(bx)\,dx = -\frac{2b^{-1}}{\pi}\left[K_{2\nu}(2\sqrt{ab}) - \frac{\pi}{2} N_{2\nu}(2\sqrt{ab})\right]$$

$$\left[a > 0,\ b > 0,\ |\operatorname{Re}\nu| < \frac{1}{2}\right].$$ ET II 62(37)a

5. $$\int_0^\infty N_\nu\left(\frac{a}{x}\right) N_\nu(bx)\,dx = -b^{-1} J_{2\nu}(2\sqrt{ab})$$

$$\left[a > 0,\ b > 0,\ |\operatorname{Re}\nu| < \frac{1}{2}\right].$$ ET II 110(14)

6. $$\int_0^\infty N_\nu\left(\frac{a}{x}\right) K_\nu(bx)\,dx = -b^{-1} e^{\frac{1}{2}\nu\pi i} K_{2\nu}\left(2e^{\frac{1}{4}\pi i}\sqrt{ab}\right) -$$

$$- b^{-1} e^{-\frac{1}{2}\nu\pi i} K_{2\nu}\left(2e^{-\frac{1}{4}\pi i}\sqrt{ab}\right)$$

$$\left[a > 0,\ \operatorname{Re} b > 0,\ |\operatorname{Re}\nu| < \frac{5}{2}\right].$$ ET II 143(37)

7. $$\int_0^\infty K_\nu\left(\frac{a}{x}\right) N_\nu(bx)\,dx = -2b^{-1}\left[\sin\left(\frac{3\nu\pi}{2}\right)\operatorname{ker}_{2\nu}(2\sqrt{ab}) + \right.$$

$$\left. + \cos\left(\frac{3\nu\pi}{2}\right)\operatorname{kei}_{2\nu}(2\sqrt{ab})\right] \qquad \left[\operatorname{Re} a > 0,\ b > 0,\ |\operatorname{Re}\nu| < \frac{1}{2}\right].$$ ET II 113(28)

8. $$\int_0^\infty K_\nu\left(\frac{a}{x}\right) K_\nu(bx)\,dx = \pi b^{-1} K_{2\nu}(2\sqrt{ab})$$

$$[\operatorname{Re} a > 0,\ \operatorname{Re} b > 0].$$ ET II 146(54)

6.515

1. $$\int_0^\infty J_\mu\left(\frac{a}{x}\right) N_\mu\left(\frac{a}{x}\right) K_0(bx)\,dx =$$

$$= -2b^{-1} J_{2\mu}(2\sqrt{ab}) K_{2\mu}(2\sqrt{ab})$$

$$[a > 0,\ \operatorname{Re} b > 0].$$ ET II 143(42)

2. $$\int_0^\infty \left[K_\mu\left(\frac{a}{x}\right)\right]^2 K_0(bx)\,dx =$$

$$= 2\pi b^{-1} K_{2\mu}\left(2e^{\frac{1}{4}\pi i}\sqrt{ab}\right) K_{2\mu}\left(2e^{-\frac{1}{4}\pi i}\sqrt{ab}\right)$$

$$[\operatorname{Re} a > 0,\ \operatorname{Re} b > 0].$$ ET II 147(59)

3. $\int_0^\infty H_\mu^{(1)}\left(\frac{a^2}{x}\right) H_\mu^{(2)}\left(\frac{a^2}{x}\right) J_0(bx)\,dx =$

$$= 16\pi^{-2} b^{-1} \cos \mu\pi K_{2\mu}\left(2e^{\frac{1}{4}\pi i} a\sqrt{b}\right) K_{2\mu}\left(2e^{-\frac{1}{4}\pi i} a\sqrt{b}\right)$$

$\left[|\arg a| < \frac{\pi}{4},\ b > 0,\ |\operatorname{Re}\mu| < \frac{1}{4}\right].$ ET II 17(36)

6.516

1. $\int_0^\infty J_{2\nu}(a\sqrt{x}) J_\nu(bx)\,dx = b^{-1} J_\nu\left(\frac{a^2}{4b}\right)$

$\left[a > 0,\ b > 0,\ \operatorname{Re}\nu > -\frac{1}{2}\right].$ ET II 58(16)

2. $\int_0^\infty J_{2\nu}(a\sqrt{x}) N_\nu(bx)\,dx = -b^{-1}\mathbf{H}_\nu\left(\frac{a^2}{4b}\right)$

$\left[a > 0,\ b > 0,\ \operatorname{Re}\nu > -\frac{1}{2}\right].$ ET II 111(18)

3. $\int_0^\infty J_{2\nu}(a\sqrt{x}) K_\nu(bx)\,dx = \frac{\pi}{2} b^{-1}\left[I_\nu\left(\frac{a^2}{4b}\right) - \mathbf{L}_\nu\left(\frac{a^2}{4b}\right)\right]$

$\left[\operatorname{Re} b > 0,\ \operatorname{Re}\nu > -\frac{1}{2}\right].$ ET II 144(45)

4. $\int_0^\infty N_{2\nu}(a\sqrt{x}) J_\nu(bx)\,dx = 2\sec(\nu\pi)\, b^{-1} \times$

$$\times\left[\frac{1}{2}\cos(\nu\pi) N_\nu\left(\frac{a^2}{4b}\right) - N_{-\nu}\left(\frac{a^2}{4b}\right) + \mathbf{H}_{-\nu}\left(\frac{a^2}{4b}\right)\right]$$

$\left[a > 0,\ b > 0,\ \operatorname{Re}\nu > -\frac{1}{2}\right].$ ET II 62(39)

5. $\int_0^\infty N_{2\nu}(a\sqrt{x}) N_\nu(bx)\,dx =$

$$= \frac{b^{-1}}{2}\left[\sec(\nu\pi) J_{-\nu}\left(\frac{a^2}{4b}\right) + \operatorname{cosec}(\nu\pi)\mathbf{H}_{-\nu}\left(\frac{a^2}{4b}\right) - \right.$$

$$\left. - 2\operatorname{ctg}(2\nu\pi)\,\mathbf{H}_\nu\left(\frac{a^2}{4b}\right)\right]$$

$\left[a > 0,\ b > 0,\ |\operatorname{Re}\nu| < \frac{1}{2}\right].$ ET II 111(19)

6. $\int_0^\infty N_{2\nu}(a\sqrt{x}) K_\nu(bx)\,dx =$

$$= \frac{\pi b^{-1}}{2}\left[\operatorname{cosec}(2\nu\pi)\,\mathbf{L}_{-\nu}\left(\frac{a^2}{4b}\right) - \operatorname{ctg}(2\nu\pi)\,\mathbf{L}_\nu\left(\frac{a^2}{4b}\right) - \right.$$

$$\left. - \operatorname{tg}(\nu\pi)\, I_\nu\left(\frac{a^2}{4b}\right) - \frac{\sec(\nu\pi)}{\pi} K_\nu\left(\frac{a^2}{4b}\right)\right]$$

$\left[\operatorname{Re} b > 0,\ |\operatorname{Re}\nu| < \frac{1}{2}\right].$ ET II 144(46)

7. $$\int_0^\infty K_{2\nu}(a\sqrt{x})\, J_\nu(bx)\, dx = \frac{1}{4}\pi b^{-1}\sec(\nu\pi)\left[\mathbf{H}_{-\nu}\left(\frac{a^2}{4b}\right) - N_{-\nu}\left(\frac{a^2}{4b}\right)\right]$$

$$\left[\operatorname{Re} a > 0,\ b > 0,\ \operatorname{Re}\nu > -\frac{1}{2}\right].$$ ET II 70(22)

8. $$\int_0^\infty K_{2\nu}(a\sqrt{x})\, N_\nu(bx)\, dx =$$

$$= -\frac{1}{4}\pi b^{-1}\left[\sec(\nu\pi) J_{-\nu}\left(\frac{a^2}{4b}\right) - \operatorname{cosec}(\nu\pi)\mathbf{H}_{-\nu}\left(\frac{a^2}{4b}\right) + {} \right.$$
$$\left. {} + 2\operatorname{cosec}(2\nu\pi)\mathbf{H}_\nu\left(\frac{a^2}{4b}\right)\right]$$

$$\left[\operatorname{Re} a > 0,\ b > 0,\ |\operatorname{Re}\nu| < \frac{1}{2}\right].$$ ET II 114(34)

9. $$\int_0^\infty K_{2\nu}(a\sqrt{x})\, K_\nu(bx)\, dx =$$

$$= \frac{\pi b^{-1}}{4\cos(\nu\pi)}\left\{K_\nu\left(\frac{a^2}{4b}\right) + \frac{\pi}{2\sin(\nu\pi)}\left[\mathbf{L}_{-\nu}\left(\frac{a^2}{4b}\right) - \mathbf{L}_\nu\left(\frac{a^2}{4b}\right)\right]\right\}$$

$$\left[\operatorname{Re} b > 0,\ |\operatorname{Re}\nu| < \frac{1}{2}\right].$$ ET II 147(63)

10. $$\int_0^\infty I_{2\nu}(a\sqrt{x})\, K_\nu(bx)\, dx = \frac{\pi b^{-1}}{2}\left[I_\nu\left(\frac{a^2}{4b}\right) + \mathbf{L}_\nu\left(\frac{a^2}{4b}\right)\right]$$

$$\left[\operatorname{Re} b > 0,\ \operatorname{Re}\nu > -\frac{1}{2}\right].$$ ET II 147(60)

6.517 $$\int_0^z J_0\left(\sqrt{z^2 - x^2}\right) dx = \sin z.$$ MO 48

6.518 $$\int_0^\infty K_{2\nu}(2z \operatorname{sh} x)\, dx = \frac{\pi^2}{8\cos\nu\pi}\left(J_\nu^2(z) + N_\nu^2(z)\right)$$

$$\left[\operatorname{Re} z > 0,\quad -\frac{1}{2} < \operatorname{Re}\nu < \frac{1}{2}\right].$$ MO 45

6.519

1. $$\int_0^{\frac{\pi}{2}} J_{2\nu}(2z\cos x)\, dx = \frac{\pi}{2} J_\nu^2(z) \qquad \left[\operatorname{Re}\nu > -\frac{1}{2}\right].$$ WH

2. $$\int_0^{\frac{\pi}{2}} J_{2\nu}(2z\sin x)\, dx = \frac{\pi}{2} J_\nu^2(z) \qquad \left[\operatorname{Re}\nu > -\frac{1}{2}\right].$$ WA 42(1)a

6.52 Bessel functions combined with x and x^2

6.521

1. $$\int_0^1 xJ_\nu(\alpha x)\, J_\nu(\beta x)\, dx = 0 \quad [\alpha \neq \beta];$$
$$= \frac{1}{2}\{J_{\nu+1}(\alpha)\}^2 \quad [\alpha = \beta]$$
$$[J_\nu(\alpha) = J_\nu(\beta) = 0, \quad \nu > -1].$$
WH

2. $$\int_0^\infty xK_\nu(ax)\, J_\nu(bx)\, dx = \frac{b^\nu}{a^\nu (b^2 + a^2)}$$
$$[\operatorname{Re} a > 0,\ b > 0,\ \operatorname{Re} \nu > -1].$$
ET II 63(2)

3. $$\int_0^\infty xK_\nu(ax)\, K_\nu(bx)\, dx = \frac{\pi (ab)^{-\nu} (a^{2\nu} - b^{2\nu})}{2 \sin(\nu\pi)(a^2 - b^2)}$$
$$[|\operatorname{Re} \nu| < 1,\ \operatorname{Re}(a + b) > 0].$$
ET II 145(48)

4. $$\int_0^a xJ_\nu(\lambda x)\, K_\nu(\mu x)\, dx = (\mu^2 + \lambda^2)^{-1}\left[\left(\frac{\lambda}{\mu}\right)^\nu + \lambda a J_{\nu+1}(\lambda a)\, K_\nu(\mu a) - \right.$$
$$\left. - \mu a J_\nu(\lambda a)\, K_{\nu+1}(\mu a)\right] \quad [\operatorname{Re} \nu > -1].$$
ET II 367(26)

6.522

1. $$\int_0^\infty x\,[J_\mu(ax)]^2\, K_\nu(bx)\, dx = \Gamma\left(\mu + \frac{1}{2}\nu + 1\right)\Gamma\left(\mu - \frac{1}{2}\nu + 1\right) b^{-2} \times$$
$$\times (1 + 4a^2b^{-2})^{-\frac{1}{2}} P_{\frac{1}{2}\nu}^{\mu}\left[(1 + 4a^2b^{-2})^{\frac{1}{2}}\right] P_{\frac{1}{2}\nu-1}^{-\mu}\left[(1 + 4a^2b^{-2})^{\frac{1}{2}}\right]$$
$$[\operatorname{Re} b > 2|\operatorname{Im} a|,\ 2\operatorname{Re}\mu > |\operatorname{Re}\nu| - 2].$$
ET II 138(19)

2. $$\int_0^\infty x\,[K_\mu(ax)]^2\, J_\nu(bx)\, dx = \frac{2e^{2\mu\pi i}\,\Gamma\left(1 + \frac{1}{2}\nu + \mu\right)}{b\,(4a^2 + b^2)^{\frac{1}{2}}\,\Gamma\left(\frac{1}{2}\nu - \mu\right)} \times$$
$$\times Q_{\frac{1}{2}\nu}^{-\mu}\left[(1 + 4a^2b^{-2})^{\frac{1}{2}}\right] Q_{\frac{1}{2}\nu-1}^{-\mu}\left[(1 + 4a^2b^{-2})^{\frac{1}{2}}\right]$$
$$\left[b > 0,\quad \operatorname{Re} a > 0,\ \operatorname{Re}\left(\frac{1}{2}\nu \pm \mu\right) > -1\right].$$
ET II 66(27)a

3. $$\int_0^\infty xK_0(ax)\, J_\nu(bx)\, J_\nu(cx)\, dx = r_1^{-1} r_2^{-1} (r_2 - r_1)^\nu (r_2 + r_1)^{-\nu},$$
$$r_1 = [a^2 + (b - c)^2]^{\frac{1}{2}}, \quad r_2 = [a^2 + (b + c)^2]^{\frac{1}{2}}$$
$$[c > 0,\ \operatorname{Re} \nu > -1,\ \operatorname{Re} a > |\operatorname{Im} b|].$$
ET II 63(6)

4. $$\int_0^\infty xI_0(ax)K_0(bx)J_0(cx)\,dx = (a^4+b^4+c^4-2a^2b^2+2a^2c^2+2b^2c^2)^{-\frac{1}{2}}$$

$[\operatorname{Re} b > \operatorname{Re} a, \quad c > 0]$. ET II 16(27)

5. $$\int_0^\infty xJ_0(ax)K_0(bx)J_0(cx)\,dx = (a^4+b^4+c^4-2a^2c^2+2a^2b^2+2b^2c^2)^{-\frac{1}{2}}$$

$[\operatorname{Re} b > |\operatorname{Im} a|, \quad c > 0]$. ET II 15(25)

6. $$\int_0^\infty xJ_0(ax)N_0(ax)J_0(bx)\,dx =$$

$$= 0 \qquad [0 < b < 2a];$$

$$= -2\pi^{-1}b^{-1}[b^2-4a^2]^{-\frac{1}{2}} \qquad [0 < 2a < b < \infty].$$

ET II 15(21)

7. $$\int_0^\infty xJ_\mu(ax)J_{\mu+1}(ax)K_\nu(bx)\,dx =$$

$$= \Gamma\left(\mu+\frac{3+\nu}{2}\right)\Gamma\left(\mu+\frac{3-\nu}{2}\right)b^{-2}(1+4a^2b^{-2})^{\frac{1}{2}}\times$$

$$\times P^{-\mu}_{\frac{1}{2}\nu-\frac{1}{2}}[(1+4a^2b^{-2})^{\frac{1}{2}}]\,P^{-\mu-1}_{\frac{1}{2}\nu-\frac{1}{2}}[(1+4a^2b^{-2})^{\frac{1}{2}}]$$

$[\operatorname{Re} b > 2|\operatorname{Im} a|, \quad 2\operatorname{Re}\mu > |\operatorname{Re}\nu| - 3]$. ET II 138(20)

8. $$\int_0^\infty xK_{\mu-\frac{1}{2}}(ax)K_{\mu+\frac{1}{2}}(ax)J_\nu(bx)\,dx =$$

$$= -\frac{2e^{2\mu\pi i}\Gamma\left(\frac{1}{2}\nu+\mu+1\right)}{b\Gamma\left(\frac{1}{2}\nu-\mu\right)(b^2+4a^2)^{\frac{1}{2}}}\,Q^{-\mu+\frac{1}{2}}_{\frac{1}{2}\nu-\frac{1}{2}}[(1+4a^2b^{-2})^{\frac{1}{2}}]\times$$

$$\times Q^{-\mu-\frac{1}{2}}_{\frac{1}{2}\nu-\frac{1}{2}}[(1+4a^2b^{-2})^{\frac{1}{2}}]$$

$\left[b > 0,\ \operatorname{Re} a > 0,\ \operatorname{Re}\nu > -1,\ |\operatorname{Re}\mu| < 1+\frac{1}{2}\operatorname{Re}\nu\right]$. ET II 67(29)a

9. $$\int_0^\infty xI_{\frac{1}{2}\nu}(ax)K_{\frac{1}{2}}(ax)J_\nu(bx)\,dx = b^{-1}(b^2+4a^2)^{-\frac{1}{2}}$$

$[b > 0,\ \operatorname{Re} a > 0,\ \operatorname{Re}\nu > -1]$. ET II 65(16)

10. $$\int_0^\infty xJ_{\frac{1}{2}\nu}(ax)N_{\frac{1}{2}\nu}(ax)J_\nu(bx)\,dx =$$

$$= 0 \qquad [a > 0,\ \operatorname{Re}\nu > -1;\ 0 < b < 2a];$$

$$= -2\pi^{-1}b^{-1}(b^2-4a^2)^{-\frac{1}{2}} \qquad [a > 0,\ \operatorname{Re}\nu > -1,\ 2a < b < \infty].$$

ET II 55(48)

11. $\int_0^\infty xJ_{\frac{1}{2}(\nu+n)}(ax)\,J_{\frac{1}{2}(\nu-n)}(ax)\,J_\nu(bx)\,dx=$

$= 2\pi^{-1}b^{-1}(4a^2-b^2)^{-\frac{1}{2}}T_n\left(\frac{b}{2a}\right)$ $[a>0,\ \operatorname{Re}\nu>-1,\ 0<b<2a]$;

$=0$ $[a>0,\ \operatorname{Re}\nu>-1,\ 2a<b<\infty]$.

ET II 52(32)

12. $\int_0^\infty xI_{\frac{1}{2}(\nu-\mu)}(ax)\,K_{\frac{1}{2}(\nu+\mu)}(ax)\,J_\nu(bx)\,dx=$

$=2^{-\mu}a^{-\mu}b^{-1}(b^2+4a^2)^{-\frac{1}{2}}[b+(b^2+4a^2)^{\frac{1}{2}}]^\mu$

$[b>0,\ \operatorname{Re}a>0,\ \operatorname{Re}\nu>-1,\ \operatorname{Re}(\nu-\mu)>-2]$. ET II 66(23)

13. $\int_0^\infty xJ_\mu(xa\sin\varphi)\,K_{\nu-\mu}(ax\cos\varphi\cos\psi)\,J_\nu(xa\sin\psi)\,dx=$

$=\frac{(\sin\varphi)^\mu(\sin\psi)^\nu(\cos\varphi)^{\nu-\mu}(\cos\psi)^{\mu-\nu}}{a^2(1-\sin^2\varphi\sin^2\psi)}$

$\left[a>0,\ 0<\varphi,\ \psi<\frac{\pi}{2},\ \operatorname{Re}\mu>-1,\ \operatorname{Re}\nu>-1\right]$. ET II 64(10)

14. $\int_0^\infty xJ_\mu(xa\sin\varphi\cos\psi)\,J_{\nu-\mu}(ax)\,J_\nu(xa\cos\varphi\sin\psi)\,dx=$

$=2\pi^{-1}a^{-2}\sin(\mu\pi)(\sin\varphi)^\mu(\sin\psi)^\nu(\cos\varphi)^{-\nu}(\cos\psi)^{-\mu}\times$

$\times[\cos(\varphi+\psi)\cos(\varphi-\psi)]^{-1}$

$\left[a>0,\ 0<\varphi,\ \psi<\frac{1}{2}\pi,\ \operatorname{Re}\nu>-1\right]$. ET II 54(39)

6.523 $\int_0^\infty x[2\pi^{-1}K_0(ax)-N_0(ax)]\,K_0(bx)\,dx=$

$=2\pi^{-1}[(a^2+b^2)^{-1}+(b^2-a^2)^{-1}]\ln\frac{b}{a}$

$[\operatorname{Re}b>|\operatorname{Im}a|,\ \operatorname{Re}(a+b)>0]$. ET II 145(50)

6.524

1. $\int_0^\infty xJ_\nu^2(ax)\,J_\nu(bx)\,N_\nu(bx)\,dx=$

$=0$ $\left[0<a<b,\ \operatorname{Re}\nu>-\frac{1}{2}\right]$;

$=-(2\pi ab)^{-1}$ $\left[0<b<a,\ \operatorname{Re}\nu>-\frac{1}{2}\right]$. ET II 352(14)

2. $\int_0^\infty x[J_0(ax)\,K_0(bx)]^2\,dx=\frac{\pi}{8ab}-\frac{1}{4ab}\arcsin\left(\frac{b^2-a^2}{b^2+a^2}\right)$

$[a>0,\ b>0]$. ET II 373(9)

6.525

1. $$\int_0^\infty x^2 J_1(ax) K_0(bx) J_0(cx)\,dx = 2a\,(a^2+b^2-c^2)\left[(a^2+b^2+c^2)^2-4a^2c^2\right]^{-\frac{3}{2}}$$
$[c>0,\ \operatorname{Re} b \geqslant |\operatorname{Im} a|,\ \operatorname{Re} a > 0]$. ET II 15(26)

2. $$\int_0^\infty x^2 I_0(ax) K_1(bx) J_0(cx)\,dx = 2b\,(b^2+c^2-a^2)\left[(a^2+b^2+c^2)^2-4a^2b^2\right]^{-\frac{3}{2}}.$$ ET II 16(28)

6.526

1. $$\int_0^\infty x J_{\frac{1}{2}\nu}(ax^2) J_\nu(bx)\,dx = (2a)^{-1} J_{\frac{1}{2}\nu}\left(\frac{b^2}{4a}\right)$$
$[a>0,\ b>0,\ \operatorname{Re}\nu > -1]$. ET II 56(1)

2. $$\int_0^\infty x J_{\frac{1}{2}\nu}(ax^2) N_\nu(bx)\,dx = (4a)^{-1}\left[N_{\frac{1}{2}\nu}\left(\frac{b^2}{4a}\right) - \operatorname{tg}\left(\frac{\nu\pi}{2}\right) J_{\frac{1}{2}\nu}\left(\frac{b^2}{4a}\right) + \sec\left(\frac{\nu\pi}{2}\right) \mathbf{H}_{-\frac{1}{2}\nu}\left(\frac{b^2}{4a}\right)\right]$$
$[a>0,\ b>0,\ \operatorname{Re}\nu > -1]$. ET II 109(9)

3. $$\int_0^\infty x J_{\frac{1}{2}\nu}(ax^2) K_\nu(bx)\,dx = \frac{\pi}{8a\cos\left(\frac{\nu\pi}{2}\right)}\left[\mathbf{H}_{-\frac{1}{2}\nu}\left(\frac{b^2}{4a}\right) - N_{-\frac{1}{2}\nu}\left(\frac{b^2}{4a}\right)\right]$$
$[a>0,\ \operatorname{Re} b>0,\ \operatorname{Re}\nu > -1]$. ET II 140(27)

4. $$\int_0^\infty x N_{\frac{1}{2}\nu}(ax^2) J_\nu(bx)\,dx = -(2a)^{-1}\mathbf{H}_{\frac{1}{2}\nu}\left(\frac{b^2}{4a}\right)$$
$[a>0,\ b>0,\ \operatorname{Re}\nu > -1]$. ET II 61(35)

5. $$\int_0^\infty x N_{\frac{1}{2}\nu}(ax^2) K_\nu(bx)\,dx = \frac{\pi}{4a\sin(\nu\pi)}\left[\cos\left(\frac{\nu\pi}{2}\right)\mathbf{H}_{-\frac{1}{2}\nu}\left(\frac{b^2}{4a}\right) - \sin\left(\frac{\nu\pi}{2}\right) J_{-\frac{1}{2}\nu}\left(\frac{b^2}{4a}\right) - \mathbf{H}_{\frac{1}{2}\nu}\left(\frac{b^2}{4a}\right)\right]$$
$[a>0,\ \operatorname{Re} b>0,\ |\operatorname{Re}\nu| < 1]$. ET II 141(28)

6. $$\int_0^\infty x K_{\frac{1}{2}\nu}(ax^2) J_\nu(bx)\,dx = \frac{\pi}{4a}\left[I_{\frac{1}{2}\nu}\left(\frac{b^2}{4a}\right) - \mathbf{L}_{\frac{1}{2}\nu}\left(\frac{b^2}{4a}\right)\right]$$
$[\operatorname{Re} a>0,\ b>0,\ \operatorname{Re}\nu > -1]$. ET II 68(9)

7. $$\int_0^\infty xK_{\frac{1}{2}\nu}(ax^2)N_\nu(bx)\,dx =$$
$$= \frac{\pi}{4a}\left[\operatorname{cosec}(\nu\pi)\,\mathbf{L}_{-\frac{1}{2}\nu}\left(\frac{b^2}{4a}\right) - \operatorname{ctg}(\nu\pi)\,\mathbf{L}_{\frac{1}{2}\nu}\left(\frac{b^2}{4a}\right) - \right.$$
$$\left. - \operatorname{tg}\left(\frac{\nu\pi}{2}\right) I_{\frac{1}{2}\nu}\left(\frac{b^2}{4a}\right) - \frac{1}{\pi}\sec\left(\frac{\nu\pi}{2}\right) K_{\frac{1}{2}\nu}\left(\frac{b^2}{4a}\right)\right]$$

$[\operatorname{Re} a > 0,\ b > 0,\ |\operatorname{Re}\nu| < 1]$. ET II 112(25)

8. $$\int_0^\infty xK_{\frac{1}{2}\nu}(ax^2)K_\nu(bx)\,dx =$$
$$= \frac{\pi}{8a}\left\{\sec\left(\frac{\nu\pi}{2}\right) K_{\frac{1}{2}\nu}\left(\frac{b^2}{4a}\right) + \right.$$
$$\left. + \pi\operatorname{cosec}(\nu\pi)\left[\mathbf{L}_{-\frac{1}{2}\nu}\left(\frac{b^2}{4a}\right) - \mathbf{L}_{\frac{1}{2}\nu}\left(\frac{b^2}{4a}\right)\right]\right\}$$

$[\operatorname{Re} a > 0,\ |\operatorname{Re}\nu| < 1]$. ET II 146(52)

6.527

1. $$\int_0^\infty x^2 J_{2\nu}(2ax)\,J_{\nu-\frac{1}{2}}(x^2)\,dx = \frac{1}{2}aJ_{\nu+\frac{1}{2}}(a^2)$$

$\left[a > 0,\ \operatorname{Re}\nu > -\frac{1}{2}\right]$. ET II 355(33)

2. $$\int_0^\infty x^2 J_{2\nu}(2ax)\,J_{\nu+\frac{1}{2}}(x^2)\,dx = \frac{1}{2}aJ_{\nu-\frac{1}{2}}(a^2)$$

$[a > 0,\ \operatorname{Re}\nu > -2]$. ET II 355(35)

3. $$\int_0^\infty x^2 J_{2\nu}(2ax)\,N_{\nu+\frac{1}{2}}(x^2)\,dx = -\frac{1}{2}a\mathbf{H}_{\nu-\frac{1}{2}}(a^2)$$

$[a > 0,\ \operatorname{Re}\nu > -2]$. ET II 355(36)

6.528

$$\int_0^\infty xK_{\frac{1}{4}\nu}\left(\frac{x^2}{4}\right) I_{\frac{1}{4}\nu}\left(\frac{x^2}{4}\right) J_\nu(bx)\,dx = K_{\frac{1}{4}\nu}\left(\frac{b^2}{4}\right) I_{\frac{1}{4}\nu}\left(\frac{b^2}{4}\right)$$

$[b > 0,\ \nu > -1]$. MO 183a

6.529

1. $$\int_0^\infty xJ_\nu(2\sqrt{ax})\,K_\nu(2\sqrt{ax})\,J_\nu(bx)\,dx = \frac{1}{2}b^{-2}e^{-\frac{2a}{b}}$$

$[\operatorname{Re} a > 0,\ b > 0,\ \operatorname{Re}\nu > -1]$ ET II 70(23)

2. $$\int_0^a x J_\lambda(2x) I_\lambda(2x) J_\mu\left(2\sqrt{a^2-x^2}\right) I_\mu\left(2\sqrt{a^2-x^2}\right) dx =$$

$$= \frac{a^{2\lambda+2\mu+2}}{2\Gamma(\lambda+1)\Gamma(\mu+1)\Gamma(\lambda+\mu+2)} \times$$

$$\times {}_1F_4\left(\frac{\lambda+\mu+1}{2};\ \lambda+1,\ \mu+1,\ \lambda+\mu+1,\ \frac{\lambda+\mu+3}{2};\ -a^4\right)$$

$[\operatorname{Re}\lambda > -1,\ \operatorname{Re}\mu > -1]$. ET II 376(31)

6.53-6.54 Combinations of Bessel functions and rational functions

6.531

1. $$\int_0^\infty \frac{N_\nu(bx)}{x+a} dx = \frac{\pi}{\sin(\pi\nu)}[\mathbf{E}_\nu(ab) + N_\nu(ab)] +$$

$$+ 2\operatorname{ctg}(\pi\nu)[\mathbf{J}_\nu(ab) - J_\nu(ab)]$$

$\left[b > 0,\ |\arg a| < \pi,\ |\operatorname{Re}\nu| < 1,\ \nu \neq 0,\ \pm\frac{1}{2}\right]$. ET II 97(5)

2. $$\int_0^\infty \frac{N_\nu(bx)}{x-a} dx = \pi\{\operatorname{ctg}(\nu\pi)[N_\nu(ab) + \mathbf{E}_\nu(ab)] +$$

$$+ \mathbf{J}_\nu(ab) + 2[\operatorname{ctg}(\nu\pi)]^2[\mathbf{J}_\nu(ab) - J_\nu(ab)]\}$$

$[b > 0,\ a > 0,\ |\operatorname{Re}\nu| < 1]$. ET II 98(9)

3. $$\int_0^\infty \frac{K_\nu(bx)}{x+a} dx = \frac{\pi^2}{2}[\operatorname{cosec}(\nu\pi)]^2[I_\nu(ab) +$$

$$+ I_{-\nu}(ab) - e^{-\frac{1}{2}i\nu\pi}\mathbf{J}_\nu(iab) - e^{\frac{1}{2}i\nu\pi}\mathbf{J}_{-\nu}(iab)]$$

$[\operatorname{Re} b > 0,\ |\arg a| < \pi,\ |\operatorname{Re}\nu| < 1]$. ET II 128(5)

6.532

1. $$\int_0^\infty \frac{J_\nu(x)}{x^2+a^2} dx = \frac{\pi[\mathbf{J}_\nu(a) - J_\nu(a)]}{a\sin(\nu\pi)}$$

$[\operatorname{Re} a > 0,\ \operatorname{Re}\nu > -1]$. ET II 340(2)

2. $$\int_0^\infty \frac{N_\nu(bx)}{x^2+a^2} dx = \frac{1}{\cos\frac{\nu\pi}{2}}\left[-\frac{\pi}{2a}\operatorname{tg}\left(\frac{\nu\pi}{2}\right) I_\nu(ab) - \frac{1}{a}K_\nu(ab) +\right.$$

$$\left. + \frac{b\sin\left(\frac{\nu\pi}{2}\right)}{1-\nu^2}\, {}_1F_2\left(1;\ \frac{3-\nu}{2},\ \frac{3+\nu}{2};\ \frac{a^2b^2}{4}\right)\right]$$

$[b > 0,\ \operatorname{Re} a > 0,\ |\operatorname{Re}\nu| < 1]$. ET II 99(13)

3. $$\int_0^\infty \frac{N_\nu(bx)}{x^2-a^2}\,dx = \frac{\pi}{2a}\left\{J_\nu(ab) + \operatorname{tg}\left(\frac{\nu\pi}{2}\right)\left\{\operatorname{tg}\left(\frac{\nu\pi}{2}\right)[\mathbf{J}_\nu(ab) - J_\nu(ab)] - \right.\right.$$
$$\left.\left. - \mathbf{E}_\nu(ab) - N_\nu(ab)\right\}\right\}$$
$[b > 0,\ a > 0,\ |\operatorname{Re}\nu| < 1]$. ET II 101(21)

4. $$\int_0^\infty \frac{xJ_0(ax)}{x^2+k^2}\,dx = K_0(ak) \quad [a > 0,\ \operatorname{Re} k > 0].$$ WA 466(5)

5. $$\int_0^\infty \frac{N_0(ax)}{x^2+k^2}\,dx = -\frac{K_0(ak)}{k} \quad [a > 0,\ \operatorname{Re} k > 0].$$ WA 466(6)

6. $$\int_0^\infty \frac{J_0(ax)}{x^2+k^2}\,dx = \frac{\pi}{2k}[I_0(ak) - \mathbf{L}_0(ak)] \quad [a > 0,\ \operatorname{Re} k > 0].$$ WA 467(7)

6.533

1. $$\int_0^z J_p(x)\,J_q(z-x)\frac{dx}{x} = \frac{J_{p+q}(z)}{p} \quad [\operatorname{Re} p > 0,\ \operatorname{Re} q > -1].$$ WA 415(3)

2. $$\int_0^z \frac{J_p(x)}{x}\,\frac{J_q(z-x)}{z-x}\,dx = \left(\frac{1}{p} + \frac{1}{q}\right)\frac{J_{p+q}(z)}{z}$$
$[\operatorname{Re} p > 0.\quad \operatorname{Re} q > 0]$. WA 415(5)

3. $$\int_0^\infty [J_0(ax)-1]J_1(bx)\frac{dx}{x^2} = \frac{-b}{4}\left[1 + 2\ln\frac{a}{b}\right] \quad [0 < b < a];$$
$$= -\frac{a^2}{4b} \quad [0 < a < b].$$ ET II 21(28)a

4. $$\int_0^\infty [1 - J_0(ax)]\,J_0(bx)\frac{dx}{x} = 0 \quad [0 < a < b];$$
$$= \ln\frac{a}{b} \quad [0 < b < a].$$ ET II 14(16)

6.534 $$\int_0^\infty \frac{x^3J_0(x)}{x^4-a^4}\,dx = \frac{1}{2}K_0(a) - \frac{1}{4}\pi N_0(a) \quad [a > 0].$$ ET II 340(5)

6.535 $$\int_0^\infty \frac{x}{x^2+a^2}[J_\nu(x)]^2\,dx = I_\nu(a)K_\nu(a) \quad [\operatorname{Re} a > 0,\ \operatorname{Re}\nu > -1].$$ ET II 342(26)

6.536 $$\int_0^\infty \frac{x^3J_0(bx)}{x^4+a^4}\,dx = \operatorname{ker}(ab) \quad \left[b > 0,\ |\arg a| < \frac{1}{4}\pi\right].$$ ET II 8(9), MO 46a

6.537 $$\int_0^\infty \frac{xJ_0(bx)}{x^4+a^4}\,dx = -\frac{1}{a^2}\operatorname{kei}(ab) \quad \left[b > 0,\ |\arg a| < \frac{\pi}{4}\right].$$ MO 46a

6.538

1. $$\int_0^\infty J_1(ax) J_1(bx) \frac{dx}{x^2} = \frac{a+b}{\pi}\left[E\left(\frac{2i\sqrt{ab}}{|b-a|}\right) - K\left(\frac{2i\sqrt{ab}}{|b-a|}\right)\right]$$
$[a > 0, \ b > 0].$ ET II 21(30)

2. $$\int_0^\infty x^{-1} J_{\nu+2n+1}(x) J_{\nu+2m+1}(x)\, dx = 0 \quad [m \neq n, \ \nu > -1];$$
$$= (4n+2\nu+2)^{-1} \quad [m = n, \ \nu > -1].$$ EH II 64

6.539

1. $$\int_a^b \frac{dx}{x[J_\nu(x)]^2} = \frac{\pi}{2}\left[\frac{N_\nu(b)}{J_\nu(b)} - \frac{N_\nu(a)}{J_\nu(a)}\right].$$ ET II 338(41)

2. $$\int_a^b \frac{dx}{x[N_\nu(x)]^2} = \frac{\pi}{2}\left[\frac{J_\nu(a)}{N_\nu(a)} - \frac{J_\nu(b)}{N_\nu(b)}\right].$$ ET II 339(49)

3. $$\int_a^b \frac{dx}{xJ_\nu(x) N_\nu(x)} = \frac{\pi}{2} \ln\left[\frac{J_\nu(a) N_\nu(b)}{J_\nu(b) N_\nu(a)}\right].$$ ET II 339(50)

6.541

1. $$\int_0^\infty x J_\nu(ax) J_\nu(bx) \frac{dx}{x^2+c^2} =$$
$$= I_\nu(bc) K_\nu(ac) \quad [0 < b < a, \ \operatorname{Re} c > 0, \ \operatorname{Re} \nu > -1];$$
$$= I_\nu(ac) K_\nu(bc) \quad [0 < a < b, \ \operatorname{Re} c > 0, \ \operatorname{Re} \nu > -1].$$
ET II 49(10)

2. $$\int_0^\infty x^{1-2n} J_\nu(ax) J_\nu(bx) \frac{dx}{x^2+c^2} =$$
$$= (-1)^n c^{-2n} I_\nu(bc) K_\nu(ac) \quad [0 < b < a, \ \operatorname{Re} c > 0, \ \operatorname{Re} \nu > n-1, \ n = 0, 1, \ldots];$$
$$= (-1)^n c^{-2n} I_\nu(ac) K_\nu(bc) \quad [0 < a < b, \ \operatorname{Re} c > 0, \ \operatorname{Re} \nu > n-1, \ n = 0, 1, \ldots].$$
ET II 49(11)

6.542 $$\int_0^\infty \frac{J_\nu(ax) N_\nu(bx) - J_\nu(bx) N_\nu(ax)}{x\{[J_\nu(bx)]^2 + [N_\nu(bx)]^2\}}\, dx =$$
$$= -\frac{\pi}{2}\left(\frac{b}{a}\right)^\nu \quad [0 < b < a].$$ ET II 352(16)

6.543 $$\int_0^\infty J_\mu(bx) \left\{\cos\left[\frac{1}{2}(\nu-\mu)\pi\right] J_\nu(ax) - \right.$$
$$\left. - \sin\left[\frac{1}{2}(\nu-\mu)\pi\right] N_\nu(ax)\right\} \frac{x\, dx}{x^2+r^2} = I_\mu(br) K_\nu(ar)$$
$[\operatorname{Re} r > 0, \ a \geqslant b > 0, \ \operatorname{Re} \mu > |\operatorname{Re} \nu| - 2].$ WA 471(5)

6.544

1. $$\int_0^\infty J_\nu\left(\frac{a}{x}\right) N_\nu\left(\frac{x}{b}\right) \frac{dx}{x^2} = -\frac{1}{a}\left[\frac{2}{\pi} K_{2\nu}\left(\frac{2\sqrt{a}}{\sqrt{b}}\right) - N_{2\nu}\left(\frac{2\sqrt{a}}{\sqrt{b}}\right)\right]$$
$$\left[a > 0,\ b > 0,\ |\operatorname{Re}\nu| < \frac{1}{2}\right].$$ EI II 357(47)

2. $$\int_0^\infty J_\nu\left(\frac{a}{x}\right) J_\nu\left(\frac{x}{b}\right) \frac{dx}{x^2} = \frac{1}{a} J_{2\nu}\left(\frac{2\sqrt{a}}{\sqrt{b}}\right)$$
$$\left[a > 0,\ b > 0,\ \operatorname{Re}\nu > -\frac{1}{2}\right].$$ ET II 57(10)

3. $$\int_0^\infty J_\nu\left(\frac{a}{x}\right) K_\nu\left(\frac{x}{b}\right) \frac{dx}{x^2} = \frac{1}{a} e^{\frac{1}{2} i\nu\pi} K_{2\nu}\left(\frac{2\sqrt{a}}{\sqrt{b}} e^{\frac{1}{4} i\pi}\right) + \frac{1}{a} e^{-\frac{1}{2} i\nu\pi} K_{2\nu}\left(\frac{2\sqrt{a}}{\sqrt{b}} e^{-\frac{1}{4} i\pi}\right)$$
$$\left[\operatorname{Re} b > 0,\ a > 0,\ |\operatorname{Re}\nu| < \frac{1}{2}\right].$$ ET II 142(32)

4. $$\int_0^\infty N_\nu\left(\frac{a}{x}\right) J_\nu\left(\frac{x}{b}\right) \frac{dx}{x^2} = \frac{2}{a\pi}\left[K_{2\nu}\left(\frac{2\sqrt{a}}{\sqrt{b}}\right) + \frac{\pi}{2} N_{2\nu}\left(\frac{2\sqrt{a}}{\sqrt{b}}\right)\right]$$
$$\left[a > 0,\ b > 0,\ |\operatorname{Re}\nu| < \frac{1}{2}\right].$$ ET II 62(38)

5. $$\int_0^\infty N_\nu\left(\frac{a}{x}\right) K_\nu\left(\frac{x}{b}\right) \frac{dx}{x^2} = \frac{1}{a}\left[e^{\frac{1}{2} i(\nu+1)\pi} K_{2\nu}\left(\frac{2\sqrt{a}}{\sqrt{b}} e^{\frac{1}{4} i\pi}\right) + e^{-\frac{1}{2} i(\nu+1)\pi} K_{2\nu}\left(\frac{2\sqrt{a}}{\sqrt{b}} e^{-\frac{1}{4} i\pi}\right)\right]$$
$$\left[\operatorname{Re} b > 0,\ a > 0,\ |\operatorname{Re}\nu| < \frac{1}{2}\right].$$ ET II 143(38)

6. $$\int_0^\infty K_\nu\left(\frac{a}{x}\right) J_\nu\left(\frac{x}{b}\right) \frac{dx}{x^2} = \frac{i}{a}\left[e^{\frac{1}{2}\nu\pi i} K_{2\nu}\left(e^{\frac{1}{4}\pi i} \frac{2\sqrt{a}}{\sqrt{b}}\right) - e^{-\frac{1}{2}\nu\pi i} K_{2\nu}\left(e^{-\frac{1}{4}\pi i} \frac{2\sqrt{a}}{\sqrt{b}}\right)\right]$$
$$\left[\operatorname{Re} a > 0,\ b > 0,\ |\operatorname{Re}\nu| < \frac{5}{2}\right].$$ ET II 70(19)

7. $$\int_0^\infty K_\nu\left(\frac{a}{x}\right) N_\nu\left(\frac{x}{b}\right) \frac{dx}{x^2} = \frac{2}{a}\left[\sin\left(\frac{3}{2}\pi\nu\right) \operatorname{kei}_{2\nu}\left(\frac{2\sqrt{a}}{\sqrt{b}}\right) - \cos\left(\frac{3}{2}\pi\nu\right) \operatorname{ker}_{2\nu}\left(\frac{2\sqrt{a}}{\sqrt{b}}\right)\right]$$
$$\left[\operatorname{Re} a > 0,\ b > 0,\ |\operatorname{Re}\nu| < \frac{5}{2}\right].$$ ET II 113(29)

8. $$\int_0^\infty K_\nu\left(\frac{a}{x}\right) K_\nu\left(\frac{x}{b}\right) \frac{dx}{x^2} = \frac{\pi}{a} K_{2\nu}\left(\frac{2\sqrt{a}}{\sqrt{b}}\right)$$
$$[\operatorname{Re} a > 0,\ \operatorname{Re} b > 0].$$ ET II 146(55)

6.55 Combinations of Bessel functions and algebraic functions

6.551

1. $$\int_0^1 x^{\frac{1}{2}} J_\nu(xy)\,dx = \sqrt{2}\,y^{-\frac{3}{2}} \frac{\Gamma\left(\frac{3}{4}+\frac{1}{2}\nu\right)}{\Gamma\left(\frac{1}{4}+\frac{1}{2}\nu\right)} + $$
$$+ y^{-\frac{1}{2}}\left[\left(\nu-\frac{1}{2}\right) + J_\nu(y) S_{-\frac{1}{2},\,\nu-1}(y) - J_{\nu-1}(y) S_{\frac{1}{2},\,\nu}(y)\right]$$
$$\left[y > 0,\ \operatorname{Re}\nu > -\frac{3}{2}\right].$$
ET II 21(1)

2. $$\int_1^\infty x^{\frac{1}{2}} J_\nu(xy)\,dx = y^{-\frac{1}{2}}\left[J_{\nu-1}(y) S_{\frac{1}{2},\,\nu}(y) + \right.$$
$$\left. + \left(\frac{1}{2}-\nu\right) J_\nu(y) S_{-\frac{1}{2},\,\nu-1}(y)\right] \qquad [y > 0].$$
ET II 22(2)

6.552

1. $$\int_0^\infty J_\nu(xy) \frac{dx}{(x^2+a^2)^{\frac{1}{2}}} = I_{\frac{1}{2}\nu}\left(\frac{1}{2}ay\right) K_{\frac{1}{2}\nu}\left(\frac{1}{2}ay\right)$$
$$[\operatorname{Re} a > 0,\ y > 0,\ \operatorname{Re}\nu > -1].$$
ET II 23(11), WA 477(3), MO 44

2. $$\int_0^\infty N_\nu(xy) \frac{dx}{(x^2+a^2)^{\frac{1}{2}}} = -\frac{1}{\pi}\sec\left(\frac{1}{2}\nu\pi\right) K_{\frac{1}{2}\nu}\left(\frac{1}{2}ay\right) \times$$
$$\times \left[K_{\frac{1}{2}\nu}\left(\frac{1}{2}ay\right) + \pi\sin\left(\frac{1}{2}\nu\pi\right) I_{\frac{1}{2}\nu}\left(\frac{1}{2}ay\right)\right]$$
$$[y > 0,\ \operatorname{Re} a > 0,\ |\operatorname{Re}\nu| < 1].$$
ET II 100(18)

3. $$\int_0^\infty K_\nu(xy) \frac{dx}{(x^2+a^2)^{\frac{1}{2}}} = \frac{\pi^2}{8}\sec\left(\frac{1}{2}\nu\pi\right) \times$$
$$\times \left\{\left[J_{\frac{1}{2}\nu}\left(\frac{1}{2}ay\right)\right]^2 + \left[N_{\frac{1}{2}\nu}\left(\frac{1}{2}ay\right)\right]^2\right\}$$
$$[\operatorname{Re} a > 0.\ \operatorname{Re} y > 0,\ |\operatorname{Re}\nu| < 1].$$
ET II 128(6)

4. $$\int_0^1 J_\nu(xy) \frac{dx}{(1-x^2)^{\frac{1}{2}}} = \frac{\pi}{2}\left[J_{\frac{1}{2}\nu}\left(\frac{1}{2}y\right)\right]^2$$
$$[y > 0,\ \operatorname{Re}\nu > -1].$$
ET II 24(22)a

5. $$\int_0^1 N_0(xy) \frac{dx}{(1-x^2)^{\frac{1}{2}}} = \frac{\pi}{2} J_0\left(\frac{1}{2}y\right) N_0\left(\frac{1}{2}y\right)$$
$$[y > 0].$$
ET II 102(26)a

6. $\int\limits_{1}^{\infty} J_{\nu}(xy)\frac{dx}{(x^2-1)^{\frac{1}{2}}} = -\frac{\pi}{2} J_{\frac{1}{2}\nu}\left(\frac{1}{2}y\right) N_{\frac{1}{2}\nu}\left(\frac{1}{2}y\right)$

$[y > 0].$ ET II 24(23)a

7. $\int\limits_{1}^{\infty} N_{\nu}(xy)\frac{dx}{(x^2-1)^{\frac{1}{2}}} = \frac{\pi}{4}\left\{\left[J_{\frac{1}{2}\nu}\left(\frac{1}{2}y\right)\right]^2 - \left[N_{\frac{1}{2}\nu}\left(\frac{1}{2}y\right)\right]^2\right\}$

$[y > 0].$ ET II 102(27)

6.553 $\int\limits_{0}^{\infty} x^{-\frac{1}{2}} I_{\nu}(x) K_{\nu}(x) K_{\mu}(2x)\,dx =$

$$= \frac{\Gamma\left(\frac{1}{4}+\frac{1}{2}\mu\right)\Gamma\left(\frac{1}{4}-\frac{1}{2}\mu\right)\Gamma\left(\frac{1}{4}+\nu+\frac{1}{2}\mu\right)\Gamma\left(\frac{1}{4}+\nu-\frac{1}{2}\mu\right)}{4\Gamma\left(\frac{3}{4}+\nu+\frac{1}{2}\mu\right)\Gamma\left(\frac{3}{4}+\nu-\frac{1}{2}\mu\right)}$$

$\left[|\operatorname{Re}\mu| < \frac{1}{2},\ 2\operatorname{Re}\nu > |\operatorname{Re}\mu| - \frac{1}{2}\right].$ ET II 372(2)

6.554

1. $\int\limits_{0}^{\infty} xJ_0(xy)\frac{dx}{(a^2+x^2)^{\frac{1}{2}}} = y^{-1}e^{-ay}$ $[y > 0,\ \operatorname{Re} a > 0].$ ET II 7(4)

2. $\int\limits_{0}^{1} xJ_0(xy)\frac{dx}{(1-x^2)^{\frac{1}{2}}} = y^{-1}\sin y$ $[y > 0].$ ET II 7(5)a

3. $\int\limits_{1}^{\infty} xJ_0(xy)\frac{dx}{(x^2-1)^{\frac{1}{2}}} = y^{-1}\cos y$ $[y > 0].$ ET II 7(6)a

4. $\int\limits_{0}^{\infty} xJ_0(xy)\frac{dx}{(x^2+a^2)^{\frac{3}{2}}} = a^{-1}e^{-ay}$ $[y > 0,\ \operatorname{Re} a > 0].$ ET II 7(7)a

5. $\int\limits_{0}^{\infty} \frac{xJ_0(ax)}{\sqrt{x^4+4k^4}}\,dx = K_0(ak)J_0(ak)$ $[a > 0,\ k > 0].$ WA 473(1)

6.555 $\int\limits_{0}^{\infty} x^{\frac{1}{2}} J_{2\nu-1}\left(ax^{\frac{1}{2}}\right) N_{\nu}(xy)\,dx = -\frac{a}{2y^2}\mathbf{H}_{\nu-1}\left(\frac{a^2}{4y}\right)$

$\left[a > 0,\ y > 0,\ \operatorname{Re}\nu > -\frac{1}{2}\right].$ ET II 111(17)

6.556 $\int\limits_{0}^{\infty} J_{\nu}\left[a(x^2+1)^{\frac{1}{2}}\right]\frac{dx}{\sqrt{x^2+1}} = -\frac{\pi}{2} J_{\frac{1}{2}\nu}\left(\frac{a}{2}\right) N_{\frac{1}{2}\nu}\left(\frac{a}{2}\right)$

$[\operatorname{Re}\nu > -1,\ a > 0].$ MO 46

6.56-6.58 Combinations of Bessel functions and powers

6.561

1. $$\int_0^1 x^\nu J_\nu(ax)\,dx = 2^{\nu-1}a^{-\nu}\pi^{\frac{1}{2}}\Gamma\left(\nu+\frac{1}{2}\right)\times \times [J_\nu(a)\,\mathbf{H}_{\nu-1}(a)-\mathbf{H}_\nu(a)\,J_{\nu-1}(a)] \quad \left[\operatorname{Re}\nu > -\frac{1}{2}\right].$$ ET II 333(2)a

2. $$\int_0^1 x^\nu N_\nu(ax)\,dx = 2^{\nu-1}a^{-\nu}\pi^{\frac{1}{2}}\Gamma\left(\nu+\frac{1}{2}\right)\times \times [N_\nu(a)\,\mathbf{H}_{\nu-1}(a)-\mathbf{H}_\nu(a)\,N_{\nu-1}(a)] \quad \left[\operatorname{Re}\nu > -\frac{1}{2}\right].$$ ET II 338(43)a

3. $$\int_0^1 x^\nu I_\nu(ax)\,dx = 2^{\nu-1}a^{-\nu}\pi^{\frac{1}{2}}\Gamma\left(\nu+\frac{1}{2}\right)\times \times [I_\nu(a)\,\mathbf{L}_{\nu-1}(a)-\mathbf{L}_\nu(a)\,I_{\nu-1}(a)] \quad \left[\operatorname{Re}\nu > -\frac{1}{2}\right].$$ ET II 364(2)a

4. $$\int_0^1 x^\nu K_\nu(ax)\,dx = 2^{\nu-1}a^{-\nu}\pi^{\frac{1}{2}}\Gamma\left(\nu+\frac{1}{2}\right)\times \times [K_\nu(a)\,\mathbf{L}_{\nu-1}(a)+\mathbf{L}_\nu(a)\,K_{\nu-1}(a)] \quad \left[\operatorname{Re}\nu > -\frac{1}{2}\right].$$ ET II 367(21)a

5. $$\int_0^1 x^{\nu+1} J_\nu(ax)\,dx = a^{-1}J_{\nu+1}(a) \qquad [\operatorname{Re}\nu > -1].$$ ET II 333(3)a

6. $$\int_0^1 x^{\nu+1} N_\nu(ax)\,dx = a^{-1}N_{\nu+1}(a) + 2^{\nu+1}a^{-\nu-2}\Gamma(\nu+1) \qquad [\operatorname{Re}\nu > -1].$$ ET II 339(44)a

7. $$\int_0^1 x^{\nu+1} I_\nu(ax)\,dx = a^{-1}I_{\nu+1}(a) \qquad [\operatorname{Re}\nu > -1].$$ ET II 365(3)a

8. $$\int_0^1 x^{\nu+1} K_\nu(ax)\,dx = 2^{\nu}a^{-\nu-2}\Gamma(\nu+1) - a^{-1}K_{\nu+1}(a) \qquad [\operatorname{Re}\nu > -1].$$ ET II 367(22)a

9. $$\int_0^1 x^{1-\nu} J_\nu(ax)\,dx = \frac{a^{\nu-2}}{2^{\nu-1}\Gamma(\nu)} - a^{-1}J_{\nu-1}(a).$$ ET II 333(4)a

10. $$\int_0^1 x^{1-\nu} N_\nu(ax)\,dx = \frac{a^{\nu-2}\operatorname{ctg}(\nu\pi)}{2^{\nu-1}\Gamma(\nu)} - a^{-1}N_{\nu-1}(a) \qquad [\operatorname{Re}\nu < 1].$$ ET II 339(45)a

11. $\int_0^1 x^{1-\nu} I_\nu(ax)\,dx = a^{-1} I_{\nu-1}(a) - \frac{a^{\nu-2}}{2^{\nu-1}\Gamma(\nu)}$. ET II 365(4)a

12. $\int_0^1 x^{1-\nu} K_\nu(ax)\,dx = 2^{-\nu} a^{\nu-2}\Gamma(1-\nu) - a^{-1} K_{\nu-1}(a)$

$[\operatorname{Re}\nu < 1]$. ET II 367(23)a

13. $\int_0^1 x^\mu J_\nu(ax)\,dx = a^{-\mu-1}\Bigg[(\nu+\mu-1)\, a J_\nu(a) +$

$+ S_{\mu-1,\,\nu-1}(a) - a J_{\nu-1}(a)\, S_{\mu,\,\nu}(a) + 2^\mu \frac{\Gamma\left(\frac{1}{2} + \frac{1}{2}\mu + \frac{1}{2}\quad\right)}{\Gamma\left(\frac{1}{2}\nu + \frac{1}{2} - \frac{1}{2}\mu\right)}\Bigg]$

$[a > 0,\ \operatorname{Re}(\mu+\nu) > -1]$. ET II 22(8)a

14. $\int_0^\infty x^\mu J_\nu(ax)\,dx = 2^\mu a^{-\mu-1} \frac{\Gamma\left(\frac{1}{2} + \frac{1}{2}\nu + \frac{1}{2}\mu\right)}{\Gamma\left(\frac{1}{2} + \frac{1}{2}\nu - \frac{1}{2}\mu\right)}$

$\left[-\operatorname{Re}\nu - 1 < \operatorname{Re}\mu < \frac{1}{2},\ a > 0\right]$. EH II 49(19)

15. $\int_0^\infty x^\mu N_\nu(ax)\,dx = 2^\mu \operatorname{ctg}\left[\frac{1}{2}(\nu+1-\mu)\pi\right] a^{-\mu-1} \frac{\Gamma\left(\frac{1}{2} + \frac{1}{2}\nu + \frac{1}{2}\mu\right)}{\Gamma\left(\frac{1}{2} + \frac{1}{2}\nu - \frac{1}{2}\mu\right)}$

$\left[|\operatorname{Re}\nu| - 1 < \mu < \frac{1}{2},\ a > 0\right]$. ET II 97(3)a

16. $\int_0^\infty x^\mu K_\nu(ax)\,dx = 2^{\mu-1} a^{-\mu-1} \Gamma\left(\frac{1+\mu+\nu}{2}\right)\Gamma\left(\frac{1+\mu-\nu}{2}\right)$

$[\operatorname{Re}(\mu+1\pm\nu) > 0,\ \operatorname{Re} a > 0]$. EH II 51(27)

17. $\int_0^\infty \frac{J_\nu(ax)}{x^{\nu-q}}\,dx = \frac{\Gamma\left(\frac{1}{2}q + \frac{1}{2}\right)}{2^{\nu-q} a^{q-\nu+1} \Gamma\left(\nu - \frac{1}{2}q + \frac{1}{2}\right)}$

$\left[-1 < \operatorname{Re} q < \operatorname{Re}\nu - \frac{1}{2}\right]$. WA 428(1), KU 144(5)

18. $\int_0^\infty \frac{N_\nu(x)}{x^{\nu-\mu}}\,dx = \frac{\Gamma\left(\frac{1}{2} + \frac{1}{2}\mu\right)\Gamma\left(\frac{1}{2} + \frac{1}{2}\mu - \nu\right)\sin\left(\frac{1}{2}\mu - \nu\right)\pi}{2^{\nu-\mu}\pi}$

$\left[|\operatorname{Re}\nu| < \operatorname{Re}(1+\mu-\nu) < \frac{3}{2}\right]$. WA 430(5)

6.562

1. $$\int_0^\infty x^\mu N_\nu(bx)\frac{dx}{x+a} =$$
$$= (2a)^\mu \pi^{-1}\left\{\sin\left[\frac{1}{2}\pi(\mu-\nu)\right]\Gamma\left[\frac{1}{2}(\mu+\nu+1)\right]\times\right.$$
$$\times\Gamma\left[\frac{1}{2}(1+\mu-\nu)\right]S_{-\mu,\nu}(ab) - 2\cos\left[\frac{1}{2}\pi(\mu-\nu)\right]\times$$
$$\left.\times\Gamma\left(1+\frac{1}{2}\mu+\frac{1}{2}\nu\right)\Gamma\left(1+\frac{1}{2}\mu-\frac{1}{2}\nu\right)S_{-\mu-1,\nu}(ab)\right\}$$
$$\left[b>0,\ |\arg a|<\pi,\ \operatorname{Re}(\mu\pm\nu)>-1,\ \operatorname{Re}\mu<\frac{3}{2}\right].$$
ET II 98(8)

2. $$\int_0^\infty \frac{x^\nu J_\nu(ax)}{x+k}dx = \frac{\pi k^\nu}{2\cos\nu\pi}[\mathbf{H}_{-\nu}(ak) - N_{-\nu}(ak)]$$
$$\left[-\frac{1}{2}<\operatorname{Re}\nu<\frac{3}{2},\ a>0,\ |\arg k|<\pi\right].$$ WA 479(7)

3. $$\int_0^\infty x^\mu K_\nu(bx)\frac{dx}{x+a} =$$
$$= 2^{\mu-2}\Gamma\left[\frac{1}{2}(\mu+\nu)\right]\Gamma\left[\frac{1}{2}(\mu-\nu)\right]b^{-\mu}\times$$
$$\times {}_1F_2\left(1;\ 1-\frac{\mu+\nu}{2},\ 1-\frac{\mu-\nu}{2};\ \frac{a^2b^2}{4}\right)-$$
$$-2^{\mu-3}\Gamma\left[\frac{1}{2}(\mu-\nu-1)\right]\Gamma\left[\frac{1}{2}(\mu+\nu-1)\right]ab^{1-\mu}\times$$
$$\times {}_1F_2\left(1;\ \frac{3-\mu-\nu}{2},\ \frac{3-\mu+\nu}{2};\ \frac{a^2b^2}{4}\right)-$$
$$-\pi a^\mu \operatorname{cosec}[\pi(\mu-\nu)]\{K_\nu(ab)+\pi\cos(\mu\pi)\operatorname{cosec}[\pi(\nu+\mu)]I_\nu(ab)\}$$
$$[\operatorname{Re} b>0,\quad |\arg a|<\pi,\quad \operatorname{Re}\mu>|\operatorname{Re}\nu|-1].$$ ET II 127(4)

6.563 $$\int_0^\infty x^{\varrho-1}J_\nu(bx)\frac{dx}{(x+a)^{1+\mu}} = \frac{\pi a^{\varrho-\mu-1}}{\sin[(\varrho+\nu-\mu)\pi]\Gamma(\mu+1)}\times$$
$$\times\left\{\sum_{m=0}^\infty \frac{(-1)^m\left(\frac{1}{2}ab\right)^{\nu+2m}\Gamma(\varrho+\nu+2m)}{m!\,\Gamma(\nu+m+1)\Gamma(\varrho+\nu-\mu+2m)}-\right.$$
$$\left.-\sum_{m=0}^\infty \frac{\left(\frac{1}{2}ab\right)^{\mu+1-\varrho+m}\Gamma(\mu+m+1)}{m!\,\Gamma\left[\frac{1}{2}(\mu+\nu-\varrho+m+3)\right]}\,\frac{\sin\left[\frac{1}{2}(\varrho+\nu-\mu-m)\pi\right]}{\Gamma\left[\frac{1}{2}(\mu-\nu-\varrho+m+3)\right]}\right\}$$
$$\left[b>0,\ |\arg a|<\pi,\ \operatorname{Re}(\varrho+\nu)>0,\ \operatorname{Re}(\varrho-\mu)<\frac{5}{2}\right].$$
ET II 23(10), WA 479

6.564

1. $$\int_0^\infty x^{\nu+1} J_\nu(bx) \frac{dx}{\sqrt{x^2+a^2}} = \sqrt{\frac{2}{\pi b}}\, a^{\nu+\frac{1}{2}} K_{\nu+\frac{1}{2}}(ab)$$

$\left[\operatorname{Re} a > 0, \quad b > 0, \quad -1 < \operatorname{Re}\nu < \frac{1}{2}\right].$ ET II 23(15)

2. $$\int_0^\infty x^{1-\nu} J_\nu(bx) \frac{dx}{\sqrt{x^2+a^2}} = \sqrt{\frac{\pi}{2b}}\, a^{\frac{1}{2}-\nu} \left[I_{\nu-\frac{1}{2}}(ab) - \mathbf{L}_{\nu-\frac{1}{2}}(ab)\right]$$

$\left[\operatorname{Re} a > 0, \quad b > 0, \quad \operatorname{Re}\nu > -\frac{1}{2}\right].$ ET II 23(16)

6.565

1. $$\int_0^\infty x^{-\nu} (x^2+a^2)^{-\nu-\frac{1}{2}} J_\nu(bx)\, dx = 2^\nu a^{-2\nu} b^\nu \frac{\Gamma(\nu+1)}{\Gamma(2\nu+1)} I_\nu\left(\frac{ab}{2}\right) K_\nu\left(\frac{ab}{2}\right)$$

$\left[\operatorname{Re} a > 0, \quad b > 0, \quad \operatorname{Re}\nu > -\frac{1}{2}\right].$ WA 477(4), ET II 23(17)

2. $$\int_0^\infty x^{\nu+1} (x^2+a^2)^{-\nu-\frac{1}{2}} J_\nu(bx)\, dx = \frac{\sqrt{\pi}\, b^{\nu-1}}{2^\nu e^{ab}\, \Gamma\left(\nu+\frac{1}{2}\right)}$$

$\left[\operatorname{Re} a > 0, \; b > 0, \; \operatorname{Re}\nu > -\frac{1}{2}\right].$ ET II 24(18)

3. $$\int_0^\infty x^{\nu+1} (x^2+a^2)^{-\nu-\frac{3}{2}} J_\nu(bx)\, dx = \frac{b^\nu \sqrt{\pi}}{2^{\nu+1} a e^{ab}\, \Gamma\left(\nu+\frac{3}{2}\right)}$$

$[\operatorname{Re} a > 0, \; b > 0, \; \operatorname{Re}\nu > -1].$ ET II 24(19)

4. $$\int_0^\infty \frac{J_\nu(bx)\, x^{\nu+1}}{(x^2+a^2)^{\mu+1}}\, dx = \frac{a^{\nu-\mu} b^\mu}{2^\mu\, \Gamma(\mu+1)} K_{\nu-\mu}(ab)$$

$\left[-1 < \operatorname{Re}\nu < \operatorname{Re}\left(2\mu+\frac{3}{2}\right), \; a > 0, \; b > 0\right].$ MO 43

5. $$\int_0^\infty x^{\nu+1} (x^2+a^2)^\mu N_\nu(bx)\, dx = 2^{\nu-1} \pi^{-1} a^{2\mu+2} (1+\mu)^{-1} \Gamma(\nu) b^{-\nu} \times$$
$$\times {}_1F_2\left(1;\; 1-\nu,\; 2+\mu;\; \frac{a^2b^2}{4}\right) - 2^\mu a^{\mu+\nu+1} [\sin(\nu\pi)]^{-1} \times$$
$$\times \Gamma(\mu+1)\, b^{-1-\mu} \left[I_{\mu+\nu+1}(ab) - 2\cos(\mu\pi) K_{\mu+\nu+1}(ab)\right]$$

$[b > 0, \; \operatorname{Re} a > 0, \; -1 < \operatorname{Re}\nu < -2\operatorname{Re}\mu].$ ET II 100(19)

6. $$\int_0^\infty x^{1-\nu} (x^2+a^2)^\mu N_\nu(bx)\, dx = 2^\mu a^{\mu-\nu+1} b^{-1-\mu} \left\{\frac{\cos(\nu\pi)}{\pi} \Gamma(\mu+1) \times\right.$$
$$\left.\times \Gamma(\nu) I_{\nu-\mu-1}(ab) - 2\operatorname{cosec}(\nu\pi) [\Gamma(-\mu)]^{-1} K_{\nu-\mu-1}(ab)\right\} -$$
$$- \frac{a^{2\mu+2} \operatorname{ctg}(\nu\pi)\, b^\nu}{2^{\nu+1} (\mu+1)\, \Gamma(\nu+1)} {}_1F_2\left(1;\; \nu+1,\; \mu+2;\; \frac{a^2b^2}{4}\right)$$

$\left[b > 0, \; \operatorname{Re} a > 0, \; \frac{1}{2} + 2\operatorname{Re}\mu < \operatorname{Re}\nu < 1\right].$ ET II 100(20)

7. $$\int_0^\infty x^{1+\nu}(x^2+a^2)^\mu K_\nu(bx)\,dx =$$
$$= 2^\nu \Gamma(\nu+1) a^{\nu+\mu+1} b^{-1-\mu} S_{\mu-\nu,\,\mu+\nu+1}(ab)$$
$$[\operatorname{Re} a>0, \quad \operatorname{Re} b>0, \quad \operatorname{Re}\nu>-1].$$ ET II 128(8)

8. $$\int_0^\infty \frac{x^{\varrho-1} J_\nu(ax)}{(x^2+k^2)^{\mu+1}}\,dx = \frac{a^\nu k^{\varrho+\nu-2\mu-2}\,\Gamma\left(\frac{1}{2}\varrho+\frac{1}{2}\nu\right)\Gamma\left(\mu+1-\frac{1}{2}\varrho-\frac{1}{2}\nu\right)}{2^{\nu+1}\,\Gamma(\mu+1)\,\Gamma(\nu+1)}\times$$
$$\times\, {}_1F_2\left(\frac{\varrho+\nu}{2};\ \frac{\varrho+\nu}{2}-\mu,\ \nu+1;\ \frac{a^2k^2}{4}\right)+$$
$$+\frac{a^{2\mu+2-\varrho}\,\Gamma\left(\frac{1}{2}\nu+\frac{1}{2}\varrho-\mu-1\right)}{2^{2\mu+3-\varrho}\,\Gamma\left(\mu+2+\frac{1}{2}\nu-\frac{1}{2}\varrho\right)}\times$$
$$\times\, {}_1F_2\left(\mu+1;\ \mu+2+\frac{\nu-\varrho}{2},\ \mu+2-\frac{\nu+\varrho}{2};\ \frac{a^2k^2}{4}\right)$$
$$\left[a>0, \quad -\operatorname{Re}\nu<\operatorname{Re}\varrho<2\operatorname{Re}\mu+\frac{7}{2}\right].$$ WA 477(1)

6.566

1. $$\int_0^\infty x^\mu N_\nu(bx)\frac{dx}{x^2+a^2} = 2^{\mu-2}\pi^{-1} b^{1-\mu}\times$$
$$\times\cos\left[\frac{\pi}{2}(\mu-\nu+1)\right]\Gamma\left(\frac{1}{2}\mu+\frac{1}{2}\nu-\frac{1}{2}\right)\Gamma\left(\frac{1}{2}\mu-\frac{1}{2}\nu-\frac{1}{2}\right)\times$$
$$\times\, {}_1F_2\left(1;\ 2-\frac{\mu+1+\nu}{2},\ 2-\frac{\mu+1-\nu}{2};\ \frac{a^2b^2}{4}\right)-$$
$$-\frac{1}{2}\pi a^{\mu-1}\operatorname{cosec}\left[\frac{\pi}{2}(\mu+\nu+1)\right]\operatorname{ctg}\left[\frac{\pi}{2}(\mu-\nu+1)\right] I_\nu(ab)-$$
$$-a^{\mu-1}\operatorname{cosec}\left[\frac{\pi}{2}(\mu-\nu+1)\right] K_\nu(ab)$$
$$\left[b>0, \ \operatorname{Re} a>0, \ |\operatorname{Re}\nu|-1<\operatorname{Re}\mu<\frac{5}{2}\right].$$ ET II 100(17)

2. $$\int_0^\infty x^{\nu+1} J_\nu(ax)\frac{dx}{x^2+b^2} = b^\nu K_\nu(ab)$$
$$\left[a>0, \ \operatorname{Re} b>0, \ -1<\operatorname{Re}\nu<\frac{3}{2}\right].$$ EH II 96(58)

3. $$\int_0^\infty x^\nu K_\nu(ax)\frac{dx}{x^2+b^2} = \frac{\pi^2 b^{\nu-1}}{4\cos\nu\pi}\left[\mathbf{H}_{-\nu}(ab)-N_{-\nu}(ab)\right]$$
$$\left[a>0, \ \operatorname{Re} b>0, \ \operatorname{Re}\nu>-\frac{1}{2}\right].$$ WA 468(9)

4. $$\int_0^\infty x^{-\nu} K_\nu(ax)\frac{dx}{x^2+b^2} = \frac{\pi^2}{4b^{\nu+1}\cos\nu\pi}\left[\mathbf{H}_\nu(ab)-N_\nu(ab)\right]$$
$$\left[a>0, \ \operatorname{Re} b>0, \ \operatorname{Re}\nu<\frac{1}{2}\right].$$ WA 468(10)

5. $\int_0^\infty x^{-\nu} J_\nu(ax) \frac{dx}{x^2+b^2} = \frac{\pi}{2b^{\nu+1}} [I_\nu(ab) - \mathbf{L}_\nu(ab)]$

$\left[a > 0, \ \operatorname{Re} b > 0, \ \operatorname{Re} \nu > -\frac{5}{2}\right].$ WA 468(11)

6.567

1. $\int_0^1 x^{\nu+1} (1-x^2)^\mu J_\nu(bx)\, dx = 2^\mu \Gamma(\mu+1) b^{-(\mu+1)} J_{\nu+\mu+1}(b)$

$[b > 0, \ \operatorname{Re} \nu > -1, \ \operatorname{Re} \mu > -1].$ ET II 26(33)a

2. $\int_0^1 x^{\nu+1} (1-x^2)^\mu N_\nu(bx)\, dx = b^{-(\mu+1)} [2^\mu \Gamma(\mu+1) N_{\mu+\nu+1}(b) +$

$+ 2^{\nu+1} \pi^{-1} \Gamma(\nu+1) S_{\mu-\nu, \mu+\nu+1}(b)]$

$[b > 0, \ \operatorname{Re} \mu > -1, \ \operatorname{Re} \nu > -1].$ ET II 103(35)a

3. $\int_0^1 x^{1-\nu} (1-x^2)^\mu J_\nu(bx)\, dx = \frac{2^{1-\nu} s_{\nu+\mu, \mu-\nu+1}(b)}{b^{\mu+1} \Gamma(\nu)}$

$[b > 0, \ \operatorname{Re} \mu > -1].$ ET II 25(31)a

4. $\int_0^1 x^{1-\nu} (1-x^2)^\mu N_\nu(bx)\, dx = b^{-(\mu+1)} [2^{1-\nu} \pi^{-1} \cos(\nu\pi) \Gamma(1-\nu) \times$

$\times s_{\mu+\nu, \mu-\nu+1}(b) - 2^\mu \operatorname{cosec}(\nu\pi) \Gamma(\mu+1) J_{\mu-\nu+1}(b)]$

$[b > 0, \ \operatorname{Re} \mu > -1, \ \operatorname{Re} \nu < 1].$ ET II 104(37)a

5. $\int_0^1 x^{1-\nu} (1-x^2)^\mu K_\nu(bx)\, dx = 2^{-\nu-2} b^\nu (\mu+1)^{-1} \Gamma(-\nu) \times$

$\times {}_1F_2\left(1; \ \nu+1, \ \mu+2; \ \frac{b^2}{4}\right) + \pi 2^{\mu-1} b^{-(\mu+1)} \operatorname{cosec}(\nu\pi) \times$

$\times \Gamma(\mu+1) I_{\mu-\nu+1}(b) \quad [\operatorname{Re} \mu > -1, \ \operatorname{Re} \nu < 1].$ ET II 129(12)a

6. $\int_0^1 x^{1-\nu} J_\nu(bx) \frac{dx}{\sqrt{1-x^2}} = \sqrt{\frac{\pi}{2b}} \mathbf{H}_{\nu-\frac{1}{2}}(b) \qquad [b > 0].$ ET II 24(24)a

7. $\int_0^1 x^{1+\nu} N_\nu(bx) \frac{dx}{\sqrt{1-x^2}} = \sqrt{\frac{\pi}{2b}} \operatorname{cosec}(\nu\pi) [\cos(\nu\pi) J_{\nu+\frac{1}{2}}(b) - \mathbf{H}_{-\nu-\frac{1}{2}}(b)]$

$[b > 0, \ \operatorname{Re} \nu > -1].$ ET II 102(28)a

8. $\int_0^1 x^{1-\nu} N_\nu(bx) \frac{dx}{\sqrt{1-x^2}} = \sqrt{\frac{\pi}{2b}} \{\operatorname{ctg}(\nu\pi) [\mathbf{H}_{\nu-\frac{1}{2}}(b) - N_{\nu-\frac{1}{2}}(b)] - J_{\nu-\frac{1}{2}}(b)\}$

$[b > 0, \ \operatorname{Re} \nu < 1].$ ET II 102(30)a

9. $\int_0^1 x^\nu (1-x^2)^{\nu-\frac{1}{2}} J_\nu(bx)\, dx = 2^{\nu-1} \sqrt{\pi}\, b^{-\nu} \Gamma\left(\nu+\frac{1}{2}\right) \left[J_\nu\left(\frac{b}{2}\right)\right]^2$

$\left[b > 0, \ \operatorname{Re} \nu > -\frac{1}{2}\right].$ ET II 24(25)a

10. $$\int_0^1 x^\nu (1-x^2)^{\nu-\frac{1}{2}} N_\nu(bx)\,dx = 2^{\nu-1}\sqrt{\pi}\, b^{-\nu}\Gamma\left(\nu+\frac{1}{2}\right) J_\nu\left(\frac{b}{2}\right) N_\nu\left(\frac{b}{2}\right)$$
$$\left[b>0,\ \operatorname{Re}\nu > -\frac{1}{2}\right].$$ ET II 102(31)a

11. $$\int_0^1 x^\nu (1-x^2)^{\nu-\frac{1}{2}} K_\nu(bx)\,dx = 2^{\nu-1}\sqrt{\pi}\, b^{-\nu}\Gamma\left(\nu+\frac{1}{2}\right) I_\nu\left(\frac{b}{2}\right) K_\nu\left(\frac{b}{2}\right)$$
$$\left[\operatorname{Re}\nu > -\frac{1}{2}\right].$$ ET II 129(10)a

12. $$\int_0^1 x^\nu (1-x^2)^{\nu-\frac{1}{2}} I_\nu(bx)\,dx = 2^{-\nu-1}\sqrt{\pi}\, b^{-\nu}\Gamma\left(\nu+\frac{1}{2}\right) \left[I_\nu\left(\frac{b}{2}\right)\right]^2$$ ET II 365(5)a

13. $$\int_0^1 x^{\nu+1} (1-x^2)^{-\nu-\frac{1}{2}} J_\nu(bx)\,dx = 2^{-\nu}\frac{b^{\nu-1}}{\sqrt{\pi}}\Gamma\left(\frac{1}{2}-\nu\right)\sin b$$
$$\left[b>0,\ |\operatorname{Re}\nu| < \frac{1}{2}\right].$$ ET II 25(27)a

14. $$\int_1^\infty x^\nu (x^2-1)^{\nu-\frac{1}{2}} N_\nu(bx)\,dx = 2^{\nu-2}\sqrt{\pi}\, b^{-\nu}\Gamma\left(\nu+\frac{1}{2}\right)\times$$
$$\times\left[J_\nu\left(\frac{b}{2}\right) J_{-\nu}\left(\frac{b}{2}\right) - N_\nu\left(\frac{b}{2}\right) N_{-\nu}\left(\frac{b}{2}\right)\right]$$
$$\left[|\operatorname{Re}\nu| < \frac{1}{2},\quad b>0\right].$$ ET II 103(32)a

15. $$\int_1^\infty x^\nu (x^2-1)^{\nu-\frac{1}{2}} K_\nu(bx)\,dx = \frac{2^{\nu-1}}{\sqrt{\pi}} b^{-\nu}\Gamma\left(\nu+\frac{1}{2}\right)\left[K_\nu\left(\frac{b}{2}\right)\right]^2$$
$$\left[\operatorname{Re} b>0,\quad \operatorname{Re}\nu > -\frac{1}{2}\right].$$ ET II 129(11)a

16. $$\int_1^\infty x^{-\nu} (x^2-1)^{-\nu-\frac{1}{2}} J_\nu(bx)\,dx = -2^{-\nu-1}\sqrt{\pi}\, b^{\nu}\Gamma\left(\frac{1}{2}-\nu\right) J_\nu\left(\frac{b}{2}\right) N_\nu\left(\frac{b}{2}\right)$$
$$\left[b>0,\quad |\operatorname{Re}\nu| < \frac{1}{2}\right].$$ ET II 25(26)a

17. $$\int_1^\infty x^{-\nu+1} (x^2-1)^{\nu-\frac{1}{2}} J_\nu(bx)\,dx = \frac{2^{-\nu}}{\sqrt{\pi}} b^{-\nu-1}\Gamma\left(\frac{1}{2}+\nu\right)\cos b$$
$$\left[b>0,\quad |\operatorname{Re}\nu| < \frac{1}{2}\right].$$ ET II 25(28)

6.568

1. $$\int_0^\infty x^\nu N_\nu(bx)\frac{dx}{x^2-a^2}=\frac{\pi}{2}a^{\nu-1}J_\nu(ab)$$
$$\left[a>0,\quad b>0,\quad -\frac{1}{2}<\operatorname{Re}\nu<\frac{5}{2}\right].$$ ET II 101(22)

2. $$\int_0^\infty x^\mu N_\nu(bx)\frac{dx}{x^2-a^2}=$$
$$=\frac{\pi}{2}a^{\mu-1}J_\nu(ab)+2^\mu\pi^{-1}a^{\mu-1}\cos\left[\frac{\pi}{2}(\mu-\nu+1)\right]\times$$
$$\times\Gamma\left(\frac{\mu-\nu+1}{2}\right)\Gamma\left(\frac{\mu+\nu+1}{2}\right)S_{-\mu,\nu}(ab)$$
$$\left[a>0,\quad b>0,\quad |\operatorname{Re}\nu|-1<\operatorname{Re}\mu<\frac{5}{2}\right].$$ ET II (101)(25)

6.569 $$\int_0^1 x^\lambda(1-x)^{\mu-1}J_\nu(ax)\,dx=\frac{\Gamma(\mu)\,\Gamma(1+\lambda+\nu)\,2^{-\nu}a^\nu}{\Gamma(\nu+1)\,\Gamma(1+\lambda+\mu+\nu)}\times$$
$$\times{}_2F_3\left(\frac{\lambda+1+\nu}{2},\ \frac{\lambda+2+\nu}{2};\ \nu+1,\ \frac{\lambda+1+\mu+\nu}{2},\ \frac{\lambda+2+\mu+\nu}{2};\ -\frac{a^2}{4}\right)$$
$$[\operatorname{Re}\mu>0,\quad \operatorname{Re}(\lambda+\nu)>-1].$$ ET II 193(56)a

6.571

1. $$\int_0^\infty\left[(x^2+a^2)^{\frac{1}{2}}\pm x\right]^\mu J_\nu(bx)\frac{dx}{\sqrt{x^2+a^2}}=a^\mu I_{\frac{1}{2}(\nu\mp\mu)}\left(\frac{ab}{2}\right)K_{\frac{1}{2}(\nu\pm\mu)}\left(\frac{ab}{2}\right)$$
$$\left[\operatorname{Re}a>0,\quad b>0,\quad \operatorname{Re}\nu>-1,\ \operatorname{Re}\mu<\frac{3}{2}\right].$$ ET II 26(38)

2. $$\int_0^\infty\left[(x^2+a^2)^{\frac{1}{2}}-x\right]^\mu N_\nu(bx)\frac{dx}{\sqrt{x^2+a^2}}=$$
$$=a^\mu\left[\operatorname{ctg}(\nu\pi)\,I_{\frac{1}{2}(\mu+\nu)}\left(\frac{ab}{2}\right)K_{\frac{1}{2}(\mu-\nu)}\left(\frac{ab}{2}\right)-\right.$$
$$\left.-\operatorname{cosec}(\nu\pi)\,I_{\frac{1}{2}(\mu-\nu)}\left(\frac{ab}{2}\right)K_{\frac{1}{2}(\mu+\nu)}\left(\frac{ab}{2}\right)\right]$$
$$\left[\operatorname{Re}a>0,\quad b>0,\quad \operatorname{Re}\mu>-\frac{3}{2},\quad |\operatorname{Re}\nu|<1\right].$$ ET II 104(40)

3. $$\int_0^\infty\left[(x^2+a^2)^{\frac{1}{2}}+x\right]^\mu K_\nu(bx)\frac{dx}{\sqrt{x^2+a^2}}=$$
$$=\frac{\pi^2}{4}a^\mu\operatorname{cosec}(\nu\pi)\left[J_{\frac{1}{2}(\nu-\mu)}\left(\frac{ab}{2}\right)N_{-\frac{1}{2}(\nu+\mu)}\left(\frac{ab}{2}\right)-\right.$$
$$\left.-N_{\frac{1}{2}(\nu-\mu)}\left(\frac{ab}{2}\right)J_{-\frac{1}{2}(\nu+\mu)}\left(\frac{ab}{2}\right)\right]$$
$$[\operatorname{Re}a>0,\quad \operatorname{Re}b>0].$$ ET II 130(15)

6.572

1. $$\int_0^\infty x^{-\mu}[(x^2+a^2)^{\frac{1}{2}}+a]^\mu J_\nu(bx)\frac{dx}{\sqrt{x^2+a^2}}=$$
$$=\frac{\Gamma\left(\frac{1+\nu-\mu}{2}\right)}{ab\Gamma(\nu+1)}W_{\frac{1}{2}\mu,\frac{1}{2}\nu}(ab)M_{-\frac{1}{2}\mu,\frac{1}{2}\nu}(ab)$$
$[\operatorname{Re} a>0, \quad b>0, \quad \operatorname{Re}(\nu-\mu)>-1]$. ET II 26(40)

2. $$\int_0^\infty x^{-\mu}[(x^2+a^2)^{\frac{1}{2}}+a]^\mu K_\nu(bx)\frac{dx}{\sqrt{x^2+a^2}}=$$
$$=\frac{\Gamma\left(\frac{1+\nu-\mu}{2}\right)\Gamma\left(\frac{1-\nu-\mu}{2}\right)}{2ab}W_{\frac{1}{2}\mu,\frac{1}{2}\nu}(iab)W_{\frac{1}{2}\mu,\frac{1}{2}\nu}(-iab)$$
$[\operatorname{Re} a>0, \quad \operatorname{Re} b>0, \quad \operatorname{Re}\mu+|\operatorname{Re}\nu|<1]$. ET II 130(18), BU 87(6a)

3. $$\int_0^\infty x^{-\mu}[(x^2+a^2)^{\frac{1}{2}}-a]^\mu N_\nu(bx)\frac{dx}{\sqrt{x^2+a^2}}=$$
$$=-\frac{1}{ab}W_{-\frac{1}{2}\mu,\frac{1}{2}\nu}(ab)\left\{\frac{\Gamma\left(\frac{1+\nu+\mu}{2}\right)}{\Gamma(\nu+1)}\operatorname{tg}\left(\frac{\nu-\mu}{2}\pi\right)M_{\frac{1}{2}\mu,\frac{1}{2}\nu}(ab)+\right.$$
$$\left.+\sec\left(\frac{\nu-\mu}{2}\pi\right)W_{\frac{1}{2}\mu,\frac{1}{2}\nu}(ab)\right\}$$
$\left[\operatorname{Re} a>0, \quad b>0, \quad |\operatorname{Re}\nu|<\frac{1}{2}+\frac{1}{2}\operatorname{Re}\mu\right]$. ET II 105(42)

6.573

1. $$\int_0^\infty x^{\nu-M+1}J_\nu(bx)\prod_{i=1}^{k}J_{\mu_i}(a_ix)\,dx=0, \qquad M=\sum_{i=1}^{k}\mu_i$$
$\left[a_i>0, \quad \sum_{i=1}^{k}a_i<b<\infty, \quad -1<\operatorname{Re}\nu<\operatorname{Re}M+\frac{1}{2}k-\frac{1}{2}\right]$.
ET II 54(42)

2. $$\int_0^\infty x^{\nu-M-1}J_\nu(bx)\prod_{i=1}^{k}J_{\mu_i}(a_ix)\,dx=$$
$$=2^{\nu-M-1}b^{-\nu}\Gamma(\nu)\prod_{i=1}^{k}\frac{a_i^{\mu_i}}{\Gamma(1+\mu_i)}, \qquad M=\sum_{i=1}^{k}\mu_i$$
$\left[a_i>0, \quad \sum_{i=1}^{k}a_i<b<\infty, \quad 0<\operatorname{Re}\nu<\operatorname{Re}M+\frac{1}{2}k+\frac{3}{2}\right]$.

WA 460(16)a, ET II 54(43)

6.574

1. $\int\limits_0^\infty J_\nu(\alpha t) J_\mu(\beta t) t^{-\lambda} dt =$

$$= \frac{\alpha^\nu \Gamma\left(\frac{\nu+\mu-\lambda+1}{2}\right)}{2^\lambda \beta^{\nu-\lambda+1} \Gamma\left(\frac{-\nu+\mu+\lambda+1}{2}\right) \Gamma(\nu+1)} \times$$
$$\times F\left(\frac{\nu+\mu-\lambda+1}{2}, \frac{\nu-\mu-\lambda+1}{2}; \nu+1; \frac{\alpha^2}{\beta^2}\right)$$

$[\operatorname{Re}(\nu+\mu-\lambda+1) > 0, \quad \operatorname{Re}\lambda > -1, \quad 0 < \alpha < \beta]$. WA 439(2)a, MO 49

If we reverse the positions of ν and μ and at the same time reverse the positions of α and β, the function on the right hand side of this equation will change. Thus, the right hand side represents a function of $\frac{\alpha}{\beta}$ that is not analytic at $\frac{\alpha}{\beta} = 1$.

For $\alpha = \beta$, we have the following equation

2. $\int\limits_0^\infty J_\nu(\alpha t) J_\mu(\alpha t) t^{-\lambda} dt =$

$$= \frac{\alpha^{\lambda-1} \Gamma(\lambda) \Gamma\left(\frac{\nu+\mu-\lambda+1}{2}\right)}{2^\lambda \Gamma\left(\frac{-\nu+\mu+\lambda+1}{2}\right) \Gamma\left(\frac{\nu+\mu+\lambda+1}{2}\right) \Gamma\left(\frac{\nu-\mu+\lambda+1}{2}\right)}$$

$[\operatorname{Re}(\nu+\mu+1) > \operatorname{Re}\lambda > 0, \quad \alpha > 0]$. MO 49, WA 441(2)a

3. $\int\limits_0^\infty J_\nu(\alpha t) J_\mu(\beta t) t^{-\lambda} dt =$

$$= \frac{\beta^\mu \Gamma\left(\frac{\nu+\mu-\lambda+1}{2}\right)}{2^\lambda \alpha^{\mu-\lambda+1} \Gamma\left(\frac{\nu-\mu+\lambda+1}{2}\right) \Gamma(\mu+1)} \times$$
$$\times F\left(\frac{\nu+\mu-\lambda+1}{2}, \frac{-\nu+\mu-\lambda+1}{2}; \mu+1; \frac{\beta^2}{\alpha^2}\right)$$

$[\operatorname{Re}(\nu+\mu-\lambda+1) > 0, \quad \operatorname{Re}\lambda > -1, \quad 0 < \beta < \alpha]$. MO 50, WA 440(3)a

If $\mu-\nu+\lambda+1$ (or $\nu-\mu+\lambda+1$) is a negative integer, the right hand side of equation 6.574 1. (or 6.574 3.) vanishes. The cases in which the hypergeometric function F in 6.574 3. (or 6.574 1.) can be reduced to an elementary function are then especially important.

6.575

1. $\int\limits_0^\infty J_{\nu+1}(\alpha t) J_\mu(\beta t) t^{\mu-\nu} dt = 0 \qquad [\alpha < \beta];$

$$= \frac{(\alpha^2-\beta^2)^{\nu-\mu}\beta^\mu}{2^{\nu-\mu}\alpha^{\nu+1}\Gamma(\nu-\mu+1)} \qquad [\alpha \geqslant \beta]$$

$[\operatorname{Re}\mu > \operatorname{Re}(\nu+1) > 0]$. MO 51

2. $$\int_0^\infty \frac{J_\nu(x) J_\mu(x)}{x^{\nu+\mu}} dx = \frac{\sqrt{\pi}\,\Gamma(\nu+\mu)}{2^{\nu+\mu}\Gamma\left(\nu+\mu+\frac{1}{2}\right)\Gamma\left(\nu+\frac{1}{2}\right)\Gamma\left(\mu+\frac{1}{2}\right)}$$

$[\operatorname{Re}(\nu+\mu) > 0]$. KU 147(17), WA 434(1)

6.576

1. $$\int_0^\infty x^{\mu-\nu+1} J_\mu(x) K_\nu(x)\, dx = \frac{1}{2}\Gamma(\mu-\nu+1)$$

$[\operatorname{Re}\mu > -1, \quad \operatorname{Re}(\mu-\nu) > -1]$. ET II 370(47)

2. $$\int_0^\infty x^{-\lambda} J_\nu(ax) J_\nu(bx)\, dx = \frac{a^\nu b^\nu \Gamma\left(\nu+\frac{1-\lambda}{2}\right)}{2^\lambda (a+b)^{2\nu-\lambda+1}\Gamma(\nu+1)\Gamma\left(\frac{1+\lambda}{2}\right)} \times F\left[\nu+\frac{1-\lambda}{2}, \nu+\frac{1}{2}; 2\nu+1; \frac{4ab}{(a+b)^2}\right]$$

$[a > 0, \quad b > 0, \quad 2\operatorname{Re}\nu + 1 > \operatorname{Re}\lambda > -1]$. ET II 47(4)

3. $$\int_0^\infty x^{-\lambda} K_\mu(ax) J_\nu(bx)\, dx = \frac{b^\nu \Gamma\left(\frac{\nu-\lambda+\mu+1}{2}\right)\Gamma\left(\frac{\nu-\lambda-\mu+1}{2}\right)}{2^{\lambda+1} a^{\nu-\lambda+1}\Gamma(1+\nu)} \times F\left(\frac{\nu-\lambda+\mu+1}{2}, \frac{\nu-\lambda-\mu+1}{2}; \nu+1; -\frac{b^2}{a^2}\right)$$

$[\operatorname{Re}(a \pm ib) > 0, \quad \operatorname{Re}(\nu-\lambda+1) > |\operatorname{Re}\mu|]$.

EH II 52(31), ET II 63(4), WA 449(1)

4. $$\int_0^\infty x^{-\lambda} K_\mu(ax) K_\nu(bx)\, dx = \frac{2^{-2-\lambda} a^{-\nu+\lambda-1} b^\nu}{\Gamma(1-\lambda)} \Gamma\left(\frac{1-\lambda+\mu+\nu}{2}\right)\Gamma\left(\frac{1-\lambda-\mu+\nu}{2}\right) \times \Gamma\left(\frac{1-\lambda+\mu-\nu}{2}\right)\Gamma\left(\frac{1-\lambda-\mu-\nu}{2}\right) \times F\left(\frac{1-\lambda+\mu+\nu}{2}, \frac{1-\lambda-\mu+\nu}{2}; 1-\lambda; 1-\frac{b^2}{a^2}\right)$$

$[\operatorname{Re}(a+b) > 0, \operatorname{Re}\lambda < 1 - |\operatorname{Re}\mu| - |\operatorname{Re}\nu|]$. ET II 145(49), EH II 93(36)

5. $$\int_0^\infty x^{-\lambda} K_\mu(ax) I_\nu(bx)\, dx = \frac{b^\nu \Gamma\left(\frac{1}{2}-\frac{1}{2}\lambda+\frac{1}{2}\mu+\frac{1}{2}\nu\right)\Gamma\left(\frac{1}{2}-\frac{1}{2}\lambda-\frac{1}{2}\mu+\frac{1}{2}\nu\right)}{2^{\lambda+1}\Gamma(\nu+1) a^{-\lambda+\nu+1}} \times F\left(\frac{1}{2}-\frac{1}{2}\lambda+\frac{1}{2}\mu+\frac{1}{2}\nu, \frac{1}{2}-\frac{1}{2}\lambda-\frac{1}{2}\mu+\frac{1}{2}\nu; \nu+1; \frac{b^2}{a^2}\right)$$

$[\operatorname{Re}(\nu+1-\lambda \pm \mu) > 0, \ a > b]$. EH II 93(35)

6. $\int_0^\infty x^{-\lambda} N_\mu(ax) J_\nu(bx)\, dx = \frac{2}{\pi} \sin \frac{\pi(\nu-\mu-\lambda)}{2} \int_0^\infty x^{-\lambda} K_\mu(ax) I_\nu(bx)\, dx$

$[a > b,\ \operatorname{Re}(\nu-\lambda+1 \pm \mu) > 0]$; (see 6.576 5.). EH II 93(37)

7. $\int_0^\infty x^{\mu+\nu+1} J_\mu(ax) K_\nu(bx)\, dx = 2^{\mu+\nu} a^\mu b^\nu \frac{\Gamma(\mu+\nu+1)}{(a^2+b^2)^{\mu+\nu+1}}$

$[\operatorname{Re}\mu > |\operatorname{Re}\nu| - 1,\ \operatorname{Re} b > |\operatorname{Im} a|]$. ET 137(16), EH II 93(36), B 449(2)

6.577

1. $\int_0^\infty x^{\nu-\mu+1+2n} J_\mu(ax) J_\nu(bx) \frac{dx}{x^2+c^2} = (-1)^n c^{\nu-\mu+2n} I_\mu(ac) K_\nu(bc)$

$[a > 0,\ b > a,\ \operatorname{Re} c > 0,\ 1 + \operatorname{Re}\mu - 2n > \operatorname{Re}\nu > -1-n,\ n \geqslant 0$ an integer$]$. ET II 49(13)

2. $\int_0^\infty x^{\mu-\nu+1+2n} J_\mu(ax) J_\nu(bx) \frac{dx}{x^2+c^2} = (-1)^n c^{\mu-\nu+2n} I_\nu(bc) K_\mu(ac)$

$[b > 0,\ a > b,\ \operatorname{Re}\nu - 2n + 1 > \operatorname{Re}\mu > -n-1,\ n \geqslant 0$ an integer$]$. ET II 49(15)

6.578

1. $\int_0^\infty x^{\varrho-1} J_\lambda(ax) J_\mu(bx) J_\nu(cx)\, dx =$

$$= \frac{2^{\varrho-1} a^\lambda b^\mu c^{-\lambda-\mu-\varrho} \Gamma\left(\frac{\lambda+\mu+\nu+\varrho}{2}\right)}{\Gamma(\lambda+1)\Gamma(\mu+1)\Gamma\left(1-\frac{\lambda+\mu-\nu+\varrho}{2}\right)} \times$$

$$\times F_4\left(\frac{\lambda+\mu-\nu+\varrho}{2}, \frac{\lambda+\mu+\nu+\varrho}{2}; \lambda+1, \mu+1; \frac{a^2}{c^2}, \frac{b^2}{c^2}\right)$$

$\left[\operatorname{Re}(\lambda+\mu+\nu+\varrho) > 0,\ \operatorname{Re}\varrho < \frac{5}{2},\ a > 0,\ b > 0,\ c > 0,\ c > a+b\right]$. ET II 351(9)

2. $\int_0^\infty x^{\varrho-1} J_\lambda(ax) J_\mu(bx) K_\nu(cx)\, dx =$

$$= \frac{2^{\varrho-2} a^\lambda b^\mu c^{-\varrho-\lambda-\mu}}{\Gamma(\lambda+1)\Gamma(\mu+1)} \Gamma\left(\frac{\varrho+\lambda+\mu-\nu}{2}\right) \Gamma\left(\frac{\varrho+\lambda+\mu+\nu}{2}\right) \times$$

$$\times F_4\left(\frac{\varrho+\lambda+\mu-\nu}{2}, \frac{\varrho+\lambda+\mu+\nu}{2}; \lambda+1, \mu+1; -\frac{a^2}{c^2}, -\frac{b^2}{c^2}\right)$$

$[\operatorname{Re}(\varrho+\lambda+\mu) > |\operatorname{Re}\nu|,\ \operatorname{Re} c > |\operatorname{Im} a| + |\operatorname{Im} b|]$. ET II 373(8)

3. $\int_0^\infty x^{\lambda-\mu-\nu+1} J_\nu(ax) J_\mu(bx) J_\lambda(cx)\, dx = 0$

$\left[\operatorname{Re}\lambda > -1,\ \operatorname{Re}(\lambda-\mu-\nu) < \frac{1}{2},\ c > b > 0,\ 0 < a < c-b\right]$. ET II 53(36)

4. $$\int_0^\infty x^{\lambda-\mu-\nu-1} J_\nu(ax) J_\mu(bx) J_\lambda(cx)\,dx = \frac{2^{\lambda-\mu-\nu-1} a^\nu b^\mu \Gamma(\lambda)}{c^\lambda \Gamma(\mu+1)\Gamma(\nu+1)}$$

$$\left[\operatorname{Re}\lambda > 0,\ \operatorname{Re}(\lambda-\mu-\nu) < \tfrac{5}{2},\ c > b > 0,\ 0 < a < c-b\right].$$

ET II 53(37)

5 $$\int_0^\infty x^{1+\mu} N_\mu(ax) J_\nu(bx) J_\nu(cx)\,dx = 0 \quad [0 < b < c,\ 0 < a < c-b].$$

ET II 352(13)

6. $$\int_0^\infty x^{\mu+1} K_\mu(ax) J_\nu(bx) J_\nu(cx)\,dx = \frac{1}{\sqrt{2\pi}} a^\mu b^{-\mu-1} c^{-\mu-1} e^{-\left(\mu+\frac{1}{2}\right)\pi i} \times$$

$$\times (u^2-1)^{-\frac{1}{2}\mu-\frac{1}{4}} Q_{\nu-\frac{1}{2}}^{\mu+\frac{1}{2}}(u), \quad 2bcu = a^2+b^2+c^2$$

$$[\operatorname{Re} a > |\operatorname{Im} b|,\ c > 0,\ \operatorname{Re}\nu > -1,\ \operatorname{Re}(\mu+\nu) > -1].$$

WA 452(2), ET II 64(12)

7. $$\int_0^\infty x^{\mu+1} I_\nu(ax) K_\mu(bx) J_\nu(cx)\,dx =$$

$$= \frac{1}{\sqrt{2\pi}} a^{-\mu-1} b^\mu c^{-\mu-1} e^{-\left(\mu-\frac{1}{2}\nu+\frac{1}{4}\right)\pi i} (v^2+1)^{-\frac{1}{2}\mu-\frac{1}{4}} Q_{\nu-\frac{1}{2}}^{\mu+\frac{1}{2}}(iv),$$

$$2acv = b^2-a^2+c^2$$

$$[\operatorname{Re} b > |\operatorname{Re} a|,\ c > 0,\ \operatorname{Re}\nu > -1,\ \operatorname{Re}(\mu+\nu) > -1].$$ ET II 66(22)

8. $$\int_0^\infty x^{1-\mu} J_\mu(ax) J_\nu(bx) J_\nu(cx)\,dx =$$

$$= \frac{c^{\mu-1}(\operatorname{sh} u)^{\mu-\frac{1}{2}}}{\sqrt{\frac{1}{2}\pi^3}\, a^\mu b^{1-\mu}} e^{\left(\mu-\frac{1}{2}\right)\pi i} \sin[(\mu-\nu)\pi] Q_{\nu-\frac{1}{2}}^{\frac{1}{2}-\mu}(\operatorname{ch} u),$$

$$2bc \operatorname{ch} u = a^2-b^2-c^2$$

$$\left[\operatorname{Re}\nu > -1,\ \operatorname{Re}\mu > -\tfrac{1}{2},\ 0 < c < a-b,\ b > 0\right];$$

$$= \frac{b^{\mu-1} c^{\mu-1}}{\sqrt{2\pi}\, a^\mu} (\sin v)^{\mu-\frac{1}{2}} \mathrm{P}_{\nu-\frac{1}{2}}^{\frac{1}{2}-\mu}(\cos v),\ 2bc\cos v = b^2+c^2-a^2$$

$$\left[\operatorname{Re}\nu > -1,\ \operatorname{Re}\mu > -\tfrac{1}{2},\ |a-b| < c < a+b,\ a > 0,\ b > 0\right];$$

$$= 0 \left[\operatorname{Re}\nu > -1,\ \operatorname{Re}\mu > -\tfrac{1}{2},\ 0 < c < b-a \text{ or } a+b < c < \infty,\ a > 0,\ b > 0\right].$$

ET II 52(34)

9. $$\int_0^\infty J_\nu(ax) J_\nu(bx) J_\nu(cx) x^{1-\nu}\,dx = \frac{2^{\nu-1}\Delta^{2\nu-1}}{(abc)^\nu \Gamma\left(\nu+\frac{1}{2}\right)\Gamma\left(\frac{1}{2}\right)},$$

where Δ is the area of a triangle whose sides are a, b, and c. In the case in which the segments whose lengths are a, b, and c cannot form a triangle, the value of the integral is zero $\left[\operatorname{Re}\nu > -\frac{1}{2}\right]$.

MO 52, WA 451(3)

10. $$\int_0^\infty x^{\nu+1} K_\mu(ax) K_\mu(bx) J_\nu(cx)\,dx =$$

$$= \frac{\sqrt{\pi}\, c^\nu \Gamma(\nu+\mu+1)\Gamma(\nu-\mu+1)}{2^{\frac{3}{2}}(ab)^{\nu+1}(u^2-1)^{\frac{1}{2}\nu+\frac{1}{4}}} P_{\mu-\frac{1}{2}}^{-\nu-\frac{1}{2}}(u),$$

$$2abu = a^2+b^2+c^2$$

$[\operatorname{Re} a > 0,\ \operatorname{Re} b > 0,\ c > 0,\ \operatorname{Re}(\nu \pm \mu) > -1,\ \operatorname{Re}\nu > -1]$. ET II 67(30)

11. $$\int_0^\infty x^{\nu+1} K_\mu(ax) I_\mu(bx) J_\nu(cx)\,dx = \frac{(ab)^{-\nu-1} c^\nu e^{-\left(\nu+\frac{1}{2}\right)\pi i} Q_{\mu-\frac{1}{2}}^{\nu+\frac{1}{2}}(u)}{\sqrt{2\pi}\,(u^2-1)^{\frac{1}{2}\nu+\frac{1}{4}}},$$

$$2abu = a^2+b^2+c^2$$

$[\operatorname{Re} a > |\operatorname{Re} b|,\ c > 0,\ \operatorname{Re}\nu > -1,\ \operatorname{Re}(\mu+\nu) > -1]$. ET II 66(24)

12. $$\int_0^\infty x^{\nu+1}[J_\nu(ax)]^2 N_\nu(bx)\,dx =$$

$$= 0 \qquad \left[a > 0,\ 0 < b < 2a,\ |\operatorname{Re}\nu| < \frac{1}{2}\right];$$

$$= \frac{2^{3\nu+1} a^{2\nu} b^{-\nu-1}}{\sqrt{\pi}\,\Gamma\left(\frac{1}{2}-\nu\right)}(b^2-4a^2)^{-\nu-\frac{1}{2}}$$

$\left[a > 0,\ 2a < b < \infty,\ |\operatorname{Re}\nu| < \frac{1}{2}\right]$. ET II 109(3)

13. $$\int_0^\infty x^{\nu+1} J_\nu(ax) N_\nu(ax) J_\nu(bx)\,dx =$$

$$= 0 \qquad \left[a > 0,\ |\operatorname{Re}\nu| < \frac{1}{2},\ 0 < b < 2a\right];$$

$$= -\frac{2^{3\nu+1} a^{2\nu} b^{-\nu-1}}{\sqrt{\pi}\,\Gamma\left(\frac{1}{2}-\nu\right)}(b^2-4a^2)^{-\nu-\frac{1}{2}}$$

$\left[a > 0,\ 2a < b < \infty,\ |\operatorname{Re}\nu| < \frac{1}{2}\right]$. ET II 55(49)

14. $$\int_0^\infty x^{\nu+1} J_\mu(xa\sin\psi)\, J_\nu(xa\sin\varphi)\, K_\mu(xa\cos\varphi\cos\psi)\, dx =$$

$$= \frac{2^\nu \Gamma(\mu+\nu+1)(\sin\varphi)^\nu \left(\cos\frac{\alpha}{2}\right)^{2\nu+1}}{a^{\nu+2}(\cos\psi)^{2\nu+2}} P_\nu^{-\mu}(\cos\alpha), \quad \operatorname{tg}\frac{1}{2}\alpha = \operatorname{tg}\psi\cos\varphi$$

$$\left[a>0,\ \frac{\pi}{2}>\varphi>0,\ 0<\psi<\frac{\pi}{2},\ \operatorname{Re}\nu>-1,\ \operatorname{Re}(\mu+\nu)>-1\right].$$

ET II 64(11)

15. $$\int_0^\infty x^{\nu+1} J_\nu(ax)\, K_\nu(bx)\, J_\nu(cx)\, dx = \frac{2^{3\nu}(abc)^\nu \Gamma\left(\nu+\frac{1}{2}\right)}{\sqrt{\pi}\,[(a^2+b^2+c^2)^2-4a^2c^2]^{\nu+\frac{1}{2}}}$$

$$\left[\operatorname{Re} b > |\operatorname{Im} a|,\ c>0,\ \operatorname{Re}\nu > -\frac{1}{2}\right].$$ ET II 63(8)

16. $$\int_0^\infty x^{\nu+1} I_\nu(ax)\, K_\nu(bx)\, J_\nu(cx)\, dx = \frac{2^{3\nu}(abc)^\nu \Gamma\left(\nu+\frac{1}{2}\right)}{\sqrt{\pi}\,[(b^2-a^2+c^2)^2+4a^2c^2]^{\nu+\frac{1}{2}}}$$

$$\left[\operatorname{Re} b > \operatorname{Re} a,\ c>0,\ \operatorname{Re}\nu > -\frac{1}{2}\right].$$ ET II 65(18)

6.579

1. $$\int_0^\infty x^{2\nu+1} J_\nu(ax)\, N_\nu(ax)\, J_\nu(bx)\, N_\nu(bx)\, dx =$$

$$= \frac{a^{2\nu}\Gamma(3\nu+1)}{2\pi b^{4\nu+2}\Gamma\left(\frac{1}{2}-\nu\right)\Gamma\left(2\nu+\frac{3}{2}\right)} \times$$

$$\times F\left(\nu+\frac{1}{2},\ 3\nu+1;\ 2\nu+\frac{3}{2};\ \frac{a^2}{b^2}\right)$$

$$\left[0<a<b,\ -\frac{1}{3}<\operatorname{Re}\nu<\frac{1}{2}\right].$$ EH II 94(45), ET II 352(15)

2. $$\int_0^\infty x^{2\nu+1} J_\nu(ax)\, K_\nu(ax)\, J_\nu(bx)\, K_\nu(bx)\, dx =$$

$$= \frac{2^{\nu-3}a^{2\nu}\Gamma\left(\frac{\nu+1}{2}\right)\Gamma\left(\nu+\frac{1}{2}\right)\Gamma\left(\frac{3\nu+1}{2}\right)}{\sqrt{\pi}\, b^{4\nu+2}\Gamma(\nu+1)} \times$$

$$\times F\left(\nu+\frac{1}{2},\ \frac{3\nu+1}{2};\ 2\nu+1;\ 1-\frac{a^4}{b^4}\right)$$

$$\left[0<a<b,\ \operatorname{Re}\nu > -\frac{1}{3}\right].$$ ET II 373(10)

3. $$\int_0^\infty x^{1-2\nu}[J_\nu(x)]^4\, dx = \frac{\Gamma(\nu)\Gamma(2\nu)}{2\pi\left[\Gamma\left(\nu+\frac{1}{2}\right)\right]^2\Gamma(3\nu)}$$

$$[\operatorname{Re}\nu > 0].$$ ET II 342(25)

4. $$\int_0^\infty x^{1-2\nu}[J_\nu(ax)]^2[J_\nu(bx)]^2\,dx = \frac{a^{2\nu-1}\Gamma(\nu)}{2\pi b\Gamma\left(\nu+\frac{1}{2}\right)\Gamma\left(2\nu+\frac{1}{2}\right)} F\left(\nu, \frac{1}{2}-\nu;\ 2\nu+\frac{1}{2};\ \frac{a^2}{b^2}\right).$$

ET II 351(10)

6.581

1. $$\int_0^a x^{\lambda-1}J_\mu(x)J_\nu(a-x)\,dx = 2^\lambda \sum_{m=0}^\infty \frac{(-1)^m\Gamma(\lambda+\mu+m)\Gamma(\lambda+m)}{m!\,\Gamma(\lambda)\Gamma(\mu+m+1)} J_{\lambda+\mu+\nu+2m}(a)$$

$[\operatorname{Re}(\lambda+\mu)>0,\ \operatorname{Re}\nu>-1]$. ET II 354(25)

2. $$\int_0^a x^{\lambda-1}(a-x)^{-1}J_\mu(x)J_\nu(a-x)\,dx = \frac{2^\lambda}{a\nu}\sum_{m=0}^\infty \frac{(-1)^m\Gamma(\lambda+\mu+m)\Gamma(\lambda+m)}{m!\,\Gamma(\lambda)\Gamma(\mu+m+1)}(\lambda+\mu+\nu+2m)J_{\lambda+\mu+\nu+2m}(a)$$

$[\operatorname{Re}(\lambda+\mu)>0,\ \operatorname{Re}\nu>0]$. ET II 354(27)

3. $$\int_0^a x^\mu (a-x)^\nu J_\mu(x)J_\nu(a-x)\,dx = \frac{\Gamma\left(\mu+\frac{1}{2}\right)\Gamma\left(\nu+\frac{1}{2}\right)}{\sqrt{2\pi}\,\Gamma(\mu+\nu+1)} a^{\mu+\nu+\frac{1}{2}} J_{\mu+\nu+\frac{1}{2}}(a)$$

$\left[\operatorname{Re}\mu>-\frac{1}{2},\ \operatorname{Re}\nu>-\frac{1}{2}\right]$. ET II 354(28), EH II 46(6)

4. $$\int_0^a x^\mu (a-x)^{\nu+1} J_\mu(x)J_\nu(a-x)\,dx = \frac{\Gamma\left(\mu+\frac{1}{2}\right)\Gamma\left(\nu+\frac{3}{2}\right)}{\sqrt{2\pi}\,\Gamma(\mu+\nu+2)} a^{\mu+\nu+\frac{3}{2}} J_{\mu+\nu+\frac{1}{2}}(a)$$

$\left[\operatorname{Re}\nu>-1,\ \operatorname{Re}\mu>-\frac{1}{2}\right]$. ET II 354(29)

5. $$\int_0^a x^\mu (a-x)^{-\mu-1} J_\mu(x)J_\nu(a-x)\,dx = \frac{2^\mu\Gamma\left(\mu+\frac{1}{2}\right)\Gamma(\nu-\mu)}{\sqrt{\pi}\,\Gamma(\mu+\nu+1)} a^\mu J_\nu(a) \quad \left[\operatorname{Re}\nu>\operatorname{Re}\mu>-\frac{1}{2}\right].$$

ET II 355(30)

6.582 $$\int_0^\infty x^{\mu-1}|x-b|^{-\mu}K_\mu(|x-b|)\,K_\nu(x)\,dx = \frac{1}{\sqrt{\pi}}(2b)^{-\mu}\Gamma\left(\frac{1}{2}-\mu\right)\Gamma(\mu+\nu)\,\Gamma(\mu-\nu)\,K_\nu(b)$$
$$\left[b>0,\ \operatorname{Re}\mu<\frac{1}{2},\ \operatorname{Re}\mu>|\operatorname{Re}\nu|\right].$$ ET II 374(14)

6.583 $$\int_0^\infty x^{\mu-1}(x+b)^{-\mu}K_\mu(x+b)\,K_\nu(x)\,dx = \frac{\sqrt{\pi}\,\Gamma(\mu+\nu)\,\Gamma(\mu-\nu)}{2^\mu b^\mu\Gamma\left(\mu+\frac{1}{2}\right)}K_\nu(b)$$
$$[|\arg b|<\pi,\ \operatorname{Re}\mu>|\operatorname{Re}\nu|].$$ ET II 374(15)

6.584

1. $$\int_0^\infty \frac{x^{\varrho-1}\left[H_\nu^{(1)}(ax)-e^{\varrho\pi i}H_\nu^{(1)}(axe^{\pi i})\right]}{(x^2-r^2)^{m+1}}dx = \frac{\pi i}{m!}\left(\frac{d}{dr^2}\right)^m\left[r^{\varrho-2}H_\nu^{(1)}(ar)\right]$$
$$\left[m=0,1,2,\ldots,\ \operatorname{Im}r>0,\ a>0,\ |\operatorname{Re}\nu|<\operatorname{Re}\varrho<2m+\frac{7}{2}\right].$$ WA 465

2. $$\int_0^\infty \left[\cos\frac{1}{2}(\varrho-\nu)\pi J_\nu(ax)+\sin\frac{1}{2}(\varrho-\nu)\pi\cdot N_\nu(ax)\right]\frac{x^{\varrho-1}}{(x^2+k^2)^{m+1}}dx = \frac{(-1)^{m+1}}{2^m\cdot m!}\left(\frac{d}{k\,dk}\right)^m\left[k^{\varrho-2}K_\nu(ak)\right]$$
$$\left[m=0,1,2,\ldots,\ \operatorname{Re}k>0,\ a>0,\ |\operatorname{Re}\nu|<\operatorname{Re}\varrho<2m+\frac{7}{2}\right].$$ WA 466(2)

3. $$\int_0^\infty \{\cos\nu\pi\,J_\nu(ax)-\sin\nu\pi\,N_\nu(ax)\}\frac{x^{1-\nu}\,dx}{(x^2+k^2)^{m+1}} = \frac{a^mK_{\nu+m}(ak)}{2^m\cdot m!\,k^{\nu+m}}$$
$$\left[m=0,1,2,\ldots,\ \operatorname{Re}k>0,\ a>0,\ -2m-\frac{3}{2}<\operatorname{Re}\nu<1\right].$$ WA 466(3)

4. $$\int_0^\infty \left\{\cos\left[\left(\frac{1}{2}\varrho-\frac{1}{2}\nu-\mu\right)\pi\right]J_\nu(ax)+\sin\left[\left(\frac{1}{2}\varrho-\frac{1}{2}\nu-\mu\right)\pi\right]N_\nu(ax)\right\}\frac{x^{\varrho-1}}{(x^2+k^2)^{\mu+1}}dx =$$
$$= \frac{\pi k^{\varrho-2\mu-2}}{2\sin\nu\pi\cdot\Gamma(\mu+1)}\left[\frac{\left(\frac{1}{2}ak\right)^\nu\Gamma\left(\frac{1}{2}\varrho+\frac{1}{2}\nu\right)}{\Gamma(\nu+1)\,\Gamma\left(\frac{1}{2}\varrho+\frac{1}{2}\nu-\mu\right)}\times {}_1F_2\left(\frac{\varrho+\nu}{2};\ \frac{\varrho+\nu}{2}-\mu,\ \nu+1;\ \frac{a^2k^2}{4}\right)-\right.$$
$$\left.-\frac{\left(\frac{1}{2}ak\right)^{-\nu}\Gamma\left(\frac{1}{2}\varrho-\frac{1}{2}\nu\right)}{\Gamma(1-\nu)\,\Gamma\left(\frac{1}{2}\varrho-\frac{1}{2}\nu-\mu\right)}{}_1F_2\left(\frac{\varrho-\nu}{2};\ \frac{\varrho-\nu}{2}-\mu,\ 1-\nu;\ \frac{a^2k^2}{4}\right)\right]$$
$$\left[a>0,\ \operatorname{Re}k>0,\ |\operatorname{Re}\nu|<\operatorname{Re}\varrho<2\operatorname{Re}\mu+\frac{7}{2}\right].$$ WA 470(1)

5. $$\int_0^\infty \left[\prod_{j,n} J_{\mu_j}(b_n x)\right] \left\{\cos\left[\frac{1}{2}\left(\varrho + \sum \mu_j - \nu\right)\pi\right] J_\nu(ax) + \sin\left[\frac{1}{2}\left(\varrho + \sum_j \mu_j - \nu\right)\pi\right] N_\nu(ax)\right\} \frac{x^{\varrho-1}}{x^2+k^2}\,dx = -\left[\prod_{j,n} I_{\mu_j}(b_n k)\right] K_\nu(ak)\,k^{\varrho-2}$$

$\left[\operatorname{Re} k > 0,\ a > \sum_n |\operatorname{Re} b_n|,\ \operatorname{Re}\left(\varrho + \sum \mu_j\right) > |\operatorname{Re}\nu|\right].$ WA 472(9)

6.59 Combinations of powers and Bessel functions of more complicated arguments

6.591

1. $$\int_0^\infty x^{2\nu+\frac{1}{2}} J_{\nu+\frac{1}{2}}\left(\frac{a}{x}\right) K_\nu(bx)\,dx = \sqrt{2\pi}\, b^{-\nu-1} a^{\nu+\frac{1}{2}} J_{1+2\nu}\left(\sqrt{2ab}\right) K_{1+2\nu}\left(\sqrt{2ab}\right)$$
$[a > 0,\ \operatorname{Re} b > 0,\ \operatorname{Re}\nu > -1].$ ET II 142(35)

2. $$\int_0^\infty x^{2\nu+\frac{1}{2}} N_{\nu+\frac{1}{2}}\left(\frac{a}{x}\right) K_\nu(bx)\,dx = \sqrt{2\pi}\, b^{-\nu-1} a^{\nu+\frac{1}{2}} N_{2\nu+1}\left(\sqrt{2ab}\right) K_{2\nu+1}\left(\sqrt{2ab}\right)$$
$[a > 0,\ \operatorname{Re} b > 0,\ \operatorname{Re}\nu > -1].$ ET II 143(41)

3. $$\int_0^\infty x^{2\nu+\frac{1}{2}} K_{\nu+\frac{1}{2}}\left(\frac{a}{x}\right) K_\nu(bx)\,dx = \sqrt{2\pi}\, b^{-\nu-1} a^{\nu+\frac{1}{2}} K_{2\nu+1}\left(e^{\frac{1}{4}i\pi}\sqrt{2ab}\right) K_{2\nu+1}\left(e^{-\frac{1}{4}i\pi}\sqrt{2ab}\right)$$
$[\operatorname{Re} a > 0,\ \operatorname{Re} b > 0].$ ET II 146(56)

4. $$\int_0^\infty x^{-2\nu+\frac{1}{2}} J_{\nu-\frac{1}{2}}\left(\frac{a}{x}\right) K_\nu(bx)\,dx = \sqrt{2\pi}\, b^{\nu-1} a^{\frac{1}{2}-\nu} K_{2\nu-1}\left(\sqrt{2ab}\right) \times \left[\sin(\nu\pi) J_{2\nu-1}\left(\sqrt{2ab}\right) + \cos(\nu\pi) N_{2\nu-1}\left(\sqrt{2ab}\right)\right]$$
$[a > 0,\ \operatorname{Re} b > 0,\ \operatorname{Re}\nu < 1].$ ET II 142(34)

5. $$\int_0^\infty x^{-2\nu+\frac{1}{2}} N_{\nu-\frac{1}{2}}\left(\frac{a}{x}\right) K_\nu(bx)\,dx = -\sqrt{\frac{\pi}{2}}\, b^{\nu-1} a^{\frac{1}{2}-\nu} \sec(\nu\pi) K_{2\nu-1}\left(\sqrt{2ab}\right) \times \left[J_{2\nu-1}\left(\sqrt{2ab}\right) - J_{1-2\nu}\left(\sqrt{2ab}\right)\right]$$
$[a > 0,\ \operatorname{Re}\nu < 1].$ ET II 143(40)

6. $$\int_0^\infty x^{-2\nu+\frac{1}{2}} J_{\frac{1}{2}-\nu}\left(\frac{a}{x}\right) J_\nu(bx)\,dx =$$
$$= -\frac{1}{2} i \operatorname{cosec}(2\nu\pi)\, b^{\nu-1} a^{\frac{1}{2}-\nu} [e^{2\nu\pi i} J_{1-2\nu}(u) J_{2\nu-1}(v) - e^{-2\nu\pi i} J_{2\nu-1}(u) J_{1-2\nu}(v)],$$
$$u = \left(\frac{1}{2} ab\right)^{\frac{1}{2}} e^{\frac{1}{4}\pi i}; \quad v = \left(\frac{1}{2} ab\right)^{\frac{1}{2}} e^{-\frac{1}{4}\pi i}$$
$$\left[a > 0,\ b > 0,\ -\frac{1}{2} < \operatorname{Re}\nu < 3\right].$$
ET II 58(12)

7 $$\int_0^\infty x^{-2\nu+\frac{1}{2}} K_{\nu-\frac{1}{2}}\left(\frac{a}{x}\right) N_\nu(bx)\,dx =$$
$$= \sqrt{2\pi}\, b^{\nu-1} a^{\frac{1}{2}-\nu} N_{2\nu-1}(\sqrt{2ab})\, K_{2\nu-1}(\sqrt{2ab})$$
$$\left[b > 0,\ \operatorname{Re} a > 0,\ \operatorname{Re}\nu > \frac{1}{6}\right].$$
ET II 113(30)

8. $$\int_0^\infty x^{\varrho-1} J_\mu(ax) J_\nu\left(\frac{b}{x}\right) dx = \frac{a^{\nu-\varrho} b^\nu \Gamma\left(\frac{1}{2}\mu + \frac{1}{2}\varrho - \frac{1}{2}\nu\right)}{2^{2\nu-\varrho+1}\Gamma(\nu+1)\Gamma\left(\frac{1}{2}\mu + \frac{1}{2}\nu - \frac{1}{2}\varrho + 1\right)} \times$$
$$\times {}_0F_3\left(\nu+1,\ \frac{\nu-\mu-\varrho}{2}+1,\ \frac{\nu+\mu-\varrho}{2}+1;\ \frac{a^2b^2}{16}\right) +$$
$$+ \frac{a^\mu b^{\mu+\varrho}\Gamma\left(\frac{1}{2}\nu - \frac{1}{2}\mu - \frac{1}{2}\varrho\right)}{2^{2\mu+\varrho+1}\Gamma(\mu+1)\Gamma\left(\frac{1}{2}\mu + \frac{1}{2}\nu + \frac{1}{2}\varrho + 1\right)} \times$$
$$\times {}_0F_3\left(\mu+1,\ \frac{\mu-\nu+\varrho}{2}+1,\ \frac{\nu+\mu+\varrho}{2}+1;\ \frac{a^2b^2}{16}\right)$$
$$\left[a > 0,\ b > 0,\ -\operatorname{Re}\left(\mu+\frac{3}{2}\right) < \operatorname{Re}\varrho < \operatorname{Re}\left(\nu+\frac{3}{2}\right)\right].$$
WA 480(1)

6.592

1. $$\int_0^1 x^\lambda (1-x)^{\mu-1} N_\nu(a\sqrt{x})\,dx =$$
$$= 2^{-\nu} a^\nu \operatorname{ctg}(\nu\pi) \frac{\Gamma(\mu)\Gamma\left(\lambda+1+\frac{1}{2}\nu\right)}{\Gamma(1+\nu)\Gamma\left(\lambda+1+\mu+\frac{1}{2}\nu\right)} \times$$
$$\times {}_1F_2\left(\lambda+1+\frac{1}{2}\nu;\ 1+\nu,\ \lambda+1+\mu+\frac{1}{2}\nu;\ -\frac{a^2}{4}\right) -$$
$$- 2^\nu a^{-\nu} \operatorname{cosec}(\nu\pi) \frac{\Gamma(\mu)\Gamma\left(\lambda+1-\frac{1}{2}\nu\right)}{\Gamma(1-\nu)\Gamma\left(\lambda+1+\mu-\frac{1}{2}\nu\right)} \times$$
$$\times {}_1F_2\left(\lambda-\frac{1}{2}\nu+1;\ 1-\nu,\ \lambda+1+\mu-\frac{1}{2}\nu;\ -\frac{a^2}{4}\right)$$
$$\left[\operatorname{Re}\lambda > -1 + \frac{1}{2}|\operatorname{Re}\nu|,\ \operatorname{Re}\mu > 0\right].$$
ET II 197(76)a

2. $$\int_0^1 x^{\lambda}(1-x)^{\mu-1} K_{\nu}(a\sqrt{x})\,dx =$$

$$= 2^{\nu-1}a^{-\nu}\frac{\Gamma(\nu)\Gamma(\mu)\Gamma\left(\lambda+1-\frac{1}{2}\nu\right)}{\Gamma\left(\lambda+1+\mu-\frac{1}{2}\nu\right)}\times$$

$$\times {}_1F_2\left(\lambda+1-\frac{1}{2}\nu;\ 1-\nu,\ \lambda+1+\mu-\frac{1}{2}\nu;\ \frac{a^2}{4}\right)+$$

$$+2^{1-\nu}a^{\nu}\frac{\Gamma(-\nu)\Gamma\left(\lambda+1+\frac{1}{2}\nu\right)\Gamma(\mu)}{\Gamma\left(\lambda+1+\mu+\frac{1}{2}\nu\right)}\times$$

$$\times {}_1F_2\left(\lambda+1+\frac{1}{2}\nu;\ 1+\nu,\ \lambda+1+\mu+\frac{1}{2}\nu;\ \frac{a^2}{4}\right)$$

$$\left[\operatorname{Re}\lambda > -1+\frac{1}{2}|\operatorname{Re}\nu|,\ \operatorname{Re}\mu > 0\right].$$ ET II 198(87)a

3. $$\int_1^{\infty} x^{\lambda}(x-1)^{\mu-1} J_{\nu}(a\sqrt{x})\,dx =$$

$$= 2^{2\lambda}a^{-2\lambda}G_{13}^{20}\left(\frac{a^2}{4}\middle|\begin{matrix}0\\ -\mu,\ \lambda+\frac{1}{2}\nu,\ \lambda-\frac{1}{2}\nu\end{matrix}\right)\Gamma(\mu)$$

$$\left[a>0,\quad 0<\operatorname{Re}\mu<\frac{1}{4}-\operatorname{Re}\lambda\right].$$ ET II 205(36)a

4. $$\int_1^{\infty} x^{\lambda}(x-1)^{\mu-1} K_{\nu}(a\sqrt{x})\,dx =$$

$$= \Gamma(\mu)\,2^{2\lambda-1}a^{-2\lambda}G_{13}^{30}\left(\frac{a^2}{4}\middle|\begin{matrix}0\\ -\mu,\ \frac{1}{2}\nu+\lambda,\ -\frac{1}{2}\nu+\lambda\end{matrix}\right)$$

$$[\operatorname{Re}a>0,\quad \operatorname{Re}\mu>0].$$ ET II 209(60)a

5. $$\int_0^1 x^{-\frac{1}{2}}(1-x)^{-\frac{1}{2}} J_{\nu}(a\sqrt{x})\,dx = \pi\left[J_{\frac{1}{2}\nu}\left(\frac{1}{2}a\right)\right]^2$$

$$[\operatorname{Re}\nu > -1].$$ ET II 194(59)a

6. $$\int_0^1 x^{-\frac{1}{2}}(1-x)^{-\frac{1}{2}} I_{\nu}(a\sqrt{x})\,dx = \pi\left[I_{\frac{1}{2}\nu}\left(\frac{1}{2}a\right)\right]^2$$

$$[\operatorname{Re}\nu > -1].$$ ET II 197(79)

7. $$\int_0^1 x^{-\frac{1}{2}}(1-x)^{-\frac{1}{2}} K_{\nu}(a\sqrt{x})\,dx =$$

$$= \frac{\sqrt{\pi}}{2}\sec(\nu\pi)\left[I_{\frac{\nu}{2}}\left(\frac{a}{2}\right)+I_{-\frac{\nu}{2}}\left(\frac{a}{2}\right)\right]K_{\frac{\nu}{2}}\left(\frac{a}{2}\right)$$

$$[|\operatorname{Re}\nu|<1].$$ ET II 198(85)a

8. $$\int_1^\infty x^{-\frac{1}{2}}(x-1)^{-\frac{1}{2}} K_\nu(a\sqrt{x})\,dx = \left[K_{\frac{\nu}{2}}\left(\frac{a}{2}\right)\right]^2$$
$$[\operatorname{Re} a > 0].$$ ET II 208(56)a

9. $$\int_0^1 x^{-\frac{1}{2}}(1-x)^{-\frac{1}{2}} N_\nu(a\sqrt{x})\,dx =$$
$$= \pi\left\{\operatorname{ctg}(\nu\pi)\left[J_{\frac{\nu}{2}}\left(\frac{a}{2}\right)\right]^2 - \operatorname{cosec}(\nu\pi)\left[J_{-\frac{\nu}{2}}\left(\frac{a}{2}\right)\right]^2\right\}$$
$$[|\operatorname{Re}\nu| < 1].$$ ET II 195(68)a

10. $$\int_1^\infty x^{-\frac{1}{2}\nu}(x-1)^{\mu-1} J_\nu(a\sqrt{x})\,dx = \Gamma(\mu)\, 2^\mu a^{-\mu} J_{\nu-\mu}(a)$$
$$\left[a > 0,\ 0 < \operatorname{Re}\mu < \frac{1}{2}\operatorname{Re}\nu + \frac{3}{4}\right].$$ ET II 205(34)a

11. $$\int_1^\infty x^{-\frac{1}{2}\nu}(x-1)^{\mu-1} J_{-\nu}(a\sqrt{x})\,dx =$$
$$= \Gamma(\mu)\, 2^\mu a^{-\mu}\left[\cos(\nu\pi) J_{\nu-\mu}(a) - \sin(\nu\pi) N_{\nu-\mu}(a)\right]$$
$$\left[a > 0,\ 0 < \operatorname{Re}\mu < \frac{1}{2}\operatorname{Re}\nu + \frac{3}{4}\right].$$ ET II 205(35)a

12. $$\int_1^\infty x^{-\frac{1}{2}\nu}(x-1)^{\mu-1} K_\nu(a\sqrt{x})\,dx = \Gamma(\mu)\, 2^\mu a^{-\mu} K_{\nu-\mu}(a)$$
$$[\operatorname{Re} a > 0, \quad \operatorname{Re}\mu > 0].$$ ET II 209(59)a

13. $$\int_1^\infty x^{-\frac{1}{2}\nu}(x-1)^{\mu-1} N_\nu(a\sqrt{x})\,dx = 2^\mu a^{-\mu} N_{\nu-\mu}(a)\,\Gamma(\mu)$$
$$\left[a > 0,\ 0 < \operatorname{Re}\mu < \frac{1}{2}\operatorname{Re}\nu + \frac{3}{4}\right].$$ ET II 206(40)a

14. $$\int_1^\infty x^{-\frac{1}{2}\nu}(x-1)^{\mu-1} H_\nu^{(1)}(a\sqrt{x})\,dx = 2^\mu a^{-\mu} H_{\nu-\mu}^{(1)}(a)\,\Gamma(\mu)$$
$$[\operatorname{Re}\mu > 0, \quad \operatorname{Im} a > 0].$$ ET II 206(45)a

15. $$\int_1^\infty x^{-\frac{1}{2}\nu}(x-1)^{\mu-1} H_\nu^{(2)}(a\sqrt{x})\,dx = 2^\mu a^{-\mu} H_{\nu-\mu}^{(2)}(a)\,\Gamma(\mu)$$
$$[\operatorname{Re}\mu > 0, \quad \operatorname{Im} a < 0].$$ ET II 207(48)a

16. $$\int_0^1 x^{-\frac{1}{2}\nu}(1-x)^{\mu-1} J_\nu(a\sqrt{x})\,dx =$$
$$= \frac{2^{2-\nu} a^{-\mu}}{\Gamma(\nu)} s_{\mu+\nu-1,\,\mu-\nu}(a) \qquad [\operatorname{Re}\mu > 0].$$ ET II 194(64)a

17. $\int\limits_0^1 x^{-\frac{1}{2}\nu}(1-x)^{\mu-1}N_\nu(a\sqrt{x})\,dx =$

$$= \frac{2^{2-\nu}a^{-\mu}\operatorname{ctg}(\nu\pi)}{\Gamma(\nu)} s_{\mu+\nu-1,\,\mu-\nu}(a) -$$

$$- 2^{\mu}a^{-\mu}\operatorname{cosec}(\nu\pi)\,J_{\mu-\nu}(a)\,\Gamma(\mu)$$

$[\operatorname{Re}\mu > 0, \quad \operatorname{Re}\nu < 1].$ ET II 196(75)a

6.593

1. $\int\limits_0^\infty \sqrt{x}J_{2\nu-1}(a\sqrt{x})\,J_\nu(bx)\,dx = \frac{1}{2}ab^{-2}J_{\nu-1}\left(\frac{a^2}{4b}\right)$

$\left[b > 0, \quad \operatorname{Re}\nu > -\frac{1}{2}\right].$ ET II 58(15)

2. $\int\limits_0^\infty \sqrt{x}J_{2\nu-1}(a\sqrt{x})\,K_\nu(bx)\,dx =$

$$= \frac{\pi a}{4b^2}\left[I_{\nu-1}\left(\frac{a^2}{4b}\right) - \mathbf{L}_{\nu-1}\left(\frac{a^2}{4b}\right)\right]$$

$\left[\operatorname{Re}b > 0, \quad \operatorname{Re}\nu > -\frac{1}{2}\right].$ ET II 144(44)

6.594

1. $\int\limits_0^\infty x^\nu I_{2\nu-1}(a\sqrt{x})J_{2\nu-1}(a\sqrt{x})\,K_\nu(bx)\,dx =$

$$= \sqrt{\pi}\,2^{-\nu}a^{2\nu-1}b^{-2\nu-\frac{1}{2}}J_{\nu-\frac{1}{2}}\left(\frac{a^2}{2b}\right)$$

$[\operatorname{Re}b > 0, \quad \operatorname{Re}\nu > 0].$ ET II 148(65)

2. $\int\limits_0^\infty x^\nu I_{2\nu-1}(a\sqrt{x})\,N_{2\nu-1}(a\sqrt{x})\,K_\nu(bx)\,dx =$

$$= \sqrt{\pi}\,2^{-\nu-1}a^{2\nu-1}b^{-2\nu-\frac{1}{2}}\operatorname{cosec}(\nu\pi)\left[\mathbf{H}_{\frac{1}{2}-\nu}\left(\frac{a^2}{2b}\right) + \right.$$

$$\left. + \cos(\nu\pi)\,J_{\nu-\frac{1}{2}}\left(\frac{a^2}{2b}\right) + \sin(\nu\pi)\,N_{\nu-\frac{1}{2}}\left(\frac{a^2}{2b}\right)\right]$$

$[\operatorname{Re}b > 0, \quad \operatorname{Re}\nu > 0].$ ET II 148(66)

3. $\int\limits_0^\infty x^\nu J_{2\nu-1}(a\sqrt{x})\,K_{2\nu-1}(a\sqrt{x})\,K_\nu(bx)\,dx =$

$$= \pi^2 2^{-\nu-2}a^{2\nu-1}b^{-2\nu-\frac{1}{2}}\operatorname{cosec}(\nu\pi)\left[\mathbf{H}_{\frac{1}{2}-\nu}\left(\frac{a^2}{2b}\right) - N_{\frac{1}{2}-\nu}\left(\frac{a^2}{2b}\right)\right]$$

$[\operatorname{Re}b > 0, \quad \operatorname{Re}\nu > 0].$ ET II 148(67)

6.595

1. $$\int_0^\infty x^{\nu+1} J_\nu(cx) \prod_{i=1}^n z_i^{-\mu_i} J_{\mu_i}(a_i z_i)\, dx = 0,$$

$$z_i = \sqrt{x^2+b_i^2} \qquad \left[a_i > 0, \quad \operatorname{Re} b_i > 0, \quad \sum_{i=1}^n a_i < c;\right.$$

$$\left.\operatorname{Re}\left(\frac{1}{2}n + \sum_{i=1}^n \mu_i - \frac{1}{2}\right) > \operatorname{Re}\nu > -1\right].$$ EH II 52(33), ET II 60(26)

2. $$\int_0^\infty x^{\nu-1} J_\nu(cx) \prod_{i=1}^n z_i^{-\mu_i} J_{\mu_i}(a_i z_i)\, dx = 2^{\nu-1}\Gamma(\nu) c^{-\nu} \prod_{i=1}^n \left[b_i^{-\mu_i} J_{\mu_i}(a_i b_i)\right],$$

$$z_i = \sqrt{x^2+b_i^2} \left[a_i > 0, \quad \operatorname{Re} b_i > 0, \sum_{i=1}^n a_i < c, \operatorname{Re}\left(\frac{1}{2}n + \sum_{i=1}^n \mu_i + \frac{3}{2}\right) > \operatorname{Re}\nu > 0\right]$$

EH II 52(34), ET II 60(27)

6.596

1. $$\int_0^\infty J_\nu\left(\alpha\sqrt{x^2+z^2}\right) \frac{x^{2\mu+1}}{\sqrt{(x^2+z^2)^\nu}}\, dx = \frac{2^\mu \Gamma(\mu+1)}{\alpha^{\mu+1} z^{\nu-\mu-1}} J_{\nu-\mu-1}(\alpha z)$$

$$\left[\alpha > 0, \quad \operatorname{Re}\left(\frac{1}{2}\nu - \frac{1}{4}\right) > \operatorname{Re}\mu > -1\right].$$ WA 457(5)

2. $$\int_0^\infty \frac{J_\nu\left(\alpha\sqrt{t^2+1}\right)}{\sqrt{t^2+1}}\, dt = -\frac{\pi}{2} J_{\frac{\nu}{2}}\left(\frac{\alpha}{2}\right) N_{\frac{\nu}{2}}\left(\frac{\alpha}{2}\right)$$

$$[\operatorname{Re}\nu > -1, \quad \alpha > 0].$$ MO 46

3. $$\int_0^\infty K_\nu\left(\alpha\sqrt{x^2+z^2}\right) \frac{x^{2\mu+1}}{\sqrt{(x^2+z^2)^\nu}}\, dx = \frac{2^\mu \Gamma(\mu+1)}{\alpha^{\mu+1} z^{\nu-\mu-1}} K_{\nu-\mu-1}(\alpha z)$$

$$[\alpha > 0, \quad \operatorname{Re}\mu > -1].$$ WA 457(6)

4. $$\int_0^\infty J_\nu(\beta x) \frac{J_{\mu-1}\left\{\alpha\sqrt{x^2+z^2}\right\}}{(x^2+z^2)\sqrt{(x^2+z^2)^\mu}} x^{\nu+1}\, dx = \frac{\alpha^{\mu-1} z^\nu}{2^{\mu-1}\Gamma(\mu)} K_\nu(\beta z)$$

$$[\alpha < \beta, \quad \operatorname{Re}(\mu+2) > \operatorname{Re}\nu > -1].$$ WA 459(11)a, ET II 59(19)

5. $$\int_0^\infty J_\nu(\beta x) \frac{J_\mu\left\{\alpha\sqrt{x^2+z^2}\right\}}{\sqrt{(x^2+z^2)^\mu}} x^{\nu-1}\, dx = \frac{2^{\nu-1}\Gamma(\nu)}{\beta^\nu} \frac{J_\mu(\alpha z)}{z^\mu}$$

$$[\operatorname{Re}(\mu+2) > \operatorname{Re}\nu > 0, \quad \beta > \alpha > 0].$$ WA 459(12)

6. $$\int_0^\infty J_\nu(\beta x)\frac{J_\mu(\alpha\sqrt{x^2+z^2})}{\sqrt{(x^2+z^2)^\mu}}x^{\nu+1}\,dx=0 \qquad [0<\alpha<\beta];$$
$$=\frac{\beta^\nu}{\alpha^\mu}\left\{\frac{\sqrt{\alpha^2-\beta^2}}{z}\right\}^{\mu-\nu-1}J_{\mu-\nu-1}\{z\sqrt{\alpha^2-\beta^2}\} \qquad [\alpha>\beta>0];$$
$$[\operatorname{Re}\mu>\operatorname{Re}\nu>-1].$$
WA 455(1)

7. $$\int_0^\infty J_\nu(\beta x)\frac{K_\mu(\alpha\sqrt{x^2+z^2})}{\sqrt{(x^2+z^2)^\mu}}x^{\nu+1}\,dx=$$
$$=\frac{\beta^\nu}{\alpha^\mu}\left(\frac{\sqrt{\alpha^2+\beta^2}}{z}\right)^{\mu-\nu-1}K_{\mu-\nu-1}(z\sqrt{\alpha^2+\beta^2})$$
$$\left[\alpha>0,\ \beta>0,\ \operatorname{Re}\nu>-1,\ |\arg z|<\frac{\pi}{2}\right].$$
KU 151(31), WA 456(2)

8. $$\int_0^\infty J_\nu(\beta t)\frac{K_\mu(\alpha\sqrt{t^2-y^2})}{\sqrt{(t^2-y^2)^\mu}}t^{\nu+1}\,dt=\frac{\pi}{2}\frac{\beta^\nu}{\alpha^\mu}\left\{\frac{\sqrt{\alpha^2+\beta^2}}{y}\right\}^{\mu-\nu-1}\times$$
$$\times\exp\left[-\frac{\pi}{2}\left(\mu-\nu-\frac{1}{2}\right)\right]\{J_{\mu-\nu-1}[y\sqrt{\alpha^2+\beta^2}]-iN_{\mu-\nu-1}[y\sqrt{\alpha^2+\beta^2}]\}$$

[$\operatorname{Re}\mu<1$. Here, it is assumed that the integration contour does not contain the singularity $t=y$, which can be excluded by going *upwards* around it, and that the sign of $\sqrt{t^2-y^2}$ is chosen in such a way that the expression in question is positive for $t>y$; $\alpha>0$, $\beta>0$, $y>0$].

9. $$\int_0^\infty J_\nu(ux)K_\mu(v\sqrt{x^2-y^2})(x^2-y^2)^{-\frac{\mu}{2}}x^{\nu+1}\,dx=$$
$$=\frac{\pi}{2}\exp\left[-i\pi\left(\mu-\nu-\frac{1}{2}\right)\right]\cdot\frac{u^\nu}{v^\mu}\cdot\left[\frac{\sqrt{u^2+v^2}}{y}\right]^{\mu-\nu-1}\times$$
$$\times H^{(2)}_{\mu-\nu-1}(y\sqrt{u^2+v^2})$$
$$\left[\operatorname{Re}\mu<1,\ \operatorname{Re}\nu>-1,\ u>0,\ v>0;\ \arg\sqrt{x^2-y^2}=0 \text{ for } x>y;\right.$$
$$\left.\text{if } x<y, \text{ then } \arg(x^2-y^2)^\sigma=\pi\sigma, \text{ where } \sigma=\frac{1}{2} \text{ or } \sigma=-\frac{\mu}{2}\right].$$
MO 43

10. $$\int_0^\infty J_\nu(ux)H^{(2)}_\mu(v\sqrt{x^2+y^2})(x^2+y)^{-\frac{\mu}{2}}x^{\nu+1}\,dx=$$
$$=\frac{u^\nu}{v^\mu}\left[\frac{\sqrt{v^2-u^2}}{y}\right]^{\mu-\nu-1}H^{(2)}_{\mu-\nu-1}(y\sqrt{v^2-u^2}) \quad [u<v]$$
$$\left[\operatorname{Re}\mu<\operatorname{Re}\nu,\ \operatorname{Re}\nu>-1,\ u>0,\ v>0,\ y>0;\ \arg\sqrt{v^2-u^2}=0\right.$$
$$\text{for } v>u,\ \arg(v^2-u^2)^\sigma=-\pi\sigma \text{ for } v<u,$$
$$\left.\text{where } \sigma=\frac{1}{2} \text{ or } \sigma=\frac{\mu-\nu-1}{2}\right].$$
MO 43

11. $$\int_0^\infty J_\nu(\beta x)J_\mu(\alpha\sqrt{x^2+z^2})J_\mu(\gamma\sqrt{x^2+z^2})\frac{x^{\nu-1}}{(x^2+z^2)^\mu}\,dx=$$
$$=\frac{2^{\nu-1}\Gamma(\nu)}{\beta^\nu}\frac{J_\mu(\alpha z)}{z^\mu}\frac{J_\mu(\gamma z)}{z^\mu}$$
$$\left[\alpha>0;\ \beta>\alpha+\gamma;\ \gamma>0,\ \operatorname{Re}\left(2\mu+\frac{5}{2}\right)>\operatorname{Re}\nu>0\right].$$
WA 459(14)

12. $\int_0^\infty J_\nu(\beta t) \prod_{k=1}^{n} J_\mu(\alpha_k \sqrt{t^2+x^2}) \sqrt{(t^2+x^2)^{-n\mu}}\, t^{\nu-1}\, dt =$

$$= 2^{\nu-1}\beta^{-\nu}\Gamma(\nu) \prod_{k=1}^{n} [x^{-\mu} J_\mu(\alpha_k x)]$$

$\left[x > 0,\ \alpha_1 > 0.\ \alpha_2 > 0,\ \ldots,\ \alpha_n > 0,\ \beta > \sum_{k=1}^{n} \alpha_k;\right.$

$\left. \operatorname{Re}\left(n\mu + \frac{1}{2}n + \frac{1}{2}\right) > \operatorname{Re}\nu > 0\right].$ MO 43

13. $\int_0^\infty \frac{J_\nu^2(\sqrt{a^2+x^2})}{(a^2+x^2)^\nu} x^{2\nu-2}\, dx = \frac{\Gamma\left(\nu-\frac{1}{2}\right)}{2a^{\nu+1}\sqrt{\pi}} \mathbf{H}_\nu(2a) \left[\operatorname{Re}\nu > \frac{1}{2}\right].$ WA 457(8)

6.597 $\int_0^\infty t^{\nu+1} J_\mu[b(t^2+y^2)^{\frac{1}{2}}](t^2+y^2)^{-\frac{1}{2}\mu}(t^2+\beta^2)^{-1} J_\nu(at)\, dt =$

$$= \beta^\nu J_\mu[b(y^2-\beta^2)^{\frac{1}{2}}](y^2-\beta^2)^{-\frac{1}{2}\mu} K_\nu(a\beta)$$

$[a \geqslant b,\ \operatorname{Re}\beta > 0,\ -1 < \operatorname{Re}\nu < 2 + \operatorname{Re}\mu].$ EH II 95(56)

6.598 $\int_0^1 x^{\frac{\mu}{2}}(1-x)^{\frac{\nu}{2}} J_\mu(a\sqrt{x}) J_\nu(b\sqrt{1-x})\, dx =$

$$= 2a^\mu b^\nu (a^2+b^2)^{-\frac{1}{2}(\nu+\mu+1)} J_{\nu+\mu+1}\left(\sqrt{a^2+b^2}\right)$$

$[\operatorname{Re}\nu > -1,\ \operatorname{Re}\mu > -1].$ EH II 46a

6.61 Combinations of Bessel functions and exponentials

6.611

1. $\int_0^\infty e^{-\alpha x} J_\nu(\beta x)\, dx = \frac{\beta^{-\nu}[\sqrt{\alpha^2+\beta^2}-\alpha]^\nu}{\sqrt{\alpha^2+\beta^2}}$

$[\operatorname{Re}\nu > -1,\ \operatorname{Re}(\alpha \pm i\beta) > 0].$ EH II 49(18), WA 422(8)

2. $\int_0^\infty e^{-\alpha x} N_\nu(\beta x)\, dx = (\alpha^2+\beta^2)^{-\frac{1}{2}} \operatorname{cosec}(\nu\pi) \times$

$$\times \{\beta^\nu[(\alpha^2+\beta^2)^{\frac{1}{2}}+\alpha]^{-\nu}\cos(\nu\pi) - \beta^{-\nu}[(\alpha^2+\beta^2)^{\frac{1}{2}}+\alpha]^\nu\}$$

$[\operatorname{Re}\alpha > 0,\ \beta > 0,\ |\operatorname{Re}\nu| < 1].$ MO 179, ET II 105(1)

3. $\int_0^\infty e^{-\alpha x} K_\nu(\beta x)\, dx = \frac{\pi}{\beta \sin(\nu\pi)} \frac{\sin(\nu\theta)}{\sin\theta}$

$\left[\cos\theta = \frac{\alpha}{\beta};\ \theta \to \frac{\pi}{2} \text{ for } \beta \to \infty\right];$ ET II 131(22)

$$= \frac{\pi \operatorname{cosec}(\nu\pi)}{2\sqrt{\alpha^2-\beta^2}} \left[\beta^{-\nu}\left(\alpha+\sqrt{\alpha^2-\beta^2}\right)^\nu - \beta^\nu\left(\sqrt{\alpha^2-\beta^2}+\alpha\right)^{-\nu}\right]$$

$[|\operatorname{Re}\nu| < 1,\ \operatorname{Re}(\alpha+\beta) > 0].$ ET I 197(24), MO 180

4. $$\int_0^\infty e^{-\alpha x} I_\nu(\beta x)\,dx = \frac{\beta^\nu}{\sqrt{\alpha^2-\beta^2}\,(\alpha+\sqrt{\alpha^2-\beta^2})^\nu}$$

$[\operatorname{Re}\nu > -1,\ \operatorname{Re}\alpha > |\operatorname{Re}\beta|]$. MO 180, ET I 195(1)

5. $$\int_0^\infty e^{-\alpha x} H_\nu^{(1,2)}(\beta x)\,dx =$$

$$= \frac{(\sqrt{\alpha^2+\beta^2}-\alpha)^\nu}{\beta^\nu \sqrt{\alpha^2+\beta^2}} \left\{1 \pm \frac{i}{\sin(\nu\pi)}\left[\cos(\nu\pi) - \frac{(\alpha+\sqrt{\alpha^2+\beta^2})^{2\nu}}{\beta^{2\nu}}\right]\right\}$$

$[-1 < \operatorname{Re}\nu < 1$; a plus sign corresponds to the function $H_\nu^{(1)}$, a minus sign to the function $H_\nu^{(2)}]$. MO 180, ET I 188(54, 55)

6. $$\int_0^\infty e^{-\alpha x} H_0^{(1)}(\beta x)\,dx = \frac{1}{\sqrt{\alpha^2+\beta^2}}\left\{1 - \frac{2i}{\pi}\ln\left[\frac{\alpha}{\beta} + \sqrt{1+\left(\frac{\alpha}{\beta}\right)^2}\right]\right\}$$

$[\operatorname{Re}\alpha > |\operatorname{Im}\beta|]$. MO 180, ET I 188(52)

7. $$\int_0^\infty e^{-\alpha x} H_0^{(2)}(\beta x)\,dx = \frac{1}{\sqrt{\alpha^2+\beta^2}}\left\{1 + \frac{2i}{\pi}\ln\left[\frac{\alpha}{\beta} + \sqrt{1+\left(\frac{\alpha}{\beta}\right)^2}\right]\right\}$$

$[\operatorname{Re}\alpha > |\operatorname{Im}\beta|]$. MO 180, ET I 188(53)

8. $$\int_0^\infty e^{-\alpha x} N_0(\beta x)\,dx = \frac{-2}{\pi\sqrt{\alpha^2+\beta^2}}\ln\frac{\alpha+\sqrt{\alpha^2+\beta^2}}{\beta}$$

$[\operatorname{Re}\alpha > |\operatorname{Im}\beta|]$. MO 47, ET I 187(44)

9. $$\int_0^\infty e^{-\alpha x} K_0(\beta x)\,dx = \frac{\arccos\frac{\alpha}{\beta}}{\sqrt{\beta^2-\alpha^2}}$$

$[0 < \alpha < \beta,\ \operatorname{Re}(\alpha+\beta) > 0]$; WA 424, ET II 131(22)

$$= \frac{1}{\sqrt{\alpha^2-\beta^2}}\ln\left(\frac{\alpha}{\beta} + \sqrt{\frac{\alpha^2}{\beta^2}-1}\right) \quad [0 \leqslant \beta < \alpha,\ \operatorname{Re}(\alpha+\beta) > 0].$$ MO 48

6.612

1. $$\int_0^\infty e^{-2\alpha x} J_0(x) N_0(x)\,dx = \frac{\boldsymbol{K}\left[\alpha(\alpha^2+1)^{-\frac{1}{2}}\right]}{\pi(\alpha^2+1)^{\frac{1}{2}}}$$

$[\operatorname{Re}\alpha > 0]$. ET II 347(58)

2. $$\int_0^\infty e^{-2\alpha x} I_0(x) K_0(x)\,dx =$$

$$= \frac{1}{2}\boldsymbol{K}\left[(1-\alpha^2)^{\frac{1}{2}}\right] \quad [0 < \alpha < 1];$$

$$= \frac{1}{2\alpha}\boldsymbol{K}\left[\left(1-\frac{1}{\alpha^2}\right)^{\frac{1}{2}}\right] \quad [1 < \alpha < \infty].$$ ET II 370(48)

3. $\int_0^\infty e^{-\alpha x} J_\nu(\beta x) J_\nu(\gamma x)\, dx =$

$$= \frac{1}{\pi\sqrt{\gamma\beta}} Q_{\nu-\frac{1}{2}}\left(\frac{\alpha^2+\beta^2+\gamma^2}{2\beta\gamma}\right)$$

$\left[\operatorname{Re}\alpha > \operatorname{Im}\beta > 0, \quad \gamma > 0, \ \operatorname{Re}\nu > -\frac{1}{2}\right]$. WA 426(2), ET II 50(17)

4. $\int_0^\infty e^{-\alpha x} [J_0(\beta x)]^2\, dx = \frac{2}{\pi\sqrt{\alpha^2+4\beta^2}} \boldsymbol{K}\left(\frac{2\beta}{\sqrt{\alpha^2+4\beta^2}}\right)$. MO 178

5. $\int_0^\infty e^{-2\alpha x} J_1^2(\beta x)\, dx = \frac{(2\alpha^2+\beta^2)\boldsymbol{K}\left(\frac{\beta}{\sqrt{\alpha^2+\beta^2}}\right) - 2(\alpha^2+\beta^2)\boldsymbol{E}\left(\frac{\beta}{\sqrt{\alpha^2+\beta^2}}\right)}{\pi\beta^2\sqrt{\alpha^2+\beta^2}}$.

WA 428(3)

6.613 $\int_0^\infty e^{-x^2} J_{\nu+\frac{1}{2}}\left(\frac{x^2}{2}\right) dx = \frac{\Gamma(\nu+1)}{\sqrt{\pi}} D_{-\nu-1}\left(ze^{\frac{\pi}{4}i}\right) D_{-\nu-1}\left(ze^{-\frac{\pi i}{4}}\right)$

$[\operatorname{Re}\nu > -1]$. MO 122

6.614

1. $\int_0^\infty e^{-\alpha x} J_\nu(\beta\sqrt{x})\, dx =$

$$= \frac{\beta}{4}\sqrt{\frac{\pi}{\alpha^3}} \exp\left(-\frac{\beta^2}{8\alpha}\right)\left[I_{\frac{1}{2}(\nu-1)}\left(\frac{\beta^2}{8\alpha}\right) - I_{\frac{1}{2}(\nu+1)}\left(\frac{\beta^2}{8\alpha}\right)\right].$$ MO 178

2. $\int_0^\infty e^{-\alpha x} N_{2\nu}(2\sqrt{\beta x})\, dx =$

$$= \frac{e^{-\frac{1}{2}\frac{\beta}{\alpha}}}{\sqrt{\alpha\beta}}\left\{\operatorname{ctg}(\nu\pi)\frac{\Gamma(\nu+1)}{\Gamma(2\nu+1)} M_{\frac{1}{2},\nu}\left(\frac{\beta}{\alpha}\right) - \operatorname{cosec}(\nu\pi) W_{\frac{1}{2},\nu}\left(\frac{\beta}{\alpha}\right)\right\}$$

$[\operatorname{Re}\alpha > 0, \ |\operatorname{Re}\nu| < 1]$. ET I 188(50)a

3. $\int_0^\infty e^{-\alpha x} I_{2\nu}(2\sqrt{\beta x})\, dx = \frac{e^{\frac{1}{2}\frac{\beta}{\alpha}}}{\sqrt{\alpha\beta}} \frac{\Gamma(\nu+1)}{\Gamma(2\nu+1)} M_{-\frac{1}{2},\nu}\left(\frac{\beta}{\alpha}\right)$

$[\operatorname{Re}\alpha > 0, \ \operatorname{Re}\nu > -1]$. ET I 197(20)a

4. $\int_0^\infty e^{-\alpha x} K_{2\nu}(2\sqrt{\beta x})\, dx = \frac{e^{\frac{1}{2}\frac{\beta}{\alpha}}}{2\sqrt{\alpha\beta}} \Gamma(\nu+1)\Gamma(1-\nu) W_{-\frac{1}{2},\nu}\left(\frac{\beta}{\alpha}\right)$

$[\operatorname{Re}\alpha > 0, \ |\operatorname{Re}\nu| < 1]$. ET I 199(37)a

5. $\int_0^\infty e^{-\alpha x} K_1(\beta\sqrt{x})\, dx =$

$$= \frac{\beta}{8}\sqrt{\frac{\pi}{\alpha^3}} \exp\left(\frac{\beta^2}{8\alpha}\right)\left[K_1\left(\frac{\beta^2}{8\alpha}\right) - K_0\left(\frac{\beta^2}{8\alpha}\right)\right].$$ MO 181

6.615 $$\int_0^\infty e^{-\alpha x} J_\nu(2\beta\sqrt{x})\, J_\nu(2\gamma\sqrt{x})\, dx = \frac{1}{\alpha} I_\nu\left(\frac{2\beta\gamma}{\alpha}\right) \exp\left(-\frac{\beta^2+\gamma^2}{\alpha}\right)$$

$[\operatorname{Re}\nu > -1]$. MO 178

6.616

1. $$\int_0^\infty e^{-\alpha x} J_0(\beta\sqrt{x^2+2\gamma x})\, dx = \frac{1}{\sqrt{\alpha^2+\beta^2}} \exp[\gamma(\alpha - \sqrt{\alpha^2+\beta^2})].$$ MO 179

2. $$\int_1^\infty e^{-\alpha x} J_0(\beta\sqrt{x^2-1})\, dx = \frac{1}{\sqrt{\alpha^2+\beta^2}} \exp(-\sqrt{\alpha^2+\beta^2}).$$ MO 179

3. $$\int_{-\infty}^\infty e^{itx} H_0^{(1)}(r\sqrt{\alpha^2-t^2})\, dt = -2i \frac{e^{i\alpha\sqrt{r^2+x^2}}}{\sqrt{r^2+x^2}}$$

$[0 \leqslant \arg\sqrt{\alpha^2-t^2} < \pi,\ 0 \leqslant \arg\alpha < \pi;\ r$ and x are real$]$. MO 49

4. $$\int_{-\infty}^\infty e^{-itx} H_0^{(2)}(r\sqrt{\alpha^2-t^2})\, dt = 2i \frac{e^{-i\alpha\sqrt{r^2+x^2}}}{\sqrt{r^2+x^2}}$$

$[-\pi < \arg\sqrt{\alpha^2-t^2} \leqslant 0,\ -\pi < \arg\alpha \leqslant 0,\ r$ and x are real$]$. MO 49

6.617

1. $$\int_0^\infty K_{q-p}(2z \operatorname{sh} x)\, e^{(p+q)x}\, dx = \frac{\pi^2}{4\sin[(p-q)\pi]}[J_p(z) N_q(z) - J_q(z) N_p(z)]$$

$[\operatorname{Re} z > 0,\ -1 < \operatorname{Re}(p-q) < 1]$. MO 44

2. $$\int_0^\infty K_0(2z \operatorname{sh} x)\, e^{-2px}\, dx = -\frac{\pi}{4}\left\{J_p(z)\frac{\partial N_p(z)}{\partial p} - N_p(z)\frac{\partial J_p(z)}{\partial p}\right\}$$

$[\operatorname{Re} z > 0]$. MO 44

6.618

1. $$\int_0^\infty e^{-\alpha x^2} J_\nu(\beta x)\, dx = \frac{\sqrt{\pi}}{2\sqrt{\alpha}} \exp\left(-\frac{\beta^2}{8\alpha}\right) I_{\frac{1}{2}\nu}\left(\frac{\beta^2}{8\alpha}\right)$$

$[\operatorname{Re}\alpha > 0,\ \beta > 0,\ \operatorname{Re}\nu > -1]$. WA 432(5), ET II 29(8)

2. $$\int_0^\infty e^{-\alpha x^2} N_\nu(\beta x)\, dx = -\frac{\sqrt{\pi}}{2\sqrt{\alpha}} \exp\left(-\frac{\beta^2}{8\alpha}\right) \times$$
$$\times \left[\operatorname{tg}\frac{\nu\pi}{2} I_{\frac{1}{2}\nu}\left(\frac{\beta^2}{8\alpha}\right) + \frac{1}{\pi}\sec\left(\frac{\nu\pi}{2}\right) K_{\frac{1}{2}\nu}\left(\frac{\beta^2}{8\alpha}\right)\right]$$

$[\operatorname{Re}\alpha > 0,\ \beta > 0,\ |\operatorname{Re}\nu| < 1]$. WA 432(6), ET II 106(3)

3. $$\int_0^\infty e^{-\alpha x^2} K_\nu(\beta x)\, dx = \frac{1}{4}\sec\left(\frac{\nu\pi}{2}\right) \frac{\sqrt{\pi}}{\sqrt{\alpha}} \exp\left(\frac{\beta^2}{8\alpha}\right) K_{\frac{1}{2}\nu}\left(\frac{\beta^2}{8\alpha}\right)$$

$[\operatorname{Re}\alpha > 0,\ |\operatorname{Re}\nu| < 1]$. EH II 51(28), ET II 132(24)

4. $$\int_0^\infty e^{-\alpha x^2} I_\nu(\beta x)\,dx = \frac{\sqrt{\pi}}{2\sqrt{\alpha}} \exp\left(\frac{\beta^2}{8\alpha}\right) I_{\frac{1}{2}\nu}\left(\frac{\beta^2}{8\alpha}\right)$$
$[\operatorname{Re}\nu > -1,\ \operatorname{Re}\alpha > 0]$. EH II 92(27)

5. $$\int_0^\infty e^{-\alpha x^2} J_\mu(\beta x) J_\nu(\beta x)\,dx = 2^{-\nu-\mu-1} \alpha^{-\frac{\nu+\mu+1}{2}} \beta^{\nu+\mu} \frac{\Gamma\left(\frac{\mu+\nu+1}{2}\right)}{\Gamma(\mu+1)\Gamma(\nu+1)} \times$$
$$\times {}_3F_3\left(\frac{\nu+\mu+1}{2}, \frac{\nu+\mu+2}{2}, \frac{\nu+\mu+1}{2}, \mu+1, \nu+1, \nu+\mu+1; -\frac{\beta^2}{\alpha}\right)$$
$[\operatorname{Re}(\nu+\mu) > -1,\ \operatorname{Re}\alpha > 0]$. EH II 50(21)a

6.62-6.63 Combinations of Bessel functions, exponentials, and powers

6.621

1. $$\int_0^\infty e^{-\alpha x} J_\nu(\beta x) x^{\mu-1}\,dx =$$
$$= \frac{\left(\frac{\beta}{2\alpha}\right)^\nu \Gamma(\nu+\mu)}{\alpha^\mu \Gamma(\nu+1)} F\left(\frac{\nu+\mu}{2}, \frac{\nu+\mu+1}{2}; \nu+1; -\frac{\beta^2}{\alpha^2}\right);$$ WA 421(2)
$$= \frac{\left(\frac{\beta}{2\alpha}\right)^\nu \Gamma(\nu+\mu)}{\alpha^\mu \Gamma(\nu+1)} \left(1+\frac{\beta^2}{\alpha^2}\right)^{\frac{1}{2}-\mu} \times$$
$$\times F\left(\frac{\nu-\mu+1}{2}, \frac{\nu-\mu}{2}+1; \nu+1; -\frac{\beta^2}{\alpha^2}\right);$$ WA 421(3)
$$= \frac{\left(\frac{\beta}{2}\right)^\nu \Gamma(\nu+\mu)}{\sqrt{(\alpha^2+\beta^2)^{\nu+\mu}}\,\Gamma(\nu+1)} F\left(\frac{\nu+\mu}{2}, \frac{1-\mu+\nu}{2}; \nu+1; \frac{\beta^2}{\alpha^2+\beta^2}\right)$$
$[\operatorname{Re}(\nu+\mu) > 0,\ \operatorname{Re}(\alpha+i\beta) > 0,\ \operatorname{Re}(\alpha-i\beta) > 0]$; WA 421(3)
$$= (\alpha^2+\beta^2)^{-\frac{1}{2}\mu} \Gamma(\nu+\mu) P_{\mu-1}^{-\nu}\left[\alpha(\alpha^2+\beta^2)^{-\frac{1}{2}}\right]$$
$[\alpha > 0,\ \beta > 0,\ \operatorname{Re}(\nu+\mu) > 0]$. ET II 29(6)

2. $$\int_0^\infty e^{-\alpha x} N_\nu(\beta x) x^{\mu-1}\,dx =$$
$$= \operatorname{ctg}\nu\pi \frac{\left(\frac{\beta}{2}\right)^\nu \Gamma(\nu+\mu)}{\sqrt{(\alpha^2+\beta^2)^{\nu+\mu}}\,\Gamma(\nu+1)} F\left(\frac{\nu+\mu}{2}, \frac{\nu-\mu+1}{2}; \nu+1; \frac{\beta^2}{\alpha^2+\beta^2}\right) -$$
$$- \operatorname{cosec}\nu\pi \frac{\left(\frac{\beta}{2}\right)^{-\nu} \Gamma(\mu-\nu)}{\sqrt{(\alpha^2+\beta^2)^{\mu-\nu}}\,\Gamma(1-\nu)} F\left(\frac{\mu-\nu}{2}, \frac{1-\nu-\mu}{2}; 1-\nu; \frac{\beta^2}{\alpha^2+\beta^2}\right)$$
$[\operatorname{Re}\mu \geqslant |\operatorname{Re}\nu|,\ \operatorname{Re}(\alpha \pm i\beta) > 0]$; WA 421(4)
$$= -\frac{2}{\pi} \Gamma(\nu+\mu)(\beta^2+\alpha^2)^{-\frac{1}{2}\mu} Q_{\mu-1}^{-\nu}\left[\alpha(\alpha^2+\beta^2)^{-\frac{1}{2}}\right]$$
$[\alpha > 0,\ \beta > 0,\ \operatorname{Re}\mu > |\operatorname{Re}\nu|]$. ET II 105(2)

3. $$\int_0^\infty x^{\mu-1} e^{-\alpha x} K_\nu(\beta x)\, dx =$$

$$= \frac{\sqrt{\pi}\,(2\beta)^\nu}{(\alpha+\beta)^{\mu+\nu}} \frac{\Gamma(\mu+\nu)\,\Gamma(\mu-\nu)}{\Gamma\left(\mu+\frac{1}{2}\right)} F\left(\mu+\nu, \nu+\frac{1}{2};\ \mu+\frac{1}{2};\ \frac{\alpha-\beta}{\alpha+\beta}\right)$$

$[\operatorname{Re}\mu > |\operatorname{Re}\nu|,\ \operatorname{Re}(\alpha+\beta) > 0]$.

ET II 131(23)a, EH II 50(26)

4. $$\int_0^\infty x^{m+1} e^{-\alpha x} J_\nu(\beta x)\, dx = (-1)^{m+1} \beta^{-\nu} \frac{d^{m+1}}{d\alpha^{m+1}} \left[\frac{(\sqrt{\alpha^2+\beta^2}-\alpha)^\nu}{\sqrt{\alpha^2+\beta^2}} \right]$$

$[\beta > 0,\ \operatorname{Re}\nu > -m-2]$. ET II 28(3)

6.622

1. $$\int_0^\infty (J_0(x) - e^{-\alpha x}) \frac{dx}{x} = \ln 2\alpha \quad [\alpha > 0].$$ NT 66(13)

2. $$\int_0^\infty \frac{e^{i(u+x)}}{u+x} J_0(x)\, dx = \frac{\pi}{2} i H_0^{(1)}(u).$$ MO 44

3. $$\int_0^\infty e^{-x \operatorname{ch}\alpha} I_p(x) \frac{dx}{\sqrt{x}} = \sqrt{\frac{2}{\pi}}\, Q_{p-\frac{1}{2}}(\operatorname{ch}\alpha).$$ WA 424(5)

6.623

1. $$\int_0^\infty e^{-\alpha x} J_\nu(\beta x)\, x^\nu\, dx = \frac{(2\beta)^\nu\, \Gamma\left(\nu+\frac{1}{2}\right)}{\sqrt{\pi}\,(\alpha^2+\beta^2)^{\nu+\frac{1}{2}}}$$

$\left[\operatorname{Re}\nu > -\frac{1}{2},\ \operatorname{Re}\alpha > |\operatorname{Im}\beta|\right]$. WA 422(5)

2. $$\int_0^\infty e^{-\alpha x} J_\nu(\beta x)\, x^{\nu+1}\, dx = \frac{2\alpha\,(2\beta)^\nu\, \Gamma\left(\nu+\frac{3}{2}\right)}{\sqrt{\pi}\,(\alpha^2+\beta^2)^{\nu+\frac{3}{2}}}$$

$[\operatorname{Re}\nu > -1,\ \operatorname{Re}\alpha > |\operatorname{Im}\beta|]$. WA 422(6)

3. $$\int_0^\infty e^{-\alpha x} J_\nu(\beta x) \frac{dx}{x} = \frac{(\sqrt{\alpha^2+\beta^2}-\alpha)^\nu}{\nu\beta^\nu}$$

$[\operatorname{Re}\nu > 0;\ \operatorname{Re}\alpha > |\operatorname{Im}\beta|]$ (cf. **6.611** 1.). WA 422(7)

6.624

1. $$\int_0^\infty x e^{-\alpha x} K_0(\beta x)\, dx = \frac{1}{\alpha^2-\beta^2} \left\{ \frac{\alpha}{\sqrt{\alpha^2-\beta^2}} \ln\left[\frac{\alpha}{\beta} + \sqrt{\left(\frac{\alpha}{\beta}\right)^2 - 1}\right] - 1 \right\}.$$

MO 181

2. $$\int_0^\infty \sqrt{x}\, e^{-\alpha x} K_{\pm\frac{1}{2}}(\beta x)\, dx = \sqrt{\frac{\pi}{2\beta}}\, \frac{1}{\alpha+\beta}.$$ MO 181

3. $$\int_0^\infty e^{-tz}\left(z^2-1\right)^{-\frac{1}{2}} K_\mu(t)\, t^\nu\, dt = \frac{\Gamma(\nu-\mu+1)}{\left(z^2-1\right)^{-\frac{1}{2}(\nu+1)}} e^{-i\mu\pi} Q_\nu^\mu(z)$$

$[\operatorname{Re}(\nu \pm \mu) > -1]$. EH II 57(7)

4. $$\int_0^\infty e^{-tz}\left(z^2-1\right)^{-\frac{1}{2}} I_{-\mu}(t)\, t^\nu\, dt = \frac{\Gamma(-\nu-\mu)}{\left(z^2-1\right)^{\frac{1}{2}\nu}} P_\nu^\mu(z) \qquad [\operatorname{Re}(\nu+\mu) < 0].$$

EH II 57(8)

5. $$\int_0^\infty e^{-tz}\left(z^2-1\right)^{-\frac{1}{2}} I_\mu(t)\, t^\nu\, dt = \frac{\Gamma(\nu+\mu+1)}{\left(z^2-1\right)^{-\frac{1}{2}(\nu+1)}} P_\nu^{-\mu}(z) \quad [\operatorname{Re}(\nu+\mu) > -1].$$

EH II 57(9)

6. $$\int_0^\infty e^{-t\cos\theta} J_\mu(t\sin\theta)\, t^\nu\, dt = \Gamma(\nu+\mu+1)\, \mathrm{P}_\nu^{-\mu}(\cos\theta)$$

$$\left[\operatorname{Re}(\nu+\mu) > -1,\ 0 \leqslant \theta < \frac{1}{2}\pi\right].$$ EH II 57(10)

7. $$\int_0^\infty \frac{J_\nu(bx)\, x^\nu}{e^{\pi x}-1}\, dx = \frac{(2b)^\nu\, \Gamma\left(\nu+\frac{1}{2}\right)}{\sqrt{\pi}} \sum_{n=1}^\infty \frac{1}{\left(n^2\pi^2+b^2\right)^{\nu+\frac{1}{2}}}$$

$[\operatorname{Re}\nu > 0,\ |\operatorname{Im} b| < \pi]$. WA 423(9)

6.625

1. $$\int_0^1 x^{\lambda-\nu-1}(1-x)^{\mu-1} e^{\pm i\alpha x} J_\nu(\alpha x)\, dx =$$

$$= \frac{2^{-\nu}\alpha^\nu\, \Gamma(\lambda)\, \Gamma(\mu)}{\Gamma(\lambda+\mu)\, \Gamma(\nu+1)}\, {}_2F_2\left(\lambda, \nu+\frac{1}{2}; \lambda+\mu, 2\nu+1; \pm 2i\alpha\right)$$

$[\operatorname{Re}\lambda > 0,\ \operatorname{Re}\mu > 0]$. ET II 194(58)a

2. $$\int_0^1 x^\nu (1-x)^{\mu-1} e^{\pm i\alpha x} J_\nu(\alpha x)\, dx =$$

$$= \frac{(2\alpha)^\nu\, \Gamma(\mu)\, \Gamma\left(\nu+\frac{1}{2}\right)}{\sqrt{\pi}\, \Gamma(\mu+2\nu+1)}\, {}_1F_1\left(\nu+\frac{1}{2};\ \mu+2\nu+1;\ \pm 2i\alpha\right)$$

$$\left[\operatorname{Re}\mu > 0,\ \operatorname{Re}\nu > -\frac{1}{2}\right].$$ ET II 194(57)a

3 $$\int_0^1 x^\nu (1-x)^{\mu-1} e^{\pm \alpha x} I_\nu(\alpha x)\, dx =$$

$$= \frac{(2\alpha)^\nu\, \Gamma\left(\nu+\frac{1}{2}\right) \Gamma(\mu)}{\sqrt{\pi}\, \Gamma(\mu+2\nu+1)}\, {}_1F_1\left(\nu+\frac{1}{2};\ \mu+2\nu+1;\ \pm 2\alpha\right)$$

$$\left[\operatorname{Re}\mu > 0,\ \operatorname{Re}\nu > -\frac{1}{2}\right].$$ BU 9(16a), ET II 197(77)a

4. $\int_0^1 x^{\lambda-1}(1-x)^{\mu-1} e^{\pm \alpha x} I_\nu(\alpha x)\, dx =$

$$= \frac{\left(\frac{1}{2}\alpha\right)^\nu \Gamma(\lambda+\nu)\Gamma(\mu)}{\Gamma(\nu+1)\Gamma(\lambda+\mu+\nu)} {}_2F_2\left(\nu+\frac{1}{2},\ \lambda+\nu;\ 2\nu+1,\ \mu+\lambda+\nu;\ \pm 2\alpha\right)$$

$[\operatorname{Re}\mu > 0,\ \operatorname{Re}(\lambda+\nu) > 0]$. ET II 197(78)a

5. $\int_0^1 x^{\mu-\varkappa}(1-x)^{2\varkappa-1} I_{\mu-\varkappa}\left(\frac{1}{2}xz\right) e^{-\frac{1}{2}xz}\, dx =$

$$= \frac{\Gamma(2\varkappa)}{\sqrt{\pi}\,\Gamma(1+2\mu)} e^{\frac{z}{2}} z^{-\varkappa-\frac{1}{2}} M_{\varkappa,\mu}(z)$$

$\left[\operatorname{Re}\left(\varkappa-\frac{1}{2}-\mu\right) < 0,\ \operatorname{Re}\varkappa > 0\right]$ BU 129(14a)

6. $\int_1^\infty x^{-\lambda}(x-1)^{\mu-1} e^{-\alpha x} I_\nu(\alpha x)\, dx = \frac{(2\alpha)^\lambda \Gamma(\mu)}{\sqrt{\pi}} G_{23}^{21}\left(2\alpha \middle| \begin{matrix} \frac{1}{2}-\lambda,\ 0 \\ -\mu,\ \nu-\lambda,\ -\nu-\lambda \end{matrix}\right)$

$\left[0 < \operatorname{Re}\mu < \frac{1}{2}+\operatorname{Re}\lambda,\quad \operatorname{Re}\alpha > 0\right]$. ET II 207(50)a

7. $\int_1^\infty x^{-\lambda}(x-1)^{\mu-1} e^{-\alpha x} K_\nu(\alpha x)\, dx =$

$$= \Gamma(\mu)\sqrt{\pi}(2\alpha)^\lambda G_{23}^{30}\left(2\alpha \middle| \begin{matrix} 0,\ \frac{1}{2}-\lambda \\ -\mu,\ \nu-\lambda,\ -\nu-\lambda \end{matrix}\right)$$

$[\operatorname{Re}\mu > 0,\ \operatorname{Re}\alpha > 0]$. ET II 208(55)a

8. $\int_1^\infty x^{-\nu}(x-1)^{\mu-1} e^{-\alpha x} I_\nu(\alpha x)\, dx =$

$$= \frac{(2\alpha)^{\nu-\mu}\Gamma\left(\frac{1}{2}-\mu+\nu\right)\Gamma(\mu)}{\sqrt{\pi}\,\Gamma(1-\mu+2\nu)} {}_1F_1\left(\frac{1}{2}-\mu+\nu;\ 1-\mu+2\nu;\ -2\alpha\right)$$

$\left[0 < \operatorname{Re}\mu < \frac{1}{2}+\operatorname{Re}\nu,\ \operatorname{Re}\alpha > 0\right]$. ET II 207(49)a

9. $\int_1^\infty x^{-\nu}(x-1)^{\mu-1} e^{-\alpha x} K_\nu(\alpha x)\, dx = \sqrt{\pi}\,\Gamma(\mu)(2\alpha)^{-\frac{1}{2}\mu-\frac{1}{2}} e^{-\alpha} W_{-\frac{1}{2}\mu,\ \nu-\frac{1}{2}\mu}(2\alpha)$

$[\operatorname{Re}\mu > 0.\quad \operatorname{Re}\alpha > 0]$. ET II 208(53)a

10. $\int_1^\infty x^{-\mu-\frac{1}{2}}(x-1)^{\mu-1} e^{-\alpha x} K_\nu(\alpha x)\, dx = \sqrt{\pi}\,\Gamma(\mu)(2\alpha)^{-\frac{1}{2}} e^{-\alpha} W_{-\mu,\nu}(2\alpha)$

$[\operatorname{Re}\mu > 0,\qquad \operatorname{Re}\alpha > 0]$. ET II 207(51)a

6.626

1. $$\int_0^\infty x^{\lambda-1} e^{-\alpha x} J_\mu(\beta x) J_\nu(\gamma x)\, dx =$$

$$= \frac{\beta^\mu \gamma^\nu}{\Gamma(\nu+1)} 2^{-\nu-\mu} \alpha^{-\lambda-\mu-\nu} \sum_{m=0}^\infty \frac{\Gamma(\lambda+\mu+\nu+2m)}{m!\,\Gamma(\mu+m+1)} \times$$

$$\times F\left(-m, \ -\mu-m; \ \nu+1; \ \frac{\gamma^2}{\beta^2}\right) \left(-\frac{\beta^2}{4\alpha^2}\right)^m$$

$[\operatorname{Re}(\lambda+\mu+\nu) > 0, \quad \operatorname{Re}(\alpha \pm i\beta \pm i\gamma) > 0]$. EH II 48(15)

2. $$\int_0^\infty e^{-2\alpha x} J_\nu(\beta x) J_\mu(\beta x) x^{\nu+\mu}\, dx =$$

$$= \frac{\Gamma\left(\nu+\mu+\frac{1}{2}\right) \beta^{\nu+\mu}}{\sqrt{\pi^3}} \int_0^{\frac{\pi}{2}} \frac{\cos^{\nu+\mu}\varphi \cos(\nu-\mu)\varphi}{(\alpha^2+\beta^2\cos^2\varphi)^{\nu+\mu}\sqrt{\alpha^2+\beta^2\cos^2\varphi}}\, d\varphi$$

$\left[\operatorname{Re}\alpha > |\operatorname{Im}\beta|, \qquad \operatorname{Re}(\nu+\mu) > -\frac{1}{2}\right]$. WA 427(1)

3. $$\int_0^\infty e^{-2\alpha x} J_0(\beta x) J_1(\beta x)\, x\, dx = \frac{\boldsymbol{K}\left(\frac{\beta}{\sqrt{\alpha^2+\beta^2}}\right) - \boldsymbol{E}\left(\frac{\beta}{\sqrt{\alpha^2+\beta^2}}\right)}{2\pi\beta\sqrt{\alpha^2+\beta^2}}.$$ WA 427(2)

4. $$\int_0^\infty e^{-2\alpha x} I_0(\beta x) I_1(\beta x)\, x\, dx = \frac{1}{2\pi\beta}\left\{\frac{\alpha}{\alpha^2-\beta^2} \boldsymbol{E}\left(\frac{\beta}{\alpha}\right) - \frac{1}{\alpha} \boldsymbol{K}\left(\frac{\beta}{\alpha}\right)\right\}$$

$[\operatorname{Re}\alpha > \operatorname{Re}\beta]$. WA 428(5)

6.627 $$\int_0^\infty \frac{(\sqrt{x})^{-1}}{x+a} e^{-x} K_\nu(x)\, dx = \frac{\pi e^a K_\nu(a)}{\sqrt{a}\cos(\nu\pi)}$$

$\left[|\arg a| < \pi, \quad |\operatorname{Re}\nu| < \frac{1}{2}\right]$. ET II 368(29)

6.628

1. $$\int_0^\infty e^{-x\cos\beta} J_{-\nu}(x\sin\beta) x^\mu\, dx = \Gamma(\mu-\nu+1) \operatorname{P}_\mu^\nu(\cos\beta)$$

$\left[0 < \beta < \frac{\pi}{2}, \quad \operatorname{Re}(\mu-\nu) > -1\right]$. WA 424(3), WH

2. $$\int_0^\infty e^{-x\cos\beta} N_\nu(x\sin\beta) x^\mu\, dx =$$

$$= -\frac{\sin\mu\pi}{\sin(\mu+\nu)\pi} \frac{\Gamma(\mu-\nu+1)}{\pi} [Q_\mu^\nu(\cos\beta + 0\cdot i) e^{\frac{1}{2}\nu\pi i} +$$

$$+ Q_\mu^\nu(\cos\beta - 0\cdot i) e^{-\frac{1}{2}\nu\pi i}]$$

$\left[\operatorname{Re}(\mu+\nu) > -1, \quad 0 < \beta < \frac{\pi}{2}\right]$. WA 424(4)

3. $$\int_0^1 e^{\frac{xu}{2}} (1-x)^{2\nu-1} x^{\mu-\nu} J_{\mu-\nu}\left(\frac{ixu}{2}\right) dx =$$

$$= 2^{2(\nu-\mu)} e^{\frac{\pi}{2}(\mu-\nu)i} \frac{\mathrm{B}(2\nu, 2\mu-2\nu+1)}{\Gamma(\mu-\nu+1)} \frac{e^{\frac{u}{2}}}{u^{\nu+\frac{1}{2}}} M_{\nu,\mu}(u).$$ MO 118a

4. $$\int_0^\infty e^{-x \operatorname{ch} \alpha} I_\nu (x \operatorname{sh} \alpha) x^\mu dx = \Gamma(\nu+\mu+1) P_\mu^{-\nu}(\operatorname{ch} \alpha)$$

$[\operatorname{Re} \mu > -2]$. WA 423(1)

5. $$\int_0^\infty e^{-x \operatorname{ch} \alpha} K_\nu (x \operatorname{sh} \alpha) x^\mu dx = \frac{\sin \mu\pi}{\sin(\nu+\mu)\pi} \Gamma(\mu-\nu+1) Q_\mu^\nu (\operatorname{ch} \alpha)$$

$[\operatorname{Re}(\mu+1) > |\operatorname{Re} \nu|]$. WA 423(2)

6. $$\int_0^\infty e^{-x \operatorname{ch} \alpha} I_\nu (x) x^{\mu-1} dx = \frac{\cos \nu\pi}{\sin(\mu+\nu)\pi} \frac{Q_{\nu-\frac{1}{2}}^{\mu-\frac{1}{2}}(\operatorname{ch} \alpha)}{\sqrt{\frac{\pi}{2}} (\operatorname{sh} \alpha)^{\mu-\frac{1}{2}}}$$

$[\operatorname{Re}(\mu+\nu) > 0, \quad \operatorname{Re}(\operatorname{ch} \alpha) > 1]$. WA 424(6)

7. $$\int_0^\infty e^{-x \operatorname{ch} \alpha} K_\nu (x) x^{\mu-1} dx = \sqrt{\frac{\pi}{2}} \Gamma(\mu-\nu) \Gamma(\mu+\nu) \frac{P_{\nu-\frac{1}{2}}^{\frac{1}{2}-\mu}(\operatorname{ch} \alpha)}{(\operatorname{sh} \alpha)^{\mu-\frac{1}{2}}}$$

$[\operatorname{Re} \mu > |\operatorname{Re} \nu|, \ \operatorname{Re}(\operatorname{ch} \alpha) > -1]$. WA 424(7)

6.629 $$\int_0^\infty (\sqrt{x})^{-1} e^{-x\alpha \cos \varphi \cos \psi} J_\mu (\alpha x \sin \varphi) J_\nu (\alpha x \sin \psi) dx =$$

$$= \Gamma\left(\mu+\nu+\frac{1}{2}\right) \alpha^{-\frac{1}{2}} \mathrm{P}_{\nu-\frac{1}{2}}^{-\mu}(\cos \varphi) \mathrm{P}_{\mu-\frac{1}{2}}^{-\nu}(\cos \psi)$$

$\left[\alpha > 0, \quad 0 < \varphi, \quad \psi < \frac{\pi}{2}, \quad \operatorname{Re}(\mu+\nu) > -\frac{1}{2}\right]$. ET II 50(19)

6.631

1. $$\int_0^\infty x^\mu e^{-\alpha x^2} J_\nu (\beta x) dx = \frac{\beta^\nu \Gamma\left(\frac{1}{2}\nu + \frac{1}{2}\mu + \frac{1}{2}\right)}{2^{\nu+1} \alpha^{\frac{1}{2}(\mu+\nu+1)} \Gamma(\nu+1)} {}_1F_1\left(\frac{\nu+\mu+1}{2}; \ \nu+1; \ -\frac{\beta^2}{4\alpha}\right);$$

BU 8(15)

$$= \frac{\Gamma\left(\frac{1}{2}\nu + \frac{1}{2}\mu + \frac{1}{2}\right)}{\beta \alpha^{\frac{1}{2}\mu} \Gamma(\nu+1)} \exp\left(-\frac{\beta^2}{8\alpha}\right) M_{\frac{1}{2}\mu, \frac{1}{2}\nu}\left(\frac{\beta^2}{4\alpha}\right)$$

$[\operatorname{Re} \alpha > 0, \quad \operatorname{Re}(\mu+\nu) > -1,$

EH II 50(22), ET II 30(14), BU 14(13b)

2. $$\int_0^\infty x^\mu e^{-\alpha x^2} N_\nu(\beta x)\,dx = -\alpha^{-\frac{1}{2}\mu}\beta^{-1}\sec\left(\frac{\nu-\mu}{2}\pi\right)\exp\left(-\frac{\beta^2}{8\alpha}\right)\times$$
$$\times\left\{\frac{\Gamma\left(\frac{1}{2}+\frac{1}{2}\mu+\frac{1}{2}\nu\right)}{\Gamma(1+\nu)}\sin\left(\frac{\nu-\mu}{2}\pi\right)M_{\frac{1}{2}\mu,\frac{1}{2}\nu}\left(\frac{\beta^2}{4\alpha}\right)+\right.$$
$$\left.+W_{\frac{1}{2}\mu,\frac{1}{2}\nu}\left(\frac{\beta^2}{4\alpha}\right)\right\}$$
$$[\operatorname{Re}\alpha>0,\quad \operatorname{Re}\mu>|\operatorname{Re}\nu|-1,\quad \beta>0].$$ ET II 106(4)

3 $$\int_0^\infty x^\mu e^{-\alpha x^2} K_\nu(\beta x)\,dx = \frac{1}{2}\alpha^{-\frac{1}{2}\mu}\beta^{-1}\times$$
$$\times\Gamma\left(\frac{1+\nu+\mu}{2}\right)\Gamma\left(\frac{1-\nu+\mu}{2}\right)\exp\left(\frac{\beta^2}{8\alpha}\right)W_{-\frac{1}{2}\mu,\frac{1}{2}\nu}\left(\frac{\beta^2}{4\alpha}\right)$$
$$[\operatorname{Re}\mu>|\operatorname{Re}\nu|-1].$$ ET II 132(25)

4. $$\int_0^\infty x^{\nu+1} e^{-\alpha x^2} J_\nu(\beta x)\,dx = \frac{\beta^\nu}{(2\alpha)^{\nu+1}}\exp\left(-\frac{\beta^2}{4\alpha}\right)$$
$$[\operatorname{Re}\alpha>0,\quad \operatorname{Re}\nu>-1].$$ WA 43(4), ET II 29(10)

5. $$\int_0^\infty x^{\nu-1} e^{-\alpha x^2} J_\nu(\beta x)\,dx = 2^{\nu-1}\beta^{-\nu}\gamma\left(\nu,\ \frac{\beta^2}{4\alpha}\right)$$
$$[\operatorname{Re}\alpha>0,\quad \operatorname{Re}\nu>0].$$ ET II 30(11)

6. $$\int_0^\infty x^{\nu+1} e^{\pm i\alpha x^2} J_\nu(\beta x)\,dx = \frac{\beta^\nu}{(2\alpha)^{\nu+1}}\exp\left[\pm i\left(\frac{\nu+1}{2}\pi-\frac{\beta^2}{4\alpha}\right)\right]$$
$$\left[\alpha>0,\quad -1<\operatorname{Re}\nu<\frac{1}{2},\quad \beta>0\right].$$ ET II 30(12)

7. $$\int_0^\infty x e^{-\alpha x^2} J_\nu(\beta x)\,dx = \frac{\sqrt{\pi}\,\beta}{8\alpha^{\frac{3}{2}}}\exp\left(-\frac{\beta^2}{8\alpha}\right)\left[I_{\frac{1}{2}\nu-\frac{1}{2}}\left(\frac{\beta^2}{8\alpha}\right)-I_{\frac{1}{2}\nu+\frac{1}{2}}\left(\frac{\beta^2}{8\alpha}\right)\right]$$
$$[\operatorname{Re}\alpha>0,\quad \operatorname{Re}\nu>-2].$$ ET II 29(9)

8. $$\int_0^1 x^{n+1} e^{-\alpha x^2} I_n(2\alpha x)\,dx = \frac{1}{4\alpha}\left[e^\alpha - e^{-\alpha}\sum_{r=-n}^{n} I_r(2\alpha)\right]$$
$$[n=0,\ 1,\ \ldots].$$ ET II 365(8)a

9. $$\int_1^\infty x^{1-n} e^{-\alpha x^2} I_n(2\alpha x)\,dx = \frac{1}{4\alpha}\left[e^\alpha - e^{-\alpha}\sum_{r=1-n}^{n-1} I_r(2\alpha)\right]$$
$$[n=1,\ 2,\ \ldots].$$ ET II 367(20)a

10. $\int_0^\infty e^{-x^2} x^{2n+\mu+1} J_\mu (2x\sqrt{z})\, dx = \frac{n!}{2} e^{-z} z^{\frac{1}{2}\mu} L_n^\mu (z)$

$[n = 0, 1, \ldots; \quad n + \operatorname{Re}\mu > -1]$. BU 135(5)

6.632 $\int_0^\infty x^{-\frac{1}{2}} \exp[-(x^2+a^2-2ax\cos\varphi)^{\frac{1}{2}}] [x^2+a^2-2ax\cos\varphi]^{-\frac{1}{2}} K_\nu(x)\, dx =$

$$= \pi a^{-\frac{1}{2}} \sec(\nu\pi) P_{\nu-\frac{1}{2}}(-\cos\varphi) K_\nu(a)$$

$\left[|\arg a| + |\operatorname{Re}\varphi| < \pi, \quad |\operatorname{Re}\nu| < \frac{1}{2} \right]$. ET II 368(32)

6.633

1. $\int_0^\infty x^{\lambda+1} e^{-\alpha x^2} J_\mu(\beta x) J_\nu(\gamma x)\, dx =$

$$= \frac{\beta^\mu \gamma^\nu \alpha^{-\frac{\mu+\nu+\lambda+2}{2}}}{2^{\nu+\mu+1}\Gamma(\nu+1)} \sum_{m=0}^\infty \frac{\Gamma\left(m+\frac{1}{2}\nu+\frac{1}{2}\mu+\frac{1}{2}\lambda+1\right)}{m!\,\Gamma(m+\mu+1)} \left(-\frac{\beta^2}{4\alpha}\right)^m \times$$

$$\times F\left(-m, \ -\mu-m; \ \nu+1; \ \frac{\gamma^2}{\beta^2}\right)$$

$[\operatorname{Re}\alpha > 0, \quad \operatorname{Re}(\mu+\nu+\lambda) > -2, \quad \beta > 0, \quad \gamma > 0]$.

EH II 49(20)a, ET II 51(24)a

2. $\int_0^\infty e^{-\varrho^2 x^2} J_p(\alpha x) J_p(\beta x)\, x\, dx = \frac{1}{2\varrho^2} \exp\left(-\frac{\alpha^2+\beta^2}{4\varrho^2}\right) I_p\left(\frac{\alpha\beta}{2\varrho^2}\right)$

$[\operatorname{Re} p > -1, \ |\arg\varrho| < \frac{\pi}{4}, \ \alpha > 0, \ \beta > 0]$. KU 146(16)a, WA 433(1)

3. $\int_0^\infty x^{2\nu+1} e^{-\alpha x^2} J_\nu(x) N_\nu(x)\, dx = -\frac{1}{2\sqrt{\pi}} \alpha^{-\frac{3}{2}\nu-\frac{1}{2}} \exp\left(-\frac{1}{2\alpha}\right) W_{\frac{1}{2}\nu, \frac{1}{2}\nu}\left(\frac{1}{\alpha}\right)$

$\left[\operatorname{Re}\alpha > 0, \quad \operatorname{Re}\nu > -\frac{1}{2}\right]$. ET II 347(59)

4. $\int_0^\infty x e^{-\alpha x^2} I_\nu(\beta x) J_\nu(\gamma x)\, dx = \frac{1}{2\alpha} \exp\left(\frac{\beta^2-\gamma^2}{4\alpha}\right) J_\nu\left(\frac{\beta\gamma}{2\alpha}\right)$

$[\operatorname{Re}\alpha > 0, \quad \operatorname{Re}\nu > -1]$. ET II 63(1)

5. $\int_0^\infty x^{\lambda-1} e^{-\alpha x^2} J_\mu(\beta x) J_\nu(\beta x)\, dx =$

$$= 2^{-\nu-\mu-1} \alpha^{-\frac{1}{2}(\nu+\lambda+\mu)} \beta^{\nu+\mu} \frac{\Gamma\left(\frac{1}{2}\lambda+\frac{1}{2}\mu+\frac{1}{2}\nu\right)}{\Gamma(\mu+1)\Gamma(\nu+1)} \times$$

$$\times {}_3F_3\left[\frac{\nu}{2}+\frac{\mu}{2}+\frac{1}{2}, \ \frac{\nu}{2}+\frac{\mu}{2}+1, \ \frac{\nu+\mu+\lambda}{2}; \ \mu+1, \ \nu+1, \ \mu+\nu+1; \ -\frac{\beta^2}{\alpha}\right]$$

$[\operatorname{Re}(\nu+\lambda+\mu) > 0, \ \operatorname{Re}\alpha > 0]$. WA 434, EH II 50(21)

6.634 $$\int_0^\infty xe^{-\frac{x^2}{2a}}\left[I_\nu(x)+I_{-\nu}(x)\right]K_\nu(x)\,dx = ae^aK_\nu(a)$$

$[\operatorname{Re} a>0,\quad -1<\operatorname{Re}\ \nu<1]$. ET II 371(49)

6.635

1. $$\int_0^\infty x^{-1}e^{-\frac{\alpha}{x}}J_\nu(\beta x)\,dx = 2J_\nu\left(\sqrt{2\alpha\beta}\right)K_\nu\left(\sqrt{2\alpha\beta}\right)$$ $[\operatorname{Re}\alpha>0,\ \beta>0]$.

ET II 30(15)

2. $$\int_0^\infty x^{-1}e^{-\frac{\alpha}{x}}N_\nu(\beta x)\,dx = 2N_\nu\left(\sqrt{2\alpha\beta}\right)K_\nu\left(\sqrt{2\alpha\beta}\right)$$ $[\operatorname{Re}\alpha>0,\ \beta>0]$.

ET II 106(5)

3. $$\int_0^\infty x^{-1}e^{-\frac{\alpha}{x}-\beta x}J_\nu(\gamma x)\,dx =$$

$$= 2J_\nu\left\{\sqrt{2\alpha}\left[\sqrt{\beta^2+\gamma^2}-\beta\right]^{\frac{1}{2}}\right\}K_\nu\left\{\sqrt{2\alpha}\left[\sqrt{\beta^2+\gamma^2}+\beta\right]^{\frac{1}{2}}\right\}$$

$[\operatorname{Re}\alpha>0,\ \operatorname{Re}\beta>0,\ \gamma>0]$. ET II 30(16)

6.636 $$\int_0^\infty x^{-\frac{1}{2}}e^{-\alpha\sqrt{x}}J_\nu(\beta x)\,dx =$$

$$= \frac{\sqrt{2}}{\sqrt{\pi\beta}}\Gamma\left(\nu+\frac{1}{2}\right)D_{-\nu-\frac{1}{2}}\left(2^{-\frac{1}{2}}\alpha e^{\frac{1}{4}\pi i}\beta^{-\frac{1}{2}}\right)D_{-\nu-\frac{1}{2}}\left(2^{-\frac{1}{2}}\alpha e^{-\frac{1}{4}\pi i}\beta^{-\frac{1}{2}}\right)$$

$\left[\operatorname{Re}\alpha>0,\quad \beta>0,\ \operatorname{Re}\nu>-\frac{1}{2}\right]$. ET II 30(17)

6.637

1. $$\int_0^\infty (\beta^2+x^2)^{-\frac{1}{2}}\exp\left[-\alpha(\beta^2+x^2)^{\frac{1}{2}}\right]J_\nu(\gamma x)\,dx =$$

$$= I_{\frac{1}{2}\nu}\left\{\frac{1}{2}\beta\left[(\alpha^2+\gamma^2)^{\frac{1}{2}}-\alpha\right]\right\}K_{\frac{1}{2}\nu}\left\{\frac{1}{2}\beta\left[(\alpha^2+\gamma^2)^{\frac{1}{2}}+\alpha\right]\right\}$$

$[\operatorname{Re}\alpha>0,\ \operatorname{Re}\beta>0,\ \gamma>0,\ \operatorname{Re}\nu>-1]$. ET II 31(20)

2. $$\int_0^\infty (\beta^2+x^2)^{-\frac{1}{2}}\exp\left[-\alpha(\beta^2+x^2)^{\frac{1}{2}}\right]N_\nu(\gamma x)\,dx =$$

$$= -\sec\left(\frac{\nu\pi}{2}\right)K_{\frac{1}{2}\nu}\left\{\frac{1}{2}\beta\left[(\alpha^2+\gamma^2)^{\frac{1}{2}}+\alpha\right]\right\}\times$$

$$\times\left(\frac{1}{\pi}K_{\frac{1}{2}\nu}\left\{\frac{1}{2}\beta\left[(\alpha^2+\gamma^2)^{\frac{1}{2}}+\alpha\right]\right\}+\right.$$

$$\left.+\sin\left(\frac{\nu\pi}{2}\right)I_{\frac{1}{2}\nu}\left\{\frac{1}{2}\beta\left[(\alpha^2+\gamma^2)^{\frac{1}{2}}-\alpha\right]\right\}\right)$$

$[\operatorname{Re}\alpha>0,\ \operatorname{Re}\beta>0,\ \gamma>0,\ |\operatorname{Re}\nu|<1]$. ET II 106(6)

3. $\int\limits_0^\infty (x^2+\beta^2)^{-\frac{1}{2}} \exp\left[-\alpha (x^2+\beta^2)^{\frac{1}{2}}\right] K_\nu(\gamma x)\, dx =$

$$= \frac{1}{2} \sec\left(\frac{\nu\pi}{2}\right) K_{\frac{1}{2}\nu} \left\{\frac{1}{2}\beta\left[\alpha+(\alpha^2-\gamma^2)^{\frac{1}{2}}\right]\right\} K_{\frac{1}{2}\nu} \left\{\frac{1}{2}\beta\left[\alpha-(\alpha^2-\gamma^2)^{\frac{1}{2}}\right]\right\}$$

$[\operatorname{Re}\alpha > 0,\ \operatorname{Re}\beta > 0,\ \operatorname{Re}(\gamma+\beta) > 0,\ |\operatorname{Re}\nu| < 1]$. ET II 132(26)

6.64 Combinations of Bessel functions of more complicated arguments, exponentials, and powers

6.641 $\int\limits_0^\infty \sqrt{x}\, e^{-\alpha x} J_{\pm\frac{1}{4}}(x^2)\, dx =$

$$= \frac{\sqrt{\pi\alpha}}{4}\left[\mathbf{H}_{\mp\frac{1}{4}}\left(\frac{\alpha^2}{4}\right) - N_{\mp\frac{1}{4}}\left(\frac{\alpha^2}{4}\right)\right].$$ MI 42

6.642

1. $\int\limits_0^\infty x^{-1} e^{-\alpha x} N_\nu\left(\frac{2}{x}\right) dx = N_\nu(\sqrt{\alpha})\, K_\nu(\sqrt{\alpha}).$ MI 44

2. $\int\limits_0^\infty x^{-1} e^{-\alpha x} H_\nu^{(1,2)}\left(\frac{2}{x}\right) dx = H_\nu^{(1,2)}(\sqrt{\alpha})\, K_\nu(\sqrt{\alpha}).$

MI 44, EH II 91(26)

6.643

1. $\int\limits_0^\infty x^{\mu-\frac{1}{2}} e^{-\alpha x} J_{2\nu}(2\beta\sqrt{x})\, dx = \frac{\Gamma\left(\mu+\nu+\frac{1}{2}\right)}{\beta\Gamma(2\nu+1)} e^{-\frac{\beta^2}{2\alpha}} \alpha^{-\mu} M_{\mu,\nu}\left(\frac{\beta^2}{\alpha}\right)$

$\left[\operatorname{Re}\left(\mu+\nu+\frac{1}{2}\right) > 0\right]$, (cf. **6.631** 1.). BU 14(13a), MI 42a

2. $\int\limits_0^\infty x^{\mu-\frac{1}{2}} e^{-\alpha x} I_{2\nu}(2\beta\sqrt{x})\, dx =$

$$= \frac{\Gamma\left(\mu+\nu+\frac{1}{2}\right)}{\Gamma(2\nu+1)} \beta^{-1} e^{\frac{\beta^2}{2\alpha}} \alpha^{-\mu} M_{-\mu,\nu}\left(\frac{\beta^2}{\alpha}\right)$$

$\left[\operatorname{Re}\left(\mu+\nu+\frac{1}{2}\right) > 0\right]$. MI 45

3. $\int\limits_0^\infty x^{\mu-\frac{1}{2}} e^{-\alpha x} K_{2\nu}(2\beta\sqrt{x})\, dx =$

$$= \frac{\Gamma\left(\mu+\nu+\frac{1}{2}\right)\Gamma\left(\mu-\nu+\frac{1}{2}\right)}{2\beta} e^{\frac{\beta^2}{2\alpha}} \alpha^{-\mu} W_{-\mu,\nu}\left(\frac{\beta^2}{\alpha}\right)$$

$\left[\operatorname{Re}\left(\mu+\nu+\frac{1}{2}\right) > 0\right]$, (cf. **6.631** 3.). MI 47a

4. $$\int_0^\infty x^{n+\frac{1}{2}\nu} e^{-\alpha x} J_\nu(2\beta\sqrt{x})\,dx = n!\beta^\nu e^{-\frac{\beta^2}{\alpha}} \alpha^{-n-\nu-1} L_n^\nu\left(\frac{\beta^2}{\alpha}\right)$$

$[n+\nu > -1]$. MO 178a

5. $$\int_0^\infty x^{-\frac{1}{2}} e^{-\alpha x} N_{2\nu}(\beta\sqrt{x})\,dx =$$

$$= -\sqrt{\frac{\pi}{\alpha}} \frac{\exp\left(-\frac{\beta^2}{8\alpha}\right)}{\cos(\nu\pi)} \left[\sin(\nu\pi) I_\nu\left(\frac{\beta^2}{8\alpha}\right) + \frac{1}{\pi} K_\nu\left(\frac{\beta^2}{8\alpha}\right)\right]$$

$\left[|\operatorname{Re}\nu| < \frac{1}{2}\right]$. MI 44

6. $$\int_0^\infty x^{\frac{1}{2}m} e^{-\alpha x} K_m(2\sqrt{x})\,dx =$$

$$= \frac{\Gamma(m+1)}{2\alpha} \left(\frac{1}{\alpha}\right)^{\frac{1}{2}m-\frac{1}{2}} e^{\frac{1}{2\alpha}} W_{-\frac{1}{2}(m+1),\,-\frac{1}{2}m}\left(\frac{1}{\alpha}\right).$$ MI 48a

6.644 $$\int_0^\infty e^{-\beta x} J_{2\nu}(2a\sqrt{x}) J_\nu(bx)\,dx =$$

$$= \exp\left(-\frac{a^2\beta}{\beta^2+b^2}\right) J_\nu\left(\frac{a^2 b}{\beta^2+b^2}\right) \frac{1}{\sqrt{\beta^2+b^2}}$$

$\left[\operatorname{Re}\beta > 0,\ b > 0,\ \operatorname{Re}\nu > -\frac{1}{2}\right]$. ET II 58(17)

6.645

1. $$\int_1^\infty (x^2-1)^{-\frac{1}{2}} e^{-\alpha x} J_\nu(\beta\sqrt{x^2-1})\,dx =$$

$$= I_{\frac{1}{2}\nu}\left[\frac{1}{2}\left(\sqrt{\alpha^2+\beta^2}-\alpha\right)\right] K_{\frac{1}{2}\nu}\left[\frac{1}{2}\left(\sqrt{\alpha^2+\beta^2}+\alpha\right)\right].$$ MO 179a

2. $$\int_1^\infty (x^2-1)^{\frac{1}{2}\nu} e^{-\alpha x} J_\nu(\beta\sqrt{x^2-1})\,dx =$$

$$= \sqrt{\frac{2}{\pi}} \beta^\nu (\alpha^2+\beta^2)^{-\frac{1}{2}\nu-\frac{1}{4}} K_{\nu+\frac{1}{2}}\left(\sqrt{\alpha^2+\beta^2}\right).$$ MO 179a

6.646

1. $$\int_1^\infty \left(\frac{x-1}{x+1}\right)^{\frac{1}{2}\nu} e^{-\alpha x} J_\nu(\beta\sqrt{x^2-1})\,dx =$$

$$= \frac{\exp\left(-\sqrt{\alpha^2+\beta^2}\right)}{\sqrt{\alpha^2+\beta^2}} \left(\frac{\beta}{\alpha+\sqrt{\alpha^2+\beta^2}}\right)^\nu$$ $[\operatorname{Re}\nu > -1]$.

EF 89(52), MO 179

2. $$\int_1^\infty \left(\frac{x-1}{x+1}\right)^{\frac{1}{2}\nu} e^{-\alpha x} I_\nu\left(\beta\sqrt{x^2-1}\right) dx = \frac{\exp\left(-\sqrt{\alpha^2-\beta^2}\right)}{\sqrt{\alpha^2-\beta^2}}\left(\frac{\beta}{\alpha+\sqrt{\alpha^2-\beta^2}}\right)^\nu \quad [\operatorname{Re}\nu > -1,\ \alpha > \beta].$$ MO 180

3. $$\int_1^\infty \left(\frac{x-1}{x+1}\right)^{\frac{1}{2}\nu} e^{-\alpha x} K_\nu\left(\beta\sqrt{x^2-1}\right) dx = \frac{\pi\exp\left(-\sqrt{\alpha^2-\beta^2}\right)}{2\sqrt{\alpha^2-\beta^2}\sin(\nu\pi)}\left[\left(\frac{\alpha+\sqrt{\alpha^2-\beta^2}}{\beta}\right)^\nu - \left(\frac{\beta}{\alpha+\sqrt{\alpha^2-\beta^2}}\right)^\nu\right]$$
$$[|\operatorname{Re}\nu| < 1,\ \alpha+\beta > 0].$$ ME 39a

6.647

1. $$\int_0^\infty x^{-\lambda-\frac{1}{2}}(\beta+x)^{\lambda-\frac{1}{2}} e^{-\alpha x} K_{2\mu}\left[\sqrt{x(\beta+x)}\right] dx = \frac{1}{\beta} e^{\frac{1}{2}\alpha\beta}\,\Gamma\left(\frac{1}{2}-\lambda+\mu\right)\Gamma\left(\frac{1}{2}-\lambda-\mu\right) W_{\lambda,\mu}(z_1)\, W_{\lambda,\mu}(z_2),$$
$$z_1 = \frac{1}{2}\beta\left(\alpha+\sqrt{\alpha^2-1}\right),$$
$$z_2 = \frac{1}{2}\beta\left(\alpha-\sqrt{\alpha^2-1}\right),$$
$$\left[|\arg\beta| < \pi,\ \operatorname{Re}\alpha > -1,\ \operatorname{Re}\lambda + |\operatorname{Re}\mu| < \frac{1}{2}\right].$$ ET II 377(37)

2. $$\int_0^\infty (\alpha+x)^{-\frac{1}{2}} x^{-\frac{1}{2}} e^{-x\operatorname{ch} t} K_\nu\left[\sqrt{x(\alpha+x)}\right] dx = \frac{1}{2}\sec\left(\frac{\nu\pi}{2}\right) e^{\frac{1}{2}\alpha\operatorname{ch} t} K_{\frac{1}{2}\nu}\left(\frac{1}{4}\alpha e^t\right) K_{\frac{1}{2}\nu}\left(\frac{1}{4}\alpha e^{-t}\right)$$
$$[-1 < \operatorname{Re}\nu < 1].$$ ET II 377(36)

3. $$\int_0^\alpha x^{\lambda-\frac{1}{2}}(\alpha-x)^{-\lambda-\frac{1}{2}} e^{-x\operatorname{sh} t} I_{2\mu}\left[\sqrt{x(\alpha-x)}\right] dx = \frac{2\Gamma\left(\frac{1}{2}+\lambda+\mu\right)\Gamma\left(\frac{1}{2}-\lambda+\mu\right)}{\alpha\left[\Gamma(2\mu+1)\right]^2} M_{\lambda,\mu}\left(\frac{1}{2}\alpha e^t\right) M_{-\lambda,\mu}\left(\frac{1}{2}\alpha e^{-t}\right)$$
$$\left[\operatorname{Re}\mu > |\operatorname{Re}\lambda| - \frac{1}{2}\right].$$ ET II 377(32)

6.648 $$\int_{-\infty}^\infty e^{\varrho x}\left(\frac{\alpha+\beta e^x}{\alpha e^x+\beta}\right) K_{2\nu}\left[(\alpha^2+\beta^2+2\alpha\beta\operatorname{ch} x)^{\frac{1}{2}}\right] dx = 2K_{\nu+\varrho}(\alpha)\, K_{\nu-\varrho}(\beta)$$
$$[\operatorname{Re}\alpha > 0,\ \operatorname{Re}\beta > 0].$$ ET II 379(45)

6.649

1. $$\int_0^\infty K_{\mu-\nu}(2z\operatorname{sh} x)\, e^{(\nu+\mu)x} dx = \frac{\pi^2}{4\sin[(\nu-\mu)\pi]}\left[J_\nu(z) N_\mu(z) - J_\mu(z) N_\nu(z)\right]$$
$$[\operatorname{Re} z > 0,\ -1 < \operatorname{Re}(\nu-\mu) < 1].$$ MO 44

2. $$\int_0^\infty J_{\nu+\mu}(2x \operatorname{sh} t)\, e^{(\nu-\mu)t}\, dt = K_\nu(x)\, I_\mu(x)$$

$$\left[\operatorname{Re}(\nu-\mu) < \frac{3}{2},\ \operatorname{Re}(\nu+\mu) > -1,\ x > 0\right].$$ EH II 97(68)

3. $$\int_0^\infty N_{\nu-\mu}(2x \operatorname{sh} t)\, e^{-(\nu+\mu)t}\, dt =$$

$$= \frac{1}{\sin[\pi(\mu-\nu)]} \{I_\mu(x)\, K_\nu(x) - \cos[(\nu-\mu)\pi]\, I_\nu(x)\, K_\mu(x)\}$$

$$\left[|\operatorname{Re}(\nu-\mu)| < 1,\ \operatorname{Re}(\nu+\mu) > -\frac{1}{2},\ x > 0\right].$$ EH II 97(73)

4. $$\int_0^\infty K_0(2z \operatorname{sh} x)\, e^{-2\nu x}\, dx = -\frac{\pi}{4}\left\{J_\nu(z)\frac{\partial N_\nu(z)}{\partial \nu} - N_\nu(z)\frac{\partial J_\nu(z)}{\partial \nu}\right\}.$$

6.65 Combinations of Bessel and exponential functions of more complicated arguments and powers

6.651

1. $$\int_0^\infty x^{\lambda+\frac{1}{2}} e^{-\frac{1}{4}\alpha^2x^2} I_\mu\left(\frac{1}{4}\alpha^2x^2\right) J_\nu(\beta x)\, dx =$$

$$= \frac{1}{\sqrt{2\pi}} 2^{\lambda+1} \beta^{-\lambda-\frac{3}{2}} G_{23}^{21}\left(\frac{\beta^2}{2\alpha^2}\,\middle|\, \begin{matrix} 1-\mu,\ 1+\mu \\ h,\ \frac{1}{2},\ k \end{matrix}\right),$$

$$h = \frac{3}{4} + \frac{1}{2}\lambda + \frac{1}{2}\nu,$$

$$k = \frac{3}{4} + \frac{1}{2}\lambda - \frac{1}{2}\nu$$

$$\left[|\arg\alpha| < \frac{\pi}{4},\ \beta > 0,\ -\frac{3}{2} - \operatorname{Re}(2\mu+\nu) < \operatorname{Re}\lambda < 0\right].$$ ET II 68(8)

2. $$\int_0^\infty x^{\lambda+\frac{1}{2}} e^{-\frac{1}{4}\alpha^2x^2} K_\mu\left(\frac{1}{4}\alpha^2x^2\right) J_\nu(\beta x)\, dx =$$

$$= \sqrt{\frac{\pi}{2}}\, 2^{\lambda+1} \beta^{-\lambda-\frac{3}{2}} G_{23}^{12}\left(\frac{\beta^2}{2\alpha^2}\,\middle|\, \begin{matrix} 1-\mu,\ 1+\mu \\ h,\ \frac{1}{2},\ k \end{matrix}\right),$$

$$h = \frac{3}{4} + \frac{1}{2}\lambda + \frac{1}{2}\nu,$$

$$k = \frac{3}{4} + \frac{1}{2}\lambda - \frac{1}{2}\nu$$

$$\left[|\arg\alpha| < \frac{\pi}{4},\ \operatorname{Re}(\lambda+\nu\pm 2\mu) > -\frac{3}{2}\right].$$ ET II 69(15)

3. $$\int_0^\infty x^{2\mu-\nu+1}e^{-\frac{1}{4}\alpha x^2} I_\mu\left(\frac{1}{4}\alpha x^2\right) J_\nu(\beta x)\,dx =$$
$$= 2^{2\mu-\nu+\frac{1}{2}}(\pi\alpha)^{-\frac{1}{2}}\Gamma\left(\frac{1}{2}+\mu\right)\frac{\beta^{\nu-2\mu-1}}{\Gamma\left(\frac{1}{2}-\mu+\nu\right)}\times$$
$$\times {}_1F_1\left(\frac{1}{2}+\mu;\ \frac{1}{2}-\mu+\nu;\ -\frac{\beta^2}{2\alpha}\right)$$
$$\left[\operatorname{Re}\alpha > 0,\ \beta > 0,\ \operatorname{Re}\nu > 2\operatorname{Re}\mu + \frac{1}{2} > -\frac{1}{2}\right].$$ ET II 68(6)

4. $$\int_0^\infty x^{2\mu+\nu+1}e^{-\frac{1}{4}\alpha^2 x^2} K_\mu\left(\frac{1}{4}\alpha^2 x^2\right) J_\nu(\beta x)\,dx =$$
$$= \sqrt{\pi}\, 2^\mu \alpha^{-2\mu-2\nu-2}\beta^\nu \frac{\Gamma(1+2\mu+\nu)}{\Gamma\left(\mu+\nu+\frac{3}{2}\right)}\times$$
$$\times {}_1F_1\left(1+2\mu+\nu;\ \mu+\nu+\frac{3}{2};\ -\frac{\beta^2}{2\alpha^2}\right)$$
$$\left[|\arg\alpha| < \frac{1}{4}\pi,\ \operatorname{Re}\nu > -1,\ \operatorname{Re}(2\mu+\nu) > -1,\ \beta > 0\right].$$ ET II 69(13)

5. $$\int_0^\infty x^{2\mu+\nu+1}e^{-\frac{1}{2}\alpha x^2} I_\mu\left(\frac{1}{2}\alpha x^2\right) K_\nu(\beta x)\,dx =$$
$$= \frac{2^{\mu-\frac{1}{2}}}{\sqrt{\pi}}\beta^{-\mu-\frac{3}{2}}\alpha^{-\frac{1}{2}\mu-\frac{1}{2}\nu-\frac{1}{4}}\Gamma(2\mu+\nu+1)\,\Gamma\left(\mu+\frac{1}{2}\right)\exp\left(\frac{\beta^2}{8\alpha}\right)W_{k,m}\left(\frac{\beta^2}{4\alpha}\right),$$
$$2k = -3\mu-\nu-\frac{1}{2},$$
$$2m = \mu+\nu+\frac{1}{2}$$
$$\left[\operatorname{Re}\alpha > 0,\ \operatorname{Re}\mu > -\frac{1}{2},\ \operatorname{Re}(2\mu+\nu) > -1\right].$$ ET II 146(53)

6. $$\int_0^\infty xe^{-\frac{1}{4}\alpha x^2} J_{\frac{1}{2}\nu}\left(\frac{1}{4}\beta x^2\right) J_\nu(\gamma x)\,dx =$$
$$= 2(\alpha^2+\beta^2)^{-\frac{1}{2}}\exp\left(-\frac{\alpha\gamma^2}{\alpha^2+\beta^2}\right)J_{\frac{1}{2}\nu}\left(\frac{\beta\gamma^2}{\alpha^2+\beta^2}\right)$$
$$[\gamma > 0,\ \operatorname{Re}\alpha > |\operatorname{Im}\beta|,\ \operatorname{Re}\nu > -1].$$ ET II 56(2)

7. $$\int_0^\infty xe^{-\frac{1}{4}\alpha x^2} I_{\frac{1}{2}\nu}\left(\frac{1}{4}\alpha x^2\right) J_\nu(\beta x)\,dx = \left(\frac{1}{2}\pi\alpha\right)^{-\frac{1}{2}}\beta^{-1}\exp\left(-\frac{\beta^2}{2\alpha}\right)$$
$$[\operatorname{Re}\alpha > 0,\ \beta > 0,\ \operatorname{Re}\nu > -1].$$ ET II 67(3)

8. $$\int_0^\infty x^{1-\nu}e^{-\frac{1}{4}\alpha^2 x^2} I_\nu\left(\frac{1}{4}\alpha^2 x^2\right) J_\nu(\beta x)\,dx =$$
$$= \sqrt{\frac{2}{\pi}}\frac{\beta^{\nu-1}}{\alpha}\exp\left(-\frac{\beta^2}{4\alpha^2}\right)D_{-2\nu}\left(\frac{\beta}{\alpha}\right)$$
$$\left[|\arg\alpha| < \frac{1}{4}\pi,\ \beta > 0,\ \operatorname{Re}\nu > -\frac{1}{2}\right].$$ ET II 67(1)

9. $$\int_0^\infty x^{-\nu-1}e^{-\frac{1}{4}\alpha^2x^2}I_{\nu+1}\left(\frac{1}{4}\alpha^2x^2\right)J_\nu(\beta x)\,dx = \sqrt{\frac{2}{\pi}}\beta^\nu\exp\left(-\frac{\beta^2}{4\alpha^2}\right)D_{-2\nu-3}\left(\frac{\beta}{\alpha}\right)$$

$$\left[|\arg\alpha| < \frac{1}{4}\pi,\ \operatorname{Re}\nu > -1,\ \beta > 0\right].$$ ET II 67(2)

6.652 $$\int_0^\infty x^{2\nu}e^{-\left(\frac{x^2}{8}+\alpha x\right)}I_\nu\left(\frac{x^2}{8}\right)dx = \frac{\Gamma(4\nu+1)}{2^{4\nu}\Gamma(\nu+1)}\frac{e^{\frac{\alpha^2}{2}}}{\alpha^{\nu+1}}W_{-\frac{3}{2}\nu,\frac{1}{2}\nu}(\alpha^2)$$

$$\left[\operatorname{Re}\left(\nu+\frac{1}{4}\right) > 0\right].$$ MI 45

6.653

1. $$\int_0^\infty \exp\left[-\frac{1}{2}x - \frac{1}{2x}(a^2+b^2)\right]I_\nu\left(\frac{ab}{x}\right)\frac{dx}{x} =$$

$$= 2I_\nu(a)K_\nu(b) \qquad [0 < a < b];$$
$$= 2K_\nu(a)I_\nu(b) \qquad [0 < b < a]$$

$$[\operatorname{Re}\nu > -1].$$ WA 482(2)a, EH II 53(37), WA 482(3)a

2. $$\int_0^\infty \exp\left[-\frac{1}{2}x - \frac{1}{2x}(z^2+w^2)\right]K_\nu\left(\frac{zw}{x}\right)\frac{dx}{x} = 2K_\nu(z)K_\nu(w)$$

$$\left[|\arg z| < \pi,\ |\arg w| < \pi,\ |\arg(z+w)| < \frac{1}{4}\pi\right].$$ WA 483(1), EH II 53(36)

6.654 $$\int_0^\infty x^{-\frac{1}{2}}e^{-\frac{\beta^2}{8x}-\alpha x}K_\nu\left(\frac{\beta^2}{8x}\right)dx = \sqrt{4\pi}\alpha^{-\frac{1}{2}}K_{2\nu}(\beta\sqrt{\alpha}).$$ ME 39

6.655 $$\int_0^\infty x(\beta^2+x^2)^{-\frac{1}{2}}\exp\left(-\frac{\alpha^2\beta}{\beta^2+x^2}\right)J_\nu\left(\frac{\alpha^2 x}{\beta^2+x^2}\right)J_\nu(\gamma x)\,dx = \gamma^{-1}e^{-\beta\gamma}J_{2\nu}(2\alpha\sqrt{\gamma})$$

$$\left[\operatorname{Re}\beta > 0,\ \gamma > 0,\ \operatorname{Re}\nu > -\frac{1}{2}\right].$$ ET II 58(14)

6.656

1. $$\int_0^\infty e^{-(\xi-z)\operatorname{ch}t}J_{2\nu}\left[2(z\xi)^{\frac{1}{2}}\operatorname{sh}t\right]dt = I_\nu(z)K_\nu(\xi)$$

$$\left[\operatorname{Re}\nu > -\frac{1}{2},\ \operatorname{Re}(\xi-z) > 0\right].$$ EH II 98(78)

2. $$\int_0^\infty e^{-(\xi+z)\operatorname{ch}t}K_{2\nu}\left[2(z\xi)^{\frac{1}{2}}\operatorname{sh}t\right]dt = \frac{1}{2}K_\nu(z)K_\nu(\xi)\sec(\nu\pi)$$

$$\left[|\operatorname{Re}\nu| < \frac{1}{2},\ \operatorname{Re}(z^{\frac{1}{2}}+\xi^{\frac{1}{2}})^2 \geqslant 0\right].$$ EH II 98(79)

6.66 Combinations of Bessel, hyperbolic, and exponential functions

Bessel and hyperbolic functions

6.661

1. $$\int_0^\infty \operatorname{sh}(ax) K_\nu(bx)\,dx = \frac{\pi}{2} \frac{\operatorname{cosec}\left(\frac{\nu\pi}{2}\right) \sin\left[\nu \arcsin\left(\frac{a}{b}\right)\right]}{\sqrt{b^2-a^2}}$$

$[\operatorname{Re} b > |\operatorname{Re} a|,\ |\operatorname{Re} \nu| < 2]$. ET II 133(32)

2. $$\int_0^\infty \operatorname{ch}(ax) K_\nu(bx)\,dx = \frac{\pi \cos\left[\nu \arcsin\left(\frac{a}{b}\right)\right]}{2\sqrt{b^2-a^2}\cos\left(\frac{\nu\pi}{2}\right)}$$

$[\operatorname{Re} b > |\operatorname{Re} a|,\ |\operatorname{Re} \nu| < 1]$. ET II 134(33)

6.662

1. $$\int_0^\infty \operatorname{ch}(\beta x) K_0(\alpha x) J_0(\gamma x)\,dx = \frac{\boldsymbol{K}(k)}{\sqrt{u+v}},$$

$$u = \frac{1}{2}\{[(\alpha^2+\beta^2+\gamma^2)^2 - 4\alpha^2\beta^2]^{\frac{1}{2}} + \alpha^2 - \beta^2 - \gamma^2\},$$

$$v = \frac{1}{2}\{[(\alpha^2+\beta^2+\gamma^2)^2 - 4\alpha^2\beta^2]^{\frac{1}{2}} - \alpha^2 + \beta^2 + \gamma^2\},$$

$$k^2 = v(u+v)^{-1}$$

$[\operatorname{Re} \alpha > |\operatorname{Re} \beta|,\ \gamma > 0]$. ET II 15(23)

2. $$\int_0^\infty \operatorname{sh}(\beta x) K_1(\alpha x) J_0(\gamma x)\,dx =$$

$$= \alpha^{-1}\left[u\boldsymbol{E}(k) - \boldsymbol{K}(k) E(u) + \frac{\boldsymbol{K}(k) \operatorname{sn} u \operatorname{dn} u}{\operatorname{cn} u}\right],$$

$$\operatorname{cn}^2 u = 2\gamma^2\{[(\alpha^2+\beta^2+\gamma^2)^2 - 4\alpha^2\beta^2]^{\frac{1}{2}} - \alpha^2 + \beta^2 + \gamma^2\}^{-1},$$

$$k^2 = \frac{1}{2}\{1 - (\alpha^2-\beta^2-\gamma^2)[(\alpha^2+\beta^2+\gamma^2)^2 - 4\alpha^2\beta^2]^{-\frac{1}{2}}\}$$

$[\operatorname{Re} \alpha > |\operatorname{Re} \beta|,\ \gamma > 0]$. ET II 15(24)

6.663

1. $$\int_0^\infty K_{\nu\pm\mu}(2z \operatorname{ch} t) \operatorname{ch}[(\mu \mp \nu) t]\,dt = \frac{1}{2} K_\mu(z) K_\nu(z)$$

$[\operatorname{Re} z > 0]$. WA 484(1), EH II 54(39)

2. $$\int_0^\infty N_{\mu+\nu}(2z \operatorname{ch} t) \operatorname{ch}[(\mu-\nu) t]\,dt = \frac{\pi}{4}[J_\mu(z) J_\nu(z) - N_\mu(z) N_\nu(z)]$$

$[z > 0]$. EH II 96(64)

3. $\int_0^\infty J_{\mu+\nu}(2z \operatorname{ch} t) \operatorname{ch}[(\mu-\nu)t]\,dt = -\frac{\pi}{4}[J_\mu(z)N_\nu(z) + J_\nu(z)N_\mu(z)]$

$[z > 0].$ EH II 97(65)

4. $\int_0^\infty J_{\mu+\nu}(2z \operatorname{sh} t) \operatorname{ch}[(\mu-\nu)t]\,dt = \frac{1}{2}[I_\nu(z)K_\mu(z) + I_\mu(z)K_\nu(z)]$

$\left[\operatorname{Re}(\nu+\mu) > -1,\ |\operatorname{Re}(\mu-\nu)| < \frac{3}{2},\ z > 0\right].$ EH II 97(71)

5. $\int_0^\infty J_{\mu+\nu}(2z \operatorname{sh} t) \operatorname{sh}[(\mu-\nu)t]\,dt = \frac{1}{2}[I_\nu(z)K_\mu(z) - I_\mu(z)K_\nu(z)]$

$\left[\operatorname{Re}(\nu+\mu) > -1,\ |\operatorname{Re}(\mu-\nu)| < \frac{3}{2},\ z > 0\right].$ EH II 97(72)

6.664

1. $\int_0^\infty J_0(2z \operatorname{sh} t) \operatorname{sh}(2\nu t)\,dt = \frac{\sin(\nu\pi)}{\pi}[K_\nu(z)]^2$

$\left[|\operatorname{Re}\nu| < \frac{3}{4},\ z > 0\right].$ EH II 97(69)

2. $\int_0^\infty N_0(2z \operatorname{sh} t) \operatorname{ch}(2\nu t)\,dt = -\frac{\cos(\nu\pi)}{\pi}[K_\nu(z)]^2$

$\left[|\operatorname{Re}\nu| < \frac{3}{4},\ z > 0\right].$ EH II 97(70)

3. $\int_0^\infty N_0(2z \operatorname{sh} t) \operatorname{sh}(2\nu t)\,dt =$

$= \frac{1}{\pi}\left[I_\nu(z)\frac{\partial K_\nu(z)}{\partial\nu} - K_\nu(z)\frac{\partial I_\nu(z)}{\partial\nu}\right] - \frac{1}{\pi}\cos(\nu\pi)[K_\nu(z)]^2$

$\left[|\operatorname{Re}\nu| < \frac{3}{4},\ z > 0\right].$ EH II 97(75)

4. $\int_0^\infty K_0(2z \operatorname{sh} t) \operatorname{ch} 2\nu t\,dt = \frac{\pi^2}{8}\{J_\nu^2(z) + N_\nu^2(z)\}$ $[\operatorname{Re} z > 0].$ MO 44

5. $\int_0^\infty K_{2\mu}(z \operatorname{sh} 2t) \operatorname{cth}^{2\nu} t\,dt =$

$= \frac{1}{4z}\Gamma\left(\frac{1}{2}+\mu-\nu\right)\Gamma\left(\frac{1}{2}-\mu-\nu\right)W_{\nu,\mu}(iz)W_{\nu,\mu}(-iz)$

$\left[|\arg z| \leqslant \frac{\pi}{2},\ |\operatorname{Re}\mu| + \operatorname{Re}\nu < \frac{1}{2}\right].$ MO 119

6. $\int_0^\infty \operatorname{ch}(2\mu x) K_{2\nu}(2a \operatorname{ch} x)\,dx = \frac{1}{2}K_{\mu+\nu}(a)K_{\mu-\nu}(a)$ $[\operatorname{Re} a > 0].$

ET II 378(42)

6.665 $\int_0^\infty \operatorname{sech} x \operatorname{ch}(2\lambda x) I_{2\mu}(a \operatorname{sech} x)\,dx =$

$$= \frac{\Gamma\left(\frac{1}{2}+\lambda+\mu\right)\Gamma\left(\frac{1}{2}-\lambda+\mu\right)}{2a\,[\Gamma(2\mu+1)]^2} M_{\lambda,\,\mu}(a) M_{-\lambda,\,\mu}(a)$$

$$\left[|\operatorname{Re}\lambda| - \operatorname{Re}\mu < \frac{1}{2}\right].$$ ET II 378(43)

Bessel, hyperbolic, and algebraic functions

6.666 $\int_0^\infty x^{\nu+1} \operatorname{sh}(\alpha x) \operatorname{cosech} \pi x\, J_\nu(\beta x)\,dx =$

$$= \frac{2}{\pi} \sum_{n=1}^{\infty} (-1)^{n-1} n^{\nu+1} \sin(n\alpha) K_\nu(n\beta)$$

$$[|\operatorname{Re}\alpha| < \pi, \ \operatorname{Re}\nu > -1].$$ ET II 41(3), WA 469(12)

6.667

1. $\int_0^a y^{-1} \operatorname{ch}(y \operatorname{sh} t) I_{2\nu}(x)\,dx = \frac{\pi}{2} I_\nu(ae^t) I_\nu(ae^{-t}),$

$$y = (a^2 - x^2)^{\frac{1}{2}} \quad \left[\operatorname{Re}\nu > -\frac{1}{2}\right].$$ ET II 365(10)

2. $\int_0^a y^{-1} \operatorname{ch}(y \operatorname{sh} t) K_{2\nu}(x)\,dx =$

$$= \frac{\pi^2}{4} \operatorname{cosec}(\nu\pi)\,[I_{-\nu}(ae^t) I_{-\nu}(ae^{-t}) - I_\nu(ae^t) I_\nu(ae^{-t})],$$

$$y = (a^2 - x^2)^{\frac{1}{2}} \quad \left[|\operatorname{Re}\nu| < \frac{1}{2}\right].$$ ET II 367(25)

Exponential, hyperbolic, and Bessel functions

6.668

1. $\int_0^\infty e^{-\alpha x} \operatorname{sh}(\beta x) J_0(\gamma x)\,dx = (\alpha\beta)^{\frac{1}{2}} r_1^{-1} r_2^{-1} (r_2 - r_1)^{\frac{1}{2}} (r_2 + r_1)^{-\frac{1}{2}},$

$$r_1 = [\gamma^2 + (\beta - \alpha)^2]^{\frac{1}{2}}, \quad r_2 = [\gamma^2 + (\beta + \alpha)^2]^{\frac{1}{2}}$$

$$[\operatorname{Re}\alpha > |\operatorname{Re}\beta|, \ \gamma > 0].$$ ET II 12(52)

2. $\int_0^\infty e^{-\alpha x} \operatorname{ch}(\beta x) J_0(\gamma x)\,dx = (\alpha\beta)^{\frac{1}{2}} r_1^{-1} r_2^{-1} (r_2 + r_1)^{\frac{1}{2}} (r_2 - r_1)^{-\frac{1}{2}},$

$$r_1 = [\gamma^2 + (\beta - \alpha)^2]^{\frac{1}{2}}, \quad r_2 = [\gamma^2 + (\beta + \alpha)^2]^{\frac{1}{2}}$$

$$[\operatorname{Re}\alpha > |\operatorname{Re}\beta|, \ \gamma > 0].$$ ET II 12(54)

6.669

1. $$\int_0^\infty \left[\operatorname{cth}\left(\frac{1}{2}x\right)\right]^{2\lambda} e^{-\beta \operatorname{ch} x} J_{2\mu}(\alpha \operatorname{sh} x)\,dx =$$

$$= \frac{\Gamma\left(\frac{1}{2}-\lambda+\mu\right)}{\alpha\Gamma(2\mu+1)} M_{-\lambda,\,\mu}\left[(\alpha^2+\beta^2)^{\frac{1}{2}}-\beta\right] W_{\lambda,\,\mu}\left[(\alpha^2+\beta^2)^{\frac{1}{2}}+\beta\right]$$

$$\left[\operatorname{Re}\beta > |\operatorname{Re}\alpha|,\ \operatorname{Re}(\mu-\lambda) > -\frac{1}{2}\right].$$ BU 86(5b)a, ET II 363(34)

2. $$\int_0^\infty \left[\operatorname{cth}\left(\frac{1}{2}x\right)\right]^{2\lambda} e^{-\beta \operatorname{ch} x} N_{2\mu}(\alpha \operatorname{sh} x)\,dx =$$

$$= -\frac{\sec[(\mu+\lambda)\pi]}{\alpha} W_{\lambda,\,\mu}\left(\sqrt{\alpha^2+\beta^2}+\beta\right) W_{-\lambda,\,\mu}\left(\sqrt{\alpha^2+\beta^2}-\beta\right) -$$

$$-\frac{\operatorname{tg}[(\mu+\lambda)\pi]\,\Gamma\left(\frac{1}{2}-\lambda+\mu\right)}{\alpha\Gamma(2\mu+1)} W_{\lambda,\,\mu}\left(\sqrt{\alpha^2+\beta^2}+\beta\right) M_{-\lambda,\,\mu}\left(\sqrt{\alpha^2+\beta^2}-\beta\right)$$

$$\left[\operatorname{Re}\beta > |\operatorname{Re}\alpha|,\ \operatorname{Re}\lambda < \frac{1}{2}-|\operatorname{Re}\mu|\right].$$ ET II 363(35)

3. $$\int_0^\infty e^{-\frac{1}{2}(a_1+a_2)t \operatorname{ch} x}\left[\operatorname{cth}\left(\frac{1}{2}x\right)\right]^{2\nu} K_{2\mu}\left(t\sqrt{a_1a_2}\operatorname{sh} x\right)dx =$$

$$= \frac{\Gamma\left(\frac{1}{2}+\mu-\nu\right)\Gamma\left(\frac{1}{2}-\mu-\nu\right)}{2t\sqrt{a_1a_2}} W_{\nu,\,\mu}(a_1t)\, W_{\nu,\,\mu}(a_2t)$$

$$\left[\operatorname{Re}\nu < \operatorname{Re}\frac{1\pm 2\mu}{2},\ \operatorname{Re}\left[t\left(\sqrt{a_1}+\sqrt{a_2}\right)^2\right] > 0\right].$$ BU 85(4a)

4. $$\int_0^\infty e^{-\frac{1}{2}(a_1+a_2)t \operatorname{ch} x}\left[\operatorname{cth}\left(\frac{x}{2}\right)\right]^{2\nu} I_{2\mu}\left(t\sqrt{a_1a_2}\operatorname{sh} x\right)dx =$$

$$= \frac{\Gamma\left(\frac{1}{2}+\mu-\nu\right)}{t\sqrt{a_1a_2}\,\Gamma(1+2\mu)} W_{\nu,\,\mu}(a_1t)\, M_{\nu,\,\mu}(a_2t)$$

$$\left[\operatorname{Re}\left(\frac{1}{2}+\mu-\nu\right) > 0,\ \operatorname{Re}\mu > 0,\ a_1 > a_2\right].$$ BU 86(5c)

5. $$\int_{-\infty}^\infty e^{2\nu s-\frac{x-y}{2}\operatorname{th} s} I_{2\mu}\left(\frac{\sqrt{xy}}{\operatorname{ch} s}\right)\frac{ds}{\operatorname{ch} s} =$$

$$= \frac{\Gamma\left(\frac{1}{2}+\mu+\nu\right)\Gamma\left(\frac{1}{2}+\mu-\nu\right)}{\sqrt{xy}\,[\Gamma(1+2\mu)]^2} M_{\nu,\,\mu}(x)\, M_{-\nu,\,\mu}(y)$$

$$\left[\operatorname{Re}\left(\pm\nu+\frac{1}{2}+\mu\right) > 0\right].$$ BU 83(3a)a

6. $\int\limits_{-\infty}^{\infty} e^{2\nu s-\frac{x+y}{2}\operatorname{th} s} J_{2\mu}\left(\frac{\sqrt{xy}}{\operatorname{ch} s}\right)\frac{ds}{\operatorname{ch} s} =$

$$=\frac{\Gamma\left(\frac{1}{2}+\mu+\nu\right)\Gamma\left(\frac{1}{2}+\mu-\nu\right)}{\sqrt{xy}\,[\Gamma(1+2\mu)]^2} M_{\nu,\mu}(x) M_{\nu,\mu}(y)$$

$\left[\operatorname{Re}\left(\mp\nu+\frac{1}{2}+\mu\right)>0\right].$ BU 84(3b)a

6.67-6.68 Combinations of Bessel and trigonometric functions

6.671

1. $\int\limits_0^{\infty} J_\nu(\alpha x)\sin\beta x\,dx=\frac{\sin\left(\nu\arcsin\frac{\beta}{\alpha}\right)}{\sqrt{\alpha^2-\beta^2}}$ $[\beta<\alpha];$

$=\infty$ or 0 $[\beta=\alpha];$

$=\frac{\alpha^\nu\cos\frac{\nu\pi}{2}}{\sqrt{\beta^2-\alpha^2}\,(\beta+\sqrt{\beta^2-\alpha^2})^\nu}$ $[\beta>\alpha].$

$[\operatorname{Re}\nu>-2].$ WA 444(4)

2. $\int\limits_0^{\infty} J_\nu(\alpha x)\cos\beta x\,dx=\frac{\cos\left(\nu\arcsin\frac{\beta}{\alpha}\right)}{\sqrt{\alpha^2-\beta^2}}$ $[\beta<\alpha];$

$=\infty$ or 0 $[\beta=\alpha];$

$=\frac{-\alpha^\nu\sin\frac{\nu\pi}{2}}{\sqrt{\beta^2-\alpha^2}\,(\beta+\sqrt{\beta^2-\alpha^2})^\nu}$ $[\beta>\alpha].$

$[\operatorname{Re}\nu>-1].$ WA 444(5)

3. $\int\limits_0^{\infty} N_\nu(ax)\sin(bx)\,dx=\operatorname{ctg}\left(\frac{\nu\pi}{2}\right)(a^2-b^2)^{-\frac{1}{2}}\sin\left[\nu\arcsin\left(\frac{b}{a}\right)\right]$

$[0<b<a,\ |\operatorname{Re}\nu|<2];$

$$=\frac{1}{2}\operatorname{cosec}\left(\frac{\nu\pi}{2}\right)(b^2-a^2)^{-\frac{1}{2}}\{a^{-\nu}\cos(\nu\pi)[b-(b^2-a^2)^{\frac{1}{2}}]^\nu-a^\nu[b-(b^2-a^2)^{\frac{1}{2}}]^{-\nu}\}$$

$[0<a<b,\ |\operatorname{Re}\nu|<2].$ ET I 103(33)

4. $\int\limits_0^{\infty} N_\nu(ax)\cos(bx)\,dx=-\frac{\operatorname{tg}\left(\frac{\nu\pi}{2}\right)}{(a^2-b^2)^{\frac{1}{2}}}\cos\left[\nu\arcsin\left(\frac{b}{a}\right)\right]$

$[0<b<a,\ |\operatorname{Re}\nu|<1];$

$$=-\sin\left(\frac{\nu\pi}{2}\right)(b^2-a^2)^{-\frac{1}{2}}\{a^{-\nu}[b-(b^2-a^2)^{\frac{1}{2}}]^\nu+\operatorname{ctg}(\nu\pi)+$$

$$+a^\nu[b-(b^2-a^2)^{\frac{1}{2}}]^{-\nu}\operatorname{cosec}(\nu\pi)\}$$ $[0<a<b,\ |\operatorname{Re}\nu|<1].$

ET I 47(29)

5. $$\int_0^\infty K_\nu(ax)\sin(bx)\,dx =$$

$$= \frac{1}{4}\pi a^{-\nu}\operatorname{cosec}\left(\frac{\nu\pi}{2}\right)(a^2+b^2)^{-\frac{1}{2}}\{[(b^2+a^2)^{\frac{1}{2}}+b]^\nu - [(b^2+a^2)^{\frac{1}{2}}-b]^\nu\}$$

$[\operatorname{Re} a > 0,\ b > 0,\ |\operatorname{Re}\nu| < 2,\ \nu \neq 0]$. ET I 105(48)

6. $$\int_0^\infty K_\nu(ax)\cos(bx)\,dx =$$

$$= \frac{\pi}{4}(b^2+a^2)^{-\frac{1}{2}}\sec\left(\frac{\nu\pi}{2}\right)\{a^{-\nu}[b+(b^2+a^2)^{\frac{1}{2}}]^\nu + a^\nu[b+(b^2+a^2)^{\frac{1}{2}}]^{-\nu}\}$$

$[\operatorname{Re} a > 0,\ b > 0,\ |\operatorname{Re}\nu| < 1]$. ET I 49(40)

7. $$\int_0^\infty J_0(ax)\sin(bx)\,dx = 0 \quad [0 < b < a];$$

$$= \frac{1}{\sqrt{b^2-a^2}} \quad [0 < a < b].$$ ET I 99(1)

8. $$\int_0^\infty J_0(ax)\cos(bx)\,dx = \frac{1}{\sqrt{a^2-b^2}} \quad [0 < b < a];$$

$$= \infty \quad [a = b];$$

$$= 0 \quad [0 < a < b].$$ ET I 43(1)

9. $$\int_0^\infty J_{2n+1}(ax)\sin(bx)\,dx =$$

$$= (-1)^n \frac{1}{\sqrt{a^2-b^2}} T_{2n+1}\left(\frac{b}{a}\right) \quad [0 < b < a];$$

$$= 0 \quad [0 < a < b].$$ ET I 99(2)

10. $$\int_0^\infty J_{2n}(ax)\cos(bx)\,dx =$$

$$= (-1)^n \frac{1}{\sqrt{a^2-b^2}} T_{2n}\left(\frac{b}{a}\right) \quad [0 < b < a];$$

$$= 0 \quad [0 < a < b].$$ ET I 43(2)

11. $$\int_0^\infty N_0(ax)\sin(bx)\,dx = \frac{2\arcsin\left(\frac{b}{a}\right)}{\pi\sqrt{a^2-b^2}} \quad [0 < b < a];$$

$$= \frac{2}{\pi}\frac{1}{\sqrt{b^2-a^2}}\ln\left[\frac{b}{a} - \sqrt{\frac{b^2}{a^2}-1}\right] \quad [0 < a < b].$$ ET I 103(31)

12. $$\int_0^\infty N_0(ax)\cos(bx)\,dx = 0 \quad [0 < b < a];$$

$$= -\frac{1}{\sqrt{b^2-a^2}} \quad [0 < a < b].$$ ET I 47(28)

13. $\int_0^\infty K_0(\beta x)\sin\alpha x\,dx = \frac{1}{\sqrt{\alpha^2+\beta^2}}\ln\left(\frac{\alpha}{\beta}+\sqrt{\frac{\alpha^2}{\beta^2}+1}\right)$

$[\alpha > 0,\ \beta > 0].$ WA 425(11)a, MO 48

14 $\int_0^\infty K_0(\beta x)\cos\alpha x\,dx = \frac{\pi}{2\sqrt{\alpha^2+\beta^2}}$

[α and β are real; $\beta > 0$]. WA 425(10)a, MO 48

6.672

1. $\int_0^\infty J_\nu(ax)J_\nu(bx)\sin(cx)\,dx =$

$= 0 \quad [\operatorname{Re}\nu > -1,\ 0 < c < b-a,\ 0 < a < b];$

$= \frac{1}{2\sqrt{ab}} P_{\nu-\frac{1}{2}}\left(\frac{b^2+a^2-c^2}{2ab}\right) \quad [\operatorname{Re}\nu > -1,\ b-a < c < b+a,\ 0 < a < b];$

$= -\frac{\cos(\nu\pi)}{\pi\sqrt{ab}} Q_{\nu-\frac{1}{2}}\left(-\frac{b^2+a^2-c^2}{2ab}\right) \quad [\operatorname{Re}\nu > -1,\ b+a < c,\ 0 < a < b]$

ET I 102(27)

2. $\int_0^\infty J_\nu(x)J_{-\nu}(x)\cos(bx)\,dx =$

$= \frac{1}{2}P_{\nu-\frac{1}{2}}\left(\frac{1}{2}b^2-1\right) \quad [0 < b < 2];$

$= 0 \quad [2 < b].$ ET I 46(21)

3. $\int_0^\infty K_\nu(ax)K_\nu(bx)\cos(cx)\,dx =$

$= \frac{\pi^2}{4\sqrt{ab}}\sec(\nu\pi)\,P_{\nu-\frac{1}{2}}[(a^2+b^2+c^2)(2ab)^{-1}]$

$\left[\operatorname{Re}(a+b) > 0,\ c > 0,\ |\operatorname{Re}\nu| < \frac{1}{2}\right].$ ET I 50(51)

4. $\int_0^\infty K_\nu(ax)I_\nu(bx)\cos(cx)\,dx = \frac{1}{2\sqrt{ab}}Q_{\nu-\frac{1}{2}}\left(\frac{a^2+b^2+c^2}{2ab}\right)$

$\left[\operatorname{Re}a > |\operatorname{Re}b|,\ c > 0,\ \operatorname{Re}\nu > -\frac{1}{2}\right].$ ET I 49(47)

5. $\int_0^\infty \sin(2ax)[J_\nu(x)]^2\,dx =$

$= \frac{1}{2}P_{\nu-\frac{1}{2}}(1-2a^2) \quad [0 < a < 1,\ \operatorname{Re}\nu > -1];$

$= \frac{1}{\pi}\cos(\nu\pi)\,Q_{\nu-\frac{1}{2}}(2a^2-1) \quad [a > 1,\ \operatorname{Re}\nu > -1].$ ET II 343(30)

6. $\int\limits_0^\infty \cos(2ax)[J_\nu(x)]^2\,dx =$

$$= \frac{1}{\pi} Q_{\nu-\frac{1}{2}}(1-2a^2) \qquad \left[0 < a < 1, \ \operatorname{Re}\nu > -\frac{1}{2}\right];$$

$$= -\frac{1}{\pi}\sin(\nu\pi)\, Q_{\nu-\frac{1}{2}}(2a^2-1) \qquad \left[a > 1, \ \operatorname{Re}\nu > -\frac{1}{2}\right].$$ ET II 344(32)

7. $\int\limits_0^\infty \sin(2ax)\, J_0(x)\, N_0(x)\,dx = 0 \qquad [0 < a < 1];$

$$= -\frac{\boldsymbol{K}\left[(1-a^{-2})^{\frac{1}{2}}\right]}{\pi a} \qquad [a > 1].$$ ET II 348(60)

8. $\int\limits_0^\infty K_0(ax)\, I_0(bx)\cos(cx)\,dx = \frac{1}{\sqrt{c^2+(a+b)^2}}\, \boldsymbol{K}\left\{\frac{\sqrt{2ab}}{\sqrt{c^2+(a+b)^2}}\right\}$

$$[\operatorname{Re} a > |\operatorname{Re} b|, \ c > 0].$$ ET I 49(46)

9. $\int\limits_0^\infty \cos(2ax)\, J_0(x)\, N_0(x)\,dx =$

$$= -\frac{1}{\pi}\boldsymbol{K}(a) \qquad [0 < a < 1];$$

$$= -\frac{1}{\pi a}\boldsymbol{K}\left(\frac{1}{a}\right) \qquad [a > 1].$$ ET II 348(61)

10. $\int\limits_0^\infty \cos(2ax)[N_0(x)]^2\,dx =$

$$= \frac{1}{\pi}\boldsymbol{K}\left(\sqrt{1-a^2}\right) \qquad [0 < a < 1];$$

$$= \frac{2}{\pi a}\boldsymbol{K}\left(\sqrt{1-\frac{1}{a^2}}\right) \qquad [a > 1].$$ ET II 348(62)

6.673

1. $\int\limits_0^\infty \left[J_\nu(ax)\cos\left(\frac{\nu\pi}{2}\right) - N_\nu(ax)\sin\left(\frac{\nu\pi}{2}\right)\right]\sin(bx)\,dx = 0$

$$[0 < b < a, \ |\operatorname{Re}\nu| < 2];$$

$$= \frac{1}{2a^\nu\sqrt{b^2-a^2}}\left\{\left[b+(b^2-a^2)^{\frac{1}{2}}\right]^\nu + \left[b-(b^2-a^2)^{\frac{1}{2}}\right]^\nu\right\}$$

$$[0 < a < b, \ |\operatorname{Re}\nu| < 2].$$ ET I 104(39)

2. $\int\limits_0^\infty \left[N_\nu(ax)\cos\left(\frac{\nu\pi}{2}\right) + J_\nu(ax)\sin\left(\frac{\nu\pi}{2}\right)\right]\cos(bx)\,dx = 0$

$$[0 < b < a, \ |\operatorname{Re}\nu| < 1];$$

$$= -\frac{1}{2a^\nu\sqrt{b^2-a^2}}\left\{\left[b+(b^2-a^2)^{\frac{1}{2}}\right]^\nu + \left[b-(b^2-a^2)^{\frac{1}{2}}\right]^\nu\right\}$$

$$[0 < a < b, \ |\operatorname{Re}\nu| < 1].$$ ET I 48(32)

6.674

1. $$\int_0^a \sin(a-x) J_\nu(x)\,dx = aJ_{\nu+1}(a) - 2\nu \sum_{n=0}^{\infty} (-1)^n J_{\nu+2n+2}(a) \qquad [\operatorname{Re}\nu > -1].$$ ET II 334(12)

2. $$\int_0^a \cos(a-x) J_\nu(x)\,dx = aJ_\nu(a) - 2\nu \sum_{n=0}^{\infty} (-1)^n J_{\nu+2n+1}(a) \qquad [\operatorname{Re}\nu > -1].$$ ET II 336(23)

3. $$\int_0^a \sin(a-x) J_{2n}(x)\,dx = aJ_{2n+1}(a) + (-1)^n 2n\left[\cos a - J_0(a) - 2\sum_{m=1}^{n} (-1)^m J_{2m}(a)\right] \qquad [n = 0, 1, 2, \ldots].$$ ET II 334(10)

4. $$\int_0^a \cos(a-x) J_{2n}(x)\,dx = aJ_{2n}(a) - (-1)^n 2n\left[\sin a - 2\sum_{m=0}^{n-1} (-1)^m J_{2m+1}(a)\right] \qquad [n = 0, 1, 2, \ldots].$$ ET II 335(21)

5. $$\int_0^a \sin(a-x) J_{2n+1}(x)\,dx = aJ_{2n+2}(a) + (-1)^n (2n+1)\left[\sin a - 2\sum_{m=0}^{n} (-1)^m J_{2m+1}(a)\right] \qquad [n = 0, 1, 2, \ldots].$$ ET II 334(11)

6. $$\int_0^a \cos(a-x) J_{2n+1}(x)\,dx = aJ_{2n+1}(a) + (-1)^n (2n+1)\left[\cos a - J_0(a) - 2\sum_{m=1}^{n} (-1)^m J_{2m}(a)\right] \qquad [n = 0, 1, 2, \ldots].$$ ET II 336(22)

7. $$\int_0^z \sin(z-x) J_0(x)\,dx = zJ_1(z).$$ WA 415(2)

8. $$\int_0^z \cos(z-x) J_0(x)\,dx = zJ_0(z).$$ WA 415(1)

6.675

1. $$\int_0^\infty J_\nu(a\sqrt{x})\sin(bx)\,dx = \frac{a\sqrt{\pi}}{4b^{\frac{3}{2}}}\left[\cos\left(\frac{a^2}{8b}-\frac{\nu\pi}{4}\right)J_{\frac{1}{2}\nu-\frac{1}{2}}\left(\frac{a^2}{8b}\right)- \sin\left(\frac{a^2}{8b}-\frac{\nu\pi}{4}\right)J_{\frac{1}{2}\nu+\frac{1}{2}}\left(\frac{a^2}{8b}\right)\right]$$

   $[a > 0, \ b > 0, \ \operatorname{Re}\nu > -4]$. ET I 110(23)

2. $$\int_0^\infty J_\nu(a\sqrt{x})\cos(bx)\,dx = -\frac{a\sqrt{\pi}}{4b^{\frac{3}{2}}}\left[\sin\left(\frac{a^2}{8b}-\frac{\nu\pi}{4}\right)J_{\frac{1}{2}\nu-\frac{1}{2}}\left(\frac{a^2}{8b}\right)+ \cos\left(\frac{a^2}{8b}-\frac{\nu\pi}{4}\right)J_{\frac{1}{2}\nu+\frac{1}{2}}\left(\frac{a^2}{8b}\right)\right]$$

   $[a > 0, \ b > 0, \ \operatorname{Re}\nu > -2]$. ET I 53(22)a

3. $$\int_0^\infty J_0(a\sqrt{x})\sin(bx)\,dx = \frac{1}{b}\cos\left(\frac{a^2}{4b}\right)$$

   $[a > 0, \ b > 0]$. ET I 110(22)

4. $$\int_0^\infty J_0(a\sqrt{x})\cos(bx)\,dx = \frac{1}{b}\sin\left(\frac{a^2}{4b}\right)$$

   $[a > 0, \ b > 0]$. ET I 53(21)

6.676

1. $$\int_0^\infty J_\nu(a\sqrt{x})J_\nu(b\sqrt{x})\sin(cx)\,dx = \frac{1}{c}J_\nu\left(\frac{ab}{2c}\right)\cos\left(\frac{a^2+b^2}{4c}-\frac{\nu\pi}{2}\right)$$

   $[a > 0, \ b > 0, \ c > 0, \ \operatorname{Re}\nu > -2]$. ET I 111(29)a

2. $$\int_0^\infty J_\nu(a\sqrt{x})J_\nu(b\sqrt{x})\cos(cx)\,dx = \frac{1}{c}J_\nu\left(\frac{ab}{2c}\right)\sin\left(\frac{a^2+b^2}{4c}-\frac{\nu\pi}{2}\right)$$

   $[a > 0, \ b > 0, \ c > 0, \ \operatorname{Re}\nu > -1]$. ET I 54(27)

3. $$\int_0^\infty J_0(a\sqrt{x})K_0(a\sqrt{x})\sin(bx)\,dx = \frac{1}{2b}K_0\left(\frac{a^2}{2b}\right)$$

   $[\operatorname{Re} a > 0, \ b > 0]$. ET I 111(31)

4. $\int_0^\infty J_0\left(\sqrt{ax}\right) K_0\left(\sqrt{ax}\right) \cos(bx)\, dx =$

$$= \frac{\pi}{4b}\left[I_0\left(\frac{a}{2b}\right) - \mathbf{L}_0\left(\frac{a}{2b}\right)\right]$$

$[\operatorname{Re} a > 0,\ b > 0]$. ET I 54(29)

5. $\int_0^\infty K_0\left(\sqrt{ax}\right) N_0\left(\sqrt{ax}\right) \cos(bx)\, dx = -\frac{1}{2b} K_0\left(\frac{a}{2b}\right)$

$[\operatorname{Re}\sqrt{a} > 0,\ b > 0]$. ET I 54(30)

6. $\int_0^\infty K_0\left(\sqrt{ax}\, e^{\frac{1}{4}\pi i}\right) K_0\left(\sqrt{ax}\, e^{-\frac{1}{4}\pi i}\right) \cos(bx)\, dx =$

$$= \frac{\pi^2}{8b}\left[\mathbf{H}_0\left(\frac{a}{2b}\right) - N_0\left(\frac{a}{2b}\right)\right]$$

$[\operatorname{Re} a > 0,\ b > 0]$. ET I 54(31)

6.677

1. $\int_a^\infty J_0\left(b\sqrt{x^2 - a^2}\right) \sin(cx)\, dx =$

$$= 0 \qquad [0 < c < b];$$

$$= \frac{\cos\left(a\sqrt{c^2 - b^2}\right)}{\sqrt{c^2 - b^2}} \qquad [0 < b < c].$$ ET I 113(47)

2. $\int_a^\infty J_0\left(b\sqrt{x^2 - a^2}\right) \cos(cx)\, dx = \frac{\exp\left(-a\sqrt{b^2 - c^2}\right)}{\sqrt{b^2 - c^2}} \qquad [0 < c < b];$

$$= \frac{-\sin\left(a\sqrt{c^2 - b^2}\right)}{\sqrt{c^2 - b^2}} \qquad [0 < b < c].$$ ET I 57(48)a

3. $\int_0^\infty J_0\left(\alpha\sqrt{x^2 + z^2}\right) \cos\beta x\, dx = \frac{\cos z\sqrt{\alpha^2 - \beta^2}}{\sqrt{\alpha^2 - \beta^2}} \qquad [0 < \beta < \alpha,\ z > 0];$

$$= 0 \qquad [0 < \alpha \leqslant \beta,\ z > 0].$$ MO 47a

4. $\int_0^\infty N_0\left(\alpha\sqrt{x^2 + z^2}\right) \cos\beta x\, dx = \frac{1}{\sqrt{\alpha^2 - \beta^2}} \sin\left(z\sqrt{\alpha^2 - \beta^2}\right)$

$[0 < \beta < \alpha,\ z > 0];$

$$= -\frac{1}{\sqrt{\beta^2 - \alpha^2}} \exp\left(-z\sqrt{\beta^2 - \alpha^2}\right)$$

$[0 < \alpha < \beta,\ z > 0]$. MO 47a

5. $\int_0^\infty K_0\left[\alpha\sqrt{x^2 + \beta^2}\right] \cos(\gamma x)\, dx = \frac{\pi}{2\sqrt{\alpha^2 + \gamma^2}} \exp\left(-\beta\sqrt{\alpha^2 + \gamma^2}\right)$

$[\operatorname{Re}\alpha > 0,\ \operatorname{Re}\beta > 0,\ \gamma > 0]$. ET I 56(43)

6. $$\int_0^a J_0\left(b\sqrt{a^2-x^2}\right)\cos(cx)\,dx = \frac{\sin\left(a\sqrt{b^2+c^2}\right)}{\sqrt{b^2+c^2}}$$
$[b>0]$. MO 48a, ET I 57(47)

7. $$\int_0^\infty J_0\left(b\sqrt{x^2-a^2}\right)\cos(cx)\,dx =$$
$$= \frac{\operatorname{ch}\left(a\sqrt{b^2-c^2}\right)}{\sqrt{b^2-c^2}} \qquad [0<c<b,\ a>0];$$
$$= 0 \qquad [0<b<c,\ a>0].$$
ET I 57(49)

8. $$\int_0^\infty H_0^{(1)}\left(\alpha\sqrt{\beta^2-x^2}\right)\cos(\gamma x)\,dx = -i\,\frac{\exp\left(i\beta\sqrt{\alpha^2+\gamma^2}\right)}{\sqrt{\alpha^2+\gamma^2}}$$
$\left[\pi > \arg\sqrt{\beta^2-x^2} \geqslant 0,\ \alpha>0,\ \gamma>0\right]$. ET I 59(59)

9. $$\int_0^\infty H_0^{(2)}\left(\alpha\sqrt{\beta^2-x^2}\right)\cos(\gamma x)\,dx = \frac{i\exp\left(-i\beta\sqrt{\alpha^2+\gamma^2}\right)}{\sqrt{\alpha^2+\gamma^2}}$$
$\left[-\pi < \arg\sqrt{\beta^2-x^2} \leqslant 0,\ \alpha>0,\ \gamma>0\right]$. ET I 58(58)

6.678 $$\int_0^\infty \left[K_0\left(2\sqrt{x}\right)+\frac{\pi}{2}N_0\left(2\sqrt{x}\right)\right]\sin(bx)\,dx = \frac{\pi}{2b}\sin\left(\frac{1}{b}\right)$$
$[b>0]$. ET I 111(34)

6.679

1. $$\int_0^\infty J_{2\nu}\left[2b\operatorname{sh}\left(\frac{x}{2}\right)\right]\sin(bx)\,dx = -i\left[I_{\nu-ib}(a)K_{\nu+ib}(a) - I_{\nu+ib}(a)K_{\nu-ib}(a)\right]$$
$[a>0,\ b>0,\ \operatorname{Re}\nu>-1]$. ET I 115(59)

2. $$\int_0^\infty J_{2\nu}\left[2a\operatorname{sh}\left(\frac{x}{2}\right)\right]\cos(bx)\,dx =$$
$$= I_{\nu-ib}(a)K_{\nu+ib}(a) + I_{\nu+ib}(a)K_{\nu-ib}(a)$$
$\left[a>0,\ b>0,\ \operatorname{Re}\nu>-\frac{1}{2}\right]$. ET I 59(64)

3. $$\int^\infty J_{2\nu}\left[2a\operatorname{ch}\left(\frac{x}{2}\right)\right]\cos(bx)\,dx =$$
$$= -\frac{\pi}{2}\left[J_{\nu+ib}(a)N_{\nu-ib}(a) + J_{\nu-ib}(a)N_{\nu+ib}(a)\right].$$
ET I 59(63)

4. $$\int_0^\infty J_0\left[2a\operatorname{sh}\left(\frac{x}{2}\right)\right]\sin(bx)\,dx =$$
$$= \frac{2}{\pi}\operatorname{sh}(\pi b)\left[K_{ib}(a)\right]^2 \qquad [a>0,\ b>0].$$
ET I 115(58)

5. $\int_0^\infty J_0\left[2a \operatorname{sh}\left(\frac{x}{2}\right)\right] \cos(bx)\,dx =$

$= [I_{ib}(a) + I_{-ib}(a)]\,K_{ib}(a) \qquad [a > 0,\ b > 0].$ ET I 59(62)

6. $\int_0^\infty N_0\left[2a \operatorname{sh}\left(\frac{x}{2}\right)\right] \cos(bx)\,dx =$

$= -\frac{2}{\pi} \operatorname{ch}(\pi b)\,[K_{ib}(a)]^2 \qquad [a > 0,\ b > 0].$ ET I 59(65)

7. $\int_0^\infty K_0\left[2a \operatorname{sh}\left(\frac{x}{2}\right)\right] \cos(bx)\,dx =$

$= \frac{\pi^2}{4}\{[J_{ib}(a)]^2 + [N_{ib}(a)]^2\} \qquad [\operatorname{Re} a > 0,\ b > 0].$ ET I 59(66)

6.681

1. $\int_0^{\frac{\pi}{2}} \cos(2\mu x)\,J_{2\nu}(2a\cos x)\,dx = \frac{\pi}{2} J_{\nu+\mu}(a)\,J_{\nu-\mu}(a)$

$\left[\operatorname{Re}\nu > -\frac{1}{2}\right].$ ET II 361(23)

2. $\int_0^{\frac{\pi}{2}} \cos(2\mu x)\,N_{2\nu}(2a\cos x)\,dx =$

$= \frac{\pi}{2}[\operatorname{ctg}(2\nu\pi)\,J_{\nu+\mu}(a)\,J_{\nu-\mu}(a) - \operatorname{cosec}(2\nu\pi)\,J_{\mu-\nu}(a)\,J_{-\mu-\nu}(a)]$

$\left[|\operatorname{Re}\nu| < \frac{1}{2}\right].$ ET II 361(24)

3. $\int_0^{\frac{\pi}{2}} \cos(2\mu x)\,I_{2\nu}(2a\cos x)\,dx = \frac{\pi}{2} I_{\nu-\mu}(a)\,I_{\nu+\mu}(a)$

$\left[\operatorname{Re}\nu > -\frac{1}{2}\right].$ ET I 59(61)

4. $\int_0^{\frac{\pi}{2}} \cos(\nu x)\,K_\nu(2a\cos x)\,dx = \frac{\pi}{2} I_0(a)\,K_\nu(a)$

$[\operatorname{Re}\nu < 1].$ WA 484(3)

5. $\int_0^{\pi} J_0(2z\cos x)\cos 2nx\,dx = (-1)^n \pi J_n^2(z).$ MO 45

6. $\int_0^{\pi} J_0(2z\sin x)\cos 2nx\,dx = \pi J_n^2(z).$ WA 43(3), MO 45

7. $\int_0^{\frac{\pi}{2}} \cos(2nx)\,N_0(2a\sin x)\,dx = \frac{\pi}{2} J_n(a)\,N_n(a)$

$[n = 0,\ 1,\ 2,\ \ldots].$ ET II 360(16)

8. $$\int_0^{\pi} \sin(2\mu x) J_{2\nu}(2a\sin x)\,dx = \pi\sin(\mu\pi) J_{\nu-\mu}(a) J_{\nu+\mu}(a) \qquad [\operatorname{Re}\nu > -1].$$ ET II 360(13)

9. $$\int_0^{\pi} \cos(2\mu x) J_{2\nu}(2a\sin x)\,dx = \pi\cos(\mu\pi) J_{\nu-\mu}(a) J_{\nu+\mu}(a) \qquad \left[\operatorname{Re}\nu > -\tfrac{1}{2}\right].$$ ET II 360(14)

10. $$\int_0^{\frac{\pi}{2}} J_{\nu+\mu}(2z\cos x)\cos[(\nu-\mu)x]\,dx = \frac{\pi}{2} J_\nu(z) J_\mu(z) \qquad [\operatorname{Re}(\nu+\mu) > -1].$$ MO 42

11. $$\int_0^{\frac{\pi}{2}} \cos[(\mu-\nu)x] I_{\mu+\nu}(2a\cos x)\,dx = \frac{\pi}{2} I_\mu(a) I_\nu(a) \qquad [\operatorname{Re}(\mu+\nu) > -1].$$ WA 484(2), ET II 378(39)

12. $$\int_0^{\frac{\pi}{2}} \cos[(\mu-\nu)x] K_{\mu+\nu}(2a\cos x)\,dx = \frac{\pi}{2}\operatorname{cosec}[(\mu+\nu)\pi][I_{-\mu}(a) I_{-\nu}(a) - I_\mu(a) I_\nu(a)] \qquad [|\operatorname{Re}(\mu+\nu)| < 1].$$ ET II 378(40)

13. $$\int_0^{\frac{\pi}{2}} K_{\nu-m}(2a\cos x)\cos[(m+\nu)x]\,dx = (-1)^m \frac{\pi}{2} I_m(a) K_\nu(a) \qquad [|\operatorname{Re}(\nu-m)| < 1].$$ WA 485(4)

6.682

1. $$\int_0^{\frac{\pi}{2}} J_{\nu-\frac{1}{2}}(x\sin t)\sin^{\nu+\frac{1}{2}} t\,dt = \sqrt{\frac{\pi}{2x}} J_\nu(x)$$

[ν may be zero, a natural number, one half, or a natural number plus one half; $x > 0$]. MO 42a

2. $$\int_0^{\frac{\pi}{2}} J_\nu(z\sin x)\sin^\nu x\cos^{2\nu} x\,dx = 2^{\nu-1}\sqrt{\pi}\,\Gamma\left(\nu+\frac{1}{2}\right) z^{-\nu} J_\nu^2\left(\frac{z}{2}\right) \qquad \left[\operatorname{Re}\nu > -\tfrac{1}{2}\right].$$ MO 42a

6.683

1. $$\int_0^{\frac{\pi}{2}} J_\nu(z\sin x) I_\mu(z\cos x) \operatorname{tg}^{\nu+1} x\, dx = \frac{\left(\frac{z}{2}\right)^\nu \Gamma\left(\frac{\mu-\nu}{2}\right)}{\Gamma\left(\frac{\mu+\nu}{2}+1\right)} J_\mu(z)$$
$[\operatorname{Re}\nu > \operatorname{Re}\mu > -1]$. WA 407(4)

2. $$\int_0^{\frac{\pi}{2}} J_\nu(z_1\sin x) J_\mu(z_2\cos x) \sin^{\nu+1} x \cos^{\mu+1} x\, dx =$$
$$= \frac{z_1^\nu z_2^\mu J_{\nu+\mu+1}\left(\sqrt{z_1^2+z_2^2}\right)}{\sqrt{(z_1^2+z_2^2)^{\nu+\mu+1}}}$$
$[\operatorname{Re}\nu > -1,\ \operatorname{Re}\mu > -1]$. WA 410(1)

3. $$\int_0^{\frac{\pi}{2}} J_\nu(z\cos^2 x) J_\mu(z\sin^2 x) \sin x \cos x\, dx =$$
$$= \frac{1}{z} \sum_{k=0}^{\infty} (-1)^k J_{\nu+\mu+2k+1}(z)$$
$[\operatorname{Re}\nu > -1,\ \operatorname{Re}\mu > -1]$

(see also 6.513 6.). WA 414(1)

4. $$\int_0^{\frac{\pi}{2}} J_\mu(z\sin\theta)(\sin\theta)^{1-\mu}(\cos\theta)^{2\nu+1}\, d\theta =$$
$$= \frac{s_{\mu+\nu,\ \nu-\mu+1}(z)}{2^{\mu-1} z^{\nu+1}\Gamma(\mu)}$$
$[\operatorname{Re}\nu > -1]$. WA 407(2)

5. $$\int_0^{\frac{\pi}{2}} J_\mu(z\sin\theta)(\sin\theta)^{1-\mu}\, d\theta = \frac{\mathbf{H}_{\mu-\frac{1}{2}}(z)}{\sqrt{\frac{2z}{\pi}}}.$$ WA 407(3)

6. $$\int_0^{\frac{\pi}{2}} J_\mu(a\sin\theta)(\sin\theta)^{\mu+1}(\cos\theta)^{2\varrho+1}\, d\theta = 2^\varrho\Gamma(\varrho+1)\, a^{-\varrho-1} J_{\varrho+\mu+1}(a)$$
$[\operatorname{Re}\varrho > -1,\ \operatorname{Re}\mu > -1]$. WA 406(1), EH II 46(5)

7. $$\int_0^{\frac{\pi}{2}} J_\nu(2z\sin\theta)(\sin\theta)^\nu(\cos\theta)^{2\nu}\, d\theta =$$
$$= \frac{1}{2}\sum_{m=0}^{\infty} \frac{(-1)^m z^{\nu+2m}\Gamma\left(\nu+m+\frac{1}{2}\right)\Gamma\left(\nu+\frac{1}{2}\right)}{m!\,\Gamma(\nu+m+1)\,\Gamma(2\nu+m+1)};$$
$$= \frac{1}{2} z^{-\nu}\sqrt{\pi}\,\Gamma\left(\nu+\frac{1}{2}\right)[J_\nu(z)]^2 \qquad \left[\operatorname{Re}\nu > -\frac{1}{2}\right].$$ EH II 47(10)

8. $\int_0^{\frac{\pi}{2}} J_\nu(z\sin\theta)(\sin\theta)^{\nu+1}(\cos\theta)^{-2\nu}\,d\theta = 2^{-\nu}\frac{z^{\nu-1}}{\sqrt{\pi}}\Gamma\left(\frac{1}{2}-\nu\right)\sin z$

$\left[-1 < \operatorname{Re}\nu < \frac{1}{2}\right].$ EH II 68(39)

9. $\int_0^{\frac{\pi}{2}} J_\nu(z\sin^2\theta)J_\nu(z\cos^2\theta)(\sin\theta)^{2\nu+1}(\cos\theta)^{2\nu+1}\,d\theta =$

$$= \frac{\Gamma\left(\frac{1}{2}+\nu\right)J_{2\nu+\frac{1}{2}}(z)}{2^{2\nu+\frac{3}{2}}\Gamma(\nu+1)\sqrt{z}} \qquad \left[\operatorname{Re}\nu > -\frac{1}{2}\right].$$ WA 409(1)

10 $\int_0^{\frac{\pi}{2}} J_\mu(z\sin^2\theta)J_\nu(z\cos^2\theta)\sin^{2\mu+1}\theta\cos^{2\nu+1}\theta\,d\theta =$

$$= \frac{\Gamma\left(\mu+\frac{1}{2}\right)\Gamma\left(\nu+\frac{1}{2}\right)J_{\mu+\nu+\frac{1}{2}}(z)}{2\sqrt{\pi}\Gamma(\mu+\nu+1)\sqrt{2z}}$$

$\left[\operatorname{Re}\mu > -\frac{1}{2},\ \operatorname{Re}\nu > -\frac{1}{2}\right].$ WA 417(1)

6.684

1. $\int_0^{\pi} (\sin x)^{2\nu}\frac{J_\nu\left(\sqrt{\alpha^2+\beta^2-2\alpha\beta\cos x}\right)}{\left(\sqrt{\alpha^2+\beta^2-2\alpha\beta\cos x}\right)^\nu}\,dx =$

$$= 2^\nu\sqrt{\pi}\Gamma\left(\nu+\frac{1}{2}\right)\frac{J_\nu(\alpha)}{\alpha^\nu}\frac{J_\nu(\beta)}{\beta^\nu} \qquad \left[\operatorname{Re}\nu > -\frac{1}{2}\right].$$ ET II 362(27)

2. $\int_0^{\pi} (\sin x)^{2\nu}\frac{N_\nu\left(\sqrt{\alpha^2+\beta^2-2\alpha\beta\cos x}\right)}{\left(\sqrt{\alpha^2+\beta^2-2\alpha\beta\cos x}\right)^\nu}\,dx =$

$$= 2^\nu\sqrt{\pi}\Gamma\left(\nu+\frac{1}{2}\right)\frac{J_\nu(\alpha)}{\alpha^\nu}\frac{N_\nu(\beta)}{\beta^\nu}$$

$\left[|\alpha| < |\beta|,\ \operatorname{Re}\nu > -\frac{1}{2}\right].$ ET II 362(28)

6.685 $\int_0^{\frac{\pi}{2}} \sec x\cos(2\lambda x)K_{2\mu}(a\sec x)\,dx = \frac{\pi}{2a}W_{\lambda,\mu}(a)W_{-\lambda,\mu}(a) \qquad [\operatorname{Re}a > 0].$

ET II 378(41)

6.686

1. $\int_0^{\infty} \sin(ax^2)J_\nu(bx)\,dx = -\frac{\sqrt{\pi}}{2\sqrt{a}}\sin\left(\frac{b^2}{8a}-\frac{\nu+1}{4}\pi\right)J_{\frac{1}{2}\nu}\left(\frac{b^2}{8a}\right)$

$[a > 0,\ b > 0,\ \operatorname{Re}\nu > -3].$ ET II 34(13)

2. $\int_0^\infty \cos(ax^2) J_\nu(bx)\,dx = \frac{\sqrt{\pi}}{2\sqrt{a}} \cos\left(\frac{b^2}{8a} - \frac{\nu+1}{4}\pi\right) J_{\frac{1}{2}\nu}\left(\frac{b^2}{8a}\right)$

$[a > 0,\ b > 0,\ \operatorname{Re}\nu > -1].$ ET II 38(38)

3. $\int_0^\infty \sin(ax^2) N_\nu(bx)\,dx = -\frac{\sqrt{\pi}}{4\sqrt{a}} \sec\left(\frac{\nu\pi}{2}\right) \times$

$\times \left[\cos\left(\frac{b^2}{8a} - \frac{3\nu+1}{4}\pi\right) J_{\frac{1}{2}\nu}\left(\frac{b^2}{8a}\right) - \right.$

$\left. - \sin\left(\frac{b^2}{8a} + \frac{\nu-1}{4}\pi\right) N_{\frac{1}{2}\nu}\left(\frac{b^2}{8a}\right)\right]$

$[a > 0,\ b > 0,\ -3 < \operatorname{Re}\nu < 3].$ ET II 107(7)

4. $\int_0^\infty \cos(ax^2) N_\nu(bx)\,dx = \frac{\sqrt{\pi}}{4\sqrt{a}} \sec\left(\frac{\nu\pi}{2}\right) \left[\sin\left(\frac{b^2}{8a} - \frac{3\nu+1}{4}\pi\right) J_{\frac{1}{2}\nu}\left(\frac{b^2}{8a}\right) + \right.$

$\left. + \cos\left(\frac{b^2}{8a} + \frac{\nu-1}{4}\pi\right) N_{\frac{1}{2}\nu}\left(\frac{b^2}{8a}\right)\right]$

$[a > 0,\ b > 0,\ -1 < \operatorname{Re}\nu < 1].$ ET II 107(8)

5. $\int_0^\infty \sin(ax^2) J_1(bx)\,dx = \frac{1}{b} \sin\frac{b^2}{4a} \qquad [a > 0,\ b > 0].$ ET II 19(16)

6. $\int_0^\infty \cos(ax^2) J_1(bx)\,dx = \frac{2}{b} \sin^2\left(\frac{b^2}{8a}\right) \qquad [a > 0,\ b > 0].$ ET II 20(20)

7. $\int_0^\infty \sin^2(ax^2) J_1(bx)\,dx = \frac{1}{2b} \cos\left(\frac{b^2}{8a}\right) \qquad [a > 0,\ b > 0].$ ET II 19(17)

6.687 $\int_0^\infty \cos\left(\frac{x^2}{2a}\right) K_{2\nu}\left(xe^{i\frac{\pi}{4}}\right) K_{2\nu}\left(xe^{-i\frac{\pi}{4}}\right) dx =$

$= \frac{\Gamma\left(\frac{1}{4}+\nu\right) \Gamma\left(\frac{1}{4}-\nu\right) \sqrt{\pi}}{8\sqrt{a}} W_{\frac{1}{4},\nu}\left(ae^{i\frac{\pi}{2}}\right) W_{\frac{1}{4},\nu}\left(ae^{-i\frac{\pi}{2}}\right)$

$\left[a > 0,\ |\operatorname{Re}\nu| < \frac{1}{4}\right].$ ET II 372(1)

6.688

1. $\int_0^{\frac{\pi}{2}} J_\nu(\mu z \sin t) \cos(\mu x \cos t)\,dt =$

$= \frac{\pi}{2} J_{\frac{\nu}{2}}\left(\mu \frac{\sqrt{x^2+z^2}+x}{2}\right) J_{\frac{\nu}{2}}\left(\mu \frac{\sqrt{x^2+z^2}-x}{2}\right)$

$[\operatorname{Re}\nu > -1,\ \operatorname{Re} z > 0].$ MO 46

2. $\int_0^{\frac{\pi}{2}} (\sin x)^{\nu+1} \cos(\beta \cos x) J_\nu(\alpha \sin x)\, dx =$

$= 2^{-\frac{1}{2}} \sqrt{\pi}\, \alpha^\nu (\alpha^2+\beta^2)^{-\frac{1}{2}\nu-\frac{1}{4}} J_{\nu+\frac{1}{2}} [(\alpha^2+\beta^2)^{\frac{1}{2}}]$ $[\operatorname{Re} \nu > -1]$. ET II 361(19)

3. $\int_0^{\frac{\pi}{2}} \cos[(z-\zeta)\cos\theta] J_{2\nu}[2\sqrt{z\zeta}\sin\theta]\, d\theta = \frac{\pi}{2} J_\nu(z) J_\nu(\zeta)$

$\left[\operatorname{Re} \nu > -\frac{1}{2}\right]$. EH II 47(8)

6.69-6.74 Combinations of Bessel and trigonometric functions and powers

6.691 $\int_0^\infty x \sin(bx) K_0(ax)\, dx = \frac{\pi b}{2} (a^2+b^2)^{-\frac{3}{2}}$

$[\operatorname{Re} a > 0,\ b > 0]$. ET I 105(47)

6.692

1. $\int_0^\infty x K_\nu(ax) I_\nu(bx) \sin(cx)\, dx =$

$= \frac{1}{2} (ab)^{-\frac{3}{2}} c\, (u^2-1)^{-\frac{1}{2}} Q^1_{\nu-\frac{1}{2}}(u), \qquad u = (2ab)^{-1}(a^2+b^2+c^2)$

$\left[\operatorname{Re} a > |\operatorname{Re} b|,\ c > 0,\ \operatorname{Re} \nu > -\frac{3}{2}\right]$. ET I 106(54)

2. $\int_0^\infty x K_\nu(ax) K_\nu(bx) \sin(cx)\, dx =$

$= \frac{\pi}{4} (ab)^{-\frac{3}{2}} c\, (u^2-1)^{-\frac{1}{2}} \Gamma\left(\frac{3}{2}+\nu\right) \Gamma\left(\frac{3}{2}-\nu\right) P^{-1}_{\nu-\frac{1}{2}}(u),$

$u = (2ab)^{-1}(a^2+b^2+c^2)$

$\left[\operatorname{Re}(a+b) > 0,\ c > 0.\ |\operatorname{Re} \nu| < \frac{3}{2}\right]$. ET I 107(61)

6.693

1. $\int_0^\infty J_\nu(\alpha x) \sin \beta x \frac{dx}{x} = \frac{1}{\nu} \sin\left(\nu \arcsin \frac{\beta}{\alpha}\right)$ $[\beta \leqslant \alpha]$

$= \dfrac{\alpha^\nu \sin \frac{\nu\pi}{2}}{\nu\left(\beta + \sqrt{\beta^2-\alpha^2}\right)^\nu}$ $[\beta \geqslant \alpha]$

$[\operatorname{Re} \nu > -1]$. WA 443(2)

2. $\int_0^\infty J_\nu(\alpha x) \cos \beta x \frac{dx}{x} = \frac{1}{\nu} \cos\left(\nu \arcsin \frac{\beta}{\alpha}\right)$ $[\beta \leqslant \alpha]$

$= \dfrac{\alpha^\nu \cos \frac{\nu\pi}{2}}{\nu\left(\beta + \sqrt{\beta^2-\alpha^2}\right)^\nu}$ $[\beta \geqslant \alpha]$

$[\operatorname{Re} \nu > 0]$. WA 443(3)

3. $\int_0^\infty N_\nu(ax)\sin(bx)\frac{dx}{x} = -\frac{1}{\nu}\operatorname{tg}\left(\frac{\nu\pi}{2}\right)\sin\left[\nu\arcsin\left(\frac{b}{a}\right)\right]$ $[0 < b < a, \ |\operatorname{Re}\nu| < 1];$

$= \frac{1}{2\nu}\sec\left(\frac{\nu\pi}{2}\right)\left\{a^{-\nu}\cos(\nu\pi)\left[b-(b^2-a^2)^{\frac{1}{2}}\right]^\nu - a^\nu\left[b-(b^2-a^2)^{\frac{1}{2}}\right]^{-\nu}\right\}$ $[0 < a < b, \ |\operatorname{Re}\nu| < 1].$ ET I 103(35)

4. $\int_0^\infty J_\nu(ax)\sin(bx)\frac{dx}{x^2} = \frac{\sqrt{a^2-b^2}\sin\left[\nu\arcsin\left(\frac{b}{a}\right)\right]}{\nu^2-1} - \frac{b\cos\left[\nu\arcsin\left(\frac{b}{a}\right)\right]}{\nu(\nu^2-1)}$ $[0 < b < a, \ \operatorname{Re}\nu > 0];$

$= \frac{-a^\nu\cos\left(\frac{\nu\pi}{2}\right)\left[b+\nu\sqrt{b^2-a^2}\right]}{\nu(\nu^2-1)\left[b+\sqrt{b^2-a^2}\right]^\nu}$ $[0 < a < b, \ \operatorname{Re}\nu > 0].$ ET I 99(6)

5. $\int_0^\infty J_\nu(ax)\cos(bx)\frac{dx}{x^2} =$

$= \frac{a\cos\left[(\nu-1)\arcsin\left(\frac{b}{a}\right)\right]}{2\nu(\nu-1)} + \frac{a\cos\left[(\nu+1)\arcsin\left(\frac{b}{a}\right)\right]}{2\nu(\nu+1)}$ $[0 < b < a, \ \operatorname{Re}\nu > 1];$

$= \frac{a^\nu\sin\left(\frac{\nu\pi}{2}\right)}{2\nu(\nu-1)\left[b+\sqrt{b^2-a^2}\right]^{\nu-1}} - \frac{a^{\nu+2}\sin\left(\frac{\nu\pi}{2}\right)}{2\nu(\nu+1)\left[b+\sqrt{b^2-a^2}\right]^{\nu+1}}$ $[0 < a < b, \ \operatorname{Re}\nu > 1].$ ET I 44(6)

6. $\int_0^\infty J_0(\alpha x)\sin x\frac{dx}{x} = \frac{\pi}{2}$ $[0 < \alpha < 1];$

$= \operatorname{arccosec}\alpha$ $[\alpha > 1].$ WH

7. $\int_0^\infty J_0(x)\sin\beta x\frac{dx}{x} = \frac{\pi}{2}$ $[\beta > 1];$

$= \arcsin\beta$ $[\beta^2 < 1];$

$= -\frac{\pi}{2}$ $[\beta < -1].$

8. $\int_0^\infty [J_0(x) - \cos\alpha x]\frac{dx}{x} = \ln 2\alpha.$ NT 66(13)

9. $\int_0^z J_\nu(x)\sin(z-x)\frac{dx}{x} = \frac{2}{\nu}\sum_{k=0}^\infty(-1)^k J_{\nu+2k+1}(z)$ $[\operatorname{Re}\nu > 0].$ WA 416(4)

10. $$\int_0^z J_\nu(x)\cos(z-x)\frac{dx}{x}=\frac{1}{\nu}J_\nu(z)+\frac{2}{\nu}\sum_{k=1}^{\infty}(-1)^k J_{\nu+2k}(z)$$
$[\operatorname{Re}\nu>0]$. WA 416(5)

6.694 $$\int_0^\infty\left[\frac{J_1(ax)}{x}\right]^2\sin(bx)\,dx=\frac{1}{2}b-\left(\frac{4a}{3\pi}\right)\left[\left(1+\frac{b^2}{4a^2}\right)\boldsymbol{E}\left(\frac{b}{2a}\right)+\right.$$
$$\left.-\left(1-\frac{b^2}{4a^2}\right)\boldsymbol{K}\left(\frac{b}{2a}\right)\right]\qquad[0<b\leqslant 2a].$$ ET I 102(22)

$$=\frac{1}{2}b-\frac{2b}{3\pi}\left[2\boldsymbol{E}\left(\frac{2a}{b}\right)-\left(1-\frac{4a^2}{b^2}\right)\boldsymbol{K}\left(\frac{2a}{b}\right)\right]\qquad[0<2a\leqslant b].$$

6.695

1. $$\int_0^\infty\frac{\sin\alpha x}{\beta^2+x^2}J_0(ux)\,dx=\frac{\operatorname{sh}\alpha\beta}{\beta}K_0(\beta u)\qquad[\alpha>0,\ \operatorname{Re}\beta>0,\ u>\alpha].$$ MO 46

2. $$\int_0^\infty\frac{\cos\alpha x}{\beta^2+x^2}J_0(ux)\,dx=\frac{\pi}{2}\frac{e^{-\alpha\beta}}{\beta}I_0(\beta u)$$
$[\alpha>0,\ \operatorname{Re}\beta>0,\ -\alpha<u<\alpha]$. MO 46

3. $$\int_0^\infty\frac{x}{x^2+\beta^2}\sin(\alpha x)J_0(\gamma x)\,dx=\frac{\pi}{2}e^{-\alpha\beta}I_0(\gamma\beta)$$
$[\alpha>0,\ \operatorname{Re}\beta>0,\ 0<\gamma<\alpha]$. ET II 10(36)

4. $$\int_0^\infty\frac{x}{x^2+\beta^2}\cos(\alpha x)J_0(\gamma x)\,dx=\operatorname{ch}(\alpha\beta)K_0(\beta\gamma)$$
$[\alpha>0,\ \operatorname{Re}\beta>0,\ \alpha<\gamma]$. ET II 11(45)

6.696 $$\int_0^\infty[1-\cos(\alpha x)]J_0(\beta x)\frac{dx}{x}=$$
$$=\operatorname{Arch}\left(\frac{\alpha}{\beta}\right)\qquad[0<\beta<\alpha];$$
$$=0\qquad[0<\alpha<\beta].$$ ET II 11(43)

6.697

1. $$\int_{-\infty}^\infty\frac{\sin[\alpha(x+\beta)]}{x+\beta}J_0(x)\,dx=2\int_0^\alpha\frac{\cos\beta u}{\sqrt{1-u^2}}du\qquad[0\leqslant\alpha\leqslant 1];$$ WA 463(2)
$$=\pi J_0(\beta)\qquad[1\leqslant\alpha<\infty].$$ WA 463(1), ET II 345(42)

2. $$\int_0^\infty\frac{\sin(x+t)}{x+t}J_0(t)\,dt=\frac{\pi}{2}J_0(x)\qquad[x>0].$$ WA 475(4)

3. $$\int_0^\infty\frac{\cos(x+t)}{x+t}J_0(t)\,dt=-\frac{\pi}{2}N_0(x)\qquad[x>0].$$ WA 475(5)

4. $$\int_{-\infty}^{\infty} \frac{|x|}{x+\beta} \sin[\alpha(x+\beta)] J_0(bx)\,dx = 0$$

$[0 \leqslant \alpha < b]$. WA 464(5), ET II 345(43)a

5. $$\int_{-\infty}^{\infty} \frac{\sin[\alpha(x+\beta)]}{x+\beta} [J_{n+\frac{1}{2}}(x)]^2\,dx = \pi [J_{n+\frac{1}{2}}(\beta)]^2$$

$[2 \leqslant \alpha < \infty,\ n = 0, 1, \ldots]$. ET II 346(45)

6. $$\int_{-\infty}^{\infty} \frac{\sin[\alpha(x+\beta)]}{x+\beta} J_{n+\frac{1}{2}}(x) J_{-n-\frac{1}{2}}(x)\,dx = \pi J_{n+\frac{1}{2}}(\beta) J_{-n-\frac{1}{2}}(\beta) \qquad [2 \leqslant \alpha < \infty,\ n = 0, 1, \ldots].$$

ET II 346(46)

7. $$\int_{-\infty}^{\infty} \frac{J_\mu[a(z+x)]}{(z+x)^\mu} \frac{J_\nu[a(\zeta+x)]}{(\zeta+x)^\nu}\,dx = \frac{\Gamma(\mu+\nu)\sqrt{\pi}\sqrt{\frac{2}{a}}}{\Gamma\left(\mu+\frac{1}{2}\right)\Gamma\left(\nu+\frac{1}{2}\right)} \cdot \frac{J_{\mu+\nu-\frac{1}{2}}[a(z-\zeta)]}{(z-\zeta)^{\mu+\nu-\frac{1}{2}}}$$

$[\operatorname{Re}(\mu+\nu) > 0]$. WA 463(3)

6.698

1. $$\int_0^\infty \sqrt{x}\, J_{\nu+\frac{1}{4}}(ax) J_{-\nu+\frac{1}{4}}(ax) \sin(bx)\,dx = \sqrt{\frac{2}{\pi b}} \frac{\cos\left[2\nu \arccos\left(\frac{b}{2a}\right)\right]}{\sqrt{4a^2-b^2}} \quad [0 < b < 2a];$$
$$= 0 \quad [0 < 2a < b].$$
ET I 102(26)

2. $$\int_0^\infty \sqrt{x}\, J_{\nu-\frac{1}{4}}(ax) J_{-\nu-\frac{1}{4}}(ax) \cos(bx)\,dx = \sqrt{\frac{2}{\pi b}} \frac{\cos\left[2\nu \arccos\left(\frac{b}{2a}\right)\right]}{\sqrt{4a^2-b^2}} \quad [0 < b < 2a];$$
$$= 0 \quad [0 < 2a < b].$$
ET I 46(24)

3. $$\int_0^\infty \sqrt{x}\, I_{\frac{1}{4}-\nu}\left(\frac{1}{2}ax\right) K_{\frac{1}{4}+\nu}\left(\frac{1}{2}ax\right) \sin(bx)\,dx = \sqrt{\frac{\pi}{2b}}\, a^{-2\nu} \frac{(b+\sqrt{a^2+b^2})^{2\nu}}{\sqrt{a^2+b^2}}$$

$\left[\operatorname{Re} a > 0,\ b > 0,\ \operatorname{Re}\nu < \frac{5}{4}\right]$. ET I 106(56)

4. $$\int_0^\infty \sqrt{x}\, I_{-\frac{1}{4}-\nu}\left(\frac{1}{2}ax\right) K_{-\frac{1}{4}+\nu}\left(\frac{1}{2}ax\right) \cos(bx)\,dx = \sqrt{\frac{\pi}{2b}}\, a^{-2\nu} \frac{(b+\sqrt{a^2+b^2})^{2\nu}}{\sqrt{a^2+b^2}}$$

$\left[\operatorname{Re} a > 0,\ b > 0,\ \operatorname{Re}\nu < \frac{3}{4}\right]$. ET I 50(49)

6.699

1. $$\int_0^\infty x^\lambda J_\nu(ax)\sin(bx)\,dx = 2^{1+\lambda} a^{-(2+\lambda)} b \frac{\Gamma\left(\frac{2+\lambda+\nu}{2}\right)}{\Gamma\left(\frac{\nu-\lambda}{2}\right)} \times$$
$$\times F\left(\frac{2+\lambda+\nu}{2}, \frac{2+\lambda-\nu}{2}; \frac{3}{2}; \frac{b^2}{a^2}\right)$$
$$\left[0 < b < a, \quad -\operatorname{Re}\nu - 1 < 1 + \operatorname{Re}\lambda < \frac{3}{2}\right];$$
$$= \left(\frac{1}{2}a\right)^\nu b^{-(\nu+\lambda+1)} \frac{\Gamma(\nu+\lambda+1)}{\Gamma(\nu+1)} \sin\left[\pi\left(\frac{1+\lambda+\nu}{2}\right)\right] \times$$
$$\times F\left(\frac{2+\lambda+\nu}{2}, \frac{1+\lambda+\nu}{2}; \nu+1; \frac{a^2}{b^2}\right)$$
$$\left[0 < a < b, \quad -\operatorname{Re}\nu - 1 < 1 + \operatorname{Re}\lambda < \frac{3}{2}\right].$$
ET I 100(11)

2. $$\int_0^\infty x^\lambda J_\nu(ax)\cos(bx)\,dx = \frac{2^\lambda a^{-(1+\lambda)}\Gamma\left(\frac{1+\lambda+\nu}{2}\right)}{\Gamma\left(\frac{\nu-\lambda+1}{2}\right)} \times$$
$$\times F\left(\frac{1+\lambda+\nu}{2}, \frac{1+\lambda-\nu}{2}; \frac{1}{2}; \frac{b^2}{a^2}\right)$$
$$\left[0 < b < a, \quad -\operatorname{Re}\nu < 1 + \operatorname{Re}\lambda < \frac{3}{2}\right];$$
$$= \frac{\left(\frac{a}{2}\right)^\nu b^{-(\nu+1+\lambda)}\Gamma(1+\lambda+\nu)\cos\left[\frac{\pi}{2}(1+\lambda+\nu)\right]}{\Gamma(\nu+1)} \times$$
$$\times F\left(\frac{1+\lambda+\nu}{2}, \frac{2+\lambda+\nu}{2}; \nu+1; \frac{a^2}{b^2}\right)$$
$$\left[0 < a < b, \quad -\operatorname{Re}\nu < 1 + \operatorname{Re}\lambda < \frac{3}{2}\right].$$
ET I 45(13)

3. $$\int_0^\infty x^\lambda K_\mu(ax)\sin(bx)\,dx = \frac{2^\lambda b\Gamma\left(\frac{2+\mu+\lambda}{2}\right)\Gamma\left(\frac{2+\lambda-\mu}{2}\right)}{a^{2+\lambda}} \times$$
$$\times F\left(\frac{2+\mu+\lambda}{2}, \frac{2+\lambda-\mu}{2}; \frac{3}{2}; -\frac{b^2}{a^2}\right)$$
$$[\operatorname{Re}(-\lambda \pm \mu) < 2, \quad \operatorname{Re} a > 0, \quad b > 0].$$
ET I 106(50)

4. $$\int_0^\infty x^\lambda K_\mu(ax)\cos(bx)\,dx = 2^{\lambda-1} a^{-\lambda-1}\Gamma\left(\frac{\mu+\lambda+1}{2}\right)\Gamma\left(\frac{1+\lambda-\mu}{2}\right) \times$$
$$\times F\left(\frac{\mu+\lambda+1}{2}, \frac{1+\lambda-\mu}{2}; \frac{1}{2}; -\frac{b^2}{a^2}\right)$$
$$[\operatorname{Re}(-\lambda \pm \mu) < 1, \quad \operatorname{Re} a > 0, \quad b > 0].$$
ET I 49(42)

5. $\int_0^\infty x^\nu \sin(ax) J_\nu(bx)\,dx =$

$$= \frac{\sqrt{\pi}\, 2^\nu b^\nu (a^2-b^2)^{-\nu-\frac{1}{2}}}{\Gamma\left(\frac{1}{2}-\nu\right)} \qquad \left[0 < b < a, \quad -1 < \operatorname{Re}\nu < \frac{1}{2}\right];$$

$$= 0 \qquad \left[0 < a < b, \quad -1 < \operatorname{Re}\nu < \frac{1}{2}\right].$$

ET II 32(4)

6. $\int_0^\infty x^\nu \cos(ax) J_\nu(bx)\,dx =$

$$= -2^\nu \frac{\sin(\nu\pi)}{\sqrt{\pi}} \Gamma\left(\frac{1}{2}+\nu\right) b^\nu (a^2-b^2)^{-\nu-\frac{1}{2}} \qquad \left[0 < b < a, \quad |\operatorname{Re}\nu| < \frac{1}{2}\right];$$

$$= 2^\nu \frac{b^\nu}{\sqrt{\pi}} \Gamma\left(\frac{1}{2}+\nu\right) (b^2-a^2)^{-\nu-\frac{1}{2}} \qquad \left[0 < a < b, \quad |\operatorname{Re}\nu| < \frac{1}{2}\right].$$

ET II 36(29)

7. $\int_0^\infty x^{\nu+1} \sin(ax) J_\nu(bx)\,dx =$

$$= -2^{1+\nu} a \frac{\sin(\nu\pi)}{\sqrt{\pi}} b^\nu \Gamma\left(\nu+\frac{3}{2}\right) (a^2-b^2)^{-\nu-\frac{3}{2}}$$

$$\left[0 < b < a, \quad -\frac{3}{2} < \operatorname{Re}\nu < -\frac{1}{2}\right];$$

$$= -\frac{2^{1+\nu}}{\sqrt{\pi}} a b^\nu \Gamma\left(\nu+\frac{3}{2}\right) (b^2-a^2)^{-\nu-\frac{3}{2}}$$

$$\left[0 < a < b, \quad -\frac{3}{2} < \operatorname{Re}\nu < -\frac{1}{2}\right].$$

ET II 32(3)

8. $\int_0^\infty x^{\nu+1} \cos(ax) J_\nu(bx)\,dx =$

$$= 2^{1+\nu} \sqrt{\pi}\, a b^\nu \frac{(a^2-b^2)^{-\nu-\frac{3}{2}}}{\Gamma\left(-\frac{1}{2}-\nu\right)} \qquad \left[0 < b < a, \quad -1 < \operatorname{Re}\nu < -\frac{1}{2}\right];$$

$$= 0 \qquad \left[0 < a < b, \quad -1 < \operatorname{Re}\nu < -\frac{1}{2}\right].$$

ET II 36(28)

9. $\int_0^1 x^\nu \sin(ax) J_\nu(ax)\,dx = \frac{1}{2\nu+1}\left[\sin a J_\nu(a) - \cos a J_{\nu+1}(a)\right]$

$$[\operatorname{Re}\nu > -1].$$

ET II 334(9)a

10. $\int_0^1 x^\nu \cos(ax) J_\nu(ax)\,dx = \frac{1}{2\nu+1}\left[\cos a J_\nu(a) + \sin a J_{\nu+1}(a)\right]$

$$\left[\operatorname{Re}\nu > -\frac{1}{2}\right].$$

ET II 335(20)

11. $$\int_0^\infty x^{1+\nu} K_\nu(ax) \sin(bx)\,dx = \sqrt{\pi}(2a)^\nu \Gamma\left(\frac{3}{2}+\nu\right) b(b^2+a^2)^{-\frac{3}{2}-\nu}$$
$$\left[\operatorname{Re} a > 0,\ b > 0,\ \operatorname{Re}\nu > -\frac{3}{2}\right].$$ ET I 105(49)

12. $$\int_0^\infty x^\mu K_\mu(ax) \cos(bx)\,dx = \frac{1}{2}\sqrt{\pi}(2a)^\mu \Gamma\left(\mu+\frac{1}{2}\right)(b^2+a^2)^{-\mu-\frac{1}{2}}$$
$$\left[\operatorname{Re} a > 0,\ b > 0,\ \operatorname{Re}\mu > -\frac{1}{2}\right].$$ ET I 49(41)

13. $$\int_0^\infty x^\nu N_{\nu-1}(ax) \sin(bx)\,dx =$$
$$= 0 \qquad \left[0 < b < a,\ |\operatorname{Re}\nu| < \frac{1}{2}\right];$$
$$= \frac{2^\nu \sqrt{\pi}\, a^{\nu-1} b}{\Gamma\left(\frac{1}{2}-\nu\right)}(b^2-a^2)^{-\nu-\frac{1}{2}} \qquad \left[0 < a < b,\ |\operatorname{Re}\nu| < \frac{1}{2}\right].$$
ET I 104(36)

14. $$\int_0^\infty x^\nu N_\nu(ax) \cos(bx)\,dx =$$
$$= 0 \qquad \left[0 < b < a,\ |\operatorname{Re}\nu| < \frac{1}{2}\right];$$
$$= -2^\nu \sqrt{\pi} a^\nu \frac{(b^2-a^2)^{-\nu-\frac{1}{2}}}{\Gamma\left(\frac{1}{2}-\nu\right)} \qquad \left[0 < a < b,\ |\operatorname{Re}\nu| < \frac{1}{2}\right].$$
ET I 47(30)

6.711

1. $$\int_0^\infty x^{\nu-\mu} J_\mu(ax) J_\nu(bx) \sin(cx)\,dx = 0$$
$$[0 < c < b-a,\ -1 < \operatorname{Re}\nu < 1+\operatorname{Re}\mu].$$ ET I 103(28)

2. $$\int_0^\infty x^{\nu-\mu+1} J_\mu(ax) J_\nu(bx) \cos(cx)\,dx = 0$$
$$[0 < c < b-a,\ a > 0,\ b > 0,\ -1 < \operatorname{Re}\nu < \operatorname{Re}\mu].$$ ET I 47(25)

3. $$\int_0^\infty x^{\nu-\mu-2} J_\mu(ax) J_\nu(bx) \sin(cx)\,dx = 2^{\nu-\mu-1} a^\mu b^{-\nu} \frac{c\Gamma(\nu)}{\Gamma(\mu+1)}$$
$$[0 < a,\ 0 < b,\ 0 < c < b-a,\ 0 < \operatorname{Re}\nu < \operatorname{Re}\mu+3].$$ ET I 103(29)

4. $$\int_0^\infty x^{\varrho-\mu-1} J_\mu(ax) J_\varrho(bx) \cos(cx)\,dx = 2^{\varrho-\mu-1} b^{-\varrho} a^\mu \frac{\Gamma(\varrho)}{\Gamma(\mu+1)}$$
$$[b > 0,\ a > 0,\ 0 < c < b-a,\ 0 < \operatorname{Re}\varrho < \operatorname{Re}\mu+2].$$ ET I 47(26)

5. $$\int_0^\infty x^{1-2\nu} \sin(2ax) J_\nu(x) N_\nu(x)\,dx =$$
$$= -\frac{\Gamma\left(\frac{3}{2}-\nu\right) a}{2\Gamma\left(2\nu-\frac{1}{2}\right)\Gamma(2-\nu)} F\left(\frac{3}{2}-\nu, \frac{3}{2}-2\nu; 2-\nu; a^2\right)$$
$$\left[0 < \operatorname{Re}\nu < \frac{3}{2}, \ 0 < a < 1\right].$$
ET II 348(63)

1. $$\int_0^\infty x^\nu [J_\nu(ax)\cos(ax) + N_\nu(ax)\sin(ax)]\sin(bx)\,dx =$$
$$= \frac{\sqrt{\pi}(2a)^\nu}{\Gamma\left(\frac{1}{2}-\nu\right)} (b^2+2ab)^{-\nu-\frac{1}{2}}$$
$$\left[b > 0, \ -1 < \operatorname{Re}\nu < \frac{1}{2}\right].$$
ET I 104(40)

2. $$\int_0^\infty x^\nu [N_\nu(ax)\cos(ax) - J_\nu(ax)\sin(ax)]\cos(bx)\,dx =$$
$$= -\frac{\sqrt{\pi}(2a)^\nu}{\Gamma\left(\frac{1}{2}-\nu\right)} (b^2+2ab)^{-\nu-\frac{1}{2}}.$$
ET I 48(35)

3. $$\int_0^\infty x^\nu [J_\nu(ax)\cos(ax) - N_\nu(ax)\sin(ax)]\sin(bx)\,dx = 0$$
$$\left[0 < b < 2a, \ -1 < \operatorname{Re}\nu < \frac{1}{2}\right];$$
$$= \frac{2^\nu \sqrt{\pi}\, b^\nu}{\Gamma\left(\frac{1}{2}-\nu\right)} (b^2-2ab)^{-\nu-\frac{1}{2}} \qquad \left[2a < b, \ -1 < \operatorname{Re}\nu < \frac{1}{2}\right].$$
ET I 104(41)

4. $$\int_0^\infty x^\nu [J_\nu(ax)\sin(ax) + N_\nu(ax)\cos(ax)]\cos(bx)\,dx = 0$$
$$\left[0 < b < 2a, \ |\operatorname{Re}\nu| < \frac{1}{2}\right];$$
$$= -\frac{\sqrt{\pi}(2a)^\nu}{\Gamma\left(\frac{1}{2}-\nu\right)} (b^2-2ab)^{-\nu-\frac{1}{2}} \qquad \left[0 < 2a < b, \ |\operatorname{Re}\nu| < \frac{1}{2}\right].$$
ET I 48(33)

6.713

1. $$\int_0^\infty x^{1-2\nu} \sin(2ax) \{[J_\nu(x)]^2 - [N_\nu(x)]^2\}\,dx =$$
$$= \frac{\sin(2\nu\pi)\Gamma\left(\frac{3}{2}-\nu\right)\Gamma\left(\frac{3}{2}-2\nu\right) a}{\pi\Gamma(2-\nu)} F\left(\frac{3}{2}-\nu, \frac{3}{2}-2\nu; 2-\nu; a^2\right)$$
$$\left[0 < \operatorname{Re}\nu < \frac{3}{4}, \ 0 < a < 1\right].$$
ET II 348(64)

2. $$\int_0^\infty x^{2-2\nu}\sin(2ax)\left[J_\nu(x)J_{\nu-1}(x)-N_\nu(x)N_{\nu-1}(x)\right]dx=$$
$$=-\frac{\sin(2\nu\pi)\Gamma\left(\frac{3}{2}-\nu\right)\Gamma\left(\frac{5}{2}-2\nu\right)a}{\pi\Gamma(2-\nu)}F\left(\frac{3}{2}-\nu,\ \frac{5}{2}-2\nu;\ 2-\nu;\ a^2\right)$$
$$\left[\frac{1}{2}<\operatorname{Re}\nu<\frac{5}{4},\ 0<a<1\right].$$ ET II 348(65)

3. $$\int_0^\infty x^{2-2\nu}\sin(2ax)\left[J_\nu(x)N_{\nu-1}(x)+N_\nu(x)J_{\nu-1}(x)\right]dx=$$
$$=-\frac{\Gamma\left(\frac{3}{2}-\nu\right)a}{\Gamma\left(2\nu-\frac{3}{2}\right)\Gamma(2-\nu)}F\left(\frac{3}{2}-\nu,\ \frac{5}{2}-2\nu;\ 2-\nu;\ a^2\right)$$
$$\left[\frac{1}{2}<\operatorname{Re}\nu<\frac{5}{2},\ 0<a<1\right].$$ ET II 349(66)

6.714

1. $$\int_0^\infty \sin(2ax)\left[x^\nu J_\nu(x)\right]^2dx=$$
$$=\frac{a^{-2\nu}\Gamma\left(\frac{1}{2}+\nu\right)}{2\sqrt{\pi}\Gamma(1-\nu)}F\left(\frac{1}{2}+\nu,\ \frac{1}{2};\ 1-\nu;\ a^2\right)$$
$$\left[0<a<1,\quad |\operatorname{Re}\nu|<\frac{1}{2}\right],$$
$$=\frac{a^{-4\nu-1}\Gamma\left(\frac{1}{2}+\nu\right)}{2\Gamma(1+\nu)\Gamma\left(\frac{1}{2}-2\nu\right)}F\left(\frac{1}{2}+\nu,\ \frac{1}{2}+2\nu;\ 1+\nu;\ \frac{1}{a^2}\right)$$
$$\left[a>1,\ |\operatorname{Re}\nu|<\frac{1}{2}\right].$$ ET II 343(31)

2. $$\int_0^\infty \cos(2ax)\left[x^\nu J_\nu(x)\right]^2dx=$$
$$=\frac{a^{-2\nu}\Gamma(\nu)}{2\sqrt{\pi}\Gamma\left(\frac{1}{2}-\nu\right)}F\left(\nu+\frac{1}{2},\ \frac{1}{2};\ 1-\nu;\ a^2\right)+$$
$$+\frac{\Gamma(-\nu)\Gamma\left(\frac{1}{2}+2\nu\right)}{2\pi\Gamma\left(\frac{1}{2}-\nu\right)}F\left(\frac{1}{2}+\nu,\ \frac{1}{2}+2\nu;\ 1+\nu;\ a^2\right)$$
$$\left[0<a<1,\ -\frac{1}{4}<\operatorname{Re}\nu<\frac{1}{2}\right];$$
$$=-\frac{\sin(\nu\pi)a^{-4\nu-1}\Gamma\left(\frac{1}{2}+2\nu\right)}{\Gamma(1+\nu)\Gamma\left(\frac{1}{2}-\nu\right)}F\left(\frac{1}{2}+\nu,\ \frac{1}{2}+2\nu;\ 1+\nu;\ \frac{1}{a^2}\right)$$
$$\left[a>1,\ -\frac{1}{4}<\operatorname{Re}\nu<\frac{1}{2}\right].$$ ET II 344(33)

6.715

1. $$\int_0^\infty \frac{x^\nu}{x+\beta} \sin(x+\beta) J_\nu(x)\, dx = \frac{\pi}{2} \sec(\nu\pi) \beta^\nu J_{-\nu}(\beta)$$

$$\left[|\arg\beta| < \pi,\ |\operatorname{Re}\nu| < \tfrac{1}{2} \right].$$ ET II 340(8)

2. $$\int_0^\infty \frac{x^\nu}{x+\beta} \cos(x+\beta) J_\nu(x)\, dx = -\frac{\pi}{2} \sec(\nu\pi) \beta^\nu N_{-\nu}(\beta)$$

$$\left[|\arg\beta| < \pi,\ |\operatorname{Re}\nu| < \tfrac{1}{2} \right].$$ ET II 340(9)

6.716

1. $$\int_0^a x^\lambda \sin(a-x) J_\nu(x)\, dx =$$

$$= 2a^{\lambda+1} \sum_{n=0}^\infty \frac{(-1)^n \Gamma(\nu-\lambda+2n)\, \Gamma(\nu+\lambda+1)}{\Gamma(\nu-\lambda)\, \Gamma(\nu+\lambda+3+2n)} (\nu+2n+1) J_{\nu+2n+1}(a)$$

$$[\operatorname{Re}(\lambda+\nu) > -1].$$ ET II 335(16)

2. $$\int_0^a x^\lambda \cos(a-x) J_\nu(x)\, dx = \frac{a^{\lambda+1} J_\nu(a)}{\lambda+\nu+1} +$$

$$+ 2a^{\lambda+1} \sum_{n=1}^\infty \frac{(-1)^n \Gamma(\nu-\lambda+2n-1)\, \Gamma(\nu+\lambda+1)}{\Gamma(\nu-\lambda)\, \Gamma(\nu+\lambda+2n+2)} (\nu+2n) J_{\nu+2n}(a)$$

$$[\operatorname{Re}(\lambda+\nu) > -1].$$ ET II 336(26)

6.717 $$\int_{-\infty}^\infty \frac{\sin[a(x+\beta)]}{x^\nu (x+\beta)} J_{\nu+2n}(x)\, dx = \pi\beta^{-\nu} J_{\nu+2n}(\beta)$$

$$\left[1 \leqslant a < \infty,\ n = 0, 1, 2, \ldots;\ \operatorname{Re}\nu > -\tfrac{3}{2} \right].$$ ET II 345(44)

6.718

1. $$\int_0^\infty \frac{x^\nu}{x^2+\beta^2} \sin(\alpha x) J_\nu(\gamma x)\, dx = \beta^{\nu-1} \operatorname{sh}(\alpha\beta) K_\nu(\beta\gamma)$$

$$\left[0 < \alpha \leqslant \gamma,\ \operatorname{Re}\beta > 0,\ -1 < \operatorname{Re}\nu < \tfrac{3}{2} \right].$$ ET II 33(8)

2. $$\int_0^\infty \frac{x^{\nu+1}}{x^2+\beta^2} \cos(\alpha x) J_\nu(\gamma x)\, dx = \beta^\nu \operatorname{ch}(\alpha\beta) K_\nu(\beta\gamma)$$

$$\left[0 < \alpha \leqslant \gamma,\ \operatorname{Re}\beta > 0,\ -1 < \operatorname{Re}\nu < \tfrac{1}{2} \right].$$ ET II 37(33)

3. $$\int_0^\infty \frac{x^{1-\nu}}{x^2+\beta^2}\sin(\alpha x) J_\nu(\gamma x)\,dx = \frac{\pi}{2}\beta^{-\nu}e^{-\alpha\beta}I_\nu(\beta\gamma)$$

$$\left[0<\gamma\leqslant\alpha,\ \operatorname{Re}\beta>0,\ \operatorname{Re}\nu>-\frac{1}{2}\right].$$ ET II 33(9)

4. $$\int_0^\infty \frac{x^{-\nu}}{x^2+\beta^2}\cos(\alpha x) J_\nu(\gamma x)\,dx = \frac{\pi}{2}\beta^{-\nu-1}e^{-\alpha\beta}I_\nu(\beta\gamma)$$

$$\left[0<\gamma\leqslant\alpha,\ \operatorname{Re}\beta>0,\ \operatorname{Re}\nu>-\frac{3}{2}\right].$$ ET II 37(34)

6.719

1. $$\int_0^\alpha \frac{\sin(\beta x)}{\sqrt{\alpha^2-x^2}} J_\nu(x)\,dx =$$

$$= \pi\sum_{n=0}^\infty (-1)^n J_{2n+1}(\alpha\beta) J_{\frac{1}{2}\nu+n+\frac{1}{2}}\left(\frac{1}{2}\alpha\right) J_{\frac{1}{2}\nu-n-\frac{1}{2}}(\alpha)$$

$$[\operatorname{Re}\nu>-2].$$ ET II 335(17)

2. $$\int_0^\alpha \frac{\cos(\beta x)}{\sqrt{\alpha^2-x^2}} J_\nu(x)\,dx = \frac{\pi}{2} J_0(\alpha\beta)\left[J_{\frac{1}{2}\nu}\left(\frac{1}{2}\alpha\right)\right]^2 +$$

$$+\pi\sum_{n=1}^\infty (-1)^n J_{2n}(\alpha\beta) J_{\frac{1}{2}\nu+n}\left(\frac{1}{2}\alpha\right) J_{\frac{1}{2}\nu-n}\left(\frac{1}{2}\alpha\right).$$

$$[\operatorname{Re}\nu>-1].$$ ET II 336(27)

6.721

1. $$\int_0^\infty \sqrt{x}\, J_{\frac{1}{4}}(a^2x^2)\sin(bx)\,dx = 2^{-\frac{3}{2}}a^{-2}\sqrt{\pi b}\, J_{\frac{1}{4}}\left(\frac{b^2}{4a^2}\right)$$

$$[b>0].$$ ET I 108(1)

2. $$\int_0^\infty \sqrt{x}\, J_{-\frac{1}{4}}(a^2x^2)\cos(bx)\,dx = 2^{-\frac{3}{2}}a^{-2}\sqrt{\pi b}\, J_{-\frac{1}{4}}\left(\frac{b^2}{4a^2}\right)$$

$$[b>0].$$ ET I 51(1)

3. $$\int_0^\infty \sqrt{x}\, N_{\frac{1}{4}}(a^2x^2)\sin(bx)\,dx =$$

$$= -2^{-\frac{3}{2}}\sqrt{\pi b}\, a^{-2}\mathbf{H}_{\frac{1}{4}}\left(\frac{b^2}{4a^2}\right).$$ ET I 108(7)

4. $$\int_0^\infty \sqrt{x}\, N_{-\frac{1}{4}}(a^2x^2)\cos(bx)\,dx =$$

$$= -2^{-\frac{3}{2}}\sqrt{\pi b}\, a^{-2}\mathbf{H}_{-\frac{1}{4}}\left(\frac{b^2}{4a^2}\right)$$ ET I 52(7)

5. $$\int_0^\infty \sqrt{x} K_{\frac{1}{4}}(a^2x^2) \sin(bx)\, dx =$$

$$= 2^{-\frac{5}{2}} \sqrt{\pi^3 b}\, a^{-2} \left[I_{\frac{1}{4}}\left(\frac{b^2}{4a^2}\right) - \mathbf{L}_{\frac{1}{4}}\left(\frac{b^2}{4a^2}\right) \right]$$

$$\left[|\arg a| < \frac{\pi}{4},\ b > 0 \right].$$ ET I 109(11)

6. $$\int_0^\infty \sqrt{x} K_{-\frac{1}{4}}(a^2x^2) \cos(bx)\, dx =$$

$$= 2^{-\frac{5}{2}} \sqrt{\pi^3 b}\, a^{-2} \left[I_{-\frac{1}{4}}\left(\frac{b^2}{4a^2}\right) - \mathbf{L}_{-\frac{1}{4}}\left(\frac{b^2}{4a^2}\right) \right] \qquad [b > 0].$$

ET I 52(10)

6.722

1. $$\int_0^\infty \sqrt{x} K_{\frac{1}{8}+\nu}(a^2x^2) I_{\frac{1}{8}-\nu}(a^2x^2) \sin(bx)\, dx =$$

$$= \sqrt{2\pi}\, b^{-\frac{3}{2}} \frac{\Gamma\left(\frac{5}{8} - \nu\right)}{\Gamma\left(\frac{5}{4}\right)} W_{\nu, \frac{1}{8}}\left(\frac{b^2}{8a^2}\right) M_{-\nu, \frac{1}{8}}\left(\frac{b^2}{8a^2}\right)$$

$$\left[\operatorname{Re} \nu < \frac{5}{8},\ |\arg a| < \frac{\pi}{4},\ b > 0 \right].$$ ET I 109(13)

2. $$\int_0^\infty \sqrt{x} J_{-\frac{1}{8}-\nu}(a^2x^2) J_{-\frac{1}{8}+\nu}(a^2x^2) \cos(bx)\, dx =$$

$$= \sqrt{\frac{2}{\pi}}\, b^{-\frac{3}{2}} \left[e^{-\frac{i\pi}{8}} W_{\nu, -\frac{1}{8}}\left(\frac{b^2 e^{-\frac{\pi i}{2}}}{8a^2}\right) W_{-\nu, -\frac{1}{8}}\left(\frac{b^2 e^{-\frac{\pi i}{2}}}{8a^2}\right) + \right.$$

$$\left. + e^{\frac{i\pi}{8}} W_{\nu, -\frac{1}{8}}\left(\frac{b^2 e^{\frac{\pi i}{2}}}{8a^2}\right) W_{-\nu, -\frac{1}{8}}\left(\frac{b^2 e^{\frac{\pi i}{2}}}{8a^2}\right) \right]$$

$$[b > 0].$$ ET I 52(6)

3. $$\int_0^\infty \sqrt{x} J_{\frac{1}{8}-\nu}(a^2x^2) J_{\frac{1}{8}+\nu}(a^2x^2) \sin(bx)\, dx =$$

$$= \sqrt{\frac{2}{\pi}}\, b^{-\frac{3}{2}} \left[e^{\frac{\pi i}{8}} W_{\nu, \frac{1}{8}}\left(\frac{b^2 e^{\frac{\pi i}{2}}}{8a^2}\right) W_{-\nu, \frac{1}{8}}\left(\frac{b^2 e^{\frac{\pi i}{2}}}{8a^2}\right) + \right.$$

$$\left. + e^{-\frac{\pi i}{8}} W_{\nu, \frac{1}{8}}\left(\frac{b^2 e^{-\frac{\pi i}{2}}}{8a^2}\right) W_{-\nu, \frac{1}{8}}\left(\frac{b^2 e^{-\frac{\pi i}{2}}}{8a^2}\right) \right] \qquad [b > 0].$$ ET I 108(6)

4. $$\int_0^\infty \sqrt{x} K_{\frac{1}{8}-\nu}(a^2x^2) I_{-\frac{1}{8}-\nu}(a^2x^2)\cos(bx)\,dx =$$
$$= \sqrt{2\pi}\, b^{-\frac{3}{2}} \frac{\Gamma\left(\frac{3}{8}-\nu\right)}{\Gamma\left(\frac{3}{4}\right)} W_{\nu,-\frac{1}{8}}\left(\frac{b^2}{8a^2}\right) M_{-\nu,-\frac{1}{8}}\left(\frac{b^2}{8a^2}\right)$$
$$\left[\operatorname{Re}\nu < \frac{3}{8},\ b > 0\right].$$ ET I 52(12)

6.723 $$\int_0^\infty xJ_\nu(x^2)\left[\sin(\nu\pi)J_\nu(x^2) - \cos(\nu\pi)N_\nu(x^2)\right]J_{4\nu}(4ax)\,dx =$$
$$= \frac{1}{4}J_\nu(a^2)J_{-\nu}(a^2)$$
$$[a > 0,\ \operatorname{Re}\nu > -1].$$ ET II 375(20)

6.724

1. $$\int_0^\infty x^{2\lambda}J_{2\nu}\left(\frac{a}{x}\right)\sin(bx)\,dx =$$
$$= \frac{\sqrt{\pi}\,a^{2\nu}\Gamma(\lambda-\nu+1)\,b^{2\nu-2\lambda-1}}{4^{2\nu-\lambda}\Gamma(2\nu+1)\,\Gamma\left(\nu-\lambda+\frac{1}{2}\right)}\, {}_0F_3\left(2\nu+1,\ \nu-\lambda,\ \nu-\lambda+\frac{1}{2};\ \frac{a^2b^2}{16}\right)+$$
$$+\frac{a^{2\lambda+2}\Gamma(\nu-\lambda-1)\,b}{2^{2\lambda+3}\Gamma(\nu+\lambda+2)}\, {}_0F_3\left(\frac{3}{2},\ \lambda-\nu+2,\ \lambda+\nu+2;\ \frac{a^2b^2}{16}\right)$$
$$\left[-\frac{5}{4} < \operatorname{Re}\lambda < \operatorname{Re}\nu,\ a > 0,\ b > 0\right].$$ ET I 109(15)

2. $$\int_0^\infty x^{2\lambda}J_{2\nu}\left(\frac{a}{x}\right)\cos(bx)\,dx = 4^{\lambda-2\nu}\sqrt{\pi}\,a^{2\nu}b^{2\nu-2\lambda-1}\times$$
$$\times \frac{\Gamma\left(\lambda-\nu+\frac{1}{2}\right)}{\Gamma(2\nu+1)\,\Gamma(\nu-\lambda)}\, {}_0F_3\left(2\nu+1,\ \nu-\lambda+\frac{1}{2},\ \nu-\lambda;\ \frac{a^2b^2}{16}\right)+$$
$$+4^{-\lambda-1}a^{2\lambda+1}\frac{\Gamma\left(\nu-\lambda-\frac{1}{2}\right)}{\Gamma\left(\nu+\lambda+\frac{3}{2}\right)}\, {}_0F_3\left(\frac{1}{2},\ \lambda-\nu+\frac{3}{2},\ \nu+\lambda+\frac{3}{2};\ \frac{a^2b^2}{16}\right)$$
$$\left[-\frac{3}{4} < \operatorname{Re}\lambda < \operatorname{Re}\nu - \frac{1}{2},\ a > 0,\ b > 0\right].$$ ET I 53(14)

6.725

1. $$\int_0^\infty \frac{\sin(bx)}{\sqrt{x}}J_\nu(a\sqrt{x})\,dx = -\sqrt{\frac{\pi}{b}}\sin\left(\frac{a^2}{8b}-\frac{\nu\pi}{4}-\frac{\pi}{4}\right)J_{\frac{\nu}{2}}\left(\frac{a^2}{8b}\right)$$
$$\left[\operatorname{Re}\nu > -3,\ a > 0,\ b > 0\right].$$ ET I 110(27)

2. $$\int_0^\infty \frac{\cos(bx)}{\sqrt{x}}J_\nu(a\sqrt{x})\,dx =$$
$$= \sqrt{\frac{\pi}{b}}\cos\left(\frac{a^2}{8b}-\frac{\nu\pi}{4}-\frac{\pi}{4}\right)J_{\frac{1}{2}\nu}\left(\frac{a^2}{8b}\right)$$
$$[\operatorname{Re}\nu > -1,\ a > 0,\ b > 0].$$ ET I 54(25)

3. $$\int_0^\infty x^{\frac{1}{2}\nu} J_\nu(a\sqrt{x}) \sin(bx)\,dx = 2^{-\nu} a^\nu b^{-\nu-1} \cos\left(\frac{a^2}{4b} - \frac{\nu\pi}{2}\right)$$

$$\left[-2 < \operatorname{Re}\nu < \frac{1}{2},\ a > 0,\ b > 0\right].$$ ET I 110(28)

4. $$\int_0^\infty x^{\frac{1}{2}\nu} J_\nu(a\sqrt{x}) \cos(bx)\,dx = 2^{-\nu} b^{-\nu-1} a^\nu \sin\left(\frac{a^2}{4b} - \frac{\nu\pi}{2}\right)$$

$$\left[-1 < \operatorname{Re}\nu < \frac{1}{2},\ a > 0,\ b > 0\right].$$ ET I 54(26)

6.726

1. $$\int_0^\infty x(x^2+b^2)^{-\frac{1}{2}\nu} J_\nu\left(a\sqrt{x^2+b^2}\right) \sin(cx)\,dx =$$

$$= \sqrt{\frac{\pi}{2}}\, a^{-\nu} b^{-\nu+\frac{3}{2}} c\,(a^2-c^2)^{\frac{1}{2}\nu-\frac{3}{4}} J_{\nu-\frac{3}{2}}\left(b\sqrt{a^2-c^2}\right) \qquad \left[0 < c < a,\ \operatorname{Re}\nu > \frac{1}{2}\right];$$

$$= 0 \qquad \left[0 < a < c,\ \operatorname{Re}\nu > \frac{1}{2}\right].$$ ET I 111(37)

2. $$\int_0^\infty (x^2+b^2)^{-\frac{1}{2}\nu} J_\nu\left(a\sqrt{x^2+b^2}\right) \cos(cx)\,dx =$$

$$= \sqrt{\frac{\pi}{2}}\, a^{-\nu} b^{-\nu+\frac{1}{2}} (a^2-c^2)^{\frac{1}{2}\nu-\frac{1}{4}} J_{\nu-\frac{1}{2}}\left(b\sqrt{a^2-c^2}\right) \qquad \left[0 < c < a,\ b > 0,\ \operatorname{Re}\nu > -\frac{1}{2}\right];$$

$$= 0 \qquad \left[0 < a < c,\ b > 0,\ \operatorname{Re}\nu > -\frac{1}{2}\right].$$ ET I 55(37)

3. $$\int_0^\infty x(x^2+b^2)^{\frac{1}{2}\nu} K_{\pm\nu}\left(a\sqrt{x^2+b^2}\right) \sin(cx)\,dx =$$

$$= \sqrt{\frac{\pi}{2}}\, a^\nu b^{\nu+\frac{3}{2}} c\,(a^2+c^2)^{-\frac{1}{2}\nu-\frac{3}{4}} K_{-\nu-\frac{3}{2}}\left(b\sqrt{a^2+c^2}\right)$$

$$[\operatorname{Re} a > 0,\ \operatorname{Re} b > 0,\ c > 0].$$ ET I 113(45)

4. $$\int_0^\infty (x^2+b^2)^{\mp\frac{1}{2}\nu} K_\nu\left(a\sqrt{x^2+b^2}\right) \cos(cx)\,dx =$$

$$= \sqrt{\frac{\pi}{2}}\, a^{\mp\nu} b^{\frac{1}{2}\mp\nu} (a^2+c^2)^{\pm\frac{1}{2}\nu-\frac{1}{4}} K_{\pm\nu-\frac{1}{2}}\left(b\sqrt{a^2+c^2}\right)$$

$$[\operatorname{Re} a > 0,\ \operatorname{Re} b > 0,\ c > 0].$$ ET I 56(45)

5. $$\int_0^\infty (x^2+a^2)^{-\frac{1}{2}\nu} N_\nu\left(b\sqrt{x^2+a^2}\right)\cos(cx)\,dx =$$

$$= \sqrt{\frac{a\pi}{2}}(ab)^{-\nu}(b^2-c^2)^{\frac{1}{2}\nu-\frac{1}{4}} N_{\nu-\frac{1}{2}}\left(a\sqrt{b^2-c^2}\right)$$

$$\left[0<c<b,\ a>0,\ \operatorname{Re}\nu>-\frac{1}{2}\right];$$

$$= -\sqrt{\frac{2a}{\pi}}(ab)^{-\nu}(c^2-b^2)^{\frac{1}{2}\nu-\frac{1}{4}} K_{\nu-\frac{1}{2}}\left(a\sqrt{c^2-b^2}\right)$$

$$\left[0<b<c,\ a>0,\ \operatorname{Re}\nu>-\frac{1}{2}\right].$$ ET I 56(41)

6.727

1. $$\int_0^a \frac{\sin(cx)}{\sqrt{a^2-x^2}} J_\nu\left(b\sqrt{a^2-x^2}\right)dx =$$

$$= \frac{\pi}{2} J_{\frac{1}{2}\nu}\left[\frac{a}{2}\left(\sqrt{b^2+c^2}-c\right)\right] J_{\frac{1}{2}\nu}\left[\frac{a}{2}\left(\sqrt{b^2+c^2}+c\right)\right]$$

$$[\operatorname{Re}\nu>-1,\ c>0,\ a>0].$$ ET I 113(48)

2. $$\int_a^\infty \frac{\sin(cx)}{\sqrt{x^2-a^2}} J_\nu\left(b\sqrt{x^2-a^2}\right)dx =$$

$$= \frac{\pi}{2} J_{\frac{1}{2}\nu}\left[\frac{a}{2}\left(c-\sqrt{c^2-b^2}\right)\right] J_{-\frac{1}{2}\nu}\left[\frac{a}{2}\left(c+\sqrt{c^2-b^2}\right)\right]$$

$$[0<b<c,\ a>0,\ \operatorname{Re}\nu>-1].$$ ET I 113(49)

3. $$\int_a^\infty \frac{\cos(cx)}{\sqrt{x^2-a^2}} J_\nu\left(b\sqrt{x^2-a^2}\right)dx =$$

$$= -\frac{\pi}{2} J_{\frac{1}{2}\nu}\left[\frac{a}{2}\left(c-\sqrt{c^2-b^2}\right)\right] N_{-\frac{1}{2}\nu}\left[\frac{a}{2}\left(c+\sqrt{c^2-b^2}\right)\right]$$

$$[0<b<c,\ a>0,\ \operatorname{Re}\nu>-1].$$ ET I 58(54)

4 $$\int_0^a (a^2-x^2)^{\frac{1}{2}\nu}\cos x\, I_\nu\left(\sqrt{a^2-x^2}\right)dx = \frac{\sqrt{\pi}\,a^{2\nu+1}}{2^{\nu+1}\Gamma\left(\nu+\frac{3}{2}\right)}$$

$$\left[\operatorname{Re}\nu>-\frac{1}{2}\right].$$ WA 409(2)

6.728

1. $$\int_0^\infty x\sin(ax^2) J_\nu(bx)\,dx =$$

$$= \frac{\sqrt{\pi}\,b}{8a^{\frac{3}{2}}}\left[\cos\left(\frac{b^2}{8a}-\frac{\nu\pi}{4}\right) J_{\frac{1}{2}\nu-\frac{1}{2}}\left(\frac{b^2}{8a}\right) - \right.$$

$$\left. -\sin\left(\frac{b^2}{8a}-\frac{\nu\pi}{4}\right) J_{\frac{1}{2}\nu+\frac{1}{2}}\left(\frac{b^2}{8a}\right)\right]$$

$$[a>0,\ b>0,\ \operatorname{Re}\nu>-4].$$ ET II 34(14)

2. $$\int_0^\infty x\cos(ax^2)J_\nu(bx)\,dx = \frac{\sqrt{\pi}\,b}{8a^{\frac{3}{2}}}\left[\cos\left(\frac{b^2}{8a}-\frac{\nu\pi}{4}\right)J_{\frac{1}{2}\nu+\frac{1}{2}}\left(\frac{b^2}{8a}\right)+\sin\left(\frac{b^2}{8a}-\frac{\nu\pi}{4}\right)J_{\frac{1}{2}\nu-\frac{1}{2}}\left(\frac{b^2}{8a}\right)\right]$$
$[a>0,\ b>0,\ \operatorname{Re}\nu>-2]$. ET II 38(39)

3. $$\int_0^\infty J_0(\beta x)\sin(\alpha x^2)\,x\,dx = \frac{1}{2\alpha}\cos\frac{\beta^2}{4\alpha}\qquad [\alpha>0,\ \beta>0].$$ MO 47

4. $$\int_0^\infty J_0(\beta x)\cos(\alpha x^2)\,x\,dx = \frac{1}{2\alpha}\sin\frac{\beta^2}{4\alpha}\qquad [\alpha>0,\ \beta>0].$$ MO 47

5. $$\int_0^\infty x^{\nu+1}\sin(ax^2)J_\nu(bx)\,dx = \frac{b^\nu}{2^{\nu+1}a^{\nu+1}}\cos\left(\frac{b^2}{4a}-\frac{\nu\pi}{2}\right)$$
$\left[a>0,\ b>0,\ -2<\operatorname{Re}\nu<\frac{1}{2}\right]$. ET II 34(15)

6. $$\int_0^\infty x^{\nu+1}\cos(ax^2)J_\nu(bx)\,dx = \frac{b^\nu}{2^{\nu+1}a^{\nu+1}}\sin\left(\frac{b^2}{4a}-\frac{\nu\pi}{2}\right)$$
$\left[a>0,\ b>0,\ -1<\operatorname{Re}\nu<\frac{1}{2}\right]$. ET II 38(40)

6.729

1. $$\int_0^\infty x\sin(ax^2)J_\nu(bx)J_\nu(cx)\,dx = \frac{1}{2a}\cos\left(\frac{b^2+c^2}{4a}-\frac{\nu\pi}{2}\right)J_\nu\left(\frac{bc}{2a}\right)$$
$[a>0,\ b>0,\ c>0,\ \operatorname{Re}\nu>-2]$. ET II 51(26)

2. $$\int_0^\infty x\cos(ax^2)J_\nu(bx)J_\nu(cx)\,dx = \frac{1}{2a}\sin\left(\frac{b^2+c^2}{4a}-\frac{\nu\pi}{2}\right)J_\nu\left(\frac{bc}{2a}\right)$$
$[a>0,\ b>0,\ c>0,\ \operatorname{Re}\nu>-1]$. ET II 51(27)

6.731

1. $$\int_0^\infty x\sin(ax^2)J_\nu(bx^2)J_{2\nu}(2cx)\,dx =$$
$$= \frac{1}{2\sqrt{b^2-a^2}}\sin\left(\frac{ac^2}{b^2-a^2}\right)J_\nu\left(\frac{bc^2}{b^2-a^2}\right)\qquad [0<a<b,\ \operatorname{Re}\nu>-1];$$
$$= \frac{1}{2\sqrt{a^2-b^2}}\cos\left(\frac{ac^2}{a^2-b^2}\right)J_\nu\left(\frac{bc^2}{a^2-b^2}\right)\qquad [0<b<a,\ \operatorname{Re}\nu>-1].$$
ET II 356(41)a

2. $$\int_0^\infty x\cos(ax^2)\,J_\nu(bx^2)\,J_{2\nu}(2cx)\,dx =$$

$$= \frac{1}{2\sqrt{b^2-a^2}}\cos\left(\frac{ac^2}{b^2-a^2}\right)J_\nu\left(\frac{bc^2}{b^2-a^2}\right) \quad \left[0<a<b,\ \operatorname{Re}\nu>-\tfrac{1}{2}\right];$$

$$= \frac{1}{2\sqrt{a^2-b^2}}\sin\left(\frac{ac^2}{a^2-b^2}\right)J_\nu\left(\frac{bc^2}{a^2-b^2}\right) \quad \left[0<b<a,\ \operatorname{Re}\nu>-\tfrac{1}{2}\right].$$

ET II 356(42)a

6.732 $$\int_0^\infty x^3\cos\left(\frac{x^2}{2a}\right)N_1(x)\,K_1(x)\,dx = -a^3K_0(a) \quad [a>0].$$ ET II 371(52)

6.733

1. $$\int_0^\infty \sin\left(\frac{a}{2x}\right)[\sin x J_0(x)+\cos x N_0(x)]\frac{dx}{x} = \pi J_0(\sqrt{a})\,N_0(\sqrt{a}) \quad [a>0].$$ ET II 346(51)

2. $$\int_0^\infty \cos\left(\frac{a}{2x}\right)[\sin x N_0(x)-\cos x J_0(x)]\frac{dx}{x} = \pi J_0(\sqrt{a})\,N_0(\sqrt{a}) \quad [a>0].$$ ET II 347(52)

3. $$\int_0^\infty x\sin\left(\frac{a}{2x}\right)K_0(x)\,dx = \frac{\pi a}{2}J_1(\sqrt{a})\,K_1(\sqrt{a}) \quad [a>0].$$ ET II 368(34)

4. $$\int_0^\infty x\cos\left(\frac{a}{2x}\right)K_0(x)\,dx = -\frac{\pi a}{2}N_1(\sqrt{a})\,K_1(\sqrt{a}) \quad [a>0].$$ ET II 369(35)

6.734 $$\int_0^\infty \cos(a\sqrt{x})\,K_\nu(bx)\frac{dx}{\sqrt{x}} =$$

$$= \frac{\pi}{2\sqrt{b}}\sec(\nu\pi)\left[D_{\nu-\frac{1}{2}}\left(\frac{a}{\sqrt{2b}}\right)D_{-\nu-\frac{1}{2}}\left(-\frac{a}{\sqrt{2b}}\right)+ D_{\nu-\frac{1}{2}}\left(-\frac{a}{\sqrt{2b}}\right)D_{-\nu-\frac{1}{2}}\left(\frac{a}{\sqrt{2b}}\right)\right]$$

$$\left[\operatorname{Re}b>0,\ |\operatorname{Re}\nu|<\tfrac{1}{2}\right].$$ ET II 132(27)

6.735

1. $$\int_0^\infty x^{\frac{1}{4}}\sin(2a\sqrt{x})\,J_{-\frac{1}{4}}(x)\,dx = \sqrt{\pi}\,a^{\frac{3}{2}}J_{\frac{3}{4}}(a^2) \quad [a>0].$$ ET II 341(10)

2. $$\int_0^\infty x^{\frac{1}{4}}\cos(2a\sqrt{x})\,J_{\frac{1}{4}}(x)\,dx = \sqrt{\pi}\,a^{\frac{3}{2}}J_{-\frac{3}{4}}(a^2) \quad [a>0].$$ ET II 341(12)

3. $$\int_0^\infty x^{\frac{1}{4}}\sin(2a\sqrt{x})\,J_{\frac{3}{4}}(x)\,dx = \sqrt{\pi}\,a^{\frac{3}{2}}J_{-\frac{1}{4}}(a^2) \quad [a>0].$$ ET II 341(11)

4. $$\int_0^\infty x^{\frac{1}{4}}\cos(2a\sqrt{x})\,J_{-\frac{3}{4}}(x)\,dx = \sqrt{\pi}\,a^{\frac{3}{2}}J_{\frac{1}{4}}(a^2) \quad [a>0].$$ ET II 341(13)

6.736

1. $$\int_0^\infty x^{-\frac{1}{2}} \sin x \cos(4a\sqrt{x}) J_0(x)\,dx = -2^{-\frac{3}{2}}\sqrt{\pi}\left[\cos\left(a^2-\frac{\pi}{4}\right)J_0(a^2)-\sin\left(a^2-\frac{\pi}{4}\right)N_0(a^2)\right]$$ $[a>0]$. ET II 341(18)

2. $$\int_0^\infty x^{-\frac{1}{2}} \cos x \cos(4a\sqrt{x}) J_0(x)\,dx = -2^{-\frac{3}{2}}\sqrt{\pi}\left[\sin\left(a^2-\frac{\pi}{4}\right)J_0(a^2)+\cos\left(a^2-\frac{\pi}{4}\right)N_0(a^2)\right]$$ $[a>0]$. ET II 342(22)

3. $$\int_0^\infty x^{-\frac{1}{2}} \sin x \sin(4a\sqrt{x}) J_0(x)\,dx = \sqrt{\frac{\pi}{2}}\cos\left(a^2+\frac{\pi}{4}\right)J_0(a^2) \qquad [a>0].$$ ET II 341(16)

4. $$\int_0^\infty x^{-\frac{1}{2}} \cos x \sin(4a\sqrt{x}) J_0(x)\,dx = \sqrt{\frac{\pi}{2}}\cos\left(a^2-\frac{\pi}{4}\right)J_0(a^2) \qquad [a>0].$$ ET II 342(20)

5. $$\int_0^\infty x^{-\frac{1}{2}} \sin x \cos(4a\sqrt{x}) N_0(x)\,dx = 2^{-\frac{3}{2}}\sqrt{\pi}\left[3\sin\left(a^2-\frac{\pi}{4}\right)J_0(a^2)-\cos\left(a^2-\frac{\pi}{4}\right)N_0(a^2)\right]$$ $[a>0]$. ET II 347(55)

6. $$\int_0^\infty x^{-\frac{1}{2}} \cos x \cos(4a\sqrt{x}) N_0(x)\,dx = -2^{-\frac{3}{2}}\sqrt{\pi}\left[3\cos\left(a^2-\frac{\pi}{4}\right)J_0(a^2)+\sin\left(a^2-\frac{\pi}{4}\right)N_0(a^2)\right]$$ $[a>0]$. ET II 347(56)

6.737

1. $$\int_0^\infty \frac{\sin\left(a\sqrt{x^2+b^2}\right)}{\sqrt{x^2+b^2}} J_\nu(cx)\,dx = \frac{\pi}{2} J_{\frac{1}{2}\nu}\left[\frac{b}{2}\left(a-\sqrt{a^2-c^2}\right)\right] J_{-\frac{1}{2}\nu}\left[\frac{b}{2}\left(a+\sqrt{a^2-c^2}\right)\right]$$ $[a>0,\ \operatorname{Re} b>0,\ c>0,\ a>c,\ \operatorname{Re}\nu>-1]$. ET II 35(19)

2. $$\int_0^\infty \frac{\cos\left(a\sqrt{x^2+b^2}\right)}{\sqrt{x^2+b^2}} J_\nu(cx)\,dx = -\frac{\pi}{2} J_{\frac{1}{2}\nu}\left[\frac{b}{2}\left(a-\sqrt{a^2-c^2}\right)\right] N_{-\frac{1}{2}\nu}\left[\frac{b}{2}\left(a+\sqrt{a^2-c^2}\right)\right]$$
$[a>0,\ \operatorname{Re} b>0,\ c>0,\ a>c,\ \operatorname{Re}\nu>-1]$. ET II 39(44)

3. $$\int_0^a \frac{\cos\left(b\sqrt{a^2-x^2}\right)}{\sqrt{a^2-x^2}} J_\nu(cx)\,dx = \frac{\pi}{2} J_{\frac{1}{2}\nu}\left[\frac{a}{2}\left(\sqrt{b^2+c^2}-b\right)\right] J_{\frac{1}{2}\nu}\left[\frac{a}{2}\left(\sqrt{b^2+c^2}+b\right)\right]$$
$[c>0,\ \operatorname{Re}\nu>-1]$. ET II 39(47)

4. $$\int_0^a x^{\nu+1} \frac{\cos\left(\sqrt{a^2-x^2}\right)}{\sqrt{a^2-x^2}} I_\nu(x)\,dx = \frac{\sqrt{\pi}\,a^{2\nu+1}}{2^{\nu+1}\Gamma\left(\nu+\frac{3}{2}\right)}$$
$[\operatorname{Re}\nu>-1]$. ET II 365(9)

5. $$\int_0^\infty x^{\nu+1} \frac{\sin\left(a\sqrt{b^2+x^2}\right)}{\sqrt{b^2+x^2}} J_\nu(cx)\,dx =$$
$$= \sqrt{\frac{\pi}{2}}\, b^{\frac{1}{2}+\nu} c^\nu (a^2-c^2)^{-\frac{1}{4}-\frac{1}{2}\nu} J_{-\nu-\frac{1}{2}}\left(b\sqrt{a^2-c^2}\right)$$
$\left[0<c<a,\ \operatorname{Re} b>0,\ -1<\operatorname{Re}\nu<\frac{1}{2}\right]$;
$$=0$$
$\left[0<a<c,\ \operatorname{Re} b>0,\ -1<\operatorname{Re}\nu<\frac{1}{2}\right]$. ET II 35(20)

6. $$\int_0^\infty x^{\nu+1} \frac{\cos\left(a\sqrt{x^2+b^2}\right)}{\sqrt{x^2+b^2}} J_\nu(cx)\,dx =$$
$$= -\sqrt{\frac{\pi}{2}}\, b^{\frac{1}{2}+\nu} c^\nu (a^2-c^2)^{-\frac{1}{4}-\frac{1}{2}\nu} N_{-\nu-\frac{1}{2}}\left(b\sqrt{a^2-c^2}\right)$$
$\left[0<c<a,\ \operatorname{Re} b>0,\ -1<\operatorname{Re}\nu<\frac{1}{2}\right]$;
$$= \sqrt{\frac{2}{\pi}}\, b^{\frac{1}{2}+\nu} c^\nu (c^2-a^2)^{-\frac{1}{4}-\frac{1}{2}\nu} K_{\nu+\frac{1}{2}}\left(b\sqrt{c^2-a^2}\right)$$
$\left[0<a<c,\ \operatorname{Re} b>0,\ -1<\operatorname{Re}\nu<\frac{1}{2}\right]$. ET II 39(45)

6.738

1. $$\int_0^a x^{\nu+1} \sin\left(b\sqrt{a^2-x^2}\right) J_\nu(x)\,dx = \sqrt{\frac{\pi}{2}}\, a^{\nu+\frac{3}{2}} b(1+b^2)^{-\frac{1}{2}\nu-\frac{3}{4}} J_{\nu+\frac{3}{2}}\left(a\sqrt{1+b^2}\right)$$
$[\operatorname{Re}\nu>-1]$. ET II 335(19)

2. $\int\limits_0^\infty x^{\nu+1} \cos\left(a\sqrt{x^2+b^2}\right) J_\nu(cx)\,dx =$

$$= \sqrt{\frac{\pi}{2}}\, ab^{\nu+\frac{3}{2}} c^\nu (a^2-c^2)^{-\frac{1}{2}\nu-\frac{3}{4}} \left[\cos(\pi\nu) J_{\nu+\frac{3}{2}}\left(b\sqrt{a^2-c^2}\right) - \sin(\pi\nu) N_{\nu+\frac{3}{2}}\left(b\sqrt{a^2-c^2}\right)\right]$$

$\left[0<c<a,\ \operatorname{Re} b>0,\ -1<\operatorname{Re}\nu<-\frac{1}{2}\right];$

$= 0 \qquad \left[0<a<c,\ \operatorname{Re} b>0,\ -1<\operatorname{Re}\nu<-\frac{1}{2}\right].$

ET II 39(43)

6.739 $\int\limits_0^t x^{-\frac{1}{2}} \frac{\cos\left(b\sqrt{t-x}\right)}{\sqrt{t-x}} J_{2\nu}\left(a\sqrt{x}\right) dx =$

$$= \pi J_\nu\left[\frac{\sqrt{t}}{2}\left(\sqrt{a^2+b^2}+b\right)\right] J_\nu\left[\frac{\sqrt{t}}{2}\left(\sqrt{a^2+b^2}-b\right)\right]$$

$\left[\operatorname{Re}\nu>-\frac{1}{2}\right].$ EH II 47(7)

6.741

1. $\int\limits_0^1 \frac{\cos(\mu \arccos x)}{\sqrt{1-x^2}} J_\nu(ax)\,dx = \frac{\pi}{2} J_{\frac{1}{2}(\mu+\nu)}\left(\frac{a}{2}\right) J_{\frac{1}{2}(\nu-\mu)}\left(\frac{a}{2}\right)$

$[\operatorname{Re}(\mu+\nu)>-1,\ a>0].$ ET II 41(54)

2. $\int\limits_0^1 \frac{\cos[(\nu+1)\arccos x]}{\sqrt{1-x^2}} J_\nu(ax)\,dx = \sqrt{\frac{\pi}{a}} \cos\left(\frac{a}{2}\right) J_{\nu+\frac{1}{2}}\left(\frac{a}{2}\right)$

$[\operatorname{Re}\nu>-1,\ a>0].$ ET II 40(53)

3. $\int\limits_0^1 \frac{\cos[(\nu-1)\arccos x]}{\sqrt{1-x^2}} J_\nu(ax)\,dx = \sqrt{\frac{\pi}{a}} \sin\left(\frac{a}{2}\right) J_{\nu-\frac{1}{2}}\left(\frac{a}{2}\right)$

$[\operatorname{Re}\nu>0,\ a>0].$ ET II 40(52)a

6.75 Combinations of Bessel, trigonometric, and exponential functions and powers

6.751

1. $\int\limits_0^\infty e^{-\frac{1}{2}ax} \sin(bx)\, I_0\left(\frac{1}{2}ax\right) dx = \frac{1}{\sqrt{2b}} \frac{1}{\sqrt{b^2+a^2}} \sqrt{b+\sqrt{b^2+a^2}}$

$[\operatorname{Re} a>0,\ b>0].$ ET I 105(44)

2. $\int\limits_0^\infty e^{-\frac{1}{2}ax} \cos(bx)\, I_0\left(\frac{1}{2}ax\right) dx = \frac{a}{\sqrt{2b}} \frac{1}{\sqrt{a^2+b^2}\sqrt{b+\sqrt{a^2+b^2}}}$

$[\operatorname{Re} a>0,\ b>0].$ ET I 48(38)

3. $$\int_0^\infty e^{-bx}\cos(ax)\,J_0(cx)\,dx=\frac{\left[\sqrt{(b^2+c^2-a^2)^2+4a^2b^2}+b^2+c^2-a^2\right]^{\frac{1}{2}}}{\sqrt{2}\sqrt{(b^2+c^2-a^2)^2+4a^2b^2}}$$

$[c>0]$. ET II 11(46)

6.752

1. $$\int_0^\infty e^{-ax}J_0(bx)\sin(cx)\frac{dx}{x}=\arcsin\left(\frac{2c}{\sqrt{a^2+(c+b)^2}+\sqrt{a^2+(c-b)^2}}\right)$$

$[\operatorname{Re}a>|\operatorname{Im}b|,\ c>0]$. ET I 101(17)

2. $$\int_0^\infty e^{-ax}J_1(cx)\sin(bx)\frac{dx}{x}=\frac{b}{c}(1-r),$$

$\left[b^2=\frac{c^2}{1-r^2}-\frac{a^2}{r^2},\quad c>0\right]$. ET II 19(15)

6.753

1. $$\int_0^\infty \frac{\sin(xa\sin\psi)}{x}e^{-xa\cos\varphi\cos\psi}J_\nu(xa\sin\varphi)\,dx=\nu^{-1}\left(\operatorname{tg}\frac{\varphi}{2}\right)^\nu\sin(\nu\psi)$$

$\left[\operatorname{Re}\nu>-1,\ a>0,\ 0<\varphi,\ \psi<\frac{\pi}{2}\right]$. ET II 33(10)

2. $$\int_0^\infty \frac{\cos(xa\sin\psi)}{x}e^{-xa\cos\varphi\cos\psi}J_\nu(xa\sin\varphi)\,dx=\nu^{-1}\left(\operatorname{tg}\frac{\varphi}{2}\right)^\nu\cos(\nu\psi)$$

$\left[\operatorname{Re}\nu>0,\ a>0,\ 0<\varphi,\ \psi<\frac{\pi}{2}\right]$. ET II 38(35)

3. $$\int_0^\infty x^{\nu+1}e^{-ax\cos\varphi\cos\psi}\sin(ax\sin\psi)\,J_\nu(ax\sin\varphi)\,dx=$$

$$=2^{\nu+1}\frac{\Gamma\left(\nu+\frac{3}{2}\right)}{\sqrt{\pi}}a^{-\nu-2}(\sin\varphi)^\nu(\cos^2\psi+\sin^2\psi\cos^2\varphi)^{-\nu-\frac{3}{2}}\sin\left[\left(\nu+\frac{3}{2}\right)\beta\right],$$

$$\operatorname{tg}\frac{\beta}{2}=\operatorname{tg}\psi\cos\varphi$$

$\left[a>0,\ 0<\varphi,\ \psi<\frac{\pi}{2},\ \operatorname{Re}\nu>-\frac{3}{2}\right]$. ET II 34(11)

4. $$\int_0^\infty x^{\nu+1}e^{-ax\cos\varphi\cos\psi}\cos(ax\sin\psi)\,J_\nu(ax\sin\varphi)\,dx=$$

$$=2^{\nu+1}\frac{\Gamma\left(\nu+\frac{3}{2}\right)}{\sqrt{\pi}}a^{-\nu-2}(\sin\varphi)^\nu(\cos^2\psi+\sin^2\psi\cos^2\varphi)^{-\nu-\frac{3}{2}}\cos\left[\left(\nu+\frac{3}{2}\right)\beta\right],$$

$\operatorname{tg}\frac{\beta}{2}=\operatorname{tg}\psi\cos\varphi\quad\left[a>0,\ 0<\varphi,\ \psi<\frac{\pi}{2},\ \operatorname{Re}\nu>-1\right]$.

ET II 38(36)

5. $\int\limits_0^\infty x^\nu e^{-ax\cos\varphi\cos\psi}\sin(ax\sin\psi)\,J_\nu(ax\sin\varphi)\,dx =$

$$= 2^\nu\frac{\Gamma\left(\nu+\frac{1}{2}\right)}{\sqrt{\pi}}a^{-\nu-1}(\sin\varphi)^\nu(\cos^2\psi+\sin^2\psi\cos^2\varphi)^{-\nu-\frac{1}{2}}\sin\left[\left(\nu+\frac{3}{2}\right)\beta\right],$$

$$\operatorname{tg}\frac{\beta}{2} = \operatorname{tg}\psi\cos\varphi$$

$$\left[a>0,\ 0<\varphi,\ \psi<\frac{\pi}{2},\ \operatorname{Re}\nu>-1\right].$$ ET II 34(12)

6. $\int\limits_0^\infty x^\nu e^{-ax\cos\varphi\cos\psi}\cos(ax\sin\psi)\,J_\nu(ax\sin\varphi)\,dx =$

$$= 2^\nu\frac{\Gamma\left(\nu+\frac{1}{2}\right)}{\sqrt{\pi}}a^{-\nu-1}(\sin\varphi)^\nu(\cos^2\psi+\sin^2\psi\cos^2\varphi)^{-\nu-\frac{1}{2}}\cos\left[\left(\nu+\frac{1}{2}\right)\beta\right],$$

$$\operatorname{tg}\frac{\beta}{2} = \operatorname{tg}\psi\cos\varphi$$

$$\left[a>0,\ 0<\varphi,\ \psi<\frac{\pi}{2},\ \operatorname{Re}\nu>-\frac{1}{2}\right].$$ ET II 38(37)

6.754

1. $\int\limits_0^\infty e^{-x^2}\sin(bx)\,I_0(x^2)\,dx = \frac{\sqrt{\pi}}{2^{\frac{3}{2}}}e^{-\frac{b^2}{8}}I_0\left(\frac{b^2}{8}\right) \quad [b>0].$ ET I 108(9)

2. $\int\limits_0^\infty e^{-ax}\cos(x^2)\,J_0(x^2)\,dx = \frac{1}{4}\sqrt{\frac{\pi}{2}}\left[J_0\left(\frac{a^2}{16}\right)\cos\left(\frac{a^2}{16}-\frac{\pi}{4}\right)-\right.$

$$\left.-N_0\left(\frac{a^2}{16}\right)\cos\left(\frac{a^2}{16}+\frac{\pi}{4}\right)\right] \quad [a>0].$$ MI 42

3. $\int\limits_0^\infty e^{-ax}\sin(x^2)\,J_0(x^2)\,dx = \frac{1}{4}\sqrt{\frac{\pi}{2}}\left[J_0\left(\frac{a^2}{16}\right)\sin\left(\frac{a^2}{16}-\frac{\pi}{4}\right)-\right.$

$$\left.-N_0\left(\frac{a^2}{16}\right)\sin\left(\frac{a^2}{16}+\frac{\pi}{4}\right)\right] \quad [a>0].$$ MI 42

6.755

1. $\int\limits_0^\infty x^{-\nu}e^{-x}\sin(4a\sqrt{x})\,I_\nu(x)\,dx = (2^{\frac{3}{2}}a)^{\nu-1}e^{-a^2}W_{\frac{1}{2}-\frac{3}{2}\nu,\ \frac{1}{2}-\frac{1}{2}\nu}(2a^2)$

$$[a>0,\ \operatorname{Re}\nu>0].$$ ET II 366(14)

2. $\int\limits_0^\infty x^{-\nu-\frac{1}{2}}e^{-x}\cos(4a\sqrt{x})\,I_\nu(x)\,dx = 2^{\frac{3}{2}\nu-1}a^{\nu-1}e^{-a^2}W_{-\frac{3}{2}\nu,\ \frac{1}{2}\nu}(2a^2)$

$$\left[a>0,\ \operatorname{Re}\nu>-\frac{1}{2}\right].$$ ET II 366(16)

3. $$\int_0^\infty x^{-\nu} e^x \sin(4a\sqrt{x}) K_\nu(x)\,dx =$$

$$= (2^{\frac{3}{2}} a)^{\nu-1} \pi \frac{\Gamma\left(\frac{3}{2} - 2\nu\right)}{\Gamma\left(\frac{1}{2} + \nu\right)} e^{a^2} W_{\frac{3}{2}\nu - \frac{1}{2}, \frac{1}{2} - \frac{1}{2}\nu}(2a^2)$$

$$\left[a > 0,\ 0 < \operatorname{Re}\nu < \frac{3}{4}\right].$$ ET II 369(38)

4. $$\int_0^\infty x^{-\nu-\frac{1}{2}} e^x \cos(4a\sqrt{x}) K_\nu(x)\,dx =$$

$$= 2^{\frac{3}{2}\nu - 1} \pi a^{\nu-1} \frac{\Gamma\left(\frac{1}{2} - 2\nu\right)}{\Gamma\left(\frac{1}{2} + \nu\right)} e^{a^2} W_{\frac{3}{2}\nu, -\frac{1}{2}\nu}(2a^2)$$

$$\left[a > 0,\ -\frac{1}{2} < \operatorname{Re}\nu < \frac{1}{4}\right].$$ ET II 369(42)

5. $$\int_0^\infty x^{\varrho-\frac{3}{2}} e^{-x} \sin(4a\sqrt{x}) K_\nu(x)\,dx =$$

$$= \frac{\sqrt{\pi}\, a \Gamma(\varrho+\nu) \Gamma(\varrho-\nu)}{2^{\varrho-2} \Gamma\left(\varrho + \frac{1}{2}\right)} {}_2F_2\left(\varrho+\nu, \varrho-\nu; \frac{3}{2}, \varrho + \frac{1}{2}; -2a^2\right)$$

$$[\operatorname{Re}\varrho > |\operatorname{Re}\nu|].$$ ET II 369(39)

6. $$\int_0^\infty x^{\varrho-1} e^{-x} \cos(4a\sqrt{x}) K_\nu(x)\,dx =$$

$$= \frac{\sqrt{\pi}\, \Gamma(\varrho+\nu) \Gamma(\varrho-\nu)}{2^{\varrho} \Gamma\left(\varrho + \frac{1}{2}\right)} {}_2F_2\left(\varrho+\nu, \varrho-\nu; \frac{1}{2}, \varrho + \frac{1}{2}; -2a^2\right)$$

$$[\operatorname{Re}\varrho > |\operatorname{Re}\nu|].$$ ET II 370(43)

7. $$\int_0^\infty x^{-\frac{1}{2}} e^{-x} \cos(4a\sqrt{x}) I_0(x)\,dx = \frac{1}{\sqrt{2\pi}} e^{-a^2} K_0(a^2) \quad [a > 0].$$

ET II 366(15)

8. $$\int_0^\infty x^{-\frac{1}{2}} e^{x} \cos(4a\sqrt{x}) K_0(x)\,dx = \sqrt{\frac{\pi}{2}} e^{a^2} K_0(a^2) \quad [a > 0].$$

ET II 369(40)

9. $$\int_0^\infty x^{-\frac{1}{2}} e^{-x} \cos(4a\sqrt{x}) K_0(x)\,dx = \frac{1}{\sqrt{2}} \pi^{\frac{3}{2}} e^{-a^2} I_0(a^2).$$ ET II 369(41)

6.756

1. $$\int_0^\infty x^{-\frac{1}{2}} e^{-a\sqrt{x}} \sin(a\sqrt{x}) J_\nu(bx)\,dx = \frac{i}{\sqrt{2\pi b}} \Gamma\left(\nu+\frac{1}{2}\right) D_{-\nu-\frac{1}{2}}\left(\frac{a}{\sqrt{b}}\right) \times \left[D_{-\nu-\frac{1}{2}}\left(\frac{ia}{\sqrt{b}}\right) - D_{-\nu-\frac{1}{2}}\left(-\frac{ia}{\sqrt{b}}\right)\right]$$
$[a>0,\ b>0,\ \operatorname{Re}\nu > -1]$. ET II 34(17)

2. $$\int_0^\infty x^{-\frac{1}{2}} e^{-a\sqrt{x}} \cos(a\sqrt{x}) J_\nu(bx)\,dx = \frac{1}{\sqrt{2\pi b}} \Gamma\left(\nu+\frac{1}{2}\right) D_{-\nu-\frac{1}{2}}\left(\frac{a}{\sqrt{b}}\right) \times \left[D_{-\nu-\frac{1}{2}}\left(\frac{ia}{\sqrt{b}}\right) + D_{-\nu-\frac{1}{2}}\left(-\frac{ia}{\sqrt{b}}\right)\right]$$
$\left[a>0,\ b>0,\ \operatorname{Re}\nu > -\frac{1}{2}\right]$. ET II 39(42)

3. $$\int_0^\infty x^{-\frac{1}{2}} e^{-a\sqrt{x}} \sin(a\sqrt{x}) J_0(bx)\,dx = \frac{1}{2b} a I_{\frac{1}{4}}\left(\frac{a^2}{4b}\right) K_{\frac{1}{4}}\left(\frac{a^2}{4b}\right) \quad \left[|\arg a| < \frac{\pi}{4},\ b>0\right].$$
ET II 11(40)

4. $$\int_0^\infty x^{-\frac{1}{2}} e^{-a\sqrt{x}} \cos(a\sqrt{x}) J_0(bx)\,dx = \frac{a}{2b} I_{-\frac{1}{4}}\left(\frac{a^2}{4b}\right) K_{\frac{1}{4}}\left(\frac{a^2}{4b}\right)$$
$\left[|\arg a| < \frac{\pi}{4},\ b>0\right]$. ET II 12(49)

6.757

1. $$\int_0^\infty e^{-bx} \sin[a(1-e^{-x})] J_\nu(ae^{-x})\,dx = 2\sum_{n=0}^\infty \frac{(-1)^n \Gamma(\nu-b+2n+1)\Gamma(\nu+b)}{\Gamma(\nu-b+1)\Gamma(\nu+b+2n+2)} \times (\nu+2n-1) J_{\nu+2n+1}(a)$$
$[\operatorname{Re} b > -\operatorname{Re}\nu]$. ET I 193(26)

2. $$\int_0^\infty e^{-bx} \cos[a(1-e^{-x})] J_\nu(ae^{-x})\,dx = \frac{J_\nu(a)}{\nu+b} + \sum_{n=0}^\infty 2(-1)^n \frac{\Gamma(\nu-b+2n)\Gamma(\nu+b)}{\Gamma(\nu-b+1)\Gamma(\nu+b+2n+1)} (\nu+2n) J_{\nu+2n}(a)$$
$[\operatorname{Re} b > -\operatorname{Re}\nu]$. ET I 193(27)

6.758 $$\int_{-\frac{\pi}{2}}^{\frac{\pi}{2}} e^{i(\mu-\nu)\theta} (\cos\theta)^{\nu+\mu} (\lambda z)^{-\nu-\mu} J_{\nu+\mu}(\lambda z)\,d\theta = \pi(2az)^{-\mu}(2bz)^{-\nu} J_\mu(az) J_\nu(bz);$$
$\lambda = \sqrt{2\cos\theta\,(a^2e^{i\theta} + b^2e^{-i\theta})}$ $\quad [\operatorname{Re}(\nu+\mu) > -1]$. EH II 48(12)

6.76 Combinations of Bessel, trigonometric, and hyperbolic functions

6.761 $\int_0^\infty \operatorname{ch} x \cos(2a \operatorname{sh} x) J_\nu(be^x) J_\nu(be^{-x})\,dx =$

$$= \frac{J_{2\nu}(2\sqrt{b^2-a^2})}{2\sqrt{b^2-a^2}} \qquad [0 < a < b,\ \operatorname{Re}\nu > -1];$$

$$= 0 \qquad [0 < b < a,\ \operatorname{Re}\nu > -1].$$

ET II 359(10)

6.762 $\int_0^\infty \operatorname{ch} x \sin(2a \operatorname{sh} x)[J_\nu(be^x) N_\nu(be^{-x}) - N_\nu(be^x) J_\nu(be^{-x})]\,dx =$

$$= 0 \qquad \left[0 < a < b,\ |\operatorname{Re}\nu| < \frac{1}{2}\right];$$

$$= -\frac{2}{\pi}\cos(\nu\pi)(a^2-b^2)^{-\frac{1}{2}} K_{2\nu}[2(a^2-b^2)^{\frac{1}{2}}] \qquad \left[0 < b < a,\ |\operatorname{Re}\nu| < \frac{1}{2}\right].$$

ET II 360(12)

6.763 $\int_0^\infty \operatorname{ch} x \cos(2a \operatorname{sh} x) N_\nu(be^x) N_\nu(be^{-x})\,dx =$

$$= -\frac{1}{2}(b^2-a^2)^{-\frac{1}{2}} J_{2\nu}[2(b^2-a^2)^{\frac{1}{2}}] \qquad [0 < a < b,\ |\operatorname{Re}\nu| < 1];$$

$$= \frac{2}{\pi}\cos(\nu\pi)(a^2-b^2)^{-\frac{1}{2}} K_{2\nu}[2(a^2-b^2)^{\frac{1}{2}}]$$

$$[0 < b < a,\ |\operatorname{Re}\nu| < 1].$$

ET II 360(11)

6.77 Combinations of Bessel functions and the logarithm, or arctangent

6.771
$$\int_0^\infty x^{\mu+\frac{1}{2}} \ln x\, J_\nu(ax)\,dx = \frac{2^{\mu-\frac{1}{2}}\Gamma\left(\frac{\mu+\nu}{2}+\frac{3}{4}\right)}{\Gamma\left(\frac{\nu-\mu}{2}+\frac{1}{4}\right)a^{\mu+\frac{3}{2}}} \times$$
$$\times \left[\psi\left(\frac{\mu+\nu}{2}+\frac{3}{4}\right)+\psi\left(\frac{\nu-\mu}{2}+\frac{1}{4}\right)-\ln\frac{a^2}{4}\right]$$
$$\left[a > 0,\ -\operatorname{Re}\nu - \frac{3}{2} < \operatorname{Re}\mu < 0\right].$$

ET II 32(25)

6.772

1. $\int_0^\infty \ln x J_0(ax)\,dx = -\frac{1}{a}[\ln(2a) + C].$ WA 430(4)a, ET II 10(27)

2. $\int_0^\infty \ln x J_1(ax)\,dx = -\frac{1}{a}\left[\ln\left(\frac{a}{2}\right) + C\right].$ ET II 19(11)

3. $\int_0^\infty \ln(a^2+x^2) J_1(bx)\,dx = \frac{2}{b}[K_0(ab) + \ln a].$ ET II 19(12)

4. $\int_0^\infty J_1(tx)\ln\sqrt{1+t^4}\,dt = \frac{2}{x}\operatorname{ker} x.$ MO 46

6.773 $\int_0^\infty \frac{\ln(x+\sqrt{x^2+a^2})}{\sqrt{x^2+a^2}} J_0(bx)\,dx =$

$$= \left[\frac{1}{2}K_0^2\left(\frac{ab}{2}\right) + \ln a I_0\left(\frac{ab}{2}\right)K_0\left(\frac{ab}{2}\right)\right]$$

$[a > 0,\ b > 0].$ ET II 10(28)

6.774 $\int_0^\infty \ln\frac{\sqrt{x^2+a^2}+x}{\sqrt{x^2+a^2}-x} J_0(bx)\frac{dx}{\sqrt{x^2+a^2}} = K_0^2\left(\frac{ab}{2}\right)$

$[\operatorname{Re} a > 0,\ b > 0].$ ET II 10(29)

6.775 $\int_0^\infty x\left[\ln\left(a+\sqrt{a^2+x^2}\right) - \ln x\right] J_0(bx)\,dx =$

$$= \frac{1}{b^2}(1-e^{-ab}) \qquad [\operatorname{Re} a > 0,\ b > 0].$$ ET II 12(55)

6.776 $\int_0^\infty x\ln\left(1+\frac{a^2}{x^2}\right) J_0(bx)\,dx = \frac{2}{b}\left[\frac{1}{b} - aK_1(ab)\right]$

$[\operatorname{Re} a > 0,\ b > 0].$ ET II 10(30)

6.777 $\int_0^\infty J_1(tx)\operatorname{arctg} t^2\,dt = -\frac{2}{x}\operatorname{kei} x.$ MO 46

6.78 Combinations of Bessel and other special functions

6.781 $\int_0^\infty \operatorname{si}(ax) J_0(bx)\,dx = -\frac{1}{b}\arcsin\left(\frac{b}{a}\right) \qquad [0 < b < a];$

$$= 0 \qquad [0 < a < b].$$

ET II 13(6)

6.782

1. $\int_0^\infty \operatorname{Ei}(-x) J_0\left(2\sqrt{zx}\right)dx = \frac{e^{-z}-1}{z}.$ NT 60(4)

2. $\int_0^\infty \operatorname{si}(x) J_0\left(2\sqrt{zx}\right)dx = -\frac{\sin z}{z}.$ NT 60(6)

3. $\int_0^\infty \operatorname{ci}(x) J_0\left(2\sqrt{zx}\right)dx = \frac{\cos z - 1}{z}.$ NT 60(5)

4. $\int_0^\infty \operatorname{Ei}(-x) J_1\left(2\sqrt{zx}\right)\frac{dx}{\sqrt{x}} = \frac{\operatorname{Ei}(-z) - C - \ln z}{\sqrt{z}}.$ NT 60(7)

5. $\int_0^\infty \operatorname{si}(x) J_1(2\sqrt{zx}) \frac{dx}{\sqrt{x}} = -\frac{\frac{\pi}{2} - \operatorname{si}(z)}{\sqrt{z}}.$ NT 60(9)

6. $\int_0^\infty \operatorname{ci}(z) J_1(2\sqrt{zx}) \frac{dx}{\sqrt{x}} = \frac{\operatorname{ci}(z) - C - \ln z}{\sqrt{z}}.$ NT 60(8)

7. $\int_0^\infty \operatorname{Ei}(-x) N_0(2\sqrt{zx})\, dx = \frac{C + \ln z - e^z \operatorname{Ei}(-z)}{\pi z}.$ NT 63(5)

6.783

1. $\int_0^\infty x \operatorname{si}(a^2x^2) J_0(bx)\, dx = -\frac{2}{b^2} \sin\left(\frac{b^2}{4a^2}\right)$

$[a > 0].$ ET II 13(7)a

2. $\int_0^\infty x \operatorname{ci}(a^2x^2) J_0(bx)\, dx = \frac{2}{b^2}\left[1 - \cos\left(\frac{b^2}{4a^2}\right)\right]$

$[a > 0].$ ET II 13(8)a

3. $\int_0^\infty \operatorname{ci}(a^2x^2) J_0(bx)\, dx = \frac{1}{b}\left[\operatorname{ci}\left(\frac{b^2}{4a^2}\right) + \ln\left(\frac{b^2}{4a^2}\right) + 2C\right]$

$[a > 0].$ ET II 13(9)a

4. $\int_0^\infty \operatorname{si}(a^2x^2) J_1(bx)\, dx = \frac{1}{b}\left[-\operatorname{si}\left(\frac{b^2}{4a^2}\right) - \frac{\pi}{2}\right]$

$[a > 0].$ ET II 20(25)a

6.784

1. $\int_0^\infty x^{\nu+1}[1 - \Phi(ax)] J_\nu(bx)\, dx =$

$$= a^{-\nu} \frac{\Gamma\left(\nu + \frac{3}{2}\right)}{b^2 \Gamma(\nu+2)} \exp\left(-\frac{b^2}{8a^2}\right) M_{\frac{1}{2}\nu + \frac{1}{2}, \frac{1}{2}\nu + \frac{1}{2}}\left(\frac{b^2}{4a^2}\right)$$

$\left[|\arg a| < \frac{\pi}{4},\ b > 0,\ \operatorname{Re}\nu > -1\right].$ ET II 92(22)

2. $\int_0^\infty x^\nu [1 - \Phi(ax)] J_\nu(bx)\, dx =$

$$= \frac{a^{\frac{1}{2}-\nu} \Gamma\left(\nu + \frac{1}{2}\right)}{\sqrt{2}\, b^{\frac{3}{2}} \Gamma\left(\nu + \frac{3}{2}\right)} \exp\left(-\frac{b^2}{8a^2}\right) M_{\frac{1}{2}\nu - \frac{1}{4}, \frac{1}{2}\nu + \frac{1}{4}}\left(\frac{b^2}{4a^2}\right)$$

$\left[|\arg a| < \frac{\pi}{4},\ \operatorname{Re}\nu > -\frac{1}{2},\ b > 0\right].$ ET II 92(23)

6.785 $$\int_0^\infty \frac{\exp\left(\frac{a^2}{2x}-x\right)}{x}\left[1-\Phi\left(\frac{a}{\sqrt{2x}}\right)\right] K_\nu(x)\,dx = \frac{\pi^{\frac{5}{2}}}{4}\sec(\nu\pi)\{[J_\nu(a)]^2+[N_\nu(a)]^2\}$$
$$\left[\operatorname{Re} a>0,\ |\operatorname{Re}\nu|<\frac{1}{2}\right].$$ ET II 370(46)

6.786 $$\int_0^\infty x^{\nu-2\mu+2n+2}e^{x^2}\Gamma(\mu,\ x^2)\,N_\nu(bx)\,dx = (-1)^n\frac{\Gamma\left(\frac{3}{2}-\mu+\nu+n\right)\Gamma\left(\frac{3}{2}-\mu+n\right)}{b\Gamma(1-\mu)}\times$$
$$\times\exp\left(\frac{b^2}{8}\right)W_{\mu-\frac{1}{2}\nu-n-1,\ \frac{1}{2}\nu}\left(\frac{b^2}{4}\right)$$
$$\left[n\text{ — an integer},\ b>0.\ \operatorname{Re}(\nu-\mu+n)>-\frac{3}{2},\ \operatorname{Re}(-\mu+n)>-\frac{3}{2},\ \operatorname{Re}\nu<\frac{1}{2}-2n\right].$$ ET II 108(2)

6.787 $$\int_0^\infty \frac{x^{\nu+2n-\frac{1}{2}}}{\mathrm{B}(a+x,\ a-x)}J_\nu(bx)\,dx=0$$
$$\left[\pi\leqslant b<\infty,\ -1<\operatorname{Re}\nu<2a-2n-\frac{7}{2}\right].$$ ET II 92(21)

6.79 Integration of Bessel functions with respect to the order

6.791

1. $$\int_{-\infty}^\infty K_{ix+iy}(a)\,K_{ix+iz}(b)\,dx=\pi K_{iy-iz}(a+b)$$
$$[|\arg a|+|\arg b|<\pi].$$ ET II 382(21)

2. $$\int_{-\infty}^\infty J_{\nu-x}(a)\,J_{\mu+x}(a)\,dx=J_{\mu+\nu}(2a)\qquad[\operatorname{Re}(\mu+\nu)>1].$$ ET II 379(1)

3. $$\int_{-\infty}^\infty J_{\varkappa+x}(a)\,J_{\lambda-x}(a)\,J_{\mu+x}(a)\,J_{\nu-x}(a)\,dx = \frac{\Gamma(\varkappa+\lambda+\mu+\nu+1)}{\Gamma(\varkappa+\lambda+1)\,\Gamma(\lambda+\mu+1)\,\Gamma(\mu+\nu+1)\,\Gamma(\nu+\varkappa+1)}\times$$
$$\times\,{}_4F_5\left(\frac{\varkappa+\lambda+\mu+\nu+1}{2},\ \frac{\varkappa+\lambda+\mu+\nu+1}{2},\ \frac{\varkappa+\lambda+\mu+\nu}{2}+1,\ \frac{\varkappa+\lambda+\mu+\nu}{2}+1;\right.$$
$$\left.\varkappa+\lambda+\mu+\nu+1,\ \varkappa+\lambda+1,\ \lambda+\mu+1,\ \mu+\nu+1,\ \nu+\varkappa+1;\ -4a^2\right)$$
$$[\operatorname{Re}(\varkappa+\lambda+\mu+\nu)>-1].$$ ET II 379(3)

6.792

1. $$\int_{-\infty}^{\infty} e^{\pi x} K_{ix+iy}(a) K_{ix+iz}(b)\, dx = \pi e^{-\pi z} K_{i(y-z)}(a-b)$$
$[a > b > 0]$. ET II 382(22)

2. $$\int_{-\infty}^{\infty} e^{i\varrho x} K_{\nu+ix}(\alpha) K_{\nu-ix}(\beta)\, dx =$$
$$= \pi \left(\frac{\alpha + \beta e^{\varrho}}{\alpha e^{\varrho} + \beta} \right)^{\nu} K_{2\nu}\left(\sqrt{\alpha^2 + \beta^2 + 2\alpha\beta \operatorname{ch} \varrho}\right)$$
$[|\arg \alpha| + |\arg \beta| + |\operatorname{Im} \varrho| < \pi]$. ET II 382(23)

3. $$\int_{-\infty}^{\infty} e^{(\pi-\gamma)x} K_{ix+iy}(a) K_{ix+iz}(b)\, dx = \pi e^{-\beta y - \alpha z} K_{iy-iz}(c)$$
$[0 < \gamma < \pi,\ a > 0,\ b > 0,\ c > 0,\ \alpha,\ \beta,\ \gamma$ — the angles of the triangle with sides $a,\ b,\ c]$. ET II 382(24), EH II 55(44)a

4. $$\int_{-\infty}^{\infty} e^{-cxi} H^{(2)}_{\nu-ix}(a) H^{(2)}_{\nu+ix}(b)\, dx = 2i \left(\frac{h}{k} \right)^{2\nu} H^{(2)}_{2\nu}(hk),$$
$$h = \sqrt{ae^{\frac{1}{2}c} + be^{-\frac{1}{2}c}}, \quad k = \sqrt{ae^{-\frac{1}{2}c} + be^{\frac{1}{2}c}}$$
$[a,\ b > 0,\ \operatorname{Im} c = 0]$. ET II 380(11)

5. $$\int_{-\infty}^{\infty} a^{-\mu-x} b^{-\nu+x} e^{cxi} J_{\mu+x}(a) J_{\nu-x}(b)\, dx =$$
$$= \left[\frac{2 \cos\left(\frac{c}{2}\right)}{a^2 e^{-\frac{1}{2}ci} + b^2 e^{\frac{1}{2}ci}} \right]^{\frac{1}{2}\mu + \frac{1}{2}\nu} \exp\left[\frac{c}{2} (\nu - \mu) i \right] \times$$
$$\times J_{\mu+\nu} \left\{ \left[2 \cos\left(\frac{c}{2}\right) \left(a^2 e^{-\frac{1}{2}ci} + b^2 e^{\frac{1}{2}ci} \right) \right]^{\frac{1}{2}} \right\}$$
$[b > 0,\ a > 0,\ |c| < \pi,\ \operatorname{Re}(\mu + \nu) > 1]$;
$$= 0 \qquad [a > 0,\ b > 0,\ |c| \geqslant \pi,\ \operatorname{Re}(\mu + \nu) > 1].$$
EH II 54(41), ET II 379(2)

6.793

1. $$\int_{-\infty}^{\infty} e^{-cxi} [J_{\nu-ix}(a) N_{\nu+ix}(b) + N_{\nu-ix}(a) J_{\nu+ix}(b)]\, dx =$$
$$= -2 \left(\frac{h}{k} \right)^{2\nu} J_{2\nu}(hk),$$
$$h = \sqrt{ae^{\frac{1}{2}c} + be^{-\frac{1}{2}c}}, \quad k = \sqrt{ae^{-\frac{1}{2}c} + be^{\frac{1}{2}c}}$$
$[a,\ b > 0,\ \operatorname{Im} c = 0]$. ET II 380(9)

2. $$\int_{-\infty}^{\infty} e^{-cxi}\left[J_{\nu-ix}(a)\,J_{\nu+ix}(b)-N_{\nu-ix}(a)\,N_{\nu+ix}(b)\right]dx=$$
$$=2\left(\frac{h}{k}\right)^{2\nu}N_{2\nu}(hk),$$
$$h=\sqrt{ae^{\frac{1}{2}c}+be^{-\frac{1}{2}c}},\quad k=\sqrt{ae^{-\frac{1}{2}c}+be^{\frac{1}{2}c}}$$
$[a,\ b>0,\ \operatorname{Im} c=0]$. ET II 380(10)

6.794

1. $$\int_{0}^{\infty} K_{ix}(a)\,K_{ix}(b)\operatorname{ch}\left[(\pi-\varphi)x\right]dx=$$
$$=\frac{\pi}{2}K_0\left(\sqrt{a^2+b^2-2ab\cos\varphi}\right).$$ EH II 55(42)

2. $$\int_{0}^{\infty}\operatorname{ch}\left(\frac{\pi}{2}x\right)K_{ix}(a)\,dx=\frac{\pi}{2}\qquad [a>0].$$ ET II 382(19)

3. $$\int_{0}^{\infty}\operatorname{ch}(\varrho x)\,K_{ix+\nu}(a)K_{-ix+\nu}(a)\,dx=\frac{\pi}{2}K_{2\nu}\left[2a\cos\left(\frac{\varrho}{2}\right)\right]$$
$[2|\arg a|+|\operatorname{Re}\varrho|<\pi]$. ET II 383(28)

4. $$\int_{-\infty}^{\infty}\operatorname{sech}\left(\frac{\pi}{2}x\right)J_{ix}(a)\,dx=2\sin a\qquad [a>0].$$ ET II 380(6)

5. $$\int_{-\infty}^{\infty}\operatorname{cosech}\left(\frac{\pi}{2}x\right)J_{ix}(a)\,dx=-2i\cos a\qquad [a>0].$$ ET II 380(7)

6. $$\int_{0}^{\infty}\operatorname{sech}(\pi x)\left\{[J_{ix}(a)]^2+[N_{ix}(a)]^2\right\}dx=-N_0(2a)-\mathbf{E}_0(2a)$$
$[a>0]$. ET II 380(12)

7. $$\int_{0}^{\infty}x\operatorname{sh}\left(\frac{\pi}{2}x\right)K_{ix}(a)\,dx=\frac{\pi a}{2}\qquad [a>0].$$ ET II 382(20)

8. $$\int_{0}^{\infty}x\operatorname{th}(\pi x)\,K_{ix}(\beta)\,K_{ix}(\alpha)\,dx=\frac{\pi}{2}\sqrt{\alpha\beta}\,\frac{\exp(-\beta-\alpha)}{\alpha+\beta}$$
$[|\arg\beta|<\pi,\ |\arg\alpha|<\pi]$. ET II 175(4)

9. $$\int_{0}^{\infty}x\operatorname{sh}(\pi x)\,K_{2ix}(\alpha)\,K_{ix}(\beta)\,dx=$$
$$=\frac{\pi^{\frac{3}{2}}\alpha}{2^{\frac{5}{2}}\sqrt{\beta}}\exp\left(-\beta-\frac{\alpha^2}{8\beta}\right)\quad\left[\beta>0,\ |\arg\alpha|<\frac{\pi}{4}\right].$$ ET II 175(5)

10. $\int_0^\infty \frac{x \operatorname{sh}(\pi x)}{x^2+n^2} K_{ix}(\alpha) K_{ix}(\beta)\, dx =$

$= \frac{\pi^2}{2} I_n(\beta) K_n(\alpha) \quad [0 < \beta < \alpha;\ n = 0, 1, 2, \ldots];$

$= \frac{\pi^2}{2} I_n(\alpha) K_n(\beta) \quad [0 < \alpha < \beta;\ n = 0, 1, 2, \ldots].$ ET II 176(8)

11. $\int_0^\infty x \operatorname{sh}(\pi x) K_{ix}(\alpha) K_{ix}(\beta) K_{ix}(\gamma)\, dx =$

$= \frac{\pi^2}{4} \exp\left[-\frac{\gamma}{2}\left(\frac{\alpha}{\beta} + \frac{\beta}{\alpha} + \frac{\alpha\beta}{\gamma^2}\right)\right] \quad \left[|\arg\alpha| + |\arg\beta| < \frac{\pi}{2},\ \gamma > 0\right]$

ET II 176(9)

12. $\int_0^\infty x \operatorname{sh}\left(\frac{\pi}{2} x\right) K_{\frac{1}{2} ix}(\alpha) K_{\frac{1}{2} ix}(\beta) K_{ix}(\gamma)\, dx =$

$= \frac{\pi^2\gamma}{2\sqrt{\gamma^2+4\alpha\beta}} \exp\left[-\frac{(\alpha+\beta)\sqrt{\gamma^2+4\alpha\beta}}{2\sqrt{\alpha\beta}}\right]$

$[|\arg\alpha| + |\arg\beta| < \pi,\ \gamma > 0].$ ET II 176(10)

13. $\int_0^\infty x \operatorname{sh}(\pi x) K_{\frac{1}{2} ix+\lambda}(\alpha) K_{\frac{1}{2} ix-\lambda}(\alpha) K_{ix}(\gamma)\, dx =$

$= 0 \quad [0 < \gamma < 2\alpha];$

$= \frac{\pi^2\gamma}{2^{2\lambda+1}\alpha^{2\lambda} z} \quad [(\gamma + z)^{2\lambda} + (\gamma - z)^{2\lambda}],$

$z = \sqrt{\gamma^2 - 4\alpha^2} \quad [0 < 2\alpha < \gamma].$ ET II 176(11)

6.795

1. $\int_0^\infty \cos(bx) K_{ix}(a)\, dx = \frac{\pi}{2} e^{-a \operatorname{ch} b}$

$\left[|\operatorname{Im} b| < \frac{\pi}{2},\ a > 0\right].$ EH II 55(46), ET II 175(2)

2. $\int_0^\infty J_x(ax) J_{-x}(ax) \cos(\pi x)\, dx = \frac{1}{4}(1 - a^2)^{-\frac{1}{2}} \quad [|a| < 1].$ ET II 380(4)

3. $\int_0^\infty x \sin(ax) K_{ix}(bx)\, dx = \frac{\pi b}{2} \operatorname{sh} a \exp(-b \operatorname{ch} a)$

$\left[|\operatorname{Im} a| < \frac{\pi}{2},\ b > 0\right].$ ET II 175(1)

4. $\int_{-\infty}^\infty \frac{\sin[(\nu+ix)\pi]}{n+\nu+ix} K_{\nu+ix}(a) K_{\nu-ix}(b)\, dx =$

$= \pi^2 I_n(a) K_{n+2\nu}(b) \quad [0 < a < b;\ n = 0, 1, \ldots];$

$= \pi^2 K_{n+2\nu}(a) I_n(b) \quad [0 < b < a;\ n = 0, 1, \ldots].$ ET II 382(25)

5. $$\int_0^\infty x \sin\left(\frac{1}{2}\pi x\right) K_{\frac{1}{2}ix}(a) K_{ix}(b)\, dx =$$

$$= \frac{\pi^{\frac{3}{2}} b}{\sqrt{2a}} \exp\left(-a - \frac{b^2}{8a}\right) \quad \left[|\arg a| < \frac{\pi}{2},\ b > 0\right].$$ ET II 175(6)

6.796

1. $$\int_{-\infty}^\infty \frac{e^{\frac{1}{2}\pi x} \cos(bx)}{\operatorname{sh}(\pi x)} J_{ix}(a)\, dx = -i \exp(ia \operatorname{ch} b) \quad [a > 0,\ b > 0].$$ ET II 380(8)

2. $$\int_0^\infty \cos(bx) \operatorname{ch}\left(\frac{1}{2}\pi x\right) K_{ix}(a)\, dx = \frac{\pi}{2} \cos(a \operatorname{sh} b).$$ EH II 55(47)

3. $$\int_0^\infty \sin(bx) \operatorname{sh}\left(\frac{1}{2}\pi x\right) K_{ix}(a)\, dx = \frac{\pi}{2} \sin(a \operatorname{sh} b).$$ EH II 55(48)

4. $$\int_0^\infty \cos(bx) \operatorname{ch}(\pi x) [K_{ix}(a)]^2\, dx = -\frac{\pi^2}{4} N_0\left[2a \operatorname{sh}\left(\frac{b}{2}\right)\right]$$

$$[a > 0,\ b > 0].$$ ET II 383(27)

5. $$\int_0^\infty \sin(bx) \operatorname{sh}(\pi x) [K_{ix}(a)]^2\, dx = \frac{\pi^2}{4} J_0\left[2a \operatorname{sh}\left(\frac{b}{2}\right)\right]$$

$$[a > 0,\ b > 0].$$ ET II 382(26)

6.797

1. $$\int_0^\infty x e^{\pi x} \operatorname{sh}(\pi x) \Gamma(\nu + ix) \Gamma(\nu - ix) H_{ix}^{(2)}(a) H_{ix}^{(2)}(b)\, dx =$$

$$= i 2^\nu \sqrt{\pi}\, \Gamma\left(\frac{1}{2} + \nu\right) (ab)^\nu (a+b)^{-\nu} K_\nu(a+b)$$

$$[a > 0,\ b > 0,\ \operatorname{Re} \nu > 0].$$ ET II 381(14)

2. $$\int_0^\infty x e^{\pi x} \operatorname{sh}(\pi x) \operatorname{ch}(\pi x) \Gamma(\nu + ix) \Gamma(\nu - ix) H_{ix}^{(2)}(a) H_{ix}^{(2)}(b)\, dx =$$

$$= \frac{i \pi^{\frac{3}{2}} 2^\nu}{\Gamma\left(\frac{1}{2} - \nu\right)} (b-a)^{-\nu} H_\nu^{(2)}(b-a) \quad \left[0 < a < b,\ 0 < \operatorname{Re} \nu < \frac{1}{2}\right].$$

ET II 381(15)

3. $$\int_0^\infty x e^{\pi x} \operatorname{sh}(\pi x) \Gamma\left(\frac{\nu + ix}{2}\right) \Gamma\left(\frac{\nu - ix}{2}\right) H_{ix}^{(2)}(a) H_{ix}^{(2)}(b)\, dx =$$

$$= i\pi 2^{2-\nu} (ab)^\nu (a^2 + b^2)^{-\frac{1}{2}\nu} H_\nu^{(2)}\left(\sqrt{a^2 + b^2}\right)$$

$$[a > 0,\ b > 0,\ \operatorname{Re} \nu > 0].$$ ET II 381(16)

4. $$\int_0^\infty x \operatorname{sh}(\pi x) \Gamma(\lambda + ix) \Gamma(\lambda - ix) K_{ix}(a) K_{ix}(b)\, dx =$$
$$= 2^{\nu-1} \pi^{\frac{3}{2}} (ab)^\lambda (a+b)^{-\lambda} \Gamma\left(\lambda + \frac{1}{2}\right) K_\lambda (a+b)$$
$$[|\arg a| < \pi, \ \operatorname{Re} \lambda > 0, \ b > 0].$$ ET II 176(12)

5. $$\int_0^\infty x \operatorname{sh}(2\pi x) \Gamma(\lambda + ix) \Gamma(\lambda - ix) K_{ix}(a) K_{ix}(b)\, dx =$$
$$= \frac{2^\lambda \pi^{\frac{5}{2}}}{\Gamma\left(\frac{1}{2} - \lambda\right)} \left(\frac{ab}{|b-a|}\right)^\lambda K_\lambda(|b-a|)$$
$$\left[a > 0, \ 0 < \operatorname{Re}\lambda < \frac{1}{2}, \ b > 0\right].$$ ET II 176(13)

6. $$\int_0^\infty x \operatorname{sh}(\pi x) \Gamma\left(\lambda + \frac{1}{2} ix\right) \Gamma\left(\lambda - \frac{1}{2} ix\right) K_{ix}(a) K_{ix}(b)\, dx =$$
$$= 2\pi^2 \left(\frac{ab}{2\sqrt{a^2+b^2}}\right) K_{2\lambda}\left(\sqrt{a^2+b^2}\right)$$
$$\left[|\arg a| < \frac{\pi}{2}, \ \operatorname{Re}\lambda > 0, \ b > 0\right].$$ ET II 177(14)

7. $$\int_0^\infty \frac{x \operatorname{th}(\pi x) K_{ix}(a) K_{ix}(b)}{\Gamma\left(\frac{3}{4} + \frac{1}{2} ix\right) \Gamma\left(\frac{3}{4} - \frac{1}{2} ix\right)} dx = \frac{1}{2} \sqrt{\frac{\pi ab}{a^2+b^2}} \exp\left(-\sqrt{a^2+b^2}\right)$$
$$\left[|\arg a| < \frac{\pi}{2}, \ b > 0\right],$$ (see also **7.335**). ET II 177(15)

6.8 Functions Generated by Bessel Functions

6.81 Struve functions

6.811

1. $$\int_0^\infty \mathbf{H}_\nu(bx)\, dx = -\frac{\operatorname{ctg}\left(\frac{\nu\pi}{2}\right)}{b} \quad [-2 < \operatorname{Re}\nu < 0, \ b > 0].$$ ET II 158(1)

2. $$\int_0^\infty \mathbf{H}_\nu\left(\frac{a^2}{x}\right) \mathbf{H}_\nu(bx)\, dx = -\frac{J_{2\nu}(2a\sqrt{b})}{b}$$
$$\left[a > 0, \ b > 0, \ \operatorname{Re}\nu > -\frac{3}{2}\right].$$ ET II 170(37)

3. $$\int_0^\infty \mathbf{H}_{\nu-1}\left(\frac{a^2}{x}\right) \mathbf{H}_\nu(bx) \frac{dx}{x} = -\frac{1}{a\sqrt{b}} J_{2\nu-1}(2a\sqrt{b})$$
$$\left[a > 0, \ b > 0, \ \operatorname{Re}\nu > -\frac{1}{2}\right].$$ ET II 170(38)

6.812

1. $$\int_0^\infty \frac{\mathbf{H}_1(bx)\,dx}{x^2+a^2} = \frac{\pi}{2a}\left[I_1(ab) - \mathbf{L}_1(ab)\right] \quad [\operatorname{Re} a > 0,\ b > 0].$$ ET II 158(6)

2. $$\int_0^\infty \frac{\mathbf{H}_\nu(bx)}{x^2+a^2}\,dx = -\frac{\pi}{2a\sin\left(\frac{\nu\pi}{2}\right)}\mathbf{L}_\nu(ab) + \frac{b\operatorname{ctg}\left(\frac{\nu\pi}{2}\right)}{1-\nu^2}\,{}_1F_2\left(1;\ \frac{3-\nu}{2};\ \frac{3+\nu}{2};\ \frac{a^2b^2}{2}\right)$$ $[\operatorname{Re} a > 0,\ b > 0,\ |\operatorname{Re}\nu| < 2].$ ET II 159(7)

6.813

1. $$\int_0^\infty x^{s-1}\mathbf{H}_\nu(ax)\,dx = \frac{2^{s-1}\Gamma\left(\frac{s+\nu}{2}\right)}{a^s\Gamma\left(\frac{1}{2}\nu - \frac{1}{2}s+1\right)}\operatorname{tg}\left(\frac{s+\nu}{2}\pi\right)$$ $$\left[a > 0,\quad -1-\operatorname{Re}\nu < \operatorname{Re}s < \min\left(\frac{3}{2},\ 1-\operatorname{Re}\nu\right)\right].$$ WA 429(2), ET I 335(52)

2. $$\int_0^\infty x^{-\nu-1}\mathbf{H}_\nu(x)\,dx = \frac{2^{-\nu-1}\pi}{\Gamma(\nu+1)} \quad \left[\operatorname{Re}\nu > -\frac{3}{2}\right].$$ ET II 383(2)

3. $$\int_0^\infty x^{-\mu-\nu}\mathbf{H}_\mu(x)\,\mathbf{H}_\nu(x)\,dx = \frac{2^{-\mu-\nu}\sqrt{\pi}\,\Gamma(\mu+\nu)}{\Gamma\left(\mu+\frac{1}{2}\right)\Gamma\left(\nu+\frac{1}{2}\right)\Gamma\left(\mu+\nu+\frac{1}{2}\right)}$$ $[\operatorname{Re}(\mu+\nu) > 0].$ WA 435(2), ET II 384(8)

4. $$\int_0^1 x^{\nu+1}\mathbf{H}_\nu(ax)\,dx = \frac{1}{a}\mathbf{H}_{\nu+1}(a) \quad \left[a > 0,\ \operatorname{Re}\nu > -\frac{3}{2}\right].$$ ET II 158(2)a

5. $$\int_0^1 x^{1-\nu}\mathbf{H}_\nu(ax)\,dx = \frac{a^{\nu-1}}{2^{\nu-1}\sqrt{\pi}\,\Gamma\left(\nu+\frac{1}{2}\right)} - \frac{1}{a}\mathbf{H}_{\nu-1}(a) \quad [a > 0].$$ ET II 158(3)a

6.814

1. $$\int_0^\infty \frac{x^\lambda\mathbf{H}_\nu(bx)}{(x^2+a^2)^{1-\mu}}\,dx = \frac{1}{\sqrt{2b}}\frac{a^{\lambda+2\mu-\frac{3}{2}}}{\Gamma(1-\mu)}G_{24}^{22}\left(\frac{a^2b^2}{4}\middle|\begin{matrix} l,\ m \\ l,\ m-\mu,\ h,\ k\end{matrix}\right),$$ $$h = \frac{1}{4}+\frac{\nu}{2},\quad k = \frac{1}{4}-\frac{\nu}{2},\quad l = \frac{3}{4}+\frac{\nu}{2},\quad m = \frac{3}{4}-\frac{\lambda}{2}$$ $$\left[\operatorname{Re}a > 0,\ b > 0,\ \operatorname{Re}(\lambda+\nu) > -2,\ \operatorname{Re}(\lambda+2\mu) < \frac{5}{2},\ \operatorname{Re}(\lambda+2\mu+\nu) < 2\right].$$ ET II 159(10)

2. $$\int_0^\infty \frac{x^{\nu+1}\mathbf{H}_\nu(bx)}{(x^2+a^2)^{1-\mu}}\,dx = \frac{2^{\mu-1}\pi a^{\mu+\nu}b^{-\mu}}{\Gamma(1-\mu)\cos[(\mu+\nu)\pi]}[I_{-\mu-\nu}(ab)-\mathbf{L}_{\mu+\nu}(ab)]$$

$$\left[\operatorname{Re} a>0,\ b>0,\ \operatorname{Re}\nu>-\frac{3}{2},\ \operatorname{Re}(\mu+\nu)<\frac{1}{2},\ \operatorname{Re}(2\mu+\nu)<\frac{3}{2}\right].$$

ET II 159(8)

6.815

1. $$\int_0^1 x^{\frac{1}{2}\nu}(1-x)^{\mu-1}\mathbf{H}_\nu(a\sqrt{x})\,dx = 2^\mu a^{-\mu}\Gamma(\mu)\,\mathbf{H}_{\mu+\nu}(a)$$

$$\left[\operatorname{Re}\nu>-\frac{3}{2},\ \operatorname{Re}\mu>0\right].$$

ET II 199(88)a

2. $$\int_0^1 x^{\lambda-\frac{1}{2}\nu-\frac{3}{2}}(1-x)^{\mu-1}\,\mathbf{H}_\nu(a\sqrt{x})\,dx =$$

$$= \frac{\mathrm{B}(\lambda,\mu)\,a^{\nu+1}}{2^\nu\sqrt{\pi}\,\Gamma\left(\nu+\frac{3}{2}\right)}\,{}_2F_3\left(1,\ \lambda;\ \frac{3}{2},\ \nu+\frac{3}{2},\ \lambda+\mu;\ -\frac{a^2}{4}\right)$$

$$[\operatorname{Re}\lambda>0,\ \operatorname{Re}\mu>0].$$

ET II 199(89)a

6.82 Combinations of Struve functions, exponentials, and powers

6.821

1. $$\int_0^\infty e^{-\alpha x}\mathbf{H}_{-n-\frac{1}{2}}(\beta x)\,dx = (-1)^n\beta^{n+\frac{1}{2}}\left(\alpha+\sqrt{\alpha^2+\beta^2}\right)^{-n-\frac{1}{2}}\frac{1}{\sqrt{\alpha^2+\beta^2}}$$

$$[\operatorname{Re}\alpha>|\operatorname{Im}\beta|].$$

ET II 206(6)

2. $$\int_0^\infty e^{-\alpha x}\mathbf{L}_{-n-\frac{1}{2}}(\beta x)\,dx = \beta^{n+\frac{1}{2}}\left(\alpha+\sqrt{\alpha^2-\beta^2}\right)^{-n-\frac{1}{2}}\frac{1}{\sqrt{\alpha^2-\beta^2}}$$

$$[\operatorname{Re}\alpha>|\operatorname{Re}\beta|].$$

ET II 208(26)

3 $$\int_0^\infty e^{-\alpha x}\mathbf{H}_0(\beta x)\,dx = \frac{2}{\pi}\frac{\ln\left(\frac{\sqrt{\alpha^2+\beta^2}+\beta}{\alpha}\right)}{\sqrt{\alpha^2+\beta^2}} \qquad [\operatorname{Re}\alpha>|\operatorname{Im}\beta|].$$

ET II 205(1)

4. $$\int_0^\infty e^{-\alpha x}\mathbf{L}_0(\beta x)\,dx = \frac{2}{\pi}\frac{\arcsin\left(\frac{\beta}{\alpha}\right)}{\sqrt{\alpha^2+\beta^2}} \qquad [\operatorname{Re}\alpha>|\operatorname{Re}\beta|].$$

ET II 207(18)

6.822 $$\int_0^\infty e^{(\nu+1)x}\mathbf{H}_\nu(a\,\operatorname{sh}x)\,dx =$$

$$= \sqrt{\frac{\pi}{a}}\operatorname{cosec}(\nu\pi)\left[\operatorname{sh}\left(\frac{a}{2}\right)I_{\nu+\frac{1}{2}}\left(\frac{a}{2}\right)-\operatorname{ch}\left(\frac{a}{2}\right)I_{-\nu-\frac{1}{2}}\left(\frac{a}{2}\right)\right]$$

$$[\operatorname{Re} a>0,\ -2<\operatorname{Re}\nu<0].$$

ET II 385(11)

6.823

1. $$\int_0^\infty x^\lambda e^{-ax} \mathbf{H}_\nu (bx)\,dx = \frac{b^{\nu+1}\Gamma(\lambda+\nu+2)}{2^\nu a^{\lambda+\nu+2}\sqrt{\pi}\,\Gamma\left(\nu+\frac{3}{2}\right)} \times$$
$$\times\, {}_3F_2\left(1, \frac{\lambda+\nu}{2}+1, \frac{\lambda+\nu+3}{2}; \frac{3}{2}, \nu+\frac{3}{2}; -\frac{b^2}{a^2}\right)$$
$[\operatorname{Re} a > 0,\ b > 0,\ \operatorname{Re}(\lambda+\nu) > -2]$. ET II 161(19)

2. $$\int_0^\infty x^\nu e^{-\alpha x} \mathbf{L}_\nu (\beta x)\,dx = \frac{(2\beta)^\nu \Gamma\left(\nu+\frac{1}{2}\right)}{\sqrt{\pi}\left(\sqrt{\alpha^2-\beta^2}\right)^{2\nu+1}} -$$
$$- \frac{\Gamma(2\nu+1)\left(\frac{\beta}{\alpha}\right)^\nu}{\sqrt{\frac{\pi}{2}}\,\alpha\,(\beta^2-\alpha^2)^{\frac{1}{2}\nu+\frac{1}{4}}} P_{-\nu-\frac{1}{2}}^{-\nu-\frac{1}{2}}\left(\frac{\beta}{\alpha}\right)$$
$\left[\operatorname{Re}\alpha > |\operatorname{Re}\beta|,\ \operatorname{Re}\nu > -\frac{1}{2}\right]$. ET I 209(35)a

6.824

1. $$\int_0^\infty t^\nu e^{-at} \mathbf{L}_{2\nu}(2\sqrt{t})\,dt = \frac{1}{a^{2\nu+1}} e^{\frac{1}{a}} \Phi\left(\frac{1}{\sqrt{a}}\right).$$ MI 51

2. $$\int_0^\infty t^\nu e^{-at} \mathbf{L}_{-2\nu}(\sqrt{t})\,dt =$$
$$= \frac{1}{\Gamma\left(\frac{1}{2}-2\nu\right) a^{2\nu+1}} e^{\frac{1}{a}} \gamma\left(\frac{1}{2}-2\nu, \frac{1}{a}\right).$$ MI 51

6.825 $$\int_0^\infty x^{s-1} e^{-\alpha^2 x^2} \mathbf{H}_\nu(\beta x)\,dx = \frac{\beta^{\nu+1}\Gamma\left(\frac{1}{2}+\frac{s}{2}+\frac{\nu}{2}\right)}{2^{\nu+1}\sqrt{\pi}\,\alpha^{\nu+s+1}\Gamma\left(\nu+\frac{3}{2}\right)} \times$$
$$\times\, {}_2F_2\left(1, \frac{\nu+s+1}{2}; \frac{3}{2}, \nu+\frac{3}{2}; -\frac{\beta^2}{4\alpha^2}\right)$$
$\left[\operatorname{Re} s > -\operatorname{Re}\nu - 1,\ |\arg\alpha| < \frac{\pi}{4}\right]$. ET I 335(51)a, ET II 162(20)

6.83 Combinations of Struve and trigonometric functions

6.831 $$\int_0^\infty x^{-\nu} \sin(ax)\, \mathbf{H}_\nu(bx)\,dx =$$
$$= 0 \qquad \left[0 < b < a,\ \operatorname{Re}\nu > -\frac{1}{2}\right];$$
$$= \sqrt{\pi}\, 2^{-\nu} b^{-\nu} \frac{(b^2-a^2)^{\nu-\frac{1}{2}}}{\Gamma\left(\nu+\frac{1}{2}\right)} \qquad \left[0 < a < b,\ \operatorname{Re}\nu > -\frac{1}{2}\right].$$
ET II 162(21)

6.832 $\int_0^\infty \sqrt{x}\sin(ax)\,\mathbf{H}_{\frac{1}{4}}(b^2x^2)\,dx = -2^{-\frac{3}{2}}\sqrt{\pi}\frac{\sqrt{a}}{b^2}N_{\frac{1}{4}}\left(\frac{a^2}{4b^2}\right)$

$[a>0]$. ET I 109(14)

6.84-6.85 Combinations of Struve and Bessel functions

6.841 $\int_0^\infty \mathbf{H}_{\nu-1}(ax)N_\nu(bx)\,dx =$

$= -a^{\nu-1}b^{-\nu} \quad \left[0<b<a, \quad |\operatorname{Re}\nu|<\frac{1}{2}\right];$

$= 0 \quad \left[0<a<b, \quad |\operatorname{Re}\nu|<\frac{1}{2}\right].$ ET II 114(36)

6.842 $\int_0^\infty [\mathbf{H}_0(ax)-N_0(ax)]\,J_0(bx)\,dx = \frac{4}{\pi(a+b)}\boldsymbol{K}\left[\frac{|a-b|}{a+b}\right]$

$\left[a>0, \quad b>0\right].$ ET II 15(22)

6.843

1. $\int_0^\infty J_{2\nu}(a\sqrt{x})\,\mathbf{H}_\nu(bx)\,dx = -\frac{1}{b}N_\nu\left(\frac{a^2}{4b}\right)$

$\left[a>0, \quad b>0, \quad -1<\operatorname{Re}\nu<\frac{5}{4}\right].$ ET II 164(10)

2. $\int_0^\infty K_{2\nu}(2a\sqrt{x})\,\mathbf{H}_\nu(bx)\,dx = \frac{2^\nu}{\pi b}\Gamma(\nu+1)\,S_{-\nu-1,\,\nu}\left(\frac{a^2}{b}\right)$

$[\operatorname{Re}a>0, \quad b>0, \quad \operatorname{Re}\nu>-1].$ ET II 168(27)

6.844 $\int_0^\infty \left[\cos\left(\frac{\mu-\nu}{2}\pi\right)J_\mu(a\sqrt{x})-\sin\left(\frac{\mu-\nu}{2}\pi\right)N_\mu(a\sqrt{x})\right]\times$

$\times K_\mu(a\sqrt{x})\,\mathbf{H}_\nu(bx)\,dx = \frac{1}{a^2}W_{\frac{1}{2}\nu,\,\frac{1}{2}\mu}\left(\frac{a^2}{2b}\right)W_{-\frac{1}{2}\nu,\,\frac{1}{2}\mu}\left(\frac{a^2}{2b}\right)$

$\left[|\arg a|<\frac{\pi}{4}, \quad b>0, \quad \operatorname{Re}\nu>|\operatorname{Re}\mu|-2\right].$ ET II 169(35)

6.845

1. $\int_0^\infty \left[\mathbf{H}_{-\nu}\left(\frac{a}{x}\right)-N_{-\nu}\left(\frac{a}{x}\right)\right]J_\nu(bx)\,dx = \frac{4}{\pi b}\cos(\nu\pi)\,K_{2\nu}(2\sqrt{ab})$

$\left[|\arg a|<\pi, \quad b>0, \quad |\operatorname{Re}\nu|<\frac{1}{2}\right].$ ET II 73(7)

2. $$\int_0^\infty \left[J_{-\nu}\left(\frac{a^2}{x}\right)+\sin(\nu\pi)\,\mathbf{H}_\nu\left(\frac{a^2}{x}\right)\right]\mathbf{H}_\nu(bx)\,dx = \frac{1}{b}\left[\frac{2}{\pi}K_{2\nu}(2a\sqrt{b})-N_{2\nu}(2a\sqrt{b})\right]$$

$$\left[a>0, b>0, \ -\frac{3}{2}<\operatorname{Re}\nu<0\right].$$ ET II 170(39)

6.846 $$\int_0^\infty \left[\frac{2}{\pi}K_{2\nu}(2a\sqrt{x})+N_{2\nu}(2a\sqrt{x})\right]\mathbf{H}_\nu(bx)\,dx=\frac{1}{b}J_\nu\left(\frac{a^2}{b}\right)$$

$$\left[a>0, \quad b>0, \quad |\operatorname{Re}\nu|<\frac{1}{2}\right].$$ ET II 169(30)

6.847 $$\int_0^\infty \left[\cos\frac{\nu\pi}{2}J_\nu(ax)+\sin\frac{\nu\pi}{2}\mathbf{H}_\nu(ax)\right]\frac{dx}{x^2+k^2}=\frac{\pi}{2k}\left[I_\nu(ak)-\mathbf{L}_\nu(ak)\right]$$

$$\left[a>0, \ \operatorname{Re}k>0, \ -\frac{1}{2}<\operatorname{Re}\nu<2\right].$$ ET II 384(5)a, WA 467(8)

6.848

1. $$\int_0^\infty x\left[I_\nu(ax)-\mathbf{L}_{-\nu}(ax)\right]J_\nu(bx)\,dx=\frac{2}{\pi}\left(\frac{a}{b}\right)^{\nu-1}\cos(\nu\pi)\frac{1}{a^2+b^2}$$

$$\left[\operatorname{Re}a>0, \quad b>0, \quad -1<\operatorname{Re}\nu<-\frac{1}{2}\right].$$ ET II 74(12)

2. $$\int_0^\infty x\left[\mathbf{H}_{-\nu}(ax)-N_{-\nu}(ax)\right]J_\nu(bx)\,dx=2\frac{\cos(\nu\pi)}{a^\nu\pi}b^{\nu-1}\frac{1}{a+b}$$

$$\left[|\arg a|<\pi, \quad -\frac{1}{2}<\operatorname{Re}\nu, \quad b>0\right].$$ ET II 73(5)

6.849

1. $$\int_0^\infty xK_\nu(ax)\,\mathbf{H}_\nu(bx)\,dx=a^{-\nu-1}b^{\nu+1}\frac{1}{a^2+b^2}$$

$$\left[\operatorname{Re}a>0, \quad b>0, \quad \operatorname{Re}\nu>-\frac{3}{2}\right].$$ ET II 164(12)

2. $$\int_0^\infty x\left[K_\mu(ax)\right]^2\mathbf{H}_0(bx)\,dx=-2^{-\mu-1}\pi a^{-2\mu}\frac{\left[(z+b)^{2\mu}+(z-b)^{2\mu}\right]}{bz}\sec(\mu\pi),$$

$$z=\sqrt{4a^2+b^2} \qquad \left[\operatorname{Re}a>0, \quad b>0, \quad |\operatorname{Re}\mu|<\frac{3}{2}\right].$$ ET II 166(18)

6.851

1. $$\int_0^\infty x\left\{\left[J_{\frac{1}{2}\nu}(ax)\right]^2-\left[N_{\frac{1}{2}\nu}(ax)\right]^2\right\}\mathbf{H}_\nu(bx)\,dx=$$

$$=0 \qquad \left[0<b<2a, \ -\frac{3}{2}<\operatorname{Re}\nu<0\right];$$

$$=\frac{4}{\pi b}\frac{1}{\sqrt{b^2-4a^2}} \qquad \left[0<2a<b, \ -\frac{3}{2}<\operatorname{Re}\nu<0\right].$$ ET II 164(7)

2. $\int_0^\infty x^{\nu+1} \{[J_\nu(ax)]^2 - [N_\nu(ax)]^2\} \mathbf{H}_\nu(bx)\, dx =$

$= 0 \qquad \left[0 < b < 2a, \ -\frac{3}{4} < \operatorname{Re}\nu < 0\right];$

$= \frac{2^{3\nu+2} a^{2\nu} b^{-\nu-1}}{\sqrt{\pi}\,\Gamma\left(\frac{1}{2}-\nu\right)} (b^2 - 4a^2)^{-\nu-\frac{1}{2}} \qquad \left[0 < 2a < b, \ -\frac{3}{4} < \operatorname{Re}\nu < 0\right].$

ET II 163(6)

6.852

1. $\int_0^\infty x^{1-\mu-\nu} J_\nu(x) \mathbf{H}_\mu(x)\, dx = \frac{(2\nu-1)\, 2^{-\mu-\nu}}{(\mu+\nu-1)\,\Gamma\left(\mu+\frac{1}{2}\right)\Gamma\left(\nu+\frac{1}{2}\right)}$

$\left[\operatorname{Re}\nu > \frac{1}{2}, \quad \operatorname{Re}(\mu+\nu) > 1\right].$ ET II 383(4)

2. $\int_0^\infty x^{\mu-\nu+1} N_\mu(ax) \mathbf{H}_\nu(bx)\, dx =$

$= 0 \qquad \left[0 < b < a, \ \operatorname{Re}(\nu-\mu) > 0, \ -\frac{3}{2} < \operatorname{Re}\mu < \frac{1}{2}\right];$

$= \frac{2^{1+\mu-\nu} a^\mu b^{-\nu}}{\Gamma(\nu-\mu)} (b^2 - a^2)^{\nu-\mu-1}$

$\left[0 < a < b, \ \operatorname{Re}(\nu-\mu) > 0, \ -\frac{3}{2} < \operatorname{Re}\mu < \frac{1}{2}\right].$ ET II 163(3)

3 $\int_0^\infty x^{\mu+\nu+1} K_\mu(ax) \mathbf{H}_\nu(bx)\, dx =$

$= \frac{2^{\mu+\nu+1} b^{\nu+1}}{\sqrt{\pi}\, a^{\mu+2\nu+3}} \Gamma\left(\mu+\nu+\frac{3}{2}\right) F\left(1, \ \mu+\nu+\frac{3}{2}; \ \frac{3}{2}; \ -\frac{b^2}{a^2}\right)$

$\left[\operatorname{Re} a > 0, \ b > 0, \ \operatorname{Re}\nu > -\frac{3}{2}, \ \operatorname{Re}(\mu+\nu) > -\frac{3}{2}\right].$ ET II 165(13)

6.853

1. $\int_0^\infty x^{1-\mu} [\sin(\mu\pi) J_{\mu+\nu}(ax) + \cos(\mu\pi) N_{\mu+\nu}(ax)] \ \mathbf{H}_\nu(bx)\, dx = 0$

$\left[0 < b < a, \ 1 < \operatorname{Re}\mu < \frac{3}{2}, \ \operatorname{Re}\nu > -\frac{3}{2}, \ \operatorname{Re}(\nu-\mu) < \frac{1}{2}\right];$

$= \frac{b^\nu (b^2 - a^2)^{\mu-1}}{2^{\mu-1} a^{\mu+\nu} \Gamma(\mu)}$

$\left[0 < a < b, \ 1 < \operatorname{Re}\mu < \frac{3}{2}, \ \operatorname{Re}\nu > -\frac{3}{2}, \ \operatorname{Re}(\nu-\mu) < \frac{1}{2}\right].$

ET II 163(4)

2. $$\int_0^\infty x^{\lambda+\frac{1}{2}} [I_\mu(ax) - \mathbf{L}_{-\mu}(ax)] J_\nu(bx)\, dx =$$

$$= 2^{\lambda+\frac{1}{2}} \frac{\cos(\mu\pi)}{\pi} b^{-\lambda-\frac{3}{2}} G_{33}^{22}\left(\frac{b^2}{a^2} \middle| \begin{matrix} \frac{1+\mu}{2}, & 1-\frac{\mu}{2}, & 1+\frac{\mu}{2} \\ \frac{3}{4}+\frac{\lambda+\nu}{2}, & \frac{1+\mu}{2}, & \frac{3}{4}+\frac{\lambda-\nu}{2} \end{matrix}\right)$$

$\left[\operatorname{Re} a > 0, \quad b > 0, \quad \operatorname{Re}(\mu+\nu+\lambda) > -\frac{3}{2}, \quad -\operatorname{Re}\nu - \frac{5}{2} < \operatorname{Re}(\lambda-\mu) < 1\right].$

ET II 76(21)

3. $$\int_0^\infty x^{\lambda+\frac{1}{2}} [\mathbf{H}_\mu(ax) - N_\mu(ax)] J_\nu(bx)\, dx =$$

$$= 2^{\lambda+\frac{1}{2}} \frac{\cos(\mu\pi)}{\pi^2} b^{-\lambda-\frac{3}{2}} G_{33}^{23}\left(\frac{b^2}{a^2} \middle| \begin{matrix} \frac{1-\mu}{2}, & 1-\frac{\mu}{2}, & 1+\frac{\mu}{2} \\ \frac{3}{4}+\frac{\lambda+\nu}{2}, & \frac{1-\mu}{2}, & \frac{3}{4}+\frac{\lambda-\nu}{2} \end{matrix}\right)$$

$\left[b > 0, \quad |\arg a| < \pi, \quad \operatorname{Re}(\lambda+\mu) < 1, \quad \operatorname{Re}(\lambda+\nu) + \frac{3}{2} > |\operatorname{Re}\mu|\right].$

ET II 73(6)

4. $$\int_0^\infty \sqrt{x}\, [I_{\nu-\frac{1}{2}}(ax) - \mathbf{L}_{\nu-\frac{1}{2}}(ax)] J_\nu(bx)\, dx =$$

$$= \sqrt{\frac{2}{\pi}}\, a^{\nu-\frac{1}{2}} b^{-\nu} \frac{1}{\sqrt{a^2+b^2}} \qquad \left[\operatorname{Re} a > 0, \quad b > 0, \quad |\operatorname{Re}\nu| < \frac{1}{2}\right].$$

ET II 74(11)

5. $$\int_0^\infty x^{\mu-\nu+1} [I_\mu(ax) - \mathbf{L}_\mu(ax)] J_\nu(bx)\, dx =$$

$$= \frac{2^{\mu-\nu+1} a^{\mu-1} b^{\nu-2\mu-1}}{\sqrt{\pi}\,\Gamma\left(\nu-\mu+\frac{1}{2}\right)} F\left(1, \frac{1}{2}; \nu-\mu+\frac{1}{2}; -\frac{b^2}{a^2}\right)$$

$\left[-1 < 2\operatorname{Re}\mu + 1 < \operatorname{Re}\nu + \frac{1}{2}, \quad \operatorname{Re} a > 0, \quad b > 0\right].$

ET II 74(13)

6. $$\int_0^\infty x^{\mu-\nu+1} [I_\mu(ax) - \mathbf{L}_{-\mu}(ax)] J_\nu(bx)\, dx =$$

$$= \frac{2^{\mu-\nu+1} a^{-\mu-1} b^{\nu-1}}{\Gamma\left(\frac{1}{2}-\mu\right)\Gamma\left(\frac{1}{2}+\nu\right)} F\left(1, \frac{1}{2}+\mu; \frac{1}{2}+\nu; -\frac{b^2}{a^2}\right)$$

$\left[\operatorname{Re} a > 0, \quad \operatorname{Re}\nu > -\frac{1}{2}, \quad \operatorname{Re}\mu > -1, \quad b > 0\right].$

ET II 75(18)

6.854

1. $$\int_0^\infty x\mathbf{H}_{\frac{1}{2}\nu}(ax^2)K_\nu(bx)\,dx = \frac{\Gamma\left(\frac{1}{2}\nu+1\right)}{2^{1-\frac{1}{2}\nu}a\pi} S_{-\frac{1}{2}\nu-1,\frac{1}{2}\nu}\left(\frac{b^2}{4a}\right)$$

$[a > 0, \quad \operatorname{Re} b > 0, \quad \operatorname{Re}\nu > -2]$. ET II 150(75)

2. $$\int_0^\infty x\mathbf{H}_{\frac{1}{2}\nu}(ax^2)J_\nu(bx)\,dx = -\frac{1}{2a}N_{\frac{1}{2}\nu}\left(\frac{b^2}{4a}\right)$$

$\left[a > 0, \quad b > 0, \quad -2 < \operatorname{Re}\nu < \frac{3}{2}\right]$. ET II 73(3)

6.855

1. $$\int_0^\infty x^{2\nu+\frac{1}{2}}\left[I_{\nu+\frac{1}{2}}\left(\frac{a}{x}\right) - \mathbf{L}_{\nu+\frac{1}{2}}\left(\frac{a}{x}\right)\right]J_\nu(bx)\,dx = 2^{\frac{3}{2}}\frac{a^{\nu+\frac{1}{2}}}{\sqrt{\pi}\,b^{\nu+1}}J_{2\nu+1}\left(\sqrt{2ab}\right)K_{2\nu+1}\left(\sqrt{2ab}\right)$$

$\left[\operatorname{Re} a > 0, \ b > 0, \ -1 < \operatorname{Re}\nu < \frac{1}{2}\right]$. ET II 76(22)

2. $$\int_0^\infty \left[\mathbf{H}_{-\nu-1}\left(\frac{a}{x}\right) - N_{-\nu-1}\left(\frac{a}{x}\right)\right]J_\nu(bx)\frac{dx}{x} = -\frac{4}{\pi\sqrt{ab}}\cos(\nu\pi)K_{-2\nu-1}\left(2\sqrt{ab}\right)$$

$\left[|\arg a| < \pi, \ b > 0, \ |\operatorname{Re}\nu| < \frac{1}{2}\right]$. ET II 74(8)

3. $$\int_0^\infty x^{2\nu+\frac{1}{2}}\left[\mathbf{H}_{\nu+\frac{1}{2}}\left(\frac{a}{x}\right) - N_{\nu+\frac{1}{2}}\left(\frac{a}{x}\right)\right]J_\nu(bx)\,dx = -2^{\frac{5}{2}}\pi^{-\frac{3}{2}}a^{\nu+\frac{1}{2}}b^{-\nu-1}\sin(\nu\pi)K_{2\nu+1}\left(\sqrt{2ab}\,e^{\frac{1}{4}\pi i}\right)K_{2\nu+1}\left(\sqrt{2ab}\,e^{-\frac{1}{4}\pi i}\right)$$

$\left[|\arg a| < \pi, \ b > 0, \ -1 < \operatorname{Re}\nu < -\frac{1}{6}\right]$. ET II 74(9)

6.856 $$\int_0^\infty xN_\nu\left(a\sqrt{x}\right)K_\nu\left(a\sqrt{x}\right)\mathbf{H}_\nu(bx)\,dx = \frac{1}{2b^2}\exp\left(-\frac{a^2}{2b}\right)$$

$\left[b > 0, \ |\arg a| < \frac{\pi}{4}, \ \operatorname{Re}\nu > -\frac{3}{2}\right]$. ET II 169(32)

6.857

1. $$\int_0^\infty x \exp\left(\frac{a^2x^2}{8}\right) K_{\frac{1}{2}\nu}\left(\frac{a^2x^2}{8}\right) \mathbf{H}_\nu(bx)\,dx =$$

$$= \frac{2}{\sqrt{\pi}} a^{-\frac{\nu}{2}-1} b^{\frac{\nu}{2}-1} \cos\left(\frac{\nu\pi}{2}\right) \Gamma\left(-\frac{1}{2}\nu\right) \exp\left(\frac{b^2}{2a^2}\right) W_{k,m}\left(\frac{b^2}{a^2}\right),$$

$$k = \frac{1}{4}\nu, \quad m = \frac{1}{2} + \frac{1}{4}\nu$$

$$\left[|\arg a| < \frac{3}{4}\pi, \quad b > 0, \quad -\frac{3}{2} < \operatorname{Re}\nu < 0\right].$$ ET II 167(24)

2. $$\int_0^\infty x^{\sigma-2} \exp\left(-\frac{1}{2}a^2x^2\right) K_\mu\left(\frac{1}{2}a^2x^2\right) \mathbf{H}_\nu(bx)\,dx =$$

$$= \frac{\sqrt{\pi}}{2^{\nu+2}} a^{-\nu-\sigma} b^{\nu+1} \frac{\Gamma\left(\frac{\nu+\sigma}{2}+\mu\right)\Gamma\left(\frac{\nu+\sigma}{2}-\mu\right)}{\Gamma\left(\frac{3}{2}\right)\Gamma\left(\nu+\frac{3}{2}\right)\Gamma\left(\frac{\nu+\sigma}{2}\right)} \times$$

$$\times {}_3F_3\left(1, \frac{\nu+\sigma}{2}+\mu, \frac{\nu+\sigma}{2}-\mu; \frac{3}{2}, \nu+\frac{3}{2}, \frac{\nu+\sigma}{2}; -\frac{b^2}{4a^2}\right)$$

$$\left[b > 0, \ |\arg a| < \frac{\pi}{4}, \ \operatorname{Re}(\sigma+\nu) > 2|\operatorname{Re}\mu|\right].$$ ET II 167(23)

6.86 Lommel functions

6.861

$$\int_0^\infty x^{\lambda-1} s_{\mu,\nu}(x)\,dx =$$

$$= \frac{\Gamma\left[\frac{1}{2}(1+\lambda+\mu)\right]\Gamma\left[\frac{1}{2}(1-\lambda-\mu)\right]\Gamma\left[\frac{1}{2}(1+\mu+\nu)\right]\Gamma\left[\frac{1}{2}(1+\mu-\nu)\right]}{2^{2-\lambda-\mu}\Gamma\left[\frac{1}{2}(\nu-\lambda)+1\right]\Gamma\left[1-\frac{1}{2}(\lambda+\nu)\right]}$$

$$\left[-\operatorname{Re}\mu < \operatorname{Re}\lambda + 1 < \frac{5}{2}\right].$$ ET II 385(17)

6.862

1. $$\int_0^u x^{\lambda-\frac{1}{2}\mu-\frac{1}{2}} (u-x)^{\sigma-1} s_{\mu,\nu}\left(a\sqrt{x}\right)dx =$$

$$= \Gamma(\sigma) \frac{a^{\mu+1} u^{\lambda+\sigma} \Gamma(\lambda+1)}{(\mu-\nu+1)(\mu+\nu+1)\Gamma(\lambda+\sigma+1)} \times$$

$$\times {}_2F_3\left(1, 1+\lambda; \frac{\mu-\nu+3}{2}, \frac{\mu+\nu+3}{2}, \lambda+\sigma+1; -\frac{a^2u}{4}\right)$$

$$[\operatorname{Re}\lambda > -1, \ \operatorname{Re}\sigma > 0].$$ ET II 199(92)

2. $\int\limits_u^\infty x^{\frac{1}{2}\nu}(x-u)^{\mu-1}S_{\lambda,\nu}(a\sqrt{x})\,dx =$

$$= \frac{\mathrm{B}\left[\mu, \frac{1}{2}(1-\lambda-\nu)-\mu\right]u^{\frac{1}{2}\mu+\frac{1}{2}\nu}}{a^\mu} S_{\lambda+\mu,\,\mu+\nu}(a\sqrt{u})$$

$[|\arg(a\sqrt{u})| < \pi,\ 0 < 2\operatorname{Re}\mu < 1 - \operatorname{Re}(\lambda+\nu)].$ ET II 211(71)

6.863 $\int\limits_0^\infty \sqrt{x}\,e^{-\alpha x}s_{\mu,\frac{1}{4}}\left(\frac{x^2}{2}\right)dx = 2^{-2\mu-1}\sqrt{\alpha}\,\Gamma\left(2\mu+\frac{3}{2}\right)S_{-\mu-1,\frac{1}{4}}\left(\frac{\alpha^2}{2}\right)$

$\left[\operatorname{Re}\alpha > 0,\ \operatorname{Re}\mu > -\frac{3}{4}\right].$ ET I 209(38)

6.864 $\int\limits_0^\infty \exp[(\mu+1)x]\,s_{\mu,\nu}(a\operatorname{sh}x)\,dx = 2^{\mu-2}\pi\operatorname{cosec}(\mu\pi)\,\Gamma(\varrho)\,\Gamma(\sigma)\times$

$$\times\left[I_\varrho\left(\frac{a}{2}\right)I_\sigma\left(\frac{a}{2}\right) - I_{-\varrho}\left(\frac{a}{2}\right)I_{-\sigma}\left(\frac{a}{2}\right)\right],$$

$2\varrho = \mu+\nu+1,\ 2\sigma = \mu-\nu+1\ [a>0,\ -2 < \operatorname{Re}\mu < 0].$ ET II 386(22)

6.865 $\int\limits_0^\infty \sqrt{\operatorname{sh}x}\operatorname{ch}(\nu x)\,S_{\mu,\frac{1}{2}}(a\operatorname{ch}x)\,dx =$

$$= \frac{\mathrm{B}\left(\frac{1}{4}-\frac{\mu+\nu}{2}, \frac{1}{4}-\frac{\mu-\nu}{2}\right)}{\sqrt{a}\,2^{\mu+\frac{3}{2}}} S_{\mu+\frac{1}{2},\nu}(a)$$

$\left[|\arg a| < \pi,\ \operatorname{Re}\mu + |\operatorname{Re}\nu| < \frac{1}{2}\right]$ ET II 388(31)

6.866

1. $\int\limits_0^\infty x^{-\mu-1}\cos(ax)\,s_{\mu,\nu}(x)\,dx = 0 \quad [a > 1];$

$$= 2^{\mu-\frac{1}{2}}\sqrt{\pi}\,\Gamma\left(\frac{\mu+\nu+1}{2}\right)\Gamma\left(\frac{\mu-\nu+1}{2}\right)(1-a^2)^{\frac{1}{2}\mu+\frac{1}{4}}\mathrm{P}_{\nu-\frac{1}{2}}^{-\mu-\frac{1}{2}}(a)$$

$[0 < a < 1].$ ET II 386(18)

2. $\int\limits_0^\infty x^{-\mu}\sin(ax)\,S_{\mu,\nu}(x)\,dx =$

$$= 2^{-\mu-\frac{1}{2}}\sqrt{\pi}\,\Gamma\left(1-\frac{\mu+\nu}{2}\right)\Gamma\left(1-\frac{\mu-\nu}{2}\right)(a^2-1)^{\frac{1}{2}\mu-\frac{1}{4}}\mathrm{P}_{\nu-\frac{1}{2}}^{\mu-\frac{1}{2}}(a)$$

$[a > 1,\ \operatorname{Re}\mu < 1 - |\operatorname{Re}\nu|].$ ET II 387(23)

6.867

1. $\int\limits_0^{\frac{\pi}{2}} \cos(2\mu x) S_{2\mu-1,\,2\nu}(a\cos x)\,dx =$

$$= \frac{\pi 2^{2\mu-3} a^{2\mu} \operatorname{cosec}(2\nu\pi)}{\Gamma(1-\mu-\nu)\,\Gamma(1-\mu+\nu)} \left[J_{\mu+\nu}\left(\frac{a}{2}\right) N_{\mu-\nu}\left(\frac{a}{2}\right) - J_{\mu-\nu}\left(\frac{a}{2}\right) N_{\mu+\nu}\left(\frac{a}{2}\right) \right]$$

$[\operatorname{Re}\mu > -2, \ |\operatorname{Re}\nu| < 1]$. ET II 388(29)

2. $\int\limits_0^{\frac{\pi}{2}} \cos[(\mu+1)x]\, s_{\mu,\,\nu}(a\cos x)\,dx =$

$$= 2^{\mu-2}\pi\Gamma(\varrho)\,\Gamma(\sigma)\, J_{\varrho}\left(\frac{a}{2}\right) J_{\sigma}\left(\frac{a}{2}\right),$$

$2\varrho = \mu+\nu+1, \quad 2\sigma = \mu-\nu+1 \quad [\operatorname{Re}\mu > -2]$. ET II 386(21)

6.868 $\int\limits_0^{\frac{\pi}{2}} \frac{\cos(2\mu x)}{\cos x} S_{2\mu,\,2\nu}(a\sec x)\,dx = \frac{\pi 2^{2\mu-1}}{a} W_{\mu,\,\nu}(ae^{i\frac{\pi}{2}})\, W_{\mu,\,\nu}(ae^{-i\frac{\pi}{2}})$

$[|\arg a| < \pi, \ \operatorname{Re}\mu < 1]$. ET II 388(30)

6.869

1. $\int\limits_0^{\infty} x^{1-\mu-\nu} J_{\nu}(ax)\, S_{\mu,\,-\mu-2\nu}(x)\,dx =$

$$= \frac{\sqrt{\pi}\, a^{\nu-1}\,\Gamma(1-\mu-\nu)}{2^{\mu+2\nu}\,\Gamma\left(\nu+\frac{1}{2}\right)} (a^2-1)^{\frac{1}{2}(\mu+\nu-1)} P^{\mu+\nu-1}_{\mu+\nu}(a)$$

$\left[a > 1, \ \operatorname{Re}\nu > -\frac{1}{2}, \quad \operatorname{Re}(\mu+\nu) < 1\right]$. ET II 388(28)

2. $\int\limits_0^{\infty} x^{-\mu} J_{\nu}(ax)\, s_{\nu+\mu,\,-\nu+\mu+1}(x)\,dx =$

$= 2^{\nu-1}\Gamma(\nu)\, a^{-\nu}(1-a^2)^{\mu} \quad \left[0 < a < 1, \ \operatorname{Re}\mu > -1, \ -1 < \operatorname{Re}\nu < \frac{3}{2}\right];$

$= 0 \quad \left[1 < a, \ \operatorname{Re}\mu > -1, \ -1 < \operatorname{Re}\nu < \frac{3}{2}\right]$.

ET II 92(24)

3. $\int\limits_0^{\infty} xK_{\nu}(bx)\, s_{\mu,\,\frac{1}{2}\nu}(ax^2)\,dx =$

$$= \frac{1}{4a}\Gamma\left(\mu+\frac{1}{2}\nu+1\right)\Gamma\left(\mu-\frac{1}{2}\nu+1\right) S_{-\mu-1,\,\frac{1}{2}\nu}\left(\frac{b^2}{4a}\right)$$

$\left[\operatorname{Re}\mu > \frac{1}{2}|\operatorname{Re}\nu| - 2, \quad a > 0, \quad \operatorname{Re} b > 0\right]$. ET II 151(78)

6.87 Thomson functions

6.871

1. $$\int_0^\infty e^{-\beta x} \operatorname{ber} x \, dx = \frac{\left(\sqrt{\beta^4+1}+\beta^2\right)^{\frac{1}{2}}}{\sqrt{2(\beta^4+1)}}.$$ ME 40

2. $$\int_0^\infty e^{-\beta x} \operatorname{bei} x \, dx = \frac{\left(\sqrt{\beta^4+1}-\beta^2\right)^{\frac{1}{2}}}{\sqrt{2(\beta^4+1)}}.$$ ME 40

6.872

1. $$\int_0^\infty e^{-\beta x} \operatorname{ber}_\nu (2\sqrt{x}) \, dx = \frac{1}{2\beta} \sqrt{\frac{\pi}{\beta}} \left[J_{\frac{1}{2}(\nu-1)}\left(\frac{1}{2\beta}\right) \cos\left(\frac{1}{2\beta} + \frac{3\nu\pi}{4}\right) - J_{\frac{1}{2}(\nu+1)}\left(\frac{1}{2\beta}\right) \cos\left(\frac{1}{2\beta} + \frac{3\nu+6}{4}\pi\right) \right].$$ MI 49

2. $$\int_0^\infty e^{-\beta x} \operatorname{bei}_\nu (2\sqrt{x}) \, dx = \frac{1}{2\beta} \sqrt{\frac{\pi}{\beta}} \left[J_{\frac{1}{2}(\nu-1)}\left(\frac{1}{2\beta}\right) \sin\left(\frac{1}{2\beta} + \frac{3\nu}{4}\pi\right) - J_{\frac{1}{2}(\nu+1)}\left(\frac{1}{2\beta}\right) \sin\left(\frac{1}{2\beta} + \frac{3\nu+6}{4}\pi\right) \right].$$ MI 49

3. $$\int_0^\infty e^{-\beta x} \operatorname{ber} (2\sqrt{x}) \, dx = \frac{1}{\beta} \cos\frac{1}{\beta}.$$ ME 40

4 $$\int_0^\infty e^{-\beta x} \operatorname{bei} (2\sqrt{x}) \, dx = \frac{1}{\beta} \sin\frac{1}{\beta}.$$ ME 40

5. $$\int_0^\infty e^{-\beta x} \operatorname{ker} (2\sqrt{x}) \, dx = -\frac{1}{2\beta} \left[\cos\frac{1}{\beta} \operatorname{ci}\frac{1}{\beta} + \sin\frac{1}{\beta} \operatorname{si}\frac{1}{\beta} \right].$$ MI 50

6. $$\int_0^\infty e^{-\beta x} \operatorname{kei} (2\sqrt{x}) \, dx = -\frac{1}{2\beta} \left[\sin\frac{1}{\beta} \operatorname{ci}\frac{1}{\beta} - \cos\frac{1}{\beta} \operatorname{si}\frac{1}{\beta} \right].$$ MI 50

7. $$\int_0^\infty e^{-\beta x} \operatorname{ber}_\nu (2\sqrt{x}) \operatorname{bei}_\nu (2\sqrt{x}) \, dx = \frac{1}{2\beta} J_\nu\left(\frac{2}{\beta}\right) \sin\left(\frac{2}{\beta} + \frac{3\nu\pi}{2}\right) \quad [\operatorname{Re} \nu > -1].$$ MI 49

6.873 $$\int_0^\infty \left[\operatorname{ber}_\nu^2 (2\sqrt{x}) + \operatorname{bei}_\nu^2 (2\sqrt{x})\right] e^{-\beta x} \, dx = \frac{1}{\beta} I_\nu\left(\frac{2}{\beta}\right) \quad [\operatorname{Re} \nu > -1].$$ ME 40

6.874

1. $$\int_0^\infty \frac{e^{-\beta x}}{\sqrt{x}} \operatorname{ber}_{2\nu} (2\sqrt{2x}) \, dx = \sqrt{\frac{\pi}{\beta}} J_\nu\left(\frac{1}{\beta}\right) \cos\left(\frac{1}{\beta} - \frac{3\pi}{4} + \frac{3\nu\pi}{2}\right) \quad \left[\operatorname{Re} \nu > -\frac{1}{2}\right].$$ MI 49

2. $$\int_0^\infty \frac{e^{-\beta x}}{\sqrt{x}} \operatorname{bei}_{2\nu}(2\sqrt{2x})\, dx = \sqrt{\frac{\pi}{\beta}} J_\nu\left(\frac{1}{\beta}\right) \sin\left(\frac{1}{\beta} - \frac{3\pi}{4} + \frac{3\nu\pi}{2}\right)$$

$\left[\operatorname{Re} \nu > -\frac{1}{2}\right]$. MI 49

3. $$\int_0^\infty x^{\frac{\nu}{2}} \operatorname{ber}_\nu(\sqrt{x})\, e^{-\beta x}\, dx = \frac{2^{-\nu}}{\beta^{1+\nu}} \cos\left(\frac{1}{4\beta} + \frac{3\nu\pi}{4}\right)$$

$[\operatorname{Re} \nu > -1]$. ME 40

4. $$\int_0^\infty x^{\frac{\nu}{2}} \operatorname{bei}_\nu(\sqrt{x})\, e^{-\beta x}\, dx = \frac{2^{-\nu}}{\beta^{1+\nu}} \sin\left(\frac{1}{4\beta} + \frac{3\nu\pi}{4}\right) \qquad [\operatorname{Re} \nu > -1].$$ ME 40

6.875

1. $$\int_0^\infty e^{-\beta x}\left[\operatorname{ker}(2\sqrt{x}) - \frac{1}{2} \ln x \operatorname{ber}(2\sqrt{x})\right] dx = \frac{1}{\beta}\left[\ln \beta \cos\frac{1}{\beta} + \frac{\pi}{4} \sin\frac{1}{\beta}\right].$$ MI 50

2. $$\int_0^\infty e^{-\beta x}\left[\operatorname{kei}(2\sqrt{x}) - \frac{1}{2} \ln x \operatorname{bei}(2\sqrt{x})\right] dx = \frac{1}{\beta}\left[\ln \beta \sin\frac{1}{\beta} - \frac{\pi}{4} \cos\frac{1}{\beta}\right].$$ MI 50

6.876

1. $$\int_0^\infty x \operatorname{kei} x J_1(ax)\, dx = -\frac{1}{2a} \operatorname{arctg} a^2 \qquad [a > 0].$$ ET II 21(32)

2. $$\int_0^\infty x \operatorname{ker} x J_1(ax)\, dx = \frac{1}{2a} \ln (1 + a^4)^{\frac{1}{2}} \qquad [a > 0].$$ ET II 21(33)

6.9 Mathieu Functions

Notation: $k^2 = q$. For definition of the coefficients $A_p^{(m)}$ and $B_p^{(m)}$ see 8.6

6.91 Mathieu functions

6.911

1. $$\int_0^{2\pi} \operatorname{ce}_m(z, q) \operatorname{ce}_p(z, q)\, dz = 0 \qquad [m \neq p].$$ MA

2. $$\int_0^{2\pi} [\operatorname{ce}_{2n}(z, q)]^2\, dz = 2\pi [A_0^{(2n)}]^2 + \pi \sum_{r=1}^\infty [A_{2r}^{(2n)}]^2 = \pi.$$ MA

3. $$\int_0^{2\pi} [\operatorname{ce}_{2n+1}(z, q)]^2\, dz = \pi \sum_{r=0}^\infty [A_{2r+1}^{(2n+1)}]^2 = \pi.$$ MA

4. $\int_0^{2\pi} \operatorname{se}_m(z, q)\operatorname{se}_p(z, q)\,dz = 0 \qquad [m \neq p].$ MA

5. $\int_0^{2\pi} [\operatorname{se}_{2n+1}(z, q)]^2\,dz = \pi \sum_{r=0}^{\infty} [B_{2r+1}^{(2n+1)}]^2 = \pi.$ MA

6. $\int_0^{2\pi} [\operatorname{se}_{2n+2}(z, q)]^2\,dz = \pi \sum_{r=0}^{\infty} [B_{2r+2}^{(2n+2)}]^2 = \pi.$ MA

7. $\int_0^{2\pi} \operatorname{se}_m(z, q)\operatorname{ce}_p(z, q)\,dz = 0 \qquad [m = 1, 2, \ldots;\ p = 1, 2, \ldots].$ MA

6.92 Combinations of Mathieu, hyperbolic, and trigonometric functions

6.921

1. $\int_0^{\pi} \operatorname{ch}(2k\cos u \operatorname{sh} z)\operatorname{ce}_{2n}(u, q)\,du =$

$$= \frac{\pi A_0^{(2n)}}{\operatorname{ce}_{2n}\left(\frac{\pi}{2}, q\right)}(-1)^n \operatorname{Ce}_{2n}(z, -q) \qquad [q > 0].$$ MA

2. $\int_0^{\pi} \operatorname{ch}(2k\sin u \operatorname{ch} z)\operatorname{ce}_{2n}(u, q)\,du =$

$$= \frac{\pi A_0^{(2n)}}{\operatorname{ce}_{2n}(0, q)}(-1)^n \operatorname{Ce}_{2n}(z, -q) \qquad [q > 0].$$ MA

3. $\int_0^{\pi} \operatorname{sh}(2k\sin u \operatorname{ch} z)\operatorname{se}_{2n+1}(u, q)\,du =$

$$= \frac{\pi k B_1^{(2n+1)}}{\operatorname{se}'_{2n+1}(0, q)}(-1)^n \operatorname{Ce}_{2n+1}(z, -q) \qquad [q > 0].$$ MA

4. $\int_0^{\pi} \operatorname{sh}(2k\cos u \operatorname{sh} z)\operatorname{ce}_{2n+1}(u, q)\,du =$

$$= \frac{\pi k A_1^{(2n+1)}}{\operatorname{ce}'_{2n+1}\left(\frac{\pi}{2}, q\right)}(-1)^{n+1} \operatorname{Se}_{2n+1}(z, -q) \qquad [q > 0].$$ MA

5. $\int_0^{\pi} \operatorname{sh}(2k\sin u \sin z)\operatorname{se}_{2n+1}(u, q)\,du =$

$$= \frac{\pi k B_1^{(2n+1)}}{\operatorname{se}'_{2n+1}(0, q)} \operatorname{se}_{2n+1}(z, q) \qquad [q > 0].$$ MA

6.922

1. $\int_0^{\pi} \cos u \operatorname{ch} z \cos(2k\sin u \operatorname{sh} z)\operatorname{ce}_{2n+1}(u, q)\,du =$

$$= \frac{\pi A_1^{(2n+1)}}{2\operatorname{ce}_{2n+1}(0, q)} \operatorname{Ce}_{2n+1}(z, q) \qquad [q > 0].$$ MA

2. $$\int_0^\pi \sin u \operatorname{sh} z \cos(2k \cos u \operatorname{ch} z) \operatorname{se}_{2n+1}(u, q)\, du =$$
$$= \frac{\pi B_1^{(2n+1)}}{2\operatorname{se}_{2n+1}\left(\frac{\pi}{2}, q\right)} \operatorname{Se}_{2n+1}(z, q) \qquad [q > 0].$$ MA

3. $$\int_0^\pi \sin u \operatorname{sh} z \sin(2k \cos u \operatorname{ch} z) \operatorname{se}_{2n+2}(u, q)\, du =$$
$$= -\frac{\pi k B_2^{(2n+2)}}{2\operatorname{se}'_{2n+2}\left(\frac{\pi}{2}, q\right)} \operatorname{Se}_{2n+2}(z, q) \qquad [q > 0].$$ MA

4. $$\int_0^\pi \cos u \operatorname{ch} z \sin(2k \sin u \operatorname{sh} z) \operatorname{se}_{2n+2}(u, q)\, du =$$
$$= \frac{\pi k B_2^{(2n+2)}}{2\operatorname{se}'_{2n+2}(0, q)} \operatorname{Se}_{2n+2}(z, q) \qquad [q > 0].$$ MA

5. $$\int_0^\pi \sin u \operatorname{ch} z \operatorname{ch}(2k \cos u \operatorname{sh} z) \operatorname{se}_{2n+1}(u, q)\, du =$$
$$= \frac{\pi B_1^{(2n+1)}}{2\operatorname{se}_{2n+1}\left(\frac{\pi}{2}, q\right)} (-1)^n \operatorname{Ce}_{2n+1}(z, -q) \qquad [q > 0].$$ MA

6. $$\int_0^\pi \cos u \operatorname{sh} z \operatorname{ch}(2k \sin u \operatorname{ch} z) \operatorname{ce}_{2n+1}(u, q)\, du =$$
$$= \frac{\pi A_1^{(2n+1)}}{2\operatorname{ce}_{2n+1}(0, q)} (-1)^n \operatorname{Se}_{2n+1}(z, -q) \qquad [q > 0].$$ MA

7. $$\int_0^\pi \sin u \operatorname{ch} z \operatorname{sh}(2k \cos u \operatorname{sh} z) \operatorname{se}_{2n+2}(u, q)\, du =$$
$$= \frac{\pi k B_2^{(2n+2)}}{2\operatorname{se}'_{2n+2}\left(\frac{\pi}{2}, q\right)} (-1)^{n+1} \operatorname{Se}_{2n+2}(z, -q) \qquad [q > 0].$$ MA

8. $$\int_0^\pi \cos u \operatorname{sh} z \operatorname{sh}(2k \sin u \operatorname{ch} z) \operatorname{se}_{2n+2}(u, q)\, du =$$
$$= \frac{\pi k B_2^{(2n+2)}}{2\operatorname{se}'_{2n+2}(0, q)} (-1)^n \operatorname{Se}_{2n+2}(z, -q) \qquad [q > 0].$$ MA

6.923

1. $$\int_0^\infty \sin(2k \operatorname{ch} z \operatorname{ch} u) \operatorname{sh} z \operatorname{sh} u \operatorname{Se}_{2n+1}(u, q)\, du =$$
$$= -\frac{\pi B_1^{(2n+1)}}{4\operatorname{se}_{2n+1}\left(\frac{\pi}{2}, q\right)} \operatorname{Se}_{2n+1}(z, q) \qquad [q > 0].$$ MA

2. $$\int_0^\infty \cos(2k \operatorname{ch} z \operatorname{ch} u) \operatorname{sh} z \operatorname{sh} u \operatorname{Se}_{2n+1}(u, q)\, du = -\frac{\pi B_1^{(2n+1)}}{4\operatorname{se}_{2n+1}\left(\frac{\pi}{2}, q\right)} \operatorname{Gey}_{2n+1}(z, q) \qquad [q > 0].$$ MA

3 $$\int_0^\infty \sin(2k \operatorname{ch} z \operatorname{ch} u) \operatorname{sh} z \operatorname{sh} u \operatorname{Se}_{2n+2}(u, q)\, du = -\frac{k\pi B_2^{(2n+2)}}{4\operatorname{se}'_{2n+2}\left(\frac{\pi}{2}, q\right)} \operatorname{Gey}_{2n+2}(z, q) \qquad [q > 0].$$ MA

4. $$\int_0^\infty \cos(2k \operatorname{ch} z \operatorname{ch} u) \operatorname{sh} z \operatorname{sh} u \operatorname{Se}_{2n+2}(u, q)\, du = -\frac{k\pi B_2^{(2n+2)}}{4\operatorname{se}_{2n+2}\left(\frac{\pi}{2}, q\right)} \operatorname{Se}_{2n+2}(z, q) \qquad [q > 0].$$ MA

5 $$\int_0^\infty \sin(2k \operatorname{ch} z \operatorname{ch} u) \operatorname{Ce}_{2n}(u, q)\, du = \frac{\pi A_0^{(2n)}}{2\operatorname{ce}_{2n}\left(\frac{1}{2}\pi, q\right)} \operatorname{Ce}_{2n}(z, q) \qquad [q > 0].$$ MA

6. $$\int_0^\infty \cos(2k \operatorname{ch} z \operatorname{ch} u) \operatorname{Ce}_{2n}(u, q)\, du = -\frac{\pi A_0^{(2n)}}{2\operatorname{ce}_{2n}\left(\frac{\pi}{2}, q\right)} \operatorname{Fey}_{2n}(z, q) \qquad [q > 0].$$ MA

7 $$\int_0^\infty \sin(2k \operatorname{ch} z \operatorname{ch} u) \operatorname{Ce}_{2n+1}(u, q)\, du = \frac{k\pi A_1^{(2n+1)}}{2\operatorname{ce}'_{2n+1}\left(\frac{\pi}{2}, q\right)} \operatorname{Fey}_{2n+1}(z, q) \qquad [q > 0].$$ MA

8. $$\int_0^\infty \cos(2k \operatorname{ch} z \operatorname{ch} u) \operatorname{Ce}_{2n+1}(u, q)\, du = \frac{k\pi A_1^{(2n+1)}}{2\operatorname{ce}'_{2n+1}\left(\frac{\pi}{2}, q\right)} \operatorname{Ce}_{2n+1}(z, q) \qquad [q > 0].$$ MA

6.924

1. $$\int_0^\pi \cos(2k \cos u \cos z) \operatorname{ce}_{2n}(u, q)\, du = \frac{\pi A_0^{(2n)}}{\operatorname{ce}_{2n}\left(\frac{\pi}{2}, q\right)} \operatorname{ce}_{2n}(z, q) \qquad [q > 0].$$ MA

2. $\int\limits_0^{\pi} \sin(2k\cos u\cos z)\,\mathrm{ce}_{2n+1}(u, q)\,du =$

$$= -\frac{\pi k A_1^{(2n+1)}}{\mathrm{ce}'_{2n+1}\left(\frac{\pi}{2}, q\right)}\,\mathrm{ce}_{2n+1}(z, q) \quad [q > 0].$$ MA

3. $\int\limits_0^{\pi} \cos(2k\cos u\,\mathrm{ch}\,z)\,\mathrm{ce}_{2n}(u, q)\,du =$

$$= \frac{\pi A_0^{(2n)}}{\mathrm{ce}_{2n}\left(\frac{\pi}{2}, q\right)}\,\mathrm{Ce}_{2n}(z, q) \quad [q > 0].$$ MA

4. $\int\limits_0^{\pi} \cos(2k\sin u\,\mathrm{sh}\,z)\,\mathrm{ce}_{2n}(u, q)\,du =$

$$= \frac{\pi A_0^{(2n)}}{\mathrm{ce}_{2n}(0, q)}\,\mathrm{Ce}_{2n}(z, q) \quad [q > 0].$$ MA

5. $\int\limits_0^{\pi} \sin(2k\cos u\,\mathrm{ch}\,z)\,\mathrm{ce}_{2n+1}(u, q)\,du =$

$$= -\frac{\pi k A_1^{(2n+1)}}{\mathrm{ce}'_{2n+1}\left(\frac{\pi}{2}, q\right)}\,\mathrm{Ce}_{2n+1}(z, q) \quad [q > 0].$$ MA

6. $\int\limits_0^{\pi} \sin(2k\sin u\,\mathrm{sh}\,z)\,\mathrm{se}_{2n+1}(u, q)\,du =$

$$= \frac{\pi k B_1^{(2n+1)}}{\mathrm{se}'_{2n+1}(0, q)}\,\mathrm{Se}_{2n+1}(z, q) \quad [q > 0].$$ MA

6.925 Notation: $z_1 = 2k\sqrt{\mathrm{ch}^2\xi - \sin^2\eta}$, $\mathrm{tg}\,\alpha = \mathrm{th}\,\xi\,\mathrm{tg}\,\eta$

1. $\int\limits_0^{2\pi} \sin[z_1\cos(\theta - \alpha)]\,\mathrm{ce}_{2n}(\theta, q)\,d\theta = 0.$ MA

2. $\int\limits_0^{2\pi} \cos[z_1\cos(\theta - \alpha)]\,\mathrm{ce}_{2n}(\theta, q)\,d\theta =$

$$= \frac{2\pi A_0^{(2n)}}{\mathrm{ce}_{2n}(0, q)\,\mathrm{ce}_{2n}\left(\frac{\pi}{2}, q\right)}\,\mathrm{Ce}_{2n}(\xi, q)\,\mathrm{ce}_{2n}(\eta, q).$$ MA

3. $\int\limits_0^{2\pi} \sin[z_1\cos(\theta - \alpha)]\,\mathrm{ce}_{2n+1}(\theta, q)\,d\theta =$

$$= -\frac{2\pi k A_1^{(2n+1)}}{\mathrm{ce}_{2n+1}(0, q)\,\mathrm{ce}'_{2n+1}\left(\frac{\pi}{2}, q\right)}\,\mathrm{Ce}_{2n+1}(\xi, q)\,\mathrm{ce}_{2n+1}(\eta, q).$$ MA

4. $$\int_0^{2\pi} \cos[z_1 \cos(\theta-\alpha)]\, \mathrm{ce}_{2n+1}(\theta, q)\, d\theta = 0.$$ MA

5. $$\int_0^{2\pi} \sin[z_1 \cos(\theta-\alpha)]\, \mathrm{se}_{2n+1}(\theta, q)\, d\theta =$$
$$= \frac{2\pi k B_1^{(2n+1)}}{\mathrm{se}_{2n+1}(0, q)\, \mathrm{se}_{2n+1}\left(\frac{\pi}{2}, q\right)} \mathrm{Se}_{2n+1}(\xi, q)\, \mathrm{se}_{2n+1}(\eta, q).$$ MA

6. $$\int_0^{2\pi} \cos[z_1 \cos(\theta-\alpha)]\, \mathrm{se}_{2n+1}(\theta, q)\, d\theta = 0.$$ MA

7. $$\int_0^{2\pi} \sin[z_1 \cos(\theta-\alpha)]\, \mathrm{se}_{2n+2}(\theta, q)\, d\theta = 0.$$ MA

8. $$\int_0^{2\pi} \cos[z_1 \cos(\theta-\alpha)]\, \mathrm{se}_{2n+2}(\theta, q)\, d\theta =$$
$$= \frac{2\pi k^2 B_2^{(2n+2)}}{\mathrm{se}'_{2n+2}(0, q)\, \mathrm{se}'_{2n+2}\left(\frac{\pi}{2}, q\right)} \mathrm{Se}_{2n+2}(\xi, q)\, \mathrm{se}_{2n+2}(\eta, q)$$ MA

6.926 $$\int_0^{\pi} \sin u \sin z \sin(2k \cos u \cos z)\, \mathrm{se}_{2n+2}(u, q)\, du =$$
$$= -\frac{\pi k B_2^{(2n+2)}}{2\, \mathrm{se}'_{2n+2}\left(\frac{\pi}{2}, q\right)} \mathrm{se}_{2n+2}(z, q) \quad [q > 0].$$ MA

6.93 Combinations of Mathieu and Bessel functions

6.931

1. $$\int_0^{\pi} J_0\{k[2(\cos 2u + \cos 2z)]^{\frac{1}{2}}\}\, \mathrm{ce}_{2n}(u, q)\, du =$$
$$= \frac{\pi [A_0^{(2n)}]^2}{\mathrm{ce}_{2n}(0, q)\, \mathrm{ce}_{2n}\left(\frac{\pi}{2}, q\right)} \mathrm{ce}_{2n}(z, q).$$ MA

2. $$\int_0^{2\pi} N_0\{k[2(\cos 2u + \operatorname{ch} 2z)]^{\frac{1}{2}}\}\, \mathrm{ce}_{2n}(u, q)\, du =$$
$$= \frac{2\pi [A_0^{(2n)}]^2}{\mathrm{ce}_{2n}(0, q)\, \mathrm{ce}_{2n}\left(\frac{\pi}{2}, q\right)} \mathrm{Fey}_{2n}(z, q).$$ MA

7.1-7.2 Associated Legendre Functions

7.11 Associated Legendre functions

7.111 $\int\limits_{\cos\varphi}^{1} P_\nu(x)\,dx = \sin\varphi P_\nu^{-1}(\cos\varphi).$ MO 90

7.112

1. $\int\limits_{-1}^{1} P_n^m(x) P_k^m(x)\,dx = 0 \qquad [n \neq k];$

$$= \frac{2}{2n+1}\frac{(n+m)!}{(n-m)!} \qquad [n = k].$$ SM III 185, WH

2. $\int\limits_{-1}^{1} Q_n^m(x) P_k^m(x)\,dx = (-1)^m \frac{1-(-1)^{n+k}(n+m)!}{(k-n)(k+n+1)(n-m)!}.$ EH I 171(18)

3. $\int\limits_{-1}^{1} P_\nu(x) P_\sigma(x)\,dx = \frac{2\pi \sin\pi(\sigma-\nu) + 4\sin(\pi\nu)\sin(\pi\sigma)[\psi(\nu+1)-\psi(\sigma+1)]}{\pi^2(\sigma-\nu)(\sigma+\nu+1)}$

$$[\sigma+\nu+1 \neq 0];$$ EH I 170(7)

$$= \frac{\pi^2 - 2(\sin\pi\nu)^2\psi'(\nu+1)}{\pi^2\left(\nu+\frac{1}{2}\right)} \qquad [\sigma = \nu].$$ EH I 170(9)a

4. $\int\limits_{-1}^{1} Q_\nu(x) Q_\sigma(x)\,dx =$

$$= \frac{[\psi(\nu+1)-\psi(\sigma+1)][1+\cos(\pi\sigma)\cos(\nu\pi)] - \frac{\pi}{2}\sin\pi(\nu-\sigma)}{(\sigma-\nu)(\sigma+\nu+1)}$$

$$[\sigma+\nu+1 \neq 0;\ \nu, \sigma \neq -1, -2, -3, \ldots];$$ EH I 170(11)

$$= \frac{\frac{1}{2}\pi^2 - \psi'(\nu+1)[1+(\cos\nu\pi)^2]}{2\nu+1} \qquad [\nu = \sigma,\ \nu \neq -1, -2, -3, \ldots].$$

EH I 170(12)

5. $\int\limits_{-1}^{1} P_\nu(x) Q_\sigma(x)\,dx =$

$$= \frac{1-\cos\pi(\sigma-\nu) - 2\pi^{-1}\sin(\pi\nu)\cos(\pi\sigma)[\psi(\nu+1)-\psi(\sigma+1)]}{(\nu-\sigma)(\nu+\sigma+1)}$$

$$[\mathrm{Re}\,\nu > 0,\ \mathrm{Re}\,\sigma > 0,\ \sigma \neq \nu];$$ EH I 170(13)

$$= -\frac{\sin(2\nu\pi)\psi'(\nu+1)}{\pi(2\nu+1)} \qquad [\mathrm{Re}\,\nu > 0,\ \sigma = \nu].$$ EH I 171(14)

7.113 Notation: $A=\dfrac{\Gamma\left(\frac{1}{2}+\frac{\nu}{2}\right)\Gamma\left(1+\frac{\sigma}{2}\right)}{\Gamma\left(\frac{1}{2}+\frac{\sigma}{2}\right)\Gamma\left(1+\frac{\nu}{2}\right)}$

1. $$\int_0^1 \mathrm{P}_\nu(x)\,\mathrm{P}_\sigma(x)\,dx=\frac{A\sin\frac{\pi\sigma}{2}\cos\frac{\pi\nu}{2}-A^{-1}\sin\frac{\pi\nu}{2}\cos\frac{\pi\sigma}{2}}{\frac{1}{2}\pi(\sigma-\nu)(\sigma+\nu+1)}.$$

EH I 171(15)

2. $$\int_0^1 \mathrm{Q}_\nu(x)\,\mathrm{Q}_\sigma(x)\,dx=$$
$$=\frac{\psi(\nu+1)-\psi(\sigma+1)-\frac{\pi}{2}\left[(A-A^{-1})\sin\frac{\pi(\sigma+\nu)}{2}-(A+A^{-1})\sin\frac{\pi(\sigma-\nu)}{2}\right]}{(\sigma-\nu)(\sigma+\nu+1)}$$
$[\operatorname{Re}\nu>0,\ \operatorname{Re}\sigma>0]$. EH I 171(16)

3. $$\int_0^1 \mathrm{P}_\nu(x)\,\mathrm{Q}_\sigma(x)\,dx=\frac{A^{-1}\cos\frac{\pi(\nu-\sigma)}{2}-1}{(\sigma-\nu)(\sigma+\nu+1)}\quad[\operatorname{Re}\nu>0,\ \operatorname{Re}\sigma>0].$$

EH I 171(17)

7.114

1. $$\int_1^\infty P_\nu(x)\,Q_\sigma(x)\,dx=\frac{1}{(\sigma-\nu)(\sigma+\nu+1)}$$
$[\operatorname{Re}(\sigma-\nu)>0,\ \operatorname{Re}(\sigma+\nu)>-1]$. ET II 324(19)

2. $$\int_1^\infty Q_\nu(x)\,Q_\sigma(x)\,dx=\frac{\psi(\sigma+1)-\psi(\nu+1)}{(\sigma-\nu)(\sigma+\nu+1)}$$
$[\operatorname{Re}(\nu+\sigma)>-1;\quad \sigma,\ \nu\neq-1,\ -2,\ -3,\ \ldots]$. EH I 170(5)

3. $$\int_1^\infty [Q_\nu(x)]^2\,dx=\frac{\psi'(\nu+1)}{2\nu+1}\qquad\left[\operatorname{Re}\nu>-\frac{1}{2}\right].$$ EH I 170(6)

7.115 $$\int_1^\infty Q_\nu(x)\,dx=\frac{1}{\nu(\nu+1)}\quad[\operatorname{Re}\nu>0].$$ ET II 324(18)

7.12-7.13 Combinations of associated Legendre functions and powers

7.121 $$\int_{\cos\varphi}^1 x\mathrm{P}_\nu(x)\,dx=\frac{\sin\varphi}{(\nu-1)(\nu+2)}\left[\sin\varphi\,\mathrm{P}_\nu(\cos\varphi)+\cos\varphi\,\mathrm{P}_\nu^1(\cos\varphi)\right].$$ MO 90

7.122

1. $$\int_0^1 \frac{[\mathrm{P}_n^m(x)]^2}{1-x^2}\,dx=\frac{1}{2m}\,\frac{(n+m)!}{(n-m)!}\quad[0<m\leqslant n].$$ MO 74

2. $$\int_0^1 [\mathrm{P}_\nu^\mu(x)]^2\,\frac{dx}{1-x^2}=-\frac{\Gamma(1+\mu+\nu)}{2\mu\Gamma(1-\mu+\nu)}$$
$[\operatorname{Re}\mu<0,\ \nu+\mu$ — a positive integer$]$. EH I 172(26)

3. $$\int_0^1 [P_\nu^{n-\nu}(x)]^2 \frac{dx}{1-x^2} = -\frac{n!}{2(n-\nu)\Gamma(1-n+2\nu)}$$

$[n=0, 1, 2, \ldots; \operatorname{Re}\nu > n]$ ET II 315(9)

7.123 $$\int_{-1}^1 P_n^m(x) P_n^k(x) \frac{dx}{1-x^2} = 0 \quad [0 \leqslant m \leqslant n, \ 0 \leqslant k \leqslant n; \ m \neq k].$$ MO 74

7.124 $$\int_{-1}^1 x^k (z-x)^{-1} (1-x^2)^{\frac{1}{2}m} P_n^m(x)\,dx = (+2)(z^2-1)^{\frac{1}{2}m} Q_n^m(z)\cdot z^k$$

$[m \leqslant n; \ k=0, 1, \ldots, n-m; \ z$ in the complex plane with a cut along the interval $(-1, 1)$ on the real axis$]$.

ET II 279(26)

7.125 $$\int_{-1}^1 (1-x^2)^{\frac{1}{2}m} P_k^m(x) P_l^m(x) P_n^m(x)\,dx =$$

$$= (-1)^m \pi^{-\frac{3}{2}} \frac{(k+m)!\,(l+m)!\,(n+m)!\,(s-m)!}{(k-m)!\,(l-m)!\,(n-m)!\,(s-k)!} \times$$

$$\times \frac{\Gamma\left(m+\frac{1}{2}\right)\Gamma\left(t-k+\frac{1}{2}\right)\Gamma\left(t-l+\frac{1}{2}\right)\Gamma\left(t-n+\frac{1}{2}\right)}{(s-l)!\,(s-n)!\,\Gamma\left(s+\frac{3}{2}\right)}$$

$[2s = k+l+n+m$ and $2t = k+l+n-m$ — both odd;
$l \geqslant m, \ m \leqslant k-l-m \leqslant n \leqslant k+l+m]$. ET II 280(32)

7.126

1. $$\int_0^1 P_\nu(x) x^\sigma\,dx = \frac{\sqrt{\pi}\,2^{-\sigma-1}\Gamma(1+\sigma)}{\Gamma\left(1+\frac{1}{2}\sigma-\frac{1}{2}\nu\right)\Gamma\left(\frac{1}{2}\sigma+\frac{1}{2}\nu+\frac{3}{2}\right)}$$

$[\operatorname{Re}\sigma > -1]$. EH I 171(23)

2. $$\int_0^1 x^\sigma P_\nu^m(x)\,dx = \frac{(-1)^m \pi^{\frac{1}{2}} 2^{-2m-1}\Gamma\left(\frac{1+\sigma}{2}\right)\Gamma(1+m+\nu)}{\Gamma\left(\frac{1}{2}+\frac{1}{2}m\right)\Gamma\left(\frac{3}{2}+\frac{\sigma}{2}+\frac{m}{2}\right)\Gamma(1-m+\nu)} \times$$

$$\times {}_3F_2\left(\frac{m+\nu+1}{2}, \ \frac{m-\nu}{2}, \ \frac{m}{2}+1; \ m+1, \ \frac{3+\sigma+m}{2}; \ 1\right)$$

$[\operatorname{Re}\sigma > -1; \ m=0, 1, 2, \ldots]$. ET II 313(2)

3. $$\int_0^1 x^\sigma P_\nu^\mu(x)\,dx = \frac{\pi^{\frac{1}{2}} 2^{2\mu-1}\Gamma\left(\frac{1+\sigma}{2}\right)}{\Gamma\left(\frac{1-\mu}{2}\right)\Gamma\left(\frac{3+\sigma-\mu}{2}\right)} \times$$

$$\times {}_3F_2\left(\frac{\nu-\mu+1}{2}, \ -\frac{\mu+\nu}{2}, \ 1-\frac{\mu}{2}; \ 1-\mu, \ \frac{3+\sigma-\mu}{2}; \ 1\right)$$

$[\operatorname{Re}\sigma > -1, \ \operatorname{Re}\mu < 2]$. ET II 313(3)

4. $$\int_1^\infty x^{\mu-1} Q_\nu(ax)\,dx = e^{\mu\pi i}\Gamma(\mu)\,a^{-\mu}(a^2-1)^{\frac{1}{2}\mu} Q_\nu^{-\mu}(a)$$

$[|\arg(a-1)| < \pi, \ \operatorname{Re}\mu > 0, \ \operatorname{Re}(\nu-\mu) > -1]$. ET II 325(26)

7.127 $$\int_{-1}^{1}(1+x)^{\sigma}P_{\nu}(x)\,dx=\frac{2^{\sigma+1}[\Gamma(\sigma+1)]^{2}}{\Gamma(\sigma+\nu+2)\,\Gamma(1+\sigma-\nu)}$$

$[\operatorname{Re}\sigma>-1]$. ET II 316(15)

7.128

1. $$\int_{-1}^{1}(1-x)^{-\frac{1}{2}\mu}(1+x)^{\frac{1}{2}\mu-\frac{1}{2}}(z+x)^{\mu-\frac{3}{2}}P_{\nu}^{\mu}(x)\,dx=$$

$$=-\frac{\Gamma\left(\mu-\frac{1}{2}\right)(z-1)^{\mu-\frac{1}{2}}(z+1)^{-\frac{1}{2}}}{\pi^{\frac{1}{2}}e^{2\mu\pi i}\,\Gamma(\mu+\nu)\,\Gamma(\mu-\nu-1)}\times$$

$$\times\left\{Q_{\nu}^{\mu}\left[\left(\frac{1+z}{2}\right)^{\frac{1}{2}}\right]Q_{-\nu-1}^{\mu-1}\left[\left(\frac{1+z}{2}\right)^{\frac{1}{2}}\right]+\right.$$

$$\left.+Q_{\nu}^{\mu-1}\left[\left(\frac{1+z}{2}\right)^{\frac{1}{2}}\right]Q_{-\nu-1}^{\mu}\left[\left(\frac{1+z}{2}\right)^{\frac{1}{2}}\right]\right\}$$

$\left[-\frac{1}{2}<\operatorname{Re}\mu<1,\ z\right.$ — in the complex plane with a cut along the interval $\left.(-1,\ 1)\text{ of the real axis}\right]$. ET II 317(20)

2. $$\int_{-1}^{1}(1-x)^{-\frac{1}{2}\mu}(1+x)^{\frac{1}{2}\mu-\frac{1}{2}}(z+x)^{\mu-\frac{1}{2}}P_{\nu}^{\mu}(x)\,dx=$$

$$=\frac{2e^{-2\mu\pi i}\,\Gamma\left(\frac{1}{2}+\mu\right)}{\pi^{\frac{1}{2}}\,\Gamma(\mu-\nu)\,\Gamma(\mu+\nu+1)}(z-1)^{\mu}\,Q_{\nu}^{\mu}\left[\left(\frac{1+z}{2}\right)^{\frac{1}{2}}\right]Q_{-\nu-1}^{\mu}\left[\left(\frac{1+z}{2}\right)^{\frac{1}{2}}\right]$$

$\left[-\frac{1}{2}<\operatorname{Re}\mu<1,\ z\right.$ — in the complex plane with a cut along the interval $\left.(-1,\ 1)\text{ of the real axis}\right]$ ET II 316(18)

7.129 $$\int_{-1}^{1}P_{\nu}(x)\,P_{\lambda}(x)\,(1+x)^{\lambda+\nu}\,dx=\frac{2^{\lambda+\nu+1}[\Gamma(\lambda+\nu+1)]^{4}}{[\Gamma(\lambda+1)\,\Gamma(\nu+1)]^{2}\,\Gamma(2\lambda+2\nu+2)}$$

$[\operatorname{Re}(\nu+\lambda+1)>0]$. EH I 172(30)

7.131

1. $$\int_{1}^{\infty}(x-1)^{-\frac{1}{2}\mu}(x+1)^{\frac{1}{2}\mu-\frac{1}{2}}(z+x)^{\mu-\frac{1}{2}}P_{\nu}^{\mu}(x)\,dx=$$

$$=\pi^{\frac{1}{2}}\frac{\Gamma(-\mu-\nu)\,\Gamma(1-\mu+\nu)}{\Gamma\left(\frac{1}{2}-\mu\right)}(z-1)^{\mu}\left\{P_{\nu}^{\mu}\left[\left(\frac{1+z}{2}\right)^{\frac{1}{2}}\right]\right\}^{2}$$

$[\operatorname{Re}(\mu+\nu)<0,\ \operatorname{Re}(\mu-\nu)<1,\ |\arg(z+1)|<\pi]$. ET II 321(6)

2. $$\int_1^\infty (x-1)^{-\frac{1}{2}\mu}(x+1)^{\frac{1}{2}\mu-\frac{1}{2}}(z+x)^{\mu-\frac{3}{2}} P_\nu^\mu(x)\,dx =$$

$$= \frac{\pi^{\frac{1}{2}}\Gamma(1-\mu-\nu)\,\Gamma(2-\mu+\nu)\,(z-1)^{\mu-\frac{1}{2}}(z+1)^{-\frac{1}{2}}}{\Gamma\left(\frac{3}{2}-\mu\right)} \times$$

$$\times P_\nu^\mu\left[\left(\frac{1+z}{2}\right)^{\frac{1}{2}}\right] P_\nu^{\mu-1}\left[\left(\frac{1+z}{2}\right)^{\frac{1}{2}}\right]$$

$[\mathrm{Re}\,\mu < 1,\ \mathrm{Re}(\mu+\nu) < 1,\ \mathrm{Re}(\mu-\nu) < 2,\ |\arg(1+z)| < \pi]$

ET II 321(7)

7.132

1. $$\int_{-1}^1 (1-x^2)^{\lambda-1}\,\mathrm{P}_\nu^\mu(x)\,dx =$$

$$= \frac{\pi 2^\mu \Gamma\left(\lambda+\frac{1}{2}\mu\right)\Gamma\left(\lambda-\frac{1}{2}\mu\right)}{\Gamma\left(\lambda+\frac{1}{2}\nu+1\right)\Gamma\left(\lambda-\frac{1}{2}\nu\right)\Gamma\left(-\frac{1}{2}\mu+\frac{1}{2}\nu+1\right)\Gamma\left(-\frac{1}{2}\mu-\frac{1}{2}\nu+\frac{1}{2}\right)}$$

$[2\mathrm{Re}\,\lambda > |\mathrm{Re}\,\mu|]$. ET II 316(6)

2. $$\int_1^\infty (x^2-1)^{\lambda-1} P_\nu^\mu(x)\,dx =$$

$$= \frac{2^{\mu-1}\Gamma\left(\lambda-\frac{1}{2}\mu\right)\Gamma\left(1-\lambda+\frac{1}{2}\nu\right)\Gamma\left(\frac{1}{2}-\lambda-\frac{1}{2}\nu\right)}{\Gamma\left(1-\frac{1}{2}\mu+\frac{1}{2}\nu\right)\Gamma\left(\frac{1}{2}-\frac{1}{2}\mu-\frac{1}{2}\nu\right)\Gamma\left(1-\lambda-\frac{1}{2}\mu\right)}$$

$[\mathrm{Re}\,\lambda > \mathrm{Re}\,\mu,\ \mathrm{Re}(1-2\lambda-\nu) > 0,\ \mathrm{Re}(2-2\lambda+\nu) > 0]$. ET II 320(2)

3. $$\int_1^\infty (x^2-1)^{\lambda-1} Q_\nu^\mu(x)\,dx =$$

$$= e^{\mu\pi i}\frac{\Gamma\left(\frac{1}{2}+\frac{1}{2}\nu+\frac{1}{2}\mu\right)\Gamma\left(1-\lambda+\frac{1}{2}\nu\right)\Gamma\left(\lambda+\frac{1}{2}\mu\right)\Gamma\left(\lambda-\frac{1}{2}\mu\right)}{2^{2\lambda-\mu}\Gamma\left(1+\frac{1}{2}\nu-\frac{1}{2}\mu\right)\Gamma\left(\frac{1}{2}+\lambda+\frac{1}{2}\nu\right)}$$

$[|\mathrm{Re}\,\mu| < 2\mathrm{Re}\,\lambda < \mathrm{Re}\,\nu+2]$. ET II 324(23)

4. $$\int_0^1 x^\sigma (1-x^2)^{-\frac{1}{2}\mu}\,\mathrm{P}_\nu^\mu(x)\,dx =$$

$$= \frac{2^{\mu-1}\Gamma\left(\frac{1}{2}+\frac{1}{2}\sigma\right)\Gamma\left(1+\frac{1}{2}\sigma\right)}{\Gamma\left(1+\frac{1}{2}\sigma-\frac{1}{2}\nu-\frac{1}{2}\mu\right)\Gamma\left(\frac{1}{2}\sigma+\frac{1}{2}\nu-\frac{1}{2}\mu+\frac{3}{2}\right)}$$

$[\mathrm{Re}\,\mu < 1,\quad \mathrm{Re}\,\sigma > -1]$. EH I 172(24)

5. $$\int_0^1 x^\sigma (1-x^2)^{\frac{1}{2}m} P_\nu^m(x)\,dx =$$
$$= \frac{(-1)^m 2^{-m-1}\Gamma\left(\frac{1}{2}+\frac{1}{2}\sigma\right)\Gamma\left(1+\frac{1}{2}\sigma\right)\Gamma(1+m+\nu)}{\Gamma(1-m+\nu)\Gamma\left(1+\frac{1}{2}\sigma+\frac{1}{2}m-\frac{1}{2}\nu\right)\Gamma\left(\frac{3}{2}+\frac{1}{2}\sigma+\frac{1}{2}m+\frac{1}{2}\nu\right)}$$
[$\mathrm{Re}\,\sigma > -1$, m is a positive integer]. EH I 172(25), ET II 313(4)

6. $$\int_0^1 x^\sigma (1-x^2)^\eta P_\nu^\mu(x)\,dx = \frac{2^{\mu-1}\Gamma\left(1+\eta-\frac{1}{2}\mu\right)\Gamma\left(\frac{1}{2}+\frac{1}{2}\sigma\right)}{\Gamma(1-\mu)\Gamma\left(\frac{3}{2}+\eta+\frac{1}{2}\sigma-\frac{1}{2}\mu\right)}\times$$
$$\times {}_3F_2\left(\frac{\nu-\mu+1}{2},\ -\frac{\mu+\nu}{2},\ 1+\eta-\frac{\mu}{2};\ 1-\mu,\ \frac{3+\sigma-\mu}{2}+\eta;\ 1\right)$$
$$\left[\mathrm{Re}\left(\eta-\frac{1}{2}\mu\right) > -1,\ \mathrm{Re}\,\sigma > -1\right].$$ ET II 314(6)

7. $$\int_1^\infty x^{-\varrho}(x^2-1)^{-\frac{1}{2}\mu} P_\nu^\mu(x)\,dx = \frac{2^{\varrho+\mu-2}\Gamma\left(\frac{\varrho+\mu+\nu}{2}\right)\Gamma\left(\frac{\varrho+\mu-\nu-1}{2}\right)}{\sqrt{\pi}\,\Gamma(\varrho)}$$
[$\mathrm{Re}\,\mu < 1$, $\mathrm{Re}(\varrho+\mu+\nu) > 0$, $\mathrm{Re}(\varrho+\mu-\nu) > 1$]. ET II 320(3)

7.133

1. $$\int_u^\infty Q_\nu(x)(x-u)^{\mu-1}\,dx = \Gamma(\mu)\,e^{\mu\pi i}(u^2-1)^{\frac{1}{2}\mu} Q_\nu^{-\mu}(u)$$
[$|\arg(u-1)| < \pi$, $0 < \mathrm{Re}\,\mu < 1+\mathrm{Re}\,\nu$]. MO 90a

2. $$\int_u^\infty (x^2-1)^{\frac{1}{2}\lambda} Q_\nu^{-\lambda}(x)(x-u)^{\mu-1}\,dx = \Gamma(\mu)\,e^{\mu\pi i}(u^2-1)^{\frac{1}{2}\lambda+\frac{1}{2}\mu} Q_\nu^{-\lambda-\mu}(u)$$
[$|\arg(u-1)| < \pi$, $0 < \mathrm{Re}\,\mu < 1+\mathrm{Re}(\nu-\lambda)$]. ET II 204(30)

7.134

1. $$\int_1^\infty (x-1)^{\lambda-1}(x^2-1)^{\frac{1}{2}\mu} P_\nu^\mu(x)\,dx = \frac{2^{\lambda+\mu}\Gamma(\lambda)\Gamma(-\lambda-\mu-\nu)\Gamma(1-\lambda-\mu+\nu)}{\Gamma(1-\mu+\nu)\Gamma(-\mu-\nu)\Gamma(1-\lambda-\mu)}$$
[$\mathrm{Re}\,\lambda > 0$, $\mathrm{Re}(\lambda+\mu+\nu) < 0$, $\mathrm{Re}(\lambda+\mu-\nu) < 1$]. ET II 321(4)

2. $$\int_1^\infty (x-1)^{\lambda-1}(x^2-1)^{-\frac{1}{2}\mu} P_\nu^\mu(x)\,dx =$$
$$= -\frac{2^{\lambda-\mu}\sin\pi\nu\,\Gamma(\lambda-\mu)\Gamma(-\lambda+\mu-\nu)\Gamma(1-\lambda+\mu+\nu)}{\pi\Gamma(1-\lambda)}$$
[$\mathrm{Re}(\lambda-\mu) > 0$, $\mathrm{Re}(\mu-\lambda-\nu) > 0$, $\mathrm{Re}(\mu-\lambda+\nu) > -1$]. ET II 321(5)

7.135

1. $$\int_{-1}^1 (1-x^2)^{-\frac{1}{2}\mu}(z-x)^{-1} P_{\mu+n}^\mu(x)\,dx = 2e^{-i\mu\pi}(z^2-1)^{-\frac{1}{2}\mu} Q_{\mu+n}^\mu(z)$$
[$n=0, 1, 2, \ldots$, $\mathrm{Re}\,\mu+n > -1$, z–in the complex plane with a cut along the interval $(-1, 1)$ of the real axis]. ET II 316(17)

2. $$\int_1^\infty (x-1)^{\lambda-1} (x^2-1)^{\frac{1}{2}\mu} (x+z)^{-\varrho} P_\nu^\mu(x)\, dx =$$

$$= \frac{2^{\lambda+\mu-\varrho}\Gamma(\lambda-\varrho)\Gamma(\varrho-\lambda-\mu-\nu)\Gamma(\varrho-\lambda-\mu+\nu+1)}{\Gamma(1-\mu+\nu)\Gamma(-\mu-\nu)\Gamma(1+\varrho-\lambda-\mu)} \times$$

$$\times {}_3F_2\left(\varrho,\ \varrho-\lambda-\mu-\nu,\ \varrho-\lambda-\mu+\nu+1;\ \ \varrho-\lambda+1,\ \varrho-\lambda-\mu+1; \frac{1+z}{2}\right) +$$

$$+ \frac{\Gamma(\varrho-\lambda)\Gamma(\lambda)}{\Gamma(\varrho)\Gamma(1-\mu)} 2^\mu (z+1)^{\lambda-\varrho} {}_3F_2\left(\lambda,\ -\mu-\nu,\ 1-\mu+\nu;\ 1-\mu,\ 1-\varrho+\lambda; \frac{1+z}{2}\right)$$

$[\operatorname{Re}\lambda > 0,\ \operatorname{Re}(\varrho-\lambda-\mu-\nu) > 0,\ \ \operatorname{Re}(\varrho-\lambda-\mu+\nu+1) > 0$
$|\arg(z+1)| < \pi].$ ET II 322(9)

3. $$\int_1^\infty (x-1)^{\lambda-1} (x^2-1)^{-\frac{1}{2}\mu} (x+z)^{-\varrho} P_\nu^\mu(x)\, dx =$$

$$= -\frac{\sin(\nu\pi)\Gamma(\lambda-\mu-\varrho)\Gamma(\varrho-\lambda+\mu-\nu)\Gamma(\varrho-\lambda+\mu+\nu+1)}{2^{\varrho-\lambda+\mu}\pi\Gamma(1+\varrho-\lambda)} \times$$

$$\times {}_3F_2\left(\varrho,\ \varrho-\lambda+\mu-\nu,\ \varrho-\lambda+\mu+\nu+1;\ 1+\varrho-\lambda,\ \ 1+\varrho-\lambda+\mu; \frac{1+z}{2}\right) +$$

$$+ \frac{\Gamma(\lambda-\mu)\Gamma(\varrho-\lambda+\mu)}{\Gamma(\varrho)\Gamma(1-\mu)} (z+1)^{\lambda-\varrho-\mu} \times$$

$$\times {}_3F_2\left(\lambda-\mu,\ -\nu,\ \nu+1;\ 1+\lambda-\mu-\varrho,\ 1-\mu; \frac{1+z}{2}\right)$$

$[\operatorname{Re}(\lambda-\mu) > 0,\ \ \operatorname{Re}(\varrho-\lambda+\mu-\nu) > 0,\ \operatorname{Re}(\varrho-\lambda+\mu+\nu+1) > 0,$
$|\arg(z+1)| < \pi].$ ET II 322(10)

7.136

1. $$\int_{-1}^1 (1-x^2)^{\lambda-1} (1-a^2x^2)^{\frac{1}{2}\mu} \mathrm{P}_\nu(ax)\, dx =$$

$$= \frac{\pi 2^\mu \Gamma(\lambda)}{\Gamma\left(\frac{1}{2}+\lambda\right)\Gamma\left(\frac{1}{2}-\frac{1}{2}\mu-\frac{1}{2}\nu\right)\Gamma\left(1-\frac{1}{2}\mu+\frac{1}{2}\nu\right)} \times$$

$$\times {}_2F_1\left(-\frac{\mu+\nu}{2},\ \frac{1-\mu+\nu}{2};\ \frac{1}{2}+\lambda;\ a^2\right)$$

$[\operatorname{Re}\lambda > 0,\ \ \ -1 < a < 1].$ ET II 318(31)

2. $$\int_1^\infty (x^2-1)^{\lambda-1} (a^2x^2-1)^{\frac{1}{2}\mu} P_\nu^\mu(ax)\, dx =$$

$$= \frac{\Gamma(\lambda)\Gamma\left(1-\lambda-\frac{1}{2}\mu+\frac{1}{2}\nu\right)\Gamma\left(\frac{1}{2}-\lambda-\frac{1}{2}\mu-\frac{1}{2}\nu\right)}{\Gamma\left(1-\frac{1}{2}\mu+\frac{1}{2}\nu\right)\Gamma\left(\frac{1}{2}-\frac{1}{2}\nu-\frac{1}{2}\mu\right)\Gamma(1-\lambda-\mu)} \times$$

$$\times 2^{\mu-1} a^{\mu-\nu-1} {}_2F_1\left(\frac{1-\mu+\nu}{2},\ 1-\lambda-\frac{\mu-\nu}{2};\ 1-\lambda-\mu;\ 1-\frac{1}{a^2}\right)$$

$[\operatorname{Re} a > 0,\ \operatorname{Re}\lambda > 0,\ \operatorname{Re}(\nu-\mu-2\lambda) > -2,\ \ \operatorname{Re}(2\lambda+\mu+\nu) < 1].$

ET II 325(25)

3. $$\int_1^\infty (x^2-1)^{\lambda-1}(a^2x^2-1)^{-\frac{1}{2}\mu} Q_\nu^\mu(ax)\,dx =$$

$$= \frac{\Gamma\left(\frac{\mu+\nu+1}{2}\right)\Gamma(\lambda)\Gamma\left(1-\lambda+\frac{\mu+\nu}{2}\right)2^{\mu-2}e^{\mu\pi i}a^{-\mu-\nu-1}}{\Gamma\left(\nu+\frac{3}{2}\right)} \times$$

$$\times {}_2F_1\left(\frac{\mu+\nu+1}{2},\ 1-\lambda+\frac{\mu+\nu}{2};\ \nu+\frac{3}{2};\ a^{-2}\right)$$

$[|\arg(a-1)| < \pi,\ \operatorname{Re}\lambda > 0,\ \operatorname{Re}(2\lambda-\mu-\nu) < 2]$. ET II 325(27)

7.137

1. $$\int_1^\infty x^{-\frac{1}{2}\mu-\frac{1}{2}}(x-1)^{-\mu-\frac{1}{2}}(1+ax)^{\frac{1}{2}\mu} Q_\nu^\mu(1+2ax)\,dx =$$

$$= \pi^{-\frac{1}{2}} e^{-\mu\pi i}\Gamma\left(\frac{1}{2}-\mu\right) a^{\frac{1}{2}\mu}\left\{Q_\nu^\mu\left[(1+a)^{\frac{1}{2}}\right]\right\}^2$$

$\left[|\arg a| < \pi,\quad \operatorname{Re}\mu < \frac{1}{2},\quad \operatorname{Re}(\mu+\nu) > -1\right]$. ET II 325(28)

2. $$\int_1^\infty x^{-\frac{1}{2}\mu-\frac{1}{2}}(x-1)^{-\mu-\frac{3}{2}}(1+ax)^{\frac{1}{2}\mu} Q_\nu^\mu(1+2ax)\,dx =$$

$$= -\pi^{-\frac{1}{2}} e^{-\mu\pi i}\Gamma\left(-\mu-\frac{1}{2}\right) a^{\frac{1}{2}\mu+\frac{1}{2}}(1+a^2)^{-\frac{1}{2}} Q_\nu^{\mu+1}\left[(1+a)^{\frac{1}{2}}\right] Q_\nu^\mu\left[(1+a)^{\frac{1}{2}}\right]$$

$\left[|\arg a| < \pi,\ \operatorname{Re}\mu < -\frac{1}{2},\ \operatorname{Re}(\mu+\nu+2) > 0\right]$. ET II 326(29)

3. $$\int_0^1 x^{-\frac{1}{2}\mu-\frac{1}{2}}(1-x)^{-\mu-\frac{1}{2}}(1+ax)^{\frac{1}{2}\mu} P_\nu^\mu(1+2ax)\,dx =$$

$$= \pi^{\frac{1}{2}}\Gamma\left(\frac{1}{2}-\mu\right) a^{\frac{1}{2}\mu}\left\{P_\nu^\mu\left[(1+a)^{\frac{1}{2}}\right]\right\}^2$$

$\left[\operatorname{Re}\mu < \frac{1}{2},\ |\arg a| < \pi\right]$. ET II 319(32)

4. $$\int_0^1 x^{-\frac{1}{2}\mu-\frac{1}{2}}(1-x)^{-\mu-\frac{3}{2}}(1+ax)^{\frac{1}{2}\mu} P_\nu^\mu(1+2ax)\,dx =$$

$$= \pi^{\frac{1}{2}}\Gamma\left(-\frac{1}{2}-\mu\right) a^{\frac{1}{2}\mu+\frac{1}{2}} P_\nu^{\mu+1}\left[(1+a)^{\frac{1}{2}}\right] P_\nu^\mu\left[(1+a)^{\frac{1}{2}}\right]$$

$\left[\operatorname{Re}\mu < -\frac{1}{2},\ |\arg a| < \pi\right]$. ET II 319(33)

5. $$\int_0^1 x^{\frac{1}{2}\mu-\frac{1}{2}}(1-x)^{\mu-\frac{1}{2}}(1+ax)^{-\frac{1}{2}\mu} P_\nu^\mu(1+2ax)\,dx =$$

$$= \pi^{\frac{1}{2}}\Gamma\left(\frac{1}{2}+\mu\right) a^{-\frac{1}{2}\mu} P_\nu^\mu\left[(1+a)^{\frac{1}{2}}\right] P_\nu^{-\mu}\left[(1+a)^{\frac{1}{2}}\right]$$

$\left[\operatorname{Re}\mu > -\frac{1}{2},\quad |\arg a| < \pi\right]$. ET II 319(34)

6. $$\int_0^1 x^{\frac{1}{2}\mu-\frac{1}{2}}(1-x)^{\mu-\frac{3}{2}}(1+ax)^{-\frac{1}{2}\mu}P_\nu^\mu(1+2ax)\,dx =$$
$$= \frac{1}{2}\pi^{\frac{1}{2}}\Gamma\left(\mu-\frac{1}{2}\right)a^{\frac{1}{2}-\frac{1}{2}\mu}(1+a)^{-\frac{1}{2}}\{P_\nu^{1-\mu}[(1+a)^{\frac{1}{2}}]P_\nu^\mu[(1+a)^{\frac{1}{2}}]+$$
$$+(\mu+\nu)(1-\mu+\nu)P_\nu^{-\mu}[(1+a)^{\frac{1}{2}}]P_\nu^\mu[(1+a)^{\frac{1}{2}}]\}$$
$$\left[\operatorname{Re}\mu>\frac{1}{2},\quad |\arg a|<\pi\right].$$ ET II 319(35)

7. $$\int_0^1 x^{-\frac{\mu}{2}-\frac{1}{2}}(1-x)^{-\mu-\frac{1}{2}}(1+ax)^{\frac{1}{2}\mu}Q_\nu^\mu(1+2ax)\,dx =$$
$$= \pi^{\frac{1}{2}}\Gamma\left(\frac{1}{2}-\mu\right)a^{\frac{1}{2}\mu}P_\nu^\mu[(1+a)^{\frac{1}{2}}]Q_\nu^\mu[(1+a)^{\frac{1}{2}}]$$
$$\left[\operatorname{Re}\mu<\frac{1}{2},\quad |\arg a|<\pi\right].$$ ET II 320(38)

8. $$\int_0^1 x^{-\frac{\mu}{2}-\frac{1}{2}}(1-x)^{-\mu-\frac{3}{2}}(1+ax)^{\frac{1}{2}\mu}Q_\nu^\mu(1+2ax)\,dx =$$
$$= \frac{1}{2}\pi^{\frac{1}{2}}\Gamma\left(-\mu-\frac{1}{2}\right)(1+a)^{-\frac{1}{2}}a^{\frac{1}{2}\mu+\frac{1}{2}}\times$$
$$\times\{P_\nu^{\mu+1}[(1+a)^{\frac{1}{2}}]Q_\nu^\mu[(1+a)^{\frac{1}{2}}]+P_\nu^\mu[(1+a)^{\frac{1}{2}}]Q_\nu^{\mu+1}[(1+a)^{\frac{1}{2}}]\}$$
$$\left[\operatorname{Re}\mu<-\frac{1}{2},\quad |\arg a|<\pi\right].$$ ET II 320(39)

9. $$\int_0^y (y-x)^{\mu-1}\left[x\left(1+\frac{1}{2}\gamma x\right)\right]^{-\frac{1}{2}\lambda}P_\nu^\lambda(1+\gamma x)\,dx =$$
$$= \Gamma(\mu)\left(\frac{2}{\gamma}\right)^{\frac{1}{2}\mu}\left[y\left(1+\frac{1}{2}\gamma y\right)\right]^{\frac{1}{2}\mu-\frac{1}{2}\lambda}P_\nu^{\lambda-\mu}(1+\gamma y)$$
$$\left[\operatorname{Re}\lambda<1,\quad \operatorname{Re}\mu>0,\quad |\arg\gamma y|<\pi\right].$$ ET II 193(52)

10. $$\int_0^y (y-x)^{\mu-1}x^{\sigma+\frac{1}{2}\lambda-1}\left(1+\frac{1}{2}\gamma x\right)^{-\frac{1}{2}\lambda}P_\nu^\lambda(1+\gamma x)\,dx =$$
$$= \frac{\left(\frac{\gamma}{2}\right)^{-\frac{1}{2}\lambda}\Gamma(\sigma)\Gamma(\mu)y^{\sigma+\mu-1}}{\Gamma(1-\lambda)\Gamma(\sigma+\mu)}\times$$
$$\times {}_3F_2\left(-\nu,\ 1+\nu,\ \sigma;\ 1-\lambda,\ \sigma+\mu;\ -\frac{1}{2}\gamma y\right)$$
$$[\operatorname{Re}\sigma>0,\ \operatorname{Re}\mu>0,\ |\gamma y|<1].$$ ET II 193(53)

11. $\int_0^y (y-x)^{\mu-1}[x(1-x)]^{-\frac{1}{2}\lambda} P_\nu^\lambda(1-2x)\,dx =$

$$= \Gamma(\mu)[y(1-y)]^{\frac{1}{2}\mu-\frac{1}{2}\lambda} P_\nu^{\lambda-\mu}(1-2y)$$

$[\operatorname{Re}\lambda < 1,\ \operatorname{Re}\mu > 0,\ 0 < y < 1]$. ET II 193(54)

12. $\int_0^y (y-x)^{\mu-1} x^{\sigma+\frac{1}{2}\lambda-1}(1-x)^{-\frac{1}{2}\lambda} P_\nu^\lambda(1-2x)\,dx =$

$$= \frac{\Gamma(\mu)\Gamma(\sigma)\, y^{\sigma+\mu-1}}{\Gamma(\sigma+\mu)\Gamma(1-\lambda)}\, {}_3F_2(-\nu,\ 1+\nu,\ \sigma;\ 1-\lambda,\ \sigma+\mu;\ y)$$

$[\operatorname{Re}\sigma > 0,\ \operatorname{Re}\mu > 0,\ 0 < y < 1]$. ET II 193(155)

7.138 $\int_0^\infty (a+x)^{-\mu-\nu-2} P_\mu\left(\frac{a-x}{a+x}\right) P_\nu\left(\frac{a-x}{a+x}\right) dx =$

$$= \frac{a^{-\mu-\nu-1}[\Gamma(\mu+\nu+1)]^4}{[\Gamma(\mu+1)\Gamma(\nu+1)]^2\,\Gamma(2\mu+2\nu+2)}$$

$[|\arg a| < \pi,\quad \operatorname{Re}(\mu+\nu) > -1]$. ET II 326(3)

7.14 Combinations of associated Legendre functions, exponentials, and powers

7.141

1. $\int_1^\infty e^{-ax}(x-1)^{\lambda-1}(x^2-1)^{\frac{1}{2}\mu} P_\nu^\mu(x)\,dx =$

$$= \frac{a^{-\lambda-\mu}e^{-a}}{\Gamma(1-\mu+\nu)\Gamma(-\mu-\nu)} G_{23}^{31}\left(2a \middle| \begin{matrix} 1+\mu,\ 1 \\ \lambda+\mu,\ -\nu,\ 1+\nu \end{matrix}\right)$$

$[\operatorname{Re} a > 0,\ \operatorname{Re}\lambda > 0]$. ET II 323(13)

2. $\int_1^\infty e^{-ax}(x-1)^{\lambda-1}(x^2-1)^{\frac{1}{2}\mu} Q_\nu^\mu(x)\,dx =$

$$= \frac{\Gamma(\nu+\mu+1)\,e^{\mu\pi i}}{2\Gamma(\nu-\mu+1)}\, a^{-\lambda-\mu}e^{-a} G_{23}^{22}\left(2a \middle| \begin{matrix} 1+\mu,\ 1 \\ \lambda+\mu,\ \nu+1,\ -\nu \end{matrix}\right)$$

$[\operatorname{Re} a > 0,\quad \operatorname{Re}\lambda > 0,\quad \operatorname{Re}(\lambda+\mu) > 0]$. ET II 325(24)

3. $\int_1^\infty e^{-ax}(x-1)^{\lambda-1}(x^2-1)^{-\frac{1}{2}\mu} P_\nu^\mu(x)\,dx =$

$$= -\pi^{-1}\sin(\nu\pi)\, a^{\mu-\lambda}e^{-a} G_{23}^{31}\left(2a \middle| \begin{matrix} 1,\ 1-\mu \\ \lambda-\mu,\ 1+\nu,\ -\nu \end{matrix}\right).$$

$[\operatorname{Re} a > 0,\ \operatorname{Re}(\lambda-\mu) > 0]$. ET II 323(15)

4. $\int_1^\infty e^{-ax}(x-1)^{\lambda-1}(x^2-1)^{-\frac{1}{2}\mu} Q_\nu^\mu(x)\,dx =$

$$= \frac{1}{2}e^{\mu\pi i} a^{\mu-\lambda}e^{-a} G_{23}^{22}\left(2a \middle| \begin{matrix} 1-\mu,\ 1 \\ \lambda-\mu,\ \nu+1,\ -\nu \end{matrix}\right)$$

$[\operatorname{Re} a > 0,\quad \operatorname{Re}\lambda > 0,\quad \operatorname{Re}(\lambda-\mu) > 0]$. ET II 323(14)

5. $$\int_1^\infty e^{-ax}(x^2-1)^{-\frac{1}{2}\mu} P_\nu^\mu(x)\,dx = 2^{\frac{1}{2}}\pi^{-\frac{1}{2}} a^{\mu-\frac{1}{2}} K_{\nu+\frac{1}{2}}(a)$$

$[\operatorname{Re} a > 0, \quad \operatorname{Re}\mu < 1]$. ET II 323(11), MO 90

7.142 $$\int_1^\infty e^{-\frac{1}{2}ax}\left(\frac{x+1}{x-1}\right)^{\frac{1}{2}\mu} P^\mu_{\nu-\frac{1}{2}}(x)\,dx = \frac{2}{a} W_{\mu,\nu}(a)$$

$\left[\operatorname{Re}\mu < 1, \ \nu - \frac{1}{2} \neq 0, \ \pm 1, \ \pm 2, \ldots\right]$. BU 79(34), MO 118

7.143

1. $$\int_0^\infty [x(1+x)]^{-\frac{1}{2}\mu} e^{-\beta x} P_\nu^\mu(1+2x)\,dx =$$

$$= \frac{\beta^{\mu-\frac{1}{2}}}{\sqrt{\pi}} e^{\frac{1}{2}\beta} K_{\nu+\frac{1}{2}}\left(\frac{\beta}{2}\right)$$ $[\operatorname{Re}\mu < 1, \ \operatorname{Re}\beta > 0]$. ET I 179(1)

2. $$\int_0^\infty \left(1+\frac{1}{x}\right)^{\frac{1}{2}\mu} e^{-\beta x} P_\nu^\mu(1+2x)\,dx = \frac{e^{\frac{1}{2}\beta}}{\beta} W_{\mu,\nu+\frac{1}{2}}(\beta)$$

$[\operatorname{Re}\mu < 1, \quad \operatorname{Re}\beta > 0]$. ET I 179(2)

7.144

1. $$\int_0^\infty e^{-\beta x} x^{\lambda+\frac{1}{2}\mu-1} (x+2)^{\frac{1}{2}\mu} Q_\nu^\mu(1+x)\,dx =$$

$$= \frac{\Gamma(\nu+\mu+1)}{\Gamma(\nu-\mu+1)}\left\{\frac{\sin(\nu\pi)}{2\beta^{\lambda+\mu}\sin(\mu\pi)} E(-\nu, \ \nu+1, \ \lambda+\mu : \mu+1 : 2\beta) - \right.$$

$$\left. - \frac{\sin[(\mu+\nu)\pi]}{2^{1-\mu}\beta^\lambda \sin(\mu\pi)} E(\nu-\mu+1, \ -\nu-\mu, \ \lambda : 1-\mu : 2\beta)\right\}$$

$[\operatorname{Re}\beta > 0, \quad \operatorname{Re}\lambda > 0, \quad \operatorname{Re}(\lambda+\mu) > 0]$. ET I 181(16)

2. $$\int_0^\infty e^{-\beta x} x^{\lambda-\frac{1}{2}\mu-1} (x+2)^{\frac{1}{2}\mu} Q_\nu^\mu(1+x)\,dx =$$

$$= -\frac{\sin(\nu\pi)}{2\beta^{\lambda-\mu}\sin(\mu\pi)} E(-\nu, \ \nu+1, \ \lambda-\mu : 1-\mu : 2\beta) -$$

$$- \frac{\sin[(\mu-\nu)\pi]}{2^{1+\mu}\beta^\lambda \sin(\mu\pi)} E(\mu+\nu+1, \ \mu-\nu, \ \lambda : 1+\mu : 2\beta)$$

$[\operatorname{Re}\beta > 0, \quad \operatorname{Re}\lambda > 0, \quad \operatorname{Re}(\lambda-\mu)] > 0$. ET I 181(17)

7.145

1. $$\int_0^\infty \frac{e^{-\beta x}}{1+x} \mathrm{P}_\nu\left[\frac{1}{(1+x)^2} - 1\right] dx = \frac{e^\beta}{\beta} W_{\nu+\frac{1}{2},0}(\beta) W_{-\nu-\frac{1}{2},0}(\beta)$$

$[\operatorname{Re}\beta > 0]$. ET I 180(6)

2. $$\int_0^\infty x^{-1} e^{-\beta x} Q_{-\frac{1}{2}}(1+2x^{-2})\,dx = \frac{\pi^2}{8}\left\{\left[J_0\left(\frac{1}{2}\beta\right)\right]^2 + \left[N_0\left(\frac{1}{2}\beta\right)\right]^2\right\}$$

$[\operatorname{Re}\beta > 0]$. ET II 327(5)

3. $$\int_0^\infty x^{-1} e^{-ax} Q_\nu(1+2x^{-2})\,dx = \frac{1}{2}[\Gamma(\nu+1)]^2 a^{-1} W_{-\nu-\frac{1}{2},0}(ai) W_{-\nu-\frac{1}{2},0}(-ai)$$

$[\operatorname{Re} a > 0, \quad \operatorname{Re}\nu > -1]$. ET II 327(6)

7.146

1. $$\int_0^\infty x^{-\frac{1}{2}\mu} e^{-\beta x} P_\nu^\mu(\sqrt{1+x})\,dx = 2^\mu \beta^{\frac{1}{2}\mu-\frac{5}{4}} e^{\frac{\beta}{2}} W_{\frac{1}{2}\mu+\frac{1}{4},\ \frac{1}{2}\nu+\frac{1}{4}}(\beta)$$

$[\operatorname{Re}\mu < 1, \quad \operatorname{Re}\beta > 0]$. ET I 180(7)

2. $$\int_0^\infty x^{-\frac{1}{2}\mu} \frac{e^{-\beta x}}{\sqrt{1+x}} P_\nu^\mu(\sqrt{1+x})\,dx = 2^\mu \beta^{\frac{1}{2}\mu-\frac{3}{4}} e^{\frac{1}{2}\beta} W_{\frac{1}{2}\mu-\frac{1}{4},\ \frac{1}{2}\nu+\frac{1}{4}}(\beta)$$

$[\operatorname{Re}\mu < 1, \quad \operatorname{Re}\beta > 0]$. ET I 180(8)a

3. $$\int_0^\infty \sqrt{x} e^{-\beta x} P_\nu^{\frac{1}{4}}(\sqrt{1+x^2}) P_\nu^{-\frac{1}{4}}(\sqrt{1+x^2})\,dx =$$

$$= \frac{1}{2}\sqrt{\frac{\pi}{2\beta}} H_{\nu+\frac{1}{2}}^{(1)}\left(\frac{1}{2}\beta\right) H_{\nu+\frac{1}{2}}^{(2)}\left(\frac{1}{2}\beta\right) \quad [\operatorname{Re}\beta > 0].$$ ET I 180(9)

7.147 $$\int_0^\infty x^{\lambda-1}(x^2+a^2)^{\frac{1}{2}\nu} e^{-\beta x} \mathrm{P}_\nu^\mu\left[\frac{x}{(x^2+a^2)^{\frac{1}{2}}}\right]dx =$$

$$= \frac{2^{-\nu-2} a^{\lambda+\nu}}{\pi\Gamma(-\mu-\nu)} G_{24}^{32}\left(\frac{a^2\beta^2}{4}\middle| \begin{matrix} 1-\frac{\lambda}{2}, & \frac{1-\lambda}{2} & & \\ 0, & \frac{1}{2}, & -\frac{\lambda+\mu+\nu}{2}, & -\frac{\lambda-\mu+\nu}{2} \end{matrix}\right)$$

$[a > 0, \quad \operatorname{Re}\beta > 0, \quad \operatorname{Re}\lambda > 0]$. ET II 327(7)

7.148 $$\int_{-1}^1 (1-x)^{-\frac{1}{2}\mu}(1+x)^{\frac{1}{2}\mu+\nu-1} \exp\left(-\frac{1-x}{1+x}y\right) \mathrm{P}_\nu^\mu(x)\,dx =$$

$$= 2^\nu y^{\frac{1}{2}\mu+\nu-\frac{1}{2}} e^{\frac{1}{2}y} W_{\frac{1}{2}\mu-\nu-\frac{1}{2},\ \frac{1}{2}\mu}(y) \qquad [\operatorname{Re} y > 0].$$ ET II 317(21)

7.149 $$\int_1^\infty (\alpha^2+\beta^2+2\alpha\beta x)^{-\frac{1}{2}} \exp[-(\alpha^2+\beta^2+2\alpha\beta x)^{\frac{1}{2}}] P_\nu(x)\,dx =$$

$$= 2\pi^{-1}(\alpha\beta)^{-\frac{1}{2}} K_{\nu+\frac{1}{2}}(\alpha) K_{\nu+\frac{1}{2}}(\beta) \qquad [\operatorname{Re}\alpha > 0, \ \operatorname{Re}\beta > 0].$$ ET II 323(16)

7.15 Combinations of associated Legendre and hyperbolic functions

7.151

1. $$\int_0^\infty (\operatorname{sh} x)^{\alpha-1} P_\nu^{-\mu}(\operatorname{ch} x)\,dx =$$

$$= \frac{2^{-1-\mu}\Gamma\left(\frac{1}{2}\alpha+\frac{1}{2}\mu\right)\Gamma\left(\frac{1}{2}\nu-\frac{1}{2}\alpha+1\right)\Gamma\left(\frac{1}{2}-\frac{1}{2}\alpha-\frac{1}{2}\nu\right)}{\Gamma\left(\frac{1}{2}\mu+\frac{1}{2}\nu+1\right)\Gamma\left(\frac{1}{2}+\frac{1}{2}\mu-\frac{1}{2}\nu\right)\Gamma\left(1+\frac{1}{2}\mu-\frac{1}{2}\alpha\right)}$$

$[\operatorname{Re}(\alpha+\mu)>0,\ \operatorname{Re}(\nu-\alpha+2)>0\ \ \operatorname{Re}(1-\alpha-\nu)>0]$. EH I 172(28)

2. $$\int_0^\infty (\operatorname{sh} x)^{\alpha-1} Q_\nu^\mu(\operatorname{ch} x)\,dx =$$

$$= \frac{e^{i\mu\pi}2^{\mu-\alpha}\Gamma\left(\frac{1}{2}+\frac{1}{2}\nu+\frac{1}{2}\mu\right)\Gamma\left(1+\frac{1}{2}\nu-\frac{1}{2}\alpha\right)}{\Gamma\left(1+\frac{1}{2}\nu-\frac{1}{2}\mu\right)\Gamma\left(\frac{1}{2}+\frac{1}{2}\nu+\frac{1}{2}\alpha\right)} \times$$

$$\times \Gamma\left(\frac{1}{2}\alpha+\frac{1}{2}\mu\right)\Gamma\left(\frac{1}{2}\alpha-\frac{1}{2}\mu\right)$$

$[\operatorname{Re}(\alpha\pm\mu)>0,\quad \operatorname{Re}(\nu-\alpha+2)>0]$. EH I 172(29)

7.152 $$\int_0^\infty e^{-\alpha x}\operatorname{sh}^{2\mu}\left(\frac{1}{2}x\right)P_{2n}^{-2\mu}\left[\operatorname{ch}\left(\frac{1}{2}x\right)\right]dx =$$

$$= \frac{\Gamma\left(2\mu+\frac{1}{2}\right)\Gamma(\alpha-n-\mu)\Gamma\left(\alpha+n-\mu+\frac{1}{2}\right)}{4^\mu\sqrt{\pi}\,\Gamma(\alpha+n+\mu+1)\Gamma\left(\alpha-n+\mu+\frac{1}{2}\right)}$$

$\left[\operatorname{Re}\alpha>n+\operatorname{Re}\mu,\quad \operatorname{Re}\mu>-\frac{1}{4}\right]$. ET I 181(15)

7.16 Combinations of associated Legendre functions, powers, and trigonometric functions

7.161

1. $$\int_0^1 x^{\lambda-1}(1-x^2)^{-\frac{1}{2}\mu}\sin(ax)\,\mathrm{P}_\nu^\mu(x)\,dx =$$

$$= \frac{\pi^{\frac{1}{2}}2^{\mu-\lambda-1}\Gamma(\lambda+1)\,a}{\Gamma\left(1+\frac{\lambda-\mu-\nu}{2}\right)\Gamma\left(\frac{3+\lambda-\mu+\nu}{2}\right)} \times$$

$$\times {}_2F_3\left(\frac{1+\lambda}{2},\ 1+\frac{\lambda}{2};\ \frac{3}{2},\ 1+\frac{\lambda-\mu-\nu}{2},\ \frac{3+\lambda-\mu+\nu}{2};\ -\frac{a^2}{4}\right)$$

$[\operatorname{Re}\lambda>-1,\quad \operatorname{Re}\mu<1]$. ET II 314(7)

2. $\int_0^1 x^{\lambda-1}(1-x^2)^{-\frac{1}{2}\mu}\cos(ax)\,\mathrm{P}_\nu^\mu(x)\,dx =$

$$= \frac{\pi^{\frac{1}{2}}2^{\mu-\lambda}\Gamma(\lambda)}{\Gamma\left(1+\frac{\lambda-\mu+\nu}{2}\right)\Gamma\left(\frac{1+\lambda-\mu-\nu}{2}\right)}\times$$
$$\times\,{}_2F_3\left(\frac{\lambda}{2},\ \frac{\lambda+1}{2};\ \frac{1}{2},\ \frac{1+\lambda-\mu-\nu}{2},\ 1+\frac{\lambda-\mu+\nu}{2};\ -\frac{a^2}{4}\right)$$

$[\operatorname{Re}\lambda > 0,\quad \operatorname{Re}\mu < 1].$ ET II 314(8)

3. $\int_0^\infty (x^2-1)^{\frac{1}{2}\mu}\sin(ax)\,P_\nu^\mu(x)\,dx =$

$$= \frac{2^\mu\pi^{\frac{1}{2}}a^{-\mu-\frac{1}{2}}}{\Gamma\left(\frac{1}{2}-\frac{1}{2}\mu-\frac{1}{2}\nu\right)\Gamma\left(1-\frac{1}{2}\mu+\frac{1}{2}\nu\right)}S_{\mu+\frac{1}{2},\,\nu+\frac{1}{2}}(a)$$

$\left[a > 0,\quad \operatorname{Re}\mu < \frac{3}{2},\quad \operatorname{Re}(\mu+\nu) < 1\right].$ ET II 320(1)

7.162

1. $\int_a^\infty P_\nu(2x^2a^{-2}-1)\sin(bx)\,dx =$

$$= -\frac{\pi a}{4\cos(\nu\pi)}\left\{\left[J_{\nu+\frac{1}{2}}\left(\frac{ab}{2}\right)\right]^2-\left[J_{-\nu-\frac{1}{2}}\left(\frac{ab}{2}\right)\right]^2\right\}$$

$[a > 0,\ b > 0,\ -1 < \operatorname{Re}\nu < 0].$ ET II 326(1)

2. $\int_a^\infty P_\nu(2x^2a^{-2}-1)\cos(bx)\,dx =$

$$= -\frac{\pi}{4}a\left[J_{\nu+\frac{1}{2}}\left(\frac{ab}{2}\right)J_{-\nu-\frac{1}{2}}\left(\frac{ab}{2}\right)-N_{\nu+\frac{1}{2}}\left(\frac{ab}{2}\right)N_{-\nu-\frac{1}{2}}\left(\frac{ab}{2}\right)\right]$$

$[a > 0,\ b > 0,\ -1 < \operatorname{Re}\nu < 0].$ ET II 326(2)

3. $\int_0^\infty (x^2+2)^{-\frac{1}{2}}\sin(ax)\,P_\nu^{-1}(x^2+1)\,dx = 2^{-\frac{1}{2}}\pi^{-1}a\sin(\nu\pi)\left[K_{\nu+\frac{1}{2}}(2^{-\frac{1}{2}}a)\right]^2$

$[a > 0,\quad -2 < \operatorname{Re}\nu < 1].$ ET I 98(22)

4. $\int_0^\infty (x^2+2)^{-\frac{1}{2}}\sin(ax)\,Q_\nu^1(x^2+1)\,dx = -2^{-\frac{3}{2}}\pi a K_{\nu+\frac{1}{2}}(2^{-\frac{1}{2}}a)\,I_{\nu+\frac{1}{2}}(2^{-\frac{1}{2}}a)$

$\left[a > 0,\quad \operatorname{Re}\nu > -\frac{3}{2}\right].$ ET 98(23)

5. $\int_0^\infty \cos(ax)\,P_\nu(1+x^2)\,dx = -\frac{\sqrt{2}}{\pi}\sin(\nu\pi)\left[K_{\nu+\frac{1}{2}}\left(\frac{a}{\sqrt{2}}\right)\right]^2$

$[a > 0,\quad -1 < \operatorname{Re}\nu < 0].$ ET I 42(23)

6. $\int_0^\infty \cos(ax)\,Q_\nu(1+x^2)\,dx = \frac{\pi}{\sqrt{2}}K_{\nu+\frac{1}{2}}\left(\frac{a}{\sqrt{2}}\right)I_{\nu+\frac{1}{2}}\left(\frac{a}{\sqrt{2}}\right)$

$[a > 0,\quad \operatorname{Re}\nu > -1].$ ET I 42(24)

7. $\int_0^1 \cos(ax)\,P_\nu(2x^2-1)\,dx = \frac{\pi}{2} J_{\nu+\frac{1}{2}}\left(\frac{a}{2}\right) J_{-\nu-\frac{1}{2}}\left(\frac{a}{2}\right)$

$[a > 0]$. ET I 42(25)

7.163

1. $\int_a^\infty (x^2-a^2)^{\frac{1}{2}\nu-\frac{1}{4}} \sin(bx)\,P_0^{\frac{1}{2}-\nu}(ax^{-1})\,dx = b^{-\nu-\frac{1}{2}} \cos\left(ab - \frac{\nu\pi}{2} + \frac{\pi}{4}\right)$

$\left[a > 0, \quad |\operatorname{Re}\nu| < \frac{1}{2}\right]$. ET I 98(24)

2. $\int_0^1 x^{-1} \cos(ax)\,P_\nu(2x^{-2}-1)\,dx =$

$= -\frac{1}{2}\pi \operatorname{cosec}(\nu\pi)\,{}_1F_1(\nu+1;\ 1;\ ai)\,{}_1F_1(\nu+1;\ 1;\ -ai)$

$[a > 0, \quad -1 < \operatorname{Re}\nu < 0]$. ET II 327(4)

7.164

1. $\int_0^\infty x^{\frac{1}{2}} \sin(bx)\left[P_\nu^{-\frac{1}{4}}\left(\sqrt{1+a^2x^2}\right)\right]^2 dx =$

$= \frac{\sqrt{\frac{2}{\pi}}\,a^{-1}b^{-\frac{1}{2}}}{\Gamma\left(\frac{5}{4}+\nu\right)\Gamma\left(\frac{1}{4}-\nu\right)} \left[K_{\nu+\frac{1}{2}}\left(\frac{b}{2a}\right)\right]^2$

$\left[\operatorname{Re}a > 0, \quad b > 0, \quad -\frac{5}{4} < \operatorname{Re}\nu < \frac{1}{4}\right]$. ET II 327(8)

2. $\int_0^\infty x^{\frac{1}{2}} \sin(bx)\,P_\nu^{-\frac{1}{4}}\left(\sqrt{1+a^2x^2}\right) Q_\nu^{-\frac{1}{4}}\left(\sqrt{1+a^2x^2}\right) dx =$

$= \frac{\sqrt{\frac{\pi}{2}}\,e^{-\frac{1}{4}\pi i}\,\Gamma\left(\nu+\frac{5}{4}\right)}{ab^{\frac{1}{2}}\Gamma\left(\nu+\frac{3}{4}\right)} I_{\nu+\frac{1}{2}}\left(\frac{b}{2a}\right) K_{\nu+\frac{1}{2}}\left(\frac{b}{2a}\right)$

$\left[\operatorname{Re}a > 0, \quad b > 0, \quad \operatorname{Re}\nu > -\frac{5}{4}\right]$. ET II 327(9)

3. $\int_0^\infty x^{\frac{1}{2}} \sin(bx)\,P_\nu^{-\frac{1}{4}}\left(\sqrt{1+a^2x^2}\right) P_{\nu-1}^{-\frac{1}{4}}\left(\sqrt{1+a^2x^2}\right) \frac{dx}{\sqrt{1+a^2x^2}} =$

$= \frac{a^{-2}b^{\frac{1}{2}}}{\sqrt{2\pi}\,\Gamma\left(\frac{5}{4}+\nu\right)\Gamma\left(\frac{5}{4}-\nu\right)} K_{\nu-\frac{1}{2}}\left(\frac{b}{2a}\right) K_{\nu+\frac{1}{2}}\left(\frac{b}{2a}\right)$

$\left[\operatorname{Re}a > 0, \quad b > 0, \quad -\frac{5}{4} < \operatorname{Re}\nu < \frac{5}{4}\right]$. ET II 328(10)

4. $$\int_0^\infty x^{\frac{1}{2}} \sin(bx) P_\nu^{\frac{1}{4}}(\sqrt{1+a^2x^2}) P_\nu^{-\frac{3}{4}}(\sqrt{1+a^2x^2}) \frac{dx}{\sqrt{1+a^2x^2}} =$$

$$= \frac{a^{-2}b^{\frac{1}{2}}}{\sqrt{2\pi}\,\Gamma\left(\frac{7}{4}+\nu\right)\Gamma\left(\frac{3}{4}-\nu\right)} \left[K_{\nu+\frac{1}{2}}\left(\frac{b}{2a}\right)\right]^2$$

$$\left[\operatorname{Re} a > 0, \quad b > 0, \quad -\frac{7}{4} < \operatorname{Re}\nu < \frac{3}{4}\right].$$ ET II 328(11)

5. $$\int_0^\infty x^{\frac{1}{2}} \cos(bx) [P_\nu^{\frac{1}{4}}(\sqrt{1+a^2x^2})]^2 \, dx =$$

$$= \frac{a^{-1}\left(\frac{\pi b}{2}\right)^{-\frac{1}{2}}}{\Gamma\left(\frac{3}{4}+\nu\right)\Gamma\left(-\frac{1}{4}-\nu\right)} \left[K_{\nu+\frac{1}{2}}\left(\frac{b}{2a}\right)\right]^2$$

$$\left[\operatorname{Re} a > 0, \quad b > 0, \quad -\frac{3}{4} < \operatorname{Re}\nu < -\frac{1}{4}\right].$$ ET II 328(12)

6. $$\int_0^\infty x^{\frac{1}{2}} \cos(bx) P_\nu^{\frac{1}{4}}(\sqrt{1+a^2x^2}) Q_\nu^{\frac{1}{4}}(\sqrt{1+a^2x^2}) \, dx =$$

$$= \frac{\sqrt{\frac{\pi}{2}}\, e^{\frac{1}{4}\pi i}\, \Gamma\left(\nu+\frac{3}{4}\right)}{a b^{\frac{1}{2}} \Gamma\left(\nu+\frac{5}{4}\right)} I_{\nu+\frac{1}{2}}\left(\frac{b}{2a}\right) K_{\nu+\frac{1}{2}}\left(\frac{b}{2a}\right)$$

$$\left[\operatorname{Re} a > 0, \quad b > 0, \quad \operatorname{Re}\nu > -\frac{3}{4}\right].$$ ET II 328(13)

7. $$\int_0^\infty x^{\frac{1}{2}} \cos(bx) P_\nu^{-\frac{1}{4}}(\sqrt{1+a^2x^2}) P_\nu^{\frac{3}{4}}(\sqrt{1+a^2x^2}) \frac{dx}{\sqrt{1+a^2x^2}} =$$

$$= \frac{a^{-2}b^{\frac{1}{2}}}{\sqrt{2\pi}\,\Gamma\left(\frac{5}{4}+\nu\right)\Gamma\left(\frac{1}{4}-\nu\right)} \left[K_{\nu+\frac{1}{2}}\left(\frac{b}{2a}\right)\right]^2$$

$$\left[\operatorname{Re} a > 0, \quad b > 0, \quad -\frac{5}{4} < \operatorname{Re}\nu < \frac{1}{4}\right].$$ ET II 328(14)

8. $$\int_0^\infty x^{\frac{1}{2}} \cos(bx) P_\nu^{\frac{1}{4}}(\sqrt{1+a^2x^2}) P_{\nu-1}^{\frac{1}{4}}(\sqrt{1+a^2x^2}) \frac{dx}{\sqrt{1+a^2x^2}} =$$

$$= \frac{a^{-2}b^{\frac{1}{2}}}{\sqrt{2\pi}\,\Gamma\left(\frac{3}{4}+\nu\right)\Gamma\left(\frac{3}{4}-\nu\right)} K_{\nu-\frac{1}{2}}\left(\frac{b}{2a}\right) K_{\nu+\frac{1}{2}}\left(\frac{b}{2a}\right)$$

$$\left[\operatorname{Re} a > 0, \quad b > 0, \quad |\operatorname{Re}\nu| < \frac{3}{4}\right].$$ ET II 329(15)

7.165 $\int\limits_0^\infty \cos(ax) P_\nu(\operatorname{ch} x)\,dx =$

$$= -\frac{\sin(\nu\pi)}{4\pi^2}\Gamma\left(\frac{1+\nu+ia}{2}\right)\Gamma\left(\frac{1+\nu-ia}{2}\right)\Gamma\left(-\frac{\nu+ia}{2}\right)\Gamma\left(-\frac{\nu-ia}{2}\right)$$

$[a>0, \quad -1<\operatorname{Re}\nu<0]$. ET II 329(18)

7.166 $\int\limits_0^\pi \mathrm{P}_\nu^{-\mu}(\cos\varphi)\sin^{\alpha-1}\varphi\,d\varphi =$

$$= \frac{2^{-\mu}\pi\Gamma\left(\frac{1}{2}\alpha+\frac{1}{2}\mu\right)\Gamma\left(\frac{1}{2}\alpha-\frac{1}{2}\mu\right)}{\Gamma\left(\frac{1}{2}+\frac{1}{2}\alpha+\frac{1}{2}\nu\right)\Gamma\left(\frac{1}{2}\alpha-\frac{1}{2}\nu\right)\Gamma\left(\frac{1}{2}\mu+\frac{1}{2}\nu+1\right)\Gamma\left(\frac{1}{2}\mu-\frac{1}{2}\nu+\frac{1}{2}\right)}$$

$[\operatorname{Re}(\alpha\pm\mu)>0]$. MO 90, EH I 172(27)

7.167 $\int\limits_0^a \mathrm{P}_\nu^{-\mu}(\cos x)\,\mathrm{P}_\nu^{-\eta}[\cos(a-x)]\left[\frac{\sin(a-x)}{\sin x}\right]^\eta \frac{dx}{\sin x} =$

$$= \frac{2^\eta\Gamma(\mu-\eta)\Gamma\left(\eta+\frac{1}{2}\right)(\sin a)^\eta}{\sqrt{\pi}\,\Gamma(\eta+\mu+1)}\mathrm{P}_\nu^{-\mu}(\cos a)$$

$\left[\operatorname{Re}\mu>\operatorname{Re}\eta>-\frac{1}{2}\right]$. ET II 329(16)

7.17 A combination of an associated Legendre function and the probability integral

7.171 $\int\limits_1^\infty (x^2-1)^{-\frac{1}{2}\mu}\exp(a^2x^2)[1-\Phi(ax)]P_\nu^\mu(x)\,dx =$

$$= \pi^{-1}2^{\mu-1}\Gamma\left(\frac{1+\mu+\nu}{2}\right)\Gamma\left(\frac{\mu-\nu}{2}\right)a^{\mu-\frac{3}{2}}e^{\frac{a^2}{2}}W_{\frac{1}{4}-\frac{1}{2}\mu,\ \frac{1}{4}+\frac{1}{2}\nu}(a^2)$$

$[\operatorname{Re}a>0,\ \operatorname{Re}\mu<1,\ \operatorname{Re}(\mu+\nu)>-1,\ \operatorname{Re}(\mu-\nu)>0]$. ET II 324(17)

7.18 Combinations of associated Legendre and Bessel functions

7.181

1. $\int\limits_1^\infty P_{\nu-\frac{1}{2}}(x)\,x^{\frac{1}{2}}N_\nu(ax)\,dx =$

$$= 2^{-\frac{1}{2}}a^{-1}\left[\cos\left(\frac{1}{2}a\right)J_\nu\left(\frac{1}{2}a\right)-\sin\left(\frac{1}{2}a\right)N_\nu\left(\frac{1}{2}a\right)\right]$$

$\left[a>0, \quad \operatorname{Re}\nu<\frac{1}{2}\right]$. ET II 108(3)a

2. $\int\limits_1^\infty P_{\nu-\frac{1}{2}}(x)\,x^{\frac{1}{2}}J_\nu(ax)\,dx =$

$$= -\frac{1}{\sqrt{2a}}\left[\cos\left(\frac{1}{2}a\right)N_\nu\left(\frac{1}{2}a\right)+\sin\left(\frac{1}{2}a\right)J_\nu\left(\frac{1}{2}a\right)\right]$$

$\left[|\operatorname{Re}\nu|<\frac{1}{2}\right]$. ET II 344(36)a

7.182

1. $$\int_1^\infty x^\nu (x^2-1)^{\frac{1}{2}\lambda-\frac{1}{2}} P_\lambda^{\lambda-1}(x) J_\nu(ax)\,dx = \frac{2^{\lambda+\nu}a^{-\lambda}\Gamma\left(\frac{1}{2}+\nu\right)}{\pi^{\frac{1}{2}}\Gamma(1-\lambda)} S_{\lambda-\nu,\,\lambda+\nu}(a)$$

$$\left[a>0, \quad \operatorname{Re}\nu<\frac{5}{2}, \quad \operatorname{Re}(2\lambda+\nu)<\frac{3}{2}\right].$$ ET II 345(38)a

2. $$\int_1^\infty x^{\frac{1}{2}-\mu}(x^2-1)^{-\frac{1}{2}\mu} P^\mu_{\nu-\frac{1}{2}}(x) J_\nu(ax)\,dx =$$

$$= -2^{-\frac{3}{2}}\pi^{\frac{1}{2}} a^{\mu-\frac{1}{2}}\left[J_{\mu-\frac{1}{2}}\left(\frac{1}{2}a\right)N_\nu\left(\frac{1}{2}a\right)+N_{\mu-\frac{1}{2}}\left(\frac{1}{2}a\right)J_\nu\left(\frac{1}{2}a\right)\right]$$

$$\left[-\frac{1}{4}<\operatorname{Re}\mu<1,\ a>0,\ |\operatorname{Re}\nu|<\frac{1}{2}+2\operatorname{Re}\mu\right].$$ ET II 344(37)a

3. $$\int_1^\infty x^{\frac{1}{2}-\mu}(x^2-1)^{-\frac{1}{2}\mu} P^\mu_{\nu-\frac{1}{2}}(x) N_\nu(ax)\,dx =$$

$$= 2^{-\frac{3}{2}}\pi^{\frac{1}{2}} a^{\mu-\frac{1}{2}}\left[J_\nu\left(\frac{1}{2}a\right)J_{\mu-\frac{1}{2}}\left(\frac{1}{2}a\right)-N_\nu\left(\frac{1}{2}a\right)N_{\mu-\frac{1}{2}}\left(\frac{1}{2}a\right)\right]$$

$$\left[-\frac{1}{4}<\operatorname{Re}\mu<1,\ a>0,\ \operatorname{Re}(2\mu-\nu)>-\frac{1}{2}\right].$$ ET II 349(67)a

4. $$\int_0^1 x^{\frac{1}{2}-\mu}(1-x^2)^{-\frac{1}{2}\mu} \mathrm{P}^\mu_\nu(x) J_{\nu+\frac{1}{2}}(ax)\,dx =$$

$$= \sqrt{\frac{\pi}{2}}\, a^{\mu-\frac{1}{2}} J_{\frac{1}{2}-\mu}\left(\frac{1}{2}a\right) J_{\nu+\frac{1}{2}}\left(\frac{1}{2}a\right)$$

$$[\operatorname{Re}\mu<1,\ \operatorname{Re}(\mu-\nu)<2].$$ ET II 337(33)a

5. $$\int_1^\infty x^{\frac{1}{2}-\mu}(x^2-1)^{-\frac{1}{2}\mu} P^\mu_{\nu-\frac{1}{2}}(x) K_\nu(ax)\,dx =$$

$$= (2\pi)^{-\frac{1}{2}} a^{\mu-\frac{1}{2}} K_\nu\left(\frac{1}{2}a\right) K_{\mu-\frac{1}{2}}\left(\frac{1}{2}a\right) \quad [\operatorname{Re}\mu<1,\ \operatorname{Re}a>0]$$ ET II 135(5)a

6. $$\int_1^\infty x^{\mu+\frac{1}{2}}(x^2-1)^{-\frac{1}{2}\mu} P^\mu_{\nu-\frac{1}{2}}(x) K_\nu(ax)\,dx = \sqrt{\frac{\pi}{2}}\, a^{-\frac{3}{2}} e^{-\frac{1}{2}a} W_{\mu,\nu}(a)$$

$$[\operatorname{Re}\mu<1,\ \operatorname{Re}a>0].$$ ET II 135(3)a

7. $$\int_1^\infty x^{\mu-\frac{3}{2}}(x^2-1)^{-\frac{1}{2}\mu} P^\mu_{\nu-\frac{1}{2}}(x) K_\nu(ax)\,dx = \sqrt{\frac{\pi}{2}}\, a^{-\frac{1}{2}} e^{-\frac{1}{2}a} W_{\mu-1,\nu}(a)$$

$$[\operatorname{Re}\mu<1,\ \operatorname{Re}a>0].$$ ET II 135(4)a

8. $$\int_1^\infty x^{\mu-\frac{1}{2}}(x^2-1)^{-\frac{1}{2}\mu} P^{\mu}_{\nu-\frac{3}{2}}(x) K_\nu(ax)\,dx = \sqrt{\frac{\pi}{2}}\, a^{-1} e^{-\frac{1}{2}a} W_{\mu-\frac{1}{2},\,\nu-\frac{1}{2}}(a)$$

$[\operatorname{Re}\mu < 1]$. ET II 135(6)a

9. $$\int_1^\infty x^{\frac{1}{2}}(x^2-1)^{\frac{1}{2}\nu-\frac{1}{4}} P_\mu^{\frac{1}{2}-\nu}(2x^2-1) K_\nu(ax)\,dx = \pi^{-\frac{1}{2}} a^{-\nu} 2^{\nu-1}\left[K_{\mu+\frac{1}{2}}\left(\frac{a}{2}\right)\right]^2$$

$\left[\operatorname{Re}\nu > -\frac{1}{2},\ \operatorname{Re}a > 0\right]$. ET II 136(11)a

10. $$\int_1^\infty x^{\frac{1}{2}}(x^2-1)^{\frac{1}{2}\nu-\frac{1}{4}} P_\mu^{\frac{1}{2}-\nu}(2x^2-1) N_\nu(ax)\,dx =$$

$$= \pi^{\frac{1}{2}} 2^{\nu-2} a^{-\nu}\left[J_{\mu+\frac{1}{2}}\left(\frac{a}{2}\right) J_{-\mu-\frac{1}{2}}\left(\frac{a}{2}\right) - N_{\mu+\frac{1}{2}}\left(\frac{a}{2}\right) N_{-\mu-\frac{1}{2}}\left(\frac{a}{2}\right)\right]$$

$\left[\operatorname{Re}\nu > -\frac{1}{2},\ a > 0,\ \operatorname{Re}\nu + |2\operatorname{Re}\mu + 1| < \frac{3}{2}\right]$. ET II 108(5)a

11. $$\int_1^\infty x^{\frac{1}{2}}(x^2-1)^{\frac{1}{2}\nu-\frac{1}{4}} P_\mu^{\frac{1}{2}-\nu}(2x^2-1) J_\nu(ax)\,dx =$$

$$= -2^{\nu-2} a^{-\nu} \pi^{\frac{1}{2}} \sec(\mu\pi)\left\{\left[J_{\mu+\frac{1}{2}}\left(\frac{a}{2}\right)\right]^2 - \left[J_{-\mu-\frac{1}{2}}\left(\frac{a}{2}\right)\right]^2\right\}$$

$\left[\operatorname{Re}\nu > -\frac{1}{2},\ a > 0,\ \operatorname{Re}\nu - \frac{3}{2} < 2\operatorname{Re}\mu < \frac{1}{2} - \operatorname{Re}\nu\right]$. ET II 345(39)a

12. $$\int_1^\infty x(x^2-1)^{-\frac{1}{2}\nu} P^\nu_\mu(2x^2-1) K_\nu(ax)\,dx = 2^{-\nu} a^{\nu-1} K_{\mu+1}(a)$$

$[\operatorname{Re}a > 0,\ \operatorname{Re}\nu < 1]$. ET II 136(10)a

13 $$\int_0^\infty x(x^2+a^2)^{\frac{1}{2}\nu} P^\nu_\mu(1+2x^2a^{-2}) K_\nu(xy)\,dx = 2^{-\nu} a y^{-\nu-1} S_{2\nu,\,2\mu+1}(ay)$$

$[\operatorname{Re}a > 0,\ \operatorname{Re}y > 0,\ \operatorname{Re}\nu < 1]$. ET II 135(7)

14. $$\int_0^\infty x(x^2+a^2)^{\frac{1}{2}\nu}\left[(\mu-\nu) P^\nu_\mu(1+2x^2a^{-2}) +\right.$$

$$\left. + (\mu+\nu) P^\nu_{-\mu}(1+2x^2a^{-2})\right] K_\nu(xy)\,dx = 2^{1-\nu}\mu y^{-\nu-2} S_{2\nu+1,\,2\mu}(ay)$$

$[\operatorname{Re}a > 0,\ \operatorname{Re}y > 0,\ \operatorname{Re}\nu < 1]$. ET II 136(8)

15. $$\int_0^\infty x(x^2+a^2)^{\frac{1}{2}\nu-1}\left[P^\nu_\mu(1+2x^2a^{-2}) +\right.$$

$$\left. + P^\nu_{-\mu}(1+2x^2a^{-2})\right] K_\nu(xy)\,dx = 2^{1-\nu} y^{-\nu} S_{2\nu-1,\,2\mu}(ay)$$

$[\operatorname{Re}a > 0,\ \operatorname{Re}y > 0,\ \operatorname{Re}\nu < 1]$. ET II 136(9)

16. $$\int_0^\infty x^{\frac{1}{2}}(x^2+2)^{-\frac{1}{2}\nu-\frac{1}{4}} P_\mu^{-\nu-\frac{1}{2}}(x^2+1) J_\nu(xy)\,dx = \frac{y^{-\frac{1}{2}} 2^{\frac{1}{2}-\nu} \pi^{-\frac{1}{2}} \left[K_{\mu+\frac{1}{2}}\left(2^{-\frac{1}{2}} y\right)\right]^2}{\Gamma\left(\nu+\mu+\frac{3}{2}\right)\Gamma\left(\nu-\mu+\frac{1}{2}\right)}$$

$$\left[-\frac{3}{2}-\operatorname{Re}\nu < \operatorname{Re}\mu < \operatorname{Re}\nu+\frac{1}{2},\ y>0\right].$$ ET II 44(1)

17. $$\int_0^\infty x^{\frac{1}{2}}(x^2+2)^{-\frac{1}{2}\nu-\frac{1}{4}} Q_\mu^{\nu+\frac{1}{2}}(x^2+1) J_\nu(xy)\,dx =$$

$$= 2^{-\nu-\frac{1}{2}} \pi^{\frac{1}{2}} e^{\left(\nu+\frac{1}{2}\right)\pi i} y^\nu K_{\mu+\frac{1}{2}}\left(2^{-\frac{1}{2}} y\right) I_{\mu+\frac{1}{2}}\left(2^{-\frac{1}{2}} y\right)$$

$$\left[\operatorname{Re}\nu > -1,\ \operatorname{Re}(2\mu+\nu) > -\frac{5}{2},\ y>0\right].$$ ET II 46(12)

7.183 $$\int_0^\infty x^{1-\mu}(1+a^2x^2)^{-\frac{1}{2}\mu-\frac{1}{4}} Q_{\nu-\frac{1}{2}}^{\mu+\frac{1}{2}}(\pm iax) J_\nu(xy)\,dx =$$

$$= i(2\pi)^{\frac{1}{2}} e^{i\pi\left(\mu \mp \frac{1}{2}\nu \mp \frac{1}{4}\right)} a^{-1} y^{\mu-1} I_\nu\left(\frac{1}{2}a^{-1}y\right) K_\mu\left(\frac{1}{2}a^{-1}y\right)$$

$$\left[-\frac{3}{4}-\frac{1}{2}\operatorname{Re}\nu < \operatorname{Re}\mu < 1+\operatorname{Re}\nu,\ y>0,\ \operatorname{Re}a>0\right].$$ ET II 46(11)

7.184

1. $$\int_1^\infty x^{\frac{1}{2}}(x^2-1)^{\frac{1}{2}\mu-\frac{1}{4}} \mathrm{P}_{-\frac{1}{2}+\nu}^{\frac{1}{2}-\mu}(x^{-1}) J_\nu(xa)\,dx =$$

$$= 2^{\frac{1}{2}} a^{-1-\mu} \pi^{-\frac{1}{2}} \cos\left[a+\frac{1}{2}(\nu-\mu)\pi\right]$$

$$\left[|\operatorname{Re}\mu| < \frac{1}{2},\ \operatorname{Re}\nu > -1,\ a>0\right].$$ ET II 44(2)a

2. $$\int_1^\infty x^{-\nu}(x^2-1)^{\frac{1}{4}-\frac{1}{2}\nu} \mathrm{P}_\mu^{\nu-\frac{1}{2}}(2x^{-2}-1) K_\nu(ax)\,dx =$$

$$= \pi^{\frac{1}{2}} 2^{-\nu} a^{-2+\nu} W_{\mu+\frac{1}{2},\ \nu-\frac{1}{2}}(a) W_{-\mu-\frac{1}{2},\ \nu-\frac{1}{2}}(a)$$

$$\left[\operatorname{Re}\nu < \frac{3}{2},\ a>0\right].$$ ET II 370(45)a

3. $$\int_0^\infty x^\nu (1+x^2)^{\frac{1}{4}+\frac{\nu}{2}} Q_\mu^{\nu+\frac{1}{2}}\left(1+\frac{2}{x^2}\right) J_\nu(ax)\,dx =$$

$$= -ie^{i\pi\nu} \pi^{-\frac{1}{2}} 2^\nu a^{-\nu-2} \left[\Gamma\left(\frac{3}{2}+\mu+\nu\right)\right]^2 \Gamma\left(\frac{1}{2}+\nu-\mu\right) \times$$

$$\times W_{-\mu-\frac{1}{2},\ \nu+\frac{1}{2}}(a) \left[\frac{\cos(\mu\pi)}{\Gamma(2+2\nu)} M_{\mu+\frac{1}{2},\ \nu+\frac{1}{2}}(a) + \frac{\sin(\nu\pi)}{\Gamma\left(\nu+\mu+\frac{3}{2}\right)} W_{\mu+\frac{1}{2},\ \nu+\frac{1}{2}}(a)\right]$$

$$\left[a>0,\ \operatorname{Re}(\mu+\nu) > -\frac{3}{2},\ \operatorname{Re}(\mu-\nu) < \frac{1}{2}\right].$$ ET II 46(14)

4. $$\int_0^1 x^\nu (1-x^2)^{\frac{1}{2}\nu+\frac{1}{4}} P_\mu^{-\nu-\frac{1}{2}} (2x^{-2}-1) J_\nu(xy)\,dx =$$

$$= 2^{\nu+\frac{1}{2}} y^\nu \frac{\Gamma\left(\frac{3}{2}+\mu+\nu\right)\Gamma\left(\frac{1}{2}+\nu-\mu\right)}{(2\pi)^{\frac{1}{2}}\left[\Gamma\left(\frac{3}{2}+\nu\right)\right]^2} \times$$

$$\times\, {}_1F_1\left(\nu+\mu+\frac{3}{2};\ 2\nu+2;\ iy\right) {}_1F_1\left(\nu+\mu+\frac{3}{2};\ 2\nu+2;\ -iy\right)$$

$$\left[y>0,\ -\frac{3}{2}-\operatorname{Re}\nu<\operatorname{Re}\mu<\operatorname{Re}\nu+\frac{1}{2}\right].$$ ET II 45(3)

5 $$\int_0^\infty x^{-\nu}(x^2+a^2)^{\frac{1}{4}-\frac{1}{2}\nu} Q_\mu^{\frac{1}{2}-\nu}(1+2a^2x^{-2}) K_\nu(xy)\,dx =$$

$$= ie^{-i\pi\nu}\pi^{\frac{1}{2}} 2^{-\nu-1} a^{-\nu-\frac{1}{2}} y^{\nu-2}\left[\Gamma\left(\frac{3}{2}+\mu-\nu\right)\right]^2 \times$$

$$\times W_{-\mu-\frac{1}{2},\,\nu-\frac{1}{2}}(iay)\, W_{-\mu-\frac{1}{2},\,\nu-\frac{1}{2}}(-iay)$$

$$\left[\operatorname{Re}a>0,\ \operatorname{Re}y>0,\ \operatorname{Re}\mu>-\frac{3}{2},\ \operatorname{Re}(\mu-\nu)>-\frac{3}{2}\right].$$ ET II 137(13)

6. $$\int_0^\infty x^{-\nu}(x^2+1)^{\frac{1}{4}-\frac{1}{2}\nu} Q_\mu^{\frac{1}{2}-\nu}(1+2x^{-2}) J_\nu(ax)\,dx =$$

$$= 2^{-\nu}a^{-\nu-2}\frac{ie^{-i\nu\pi}\pi^{\frac{1}{2}}\Gamma\left(\frac{3}{2}+\mu-\nu\right)}{\Gamma(2\nu)} M_{\mu+\frac{1}{2},\,\nu-\frac{1}{2}}(a)\, W_{-\mu-\frac{1}{2},\,\nu-\frac{1}{2}}(a)$$

$$\left[a>0,\ 0<\operatorname{Re}\nu<\operatorname{Re}\mu+\frac{3}{2}\right].$$ ET II 47(15)a

7 $$\int_0^\infty x^{-\nu}(x^2+a^2)^{\frac{1}{4}-\frac{1}{2}\nu} Q_{-\frac{1}{2}}^{\frac{1}{2}-\nu}(1+2a^2x^{-2}) K_\nu(xy)\,dx =$$

$$= ie^{-i\pi\nu}\pi^{\frac{3}{2}} 2^{-\nu-3} a^{\frac{1}{2}-\nu} y^{\nu-1}[\Gamma(1-\nu)]^2 \times$$

$$\times\left\{\left[J_{\nu-\frac{1}{2}}\left(\frac{ay}{2}\right)\right]^2+\left[N_{\nu-\frac{1}{2}}\left(\frac{ay}{2}\right)\right]^2\right\}$$

$$[\operatorname{Re}a>0,\ \operatorname{Re}y>0,\ \operatorname{Re}\nu<1].$$ ET II 136(12)

7.185 $$\int_0^\infty x^{\frac{1}{2}} Q_{\nu-\frac{1}{2}}[(a^2+x^2)x^{-1}] J_\nu(xy)\,dx =$$

$$= 2^{-\frac{1}{2}}\pi y^{-1}\exp\left[-\left(a^2-\frac{1}{4}\right)^{\frac{1}{2}} y\right] J_\nu\left(\frac{1}{2}y\right)$$

$$\left[\operatorname{Re}\nu>-\frac{1}{2},\ y>0\right].$$ ET II 46(10)

7.186 $\int_0^\infty x(1+x^2)^{-\nu-1} \mathrm{P}_\nu\left(\frac{1-x^2}{1+x^2}\right) J_0(xy)\,dx =$

$$= y^{2\nu}[2^\nu \Gamma(\nu+1)]^{-2} K_0(y) \qquad [\operatorname{Re}\nu > 0]. \qquad \text{ET II 13(10)}$$

7.187

1. $\int_0^\infty x P_\mu^\nu(\sqrt{1+x^2})\, K_\nu(xy)\,dx = y^{-\frac{3}{2}} S_{\nu+\frac{1}{2},\,\mu+\frac{1}{2}}(y)$

$$[\operatorname{Re}\nu < 1,\ \operatorname{Re}y > 0]. \qquad \text{ET II 137(14)}$$

2. $\int_0^\infty x\left[P_{\lambda-\frac{1}{2}}(\sqrt{1+a^2x^2})\right]^2 J_0(xy)\,dx = 2\pi^{-2}y^{-1}a^{-1}\cos(\lambda\pi)\left[K_\lambda\left(\frac{y}{2a}\right)\right]^2$

$$\left[\operatorname{Re}a > 0,\ |\operatorname{Re}\lambda| < \frac{1}{4},\ y > 0\right]. \qquad \text{ET II 13(11)}$$

3. $\int_0^\infty x(1+x^2)^{-\frac{1}{2}} P_\mu^\nu(\sqrt{1+x^2})\, K_\nu(xy)\,dx = y^{-\frac{1}{2}} S_{\nu-\frac{1}{2},\,\mu+\frac{1}{2}}(y)$

$$[\operatorname{Re}\nu < 1,\ \operatorname{Re}y > 0]. \qquad \text{ET II 137(15)}$$

4. $\int_0^\infty x P_\mu^{-\frac{1}{2}\nu}(\sqrt{1+a^2x^2})\, Q_\mu^{-\frac{1}{2}\nu}(\sqrt{1+a^2x^2})\, J_\nu(xy)\,dx =$

$$= \frac{y^{-1}e^{-\frac{1}{2}\nu\pi i}\Gamma\left(1+\mu+\frac{1}{2}\nu\right)}{a\Gamma\left(1+\mu-\frac{1}{2}\nu\right)} I_{\mu+\frac{1}{2}}\left(\frac{y}{2a}\right) K_{\mu+\frac{1}{2}}\left(\frac{y}{2a}\right)$$

$$\left[\operatorname{Re}a > 0,\ y > 0,\ \operatorname{Re}\mu > -\frac{3}{4},\ \operatorname{Re}\nu > -1\right]. \qquad \text{ET II 47(16)}$$

5. $\int_0^\infty x P_{\sigma-\frac{1}{2}}^\mu(\sqrt{1+a^2x^2})\, Q_{\sigma-\frac{1}{2}}^\mu(\sqrt{1+a^2x^2})\, J_0(xy)\,dx =$

$$= y^{-2}e^{\mu\pi i}\frac{\Gamma\left(\frac{1}{2}+\sigma-\mu\right)}{\Gamma(1+2\sigma)} W_{\mu,\sigma}\left(\frac{y}{a}\right) M_{-\mu,\sigma}\left(\frac{y}{a}\right)$$

$$\left[\operatorname{Re}a > 0,\ y > 0,\ \operatorname{Re}\sigma > -\frac{1}{4},\ \operatorname{Re}\mu < 1\right]. \qquad \text{ET II 14(15)}$$

6. $\int_0^\infty x P_{\sigma-\frac{1}{2}}^\mu(\sqrt{1+a^2x^2})\, P_{\sigma-\frac{1}{2}}^{-\mu}(\sqrt{1+a^2x^2})\, J_0(xy)\,dx =$

$$= 2\pi^{-1}y^{-2}\cos(\sigma\pi)\, W_{\mu,\sigma}\left(\frac{y}{a}\right) W_{-\mu,\sigma}\left(\frac{y}{a}\right)$$

$$\left[\operatorname{Re}a > 0,\ y > 0,\ |\operatorname{Re}\sigma| < \frac{1}{4}\right]. \qquad \text{ET II 14(14)}$$

7. $\int_0^\infty x\left\{P_{\sigma-\frac{1}{2}}^\mu(\sqrt{1+a^2x^2})\right\}^2 J_0(xy)\,dx =$

$$= -i\pi^{-1}y^{-2}W_{\mu,\sigma}\left(\frac{y}{a}\right)\left[W_{\mu,\sigma}\left(e^{\pi i}\frac{y}{a}\right) - W_{\mu,\sigma}\left(e^{-\pi i}\frac{y}{a}\right)\right]$$

$$\left[\operatorname{Re}a > 0,\ y > 0,\ |\operatorname{Re}\sigma| < \frac{1}{4},\ \operatorname{Re}\mu < 1\right]. \qquad \text{ET II 14(13)}$$

8. $$\int_0^\infty x(1+a^2x^2)^{-\frac{1}{2}} P_\mu^{-\frac{1}{2}-\frac{1}{2}\nu}(\sqrt{1+a^2x^2})\, P_\mu^{\frac{1}{2}-\frac{1}{2}\nu}(\sqrt{1+a^2x^2})\, J_\nu(xy)\,dx =$$
$$= \frac{\left[K_{\mu+\frac{1}{2}}\left(\frac{y}{2a}\right)\right]^2}{\pi a^2 \Gamma\left(\frac{\nu}{2}+\mu+\frac{3}{2}\right)\Gamma\left(\frac{\nu}{2}-\mu+\frac{1}{2}\right)}$$
$$\left[\operatorname{Re} a > 0,\ y > 0,\ -\frac{5}{4} < \operatorname{Re}\mu < \frac{1}{4}\right].$$ ET II 46(9)

9. $$\int_0^\infty x\{P_\mu^{-\frac{1}{2}\nu}(\sqrt{1+a^2x^2})\}^2 J_\nu(xy)\,dx = \frac{2\left[K_{\mu+\frac{1}{2}}\left(\frac{y}{2a}\right)\right]^2 y^{-1}}{\pi a \Gamma\left(1+\mu+\frac{1}{2}\nu\right)\Gamma\left(\frac{1}{2}\nu-\mu\right)}$$
$$\left[\operatorname{Re} a > 0,\ y > 0,\ -\frac{3}{4} < \operatorname{Re}\mu < -\frac{1}{4},\ \operatorname{Re}\nu > -1\right].$$ ET II 45(7)

10. $$\int_0^\infty x(1+a^2x^2)^{-\frac{1}{2}} P_\mu^{-\frac{1}{2}\nu}(\sqrt{1+a^2x^2})\, P_{\mu+1}^{-\frac{1}{2}\nu}(\sqrt{1+a^2x^2})\, J_\nu(xy)\,dx =$$
$$= \frac{K_{\mu+\frac{1}{2}}\left(\frac{y}{2a}\right) K_{\mu+\frac{3}{2}}\left(\frac{y}{2a}\right)}{\pi a^2 \Gamma\left(2+\frac{1}{2}\nu+\mu\right)\Gamma\left(\frac{1}{2}\nu-\mu\right)}$$
$$\left[\operatorname{Re} a > 0,\ y > 0,\ -\frac{7}{4} < \operatorname{Re}\mu < -\frac{1}{4}\right].$$ ET II 45(8)

7.188

1. $$\int_0^\infty x(a^2+x^2)^{-\frac{1}{2}\mu} \mathrm{P}_{\mu-1}^{-\nu}\left[\frac{a}{\sqrt{a^2+x^2}}\right] J_\nu(xy)\,dx = \frac{y^{\mu-2}e^{-ay}}{\Gamma(\mu+\nu)}$$
$$\left[\operatorname{Re} a > 0,\ y > 0,\ \operatorname{Re}\nu > -1,\ \operatorname{Re}\mu > \frac{1}{2}\right].$$ ET II 45(4)

2. $$\int_0^\infty x^{\nu+1}(x^2+a^2)^{\frac{1}{2}\nu} P_\nu\left(\frac{x^2+2a^2}{2a\sqrt{x^2+a^2}}\right) J_\nu(xy)\,dx =$$
$$= \frac{(2a)^{\nu+1}y^{-\nu-1}}{\pi\Gamma(-\nu)}\left[K_{\nu+\frac{1}{2}}\left(\frac{ya}{2}\right)\right]^2$$
$$[\operatorname{Re} a > 0,\ -1 < \operatorname{Re}\nu < 0,\ y > 0].$$ ET II 45(5)

3. $$\int_0^\infty x^{1-\nu}(x^2+a^2)^{-\frac{1}{2}\nu} P_{\nu-1}\left(\frac{x^2+2a^2}{2a\sqrt{x^2+a^2}}\right) J_\nu(xy)\,dx =$$
$$= \frac{(2a)^{1-\nu}y^{\nu-1}}{\Gamma(\nu)} I_{\nu-\frac{1}{2}}\left(\frac{ay}{2}\right) K_{\nu-\frac{1}{2}}\left(\frac{ay}{2}\right)$$
$$[\operatorname{Re} a > 0,\ y > 0,\ 0 < \operatorname{Re}\nu < 1].$$ ET II 45(6)

7.189

1. $$\int_0^\infty (a+x)^{\mu} e^{-x} P_\nu^{-2\mu}\left(1+\frac{2x}{a}\right) I_\mu(x)\,dx = 0$$

$$\left[-\frac{1}{2} < \operatorname{Re}\mu < 0, \ -\frac{1}{2} + \operatorname{Re}\mu < \operatorname{Re}\nu < -\frac{1}{2} - \operatorname{Re}\mu\right].$$ ET II 366(18)

2. $$\int_0^\infty (x+a)^{-\mu} e^{-x} P_\nu^{-2\mu}\left(1+\frac{2x}{a}\right) I_\mu(x)\,dx =$$

$$= \frac{2^{\mu-1}\Gamma\left(\mu+\nu+\frac{1}{2}\right)\Gamma\left(\mu-\nu-\frac{1}{2}\right)e^{a}}{\pi^{\frac{1}{2}}\Gamma(2\mu+\nu+1)\Gamma(2\mu-\nu)} W_{\frac{1}{2}-\mu,\frac{1}{2}+\nu}(2a)$$

$$\left[|\arg a| < \pi, \ \operatorname{Re}\mu > \left|\operatorname{Re}\nu + \frac{1}{2}\right|\right].$$ ET II 367(19)

3. $$\int_0^\infty x^{-\mu} e^{x} P_\nu^{2\mu}\left(1+\frac{2x}{a}\right) K_\mu(x+a)\,dx =$$

$$= \pi^{-\frac{1}{2}} 2^{\mu-1} \cos(\mu\pi)\,\Gamma\left(\mu+\nu+\frac{1}{2}\right)\Gamma\left(\mu-\nu+\frac{1}{2}\right) W_{\frac{1}{2}-\mu,\frac{1}{2}+\nu}(2a)$$

$$\left[|\arg a| < \pi, \ \operatorname{Re}\mu > \left|\operatorname{Re}\nu + \frac{1}{2}\right|\right].$$ ET II 373(11)

4. $$\int_0^\infty x^{-\frac{1}{2}\mu}(x+a)^{-\frac{1}{2}} e^{-x} \mathrm{P}^{\mu}_{\nu-\frac{1}{2}}\left(\frac{a-x}{a+x}\right) K_\nu(a+x)\,dx =$$

$$= \sqrt{\frac{\pi}{2}}\, a^{-\frac{1}{2}\mu}\Gamma(\mu, 2a) \qquad [a > 0, \ \operatorname{Re}\mu < 1].$$ ET II 374(12)

5. $$\int_0^\infty (\operatorname{sh} x)^{\mu+1} (\operatorname{ch} x)^{-2\mu-\frac{3}{2}} P_\nu^{-\mu}[\operatorname{ch}(2x)]\, I_{\mu-\frac{1}{2}}(a \operatorname{sech} x)\,dx =$$

$$= \frac{2^{\mu-\frac{1}{2}}\Gamma(\mu-\nu)\Gamma(\mu+\nu+1)}{\pi^{\frac{1}{2}} a^{\mu+\frac{3}{2}} [\Gamma(\mu+1)]^2} M_{\nu+\frac{1}{2},\mu}(a) M_{-\nu-\frac{1}{2},\mu}(a)$$

$$[\operatorname{Re}\mu > \operatorname{Re}\nu, \ \operatorname{Re}\mu > -\operatorname{Re}\nu - 1].$$ ET II 378(44)

7.19 Combinations of associated Legendre functions and functions generated by Bessel functions

7.191

1. $$\int_a^\infty x^{\frac{1}{2}}(x^2-a^2)^{-\frac{1}{4}-\frac{1}{2}\nu} P_\mu^{\nu+\frac{1}{2}}(2x^2a^{-2}-1)\,[\mathbf{H}_\nu(x) - N_\nu(x)]\,dx =$$

$$= 2^{-\nu-2}\pi^{\frac{1}{2}} a \operatorname{cosec}(\mu\pi)\cos(\nu\pi)\left\{\left[N_\nu\left(\frac{1}{2}a\right)\right]^2 - \left[J_\nu\left(\frac{1}{2}a\right)\right]^2\right\}$$

$$\left[-1 < \operatorname{Re}\mu < 0, \ \operatorname{Re}\nu < \frac{1}{2}\right].$$ ET II 384(6)

2. $$\int_a^\infty x^{\frac{1}{2}}(x^2-a^2)^{-\frac{1}{4}-\frac{1}{2}\nu} P_\mu^{\nu+\frac{1}{2}}(2x^2a^{-2}-1)\left[I_{-\nu}(x)-\mathbf{L}_\nu(x)\right]dx =$$
$$= 2^{-\nu-1}\pi^{\frac{1}{2}} a \operatorname{cosec}(2\mu\pi)\cos(\nu\pi)\left\{\left[I_\nu\left(\frac{1}{2}a\right)\right]^2-\left[I_{-\nu}\left(\frac{1}{2}a\right)\right]^2\right\}$$
$$\left[-1<\operatorname{Re}\mu<0,\ \operatorname{Re}\nu<\frac{1}{2}\right].$$
ET II 385(15)

7.192

1. $$\int_0^1 x^{\frac{1}{2}(\nu-\mu-1)}(1-x^2)^{\frac{1}{4}(\nu-\mu-2)} \mathrm{P}_{\nu-\frac{1}{2}}^{\frac{1}{2}(\mu-\nu+2)}(x) S_{\mu,\nu}(ax)\,dx =$$
$$= 2^{\mu-\frac{3}{2}}\pi^{\frac{1}{2}} a^{-\frac{1}{2}(\nu-\mu-1)}\Gamma\left(\frac{\mu+\nu+3}{4}\right)\Gamma\left(\frac{\mu-3\nu+3}{4}\right)\cos\left(\frac{\mu-\nu}{2}\pi\right)\times$$
$$\times\left[J_\nu\left(\frac{1}{2}a\right)N_{-\frac{1}{2}(\mu-\nu+1)}\left(\frac{1}{2}a\right)-N_\nu\left(\frac{1}{2}a\right)J_{-\frac{1}{2}(\mu-\nu+1)}\left(\frac{1}{2}a\right)\right]$$
$$[\operatorname{Re}(\mu-\nu)<0,\ a>0,\ |\operatorname{Re}(\mu+\nu)|<1,\ \operatorname{Re}(\mu-3\nu)<1].$$
ET II 387(24)a

2. $$\int_1^\infty x^{\frac{1}{2}}(x^2-1)^{-\frac{1}{2}\beta} P_\nu^\beta(x) S_{\mu,\frac{1}{2}}(ax)\,dx =$$
$$= \frac{2^{-\frac{3}{2}+\beta-\mu} a^{\beta-1}\Gamma\left(\frac{\beta-\mu+\nu}{2}+\frac{1}{4}\right)\Gamma\left(\frac{\beta-\mu-\nu}{2}\ \ \frac{1}{4}\right)}{\pi^{\frac{1}{2}}\Gamma\left(\frac{1}{2}-\mu\right)} S_{\mu-\beta+1,\ \nu+\frac{1}{2}}(a)$$
$$\left[\operatorname{Re}\beta<1,\ a>0,\ \operatorname{Re}(\mu+\nu-\beta)<-\frac{1}{2},\ \operatorname{Re}(\mu-\nu-\beta)<\frac{1}{2}\right].$$
ET II 387(25)a

7.193

1. $$\int_1^\infty x^{-\nu}(x^2-1)^{\frac{1}{4}-\frac{1}{2}\nu} \mathrm{P}_{\frac{1}{2}\mu-\frac{1}{2}\nu}^{\nu-\frac{1}{2}}(2x^{-2}-1) S_{\mu,\nu}(ax)\,dx =$$
$$= \frac{2^{\mu-\nu} a^{\nu-2}\pi^{\frac{1}{2}}\Gamma\left(\frac{3\nu-\mu-1}{2}\right)}{\Gamma\left(\frac{1+\nu-\mu}{2}\right)} W_{\varrho,\sigma}\left(ae^{i\frac{\pi}{2}}\right)W_{\varrho,\sigma}\left(ae^{-i\frac{\pi}{2}}\right);$$
$$\varrho=\frac{1}{2}(\mu+1-\nu),\quad \sigma=\nu-\frac{1}{2}$$
$$\left[\operatorname{Re}(\mu-\nu)<0,\ a>0,\ \operatorname{Re}\nu<\frac{3}{2},\ \operatorname{Re}(3\nu-\mu)>1\right].$$
ET II 387(27)a

2. $$\int_1^\infty x(x^2-1)^{-\frac{1}{2}\nu} P_\lambda^\nu(2x^2-1) S_{\mu,\nu}(ax)\,dx =$$
$$= \frac{a^{\nu-1}\Gamma\left(\frac{\nu-\mu+1}{2}+\lambda\right)\Gamma\left(\frac{\nu-\mu-1}{2}-\lambda\right)}{2\Gamma\left(\frac{1-\mu-\nu}{2}\right)\Gamma\left(\frac{1-\mu+\nu}{2}\right)} S_{\mu-\nu+1,\ 2\lambda+1}(a)$$
$$[\operatorname{Re}\nu<1,\ a>0,\ \operatorname{Re}(\mu-\nu+\lambda)<-1,\ \operatorname{Re}(\mu-\nu+\lambda)<0].$$
ET II 387(26)a

7.21 Integration of associated Legendre functions with respect to the order

7.211

1. $$\int_0^\infty \mathrm{P}_{-x-\frac{1}{2}}(\cos\theta)\,dx = \frac{1}{2}\operatorname{cosec}\left(\frac{1}{2}\theta\right) \qquad [0<\theta<\pi].$$ ET II 329(19)

2. $$\int_{-\infty}^\infty \mathrm{P}_x(\cos\theta)\,dx = \operatorname{cosec}\left(\frac{1}{2}\theta\right) \qquad [0<\theta<\pi].$$ ET II 329(20)

7.212 $$\int_0^\infty x^{-1}\operatorname{th}(\pi x)\,P_{-\frac{1}{2}+ix}(\operatorname{ch} a)\,dx = 2e^{-\frac{1}{2}a}\boldsymbol{K}(e^{-a}) \qquad [a>0].$$ ET II 330(22)

7.213 $$\int_0^\infty \frac{x\operatorname{th}(\pi x)}{a^2+x^2}P_{-\frac{1}{2}+ix}(\operatorname{ch} b)\,dx = Q_{a-\frac{1}{2}}(\operatorname{ch} b) \qquad [\operatorname{Re} a>0].$$ ET II 387(23)

7.214 $$\int_0^\infty \operatorname{sh}(\pi x)\cos(ax)\,P_{-\frac{1}{2}+ix}(b)\,dx = \frac{1}{\sqrt{2(b+\operatorname{ch} a)}}$$
$$[a>0,\ |b|<1].$$ ET I 42(27)

7.215 $$\int_0^\infty \cos(bx)\,P^{\mu}_{-\frac{1}{2}+ix}(\operatorname{ch} a)\,dx = 0 \qquad [0<a<b];$$
$$= \frac{\sqrt{\frac{\pi}{2}}(\operatorname{sh} a)^{\mu}}{\Gamma\left(\frac{1}{2}-\mu\right)(\operatorname{ch} a-\operatorname{ch} b)^{\mu+\frac{1}{2}}} \qquad [0<b<a].$$ ET II 330(21)

7.216 $$\int_0^\infty \cos(bx)\,\Gamma(\mu+ix)\,\Gamma(\mu-ix)\,P^{\frac{1}{2}-\mu}_{-\frac{1}{2}+ix}(\operatorname{ch} a)\,dx =$$
$$= \frac{\sqrt{\frac{\pi}{2}}\,\Gamma(\mu)\,(\operatorname{sh} a)^{\mu-\frac{1}{2}}}{(\operatorname{ch} a+\operatorname{ch} b)^{\mu}} \qquad [a>0,\ b>0,\ \operatorname{Re}\mu>0].$$ ET II 330(24)

7.217

1. $$\int_{-\infty}^\infty \left(\nu-\frac{1}{2}+ix\right)\Gamma\left(\frac{1}{2}-ix\right)\Gamma\left(2\nu-\frac{1}{2}+ix\right)\times$$
$$\times P^{\frac{1}{2}-\nu}_{\nu+ix-1}(\cos\theta)\,I_{\nu-\frac{1}{2}+ix}(a)\,K_{\nu-\frac{1}{2}+ix}(b)\,dx =$$
$$= \sqrt{2\pi}(\sin\theta)^{\nu-\frac{1}{2}}\left(\frac{ab}{\omega}\right)^{\nu}K_{\nu}(\omega); \quad \omega = (a^2+b^2+2ab\cos\theta)^{\frac{1}{2}}.$$ ET II 383(29)

2. $$\int_0^\infty xe^{\pi x}\operatorname{th}(\pi x)\,P_{-\frac{1}{2}+ix}(-\cos\theta)\,H^{(2)}_{ix}(ka)\,H^{(2)}_{ix}(kb)\,dx = -\frac{2(ab)^{\frac{1}{2}}}{\pi R}e^{-ikR};$$
$$R = (a^2+b^2-2ab\cos\theta)^{\frac{1}{2}}$$
$$[a>0,\ b>0,\ 0<\theta<\pi,\ \operatorname{Im} k\leqslant 0].$$ ET II 381(17)

3. $$\int_0^\infty x e^{\pi x} \operatorname{sh}(\pi x)\, \Gamma(\nu + ix)\, \Gamma(\nu - ix)\, P^{\frac{1}{2}-\nu}_{-\frac{1}{2}+ix}(-\cos\theta)\, H^{(2)}_{ix}(a) H^{(2)}_{ix}(b)\, dx =$$

$$= i(2\pi)^{\frac{1}{2}} (\sin\theta)^{\nu-\frac{1}{2}} \left(\frac{ab}{R}\right)^{\nu} H^{(2)}_{\nu}(R); \qquad R = (a^2 + b^2 - 2ab\cos\theta)^{\frac{1}{2}}$$

$[a > 0,\ b > 0,\ 0 < \theta < \pi,\ \operatorname{Re}\nu > 0]$. ET II 381(18)

4. $$\int_0^\infty x \operatorname{sh}(\pi x)\, \Gamma(\lambda + ix)\, \Gamma(\lambda - ix)\, K_{ix}(a)\, K_{ix}(b)\, P^{\frac{1}{2}-\lambda}_{-\frac{1}{2}+ix}(\beta)\, dx =$$

$$= \frac{\pi^{\frac{1}{2}}}{\sqrt{2}} \left(\frac{ab}{z}\right)^{\lambda} (\beta^2 - 1)^{\frac{1}{2}\lambda - \frac{1}{4}} K_{\lambda}(z); \qquad z = \sqrt{a^2 + b^2 + 2ab\beta}$$

$\left[|\arg a| < \frac{\pi}{2},\ |\arg(\beta - 1)| < \pi,\ \operatorname{Re}\lambda > 0\right]$. ET II 177(16)

7.22 Combinations of Legendre polynomials, rational functions, and algebraic functions

7.221

1. $$\int_{-1}^{1} P_n(x) P_m(x)\, dx = 0 \qquad [m \neq n]$$

$$= \frac{2}{2n+1} \qquad [m = n].$$ WH, EH I 170(8, 10)

2. $$\int_0^1 P_n(x) P_m(x)\, dx = \frac{1}{2n+1} \qquad [m = n];$$

$$= 0 \qquad [n - m \text{ is even},\ m \neq n];$$

$$= \frac{(-1)^{\frac{1}{2}(m+n-1)}\, m!\, n!}{2^{m+n-1}(n-m)(n+m+1)\left[\left(\frac{n}{2}\right)!\left(\frac{m-1}{2}\right)!\right]^2}$$

$[n$ — even, m — odd$]$. WH

3. $$\int_0^{2\pi} P_{2n}(\cos\varphi)\, d\varphi = 2\pi \left[\binom{2n}{n} 2^{-2n}\right]^2.$$ MO 70, EH II 183(50)

7.222

1. $$\int_{-1}^{1} x^m P_n(x)\, dx = 0 \qquad [m < n].$$

2. $$\int_{-1}^{1} (1+x)^{m+n} P_m(x) P_n(x)\, dx = \frac{2^{m+n+1}[(m+n)!]^4}{(m!\,n!)^2 (2m+2n+1)!}.$$ ET II 277(15)

3. $$\int_{-1}^{1} (1+x)^{m-n-1} P_m(x) P_n(x)\, dx = 0 \qquad [m > n].$$ ET II 278(16)

4. $$\int_{-1}^{1} (1-x^2)^n P_{2m}(x)\,dx = \frac{2n^2}{(n-m)(2m+2n+1)} \int_{-1}^{1} (1-x^2)^{n-1} P_{2m}(x)\,dx$$ $[m < n]$. WH

5. $$\int_{0}^{1} x^2 P_{n+1}(x) P_{n-1}(x)\,dx = \frac{n(n+1)}{(2n-1)(2n+1)(2n+3)}.$$ WH

7.223 $$\int_{-1}^{1} \frac{1}{z-x} \{P_n(x) P_{n-1}(z) - P_{n-1}(x) P_n(z)\}\,dx = -\frac{2}{n}.$$ WH

7.224 [z belongs to the complex plane with a discontinuity along the interval from -1 to $+1$].

1. $$\int_{-1}^{1} (z-x)^{-1} P_n(x)\,dx = 2Q_n(z).$$ ET II 277(7)

2. $$\int_{-1}^{1} x(z-x)^{-1} P_0(x)\,dx = 2Q_1(z).$$ ET II 277(8)

3. $$\int_{-1}^{1} x^{n+1}(z-x)^{-1} P_n(x)\,dx = 2z^{n+1}Q_n(z) - \frac{2^{n+1}(n!)^2}{(2n+1)!}.$$ ET II 277(9)

4. $$\int_{-1}^{1} x^m (z-x)^{-1} P_n(x)\,dx = 2z^m Q_n(z) \qquad [m \leqslant n].$$ ET II 277(10)a

5. $$\int_{-1}^{1} (z-x)^{-1} P_m(x) P_n(x)\,dx = 2P_m(z) Q_n(z) \qquad [m \leqslant n].$$ ET II 278(18)a

6. $$\int_{-1}^{1} (z-x)^{-1} P_n(x) P_{n+1}(x)\,dx = 2P_{n+1}(z) Q_n(z) - \frac{2}{n+1}.$$ ET II 278(19)

7. $$\int_{-1}^{1} x(z-x)^{-1} P_m(x) P_n(x)\,dx = 2zP_m(z) Q_n(z) \qquad [m < n].$$ ET II 278(21)

8. $$\int_{-1}^{1} x(z-x)^{-1} [P_n(x)]^2\,dx = 2zP_n(z) Q_n(z) - \frac{2}{2n+1}.$$ ET II 278(20)

7.225

1. $$\int_{-1}^{x} (x-t)^{-\frac{1}{2}} P_n(t)\,dt = \left(n+\frac{1}{2}\right)^{-1} (1+x)^{-\frac{1}{2}} [T_n(x) + T_{n+1}(x)].$$ EH II 187(43)

2. $$\int_{x}^{1} (t-x)^{-\frac{1}{2}} P_n(t)\,dt = \left(n+\frac{1}{2}\right)^{-1} (1-x)^{-\frac{1}{2}} [T_n(x) - T_{n+1}(x)].$$ EH II 187(44)

3. $$\int_{-1}^{1} (1-x)^{-\frac{1}{2}} P_n(x)\,dx = \frac{2^{\frac{3}{2}}}{2n+1}.$$ EH II 183(49)

4. $$\int_{-1}^{1} (\operatorname{ch} 2p - x)^{-\frac{1}{2}} P_n(x)\,dx = \frac{2\sqrt{2}}{2n+1} \exp[-(2n+1)p] \quad [p > 0].$$ WH

7.226

1. $$\int_{-1}^{1} (1-x^2)^{-\frac{1}{2}} P_{2m}(x)\,dx = \left[\frac{\Gamma\left(\frac{1}{2}+m\right)}{m!}\right]^2.$$ ET II 276(4)

2. $$\int_{-1}^{1} x(1-x^2)^{-\frac{1}{2}} P_{2m+1}(x)\,dx = \frac{\Gamma\left(\frac{1}{2}+m\right)\Gamma\left(\frac{3}{2}+m\right)}{m!\,(m+1)!}$$ ET II 276(5)

3. $$\int_{-1}^{1} (1+px^2)^{-m-\frac{3}{2}} P_{2m}(x)\,dx = \frac{2}{2m+1}(-p)^m (1+p)^{-m-\frac{1}{2}}$$ $[|p| < 1]$. MO 71

7.227 $$\int_{0}^{1} x(a^2+x^2)^{-\frac{1}{2}} P_n(1-2x^2)\,dx = \frac{\left[a+(a^2+1)^{\frac{1}{2}}\right]^{-2n-1}}{2n+1}$$ $[\operatorname{Re} a > 0]$. ET II 278(23)

7.23 Combinations of Legendre polynomials and powers

7.231

1. $$\int_{0}^{1} x^{\lambda} P_{2m}(x)\,dx = \frac{(-1)^m \Gamma\left(m-\frac{1}{2}\lambda\right)\Gamma\left(\frac{1}{2}+\frac{1}{2}\lambda\right)}{2\Gamma\left(-\frac{1}{2}\lambda\right)\Gamma\left(m+\frac{3}{2}+\frac{1}{2}\lambda\right)}$$ $[\operatorname{Re}\lambda > -1]$. EH II 183(51)

2. $$\int_{0}^{1} x^{\lambda} P_{2n+1}(x)\,dx = \frac{(-1)^m \Gamma\left(m+\frac{1}{2}-\frac{1}{2}\lambda\right)\Gamma\left(1+\frac{1}{2}\lambda\right)}{2\Gamma\left(\frac{1}{2}-\frac{1}{2}\lambda\right)\Gamma\left(m+2+\frac{1}{2}\lambda\right)}$$ $[\operatorname{Re}\lambda > -2]$. EH II 183(52)

7.232

1. $$\int_{-1}^{1} (1-x)^{a-1} P_m(x) P_n(x)\,dx =$$
$$= \frac{2^a \Gamma(a)\Gamma(n-a+1)}{\Gamma(1-a)\Gamma(n+a+1)}\, {}_4F_3(-m, m+1, a, a; 1, a+n+1, a-n; 1)$$ $[\operatorname{Re} a > 0]$. ET II 278(17)

2. $\int_{-1}^{1} (1-x)^{a-1} (1+x)^{b-1} P_n(x)\, dx =$

$$= \frac{2^{a+b-1}\Gamma(a)\Gamma(b)}{\Gamma(a+b)} {}_3F_2(-n, 1+n, a; 1, a+b; 1)$$

$[\operatorname{Re} a > 0, \quad \operatorname{Re} b > 0].$ ET II 276(6)

3. $\int_0^1 (1-x)^{\mu-1} P_n(1-\gamma x)\, dx = \frac{\Gamma(\mu)\, n!}{\Gamma(\mu+n+1)} P_n^{(\mu, -\mu)}(1-\gamma)$

$[\operatorname{Re} \mu > 0].$ ET II 190(37)a

4. $\int_0^1 (1-x)^{\mu-1} x^{\nu-1} P_n(1-\gamma x)\, dx =$

$$= \frac{\Gamma(\mu)\Gamma(\nu)}{\Gamma(\mu+\nu)} {}_3F_2\left(-n, n+1, \nu; 1, \mu+\nu; \frac{1}{2}\gamma\right)$$

$[\operatorname{Re} \mu > 0, \ \operatorname{Re} \nu > 0].$ ET II 190(38)

7.233 $\int_0^1 x^{2\mu-1} P_n(1-2x^2)\, dx = \frac{(-1)^n [\Gamma(\mu)]^2}{2\Gamma(\mu+n)\Gamma(\mu-n)}$

$[\operatorname{Re} \mu > 0].$ ET II 278(22)

7.24 Combinations of Legendre polynomials and other elementary functions

7.241 $\int_0^\infty P_n(1-x) e^{-ax}\, dx = e^{-a} a^n \left(\frac{1}{a}\frac{d}{da}\right)^n \left(\frac{e^a}{a}\right);$

$$= a^n \left(1 + \frac{1}{2}\frac{d}{da}\right)^n \left(\frac{1}{a^{n+1}}\right)$$

$[\operatorname{Re} a > 0].$ ET I 171(2)

7.242 $\int_0^\infty P_n(e^{-x}) e^{-ax}\, dx = \frac{(a-1)(a-2)\dots(a-n+1)}{(a+n)(a+n-2)\dots(a-n+2)}$

$[n \geqslant 2, \ \operatorname{Re} a > 0].$ ET I 171(3)

7.243

1. $\int_0^\infty P_{2n}(\operatorname{ch} x) e^{-ax}\, dx = \frac{(a^2-1^2)(a^2-3^2)\dots[a^2-(2n-1)^2]}{a(a^2-2^2)(a^2-4^2)\dots[a^2-(2n)^2]}$

$[\operatorname{Re} a > 2n].$ ET I 171(6)

2. $\int_0^\infty P_{2n+1}(\operatorname{ch} x) e^{-ax}\, dx = \frac{a(a^2-2^2)(a^2-4^2)\dots[a^2-(2n)^2]}{(a^2-1)(a^2-3^2)\dots[a^2-(2n+1)^2]}$

$[\operatorname{Re} a > 2n+1].$ ET I 171(7)

3. $\int_0^\infty P_{2n}(\cos x) e^{-ax}\, dx = \frac{(a^2+1^2)(a^2+3^2)\dots[a^2+(2n-1)^2]}{a(a^2+2^2)(a^2+4^2)\dots[a^2+(2n)^2]}$

$[\operatorname{Re} a > 0].$ ET I 171(4)

4. $$\int_0^\infty P_{2n+1}(\cos x)\, e^{-ax}\, dx = \frac{a(a^2+2^2)(a^2+4^2)\ldots[a^2+(2n)^2]}{(a^2+1^2)(a^2+3^2)\ldots[a^2+(2n+1)^2]}$$

$[\operatorname{Re} a > 0]$. ET I 171(5)

7.244

1. $$\int_0^1 P_n(1-2x^2)\sin ax\, dx = \frac{\pi}{2}\left[J_{n+\frac{1}{2}}\left(\frac{a}{2}\right)\right]^2 \qquad [a > 0].$$ ET I 94(2)

2. $$\int_0^1 P_n(1-2x^2)\cos ax\, dx = \frac{\pi}{2}(-1)^n J_{n+\frac{1}{2}}\left(\frac{a}{2}\right) J_{-n-\frac{1}{2}}\left(\frac{a}{2}\right)$$

$[a > 0]$. ET I 38(1)

7.245

1. $$\int_0^{2\pi} P_{2m+1}(\cos\theta)\cos\theta\, d\theta = \frac{\pi}{2^{4m+1}}\binom{2m}{m}\binom{2m+2}{m+1}.$$

MO 70, EH II 183(50)

2. $$\int_0^\pi P_m(\cos\theta)\sin n\theta\, d\theta =$$

$$= \frac{2(n-m+1)(n-m+3)\ldots(n+m-1)}{(n-m)(n-m+2)\ldots(n+m)}$$

$[n > m, \quad n+m$ is odd$]$;

$= 0 \quad [n \leqslant m$ or $n+m$ is even$]$. MO 71

7.246 $$\int_0^\pi P_n(1-2\sin^2 x\sin^2\theta)\sin x\, dx = \frac{2\sin(2n+1)\theta}{(2n+1)\sin\theta}.$$ MO 71

7.247 $$\int_0^1 P_{2n+1}(x)\sin ax\frac{dx}{\sqrt{x}} = (-1)^{n+1}\sqrt{\frac{\pi}{2a}}\, J_{2n+\frac{3}{2}}(a)$$

$[a > 0]$. ET I 94(1)

7.248

1. $$\int_{-1}^1 (a^2+b^2-2abx)^{-\frac{1}{2}}\sin[\lambda(a^2+b^2-2abx)^{\frac{1}{2}}]P_n(x)\, dx =$$

$$= \pi(ab)^{-\frac{1}{2}} J_{n+\frac{1}{2}}(a\lambda)\, J_{n+\frac{1}{2}}(b\lambda)$$

$[a > 0, \quad b > 0]$. ET II 277(11)

2. $$\int_{-1}^1 (a^2+b^2-2abx)^{-\frac{1}{2}}\cos[\lambda(a^2+b^2-2abx)^{\frac{1}{2}}]P_n(x)\, dx =$$

$$= -\pi(ab)^{-\frac{1}{2}} J_{n+\frac{1}{2}}(a\lambda)\, N_{n+\frac{1}{2}}(b\lambda) \qquad [0 \leqslant a \leqslant b].$$ ET II 277(12)

7.249

1. $$\int_{-1}^{1} P_n(x) \arcsin x \, dx = 0 \qquad [n \text{ — even}];$$

$$= \pi \left\{ \frac{(n-2)!!}{2^{\frac{1}{2}(n+1)} \left(\frac{n+1}{2} \right)!} \right\}^2 \qquad [n \text{ — odd}].$$ WH

2. $$P_n(x) = \frac{1}{t} \sum_{r=0}^{t-1} \left(x + \sqrt{x^2-1} \cos \frac{2\pi r}{t} \right)^n \qquad [t > n].^*$$

7.25 Combinations of Legendre polynomials and Bessel functions

7.251

1. $$\int_0^1 x P_n(1-2x^2) N_\nu(xy) \, dx = \pi^{-1} y^{-1} [S_{2n+1}(y) + \pi N_{2n+1}(y)]$$

$$[n = 0, 1, \ldots; y > 0, \nu > 0].$$ ET II 108(1)

2. $$\int_0^1 x P_n(1-2x^2) K_0(xy) \, dx = y^{-1} \left[(-1)^{n+1} K_{2n+1}(y) + \frac{i}{2} S_{2n+1}(iy) \right]$$

$$[y > 0].$$ ET II 134(1)

3. $$\int_0^1 x P_n(1-2x^2) J_0(xy) \, dx = y^{-1} J_{2n+1}(y) \qquad [y > 0].$$ ET II 13(1)

4. $$\int_0^1 x P_n(1-2x^2) [J_0(ax)]^2 \, dx = \frac{1}{2(2n+1)} \{[J_n(a)]^2 + [J_{n+1}(a)]^2\}.$$ ET II 338(39)a

5. $$\int_0^1 x P_n(1-2x^2) J_0(ax) N_0(ax) \, dx =$$

$$= \frac{1}{2(2n+1)} [J_n(a) N_n(a) + J_{n+1}(a) N_{n+1}(a)].$$ ET II 339(48)a

6. $$\int_0^1 x^2 P_n(1-2x^2) J_1(xy) \, dx = y^{-1} (2n+1)^{-1} [(n+1) J_{2n+2}(y) -$$

$$- n J_{2n}(y)] \qquad [y > 0].$$ ET II 20(23)

7. $$\int_0^1 x^{\mu-1} P_n(2x^2-1) J_\nu(ax) \, dx =$$

$$= \frac{2^{-\nu-1} a^\nu \left[\Gamma\left(\frac{1}{2}\mu + \frac{1}{2}\nu \right) \right]^2}{\Gamma(\nu+1) \Gamma\left(\frac{1}{2}\mu + \frac{1}{2}\nu + n + 1 \right) \Gamma\left(\frac{1}{2} + \frac{1}{2}\nu - n \right)} \times$$

$$\times {}_2F_3 \left(\frac{\mu+\nu}{2}, \frac{\mu+\nu}{2}; \nu+1, \frac{\mu+\nu}{2} + n + 1, \frac{\mu+\nu}{2} - n; -\frac{a^2}{4} \right)$$

$$[a > 0, \operatorname{Re}(\mu+\nu) > 0].$$ ET II 337(32)a

*I. J. Good. Proc. Camb. Philos. Soc. *51* (1955), 385-388.

7.252 $$\int_0^1 e^{-ax} P_n(1-2x) I_0(ax)\, dx = \frac{e^{-a}}{2n+1}[I_n(a) + I_{n+1}(a)]$$

$[a > 0]$. ET II 366(11)a

7.253 $$\int_0^{\frac{\pi}{2}} \sin(2x) P_n(\cos 2x) J_0(a \sin x)\, dx = a^{-1} J_{2n+1}(a).$$ ET II 361(20)

7.254 $$\int_0^1 x P_n(1-2x^2)[I_0(ax) - \mathbf{L}_0(ax)]\, dx = (-1)^n [I_{2n+1}(a) - \mathbf{L}_{2n+1}(a)]$$

$[a > 0]$. ET II 385(14)a

7.3-7.4 Orthogonal Polynomials

7.31 Combinations of Gegenbauer polynomials $C_n^\nu(x)$ and powers

7.311

1. $$\int_{-1}^1 (1-x^2)^{\nu-\frac{1}{2}} C_n^\nu(x)\, dx = 0 \qquad \left[n > 0,\ \operatorname{Re}\nu > -\frac{1}{2}\right].$$ ET II 280(1)

2. $$\int_0^1 x^{n+2\varrho}(1-x^2)^{\nu-\frac{1}{2}} C_n^\nu(x)\, dx =$$
$$= \frac{\Gamma(2\nu+n)\Gamma(2\varrho+n+1)\Gamma\left(\nu+\frac{1}{2}\right)\Gamma\left(\varrho+\frac{1}{2}\right)}{2^{n+1}\Gamma(2\nu)\Gamma(2\varrho+1)\, n!\, \Gamma(n+\nu+\varrho+1)}$$
$\left[\operatorname{Re}\varrho > -\frac{1}{2},\ \operatorname{Re}\nu > -\frac{1}{2}\right]$. ET II 280(2)

3. $$\int_{-1}^1 (1-x)^{\nu-\frac{1}{2}}(1+x)^\beta C_n^\nu(x)\, dx =$$
$$= \frac{2^{\beta+\nu+\frac{1}{2}}\Gamma(\beta+1)\Gamma\left(\nu+\frac{1}{2}\right)\Gamma(2\nu+n)\Gamma\left(\beta-\nu+\frac{3}{2}\right)}{n!\,\Gamma(2\nu)\Gamma\left(\beta-\nu-n+\frac{3}{2}\right)\Gamma\left(\beta+\nu+n+\frac{3}{2}\right)}$$
$\left[\operatorname{Re}\beta > -1, [\operatorname{Re}\nu > -\frac{1}{2}\right]$. ET II 280(3)

4. $$\int_{-1}^1 (1-x)^\alpha (1+x)^\beta C_n^\nu(x)\, dx =$$
$$= \frac{2^{\alpha+\beta+1}\Gamma(\alpha+1)\Gamma(\beta+1)\Gamma(n+2\nu)}{n!\,\Gamma(2\nu)\Gamma(\alpha+\beta+2)} \times$$
$$\times\ {}_3F_2\left(-n,\ n+2\nu,\ \alpha+1;\ \nu+\frac{1}{2},\ \alpha+\beta+2;\ 1\right)$$
$[\operatorname{Re}\alpha > -1,\ \operatorname{Re}\beta > -1]$. ET II 281(4)

7.312 In the following integrals, z belongs to the complex plane with a cut along the interval of the real axis from -1 to 1.

1. $$\int_{-1}^{1} x^m (z-x)^{-1} (1-x^2)^{\nu-\frac{1}{2}} C_n^\nu(x)\,dx =$$
$$= \frac{\pi^{\frac{1}{2}} 2^{\frac{3}{2}-\nu}}{\Gamma(\nu)} e^{-\left(\nu-\frac{1}{2}\right)\pi i} z^m (z^2-1)^{\frac{1}{2}\nu-\frac{1}{4}} Q_{n+\nu-\frac{1}{2}}^{\nu-\frac{1}{2}}(z)$$
$$\left[m \leqslant n,\ \operatorname{Re}\nu > -\frac{1}{2}\right].$$ ET II 281(5)

2. $$\int_{-1}^{1} x^{n+1} (z-x)^{-1} (1-x^2)^{\nu-\frac{1}{2}} C_n^\nu(x)\,dx =$$
$$= \frac{\pi^{\frac{1}{2}} 2^{\frac{3}{2}-\nu}}{\Gamma(\nu)} e^{-\left(\nu-\frac{1}{2}\right)\pi i} z^{n+1} (z^2-1)^{\frac{1}{2}\nu-\frac{1}{4}} Q_{n+\nu-\frac{1}{2}}^{\nu-\frac{1}{2}}(z) -$$
$$- \frac{\pi\, 2^{1-2\nu-n} n!}{\Gamma(\nu)\,\Gamma(\nu+n+1)}$$
$$\left[\operatorname{Re}\nu > -\frac{1}{2}\right].$$ ET II 281(6)

3. $$\int_{-1}^{1} (z-x)^{-1} (1-x^2)^{\nu-\frac{1}{2}} C_m^\nu(x)\, C_n^\nu(x)\,dx =$$
$$= \frac{\pi^{\frac{1}{2}} 2^{\frac{1}{2}-\nu}}{\Gamma(\nu)} e^{-\left(\nu-\frac{1}{2}\right)\pi i} (z^2-1)^{\frac{1}{2}\nu-\frac{1}{4}} C_m^\nu(z)\, Q_{n+\nu-\frac{1}{2}}^{\nu-\frac{1}{2}}(z)$$
$$\left[m \leqslant n,\ \operatorname{Re}\nu > -\frac{1}{2}\right].$$ ET II 283(17)

7.313

1. $$\int_{-1}^{1} (1-x^2)^{\nu-\frac{1}{2}} C_m^\nu(x)\, C_n^\nu(x)\,dx = 0$$
$$\left[m \neq n,\ \operatorname{Re}\nu > -\frac{1}{2}\right].$$ ET II 282(12), MO 98a, EH I 177(16)

2. $$\int_{-1}^{1} (1-x^2)^{\nu-\frac{1}{2}} [C_n^\nu(x)]^2\,dx = \frac{\pi 2^{1-2\nu}\Gamma(2\nu+n)}{n!\,(n+\nu)\,[\Gamma(\nu)]^2}$$
$$\left[\operatorname{Re}\nu > -\frac{1}{2}\right].$$ ET II 281(8), MO 98a, EH I 177(17)

7.314

1. $$\int_{-1}^{1} (1-x)^{\nu-\frac{3}{2}} (1+x)^{\nu-\frac{1}{2}} [C_n^\nu(x)]^2\,dx = \frac{\pi^{\frac{1}{2}}\Gamma\left(\nu-\frac{1}{2}\right)\Gamma(2\nu+n)}{n!\,\Gamma(\nu)\,\Gamma(2\nu)}$$
$$\left[\operatorname{Re}\nu > \frac{1}{2}\right].$$ ET II 281(9)

2. $$\int_{-1}^{1} (1-x)^{\nu-\frac{1}{2}} (1+x)^{2\nu-1} [C_n^{\nu}(x)]^2 dx = \frac{2^{3\nu-\frac{1}{2}} [\Gamma(2\nu+n)]^2 \Gamma\left(2n+\nu+\frac{1}{2}\right)}{(n!)^2 \Gamma(2\nu) \Gamma\left(3\nu+2n+\frac{1}{2}\right)}$$

$[\operatorname{Re} \nu > 0]$. ET II 282(10)

3. $$\int_{-1}^{1} (1-x)^{3\nu+2n-\frac{3}{2}} (1+x)^{\nu-\frac{1}{2}} [C_n^{\nu}(x)]^2 dx =$$

$$= \frac{\pi^{\frac{1}{2}} \left[\Gamma\left(\nu+\frac{1}{2}\right)\right]^2 \Gamma\left(\nu+2n+\frac{1}{2}\right) \Gamma(2\nu+2n) \Gamma\left(3\nu+2n-\frac{1}{2}\right)}{2^{2\nu+2n} \left[n!\, \Gamma\left(\nu+n+\frac{1}{2}\right) \Gamma(2\nu)\right]^2 \Gamma\left(2\nu+2n+\frac{1}{2}\right)}$$

$\left[\operatorname{Re} \nu > \frac{1}{6}\right]$. ET II 282(11)

4. $$\int_{-1}^{1} (1-x)^{\nu-\frac{1}{2}} (1+x)^{\nu+m-n-\frac{3}{2}} C_m^{\nu}(x) C_n^{\nu}(x) dx =$$

$$= (-1)^m \frac{2^{2-2\nu-m+n} \pi^{\frac{3}{2}} \Gamma(2\nu+n)}{m!\,(n-m)!\,[\Gamma(\nu)]^2 \Gamma\left(\frac{1}{2}+\nu+m\right)} \times$$

$$\times \frac{\Gamma\left(\nu-\frac{1}{2}+m-n\right) \Gamma\left(\frac{1}{2}-\nu+m-n\right)}{\Gamma\left(\frac{1}{2}-\nu-n\right) \Gamma\left(\frac{1}{2}+m-n\right)}$$

$\left[\operatorname{Re} \nu > -\frac{1}{2};\ n \geqslant m\right]$. ET II 282(13)a

5. $$\int_{-1}^{1} (1-x)^{2\nu-1} (1+x)^{\nu-\frac{1}{2}} C_m^{\nu}(x) C_n^{\nu}(x) dx =$$

$$= \frac{2^{3\nu-\frac{1}{2}} \Gamma\left(\nu+\frac{1}{2}\right) \Gamma(2\nu+m) \Gamma(2\nu+n)}{m!\,n!\, \Gamma(2\nu) \Gamma\left(\frac{1}{2}-\nu\right)} \times$$

$$\times \frac{\Gamma\left(\nu+\frac{1}{2}+m+n\right) \Gamma\left(\frac{1}{2}-\nu+n-m\right)}{\Gamma\left(\nu+\frac{1}{2}+n-m\right) \Gamma\left(3\nu+\frac{1}{2}+m+n\right)}$$

$[\operatorname{Re} \nu > 0]$. ET II 282(14)

6. $$\int_{-1}^{1} (1-x)^{\nu-\frac{1}{2}} (1+x)^{3\nu+m+n-\frac{3}{2}} C_m^{\nu}(x) C_n^{\nu}(x) dx =$$

$$= \frac{2^{4\nu+m+n-1} \left[\Gamma\left(\nu+\frac{1}{2}\right) \Gamma(2\nu+m+n)\right]^2}{\Gamma\left(\nu+m+\frac{1}{2}\right) \Gamma\left(\nu+n+\frac{1}{2}\right) \Gamma(2\nu+m)} \times$$

$$\times \frac{\Gamma\left(\nu+m+n+\frac{1}{2}\right) \Gamma\left(3\nu+m+n-\frac{1}{2}\right)}{\Gamma(2\nu+n) \Gamma(4\nu+2m+2n)}$$

$\left[\operatorname{Re} \nu > \frac{1}{6}\right]$. ET II 282(15)

7. $$\int_{-1}^{1} (1-x)^{\alpha}(1+x)^{\nu-\frac{1}{2}} C_m^{\mu}(x) C_n^{\nu}(x)\, dx =$$

$$= \frac{2^{\alpha+\nu+\frac{1}{2}} \Gamma(\alpha+1)\Gamma\left(\nu+\frac{1}{2}\right)\Gamma\left(\nu-\alpha+n-\frac{1}{2}\right)}{m!n!\,\Gamma\left(\nu-\alpha-\frac{1}{2}\right)\Gamma\left(\nu-\alpha+n+\frac{3}{2}\right)} \frac{\Gamma(2\mu+m)\Gamma(2\nu+n)}{\Gamma(2\mu)\Gamma(2\nu)} \times$$

$$\times {}_4F_3\left(-m,\ m+2\mu,\ \alpha+1,\ \alpha-\nu+\frac{3}{2};\ \mu+\frac{1}{2},\ \nu+\alpha+n+\frac{3}{2},\ \alpha-\nu-n+\frac{3}{2};\ 1\right)$$

$$\left[\operatorname{Re}\alpha > -1,\ \operatorname{Re}\nu > -\frac{1}{2}\right].$$ ET II 283(16)

7.315 $$\int_{-1}^{1} (1-x^2)^{\frac{1}{2}\nu-1} C_{2n}^{\nu}(ax)\, dx = \frac{\pi^{\frac{1}{2}}\Gamma\left(\frac{1}{2}\nu\right)}{\Gamma\left(\frac{1}{2}\nu+\frac{1}{2}\right)} C_n^{\frac{1}{2}\nu}(2a^2-1)$$

$$[\operatorname{Re}\nu > 0].$$ ET II 283(19)

7.316 $$\int_{-1}^{1} (1-x^2)^{\nu-1} C_n^{\nu}(\cos\alpha\cos\beta + x\sin\alpha\sin\beta)\, dx =$$

$$= \frac{2^{2\nu-1} n!\,[\Gamma(\nu)]^2}{\Gamma(2\nu+n)} C_n^{\nu}(\cos\alpha)\, C_n^{\nu}(\cos\beta)$$

$$[\operatorname{Re}\nu > 0].$$ ET II 283(20)

7.317

1. $$\int_0^1 (1-x)^{\mu-1} x^{\lambda-\frac{1}{2}} C_n^{\lambda}(1-\gamma x)\, dx = \frac{\Gamma(2\lambda+n)\Gamma\left(\lambda+\frac{1}{2}\right)\Gamma(\mu)}{\Gamma(2\lambda)\Gamma\left(\lambda+\mu+n+\frac{1}{2}\right)} P_n^{(\alpha,\beta)}(1-\gamma),$$

$$\alpha = \lambda+\mu-\frac{1}{2},\quad \beta = \lambda-\mu-\frac{1}{2}$$

$$\left[\operatorname{Re}\lambda > -1,\ \lambda \neq 0,\ -\frac{1}{2},\ \operatorname{Re}\mu > 0\right].$$ ET II 190(39)a

2. $$\int_0^1 (1-x)^{\mu-1} x^{\nu-1} C_n^{\lambda}(1-\gamma x)\, dx = \frac{\Gamma(2\lambda+n)\Gamma(\mu)\Gamma(\nu)}{n!\,\Gamma(2\lambda)\Gamma(\mu+\nu)} \times$$

$$\times {}_3F_2\left(-n,\ n+2\lambda,\ \nu;\ \lambda+\frac{1}{2},\ \mu+\nu;\ \frac{\gamma}{2}\right)$$

$$[2\lambda \neq 0,\ -1,\ -2,\ \ldots,\ \operatorname{Re}\mu > 0,\ \operatorname{Re}\nu > 0].$$ ET II 191(40)a

7.318 $$\int_0^1 x^{2\nu}(1-x^2)^{\sigma-1} C_n^{\nu}(1-x^2 y)\, dx =$$

$$= \frac{\Gamma(2\nu+n)\Gamma\left(\nu+\frac{1}{2}\right)\Gamma(\sigma)}{2\Gamma(2\nu)\Gamma\left(n+\nu+\sigma+\frac{1}{2}\right)} P_n^{(\alpha,\beta)}(1-y),$$

$$\alpha = \nu+\sigma-\frac{1}{2},\quad \beta = \nu-\sigma-\frac{1}{2}$$

$$\left[\operatorname{Re}\nu > -\frac{1}{2},\ \operatorname{Re}\sigma > 0\right].$$ ET II 283(21)

7.319

1. $$\int_0^1 (1-x)^{\mu-1} x^{\nu-1} C_{2n}^{\lambda}\left(\gamma x^{\frac{1}{2}}\right) dx = (-1)^n \frac{\Gamma(\lambda+n)\Gamma(\mu)\Gamma(\nu)}{n!\,\Gamma(\lambda)\Gamma(\mu+\nu)} \times$$
$$\times\, {}_3F_2\left(-n,\ n+\lambda,\ \nu;\ \frac{1}{2},\ \mu+\nu;\ \gamma^2\right)$$
$$[\operatorname{Re}\mu > 0,\ \operatorname{Re}\nu > 0].$$ ET II 191(41)a

2. $$\int_0^1 (1-x)^{\mu-1} x^{\nu-1} C_{2n+1}^{\lambda}\left(\gamma x^{\frac{1}{2}}\right) dx =$$
$$= \frac{(-1)^n 2\gamma\Gamma(\mu)\Gamma(\lambda+n+1)\Gamma\left(\nu+\frac{1}{2}\right)}{n!\,\Gamma(\lambda)\Gamma\left(\mu+\nu+\frac{1}{2}\right)} \times$$
$$\times\, {}_3F_2\left(-n,\ n+\lambda+1,\ \nu+\frac{1}{2};\ \frac{3}{2},\ \mu+\nu+\frac{1}{2};\ \gamma^2\right)$$
$$\left[\operatorname{Re}\mu > 0,\ \operatorname{Re}\nu > -\frac{1}{2}\right].$$ ET II 191(42)

7.32 Combinations of the polynomials $C_n^{\nu}(x)$ and some elementary functions

7.321 $$\int_{-1}^1 (1-x^2)^{\nu-\frac{1}{2}} e^{iax} C_n^{\nu}(x)\, dx =$$
$$= \frac{\pi 2^{1-\nu} i^n \Gamma(2\nu+n)}{n!\,\Gamma(\nu)} a^{-\nu} J_{\nu+n}(a)$$
$$\left[\operatorname{Re}\nu > -\frac{1}{2}\right].$$ ET II 281(7), MO 99a

7.322 $$\int_0^{2a} [x(2a-x)]^{\nu-\frac{1}{2}} C_n^{\nu}\left(\frac{x}{a}-1\right) e^{-bx}\, dx =$$
$$= (-1)^n \frac{\pi\Gamma(2\nu+n)}{n!\,\Gamma(\nu)} \left(\frac{a}{2b}\right)^{\nu} e^{-ab} I_{\nu+n}(ab)$$
$$\left[\operatorname{Re}\nu > -\frac{1}{2}\right].$$ ET I 171(9)

7.323

1. $$\int_0^{\pi} C_n^{\nu}(\cos\varphi)(\sin\varphi)^{2\nu}\, d\varphi = 0 \qquad [n = 1, 2, 3, \ldots];$$
$$= 2^{-2\nu}\pi\Gamma(2\nu+1)[\Gamma(1+\nu)]^{-2} \qquad [n = 0].$$ EH I 177(18)

2. $$\int_0^{\pi} C_n^{\nu}(\cos\psi\cos\psi' + \sin\psi\sin\psi'\cos\varphi)(\sin\varphi)^{2\nu-1}\, d\varphi =$$
$$= 2^{2\nu-1} n!\,[\Gamma(\nu)]^2 C_n^{\nu}(\cos\psi) C_n^{\nu}(\cos\psi')[\Gamma(2\nu+n)]^{-1}$$
$$[\operatorname{Re}\nu > 0].$$ EH I 177(20)

7.324

1. $$\int_0^1 (1-x^2)^{\nu-\frac{1}{2}} C_{2n+1}^{\nu}(x) \sin ax\, dx = (-1)^n \pi \frac{\Gamma(2n+2\nu+1) J_{2n+\nu+1}(a)}{(2n+1)!\, \Gamma(\nu)(2a)^{\nu}}$$
$$\left[\operatorname{Re}\nu > -\tfrac{1}{2},\ a > 0\right].$$ ET I 94(4)

2. $$\int_0^1 (1-x^2)^{\nu-\frac{1}{2}} C_{2n}^{\nu}(x) \cos ax\, dx = \frac{(-1)^n \pi \Gamma(2n+2\nu) J_{\nu+2n}(a)}{(2n)!\, \Gamma(\nu)(2a)^{\nu}} \qquad \left[\operatorname{Re}\nu > -\tfrac{1}{2},\ a > 0\right].$$ ET I 38(3)a

7.33 Combinations of the polynomials $C_n^{\nu}(x)$ and Bessel functions. Integration of Gegenbauer functions with respect to the index

7.331

1. $$\int_1^{\infty} x^{2n+1-\nu}(x^2-1)^{\nu-2n-\frac{1}{2}} C_{2n}^{\nu-2n}\left(\frac{1}{x}\right) J_{\nu}(xy)\, dx = (-1)^n 2^{2n-\nu+1} y^{-\nu+2n-1} [(2n)!]^{-1} \Gamma(2\nu-2n) [\Gamma(\nu-2n)]^{-1} \cos y$$
$$\left[y > 0,\ 2n-\tfrac{1}{2} < \operatorname{Re}\nu < 2n+\tfrac{1}{2}\right].$$ ET II 44(10)a

2. $$\int_1^{\infty} x^{2n-\nu+2}(x^2-1)^{\nu-2n-\frac{3}{2}} C_{2n+1}^{\nu-2n-1}\left(\frac{1}{x}\right) J_{\nu}(xy)\, dx = (-1)^n 2^{2n-\nu+2} y^{-\nu+2n} \Gamma(2\nu-2n-1) \times [(2n+1)!\, \Gamma(\nu-2n-1)]^{-1} \sin y$$
$$\left[y > 0,\ 2n+\tfrac{1}{2} < \operatorname{Re}\nu < 2n+\tfrac{3}{2}\right].$$ ET II 44(11)a

7.332

1. $$\int_0^{\infty} x^{\nu+1}(x^2+\beta^2)^{-\frac{1}{2}\nu-\frac{3}{4}} C_{2n+1}^{\nu+\frac{1}{2}}\left[(x^2+\beta^2)^{-\frac{1}{2}}\beta\right] \times J_{\nu+\frac{3}{2}+2n}\left[(x^2+\beta^2)^{\frac{1}{2}} a\right] J_{\nu}(xy)\, dx =$$
$$= (-1)^n 2^{\frac{1}{2}} \pi^{-\frac{1}{2}} a^{\frac{1}{2}-\nu} y^{\nu} (a^2-y^2)^{-\frac{1}{2}} \sin\left[\beta(a^2-y^2)^{\frac{1}{2}}\right] \times C_{2n+1}^{\nu+\frac{1}{2}}\left[\left(1-\frac{y^2}{a^2}\right)^{\frac{1}{2}}\right] \qquad [0 < y < a];$$
$$= 0 \qquad [a < y < \infty]$$
$$[a > 0,\ \operatorname{Re}\beta > 0,\ \operatorname{Re}\nu > -1].$$ ET II 59(23)

2. $$\int_0^\infty x^{\nu+1}(x^2+\beta^2)^{-\frac{1}{2}\nu-\frac{3}{4}} C_{2n}^{\nu+\frac{1}{2}}\left[\beta(x^2+\beta^2)^{-\frac{1}{2}}\right]\times$$
$$\times J_{\nu+\frac{1}{2}+2n}\left[(x^2+\beta^2)^{\frac{1}{2}}a\right]J_\nu(xy)\,dx =$$
$$=(-1)^n 2^{\frac{1}{2}}\pi^{-\frac{1}{2}}a^{\frac{1}{2}-\nu}y^\nu(a^2-y^2)^{-\frac{1}{2}}\cos\left[\beta(a^2-y^2)^{\frac{1}{2}}\right]\times$$
$$\times C_{2n}^{\nu+\frac{1}{2}}\left[\left(1-\frac{y^2}{a^2}\right)^{\frac{1}{2}}\right] \qquad [0<y<a];$$
$$=0 \qquad [a<y<\infty]$$
$$[a>0,\ \operatorname{Re}\beta>0,\ \operatorname{Re}\nu>-1].$$ ET II 59(24)

7.333

1. $$\int_0^\pi (\sin x)^{\nu+1}\cos(a\cos\theta\cos x)\,C_n^{\nu+\frac{1}{2}}(\cos x)J_\nu(a\sin\theta\sin x)\,dx =$$
$$=(-1)^{\frac{n}{2}}\left(\frac{2\pi}{a}\right)^{\frac{1}{2}}(\sin\theta)^\nu C_n^{\nu+\frac{1}{2}}(\cos\theta)\,J_{\nu+\frac{1}{2}+n}(a) \qquad [n=0,2,4,\ldots];$$
$$=0 \qquad [n=1,3,5,\ldots]$$
$$[\operatorname{Re}\nu>-1].$$ WA 414(2)a

2. $$\int_0^\pi (\sin x)^{\nu+1}\sin(a\cos\theta\cos x)\,C_n^{\nu+\frac{1}{2}}(\cos x)J_\nu(a\sin\theta\sin x)\,dx =$$
$$=0 \qquad [n=0,2,4,\ldots];$$
$$=(-1)^{\frac{n-1}{2}}\left(\frac{2\pi}{a}\right)^{\frac{1}{2}}(\sin\theta)^\nu C_n^{\nu+\frac{1}{2}}(\cos\theta)\,J_{\nu+\frac{1}{2}+n}(a) \qquad [n=1,3,5,\ldots]$$
$$[\operatorname{Re}\nu>-1].$$ WA 414(3)a

7.334

1. $$\int_0^\pi (\sin x)^{2\nu}C_n^\nu(\cos x)\frac{J_\nu(\omega)}{\omega^\nu}\,dx =$$
$$=\frac{\pi\Gamma(2\nu+n)}{2^{\nu-1}n!\,\Gamma(\nu)}\frac{J_{\nu+n}(\alpha)}{\alpha^\nu}\frac{J_{\nu+n}(\beta)}{\beta^\nu},$$
$$\omega=(\alpha^2+\beta^2-2\alpha\beta\cos x)^{\frac{1}{2}} \qquad \left[n=0,1,2,\ldots;\ \operatorname{Re}\nu>-\frac{1}{2}\right].$$
ET II 362(29)

2. $$\int_0^\pi (\sin x)^{2\nu}C_n^\nu(\cos x)\frac{N_\nu(\omega)}{\omega^\nu}\,dx =$$
$$=\frac{\pi\Gamma(2\nu+n)}{2^{\nu-1}n!\,\Gamma(\nu)}\frac{J_{\nu+n}(\alpha)}{\alpha^\nu}\frac{N_{\nu+n}(\beta)}{\beta^\nu},$$
$$\omega=(\alpha^2+\beta^2-2\alpha\beta\cos x)^{\frac{1}{2}} \qquad \left[|\alpha|<|\beta|,\ \operatorname{Re}\nu>-\frac{1}{2}\right].$$
ET II 362(30)

Integration of Gegenbauer functions with respect to the index

7.335 $$\int_{c-i\infty}^{c+i\infty} [\sin(\alpha\pi)]^{-1} t^{\alpha} C_{\alpha}^{\nu}(z)\, d\alpha = -2i(1+2tz+t^2)^{-\nu}$$

$$[-2 < \operatorname{Re}\nu < c < 0, \ |\arg(z\pm 1)| < \pi].$$ EH I 178(25)

7.336 $$\int_{-\infty}^{\infty} \operatorname{sech}(\pi x)\left(\nu - \tfrac{1}{2} + ix\right) K_{\nu-\frac{1}{2}+ix}(a) I_{\nu-\frac{1}{2}+ix}(b) C^{\nu}_{-\frac{1}{2}+ix}(-\cos\varphi)\, dx =$$

$$= \frac{2^{-\nu+1}(ab)^{\nu}}{\Gamma(\nu)} \omega^{-\nu} K_{\nu}(\omega),$$

$$\omega = \sqrt{a^2+b^2-2ab\cos\varphi}.$$ EH II 55(45)

7.34 Combinations of Chebyshev polynomials and powers

7.341 $$\int_{-1}^{1} [T_n(x)]^2\, dx = 1-(4n^2-1)^{-1}.$$ ET II 271(6)

7.342 $$\int_{-1}^{1} U_n\left[x(1-y^2)^{\frac{1}{2}}(1-z^2)^{\frac{1}{2}} + yz\right] dx =$$

$$= \frac{2}{n+1} U_n(y) U_n(z) \qquad [|y| < 1, \ |z| < 1].$$ ET II 275(34)

7.343

1. $$\int_{-1}^{1} T_n(x) T_m(x) \frac{dx}{\sqrt{1-x^2}} = 0 \qquad [m \neq n];$$

$$= \frac{\pi}{2} \qquad [m = n \neq 0];$$

$$= \pi \qquad [m = n = 0].$$

MO 104

2. $$\int_{-1}^{1} \sqrt{1-x^2}\, U_n(x) U_m(x)\, dx = 0 \qquad [m \neq n \quad \text{or} \quad m = n = 0];$$

ET II 274(28)

$$= \frac{\pi}{2} \qquad [m = n \neq 0].$$

ET II 274(27), MO 105a

7.344

1. $$\int_{-1}^{1} (y-x)^{-1}(1-y^2)^{-\frac{1}{2}} T_n(y)\, dy = \pi U_{n-1}(x)$$

$$[n = 1, 2, \ldots].$$ EH II 187(47)

2. $$\int_{-1}^{1} (y-x)^{-1}(1-y^2)^{\frac{1}{2}} U_{n-1}(y)\, dy = -\pi T_n(x)$$

$$[n = 1, 2, \ldots].$$ EH II 187(48)

7.345

1. $$\int_{-1}^{1} (1-x)^{-\frac{1}{2}} (1+x)^{m-n-\frac{3}{2}} T_m(x) T_n(x)\, dx = 0 \quad [m > n].$$ ET II 272(10)

2. $$\int_{-1}^{1} (1-x)^{-\frac{1}{2}} (1+x)^{m+n-\frac{3}{2}} T_m(x) T_n(x)\, dx = \frac{\pi (2m+2n-2)!}{2^{m+n}(2m-1)!\,(2n-1)!}$$
$[m+n \neq 0]$. ET II 272(11)

3. $$\int_{-1}^{1} (1-x)^{\frac{1}{2}} (1+x)^{m+n+\frac{3}{2}} U_m(x) U_n(x)\, dx = \frac{\pi (2m+2n+2)!}{2^{m+n+2}(2m+1)!\,(2n+1)!}.$$ ET II 274(31)

4. $$\int_{-1}^{1} (1-x)^{\frac{1}{2}} (1+x)^{m-n-\frac{1}{2}} U_m(x) U_n(x)\, dx = 0 \quad [m > n].$$ ET II 274(30)

5. $$\int_{-1}^{1} (1-x)(1+x)^{\frac{1}{2}} U_m(x) U_n(x)\, dx = \frac{2^{\frac{5}{2}}(m+1)(n+1)}{\left(m+n+\frac{3}{2}\right)\left(m+n+\frac{5}{2}\right)[1-4(m-n)^2]}.$$ ET II 274(29)

6. $$\int_{-1}^{1} (1+x)^{-\frac{1}{2}} (1-x)^{\alpha-1} T_m(x) T_n(x)\, dx = \frac{\pi^{\frac{1}{2}} 2^{\alpha-\frac{1}{2}} \Gamma(\alpha) \Gamma\left(n-\alpha+\frac{1}{2}\right)}{\Gamma\left(\frac{1}{2}-\alpha\right)\Gamma\left(\alpha+n+\frac{1}{2}\right)} \times {}_4F_3\left(-m, m, \alpha, \alpha+\frac{1}{2};\ \frac{1}{2}, \alpha+n+\frac{1}{2}, \alpha-n+\frac{1}{2};\ 1\right)$$
$[\operatorname{Re}\alpha > 0]$. ET II 272(12)

7. $$\int_{-1}^{1} (1+x)^{\frac{1}{2}} (1-x)^{\alpha-1} U_m(x) U_n(x)\, dx = \frac{\pi^{\frac{1}{2}} 2^{\alpha-\frac{1}{2}} (m+1)(n+1) \Gamma(\alpha) \Gamma\left(n-\alpha+\frac{3}{2}\right)}{\Gamma\left(\frac{3}{2}-\alpha\right)\Gamma\left(\frac{3}{2}+\alpha+n\right)} \times {}_4F_3\left(-m, m+2, \alpha, \alpha-\frac{1}{2};\ \frac{3}{2}, \alpha+n+\frac{3}{2}, \alpha-n-\frac{1}{2};\ 1\right)$$
$[\operatorname{Re}\alpha > 0]$. ET II 275(32)

7.346 $$\int_{0}^{1} x^{s-1} T_n(x) \frac{dx}{\sqrt{1-x^2}} = \frac{\pi}{s 2^s \mathrm{B}\left(\frac{1}{2}+\frac{1}{2}s+\frac{1}{2}n,\ \frac{1}{2}+\frac{1}{2}s-\frac{1}{2}n\right)}$$
$[\operatorname{Re} s > 0]$. ET II 324(2)

7.347

1. $$\int_{-1}^{1} (1-x)^{\alpha}(1+x)^{\beta} T_n(x)\,dx = \frac{2^{\alpha+\beta+2n+1}(n!)^2\,\Gamma(\alpha+1)\,\Gamma(\beta+1)}{(2n)!\,\Gamma(\alpha+\beta+2)}\, {}_3F_2\left(-n,\ n,\ \alpha+1;\ \frac{1}{2},\ \alpha+\beta+2;\ 1\right)$$

$[\operatorname{Re}\alpha > -1,\ \operatorname{Re}\beta > -1]$. ET II 271(2)

2. $$\int_{-1}^{1} (1-x)^{\alpha}(1+x)^{\beta} U_n(x)\,dx = \frac{2^{\alpha+\beta+2n+2}[(n+1)!]^2\,\Gamma(\alpha+1)\,\Gamma(\beta+1)}{(2n+2)!\,\Gamma(\alpha+\beta+2)}\times {}_3F_2\left(-n,\ n+1,\ \alpha+1;\ \frac{3}{2},\ \alpha+\beta+2;\ 1\right).$$

ET II 273(22)

7.348 $$\int_{-1}^{1} (1-x^2)^{-\frac{1}{2}} U_{2n}(xz)\,dx = \pi P_n(2z^2-1) \qquad [|z|<1].$$ ET II 275(33)

7.349 $$\int_{-1}^{1} (1-x^2)^{-\frac{1}{2}} T_n(1-x^2y)\,dx = \frac{1}{2}\pi\left[P_n(1-y)+P_{n-1}(1-y)\right].$$ ET II 272(14)

7.35 Combinations of Chebyshev polynomials and some elementary functions

7.351 $$\int_0^1 x^{-\frac{1}{2}}(1-x^2)^{-\frac{1}{2}} e^{-\frac{2a}{x}} T_n(x)\,dx = \pi^{\frac{1}{2}} D_{n-\frac{1}{2}}\left(2a^{\frac{1}{2}}\right) D_{-n-\frac{1}{2}}\left(2a^{\frac{1}{2}}\right)$$

$[\operatorname{Re} a > 0]$. ET II 272(13)

7.352

1. $$\int_0^{\infty} \frac{xU_n\left[a(a^2+x^2)^{-\frac{1}{2}}\right]}{(a^2+x^2)^{\frac{1}{2}n+1}(e^{\pi x}+1)}\,dx = \frac{a^{-n}}{2n} - 2^{-n-1}\zeta\left(n+1,\ \frac{a+1}{2}\right)$$

$[\operatorname{Re} a > 0]$. ET II 275(39)

2. $$\int_0^{\infty} \frac{xU_n\left[a(a^2+x^2)^{-\frac{1}{2}}\right]}{(a^2+x^2)^{\frac{1}{2}n+1}(e^{2\pi x}-1)}\,dx = \frac{1}{2}\zeta(n+1,\ a) - \frac{a^{-n-1}}{4} - \frac{a^{-n}}{2n}$$

$[\operatorname{Re} a > 0]$. ET II 276(40)

7.353

1. $$\int_0^{\infty} (a^2+x^2)^{-\frac{1}{2}n} \operatorname{sech}\left(\frac{1}{2}\pi x\right) T_n\left[a(a^2+x^2)^{-\frac{1}{2}}\right] dx = 2^{1-2n}\left[\zeta\left(n,\ \frac{a+1}{4}\right) - \zeta\left(n,\ \frac{a+3}{4}\right)\right] = 2^{1-n}\Phi\left(-1,\ n,\ \frac{a+1}{2}\right)$$

$[\operatorname{Re} a > 0]$. ET II 273(19)

2. $$\int_0^{\infty} (a^2+x^2)^{-\frac{1}{2}n}\left[\operatorname{ch}\left(\frac{1}{2}\pi x\right)\right]^{-2} T_n\left[a(a^2+x^2)^{-\frac{1}{2}}\right] dx = \pi^{-1} n 2^{1-n}\zeta\left(n+1,\ \frac{a+1}{2}\right) \qquad [\operatorname{Re} a > 0].$$ ET II 273(20)

7.354

1. $$\int_{-1}^{1} \sin(xyz) \cos\left[(1-x^2)^{\frac{1}{2}}(1-y^2)^{\frac{1}{2}} z\right] T_{2n+1}(x)\, dx = (-1)^n \pi T_{2n+1}(y) J_{2n+1}(z).$$ ET II 271(4)

2. $$\int_{-1}^{1} \sin(xyz) \sin\left[(1-x^2)^{\frac{1}{2}}(1-y^2)^{\frac{1}{2}} z\right] U_{2n+1}(x)\, dx = (-1)^n \pi (1-y^2)^{\frac{1}{2}} U_{2n+1}(y) J_{2n+2}(z).$$ ET II 274(25)

3. $$\int_{-1}^{1} \cos(xyz) \cos\left[(1-x^2)^{\frac{1}{2}}(1-y^2)^{\frac{1}{2}} z\right] T_{2n}(x)\, dx = (-1)^n \pi T_{2n}(y) J_{2n}(z).$$ ET II 271(5)

4. $$\int_{-1}^{1} \cos(xyz) \sin\left[(1-x^2)^{\frac{1}{2}}(1-y^2)^{\frac{1}{2}} z\right] U_{2n}(x)\, dx = (-1)^n \pi (1-y^2)^{\frac{1}{2}} U_{2n}(y) J_{2n+1}(z).$$ ET II 274(24)

7.355

1. $$\int_0^1 T_{2n+1}(x) \sin ax \frac{dx}{\sqrt{1-x^2}} = (-1)^n \frac{\pi}{2} J_{2n+1}(a) \quad [a > 0].$$ ET I 94(3)a

2. $$\int_0^1 T_{2n}(x) \cos ax \frac{dx}{\sqrt{1-x^2}} = (-1)^n \frac{\pi}{2} J_{2n}(a) \qquad [a > 0].$$ ET I 38(2)a

7.36 Combinations of Chebyshev polynomials and Bessel functions

7.361 $$\int_0^1 (1-x^2)^{-\frac{1}{2}} T_n(x) J_\nu(xy)\, dx = \frac{1}{2}\pi J_{\frac{1}{2}(\nu+n)}\left(\frac{1}{2}y\right) J_{\frac{1}{2}(\nu-n)}\left(\frac{1}{2}y\right)$$ $$[y > 0, \ \operatorname{Re}\nu > -n-1].$$ ET II 42(1)

7.362 $$\int_1^\infty (x^2-1)^{-\frac{1}{2}} T_n\left(\frac{1}{x}\right) K_{2\mu}(ax)\, dx = \frac{\pi}{2a} W_{\frac{1}{2}n, \mu}(a) W_{-\frac{1}{2}n, \mu}(a)$$ $$[\operatorname{Re} a > 0].$$ ET II 366(17)a

7.37-7.38 Hermite polynomials

7.371 $$\int_0^x H_n(y)\, dy = [2(n+1)]^{-1} [H_{n+1}(x) - H_{n+1}(0)].$$ EH II 194(27)

7.372 $$\int_{-1}^{1} (1-t^2)^{\alpha-\frac{1}{2}} H_{2n}(\sqrt{x}\, t)\, dt = \frac{(-1)^n \pi^{\frac{1}{2}} (2n)!\, \Gamma\left(\alpha+\frac{1}{2}\right) L_n^\alpha(x)}{\Gamma(n+\alpha+1)}$$ $$\left[\operatorname{Re}\alpha > -\frac{1}{2}\right].$$ EH II 195(34)

7.373

1. $\int\limits_0^x e^{-y^2} H_n(y)\,dy = H_{n-1}(0) - e^{-x^2} H_{n-1}(x)$. (see 8.956). EH II 194(26)

2. $\int\limits_{-\infty}^{\infty} e^{-x^2} H_{2m}(xy)\,dx = \sqrt{\pi}\,\frac{(2m)!}{m!}\,(y^2-1)^m$. EH II 195(28)

7.374

1. $\int\limits_{-\infty}^{\infty} e^{-x^2} H_n(x) H_m(x)\,dx = 0 \qquad [m \neq n];$ SM III 567

$= 2^n \cdot n!\,\sqrt{\pi} \qquad [m = n].$ SM III 568

2. $\int\limits_{-\infty}^{\infty} e^{-2x^2} H_m(x) H_n(x)\,dx = (-1)^{\frac{1}{2}(m+n)}\, 2^{\frac{m+n-1}{2}}\, \Gamma\left(\frac{m+n+1}{2}\right)$

$[m+n \text{ is even}].$ ET II 289(10)a

3. $\int\limits_{-\infty}^{\infty} e^{-x^2} H_m(ax) H_n(x)\,dx = 0 \qquad [m < n].$ ET II 290(20)a

4. $\int\limits_{-\infty}^{\infty} e^{-x^2} H_{2m+n}(ax) H_n(x)\,dx = \sqrt{\pi}\, 2^{-m+\frac{1}{2}}\,\frac{(2m+n)!}{m!}\,(a^2-1)^m a^n.$

ET II 291(21)a

5. $\int\limits_{-\infty}^{\infty} e^{-2\alpha^2 x^2} H_m(x) H_n(x)\,dx = 2^{\frac{m+n-1}{2}}\, \alpha^{-m-n-1}\,(1-2\alpha^2)^{\frac{m+n}{2}}\, \Gamma\left(\frac{m+n+1}{2}\right) \times$

$\times\, {}_2F_1\left(-m,\ -n;\ \frac{1-m-n}{2};\ \frac{\alpha^2}{2\alpha^2-1}\right)$

$[\operatorname{Re}\alpha^2 > 0,\ m+n \text{ is even}].$ ET II 289(12)a

6. $\int\limits_{-\infty}^{\infty} e^{-(x-y)^2} H_n(x)\,dx = \pi^{\frac{1}{2}}\, y^n 2^n.$ ET II 288(2)a, EH II 195(31)

7. $\int\limits_{-\infty}^{\infty} e^{-(x-y)^2} H_m(x) H_n(x)\,dx = 2^n \pi^{\frac{1}{2}} m!\, y^{n-m} L_n^{n-m}(-2y^2)$

$[m \leqslant n].$ BU 148(15), ET II 289(13)a

8. $\int\limits_{-\infty}^{\infty} e^{-(x-y)^2} H_n(\alpha x)\,dx = \pi^{\frac{1}{2}}\,(1-\alpha^2)^{\frac{n}{2}} H_n\left[\frac{\alpha y}{(1-\alpha^2)^{\frac{1}{2}}}\right].$ ET II 290(17)a

9. $\int\limits_{-\infty}^{\infty} e^{-(x-y)^2} H_m(\alpha x) H_n(\alpha x)\,dx =$

$= \pi^{\frac{1}{2}} \sum\limits_{k=0}^{\min(m,n)} 2^k k! \binom{m}{k}\binom{n}{k}(1-\alpha^2)^{\frac{m+n}{2}-k} H_{m+n-2k}\left[\frac{\alpha y}{(1-\alpha^2)^{\frac{1}{2}}}\right].$

ET II 291(26)a

10. $\int_{-\infty}^{\infty} e^{-\frac{(x-y)^2}{2u}} H_n(x)\,dx = (2\pi u)^{\frac{1}{2}} (1-2u)^{\frac{n}{2}} H_n\left[y(1-2u)^{-\frac{1}{2}}\right]$

$\left[0 \leqslant u < \frac{1}{2}\right]$. EH II 195(30)

7.375

1. $\int_{-\infty}^{\infty} e^{-2x^2} H_k(x) H_m(x) H_n(x)\,dx =$

$= \pi^{-1} 2^{\frac{1}{2}(m+n+k-1)} \Gamma(s-k)\Gamma(s-m)\Gamma(s-n).$

$2s = k+m+n+1$ [$k+m+n$ is even]. ET II 290(14)a

2. $\int_{-\infty}^{\infty} e^{-x^2} H_k(x) H_m(x) H_n(x)\,dx = \frac{2^{\frac{m+n+k}{2}} \pi^{\frac{1}{2}} k!\, m!\, n!}{(s-k)!\,(s-m)!\,(s-n)!},$

$2s = m+n+k$ [$k+m+n$ is even]. ET II 290(15)a

7.376

1. $\int_{-\infty}^{\infty} e^{ixy} e^{-\frac{x^2}{2}} H_n(x)\,dx = (2\pi)^{\frac{1}{2}} e^{-\frac{y^2}{2}} H_n(y)\, i^n.$ MO 165a

2. $\int_0^{\infty} e^{-2\alpha x^2} x^{\nu} H_{2n}(x)\,dx = (-1)^n 2^{2n-\frac{3}{2}-\frac{1}{2}\nu} \times$

$\times \frac{\Gamma\left(\frac{\nu+1}{2}\right)\Gamma\left(n+\frac{1}{2}\right)}{\sqrt{\pi}\,\alpha^{\frac{1}{2}(\nu+1)}} F\left(-n, \frac{\nu+1}{2}; \frac{1}{2}; \frac{1}{2\alpha}\right)$

[$\operatorname{Re}\alpha > 0$, $\operatorname{Re}\nu > -1$]. BU 150(18a)

3. $\int_0^{\infty} e^{-2\alpha x^2} x^{\nu} H_{2n+1}(x)\,dx =$

$= (-1)^n 2^{2n-\frac{1}{2}\nu} \frac{\Gamma\left(\frac{\nu+1}{2}\right)\Gamma\left(n+\frac{1}{2}\right)}{\sqrt{\pi}\,\alpha^{\frac{1}{2}\nu+1}} F\left(-n, \frac{\nu}{2}+1; \frac{3}{2}; \frac{1}{2\alpha}\right)$

[$\operatorname{Re}\alpha > 0$, $\operatorname{Re}\nu > -2$]. BU 150(18b)

7.377 $\int_{-\infty}^{\infty} e^{-x^2} H_m(x+y) H_n(x+z)\,dx = 2^n \pi^{\frac{1}{2}} m!\, z^{n-m} L_m^{n-m}(-2yz)$

[$m \leqslant n$]. ET II 292(30)a

7.378 $\int_0^{\infty} x^{\alpha-1} e^{-\beta x} H_n(x)\,dx =$

$= 2^n \sum_{m=0}^{E\left(\frac{n}{2}\right)} \frac{n!\,\Gamma(\alpha+n-2m)}{m!\,(n-2m)!} (-1)^m 2^{-2m} \beta^{2m-\alpha-n}$

[$\operatorname{Re}\alpha > 0$, if n is even; $\operatorname{Re}\alpha > -1$, if n is odd; $\operatorname{Re}\beta > 0$].

ET I 172(11)a

7.379

1. $$\int_{-\infty}^{\infty} xe^{-x^2}H_{2m+1}(xy)\,dx = \pi^{\frac{1}{2}}\frac{(2m+1)!}{m!}\,y\,(y^2-1)^m.$$ EH II 195(28)

2. $$\int_{-\infty}^{\infty} x^n e^{-x^2}H_n(xy)\,dx = \pi^{\frac{1}{2}}\,n!\,P_n(y).$$ EH II 195(29)

7.381 $$\int_{-\infty}^{\infty}(x\pm ic)^{\nu}e^{-x^2}H_n(x)\,dx = 2^{n-1-\nu}\pi^{\frac{1}{2}}\frac{\Gamma\left(\frac{n-\nu}{2}\right)}{\Gamma(-\nu)}\exp\left[\pm\frac{1}{2}\pi(\nu+n)\,i\right]$$

$[c>0]$. ET II 288(3)a

7.382 $$\int_0^{\infty} x^{-1}(x^2+a^2)^{-1}e^{-x^2}H_{2n+1}(x)\,dx =$$

$$=(-2)^n(\pi)^{\frac{1}{2}}a^{-2}\left[2^n n! - (2n+1)!\,e^{\frac{1}{2}a^2}D_{-2n-2}\left(a\sqrt{2}\right)\right].$$

ET II 288(4)a

7.383

1. $$\int_0^{\infty} e^{-xp}H_{2n+1}\left(\sqrt{x}\right)dx = (-1)^n 2^n(2n+1)!!\,\pi^{\frac{1}{2}}(p-1)^n p^{-n-\frac{3}{2}}$$

$[\operatorname{Re} p>0]$. EF 151(261)a, ET I 172(12)a

2. $$\int_0^{\infty} e^{-(b-\beta)x}H_{2n+1}\left(\sqrt{(\alpha-\beta)x}\right)dx = (-1)^n\sqrt{\pi}\sqrt{\alpha-\beta}\,\frac{(2n+1)!}{n!}\,\frac{(b-\alpha)^n}{(b-\beta)^{n+\frac{3}{2}}}$$

$[\operatorname{Re}(b-\beta)>0]$. ET I 172(15)a

3. $$\int_0^{\infty}\frac{1}{\sqrt{x}}e^{-(b-\beta)x}H_{2n}\left(\sqrt{(\alpha-\beta)x}\right)dx = (-1)^n\sqrt{\pi}\,\frac{(2n)!}{n!}\,\frac{(b-\alpha)^n}{(b-\beta)^{n+\frac{1}{2}}}$$

$[\operatorname{Re}(b-\beta)>0]$. ET I 172(16)a

4. $$\int_0^{\infty} x^{a-\frac{1}{2}n-1}e^{-bx}H_n\left(\sqrt{x}\right)dx = 2^{\frac{n}{2}}\Gamma(a)\,b^{-a}{}_2F_1\left(-\frac{1}{2}n,\ \frac{1}{2}-\frac{1}{2}n;\ 1-a;\ b\right)$$

$\left[\operatorname{Re} a>\frac{1}{2}n\right.$, if n is even; $\operatorname{Re} a>\frac{1}{2}n-\frac{1}{2}$, if n is odd; $\operatorname{Re} b>0$. If a is even, only the first $1+E\left(\frac{n}{2}\right)$ terms are kept in the series for $\left.{}_2F_1\right]$. ET I 172(14)a

5. $$\int_0^{\infty} x^{-\frac{1}{2}}e^{-px}H_{2n}\left(\sqrt{x}\right)dx = (-1)^n 2^n(2n-1)!!\,\pi^{\frac{1}{2}}(p-1)^n p^{-n-\frac{1}{2}}.$$

MO 177a

7.384 $\int_0^\infty \frac{1}{\sqrt{x}} e^{-bx} \left[H_n\left(\frac{\alpha+\sqrt{x}}{\lambda}\right) + H_n\left(\frac{\alpha-\sqrt{x}}{\lambda}\right)\right] dx =$

$$= \sqrt{\frac{2\pi}{b}} (1-\lambda^{-2}b^{-1})^{\frac{n}{2}} H_n\left(\frac{\alpha}{\sqrt{\lambda^2-\frac{1}{b}}}\right) \quad [\operatorname{Re} b > 0].$$ ET I 173(17)a

7.385

1. $\int_0^\infty \frac{e^{-bx}}{\sqrt{e^x-1}} H_{2n}\left[\sqrt{s(1-e^{-x})}\right] dx =$

$$= (-1)^n 2^{2n} \sqrt{\pi} \frac{(2n)!\,\Gamma\left(b+\frac{1}{2}\right)}{\Gamma(n+b+1)} L_n^b(s)$$

$$\left[\operatorname{Re} b > -\frac{1}{2}\right].$$ ET I 174(23)a

2. $\int_0^\infty e^{-bx} H_{2n+1}\left[\sqrt{s}\sqrt{1-e^{-x}}\right] dx = (-1)^n 2^{2n} \sqrt{\pi s} \frac{(2n+1)!\,\Gamma(b)}{\Gamma\left(n+b+\frac{3}{2}\right)} L_n^b(s)$

$$[\operatorname{Re} b > 0].$$ ET I 174(24)a

7.386 $\int_0^\infty x^{-\frac{n+1}{2}} e^{-\frac{q^2}{4x}} H_n\left(\frac{q}{2\sqrt{x}}\right) e^{-px} dx = 2^n \pi^{\frac{1}{2}} p^{\frac{n-1}{2}} e^{-q\sqrt{p}}.$ EF 129(117)

7.387

1. $\int_0^\infty e^{-x^2} \operatorname{sh}(\sqrt{2}\beta x) H_{2n+1}(x)\, dx = 2^{n-\frac{1}{2}} \pi^{\frac{1}{2}} \beta^{2n+1} e^{\frac{1}{2}\beta^2}.$ ET II 289(7)a

2. $\int_0^\infty e^{-x^2} \operatorname{ch}(\sqrt{2}\beta x) H_{2n}(x)\, dx = 2^{n-1} \pi^{\frac{1}{2}} \beta^{2n} e^{\frac{1}{2}\beta^2}.$ ET II 289(8)a

7.388

1. $\int_0^\infty e^{-x^2} \sin(\sqrt{2}\beta x) H_{2n+1}(x)\, dx = (-1)^n 2^{n-\frac{1}{2}} \pi^{\frac{1}{2}} \beta^{2n+1} e^{-\frac{1}{2}\beta^2}$

ET II 288(5)a

2. $\int_0^\infty e^{-x^2} \sin(\sqrt{2}\beta x) H_{2n+1}(ax)\, dx =$

$$= (-1)^n 2^{-1} \pi^{\frac{1}{2}} (a^2-1)^{n+\frac{1}{2}} e^{-\frac{1}{2}\beta^2} H_{2n+1}\left(\frac{a\beta}{\sqrt{2}(a^2-1)^{\frac{1}{2}}}\right).$$ ET II 290(18)a

3. $\int_0^\infty e^{-x^2} \cos(\sqrt{2}\beta x) H_{2n}(x)\, dx = (-1)^n 2^{n-1} \pi^{\frac{1}{2}} \beta^{2n} e^{-\frac{1}{2}\beta^2}.$ ET II 289(6)a

4. $\int_0^\infty e^{-x^2} \cos(\sqrt{2}\beta x) H_{2n}(ax)\, dx = 2^{-1} \pi^{\frac{1}{2}} (1-a^2)^n e^{-\frac{1}{2}\beta^2} H_{2n}\left[\frac{a\beta}{\sqrt{2}(a^2-1)^{\frac{1}{2}}}\right].$

ET II 290(19)a

5. $$\int_0^\infty e^{-y^2}[H_n(y)]^2\cos(\sqrt{2}\,\beta y)\,dy=\pi^{\frac{1}{2}}2^{n-1}n!\,L_n(\beta^2).$$ EH II 195(33)

6. $$\int_0^\infty e^{-x^2}\sin(bx)\,H_n(x)\,H_{n+2m+1}(x)\,dx=$$
$$=2^n(-1)^m\sqrt{\frac{\pi}{2}}\,n!\,b^{2m}e^{-\frac{b^2}{4}}L_n^{2m+1}\left(\frac{b^2}{2}\right)$$
$[b>0]$. ET I 39(11)a

7. $$\int_0^\infty e^{-x^2}\cos(bx)\,H_n(x)\,H_{n+2m}(x)\,dx=$$
$$=2^{n-\frac{1}{2}}\sqrt{\frac{\pi}{2}}\,n!\,(-1)^m b^{2m}e^{-\frac{b^2}{4}}L_n^{2m}\left(\frac{b^2}{2}\right)$$
$[b>0]$. ET I 39(11)a

7.389 $$\int_0^\pi(\cos x)^n H_{2n}\left[a(1-\sec x)^{\frac{1}{2}}\right]dx=2^{-n}(-1)^n\pi\frac{(2n)!}{(n!)^2}[H_n(a)]^2.$$ ET II 292(31)

7.39 Jacobi polynomials

7.391

1. $$\int_{-1}^{1}(1-x)^\alpha(1+x)^\beta P_n^{(\alpha,\beta)}(x)\,P_m^{(\alpha,\beta)}(x)\,dx=$$
$=0$ $[m\neq n,\ \operatorname{Re}\alpha>-1,\ \operatorname{Re}\beta>-1]$;
$$=\frac{2^{\alpha+\beta+1}\Gamma(\alpha+n+1)\,\Gamma(\beta+n+1)}{n!\,(\alpha+\beta+1+2n)\,\Gamma(\alpha+\beta+n+1)}$$ $[m=n,\ \operatorname{Re}\alpha>-1,\ \operatorname{Re}\beta>-1]$. ET II 285(5, 9)

2. $$\int_{-1}^{1}(1-x)^\varrho(1+x)^\sigma P_n^{(\alpha,\beta)}(x)\,dx=\frac{2^{\varrho+\sigma+1}\Gamma(\varrho+1)\,\Gamma(\sigma+1)\,\Gamma(n+1+\alpha)}{n!\,\Gamma(\varrho+\sigma+2)\,\Gamma(1+\alpha)}\times$$
$$\times\,{}_3F_2(-n,\ \alpha+\beta+n+1,\ \varrho+1;\ \alpha+1,\ \varrho+\sigma+2;\ 1)$$
$[\operatorname{Re}\varrho>-1,\ \operatorname{Re}\sigma>-1]$. ET II 284(3)

3. $$\int_{-1}^{1}(1-x)^\alpha(1+x)^\sigma P_n^{(\alpha,\beta)}(x)\,dx=\frac{2^{\alpha+\sigma+1}\Gamma(\sigma+1)\,\Gamma(\alpha+n+1)\,\Gamma(\sigma-\beta+1)}{\Gamma(\sigma-\beta-n+1)\,\Gamma(\alpha+\sigma+n+2)}$$
$[\operatorname{Re}\alpha>-1,\ \operatorname{Re}\sigma>-1]$. ET II 284(1)

4. $$\int_{-1}^{1}(1-x)^\varrho(1+x)^\beta P_n^{(\alpha,\beta)}(x)\,dx=\frac{2^{\beta+\varrho+1}\Gamma(\varrho+1)\,\Gamma(\beta+n+1)\,\Gamma(\alpha-\varrho+n)}{n!\,\Gamma(\alpha-\varrho)\,\Gamma(\beta+\varrho+n+2)}$$
$[\operatorname{Re}\varrho>-1,\ \operatorname{Re}\beta>-1]$. ET II 284(2)

5. $$\int_{-1}^{1} (1-x)^{\alpha-1}(1+x)^{\beta}[P_n^{(\alpha,\beta)}(x)]^2\,dx = \frac{2^{\alpha+\beta}\Gamma(\alpha+n+1)\Gamma(\beta+n+1)}{n!\,\alpha\Gamma(\alpha+\beta+n+1)}$$

$[\operatorname{Re}\alpha > 0,\ \operatorname{Re}\beta > -1]$. ET II 285(6)

6. $$\int_{-1}^{1} (1-x)^{2\alpha}(1+x)^{\beta}[P_n^{(\alpha,\beta)}(x)]^2\,dx =$$

$$= \frac{2^{4\alpha+\beta+1}\Gamma\left(\alpha+\frac{1}{2}\right)[\Gamma(\alpha+n+1)]^2\Gamma(\beta+2n+1)}{\sqrt{\pi}\,(n!)^2\Gamma(\alpha+1)\Gamma(2\alpha+\beta+2n+2)}$$

$\left[\operatorname{Re}\alpha > -\frac{1}{2},\ \operatorname{Re}\beta > -1\right]$. ET II 285(7)

7. $$\int_{-1}^{1} (1-x)^{\varrho}(1+x)^{\beta}P_n^{(\alpha,\beta)}(x)P_n^{(\varrho,\beta)}(x)\,dx =$$

$$= \frac{2^{\varrho+\beta+1}\Gamma(\varrho+n+1)\Gamma(\beta+n+1)\Gamma(\alpha+\beta+2n+1)}{n!\,\Gamma(\beta+\varrho+2n+2)\Gamma(\alpha+\beta+n+1)}$$

$[\operatorname{Re}\varrho > -1,\ \operatorname{Re}\beta > -1]$. ET II 285(10)

8. $$\int_{-1}^{1} (1-x)^{\varrho-1}(1+x)^{\beta}P_n^{(\alpha,\beta)}(x)P_n^{(\varrho,\beta)}(x)\,dx =$$

$$= \frac{2^{\varrho+\beta}\Gamma(\alpha+n+1)\Gamma(\beta+n+1)\Gamma(\varrho)}{n!\,\Gamma(\alpha+1)\Gamma(\varrho+\beta+n+1)}$$ $[\operatorname{Re}\beta > -1,\ \operatorname{Re}\varrho > 0]$. ET II 286(11)

9. $$\int_{-1}^{1} (1-x)^{\alpha}(1+x)^{\sigma}P_n^{(\alpha,\beta)}(x)P_m^{(\alpha,\sigma)}(x)\,dx =$$

$$= \frac{2^{\alpha+\sigma+1}\Gamma(\alpha+n+1)\Gamma(\alpha+\beta+m+n+1)\Gamma(\sigma+m+1)\Gamma(\sigma-\beta+1)}{m!\,(n-m)!\,\Gamma(\alpha+\beta+n+1)\Gamma(\alpha+\sigma+m+n+2)\Gamma(\sigma-\beta+m+1)}$$

$[\operatorname{Re}\alpha > -1,\ \operatorname{Re}\sigma > -1]$. ET II 286(12)

10. $$\int_{-1}^{1} (1-x)^{\varrho}(1+x)^{\beta}P_n^{(\alpha,\beta)}(x)P_m^{(\varrho,\beta)}(x)\,dx =$$

$$= \frac{2^{\beta+\varrho+1}\Gamma(\alpha+\beta+m+n+1)\Gamma(\beta+n+1)\Gamma(\varrho+m+1)}{n!\,(n-m)!\,\Gamma(\alpha+\beta+n+1)\Gamma(\beta+\varrho+m+n+2)}\,\frac{\Gamma(\varrho-\alpha-m+n)}{\Gamma(\varrho-\alpha)}$$

$[\operatorname{Re}\beta > -1,\ \operatorname{Re}\varrho > -1]$. ET II 287(16)

11. $$\int_{0}^{x} (1-y)^{\alpha}(1+y)^{\beta}P_n^{(\alpha,\beta)}(y)\,dy = \frac{1}{2n}[P_{n-1}^{(\alpha+1,\beta+1)}(0) -$$

$$- (1-x)^{\alpha+1}(1+x)^{\beta+1}P_{n-1}^{(\alpha+1,\beta+1)}(x)].$$ EH II 173(38)

7.392

1. $$\int_{0}^{1} x^{\lambda-1}(1-x)^{\mu-1}P_n^{(\alpha,\beta)}(1-\gamma x)\,dx =$$

$$= \frac{\Gamma(\alpha+n+1)\Gamma(\lambda)\Gamma(\mu)}{n!\,\Gamma(\alpha+1)\Gamma(\lambda+\mu)}\,{}_3F_2\left(-n,\ n+\alpha+\beta+1,\ \lambda;\ \alpha+1,\ \lambda+\mu;\ \frac{1}{2}\gamma\right)$$

$[\operatorname{Re}\lambda > 0,\ \operatorname{Re}\mu > 0]$. ET II 192(46)a

2. $$\int_0^1 x^{\lambda-1}(1-x)^{\mu-1} P_n^{(\alpha,\beta)}(\gamma x-1)\,dx =$$
$$=(-1)^n \frac{\Gamma(\beta+n+1)\Gamma(\lambda)\Gamma(\mu)}{n!\,\Gamma(\beta+1)\Gamma(\lambda+\mu)}\, {}_3F_2\left(-n,\ n+\alpha+\beta+1,\ \lambda;\ \beta+1,\ \lambda+\mu;\ \frac{1}{2}\gamma\right)$$
$[\operatorname{Re}\lambda > 0,\ \operatorname{Re}\mu > 0]$. ET II 192(47)a

3. $$\int_0^1 x^{\alpha}(1-x)^{\mu-1} P_n^{(\alpha,\beta)}(1-\gamma x)\,dx = \frac{\Gamma(\alpha+n+1)\Gamma(\mu)}{\Gamma(\alpha+\mu+n+1)} P_n^{(\alpha+\mu,\,\beta-\mu)}(1-\gamma)$$
$[\operatorname{Re}\alpha > -1,\ \operatorname{Re}\mu > 0]$. ET II 191(43)a

4. $$\int_0^1 x^{\beta}(1-x)^{\mu-1} P_n^{(\alpha,\beta)}(\gamma x-1)\,dx = \frac{\Gamma(\beta+n+1)\Gamma(\mu)}{\Gamma(\beta+\mu+n+1)} P_n^{(\alpha-\mu,\,\beta+\mu)}(\gamma-1)$$
$[\operatorname{Re}\beta > -1,\ \operatorname{Re}\mu > 0]$. ET II 191(44)a

7.393

1. $$\int_0^1 (1-x^2)^{\nu}\sin bx\, P_{2n+1}^{(\nu,\nu)}(x)\,dx = \frac{(-1)^n\sqrt{\pi}\,\Gamma(2n+\nu+2)\,J_{2n+\nu+\frac{3}{2}}(b)}{2^{\frac{1}{2}-\nu}(2n+1)!\,b^{\nu+\frac{1}{2}}}$$
$[b > 0,\ \operatorname{Re}\nu > -1]$. ET I 94(5)

2. $$\int_0^1 (1-x^2)^{\nu}\cos bx\, P_{2n}^{(\nu,\nu)}(x)\,dx = \frac{(-1)^n 2^{\nu-\frac{1}{2}}\sqrt{\pi}\,\Gamma(2n+\nu-1)\,J_{2n+\nu+\frac{1}{2}}(b)}{(2n)!\,b^{\nu+\frac{1}{2}}}$$
$[b > 0,\ \operatorname{Re}\nu > -1]$. ET I 38(4)

7.41-7.42 Laguerre polynomials

7.411

1. $$\int_0^t L_n(x)\,dx = L_n(t) - L_{n+1}(t).$$ MO 110

2. $$\int_0^t L_n^{\alpha}(x)\,dx = L_n^{\alpha}(t) - L_{n+1}^{\alpha}(t) - \binom{n+\alpha}{n} + \binom{n+1+\alpha}{n+1}.$$ EH II 189(16)a

3. $$\int_0^t L_{n-1}^{\alpha+1}(x)\,dx = -L_n^{\alpha}(t) + \binom{n+\alpha}{n}.$$ EH II 189(15)a

4. $$\int_0^t L_m(x)L_n(t-x)\,dx = L_{m+n}(t) - L_{m+n+1}(t).$$ EH II 191(31)

5. $$\sum_{k=0}^{\infty}\left[\int_0^t L_k(x)\,dx\right]^2 = e^t - 1 \quad [t \geqslant 0].$$ MO 110

7.412

1. $$\int_0^1 (1-x)^{\mu-1} x^{\alpha} L_n^{\alpha}(ax)\,dx = \frac{\Gamma(\alpha+n+1)\,\Gamma(\mu)}{\Gamma(\alpha+\mu+n+1)} L_n^{\alpha+\mu}(a)$$

$[\operatorname{Re}\alpha > -1,\ \operatorname{Re}\mu > 0]$. EH II 191(30)a, BU 129(14c)

2. $$\int_0^1 (1-x)^{\mu-1} x^{\lambda-1} L_n^{\alpha}(\beta x)\,dx =$$
$$= \frac{\Gamma(\alpha+n+1)\,\Gamma(\lambda)\,\Gamma(\mu)}{n!\,\Gamma(\alpha+1)\,\Gamma(\lambda+\mu)}\, {}_2F_2(-n, \lambda; \alpha+1, \lambda+\mu: \beta)$$

$[\operatorname{Re}\lambda > 0,\ \operatorname{Re}\mu > 0]$. ET II 192(50)a

7.413 $$\int_0^1 x^{\alpha}(1-x)^{\beta} L_m^{\alpha}(xy)\, L_n^{\beta}[(1-x)y]\,dx =$$
$$= \frac{(m+n)!\,\Gamma(\alpha+m+1)\,\Gamma(\beta+n+1)}{m!\,n!\,\Gamma(\alpha+\beta+m+n+2)} L_{m+n}^{\alpha+\beta+1}(y)$$

$[\operatorname{Re}\alpha > -1, \operatorname{Re}\beta > -1]$. ET II 293(7)

7.414

1. $$\int_y^\infty e^{-x} L_n^{\alpha}(x)\,dx = e^{-y}[L_n^{\alpha}(y) - L_{n-1}^{\alpha}(y)].$$ EH II 191(29)

2. $$\int_0^\infty e^{-bx} L_n(\lambda x)\, L_n(\mu x)\,dx = \frac{(b-\lambda-\mu)^n}{b^{n+1}} P_n\left[\frac{b^2-(\lambda+\mu)b+2\lambda\mu}{b(b-\lambda-\mu)}\right]$$

$[\operatorname{Re} b > 0]$. ET I 175(34)

3. $$\int_0^\infty e^{-x} x^{\alpha} L_n^{\alpha}(x)\, L_m^{\alpha}(x)\,dx =$$

$= 0$ $[m \neq n,\ \operatorname{Re}\alpha > -1]$; BU 115(8), ET II 293(3)

$= \frac{\Gamma(\alpha+n+1)}{n!}$ $[m = n,\ \operatorname{Re}\alpha > 0]$. BU 115(8), ET II 292(2)

4. $$\int_0^\infty e^{-bx} x^{\alpha} L_n^{\alpha}(\lambda x)\, L_m^{\alpha}(\mu x)\,dx = \frac{\Gamma(m+n+\alpha+1)}{m!\,n!} \frac{(b-\lambda)^n (b-\mu)^m}{b^{m+n+\alpha+1}} \times$$
$$\times F\left[-m, -n; -m-n-\alpha; \frac{b(b-\lambda-\mu)}{(b-\lambda)(b-\mu)}\right]$$

$[\operatorname{Re}\alpha > -1,\ \operatorname{Re} b > 0]$. ET I 175(35)

5. $$\int_0^\infty e^{-bx} L_n^{a}(x)\,dx = \sum_{m=0}^{n} \binom{a+m-1}{m} \frac{(b-1)^{n-m}}{b^{n-m+1}}$$ $[\operatorname{Re} b > 0]$. ET I 174(27)

6. $$\int_0^\infty e^{-bx} L_n(x)\,dx = (b-1)^n\, b^{-n-1}$$ $[\operatorname{Re} b > 0]$. ET I 174(25)

7. $$\int_0^\infty e^{-st} t^{\beta} L_n^{\alpha}(t)\,dt = \frac{\Gamma(\beta+1)\,\Gamma(\alpha+n+1)}{n!\,\Gamma(\alpha+1)} s^{-\beta-1} F\left(-n, \beta+1; \alpha+1; \frac{1}{s}\right)$$

$[\operatorname{Re}\beta > -1,\ \operatorname{Re} s > 0]$. BU 119(4b), EH II 191(133)

8. $$\int_0^\infty e^{-st} t^\alpha L_n^\alpha(t)\,dt = \frac{\Gamma(\alpha+n+1)(s-1)^n}{n!\, s^{\alpha+n+1}}$$

$[\operatorname{Re}\alpha > -1,\ \operatorname{Re}s > 0]$. EH II 191(32), MO 176a

9. $$\int_0^\infty e^{-x} x^{\alpha+\beta} L_m^\alpha(x) L_n^\beta(x)\,dx = (-1)^{m+n}\binom{\alpha+m}{n}\binom{\beta+n}{m}$$

$[\operatorname{Re}(\alpha+\beta) > -1]$. ET II 293(4)

10. $$\int_0^\infty e^{-bx} x^{2a}[L_n^a(x)]^2\,dx = \frac{2^{2a}\Gamma\left(a+\frac{1}{2}\right)\Gamma\left(n+\frac{1}{2}\right)}{\pi (n!)^2 b^{2a+1}} \times$$
$$\times F\left(-n, a+\frac{1}{2};\ \frac{1}{2}-n;\ \left(1-\frac{2}{b}\right)^2\right)$$

$\left[\operatorname{Re}a > -\frac{1}{2},\ \operatorname{Re}b > 0\right]$. ET I 174(30)

11. $$\int_0^\infty e^{-x} x^{\gamma-1} L_n^\mu(x)\,dx = \frac{\Gamma(\gamma)\Gamma(1+\mu+n-\gamma)}{n!\,\Gamma(1+\mu-\gamma)}$$ $[\operatorname{Re}\gamma > 0]$. BU 120(4b)

12. $$\int_0^\infty e^{-x\left(s+\frac{a_1+a_2}{2}\right)} x^{\mu+\beta} L_k^\mu(a_1 x) L_k^\mu(a_2 x)\,dx =$$
$$= \frac{\Gamma(1+\mu+\beta)\Gamma(1+\mu+k)}{k!\,k!\,\Gamma(1+\mu)}\left\{\frac{d^k}{dh^k}\left[\frac{F\left(\frac{1+\mu+\beta}{2}, 1+\frac{\mu+\beta}{2};\ 1+\mu;\ \frac{A^2}{B^2}\right)}{(1-h)^{1+\mu} B^{1+\mu+\beta}}\right]\right\}_{h=0},$$
$$A^2 = \frac{4a_1a_2h}{(1-h)^2};\quad B = s + \frac{a_1+a_2}{2}\frac{1+h}{1-h}$$

$\left[\operatorname{Re}\left(s+\frac{a_1+a_2}{2}\right) > 0,\ a_1 > 0,\ a_2 > 0,\ \operatorname{Re}(\mu+\beta) > -1\right]$.

BU 142(19)

13 $$\int_0^\infty e^{-x\left(s+\frac{a_1+a_2}{2}\right)} x^\mu L_k^\mu(a_1 x) L_k^\mu(a_2 x)\,dx = \frac{\Gamma(1+\mu+k)}{b_0^{1+\mu+k}}\cdot\frac{b_2^k}{k!}\cdot P_k^{(\mu,0)}\left(\frac{b_1^2}{b_0 b_2}\right),$$
$$b_0 = s+\frac{a_1+a_2}{2},\quad b_1^2 = b_0 b_2 + 2a_1a_2,\quad b_2 = s - \frac{a_1+a_2}{2}$$

$\left[\operatorname{Re}\mu > -1,\ \operatorname{Re}\left(s+\frac{a_1+a_2}{2}\right) > 0\right]$. BU 144(22)

7.415 $$\int_0^1 (1-x)^{\mu-1} x^{\lambda-1} e^{-\beta x} L_n^\alpha(\beta x)\,dx =$$
$$= \frac{\Gamma(\alpha+n+1)}{n!\,\Gamma(\alpha+1)} \mathrm{B}(\lambda,\mu)\, {}_2F_2(\alpha+n+1, \lambda;\ \alpha+1,\ \lambda+\mu;\ -\beta)$$

$[\operatorname{Re}\lambda > 0,\ \operatorname{Re}\mu > 0]$. ET II 193(51)a

7.416 $\int\limits_{-\infty}^{\infty} x^{m-n} \exp\left[-\frac{1}{2}(x-y)^2\right] L_n^{m-n}(x^2)\,dx =$

$$= \frac{(2\pi)^{\frac{1}{2}}}{n!} i^{n-m} 2^{-\frac{n+m}{2}} H_n\left(\frac{iy}{\sqrt{2}}\right) H_m\left(\frac{iy}{\sqrt{2}}\right).$$

BU 149(15b)a, ET II 293(8)a

7.417

1 $\int\limits_0^{\infty} x^{\nu-2n-1} e^{-ax} \sin(bx) L_{2n}^{\nu-2n-1}(ax)\,dx =$

$$= (-1)^n i\Gamma(\nu) \frac{b^{2n}[(a-ib)^{-\nu}-(a+ib)^{-\nu}]}{2(2n)!}$$

$[b > 0, \quad \operatorname{Re} a > 0, \quad \operatorname{Re}\nu > 2n].$ ET I 95(12)

2. $\int\limits_0^{\infty} x^{\nu-2n-2} e^{-ax} \sin(bx) L_{2n+1}^{\nu-2n-2}(ax)\,dx =$

$$= (-1)^{n+1}\Gamma(\nu) \frac{b^{2n+1}[(a+ib)^{-\nu}+(a-ib)^{-\nu}]}{2(2n+1)!}$$

$[b > 0, \quad \operatorname{Re} a > 0, \quad \operatorname{Re}\nu > 2n+1].$ ET I 95(13)

3. $\int\limits_0^{\infty} x^{\nu-2n} e^{-ax} \cos(bx) L_{2n-1}^{\nu-2n}(ax)\,dx =$

$$= i(-1)^{n+1}\Gamma(\nu) \frac{b^{2n-1}[(a-ib)^{-\nu}-(a+ib)^{-\nu}]}{2(2n-1)!}$$

$[b > 0, \quad \operatorname{Re} a > 0, \quad \operatorname{Re}\nu > 2n-1].$ ET I 39(12)

4. $\int\limits_0^{\infty} x^{\nu-2n-1} e^{-ax} \cos(bx) L_{2n}^{\nu-2n-1}(ax)\,dx =$

$$= (-1)^n \Gamma(\nu) \frac{b^{2n}[(a+ib)^{-\nu}+(a-ib)^{-\nu}]}{2(2n)!}$$

$[b > 0, \quad \operatorname{Re}\nu > 2n, \quad \operatorname{Re} a > 0].$ ET I 39(13)

7.418

1. $\int\limits_0^{\infty} e^{-\frac{1}{2}x^2} \sin(bx) L_n(x^2)\,dx = (-1)^n \frac{i}{2} n! \frac{1}{\sqrt{2\pi}} \{[D_{-n-1}(ib)]^2 - [D_{-n-1}(-ib)]^2\}$

$[b > 0].$ ET I 95(14)

2. $\int\limits_0^{\infty} e^{-\frac{1}{2}x^2} \cos(bx) L_n(x^2)\,dx = \sqrt{\frac{\pi}{2}} (n!)^{-1} e^{-\frac{1}{2}b^2} 2^{-n} \left[H_n\left(\frac{b}{\sqrt{2}}\right)\right]^2$

$[b > 0].$ ET I 39(14)

3. $\int\limits_0^{\infty} x^{2n+1} e^{-\frac{1}{2}x^2} \sin(bx) L_n^{n+\frac{1}{2}}\left(\frac{1}{2}x^2\right) dx =$

$$= \sqrt{\frac{\pi}{2}} b^{2n+1} e^{-\frac{1}{2}b^2} L_n^{n+\frac{1}{2}}\left(\frac{b^2}{2}\right) \qquad [b > 0].$$ ET I 95(15)

4. $\int_0^\infty x^{2n} e^{-\frac{1}{2}x^2} \cos(bx) L_n^{n-\frac{1}{2}}\left(\frac{1}{2}x^2\right) dx = \sqrt{\frac{\pi}{2}}\, b^{2n} e^{-\frac{1}{2}b^2} L_n^{n+\frac{1}{2}}\left(\frac{1}{2}b^2\right)$

$[b > 0]$. ET I 39(13)

5 $\int_0^\infty x e^{-\frac{1}{2}x^2} L_n^{\alpha}\left(\frac{1}{2}x^2\right) L_n^{\frac{1}{2}-\alpha}\left(\frac{1}{2}x^2\right) \sin(xy)\, dx =$

$= \left(\frac{\pi}{2}\right)^{\frac{1}{2}} y e^{-\frac{1}{2}y^2} L_n^{\alpha}\left(\frac{1}{2}y^2\right) L_n^{\frac{1}{2}-\alpha}\left(\frac{1}{2}y^2\right).$ ET II 294(11)

6. $\int_0^\infty e^{-\frac{1}{2}x^2} L_n^{\alpha}\left(\frac{1}{2}x^2\right) L_n^{-\frac{1}{2}-\alpha}\left(\frac{1}{2}x^2\right) \cos(xy)\, dx =$

$= \left(\frac{\pi}{2}\right)^{\frac{1}{2}} e^{-\frac{1}{2}y^2} L_n^{\alpha}\left(\frac{1}{2}y^2\right) L_n^{-\alpha-\frac{1}{2}}\left(\frac{1}{2}y^2\right).$ ET II 294(12)

7.419 $\int_0^\infty x^{n+2\nu-\frac{1}{2}} \exp[-(1+a)x] L_n^{2\nu}(ax) K_\nu(x)\, dx =$

$= \frac{\pi^{\frac{1}{2}} \Gamma\left(n+\nu+\frac{1}{2}\right) \Gamma\left(n+3\nu+\frac{1}{2}\right)}{2^{n+2\nu+\frac{1}{2}} n!\, \Gamma(2\nu+1)} F\left(n+\nu+\frac{1}{2}, n+3\nu+\frac{1}{2}; 2\nu+1; -\frac{1}{2}a\right)$

$\left[\operatorname{Re} a > -2, \quad \operatorname{Re}(n+\nu) > -\frac{1}{2}, \quad \operatorname{Re}(n+3\nu) > -\frac{1}{2}\right].$ ET II 370(44)

7.421

1. $\int_0^\infty x e^{-\frac{1}{2}\alpha x^2} L_n\left(\frac{1}{2}\beta x^2\right) J_0(xy)\, dx = \frac{(\alpha-\beta)^n}{\alpha^{n+1}} e^{-\frac{1}{2\alpha}y^2} L_n\left[\frac{\beta y^2}{2\alpha(\beta-\alpha)}\right]$

$[y > 0, \quad \operatorname{Re}\alpha > 0]$. ET II 13(4)a

2. $\int_0^\infty x e^{-x^2} L_n(x^2) J_0(xy)\, dx = \frac{2^{-2n-1}}{n!} y^{2n} e^{-\frac{1}{4}y^2}.$ ET II 13(5)

3. $\int_0^\infty x^{2n+\nu+1} e^{-\frac{1}{2}x^2} L_n^{\nu+n}\left(\frac{1}{2}x^2\right) J_\nu(xy)\, dx = y^{2n+\nu} e^{-\frac{1}{2}y^2} L_n^{\nu+n}\left(\frac{1}{2}y^2\right)$

$[y > 0, \quad \operatorname{Re}\nu > -1]$. MO 183

4. $\int_0^\infty x^{\nu+1} e^{-\beta x^2} L_n^{\nu}(\alpha x^2) J_\nu(xy)\, dx =$

$= 2^{-\nu-1} \beta^{-\nu-n-1} (\beta-\alpha)^n y^\nu e^{-\frac{y^2}{4\beta}} L_n^{\nu}\left[\frac{\alpha y^2}{4\beta(\alpha-\beta)}\right].$ ET II 43(5)

5. $\int_0^\infty e^{-\frac{1}{2q}x^2} x^{\nu+1} L_n^{\nu}\left[\frac{x^2}{2q(1-q)}\right] J_\nu(xy)\, dx =$

$= \frac{q^{n+\nu+1}}{(q-1)^n} e^{-\frac{qy^2}{2}} y^\nu L_n^{\nu}\left(\frac{y^2}{2}\right) \qquad [\nu > 0].$ MO 183

7.422

1. $$\int_0^\infty x^{\nu+1}e^{-\beta x^2}[L_n^{\frac{1}{2}\nu}(\alpha x^2)]^2 J_\nu(xy)\,dx = \frac{y^\nu}{\pi n!}\Gamma\left(n+1+\frac{1}{2}\nu\right)(2\beta)^{-\nu-1}e^{-\frac{y^2}{4\beta}}\times$$
$$\times\sum_{l=0}^{n}\frac{(-1)^l\,\Gamma\left(n-l+\frac{1}{2}\right)\Gamma\left(l+\frac{1}{2}\right)}{\Gamma\left(l+1+\frac{1}{2}\nu\right)(n-l)!}\left(\frac{2\alpha-\beta}{\beta}\right)^{2l}L_{2l}^{\nu}\left[\frac{\alpha y^2}{2\beta(2\alpha-\beta)}\right]$$
$[y>0, \quad \operatorname{Re}\beta>0, \quad \operatorname{Re}\nu>-1]$. ET II 43(7)

2. $$\int_0^\infty x^{\nu+1}e^{-\alpha x^2}L_m^{\nu-\sigma}(\alpha x^2)\,L_n^{\sigma}(\alpha x^2)\,J_\nu(xy)\,dx =$$
$$=(-1)^{m+n}(2\alpha)^{-\nu-1}y^\nu e^{-\frac{y^2}{4\alpha}}L_n^{\sigma-m+n}\left(\frac{y^2}{4\alpha}\right)L_m^{\nu-\sigma+m-n}\left(\frac{y^2}{4\alpha}\right)$$
$[y>0, \quad \operatorname{Re}\alpha>0, \quad \operatorname{Re}\nu>-1]$. ET II 43(8)

7.423

1. $$\int_0^\infty e^{-\frac{1}{2}x^2}L_n\left(\frac{1}{2}x^2\right)H_{2n+1}\left(\frac{x}{2\sqrt{2}}\right)\sin(xy)\,dx =$$
$$=\left(\frac{\pi}{2}\right)^{\frac{1}{2}}e^{-\frac{1}{2}y^2}L_n\left(\frac{1}{2}y^2\right)H_{2n+1}\left(\frac{y}{2\sqrt{2}}\right).$$ ET II 294(13)a

2. $$\int_0^\infty e^{-\frac{1}{2}x^2}L_n\left(\frac{1}{2}x^2\right)H_{2n}\left(\frac{x}{2\sqrt{2}}\right)\cos(xy)\,dx =$$
$$=\left(\frac{\pi}{2}\right)^{\frac{1}{2}}e^{-\frac{1}{2}y^2}L_n\left(\frac{1}{2}y^2\right)H_{2n}\left(\frac{y}{2\sqrt{2}}\right).$$ ET II 294(14)a

7.5 Hypergeometric Functions

7.51 Combinations of hypergeometric functions and powers

7.511 $$\int_0^\infty F(a, b; c; -z)\,z^{-s-1}\,dz = \frac{\Gamma(a+s)\,\Gamma(b+s)\,\Gamma(c)\,\Gamma(-s)}{\Gamma(a)\,\Gamma(b)\,\Gamma(c+s)}$$
$[c\neq 0, -1, -2, \ldots, \operatorname{Re}s<0, \operatorname{Re}(a+s)>0, \operatorname{Re}(b+s)>0]$.

7.512

EH I 79(4)

1. $$\int_0^1 x^{\alpha-\gamma}(1-x)^{\gamma-\beta-1}F(\alpha, \beta; \gamma; x)\,dx =$$
$$=\frac{\Gamma\left(1+\frac{\alpha}{2}\right)\Gamma(\gamma)\,\Gamma(\alpha-\gamma+1)\,\Gamma\left(\gamma-\frac{\alpha}{2}-\beta\right)}{\Gamma(1+\alpha)\,\Gamma\left(1+\frac{\alpha}{2}-\beta\right)\Gamma\left(\gamma-\frac{\alpha}{2}\right)}$$
$\left[\operatorname{Re}\alpha+1>\operatorname{Re}\gamma>\operatorname{Re}\beta, \quad \operatorname{Re}\left(\gamma-\frac{\alpha}{2}-\beta\right)>0\right]$.

ET II 398(1)

2. $$\int_0^1 x^{\varrho-1}(1-x)^{\beta-\gamma-n}F(-n, \beta; \gamma; x)\,dx = \frac{\Gamma(\gamma)\Gamma(\varrho)\Gamma(\beta-\gamma+1)\Gamma(\gamma-\varrho+n)}{\Gamma(\gamma+n)\Gamma(\gamma-\varrho)\Gamma(\beta-\gamma+\varrho+1)}$$

$[n=0, 1, 2, \ldots; \quad \operatorname{Re}\varrho > 0, \quad \operatorname{Re}(\beta-\gamma) > n-1]$. ET II 398(2)

3. $$\int_0^1 x^{\varrho-1}(1-x)^{\beta-\varrho-1}F(\alpha, \beta; \gamma; x)\,dx = \frac{\Gamma(\gamma)\Gamma(\varrho)\Gamma(\beta-\varrho)\Gamma(\gamma-\alpha-\varrho)}{\Gamma(\beta)\Gamma(\gamma-\alpha)\Gamma(\gamma-\varrho)}$$

$[\operatorname{Re}\varrho > 0, \quad \operatorname{Re}(\beta-\varrho) > 0, \quad \operatorname{Re}(\gamma-\alpha-\varrho) > 0]$. ET II 399(3)

4. $$\int_0^1 x^{\gamma-1}(1-x)^{\varrho-1}F(\alpha, \beta; \gamma; x)\,dx = \frac{\Gamma(\gamma)\Gamma(\varrho)\Gamma(\gamma+\varrho-\alpha-\beta)}{\Gamma(\gamma+\varrho-\alpha)\Gamma(\gamma+\varrho-\beta)}$$

$[\operatorname{Re}\gamma > 0, \quad \operatorname{Re}\varrho > 0, \quad \operatorname{Re}(\gamma+\varrho-\alpha-\beta) > 0]$. ET II 399(4)

5. $$\int_0^1 x^{\varrho-1}(1-x)^{\sigma-1}F(\alpha, \beta; \gamma; x)\,dx = \frac{\Gamma(\varrho)\Gamma(\sigma)}{\Gamma(\varrho+\sigma)}\,{}_3F_2(\alpha, \beta, \varrho; \gamma, \varrho+\sigma; 1)$$

$[\operatorname{Re}\varrho > 0, \quad \operatorname{Re}\sigma > 0, \quad \operatorname{Re}(\gamma+\sigma-\alpha-\beta) > 0]$. ET II 399(5)

6. $$\int_0^1 x^{\lambda-1}(1-x)^{\beta-\lambda-1}F\left(\alpha, \beta; \lambda; \frac{zx}{b}\right)dx = \mathrm{B}(\lambda, \beta-\lambda)\,F\left(\alpha, b; \beta; \frac{z}{b}\right).$$

BU 9

7. $$\int_0^1 x^{\gamma-1}(1-x)^{\delta-\gamma-1}F(\alpha, \beta; \gamma; xz)\,F(\delta-\alpha, \delta-\beta; \delta-\gamma; (1-x)\zeta)\,dx =$$
$$= \frac{\Gamma(\gamma)\Gamma(\delta-\gamma)}{\Gamma(\delta)}(1-\zeta)^{2\alpha-\delta}F(\alpha, \beta; \delta; z+\zeta-z\zeta)$$

$[0 < \operatorname{Re}\gamma < \operatorname{Re}\delta, \quad |\arg(1-z)| < \pi, \quad |\arg(1-\zeta)| < \pi]$.

ET II 400(11)

8. $$\int_0^1 x^{\gamma-1}(1-x)^{\varepsilon-1}(1-xz)^{-\delta}F(\alpha, \beta; \gamma; xz)\,F\left[\delta, \beta-\gamma; \varepsilon; \frac{(1-x)z}{(1-xz)}\right]dx =$$
$$= \frac{\Gamma(\gamma)\Gamma(\varepsilon)}{\Gamma(\gamma+\varepsilon)}F(\alpha+\delta, \beta; \gamma+\varepsilon; z)$$

$[\operatorname{Re}\gamma > 0, \quad \operatorname{Re}\varepsilon > 0, \; |\arg(z-1)| < \pi]$. ET II 400(12), EH I 78(3)

9. $$\int_0^1 x^{\gamma-1}(1-x)^{\varrho-1}(1-zx)^{-\sigma}F(\alpha, \beta; \gamma; x)\,dx =$$
$$= \frac{\Gamma(\gamma)\Gamma(\varrho)\Gamma(\gamma+\varrho-\alpha-\beta)}{\Gamma(\gamma+\varrho-\alpha)\Gamma(\gamma+\varrho-\beta)}(1-z)^{\sigma} \times$$
$$\times\,{}_3F_2\left(\varrho, \sigma, \gamma+\varrho-\alpha-\beta; \gamma+\varrho-\alpha, \gamma+\varrho-\beta; \frac{z}{z-1}\right)$$

$[\operatorname{Re}\gamma > 0, \; \operatorname{Re}\varrho > 0, \; \operatorname{Re}(\gamma+\varrho-\alpha-\beta) > 0, \quad |\arg(1-z)| < \pi]$.

ET II 399(6)

10. $$\int_0^\infty x^{\gamma-1}(x+z)^{-\sigma}F(\alpha, \beta; \gamma; -x)\,dx = \frac{\Gamma(\gamma)\Gamma(\alpha-\gamma+\sigma)\Gamma(\beta-\gamma+\sigma)}{\Gamma(\sigma)\Gamma(\alpha+\beta-\gamma+\sigma)} \times$$
$$\times F(\alpha-\gamma+\sigma, \beta-\gamma+\sigma; \alpha+\beta-\gamma+\sigma; 1-z)$$

$[\operatorname{Re}\gamma > 0, \quad \operatorname{Re}(\alpha-\gamma+\sigma) > 0, \quad \operatorname{Re}(\beta-\gamma+\sigma) > 0, \; |\arg z| < \pi]$. ET II 400(10)

11. $\int_0^1 (1-x)^{\mu-1} x^{\nu-1} {}_pF_q(a_1, \ldots, a_p; \nu, b_2, \ldots, b_q; ax)\, dx =$

$$= \frac{\Gamma(\mu)\Gamma(\nu)}{\Gamma(\mu+\nu)} {}_pF_q(a_1, \ldots, a_p; \mu+\nu, b_2, \ldots, b_q; a)$$

$[\operatorname{Re}\mu > 0, \quad \operatorname{Re}\nu > 0, \quad p \leqslant q+1; \quad \text{if } p = q+1, \text{ then } |a| < 1].$

ET II 200(94)

12. $\int_0^1 (1-x)^{\mu-1} x^{\nu-1} {}_pF_q(a_1, \ldots, a_p; b_1, \ldots, b_q; ax)\, dx =$

$$= \frac{\Gamma(\mu)\Gamma(\nu)}{\Gamma(\mu+\nu)} {}_{p+1}F_{q+1}(\nu, a_1, \ldots, a_p; \mu+\nu, b_1, \ldots, b_q; a)$$

$[\operatorname{Re}\mu > 0, \operatorname{Re}\nu > 0, p \leqslant q+1, \text{ if } p = q+1, \text{ then } |a| < 1].$

ET II 200(95)

7.513 $\int_0^1 x^{s-1}(1-x^2)^{\nu} F(-n, a; b; x^2)\, dx =$

$$= \frac{1}{2} \mathrm{B}\left(\nu+1, \frac{s}{2}\right) {}_3F_2\left(-n, a, \frac{s}{2}; b, \nu+1+\frac{s}{2}; 1\right)$$

$[\operatorname{Re} s > 0, \quad \operatorname{Re}\nu > -1].$ ET I 336(4)

7.52 Combinations of hypergeometric functions and exponentials

7.521 $\int_0^\infty e^{-st} {}_pF_q(a_1, \ldots, a_p; b_1, \ldots, b_q, t)\, dt =$

$$= \frac{1}{s} {}_{p+1}F_q(1, a_1, \ldots, a_p; b_1, \ldots, b_q, s^{-1}) \qquad [p \leqslant q].$$

EH I 192

7.522

1. $\int_0^\infty e^{-\lambda x} x^{\gamma-1} {}_2F_1(\alpha, \beta; \delta; -x)\, dx = \frac{\Gamma(\delta)\lambda^{-\gamma}}{\Gamma(\alpha)\Gamma(\beta)} E(\alpha, \beta, \gamma: \delta : \lambda)$

$[\operatorname{Re}\lambda > 0, \operatorname{Re}\gamma > 0].$ EH I 205(10)

2. $\int_0^\infty e^{-bx} x^{a-1} F\left(\frac{1}{2}+\nu, \frac{1}{2}-\nu; a; -\frac{x}{2}\right) dx = \frac{1}{\sqrt{\pi}} \Gamma(a)(2b)^{\frac{1}{2}-a} K_\nu(b)$

$[\operatorname{Re} a > 0, \operatorname{Re} b > 0].$ ET I 212(1)

3. $\int_0^\infty e^{-bx} x^{\gamma-1} F(2\alpha, 2\beta; \gamma; -\lambda x)\, dx =$

$$= \Gamma(\gamma) b^{-\gamma} \left(\frac{b}{\lambda}\right)^{\alpha+\beta-\frac{1}{2}} e^{\frac{b}{2\lambda}} W_{\frac{1}{2}-\alpha-\beta, \alpha-\beta}\left(\frac{b}{2\lambda}\right)$$

$[\operatorname{Re} b > 0, \operatorname{Re}\gamma > 0, |\arg\lambda| < \pi].$ BU 78(30), ET I 212(4)

4. $\int_0^\infty e^{-xt} t^{b-1} F(a, a-c+1; b; -t)\, dt = x^{b-a}\Gamma(b)\Psi(a, c; x)$

$[\operatorname{Re} b > 0, \operatorname{Re} x > 0].$ EH I 273(11)

5. $$\int_0^\infty e^{-x}x^{s-1}{}_pF_q(a_1, \ldots, a_p, b_1, \ldots, b_q; ax)\,dx =$$
$$= \Gamma(s)\,{}_{p+1}F_q(s, a_1, \ldots, a_p; b_1, \ldots, b_q; a)$$
$[p < q, \operatorname{Re} s > 0]$. ET I 337(11)

6. $$\int_0^\infty x^{\beta-1}e^{-\mu x}{}_2F_2(-n, n+1; 1, \beta; x)\,dx = \Gamma(\beta)\,\mu^{-\beta}P_n\left(1 - \frac{2}{\mu}\right)$$
$[\operatorname{Re}\mu > 0, \operatorname{Re}\beta > 0]$. ET I 218(6)

7. $$\int_0^\infty x^{\beta-1}e^{-\mu x}{}_2F_2\left(-n, n; \beta, \frac{1}{2}; x\right)dx = \Gamma(\beta)\,\mu^{-\beta}\cos\left[2n\arcsin\left(\frac{1}{\sqrt{\mu}}\right)\right]$$
$[\operatorname{Re}\mu > 0, \operatorname{Re}\beta > 0]$. ET I 218(7)

8. $$\int_0^\infty x^{\varrho_n-1}e^{-\mu x}{}_mF_n(a_1, \ldots, a_m; \varrho_1, \ldots, \varrho_n; \lambda x)\,dx =$$
$$= \Gamma(\varrho_n)\,\mu^{-\varrho_n}{}_mF_{n-1}\left(a_1, \ldots, a_m; \varrho_1, \ldots, \varrho_{n-1}; \frac{\lambda}{\mu}\right)$$
$[m \leqslant n;\ \operatorname{Re}\varrho_n > 0,\ \operatorname{Re}\mu > 0,$ if $m < n;\ \operatorname{Re}\mu > \operatorname{Re}\lambda,$ if $m = n]$.
ET I 219(16)a

9. $$\int_0^\infty x^{\sigma-1}e^{-\mu x}{}_mF_n(a_1, \ldots, a_m; \varrho_1, \ldots, \varrho_n; \lambda x)\,dx =$$
$$= \Gamma(\sigma)\,\mu^{-\sigma}{}_{m+1}F_n\left(a_1, \ldots, a_m, \sigma; \varrho_1, \ldots, \varrho_n; \frac{\lambda}{\mu}\right)$$
$[m \leqslant n,\ \operatorname{Re}\sigma > 0,\ \operatorname{Re}\mu > 0,$ if $m < n;\ \operatorname{Re}\mu > \operatorname{Re}\lambda,$ if $m = n]$.
ET I 219(17)

7.523 $$\int_0^1 x^{\gamma-1}(1-x)^{\varrho-1}e^{-xz}F(\alpha, \beta; \gamma; x)\,dx =$$
$$= \frac{\Gamma(\gamma)\,\Gamma(\varrho)\,\Gamma(\gamma+\varrho-\alpha-\beta)}{\Gamma(\gamma+\varrho-\alpha)\,\Gamma(\gamma+\varrho-\beta)}\,e^{-z}{}_2F_2(\varrho, \gamma+\varrho-\alpha-\beta; \gamma+\varrho-\alpha, \gamma+\varrho-\beta; z)$$
$[\operatorname{Re}\gamma > 0,\ \operatorname{Re}\varrho > 0,\ \operatorname{Re}(\gamma+\varrho-\alpha-\beta) > 0]$. ET II 400(8)

7.524

1. $$\int_0^\infty e^{-\lambda x}F\left(\alpha, \beta; \frac{1}{2}; -x^2\right)dx = \lambda^{\alpha+\beta-1}\,S_{1-\alpha-\beta,\,\alpha-\beta}(\lambda)$$
$[\operatorname{Re}\lambda > 0]$. ET II 401(13)

2. $$\int_0^\infty e^{-st}{}_pF_q(a_1, \ldots, a_p; b_1, \ldots, b_q; t^2)\,dt =$$
$$= s^{-1}{}_{p+2}F_q\left(a_1, \ldots, a_p, 1, \frac{1}{2}; b_1, \ldots, b_q; \frac{4}{s^2}\right) \quad [p < q].$$
MO 176

3. $$\int_0^\infty e^{-st}{}_0F_q\left(\frac{1}{q}, \frac{2}{q}, \ldots, \frac{q-1}{q}, 1; \frac{t^q}{q^q}\right)dt = s^{-1}\exp(s^{-q}).$$
MO 176

7.525

1. $$\int_0^\infty x^{\sigma-1} e^{-\mu x} {}_mF_n[a_1, \ldots, a_m; \varrho_1, \ldots, \varrho_n; (\lambda x)^k]\, dx =$$
$$= \Gamma(\sigma)\mu^{-\sigma} {}_{m+k}F_n\left[a_1, \ldots, a_m, \frac{\sigma}{k}, \frac{\sigma+1}{k}, \ldots, \frac{\sigma+k-1}{k}; \varrho_1, \ldots, \varrho_n; \left(\frac{k\lambda}{\mu}\right)^k\right]$$
$[m+k \leqslant n+1,\ \operatorname{Re}\sigma > 0;\ \operatorname{Re}\mu > 0,\quad \text{if}\ m+k \leqslant n;$
$\operatorname{Re}(\mu + k\lambda e^{\frac{2\pi ri}{k}}) > 0;\ r = 0, 1, \ldots, k-1\ \text{for}\ m+k = n+1].$ ET I 220(19)

2. $$\int_0^\infty x e^{-\lambda x} F\left(\alpha, \beta; \frac{3}{2}; -x^2\right) dx = \lambda^{\alpha+\beta-2} S_{1-\alpha-\beta,\, \alpha-\beta}(\lambda)$$
$[\operatorname{Re}\lambda > 0].$ ET II 401(14)

7.526

1. $$\int_{\gamma-i\infty}^{\gamma+i\infty} e^{st} s^{-b} F\left(a, b; a+b-c+1; 1-\frac{1}{s}\right) ds =$$
$$= 2\pi i \frac{\Gamma(a+b-c+1)}{\Gamma(b)\Gamma(b-c+1)} t^{b-1} \Psi(a; c; t)$$
$\left[\operatorname{Re} b > 0,\ \operatorname{Re}(b-c) > -1,\ \gamma > \frac{1}{2}\right].$ EH I 273(12)

2. $$\int_0^\infty e^{-t} t^{\gamma-1} (x+t)^{-\alpha} (y+t)^{-\alpha'} F\left[a, a'; \gamma; \frac{t(x+y+t)}{(x+t)(y+t)}\right] dt =$$
$$= \Gamma(\gamma)\Psi(a, c; x)\Psi(a', c; y),$$
$\gamma = a + a' - c + 1\quad [\operatorname{Re}\gamma > 0,\ xy \neq 0].$ EH I 287(21)

3. $$\int_0^\infty x^{\gamma-1}(x+y)^{-\alpha}(x+z)^{-\beta} e^{-x} F\left[\alpha, \beta; \gamma; \frac{x(x+y+z)}{(x+y)(x+z)}\right] dx =$$
$$= \Gamma(\gamma)(zy)^{-\frac{1}{2}-\mu} e^{\frac{y+z}{2}} W_{\nu,\mu}(y) W_{\lambda,\mu}(z),$$
$2\nu = 1 - \alpha + \beta - \gamma;\ 2\lambda = 1 + \alpha - \beta - \gamma;\ 2\mu = \alpha + \beta - \gamma$
$[\operatorname{Re}\gamma > 0,\ |\arg y| < \pi,\ |\arg z| < \pi].$ ET II 401(15)

7.527

1. $$\int_0^\infty (1-e^{-x})^{\lambda-1} e^{-\mu x} F(\alpha, \beta; \gamma; \delta e^{-x})\, dx = \mathrm{B}(\mu, \lambda)\, {}_3F_2(\alpha, \beta, \mu; \gamma, \mu+\lambda; \delta)$$
$[\operatorname{Re}\lambda > 0,\ \operatorname{Re}\mu > 0,\ |\arg(1-\delta)| < \pi].$ ET I 213(9)

2. $$\int_0^\infty (1-e^{-x})^{\mu} e^{-\alpha x} F(-n, \mu+\beta+n; \beta; e^{-x})\, dx =$$
$$= \frac{\mathrm{B}(\alpha, \mu+n+1)\,\mathrm{B}(\alpha, \beta+n-\alpha)}{\mathrm{B}(\alpha, \beta-\alpha)}$$
$[\operatorname{Re}\alpha > 0,\ \operatorname{Re}\mu > -1].$ ET I 213(10)

3. $\int_0^\infty (1-e^{-x})^{\gamma-1} e^{-\mu x} F(\alpha, \beta; \gamma; 1-e^{-x})\, dx = \frac{\Gamma(\mu)\,\Gamma(\gamma-\alpha-\beta+\mu)\,\Gamma(\gamma)}{\Gamma(\gamma-\alpha+\mu)\,\Gamma(\gamma-\beta+\mu)}$

$[\operatorname{Re}\mu > 0,\ \operatorname{Re}\mu > \operatorname{Re}(\alpha+\beta-\gamma),\ \operatorname{Re}\gamma > 0].$ ET I 213(11)

4. $\int_0^\infty (1-e^{-x})^{\gamma-1} e^{-\mu x} F[\alpha, \beta; \gamma; \delta(1-e^{-x})]\, dx = \mathrm{B}(\mu, \gamma) F(\alpha, \beta; \mu+\gamma; \delta)$

$[\operatorname{Re}\mu > 0,\ \operatorname{Re}\gamma > 0,\ |\arg(1-\delta)| < \pi].$ ET I 213(12)

7.53 Hypergeometric and trigonometric functions

7.531

1. $\int_0^\infty x \sin\mu x\, F\left(\alpha, \beta; \frac{3}{2}; -c^2x^2\right) dx = 2^{-\alpha-\beta+1} \pi c^{-\alpha-\beta} \mu^{\alpha+\beta-2} \frac{K_{\alpha-\beta}\left(\frac{\mu}{c}\right)}{\Gamma(\alpha)\,\Gamma(\beta)}$

$\left[\mu > 0,\ \operatorname{Re}\alpha > \frac{1}{2},\ \operatorname{Re}\beta > \frac{1}{2}\right].$ ET I 115(6)

2. $\int_0^\infty \cos\mu x\, F\left(\alpha, \beta; \frac{1}{2}; -c^2x^2\right) dx = 2^{-\alpha-\beta+1} \pi c^{-\alpha-\beta} \mu^{\alpha+\beta-1} \frac{K_{\alpha-\beta}\left(\frac{\mu}{c}\right)}{\Gamma(\alpha)\,\Gamma(\beta)}$

$[\mu > 0,\ \operatorname{Re}\alpha > 0,\ \operatorname{Re}\beta > 0,\ c > 0].$ ET I 61(9)

7.54 Combinations of hypergeometric and Bessel functions

7.541 $\int_0^\infty x^{\alpha+\beta-2\nu-1}(x+1)^{-\nu} e^{xz} K_\nu[(x+1)z]\, F(\alpha, \beta; \alpha+\beta-2\nu; -x)\, dx =$

$= \pi^{-\frac{1}{2}} \cos(\nu\pi)\, \Gamma\left(\frac{1}{2}-\alpha+\nu\right) \Gamma\left(\frac{1}{2}-\beta+\nu\right) \Gamma(\gamma) \times$

$\times (2z)^{-\frac{1}{2}-\frac{1}{2}\gamma} W_{\frac{1}{2}\gamma, \frac{1}{2}(\beta-\alpha)}(2z), \qquad \gamma = \alpha+\beta-2\nu$

$\left[\operatorname{Re}(\alpha+\beta-2\nu) > 0,\ \operatorname{Re}\left(\frac{1}{2}-\alpha+\nu\right) > 0,\ \operatorname{Re}\left(\frac{1}{2}-\beta+\nu\right) > 0,\ |\arg z| < \frac{3\pi}{2}\right].$ ET II 401(16)

7.542

1. $\int_0^\infty x^{\sigma-1} {}_pF_{p-1}(a_1, \ldots, a_p; b_1, \ldots, b_{p-1}; -\lambda x^2) N_\nu(xy)\, dx =$

$= \frac{\Gamma(b_1)\ldots\Gamma(b_{p-1})}{2\lambda^{\frac{1}{2}\sigma}\,\Gamma(a_1)\ldots\Gamma(a_p)} G^{p+2,\,1}_{p+2,\,p+3}\left(\frac{y^2}{4\lambda}\middle| \begin{matrix} b_0^*, \ldots, b_{p-1}^*, l \\ h, k, a_1^*, \ldots, a_p^*, l \end{matrix}\right),$

$a_j^* = a_j - \frac{\sigma}{2},\ j = 1, \ldots, p;\ b_0^* = 1 - \frac{\sigma}{2};\ b_j^* = b_j - \frac{\sigma}{2},\ j = 1, \ldots, p-1;$

$h = \frac{\nu}{2},\ k = -\frac{\nu}{2},\ l = -\frac{1+\nu}{2}$ $\left[|\arg\lambda| < \pi,\ \operatorname{Re}\sigma > |\operatorname{Re}\nu|,\ \operatorname{Re} a_j > \frac{1}{2}\operatorname{Re}\sigma - \frac{3}{4},\ y > 0\right].$ ET II 118(53)

2. $\int_0^\infty x^{\sigma-1}{}_pF_p(a_1, \ldots, a_p; b_1, \ldots, b_p; -\lambda x^2) N_\nu(xy)\, dx =$

$$= \frac{\Gamma(b_1)\ldots\Gamma(b_p)}{2\lambda^{\frac{1}{2}\sigma}\Gamma(a_1)\ldots\Gamma(a_p)} G_{p+2,\, p+3}^{p+2,\, 1}\left(\frac{y^2}{4\lambda}\middle|\begin{matrix} b_0^*, \ldots, b_p^*, l \\ h, k, a_1^*, \ldots, a_p^*, l\end{matrix}\right),$$

$b_0^* = 1 - \frac{\sigma}{2}$; $a_j^* = a_j - \frac{\sigma}{2}$, $b_j^* = b_j - \frac{\sigma}{2}$; $j = 1, \ldots, p$; $h = \frac{\nu}{2}$, $k = -\frac{\nu}{2}$, $l = -\frac{1+\nu}{2}$ $\left[\operatorname{Re}\lambda > 0,\ \operatorname{Re}\sigma > |\operatorname{Re}\nu|,\ \operatorname{Re} a_j > \frac{1}{2}\operatorname{Re}\sigma - \frac{3}{4},\ y > 0\right].$ ET II 119(54)

3. $\int_0^\infty x^{\sigma-1}{}_pF_q(a_1, \ldots, a_p; b_1, \ldots, b_q; -\lambda x^2) N_\nu(xy)\, dx =$

$$= -\pi^{-1}2^{\sigma-1}y^{-\sigma}\cos\left[\frac{\pi}{2}(\sigma-\nu)\right]\Gamma\left(\frac{\sigma+\nu}{2}\right)\Gamma\left(\frac{\sigma-\nu}{2}\right)\times$$
$$\times {}_{p+2}F_q\left(a_1, \ldots, a_p, \frac{\sigma+\nu}{2}, \frac{\sigma-\nu}{2}; b_1, \ldots, b_q; -\frac{4\lambda}{y^2}\right)$$
$[y > 0,\ p \leqslant q-1,\ \operatorname{Re}\sigma > |\operatorname{Re}\nu|].$ ET II 119(55)

4. $\int_0^\infty x^{\sigma-1}{}_pF_q(a_1, \ldots, a_p; b_1, \ldots, b_q; -\lambda x^2) K_\nu(xy)\, dx =$

$$= 2^{\sigma-2}y^{-\sigma}\Gamma\left(\frac{\sigma+\nu}{2}\right)\Gamma\left(\frac{\sigma-\nu}{2}\right)\times$$
$$\times {}_{p+2}F_q\left(a_1, \ldots, a_p, \frac{\sigma+\nu}{2}, \frac{\sigma-\nu}{2}; b_1, \ldots, b_q; \frac{4\lambda}{y^2}\right)$$
$[\operatorname{Re} y > 0,\ p \leqslant q-1,\ \operatorname{Re}\sigma > |\operatorname{Re}\nu|].$ ET II 153(88)

5. $\int_0^\infty x^{2\varrho}{}_pF_p(a_1, \ldots, a_p; b_1, \ldots, b_p; -\lambda x^2) J_\nu(xy)\, dx =$

$$= \frac{2^{2\varrho}\Gamma(b_1)\ldots\Gamma(b_p)}{y^{2\varrho+1}\Gamma(a_1)\ldots\Gamma(a_p)} G_{p+1,\, p+2}^{p+1,\, 1}\left(\frac{y^2}{4\lambda}\middle|\begin{matrix} 1, b_1, \ldots, b_p \\ h, a_1, \ldots, a_p, k\end{matrix}\right),$$
$$h = \frac{1}{2} + \varrho + \frac{1}{2}\nu,\quad k = \frac{1}{2} + \varrho - \frac{1}{2}\nu,$$
$\left[y > 0,\ \operatorname{Re}\lambda > 0,\ -1 - \operatorname{Re}\nu < 2\operatorname{Re}\varrho < \frac{1}{2} + 2\operatorname{Re} a_r,\ r = 1, \ldots, p\right].$

ET II 91(18)

6. $\int_0^\infty x^{2\varrho}{}_{m+1}F_m(a_1, \ldots, a_{m+1}; b_1, \ldots, b_m; -\lambda^2x^2) J_\nu(xy)\, dx =$

$$= \frac{2^{2\varrho}\Gamma(b_1)\ldots\Gamma(b_m)\, y^{-2\varrho-1}}{\Gamma(a_1)\ldots\Gamma(a_{m+1})} G_{m+1,\, m+3}^{m+2,\, 1}\left(\frac{y^2}{4\lambda^2}\middle|\begin{matrix} 1, b_1, \ldots, b_m \\ h, a_1, \ldots, a_{m+1}, k\end{matrix}\right),$$
$$h = \frac{1}{2} + \varrho + \frac{1}{2}\nu,\quad k = \frac{1}{2} + \varrho - \frac{1}{2}\nu,$$
$\left[y > 0,\ \operatorname{Re}\lambda > 0,\ \operatorname{Re}(2\varrho+\nu) > -1,\ \operatorname{Re}(\varrho - a_r) < \frac{1}{4}; r = 1, \ldots, m+1\right].$

ET II 91(19)

7. $$\int_0^\infty x^\delta F(\alpha, \beta; \gamma; -\lambda^2x^2) J_\nu(xy)\,dx =$$
$$= \frac{2^\delta\Gamma(\gamma)}{\Gamma(\alpha)\Gamma(\beta)} y^{-\delta-1} G_{24}^{22}\left(\frac{y^2}{4\lambda^2}\left|\begin{matrix}1-\alpha, 1-\beta \\ \frac{1+\delta+\nu}{2}, 0, 1-\gamma, \frac{1+\delta-\nu}{2}\end{matrix}\right.\right)$$
$$\left[y>0,\ \operatorname{Re}\lambda>0,\ -1-\operatorname{Re}\nu-2\min(\operatorname{Re}\alpha, \operatorname{Re}\beta)<\operatorname{Re}\delta<-\tfrac{1}{2}\right].$$

ET II 82(9)

8. $$\int_0^\infty x^\delta F(\alpha, \beta; \gamma; -\lambda^2x^2) J_\nu(xy)\,dx =$$
$$= \frac{2^\delta y^{-\delta-1}\Gamma(\gamma)}{\Gamma(\alpha)\Gamma(\beta)} G_{24}^{31}\left(\frac{y^2}{4\lambda^2}\left|\begin{matrix}1, \gamma \\ \frac{1+\delta+\nu}{2}, \alpha, \beta, \frac{1+\delta-\nu}{2}\end{matrix}\right.\right)$$
$$\left[y>0,\ \operatorname{Re}\lambda>0,\ -\operatorname{Re}\nu-1<\operatorname{Re}\delta<2\max(\operatorname{Re}\alpha, \operatorname{Re}\beta)-\tfrac{1}{2}\right].$$

ET II 81(6)

9. $$\int_0^\infty x^{\nu+1} F(\alpha, \beta; \gamma; -\lambda^2x^2) J_\nu(xy)\,dx =$$
$$= \frac{2^{\nu+1}\Gamma(\gamma)}{\Gamma(\alpha)\Gamma(\beta)} y^{-\nu-2} G_{13}^{30}\left(\frac{y^2}{4\lambda^2}\left|\begin{matrix}\gamma \\ \nu+1, \alpha, \beta\end{matrix}\right.\right)$$
$$\left[y>0,\ \operatorname{Re}\lambda>0,\ -1<\operatorname{Re}\nu<2\max(\operatorname{Re}\alpha, \operatorname{Re}\beta)-\tfrac{3}{2}\right].$$

ET II 81(5)

10. $$\int_0^\infty x^{\nu+1} F(\alpha, \beta; \nu+1; -\lambda^2x^2) J_\nu(xy)\,dx =$$
$$= \frac{2^{\nu-\alpha-\beta+2}\Gamma(\nu+1)}{\lambda^{\alpha+\beta}\Gamma(\alpha)\Gamma(\beta)} y^{\alpha+\beta-\nu-2} K_{\alpha-\beta}\left(\frac{y}{\lambda}\right)$$
$$\left[y>0,\ \operatorname{Re}\lambda>0,\ -1<\operatorname{Re}\nu<2\max(\operatorname{Re}\alpha, \operatorname{Re}\beta)-\tfrac{3}{2}\right].$$

ET II 81(3)

11. $$\int_0^\infty x^{\nu+1} F(\alpha, \beta; \nu+1; -\lambda^2x^2) K_\nu(xy)\,dx =$$
$$= 2^{\nu+1}\lambda^{-\alpha-\beta} y^{\alpha+\beta-\nu-2}\Gamma(\nu+1) S_{1-\alpha-\beta, \alpha-\beta}\left(\frac{y}{\lambda}\right)$$
$$[\operatorname{Re} y>0,\ \operatorname{Re}\lambda>0,\ \operatorname{Re}\nu>-1].$$

ET II 152(86)

12. $$\int_0^\infty x^{\nu+1} F\left(\alpha, \beta; \frac{\beta+\nu}{2}+1; -\lambda^2x^2\right) J_\nu(xy)\,dx =$$
$$= \frac{\Gamma\left(\frac{\beta+\nu+2}{2}\right) y^{\beta-1}\lambda^{-\nu-\beta-1}}{\pi^{\frac{1}{2}}\Gamma(\alpha)\Gamma(\beta)2^{\beta-1}}\left[K_{\frac{1}{2}(\nu-\beta+1)}\left(\frac{y}{2\lambda}\right)\right]^2$$
$$\left[y>0,\ -1<\operatorname{Re}\nu<2\max(\operatorname{Re}\alpha, \operatorname{Re}\beta)-\tfrac{3}{2}\right].$$

ET II 81(4)

13. $\int\limits_0^\infty x^{\sigma+\frac{1}{2}} F(\alpha, \beta; \gamma; -\lambda^2 x^2) N_\nu(xy)\,dx =$

$$= \frac{\lambda^{-\sigma-1} y^{-\frac{1}{2}} \Gamma(\gamma)}{\sqrt{2}\,\Gamma(\alpha)\Gamma(\beta)} G_{35}^{41}\left(\frac{y^2}{4\lambda^2}\middle|\begin{matrix}1-p, \gamma-p, l \\ h, k, \alpha-p, \beta-p, l\end{matrix}\right),$$

$$h = \frac{1}{4} + \frac{1}{2}\nu,\quad k = \frac{1}{4} - \frac{1}{2}\nu,\quad l = -\frac{1}{4} - \frac{1}{2}\nu,\quad p = \frac{1}{2} + \frac{1}{2}\sigma$$

$\left[y > 0,\ \operatorname{Re}\lambda > 0,\ \operatorname{Re}\sigma > |\operatorname{Re}\nu| - \frac{3}{2},\ \operatorname{Re}\sigma < 2\operatorname{Re}\alpha,\ \operatorname{Re}\sigma < 2\operatorname{Re}\beta\right].$

ET II 118(52)

14. $\int\limits_0^\infty x^{\nu+2} F\left(\frac{1}{2}, \frac{1}{2}-\nu; \frac{3}{2}; -\lambda^2 x^2\right) N_\nu(xy)\,dx =$

$$= \frac{2^\nu y^{-\nu-1}}{\pi^{\frac{1}{2}} \lambda^2 \Gamma\left(\frac{1}{2}-\nu\right)} K_\nu\left(\frac{y}{2\lambda}\right) K_{\nu+1}\left(\frac{y}{2\lambda}\right)$$

$\left[y > 0,\ \operatorname{Re}\lambda > 0,\ -\frac{3}{2} < \operatorname{Re}\nu < -\frac{1}{2}\right].$ ET II 117(49)

15. $\int\limits_0^\infty x^{\nu+2} F\left(1, 2\nu+\frac{3}{2}; \nu+2; -\lambda^2 x^2\right) N_\nu(xy)\,dx =$

$$= \pi^{-\frac{1}{2}} 2^{-\nu} \lambda^{-2\nu-3} y^\nu \frac{\Gamma(\nu+2)}{\Gamma\left(2\nu+\frac{3}{2}\right)} \left[K_\nu\left(\frac{y}{2\lambda}\right)\right]^2$$

$\left[y > 0,\ \operatorname{Re}\lambda > 0,\ -\frac{1}{2} < \operatorname{Re}\nu < \frac{1}{2}\right].$ ET II 117(50)

16. $\int\limits_0^\infty x^{\nu+2} F\left(1, \mu+\nu+\frac{3}{2}; \frac{3}{2}; -\lambda^2 x^2\right) N_\nu(xy)\,dx =$

$$= \frac{\pi^{\frac{1}{2}} 2^{-\mu-\nu-1} \lambda^{-\mu-2\nu-3} y^{\mu+\nu}}{\Gamma\left(\mu+\nu+\frac{3}{2}\right)} K_\mu\left(\frac{y}{\lambda}\right)$$

$\left[y > 0,\ \operatorname{Re}\lambda > 0,\ -\frac{3}{2} < \operatorname{Re}\nu < \frac{1}{2},\ \operatorname{Re}(2\mu+\nu) > -\frac{3}{2}\right].$

ET II 118(51)

17. $\int\limits_0^\infty x^{2\alpha+\nu} F\left(\alpha-\nu-\frac{1}{2}, \alpha; 2\alpha; -\lambda^2 x^2\right) J_\nu(xy)\,dx =$

$$= \frac{i\Gamma\left(\frac{1}{2}+\alpha\right)\Gamma\left(\frac{1}{2}+\alpha+\nu\right)}{\pi 2^{1-\nu-2\alpha} \lambda^{2\alpha-1} y^{\nu+2}} W_{\frac{1}{2}-\alpha,\, -\frac{1}{2}-\nu}\left(\frac{y}{\lambda}\right) \times$$

$$\times \left[W_{\frac{1}{2}-\alpha,\, -\frac{1}{2}-\nu}\left(e^{-i\pi}\frac{y}{\lambda}\right) - W_{\frac{1}{2}-\alpha,\, -\frac{1}{2}-\nu}\left(e^{i\pi}\frac{y}{\lambda}\right)\right]$$

$\left[y > 0,\ \operatorname{Re}\lambda > 0,\ \operatorname{Re}\nu < -\frac{1}{2},\ \operatorname{Re}(\alpha+\nu) > -\frac{1}{2}\right].$ ET II 80(1)

18. $$\int_0^\infty x^{2\alpha-\nu} F\left(\nu+\alpha-\frac{1}{2},\ \alpha;\ 2\alpha;\ -\lambda^2 x^2\right) J_\nu(xy)\,dx =$$
$$= \frac{2^{2\alpha-\nu}\Gamma\left(\frac{1}{2}+\alpha\right) y^{\nu-2}}{\lambda^{2\alpha-1}\Gamma(2\nu)} M_{\alpha-\frac{1}{2},\ \nu-\frac{1}{2}}\left(\frac{y}{\lambda}\right) W_{\frac{1}{2}-\alpha,\ \nu-\frac{1}{2}}\left(\frac{y}{\lambda}\right).$$

ET II 80(2)

7.543

1. $$\int_0^\infty x^{-2\alpha-1} F\left(\frac{1}{2}+\alpha,\ 1+\alpha;\ 1+2\alpha;\ -\frac{4\lambda^2}{x^2}\right) J_\nu(xy)\,dx =$$
$$= \lambda^{-2\alpha} I_{\frac{1}{2}\nu+\alpha}(\lambda y) K_{\frac{1}{2}\nu-\alpha}(\lambda y)$$
$$\left[y>0,\ \operatorname{Re}\lambda>0,\ \operatorname{Re}\nu>-1,\ \operatorname{Re}\alpha>-\frac{1}{2}\right].$$ ET II 81(7)

2. $$\int_0^\infty x^{\nu+1-4\alpha} F\left(\alpha,\ \alpha+\frac{1}{2};\ \nu+1;\ -\frac{\lambda^2}{x^2}\right) J_\nu(xy)\,dx =$$
$$= \frac{\Gamma(\nu)}{\Gamma(2\alpha)} 2^\nu \lambda^{1-2\alpha} y^{2\alpha-\nu-1} I_\nu\left(\frac{1}{2}\lambda y\right) K_{2\alpha-\nu-1}\left(\frac{1}{2}\lambda y\right)$$
$$\left[y>0,\ \operatorname{Re}\lambda>0,\ \operatorname{Re}\alpha-1<\operatorname{Re}\nu<4\operatorname{Re}\alpha-\frac{3}{2}\right].$$ ET II 81(8)

7.544 $$\int_0^\infty x^{\nu+1}(1+x)^{-2\alpha} F\left[\alpha,\ \nu+\frac{1}{2};\ 2\nu+1;\ \frac{4x}{(1+x)^2}\right] J_\nu(xy)\,dx =$$
$$= \frac{\Gamma(\nu+1)\Gamma(\nu-\alpha+1)}{\Gamma(\alpha)} 2^{2\nu-2\alpha+1} y^{2(\alpha-\nu-1)} J_\nu(y)$$
$$\left[y>0,\ -1<\operatorname{Re}\nu<2\operatorname{Re}\alpha-\frac{3}{2}\right].$$ ET II 82(10)

7.6 Degenerate Hypergeometric Functions

7.61 Combinations of degenerate hypergeometric functions and powers

7.611

1. $$\int_0^\infty x^{-1} W_{k,\mu}(x)\,dx = \frac{\pi^{\frac{3}{2}} 2^k \sec(\mu\pi)}{\Gamma\left(\frac{3}{4}-\frac{1}{2}k+\frac{1}{2}\mu\right)\Gamma\left(\frac{3}{4}-\frac{1}{2}k-\frac{1}{2}\mu\right)}$$
$$\left[|\operatorname{Re}\mu|<\frac{1}{2}\right].$$ ET II 406(22)

2. $$\int_0^\infty x^{-1} M_{k,\mu}(x) W_{\lambda,\mu}(x)\,dx = \frac{\Gamma(2\mu+1)}{(k-\lambda)\Gamma\left(\frac{1}{2}+\mu-\lambda\right)}$$
$$\left[\operatorname{Re}\mu>-\frac{1}{2},\ \operatorname{Re}(k-\lambda)>0\right].$$ BU 116(11), ET II 409(39)

3. $$\int_0^\infty x^{-1} W_{k,\mu}(x) W_{\lambda,\mu}(x)\,dx =$$
$$= \frac{1}{(k-\lambda)\sin(2\mu\pi)} \left[\frac{1}{\Gamma\left(\frac{1}{2}-k+\mu\right)\Gamma\left(\frac{1}{2}-\lambda-\mu\right)} - \right.$$
$$\left. - \frac{1}{\Gamma\left(\frac{1}{2}-k-\mu\right)\Gamma\left(\frac{1}{2}-\lambda+\mu\right)} \right] \qquad \left[|\operatorname{Re}\mu| < \frac{1}{2} \right].$$

BU 116(12), ET II 409(40)

4. $$\int_0^\infty \{W_{\varkappa,\mu}(z)\}^2 \frac{dz}{z} = \frac{\pi}{\sin 2\pi\mu} \frac{\psi\left(\frac{1}{2}+\mu-\varkappa\right) - \psi\left(\frac{1}{2}-\mu-\varkappa\right)}{\Gamma\left(\frac{1}{2}+\mu-\varkappa\right)\Gamma\left(\frac{1}{2}-\mu-\varkappa\right)}$$
$$\left[|\operatorname{Re}\mu| < \frac{1}{2} \right].$$

BU 117(12a)

5. $$\int_0^\infty \frac{1}{z} [W_{\varkappa,0}(z)]^2\,dz = \frac{\psi'\left(\frac{1}{2}-\varkappa\right)}{\left[\Gamma\left(\frac{1}{2}-\varkappa\right)\right]^2}.$$

BU 117(12b)

6. $$\int_0^\infty x^{\varrho-1} W_{k,\mu}(x) W_{-k,\mu}(x)\,dx =$$
$$= \frac{\Gamma(\varrho+1)\Gamma\left(\frac{1}{2}\varrho+\frac{1}{2}+\mu\right)\Gamma\left(\frac{1}{2}\varrho+\frac{1}{2}-\mu\right)}{2\Gamma\left(1+\frac{1}{2}\varrho+k\right)\Gamma\left(1+\frac{1}{2}\varrho-k\right)}$$
$$[\operatorname{Re}\varrho > 2|\operatorname{Re}\mu| - 1].$$

ET II 409(41)

7. $$\int_0^\infty x^{\varrho-1} W_{k,\mu}(x) W_{\lambda,\nu}(x)\,dx = \frac{\Gamma(1+\mu+\nu+\varrho)\Gamma(1-\mu+\nu+\varrho)\Gamma(-2\nu)}{\Gamma\left(\frac{1}{2}-\lambda-\nu\right)\Gamma\left(\frac{3}{2}-k+\nu+\varrho\right)} \times$$
$$\times {}_3F_2\left(1+\mu+\nu+\varrho,\ 1-\mu+\nu+\varrho,\ \frac{1}{2}-\lambda+\nu;\ 1+2\nu,\ \frac{3}{2}-k+\nu+\varrho;\ 1\right) +$$
$$+ \frac{\Gamma(1+\mu-\nu+\varrho)\Gamma(1-\mu-\nu+\varrho)\Gamma(2\nu)}{\Gamma\left(\frac{1}{2}-\lambda+\nu\right)\Gamma\left(\frac{3}{2}-k-\nu+\varrho\right)} \times$$
$$\times {}_3F_2\left(1+\mu-\nu+\varrho,\ 1-\mu-\nu+\varrho,\ \frac{1}{2}-\lambda-\nu;\ 1-2\nu,\ \frac{3}{2}-k-\nu+\varrho;\ 1\right)$$
$$[|\operatorname{Re}\mu| + |\operatorname{Re}\nu| < \operatorname{Re}\varrho + 1].$$

ET II 410(42)

7.612

1. $$\int_0^\infty t^{b-1}\,{}_1F_1(a; c; -t)\,dt = \frac{\Gamma(b)\Gamma(c)\Gamma(a-b)}{\Gamma(a)\Gamma(c-b)} \qquad [0 < \operatorname{Re} b < \operatorname{Re} a].$$

EH I 285(10)

2. $$\int_0^\infty t^{b-1}\,\Psi(a, c; t)\,dt = \frac{\Gamma(b)\Gamma(a-b)\Gamma(b-c+1)}{\Gamma(a)\Gamma(a-c+1)}$$
$$[0 < \operatorname{Re} b < \operatorname{Re} a, \quad \operatorname{Re} c < \operatorname{Re} b + 1].$$

EH I 285(11)

7.613

1. $$\int_0^t x^{\gamma-1}(t-x)^{c-\gamma-1}\, {}_1F_1(a;\ \gamma;\ x)\,dx = t^{c-1}\frac{\Gamma(\gamma)\,\Gamma(c-\gamma)}{\Gamma(c)}\, {}_1F_1(a;\ c;\ t)$$
$$[\operatorname{Re} c > \operatorname{Re}\gamma > 0].$$ BU 9(16)a, EH I 271(16)

2. $$\int_0^t x^{\beta-1}(t-x)^{\gamma-1}\, {}_1F_1(t;\ \beta;\ x)\,dx = \frac{\Gamma(\beta)\,\Gamma(\gamma)}{\Gamma(\beta+\gamma)}\, t^{\beta+\gamma-1}\, {}_1F_1(t;\ \beta+\gamma;\ t)$$
$$[\operatorname{Re}\beta > 0,\ \operatorname{Re}\gamma > 0].$$ ET II 401(1)

3. $$\int_0^1 x^{\lambda-1}(1-x)^{2\mu-\lambda}\, {}_1F_1\left(\frac{1}{2}+\mu-\nu;\ \lambda;\ xz\right)dx =$$
$$= \mathrm{B}(\lambda,\ 1+2\mu-\lambda)\, e^{\frac{1}{2}z}\, z^{-\frac{1}{2}-\mu}\, M_{\nu,\mu}(z)$$
$$[\operatorname{Re}\lambda > 0,\ \operatorname{Re}(2\mu-\lambda) > -1].$$ BU 14(14)

4. $$\int_0^t x^{\beta-1}(t-x)^{\delta-1}\, {}_1F_1(t;\ \beta;\ x)\, {}_1F_1(\gamma;\ \delta;\ t-x)\,dx =$$
$$= \frac{\Gamma(\beta)\,\Gamma(\delta)}{\Gamma(\beta+\delta)}\, t^{\beta+\delta-1}\, {}_1F_1(t+\gamma;\ \beta+\delta;\ t)$$
$$[\operatorname{Re}\beta > 0,\ \operatorname{Re}\delta > 0].$$ ET II 402(2), EH I 271(15)

5. $$\int_0^t x^{\mu-\frac{1}{2}}(t-x)^{\nu-\frac{1}{2}} M_{k,\mu}(x)\, M_{\lambda,\nu}(t-x)\,dx =$$
$$= \frac{\Gamma(2\mu+1)\,\Gamma(2\nu+1)}{\Gamma(2\mu+2\nu+2)}\, t^{\mu+\nu} M_{k+\lambda,\ \mu+\nu+\frac{1}{2}}(t)$$
$$\left[\operatorname{Re}\mu > -\frac{1}{2},\ \operatorname{Re}\nu > -\frac{1}{2}\right].$$ BU 128(14), ET II 402(7)

6. $$\int_0^1 x^{\beta-1}(1-x)^{\sigma-\beta-1}\, {}_1F_1(\alpha;\ \beta;\ \lambda x)\, {}_1F_1[\sigma-\alpha;\ \sigma-\beta;\ \mu(1-x)]\,dx =$$
$$= \frac{\Gamma(\beta)\,\Gamma(\sigma-\beta)}{\Gamma(\sigma)}\, e^{\lambda}\, {}_1F_1(\alpha;\ \sigma;\ \mu-\lambda)$$
$$[0 < \operatorname{Re}\beta < \operatorname{Re}\sigma].$$ ET II 402(3)

7.62-7.63 Combinations of degenerate hypergeometric functions and exponentials

7.621

1. $$\int_0^\infty e^{-st}\, t^{\alpha} M_{\mu,\nu}(t)\,dt = \frac{\Gamma\left(\alpha+\nu+\frac{3}{2}\right)}{\left(\frac{1}{2}+s\right)^{\alpha+\nu+\frac{3}{2}}} \times$$
$$\times F\left(\alpha+\nu+\frac{3}{2},\ -\mu+\nu+\frac{1}{2};\ 2\nu+1;\ \frac{2}{2s+1}\right)$$
$$\left[\operatorname{Re}\left(\alpha+\mu+\frac{3}{2}\right) > 0,\ \operatorname{Re} s > \frac{1}{2}\right].$$ BU 118(1), MO 176a, EH I 270(12)a

2. $\int\limits_0^\infty e^{-st} t^{\mu-\frac{1}{2}} M_{\lambda,\mu}(qt)\,dt =$

$$= q^{\mu+\frac{1}{2}} \Gamma(2\mu+1) \left(s-\frac{1}{2}q\right)^{\lambda-\mu-\frac{1}{2}} \left(s+\frac{1}{2}q\right)^{-\lambda-\mu-\frac{1}{2}}$$
$$\left[\operatorname{Re}\mu > -\frac{1}{2},\ \operatorname{Re}s > \frac{|\operatorname{Re}q|}{2}\right].$$

BU 119(4c), MO176a, EH I 271(13)a

3. $\int\limits_0^\infty e^{-st} t^{\alpha} W_{\lambda,\mu}(qt)\,dt =$

$$= \frac{\Gamma\left(\alpha+\mu+\frac{3}{2}\right)\Gamma\left(\alpha-\mu+\frac{3}{2}\right) q^{\mu+\frac{1}{2}}}{\Gamma(\alpha-\lambda+2)} \left(s+\frac{1}{2}q\right)^{-\alpha-\mu-\frac{3}{2}} \times$$
$$\times F\left(\alpha+\mu+\frac{3}{2},\ \mu-\lambda+\frac{1}{2};\ \alpha-\lambda+2;\ \frac{2s-q}{2s+q}\right)$$
$$\left[\operatorname{Re}\left(\alpha\pm\mu+\frac{3}{2}\right) > 0,\ \operatorname{Re}s > -\frac{q}{2},\ q > 0\right].$$

EH I 271(14)a, BU 121(6), MO 176

4. $\int\limits_0^\infty e^{-st} t^{b-1}\, {}_1F_1(a;\ c;\ kt)\,dt = \Gamma(b) s^{-b} F(a,\ b;\ c;\ ks^{-1}) \qquad [|s| > |k|];$

$$= \Gamma(b)(s-k)^{-b} F\left(c-a,\ b;\ c;\ \frac{k}{k-s}\right) \qquad [|s-k| > |k|];$$
$$[\operatorname{Re}b > 0,\quad \operatorname{Re}s > \max(0,\ \operatorname{Re}k)].$$

EH I 269(5)

5. $\int\limits_0^\infty t^{c-1}\, {}_1F_1(a;\ c;\ t) e^{-st}\,dt = \Gamma(c) s^{-c} (1-s^{-1})^{-a}$

$$[\operatorname{Re}c > 0,\ \operatorname{Re}s > 1].$$

EH I 270(6)

6. $\int\limits_0^\infty t^{b-1} \Psi(a,\ c;\ t) e^{-st}\,dt =$

$$= \frac{\Gamma(b)\Gamma(b-c+1)}{\Gamma(a+b-c+1)} F(b,\ b-c+1;\ a+b-c+1;\ 1-s)$$
$$[\operatorname{Re}b > 0,\ \operatorname{Re}c < \operatorname{Re}b+1,\ |1-s| < 1];$$
$$= \frac{\Gamma(b)\Gamma(b-c+1)}{\Gamma(a+b-c+1)} s^{-b} F(a,\ b;\ a+b-c+1;\ 1-s^{-1})$$
$$\left[\operatorname{Re}s > \frac{1}{2}\right].$$

EH I 270(7)

7. $\int\limits_0^\infty e^{-\frac{b}{2}x} x^{\nu-1} M_{\varkappa,\mu}(bx)\,dx = \frac{\Gamma(1+2\mu)\Gamma(\varkappa-\nu)\Gamma\left(\frac{1}{2}+\mu+\nu\right)}{\Gamma\left(\frac{1}{2}+\mu+\varkappa\right)\Gamma\left(\frac{1}{2}+\mu-\nu\right)} b^{\nu}$

$$\left[\operatorname{Re}\left(\nu+\frac{1}{2}+\mu\right) > 0,\ \operatorname{Re}(\varkappa-\nu) > 0\right].$$

BU 119(3)a, ET I 215(11)a

8. $$\int_0^\infty e^{-sx} M_{\varkappa,\mu}(x) \frac{dx}{x} = \frac{2\Gamma(1+2\mu) e^{-i\pi\varkappa}}{\Gamma\left(\frac{1}{2}+\mu+\varkappa\right)} \left(\frac{s-\frac{1}{2}}{s+\frac{1}{2}}\right)^{\frac{\varkappa}{2}} Q^{\varkappa}_{\mu-\frac{1}{2}}(2s)$$

$$\left[\operatorname{Re}\left(\frac{1}{2}+\mu\right) > 0,\ \operatorname{Re} s > \frac{1}{2}\right].$$ BU 119(4a)

9. $$\int_0^\infty e^{-sx} W_{\varkappa,\mu}(x) \frac{dx}{x} = \frac{\pi}{\cos\left(\frac{\pi\mu}{2}\right)} \left(\frac{s-\frac{1}{2}}{s+\frac{1}{2}}\right)^{\frac{\varkappa}{2}} P^{\varkappa}_{\mu-\frac{1}{2}}(2s)$$

$$\left[\operatorname{Re}\left(\frac{1}{2}\pm\mu\right) > 0,\ \operatorname{Re} s > -\frac{1}{2}\right].$$ BU 121(7)

10. $$\int_0^\infty x^{k+2\mu-1} e^{-\frac{3}{2}x} W_{k,\mu}(x)\, dx = \frac{\Gamma\left(k+\mu+\frac{1}{2}\right)\Gamma\left[\frac{1}{4}(2k+6\mu+5)\right]}{\left(k+3\mu+\frac{1}{2}\right)\Gamma\left[\frac{1}{4}(2\mu-2k+3)\right]}$$

$$\left[\operatorname{Re}(k+\mu) > -\frac{1}{2},\ \operatorname{Re}(k+3\mu) > -\frac{1}{2}\right].$$

BU 122(8a), ET II 406(23)

11. $$\int_0^\infty e^{-\frac{1}{2}x} x^{\nu-1} W_{\varkappa,\mu}(x)\, dx = \frac{\Gamma\left(\nu+\frac{1}{2}-\mu\right)\Gamma\left(\nu+\frac{1}{2}+\mu\right)}{\Gamma(\nu-\varkappa+1)}$$

$$\left[\operatorname{Re}\left(\nu+\frac{1}{2}\pm\mu\right) > 0\right].$$ BU 122(8b)

12. $$\int_0^\infty e^{\frac{1}{2}x} x^{\nu-1} W_{\varkappa,\mu}(x)\, dx = \Gamma(-\varkappa-\mu) \frac{\Gamma\left(\frac{1}{2}+\mu+\nu\right)\Gamma\left(\frac{1}{2}-\mu+\nu\right)}{\Gamma\left(\frac{1}{2}-\mu-\varkappa\right)\Gamma\left(\frac{1}{2}+\mu-\varkappa\right)}$$

$$\left[\operatorname{Re}\left(\nu+\frac{1}{2}\pm\mu\right) > 0,\ \operatorname{Re}(\varkappa+\nu) < 0\right].$$ BU 122(8c)a

7.622

1. $$\int_0^\infty e^{-st} t^{c-1} {}_1F_1(a;\, c;\, t)\, {}_1F_1(\alpha;\, c;\, \lambda t)\, dt =$$

$$= \Gamma(c)(s-1)^{-a}(s-\lambda)^{-\alpha} s^{a+\alpha-c} F\left[a,\, \alpha;\, c;\, \lambda(s-1)^{-1}(s-\lambda)^{-1}\right]$$

$$[\operatorname{Re} c > 0,\ \operatorname{Re} s > \operatorname{Re}\lambda + 1].$$ EH I 287(22)

2. $$\int_0^\infty e^{-t} t^{\varrho} {}_1F_1(a;\, c;\, t)\, \Psi(a';\, c';\, \lambda t)\, dt = C \frac{\Gamma(c)\Gamma(\beta)}{\Gamma(\gamma)} \lambda^{\sigma} F(c-a,\, \beta;\, \gamma;\, 1-\lambda^{-1}),$$

$$\varrho = c-1,\ \sigma = -c,\ \beta = c-c'+1,\ \gamma = c-a+a'-c'+1,\ C = \frac{\Gamma(a'-a)}{\Gamma(a')},$$

or

$$\varrho = c+c'-2,\ \sigma = 1-c-c',\ \beta = c+c'-1,\ \gamma = a'-a+c,\ C = \frac{\Gamma(a'-a-c'+1)}{\Gamma(a'-c'+1)}.$$

EH I 287(24)

3. $$\int_0^\infty x^{\nu-1}e^{-bx}M_{\lambda_1,\mu_1-\frac{1}{2}}(a_1x)\ldots M_{\lambda_n,\mu_n-\frac{1}{2}}(a_nx)\,dx=$$
$$=a_1^{\mu_1}\ldots a_n^{\mu_n}(b+A)^{-\nu-M}\,\Gamma(\nu+M)\times$$
$$\times F_A\left(\nu+M;\ \mu_1-\lambda_1,\ \ldots,\ \mu_n-\lambda_n;\ 2\mu_1,\ \ldots,\ 2\mu_n:\frac{a_1}{b+A},\ \ldots,\ \frac{a_n}{b+A}\right),$$
$$M=\mu_1+\ldots+\mu_n,\quad A=\frac{1}{2}(a_1+\ldots+a_n)$$
$$\left[\operatorname{Re}(\nu+M)>0,\ \operatorname{Re}\left(b\pm\frac{1}{2}a_1\pm\ldots\pm\frac{1}{2}a_n\right)>0\right].$$ ET I 216(14)

7.623

1. $$\int_0^\infty e^{-x}x^{c+n-1}(x+y)^{-1}{}_1F_1(a;\ c;\ x)\,dx=$$
$$=(-1)^n\Gamma(c)\,\Gamma(1-a)\,y^{c+n-1}\Psi(c-a,\ c;\ y)$$
$$[-\operatorname{Re}c<n<1-\operatorname{Re}a,\ n=0,\ 1,\ 2,\ \ldots,\ |\arg y|<\pi].$$ EH I 285(16)

2. $$\int_0^t x^{-1}(t-x)^{k-1}e^{\frac{1}{2}(t-x)}M_{k,\mu}(x)\,dx=\frac{\Gamma(k)\,\Gamma(2\mu+1)}{\Gamma\left(k+\mu+\frac{1}{2}\right)}\pi^{\frac{1}{2}}t^{k-\frac{1}{2}}I_\mu\left(\frac{1}{2}t\right)$$
$$\left[\operatorname{Re}k>0,\ \operatorname{Re}\mu>-\frac{1}{2}\right].$$ ET II 402(5)

3. $$\int_0^t x^{k-1}(t-x)^{\lambda-1}e^{\frac{1}{2}(t-x)}M_{k+\lambda,\mu}(x)\,dx=\frac{\Gamma(\lambda)\,\Gamma\left(k+\mu+\frac{1}{2}\right)t^{k+\lambda-1}}{\Gamma\left(k+\lambda+\mu+\frac{1}{2}\right)}M_{k,\mu}(t)$$
$$\left[\operatorname{Re}(k+\mu)>-\frac{1}{2},\ \operatorname{Re}\lambda>0\right].$$ ET II 402(6)

4. $$\int_0^t x^{-k-\lambda-1}(t-x)^{\lambda-1}e^{\frac{1}{2}x}W_{k,\mu}(x)\,dx=$$
$$=\frac{\Gamma(\lambda)\,\Gamma\left(\frac{1}{2}-k-\lambda+\mu\right)\Gamma\left(\frac{1}{2}-k-\lambda-\mu\right)}{t^{k+1}\Gamma\left(\frac{1}{2}-k+\mu\right)\Gamma\left(\frac{1}{2}-k-\mu\right)}W_{k+\lambda,\mu}(t)$$
$$\left[\operatorname{Re}\lambda>0,\ \operatorname{Re}(k+\lambda)<\frac{1}{2}-|\operatorname{Re}\mu|\right].$$ ET II 405(21)

5. $$\int_1^\infty (x-1)^{\mu-1}x^{\lambda-\frac{1}{2}}e^{\frac{1}{2}ax}W_{k,\lambda}(ax)\,dx=$$
$$=\frac{\Gamma(\mu)\Gamma\left(\frac{1}{2}-k-\lambda-\mu\right)}{\Gamma\left(\frac{1}{2}-k-\lambda\right)}a^{-\frac{1}{2}\mu}e^{\frac{1}{2}a}W_{k+\frac{1}{2}\mu,\lambda+\frac{1}{2}\mu}(a)$$
$$\left[|\arg(a)|<\frac{3}{2}\pi,\ 0<\operatorname{Re}\mu<\frac{1}{2}-\operatorname{Re}(k+\lambda)\right].$$ ET II 211(72)a

6. $$\int_1^\infty (x-1)^{\mu-1} x^{\lambda-\frac{1}{2}} e^{-\frac{1}{2}ax} W_{k,\lambda}(ax)\,dx = a^{-\frac{1}{2}\mu}\Gamma(\mu)\,e^{-\frac{1}{2}a} W_{k-\frac{1}{2}\mu,\,\lambda-\frac{1}{2}\mu}(a)$$

$[\operatorname{Re}\mu > 0,\ \operatorname{Re} a > 0]$. ET II 211(74)a

7 $$\int_1^\infty (x-1)^{\mu-1} x^{k-\mu-1} e^{-\frac{1}{2}ax} W_{k,\lambda}(ax)\,dx = \Gamma(\mu)\,e^{-\frac{1}{2}a} W_{k-\mu,\lambda}(a)$$

$[\operatorname{Re}\mu > 0,\ \operatorname{Re} a > 0]$. ET II 211(73)a

8. $$\int_0^1 (1-x)^{\mu-1} x^{k-\mu-1} e^{-\frac{1}{2}ax} W_{k,\lambda}(ax)\,dx = \Gamma(\mu)\,e^{-\frac{1}{2}a} \sec[(k-\mu-\lambda)\pi] \times$$

$$\times \left\{\sin(\mu\pi)\,\frac{\Gamma\left(k-\mu+\lambda+\frac{1}{2}\right)}{\Gamma(2\lambda+1)} M_{k-\mu,\lambda}(a) + \cos[(k-\lambda)\pi]\,W_{k-\mu,\lambda}(a)\right\}$$

$\left[0 < \operatorname{Re}\mu < \operatorname{Re} k - |\operatorname{Re}\lambda| + \frac{1}{2}\right]$. ET II 200(93)a

7.624

1. $$\int_0^\infty x^{\varrho-1}\left[x^{\frac{1}{2}} + (a+x)^{\frac{1}{2}}\right]^{2\sigma} e^{-\frac{1}{2}x} M_{k,\mu}(x)\,dx =$$

$$= \frac{-\sigma\Gamma(2\mu+1)\,a^\sigma}{\pi^{\frac{1}{2}}\Gamma\left(\frac{1}{2}+k+\mu\right)} G^{23}_{34}\left(a\,\middle|\,\begin{matrix}\frac{1}{2},\ 1,\ 1-k+\varrho\\ \frac{1}{2}+\mu+\varrho,\ -\sigma,\ \sigma,\ \frac{1}{2}-\mu+\varrho\end{matrix}\right)$$

$\left[|\arg a| < \pi,\ \operatorname{Re}(\mu+\varrho) > -\frac{1}{2},\ \operatorname{Re}(k-\varrho-\sigma) > 0\right]$. ET II 403(8)

2. $$\int_0^\infty x^{\varrho-1}\left[x^{\frac{1}{2}} + (a+x)^{\frac{1}{2}}\right]^{2\sigma} e^{-\frac{1}{2}x} W_{k,\mu}(x)\,dx =$$

$$= -\pi^{-\frac{1}{2}}\sigma a^\sigma G^{32}_{34}\left(a\,\middle|\,\begin{matrix}\frac{1}{2},\ 1,\ 1-k+\varrho\\ \frac{1}{2}+\mu+\varrho,\ \frac{1}{2}-\mu+\varrho,\ -\sigma,\ \sigma\end{matrix}\right)$$

$\left[|\arg a| < \pi,\ \operatorname{Re}\varrho > |\operatorname{Re}\mu| - \frac{1}{2}\right]$. ET II 406(24)

3 $$\int_0^\infty x^{\varrho-1}\left[x^{\frac{1}{2}} + (a+x)^{\frac{1}{2}}\right]^{2\sigma} e^{\frac{1}{2}x} W_{k,\mu}(x)\,dx =$$

$$= -\frac{\sigma\pi^{-\frac{1}{2}}a^\sigma}{\Gamma\left(\frac{1}{2}-k+\mu\right)\Gamma\left(\frac{1}{2}-k-\mu\right)} G^{33}_{34}\left(a\,\middle|\,\begin{matrix}\frac{1}{2},\ 1,\ 1+k+\varrho\\ \frac{1}{2}+\mu+\varrho,\ \frac{1}{2}-\mu+\varrho,\ -\sigma,\ \sigma\end{matrix}\right)$$

$\left[|\arg a| < \pi,\ \operatorname{Re}\varrho > |\operatorname{Re}\mu| - \frac{1}{2},\ \operatorname{Re}(k+\varrho+\sigma) < 0\right]$. ET II 406(25)

4. $\int_0^\infty x^{\varrho-1}(a+x)^{-\frac{1}{2}}[x^{\frac{1}{2}}+(a+x)^{\frac{1}{2}}]^{2\sigma}e^{-\frac{1}{2}x}M_{k,\mu}(x)\,dx =$

$$= \frac{\Gamma(2\mu+1)\,a^\sigma}{\pi^{\frac{1}{2}}\Gamma\left(\frac{1}{2}+k+\mu\right)} G_{34}^{23}\left(a\,\middle|\,\begin{matrix}0,\ \frac{1}{2},\ \frac{1}{2}-k-\varrho\\ -\sigma,\ \varrho+\mu,\ \varrho-\mu,\ \sigma\end{matrix}\right)$$

$\left[|\arg a|<\pi,\ \mathrm{Re}(\varrho+\mu)>-\frac{1}{2},\ \mathrm{Re}(k-\varrho-\sigma)>-\frac{1}{2}\right]$. ET II 403(9)

5. $\int_0^\infty x^{\varrho-1}(a+x)^{-\frac{1}{2}}[x^{\frac{1}{2}}+(a+x)^{\frac{1}{2}}]^{2\sigma}e^{\frac{1}{2}x}W_{k,\mu}(x)\,dx =$

$$= \frac{\pi^{-\frac{1}{2}}a^\sigma}{\Gamma\left(\frac{1}{2}-k+\mu\right)\Gamma\left(\frac{1}{2}-k-\mu\right)} G_{34}^{33}\left(a\,\middle|\,\begin{matrix}0,\ \frac{1}{2},\ \frac{1}{2}+k+\varrho\\ -\sigma,\ \varrho+\mu,\ \varrho-\mu,\ \sigma\end{matrix}\right)$$

$\left[|\arg a|<\pi,\ \mathrm{Re}\,\varrho>|\mathrm{Re}\,\mu|-\frac{1}{2},\ \mathrm{Re}(k+\varrho+\sigma)<\frac{1}{2}\right]$. ET II 406(26)

6. $\int_0^\infty x^{\varrho-1}(a+x)^{-\frac{1}{2}}[x^{\frac{1}{2}}+(a+x)^{\frac{1}{2}}]^{2\sigma}e^{-\frac{1}{2}x}W_{k,\mu}(x)\,dx =$

$$= \pi^{-\frac{1}{2}}a^\sigma G_{34}^{32}\left(a\,\middle|\,\begin{matrix}0,\ \frac{1}{2},\ \frac{1}{2}-k+\varrho\\ -\sigma,\ \varrho+\mu,\ \varrho-\mu,\ \sigma\end{matrix}\right) \quad \left[|\arg a|<\pi,\ \mathrm{Re}\,\varrho>|\mathrm{Re}\,\mu|-\frac{1}{2}\right].$$

ET II 406(27)

7.625

1. $\int_0^\infty x^{\varrho-1}\exp\left[-\frac{1}{2}(\alpha+\beta)x\right]M_{k,\mu}(\alpha x)W_{\lambda,\nu}(\beta x)\,dx =$

$$= \frac{\Gamma(1+\mu+\nu+\varrho)\,\Gamma(1+\mu-\nu+\varrho)}{\Gamma\left(\frac{3}{2}-\lambda+\mu+\varrho\right)}\alpha^{\mu+\frac{1}{2}}\beta^{-\mu-\varrho-\frac{1}{2}}\times$$

$$\times {}_3F_2\left(\frac{1}{2}+k+\mu,\ 1+\mu+\nu+\varrho,\ 1+\mu-\nu+\varrho;\ 2\mu+1,\ \frac{3}{2}-\lambda+\mu+\varrho;\ -\frac{\alpha}{\beta}\right)$$

$[\mathrm{Re}\,\alpha>0,\ \mathrm{Re}\,\beta>0,\ \mathrm{Re}(\varrho+\mu)>|\mathrm{Re}\,\nu|-1]$. ET II 410(43)

2. $\int_0^\infty x^{\varrho-1}\exp\left[\frac{1}{2}(\alpha+\beta)x\right]W_{k,\mu}(\alpha x)W_{\lambda,\nu}(\beta x)\,dx =$

$$= \beta^{-\varrho}\left[\Gamma\left(\frac{1}{2}-k+\mu\right)\Gamma\left(\frac{1}{2}-k-\mu\right)\Gamma\left(\frac{1}{2}-\lambda+\nu\right)\Gamma\left(\frac{1}{2}-\lambda-\nu\right)\right]^{-1}\times$$

$$\times G_{33}^{33}\left(\frac{\beta}{\alpha}\,\middle|\,\begin{matrix}\frac{1}{2}+\mu,\ \frac{1}{2}-\mu,\ 1+\lambda+\varrho\\ \frac{1}{2}+\nu+\varrho,\ \frac{1}{2}-\nu+\varrho,\ -k\end{matrix}\right)$$

$[|\mathrm{Re}\,\mu|+|\mathrm{Re}\,\nu|<\mathrm{Re}\,\varrho+1,\ \mathrm{Re}(k+\lambda+\varrho)<0]$. ET II 410(44)a

3. $$\int_0^\infty x^{\varrho-1} \exp\left[-\frac{1}{2}(\alpha+\beta)x\right] W_{k,\mu}(\alpha x) W_{\lambda,\nu}(\beta x)\,dx =$$

$$= \beta^{-\varrho} G_{33}^{22}\left(\frac{\beta}{\alpha}\left|\begin{matrix}\frac{1}{2}+\mu, \frac{1}{2}-\nu, 1-\lambda+\varrho \\ \frac{1}{2}+\nu+\varrho, \frac{1}{2}-\nu+\varrho, k\end{matrix}\right.\right)$$

$[\mathrm{Re}\,(\alpha+\beta) > 0, \ |\mathrm{Re}\,\mu| + |\mathrm{Re}\,\nu| < \mathrm{Re}\,\varrho + 1].$ ET II 411(46)

4. $$\int_0^\infty x^{\varrho-1} \exp\left[-\frac{1}{2}(\alpha-\beta)x\right] W_{k,\mu}(\alpha x) W_{\lambda,\nu}(\beta x)\,dx =$$

$$= \beta^{-\varrho}\left[\Gamma\left(\frac{1}{2}-\lambda+\nu\right)\Gamma\left(\frac{1}{2}-\lambda-\nu\right)\right]^{-1} \times$$

$$\times G_{33}^{23}\left(\frac{\beta}{\alpha}\left|\begin{matrix}\frac{1}{2}+\mu, \frac{1}{2}-\mu, 1+\lambda+\varrho \\ \frac{1}{2}+\nu+\varrho, \frac{1}{2}-\nu+\varrho, k\end{matrix}\right.\right)$$

$[\mathrm{Re}\,\alpha > 0, \ |\mathrm{Re}\,\mu| + |\mathrm{Re}\,\nu| < \mathrm{Re}\,\varrho + 1].$ ET II 411(45)

7.626

1. $$\int_0^1 \left[\frac{k}{x} - \frac{1}{4}(\xi+\eta)\right] \exp\left[-\frac{1}{2}(\xi+\eta)x\right] x^c \times$$

$$\times {}_1F_1(a;\ c;\ \xi x)\, {}_1F_1(a;\ c;\ \eta x)\,dx$$

$$= 0 \qquad [\xi \neq \eta, \ \mathrm{Re}\,c > 0];$$

$$= \frac{a}{\xi} e^{-\xi} [{}_1F_1(a+1;\ c;\ \xi)]^2 \qquad [\xi = \eta, \ \mathrm{Re}\,c > 0]$$

[where ξ and η are any two zeros of the function ${}_1F_1(a;\ c;\ x)$].

EH I 285

2. $$\int_1^\infty \left[\frac{k}{x} - \frac{1}{4}(\xi+\eta)\right] e^{-\frac{1}{2}(\xi+\eta)x} x^c \Psi(a,\ c;\ \xi x)\, \Psi(a,\ c;\ \eta x)\,dx =$$

$$= 0 \qquad [\xi \neq \eta];$$

$$= -\xi^{-1} e^{-\xi} [\Psi(a-1,\ c;\ \xi)]^2 \qquad [\xi = \eta]$$

[where ξ and η are any two zeros of the function $\Psi(a,\ c;\ x)$].

EH I 286

7.627

1. $$\int_0^\infty x^{2\lambda-1}(a+x)^{-\mu-\frac{1}{2}} e^{\frac{1}{2}x} W_{k,\mu}(a+x)\,dx =$$

$$= \frac{\Gamma(2\lambda)\,\Gamma\left(\frac{1}{2}-k+\mu-2\lambda\right)}{\Gamma\left(\frac{1}{2}-k+\mu\right)} a^{\lambda-\mu-\frac{1}{2}} W_{k+\lambda,\mu-\lambda}(a)$$

$\left[|\arg a| < \pi, \ 0 < 2\,\mathrm{Re}\,\lambda < \frac{1}{2} - \mathrm{Re}\,(k+\mu)\right].$ ET II 411(50)

2. $$\int_0^\infty x^{2\lambda-1}(a+x)^{-\mu-\frac{1}{2}}e^{-\frac{1}{2}x}M_{k,\mu}(a+x)\,dx =$$

$$= \frac{\Gamma(2\lambda)\Gamma(2\mu+1)\Gamma\left(k+\mu-2\lambda+\frac{1}{2}\right)}{\Gamma\left(k+\mu+\frac{1}{2}\right)\Gamma(1-2\lambda+2\mu)}a^{\lambda-\mu-\frac{1}{2}}M_{k-\lambda,\mu-\lambda}(a)$$

$$\left[\operatorname{Re}\lambda > 0,\ \operatorname{Re}(k+\mu-2\lambda) > -\frac{1}{2}\right].$$ ET II 405(20)

3. $$\int_0^\infty x^{2\lambda-1}(a+x)^{-\mu-\frac{1}{2}}e^{-\frac{1}{2}x}W_{k,\mu}(a+x)\,dx =$$

$$= \Gamma(2\lambda)a^{\lambda-\mu-\frac{1}{2}}W_{k-\lambda,\mu-\lambda}(a) \quad [|\arg a| < \pi,\ \operatorname{Re}\lambda > 0].$$ ET II 411(47)

4. $$\int_0^\infty x^{\lambda-1}(a+x)^{k-\lambda-1}e^{-\frac{1}{2}x}W_{k,\mu}(a+x)\,dx = \Gamma(\lambda)a^{k-1}W_{k-\lambda,\mu}(a)$$

$$[|\arg a| < \pi,\ \operatorname{Re}\lambda > 0].$$ ET II 411(48)

5. $$\int_0^\infty x^{\varrho-1}(a+x)^{-\sigma}e^{-\frac{1}{2}x}W_{k,\mu}(a+x)\,dx =$$

$$= \Gamma(\varrho)a^{\varrho}e^{\frac{1}{2}a}G_{23}^{30}\left(a\,\middle|\,\begin{matrix}0, & 1-k-\sigma & \\ -\varrho, & \frac{1}{2}+\mu-\sigma, & \frac{1}{2}-\mu-\sigma\end{matrix}\right)$$

$$[|\arg a| < \pi,\quad \operatorname{Re}\varrho > 0].$$ ET II 411(49)

6. $$\int_0^\infty x^{\varrho-1}(a+x)^{-\sigma}e^{\frac{1}{2}x}W_{k,\mu}(a+x)\,dx =$$

$$= \frac{\Gamma(\varrho)a^{\varrho}e^{-\frac{1}{2}a}}{\Gamma\left(\frac{1}{2}-k+\mu\right)\Gamma\left(\frac{1}{2}-k-\mu\right)}G_{23}^{31}\left(a\,\middle|\,\begin{matrix}k-\sigma+1, & 0 & \\ -\varrho, & \frac{1}{2}+\mu-\sigma, & \frac{1}{2}-\mu-\sigma\end{matrix}\right)$$

$$[|\arg a| < \pi,\ 0 < \operatorname{Re}\varrho < \operatorname{Re}(\sigma-k)].$$ ET II 412(51)

7. $$\int_0^\infty e^{-\frac{1}{2}(a+x)}\frac{(a+x)^{2\varkappa-1}}{(ax)^{\varkappa}}W_{\varkappa,\mu}(x)\frac{dx}{x} =$$

$$= \frac{\Gamma\left(\frac{1}{2}-\mu-\varkappa\right)\Gamma\left(\frac{1}{2}+\mu-\varkappa\right)}{a\Gamma(1-2\varkappa)}W_{\varkappa,\mu}(a)$$

$$\left[\operatorname{Re}\left(\frac{1}{2}\pm\mu-\varkappa\right) > 0\right].$$ BU 126(7a)

8. $$\int_0^\infty e^{-\frac{1}{2}x}x^{\gamma+\alpha-1}M_{\varkappa,\mu}(x)\frac{dx}{(x+a)^{\alpha}} =$$

$$= \frac{\Gamma(1+2\mu)\Gamma\left(\frac{1}{2}+\mu+\gamma\right)\Gamma(\varkappa-\gamma)}{\Gamma\left(\frac{1}{2}+\mu-\gamma\right)\Gamma\left(\frac{1}{2}+\mu+\varkappa\right)}{}_2F_2\left(\alpha,\ \varkappa-\gamma;\frac{1}{2}+\mu-\gamma,\frac{1}{2}-\mu-\gamma;\ a\right)+$$

$$+\frac{\Gamma\left(\alpha+\gamma+\frac{1}{2}+\mu\right)\Gamma\left(-\gamma-\frac{1}{2}-\mu\right)}{\Gamma(\alpha)}a^{\gamma+\frac{1}{2}+\mu}\times$$

$$\times {}_2F_2\left(\alpha+\gamma+\mu+\frac{1}{2},\ \varkappa+\mu+\frac{1}{2};\ 1+2\mu,\ \frac{3}{2}+\mu+\gamma;\ a\right)$$

$$\left[\operatorname{Re}\left(\gamma+\alpha+\frac{1}{2}+\mu\right)>0,\ \operatorname{Re}(\gamma-\varkappa)<0\right]. \qquad \text{BU 126(8)a}$$

9. $$\int_0^\infty e^{-\frac{1}{2}x}x^{n+\mu+\frac{1}{2}}M_{\varkappa,\mu}(x)\frac{dx}{x+a}=$$

$$=(-1)^{n+1}a^{n+\mu+\frac{1}{2}}e^{\frac{1}{2}a}\Gamma(1+2\mu)\Gamma\left(\frac{1}{2}-\mu+\varkappa\right)W_{-\varkappa,\mu}(a)$$

$$\left[n=0,\ 1,\ 2,\ \ldots,\ \operatorname{Re}\left(\mu+1+\frac{n}{2}\right)>0,\ \operatorname{Re}\left(\varkappa-\mu-\frac{1}{2}\right)<n,\ |\arg a|<\pi\right]$$

BU 127(10a)a

7.628

1. $$\int_0^\infty e^{-st}e^{-t^2}t^{2c-2}{}_1F_1(a;\ c;\ t^2)\,dt=$$

$$=2^{1-2c}\Gamma(2c-1)\Psi\left(c-\frac{1}{2},\ a+\frac{1}{2};\ \frac{1}{4}s^2\right)$$

$$\left[\operatorname{Re}c>\frac{1}{2},\ \operatorname{Re}s>0\right]. \qquad \text{EH I 270(11)}$$

2. $$\int_0^\infty t^{2\nu-1}e^{-\frac{1}{2a}t^2}e^{-st}M_{-3\nu,\nu}\left(\frac{t^2}{a}\right)dt=$$

$$=\frac{1}{2\sqrt{\pi}}\Gamma(4\nu+1)a^{-\nu}s^{-4\nu}e^{\frac{1}{8}as^2}K_{2\nu}\left(\frac{as^2}{8}\right)$$

$$\left[\operatorname{Re}a>0,\operatorname{Re}\nu>-\frac{1}{4},\ \operatorname{Re}s>0\right]. \qquad \text{ET I 215(12)}$$

3. $$\int_0^\infty t^{2\mu-1}e^{-\frac{1}{2a}t^2}e^{-st}M_{\lambda,\mu}\left(\frac{t^2}{a}\right)dt=$$

$$=2^{-3\mu-\lambda}\Gamma(4\mu+1)a^{\frac{1}{2}(\lambda+\mu-1)}s^{\lambda-\mu-1}e^{\frac{as^2}{8}}W_{-\frac{1}{2}(\lambda+3\mu),\ \frac{1}{2}(\lambda-\mu)}\left(\frac{as^2}{4}\right)$$

$$\left[\operatorname{Re}a>0,\ \operatorname{Re}\mu>-\frac{1}{4},\ \operatorname{Re}s>0\right]. \qquad \text{ET I 215(13)}$$

7.629

1. $$\int_0^\infty t^k\exp\left(\frac{a}{2t}\right)e^{-st}W_{k,\mu}\left(\frac{a}{t}\right)dt=$$

$$=2^{1-2k}\sqrt{a}s^{-k-\frac{1}{2}}S_{2k,2\mu}(2\sqrt{as})$$

$$\left[|\arg a|<\pi,\ \operatorname{Re}(k\pm\mu)>-\frac{1}{2},\ \operatorname{Re}s>0\right]. \qquad \text{ET I 217(21)}$$

2. $$\int_0^\infty t^{-k} \exp\left(-\frac{a}{2t}\right) e^{-st} W_{k,\mu}\left(\frac{a}{t}\right) dt = 2\sqrt{a}\, s^{k-\frac{1}{2}} K_{2\mu}(2\sqrt{as})$$

$[\operatorname{Re} a > 0,\ \operatorname{Re} s > 0].$ ET I 217(22)

7.631

1. $$\int_0^\infty x^{\varrho-1} \exp\left[\frac{1}{2}(\alpha^{-1}x - \beta x^{-1})\right] W_{k,\mu}(\alpha^{-1}x) W_{\lambda,\nu}(\beta x^{-1})\, dx =$$

$$= \beta^\varrho \left[\Gamma\left(\frac{1}{2} - k + \mu\right)\Gamma\left(\frac{1}{2} - k - \mu\right)\right]^{-1} \times$$

$$\times G_{24}^{41}\left(\frac{\beta}{\alpha}\left|\begin{matrix} 1+k,\ 1-\lambda-\varrho \\ \frac{1}{2}+\mu,\ \frac{1}{2}-\mu,\ \frac{1}{2}+\nu-\varrho,\ \frac{1}{2}-\nu-\varrho \end{matrix}\right.\right)$$

$\left[|\arg\alpha| < \frac{3}{2}\pi,\ \operatorname{Re}\beta > 0,\ \operatorname{Re}(k+\varrho) < -|\operatorname{Re}\nu| - \frac{1}{2}\right].$

ET II 412(55)

2. $$\int_0^\infty x^{\varrho-1} \exp\left[\frac{1}{2}(\alpha^{-1}x + \beta x^{-1})\right] W_{k,\mu}(\alpha^{-1}x) W_{\lambda,\nu}(\beta x^{-1})\, dx =$$

$$= \beta^\varrho \left[\Gamma\left(\frac{1}{2} - k + \mu\right)\Gamma\left(\frac{1}{2} - k - \mu\right)\Gamma\left(\frac{1}{2} - \lambda + \nu\right)\Gamma\left(\frac{1}{2} - \lambda - \nu\right)\right]^{-1} \times$$

$$\times G_{24}^{42}\left(\frac{\beta}{\alpha}\left|\begin{matrix} 1+k,\ 1+\lambda-\varrho \\ \frac{1}{2}+\mu,\ \frac{1}{2}-\mu,\ \frac{1}{2}+\nu-\varrho,\ \frac{1}{2}-\nu-\varrho \end{matrix}\right.\right)$$

$\left[|\arg\alpha| < \frac{3}{2}\pi,\ |\arg\beta| < \frac{3}{2}\pi,\ \operatorname{Re}(\lambda-\varrho) < \frac{1}{2} - |\operatorname{Re}\mu|,\right.$

$\left.\operatorname{Re}(k+\varrho) < \frac{1}{2} - |\operatorname{Re}\nu|\right].$ ET II 412(57)

3. $$\int_0^\infty x^{\varrho-1} \exp\left[-\frac{1}{2}(\alpha^{-1}x + \beta x^{-1})\right] W_{k,\mu}(\alpha^{-1}x) W_{\lambda,\nu}(\beta x^{-1})\, dx =$$

$$= \beta^\varrho G_{24}^{40}\left(\frac{\beta}{\alpha}\left|\begin{matrix} 1-k,\ 1-\lambda-\varrho \\ \frac{1}{2}+\mu,\ \frac{1}{2}-\mu,\ \frac{1}{2}+\nu-\varrho,\ \frac{1}{2}-\nu-\varrho \end{matrix}\right.\right)$$

$[\operatorname{Re}\alpha > 0,\ \operatorname{Re}\beta > 0].$ ET II 412(54)

7.632 $$\int_0^\infty e^{-st}(e^t - 1)^{\mu-\frac{1}{2}} \exp\left(-\frac{1}{2}\lambda e^t\right) M_{k,\mu}(\lambda e^t - \lambda)\, dt =$$

$$= \frac{\Gamma(2\mu+1)\Gamma\left(\frac{1}{2}+k-\mu+s\right)}{\Gamma(s+1)} W_{-k-\frac{1}{2}s,\ \mu-\frac{1}{2}s}(\lambda)$$

$\left[\operatorname{Re}\mu > -\frac{1}{2},\ \operatorname{Re} s > \operatorname{Re}(\mu-k) - \frac{1}{2}\right].$ ET I 216(15)

7.64 Combinations of degenerate hypergeometric and trigonometric functions

7.641 $$\int_0^\infty \cos(ax)\,{}_1F_1(\nu+1;\ 1;\ ix)\,{}_1F_1(\nu+1;\ 1;\ -ix)\,dx =$$
$$= -a^{-1}\sin(\nu\pi)\,P_\nu(2a^{-2}-1) \qquad [0<a<1];$$
$$= 0 \qquad [1<a<\infty]$$
$$[-1<\operatorname{Re}\nu<0].$$ ET II 402(4)

7.642 $$\int_0^\infty \cos(2xy)\,{}_1F_1(a;\ c;\ -x^2)\,dx =$$
$$=\frac{1}{2}\pi^{\frac{1}{2}}\frac{\Gamma(c)}{\Gamma(a)}\,y^{2a-1}e^{-y^2}\,\Psi\left(c-\frac{1}{2},\ a+\frac{1}{2};\ y^2\right).$$ EH I 285(12)

7.643

1. $$\int_0^\infty x^{4\nu}e^{-\frac{1}{2}x^2}\sin(bx)\,{}_1F_1\left(\frac{1}{2}-2\nu;\ 2\nu+1;\ \frac{1}{2}x^2\right)dx =$$
$$=\sqrt{\frac{\pi}{2}}\,b^{4\nu}e^{-\frac{1}{2}b^2}\,{}_1F_1\left(\frac{1}{2}-2\nu;\ 1+2\nu;\ \frac{1}{2}b^2\right)$$
$$\left[b>0,\ \operatorname{Re}\nu>-\frac{1}{4}\right].$$ ET I 115(5)

2. $$\int_0^\infty x^{2\nu-1}e^{-\frac{1}{4}x^2}\sin(bx)\,M_{3\nu,\nu}\left(\frac{1}{2}x^2\right)dx=\sqrt{\frac{\pi}{2}}\,b^{2\nu-1}e^{-\frac{1}{4}b^2}M_{3\nu,\nu}\left(\frac{1}{2}b^2\right)$$
$$\left[b>0,\ \operatorname{Re}\nu>-\frac{1}{4}\right].$$ ET I 116(10)

3. $$\int_0^\infty x^{-2\nu-1}e^{\frac{1}{4}x^2}\cos(bx)\,W_{3\nu,\nu}\left(\frac{1}{2}x^2\right)dx=\sqrt{\frac{\pi}{2}}\,b^{-2\nu-1}e^{\frac{1}{4}b^2}W_{3\nu,\nu}\left(\frac{1}{2}b^2\right)$$
$$\left[\operatorname{Re}\nu<\frac{1}{4},\ b>0\right].$$ ET I 61(7)

4. $$\int_0^\infty x^{-2\nu}e^{\frac{1}{4}x^2}\sin(bx)\,W_{3\nu-1,\nu}\left(\frac{1}{2}x^2\right)dx=$$
$$=\sqrt{\frac{\pi}{2}}\,b^{-2\nu}e^{\frac{1}{4}b^2}W_{3\nu-1,\nu}\left(\frac{1}{2}b^2\right)$$
$$\left[\operatorname{Re}\nu<\frac{1}{2},\ b>0\right].$$ ET I 116(9)

7.644

1. $$\int_0^\infty x^{-\mu-\frac{1}{2}}e^{-\frac{1}{2}x}\sin(2ax^{\frac{1}{2}})\,M_{k,\mu}(x)\,dx=$$
$$=\pi^{\frac{1}{2}}a^{k+\mu-1}\frac{\Gamma(3-2\mu)}{\Gamma\left(\frac{1}{2}+k+\mu\right)}\exp\left(-\frac{a^2}{2}\right)W_{\varrho,\sigma}(a^2),$$
$$2\varrho=k-3\mu+1,\quad 2\sigma=k+\mu-1$$
$$[a>0,\ \operatorname{Re}(k+\mu)>0].$$ ET II 403(10)

2. $$\int_0^\infty x^{\varrho-1}\sin\left(cx^{\frac{1}{2}}\right)e^{-\frac{1}{2}x}W_{k,\mu}(x)\,dx=\frac{c\Gamma(1+\mu+\varrho)\Gamma(1-\mu+\varrho)}{\Gamma\left(\frac{3}{2}-k+\varrho\right)}\times$$
$$\times{}_2F_2\left(1+\mu+\varrho,\ 1-\mu+\varrho;\ \frac{3}{2},\ \frac{3}{2}-k+\varrho;\ -\frac{c^2}{4}\right)$$
$[\operatorname{Re}\varrho>|\operatorname{Re}\mu|-1]$. ET II 407(28)

3. $$\int_0^\infty x^{\varrho-1}\sin\left(cx^{\frac{1}{2}}\right)e^{\frac{1}{2}x}W_{k,\mu}(x)\,dx=$$
$$=\frac{\pi^{\frac{1}{2}}}{\Gamma\left(\frac{1}{2}-k+\mu\right)\Gamma\left(\frac{1}{2}-k-\mu\right)}G_{23}^{22}\left(\frac{c^2}{4}\middle|\begin{matrix}\frac{1}{2}+\mu-\varrho,\ \frac{1}{2}-\mu-\varrho\\ \frac{1}{2},\ -k-\varrho,\ 0\end{matrix}\right)$$
$\left[c>0,\ \operatorname{Re}\varrho>|\operatorname{Re}\mu|-1,\ \operatorname{Re}(k+\varrho)<\frac{1}{2}\right]$. ET II 407(29)

4. $$\int_0^\infty x^{\varrho-1}\cos\left(cx^{\frac{1}{2}}\right)e^{-\frac{1}{2}x}W_{k,\mu}(x)\,dx=\frac{\Gamma\left(\frac{1}{2}+\mu+\varrho\right)\Gamma\left(\frac{1}{2}-\mu+\varrho\right)}{\Gamma(1-k+\varrho)}\times$$
$$\times{}_2F_2\left(\frac{1}{2}+\mu+\varrho,\ \frac{1}{2}-\mu+\varrho;\ \frac{1}{2},\ 1-k+\varrho;\ -\frac{c^2}{4}\right)$$
$\left[\operatorname{Re}\varrho>|\operatorname{Re}\mu|-\frac{1}{2}\right]$. ET II 407(30)

5. $$\int_0^\infty x^{\varrho-1}\cos\left(cx^{\frac{1}{2}}\right)e^{\frac{1}{2}x}W_{k,\mu}(x)\,dx=$$
$$=\frac{\pi^{\frac{1}{2}}}{\Gamma\left(\frac{1}{2}-k+\mu\right)\Gamma\left(\frac{1}{2}-k-\mu\right)}G_{23}^{22}\left(\frac{c^2}{4}\middle|\begin{matrix}\frac{1}{2}+\mu-\varrho,\ \frac{1}{2}-\mu-\varrho\\ 0,\ -k-\varrho,\ \frac{1}{2}\end{matrix}\right)$$
$\left[c>0,\ \operatorname{Re}\varrho>|\operatorname{Re}\mu|-\frac{1}{2},\ \operatorname{Re}(k+\varrho)<\frac{1}{2}\right]$. ET II 407(31)

7.65 Combinations of degenerate hypergeometric functions and Bessel functions

7.651

1. $$\int_0^\infty J_\nu(xy)M_{-\frac{1}{2}\mu,\frac{1}{2}\nu}(ax)W_{\frac{1}{2}\mu,\frac{1}{2}\nu}(ax)\,dx=$$
$$=ay^{-\mu-1}\frac{\Gamma(\nu+1)}{\Gamma\left(\frac{1}{2}-\frac{1}{2}\mu+\frac{1}{2}\nu\right)}\left[a+(a^2+y^2)^{\frac{1}{2}}\right]^\mu(a^2+y^2)^{-\frac{1}{2}}$$
$\left[y>0,\ \operatorname{Re}\nu>-1,\ \operatorname{Re}\mu<\frac{1}{2},\ \operatorname{Re}a>0\right]$. ET II 85(19)

2. $$\int_0^\infty M_{k,\frac{1}{2}\nu}(-iax)M_{-k,\frac{1}{2}\nu}(-iax)J_\nu(xy)\,dx=$$
$$=\frac{ae^{-\frac{1}{2}(\nu+1)\pi i}[\Gamma(1+\nu)]^2}{\Gamma\left(\frac{1}{2}+k+\frac{1}{2}\nu\right)\Gamma\left(\frac{1}{2}-k+\frac{1}{2}\nu\right)}y^{-1-2k}\times$$

$$\times (a^2-y^2)^{-\frac{1}{2}}\{[a+(a^2-y^2)^{\frac{1}{2}}]^{2k}+[a-(a^2-y^2)^{\frac{1}{2}}]^{2k}\} \qquad [0<y<a];$$
$$=0 \qquad [a<y<\infty]$$
$$\left[a>0,\ \operatorname{Re}\nu>-1,\ |\operatorname{Re}k|<\frac{1}{4}\right]. \qquad \text{ET II 85(18)}$$

7.652 $$\int_0^\infty M_{-\mu,\frac{1}{2}\nu}\{a[(b^2+x^2)^{\frac{1}{2}}-b]\}W_{\mu,\frac{1}{2}\nu}\{a[(b^2+x^2)^{\frac{1}{2}}+b]\}J_\nu(xy)\,dx=$$
$$=\frac{ay^{-2\mu-1}\Gamma(1+\nu)[(a^2+y^2)^{\frac{1}{2}}+a]^{2\mu}}{\Gamma\left(\frac{1}{2}+\frac{1}{2}\nu-\mu\right)(a^2+y^2)^{\frac{1}{2}}}\exp[-b(a^2+y^2)^{\frac{1}{2}}]$$
$$\left[y>0,\ \operatorname{Re}\nu>-1,\ \operatorname{Re}\mu<\frac{1}{4},\ \operatorname{Re}a>0,\ \operatorname{Re}b>0\right]. \qquad \text{ET II 87(29)}$$

7.66 Combinations of degenerate hypergeometric functions, Bessel functions, and powers

7.661

1. $$\int_0^\infty x^{-1}W_{k,\mu}(ax)M_{-k,\mu}(ax)J_0(xy)\,dx=$$
$$=e^{-ik\pi}\frac{\Gamma(1+2\mu)}{\Gamma\left(\frac{1}{2}+\mu+k\right)}P^k_{\mu-\frac{1}{2}}\left[\left(1+\frac{y^2}{a^2}\right)^{\frac{1}{2}}\right]Q^k_{\mu-\frac{1}{2}}\left[\left(1+\frac{y^2}{a^2}\right)^{\frac{1}{2}}\right]$$
$$\left[y>0,\ \operatorname{Re}a>0,\ \operatorname{Re}\mu>-\frac{1}{2},\ \operatorname{Re}k<\frac{3}{4}\right]. \qquad \text{ET II 18(44)}$$

2. $$\int_0^\infty x^{-1}W_{k,\mu}(ax)W_{-k,\mu}(ax)J_0(xy)\,dx=$$
$$=\frac{1}{2}\pi\cos(\mu\pi)P^k_{\mu-\frac{1}{2}}\left[\left(1+\frac{y^2}{a^2}\right)^{\frac{1}{2}}\right]P^{-k}_{\mu-\frac{1}{2}}\left[\left(1+\frac{y^2}{a^2}\right)^{\frac{1}{2}}\right]$$
$$\left[y>0,\ \operatorname{Re}a>0,\ |\operatorname{Re}\mu|<\frac{1}{2}\right]. \qquad \text{ET II 18(45)}$$

3. $$\int_0^\infty x^{2\mu-\nu}W_{k,\mu}(ax)M_{-k,\mu}(ax)J_\nu(xy)\,dx=$$
$$=2^{2\mu-\nu+2k}a^{2k}y^{\nu-2\mu-2k-1}\frac{\Gamma(2\mu+1)}{\Gamma\left(\nu-k-\mu+\frac{1}{2}\right)}\times$$
$$\times {}_3F_2\left(\frac{1}{2}-k,\ 1-k,\ \frac{1}{2}-k+\mu;\ 1-2k,\ \frac{1}{2}-k-\mu+\nu;\ -\frac{y^2}{a^2}\right)$$
$$\left[y>0,\ \operatorname{Re}\mu>-\frac{1}{2},\ \operatorname{Re}a>0,\ \operatorname{Re}(2\mu+2k-\nu)<\frac{1}{2}\right]. \qquad \text{ET II 85(20)}$$

4. $$\int_0^\infty x^{2\varrho-\nu} W_{k,\mu}(iax) W_{k,\mu}(-iax) J_\nu(xy)\,dx =$$

$$= 2^{2\varrho-\nu} y^{\nu-2\varrho-1} \pi^{-\frac{1}{2}} \left[\Gamma\left(\frac{1}{2}-k+\mu\right) \Gamma\left(\frac{1}{2}-k-\mu\right) \right]^{-1} \times$$

$$\times G_{44}^{24}\left(\frac{y^2}{a^2} \middle| \begin{matrix} \frac{1}{2},\ 0,\ \frac{1}{2}-\mu,\ \frac{1}{2}+\mu \\ \varrho+\frac{1}{2},\ -k,\ k,\ \varrho-\nu+\frac{1}{2} \end{matrix} \right)$$

$$\left[y > 0,\ \operatorname{Re} a > 0,\ \operatorname{Re}\varrho > |\operatorname{Re}\mu| - 1,\ \operatorname{Re}(2\varrho + 2k - \nu) < \frac{1}{2} \right].$$

ET II 86(23)a

5. $$\int_0^\infty x^{2\varrho-\nu} W_{k,\mu}(ax) M_{-k,\mu}(ax) J_\nu(xy)\,dx =$$

$$= \frac{2^{2\varrho-\nu}\Gamma(2\mu+1)}{\pi^{\frac{1}{2}}\Gamma\left(\frac{1}{2}-k+\mu\right)} y^{\nu-2\varrho-1} G_{44}^{23}\left(\frac{y^2}{a^2} \middle| \begin{matrix} \frac{1}{2},\ 0,\ \frac{1}{2}-\mu,\ \frac{1}{2}+\mu \\ \varrho+\frac{1}{2},\ -k,\ k,\ \varrho-\nu+\frac{1}{2} \end{matrix} \right)$$

$$\left[y > 0,\ \operatorname{Re} a > 0,\ \operatorname{Re}\varrho > -1,\ \operatorname{Re}(\varrho+\mu) > -1), \right.$$
$$\left. \operatorname{Re}(2\varrho + 2k + \nu) < \frac{1}{2} \right].$$

ET II 86(21)a

6. $$\int_0^\infty x^{2\varrho-\nu} W_{k,\mu}(ax) W_{-k,\mu}(ax) J_\nu(xy)\,dx =$$

$$= \frac{\Gamma(\varrho+1+\mu)\Gamma(\varrho+1-\mu)\Gamma(2\varrho+2)}{\Gamma\left(\frac{3}{2}+k+\varrho\right)\Gamma\left(\frac{3}{2}-k+\varrho\right)\Gamma(1+\nu)} y^\nu 2^{-\nu-1} a^{-2\varrho-1} \times$$

$$\times {}_4F_3\left(\varrho+1,\ \varrho+\frac{3}{2},\ \varrho+1+\mu,\ \varrho+1-\mu;\ \frac{3}{2}+k+\varrho,\ \frac{3}{2}-k+\varrho,\ 1+\nu;\ -\frac{y^2}{a^2}\right)$$

$$[y > 0,\ \operatorname{Re}\varrho > |\operatorname{Re}\mu| - 1,\ \operatorname{Re} a > 0].$$

ET II 86(22)a

7.662

1. $$\int_0^\infty x^{-1} M_{-\mu,\frac{1}{4}\nu}\left(\frac{1}{2}x^2\right) W_{\mu,\frac{1}{4}\nu}\left(\frac{1}{2}x^2\right) J_\nu(xy)\,dx =$$

$$= \frac{\Gamma\left(1+\frac{1}{2}\nu\right)}{\Gamma\left(\frac{1}{2}+\frac{1}{4}\nu-\mu\right)} I_{\frac{1}{4}\nu-\mu}\left(\frac{1}{4}y^2\right) K_{\frac{1}{4}\nu+\mu}\left(\frac{1}{4}y^2\right)$$

$$[y > 0,\ \operatorname{Re}\nu > -1].$$

ET II 86(24)

2. $\int_0^\infty x^{-1} M_{\alpha-\beta,\frac{1}{4}\nu-\gamma}\left(\frac{1}{2}x^2\right) W_{\alpha+\beta,\frac{1}{4}\nu+\gamma}\left(\frac{1}{2}x^2\right) J_\nu(xy)\,dx =$

$$= \frac{\Gamma\left(1+\frac{1}{2}\nu-2\gamma\right)}{\Gamma\left(1+\frac{1}{2}\nu-2\beta\right)} y^{-2} M_{\alpha-\gamma,\frac{1}{4}\nu-\beta}\left(\frac{1}{2}y^2\right) W_{\alpha+\gamma,\frac{1}{4}\nu+\beta}\left(\frac{1}{2}y^2\right)$$

$\left[y>0,\ \operatorname{Re}\beta<\frac{1}{8},\ \operatorname{Re}\nu>-1,\quad \operatorname{Re}(\nu-4\gamma)>-2\right].$ ET II 86(25)

3. $\int_0^\infty x^{-1} M_{k,0}(iax^2) M_{k,0}(-iax^2) K_0(xy)\,dx =$

$$= \frac{\pi}{16}\left\{\left[J_k\left(\frac{y^2}{8a}\right)\right]^2 + \left[N_k\left(\frac{y^2}{8a}\right)\right]^2\right\}$$

$[a>0].$ ET II 152(83)

4. $\int_0^\infty x^{-1} M_{k,\mu}(iax^2) M_{k,\mu}(-iax^2) K_0(xy)\,dx =$

$$= ay^{-2}[\Gamma(2\mu+1)]^2 W_{-\mu,k}\left(\frac{iy^2}{4a}\right) W_{-\mu,k}\left(-\frac{iy^2}{4a}\right)$$

$\left[a>0,\ \operatorname{Re} y>0,\ \operatorname{Re}\mu>-\frac{1}{2}\right].$ ET II 152(84)

7.663

1. $\int_0^\infty x^{2\varrho} {}_1F_1(a;\ b;\ -\lambda x^2) J_\nu(xy)\,dx =$

$$= \frac{2^{2\varrho}\Gamma(b)}{\Gamma(a)\, y^{2\varrho+1}} G_{23}^{21}\left(\frac{y^2}{4\lambda}\left|\begin{matrix} 1, & b \\ \frac{1}{2}+\varrho+\frac{1}{2}\nu, & a, & \frac{1}{2}+\varrho-\frac{1}{2}\nu\end{matrix}\right.\right)$$

$\left[y>0,\ -1-\operatorname{Re}\nu<2\operatorname{Re}\varrho<\frac{1}{2}+2\operatorname{Re} a,\ \operatorname{Re}\lambda>0\right].$ ET II 88(6)

2. $\int_0^\infty x^{\nu+1} {}_1F_1\left(2a-\nu;\ a+1;\ -\frac{1}{2}x^2\right) J_\nu(xy)\,dx =$

$$= \frac{2^{\nu-a+\frac{1}{2}}\Gamma(a+1)}{\pi^{\frac{1}{2}}\Gamma(2a-\nu)} y^{2a-\nu-1} e^{-\frac{1}{4}y^2} K_{a-\nu-\frac{1}{2}}\left(\frac{1}{4}y^2\right)$$

$\left[y>0,\ \operatorname{Re}\nu>-1,\ \operatorname{Re}(4a-3\nu)>\frac{1}{2}\right].$ ET II 87(1)

3. $\int_0^\infty x^a {}_1F_1\left(a;\ \frac{1+a+\nu}{2};\ -\frac{1}{2}x^2\right) J_\nu(xy)\,dx =$

$$= y^{a-1} {}_1F_1\left(a;\ \frac{1+a+\nu}{2};\ -\frac{y^2}{2}\right)$$

$\left[y>0,\ \operatorname{Re} a>-\frac{1}{2},\ \operatorname{Re}(a+\nu)>-1\right].$ ET II 87(2)

4. $$\int_0^\infty x^{\nu+1-2a}{}_1F_1\left(a; 1+\nu-a; \; -\frac{1}{2}x^2\right) J_\nu(xy)\,dx =$$

$$= \frac{\pi^{\frac{1}{2}}\Gamma(1+\nu-a)}{\Gamma(a)} 2^{-2a+\nu+\frac{1}{2}} y^{2a-\nu-1} e^{-\frac{1}{4}y^2} I_{a-\frac{1}{2}}\left(\frac{1}{4}y^2\right)$$

$$\left[y>0, \; \operatorname{Re} a - 1 < \operatorname{Re}\nu < 4\operatorname{Re} a - \frac{1}{2}\right].$$ ET II 87(3)

5. $$\int_0^\infty x\,{}_1F_1(\lambda; 1; -x^2) J_0(xy)\,dx = [2^{2\lambda-1}\Gamma(\lambda)]^{-1} y^{2\lambda-2} e^{-\frac{1}{4}y^2}$$

$$[y>0, \; \operatorname{Re}\lambda>0].$$ ET II 18(46)

6. $$\int_0^\infty x^{\nu+1}{}_1F_1(a; \; b; \; -\lambda x^2) J_\nu(xy)\,dx =$$

$$= \frac{2^{1-a}\Gamma(b)}{\Gamma(a)\lambda^{\frac{1}{2}a+\frac{1}{2}\nu}} y^{a-2} e^{-\frac{y^2}{8\lambda}} W_{k,\mu}\left(\frac{y^2}{4\lambda}\right),$$

$$2k = a-2b+\nu+2, \; 2\mu = a-\nu-1$$

$$\left[y>0, \; -1<\operatorname{Re}\nu<2\operatorname{Re} a-\frac{1}{2}, \; \operatorname{Re}\lambda>0\right].$$ ET II 88(4)

7. $$\int_0^\infty x^{2b-\nu-1}{}_1F_1(a; \; b; \; -\lambda x^2) J_\nu(xy)\,dx =$$

$$= \frac{2^{2b-2a-\nu-1}\Gamma(b)}{\Gamma(a-b+\nu+1)} \lambda^{-a} y^{2a-2b+\nu}{}_1F_1\left(a; \; 1+a-b+\nu; \; -\frac{y^2}{4\lambda}\right)$$

$$\left[y>0, \; 0<\operatorname{Re} b<\frac{3}{4}+\operatorname{Re}\left(a+\frac{1}{2}\nu\right), \; \operatorname{Re}\lambda>0\right].$$ ET II 88(5)

7.664

1. $$\int_0^\infty x W_{\frac{1}{2}\nu,\mu}\left(\frac{a}{x}\right) W_{-\frac{1}{2}\nu,\mu}\left(\frac{a}{x}\right) K_\nu(xy)\,dx =$$

$$= 2ay^{-1} K_{2\mu}\left[(2ay)^{\frac{1}{2}} e^{\frac{1}{4}i\pi}\right] K_{2\mu}\left[(2ay)^{\frac{1}{2}} e^{-\frac{1}{4}i\pi}\right]$$

$$[\operatorname{Re} y>0, \quad \operatorname{Re} a>0].$$ ET II 152(85)

2. $$\int_0^\infty x W_{\frac{1}{2}\nu,\mu}\left(\frac{2}{x}\right) W_{-\frac{1}{2}\nu,\mu}\left(\frac{2}{x}\right) J_\nu(xy)\,dx =$$

$$= -4y^{-1}\left\{\sin\left[\left(\mu-\frac{1}{2}\nu\right)\pi\right] J_{2\mu}(2y^{\frac{1}{2}}) + \right.$$

$$\left. + \cos\left[\left(\mu-\frac{1}{2}\nu\right)\pi\right] N_{2\mu}(2y^{\frac{1}{2}})\right\} K_{2\mu}(2y^{\frac{1}{2}})$$

$$[y>0, \; \operatorname{Re}(\nu\pm 2\mu)>-1].$$ ET II 87(27)

3. $$\int_0^\infty x W_{\frac{1}{2}\nu,\mu}\left(\frac{2}{x}\right) W_{-\frac{1}{2}\nu,\mu}\left(\frac{2}{x}\right) N_\nu(xy)\,dx =$$
$$= 4y^{-1}\left\{\cos\left[\left(\mu-\frac{1}{2}\nu\right)\pi\right] J_{2\mu}(2y^{\frac{1}{2}}) - \right.$$
$$\left. - \sin\left[\left(\mu-\frac{1}{2}\nu\right)\pi\right] N_{2\mu}(2y^{\frac{1}{2}})\right\} K_{2\mu}(2y^{\frac{1}{2}})\}$$
$$\left[y>0,\ |\operatorname{Re}\mu|<\frac{1}{4}\right].$$ ET II 117(48)

4 $$\int_0^\infty x W_{-\frac{1}{2}\nu,\mu}\left(\frac{2}{x}\right) M_{\frac{1}{2}\nu,\mu}\left(\frac{2}{x}\right) J_\nu(xy)\,dx =$$
$$= \frac{4\Gamma(1+2\mu)\,y^{-1}}{\Gamma\left(\frac{1}{2}+\frac{1}{2}\nu+\mu\right)} J_{2\mu}(2y^{\frac{1}{2}}) K_{2\mu}(2y^{\frac{1}{2}})$$
$$\left[y>0,\ \operatorname{Re}\nu>-1,\ \operatorname{Re}\mu>-\frac{1}{4}\right].$$ ET II 86(26)

5. $$\int_0^\infty x W_{-\frac{1}{2}\nu,\mu}\left(\frac{ia}{x}\right) W_{-\frac{1}{2}\nu,\mu}\left(-\frac{ia}{x}\right) J_\nu(xy)\,dx =$$
$$= 4ay^{-1}\left[\Gamma\left(\frac{1}{2}+\mu+\frac{1}{2}\nu\right)\Gamma\left(\frac{1}{2}-\mu+\frac{1}{2}\nu\right)\right]^{-1} K_\mu[(2iay)^{\frac{1}{2}}]\, K_\mu[(-2iay)^{\frac{1}{2}}]$$
$$\left[y>0,\ \operatorname{Re}a>0,\ |\operatorname{Re}\mu|<\frac{1}{2},\ \operatorname{Re}\nu>-1\right].$$ ET II 87(28)

7.665

1. $$\int_0^\infty x^{-\frac{1}{2}} J_\nu(ax^{\frac{1}{2}}) K_{\frac{1}{2}\nu-\mu}\left(\frac{1}{2}x\right) M_{k,\mu}(x)\,dx =$$
$$= \frac{\Gamma(2\mu+1)}{a\Gamma\left(k+\frac{1}{2}\nu+1\right)} W_{\frac{1}{2}(k-\mu),\,\frac{1}{2}k-\frac{1}{4}\nu}\left(\frac{a^2}{2}\right) M_{\frac{1}{2}(k+\mu),\,\frac{1}{2}k+\frac{1}{4}\nu}\left(\frac{a^2}{2}\right)$$
$$\left[a>0,\ \operatorname{Re}k>-\frac{1}{4},\ \operatorname{Re}\mu>-\frac{1}{2},\ \operatorname{Re}\nu>-1\right].$$ ET II 405(18)

2. $$\int_0^\infty x^{\frac{1}{2}c+\frac{1}{2}c'-1}\,\Psi(a,\ c;\ x)\,{}_1F_1(a';\ c';\ -x)\,J_{c+c'-2}[2(xy)^{\frac{1}{2}}]\,dx =$$
$$= \frac{\Gamma(c')}{\Gamma(a+a')} y^{\frac{1}{2}c+\frac{1}{2}c'-1}\,\Psi(c'-a',\ c+c'-a-a';\ y)\,{}_1F_1(a';\ a+a';\ -y)$$
$$\left[\operatorname{Re}c'>0,\ 1<\operatorname{Re}(c+c')<2\operatorname{Re}(a+a')+\frac{1}{2}\right].$$ EH I 287(23)

7.666 $$\int_0^\infty x^{\frac{1}{2}c-\frac{1}{2}}\,{}_1F_1(a;\ c;\ -2x^{\frac{1}{2}})\,\Psi(a,\ c;\ 2x^{\frac{1}{2}})\,J_{c-1}[2(xy)^{\frac{1}{2}}]\,dx =$$
$$= 2^{-c}\frac{\Gamma(c)}{\Gamma(a)} y^{a-\frac{1}{2}c-\frac{1}{2}}[1+(1+y)^{\frac{1}{2}}]^{c-2a}(1+y)^{-\frac{1}{2}}$$
$$\left[\operatorname{Re}c>2,\ \operatorname{Re}(c-2a)<\frac{1}{2}\right].$$ EH I 285(13)

7.67 Combinations of degenerate hypergeometric functions, Bessel functions, exponentials, and powers

7.671

1. $$\int_0^\infty x^{k-\frac{3}{2}} \exp\left[-\frac{1}{2}(a+1)x\right] K_\nu\left(\frac{1}{2}ax\right) M_{k,\nu}(x)\,dx =$$

$$= \frac{\pi^{\frac{1}{2}}\Gamma(k)\Gamma(k+2\nu)}{a^{k+\nu}\Gamma\left(k+\nu+\frac{1}{2}\right)}\,{}_2F_1(k,\ k+2\nu;\ 2\nu+1;\ -a^{-1})$$

$[\operatorname{Re} a > 0,\ \operatorname{Re} k > 0,\ \operatorname{Re}(k+2\nu) > 0].$ ET II 405(17)

2. $$\int_0^\infty x^{-k-\frac{3}{2}} \exp\left[-\frac{1}{2}(a-1)x\right] K_\mu\left(\frac{1}{2}ax\right) W_{k,\mu}(x)\,dx =$$

$$= \frac{\pi\Gamma(-k)\Gamma(2\mu-k)\Gamma(-2\mu-k)}{\Gamma\left(\frac{1}{2}-k\right)\Gamma\left(\frac{1}{2}+\mu-k\right)\Gamma\left(\frac{1}{2}-\mu-k\right)} \times$$

$$\times 2^{2k+1}a^{k-\nu}{}_2F_1(-k,\ 2\mu-k;\ -2k;\ 1-a^{-1})$$

$[\operatorname{Re} a > 0,\ \operatorname{Re} k < 2\operatorname{Re}\mu < -\operatorname{Re} k].$ ET II 408(36)

7.672

1. $$\int_0^\infty x^{2\varrho} e^{-\frac{1}{2}ax^2} M_{k,\mu}(ax^2) J_\nu(xy)\,dx =$$

$$= \frac{\Gamma(2\mu+1)}{\Gamma\left(\mu+k+\frac{1}{2}\right)} 2^{2\varrho} y^{-2\varrho-1} G_{23}^{21}\left(\frac{y^2}{4a}\middle|\begin{matrix}\frac{1}{2}-\mu,\ \frac{1}{2}+\mu \\ \frac{1}{2}+\varrho+\frac{1}{2}\nu,\ k,\ \frac{1}{2}+\varrho-\frac{1}{2}\nu\end{matrix}\right)$$

$\left[y > 0,\ -1-\operatorname{Re}\left(\frac{1}{2}\nu+\mu\right) < \operatorname{Re}\varrho < \operatorname{Re} k - \frac{1}{4},\ \operatorname{Re} a > 0\right].$

ET II 83(10)

2. $$\int_0^\infty x^{2\varrho} e^{-\frac{1}{2}ax^2} W_{k,\mu}(ax^2) J_\nu(xy)\,dx =$$

$$= \frac{\Gamma\left(1+\mu+\frac{1}{2}\nu+\varrho\right)\Gamma\left(1-\mu+\frac{1}{2}\nu+\varrho\right)2^{-\nu-1}}{\Gamma(\nu+1)\Gamma\left(\frac{3}{2}-k+\frac{1}{2}\nu+\varrho\right)} a^{-\frac{1}{2}\nu-\varrho-\frac{1}{2}} y^\nu \times$$

$$\times {}_2F_2\left(\lambda+\mu,\ \lambda-\mu;\ \nu+1,\ \frac{1}{2}-k+\lambda;\ -\frac{y^2}{4a}\right),$$

$$\lambda = 1+\frac{1}{2}\nu+\varrho$$

$\left[y > 0,\ \operatorname{Re} a > 0,\ \operatorname{Re}\left(\varrho \pm \mu + \frac{1}{2}\nu\right) > -1\right].$ ET II 85(16)

3. $$\int_0^\infty x^{2\varrho} e^{\frac{1}{2}ax^2} W_{k,\mu}(ax^2) J_\nu(xy)\,dx = \frac{2^{2\varrho} y^{-2\varrho-1}}{\Gamma\left(\frac{1}{2}+\mu-k\right)\Gamma\left(\frac{1}{2}-\mu-k\right)} \times$$

$$\times G_{23}^{22}\left(\frac{y^2}{4a}\middle|\begin{matrix}\frac{1}{2}-\mu, & \frac{1}{2}+\mu \\ \frac{1}{2}+\varrho+\frac{1}{2}\nu, & -k, & \frac{1}{2}+\varrho-\frac{1}{2}\nu\end{matrix}\right)$$

$$\left[y>0,\ |\arg a|<\pi,\quad -1-\operatorname{Re}\left(\frac{1}{2}\nu\pm\mu\right)<\operatorname{Re}\varrho<-\frac{1}{4}-\operatorname{Re}k\right].$$

ET II 85(17)

4. $$\int_0^\infty x^{2\lambda+\frac{1}{2}} e^{-\frac{1}{4}x^2} M_{k,\mu}\left(\frac{1}{2}x^2\right) N_\nu(xy)\,dx =$$

$$= \frac{2^\lambda y^{-\frac{1}{2}} \Gamma(2\mu+1)}{\Gamma\left(\frac{1}{2}+k+\mu\right)} G_{34}^{31}\left(\frac{y^2}{2}\middle|\begin{matrix}-\mu-\lambda, & \mu-\lambda, & l \\ h, & \varkappa, & k-\lambda-\frac{1}{2}, & l\end{matrix}\right),$$

$$h=\frac{1}{4}+\frac{1}{2}\nu,\quad \varkappa=\frac{1}{4}-\frac{1}{2}\nu,\quad l=-\frac{1}{4}-\frac{1}{2}\nu$$

$$\left[y>0,\quad \operatorname{Re}(k-\lambda)>0,\quad \operatorname{Re}(2\lambda+2\mu\pm\nu)>-\frac{5}{2}\right].$$ ET II 116(45)

5. $$\int_0^\infty x^{2\lambda+\frac{1}{2}} e^{\frac{1}{4}x^2} W_{k,\mu}\left(\frac{1}{2}x^2\right) N_\nu(xy)\,dx =$$

$$= 2^\lambda\left[\Gamma\left(\frac{1}{2}-k+\mu\right)\Gamma\left(\frac{1}{2}-k-\mu\right)\right]^{-1} \times$$

$$\times G_{34}^{32}\left(\frac{y^2}{2}\middle|\begin{matrix}-\mu-\lambda, & \mu-\lambda, & l \\ h, & \varkappa & -\frac{1}{2}-k-\lambda, & l\end{matrix}\right) y^{-\frac{1}{2}},$$

$$h=\frac{1}{4}+\frac{1}{2}\nu,\quad \varkappa=\frac{1}{4}-\frac{1}{2}\nu,\quad l=-\frac{1}{4}-\frac{1}{2}\nu$$

$$\left[y>0,\quad \operatorname{Re}(k+\lambda)<0,\quad \operatorname{Re}(2\lambda\pm2\mu\pm\nu)>-\frac{5}{2}\right].$$ ET II 117(47)

6. $$\int_0^\infty x^{-\frac{1}{2}} e^{-\frac{1}{2}x^2} M_{\frac{1}{2}\nu-\frac{1}{4},\ \frac{1}{2}\nu+\frac{1}{4}}(x^2) J_\nu(xy)\,dx =$$

$$= (2\nu+1)\,2^{-\nu} y^{\nu-1}\left[1-\Phi\left(\frac{1}{2}y\right)\right]$$

$$\left[y>0,\quad \operatorname{Re}\nu>-\frac{1}{2}\right].$$ ET II 82(1)

7. $$\int_0^\infty x^{-1} e^{-\frac{1}{2}x^2} M_{\frac{1}{2}\nu+\frac{1}{2},\ \frac{1}{2}\nu+\frac{1}{2}}(x^2) J_\nu(xy)\,dx =$$

$$= \frac{\Gamma(\nu+2)\,y^\nu}{\Gamma\left(\nu+\frac{3}{2}\right)2^\nu}\left[1-\Phi\left(\frac{1}{2}y\right)\right]$$

$$[y>0,\quad \operatorname{Re}\nu>-1].$$ ET II 82(2)

8. $$\int_0^\infty e^{-\frac{1}{4}x^2} M_{k,\frac{1}{2}\nu}\left(\frac{1}{2}x^2\right) J_\nu(xy)\,dx = \frac{2^{-k}\Gamma(\nu+1)}{\Gamma\left(k+\frac{1}{2}\nu+\frac{1}{2}\right)} y^{2k-1} e^{-\frac{1}{2}y^2}$$

$$\left[y>0, \quad \operatorname{Re}\nu>-1, \quad \operatorname{Re}k<\frac{1}{2}\right].$$ ET II 83(7)

9. $$\int_0^\infty x^{\nu-2\mu} e^{-\frac{1}{4}x^2} M_{k,\mu}\left(\frac{1}{2}x^2\right) J_\nu(xy)\,dx =$$

$$= 2^{\frac{1}{2}\left(\frac{1}{2}-k-3\mu+\nu\right)} \frac{\Gamma(2\mu+1)}{\Gamma\left(\mu+k+\frac{1}{2}\right)} y^{k+\mu-\frac{3}{2}} e^{-\frac{1}{4}y^2} W_{\alpha,\beta}\left(\frac{1}{2}y^2\right),$$

$$2\alpha = k-3\mu+\nu+\frac{1}{2}, \qquad 2\beta = k+\mu-\nu-\frac{1}{2}$$

$$\left[y>0, \quad -1<\operatorname{Re}\nu<2\operatorname{Re}(k+\mu)-\frac{1}{2}\right].$$ ET II 83(9)

10. $$\int_0^\infty x^{\nu-2\mu} e^{-\frac{1}{4}x^2} W_{k,\pm\mu}\left(\frac{1}{2}x^2\right) J_\nu(xy)\,dx =$$

$$= \frac{\Gamma(1+\nu-2\mu)}{\Gamma(1+2\beta)} 2^{\beta-\mu} y^{k+\mu-\frac{3}{2}} e^{-\frac{1}{4}y^2} M_{\alpha,\beta}\left(\frac{1}{2}y^2\right),$$

$$2\alpha = \frac{1}{2}+k+\nu-3\mu, \qquad 2\beta = \frac{1}{2}-k+\nu-\mu$$

$$[y>0, \quad \operatorname{Re}\nu>-1, \quad \operatorname{Re}(\nu-2\mu)>-1].$$ ET II 84(14)

11. $$\int_0^\infty x^{\nu-2\mu} e^{\frac{1}{4}x^2} W_{k,\pm\mu}\left(\frac{1}{2}x^2\right) J_\nu(xy)\,dx =$$

$$= \frac{\Gamma(1+\nu-2\mu)}{\Gamma\left(\frac{1}{2}+\mu-k\right)} 2^{\frac{1}{2}\left(\frac{1}{2}+k-3\mu+\nu\right)} y^{\mu-k-\frac{3}{2}} e^{\frac{1}{4}y^2} W_{\alpha,\beta}\left(\frac{1}{2}y^2\right),$$

$$2\alpha = k+3\mu-\nu-\frac{1}{2}, \qquad 2\beta = k-\mu+\nu+\frac{1}{2}$$

$$\left[y>0, \ \operatorname{Re}\nu>-1, \ \operatorname{Re}(\nu-2\mu)>-1, \operatorname{Re}\left(k-\mu+\frac{1}{2}\nu\right)<-\frac{1}{4}\right].$$

ET II 84(15)

12. $$\int_0^\infty x^{2\mu-\nu} e^{-\frac{1}{4}x^2} M_{k,\mu}\left(\frac{1}{2}x^2\right) J_\nu(xy)\,dx =$$

$$= \frac{\Gamma(2\mu+1)}{\Gamma\left(\frac{1}{2}+k-\mu+\nu\right)} 2^{\frac{1}{2}\left(\frac{1}{2}-k+3\mu-\nu\right)} y^{k-\mu-\frac{3}{2}} e^{-\frac{1}{4}y^2} M_{\alpha,\beta}\left(\frac{1}{2}y^2\right),$$

$$2\alpha = \frac{1}{2}+k+3\mu-\nu, \quad 2\beta = -\frac{1}{2}+k-\mu+\nu$$

$$\left[y>0, \quad -\frac{1}{2}<\operatorname{Re}\mu<\operatorname{Re}\left(k+\frac{1}{2}\nu\right)-\frac{1}{4}\right].$$ ET II 83(8)

13. $$\int_0^\infty x^{2\mu-\nu} e^{-\frac{1}{4}x^2} M_{k,\mu}\left(\frac{1}{2}x^2\right) N_\nu(xy)\,dx = \pi^{-1}2^{\mu+\beta} y^{k-\mu-\frac{3}{2}} e^{-\frac{1}{4}y^2}\Gamma(2\mu+1)\times$$
$$\times\Gamma\left(\frac{1}{2}-k-\mu\right)\left\{\cos[(\nu-2\mu)\pi]\frac{\Gamma(2\mu-\nu-1)}{\Gamma(2\beta+1)}M_{\alpha,\beta}\left(\frac{1}{2}y^2\right)-\right.$$
$$\left.-\sin[(\nu+k-\mu)\pi]\,W_{\alpha,\beta}\left(\frac{1}{2}y^2\right)\right\},$$
$$2\alpha=3\mu-\nu+k+\frac{1}{2},\qquad 2\beta=\mu-\nu-k+\frac{1}{2}$$
$$\left[y>0,\quad -1<2\operatorname{Re}\mu<\operatorname{Re}(2k+\nu)+\frac{1}{2},\quad \operatorname{Re}(2\mu-\nu)>-1\right].$$

ET II 116(44)

14. $$\int_0^\infty x^{2\mu+\nu} e^{-\frac{1}{4}x^2} M_{k,\mu}\left(\frac{1}{2}x^2\right) N_\nu(xy)\,dx = \pi^{-1}2^{\mu+\beta} y^{k-\mu-\frac{3}{2}}\Gamma(2\mu+1)\times$$
$$\times\Gamma\left(\frac{1}{2}-\mu-k\right)e^{-\frac{1}{4}y^2}\left\{\cos(2\mu\pi)\frac{\Gamma(2\mu+\nu+1)}{\Gamma\left(\mu+\nu-k+\frac{3}{2}\right)}M_{\alpha,\beta}\left(\frac{1}{2}y^2\right)+\right.$$
$$\left.+\sin[(\mu-k)\pi]\,W_{\alpha,\beta}\left(\frac{1}{2}y^2\right)\right\},$$
$$2\alpha=3\mu+\nu+k+\frac{1}{2},\qquad 2\beta=\mu+\nu-k+\frac{1}{2}$$
$$\left[y>0,\ -1<2\operatorname{Re}\mu<\operatorname{Re}(2k-\nu)+\frac{1}{2},\ \operatorname{Re}(2\mu+\nu)>-1\right].$$

ET II 116(43)

15. $$\int_0^\infty x^{2\mu+\nu} e^{-\frac{1}{2}ax^2} M_{k,\mu}(ax^2)\,K_\nu(xy)\,dx = 2^{\mu-k-\frac{1}{2}} a^{\frac{1}{4}-\frac{1}{2}(\mu+\nu+k)} y^{k-\mu-\frac{3}{2}}\times$$
$$\times\Gamma(2\mu+1)\Gamma(2\mu+\nu+1)\exp\left(\frac{y^2}{8a}\right)W_{\varkappa,m}\left(\frac{y^2}{4a}\right),$$
$$2\varkappa=-3\mu-\nu-k-\frac{1}{2},\quad 2m=\mu+\nu-k+\frac{1}{2}$$
$$\left[\operatorname{Re}y>0,\quad \operatorname{Re}a>0,\quad \operatorname{Re}\mu>-\frac{1}{2},\quad \operatorname{Re}(2\mu+\nu)>-1\right].$$

ET II 152(82)

7.673

1 $$\int_0^\infty e^{-\frac{1}{2}ax} x^{\frac{1}{2}(\mu-\nu-1)} M_{\varkappa,\frac{1}{2}\mu}(ax)\,J_\nu(2\sqrt{bx})\,dx=$$
$$=\left(\frac{b}{a}\right)^{\frac{\varkappa-1}{2}-\frac{1+\mu}{4}} a^{-\frac{1}{2}(\mu+1-\nu)}\Gamma(1+\mu)\,e^{-\frac{b}{2a}}\frac{1}{\Gamma\left(1+\frac{\varkappa+\nu}{2}-\frac{1+\mu}{4}\right)}\times$$
$$\times M_{\frac{1}{2}(\varkappa-\nu-1)+\frac{3}{4}(1+\mu),\ \frac{\varkappa+\nu}{2}-\frac{1+\mu}{4}}\left(\frac{b}{a}\right)+$$
$$\left[\operatorname{Re}(1+\mu)>0,\ \operatorname{Re}\left(\varkappa+\frac{\nu-\mu}{2}\right)>-\frac{3}{4},\ \operatorname{Im}b=0\right].$$ BU 128(12)a

2. $$\int_0^\infty e^{\frac{1}{2}ax} x^{\frac{1}{2}(\nu-1\mp\mu)} W_{\varkappa, \frac{1}{2}\mu}(ax) J_\nu(2\sqrt{bx})\, dx =$$
$$= a^{-\frac{1}{2}(\nu+1\mp\mu)} \frac{\Gamma(\nu+1\mp\mu)\, e^{\frac{b}{2a}}}{\Gamma\left(\frac{1\pm\mu}{2}-\varkappa\right)} \left(\frac{a}{b}\right)^{\frac{1}{2}(\varkappa+1)+\frac{1}{4}(1\mp\mu)} \times$$
$$\times W_{\frac{1}{2}(\varkappa+1-\nu)-\frac{3}{4}(1\mp\mu),\ \frac{1}{2}(\varkappa+\nu)+\frac{1}{4}(1\mp\mu)}\left(\frac{b}{a}\right)$$
$$\left[\operatorname{Re}\left(\frac{\nu\mp\mu}{2}+\varkappa\right) < \frac{3}{4},\ \operatorname{Re}\nu > -1\right].$$ BU 128(13)

7.674

1. $$\int_0^\infty x^{\varrho-1} e^{-\frac{1}{2}x} J_{\lambda+\nu}(ax^{\frac{1}{2}}) J_{\lambda-\nu}(ax^{\frac{1}{2}}) W_{k,\mu}(x)\, dx =$$
$$= \frac{\left(\frac{1}{2}a\right)^{2\lambda} \Gamma\left(\frac{1}{2}+\lambda+\mu+\varrho\right)\Gamma\left(\frac{1}{2}+\lambda-\mu+\varrho\right)}{\Gamma(1+\lambda+\nu)\,\Gamma(1+\lambda-\nu)\,\Gamma(1+\lambda-k+\varrho)} \times$$
$$\times {}_4F_4\left(1+\lambda,\ \frac{1}{2}+\lambda,\ \frac{1}{2}+\lambda+\mu+\varrho,\ \frac{1}{2}+\lambda-\mu+\varrho;\ 1+\lambda+\nu,\right.$$
$$\left. 1+\lambda-\nu,\ 1+2\lambda,\ 1+\lambda-k+\varrho;\ -a^2\right)$$
$$\left[|\operatorname{Re}\mu| < \operatorname{Re}(\lambda+\varrho)+\frac{1}{2}\right].$$ ET II 409(37)

2. $$\int_0^\infty x^{\varrho-1} e^{-\frac{1}{2}x} I_{\lambda+\nu}(ax^{\frac{1}{2}}) K_{\lambda-\nu}(ax^{\frac{1}{2}}) W_{k,\mu}(x)\, dx =$$
$$= \frac{\pi^{-\frac{1}{2}}}{2} G_{45}^{24}\left(a^2 \middle| \begin{matrix} 0,\ \frac{1}{2},\ \frac{1}{2}+\mu-\varrho,\ \frac{1}{2}-\mu-\varrho \\ \lambda,\ \nu,\ -\lambda,\ -\nu,\ k-\varrho \end{matrix}\right)$$
$$\left[|\operatorname{Re}\mu| < \operatorname{Re}(\lambda+\varrho)+\frac{1}{2},\ |\operatorname{Re}\mu| < \operatorname{Re}(\nu+\varrho)+\frac{1}{2}\right].$$ ET II 409(38)

Combinations of Struve functions and degenerate hypergeometric functions

7.675

1. $$\int_0^\infty x^{2\lambda+\frac{1}{2}} e^{-\frac{1}{4}x^2} M_{k,\mu}\left(\frac{1}{2}x^2\right) \mathbf{H}_\nu(xy)\, dx =$$
$$= \frac{2^{-\lambda}\Gamma(2\mu+1)}{y^{\frac{1}{2}}\Gamma\left(\frac{1}{2}+k+\mu\right)} G_{34}^{22}\left(\frac{y^2}{2} \middle| \begin{matrix} l,\ -\mu-\lambda,\ \mu-\lambda \\ l,\ k-\lambda-\frac{1}{2},\ h,\ \varkappa \end{matrix}\right),$$
$$h = \frac{1}{4}+\frac{1}{2}\nu, \quad \varkappa = \frac{1}{4}-\frac{1}{2}\nu, \quad l = \frac{3}{4}+\frac{1}{2}\nu$$
$$\left[\operatorname{Re}(2\lambda+2\mu+\nu) > -\frac{7}{2},\quad \operatorname{Re}(k-\lambda) > 0,\quad y > 0,\right.$$
$$\left.\operatorname{Re}(2\lambda-2k+\nu) < -\frac{1}{2}\right].$$ ET II 171(42)

2. $$\int_0^\infty x^{2\lambda+\frac{1}{2}} e^{-\frac{1}{4}x^2} W_{k,\mu}\left(\frac{1}{2}x^2\right) \mathbf{H}_\nu(xy)\,dx =$$
$$= 2^{\frac{1}{4}-\lambda-\frac{1}{2}\nu} \pi^{-\frac{1}{2}} y^{\nu+1} \frac{\Gamma\left(\frac{7}{4}+\frac{1}{2}\nu+\lambda+\mu\right)\Gamma\left(\frac{7}{4}+\frac{1}{2}\nu+\lambda-\mu\right)}{\Gamma\left(\nu+\frac{3}{2}\right)\Gamma\left(\frac{9}{4}+\lambda-k-\frac{1}{2}\nu\right)} \times$$
$$\times {}_3F_3\left(1, \frac{7}{4}+\frac{\nu}{2}+\lambda+\mu, \frac{7}{4}+\frac{\nu}{2}+\lambda-\mu; \frac{3}{2}, \nu+\frac{3}{2}, \frac{9}{4}+\lambda-k+\frac{\nu}{2}; -\frac{y^2}{2}\right)$$
$$\left[\operatorname{Re}(2\lambda+\nu) > 2|\operatorname{Re}\mu| - \frac{7}{2},\ y > 0\right].$$
ET II 171(43)

3. $$\int_0^\infty x^{2\lambda+\frac{1}{2}} e^{\frac{1}{4}x^2} W_{k,\mu}\left(\frac{1}{2}x^2\right) \mathbf{H}_\nu(xy)\,dx =$$
$$= \left[2^\lambda \Gamma\left(\frac{1}{2}-k+\mu\right)\Gamma\left(\frac{1}{2}-k-\mu\right)\right]^{-1} y^{-\frac{1}{2}} \times$$
$$\times G^{23}_{34}\left(\frac{y^2}{2}\,\middle|\,\begin{matrix} l, -\mu-\lambda, \mu-\lambda \\ l, -k-\lambda-\frac{1}{2}, h, \varkappa \end{matrix}\right),$$
$$h = \frac{1}{4}+\frac{1}{2}\nu,\quad \varkappa = \frac{1}{4}-\frac{1}{2}\nu,\quad l = \frac{3}{4}+\frac{1}{2}\nu$$
$$\left[y > 0,\ \operatorname{Re}(2\lambda+\nu) > 2|\operatorname{Re}\mu| - \frac{7}{2},\right.$$
$$\left.\operatorname{Re}(2k+2\lambda+\nu) < -\frac{1}{2},\ \operatorname{Re}(k+\lambda) < 0\right].$$
ET II 172(46)a

4. $$\int_0^\infty e^{\frac{1}{2}x^2} W_{-\frac{1}{2}\nu-\frac{1}{2}, \frac{1}{2}\nu}(x^2)\, \mathbf{H}_\nu(xy)\,dx =$$
$$= 2^{-\nu-1} y^\nu \pi e^{\frac{1}{4}y^2}\left[1-\Phi\left(\frac{y}{2}\right)\right] \quad [y > 0,\ \operatorname{Re}\nu > -1].$$
ET II 171(44)

7.68 Combinations of degenerate hypergeometric functions and other special functions

Combinations of degenerate hypergeometric functions and associated Legendre functions

7.681

1. $$\int_0^\infty x^{-\frac{1}{2}}(a+x)^\mu e^{-\frac{1}{2}x} P_\nu^{-2\mu}\left(1+2\frac{x}{a}\right) M_{k,\mu}(x)\,dx =$$
$$= -\frac{\sin(\nu\pi)}{\pi\Gamma(k)}\Gamma(2\mu+1)\Gamma\left(k-\mu+\nu+\frac{1}{2}\right)\times$$
$$\times \Gamma\left(k-\mu-\nu-\frac{1}{2}\right) e^{\frac{1}{2}a} W_{\varrho,\sigma}(a),$$
$$\varrho = \frac{1}{2}-k+\mu,\quad \sigma = \frac{1}{2}+\nu$$
$$\left[|\arg a| < \pi,\ \operatorname{Re}\mu > -\frac{1}{2},\ \operatorname{Re}(k-\mu) > \left|\operatorname{Re}\nu+\frac{1}{2}\right|\right].$$
ET II 403(11)

2. $$\int_0^\infty x^{-\frac{1}{2}}(a+x)^{-\mu}e^{-\frac{1}{2}x}P_\nu^{-2\mu}\left(1+2\frac{x}{a}\right)M_{k,\mu}(x)\,dx =$$

$$=\frac{\Gamma(2\mu+1)\,\Gamma\left(k+\mu+\nu+\frac{1}{2}\right)\Gamma\left(k+\mu-\nu-\frac{1}{2}\right)e^{\frac{1}{2}a}}{\Gamma\left(k+\mu+\frac{1}{2}\right)\Gamma(2\mu+\nu+1)\,\Gamma(2\mu-\nu)}W_{\frac{1}{2}-k-\mu,\ \frac{1}{2}+\nu}(a)$$

$\left[|\arg a|<\pi,\ \operatorname{Re}\mu>-\frac{1}{2},\ \operatorname{Re}(k+\mu)>\left|\operatorname{Re}\nu+\frac{1}{2}\right|\right].$ ET II 403(12)

3. $$\int_0^\infty x^{-\frac{1}{2}-\frac{1}{2}\mu-\nu}(a+x)^{\frac{1}{2}\mu}e^{-\frac{1}{2}x}P^\mu_{k+\nu-\frac{3}{2}}\left(1+2\frac{x}{a}\right)W_{k,\nu}(x)\,dx =$$

$$=\frac{\Gamma(1-\mu-2\nu)}{\Gamma\left(\frac{3}{2}-k-\mu-\nu\right)}a^{-\frac{1}{4}+\frac{1}{2}k-\frac{1}{2}\nu}e^{\frac{1}{2}a}W_{\varrho,\sigma}(a),$$

$$2\varrho=\frac{1}{2}+2\mu+\nu-k,\quad 2\sigma=k+3\nu-\frac{3}{2}$$

$[|\arg a|<\pi,\ \operatorname{Re}\mu<1,\ \operatorname{Re}(\mu+2\nu)<1].$ ET II 407(32)

4. $$\int_0^\infty x^{-\frac{1}{2}-\frac{1}{2}\mu-\nu}(a+x)^{-\frac{1}{2}\mu}e^{-\frac{1}{2}x}P^\mu_{k+\mu+\nu-\frac{3}{2}}\left(1+2\frac{x}{a}\right)W_{k,\nu}(x)\,dx =$$

$$=\frac{\Gamma(1-\mu-2\nu)}{\Gamma\left(\frac{3}{2}-k-\mu-\nu\right)}a^{-\frac{1}{2}+\frac{1}{2}k-\frac{1}{2}\nu}e^{\frac{1}{2}a}W_{\varrho,\sigma}(a),$$

$$2\varrho=\frac{1}{2}-k+\nu,\quad 2\sigma=k+2\mu+3\nu-\frac{3}{2}$$

$[|\arg a|<\pi,\ \operatorname{Re}\mu<1,\ \operatorname{Re}(\mu+2\nu)<1].$ ET II 408(33)

5. $$\int_0^\infty x^{\mu-\frac{1}{4}k-\frac{1}{2}\nu-\frac{1}{2}}(a+x)^{\frac{1}{2}\nu}e^{-\frac{1}{2}x}Q^\nu_{\mu-k+\frac{3}{2}}\left(1+2\frac{x}{a}\right)M_{k,\mu}(x)\,dx =$$

$$=\frac{e^{\nu\pi i}\,\Gamma(1+2\mu-\nu)\,\Gamma(1+2\mu)\,\Gamma\left(\frac{5}{2}-k+\mu+\nu\right)}{2\Gamma\left(\frac{1}{2}+k+\mu\right)}a^{\frac{1}{4}(k+2\mu-2\nu+5)}e^{\frac{1}{2}a}W_{\varrho,\sigma}(a),$$

$$2\varrho=\frac{1}{2}-k-\mu+2\nu,\quad 2\sigma=k-3\mu-\frac{3}{2}$$

$\left[|\arg a|<\pi,\ \operatorname{Re}\mu>-\frac{1}{2},\ \operatorname{Re}(2\mu-\nu)>-1\right].$ ET II 404(14)

7.682

1. $$\int_0^\infty x^{-\frac{1}{2}}e^{-\frac{1}{2}x}P_\nu^{-2\mu}\left[\left(1+\frac{x}{a}\right)^{\frac{1}{2}}\right]M_{k,\mu}(x)\,dx =$$

$$=\frac{\Gamma(2\mu+1)\,\Gamma\left(k+\frac{1}{2}\nu\right)\Gamma\left(k-\frac{1}{2}\nu-\frac{1}{2}\right)e^{\frac{1}{2}a}}{2^{2\mu}a^{\frac{1}{4}}\Gamma\left(k+\mu+\frac{1}{2}\right)\Gamma\left(\mu+\frac{1}{2}\nu+\frac{1}{2}\right)\Gamma\left(\mu-\frac{1}{2}\nu\right)}W_{\frac{3}{4}-k,\ \frac{1}{4}+\frac{1}{2}\nu}(a)$$

$\left[|\arg a|<\pi,\ \operatorname{Re}k>\frac{1}{2}\operatorname{Re}\nu-\frac{1}{2},\ \operatorname{Re}k>-\frac{1}{2}\operatorname{Re}\nu\right].$ ET II 404(13)

2. $$\int_0^\infty x^{\frac{1}{2}(k+\mu+\nu)-1}(a+x)^{-\frac{1}{2}}e^{-\frac{1}{2}x}Q_{k-\mu-\nu-1}^{1-k+\mu-\nu}\left[\left(1+\frac{x}{a}\right)^{\frac{1}{2}}\right]M_{k,\mu}(x)\,dx =$$

$$= e^{(1-k+\mu-\nu)\pi i}2^{\mu-k-\nu}a^{\frac{1}{2}(k+\mu-1)}\times$$

$$\times\frac{\Gamma\left(\frac{1}{2}-\nu\right)\Gamma(1+2\mu)\Gamma(k+\mu+\nu)}{\Gamma\left(k+\mu+\frac{1}{2}\right)}e^{\frac{1}{2}a}W_{\varrho,\sigma}(a),$$

$\varrho=\frac{1}{2}-k-\frac{1}{2}\nu,\quad \sigma=\mu+\frac{1}{2}\nu\qquad \left[|\arg a|<\pi,\ \operatorname{Re}\mu>-\frac{1}{2},\ \operatorname{Re}(k+\mu+\nu)>0\right].$ ET II 404(15)

3. $$\int_0^\infty x^{\nu-\frac{1}{2}}e^{-\frac{1}{2}x}Q_{2k-2\nu-3}^{2\mu-2\nu}\left[\left(1+\frac{x}{a}\right)^{\frac{1}{2}}\right]M_{k,\mu}(x)\,dx =$$

$$= e^{2(\mu-\nu)\pi i}\,2^{2\mu-2\nu-1}a^{\frac{1}{2}(k+\mu-1)}e^{\frac{1}{2}a}\times$$

$$\times\frac{\Gamma(2\mu+1)\Gamma(\nu+1)\Gamma\left(k+\mu-2\nu-\frac{1}{2}\right)}{\Gamma\left(k+\mu+\frac{1}{2}\right)}W_{\varrho,\sigma}(a),$$

$$2\varrho=1-k+\mu-2\nu,\quad 2\sigma=k-\mu-2\nu-2$$

$\left[|\arg a|<\pi,\quad \operatorname{Re}\mu>-\frac{1}{2},\ \operatorname{Re}\nu>-1,\ \operatorname{Re}(k+\mu-2\nu)>\frac{1}{2}\right].$ ET II 404(16)

4. $$\int_0^\infty x^{-\frac{1}{2}-\frac{1}{2}\mu-\nu}e^{-\frac{1}{2}x}P_{2k+\mu+2\nu-3}^{\mu}\left[\left(1+\frac{x}{a}\right)^{\frac{1}{2}}\right]W_{k,\nu}(x)\,dx =$$

$$=\frac{2^{\mu}\Gamma(1-\mu-2\nu)}{\Gamma\left(\frac{3}{2}-k-\mu-\nu\right)}a^{-\frac{1}{2}+\frac{1}{2}k-\frac{1}{2}\nu}e^{\frac{1}{2}a}W_{\varrho,\sigma}(a),$$

$$2\varrho=1-k+\mu+\nu,\quad 2\sigma=k+\mu+3\nu-2$$

$[|\arg a|<\pi,\quad \operatorname{Re}\mu<1,\quad \operatorname{Re}(\mu+2\nu)<1].$ ET II 408(34)

5. $$\int_0^\infty x^{-\frac{1}{2}-\frac{1}{2}\mu-}(a+x)^{-\frac{1}{2}}e^{-\frac{1}{2}x}P_{2k+\mu+2\nu-2}^{\mu}\left[\left(1+\frac{x}{a}\right)^{\frac{1}{2}}\right]W_{k,\nu}(x)\,dx =$$

$$=\frac{2^{\mu}\Gamma(1-\mu-2\nu)}{\Gamma\left(\frac{3}{2}-k-\mu-\nu\right)}a^{-\frac{1}{2}+\frac{1}{2}k-\frac{1}{2}\nu}e^{\frac{1}{2}a}W_{\varrho,\sigma}(a),$$

$$2\varrho=\mu+\nu-k,\ 2\sigma=k+\mu+3\nu-1$$

$[|\arg a|<\pi,\ \operatorname{Re}\mu>0,\ \operatorname{Re}\nu>0].$ ET II 408(35)

A combination of degenerate hypergeometric functions and orthogonal polynomials

7.683

$$\int_0^1 e^{-\frac{1}{2}ax}x^{\alpha}(1-x)^{\frac{\mu-\alpha}{2}-1}L_n^{\alpha}(ax)\,M_{\varkappa-\frac{1+\alpha}{2},\,\frac{\mu-\alpha-1}{2}}[a(1-x)]\,dx =$$

$$=\frac{\Gamma(\mu-\alpha)}{\Gamma(1+\mu)}\frac{\Gamma(1+n+\alpha)}{n!}a^{-\frac{1+\alpha}{2}}M_{\varkappa+n,\frac{\mu}{2}}(a)$$

$[\operatorname{Re}\alpha>-1,\ \operatorname{Re}(\mu-\alpha)>0,\ n=0,1,2,\ldots].$ BU 129(14b)

A combination of hypergeometric and degenerate hypergeometric functions

7.684 $$\int_0^\infty x^{\varrho-1} e^{-\frac{1}{2}x} M_{\gamma+\varrho,\, \beta+\varrho+\frac{1}{2}}(x)\, {}_2F_1\left(\alpha, \beta; \gamma; -\frac{\lambda}{x}\right) dx =$$
$$= \frac{\Gamma(\alpha+\beta+2\varrho)\,\Gamma(2\beta+2\varrho)\,\Gamma(\gamma)}{\Gamma(\beta)\,\Gamma(\beta+\gamma+2\varrho)} \lambda^{\frac{1}{2}\beta+\varrho-\frac{1}{2}} e^{\frac{1}{2}\lambda} W_{k,\mu}(\lambda);$$
$$k = \frac{1}{2} - \alpha - \frac{1}{2}\beta - \varrho, \quad \mu = \frac{1}{2}\beta + \varrho$$
$$[|\arg\lambda| < \pi, \ \operatorname{Re}(\beta+\varrho) > 0, \ \operatorname{Re}(\alpha+\beta+2\varrho) > 0, \ \operatorname{Re}\gamma > 0].$$
ET II 405(19)

7.69 Integration of degenerate hypergeometric functions with respect to the index

7.691 $$\int_{-\infty}^{\infty} \operatorname{sech}(\pi x)\, W_{ix,0}(\alpha)\, W_{-ix,0}(\beta)\, dx =$$
$$= 2\frac{(\alpha\beta)^{\frac{1}{2}}}{\alpha+\beta} \exp\left[-\frac{1}{2}(\alpha+\beta)\right].$$
ET II 414(61)

7.692 $$\int_{-i\infty}^{i\infty} \Gamma(-a)\,\Gamma(c-a)\,\Psi(a, c; x)\,\Psi(c-a, c; y)\, da =$$
$$= 2\pi i\, \Gamma(c)\, \Psi(c, 2c; x+y).$$
EH I 285(15)

7.693

1. $$\int_{-\infty}^{\infty} \Gamma(ix)\,\Gamma(2k+ix)\, W_{k+ix,\, k-\frac{1}{2}}(\alpha)\, W_{-k-ix,\, k-\frac{1}{2}}(\beta)\, dx =$$
$$= 2\pi^{\frac{1}{2}}\, \Gamma(2k)\, (\alpha\beta)^k\, (\alpha+\beta)^{\frac{1}{2}-2k} K_{2k-\frac{1}{2}}\left(\frac{\alpha+\beta}{2}\right).$$
ET II 414(62)

2. $$\int_{-i\infty}^{i\infty} \Gamma\left(\frac{1}{2}+\nu+\mu+x\right)\Gamma\left(\frac{1}{2}+\nu+\mu-x\right) \times$$
$$\times \Gamma\left(\frac{1}{2}+\nu-\mu+x\right)\Gamma\left(\frac{1}{2}+\nu-\mu-x\right) M_{\mu+ix,\,\nu}(\alpha)\, M_{\mu-ix,\,\nu}(\beta)\, dx =$$
$$= \frac{2\pi\,(\alpha\beta)^{\nu+\frac{1}{2}}\,[\Gamma(2\nu+1)]^2\,\Gamma(2\nu+2\mu+1)\,\Gamma(2\nu-2\mu+1)}{(\alpha+\beta)^{2\nu+1}\,\Gamma(4\nu+2)} M_{2\mu,\, 2\nu+\frac{1}{2}}(\alpha+\beta)$$
$$\left[\operatorname{Re}\nu > |\operatorname{Re}\mu| - \frac{1}{2}\right].$$
ET II 413(59)

7.694 $$\int_{-\infty}^{\infty} e^{-2\varrho x i}\, \Gamma\left(\frac{1}{2}+\nu+ix\right)\Gamma\left(\frac{1}{2}+\nu-ix\right) M_{ix,\nu}(\alpha)\, M_{ix,\nu}(\beta)\, dx =$$
$$= \frac{2\pi\,(\alpha\beta)^{\frac{1}{2}}}{\operatorname{ch}\varrho} \exp\left[-(\alpha+\beta)\operatorname{th}\varrho\right] J_{2\nu}\left(\frac{2\alpha^{\frac{1}{2}}\beta^{\frac{1}{2}}}{\operatorname{ch}\varrho}\right)$$
$$\left[|\operatorname{Im}\varrho| < \frac{1}{2}\pi, \ \operatorname{Re}\nu > -\frac{1}{2}\right].$$
ET II 414(60)

7.7 Parabolic-Cylinder Functions*

7.71 Parabolic-cylinder functions

7.711

1. $$\int_{-\infty}^{\infty} D_n(x) D_m(x)\, dx = 0 \qquad [m \neq n];$$ WH

$$= n!\,(2\pi)^{\frac{1}{2}} \qquad [m = n].$$ WH

2. $$\int_0^{\infty} D_\mu(\pm t) D_\nu(t)\, dt =$$

$$= \frac{\pi 2^{\frac{1}{2}(\mu+\nu+1)}}{\mu-\nu}\left[\frac{1}{\Gamma\left(\frac{1}{2}-\frac{1}{2}\mu\right)\Gamma\left(-\frac{1}{2}\nu\right)} \mp \frac{1}{\Gamma\left(\frac{1}{2}-\frac{1}{2}\nu\right)\Gamma\left(-\frac{1}{2}\mu\right)}\right]$$

[when the lower sign is taken, $\operatorname{Re}\mu > \operatorname{Re}\nu$].

BU 11 117(13a), EH II 122(21)

3. $$\int_0^{\infty} [D_\nu(t)]^2\, dt = \pi^{\frac{1}{2}} 2^{-\frac{3}{2}} \frac{\psi\left(\frac{1}{2}-\frac{1}{2}\nu\right) - \psi\left(-\frac{1}{2}\nu\right)}{\Gamma(-\nu)}.$$

BU 117(13b)a, EH II 122(22)a

7.72 Combinations of parabolic-cylinder functions, powers, and exponentials

7.721

1. $$\int_{-\infty}^{\infty} e^{-\frac{1}{4}x^2}(x-z)^{-1} D_n(x)\, dx = \pm\, i e^{\mp n\pi i}(2\pi)^{\frac{1}{2}} n!\, e^{-\frac{1}{4}z^2} D_{-n-1}(\mp iz)$$

[The upper or lower sign is taken according as the imaginary part of z is positive or negative]. WH

2. $$\int_1^{\infty} x^\nu (x-1)^{\frac{1}{2}\mu-\frac{1}{2}\nu-1} \exp\left[-\frac{(x-1)^2 a^2}{4}\right] D_\mu(ax)\, dx =$$

$$= 2^{\mu-\nu-2} a^{\frac{\mu}{2}-\frac{\nu}{2}-1}\, \Gamma\left(\frac{\mu-\nu}{2}\right) D_\nu(a)$$

[$\operatorname{Re}(\mu-\nu) > 0$]. ET II 395(4)a

7.722

1. $$\int_0^{\infty} e^{-\frac{3}{4}x^2} x^\nu D_{\nu+1}(x)\, dx = 2^{-\frac{1}{2}-\frac{1}{2}\nu}\, \Gamma(\nu+1) \sin\frac{1}{4}(1-\nu)\pi$$

[$\operatorname{Re}\nu > -1$]. WH

2. $$\int_0^{\infty} e^{-\frac{1}{4}x^2} x^{\mu-1} D_{-\nu}(x)\, dx = \frac{\pi^{\frac{1}{2}} 2^{-\frac{1}{2}\mu-\frac{1}{2}\nu}\,\Gamma(\mu)}{\Gamma\left(\frac{1}{2}\mu+\frac{1}{2}\nu+\frac{1}{2}\right)}$$

[$\operatorname{Re}\mu > 0$]. EH II 122(20)

*See Whitaker, E. T., & Watson, G. N., *Modern Analysis*, Cambridge University Press 1952, page 437 for definition.

3. $$\int_0^\infty e^{-\frac{3}{4}x^2} x^\nu D_{\nu-1}(x)\,dx = 2^{-\frac{1}{2}\nu-1}\Gamma(\nu)\sin\frac{1}{4}\pi\nu$$

$[\operatorname{Re}\nu > -1]$. ET II 395(2)

7.723

1. $$\int_0^\infty e^{-\frac{1}{4}x^2} x^\nu (x^2+y^2)^{-1} D_\nu(x)\,dx = \left(\frac{\pi}{2}\right)^{\frac{1}{2}}\Gamma(\nu+1)\,y^{\nu-1}e^{\frac{1}{4}y^2}D_{-\nu-1}(y)$$

$[\operatorname{Re} y > 0,\ \operatorname{Re}\nu > -1]$. EH II 121(18)a, ET II 396(6)a

2. $$\int_0^\infty e^{-\frac{1}{4}x^2} x^{\nu-1}(x^2+y^2)^{-\frac{1}{2}} D_\nu(x)\,dx = y^{\nu-1}\Gamma(\nu)\,e^{\frac{1}{4}y^2}D_{-\nu}(y)$$

$[\operatorname{Re} y > 0,\quad \operatorname{Re}\nu > 0]$. ET II 396(7)

3. $$\int_0^1 x^{2\nu-1}(1-x^2)^{\lambda-1}e^{\frac{a^2x^2}{4}}D_{-2\lambda-2\nu}(ax)\,dx = \frac{\Gamma(\lambda)\,\Gamma(2\nu)}{\Gamma(2\lambda+2\nu)}2^{\lambda-1}e^{\frac{a^2}{4}}D_{-2\nu}(a)$$

$[\operatorname{Re}\lambda > 0,\ \operatorname{Re}\nu > 0]$. ET II 395(3)a

7.724 $$\int_{-\infty}^\infty e^{-\frac{(x-y)^2}{2\mu}}e^{\frac{1}{4}x^2}D_\nu(x)\,dx =$$

$$= (2\pi\mu)^{\frac{1}{2}}(1-\mu)^{\frac{1}{2}\nu}e^{\frac{y^2}{4-4\mu}}D_\nu\left[y(1-\mu)^{-\frac{1}{2}}\right]$$

$[0 < \operatorname{Re}\mu < 1]$. EH II 121(15)

7.725

1. $$\int_0^\infty e^{-pt}(2t)^{\frac{\nu-1}{2}}e^{-\frac{t}{2}}D_{-\nu-2}(\sqrt{2t})\,dt =$$

$$= \left(\frac{\pi}{2}\right)^{\frac{1}{2}}\frac{(\sqrt{p+1}-1)^{\nu+1}}{(\nu+1)\,p^{\nu+1}}$$ $[\operatorname{Re}\nu > -1]$. MO 175

2. $$\int_0^\infty e^{-pt}(2t)^{\frac{\nu-1}{2}}e^{-\frac{t}{2}}D_{-\nu}(\sqrt{2t})\,dt =$$

$$= \left(\frac{\pi}{2}\right)^{\frac{1}{2}}\frac{(\sqrt{p+1}-1)^{\nu}}{p^\nu\sqrt{p+1}}$$ $[\operatorname{Re}\nu > -1]$. MO 175

3. $$\int_0^\infty e^{-bx}D_{2n+1}(\sqrt{2x})\,dx = (-2)^n\,\Gamma\left(n+\frac{3}{2}\right)\left(b-\frac{1}{2}\right)^n\left(b+\frac{1}{2}\right)^{-n-\frac{3}{2}}$$

$\left[\operatorname{Re} b > -\frac{1}{2}\right]$. ET I 210(3)

4. $$\int_0^\infty (\sqrt{x})^{-1}e^{-bx}D_{2n}(\sqrt{2x})\,dx =$$

$$= (-2)^n\,\Gamma\left(n+\frac{1}{2}\right)\left(b-\frac{1}{2}\right)^n\left(b+\frac{1}{2}\right)^{-n-\frac{1}{2}}$$

$\left[\operatorname{Re} b > -\frac{1}{2}\right]$. ET I 210(5)

5. $$\int_0^\infty x^{-\frac{1}{2}(\nu+1)} e^{-sx} D_\nu(\sqrt{x})\,dx = \sqrt{\pi}\left(1+\sqrt{\frac{1}{2}+2s}\right)^\nu \frac{1}{\sqrt{\frac{1}{4}+s}}$$

$$\left[\operatorname{Re} s > -\frac{1}{4},\quad \operatorname{Re}\nu < 1\right].$$ ET I 210(7)

6. $$\int_0^\infty e^{-zt} t^{-1+\frac{\beta}{2}} D_{-\nu}\left[2(kt)^{\frac{1}{2}}\right] dt =$$

$$= \frac{2^{1-\beta-\frac{\nu}{2}} \pi^{\frac{1}{2}} \Gamma(\beta)}{\Gamma\left(\frac{1}{2}\nu+\frac{1}{2}\beta+\frac{1}{2}\right)} (z+k)^{-\frac{\beta}{2}} F\left(\frac{\nu}{2}, \frac{\beta}{2}; \frac{\nu+\beta+1}{2}; \frac{z-k}{z+k}\right)$$

$$\left[\operatorname{Re}(z+k) > 0,\ \operatorname{Re}\frac{z}{k} > 0\right].$$ EH II 121(11)

7.726 $$\int_{-\infty}^\infty e^{ixy - \frac{(1+\lambda)x^2}{4}} D_\nu\left[x(1-\lambda)^{\frac{1}{2}}\right] dx = (2\pi)^{\frac{1}{2}} \lambda^{\frac{1}{2}\nu} e^{-\frac{(1+\lambda)y^2}{4\lambda}} D_\nu\left[i(\lambda^{-1}-1)^{\frac{1}{2}} y\right]$$

$$[\operatorname{Re}\lambda > 0].$$ EH II 121(16)

7.727 $$\int_0^\infty \frac{e^{\frac{1}{2}x} e^{-bx}}{(e^x-1)^{\mu+\frac{1}{2}}} \exp\left(-\frac{a}{1-e^{-x}}\right) D_{2\mu}\left(\frac{2\sqrt{a}}{\sqrt{1-e^{-x}}}\right) dx =$$

$$= e^{-a} 2^{b+\mu} \Gamma(b+\mu) D_{-2b}(2\sqrt{a})$$

$$[\operatorname{Re} a > 0,\ \operatorname{Re} b > -\operatorname{Re}\mu].$$ ET I 211(13)

7.728 $$\int_0^\infty (2t)^{-\frac{\nu}{2}} e^{-pt} e^{-\frac{q^2}{8t}} D_{\nu-1}\left(\frac{q}{\sqrt{2t}}\right) dt = \left(\frac{\pi}{2}\right)^{\frac{1}{2}} p^{\frac{1}{2}\nu-1} e^{-q\sqrt{p}}.$$ MO 175

7.73 Combinations of parabolic-cylinder and hyperbolic functions

7.731

1. $$\int_0^\infty \operatorname{ch}(2\mu x) \exp\left[-(a \operatorname{sh} x)^2\right] D_{2k}(2a \operatorname{ch} x)\,dx = 2^{k-\frac{3}{2}} \pi^{\frac{1}{2}} a^{-1} W_{k,\mu}(2a^2)$$

$$[\operatorname{Re} a^2 > 0].$$ ET II 398(20)

2. $$\int_0^\infty \operatorname{ch}(2\mu x) \exp\left[(a \operatorname{sh} x)^2\right] D_{2k}(2a \operatorname{ch} x)\,dx =$$

$$= \frac{\Gamma(\mu-k)\Gamma(-\mu-k)}{2^{k+\frac{5}{2}} a \Gamma(-2k)} W_{k+\frac{1}{2},\mu}(2a^2)$$

$$\left[|\arg a| < \frac{3\pi}{4},\ \operatorname{Re} k + |\operatorname{Re}\mu| < 0\right].$$ ET II 398(21)

7.74 Combinations of parabolic-cylinder and trigonometric functions

7.741

1. $$\int_0^\infty \sin(bx)\{[D_{-n-1}(ix)]^2 - [D_{-n-1}(-ix)]^2\}\,dx = $$
$$= (-1)^{n+1}\frac{i}{n!}\pi\sqrt{2\pi}e^{-\frac{1}{2}b^2}L_n(b^2) \qquad [b>0].$$ ET I 115(3)

2. $$\int_0^\infty e^{-\frac{1}{4}x^2}\sin(bx)\,D_{2n+1}(x)\,dx = (-1)^n\sqrt{\frac{\pi}{2}}\,b^{2n+1}e^{-\frac{1}{2}b^2} \qquad [b>0].$$ ET I 115(1)

3. $$\int_0^\infty e^{-\frac{1}{4}x^2}\cos(bx)\,D_{2n}(x)\,dx = (-1)^n\sqrt{\frac{\pi}{2}}\,b^{2n}e^{-\frac{1}{2}b^2} \qquad [b>0].$$ ET I 60(2)

4. $$\int_0^\infty e^{-\frac{1}{4}x^2}\sin(bx)\left[D_{2\nu-\frac{1}{2}}(x) - D_{2\nu-\frac{1}{2}}(-x)\right]dx =$$
$$= \sqrt{2\pi}\sin\left[\left(\nu-\frac{1}{4}\right)\pi\right]b^{2\nu-\frac{1}{2}}e^{-\frac{1}{2}b^2} \qquad \left[\operatorname{Re}\nu > \frac{1}{4},\ b>0\right].$$ ET I 115(2)

5. $$\int_0^\infty e^{-\frac{1}{2}x^2}\cos(bx)\left[D_{2\nu-\frac{1}{2}}(x) + D_{2\nu-\frac{1}{2}}(-x)\right]dx =$$
$$= \frac{2^{\frac{1}{4}-2\nu}\sqrt{\pi}\,b^{2\nu-\frac{1}{2}}e^{-\frac{1}{4}b^2}}{\operatorname{cosec}\left[\left(\nu+\frac{1}{4}\right)\pi\right]} \qquad \left[\operatorname{Re}\nu > \frac{1}{4},\ b>0\right].$$ ET I 61(4)

7.742

1. $$\int_0^\infty x^{2\varrho-1}\sin(ax)\,e^{-\frac{x^2}{4}}D_{2\nu}(x)\,dx =$$
$$= 2^{\nu-\varrho-\frac{1}{2}}\pi^{\frac{1}{2}}a\frac{\Gamma(2\varrho+1)}{\Gamma(\varrho-\nu+1)}\,{}_2F_2\left(\varrho+\frac{1}{2},\ \varrho+1;\ \frac{3}{2},\ \varrho-\nu+1;\ -\frac{a^2}{2}\right) \qquad \left[\operatorname{Re}\varrho > -\frac{1}{2}\right].$$ ET II 396(8)

2. $$\int_0^\infty x^{2\varrho-1}\sin(ax)\,e^{\frac{x^2}{4}}D_{2\nu}(x)\,dx = \frac{2^{\varrho-\nu-2}}{\Gamma(-2\nu)}G_{23}^{22}\left(\frac{a^2}{2}\middle|\begin{matrix}\frac{1}{2}-\varrho,\ 1-\varrho\\ -\varrho-\nu,\ \frac{1}{2},\ 0\end{matrix}\right)$$
$$\left[a>0,\ \operatorname{Re}\varrho > -\frac{1}{2},\ \operatorname{Re}(\varrho+\nu) < \frac{1}{2}\right].$$ ET II 396(9)

3. $$\int_0^\infty x^{2\varrho-1}\cos(ax)e^{-\frac{x^2}{4}}D_{2\nu}(x)\,dx = \frac{2^{\nu-\varrho}\Gamma(2\varrho)\pi^{\frac{1}{2}}}{\Gamma\left(\varrho-\nu+\frac{1}{2}\right)}\,{}_2F_2\left(\varrho,\ \varrho+\frac{1}{2};\ \frac{1}{2},\ \varrho-\nu+\frac{1}{2};\ -\frac{a^2}{2}\right)$$

$[\operatorname{Re}\varrho > 0]$. ET II 396(10)a

4. $$\int_0^\infty x^{2\varrho-1}\cos(ax)\,e^{\frac{x^2}{4}}D_{2\nu}(x)\,dx = \frac{2^{\varrho-\nu-2}}{\Gamma(-2\nu)}G_{23}^{22}\left(\frac{a^2}{2}\,\middle|\,\begin{matrix}\frac{1}{2}-\varrho,\ 1-\varrho\\ -\varrho-\nu,\ 0,\ \frac{1}{2}\end{matrix}\right)$$

$\left[a > 0,\ \operatorname{Re}\varrho > 0,\ \operatorname{Re}(\varrho+\nu) < \frac{1}{2}\right]$. ET II 396(11)

7.743 $$\int_0^{\frac{\pi}{2}}(\cos x)^{-\mu-2}(\sin x)^{-\nu}D_\nu(a\sin x)D_\mu(a\cos x)\,dx = -\left(\frac{1}{2}\pi\right)^{\frac{1}{2}}(1+\mu)^{-1}D_{\mu+\nu+1}(a)$$

$[\operatorname{Re}\nu < 1,\ \operatorname{Re}\mu < -1]$. ET II 397(19)

7.744

1. $$\int_0^\infty \sin(bx)\left[D_{-\nu-\frac{1}{2}}(\sqrt{2x}) - D_{-\nu-\frac{1}{2}}(-\sqrt{2x})\right]D_{\nu-\frac{1}{2}}(\sqrt{2x})\,dx = -\sqrt{2\pi}\sin\left[\left(\frac{1}{4}+\frac{1}{2}\nu\right)\pi\right]b^{-\nu-\frac{1}{2}}\frac{(1+\sqrt{1+b^2})^\nu}{\sqrt{1+b^2}}$$

$[b > 0]$. ET I 115(4)

2 $$\int_0^\infty \cos(bx)\left[D_{-2\nu-\frac{1}{2}}(\sqrt{2x}) + D_{-2\nu-\frac{1}{2}}(-\sqrt{2x})\right]D_{2\nu-\frac{1}{2}}(\sqrt{2x})\,dx = -\frac{\sqrt{\pi}\sin\left[\left(\nu-\frac{1}{4}\right)\pi\right](1+\sqrt{1+b^2})^{2\nu}}{\sqrt{1+b^2}\,b^{2\nu+\frac{1}{2}}}$$

$[b > 0]$. ET I 60(3)

7.75 Combinations of parabolic-cylinder and Bessel functions

7.751

1. $$\int_0^\infty [D_n(ax)]^2 J_1(xy)\,dx = (-1)^{n-1}y^{-1}\left[D_n\left(\frac{y}{a}\right)\right]^2$$

$[y > 0]$. ET II 20(24)

2. $$\int_0^\infty J_0(xy) D_n(ax) D_{n+1}(ax)\,dx = (-1)^n y^{-1} D_n\left(\frac{y}{a}\right) D_{n+1}\left(\frac{y}{a}\right)$$

$$\left[y>0,\ |\arg a| < \frac{1}{4}\pi\right]$$ ET II 17(42)

3. $$\int_0^\infty J_0(xy) D_\nu(x) D_{\nu+1}(x)\,dx =$$

$$= 2^{-1} y^{-1} [D_\nu(-y) D_{\nu+1}(y) - D_{\nu+1}(-y) D_\nu(y)].$$ ET II 397(17)a

7.752

1. $$\int_0^\infty x^\nu e^{-\frac{1}{4}x^2} D_{2\nu-1}(x) J_\nu(xy)\,dx =$$

$$= -\frac{1}{2}\sec(\nu\pi)\, y^{\nu-1} e^{-\frac{1}{4}y^2} [D_{2\nu-1}(y) - D_{2\nu-1}(-y)]$$

$$\left[y>0,\ \operatorname{Re}\nu > -\frac{1}{2}\right].$$ ET II 76(1), MO 183

2. $$\int_0^\infty x^\nu e^{\frac{1}{4}x^2} D_{2\nu-1}(x) J_\nu(xy)\,dx = 2^{\frac{1}{2}-\nu}\pi \sin(\nu\pi)\, y^{-\nu}\Gamma(2\nu)\, e^{\frac{1}{4}y^2} K_\nu\left(\frac{1}{4}y^2\right)$$

$$\left[y>0,\ -\frac{1}{2} < \operatorname{Re}\nu < \frac{1}{2}\right].$$ ET II 77(4)

3. $$\int_0^\infty x^{\nu+1} e^{-\frac{1}{4}x^2} D_{2\nu}(x) J_\nu(xy)\,dx =$$

$$= \frac{1}{2}\sec(\nu\pi)\, y^{\nu-1} e^{-\frac{1}{4}y^2} [D_{2\nu+1}(y) - D_{2\nu+1}(-y)]$$

$$[y>0,\ \operatorname{Re}\nu > -1].$$ ET II 78(13)

4. $$\int_0^\infty x^\nu e^{-\frac{1}{4}x^2} D_{2\nu+1}(x) J_\nu(xy)\,dx =$$

$$= \frac{1}{2}\sec(\nu\pi)\, e^{-\frac{1}{4}y^2} y^\nu [D_{2\nu}(y) + D_{2\nu}(-y)]$$

$$\left[y>0,\ \operatorname{Re}\nu > -\frac{1}{2}\right].$$ ET II 77(5)

5 $$\int_0^\infty x^{\nu+1} e^{-\frac{1}{4}x^2} D_{2\nu+2}(x) J_\nu(xy)\,dx =$$

$$= -\frac{1}{2}\sec(\nu\pi)\, y^\nu e^{-\frac{1}{4}y^2} [D_{2\nu+2}(y) + D_{2\nu+2}(-y)]$$

$$[\operatorname{Re}\nu > -1,\ y>0].$$ ET II 78(16)

6. $$\int_0^\infty x^{\nu+1} e^{\frac{1}{4}x^2} D_{2\nu+2}(x) J_\nu(xy)\,dx =$$

$$= \pi^{-1}\sin(\nu\pi)\,\Gamma(2\nu+3)\, y^{-\nu-2} e^{\frac{1}{4}y^2} K_{\nu+1}\left(\frac{1}{4}y^2\right)$$

$$\left[y>0,\ -1 < \operatorname{Re}\nu < -\frac{5}{6}\right].$$ ET II 78(19)

7. $$\int_0^\infty x^\nu e^{-\frac{1}{4}x^2} D_{-2\nu}(x) J_\nu(xy)\,dx = 2^{-\frac{1}{2}}\pi^{\frac{1}{2}} y^{-\nu} e^{-\frac{1}{4}y^2} I_\nu\left(\frac{1}{4}y^2\right)$$

$$\left[y > 0,\ \operatorname{Re}\nu > -\frac{1}{2}\right].$$ ET II 77(8)

8. $$\int_0^\infty x^\nu e^{\frac{1}{4}x^2} D_{-2\nu}(x) J_\nu(xy)\,dx = y^{\nu-1} e^{\frac{1}{4}y^2} D_{-2\nu}(y)$$

$$\left[\operatorname{Re}\nu > -\frac{1}{2},\ y > 0\right].$$ ET II 77(9), EH II 121(17)

9. $$\int_0^\infty x^\nu e^{\frac{1}{4}x^2} D_{-2\nu-2}(x) J_\nu(xy)\,dx = (2\nu+1)^{-1} y^\nu e^{\frac{1}{4}y^2} D_{-2\nu-1}(y)$$

$$\left[y > 0,\ \operatorname{Re}\nu > -\frac{1}{2}\right].$$ ET II 77(10)

10. $$\int_0^\infty x^\nu e^{-\frac{1}{4}a^2x^2} D_{2\mu}(ax) J_\nu(xy)\,dx =$$

$$= \frac{2^{\mu-\frac{1}{2}}\Gamma\left(\nu+\frac{1}{2}\right) y^\nu}{\Gamma(\nu-\mu+1)\, a^{1+2\nu}}\, {}_1F_1\left(\nu+\frac{1}{2};\ \nu-\mu+1;\ -\frac{y^2}{2a^2}\right)$$

$$\left[y > 0,\ |\arg a| < \frac{1}{4}\pi,\ \operatorname{Re}\nu > -\frac{1}{2}\right].$$ ET II 77(11)

11. $$\int_0^\infty x^\nu e^{\frac{1}{4}a^2x^2} D_{2\mu}(ax) J_\nu(xy)\,dx = \frac{\Gamma\left(\frac{1}{2}+\nu\right) a^{2k} 2^{m+\mu}}{\Gamma\left(\frac{1}{2}-\mu\right) y^{\mu+\frac{3}{2}}} e^{\frac{y^2}{4a^2}} W_{k,m}\left(\frac{y^2}{4a^2}\right),$$

$$2k = \frac{1}{2}+\mu-\nu, \quad 2m = \frac{1}{2}+\mu+\nu$$

$$\left[y > 0,\ |\arg a| < \frac{1}{4}\pi,\ -\frac{1}{2} < \operatorname{Re}\nu < \operatorname{Re}\left(\frac{1}{2}-2\mu\right)\right].$$ ET II 78(12)

12. $$\int_0^\infty x^{\nu+1} e^{-\frac{1}{4}a^2x^2} D_{2\mu}(ax) J_\nu(xy)\,dx =$$

$$= \frac{2^\mu \Gamma\left(\nu+\frac{3}{2}\right) y^\nu}{\Gamma\left(\nu-\mu+\frac{3}{2}\right) a^{2\nu+2}}\, {}_1F_1\left(\nu+\frac{3}{2};\ \nu-\mu+\frac{3}{2};\ -\frac{y^2}{2a^2}\right)$$

$$\left[y > 0,\ |\arg a| < \frac{1}{4}\pi,\ \operatorname{Re}\nu > -1\right].$$ ET II 79(23)

13. $$\int_0^\infty x^{\nu+1} e^{\frac{1}{4}a^2x^2} D_{2\mu}(ax) J_\nu(xy)\,dx =$$

$$= \frac{\Gamma\left(\frac{3}{2}+\nu\right) 2^{\frac{1}{2}+m+\mu} a^{2k+1}}{\Gamma(-\mu)\, y^{\mu+2}} e^{\frac{y^2}{4a^2}} W_{k,m}\left(\frac{y^2}{2a^2}\right),$$

$$2k = \mu-\nu-1, \quad 2m = \mu+\nu+1$$

$$\left[y > 0,\ |\arg a| < \frac{3}{4}\pi,\ -1 < \operatorname{Re}\nu < -\frac{1}{2} - 2\operatorname{Re}\mu\right].$$ ET II 79(24)

14. $\int\limits_0^\infty x^{\lambda+\frac{1}{2}} e^{\frac{1}{4}a^2x^2} D_\mu(ax) J_\nu(xy)\,dx =$

$$= \frac{2^{\lambda-\frac{1}{2}\mu}\pi^{-\frac{1}{2}}}{\Gamma(-\mu)\,y^{\lambda+\frac{3}{2}}} G_{23}^{22}\left(\frac{y^2}{2a^2}\middle|\begin{matrix}\frac{1}{2},\ 1\\ \frac{3}{4}+\frac{\lambda+\nu}{2},\ -\frac{\mu}{2},\ \frac{3}{4}+\frac{\lambda-\nu}{2}\end{matrix}\right)$$

$\left[y>0,\ |\arg a|<\frac{3}{4}\pi,\ \operatorname{Re}\mu<-\operatorname{Re}\lambda<\operatorname{Re}\nu+\frac{3}{2}\right]$. ET II 80(26)

15. $\int\limits_0^\infty x^{\nu+1} e^{\frac{1}{4}x^2} D_{-2\nu-1}(x) J_\nu(xy)\,dx = (2\nu+1)\,y^{\nu-1} e^{\frac{1}{4}y^2} D_{-2\nu-2}(y)$

$\left[y>0,\ \operatorname{Re}\nu>-\frac{1}{2}\right]$. ET II 79(20)

16 $\int\limits_0^\infty x^{\nu+1} e^{-\frac{1}{4}x^2} D_{-2\nu-3}(x) J_\nu(xy)\,dx = 2^{-\frac{1}{2}}\pi^{\frac{1}{2}} y^{-\nu-2} e^{-\frac{1}{4}y^2} I_{\nu+1}\left(\frac{1}{4}y^2\right)$

$[y>0,\ \operatorname{Re}\nu>-1]$. ET II 79(21)

17. $\int\limits_0^\infty x^{\nu+1} e^{\frac{1}{4}x^2} D_{-2\nu-3}(x) J_\nu(xy)\,dx = y^\nu e^{\frac{1}{4}y^2} D_{-2\nu-3}(y)$

$[y>0,\ \operatorname{Re}\nu>-1]$. ET II 79(22)

18. $\int\limits_0^\infty x^\nu e^{\frac{1}{4}a^2x^2} D_{\frac{1}{2}\nu-\frac{1}{2}}(ax) N_\nu(xy)\,dx =$

$$= -\pi^{-1} 2^{\frac{3}{4}\nu+\frac{3}{4}} a^{-\nu} y^{-1} \Gamma(\nu+1)\, e^{\frac{y^2}{4a^2}} W_{-\frac{1}{2}\nu-\frac{1}{2},\ \frac{1}{2}\nu}\left(\frac{y^2}{2a^2}\right)$$

$\left[y>0,\ |\arg a|<\frac{3}{4}\pi,\ -\frac{1}{2}<\operatorname{Re}\nu<\frac{2}{3}\right]$. ET II 115(39)

7.753

1. $\int\limits_0^\infty x^{\nu-\frac{1}{2}} e^{-(x+a)^2} I_{\nu-\frac{1}{2}}(2ax)\, D_\nu(2x)\,dx = \frac{1}{2}\pi^{-\frac{1}{2}}\Gamma(\nu)\, a^{\nu-\frac{1}{2}} D_{-\nu}(2a)$

$[\operatorname{Re}a>0,\ \operatorname{Re}\nu>0]$. ET II 397(12)

2. $\int\limits_0^\infty x^{\nu-\frac{3}{2}} e^{-(x+a)^2} I_{\nu-\frac{3}{2}}(2ax)\, D_\nu(2x)\,dx = \frac{1}{2}\pi^{-\frac{1}{2}}\Gamma(\nu)\, a^{\nu-\frac{3}{2}} D_{-\nu}(2a)$

$[\operatorname{Re}a>0,\ \operatorname{Re}\nu>1]$. ET II 397(13)

7.754

1. $$\int_0^\infty x^\nu e^{-\frac{1}{4}x^2}\{[1 \mp 2\cos(\nu\pi)]\, D_{2\nu-1}(x) - D_{2\nu-1}(-x)\}\, J_\nu(xy)\, dx =$$
$$= \pm\, y^{\nu-1} e^{-\frac{1}{4}y^2}\{[1 \mp 2\cos(\nu\pi)]\, D_{2\nu-1}(y) - D_{2\nu-1}(-y)\}$$
$$\left[y > 0,\ \operatorname{Re}\nu > -\frac{1}{2}\right].$$ ET II 76(2, 3)

2. $$\int_0^\infty x^\nu e^{-\frac{1}{4}x^2}\{[1 \mp 2\cos(\nu\pi)]\, D_{2\nu+1}(x) - D_{2\nu+1}(-x)\}\, J_\nu(xy)\, dx =$$
$$= \mp\, y^\nu e^{-\frac{1}{4}y^2}\{[1 \mp 2\cos(\nu\pi)]\, D_{2\nu}(y) + D_{2\nu}(-y)\}$$
$$\left[y > 0,\ \operatorname{Re}\nu > -\frac{1}{2}\right].$$ ET II 77(6, 7)

3. $$\int_0^\infty x^{\nu+1} e^{-\frac{1}{4}x^2}\{[1 \pm 2\cos(\nu\pi)]\, D_{2\nu}(x) + D_{2\nu}(-x)\}\, J_\nu(xy)\, dx =$$
$$= \pm\, y^{\nu-1} e^{-\frac{1}{4}y^2}\{[1 \pm 2\cos(\nu\pi)]\, D_{2\nu+1}(y) - D_{2\nu+1}(-y)\}$$
$$[y > 0,\ \operatorname{Re}\nu > -1].$$ ET II 78(14, 15)

4 $$\int_0^\infty x^{\nu+1} e^{-\frac{1}{4}x^2}\{[1 \mp 2\cos(\nu\pi)]\, D_{2\nu+2}(x) + D_{2\nu+2}(-x)\}\, J_\nu(xy)\, dx =$$
$$= \pm\, y^\nu e^{-\frac{1}{4}y^2}\{[1 \mp 2\cos(\nu\pi)]\, D_{2\nu+2}(y) + D_{2\nu+2}(-y)\}$$
$$[y > 0,\ \operatorname{Re}\nu > -1].$$ ET II 78(17, 18)

7.755

1. $$\int_0^\infty x^{-\frac{1}{2}} D_\nu\left(a^{\frac{1}{2}} x^{\frac{1}{2}}\right) D_{-\nu-1}\left(a^{\frac{1}{2}} x^{\frac{1}{2}}\right) J_0(xy)\, dx =$$
$$= 2^{-\frac{3}{2}} \pi a^{-\frac{1}{2}} P_{-\frac{1}{4}}^{\frac{1}{2}\nu+\frac{1}{4}}\left[\left(1+\frac{4y^2}{a^2}\right)^{\frac{1}{2}}\right] P_{-\frac{1}{4}}^{-\frac{1}{2}\nu-\frac{1}{4}}\left[\left(1+\frac{4y^2}{a^2}\right)^{\frac{1}{2}}\right]$$
$$[y > 0,\ \operatorname{Re} a > 0].$$ ET II 17(43)

2. $$\int_0^\infty x^{\frac{1}{2}} D_{-\frac{1}{2}-\nu}\left(a e^{\frac{1}{4}\pi i} x^{\frac{1}{2}}\right) D_{-\frac{1}{2}-\nu}\left(a e^{-\frac{1}{4}\pi i} x^{\frac{1}{2}}\right) J_\nu(xy)\, dx =$$
$$= 2^{-\nu} \pi^{\frac{1}{2}} y^{-\nu-1} (a^2 + 2y)^{-\frac{1}{2}} \left[\Gamma\left(\nu + \frac{1}{2}\right)\right]^{-1} [(a^2 + 2y)^{\frac{1}{2}} - a]^{2\nu}$$
$$\left[y > 0,\ \operatorname{Re} a > 0,\ \operatorname{Re}\nu > -\frac{1}{2}\right].$$ ET II 80(27)

3. $$\int_0^\infty D_{-\frac{1}{2}-\nu}\left(ae^{\frac{1}{4}\pi i}x^{-\frac{1}{2}}\right)D_{-\frac{1}{2}-\nu}\left(ae^{-\frac{1}{4}\pi i}x^{-\frac{1}{2}}\right)J_\nu(xy)\,dx =$$
$$= 2^{\frac{1}{2}}\pi^{\frac{1}{2}}y^{-1}\left[\Gamma\left(\nu+\frac{1}{2}\right)\right]^{-1}\exp\left[-a(2y)^{\frac{1}{2}}\right]$$
$$\left[y > 0,\ \operatorname{Re} a > 0,\ \operatorname{Re}\nu > -\frac{1}{2}\right].$$
ET II 80(28)a

4. $$\int_0^\infty x^{\frac{1}{2}}D_{\nu-\frac{1}{2}}\left(ax^{-\frac{1}{2}}\right)D_{-\nu-\frac{1}{2}}\left(ax^{-\frac{1}{2}}\right)N_\nu(xy)\,dx =$$
$$= y^{-\frac{3}{2}}\exp\left(-ay^{\frac{1}{2}}\right)\sin\left[ay^{\frac{1}{2}}-\frac{1}{2}\left(\nu-\frac{1}{2}\right)\pi\right].$$
$$\left[y > 0,\ |\arg a| < \frac{1}{4}\pi\right].$$
ET II 115(40)

5. $$\int_0^\infty x^{\frac{1}{2}}D_{\nu-\frac{1}{2}}\left(ax^{-\frac{1}{2}}\right)D_{-\nu-\frac{1}{2}}\left(ax^{-\frac{1}{2}}\right)K_\nu(xy)\,dx = 2^{-1}y^{-\frac{3}{2}}\pi\exp\left[-a(2y)^{\frac{1}{2}}\right]$$
$$\left[\operatorname{Re} y > 0,\ |\arg a| < \frac{1}{4}\pi\right].$$
ET II 151(81)

Combinations of parabolic-cylinder and Struve functions

7.756 $$\int_0^\infty x^{-\nu}e^{-\frac{1}{4}x^2}\left[D_\mu(x)-D_\mu(-x)\right]\mathbf{H}_\nu(xy)\,dx =$$
$$= \frac{2^{\frac{3}{2}}\Gamma\left(\frac{1}{2}\mu+\frac{1}{2}\right)}{\Gamma\left(\frac{1}{2}\mu+\nu+1\right)}y^{\mu+\nu}\sin\left(\frac{1}{2}\mu\pi\right){}_1F_1\left(\frac{1}{2}\mu+\frac{1}{2};\frac{1}{2}\mu+\nu+1:\ -\frac{1}{2}y^2\right)$$
$$\left[y > 0,\ \operatorname{Re}(\mu+\nu) > -\frac{3}{2},\ \operatorname{Re}\mu > -1\right].$$
ET II 171(41)

7.76 Combinations of parabolic-cylinder functions and degenerate hypergeometric functions

7.761

1. $$\int_0^\infty e^{\frac{1}{4}t^2}t^{2c-1}D_{-\nu}(t)\,{}_1F_1\left(a;\ c;\ -\frac{1}{2}pt^2\right)dt =$$
$$= \frac{\pi^{\frac{1}{2}}}{2^{c+\frac{1}{2}\nu}}\frac{\Gamma(2c)\,\Gamma\left(\frac{1}{2}\nu-c+a\right)}{\Gamma\left(\frac{1}{2}\nu\right)\Gamma\left(a+\frac{1}{2}+\frac{1}{2}\nu\right)}F\left(a,\ c+\frac{1}{2};\ a+\frac{1}{2}+\frac{1}{2}\nu;\ 1-p\right)$$
$$\left[|1-p| < 1,\ \operatorname{Re} c > 0,\ \operatorname{Re}\nu > 2\operatorname{Re}(c-a)\right].$$
EH II 121(12)

2. $$\int_0^\infty e^{\frac{1}{4}t^2} t^{2c-2} D_{-\nu}(t)\, {}_1F_1\left(a;\ c;\ -\frac{1}{2}pt^2\right) dt =$$

$$= \frac{\pi^{\frac{1}{2}}}{2^{+\frac{1}{2}\nu-\frac{1}{2}}} \frac{\Gamma(2c-1)\,\Gamma\left(\frac{1}{2}\nu+\frac{1}{2}-c+a\right)}{\Gamma\left(\frac{1}{2}+\frac{1}{2}\nu\right)\Gamma\left(a+\frac{1}{2}\nu\right)} F\left(a,\ c-\frac{1}{2};\ a+\frac{1}{2}\nu;\ 1-p\right)$$

$\left[|1-p|<1,\ \operatorname{Re} c>\frac{1}{2},\ \operatorname{Re}\nu>2\operatorname{Re}(c-a)-1\right]$. EH II 121(13)

7.77 Integration of a parabolic-cylinder function with respect to the index

7.771 $$\int_0^\infty \cos(ax)\, D_{x-\frac{1}{2}}(\beta)\, D_{-x-\frac{1}{2}}(\beta)\, dx =$$

$$= \frac{1}{2}\left(\frac{\pi}{\cos a}\right)^{\frac{1}{2}} \exp\left(-\frac{\beta^2 \cos a}{2}\right) \qquad \left[|a|<\frac{1}{2}\pi\right];$$

$$= 0 \qquad \left[|a|>\frac{1}{2}\pi\right].$$ ET II 298(22)

7.772

1. $$\int_{-\frac{1}{2}-i\infty}^{-\frac{1}{2}+i\infty} \left[\frac{\left(\operatorname{tg}\frac{1}{2}\varphi\right)^\nu}{\cos\frac{1}{2}\varphi} D_\nu\left(-e^{\frac{1}{4}i\pi}\xi\right) D_{-\nu-1}\left(e^{\frac{1}{4}i\pi}\eta\right) + \right.$$

$$\left. + \frac{\left(\operatorname{ctg}\frac{1}{2}\varphi\right)^\nu}{\sin\frac{1}{2}\varphi} D_{-\nu-1}\left(e^{\frac{1}{4}i\pi}\xi\right) D_\nu\left(-e^{\frac{1}{4}i\pi}\eta\right)\right] \frac{d\nu}{\sin\nu\pi} =$$

$$= -2i(2\pi)^{\frac{1}{2}} \exp\left[-\frac{1}{4}i(\xi^2-\eta^2)\cos\varphi - \frac{1}{2}i\xi\eta\sin\varphi\right].$$

EH II 125(7)

2. $$\int_{-\frac{1}{2}-i\infty}^{-\frac{1}{2}+i\infty} \frac{\left(\operatorname{tg}\frac{1}{2}\varphi\right)^\nu}{\cos\frac{1}{2}\varphi} D_\nu\left(-e^{\frac{1}{4}i\pi}\zeta\right) D_{-\nu-1}\left(e^{\frac{1}{4}i\pi}\eta\right) \frac{d\nu}{\sin\nu\pi} =$$

$$= -2i\, D_0\left[e^{\frac{1}{4}i\pi}\left(\zeta\cos\frac{1}{2}\varphi + \eta\sin\frac{1}{2}\varphi\right)\right] \times$$

$$\times D_{-1}\left[e^{\frac{1}{4}i\pi}\left(\eta\cos\frac{1}{2}\varphi - \zeta\sin\frac{1}{2}\varphi\right)\right].$$ EH II 125(8)

7.773

1. $$\int_{c-i\infty}^{c+i\infty} D_\nu(z)\, t^\nu\, \Gamma(-\nu)\, d\nu = 2\pi i e^{-\frac{1}{4}z^2 - zt - \frac{1}{2}t^2}$$

$$\left[c<0,\ |\arg t|<\frac{\pi}{4}\right].$$ EH II 126(10)

2. $$\int_{c-i\infty}^{c+i\infty} [D_\nu(x) D_{-\nu-1}(iy) + D_\nu(-x) D_{-\nu-1}(-iy)] \frac{t^{-\nu-1}\,d\nu}{\sin(-\nu\pi)} =$$
$$= \frac{2\pi i}{\left(\frac{\pi}{2}\right)^{\frac{1}{2}}} (1+t^2)^{-\frac{1}{2}} \exp\left[\frac{1}{4}\frac{1-t^2}{1+t^2}(x^2+y^2) + i\frac{txy}{1+t^2}\right]$$
$$\left[-1 < c < 0,\ |\arg t| < \frac{1}{2}\pi\right].$$ EH II 126(11)

7.774 $$\int_{c-i\infty}^{c+i\infty} D_\nu[k^{\frac{1}{2}}(1+i)\xi] D_{-\nu-1}[k^{\frac{1}{2}}(1+i)\eta] \Gamma\left(-\frac{1}{2}\nu\right) \Gamma\left(\frac{1}{2}+\frac{1}{2}\nu\right) d\nu =$$
$$= 2^{\frac{1}{2}} \pi^2 H_0^{(2)}\left[\frac{1}{2}k(\xi^2+\eta^2)\right]$$
$$[-1 < c < 0,\quad \mathrm{Re}\, ik \geqslant 0].$$ EH II 125(9)

7.8 Meijer's and MacRobert's Functions (G and E)

7.81 Combinations of the functions G and E and the elementary functions

7.811

1. $$\int_0^\infty G_{p,q}^{m,n}\left(\eta x \left|\begin{matrix} a_1, \dots, a_p \\ b_1, \dots, b_q \end{matrix}\right.\right) G_{\sigma,\tau}^{\mu,\nu}\left(\omega x \left|\begin{matrix} c_1, \dots, c_\sigma \\ d, \dots, d_\tau \end{matrix}\right.\right) dx =$$
$$= \frac{1}{\eta} G_{q+\sigma,p+\tau}^{n+\mu,m+\nu}\left(\frac{\omega}{\eta} \left|\begin{matrix} -b_1, \dots, -b_m, c_1, \dots, c_\sigma, -b_{m+1}, \dots, -b_q \\ -a_1, \dots, -a_n, d_1, \dots, d_\tau, -a_{n+1}, \dots, -a_p \end{matrix}\right.\right)$$
$[m, n, p, q, \mu, \nu, \sigma, \tau$ — are integers; $1 \leqslant n \leqslant p < q < p+\tau-\sigma$,
$\frac{1}{2}p + \frac{1}{2}q - n < m \leqslant q$, $0 \leqslant \nu \leqslant \sigma$, $\frac{1}{2}\sigma + \frac{1}{2}\tau - \nu < \mu \leqslant \tau$;
$\mathrm{Re}(b_j + d_k) > -1 \quad (j = 1, \dots, m;\ k = 1, \dots, \mu)$,
$\mathrm{Re}(a_j + c_k) < 1 \quad (j = 1, \dots, n;\ k = 1, \dots, \tau)$;

must not be integers:

$b_j - b_k \quad (j = 1, \dots, m; k = 1, \dots, m;\ j \neq k)$,
$a_j - a_k \quad (j = 1, \dots, n; k = 1, \dots, n;\ j \neq k)$,
$d_j - d_k \quad (j = 1, \dots, \mu; k = 1, \dots, \mu;\ j \neq k)$,
$a_j + d_k \quad (j = 1, \dots, n; k = 1, \dots, n)$;

must not be positive integers:

$a_j - b_k \quad (j = 1, \dots, n;\ k = 1, \dots, m)$,
$c_j - d_k \quad (j = 1, \dots, \nu;\ k = 1, \dots, \mu)$;
$\omega \neq 0$, $\eta \neq 0$, $|\arg \eta| < \left(m + n - \frac{1}{2}p - \frac{1}{2}q\right)\pi$,
$|\arg \omega| < \left(\mu + \nu - \frac{1}{2}\sigma - \frac{1}{2}\tau\right)\pi\Big]$.

Formula **7.811** 1 also holds for four sets of restrictions. See C. C. Meijer, Neue Integraldarstellungen für Whittakersche Funktionen, Nederl. Akad. Wetensch. Proc. 44 (1941), 82–92.

ET II 422(14)

2. $$\int_0^1 x^{\varrho-1}(1-x)^{\sigma-1} G_{pq}^{mn}\left(\alpha x \middle| \begin{matrix} a_1, \ldots, a_p \\ b_1, \ldots, b_q \end{matrix}\right) dx =$$

$$= \Gamma(\sigma) G_{p+1,\,q+1}^{m,\,n+1}\left(\alpha \middle| \begin{matrix} 1-\varrho, a_1, \ldots, a_p \\ b_1, \ldots, b_q, 1-\varrho-\sigma \end{matrix}\right)$$

$$\Big[(p+q) < 2(m+n),\ |\arg\alpha| < \left(m+n-\tfrac{1}{2}p-\tfrac{1}{2}q\right)\pi,$$
$$\operatorname{Re}(\varrho+b_j) > 0;\ j=1, \ldots, m;\ \operatorname{Re}\sigma > 0,$$

either

$$p+q \leqslant 2(m+n),\quad |\arg\alpha| \leqslant \left(m+n-\tfrac{1}{2}\varrho-\tfrac{1}{2}q\right)\pi,$$
$$\operatorname{Re}(\varrho+b_j) > 0;\ j=1, \ldots, m;\ \operatorname{Re}\sigma > 0,$$
$$\operatorname{Re}\left[\sum_{j=1}^{p} a_j - \sum_{j=1}^{q} b_j + (p-q)\left(\varrho-\tfrac{1}{2}\right)\right] > -\tfrac{1}{2},$$

or

$$p < q \ (\text{or}\quad p \leqslant q \ \text{ for } |\alpha| < 1),$$
$$\operatorname{Re}(p+b_j) > 0;\ j=1, \ldots, m;\ \operatorname{Re}\sigma > 0\Big].$$ ET II 417(1)

3. $$\int_1^\infty x^{-\varrho}(x-1)^{\sigma-1} G_{pq}^{mn}\left(\alpha x \middle| \begin{matrix} a_1, \ldots, a_p \\ b_1, \ldots, b_q \end{matrix}\right) dx =$$

$$= \Gamma(\sigma) G_{p+1,\,q+1}^{m+1,\,n}\left(\alpha \middle| \begin{matrix} a_1, \ldots, a_p, \varrho \\ \varrho-\sigma, b_1, \ldots, b_q \end{matrix}\right)$$

$$\Big[p+q < 2(m+n),\ |\arg\alpha| < \left(m+n-\tfrac{1}{2}p-\tfrac{1}{2}q\right)\pi,$$
$$\operatorname{Re}(\varrho-\sigma-a_j) > -1;\ j=1, \ldots, n;\ \operatorname{Re}\sigma > 0,$$

either

$$p+q \leqslant 2(m+n),\quad |\arg\alpha| \leqslant \left(m+n-\tfrac{1}{2}p-\tfrac{1}{2}q\right)\pi,$$
$$\operatorname{Re}(\varrho-\sigma-a_j) > -1;\ j=1, \ldots, n;\ \operatorname{Re}\sigma > 0,$$
$$\operatorname{Re}\left[\sum_{j=1}^{p} a_j - \sum_{j=1}^{q} b_j + (q-p)\left(\varrho-\sigma+\tfrac{1}{2}\right)\right] > -\tfrac{1}{2},$$

or

$$q < p \ (\text{or}\quad q \leqslant p \ \text{ for } |\alpha| > 1),\ \operatorname{Re}(\varrho-\sigma-a_j) > -1;$$
$$j=1, \ldots, n;\ \operatorname{Re}\sigma > 0\Big].$$ ET II 417(2)

4. $$\int_0^\infty x^{\varrho-1} G_{pq}^{mn}\left(\alpha x \middle| \begin{matrix} a_1, \ldots, a_p \\ b_1, \ldots, b_q \end{matrix}\right) dx = \frac{\prod\limits_{j=1}^{m} \Gamma(b_j+\varrho) \prod\limits_{j=1}^{n} \Gamma(1-a_j-\varrho)}{\prod\limits_{j=m+1}^{q} \Gamma(1-b_j-\varrho) \prod\limits_{j=n+1}^{p} \Gamma(a_j+\varrho)} \alpha^{-\varrho}$$

$$\Big[p+q < 2(m+n),\ |\arg\alpha| < \left(m+n-\tfrac{1}{2}p-\tfrac{1}{2}q\right)\pi,$$
$$-\min_{1 \leqslant j \leqslant m} \operatorname{Re} b_j < \operatorname{Re}\varrho < 1 - \max_{1 \leqslant j \leqslant n} \operatorname{Re} a_j\Big].$$ ET II 418(3)a, ET I 337(14)

5. $$\int_0^\infty x^{\varrho-1}(x+\beta)^{-\sigma} G_{pq}^{mn}\left(\alpha x \,\middle|\, \begin{matrix} a_1, \ldots, a_p \\ b_1, \ldots, b_q \end{matrix}\right) dx =$$
$$= \frac{\beta^{\varrho-\sigma}}{\Gamma(\sigma)} G_{p+1,\, q+1}^{m+1,\, n+1}\left(\alpha\beta \,\middle|\, \begin{matrix} 1-\varrho, a_1, \ldots, a_p \\ \sigma-\varrho, b_1, \ldots, b_q \end{matrix}\right)$$
$$\left[p+q < 2(m+n),\ |\arg\alpha| < \left(m+n-\tfrac{1}{2}p-\tfrac{1}{2}q\right)\pi,\ |\arg\beta| < \pi\right.$$
$$\operatorname{Re}(\varrho+b_j) > 0,\ j=1, \ldots, m,\ \operatorname{Re}(\varrho-\sigma+a_j) < 1,\ j=1, \ldots, n,$$

either

$$p \leqslant q,\ p+q \leqslant 2(m+n),\ |\arg\alpha| \leqslant \left(m+n-\tfrac{1}{2}p-\tfrac{1}{2}q\right)\pi,\ |\arg\beta| < \pi$$
$$\operatorname{Re}(\varrho+b_j) > 0,\ j=1, \ldots, m,\ \operatorname{Re}(\varrho-\sigma+a_j) < 1,\ j=1, \ldots, n,$$
$$\operatorname{Re}\left[\sum_{j=1}^{p} a_j - \sum_{=1}^{q} b_j - (q-p)\left(\varrho-\sigma-\tfrac{1}{2}\right)\right] > 1,$$

or

$$p \geqslant q,\ p+q \leqslant 2(m+n),\ |\arg\alpha| \leqslant \left(m+n-\tfrac{1}{2}p-\tfrac{1}{2}q\right)\pi,\ |\arg\beta| < \pi,$$
$$\operatorname{Re}(\varrho+b_j) > 0,\ j=1, \ldots, m,\ \operatorname{Re}(\varrho-\sigma+a_j) < 1,\ j=1, \ldots, n,$$
$$\left.\operatorname{Re}\left[\sum_{j=1}^{p} a_j - \sum_{j=1}^{q} b_j + (p-q)\left(\varrho-\tfrac{1}{2}\right)\right] > 1\right].$$ ET II 418(4)

7.812

1. $$\int_0^1 x^{\beta-1}(1-x)^{\gamma-\beta-1} E\left(a_1, \ldots, a_p : \varrho_1, \ldots, \varrho_q : \frac{z}{x^m}\right) dx =$$
$$= \Gamma(\gamma-\beta)\, m^{\beta-\gamma} E(a_1, \ldots, a_{p+m} : \varrho_1, \ldots, \varrho_{q+m} : z),$$
$$a_{p+k} = \frac{\beta+k-1}{m},\quad \varrho_{q+k} = \frac{\gamma+k-1}{m},\quad k=1, \ldots, m$$
$$[\operatorname{Re}\gamma > \operatorname{Re}\beta > 0,\ m=1, 2, \ldots].$$ ET II 414(2)

2 $$\int_0^\infty x^{\varrho-1}(1+x)^{-\sigma} E[a_1, \ldots, a_p : \varrho_1, \ldots, \varrho_q : (1+x)z]\, dx =$$
$$= \Gamma(\varrho) E(a_1, \ldots, a_p, \sigma-\varrho : \varrho_1, \ldots, \varrho_q, \sigma : z)$$
$$[\operatorname{Re}\sigma > \operatorname{Re}\varrho > 0].$$ ET II 415(3)

3. $$\int_0^\infty (1+x)^{-\beta} x^{s-1} G_{p,\, q}^{m,\, n}\left(\frac{ax}{1+x} \,\middle|\, \begin{matrix} a_1, \ldots, a_p \\ b_1, \ldots, b_q \end{matrix}\right) dx =$$
$$= \Gamma(\beta-s)\, G_{p+1,\, q+1}^{m,\, n+1}\left(a \,\middle|\, \begin{matrix} 1-s, a_1, \ldots, a_p \\ b_1, \ldots, b_q, 1-\beta \end{matrix}\right)$$
$$\left[-\min \operatorname{Re} b_k < \operatorname{Re} s < \operatorname{Re}\beta,\ 1 \leqslant k \leqslant m;\ (p+q) < 2(m+n),\right.$$
$$\left.|\arg a| < \left(m+n-\tfrac{1}{2}p-\tfrac{1}{2}q\right)\pi\right].$$ ET I 338(19)

7.813

1. $$\int_0^\infty x^{-\varrho} e^{-\beta x} G_{pq}^{mn}\left(\alpha x \left| \begin{matrix} a_1, \ldots, a_p \\ b_1, \ldots, b_q \end{matrix} \right.\right) dx = \beta^{\varrho-1} G_{p+1,q}^{m,n+1}\left(\frac{\alpha}{\beta} \left| \begin{matrix} \varrho, a_1, \ldots, a_p \\ b_1, \ldots, b_q \end{matrix} \right.\right)$$

$$\left[p+q<2(m+n),\ |\arg\alpha|<\left(m+n-\frac{1}{2}p-\frac{1}{2}q\right)\pi,\right.$$
$$\left.|\arg\beta|<\frac{1}{2}\pi,\ \operatorname{Re}(b_j-\varrho)>-1,\ j=1,\ldots,m\right]$$

ET II 419(5)

2. $$\int_0^\infty e^{-\beta x} G_{pq}^{mn}\left(\alpha x^2 \left| \begin{matrix} a_1, \ldots, a_p \\ b_1, \ldots, b_q \end{matrix} \right.\right) dx = \pi^{-\frac{1}{2}} \beta^{-1} G_{p+2,q}^{m,n+2}\left(\frac{4\alpha}{\beta^2} \left| \begin{matrix} 0, \frac{1}{2}, a_1, \ldots, a_p \\ b_1, \ldots, b_q \end{matrix} \right.\right)$$

$$\left[p+q<2(m+n),\ |\arg\alpha|<\left(m+n-\frac{1}{2}p-\frac{1}{2}q\right)\pi,\right.$$
$$\left.|\arg\beta|<\frac{1}{2}\pi,\ \operatorname{Re} b_j>-\frac{1}{2};\ j=1,\ldots,m\right].$$

ET II 419(6)

7.814

1. $$\int_0^\infty x^{\beta-1} e^{-x} E(a_1, \ldots, a_p : \varrho_1, \ldots, \varrho_q : xz)\, dx =$$

$$= \pi \operatorname{cosec}(\beta\pi)\left[E(a_1, \ldots, a_p : 1-\beta, \varrho_1, \ldots, \varrho_q : e^{\pm i\pi} z) -\right.$$
$$\left.- z^{-\beta} E(a_1+\beta, \ldots, a_p+\beta : 1+\beta, \varrho_1+\beta, \ldots, \varrho_q+\beta : e^{\pm i\pi} z)\right]$$

[$p \geqslant q+1$, $\operatorname{Re}(a_r+\beta)>0$, $r=1, \ldots, p$, $|\arg z|<\pi$. The formula holds also for $p<q+1$, provided the integral converges].

ET II 415(4)

2. $$\int_0^\infty x^{\beta-1} e^{-x} E(a_1, \ldots, a_p : \varrho_1, \ldots, \varrho_q : x^{-m} z)\, dx =$$

$$= (2\pi)^{\frac{1}{2}-\frac{1}{2}m} m^{\beta-\frac{1}{2}} E(a_1, \ldots, a_{p+m} : \varrho_1, \ldots, \varrho_q : m^{-m} z)$$

$$\left[\operatorname{Re}\beta>0,\ a_{p+k}=\frac{\beta+k-1}{m},\ k=1, \ldots, m;\ m=1, 2, \ldots\right].$$

ET II 415(5)

7.815

1. $$\int_0^\infty \sin(cx)\, G_{pq}^{mn}\left(\alpha x^2 \left| \begin{matrix} a_1, \ldots, a_p \\ b_1, \ldots, b_q \end{matrix} \right.\right) dx =$$

$$= \pi^{\frac{1}{2}} c^{-1} G_{p+2,q}^{m,n+1}\left(\frac{4\alpha}{c^2} \left| \begin{matrix} 0, a_1, \ldots, a_p, \frac{1}{2} \\ b_1, \ldots, b_q \end{matrix} \right.\right)$$

$$\left[p+q<2(m+n),\ |\arg\alpha|<\left(m+n-\frac{1}{2}p-\frac{1}{2}q\right)\pi,\right.$$
$$\left.c>0,\ \operatorname{Re} b_j>-1,\ j=1, 2, \ldots, m,\ \operatorname{Re} a_j<\frac{1}{2},\ j=1, \ldots, n\right].$$

ET II 420(7)

2. $$\int_0^\infty \cos(cx)\, G_{pq}^{mn}\left(\alpha x^2 \middle| \begin{matrix} a_1, \ldots, a_p \\ b_1, \ldots, b_q \end{matrix}\right) dx =$$

$$= \pi^{\frac{1}{2}} c^{-1} G_{p+2,q}^{m,n+1}\left(\frac{4\alpha}{c^2} \middle| \begin{matrix} \frac{1}{2}, a_1, \ldots, a_p, 0 \\ b_1, \ldots, b_q \end{matrix}\right)$$

$$\left[p+q < 2(m+n),\ |\arg\alpha| < \left(m+n-\frac{1}{2}p-\frac{1}{2}q\right)\pi,\right.$$
$$\left. c>0,\ \operatorname{Re} b_j > -\frac{1}{2},\ j=1, \ldots, m,\ \operatorname{Re} a_j < \frac{1}{2},\ j=1, \ldots, n\right].$$

ET II 420(8)

7.82 Combinations of the functions *G* and *E* and Bessel functions

7.821

1. $$\int_0^\infty x^{-\varrho} J_\nu(2\sqrt{x})\, G_{pq}^{mn}\left(\alpha x \middle| \begin{matrix} a_1, \ldots, a_p \\ b_1, \ldots, b_q \end{matrix}\right) dx =$$

$$= G_{p+2,q}^{m,n+1}\left(\alpha \middle| \begin{matrix} \varrho-\frac{1}{2}\nu, a_1, \ldots, a_p, \varrho+\frac{1}{2}\nu \\ b_1, \ldots, b_q \end{matrix}\right)$$

$$\left[p+q < 2(m+n),\ |\arg\alpha| < \left(m+n-\frac{1}{2}p-\frac{1}{2}q\right)\pi,\right.$$
$$\left. -\frac{3}{4} + \max_{1\leqslant j\leqslant n} \operatorname{Re} a_j < \operatorname{Re}\varrho < 1 + \frac{1}{2}\operatorname{Re}\nu + \min_{1\leqslant j\leqslant m} \operatorname{Re} b_j\right].$$

ET II 420(9)

2. $$\int_0^\infty x^{-\varrho} N_\nu(2\sqrt{x})\, G_{pq}^{mn}\left(\alpha x \middle| \begin{matrix} a_1, \ldots, a_p \\ b_1, \ldots, b_q \end{matrix}\right) dx =$$

$$= G_{p+3,q+1}^{m,n+2}\left(\alpha \middle| \begin{matrix} \varrho-\frac{1}{2}\nu, \varrho+\frac{1}{2}\nu, a_1, \ldots, a_p, \varrho+\frac{1}{2}+\frac{1}{2}\nu \\ b_1, \ldots, b_q, \varrho+\frac{1}{2}+\frac{1}{2}\nu \end{matrix}\right)$$

$$\left[p+q < 2(m+n),\ |\arg\alpha| < \left(m+n-\frac{1}{2}p-\frac{1}{2}q\right)\pi,\right.$$
$$\left. -\frac{3}{4} + \max_{1\leqslant j\leqslant n} \operatorname{Re} a_j < \operatorname{Re}\varrho < \min_{1\leqslant j\leqslant m} \operatorname{Re} b_j + \frac{1}{2}|\operatorname{Re}\nu| + 1\right]$$

ET II 420(10)

3. $$\int_0^\infty x^{-\varrho} K_\nu(2\sqrt{x})\, G_{pq}^{mn}\left(\alpha x \middle| \begin{matrix} a_1, \ldots, a_p \\ b_1, \ldots, b_q \end{matrix}\right) dx =$$

$$= \frac{1}{2} G_{p+2,q}^{m,n+2}\left(\alpha \middle| \begin{matrix} \varrho-\frac{1}{2}\nu, \varrho+\frac{1}{2}\nu, a_1, \ldots, a_p \\ b_1, \ldots, b_q \end{matrix}\right)$$

$$\left[p+q < 2(m+n),\ |\arg\alpha| < \left(m+n-\frac{1}{2}p-\frac{1}{2}q\right)\pi,\right.$$
$$\left. \operatorname{Re}\varrho < 1 - \frac{1}{2}|\operatorname{Re}\nu| + \min_{1\leqslant j\leqslant m} \operatorname{Re} b_j\right].$$

ET II 421(11)

7.822

1. $$\int_0^\infty x^{2\varrho} J_\nu(xy) G_{pq}^{mn}\left(\lambda x^2 \middle| \begin{matrix} a_1, \ldots, a_p \\ b_1, \ldots, b_q \end{matrix}\right) dx =$$

$$= \frac{2^{2\varrho}}{y^{2\varrho+1}} G_{p+2,q}^{m,n+1}\left(\frac{4\lambda}{y^2} \middle| \begin{matrix} h, a_1, \ldots, a_p, k \\ b_1, \ldots, b_q \end{matrix}\right),$$

$$h = \frac{1}{2} - \varrho - \frac{1}{2}\nu, \quad k = \frac{1}{2} - \varrho + \frac{1}{2}\nu$$

$$\left[p + q < 2(m+n), \ |\arg\lambda| < \left(m + n - \frac{1}{2}p - \frac{1}{2}q\right)\pi,\right.$$

$$\operatorname{Re}\left(b_j + \varrho + \frac{1}{2}\nu\right) > -\frac{1}{2}, \ j = 1, 2, \ldots, m,$$

$$\left.\operatorname{Re}(a_j + \varrho) < \frac{3}{4}, \ j = 1, \ldots, n, \ y > 0\right].$$ ET II 91(20)

2. $$\int_0^\infty x^{\frac{1}{2}} N_\nu(xy) G_{pq}^{mn}\left(\lambda x^2 \middle| \begin{matrix} a_1, \ldots, a_p \\ b_1, \ldots, b_q \end{matrix}\right) dx =$$

$$= (2\lambda)^{-\frac{1}{2}} y^{-\frac{1}{2}} G_{q+1,p+3}^{n+2,m}\left(\frac{y^2}{4\lambda} \middle| \begin{matrix} \frac{1}{2} - b_1, \ldots, \frac{1}{2} - b_q, l \\ h, k, \frac{1}{2} - a_1, \ldots, \frac{1}{2} - a_p, l \end{matrix}\right)$$

$$h = \frac{1}{4} + \frac{1}{2}\nu, \quad k = \frac{1}{4} - \frac{1}{2}\nu, \quad l = -\frac{1}{4} - \frac{1}{2}\nu$$

$$\left[p + q < 2(m+n), \ |\arg\lambda| < \left(m + n - \frac{1}{2}p - \frac{1}{2}q\right)\pi. \ y > 0,\right.$$

$$\left.\operatorname{Re} a_j < 1, \ j = 1, \ldots, n, \ \operatorname{Re}\left(b_j \pm \frac{1}{2}\nu\right) > -\frac{3}{4}, \ j = 1, \ldots, m\right].$$

ET II 119(56)

3. $$\int_0^\infty x^{\frac{1}{2}} K_\nu(xy) G_{pq}^{mn}\left(\lambda x^2 \middle| \begin{matrix} a_1, \ldots, a_p \\ b_1, \ldots, b_q \end{matrix}\right) dx =$$

$$= 2^{-\frac{3}{2}} \lambda^{-\frac{1}{2}} y^{-\frac{1}{2}} G_{q,p+2}^{n+2,m}\left(\frac{y^2}{4\lambda} \middle| \begin{matrix} \frac{1}{2} - b_1, \ldots, \frac{1}{2} - b_q \\ h, k, \frac{1}{2} - a_1, \ldots, \frac{1}{2} - a_p \end{matrix}\right),$$

$$h = \frac{1}{4} + \frac{1}{2}\nu, \quad k = \frac{1}{4} - \frac{1}{2}\nu$$

$$\left[\operatorname{Re} y > 0, \ p + q < 2(m+n), \ |\arg\lambda| < \left(m + n - \frac{1}{2}p - \frac{1}{2}q\right)\pi,\right.$$

$$\left.\operatorname{Re} b_j > \frac{1}{2}|\operatorname{Re}\nu| - \frac{3}{4}, \ j = 1, \ldots, m\right].$$ ET II 153(90)

7.823

1. $$\int_0^\infty x^{\beta-1} J_\nu(x)\, E(a_1, \ldots, a_p : \varrho_1, \ldots, \varrho_q : x^{-2m} z)\, dx =$$
$$= (2\pi)^{-m} (2m)^{\beta-1} \left\{ \exp\left[\tfrac{1}{2}\pi(\beta-\nu-1)i\right] \times \right.$$
$$\times E\left[a_1, \ldots, a_{p+2m} : \varrho_1, \ldots, \varrho_q : (2m)^{-2m} z e^{-m\pi i}\right] +$$
$$+ \exp\left[-\tfrac{1}{2}\pi(\beta-\nu-1)i\right] \times$$
$$\left. \times E\left[a_1, \ldots, a_{p+2m} : \varrho_1, \ldots, \varrho_q : (2m)^{-2m} z e^{m\pi i}\right] \right\},$$
$$a_{p+k} = \frac{\beta+\nu+2k-2}{2m}, \quad a_{p+m+k} = \frac{\beta-\nu+2k-2}{2m},$$
$$m = 1, 2, \ldots; \quad k = 1, \ldots, m$$
$$\left[\operatorname{Re}(\beta+\nu) > 0, \ \operatorname{Re}(2a_r m - \beta) > -\tfrac{3}{2}, \ r = 1, \ldots, p\right]$$

ET II 415(7)

2. $$\int_0^\infty x^{\beta-1} K_\nu(x)\, E(a_1, \ldots, a_p : \varrho_1, \ldots, \varrho_q : x^{-2m} z)\, dx =$$
$$= (2\pi)^{1-m} 2^{\beta-2} m^{\beta-1} E\left[a_1, \ldots, a_{p+2m} : \varrho_1, \ldots, \varrho_q : (2m)^{-2m} z\right],$$
$$a_{p+k} = \frac{\beta+\nu+2k-2}{2m}, \quad a_{p+m+k} = \frac{\beta-\nu+2k-2}{2m}, \quad k = 1, 2, \ldots, m$$
$$[\operatorname{Re}\beta > |\operatorname{Re}\nu|, \quad m = 1, 2, \ldots].$$

ET II 416(8)

7.824

1. $$\int_0^\infty x^{\frac{1}{2}} \mathbf{H}_\nu(xy)\, G_{pq}^{mn}\left(\lambda x^2 \middle| \begin{matrix} a_1, \ldots, a_p \\ b_1, \ldots, b_q \end{matrix}\right) dx =$$
$$= (2\lambda y)^{-\frac{1}{2}} G_{q+1,\, p+3}^{n+1,\, m+1}\left(\frac{y^2}{4\lambda} \middle| \begin{matrix} l, \frac{1}{2}-b_1, \ldots, \frac{1}{2}-b_q \\ l, \frac{1}{2}-a_1, \ldots, \frac{1}{2}-a_p, h, k \end{matrix}\right)$$
$$h = \frac{1}{4} + \frac{\nu}{2}, \quad k = \frac{1}{4} - \frac{\nu}{2}, \quad l = \frac{3}{4} + \frac{\nu}{2}$$
$$\left[p+q < 2(m+n), \ |\arg\lambda| < \left(m+n-\tfrac{1}{2}p-\tfrac{1}{2}q\right)\pi, \ y > 0, \right.$$
$$\operatorname{Re} a_j < \min\left(1, \tfrac{3}{4} - \tfrac{1}{2}\nu\right), \ j = 1, \ldots, n,$$
$$\left. \operatorname{Re}(2b_j + \nu) > -\tfrac{5}{2}, \ j = 1, \ldots, m\right].$$

ET II 172(47)

2. $$\int_0^\infty x^{-\varrho} \mathbf{H}_\nu(2\sqrt{x})\, G_{pq}^{mn}\left(\alpha x \,\middle|\, \begin{matrix} a_1, \ldots, a_p \\ b_1, \ldots, b_q \end{matrix}\right) dx =$$

$$= G_{p+3,\,q+1}^{m+1,\,n+1}\left(\alpha \,\middle|\, \begin{matrix} \varrho-\frac{1}{2}-\frac{1}{2}\nu,\, a_1, \ldots, a_p,\, \varrho+\frac{1}{2}\nu,\, \varrho-\frac{1}{2}\nu \\ \varrho-\frac{1}{2}-\frac{1}{2}\nu,\, b_1, \ldots, b_q \end{matrix}\right)$$

$$\left[p+q<2(m+n),\ |\arg\alpha|<\left(m+n-\frac{1}{2}p-\frac{1}{2}q\right)\pi,\right.$$
$$\left.\max\left(-\frac{3}{4},\ \operatorname{Re}\frac{\nu-1}{2}\right)+\max_{1\leqslant j\leqslant n}\operatorname{Re}a_j<\operatorname{Re}\varrho<\min_{1\leqslant j\leqslant m}\operatorname{Re}b_j+\frac{1}{2}\operatorname{Re}\nu+\frac{3}{2}\right]$$

ET II 421(12)

7.83 Combinations of the functions G and E and other special functions

7.831 $$\int_1^\infty x^{-\varrho}(x-1)^{\sigma-1} F(k+\sigma-\varrho,\ \lambda+\sigma-\varrho;\ \sigma;\ 1-x)\times$$

$$\times G_{pq}^{mn}\left(\alpha x \,\middle|\, \begin{matrix} a_1, \ldots, a_p \\ b_1, \ldots, b_q \end{matrix}\right) dx = \Gamma(\sigma)\, G_{p+2,\,q+2}^{m+2,\,n}\left(\alpha \,\middle|\, \begin{matrix} a_1, \ldots, a_p,\, k+\lambda+\sigma-\varrho,\, \varrho \\ k,\, \lambda,\, b_1, \ldots, b_q \end{matrix}\right)$$

$$\left[p+q<2(m+n),\ |\arg\alpha|<\left(m+n-\frac{1}{2}p-\frac{1}{2}q\right)\pi,\right.$$
$$\operatorname{Re}\sigma>0,\ \operatorname{Re}k\geqslant\operatorname{Re}\lambda>\operatorname{Re}a_j-1,\ j=1,\ \ldots,\ n,$$

or

$$p+q\leqslant 2(m+n),\quad |\arg\alpha|\leqslant\left(m+n-\frac{1}{2}p-\frac{1}{2}q\right)\pi,$$
$$\operatorname{Re}\sigma>0,\quad \operatorname{Re}k\geqslant\operatorname{Re}\lambda>\operatorname{Re}a_j-1,\quad j=1,\ \ldots,\ n,$$
$$\operatorname{Re}\left[\sum_{j=1}^{p}a_j-\sum_{j=1}^{q}b_j+(q-p)\left(k+\frac{1}{2}\right)\right]>-\frac{1}{2},$$
$$\left.\operatorname{Re}\left[\sum_{j=1}^{p}a_j-\sum_{j=1}^{q}b_j+(q-p)\left(\lambda+\frac{1}{2}\right)\right]>-\frac{1}{2}\right].$$

ET II 421(13)

7.832 $$\int_0^\infty x^{\beta-1}e^{-\frac{1}{2}x}W_{\varkappa,\,\mu}(x)\,E(a_1,\ \ldots,\ a_p:\varrho_1,\ \ldots,\ \varrho_q:x^{-m}z)\,dx=$$

$$=(2\pi)^{\frac{1}{2}-\frac{1}{2}m}m^{\beta+\varkappa-\frac{1}{2}}E(a_1,\ \ldots,\ a_{p+2m}:\varrho_1,\ \ldots,\ \varrho_{q+m}:m^{-m}z),$$
$$a_{p+k}=\frac{\beta+k+\mu-\frac{1}{2}}{m},\quad a_{p+m+k}=\frac{\beta-\mu+k-\frac{1}{2}}{m},$$
$$\varrho_{q+k}=\frac{\beta-\varkappa+k}{m},\quad k=1,\ \ldots,\ m$$
$$\left[\operatorname{Re}\beta>|\operatorname{Re}\mu|-\frac{1}{2},\quad m=1,\ 2,\ \ldots\right].$$

ET II 416(10)

8-9 SPECIAL FUNCTIONS

8.1 Elliptic Integrals and Functions

8.11 Elliptic integrals

8.110

1. Every integral of the form $\int R(x, \sqrt{P(x)})\, dx$, where $P(x)$ is a third- or fourth-degree polynomial, can be reduced to a linear combination of integrals leading to elementary functions and the following three integrals:

$$\int \frac{dx}{\sqrt{(1-x^2)(1-k^2x^2)}}, \quad \int \frac{\sqrt{1-k^2x^2}}{\sqrt{1-x^2}}\, dx, \quad \int \frac{dx}{(1+nx^2)\sqrt{(1-x^2)(1-k^2x^2)}},$$

which are called respectively *elliptic integrals of the first, second, and third kind in the Legendre normal form.* The results of this reduction for the more frequently encountered integrals are given in formulas **3.13 — 3.17**. The number k is called the *modulus** of these integrals, the number $k' = \sqrt{1-k^2}$ is called the complementary modulus, and the number n is called the parameter of the integral of the third kind.

FI II 97-106

2. By means of the substitution $x = \sin\varphi$, elliptic integrals can be reduced to the normal trigonometric form

$$\int \frac{d\varphi}{\sqrt{1-k^2\sin^2\varphi}}, \quad \int \sqrt{1-k^2\sin^2\varphi}\, d\varphi, \quad \int \frac{d\varphi}{(1+n\sin^2\varphi)\sqrt{1-k^2\sin^2\varphi}}.$$

FI II 106

The results of reducing integrals of trigonometric functions to normal form are given in **2.58 — 2.62**.

3. Elliptic integrals from 0 to $\frac{\pi}{2}$ are called *complete elliptic integrals.*

8.111 Notations:

1. $\Delta\varphi = \sqrt{1-k^2\sin^2\varphi}; \quad k' = \sqrt{1-k^2}; \quad k^2 < 1.$

2. The elliptic integral of the first kind:

$$F(\varphi, k) = \int_0^{\varphi} \frac{d\alpha}{\sqrt{1-k^2\sin^2\alpha}} = \int_0^{\sin\varphi} \frac{dx}{\sqrt{(1-x^2)(1-k^2x^2)}}.$$

---

*The quantity k is sometimes called the *module* of the functions.

3. The elliptic integral of the second kind:

$$E(\varphi, k) = \int_0^{\varphi} \sqrt{1-k^2\sin^2\alpha}\,d\alpha = \int_0^{\sin\varphi} \frac{\sqrt{1-k^2x^2}}{\sqrt{1-x^2}}\,dx.$$ FI II 135

4. The elliptic integral of the third kind:

$$\Pi(\varphi, n, k) = \int_0^{\varphi} \frac{d\alpha}{(1+n\sin^2\alpha)\sqrt{1-k^2\sin^2\alpha}} = \int_0^{\sin\varphi} \frac{dx}{(1+nx^2)\sqrt{(1-x^2)(1-k^2x^2)}}.$$ SI 13

5. $$D(\varphi, k) = \frac{F(\varphi, k) - E(\varphi, k)}{k^2} = \int_0^{\varphi} \frac{\sin^2\alpha\,d\alpha}{\sqrt{1-k^2\sin^2\alpha}} = \int_0^{\sin\varphi} \frac{x^2\,dx}{\sqrt{(1-x^2)(1-k^2x^2)}}.$$

8.112 Complete elliptic integrals

1. $\boldsymbol{K}(k) = F\left(\frac{\pi}{2}, k\right) = \boldsymbol{K}'(k')$.

2. $\boldsymbol{E}(k) = E\left(\frac{\pi}{2}, k\right) = \boldsymbol{E}'(k')$.

3. $\boldsymbol{K}'(k) = F\left(\frac{\pi}{2}, k'\right) = \boldsymbol{K}(k')$.

4. $\boldsymbol{E}'(k) = E\left(\frac{\pi}{2}, k'\right) = \boldsymbol{E}(k')$.

5. $\boldsymbol{D} = D\left(\frac{\pi}{2}, k\right) = \frac{\boldsymbol{K}-\boldsymbol{E}}{k^2}$.

In writing complete elliptic integrals, the modulus k, which acts as an independent variable, is often omitted and we write

$$\boldsymbol{K}(\equiv \boldsymbol{K}(k)), \quad \boldsymbol{K}'(\equiv \boldsymbol{K}'(k)), \quad \boldsymbol{E}(\equiv \boldsymbol{E}(k)), \quad \boldsymbol{E}'(\equiv \boldsymbol{E}'(k)).$$

Series representations

8.113

1. $$\boldsymbol{K} = \frac{\pi}{2}\left\{1 + \left(\frac{1}{2}\right)^2 k^2 + \left(\frac{1\cdot 3}{2\cdot 4}\right)^2 k^4 + \ldots + \left[\frac{(2n-1)!!}{2^n n!}\right]^2 k^{2n} + \ldots\right\} = \frac{\pi}{2} F\left(\frac{1}{2}, \frac{1}{2}; 1; k^2\right).$$ FI II 487, WH

2. $$\boldsymbol{K} = \frac{\pi}{1+k'}\left\{1 + \left(\frac{1}{2}\right)^2\left(\frac{1-k'}{1+k'}\right)^2 + \left(\frac{1\cdot 3}{2\cdot 4}\right)^2\left(\frac{1-k'}{1+k'}\right)^4 + \ldots + \left[\frac{(2n-1)!!}{2^n n!}\right]^2\left(\frac{1-k'}{1+k'}\right)^{2n} + \ldots\right\}.$$ DW

3. $$\boldsymbol{K} = \ln\frac{4}{k'} + \left(\frac{1}{2}\right)^2\left(\ln\frac{4}{k'} - \frac{2}{1\cdot 2}\right)k'^2 + \left(\frac{1\cdot 3}{2\cdot 4}\right)^2\left(\ln\frac{4}{k'} - \frac{2}{1\cdot 2} - \frac{2}{3\cdot 4}\right)k'^4 + \left(\frac{1\cdot 3\cdot 5}{2\cdot 4\cdot 6}\right)^2\left(\ln\frac{4}{k'} - \frac{2}{1\cdot 2} - \frac{2}{3\cdot 4} - \frac{2}{5\cdot 6}\right)k'^6 + \ldots$$ DW

See also 8.197 1., 8.197 2.

8.114

1. $$\boldsymbol{E} = \frac{\pi}{2}\left\{1 - \frac{1}{2^2}k^2 - \frac{1^2\cdot 3}{2^2\cdot 4^2}k^4 - \ldots - \left[\frac{(2n-1)!!}{2^n n!}\right]^2\frac{k^{2n}}{2n-1} - \ldots\right\} = \frac{\pi}{2} F\left(-\frac{1}{2}, \frac{1}{2}; 1, k^2\right).$$ FI II 487

2. $E=\frac{(1+k')\pi}{4}\left\{1+\frac{1}{2^2}\left(\frac{1-k'}{1+k'}\right)^2+\frac{1^2}{2^2\cdot4^2}\left(\frac{1-k'}{1+k'}\right)^4+\ldots\right.$
$\left.\ldots+\left[\frac{(2n-3)!!}{2^n n!}\right]^2\left(\frac{1-k'}{1+k'}\right)^{2n}+\ldots\right\}$ DW

3. $E=1+\frac{1}{2}\left(\ln\frac{4}{k'}-\frac{1}{1\cdot2}\right)k'^2+\frac{1^2\cdot3}{2^2\cdot4}\left(\ln\frac{4}{k'}-\frac{2}{1\cdot2}-\frac{1}{3\cdot4}\right)k'^4+$
$+\frac{1^2\cdot3^2\cdot5}{2^2\cdot4^2\cdot6}\left(\ln\frac{4}{k'}-\frac{2}{1\cdot2}-\frac{2}{3\cdot4}-\frac{1}{5\cdot6}\right)k'^6+\ldots$ DW

8.115 $D=\pi\left\{\frac{1}{1}\left(\frac{1}{2}\right)^2+\frac{2}{3}\left(\frac{1\cdot3}{2\cdot4}\right)^2k^2+\ldots+\right.$
$\left.+\frac{n}{2n-1}\left[\frac{(2n-1)!!}{2^n n!}\right]^2k^{2(n-1)}+\ldots\right\}.$ ZH 43(158)

8.116 $$\int_0^{\frac{\pi}{2}}\frac{\sqrt{1-k^2\sin^2\varphi}}{1-n^2\sin^2\varphi}d\varphi=\sqrt{n'^2-k'^2}\left(\frac{\arccos\frac{1}{n'}}{n'\sqrt{n'^2-1}}+R\right),$$

where $R=\frac{k'^2}{2}\left(p+\frac{1}{2}\right)\frac{1}{n'^3}+\frac{k'^4}{16}\left[-1+\left(p+\frac{1}{4}\right)\frac{1}{n'^3}\left(1+\frac{6}{n'^2}\right)\right]+$
$+\frac{k'^6}{16}\left[-\frac{7}{16}-\frac{1}{n'^2}+\left(p+\frac{1}{6}\right)\frac{1}{n'^3}\left(\frac{3}{8}+\frac{1}{n'^2}+\frac{5}{n'^4}\right)\right]+$
$+\frac{15k'^8}{256}\left[-\frac{37}{144}-\frac{21}{40n'^2}-\frac{1}{n'^4}+\right.$
$\left.+\left(p+\frac{1}{8}\right)\frac{1}{n'^3}\left(\frac{5}{24}+\frac{9}{20n'^2}+\frac{1}{n'^4}+\frac{14}{3n'^6}\right)\right]+\ldots,$

$p=\ln\frac{4}{k'},\quad k'=4e^{-p},\quad k'^2=1-k^2,\quad n'^2=1-n^2.$ ZH 44(163)

Trigonometric series

8.117 For *small* values of k and φ, we may use the series

1. $F(\varphi,k)=\frac{2}{\pi}K\varphi-\sin\varphi\cos\varphi\left(a_0+\frac{2}{3}a_1\sin^2\varphi+\frac{2\cdot4}{3\cdot5}a_2\sin^4\varphi+\ldots\right),$

where

$$a_0=\frac{2}{\pi}K-1;\quad a_n=a_{n-1}-\left[\frac{(2n-1)!!}{2^n n!}\right]^2k^{2n}.$$ ZH 10(19)

2. $E(\varphi,k)=\frac{2}{\pi}E\varphi-\sin\varphi\cos\varphi\left(b_0+\frac{2}{3}b_1\sin^2\varphi+\frac{2\cdot4}{3\cdot5}b_2\sin^4\varphi+\ldots\right),$

where

$$b_0=1-\frac{2}{\pi}E,\quad b_n=b_{n-1}-\left[\frac{(2n-1)!!}{2^n n!}\right]^2\frac{k^{2n}}{2n-1}.$$ ZH 27(86)

8.118 For k close to 1, we may use the series

1. $F(\varphi,k)=\frac{2}{\pi}K'\ln\operatorname{tg}\left(\frac{\varphi}{2}+\frac{\pi}{4}\right)-$
$-\frac{\operatorname{tg}\varphi}{\cos\varphi}\left(a_0'-\frac{2}{3}a_1'\operatorname{tg}^2\varphi+\frac{2\cdot4}{3\cdot5}a_2'\operatorname{tg}^4\varphi-\ldots\right),$

where

$$a_0'=\frac{2}{\pi}K'-1;\quad a_n'=a_{n-1}-\left[\frac{(2n-1)!!}{2^n n!}\right]^2k'^{2n}.$$ ZH 10(23)

2. $E(\varphi, k) = \frac{2}{\pi} \boldsymbol{E}' \ln \operatorname{tg}\left(\frac{\varphi}{2} + \frac{\pi}{4}\right) +$

$$+ \frac{\operatorname{tg}\varphi}{\cos\varphi}\left(b_0' - \frac{2}{3} b' \operatorname{tg}^2\varphi + \frac{2\cdot 4}{3\cdot 5} b_2' \operatorname{tg}^4\varphi - \ldots\right),$$

where

$$b_0' = \frac{2}{\pi} \boldsymbol{E}' - 1, \quad b_n' = b_{n-1}' - \left[\frac{(2n-1)!!}{2^n n!}\right]^2 \frac{k'^{2n}}{2n-1}.$$ ZH 27(90)

For the expansion of complete elliptic integrals in Legendre polynomials, see 8.928.

8.119 Representation in the form of an infinite product:

1. $\boldsymbol{K}(k) = \frac{\pi}{2} \prod_{n=1}^{\infty} (1 + k_n),$

where

$$k_n = \frac{1 - \sqrt{1 - k_{n-1}^2}}{1 + \sqrt{1 - k_{n-1}^2}}; \quad k_0 = k.$$ FI II 166

See also 8.197.

8.12 Functional relations between elliptic integrals

8.121

1. $F(-\varphi, k) = -F(\varphi, k).$ JA
2. $E(-\varphi, k) = -E(\varphi, k).$ JA
3. $F(n\pi \pm \varphi, k) = 2n\boldsymbol{K}(k) \pm F(\varphi, k).$ JA
4. $E(n\pi \pm \varphi, k) = 2n\boldsymbol{E}(k) \pm E(\varphi, k).$ JA

8.122 $\boldsymbol{E}(k)\boldsymbol{K}'(k) + \boldsymbol{E}'(k)\boldsymbol{K}(k) - \boldsymbol{K}(k)\boldsymbol{K}'(k) = \frac{\pi}{2}.$ FI II 691, 791

8.123

1. $\frac{\partial F}{\partial k} = \frac{1}{k'^2}\left(\frac{E - k'^2 F}{k} - \frac{k \sin\varphi \cos\varphi}{\sqrt{1 - k^2 \sin^2\varphi}}\right).$ MO 138, BY (710.07)

2. $\frac{d\boldsymbol{K}(k)}{dk} = \frac{\boldsymbol{E}(k)}{kk'^2} - \frac{\boldsymbol{K}(k)}{k}.$ FI II 691

3. $\frac{\partial E}{\partial k} = \frac{E - F}{k}.$ MO 138

4. $\frac{d\boldsymbol{E}(k)}{dk} = \frac{\boldsymbol{E}(k) - \boldsymbol{K}(k)}{k}.$ FI II 690

8.124

1. The functions $\boldsymbol{K}$ and $\boldsymbol{K}'$ satisfy the equation

$$\frac{d}{dk}\left\{kk'^2 \frac{du}{dk}\right\} - ku = 0.$$ WH

2. The functions $\boldsymbol{E}$ and $\boldsymbol{E}' - \boldsymbol{K}'$ satisfy the equation

$$k'^2 \frac{d}{dk}\left(k \frac{du}{dk}\right) + ku = 0.$$ WH

Transformation formulas

8.125

1. $F\left(\psi, \frac{1-k'}{1+k'}\right) = (1+k') F(\varphi, k)$ MO 130

2. $E\left(\psi, \frac{1-k'}{1+k'}\right) = \frac{2}{1+k'} [E(\varphi, k) + k' F(\varphi, k)] - \frac{1-k'}{1+k'} \sin\psi$ $\quad [\operatorname{tg}(\psi - \varphi) = k' \operatorname{tg}\varphi]$. MO 131

3. $F\left(\psi, \frac{2\sqrt{k}}{1+k}\right) = (1+k) F(\varphi, k)$.

4. $E\left(\psi, \frac{2\sqrt{k}}{1+k}\right) = \frac{1}{1+k}\left[2E(\varphi, k) - k'^2 F(\varphi, k) + 2k \frac{\sin\varphi\cos\varphi}{1+k\sin^2\varphi}\sqrt{1-k^2\sin^2\varphi}\right]$ $\quad \left[\sin\psi = \frac{(1+k)\sin\varphi}{1+k\sin^2\varphi}\right]$. MO 131

8.126 In particular,

1. $\boldsymbol{K}\left(\frac{1-k'}{1+k'}\right) = \frac{1+k'}{2} \boldsymbol{K}(k)$. MO 130

2. $\boldsymbol{E}\left(\frac{1-k'}{1+k'}\right) = \frac{1}{1+k'} [\boldsymbol{E}(k) + k' \boldsymbol{K}(k)]$. MO 130

3. $\boldsymbol{K}\left(\frac{2\sqrt{k}}{1+k}\right) = (1+k) \boldsymbol{K}(k)$. MO 130

4. $\boldsymbol{E}\left(\frac{2\sqrt{k}}{1+k}\right) = \frac{1}{1+k} [2\boldsymbol{E}(k) - k'^2 \boldsymbol{K}(k)]$. MO 130

8.127

k_1	$\sin\varphi_1$	$\cos\varphi_1$	$F(\varphi_1, k_1)$	$E(\varphi_1, k_1)$
$i\frac{k}{k'}$	$k'\frac{\sin\varphi}{\Delta\varphi}$	$\frac{\cos\varphi}{\Delta\varphi}$	$k' F(\varphi, k)$	$\frac{1}{k'}\left[E(\varphi, k) - \frac{k^2\sin\varphi\cos\varphi}{\Delta\varphi}\right]$
k'	$-i\operatorname{tg}\varphi$	$\sec\varphi$	$-iF(\varphi, k)$	$i[E(\varphi, k) - F(\varphi, k) - \Delta\varphi\operatorname{tg}\varphi]$
$\frac{1}{k}$	$k\sin\varphi$	$\Delta\varphi$	$kF(\varphi, k)$	$\frac{1}{k}[E(\varphi, k) - k'^2 F(\varphi, k)]$
$\frac{1}{k'}$	$-ik'\operatorname{tg}\varphi$	$\frac{\Delta\varphi}{\cos\varphi}$	$-ik' F(\varphi, k)$	$\frac{i}{k'}[E(\varphi, k) - k'^2 F(\varphi, k) - \Delta\varphi\operatorname{tg}\varphi]$
$\frac{k'}{ik}$	$\frac{-ik\sin\varphi}{\Delta\varphi}$	$\frac{1}{\Delta\varphi}$	$-ikF(\varphi, k)$	$\frac{i}{k}\left[E(\varphi, k) - F(\varphi, k) - \frac{k^2\sin\varphi\cos\varphi}{\Delta\varphi}\right]$

(see 8.111 1.). MO 131

8.128 In particular,

1. $\boldsymbol{K}\left(i\frac{k}{k'}\right) = k' \boldsymbol{K}(k)$. MO 130

2. $\boldsymbol{K}'\left(i\frac{k}{k'}\right) = k' [\boldsymbol{K}(k') - i\boldsymbol{K}(k)]$. MO 130

3. $\boldsymbol{K}\left(\frac{1}{k}\right) = k\boldsymbol{K}(k) + i\boldsymbol{K}'(k)$. MO 130

For integrals of elliptic integrals, see **6.11 – 6.15**. For indefinite integrals of complete elliptic integrals, see **5.11**.

8.129 Special values:

1. $$K\left(\sin\frac{\pi}{4}\right)=K\left(\frac{\sqrt{2}}{2}\right)=K'\left(\frac{\sqrt{2}}{2}\right)=\sqrt{2}\int_0^1\frac{dt}{\sqrt{1-t^4}}=$$
$$=\frac{1}{4\sqrt{\pi}}\left[\Gamma\left(\frac{1}{4}\right)\right]^2.$$ MO 130

2. $K'(\sqrt{2}-1)=\sqrt{2}\,K(\sqrt{2}-1)$. MO 130

3. $K'\left(\sin\frac{\pi}{12}\right)=\sqrt{3}\,K\left(\sin\frac{\pi}{12}\right)$. MO 130

4. $K'\left(\operatorname{tg}^2\frac{\pi}{8}\right)=K'\left(\frac{2-\sqrt{2}}{2+\sqrt{2}}\right)=2K\left(\operatorname{tg}^2\frac{\pi}{8}\right)$. MO 130

8.13 Elliptic functions

8.130 Definition and general properties.

1. A rational function $f(z)$ of a complex variable is said to be elliptic if it has two periods $2\omega_1$ and $2\omega_2$, that is

$$f(z+2m\omega_1+2n\omega_2)=f(z) \qquad [m,\ n \text{ integers}].$$

The ratio of the periods of an analytic function cannot be a real number. For an elliptic function $f(z)$, the z-plane can be partitioned into parallelograms – the period parallelograms – the vertices of which are the points $z_0+2m\omega_1+2n\omega_2$. At corresponding points of these parallelograms, the function $f(z)$ has the same value. ZH 117, SI 299

2. Suppose that α is the angle between the sides a and b of one of the period parallelograms. Then,

$$\tau=\frac{\omega_1}{\omega_2}=\frac{a}{b}e^{i\alpha}, \qquad q=e^{i\pi\tau}=e^{-\frac{a}{b}\pi\sin\alpha}\left[\cos\left(\frac{a}{b}\pi\cos\alpha\right)+i\sin\left(\frac{a}{b}\pi\cos\alpha\right)\right].$$

3. The *derivative* of an elliptic function is also an elliptic function with the same periods. SM III 598

4. A nonconstant elliptic function has a finite number of poles in a period parallelogram: it can have no more than two simple and one second-order pole in such a parallelogram. Suppose that these poles lie at the points $a_1, a_2, \ldots, a_n$ and that their orders are $\alpha_1, \alpha_2, \ldots, \alpha_n$. Suppose that the zeros of an analytic function that occur in a single parallelogram are $b_1, b_2, \ldots, b_m$ and that the orders of the zeros are $\beta_1, \beta_2, \ldots, \beta_m$, respectively. Then,

$$\gamma=\alpha_1+\alpha_2+\ldots+\alpha_n=\beta_1+\beta_2+\ldots+\beta_m.$$ ZH 118

The number γ representing this sum is called the *order* of the elliptic function.

5. The sum of the residues of an elliptic function with respect to all the poles belonging to a period parallelogram is equal to zero.

6. The difference between the sum of all the zeros and the sum of all the poles of an elliptic function that are located in a period parallelogram is equal to one of its periods.

7. Every two elliptic functions with the same periods are related by an algebraic relationship. GO II 151

8. A single-valued function cannot have more than two periods. GO II 147

9. An elliptic function of order γ assumes *an arbitrary value* γ times in a period parallelogram.
SM 601, SI 301

8.14 Jacobian elliptic functions

8.141 Consider the upper limit φ of the integral

$$u=\int_0^{\varphi}\frac{d\alpha}{\sqrt{1-k^2\sin^2\alpha}}$$

as a function of u. Using the notation

$$\varphi=\operatorname{am}u$$

we call this upper limit the *amplitude*. The quantity u is called the *argument*, and its dependence on φ is written

$$u=\arg\varphi.$$

8.142 The amplitude is an *infinitely-many-valued* function of u and has a period of $4\boldsymbol{K}i$. The *branch points* of the amplitude correspond to the values of the argument

$$u=2m\boldsymbol{K}+(2n+1)\boldsymbol{K}'i,$$ ZH 67-69

where m and n are arbitrary integers (see also 8.151).

8.143 The first two of the following functions

$$\operatorname{sn}u=\sin\varphi=\sin\operatorname{am}u,\quad \operatorname{cn}u=\cos\varphi=\cos\operatorname{am}u,$$

$$\operatorname{dn}u=\Delta\varphi=\sqrt{1-k^2\sin^2\varphi}=\frac{d\varphi}{du}$$

are called, respectively, the *sine-amplitude* and the *cosine-amplitude* while the third may be called the *delta amplitude*. All these elliptic functions were exhibited by Jacobi and they bear his name.
SI 16

The Jacobian elliptic functions are *doubly-periodic* functions and have *two simple poles* in a period parallelogram.
ZH 69

8.144

$$\left.\begin{aligned}&1.\quad u=\int_0^{\operatorname{sn}u}\frac{dt}{\sqrt{(1-t^2)(1-k^2t^2)}}\\&2.\quad u=\int_1^{\operatorname{cn}u}\frac{dt}{\sqrt{(1-t^2)(k'^2+k^2t^2)}}\,.\\&3.\quad u=\int_1^{\operatorname{dn}u}\frac{dt}{\sqrt{(1-t^2)(t^2-k'^2)}}\,.\end{aligned}\right\}$$ SI 21(23)

8.145 Power series representations:

1. $$\operatorname{sn}u=u-\frac{1+k^2}{3!}u^3+\frac{1+14k^2+k^4}{5!}u^5-\frac{1+135k^2+135k^4+k^6}{7!}u^7+$$
$$+\frac{1+1228k^2+5478k^4+1228k^6+k^8}{9!}u^9-\dots\qquad[|u|<|\boldsymbol{K}'|].$$ ZH 81(97)

2. $$\operatorname{cn} u = 1 - \frac{1}{2!}u^2 + \frac{1+4k^2}{4!}u^4 - \frac{1+44k^2+16k^4}{6!}u^6 + \\ + \frac{1+408k^2+912k^4+64k^6}{8!}u^8 - \ldots \quad [|u| < |\boldsymbol{K}'|].$$ ZH 81(98)

3. $$\operatorname{dn} u = 1 - \frac{k^2}{2!}u^2 + \frac{k^2(4+k^2)}{4!}u^4 - \frac{k^2(16+44k^2+k^4)}{6!}u^6 + \\ + \frac{k^2(64+912k^2+408k^4+k^6)}{8!}u^8 - \ldots \quad [|u| < |\boldsymbol{K}'|].$$ ZH 81(99)

4. $$\operatorname{am} u = u - \frac{k^2}{3!}u^3 + \frac{k^2(4+k^2)}{5!}u^5 - \frac{k^2(16+44k^2+k^4)}{7!}u^7 + \\ + \frac{k^2(64+912k^2+408k^4+k^6)}{9!}u^9 - \ldots \quad [|u| < |\boldsymbol{K}'|].$$ LA 380(4)

8.146 Representation as a trigonometric series or a product $\left(q = e^{-\frac{\pi \boldsymbol{K}'}{\boldsymbol{K}}}\right)$*:

1. $$\operatorname{sn} u = \frac{2\pi}{k\boldsymbol{K}} \sum_{n=1}^{\infty} \frac{q^{n-\frac{1}{2}}}{1-q^{2n-1}} \sin(2n-1)\frac{\pi u}{2\boldsymbol{K}}.$$ WH, ZH 84(108)

2. $$\operatorname{cn} u = \frac{2\pi}{k\boldsymbol{K}} \sum_{n=1}^{\infty} \frac{q^{n-\frac{1}{2}}}{1+q^{2n-1}} \cos(2n-1)\frac{\pi u}{2\boldsymbol{K}}.$$ WH, ZH 84(109)

3. $$\operatorname{dn} u = \frac{\pi}{2\boldsymbol{K}} + \frac{2\pi}{\boldsymbol{K}} \sum_{n=1}^{\infty} \frac{q^n}{1+q^{2n}} \cos\frac{n\pi u}{\boldsymbol{K}}.$$ WH, ZH 84(110)

4. $$\operatorname{am} u = \frac{\pi u}{2\boldsymbol{K}} + 2\sum_{n=1}^{\infty} \frac{1}{n}\frac{q^n}{1+q^{2n}} \sin\frac{n\pi u}{\boldsymbol{K}}.$$ WH

5. $$\frac{1}{\operatorname{sn} u} = \frac{\pi}{2\boldsymbol{K}}\left[\frac{1}{\sin\frac{\pi u}{2\boldsymbol{K}}} + 4\sum_{n=1}^{\infty} \frac{q^{2n-1}}{1-q^{2n-1}} \sin(2n-1)\frac{\pi u}{2\boldsymbol{K}}\right].$$ LA 369(3)

6. $$\frac{1}{\operatorname{cn} u} = \frac{\pi}{2k'\boldsymbol{K}}\left[\frac{1}{\cos\frac{\pi u}{2\boldsymbol{K}}} + 4\sum_{n=1}^{\infty} (-1)^n \frac{q^{2n-1}}{1+q^{2n-1}} \cos(2n-1)\frac{\pi u}{2\boldsymbol{K}}\right].$$ LA 369(3)

7. $$\frac{1}{\operatorname{dn} u} = \frac{\pi}{2k'\boldsymbol{K}}\left[1 + 4\sum_{n=1}^{\infty} (-1)^n \frac{q^n}{1+q^{2n}} \cos\frac{n\pi u}{\boldsymbol{K}}\right].$$ LA 369(3)

8. $$\frac{\operatorname{sn} u}{\operatorname{cn} u} = \frac{\pi}{2k'\boldsymbol{K}}\left[\operatorname{tg}\frac{\pi u}{2\boldsymbol{K}} + 4\sum_{n=1}^{\infty} (-1)^n \frac{q^{2n}}{1+q^{2n}} \sin\frac{n\pi u}{\boldsymbol{K}}\right].$$ LA 369(4)

9. $$\frac{\operatorname{sn} u}{\operatorname{dn} u} = -\frac{2\pi}{kk'\boldsymbol{K}} \sum_{n=1}^{\infty} (-1)^n \frac{q^{n-\frac{1}{2}}}{1+q^{2n-1}} \sin(2n-1)\frac{\pi u}{2\boldsymbol{K}}.$$ LA 369(4)

---

*The expansions 1—22 are valid in every strip of the form $\left|\operatorname{Im}\frac{\pi u}{2\boldsymbol{K}}\right| < \frac{1}{2}\pi \operatorname{Im}\tau$. The expansions 23—25 are valid in an arbitrary bounded portion of u.

10. $\frac{\operatorname{cn} u}{\operatorname{sn} u}=\frac{\pi}{2K}\left[\operatorname{ctg}\frac{\pi u}{2K}-4\sum_{n=1}^{\infty}\frac{q^{2n}}{1+q^{2n}}\sin\frac{\pi n u}{K}\right].$ LA 369(5)

11. $\frac{\operatorname{cn} u}{\operatorname{dn} u}=-\frac{2\pi}{kK}\sum_{n=1}^{\infty}(-1)^n\frac{q^{n-\frac{1}{2}}}{1-q^{2n-1}}\cos(2n-1)\frac{\pi u}{2K}.$ LA 369(5)

12. $\frac{\operatorname{dn} u}{\operatorname{sn} u}=\frac{\pi}{2K}\left[\frac{1}{\sin\frac{\pi u}{2K}}-4\sum_{n=1}^{\infty}\frac{q^{2n-1}}{1+q^{2n-1}}\sin(2n-1)\frac{\pi u}{2K}\right].$ LA 369(6)

13. $\frac{\operatorname{dn} u}{\operatorname{cn} u}=\frac{\pi}{2K}\left[\frac{1}{\cos\frac{\pi u}{2K}}-4\sum_{n=1}^{\infty}(-1)^n\frac{q^{2n-1}}{1-q^{2n-1}}\cos(2n-1)\frac{\pi u}{2K}\right].$ LA 369(6)

14. $\frac{\operatorname{cn} u\operatorname{dn} u}{\operatorname{sn} u}=\frac{\pi}{2K}\left[\operatorname{ctg}\frac{\pi u}{2K}-4\sum_{n=1}^{\infty}\frac{q^n}{1+q^n}\sin\frac{n\pi u}{K}\right].$ LA 369(7)

15. $\frac{\operatorname{sn} u\operatorname{dn} u}{\operatorname{cn} u}=\frac{\pi}{2K}\left\{\operatorname{tg}\frac{\pi u}{2K}+4\sum_{n=1}^{\infty}\frac{q^n}{1+(-1)^n q^n}\sin\frac{n\pi u}{K}\right\}.$ LA 369(7)

16. $\frac{\operatorname{sn} u\operatorname{cn} u}{\operatorname{dn} u}=\frac{4\pi^2}{k^2K}\sum_{n=1}^{\infty}\frac{q^{2n-1}}{1-q^{2(2n-1)}}\sin(2n-1)\frac{\pi u}{K}.$ LA 369(7)

17. $\frac{\operatorname{sn} u}{\operatorname{cn} u\operatorname{dn} u}=\frac{\pi}{2(1-k^2)K}\left[\operatorname{tg}\frac{\pi u}{2K}+4\sum_{n=1}^{\infty}(-1)^n\frac{q^n}{1-q^n}\sin\frac{n\pi u}{K}\right].$ LA 369(8)

18. $\frac{\operatorname{cn} u}{\operatorname{sn} u\operatorname{dn} u}=\frac{\pi}{2K}\left[\operatorname{ctg}\frac{\pi u}{2K}-4\sum_{n=1}^{\infty}\frac{(-1)^n q^n}{1+(-1)^n q^n}\sin\frac{n\pi u}{K}\right].$ LA 369(8)

19. $\frac{\operatorname{dn} u}{\operatorname{sn} u\operatorname{cn} u}=\frac{\pi}{K}\left[\frac{1}{\sin\frac{\pi u}{K}}+4\sum_{n=1}^{\infty}\frac{q^{2(2n-1)}}{1-q^{2(2n-1)}}\sin(2n-1)\frac{\pi u}{K}\right].$ LA 369(8)

20. $\ln\operatorname{sn} u=\ln\frac{2K}{\pi}+\ln\sin\frac{\pi u}{2K}-4\sum_{n=1}^{\infty}\frac{1}{n}\frac{q^n}{1+q^n}\sin^2\frac{n\pi u}{2K}.$ LA 369(2)

21. $\ln\operatorname{cn} u=\ln\cos\frac{\pi u}{2K}-4\sum_{n=1}^{\infty}\frac{1}{n}\frac{q^n}{1+(-1)^n q^n}\sin^2\frac{n\pi u}{2K}.$ LA 369(2)

22. $\ln\operatorname{dn} u=-8\sum_{n=1}^{\infty}\frac{1}{2n-1}\frac{q^{2n-1}}{1-q^{2(2n-1)}}\sin^2(2n-1)\frac{\pi u}{2K}.$ LA 369(2)

23. $\operatorname{sn} u=\frac{2\sqrt[4]{q}}{\sqrt{k}}\sin\frac{\pi u}{2K}\prod_{n=1}^{\infty}\frac{1-2q^{2n}\cos\frac{\pi u}{K}+q^{4n}}{1-2q^{2n-1}\cos\frac{\pi u}{K}+q^{4n-2}}.$ ZH 86(145)

24. $\operatorname{cn} u=\frac{2\sqrt{k'}\sqrt[4]{q}}{\sqrt{k}}\cos\frac{\pi u}{2K}\prod_{n=1}^{\infty}\frac{1+2q^{2n}\cos\frac{\pi u}{K}+q^{4n}}{1-2q^{2n-1}\cos\frac{\pi u}{K}+q^{4n-2}}.$ ZH 86(146)

25. $$\operatorname{dn} u = \sqrt{k'} \prod_{n=1}^{\infty} \frac{1+2q^{2n-1}\cos\frac{\pi u}{K}+q^{4n-2}}{1-2q^{2n-1}\cos\frac{\pi u}{K}+q^{4n-2}}.$$ ZH 86(147)

26. $$\operatorname{sn}^2 u = \sum_{n=0}^{\infty} \left[\frac{1+k^2}{2k^3} - \frac{(2n+1)^2}{2k^3}\frac{\pi^2}{4K^2}\right] \frac{2\pi q^{n+\frac{1}{2}} \sin(2n+1)\frac{\pi u}{2K}}{K(1-q^{2n+1})}$$
$$\left[\left|\operatorname{Im}\frac{u}{2K}\right| < \operatorname{Im}\tau\right].$$ MO 147

27. $$\frac{1}{\operatorname{sn}^2 u} = \frac{\pi^2}{4K^2}\operatorname{cosec}^2\frac{\pi u}{2K} + \frac{K-E}{K} - \frac{2\pi^2}{K^2}\sum_{n=1}^{\infty}\frac{nq^{2n}\cos\frac{n\pi u}{K}}{1-q^{2n}}$$
$$\left[\left|\operatorname{Im}\frac{u}{2K}\right| < \frac{1}{2}\operatorname{Im}\tau\right].$$ MO 148

8.147

1. $$\operatorname{sn} u = \frac{\pi}{2kK}\sum_{n=-\infty}^{\infty}\frac{1}{\sin\frac{\pi}{2K}[u-(2n-1)iK']}.$$ MO 149

2. $$\operatorname{cn} u = \frac{\pi i}{2kK}\sum_{n=-\infty}^{\infty}\frac{(-1)^n}{\sin\frac{\pi}{2K}[u-(2n-1)iK']}.$$ MO 150

3. $$\operatorname{dn} u = \frac{\pi i}{2K}\sum_{n=-\infty}^{\infty}\frac{(-1)^n}{\operatorname{tg}\frac{\pi}{2K}[u-(2n-1)iK']}.$$ MO 150

8.148 The Weierstrass expansions of the functions $\operatorname{sn} u$, $\operatorname{cn} u$, $\operatorname{dn} u$:

$$\operatorname{sn} u = \frac{B}{A}, \quad \operatorname{cn} u = \frac{C}{A}, \quad \operatorname{dn} u = \frac{D}{A},$$

where

$$A = 1 - \sum_{n=1}^{\infty}(-1)^{n+1} a_{n+1}\frac{u^{2n+2}}{(2n+2)!}$$

$$[a_2 = 2k^2, \quad a_3 = 8(k^2+k^4), \quad a_4 = 32(k^2+k^6)+68k^4, \; a_5 = 128(k^2+k^8)+ \\ +480(k^4+k^6), \quad a_6 = 512(k^2+k^{10})+3008(k^4+k^8)+5400k^6, \; \ldots]$$

$$B = \sum_{n=0}^{\infty}(-1)^n b_n\frac{u^{2n+1}}{(2n+1)!}$$

$$[b_0 = 1, \; b_1 = 1+k^2, \; b_2 = 1+k^4+4k^2, \; b_3 = 1+k^6+9(k^2+k^4), \\ b_4 = 1+k^8+16(k^2+k^6)-6k^4, \quad b_5 = 1+k^{10}+25(k^2+k^8)-494(k^4+k^6), \\ b_6 = 1+k^{12}+36(k^2+k^{10})-5781(k^4+k^8)-12184k^6, \; \ldots].$$

$$C = \sum_{n=0}^{\infty}(-1)^n c_n\frac{u^{2n}}{(2n)!}$$

$$[c_0 = 1, \quad c_1 = 1, \quad c_2 = 1+2k^2, \quad c_3 = 1+6k^2+8k^4, \quad c_4 = 1+12k^2+60k^4+32k^6, \\ c_5 = 1+20k^2+348k^4+448k^6+128k^8, \\ c_6 = 1+30k^2+2372k^4+4600k^6+2880k^8+512k^{10}, \; \ldots].$$

$$D=\sum_{n=0}^{\infty}(-1)^n d_n \frac{u^{2n}}{(2n)!}$$

$[d_0=1,\ d_1=k^2,\ d_2=2k^2+k^4,\ d_3=8k^2+6k^4+k^6,\ d_4=32k^2+60k^4+12k^6+k^8,$
$d_5=128k^2+448k^4+348k^6+20k^8+k^{10},$
$d_6=512k^2+2880k^4+4600k^6+2372k^8+30k^{10}+k^{12},\ \ldots].$

ZH 82-83(105, 106, 107)

8.15 Properties of Jacobian elliptic functions and functional relationships between them

8.151 The periods, zeros, poles, and residues of Jacobian elliptic functions:

1.

	Periods	Zeros	Poles	Residues
$\operatorname{sn} u$	$4m\boldsymbol{K}+2n\boldsymbol{K}'i$	$2m\boldsymbol{K}+2n\boldsymbol{K}'i$	$2m\boldsymbol{K}+(2n+1)\boldsymbol{K}'i$	$(-1)^m\frac{1}{k}$
$\operatorname{cn} u$	$4m\boldsymbol{K}+2n(\boldsymbol{K}+\boldsymbol{K}'i)$	$(2m+1)\boldsymbol{K}+2n\boldsymbol{K}'i$	$2m\boldsymbol{K}+(2n+1)\boldsymbol{K}'i$	$(-1)^{m-1}\frac{i}{k}$
$\operatorname{dn} u$	$2m\boldsymbol{K}+4n\boldsymbol{K}'i$	$(2m+1)\boldsymbol{K}+(2n+1)\boldsymbol{K}'i$	$2m\boldsymbol{K}+(2n+1)\boldsymbol{K}'i$	$(-1)^{n-1}i$

SM 630, ZH 69-72

2.

$u^*=u+\boldsymbol{K}$	$u+i\boldsymbol{K}$	$u+\boldsymbol{K}+i\boldsymbol{K}'$	$u+2\boldsymbol{K}$	$u+2i\boldsymbol{K}'$	$u+2\boldsymbol{K}+2i\boldsymbol{K}'$
$\operatorname{sn} u^*=\frac{\operatorname{cn} u}{\operatorname{dn} u}$	$\frac{1}{k\operatorname{sn} u}$	$\frac{1}{k}\frac{\operatorname{dn} u}{\operatorname{cn} u}$	$-\operatorname{sn} u$	$\operatorname{sn} u$	$-\operatorname{sn} u$
$\operatorname{cn} u^*=-k'\frac{\operatorname{sn} u}{\operatorname{dn} u}$	$-\frac{i}{k}\frac{\operatorname{dn} u}{\operatorname{sn} u}$	$-\frac{ik'}{k\operatorname{cn} u}$	$-\operatorname{cn} u$	$-\operatorname{cn} u$	$\operatorname{cn} u$
$\operatorname{dn} u^*=k'\frac{1}{\operatorname{dn} u}$	$-i\frac{\operatorname{cn} u}{\operatorname{sn} u}$	$ik'\frac{\operatorname{sn} u}{\operatorname{cn} u}$	$\operatorname{dn} u$	$-\operatorname{dn} u$	$-\operatorname{dn} u$

SM 630

3.

$u^*=0$	$-u$	$\frac{1}{2}\boldsymbol{K}$	$\frac{1}{2}(\boldsymbol{K}+i\boldsymbol{K}')$	$\frac{1}{2}i\boldsymbol{K}'$	$u+2m\boldsymbol{K}+2n\boldsymbol{K}'i$
$\operatorname{sn} u^*=0$	$-\operatorname{sn} u$	$\frac{1}{\sqrt{1+k'}}$	$\frac{\sqrt{1+k}+i\sqrt{1-k}}{\sqrt{2k}}$	$\frac{i}{\sqrt{k}}$	$(-1)^m\operatorname{sn} u$
$\operatorname{cn} u^*=1$	$\operatorname{cn} u$	$\frac{\sqrt{k'}}{\sqrt{1+k'}}$	$\frac{(1-i)\sqrt{k'}}{\sqrt{2k}}$	$\frac{\sqrt{1+k}}{\sqrt{k}}$	$(-1)^{m+n}\operatorname{cn} u$
$\operatorname{dn} u^*=1$	$\operatorname{dn} u$	$\sqrt{k'}$	$\frac{\sqrt{k'}(\sqrt{1+k'}-i\sqrt{1-k'})}{\sqrt{2}}$	$\sqrt{1+k}$	$(-1)^n\operatorname{dn} u$

SI 19, SI 18(13), WH, WH, WH, WH

8.152 Transformation formulas

u_1	k_1	$\mathrm{sn}(u_1, k_1)$	$\mathrm{cn}(u_1, k_1)$	$\mathrm{dn}(u_1, k_1)$
ku	$\frac{1}{k}$	$k\,\mathrm{sn}(u, k)$	$\mathrm{dn}(u, k)$	$\mathrm{cn}(u, k)$
iu	k'	$i\frac{\mathrm{sn}(u, k)}{\mathrm{cn}(u, k)}$	$\frac{1}{\mathrm{cn}(u, k)}$	$\frac{\mathrm{dn}(u, k)}{\mathrm{cn}(u, k)}$
$k'u$	$i\frac{k}{k'}$	$k'\frac{\mathrm{sn}(u, k)}{\mathrm{dn}(u, k)}$	$\frac{\mathrm{cn}(u, k)}{\mathrm{dn}(u, k)}$	$\frac{1}{\mathrm{dn}(u, k)}$
iku	$i\frac{k'}{k}$	$ik\frac{\mathrm{sn}(u, k)}{\mathrm{dn}(u, k)}$	$\frac{1}{\mathrm{dn}(u, k)}$	$\frac{\mathrm{cn}(u, k)}{\mathrm{dn}(u, k)}$
$ik'u$	$\frac{1}{k'}$	$ik'\frac{\mathrm{sn}(u, k)}{\mathrm{cn}(u, k)}$	$\frac{\mathrm{dn}(u, k)}{\mathrm{cn}(u, k)}$	$\frac{1}{\mathrm{cn}(u, k)}$
$(1+k)u$	$\frac{2\sqrt{k}}{1+k}$	$\frac{(1+k)\,\mathrm{sn}(u, k)}{1+k\,\mathrm{sn}^2(u, k)}$	$\frac{\mathrm{cn}(u, k)\,\mathrm{dn}(u, k)}{1+k\,\mathrm{sn}^2(u, k)}$	$\frac{1-k\,\mathrm{sn}^2(u, k)}{1+k\,\mathrm{sn}^2(u, k)}$
$(1+k')u$	$\frac{1-k'}{1+k'}$	$(1+k')\frac{\mathrm{sn}(u, k)\,\mathrm{cn}(u, k)}{\mathrm{dn}(u, k)}$	$\frac{1-(1+k')\,\mathrm{sn}^2(u, k)}{\mathrm{dn}(u, k)}$	$\frac{1-(1-k')\,\mathrm{sn}^2(u, k)}{\mathrm{dn}(u, k)}$
$\frac{(1+\sqrt{k'})^2}{2}u$	$\left(\frac{1-\sqrt{k'}}{1+\sqrt{k'}}\right)^2$	$\frac{k^2\,\mathrm{sn}(u, k)\,\mathrm{cn}(u, k)}{\sqrt{k_1}\,[1+\mathrm{dn}(u, k)]\,[k'+\mathrm{dn}(u, k)]}$	$\frac{\mathrm{dn}(u, k)-\sqrt{k'}}{1-\sqrt{k'}}\times \sqrt{\frac{2(1+k')}{[1+\mathrm{dn}(u, k)]\,[k'+\mathrm{dn}(u, k)]}}$	$\frac{\sqrt{1+k_1}\,(\mathrm{dn}(u, k)+\sqrt{k'})}{\sqrt{[1+\mathrm{dn}(u, k)]\,[k'+\mathrm{dn}(u, k)]}}$

JA

8.153

1. $\operatorname{sn}(iu, k) = i\dfrac{\operatorname{sn}(u, k')}{\operatorname{cn}(u, k')}$. SI 50(64)

2. $\operatorname{cn}(iu, k) = \dfrac{1}{\operatorname{cn}(u, k')}$. SI 50(65)

3. $\operatorname{dn}(iu, k) = \dfrac{\operatorname{dn}(u, k')}{\operatorname{cn}(u, k')}$. SI 50(65)

4. $\operatorname{sn}(u, k) = k^{-1}\operatorname{sn}(ku, k^{-1})$.

5. $\operatorname{cn}(u, k) = \operatorname{dn}(ku, k^{-1})$.

6. $\operatorname{dn}(u, k) = \operatorname{cn}(ku, k^{-1})$.

7. $\operatorname{sn}(u, ik) = \dfrac{1}{\sqrt{1+k^2}}\dfrac{\operatorname{sn}\left(u\sqrt{1+k^2},\ k(1+k^2)^{-\frac{1}{2}}\right)}{\operatorname{dn}\left(u\sqrt{1+k^2},\ k(1+k^2)^{-1/2}\right)}$.

8. $\operatorname{cn}(u, ik) = \dfrac{\operatorname{cn}\left(u(1+k^2)^{\frac{1}{2}},\ k(1+k^2)^{-\frac{1}{2}}\right)}{\operatorname{dn}\left(u(1+k^2)^{1/2},\ k(1+k^2)^{-1/2}\right)}$.

9. $\operatorname{dn}(u, ik) = \dfrac{1}{\operatorname{dn}\left(u(1+k^2)^{1/2},\ k(1+k^2)^{-1/2}\right)}$.

Functional relations

8.154

1. $\operatorname{sn}^2 u = \dfrac{1-\operatorname{cn} 2u}{1+\operatorname{dn} 2u}$. MO 146

2. $\operatorname{cn}^2 u = \dfrac{\operatorname{cn} 2u + \operatorname{dn} 2u}{1+\operatorname{dn} 2u}$. MO 146

3. $\operatorname{dn}^2 u = \dfrac{\operatorname{dn} 2u + k^2 \operatorname{cn} 2u + k'^2}{1+\operatorname{dn} 2u}$. MO 146

4. $\operatorname{sn}^2 u + \operatorname{cn}^2 u = 1$. SI 16(9)

5. $\operatorname{dn}^2 u + k^2 \operatorname{sn}^2 u = 1$. SI 16(9)

8.155

1. $\dfrac{1-\operatorname{dn} 2u}{1+\operatorname{dn} 2u} = k^2\dfrac{\operatorname{sn}^2 u \operatorname{cn}^2 u}{\operatorname{dn}^2 u}$. MO 146

2. $\dfrac{1-\operatorname{cn} 2u}{1+\operatorname{cn} 2u} = \dfrac{\operatorname{sn}^2 u \operatorname{dn}^2 u}{\operatorname{cn}^2 u}$. MO 146

8.156

1. $\operatorname{sn}(u \pm v) = \dfrac{\operatorname{sn} u \operatorname{cn} v \operatorname{dn} v \pm \operatorname{sn} v \operatorname{cn} u \operatorname{dn} u}{1-k^2 \operatorname{sn}^2 u \operatorname{sn}^2 v}$. SI 46(56)

2. $\operatorname{cn}(u \pm v) = \dfrac{\operatorname{cn} u \operatorname{cn} v \mp \operatorname{sn} u \operatorname{sn} v \operatorname{dn} u \operatorname{dn} v}{1-k^2 \operatorname{sn}^2 u \operatorname{sn}^2 v}$. SI 46(57)

3. $\operatorname{dn}(u \pm v) = \dfrac{\operatorname{dn} u \operatorname{dn} v \mp k^2 \operatorname{sn} u \operatorname{sn} v \operatorname{cn} u \operatorname{cn} v}{1-k^2 \operatorname{sn}^2 u \operatorname{sn}^2 v}$. SI 46(58)

8.157

1. $\operatorname{sn}\frac{u}{2}=\pm\frac{1}{k}\sqrt{\frac{1-\operatorname{dn}u}{1+\operatorname{cn}u}}=\pm\sqrt{\frac{1-\operatorname{cn}u}{1+\operatorname{dn}u}}$. SI 47(61), SI 67(15)

2. $\operatorname{cn}\frac{u}{2}=\pm\sqrt{\frac{\operatorname{cn}u+\operatorname{dn}u}{1+\operatorname{dn}u}}=\pm\frac{k}{k}\sqrt{\frac{1-\operatorname{dn}u}{\operatorname{dn}u-\operatorname{cn}u}}$. SI 48(62), SI 67(16)

3. $\operatorname{dn}\frac{u}{2}=\pm\sqrt{\frac{\operatorname{cn}u+\operatorname{dn}u}{1+\operatorname{cn}u}}=\pm k'\sqrt{\frac{1-\operatorname{cn}u}{\operatorname{dn}u-\operatorname{cn}u}}$. SI 48(63), SI 67(17)

8.158

1. $\frac{d}{du}\operatorname{sn}u=\operatorname{cn}u\operatorname{dn}u.$

2. $\frac{d}{du}\operatorname{cn}u=-\operatorname{sn}u\operatorname{dn}u$

3. $\frac{d}{du}\operatorname{dn}u=-k^2\operatorname{sn}u\operatorname{cn}u.$

SI 21(21)

8.159 Jacobian elliptic functions are solutions of the following differential equations:

1. $\frac{d}{du}\operatorname{sn}u=\sqrt{(1-\operatorname{sn}^2u)(1-k^2\operatorname{sn}^2u)}.$

2. $\frac{d}{du}\operatorname{cn}u=-\sqrt{(1-\operatorname{cn}^2u)(k'^2+k^2\operatorname{cn}^2u)}.$

3. $\frac{d}{du}\operatorname{dn}u=-\sqrt{(1-\operatorname{dn}^2u)(\operatorname{dn}^2u-k'^2)}.$

SI 21(22)

For the indefinite integrals of Jacobi's elliptic functions, see 5.13.

8.16 The Weierstrass function $\wp(u)$

8.160 The Weierstrass elliptic function $\wp(u)$ is defined by

1. $\wp(u)=\frac{1}{u^2}+\sum_{m,n}{}'\left\{\frac{1}{(u-2m\omega_1-2n\omega_2)^2}-\frac{1}{(2m\omega_1+2n\omega_2)^2}\right\},$ SI 307(6)

where the symbol $\sum'$ means that the summation is made over all combinations of integers m and n except for the combination $m=n=0$; $2\omega_1$ and $2\omega_2$ are the periods of the function $\wp(u)$. Obviously,

2. $\wp(u+2m\omega_1+2n\omega_2)=\wp(u)$ and $\operatorname{Im}\left(\frac{\omega_1}{\omega_2}\right)\neq 0,$

3. $\frac{d}{du}\wp(u)=-2\sum_{m,n}\frac{1}{(u-2m\omega_1-2n\omega_2)^3},$

where the summation is made over all integral values of m and n.

The series 8.160 1. and 8.160 3. converge everywhere except at the poles, that is, at the points $2m\omega_1+2n\omega_2$ (where m and n are integers).

4. The function $\wp(u)$ is a *second-order periodic function* and has *one second-order pole* in a period parallelogram.

SI 306

8.161 The function $\wp(u)$ satisfies the differential equation

1. $$\left[\frac{d}{du}\wp(u)\right]^2 = 4\wp^3(u) - g_2\wp(u) - g_3,$$ SI 142, 310, WH

where

2. $$g_2 = 60 \sum_{m,\,n}{}' (m\omega_1 + n\omega_2)^{-4}; \quad g_3 = 140 \sum_{m,\,n}{}' (m\omega_1 + n\omega_2)^{-6}.$$

WH, SI 310

The functions g_2 and g_3 are called the *invariants* of the function $\wp(u)$.

8.162 $$u = \int_{\wp(u)}^{\infty} \frac{dz}{\sqrt{4z^3 - g_2 z - g_3}} = \int_{\wp(u)}^{\infty} \frac{dz}{\sqrt{4(z-e_1)(z-e_2)(z-e_3)}},$$

where e_1, e_2, and e_3 are the roots of the equation $4z^3 - g_2 z - g_3 = 0$; that is,

$$e_1 + e_2 + e_3 = 0, \quad e_1 e_2 + e_2 e_3 + e_3 e_1 = -\frac{g_2}{4}, \quad e_1 e_2 e_3 = \frac{g_3}{4}.$$ SI 142, 143, 144

8.163 $\wp(\omega_1) = e_1$, $\wp(\omega_1 + \omega_2) = e_2$, $\wp(\omega_2) = e_3$. Here, it is assumed that if e_1, e_2, and e_3 lie on a straight line in the complex plane, e_2 lies between e_1 and e_3.

8.164 The number $\Delta = g_2^3 - 27g_3^2$ is called the *discriminant* of the function $\wp(u)$. If $\Delta > 0$, all roots e_1, e_2, and e_3 of the equation $4z^3 - g_2 z - g_3 = 0$ (where g_2 and g_3 are real numbers) are *real*. In this case, the roots e_1, e_2, and e_3 are numbered in such a way that $e_1 > e_2 > e_3$.

1. If $\Delta > 0$, then

$$\omega_1 = \int_{e_1}^{\infty} \frac{dz}{\sqrt{4z^3 - g_2 z - g_3}}, \quad \omega_2 = i \int_{-\infty}^{e_3} \frac{dz}{\sqrt{g_3 + g_2 z - 4z^3}},$$

where ω_1 is real and ω_2 is a purely imaginary number. Here, the values of the radical in the integrand are chosen in such a way that ω_1 and $\frac{\omega_2}{i}$ will be positive.

2. If $\Delta < 0$, the root e_2 of the equation $4z^3 - g_2 z - g_3 = 0$ is *real* and the remaining two roots (e_1 and e_3) are *complex conjugates*. Suppose that $e_1 = \alpha + i\beta$, and $e_3 = \alpha - i\beta$. In this case, it is convenient to take

$$\omega' = \int_{e_1}^{\infty} \frac{dz}{\sqrt{4z^3 - g_2 z - g_3}} \quad \text{and} \quad \omega'' = \int_{e_3}^{\infty} \frac{dz}{\sqrt{4z^3 - g_2 z - g_3}}.$$

as basic semiperiods.

In the first integral, the integration is taken over a path lying entirely in the upper half-plane and in the second over a path lying entirely in the lower-half plane.

SI 151(21, 22)

8.165 Series representation:

1. $$\wp(u) = \frac{1}{u^2} + \frac{g_2 u^2}{4\cdot 5} + \frac{g_3 u^4}{4\cdot 7} + \frac{g_2^2 u^6}{2^4\cdot 3\cdot 5^2} + \frac{3 g_2 g_3 u^8}{2^4\cdot 5\cdot 7\cdot 11} + \dots$$ WH

8.166 Functional relations

1. $\wp(u)=\wp(-u),\ \wp'(u)=-\wp'(-u).$

2. $\wp(u+v)=-\wp(u)-\wp(v)+\frac{1}{4}\left[\frac{\wp'(u)-\wp'(v)}{\wp(u)-\wp(v)}\right]^2.$ SI 163(32)

8.167 $\wp(u;\ g_2,\ g_3)=\mu^2\wp\left(\mu u;\ \frac{g_2}{\mu^4},\ \frac{g_3}{\mu^6}\right)$ (the formula for homogeneity). SI 149(13)

The special case: $\mu=i$.

1. $\wp(u;\ g_2,\ g_3)=-\wp(iu;\ g_2,\ -g_3).$

8.168 An arbitrary elliptic function can be expressed in terms of the elliptic function $\wp(u)$ having the same periods as the original function and its derivative $\wp'(u)$. This expression is rational with respect to $\wp(u)$ and linear with respect to $\wp'(u)$.

8.169 A connection with the Jacobian elliptic functions. For $\Delta>0$ (see 8.164 1.).

1. $$\wp\left(\frac{u}{\sqrt{e_1-e_3}}\right)=e_1+(e_1-e_3)\frac{\mathrm{cn}^2(u;k)}{\mathrm{sn}^2(u;k)};$$

$$=e_2+(e_1-e_3)\frac{\mathrm{dn}^2(u;k)}{\mathrm{sn}^2(u;k)};$$

$$=e_3+(e_1-e_3)\frac{1}{\mathrm{sn}^2(u,k)};$$

SI 145(5), ZH 120 (197-199)a

2. $\omega_1=\frac{K}{\sqrt{e_1-e_3}},\quad \omega_2=\frac{iK'}{\sqrt{e_1-e_3}},$ SI 154(29)

where

3. $k=\sqrt{\frac{e_2-e_3}{e_1-e_3}},\quad k'=\sqrt{\frac{e_1-e_2}{e_1-e_3}}.$ SI 145(7)

For $\Delta<0$ (see 8.164 2.)

4. $\wp\left(\frac{u}{\sqrt[4]{9\alpha^2+\beta^2}}\right)=e_2+\sqrt{9\alpha^2+\beta^2}\,\frac{1+\mathrm{cn}(2u;k)}{1-\mathrm{cn}(2u;k)};$ SI 147(12)

5. $\omega'=\frac{K-iK'}{2\sqrt{9\alpha^2+\beta^2}},\quad \omega''=\frac{K+iK'}{\sqrt[4]{9\alpha^2+\beta^2}},$ SI 153(28)

where

6. $k=\sqrt{\frac{1}{2}-\frac{3e_2}{\sqrt{9\alpha^2+\beta^2}}};\quad k'=\sqrt{\frac{1}{2}+\frac{3e_2}{\sqrt{9\alpha^2+\beta^2}}}.$ SI 147

For $\Delta=0$, all the roots e_1, e_2, and e_3 are real and if $g_2g_3\neq 0$, two of them are equal to each other.

If $e_1=e_2\neq e_3$, then

7. $\wp(u)=\frac{3g_3}{g_2}-\frac{9g_3}{2g_2}\,\mathrm{cth}^2\left(u\sqrt{-\frac{9g_3}{2g_2}}\right).$ SI 148

If $e_1\neq e_2=e_3$, then

8. $\wp(u)=-\frac{3g_3}{2g_2}+\frac{9g_3}{2g_2}\,\frac{1}{\sin^2\left(u\sqrt{\frac{9g_3}{2g_2}}\right)}.$ SI 149

If $g_2=g_3=0$, then $e_1=e_2=e_3=0$, and

9. $\wp(u)=\frac{1}{u^2}.$ SI 149

8.17 The functions $\zeta(u)$ and $\sigma(u)$

8.171 Definitions:

1. $\zeta(u)=\frac{1}{u}-\int\limits_0^u\left(\wp(z)-\frac{1}{z^2}\right)dz.$ SI 181(45)

2. $\sigma(u)=u\exp\left\{\int\limits_0^u\left(\zeta(z)-\frac{1}{z}\right)dz\right\}.$ SI 181(46)

8.172 Series and infinite-product representation

1. $\zeta(u)=\frac{1}{u}+\sum{}'\left(\frac{1}{u-2m\omega_1-2n\omega_2}+\frac{1}{2m\omega_1+2n\omega_2}+\frac{u}{(2m\omega_1+2n\omega_2)^2}\right).$ SI 307(8)

2. $\sigma(u)=u\prod{}'\left(1-\frac{u}{2m\omega_1+2n\omega_2}\right)\exp\left\{\frac{u}{2m\omega_1+2n\omega_2}+\frac{u^2}{2(2m\omega_1+2n\omega_2)^2}\right\}.$ SI 308(9)

8.173

1. $\zeta(u)=u-\frac{g_2u^3}{2^2\cdot3\cdot5}-\frac{g_3u^5}{2^2\cdot5\cdot7}-\frac{g_2^2u^7}{2^4\cdot3\cdot5^2\cdot7}-\frac{3g_2g_3u^9}{2^4\cdot5\cdot7\cdot9\cdot11}-\dots$ SI 181(49)

2. $\sigma(u)=u-\frac{g_2u^5}{2^4\cdot3\cdot5}-\frac{g_3u^7}{2^3\cdot3\cdot5\cdot7}-\frac{g_2^2u^9}{2^9\cdot3^2\cdot5\cdot7}-\frac{g_2g_3u^{11}}{2^7\cdot3^2\cdot5^2\cdot7\cdot11}-\dots$ SI 181(50)

8.174 $\zeta(u)=\frac{\zeta(\omega_1)}{\omega_1}u+\frac{\pi}{2\omega_1}\operatorname{ctg}\frac{\pi u}{2\omega_1}+\frac{\pi}{2\omega_1}\sum_{n=1}^{\infty}\left\{\operatorname{ctg}\left(\frac{\pi u}{2\omega_1}+m\pi\frac{\omega_2}{\omega_1}\right)+\right.$

$\left.+\operatorname{ctg}\left(\frac{\pi u}{2\omega_1}-m\pi\frac{\omega_2}{\omega_1}\right)\right\};$ MO 154

$=\frac{\zeta(\omega_1)}{\omega_1}u+\frac{\pi}{2\omega_1}\operatorname{ctg}\frac{\pi u}{2\omega_1}+\frac{2\pi}{\omega_1}\sum_{n=1}^{\infty}\frac{q^{2n}}{1-q^{2n}}\sin\frac{\pi nu}{\omega_1}$ MO 155

Functional relations and properties

8.175 $\zeta(u)=-\zeta(-u),\quad \sigma(u)=-\sigma(-u).$ SI 181

8.176

1. $\zeta(u+2\omega_1)=\zeta(u)+2\zeta(\omega_1).$ SI 184(57)
2. $\zeta(u+2\omega_2)=\zeta(u)+2\zeta(\omega_2).$ SI 184(57)
3. $\sigma(u+2\omega_1)=-\sigma(u)\exp\{2(u+\omega_1)\zeta(\omega_1)\}.$ SI 185(60)
4. $\sigma(u+2\omega_2)=-\sigma(u)\exp\{2(u+\omega_2)\zeta(\omega_2)\}.$ SI 185(60)
5. $\omega_2\zeta(\omega_1)-\omega_1\zeta(\omega_2)=\frac{\pi}{2}i.$ SI 186(62)

8.177

1. $\zeta(u+v)-\zeta(u)-\zeta(v)=\frac{1}{2}\frac{\wp'(u)-\wp'(v)}{\wp(u)-\wp(v)}.$ SI 182(53)

2. $\wp(u)-\wp(v)=-\frac{\sigma(u-v)\sigma(u+v)}{\sigma^2(u)\sigma^2(v)}.$ SI 183(54)

3. $\zeta(u-v)+\zeta(u+v)-2\zeta(u)=\frac{\wp'(u)}{\wp(u)-\wp(v)}.$ SI 182(51)

8.178

1. $\zeta(u;\ \omega_1,\ \omega_2) = t\zeta(tu;\ t\omega_1,\ t\omega_2)$. MO 154

2. $\sigma(u;\ \omega_1,\ \omega_2) = t^{-1}\sigma(tu;\ t\omega_1,\ t\omega_2)$. MO 156

For the (indefinite) integrals of Weierstrass elliptic functions, see 5.14.

8.18-8.19 Theta functions

8.180 *Theta functions* are defined as the sums (for $|q| < 1$) of the following series:

1. $$\vartheta_4(u) = \sum_{n=-\infty}^{\infty} (-1)^n q^{n^2} e^{2nui} = 1 + 2\sum_{n=1}^{\infty} (-1)^n q^{n^2} \cos 2nu.$$ WH

2. $$\vartheta_1(u) = \frac{1}{i}\sum_{n=-\infty}^{\infty} (-1)^n q^{\left(n+\frac{1}{2}\right)^2} e^{(2n+1)ui} =$$
$$= 2\sum_{n=1}^{\infty} (-1)^{n+1} q^{\left(n-\frac{1}{2}\right)^2} \sin(2n-1)u.$$ WH

3. $$\vartheta_2(u) = \sum_{n=-\infty}^{\infty} q^{\left(n+\frac{1}{2}\right)^2} e^{(2n+1)ui} = 2\sum_{n=1}^{\infty} q^{\left(n-\frac{1}{2}\right)^2} \cos(2n-1)u.$$ WH

4. $$\vartheta_3(u) = \sum_{n=-\infty}^{\infty} q^{n^2} e^{2nui} = 1 + 2\sum_{n=1}^{\infty} q^{n^2} \cos 2nu.$$ WH

The notations $\vartheta(u, q)$, $\vartheta(u\,|\,\tau)$, where τ and q are related by $q = e^{i\pi\tau}$, are also used.

8.181 Representation of theta functions in terms of infinite products

1. $$\vartheta_4(u) = \prod_{n=1}^{\infty} (1 - 2q^{2n-1}\cos 2u + q^{2(2n-1)})(1 - q^{2n}).$$ SI 200(9), ZH 90(9)

2. $$\vartheta_3(u) = \prod_{n=1}^{\infty} (1 + 2q^{2n-1}\cos 2u + q^{2(2n-1)})(1 - q^{2n}).$$ SI 200(9), ZH 90(9)

3. $$\vartheta_1(u) = 2\sqrt[4]{q}\sin u \prod_{n=1}^{\infty} (1 - 2q^{2n}\cos 2u + q^{4n})(1 - q^{2n}).$$ SI 200(9), ZH 90(9)

4. $$\vartheta_2(u) = 2\sqrt{q}\cos u \prod_{n=1}^{\infty} (1 + 2q^{2n}\cos 2u + q^{4n})(1 - q^{2n}).$$ SI 200(9), ZH 90(9)

Functional relations and properties

8.182 Quasiperiodicity. Suppose that $q = e^{\pi\tau i}$ ($\operatorname{Im}\tau > 0$). Then, theta functions that are periodic functions of u are called *quasiperiodic functions* of τ and u. This property follows from the equations

1. $\vartheta_4(u + \pi) = \vartheta_4(u)$. SI 200(10)

2. $\vartheta_4(u + \tau\pi) = -\dfrac{1}{q} e^{-2iu}\vartheta_4(u)$. SI 200(10)

3. $\vartheta_1(u+\pi) = -\vartheta_1(u).$ SI 200(10)

4. $\vartheta_1(u+\tau\pi) = -\frac{1}{q} e^{-2iu} \vartheta_1(u).$ SI 200(10)

5. $\vartheta_2(u+\pi) = -\vartheta_2(u).$ SI 200(10)

6. $\vartheta_2(u+\tau\pi) = \frac{1}{q} e^{-2iu} \vartheta_2(u).$ SI 200(10)

7. $\vartheta_3(u+\pi) = \vartheta_3(u).$ SI 200(10)

8. $\vartheta_3(u+\tau\pi) = \frac{1}{q} e^{-2iu} \vartheta_3(u).$ SI 200(10)

8.183

1. $\vartheta_4\left(u+\frac{1}{2}\pi\right) = \vartheta_3(u).$ WH

2. $\vartheta_1\left(u+\frac{1}{2}\pi\right) = \vartheta_2(u).$ WH

3. $\vartheta_2\left(u+\frac{1}{2}\pi\right) = -\vartheta_1(u).$ WH

4. $\vartheta_3\left(u+\frac{1}{2}\pi\right) = \vartheta_4(u).$ WH

5. $\vartheta_4\left(u+\frac{1}{2}\pi\tau\right) = iq^{-\frac{1}{4}} e^{-iu} \vartheta_1(u).$ WH

6. $\vartheta_1\left(u+\frac{1}{2}\pi\tau\right) = iq^{-\frac{1}{4}} e^{-iu} \vartheta_4(u).$ WH

7. $\vartheta_2\left(u+\frac{1}{2}\pi\tau\right) = q^{-\frac{1}{4}} e^{-iu} \vartheta_3(u).$ WH

8. $\vartheta_3\left(u+\frac{1}{2}\pi\tau\right) = q^{-\frac{1}{4}} e^{-iu} \vartheta_2(u).$ WH

8.184 Even and odd theta functions

1. $\vartheta_1(-u) = -\vartheta_1(u)$ WH
2. $\vartheta_2(-u) = \vartheta_2(u)$ WH
3. $\vartheta_3(-u) = \vartheta_3(u)$ WH
4. $\vartheta_4(-u) = \vartheta_4(u).$ WH

8.185 $\vartheta_4^4(u) + \vartheta_2^4(u) = \vartheta_1^4(u) + \vartheta_3^4(u).$ WH

8.186 Considering the theta functions as functions of two independent variables u and τ, we have

$$\pi i \frac{\partial^2 \vartheta_k(u \mid \tau)}{\partial u^2} + 4 \frac{\partial \vartheta_k(u \mid \tau)}{\partial \tau} = 0 \qquad [k = 1, 2, 3, 4].$$ WH

8.187 We denote the partial derivatives of the theta functions with respect to u by a prime and consider them as functions of the single argument u. Then,

1. $\vartheta_1'(0) = \vartheta_2(0)\vartheta_3(0)\vartheta_4(0).$ WH

2. $\frac{\vartheta_1'''(0)}{\vartheta_1'(0)} = \frac{\vartheta_2''(0)}{\vartheta_2(0)} + \frac{\vartheta_3''(0)}{\vartheta_3(0)} + \frac{\vartheta_4''(0)}{\vartheta_4(0)}.$ WH

8.188 $\vartheta_1(u)\vartheta_2(u)\vartheta_3(u)\vartheta_4(u) = \frac{1}{2}\vartheta_1(2u)\vartheta_2(0)\vartheta_3(0)\vartheta_4(0).$ WH

8.189 The zeros of the theta functions:

1. $\vartheta_4(u)=0$ for $u=2m\frac{\pi}{2}+(2n-1)\frac{\pi\tau}{2}$. SI 201

2. $\vartheta_1(u)=0$ for $u=2m\frac{\pi}{2}+2u\frac{\pi\tau}{2}$. SI 201

3. $\vartheta_2(u)=0$ for $u=(2m-1)\frac{\pi}{2}+2n\frac{\pi\tau}{2}$. SI 201

4. $\vartheta_3(u)=0$ for $u=(2m-1)\frac{\pi}{2}+(2n-1)\frac{\pi\tau}{2}$ SI 201

[m and n — integers].

For integrals of theta functions, see 6.16.

8.191 Connections with the Jacobian elliptic functions:

For $\tau=i\frac{K'}{K}$, i.e. for $q=\exp\left(-\pi\frac{K'}{K}\right)$,

1. $$\operatorname{sn} u=\frac{1}{\sqrt{k}}\frac{\vartheta_1\left(\frac{\pi u}{2K}\right)}{\vartheta_4\left(\frac{\pi u}{2K}\right)}=\frac{1}{\sqrt{k}}\frac{H(u)}{\Theta(u)}.$$ SI 206(22), SI 209(35)

2. $$\operatorname{cn} u=\sqrt{\frac{k'}{k}}\frac{\vartheta_2\left(\frac{\pi u}{2K}\right)}{\vartheta_4\left(\frac{\pi u}{2K}\right)}=\sqrt{\frac{k'}{k}}\frac{H_1(u)}{\Theta(u)}.$$ SI 207(23), SI 209(35)

3. $$\operatorname{dn} u=\sqrt{k'}\frac{\vartheta_3\left(\frac{\pi u}{2K}\right)}{\vartheta_4\left(\frac{\pi u}{2K}\right)}=\sqrt{k'}\frac{\Theta_1(u)}{\Theta(u)}.$$ SI 207(24), SI 209(35)

8.192 Series representation of the functions H, H_1, Θ, Θ_1.

1. $$\Theta(u)=\vartheta_4\left(\frac{\pi u}{2K}\right)=1+2\sum_{n=1}^{\infty}(-1)^n q^{n^2}\cos\frac{n\pi u}{K}.$$ SI 207(25), SI 212(42)

2. $$H(u)=\vartheta_1\left(\frac{\pi u}{2K}\right)=2\sum_{n=1}^{\infty}(-1)^{n+1}\sqrt[4]{q^{(2n+1)^2}}\sin(2n-1)\frac{\pi u}{2K}.$$ SI 207(25), SI 212(43)

3. $$\Theta_1(u)=\vartheta_3\left(\frac{\pi u}{2K}\right)=1+2\sum_{n=1}^{\infty}q^{n^2}\cos\frac{n\pi u}{K}.$$ SI 207(25), SI 212(45)

4. $$H_1(u)=\vartheta_2\left(\frac{\pi u}{2K}\right)=2\sum_{n=1}^{\infty}\sqrt[4]{q^{(2n-1)^2}}\cos(2n-1)\frac{\pi u}{2K}.$$ SI 207(25), SI 212(44)

In formulas 8.192 $q=\exp\left(-\pi\frac{K'}{K}\right)$.

8.193 Connections with the Weierstrass elliptic functions

1. $$\wp(u)=e_1+\left[\frac{H_1(u\sqrt{\lambda})H'(0)}{H_1(0)H(u\sqrt{\lambda})}\right]^2\lambda=e_2+\left[\frac{\Theta_1(u\sqrt{\lambda})H'(0)}{\Theta_1(0)H(u\sqrt{\lambda})}\right]^2\lambda=$$
$$=e_3+\left[\frac{\Theta(u\sqrt{\lambda})H'(0)}{\Theta(0)H(u\sqrt{\lambda})}\right]^2\lambda.$$ SI 235(77, 78)

2. $\zeta(u) = \frac{\eta_1 u}{\omega_1} + \sqrt{\lambda}\frac{H'(u\sqrt{\lambda})}{H(u\sqrt{\lambda})}$. SI 234(73)

3. $\sigma(u) = \frac{1}{\sqrt{\lambda}}\exp\left(\frac{\eta_1 u^2}{2\omega_1}\right)\frac{H(u\sqrt{\lambda})}{H'(0)}$, SI 234(72)

where

$$\lambda = e_1 - e_3;\quad \eta_1 = \zeta(\omega_1) = -\frac{\omega_1\lambda}{3}\frac{H'''(0)}{H'(0)}.$$ SI 236

8.194 The connection with elliptic integrals:

1. $E(u,k) = u - u\frac{\Theta''(0)}{\Theta(0)} + \frac{\Theta'(u)}{\Theta(u)}$. SI 228(65)

2. $\Pi(u, -k^2\sin^2 a, k) = \int\limits_0^u \frac{d\varphi}{1 - k^2\sin^2 a\,\mathrm{sn}^2\varphi} =$

$$= u + \frac{\mathrm{sn}\,a}{\mathrm{cn}\,a\,\mathrm{dn}\,a}\left[\frac{\Theta'(a)}{\Theta(a)}u + \frac{1}{2}\ln\frac{\Theta(u-a)}{\Theta(u+a)}\right].$$ SI 232(69)

q-series and products $\left[q = \exp\left(-\pi\frac{\boldsymbol{K}'}{\boldsymbol{K}}\right)\right]$

8.195 $\frac{\pi}{2}\left[1 + 2\sum_{n=1}^{\infty} q^{n^2}\right]^2 = \boldsymbol{K} = \frac{\pi}{2}\Theta^2(\boldsymbol{K})$ (cf. 8.197 1.). SI 219

8.196 $\boldsymbol{E} = \boldsymbol{K} - \boldsymbol{K}\frac{\Theta''(0)}{\Theta(0)} = \boldsymbol{K} - \frac{2\pi^2}{\boldsymbol{K}}\frac{\sum_{n=1}^{\infty}(-1)^{n+1}n^2 q^{n^2}}{1 + 2\sum_{n=1}^{\infty}(-1)^n q^{n^2}}$. SI 230(67)

8.197

1. $1 + 2\sum_{n=1}^{\infty} q^{n^2} = \sqrt{\frac{2\boldsymbol{K}}{\pi}} = \vartheta_3(0)$ (cf. 8.195). WH

2. $\sum_{n=1}^{\infty} q^{\left(\frac{2n-1}{2}\right)^2} = \sqrt{\frac{k\boldsymbol{K}}{2\pi}} = \frac{1}{2}\vartheta_2(0)$. WH

3. $4\sqrt{q}\prod_{n=1}^{\infty}\left(\frac{1+q^{2n}}{1+q^{2n-1}}\right)^4 = k$. SI 206(17, 18)

4. $\prod_{n=1}^{\infty}\left(\frac{1-q^{2n-1}}{1+q^{2n-1}}\right)^4 = k'$ SI 206(19, 20)

5. $2\sqrt[4]{q}\prod_{n=1}^{\infty}\left(\frac{1-q^{2n}}{1-q^{2n-1}}\right)^2 = 2\sqrt{k}\,\frac{\boldsymbol{K}}{\pi}$. WH

6. $\prod_{n=1}^{\infty}\left(\frac{1-q^{2n}}{1+q^{2n}}\right)^2 = 2\sqrt{k'}\,\frac{\boldsymbol{K}}{\pi}$. WH

8.198

1. $\lambda=\frac{1}{2}\frac{1-\sqrt{k'}}{1+\sqrt{k'}}=\frac{\sum\limits_{n=0}^{\infty} q^{(2n+1)^2}}{1+2\sum\limits_{n=1}^{\infty} q^{4n^2}}$ $\left[\text{for } 0<k<1, \text{ we have } 0<\lambda<\frac{1}{2}\right]$ WH

The series

2. $q=\lambda+2\lambda^5+15\lambda^9+150\lambda^{13}+1707\lambda^{17}+\ldots$ is used to determine q from the given modulus k. WH

8.2 The Exponential-Integral Function and Functions Generated by It

8.21 The exponential-integral function Ei (x)

8.211

1. $\operatorname{Ei}(x)=-\int\limits_{-x}^{\infty}\frac{e^{-t}}{t}dt=\int\limits_{-\infty}^{x}\frac{e^{t}}{t}dt=\operatorname{li}(e^{x})$ $[x<0]$.

2. $\operatorname{Ei}(x)=-\lim\limits_{\varepsilon\to+0}\left[\int\limits_{-x}^{-\varepsilon}\frac{e^{-t}}{t}dt+\int\limits_{\varepsilon}^{\infty}\frac{e^{-t}}{t}dt\right]$ $[x>0]$.

8.212

1. $\operatorname{Ei}(-x)=C+\ln x+\int\limits_{0}^{x}\frac{e^{-t}-1}{t}dt$ $[x>0]$; NT 11(1)

$=C+e^{-x}\ln x+\int\limits_{0}^{x}e^{-t}\ln t\,dt$ $[x>0]$. NT 11(10)

2. $\operatorname{Ei}(x)=e^{x}\left[\frac{1}{x}+\int\limits_{0}^{\infty}\frac{e^{-t}\,dt}{(x-t)^2}\right]$ $[x>0]$ (cf. 8.211 1.).

3. $\operatorname{Ei}(-x)=e^{-x}\left[-\frac{1}{x}+\int\limits_{0}^{\infty}\frac{e^{-t}\,dt}{(x+t)^2}\right]$ $[x>0]$ (cf. 8.211 1.).

LA 281(28)

4. $\operatorname{Ei}(\pm x)=\pm e^{\pm x}\int\limits_{0}^{1}\frac{dt}{x\pm\ln t}$ $[x>0]$ (cf. 8.211 1.).

5. $\operatorname{Ei}(\pm xy)=\pm e^{\pm xy}\int\limits_{0}^{\infty}\frac{e^{-xt}}{y\mp t}dt$ $[\operatorname{Re} y>0,\ x>0]$. NT 19(11)

6. $\operatorname{Ei}(\pm x)=-e^{\pm x}\int\limits_{0}^{\infty}\frac{e^{-it}}{t\pm ix}dt$ $[x>0]$. NT 23(2, 3)

7. $\mathrm{Ei}(xy) = e^{xy}\int_0^1 \frac{t^{y-1}}{x+\ln t}\,dt;$ LA 282(44)a

$= x^{-1}e^{xy}\left[\int_0^1 \frac{t^{x-1}}{(y+\ln t)^2}\,dt + y^{-1}\right] \quad [x>0,\ y>0].$ LA 283(46)a

8. $\mathrm{Ei}(-xy) = -e^{-xy}\int_0^1 \frac{t^{y-1}}{x-\ln t}\,dt;$ LA 282(45)a

$= x^{-1}e^{-xy}\left[\int_0^1 \frac{t^{x-1}}{(y-\ln t)^2}\,dt - y^{-1}\right] \quad [x>0,\ y>0].$ LA 283(47)a

9. $\mathrm{Ei}(x) = e^{x}\int_1^\infty \frac{1}{x-\ln t}\frac{dt}{t^2} \quad [x>0].$ LA 283(48)

10. $\mathrm{Ei}(-x) = -e^{-x}\int_1^\infty \frac{1}{x+\ln t}\frac{dt}{t^2} \quad [x>0].$ LA 283(48)

11. $\mathrm{Ei}(-x) = -e^{-x}\int_0^\infty \frac{t\cos t + x\sin t}{t^2+x^2}\,dt \quad [x>0].$ NT 23(6)

12. $\mathrm{Ei}(-x) = -e^{-x}\int_0^\infty \frac{t\cos t - x\sin t}{t^2+x^2}\,dt \quad [x<0].$ NT 23(6)

13. $\mathrm{Ei}(-x) = \frac{2}{\pi}\int_0^\infty \frac{\cos t}{t}\operatorname{arctg}\frac{t}{x}\,dt \quad [\mathrm{Re}\,x>0].$ NT 25(13)

14. $\mathrm{Ei}(-x) = \frac{2e^{-x}}{\pi}\int_0^\infty \frac{x\cos t - t\sin t}{t^2+x^2}\ln t\,dt \quad [x>0].$ NT 26(7)

15. $\mathrm{Ei}(x) = 2\ln x - \frac{2e^{x}}{\pi}\int_0^\infty \frac{x\cos t + t\sin t}{t^2+x^2}\ln t\,dt \quad [x>0].$ NT 27(8)

16. $\mathrm{Ei}(-x) = -x\int_1^\infty e^{-tx}\ln t\,dt \quad [x>0].$ NT 32(12)

See also **3.327**, **3.881** 8., **3.916** 2. and 3., **4.326** 1., **4.326** 2., **4.331** 2., **4.351** 3., **4.425** 3., **4.581**.

For integrals of the exponential-integral function, see **6.22**–**6.23**, **6.78**.

Series and asymptotic representations

8.213

1. $\mathrm{li}(x) = C + \ln(-\ln x) + \sum_{k=1}^{\infty}\frac{(\ln x)^k}{k\cdot k!} \quad [0<x<1].$ NT 3(9)

2. $\mathrm{li}(x) = C + \ln\ln x + \sum_{k=1}^{\infty}\frac{(\ln x)^k}{k\cdot k!} \quad [x>1].$ NT 3(10)

8.214

1. $$\mathrm{Ei}(x) = C + \ln(-x) + \sum_{k=1}^{\infty} \frac{x^k}{k \cdot k!} \qquad [x < 0].$$

2. $$\mathrm{Ei}(x) = C + \ln x + \sum_{k=1}^{\infty} \frac{x^k}{k \cdot k!} \qquad [x > 0].$$

3. $$\mathrm{Ei}(x) - \mathrm{Ei}(-x) = 2x \sum_{k=0}^{\infty} \frac{x^{2k}}{(2k+1)(2k+1)!} \qquad [x > 0].$$ NT 39(13)

8.215 $$\mathrm{Ei}(-x) = e^{-x} \sum_{k=1}^{n} (-1)^k \frac{(k-1)!}{x^k} + R_n,$$

where

$$|R_n| < \frac{n!}{|x|^{n+1} \cos\frac{\varphi}{2}}, \qquad x = |x| e^{i\varphi}, \ \varphi^2 < \pi^2.$$ NT 37(9)

8.216 $$\mathrm{Ei}(nx) - \mathrm{Ei}(-nx) = e^{nx'} \left(\frac{1}{nx} + \frac{1}{n^2x^2} + \frac{k_n}{n^3x^3} \right),$$

where

$x' = x \operatorname{sign} \operatorname{Re}(x)$, $k_n = O(n^0)$, and n large. NT 39(15)

8.217 Functional relations:

1. $$e^{x'} \mathrm{Ei}(-x') - e^{-x'} \mathrm{Ei}(x') = -2 \int_0^{\infty} \frac{x' \sin t}{t^2+x^2}\, dt =$$

$$= \frac{4}{\pi} \int_0^{\infty} \frac{x' \cos t}{t^2+x^2} \ln t\, dt - 2e^{-x'} \ln x' \qquad [x' = x \operatorname{sign} \operatorname{Re} x].$$ NT 24(11) NT 27(9)

2. $$e^{x'} \mathrm{Ei}(-x') + e^{-x'} \mathrm{Ei}(x') = -2 \int_0^{\infty} \frac{t \cos t}{t^2+x^2}\, dt =$$

$$= 2e^{-x'} \ln x' - \frac{4}{\pi} \int_0^{\infty} \frac{t \sin t}{t^2+x^2} \ln t\, dt \qquad [x' = x \operatorname{sign} \operatorname{Re} x].$$

NT 24(10), NT 27(10)

3. $$\mathrm{Ei}(-x) - \mathrm{Ei}\left(-\frac{1}{x}\right) = \frac{2}{\pi} \int_0^{\infty} \frac{\cos t}{t} \operatorname{arctg} \frac{t\left(x - \frac{1}{x}\right)}{1+t^2}\, dt$$

$[\operatorname{Re} x > 0]$. NT 25(14)

4. $\mathrm{Ei}(-\alpha x)\,\mathrm{Ei}(-\beta x) - \ln(\alpha\beta)\,\mathrm{Ei}[-(\alpha+\beta)x] =$

$$= e^{-(\alpha+\beta)x} \int_0^\infty \frac{e^{-tx} \ln[(\alpha+t)(\beta+t)]}{t+\alpha+\beta}\, dt.$$ NT 32(9)

See also **3.723** 1. and 5., **3.742** 2. and 4., **3.824** 4., **4.573** 2..

For a connection with a degenerate hypergeometric function, see **9.237**.

For integrals of the exponential-integral function, see **5.21**, **5.22**, **5.23**, **6.22**, and **6.23**.

8.218 Two numerical values:

1. $\mathrm{Ei}(-1) = -0.219\ 383\ 934\ 395\ 520\ 273\ 665 \ldots$ NT 89

2. $\mathrm{Ei}(1) = 1.895\ 117.816\ 355\ 936\ 755\ 478 \ldots$ NT 89

8.22 The hyperbolic-sine-integral shi x and the hyperbolic-cosine-integral chi x

8.221

1. $$\mathrm{shi}\, x = \int_0^x \frac{\mathrm{sh}\, t}{t}\, dt = -i\left[\frac{\pi}{2} + \mathrm{si}(ix)\right]$$ (see 8.230 1.). EH II 146(17)

2. $$\mathrm{chi}\, x = C + \ln x + \int_0^x \frac{\mathrm{ch}\, t - 1}{t}\, dt.$$ EH II 146(18)

8.23 The sine integral and the cosine integral: si (x) and ci (x)

8.230

1. $$\mathrm{si}(x) = -\int_x^\infty \frac{\sin t}{t}\, dt = -\frac{\pi}{2} + \int_0^x \frac{\sin t}{t}\, dt.$$ NT 11(3)

2. $$\mathrm{ci}(x) = -\int_x^\infty \frac{\cos t}{t}\, dt = C + \ln x + \int_0^x \frac{\cos t - 1}{t}\, dt.$$ NT 11(2)

8.231

1. $$\mathrm{si}(xy) = -\int_x^\infty \frac{\sin ty}{t}\, dt.$$ NT 18(7)

2. $$\mathrm{ci}(xy) = -\int_x^\infty \frac{\cos ty}{t}\, dt.$$ NT 18(6)

3. $$\mathrm{si}(x) = -\int_0^{\frac{\pi}{2}} e^{-x\cos t} \cos(x \sin t)\, dt.$$ NT 13(26)

8.232

1. $$\mathrm{si}(x) = -\frac{\pi}{2} + \sum_{k=1}^\infty \frac{(-1)^{k+1} x^{2k-1}}{(2k-1)(2k-1)!}.$$ NT 7(4)

2. $\operatorname{ci}(x) = C + \ln(x) + \sum_{k=1}^{\infty} (-1)^k \frac{x^{2k}}{2k(2k)!}$. NT 7(3)

8.233

1. $\operatorname{ci}(x) \pm i \operatorname{si}(x) = \operatorname{Ei}(\pm ix)$. NT 6a

2. $\operatorname{ci}(x) - \operatorname{ci}(xe^{\pm\pi i}) = \mp \pi i$. NT 7(5)

3. $\operatorname{si}(x) + \operatorname{si}(-x) = -\pi$. NT 7(7)

8.234

1. $\operatorname{Ei}(-x) - \operatorname{ci}(x) = \int_0^{\frac{\pi}{2}} e^{-x\cos\varphi} \sin(x\sin\varphi)\, d\varphi$. NT 13(27)

2. $[\operatorname{ci}(x)]^2 + [\operatorname{si}(x)]^2 = -2 \int_0^{\frac{\pi}{2}} \frac{\exp(-x \operatorname{tg}\varphi) \ln\cos\varphi}{\sin\varphi\cos\varphi}\, d\varphi$

$[\operatorname{Re} x > 0]$ (see also 4.366). NT 32(11)

See also 3.341, 3.351 1. and 2., 3.354 1 and 2., 3.721 2. and 3., 3.722 1., 3., 5. and 7., 3.723 8. and 11., 4.338 1., 4.366 1..

8.235

1. $\lim_{x\to+\infty} (x^{\varrho} \operatorname{si}(x)) = 0, \quad \lim_{x\to+\infty} (x^{\varrho} \operatorname{ci}(x)) = 0 \quad [\varrho < 1]$. NT 38(5)

2. $\lim_{x\to-\infty} \operatorname{si}(x) = -\pi, \quad \lim_{x\to-\infty} \operatorname{ci}(x) = \pm \pi i$. NT 38(6)

For integrals of the sine integral and cosine integral, see 6.24–6.26, 6.781, 6.782, and 6.783.

For indefinite integrals of the sine-integral and cosine-integral, see 5.3.

8.24 The logarithm-integral $\operatorname{li}(x)$

8.240

1. $\operatorname{li}(x) = \int_0^x \frac{dt}{\ln t} = \operatorname{Ei}(\ln x) \quad [x < 1]$. JA

2. $\operatorname{li}(x) = \lim_{\varepsilon\to 0} \left[\int_0^{1-\varepsilon} \frac{dt}{\ln t} + \int_{1+\varepsilon}^{x} \frac{dt}{\ln t} \right] = \operatorname{Ei}(\ln x) \quad [x > 1]$. JA

3. $\operatorname{li}\{\exp(-xe^{\pm\pi i})\} = \operatorname{Ei}(-xe^{\pm i\pi}) = \operatorname{Ei}(x \mp i0) = \operatorname{Ei}(x) \pm i\pi =$
$= \operatorname{li}(e^x) \pm i\pi \quad [x > 0]$. JA, NT 2(6)

Integral representations

8.241

1. $\operatorname{li}(x) = \int_{-\infty}^{\ln x} \frac{e^t}{t}\, dt = x \ln\ln\frac{1}{x} - \int_{-\ln x}^{\infty} e^{-t} \ln t\, dt \quad [x > 1]$. LA 281(33)

2. $\operatorname{li}(x) = x \int_0^1 \frac{dt}{\ln x + \ln t}$; LA 280(22)

$$= \frac{x}{\ln x} + x \int_0^1 \frac{dt}{(\ln x + \ln t)^2};$$ LA 280(29)

$$= x \int_1^\infty \frac{1}{\ln x - \ln t} \frac{dt}{t^2} \qquad [x < 1].$$ LA 280(30)

3. $\operatorname{li}(a^x) = \frac{1}{\ln a} \int_{-\infty}^x \frac{a^t}{t} dt \qquad [x > 0].$

For integrals of the logarithm integral, see **6.21**

8.25 The probability integral and Fresnel integrals $\Phi(x)$, $S(x)$ and $C(x)$

8.250 Definition:

1. $\Phi(x) = \frac{2}{\sqrt{\pi}} \int_0^x e^{-t^2} dt.$

2. $S(x) = \frac{2}{\sqrt{2\pi}} \int_0^x \sin t^2 dt.$

3. $C(x) = \frac{2}{\sqrt{2\pi}} \int_0^x \cos t^2 dt.$

Integral representations

8.251

1. $\Phi(x) = \frac{1}{\sqrt{\pi}} \int_0^{x^2} \frac{e^{-t}}{\sqrt{t}} dt$ (see also **3.361** 1.).

2. $S(x) = \frac{1}{\sqrt{2\pi}} \int_0^{x^2} \frac{\sin t}{\sqrt{t}} dt.$

3. $C(x) = \frac{1}{\sqrt{2\pi}} \int_0^{x^2} \frac{\cos t}{\sqrt{t}} dt.$

8.252

1. $\Phi(xy) = \frac{2y}{\sqrt{\pi}} \int_0^x e^{-t^2y^2} dt.$

2. $S(xy) = \frac{2y}{\sqrt{2\pi}} \int_0^x \sin(t^2y^2) dt.$

3. $C(xy) = \frac{2y}{\sqrt{2\pi}} \int_0^x \cos(t^2y^2) dt.$

4. $$\left.\begin{aligned}\Phi(xy) &= 1 - \frac{2}{\sqrt{\pi}} e^{-x^2y^2} \int_0^\infty \frac{e^{-t^2y^2} ty\, dt}{\sqrt{t^2+x^2}} \\ &= 1 - \frac{2x}{\pi} e^{-x^2y^2} \int_0^\infty \frac{e^{-t^2y^2}\, dt}{t^2+x^2}\end{aligned}\right\} [\operatorname{Re} y^2 > 0].$$ NT 19(11)a, NT 19(13)a

5. $$\Phi\left(\frac{-y}{2xi}\right) - \Phi\left(\frac{y}{2xi}\right) = \frac{4xie^{\frac{y^2}{4x^2}}}{\sqrt{\pi}} \int_0^\infty e^{-t^2x^2} \sin(ty)\, dt \qquad [\operatorname{Re} x^2 > 0].$$ NT 28(3)a

6. $$\Phi\left(\frac{y}{2x}\right) = 1 - \frac{2}{\sqrt{\pi}} xe^{-\frac{y^2}{4x^2}} \int_0^\infty e^{-t^2x^2-ty}\, dt \qquad [\operatorname{Re} x^2 > 0].$$ NT 27(1)a

See also 3.322, 3.362 2., 3.363, 3.468, 3.897, 6.511 4. and 5.

8.253 Series representations:

1. $$\Phi(x) = \frac{2}{\sqrt{\pi}} \sum_{k=1}^{\infty} (-1)^{k+1} \frac{x^{2k-1}}{(2k-1)(k-1)!};$$ NT 7(9)a

$$= \frac{2}{\sqrt{\pi}} e^{-x^2} \sum_{k=0}^{\infty} \frac{2^k x^{2k+1}}{(2k+1)!!}.$$ NT 10(11)a

2. $$S(x) = \frac{2}{\sqrt{2\pi}} \sum_{k=0}^{\infty} \frac{(-1)^k x^{4k+3}}{(2k+1)!\,(4k+3)};$$ NT 8(14)a

$$= \frac{2}{\sqrt{2\pi}} \left\{\sin x^2 \sum_{k=0}^{\infty} \frac{(-1)^k 2^{2k} x^{4k+1}}{(4k+1)!!} - \cos x^2 \sum_{k=0}^{\infty} \frac{(-1)^k 2^{2k+1} x^{4k+3}}{(4k+3)!!}\right\}.$$ NT 10(13)a

3. $$C(x) = \frac{2}{\sqrt{2\pi}} \sum_{k=0}^{\infty} \frac{(-1)^k x^{4k+1}}{(2k)!\,(4k+1)};$$ NT 8(13)a

$$= \frac{2}{\sqrt{2\pi}} \left\{\sin x^2 \sum_{k=0}^{\infty} \frac{(-1)^k 2^{2k+1} x^{4k+3}}{(4k+3)!!} + \cos x^2 \sum_{k=0}^{\infty} \frac{(-1)^k 2^{2k} x^{4k+1}}{(4k+1)!!}\right\}$$ NT 10(12)a

For the expansions in Bessel functions, see 8.515 2., 8.515 3.

Asymptotic representations

8.254 $$\Phi(\sqrt{x}) = 1 - \frac{1}{\pi} e^{-x} \sum_{k=0}^{n-1} \frac{(-1)^k \Gamma\left(k+\frac{1}{2}\right)}{x^{k+\frac{1}{2}}} + \frac{e^{-x}}{\pi} R_n,$$

where $$|R_n| < \frac{\Gamma\left(n+\frac{1}{2}\right)}{|x|^{n+\frac{1}{2}} \cos\frac{\varphi}{2}}, \quad x = |x| e^{i\varphi} \text{ и } \varphi^2 < \pi^2.$$ NT 37(10)a

8.255

1. $S(x)=\frac{1}{2}-\frac{1}{\sqrt{2\pi x}}\cos x^2+O\left(\frac{1}{x^2}\right) \quad [x\to\infty].$ MO 127a

2. $C(x)=\frac{1}{2}+\frac{1}{\sqrt{2\pi}\,x}\sin x^2+O\left(\frac{1}{x^2}\right) \quad [x\to\infty].$ MO 127a

8.256 Functional relations:

1. $$C(z)+iS(z)=\sqrt{\frac{i}{2}}\Phi\left(\frac{z}{\sqrt{i}}\right)=\frac{2}{\sqrt{2\pi}}\int_0^z e^{it^2}dt.$$

2. $$C(z)-iS(z)=\frac{1}{\sqrt{2i}}\Phi(z\sqrt{i})=\frac{2}{\sqrt{2\pi}}\int_0^z e^{-it^2}\,dt.$$

3. $$[\cos u^2C(u)+\sin u^2S(u)]=$$
$$=\frac{1}{2}[\cos u^2+\sin u^2]-\sqrt{\frac{2}{\pi}}\int_0^\infty e^{-2ut}\sin t^2\,dt \qquad [\operatorname{Re} u>0].$$
NT 28(6)a

4. $$[\cos u^2\,S(u)-\sin u^2C(u)]=$$
$$=\frac{1}{2}[\cos u^2-\sin u^2]-\sqrt{\frac{2}{\pi}}\int_0^\infty e^{-2ut}\cos t^2\,dt \qquad [\operatorname{Re} u>0].$$
NT 28(5)a

5. $$\left[C(x)-\frac{1}{2}\right]^2+\left[S(x)-\frac{1}{2}\right]^2=\frac{2}{\pi}\int_0^{\frac{\pi}{2}}\frac{\exp(-x^2\operatorname{tg}\varphi)\sin\frac{\varphi}{2}\sqrt{\cos\varphi}}{\sin 2\varphi}\,d\varphi.$$

See also 6.322. NT 33(18)a

For a connection with a degenerate hypergeometric function, see 9.236.

For a connection with a parabolic-cylinder function, see 9.254.

8.257

1. $\lim\limits_{x\to+\infty}\left(x^{\varrho}\left[S(x)-\frac{1}{2}\right]\right)=0 \quad [\varrho<1].$ NT 38(11)

2. $\lim\limits_{x\to+\infty}\left(x^{\varrho}\left[C(x)-\frac{1}{2}\right]\right)=0 \quad [\varrho<1].$ NT 38(11)

3. $\lim\limits_{x\to+\infty}S(x)=\frac{1}{2}.$ NT 38(12)a

4. $\lim\limits_{x\to+\infty}C(x)=\frac{1}{2}.$ NT 38(12)a

For integrals of the probability integral, see 6.28 – 6.31.

For integrals of Fresnel's sine-integral and cosine-integral, see 6.32.

8.26 Lobachevskiy's function $L(x)$

8.260 Definition:

$$L(x) = -\int_0^x \ln \cos t \, dt.$$ LO III 184(10)

For integral representations of the function $L(x)$, see also 3.531 8., 3.532 2., 3.533, and 4.224.

8.261 Representation in the form of a series:

$$L(x) = x \ln 2 - \frac{1}{2} \sum_{k=1}^{\infty} (-1)^{k-1} \frac{\sin 2kx}{k^2}.$$ LO III 185(11)

8.262 Functional relationships:

1. $L(-x) = -L(x) \quad \left[-\frac{\pi}{2} \leqslant x \leqslant \frac{\pi}{2}\right].$ LO III 185(13)

2. $L(\pi - x) = \pi \ln 2 - L(x).$ LO III 286

3. $L(\pi + x) = \pi \ln 2 + L(x).$ LO III 286

4. $L(x) - L\left(\frac{\pi}{2} - x\right) = \left(x - \frac{\pi}{4}\right) \ln 2 - \frac{1}{2} L\left(\frac{\pi}{2} - 2x\right) \quad \left[0 \leqslant x < \frac{\pi}{4}\right].$ LO III 186(14)

8.3 Euler's Integrals of the First and Second Kinds and Functions Generated by Them

8.31 The gamma function (Euler's integral of the second kind): $\Gamma(z)$

8.310 Definition:

1. $\Gamma(z) = \int_0^\infty e^{-t} t^{z-1} \, dt \quad [\operatorname{Re} z > 0].$ (Euler). FI II 777(6)

Generalization:

2. $\Gamma(z) = -\frac{1}{2i \sin \pi z} \int_C (-t)^{z-1} e^{-t} \, dt$

for z not an integer.

The contour C is shown in the drawing. WH

$\Gamma(z)$ is a fractional analytic function z with simple poles at the points $z = -l$ (for $l = 0, 1, 2, \ldots$) to which correspond the residues $\frac{(-1)^l}{l!}$. $\Gamma(z)$ satisfies the relation $\Gamma(1) = 1$. WH, MO 1

Integral representations

8.311 $\Gamma(z)=\frac{1}{e^{2\pi iz}-1}\int\limits_{\infty}^{(0+)} e^{-t}t^{z-1}\,dt.$ MO 2

8.312

1. $\Gamma(z)=\int\limits_0^1\left(\ln\frac{1}{t}\right)^{z-1}dt \quad [\operatorname{Re} z>0].$ FI II 778

2. $\Gamma(z)=x^z\int\limits_0^\infty e^{-xt}t^{z-1}\,dt \quad [\operatorname{Re} z>0,\ \operatorname{Re} x>0].$ FI II 779(8)

3. $\Gamma(z)=\frac{2a^z e^a}{\sin\pi z}\int\limits_0^\infty e^{-at^2}(1+t^2)^{z-\frac{1}{2}}\cos[2at+(2z-1)\operatorname{arctg} t]\,dt$

$[a>0].$ WH

4. $\Gamma(z)=\frac{1}{2\sin\pi z}\int\limits_0^\infty e^{-t^2}t^{z-1}(1+t^2)^{\frac{z}{2}}\{3\sin[t+z\operatorname{arcctg}(-t)]+$

$+\sin[t+(z-2)\operatorname{arcctg}(-t)]\}\,dt$

[arcctg denotes an obtuse angle]. WH

5. $\Gamma(y)=x^y e^{-i\beta y}\int\limits_0^\infty t^{y-1}\exp(-xte^{-i\beta})\,dt$

$\left[x,\ y,\ \beta \text{ real},\ x>0,\ y>0,\ |\beta|<\frac{\pi}{2}\right].$ MO 8

6. $\Gamma(z)=\frac{b^z}{2\sin\pi z}\int\limits_{-\infty}^\infty e^{bti}(it)^{z-1}\,dt \quad [b>0,\ 0<\operatorname{Re} z<1].$ NH 154(3)

7. $\Gamma(z)=\frac{(\sqrt{a^2+b^2})^z}{\cos\left(z\operatorname{arctg}\frac{b}{a}\right)}\int\limits_0^\infty e^{-at}\cos(bt)\,t^{z-1}\,dt;$ NH 152(1)a

$=\frac{(\sqrt{a^2+b^2})^z}{\sin\left(z\operatorname{arctg}\frac{b}{a}\right)}\int\limits_0^\infty e^{-at}\sin(bt)\,t^{z-1}\,dt$ NH 152(2)

$[a>0,\ b\geqslant 0,\ \operatorname{Re} z>0].$

8. $\Gamma(z)=\frac{b^z}{\cos\frac{\pi z}{2}}\int\limits_0^\infty\cos(bt)\,t^{z-1}\,dt;$ NH 152(4)

$=\frac{b^z}{\sin\frac{\pi z}{2}}\int\limits_0^\infty\sin(bt)\,t^{z-1}\,dt$ NH 152(5)

$[b>0,\ 0<\operatorname{Re} z<1].$

9. $\Gamma(z)=\int\limits_0^\infty e^{-t}(t-z)\,t^{z-1}\ln t\,dt;$ NH 173(7)

10. $\Gamma(z)=\int\limits_{-\infty}^\infty\exp(zt-e^t)\,dt$ NH 145(14)

$[\operatorname{Re} z>0].$

11. $\Gamma(x)\cos ax = \lambda^x \int_0^\infty t^{x-1} e^{-\lambda t \cos\alpha} \cos(\lambda t \sin\alpha)\, dt;$

12. $\Gamma(x)\sin ax = \lambda^x \int_0^\infty t^{x-1} e^{-\lambda t \cos\alpha} \sin(\lambda t \sin\alpha)\, dt$

$$\left[\lambda > 0,\ x > 0,\ -\frac{\pi}{2} < \alpha < \frac{\pi}{2}\right].$$ WH

13. $\Gamma(-z) = \int_0^\infty \left[\frac{e^{-t} - \sum_{k=0}^{n} (-1)^k \frac{t^k}{k!}}{t^{z+1}}\right] dt \quad [n = E(\operatorname{Re} z)].$ MO 2

8.313 $\Gamma\left(\frac{z+1}{v}\right) = v u^{\frac{z+1}{v}} \int_0^\infty \exp(-ut^v)\, t^z\, dt$

$$[\operatorname{Re} u > 0,\ \operatorname{Re} v > 0,\ \operatorname{Re} z > -1].$$ JA, MO 7a

8.314 $\Gamma(z) = \int_1^\infty e^{-t} t^{z-1}\, dt + \sum_{k=0}^\infty \frac{(-1)^k}{k!\,(z+k)}.$ MO 2

8.315

1. $\frac{1}{\Gamma(z)} = \frac{i}{2\pi} \int_C (-t)^{-z} e^{-t}\, dt$

for z, not an integer.
This curve is shown in the drawing accompanying 8.310 2.

WH

2. $\frac{e^{ab} b^{1-z}}{2\pi} \int_{-\infty}^\infty \frac{e^{bti}}{(a+it)^z}\, dt = \frac{1}{\Gamma(z)} \quad \text{for} \quad b > 0;$

$= 0 \quad \text{for} \quad b < 0$

$$\left[a > 0,\ \operatorname{Re} z > 0,\ -\frac{\pi}{2} < \arg(a+it) < \frac{\pi}{2}\right].$$ NH 155(8), MO 7

3. $\frac{1}{\Gamma(z)} = a^{1-z} \frac{e^a}{\pi} \int_0^{\frac{\pi}{2}} \cos(a \operatorname{tg}\theta - z\theta) \cos^{z-2}\theta\, d\theta \quad [\operatorname{Re} z > 1].$ NH 157(14)

See also 3.324 2., 3.326, 3.328, 3.381 4., 3.382 2., 3.389 2., 3.433, 3.434, 3.478 1., 3.551 1., 2., 3.827 1., 4.267 7., 4.272, 4.353 1., 4.369 1., 6.214, 6.223, 6.246, 6.281.

8.32 Representation of the gamma function as series and products

8.321 Representation in the form of a series:

1. $\Gamma(z+1) = \sum_{k=0}^\infty c_k z^k$

$$\left[c_0 = 1,\ c_{n+1} = \frac{\sum_{k=0}^{n} (-1)^{k+1} s_{k+1} c_{n-k}}{n+1};\quad s_1 = \boldsymbol{C},\ s_n = \zeta(n) \text{ for } n \geqslant 2,\ \operatorname{Re} z > 0\right].$$

NH 40(1, 3)

2. $$\frac{1}{\Gamma(z+1)} = \sum_{k=0}^{\infty} d_k z^k$$

$$\left[d_0 = 1,\ d_{n+1} = \frac{\sum_{k=0}^{n} (-1)^k s_{k+1} d_{n-k}}{n+1};\ s_1 = C,\ s_n = \zeta(n) \text{ for } n \geqslant 2 \right].$$

NH 41(4, 6)

Infinite-product representation

8.322 $$\Gamma(z) = e^{-Cz} \frac{1}{z} \prod_{k=1}^{\infty} \frac{e^{\frac{z}{k}}}{1+\frac{z}{k}} \quad [\operatorname{Re} z > 0];$$ SM 269

$$= \frac{1}{z} \prod_{k=1}^{\infty} \frac{\left(1+\frac{1}{k}\right)^z}{1+\frac{z}{k}} \quad [\operatorname{Re} z > 0];$$ WH

$$= \lim_{n\to\infty} \frac{n^z}{z} \prod_{k=1}^{n} \frac{k}{z+k} \quad [\operatorname{Re} z > 0].$$ SM 267(130)

8.323 $$\Gamma(z) = 2z^z e^{-z} \prod_{k=1}^{\infty} \sqrt[2^k]{\mathrm{B}\left(2^{k-1} z, \frac{1}{2}\right)}.$$ NH 98(12)

8.324 $$\Gamma(1+z) = 4^z \prod_{k=1}^{\infty} \frac{\Gamma\left(\frac{1}{2}+\frac{z}{2^k}\right)}{\Gamma\left(\frac{1}{2}\right)}.$$ MO 3

8.325

1. $$\frac{\Gamma(\alpha)\Gamma(\beta)}{\Gamma(\alpha+\gamma)\Gamma(\beta-\gamma)} = \prod_{k=0}^{\infty} \left[\left(1+\frac{\gamma}{\alpha+k}\right)\left(1-\frac{\gamma}{\beta+k}\right)\right].$$ NH 62(2)

2. $$\frac{e^{Cx}\Gamma(z+1)}{\Gamma(z-x+1)} = \prod_{k=1}^{\infty} \left[\left(1-\frac{x}{z+k}\right) e^{\frac{x}{k}}\right] \quad [z \neq 0, -1, -2, \ldots;\ \operatorname{Re} z > 0, \operatorname{Re}(z-x) > 0].$$

3. $$\frac{\Gamma\left(\frac{1}{2}\right)}{\Gamma\left(1+\frac{z}{2}\right)\Gamma\left(\frac{1}{2}-\frac{z}{2}\right)} = \prod_{k=1}^{\infty} \left(1-\frac{z}{2k-1}\right)\left(1+\frac{z}{2k}\right).$$ MO 2

8.326

1. $$\frac{\frac{[\Gamma(x)]^2}{\Gamma(2x)}}{\mathrm{B}(x+iy, x-iy)} = \left|\frac{\Gamma(x)}{\Gamma(x+iy)}\right|^2 = \prod_{k=0}^{\infty} \left(1+\frac{y^2}{(x+k)^2}\right)$$

$[x, y$ real, $x \neq 0, -1, -2, \ldots]$. LO V, NH 63(4)

2. $$\frac{\Gamma(x+iy)}{\Gamma(x)} = \frac{xe^{-iCy}}{x+iy} \prod_{n=1}^{\infty} \frac{\exp\left(\frac{iy}{n}\right)}{1+\frac{iy}{x+n}} \quad [x, y \text{ real}, x \neq 0, -1, -2, \ldots].$$ MO 2

8.327 Asymptotic representation for large values of $|z|$:

$$\Gamma(z) = z^{z-\frac{1}{2}} e^{-z} \sqrt{2\pi} \left\{1 + \frac{1}{12z} + \frac{1}{288z^2} - \frac{139}{51840z^3} - \frac{571}{2488320z^4} + O(z^{-5})\right\}$$

$[|\arg z| < \pi]$. WH

For z real and positive, the remainder of the series is less than the last term that is retained.

8.328

1. $$\lim_{|y|\to\infty} |\Gamma(x+iy)| e^{\frac{\pi}{2}|y|} |y|^{\frac{1}{2}-x} = \sqrt{2\pi} \quad [x \text{ and } y \text{ are real}].$$ MO 6

2. $$\lim_{|z|\to\infty} \frac{\Gamma(z+a)}{\Gamma(z)} e^{-a \ln z} = 1.$$ MO 6

8.33 Functional relations involving the gamma function

8.331 $\Gamma(x+1) = x\Gamma(x)$.

8.332

1. $$|\Gamma(iy)|^2 = \frac{\pi}{y \operatorname{sh} \pi y}$$ MO 3

2. $$\left|\Gamma\left(\frac{1}{2}+iy\right)\right|^2 = \frac{\pi}{\operatorname{ch} \pi y}$$

$[y$ is real$]$.

3. $$\Gamma(1+ix)\Gamma(1-ix) = \frac{\pi x}{\operatorname{sh} x\pi} \quad [x \text{ is real}].$$ LO V

4. $$\Gamma(1+x+iy)\Gamma(1-x+iy)\Gamma(1+x-iy)\Gamma(1-x-iy) = \frac{2\pi^2(x^2+y^2)}{\operatorname{sh} 2y\pi - \cos 2x\pi} \quad [x \text{ and } y \text{ are real}].$$ LO V

8.333 $$[\Gamma(n+1)]^n = G(n+1) \prod_{k=1}^{n} k^k,$$

where n is a natural number and

$$G(z+1) = (2\pi)^{\frac{z}{2}} \exp\left[-\frac{z(z+1)}{2} - \frac{C}{2} z^2\right] \prod_{n=1}^{\infty} \left\{\left(1+\frac{z}{n}\right)^n \exp\left(-z+\frac{z^2}{2n}\right)\right\}.$$

WH

8.334

1. $$\prod_{k=1}^{n} \frac{1}{\Gamma\left(-z \exp \frac{2\pi k i}{n}\right)} = -z^n \prod_{k=1}^{\infty} \left[1 - \left(\frac{z}{k}\right)^n\right] \quad [n = 2, 3, \ldots].$$ MO 2

2. $$\Gamma\left(\frac{1}{2}+x\right)\Gamma\left(\frac{1}{2}-x\right) = \frac{\pi}{\cos \pi x}.$$

3. $$\Gamma(1-x)\Gamma(x) = \frac{\pi}{\sin \pi x}.$$ FI II 430

8.335 $$\Gamma(nx) = (2\pi)^{\frac{1-n}{2}} n^{nx-\frac{1}{2}} \prod_{k=0}^{n-1} \Gamma\left(x+\frac{k}{n}\right) \quad [\text{product theorem}].$$

FI II 782a, WH

Special cases

1. $\Gamma(2x)=\frac{2^{2x-1}}{\sqrt{\pi}}\Gamma(x)\Gamma\left(x+\frac{1}{2}\right)$ [doubling formula].

2. $\Gamma(3x)=\frac{3^{3x-\frac{1}{2}}}{2\pi}\Gamma(x)\Gamma\left(x+\frac{1}{3}\right)\Gamma\left(x+\frac{2}{3}\right)$.

3. $\prod_{k=1}^{n-1}\Gamma\left(\frac{k}{n}\right)\Gamma\left(1-\frac{k}{n}\right)=\frac{(2\pi)^{n-1}}{n}$. WH

8.336 $\Gamma\left(-\frac{yz+xi}{2y}\right)\Gamma(1+z)=(2i)^{z+1}y\Gamma\left(1+\frac{yz-xi}{2y}\right)\int_0^\infty e^{-tx}\sin^z(ty)\,dt$

$[\mathrm{Re}(yi)>0,\ \mathrm{Re}(x-yzi)>0]$. NH 133(10)

For a connection with the psi function, see 8.361 1.
For a connection with the beta function, see 8.384 1.
For integrals of the gamma function, see 8.412 4., 8.414, 9.223, 9.242 3., 9.242 4.

8.337

1. $[\Gamma'(x)]^2<\Gamma(x)\Gamma''(x)$ $[x>0]$. MO 1
2. For $x>0$, $\min\Gamma(1+x)=0.88560\ldots$ is attained when $x=0.46163\ldots$ JA

Particular values

8.338

1. $\Gamma(1)=\Gamma(2)=1$.
2. $\Gamma\left(\frac{1}{2}\right)=\sqrt{\pi}$.
3. $\Gamma\left(-\frac{1}{2}\right)=-2\sqrt{\pi}$.
4. $\left[\Gamma\left(\frac{1}{4}\right)\right]^4=16\pi^2\prod_{k=1}^{\infty}\frac{(4k-1)^2[(4k+1)^2-1]}{[(4k-1)^2-1](4k+1)^2}$. MO 1a
5. $\prod_{k=1}^{8}\Gamma\left(\frac{k}{3}\right)=\frac{640}{3^6}\left(\frac{\pi}{\sqrt{3}}\right)^3$. WH

8.339 For n a natural number

1. $\Gamma(n)=(n-1)!$
2. $\Gamma\left(n+\frac{1}{2}\right)=\frac{\sqrt{\pi}}{2^n}(2n-1)!!$
3. $\Gamma\left(\frac{1}{2}-n\right)=(-1)^n\frac{2^n\sqrt{\pi}}{(2n-1)!!}$.
4. $\frac{\Gamma\left(p+n+\frac{1}{2}\right)}{\Gamma\left(p-n+\frac{1}{2}\right)}=\frac{(4p^2-1^2)(4p^2-3^2)\ldots[4p^2-(2n-1)^2]}{2^{2n}}$. WA 221

8.34 The logarithm of the gamma function

8.341 Integral representation:

1 $$\ln \Gamma(z) = \left(z - \frac{1}{2}\right) \ln z - z + \frac{1}{2} \ln 2\pi + \int_0^\infty \left(\frac{1}{2} - \frac{1}{t} + \frac{1}{e^t - 1}\right) \frac{e^{-tz}}{t} dt$$

$[\operatorname{Re} z > 0]$. WH

2. $$\ln \Gamma(z) = z \ln z - z - \frac{1}{2} \ln z + \ln \sqrt{2\pi} + 2 \int_0^\infty \frac{\operatorname{arctg} \frac{t}{z}}{e^{2\pi t} - 1} dt$$

$\left[\operatorname{Re} z > 0 \text{ and } \operatorname{arctg} w = \int_0^w \frac{du}{1+u^2}\right.$ is taken over a rectangular path in the w-plane$\left.\right]$ WH

3. $$\ln \Gamma(z) = \int_0^\infty \left\{\frac{e^{-zt} - e^{-t}}{1 - e^{-t}} + (z-1) e^{-t}\right\} \frac{dt}{t} \qquad [\operatorname{Re} z > 0].$$ WH

4. $$\ln \Gamma(z) = \int_0^\infty \left\{(z-1) e^{-t} + \frac{(1+t)^{-z} - (1+t)^{-1}}{\ln(1+t)}\right\} \frac{dt}{t} \qquad [\operatorname{Re} z > 0].$$ WH

5. $$\ln \Gamma(x) = \frac{\ln \pi - \ln \sin \pi x}{2} + \frac{1}{2} \int_0^\infty \left\{\frac{\operatorname{sh}\left(\frac{1}{2} - x\right) t}{\operatorname{sh} \frac{t}{2}} - (1 - 2x) e^{-t}\right\} \frac{dt}{t}$$

$[0 < x < 1]$. WH

6. $$\ln \Gamma(z) = \int_0^1 \left\{\frac{t^z - t}{t - 1} - t(z-1)\right\} \frac{dt}{t \ln t} \qquad [\operatorname{Re} z > 0].$$ WH

7 $$\ln \Gamma(z) = \int_0^\infty \left[(z-1) e^{-t} + \frac{e^{-tz} - e^{-t}}{1 - e^{-t}}\right] \frac{dt}{t} \qquad [\operatorname{Re} z > 0].$$ NH 187(7)

See also **3.427** 9., **3.554** 5.

8.342 Series representations:

1. $$\ln \Gamma(z+1) = \frac{1}{2}\left[\ln\left(\frac{\pi z}{\sin \pi z}\right) - \ln \frac{1+z}{1-z}\right] + (1 - C) z + \sum_{k=1}^\infty \frac{1 - \zeta(2k+1)}{2k+1} z^{2k+1} = -Cz + \sum_{k=2}^\infty (-1)^k \frac{z^k}{k} \zeta(k) \qquad [|z| < 1].$$

NH 38(16, 12)

2. $$\ln \Gamma(1+x) = \frac{1}{2} \ln \frac{\pi x}{\sin \pi x} - Cx - \sum_{n=1}^\infty \frac{x^{2n+1}}{2n+1} \{\zeta(2n+1)\} \qquad [|x| < 1].$$

NH 38(14)

8.343

1. $$\ln \Gamma(x) = \ln \sqrt{2\pi} + \sum_{n=1}^{\infty} \left\{ \frac{1}{2n} \cos 2n\pi x + \frac{1}{n\pi} (C + \ln 2n\pi) \sin 2n\pi x \right\}$$

$[0 < x < 1]$. FI III 558

2. $$\ln \Gamma(z) = z \ln z - z - \frac{1}{2} \ln z + \ln \sqrt{2\pi} + \\ + \frac{1}{2} \sum_{m=1}^{\infty} \frac{m}{(m+1)(m+2)} \sum_{n=1}^{\infty} \frac{1}{(z+n)^{m+1}}$$

$[|\arg z| < \pi]$. MO 9

8.344 Asymptotic expansion for large values of $|z|$:

$$\ln \Gamma(z) = z \ln z - z - \frac{1}{2} \ln z + \ln \sqrt{2\pi} + \sum_{k=1}^{n-1} \frac{B_{2k}}{2k(2k-1) z^{2k-1}} + R_n(z),$$

where

$$|R_n(z)| < \frac{|B_{2n}|}{2n(2n-1)|z|^{2n-1} \cos^{2n-1}\left(\frac{1}{2} \arg z\right)}.$$

MO 5

For integrals of $\ln \Gamma(x)$, see **6.44**.

8.35 The incomplete gamma function

8.350 Definition:

1. $$\gamma(\alpha, x) = \int_0^x e^{-t} t^{\alpha-1}\, dt \quad [\operatorname{Re} \alpha > 0].$$

EH II 133(1), NH 1(1)

2. $$\Gamma(\alpha, x) = \int_x^{\infty} e^{-t} t^{\alpha-1}\, dt.$$

EH II 133(2), NH 2(2), LE 339

8.351

1. $\gamma^*(\alpha, x) = \frac{x^{-\alpha}}{\Gamma(\alpha)} \gamma(\alpha, x)$ is an analytic function with respect to α and x.

EH II 133(5)

2. Another definition of $\gamma(\alpha, x)$, that is also suitable for the case $\operatorname{Re} \alpha \leqslant 0$:

$$\gamma(\alpha, x) = \frac{x^{\alpha}}{\alpha} e^{-x} \Phi(1, 1+\alpha; x) = \frac{x^{\alpha}}{\alpha} \Phi(\alpha, 1+\alpha; -x).$$

EH II 133(3)

3. For fixed x, $\Gamma(\alpha, x)$ is an entire function of α. For nonintegral α, $\Gamma(\alpha, x)$ is a multiple-valued function of x with a branch point at $x = 0$.

4. A second definition of $\Gamma(\alpha, x)$:

$$\Gamma(\alpha, x) = x^{\alpha} e^{-x} \Psi(1, 1+\alpha; x) = e^{-x} \Psi(1-\alpha, 1-\alpha; x).$$

EH II 133(4)

8.352 Special cases:

1. $$\gamma(1+n, x) = n! \left[1 - e^{-x} \left(\sum_{m=0}^{n} \frac{x^m}{m!} \right) \right]$$

$[n = 0, 1, \ldots]$. EH II 136(17, 16), NH 6(11)

2. $\Gamma(1+n, x) = n!\, e^{-x} \sum_{m=0}^{n} \frac{x^m}{m!} \qquad [n = 0, 1, \ldots].$ EH II 136(16, 18)

3. $\Gamma(-n, x) = \frac{(-1)^n}{n!}\left[\Gamma(0, x) - e^{-x} \sum_{m=0}^{n-1} (-1)^m \frac{m!}{x^{m+1}}\right]$

$[n = 1, 2, \ldots].$ EH II 137(20), NH 4(4)

8.353 Integral representations:

1. $\gamma(\alpha, x) = x^{\alpha} \operatorname{cosec} \pi\alpha \int_0^{\pi} e^{x \cos\theta} \cos(\alpha\theta + x \sin\theta)\, d\theta$

$[x \neq 0, \operatorname{Re}\alpha > 0, \alpha \neq 1, 2, \ldots].$ EH II 137(2)

2. $\gamma(\alpha, x) = x^{\frac{1}{2}\alpha} \int_0^{\infty} e^{-t} t^{\frac{1}{2}\alpha - 1} J_{\alpha}(2\sqrt{xt})\, dt \qquad [\operatorname{Re}\alpha > 0].$ EH II 138(4)

3. $\Gamma(\alpha, x) = \frac{e^{-x} x^{\alpha}}{\Gamma(1-\alpha)} \int_0^{\infty} \frac{e^{-t} t^{-\alpha}}{x+t}\, dt$

$[\operatorname{Re}\alpha < 1, x > 0].$ EH II 137(3), NH 19(12)

4. $\Gamma(\alpha, x) = \frac{2x^{\frac{1}{2}\alpha} e^{-x}}{\Gamma(1-\alpha)} \int_0^{\infty} e^{-t} t^{-\frac{1}{2}\alpha} K_{\alpha}[2\sqrt{xt}]\, dt$

$[\operatorname{Re}\alpha < 1].$ EH II 138(5)

5. $\Gamma(\alpha, xy) = y^{\alpha} e^{-xy} \int_0^{\infty} e^{-ty} (t+x)^{\alpha-1}\, dt$

$[\operatorname{Re} y > 0, \quad x > 0, \quad \operatorname{Re}\alpha > 1].$ (See also 3.936 5., 3.944 1.—4.)

NH 19(10)

For integrals of the gamma function, see 6.45.

8.354 Series representations:

1. $\gamma(\alpha, x) = \sum_{n=0}^{\infty} \frac{(-1)^n x^{\alpha+n}}{n!\,(\alpha+n)}.$ EH II 135(4)

2. $\Gamma(\alpha, x) = \Gamma(\alpha) - \sum_{n=0}^{\infty} \frac{(-1)^n x^{\alpha+n}}{n!\,(\alpha+n)}$

$[\alpha \neq 0, -1, -2, \ldots].$ EH II 135(5), LE 340(2)

3. $\Gamma(\alpha, x) - \Gamma(\alpha, x+y) = \gamma(\alpha, x+y) - \gamma(\alpha, x) =$

$= e^{-x} x^{\alpha-1} \sum_{k=0}^{\infty} \frac{(-1)^k [1 - e^{-y} e_k(y)]\, \Gamma(1-\alpha+k)}{x^k\, \Gamma(1-\alpha)}, \quad e_k(x) = \sum_{m=0}^{k} \frac{x^m}{m!}$

$[|y| < |x|].$ EH II 139(2)

4. $\gamma(\alpha, x) = \Gamma(\alpha)\, e^{-x} x^{\frac{1}{2}\alpha} \sum_{n=0}^{\infty} x^{\frac{1}{2}n} I_{n+\alpha}(2\sqrt{x}) \sum_{m=0}^{n} \frac{(-1)^m}{m!}$

$[x \neq 0, \alpha \neq 0, -1, -2, \ldots].$ EH II 139(3)

5. $\Gamma(\alpha, x) = e^{-x}x^{\alpha} \sum_{n=0}^{\infty} \frac{L_n^{\alpha}(x)}{n+1} \quad [x > 0].$ EH II 140(5)

8.355 $\Gamma(\alpha, x)\gamma(\alpha, y) = e^{-x-y}(xy)^{\alpha} \sum_{n=0}^{\infty} \frac{n!\,\Gamma(\alpha)}{(n+1)\Gamma(\alpha+n+1)} L_n^{\alpha}(x) L_n^{\alpha}(y)$

$[y > 0,\ x \geqslant y,\ \alpha \neq 0,\ -1, \dots].$ EH II 139(4)

8.356 Functional relations:

1. $\gamma(\alpha+1, x) = \alpha\gamma(\alpha, x) - x^{\alpha}e^{-x}.$ EH II 134(2)

2. $\Gamma(\alpha+1, x) = \alpha\Gamma(\alpha, x) + x^{\alpha}e^{-x}.$ EH II 134(3)

3. $\Gamma(\alpha, x) + \gamma(\alpha, x) = \Gamma(\alpha).$ EH II 134(1)

4. $\frac{d\gamma(\alpha, x)}{dx} = -\frac{d\Gamma(\alpha, x)}{dx} = x^{\alpha-1}e^{-x}.$ EH II 135(8)

5. $\frac{\Gamma(\alpha+n, x)}{\Gamma(\alpha+n)} = \frac{\Gamma(\alpha, x)}{\Gamma(\alpha)} + e^{-x} \sum_{s=0}^{n-1} \frac{x^{\alpha+s}}{\Gamma(\alpha+s+1)}.$ NH 4(3)

6. $\Gamma(\alpha)\Gamma(\alpha+n, x) - \Gamma(\alpha+n)\Gamma(\alpha, x) =$

$= \Gamma(\alpha+n)\gamma(\alpha, x) - \Gamma(\alpha)\Gamma(\alpha+n, x).$ NH 5

8.357 Asymptotic representation for large values of $|x|$:

$$\Gamma(\alpha, x) = x^{\alpha-1}e^{-x}\left[\sum_{m=0}^{M-1} \frac{(-1)^m\,\Gamma(1-\alpha+m)}{x^m\,\Gamma(1-\alpha)} + O(|x|^{-M})\right]$$

$$\left[|x| \to \infty,\ -\frac{3\pi}{2} < \arg x < \frac{3\pi}{2},\ M = 1, 2, \dots\right].$$

EH II 135(6), NH 37(7), LE 340(3)

8.358 Representation as a continued fraction:

$$\Gamma(\alpha, x) = \cfrac{e^{-x}x^{\alpha}}{x + \cfrac{1-\alpha}{1 + \cfrac{1}{x + \cfrac{2-\alpha}{1 + \cfrac{2}{x + \cfrac{3-\alpha}{1 + \dots}}}}}}$$

EH II 136(13), NH 42(9)

8.359 Relationships with other functions:

1. $\Gamma(0, x) = -\mathrm{Ei}(-x).$ EH II 143(1)

2. $\Gamma\left(0, \ln\frac{1}{x}\right) = -\mathrm{li}(x).$ EH II 143(2)

3. $\Gamma\left(\frac{1}{2}, x^2\right) = \sqrt{\pi} - \sqrt{\pi}\,\Phi(x).$ EH II 147(2)

4. $\gamma\left(\frac{1}{2}, x^2\right) = \sqrt{\pi}\,\Phi(x).$ EH II 147(1)

8.36 The psi function $\psi(x)$

8.360 Definition:

$$\psi(x) = \frac{d}{dx} \ln \Gamma(x).$$

8.361 Integral representations:

1. $$\psi(z) = \frac{d \ln \Gamma(z)}{dz} = \int_0^\infty \left(\frac{e^{-t}}{t} - \frac{e^{-zt}}{1-e^{-t}} \right) dt$$
$[\operatorname{Re} z > 0]$. NH 183(1), WH

2. $$\psi(z) = \int_0^\infty \left\{ e^{-t} - \frac{1}{(1+t)^z} \right\} \frac{dt}{t} \quad [\operatorname{Re} z > 0].$$ NH 184(7), WH

3. $$\psi(z) = \ln z - \frac{1}{2z} - 2 \int_0^\infty \frac{t\, dt}{(t^2+z^2)(e^{2\pi t}-1)} \quad [\operatorname{Re} z > 0].$$ WH

4. $$\psi(z) = \int_0^1 \left(\frac{1}{-\ln t} - \frac{t^{z-1}}{1-t} \right) dt \quad [\operatorname{Re} z > 0].$$ WH

5. $$\psi(z) = \int_0^\infty \frac{e^{-t} - e^{-zt}}{1-e^{-t}}\, dt - C,$$ WH

6. $$\psi(z) = \int_0^\infty \{(1+t)^{-1} - (1+t)^{-z}\} \frac{dt}{t} - C,$$ WH

7. $$\psi(z) = \int_0^1 \frac{t^{z-1} - 1}{t-1}\, dt - C$$ FI II 796, WH

$[\operatorname{Re} z > 0]$ (for 5.–7.)

8. $$\psi(z) = \ln z + \int_0^\infty e^{-tz} \left[\frac{1}{t} - \frac{1}{1-e^{-t}} \right] dt \quad \left[|\arg z| < \frac{\pi}{2} \right].$$ MO 4

See also **3.244** 3., **3.311** 6., **3.317** 1., **3.457**, **3.458** 2., **3.471** 14., **4.253** 1. and 6., **4.275** 2., **4.281** 4., **4.482** 5.
For integrals of the psi function, see **6.46**, **6.47**.

Series representation

8.362

1. $$\psi(x) = -C - \sum_{k=0}^\infty \left(\frac{1}{x+k} - \frac{1}{k+1} \right);$$ FI II 799(26), KU 26(1)

$$= -C - \frac{1}{x} + x \sum_{k=1}^\infty \frac{1}{k(x+k)}.$$ FI II 495

2. $$\psi(x) = \ln x - \sum_{k=0}^\infty \left[\frac{1}{z+k} - \ln\left(1 + \frac{1}{z+k}\right) \right].$$ MO 4

3. $\psi(x) = -C + \frac{\pi^2}{6}(x-1) - (x-1)\sum_{k=1}^{\infty}\left(\frac{1}{k+1} - \frac{1}{x+k}\right)\sum_{n=0}^{k-1}\frac{1}{x+n}.$

NH 54(12)

8.363

1. $\psi(x+1) = -C + \sum_{k=2}^{\infty}(-1)^k \zeta(k) x^{k-1}.$ NH 37(5)

2. $\psi(x+1) = \frac{1}{2x} - \frac{\pi}{2}\operatorname{ctg}\pi x - \frac{x^2}{1-x^2} - C + \sum_{k=1}^{\infty}[1 - \zeta(2k+1)]\, x^{2k}.$

NH 38(10)

3. $\psi(x) - \psi(y) = \sum_{k=0}^{\infty}\left(\frac{1}{y+k} - \frac{1}{x+k}\right)$ (see also **3.219, 3.231** 5., **3.311** 7., **3.688** 20., **4.253** 1., **4.295** 37.). NH 99(3)

4. $\psi(x+iy) - \psi(x-iy) = \sum_{k=0}^{\infty}\frac{2yi}{y^2+(x+k)^2}.$

5. $\psi\left(\frac{p}{q}\right) = -C + \sum_{k=0}^{\infty}\left(\frac{1}{k+1} - \frac{q}{p+kq}\right)$ (see also **3.244** 3.). NH 29(1)

6. $\psi\left(\frac{p}{q}\right) = -C - \ln q - \frac{\pi}{2}\operatorname{ctg}\frac{p\pi}{q} + 2\sum_{k=1}^{E\left(\frac{q+1}{2}\right)-1}\left[\cos\frac{2kp\pi}{q}\ln\sin\frac{k\pi}{q}\right]$

$[q = 2, 3, \ldots, p = 1, 2, \ldots, q-1].$ MO 4, EH I 19(29)

7. $\psi\left(\frac{p}{q}\right) - \psi\left(\frac{p-1}{q}\right) = q\sum_{n=2}^{\infty}\sum_{k=0}^{\infty}\frac{1}{(p+kq)^n - 1}.$ NH 59(3)

8. $\psi^{(n)}(x) = (-1)^{n+1} n! \sum_{k=0}^{\infty}\frac{1}{(x+k)^{n+1}}.$ NH 37(1)

Infinite-product representation

8.364

1. $e^{\psi(x)} = x\prod_{k=0}^{\infty}\left(1 + \frac{1}{x+k}\right)e^{-\frac{1}{x+k}}.$ NH 65(12)

2. $e^{y\psi(x)} = \frac{\Gamma(x+y)}{\Gamma(x)}\prod_{k=0}^{\infty}\left(1 + \frac{y}{x+k}\right)e^{-\frac{y}{x+k}}$ NH 65(11)

See also 8.37.
For a connection with Riemann's zeta function, see **9.533** 2.
For a connection with the gamma function, see **4.325** 12. and **4.352** 1.
For a connection with the beta function, see **4.253** 1.

For series of psi functions, see **8.403** 2., **8.446**, and **8.447** 3. (Bessel functions) **8.761** (derivatives of associated Legendre functions with respect to the degree), **9.153**, **9.154** (hypergeometric function), **9.238** (degenerate hypergeometric function).

For integrals containing psi functions, see **6.46**—**6.47**.

8.365 Functional relations:

1. $\psi(x+1)=\psi(x)+\frac{1}{x}$. JA
2. $\psi\left(\frac{x+1}{2}\right)-\psi\left(\frac{x}{2}\right)=2\beta(x)$ (cf. **8.37** 0).
3. $\psi(x+n)=\psi(x)+\sum_{k=0}^{n-1}\frac{1}{x+k}$. GA 154(64)a
4. $\psi(n+1)=-C+\sum_{k=1}^{n}\frac{1}{k}$. MO 4
5. $\lim_{n\to\infty}[\psi(z+n)-\ln n]=0$. MO 3
6. $\psi(nz)=\frac{1}{n}\sum_{k=0}^{n-1}\psi\left(z+\frac{k}{n}\right)+\ln n \quad [n=2, 3, 4, \ldots]$. MO 3
7. $\psi(x-n)=\psi(x)-\sum_{k=1}^{n}\frac{1}{x-k}$.
8. $\psi(1-z)=\psi(z)+\pi\operatorname{ctg}\pi z$. GA 155(68)a
9. $\psi\left(\frac{1}{2}+z\right)=\psi\left(\frac{1}{2}-z\right)+\pi\operatorname{tg}\pi z$. JA
10. $\psi\left(\frac{3}{4}-n\right)=\psi\left(\frac{1}{4}+n\right)+\pi \quad [n \text{— a natural number}]$.

8.366 Particular values

1. $\psi(1)=-C$ (cf. **8.367** 1.).
2. $\psi\left(\frac{1}{2}\right)=-C-2\ln 2=-1.963\,510\,026\ldots$ GA 155a
3. $\psi\left(\frac{1}{2}\pm n\right)=-C+2\left[\sum_{k=1}^{n}\frac{1}{2k-1}-\ln 2\right]$. JA
4. $\psi\left(\frac{1}{4}\right)=-C-\frac{\pi}{2}-3\ln 2$. GA 157a
5. $\psi\left(\frac{3}{4}\right)=-C+\frac{\pi}{2}-3\ln 2$. GA 157a
6. $\psi\left(\frac{1}{3}\right)=-C-\frac{\pi}{2}\sqrt{\frac{1}{3}}-\frac{3}{2}\ln 3$. GA 157a
7. $\psi\left(\frac{2}{3}\right)=-C+\frac{\pi}{2}\sqrt{\frac{1}{3}}-\frac{3}{2}\ln 3$. GA 157a
8. $\psi'(1)=\frac{\pi^2}{6}=1.644\,934\,067\ldots$ JA
9. $\psi'\left(\frac{1}{2}\right)=\frac{\pi^2}{2}=4.934\,802\,201\ldots$ JA

10. $\psi'(-n)=\infty$

11. $\psi'(n)=\frac{\pi^2}{6}-\sum_{k=1}^{n-1}\frac{1}{k^2}$

12. $\psi'\left(\frac{1}{2}+n\right)=\frac{\pi^2}{2}-4\sum_{k=1}^{n}\frac{1}{(2k-1)^2}$

13. $\psi'\left(\frac{1}{2}-n\right)=\frac{\pi^2}{2}+4\sum_{k=1}^{n}\frac{1}{(2k-1)^2}$

[n — a natural number]. JA

8.367 Euler's constant:

1. $C=-\psi(1)=0.577\,215\,664\,90\ldots$ FI II 319, 795

2. $C=\lim_{n\to\infty}\left[\sum_{k=1}^{n-1}\frac{1}{k}-\ln n\right].$ FI II 801a

3. $C=\lim_{x\to 1+0}\left[\zeta(x)-\frac{1}{x-1}\right].$ FI II 804

Integral representations:

4. $C=-\int_0^\infty e^{-t}\ln t\,dt.$ FI II 807

5. $C=-\int_0^1 \ln\left(\ln\frac{1}{t}\right)dt.$ FI II 807

6. $C=\int_0^1\left[\frac{1}{\ln t}+\frac{1}{1-t}\right]dt.$ DW

7. $C=-\int_0^\infty\left[\cos t-\frac{1}{1+t}\right]\frac{dt}{t}.$ MO 10

8. $C=1-\int_0^\infty\left[\frac{\sin t}{t}-\frac{1}{1+t}\right]\frac{dt}{t}.$ MO 10

9. $C=-\int_0^\infty\left[e^{-t}-\frac{1}{1+t}\right]\frac{dt}{t}.$ FI II 795, 802

10. $C=-\int_0^\infty\left[e^{-t}-\frac{1}{1+t^2}\right]\frac{dt}{t}.$ DW, MO 10

11. $C=\int_0^\infty\left[\frac{1}{e^t-1}-\frac{1}{te^t}\right]dt.$ DW

12. $C=\int_0^1(1-e^{-t})\frac{dt}{t}-\int_1^\infty\frac{e^{-t}}{t}dt.$ FI II 802

See also 8.361 5. — 8.361 7., 3.311 6., 3.435 3. and 4., 3.476 2., 3.481 1. and 2., 3.951 10., 4.283 9., 4.331 1., 4.421 1., 4.424 1., 4.553, 4.572, 6.234, 6.264 1., 6.468.

13. Asymptotic expansions

$$C=\sum_{k=1}^{n-1}\frac{1}{k}-\ln n+\frac{1}{2n}+\frac{1}{12n^2}-\frac{1}{120n^4}+\frac{1}{252n^6}-\frac{1}{240n^8}+\dots$$

$$\dots+\frac{B_{2r}}{2r}\frac{1}{n^{2r}}+\frac{B_{2r+2}}{2(r+1)}\frac{\theta}{n^{2r+2}}\qquad [0<\theta<1].$$ FI II 827

8.37 The function $\beta(x)$

Definition:

8.370 $$\beta(x)=\frac{1}{2}\left[\psi\left(\frac{x+1}{2}\right)-\psi\left(\frac{x}{2}\right)\right].$$ NH 16(13)

8.371 Integral representations:

1. $$\beta(x)=\int_0^\infty\frac{t^{x-1}}{1+t}\,dt\qquad [\operatorname{Re}x>0].$$ WH

2. $$\beta(x)=\int_0^\infty\frac{e^{-xt}}{1+e^{-t}}\,dt\qquad [\operatorname{Re}x>0].$$ MO 4

3. $$\beta\left(\frac{x+1}{2}\right)=\int_0^\infty\frac{e^{-xt}}{\operatorname{ch}t}\,dt\qquad [\operatorname{Re}x>0]$$ (cf. 8.371 1.).

See also **3.241** 1., **3.251** 7., **3.522** 2. and 4., **3.623** 2. and 3., **4.282** 2., **4.389** 3., **4.532** 1. and 3.

Series representation

8.372

1. $$\beta(x)=\sum_{k=0}^\infty\frac{(-1)^k}{x+k}.$$ NH 37, 101(1)

2. $$\beta(x)=\sum_{k=0}^\infty\frac{1}{(x+2k)(x+2k+1)}.$$ NH 101(2)

3. $$\beta(x)=\frac{1}{2}\sum_{k=0}^\infty\frac{k!}{x(x+1)\dots(x+k)}\frac{1}{2^k}.$$ NH 246(7)

8.373

1. $$\beta(x+1)=\sum_{k=1}^\infty(-1)^{k+1}(1-2^{1-k})\zeta(k)x^{k-1}.$$ NH 37(5)

2. $$\beta(x+1)=\frac{1}{2x}-\frac{1}{2\sin\pi x}+\frac{1}{1-x^2}-\sum_{k=0}^\infty[1-(1-2^{-2k})\zeta(2k+1)]x^{2k}.$$

NH 38(11)

8.374 $$\beta^{(n)}(x)=(-1)^n n!\sum_{k=0}^\infty\frac{(-1)^k}{(x+k)^{n+1}}.$$ NH 37(2)

8.375 Representation in the form of a finite sum:

1. $$\beta\left(\frac{p}{q}\right)=\frac{\pi}{2\sin\frac{p\pi}{q}}-\sum_{k=0}^{E\left(\frac{q-1}{2}\right)}\cos\frac{p(2k+1)\pi}{q}\ln\left(2-2\cos\frac{(2k+1)\pi}{q}\right)$$

$[q=2, 3, \ldots, p=1, 2, 3, \ldots]$ (see also 8.362 5.—7.). NH 23(9)

2. $$\beta(n)=(-1)^{n+1}\ln 2+\sum_{k=1}^{n-1}\frac{(-1)^{k+n+1}}{k}.$$

Functional relations

8.376 $$\sum_{k=0}^{2n}(-1)^k\beta\left(\frac{x+k}{2n+1}\right)=(2n+1)\beta(x).$$ NH 19

8.377 $$\sum_{k=1}^{n}\beta(2^k x)=\psi(2^n x)-\psi(x)-n\ln 2.$$ NH 20(10)

8.38 *The beta function (Euler's integral of the first kind):* $\mathrm{B}(x, y)$

Integral representation

8.380

1. $$\mathrm{B}(x, y)=\int_0^1 t^{x-1}(1-t)^{y-1}\,dt\ ^{*});$$ FI II 774(1)

$$=2\int_0^1 t^{2x-1}(1-t^2)^{y-1}\,dt \qquad [\operatorname{Re}x>0, \operatorname{Re}y>0].$$

2. $$\mathrm{B}(x, y)=2\int_0^{\frac{\pi}{2}}\sin^{2x-1}\varphi\cos^{2y-1}\varphi\,d\varphi \qquad [\operatorname{Re}x>0, \operatorname{Re}y>0].$$ KU 10

3. $$\mathrm{B}(x, y)=\int_0^{\infty}\frac{t^{x-1}}{(1+t)^{x+y}}\,dt=2\int_0^{\infty}\frac{t^{2x-1}}{(1+t^2)^{x+y}}\,dt \qquad [\operatorname{Re}x>0, \operatorname{Re}y>0].$$ FI II 775

4. $$\mathrm{B}(x, y)=2^{2-y-x}\int_{-1}^{1}\frac{(1+t)^{2x-1}(1-t)^{2y-1}}{(1+t^2)^{x+y}}\,dt \qquad [\operatorname{Re}x>0, \operatorname{Re}y>0].$$ MO 7

5. $$\mathrm{B}(x, y)=\int_0^1\frac{t^{x-1}+t^{y-1}}{(1+t)^{x+y}}\,dt=\int_1^{\infty}\frac{t^{x-1}+t^{y-1}}{(1+t)^{x+y}}\,dt$$

6. $$\mathrm{B}(x, y)=\frac{1}{2^{x+y-1}}\int_0^1[(1+t)^{x-1}(1-t)^{y-1}+(1+t)^{y-1}(1-t)^{x-1}]\,dt$$

$[\operatorname{Re}x>0, \operatorname{Re}y>0]$. BI ((1))(15)

---

*This equation is used as the definition of the function $\mathrm{B}(x, y)$.

7. $\mathrm{B}(x, y) = z^y (1+z)^x \int_0^1 \frac{t^{x-1}(1-t)^{y-1}}{(t+z)^{x+y}}$ $[\operatorname{Re} x > 0, \operatorname{Re} y > 0,$ WH

8. $\mathrm{B}(x, y) = z^y (1+z)^x \int_0^{\frac{\pi}{2}} \frac{\cos^{2x-1}\varphi \sin^{2y-1}\varphi}{(z+\cos^2\varphi)^{x+y}} d\varphi$ $0 > z > -1, \operatorname{Re}(x+y) < 1]$. NH 163(8)

See also **3.196** 3., **3.198**, **3.199**, **3.215**, **3.238** 3., **3.251** 1.—3., 11., **3.253**, **3.312** 1., **3.512** 1. and 2., **3.541** 1., **3.542** 1., **3.621** 5., **3.623** 1., **3.631** 1., 8., 9., **3.632** 2., **3.633** 1., 4., **3.634** 1., 2., **3.637**, **3.642** 1., **3.667** 8., **3.681** 2.

9. $\mathrm{B}(x, x) = \frac{1}{2^{2x-2}} \int_0^1 (1-t^2)^{x-1} dt = \frac{1}{2^{2x-1}} \int_0^1 \frac{(1-t)^{x-1}}{\sqrt{t}} dt.$

See **8.384** 4., **8.382** 3., and also **3.621** 1., **3.642** 2., **3.665** 1., **3.821** 6., **3.839** 6.

10. $\mathrm{B}(x+y, x-y) = 4^{1-x} \int_0^\infty \frac{\operatorname{ch} 2yt}{\operatorname{ch}^{2x} t} dt$ $[\operatorname{Re} x > |\operatorname{Re} y|, \operatorname{Re} x > 0]$. MO 9

11. $\mathrm{B}\left(x, \frac{y}{z}\right) = z \int_0^1 (1-t^z)^{x-1} t^{y-1} dt$ $\left[\operatorname{Re} z > 0, \operatorname{Re} \frac{y}{z} > 0, \operatorname{Re} x > 0\right]$. FI II 787a

8.381

1. $\int_{-\infty}^{\infty} \frac{dt}{(a+it)^x (b-it)^y} = \frac{2\pi (a+b)^{1-x-y}}{(x+y-1)\,\mathrm{B}(x, y)}$

2. $\int_{-\infty}^{\infty} \frac{dt}{(a-it)^x (b-it)^y} = 0$

$[a > 0, b > 0; x$ and y are real, $x+y > 1]$. MO 7

3. $\mathrm{B}(x+iy, x-iy) = 2^{1-2x} \alpha e^{-2i\gamma y} \int_{-\infty}^{\infty} \frac{e^{2i\alpha y t}\, dt}{\operatorname{ch}^{2x}(\alpha t - \gamma)}$

$[y, \alpha, \gamma$ are real, $\alpha > 0; \operatorname{Re} x > 0]$. MO 8a

For an integral representation of $\ln \mathrm{B}(x, y)$, see **3.428** 7.

4. $\frac{1}{\mathrm{B}(x, y)} = \frac{2^{x+y-1}(x+y-1)}{\pi} \int_0^{\frac{\pi}{2}} \cos[(x-y)t] \cos^{x+y-2} t\, dt;$ NH 158(5)a

$= \frac{2^{x+y-2}(x+y-1)}{\pi \cos\left[(x-y)\frac{\pi}{2}\right]} \int_0^{\pi} \cos[(x-y)t] \sin^{x+y-2} t\, dt;$ NH 159(8)a

$= \frac{2^{x+y-2}(x+y-1)}{\pi \sin\left[(x-y)\frac{\pi}{2}\right]} \int_0^{\pi} \sin[(x-y)t] \sin^{x+y-2} t\, dt.$ NH 159(9)a

Series representation

8.382

1. $$\mathrm{B}(x, y)=\frac{1}{y}\sum_{n=0}^{\infty}(-1)^n\frac{y(y-1)\ldots(y-n)}{n!\,(x+n)} \qquad [y>0].$$ WH

2. $$\ln \mathrm{B}\left(\frac{1+x}{2}, \frac{1}{2}\right)=\ln\sqrt{2\pi}+\frac{1}{2}\left[\ln\left(\frac{\operatorname{tg}\frac{\pi x}{2}}{x}\right)-\ln\left(\frac{1+x}{1-x}\right)\right]+$$
$$+\sum_{k=0}^{\infty}\frac{1-(1-2^{-2k})\,\zeta(2k+1)}{2k+1}x^{2k+1} \qquad [|x|<2].$$ NH 39(17)

3. $$\mathrm{B}\left(z, \frac{1}{2}\right)=\sum_{k=1}^{\infty}\frac{(2k-1)!!}{2^k k!}\frac{1}{z+k}+\frac{1}{z}$$ (see also 8.384 and 8.380 9.).

WH

8.383 Infinite-product representation:

$$(x+y+1)\,\mathrm{B}(x+1, y+1)=\prod_{k=1}^{\infty}\frac{k(x+y+k)}{(x+k)(y+k)} \qquad [x, y\neq -1, -2, \ldots].$$ MO 2

8.384 Functional relations involving the beta function:

1. $$\mathrm{B}(x, y)=\frac{\Gamma(x)\,\Gamma(y)}{\Gamma(x+y)}=\mathrm{B}(y, x).$$ FI II 779

2. $$\mathrm{B}(x, y)\,\mathrm{B}(x+y, z)=\mathrm{B}(y, z)\,\mathrm{B}(y+z, x).$$ MO 6

3. $$\sum_{k=0}^{\infty}\mathrm{B}(x, y+k)=\mathrm{B}(x-1, y).$$ WH

4. $$\mathrm{B}(x, x)=2^{1-2x}\mathrm{B}\left(\frac{1}{2}, x\right)$$ (see also 8.380 9. and 8.382 3.). FI II 784

5. $$\mathrm{B}(x, x)\,\mathrm{B}\left(x+\frac{1}{2}, x+\frac{1}{2}\right)=\frac{\pi}{2^{4x-1}x}.$$ WH

6. $$\frac{1}{\mathrm{B}(n, m)}=m\binom{n+m-1}{n-1}=n\binom{n+m-1}{m-1}$$
[m and n — natural numbers].

For a connection with the psi function, see 4.253 1.

8.39 The incomplete beta function $\mathrm{B}_x(p, q)$

8.391 $$\mathrm{B}_x(p, q)=\int_0^x t^{p-1}(1-t)^{q-1}\,dt=\frac{x^p}{p}\,{}_2F_1(p, 1-q; p+1; x).$$ ET I 373

8.392 $$I_x(p, q)=\frac{\mathrm{B}_x(p, q)}{\mathrm{B}(p, q)}.$$ ET II 429

8.4-8.5 Bessel Functions and Functions Associated with Them

8.40 Definitions

8.401. Bessel functions $Z_\nu(z)$ are solutions of the differential equation

$$\frac{d^2Z_\nu}{dz^2}+\frac{1}{z}\frac{dZ_\nu}{dz}+\left(1-\frac{\nu^2}{z^2}\right)Z_\nu=0.$$ KU 37(1)

Special types of Bessel functions are what are called Bessel functions of the first kind $J_\nu(z)$, Bessel functions of the second kind $N_\nu(z)$ (also called Neumann functions), and Bessel functions of the third kind $H_\nu^{(1)}(z)$ and $H_\nu^{(2)}(z)$ (also called Hankel's functions).

8.402 $$J_\nu(z)=\frac{z^\nu}{2^\nu}\sum_{k=0}^{\infty}(-1)^k\frac{z^{2k}}{2^{2k}k!\,\Gamma(\nu+k+1)}\quad[|\arg z|<\pi].$$ KU 55(1)

8.403

1. $$N_\nu(z)=\frac{1}{\sin\nu\pi}[\cos\nu\pi J_\nu(z)-J_{-\nu}(z)]\quad[\text{for nonintegral } \nu,\ |\arg z|<\pi].$$ KU 41(3)

2. $$\pi N_n(z)=2J_n(z)\ln\frac{z}{2}-\sum_{k=0}^{n-1}\frac{(n-k-1)!}{k!}\left(\frac{z}{2}\right)^{2k-n}-$$
$$-\sum_{k=0}^{\infty}(-1)^k\frac{1}{k!\,(k+n)!}\left(\frac{z}{2}\right)^{n+2k}[\psi(k+1)+\psi(k+n+1)];$$ KU 43(10)
$$=2J_n(z)\left(\ln\frac{z}{2}+C\right)-\sum_{k=0}^{n-1}\frac{(n-k-1)!}{k!}\left(\frac{z}{2}\right)^{2k-n}-$$
$$-\left(\frac{z}{2}\right)^n\frac{1}{n!}\sum_{k=1}^{n}\frac{1}{k}-\sum_{k=1}^{\infty}\frac{(-1)^k\left(\frac{z}{2}\right)^{n+2k}}{k!\,(k+n)!}\left[\sum_{m=1}^{n+k}\frac{1}{m}+\sum_{m=1}^{k}\frac{1}{m}\right]$$
$[n+1$ — a natural number, $|\arg z|<\pi]$. KU 44, WA 75(3)a

8.404

1. $N_{-n}(z)=(-1)^nN_n(z)$
2. $J_{-n}(z)=(-1)^nJ_n(z)$

$[n$ — a natural number$]$. KU 41(2)

8.405

1. $H_\nu^{(1)}(z)=J_\nu(z)+iN_\nu(z).$
2. $H_\nu^{(2)}(z)=J_\nu(z)-iN_\nu(z).$

KU 44(1)

In all relationships that hold for an arbitrary Bessel function $Z_\nu(z)$, that is, for the functions $J_\nu(z)$, $N_\nu(z)$, and linear combinations of them, for example, $H_\nu^{(1)}(z)$ and $H_\nu^{(2)}(z)$, we shall write simply the letter Z instead of the letters J, N, $H^{(1)}$, and $H^{(2)}$.

Bessel functions of imaginary argument $I_\nu(z)$ and $K_\nu(z)$

8.406

1. $I_\nu(z) = e^{-\frac{\pi}{2}\nu i} J_\nu\left(e^{\frac{\pi}{2}i} z\right) \quad \left[-\pi < \arg z \leqslant \frac{\pi}{2}\right].$ WA 92

2. $I_\nu(z) = e^{\frac{3}{2}\pi\nu i} J_\nu\left(e^{-\frac{3}{2}\pi i} z\right) \quad \left[\frac{\pi}{2} < \arg z \leqslant \pi\right].$ WA 92

For integral ν,

3. $I_n(z) = i^{-n} J_n(iz).$ KU 46(1)

8.407

1. $K_\nu(z) = \frac{\pi i}{2} e^{\frac{\pi}{2}\nu i} H_\nu^{(1)}(iz).$

2. $K_\nu(z) = \frac{\pi i}{2} e^{-\frac{\pi}{2}\nu i} H_{-\nu}^{(1)}(iz).$ WA 92(8)

For the differential equation defining these functions, see 8.494.

8.41 Integral representations of the functions $J_\nu(z)$ and $N_\nu(z)$

8.411

1. $J_n(z) = \frac{1}{2\pi} \int_{-\pi}^{\pi} e^{-ni\theta + iz\sin\theta}\, d\theta;$

$$= \frac{1}{\pi} \int_0^{\pi} \cos(n\theta - z\sin\theta)\, d\theta \quad [n \text{ — a natural number}].$$ WH

2. $J_{2n}(z) = \frac{1}{\pi} \int_0^{\pi} \cos 2n\theta \cos(z\sin\theta)\, d\theta = \frac{2}{\pi} \int_0^{\frac{\pi}{2}} \cos 2n\theta \cos(z\sin\theta)\, d\theta$

$[n \text{ — an integer}].$ WA 30(7)

3. $J_{2n+1}(z) = \frac{1}{\pi} \int_0^{\pi} \sin(2n+1)\theta \sin(z\sin\theta)\, d\theta =$

$$= \frac{2}{\pi} \int_0^{\frac{\pi}{2}} \sin(2n+1)\theta \sin(z\sin\theta)\, d\theta \quad [n \text{ — an integer}].$$ WA 30(6)

4. $J_\nu(z) = 2 \frac{\left(\frac{z}{2}\right)^\nu}{\Gamma\left(\nu+\frac{1}{2}\right)\Gamma\left(\frac{1}{2}\right)} \int_0^{\frac{\pi}{2}} \sin^{2\nu}\theta \cos(z\cos\theta)\, d\theta \quad \left[\operatorname{Re}\nu > -\frac{1}{2}\right].$ WH

5. $J_\nu(z) = \frac{\left(\frac{z}{2}\right)^\nu}{\Gamma\left(\nu+\frac{1}{2}\right)\Gamma\left(\frac{1}{2}\right)} \int_0^{\pi} \sin^{2\nu}\theta \cos(z\cos\theta)\, d\theta \quad \left[\operatorname{Re}\nu > -\frac{1}{2}\right].$

6. $$J_\nu(z) = \frac{\left(\frac{z}{2}\right)^\nu}{\Gamma\left(\nu+\frac{1}{2}\right)\Gamma\left(\frac{1}{2}\right)} \int_{-\frac{\pi}{2}}^{\frac{\pi}{2}} \cos(z \sin\theta) \cos^{2\nu}\theta \, d\theta \qquad \left[\operatorname{Re}\nu > -\frac{1}{2}\right].$$

74 KU 65(5), WA 35(4)a

7. $$J_\nu(z) = \frac{\left(\frac{z}{2}\right)^\nu}{\Gamma\left(\nu+\frac{1}{2}\right)\Gamma\left(\frac{1}{2}\right)} \int_0^\pi e^{\pm iz\cos\varphi} \sin^{2\nu}\varphi \, d\varphi \quad \left[\operatorname{Re}\left(\nu+\frac{1}{2}\right) > 0\right].$$

WH

8. $$J_\nu(z) = \frac{\left(\frac{z}{2}\right)^\nu}{\Gamma\left(\nu+\frac{1}{2}\right)\Gamma\left(\frac{1}{2}\right)} \int_{-1}^{1} (1-t^2)^{\nu-\frac{1}{2}} \cos zt \, dt \qquad \left[\operatorname{Re}\nu > -\frac{1}{2}\right].$$

KU 65(6), WH

9. $$J_\nu(x) = 2\frac{\left(\frac{x}{2}\right)^{-\nu}}{\Gamma\left(\frac{1}{2}-\nu\right)\Gamma\left(\frac{1}{2}\right)} \int_1^\infty \frac{\sin xt}{(t^2-1)^{\nu+\frac{1}{2}}} dt$$

$$\left[-\frac{1}{2} < \operatorname{Re}\nu < \frac{1}{2},\ x > 0\right].$$ MO 37

10. $$J_\nu(z) = \frac{\left(\frac{z}{2}\right)^\nu}{\Gamma\left(\nu+\frac{1}{2}\right)\Gamma\left(\frac{1}{2}\right)} \int_{-1}^{1} e^{izt} (1-t^2)^{\nu-\frac{1}{2}} \, dt \qquad \left[\operatorname{Re}\nu > -\frac{1}{2}\right].$$

WA 34(3)

11. $$J_\nu(x) = \frac{2}{\pi} \int_0^\infty \sin\left(x \operatorname{ch} t - \frac{\nu\pi}{2}\right) \operatorname{ch} \nu t \, dt.$$ WA 199(12)

12. $$J_\nu(z) = \frac{2^{\nu+1} z^\nu}{\Gamma\left(\nu+\frac{1}{2}\right)\Gamma\left(\frac{1}{2}\right)} \int_0^{\frac{\pi}{2}} \frac{\cos^{\nu-\frac{1}{2}}\theta \sin\left(z - \nu\theta + \frac{1}{2}\theta\right)}{\sin^{2\nu+1}\theta} e^{-2z\operatorname{ctg}\theta} \, d\theta$$

$$\left[|\arg z| < \frac{\pi}{2},\ \operatorname{Re}\left(\nu+\frac{1}{2}\right) > 0\right].$$ WH

13. $$J_\nu(z) = \frac{1}{\pi} \int_0^\pi \cos(\nu\theta - z\sin\theta) \, d\theta - \frac{\sin\nu\pi}{\pi} \int_0^\infty e^{-\nu\theta - z\operatorname{sh}\theta} \, d\theta$$

[ν — arbitrary, $\operatorname{Re} z > 0$]. WA 195(4)

14. $$J_\nu(z) = \frac{e^{\pm\nu\pi i}}{\pi} \left[\int_0^\pi \cos(\nu\theta + z\sin\theta) \, d\theta - \sin\nu\pi \int_0^\infty e^{-\nu\theta + z\operatorname{sh}\theta} \, d\theta\right]$$

[for $\frac{\pi}{2} < |\arg z| < \pi$, with the upper sign taken for $\arg z > \frac{\pi}{2}$ and the lower sign taken for $\arg z < -\frac{\pi}{2}$]. WH

8.412

1. $$J_\nu(z)=\frac{1}{2\pi i}\int_{-\infty}^{(0+)} t^{-\nu-1}\exp\left[\frac{z}{2}\left(t-\frac{1}{t}\right)\right]dt \qquad \left[|\arg z|<\frac{\pi}{2}\right].$$
WH, WA 195(2)

2. $$J_\nu(z)=\frac{z^\nu}{2^{\nu+1}\pi i}\int_{-\infty}^{(0+)} t^{-\nu-1}\exp\left(t-\frac{z^2}{4t}\right)dt.$$
WA 195(1)

3. $$J_\nu(z)=\frac{z}{2^{\nu+1}\pi i}\sum_{k=1}^{\infty}\frac{(-1)^k z^{2k}}{2^{2k}k!}\int_{-\infty}^{(0+)} e^t t^{-\nu-k-1}\,dt.$$
WA 195(1)

4. $$J_\nu(x)=\frac{1}{2\pi i}\int_{-i\infty}^{i\infty}\frac{\Gamma(-t)}{\Gamma(\nu+t+1)}\left(\frac{x}{2}\right)^{\nu+2t}dt \qquad [\operatorname{Re}\nu\geqslant 0,\ x>0].$$
WA 214(7)

5. $$J_\nu(z)=\frac{\Gamma\left(\frac{1}{2}-\nu\right)\left(\frac{z}{2}\right)^\nu}{2\pi i\,\Gamma\left(\frac{1}{2}\right)}\int_A^{(1+,\,-1-)}(t^2-1)^{\nu-\frac{1}{2}}\cos(zt)\,dt$$

$\left[\left\{\Gamma\left(\frac{1}{2}-\nu\right)\right\}^{-1}\neq 0;\right.$ The point A falls to the right of the point $t=1$, and $\arg(t-1)=\arg(t+1)=0$ at the point $\left.A\right]$. WH

6. $$J_\nu(z)=\frac{1}{2\pi}\int_{-\pi+\infty i}^{\pi+\infty i} e^{-iz\sin\theta+i\nu\theta}\,d\theta \qquad [\operatorname{Re} z>0].$$

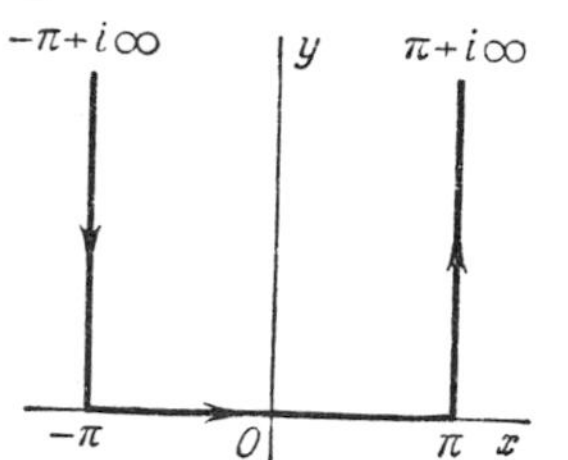

The path of integration is shown in the drawing.

8.413 $$\frac{J_\nu(\sqrt{z^2-\zeta^2})}{(z^2-\zeta^2)^{\frac{\nu}{2}}}=\frac{1}{\pi(z+\zeta)^\nu}\left\{\int_0^\infty e^{\zeta\cos t}\cos(z\sin t-\nu t)\,dt-\right.$$
$$\left.-\sin\nu\pi\int_0^\infty \exp(-z\operatorname{sh} t-\zeta\operatorname{ch} t-\nu t)\,dt\right\} \qquad [\operatorname{Re}(z+\zeta)>0].$$
MO 40

8.414 $$\int_{2x}^{\infty}\frac{J_0(t)}{t}\,dt=\frac{1}{4\pi}\int_{-\frac{1}{2}-i\infty}^{-\frac{1}{2}+i\infty}\frac{\Gamma(-t)}{t\Gamma(1+t)}x^{2t}\,dt \qquad [x>0].$$
MO 41

See 3.715 2., 9., 10., 13., 14., 19.—21., 3.865 1., 2., 4., 3.996 4. For an integral representation of $J_0(z)$ see 3.714 2., 3.753 2., 3., and 4.124. For an integral representation of $J_1(z)$ see 3.697, 3.711, 3.752 2., and 3.753 5.

8.415

1. $$N_0(x)=\frac{4}{\pi^2}\int_0^1\frac{\arcsin t}{\sqrt{1-t^2}}\sin(xt)\,dt-\frac{4}{\pi^2}\int_1^\infty\frac{\ln(t+\sqrt{t^2-1})}{\sqrt{t^2-1}}\sin(xt)\,dt$$
$$[x>0].$$
MO 37

2. $$N_\nu(x) = -2\frac{\left(\frac{x}{2}\right)^{-\nu}}{\Gamma\left(\frac{1}{2}-\nu\right)\Gamma\left(\frac{1}{2}\right)}\int_1^\infty \frac{\cos xt}{(t^2-1)^{\nu+\frac{1}{2}}}\,dt$$
$$\left[-\frac{1}{2} < \operatorname{Re}\nu < \frac{1}{2},\quad x > 0\right].$$ KU 89(28)a, MO 38

3. $$N_\nu(x) = -\frac{2}{\pi}\int_0^\infty \cos\left(x \operatorname{ch} t - \frac{\nu\pi}{2}\right)\operatorname{ch}\nu t\,dt \qquad [-1 < \operatorname{Re}\nu < 1,\ x > 0].$$ WA 199(13)

4. $$N_\nu(z) = \frac{1}{\pi}\int_0^\pi \sin(z\sin\theta - \nu\theta)\,d\theta - $$
$$-\frac{1}{\pi}\int_0^\infty (e^{\nu t} + e^{-\nu t}\cos\nu\pi)\,e^{-z\operatorname{sh} t}\,dt \qquad [\operatorname{Re} z > 0].$$ WA 197(1)

5. $$N_\nu(z) = \frac{2\left(\frac{z}{2}\right)^\nu}{\Gamma\left(\nu+\frac{1}{2}\right)\Gamma\left(\frac{1}{2}\right)}\left[\int_0^{\frac{\pi}{2}} \sin(z\sin\theta)\cos^{2\nu}\theta\,d\theta - \right.$$
$$\left. -\int_0^\infty e^{-z\operatorname{sh}\theta}\operatorname{ch}^{2\nu}\theta\,d\theta\right] \qquad \left[\operatorname{Re}\nu > -\frac{1}{2},\quad \operatorname{Re} z > 0\right].$$ WA 181(5)a

6. $$N_\nu(z) = -\frac{2^{\nu+1}z^\nu}{\Gamma\left(\nu+\frac{1}{2}\right)\Gamma\left(\frac{1}{2}\right)}\int_0^{\frac{\pi}{2}} \frac{\cos^{\nu-\frac{1}{2}}\theta\cos\left(z - \nu\theta + \frac{1}{2}\theta\right)}{\sin^{2\nu+1}\theta}\,e^{-2z\operatorname{ctg}\theta}\,d\theta$$
$$\left[|\arg z| < \frac{\pi}{2},\qquad \operatorname{Re}\left(\nu+\frac{1}{2}\right) > 0\right].$$ WA 186(8)

For an integral representation of $N_0(z)$, see 3.714 3., 3.753 4., 3.864. See also 3.865 3.

8.42 Integral representations of the functions $H_\nu^{(1)}(z)$ and $H_\nu^{(2)}(z)$

8.421

1. $$H_\nu^{(1)}(x) = \frac{e^{-\frac{\nu\pi i}{2}}}{\pi i}\int_{-\infty}^\infty e^{ix\operatorname{ch} t - \nu t}\,dt = $$
$$= \frac{2e^{-\frac{\nu\pi i}{2}}}{\pi i}\int_0^\infty e^{ix\operatorname{ch} t}\operatorname{ch}\nu t\,dt$$ WA 199(10)

2. $$H_\nu^{(2)}(x) = -\frac{e^{\frac{\nu\pi i}{2}}}{\pi i}\int_{-\infty}^\infty e^{-ix\operatorname{ch} t - \nu t}\,dt = $$
$$= -\frac{2e^{\frac{\nu\pi i}{2}}}{\pi i}\int_0^\infty e^{-ix\operatorname{ch} t}\operatorname{ch}\nu t\,dt$$ WA 199(11)

$[-1 < \operatorname{Re}\nu < 1,\ x > 0]$ (applies to 1 and 2).

3. $$H_\nu^{(1)}(z) = -\frac{2^{\nu+1} i z^\nu}{\Gamma\left(\nu+\frac{1}{2}\right)\Gamma\left(\frac{1}{2}\right)} \int_0^{\frac{\pi}{2}} \frac{\cos^{\nu-\frac{1}{2}} t\, e^{i\left(z-\nu t+\frac{t}{2}\right)}}{\sin^{2\nu+1} t} \exp(-2z \operatorname{ctg} t)\, dt$$

$$\left[\operatorname{Re}\nu > -\frac{1}{2}, \quad \operatorname{Re} z > 0\right].$$ WA 186(5)

4. $$H_\nu^{(2)}(z) = \frac{2^{\nu+1} i z^\nu}{\Gamma\left(\nu+\frac{1}{2}\right)\Gamma\left(\frac{1}{2}\right)} \int_0^{\frac{\pi}{2}} \frac{\cos^{\nu-\frac{1}{2}} t\, e^{-i\left(z-\nu t+\frac{t}{2}\right)}}{\sin^{2\nu+1} t} \exp(-2z \operatorname{ctg} t)\, dt$$

$$\left[\operatorname{Re}\nu > -\frac{1}{2}, \quad \operatorname{Re} z > 0\right].$$ WA 186(6)

5. $$H_\nu^{(1)}(x) = -\frac{2i\left(\frac{x}{2}\right)^{-\nu}}{\sqrt{\pi}\,\Gamma\left(\frac{1}{2}-\nu\right)} \int_1^\infty \frac{e^{ixt}}{(t^2-1)^{\nu+\frac{1}{2}}}\, dt$$

$$\left[-\frac{1}{2} < \operatorname{Re}\nu < \frac{1}{2}, \quad x > 0\right].$$ WA 187(1)

6. $$H_\nu^{(2)}(x) = \frac{2i\left(\frac{x}{2}\right)^{-\nu}}{\sqrt{\pi}\,\Gamma\left(\frac{1}{2}-\nu\right)} \int_1^\infty \frac{e^{-ixt}}{(t^2-1)^{\nu+\frac{1}{2}}}\, dt$$

$$\left[-\frac{1}{2} < \operatorname{Re}\nu < \frac{1}{2}, \quad x > 0\right].$$ WA 187(2)

7. $$H_\nu^{(1)}(z) = -\frac{i}{\pi} e^{-\frac{1}{2} i\nu\pi} \int_0^\infty \exp\left[\frac{1}{2} iz\left(t+\frac{1}{t}\right)\right] t^{-\nu-1}\, dt$$

$$[0 < \arg z < \pi; \quad \text{or} \quad \arg z = 0 \text{ and } -1 < \operatorname{Re}\nu < 1].$$ MO 38

8. $$H_\nu^{(1)}(xz) = -\frac{i}{\pi} e^{-\frac{1}{2} i\nu\pi} z^\nu \int_0^\infty \exp\left[\frac{1}{2} ix\left(t+\frac{z^2}{t}\right)\right] t^{-\nu-1}\, dt$$

$$\left[0 < \arg z < \frac{\pi}{2}, \quad x > 0, \quad \operatorname{Re}\nu > -1; \quad \text{or} \quad \arg z = \frac{\pi}{2}, \quad x > 0 \text{ and } -1 < \operatorname{Re}\nu < 1\right].$$ MO 38

9. $$H_\nu^{(1)}(xz) = \sqrt{\frac{2}{\pi z}}\, \frac{x^\nu \exp\left[i\left(xz - \frac{\pi}{2}\nu - \frac{\pi}{4}\right)\right]}{\Gamma\left(\nu+\frac{1}{2}\right)} \int_0^\infty \left(1+\frac{it}{2z}\right)^{\nu-\frac{1}{2}} t^{\nu-\frac{1}{2}} e^{-xt}\, dt$$

$$\left[\operatorname{Re}\nu > -\frac{1}{2}, \quad -\frac{\pi}{2} < \arg z < \frac{3}{2}\pi, \quad x > 0\right].$$ MO 39

10. $$H_\nu^{(1)}(z) = \frac{-2ie^{-i\nu\pi}\left(\frac{z}{2}\right)^\nu}{\sqrt{\pi}\,\Gamma\left(\nu+\frac{1}{2}\right)} \int_0^\infty e^{iz \operatorname{ch} t} \operatorname{sh}^{2\nu} t\, dt$$

$$\left[0 < \arg z < \pi, \; \operatorname{Re}\nu > -\frac{1}{2} \quad \text{or} \quad \arg z = 0 \text{ and } -\frac{1}{2} < \operatorname{Re}\nu < \frac{1}{2}\right].$$

MO 38

11. $H_0^{(1)}(x) = -\frac{i}{\pi}\int_{-\infty}^{\infty} \frac{\exp(i\sqrt{x^2+t^2})}{\sqrt{x^2+t^2}}\,dt \quad [x > 0].$ MO 38

8.422

1. $H_\nu^{(1)}(z) = \frac{\Gamma\left(\frac{1}{2}-\nu\right)\left(\frac{z}{2}\right)^\nu}{\pi i \Gamma\left(\frac{1}{2}\right)} \int_{1+\infty i}^{(1+)} e^{izt}(t^2-1)^{\nu-\frac{1}{2}}\,dt \quad [-\pi < \arg z < 2\pi].$ WA 183(4)

2. $H_\nu^{(2)}(z) = \frac{\Gamma\left(\frac{1}{2}-\nu\right)\left(\frac{z}{2}\right)^\nu}{\pi i \Gamma\left(\frac{1}{2}\right)} \times$

$\times \int_{-1+\infty i}^{(-1-)} e^{izt}(t^2-1)^{\nu-\frac{1}{2}}\,dt$

$[-2\pi < \arg z < \pi].$

The paths of integration are shown in the drawing.

8.423

1. $H_\nu^{(1)}(z) = -\frac{1}{\pi}\int_{-\infty i}^{-\pi+\infty i} e^{-iz\sin\theta+i\nu\theta}\,d\theta \quad [\operatorname{Re} z > 0].$ WA 197(2)a

2. $H_\nu^{(2)}(z) = -\frac{1}{\pi}\int_{\pi+\infty i}^{-\infty i} e^{-iz\sin\theta+i\nu\theta}\,d\theta \quad [\operatorname{Re} z > 0].$ WA 197(3)a

The path of integration for formula 1 is shown in the left hand drawing and for formula 2 in the right hand drawing.

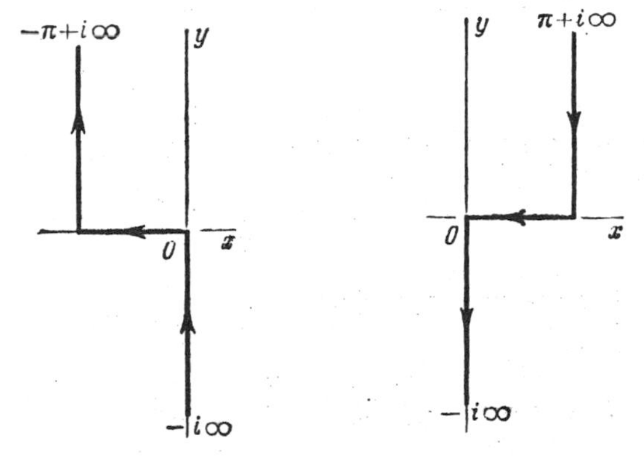

8.424

1. $H_\nu^{(1)}(z)\,J_\nu(\zeta) = \frac{1}{\pi i}\int_0^{\gamma+i\infty} \exp\left[\frac{1}{2}\left(t - \frac{z^2+\zeta^2}{t}\right)\right] I_\nu\left(\frac{z\zeta}{t}\right)\frac{dt}{t}$

2. $H_\nu^{(2)}(z)\,J_\nu(\zeta) = \frac{i}{\pi}\int_0^{\gamma-i\infty} \exp\left[\frac{1}{2}\left(t - \frac{z^2+\zeta^2}{t}\right)\right] I_\nu\left(\frac{z\zeta}{t}\right)\frac{dt}{t}$

$[\gamma > 0, \quad \operatorname{Re}\nu > -1, \quad |\zeta| < |z|].$ MO 45

8.43 Integral representations of the functions $I_\nu(z)$ and $K_\nu(z)$

The function $I_\nu(z)$

8.431

1. $$I_\nu(z) = \frac{\left(\frac{z}{2}\right)^\nu}{\Gamma\left(\nu+\frac{1}{2}\right)\Gamma\left(\frac{1}{2}\right)} \int_{-1}^{1} (1-t^2)^{\nu-\frac{1}{2}} e^{\pm zt}\, dt$$

2. $$I_\nu(z) = \frac{\left(\frac{z}{2}\right)^\nu}{\Gamma\left(\nu+\frac{1}{2}\right)\Gamma\left(\frac{1}{2}\right)} \int_{-1}^{1} (1-t^2)^{\nu-\frac{1}{2}} \operatorname{ch} zt\, dt$$

3. $$I_\nu(z) = \frac{\left(\frac{z}{2}\right)^\nu}{\Gamma\left(\nu+\frac{1}{2}\right)\Gamma\left(\frac{1}{2}\right)} \int_{0}^{\pi} e^{\pm z\cos\theta} \sin^{2\nu}\theta\, d\theta$$

4. $$I_\nu(z) = \frac{\left(\frac{z}{2}\right)^\nu}{\Gamma\left(\nu+\frac{1}{2}\right)\Gamma\left(\frac{1}{2}\right)} \int_{0}^{\pi} \operatorname{ch}(z\cos\theta) \sin^{2\nu}\theta\, d\theta$$

$\left[\operatorname{Re}\left(\nu+\frac{1}{2}\right)>0\right]$. WA 94(9)

5. $$I_\nu(z) = \frac{1}{\pi}\int_0^\pi e^{z\cos\theta}\cos\nu\theta\, d\theta - \frac{\sin\nu\pi}{\pi}\int_0^\infty e^{-z\operatorname{ch} t-\nu t}\, dt$$

$\left[|\arg z| \leqslant \frac{\pi}{2},\quad \operatorname{Re}\nu>0\right]$. WA 201(4)

See also 3.383 2., 3.387 1., 3.471 6., 3.714 5.

For an integral representation of $I_0(z)$ and $I_1(z)$, see 3.366 1., 3.534, 3.856 6.

The function $K_\nu(z)$

8.432

1. $$K_\nu(z) = \int_0^\infty e^{-z\operatorname{ch} t}\operatorname{ch}\nu t\, dt$$

$\left[|\arg z| < \frac{\pi}{2}\quad \text{or}\quad \operatorname{Re} z=0 \text{ and } \nu=0\right]$. MO 39

2. $$K_\nu(z) = \frac{\left(\frac{z}{2}\right)^\nu\Gamma\left(\frac{1}{2}\right)}{\Gamma\left(\nu+\frac{1}{2}\right)}\int_0^\infty e^{-z\operatorname{ch} t}\operatorname{sh}^{2\nu} t\, dt$$

$\left[\operatorname{Re}\nu>-\frac{1}{2},\ \operatorname{Re} z>0;\quad \text{or}\quad \operatorname{Re} z=0 \text{ and } -\frac{1}{2}<\operatorname{Re}\nu<\frac{1}{2}\right]$.

WA 190(5), WH

3. $$K_\nu(z) = \frac{\left(\frac{z}{2}\right)^\nu\Gamma\left(\frac{1}{2}\right)}{\Gamma\left(\nu+\frac{1}{2}\right)}\int_1^\infty e^{-zt}(t^2-1)^{\nu-\frac{1}{2}}\, dt$$

$\left[\operatorname{Re}\left(\nu+\frac{1}{2}\right)>0,\ |\arg z|<\frac{\pi}{2};\quad \text{or}\quad \operatorname{Re} z=0 \text{ and } \nu=0\right]$.

WA 190(4)

4. $K_\nu(x) = \frac{1}{\cos\frac{\nu\pi}{2}} \int_0^\infty \cos(x \operatorname{sh} t) \operatorname{ch} \nu t \, dt$

$[x > 0, \quad -1 < \operatorname{Re} \nu < 1]$. WA 202(13)

5. $K_\nu(xz) = \frac{\Gamma\left(\nu+\frac{1}{2}\right)(2z)^\nu}{x^\nu \Gamma\left(\frac{1}{2}\right)} \int_0^\infty \frac{\cos xt \, dt}{(t^2+z^2)^{\nu+\frac{1}{2}}}$

$\left[\operatorname{Re}\left(\nu+\frac{1}{2}\right) \geqslant 0, \ x > 0, \ |\arg z| < \frac{\pi}{2}\right]$. WA 191(1)

6. $K_\nu(z) = \frac{1}{2}\left(\frac{z}{2}\right)^\nu \int_0^\infty \frac{e^{-t-\frac{z^2}{4t}} \, dt}{t^{\nu+1}}$ $\left[|\arg z| < \frac{\pi}{2}, \ \operatorname{Re} z^2 > 0\right]$.

WA 203(15)

7. $K_\nu(xz) = \frac{z^\nu}{2} \int_0^\infty \exp\left[-\frac{x}{2}\left(t+\frac{z^2}{t}\right)\right] t^{-\nu-1} \, dt$

$\left[|\arg z| < \frac{\pi}{4} \text{ or } |\arg z| = \frac{\pi}{4} \text{ and } \operatorname{Re} \nu < 1\right]$. MO 39

8. $K_\nu(xz) = \sqrt{\frac{\pi}{2z}} \frac{x^\nu e^{-xz}}{\Gamma\left(\nu+\frac{1}{2}\right)} \int_0^\infty e^{-xt} t^{\nu-\frac{1}{2}} \left(1+\frac{t}{2z}\right)^{\nu-\frac{1}{2}} dt$

$\left[|\arg z| < \pi, \ \operatorname{Re} \nu > -\frac{1}{2}, \ x > 0\right]$. MO 39

9. $K_\nu(xz) = \frac{\sqrt{\pi}}{\Gamma\left(\nu+\frac{1}{2}\right)} \left(\frac{x}{2z}\right)^\nu \int_0^\infty \frac{\exp\left(-x\sqrt{t^2+z^2}\right)}{\sqrt{t^2+z^2}} t^{2\nu} \, dt$

$\left[\operatorname{Re} \nu > -\frac{1}{2}, \ \operatorname{Re} z > 0, \ \operatorname{Re} \sqrt{t^2+z^2} > 0, \ x > 0\right]$. MO 39

See also **3.337** 4., **3.383** 3., **3.387** 3., 6., **3.388** 2., **3.389** 4., **3.391**, **3.395** 1., **3.471** 9., **3.483**, **3.547** 2., **3.856**, **3.871** 3.,4., **7.141** 5.

8.433 $K_{\frac{1}{3}}\left(\frac{2x\sqrt{x}}{3\sqrt{3}}\right) = \frac{3}{\sqrt{x}} \int_0^\infty \cos(t^3+xt) \, dt$. KU 98(31), WA 211(2)

For an integral representation of $K_0(z)$, see **3.754** 2., **3.864**, **4.343**, **4.356**, **4.367**..

8.44 Series representation

The function $J_\nu(z)$

8.440 $J_\nu(z) = \left(\frac{z}{2}\right)^\nu \sum_{k=0}^\infty \frac{(-1)^k}{k!\,\Gamma(\nu+k+1)} \left(\frac{z}{2}\right)^{2k}$ $[|\arg z| < \pi]$.

8.441 Special cases:

1. $J_0(z) = \sum_{k=0}^\infty (-1)^k \frac{z^{2k}}{2^{2k}(k!)^2}$.

2. $J_1(z) = -J_0'(z) = \frac{z}{2}\sum_{k=0}^{\infty}\frac{(-1)^k x^{2k}}{2^{2k}k!\,(k+1)!}$.

3. $J_{\frac{1}{3}}(z) = \frac{\sqrt[3]{\frac{z}{2}}}{\Gamma\left(\frac{4}{3}\right)}\sum_{k=0}^{\infty}(-1)^k\frac{(z\sqrt{3})^{2k}}{2^{2k}k!\cdot 1\cdot 4\cdot 7\cdot\ldots\cdot(3k+1)}$.

4. $J_{-\frac{1}{3}}(z) = \frac{1}{\Gamma\left(\frac{2}{3}\right)}\sqrt[3]{\frac{2}{z}}\left\{1+\sum_{k=1}^{\infty}(-1)^k\frac{(z\sqrt{3})^{2k}}{2^{2k}k!\cdot 2\cdot 5\cdot 8\cdot\ldots\cdot(3k-1)}\right.$.

For the expansion of $J_\nu(z)$ in Laguerre polynomials, see 8.975 3.

8.442

1. $J_\nu(z)J_\mu(z) = \sum_{k=0}^{\infty}\frac{(-1)^k\left(\frac{z}{2}\right)^{\nu+\mu+2k}\Gamma(\nu+\mu+2k+1)}{\Gamma(\nu+\mu+k+1)\,\Gamma(\nu+k+1)\,\Gamma(\mu+k+1)}$.

If $\nu^2 \neq \mu^2$, then 2ν, 2μ, and $2(\nu+\mu)$ cannot be negative integers in this formula. If $\nu=\mu$, then 2ν cannot be a negative integer. If $\nu=-\mu$, then ν cannot be a negative integer.

WA 161(5)

2. $J_\nu(az)J_\mu(bz) = \frac{\left(\frac{az}{2}\right)^\nu\left(\frac{bz}{2}\right)^\mu}{\Gamma(\mu+1)}\times$

$\times\sum_{k=0}^{\infty}\frac{(-1)^k\left(\frac{az}{2}\right)^{2k}F\left(-k,\ -\nu-\kappa;\ \mu+1;\ \frac{b^2}{a^2}\right)}{k!\,\Gamma(\nu+k+1)}$. MO 28

The function $N_\nu(z)$

8.443 $N_\nu(z) = \frac{1}{\sin\nu\pi}\left\{\cos\nu\pi\left(\frac{z}{2}\right)^\nu\sum_{k=0}^{\infty}(-1)^k\frac{z^{2k}}{2^{2k}k!\,\Gamma(\nu+k+1)}-\right.$

$\left.-\left(\frac{z}{2}\right)^{-\nu}\sum_{k=0}^{\infty}(-1)^k\frac{z^{2k}}{2^{2k}k!\,\Gamma(k-\nu+1)}\right\}$ [$\nu\neq$ an integer]

(cf. 8.403 1.).

For $\nu+1$ a natural number, see 8.403 2.; for ν a negative integer see 8.404 1

8.444 Special cases,

1. $\pi N_0(z) = 2J_0(z)\left(\ln\frac{z}{2}+C\right)-2\sum_{k=1}^{\infty}\frac{(-1)^k}{(k!)^2}\left(\frac{z}{2}\right)^{2k}\sum_{m=1}^{k}\frac{1}{m}$. KU 44

2. $\pi N_1(z) = 2J_1(z)\left(\ln\frac{z}{2}+C\right)-$

$-\frac{2}{z}-\sum_{k=1}^{\infty}\frac{(-1)^{k+1}\left(\frac{z}{2}\right)^{2k-1}}{k!\,(k-1)!}\left\{\begin{matrix}\times\left[\sum_{r=1}^{n}\frac{1}{r}\right]\text{ when } k=1\\ \times\left[2\sum_{m=1}^{k-1}\frac{1}{m}+\frac{1}{k}\right]\text{ when } k\neq 1\end{matrix}\right\}$.

The functions $I_\nu(z)$ and $K_n(z)$

8.445 $$I_\nu(z) = \sum_{k=0}^{\infty} \frac{1}{k!\,\Gamma(\nu+k+1)} \left(\frac{z}{2}\right)^{\nu+2k}.$$ WH

8.446 $$K_n(z) = \frac{1}{2} \sum_{k=0}^{n-1} (-1)^k \frac{(n-k-1)!}{k! \left(\frac{z}{2}\right)^{n-2k}} +$$

$$+ (-1)^{n+1} \sum_{k=0}^{\infty} \frac{\left(\frac{z}{2}\right)^{n+2k}}{k!\,(n+k)!} \left[\ln\frac{z}{2} - \frac{1}{2}\psi(k+1) - \frac{1}{2}\psi(n+k+1) \right].$$

WA 95(15)

$$= (-1)^{n+1} I_n(z) \ln\frac{Ze^C}{2} + \frac{1}{2}(-1)^n \sum_{l=0}^{\infty} \frac{\left(\frac{z}{2}\right)^{n+2l}}{l!\,(n+l)!} \left(\sum_{k=1}^{l} \frac{1}{k} + \sum_{k=1}^{n+l} \frac{1}{k} \right) +$$

$$+ \frac{1}{2} \sum_{l=0}^{n-1} \frac{(-1)^l (n-l-1)!}{l!} \left(\frac{z}{2}\right)^{2l-n} \qquad [n+1 \text{ — a natural number}].$$

[Omit the term $\sum_{k=1}^{l} \frac{1}{k}$ in line 3 when $l = 0$] MO 29

8.447 Special cases:

1. $$I_0(z) = \sum_{k=0}^{\infty} \frac{\left(\frac{z}{2}\right)^{2k}}{(k!)^2}.$$

2. $$I_1(z) = I_0'(z) = \sum_{k=0}^{\infty} \frac{\left(\frac{z}{2}\right)^{2k+1}}{k!\,(k+1)!}.$$

3. $$K_0(z) = -\ln\frac{z}{2} I_0(z) + \sum_{k=0}^{\infty} \frac{z^{2k}}{2^{2k}(k!)^2} \psi(k+1).$$ WA 95(14)

8.45 Asymptotic expansions of Bessel functions

8.451 For large values of $|z|$ *)

1. $$J_{\pm\nu}(z) = \sqrt{\frac{2}{\pi z}} \left\{ \cos\left(z \mp \frac{\pi}{2}\nu - \frac{\pi}{4}\right) \times \right.$$

$$\times \left[\sum_{k=0}^{n-1} \frac{(-1)^k}{(2z)^{2k}} \frac{\Gamma\left(\nu+2k+\frac{1}{2}\right)}{(2k)!\,\Gamma\left(\nu-2k+\frac{1}{2}\right)} + R_1 \right] -$$

$$\left. - \sin\left(z \mp \frac{\pi}{2}\nu - \frac{\pi}{4}\right) \left[\sum_{k=0}^{n-1} \frac{(-1)^k}{(2z)^{2k+1}} \frac{\Gamma\left(\nu+2k+\frac{3}{2}\right)}{(2k+1)!\,\Gamma\left(\nu-2k-\frac{1}{2}\right)} + R_2 \right] \right\}$$

$[|\arg z| < \pi]$ (see 8.339 4.). WA 222(1, 3)

*An estimate of the remainders in formulas 8.451 is given in 8.451 7. and 8.451 8.

2. $$N_{\pm\nu}(z)=\sqrt{\frac{2}{\pi z}}\left\{\sin\left(z\mp\frac{\pi}{2}\nu-\frac{\pi}{4}\right)\times\right.$$
$$\times\left[\sum_{k=0}^{n-1}\frac{(-1)^k}{(2z)^{2k}}\frac{\Gamma\left(\nu+2k+\frac{1}{2}\right)}{(2k)!\,\Gamma\left(\nu-2k+\frac{1}{2}\right)}+R_1\right]+$$
$$\left.+\cos\left(z\mp\frac{\pi}{2}\nu-\frac{\pi}{4}\right)\left[\sum_{k=0}^{n-1}\frac{(-1)^k}{(2z)^{2k+1}}\frac{\Gamma\left(\nu+2k+\frac{3}{2}\right)}{(2k+1)!\,\Gamma\left(\nu-2k-\frac{1}{2}\right)}+R_2\right]\right\}$$

$[|\arg z|<\pi]$ (see 8.339 4.). WA 222(2, 4, 5)

3. $$H_\nu^{(1)}(z)=\sqrt{\frac{2}{\pi z}}e^{i\left(z-\frac{\pi}{2}\nu-\frac{\pi}{4}\right)}\times$$
$$\times\left[\sum_{k=0}^{n-1}\frac{(-1)^k}{(2iz)^k}\frac{\Gamma\left(\nu+k+\frac{1}{2}\right)}{k!\,\Gamma\left(\nu-k+\frac{1}{2}\right)}+\theta_1\frac{(-1)^n}{(2iz)^n}\frac{\Gamma\left(\nu+n+\frac{1}{2}\right)}{k!\,\Gamma\left(\nu-n+\frac{1}{2}\right)}\right]$$

$\left[\operatorname{Re}\nu>-\frac{1}{2},\ |\arg z|<\pi\right]$ (see 8.339 4.). WA 221(5)

4. $$H_\nu^{(2)}(z)=\sqrt{\frac{2}{\pi z}}e^{-i\left(z-\frac{\pi}{2}\nu-\frac{\pi}{4}\right)}\times$$
$$\times\left[\sum_{k=0}^{n-1}\frac{1}{(2iz)^k}\frac{\Gamma\left(\nu+k+\frac{1}{2}\right)}{k!\,\Gamma\left(\nu-k+\frac{1}{2}\right)}+\theta_2\frac{1}{(2iz)^n}\frac{\Gamma\left(\nu+n+\frac{1}{2}\right)}{n!\,\Gamma\left(\nu-n+\frac{1}{2}\right)}\right]$$

$\left[\operatorname{Re}\nu>-\frac{1}{2},\ |\arg z|<\pi\right]$ (see 8.339 4.). WA 221(6)

For indices of the form $\nu=\frac{2n-1}{2}$ (where n is a natural number), the series 8.451 terminate. In this case, the closed formulas 8.46 are valid for all values.

5. $$I_\nu(z)\sim\frac{e^z}{\sqrt{2\pi z}}\sum_{k=0}^{\infty}\frac{(-1)^k}{(2z)^k}\frac{\Gamma\left(\nu+k+\frac{1}{2}\right)}{k!\,\Gamma\left(\nu-k+\frac{1}{2}\right)}+$$
$$+\frac{\exp\left[-z\pm\left(\nu+\frac{1}{2}\right)\pi i\right]}{\sqrt{2\pi z}}\sum_{k=0}^{\infty}\frac{1}{(2z)^k}\frac{\Gamma\left(\nu+k+\frac{1}{2}\right)}{k!\,\Gamma\left(\nu-k+\frac{1}{2}\right)}.$$

[The sign $+$ is taken for $-\frac{\pi}{2}<\arg z<\frac{3}{2}\pi$, the sign $-$ for $-\frac{3}{2}\pi<\arg z<$ $<\frac{\pi}{2}$*)] (see 8.339 4.).

WA 226(2, 3)

---

*The contradiction that this condition contains at first glance is explained by the so-called Stokes phenomenon (see Watson, G.N., *A Treatise on the Theory of Bessel Functions*, 2nd Edition, Cambridge Univ. Press, 1944, page 201).

6. $$K_\nu(z)=\sqrt{\frac{\pi}{2z}}e^{-z}\left[\sum_{k=0}^{n-1}\frac{1}{(2z)^k}\frac{\Gamma\left(\nu+k+\frac{1}{2}\right)}{k!\Gamma\left(\nu-k+\frac{1}{2}\right)}+\theta_3\frac{\Gamma\left(\nu+n+\frac{1}{2}\right)}{(2z)^n\, n!\,\Gamma\left(\nu-n+\frac{1}{2}\right)}\right]$$

(see 8.339 4.). WA 231, 245(9)

An estimate of the remainders of the asymptotic series in formulas 8.451:

7. $$|R_1|<\left|\frac{\Gamma\left(\nu+2n+\frac{1}{2}\right)}{(2z)^{2n}(2n)!\,\Gamma\left(\nu-2n+\frac{1}{2}\right)}\right| \qquad \left[n>\frac{\nu}{2}-\frac{1}{4}\right].$$ WA 231

8. $$|R_2|<\left|\frac{\Gamma\left(\nu+2n+\frac{3}{2}\right)}{(2z)^{2n+1}(2n+1)!\,\Gamma\left(\nu-2n-\frac{1}{2}\right)}\right| \qquad \left[n\geqslant\frac{\nu}{2}-\frac{3}{4}\right].$$ WA 231

For $-\frac{\pi}{2}<\arg z<\frac{3}{2}\pi$, ν real, and $n+\frac{1}{2}>|\nu|$

$$|\theta_1|<1,\ \text{if}\ \operatorname{Im} z\geqslant 0;\ |\theta_1|<|\sec(\arg z)|,\ \text{if}\ \operatorname{Im} z\leqslant 0.$$ WA 245

For $-\frac{3}{2}\pi<\arg z<\frac{\pi}{2}$, ν real, and $n+\frac{1}{2}>|\nu|$

$$|\theta_2|<1,\ \text{if}\ \operatorname{Im} z\leqslant 0;\ |\theta_2|<|\sec(\arg z)|,\ \text{if}\ \operatorname{Im} z\geqslant 0.$$ WA 246

For ν real,

$$|\theta_3|<1 \text{ and } \operatorname{Re}\theta_3\geqslant 0,\ \text{if}\ \operatorname{Re} z\geqslant 0;$$
$$|\theta_3|<|\operatorname{cosec}(\arg z)|,\ \text{if}\ \operatorname{Re} z<0.$$ WA 245

For ν and z real and $n\geqslant\nu-\frac{1}{2}$,

$$0\leqslant|\theta_3|\leqslant 1.$$ WA 231

In particular, it follows from 8.451 7. and 8.451 8. that for real positive values of z and ν, the errors $|R_1|$ and $|R_2|$ are less than the absolute value of the first discarded term. For values of $|\arg z|$ close to π, the series 8.451 1. and 8.451 2. may not be suitable for calculations. In particular, the error for $|\arg z|>\pi$ can be greater in absolute value than the first discarded term.

"Approximation by tangents"

8.452 For large values of the index (where the argument is less than the index).

Suppose that $x>0$ and $\nu>0$. Let us set $\frac{\nu}{x}=\operatorname{ch}\alpha$. Then, for large values of ν, the following expansions are valid:

1. $$J_\nu\left(\frac{\nu}{\operatorname{ch}\alpha}\right)\sim\frac{\exp(\nu\operatorname{th}\alpha-\nu\alpha)}{\sqrt{2\nu\pi\operatorname{th}\alpha}}\left\{1+\frac{1}{\nu}\left(\frac{1}{8}\operatorname{cth}\alpha-\frac{5}{24}\operatorname{cth}^3\alpha\right)+\right.$$
$$\left.+\frac{1}{\nu^2}\left(\frac{9}{128}\operatorname{cth}^2\alpha-\frac{231}{576}\operatorname{cth}^4\alpha+\frac{1155}{3456}\operatorname{cth}^6\alpha\right)+\dots\right\}.$$ WA 269(3)

2. $$N_\nu\left(\frac{\nu}{\operatorname{ch}\alpha}\right)\sim\frac{\exp(\nu\alpha-\nu\operatorname{th}\alpha)}{\sqrt{\frac{\pi}{2}\nu\operatorname{th}\alpha}}\left\{1-\frac{1}{\nu}\left(\frac{1}{8}\operatorname{cth}\alpha-\frac{5}{24}\operatorname{cth}^3\alpha\right)+\right.$$
$$\left.+\frac{1}{\nu^2}\left(\frac{9}{128}\operatorname{cth}^2\alpha-\frac{231}{576}\operatorname{cth}^4\alpha+\frac{1155}{3456}\operatorname{cth}^6\alpha\right)+\dots\right\}.$$ WA 270(5)

8.453 For large values of the index (where the argument is greater than the index).

Suppose that $x>0$ and $\nu>0$. Let us set $\frac{\nu}{x}=\cos\beta$. Then, for large values of ν, the following expansions are valid:

1. $$J_\nu(\nu\sec\beta)\sim\sqrt{\frac{2}{\nu\pi\operatorname{tg}\beta}}\left\{\left[1-\frac{1}{\nu^2}\left(\frac{9}{128}\operatorname{ctg}^2\beta+\frac{231}{576}\operatorname{ctg}^4\beta+\frac{1155}{3456}\operatorname{ctg}^6\beta\right)+\ldots\right]\cos\left(\nu\operatorname{tg}\beta-\nu\beta-\frac{\pi}{4}\right)+\left[\frac{1}{\nu}\left(\frac{1}{8}\operatorname{ctg}\beta+\frac{5}{24}\operatorname{ctg}^3\beta\right)-\ldots\right]\sin\left(\nu\operatorname{tg}\beta-\nu\beta-\frac{\pi}{4}\right)\right\}.$$

WA 271(4)

2. $$N_\nu(\nu\sec\beta)\sim\sqrt{\frac{2}{\nu\pi\operatorname{tg}\beta}}\left\{\left[1-\frac{1}{\nu^2}\left(\frac{9}{128}\operatorname{ctg}^2\beta+\frac{231}{576}\operatorname{ctg}^4\beta+\frac{1155}{3456}\operatorname{ctg}^6\beta\right)+\ldots\right]\sin\left(\nu\operatorname{tg}\beta-\nu\beta-\frac{\pi}{4}\right)-\left[\frac{1}{\nu}\left(\frac{1}{8}\operatorname{ctg}\beta+\frac{5}{24}\operatorname{ctg}^3\beta\right)-\ldots\right]\cos\left(\nu\operatorname{tg}\beta-\nu\beta-\frac{\pi}{4}\right)\right\}.$$

WA 271(5)

3. $$H_\nu^{(1)}(\nu\sec\beta)\sim\frac{\exp\left[\nu i(\operatorname{tg}\beta-\beta)-\frac{\pi}{4}i\right]}{\sqrt{\frac{\pi}{2}\nu\operatorname{tg}\beta}}\left\{1-\frac{i}{\nu}\left(\frac{1}{8}\operatorname{ctg}\beta+\frac{5}{24}\operatorname{ctg}^3\beta\right)-\frac{1}{\nu^2}\left(\frac{9}{128}\operatorname{ctg}^2\beta+\frac{231}{576}\operatorname{ctg}^4\beta+\frac{1155}{3456}\operatorname{ctg}^6\beta\right)+\ldots\right\}.$$ WA 271(1)

4. $$H_\nu^{(2)}(\nu\sec\beta)\sim\frac{\exp\left[-\nu i(\operatorname{tg}\beta-\beta)+\frac{\pi}{4}i\right]}{\sqrt{\frac{\pi}{2}\nu\operatorname{tg}\beta}}\left\{1+\frac{i}{\nu}\left(\frac{1}{8}\operatorname{ctg}\beta+\frac{5}{24}\operatorname{ctg}^3\beta\right)-\frac{1}{\nu^2}\left(\frac{9}{128}\operatorname{ctg}^2\beta+\frac{231}{576}\operatorname{ctg}^4\beta+\frac{1155}{3456}\operatorname{ctg}^6\beta\right)+\ldots\right\}.$$ WA 271(2)

Formulas **8.453** are not valid when $|x-\nu|$ is of a size comparable to $x^{\frac{1}{3}}$. For arbitrary small (and also large) values of $|x-\nu|$, we may use the following formulas:

8.454 Suppose that $x>0$ and $\nu>0$, we set

$$w=\sqrt{\frac{x^2}{\nu^2}-1};$$

Then,

1. $$H_\nu^{(1)}(x)=\frac{w}{\sqrt{3}}\exp\left\{\left[\frac{\pi}{6}+\nu\left(w-\frac{w^3}{3}-\operatorname{arctg}w\right)\right]i\right\}H_{\frac{1}{3}}^{(1)}\left(\frac{\nu}{3}w^3\right)+O\left(\frac{1}{|\nu|}\right).$$

2. $$H_\nu^{(2)}(x)=\frac{w}{\sqrt{3}}\exp\left\{\left[-\frac{\pi}{6}-\nu\left(w-\frac{w^3}{3}-\operatorname{arctg}w\right)\right]i\right\}H_{\frac{1}{3}}^{(2)}\left(\frac{\nu}{3}w^3\right)+O\left(\frac{1}{|\nu|}\right).$$

MO 34

The absolute value of the error $O\left(\frac{1}{|\nu|}\right)$ is then less than $24\sqrt{2}\left|\frac{1}{\nu}\right|$.

8.455 For x real and ν a natural number ($\nu = n$), if $n \gg 1$, the following approximations are valid:

1. $$J_n(x) \approx \frac{1}{\pi}\sqrt{\frac{2(n-x)}{3x}}\,K_{\frac{1}{3}}\left\{\frac{[2(n-x)]^{\frac{3}{2}}}{3\sqrt{x}}\right\} \quad [n > x] \qquad \text{(see also 8.433);}$$

WA 276(1)

$$\approx \frac{1}{2}e^{\frac{2}{3}\pi i}\sqrt{\frac{2(n-x)}{3x}}\,H^{(1)}_{\frac{1}{3}}\left\{\frac{i}{3}\frac{[2(n-x)]^{\frac{3}{2}}}{\sqrt{x}}\right\} \quad [n > x]; \qquad \text{MO 34}$$

$$\approx \frac{1}{\sqrt{3}}\sqrt{\frac{2(x-n)}{3x}}\left\{J_{\frac{1}{3}}\left[\frac{\{2(x-n)\}^{\frac{3}{2}}}{3\sqrt{x}}\right] + J_{-\frac{1}{3}}\left[\frac{\{2(x-n)\}^{\frac{3}{2}}}{3\sqrt{x}}\right]\right\}$$

(see also 8.441 3., 8.441 4.). WA 276(2)

2. $$N_n(x) \approx -\sqrt{\frac{2(x-n)}{3x}}\left\{J_{-\frac{1}{3}}\left[\frac{\{2(x-n)\}^{\frac{3}{2}}}{3\sqrt{x}}\right] - J_{\frac{1}{3}}\left[\frac{\{2(x-n)\}^{\frac{3}{2}}}{3\sqrt{x}}\right]\right\}$$

$[x > n]$. WA 276(3)

An estimate of the error in formulas 8.455 has not yet been achieved.

8.456 $$J_\nu^2(z) + N_\nu^2(z) \approx \frac{2}{\pi z}\sum_{k=0}^{\infty}\frac{(2k-1)!!}{2^k z^{2k}}\,\frac{\Gamma\left(\nu+k+\frac{1}{2}\right)}{k!\,\Gamma\left(\nu-k+\frac{1}{2}\right)}$$

$[|\arg z| < \pi]$ (see also 8.479 1.). WA 250(5)

8.457 $$J_\nu^2(x) + J_{\nu+1}^2(x) \approx \frac{2}{\pi x} \quad [x \gg |\nu|]. \qquad \text{WA 223}$$

8.46 Bessel functions of order equal to an integer plus one-half

The function $J_\nu(z)$

8.461

1. $$J_{n+\frac{1}{2}}(z) = \sqrt{\frac{2}{\pi z}}\left\{\sin\left(z - \frac{\pi}{2}n\right)\sum_{k=0}^{E\left(\frac{n}{2}\right)}\frac{(-1)^k(n+2k)!}{(2k)!\,(n-2k)!\,(2z)^{2k}} + \right.$$

$$\left. + \cos\left(z - \frac{\pi}{2}n\right)\sum_{k=0}^{E\left(\frac{n-1}{2}\right)}\frac{(-1)^k(n+2k+1)!}{(2k+1)!\,(n-2k-1)!\,(2z)^{2k+1}}\right\}$$

[$n+1$ — is a natural number] (cf. 8.451 1.). KU 59(6), WA 66(2)

2. $$J_{-n-\frac{1}{2}}(z)=\sqrt{\frac{2}{\pi z}}\left\{\cos\left(z+\frac{\pi}{2}n\right)\sum_{k=0}^{E\left(\frac{n}{2}\right)}\frac{(-1)^k(n+2k)!}{(2k)!\,(n-2k)!\,(2z)^{2k}}-\right.$$
$$\left.-\sin\left(z+\frac{\pi}{2}n\right)\sum_{k=0}^{E\left(\frac{n-1}{2}\right)}\frac{(-1)^k(n+2k+1)!}{(2k+1)!\,(n-2k-1)!\,(2z)^{2k+1}}\right\}$$
[$n+1$ is a natural number] (cf. 8.451 1.). KU 58(7), WA 67(5)

8.462

1. $$J_{n+\frac{1}{2}}(z)=\frac{1}{\sqrt{2\pi z}}\left\{e^{iz}\sum_{k=0}^{n}\frac{i^{-n+k-1}(n+k)!}{k!\,(n-k)!\,(2z)^k}+e^{-iz}\sum_{k=0}^{n}\frac{(-i)^{-n+k-1}(n+k)!}{k!\,(n-k)!\,(2z)^k}\right\}$$
[$n+1$ is a natural number]. KU 59(6), WA 66(1)

2. $$J_{-n-\frac{1}{2}}(z)=\frac{1}{\sqrt{2\pi z}}\left\{e^{iz}\sum_{k=0}^{n}\frac{i^{n+k}(n+k)!}{k!\,(n-k)!\,(2z)^k}+e^{-iz}\sum_{k=0}^{n}\frac{(-i)^{n+k}(n+k)!}{k!\,(n-k)!\,(2z)^k}\right\}$$
[$n+1$ is a natural number]. KU 59(7), WA 67(4)

8.463

1. $$J_{n+\frac{1}{2}}(z)=(-1)^n z^{n+\frac{1}{2}}\sqrt{\frac{2}{\pi}}\frac{d^n}{(z\,dz)^n}\left(\frac{\sin z}{z}\right).$$ KU 58(4)

2. $$J_{-n-\frac{1}{2}}(z)=z^{n+\frac{1}{2}}\sqrt{\frac{2}{\pi}}\frac{d^n}{(z\,dz)^n}\left(\frac{\cos z}{z}\right).$$ KU 58(5)

8.464 Special cases:

1. $$J_{\frac{1}{2}}(z)=\sqrt{\frac{2}{\pi z}}\sin z.$$ DW

2. $$J_{-\frac{1}{2}}(z)=\sqrt{\frac{2}{\pi z}}\cos z.$$ DW

3. $$J_{\frac{3}{2}}(z)=\sqrt{\frac{2}{\pi z}}\left(\frac{\sin z}{z}-\cos z\right).$$ DW

4. $$J_{-\frac{3}{2}}(z)=\sqrt{\frac{2}{\pi z}}\left(-\sin z-\frac{\cos z}{z}\right).$$ DW

5. $$J_{\frac{5}{2}}(z)=\sqrt{\frac{2}{\pi z}}\left\{\left(\frac{3}{z^2}-1\right)\sin z-\frac{3}{z}\cos z\right\}.$$ DW

6. $$J_{-\frac{5}{2}}(z)=\sqrt{\frac{2}{\pi z}}\left\{\frac{3}{z}\sin z+\left(\frac{3}{z^2}-1\right)\cos z\right\}.$$ DW

The function $N_{n+\frac{1}{2}}(z)$

8.465

1. $$N_{n+\frac{1}{2}}(z)=(-1)^{n-1}J_{-n-\frac{1}{2}}(z).$$ JA

2. $$N_{-n-\frac{1}{2}}(z)=(-1)^n J_{n+\frac{1}{2}}(z).$$ JA

The functions $H^{(1,2)}_{n+\frac{1}{2}}(z)$, $I_{n+\frac{1}{2}}(z)$, $K_{n+\frac{1}{2}}(z)$

8.466

1. $$H^{(1)}_{n-\frac{1}{2}}(z)=\sqrt{\frac{2}{\pi z}}\, i^{-n} e^{iz} \sum_{k=0}^{n-1}(-1)^k \frac{(n+k-1)!}{k!\,(n-k-1)!}\frac{1}{(2iz)^k}$$ (cf. 8.451 3.).

2. $$H^{(2)}_{n-\frac{1}{2}}(z)=\sqrt{\frac{2}{\pi z}}\, i^{n} e^{-iz} \sum_{k=0}^{n-1}\frac{(n+k-1)!}{k!\,(n-k-1)!}\frac{1}{(2iz)^k}$$ (cf. 8.451 4.).

8.467 $$I_{\pm\left(n+\frac{1}{2}\right)}(z)=\frac{1}{\sqrt{2\pi z}}\left[e^z\sum_{k=0}^{n}\frac{(-1)^k(n+k)!}{k!\,(n-k)!\,(2z)^k}\pm(-1)^{n+1}e^{-z}\sum_{k=0}^{n}\frac{(n+k)!}{k!\,(n-k)!\,(2z)^k}\right]$$ (cf. 8.451 5.). KU 60a

8.468 $$K_{n+\frac{1}{2}}(z)=\sqrt{\frac{\pi}{2z}}\,e^{-z}\sum_{k=0}^{n}\frac{(n+k)!}{k!\,(n-k)!\,(2z)^k}$$ (cf. 8.451 6.). KU 60

8.469 Special cases:

1. $N_{\frac{1}{2}}(z)=-\sqrt{\frac{2}{\pi z}}\cos z.$

2. $N_{-\frac{1}{2}}(z)=\sqrt{\frac{2}{\pi z}}\sin z.$

3. $K_{\pm\frac{1}{2}}(z)=\sqrt{\frac{\pi}{2z}}\,e^{-z}.$ WA 95(13)

4. $H^{(1)}_{\frac{1}{2}}(z)=\sqrt{\frac{2}{\pi z}}\,\frac{e^{iz}}{i}.$ MO 27

5. $H^{(2)}_{\frac{1}{2}}(z)=\sqrt{\frac{2}{\pi z}}\,\frac{e^{-iz}}{-i}.$ MO 27

6. $H^{(1)}_{-\frac{1}{2}}(z)=\sqrt{\frac{2}{\pi z}}\,e^{iz}.$ MO 27

7. $H^{(2)}_{-\frac{1}{2}}(z)=\sqrt{\frac{2}{\pi z}}\,e^{-iz}.$ MO 27

8.47-8.48 Functional relations

8.471 Recursion formulas:

1. $zZ_{\nu-1}(z)+zZ_{\nu+1}(z)=2\nu Z_\nu(z).$ KU 56(13), WA 56(1), WA 79(1), WA 88(3)

2. $Z_{\nu-1}(z)-Z_{\nu+1}(z)=2\frac{d}{dz}Z_\nu(z).$ KU 56(12), WA 56(2), WA 79(2), WA 88(4)

Sonine and Nielsen, in their construction of the theory of Bessel functions, defined Bessel functions as analytic functions of z that satisfy the recursion relations 8.471.

8.472 Consequences of the recursion formulas:

1. $z\frac{d}{dz}Z_\nu(z)+\nu Z_\nu(z)=zZ_{\nu-1}(z).$ KU 56(11), WA 56(3), WA 79(3), WA 88(5)

2. $z\frac{d}{dz}Z_\nu(z) - \nu Z_\nu(z) = -zZ_{\nu+1}(z).$ KU 56(10), WA 56(4), WA 79(4), WA 88(6)

3. $\left(\frac{d}{z\,dz}\right)^m (z^\nu Z_\nu(z)) = z^{\nu-m} Z_{\nu-m}(z).$ KU 56(8), WA 57(5), WA 89(9)

4. $\left(\frac{d}{z\,dz}\right)^m (z^{-\nu} Z_\nu(z)) = (-1)^m z^{-\nu-m} Z_{\nu+m}(z).$

WA 89(10), Ku 55(5), WA 57(6)

5. $Z_{-n}(z) = (-1)^n Z_n(z)$ [n is a natural number]. (cf. 8.404).

8.473 Special cases:

1. $J_2(z) = \frac{2}{z} J_1(z) - J_0(z).$

2. $N_2(z) = \frac{2}{z} N_1(z) - N_0(z).$

3. $H_2^{(1,2)}(z) = \frac{2}{z} H_1^{(1,2)}(z) - H_0^{(1,2)}(z).$

4. $\frac{d}{dz} J_0(z) = -J_1(z).$

5. $\frac{d}{dz} N_0(z) = -N_1(z).$

6. $\frac{d}{dz} H_0^{(1,2)}(z) = -H_1^{(1,2)}(z).$

8.474 Each of the pairs of functions $J_\nu(z)$ and $J_{-\nu}(z)$ (for $\nu \neq 0, \pm 1, \pm 2, \ldots$), $J_\nu(z)$ and $N_\nu(z)$, and $H_\nu^{(1)}(z)$ and $H_\nu^{(2)}(z)$, which are solutions of equation 8.401, and also the pair $I_\nu(z)$ and $K_\nu(z)$ is a pair of linearly independent functions. The Wronskians of these pairs are, respectively,

$$\frac{2}{\pi z}\sin\nu\pi, \quad \frac{2}{\pi z}, \quad -\frac{4i}{\pi z}, \quad -\frac{1}{z}.$$

KU 52(10, 11, 12), WA 90(1, 4)

8.475 The functions $J_\nu(z)$, $N_\nu(z)$, $H_\nu^{(1,2)}(z)$, $I_\nu(z)$, $K_\nu(z)$ with the exception of $J_n(z)$ for n an integer are *non-single-valued*: $z=0$ is a branch point for these functions. The branches of these functions that lie on opposite sides of the cut $(-\infty, 0)$ are connected by the relations

8.476

1. $J_\nu(e^{m\pi i}z) = e^{m\nu\pi i} J_\nu(z).$ WA 90(1)

2. $N_\nu(e^{m\pi i}z) = e^{-m\nu\pi i} N_\nu(z) + 2i\sin m\nu\pi \operatorname{ctg}\nu\pi J_\nu(z).$ WA 90(3)

3. $N_{-\nu}(e^{m\pi i}z) = e^{-m\nu\pi i} N_{-\nu}(z) + 2i\sin m\nu\pi \operatorname{cosec}\nu\pi\, J_\nu(z).$ WA 90(4)

4. $I_\nu(e^{m\pi i}z) = e^{m\nu\pi i} I_\nu(z).$ WA 95(17)

5. $K_\nu(e^{m\pi i}z) = e^{-m\nu\pi i} K_\nu(z) - i\pi\frac{\sin m\nu\pi}{\sin\nu\pi} I_\nu(z)$ [ν not an integer].

WA 95(18)

6. $H_\nu^{(1)}(e^{m\pi i}z) = e^{-m\nu\pi i} H_\nu^{(1)}(z) - 2e^{-\nu\pi i}\frac{\sin m\nu\pi}{\sin\nu\pi} J_\nu(z) =$

$$= \frac{\sin(1-m)\nu\pi}{\sin\nu\pi} H_\nu^{(1)}(z) - e^{-\nu\pi i}\frac{\sin m\nu\pi}{\sin\nu\pi} H_\nu^{(2)}(z).$$ WA 95(5)

7. $H_{\nu}^{(2)}(e^{m\pi i}z) = e^{-m\nu\pi i} H_{\nu}^{(2)}(z) + 2e^{\nu\pi i} \frac{\sin m\nu\pi}{\sin \nu\pi} J_{\nu}(z) =$

$= \frac{\sin(1+m)\nu\pi}{\sin \nu\pi} H_{\nu}^{(2)}(z) + e^{\nu\pi i} \frac{\sin m\nu\pi}{\sin \nu\pi} H_{\nu}^{(1)}(z)$ WA 90(6)

[m — an integer].

8. $H_{\nu}^{(1)}(e^{i\pi}z) = -H_{-\nu}^{(2)}(z) = -e^{-i\pi\nu} H_{\nu}^{(2)}(z).$ MO 26

9. $H_{\nu}^{(2)}(e^{-i\pi}z) = -H_{-\nu}^{(1)}(z) = -e^{i\pi\nu} H_{\nu}^{(1)}(z).$ MO 26

10. $\overline{H_{\nu}^{(2)}}(z) = H_{\nu}^{(1)}(\bar{z}).$ MO 26

8.477

1. $J_{\nu}(z) N_{\nu+1}(z) - J_{\nu+1}(z) N_{\nu}(z) = -\frac{2}{\pi z}.$ WA 91(12)

2. $I_{\nu}(z) K_{\nu+1}(z) + J_{\nu+1}(z) K_{\nu}(z) = \frac{1}{z}.$ WA 95(20)

See also 3.864.
For a connection with Legendre functions, see 8.722.
For a connection with the polynomials $C_n^{\lambda}(t)$, see 8.936 4.
For a connection with a degenerate hypergeometric function, see 9.235.

8.478 For $\nu > 0$ and $x > 0$, the product

$$x[J_{\nu}^2(x) + N_{\nu}^2(x)],$$

considered as a function of x, decreases monotonically, if $\nu > \frac{1}{2}$ and increases monotonically if $0 < \nu < \frac{1}{2}$. MO 35

8.479

1. $\frac{1}{\sqrt{x^2 - \nu^2}} > \frac{\pi}{2} [J_{\nu}^2(x) + N_{\nu}^2(x)] \geqslant \frac{1}{x} \qquad \left[x \geqslant \nu \geqslant \frac{1}{2}\right]$

(see also 6.518, 6.664 4., 8.456). MO 35

2. $|J_n(nz)| \leqslant 1 \qquad \left[\left|\frac{z \exp\sqrt{1-z^2}}{1+\sqrt{1-z^2}}\right| < 1, n \text{ — a natural number}\right].$ MO 35

Relations between Bessel functions of the first, second, and third kinds

8.481 $J_{\nu}(z) = \frac{N_{-\nu}(z) - N_{\nu}(z)\cos\nu\pi}{\sin\nu\pi} = H_{\nu}^{(1)}(z) - iN_{\nu}(z) =$

$= H_{\nu}^{(2)}(z) + iN_{\nu}(z) = \frac{1}{2}(H_{\nu}^{(1)}(z) + H_{\nu}^{(2)}(z))$

(cf. 8.403 1., 8.405). WA 89(1), JA

8.482 $N_{\nu}(z) = \frac{J_{\nu}(z)\cos\nu\pi - J_{-\nu}(z)}{\sin\nu\pi} = iJ_{\nu}(z) - iH_{\nu}^{(1)}(z) =$

$= iH_{\nu}^{(2)}(z) - iJ_{\nu}(z) = \frac{i}{2}(H_{\nu}^{(2)}(z) - H_{\nu}^{(1)}(z))$

(cf. 8.403 1., 8.405). WA 89(3), JA

8.483

1. $H_{\nu}^{(1)}(z) = \frac{J_{-\nu}(z) - e^{-\nu\pi i} J_{\nu}(z)}{i \sin \nu\pi} = \frac{N_{-\nu}(z) - e^{-\nu\pi i} N_{\nu}(z)}{\sin \nu\pi} = J_{\nu}(z) + iN_{\nu}(z).$ WA 89(5)

2. $H_{\nu}^{(2)}(z) = \frac{e^{\nu\pi i} J_{\nu}(z) - J_{-\nu}(z)}{i \sin \nu\pi} = \frac{N_{-\nu}(z) - e^{\nu\pi i} N_{\nu}(z)}{\sin \nu\pi} = J_{\nu}(z) - iN_{\nu}(z)$ (cf. 8.405). WA 89(6)

8.484

1. $H_{-\nu}^{(1)}(z) = e^{\nu\pi i} H_{\nu}^{(1)}(z).$ WA 89(7)
2. $H_{-\nu}^{(2)}(z) = e^{-\nu\pi i} H_{\nu}^{(2)}(z).$ WA 89(7)

8.485 $K_{\nu}(z) = \frac{\pi}{2} \frac{I_{-\nu}(z) - I_{\nu}(z)}{\sin \nu\pi}$ [ν not an integer] (see also 8.407) WA 92(6)

8.486 Recursion formulas for the functions $I_{\nu}(z)$ and $K_{\nu}(z)$ and their consequences:

1. $zI_{\nu-1}(z) - zI_{\nu+1}(z) = 2\nu I_{\nu}(z).$ WA 93(1)
2. $I_{\nu-1}(z) + I_{\nu+1}(z) = 2\frac{d}{dz} I_{\nu}(z).$ WA 93(2)
3. $z\frac{d}{dz} I_{\nu}(z) + \nu I_{\nu}(z) = zI_{\nu-1}(z).$ WA 93(3)
4. $z\frac{d}{dz} I_{\nu}(z) - \nu I_{\nu}(z) = zI_{\nu+1}(z).$ WA 93(4)
5. $\left(\frac{d}{z\,dz}\right)^m \{z^{\nu} I_{\nu}(z)\} = z^{\nu-m} I_{\nu-m}(z).$ WA 93(5)
6. $\left(\frac{d}{z\,dz}\right)^m \{z^{-\nu} I_{\nu}(z)\} = z^{-\nu-m} I_{\nu+m}(z).$ WA 93(6)
7. $I_{-n}(z) = I_n(z)$ [n — a natural number]. WA 93(8)
8. $I_2(z) = -\frac{2}{z} I_1(z) + I_0(z).$
9. $\frac{d}{dz} I_0(z) = I_1(z).$ WA 93(7)
10. $zK_{\nu-1}(z) - zK_{\nu+1}(z) = -2\nu K_{\nu}(z).$ WA 93(1)
11. $K_{\nu-1}(z) + K_{\nu+1}(z) = -2\frac{d}{dz} K_{\nu}(z).$ WA 93(2)
12. $z\frac{d}{dz} K_{\nu}(z) + \nu K_{\nu}(z) = -zK_{\nu-1}(z).$ WA 93(3)
13. $z\frac{d}{dz} K_{\nu}(z) - \nu K_{\nu}(z) = -zK_{\nu+1}(z).$ WA 93(4)
14. $\left(\frac{d}{z\,dz}\right)^m \{z^{\nu} K_{\nu}(z)\} = (-1)^m z^{\nu-m} K_{\nu-m}(z).$ WA 93(5)
15. $\left(\frac{d}{z\,dz}\right)^m \{z^{-\nu} K_{\nu}(z)\} = (-1)^m z^{-\nu-m} K_{\nu+m}(z).$ WA 93(6)
16. $K_{-\nu}(z) = K_{\nu}(z).$ WA 93(8)
17. $K_2(z) = \frac{2}{z} K_1(z) + K_0(z).$
18. $\frac{d}{dz} K_0(z) = -K_1(z).$ WA 93(7)

8.487 Continuity with respect to the order*:

1. $\lim\limits_{\nu \to n} N_\nu(z) = N_n(z)$ WA 76

2. $\lim\limits_{\nu \to n} H_\nu^{(1,2)}(z) = H_n^{(1,2)}(z)$ [n — an integer]. WA 183

3. $\lim\limits_{\nu \to n} K_\nu(z) = K_n(z)$ WA 92

8.49 Differential equations leading to Bessel functions

See also 8.401

8.491

1. $\frac{1}{z}\frac{d}{dz}(zu') + \left(\beta^2 - \frac{\nu^2}{z^2}\right)u = 0, \qquad u = Z_\nu(\beta z).$ JA

2. $\frac{1}{z}\frac{d}{dz}(zu') + \left[(\beta\gamma z^{\gamma-1})^2 - \left(\frac{\nu\gamma}{z}\right)^2\right]u = 0, \qquad u = Z_\nu(\beta z^\gamma).$ JA

3. $u'' + \frac{1-2\alpha}{z}u' + \left[(\beta\gamma z^{\gamma-1})^2 - \frac{\alpha^2 - \nu\gamma^2}{z^2}\right]u = 0, \qquad u = z^\alpha Z_\nu(\beta z^\gamma).$ JA

4. $u'' + \left[(\beta\gamma z^{\gamma-1})^2 - \frac{4\nu^2\gamma^2 - 1}{4z^2}\right]u = 0, \qquad u = \sqrt{z}\,Z_\nu(\beta z^\gamma).$ JA

5. $u'' + \left(\beta^2 - \frac{4\nu^2 - 1}{4z^2}\right)u = 0, \qquad u = \sqrt{z}Z_\nu(\beta z).$ JA

6. $u'' + \frac{1-2\alpha}{z}u' + \left(\beta^2 + \frac{\alpha^2 - \nu^2}{z^2}\right)u = 0, \qquad u = z^\alpha Z_\nu(\beta z).$ JA

7. $u'' + bz^m u = 0, \quad u = \sqrt{z}\,Z_{\frac{1}{m+2}}\left(\frac{2\sqrt{b}}{m+2}z^{\frac{m+2}{2}}\right).$ JA 111(5)

8. $u'' + \frac{1}{z}u' + 4\left(z^2 - \frac{\nu^2}{z^2}\right)u = 0, \qquad u = Z_\nu(z^2).$ WA 111(6)

9. $u'' + \frac{1}{z}u' + \frac{1}{4z}\left(1 - \frac{\nu^2}{z}\right)u = 0, \qquad u = Z_\nu(\sqrt{z}).$ WA 111(7)

10. $u'' + \frac{1-\nu}{z}u' + \frac{1}{4}\frac{u}{z} = 0, \quad u = z^{\frac{\nu}{2}}Z_\nu(\sqrt{z}).$ WA 111(9)a

11. $u'' + \beta^2\gamma^2 z^{2\beta-2}u = 0, \quad u = z^{\frac{1}{2}}Z_{\frac{1}{2\beta}}(\gamma z^\beta).$ WA 110(3)

12. $z^2u'' + (2\alpha - 2\beta\nu + 1)zu' + [\beta^2\gamma^2 z^{2\beta} + \alpha(\alpha - 2\beta\nu)]u = 0,$
$u = z^{\beta\nu-\alpha}Z_\nu(\gamma z^\beta).$ WA 112(21)

8.492

1. $u'' + (e^{2z} - \nu^2)u = 0, \quad u = Z_\nu(e^z).$ WA 112(22)

2. $u'' + \frac{e^{\frac{2}{z}} - \nu^2}{z^4}u = 0, \quad u = zZ_\nu(e^{\frac{1}{z}}).$ WA 112(22)

*The continuity of the functions $J_\nu(z)$ and $I_\nu(z)$ follows directly from the series representations of these functions.

8.493

1. $u'' + \left(\frac{1}{z} - 2\operatorname{tg} z\right) u' - \left(\frac{\nu^2}{z^2} + \frac{\operatorname{tg} z}{z}\right) u = 0, \quad u = \sec z \, Z_\nu(z).$ JA

2. $u'' + \left(\frac{1}{z} + 2\operatorname{ctg} z\right) u' - \left(\frac{\nu^2}{z^2} - \frac{\operatorname{ctg} z}{z}\right) u = 0, \quad u = \operatorname{cosec} z \, Z_\nu(z).$ JA

8.494

1. $u'' + \frac{1}{z} u' - \left(1 + \frac{\nu^2}{z^2}\right) u = 0, \quad u = Z_\nu(iz) = C_1 I_\nu(z) + C_2 K_\nu(z).$ JA

2. $u'' + \frac{1}{z} u' - \left[\frac{1}{z} + \left(\frac{\nu}{2z}\right)^2\right] u = 0, \quad u = Z_\nu(2i\sqrt{z}).$ JA

3. $u'' + u' + \frac{1}{z^2}\left(\frac{1}{4} - \nu^2\right) u = 0, \quad u = \sqrt{z}\, e^{-\frac{z}{2}} Z_\nu\left(\frac{iz}{2}\right).$ JA

4. $u'' + \left(\frac{2\nu+1}{z} - k\right) u' - \frac{2\nu+1}{2z} k u = 0, \quad u = z^{-\nu} e^{\frac{1}{2}kz} Z_\nu\left(\frac{ikz}{2}\right).$ JA

5. $u'' + \frac{1-\nu}{z} u' - \frac{1}{4}\frac{u}{z} = 0, \quad u = z^{\frac{\nu}{2}} Z_\nu(i\sqrt{z}).$ WA 111(8)

6. $u'' \pm \frac{u}{\sqrt{z}} = 0, \quad u = \sqrt{z}\, Z_{\frac{2}{3}}\left(\frac{4}{3} z^{\frac{3}{4}}\right), \quad \sqrt{z}\, Z_{\frac{2}{3}}\left(\frac{4}{3} i z^{\frac{3}{4}}\right).$ WA 111(10)

7. $u'' \pm zu = 0, \quad u = \sqrt{z}\, Z_{\frac{1}{3}}\left(\frac{2}{3} z^{\frac{3}{2}}\right), \quad \sqrt{z}\, Z_{\frac{1}{3}}\left(\frac{2}{3} i z^{\frac{3}{2}}\right).$ WA 111(10)

8. $u'' - \left(c^2 + \frac{\nu(\nu+1)}{z^2}\right) u = 0, \quad u = \sqrt{z}\, Z_{\nu+\frac{1}{2}}(icz).$ WA 108(1)

9. $u'' - \frac{2\nu}{z} u' - c^2 u = 0, \quad u = z^{\nu+\frac{1}{2}} Z_{\nu+\frac{1}{2}}(icz).$ WA 109(3, 4)

10. $u'' - c^2 z^{2\nu-2} u = 0, \quad u = \sqrt{z}\, Z_{\frac{1}{2\nu}}\left(i \frac{c}{\nu} z^\nu\right).$ WA 109(5, 6)

8.495

1. $u'' + \frac{1}{z} u' + \left(i - \frac{\nu^2}{z^2}\right) u = 0, \quad u = Z_\nu(z\sqrt{i}).$ JA

2. $u'' + \left(\frac{1}{z} \mp 2i\right) u' - \left(\frac{\nu^2}{z^2} \pm \frac{i}{z}\right) u = 0, \quad u = e^{\pm iz} Z_\nu(z).$ JA

3. $u'' + \frac{1}{z} u' + s e^{i\alpha} u = 0, \quad u = Z_0\left(\sqrt{s}\, z e^{\frac{i}{2}\alpha}\right).$ JA

4. $u'' + \left(s e^{i\alpha} + \frac{1}{4z^2}\right) u = 0, \quad u = \sqrt{z}\, Z_0\left(\sqrt{s} z e^{\frac{i}{2}\alpha}\right).$ JA

8.496

1. $\frac{d^2}{dz^2}\left(z^4 \frac{d^2 u}{dz^2}\right) - z^2 u = 0, \quad u = \frac{1}{z}\{Z_2(2\sqrt{z}) + \overline{Z}_2(2i\sqrt{z})\}.$ WA 122(7)

2. $\frac{d^2}{dz^2}\left(z^{\frac{16}{5}} \frac{d^2 u}{dz^2}\right) - z^{\frac{8}{5}} u = 0, \quad u = z^{-\frac{7}{10}}\left\{Z_{\frac{5}{6}}\left(\frac{5}{3} z^{\frac{3}{5}}\right) + \overline{Z}_{\frac{5}{6}}\left(\frac{5}{3} i z^{\frac{3}{5}}\right)\right\}.$

WA 122(8)

3. $\frac{d^2}{dz^2}\left(z^{12}\frac{d^2u}{dz^2}\right)-z^6u=0, \quad u=z^{-4}\{Z_{10}(2z^{-\frac{1}{2}})+\overline{Z}_{10}(2iz^{-\frac{1}{2}})\}.$ WA 122(9)

4. $\frac{d^4u}{dz^4}+\frac{2}{z}\frac{d^3u}{dz^3}-\frac{2\nu^2+1}{z^2}\frac{d^2u}{dz^2}+\frac{2\nu^2+1}{z^3}\frac{du}{dz}+\left(\frac{\nu^4-4\nu^2}{z^4}-1\right)u=0,$

$u=A_1J_\nu(z)+A_2N_\nu(z)+A_3I_\nu(z)+A_4K_\nu(z),$ where A_1, A_2, A_3, A_4 — are constants MO 29

8.51-8.52 *Series of Bessel functions*

8.511 Generating function for Bessel functions:

1. $\exp\frac{1}{2}\left(t-\frac{1}{t}\right)z=J_0(z)+\sum_{k=1}^{\infty}[t^k+(-t)^{-k}]J_k(z)=\sum_{k=-\infty}^{\infty}J_k(z)t^k$ $[|z|<|t|].$ KU 119(12)

2. $\exp\left(t-\frac{1}{t}\right)z=\left\{\sum_{k=-\infty}^{\infty}t^kJ_k(z)\right\}\left\{\sum_{m=-\infty}^{\infty}t^mJ_m(z)\right\}.$ WA 40

3. $\exp(\pm iz\sin\varphi)=J_0(z)+2\sum_{k=1}^{\infty}J_{2k}(z)\cos 2k\varphi\,\pm$

$\pm\, 2i\sum_{k=1}^{\infty}J_{2k+1}(z)\sin(2k+1)\varphi.$ KU 120(13)

4. $\exp(iz\cos\varphi)=\sqrt{\frac{\pi}{2z}}\sum_{k=0}^{\infty}(2k+1)i^kJ_{k+\frac{1}{2}}(z)P_k(\cos\varphi);$ WA 401(1)

$=\sum_{k=-\infty}^{\infty}i^kJ_k(z)e^{ik\varphi};$ MO 27

$=J_0(z)+2\sum_{k=1}^{\infty}i^kJ_k(z)\cos k\varphi.$ MO 27

5 $\sqrt{\frac{i}{\pi}}e^{iz\cos 2\varphi}\int_{-\infty}^{\sqrt{2z}\cos\varphi}e^{-it^2}dt=\frac{1}{2}J_0(z)+\sum_{k=1}^{\infty}e^{\frac{1}{4}k\pi i}J_{\frac{k}{2}}(z)\cos k\varphi.$ MO 28

The series $\Sigma J_k(z)$

8.512

1. $J_0(z)+2\sum_{k=1}^{\infty}J_{2k}(z)=1.$ WA 44

2 $\sum_{k=0}^{\infty}\frac{(n+2k)(n+k-1)!}{k!}J_{n+2k}(z)=\left(\frac{z}{2}\right)^n.$ WA 45

3. $\sum_{k=0}^{\infty}\frac{(4k+1)(2k-1)!!}{2^kk!}J_{2k+\frac{1}{2}}(z)=\sqrt{2z/\pi}.$

8.513

1. $\sum_{k=1}^{\infty} (2k)^{2p} J_{2k}(z) = \sum_{k=0}^{p} Q_{2k}^{(2p)} z^{2k} \quad [p = 1, 2, 3, \ldots].$ WA 46(1)

2. $\sum_{k=0}^{\infty} (2k+1)^{2p+1} J_{2k+1}(z) = \sum_{k=0}^{p} Q_{2k+1}^{(2p+1)} z^{2k+1} \quad [p = 0, 1, 2, 3, \ldots].$ WA 46(2)

$$\left[\text{In formulas } 8.513 \quad Q_k^{(p)} = \sum_{m=0}^{E\left(\frac{k-1}{2}\right)} \frac{(-1)^m \binom{k}{m} (k-2m)^p}{2^k k!} \right].$$

In particular:

3. $\sum_{k=0}^{\infty} (2k+1)^3 J_{2k+1}(z) = \frac{1}{2}(z + z^3).$ WA 47(4)

4. $\sum_{k=1}^{\infty} (2k)^2 J_{2k}(z) = \frac{1}{2} z^2.$ WA 47(4)

5. $\sum_{k=1}^{\infty} 2k(2k+1)(2k+2) J_{2k+1}(z) = \frac{1}{2} z^3.$ WA 47(4)

8.514

1. $\sum_{k=0}^{\infty} (-1)^k J_{2k+1}(z) = \frac{\sin z}{2}.$ WH

2. $J_0(z) + 2 \sum_{k=1}^{\infty} (-1)^k J_{2k}(z) = \cos z.$ WH

3. $\sum_{k=1}^{\infty} (-1)^{k+1} (2k)^2 J_{2k}(z) = \frac{z \sin z}{2}.$ WA 32(9)

4. $\sum_{k=0}^{\infty} (-1)^k (2k+1)^2 J_{2k+1}(z) = \frac{z \cos z}{2}.$ WA 32(10)

5. $J_0(z) + 2 \sum_{k=1}^{\infty} J_{2k}(z) \cos 2k\theta = \cos(z \sin\theta).$ KU 120(14), WA 32

6. $\sum_{k=0}^{\infty} J_{2k+1}(z) \sin(2k+1)\theta = \frac{\sin(z \sin\theta)}{2}.$ KU 120(15), WA 32

7. $\sum_{k=0}^{\infty} J_{2k+1}(x) = \frac{1}{2} \int_0^x J_0(t)\, dt \quad [x \text{ real}].$ WA 638

8.515

1. $\sum_{k=0}^{\infty} \frac{(-1)^k t^k}{k!} \left(\frac{2z+t}{2z}\right)^k J_{\nu+k}(z) = \left(\frac{z}{z+t}\right)^{\nu} J_{\nu}(z+t).$ AD (9140)

2. $\sum_{k=1}^{\infty} J_{2k-\frac{1}{2}}(x^2) = S(x).$ MO 127a

3. $\sum_{k=0}^{\infty} J_{2k+\frac{1}{2}}(x^2) = C(x).$ MO 127a

8.516 $\sum_{k=0}^{\infty} \frac{(2n+2k)(2n+k-1)!}{k!} J_{2n+2k}(2z\sin\theta) = (z\sin\theta)^{2n}$ WA 47

The series $\Sigma a_k J_k(kx)$ and $\Sigma a_k J'_k(kx)$

8.517

1. $\sum_{k=1}^{\infty} J_k(kz) = \frac{z}{2(1-z)}$ WA 615(1)

2. $\sum_{k=1}^{\infty} (-1)^k J_k(kz) = -\frac{z}{2(1+z)}$ WA 622(1)

3. $\sum_{k=1}^{\infty} J_{2k}(2kz) = \frac{z^2}{2(1-z^2)}$ MO 58

$\left[\left|\frac{z\exp\sqrt{1-z^2}}{1+\sqrt{1-z^2}}\right| < 1\right].$

8.518

1. $\sum_{k=1}^{\infty} \frac{J'_k(kx)}{k} = \frac{1}{2} + \frac{x}{4}$ $[0 \leqslant x < 1].$ MO 58

2. $\sum_{k=1}^{\infty} (-1)^{k-1} \frac{J'_k(kx)}{k} = \frac{1}{2} - \frac{x}{4}$ $[0 \leqslant x < 1].$ MO 58

3. $\sum_{k=1}^{\infty} kJ'_k(kx) = \frac{1}{2(1-x)^2}$ $[0 \leqslant x < 1].$ MO 58

4. $\sum_{k=1}^{\infty} (-1)^{k-1} J'_k(kx)k = \frac{1}{2(1+x)^2}$ $[0 \leqslant x < 1].$ MO 58

The series $\Sigma a_k J_0(kx)$

8.519 If, on the interval $[0 \leqslant x \leqslant \pi]$, a function $f(x)$ possesses a continuous derivative with respect to x that is of bounded variation, then

1. $f(x) = \frac{a_0}{2} + \sum_{k=1}^{\infty} a_k J_0(kx)$ $[0 < x < \pi],$

where

2. $a_0 = 2f(0) + \frac{2}{\pi} \int_0^{\pi} du \int_0^{\frac{\pi}{2}} uf'(u\sin\varphi)\,d\varphi.$

3. $a_n = \frac{2}{\pi} \int_0^{\pi} du \int_0^{\frac{\pi}{2}} uf'(u\sin\varphi)\cos nu\,d\varphi.$ WH

8.521 Examples:

1. $$\sum_{k=1}^{\infty} J_0(kx) = -\frac{1}{2} + \frac{1}{x} + 2\sum_{m=1}^{n} \frac{1}{\sqrt{x^2 - 4m^2\pi^2}} \qquad [2n\pi < x < 2(n+1)\pi].$$

MO 59

2. $$\sum_{k=1}^{\infty} (-1)^{k+1} J_0(kx) = \frac{1}{2} \qquad [0 < x < \pi].$$ KU 124(12)

3. $$\sum_{k=1}^{\infty} \frac{1}{(2k-1)^2} J_0\{(2k-1)x\} = \frac{\pi^2}{8} - \frac{|x|}{2} \qquad [-\pi < x < \pi];$$ KU 124

$$= \frac{\pi^2}{8} + \sqrt{x^2 - \pi^2} - \frac{x}{2} - \pi \arccos \frac{\pi}{x} \qquad [\pi < x < 2\pi].$$ MO 59

4. $$\sum_{k=1}^{\infty} e^{-kz} J_0\left(k\sqrt{x^2 + y^2}\right) =$$

$$= \frac{1}{r} - \frac{1}{2} + \sum_{k=1}^{\infty} \left\{ \frac{1}{\sqrt{(2ki\pi + z)^2 + x^2 + y^2}} + \frac{1}{\sqrt{(2ki\pi - z)^2 + x^2 + y^2}} \right\};$$

$$= \frac{1}{r} - \frac{1}{2} + \sum_{k=1}^{\infty} \frac{1}{(2k)!} B_{2k} r^{2k-1} P_{2k-1}\left(\frac{z}{r}\right) \qquad [0 < r < 2\pi],$$ MO 59

where $r = \sqrt{x^2 + y^2 + z^2}$ and where the radical indicates the square root with a positive real part. In formula 8.521 4., the first equation holds when x and y are real and $\operatorname{Re} z > 0$; the second equation holds when x, y, and z are all real.

The series $\sum a_k Z_0(kx) \sin kx$ and $\sum a_k Z_0(kx) \cos kx$

8.522

1. $$\sum_{k=1}^{\infty} J_0(kx) \cos kxt = -\frac{1}{2} + \sum_{l=1}^{m} \frac{1}{\sqrt{x^2 - (2\pi l + tx)^2}} +$$

$$+ \frac{1}{x\sqrt{1-t^2}} + \sum_{l=1}^{n} \frac{1}{\sqrt{x^2 - (2\pi l - tx)^2}}.$$ MO 59

2. $$\sum_{k=1}^{\infty} J_0(kx) \sin kxt = \frac{1}{2\pi} \left\{ \sum_{l=1}^{n} \frac{1}{l} - \sum_{l=1}^{m} \frac{1}{l} \right\} +$$

$$+ \sum_{l=m+1}^{\infty} \left\{ \frac{1}{\sqrt{(2\pi l + tx)^2 - x^2}} - \frac{1}{2\pi l} \right\} - \sum_{l=n+1}^{\infty} \left\{ \frac{1}{\sqrt{(2\pi l - tx)^2 - x^2}} - \frac{1}{2\pi l} \right\}.$$

MO 59

3. $$\sum_{k=1}^{\infty} N_0(kx)\cos kxt = -\frac{1}{\pi}\left(C+\ln\frac{x}{4\pi}\right)+\frac{1}{2\pi}\left\{\sum_{l=1}^{m}\frac{1}{l}+\sum_{l=1}^{n}\frac{1}{l}\right\}-$$
$$-\sum_{l=m+1}^{\infty}\left\{\frac{1}{\sqrt{(2\pi l+tx)^2-x^2}}-\frac{1}{2\pi l}\right\}-\sum_{l=n+1}^{\infty}\left\{\frac{1}{\sqrt{(2\pi l-tx)^2-x^2}}-\frac{1}{2\pi l}\right\}.$$ MO 60

In formulas 8.522, $x>0$, $0\leqslant t<1$, $2\pi m<x(1-t)<2(m+1)\pi$, $2n\pi<x(1+t)<2(n+1)\pi$, $m+1$ and $n+1$ are natural numbers.

8.523

1. $$\sum_{k=1}^{\infty}(-1)^k J_0(kx)\cos kxt = -\frac{1}{2}+\sum_{l=1}^{m}\frac{1}{\sqrt{x^2-[(2l-1)\pi+tx]^2}}+$$
$$+\sum_{l=1}^{n}\frac{1}{\sqrt{x^2-[(2l-1)\pi-tx]^2}}.$$ MO 60

2. $$\sum_{k=1}^{\infty}(-1)^k J_0(kx)\sin kxt = \frac{1}{2\pi}\left\{\sum_{l=1}^{n}\frac{1}{l}-\sum_{l=1}^{m}\frac{1}{l}\right\}+$$
$$+\sum_{l=m+1}^{\infty}\left\{\frac{1}{\sqrt{[(2l-1)\pi+tx]^2-x^2}}-\frac{1}{2l\pi}\right\}-$$
$$-\sum_{l=n+1}^{\infty}\left\{\frac{1}{\sqrt{[(2l-1)\pi-tx]^2-x^2}}-\frac{1}{2l\pi}\right\}.$$ MO 60

3. $$\sum_{k=1}^{\infty}(-1)^k N_0(kx)\cos kxt = -\frac{1}{\pi}\left(C+\ln\frac{x}{4\pi}\right)+$$
$$+\frac{1}{2\pi}\left\{\sum_{l=1}^{m}\frac{1}{l}+\sum_{l=1}^{n}\frac{1}{l}\right\}-\sum_{l=m+1}^{\infty}\left\{\frac{1}{\sqrt{[(2l-1)\pi+tx]^2-x^2}}-\frac{1}{2l\pi}\right\}-$$
$$-\sum_{l=n+1}^{\infty}\left\{\frac{1}{\sqrt{[(2l-1)\pi-tx]^2-x^2}}-\frac{1}{2l\pi}\right\}.$$ MO 60

In formulas 8.523, $x>0$, $0\leqslant t<1$, $(2m-1)\pi<x(1-t)<(2m+1)\pi$, $(2n-1)\pi<x(1+t)<(2n+1)\pi$, m and n are natural numbers.

8.524

1. $$\sum_{k=1}^{\infty} J_0(kx)\cos kxt = -\frac{1}{2}+\sum_{l=m+1}^{n}\frac{1}{\sqrt{x^2-(2l\pi-tx)^2}}.$$ MO 60

2. $$\sum_{k=1}^{\infty} J_0(kx)\sin kxt = \sum_{l=0}^{m}\frac{1}{\sqrt{(2l\pi-tx)^2-x^2}}+$$
$$+\sum_{l=1}^{\infty}\left\{\frac{1}{\sqrt{(2l\pi+tx)^2-x^2}}-\frac{1}{2l\pi}\right\}-$$
$$-\sum_{l=n+1}^{\infty}\left\{\frac{1}{\sqrt{(2l\pi-tx)^2-x^2}}-\frac{1}{2l\pi}\right\}+\frac{1}{2\pi}\sum_{l=1}^{n}\frac{1}{l}.$$ MO 60

3. $$\sum_{k=1}^{\infty} N_0(kx)\cos kxt = -\frac{1}{\pi}\left(C+\ln\frac{x}{4\pi}\right) - \sum_{l=0}^{m} \frac{1}{\sqrt{(2\pi l - tx)^2 - x^2}} + \\ + \frac{1}{2\pi}\sum_{l=1}^{n}\frac{1}{l} - \sum_{l=1}^{\infty}\left\{\frac{1}{\sqrt{(2l\pi+tx)^2-x^2}} - \frac{1}{2l\pi}\right\} - \\ - \sum_{l=n+1}^{\infty}\left\{\frac{1}{\sqrt{(2l\pi-tx)^2-x^2}} - \frac{1}{2l\pi}\right\}.$$ MO 61

In formulas 8.524, $x > 0$, $t > 1$, $2m\pi < x(t-1) < 2(m+1)\pi$, $2n\pi < x(t+1) < 2(n+1)\pi$, $m+1$ and $n+1$ are natural numbers.

8.525

1. $$\sum_{k=1}^{\infty}(-1)^k J_0(kx)\cos kxt = -\frac{1}{2} + \sum_{l=m+1}^{n}\frac{1}{\sqrt{x^2-[(2l-1)\pi-tx]^2}}.$$ MO 61

2. $$\sum_{k=1}^{\infty}(-1)^k J_0(kx)\sin kxt = \sum_{l=1}^{m}\frac{1}{\sqrt{[(2l-1)\pi-tx]^2-x^2}} + \frac{1}{2\pi}\sum_{l=1}^{n}\frac{1}{l} + \\ + \sum_{l=1}^{\infty}\left\{\frac{1}{\sqrt{[(2l-1)\pi+tx]^2-x^2}} - \frac{1}{2l\pi}\right\} - \\ - \sum_{l=n+1}^{\infty}\left\{\frac{1}{\sqrt{[(2l-1)\pi-tx]^2-x^2}} - \frac{1}{2l\pi}\right\}.$$ MO 61

3. $$\sum_{k=1}^{\infty}(-1)^k N_0(kx)\cos kxt = -\frac{1}{\pi}\left(C+\ln\frac{x}{4\pi}\right) + \frac{1}{2\pi}\sum_{l=1}^{n}\frac{1}{l} - \\ - \sum_{l=1}^{m}\frac{1}{\sqrt{[(2l-1)\pi-tx]^2-x^2}} - \sum_{l=1}^{\infty}\left\{\frac{1}{\sqrt{[(2l-1)\pi+tx]^2-x^2}} - \frac{1}{2l\pi}\right\} - \\ - \sum_{l=n+1}^{\infty}\left\{\frac{1}{\sqrt{[(2l-1)\pi-tx]^2-x^2}} - \frac{1}{2l\pi}\right\}.$$ MO 61

In formulas 8.525, $x > 0$, $t > 1$, $(2m-1)\pi < x(t-1) < (2m+1)\pi$, $(2n-1)\pi < x(t+1) < (2n+1)\pi$, m and n are natural numbers.

8.526

1. $$\sum_{k=1}^{\infty} K_0(kx)\cos kxt = \frac{1}{2}\left(C+\ln\frac{x}{4\pi}\right) + \frac{\pi}{2x\sqrt{1+t^2}} + \\ + \frac{\pi}{2}\sum_{l=1}^{\infty}\left\{\frac{1}{\sqrt{x^2+(2l\pi-tx)^2}} - \frac{1}{2l\pi}\right\} + \frac{\pi}{2}\sum_{l=1}^{\infty}\left\{\frac{1}{\sqrt{x^2+(2l\pi+tx)^2}} - \frac{1}{2l\pi}\right\}.$$ MO 61

2. $$\sum_{k=1}^{\infty}(-1)^k K_0(kx)\cos kxt = \frac{1}{2}\left(C+\ln\frac{x}{4\pi}\right)+$$
$$+\frac{\pi}{2}\sum_{l=1}^{\infty}\left\{\frac{1}{\sqrt{x^2+[(2l-1)\pi-xt]^2}}-\frac{1}{2l\pi}\right\}+$$
$$+\frac{\pi}{2}\sum_{l=1}^{\infty}\left\{\frac{1}{\sqrt{x^2+[(2l-1)\pi+xt]^2}}-\frac{1}{2l\pi}\right\}$$

[$x>0$, t real], (see also 8.66). MO 62

8.53 Expansion in products of Bessel functions

"Summation theorems"

8.530 Suppose that $r>0$, $\varrho>0$, $\varphi>0$, and $R=\sqrt{r^2+\varrho^2-2r\varrho\cos\varphi}$; that is, suppose that r, ϱ, and R are the sides of a triangle such that the angle between the sides r and ϱ is equal to φ. Suppose also that $\varrho<r$ and that ψ is the angle opposite the side ϱ, so that

1. $$0<\psi<\frac{\pi}{2},\quad e^{2i\psi}=\frac{r-\varrho e^{-i\varphi}}{r-\varrho e^{i\varphi}}.$$

When these conditions are satisfied, we have the "summation theorem" for Bessel functions:

2. $$e^{i\nu\psi}Z_\nu(mR)=\sum_{k=-\infty}^{\infty}J_k(m\varrho)Z_{\nu+k}(mr)e^{ik\varphi}$$

[m — an arbitrary complex number]. WA 394(6)

For $Z_\nu=J_\nu$ and ν an integer, the restriction $\varrho<r$ is superfluous.

MO 31

8.531 Special cases:

1 $$J_0(mR)=J_0(m\varrho)J_0(mr)+2\sum_{k=1}^{\infty}J_k(m\varrho)J_k(mr)\cos k\varphi.$$ WA 391(1)

2. $$H_0^{(1,2)}(mR)=J_0(m\varrho)H_0^{(1,2)}(mr)+2\sum_{k=1}^{\infty}J_k(m\varrho)H_k^{(1,2)}(mr)\cos k\varphi.$$ MO 31

3. $$J_0(z\sin\alpha)=J_0^2\left(\frac{z}{2}\right)+2\sum_{k=1}^{\infty}J_k^2\left(\frac{z}{2}\right)\cos 2k\alpha;$$
$$=\sqrt{\frac{2\pi}{z}}\sum_{k=0}^{\infty}\left(2k+\frac{1}{2}\right)\frac{(2k-1)!!}{2^k k!}J_{2k+\frac{1}{2}}(z)P_{2k}(\cos\alpha).$$ MO 31

8.532 The term "summation theorem" is also applied to the formula

1. $$\frac{Z_\nu(mR)}{R^\nu}=2^\nu m^{-\nu}\Gamma(\nu)\sum_{k=0}^{\infty}(\nu+k)\frac{J_{\nu+k}(m\varrho)}{\varrho^\nu}\frac{Z_{\nu+k}(mr)}{r^\nu}C_k^\nu(\cos\varphi).$$

[$\nu\neq-1,-2,-3,\ldots$; the conditions on r, ϱ, R, φ, and m are the same as in formula 8.530; for $Z_\nu=J_\nu$ and ν an integer, formula 8.532 1 is valid for arbitrary r, ϱ, and φ]. WA 398(4)

8.533 Special cases:

1. $$\frac{e^{imR}}{R}=\frac{\pi i}{2\sqrt{r\varrho}}\sum_{k=0}^{\infty}(2k+1)J_{k+\frac{1}{2}}(m\varrho)H^{(1)}_{k+\frac{1}{2}}(mr)P_k(\cos\varphi).$$ MO 31

2. $$\frac{e^{-imR}}{R}=-\frac{\pi i}{2\sqrt{r\varrho}}\sum_{k=0}^{\infty}(2k+1)J_{k+\frac{1}{2}}(m\varrho)H^{(2)}_{k+\frac{1}{2}}(mr)P_k(\cos\varphi).$$ MO 31

8.534 A degenerate addition theorem $(r\to\infty)$:

$$e^{im\varrho\cos\varphi}=\sqrt{\frac{\pi}{2m\varrho}}\sum_{k=0}^{\infty}i^k(2k+1)J_{k+\frac{1}{2}}(m\varrho)P_k(\cos\varphi);$$ **WA 401(1)**

$$=2^{\nu}\Gamma(\nu)\sum_{k=0}^{\infty}(\nu+k)\,i^k(m\varrho)^{-\nu}J_{\nu+k}(m\varrho)C_k^{\nu}(\cos\varphi)$$ WA 401(2)

$$[\nu\neq 0,\ -1,\ -2,\ \ldots].$$

8.535 The term "product theorem" is also applied to the formula

$$Z_\nu(\lambda z)=\lambda^\nu\sum_{k=0}^{\infty}\frac{1}{k!}Z_{\nu+k}(z)\left[\frac{1-\lambda^2}{2}z\right]^k \qquad [|1-\lambda|^2<1].$$

For $Z_\nu=J_\nu$, it is valid for all values of λ and z. MO 32

8.536

1. $$\sum_{k=0}^{\infty}\frac{(2n+2k)(2n+k-1)!}{k!}J^2_{n+k}(z)=\frac{(2n)!}{(n!)^2}\left(\frac{z}{2}\right)^{2n} \qquad [n>0].$$ WA 47(1)

2. $$2\sum_{k=n}^{\infty}\frac{k\Gamma(n+k)}{\Gamma(k-n+1)}J^2_k(z)=\frac{(2n)!}{(n!)^2}\left(\frac{z}{2}\right)^{2n} \qquad [n>0].$$ WA 47(2)

3. $$J_0^2(z)+2\sum_{k=1}^{\infty}J_k^2(z)=1.$$ WA 41(3)

8.537 $$\sum_{k=-\infty}^{\infty}Z_{\nu-k}(t)J_k(z)=Z_\nu(z+t) \qquad [|z|<|t|].$$ WA 158(2)

In particular:

$$\sum_{k=-\infty}^{\infty}J_k(z)J_{n-k}(z)=J_n(2z).$$ WA 41

8.538

1. $$\sum_{k=-\infty}^{\infty}(-1)^kJ_{-\nu+k}(t)J_k(z)=J_{-\nu}(z+t) \qquad [|z|<|t|].$$ WA 159

2. $$\sum_{k=-\infty}^{\infty}Z_{\nu+k}(t)J_k(z)=Z_\nu(t-z) \qquad [|z|<|t|].$$ WA 159(5)

8.54 The zeros of Bessel functions

8.541 For arbitrary real ν, the function $J_\nu(z)$ has infinitely many real zeros. For $\nu>-1$, all its zeros are real.

WA 526, 530

A Bessel function $Z_\nu(z)$ has no multiple zeros except possibly the coordinate origin.

WA 528

8.542 All zeros of the function $N_0(z)$ with positive real parts are real. WA 531

8.543 If $-(2s+2)<\nu<-(2s+1)$, where s is a natural number or 0, then $J_\nu(z)$ has exactly $4s+2$ complex roots, two of which are purely imaginary. If $-(2s+1)<\nu<-2s$, where s is a natural number, then the function $J_\nu(z)$ has exactly $4s$ complex zeros none of which are purely imaginary. WA 532

8.544 If x_ν and x'_ν are, respectively, the smallest positive zeros of the functions $J_\nu(z)$ and $J'_\nu(z)$ for $\nu>0$, then $x_\nu>\nu$ and $x'_\nu>\nu$. Suppose also that y_ν is the smallest positive zero of the function $N_\nu(z)$. Then, $x_\nu<y_\nu<x'_\nu$.

WA 534, 536

Suppose that $z_{\nu, m}$ (for $m=1, 2, 3, \ldots$) are the zeros of the function $z^{-\nu}J_\nu(z)$, numbered in order of the absolute value of their real parts. Here, we assume that $\nu\neq-1, -2, -3, \ldots$. Then, for arbitrary z,

$$J_\nu(z)=\frac{\left(\frac{z}{2}\right)^\nu}{\Gamma(\nu+1)}\prod_{m=1}^{\infty}\left(1-\frac{z^2}{z_{\nu, m}^2}\right). \quad \text{WA 550}$$

8.545 The number of zeros of the function $z^{-\nu}J_\nu(z)$ that occur between the imaginary axis and the line on which

$$\operatorname{Re} z=m\pi+\left[\frac{1}{2}\operatorname{Re}\nu+\frac{1}{4}\right]\pi, \quad \textbf{WA 548}$$

is exactly m.

8.546 For $\nu\geqslant0$, the number of zeros of the function $K_\nu(z)$ that occur in the region $\operatorname{Re} z<0, |\arg z|<\pi$ is equal to the even number closest to $\nu-\frac{1}{2}$. WA 562

8.547 Large zeros of the functions $J_\nu(z)\cos\alpha-N_\nu(z)\sin\alpha$, where ν and α are real numbers, are given by the asymptotic expansion

$$x_{\nu, m}\sim\left(m+\frac{1}{2}\nu-\frac{1}{4}\right)\pi-\alpha-\frac{4\nu^2-1}{8\left[\left(m+\frac{1}{2}\nu-\frac{1}{4}\right)\pi-\alpha\right]}-$$

$$-\frac{(4\nu^2-1)(28\nu^2-31)}{384\left[\left(m+\frac{1}{2}\nu-\frac{1}{4}\right)\pi-\alpha\right]^3}-\ldots \quad \text{KU 109(24), WA 558}$$

8.548 In particular, large zeros of the function $J_0(z)$ are given by the expansion

$$x_{0, m}\sim\frac{\pi}{4}(4m-1)+\frac{1}{2\pi(4m-1)}-\frac{31}{6\pi^3(4m-1)^3}+\frac{3779}{15\pi^5(4m-1)^5}-\ldots$$

KU 109(25), WA 556

This series is suitable for calculating all (except the smallest x_{01}) zeros of the function $J_0(z)$ correctly to at least five digits.

8.549 To calculate the roots $x_{\nu, m}$ of the function $J_\nu(z)$ of smallest absolute value, we may use the identity

$$\sum_{m=1}^{\infty}\frac{1}{x_{\nu,m}^{16}}=\frac{429\nu^5+7640\nu^4+53752\nu^3+185430\nu^2+311387\nu+202738}{2^{16}(\nu+1)^8(\nu+2)^4(\nu+3)^2(\nu+4)^2(\nu+5)(\nu+6)(\nu+7)(\nu+8)}.$$

KU 112(27)a, WA 554

8.55 Struve functions

8.550 Definitions:

1. $$\mathbf{H}_\nu(z) = \sum_{m=0}^{\infty} (-1)^m \frac{\left(\frac{z}{2}\right)^{2m+\nu+1}}{\Gamma\left(m+\frac{3}{2}\right)\Gamma\left(\nu+m+\frac{3}{2}\right)}.$$ WA 358(2)

2. $$\mathbf{L}_\nu(z) = -ie^{-i\nu\frac{\pi}{2}}\mathbf{H}_\nu\left(ze^{i\frac{\pi}{2}}\right) = \sum_{m=0}^{\infty} \frac{\left(\frac{z}{2}\right)^{2m+\nu+1}}{\Gamma\left(m+\frac{3}{2}\right)\Gamma\left(\nu+m+\frac{3}{2}\right)}.$$ WA 360(11)

8.551 Integral representations:

1. $$\mathbf{H}_\nu(z) = \frac{2\left(\frac{z}{2}\right)^\nu}{\sqrt{\pi}\,\Gamma\left(\nu+\frac{1}{2}\right)} \int_0^1 (1-t^2)^{\nu-\frac{1}{2}} \sin zt\, dt = \frac{2\left(\frac{z}{2}\right)^\nu}{\sqrt{\pi}\,\Gamma\left(\nu+\frac{1}{2}\right)} \int_0^{\frac{\pi}{2}} \sin(z\cos\varphi)\,(\sin\varphi)^{2\nu}\, d\varphi \quad \left[\operatorname{Re}\nu > -\frac{1}{2}\right].$$ WA 358(1)

2. $$\mathbf{L}_\nu(z) = \frac{2\left(\frac{z}{2}\right)^\nu}{\sqrt{\pi}\,\Gamma\left(\nu+\frac{1}{2}\right)} \int_0^{\frac{\pi}{2}} \operatorname{sh}(z\cos\varphi)\,(\sin\varphi)^{2\nu}\, d\varphi \quad \left[\operatorname{Re}\nu > -\frac{1}{2}\right].$$ WA 360(11)

8.552 Special cases:

1. $$\mathbf{H}_n(z) = \frac{1}{\pi} \sum_{m=0}^{E\left(\frac{n-1}{2}\right)} \frac{\Gamma\left(m+\frac{1}{2}\right)\left(\frac{z}{2}\right)^{n-2m-1}}{\Gamma\left(n+\frac{1}{2}-m\right)} - \mathbf{E}_n(z) \quad [n = 1, 2, \ldots].$$ EH II 40(66), WA 367(1)

2. $$\mathbf{H}_{-n}(z) = (-1)^{n+1}\frac{1}{\pi} \sum_{m=0}^{E\left(\frac{n-1}{2}\right)} \frac{\Gamma\left(n-m-\frac{1}{2}\right)\left(\frac{z}{2}\right)^{-n+2m+1}}{\Gamma\left(m+\frac{3}{2}\right)} - \mathbf{E}_{-n}(z) \quad [n = 1, 2, \ldots].$$ EH II 40(67), WA 367(2)

3. $$\mathbf{H}_{n+\frac{1}{2}}(z) = N_{n+\frac{1}{2}}(z) + \frac{1}{\pi} \sum_{m=0}^{n} \frac{\Gamma\left(m+\frac{1}{2}\right)\left(\frac{z}{2}\right)^{-2m+n-\frac{1}{2}}}{\Gamma(n+1-m)} \quad [n = 0, 1, \ldots].$$ EH II 39(64)

4. $$\mathbf{H}_{-\left(n+\frac{1}{2}\right)}(z) = (-1)^n J_{n+\frac{1}{2}}(z) \quad [n = 0, 1, \ldots].$$ EH II 39(65)

5. $$\mathbf{L}_{-\left(n+\frac{1}{2}\right)}(z) = I_{n+\frac{1}{2}}(z) \quad [n = 0, 1, \ldots].$$ EH II 39(65)

6. $\mathbf{H}_{\frac{1}{2}}(z) = \frac{\sqrt{2}}{\sqrt{\pi z}}(1 - \cos z).$ EH II39, WA 364(3)

7. $\mathbf{H}_{\frac{3}{2}}(z) = \left(\frac{z}{2\pi}\right)^{\frac{1}{2}}\left(1 + \frac{2}{z^2}\right) - \left(\frac{2}{\pi z}\right)^{\frac{1}{2}}\left(\sin z + \frac{\cos z}{z}\right).$ WA 364(3)

8.553 Functional relations:

1. $\mathbf{H}_\nu(ze^{im\pi}) = e^{i\pi(\nu+1)m}\mathbf{H}_\nu(z) \qquad [m = 1, 2, 3, \ldots].$ WA 362(5)

2. $\frac{d}{dz}[z^\nu \mathbf{H}_\nu(z)] = z^\nu \mathbf{H}_{\nu-1}(z).$ WA 358

3. $\frac{d}{dz}[z^{-\nu}\mathbf{H}_\nu(z)] = 2^{-\nu}\pi^{-\frac{1}{2}}\left[\Gamma\left(\nu + \frac{3}{2}\right)\right]^{-1} - z^{-\nu}\mathbf{H}_{\nu+1}(z).$ WA 359

4. $\mathbf{H}_{\nu-1}(z) + \mathbf{H}_{\nu+1}(z) = 2\nu z^{-1}\mathbf{H}_\nu(z) + \pi^{-\frac{1}{2}}\left(\frac{z}{2}\right)^\nu\left[\Gamma\left(\nu + \frac{3}{2}\right)\right]^{-1}.$ WA 359(5)

5. $\mathbf{H}_{\nu-1}(z) - \mathbf{H}_{\nu+1}(z) = 2\mathbf{H}'_\nu(z) - \pi^{-\frac{1}{2}}\left(\frac{z}{2}\right)^\nu\left[\Gamma\left(\nu + \frac{3}{2}\right)\right]^{-1}.$ WA 359(6)

8.554 Asymptotic representations:

$$\mathbf{H}_\nu(\xi) = N_\nu(\xi) + \frac{1}{\pi}\sum_{m=0}^{p-1}\frac{\Gamma\left(m + \frac{1}{2}\right)\left(\frac{\xi}{2}\right)^{-2m+\nu-1}}{\Gamma\left(\nu + \frac{1}{2} - m\right)} + O(|\xi|^{\nu-2p-1})$$

$[|\arg\xi| < \pi].$ EH II 39(63), WA 363(2)

For the asymptotic representation of $N_\nu(\xi)$, see 8.451 2.

8.555 The differential equation for Struve functions:

$$z^2y'' + zy' + (z^2 - \nu^2)y = \frac{1}{\sqrt{\pi}} \cdot \frac{4\left(\frac{z}{2}\right)^{\nu+1}}{\Gamma\left(\nu + \frac{1}{2}\right)}.$$ WA 359(10)

8.56 Thomson functions and their generalizations: $\operatorname{ber}_\nu(z)$, $\operatorname{bei}_\nu(z)$, $\operatorname{her}_\nu(z)$, $\operatorname{hei}_\nu(z)$, $\operatorname{ker}(z)$, $\operatorname{kei}(z)$

8.561

1. $\operatorname{ber}_\nu(z) + i\operatorname{bei}_\nu(z) = J_\nu(ze^{\frac{3}{4}\pi i}).$

2. $\operatorname{ber}_\nu(z) - i\operatorname{bei}_\nu(z) = J_\nu(ze^{-\frac{3}{4}\pi i}).$

WA 96(6)

8.562

1. $\operatorname{her}_\nu(z) + i\operatorname{hei}_\nu(z) = H_\nu^{(1)}(ze^{\frac{3}{4}\pi i})$

2. $\operatorname{her}_\nu(z) - i\operatorname{hei}_\nu(z) = H_\nu^{(1)}(ze^{-\frac{3}{4}\pi i})$

(see also 8.567). WA 96(7)

8.563

1. $\operatorname{ber}_0(z) \equiv \operatorname{ber}(z); \quad \operatorname{bei}_0(z) \equiv \operatorname{bei}(z).$

2. $\operatorname{ker}(z) \equiv -\frac{\pi}{2}\operatorname{hei}_0(z); \quad \operatorname{kei}(z) \equiv \frac{\pi}{2}\operatorname{her}_0(z).$ WA 96(8)

For integral representations, see 6.251, 6.536, 6.537, 6.772 4., 6.777.

Series representation

8.564

1. $$\operatorname{ber}(z) = \sum_{k=0}^{\infty} \frac{(-1)^k z^{4k}}{2^{4k}[(2k)!]^2}.$$ WA 96(3)

2. $$\operatorname{bei}(z) = \sum_{k=0}^{\infty} \frac{(-1)^k z^{4k+2}}{2^{4k+2}[(2k+1)!]^2}.$$ WA 96(4)

3. $$\operatorname{ker}(z) = \left(\ln\frac{2}{z} - C\right)\operatorname{ber}(z) + \frac{\pi}{4}\operatorname{bei}(z) + \sum_{k=1}^{\infty}(-1)^k \frac{z^{4k}}{2^{4k}[(2k)!]^2}\sum_{m=1}^{2k}\frac{1}{m}.$$ WA 96(9)a, DW

4. $$\operatorname{kei}(z) = \left(\ln\frac{2}{z} - C\right)\operatorname{bei}(z) - \frac{\pi}{4}\operatorname{ber}(z) + \sum_{k=0}^{\infty}(-1)^k \frac{z^{4k+2}}{2^{4k+2}[(2k+1)!]^2}\sum_{m=1}^{2k+1}\frac{1}{m}.$$ WA 96(10)a, DW

8.565 $$\operatorname{ber}_\nu^2(z) + \operatorname{bei}_\nu^2(z) = \sum_{k=0}^{\infty} \frac{\left(\frac{z}{2}\right)^{2\nu+4k}}{k!\,\Gamma(\nu+k+1)\,\Gamma(\nu+2k+1)}.$$ WA 163(6)

Asymptotic representation

8.566

1. $$\operatorname{ber}(z) = \frac{e^{\alpha(z)}}{\sqrt{2\pi z}}\cos\beta(z) \qquad \left[|\arg z| < \frac{\pi}{4}\right].$$ WA 227(1)

2. $$\operatorname{bei}(z) = \frac{e^{\alpha(z)}}{\sqrt{2\pi z}}\sin\beta(z) \qquad \left[|\arg z| < \frac{\pi}{4}\right].$$ WA 227(1)

3. $$\operatorname{ker}(z) = \sqrt{\frac{\pi}{2z}}\, e^{\alpha(-z)}\cos\beta(-z) \qquad \left[|\arg z| < \frac{5}{4}\pi\right].$$ WA 227(2)

4. $$\operatorname{kei}(z) = \sqrt{\frac{\pi}{2z}}\, e^{\alpha(-z)}\sin\beta(-z) \qquad \left[|\arg z| < \frac{5}{4}\pi\right],$$ WA 227(2)

where

$$\alpha(z) \sim \frac{z}{\sqrt{2}} + \frac{1}{8z\sqrt{2}} - \frac{25}{384z^3\sqrt{2}} - \frac{13}{128z^4} - \ldots,$$

$$\beta(z) \sim \frac{z}{\sqrt{2}} - \frac{\pi}{8} - \frac{1}{8z\sqrt{2}} - \frac{1}{16z^2} - \frac{25}{384z^3\sqrt{2}} + \ldots.$$

8.567 Functional relations

1. $\operatorname{ker}(z) + i \operatorname{kei}(z) = K_0(z\sqrt{i})$
2. $\operatorname{ker}(z) - i \operatorname{kei}(z) = K_0(z\sqrt{-i})$

(see 8.562). WA 96(5), DW

For integrals of Thomson's functions, see 6.87.

8.57 Lommel functions

8.570 Definitions of the Lommel functions $s_{\mu,\nu}(z)$ and $S_{\mu,\nu}(z)$:

1. $$s_{\mu,\nu}(z) = \sum_{m=0}^{\infty} \frac{(-1)^m z^{\mu+1+2m}}{[(\mu+1)^2-\nu^2][(\mu+3)^2-\nu^2]\dots[(\mu+2m+1)^2-\nu^2]};$$

$$= z^{\mu-1} \sum_{m=0}^{\infty} \frac{(-1)^m \left(\frac{z}{2}\right)^{2m+2} \Gamma\left(\frac{1}{2}\mu-\frac{1}{2}\nu+\frac{1}{2}\right)\Gamma\left(\frac{1}{2}\mu+\frac{1}{2}\nu+\frac{1}{2}\right)}{\Gamma\left(\frac{1}{2}\mu-\frac{1}{2}\nu+m+\frac{3}{2}\right)\Gamma\left(\frac{1}{2}\mu+\frac{1}{2}\nu+m+\frac{3}{2}\right)}$$

[$\mu \pm \nu$ is not a negative odd integer]. EH II 40(69), WA 377(2)

2. $$S_{\mu,\nu}(z) = s_{\mu,\nu}(z) + \left[2^{\mu-1}\Gamma\left(\frac{1}{2}\mu-\frac{1}{2}\nu+\frac{1}{2}\right)\Gamma\left(\frac{1}{2}\mu+\frac{1}{2}\nu+\frac{1}{2}\right)\right] \times$$
$$\times \frac{\cos\left[\frac{1}{2}(\mu-\nu)\pi\right] J_{-\nu}(z) - \cos\left[\frac{1}{2}(\mu+\nu)\pi\right] J_{\nu}(z)}{\sin \nu\pi}$$

[$\mu \pm \nu$ is a positive odd integer, ν is an odd integer];

EH II 40(71), WA 379(2)

$$= s_{\mu,\nu}(z) + 2^{\mu-1}\Gamma\left(\frac{1}{2}\mu-\frac{1}{2}\nu+\frac{1}{2}\right)\Gamma\left(\frac{1}{2}\mu+\frac{1}{2}\nu+\frac{1}{2}\right) \times$$
$$\times \left\{\sin\left[\frac{1}{2}(\mu-\nu)\pi\right] J_{\nu}(z) - \cos\left[\frac{1}{2}(\mu-\nu)\pi\right] N_{\nu}(z)\right\}$$

[$\mu \pm \nu$ is a positive odd integer, ν is an integer].

EH II 41(71), WA 379(3)

Integral representations

8.571 $$s_{\mu,\nu}(z) = \frac{\pi}{2}\left[N_\nu(z)\int_0^z z^\mu J_\nu(z)\,dz - J_\nu(z)\int_0^z z^\mu N_\nu(z)\,dz\right].$$ WA 378(9)

8.572. $$s_{\mu,\nu}(z) = 2^\mu \left(\frac{z}{2}\right)^{\frac{1}{2}(1+\nu+\mu)} \Gamma\left(\frac{1}{2}+\frac{1}{2}\mu-\frac{1}{2}\nu\right) \times$$
$$\times \int_0^{\frac{\pi}{2}} J_{\frac{1}{2}(1+\mu-\nu)}(z\sin\theta)(\sin\theta)^{\frac{1}{2}(1+\nu-\mu)}(\cos\theta)^{\nu+\mu}\,d\theta$$

[$\operatorname{Re}(\nu+\mu+1) > 0$]. EH II 42(86)

8.573 Special cases:

1. $S_{1,2n}(z) = zO_{2n}(z)$. WA 382(1)
2. $S_{0,2n+1}(z) = \frac{z}{2n+1} O_{2n+1}(z)$. WA 382(1)

3. $S_{-1,\,2n}(z)=\frac{1}{4n}S_{2n}(z).$ WA 382(2)

4. $S_{0,\,2n+1}(z)=\frac{1}{2}S_{2n+1}(z).$ WA 382(2)

5. $s_{\nu,\nu}(z)=\Gamma\left(\nu+\frac{1}{2}\right)\sqrt{\pi}\,2^{\nu-1}\mathbf{H}_\nu(z).$ EH II 42(84)

6. $S_{\nu,\nu}(z)=[\mathbf{H}_\nu(z)-N_\nu(z)]\,2^{\nu-1}\sqrt{\pi}\,\Gamma\left(\nu+\frac{1}{2}\right).$ EH II 42(84)

8.574 Connections with other special functions:

1. $\mathbf{J}_\nu(z)=\frac{1}{\pi}\sin(\nu\pi)\,[s_{0,\nu}(z)-\nu s_{-1,\nu}(z)].$ EH II 41(82)

2. $\mathbf{E}_\nu(z)=-\frac{1}{\pi}\,[(1+\cos\nu\pi)\,s_{0,\nu}(z)+\nu(1-\cos\nu\pi)\,s_{-1,\nu}(z)].$ EH II 42(83)

A connection with a hypergeometric function

3. $s_{\mu,\nu}(z)=\frac{z^{\mu+1}}{(\mu-\nu+1)(\mu+\nu+1)}\,{}_1F_2\left(1;\ \frac{\mu-\nu+3}{2},\ \frac{\mu+\nu+3}{2};\ -\frac{z^2}{4}\right).$

EH II 40(69), WA 378(10)

8.575 Functional relations:

1. $s_{\mu+2,\,\nu}(z)=z^{\mu+1}-[(\mu+1)^2-\nu^2]\,s_{\mu,\,\nu}(z).$ EH II 41(73), WA 380(1)

2. $s'_{\mu,\,\nu}(z)+\left(\frac{\nu}{z}\right)s_{\mu,\,\nu}(z)=(\mu+\nu-1)\,s_{\mu-1,\,\nu-1}(z).$

EH II 41(74), WA 380(2)

3. $s'_{\mu,\,\nu}(z)-\left(\frac{\nu}{z}\right)s_{\mu,\,\nu}(z)=(\mu-\nu-1)\,s_{\mu-1,\,\nu+1}(z).$

EH II 41(75), WA 380(3)

4. $\left(2\frac{\nu}{z}\right)s_{\mu,\,\nu}(z)=(\mu+\nu-1)\,s_{\mu-1,\,\nu-1}(z)-(\mu-\nu-1)\,s_{\mu-1,\,\nu+1}(z).$

EH II 41(76), WA 380(4)

5. $2s_{\mu,\,\nu}(z)=(\mu+\nu-1)\,s_{\mu-1,\,\nu-1}(z)+(\mu-\nu-1)\,s_{\mu-1,\,\nu+1}(z).$

EH II 41(77), WA 380(5)

In formulas 8.575 1.—5., $s_{\mu,\,\nu}(z)$ can be replaced with $S_{\mu,\,\nu}(z)$.

8.576 Asymptotic expansion of $S_{\mu,\,\nu}(z)$. In the case in which $\mu\pm\nu$ is not a positive odd integer, the following asymptotic expansion is valid for $S_{\mu,\,\nu}(z)$:

$$S_{\mu,\,\nu}(z)=z^{\mu-1}\sum_{m=0}^{p-1}\frac{(-1)^m\,\Gamma\left(\frac{1}{2}-\frac{1}{2}\mu+\frac{1}{2}\nu+m\right)}{\left(\frac{z}{2}\right)^m\Gamma\left(\frac{1}{2}-\frac{1}{2}\mu+\frac{1}{2}\nu\right)}\,\frac{\Gamma\left(\frac{1}{2}-\frac{1}{2}\mu-\frac{1}{2}\nu+m\right)}{\Gamma\left(\frac{1}{2}-\frac{1}{2}\mu-\frac{1}{2}\nu\right)}+$$
$$+O(z^{\mu-2p}).$$ WA 385

8.577 Lommel functions satisfy the following differential equation:

$$z^2w''+zw'+(z^2-\nu^2)\,w=z^{\mu+1}.$$ WA 377(1), EH II 40(68)

8.578 Lommel functions of two variables $U_\nu(w, z)$, $V_\nu(w, z)$:

Definition

1. $U_\nu(w, z) = \sum_{m=0}^{\infty} (-1)^m \left(\frac{w}{z}\right)^{\nu+2m} J_{\nu+2m}(z).$ EH II 42(87), WA 591(5)

2. $V_\nu(w, z) = \cos\left[\frac{1}{2}\left(w + \frac{z^2}{w} + \nu\pi\right)\right] + U_{-\nu+2}(w, z).$

EH II 42(88), WA 591(6)

Particular values:

3. $U_0(z, z) = V_0(z, z) = \frac{1}{2}\{J_0(z) + \cos z\}.$ WA 591(9)

4. $U_1(z, z) = -V_1(z, z) = \frac{1}{2}\sin z.$ WA 591(10)

5. $U_{2n}(z, z) = V_{2n}(z, z) = \frac{(-1)^n}{2}\left\{\cos z - \sum_{m=0}^{n-1} (-1)^m \varepsilon_{2m} J_{2m}(z)\right\}$

$[n \geqslant 1], \quad \varepsilon_m = \begin{cases} 2, & m > 0, \\ 1, & m = 0. \end{cases}$ WA 591(11)

6. $U_{2n+1}(z, z) = -V_{2n+1}(z, z) = \frac{(-1)^n}{2}\left\{\sin z - \sum_{m=0}^{n-1} (-1)^m \varepsilon_{2m+1} J_{2m+1}(z)\right\},$

$[n \geqslant 0], \quad \varepsilon_m = \begin{cases} 2, & m > 0, \\ 1, & m = 0. \end{cases}$ WA 591(12)

7. $V_n(w, z) = (-1)^n U_n\left(\frac{z^2}{w}, z\right).$

8. $U_\nu(w, 0) = \frac{\left(\frac{w}{2}\right)^{\frac{1}{2}}}{\Gamma(\nu-1)} s_{\nu-\frac{3}{2}, \frac{1}{2}}\left(\frac{w}{2}\right).$ WA 593(9)

9. $V_{-\nu+2}(w, 0) = \frac{\left(\frac{w}{2}\right)^{\frac{1}{2}}}{\Gamma(\nu-1)} S_{\nu-\frac{3}{2}, \frac{1}{2}}\left(\frac{w}{2}\right).$ WA 593(10)

8.579 Functional relations:

1. $2\frac{\partial}{\partial w} U_\nu(w, z) = U_{\nu-1}(w, z) + \left(\frac{z}{w}\right)^2 U_{\nu+1}(w, z).$ WA 593(2)

2. $2\frac{\partial}{\partial w} V_\nu(w, z) = V_{\nu+1}(w, z) + \left(\frac{z}{w}\right)^2 V_{\nu-1}(w, z).$ WA 593(4)

3. The function $U_\nu(w, z)$ is a particular solution of the differential equation

$$\frac{\partial^2 U}{\partial z^2} - \frac{1}{z}\frac{\partial U}{\partial z} + \frac{z^2 U}{w^2} = \left(\frac{w}{z}\right)^{\nu-2} J_\nu(z).$$ WA 592(2)

4. The function $V_\nu(w, z)$ is a particular solution of the differential equation

$$\frac{\partial^2 V}{\partial z^2} - \frac{1}{z}\frac{\partial V}{\partial z} + \frac{z^2 V}{w^2} = \left(\frac{w}{z}\right)^{-\nu} J_{-\nu+2}(z).$$ WA 592(3)

8.58 Anger and Weber functions $\mathbf{J}_\nu(z)$ and $\mathbf{E}_\nu(z)$

8.580 Definitions:

1. The Anger function $\mathbf{J}_\nu(z)$:

$$\mathbf{J}_\nu(z) = \frac{1}{\pi}\int_0^\pi \cos(\nu\theta - z\sin\theta)\,d\theta.$$ WA 336(1), EH II 35(32)

2. The Weber function $\mathbf{E}_\nu(z)$:

$$\mathbf{E}_\nu(z) = \frac{1}{\pi}\int_0^\pi \sin(\nu\theta - z\sin\theta)\,d\theta.$$ WA 336(2), EH II 35(32)

8.581 Series representations:

1. $$\mathbf{J}_\nu(z) = \cos\frac{\nu\pi}{2}\sum_{n=0}^{\infty}\frac{(-1)^n\left(\frac{z}{2}\right)^{2n}}{\Gamma\left(n+1+\frac{1}{2}\nu\right)\Gamma\left(n+1-\frac{1}{2}\nu\right)} + \sin\frac{\nu\pi}{2}\sum_{n=0}^{\infty}\frac{(-1)^n\left(\frac{z}{2}\right)^{2n+1}}{\Gamma\left(n+\frac{3}{2}+\frac{1}{2}\nu\right)\Gamma\left(n+\frac{3}{2}-\frac{1}{2}\nu\right)}.$$ EH II 36(36), WA 337(3)

2. $$\mathbf{E}_\nu(z) = \sin\frac{\nu\pi}{2}\sum_{n=0}^{\infty}\frac{(-1)^n\left(\frac{z}{2}\right)^{2n}}{\Gamma\left(n+1+\frac{1}{2}\nu\right)\Gamma\left(n+1-\frac{1}{2}\nu\right)} - \cos\frac{\nu\pi}{2}\sum_{n=0}^{\infty}\frac{(-1)^n\left(\frac{z}{2}\right)^{2n+1}}{\Gamma\left(n+\frac{3}{2}+\frac{1}{2}\nu\right)\Gamma\left(n+\frac{3}{2}-\frac{1}{2}\nu\right)}.$$ EH II 36(40), WA 338(4)

8.582 Functional relations:

1. $2\mathbf{J}_\nu'(z) = \mathbf{J}_{\nu-1}(z) - \mathbf{J}_{\nu+1}(z).$ EH II 36(41), WA 340(2)

2. $2\mathbf{E}_\nu'(z) = \mathbf{E}_{\nu-1}(z) - \mathbf{E}_{\nu+1}(z).$ EH II 36(42), WA 340(6)

3. $\mathbf{J}_{\nu-1}(z) + \mathbf{J}_{\nu+1}(z) = 2\nu z^{-1}\mathbf{J}_\nu(z) - 2(\pi z)^{-1}\sin(\nu\pi).$ EH II 36(43), WA 340(1)

4. $\mathbf{E}_{\nu-1}(z) + \mathbf{E}_{\nu+1}(z) = 2\nu z^{-1}\mathbf{E}_\nu(z) - 2(\pi z)^{-1}(1-\cos\nu\pi).$ EH II 36(44), WA 340(5)

8.583 Asymptotic expansions:

1. $$\mathbf{J}_\nu(z) = J_\nu(z) + \frac{\sin \nu\pi}{\pi z}\left[\sum_{n=0}^{p-1}(-1)^n 2^{2n}\frac{\Gamma\left(n+\frac{1+\nu}{2}\right)}{\Gamma\left(\frac{1+\nu}{2}\right)}\frac{\Gamma\left(n+\frac{1-\nu}{2}\right)}{\Gamma\left(\frac{1-\nu}{2}\right)}z^{-2n} + O(|z|^{-2p}) + \nu\sum_{n=0}^{p-1}(-1)^n 2^{2n}\frac{\Gamma\left(n+1+\frac{1}{2}\nu\right)\Gamma\left(n+1-\frac{1}{2}\nu\right)}{\Gamma\left(1+\frac{1}{2}\nu\right)\Gamma\left(1-\frac{1}{2}\nu\right)}z^{-2n-1} + \nu O(|z|^{-2p-1})\right] \quad [|\arg z| < \pi].$$

EH II 37(47), WA 344(1)

2. $$\mathbf{E}_\nu(z) = -N_\nu(z) - \frac{1+\cos(\nu\pi)}{\pi z}\left[\sum_{n=0}^{p-1}(-1)^n 2^{2n}\frac{\Gamma\left(n+\frac{1+\nu}{2}\right)\Gamma\left(n+\frac{1-\nu}{2}\right)}{\Gamma\left(\frac{1+\nu}{2}\right)\Gamma\left(\frac{1-\nu}{2}\right)}z^{-2n} + O(|z|^{-2p})\right] - \frac{\nu(1-\cos\nu\pi)}{z\pi} \times \left[\sum_{n=0}^{p-1}(-1)^n 2^{2n}\frac{\Gamma\left(n+1+\frac{1}{2}\nu\right)\Gamma\left(n+1-\frac{1}{2}\nu\right)}{\Gamma\left(1+\frac{1}{2}\nu\right)\Gamma\left(1-\frac{1}{2}\nu\right)}z^{-2n-1} + O(|z|^{-2p-1})\right].$$

WA 344(2), EH II 37(48)

For the asymptotic expansion of $J_\nu(z)$ and $N_\nu(z)$, see 8.451.

8.584 The Anger and Weber functions satisfy the differential equation

$$y'' + z^{-1}y' + \left(1 - \frac{\nu^2}{z^2}\right)y = f(\nu, z),$$

where $f(\nu, z) = \frac{z-\nu}{\pi z^2}\sin\nu\pi$ for $\mathbf{J}_\nu(z)$ WA 341(9), EH II 37(44)

and $f(\nu, z) = -\frac{1}{\pi z^2}[z + \nu + (z-\nu)\cos\nu\pi]$ for $\mathbf{E}_\nu(z)$

EH II 37(45), WA 341(10)

8.59 Neumann's and Schläfli's polynomials: $O_n(z)$ and $S_n(z)$

8.590 Definition of Neumann's polynomials

1. $$O_n(z) = \frac{1}{4}\sum_{m=0}^{E\left(\frac{n}{2}\right)}\frac{n(n-m-1)!}{m!}\left(\frac{z}{2}\right)^{2m-n-1} \quad [n \geqslant 1].$$

WA 299(2), EH II 33(6)

2. $O_{-n}(z) = (-1)^n O_n(z) \quad [n \geqslant 1].$ WA 303(8)

3. $O_0(z) = \frac{1}{z}.$ WA 299(3), EH II 33(7)

4. $O_1(z) = \frac{1}{z^2}.$ EH II 33(7)

5. $O_2(z) = \frac{1}{z} + \frac{4}{z^3}$. EH II 33(7)

In general, $O_n(z)$ is a polynomial in z^{-1} of degree $n+1$.

8.591 Functional relations:

1. $O_0'(z) = -O_1(z)$. EH II 33(9), WA 301(3)

2. $2O_n'(z) = O_{n-1}(z) - O_{n+1}(z) \quad [n \geqslant 1]$. EH II 33(10), WA 301(2)

3. $$(n-1)O_{n+1}(z) + (n+1)O_{n-1}(z) - 2z^{-1}(n^2-1)O_n(z) = \\ = 2nz^{-1}\left(\sin n\frac{\pi}{2}\right)^2 \quad [n \geqslant 1].$$ EH II 33(11), WA 301(1)

4. $$nzO_{n-1}(z) - (n^2-1)O_n(z) = (n-1)zO_n'(z) + n\left(\sin n\frac{\pi}{2}\right)^2.$$ EH II 33(12), WA 303(4)

5. $$nzO_{n+1}(z) - (n^2-1)O_n(z) = -(n+1)zO_n'(z) + n\left(\sin n\frac{\pi}{2}\right)^2.$$ EH II 33(13), WA 303(5)a

8.592 The generating function:

$$\frac{1}{z-\xi} = J_0(\xi)z^{-1} + 2\sum_{n=1}^{\infty} J_n(\xi)O_n(z) \qquad [|\xi| < |z|].$$ EH II 32(1), WA 298(1)

8.593 The integral representation:

$$O_n(z) = \int_0^{\infty} \frac{[u+\sqrt{u^2+z^2}]^n + [u-\sqrt{u^2+z^2}]^n}{2z^{n+1}} e^{-u}\,du.$$

See also 3.547 6., 8., 3.549 1., 2. EH II 32(3), WA 305(1)

8.594 The inequality

$$|O_n(z)| \leqslant 2^{n-1}n!\,|z|^{-n-1}e^{\frac{1}{4}|z|^2} \qquad [n > 1].$$ EH II 33(8), WA 300(8)

8.595 Neumann's polynomial $O_n(z)$ satisfies the differential equation

$$z^2\frac{d^2y}{dz^2} + 3z\frac{dy}{dz} + (z^2+1-n^2)y = z\left(\cos n\frac{\pi}{2}\right)^2 + n\left(\sin n\frac{\pi}{2}\right)^2.$$ EH II 33(14), WA 303(1)

8.596 Schläfli's polynomials $S_n(z)$. These are the functions that satisfy the formulas

1. $S_0(z) = 0$. EH II 34(18), WA 312(2)

2. $$S_n(z) = \frac{1}{n}\left[2zO_n(z) - 2\left(\cos n\frac{\pi}{2}\right)^2\right] \qquad [n \geqslant 1];$$ EH II 34(19), WA 312(3)

$$= \sum_{m=0}^{E\left(\frac{n}{2}\right)} \frac{(n-m-1)!}{m!}\left(\frac{z}{2}\right)^{2m-n} \qquad [n \geqslant 1].$$ EH II 34(18)

3. $S_{-n}(z) = (-1)^{n+1}S_n(z)$. WA 313(6)

8.597 Functional relations:

WA 313(7)

1. $S_{n-1}(z)+S_{n+1}(z)=4O_n(z)$.

Other functional relations may be obtained from 8.591 by replacing $O_n(z)$ with the expression for $S_n(z)$ given by 8.596 2.

8.6 Mathieu Functions

8.60 Mathieu's equation

$$\frac{d^2y}{dz^2}+(a-2k^2\cos 2z)\,y=0, \quad k^2=q.$$ MA

8.61 Periodic Mathieu functions

8.610 In general, Mathieu's equation 8.60 does not have periodic solutions. If k is a real number, there exist infinitely many *eigenvalues* a, not identically equal to zero, corresponding to the periodic solutions

$$y(z)=y(2\pi+z),$$

If k is nonzero, there are no other linearly independent periodic solutions. Periodic solutions of Mathieu's equations are called *Mathieu's periodic functions* or *Mathieu functions of the first kind*, or, more simply, *Mathieu functions*.

8.611 Mathieu's equation has four series of distinct periodic solutions:

1. $\mathrm{ce}_{2n}(z,\ q)=\sum_{r=0}^{\infty} A_{2r}^{(2n)}\cos 2rz.$ MA

2. $\mathrm{ce}_{2n+1}(z,\ q)=\sum_{r=0}^{\infty} A_{2r+1}^{(2n+1)}\cos(2r+1)\,z.$ MA

3. $\mathrm{se}_{2n+1}(z,\ q)=\sum_{r=0}^{\infty} B_{2r+1}^{(2n+1)}\sin(2r+1)\,z.$ MA

4. $\mathrm{se}_{2n+2}(z,\ q)=\sum_{r=0}^{\infty} B_{2r+2}^{(2n+2)}\sin(2r+2)\,z.$ MA

5. The coefficients A and B depend on q. The eigenvalues a of the functions ce_{2n}, ce_{2n+1}, se_{2n}, se_{2n+1} are denoted by a_{2n}, a_{2n+1}, b_{2n}, b_{2n+1}.

8.612 The solutions of Mathieu's equation are normalized so that

$$\int_0^{2\pi} y^2\,dx=\pi.$$ MO 65

8.613

1. $\lim_{q\to 0}\mathrm{ce}_0(x)=\frac{1}{\sqrt{2}}$.

2. $\lim_{q\to 0}\mathrm{ce}_n(x)=\cos nx \quad [n\neq 0]$.

3. $\lim_{q\to 0}\mathrm{se}_n(x)=\sin nx$. MO 65

8.62 Recursion relations for the coefficients $A_{2r}^{(2n)}$, $A_{2r+1}^{(2n+1)}$, $B_{2r+1}^{(2n+1)}$, $B_{2r+2}^{(2n+2)}$

8.621

1. $aA_0^{(2n)} - qA_2^{(2n)} = 0.$ MA

2. $(a-4)A_2^{(2n)} - q(A_4^{(2n)} + 2A_0^{(2n)}) = 0.$ MA

3. $(a-4r^2)A_{2r}^{(2n)} - q(A_{2r+2}^{(2n)} + A_{2r-2}^{(2n)}) = 0 \quad [r \geqslant 2].$ MA

8.622

1. $(a-1-q)A_1^{(2n+1)} - qA_3^{(2n+1)} = 0.$ MA

2. $[a-(2r+1)^2]A_{2r+1}^{(2n+1)} - q(A_{2r+3}^{(2n+1)} + A_{2r-1}^{(2n+1)}) = 0 \quad [r \geqslant 1].$ MA

8.623

1. $(a-1+q)B_1^{(2n+1)} - qB_3^{(2n+1)} = 0.$ MA

2. $[a-(2r+1)^2]B_{2r+1}^{(2n+1)} - q(B_{2r+3}^{(2n+1)} + B_{2r-1}^{(2n+1)}) = 0 \quad [r \geqslant 1].$ MA

8.624

1. $(a-4)B_2^{(2n+2)} - qB_4^{(2n+2)} = 0.$ MA

2. $(a-4r^2)B_{2r}^{(2n+2)} - q(B_{2r+2}^{(2n+2)} - B_{2r-2}^{(2n+2)}) = 0 \quad [r \geqslant 2].$ MA

8.625 We can determine the coefficients A and B from equations 8.612, 8.613 and 8.621 – 8.624 provided a is known. Suppose, for example, that we need to determine the coefficients $A_{2r}^{(2n)}$ for the function $\mathrm{ce}_{2n}(z, q)$. From the recursion formulas, we have

1.
$$\begin{vmatrix} a & -q & 0 & 0 & 0 & \cdots \\ -2q & a-4 & -q & 0 & 0 & \cdots \\ 0 & -q & a-16 & -q & 0 & \cdots \\ 0 & 0 & -q & a-36 & -q & \cdots \\ 0 & 0 & 0 & -q & a-64 & \cdots \\ \cdots & \cdots & \cdots & \cdots & \cdots & \cdots \end{vmatrix} = 0.$$ ST

For given q in equation 8.625 1., we may determine the eigenvalues

2. $a = a_0, a_2, a_4, \ldots \quad [|a_0| \leqslant |a_2| \leqslant |a_4| \leqslant \ldots].$

If we now set $a = a_{2n}$, we can determine the coefficients $A_{2r}^{(2n)}$ from the recursion formulas 8.621 up to a proportionality coefficient. This coefficient is determined from the formula

3. $2[A_0^{(2n)}]^2 + \sum_{r=1}^{\infty} [A_{2r}^{(2n)}]^2 = 1,$ MA

which follows from the conditions of normalization.

8.63 Mathieu functions with a purely imaginary argument

8.630 If, in equation 8.60, we replace z with iz, we arrive at the differential equation

1. $\frac{d^2y}{dz^2} + (-a + 2q\,\mathrm{ch}\,2x)\,y = 0.$

We can find the solutions of this equation if we replace the argument z with iz in the functions $ce_n(z, q)$ and $se_n(z, q)$. The functions obtained in this way are called *associated Mathieu functions of the first kind* and are denoted as follows:

2. $Ce_{2n}(z, q)$, $Ce_{2n+1}(z, q)$, $Se_{2n+1}(z, q)$, $Se_{2n+2}(z, q)$.

8.631

1. $Ce_{2n}(z, q) = \sum_{r=0}^{\infty} A_{2r}^{(2n)} \operatorname{ch} 2rz.$ MA

2. $Ce_{2n+1}(z, q) = \sum_{r=0}^{\infty} A_{2r+1}^{(2n+1)} \operatorname{ch}(2r+1) z.$ MA

3. $Se_{2n+1}(z, q) = \sum_{r=0}^{\infty} B_{2r+1}^{(2n+1)} \operatorname{sh}(2r+1) z.$ MA

4. $Se_{2n+2}(z, q) = \sum_{r=0}^{\infty} B_{2r+2}^{(2n+2)} \operatorname{sh}(2r+2) z.$ MA

8.64 Nonperiodic solutions of Mathieu's equation

Along with each periodic solution of equation 8.60, there exists a second nonperiodic solution that is linearly independent. The nonperiodic solutions are denoted as follows:

$$fe_{2n}(z, q), \quad fe_{2n+1}(z, q), \quad ge_{2n+1}(z, q), \quad ge_{2n+2}(z, q).$$

Analogously, the second solutions of equation 8.630 1. are denoted by

$$Fe_{2n}(z, q), \quad Fe_{2n+1}(z, q), \quad Ge_{2n+1}(z, q), \quad Ge_{2n+2}(z, q).$$

8.65 Mathieu functions for negative q

8.651 If we replace the argument z in equation 8.60 with $+\left(\frac{\pi}{2} \pm z\right)$, we get the equation

$$\frac{d^2y}{dz^2} + (a + 2q \cos 2z) y = 0. \quad \text{MA}$$

This equation has the following solutions:

8.652

1. $ce_{2n}(z, -q) = (-1)^n ce_{2n}\left(\frac{1}{2}\pi - z, q\right).$ MA

2. $ce_{2n+1}(z, -q) = (-1)^n se_{2n+1}\left(\frac{1}{2}\pi - z, q\right).$ MA

3. $se_{2n+1}(z, -q) = (-1)^n ce_{2n+1}\left(\frac{1}{2}\pi - z, q\right).$ MA

4. $se_{2n+2}(z, -q) = (-1)^n se_{2n+2}\left(\frac{1}{2}\pi - z, q\right).$ MA

5. $fe_{2n}(z, -q) = (-1)^{n+1} fe_{2n}\left(\frac{1}{2}\pi - z, q\right).$ MA

6. $fe_{2n+1}(z, -q) = (-1)^n ge_{2n+1}\left(\frac{1}{2}\pi - z, q\right).$ MA

7. $ge_{2n+1}(z, -q) = (-1)^n fe_{2n+1}\left(\frac{1}{2}\pi - z, q\right).$ MA

8. $\text{ge}_{2n+2}(z, -q) = (-1)^n \text{ge}_{2n+2}\left(\frac{1}{2}\pi - z, q\right)$. MA

8.653 Analogously, if we replace z with $\frac{\pi}{2}i + z$ in equation 8.630 1. we get the equation

$$\frac{d^2y}{dz^2} - (a + 2q \operatorname{ch} z)\, y = 0.$$

It has the following solutions:

8.654

1. $\text{Ce}_{2n}(z, -q) = (-1)^n \text{Ce}_{2n}\left(\frac{\pi}{2}i + z, q\right)$. MA

2. $\text{Ce}_{2n+1}(z, -q) = (-1)^{n+1} i\, \text{Se}_{2n+1}\left(\frac{1}{2}\pi i + z, q\right)$. MA

3. $\text{Se}_{2n+1}(z, -q) = (-1)^{n+1} i\, \text{Ce}_{2n+1}\left(\frac{\pi}{2}i + z, q\right)$. MA

4. $\text{Se}_{2n+2}(z, -q) = (-1)^{n+1} \text{Se}_{2n+2}\left(\frac{\pi}{2}i + z, q\right)$. MA

5. $\text{Fe}_{2n}(z, -q) = (-1)^n \text{Fe}_{2n}\left(\frac{1}{2}\pi i + z, q\right)$. MA

6. $\text{Fe}_{2n+1}(z, -q) = (-1)^{n+1} i\, \text{Ge}_{2n+1}\left(\frac{\pi}{2}i + z, q\right)$. MA

7. $\text{Ge}_{2n+1}(z, -q) = (-1)^{n+1} i \text{Fe}_{2n+1}\left(\frac{\pi}{2}i + z, q\right)$. MA

8. $\text{Ge}_{2n+2}(z, -q) = (-1)^{n+1} \text{Ge}_{2n+2}\left(\frac{\pi}{2}i + z, q\right)$. MA

8.66 Representation of Mathieu functions as series of Bessel functions

8.661

1. $$\text{ce}_{2n}(z, q) = \frac{\text{ce}_{2n}\left(\frac{\pi}{2}, q\right)}{A_0^{(2n)}} \sum_{r=0}^{\infty} (-1)^r A_{2r}^{(2n)} J_{2r}(2k \cos z);$$ MA

$$= \frac{\text{ce}_{2n}(0, q)}{A_0^{(2n)}} \sum_{r=0}^{\infty} (-1)^r A_{2r}^{(2n)} I_{2r}(2k \sin z).$$ MA

2. $$\text{ce}_{2n+1}(z, q) = -\frac{\text{ce}'_{2n+1}\left(\frac{\pi}{2}, q\right)}{kA_1^{(2n+1)}} \sum_{r=0}^{\infty} (-1)^r A_{2r+1}^{(2n+1)} J_{2r+1}(2k \cos z);$$

MA

$$= \frac{\text{ce}_{2n+1}(0, q)}{kA_1^{(2n+1)}} \operatorname{ctg} z \sum_{r=0}^{\infty} (-1)^r (2r+1) A_{2r+1}^{(2n+1)} I_{2r+1}(2k \sin z).$$

MA

3. $$\mathrm{se}_{2n+1}(z, q) = \frac{\mathrm{se}_{2n+1}\left(\frac{\pi}{2}, q\right)}{kB_1^{(2n+1)}} \operatorname{tg} z \times$$

$$\times \sum_{r=0}^{\infty} (-1)^r (2r+1) B_{2r+1}^{(2n+1)} J_{2r+1}(2k \cos z);$$ MA

$$= \frac{\mathrm{se}'_{2n+1}(0, q)}{kB_1^{(2n+1)}} \sum_{r=0}^{\infty} (-1)^r B_{2r+1}^{(2n+1)} I_{2r+1}(2k \sin z).$$ MA

4. $$\mathrm{se}_{2n+2}(z, q) = \frac{-\mathrm{se}'_{2n+2}\left(\frac{\pi}{2}, q\right)}{k^2 B_2^{(2n+2)}} \operatorname{tg} z \times$$

$$\times \sum_{r=0}^{\infty} (-1)^r (2r+2) B_{2r+2}^{(2n+2)} J_{2r+2}(2k \cos z);$$ MA

$$= \frac{\mathrm{se}'_{2n+2}(0, q)}{k^2 B_2^{(2n+2)}} \operatorname{ctg} z \sum_{r=0}^{\infty} (-1)^r (2r+2) B_{2r+2}^{(2n+2)} I_{2r+2}(2k \sin z).$$ MA

8.662

1. $$\mathrm{fe}_{2n}(z, q) = -\frac{\pi \mathrm{fe}'_{2n}(0, q)}{2\,\mathrm{ce}_{2n}\left(\frac{\pi}{2}, q\right)} \sum_{r=0}^{\infty} (-1)^r A_{2r}^{(2n)} \operatorname{Im}\left[J_r(ke^{iz}) N_r(ke^{-iz})\right].$$

MA

2. $$\mathrm{fe}_{2n+1}(z, q) = \frac{\pi k \mathrm{fe}'_{2n+1}(0, q)}{2\mathrm{ce}'_{2n+1}\left(\frac{\pi}{2}, q\right)} \times$$

$$\times \sum_{r=0}^{\infty} (-1)^r A_{2r+1}^{(2n+1)} \operatorname{Im}\left[J_r(ke^{iz}) N_{r+1}(ke^{-iz}) + J_{r+1}(ke^{iz}) N_r(ke^{-iz})\right].$$ MA

3. $$\mathrm{ge}_{2n+1}(z, q) = -\frac{\pi k\, \mathrm{ge}_{2n+1}(0, q)}{2\mathrm{se}_{2n+1}\left(\frac{\pi}{2}, q\right)} \times$$

$$\times \sum_{r=0}^{\infty} (-1)^r B_{2r+1}^{(2n+1)} \operatorname{Re}\left[J_r(ke^{iz}) N_{r+1}(ke^{-iz}) - J_{r+1}(ke^{iz}) N_r(ke^{-iz})\right].$$ MA

4. $$\mathrm{ge}_{2n+2}(z, q) = -\frac{\pi k^2\, \mathrm{ge}_{2n+2}(0, q)}{2\mathrm{se}'_{2n+2}\left(\frac{1}{2}\pi, q\right)} \times$$

$$\times \sum_{r=0}^{\infty} (-1)^r \operatorname{Re}\left[J_k(ke^{iz}) N_{r+2}(ke^{-iz}) - J_{r+2}(ke^{iz}) N_r(ke^{-iz})\right].$$ MA

The expansions of the functions Fe_n and Ge_n as series of the functions N_ν are denoted, respectively, by Fey_n and Gey_n and the expansions of these functions as series of the functions K_ν are denoted, respectively, by Fek_n and Gek_n.

8.663

1. $$\operatorname{Fey}_{2n}(z,q)=\frac{\operatorname{ce}_{2n}(0,q)}{A_0^{(2n)}}\sum_{r=0}^{\infty}A_{2r}^{(2n)}N_{2r}(2k\operatorname{sh}z),$$

$$k^2=q\qquad[|\operatorname{sh}z|>1,\quad\operatorname{Re}z>0];$$ MA

$$=\frac{\operatorname{ce}_{2n}\left(\frac{\pi}{2},q\right)}{A_0^{(2n)}}\sum_{r=0}^{\infty}(-1)^rA_{2r}^{(2n)}N_{2r}(2k\operatorname{ch}z)$$

$$[|\operatorname{ch}z|>1];$$ MA

$$=\frac{\operatorname{ce}_{2n}(0,q)\operatorname{ce}_{2n}\left(\frac{\pi}{2},q\right)}{[A_0^{(2n)}]^2}\sum_{r=0}^{\infty}(-1)^rA_{2r}^{(2n)}J_r(ke^{-z})N_r(ke^z).$$

MA

2. $$\operatorname{Fey}_{2n+1}(z,q)=\frac{\operatorname{ce}_{2n+1}(0,q)\operatorname{cth}z}{kA_1^{(2n+1)}}\sum_{r=0}^{\infty}(2r+1)A_{2r+1}^{(2n+1)}N_{2r+1}(2k\operatorname{sh}z),$$

$$k^2=q\quad[|\operatorname{sh}z|>1,\quad\operatorname{Re}z>0];$$ MA

$$=-\frac{\operatorname{ce}'_{2n+1}\left(\frac{\pi}{2},q\right)}{kA_1^{(2n+1)}}\sum_{r=0}^{\infty}(-1)^rA_{2r+1}^{(2n+1)}N_{2r+1}(2k\operatorname{ch}z)$$

$$[|\operatorname{ch}z|>1];$$ MA

$$=-\frac{\operatorname{ce}_{2n+1}(0,q)\operatorname{ce}'_{2n+1}\left(\frac{\pi}{2},q\right)}{k[A_1^{(2n+1)}]^2}\times$$

$$\times\sum_{r=0}^{\infty}(-1)^rA_{2r+1}^{(2n+1)}[J_r(ke^{-z})N_{r+1}(ke^z)+J_{r+1}(ke^{-z})N_r(ke^z)].$$ MA

3. $$\operatorname{Gey}_{2n+1}(z,q)=\frac{\operatorname{se}'_{2n+1}(0,q)}{kB_1^{(2n+1)}}\sum_{r=0}^{\infty}B_{2r+1}^{(2n+1)}N_{2r+1}(2k\operatorname{sh}z)$$

$$[|\operatorname{sh}z|>1,\quad\operatorname{Re}z>0];$$ MA

$$=\frac{\operatorname{se}_{2n+1}\left(\frac{\pi}{2},q\right)}{kB_1^{(2n+1)}}\operatorname{th}z\sum_{r=0}^{\infty}(-1)^r(2r+1)B_{2r+1}^{(2n+1)}N_{2r+1}(2k\operatorname{ch}z)$$

$$[|\operatorname{ch}z|>1];$$ MA

$$=\frac{\operatorname{se}_{2n+1}(0,q)\operatorname{se}_{2n+1}\left(\frac{\pi}{2},q\right)}{k[B_1^{(2n+1)}]^2}\times$$

$$\times\sum_{r=0}^{\infty}(-1)^rB_{2r+1}^{(2n+1)}[J_r(ke^{-z})N_{r+1}(ke^z)-J_{r+1}(ke^{-z})N_r(ke^z)].$$ MA

4. $$\operatorname{Gey}_{2n+2}(z,q)=\frac{\operatorname{se}'_{2n+2}(0,q)}{k^2B_2^{(2n+2)}}\operatorname{cth}z\sum_{r=0}^{\infty}(2r+2)B_{2r+2}^{(2n+2)}N_{2r+2}(2k\operatorname{sh}z)$$

$$[|\operatorname{sh}z|>1,\quad\operatorname{Re}z>0];$$ MA

$$= -\frac{\mathrm{se}'_{2n+2}\left(\frac{\pi}{2}, q\right)}{k^2 B_2^{(2n+2)}} \operatorname{th} z \times$$

$$\times \sum_{r=0}^{\infty} (-1)^r (2r+2) B_{2r+2}^{(2n+2)} N_{2r+2}(2k \operatorname{ch} z) \qquad [|\operatorname{ch} z| > 1]; \qquad \text{MA}$$

$$= \frac{\mathrm{se}'_{2n+2}(0, q)\, \mathrm{se}'_{2n+2}\left(\frac{\pi}{2}, q\right)}{k^2 [B_2^{(2n+2)}]^2} \times$$

$$\times \sum_{r=0}^{\infty} (-1)^r B_{2r+2}^{(2n+2)} [J_r(ke^{-z}) N_{r+2}(ke^z) - J_{r+2}(ke^{-z}) N_r(ke^z)]. \qquad \text{MA}$$

8.664

1. $$\mathrm{Fek}_{2n}(z, q) = \frac{\mathrm{ce}_{2n}(0, q)}{\pi A_0^{(2n)}} \sum_{r=0}^{\infty} (-1)^r A_{2r}^{(2n)} K_{2r}(-2ik \operatorname{sh} z),$$
$$k^2 = q \qquad [|\operatorname{sh} z| > 1, \quad \operatorname{Re} z > 0]. \qquad \text{MA}$$

2. $$\mathrm{Fek}_{2n+1}(z, q) = \frac{\mathrm{ce}_{2n+1}(0, q)}{\pi k A_1^{(2n+1)}} \operatorname{cth} z \sum_{r=0}^{\infty} (-1)^r (2r+1) A_{2r+1}^{(2n+1)} K_{2r+1}(-2ik \operatorname{sh} z),$$
$$k^2 = q \qquad [|\operatorname{sh} z| > 1, \quad \operatorname{Re} z > 0]. \qquad \text{MA}$$

3. $$\mathrm{Gek}_{2n+1}(z, q) = \frac{\mathrm{se}_{2n+1}\left(\frac{\pi}{2}, q\right)}{\pi k B_1^{(2n+1)}} \operatorname{th} z \sum_{r=0}^{\infty} (2r+1) B_{2r+1}^{(2n+1)} K_{2r+1}(-2ik \operatorname{ch} z). \qquad \text{MA}$$

4. $$\mathrm{Gek}_{2n+2}(z, q) = \frac{\mathrm{se}'_{2n+2}\left(\frac{\pi}{2}, q\right)}{\pi k^2 B_2^{(2n+2)}} \operatorname{th} z \sum_{r=0}^{\infty} (2r+2) B_{2r+2}^{(2n+2)} K_{2r+2}(-2ik \operatorname{ch} z). \qquad \text{MA}$$

8.67 The general theory

If $i\mu$ is not an integer, the general solution of equation 8.60 can be found in the form

8.671

1. $$y = Ae^{\mu z} \sum_{r=-\infty}^{\infty} c_{2r} e^{2rzi} + Be^{-\mu z} \sum_{r=-\infty}^{\infty} c_{2r} e^{-2rzi}. \qquad \text{MA}$$

The coefficients c_{2r} can be determined from the homogenous system of linear algebraic equations

2. $c_{2r} + \xi_{2r}(c_{2r+2} + c_{2r-2}) = 0, \; r = \ldots, -2, -1, 0, 1, 2, \ldots,$ MA

where

$$\xi_{2r} = \frac{q}{(2r - i\mu)^2 - a}.$$

The condition that this system be compatible yields an equation that μ must satisfy:

$$3.\quad \Delta(i\mu)=\begin{vmatrix} \cdot & \cdot & \cdot & \cdot & \cdot & \cdot & \cdot & \cdot \\ \cdot & \xi_{-4} & 1 & \xi_{-4} & 0 & 0 & 0 & 0 \cdot \\ \cdot & 0 & \xi_{-2} & 1 & \xi_{-2} & 0 & 0 & 0 \cdot \\ \cdot & 0 & 0 & \xi_0 & 1 & \xi_0 & 0 & 0 \cdot \\ \cdot & 0 & 0 & 0 & \xi_2 & 1 & \xi_2 & 0 \cdot \\ \cdot & \cdot & \cdot & \cdot & \cdot & \cdot & \cdot & \cdot \end{vmatrix}=0. \qquad \text{MA}$$

This equation can also be written in the form

4. $\operatorname{ch}\mu\pi=1-2\Delta(0)\sin^2\left(\frac{\pi\sqrt{a}}{2}\right)$, where $\Delta(0)$ is the value that is assumed by the determinant of the preceding article if we set $\mu=0$ in the expressions for ξ_{2r}.

5. If the pair (a, q) is such that $|\operatorname{ch}\mu\pi|<1$, then $\mu=i\beta$, $\operatorname{Im}\beta=0$, and the solution 8.671 1. is bounded on the real axis.

6. If $|\operatorname{ch}\mu\pi|>1$, μ may be real or complex and the solution **8.671** 1. will not be bounded on the real axis.

7. If $\operatorname{ch}\mu\pi=\pm 1$, $i\mu$ will be an integer. In this case, one of the solutions will be of period π or 2π (depending on whether n is even or odd). The second solution is nonperiodic (see 8.61 and 8.64).

8.7-8.8 Associated Legendre Functions

8.70 Introduction

8.700 An *associated Legendre function* is a solution of the differential equation

$$1.\quad (1-z^2)\frac{d^2u}{dz^2}-2z\frac{du}{dz}+\left[\nu(\nu+1)-\frac{\mu^2}{1-z^2}\right]u=0,$$

in which ν and μ are arbitrary complex constants.

This equation is a special case of (Riemann's) hypergeometric equation (see **9.151**). The points

$$+1, \quad -1, \quad \infty$$

are, in general, its *singular points*, specifically, its ordinary branch points.

We are interested, on the one hand, in solutions of the equation that correspond to real values of the independent variable z that lie in the interval $[-1, 1]$ and, on the other hand, in solutions corresponding to an arbitrary complex number z such that $\operatorname{Re} z>1$. These are multiple-valued in the z-plane. To separate these functions into single-valued branches, we make a cut along the real axis from $-\infty$ to $+1$. We are also interested in those solutions of equation 8.700 1. for which ν or μ or both are integers. Of especial significance is the case in which $\mu=0$.

8.701 In connection with this, we shall use the following notations:

The letter z will denote *an arbitrary complex variable*; the letter x will denote a *real* variable that varies over the interval $[-1, +1]$. We shall sometimes set $x=\cos\varphi$, where φ is a real number.

We shall use the symbols $P_\nu^\mu(z)$, $Q_\nu^\mu(z)$ to denote those solutions of equation 8.700 1., that are single-valued and regular for $|z| < 1$ and, in particular, uniquely determined for $z = x$.

We shall use the symbols $\mathrm{P}_\nu^\mu(z)$, $\mathrm{Q}_\nu^\mu(z)$ to denote those solutions of equation 8.700 1. that are single-valued and regular for $\operatorname{Re} z > 1$. When these functions cannot be unrestrictedly extended without violating their single-valuedness we make a cut along the real axis to the left of the point $z = 1$. The values of the functions $P_\nu^\mu(z)$ and $Q_\nu^\mu(z)$ on the upper and lower boundaries of that portion of the cut lying between the points -1 and $+1$ are denoted respectively by

$$P_\nu^\mu(x \pm i0), \qquad Q_\nu^\mu(x \pm i0).$$

The letters n and m denote natural numbers or zero. The letters ν and μ denote arbitrary complex numbers unless the contrary is stated.

The upper index will be omitted when it is equal to zero. That is, we set

$$P_\nu^0(z) = P_\nu(z), \quad Q_\nu^0(z) = Q(z), \quad \mathrm{P}_\nu^0(z) = \mathrm{P}_\nu(z), \quad \mathrm{P}_\nu^0(z) = \mathrm{Q}_\nu(z).$$

The *linearly independent* functions

8.702 $$P_\nu^\mu(z) = \frac{1}{\Gamma(1-\mu)}\left(\frac{z+1}{z-1}\right)^{\frac{\mu}{2}} F\left(-\nu, \nu+1; 1-\mu; \frac{1-z}{2}\right)$$

$\left[\arg\frac{z+1}{z-1} = 0, \text{ if } z \text{ is real and greater than } 1\right]$ and MO 80, WH

8.703 $$Q_\nu^\mu(z) = \frac{e^{\mu\pi i}\Gamma(\nu+\mu+1)\Gamma\left(\frac{1}{2}\right)}{2^{\nu+1}\Gamma\left(\nu+\frac{3}{2}\right)}(z^2-1)^{\frac{\mu}{2}} z^{-\nu-\mu-1} \times$$
$$\times F\left(\frac{\nu+\mu+2}{2}, \frac{\nu+\mu+1}{2}; \nu+\frac{3}{2}; \frac{1}{z^2}\right)$$

[$\arg(z^2-1) = 0$ when z is real and greater than 1; $\arg z = 0$ when z is real and greater than zero] which are solutions of the differential equation 8.700 1., are called *associated Legendre functions* (or *spherical functions*) *of the first* and *second kinds* respectively. They are uniquely defined, respectively, in the intervals $|1-z| < 2$ and $|z| > 1$ with the portion of the real axis that lies between $-\infty$ and $+1$ excluded. They can be extended by means of hypergeometric series to the entire z-plane where the above-mentioned cut was made. These expressions for $P_\nu^\mu(z)$ and $Q_\nu^\mu(z)$ lose their meaning when $1-\mu$ and $\nu+\frac{3}{2}$ are nonpositive integers respectively. MO 80

When z is a real number lying on the interval $[-1, +1]$, so that $(z = x = \cos\varphi)$, we take the following functions as linearly independent solutions of the equation

8.704 $$\mathrm{P}_\nu^\mu(x) = \frac{1}{2}\left[e^{\frac{1}{2}\mu\pi i} P_\nu^\mu(\cos\varphi + i0) + e^{-\frac{1}{2}\mu\pi i} P_\nu^\mu(\cos\varphi - i0)\right];$$ EH I 143(1)

$$= \frac{1}{\Gamma(1-\mu)}\left(\frac{1+x}{1-x}\right)^{\frac{\mu}{2}} F\left(-\nu, \nu+1; 1-\mu; \frac{1-x}{2}\right).$$ EH I 143(6)

8.705 $Q_\nu^\mu(x) = \frac{1}{2} e^{-\mu\pi i}\left[e^{-\frac{1}{2}\mu\pi i} Q_\nu^\mu(x+i0) + e^{\frac{1}{2}\mu\pi i} Q_\nu^\mu(x-i0)\right];$ EH I 143(2)

$$= \frac{\pi}{2\sin\mu\pi}\left[P_\nu^\mu(x)\cos\mu\pi - \frac{\Gamma(\nu+\mu+1)}{\Gamma(\nu-\mu+1)} P_\nu^{-\mu}(x)\right] \quad \text{(cf. 8.732 5.)}$$

If $\mu = \pm m$ is an integer, the last equation loses its meaning. In this case, we get the following formulas by passing to the limit:

8.706

1. $Q_\nu^m(x) = (-1)^m (1-x^2)^{\frac{m}{2}} \frac{d^m}{dx^m} Q_\nu(x)$ (cf. 8.752 1). EH I 149(7)

2. $Q_\nu^{-m}(x) = (-1)^m \frac{\Gamma(\nu-m+1)}{\Gamma(\nu+m+1)} Q_\nu^m(x).$ EH I 144(18)

The functions $Q_\nu^\mu(z)$ are not defined when $\nu+\mu$ is equal to a negative integer. Therefore, we must exclude the cases when $\nu+\mu = -1, -2, -3, \ldots$ for these formulas.

The functions

$$P_\nu^{\pm\mu}(\pm z), \quad Q_\nu^{\pm\mu}(\pm z), \quad P_{-\nu-1}^{\pm\mu}(\pm z), \quad Q_{-\nu-1}^{\pm\mu}(\pm z).$$

are *linearly independent solutions* of the differential equation for $\nu+\mu \neq 0, \pm 1, \pm 2, \ldots$.

8.707 Nonetheless, two linearly independent solutions can always be found. Specifically, for $\nu \pm \mu$ not an integer, the differential equation 8.700 1. has the following solutions:

1. $P_\nu^{\pm\mu}(\pm z), \quad Q_\nu^{\pm\mu}(\pm z), \quad P_{-\nu-1}^{\pm\mu}(\pm z), \quad Q_{-\nu-1}^{\pm\mu}(\pm z)$

respectively, for $z = x = \cos\varphi$,

2. $P_\nu^{\pm\mu}(\pm x), \quad Q_\nu^{\pm\mu}(\pm x), \quad P_{-\nu-1}^{\pm\mu}(\pm x), \quad Q_{-\nu-1}^{\pm\mu}(\pm x).$

If $\nu \pm \mu$ is not an integer, the solutions

3. $P_\nu^\mu(z)$, $Q_\nu^\mu(z)$, respectively, and $P_\nu^\mu(x)$, $Q_\nu^\mu(x)$

are linearly independent. If $\nu \pm \mu$ is an integer but μ itself is not an integer, the following functions are linearly independent solutions of equation 8.700 1.:

4. $P_\nu^\mu(z)$, $P_\nu^{-\mu}(z)$, respectively, and $P_\nu^\mu(x)$, $P_\nu^{-\mu}(x)$.

If $\mu = \pm m$, $\nu = n$, or $\nu = -n-1$, the following functions are linearly independent solutions of equation 8.700 1.for $n \geqslant m$:

5. $P_n^m(z)$, $Q_n^m(z)$, respectively, and $P_n^m(x)$, $Q_n^m(x)$,

and for $n < m$, the following functions will be linearly independent solutions

6. $P_n^{-m}(z)$, $Q_n^m(z)$, respectively, and $P_n^{-m}(x)$, $Q_n^m(x)$.

8.71 Integral representations

8.711

1. $$P_\nu^{-\mu}(z) = \frac{(z^2-1)^{\frac{\mu}{2}}}{2^\mu \sqrt{\pi}\,\Gamma\left(\mu+\frac{1}{2}\right)} \int_{-1}^{1} \frac{(1-t^2)^{\mu-\frac{1}{2}}}{(z+t\sqrt{z^2-1})^{\mu-\nu}} dt$$

$\left[\operatorname{Re}\mu > -\frac{1}{2}, \ |\arg(z\pm 1)| < \pi\right].$ MO 88

2. $P_\nu^m(z) = \frac{(\nu+1)(\nu+2)\dots(\nu+m)}{\pi} \int_0^\pi \left[z+\sqrt{z^2-1}\cos\varphi\right]^\nu \cos m\varphi \, d\varphi;$

$= (-1)^m \frac{\nu(\nu-1)\dots(\nu-m+1)}{\pi} \int_0^\pi \frac{\cos m\varphi \, d\varphi}{\left[z+\sqrt{z^2-1}\cos\varphi\right]^{\nu+1}}$

$\left[|\arg z| < \frac{\pi}{2},\ \arg\left(z+\sqrt{z^2-1}\cos\varphi\right) = \arg z \quad \text{for} \quad \varphi = \frac{\pi}{2}\right]$

(cf. 8.822 1.). SM 483(15), WH

3. $Q_\nu^\mu(z) = \sqrt{\pi} \frac{e^{\mu\pi i}\Gamma(\nu+\mu+1)}{2^\mu \Gamma\left(\mu+\frac{1}{2}\right)\Gamma(\nu-\mu+1)} (z^2-1)^{\frac{\mu}{2}} \int_0^\infty \frac{\operatorname{sh}^{2\mu} t \, dt}{(z+\sqrt{z^2-1}\operatorname{ch} t)^{\nu+\mu+1}}$

$[\operatorname{Re}(\nu\pm\mu) > -1,\ |\arg(z\pm 1) < \pi]$ (cf. 8.822 2.). MO 88

4. $Q_\nu^\mu(z) = \frac{e^{\mu\pi i}\Gamma(\nu+1)}{\Gamma(\nu-\mu+1)} \int_0^\infty \frac{\operatorname{ch} \mu t \, dt}{(z+\sqrt{z^2-1}\operatorname{ch} t)^{\nu+1}}$

$[\operatorname{Re}(\nu+\mu) > -1,\ \nu \neq -1, -2, -3, \dots, |\arg(z\pm 1)| < \pi].$

WH, MO 88

5. $\int_{-1}^{1} P_l^2(x) P_l^0(x) dx = -\frac{l!}{(l-2)!}\,\frac{1}{2l+1} = -\frac{l(l-1)}{2l+1}.$

8.712 $Q_\nu^\mu(z) = \frac{e^{\mu\pi i}\Gamma(\nu+\mu+1)}{2^{\nu+1}\Gamma(\nu+1)} (z^2-1)^{+\frac{\mu}{2}} \int_{-1}^{1} (1-t^2)^\nu (z-t)^{-\nu-\mu-1} dt$

$[\operatorname{Re}(\nu+\mu) > -1,\ \operatorname{Re}\mu > -1,\ |\arg(z\pm 1)| < \pi]$ (cf. 8.821 2.).

MO 88a, EH I 155(5)a

8.713

1. $Q_\nu^\mu(z) = \frac{e^{\mu\pi i}\Gamma\left(\mu+\frac{1}{2}\right)}{\sqrt{2\pi}} (z^2-1)^{\frac{\mu}{2}} \times$

$\times \left\{ \int_0^\pi \frac{\cos\left(\nu+\frac{1}{2}\right) t \, dt}{(z-\cos t)^{\mu+\frac{1}{2}}} - \cos\nu\pi \int_0^\infty \frac{e^{-\left(\nu+\frac{1}{2}\right)t} dt}{(z+\operatorname{ch} t)^{\mu+\frac{1}{2}}} \right\}$

$\left[\operatorname{Re}\mu > -\frac{1}{2},\ \operatorname{Re}(\nu+\mu) > -1,\ |\arg(z\pm 1)| < \pi\right].$ MO 89

2. $P_\nu^{-\mu}(z) = \frac{(z^2-1)^{\frac{\mu}{2}}}{2^\nu \Gamma(\mu-\nu)\Gamma(\nu+1)} \int_0^\infty \frac{\operatorname{sh}^{2\nu+1} t}{(z+\operatorname{ch} t)^{\nu+\mu+1}} dt$

$[\operatorname{Re} z > -1,\ |\arg(z\pm 1)| < \pi,\ \operatorname{Re}(\nu+1) > 0,\ \operatorname{Re}(\mu-\nu) > 0].$ MO 89

3. $P_\nu^{-\mu}(z) = \sqrt{\frac{2}{\pi}} \frac{\Gamma\left(\mu+\frac{1}{2}\right)(z^2-1)^{\frac{\mu}{2}}}{\Gamma(\nu+\mu+1)\Gamma(\mu-\nu)} \int_0^\infty \frac{\operatorname{ch}\left(\nu+\frac{1}{2}\right) t \, dt}{(z+\operatorname{ch} t)^{\mu+\frac{1}{2}}}$

$[\operatorname{Re} z > -1,\ |\arg(z\pm 1)| < \pi,\ \operatorname{Re}(\nu+\mu) > -1,\ \operatorname{Re}(\mu-\nu) > 0].$ MO 89

8.714

1. $$P_\nu^\mu(\cos\varphi)=\sqrt{\frac{2}{\pi}}\frac{\sin^\mu\varphi}{\Gamma\left(\frac{1}{2}-\mu\right)}\int_0^\varphi\frac{\cos\left(\nu+\frac{1}{2}\right)t\,dt}{(\cos t-\cos\varphi)^{\mu+\frac{1}{2}}}$$
$$\left[0<\varphi<\pi,\ \operatorname{Re}\mu<\frac{1}{2}\right];\qquad \text{(cf. 8.823)}$$ MO 87

2. $$P_\nu^{-\mu}(\cos\varphi)=\frac{\Gamma(2\mu+1)\sin^\mu\varphi}{2^\mu\Gamma(\mu+1)\Gamma(\nu+\mu+1)\Gamma(\mu-\nu)}\int_0^\infty\frac{t^{\nu+\mu}\,dt}{(1+2t\cos\varphi+t^2)^{\mu+\frac{1}{2}}}$$
$$[\operatorname{Re}(\nu+\mu)>-1,\ \operatorname{Re}(\mu-\nu)>0].$$ MO 89

3. $$Q_\nu^\mu(\cos\varphi)=\frac{1}{2^{\mu+1}}\frac{\Gamma(\nu+\mu+1)}{\Gamma(\nu-\mu+1)}\frac{\sin^\mu\varphi}{\Gamma\left(\mu+\frac{1}{2}\right)}\times$$
$$\times\int_0^\infty\left[\frac{\operatorname{sh}^{2\mu}t}{(\cos\varphi+i\sin\varphi\operatorname{ch}t)^{\nu+\mu+1}}+\frac{\operatorname{sh}^{2\mu}t}{(\cos\varphi-i\sin\varphi\operatorname{ch}t)^{\nu+\mu+1}}\right]dt$$
$$\left[\operatorname{Re}(\nu+\mu+1)>0,\ \operatorname{Re}(\nu-\mu+1)>0,\ \operatorname{Re}\mu>-\frac{1}{2}\right].$$ MO 89

4. $$P_\nu^\mu(\cos\varphi)=\frac{i}{2^\mu}\frac{\Gamma(\nu+\mu+1)}{\Gamma(\nu-\mu+1)}\frac{\sin^\mu\varphi}{\Gamma\left(\mu+\frac{1}{2}\right)}\times$$
$$\times\int_0^\infty\left[\frac{\operatorname{sh}^{2\mu}t}{(\cos\varphi+i\sin\varphi\operatorname{ch}t)^{\nu+\mu+1}}-\frac{\operatorname{sh}^{2\mu}t}{(\cos\varphi-i\sin\varphi\operatorname{ch}t)^{\nu+\mu+1}}\right]dt$$
$$\left[\operatorname{Re}(\nu\pm\mu+1)>0,\ \operatorname{Re}\mu>-\frac{1}{2}\right].$$ MO 89

8.715

1. $$P_\nu^\mu(\operatorname{ch}\alpha)=\frac{\sqrt{2}\operatorname{sh}^\mu\alpha}{\sqrt{\pi}\Gamma\left(\frac{1}{2}-\mu\right)}\int_0^\alpha\frac{\operatorname{ch}\left(\nu+\frac{1}{2}\right)t\,dt}{(\operatorname{ch}\alpha-\operatorname{ch}t)^{\mu+\frac{1}{2}}}\qquad\left[\alpha>0,\ \operatorname{Re}\mu<\frac{1}{2}\right].$$ MO 87

2. $$Q_\nu^\mu(\operatorname{ch}\alpha)=\sqrt{\frac{\pi}{2}}\frac{e^{\mu\pi i}\operatorname{sh}^\mu\alpha}{\Gamma\left(\frac{1}{2}-\mu\right)}\int_\alpha^\infty\frac{e^{-\left(\nu+\frac{1}{2}\right)t}\,dt}{(\operatorname{ch}t-\operatorname{ch}\alpha)^{\mu+\frac{1}{2}}}$$
$$\left[\alpha>0,\ \operatorname{Re}\mu<\frac{1}{2},\ \operatorname{Re}(\nu+\mu)>-1\right].$$ MO 87

See also 3.277 1., 4., 5., 7., 3.318, 3.516 3., 3.518 1., 2., 3.542 2., 3.663 1., 3.894, 3.988 3., 6.622 3., 6.628 1., 4. — 7., and also 8.742.

8.72 Asymptotic series for large values of $|\nu|$

8.721 For real values of μ, $|\nu|\gg 1$, $|\nu|\gg|\mu|$, $|\arg\nu|<\pi$, we have:

1. $P_\nu^\mu(\cos\varphi)=$
$$=\frac{2}{\sqrt{\pi}}\Gamma(\nu+\mu+1)\sum_{k=0}^\infty\frac{\Gamma\left(\mu+k+\frac{1}{2}\right)}{\Gamma\left(\mu-k+\frac{1}{2}\right)}\frac{\cos\left[\left(\nu+k+\frac{1}{2}\right)\varphi-\frac{\pi}{4}(2k+1)+\frac{\mu\pi}{2}\right]}{k!\,\Gamma\left(\nu-k+\frac{3}{2}\right)(2\sin\varphi)^{k+\frac{1}{2}}}$$
$$\left[\nu+\mu\neq-1,-2,-3,\ \ldots;\ \nu\neq-\frac{3}{2},-\frac{5}{2},-\frac{7}{2}\ldots;\ \text{for }\frac{\pi}{6}<\varphi<\frac{5\pi}{6}\right.$$

this series also converges for complex values of ν and μ. In the remaining cases, it is an asymptotic expansion

for $|\nu| \gg |\mu|$, $|\nu| \gg 1$, if $\nu > 0$, $\mu > 0$ and $0 < \varepsilon \leqslant \varphi \leqslant \pi - \varepsilon \Big]$. MO 92

2. $$Q_\nu^\mu(\cos\varphi) = \sqrt{\pi}\,\Gamma(\nu+\mu+1) \times$$

$$\times \sum_{k=0}^{\infty} (-1)^k \frac{\Gamma\left(\mu+k+\frac{1}{2}\right)}{\Gamma\left(\mu-k+\frac{1}{2}\right)} \frac{\cos\left[\left(\nu+k+\frac{1}{2}\right)\varphi + \frac{\pi}{4}(2k+1) + \frac{\mu\pi}{2}\right]}{k!\,\Gamma\left(\nu-k+\frac{3}{2}\right)(2\sin\varphi)^{k+\frac{1}{2}}}$$

$\Big[\nu+\mu \neq -1, -2, -3, \ldots; \nu \neq -\frac{3}{2}, -\frac{5}{2}, -\frac{7}{2}, \ldots$; for $\frac{\pi}{6} < \varphi < \frac{5}{6}\pi$

this series also converges for complex values of ν and μ. In the remaining cases, it is an asymptotic expansion

for $|\nu| \gg |\mu|$, $|\nu| \gg 1$, if $\nu > 0$, $\mu > 0$, $0 < \varepsilon \leqslant \varphi \leqslant \pi - \varphi \Big]$.

EH I 147(6), MO 92

3. $$P_\nu^\mu(\cos\varphi) = \frac{2}{\sqrt{\pi}} \frac{\Gamma(\nu+\mu+1)}{\Gamma\left(\nu+\frac{3}{2}\right)} \frac{\cos\left[\left(\nu+\frac{1}{2}\right)\varphi - \frac{\pi}{4} + \frac{\mu\pi}{2}\right]}{\sqrt{2\sin\varphi}} \left[1 + O\left(\frac{1}{\nu}\right)\right]$$

$$\left[0 < \varepsilon \leqslant \varphi \leqslant \pi - \varepsilon,\ |\nu| \gg \frac{1}{\varepsilon}\right].$$ MO 92

For $\nu > 0$, $\mu > 0$ and $\nu > \mu$, it follows from formulas 8.721 1. and 8.721 2. that

4. $$\nu^{-\mu} P_\nu^\mu(\cos\varphi) = \sqrt{\frac{2}{\nu\pi\sin\varphi}} \cos\left[\left(\nu+\frac{1}{2}\right)\varphi - \frac{\pi}{4} + \frac{\mu\pi}{2}\right] + O\left(\frac{1}{\sqrt{\nu^3}}\right).$$

5. $$\nu^{-\mu} Q_\nu^\mu(\cos\varphi) = \sqrt{\frac{\pi}{2\nu\sin\varphi}} \cos\left[\left(\nu+\frac{1}{2}\right)\varphi + \frac{\pi}{4} + \frac{\mu\pi}{2}\right] + O\left(\frac{1}{\sqrt{\nu^3}}\right)$$

$$\left[0 < \varepsilon \leqslant \varphi \leqslant \pi - \varepsilon;\ \nu \gg \frac{1}{\varepsilon}\right].$$ MO 92

8.722 If φ is sufficiently close to 0 or π that $\nu\varphi$ or $\nu(\pi-\varphi)$ is small in comparison with 1, the asymptotic formulas 8.721 become unsuitable. In this case, the following asymptotic representation is applicable for $\mu \geqslant 0$, $\nu \gg 1$, and *small* values of φ:

1. $$\left[\left(\nu+\frac{1}{2}\right)\cos\frac{\varphi}{2}\right]^\mu P_\nu^{-\mu}(\cos\varphi) =$$

$$= J_\mu(\eta) + \sin^2\frac{\varphi}{2}\left[\frac{J_{\mu+1}(\eta)}{2\eta} - J_{\mu+2}(\eta) + \frac{\eta}{6} J_{\mu+3}(\eta)\right] + O\left(\sin^4\frac{\varphi}{2}\right),$$

where $\eta = (2\nu+1)\sin\frac{\varphi}{2}$. In particular, it follows that

2. $$\lim_{\nu\to\infty} \nu^\mu P_\nu^{-\mu}\left(\cos\frac{x}{\nu}\right) = J_\mu(x) \quad [x \geqslant 0, \mu \geqslant 0].$$ MO 93

8.723 We can see how the functions $P_\nu^\mu(z$ and $Q_\nu^\mu(z)$ behave for large $|\nu|$ and real values of $z > \frac{3}{2\sqrt{2}}$:

1. $$P_\nu^\mu(\operatorname{ch}\alpha) = \frac{2^\mu}{\sqrt{\pi}} \left\{ \frac{\Gamma\left(-\nu-\frac{1}{2}\right)}{\Gamma(-\nu-\mu)} \frac{e^{(\mu-\nu)\alpha}\operatorname{sh}^\mu\alpha}{(e^{2\alpha}-1)^{\mu+\frac{1}{2}}} \times \right.$$
$$\times F\left(\mu+\frac{1}{2}, -\mu+\frac{1}{2}; \nu+\frac{3}{2}; \frac{1}{1-e^{2\alpha}}\right) +$$
$$\left. + \frac{\Gamma\left(\nu+\frac{1}{2}\right)}{\Gamma(\nu-\mu+1)} \frac{e^{(\nu+\mu+1)\alpha}\operatorname{sh}^\mu\alpha}{(e^{2\alpha}-1)^{\mu+\frac{1}{2}}} F\left(\mu+\frac{1}{2}, -\mu+\frac{1}{2}; -\nu+\frac{1}{2}; \frac{1}{1-e^{2\alpha}}\right)\right\}$$
$$\left[\nu \neq \pm\frac{1}{2}, \pm\frac{3}{2}, \pm\frac{5}{2}, \ldots; \alpha > \frac{1}{2}\ln 2\right]$$ MO 94

2 $$Q_\nu^\mu(\operatorname{ch}\alpha) = e^{\mu\pi i} 2^\mu \sqrt{\pi} \frac{\Gamma(\nu+\mu+1)}{\Gamma\left(\nu+\frac{3}{2}\right)} \frac{e^{-(\nu+\mu+1)\alpha}}{(1-e^{-2\alpha})^{\mu+\frac{1}{2}}} \operatorname{sh}^\mu\alpha \times$$
$$\times F\left(\mu+\frac{1}{2}, -\mu+\frac{1}{2}; \nu+\frac{3}{2}; \frac{1}{1-e^{2\alpha}}\right)$$
$$\left[\mu+\nu+1 \neq 0, -1, -2, \ldots; \alpha > \frac{1}{2}\ln 2\right].$$ MO 94

See also 8.776

8.724 The inequalities

1. $$|P_\nu^{\pm\mu}(\cos\varphi)| < \sqrt{\frac{8}{\nu\pi}} \frac{\Gamma(\nu\pm\mu+1)}{\Gamma(\nu+1)} \frac{1}{\sin^{\mu+\frac{1}{2}}\varphi},$$

2. $$|Q_\nu^{\pm\mu}(\cos\varphi)| < \sqrt{\frac{2\pi}{\nu}} \frac{\Gamma(\nu\pm\mu+1)}{\Gamma(\nu+1)} \frac{1}{\sin^{\mu+\frac{1}{2}}\varphi},$$

3. $$|P_\nu^{\pm m}(\cos\varphi)| < \frac{2}{\sqrt{\nu\pi}} \frac{\Gamma(\nu\pm m+1)}{\Gamma(\nu+1)} \frac{1}{\sin^{m+\frac{1}{2}}\varphi},$$

4. $$|Q_\nu^{\pm m}(\cos\varphi)| < \sqrt{\frac{\pi}{\nu}} \frac{\Gamma(\nu\pm m+1)}{\Gamma(\nu+1)} \frac{1}{\sin^{m+\frac{1}{2}}\varphi}$$

[ν and μ are arbitrary real numbers satisfying the inequalities $\nu \geqslant 1$, $\nu-\mu+1 > 0$, $\mu \geqslant 0$]. MO 91-92

8.73-8.74 Functional relations

8.731

1. $$(z^2-1)\frac{dP_\nu^\mu(z)}{dz} = (\nu-\mu+1)P_{\nu+1}^\mu(z) - (\nu+1)zP_\nu^\mu(z)$$
(cf. 8.832 1., 8.914 2.). EH I 161(10), MO 81

2. $$(2\nu+1)zP_\nu^\mu(z) = (\nu-\mu+1)P_{\nu+1}^\mu(z) + (\nu+\mu)P_{\nu-1}^\mu(z)$$
(cf. 8.832 2., 8.914 1.). EH I 160(2), MO 81

3. $$P_\nu^{\mu+2}(z) + 2(\mu+1)\frac{z}{\sqrt{z^2-1}}P_\nu^{\mu+1}(z) = (\nu-\mu)(\nu+\mu+1)P_\nu^\mu(z).$$

MO 82, EH I 160(1)

4. $P_{\nu+1}^{\mu}(z) - P_{\nu-1}^{\mu}(z) = (2\nu+1)\sqrt{z^2-1}\, P_{\nu}^{\mu-1}(z).$ EH I 160(3), MO 82

5. $P_{-\nu-1}^{\mu}(z) = P_{\nu}^{\mu}(z)$ (cf. 8.820, 8.832 4.). EH I 140(1), MO 82

8.732

1 $$(z^2-1)\frac{dQ_{\nu}^{\mu}(z)}{dz} = (\nu-\mu+1)\,Q_{\nu+1}^{\mu}(z) - (\nu+1)\,zQ_{\nu}^{\mu}(z)$$ (cf. 8.832 3.). MO 82

2. $$(2\nu+1)\,zQ_{\nu}^{\mu}(z) = (\nu-\mu+1)\,Q_{\nu+1}^{\mu}(z) + (\nu+\mu)\,zQ_{\nu-1}^{\mu}(z)$$ (cf. 8.832 4.). MO 82

3. $$Q_{\nu}^{\mu+2}(z) + 2(\mu+1)\frac{z}{\sqrt{z^2-1}}Q_{\nu}^{\mu+1}(z) = (\nu-\mu)(\nu+\mu+1)\,Q_{\nu}^{\mu}(z).$$ MO 82

4. $$Q_{\nu-1}^{\mu}(z) - Q_{\nu+1}^{\mu}(z) = -(2\nu+1)\sqrt{z^2-1}\,Q_{\nu}^{\mu-1}(z).$$ MO 82a

5. $$e^{-\mu\pi i}Q_{\nu}^{\mu}(x \pm i0) = e^{\pm\frac{1}{2}\mu\pi i}\left[\mathrm{Q}_{\nu}^{\mu}(x) \mp i\frac{\pi}{2}\mathrm{P}_{\nu}^{\mu}(x)\right].$$ MO 83

8.733

1. $$(1-x^2)\frac{d\mathrm{P}_{\nu}^{\mu}(x)}{dx} = (\nu+1)\,x\mathrm{P}_{\nu}^{\mu}(x) - (\nu-\mu+1)\,\mathrm{P}_{\nu+1}^{\mu}(x)$$ (cf. 8.731 1.);
$$= -\nu x\mathrm{P}_{\nu}^{\mu}(x) + (\nu+\mu)\,\mathrm{P}_{\nu-1}^{\mu}(x);$$
$$= -\sqrt{1-x^2}\,\mathrm{P}_{\nu}^{\mu+1}(x) - \mu x\mathrm{P}_{\nu}^{\mu}(x);$$
$$= (\nu-\mu+1)(\nu+\mu)\sqrt{1-x^2}\,\mathrm{P}_{\nu}^{\mu-1}(x) + \mu x\mathrm{P}_{\nu}^{\mu}(x).$$ MO 82

2 $$(2\nu+1)\,x\mathrm{P}_{\nu}^{\mu}(x) = (\nu-\mu+1)\,\mathrm{P}_{\nu+1}^{\mu}(x) + (\nu+\mu)\,\mathrm{P}_{\nu-1}^{\mu}(x)$$ (cf. 8.731 2.). MO 82

3. $$\mathrm{P}_{\nu}^{\mu+2}(x) + 2(\mu+1)\frac{x}{\sqrt{1-x^2}}\mathrm{P}_{\nu}^{\mu+1}(x) + (\nu-\mu)(\nu+\mu+1)\,\mathrm{P}_{\nu}^{\mu}(x) = 0$$ (cf. 8.731 3.). MO 82

4. $$\mathrm{P}_{\nu-1}^{\mu}(x) - \mathrm{P}_{\nu+1}^{\mu}(x) = (2\nu+1)\sqrt{1-x^2}\,\mathrm{P}_{\nu}^{\mu-1}(x)$$ (cf. 8.731 4.). MO 82

5. $\mathrm{P}_{-\nu-1}^{\mu}(x) = \mathrm{P}_{\nu}^{\mu}(x)$ (cf. 8.731 5.).

8.734

1. $(\nu+\mu+1)\,zQ_{\nu}^{\mu}(z) + \sqrt{z^2-1}\,Q_{\nu}^{\mu+1}(z) = (\nu-\mu+1)\,Q_{\nu+1}^{\mu}(z).$ MO 82

2. $(\nu+\mu)\,Q_{\nu-1}^{\mu}(z) + \sqrt{z^2-1}\,Q_{\nu}^{\mu+1}(z) = (\nu-\mu)\,zQ_{\nu}^{\mu}(z).$ MO 82

3. $Q_{\nu-1}^{\mu}(z) - zQ_{\nu}^{\mu}(z) = -(\nu-\mu+1)\sqrt{z^2-1}\,Q_{\nu}^{\mu-1}(z).$ MO 82

4. $zQ_{\nu}^{\mu}(z) - Q_{\nu+1}^{\mu}(z) = -(\nu+\mu)\sqrt{z^2-1}\,Q_{\nu}^{\mu-1}(z).$ MO 82

5. $$(\nu+\mu)(\nu+\mu+1)\,Q_{\nu-1}^{\mu}(z) + (2\nu+1)\sqrt{z^2-1}\,Q_{\nu}^{\mu+1}(z) = (\nu-\mu)(\nu-\mu+1)\,Q_{\nu+1}^{\mu}(z).$$ MO 82

8.735

1. $(\nu+\mu+1)\,x\,\mathrm{P}_{\nu}^{\mu}(x) + \sqrt{1-x^2}\,\mathrm{P}_{\nu}^{\mu+1}(x) = (\nu-\mu+1)\,\mathrm{P}_{\nu+1}^{\mu}(x).$ MO 83

2. $(\nu-\mu)\,x\,\mathrm{P}_{\nu}^{\mu}(x) - (\nu+\mu)\,\mathrm{P}_{\nu-1}^{\mu}(x) = \sqrt{1-x^2}\,\mathrm{P}_{\nu}^{\mu+1}(x).$ MO 83

3. $\mathrm{P}_{\nu-1}^{\mu}(x) - x\,\mathrm{P}_{\nu}^{\mu}(x) = (\nu-\mu+1)\sqrt{1-x^2}\,\mathrm{P}_{\nu}^{\mu-1}(x).$ MO 83

4. $x\,\mathrm{P}_{\nu}^{\mu}(x) - \mathrm{P}_{\nu+1}^{\mu}(x) = (\nu+\mu)\sqrt{1-x^2}\,\mathrm{P}_{\nu}^{\mu-1}(x).$ MO 83

5. $(\nu-\mu)(\nu-\mu+1)\,\mathrm{P}_{\nu+1}^{\mu}(x) = (\nu+\mu)(\nu+\mu+1)\,\mathrm{P}_{\nu-1}^{\mu}(x) + {}$
$+ (2\nu+1)\sqrt{1-x^2}\,\mathrm{P}_{\nu}^{\mu+1}(x).$ MO 83

8.736

1. $P_{\nu}^{-\mu}(z) = \frac{\Gamma(\nu-\mu+1)}{\Gamma(\nu+\mu+1)}\left[P_{\nu}^{\mu}(z) - \frac{2}{\pi}e^{-\mu\pi i}\sin\mu\pi Q_{\nu}^{\mu}(z)\right].$ MO 83

2. $P_{\nu}^{\mu}(-z) = e^{\nu\pi i}P_{\nu}^{\mu}(z) - \frac{2}{\pi}\sin[(\nu+\mu)\pi]\,e^{-\mu\pi i}Q_{\nu}^{\mu}(z)$
$[\operatorname{Im} z < 0]$ (cf. 8.833 1.). MO 83

3. $P_{\nu}^{\mu}(-z) = e^{-\nu\pi i}P_{\nu}^{\mu}(z) - \frac{2}{\pi}\sin[(\nu+\mu)\pi]\,e^{-\mu\pi i}Q_{\nu}^{\mu}(z)$
$[\operatorname{Im} z > 0]$ (cf. 8.833 2.). MO 83

4. $Q_{\nu}^{-\mu}(z) = e^{-2\mu\pi i}\frac{\Gamma(\nu-\mu+1)}{\Gamma(\nu+\mu+1)}Q_{\nu}^{\mu}(z).$ MO 82

5. $Q_{\nu}^{\mu}(-z) = -e^{-\nu\pi i}Q_{\nu}^{\mu}(z)$ $[\operatorname{Im} z < 0]$ (cf. 8.833 3.). MO 82

6. $Q_{\nu}^{\mu}(-z) = -e^{\nu\pi i}Q_{\nu}^{\mu}(z)$ $[\operatorname{Im} z > 0]$ (cf. 8.833 4.). MO 82

7. $Q_{\nu}^{\mu}(z)\sin[(\nu+\mu)\pi] - Q_{-\nu-1}^{\mu}(z)\sin[(\nu-\mu)\pi] = \pi e^{\mu\pi i}\cos\mu\pi\,P_{\nu}^{\mu}(z).$
MO 83

8.737

1. $\mathrm{P}_{\nu}^{-\mu}(x) = \frac{\Gamma(\nu-\mu+1)}{\Gamma(\nu+\mu+1)}\left[\cos\mu\pi\,\mathrm{P}_{\nu}^{\mu}(x) - \frac{2}{\pi}\sin(\mu\pi)\,\mathrm{Q}_{\nu}^{\mu}(x)\right].$ MO 84

2. $\mathrm{P}_{\nu}^{\mu}(-x) = \cos[(\nu+\mu)\pi]\,\mathrm{P}_{\nu}^{\mu}(x) - \frac{2}{\pi}\sin[(\nu+\mu)\pi]\,\mathrm{Q}_{\nu}^{\mu}(x).$ MO 84

3. $\mathrm{Q}_{\nu}^{\mu}(-x) = -\cos[(\nu+\mu)\pi]\,\mathrm{Q}_{\nu}^{\mu}(x) - \frac{\pi}{2}\sin[(\nu+\mu)\pi]\,\mathrm{P}_{\nu}^{\mu}(x).$
MO 83, EH I 144(15)

4. $\mathrm{Q}_{-\nu-1}^{\mu}(x) = \frac{\sin[(\nu+\mu)\pi]}{\sin[(\nu-\mu)\pi]}\mathrm{Q}_{\nu}^{\mu}(x) - \frac{\pi\cos\nu\pi\cos\mu\pi}{\sin[(\nu-\mu)\pi]}\mathrm{P}_{\nu}^{\mu}(x).$ MO 84

8.738

1. $Q_{\nu}^{\mu}(i\operatorname{ctg}\varphi) = \exp\left[i\pi\left(\mu - \frac{\nu+1}{2}\right)\right]\sqrt{\pi}\,\Gamma(\nu+\mu+1)\times$
$\times\sqrt{\frac{1}{2}\sin\varphi}\,P_{-\mu-\frac{1}{2}}^{-\nu-\frac{1}{2}}(\cos\varphi)$ $\left[0<\varphi<\frac{\pi}{2}\right].$ MO 83

2. $P_{\nu}^{\mu}(i\operatorname{ctg}\varphi) = \sqrt{\frac{2}{\pi}}\exp\left[i\pi\left(\nu+\frac{1}{2}\right)\right]\frac{\sqrt{\sin\varphi}}{\Gamma(-\nu-\mu)}Q_{-\mu-\frac{1}{2}}^{-\nu-\frac{1}{2}}(\cos\varphi - i0)$
$\left[0<\varphi<\frac{\pi}{2}\right].$ MO 83

8.739 $e^{-\mu\pi i}Q_{\nu}^{\mu}(\operatorname{ch}\alpha) = \frac{\sqrt{\pi}\,\Gamma(\nu+\mu+1)}{\sqrt{2\operatorname{sh}\alpha}}P_{-\mu-\frac{1}{2}}^{-\nu-\frac{1}{2}}(\operatorname{cth}\alpha)$ $[\operatorname{Re}(\operatorname{ch}\alpha) > 0].$ MO 83

8.741

1. $\mathrm{P}_{\nu}^{-\mu}(x)\frac{d\mathrm{P}_{\nu}^{\mu}(x)}{dx} - \mathrm{P}_{\nu}^{\mu}(x)\frac{d\mathrm{P}_{\nu}^{-\mu}(x)}{dx} = \frac{2\sin\mu\pi}{\pi(1-x^2)}.$ MO 83

2. $$P_\nu^\mu(x)\frac{dQ_\nu^\mu(x)}{dx} - Q_\nu^\mu(x)\frac{dP_\nu^\mu(x)}{dx} = \frac{2^{2\mu}}{1-x^2}\frac{\Gamma\left(\frac{\nu+\mu+1}{2}\right)\Gamma\left(\frac{\nu+\mu}{2}+1\right)}{\Gamma\left(\frac{\nu-\mu+1}{2}\right)\Gamma\left(\frac{\nu-\mu}{2}+1\right)}.$$

MO 83

8.742

1. $$\frac{\Gamma(\nu-\mu+1)}{\Gamma(\nu+\mu+1)}\left\{\cos\mu\pi\, P_\nu^\mu(\cos\varphi) - \frac{2}{\pi}\sin\mu\pi\, Q_\nu^\mu(\cos\varphi)\right\} =$$
$$= \sqrt{\frac{2}{\pi}}\frac{\operatorname{cosec}^\mu\varphi}{\Gamma\left(\mu+\frac{1}{2}\right)}\int_0^\varphi \frac{\cos\left(\nu+\frac{1}{2}\right)t\, dt}{(\cos t - \cos\varphi)^{\frac{1}{2}-\mu}} \quad \left[\operatorname{Re}\mu > -\frac{1}{2}\right].$$ MO 88

2. $$\frac{\Gamma(\nu-\mu+1)}{\Gamma(\nu+\mu+1)}\left\{\cos\nu\pi\, P_\nu^\mu(\cos\varphi) - \frac{2}{\pi}\sin\nu\pi\, Q_\nu^\mu(\cos\varphi)\right\} =$$
$$= \sqrt{\frac{2}{\pi}}\frac{\operatorname{cosec}^\mu\varphi}{\Gamma\left(\mu+\frac{1}{2}\right)}\int_\varphi^\pi \frac{\cos\left[\left(\nu+\frac{1}{2}\right)(t-\pi)\right] dt}{(\cos\varphi - \cos t)^{\frac{1}{2}-\mu}} \quad \left[\operatorname{Re}\mu > -\frac{1}{2}\right].$$

MO 88

3. $$P_\nu^\mu(\cos\varphi)\cos(\nu+\mu)\pi - \frac{2}{\pi}Q_\nu^\mu(\cos\varphi)\sin(\nu+\mu)\pi =$$
$$= \sqrt{\frac{2}{\pi}}\frac{\sin^\mu\varphi}{\Gamma\left(\frac{1}{2}-\mu\right)}\int_\varphi^\pi \frac{\cos\left[\left(\nu+\frac{1}{2}\right)(t-\pi)\right] dt}{(\cos\varphi - \cos t)^{\mu+\frac{1}{2}}}$$
$$\left[\operatorname{Re}\mu < \frac{1}{2}\right].$$ MO 88

4. $$\cos\mu\pi\, P_\nu^\mu(\cos\varphi) - \frac{2}{\pi}\sin\mu\pi\, Q_\nu^\mu(\cos\varphi) =$$
$$= \frac{1}{2^\mu\sqrt{\pi}}\frac{\Gamma(\nu+\mu+1)}{\Gamma(\nu-\mu+1)}\frac{\sin^\mu\varphi}{\Gamma\left(\mu+\frac{1}{2}\right)}\int_0^\pi \frac{\sin^{2\mu}t\, dt}{(\cos\varphi \pm i\sin\varphi\cos t)^{\nu-\mu}}$$
$$\left[\operatorname{Re}\mu > -\frac{1}{2},\ 0 < \varphi < \pi\right].$$ MO 38

For integrals of Legendre functions, see **7.11 — 7.21**.

8.75 Special cases and particular values

Special cases

8.751

1. $$P_\nu^m(x) = (-1)^m\frac{\Gamma(\nu+m+1)(1-x^2)^{\frac{m}{2}}}{2^m\Gamma(\nu-m+1)\, m!}F\left(m-\nu, m+\nu+1; m+1; \frac{1-x}{2}\right).$$

MO 84

2. $$P_\nu^m(z) = \frac{\Gamma(\nu+m+1)(z^2-1)^{\frac{m}{2}}}{2^m m!\,\Gamma(\nu-m+1)}F\left(m-\nu, m+\nu+1; m+1; \frac{1-z}{2}\right).$$

MO 84

3. $$Q^{\mu}_{-n-\frac{3}{2}}(z) = \frac{e^{\mu\pi i}\Gamma\left(\mu+n+\frac{3}{2}\right)}{2^{n+\frac{3}{2}}(n+1)!} \times$$
$$\times (z^2-1)^{\frac{\mu}{2}} z^{2n-\mu+\frac{3}{2}} F\left(\frac{\mu+n+\frac{5}{2}}{2}, \frac{\mu+n+\frac{3}{2}}{2}; n+2; \frac{1}{z^2}\right).$$ MO 84

8.752

1. $$\mathrm{P}_{\nu}^{m}(x) = (-1)^m (1-x^2)^{\frac{m}{2}} \frac{d^m}{dx^m} \mathrm{P}_{\nu}(x).$$ WH, MO 84, EH I 148(6)

2. $$\mathrm{P}_{\nu}^{-m}(x) = (-1)^m \frac{\Gamma(\nu-m+1)}{\Gamma(\nu+m+1)} \mathrm{P}_{\nu}^{m}(x) =$$
$$= (1-x^2)^{-\frac{m}{2}} \int_x^1 \dots \int_x^1 \mathrm{P}_{\nu}(x)\,(dx)^m. \quad [m \geqslant 1]$$ HO 99a, MO 85, EH I 149(10)a

3. $$P_{\nu}^{-m}(z) = (z^2-1)^{-\frac{m}{2}} \int_1^z \dots \int_1^z P_{\nu}(z)\,(dz)^m. \quad [m \geqslant 1]$$ MO 85, EH I 149(8)

4. $$Q_{\nu}^{m}(z) = (z^2-1)^{\frac{m}{2}} \frac{d^m}{dz^m} Q_{\nu}(z).$$ WH, MO 85, EH I 148(5)

5. $$Q_{\nu}^{-m}(z) = (-1)^m (z^2-1)^{-\frac{m}{2}} \int_z^{\infty} \dots \int_z^{\infty} Q_{\nu}(z)\,(dz)^m.$$ $[m \geqslant 1]$ MO 85, EH I 149(9)

Special values of the indices

8.753

1. $$\mathrm{P}_0^{\mu}(\cos\varphi) = \frac{1}{\Gamma(1-\mu)} \operatorname{ctg}^{\mu}\frac{\varphi}{2}.$$ MO 84

2. $$\mathrm{P}_{\nu}^{-1}(\cos\varphi) = -\frac{1}{\nu(\nu+1)} \frac{d\mathrm{P}_{\nu}(\cos\varphi)}{d\varphi}.$$ MO 84

3. $P_n^m(z) \equiv 0$, $\mathrm{P}_n^m(x) \equiv 0$ for $m > n$. MO 85

8.754

1. $$P^{\frac{1}{2}}_{\nu-\frac{1}{2}}(\operatorname{ch}\alpha) = \sqrt{\frac{2}{\pi \operatorname{sh}\alpha}} \operatorname{ch}\nu\alpha.$$ MO 85

2. $$\mathrm{P}^{\frac{1}{2}}_{\nu-\frac{1}{2}}(\cos\varphi) = \sqrt{\frac{2}{\pi\sin\varphi}} \cos\nu\varphi.$$ MO 85

3. $$\mathrm{P}^{-\frac{1}{2}}_{\nu-\frac{1}{2}}(\cos\varphi) = \sqrt{\frac{2}{\pi\sin\varphi}} \frac{\sin\nu\varphi}{\nu}.$$ MO 85

4. $$Q^{\frac{1}{2}}_{\nu-\frac{1}{2}}(\operatorname{ch}\alpha) = i\sqrt{\frac{\pi}{2\operatorname{sh}\alpha}} e^{-\nu\alpha}.$$ MO 85

8.755

1. $$\mathrm{P}_{\nu}^{-\nu}(\cos\varphi) = \frac{1}{\Gamma(1+\nu)} \left(\frac{\sin\varphi}{2}\right)^{\nu}.$$ MO 85

2. $$P_{\nu}^{-\nu}(\operatorname{ch}\alpha) = \frac{1}{\Gamma(1+\nu)} \left(\frac{\operatorname{sh}\alpha}{2}\right)^{\nu}.$$ MO 85

Special values of Legendre functions

8.756

1. $$P_\nu^\mu(0) = \frac{2^\mu \sqrt{\pi}}{\Gamma\left(\frac{\nu-\mu}{2}+1\right)\Gamma\left(\frac{-\nu-\mu+1}{2}\right)}.$$ MO 84

2. $$\frac{dP_\nu^\mu(0)}{dx} = \frac{2^{\mu+1}\sin\frac{1}{2}(\nu+\mu)\pi\,\Gamma\left(\frac{\nu+\mu}{2}+1\right)}{\sqrt{\pi}\,\Gamma\left(\frac{\nu-\mu+1}{2}\right)}.$$ MO 84

3. $$Q_\nu^\mu(0) = -2^{\mu-1}\sqrt{\pi}\sin\frac{1}{2}(\nu+\mu)\pi\,\frac{\Gamma\left(\frac{\nu+\mu+1}{2}\right)}{\Gamma\left(\frac{\nu-\mu}{2}+1\right)}.$$ MO 84

4. $$\frac{dQ_\nu^\mu(0)}{dx} = 2^\mu\sqrt{\pi}\cos\frac{1}{2}(\nu+\mu)\pi\,\frac{\Gamma\left(\frac{\nu+\mu}{2}+1\right)}{\Gamma\left(\frac{\nu-\mu+1}{2}\right)}.$$ MO 84

8.76 Derivatives with respect to the order

8.761 $$\frac{\partial P_\nu^{-\mu}(x)}{\partial\nu} =$$
$$= \frac{1}{\Gamma(\mu+1)}\left(\frac{1-x}{1+x}\right)^{\frac{\mu}{2}}\sum_{n=1}^{\infty}\frac{(-\nu)(1-\nu)\dots(n-1-\nu)(\nu+1)(\nu+2)\dots(\nu+n)}{(\mu+1)(\mu+2)\dots(\mu+n)\,1\cdot 2\dots n}\times$$
$$\times[\psi(\nu+n+1)-\psi(\nu-n+1)]\left(\frac{1-x}{2}\right)^n$$
$$[\nu \neq 0, \pm 1, \pm 2, \dots; \operatorname{Re}\mu > -1].$$ MO 94

8.762

1. $$\left[\frac{\partial P_\nu(\cos\varphi)}{\partial\nu}\right]_{\nu=0} = 2\ln\cos\frac{\varphi}{2}.$$ MO 94

2. $$\left[\frac{\partial P_\nu^{-1}(\cos\varphi)}{\partial\nu}\right]_{\nu=0} = -\operatorname{tg}\frac{\varphi}{2} - 2\operatorname{ctg}\frac{\varphi}{2}\ln\cos\frac{\varphi}{2}.$$ MO 94

3. $$\left[\frac{\partial P_\nu^{-1}(\cos\varphi)}{\partial\nu}\right]_{\nu=1} = -\frac{1}{2}\operatorname{tg}\frac{\varphi}{2}\sin^2\frac{\varphi}{2} + \sin\varphi\ln\cos\frac{\varphi}{2}.$$ MO 94

For a connection with the polynomials $C_n^\lambda(x)$, see 8.936.
For a connection with a hypergeometric function, see 8.77.

8.77 Series representation

For a representation in the form of a series, see 8.721. It is also possible to represent associated Legendre functions in the form of a series by expressing them in terms of a hypergeometric function.

8.771

1. $$P_\nu^\mu(z) = \left(\frac{z+1}{z-1}\right)^{\frac{\mu}{2}}\frac{1}{\Gamma(1-\mu)}F\left(-\nu, \nu+1; 1-\mu; \frac{1-z}{2}\right).$$ MO 15

2. $$Q_\nu^\mu(z) = \frac{e^{\mu\pi i}}{2^{\nu+1}} \frac{\Gamma(\nu+\mu+1)}{\Gamma\left(\nu+\frac{3}{2}\right)} \frac{\Gamma\left(\frac{1}{2}\right)(z^2-1)^{\frac{\mu}{2}}}{z^{\nu+\mu+1}} \times$$
$$\times F\left(\frac{\nu+\mu}{2}+1, \frac{\nu+\mu+1}{2}; \nu+\frac{3}{2}; \frac{1}{z^2}\right).$$ MO 15

See also 8.702, 8.703, 8.704, 8.723, 8.751, 8.772.

The analytic continuation for $|z| \gg 1$

The formulas are consequences of theorems on the analytic continuation of hypergeometric series (see 9.154 and 9.155):

8.772

1. $$P_\nu^\mu(z) = \frac{\sin(\nu+\mu)\pi\Gamma(\nu+\mu+1)}{2^{\nu+1}\sqrt{\pi}\cos\nu\pi\Gamma\left(\nu+\frac{3}{2}\right)} \times$$
$$\times (z^2-1)^{\frac{\mu}{2}} z^{-\nu-\mu-1} F\left(\frac{\nu+\mu}{2}+1, \frac{\nu+\mu+1}{2}; \nu+\frac{3}{2}; \frac{1}{z^2}\right)+$$
$$+\frac{2^\nu\Gamma\left(\nu+\frac{1}{2}\right)}{\sqrt{\pi}\Gamma(\nu-\mu+1)}(z^2-1)^{\frac{\mu}{2}} z^{\nu-\mu} F\left(\frac{\mu-\nu+1}{2}, \frac{\mu-\nu}{2}; \frac{1}{2}-\nu; \frac{1}{z^2}\right)$$
$[2\nu \neq \pm 1, \pm 3, \pm 5, \ldots; |z| > 1; |\arg(z \pm 1)| < \pi]$. MO 85

2. $$P_\nu^\mu(z) = \frac{\Gamma\left(-\nu-\frac{1}{2}\right)(z^2-1)^{-\frac{\nu+1}{2}}}{2^{\nu+1}\sqrt{\pi}\Gamma(-\nu-\mu)} \times$$
$$\times F\left(\frac{\nu-\mu+1}{2}, \frac{\nu+\mu+1}{2}; \nu+\frac{3}{2}; \frac{1}{1-z^2}\right)+$$
$$+\frac{2^\nu\Gamma\left(\nu+\frac{1}{2}\right)}{\sqrt{\pi}\Gamma(\nu-\mu+1)}(z^2-1)^{\frac{\nu}{2}} F\left(\frac{\mu-\nu}{2}, -\frac{\mu+\nu}{2}; \frac{1}{2}-\nu; \frac{1}{1-z^2}\right)$$
$[2\nu \neq \pm 1, \pm 3, \pm 5; \ldots; |1-z^2| > 1; |\arg(z \pm 1)| < \pi]$. MO 85

3. $$P_\nu^\mu(z) = \frac{1}{\Gamma(1-\mu)}\left(\frac{z-1}{z+1}\right)^{-\frac{\mu}{2}}\left(\frac{z+1}{2}\right)^{-\nu} \times$$
$$\times F\left(-\nu, -\nu-\mu; 1-\mu; \frac{z-1}{z+1}\right) \qquad \left[\left|\frac{z-1}{z+1}\right| < 1\right]$$ MO 86

8.773

1. $$Q_\nu^\mu(z) = e^{\mu\pi i}\frac{\sqrt{\pi}\Gamma(\nu+\mu+1)}{2^{\nu+1}\Gamma\left(\nu+\frac{3}{2}\right)}(z^2-1)^{-\frac{\nu+1}{2}} \times$$
$$\times F\left(\frac{\nu+\mu+1}{2}, \frac{\nu-\mu+1}{2}; \nu+\frac{3}{2}; \frac{1}{1-z^2}\right)$$
$[\nu+\mu \neq -1, -2, -3, \ldots; |\arg(z \pm 1)| < \pi; |1-z^2| > 1]$. MO 86

2. $$Q_\nu^\mu(z) = \frac{1}{2}e^{\mu\pi i}\left\{\Gamma(\mu)\left(\frac{z+1}{z-1}\right)^{\frac{\mu}{2}} F\left(-\nu, \nu+1; 1-\mu; \frac{1-z}{2}\right)+\right.$$
$$\left.+\frac{\Gamma(-\mu)\Gamma(\nu+\mu+1)}{\Gamma(\nu-\mu+1)}\left(\frac{z-1}{z+1}\right)^{\frac{\mu}{2}} F\left(-\nu, \nu+1; 1+\mu; \frac{1-z}{2}\right)\right\}$$
$[|\arg(z \pm 1)| < \pi, |1-z| < 2]$. MO 86

8.774 $$P_{\nu}^{\mu}(i\operatorname{ctg}\varphi)=\sqrt{\frac{\sin\varphi}{2\pi}}\frac{\Gamma\left(-\nu-\frac{1}{2}\right)}{\Gamma(-\nu-\mu)}\times$$
$$\times e^{-i(\nu+1)\frac{\pi}{2}}\left(\operatorname{tg}\frac{\varphi}{2}\right)^{\nu+\frac{1}{2}}F\left(\frac{1}{2}+\mu,\ \frac{1}{2}-\mu;\ \nu+\frac{3}{2};\ \sin^2\frac{\varphi}{2}\right)+$$
$$+\sqrt{\frac{\sin\varphi}{2\pi}}\frac{\Gamma\left(\nu+\frac{1}{2}\right)}{\Gamma(\nu-\mu+1)}e^{i\nu\frac{\pi}{2}}\left(\operatorname{ctg}\frac{\varphi}{2}\right)^{\nu+\frac{1}{2}}F\left(\frac{1}{2}+\mu,\ \frac{1}{2}-\mu;\ \frac{1}{2}-\nu;\ \sin^2\frac{\varphi}{2}\right)$$
$$\left[2\nu\neq\pm1,\ \pm3,\ \pm5,\ \ldots;\ 0<\varphi<\frac{\pi}{2}\right].$$ MO 86

8.775

1. $$\mathrm{P}_{\nu}^{\mu}(x)=\frac{2^{\mu}\cos\frac{1}{2}(\nu+\mu)\pi\Gamma\left(\frac{\nu+\mu+1}{2}\right)}{\sqrt{\pi}\,\Gamma\left(\frac{\nu-\mu}{2}+1\right)}\times$$
$$\times(1-x^2)^{\frac{\mu}{2}}F\left(\frac{\nu+\mu+1}{2},\ \frac{\mu-\nu}{2};\ \frac{1}{2};\ x^2\right)+$$
$$+\frac{2^{\mu+1}}{\sqrt{\pi}}\frac{\sin\frac{1}{2}(\nu+\mu)\pi\Gamma\left(\frac{\nu+\mu}{2}+1\right)}{\Gamma\left(\frac{\nu-\mu+1}{2}\right)}x(1-x^2)^{\frac{\mu}{2}}F\left(\frac{\nu+\mu}{2}+1,\ \frac{-\nu+\mu+1}{2};\ \frac{3}{2};\ x^2\right).$$
MO 87

2. $$\mathrm{Q}_{\nu}^{\mu}(x)=-\frac{\sqrt{\pi}}{2^{1-\mu}}\frac{\sin\frac{1}{2}(\nu+\mu)\pi\Gamma\left(\frac{\nu+\mu+1}{2}\right)}{\Gamma\left(\frac{\nu-\mu}{2}+1\right)}\times$$
$$\times(1-x^2)^{\frac{\mu}{2}}F\left(\frac{\nu+\mu+1}{2},\ \frac{\mu-\nu}{2};\ \frac{1}{2};\ x^2\right)+$$
$$+2^{\mu}\sqrt{\pi}\frac{\cos\frac{1}{2}(\nu+\mu)\pi\Gamma\left(\frac{\nu+\mu}{2}+1\right)}{\Gamma\left(\frac{\nu-\mu+1}{2}\right)}x(1-x^2)^{\frac{\mu}{2}}F\left(\frac{\nu+\mu}{2}+1,\ \frac{\mu-\nu+1}{2};\ \frac{3}{2};x^2\right).$$
MO 87

8.776 For $|z|\gg1$

1. $$P_{\nu}^{\mu}(z)=\left\{\frac{2^{\nu}\Gamma\left(\nu+\frac{1}{2}\right)}{\sqrt{\pi}\,\Gamma(\nu-\mu+1)}z^{\nu}+\frac{\Gamma\left(-\nu-\frac{1}{2}\right)}{2^{\nu+1}\sqrt{\pi}\,\Gamma(-\nu-\mu)}z^{-\nu-1}\right\}\left(1+O\left(\frac{1}{z^2}\right)\right)$$
$$[2\nu\neq\pm1,\ \pm3,\ \pm5,\ \ldots,\ |\arg z|<\pi].$$ MO 87

2. $$Q_{\nu}^{\mu}(z)=\sqrt{\pi}\,\frac{e^{\mu\pi i}}{2^{\nu+1}}\frac{\Gamma(\mu+\nu+1)}{\Gamma\left(\nu+\frac{3}{2}\right)}z^{-\nu-1}\left(1+O\left(\frac{1}{z^2}\right)\right)$$
$$[2\nu\neq-3,\ -5,\ -7,\ \ldots;\ |\arg z|<\pi].$$ MO 87

8.777 Set $\zeta=z+\sqrt{z^2-1}$. The variable ζ is uniquely defined by this equation on the entire z-plane in which a cut is made from $-\infty$ to $+1$. Here, we are considering that branch of the variable ζ for which values of ζ exceeding 1 correspond to real values of z exceeding 1. In this case,

1. $$P_\nu^\mu(z) = \frac{2^\mu \Gamma\left(-\nu-\frac{1}{2}\right)}{\sqrt{\pi}\,\Gamma(-\nu-\mu)} \frac{(z^2-1)^{\frac{\mu}{2}}}{\zeta^{\nu+\mu+1}} F\left(\frac{1}{2}+\mu,\ \nu+\mu+1;\ \nu+\frac{3}{2};\ \frac{1}{\zeta^2}\right) + \frac{2^\mu}{\sqrt{\pi}} \frac{\Gamma\left(\nu+\frac{1}{2}\right)}{\Gamma(\nu-\mu+1)} \frac{(z^2-1)^{\frac{\mu}{2}}}{\zeta^{\mu-\nu}} F\left(\frac{1}{2}+\mu,\ \mu-\nu;\ \frac{1}{2}-\nu;\ \frac{1}{\zeta^2}\right)$$
$[2\nu \neq \pm 1,\ \pm 3,\ \pm 5,\ \ldots;\ |\arg(z-1)| < \pi]$. MO 86

2. $$Q_\nu^\mu(z) = 2^\mu e^{\mu\pi i} \sqrt{\pi} \frac{\Gamma(\nu+\mu+1)}{\Gamma\left(\nu+\frac{3}{2}\right)} \frac{(z^2-1)^{\frac{\mu}{2}}}{\zeta^{\nu+\mu+1}} \times F\left(\frac{1}{2}+\mu,\ \nu+\mu+1;\ \nu+\frac{3}{2};\ \frac{1}{\zeta^2}\right)$$
$[|\arg(z-1)| < \pi]$. MO 86

8.78 The zeros of associated Legendre functions

8.781 The function $P_\nu^{-\mu}(\cos\varphi)$, considered as a function of ν has infinitely many zeros for $\mu \geqslant 0$. These are all simple and real. If a number ν_0 is a zero of the function $P_\nu^{-\mu}(\cos\varphi)$, the number $-\nu_0 - 1$ is also a zero of this function. MO 91

8.782 If ν and μ are both real and $\mu \leqslant 0$, or if ν and μ are integers, the function $P_\nu^\mu(t)$ has no *real* zeros exceeding 1. If ν and μ are both real with $\nu < \mu < 0$, the function $P_\nu^\mu(t)$ has no real zeros exceeding 1 when $\sin\mu\pi \sin(\mu-\nu)\pi > 0$, but does have one such zero when $\sin\mu\pi\sin(\mu-\nu)\pi < 0$. Finally, if $\mu \leqslant \nu$, the function $P_\nu^\mu(t)$ has no zeros exceeding 1 for $E(\mu)$ even but does have one zero for $E(\mu)$ odd.

8.783 If $\nu > -\frac{3}{2}$ and $\nu+\mu+1 > 0$, the function $Q_\nu^\mu(t)$ has no real zeros exceeding 1. MO 91

8.784 The function $P_{-\frac{1}{2}+i\lambda}(z)$ has infinitely many zeros for real λ. All these zeros are *real* and *greater than unity*.

8.785 For n a natural number, the function $P_n(x)$ has exactly n real zeros which lie in the closed interval $-1, +1$.

8.786 The function $Q_n(z)$ has no zeros for which $|\arg(z-1)| < \pi$ if n is a natural number. The function $Q_n(\cos\varphi)$ has exactly $n+1$ zeros in the interval $0 \leqslant \varphi \leqslant \pi$. MO 91

8.787 The following approximate formula can be used to calculate the values of ν for which the equation $P_\nu^{-\mu}(\cos\varphi)=0$ holds for given small values of φ:

$$\nu+\frac{1}{2}=\frac{j_\mu}{2\sin\frac{\varphi}{2}}\left\{1-\frac{\sin^2\frac{\varphi}{2}}{6}\left(1-\frac{4\mu^2-1}{j_\mu^2}\right)+O\left(\sin^4\frac{\varphi}{2}\right)\right\};$$ MO 93

Here, j_μ denotes an arbitrary nonzero root of the equation $J_\mu(z)=0$ (for $\mu\geqslant 0$). If φ is close to π then, instead of this formula, we can use the following formulas:

1. $$\nu\approx\mu+k+\frac{\Gamma(2\mu+k+1)}{\Gamma(\mu)\Gamma(\mu+1)\Gamma(k+1)}\left(\frac{\pi-\varphi}{3}\right)^{2\mu}$$
$[\mu>0;\ k=0,\ 1,\ 2,\ \ldots]$. MO 93

2. $$\nu\approx k+\frac{1}{2\ln\left(\frac{2}{\pi-\varphi}\right)}\qquad [\mu=0,\ k=0,\ 1,\ 2,\ \ldots].$$ MO 93

8.79 Series of associated Legendre functions

8.791

1. $$\frac{1}{z-t}=\sum_{k=0}^{\infty}(2k+1)P_k(t)Q_k(z)\qquad \left[\left|t+\sqrt{t^2-1}\right|<\left|z+\sqrt{z^2-1}\right|\right];$$

Here, t must lie inside an ellipse passing through the point z with foci at the points ± 1.

2. $$\frac{1}{\sqrt{1-2tz+t^2}}\ln\frac{z-t+\sqrt{1-2tz+t^2}}{\sqrt{z^2-1}}=\sum_{k=0}^{\infty}t^kQ_k(z)$$
$[\operatorname{Re} z>1,\ |t|<1]$. MO 78

8.792 $$P_\nu^{-\alpha}(\cos\varphi)P_\nu^{-\beta}(\cos\psi)=$$
$$=\frac{\sin\nu\pi}{\pi}\sum_{k=1}^{\infty}(-1)^k\left[\frac{1}{\nu-k}-\frac{1}{\nu+k+1}\right]P_k^{-\alpha}(\cos\varphi)P_k^{-\beta}(\cos\psi)$$
$[\alpha\geqslant 0,\ \beta\geqslant 0,\ \nu \text{ real},\ -\pi<\varphi\pm\psi<\pi]$. MO 94

8.793 $$P_\nu^{-\mu}(\cos\varphi)=\frac{\sin\nu\pi}{\pi}\sum_{k=0}^{\infty}(-1)^k\left(\frac{1}{\nu-k}-\frac{1}{\nu+k+1}\right)P_k^{-\mu}(\cos\varphi)$$
$[\mu\geqslant 0,\ 0<\varphi<\pi]$. MO 94

Addition theorems

8.794

1. $$P_\nu(\cos\psi_1\cos\psi_2+\sin\psi_1\sin\psi_2\cos\varphi)=$$
$$=P_\nu(\cos\psi_1)P_\nu(\cos\psi_2)+2\sum_{k=1}^{\infty}(-1)^kP_\nu^{-k}(\cos\psi_1)P_\nu^k(\cos\psi_2)\cos k\varphi;$$
$$=P_\nu(\cos\psi_1)P_\nu(\cos\psi_2)+2\sum_{k=1}^{\infty}\frac{\Gamma(\nu-k+1)}{\Gamma(\nu+k+1)}P_\nu^k(\cos\psi_1)P_\nu^k(\cos\psi_2)\cos k\varphi$$
$[0\leqslant\psi_1<\pi,\ 0\leqslant\psi_2<\pi,\ \psi_1+\psi_2<\pi;\ \varphi \text{ real}]$
(cf. 8.814, 8.844 1.). MO 90

2. $$Q_\nu(\cos\psi_1\cos\psi_2+\sin\psi_1\sin\psi_2\cos\varphi)=$$
$$=P_\nu(\cos\psi_1)Q_\nu(\cos\psi_2)+2\sum_{k=1}^{\infty}(-1)^k P_\nu^{-k}(\cos\psi_1)Q_\nu^k(\cos\psi_2)\cos k\varphi$$
$$\left[0<\psi_1<\frac{\pi}{2},\ 0<\psi_2<\pi,\ 0<\psi_1+\psi_2<\pi;\ \varphi \text{ real}\right]$$
(cf. 8.844 3.). MO 90

8.795

1. $$P_\nu\left(z_1z_2-\sqrt{z_1^2-1}\sqrt{z_2^2-1}\cos\varphi\right)=$$
$$=P_\nu(z_1)P_\nu(z_2)+2\sum_{k=1}^{\infty}(-1)^k P_\nu^k(z_1)P_\nu^{-k}(z_2)\cos k\varphi$$
$[\operatorname{Re} z_1>0,\ \operatorname{Re} z_2>0,\ |\arg(z_1-1)|<\pi,\ |\arg(z_2-1)|<\pi]$. MO 91

2. $$Q_\nu\left(x_1x_2-\sqrt{x_1^2-1}\sqrt{x_2^2-1}\cos\varphi\right)=$$
$$=P_\nu(x_1)Q_\nu(x_2)+2\sum_{k=1}^{\infty}(-1)^k P_\nu^{-k}(x_1)Q_\nu^k(x_2)\cos k\varphi\cdot$$
$[1<x_1<x_2,\ \nu\neq-1,\ -2,\ -3,\ \ldots;\ \varphi \text{ real}]$ MO 91

3. $$Q_n\left(x_1x_2+\sqrt{x_1^2+1}\sqrt{x_2^2+1}\operatorname{ch}\alpha\right)=$$
$$=\sum_{k=n+1}^{\infty}\frac{1}{(k-n-1)!\,(k+n)!}Q_n^k(ix_1)Q_n^k(ix_2)e^{-k\alpha}\quad [x_1>0,\ x_2>0;\ \alpha>0].$$
MO 91

8.796

$$P_\nu(-\cos\psi_1\cos\psi_2-\sin\psi_1\sin\psi_2\cos\varphi)=P_\nu(-\cos\psi_1)P_\nu(\cos\psi_2)+$$
$$+2\sum_{k=1}^{\infty}(-1)^k\frac{\Gamma(\nu+k+1)}{\Gamma(\nu-k+1)}P_\nu^{-k}(-\cos\psi_1)P_\nu^{-k}(\cos\psi_2)\cos k\varphi$$
$[0<\psi_2<\psi_1<\pi;\ \varphi \text{ real}]$ (cf. 8.844 2.). MO 91

See also 8.934 3.

8.81 Associated Legendre functions with integral indices

8.810 For *integral* values of ν and μ, the differential equation 8.700 1. (with $|\nu|>|\mu|$) has a simple solution in the real domain, namely:

$$u=P_n^m(x)=(-1)^m(1-x^2)^{\frac{m}{2}}\frac{d^m}{dx^m}P_n(x).$$

The functions $P_n^m(x)$ are called *associated Legendre functions* (or *spherical functions*) *of the first kind*. The number n is called the *degree* and the number m is called the *order* of the function $P_n^m(x)$. The functions

$$\cos m\vartheta P_n^m(\cos\varphi),\qquad \sin m\vartheta P_n^m(\cos\varphi),$$

which depend on the angles φ and ϑ, are also called Legendre functions of the first kind, or, more specifically, *tesseral harmonics* for $m<n$ and *sectoral harmonics* for $m=n$. These last functions are periodic with respect to the angles

φ and ϑ. Their periods are respectively π and 2π. They are single-valued and continuous everywhere on the surface of the unit sphere $x_1^2+x_2^2+x_3^2=1$ (where $x_1=\sin\varphi\cos\vartheta$, $x_2=\sin\varphi\sin\vartheta$, $x_3=\cos\varphi$) and they are solutions of the differential equation

$$\frac{1}{\sin\varphi}\frac{\partial}{\partial\varphi}\left(\sin\varphi\frac{\partial Y}{\partial\varphi}\right)+\frac{1}{\sin^2\varphi}\frac{\partial^2 Y}{\partial\vartheta^2}+n(n+1)Y=0.$$

8.811 The integral equation:

$$\mathrm{P}_n^m(\cos\varphi)=\frac{(-1)^m(n+m)!}{\Gamma\left(m+\frac{1}{2}\right)(n-m)!}\sqrt{\frac{2}{\pi}}\sin^{-m}\varphi\times$$
$$\times\int_0^{\varphi}(\cos t-\cos\varphi)^{m-\frac{1}{2}}\cos\left(n+\frac{1}{2}\right)t\,dt.$$
MO 75

8.812 The series representation:

$$\mathrm{P}_n^m(x)=\frac{(-1)^m(n+m)!}{2^m m!(n-m)!}(1-x^2)^{\frac{m}{2}}\left\{1-\frac{(n-m)(m+n+1)}{1!(m+1)}\frac{1-x}{2}+\right.$$
$$\left.+\frac{(n-m)(n-m+1)(m+n+1)(m+n+2)}{2!(m+1)(m+2)}\left(\frac{1-x}{2}\right)^2-\dots\right\};$$
MO 73

$$=\frac{(-1)^m(2n-1)!!}{(n-m)!}(1-x^2)^{\frac{m}{2}}\left\{x^{n-m}-\frac{(n-m)(n-m-1)}{2(2n-1)}x^{n-m-2}+\right.$$
$$\left.+\frac{(n-m)(n-m-1)(n-m-2)(n-m-3)}{2\cdot 4(2n-1)(2n-3)}x^{n-m-4}-\dots\right\};$$
MO 73

$$=\frac{(-1)^m(2n-1)!!}{(n-m)!}(1-x^2)^{\frac{m}{2}}x^{n-m}F\left(\frac{m-n}{2},\frac{n-m+1}{2};\frac{1}{2}-n;\frac{1}{x^2}\right).$$
MO 73

8.813 Special cases:

1. $\mathrm{P}_1^1(x)=-(1-x^2)^{\frac{1}{2}}=-\sin\varphi.$ MO 73

2. $\mathrm{P}_2^1(x)=-3(1-x^2)^{\frac{1}{2}}x=-\frac{3}{2}\sin 2\varphi.$ MO 73

3. $\mathrm{P}_2^2(x)=3(1-x^2)=\frac{3}{2}(1-\cos 2\varphi).$ MO 73

4. $\mathrm{P}_3^1(x)=-\frac{3}{2}(1-x^2)^{\frac{1}{2}}(5x^2-1)=-\frac{3}{8}(\sin\varphi+5\sin 3\varphi).$ MO 73

5. $\mathrm{P}_3^2(x)=15(1-x^2)x=\frac{15}{4}(\cos\varphi-\cos 3\varphi).$ MO 73

6. $\mathrm{P}_3^3(x)=-15(1-x^2)^{\frac{3}{2}}=-\frac{15}{4}(3\sin\varphi-\sin 3\varphi).$ MO 73

Functional relations

For recursion formulas, see **8.731**.

8.814 $P_n(\cos\varphi_1\cos\varphi_2+\sin\varphi_1\sin\varphi_2\cos\Theta)=$

$$=P_n(\cos\varphi_1)P_n(\cos\varphi_2)+2\sum_{m=1}^{n}\frac{(n-m)!}{(n+m)!}\mathrm{P}_n^m(\cos\varphi_1)\mathrm{P}_n^m(\cos\varphi_2)\cos m\Theta$$

("addition theorem"). MO 74

8.815 If

$$Y_{n_1}(\varphi, \vartheta) = a_0 P_{n_1}(\cos\varphi) + \sum_{m=1}^{n_1} (a_m \cos m\vartheta + b_m \sin m\vartheta) P_{n_1}^m(\cos\varphi),$$

$$Z_{n_2}(\varphi, \vartheta) = \alpha_0 P_{n_2}(\cos\varphi) + \sum_{m=1}^{n_2} (\alpha_m \cos m\vartheta + \beta_m \sin m\vartheta) P_{n_2}^m(\cos\varphi),$$

then

$$\int_0^{2\pi} d\vartheta \int_0^{\pi} \sin\varphi \, d\varphi \, Y_{n_1}(\varphi, \vartheta) Z_{n_2}(\varphi, \vartheta) = 0,$$

$$\int_0^{2\pi} d\vartheta \int_0^{\pi} \sin\varphi \, d\varphi \, Y_n(\varphi, \vartheta) P_n[\cos\varphi\cos\psi + \sin\varphi\sin\psi\cos(\vartheta - \theta)] =$$

$$= \frac{4\pi}{2n+1} Y_n(\psi, \theta).$$ MO 75

8.816 $$(\cos\varphi + i\sin\varphi\cos\vartheta)^n = P_n(\cos\varphi) +$$

$$+ 2\sum_{m=1}^{n} (-1)^m \frac{n!}{(n+m)!} \cos m\vartheta P_n^m(\cos\varphi).$$ MO 75

For integrals of the functions, $P_n^m(x)$, see **7.112 1.**, **7.122 1.**

8.82-8.83 Legendre functions

8.820 The differential equation

$$\frac{d}{dz}\left[(1-z^2)\frac{du}{dz}\right] + \nu(\nu+1)u = 0 \qquad \text{(cf. 8.700 1.)},$$

where the parameter ν can be an arbitrary number, has the following two linearly independent solutions:

1. $$P_\nu(z) = F\left(-\nu, \nu+1; 1; \frac{1-z}{2}\right).$$

2. $$Q_\nu(z) = \frac{\Gamma(\nu+1)\Gamma\left(\frac{1}{2}\right)}{2^{\nu+1}\Gamma\left(\nu+\frac{3}{2}\right)} z^{-\nu-1} F\left(\frac{\nu+2}{2}, \frac{\nu+1}{2}; \frac{2\nu+3}{2}; \frac{1}{z^2}\right).$$

SM 518(137)

The functions $P_\nu(z)$ and $Q_\nu(z)$ are called *Legendre functions of the first* and *second kind* respectively. If ν is not an integer, the function $P_\nu(z)$ has *singularities* at $z = -1$ and $z = \infty$. However, if $\nu = n = 0, 1, 2, \ldots$, the function $P_\nu(z)$ becomes the *Legendre polynomial* $P_n(z)$ (see **8.91**). For $\nu = -n = -1, -2, \ldots$, we have

$$P_{-n-1}(z) = P_n(z).$$

3. If $\nu \neq 0, 1, 2, \ldots$, the function $Q_\nu(z)$ has singularities at the points $z = \pm 1$ and $z = \infty$. These points are branch points of the function. On the other hand, if $\nu = n = 0, 1, 2, \ldots$, the function $Q_n(z)$ is single-valued for $|z| > 1$ and regular for $z = \infty$.

4. In the right half-plane,

$$P_\nu(z)=\left(\frac{1+z}{2}\right)^\nu F\left(-\nu,\ -\nu;\ 1;\ \frac{z-1}{z+1}\right) \quad [\operatorname{Re} z>0].$$

5. The function $P_\nu(z)$ is uniquely determined by equations 8.820 1. and 8.820 4. within a circle of radius 2 with its center at the point $z=1$ in the right half-plane.

For $z=x=\cos\varphi$, a solution of equation 8.820 is the function

6. $\mathrm{P}_\nu(x)=P_\nu(\cos\varphi)=F\left(-\nu,\ \nu+1;\ 1;\ \sin^2\frac{\varphi}{2}\right);$

In general,

7. $P_\nu(z)=P_{-\nu-1}(z)=\mathrm{P}_\nu(x)=\mathrm{P}_{-\nu-1}(x),\quad$ for $z=x$.

8. The function $Q_\nu(z)$ for $|z|>1$ is uniquely determined by equation 8.820 2. everywhere in the z-plane in which a cut is made from the point $z=-\infty$ to the point $z=1$. By means of a hypergeometric series, the function can be continued analytically inside the unit circle. On the cut $(-1\leqslant x\leqslant +1)$ of the real axis, the function $Q_\nu(x)$ is determined by the equation

9. $\mathrm{Q}_\nu(x)=\frac{1}{2}\left[Q_\nu(x+i0)+Q_\nu(x-i0)\right].$ HO 52(53), WH

Integral representations

8.821

1. $$P_\nu(z)=\frac{1}{2\pi i}\int_A^{(1+,\ z+)}\frac{(t^2-1)^\nu}{2^\nu(t-z)^{\nu+1}}\,dt.$$

Here, A is a point on the real axis to the right of the point $t=1$ and to the right of z if z is real. At the point A, we set

$$\arg(t-1)=\arg(t+1)=0 \quad \text{and} \quad [|\arg(t-z)|<\pi].$$ WH

2. $$Q_\nu(z)=\frac{1}{4i\sin\nu\pi}\int_A^{(1-,\ 1+)}\frac{(t^2-1)^\nu}{2^\nu(z-t)^{\nu+1}}\,dt.$$

[ν is not an integer; the point A is at the end of the major axis of an ellipse to the right of $t=1$ drawn in the t-plane with foci at the points ± 1 and with a minor axis sufficiently small that the point z lies outside it. The contour begins at the point A, follows the path $(1-,\ -1+)$ and returns to A; $|\arg z|\leqslant\pi$ and $|\arg(z-t)|\to\arg z$ as $t\to 0$ on the contour; $\arg(t+1)=\arg(t-1)=0$ at the point A; z does not lie on the real axis between -1 and 1.]

For $\nu=n$ an integer,

3. $$Q_n(z)=\frac{1}{2^{n+1}}\int_{-1}^{1}(1-t^2)^n(z-t)^{-n-1}\,dt.$$ SM 517(134), WH

8.822

1. $$P_\nu(z)=\frac{1}{\pi}\int_0^\pi\frac{d\varphi}{\left(z+\sqrt{z^2-1}\cos\varphi\right)^{\nu+1}}=\frac{1}{\pi}\int_0^\pi\left(z+\sqrt{z^2-1}\cos\varphi\right)^\nu d\varphi$$

$$\left[\operatorname{Re} z>0 \text{ and } \arg\left\{z+\sqrt{z^2-1}\cos\varphi\right\}=\arg z \text{ for } \varphi=\frac{\pi}{2}\right].$$

WH

2. $Q_\nu(z)=\int\limits_0^\infty \frac{d\varphi}{(z+\sqrt{z^2-1}\operatorname{ch}\varphi)^{\nu+1}}$, [$\operatorname{Re}\nu>-1$; if ν is not an integer, $\{(z+\sqrt{z^2-1})\operatorname{ch}\varphi\}$ for $\varphi=0$ has its principal value]. WH

8.823 $P_\nu(\cos\theta)=\frac{2}{\pi}\int\limits_0^\theta \frac{\cos\left(\nu+\frac{1}{2}\right)\varphi}{\sqrt{2(\cos\varphi-\cos\theta)}}\,d\varphi.$ WH

8.824 $Q_n(z)=2^n n!\int\limits_z^\infty \ldots \int\limits_z^\infty \frac{(dz)^{n+1}}{(z^2-1)^{n+1}}=2^n\int\limits_z^\infty \frac{(t-z)^n}{(t^2-1)^{n+1}}\,dt;$

$$=\frac{(-1)^n}{(2n-1)!!}\frac{d^n}{dz^n}\left[(z^2-1)^n\int\limits_z^\infty \frac{dt}{(t^2-1)^{n+1}}\right]$$

[$\operatorname{Re} z>1$]. WH, MO 78

8.825 $Q_n(z)=\frac{1}{2}\int\limits_{-1}^{1}\frac{P_n(t)}{z-t}\,dt$ [$|\arg(z-1)|<\pi$]. WH, MO 78

See also 6.622 3., 8.842.

8.826 Fourier series:

1. $P_n(\cos\varphi)=\frac{2^{n+2}}{\pi}\frac{n!}{(2n+1)!!}\left[\sin(n+1)\varphi+\frac{1}{1}\frac{n+1}{2n+3}\sin(n+3)\varphi+\right.$

$\left.+\frac{1\cdot3\,(n+1)(n+2)}{1\cdot2\,(2n+3)(2n+5)}\sin(n+5)\varphi+\ldots\right]$ [$0<\varphi<\pi$]. MO 79

2. $Q_n(\cos\varphi)=2^{n+1}\frac{n!}{(2n+1)!!}\left[\cos(n+1)\varphi+\frac{1}{1}\frac{n+1}{2n+3}\cos(n+3)\varphi+\right.$

$\left.+\frac{1\cdot3}{1\cdot2}\frac{(n+1)(n+2)}{(2n+3)(2n+5)}\cos(n+5)\varphi+\ldots\right]$ [$0<\varphi<\pi$]. MO 79

The expressions for Legendre functions in terms of a hypergeometric function (see 8.820) provide other series representations of these functions.

Special cases and particular values

8.827

1. $Q_0(x)=\frac{1}{2}\ln\frac{1+x}{1-x}=\operatorname{Arth} x.$ JA

2. $Q_1(x)=\frac{x}{2}\ln\frac{1+x}{1-x}-1.$ JA

3. $Q_2(x)=\frac{1}{4}(3x^2-1)\ln\frac{1+x}{1-x}-\frac{3}{2}x.$ JA

4. $Q_3(x)=\frac{1}{4}(5x^3-3x)\ln\frac{1+x}{1-x}-\frac{5}{2}x^2+\frac{2}{3}.$ JA

5. $Q_4(x)=\frac{1}{16}(35x^4-30x^2+3)\ln\frac{1+x}{1-x}-\frac{35}{8}x^3+\frac{55}{24}x.$ JA

6. $Q_5(x)=\frac{1}{16}(63x^5-70x^3+15x)\ln\frac{1+x}{1-x}-\frac{63}{8}x^4+\frac{49}{8}x^2-\frac{8}{15}.$ JA

8.828

1. $\mathrm{P}_\nu(1) = 1.$ MO 79

2. $\mathrm{P}_\nu(0) = -\frac{1}{2}\frac{\sin\nu\pi}{\sqrt{\pi^3}}\Gamma\left(\frac{\nu+1}{2}\right)\Gamma\left(-\frac{\nu}{2}\right).$ MO 79

8.829 $\mathrm{Q}_\nu(0) = \frac{1}{4\sqrt{\pi}}(1-\cos\nu\pi)\Gamma\left(\frac{\nu+1}{2}\right)\Gamma\left(-\frac{\nu}{2}\right).$ MO 79

Functional relationships

8.831

1. $\mathrm{Q}_\nu(x) = \frac{\pi}{2\sin\nu\pi}[\cos\nu\pi P_\nu(x) - P_\nu(-x)] \qquad [\nu \neq 0, \pm 1, \pm 2, \ldots].$ MO 76

2. $\mathrm{Q}_n(x) = \frac{1}{2}P_n(x)\ln\frac{1+x}{1-x} - W_{n-1}(x) \qquad [n = 0, 1, 2, \ldots],$

where

3. $W_{n-1}(x) = \sum_{k=0}^{2E\left(\frac{n-1}{2}\right)} \frac{2(n-2k)-1}{(2k+1)(n-k)} P_{n-2k-1}(x) = \sum_{k=1}^{n} \frac{1}{k} P_{k-1}(x) P_{n-k}(x)$

and

$W_{-1}(x) \equiv 0$ (see also 8.839). SM 516(131), MO 76

4. $\sum_{k=0}^{\infty}(-1)^k\left(\frac{1}{\nu-k} - \frac{1}{\nu+k+1}\right)P_k(\cos\varphi) = \frac{\pi}{\sin\nu\pi}\mathrm{P}_\nu(\cos\varphi)$

$[\nu \text{ not an integer}; \ 0 \leqslant \varphi < \pi].$ MO 77

5. $\sum_{k=0}^{\infty}(-1)^k\left(\frac{1}{\nu-k} - \frac{1}{\nu+k+1}\right)P_k(\cos\varphi)P_k(\cos\psi) =$

$= \frac{\pi}{\sin\nu\pi}\mathrm{P}_\nu(\cos\varphi)\mathrm{P}_\nu(\cos\psi)$

$[\nu \text{ not an integer}, \ -\pi < \varphi+\psi < \pi, \ -\pi < \varphi-\psi < \pi].$ MO 77

See also 8.521 4.

8.832

1. $(z^2-1)\frac{d}{dz}P_\nu(z) = (\nu+1)[P_{\nu+1}(z) - zP_\nu(z)].$ WH

2. $(2\nu+1)zP_\nu(z) = (\nu+1)P_{\nu+1}(z) + \nu P_{\nu-1}(z).$ WH

3. $(z^2-1)\frac{d}{dz}Q_\nu(z) = (\nu+1)[Q_{\nu+1}(z) - zQ_\nu(z)].$ WH

4. $(2\nu+1)zQ_\nu(z) = (\nu+1)Q_{\nu+1}(z) + \nu Q_{\nu-1}(z).$ WH

8.833

1. $P_\nu(-z) = e^{\nu\pi i}P_\nu(z) - \frac{2}{\pi}\sin\nu\pi Q_\nu(z) \qquad [\mathrm{Im}\, z < 0].$ MO 77

2. $P_\nu(-z) = e^{-\nu\pi i}P_\nu(z) - \frac{2}{\pi}\sin\nu\pi Q_\nu(z) \qquad [\mathrm{Im}\, z > 0].$ MO 77

3. $Q_\nu(-z) = -e^{-\nu\pi i} Q_\nu(z) \qquad [\operatorname{Im} z < 0].$ MO 77

4. $Q_\nu(-z) = -e^{\nu\pi i} Q_\nu(z) \qquad [\operatorname{Im} z > 0].$ MO 77

8.834

1. $Q_\nu(x \pm i0) = \mathrm{Q}_\nu(x) \mp \frac{\pi i}{2} \mathrm{P}_\nu(x).$ MO 77

2. $Q_n(z) = \frac{1}{2} P_n(z) \ln \frac{z+1}{z-1} - W_{n-1}(z)$ (see 8.831 3.). MO 77

8.835

1. $Q_\nu(z) - Q_{-\nu-1}(z) = \pi \operatorname{ctg} \nu\pi \, \mathrm{P}_\nu(z) \qquad [\sin \nu\pi \neq 0].$ MO 77

2. $\mathrm{Q}_{-\nu-1}(\cos\varphi) = \mathrm{Q}_\nu(\cos\varphi) - \pi \operatorname{ctg} \nu\pi \, \mathrm{P}_\nu(\cos\varphi) \qquad [\sin \nu\pi \neq 0].$ MO 77

3. $\mathrm{Q}_\nu(-\cos\varphi) = -\cos\nu\pi \, \mathrm{Q}_\nu(\cos\varphi) + \frac{\pi}{2} \sin\nu\pi \, \mathrm{P}_\nu(\cos\varphi).$ MO 77

8.836

1. $Q_n(z) = \frac{1}{2^n n!} \frac{d^n}{dz^n} \left[(z^2-1)^n \ln \frac{z+1}{z-1} \right] - \frac{1}{2} P_n(z) \ln \frac{z+1}{z-1}.$ MO 79

2. $Q_n(x) = \frac{1}{2^n n!} \frac{d^n}{dx^n} \left[(x^2-1)^n \ln \frac{1+x}{1-x} \right] - \frac{1}{2} P_n(x) \ln \frac{1+x}{1-x}.$ MO 79

8.837

1. $\mathrm{P}_\nu(x) = \mathrm{P}_\nu(\cos\varphi) = F\left(-\nu, \nu+1; 1; \sin^2\frac{\varphi}{2}\right)$ (cf. 8.820 6.). MO 76

2. $$P_\nu(z) = \frac{\operatorname{tg}\nu\pi}{2^{\nu+1}\sqrt{\pi}} \frac{\Gamma(\nu+1)}{\Gamma\left(\nu+\frac{3}{2}\right)} z^{-\nu-1} F\left(\frac{\nu}{2}+1, \frac{\nu+1}{2}; \nu+\frac{3}{2}; \frac{1}{z^2}\right) +$$
$$+ \frac{2^\nu}{\sqrt{\pi}} \frac{\Gamma\left(\nu+\frac{1}{2}\right)}{\Gamma(\nu+1)} z^\nu F\left(\frac{1-\nu}{2}, -\frac{\nu}{2}; \frac{1}{2}-\nu; \frac{1}{z^2}\right).$$ MO 78

See also 8.820.

For integrals of Legendre functions, sec 7.1—7.2.

8.838 Inequalities:

1. $|\mathrm{P}_\nu(\cos\varphi) - \mathrm{P}_{\nu+2}(\cos\varphi)| \leqslant 2C_0 \sqrt{\frac{1}{\nu\pi}}.$ MO 78

2. $|\mathrm{Q}_\nu(\cos\varphi) - \mathrm{Q}_{\nu+2}(\cos\varphi)| < C_0 \sqrt{\frac{\pi}{\nu}}.$ MO 78

$[0 \leqslant \varphi \leqslant \pi,\ \nu > 1,\ C_0$ is a number that does not depend on the values of ν or $\varphi]$.

With regard to the zeros of Legendre functions of the second kind, see 8.784, 8.785, and 8.786. For the expansion of Legendre functions in series of associated Legendre functions, see 8.794, 8.795, and 8.796.

8.839 A differential equation leading to the functions $W_{n-1}(x)$ (see 8.831 3.):

$$(1-x^2)\frac{d^2W_{n-1}}{dx^2} - 2x\frac{dW_{n-1}}{dx} + (n+1)nW_{n-1} = 2\frac{dP_n(x)}{dx}.$$ MO 76

8.84 Conical functions

8.840 Let us set

$$\nu = -\frac{1}{2} + i\lambda,$$

where λ is a real parameter, in the defining differential equation 8.700 1. for associated Legendre functions. We then obtain the differential equation of the so-called conical functions. A conical function is a special case of the associated Legendre function. However, the Legendre functions

$$P_{-\frac{1}{2}+i\lambda}(x), \quad Q_{-\frac{1}{2}+i\lambda}(x)$$

have certain peculiarities that make us distinguish them as a special class – the class of conical functions. The most important of these peculiarities is the following

8.841 The functions

$$P_{-\frac{1}{2}+i\lambda}(\cos\varphi) = 1 + \frac{4\lambda^2+1^2}{2^2}\sin^2\frac{\varphi}{2} + \frac{(4\lambda^2+1^2)(4\lambda^2+3^2)}{2^2 4^2}\sin^4\frac{\varphi}{2} + \dots$$

are real for real values of φ. Also,

$$P_{-\frac{1}{2}+i\lambda}(x) \equiv P_{-\frac{1}{2}-i\lambda}(x). \qquad \text{MO 95}$$

8.842 Integral representations:

1. $$P_{-\frac{1}{2}+i\lambda}(\cos\varphi) = \frac{2}{\pi}\int_0^{\varphi} \frac{\operatorname{ch}\lambda u\,du}{\sqrt{2(\cos u - \cos\varphi)}} = \frac{2}{\pi}\operatorname{ch}\lambda\pi \int_0^{\infty} \frac{\cos\lambda u\,du}{\sqrt{2(\cos\varphi + \operatorname{ch} u)}}. \qquad \text{MO 95}$$

2. $$Q_{-\frac{1}{2}\mp\lambda i}(\cos\varphi) = \pm i\operatorname{sh}\lambda\pi \int_0^{\infty} \frac{\cos\lambda u\,du}{\sqrt{2(\operatorname{ch} u + \cos\varphi)}} + \int_0^{\infty} \frac{\operatorname{ch}\lambda u\,du}{\sqrt{2(\operatorname{ch} u - \cos\varphi)}}. \qquad \text{MO 95}$$

Functional relations

(see also 8.73)

8.843 $$P_{-\frac{1}{2}+i\lambda}(-\cos\varphi) = \frac{\operatorname{ch}\lambda\pi}{\pi}\left[Q_{-\frac{1}{2}+i\lambda}(\cos\varphi) + Q_{-\frac{1}{2}-i\lambda}(\cos\varphi)\right]. \qquad \text{MO 95}$$

8.844

1 $$P_{-\frac{1}{2}+i\lambda}(\cos\psi\cos\vartheta + \sin\psi\sin\vartheta\cos\varphi) =$$
$$= P_{-\frac{1}{2}+i\lambda}(\cos\psi)\,P_{-\frac{1}{2}+i\lambda}(\cos\vartheta) +$$
$$+ 2\sum_{k=1}^{\infty} \frac{(-1)^k 2^{2k} P^k_{-\frac{1}{2}+i\lambda}(\cos\psi)\,P^k_{-\frac{1}{2}+i\lambda}(\cos\vartheta)\cos m\varphi}{(4\lambda^2+1^2)(4\lambda^2+3^2)\dots[4\lambda^2+(2k-1)^2]}$$

$$\left[0<\vartheta<\frac{\pi}{2},\ 0<\psi<\pi,\ 0<\psi+\vartheta<\pi\right] \qquad \text{(cf. 8.794 1.).} \qquad \text{MO 95}$$

2. $$P_{-\frac{1}{2}+i\lambda}(-\cos\psi\cos\vartheta - \sin\psi\sin\vartheta\cos\varphi) =$$
$$= P_{-\frac{1}{2}+i\lambda}(\cos\psi)\,P_{-\frac{1}{2}+i\lambda}(-\cos\vartheta) +$$
$$+ 2\sum_{k=1}^{\infty} \frac{(-1)^k 2^{2k} P^k_{-\frac{1}{2}+i\lambda}(\cos\psi)\,P^k_{-\frac{1}{2}+i\lambda}(-\cos\vartheta)\cos m\varphi}{(4\lambda^2+1)(4\lambda^2+3^2)\dots[4\lambda^2+(2k-1)^2]}$$

$$\left[0<\psi<\frac{\pi}{2}<\vartheta,\ \psi+\vartheta<\pi\right] \qquad \text{(cf. 8.796).} \qquad \text{MO 95}$$

3. $Q_{-\frac{1}{2}+i\lambda}(\cos\psi\cos\vartheta+\sin\psi\sin\vartheta\cos\varphi)=P_{-\frac{1}{2}+i\lambda}(\cos\psi)Q_{-\frac{1}{2}+i\lambda}(\cos\vartheta)+$

$$+2\sum_{k=1}^{\infty}\frac{(-1)^k 2^{2k}P^k_{-\frac{1}{2}+i\lambda}(\cos\psi)\,Q^k_{-\frac{1}{2}+i\lambda}(\cos\vartheta)\cos k\varphi}{(4\lambda^2+1)(4\lambda^2+3^2)\ldots[4\lambda^2+(2k-1)^2]}$$

$\left[0<\psi<\frac{\pi}{2}<\vartheta,\ \psi+\vartheta<\pi\right]$ (cf. 8.794 2.). MO 96

Regarding the zeros of conical functions, see 8.784.

8.85 Toroidal functions*

8.850 Solutions of the differential equation

1. $$\frac{d^2u}{d\eta^2}+\frac{\operatorname{ch}\eta}{\operatorname{sh}\eta}\frac{du}{d\eta}-\left(n^2-\frac{1}{4}+\frac{m^2}{\operatorname{sh}^2\eta}\right)u=0,$$

are called toroidal functions. They are equivalent (under a coordinate transformation) to associated Legendre functions. In particular, the functions

$$P^m_{n-\frac{1}{2}}(\operatorname{ch}\eta),\quad Q^m_{n-\frac{1}{2}}(\operatorname{sh}\eta).$$ MO 96

are solutions of equation 8.850 1.

The following formulas, obtained from the formulas obtained earlier for associated Legendre functions, are valid for toroidal functions:

8.851 Integral representations:

1. $P^m_{n-\frac{1}{2}}(\operatorname{ch}\eta)=$

$$=\frac{\Gamma\left(n+m+\frac{1}{2}\right)}{\Gamma\left(n-m+\frac{1}{2}\right)}\frac{(\operatorname{sh}\eta)^m}{2^m\sqrt{\pi}\,\Gamma\left(m+\frac{1}{2}\right)}\int_0^{\pi}\frac{\sin^{2m}\varphi\,d\varphi}{(\operatorname{ch}\eta+\operatorname{sh}\eta\cos\varphi)^{n+m+\frac{1}{2}}}=$$

$$=\frac{(-1)^m}{2\pi}\frac{\Gamma\left(n+\frac{1}{2}\right)}{\Gamma\left(n-m+\frac{1}{2}\right)}\int_0^{2\pi}\frac{\cos m\varphi\,d\varphi}{(\operatorname{ch}\eta+\operatorname{sh}\eta\cos\varphi)^{n+\frac{1}{2}}}$$ MO 96

2. $$Q^m_{n-\frac{1}{2}}(\operatorname{ch}\eta)=(-1)^m\frac{\Gamma\left(n+\frac{1}{2}\right)}{\Gamma\left(n-m+\frac{1}{2}\right)}\int_0^{\infty}\frac{\operatorname{ch}mt\,dt}{(\operatorname{ch}\eta+\operatorname{sh}\eta\operatorname{ch}t)^{n+\frac{1}{2}}};\qquad [n\geqslant m]$$

$$=(-1)^m\frac{\Gamma\left(n+m+\frac{1}{2}\right)}{\Gamma\left(n+\frac{1}{2}\right)}\int_0^{\ln\operatorname{cth}\frac{\eta}{2}}(\operatorname{ch}\eta-\operatorname{sh}\eta\operatorname{ch}t)^{n-\frac{1}{2}}\operatorname{ch}mt\,dt.$$ MO 96

8.852 Functional relations:

1. $$Q^m_{n-\frac{1}{2}}(\operatorname{ch}\eta)=(-1)^m\frac{2^m\Gamma\left(n+m+\frac{1}{2}\right)\sqrt{\pi}}{\Gamma(n+1)}\operatorname{sh}^m\eta\, e^{-\left(n+m+\frac{1}{2}\right)\eta}\times$$

$$\times F\left(m+\frac{1}{2},\ n+m+\frac{1}{2};\ n+1;\ e^{-2\eta}\right).$$ MO 96

*Sometimes called *torus functions*.

2. $$P_{n-\frac{1}{2}}^{-m}(\operatorname{ch}\eta) = \frac{2^{-m}}{\Gamma(m+1)}(1-e^{-2\eta})^m e^{-\left(n+\frac{1}{2}\right)\eta} \times$$
$$\times F\left(m+\frac{1}{2},\ n+m+\frac{1}{2};\ 2m+1;\ 1-e^{-2\eta}\right).$$ MO 96

8.853 An asymptotic representation $P_{n-\frac{1}{2}}(\operatorname{ch}\eta)$ for large values of n:

$$P_{n-\frac{1}{2}}(\operatorname{ch}\eta) = \frac{\Gamma(n)\, e^{\left(n-\frac{1}{2}\right)\eta}}{\sqrt{\pi}\,\Gamma\left(n+\frac{1}{2}\right)} \times$$
$$\times \left[\frac{2\Gamma^2\left(n+\frac{1}{2}\right)}{\pi n!\,\Gamma(n)} \ln(4e^{\eta})\, e^{-2n\eta} F\left(\frac{1}{2},\ n+\frac{1}{2};\ n+1;\ e^{-2\eta}\right) + A + B\right],$$

where

$$A = 1 + \frac{1}{2^2}\,\frac{1\cdot(2n-1)}{1\cdot(n-1)}\, e^{-2\eta} + \frac{1}{2^4}\,\frac{1\cdot3\cdot(2n-1)(2n-3)}{1\cdot2\cdot(n-1)(n-2)}\, e^{-4\eta} + \ldots$$
$$\ldots + \frac{1}{2^{2n-2}}\left(\frac{(2n-1)!!}{(n-1)!}\right)^2 e^{-2(n-1)\eta}.$$

$$B = \frac{\Gamma\left(n+\frac{1}{2}\right)}{\sqrt{\pi^3}\,\Gamma(n)} \sum_{k=1}^{\infty} \frac{\Gamma\left(k+\frac{1}{2}\right)\Gamma\left(n+k+\frac{1}{2}\right)}{\Gamma(n+k+1)\,\Gamma(k+1)} \times$$
$$\times \left(u_{n+k} + u_k - v_{n+k-\frac{1}{2}} - v_{k-\frac{1}{2}}\right) e^{-2(n+k)\eta};$$

Here,

$$u_r = \sum_{s=1}^{r} \frac{1}{s},\quad v_{r-\frac{1}{2}} = \sum_{s=1}^{r} \frac{2}{2s-1} \qquad [r \text{ — a natural number}].$$ MO 97

8.9 Orthogonal Polynomials

8.90 Introduction

8.901 Suppose that $w(x)$ is a nonnegative real function of a real variable x. Let (a, b) be a fixed interval on the x-axis. Let us suppose further that, for $n = 0, 1, 2, \ldots$, the integral

$$\int_a^b x^n w(x)\, dx$$

exists and that the integral

$$\int_a^b w(x)\, dx$$

is positive. In this case, there exists a sequence of polynomials $p_0(x), p_1(x), \ldots, p_n(x), \ldots$, that is uniquely determined by the following conditions:

1. $p_n(x)$ is a polynomial of degree n and the coefficient of x^n in this polynomial is positive.

2. The polynomials $p_0(x),\ p_1(x),\ \ldots$ are orthonormal; that is,

$$\int_a^b p_n(x)\, p_m(x)\, w(x)\, dx = \begin{cases} 0 & \text{for } n \neq m, \\ 1 & \text{for } n = m. \end{cases}$$

We say that the polynomials $p_n(x)$ constitute *a system of orthogonal polynomials on the interval* (a, b) *with the weight function* $w(x)$.

8.902 If q_n is the coefficient of x^n in the polynomial $p_n(x)$, then

1. $$\sum_{k=0}^{n} p_k(x)\, p_k(y) = \frac{q_n}{q_{n+1}} \frac{p_{n+1}(x)\, p_n(y) - p_n(x)\, p_{n+1}(y)}{x-y}$$

(Darboux-Christoffel formula) EH II 159(10)

2. $$\sum_{k=0}^{n} [p_k(x)]^2 = \frac{q_n}{q_{n+1}} [p_n(x)\, p'_{n+1}(x) - p'_n(x)\, p_{n+1}(x)].$$ EH II 159(11)

8.903 Between any three consecutive orthogonal polynomials, there is a dependence

$$p_n(x) = (A_n x + B_n)\, p_{n-1}(x) - C_n p_{n-2}(x) \quad [n = 2, 3, 4, \ldots].$$

In this formula, A_n, B_n, and C_n are constants and

$$A_n = \frac{q_n}{q_{n-1}}, \qquad C_n = \frac{q_n q_{n-2}}{q_{n-1}^2}.$$ MO 102

8.904 Examples of normalized systems of orthogonal polynomials:

Notation and name	Interval	Weight
$\left(n+\frac{1}{2}\right)^{\frac{1}{2}} P_n(x)$, see 8.91	$(-1, +1)$	1
$2^\lambda \Gamma(\lambda) \left[\frac{(n+\lambda)\, n!}{2\pi\Gamma(2\lambda+n)}\right]^{\frac{1}{2}} C_n^\lambda(x)$, see 8.93	$(-1, +1)$	$(1-x^2)^{\lambda-\frac{1}{2}}$
$\sqrt{\frac{\varepsilon_n}{\pi}} T_n(x)$, $\varepsilon_0 = 1, \varepsilon_n = 2$ for $n = 1, 2, 3, \ldots$, see 8.94	$(-1, +1)$	$(1-x^2)^{-\frac{1}{2}}$
$2^{-\frac{n}{2}} \pi^{-\frac{1}{4}} (n!)^{-\frac{1}{2}} H_n(x)$, see 8.95	$(-\infty, \infty)$	e^{-x^2}
$\left[\frac{\Gamma(n+1)\Gamma(\alpha+\beta+1+n)(\alpha+\beta+1+2n)}{\Gamma(\alpha+1+n)\Gamma(\beta+1+n)\, 2^{\alpha+\beta+1}}\right]^{\frac{1}{2}} P_n^{(\alpha, \beta)}(x)$, see 8.96	$(-1, +1)$	$(1-x)^\alpha (1+x)^\beta$
$\left[\frac{\Gamma(n+1)}{\Gamma(\alpha+n+1)}\right]^{\frac{1}{2}} L_n^\alpha(x)$, see 8.97	$(0, \infty)$	$x^\alpha e^{-x}$

Cf. 7.221 1., 7.313, 7.343, 7.374 1., 7.391 1., 7.414 3.

8.91 Legendre polynomials

8.910 Definition. The Legendre polynomials $P_n(z)$ are polynomials satisfying equation 8.700 1. with $\mu=0$ and $\nu=n$: that is, they satisfy the equation

1. $$(1-z^2)\frac{d^2u}{dz^2}-2z\frac{du}{dz}+n(n+1)u=0.$$

This equation has a polynomial solution if, and only if, n is an integer. Thus, Legendre polynomials constitute a special type of associated Legendre function.
Legendre polynomials of degree n are of the form

2. $$P_n(z)=\frac{1}{2^n n!}\frac{d^n}{dz^n}(z^2-1)^n.$$

8.911 Legendre polynomials written in expanded form:

1. $$P_n(z)=\frac{1}{2^n}\sum_{k=0}^{E\left(\frac{n}{2}\right)}\frac{(-1)^k(2n-2k)!}{k!\,(n-k)!\,(n-2k)!}z^{n-2k}=$$
$$=\frac{(2n)!}{2n\,(n!)^2}\left(z^n-\frac{n(n-1)}{2(2n-1)}z^{n-2}+\frac{n(n-1)(n-2)(n-3)}{2\cdot 4(2n-1)(2n-3)}z^{n-4}-\dots\right);$$
$$=\frac{(2n-1)!!}{n!}z^nF\left(-\frac{n}{2},\frac{1-n}{2};\frac{1}{2}-n;\frac{1}{z^2}\right).$$ HO 13, AD(9001), MO 69

2. $$P_{2n}(z)=(-1)^n\frac{(2n-1)!!}{2^n n!}\left(1-\frac{2n(2n+1)}{2!}z^2+\right.$$
$$\left.+\frac{2n(2n-2)(2n+1)(2n+3)}{4!}z^4-\dots\right);$$
$$=(-1)^n\frac{(2n-1)!!}{2^n n!}F\left(-n,n+\frac{1}{2};\frac{1}{2};z^2\right).$$ AD(9002), MO 69

3. $$P_{2n+1}(z)=(-1)^n\frac{(2n+1)!!}{2^n n!}\left(z-\frac{2n(2n+3)}{3!}z^3+\right.$$
$$\left.+\frac{2n(2n-2)(2n+3)(2n+5)}{5!}z^5-\dots\right);$$
$$=(-1)^n\frac{(2n+1)!!}{2^n n!}zF\left(-n,n+\frac{3}{2};\frac{3}{2};z^2\right).$$ AD(9002), MO 69

4. $$P_n(\cos\varphi)=\frac{(2n-1)!!}{2^n n!}\left(\cos n\varphi+\frac{1}{1}\frac{n}{2n-1}\cos(n-2)\varphi+\right.$$
$$+\frac{1\cdot 3}{1\cdot 2}\frac{n(n-1)}{(2n-1)(2n-3)}\cos(n-4)\varphi+$$
$$\left.+\frac{1\cdot 3\cdot 5}{1\cdot 2\cdot 3}\frac{n(n-1)(n-2)}{(2n-1)(2n-3)(2n-5)}\cos(n-6)\varphi-\dots\right).$$ WH

5. $$P_{2n}(\cos\varphi)=(-1)^n\frac{(2n-1)!!}{2^n n!}\left\{\sin^{2n}\varphi-\frac{(2n)^2}{2!}\sin^{2n-2}\varphi\cos^2\varphi+\dots\right.$$
$$\left.\dots+(-1)^n\frac{2^n n!}{(2n-1)!!}\cos^{2n}\varphi\right\}.$$ AD (9011)

6. $$P_{2n+1}(\cos\varphi)=(-1)^n\frac{(2n+1)!!}{2^n n!}\cos\varphi\left\{\sin^{2n}\varphi-\frac{(2n)^2}{3!}\sin^{2n-2}\varphi\cos^2\varphi+\dots\right.$$
$$\left.\dots+(-1)^n\frac{2^n n!}{(2n+1)!!}\cos^{2n}\varphi\right\}.$$ AD(9012)

7. $$P_n(z)=\sum_{k=0}^{n}\frac{(-1)^k(n+k)!}{(n-k)!\,(k!)^2\,2^{k+1}}\left[(1-z)^k+(-1)^n(1+z)^k\right].$$ WH

8.912 Special cases:

1. $P_0(x) = 1.$ JA
2. $P_1(x) = x = \cos\varphi.$ JA
3. $P_2(x) = \frac{1}{2}(3x^2 - 1) = \frac{1}{4}(3\cos 2\varphi + 1).$ JA
4. $P_3(x) = \frac{1}{2}(5x^3 - 3x) = \frac{1}{8}(5\cos 3\varphi + 3\cos\varphi).$ JA
5. $P_4(x) = \frac{1}{8}(35x^4 - 30x^2 + 3) = \frac{1}{64}(35\cos 4\varphi + 20\cos 2\varphi + 9).$ JA
6. $P_5(x) = \frac{1}{8}(63x^5 - 70x^3 + 15x) = \frac{1}{128}(63\cos 5\varphi + 35\cos 3\varphi + 30\cos\varphi).$ JA

8.913 Integral representation:

$$P_n(\cos\varphi) = \frac{2}{\pi}\int_{\varphi}^{\pi} \frac{\sin\left(n+\frac{1}{2}\right)t}{\sqrt{2(\cos\varphi - \cos t)}}\,dt.$$ WH

See also **3.611** 3., **3.661** 3., 4.

Functional relations

8.914 Recurrence formulas:

1. $(n+1)P_{n+1}(z) - (2n+1)zP_n(z) + nP_{n-1}(z) = 0$ WH
2. $(z^2 - 1)\frac{dP_n}{dz} = n[zP_n(z) - P_{n-1}(z)] = \frac{n(n+1)}{2n+1}[P_{n+1}(z) - P_{n-1}(z)].$ WH

8.915

1. $\sum_{k=0}^{n}(2k+1)P_k(x)P_k(y) = (n+1)\frac{P_n(x)P_{n+1}(y) - P_n(y)P_{n+1}(x)}{y-x}.$ MO 70
2. $\sum_{k=0}(2n - 4k - 1)P_{n-2k-1}(z) = P'_n(z)$ (summation theorem) MO 70

[The summation is cut off at the first term with a negative subscript].

3. $\sum_{k=0}(2n - 4k - 3)P_{n-2k-2}(z) = zP'_n(z) - nP_n(z)$ SM 491(42), WH

[The summation is cut off at the first term with a negative subscript].

4. $\sum_{k=1}^{E\left(\frac{n}{2}\right)}(2n - 4k + 1)[k(2n - 2k + 1) - 2]P_{n-2k}(z) =$

$$= z^2P''_n(z) - n(n-1)P_n(z).$$ WH

5. $\sum_{k=0}^{m}\frac{a_{m-k}a_k a_{n-k}}{a_{n+m-k}}\left(\frac{2n+2m-4k+1}{2n+2m-2k+1}\right)P_{n+m-2k}(z) = P_n(z)P_m(z)$

$$\left[a_k = \frac{(2k-1)!!}{k!}, \quad m \leqslant n\right].$$ AD (9036)

8.916

1. $P_n(\cos\varphi) = \frac{(2n-1)!!}{2^n n!} e^{\mp in\varphi} F\left(\frac{1}{2}, -n; \frac{1}{2}-n; e^{\pm 2i\varphi}\right).$ MO 69

2. $P_n(\cos\varphi) = F\left(n+1, -n; 1; \sin^2\frac{\varphi}{2}\right).$ MO 69

3. $P_n(\cos\varphi) = (-1)^n F\left(n+1, -n; 1; \cos^2\frac{\varphi}{2}\right).$ WH

4. $P_n(\cos\varphi) = \cos^n\varphi F\left(-\frac{1}{2}n, \frac{1}{2}-\frac{1}{2}n; 1; -\operatorname{tg}^2\varphi\right).$ HO 23

5. $P_n(\cos\varphi) = \cos^{2n}\frac{\varphi}{2} F\left(-n, -n; 1; -\operatorname{tg}^2\frac{\varphi}{2}\right).$ HO 23, 29, WH

See also 8.911 1., 8.911 2., 8.911 3. For a connection with other functions, see 8.936 3., 8.836, 8.962 2. For integrals of Legendre polynomials, see 7.22.—7.25. For the zeros of Legendre polynomials, see 8.785.

8.917 Inequalities:

1. For $x > 1$ $P_0(x) < P_1(x) < P_2(x) < \ldots < P_n(x) < \ldots$ MO 71

2. For $x > -1$ $P_0(x) + P_1(x) + \ldots + P_n(x) > 0.$ MO 71

3. $[P_n(\cos\varphi)]^2 > \frac{\sin(2n+1)\varphi}{(2n+1)\sin\varphi}$ $[0 < \varphi < \pi].$ MO 71

4. $\sqrt{n\sin\varphi}\,|P_n(\cos\varphi)| \leqslant 1.$ MO 71

5. $|P_n(\cos\varphi)| \leqslant 1.$ WH

8.92 Series of Legendre polynomials

8.921 The generating function:

$$\frac{1}{\sqrt{1-2tz+t^2}} = \sum_{k=0}^{\infty} t^k P_k(z) \qquad [|t| < \min|z \pm \sqrt{z^2-1}|];$$

SM 489(31), WH

$$= \sum_{k=0}^{\infty} \frac{1}{t^{k+1}} P_k(z) \qquad [|t| > \max|z \pm \sqrt{z^2-1}|].$$ MO 70

8.922

1. $$z^{2n} = \frac{1}{2n+1} P_0(z) + \sum_{k=1}^{\infty} (4k+1) \frac{2n(2n-2)\ldots(2n-2k+2)}{(2n+1)(2n+3)\ldots(2n+2k+1)} P_{2k}(z).$$

MO 72

2. $$z^{2n+1} = \frac{3}{2n+3} P_1(z) + \sum_{k=1}^{\infty} (4k+3) \frac{2n(2n-2)\ldots(2n-2k+2)}{(2n+3)(2n+5)\ldots(2n+2k+3)} P_{2k+1}(z).$$

MO 72

3. $$\frac{1}{\sqrt{1-x^2}} = \frac{\pi}{2} \sum_{k=0}^{\infty} (4k+1) \left\{\frac{(2k-1)!!}{2^k k!}\right\}^2 P_{2k}(x) \qquad [|x| < 1, \ (-1)!! \equiv 1].$$

MO 72, LA 385(15)

4. $$\frac{x}{\sqrt{1-x^2}}=\frac{\pi}{2}\sum_{k=0}^{\infty}(4k+3)\frac{(2k-1)!!\,(2k+1)!!}{2^{2k+1}k!\,(k+1)!}P_{2k+1}(x)$$

$$[|x|<1,\ (-1)!!\equiv 1].$$ LA 385(17)

5. $$\sqrt{1-x^2}=\frac{\pi}{2}\left\{\frac{1}{2}-\sum_{k=1}^{\infty}(4k+1)\frac{(2k-3)!!\,(2k-1)!!}{2^{2k+1}k!\,(k+1)!}P_{2k}(x)\right\}$$

$$[|x|<1,\ (-1)!!\equiv 1].$$ LA 385(18)

8.923 $$\arcsin x=\frac{\pi}{2}\sum_{k=0}^{\infty}\left\{\frac{(2k-1)!!}{2^k k!}\right\}^2[P_{2k+1}(x)-P_{2k-1}(x)]$$

$$[|x|<1,\ (-1)!!\equiv 1].$$ WH

8.924

1. $$-\frac{1+\cos n\pi}{2(n^2-1)}P_0(\cos\theta)-$$

$$-\frac{1+\cos n\pi}{2}\sum_{k=0}^{\infty}\frac{(4k+5)\,n^2(n^2-2^2)\dots[n^2-(2k)^2]}{(n^2-1^2)(n^2-3^2)\dots[n^2-(2k+3)^2]}P_{2k+2}(\cos\theta)-$$

$$-\frac{3(1-\cos n\pi)}{2(n^2-2^2)}P_1(\cos\theta)-$$

$$-\frac{1-\cos n\pi}{2}\sum_{k=1}^{\infty}\frac{(4k+3)(n^2-1^2)\dots[n^2-(2k-1)^2]}{(n^2-2^2)(n^2-4^2)\dots[n^2-(2k+2)^2]}P_{2k+1}(\cos\theta)=\cos n\theta.$$

AD (9062.1)

2. $$\frac{-\sin n\pi}{2(n^2-1)}P_0(\cos\theta)-$$

$$-\frac{\sin n\pi}{2}\sum_{k=0}^{\infty}\frac{(4k+5)\,n^2(n^2-2^2)\dots[n^2-(2k)^2]}{(n^2-1^2)(n^2-3^2)\dots[n^2-(2k+3)^2]}P_{2k+2}(\cos\theta)+$$

$$+\frac{3\sin n\pi}{2(n^2-2^2)}P_1(\cos\theta)+$$

$$+\frac{\sin n\pi}{2}\sum_{k=1}^{\infty}\frac{(4k+3)(n^2-1^2)(n^2-3^2)\dots[n^2-(2k-1)^2]}{(n^2-2^2)(n^2-4^2)\dots[n^2-(2k+2)^2]}P_{2k+1}(\cos\theta)=\sin n\theta.$$

AD (9060.2)

3. $$\frac{2^{n-1}n!}{(2n-1)!!}P_n(\cos\theta)+$$

$$-n\sum_{k=1}^{E[n/2]}(2n-4k+1)\frac{2^{n-2k-1}(n-k-1)!\,(2k-3)!!}{(2n-2k+1)!!\,k!}P_{n-2k}(\cos\theta)=\cos n\theta.$$

AD (9061.1)

4. $$\frac{(2n-1)!!\,P_{n-1}(\cos\theta)}{2^{n-1}(n-1)!}-$$

$$-\frac{n}{2^{n+1}}\sum_{k=0}^{\infty}\frac{(2n+2k-1)!!\,(2k-1)!!\,(2n+4k+3)}{2^{2k}(n+k+1)!\,(k+1)!}P_{n+2k+1}(\cos\theta)=\frac{4\sin n\theta}{\pi}.$$

AD (9061.2)

8.925

1. $$\sum_{k=1}^{\infty} \frac{4k-1}{2^{2k}(2k-1)^2}\left[\frac{(2k-1)!!}{k!}\right]^2 P_{2k-1}(\cos\theta) = 1 - \frac{2\theta}{\pi}.$$

2. $$\sum_{k=1}^{\infty} \frac{4k+1}{2^{2k+1}(2k-1)(k+1)}\left[\frac{(2k-1)!!}{k!}\right]^2 P_{2k}(\cos\theta) = \frac{1}{2} - \frac{2\sin\theta}{\pi}.$$

AD (9062.2)

3. $$\sum_{k=1}^{\infty} \frac{k(4k-1)}{2^{2k-1}(2k-1)}\left[\frac{(2k-1)!!}{k!}\right]^2 P_{2k-1}(\cos\theta) = \frac{2\operatorname{ctg}\theta}{\pi}.$$ AD (9062.3)

4. $$\sum_{k=1}^{\infty} \frac{4k+1}{2^{2k}}\left[\frac{(2k-1)!!}{k!}\right]^2 P_{2k}(\cos\theta) = \frac{2}{\pi\sin\theta} - 1.$$ AD (9062.4)

8.926

1. $$\sum_{n=1}^{\infty} \frac{1}{n} P_n(\cos\theta) = \ln\frac{2\operatorname{tg}\frac{\pi-\theta}{4}}{\sin\theta} = -\ln\sin\frac{\theta}{2} - \ln\left(1+\sin\frac{\theta}{2}\right).$$

AD (9063.2)

2. $$\sum_{n=1}^{\infty} \frac{1}{n+1} P_n(\cos\theta) = \ln\frac{1+\sin\frac{\theta}{2}}{\sin\frac{\theta}{2}} - 1.$$ AD (9063.1)

8.927 $$\sum_{k=0}^{\infty} \cos\left(k+\frac{1}{2}\right)\beta P_k(\cos\varphi) = \frac{1}{\sqrt{2(\cos\beta-\cos\varphi)}} \qquad [0 \leqslant \beta < \varphi < \pi];$$
$$= 0 \quad [0 < \varphi < \beta < \pi].$$ MO 72

8.928

1. $$\sum_{n=1}^{\infty} \frac{(-1)^n (4n+1)[(2n-1)!!]^3}{2^{3n}(n!)^3} P_{2n}(\cos\theta) = \frac{4\boldsymbol{K}(\sin\theta)}{\pi^2} - 1.$$ AD (9064.1)

2. $$\sum_{n=1}^{\infty} (-1)^{n+1} \frac{(4n+1)[(2n-1)!!]^3}{(2n-1)(2n+2)2^{3n}(n!)^3} P_{2n}(\cos\theta) = \frac{4\boldsymbol{E}(\sin\theta)}{\pi^2} - \frac{1}{2}.$$ AD (9064.2)

For series of products of Bessel functions and Legendre polynomials, see 8.511 4., 8.531 3., 8.533 1., 8.543 2., and 8.534.

8.93 Gegenbauer polynomials $C_n^\lambda(t)$

8.930 Definition. The polynomials $C_n^\lambda(t)$ of degree n are the coefficients of α^n in the power-series expansion of the function

$$(1 - 2t\alpha + \alpha^2)^{-\lambda} = \sum_{n=0}^{\infty} C_n^\lambda(t)\,\alpha^n.$$ WH

Thus, the polynomials $C_n^\lambda(t)$ are a *generalization of the Legendre polynomials.*

8.931 Integral representation:

$$C_n^\lambda(t)=\frac{1}{\sqrt{\pi}}\,\frac{\Gamma(2\lambda+n)}{n!\,\Gamma(2\lambda)}\,\frac{\Gamma\left(\frac{2\lambda+1}{2}\right)}{\Gamma(\lambda)}\int_0^\pi\left(t+\sqrt{t^2-1}\cos\varphi\right)^n\sin^{2\lambda-1}\varphi\,d\varphi. \qquad \text{MO 99}$$

See also 3.252 11., 3.663 2., 3.664 4.

Functional relations

8.932 Expressions in terms of hypergeometric functions:

1. $$C_n^\lambda(t)=\frac{\Gamma(2\lambda+n)}{\Gamma(n+1)\,\Gamma(2\lambda)}F\left(2\lambda+n,\ -n;\ \lambda+\frac{1}{2};\ \frac{1-t}{2}\right)^{*}; \qquad \text{MO 97}$$
$$=\frac{2^n\Gamma(\lambda+n)}{n!\,\Gamma(\lambda)}t^nF\left(-\frac{n}{2},\ \frac{1-n}{2};\ 1-\lambda-n;\ \frac{1}{t^2}\right). \qquad \text{MO 99}$$

2. $$C_{2n}^\lambda(t)=\frac{(-1)^n}{(\lambda+n)\,\mathrm{B}(\lambda,\ n+1)}F\left(-n,\ n+\lambda;\ \frac{1}{2};\ t^2\right). \qquad \text{MO 99}$$

3. $$C_{2n+1}^\lambda(t)=\frac{(-1)^n\,2t}{\mathrm{B}(\lambda,\ n+1)}F\left(-n,\ n+\lambda+1;\ \frac{3}{2};\ t^2\right). \qquad \text{MO 99}$$

8.933 Recursion formulas:

1. $(n+2)\,C_{n+2}^\lambda(t)=2(\lambda+n+1)\,tC_{n+1}^\lambda(t)-(2\lambda+n)\,C_n^\lambda(t).$ MO 98
2. $nC_n^\lambda(t)=2\lambda\left[tC_{n-1}^{\lambda+1}(t)-C_{n-2}^{\lambda+1}(t)\right].$ WH
3. $(2\lambda+n)\,C_n^\lambda(t)=2\lambda\left[C_n^{\lambda+1}(t)-tC_{n-1}^{\lambda+1}(t)\right].$ WH
4. $nC_n^\lambda(t)=(2\lambda+n-1)\,tC_{n-1}(t)-2\lambda(1-t^2)\,C_{n-2}^{\lambda-1}(t).$ WH

8.934

1. $$C_n^\lambda(t)=\frac{(-1)^n}{2^n}\,\frac{\Gamma(2\lambda+n)\,\Gamma\left(\frac{2\lambda+1}{2}\right)}{\Gamma(2\lambda)\,\Gamma\left(\frac{2\lambda+1}{2}+n\right)}\,\frac{(1-t^2)^{\frac{1}{2}-\lambda}}{n!}\,\frac{d^n}{dt^n}\left[(1-t^2)^{\lambda+n-\frac{1}{2}}\right]. \qquad \text{WH}$$

2. $$C_n^\lambda(\cos\varphi)=\sum_{\substack{k,\ l=0\\ k+l=n}}^{n}\frac{\Gamma(\lambda+k)\,\Gamma(\lambda+l)}{k!\,l!\,[\Gamma(\lambda)]^2}\cos(k-l)\,\varphi. \qquad \text{MO 99}$$

3. $$C_n^\lambda(\cos\psi\cos\vartheta+\sin\psi\sin\vartheta\cos\varphi)=$$
$$=\frac{\Gamma(2\lambda-1)}{[\Gamma(\lambda)]^2}\sum_{k=0}^{n}\frac{2^{2k}(n-k)!\,[\Gamma(\lambda+k)]^2}{\Gamma(2\lambda+n+k)}(2\lambda+2k-1)\sin^k\psi\sin^k\vartheta\times$$
$$\times C_{n-k}^{\lambda+k}(\cos\psi)\,C_{n-k}^{\lambda+k}(\cos\vartheta)\,C_k^{\lambda-\frac{1}{2}}(\cos\varphi)$$
$$\left[\psi,\ \vartheta,\ \varphi\ \text{real};\ \lambda\neq\frac{1}{2}\right]\ [\text{"summation theorem"}]$$
(see also 8.794 – 8.796). WH

4. $$\lim_{\lambda\to 0}\Gamma(\lambda)\,C_n^\lambda(\cos\varphi)=\frac{2\cos n\varphi}{n}. \qquad \text{MO 98}$$

For orthogonality, see 8.904, 7.313.

---

*This equation is used for defining the generalized functions $C_n^\lambda(t)$, where the subscript n can be an arbitrary number.

8.935 Derivatives:

1. $$\frac{d^k}{dt^k} C_n^\lambda(t) = 2^k \frac{\Gamma(\lambda+k)}{\Gamma(\lambda)} C_{n-k}^{\lambda+k}(t).$$ MO 99

In particular,

2. $$\frac{dC_n^\lambda(t)}{dt} = 2\lambda C_{n-1}^{\lambda+1}(t).$$ WH

For integrals of the polynomials $C_n^\lambda(x)$ see **7.31 – 7.33.**

8.936 Connections with other functions:

1. $$C_n^\lambda(t) = \frac{\Gamma(2\lambda+n)\Gamma\left(\lambda+\frac{1}{2}\right)}{\Gamma(2\lambda)\Gamma(n+1)} \left\{\frac{1}{4}(t^2-1)\right\}^{\frac{1}{4}-\frac{\lambda}{2}} P_{\lambda+n-\frac{1}{2}}^{\frac{1}{2}-\lambda}(t).$$ MO 98

2. $$C_{n-m}^{m+\frac{1}{2}}(t) = \frac{1}{(2m-1)!!} \frac{d^m P_n(t)}{dt^m} = (-1)^m \frac{(1-t^2)^{-\frac{m}{2}} m!\, 2^m}{(2m)!} P_n^m(t)$$

$[m+1$ a natural number$]$. MO 98, WH

3. $$C_n^{\frac{1}{2}}(t) = P_n(t).$$

4. $$J_{\lambda-\frac{1}{2}}(r\sin\vartheta\sin\alpha)(r\sin\vartheta\sin\alpha)^{-\lambda+\frac{1}{2}} e^{-ir\cos\vartheta\cos\alpha} =$$

$$= \sqrt{2}\frac{\Gamma(\lambda)}{\Gamma\left(\lambda+\frac{1}{2}\right)} \sum_{k=0}^{\infty} (\lambda+k)\, i^{-k} \frac{J_{\lambda+k}(r)\, C_k^\lambda(\cos\vartheta)\, C_k^\lambda(\cos\alpha)}{r^\lambda C_k^\lambda(1)}.$$ MO 99

5. $$\lim_{\lambda\to\infty} \lambda^{-\frac{n}{2}} C_n^{\frac{\lambda}{2}}\left(t\sqrt{\frac{2}{\lambda}}\right) = \frac{2^{-\frac{n}{2}}}{n!} H_n(t).$$ MO 99a

See also **8.932.**

8.937 Special cases and particular values:

1. $$C_n^1(\cos\varphi) = \frac{\sin(n+1)\varphi}{\sin\varphi}.$$ MO 99

2. $$C_0^0(\cos\varphi) = 1.$$ MO 98

3 $$C_0^\lambda(t) \equiv 1.$$ MO 98

4. $$C_n^\lambda(1) = \binom{2\lambda+n-1}{n}.$$ MO 98

8.938 A differential equation leading to the polynomials $C_n^\lambda(t)$:

$$y'' + \frac{(2\lambda+1)t}{t^2-1} y' - \frac{n(2\lambda+n)}{t^2-1} y = 0 \qquad \text{(cf. 9.174).}$$ WH

For series of products of Bessel functions and the polynomials $C_n^\lambda(x)$, see **8.532, 8.534.**

8.94 The Chebyshev polynomials $T_n(x)$ and $U_n(x)$

8.940 Definition

1. Chebyshev's polynomials of the first kind

$$T_n(x) = \cos(n \arccos x) = \frac{1}{2}\left[\left(x + i\sqrt{1-x^2}\right)^n + \left(x - i\sqrt{1-x^2}\right)^n\right] =$$
$$= x^n - \binom{n}{2} x^{n-2}(1-x^2) + \binom{n}{4} x^{n-4}(1-x^2)^2 - \binom{n}{6} x^{n-6}(1-x^2)^6 + \dots$$

NA 66, 71

2. Chebyshev's polynomials of the second kind:

$$U_n(x) = \frac{\sin[(n+1)\arccos x]}{\sin[\arccos x]} =$$
$$= \frac{1}{2i\sqrt{1-x^2}}\left[\left(x + i\sqrt{1-x^2}\right)^{n+1} - \left(x - i\sqrt{1-x^2}\right)^{n+1}\right] =$$
$$= \binom{n+1}{1} x^n - \binom{n+1}{3} x^{n-2}(1-x^2) + \binom{n+1}{5} x^{n-4}(1-x^2)^2 - \dots$$

Functional relations

8.941 Recursion formulas:

1. $T_{n+1}(x) - 2xT_n(x) + T_{n-1}(x) = 0.$ NA 358
2. $U_{n+1}(x) - 2xU_n(x) + U_{n-1}(x) = 0.$
3. $T_n(x) = U_n(x) - xU_{n-1}(x).$ EH II 184(3)
4. $(1-x^2)U_{n-1}(x) = xT_n(x) - T_{n+1}(x).$ EH II 184(4)

For the orthogonality, see **7.343** and **8.904**.

8.942 Relations with other functions:

1. $T_n(x) = F\left(n, -n; \frac{1}{2}; \frac{1-x}{2}\right).$ MO 104

2. $T_n(x) = (-1)^n \frac{\sqrt{1-x^2}}{(2n-1)!!} \frac{d^n}{dx^n}(1-x^2)^{n-\frac{1}{2}}.$ MO 104

3. $U_n(x) = \frac{(-1)^n (n+1)}{\sqrt{1-x^2}\,(2n+1)!!} \frac{d^n}{dx^n}(1-x^2)^{n+\frac{1}{2}}.$ EH II 185(15)

See also **8.962** 3.

8.943 Special cases:

1. $T_0(x) = 1.$
2. $T_1(x) = x.$
3. $T_2(x) = 2x^2 - 1.$
4. $T_3(x) = 4x^3 - 3x.$
5. $T_4(x) = 8x^4 - 8x^2 + 1.$
6. $T_5(x) = 16x^5 - 20x^3 + 5x.$
7. $U_0(x) = 1.$
8. $U_1(x) = 2x.$
9. $U_2(x) = 4x^2 - 1.$
10. $U_3(x) = 8x^3 - 4x$
11. $U_4(x) = 16x^4 - 12x^2 + 1.$

8.944 Particular values:

1. $T_n(1) = 1.$
2. $T_n(-1) = (-1)^n.$

3. $T_{2n}(0)=(-1)^n$.
4. $T_{2n+1}(0)=0$
5. $U_{2n+1}(0)=0$.
6. $U_{2n}(0)=(-1)^n$.

8.945 The generating function:

1. $$\frac{1-t^2}{1-2tx+t^2}=T_0(x)+2\sum_{k=1}^{\infty}T_k(x)t^k.$$ MO 104

2. $$\frac{1}{1-2tx+t^2}=\sum_{k=0}^{\infty}U_k(x)t^k.$$ MO 104a, EH II 186(31)

8.946 Zeros. The polynomials $T_n(x)$ and $U_n(x)$ only have real simple zeros. All these zeros lie in the interval $(-1, +1)$.

8.947 The functions $T_n(x)$ and $\sqrt{1-x^2}\,U_{n-1}(x)$ are two linearly independent solutions of the differential equation

$$(1-x^2)\frac{d^2y}{dx^2}-x\frac{dy}{dx}+n^2y=0.$$ NA 69(58)

8.948 Of all polynomials of degree n with leading coefficient equal to 1, the one that deviates the least from zero on the interval $[-1, +1]$ is the polynomial $2^{-n+1}T_n(x)$. NA 63

8.95 The Hermite polynomials $H_n(x)$

8.950 Definition

1. $$H_n(x)=(-1)^n e^{x^2}\frac{d^n}{dx^n}(e^{-x^2})$$ SM 567(14)

or

2. $$H_n(x)=2^n x^n-2^{n-1}\binom{n}{2}x^{n-2}+2^{n-2}\cdot 1\cdot 3\cdot\binom{n}{4}x^{n-4}-$$
$$-2^{n-3}\cdot 1\cdot 3\cdot 5\cdot\binom{n}{6}x^{n-6}+\dots$$ MO 105a

8.951 The integral representation:

$$H_n(x)=\frac{2^n}{\sqrt{\pi}}\int_{-\infty}^{\infty}(x+it)^n e^{-t^2}\,dt.$$ MO 106a

Functional relations

8.952 Recursion formulas:

1. $$\frac{dH_n(x)}{dx}=2nH_{n-1}(x).$$ SM 569(22)
2. $$H_{n+1}(x)=2xH_n(x)-2nH_{n-1}(x).$$ SM 570(23)

For the orthogonality, see 7.374 1. and 8.904.

8.953 The connection with other functions:

1. $$H_{2n}(x)=(-1)^n\frac{(2n)!}{n!}\Phi\left(-n, \frac{1}{2}; x^2\right).$$ MO 106a
2. $$H_{2n+1}(x)=(-1)^n 2\frac{(2n+1)!}{n!}x\,\Phi\left(-n, \frac{3}{2}; x^2\right).$$ MO 106a

For a connection with the polynomials $C_n^\lambda(x)$, see 8.936 5.
For a connection with the Laguerre polynomials, see 8.972 2. and 8.972 3
For a connection with functions of a parabolic cylinder, see 9.253.

8.954 Inequalities:

$$|H_n(x)| \leqslant 2^{\frac{n}{2} - E\left(\frac{n}{2}\right)} \frac{n!}{\left[E\left(\frac{n}{2}\right)\right]!} e^{2x\sqrt{E\left(\frac{n}{2}\right)}} \qquad [x > 0].$$ MO 106a

8.955 Asymptotic representation:

1. $$H_{2n}(x) = (-1)^n 2^n (2n-1)!!\, e^{\frac{x^2}{2}} \left[\cos(\sqrt{4n+1}\,x) + O\left(\frac{1}{\sqrt[4]{n}}\right)\right]$$ SM 579

2. $$H_{2n+1}(x) = (-1)^n 2^{n+\frac{1}{2}} (2n-1)!! \sqrt{2n+1}\, e^{\frac{x^2}{2}} \left[\sin(\sqrt{4n+3}\,x) + O\left(\frac{1}{\sqrt[4]{n}}\right)\right].$$

SM 579

8.956 Special cases and particular values:

1. $H_0(x) = 1.$
2. $H_1(x) = 2x$
3. $H_2(x) = 4x^2 - 2.$
4. $H_3(x) = 8x^3 - 12x.$
5. $H_4(x) = 16x^4 - 48x^2 + 12.$
6. $H_{2n}(0) = (-1)^n 2^n (2n-1)!!.$ SM 570(24)
7. $H_{2n+1}(0) = 0.$

Series of Hermite polynomials

8.957 The generating function:

1 $$\exp(-t^2 + 2tx) = \sum_{k=0}^{\infty} \frac{t^k}{k!} H_k(x).$$ SM 569(21)

2. $$\frac{1}{e} \operatorname{sh} 2x = \sum_{k=0}^{\infty} \frac{1}{(2k+1)!} H_{2k+1}(x)$$ MO 106a

3. $$\frac{1}{e} \operatorname{ch} 2x = \sum_{k=0}^{\infty} \frac{1}{(2k)!} H_{2k}(x).$$ MO 106a

4. $$e \sin 2x = \sum_{k=0}^{\infty} (-1)^k \frac{1}{(2k+1)!} H_{2k+1}(x).$$ MO 106a

5. $$e \cos 2x = \sum_{k=0}^{\infty} (-1)^k \frac{1}{(2k)!} H_{2k}(x).$$ MO 106a

8.958 "The summation theorem":

1. $$\frac{\left(\sum_{k=1}^{r} a_k^2\right)^{\frac{n}{2}}}{n!} H_n\left(\frac{\sum_{k=1}^{r} a_k x_k}{\sqrt{\sum_{k=1}^{r} a_k^2}}\right) = \sum_{m_1+m_2+\ldots+m_r=n} \prod_{k=1}^{r} \left\{\frac{a_k^{m_k}}{m_k!} H_{m_k}(x_k)\right\}.$$

MO 106a

2. A special case:

$$2^{\frac{n}{2}} H_n(x+y) = \sum_{k=0}^{n} \binom{n}{k} H_{n-k}\left(x\sqrt{2}\right) H_k\left(y\sqrt{2}\right)$$ MO 107a

8.959 Hermite polynomials satisfy the differential equation

1. $$\frac{d^2u_n}{dx^2} - 2x\frac{du_n}{dx} + 2nu_n = 0;$$ SM 566(9)

A second solution of this differential equation is provided by the functions:

2. $$u_{2n} = (-1)^n Ax\Phi\left(\frac{1}{2}-n;\ \frac{3}{2};\ x^2\right),$$

3. $$u_{2n+1} = (-1)^n B\Phi\left(-\frac{1}{2}-n;\ \frac{1}{2};\ x^2\right)$$

[A and B – arbitrary constants]. MO 107

8.96 Jacobi's polynomials

8.960 Definition

$$P_n^{(\alpha,\beta)}(x) = \frac{(-1)^n}{2^n n!}(1-x)^{-\alpha}(1+x)^{-\beta}\frac{d^n}{dx^n}\left[(1-x)^{\alpha+n}(1+x)^{\beta+n}\right];$$

EH II 169(10), CO

$$= \frac{1}{2^n}\sum_{m=0}^{n}\binom{n+\alpha}{m}\binom{n+\beta}{n-m}(x-1)^{n-m}(x+1)^m.$$

EH II 169(2)

8.961 Functional relations:

1. $$P_n^{(\alpha,\beta)}(-x) = (-1)^n P_n^{(\beta,\alpha)}(x).$$ EH II 169(13)

2. $$2(n+1)(n+\alpha+\beta+1)(2n+\alpha+\beta)P_{n+1}^{(\alpha,\beta)}(x) =$$
$$= (2n+\alpha+\beta+1)[(2n+\alpha+\beta)(2n+\alpha+\beta+2)x+\alpha^2-\beta^2]P_n^{(\alpha,\beta)}(x) -$$
$$-2(n+\alpha)(n+\beta)(2n+\alpha+\beta+2)P_{n-1}^{(\alpha,\beta)}(x).$$ EH II 169(11)

3. $$(2n+\alpha+\beta)(1-x^2)\frac{d}{dx}P_n^{(\alpha,\beta)}(x) =$$
$$= n[(\alpha-\beta)-(2n+\alpha+\beta)x]P_n^{(\alpha,\beta)}(x) +$$
$$+2(n+\alpha)(n+\beta)P_{n-1}^{(\alpha,\beta)}(x).$$ EH II 170(15)

4. $$\frac{d^m}{dx^m}\left[P_n^{(\alpha,\beta)}(x)\right] = \frac{1}{2^m}\frac{\Gamma(n+m+\alpha+\beta+1)}{\Gamma(n+\alpha+\beta+1)}P_{n-m}^{(\alpha+m,\beta+m)}(x)$$

$[m = 1, 2, \ldots, n]$. EH II 170(17)

5. $\left(n+\frac{1}{2}\alpha+\frac{1}{2}\beta+1\right)(1-x)P_n^{(\alpha+1,\beta)}(x)=$
$$=(n+\alpha+1)P_n^{(\alpha,\beta)}(x)-(n+1)P_{n+1}^{(\alpha,\beta)}(x).$$ EH II 173(32)

6. $\left(n+\frac{1}{2}\alpha+\frac{1}{2}\beta+1\right)(1+x)P_n^{(\alpha,\beta+1)}(x)=$
$$=(n+\beta+1)P_n^{(\alpha,\beta)}(x)+(n+1)P_{n+1}^{(\alpha,\beta)}(x).$$ EH II 173(33)

7. $(1-x)P_n^{(\alpha+1,\beta)}(x)+(1+x)P_n^{(\alpha,\beta+1)}(x)=2P_n^{(\alpha,\beta)}(x).$ EH II 173(34)

8. $(2n+\alpha+\beta)P_n^{(\alpha-1,\beta)}(x)=(n+\alpha+\beta)P_n^{(\alpha,\beta)}(x)-(n+\beta)P_{n-1}^{(\alpha,\beta)}(x).$ EH II 173(35)

9. $(2n+\alpha+\beta)P_n^{(\alpha,\beta-1)}(x)=(n+\alpha+\beta)P_n^{(\alpha,\beta)}(x)+(n+\alpha)P_{n-1}^{(\alpha,\beta)}(x).$ EH II 173(36)

10. $P_n^{(\alpha,\beta-1)}(x)-P_n^{(\alpha-1,\beta)}(x)=P_{n-1}^{(\alpha,\beta)}(x).$ EH II 173(37)

8.962 Connections with other functions:

1. $P_n^{(\alpha,\beta)}(x)=\frac{(-1)^n\Gamma(n+1+\beta)}{n!\,\Gamma(1+\beta)}F\left(n+\alpha+\beta+1,\ -n;\ 1+\beta;\ \frac{1+x}{2}\right);$ CO, EH II 170(16)

$$=\frac{\Gamma(n+1+\alpha)}{n!\,\Gamma(1+\alpha)}F\left(n+\alpha+\beta+1,\ -n;\ 1+\alpha;\ \frac{1-x}{2}\right);$$ EH II 170(16)

$$=\frac{\Gamma(n+1+\alpha)}{n!\,\Gamma(1+\alpha)}\left(\frac{1+x}{2}\right)^n F\left(-n,\ -n-\beta;\ \alpha+1;\ \frac{x-1}{x+1}\right);$$ EH II 170(16)

$$=\frac{\Gamma(n+1+\beta)}{n!\,\Gamma(1+\beta)}\left(\frac{x-1}{2}\right)^n F\left(-n,\ -n-\alpha;\ \beta+1;\ \frac{x+1}{x-1}\right).$$ EH II 170(16)

2. $P_n(x)=P_n^{(0,0)}(x).$ CO, EH II 179(3)

3. $T_n(x)=\frac{2^{2n}(n!)^2}{(2n)!}P_n^{\left(-\frac{1}{2},-\frac{1}{2}\right)}(x).$ CO, EH II 184(5)a

4. $C_n^{\nu}(x)=\frac{\Gamma(n+2\nu)\Gamma\left(\nu+\frac{1}{2}\right)}{\Gamma(2\nu)\Gamma\left(n+\nu+\frac{1}{2}\right)}P_n^{\left(\nu-\frac{1}{2},\ \nu-\frac{1}{2}\right)}(x).$ MO 108a, EH II 174(4)

8.963 The generating function:

$$\sum_{n=0}^{\infty}P_n^{(\alpha,\beta)}(x)z^n=2^{\alpha+\beta}R^{-1}(1-z+R)^{-\alpha}(1+z+R)^{-\beta},$$
$$R=\sqrt{1-2xz+z^2}\qquad [|z|<1].$$ EH II 172(29)

8.964 The Jacobi polynomials constitute the *unique* rational solution of the differential (hypergeometric) equation

$$(1-x^2)y''+[\beta-\alpha-(\alpha+\beta+2)x]y'+n(n+\alpha+\beta+1)y=0.$$ EH II 169(14)

8.965 Asymptotic representation

$$P_n^{(\alpha,\beta)}(\cos\theta)=\frac{\cos\left\{\left[n+\frac{1}{2}(\alpha+\beta+1)\right]\theta-\left(\frac{1}{2}\alpha+\frac{1}{4}\right)\pi\right\}}{\sqrt{\pi n}\left(\sin\frac{1}{2}\theta\right)^{\alpha+\frac{1}{2}}\left(\cos\frac{1}{2}\theta\right)^{\beta+\frac{1}{2}}}+O\left(n^{-\frac{3}{2}}\right)$$

$$[\operatorname{Im}\alpha=\operatorname{Im}\beta=0,\ 0<\theta<\pi].$$ EH II 198(10)

8.966 A limit relationship:

$$\lim_{n\to\infty}\left[n^{-\alpha}P_n^{(\alpha,\beta)}\left(\cos\frac{z}{n}\right)\right]=\left(\frac{z}{2}\right)^{-\alpha}J_\alpha(z).$$ EH II 173(41)

8.967 If $\alpha>-1$ and $\beta>-1$, all the zeros of the polynomial $P_n^{(\alpha,\beta)}(x)$ are simple and they lie in the interval $(-1,1)$.

8.97 The Laguerre polynomials

8.970 Definition.

1. $$L_n^\alpha(x)=\frac{1}{n!}e^x x^{-\alpha}\frac{d^n}{dx^n}\left(e^{-x}x^{n+\alpha}\right);$$ EH II 188(5), MO 108

$$=\sum_{m=0}^{n}(-1)^m\binom{n+\alpha}{n-m}\frac{x^m}{m!}.$$ MO 109, EH II 188(7)

2. $$L_n^0(x)=L_n(x).$$ ET I 369

8.971 Functional relations:

1. $$\frac{d}{dx}\left[L_n^\alpha(x)-L_{n+1}^\alpha(x)\right]=L_n^\alpha(x).$$ EH II 189(16)

2. $$\frac{d}{dx}L_n^\alpha(x)=-L_{n-1}^{\alpha+1}(x).$$ EH II 189(15), SM 575(42)a

3. $$x\frac{d}{dx}L_n^\alpha(x)=nL_n^\alpha(x)-(n+\alpha)L_{n-1}^\alpha(x);$$

$$=(n+1)L_{n+1}^\alpha(x)-(n+\alpha+1-x)L_n^\alpha(x).$$

EH II 189(12), MO 109

4. $$xL_n^{\alpha+1}(x)=(n+\alpha+1)L_n^\alpha(x)-(n+1)L_{n+1}^\alpha(x);$$

$$=(n+\alpha)L_{n-1}^\alpha(x)-(n-x)L_n^\alpha(x).$$

SM 575(43)a, EH II 190(23)

5. $$L_n^{\alpha-1}(x)=L_n^\alpha(x)-L_{n-1}^\alpha(x).$$ SM 575(44)a, EH II 190(24)

6. $$(n+1)L_{n+1}^\alpha(x)-(2n+\alpha+1-x)L_n^\alpha(x)+(n+\alpha)L_{n-1}^\alpha(x)=0$$

$$[n=1,2,\ldots].$$ MO 109, EH II 190(25, 24)

8.972 Connections with other functions:

1. $$L_n^\alpha(x)=\binom{n+\alpha}{n}\Phi(-n,\alpha+1;x).$$ MO 109, FI II 189(14)

2. $$H_{2n}(x)=(-1)^n 2^{2n}n!\,L_n^{-\frac{1}{2}}(x^2).$$ EH II 193(2), SM 576(47)

3. $$H_{2n+1}(x)=(-1)^n 2^{2n+1}n!\,xL_n^{\frac{1}{2}}(x^2).$$ EH II 193(3), SM 577(48)

8.973 Special cases:

1. $L_0^{\alpha}(x)=1$ EH II 188(6)

2. $L_1^{\alpha}(x)=\alpha+1-x.$ EH II 188(6)

3. $L_n^{\alpha}(0)=\binom{n+\alpha}{n}.$ EH II 189(13)

4. $L_n^{-n}(x)=(-1)^n\frac{x^n}{n!}.$ MO 109

5. $L_1(x)=1-x.$

6. $L_2(x)=1-2x+\frac{x^2}{2}.$ MO 109

8.974 Finite sums:

1. $$\sum_{m=0}^{n}\frac{m!}{\Gamma(m+\alpha+1)}L_m^{\alpha}(x)L_m^{\alpha}(y)=$$
$$=\frac{(n+1)!}{\Gamma(n+\alpha+1)(x-y)}[L_n^{\alpha}(x)L_{n+1}^{\alpha}(y)-L_{n+1}^{\alpha}(x)L_n^{\alpha}(y)]$$ EH II 188(9)

2. $$\sum_{m=0}^{n}\frac{\Gamma(\alpha-\beta+m)}{\Gamma(\alpha-\beta)\,m!}L_{n-m}^{\beta}(x)=L_n^{\alpha}(x).$$ MO 110, EH II 192(39)

3. $$\sum_{m=0}^{n}L_m^{\alpha}(x)=L_n^{\alpha+1}(x).$$ EH II 192(38)

4. $$\sum_{m=0}^{n}L_m^{\alpha}(x)L_{n-m}^{\beta}(x)=L_n^{\alpha+\beta+1}(x+y).$$ EH II 192(41)

8.975 Arbitrary functions:

1. $$(1-z)^{-\alpha-1}\exp\frac{xz}{z-1}=\sum_{n=0}^{\infty}L_n^{\alpha}(x)z^n\quad[|z|<1].$$ EH II 189(17), MO 109

2. $$e^{-xz}(1+z)^{\alpha}=\sum_{n=0}^{\infty}L_n^{\alpha-n}(x)z^n\qquad[|z|<1].$$ MO 110, EH II 189(19)

3. $$J_{\alpha}\left(2\sqrt{xz}\right)e^{z}(xz)^{-\frac{1}{2}\alpha}=\sum_{n=0}^{\infty}\frac{z^n}{\Gamma(n+\alpha+1)}L_n^{\alpha}(x)$$
$$[\alpha>-1].$$ EH II 189(18), MO 109

8.976 Other series of Laguerre polynomials:

1. $$\sum_{n=0}^{\infty}n!\frac{L_n^{\alpha}(x)L_n^{\alpha}(y)z^n}{\Gamma(n+\alpha+1)}=\frac{(xyz)^{-\frac{1}{2}\alpha}}{1-z}\exp\left(-z\frac{x+y}{1-z}\right)I_{\alpha}\left(2\frac{\sqrt{xyz}}{1-z}\right)$$
$$[|z|<1].$$ EH II 189(20)

2. $$\sum_{n=0}^{\infty}\frac{L_n^{\alpha}(x)}{n+1}=e^{x}x^{-\alpha}\Gamma(\alpha,x)\qquad[\alpha>-1,\ x>0].$$ EH II 215(19)

3. $$[L_n^\alpha(x)]^2 = \frac{\Gamma(1+\alpha+n)}{n!}\sum_{k=0}^{\infty}\frac{(2n-2k)!\,(2k)!}{\Gamma(1+\alpha-k)}\frac{L_{2k}^{2\alpha}(2x)}{(n-k)!}.$$ MO 110

4. $$L_n^\alpha(x)\,L_n^\alpha(y) = \frac{\Gamma(1+\alpha+n)}{n!}\sum_{k=0}^{\infty}\frac{L_{n-k}^{\alpha+2k}(x+y)}{\Gamma(1+\alpha+k)}\frac{(xy)^k}{k!}.$$

MO 110, EH II 192(42)

8.977 Summation theorems:

1. $$L_n^{\alpha_1+\alpha_2+\ldots+\alpha_k+k-1}(x_1+x_2+\ldots+x_k) = \sum_{(i_1+i_2+\ldots+i_k=n)} L_{i_1}^{\alpha_1}(x_1)\,L_{i_2}^{\alpha_2}(x_2)\ldots L_{i_k}^{\alpha_k}(x_k).$$ MO 110

2. $$L_n^\alpha(x+y) = e^y\sum_{k=0}^{\infty}\frac{(-1)^k}{k!}\,y^k L_n^{\alpha+k}(x).$$ MO 110

8.978 Limit relations and asymptotic behavior:

1. $$L_n^\alpha(x) = \lim_{\beta\to\infty} P_n^{(\alpha,\,\beta)}\left(1-\frac{2x}{\beta}\right).$$ EH II 191(35)

2. $$\lim_{n\to\infty}\left[n^{-\alpha}L_n^\alpha\left(\frac{x}{n}\right)\right] = x^{-\frac{1}{2}\alpha}J_\alpha(2\sqrt{x}).$$ EH II 191(36)

3. $$L_n^\alpha(x) = \frac{1}{\sqrt{\pi}}e^{\frac{1}{2}x}x^{-\frac{1}{2}\alpha-\frac{1}{4}}n^{\frac{1}{2}\alpha-\frac{1}{4}}\cos\left[2\sqrt{nx}-\frac{\alpha\pi}{2}-\frac{\pi}{4}\right]+O\left(n^{\frac{1}{2}\alpha-\frac{3}{4}}\right)$$

$[\operatorname{Im}\alpha = 0,\ x > 0]$. EH II 199(1)

8.979 Laguerre polynomials satisfy the following differential equation:

$$x\frac{d^2u}{dx^2}+(\alpha-x+1)\frac{du}{dx}+nu = 0.$$

EH II 188(10), SM 574(34)

9.1 Hypergeometric Functions

9.10 Definition

9.100 A *hypergeometric series* is a series of the form

$$F(\alpha,\beta;\gamma;z) = 1+\frac{\alpha\cdot\beta}{\gamma\cdot 1}z+\frac{\alpha(\alpha+1)\beta(\beta+1)}{\gamma(\gamma+1)\cdot 1\cdot 2}z^2+ \frac{\alpha(\alpha+1)(\alpha+2)\beta(\beta+1)(\beta+2)}{\gamma(\gamma+1)(\gamma+2)\cdot 1\cdot 2\cdot 3}z^3+\ldots$$

9.101 A hypergeometric series terminates if α or β is equal to a negative integer or to zero. For $\gamma = -n$ $(n = 0, 1, 2, \ldots)$, the hypergeometric series is indeterminate if neither α nor β is equal to $-m$ (where $m < n$ and m is a natural number). However,

1. $$\lim_{\gamma\to -n}\frac{F(\alpha,\beta;\gamma;z)}{\Gamma(\gamma)} = \frac{\alpha(\alpha+1)\ldots(\alpha+n)\beta(\beta+1)\ldots(\beta+n)}{(n+1)!}z^{n+1}F(\alpha+n+1,\ \beta+n+1;\ n+2;\ z).$$

EH I 62(16)

9.102 If we exclude these values of the parameters α, β, γ, a hypergeometric series converges in the unit circle $|z| < 1$. F then has a branch point at $z = 1$. Then we have the following conditions for convergence on the unit circle:

1. $1 > \operatorname{Re}(\alpha+\beta-\gamma) \geqslant 0$. The series converges throughout the entire unit circle except at the point $z = 1$.
2. $\operatorname{Re}(\alpha+\beta-\gamma) < 0$. The series converges (absolutely) throughout the entire unit circle.
3. $\operatorname{Re}(\alpha+\beta-\gamma) \geqslant 1$. The series diverges on the entire unit circle.

FI II 410, WH

9.11 Integral representations

9.111 $$F(\alpha, \beta; \gamma; z) = \frac{1}{\mathrm{B}(\beta, \gamma-\beta)} \int_0^1 t^{\beta-1} (1-t)^{\gamma-\beta-1} (1-tz)^{-\alpha} dt$$

$[\operatorname{Re}\gamma > \operatorname{Re}\beta > 0]$. WH

9.112 $$F(p, n+p; n+1; z^2) = \frac{z^{-n}}{2\pi} n\mathrm{B}(p, n) \int_0^{2\pi} \frac{\cos nt \, dt}{(1-2z\cos t+z^2)^p}$$

$[n = 0, 1, 2, \ldots; \operatorname{Re} p > 0]$. WH, MO 16

9.113 $$F(\alpha, \beta; \gamma; z) = \frac{\Gamma(\gamma)}{\Gamma(\alpha)\Gamma(\beta)} \frac{1}{2\pi i} \int_{-\infty i}^{\infty i} \frac{\Gamma(\alpha+t)\Gamma(\beta+t)\Gamma(-t)}{\Gamma(\gamma+t)} (-z)^t \, dt,$$

Here, $|\arg(-z)| < \pi$ and the path of integration is chosen in such a way that the poles of the functions $\Gamma(\alpha+t)$ and $\Gamma(\beta+t)$ lie to the left of the path of integration and the poles of the function $\Gamma(-t)$ lie to the right of it.

9.114 $$F\left(-m, -\frac{p+m}{2}; 1-\frac{p+m}{2}; -1\right) = \frac{(-2)^m (p+m)}{\sin p\pi} \int_0^{\pi} \cos^m \varphi \cos p\varphi \, d\varphi$$

$[m+1$ a natural number; $p \neq 0, \pm 1, \ldots]$.

EH I 80(8), MO 16

See also 3.194 1., 2., 5., 3.196 1., 3.197 6., 9., 3.259 3., 3.312 3., 3.518 4.—6., 3.665 2., 3.671 1., 2., 3.681 1., 3.984 7.

9.12 The representation of elementary functions in terms of a hypergeometric function

9.121

1. $F(-n, \beta; \beta; -z) = (1+z)^n$ [β arbitrary]

EH I 101(4), GA 127 Ia

2. $$F\left(-\frac{n}{2}, -\frac{n-1}{2}; \frac{1}{2}; \frac{z^2}{t^2}\right) = \frac{(t+z)^n + (t-z)^n}{2t^n}.$$ GA 127 II

3. $$\lim_{\omega\to\infty} F\left(-n, \omega; 2\omega; -\frac{z}{t}\right) = \left(1+\frac{z}{2t}\right)^n.$$ GA 127 IIIa

4. $$F\left(-\frac{n-1}{2}, -\frac{n-2}{2}; \frac{3}{2}; \frac{z^2}{t^2}\right) = \frac{(t+z)^n - (t-z)^n}{2nzt^{n-1}}.$$ GA 127 IV

5. $F\left(1-n,\ 1;\ 2;\ -\frac{z}{t}\right)=\frac{(t+z)^n-t^n}{nzt^{n-1}}.$ GA 127 V

6. $F(1,\ 1;\ 2;\ -z)=\frac{\ln(1+z)}{z}.$ GA 127 VI

7. $F\left(\frac{1}{2},\ 1;\ \frac{3}{2};\ z^2\right)=\frac{\ln\frac{1+z}{1-z}}{2z}.$ GA 127 VII

8. $\lim_{k\to\infty} F\left(1,\ k;\ 1;\ \frac{z}{k}\right)=1+z\lim_{k\to\infty} F\left(1,\ k;\ 2;\ \frac{z}{k}\right)=$

$=1+z+\frac{z^2}{2}\lim_{k\to\infty} F\left(1,\ k;\ 3;\ \frac{z}{k}\right)=\ldots=e^z.$ GA 127 VIII

9. $\lim_{\substack{k\to\infty \\ k'\to\infty}} F\left(k,\ k';\ \frac{1}{2};\ \frac{z^2}{4kk'}\right)=\frac{e^z+e^{-z}}{2}=\operatorname{ch} z.$ GA 127 IX

10. $\lim_{\substack{k\to\infty \\ k'\to\infty}} F\left(k,\ k';\ \frac{3}{2};\ \frac{z^2}{4kk'}\right)=\frac{e^z-e^{-z}}{2z}=\frac{\operatorname{sh} z}{z}.$ GA 127 X

11. $\lim_{\substack{k\to\infty \\ k'\to\infty}} F\left(k,\ k';\ \frac{3}{2};\ -\frac{z^2}{4kk'}\right)=\frac{\sin z}{z}.$ GA 127 XI

12. $\lim_{\substack{k\to\infty \\ k'\to\infty}} F\left(k,\ k';\ \frac{1}{2};\ -\frac{z^2}{4kk'}\right)=\cos z.$ GA 127 XII

13. $F\left(\frac{1}{2},\ \frac{1}{2};\ \frac{3}{2};\ \sin^2 z\right)=\frac{z}{\sin z}.$ GA 127 XIII

14. $F\left(1,\ 1;\ \frac{3}{2};\ \sin^2 z\right)=\frac{z}{\sin z\cos z}.$ GA 127 XIV

15. $F\left(\frac{1}{2},\ 1;\ \frac{3}{2};\ -\operatorname{tg}^2 z\right)=\frac{z}{\operatorname{tg} z}.$ GA 127 XV

16. $F\left(\frac{n+1}{2},\ -\frac{n-1}{2};\ \frac{3}{2};\ \sin^2 z\right)=\frac{\sin nz}{n\sin z}.$ GA 127 XVI

17. $F\left(\frac{n+2}{2},\ -\frac{n-2}{2};\ \frac{3}{2};\ \sin^2 z\right)=\frac{\sin nz}{n\sin z\cos z}.$ GA 127 XVII

18. $F\left(-\frac{n-2}{2},\ -\frac{n-1}{2};\ \frac{3}{2};\ -\operatorname{tg}^2 z\right)=\frac{\sin nz}{n\sin z\cos^{n-1} z}.$ GA 127 XVIII

19. $F\left(\frac{n+2}{2},\ \frac{n+1}{2};\ \frac{3}{2};\ -\operatorname{tg}^2 z\right)=\frac{\sin nz\cos^{n+1} z}{n\sin z}.$ GA 127 XIX

20. $F\left(\frac{n}{2},\ -\frac{n}{2};\ \frac{1}{2};\ \sin^2 z\right)=\cos nz.$ EH I 101(11), GA 127 XX

21. $F\left(\frac{n+1}{2},\ -\frac{n-1}{2};\ \frac{1}{2};\ \sin^2 z\right)=\frac{\cos nz}{\cos z}.$ EH I 101(11), GA 127 XXI

22. $F\left(-\frac{n}{2},\ -\frac{n-1}{2};\ \frac{1}{2};\ -\operatorname{tg}^2 z\right)=\frac{\cos nz}{\cos^n z}.$

EH I 101(11), GA 127 XXII

23. $F\left(\frac{n+1}{2},\ \frac{n}{2};\ \frac{1}{2};\ -\operatorname{tg}^2 z\right)=\cos nz\cos^n z.$ GA 127 XXIII

24. $F\left(\frac{1}{2},\ 1;\ 2;\ 4z(1-z)\right)=\frac{1}{1-z}$ $\left[|z|\leqslant\frac{1}{2};\ |z(1-z)|\leqslant\frac{1}{4}\right].$

25. $F\left(\frac{1}{2},\ 1;\ 1;\ \sin^2 z\right)=\sec z.$

26. $F\left(\frac{1}{2}, \frac{1}{2}; \frac{3}{2}; z^2\right) = \frac{\arcsin z}{z}$ (cf. 9.121 13.).

27. $F\left(\frac{1}{2}, 1; \frac{3}{2}; -z^2\right) = \frac{\operatorname{arctg} z}{z}$ (cf. 9.121 15.).

28. $F\left(\frac{1}{2}, \frac{1}{2}; \frac{3}{2}; -z^2\right) = \frac{\operatorname{Arsh} z}{z}$ (cf. 9.121 26.).

29. $F\left(\frac{1+n}{2}, \frac{1-n}{2}; \frac{3}{2}; z^2\right) = \frac{\sin(n \arcsin z)}{nz}$ (cf. 9.121 16.).

30. $F\left(1+\frac{n}{2}, 1-\frac{n}{2}; \frac{3}{2}; z^2\right) = \frac{\sin(n \arcsin z)}{nz\sqrt{1-z^2}}$ (cf. 9.121 17.).

31. $F\left(\frac{n}{2}, -\frac{n}{2}; \frac{1}{2}; z^2\right) = \cos(n \arcsin z)$ (cf. 9.121 20.).

32. $F\left(\frac{1+n}{2}, \frac{1-n}{2}; \frac{1}{2}; z^2\right) = \frac{\cos(n \arcsin z)}{\sqrt{1-z^2}}$ (cf. 9.121 21.).

The representation of special functions in terms of a hypergeometric function

For complete elliptic integrals, see 8.113 1. and 8.114 1.;
for integrals of Bessel functions, see 6.574 1., 3., 6.576 2. – 5., 6.621 1. – 3.;
for Legendre polynomials, 8.911 and 8.916. (All these hypergeometric series terminate; that is, these series are finite sums);
for Legendre functions, see 8.820 and 8.837;
for associated Legendre functions, see 8.702, 8.703, 8.751, 8.77, 8.852, and 8.853;
for Chebyshev polynomials, see 8.942 1.;
for Jacobi's polynomials, see 8.962;
for Gegenbauer polynomials $C_n^\lambda(x)$, see 8.932;
for integrals of parabolic-cylinder functions, see 7.726 6.

9.122 Particular values:

1. $F(\alpha, \beta; \gamma; 1) = \frac{\Gamma(\gamma)\Gamma(\gamma-\alpha-\beta)}{\Gamma(\gamma-\alpha)\Gamma(\gamma-\beta)}$

$[\operatorname{Re}\gamma > \operatorname{Re}(\alpha+\beta)]$ GA 147(48), FI II 793

2. $F(\alpha, \beta; \gamma; 1) = F(-\alpha, -\beta; \gamma-\alpha-\beta; 1)$ $[\operatorname{Re}\gamma > 0]$; GA 148(49)

$= \frac{1}{F(-\alpha, \beta; \gamma-\alpha; 1)}$ $[\operatorname{Re}(\gamma-\beta) > 0]$; GA 148(50)

$= \frac{1}{F(\alpha, -\beta; \gamma-\beta; 1)}$ $[\operatorname{Re}(\gamma-\alpha) > 0]$. GA 148(51)

3. $F\left(1, 1; \frac{3}{2}; \frac{1}{2}\right) = \frac{\pi}{2}$.

9.13 Transformation formulas and the analytic continuation of functions defined by hypergeometric series

9.130 The series $F(\alpha, \beta; \gamma; z)$ defines an analytic function that, speaking generally, has singularities at the points $z = 0$, 1, and ∞. (In the general case, there are branch points.) We make a cut in the z-plane along the real axis from

$z=1$ to $z=\infty$; that is, we require that $|\arg(-z)|<\pi$ for $|z|\geqslant 1$. Then, the series $F(\alpha, \beta; \gamma; z)$ will, in the cut plane, yield a single-valued analytic continuation which we can obtain by means of the formulas below (provided $\gamma+1$ is not a natural number and $\alpha-\beta$ and $\gamma-\alpha-\beta$ are not integers). These formulas make it possible to calculate the values of F in the given region even in the case in which $|z|>1$. There are other closely related transformation formulas that can also be used to get the analytic continuation when the corresponding relationships hold between α, β, γ.

Transformation formulas

9.131

1. $$F(\alpha, \beta; \gamma; z)=(1-z)^{-\alpha} F\left(\alpha, \gamma-\beta; \gamma; \frac{z}{z-1}\right);$$ GA 218(91)

$$=(1-z)^{-\beta} F\left(\beta, \gamma-\alpha; \gamma; \frac{z}{z-1}\right);$$ GA 218(92)

$$=(1-z)^{\gamma-\alpha-\beta} F(\gamma-\alpha, \gamma-\beta; \gamma; z).$$

2. $$F(\alpha, \beta; \gamma; z)=\frac{\Gamma(\gamma)\Gamma(\gamma-\alpha-\beta)}{\Gamma(\gamma-\alpha)\Gamma(\gamma-\beta)} F(\alpha, \beta; \alpha+\beta-\gamma+1; 1-z)+$$
$$+(1-z)^{\gamma-\alpha-\beta}\frac{\Gamma(\gamma)\Gamma(\alpha+\beta-\gamma)}{\Gamma(\alpha)\Gamma(\beta)} F(\gamma-\alpha, \gamma-\beta; \gamma-\alpha-\beta+1; 1-z).$$

EH I 94, MO 13

9.132

1. $$F(\alpha, \beta; \gamma; z)=(1-z)^{-\alpha}\frac{\Gamma(\gamma)\Gamma(\beta-\alpha)}{\Gamma(\beta)\Gamma(\gamma-\alpha)} F\left(\alpha, \gamma-\beta; \alpha-\beta+1; \frac{1}{1-z}\right)+$$
$$+(1-z)^{-\beta}\frac{\Gamma(\gamma)\Gamma(\alpha-\beta)}{\Gamma(\alpha)\Gamma(\gamma-\beta)} F\left(\beta, \gamma-\alpha; \beta-\alpha+1; \frac{1}{1-z}\right).$$ MO 13

2. $$F(\alpha, \beta; \gamma; z)=$$
$$=\frac{\Gamma(\gamma)\Gamma(\beta-\alpha)}{\Gamma(\beta)\Gamma(\gamma-\alpha)}(-1)^{\alpha} z^{-\alpha} F\left(\alpha, \alpha+1-\gamma; \alpha+1-\beta; \frac{1}{z}\right)+$$
$$+\frac{\Gamma(\gamma)\Gamma(\alpha-\beta)}{\Gamma(\alpha)\Gamma(\gamma-\beta)}(-1)^{\beta} z^{-\beta} F\left(\beta, \beta+1-\gamma; \beta+1-\alpha; \frac{1}{z}\right).$$

GA 220(93)

9.133 $$F\left(2\alpha, 2\beta; \alpha+\beta+\frac{1}{2}; z\right)=F\left[\alpha, \beta; \alpha+\beta+\frac{1}{2}; 4z(1-z)\right]$$
$$\left[|z|\leqslant\frac{1}{2}, |z(1-z)|\leqslant\frac{1}{4}\right].$$ WH

9.134

1. $$F(\alpha, \beta; 2\beta; z)=\left(1-\frac{z}{2}\right)^{-\alpha} F\left[\frac{\alpha}{2}, \frac{\alpha+1}{2}; \beta+\frac{1}{2}; \left(\frac{z}{2-z}\right)^{2}\right].$$

MO 13, EH I 111(4)

2. $$F(2\alpha, 2\alpha+1-\gamma; \gamma; z)=(1+z)^{-2\alpha} F\left(\alpha, \alpha+\frac{1}{2}; \gamma; \frac{4z}{(1+z)^{2}}\right).$$

GA 225(100)

3. $$F\left(\alpha, \alpha+\frac{1}{2}-\beta; \beta+\frac{1}{2}; z^{2}\right)=(1+z)^{-2\alpha} F\left(\alpha, \beta; 2\beta; \frac{4z}{(1+z)^{2}}\right).$$

GA 225(101)

9.135 $F\left(\alpha,\ \beta;\ \alpha+\beta+\frac{1}{2};\ \sin^2\varphi\right)=F\left(2\alpha,\ 2\beta;\ \alpha+\beta+\frac{1}{2};\ \sin^2\frac{\varphi}{2}\right)$

$$\left[x=\sin^2\frac{\varphi}{2}\ \text{real};\ \frac{1-\sqrt{2}}{2}<x<\frac{1}{2}\right].$$ MO 13

9.136 We set

$$A=\frac{\Gamma\left(\alpha+\beta+\frac{1}{2}\right)\Gamma\left(\frac{1}{2}\right)}{\Gamma\left(\alpha+\frac{1}{2}\right)\Gamma\left(\beta+\frac{1}{2}\right)},\qquad B=\frac{\Gamma\left(\alpha+\beta+\frac{1}{2}\right)\Gamma\left(-\frac{1}{2}\right)}{\Gamma(\alpha)\,\Gamma(\beta)};$$

then

1. $$F\left(2\alpha,\ 2\beta;\ \alpha+\beta+\frac{1}{2};\ \frac{1-\sqrt{z}}{2}\right)=$$
$$=AF\left(\alpha,\ \beta;\frac{1}{2};\ z\right)+B\sqrt{z}F\left(\alpha+\frac{1}{2},\ \beta+\frac{1}{2};\ \frac{3}{2};\ z\right).$$ GA 227(106)

2. $$F\left(2\alpha,\ 2\beta;\ \alpha+\beta+\frac{1}{2};\ \frac{1+\sqrt{z}}{2}\right)=$$
$$=AF\left(\alpha,\ \beta;\frac{1}{2};\ z\right)-B\sqrt{z}F\left(\alpha+\frac{1}{2},\ \beta+\frac{1}{2};\ \frac{3}{2};\ z\right).$$ GA 228(107)

3. $$\frac{\left(\alpha-\frac{1}{2}\right)\left(\beta-\frac{1}{2}\right)}{\alpha+\beta-\frac{1}{2}}A\sqrt{z}F\left(\alpha,\ \beta;\frac{3}{2};\ z\right)=$$
$$=F\left(2\alpha-1,\ 2\beta-1;\ \alpha+\beta-\frac{1}{2};\ \frac{1+\sqrt{z}}{2}\right)-$$
$$-F\left(2\alpha-1, 2\beta-1;\ \alpha+\beta-\frac{1}{2};\ \frac{1-\sqrt{z}}{2}\right).$$ GA 229(110)

9.137 Gauss' recursion functions:

1. $\gamma[\gamma-1-(2\gamma-\alpha-\beta-1)z]F(\alpha,\ \beta;\ \gamma;\ z)+$
$+(\gamma-\alpha)(\gamma-\beta)zF(\alpha,\beta;\gamma+1;z)+\gamma(\gamma-1)(z-1)F(\alpha,\beta;\gamma-1;z)=0.$
2. $(2\alpha-\gamma-\alpha z+\beta z)F(\alpha,\ \beta;\ \gamma;\ z)+(\gamma-\alpha)F(\alpha-1,\ \beta;\ \gamma;\ z)+$
$+\alpha(z-1)F(\alpha+1,\ \beta;\ \gamma;\ z)=0.$
3. $(2\beta-\gamma-\beta z+\alpha z)F(\alpha,\ \beta;\ \gamma;\ z)+(\gamma-\beta)F(\alpha,\ \beta-1;\ \gamma;\ z)+$
$+\beta(z-1)F(\alpha,\ \beta+1;\ \gamma;\ z)=0.$
4. $\gamma F(\alpha,\ \beta-1;\ \gamma;\ z)-\gamma F(\alpha-1,\ \beta;\ \gamma;\ z)+(\alpha-\beta)zF(\alpha,\ \beta;\ \gamma+1;\ z)=0.$
5. $\gamma(\alpha-\beta)F(\alpha,\ \beta;\ \gamma;\ z)-\alpha(\gamma-\beta)F(\alpha+1,\ \beta;\ \gamma+1;\ z)+$
$+\beta(\gamma-\alpha)F(\alpha,\ \beta+1;\ \gamma+1;\ z)=0.$
6. $\gamma(\gamma+1)F(\alpha,\ \beta;\ \gamma;\ z)-\gamma(\gamma+1)F(\alpha,\ \beta;\ \gamma+1;\ z)-$
$-\alpha\beta zF(\alpha+1,\ \beta+1;\ \gamma+2;\ z)=0.$
7. $\gamma F(\alpha,\ \beta;\ \gamma;\ z)-(\gamma-\alpha)F(\alpha,\ \beta+1;\ \gamma+1;\ z)-$
$-\alpha(1-z)F(\alpha+1,\ \beta+1;\ \gamma+1;\ z)=0.$
8. $\gamma F(\alpha,\ \beta;\ \gamma;\ z)+(\beta-\gamma)F(\alpha+1,\ \beta;\ \gamma+1;\ z)-$
$-\beta(1-z)F(\alpha+1,\ \beta+1;\ \gamma+1;\ z)=0.$

9. $\gamma(\gamma-\beta z-\alpha)F(\alpha, \beta; \gamma; z)-\gamma(\gamma-\alpha)F(\alpha-1, \beta; \gamma; z)+$
$+\alpha\beta z(1-z)F(\alpha+1, \beta+1; \gamma+1; z)=0.$

10. $\gamma(\gamma-\alpha z-\beta)F(\alpha, \beta; \gamma; z)-\gamma(\gamma-\beta)F(\alpha, \beta-1; \gamma; z)+$
$+\alpha\beta z(1-z)F(\alpha+1, \beta+1; \gamma+1; z)=0.$

11. $\gamma F(\alpha, \beta; \gamma; z)-\gamma F(\alpha, \beta+1; \gamma; z)+\alpha z F(\alpha+1, \beta+1; \gamma+1; z)=0.$

12. $\gamma F(\alpha, \beta; \gamma; z)-\gamma F(\alpha+1, \beta; \gamma; z)+\beta z F(\alpha+1, \beta+1; \gamma+1; z)=0.$

13. $\gamma[\alpha-(\gamma-\beta)z]F(\alpha, \beta; \gamma; z)-\alpha\gamma(1-z)F(\alpha+1, \beta; \gamma; z)+$
$+(\gamma-\alpha)(\gamma-\beta)zF(\alpha, \beta; \gamma+1; z)=0.$

14. $\gamma[\beta-(\gamma-\alpha)z]F(\alpha, \beta; \gamma; z)-\beta\gamma(1-z)F(\alpha, \beta+1; \gamma; z)+$
$+(\gamma-\alpha)(\gamma-\beta)zF(\alpha, \beta; \gamma+1; z)=0.$

15. $\gamma(\gamma+1)F(\alpha, \beta; \gamma; z)-\gamma(\gamma+1)F(\alpha, \beta+1; \gamma+1; z)+$
$+\alpha(\gamma-\beta)zF(\alpha+1, \beta+1; \gamma+2; z)=0.$

16. $\gamma(\gamma+1)F(\alpha, \beta; \gamma; z)-\gamma(\gamma+1)F(\alpha+1, \beta; \gamma+1; z)+$
$+\beta(\gamma-\alpha)zF(\alpha+1, \beta+1; \gamma+2; z)=0$

17. $\gamma F(\alpha, \beta; \gamma; z)-(\gamma-\beta)F(\alpha, \beta; \gamma+1; z)-\beta F(\alpha, \beta+1; \gamma+1; z)=0.$

18 $\gamma F(\alpha, \beta; \gamma; z)-(\gamma-\alpha)F(\alpha, \beta; \gamma+1; z)-\alpha F(\alpha+1, \beta; \gamma+1; z)=0.$

MO 13-14

9.14 A generalized hypergeometric series

The series

1. $${}_pF_q(\alpha_1, \alpha_2, \ldots, \alpha_p; \beta_1, \beta_2, \ldots, \beta_q; z)=\sum_{k=0}^{\infty}\frac{(\alpha_1)_k(\alpha_2)_k\ldots(\alpha_p)_k}{(\beta_1)_k(\beta_2)_k\ldots(\beta_q)_k}\frac{z^k}{k!}$$

is called a *generalized hypergeometric series* (see also 9.210). MO 14

2. ${}_2F_1(\alpha, \beta; \gamma; z)\equiv F(\alpha, \beta; \gamma; z).$ MO 15

For integral representations, see 3.254 2., 3.259 2., and 3.478 3.

9.15 The hypergeometric differential equation

9.151 A hypergeometric series is one of the solutions of the differential equation

$$z(1-z)\frac{d^2u}{dz^2}+[\gamma-(\alpha+\beta+1)z]\frac{du}{dz}-\alpha\beta u=0,$$ WH

which is called the *hypergeometric equation.*

The solution of the hypergeometric differential equation

9.152 The hypergeometric differential equation 9.151 possesses *two linearly independent solutions*. These solutions have analytic continuations to the entire z-plane except possibly for the three points 0, 1, and ∞. Generally speaking, the points $z=0, 1, \infty$ are branch points of at least one of the branches of each solution of the hypergeometric differential equation. The ratio $w(z)$ of two linearly independent solutions satisfies the differential equation

$$2\frac{w'''}{w'}-3\left(\frac{w''}{w'}\right)^2=\frac{1-a_1^2}{z^2}+\frac{1-a_2^2}{(z-1)^2}+\frac{a_1^2+a_2^2-a_3^2-1}{z(z-1)},$$

where

$$a_1^2=(1-\gamma)^2,\quad a_2^2=(\gamma-\alpha-\beta)^2,\quad a_3^2=(\alpha-\beta)^2.$$

If α, β, γ are real, the function $w(z)$ maps the upper ($\operatorname{Im} z>0$) or the lower ($\operatorname{Im} z<0$) half-plane onto a curvilinear triangle whose angles are πa_1, πa_2, πa_3. The vertices of this triangle are the images of the points $z=0$, $z=1$, and $z=\infty$.

9.153 *Within the unit circle* $|z|<1$, the linearly independent solutions $u_1(z)$ and $u_2(z)$ of the hypergeometric differential equation are given by the following formulas:

1. If γ is not an integer,

$$u_1=F(\alpha, \beta; \gamma; z),$$
$$u_2=z^{1-\gamma}F(\alpha-\gamma+1, \beta-\gamma+1; 2-\gamma; z).$$

2. If $\gamma=1$, then

$$u_1=F(\alpha, \beta; 1; z),$$
$$u_2=F(\alpha, \beta; 1; z)\ln z+$$
$$+\sum_{k=1}^{\infty} z^k\frac{(\alpha)_k(\beta)_k}{(k!)^2}\{\psi(\alpha+k)-\psi(\alpha)+\psi(\beta+k)-\psi(\beta)-2\psi(k+1)+2\psi(1)\}$$

(see **9.14** 2.).

3. If $\gamma=m+1$ (where m is a natural number), and if neither α nor β is a positive number not exceeding m, then

$$u_1=F(\alpha, \beta; m+1; z),$$
$$u_2=F(\alpha, \beta; m+1; z)\ln z+$$
$$+\sum_{k=1}^{\infty} z^k\frac{(\alpha)_k(\beta)_k}{(1+m)_k}\{h(k)-h(0)\}-\sum_{k=1}^{m}\frac{(k-1)!\,(-m)_k}{(1-\alpha)_k(1-\beta)_k}z^{-k}$$

(see **9.14** 2.),

where

$$h(n)=\psi(\alpha+n)+\psi(\beta+n)-\psi(m+1+n)-\psi(n+1)$$

[$n+1$ a natural number].

4. Suppose that $\gamma=m+1$ (where m is a natural number) and that α or β is equal to $m'+1$, where $0\leqslant m'<m$. Then, for example, for $\alpha=m'+1$, we obtain

$$u_1=F(1+m', \beta; 1+m; z),$$
$$u_2=z^{-m}F(1+m'-m, \beta-m; 1-m; z).$$

In this case, u_2 is a polynomial in z^{-1}.

5. If $\gamma = 1 - m$ (where m is a natural number) and if α and β are both different from the numbers $0, \; -1, \; -2, \; \ldots, \; 1-m$, then

$$u_1 = z^m F(\alpha+m, \; \beta+m; \; 1+m; \; z),$$

$$u_2 = z^m F(\alpha+m, \; \beta+m; \; 1+m; \; z)\ln z + $$

$$+ z^m \sum_{k=1}^{\infty} z^k \frac{(\alpha+m)_k (\beta+m)_k}{(1+m)_k \, k!} \{h^*(k) - h^*(0)\} -$$

$$- \sum_{k=1}^{\infty} \frac{(k-1)! \, (-m)_k}{(1-\alpha-m)_k \, (1-\beta-m)_k} z^{m-n} \qquad \text{(see } \mathbf{9.14} \; 2.\text{)},$$

where

$$h^*(n) = \psi(\alpha+m+n) + \psi(\beta+m+n) - \psi(1+m+n) - \psi(1+n).$$

We note that

$$\psi(\alpha+n) - \psi(\alpha) = \frac{1}{\alpha} + \frac{1}{\alpha+1} + \ldots + \frac{1}{\alpha+n-1} \qquad \text{(cf. 8.365 3.)}$$

and that, for $\alpha = -\lambda$, where λ is a natural number or zero and $n = \lambda+1, \lambda+2, \ldots$. the expression

$$(\alpha)_k [\psi(\alpha+n) - \psi(\alpha)]$$

in formulas **9.153** 2. — 5. should be replaced with the expression

$$(-1)^{\lambda} \lambda! \, (n-\lambda-1)!.$$

6. Suppose that $\gamma = 1-m$ (where m is a natural number) and that α or β is an integer $(-m')$, where m' is one of the following numbers: $0, \; 1, \; \ldots, \; m-1$. Suppose, for example, that $\alpha = -m'$. Then,

$$u_1 = F(-m', \; \beta; \; 1-m; \; z),$$

$$u_2 = F(-m'+m, \; \beta+m; \; 1+m; \; z). \qquad \text{MO 18}$$

7. For $\gamma = \frac{1}{2}(\alpha+\beta+1)$

$$u_1 = F\left(\alpha, \; \beta; \; \frac{1}{2}(\alpha+\beta+1); \; z\right),$$

$$u_2 = F\left(\alpha, \; \beta; \; \frac{1}{2}(\alpha+\beta+1); \; 1-z\right)$$

are two linearly independent solutions of the hypergeometric differential equation provided α, β and γ are not zero or negative integers. MO 17-19

The analytic continuation of a solution that is regular at the point $z=0$

9.154 Formulas **9.153** make possible the analytic continuation, by means of the hypergeometric series, of the function $F(\alpha, \; \beta; \; \gamma; \; z)$ defined inside the circle $|z| < 1$ to the region $|z| > 1$, $|\arg(-z)| < \pi$. Here, it is assumed that $\alpha - \beta$ is not an integer. In the event that $\alpha - \beta$ is an integer (for example, if $\beta = \alpha + m$, where m is a natural number), then, for $|z| > 1$, $|\arg(-z)| < \pi$ we have:

1. $$\frac{\Gamma(\alpha)\Gamma(\alpha+m)}{\Gamma(\gamma)} F(\alpha, \alpha+m; \gamma; z) =$$

$$= \frac{\sin\pi(\gamma-\alpha)}{\pi}\left\{\sum_{k=0}^{m-1}\frac{\Gamma(\alpha+k)\Gamma(1-\gamma+\alpha+k)\Gamma(m-k)}{k!}(-z)^{-\alpha-k}+\right.$$

$$\left.+(-z)^{-\alpha-m}\sum_{k=0}^{\infty}\frac{\Gamma(\alpha+m+k)\Gamma(1-\gamma+\alpha+m+k)}{k!\,(k+m)!}g(k)\,z^{-k}\right\},$$

where

2. $$g(n) = \ln(-z) + \pi\operatorname{ctg}\pi(\gamma-\alpha) + \psi(n+1) + \psi(n+m+1) -$$
$$- \psi(\alpha+m+n) - \psi(1-\gamma+\alpha+m+n).$$

For $m = 0$, we should set $\sum_{k=0}^{m-1} = 0$.

9.155 This formula loses its meaning when α, γ or $\alpha-\gamma+1$ is equal to one of the numbers $0, -1, -2, \ldots$. In this last case, we have:

1. If α is a nonpositive integer and γ is not an integer, $F(\alpha, \alpha+m; \gamma; z)$ is a polynomial in z.

2. Suppose that γ is a nonpositive integer and that α is not an integer. We then set $\gamma = -\lambda$, where $\lambda = 0, 1, 2, \ldots$. Then,

$$\frac{\Gamma(\alpha+\lambda+1)\Gamma(\alpha+\lambda+m+1)}{\Gamma(\lambda+2)} z^{\lambda+1} F(\alpha+\lambda+1, \alpha+\lambda+m+1; \lambda+2; z)$$

is a solution of the hypergeometric equation that is regular at the point $z=0$. This solution is equal to the right hand member of formula **9.154** 1. if we replace γ with λ in this equation and in formula **9.154** 2.

3. If $\alpha-\gamma+1$ is a nonpositive integer and if α and γ are not themselves integers, we may use the formula

$$F(\alpha, \alpha+m; \gamma; z) = (1-z)^{\gamma-2\alpha-m} F(\gamma-\alpha-m, \gamma-\alpha; \gamma; z)$$

and apply formula **9.154** 1. to its right hand member provided $\gamma-\alpha-m > 0$. However, if $\alpha-\gamma-m \leqslant 0$, the right member of this expression is a polynomial taken to the $(1-z)$th power.

4. If α, β, and γ are integers, the hypergeometric differential equation always has a solution that is regular for $z=0$ and that is of the form

$$R_1(z) + \ln(1-z)R_2(z),$$

where $R_1(z)$ and $R_2(z)$ are rational functions of z. To get a solution of this form, we need to apply formulas **9.137** 1. – **9.137** 3. to the function $F(\alpha, \beta; \gamma; z)$. However, if $\gamma = -\lambda$, where $\lambda+1$ is a natural number, formulas **9.137** 1. and **9.137** 2. should be applied not to $F(\alpha, \beta; \gamma; z)$ but to the function $z^{\lambda+1}F(\alpha+\lambda+1, \beta+\lambda+1; \lambda+2, z)$.

By successive applications of these formulas, we can reduce the positive values of the parameters to the pair, unity and zero. Furthermore, we can obtain the desired form of the solution from the formulas

$$F(1, 1; 2; z) = -z^{-1}\ln(1-z),$$
$$F(0, \beta; \gamma; z) = F(\alpha, 0; \gamma; z) = 1$$

9.16 Riemann's differential equation

9.160 The hypergeometric differential equation is a particular case of Riemann's differential equation

1. $$\frac{d^2u}{dz^2}+\left[\frac{1-\alpha-\alpha'}{z-a}+\frac{1-\beta-\beta'}{z-b}+\frac{1-\gamma-\gamma'}{z-c}\right]\frac{du}{dz}+\left[\frac{\alpha\alpha'(a-b)(a-c)}{z-a}+\frac{\beta\beta'(b-c)(b-a)}{z-b}+\frac{\gamma\gamma'(c-a)(c-b)}{z-c}\right]\frac{u}{(z-a)(z-b)(z-c)}=0.$$ WH

The coefficients of this equation have poles at the points a, b, and c, and the numbers α, α'; β, β'; γ, γ' are called the indices corresponding to these poles. The indices α, α'; β, β'; γ, γ' are related by the following equation:

$$\alpha+\alpha'+\beta+\beta'+\gamma+\gamma'-1=0.$$ WH

2. The differential equations **9.160** 1. are written diagramatically as follows:

3. $$u=P\begin{Bmatrix} a & b & c & \\ \alpha & \beta & \gamma & z \\ \alpha' & \beta' & \gamma' & \end{Bmatrix}.$$

The singular points of the equation appear in the first row in this scheme, the indices corresponding to them appear beneath them, and the independent variable appears in the fourth column. WH

9.161 The two following transformation formulas are valid for Riemann's P-equation:

1. $$\left(\frac{z-a}{z-b}\right)^k\left(\frac{z-c}{z-b}\right)^l P\begin{Bmatrix} a & b & c & \\ \alpha & \beta & \gamma & z \\ \alpha' & \beta' & \gamma' & \end{Bmatrix}=P\begin{Bmatrix} a & b & c & \\ \alpha+k & \beta-k-l & \gamma+l & z \\ \alpha'+k & \beta'-k-l & \gamma'+l & \end{Bmatrix}.$$ WH

2. $$P\begin{Bmatrix} a & b & c & \\ \alpha & \beta & \gamma & z \\ \alpha' & \beta' & \gamma' & \end{Bmatrix}=P\begin{Bmatrix} a_1 & b_1 & c_1 & \\ \alpha & \beta & \gamma & z_1 \\ \alpha' & \beta' & \gamma' & \end{Bmatrix}.$$ WH

The first of these formulas means that if

$$u=P\begin{Bmatrix} a & b & c & \\ \alpha & \beta & \gamma & z \\ \alpha' & \beta' & \gamma' & \end{Bmatrix},$$

the function

$$u_1=\left(\frac{z-a}{z-b}\right)^k\left(\frac{z-c}{z-b}\right)^l u$$

satisfies a second-order differential equation having the same singular points as equation **9.161** 2. and indices equal to $\alpha+k$, $\alpha'+k$; $\beta-k-l$, $\beta'-k-l$; $\gamma+l$, $\gamma'+l$. The second transformation formula converts a differential equation with singularities at the points a, b, and c, indices α, α'; β, β'; γ, γ', and an independent variable z into a differential equation with the same indices, singular points a_1, b_1, and c_1, and independent variable z_1. The variable z_1 is connected with the variable z by the fractional transformation

$$z=\frac{Az_1+B}{Cz_1+D} \qquad [AD-BC\neq 0].$$

The same transformation connects the points a_1, b_1, and c_1 with the points a, b, and c.

WH, MO 20

9.162 By successive application of the two transformation formulas 9.161 1. and 9.161 2., we can convert Riemann's differential equation into the hypergeometric differential equation. Thus, the solution of Riemann's differential equation can be expressed in terms of a hypergeometric function.

For $k=-\alpha$, $l=-\gamma$, and $z_1=\frac{(z-a)(c-b)}{(z-b)(c-a)}$, we have

$$1.\quad u=P\begin{Bmatrix} a & b & c & \\ \alpha & \beta & \gamma & z \\ \alpha' & \beta' & \gamma' & \end{Bmatrix}=\left(\frac{z-a}{z-b}\right)^{\alpha}\left(\frac{z-c}{z-b}\right)^{\gamma}P\begin{Bmatrix} a & b & c & \\ 0 & \beta+\alpha+\gamma & 0 & z \\ \alpha'-\alpha & \beta'+\alpha+\gamma & \gamma'-\gamma & \end{Bmatrix}=$$

$$=\left(\frac{z-a}{z-b}\right)^{\alpha}\left(\frac{z-c}{z-b}\right)^{\gamma}P\begin{Bmatrix} 0 & \infty & 1 & \\ 0 & \beta+\alpha+\gamma & 0 & \frac{(z-a)(c-b)}{(z-b)(c-a)} \\ \alpha'-\alpha & \beta'+\alpha+\gamma & \gamma'-\gamma & \end{Bmatrix}.$$

MO 23

Thus, this solution can be expressed as a hypergeometric series as follows:

$$2.\quad u=\left(\frac{z-a}{z-b}\right)^{\alpha}\left(\frac{z-c}{z-b}\right)^{\gamma}F\left(\alpha+\beta+\gamma,\ \alpha+\beta'+\gamma;\ 1+\alpha-\alpha';\ \frac{(z-a)(c-b)}{(z-b)(c-a)}\right).$$

If the constants a, b, c; α, α'; β, β'; γ, γ' are permuted in a suitable manner, Riemann's equation remains unchanged. Thus, we obtain a set of 24 solutions of differential equations having the following form (provided none of the differences $\alpha-\alpha'$, $\beta-\beta'$, $\gamma-\gamma'$ are integers):

WH, MO 23

9.163

$$1.\quad u_1=\left(\frac{z-a}{z-b}\right)^{\alpha}\left(\frac{z-c}{z-b}\right)^{\gamma}F\left\{\alpha+\beta+\gamma,\ \alpha+\beta'+\gamma;\ 1+\alpha-\alpha';\ \frac{(c-b)(z-a)}{(c-a)(z-b)}\right\}.$$

$$2.\quad u_2=\left(\frac{z-a}{z-b}\right)^{\alpha'}\left(\frac{z-c}{z-b}\right)^{\gamma}F\left\{\alpha'+\beta+\gamma,\ \alpha'+\beta'+\gamma;\ 1+\alpha'-\alpha;\ \frac{(c-b)(z-a)}{(c-a)(z-b)}\right\}.$$

$$3.\quad u_3=\left(\frac{z-a}{z-b}\right)^{\alpha}\left(\frac{z-c}{z-b}\right)^{\gamma'}F\left\{\alpha+\beta+\gamma',\ \alpha+\beta'+\gamma';\ 1+\alpha-\alpha';\ \frac{(c-b)(z-a)}{(c-a)(z-b)}\right\}.$$

$$4.\quad u_4=\left(\frac{z-a}{z-b}\right)^{\alpha'}\left(\frac{z-c}{z-b}\right)^{\gamma'}F\left\{\alpha'+\beta+\gamma',\ \alpha'+\beta'+\gamma';\ 1+\alpha'-\alpha;\ \frac{(c-b)(z-a)}{(c-a)(z-b)}\right\}.$$

9.164

$$1.\quad u_5=\left(\frac{z-b}{z-c}\right)^{\beta}\left(\frac{z-a}{z-c}\right)^{\alpha}F\left\{\beta+\gamma+\alpha\ \ \beta+\gamma'+\alpha;\ 1+\beta-\beta';\ \frac{(a-c)(z-b)}{(a-b)(z-c)}\right\}.$$

$$2\quad u_6=\left(\frac{z-b}{z-c}\right)^{\beta'}\left(\frac{z-a}{z-c}\right)^{\alpha}F\left\{\beta'+\gamma+\alpha,\ \beta'+\gamma'+\alpha;\ 1+\beta'-\beta;\ \frac{(a-c)(z-b)}{(a-b)(z-c)}\right\}.$$

$$3.\quad u_7=\left(\frac{z-b}{z-c}\right)^{\beta}\left(\frac{z-a}{z-c}\right)^{\alpha'}F\left\{\beta+\gamma+\alpha',\ \beta+\gamma'+\alpha';\ 1+\beta-\beta';\ \frac{(a-c)(z-b)}{(a-b)(z-c)}\right\}.$$

$$4.\quad u_8=\left(\frac{z-b}{z-c}\right)^{\beta'}\left(\frac{z-a}{z-c}\right)^{\alpha'}F\left\{\beta'+\gamma+\alpha',\ \beta'+\alpha'+\gamma';\ 1+\beta'-\beta;\ \frac{(a-c)(z-b)}{(a-b)(z-c)}\right\}.$$

9.165

1. $u_9 = \left(\frac{z-c}{z-a}\right)^{\gamma} \left(\frac{z-b}{z-a}\right)^{\beta} F\left\{\gamma+\alpha+\beta,\ \gamma+\alpha'+\beta;\ 1+\gamma-\gamma';\ \frac{(b-a)(z-c)}{(b-c)(z-a)}\right\}$

2. $u_{10} = \left(\frac{z-c}{z-a}\right)^{\gamma'} \left(\frac{z-b}{z-a}\right)^{\beta} F\left\{\gamma'+\alpha+\beta,\ \gamma'+\alpha'+\beta;\ 1+\gamma'-\gamma;\ \frac{(b-a)(z-c)}{(b-c)(z-a)}\right\}.$

3. $u_{11} = \left(\frac{z-c}{z-a}\right)^{\gamma} \left(\frac{z-b}{z-a}\right)^{\beta'} F\left\{\gamma+\alpha+\beta',\ \gamma+\alpha'+\beta';\ 1+\gamma-\gamma';\ \frac{(b-a)(z-c)}{(b-c)(z-a)}\right\}.$

4. $u_{12} = \left(\frac{z-c}{z-a}\right)^{\gamma'} \left(\frac{z-b}{z-a}\right)^{\beta'} F\left\{\gamma'+\alpha+\beta',\ \gamma'+\alpha'+\beta';\ 1+\gamma'-\gamma;\ \frac{(b-a)(z-c)}{(b-c)(z-a)}\right\}.$

9.166

1. $u_{13} = \left(\frac{z-a}{z-c}\right)^{\alpha} \left(\frac{z-b}{z-c}\right)^{\beta} F\left\{\alpha+\gamma+\beta,\ \alpha+\gamma'+\beta;\ 1+\alpha-\alpha';\ \frac{(b-c)(z-a)}{(b-a)(z-c)}\right\}.$

2. $u_{14} = \left(\frac{z-a}{z-c}\right)^{\alpha'} \left(\frac{z-b}{z-c}\right)^{\beta} F\left\{\alpha'+\gamma+\beta,\ \alpha'+\gamma'+\beta;\ 1+\alpha'-\alpha;\ \frac{(b-c)(z-a)}{(b-a)(z-c)}\right\}.$

3. $u_{15} = \left(\frac{z-a}{z-c}\right)^{\alpha} \left(\frac{z-b}{z-c}\right)^{\beta'} F\left\{\alpha+\gamma+\beta',\ \alpha+\gamma'+\beta';\ 1+\alpha-\alpha';\ \frac{(b-c)(z-a)}{(b-a)(z-c)}\right\}.$

4. $u_{16} = \left(\frac{z-a}{z-c}\right)^{\alpha'} \left(\frac{z-b}{z-c}\right)^{\beta'} F\left\{\alpha'+\gamma+\beta',\ \alpha'+\gamma'+\beta';\ 1+\alpha'-\alpha;\ \frac{(b-c)(z-a)}{(b-a)(z-c)}\right\}.$

9.167

1. $u_{17} = \left(\frac{z-c}{z-b}\right)^{\gamma} \left(\frac{z-a}{z-b}\right)^{\alpha} F\left\{\gamma+\beta+\alpha,\ \gamma+\beta'+\alpha;\ 1+\gamma-\gamma';\ \frac{(a-b)(z-c)}{(a-c)(z-b)}\right\}.$

2. $u_{18} = \left(\frac{z-c}{z-b}\right)^{\gamma'} \left(\frac{z-a}{z-b}\right)^{\alpha} F\left\{\gamma'+\beta+\alpha,\ \gamma'+\beta'+\alpha;\ 1+\gamma'-\gamma;\ \frac{(a-b)(z-c)}{(a-c)(z-b)}\right\}.$

3. $u_{19} = \left(\frac{z-c}{z-b}\right)^{\gamma} \left(\frac{z-a}{z-b}\right)^{\alpha'} F\left\{\gamma+\beta+\alpha',\ \gamma+\beta'+\alpha';\ 1+\gamma-\gamma';\ \frac{(a-b)(z-c)}{(a-c)(z-b)}\right\}.$

4. $u_{20} = \left(\frac{z-c}{z-b}\right)^{\gamma'} \left(\frac{z-a}{z-b}\right)^{\alpha'} F\left\{\gamma'+\beta+\alpha',\ \gamma'+\beta'+\alpha';\ 1+\gamma'-\gamma;\ \frac{(a-b)(z-c)}{(a-c)(z-b)}\right\}.$

9.168

1. $u_{21} = \left(\frac{z-b}{z-a}\right)^{\beta} \left(\frac{z-c}{z-a}\right)^{\gamma} F\left\{\beta+\alpha+\gamma,\ \beta+\alpha'+\gamma;\ 1+\beta-\beta';\ \frac{(c-a)(z-b)}{(c-b)(z-a)}\right\}.$

2. $u_{22} = \left(\frac{z-b}{z-a}\right)^{\beta'} \left(\frac{z-c}{z-a}\right)^{\gamma} F\left\{\beta'+\alpha+\gamma,\ \beta'+\alpha'+\gamma;\ 1+\beta'-\beta;\ \frac{(c-a)(z-b)}{(c-b)(z-a)}\right\}.$

3. $u_{23} = \left(\frac{z-b}{z-a}\right)^{\beta} \left(\frac{z-c}{z-a}\right)^{\gamma'} F\left\{\beta+\alpha+\gamma',\ \beta+\alpha'+\gamma';\ 1+\beta-\beta';\ \frac{(c-a)(z-b)}{(c-b)(z-a)}\right\}.$

4. $u_{24} = \left(\frac{z-b}{z-a}\right)^{\beta'} \left(\frac{z-c}{z-a}\right)^{\gamma'} F\left\{\beta'+\alpha+\gamma',\ \beta'+\alpha'+\gamma';\ 1+\beta'-\beta;\ \frac{(c-a)(z-b)}{(c-b)(z-a)}\right\}.$

WH

9.17 Representation of certain second-order differential equations by means of a Riemann scheme

9.171 The hypergeometric equation (see 9.151);

$$u = P\left\{\begin{matrix} 0 & \infty & 1 & \\ 0 & \alpha & 0 & z \\ 1-\gamma & \beta & \gamma-\alpha-\beta & \end{matrix}\right\}. \qquad \text{WH}$$

9.172 The associated Legendre's equation defining the functions $P_n^m(z)$ for n and m integers (see 8.700 1.):

1. $$u = P\left\{\begin{matrix} 0 & \infty & 1 & \\ \frac{1}{2}m & n+1 & \frac{1}{2}m & \frac{1-z}{2} \\ -\frac{1}{2}m & -n & -\frac{1}{2}m & \end{matrix}\right\}. \qquad \text{WH}$$

2. $$u = P\left\{\begin{matrix} 0 & \infty & 1 & \\ -\frac{1}{2}n & \frac{1}{2}m & 0 & \frac{1}{1-z^2} \\ \frac{n+1}{2} & -\frac{1}{2}m & \frac{1}{2} & \end{matrix}\right\}. \qquad \text{WH}$$

9.173 The function $P_n^m\left(1-\frac{z^2}{2n^2}\right)$ satisfies the equation

$$u = P\left\{\begin{matrix} 4n^2 & \infty & 0 & \\ \frac{1}{2}m & n+1 & \frac{1}{2}m & z^2 \\ -\frac{1}{2}m & -n & -\frac{1}{2}m & \end{matrix}\right\}. \qquad \text{WH}$$

The function $J_m(z)$ satisfies the limiting form of this equation obtained as $n \to \infty$.

9.174 The equation defining the polynomials $C_n^\lambda(z)$ (see 8.938):

$$u = P\left\{\begin{matrix} -1 & \infty & 1 & \\ \frac{1}{2}-\lambda & n+2\lambda & \frac{1}{2}-\lambda & z \\ 0 & -n & 0 & \end{matrix}\right\}. \qquad \text{WH}$$

9.175 Bessel's equation (see 8.401) is the limiting form of the equations:

1. $$u = P\left\{\begin{matrix} 0 & \infty & c & \\ n & ic & \frac{1}{2}+ic & z \\ -n & -ic & \frac{1}{2}-ic & \end{matrix}\right\}, \qquad \text{WH}$$

2. $u = e^{iz} P \begin{Bmatrix} 0 & \infty & c & \\ n & \frac{1}{2} & 0 & z \\ -n & \frac{3}{2} - 2ic & 2ic - 1 & \end{Bmatrix},$ WH

3. $u = P \begin{Bmatrix} 0 & \infty & c^2 & \\ \frac{1}{2} n & \frac{1}{2}(c-n) & 0 & z^2 \\ -\frac{1}{2} n & -\frac{1}{2}(c+n) & n+1 & \end{Bmatrix},$ WH

as $c \to \infty$.

9.18 Hypergeometric functions of two variables

9.180

1. $$F_1(\alpha, \beta, \beta'; \gamma; x, y) = \sum_{m=0}^{\infty} \sum_{n=0}^{\infty} \frac{(\alpha)_{m+n} (\beta)_m (\beta')_n}{(\gamma)_{m+n} m!\, n!} x^m y^n.$$

EH I 224(6), AK 14(11)

Region of convergence

$$|x| < 1, \quad |y| < 1.$$ AK 16

2. $$F_2(\alpha, \beta, \beta', \gamma, \gamma'; x, y) = \sum_{m=0}^{\infty} \sum_{n=0}^{\infty} \frac{(\alpha)_{m+n} (\beta)_m (\beta')_n}{(\gamma)_m (\gamma')_n m!\, n!} x^m y^n.$$

EH I 224(7), AK 14(12)

Region of convergence

$$|x| + |y| < 1.$$ AK 17

3. $$F_3(\alpha, \alpha', \beta, \beta', \gamma; x, y) = \sum_{m=0}^{\infty} \sum_{n=0}^{\infty} \frac{(\alpha)_m (\alpha')_n (\beta)_m (\beta')_n}{(\gamma)_{m+n}\, m!\, n!} x^m y^n.$$

EH I 224(8), AK 14(13)

Region of convergence

$$|x| < 1, \quad |y| < 1.$$ AK 17

4. $$F_4(\alpha, \beta, \gamma, \gamma'; x, y) = \sum_{m=0}^{\infty} \sum_{n=0}^{\infty} \frac{(\alpha)_{m+n} (\beta)_{m+n}}{(\gamma)_m (\gamma')_n m!\, n!} x^m y^n.$$

EH I 224(9), AK 14(14)

Region of convergence

$$|\sqrt{x}| + |\sqrt{y}| < 1.$$ AK 18

9.181 The functions F_1, F_2, F_3, and F_4 satisfy the following systems of partial differential equations for z:

1. System of equations for $z=F_1$:

$$x(1-x)\frac{\partial^2 z}{\partial x^2}+y(1-x)\frac{\partial^2 z}{\partial x\,\partial y}+$$
$$+[\gamma-(\alpha+\beta+1)x]\frac{\partial z}{\partial x}-\beta y\frac{\partial z}{\partial y}-\alpha\beta z=0,$$
$$y(1-y)\frac{\partial^2 z}{\partial y^2}+x(1-y)\frac{\partial^2 z}{\partial x\,\partial y}+$$
$$+[\gamma-(\alpha+\beta'+1)y]\frac{\partial z}{\partial y}-\beta' x\frac{\partial z}{\partial x}-\alpha\beta' z=0.$$

EH I 233(9)

2. System of equations for $z=F_2$:

$$x(1-x)\frac{\partial^2 z}{\partial x^2}-xy\frac{\partial^2 z}{\partial x\,\partial y}+[\gamma-(\alpha+\beta+1)x]\frac{\partial z}{\partial x}-$$
$$-\beta y\frac{\partial z}{\partial y}-\alpha\beta z=0,$$
$$y(1-y)\frac{\partial^2 z}{\partial y^2}-xy\frac{\partial^2 z}{\partial x\,\partial y}+[\gamma'-(\alpha+\beta'+1)y]\frac{\partial z}{\partial y}-$$
$$-\beta' x\frac{\partial z}{\partial x}-\alpha\beta' z=0.$$

EH I 234(10)

3. System of equations for $z=F_3$:

$$x(1-x)\frac{\partial^2 z}{\partial x^2}+y\frac{\partial^2 z}{\partial x\,\partial y}+$$
$$+[\gamma-(\alpha+\beta+1)x]\frac{\partial z}{\partial x}-\alpha\beta z=0,$$
$$y(1-y)\frac{\partial^2 z}{\partial y^2}+x\frac{\partial^2 z}{\partial x\,\partial y}+$$
$$+[\gamma-(\alpha'+\beta'+1)y]\frac{\partial z}{\partial y}-\alpha'\beta' z=0.$$

EH I 234(11)

4. System of equations for $z=F_4$:

$$x(1-x)\frac{\partial^2 z}{\partial x^2}-y^2\frac{\partial^2 z}{\partial y^2}-2xy\frac{\partial^2 z}{\partial x\,\partial y}+$$
$$+[\gamma-\ \alpha+\beta+1)x]\frac{\partial z}{\partial x}-(\alpha+\beta+1)y\frac{\partial z}{\partial y}-\alpha\beta z=0,$$
$$y(1-y)\frac{\partial^2 z}{\partial y^2}-x^2\frac{\partial^2 z}{\partial x^2}-2xy\frac{\partial^2 z}{\partial x\,\partial y}+$$
$$+[\gamma'-(\alpha+\beta+1)y]\frac{\partial z}{\partial y}-(\alpha+\beta+1)x\frac{\partial z}{\partial x}-\alpha\beta z=0.$$

EH I 234(12) AK 44

9.182 For certain relationships between the parameters and the argument, hypergeometric functions of two variables can be expressed in terms of hypergeometric functions of a single variable or in terms of elementary functions:

1. $$F_1(\alpha,\ \beta,\ \beta',\ \beta+\beta';\ x,\ y)=(1-y)^{-\alpha}F\left(\alpha,\ \beta;\ \beta+\beta';\ \frac{x-y}{1-y}\right).$$

EH I 238(1), AK 24(28)

2. $$F_2(\alpha,\ \beta,\ \beta',\ \beta,\ \gamma';\ x,\ y)=(1-x)^{-\alpha}F\left(\alpha,\ \beta';\ \gamma';\ \frac{y}{1-x}\right).$$

EH I 238(2), AK 23

3. $F_2(\alpha, \beta, \beta', \alpha, \alpha; x, y) = (1-x)^{-\beta}(1-y)^{-\beta'} F\left[\beta, \beta'; \alpha; \frac{xy}{(1-x)(1-y)}\right].$

EH I 238(3)

4. $F_3(\alpha, \gamma-\alpha, \beta, \gamma-\beta, \gamma; x, y) = (1-y)^{\alpha+\beta-\gamma} F(\alpha, \beta; \gamma; x+y-xy).$

EH I 238(4), AK 25(35)

5. $F_4[\alpha, \gamma+\gamma'-\alpha-1, \gamma, \gamma'; x(1-y), y(1-x)] =$
$= F(\alpha, \gamma+\gamma'-\alpha-1; \gamma; x) F(\alpha, \gamma+\gamma'-\alpha-1; \gamma'; y).$ EH I 238(5)

6. $F_4\left[\alpha, \beta, \alpha, \beta; -\frac{x}{(1-x)(1-y)}, \frac{-y}{(1-x)(1-y)}\right] = \frac{(1-x)^{\beta}(1-y)^{\alpha}}{(1-xy)}.$

EH I 238(6)

7. $F_4\left[\alpha, \beta, \beta, \beta; -\frac{x}{(1-x)(1-y)}, -\frac{y}{(1-x)(1-y)}\right] =$
$= (1-x)^{\alpha}(1-y)^{\alpha} F(\alpha, 1+\alpha-\beta; \beta; xy).$ EH I 238(7)

8. $F_4\left[\alpha, \beta, 1+\alpha-\beta, \beta; -\frac{x}{(1-x)(1-y)}, -\frac{y}{(1-x)(1-y)}\right] =$
$= (1-y)^{\alpha} F\left[\alpha, \beta; 1+\alpha-\beta; -\frac{x(1-y)}{1-x}\right].$ EH I 238(8)

9. $F_4\left(\alpha, \alpha+\frac{1}{2}, \gamma, \frac{1}{2}; x, y\right) =$
$= \frac{1}{2}(1+\sqrt{y})^{-2\alpha} F\left(\alpha, \alpha+\frac{1}{2}; \gamma; \frac{x}{(1+\sqrt{y})^2}\right) +$
$+ \frac{1}{2}(1-\sqrt{y})^{-2\alpha} F\left(\alpha, \alpha+\frac{1}{2}; \gamma; \frac{x}{(1-\sqrt{y})^2}\right).$ AK 23

10. $F_1(\alpha, \beta, \beta', \gamma; x, 1) = \frac{\Gamma(\gamma)\Gamma(\gamma-\alpha-\beta')}{\Gamma(\gamma-\alpha)\Gamma(\gamma-\beta')} F(\alpha, \beta; \gamma-\beta'; x).$

EH I 239(10), AK 22(23)

11. $F_1(\alpha, \beta, \beta', \gamma; x, x) = F(\alpha, \beta+\beta'; \gamma; x).$ EH I 239(11), AK 23(25)

9.183 Functional relations between hypergeometric functions of two variables:

1. $F_1(\alpha, \beta, \beta', \gamma; x, y) =$
$= (1-x)^{-\beta}(1-y)^{-\beta'} F_1\left(\gamma-\alpha, \beta, \beta', \gamma; \frac{x}{x-1}, \frac{y}{y-1}\right);$ EH I 239(1)
$= (1-x)^{-\alpha} F_1\left(\alpha, \gamma-\beta-\beta', \beta', \gamma; \frac{x}{x-1}, \frac{y-x}{1-x}\right);$ EH I 239(2)
$= (1-y)^{-\alpha} F_1\left(\alpha, \beta, \gamma-\beta-\beta', \gamma; \frac{y-x}{y-1}, \frac{y}{y-1}\right);$ EH I 239(3)
$= (1-x)^{\gamma-\alpha-\beta}(1-y)^{-\beta'} F_1\left(\gamma-\alpha, \gamma-\beta-\beta', \beta', \gamma; x, \frac{x-y}{1-y}\right);$

EH I 240(4)

$= (1-x)^{-\beta}(1-y)^{\gamma-\alpha-\beta'} F_1\left(\gamma-\alpha, \beta, \gamma-\beta-\beta', \gamma; \frac{x-y}{x-1}, y\right).$

EH I 240(5), AK 30(5)

2. $F_2(\alpha, \beta, \beta'; \gamma, \gamma'; x, y) =$

$$= (1-x)^{-\alpha} F_2\left(\alpha, \gamma-\beta, \beta', \gamma, \gamma'; \frac{x}{x-1}, \frac{y}{1-x}\right);$$ EH I 240(6)

$$= (1-y)^{-\alpha} F_2\left(\alpha, \beta, \gamma'-\beta', \gamma, \gamma'; \frac{x}{1-y}, \frac{y}{y-1}\right);$$ EH I 240(7)

$$= (1-x-y)^{-\alpha} F_2\left(\alpha, \gamma-\beta, \gamma'-\beta', \gamma, \gamma'; \frac{x}{x+y-1}, \frac{y}{x+y-1}\right)$$

EH I 240(8), AK 32(6)

3. $F_4(\alpha, \beta, \gamma, \gamma'; x, y) =$

$$= \frac{\Gamma(\gamma')\Gamma(\beta-\alpha)}{\Gamma(\gamma'-\alpha)\Gamma(\beta)}(-y)^{-\alpha} F_4\left(\alpha, \alpha+1-\gamma', \gamma, \alpha+1-\beta; \frac{x}{y}, \frac{1}{y}\right) +$$
$$+ \frac{\Gamma(\gamma')\Gamma(\alpha-\beta)}{\Gamma(\gamma'-\beta)\Gamma(\alpha)}(-y)^{\beta} F_4\left(\beta+1-\gamma', \beta, \gamma, \beta+1-\alpha; \frac{x}{y}, \frac{1}{y}\right).$$

EH I 240(9), AK 26(37)

9.184 Integral representations:

Double integrals of the Euler type

1. $$F_1(\alpha, \beta, \beta', \gamma; x, y) = \frac{\Gamma(\gamma)}{\Gamma(\beta)\Gamma(\beta')\Gamma(\gamma-\beta-\beta')} \times$$
$$\times \iint\limits_{\substack{u \geqslant 0,\ v \geqslant 0 \\ u+v \leqslant 1}} u^{\beta-1} v^{\beta'-1} (1-u-v)^{\gamma-\beta-\beta'-1} (1-ux-vy)^{-\alpha}\, du\, dv$$
$[\operatorname{Re}\beta > 0, \operatorname{Re}\beta' > 0, \operatorname{Re}(\gamma-\beta-\beta') > 0].$ EH I 230(1), AK 28(1)

2. $$F_2(\alpha, \beta, \beta', \gamma, \gamma'; x, y) = \frac{\Gamma(\gamma)\Gamma(\gamma')}{\Gamma(\beta)\Gamma(\beta')\Gamma(\gamma-\beta)\Gamma(\gamma'-\beta')} \times$$
$$\times \int_0^1 \int_0^1 u^{\beta-1} v^{\beta'-1} (1-u)^{\gamma-\beta-1} (1-v)^{\gamma'-\beta'-1} (1-ux-vy)^{-\alpha}\, du\, dv$$
$[\operatorname{Re}\beta > 0, \operatorname{Re}\beta' > 0, \operatorname{Re}(\gamma-\beta) > 0, \operatorname{Re}(\gamma'-\beta') > 0].$

EH I 230(2), AK 28(2)

3. $$F_3(\alpha, \alpha', \beta, \beta', \gamma; x, y) = \frac{\Gamma(\gamma)}{\Gamma(\beta)\Gamma(\beta')\Gamma(\gamma-\beta-\beta')} \times$$
$$\times \iint\limits_{\substack{u \geqslant 0,\ v \geqslant 0 \\ u+v \leqslant 1}} u^{\beta-1} v^{\beta'-1} (1-u-v)^{-\gamma-\beta-\beta'-1} (1-ux)^{-\alpha} (1-vy)^{-\alpha'}\, du\, dv$$
$[\operatorname{Re}\beta > 0, \operatorname{Re}\beta' > 0, \operatorname{Re}(\gamma-\beta-\beta') > 0].$ EH I 230(3), AK 28(3)

4. $F_4[\alpha, \beta, \gamma, \gamma'; x(1-y), y(1-x)] =$
$$= \frac{\Gamma(\gamma)\Gamma(\gamma')}{\Gamma(\alpha)\Gamma(\beta)\Gamma(\gamma-\alpha)\Gamma(\gamma'-\beta)} \int_0^1 \int_0^1 u^{\alpha-1} v^{\beta-1} (1-u)^{\gamma-\alpha-1} (1-v)^{\gamma'-\beta-1} \times$$
$$\times (1-ux)^{\alpha-\gamma-\gamma'+1} (1-vy)^{\beta-\gamma-\gamma'+1} (1-ux-vy)^{\gamma+\gamma'-\alpha-\beta-1}\, du\, dv$$
$[\operatorname{Re}\alpha > 0, \operatorname{Re}\beta > 0, \operatorname{Re}(\gamma-\alpha) > 0, \operatorname{Re}(\gamma'-\beta) > 0].$ EH I 230(4)

Integrals of the Mellin-Barnes type

9.185 The functions F_1, F_2, F_3 and F_4 can be represented by means of double integrals of the following form:

$$F(x, y) = \frac{\Gamma(\gamma)}{\Gamma(\alpha)\Gamma(\beta)(2\pi i)^2} \int_{-i\infty}^{i\infty} \int_{-i\infty}^{i\infty} \Psi(s, t)\Gamma(-s)\Gamma(-t)(-x)^s(-y)^t\,ds\,dt.$$

$\Psi(s, t)$	$F(x, y)$
$\frac{\Gamma(\alpha+s+t)\Gamma(\beta+s)\Gamma(\beta'+t)}{\Gamma(\beta')\Gamma(\gamma+s+t)}$	$F_1(\alpha, \beta, \beta', \gamma; x, y)$
$\frac{\Gamma(\alpha+s+t)\Gamma(\beta+s)\Gamma(\beta'+t)\Gamma(\gamma')}{\Gamma(\beta')\Gamma(\gamma+s)\Gamma(\gamma'+t)}$	$F_2(\alpha, \beta, \beta', \gamma, \gamma'; x, y)$
$\frac{\Gamma(\alpha+s)\Gamma(\alpha'+t)\Gamma(\beta+s)\Gamma(\beta'+t)}{\Gamma(\alpha')\Gamma(\beta')\Gamma(\gamma+s+t)}$	$F_3(\alpha, \alpha', \beta, \beta'\ \gamma; x, y)$
$\frac{\Gamma(\alpha+s+t)\Gamma(\beta+s+t)\Gamma(\gamma')}{\Gamma(\gamma+s)\Gamma(\gamma'+t)}$	$F_4(\alpha, \beta, \gamma, \gamma'; x, y)$

[α, α', β, β' may not be negative integers]. EH I 232(9-13), AK 41(33)

9.19 A hypergeometric function of several variables

$$F_A(\alpha; \beta_1, \ldots, \beta_n; \gamma_1, \ldots, \gamma_n; z_1, \ldots, z_n) =$$
$$= \sum_{m_1=0}^{\infty} \sum_{m_2=0}^{\infty} \cdots \sum_{m_n=0}^{\infty} \frac{(\alpha)_{m_1+\ldots+m_n}(\beta_1)_{m_1}\cdots(\beta_n)_{m_n}}{(\gamma_1)_{m_1}\cdots(\gamma_n)_{m_n} m_1! \ldots m_n!} z_1^{m_1} z_2^{m_2} \ldots z_n^{m_n}.$$

ET I 385

9.2 A Degenerate Hypergeometric Function

9.20 Introduction

9.201 A *degenerate hypergeometric function* is obtained by taking the limit as $c \to \infty$ in the solution of Riemann's differential equation

$$P\begin{Bmatrix} 0 & \infty & c \\ \frac{1}{2}+\mu & -c & c-\lambda & z \\ \frac{1}{2}-\mu & 0 & \lambda \end{Bmatrix}.$$ WH

9.202 The equation obtained by means of this limiting process is of the form

1. $$\frac{d^2u}{dz^2} + \frac{du}{dz} + \left(\frac{\lambda}{z} + \frac{\frac{1}{4}-\mu^2}{z^2}\right)u = 0.$$ WH

Equation 9.202 1. has the following two linearly independent solutions:

2. $z^{\frac{1}{2}+\mu} e^{-z} \Phi\left(\frac{1}{2}+\mu-\lambda,\ 2\mu+1;\ z\right)$,

3. $z^{\frac{1}{2}-\mu} e^{-z} \Phi\left(\frac{1}{2}-\mu-\lambda,\ -2\mu+1;\ z\right)$,

which are defined for all values of $\mu \neq \pm \frac{1}{2},\ \pm \frac{2}{2},\ \pm \frac{3}{2}, \ldots$

MO 111

9.21 The functions $\Phi(\alpha, \gamma; z)$ and $\Psi(\alpha, \gamma; z)$

9.210 The series

1. $$\Phi(\alpha, \gamma; z) = 1 + \frac{\alpha}{\gamma} \frac{z}{1!} + \frac{\alpha(\alpha+1)}{\gamma(\gamma+1)} \frac{z^2}{2!} + \frac{\alpha(\alpha+1)(\alpha+2)}{\gamma(\gamma+1)(\gamma+2)} \frac{z^3}{3!} + \ldots$$

is also called a *degenerate hypergeometric function*.

A second notation: $\Phi(\alpha, \gamma; z) = {}_1F_1(\alpha, \gamma; z)$.

2. $$\Psi(\alpha, \gamma; z) = \frac{\Gamma(1-\gamma)}{\Gamma(\alpha-\gamma+1)} \Phi(\alpha, \gamma; z) + \frac{\Gamma(\gamma-1)}{\Gamma(\alpha)} z^{1-\gamma} \Phi(\alpha-\gamma+1, 2-\gamma; z).$$

EH I 257(7)

9.211 Integral representation:

1. $$\Phi(\alpha, \gamma; z) = \frac{2^{1-\gamma} e^{\frac{1}{2}z}}{\mathrm{B}(\alpha, \gamma-\alpha)} \int_{-1}^{1} (1-t)^{\gamma-\alpha-1} (1+t)^{\alpha-1} e^{\frac{1}{2}zt}\, dt$$

$[0 < \operatorname{Re}\alpha < \operatorname{Re}\gamma]$. MO 114

2. $$\Phi(\alpha, \gamma; z) = \frac{1}{\mathrm{B}(\alpha, \gamma-\alpha)} z^{1-\gamma} \int_{0}^{z} e^{t} t^{\alpha-1} (z-t)^{\gamma-\alpha-1}\, dt$$

$[0 < \operatorname{Re}\alpha < \operatorname{Re}\gamma]$. MO 114

3. $$\Phi(-\nu, \alpha+1; z) = \frac{\Gamma(\alpha+1)}{\Gamma(\alpha+\nu+1)} e^{z} z^{-\frac{\alpha}{2}} \int_{0}^{\infty} e^{-t} t^{\nu+\frac{\alpha}{2}} J_{\alpha}(2\sqrt{zt})\, dt$$

$\left[\operatorname{Re}(\alpha+\nu+1) > 0,\quad |\arg z| < \frac{\pi}{2}\right]$. MO 115

4. $$\Psi(\alpha, \gamma; z) = \frac{1}{\Gamma(\alpha)} \int_{0}^{\infty} e^{-zt} t^{\alpha-1} (1+t)^{\gamma-\alpha-1}\, dt \quad [\operatorname{Re}\alpha > 0].$$

ET I 255(2)

Functional relations

9.212

1. $\Phi(\alpha, \gamma; z) = e^{z} \Phi(\gamma-\alpha, \gamma; -z)$. MO 112

2. $\frac{z}{\gamma} \Phi(\alpha+1, \gamma+1; z) = \Phi(\alpha+1, \gamma; z) - \Phi(\alpha, \gamma; z)$.

3. $\alpha\Phi(\alpha+1, \gamma+1; z) = (\alpha-\gamma)\Phi(\alpha, \gamma+1; z) + \gamma\Phi(\alpha, \gamma; z)$. MO 112

4. $$\alpha\Phi(\alpha+1, \gamma; z) = (z+2\alpha-\gamma)\Phi(\alpha, \gamma; z) + (\gamma-\alpha)\Phi(\alpha-1, \gamma; z).$$

MO 112

9.213 $\frac{d\Phi}{dz} = \frac{\alpha}{\gamma} \Phi(\alpha+1, \gamma+1; z)$. MO 112

9.214 $$\lim_{\gamma\to -n}\frac{1}{\Gamma(\gamma)}\Phi(\alpha,\ \gamma;\ z)=z^{n+1}\binom{\alpha+n}{n+1}\Phi(\alpha+n+1,\ n+2;\ z)$$
$$[n=0,\ 1,\ 2,\ \ldots].$$ MO 112

9.215

1. $\Phi(\alpha,\ \alpha;\ z)=e^z.$ MO 15

2. $\Phi(\alpha,\ 2\alpha;\ 2z)=2^{\alpha-\frac{1}{2}}\exp\left[\frac{1}{4}(1-2\alpha)\pi i\right]\Gamma\left(\alpha+\frac{1}{2}\right)e^z z^{\frac{1}{2}-\alpha}J_{\alpha-\frac{1}{2}}\left(ze^{\frac{\pi}{2}i}\right).$ MO 112

3. $\Phi\left(p+\frac{1}{2},\ 2p+1;\ 2iz\right)=\Gamma(p+1)\left(\frac{z}{2}\right)^{-p}e^{iz}J_p(z).$ MO 15

For a representation of special functions in terms of a degenerate hypergeometric function $\Phi(\alpha,\ \gamma;\ z)$, see:
for the probability integral, 9.236;
for integrals of Bessel functions, 6.631 1.;
for Hermite polynomials, 8.953 and 8.959;
for Laguerre polynomials, 8.972 1.;
for parabolic-cylinder functions, 9.240;
for the functions $M_{\lambda,\mu}(z)$, 9.220 2. and 9.220 3.;
for the functions $W_{p,q}(z)$ 9.239.

9.216 The function $\Phi(\alpha,\ \gamma;\ z)$ is a solution of the differential equation

1. $$z\frac{d^2F}{dz^2}+(\gamma-z)\frac{dF}{dz}-\alpha F=0.$$ MO 111

This equation has two linearly independent solutions:

2. $\Phi(\alpha,\ \gamma;\ z)$

3. $z^{1-\gamma}\Phi(\alpha-\gamma+1,\ 2-\gamma;\ z)$ MO 112

9.22-9.23 The Whittaker functions $M_{\lambda,\mu}(z)$ and $W_{\lambda,\mu}(z)$

9.220 If we make the change of variable $u=e^{-\frac{z}{2}}W$ in equation 9.202 1., we obtain the equation

1. $$\frac{d^2W}{dz^2}+\left(-\frac{1}{4}+\frac{\lambda}{z}+\frac{\frac{1}{4}-\mu^2}{z^2}\right)W=0.$$ MO 115

Equation 9.220 1. has the following two linearly independent solutions:

2. $$M_{\lambda,\mu}(z)=z^{\mu+\frac{1}{2}}e^{-\frac{z}{2}}\Phi\left(\mu-\lambda+\frac{1}{2},\ 2\mu+1;\ z\right).$$

3. $$M_{\lambda,-\mu}(z)=z^{-\mu+\frac{1}{2}}e^{-\frac{z}{2}}\Phi\left(-\mu-\lambda+\frac{1}{2},\ -2\mu+1;\ z\right).$$ MO 115

To obtain solutions that are also suitable for $2\mu=\pm 1,\ \pm 2,\ \ldots,$ we introduce Whittaker's function

4. $$W_{\lambda,\mu}(z)=\frac{\Gamma(-2\mu)}{\Gamma\left(\frac{1}{2}-\mu-\lambda\right)}M_{\lambda,\mu}(z)+\frac{\Gamma(2\mu)}{\Gamma\left(\frac{1}{2}+\mu-\lambda\right)}M_{\lambda,-\mu}(z),$$ WH

which, for 2μ, approaches an integer and is also a solution of equation **9.220** 1.

For the functions $M_{\lambda,\mu}(z)$ and $W_{\lambda,\mu}(z)$, $z=0$ is a branch point and $z=\infty$ is an essential singular point. Therefore, we shall examine these functions only for $|\arg z|<\pi$.

The functions $W_{\lambda,\mu}(z)$ and $W_{-\lambda,\mu}(-z)$ are linearly independent solutions of equation **9.220** 1.

Integral representations

9.221 $M_{\lambda,\mu}(z)=$

$$=\frac{z^{\mu+\frac{1}{2}}}{2^{2\mu}B\left(\mu+\lambda+\frac{1}{2},\ \mu-\lambda+\frac{1}{2}\right)}\int_{-1}^{1}(1+t)^{\mu-\lambda-\frac{1}{2}}(1-t)^{\mu+\lambda-\frac{1}{2}}e^{\frac{1}{2}zt}\,dt,$$ WH

if the integral converges. See also **6.631** 1. and **7.623** 3.

9.222

1. $$W_{\lambda,\mu}(z)=\frac{z^{\mu+\frac{1}{2}}e^{-\frac{z}{2}}}{\Gamma\left(\mu-\lambda+\frac{1}{2}\right)}\int_{0}^{\infty}e^{-zt}t^{\mu-\lambda-\frac{1}{2}}(1+t)^{\mu+\lambda-\frac{1}{2}}\,dt.$$ MO 118

2. $$W_{\lambda,\mu}(z)=\frac{z^{\lambda}e^{-\frac{z}{2}}}{\Gamma\left(\mu-\lambda+\frac{1}{2}\right)}\int_{0}^{\infty}t^{\mu-\lambda-\frac{1}{2}}e^{-t}\left(1+\frac{t}{z}\right)^{\mu+\lambda-\frac{1}{2}}dt$$

$$\left[\operatorname{Re}(\mu-\lambda)>-\frac{1}{2},\ |\arg z|<\pi\right].$$ WH

9.223 $$W_{\lambda,\mu}(z)=\frac{e^{-\frac{z}{2}}}{2\pi i}\int_{-i\infty}^{i\infty}\frac{\Gamma(u-\lambda)\,\Gamma\left(-u-\mu+\frac{1}{2}\right)\Gamma\left(-u+\mu+\frac{1}{2}\right)}{\Gamma\left(-\lambda+\mu+\frac{1}{2}\right)\Gamma\left(-\lambda-\mu+\frac{1}{2}\right)}z^{u}\,du$$

[the path of integration is chosen in such a way that the poles of the function $\Gamma(u-\lambda)$ are separated from the poles of the functions $\Gamma\left(-u-\mu+\frac{1}{2}\right)$ and $\Gamma\left(-u+\mu+\frac{1}{2}\right)$]. See also **7.142**. MO 118

9.224 $$W_{\mu,\frac{1}{2}+\mu}(z)=z^{\mu+1}e^{-\frac{1}{2}z}\int_{0}^{\infty}(1+t)^{2\mu}e^{-zt}\,dt=$$

$$=z^{-\mu}e^{\frac{1}{2}z}\int_{z}^{\infty}t^{2\mu}e^{-t}\,dt\qquad[\operatorname{Re}z>0].$$ WH

9.225

1. $W_{\lambda,\mu}(x)\,W_{-\lambda,\mu}(x)=$

$$=-x\int_{0}^{\infty}\operatorname{th}^{2\lambda}\frac{t}{2}\left\{J_{2\mu}(x\operatorname{sh}t)\sin(\mu-\lambda)\pi+N_{2\mu}(x\operatorname{sh}t)\cos(\mu-\lambda)\pi\right\}dt$$

$$\left[|\operatorname{Re}\mu|-\operatorname{Re}\lambda<\frac{1}{2};\ x>0\right].$$ MO 119

2. $$W_{\varkappa,\mu}(z_1)W_{\lambda,\mu}(z_2)=\frac{(z_1z_2)^{\mu+\frac{1}{2}}\exp\left[-\frac{1}{2}(z_1+z_2)\right]}{\Gamma(1-\varkappa-\lambda)}\times$$
$$\times\int_0^\infty e^{-t}t^{-\varkappa-\lambda}(z_1+t)^{-\frac{1}{2}+\varkappa-\mu}(z_2+t)^{-\frac{1}{2}+\lambda-\mu}\times$$
$$\times F\left(\frac{1}{2}-\varkappa+\mu,\ \frac{1}{2}-\lambda+\mu;\ 1-\varkappa-\lambda;\ \Theta\right)dt,\quad \Theta=\frac{t(z_1+z_2+t)}{(z_1+t)(z_2+t)}$$
$[z_1\neq 0,\ z_2\neq 0,\ |\arg z_1|<\pi,\ |\arg z_2|<\pi,\ \operatorname{Re}(\varkappa+\lambda)<1]$ MO 119

See also **3.334**, **3.381** 6., **3.382** 3., **3.383** 4., 8., **3.384** 3., **3.471** 2.

9.226 Series representations

$$M_{0,\mu}(z)=z^{\frac{1}{2}+\mu}\left\{1+\sum_{k=1}^{\infty}\frac{z^{2k}}{2^{4k}k!(\mu+1)(\mu+2)\dots(\mu+k)}\right\}.$$ WH

Asymptotic representations

9.227 For large values of $|z|$

$$W_{\lambda,\mu}(z)\sim e^{-\frac{z}{2}}z^{\lambda}\left(1+\sum_{k=1}^{\infty}\frac{\left[\mu^2-\left(\lambda-\frac{1}{2}\right)^2\right]\left[\mu^2-\left(\lambda-\frac{3}{2}\right)^2\right]\dots\left[\mu^2-\left(\lambda-k+\frac{1}{2}\right)^2\right]}{k!z^k}\right)$$
$[|\arg z|\leqslant\pi-\alpha<\pi]$. WH

9.228 For large values of $|\lambda|$

$$M_{\lambda,\mu}(z)\sim\frac{1}{\sqrt{\pi}}\Gamma(2\mu+1)\lambda^{-\mu-\frac{1}{4}}z^{\frac{1}{4}}\cos\left(2\sqrt{\lambda z}-\mu\pi-\frac{1}{4}\pi\right).$$ MO 118

9.229

1. $$W_{\lambda,\mu}\sim-\left(\frac{4z}{\lambda}\right)^{\frac{1}{4}}e^{-\lambda+\lambda\ln\lambda}\sin\left(2\sqrt{\lambda z}-\lambda\pi-\frac{\pi}{4}\right).$$ MO 118

2. $$W_{-\lambda,\mu}\sim\left(\frac{z}{4\lambda}\right)^{\frac{1}{4}}e^{\lambda-\lambda\ln\lambda-2\sqrt{\lambda z}}$$ MO 118

Formulas **9.228** and **9.229** are applicable for $|\lambda|\gg 1,\ |\lambda|\gg|z|,\ |\lambda|\gg|\mu|,\ z\neq 0,$ $|\arg\sqrt{z}|<\frac{3\pi}{4}$ and $|\arg\lambda|<\frac{\pi}{2}]$. MO 118

Functional relations

9.231

1. $$M_{n+\mu+\frac{1}{2},\mu}(z)=\frac{z^{\frac{1}{2}-\mu}e^{\frac{1}{2}z}}{(2\mu+1)(2\mu+2)\dots(2\mu+n)}\frac{d^n}{dz^n}(z^{n+2\mu}e^{-z})$$
$[n=0,\ 1,\ 2,\ \dots;\ 2\mu\neq-1,\ -2,\ -3,\ \dots]$. MO 117

2. $$z^{-\frac{1}{2}-\mu}M_{\lambda,\mu}(z)=(-z)^{-\frac{1}{2}-\mu}M_{-\lambda,\mu}(-z)\quad [2\mu\neq-1,\ -2,\ -3,\ \dots].$$ WH

9.232

1. $W_{\lambda,\mu}(z) = W_{\lambda,-\mu}(z)$. MO 116

2. $W_{-\lambda,\mu}(-z) = \frac{\Gamma(-2\mu)}{\Gamma\left(\frac{1}{2}-\mu+\lambda\right)} M_{-\lambda,\mu}(-z) + \frac{\Gamma(2\mu)}{\Gamma\left(\frac{1}{2}+\mu+\lambda\right)} M_{-\lambda,-\mu}(-z)$

$\left[|\arg(-z)| < \frac{3}{2}\pi\right]$. WH

9.233

1. $M_{\lambda,\mu}(z) = \frac{\Gamma(2\mu+1)}{\Gamma\left(\mu-\lambda+\frac{1}{2}\right)} e^{i\pi\lambda} W_{-\lambda,\mu}(e^{i\pi}z) +$

$+ \frac{\Gamma(2\mu+1)}{\Gamma\left(\mu+\lambda+\frac{1}{2}\right)} \exp\left[i\pi\left(\lambda-\mu-\frac{1}{2}\right)\right] W_{\lambda,\mu}(z)$

$\left[-\frac{3}{2}\pi < \arg z < \frac{\pi}{2};\ 2\mu \neq -1, -2, \ldots\right]$. MO 117

2. $M_{\lambda,\mu}(z) = \frac{\Gamma(2\mu+1)}{\Gamma\left(\mu-\lambda+\frac{1}{2}\right)} e^{-i\pi\lambda} W_{-\lambda,\mu}(e^{-i\pi}z) +$

$+ \frac{\Gamma(2\mu+1)}{\Gamma\left(\mu+\lambda+\frac{1}{2}\right)} \exp\left[-i\pi\left(\lambda-\mu-\frac{1}{2}\right)\right] W_{\lambda,\mu}(z)$

$\left[-\frac{\pi}{2} < \arg z < \frac{3}{2}\pi;\ 2\mu \neq -1, -2, \ldots\right]$. MO 117

9.234 Recursion formulas

1. $W_{\mu,\lambda}(z) = \sqrt{z}\, W_{\mu-\frac{1}{2},\lambda-\frac{1}{2}}(z) + \left(\frac{1}{2}+\lambda-\mu\right) W_{\mu-1,\lambda}(z)$. WH

2. $W_{\mu,\lambda}(z) = \sqrt{z}\, W_{\mu-\frac{1}{2},\lambda+\frac{1}{2}}(z) + \left(\frac{1}{2}-\lambda-\mu\right) W_{\mu-1,\lambda}(z)$. WH

3. $z\frac{d}{dz} W_{\lambda,\mu}(z) = \left(\lambda-\frac{1}{2}z\right) W_{\lambda,\mu}(z) - \left[\mu^2 - \left(\lambda-\frac{1}{2}\right)^2\right] W_{\lambda-1,\mu}(z)$. WH

4. $\left[\left(\mu+\frac{1-z}{2}\right) W_{\lambda,\mu}(z) - z\frac{d}{dz} W_{\lambda,\mu}(z)\right]\left(\mu+\frac{1}{2}+\lambda\right) =$

$= \left[\left(\mu+\frac{1+z}{2}\right) W_{\lambda,\mu+1}(z) + z\frac{d}{dz} W_{\lambda,\mu+1}(z)\right]\left(\mu+\frac{1}{2}-\lambda\right)$ MO 117

5. $\left(\frac{3}{2}+\lambda+\mu\right)\left(\frac{1}{2}+\lambda+\mu\right) z W_{\lambda,\mu}(z) = z(z+2\mu+1)\frac{d}{dz} W_{\lambda+1,\mu+1}(z) +$

$+ \left[\frac{1}{2}z^2 + \left(\mu-\lambda-\frac{1}{2}\right)z + 2\mu^2 + 2\mu + \frac{1}{2}\right] W_{\lambda+1,\mu+1}(z)$. MO 117

Connections with other functions

9.235

1. $M_{0,\mu}(z) = 2^{2\mu}\Gamma(\mu+1)\sqrt{z}\, I_\mu\left(\frac{z}{2}\right)$. MO 125a

2. $W_{0,\mu}(z) = \sqrt{\frac{z}{\pi}}\, K_\mu\left(\frac{z}{2}\right)$. MO 125

9.236

1. $\Phi(x) = 1 - \frac{e^{-\frac{x^2}{2}}}{\sqrt{\pi x}} W_{-\frac{1}{4}, \frac{1}{4}}(x^2) = \frac{2x}{\sqrt{\pi}} \Phi\left(\frac{1}{2}, \frac{3}{2}; -x^2\right).$ WH, MO 126

2. $\operatorname{li}(z) = -\frac{\sqrt{z}}{\sqrt{\ln \frac{1}{z}}} W_{-\frac{1}{2}, 0}(-\ln z).$ WH

3. $\Gamma(\alpha, x) = e^{-x} \Psi(1-\alpha, 1-\alpha; x).$ EH I 266(21)

4. $\gamma(\alpha, x) = \frac{x^\alpha}{\alpha} \Phi(\alpha, \alpha+1; -x).$ EH I 266(22)

9.237

1. $$W_{\lambda, \mu}(z) = \frac{(-1)^{2\mu} z^{\mu+\frac{1}{2}} e^{-\frac{1}{2}z}}{\Gamma\left(\frac{1}{2}-\mu-\lambda\right) \Gamma\left(\frac{1}{2}+\mu-\lambda\right)} \times$$

$$\times \left\{ \sum_{k=0}^{\infty} \frac{\Gamma\left(\mu+k-\lambda+\frac{1}{2}\right)}{k!\,(2\mu+k)!} z^k \left[\psi(k+1) + \psi(2\mu+k+1) - \psi\left(\mu+k-\lambda+\frac{1}{2}\right) - \ln z\right] + \right.$$

$$\left. + (-z)^{-2\mu} \sum_{k=0}^{2\mu-1} \frac{\Gamma(2\mu-k)\,\Gamma\left(k-\mu-\lambda+\frac{1}{2}\right)}{k!} (-z)^k \right\}^*$$

$\left[|\arg z| < \frac{3\pi}{2};\ 2\mu+1 \text{ is a natural number}\right].$ MO 116

2 Set $\lambda - \mu - \frac{1}{2} = l$, where $l+1$ is a natural number. Then

$$W_{l+\mu+\frac{1}{2}, \mu}(z) = (-1)^l z^{\mu+\frac{1}{2}} e^{-\frac{1}{2}z} (2\mu+1)(2\mu+2)\ldots(2\mu+l)\,\Phi(-l, 2\mu+1; z) =$$

$$= (-1)^l z^{\mu+\frac{1}{2}} e^{-\frac{1}{2}z} L_l^{2\mu}(z).$$ MO 116

9.238

1. $J_\nu(x) = \frac{2^{-\nu}}{\Gamma(\nu+1)} x^\nu e^{-ix} \Phi\left(\frac{1}{2}+\nu, 1+2\nu; 2ix\right).$ EH I 265(9)

2. $I_\nu(x) = \frac{2^{-\nu}}{\Gamma(\nu+1)} x^\nu e^{-x} \Phi\left(\frac{1}{2}+\nu, 1+2\nu; 2x\right).$ EH I 265(10)

3. $K_\nu(x) = \sqrt{\pi}\, e^{-x} (2x)^\nu \Psi\left(\frac{1}{2}+\nu, 1+2\nu; 2x\right).$ EH I 265(13)

*For $\mu=0$, the last sum is equal to zero.

9.24-9.25 Parabolic cylinder functions $D_p(z)$

9.240 $$D_p(z) = 2^{\frac{1}{4}+\frac{p}{2}} W_{\frac{1}{4}+\frac{p}{2}, -\frac{1}{4}}\left(\frac{z^2}{2}\right) z^{-\frac{1}{2}} =$$

$$= 2^{\frac{p}{2}} e^{-\frac{z^2}{4}} \left\{ \frac{\sqrt{\pi}}{\Gamma\left(\frac{1-p}{2}\right)} \Phi\left(-\frac{p}{2}, \frac{1}{2}; \frac{z^2}{2}\right) - \frac{\sqrt{2\pi} z}{\Gamma\left(-\frac{p}{2}\right)} \Phi\left(\frac{1-p}{2}, \frac{3}{2}; \frac{z^2}{2}\right) \right\}$$

MO 120a

are called *parabolic cylinder functions.*

Integral representations

9.241

1. $$D_p(z) = \frac{1}{\sqrt{\pi}} 2^{p+\frac{1}{2}} e^{-\frac{\pi}{2} p i} e^{\frac{z^2}{4}} \int_{-\infty}^{\infty} x^p e^{-2x^2+2ixz} \, dx$$

$[\operatorname{Re} p > -1;\ \text{for}\ x < 0\ \arg x^p = p\pi i]$ MO 122

2. $$D_p(z) = \frac{e^{-\frac{z^2}{4}}}{\Gamma(-p)} \int_0^{\infty} e^{-zx-\frac{x^2}{2}} x^{-p-1} \, dx \qquad [\operatorname{Re} p < 0]$$

(cf. 3.462 1.). MO 122

9.242

1. $$D_p(z) = -\frac{\Gamma(p+1)}{2\pi i} e^{-\frac{1}{4}z^2} \int_{\infty}^{(0+)} e^{-zt-\frac{1}{2}t^2} (-t)^{-p-1} \, dt \qquad [|\arg(-t)| \leqslant \pi].$$

WH

2. $$D_p(z) = 2^{\frac{1}{2}(p-1)} \frac{\Gamma\left(\frac{p}{2}+1\right)}{i\pi} \int_{-\infty}^{(-1+)} e^{\frac{1}{4}z^2 t} (1+t)^{-\frac{1}{2}p-1} (1-t)^{\frac{1}{2}(p-1)} \, dt$$

$$\left[|\arg z| < \frac{\pi}{4};\ |\arg(1+t)| \leqslant \pi \right].$$ WH

3. $$D_p(z) = \frac{1}{2\pi i} e^{-\frac{1}{4}z^2} \int_{-\infty i}^{\infty i} \frac{\Gamma\left(\frac{1}{2}t - \frac{1}{2}p\right) \Gamma(-t)}{\Gamma(-p)} \left(\sqrt{2}\right)^{t-p-2} z^t \, dt$$

$$\left[|\arg z| < \frac{3}{4}\pi;\ p \text{ is not a positive integer} \right].$$ WH

4. $$D_p(z) = \frac{1}{2\pi i} e^{-\frac{1}{4}z^2} \int_{\infty}^{(0-)} \frac{\Gamma\left(\frac{1}{2}t - \frac{1}{2}p\right) \Gamma(-t)}{\Gamma(-p)} \left(\sqrt{2}\right)^{t-p-2} z^t \, dt$$

[for all values of $\arg z$; also, the contours encircle the poles of the function $\Gamma(-t)$ but they do not encircle the poles of the function $\Gamma\left(\frac{1}{2}t - \frac{1}{2}p\right)$]. WH

9.243

1. $$D_n(z)=(-1)^\mu\left(\frac{\pi}{2}\right)^{-\frac{1}{2}}(\sqrt{n})^{n+1}e^{\frac{1}{4}z^2-\frac{1}{2}n}\left\{\int_{-\infty}^{\infty}e^{-n(t-1)^2}\frac{\cos}{\sin}(zt\sqrt{n})\,dt+\right.$$
$$\left.+\int_0^{\infty}\left[e^{\frac{1}{2}n(1-t^2)}t^n-e^{-n(t-1)^2}\right]\frac{\cos}{\sin}(zt\sqrt{n})\,dt-\int_{-\infty}^{0}e^{-n(t-1)^2}\frac{\cos}{\sin}(zt\sqrt{n})\,dt\right\}$$

WH

[n is a natural number].

2. $$D_n(z)=(-1)^\mu 2^{n+2}(2\pi)^{-\frac{1}{2}}e^{\frac{1}{4}z^2}\int_0^{\infty}t^n e^{-2t^2}\frac{\cos}{\sin}(2zt)\,dt$$

$\left[n\right.$ is a natural number, $\mu=E\left(\frac{n}{2}\right)$, and the cosine or sine is chosen according as n is even or odd$\left.\right]$. WH

9.244

1. $$D_{-p-1}[(1+i)z]=\frac{e^{-\frac{iz^2}{2}}}{2^{\frac{p-1}{2}}\Gamma\left(\frac{p+1}{2}\right)}\int_0^{\infty}\frac{e^{-ix^2z^2}x^p}{(1+x^2)^{1+\frac{p}{2}}}\,dx$$

[$\operatorname{Re} p>-1,\ \operatorname{Re} iz^2\geqslant 0$]. MO 122

2. $$D_p[(1+i)z]=\frac{2^{\frac{p+1}{2}}}{\Gamma\left(-\frac{p}{2}\right)}\int_1^{\infty}e^{-\frac{i}{2}z^2x}\frac{(x+1)^{\frac{p-1}{2}}}{(x-1)^{1+\frac{p}{2}}}\,dx$$

[$\operatorname{Re} p<0;\ \operatorname{Re} iz^2\geqslant 0$]. MO 122

See also 3.383 6., 7., 3.384 2., 6., 3.966 5., 6.

9.245

1. $$D_p(x)D_{-p-1}(x)=-\frac{1}{\sqrt{\pi}}\int_0^{\infty}\operatorname{cth}^{p+\frac{1}{2}}\frac{t}{2}\frac{1}{\sqrt{\operatorname{sh} t}}\sin\frac{x^2\operatorname{sh} t+p\pi}{2}$$

[x is real, $\operatorname{Re} p<0$]. MO 122

2. $$D_p\left(ze^{\frac{\pi}{4}i}\right)D_p\left(ze^{-\frac{\pi}{4}i}\right)=\frac{1}{\Gamma(-p)}\int_0^{\infty}\operatorname{cth}^p t\exp\left(-\frac{z^2}{2}\operatorname{sh} 2t\right)\frac{dt}{\operatorname{sh} t}$$

$\left[|\arg z|<\frac{\pi}{4};\ \operatorname{Re} p<0\right]$ MO 122

See also 6.613.

9.246 Asymptotic expansions. If $|z|\gg 1,\ |z|\gg|p|$, then

1. $$D_p(z)\sim e^{-\frac{z^2}{4}}z^p\left(1-\frac{p(p-1)}{2z^2}+\frac{p(p-1)(p-2)(p-3)}{2\cdot 4z^4}-\ldots\right)$$

$\left[|\arg z|<\frac{3}{4}\pi\right]$ MO 121

2. $$D_p(z) \sim e^{-\frac{z^2}{4}} z^p \left(1 - \frac{p(p-1)}{2z^2} + \frac{p(p-1)(p-2)(p-3)}{2 \cdot 4z^4} - \dots\right) -$$
$$- \frac{\sqrt{2\pi}}{\Gamma(-p)} e^{p\pi i} e^{\frac{z^2}{4}} z^{-p-1} \left(1 + \frac{(p+1)(p+2)}{2z^2} + \frac{p(p+1)(p+2)(p+3)(p+4)}{2 \cdot 4z^4} + \dots\right)$$
$$\left[\frac{\pi}{4} < \arg z < \frac{5}{4}\pi\right].$$ MO 121

3. $$D_p(z) \sim e^{-\frac{z^2}{4}} z^p \left(1 - \frac{p(p-1)}{2z^2} + \frac{p(p-1)(p-2)(p-3)}{2 \cdot 4z^4} - \dots\right) -$$
$$- \frac{\sqrt{2\pi}}{\Gamma(-p)} e^{-p\pi i} e^{\frac{z^2}{4}} z^{-p-1} \left(1 + \frac{(p+1)(p+2)}{2z^2} + \frac{(p+1)(p+2)(p+3)(p+4)}{2 \cdot 4z^4} + \dots\right)$$
$$\left[-\frac{\pi}{4} > \arg z > -\frac{5}{4}\pi\right]$$ MO 121

Functional relations

9.247 Recursion formulas:

1. $D_{p+1}(z) - zD_p(z) + pD_{p-1}(z) = 0.$ WH

2. $\frac{d}{dz} D_p(z) + \frac{1}{2} zD_p(z) - pD_{p-1}(z) = 0.$ WH

3. $\frac{d}{dz} D_p(z) - \frac{1}{2} zD_p(z) + D_{p+1}(z) = 0.$ MO 121

9.248 Linear relations:

1. $$D_p(z) = \frac{\Gamma(p+1)}{\sqrt{2\pi}} \left[e^{\frac{\pi}{2} pi} D_{-p-1}(iz) + e^{-\frac{\pi}{2} pi} D_{-p-1}(-iz)\right];$$

2. $$= e^{-p\pi i} D_p(-z) + \frac{\sqrt{2\pi}}{\Gamma(-p)} e^{-\frac{\pi}{2}(p+1)i} D_{-p-1}(iz);$$

3. $$= e^{p\pi i} D_p(-z) + \frac{\sqrt{2\pi}}{\Gamma(-p)} e^{\frac{\pi}{2}(p+1)i} D_{-p-1}(-iz).$$ MO 121

9.249 $$D_p[(1+i)x] + D_p[-(1+i)x] =$$
$$= \frac{2^{1+\frac{p}{2}}}{\Gamma(-p)} \exp\left[-\frac{i}{2}\left(x^2 + p\frac{\pi}{2}\right)\right] \int_0^\infty \frac{\cos xt}{t^{p+1}} e^{-\frac{i}{4}t^2} dt$$
$[x \text{ real}; \; -1 < \operatorname{Re} p < 0].$ MO 122

9.251 $D_n(z) = (-1)^n e^{\frac{z^2}{4}} \frac{d^n}{dz^n}\left(e^{-\frac{z^2}{2}}\right) \quad [n = 0, 1, 2, \dots].$ WH

9.252 $$D_p(ax+by) = \exp\frac{(bx-ay)^2}{4} \left(\frac{a}{\sqrt{a^2+b^2}}\right)^p \times$$
$$\times \sum_{k=0}^{\infty} \binom{p}{k} D_{p-k}\left(\sqrt{a^2+b^2}\, x\right) D_k\left(\sqrt{a^2+b^2}\, y\right) \left(\frac{b}{a}\right)^k$$
$[a > b > 0, \; x > 0, \; y > 0, \; \operatorname{Re} p \geqslant 0]$ ["summation theorem"] MO 124

Connections with other functions

9.253 $D_n(z) = 2^{-\frac{n}{2}} e^{-\frac{z^2}{4}} H_n\left(\frac{z}{\sqrt{2}}\right).$ MO 123a

9.254

1. $D_{-1}(z) = e^{\frac{z^2}{4}} \sqrt{\frac{\pi}{2}} \left[1 - \Phi\left(\frac{z}{\sqrt{2}}\right)\right].$ MO 123

2. $D_{-2}(z) = -e^{\frac{z^2}{4}} \sqrt{\frac{\pi}{2}} \left\{\sqrt{\frac{2}{\pi}}\, e^{-\frac{z^2}{2}} - z\left[1 - \Phi\left(\frac{z}{\sqrt{2}}\right)\right]\right\}.$ MO 123

9.255 Differential equations leading to parabolic cylinder functions:

1. $\frac{d^2u}{dz^2} + \left(p + \frac{1}{2} - \frac{z^2}{4}\right) u = 0,$

$u = D_p(z),\ D_p(-z),\ D_{-p-1}(iz),\ D_{-p-1}(-iz)$

(These four solutions are linearly dependent. See 9.248)

2. $\frac{d^2u}{dz^2} + (z^2 + \lambda) u = 0, \qquad u = D_{-\frac{1+i\lambda}{2}}[\pm(1+i) z].$

EH II 118(12, 13)a, MO 123

3. $\frac{d^2u}{dz^2} + z\frac{du}{dz} + (p+1) u = 0, \qquad u = e^{-\frac{z^2}{4}} D_p(z).$ MO 123

9.26 Degenerate hypergeometric series of two variables

9.261

1. $\Phi_1(\alpha, \beta, \gamma, x, y) = \sum_{m,\, n=0}^{\infty} \frac{(\alpha)_{m+n} (\beta)_n}{(\gamma)_{m+n}\, m!\, n!} x^m y^n$

$[|x| < 1].$ EH I 225(20)

2. $\Phi_2(\beta, \beta', \gamma, x, y) = \sum_{m,\, n=0}^{\infty} \frac{(\beta)_m (\beta')_n}{(\gamma)_{m+n}\, m!\, n!} x^m y^n.$

EH I 225(21)a, ET I 385

3. $\Phi_3(\beta, \gamma, x, y) = \sum_{m,\, n=0}^{\infty} \frac{(\beta)_m}{(\gamma)_{m+n}\, m!\, n!} x^m y^n.$ EH I 225(22)

The functions Φ_1, Φ_2, Φ_3 satisfy the following systems of partial differential equations:

9.262

1. $z = \Phi_1(\alpha, \beta, \gamma, x, y).$

$$\left\{\begin{array}{l} x(1-x)\frac{\partial^2 z}{\partial x^2} + y(1-x)\frac{\partial^2 z}{\partial x\, \partial y} + [\gamma - (\alpha+\beta+1)x]\frac{\partial z}{\partial x} - \beta y\frac{\partial z}{\partial y} - \alpha\beta z = 0, \\ y\frac{\partial^2 z}{\partial y^2} + x\frac{\partial^2 z}{\partial x\, \partial y} + (\gamma - y)\frac{\partial z}{\partial y} - x\frac{\partial z}{\partial x} - \alpha z = 0. \end{array}\right.$$

EH I 235(23)

2. $z = \Phi_2(\beta, \beta', \gamma, x, y)$

$$\begin{cases} x \dfrac{\partial^2 z}{\partial x^2} + y \dfrac{\partial^2 z}{\partial x \partial y} + (\gamma - x) \dfrac{\partial z}{\partial x} - \beta z = 0, \\ y \dfrac{\partial^2 z}{\partial y^2} + x \dfrac{\partial^2 z}{\partial x \partial y} + (\gamma - y) \dfrac{\partial z}{\partial y} - \beta' z = 0. \end{cases}$$ EH I 235(24)

3. $z = \Phi_3(\beta, \gamma, x, y)$.

$$\begin{cases} x \dfrac{\partial^2 z}{\partial x^2} + y \dfrac{\partial^2 z}{\partial x \partial y} + (\gamma - x) \dfrac{\partial z}{\partial x} - \beta z = 0, \\ y \dfrac{\partial^2 z}{\partial y^2} + x \dfrac{\partial^2 z}{\partial x \partial y} + \gamma \dfrac{\partial z}{\partial y} - z = 0. \end{cases}$$ EH I 235(25)

9.3 Meijer's G-Function

9.30 Definition

9.301 $G_{p,q}^{m,n}\left(x \,\middle|\, \begin{matrix} a_1, \ldots, a_p \\ b_1, \ldots, b_q \end{matrix}\right) =$

$$= \frac{1}{2\pi i} \int_L \frac{\prod\limits_{j=1}^{m} \Gamma(b_j - s) \prod\limits_{j=1}^{n} \Gamma(1 - a_j + s)}{\prod\limits_{j=m+1}^{q} \Gamma(1 - b_j + s) \prod\limits_{j=n+1}^{p} \Gamma(a_j - s)} x^s \, ds$$

[$0 \leqslant m \leqslant q$, $0 \leqslant n \leqslant p$, and the poles of $\Gamma(b_j - s)$ must not coincide with the poles of $\Gamma(1 - a_k + s)$ for any j and k (where $j = 1, \ldots, m$; $k = 1, \ldots, n$)]. Besides **9.301**, the following notations are also used

$$G_{pq}^{mn}\left(x \,\middle|\, \begin{matrix} a_r \\ b_s \end{matrix}\right), \quad G_{pq}^{mn}(x), \quad G(x).$$ EH I 207(1)

9.302 Three types of integration paths L in the right member of **9.301**. can be exhibited:

1) The path L runs from $-\infty$ to $+\infty$ in such a way that the poles of the functions $\Gamma(1 - a_k + s)$ lie to the left, and the poles of the functions $\Gamma(b_j - s)$ lie to the right of L (for $j = 1, 2, \ldots, m$ and $k = 1, 2, \ldots, n$). In this case, the conditions under which the integral **9.301** converges are of the form

$$p + q < 2(m + n), \quad |\arg x| < \left(m + n - \frac{1}{2} p - \frac{1}{2} q\right) \pi.$$ EH I 207(2)

2) L is a loop, beginning and ending at $+\infty$, that encircles the poles of the functions $\Gamma(b_j - s)$ (for $j = 1, 2, \ldots, m$) once in the negative direction. All the poles of the functions $\Gamma(1 - a_k + s)$ must remain outside this loop. Then, the conditions under which the integral **9.301** converges are:

$$q \geqslant 1 \text{ and either } p < q, \text{ or } p = q \text{ and } |x| < 1.$$ EH I 207(3)

3) L is a loop, beginning and ending at $-\infty$, that encircles the poles of the functions $\Gamma(1 - a_k + s)$ (for $k = 1, 2, \ldots, n$) once in the positive direction. All the poles of the functions $\Gamma(b_j - s)$ (for $j = 1, 2, \ldots, m$) must remain outside this loop.

The conditions under which the integral in **9.301** converges are

$$p \geqslant 1 \text{ and either } p > q \text{ or } p = q \text{ and } |x| > 1.$$ EH I 207(4)

The function $G_{pq}^{mn}\left(x\left|{}^{a_r}_{b_s}\right.\right)$ is analytic with respect to x; it is symmetric with respect to the parameters $a_1, \ldots, a_n$ and also with respect to $a_{n+1}, \ldots, a_p$; $b_1, \ldots, b_m$; $b_{m+1}, \ldots, b_q$.

EH I 208

9.303 If no b_j (for $j=1, 2, \ldots, n$) is different from a whole number, then, under the conditions that either $p<q$ or $p=q$ and $|x|<1$,

$$G_{pq}^{mn}\left(x\left|{}^{a_r}_{b_s}\right.\right)=\sum_{n=1}^{m}\frac{\prod\limits_{j=1}^{m}{}' \Gamma(b_j-b_h)\prod\limits_{j=1}^{n}\Gamma(1+b_h-a_j)}{\prod\limits_{j=m+1}^{q}\Gamma(1+b_h-b_j)\prod\limits_{j=n+1}^{p}\Gamma(a_j-b_h)}x^{b_h}\times$$

$$\times {}_pF_{q-1}[1+b_h-a_1, \ldots, 1+b_h-a_p;\ 1+b_h-b_1, \ldots$$
$$\ldots, *, \ldots, 1+b_h-b_q;\ (-1)^{p-m-n}x]\ *).$$

EH I 208(5)

The prime by the product symbol denotes the omission of the product when $j=h$. The asterisk under the symbol for the function ${}_pF_{q-1}$ denotes the omission of the hth parameter.

9.304 If no a_k (for $k=1, 2, \ldots, n$) is different from a whole number, then, under the conditions that $q<p$ or $q=p$ and $|x|>1$,

$$G_{pq}^{mn}\left(x\left|{}^{a_r}_{b_s}\right.\right)=\sum_{h=1}^{n}\frac{\prod\limits_{j=1}^{n}{}' \Gamma(a_h-a_j)\prod\limits_{j=1}^{m}\Gamma(b_j-a_h+1)}{\prod\limits_{j=n+1}^{p}\Gamma(a_j-a_h+1)\prod\limits_{j=m+1}^{q}\Gamma(a_h-b_j)}x^{a_h-1}\times$$

$$\times {}_qF_{p-1}[1+b_1-a_h, \ldots, 1+b_q-a_h;\ 1+a_1-a_h, \ldots$$
$$\ldots, *, \ldots, 1+a_p-a_h;\ (-1)^{q-m-n}x^{-1}]*).$$

EH I 208(6)

9.31 Functional relations

If one of the parameters a_j (for $j=1, 2, \ldots, n$) coincides with one of the parameters b_j (for $j=m+1, m+2, \ldots, q$), the order of the G-function decreases. For example,

1. $$G_{pq}^{mn}\left(x\left|\begin{matrix}a_1, \ldots, a_p\\ b_1, \ldots, b_{q-1}, a_1\end{matrix}\right.\right)=G_{p-1,q-1}^{m,n-1}\left(x\left|\begin{matrix}a_2, \ldots, a_p\\ b_1, \ldots, b_{q-1}\end{matrix}\right.\right) \quad [n, p, q\geqslant 1].$$

An analogous relationship occurs when one of the parameters b_j (for $j=1, 2, \ldots, m$) coincides with one of the a_j (for $j=n+1, \ldots, p$). In this case, it is m and not n that decreases by one unit.

EH I 209(7)

The G-function with $p>q$ can be transformed into the G-function with $p<q$ by means of the relationships:

2. $$G_{pq}^{mn}\left(x^{-1}\left|\begin{matrix}a_r\\ b_s\end{matrix}\right.\right)=G_{qp}^{nm}\left(x\left|\begin{matrix}1-b_s\\ 1-a_r\end{matrix}\right.\right).$$

EH I 209(9)

3. $$x\frac{d}{dx}G_{pq}^{mn}\left(x\left|\begin{matrix}a_r\\ b_s\end{matrix}\right.\right)=G_{pq}^{mn}\left(x\left|\begin{matrix}a_1-1, a_2, \ldots, a_p\\ b_1, \ldots, b_q\end{matrix}\right.\right)+$$
$$+(a_1-1)G_{pq}^{mn}\left(x\left|\begin{matrix}a_r\\ b_s\end{matrix}\right.\right) \quad [n\geqslant 1].$$

EH I 210(13)

9.32 A differential equation for the G-function

$G_{pq}^{mn}\left(x \middle| \begin{matrix} a_r \\ b_s \end{matrix}\right)$ satisfies the following linear qth-order differential equation

$$\left[(-1)^{p-m-n} x \prod_{j=1}^{p} \left(x \frac{d}{dx} - a_j + 1\right) - \prod_{j=1}^{q} \left(x \frac{d}{dx} - b_j\right)\right] y = 0 \qquad [p \leqslant q]$$

EH I 210(1)

9.33 Series of G-functions

$$G_{pq}^{mn}\left(\lambda x \middle| \begin{matrix} a_1, \ldots, a_p \\ b_1, \ldots, b_q \end{matrix}\right) =$$

$$= \lambda^{b_1} \sum_{r=0}^{\infty} \frac{1}{r!} (1-\lambda)^r G_{pq}^{mn}\left(x \middle| \begin{matrix} a_1, \ldots, a_p \\ b_1 + r, b_2, \ldots, b_q \end{matrix}\right)$$

$[|\lambda - 1| < 1,\ m \geqslant 1,$ if $m = 1$ and $p < q,$ λ may be arbitrary];

EH I 213(1)

$$= \lambda^{b_q} \sum_{r=0}^{\infty} \frac{1}{r!} (\lambda - 1)^r G_{pq}^{mn}\left(x \middle| \begin{matrix} a_1, \ldots, a_p \\ b_1, \ldots, b_{q-1}, b_q + r \end{matrix}\right)$$

$[m < q,\ |\lambda - 1| < 1]$; EH I 213(2)

$$= \lambda^{a_1 - 1} \sum_{r=0}^{\infty} \frac{1}{r!} \left(1 - \frac{1}{\lambda}\right)^r G_{pq}^{mn}\left(x \middle| \begin{matrix} a_1 - r, a_2, \ldots, a_p \\ b_1, \ldots, b_q \end{matrix}\right)$$

$\left[n \geqslant 1,\ \operatorname{Re} \lambda > \frac{1}{2}\right.$ (if $n = 1$ and $p > q$, then λ may be arbitrary)$\left.\right]$;

EH I 213(3)

$$= \lambda^{a_p - 1} \sum_{r=0}^{\infty} \frac{1}{r!} \left(\frac{1}{\lambda} - 1\right)^r G_{pq}^{mn}\left(x \middle| \begin{matrix} a_1, \ldots, a_{p-1}, a_p - r \\ b_1, \ldots, b_q \end{matrix}\right)$$

$\left[n < p,\ \operatorname{Re} \lambda > \frac{1}{2}\right]$. EH I 213(4)

For integrals of the G-function, see **7.8**.

9.34 Connections with other special functions

1. $J_\nu(x) x^\mu = 2^\mu G_{02}^{10}\left(\frac{1}{4} x^2 \middle| \frac{1}{2}\nu + \frac{1}{2}\mu,\ \frac{1}{2}\mu - \frac{1}{2}\nu\right).$ EH I 219(44)

2. $N_\nu(x) x^\mu = 2^\mu G_{13}^{20}\left(\frac{1}{4} x^2 \middle| \begin{matrix} \frac{1}{2}\mu - \frac{1}{2}\nu - \frac{1}{2} \\ \frac{1}{2}\mu - \frac{1}{2}\nu,\ \frac{1}{2}\mu + \frac{1}{2}\nu,\ \frac{1}{2}\mu - \frac{1}{2}\nu - \frac{1}{2} \end{matrix}\right).$

EH I 219(46)

3. $K_\nu(x) x^\mu = 2^{\mu-1} G_{02}^{20}\left(\frac{1}{4} x^2 \middle| \frac{1}{2}\mu + \frac{1}{2}\nu,\ \frac{1}{2}\mu - \frac{1}{2}\nu\right).$ EH I 219(47)

4. $K_\nu(x) = e^x \sqrt{\pi}\, G_{12}^{20}\left(2x \,\middle|\, \begin{matrix} \frac{1}{2} \\ \nu, \ -\nu \end{matrix}\right).$ EH I 219(49)

5. $\mathbf{H}_\nu(x)\, x^\mu = 2^\mu G_{13}^{11}\left(\frac{1}{4}x^2 \,\middle|\, \begin{matrix} \frac{1}{2}+\frac{1}{2}\nu+\frac{1}{2}\mu \\ \frac{1}{2}+\frac{1}{2}\nu+\frac{1}{2}\mu, \ \frac{1}{2}\mu-\frac{1}{2}\nu, \ \frac{1}{2}\mu+\frac{1}{2}\nu \end{matrix}\right).$ EH I 220(51)

6. $S_{\mu,\nu}(x) = 2^{\mu-1}\dfrac{1}{\Gamma\left(\frac{1-\mu-\nu}{2}\right)\Gamma\left(\frac{1-\mu+\nu}{2}\right)} \times$

$\times G_{13}^{31}\left(\frac{1}{4}x^2 \,\middle|\, \begin{matrix} \frac{1}{2}+\frac{1}{2}\mu \\ \frac{1}{2}+\frac{1}{2}\mu, \ \frac{1}{2}\nu, \ -\frac{1}{2}\nu \end{matrix}\right).$ EH I 220(55)

7. ${}_2F_1(a, b; c; -x) = \dfrac{\Gamma(c)\, x}{\Gamma(a)\,\Gamma(b)} G_{22}^{12}\left(x \,\middle|\, \begin{matrix} -a, \ -b \\ -1, \ -c \end{matrix}\right).$ EH I 222(74)a

8. ${}_pF_q(a_1, \ldots, a_p; b_1, \ldots, b_q; x) =$

$= \dfrac{\prod_{j=1}^{q} \Gamma(b_j)}{\prod_{j=1}^{p} \Gamma(a_j)} G_{p,\,q+1}^{1,\,p}\left(-x \,\middle|\, \begin{matrix} 1-a_1, \ \ldots, \ 1-a_p \\ 0, \ 1-b_1, \ \ldots, \ 1-b_q \end{matrix}\right);$

$= \dfrac{\prod_{j=1}^{q} \Gamma(b_j)}{\prod_{j=1}^{p} \Gamma(a_j)} G_{q+1,\,p}^{p,\,1}\left(-\frac{1}{x} \,\middle|\, \begin{matrix} 1, \ b_1, \ \ldots, \ b_q \\ a_1, \ \ldots, \ a_p \end{matrix}\right).$ EH I 215(1)

9. $W_{k,m}(x) = \dfrac{2^k \sqrt{x}\, e^{\frac{1}{2}x}}{\sqrt{2\pi}} G_{24}^{40}\left(\frac{x^2}{4} \,\middle|\, \begin{matrix} \frac{1}{4}-\frac{1}{2}k, \ \frac{3}{4}-\frac{1}{2}k \\ \frac{1}{2}+\frac{1}{2}m, \ \frac{1}{2}-\frac{1}{2}m, \ \frac{1}{2}m, \ -\frac{1}{2}m \end{matrix}\right).$ EH I 221(70)

9.4 MacRobert's E-Function

9.41 Representation by means of multiple integrals

$$E(p;\ \alpha_r : q;\ \varrho_s : x) = \frac{\Gamma(\alpha_{q+1})}{\Gamma(\varrho_1-\alpha_1)\,\Gamma(\varrho_2-\alpha_2)\ldots\Gamma(\varrho_q-\alpha_q)} \times$$

$$\times \prod_{\mu=1}^{q} \int_0^\infty \lambda_\mu^{\varrho_\mu-\alpha_\mu-1}(1+\lambda_\mu)^{-\varrho_\mu}\, d\lambda_\mu \prod_{\nu=2}^{p-q-1} \int_0^\infty e^{-\lambda_{q+\nu}} \lambda_{q+\nu}^{\alpha_{q+\nu}-1}\, d\lambda_{q+\nu} \times$$

$$\times \int_0^\infty e^{-\lambda_p} \lambda_p^{\alpha_p-1}\left[1+\frac{\lambda_{q+2}\,\lambda_{q+3}\ldots\lambda_p}{(1+\lambda_1)\ldots(1+\lambda_q)\,x}\right]^{-\alpha_{q+1}} d\lambda_p$$

[$|\arg x| < \pi$, $p \geqslant q+1$, α_r and β_s are bounded by the condition that the integrals on the right be convergent.] EH I 204(3)

9.42 Functional relations

1. $\alpha_1 x E(\alpha_1, \ldots, \alpha_p : \varrho_1, \ldots, \varrho_q : x) =$
$= xE(\alpha_1+1, \alpha_2, \ldots, \alpha_p : \varrho_1, \ldots, \varrho_q : x) +$
$+ E(\alpha_1+1, \alpha_2+1, \ldots, \alpha_p+1 : \varrho_1+1, \ldots, \varrho_q+1 : x).$ **EH I 205(7)**

2. $(\varrho_1 - 1) xE(\alpha_1, \ldots, \alpha_p : \varrho_1, \ldots, \varrho_q : x) =$
$= xE(\alpha_1, \ldots, \alpha_p : \varrho_1 - 1, \varrho_2, \ldots, \varrho_q : x) +$
$+ E(\alpha_1+1, \ldots, \alpha_p+1 : \varrho_1+1, \ldots, \varrho_q+1 : x).$ **EH I 205(9)**

3. $\frac{d}{dx} E(\alpha_1, \ldots, \alpha_p : \varrho_1, \ldots, \varrho_q : x) =$
$= x^{-2} E(\alpha_1+1, \ldots, \alpha_p+1 : \varrho_1+1, \ldots, \varrho_q+1 : x).$ **EH I 205(8)**

9.5 Riemann's Zeta Functions $\zeta(z, q)$, and $\zeta(z)$, and the Functions $\Phi(z, s, v)$ and $\xi(s)$

9.51 Definition and integral representations

9.511 $\zeta(z, q) = \frac{1}{\Gamma(z)} \int_0^\infty \frac{t^{z-1} e^{-qt}}{1-e^{-t}}\, dt;$ **WH**

$$= \frac{1}{2} q^{-z} + \frac{q^{1-z}}{z-1} + 2 \int_0^\infty (q^2+t^2)^{-\frac{z}{2}} \left[\sin\left(z \operatorname{arctg} \frac{t}{q} \right) \right] \frac{dt}{e^{2\pi t}-1}$$

$[0 < q < 1, \ \operatorname{Re} z > 1].$ **WH**

9.512 $\zeta(z, q) = -\frac{\Gamma(1-z)}{2\pi i} \int_\infty^{(0+)} \frac{(-\theta)^{z-1} e^{-q\theta}}{1-e^{-\theta}}\, d\theta.$

This equation is valid for all values of z except for $z = 1, 2, 3, \ldots$. It is assumed that the path of integration (see drawing below) does not pass through the points $2n\pi i$ (where n is a natural number).

See also 4.251 4., 4.271 1., 4., 8., 4.272 9., 12., 4.294 11.

9.513

1. $\zeta(z) = \frac{1}{(1-2^{1-z})\Gamma(z)} \int_0^\infty \frac{t^{z-1}}{e^t+1}\, dt \qquad [\operatorname{Re} z > 0].$ **WH**

2. $\zeta(z) = \frac{2^z}{(2^z-1)\Gamma(z)} \int_0^\infty \frac{t^{z-1} e^t}{e^{2t}-1}\, dt \qquad [\operatorname{Re} z > 1].$ **WH**

3. $\zeta(z) = \frac{\pi^{\frac{z}{2}}}{\Gamma\left(\frac{z}{2}\right)} \left[\frac{1}{z(z-1)} + \int_1^\infty \left(t^{\frac{1-z}{2}} + t^{\frac{z}{2}}\right) t^{-1} \sum_{k=1}^\infty e^{-k^2 \pi t}\, dt \right].$ **WH**

4. $\zeta(z) = \frac{2^{z-1}}{z-1} - 2^z \int_0^\infty (1+t^2)^{-\frac{z}{2}} \sin(z \operatorname{arc\,tg} t) \frac{dt}{e^{\pi t}+1}.$ **WH**

5. $\zeta(z) = \frac{2^{z-1}}{2^z - 1} \frac{z}{z-1} + \frac{2}{2^z - 1} \int_0^\infty \left(\frac{1}{4} + t^2\right)^{-\frac{z}{2}} \sin(z \operatorname{arctg} 2t) \frac{dt}{e^{2\pi t} - 1}$. WH

See also 3.411 1., 3.523 1., 3.527 1., 3., 4.271 8.

9.52 Representation as a series or as an infinite product

9.521

1. $\zeta(z, q) = \sum_{n=0}^{\infty} \frac{1}{(q+n)^z}$ [Re $z > 1$]. WH

2. $\zeta(z, q) = \frac{2\Gamma(1-z)}{(2\pi)^{1-z}} \left[\sin\frac{z\pi}{2} \sum_{n=1}^{\infty} \frac{\cos 2\pi qn}{n^{1-z}} + \cos\frac{z\pi}{2} \sum_{n=1}^{\infty} \frac{\sin 2\pi qn}{n^{1-z}}\right]$

[Re $z > 0$], $0 < q \leqslant 1$. WH

3. $\zeta(z, q) = \sum_{n=0}^{N} \frac{1}{(q+n)^z} - \frac{1}{(1-z)(N+q)^{z-1}} - \sum_{n=N}^{\infty} F_n(z)$,

where $F_n(z) = \frac{1}{1-z}\left(\frac{1}{(n+1+q)^{z-1}} - \frac{1}{(n+q)^{z-1}}\right) - \frac{1}{(n+1+q)^z} = z \int_n^{n+1} \frac{(t-n)\,dt}{(t+q)^{z+1}}$

[Re $z > 1$, N is a natural number]. WH

9.522

1. $\zeta(z) = \sum_{n=1}^{\infty} \frac{1}{n^z}$ [Re $z > 1$]. WH

2. $\zeta(z) = \frac{1}{1-2^{1-z}} \sum_{n=1}^{\infty} (-1)^{n+1} \frac{1}{n^z}$ [Re $z > 0$]. WH

9.523

1. $\zeta(m) = \prod \frac{1}{1-p^{-m}}$, WH

2. $\ln \zeta(z) = \sum_p \sum_{k=1}^{\infty} \frac{1}{k p^{kz}}$. WH

The product and the summation are taken over all primes p.

9.524 $\frac{\zeta'(z)}{\zeta(z)} = -\sum_{k=1}^{\infty} \frac{\Delta(k)}{k^z}$,

where $\Delta(k) = 0$ when k is not a power of a prime and $\Delta(k) = \ln p$ when k is a power of a prime p [for Re $z > 1$]. WH

9.53 Functional relations

9.531 $\zeta(-n, q) = -\frac{B'_{n+2}(q)}{(n+1)(n+2)}$ [n is a nonnegative integer]. WH

9.532 $\sum_{k=2}^{\infty} \frac{(-1)^{k-1}}{k} z^k \zeta(k, q) = \ln \frac{e^{-Cz}\Gamma(q)}{\Gamma(z+q)} - \frac{z}{q} + \sum_{k=1}^{\infty} \frac{qz}{k(q+k)}$ $[|z| < q]$. WH

9.533

1. $\lim_{z \to 1} \frac{\zeta(z, q)}{\Gamma(1-z)} = -1.$ WH

2. $\lim_{z \to 1} \left\{ \zeta(z, q) - \frac{1}{z-1} \right\} = -\psi(q).$ WH

3. $\left\{ \frac{d}{dz} \zeta(z, q) \right\}_{z=0} = \ln \Gamma(q) - \frac{1}{2} \ln 2\pi.$ WH

9.534 $\zeta(z, 1) = \zeta(z).$

9.535

1. $\zeta(z) = \frac{1}{2^z - 1} \zeta\left(z, \frac{1}{2}\right)$ $[\operatorname{Re} z > 1].$ WH

2. $2^z \Gamma(1-z) \zeta(1-z) \sin \frac{z\pi}{2} = \pi^{1-z} \zeta(z).$ WH

3. $2^{1-z} \Gamma(z) \zeta(z) \cos \frac{z\pi}{2} = \pi^z \zeta(1-z).$ WH

4. $\Gamma\left(\frac{z}{2}\right) \pi^{-\frac{z}{2}} \zeta(z) = \Gamma\left(\frac{1-z}{2}\right) \pi^{\frac{z-1}{2}} \zeta(1-z).$ WH

9.536 $\lim_{z \to 1} \left\{ \zeta(z) - \frac{1}{z-1} \right\} = C.$

9.537 Set $z = \frac{1}{2} + it$; Then, $\Xi(t) = \frac{(z-1)\Gamma\left(\frac{z}{2}+1\right)}{\sqrt{\pi^z}} \zeta(z) = \Xi(-t)$

is an even function of t with real coefficients in its expansion in powers of t^2. JA

9.54 Singular points and zeros

9.541

1. $z = 1$ is the only singular point of the function $\zeta(z, q)$. WH

2. The function $\zeta(z)$ has simple zeros at the points $-2n$, where n is a natural number. All other zeros of the function $\zeta(z)$ lie in the strip $0 \leqslant \operatorname{Re} z \leqslant 1$.

3. Riemann's hypothesis: All zeros of the function $\zeta(z)$ that lie in the strip $0 \leqslant \operatorname{Re} z \leqslant 1$ lie on the straight line $\operatorname{Re} z = \frac{1}{2}$. It has been shown that an uncountable set of zeros of the zeta function lie on this line. WH

9.542 Particular values:

1. $\zeta(2m) = \frac{2^{2m-1}\pi^{2m} |B_{2m}|}{(2m)!}$, WH

2. $\zeta(1-2m) = -\frac{B_{2m}}{2m}$, WH

3. $\zeta(-2m) = 0$ WH

[m is a natural number].

4. $\zeta'(0) = -\frac{1}{2} \ln 2\pi.$ WH

9.55 The function $\Phi(z, s, v)$

9.550 Definition:

$$\Phi(z, s, v) = \sum_{n=0}^{\infty} (v+n)^{-s} z^n$$

$$[|z| < 1, \quad v \neq 0, \ -1, \ \ldots].$$ EH I 27(1)

Functional relations

9.551 $$\Phi(z, s, v) = z^m \Phi(z, s, m+v) + \sum_{n=0}^{m-1} (v+n)^{-s} z^n$$

$$[m = 1, 2, 3, \ldots, \quad v \neq 0, \ -1, \ -2, \ \ldots].$$ EH I 27(1)

9.552 $$\Phi(z, s, v) = iz^{-v}(2\pi)^{s-1}\Gamma(1-s)\left[e^{-i\pi\frac{s}{2}}\Phi\left(e^{-2\pi iv}, 1-s, \frac{\ln z}{2\pi i}\right) - e^{i\pi\left(\frac{s}{2}+2v\right)}\Phi\left(e^{2\pi iv}, 1-s, 1-\frac{\ln z}{2\pi i}\right)\right].$$ EH I 29(7)

Series representation

9.553 $$\Phi(z, s, v) = z^{-v}\Gamma(1-s)\sum_{n=-\infty}^{\infty}(-\ln z + 2\pi ni)^{s-1} e^{2\pi nvi}$$

$$[0 < v \leqslant 1, \quad \operatorname{Re} s < 0, \quad |\arg(-\ln z + 2\pi ni)| \leqslant \pi].$$ EH I 28(6)

9.554 $$\Phi(z, m, v) = z^{-v}\left\{\sum_{n=0}^{\infty}{}' \zeta(m-n, v)\frac{(\ln z)^n}{n!} + \frac{(\ln z)^{m-1}}{(m-1)!}\left[\psi(m) - \psi(v) - \ln\left(\ln\frac{1}{z}\right)\right]\right\}^*$$

$$[m = 2, 3, 4, \ldots, \quad |\ln z| < 2\pi, \quad v \neq 0, \ -1, \ -2, \ \ldots].$$ EH I 30(9)

9.555 $$\Phi(z, -m, v) = \frac{m!}{z^v}\left(\ln\frac{1}{z}\right)^{-m-1} - \frac{1}{z^v}\sum_{r=0}^{\infty}\frac{B_{m+r+1}(v)(\ln z)^r}{r!(m+r+1)}$$

$$[|\ln z| < 2\pi].$$ EH I 30(11)

Integral representation

9.556 $$\Phi(z, s, v) = \frac{1}{\Gamma(s)}\int_0^{\infty}\frac{t^{s-1}e^{-vt}}{1-ze^{-t}}\,dt = \frac{1}{\Gamma(s)}\int_0^{\infty}\frac{t^{s-1}e^{-(v-1)t}\,dt}{e^t - z}$$

$$[\operatorname{Re} v > 0, \ \text{or} \ |z| \leqslant 1, \ z \neq 1, \ \operatorname{Re} s > 0, \ \text{or} \ z = 1, \ \operatorname{Re} s > 1].$$ EH I 27(3)

---

*The prime on the symbol $\sum$ means that the term corresponding to $n = m-1$ is omitted.

Limit relationships

9.557 $\lim_{z\to 1}(1-z)^{1-s}\Phi(z, s, v) = \Gamma(1-s)$ [$\operatorname{Re} s < 1$]. EH I 30(12)

9.558 $\lim_{z\to 1}\frac{\Phi(z, 1, v)}{-\ln(1-z)} = 1.$ EH I 30(13)

A connection with a hypergeometric function

9.559 $\Phi(z, 1, v) = v^{-1}{}_2F_1(1, v; 1+v; z)$ [$|z| < 1$]. EH I 30(10)

9.56 The function $\xi(s)$

9.561 $\xi(s) = \frac{1}{2}s(s-1)\frac{\Gamma\left(\frac{1}{2}s\right)}{\pi^{\frac{1}{2}s}}\zeta(s).$ EH III 190(10)

9.562 $\xi(1-s) = \xi(s).$ EH III 190(11)

9.6 Bernoulli Numbers and Polynomials, Euler Numbers, the Functions $\nu(x)$, $\nu(x, \alpha)$, $\mu(x, \beta)$, $\mu(x, \beta, \alpha)$, $\lambda(x, y)$

9.61 Bernoulli numbers

9.610 The numbers B_n, representing the coefficients of $\frac{t^n}{n!}$ in the expansion of the function

1. $$\frac{t}{e^t-1} = \sum_{n=0}^{\infty} B_n\frac{t^n}{n!},$$

are called *Bernoulli* numbers. Thus, the function $\frac{t}{e^t-1}$ is a generating function for the Bernoulli numbers. GE 48(57), FI II 520

9.611 Integral representations

1. $$B_{2n} = (-1)^{n-1}4n\int_0^{\infty}\frac{x^{2n-1}}{e^{2\pi x}-1}dx$$ (cf. 3.411 2., 4.). FI II 721a

2. $$B_{2n} = (-1)^{n-1}\pi^{2n}\int_0^{\infty}\frac{x^{2n}}{\operatorname{sh}^2 x}dx.$$

3. $$B_{2n} = (-1)^{n-1}\frac{2n(1-2n)}{\pi}\int_0^{\infty}x^{2n-2}\ln\left(1-e^{-2\pi x}\right)dx.$$

See also 3.523 2., 4.271 3.

Properties and functional relations

9.612 A recursion formula (symbolic notation):

$$B^n = (B+1)^n; \quad B^0 = B_0 = 1. \qquad [n \neq 1]$$ GE 49(60)

For computation, after carrying out the expansion on the right hand side, we need to convert all powers into indices, thus:

$$B_n = \sum_{k=0}^{n} \binom{n}{k} B_k, \quad B_0 = 1. \quad [n \neq 1]$$ GE 49

9.613 All the Bernoulli numbers are rational numbers.

9.614 Every number B_n can be represented in the form

$$B_n = C_n - \sum \frac{1}{k+1},$$

where C_n is an integer and the sum is taken over all $k > 0$ such that $k+1$ is a prime and k is a divisor of n. GE 64

9.615 All the Bernoulli numbers with odd index are equal to zero except that $B_1 = -\frac{1}{2}$; that is, $B_{2n+1} = 0$ for n a natural number. GE 52, FI II 521

$$B_{2n} = -\frac{1}{2n+1} + \frac{1}{2} - \sum_{k=2}^{2n-2} \frac{2n(2n-1)\dots(2n-2k+2)}{k!} B_k.$$

9.616 $B_{2n} = \frac{(-1)^{n-1}(2n)!}{2^{2n-1}\pi^{2n}} \zeta(2n).$ GE 56(79), FI II 721a

9.617 $B_{2n} = (-1)^{n-1} \frac{2(2n)!}{(2\pi)^{2n}} \frac{1}{\prod_{p=1}^{\infty} \left(1 - \frac{1}{p^{2n}}\right)}$ (cf. 9.523)

(where the product is taken over all primes p).

For a connection with Riemann's zeta function, see 9.542.

For a connection with the Euler numbers, see 9.635.

For a table of values of the Bernoulli numbers, see 9.71

9.618 An inequality (symbolic notation):

$$|(B-\theta)^n| \leqslant |B_n| \quad [0 < \theta < 1].$$ CE 337

9.62 Bernoulli polynomials

9.620 The Bernoulli polynomials $B_n(x)$ are polynomials of the form

$$B_n(x) = \sum_{k=0}^{n} \binom{n}{k} B_k x^{n-k}$$ GE 51(62)

or, symbolically,

$$B_n(x) = (B + x)^n.$$ GE 52(68)

9.621 The generating function:

$$\frac{e^{xt}}{e^t - 1} = \sum_{n=0}^{\infty} B_n(x) \frac{t^{n-1}}{n!} \quad \text{(cf. 1.213)}.$$ GE 65(89)a

9.622 Series representation:

$$B_{2n}(x) = \frac{(-1)^{n-1} 2(2n)!}{(2\pi)^{2n}} \sum_{k=1}^{\infty} \frac{\cos 2k\pi x}{k^{2n}} \quad [0 < x < 1].$$ GE 71

9.623 Functional relations and properties:

1. $B_{m+1}(n) = B_{m+1} + (m+1)\sum_{k=1}^{n-1} k^m$ [n and m are natural numbers]

(see also 0.121) GE 51(65)

2. $\Delta B_n(x) \equiv B_n(x+1) - B_n(x) = nx^{n-1}$. GE 65(90)
3. $B_n'(x) = nB_{n-1}(x)$. GE 66
4. $B_n(1-x) = (-1)^n B_n(x)$. GE 66

9.624 $B_n(mx) = m^{n-1}\sum_{k=0}^{m-1} B_n\left(x + \frac{k}{m}\right)$ ["summation theorem"]. GE 67

9.625 For n odd, the differences

$$B_n(x) - B_n$$

vanish on the interval [0, 1] only at the points 0, $\frac{1}{2}$, and 1. They change sign at the point $x = \frac{1}{2}$. For n even, these differences vanish at the end points of the interval [0, 1]. Within this interval, they do not change sign and their greatest absolute value occurs at the point $x = \frac{1}{2}$.

9.626 The polynomials

$$B_{2n}(x) - B_{2n} \quad \text{and} \quad B_{2n+2}(x) - B_{2n+2}$$

have opposite signs in the interval (0, 1). GE 87

9.627 Special cases:

1. $B_1(x) = x - \frac{1}{2}$.
2. $B_2(x) = x^2 - x + \frac{1}{6}$.
3. $B_3(x) = x^3 - \frac{3}{2}x^2 + \frac{1}{2}x$.
4. $B_4(x) = x^4 - 2x^3 + x^2 - \frac{1}{30}$.
5. $B_5(x) = x^5 - \frac{5}{2}x^4 + \frac{5}{3}x^3 - \frac{1}{6}x$

GE 70

9.628 Particular values:

1. $B_n(0) = B_n$.
2. $B_n(1) = (-1)^n B_n$.

GE 76

9.63 Euler numbers

9.630 The numbers E_n, representing the coefficients of $\frac{t^n}{n!}$ in the expansion of the function

$$\frac{1}{\operatorname{ch} t} = \sum_{n=0}^{\infty} E_n \frac{t^n}{n!},$$

are known as the *Euler numbers*. Thus, the function $\frac{1}{\operatorname{ch} t}$ is a generating function for the Euler numbers. CE 330

9.631 A recursion formula (symbolic notation):

$$(E+1)^n+(E-1)^n=0, \quad E_0=1. \qquad \text{CE 329}$$

Properties of the Euler numbers

9.632 The Euler numbers are integers.

9.633 The Euler numbers of odd index are equal to zero; the signs of two adjacent numbers of even indices are opposite; that is,

$$E_{2n+1}=0, \quad E_{4n}>0, \quad E_{4n+2}<0. \qquad \text{CE 329}$$

9.634 If α, β, γ, ... are the divisors of the number $n-m$, the difference $E_{2n}-E_{2m}$ is divisible by those of the numbers $2\alpha+1$, $2\beta+1$, $2\gamma+1$, ..., that are primes.

9.635 A connection with the Bernoulli numbers (symbolic notation):

1. $$E_{n-1}=\frac{(4B-1)^n-(4B-3)^n}{2n}. \qquad \text{CE 330}$$

2. $$B_n=\frac{n(E+1)^{n-1}}{2^n(2^n-1)}. \quad [n\leqslant z]. \qquad \text{CE 330}$$

3. $$-\left(B+\frac{1}{4}\right)^{2n+1}=\frac{2n+1}{4^{2n+1}}E_{2n}. \qquad \text{CE 341}$$

For a table of values of the Euler numbers, see 9.72.

9.64 The functions $\nu(x)$, $\nu(x, \alpha)$, $\mu(x, \beta)$, $\mu(x, \beta, \alpha)$, $\lambda(x, y)$

9.640

1. $$\nu(x)=\int_0^\infty \frac{x^t\,dt}{\Gamma(t+1)}. \qquad \text{EH III 217(1)}$$

2. $$\nu(x, \alpha)=\int_0^\infty \frac{x^{\alpha+t}\,dt}{\Gamma(\alpha+t+1)}. \qquad \text{EH III 217(1)}$$

3. $$\mu(x, \beta)=\int_0^\infty \frac{x^t t^\beta\,dt}{\Gamma(\beta+1)\Gamma(t+1)}. \qquad \text{EH III 217(2)}$$

4. $$\mu(x, \beta, \alpha)=\int_0^\infty \frac{x^{\alpha+t} t^\beta\,dt}{\Gamma(\beta+1)\Gamma(\alpha+t+1)}. \qquad \text{EH III 217(2)}$$

5. $$\lambda(x, y)=\int_0^y \frac{\Gamma(u+1)\,du}{x^u}. \qquad \text{MI 9}$$

9.7 Constants

9.71 Bernoulli numbers

$$B_0=1, \qquad B_4=-\frac{1}{30},$$

$$B_1=-\frac{1}{2}, \qquad B_6=\frac{1}{42},$$

$$B_2=\frac{1}{6}, \qquad B_8=-\frac{1}{30},$$

$$B_{10}=\frac{5}{66},\qquad B_{24}=-\frac{236\,364\,091}{2730},$$

$$B_{12}=-\frac{691}{2730},\qquad B_{26}=\frac{8\,553\,103}{6},$$

$$B_{14}=\frac{7}{6},\qquad B_{28}=-\frac{23\,749\,461\,029}{870},$$

$$B_{16}=-\frac{3617}{510},\qquad B_{30}=\frac{8\,615\,841\,276\,005}{14\,322},$$

$$B_{18}=\frac{43\,867}{798},\qquad B_{32}=-\frac{7\,709\,321\,041\,217}{510},$$

$$B_{20}=-\frac{174\,611}{330},\qquad B_{34}=\frac{2\,577\,687\,858\,367}{6}.$$

$$B_{22}=\frac{854\,513}{138},$$

9.72 Euler numbers

$$E_0=1,\qquad E_{12}=2\,702\,765,$$
$$E_2=-1,\qquad E_{14}=-199\,360\,981,$$
$$E_4=5,\qquad E_{16}=19\,391\,512\,145,$$
$$E_6=-61,\qquad E_{18}=-2\,404\,879\,675\,441,$$
$$E_8=1385,$$
$$E_{10}=-50\,521,\qquad E_{20}=370\,371\,188\,237\,525.$$

The Bernoulli and Euler numbers of odd index (with the exception of B_1) are equal to zero.

9.73 Euler's and Catalan's constants

Euler's constant

$$C=0.577\,215\,664\,901\,532\,5\ldots$$

Catalan's constant

$$G=0.915\,965\,594\ldots$$

BIBLIOGRAPHIC REFERENCES USED IN PREPARATION OF TEXT

(See page xlv for explanation of the letters preceding each bibliographic reference)

AD Adams, E. P., and Hippisley, R. L., *Smithsonian Mathematical Formulae and Tables of Elliptic Functions*, Smithsonian Institute, Washington, D. C. 1922.

AK Appel, P., and Kampe de Fériet, *Fonctions hypergéometriques et hypersphériques, polynômes d'Hermite*, Paris, 1926.

BE Bertrand, J., *Traité de calcul différentiel et de calcul intégral*, V. 2, *Calcul intégral, intégrales définies et indéfinies*, Gauthier-Villars, Paris, 1870.

BI Bieren de Haan, D., *Nouvelles tables d'intégrales définies*, Amsterdam, 1867.

BR*Bromwich, T. I'A., *An Introduction to the Theory of Infinite Sets*, Macmillan, London, 1908, 2nd edition, 1926.

BU Buchholtz, H. *Die konfluente hypergeometrische Funktion mit besonderer Berücksichtigung ihrer Ahwendungen*, Springer Verlag, Berlin, 1953.

BY Byrd, P. F. and Friedman, M. D., *Handbook of Elliptic Integrals for Engineers and Physicists*, Springer Verlag, Berlin, 1954.

CE Cesaro, Z., *Elementary Class Book of Algebraic Analysis and the Calculation of Infinite Limits*, 1st ed. ONTI, Moscow and Leningrad, 1936.

CO Courant, R. and Hilbert, D., *Methods of Mathematical Physics*, Vol. I, Interscience, New York, 1953.

DW Dwight, H. B., *Tables of Integrals and Other Mathematical Data*, Macmillan, New York, 1934.

EF Efros, A. M. and Danilevskiy, A. M., *Operatsionnoye ischisleniye i konturnyye integraly* (Operational calculus and contour integrals), GNTIU, Khar'kov, 1937.

EH Erdelyi, A. et al., *Higher Transcendental Functions*, Vols. I, II, and III, McGraw-Hill, New York, 1953-1955.

ET Erdelyi, A. et al., *Tables of Integral Transforms*, Vols. I and II, McGraw-Hill, New York, 1954.

EU Euler, L., *Introductio in Analysin Infinitorum*, Lausanne, 1748.

FI Fikhtengol'ts, G. M., *Kurs differentsial'nogo i integral'nogo ischisleniya* (Course in differential and integral calculus), Vols. I, II, and III, Gostekhizdat, Moscow and Leningrad, 1947-1949.

GA Gauss, K. F., *Werke*, Bd. III, Göttingen, 1876.

GE Gel'fond, A. O., *Ischisleniye konechnykh raznostey* (Calculus of finite differences) Part I, ONTI, Moscow and Leningrad, 1936.

GI Giunter, N. M. and Kuz'min, R. O., (eds.), *Sbornik zadach po vysshey matematike* (Collection of problems in higher mathematics), Vol. I, II, and III, Gostekhizdat, Moscow and Leningrad, 1947.

GO Goursat, E. J. B., *Cours d'Analyse*, Vol. I, Gauthier-Villars, Paris, 1923.

GU Gröbner, W. et al., *Integraltafel*, Teil I, *Unbestimmte Integrale*, Braunschweig, 1944.

GW Gröbner, W. and Hofreiter, N., *Integraltafel*, Teil II, *Bestimmte Integrale*, Springer-Verlag, Wien and Innsbruck, 1958.

HO Hobson, E. W., *The Theory of Spherical and Ellipsoidal Harmonics*, Cambridge University Press, 1931.

JA Jahnke, E. and Emde, F., *Tables of Functions with Formulas and Curves*, Dover, New York, 1943.

JE James, H. M., et al. (eds.), *Theory of Servomechanisms*, McGraw-Hill, New York, 1947.

JO Jolley, L., *Summation of Series*, Chapman and Hall, London, 1925.

* The Bibliographic Reference BR* refers to the 1908 edition of Bromwich T. I.'A., *An Introduction to the Theory of Infinite Sets*. BR refers to the 1926 edition.

KR Krechmar, V. A., *Zadachnik po algebre* (Problem book in algebra) 2nd ed. Gostekhizdat, Moscow and Leningrad, 1950.

KU Kuzmin, R. O., *Besselevy funktsii* (Bessel functions), ONTI, Moscow and Leningrad, 1935.

LA Laska, W., *Sammlung von Formeln der reinen und angewandten Mathematik*, Friedrich Viewig und Sohn, Braunschweig, 1888.

LE Legendre, A. M., *Exercices calcul intégral*, Paris, 1811.

LI Lindeman, C. E., *Examen des nouvelles tables d'intégrales définies de M. Bierens de Haan*, Amsterdam, 1867, Norstedt, Stockholm, 1891.

LO Lobachevskiy, N. I., *Poloye sobraniye sochineniy* (Complete works), Vols. I, III, and V, Gostekhizdat, Moscow and Leningrad, 1946-1951.

MA McLachlan, N., *Theory and Application of Mathieu Functions*, Oxford University Press, 1947.

ME McLachlan, N. W. and Humbert, P., *Formulaire pour le calcul symbolique*, L'Acad. des Sciences de Paris et al., Fasc. 100, 1950.

MI McLachlan, N. W., et al., *Supplément au formulaire pour le calcul symbolique*, L'Acad. des Sciences de Paris et al., Fasc. 113, 1950.

MO Magnus, W. and Oberhettinger, F., *Formeln und Sätze für die speziellen Funktionen der mathematischen Physik* Springer Verlag, Berlin, 1948.

MZ Meyer Zur Capellen, W., *Integraltafeln, Sammlung unbestimmer Integrale elementarer Funktionen*, Springer Verlag, Berlin, 1950.

NA Natanson, I. P., *Konstruktivnaya teoriya funktsiy* (Constructive theory of functions), Gostekhizdat, Moscow and Leningrad, 1949.

NH Nielsen, N., *Handbuch der Theorie der Gammafunktion*, Teubner, Leipzig, 1906.

NT Nielsen, N., *Theorie des Integrallogarithmus und verwandter Transcendenten*, Teubner, Leipzig, 1906.

NV Novoselov, S. I., *Obratnyye trigonometricheskiye funktsii, posobiye dlya uchiteley* (Inverse trigonometric functions, textbook for students), 3rd ed., Uchpedgiz, Moscow and Leningrad, 1950.

PE Peirce, B. O., *A Short Table of Integrals*, 3rd ed., Ginn and Co., Boston, 1929.

SI Sikorskiy, Yu. S., *Elementy teorii ellipticheskikh funktsiy s prilozheniyama k mekhanike* (Elements of the theory of elliptic functions with applications to mechanics), ONTI, Moscow and Leningrad, 1936.

SM Smirnov, V. I., *Kurs vysshey matematiki* (A course of higher mathematics) Vol. III, Part 2, 4th ed., Gostekhizdat, Moscow and Leningrad, 1949.

ST Strutt, M. J. O., *Lamésche, Mathieusche und Verwandte Funktionen in Physik und Technik*, Springer Verlag, Berlin, 1932.

TI Timofeyev, A. F. *Integrirovaniye funktsiy* (Integration of functions), Part I, GTTI, Moscow and Leningrad, 1933.

WA Watson, G. N., *A Treatise on the Theory of Bessel Functions*, 2nd ed., Cambridge University Press, 1944.

WH Whittaker, E. T. and Watson, G. N., *Modern Analysis*, 4th ed., Cambridge University Press, 1927, Part II, 1934.

ZH Zhuravskiy, A. M., *Spravochnik po ellipticheskim funktsiyam* (Reference book on elliptic functions), Izd. Akad. Nauk. U.S.S.R., Moscow and Leningrad, 1941.

CLASSIFIED SUPPLEMENTARY REFERENCES*

General Reference Books

Bromwich, T. I'A., *An Introduction to the Theory of Infinite Series*. Macmillan, London. 2nd ed. 1926 (Reprinted 1942)

Copson, E. T., *An Introduction to the Theory of Functions of a Complex Variable*. Oxford University Press, 1935.

Courant, R. and Hilbert, D., *Methods of Mathematical Physics*, vol. I. Interscience, New York, 1953.

Erdelyi, A. et al. *Higher Transcendental Functions*, vols. I to III, McGraw Hill, New York. 1953 to 1955.

Erdelyi, A. et al., *Tables of Integral Transforms*, vols. I and II. McGraw Hill, New York, 1955.

Fletcher, A., Miller, J. C. P., and Rosenhead, L., *An Index of Mathematical Tables*. Scientific Computing Service, London, 2nd ed., 1962.

Grobner, W. and Hofreiter, N., *Integraltafel*. I, II. Springer Verlag, Wien and Innsbruck, 1949.

Hardy, G. H., Littlewood, J. E., and Polya, G., *Inequalities*. Cambridge University Press, 2nd ed., 1952.

Hartley, H. O. and Greenwood, J. A., *Guide to Tables in Mathematical Statistics*. Princeton University Press, 1962.

Jeffreys, H. and Jeffreys, B. S., *Methods of Mathematical Physics*. Cambridge University Press, 1956.

Knopp, K., *Theory and Application of Infinite Series*. Blackie, London, 1946, Hafner, New York, 1948.

Magnus, W. and Oberhettinger, F., *Formulas and Theorems for the Special Functions of Mathematical Physics*. Chelsea, New York, 1949.

National Bureau of Standards, *Handbook of Mathematical Functions*. U. S. Government Printing Office, Washington, D. C., 1964.

Truesdell, C. *A Unified Theory of Special Functions*. Princeton University Press, New Jersey, 1948.

Whittaker, E. T. and Watson, G. N., *A Course of Modern Analysis*. Cambridge University Press, 4th ed. 1940.

Exponential Integrals, The Gamma Function and Related Functions

Artin, E., *The Gamma Function*. Holt, Rinehart, and Winston, New York, 1964.

Busbridge, I. W., *The Mathematics of Radiative Transfer*. Cambridge University Press, 1960.

Erdelyi, A. et al., *Higher Transcendental Functions*, vol. II. McGraw Hill, New York, 1953.

Erdelyi, A. et al., *Tables of Integral Transforms*, vols. I. and II. McGraw Hill, New York, 1954.

Hastings, Jr., C., *Approximations for Digital Computers*. Princeton University Press. New Jersey, 1955.

Kourganoff, V., *Basic Methods in Transfer Problems*. Oxford University Press, 1952.

Losch, F. and Schoblik, F., *Die Fakultat (Gamma-funktion) und Verwandte Funktionen*. Teubner, Leipzig, 1951.

Oberhettinger, F., *Tabellen zur Fourier Transformation*. Springer Verlag, Berlin, 1957.

Error Function and Fresnel Integrals

Erdelyi, A. et al., *Higher Transcendental Functions*, vol. II. McGraw Hill, New York, 1953.

Erdelyi, A. et al., *Tables of Integral Transforms*, vol. I. McGraw Hill, New York, 1954.

Slater, L. J., *Confluent Hypergeometric Functions*. Cambridge University Press, 1960.

Tricomi, F. G., *Funzioni ipergeometriche confluenti*. Edizioni Cremonese, Italy, 1954.

---

* Prepared by Alan Jeffrey for the English language edition.

Watson, G. N., *A Treatise on the Theory of Bessel Functions*. Cambridge University Press, 2nd ed., 1958.

Legendre and Related Functions

Erdelyi, A. et al., *Higher Transcendental Functions*, vol. I. McGraw Hill, New York, 1953.

Helfenstein, H., *Ueber eine Spezielle Lamesche Differentialgleichung*. Brunner and Bodmer, Zurich, 1950. (Bibliography).

Hobson, E. W., *The Theory of Spherical and Ellipsoidal Harmonics*. Cambridge University Press, 1931. Reprinted by Chelsea, New York, 1955.

Lense, J., *Kugelfunktionen*. Geest and Portig, Leipzig, 1950.

MacRobert, T. M., *Spherical Harmonics: An Elementary Treatise on Harmonic Functions with Applications*. Methuen, England, 1927. (Revised ed. 1947; reprinted Dover, New York, 1948).

Snow, C., *The Hypergeometric and Legendre Functions with Applications to Integral Equations of Potential Theory*. National Bureau of Standards, Washington, D. C., 2nd ed. 1952.

Stratton, J. A., Morse, P. M., Chu, L. J. and Hunter, R. A., *Elliptic Cylinder and Spheroidal Wave Functions Including Tables of Separation Constants and Coefficients*. Wiley, New York, 1941.

Bessel Functions

Bickley, W. G., *Bessel Functions and Formulae*. Cambridge University Press, 1953.

Erdelyi, A. et al., *Higher Transcendental Functions*, vols. I and II. McGraw Hill, New York, 1954.

Erdelyi, A. et al., *Tables of Integral Transforms*, vols. I and II. McGraw Hill, New York, 1954.

Gray, A., Mathews, G. B. and MacRobert, T. M., *A Treatise on Bessel Functions and Their Applications to Physics*. Macmillan, 2nd ed. 1922.

McLachlan, N. W., *Bessel Functions for Engineers*. Oxford University Press, 2nd ed. 1955.

Petiau, G., *La théorie des fonctions de Bessel*. Centre National de la Recherche Scientifique, Paris, 1955.

Relton, F. E., *Applied Bessel Functions*. Blackie, London, 1946.

Watson, G. N., *A Treatise on the Theory of Bessel Functions*. Cambridge University Press, 2nd ed. 1958.

Struve Functions

Erdelyi, A. et al., *Higher Transcendental Functions*, vol.II. McGraw Hill, New York, 1954.

Gray, A., Mathews, G. B. and MacRobert, T. M., *A Treatise on Bessel Functions and Their Applications to Physics*. Macmillan, London, 2nd ed., 1922.

Watson, G. N., *A Treatise on the Theory of Bessel Functions*. Cambridge University Press, 2nd ed. 1958.

Hypergeometric and Confluent Hypergeometric Functions

Appel, P. and Kampe de Ferriet, J., *Fonctions hypergéometriques et hypersphériques*. Gauthiers-Villars, Paris, 1926.

Bailey, W. N., *Generalized Hypergeometric Functions*. Cambridge University Press, 1935.

Buchholz, H., *Die konfluente hypergeometrische Funktion*. Springer Verlag, Berlin, 1953.

Erdelyi, A. et al., *Higher Transcendental Functions*, vol. I. McGraw Hill, New York, 1953.

Jeffreys, H. and Jeffreys, B. S., *Methods of Mathematical Physics*. Cambridge University Press, 1956.

Klein, F., *Vorlesungen über hypergeometrische Funktion*. Springer Verlag, Berlin, 1933.

Slater, L. J., *Confluent Hypergeometric Functions*. Cambridge University Press, 1960.

Snow, C., *The Hypergeometric and Legendre Functions with Applications to Integral Equations of Potential Theory*. National Bureau of Standards, Washington, D. C. 2nd ed. 1952.

Swanson, C. A. and Erdelyi, A., *Asymptotic Forms of Confluent Hypergeometric Functions*. Memoir 25, American Mathematical Society, 1957.

Tricomi, F. G., *Lezioni sulla funzioni ipergeometriche confluenti*. Gheroni, Torino, 1952.

Jacobian and Weierstrass Elliptic Functions and Related Functions

Erdelyi, A. et al., *Higher Transcendental Functions*, vol. II. McGraw Hill, New York, 1954.

Byrd, P. F. and Friedman, M. D., *Handbook of Elliptic Integrals for Engineers and Physicists*. Springer Verlag, Berlin, 1954.

Graeser, E., *Einführung in die Theorie der Elliptischen Funktionen und deren Anwendungen.* Oldenbourg, Munich, 1950.

Hancock, H., *Lectures on the Theory of Elliptic Functions*, vol. I. Dover, New York, 1958.

Neville, E. H., *Jacobian Elliptic Functions*. Oxford University Press, 1944 (2nd ed. 1951).

Oberhettinger, F. and Magnus, W., *Anwendungen der Elliptischen Funktionen in Physik und Technik*. Springer Verlag, Berlin, 1949.

Roberts, W. R. W., *Elliptic and Hyperelliptic Integrals and Allied Theory*. Cambridge University Press, 1938.

Tannery, J. and Molk, J., *Eléments de la Théorie des Fonctions Elliptiques*, 4 volumes. Gauthier-Villars, Paris, 1893–1902.

Tricomi, F. G., *Elliptische Funktionen*. Akad. Verlag, Leipzig, 1948.

Parabolic Cylinder Functions

Buchholz, H., *Die konfluente hypergeometrische Funktion*. Springer Verlag, Berlin, 1953.

Erdelyi, A. et al., *Higher Transcendental Functions*, vol. II. McGraw Hill, New York, 1954.

Orthogonal Polynomials and Functions

Bibliography on Orthogonal Polynomials. Bulletin of National Research Council No. 103, Washington, D. C., 1940.

Courant, R. and Hilbert, D., *Methods of Mathematical Physics*, vol. I. Interscience, New York, 1953.

Erdelyi, A. et al., *Higher Transcendental Functions*, vol. II. McGraw Hill, New York, 1954.

Kaczmarz, St. and Steinhaus, H., *Theorie der Orthogonalreihen*. Chelsea, New York, 1951.

Lorentz, G. G., *Berstein Polynomials*. University of Toronto Press, Toronto, 1953.

Sansone, G., *Orthogonal Functions*. Interscience, New York, 1959.

Shohat, J. A. and Tamarkin, J. D., *The Problem of Moments*. American Mathematical Society, 1943.

Szego, G., *Orthogonal Polynomials*, American Mathematical Society Colloquim Pub. No. 23, 1959.

Titchmarsh, E. C., *Eigenfunction Expansions Associated with Second Order Differential Equations*. Oxford University Press, part I (1946), part II (1958).

Tricomi, F. G., *Vorlesungen über Orthogonalreihen*. Springer Verlag, Berlin, 1955.

Riemann Zeta Function

Titchmarsh, E. C., *The Zeta Function of Riemann*. Cambridge University Press, 1930.

Titchmarsh, E. C., *The Theory of the Riemann Zeta Function*. Oxford University Press, 1951.

Probability Function

Cramer, H., *Mathematical Methods of Statistics*. Princeton University Press, 1951.

Erdelyi, A. et al., *Higher Transcendental Functions*, vols. I, II, and III. McGraw Hill, New York, 1953 to 1955.

Kendall, M. G. and Stuart, A., *The Advanced Theory of Statistics*, vol. I: *Distribution Theory*. Griffin, London, 1958.

Mathieu Functions

Erdelyi, A., *Higher Transcendental Functions*, vol. III. McGraw Hill, New York, 1955.

McLachlan, N. W., *Theory and Application of Mathieu Functions*. Oxford University Press, 1947.

Meixner, J. and Schafke, F. W., *Mathieusche Funktionen und Spheroid-funktionen mit Anwendungen auf Physikalische und Technische Probleme*. Springer Verlag, Heidelberg, 1954.

Strutt, M. J. O., *Lamesche, Mathieusche und verwandte Funktionen in Physik und Technik*. Ergeb, Math. Grenzgeb. *1*, 199-323 (1932). Reprint Edwards Bros., Ann Arbor, Michigan, 1944.

Integral Transforms

Bochner, S., *Vorlesungen über Fourische Integrale*. Akad. Verlag, Leipzig, 1932. Reprint Chelsea, New York, 1948.

Bochner, S. and Chandrasekharan, K., *Fourier Transforms*. Princeton University Press, 1949.

Carslaw, H. S. and Jaeger, J. C., *Conduction of Heat in Solids*. Oxford University Press, 1948.

Doetsch, G., *Theorie und Anwendung der Laplace-Transformation*. Springer Verlag, Berlin, 1937. (Reprinted by Dover, New York, 1943)

Doetsch, G., *Theory and Application of the Laplace-Transform*. Chelsea, New York, 1965.

Doetsch, G., *Handbuch der Laplace-Transformation*, vol. I. *Theorie*. Birkhauser Verlag, Basel, 1950.

Doetsch, G., *Handbuch der Laplace-Transformation*, vol. II. *Anwendungen*. Birkhauser Verlag, Basel, 1955.

Doetsch, G., *Handbuch der Physik, Mathematische Methoden II*, 1st ed., Berlin, 1955.

Hirschmann, J. J. and Widder, D. V., *The Convolution Transformation*. Princeton University Press, New Jersey, 1955.

Van der Pol, B. and Bremmer, H., *Operational Calculus Based on the Two Sided Laplace Transformation*. Cambridge University Press, 1950.

Oberhettinger, F., *Tabellen zur Fourier Transformation*. Springer Verlag, Berlin, 1957.

Sneddon, I. N., *Fourier Transforms*. McGraw Hill, New York, 1951.

Titchmarsh, E. C., *Introduction to the Theory of Fourier Integrals*. Oxford University Press, 1937.

Widder, D. V., *The Laplace Transform*. Princeton University Press, New Jersey, 1941.

Wiener, N., *The Fourier Integral and Certain of its Applications*. Dover, New York, 1951.

Asymptotic Expansions

De Bruijn, N. G., *Asymptotic Methods in Analysis*. North-Holland Publishing Co., Amsterdam, 1958.

Copson, E. T., *Asymptotic Expansions*. Cambridge University Press, 1965.

Erdelyi, A., *Asymptotic Expansions*. Dover, New York, 1956.

Ford, W. B., *Studies on Divergent Series and Summability*. Macmillan, New York, 1916.

Hardy, G. H., *Divergent Series*. Clarendon Press, Oxford, 1949.

Watson, G. N., *A Treatise on the Theory of Bessel Functions*. Cambridge University Press, 2nd ed., 1958.

4 5 6 7 8 9 0 H 1 I 2 J 3